IBC®

INTERNATIONAL BUILDING CODE®

CODE AND COMMENTARY
VOLUME 1

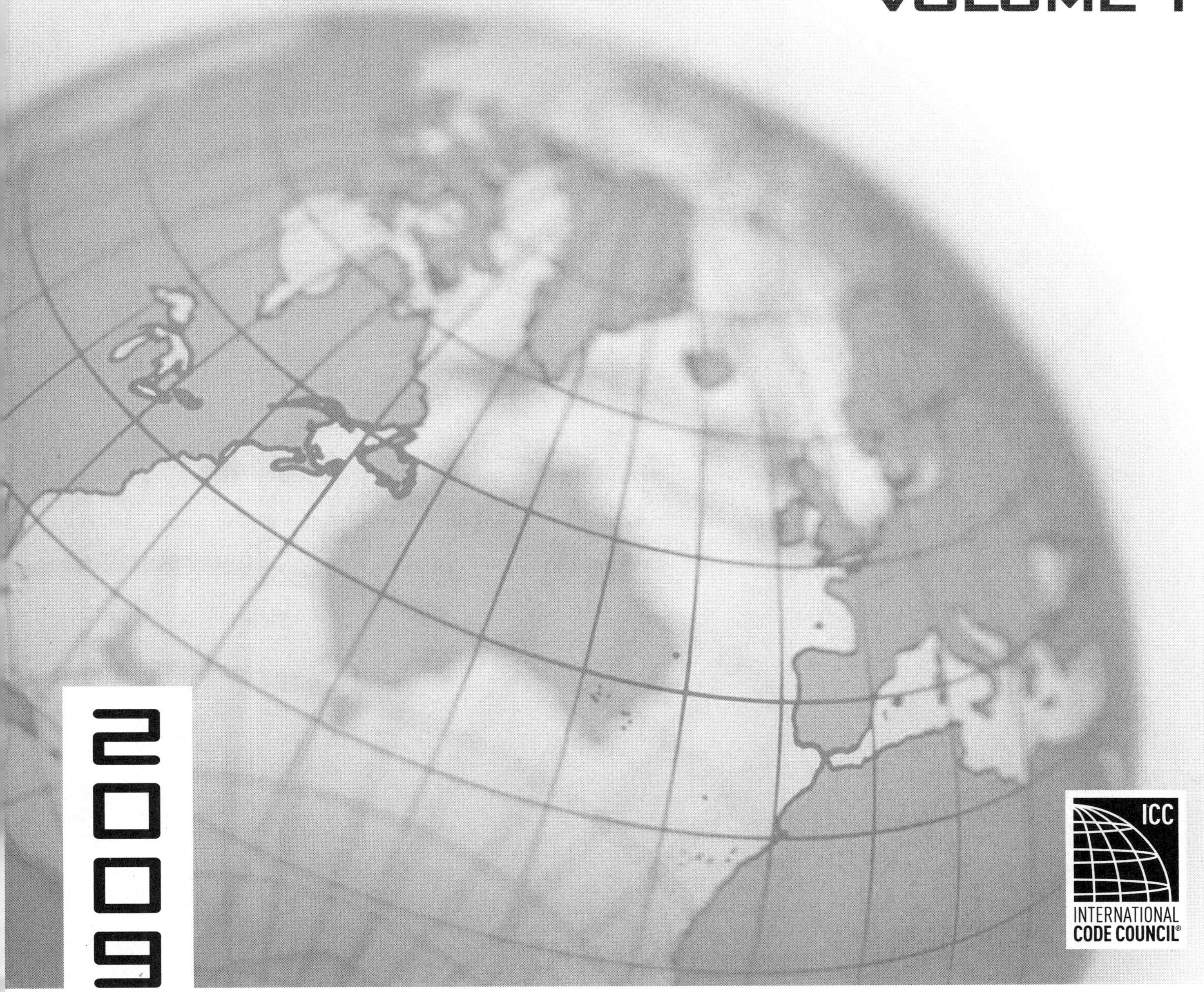

2009

ICC

INTERNATIONAL
CODE COUNCIL®

2009 International Building Code® Commentary

First Printing: January 2010
Second Printing: August 2010

ISBN : 978-1-58001-892-0

34264.T011356

PRINTED IN THE U.S.A.

PREFACE

The principal purpose of the Commentary is to provide a basic volume of knowledge and facts relating to building construction as it pertains to the regulations set forth in the 2009 *International Building Code*. The person who is serious about effectively designing, constructing and regulating buildings and structures will find the Commentary to be a reliable data source and reference to almost all components of the built environment.

As a follow-up to the *International Building Code*, we offer a companion document, the *International Building Code Commentary—Volume I*. Volume I covers Chapters 1 through 15 of the 2009 *International Building Code*. The basic appeal of the Commentary is thus: it provides in a small package and at reasonable cost thorough coverage of many issues likely to be dealt with when using the *International Building Code* — and then supplements that coverage with historical and technical background. Reference lists, information sources and bibliographies are also included.

Throughout all of this, strenuous effort has been made to keep the vast quantity of material accessible and its method of presentation useful. With a comprehensive yet concise summary of each section, the Commentary provides a convenient reference for regulations applicable to the construction of buildings and structures. In the chapters that follow, discussions focus on the full meaning and implications of the code text. Guidelines suggest the most effective method of application, and the consequences of not adhering to the code text. Illustrations are provided to aid understanding; they do not necessarily illustrate the only methods of achieving code compliance.

The format of the Commentary includes the full text of each section, table and figure in the code, followed immediately by the commentary applicable to that text. At the time of printing, the Commentary reflects the most up-to-date text of the 2009 *International Building Code*. As stated in the preface to the *International Building Code,* the content of sections in the code which begin with a letter designation (i.e., Section [F]307.1) are maintained by another code development committee. Each section's narrative includes a statement of its objective and intent, and usually includes a discussion about why the requirement commands the conditions set forth. Code text and commentary text are easily distinguished from each other. All code text is shown as it appears in the *International Building Code*, and all commentary is indented below the code text and begins with the symbol ❖.

Readers should note that the Commentary is to be used in conjunction with the *International Building Code* and not as a substitute for the code. **The Commentary is advisory only;** the code official alone possesses the authority and responsibility for interpreting the code.

Comments and recommendations are encouraged, for through your input, we can improve future editions. Please direct your comments to the Codes and Standards Development Department at the Chicago District Office.

The International Code Council would like to extend its thanks to the following individuals for their contributions to the technical content of this commentary:

Chris Marion	Gregory Cahanin
Jeff Tubbs	David Cooper
Rebecca Quinn	Dave Collins
Joann Surmar	Vickie Lovell
James Milke	John Valiulis
Richard Walke	Marcelo Hirschler
Dave Adams	Edward Keith
Zeno Martin	Phillip Samblanet
Jason Thompson	

TABLE OF CONTENTS

Chapter 1:
Scope and Administration

General Comments

This chapter contains provisions for the application, enforcement and administration of subsequent requirements of the code. In addition to establishing the scope of the code, Chapter 1 identifies which buildings and structures come under its purview. Section 101 addresses the scope of the code and references the other *International Codes*® that are mentioned elsewhere in the code. Section 102 establishes the applicability of the code and addresses existing structures.

Section 103 establishes the department of building safety and the appointment of department personnel. Section 104 outlines the duties and authority of the building official with regard to permits, inspections and right of entry. It also establishes the authority of the building official to approve alternative materials, used materials and modifications. Section 105 states when permits are required and establishes the procedures for the review of applications and the issuance of permits. Section 106 provides requirements for posting live loads greater than 50 pounds per square foot (2394 Pa) (psf). Section 107 describes the information that must be included on the construction documents submitted with the application. Section 108 authorizes the building official to issue permits for temporary structures and uses. Section 109 establishes requirements for a fee schedule. Section 110 includes inspection duties of the building official or an inspection agency that has been approved by the building official. Provisions for the issuance of certificates of occupancy are detailed in Section 111. Section 112 gives the building official the authority to approve utility connections. Section 113 establishes the board of appeals and the criteria for making applications for appeal. Administrative provisions for violations are addressed in Section 114, including provisions for unlawful acts, violation notices, prosecution and penalties. Section 115 describes procedures for stop work orders. Section 116 establishes the criteria for unsafe structures and equipment and the procedures to be followed by the building official for abatement and for notification to the responsible party.

Each state's building code enabling legislation, which is grounded within the police power of the state, is the source of all authority to enact building codes. In terms of how it is used, police power is the power of the state to legislate for the general welfare of its citizens. This power enables passage of such laws as building codes. If the state legislature has limited this power in any way, the municipality may not exceed these limitations. While the municipality may not further delegate its police power (e.g., by delegating the burden of determining code com-

pliance to the building owner, contractor or architect), it may turn over the administration of the building code to a municipal official, such as a building official, provided that sufficient criteria are given to establish clearly the basis for decisions as to whether or not a proposed building conforms to the code.

Chapter 1 is largely concerned with maintaining "due process of law" in enforcing the building performance criteria contained in the body of the code. Only through careful observation of the administrative provisions can the building official reasonably hope to demonstrate that "equal protection under the law" has been provided. While it is generally assumed that the administration and enforcement section of a code is geared toward a building official, this is not entirely true. The provisions also establish the rights and privileges of the design professional, contractor and building owner. The position of the building official is merely to review the proposed and completed work and to determine if the construction conforms to the code requirements. The design professional is responsible for the design of a safe structure. The contractor is responsible for constructing the structure in compliance with the plans.

During the course of construction, the building official reviews the activity to ascertain that the spirit and intent of the law are being met and that the safety, health and welfare of the public will be protected. As a public servant, the building official enforces the code in an unbiased, proper manner. Every individual is guaranteed equal enforcement of the provisions of the code. Furthermore, design professionals, contractors and building owners have the right of due process for any requirement in the code.

Purpose

A building code, as with any other code, is intended to be adopted as a legally enforceable document to safeguard health, safety, property and public welfare. A building code cannot be effective without adequate provisions for its administration and enforcement. The official charged with the administration and enforcement of building regulations has a great responsibility, and with this responsibility goes authority. No matter how detailed the building code may be, the building official must, to some extent, exercise his or her own judgement in determining code compliance. The building official has the responsibility to establish that the homes in which the citizens of the community reside and the buildings in which they work are designed and constructed to be structurally stable with adequate means of egress, light and ventilation and to provide a minimum acceptable level of protection to life and property from fire.

Chapter 1 contains two parts. Part 1, Scope and Application, contains all issues related to the scope and intent of the code, as well as the applicability of this code relative to other standards and laws that might also be applicable on a given building project, such as federal or state. Part 2, Administration and Enforcement, contains all issues related to the duties and powers of the building official, the issuance of permits and certificates of occupancy, and other related operational items.

PART 1—SCOPE AND APPLICATION

SECTION 101
GENERAL

101.1 Title. These regulations shall be known as the *Building Code* of [NAME OF JURISDICTION], hereinafter referred to as "this code."

❖ The purpose of this section is to identify the adopted regulations by inserting the name of the adopting jurisdiction into the code.

101.2 Scope. The provisions of this code shall apply to the construction, *alteration*, movement, enlargement, replacement, repair, equipment, use and occupancy, location, maintenance, removal and demolition of every building or structure or any appurtenances connected or attached to such buildings or structures.

> **Exception:** Detached one- and two-family *dwellings* and multiple single-family *dwellings* (*townhouses*) not more than three *stories* above *grade plane* in height with a separate *means of egress* and their accessory structures shall comply with the *International Residential Code*.

❖ This section establishes when the regulations contained in the code must be followed, whether all or in part. Something must happen (construction of a new building, modification to an existing one or allowing an existing building or structure to become unsafe) for the code to be applicable. While such activity may not be as significant as a new building, a fence is considered a structure and, therefore, its erection is within the scope of the code. The building code is not a maintenance document requiring periodic inspections that will, in turn, result in an enforcement action, although periodic inspections are addressed by the *International Fire Code*® (IFC®).

The exception mandates that detached one- and two-family dwellings and townhouses that are not more than three stories above grade and have separate means of egress are to comply with the *International Residential Code*® (IRC®) and are not required to comply with this code. This applies to all such structures, whether or not there are lot lines separating them and also to their accessory structures, such as garages and pools. Such structures four stories or more in height are beyond the scope of the IRC and must comply with the provisions of the code and its referenced codes.

101.2.1 Appendices. Provisions in the appendices shall not apply unless specifically adopted.

❖ The provisions contained in Appendices A through K are not considered part of the code and are, therefore, not enforceable unless they are specifically included in the ordinance or other adopting law or regulation of the jurisdiction. See Section 1 of the sample ordinance on page xv of the code for where the appendices to be adopted are to be specified in the adoption ordinance.

101.3 Intent. The purpose of this code is to establish the minimum requirements to safeguard the public health, safety and general welfare through structural strength, *means of egress* facilities, stability, sanitation, adequate light and ventilation, energy conservation, and safety to life and property from fire and other hazards attributed to the built environment and to provide safety to fire fighters and emergency responders during emergency operations.

❖ The intent of the code is to set forth regulations that establish the minimum acceptable level to safeguard public health, safety and welfare and to provide protection for fire fighters and emergency responders in building emergencies. The intent becomes important in the application of such sections as Sections 102, 104.11 and 114 as well as any enforcement-oriented interpretive action or judgement. Like any code, the written text is subject to interpretation. Interpretations should not be affected by economics or the potential impact on any party. The only considerations should be protection of public health, safety and welfare and emergency responder safety.

101.4 Referenced codes. The other codes listed in Sections 101.4.1 through 101.4.6 and referenced elsewhere in this code shall be considered part of the requirements of this code to the prescribed extent of each such reference.

❖ The International Code Council® (ICC®) promulgates a complete set of codes to regulate the built environment. These codes are coordinated with each other so as to avoid conflicting provisions. When the code is adopted by a jurisdiction, the codes that regulate a building's electrical, fuel gas, mechanical and plumbing systems are also included in the adoption and are considered a part of the code. The *International Property Maintenance Code*® (IPMC®) and the IFC are also referenced and enable the building official to address unsafe conditions in existing structures. Various other sections of the code also specifically refer to these codes. Note that these codes are listed in Chapter 35 and further identified by the specific year of issue. Only that edition of the code is legally adopted and any future editions are not enforceable. The issuance of new editions of all the *International Codes®* occurs concurrently and new editions of the referenced codes are adopted with each new edition of the code. Adoption is done in this manner so that there are not conflicting provisions in these codes.

101.4.1 Gas. The provisions of the *International Fuel Gas Code* shall apply to the installation of gas piping from the point of delivery, gas appliances and related accessories as covered in this code. These requirements apply to gas piping systems extending from the point of delivery to the inlet connections of appliances and the installation and operation of residential and commercial gas appliances and related accessories.

❖ The *International Fuel Gas Code®* (IFGC®) regulates gas piping and appliances and is adopted by reference from this section, as well as Section 2801.1, as the enforceable document for regulating gas systems. This section also establishes the scope of the IFGC as extending from the point of delivery to the inlet connections of each gas appliance. The "Point of delivery" is defined in the IFGC as the outlet of the service meter, regulator or shutoff valve.

101.4.2 Mechanical. The provisions of the *International Mechanical Code* shall apply to the installation, alterations, repairs and replacement of mechanical systems, including equipment, appliances, fixtures, fittings and/or appurtenances, including ventilating, heating, cooling, air-conditioning and refrigeration systems, incinerators and other energy-related systems.

❖ The *International Mechanical Code®* (IMC®) regulates all aspects of a building's mechanical systems, including ventilating, heating, cooling, air-conditioning and refrigeration systems, incinerators and other energy-related systems and is adopted by reference from this section, as well as Section 2801.1, as the enforceable document for regulating these systems.

101.4.3 Plumbing. The provisions of the *International Plumbing Code* shall apply to the installation, *alteration*, repair and replacement of plumbing systems, including equipment, appliances, fixtures, fittings and appurtenances, and where connected to a water or sewage system and all aspects of a medical gas system. The provisions of the *International Private Sewage Disposal Code* shall apply to private sewage disposal systems.

❖ The *International Plumbing Code®* (IPC®) regulates the components of a building's plumbing system, including water supply and distribution piping; sanitary and storm drainage systems; the fixtures and appliances connected thereto and medical gas and oxygen systems and is adopted by reference from this section, as well as Section 2901.1, as the enforceable document for regulating these systems. The *International Private Sewage Disposal Code®* (IPSDC®) is also adopted as the enforceable document for regulating on-site sewage disposal systems.

101.4.4 Property maintenance. The provisions of the *International Property Maintenance Code* shall apply to existing structures and premises; equipment and facilities; light, ventilation, space heating, sanitation, life and fire safety hazards; responsibilities of owners, operators and occupants; and occupancy of existing premises and structures.

❖ The applicability of the code to existing structures is set forth in Section 101.2 and Chapter 34 and is generally limited to new work or changes in use that occur in these buildings. The IPMC, however, is specifically intended to apply to existing structures and their premises and provides a jurisdiction with an enforceable document for public health, safety and welfare when occupying all buildings, including those that were constructed prior to the adoption of the current building code.

101.4.5 Fire prevention. The provisions of the *International Fire Code* shall apply to matters affecting or relating to structures, processes and premises from the hazard of fire and explosion arising from the storage, handling or use of structures, materials or devices; from conditions hazardous to life, property or public welfare in the occupancy of structures or premises; and from the construction, extension, repair, *alteration* or removal of fire suppression and alarm systems or fire hazards in the structure or on the premises from occupancy or operation.

❖ The IFC contains provisions for safeguarding structures and premises from the hazards of fire and explosion that result from: materials, substances and operations that may be present in a structure; circumstances that endanger life, property or public welfare and the modification or removal of fire suppression and alarm systems. Many of the provisions contained in the code, especially in Chapters 9 and 10, also appear in the IFC. So that all the *International Codes* contain consistent provisions, only one development committee is responsible for considering proposed changes to such provisions and that committee is identified by a letter designation in brackets that appears at the beginning of affected sections. This is described more fully in the preface to the codes. The IFC also contains provisions that are specifically applicable to existing structures and uses and, like the IPMC, provides a jurisdiction with an enforceable document for public health, safety and welfare in all buildings.

101.4.6 Energy. The provisions of the *International Energy Conservation Code* shall apply to all matters governing the design and construction of buildings for energy efficiency.

❖ The *International Energy Conservation Code®* (IECC®) contains provisions for the efficient use of energy in building construction by regulating the design of building envelopes for thermal resistance and low air leakage and the design and selection of mechanical systems for effective use of energy and is adopted by reference in this section, as well as Section 1301.1.1, as the enforceable document for regulating these systems.

SECTION 102
APPLICABILITY

102.1 General. Where there is a conflict between a general requirement and a specific requirement, the specific requirement shall be applicable. Where, in any specific case, different sections of this code specify different materials, methods of construction or other requirements, the most restrictive shall govern.

❖ In cases where the code establishes a specific requirement for a certain condition, that requirement is applicable even if it is less restrictive than a general requirement elsewhere in the code. For instance, specific requirements for certain uses and occupancies are located in Chapter 4 and take precedence over general requirements found in other chapters of the code. As an example, the requirements contained in Section 402.4 for means of egress in a covered mall building would govern over any differing requirements located in Chapter 10, whether the requirements in Section 402.4 are more or less restrictive.

The most restrictive requirement is to apply where there may be different requirements in the code for a specific issue.

102.2 Other laws. The provisions of this code shall not be deemed to nullify any provisions of local, state or federal law.

❖ In some cases, other laws enacted by the jurisdiction or the state or federal government may be applicable to a condition that is also governed by a requirement in the code. In such circumstances, the requirements of the code are in addition to the other law that is still in effect, although the building official may not be responsible for its enforcement.

102.3 Application of references. References to chapter or section numbers, or to provisions not specifically identified by number, shall be construed to refer to such chapter, section or provision of this code.

❖ In a situation where the code may make reference to a chapter or section number or to another code provision without specifically identifying its location in the code, assume that the referenced section, chapter or provision is in the code and not in a referenced code or standard.

102.4 Referenced codes and standards. The codes and standards referenced in this code shall be considered part of the requirements of this code to the prescribed extent of each such reference. Where differences occur between provisions of this code and referenced codes and standards, the provisions of this code shall apply.

❖ A referenced code, standard or portion thereof is an enforceable extension of the code as if the content of the standard were included in the body of the code. For example, Section 905.2 references NFPA 14 in its entirety for the installation of standpipe systems. In those cases when the code references only portions of a standard, the use and application of the referenced standard is limited to those portions that are specifi-

cally identified. For example, Section 412.4.6 requires that aircraft hangars must be provided with fire suppression systems as required in NFPA 409. Section 412.4.6 cannot be construed to require compliance with NFPA 409 in its entirety. It is the intent of the code to be in harmony with the referenced standards. If conflicts occur because of scope or purpose, the code text governs.

102.5 Partial invalidity. In the event that any part or provision of this code is held to be illegal or void, this shall not have the effect of making void or illegal any of the other parts or provisions.

❖ Only invalid sections of the code (as established by the court of jurisdiction) can be set aside. This is essential to safeguard the application of the code text to situations when a provision is declared illegal or unconstitutional. This section preserves the legislative action that put the legal provisions in place.

102.6 Existing structures. The legal occupancy of any structure existing on the date of adoption of this code shall be permitted to continue without change, except as is specifically covered in this code, the *International Property Maintenance Code* or the *International Fire Code*, or as is deemed necessary by the *building official* for the general safety and welfare of the occupants and the public.

❖ An existing structure is generally "grandfathered" to be considered approved with code adoption, provided that the building meets a minimum level of safety. Frequently, the criteria for this level are the regulations (or code) under which the existing building was originally constructed. If there are no previous code criteria to apply, the building official must apply those provisions that are reasonably applicable to existing buildings. A specific level of safety is dictated by provisions dealing with hazard abatement in existing buildings and maintenance provisions, as contained in this code, the IPMC and the IFC. These codes are referenced (see Sections 101.4.4 and 101.4.5) and are applicable to existing buildings. Additionally, Chapter 34 comprehensively identifies the pertinent requirements for existing buildings on which a construction operation is intended or that undergo a change of occupancy.

PART 2—ADMINISTRATION AND ENFORCEMENT

SECTION 103
DEPARTMENT OF BUILDING SAFETY

103.1 Creation of enforcement agency. The Department of Building Safety is hereby created and the official in charge thereof shall be known as the *building official*.

❖ This section creates the building department and describes its composition (see Section 110 for a discussion of the inspection duties of the department). Appendix A contains qualifications for the employees of the building department involved in the enforcement of the code. A jurisdiction can establish the qualifications outlined in Appendix A for its employees by specifically referencing Appendix A in the adopting ordinance.

The executive official in charge of the building department is named the "building official" by this section. In actuality, the person who is in charge of the department may hold a different title, such as building commissioner, building inspector or construction official. For the purpose of the code, that person is referred to as the "building official."

103.2 Appointment. The *building official* shall be appointed by the chief appointing authority of the jurisdiction.

❖ This section establishes the building official as an appointed position of the jurisdiction.

103.3 Deputies. In accordance with the prescribed procedures of this jurisdiction and with the concurrence of the appointing authority, the *building official* shall have the authority to appoint a deputy building official, the related technical officers, inspectors, plan examiners and other employees. Such employees shall have powers as delegated by the *building official*. For the maintenance of existing properties, see the *International Property Maintenance Code*.

❖ This section provides the building official with the authority to appoint other individuals to assist with the administration and enforcement of the code. These individuals would have the authority and responsibility as designated by the building official. Such appointments, however, may be exercised only with the authorization of the chief appointing authority.

SECTION 104
DUTIES AND POWERS OF BUILDING OFFICIAL

104.1 General. The *building official* is hereby authorized and directed to enforce the provisions of this code. The *building official* shall have the authority to render interpretations of this code and to adopt policies and procedures in order to clarify the application of its provisions. Such interpretations, policies and procedures shall be in compliance with the intent and purpose of this code. Such policies and procedures shall not have the effect of waiving requirements specifically provided for in this code.

❖ The duty of the building official is to enforce the code, and he or she is the "authority having jurisdiction" for all matters relating to the code and its enforcement. It is the duty of the building official to interpret the code and to determine compliance. Code compliance will not always be easy to determine and will require judgement and expertise, particularly when enforcing the provisions of Sections 104.10 and 104.11. In exercising this authority, however, the building official cannot set aside or ignore any provision of the code.

104.2 Applications and permits. The *building official* shall receive applications, review *construction documents* and issue *permits* for the erection, and *alteration*, demolition and moving of buildings and structures, inspect the premises for which such *permits* have been issued and enforce compliance with the provisions of this code.

❖ The code enforcement process is normally initiated with an application for a permit. The building official is responsible for processing applications and issuing permits for the construction or modification of buildings in accordance with the code.

104.3 Notices and orders. The *building official* shall issue all necessary notices or orders to ensure compliance with this code.

❖ An important element of code enforcement is the necessary advisement of deficiencies and corrections, which is accomplished through written notices and orders. The building official is required to issue orders to abate illegal or unsafe conditions. Section 116.3 contains additional information for these notices.

104.4 Inspections. The *building official* shall make all of the required inspections, or the *building official* shall have the authority to accept reports of inspection by *approved agencies* or individuals. Reports of such inspections shall be in writing and be certified by a responsible officer of such *approved agency* or by the responsible individual. The *building official* is authorized to engage such expert opinion as deemed necessary to report upon unusual technical issues that arise, subject to the approval of the appointing authority.

❖ The building official is required to make inspections as necessary to determine compliance with the code or to accept written reports of inspections by an approved agency. The inspection of the work in progress or accomplished is another significant element in determining code compliance. While a department does not have the resources to inspect every aspect of all work, the required inspections are those that are dictated by administrative rules and procedures based on many parameters, including available inspection resources. In order to expand the available resources for inspection purposes, the building official may approve an agency that, in his or her opinion, complies with the criteria set forth in Section 1703. When unusual, extraordinary or complex technical issues arise relative to building safety, the building official has the authority to seek the opinion and advice of experts. Since this usually involves the expenditure of funds, the approval of the jurisdiction's chief executive (or similar position) is required. A technical report from an expert requested by the building official can be used to assist in the approval process (also see Section 1704 for special inspection requirements).

104.5 Identification. The *building official* shall carry proper identification when inspecting structures or premises in the performance of duties under this code.

❖ This section requires the building official (including by definition all authorized designees) to carry identification in the course of conducting the duties of the position. This removes any question as to the purpose and authority of the inspector.

104.6 Right of entry. Where it is necessary to make an inspection to enforce the provisions of this code, or where the *building official* has reasonable cause to believe that there exists in a structure or upon a premises a condition which is contrary to or in violation of this code which makes the structure or premises unsafe, dangerous or hazardous, the *building official* is authorized to enter the structure or premises at reasonable times to inspect or to perform the duties imposed by this code, provided that if such structure or premises be occupied that credentials be presented to the occupant and entry requested. If such structure or premises is unoccupied, the *building official* shall first make a reasonable effort to locate the owner or other person having charge or control of the structure or premises and request entry. If entry is refused, the *building official* shall have recourse to the remedies provided by law to secure entry.

❖ The first part of this section establishes the right of the building official to enter the premises in order to make the permit inspections required by Section 110.3. Permit application forms typically include a statement in the certification signed by the applicant (who is the owner or owner's agent) granting the building official the authority to enter areas covered by the permit in order to enforce code provisions related to the permit. The right to enter other structures or premises is more limited. First, to protect the right of privacy, the owner or occupant must grant the building official permission before an interior inspection of the property can be conducted. Permission is not required for inspections that can be accomplished from within the public right-of-way. Second, such access may be denied by the owner or occupant. Unless the inspector has reasonable cause to believe that a violation of the code exists, access may be unattainable. Third, building officials must present proper identification (see Section 104.5) and request admittance during reasonable hours—usually the normal business hours of the establishment—to be admitted. Fourth, inspections must be aimed at securing or determining compliance with the provisions and intent of the regulations that are specifically within the established scope of the building official's authority.

Searches to gather information for the purpose of enforcing the other codes, ordinances or regulations are considered unreasonable and are prohibited by the Fourth Amendment to the U.S. Constitution. "Reasonable cause" in the context of this section must be distinguished from "probable cause," which is required to gain access to property in criminal cases. The burden of proof establishing reasonable cause may vary among jurisdictions. Usually, an inspector must show that the property is subject to inspection under the provisions of the code; that the interests of the public health, safety and welfare outweigh the individual's right to maintain privacy and that such an inspection is required solely to determine compliance with the provisions of the code.

Many jurisdictions do not recognize the concept of an administrative warrant and may require the building official to prove probable cause in order to gain access upon refusal. This burden of proof is usually more sub-stantial, often requiring the building official to stipulate in advance why access is needed (usually access is restricted to gathering evidence for seeking an indictment or making an arrest); what specific items or information is sought; its relevance to the case against the individual subject; how knowledge of the relevance of the information or items sought was obtained and how the evidence sought will be used. In all such cases, the right to privacy must always be weighed against the right of the building official to conduct an inspection to verify that public health, safety and welfare are not in jeopardy. Such important and complex constitutional issues should be discussed with the jurisdiction's legal counsel. Jurisdictions should establish procedures for securing the necessary court orders when an inspection is deemed necessary following a refusal.

104.7 Department records. The *building official* shall keep official records of applications received, *permits* and certificates issued, fees collected, reports of inspections, and notices and orders issued. Such records shall be retained in the official records for the period required for retention of public records.

❖ In keeping with the need for an efficiently conducted business practice, the building official must keep official records pertaining to permit applications, permits, fees collected, inspections, notices and orders issued. Such documentation provides a valuable resource of information if questions arise regarding the department's actions with respect to a building. The code does not require that construction documents be kept after the project is complete. It requires that other documents be kept for the length of time mandated by a jurisdiction's, or its state's, laws or administrative rules for retaining public records.

104.8 Liability. The *building official*, member of the board of appeals or employee charged with the enforcement of this code, while acting for the jurisdiction in good faith and without malice in the discharge of the duties required by this code or other pertinent law or ordinance, shall not thereby be rendered liable personally and is hereby relieved from personal liability for any damage accruing to persons or property as a result of any act or by reason of an act or omission in the discharge of official duties. Any suit instituted against an officer or employee because of an act performed by that officer or employee in the lawful discharge of duties and under the provisions of this code shall be defended by legal representative of the jurisdiction until the final termination of the proceedings. The *building official* or any subordinate shall not be liable for cost in any action, suit or proceeding that is instituted in pursuance of the provisions of this code.

❖ The building official, other department employees and members of the appeals board are not intended to be held liable for those actions performed in accordance with the code in a reasonable and lawful manner. The responsibility of the building official in this regard is subject to local, state and federal laws that may supersede this provision. This section further establishes that building officials (or subordinates) must not be liable for costs in any legal action instituted in response to the performance of lawful duties. These costs are to be borne by the state, county or municipality. The best way to be certain that the building official's action is a "lawful duty" is always to cite the applicable code section on which the enforcement action is based.

104.9 Approved materials and equipment. Materials, equipment and devices *approved* by the *building official* shall be constructed and installed in accordance with such approval.

❖ The code is a compilation of criteria with which materials, equipment, devices and systems must comply to be suitable for a particular application. The building official has a duty to evaluate such materials, equipment, devices and systems for code compliance and, when compliance is determined, approve the same for use. The materials, equipment, devices and systems must be constructed and installed in compliance with, and all conditions and limitations considered as a basis for, that approval. For example, the manufacturer's instructions and recommendations are to be followed if the approval of the material was based even in part on those instructions and recommendations. The approval authority given to the building official is a significant responsibility and is a key to code compliance. The approval process is first technical and then administrative and must be approached as such. For example, if data to determine code compliance are required, such data should be in the form of test reports or engineering analysis and not simply taken from a sales brochure.

104.9.1 Used materials and equipment. The use of used materials which meet the requirements of this code for new materials is permitted. Used equipment and devices shall not be reused unless *approved* by the *building official*.

❖ The code criteria for materials and equipment have changed over the years. Evaluation of testing and materials technology has permitted the development of new criteria that the old materials may not satisfy. As a result, used materials are required to be evaluated in the same manner as new materials. Used materials, equipment and devices must be equivalent to that required by the code if they are to be used again in a new installation.

104.10 Modifications. Wherever there are practical difficulties involved in carrying out the provisions of this code, the *building official* shall have the authority to grant modifications for individual cases, upon application of the owner or owner's representative, provided the *building official* shall first find that special individual reason makes the strict letter of this code impractical and the modification is in compliance with the intent and purpose of this code and that such modification does not lessen health, accessibility, life and fire safety, or structural requirements. The details of action granting modifications shall be recorded and entered in the files of the department of building safety.

❖ The building official may amend or make exceptions to the code as needed where strict compliance is impractical. Only the building official has authority to grant modifications. Consideration of a particular difficulty is

to be based on the application of the owner and a demonstration that the intent of the code is accomplished. This section is not intended to permit setting aside or ignoring a code provision; rather, it is intended to provide acceptance of equivalent protection. Such modifications do not, however, extend to actions that are necessary to correct violations of the code. In other words, a code violation or the expense of correcting one cannot constitute a practical difficulty.

104.11 Alternative materials, design and methods of construction and equipment. The provisions of this code are not intended to prevent the installation of any material or to prohibit any design or method of construction not specifically prescribed by this code, provided that any such alternative has been *approved*. An alternative material, design or method of construction shall be *approved* where the *building official* finds that the proposed design is satisfactory and complies with the intent of the provisions of this code, and that the material, method or work offered is, for the purpose intended, at least the equivalent of that prescribed in this code in quality, strength, effectiveness, *fire resistance*, durability and safety.

❖ The code is not intended to inhibit innovative ideas or technological advances. A comprehensive regulatory document, such as a building code, cannot envision and then address all future innovations in the industry. As a result, a performance code must be applicable to and provide a basis for the approval of an increasing number of newly developed, innovative materials, systems and methods for which no code text or referenced standards yet exist. The fact that a material, product or method of construction is not addressed in the code is not an indication that such material, product or method is intended to be prohibited. The building official is expected to apply sound technical judgement in accepting materials, systems or methods that, while not anticipated by the drafters of the current code text, can be demonstrated to offer equivalent performance. By virtue of its text, the code regulates new and innovative construction practices while addressing the relative safety of building occupants. The building official is responsible for determining if a requested alternative provides the equivalent level of protection of public health, safety and welfare as required by the code.

104.11.1 Research reports. Supporting data, where necessary to assist in the approval of materials or assemblies not specifically provided for in this code, shall consist of valid research reports from *approved* sources.

❖ When an alternative material or method is proposed for construction, it is incumbent upon the building official to determine whether this alternative is, in fact, an equivalent to the methods prescribed by the code. Reports providing evidence of this equivalency are required to be supplied by an approved source, meaning a source that the building official finds to be reliable and accurate. The ICC Evaluation Service is an example of an agency that provides research reports for alternative materials and methods.

104.11.2 Tests. Whenever there is insufficient evidence of compliance with the provisions of this code, or evidence that a material or method does not conform to the requirements of this code, or in order to substantiate claims for alternative materials or methods, the *building official* shall have the authority to require tests as evidence of compliance to be made at no expense to the jurisdiction. Test methods shall be as specified in this code or by other recognized test standards. In the absence of recognized and accepted test methods, the *building official* shall approve the testing procedures. Tests shall be performed by an *approved agency*. Reports of such tests shall be retained by the *building official* for the period required for retention of public records.

❖ To provide the basis on which the building official can make a decision regarding an alternative material or method, sufficient technical data, test reports and documentation must be provided for evaluation. If evidence satisfactory to the building official indicates that the alternative material or construction method is equivalent to that required by the code, he or she may approve it. Any such approval cannot have the effect of waiving any requirements of the code. The burden of proof of equivalence lies with the applicant who proposes the use of alternative materials or methods.

The building official must require the submission of any appropriate information and data to assist in the determination of equivalency. This information must be submitted before a permit can be issued. The type of information required includes test data in accordance with referenced standards, evidence of compliance with the referenced standard specifications and design calculations. A research report issued by an authoritative agency is particularly useful in providing the building official with the technical basis for evaluation and approval of new and innovative materials and methods of construction. The use of authoritative research reports can greatly assist the building official by reducing the time-consuming engineering analysis necessary to review these materials and methods. Failure to substantiate adequately a request for the use of an alternative is a valid reason for the building official to deny a request. Any tests submitted in support of an application must have been performed by an agency approved by the building official based on evidence that the agency has the technical expertise, test equipment and quality assurance to properly conduct and report the necessary testing. The test reports submitted to the building official must be retained in accordance with the requirements of Section 104.7.

SECTION 105
PERMITS

105.1 Required. Any owner or authorized agent who intends to construct, enlarge, alter, repair, move, demolish, or change the occupancy of a building or structure, or to erect, install, enlarge, alter, repair, remove, convert or replace any electrical, gas, mechanical or plumbing system, the installation of which is regulated by this code, or to cause any such work to be done, shall first make application to the *building official* and obtain the required *permit*.

❖ This section contains the administrative rules governing the issuance, suspension, revocation or modification of building permits. It also establishes how and by whom the application for a building permit is to be made, how it is to be processed, fees and what information it must contain or have attached to it.

In general, a permit is required for all activities that are regulated by the code or its referenced codes (see Section 101.4), and these activities cannot begin until the permit is issued, unless the activity is specifically exempted by Section 105.2. Only the owner or a person authorized by the owner can apply for the permit. Note that this section indicates a need for a permit for a change in occupancy, even if no work is contemplated. Although the occupancy of a building or portion thereof may change and the new activity is still classified in the same group, different code provisions may be applicable. The means of egress, structural loads and light and ventilation provisions are examples of requirements that are occupancy sensitive. The purpose of the permit is to cause the work to be reviewed, approved and inspected to determine compliance with the code.

105.1.1 Annual permit. In lieu of an individual *permit* for each *alteration* to an already *approved* electrical, gas, mechanical or plumbing installation, the *building official* is authorized to issue an annual *permit* upon application therefor to any person, firm or corporation regularly employing one or more qualified tradepersons in the building, structure or on the premises owned or operated by the applicant for the *permit*.

❖ In some instances, such as large buildings or industrial facilities, the repair, replacement or alteration of electrical, gas, mechanical or plumbing systems occurs on a frequent basis, and this section allows the building official to issue an annual permit for this work. This relieves both the building department and the owners of such facilities from the burden of filing and processing individual applications for this activity; however, there are restrictions on who is entitled to these permits. They can be issued only for work on a previously approved installation and only to an individual or corporation that employs persons specifically qualified in the trade for which the permit is issued. If tradespeople who perform the work involved are required to be licensed in the jurisdiction, then only those persons would be permitted to perform the work. If trade licensing is not required, then the building official needs to review and approve the qualifications of the persons who will be performing the work. The annual permit can apply only to the individual property that is owned or operated by the applicant.

105.1.2 Annual permit records. The person to whom an annual *permit* is issued shall keep a detailed record of *alterations* made under such annual *permit*. The *building official* shall have access to such records at all times or such records shall be filed with the *building official* as designated.

❖ The work performed in accordance with an annual permit must be inspected by the building official, so it is necessary to know the location of such work and when it was performed. This can be accomplished by having records of the work available to the building official either at the premises or in the official's office, as determined by the official.

105.2 Work exempt from permit. Exemptions from *permit* requirements of this code shall not be deemed to grant authorization for any work to be done in any manner in violation of the provisions of this code or any other laws or ordinances of this jurisdiction. *Permits* shall not be required for the following:

Building:

1. One-story detached accessory structures used as tool and storage sheds, playhouses and similar uses, provided the floor area does not exceed 120 square feet (11 m²).

2. Fences not over 6 feet (1829 mm) high.

3. Oil derricks.

4. Retaining walls that are not over 4 feet (1219 mm) in height measured from the bottom of the footing to the top of the wall, unless supporting a surcharge or impounding Class I, II or IIIA liquids.

5. Water tanks supported directly on grade if the capacity does not exceed 5,000 gallons (18 925 L) and the ratio of height to diameter or width does not exceed 2:1.

6. Sidewalks and driveways not more than 30 inches (762 mm) above adjacent grade, and not over any basement or *story* below and are not part of an *accessible route*.

7. Painting, papering, tiling, carpeting, cabinets, counter tops and similar finish work.

8. Temporary motion picture, television and theater stage sets and scenery.

9. Prefabricated swimming pools accessory to a Group R-3 occupancy that are less than 24 inches (610 mm) deep, do not exceed 5,000 gallons (18 925 L) and are installed entirely above ground.

10. Shade cloth structures constructed for nursery or agricultural purposes, not including service systems.

11. Swings and other playground equipment accessory to detached one- and two-family *dwellings*.

12. Window *awnings* supported by an *exterior wall* that do not project more than 54 inches (1372 mm) from the *exterior wall* and do not require additional support of Groups R-3 and U occupancies.

13. Nonfixed and movable fixtures, cases, racks, counters and partitions not over 5 feet 9 inches (1753 mm) in height.

Electrical:

Repairs and maintenance: Minor repair work, including the replacement of lamps or the connection of *approved* portable electrical equipment to *approved* permanently installed receptacles.

Radio and television transmitting stations: The provisions of this code shall not apply to electrical equipment used for radio and television transmissions, but do apply to equipment and wiring for a power supply and the installations of towers and antennas.

Temporary testing systems: A *permit* shall not be required for the installation of any temporary system required for the testing or servicing of electrical equipment or apparatus.

Gas:

1. Portable heating appliance.

2. Replacement of any minor part that does not alter approval of equipment or make such equipment unsafe.

Mechanical:

1. Portable heating appliance.

2. Portable ventilation equipment.

3. Portable cooling unit.

4. Steam, hot or chilled water piping within any heating or cooling equipment regulated by this code.

5. Replacement of any part that does not alter its approval or make it unsafe.

6. Portable evaporative cooler.

7. Self-contained refrigeration system containing 10 pounds (5 kg) or less of refrigerant and actuated by motors of 1 horsepower (746 W) or less.

Plumbing:

1. The stopping of leaks in drains, water, soil, waste or vent pipe, provided, however, that if any concealed trap, drain pipe, water, soil, waste or vent pipe becomes defective and it becomes necessary to remove and replace the same with new material, such work shall be considered as new work and a *permit* shall be obtained and inspection made as provided in this code.

2. The clearing of stoppages or the repairing of leaks in pipes, valves or fixtures and the removal and reinstallation of water closets, provided such repairs do not involve or require the replacement or rearrangement of valves, pipes or fixtures.

❖ Section 105.1 essentially requires a permit for any activity involving work on a building and its systems and other structures. This section lists those activities that are permitted to take place without first obtaining a permit from the building department. Note that in some cases, such as Items 9, 10, 11 and 12, the work is exempt only for certain occupancies. It is further the intent of the code that even though work may be exempted for permit purposes, it must still comply with the code and the owner is responsible for proper and safe construction for all work being done. Work exempted by the codes adopted by reference in Section 101.4 is also included here.

105.2.1 Emergency repairs. Where equipment replacements and repairs must be performed in an emergency situation, the *permit* application shall be submitted within the next working business day to the *building official*.

❖ This section recognizes that in some cases, emergency replacement and repair work must be done as quickly as possible, so it is not practical to take the necessary time to apply for and obtain approval. A permit for the work must be obtained the next day that the building department is open for business. Any work performed before the permit is issued must be done in accordance with the code and corrected if not approved by the building official.

105.2.2 Repairs. Application or notice to the *building official* is not required for ordinary repairs to structures, replacement of lamps or the connection of *approved* portable electrical equipment to *approved* permanently installed receptacles. Such repairs shall not include the cutting away of any wall, partition or portion thereof, the removal or cutting of any structural beam or load-bearing support, or the removal or change of any required *means of egress*, or rearrangement of parts of a structure affecting the egress requirements; nor shall ordinary repairs include *addition* to, *alteration* of, replacement or relocation of any standpipe, water supply, sewer, drainage, drain leader, gas, soil, waste, vent or similar piping, electric wiring or mechanical or other work affecting public health or general safety.

❖ This section distinguishes between what might be termed by some as repairs but are in fact alterations, wherein the code is to be applicable, and ordinary repairs, which are maintenance activities that do not require a permit.

105.2.3 Public service agencies. A *permit* shall not be required for the installation, *alteration* or repair of generation, transmission, distribution or metering or other related equipment that is under the ownership and control of public service agencies by established right.

❖ Utilities that supply electricity, gas, water, telephone, television cable, etc., do not require permits for work involving the transmission lines and metering equipment that they own and control; that is, to their point of delivery. Utilities are typically regulated by other laws that give them specific rights and authority in this area. Any equipment or appliances installed or serviced by such agencies that are not owned by them and under their full control are not exempt from a permit.

105.3 Application for permit. To obtain a *permit*, the applicant shall first file an application therefor in writing on a form furnished by the department of building safety for that purpose. Such application shall:

1. Identify and describe the work to be covered by the *permit* for which application is made.

2. Describe the land on which the proposed work is to be done by legal description, street address or similar description that will readily identify and definitely locate the proposed building or work.

3. Indicate the use and occupancy for which the proposed work is intended.

4. Be accompanied by *construction documents* and other information as required in Section 107.

5. State the valuation of the proposed work.

6. Be signed by the applicant, or the applicant's authorized agent.

7. Give such other data and information as required by the *building official*.

❖ This section requires that a written application for a permit be filed on forms provided by the building department and details the information required on the application. Permit forms will typically have sufficient space to write a very brief description of the work to be accomplished, which is sufficient for only small jobs. For larger projects, the description will be augmented by construction documents as indicated in Item 4. As required by Section 105.1, the applicant must be the owner of the property or an authorized agent of the owner, such as an engineer, architect, contractor, tenant or other. The applicant must sign the application, and permit forms typically include a statement that if the applicant is not the owner, he or she has permission from the owner to make the application.

105.3.1 Action on application. The *building official* shall examine or cause to be examined applications for *permits* and amendments thereto within a reasonable time after filing. If the application or the *construction documents* do not conform to the requirements of pertinent laws, the *building official* shall reject such application in writing, stating the reasons therefor. If the *building official* is satisfied that the proposed work conforms to the requirements of this code and laws and ordinances applicable thereto, the *building official* shall issue a *permit* therefor as soon as practicable.

❖ This section requires the building official to act with reasonable speed on a permit application. In some instances, this time period is set by state or local law. The building official must refuse to issue a permit when the application and accompanying documents do not conform to the code. In order to ensure effective communication and due process of law, the reasons for denial of an application for a permit are required to be in writing. Once the building official determines that the work described conforms with the code and other applicable laws, the permit must be issued upon payment of the fees required by Section 109.

105.3.2 Time limitation of application. An application for a *permit* for any proposed work shall be deemed to have been abandoned 180 days after the date of filing, unless such application has been pursued in good faith or a *permit* has been issued; except that the *building official* is authorized to grant one or more extensions of time for additional periods not exceeding 90 days each. The extension shall be requested in writing and justifiable cause demonstrated.

❖ Typically, an application for a permit is submitted and goes through a review process that ends with the issuance of a permit. If a permit has not been issued 180 days after the date of filing, however, the application is considered abandoned, unless the applicant was diligent in efforts to obtain the permit. The building official has the authority to extend this time limitation (in increments of 90 days), provided there is reasonable cause. This would cover delays beyond the applicant's control, such as prerequisite permits or approvals from other authorities within the jurisdiction or state. The intent of this section is to limit the time between the review process and the issuance of a permit.

105.4 Validity of permit. The issuance or granting of a *permit* shall not be construed to be a *permit* for, or an approval of, any violation of any of the provisions of this code or of any other ordinance of the jurisdiction. *Permits* presuming to give authority to violate or cancel the provisions of this code or other ordinances of the jurisdiction shall not be valid. The issuance of a *permit* based on *construction documents* and other data shall not prevent the *building official* from requiring the correction of errors in the *construction documents* and other data. The *building official* is also authorized to prevent occupancy or use of a structure where in violation of this code or of any other ordinances of this jurisdiction.

❖ This section states the fundamental premise that the permit is only a license to proceed with the work. It is not a license to violate, cancel or set aside any provisions of the code. This is significant because it means that despite any errors or oversights in the approval process, the permit applicant, not the building official, is responsible for code compliance. Also, the permit can be suspended or revoked in accordance with Section 105.6.

105.5 Expiration. Every *permit* issued shall become invalid unless the work on the site authorized by such *permit* is commenced within 180 days after its issuance, or if the work authorized on the site by such *permit* is suspended or abandoned for a period of 180 days after the time the work is commenced. The *building official* is authorized to grant, in writing, one or more extensions of time, for periods not more than 180 days each. The extension shall be requested in writing and justifiable cause demonstrated.

❖ The permit becomes invalid under two distinct situations—both based on a 180-day period. The first situation is when no work was initiated 180 days from issuance of a permit. The second situation is when the authorized work has stopped for 180 days. The person who was issued the permit should be notified, in writing, that the permit is invalid and what steps must be taken to reinstate it and restart the work. The building official has the authority to extend this time limitation (in increments of 180 days), provided the extension is requested in writing and there is reasonable cause, which typically includes events beyond the permit holder's control.

105.6 Suspension or revocation. The *building official* is authorized to suspend or revoke a *permit* issued under the provisions of this code wherever the *permit* is issued in error or on the basis of incorrect, inaccurate or incomplete information, or in violation of any ordinance or regulation or any of the provisions of this code.

❖ A permit is a license to proceed with the work. The building official, however, can suspend or revoke permits shown to be based, all or in part, on any false statement or misrepresentation of fact. A permit can also be suspended or revoked if it was issued in error, such as an omitted prerequisite approval or code violation indicated on the construction documents. An applicant may subsequently apply for a reinstatement of the permit with the appropriate corrections or modifications made to the application and construction documents.

105.7 Placement of permit. The building *permit* or copy shall be kept on the site of the work until the completion of the project.

❖ The permit, or copy thereof, is to be kept on the job site until the work is complete and made available to the building official or representative to conveniently make required entries thereon.

SECTION 106
FLOOR AND ROOF DESIGN LOADS

106.1 Live loads posted. Where the live loads for which each floor or portion thereof of a commercial or industrial building is or has been designed to exceed 50 psf (2.40 kN/m²), such design live loads shall be conspicuously posted by the owner in that part of each *story* in which they apply, using durable signs. It shall be unlawful to remove or deface such notices.

❖ This section requires that live loads be posted for most occupancies, since many of the live loads specified in Table 1607.1 exceed 50 psf (2.40 kN/m²). Where part of the floor is designed for 50 psf (2.40 kN/m²) or less and part for more than 50 psf (2.40 kN/m²), the live loads are required to be posted for those portions more than 50 psf (2.40 kN/m²). The code requires that the posting be done in the part where it applies. For example, an assembly area such as a restaurant would need to have the live load posted in the dining room.

This live load posting gives the building department easy access to the information for field verification. It also serves as a notice of the loading restriction that is stated in Section 106.3.

106.2 Issuance of certificate of occupancy. A certificate of occupancy required by Section 111 shall not be issued until the floor load signs, required by Section 106.1, have been installed.

❖ The design live load signs required by Section 106.1 need to be in place prior to the occupancy of the building for reference purposes. They serve as a record of the structural design loads for future reference, particularly when a change in occupancy is contemplated.

106.3 Restrictions on loading. It shall be unlawful to place, or cause or permit to be placed, on any floor or roof of a building, structure or portion thereof, a load greater than is permitted by this code.

❖ The loads that this section is referring to are the various structural loads specified in Chapter 16. For example, Table 1607.1 includes the minimum live loads for building design. Unless the building is designed for higher loads than specified in Table 1607.1, those values are not to be exceeded. Note that the loads in Table 1607.1 are minimum live loads. A building is permitted to be designed for higher loads, in which case the higher loads would be the limit of the actual applied loads.

SECTION 107
SUBMITTAL DOCUMENTS

107.1 General. Submittal documents consisting of *construction documents*, statement of *special inspections*, geotechnical report and other data shall be submitted in two or more sets with each *permit* application. The *construction documents* shall be prepared by a *registered design professional* where required by the statutes of the jurisdiction in which the project is to be constructed. Where special conditions exist, the *building official* is authorized to require additional *construction documents* to be prepared by a *registered design professional*.

> **Exception:** The *building official* is authorized to waive the submission of *construction documents* and other data not required to be prepared by a *registered design professional* if it is found that the nature of the work applied for is such that review of *construction documents* is not necessary to obtain compliance with this code.

❖ This section establishes the requirement to provide the building official with construction drawings, specifications and other documents that describe the structure or system for which a permit is sought (see Section 202 for a complete definition). It describes the information that must be included in the documents, who must prepare them and procedures for approving them.

A detailed description of the work for which an application is made must be submitted. When the work can be briefly described on the application form and the services of a registered design professional are not required, the building official may utilize judgement in determining the need for detailed documents. An example of work that may not involve the submission of detailed construction documents is the replacement of an existing 60-amp electrical service with a 200-amp

service. Other sections of the code also contain specific requirements for construction documents, such as Sections 1603, 1901.4, 2101.3 and 3103.2. These provisions are intended to reflect the minimum scope of information needed to determine code compliance. Although this section specifies that "one or more" sets of construction documents be submitted, note that Section 106.3.1 requires one set of approved documents to be retained by the building official and one set to be returned to the applicant, essentially requiring at least two sets of construction documents. The building official should establish a consistent policy of the number of sets required by the jurisdiction and make this information readily available to applicants.

This section also requires the building official to determine that any state professional registration laws be complied with as they apply to the preparation of construction documents.

107.2 Construction documents. *Construction documents* shall be in accordance with Sections 107.2.1 through 107.2.5.

❖ This section provides instructions regarding the information and form of construction documents.

107.2.1 Information on construction documents. *Construction documents* shall be dimensioned and drawn upon suitable material. Electronic media documents are permitted to be submitted when *approved* by the *building official*. *Construction documents* shall be of sufficient clarity to indicate the location, nature and extent of the work proposed and show in detail that it will conform to the provisions of this code and relevant laws, ordinances, rules and regulations, as determined by the *building official*.

❖ The construction documents are required to be of a quality and detail such that the building official can determine that the work conforms to the code and other applicable laws and regulations. General statements on the documents, such as "all work must comply with the *International Building Code*," are not an acceptable substitute for showing the required information. The following subsections and sections in other chapters indicated in the commentary to Sections 107.2.2 through 107.2.5 specify the detailed information that must be shown on the submitted documents. When specifically allowed by the building official, documents can be submitted in electronic form.

107.2.2 Fire protection system shop drawings. Shop drawings for the *fire protection system(s)* shall be submitted to indicate conformance to this code and the *construction documents* and shall be *approved* prior to the start of system installation. Shop drawings shall contain all information as required by the referenced installation standards in Chapter 9.

❖ Since the fire protection contractor(s) may not have been selected at the time a permit is issued for construction of a building, detailed shop drawings for fire protection systems are not available. Because they provide the information necessary to determine code compliance, as specified in the appropriate referenced standard in Chapter 9, they must be submitted and approved by the building official before the contractor can begin installing the system. For example, the professional responsible for the design of an automatic sprinkler system should determine that the water supply is adequate, but will not be able to prepare a final set of hydraulic calculations if the specific materials and pipe sizes, lengths and arrangements have not been identified. Once the installing contractor is selected, specific hydraulic calculations can be prepared. Factors, such as classification of the hazard, amount of water supply available and the density or concentration to be achieved by the system, are to be included with the submission of the shop drawings. Specific data sheets identifying sprinklers, pipe dimensions, power requirements for smoke detectors, etc., should also be included with the submission.

107.2.3 Means of egress. The *construction documents* shall show in sufficient detail the location, construction, size and character of all portions of the *means of egress* in compliance with the provisions of this code. In other than occupancies in Groups R-2, R-3, and I-1, the *construction documents* shall designate the number of occupants to be accommodated on every floor, and in all rooms and spaces.

❖ The complete means of egress system is required to be indicated on the plans to permit the building official to initiate a review and identify pertinent code requirements for each component. Additionally, requiring such information to be reflected in the construction documents requires the designer not only to become familiar with the code, but also to be aware of egress principles, concepts and purposes. The need to ensure that the means of egress leads to a public way is also a consideration during the plan review. Such an evaluation cannot be made without the inclusion of a site plan, as required by Section 107.2.5.

Information essential for determining the required capacity of the egress components (see Section 1005) and the number of egress components required from a space (see Sections 1014.1 and 1018.1) must be provided. The designer must be aware of the occupancy of a space and properly identify that information, along with its resultant occupant load, on the construction documents. In occupancies in Groups I-1, R-2 and R-3, the occupant load can be readily determined with little difference in the number so that the designation of the occupant load on the construction documents is not required.

107.2.4 Exterior wall envelope. *Construction documents* for all buildings shall describe the *exterior wall envelope* in sufficient detail to determine compliance with this code. The *construction documents* shall provide details of the *exterior wall envelope* as required, including flashing, intersections with dissimilar materials, corners, end details, control joints, intersections at roof, eaves or parapets, means of drainage, water-resistive membrane and details around openings.

The *construction documents* shall include manufacturer's installation instructions that provide supporting documentation that the proposed penetration and opening details described in the *construction documents* maintain the weather resistance of the *exterior wall envelope*. The supporting documentation shall fully describe the *exterior wall* system which was tested, where applicable, as well as the test procedure used.

❖ This section specifically identifies details of exterior wall construction that are critical to the weather resistance of the wall and requires those details to be provided on the construction documents. Where the weather resistance of the exterior wall assembly is based on tests, the submitted documentation is to describe the details of the wall envelope and the test procedure that was used. This provides the building official with the information necessary to determine code compliance.

107.2.5 Site plan. The *construction documents* submitted with the application for *permit* shall be accompanied by a site plan showing to scale the size and location of new construction and existing structures on the site, distances from *lot lines*, the established street grades and the proposed finished grades and, as applicable, flood hazard areas, floodways, and *design flood* elevations; and it shall be drawn in accordance with an accurate boundary line survey. In the case of demolition, the site plan shall show construction to be demolished and the location and size of existing structures and construction that are to remain on the site or plot. The *building official* is authorized to waive or modify the requirement for a site plan when the application for *permit* is for *alteration* or repair or when otherwise warranted.

❖ Certain code requirements are dependent on the structure's location on the lot (see Sections 506.2, 507, 705, 1025 and 1206) and the topography of the site (see Sections 1104, 1107.7.4 and 1804.3). As a result, a scaled site plan containing the data listed in this section is required to permit review for compliance. The building official can waive the requirement for a site plan when it is not required to determine code compliance, such as work involving only interior alterations or repairs.

107.2.5.1 Design flood elevations. Where *design flood* elevations are not specified, they shall be established in accordance with Section 1612.3.1.

❖ A large percentage of areas that are mapped as special flood hazard areas by the National Flood Insurance Program (NFIP) do not have either flood elevations or floodway designations (floodways are areas along riverine bodies of water that convey the bulk of floodwaters). Section 1612.3 gives the authority to the code official to require use of data which may be obtained from other sources, or to require the applicant to develop flood hazard data.

107.3 Examination of documents. The *building official* shall examine or cause to be examined the accompanying submittal documents and shall ascertain by such examinations whether the construction indicated and described is in accordance with the requirements of this code and other pertinent laws or ordinances.

❖ The requirements of this section are related to those found in Section 105.3.1 regarding the action of the building official in response to a permit application. The building official can delegate review of the construction documents to subordinates as provided for in Section 103.3.

107.3.1 Approval of construction documents. When the *building official* issues a *permit*, the *construction documents* shall be *approved*, in writing or by stamp, as "Reviewed for Code Compliance." One set of *construction documents* so reviewed shall be retained by the *building official*. The other set shall be returned to the applicant, shall be kept at the site of work and shall be open to inspection by the *building official* or a duly authorized representative.

❖ The building official must stamp or otherwise endorse as "Reviewed for Code Compliance" the construction documents on which the permit is based. One set of approved construction documents must be kept on the construction site to serve as the basis for all subsequent inspections. To avoid confusion, the construction documents on the site must be the documents that were approved and stamped. This is because inspections are to be performed with regard to the approved documents, not the code itself. Additionally, the contractor cannot determine compliance with the approved construction documents unless they are readily available. If the approved construction documents are not available, the inspection should be postponed and work on the project halted.

107.3.2 Previous approvals. This code shall not require changes in the *construction documents*, construction or designated occupancy of a structure for which a lawful *permit* has been heretofore issued or otherwise lawfully authorized, and the construction of which has been pursued in good faith within 180 days after the effective date of this code and has not been abandoned.

❖ If a permit is issued and construction proceeds at a normal pace and a new edition of the code is adopted by the legislative body, requiring that the building be constructed to conform to the new code is unreasonable. This section provides for the continuity of permits issued under previous codes, as long as such permits are being "actively prosecuted" subsequent to the effective date of the ordinance adopting this edition of the code.

107.3.3 Phased approval. The *building official* is authorized to issue a *permit* for the construction of foundations or any other part of a building or structure before the *construction documents* for the whole building or structure have been submitted, provided that adequate information and detailed statements have been filed complying with pertinent requirements of this code. The holder of such *permit* for the foundation or other parts of a building or structure shall proceed at the holder's own risk with the building operation and without assurance that a *permit* for the entire structure will be granted.

❖ The building official has the authority to issue a partial permit to allow for the practice of "fast tracking" a job. Any construction under a partial permit is "at the holder's own risk" and "without assurance that a permit for the entire structure will be granted." The building official is under no obligation to accept work or issue a complete permit in violation of the code,

ordinances or statutes simply because a partial permit had been issued. Fast tracking puts an unusual administrative and technical burden on the building official. The purpose is to proceed with construction while the design continues for other aspects of the work. Coordinating and correlating the code aspects into the project in phases requires attention to detail and project tracking so that all code issues are addressed. The coordination of these submittals is the responsibility of the registered design professional in responsible charge described in Section 107.3.4.

107.3.4 Design professional in responsible charge.

107.3.4.1 General. When it is required that documents be prepared by a *registered design professional*, the *building official* shall be authorized to require the owner to engage and designate on the building *permit* application a *registered design professional* who shall act as the *registered design professional in responsible charge*. If the circumstances require, the owner shall designate a substitute *registered design professional in responsible charge* who shall perform the duties required of the original *registered design professional in responsible charge*. The *building official* shall be notified in writing by the owner if the *registered design professional in responsible charge* is changed or is unable to continue to perform the duties.

The *registered design professional in responsible charge* shall be responsible for reviewing and coordinating submittal documents prepared by others, including phased and deferred submittal items, for compatibility with the design of the building.

❖ At the time of permit application and at various intervals during a project, the code requires detailed technical information to be submitted to the building official. This will vary depending on the complexity of the project, but typically includes the construction documents with supporting information, applications utilizing the phased approval procedure in Section 107.3.3 and reports from engineers, inspectors and testing agencies required in Chapter 17. Since these documents and reports are prepared by numerous individuals, firms and agencies, it is necessary to have a single person charged with responsibility for coordinating their submittal to the building official. This person is the point of contact for the building official for all information relating to the project. Otherwise, the building official could waste time and effort attempting to locate the source of accurate information when trying to resolve an issue such as a discrepancy in plans submitted by different designers. The requirement that the owner engage a person to act as the design professional in responsible charge is applicable to projects where the construction documents are required by law to be prepared by a registered design professional (see Section 107.1) and when required by the building official. The person employed by the owner to act as the design professional in responsible charge must be identified on the permit application, but the owner can change the designated person at any time during the course of the review process or work, provided the building official is so notified in writing.

107.3.4.2 Deferred submittals. For the purposes of this section, deferred submittals are defined as those portions of the design that are not submitted at the time of the application and that are to be submitted to the *building official* within a specified period.

Deferral of any submittal items shall have the prior approval of the *building official*. The *registered design professional in responsible charge* shall list the deferred submittals on the *construction documents* for review by the *building official*.

Documents for deferred submittal items shall be submitted to the *registered design professional in responsible charge* who shall review them and forward them to the *building official* with a notation indicating that the deferred submittal documents have been reviewed and found to be in general conformance to the design of the building. The deferred submittal items shall not be installed until the deferred submittal documents have been *approved* by the *building official*.

❖ Often, especially on larger projects, details of certain building parts are not available at the time of permit issuance because they have not yet been designed; for example, exterior cladding, prefabricated items such as trusses and stairs and the components of fire protection systems (see Section 107.2.2). The design professional in responsible charge must identify on the construction documents the items to be included in any deferred submittals. Documents required for the approval of deferred items must be reviewed by the design professional in responsible charge for compatibility with the design of the building, forwarded to the building official with a notation that this is the case and approved by the building official before installation of the items. Sufficient time must be allowed for the approval process. Note that deferred submittals differ from the phased permits described in Section 107.3.3 in that they occur after the permit for the building is issued and are not for work covered by separate permits.

107.4 Amended construction documents. Work shall be installed in accordance with the *approved construction documents*, and any changes made during construction that are not in compliance with the *approved construction documents* shall be resubmitted for approval as an amended set of *construction documents*.

❖ Any amendments to the approved construction documents must be filed before constructing the amended item. In the broadest sense, amendments include all addenda, change orders, revised drawings and marked-up shop drawings. Building officials should maintain a policy that all amendments be submitted for review. Otherwise, a significant amendment may not be submitted because of misinterpretation, resulting in an activity that is not approved and that causes a needless delay in obtaining approval of the finished work.

107.5 Retention of construction documents. One set of *approved construction documents* shall be retained by the *building official* for a period of not less than 180 days from date of completion of the permitted work, or as required by state or local laws.

❖ A set of the approved construction documents must be kept by the building official as may be required by state or local laws, but for a period of no less than 180 days after the work is complete. Questions regarding an item shown on the approved documents may arise in the period immediately following completion of the work and the documents should be available for review. See Section 104.7 for requirements to retain other records that are generated as a result of the work.

SECTION 108
TEMPORARY STRUCTURES AND USES

108.1 General. The *building official* is authorized to issue a *permit* for temporary structures and temporary uses. Such *permits* shall be limited as to time of service, but shall not be permitted for more than 180 days. The *building official* is authorized to grant extensions for demonstrated cause.

❖ In the course of construction or other activities, structures that have a limited service life are often necessary. This section contains the administrative provisions that permit such temporary structures without full compliance with the code requirements for permanently occupied structures. This section should not be confused with the scope of Section 3103, which regulates temporary structures larger than 120 square feet (11 m²) in area.

This section allows the building official to issue permits for temporary structures or uses. The applicant must specify the time period desired for the temporary structure or use, but the approval period cannot exceed 180 days. Structures or uses that are temporary but are anticipated to be in existence for more than 180 days are required to conform to code requirements for permanent structures and uses. The section also authorizes the building official to grant extensions to this time period if the applicant can provide a valid reason for the extension, which typically includes circumstances beyond the applicant's control. This provision is not intended to be used to circumvent the 180-day limitation.

108.2 Conformance. Temporary structures and uses shall conform to the structural strength, fire safety, *means of egress*, accessibility, light, ventilation and sanitary requirements of this code as necessary to ensure public health, safety and general welfare.

❖ This section prescribes those categories of the code that must be complied with, despite the fact that the structure will be removed or the use discontinued some time in the future. These criteria are essential for measuring the safety of any structure or use, temporary or permanent; therefore, the application of these criteria to a temporary structure cannot be waived.

"Structural strength" refers to the ability of the temporary structure to resist anticipated live, environmental and dead loads (see Chapter 16). It also applies to anticipated live and dead loads imposed by a temporary use in an existing structure.

"Fire safety" provisions are those required by Chapters 7, 8 and 9 invoked by virtue of the structure's size, use or location on the property.

"Means of egress" refers to full compliance with Chapter 10.

"Accessibility" refers to full compliance with Chapter 11 for making buildings accessible to physically disabled persons, a requirement that is repeated in Section 1103.1.

"Light, ventilation and sanitary" requirements are those imposed by Chapter 12 of the code or applicable sections of the IPC or IMC.

108.3 Temporary power. The *building official* is authorized to give permission to temporarily supply and use power in part of an electric installation before such installation has been fully completed and the final certificate of completion has been issued. The part covered by the temporary certificate shall comply with the requirements specified for temporary lighting, heat or power in NFPA 70.

❖ Commonly, the electrical service on most construction sites is installed and energized long before all of the wiring is completed. This procedure allows the power supply to be increased as construction demands; however, temporary permission is not intended to waive the requirements set forth in NFPA 70. Construction power from the permanent wiring of the building does not require the installation of temporary ground-fault circuit-interrupter (GFCI) protection or the assured equipment grounding program, because the building wiring installed as required by the code should be as safe for use during construction as it would be for use after completion of the building.

108.4 Termination of approval. The *building official* is authorized to terminate such *permit* for a temporary structure or use and to order the temporary structure or use to be discontinued.

❖ This section provides the building official with the necessary authority to terminate the permit for a temporary structure or use. The building official can order that a temporary structure be removed or a temporary use be discontinued if conditions of the permit have been violated or the structure or use poses an imminent hazard to the public, in which case the provisions of Section 116 become applicable. This text is important because it allows the building official to act quickly when time is of the essence in order to protect public health, safety and welfare.

SECTION 109
FEES

109.1 Payment of fees. A *permit* shall not be valid until the fees prescribed by law have been paid, nor shall an amendment to a *permit* be released until the additional fee, if any, has been paid.

❖ The code anticipates that jurisdictions will establish their own fee schedules. It is the intent that the fees collected by the department for building permit issuance, plan review and inspection be adequate to cover the costs to the department in these areas. If the department has additional duties, then its budget will need to be supplemented from the general fund. This section requires that all fees be paid prior to permit issuance or release of an amendment to a permit. Since department operations are intended to be supported by fees paid by the user of department activities, it is important that these fees are received before incurring any expense. This philosophy has resulted in some departments having fees paid prior to the performance of two areas of work: plan review and inspection.

109.2 Schedule of permit fees. On buildings, structures, electrical, gas, mechanical, and plumbing systems or *alterations* requiring a *permit*, a fee for each *permit* shall be paid as required, in accordance with the schedule as established by the applicable governing authority.

❖ The jurisdiction inserts its desired fee schedule at this location. The fees are established by law, such as in an ordinance adopting the code (see page xv of the code for a sample), a separate ordinance or legally promulgated regulation, as required by state or local law. Fee schedules are often based on a valuation of the work to be performed. This concept is based on the proposition that the valuation of a project is related to the amount of work to be expended in plan review, inspections and administering the permit, plus an excess to cover the department overhead.

To assist jurisdictions in establishing some uniformity in fees, building evaluation data are published periodically in ICC's *Building Safety Journal*.

109.3 Building permit valuations. The applicant for a *permit* shall provide an estimated *permit* value at time of application. *Permit* valuations shall include total value of work, including materials and labor, for which the *permit* is being issued, such as electrical, gas, mechanical, plumbing equipment and permanent systems. If, in the opinion of the *building official*, the valuation is underestimated on the application, the *permit* shall be denied, unless the applicant can show detailed estimates to meet the approval of the *building official*. Final building *permit* valuation shall be set by the *building official*.

❖ As indicated in Section 109.2, jurisdictions usually base their fees on the value of the work being performed. This section, therefore, requires the applicant to provide this figure, which is to include the total value of the work, including materials and labor, for which the permit is sought. If the building official believes that the value provided by the applicant is underestimated, the permit is to be denied unless the applicant can substantiate the value by providing detailed estimates of the work to the satisfaction of the building official. For the construction of new buildings, the building valuation data referred to in Section 109.2 can be used by the building official as a yardstick against which to compare the applicant's estimate.

109.4 Work commencing before permit issuance. Any person who commences any work on a building, structure, electrical, gas, mechanical or plumbing system before obtaining the necessary *permits* shall be subject to a fee established by the *building official* that shall be in addition to the required *permit* fees.

❖ The building official will incur certain costs (i.e., inspection time and administrative) when investigating and citing a person who has commenced work without having obtained a permit. The building official is, therefore, entitled to recover these costs by establishing a fee, in addition to that collected when the required permit is issued, to be imposed on the responsible party. Note that this is not a penalty, as described in Section 114.4, for which the person can also be liable.

109.5 Related fees. The payment of the fee for the construction, *alteration*, removal or demolition for work done in connection to or concurrently with the work authorized by a building *permit* shall not relieve the applicant or holder of the *permit* from the payment of other fees that are prescribed by law.

❖ The fees for a building permit may be in addition to other fees required by the jurisdiction or others for related items, such as sewer connections, water service taps, driveways and signs. It cannot be construed that the building permit fee includes these other items.

109.6 Refunds. The *building official* is authorized to establish a refund policy.

❖ This section allows for a refund of fees, which may be full or partial, typically resulting from the revocation, abandonment or discontinuance of a building project for which a permit has been issued and fees have been collected. The refund of fees should be related to the cost of enforcement services not provided because of the termination of the project. The building official, when authorizing a fee refund, is authorizing the disbursement of public funds; therefore, the request for a refund must be in writing and for good cause.

SECTION 110
INSPECTIONS

110.1 General. Construction or work for which a *permit* is required shall be subject to inspection by the *building official* and such construction or work shall remain accessible and exposed for inspection purposes until *approved*. Approval as a result of an inspection shall not be construed to be an approval of a violation of the provisions of this code or of other ordi-

nances of the jurisdiction. Inspections presuming to give authority to violate or cancel the provisions of this code or of other ordinances of the jurisdiction shall not be valid. It shall be the duty of the *permit* applicant to cause the work to remain accessible and exposed for inspection purposes. Neither the *building official* nor the jurisdiction shall be liable for expense entailed in the removal or replacement of any material required to allow inspection.

❖ The inspection function is one of the more important aspects of building department operations. This section authorizes the building official to inspect the work for which a permit has been issued and requires that the work to be inspected remain accessible to the building official until inspected and approved. Any expense incurred in removing or replacing material that conceals an item to be inspected is not the responsibility of the building official or the jurisdiction. As with the issuance of permits (see Section 105.4), approval as a result of an inspection is not a license to violate the code and an approval in violation of the code does not relieve the applicant from complying with the code and is not valid.

110.2 Preliminary inspection. Before issuing a *permit*, the *building official* is authorized to examine or cause to be examined buildings, structures and sites for which an application has been filed.

❖ The building official is granted authority to inspect the site before permit issuance. This may be necessary to verify existing conditions that impact the plan review and permit approval. This section provides the building official with the right-of-entry authority that otherwise does not occur until after the permit is issued (see Section 104.6).

110.3 Required inspections. The *building official*, upon notification, shall make the inspections set forth in Sections 110.3.1 through 110.3.10.

❖ The building official is required to verify that the building is constructed in accordance with the approved construction documents. It is the responsibility of the permit holder to notify the building official when the item is ready for inspection. The inspections that are necessary to provide such verification are listed in the following sections, with the caveat in Section 110.3.8 that inspections in addition to those listed here may be required depending on the work involved.

110.3.1 Footing and foundation inspection. Footing and foundation inspections shall be made after excavations for footings are complete and any required reinforcing steel is in place. For concrete foundations, any required forms shall be in place prior to inspection. Materials for the foundation shall be on the job, except where concrete is ready mixed in accordance with ASTM C 94, the concrete need not be on the job.

❖ It is necessary for the building official to inspect the soil upon which the footing or foundation is to be placed. This inspection also includes any reinforcing steel, concrete forms and materials to be used in the foundation, except for ready-mixed concrete that is prepared off site.

110.3.2 Concrete slab and under-floor inspection. Concrete slab and under-floor inspections shall be made after in-slab or under-floor reinforcing steel and building service equipment, conduit, piping accessories and other ancillary equipment items are in place, but before any concrete is placed or floor sheathing installed, including the subfloor.

❖ The building official must be able to inspect the soil and any required under-slab drainage, waterproofing or dampproofing material, as well as reinforcing steel, conduit, piping and other service equipment embedded in or installed below a slab prior to placing the concrete. Similarly, items installed below a floor system other than concrete must be inspected before they are concealed by the floor sheathing or subfloor.

110.3.3 Lowest floor elevation. In flood hazard areas, upon placement of the lowest floor, including the basement, and prior to further vertical construction, the elevation certification required in Section 1612.5 shall be submitted to the *building official*.

❖ Where a structure is located in a flood hazard area, as established in Section 1612.5, the building official must be provided with certification that either the lowest floor elevation (for structures located in flood hazard areas not subject to high-velocity wave action) or the elevation of the lowest horizontal structural member (for structures located in flood hazard areas subject to high-velocity wave action) is in compliance with Section 1612. This certification must be submitted prior to any construction proceeding above this level.

110.3.4 Frame inspection. Framing inspections shall be made after the roof deck or sheathing, all framing, *fireblocking* and bracing are in place and pipes, chimneys and vents to be concealed are complete and the rough electrical, plumbing, heating wires, pipes and ducts are *approved*.

❖ This section requires that the building official be able to inspect the framing members, such as studs, joists, rafters and girders and other items, such as vents and chimneys, that will be concealed by wall construction. Rough electrical work, plumbing, heating wires, pipes and ducts must have already been approved in accordance with the applicable codes prior to this inspection.

110.3.5 Lath and gypsum board inspection. Lath and gypsum board inspections shall be made after lathing and gypsum board, interior and exterior, is in place, but before any plastering is applied or gypsum board joints and fasteners are taped and finished.

> **Exception:** Gypsum board that is not part of a fire-resistance-rated assembly or a shear assembly.

❖ In order to verify that lath and gypsum board is properly attached to framing members, it is necessary for the building official to be able to conduct an inspection before the plaster or joint finish material is applied. This is required only for gypsum board that is part of either a fire-resistant assembly or a shear wall.

110.3.6 Fire- and smoke-resistant penetrations. Protection of joints and penetrations in fire-resistance-rated assemblies, *smoke barriers* and smoke partitions shall not be concealed from view until inspected and *approved*.

❖ The building official must have an opportunity to inspect joint protection required by Section 714 and penetration protection required by Section 713 for fire-resistance-rated assemblies, smoke barriers, and smoke partitions before they become concealed from view.

110.3.7 Energy efficiency inspections. Inspections shall be made to determine compliance with Chapter 13 and shall include, but not be limited to, inspections for: envelope insulation R- and U-values, fenestration U-value, duct system R-value, and HVAC and water-heating equipment efficiency.

❖ Items installed in a building that are required by the IECC to comply with certain criteria, such as insulation material, windows, HVAC and water-heating equipment, must be inspected and approved.

110.3.8 Other inspections. In addition to the inspections specified above, the *building official* is authorized to make or require other inspections of any construction work to ascertain compliance with the provisions of this code and other laws that are enforced by the department of building safety.

❖ Any item regulated by the code is subject to inspection by the building official to determine compliance with the applicable code provision, and no list can include all items in a given building. This section, therefore, gives the building official the authority to inspect any regulated items.

110.3.9 Special inspections. For *special inspections*, see Section 1704.

❖ Special inspections are to be provided by the owner for the types of work required in Section 1704. The building official is to approve special inspectors and verify that the required special inspections have been conducted. See the commentary to Section 1704 for a complete discussion of this topic.

110.3.10 Final inspection. The final inspection shall be made after all work required by the building *permit* is completed.

❖ Upon completion of the work for which the permit has been issued and before issuance of the certificate of occupancy required by Section 111.1, a final inspection is to be made. All violations of the approved construction documents and permit are to be noted and the holder of the permit is to be notified of the discrepancies.

110.4 Inspection agencies. The *building official* is authorized to accept reports of *approved* inspection agencies, provided such agencies satisfy the requirements as to qualifications and reliability.

❖ As an alternative to the building official conducting the inspection, he or she is permitted to accept inspections of and reports by approved inspection agencies. Appropriate criteria on which to base approval of inspection agencies can be found in Section 1703.

110.5 Inspection requests. It shall be the duty of the holder of the building *permit* or their duly authorized agent to notify the *building official* when work is ready for inspection. It shall be the duty of the *permit* holder to provide access to and means for inspections of such work that are required by this code.

❖ It is the responsibility of the permit holder or other authorized person, such as the contractor performing the work, to arrange for the required inspections when completed work is ready and to allow for sufficient time for the building official to schedule a visit to the site to prevent work from being concealed prior to being inspected. Access to the work to be inspected must be provided, including any special means such as a ladder.

110.6 Approval required. Work shall not be done beyond the point indicated in each successive inspection without first obtaining the approval of the *building official*. The *building official*, upon notification, shall make the requested inspections and shall either indicate the portion of the construction that is satisfactory as completed, or notify the *permit* holder or his or her agent wherein the same fails to comply with this code. Any portions that do not comply shall be corrected and such portion shall not be covered or concealed until authorized by the *building official*.

❖ This section establishes that work cannot progress beyond the point of a required inspection without the building official's approval. Upon making the inspection, the building official must either approve the completed work or notify the permit holder or other responsible party of that which does not comply with the code. Approvals and notices of noncompliance must be in writing, as required by Section 104.4, to avoid any misunderstanding as to what is required. Any item not approved cannot be concealed until it has been corrected and approved by the building official.

SECTION 111
CERTIFICATE OF OCCUPANCY

111.1 Use and occupancy. No building or structure shall be used or occupied, and no change in the existing occupancy classification of a building or structure or portion thereof shall be made, until the *building official* has issued a certificate of occupancy therefor as provided herein. Issuance of a certificate of occupancy shall not be construed as an approval of a violation of the provisions of this code or of other ordinances of the jurisdiction.

> **Exception:** Certificates of occupancy are not required for work exempt from *permits* under Section 105.2.

❖ This section establishes that a new building or structure cannot be occupied until a certificate of occupancy is issued by the building official, which reflects the conclusion of the work allowed by the building permit. Also, no change in occupancy of an existing building is permitted without first obtaining a certificate of occupancy for the new use.

The tool that the building official uses to control the uses and occupancies of various buildings and structures within the jurisdiction is the certificate of occupancy. It is unlawful to use or occupy a building or

structure unless a certificate of occupancy has been issued. Its issuance does not relieve the building owner from the responsibility for correcting any code violation that may exist.

The exception simply states that when work is not under the monitor of the building department, there is no need to deal with a certificate of occupancy.

111.2 Certificate issued. After the *building official* inspects the building or structure and finds no violations of the provisions of this code or other laws that are enforced by the department of building safety, the *building official* shall issue a certificate of occupancy that contains the following:

1. The building *permit* number.

2. The address of the structure.

3. The name and address of the owner.

4. A description of that portion of the structure for which the certificate is issued.

5. A statement that the described portion of the structure has been inspected for compliance with the requirements of this code for the occupancy and division of occupancy and the use for which the proposed occupancy is classified.

6. The name of the *building official*.

7. The edition of the code under which the *permit* was issued.

8. The use and occupancy, in accordance with the provisions of Chapter 3.

9. The type of construction as defined in Chapter 6.

10. The design *occupant load*.

11. If an *automatic sprinkler system* is provided, whether the sprinkler system is required.

12. Any special stipulations and conditions of the building *permit*.

❖ The building official is required to issue a certificate of occupancy after a successful final inspection has been completed and all deficiencies and violations have been resolved. This section lists the information that must be included on the certificate. This information is useful to both the building official and the owner because it indicates the criteria under which the structure was evaluated and approved at the time the certificate was issued. This is important when applying Chapter 34 to existing buildings.

111.3 Temporary occupancy. The *building official* is authorized to issue a temporary certificate of occupancy before the completion of the entire work covered by the *permit*, provided that such portion or portions shall be occupied safely. The *building official* shall set a time period during which the temporary certificate of occupancy is valid.

❖ The building official is permitted to issue a temporary certificate of occupancy for all or a portion of a building prior to the completion of all work. Such certification is to be issued only when the building or portion in question can be safely occupied prior to full completion.

The certification is intended to acknowledge that some building features may not be completed even though the building is safe for occupancy, or that a portion of the building can be safely occupied while work continues in another area. This provision precludes the occupancy of a building or structure that does not contain all of the required fire protection systems and means of egress. Temporary certificates should be issued only when incidental construction remains, such as site work and interior work that is not regulated by the code and exterior decoration not necessary to the integrity of the building envelope. The building official should view the issuance of a temporary certificate of occupancy as substantial an act as the issuance of the final certificate. Indeed, the issuance of a temporary certificate of occupancy offers a greater potential for conflict because once the building or structure is occupied it is very difficult to remove the occupants through legal means. The certificate must specify the time period for which it is valid.

111.4 Revocation. The *building official* is authorized to, in writing, suspend or revoke a certificate of occupancy or completion issued under the provisions of this code wherever the certificate is issued in error, or on the basis of incorrect information supplied, or where it is determined that the building or structure or portion thereof is in violation of any ordinance or regulation or any of the provisions of this code.

❖ The building official is authorized to, in writing, suspend or revoke a certificate of occupancy or completion issued under the provisions of this code wherever the certificate is issued in error, on the basis of incorrect information supplied, or where it is determined that the building or structure or portion thereof is in violation of any ordinance, regulation or any of the provisions of this code.

This section is needed to give the building official the authority to revoke a certificate of occupancy for the reasons indicated in the code text. The building official may also suspend the certificate of occupancy until all of the code violations are corrected.

SECTION 112
SERVICE UTILITIES

112.1 Connection of service utilities. No person shall make connections from a utility, source of energy, fuel or power to any building or system that is regulated by this code for which a *permit* is required, until released by the *building official*.

❖ This section establishes the authority of the building official to approve utility connections to a building for items such as water, sewer, electricity, gas and steam, and to require their disconnection when hazardous conditions or emergencies exist.

The approval of the building official is required before a connection can be made from a utility to a building system that is regulated by the code, including those referenced in Section 101.4. This includes utilities supplying water, sewer, electricity, gas and steam ser-

vices. For the protection of building occupants, including workers, such systems must have had final inspection approvals, except as allowed by Section 112.2 for temporary connections.

112.2 Temporary connection. The *building official* shall have the authority to authorize the temporary connection of the building or system to the utility source of energy, fuel or power.

❖ The building official is permitted to issue temporary authorization to make connections to the public utility system prior to the completion of all work. This acknowledges that, because of seasonal limitations, time constraints or the need for testing or partial operation of equipment, some building systems may be safely connected even though the building is not suitable for final occupancy. The temporary connection and utilization of connected equipment should be approved when the requesting permit holder has demonstrated to the building official's satisfaction that public health, safety and welfare will not be endangered.

112.3 Authority to disconnect service utilities. The *building official* shall have the authority to authorize disconnection of utility service to the building, structure or system regulated by this code and the referenced codes and standards set forth in Section 101.4 in case of emergency where necessary to eliminate an immediate hazard to life or property or when such utility connection has been made without the approval required by Section 112.1 or 112.2. The *building official* shall notify the serving utility, and wherever possible the owner and occupant of the building, structure or service system of the decision to disconnect prior to taking such action. If not notified prior to disconnecting, the owner or occupant of the building, structure or service system shall be notified in writing, as soon as practical thereafter.

❖ Disconnection of one or more of a building's utility services is the most radical method of hazard abatement available to the building official and should be reserved for cases in which all other lesser remedies have proven ineffective. Such an action must be preceded by written notice to the utility and the owner and occupants of the building. Disconnection must be accomplished within the time frame established by the building official in the notice. When the hazard to the public health, safety or welfare is so imminent as to mandate immediate disconnection, the building official has the authority and even the obligation to cause disconnection without notice. In such cases, the owner or occupants must be given written notice as soon as possible.

SECTION 113
BOARD OF APPEALS

113.1 General. In order to hear and decide appeals of orders, decisions or determinations made by the *building official* relative to the application and interpretation of this code, there shall be and is hereby created a board of appeals. The board of appeals shall be appointed by the applicable governing authority and shall hold office at its pleasure. The board shall adopt rules of procedure for conducting its business.

❖ This section provides an aggrieved party with a material interest in the decision of the building official a process to appeal such a decision before a board of appeals. This provides a forum, other than the court of jurisdiction, in which to review the building official's actions.

This section literally allows any person to appeal a decision of the building official. In practice, this section has been interpreted to permit appeals only by those aggrieved parties with a material or definitive interest in the decision of the building official. An aggrieved party may not appeal a code requirement per se. The intent of the appeal process is not to waive or set aside a code requirement; rather, it is intended to provide a means of reviewing a building official's decision on an interpretation or application of the code or to review the equivalency of protection to the code requirements. The members of the appeals board are appointed by the "governing body" of the jurisdiction, typically a council or administrator, such as a mayor or city manager, and remain members until removed from office. The board must establish procedures for electing a chairperson, scheduling and conducting meetings and administration. Note that Appendix B contains complete, detailed requirements for creating an appeals board, including number of members, qualifications and administrative procedures. Jurisdictions desiring to utilize these requirements must include Appendix B in their adopting ordinance.

113.2 Limitations on authority. An application for appeal shall be based on a claim that the true intent of this code or the rules legally adopted thereunder have been incorrectly interpreted, the provisions of this code do not fully apply or an equally good or better form of construction is proposed. The board shall have no authority to waive requirements of this code.

❖ This section establishes the grounds for an appeal, which claims that the building official has misinterpreted or misapplied a code provision. The board is not allowed to set aside any of the technical requirements of the code; however, it is allowed to consider alternative methods of compliance with the technical requirements (see Section 104.11).

113.3 Qualifications. The board of appeals shall consist of members who are qualified by experience and training to pass on matters pertaining to building construction and are not employees of the jurisdiction.

❖ It is important that the decisions of the appeals board are based purely on the technical merits involved in an appeal. It is not the place for policy or political deliberations. The members of the appeals board are, therefore, expected to have experience in building construction matters. Appendix B provides more detailed qualifications for appeals board members and can be adopted by jurisdictions desiring that level of expertise.

SECTION 114
VIOLATIONS

114.1 Unlawful acts. It shall be unlawful for any person, firm or corporation to erect, construct, alter, extend, repair, move, remove, demolish or occupy any building, structure or equipment regulated by this code, or cause same to be done, in conflict with or in violation of any of the provisions of this code.

❖ Violations of the code are prohibited and form the basis for all citations and correction notices.

114.2 Notice of violation. The *building official* is authorized to serve a notice of violation or order on the person responsible for the erection, construction, *alteration*, extension, repair, moving, removal, demolition or occupancy of a building or structure in violation of the provisions of this code, or in violation of a *permit* or certificate issued under the provisions of this code. Such order shall direct the discontinuance of the illegal action or condition and the abatement of the violation.

❖ The building official is required to notify the person responsible for the erection or use of a building found to be in violation of the code. The section that is allegedly being violated must be cited so that the responsible party can respond to the notice.

114.3 Prosecution of violation. If the notice of violation is not complied with promptly, the *building official* is authorized to request the legal counsel of the jurisdiction to institute the appropriate proceeding at law or in equity to restrain, correct or abate such violation, or to require the removal or termination of the unlawful occupancy of the building or structure in violation of the provisions of this code or of the order or direction made pursuant thereto.

❖ The building official must pursue, through the use of legal counsel of the jurisdiction, legal means to correct the violation. This is not optional.

Any extensions of time, so that the violations may be corrected voluntarily, must be for a reasonable and valid cause or the building official may be subject to criticism for "arbitrary and capricious" actions. In general, it is better to have a standard time limitation for correction of violations. Departures from this standard must be for a clear and reasonable purpose, usually stated in writing by the violator.

114.4 Violation penalties. Any person who violates a provision of this code or fails to comply with any of the requirements thereof or who erects, constructs, alters or repairs a building or structure in violation of the *approved construction documents* or directive of the *building official*, or of a *permit* or certificate issued under the provisions of this code, shall be subject to penalties as prescribed by law.

❖ Penalties for violating provisions of the code are typically contained in state law, particularly if the code is adopted at that level, and the building department must follow those procedures. If there is no such procedure already in effect, one must be established with the aid of legal counsel.

SECTION 115
STOP WORK ORDER

115.1 Authority. Whenever the *building official* finds any work regulated by this code being performed in a manner either contrary to the provisions of this code or dangerous or unsafe, the *building official* is authorized to issue a stop work order.

❖ Whenever the building official finds any work regulated by this code being performed in a manner either contrary to the provisions of this code or dangerous or unsafe, the building official is authorized to issue a stop work order.

This section provides for the suspension of work for which a permit was issued, pending the removal or correction of a severe violation or unsafe condition identified by the building official.

Normally, correction notices, issued in accordance with Section 110.6, are used to inform the permit holder of code violations. Stop work orders are issued when enforcement can be accomplished no other way or when a dangerous condition exists.

115.2 Issuance. The stop work order shall be in writing and shall be given to the owner of the property involved, or to the owner's agent, or to the person doing the work. Upon issuance of a stop work order, the cited work shall immediately cease. The stop work order shall state the reason for the order, and the conditions under which the cited work will be permitted to resume.

❖ Upon receipt of a violation notice from the building official, all construction activities identified in the notice must immediately cease, except as expressly permitted to correct the violation.

115.3 Unlawful continuance. Any person who shall continue any work after having been served with a stop work order, except such work as that person is directed to perform to remove a violation or unsafe condition, shall be subject to penalties as prescribed by law.

❖ This section states that the work in violation must terminate and that all other work, except that which is necessary to correct the violation or unsafe condition, must cease as well. As determined by the municipality or state, a penalty may be assessed for failure to comply with this section.

SECTION 116
UNSAFE STRUCTURES AND EQUIPMENT

116.1 Conditions. Structures or existing equipment that are or hereafter become unsafe, insanitary or deficient because of inadequate *means of egress* facilities, inadequate light and ventilation, or which constitute a fire hazard, or are otherwise dangerous to human life or the public welfare, or that involve illegal or improper occupancy or inadequate maintenance, shall be deemed an unsafe condition. Unsafe structures shall be taken down and removed or made safe, as the *building official* deems necessary and as provided for in this section. A vacant structure that is not secured against entry shall be deemed unsafe.

❖ This section describes the responsibility of the building official to investigate reports of unsafe structures and equipment and provides criteria for such determination.

Unsafe structures are defined as buildings or structures that are insanitary; deficient in light and ventilation or adequate exit facilities; constitute a fire hazard or are otherwise dangerous to human life.

This section establishes that unsafe buildings can result from illegal or improper occupancies. For example, prima facie evidence of an unsafe structure is an unsecured (open at door or window) vacant building. All unsafe buildings must either be demolished or made safe and secure as deemed appropriate by the building official.

116.2 Record. The *building official* shall cause a report to be filed on an unsafe condition. The report shall state the occupancy of the structure and the nature of the unsafe condition.

❖ The building official must file a report on each investigation of unsafe conditions, stating the occupancy of the structure and the nature of the unsafe condition. This report provides the basis for the notice described in Section 116.3.

116.3 Notice. If an unsafe condition is found, the *building official* shall serve on the owner, agent or person in control of the structure, a written notice that describes the condition deemed unsafe and specifies the required repairs or improvements to be made to abate the unsafe condition, or that requires the unsafe structure to be demolished within a stipulated time. Such notice shall require the person thus notified to declare immediately to the *building official* acceptance or rejection of the terms of the order.

❖ The building official must file a report on each investigation of unsafe conditions, stating the occupancy of the structure and the nature of the unsafe condition. This report provides the basis for the notice described in Section 116.3.

116.4 Method of service. Such notice shall be deemed properly served if a copy thereof is (a) delivered to the owner personally; (b) sent by certified or registered mail addressed to the owner at the last known address with the return receipt requested; or (c) delivered in any other manner as prescribed by local law. If the certified or registered letter is returned showing that the letter was not delivered, a copy thereof shall be posted in a conspicuous place in or about the structure affected by such notice. Service of such notice in the foregoing manner upon the owner's agent or upon the person responsible for the structure shall constitute service of notice upon the owner.

❖ The notice must be delivered personally to the owner. If the owner or agent cannot be located, additional procedures are established, including posting the unsafe notice on the premises in question. Such action may be considered the equivalent of personal notice; however, it may or may not be deemed by the courts as representing a "good faith" effort to notify. In addition to complying with this section, therefore, public notice through the use of newspapers and other postings in a prominent location at the government center should be used.

116.5 Restoration. The structure or equipment determined to be unsafe by the *building official* is permitted to be restored to a safe condition. To the extent that repairs, *alterations* or *additions* are made or a change of occupancy occurs during the restoration of the structure, such repairs, *alterations*, *additions* or change of occupancy shall comply with the requirements of Section 105.2.2 and Chapter 34.

❖ This section provides that unsafe structures may be restored to a safe condition. This means that the cause of the unsafe structure notice can be abated without the structure being required to comply fully with the provisions for new construction. Any work done to eliminate the unsafe condition, as well as any change in occupancy that may occur, must comply with the code.

Bibliography

The following resource materials are referenced in this chapter or are relevant to the subject matter addressed in this chapter.

IECC-09, *International Electrical Conservation Code.* Washington, DC: International Code Council, 2009.

IFC-09, *International Fire Code.* Washington, DC: International Code Council, 2009.

IFGC-09, *International Fuel Gas Code.* Washington, DC: International Code Council, 2009.

IMC-09, *International Mechanical Code.* Washington, DC: International Code Council, 2009.

IPC-09, *International Plumbing Code.* Washington, DC: International Code Council, 2009.

IPMC-09, *International Property Maintenance Code.* Washington, DC: International Code Council, 2009.

IPSDC-09, *International Private Sewage Disposal Code.* Washington, DC: International Code Council, 2009.

IRC-09, *International Residential Code.* Washington, DC: International Code Council, 2009.

Legal Aspects of Code Administration. Country Club Hills, IL: International Code Council, 2002.

NFPA 14-07, *Standpipe and Hose Systems.* Quincy, MA: National Fire Protection Association, 2007.

NFPA 70-08, *National Electrical Code.* Quincy, MA: National Fire Protection Association, 2008.

NFPA 409-04, *Standard on Aircraft.* Quincy, MA: National Fire Protection Association, 2004.

Readings in Code Administration, Volume 1: History/Philosophy/Law. Country Club Hills, IL: Building Officials and Code Administrators International, Inc., 1974.

Chapter 2:
Definitions

General Comments

All terms defined in the code are listed alphabetically in Chapter 2. The actual definitions of the terms are located as follows:

Where a term is used in more than one chapter, its definition appears in Chapter 2. Of the more than 700 words, terms and phrases defined in the code, 60 are defined in Chapter 2.

Where a term is unique or primarily pertains to a single chapter, its definition appears within that chapter. In many chapters, the second section is devoted to definitions. For example, definitions applicable to means of egress are found in Section 1002.

Where a term is unique to a single section or subsection of a chapter, its definition appears within that section or subsection. For example, definitions applicable to stages and platforms are found in Section 410.2.

Purpose

Codes, by their very nature, are technical documents. As such, literally every word, term and punctuation mark can add to or change the meaning of the intended result. This is even more so with a performance-based code where the desired result often takes on more importance than the specific words. Furthermore, the code, with its broad scope of applicability, includes terms inherent in a variety of construction disciplines. These terms often have multiple meanings depending on the context or discipline being used at the time. For these reasons, it is necessary to maintain a consensus on the specific meaning of terms contained in the code. Chapter 2 performs this function by stating clearly what specific terms mean for the purpose of the code.

SECTION 201
GENERAL

❖ This section contains language and provisions that are supplemental to the use of Chapter 2. It gives guidance to the use of the defined words relevant to tense, gender and plurality. Finally, this section provides direction on how to apply terms that are not defined in the code.

201.1 Scope. Unless otherwise expressly stated, the following words and terms shall, for the purposes of this code, have the meanings shown in this chapter.

❖ The use of words and terms in the code is governed by the provisions of this section. This includes code-defined terms as well as those terms that are not.

201.2 Interchangeability. Words used in the present tense include the future; words stated in the masculine gender include the feminine and neuter; the singular number includes the plural and the plural, the singular.

❖ While the definitions contained or referenced in Chapter 2 are to be taken literally, gender and tense are interchangeable.

201.3 Terms defined in other codes. Where terms are not defined in this code and are defined in the *International Fuel Gas Code, International Fire Code, International Mechanical Code* or *International Plumbing Code,* such terms shall have the meanings ascribed to them as in those codes.

❖ Definitions that are applicable in other *International Codes®* are applicable everywhere the term is used in the code. Definitions of terms can help in the understanding and application of code requirements.

201.4 Terms not defined. Where terms are not defined through the methods authorized by this section, such terms shall have ordinarily accepted meanings such as the context implies.

❖ Words or terms not defined within the *International Code* series are intended to be applied based on their "ordinarily accepted meanings." The intent of this statement is that a dictionary definition may suffice, provided it is in context. Often times, construction terms used throughout the code are not specifically defined in the code or even in a dictionary. In such a case, the definitions contained in the referenced standards (see Chapter 35) and published textbooks on the subject in question are good resources.

SECTION 202
DEFINITIONS

❖ This portion of the commentary addresses only those terms whose definitions appear in Chapter 2. The commentary for definitions that are located elsewhere in the code can be found in the indicated sections that contain those definitions.

AAC MASONRY. See Section 2102.1.

ACCESSIBLE. See Section 1102.1.

ACCESSIBLE MEANS OF EGRESS. See Section 1002.1.

ACCESSIBLE ROUTE. See Section 1102.1.

ACCESSIBLE UNIT. See Section 1102.1.

ACCREDITATION BODY. See Section 2302.1.

ADDITION. An extension or increase in floor area or height of a building or structure.

❖ This term is used to describe the condition when the floor area or height of an existing building or structure is increased (see Chapter 34). This term is only applicable to existing buildings, never new ones. This would include additional floor area that is added within an existing building, such as adding a new mezzanine.

ADHERED MASONRY VENEER. See Section 1402.1.

ADOBE CONSTRUCTION. See Section 2102.1.

   Adobe, stabilized. See Section 2102.1.

   Adobe, unstabilized. See Section 2102.1.

[F] AEROSOL. See Section 307.2.

   Level 1 aerosol products. See Section 307.2.

   Level 2 aerosol products. See Section 307.2.

   Level 3 aerosol products. See Section 307.2.

[F] AEROSOL CONTAINER. See Section 307.2.

AGGREGATE. See Section 1502.1.

AGRICULTURAL, BUILDING. A structure designed and constructed to house farm implements, hay, grain, poultry, livestock or other horticultural products. This structure shall not be a place of human habitation or a place of employment where agricultural products are processed, treated or packaged, nor shall it be a place used by the public.

❖ This definition is needed for the proper application of the utility and miscellaneous occupancy group and Appendix C provisions. The use of the building is quite restricted such that buildings that include habitable or public spaces are not agricultural buildings by definition.

AIR-INFLATED STRUCTURE. See Section 3102.2.

AIR-SUPPORTED STRUCTURE. See Section 3102.2.

   Double skin. See Section 3102.2.

   Single skin. See Section 3102.2.

AISLE. See Section 1002.1.

AISLE ACCESSWAY. See Section 1002.1.

[F] ALARM NOTIFICATION APPLIANCE. See Section 902.1.

[F] ALARM SIGNAL. See Section 902.1.

[F] ALARM VERIFICATION FEATURE. See Section 902.1.

ALLOWABLE STRESS DESIGN. See Section 1602.1.

ALTERATION. Any construction or renovation to an existing structure other than repair or *addition*.

❖ The code utilizes this term to reflect construction operations intended for an existing building (see Chapter 34), but not within the scope of an addition or repair (see the definitions of "Addition" and "Repair").

ALTERNATING TREAD DEVICE. See Section 1002.1.

AMBULATORY HEALTH CARE FACILITY. Buildings or portions thereof used to provide medical, surgical, psychiatric, nursing or similar care on a less than 24-hour basis to individuals who are rendered incapable of self-preservation.

❖ The code provides different requirements for outpatient clinics, ambulatory health care facilities and hospitals. Ambulatory health care facilities while still classified as a Group B occupancy has additional standards above those of a outpatient clinic primarily due to the fact that many patients are temporarily unable to respond to emergencies (see commentary, Section 423). Ambulatory health care facilities include day surgery centers and similar facilities where patients may receive fairly intensive treatment, but do not stay at the facility more than a few hours. If patients are receiving 24-hour care, such facilities would be defined as a hospital.

ANCHOR. See Section 2102.1.

ANCHOR BUILDING. See Section 402.2.

ANCHORED MASONRY VENEER. See Section 1402.1.

ANNULAR SPACE. See Section 702.1.

[F] ANNUNCIATOR. See Section 902.1.

APPROVED. Acceptable to the code official or authority having jurisdiction.

❖ As related to the process of acceptance of building installations, including materials, equipment and construction systems, this definition identifies where the ultimate authority rests. Whenever this term is used, it intends that only the enforcing authority can accept a specific installation or component as complying with the code. For the *International Building Code* the building official is identified as the code official, the person responsible for administering the code.

APPROVED AGENCY. See Section 1702.1.

APPROVED FABRICATOR. See Section 1702.1.

APPROVED SOURCE. An independent person, firm or corporation, *approved* by the *building official*, who is competent and experienced in the application of engineering principles to materials, methods or systems analyses.

❖ The building official sometimes needs to rely on evaluation reports, analyses or other types of reports that purport to validate the use of a material, system or method as complying with the code. This definition establishes that the building official needs to rely on independent, competent individuals or agencies as the source of these reports.

ARCHITECTURAL TERRA COTTA. See Section 2102.1.

AREA (for masonry). See Section 2102.1.

Bedded. See Section 2102.1.

Gross cross-sectional. See Section 2102.1.

Net cross-sectional. See Section 2102.1.

AREA, BUILDING. See Section 502.1.

AREA OF REFUGE. See Section 1002.1.

AREAWAY. A subsurface space adjacent to a building open at the top or protected at the top by a grating or *guard*.

❖ Areaways are often constructed to provide access to below-grade building services, including transformers, ventilation shafts and pipe tunnels.

ASSISTED LIVING FACILITIES. See Section 310.2, "Residential Care/Assisted living facilities."

ATRIUM. See Section 404.1.1.

ATTIC. The space between the ceiling beams of the top *story* and the roof rafters.

❖ The definition of "Attic" identifies the specific portion of a building or structure for the purposes of determining the applicability of requirements that are specific to attics, such as ventilation (see Section 1203) and draftstopping (see Section 717). Additionally, the code has access requirements (see Section 1209) and uniformly distributed live load requirements (see Table 1607.1) for attics. An attic is considered the space or area located immediately below the roof sheathing within the roof framing system of a building. Pitched roof systems, such as gabled, hip, sawtoothed or curved roofs, all create spaces between the roof sheathing and ceiling membrane, which are considered attics.

[F] AUDIBLE ALARM NOTIFICATION APPLIANCE. See Section 902.1.

AUTOCLAVED AERATED CONCRETE (AAC). See Section 2102.1.

[F] AUTOMATIC. See Section 902.1.

[F] AUTOMATIC FIRE-EXTINGUISHING SYSTEM. See Section 902.1.

[F] AUTOMATIC SMOKE DETECTION SYSTEM. See Section 902.1.

[F] AUTOMATIC SPRINKLER SYSTEM. See Section 902.1.

[F] AVERAGE AMBIENT SOUND LEVEL. See Section 902.1.

AWNING. An architectural projection that provides weather protection, identity or decoration and is wholly supported by the building to which it is attached. An *awning* is comprised of a lightweight frame structure over which a covering is attached.

❖ Similar to a canopy, an awning typically provides weather protection, signage or decoration. But unlike a canopy, an awning relies solely on the building to which it is attached for its means of support. Awnings are distinct from similar building features by the use of fabric or similar pliable materials as the covering of the frame. See Section 3105 for general requirements, Section 1607.11 for awning design loads and Section 3202 for encroachment requirements.

BACKING. See Section 1402.1.

[F] BALED COTTON. See Section 307.2.

[F] BALED COTTON, DENSELY PACKED. See Section 307.2.

BALLAST. See Section 1502.1.

[F] BARRICADE. See Section 307.2.

Artificial barricade. See Section 307.2.

Natural barricade. See Section 307.2.

BASE FLOOD. See Section 1612.2.

BASE FLOOD ELEVATION. See Section 1612.2.

BASEMENT (for other than flood loads). See Section 502.1.

BASEMENT (for flood loads). See Section 1612.2.

BEARING WALL STRUCTURE. See Section 1614.2.

BED JOINT. See Section 2102.1.

BLEACHERS. See Section 1002.1.

BOARDING HOUSE. See Section 310.2.

[F] BOILING POINT. See Section 307.2.

BOND BEAM. See Section 2102.1.

BRACED WALL LINE. See Section 2302.1.

BRACED WALL PANEL. See Section 2302.1.

BRICK. See Section 2102.1.

Calcium silicate (sand lime brick). See Section 2102.1.

Clay or shale. See Section 2102.1.

Concrete. See Section 2102.1.

BUILDING. Any structure used or intended for supporting or sheltering any use or occupancy.

❖ The code uses this term to identify those structures that provide shelter for a function or activity. See the definition for "Area, building" for situations when a single structure may be two or more "Buildings" created by fire walls.

BUILDING ELEMENT. See Section 702.1.

BUILDING LINE. The line established by law, beyond which a building shall not extend, except as specifically provided by law.

❖ This term defines the limitations or boundaries for construction of a building This line is typically established by a zoning statute or rights-of-way dedication and is not specified in the code.

BUILDING OFFICIAL. The officer or other designated authority charged with the administration and enforcement of this code, or a duly authorized representative.

❖ The statutory power to enforce the code is normally vested in a building department (or the like) of a state, county or municipality that has a designated enforcement officer termed the "building official" (see Section 103.1).

BUILT-UP ROOF COVERING. See Section 1502.1.

CABLE-RESTRAINED, AIR-SUPPORTED STRUC-TURE. See Section 3102.2.

CANOPY. A permanent structure or architectural projection of rigid construction over which a covering is attached that provides weather protection, identity or decoration, and shall be structurally independent or supported by attachment to a building on one end and by not less than one stanchion on the outer end.

❖ A canopy can be either an architectural projection from a building, or it can be independent structure. An example of the former is typically found covering an entrance walkway in front of a hotel or apartment building, perhaps a fancy restaurant. An example of the latter is a canopy built over fuel pumps at a gasoline station. The covering of the structure can be either rigid materials or fabric and membrane materials similar to an awning. If it is attached to a building, it is characterized by having supports (stanchions) at the other end. This distinguishes it from an awning that is only supported by the building. See Section 3105 for general requirements, Section 1607.11 for design loads and Section 3202 for encroachment requirements.

[F] CARBON DIOXIDE EXTINGUISHING SYSTEMS. See Section 902.1.

CAST STONE. See Section 2102.1.

[F] CEILING LIMIT. See Section 902.1.

CEILING RADIATION DAMPER. See Section 702.1.

CELL. See Section 408.1.1.

CELL (masonry). See Section 2102.1.

CELL TIER. See Section 408.1.1.

CEMENT PLASTER. See Section 2502.1.

CERAMIC FIBER BLANKET. See Section 721.1.1.

CERTIFICATE OF COMPLIANCE. See Section 1702.1.

CHILD CARE FACILITIES. See Section 308.3.1.

CHIMNEY. See Section 2102.1.

CHIMNEY TYPES. See Section 2102.1.

    High-heat appliance type. See Section 2102.1.

    Low-heat appliance type. See Section 2102.1.

    Masonry type. See Section 2102.1.

    Medium-heat appliance type. See Section 2102.1.

CIRCULATION PATH. See Section 1102.1.

[F] CLEAN AGENT. See Section 902.1.

CLEANOUT. See Section 2102.1.

CLINIC, OUTPATIENT. See Section 304.1.1.

[F] CLOSED SYSTEM. See Section 307.2.

COLLAR JOINT. See Section 2102.1.

COLLECTOR. See Section 2302.1.

COMBINATION FIRE/SMOKE DAMPER. See Section 702.1.

[F] COMBUSTIBLE DUST. See Section 307.2.

[F] COMBUSTIBLE FIBERS. See Section 307.2.

[F] COMBUSTIBLE LIQUID. See Section 307.2.

    Class II. See Section 307.2.

    Class IIIA. See Section 307.2.

    Class IIIB. See Section 307.2.

COMMON USE. See Section 1102.1.

COMMON PATH OF EGRESS TRAVEL. See Section 1002.1.

[F] COMPRESSED GAS. See Section 307.2.

COMPRESSIVE STRENGTH OF MASONRY. See Section 2102.1.

CONCRETE, CARBONATE AGGREGATE. See Section 721.1.1.

CONCRETE, CELLULAR. See Section 721.1.1.

CONCRETE, LIGHTWEIGHT AGGREGATE. See Section 721.1.1.

CONCRETE, PERLITE. See Section 721.1.1.

CONCRETE, SAND-LIGHTWEIGHT. See Section 721.1.1.

CONCRETE, SILICEOUS AGGREGATE. See Section 721.1.1.

CONCRETE, VERMICULITE. See Section 721.1.1.

CONGREGATE LIVING FACILITIES. See Section 310.2.

CONNECTOR. See Section 2102.1.

[F] CONSTANTLY ATTENDED LOCATION. See Section 902.1.

CONSTRUCTION DOCUMENTS. Written, graphic and pictorial documents prepared or assembled for describing the design, location and physical characteristics of the elements of a project necessary for obtaining a building *permit*.

❖ To determine whether or not proposed construction is in compliance with code requirements, it is necessary that sufficient information be submitted to the building official for review. This typically consists of the drawings (floor plans, elevations, sections, details, etc.), specifications and product information describing the proposed work.

CONSTRUCTION TYPES. See Section 602.

Type I. See Section 602.2.

Type II. See Section 602.2.

Type III. See Section 602.3.

Type IV. See Section 602.4.

Type V. See Section 602.5.

[F] CONTINUOUS GAS DETECTION SYSTEM. See Section 415.2.

[F] CONTROL AREA. See Section 307.2.

CONTROLLED LOW-STRENGTH MATERIAL. A self-compacted, cementitious material used primarily as a backfill in place of compacted fill.

❖ The definition provided is from ACI 229R-99. This type of material is known by many "local" names (e.g., flowable fill) and is commonly used in lieu of a compacted backfill. Requirements for its use under the code are outlined in Section 1804.6.

CONVENTIONAL LIGHT-FRAME CONSTRUCTION. See Section 2302.1.

CORRIDOR. See Section 1002.1.

CORROSION RESISTANCE. The ability of a material to withstand deterioration of its surface or its properties when exposed to its environment.

❖ There are different environments that contain different types of materials to which building construction materials are exposed. "Corrosion resistance" is not an absolute term; it is relative to the building material and where it is being used. For instance, certain types of plastic polymers might resist corrosion when used in an exterior environment, but might not be resistant to corrosion from certain chemical gases that could be present in a laboratory.

[F] CORROSIVE. See Section 307.2.

COURT. An open, uncovered space, unobstructed to the sky, bounded on three or more sides by exterior building walls or other enclosing devices.

❖ Though not specifically identified in the definition, the provisions in the code for courts (Section 1206) are only applicable to those areas created by the arrangement of exterior walls and used to provide natural light or ventilation (see Section 1206.1 and the definition of "Yard" at the end of this section). See also the definition of egress court in Section 1002.1 for a specific type of court.

COVER. See Section 2102.1.

COVERED MALL BUILDING. See Section 402.2.

Mall. See Section 402.2.

Open mall. See Section 402.2.

Open mall building. See Section 402.2.

CRIPPLE WALL. See Section 2302.1.

[F] CRYOGENIC FLUID. See Section 307.2.

DALLE GLASS. See Section 2402.1.

DAMPER. See Section 702.1.

DANGEROUS. See Section 3402.1.

[F] DAY BOX. See Section 307.2.

DEAD LOADS. See Section 1602.1.

DECORATIVE GLASS. See Section 2402.1.

[F] DECORATIVE MATERIALS. All materials applied over the building *interior finish* for decorative, acoustical or other effect (such as curtains, draperies, fabrics, streamers and surface coverings), and all other materials utilized for decorative effect (such as batting, cloth, cotton, hay, stalks, straw, vines, leaves, trees, moss and similar items), including foam plastics and materials containing foam plastics. *Decorative materials* do not include floor coverings, ordinary window shades, *interior finish* and materials 0.025 inch (0.64 mm) or less in thickness applied directly to and adhering tightly to a substrate.

❖ The significance of this definition is to provide information as to what is not regulated as decorative materials in the application of code requirements. While any dictionary would define floor coverings, window shades and wall paper as being "decorative" in a building interior, they are not considered decorative materials for the flame-resistance testing to which the code requirements are intended to apply.

DEEP FOUNDATION. See Section 1802.1.

[F] DEFLAGRATION. See Section 307.2.

[F] DELUGE SYSTEM. See Section 902.1.

DESIGN DISPLACEMENT. See Section 1908.1.1.

DESIGN EARTHQUAKE GROUND MOTION. See Section 1613.2.

DESIGN FLOOD. See Section 1612.2.

DESIGN FLOOD ELEVATION. See Section 1612.2.

DESIGN STRENGTH. See Section 1602.1.

DESIGNATED SEISMIC SYSTEM. See Section 1702.1.

[F] DETACHED BUILDING. See Section 415.2.

DETAILED PLAIN CONCRETE STRUCTURAL WALL. See Section 1908.1.1.

DETECTABLE WARNING. See Section 1102.1.

[F] DETECTOR, HEAT. See Section 902.1.

[F] DETONATION. See Section 307.2.

DETOXIFICATION FACILITY. See Section 308.3.1.

DIAPHRAGM. See Sections 1602.1 and 2302.1.

    Diaphragm, blocked. See Section 1602.1.

    Diaphragm, boundary. See Section 1602.1.

    Diaphragm, chord. See Section 1602.1.

    Diaphragm, flexible. See Section 1602.1.

    Diaphragm, rigid. See Section 1602.1.

    Diaphragm, unblocked. See Section 2302.1.

DIMENSIONS. See Section 2102.1.

    Actual. See Section 2102.1.

    Nominal. See Section 2102.1.

    Specified. See Section 2102.1.

[F] DISPENSING. See Section 307.2.

DOOR, BALANCED. See Section 1002.1.

DORMITORY. See Section 310.2.

DRAFTSTOP. See Section 702.1.

DRAG STRUT. See Section 2302.1.

DRILLED SHAFT. See Section 1802.1.

    Socketed drilled shaft. See Section 1802.1.

[F] DRY-CHEMICAL EXTINGUISHING AGENT. See Section 902.1.

DRY FLOODPROOFING. See Section 1612.2.

DURATION OF LOAD. See Section 1602.1.

DWELLING. A building that contains one or two *dwelling units* used, intended or designed to be used, rented, leased, let or hired out to be occupied for living purposes.

❖ Dwellings are buildings intended to serve as residences for one or two families. Dwellings can be owner occupied or rented. The term "dwelling," which refers to the building itself, is defined to distinguish it from the term "dwelling unit," which is a single living unit within a building. It is important to recognize that the code is not intended to regulate detached one- and two-family dwellings and townhouses that are no more than three stories in height. These dwellings are regulated by the *International Residential Code*® (IRC®) (see Section 101.2). See also the definition below for "Townhouse."

DWELLING UNIT. A single unit providing complete, independent living facilities for one or more persons, including permanent provisions for living, sleeping, eating, cooking and sanitation.

❖ A dwelling unit, as stated, is a residential unit that contains all of the necessary facilities for independent living. This provides a single, independent unit that serves a single family or single group of individuals. This terminology is used throughout the code for the determination of the application of various provisions. A dwelling unit is also distinguished from a sleeping unit which does not have all of the features of a dwelling unit and must comply with a different set of requirements (see the definition below for "Sleeping unit"). A building containing one or more dwelling units is a "dwelling" (see the definitions for "Dwelling" and "Townhouse"). A building containing three or more dwelling units is regulated as a Group R-2 occupancy. The most common term used for such a building is an apartment house or condominium. To be considered a Group R-3, the structure must have one or two dwelling units, or be subdivided by fire walls between every unit or every two units (see Section 310.1).

DWELLING UNIT OR SLEEPING UNIT, MULTISTORY. See Section 1102.1.

DWELLING UNIT OR SLEEPING UNIT, TYPE A. See Section 1102.1.

DWELLING UNIT OR SLEEPING UNIT, TYPE B. See Section 1102.1.

EGRESS COURT. See Section 1002.1.

ELEVATOR GROUP. See Section 902.1.

[F] EMERGENCY ALARM SYSTEM. See Section 902.1.

[F] EMERGENCY CONTROL STATION. See Section 415.2.

EMERGENCY ESCAPE AND RESCUE OPENING. See Section 1002.1.

[F] EMERGENCY VOICE/ALARM COMMUNICATIONS. See Section 902.1.

EMPLOYEE WORK AREA. See Section 1102.1.

EQUIPMENT PLATFORM. See Section 502.1.

ESSENTIAL FACILITIES. See Section 1602.1.

[F] EXHAUSTED ENCLOSURE. See Section 415.2.

EXISTING CONSTRUCTION. See Section 1612.2.

EXISTING STRUCTURE. See Sections 1612.2 and 3402.1.

EXIT. See Section 1002.1.

EXIT ACCESS. See Section 1002.1.

EXIT ACCESS DOORWAY. See Section 1002.1.

EXIT DISCHARGE. See Section 1002.1.

EXIT DISCHARGE, LEVEL OF. See Section 1002.1.

EXIT ENCLOSURE. See Section 1002.1.

EXIT, HORIZONTAL. See Section 1002.1.

EXIT PASSAGEWAY. See Section 1002.1.

EXPANDED VINYL WALL COVERING. See Section 802.1.

[F] EXPLOSION. See Section 307.2.

[F] EXPLOSIVE. See Section 307.2.

   High explosive. See Section 307.2.

   Low explosive. See Section 307.2.

   Mass detonating explosives. See Section 307.2.

   UN/DOTn Class 1 Explosives. See Section 307.2.

      Division 1.1. See Section 307.2.

      Division 1.2. See Section 307.2.

      Division 1.3. See Section 307.2.

      Division 1.4. See Section 307.2.

      Division 1.5. See Section 307.2.

      Division 1.6. See Section 307.2.

EXTERIOR INSULATION AND FINISH SYSTEM (EIFS). See Section 1402.1.

EXTERIOR INSULATION AND FINISH SYSTEM (EIFS) WITH DRAINAGE. See Section 1402.1.

EXTERIOR SURFACES. See Section 2502.1.

EXTERIOR WALL. See Section 1402.1.

EXTERIOR WALL COVERING. See Section 1402.1.

EXTERIOR WALL ENVELOPE. See Section 1402.1.

F RATING. See Section 702.1.

FABRIC PARTITION. See Section 1602.1.

FABRICATED ITEM. See Section 1702.1.

[F] FABRICATION AREA. See Section 415.2.

FACILITY. See Section 1102.1.

FACTORED LOAD. See Section 1602.1.

FIBER CEMENT SIDING. See Section 1402.1.

FIBER REINFORCED POLYMER. See Section 2602.1.

   Fiberglass Reinforced Polymer. See Section 2602.1.

FIBERBOARD. See Section 2302.1.

FIRE ALARM BOX, MANUAL. See Section 902.1.

[F] FIRE ALARM CONTROL UNIT. See Section 902.1.

[F] FIRE ALARM SIGNAL. See Section 902.1.

[F] FIRE ALARM SYSTEM. See Section 902.1.

FIRE AREA. See Section 902.1.

FIRE BARRIER. See Section 702.1.

[F] FIRE COMMAND CENTER. See Section 902.1.

FIRE DAMPER. See Section 702.1.

[F] FIRE DETECTOR, AUTOMATIC. See Section 902.1.

FIRE DOOR. See Section 702.1.

FIRE DOOR ASSEMBLY. See Section 702.1.

FIRE EXIT HARDWARE. See Section 1002.1.

[F] FIRE LANE. A road or other passageway developed to allow the passage of fire apparatus. A *fire lane* is not necessarily intended for vehicular traffic other than fire apparatus.

❖ The term "fire lane" is synonymous with the term "fire apparatus access road," both being a road that provides access from a fire station to a building, or portion thereof. However, it should be noted that the driving surface is not necessarily the same as that provided for a public road. The driving surface must be a surface that can be shown to adequately support the load of anticipated emergency vehicles.

FIRE PARTITION. See Section 702.1.

FIRE PROTECTION RATING. See Section 702.1.

[F] FIRE PROTECTION SYSTEM. See Section 902.1.

FIRE RESISTANCE. See Section 702.1.

FIRE-RESISTANCE RATING. See Section 702.1.

FIRE-RESISTANT JOINT SYSTEM. See Section 702.1.

[F] FIRE SAFETY FUNCTIONS. See Section 902.1.

FIRE SEPARATION DISTANCE. See Section 702.1.

FIRE WALL. See Section 702.1.

FIRE WINDOW ASSEMBLY. See Section 702.1.

FIREBLOCKING. See Section 702.1.

FIREPLACE. See Section 2102.1.

FIREPLACE THROAT. See Section 2102.1.

[F] FIREWORKS. See Section 307.2.

   Fireworks, 1.3G. See Section 307.2.

   Fireworks, 1.4G. See Section 307.2.

FIXED BASE OPERATOR (FBO). See Section 412.2.

FLAME SPREAD. See Section 802.1.

FLAME SPREAD INDEX. See Section 802.1.

[F] FLAMMABLE GAS. See Section 307.2.

[F] FLAMMABLE LIQUEFIED GAS. See Section 307.2.

[F] FLAMMABLE LIQUID. See Section 307.2.

   Class IA. See Section 307.2.

   Class IB. See Section 307.2.

   Class IC. See Section 307.2.

[F] FLAMMABLE MATERIAL. See Section 307.2.

[F] FLAMMABLE SOLID. See Section 307.2.

[F] FLAMMABLE VAPORS OR FUMES. See Section 415.2.

[F] FLASH POINT. See Section 307.2.

FLIGHT. See Section 1002.1.

FLOOD OR FLOODING. See Section 1612.2.

FLOOD DAMAGE-RESISTANT MATERIALS. See Section 1612.2.

FLOOD HAZARD AREA. See Section 1612.2.

FLOOD HAZARD AREA SUBJECT TO HIGH-VELOCITY WAVE ACTION. See Section 1612.2.

FLOOD INSURANCE RATE MAP (FIRM). See Section 1612.2.

FLOOD INSURANCE STUDY. See Section 1612.2.

FLOODWAY. See Section 1612.2.

FLOOR AREA, GROSS. See Section 1002.1.

FLOOR AREA, NET. See Section 1002.1.

FLOOR FIRE DOOR ASSEMBLY. See Section 702.1.

FLY GALLERY. See Section 410.2.

[F] FOAM-EXTINGUISHING SYSTEMS. See Section 902.1.

FOAM PLASTIC INSULATION. See Section 2602.1.

FOLDING AND TELESCOPIC SEATING. See Section 1002.1.

FOOD COURT. See Section 402.2.

FOUNDATION PIER. See Section 2102.1.

FRAME STRUCTURE. See Section 1614.2.

[F] GAS CABINET. See Section 415.2.

[F] GAS ROOM. See Section 415.2.

[F] GASEOUS HYDROGEN SYSTEM. See Section 421.2.

GLASS FIBERBOARD. See Section 721.1.1.

GLUED BUILT-UP MEMBER. See Section 2302.1.

GRADE FLOOR OPENING. A window or other opening located such that the sill height of the opening is not more than 44 inches (1118 mm) above or below the finished ground level adjacent to the opening.

❖ Openings used for emergency escape or rescue are clearly easier to use the closer they are to grade. This definition specifies that the maximum sill height above the exterior adjacent grade must be no more than 44 inches (1118 mm) for an opening to qualify as a grade floor opening. See Section 1029.2.

GRADE (LUMBER). See Section 2302.1.

GRADE PLANE. See Section 502.1.

GRANDSTAND. See Section 1002.1.

GRIDIRON. See Section 410.2.

GROSS LEASABLE AREA. See Section 402.2.

GROUTED MASONRY. See Section 2102.1.

  Grouted hollow-unit masonry. See Section 2102.1.

  Grouted multiwythe masonry. See Section 2102.1.

GUARD. See Section 1002.1.

GYPSUM BOARD. See Section 2502.1.

GYPSUM PLASTER. See Section 2502.1.

GYPSUM VENEER PLASTER. See Section 2502.1.

HABITABLE SPACE. A space in a building for living, sleeping, eating or cooking. Bathrooms, toilet rooms, closets, halls, storage or utility spaces and similar areas are not considered habitable spaces.

❖ These spaces are normally considered inhabited in the course of residential living and provide the four basic characteristics associated with it: living, sleeping, eating and cooking. All habitable spaces are considered occupiable spaces, though other occupiable spaces, such as halls or utility rooms, are not considered habitable (see the definition of "Occupiable space" in this chapter).

[F] HALOGENATED EXTINGUISHING SYSTEMS. See Section 902.1.

[F] HANDLING. See Section 307.2.

HANDRAIL. See Section 1002.1.

HARDBOARD. See Section 2302.1.

[F] HAZARDOUS MATERIALS. See Section 307.2.

[F] HAZARDOUS PRODUCTION MATERIAL (HPM). See Section 415.2.

HEAD JOINT. See Section 2102.1.

[F] HEALTH HAZARD. See Section 307.2.

HEIGHT, BUILDING. See Section 502.1.

HEIGHT, WALLS. See Section 2102.1.

HELICAL PILE. See Section 1802.1.

HELIPORT. See Section 412.2.

HELISTOP. See Section 412.2.

HIGH-RISE BUILDING. A building with an occupied floor located more than 75 feet (22 860 mm) above the lowest level of fire department vehicle access.

❖ Determining what qualifies as a high-rise building is a fairly unique measurement of height. The critical measurement is from the lowest ground location where a fire department will be able to set its fire-fighting equipment to a floor level of occupied floors (including any occupied roofs). It is not a measurement from grade plane to top of the building. The basis of the measurement is analyzing the capability of fighting a fire and rescuing occupants from the outside the building. Once past a height of 75 feet (22 860 mm) above ground level, ground based fire fighting will not be sufficient [see the commentary, Section 403 and Figure 403.1(1)].

[F] HIGHLY TOXIC. See Section 307.2.

HISTORIC BUILDINGS. Buildings that are listed in or eligible for listing in the National Register of Historic Places, or designated as historic under an appropriate state or local law (see Sections 3409 and 3411.9).

❖ The code provides a subjective exception for compliance in Sections 3409 and 3411.9 to registered historic buildings that are receiving improvements or undergoing changes of any kind. To be a historic building it need to be designated as such through a federal, state or local law. In addition, there are buildings that have been reviewed for eligibility to be listed as a national historic building. Those listed as eligible for national listing also are considered historic for the purposes of this code. Buildings that are within a historic district are not necessarily, themselves, historic buildings. The determination of their designation as historic would depend on the specifics of the listing of the historic area.

HORIZONTAL ASSEMBLY. See Section 702.1.

HOSPITALS AND MENTAL HOSPITALS. See Section 308.3.1.

HOUSING UNIT. See Section 408.1.1.

[F] HPM FLAMMABLE LIQUID. See Section 415.2.

[F] HPM ROOM. See Section 415.2.

HURRICANE-PRONE REGIONS. See Section 1609.2.

[F] HYDROGEN CUTOFF ROOM. See Section 421.2.

[F] IMMEDIATELY DANGEROUS TO LIFE AND HEALTH (IDLH). See Section 415.2.

IMPACT LOAD. See Section 1602.1.

[F] INCOMPATIBLE MATERIALS. See Section 307.2.

[F] INERT GAS. See Section 307.2.

[F] INITIATING DEVICE. See Section 902.1.

INSPECTION CERTIFICATE. See Section 1702.1.

INTENDED TO BE OCCUPIED AS A RESIDENCE. See Section 1102.1.

INTERIOR FINISH. See Section 802.1.

INTERIOR FLOOR FINISH. See Section 802.1.

[F] INTERIOR FLOOR-WALL BASE. See Section 802.1.

INTERIOR SURFACES. See Section 2502.1.

INTERIOR WALL AND CEILING FINISH. See Section 802.1.

INTERLAYMENT. See Section 1502.1.

INTUMESCENT FIRE-RESISTANT COATINGS. See Section 1702.1.

JOINT. See Section 702.1.

JURISDICTION. The governmental unit that has adopted this code under due legislative authority.

❖ The governmental unit adopting the code has the legal authority to do so under state statutes.

LABEL. An identification applied on a product by the manufacturer that contains the name of the manufacturer, the function and performance characteristics of the product or material, and the name and identification of an *approved agency* and that indicates that the representative sample of the product or material has been tested and evaluated by an *approved agency* (see Section 1703.5 and "Inspection certificate," "Manufacturer's designation" and "*Mark*").

❖ A label provides verification of testing and inspection of materials, products or assemblies (see commentary Section 1703.5). See also the commentary for definitions of "Listed," "Mark," "Labeled" and "Manufacturer's Designation" in this chapter.

LABELED. Equipment, materials or products to which has been affixed a *label*, seal, symbol or other identifying *mark* of a nationally recognized testing laboratory, inspection agency or other organization concerned with product evaluation that maintains periodic inspection of the production of the above-labeled items and whose labeling indicates either that the equipment, material or product meets identified standards or has been tested and found suitable for a specified purpose.

❖ The term is an adjective applied to equipment, materials and products that has been tested or otherwise determined to meet the intended purpose or meet a standard. The label is that of laboratory or other agency qualified to do the evaluations (see commentary, Section 1703.5). See also the commentary for definitions of "Listed," "Mark," "Label" and "Manufacturer's designation" in this chapter.

LIGHT-DIFFUSING SYSTEM. See Section 2602.1.

LIGHT-FRAME CONSTRUCTION. A type of construction whose vertical and horizontal structural elements are primarily formed by a system of repetitive wood or cold-formed steel framing members.

❖ The code uses the term "light frame" to distinguish this unique type of framing system from other structural systems. The structural integrity of light-frame construction is dependent upon numerous connections or frequent bracing. Other framing systems or terms commonly used in the building industry that are considered as light-frame construction include: "stick built," "platform frame," "western frame" and "balloon frame." Section 2210 pertains to light-frame cold-formed steel construction. Section 2308 defines a specific subcategory of light-frame construction called "Conventional light-frame construction," which is limited to wood materials.

LIGHT-TRANSMITTING PLASTIC ROOF PANELS. See Section 2602.1.

LIGHT-TRANSMITTING PLASTIC WALL PANELS. See Section 2602.1.

LIMIT STATE. See Section 1602.1.

[F] LIQUID. See Section 415.2.

[F] LIQUID STORAGE ROOM. See Section 415.2.

[F] LIQUID USE, DISPENSING AND MIXING ROOM. See Section 415.2.

LISTED. Equipment, materials, products or services included in a list published by an organization acceptable to the code official and concerned with evaluation of products or services that maintains periodic inspection of production of *listed* equipment or materials or periodic evaluation of services and whose listing states either that the equipment, material, product or service meets identified standards or has been tested and found suitable for a specified purpose.

❖ When a product is listed and labeled, it indicates that it has been tested for conformance to an applicable standard and is subject to a third-party inspection quality assurance (QA) program. The QA verifies that the minimum level of quality required by the appropriate standard is maintained. Labeling provides a readily available source of information that is useful for field inspection of installed products. The label identifies the product or material and provides other information that can be further investigated if there is any question as to its suitability for the specific installation. The labeling agency performing the third-party inspection must be approved by the building official, and the basis fro this approval may include, but is not limited to, the capacity and capability of the agency to perform the specific testing and inspection. See also the commentary for definitions of "Mark," "Label," "Labeled" and "Manufacturer's designation" in this chapter.

LIVE LOADS. See Section 1602.1.

LIVE LOADS (ROOF). See Section 1602.1.

LOAD AND RESISTANCE FACTOR DESIGN (LRFD). See Section 1602.1.

LOAD EFFECTS. See Section 1602.1.

LOAD FACTOR. See Section 1602.1.

LOADS. See Section 1602.1.

LOT. A portion or parcel of land considered as a unit.

❖ A lot is a legally recorded parcel of land, the boundaries of which are described on a deed. When code requirements are based on some element of a lot (such as yard area or lot line location), it is the physical attributes of the parcel of land that the code is addressing, not issues of ownership. Adjacent lots owned by the same party are treated as if they were owned by different parties because ownership can change at any time. However, a group of platted lots or subdivision lots could be joined together and "considered as a unit" for the purposes of the code. For example, a collection of platted lots could be used as a single building lot for the construction of a covered mall and its associated anchor buildings. Local jurisdictions may require for taxing or other purposes that the lots be legally joined, or merged, as well.

A condominium form of building ownership, whether a residential or a commercial condominium, does not create separate lots (i.e., parcels of land) and such unit owners are treated as separate tenants, not separate lot owners. The lines separating on part of a condominium from another are not lot lines but lines indicating the limits of ownership. As such, walls constructed on lines separating condominium ownership would not need to be fire (or party) walls.

LOT LINE. A line dividing one lot from another, or from a street or any public place.

❖ Lot lines are legally recorded divisions between two adjacent land parcels or lots. They are the reference point for the location of buildings for exterior separation and other code purposes (see the definition of "Lot" above).

[F] LOWER FLAMMABLE LIMIT (LFL). See Section 415.2.

LOWEST FLOOR. See Section 1612.2.

MAIN WINDFORCE-RESISTING SYSTEM. See Section 1702.1.

[F] MANUAL FIRE ALARM BOX. See Section 902.1.

MANUFACTURER'S DESIGNATION. An identification applied on a product by the manufacturer indicating that a product or material complies with a specified standard or set of rules (see also "Inspection certificate," "*Label*" and "*Mark*").

❖ This represents terminology for a manufacturer's self-certification that a product complies with a given standard (see commentary, Section 1703.4). See also the commentary for definitions of "Listed," "Mark," "Label" and "Labeled" in this chapter.

MARK. An identification applied on a product by the manufacturer indicating the name of the manufacturer and the function of a product or material (see also "Inspection certificate," "*Label*" and "Manufacturer's designation").

❖ A mark represents the manufacturer's identification placed on a product, stating who made the product and describing its function. There is, however, no certification of compliance to any particular standard and no third-party quality control (see commentary, Section 1703.4). See also the commentary for definitions of "Listed," "Label," "Labeled" and "Manufacturer's designation" in this chapter.

MARQUEE. A permanent roofed structure attached to and supported by the building and that projects into the public right-of-way.

❖ Marquees, unlike awnings, are fixed, permanent structures that justify sufficiently different requirements from those for other projections (see Section 3106 for code requirements for marquees).

MASONRY. See Section 2102.1.

    Ashlar masonry. See Section 2102.1.

    Coursed ashlar. See Section 2102.1.

    Glass unit masonry. See Section 2102.1.

    Plain masonry. See Section 2102.1.

Random ashlar. See Section 2102.1.

Reinforced masonry. See Section 2102.1.

Solid masonry. See Section 2102.1.

Unreinforced (plain) masonry. See Section 2102.1.

MASONRY UNIT. See Section 2102.1.

Clay. See Section 2102.1.

Concrete. See Section 2102.1.

Hollow. See Section 2102.1.

Solid. See Section 2102.1.

MASTIC FIRE-RESISTANT COATINGS. See Section 1702.1.

MAXIMUM CONSIDERED EARTHQUAKE GROUND MOTION. See Section 1613.2.

MEANS OF EGRESS. See Section 1002.1.

MECHANICAL-ACCESS OPEN PARKING GARAGES. See Section 406.3.2.

MECHANICAL EQUIPMENT SCREEN. See Section 1502.1.

MECHANICAL SYSTEMS. See Section 1613.2.

MEMBRANE-COVERED CABLE STRUCTURE. See Section 3102.2.

MEMBRANE-COVERED FRAME STRUCTURE. See Section 3102.2.

MEMBRANE PENETRATION. See Section 702.1.

MEMBRANE-PENETRATION FIRESTOP. See Section 702.1.

MENTAL HOSPITALS. See Section 308.3.1.

MERCHANDISE PAD. See Section 1002.1.

METAL COMPOSITE MATERIAL (MCM). See Section 1402.1.

METAL COMPOSITE MATERIAL (MCM) SYSTEM. See Section 1402.1.

METAL ROOF PANEL. See Section 1502.1.

METAL ROOF SHINGLE. See Section 1502.1.

MEZZANINE. See Section 502.1.

MICROPILE. See Section 1802.1.

MINERAL BOARD. See Section 721.1.1.

MINERAL FIBER. See Section 702.1.

MINERAL WOOL. See Section 702.1.

MODIFIED BITUMEN ROOF COVERING. See Section 1502.1.

MORTAR. See Section 2102.1.

MORTAR, SURFACE-BONDING. See Section 2102.1.

MULTILEVEL ASSEMBLY SEATING. See Section 1102.1.

[F] MULTIPLE-STATION ALARM DEVICE. See Section 902.1.

[F] MULTIPLE-STATION SMOKE ALARM. See Section 902.1.

MULTISTORY UNITS. See Section 1102.1.

NAILING, BOUNDARY. See Section 2302.1.

NAILING, EDGE. See Section 2302.1.

NAILING, FIELD. See Section 2302.1.

NATURALLY DURABLE WOOD. See Section 2302.1.

Decay resistant. See Section 2302.1.

Termite resistant. See Section 2302.1.

NOMINAL LOADS. See Section 1602.1.

NOMINAL SIZE (LUMBER). See Section 2302.1.

NONCOMBUSTIBLE MEMBRANE STRUCTURE. See Section 3102.2.

[F] NORMAL TEMPERATURE AND PRESSURE (NTP). See Section 415.2.

NOSING. See Section 1002.1.

NOTIFICATION ZONE. See Section 902.1.

[F] NUISANCE ALARM. See Section 902.1.

NURSING HOMES. See Section 308.3.1.

OCCUPANCY CATEGORY. See Section 1602.1.

OCCUPANT LOAD. See Section 1002.1.

OCCUPIABLE SPACE. A room or enclosed space designed for human occupancy in which individuals congregate for amusement, educational or similar purposes or in which occupants are engaged at labor, and which is equipped with *means of egress* and light and ventilation facilities meeting the requirements of this code.

❖ Occupiable spaces are those areas designed for human occupancy. It applies to both residential and nonresidential spaces alike. Most spaces in a building are occupiable spaces. Based on the nature of the occupancy, various code sections apply. All habitable spaces are also considered occupiable (see the definition of "Habitable space"); however, all occupiable spaces are not habitable. Additionally, some spaces are neither habitable nor occupiable. The code identifies crawl spaces, attics, penthouses and elevated platforms (mechanical or industrial equipment) as unoccupied spaces. Since the code generally states how these spaces must be accessed, but does not specifically require means of egress, they would not be occupiable spaces. If access is limited to maintenance and service personnel, it is likely that a space is not occupiable.

OPEN PARKING GARAGE. See Section 406.3.2.

[F] OPEN SYSTEM. See Section 307.2.

[F] OPERATING BUILDING. See Section 307.2.

ORDINARY PRECAST STRUCTURAL WALL. See Section 1908.1.1.

ORDINARY REINFORCED CONCRETE STRUCTURAL WALL. See Section 1908.1.1.

ORDINARY STRUCTURAL PLAIN CONCRETE WALL. See Section 1908.1.1.

[F] ORGANIC PEROXIDE. See Section 307.2.

    Class I. See Section 307.2.

    Class II. See Section 307.2.

    Class III. See Section 307.2.

    Class IV. See Section 307.2.

    Class V. See Section 307.2.

    Unclassified detonable. See Section 307.2.

ORTHOGONAL. See Section 1613.2.

OTHER STRUCTURES. See Section 1602.1.

OWNER. Any person, agent, firm or corporation having a legal or equitable interest in the property.

❖ This term defines the person or other legal entity who is responsible for a building and its compliance with the code requirements.

[F] OXIDIZER. See Section 307.2.

    Class 4. See Section 307.2.

    Class 3. See Section 307.2.

    Class 2. See Section 307.2.

    Class 1. See Section 307.2.

[F] OXIDIZING GAS. See Section 307.2.

PANEL (PART OF A STRUCTURE). See Section 1602.1.

PANIC HARDWARE. See Section 1002.1.

PARTICLEBOARD. See Section 2302.1.

PENETRATION FIRESTOP. See Section 702.1.

PENTHOUSE. See Section 1502.1.

PERMIT. An official document or certificate issued by the authority having jurisdiction which authorizes performance of a specified activity.

❖ The permit constitutes a license issued by the building official to proceed with a specific activity, such as construction of a building, in accordance with all applicable laws.

PERSON. An individual, heirs, executors, administrators or assigns, and also includes a firm, partnership or corporation, its or their successors or assigns, or the agent of any of the aforesaid.

❖ Corporations and other organizations listed in the definition are treated as persons under the law. Also, when the code provides for a penalty (see Section 114.4), the definition makes it clear that the individuals responsible for administering the activities of these various organizations are subject to these penalties.

PERSONAL CARE SERVICE. See Section 310.2.

PHOTOLUMINESCENT. See Section 1002.1.

[F] PHYSICAL HAZARD. See Section 307.2.

[F] PHYSIOLOGICAL WARNING THRESHOLD LEVEL. See Section 415.2.

PINRAIL. See Section 410.2.

PLASTIC, APPROVED. See Section 2602.1.

PLASTIC GLAZING. See Section 2602.1.

PLATFORM. See Section 410.2.

POSITIVE ROOF DRAINAGE. See Section 1502.1.

PREFABRICATED WOOD I-JOIST. See Section 2302.1.

PRESTRESSED MASONRY. See Section 2102.1.

PRIMARY FUNCTION. See Section 3402.1.

PRIMARY STRUCTURAL FRAME. The primary structural frame shall include all of the following structural members:

1. The columns;

2. Structural members having direct connections to the columns, including girders, beams, trusses and spandrels;

3. Members of the floor construction and roof construction having direct connections to the columns; and

4. Bracing members that are essential to the vertical stability of the primary structural frame under gravity loading shall be considered part of the primary structural frame whether or not the bracing member carries gravity loads.

❖ The primary structural frame and secondary members must meet different standards of design and protection as specified in Chapters 6 and 7. The definitions of these two terms spell out which elements of a structures framing system are part of the primary structural framing system essential to carrying the gravity loads of the building. Such elements are generally required have greater fire-resistance protection (see commentary, Table 601 and Section 704).

PRISM. See Section 2102.1.

PROSCENIUM WALL. See Section 410.2.

PUBLIC ENTRANCE. See Section 1102.1.

PUBLIC-USE AREAS. See Section 1102.1.

PUBLIC WAY. See Section 1002.1.

[F] PYROPHORIC. See Section 307.2.

[F] PYROTECHNIC COMPOSITION. See Section 307.2.

RAMP. See Section 1002.1.

RAMP-ACCESS OPEN PARKING GARAGES. See Section 406.3.2.

[F] RECORD DRAWINGS. See Section 902.1.

REFLECTIVE PLASTIC CORE FOIL INSULATION. An insulation material packaged in rolls, that is less than 0.5 inches thick, with at least one exterior low emittance surface (0.1 or less) and a core material containing voids or cells.

❖ This is a product distinct from foam plastic insulation as defined in Section 2602. It's unique properties dictates the need for a different testing procedure as specified in Section 2613.

REGISTERED DESIGN PROFESSIONAL. An individual who is registered or licensed to practice their respective design profession as defined by the statutory requirements of the professional registration laws of the state or jurisdiction in which the project is to be constructed.

❖ Legal qualifications for engineers and architects are established by the state having jurisdiction. Licensing and registration of engineers and architects are accomplished by written or oral examinations offered by states or by reciprocity (licensing in other states).

REGISTERED DESIGN PROFESSIONAL IN RESPONSIBLE CHARGE. A *registered design professional* engaged by the owner to review and coordinate certain aspects of the project, as determined by the *building official*, for compatibility with the design of the building or structure, including submittal documents prepared by others, deferred submittal documents and phased submittal documents.

❖ This definition refers to the registered design professional named by the owner where required by Section 107.3.4.1, which states that the owner must designate a registered design profession when required by the laws applicable to the jurisdiction in which its building is constructed. The role of the registered design professional, in responsible charge, includes the review and coordination of the following items for compatibility with a project's design requirements:

1. Submittal documents prepared by others;

2. Deferred submittal documents;

3. Phases submittal documents; and

4. Special inspection reports.

RELIGIOUS WORSHIP, PLACE OF. A building or portion thereof intended for the performance of religious services.

❖ This term has been added to the code for the purpose of making the code more broadly applicable to the worship facilities of all religions. Major religions for the world include Christianity, Islam, Hinduism, Buddhism and Judaism, which use different terms to describe the main space used for religious services. The intent in the code is for the same application for all similar types of religious facilities. The term also makes it clear that it defines the room or sanctuary for the performance of religious worship services and not retreat complexes, rectories, convents and classroom or office areas.

REPAIR. The reconstruction or renewal of any part of an existing building for the purpose of its maintenance.

❖ As indicated in Section 105.2.2, the repair of an item typically does not require a permit. This definition makes it clear that repair is limited to work on the item, and does not include complete or substantial replacement or other new work.

REROOFING. See Section 1502.1.

RESIDENTIAL AIRCRAFT HANGAR. See Section 412.2.

RESIDENTIAL CARE/ASSISTED LIVING FACILITIES. See Section 310.2.

RESISTANCE FACTOR. See Section 1602.1.

RESTRICTED ENTRANCE. See Section 1102.1.

RETRACTABLE AWNING. See Section 3105.2.

ROOF ASSEMBLY. See Section 1502.1.

ROOF COVERING. See Section 1502.1.

ROOF COVERING SYSTEM. See Section 1502.1.

ROOF DECK. See Section 1502.1.

ROOF RECOVER. See Section 1502.1.

ROOF REPAIR. See Section 1502.1.

ROOF REPLACEMENT. See Section 1502.1.

ROOF VENTILATION. See Section 1502.1.

ROOFTOP STRUCTURE. See Section 1502.1.

RUBBLE MASONRY. See Section 2102.1.

    Coursed rubble. See Section 2102.1.

    Random rubble. See Section 2102.1.

    Rough or ordinary rubble. See Section 2102.1.

RUNNING BOND. See Section 2102.1.

SALLYPORT. See Section 408.1.1.

SCISSOR STAIR. See Section 1002.1.

SCUPPER. See Section 1502.1.

SECONDARY MEMBERS. The following structural members shall be considered secondary members and not part of the primary structural frame:

1. Structural members not having direct connections to the columns;

2. Members of the floor construction not having direct connections to the columns; and

3. Bracing members other than those that are part of the *primary structural frame*.

❖ This term works in conjunction with the term "primary structural frame" to distinguish which elements of a building's structure needs to receive various levels of fire resistance protection. These requirements are found in Table 601 and Sections 704.

SEISMIC DESIGN CATEGORY. See Section 1613.2.

SEISMIC-FORCE-RESISTING SYSTEM. See Section 1613.2.

SELF-CLOSING. See Section 702.1.

SELF-LUMINOUS. See Section 1002.1.

SELF-SERVICE STORAGE FACILITY. See Section 1102.1.

[F] SERVICE CORRIDOR. See Section 415.2.

SERVICE ENTRANCE. See Section 1102.1.

SHAFT. See Section 702.1.

SHAFT ENCLOSURE. See Section 702.1.

SHALLOW FOUNDATION. See Section 1802.1.

SHEAR WALL. See Sections 2102.1 and 2302.1.

  Detailed plain masonry shear wall. See Section 2102.1.

  Intermediate prestressed masonry shear wall. See Section 2102.1.

  Intermediate reinforced masonry shear wall. See Section 2102.1.

  Ordinary plain masonry shear wall. See Section 2102.1.

  Ordinary plain prestressed masonry shear wall. See Section 2102.1.

  Ordinary reinforced masonry shear wall. See Section 2102.1.

  Perforated shear wall. See Section 2302.1.

  Perforated shear wall segment. See Section 2302.1.

  Special prestressed masonry shear wall. See Section 2102.1.

  Special reinforced masonry shear wall. See Section 2102.1.

SHELL. See Section 2102.1.

SINGLE-PLY MEMBRANE. See Section 1502.1.

[F] SINGLE-STATION SMOKE ALARM. See Section 902.1.

SITE. See Section 1102.1.

SITE CLASS. See Section 1613.2.

SITE COEFFICIENTS. See Section 1613.2.

SITE-FABRICATED STRETCH SYSTEM. See Section 802.1.

SKYLIGHT, UNIT. A factory-assembled, glazed fenestration unit, containing one panel of glazing material that allows for natural lighting through an opening in the roof assembly while preserving the weather-resistant barrier of the roof.

❖ This is a specific type of sloped glazing assembly that is factory assembled. The code contains specific provisions that are appropriate for this type of building component. Factory assembled units, as opposed to site-built skylights, can be designed, tested and rated as one component that incorporates both glazing and

framing, if applicable. The individual components of site-built glazing must be designed to resist the design loads of the codes individually, and are not usually rated as an assembly.

SKYLIGHTS AND SLOPED GLAZING. Glass or other transparent or translucent glazing material installed at a slope of 15 degrees (0.26 rad) or more from vertical. Glazing material in skylights, including unit skylights, solariums, sunrooms, roofs and sloped walls, are included in this definition.

❖ The code regulates skylights and sloped glazing since their failure could result in injury and building damage (see Section 2405 for the code requirements).

SLEEPING UNIT. A room or space in which people sleep, which can also include permanent provisions for living, eating, and either sanitation or kitchen facilities but not both. Such rooms and spaces that are also part of a *dwelling unit* are not sleeping units.

❖ This definition is included to coordinate the *Fair Housing Act Guidelines* with the code. The definition for "Sleeping unit" clarifies the differences between sleeping units and dwelling units. Some examples of sleeping units are hotel guestrooms, dormitories and boarding houses. Another example would be a studio apartment with a kitchenette (i.e., countertop microwave, sink, refrigerator). Since the cooking arrangements were not the traditional permanent appliances (i.e., a cooktop, range or oven), this configuration would be considered a sleeping unit, not a dwelling unit. As already defined in the code, a dwelling unit must contain permanent facilities for living, sleeping, eating, cooking and sanitation.

[F] SMOKE ALARM. See Section 902.1.

SMOKE BARRIER. See Section 702.1.

SMOKE COMPARTMENT. See Section 702.1.

SMOKE DAMPER. See Section 702.1.

[F] SMOKE DETECTOR. See Section 902.1.

SMOKE-DEVELOPED INDEX. See Section 802.1.

SMOKE-PROTECTED ASSEMBLY SEATING. See Section 1002.1.

SMOKEPROOF ENCLOSURE. See Section 902.1.

[F] SOLID. See Section 415.2.

SPECIAL AMUSEMENT BUILDING. See Section 411.2.

SPECIAL FLOOD HAZARD AREA. See Section 1612.2.

SPECIAL INSPECTION. See Section 1702.1.

SPECIAL INSPECTION, CONTINUOUS. See Section 1702.1.

SPECIAL INSPECTION, PERIODIC. See Section 1702.1.

SPECIAL STRUCTURAL WALL. See Section 1908.1.1.

SPECIFIED. See Section 2102.1.

SPECIFIED COMPRESSIVE STRENGTH OF MASONRY (f′_m). See Section 2102.1.

SPLICE. See Section 702.1.

SPRAYED FIRE-RESISTANT MATERIALS. See Section 1702.1.

STACK BOND. See Section 2102.1.

STAGE. See Section 410.2.

STAIR. See Section 1002.1.

STAIRWAY. See Section 1002.1.

STAIRWAY, EXTERIOR. See Section 1002.1.

STAIRWAY, INTERIOR. See Section 1002.1.

STAIRWAY, SPIRAL. See Section 1002.1.

[F] STANDPIPE SYSTEM, CLASSES OF. See Section 902.1.

   Class I system. See Section 902.1.

   Class II system. See Section 902.1.

   Class III system. See Section 902.1.

[F] STANDPIPE, TYPES OF. See Section 902.1.

   Automatic dry. See Section 902.1.

   Automatic wet. See Section 902.1.

   Manual dry. See Section 902.1.

   Manual wet. See Section 902.1.

   Semiautomatic dry. See Section 902.1.

START OF CONSTRUCTION. See Section 1612.2.

STEEL CONSTRUCTION, COLD-FORMED. See Section 2202.1.

STEEL JOIST. See Section 2202.1.

STEEL MEMBER, STRUCTURAL. See Section 2202.1.

STEEP SLOPE. A roof slope greater than two units vertical in 12 units horizontal (17-percent slope).

❖ This is the general criterion for roof slope that is used throughout the code. Slope requirements for specific roof covering materials are specified in Chapter 15.

STONE MASONRY. See Section 2102.1.

   Ashlar stone masonry. See Section 2102.1.

   Rubble stone masonry. See Section 2102.1.

[F] STORAGE, HAZARDOUS MATERIALS. See Section 415.2.

STORM SHELTER. See Section 423.2.

   Community storm shelter. See Section 423.2.

   Residential storm shelter. See Section 423.2.

STORY. That portion of a building included between the upper surface of a floor and the upper surface of the floor or roof next above (also see "Basement," "*Mezzanine*" and Section 502.1). It is measured as the vertical distance from top to top of two successive tiers of beams or finished floor surfaces and, for the topmost *story*, from the top of the floor finish to the top of the ceiling joists or, where there is not a ceiling, to the top of the roof rafters.

❖ All levels in a building that conform to this description are stories, including basements. A mezzanine is considered part of the story in which it is located. See Chapter 5 for code requirements regarding limitations on the number of stories in a building as a function of the type of construction.

STORY ABOVE GRADE PLANE. Any *story* having its finished floor surface entirely above *grade plane*, or in which the finished surface of the floor next above is:

   1. More than 6 feet (1829 mm) above *grade plane*; or

   2. More than 12 feet (3658 mm) above the finished ground level at any point.

❖ The determination of the allowed height of a building under Table 503 is based on the number of stories above grade plane. ("Grade plane," "Building height" and "Basement" are all defined in Section 502.1.) The code establishes by this definition which stories of a building are those above grade plane. Clearly, it is includes those stories which are fully above grade plane. It also includes stories which may be partially below finished ground level, but the finished floor level is more than 6 feet (1829 mm) above grade plane. It also includes those floor levels which, due to an irregular terrain, have a finished floor level more than 12 feet (3658 mm) above finished ground level at any point surrounding the building. Any building level not qualifying as a story above grade plane is, by definition in Section 502.1, a basement. See the commentary and figures in Section 502.1.

STRENGTH. See Section 2102.1.

   Design strength. See Section 2102.1.

   Nominal strength. See Sections 1602.1 and 2102.1.

   Required strength. See Sections 1602.1 and 2102.1.

STRENGTH DESIGN. See Section 1602.1.

STRUCTURAL COMPOSITE LUMBER. See Section 2302.1.

   Laminated veneer lumber (LVL). See Section 2302.1.

   Parallel strand lumber (PSL). See Section 2302.1.

STRUCTURAL GLUED-LAMINATED TIMBER. See Section 2302.1.

STRUCTURAL OBSERVATION. See Section 1702.1.

STRUCTURE. That which is built or constructed.

❖ This definition is intentionally broad so as to include within its scope, and therefore the scope of the code (see Section 101.2), everything that is built as an improvement to real property. See also the definitions for "Building" and "Area, building" for the difference between a building and structure.

SUBDIAPHRAGM. See Section 2302.1.

SUBSTANTIAL DAMAGE. See Section 1612.2.

SUBSTANTIAL IMPROVEMENT. See Section 1612.2.

SUBSTANTIAL STRUCTURAL DAMAGE. See Section 3402.1.

SUITE. See Section 1002.1.

SUNROOM. See Section 1202.1.

[F] SUPERVISING STATION. See Section 902.1.

[F] SUPERVISORY SERVICE. See Section 902.1.

[F] SUPERVISORY SIGNAL. See Section 902.1.

[F] SUPERVISORY SIGNAL-INITIATING DEVICE. See Section 902.1.

SWIMMING POOLS. See Section 3109.2.

T RATING. See Section 702.1.

TECHNICALLY INFEASIBLE. See Section 3402.1.

TENT. A structure, enclosure or shelter, with or without sidewalls or drops, constructed of fabric or pliable material supported in any manner except by air or the contents it protects.

❖ Tents can be temporary or permanent structures. When permanent, they are considered membrane-covered structures and are regulated by Section 3102. When erected as temporary enclosures, they are regulated by Section 3103.

THERMAL ISOLATION. See Section 1202.1.

THERMOPLASTIC MATERIAL. See Section 2602.1.

THERMOSETTING MATERIAL. See Section 2602.1.

THIN-BED MORTAR. See Section 2102.1.

THROUGH PENETRATION. See Section 702.1.

THROUGH-PENETRATION FIRESTOP SYSTEM. See Section 702.1.

TIE-DOWN (HOLD-DOWN). See Section 2302.1.

TIE, LATERAL. See Section 2102.1.

TIE, WALL. See Section 2102.1.

TILE. See Section 2102.1.

TILE, STRUCTURAL CLAY. See Section 2102.1.

[F] TIRES, BULK STORAGE OF. See Section 902.1.

TOWNHOUSE. A single-family *dwelling unit* constructed in a group of three or more attached units in which each unit extends from the foundation to roof and with open space on at least two sides.

❖ This specific configuration of construction is called different things in different parts of the country, such as a rowhouse. A townhouse structure which meets four

criteria is not regulated by this code but is regulated by the IRC. Those criteria are:

1. Each unit extends from foundation to roof with no vertical overlap of any parts of adjoining units;

2. Each unit must have open space on two sides (either two opposite or two adjoining sides);

3. Each unit must have a separate means of egress; and

4. The building must not exceed three stories above grade plane.

If all of these criteria are met, then according to the exception to Section 101.2, the structure is within scope of the IRC. (It should also be noted that townhouses within the IRC must be separated by a wall or walls meeting specific criteria.) If a structure does not meet these four criteria, it will need to be regulated under this code and will either be classified as a Group R-2 or Group R-3 structure, depending on how the units are separated. A building containing three or more dwelling units is regulated as a Group R-2 occupancy. To be considered a Group R-3, the structure must have one or two dwelling units, or be subdivided by fire walls between every unit or every two units (see Section 310.1 and the definition for "Area, building"). Finally, the definition of townhouse is not dependent on the presence of individual lots. A townhouse structure could be built with any number of attached units on the same lot, or it could be developed such that a property line lies at each common wall separating two units (see definition for "Lot").

[F] TOXIC. See Section 307.2.

TRANSIENT. See Section 310.2.

TRANSIENT AIRCRAFT. See Section 412.2.

TREATED WOOD. See Section 2302.1.

    Fire-retardant-treated wood. See Section 2302.1.

    Preservative-treated wood. See Section 2302.1.

TRIM. See Section 802.1.

[F] TROUBLE SIGNAL. See Section 902.1.

TYPE A UNIT. See Section 1102.1.

TYPE B UNIT. See Section 1102.1.

UNDERLAYMENT. See Section 1502.1.

[F] UNSTABLE (REACTIVE) MATERIAL. See Section 307.2.

    Class 4. See Section 307.2.

    Class 3. See Section 307.2.

    Class 2. See Section 307.2.

    Class 1. See Section 307.2.

[F] USE (MATERIAL). See Section 415.2.

VAPOR-PERMEABLE MEMBRANE. A material or covering having a permeance rating of 5 perms (52.9×10^{-10} kg/Pa · s · m²) or greater, when tested in accordance with the dessicant method using Procedure A of ASTM E 96. A vapor-permeable material permits the passage of moisture vapor.

❖ Greater demands on the building envelope due to energy considerations now dictate the need for an outer membrane that reduces wind infiltration. The membranes used in this application may need to allow vapor to pass through it, given that a vapor barrier would be needed on the inside of the wall and would be undesirable on the outside of the wall. In such cases, a vapor-permeable membrane would be used.

VAPOR RETARDER CLASS. A measure of a material or assembly's ability to limit the amount of moisture that passes through that material or assembly. Vapor retarder class shall be defined using the desiccant method of ASTM E 96 as follows:

❖ Vapor retarders are used to limit moisture intrusion into a building. The definition establishes three classes of vapor retarders based on the amount of moisture that can pass through in a given time period.

    Class I: 0.1 perm or less.

    Class II: 0.1 < perm ≤ 1.0 perm.

    Class III: 1.0 < perm ≤ 10 perm.

VEHICLE BARRIER SYSTEM. See Section 1602.1.

VEHICULAR GATE. See Section 3110.2.

VENEER. See Section 1402.1.

VENTILATION. The natural or mechanical process of supplying conditioned or unconditioned air to, or removing such air from, any space.

❖ Ventilation is the process of moving air to or from building spaces. This definition of ventilation requirements is used in this code to establish minimum levels of air movement within a building for the purposes of providing a healthful interior environment. Ventilation would include both natural (openable exterior windows and doors for wind movement) and mechanical (forced air with mechanical equipment) methods.

VINYL SIDING. See Section 1402.1.

[F] VISIBLE ALARM NOTIFICATION APPLIANCE. See Section 902.1.

WALKWAY, PEDESTRIAN. A walkway used exclusively as a pedestrian trafficway.

❖ A pedestrian walkway is an enclosed passageway external to, and not considered part of, the buildings it connects. Intended only for pedestrian use, it can be at grade, below grade or elevated above grade.

WALL. See Section 2102.1.

    Cavity wall. See Section 2102.1.

    Composite wall. See Section 2102.1.

    Dry-stacked, surface-bonded wall. See Section 2102.1.

    Masonry-bonded hollow wall. See Section 2102.1.

    Parapet wall. See Section 2102.1.

WALL, LOAD-BEARING. Any wall meeting either of the following classifications:

1. Any metal or wood stud wall that supports more than 100 pounds per linear foot (1459 N/m) of vertical load in addition to its own weight.

2. Any masonry or concrete wall that supports more than 200 pounds per linear foot (2919 N/m) of vertical load in addition to its own weight.

❖ This definition is necessary since the structural requirements and fire-resistance-rating requirements in this code vary for nonload-bearing walls and load-bearing walls. The term "load-bearing walls" are intended to refer to wall elements that support part of the structural framework of a building.

WALL, NONLOAD-BEARING. Any wall that is not a *load-bearing wall*.

❖ This definition is necessary since the structural requirements and fire-resistance-rating requirements in this code vary for nonload-bearing walls and load-bearing walls. Nonload-bearing walls do not support any portion of the building or structure except the weight of the wall itself.

WALL PIER. See Section 1908.1.1.

[F] WATER-REACTIVE MATERIAL. See Section 307.2.

    Class 3. See Section 307.2.

    Class 2. See Section 307.2.

    Class 1. See Section 307.2.

WATER-RESISTIVE BARRIER. See Section 1402.1.

WEATHER-EXPOSED SURFACES. See Section 2502.1.

WEB. See Section 2102.1.

[F] WET-CHEMICAL EXTINGUISHING SYSTEM. See Section 902.1.

WHEELCHAIR SPACE. See Section 1102.1.

WIND-BORNE DEBRIS REGION. See Section 1609.2.

WINDER. See Section 1002.1.

WIRE BACKING. See Section 2502.1.

[F] WIRELESS PROTECTION SYSTEM. See Section 902.1.

WOOD SHEAR PANEL. See Section 2302.1.

WOOD STRUCTURAL PANEL. See Section 2302.1.

    Composite panels. See Section 2302.1.

    Oriented strand board (OSB). See Section 2302.1.

    Plywood. See Section 2302.1.

[F] WORKSTATION. See Section 415.2.

WYTHE. See Section 2102.1.

YARD. An open space, other than a *court*, unobstructed from the ground to the sky, except where specifically provided by this code, on the lot on which a building is situated.

❖ This definition is used, similar to the definition of "Court," to establish the applicability of code requirements when yards are utilized for natural light or natural ventilation purposes (see Section 1206.1). Whereas a court is bounded on three or more sides with the building or structure, a yard is bounded on two or less sides by the building or structure. See also the definition of "Egress court" in Section 1002.1.

[F] ZONE. See Section 902.1.

ZONE, NOTIFICATION. See Section 902.1.

Bibliography

The following resource materials are referenced in this chapter or are relevant to the subject matter addressed in this chapter.

24 CFR, *Fair Housing Accessibility Guidelines* (FHAG). Washington, DC, Department of Housing and Urban Development, 1991.

ACI Standard 222R-99, *Controlled Low-strength Materials.* Farmington Hills, MI: American Concrete Institute, 1999.

IRC-09, *International Residential Code.* Washington, DC: International Code Council, 2009.

Chapter 3:
Use and Occupancy Classification

General Comments

Chapter 3 provides for the classification of buildings, structures and parts thereof based on the purpose or purposes for which they are used.

Section 302 identifies the groups into which all buildings, structures and parts thereof must be classified.

Sections 303 through 312 identify the occupancy characteristics of each group classification. In some sections, specific group classifications having requirements in common are collectively organized such that one term applies to all. For example, Groups A-1, A-2, A-3, A-4 and A-5 are individual groups. The general term Group A, however, includes each of these individual groups. For this reason, each specific assembly group classification is included in Section 303.

In the early years of building code development, the essence of regulatory safeguards from fire was to provide a reasonable level of protection to property. The idea was that if property was adequately protected from fire, then the building occupants would also be protected.

From this outlook on fire safety, the concept of equivalent risk has evolved in the code. This concept maintains that, in part, an acceptable level of risk against the damages of fire respective to a particular occupancy type (group) can be achieved by limiting the height and area of buildings containing such occupancies according to the building's construction type (i.e., its relative fire endurance).

The concept of equivalent risk involves three interdependent considerations: (1) the level of fire hazard associated with the specific occupancy of the facility; (2) the reduction of fire hazard by limiting the floor area(s) and the height of the building based on the fuel load (combustible contents and burnable building components) and (3) the level of overall fire resistance provided by the type of construction used for the building.

The interdependence of these fire safety considerations can be seen by first looking at Tables 601 and 602, which show the fire-resistance ratings of the principal structural elements comprising a building in relation to the five classifications for types of construction. Type I construction is the classification that generally requires the highest fire-resistance ratings for structural elements, whereas Type V construction, which is designated as a combustible type of construction, generally requires the least amount of fire-resistance-rated structural elements. If one then looks at Table 503, the relationship among group classification, allowable heights and areas and types of construction becomes apparent. Respective to each group classification, the greater the fire-resistance rating of structural elements, as represented by the type of construction, the greater the floor area and height allowances. The greater the potential fire hazards indicated as a function of the group, the lesser the height and area allowances for a particular construction type.

As a result of extensive research and advancements in fire technology, today's building codes are more comprehensive and complex regulatory instruments than they were in the earlier years of code development. While the principle of equivalent risk remains an important component in building codes, perspectives have changed and life safety is now the paramount fire issue. Even so, occupancy classification still plays a key part in organizing and prescribing the appropriate protection measures. As such, threshold requirements for fire protection and means of egress systems are based on occupancy classification (see Chapters 9 and 10).

Other sections of the code also contain requirements respective to the classification of building groups. For example, Section 102.6 deals with applicability of the code to existing structures; Section 705 deals with requirements for exterior wall fire-resistance ratings that are tied to the occupancy classification of a building and Section 803.9 contains interior finish requirements that are dependent upon the occupancy classification.

Purpose

The purpose of this chapter is to classify a building, structure or part thereof into a group based on the specific purpose for which it is designed or occupied. Throughout the code, group classifications are considered a fundamental principle in organizing and prescribing the appropriate features of construction and occupant safety requirements for buildings, especially general building limitations, means of egress, fire protection systems and interior finishes.

SECTION 301
GENERAL

301.1 Scope. The provisions of this chapter shall control the classification of all buildings and structures as to use and occupancy.

❖ As used throughout the code, the classification of an occupancy into a group is established by the requirements of this chapter. The purpose of these provisions is to provide rational criteria for the classification of various occupancies into groups based on their relative fire hazard and life safety properties. This is necessary because the code utilizes group classification as a fundamental principle for differentiating requirements in other parts of the code related to fire and life safety protection.

SECTION 302
CLASSIFICATION

302.1 General. Structures or portions of structures shall be classified with respect to occupancy in one or more of the groups listed below. A room or space that is intended to be occupied at different times for different purposes shall comply with all of the requirements that are applicable to each of the purposes for which the room or space will be occupied. Structures with multiple occupancies or uses shall comply with Section 508. Where a structure is proposed for a purpose that is not specifically provided for in this code, such structure shall be classified in the group that the occupancy most nearly resembles, according to the fire safety and relative hazard involved.

1. Assembly (see Section 303): Groups A-1, A-2, A-3, A-4 and A-5

2. Business (see Section 304): Group B

3. Educational (see Section 305): Group E

4. Factory and Industrial (see Section 306): Groups F-1 and F-2

5. High Hazard (see Section 307): Groups H-1, H-2, H-3, H-4 and H-5

6. Institutional (see Section 308): Groups I-1, I-2, I-3 and I-4

7. Mercantile (see Section 309): Group M

8. Residential (see Section 310): Groups R-1, R-2, R-3 and R-4

9. Storage (see Section 311): Groups S-1 and S-2

10. Utility and Miscellaneous (see Section 312): Group U

❖ This section requires all structures to be classified in one or more of the groups listed according to the structure's purpose and function (i.e., its occupancy). By organizing occupancies with similar fire hazard and life safety properties into groups, the code has adopted the means to differentiate occupancies such that various fire protection and life safety requirements can be rationally organized and applied. Each specific group has an individual classification. Each represents a different characteristic and level of fire hazard that requires spe-cial code provisions to lessen the associated risks. There are some group classifications that are very closely related to other specific groups and, therefore, are collectively referred to as a single group (e.g., Group F applies to Groups F-1 and F-2). In these cases, there are requirements within the code that are common to each specific group classification. These common requirements are applicable based on the reference to the collective classification. For example, the requirements of Section 1028 apply to each specific group classification listed under the term "Group A."

Example 1: Both a restaurant (Group A-2) and a church (Group A-3) are included in Group A, but they have different specific group classifications. Both Groups A-2 and A-3 are subject to the same travel distance limitations (see Table 1016.1) and corridor fire-resistance ratings (see Table 1018.1), but have different thresholds for when automatic sprinkler systems are required (see Section 903).

Buildings that contain more than one occupancy group are mixed occupancy buildings. Buildings with mixed occupancies must comply with one of the design options contained in Section 508. When within the limits of Section 508.2, accessory occupancies including incidental accessory occupancies are individually classified as provided in Section 302 but are treated as part of the main occupancy. This condition is permitted because such areas either do not represent a significant change in the performance characteristics of the structure or are otherwise appropriately protected by the provisions contained in this section. Incidental accessory occupancies, while treated as part of the main occupancy, require protection or separation from the main occupancy as specified in Section 508.2.5.

Occasionally, a building or space is intended to be occupied for completely different purposes at different times. For instance, a church hall might be used as a day care center during weekdays and as a reception hall for weddings and other similar events at other times. In these cases, the code provisions for each occupancy must be satisfied.

In cases where a structure has a purpose that is not specifically identified within any particular occupancy classification, that structure is to be classified in the group that it most closely resembles. Before an accurate classification can be made, however, a detailed description of the activities or processes taking place inside the building, the occupant load and the materials and equipment used and stored therein must be submitted to the building official. The building official must then compare this information to the various occupancy classifications, determine which one the building most closely resembles and classify the building as such.

Example 2: A designer presents the building official with a building needing an occupancy group classification. The building official is informed that the building is to be used as an indoor shooting gallery, open to the public but used mostly by police officers. After reviewing the code, he or she cannot find a specific reference to a shooting gallery in Sections 303 through 312 or in the associated tables. The building official asks the designer for additional information about the activities to be conducted in the building and is told that there will be a small sign-in booth, patron waiting/viewing area and the actual shooting area. Based on this information, the building official can determine that the most logical classification of the building is Group A-3, assembly. This classification is based on the fact that the building is used for the congregation of people for recreation. A shooting gallery is similar in many respects to a bowling center, which is classified as Group A-3 (see Figure 302.1).

SECTION 303
ASSEMBLY GROUP A

303.1 Assembly Group A. Assembly Group A occupancy includes, among others, the use of a building or structure, or a portion thereof, for the gathering of persons for purposes such as civic, social or religious functions; recreation, food or drink consumption or awaiting transportation.

Exceptions:

1. A building or tenant space used for assembly purposes with an *occupant load* of less than 50 persons shall be classified as a Group B occupancy.

2. A room or space used for assembly purposes with an *occupant load* of less than 50 persons and accessory to another occupancy shall be classified as a Group B occupancy or as part of that occupancy.

3. A room or space used for assembly purposes that is less than 750 square feet (70 m²) in area and accessory to another occupancy shall be classified as a Group B occupancy or as part of that occupancy.

4. Assembly areas that are accessory to Group E occupancies are not considered separate occupancies except when applying the assembly occupancy requirements of Chapter 11.

5. Accessory religious educational rooms and religious auditoriums with occupant loads of less than 100 are not considered separate occupancies.

Assembly occupancies shall include the following:

A-1 Assembly uses, usually with fixed seating, intended for the production and viewing of the performing arts or motion pictures including, but not limited to:

> Motion picture theaters
> Symphony and concert halls
> Television and radio studios admitting an audience
> Theaters

A-2 Assembly uses intended for food and/or drink consumption including, but not limited to:

> Banquet halls
> Night clubs
> Restaurants
> Taverns and bars

A-3 Assembly uses intended for worship, recreation or amusement and other assembly uses not classified elsewhere in Group A including, but not limited to:

> Amusement arcades
> Art galleries
> Bowling alleys
> Community halls
> Courtrooms
> Dance halls (not including food or drink consumption)
> Exhibition halls
> Funeral parlors
> Gymnasiums (without spectator seating)
> Indoor swimming pools (without spectator seating)
> Indoor tennis courts (without spectator seating)
> Lecture halls
> Libraries

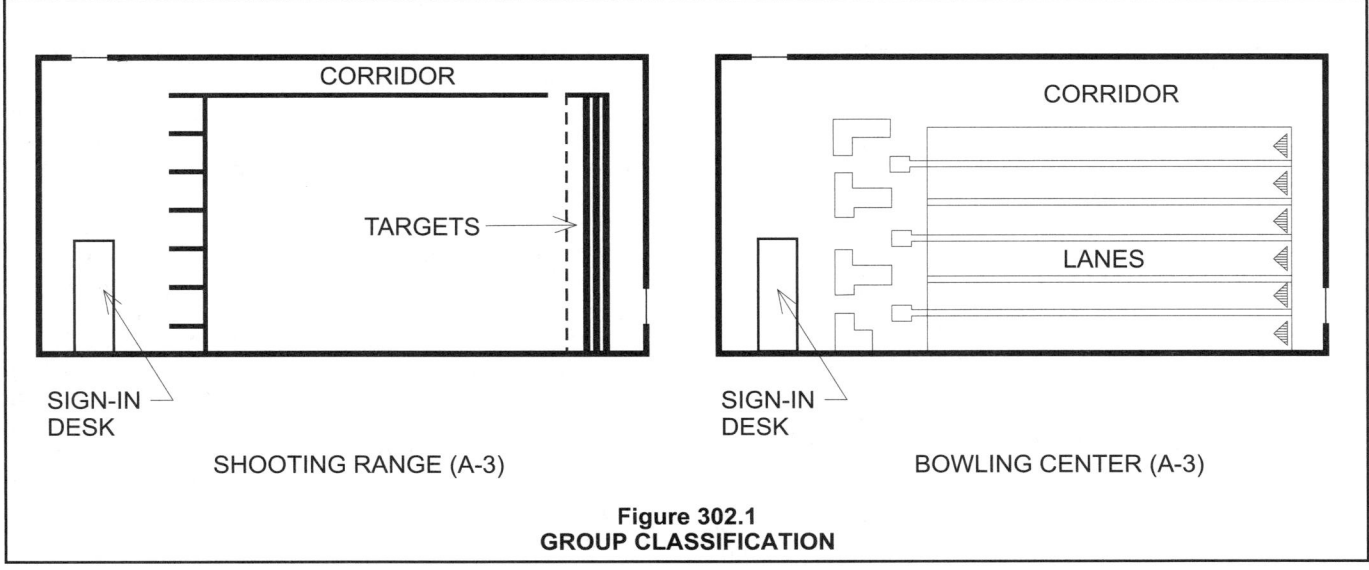

SHOOTING RANGE (A-3)

BOWLING CENTER (A-3)

Figure 302.1
GROUP CLASSIFICATION

Museums
Places of religious worship
Pool and billiard parlors
Waiting areas in transportation terminals

A-4 Assembly uses intended for viewing of indoor sporting events and activities with spectator seating including, but not limited to:

Arenas
Skating rinks
Swimming pools
Tennis courts

A-5 Assembly uses intended for participation in or viewing outdoor activities including, but not limited to:

Amusement park structures
Bleachers
Grandstands
Stadiums

❖ Because of the arrangement and density of the occupant load associated with occupancies classified in the assembly group category, the potential for multiple fatalities and injuries from fire is comparatively high. For example, no other use listed in Section 302.1 contemplates occupant loads as dense as 5 square feet (0.46 m2) per person (see Table 1004.1.1). Darkened spaces in theaters, nightclubs and the like serve to increase hazards. In sudden emergencies, the congestion caused by large numbers of people rushing to exits can cause panic conditions. For these and many other reasons, there is a relatively high degree of hazard to life safety in assembly facilities. The relative hazards of assembly occupancies are reflected in the height and area limitations of Table 503 that are, in comparison, generally more restrictive than for buildings in other group classifications.

There are five specific assembly group classifications, Groups A-1 through A-5, described in this section. Where used in the code, the general term "Group A" is intended to include all five classifications.

The fundamental characteristics of all assembly occupancies are identified in this section. Structures that are designed or occupied for assembly purposes must be placed in one of the assembly group classifications. Some exceptions to this rule are small assembly buildings, tenant spaces and assembly spaces in nonassembly buildings meeting one of the exceptions of Section 303.1.

Exception 1 recognizes that there are often small establishments that typically serve food and have a few seats that technically meet the definition of an assembly occupancy but due to the low occupant load pose a lower risk than a typical assembly occupancy. These types of buildings and tenant spaces are to be considered a Group B occupancy when they serve less than 50 people. Examples of this include small "fast food" establishments and small "mom-and-pop" restaurants or coffee shops.

Exceptions 2 and 3 address assembly rooms or spaces in nonassembly buildings. Exception 2 evaluates the occupant load (less than 50) of the assembly space, while Exception 3 evaluates the area [(less than 750 square feet (65 m2)] of the assembly space. In both cases, the purpose of the assembly space must be accessory to the principal occupancy of the structure (i.e., the activities in the assembly space are subordinate and secondary to the primary occupancy). If either the occupant load or floor area requirement is satisfied and the purpose of the assembly space is accessory to the principal occupancy, the space shall either be classified as a Group B occupancy or as part of the principal occupancy. In either case, the assembly space is not required to be less than 10 percent of the area of the story on which it is located as is specified for accessory uses in Section 508.2 (IBC Interpretation No. 20-04).

A Group E occupancy invariably contains many types of assembly spaces other than classrooms, such as auditoriums, cafeterias, gymnasiums and libraries. Therefore, Exception 4 to this section provides that accessory assembly spaces in a Group E building are not to be regulated as separate occupancies, regardless of their floor area. It is worth mentioning, for such assembly functions to be considered part of the primary Group E occupancy, the assembly functions must be ancillary and supportive to the educational operation of the building. While this exception specifically requires these assembly spaces to comply with Group A occupancy requirements specified for accessibility in Chapter 11; other code requirements specific to the use of spaces for assembly purposes must also be met. Many provisions of Chapter 10 address assembly spaces.

Places of religious worship are listed as a Group A-3 occupancy. Exception 5 permits religious education rooms and other religious auditoriums to remain in the Group A-3 occupancy even when they hold less than 100 occupants. Otherwise, they could be classified as Group E if occupied by children of ages through the 12th grade, or as Group B if the education is provided to adults. By maintaining the A-3 occupancy category for these spaces, such facilities would not need to comply with the mixed occupancy requirements of Section 508. There can be more than one room, each with an occupant load of less than 100, in the same portion of the building and this exception still applies even if the aggregate occupant load of multiple rooms exceeds 100.

The exceptions given to assembly spaces in nonassembly buildings is a practical code consideration that permits a mixed occupancy condition to exist without requiring compliance with the provisions for mixed occupancies (see Section 508) or accessory occupancies (see Section 508.2). Although the term "accessory" is used in many of these exceptions, the intent of the use of the term here is that the use of the space is related to, or part of, the main use of the space. These exceptions are not limited by the accessory use requirements found in Section 508.2.

For other code requirements, close attention needs to be paid as to whether they are based either on the Chapter 3 occupancy category, or the use of the space. For example, occupant load is determined based on the function of the space (see Table 1004.1.1), but corridor requirements for rated construction (see Table 1018.1) are based on the occupancy category being served. Sprinkler requirements in Section 903.2 are generally based on thresholds for each occupancy category. Also, please note the specific wording of each exception. For example, Exception 1 to Section 303.1 specifically says these spaces are Group B occupancies. Therefore other code requirements specific to Group B occupancies would apply. Exceptions 2 and 3 state that the small assembly spaces are Group B or the same as the occupancy to which they are accessory. Thus, the small lunchroom for 49 occupants in a factory building would be designated Group F.

Example 1: An office building, Group B, has a conference room used for staff meetings [see Figure 303.1(1)]. The occupancy of a conference room is typically considered a Group A-3. Since the occupant load of the conference room is less than 50 and its function is clearly accessory to the business area, the provisions of Exception 2 permit the room to be included in the main occupancy, Group B.

Example 2: A 749-square-foot (70 m²) assembly area is located on a mercantile floor area of 5,000 square feet (465 m²) [see Figure 303.1(2)]. While the assembly use area occupies 15 percent of the 5,000-square-foot (465 m²) floor area, it does not exceed 750 square feet (70 m²) and is not considered a separate Group A occupancy.

A-1: Some of the characteristics of Group A-1 occupancies are large, concentrated occupant loads, low lighting levels, above-normal sound levels and a moderate fuel load.

Group A-1 is characterized by two basic types of activities. The first type is one in which the facility is occupied for the production and viewing of theatrical or operatic performances. Facilities of this type ordinarily have fixed seating; a permanent raised stage; a proscenium wall and curtain; fixed or portable scenery drops; lighting devices; dressing rooms; mechanical appliances or other theatrical accessories and equipment [see Figure 303.1(3)].

The second type is one in which the structure is primarily occupied for the viewing of motion pictures. Facilities of this type ordinarily have fixed seating, no stage, a viewing screen, motion picture projection booth(s) and equipment [see Figure 303.1(4)].

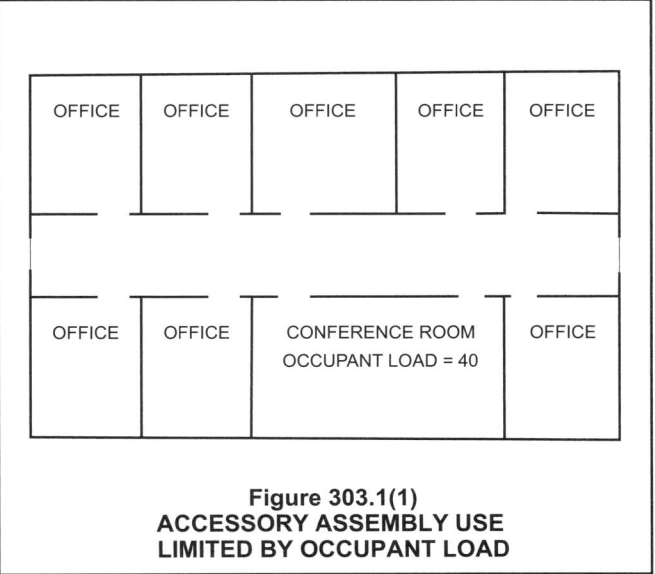

Figure 303.1(1)
ACCESSORY ASSEMBLY USE
LIMITED BY OCCUPANT LOAD

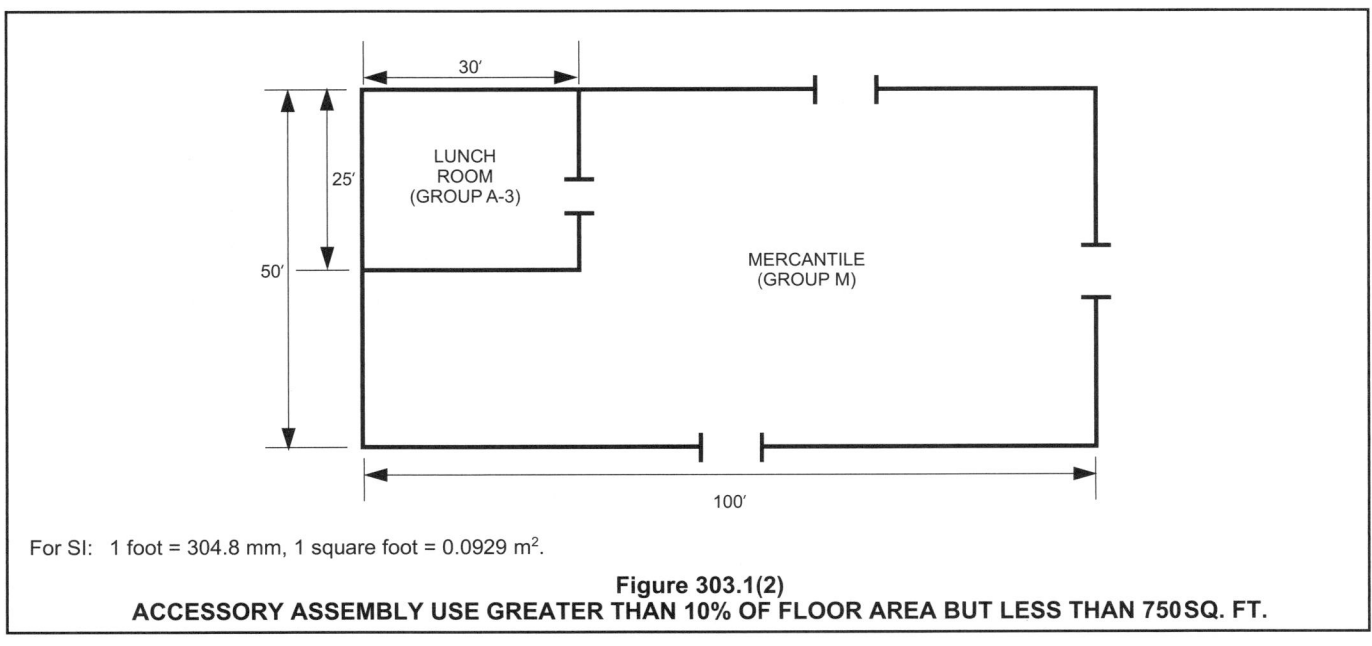

For SI: 1 foot = 304.8 mm, 1 square foot = 0.0929 m².

Figure 303.1(2)
ACCESSORY ASSEMBLY USE GREATER THAN 10% OF FLOOR AREA BUT LESS THAN 750 SQ. FT.

Group A-1 presents a significant potential life safety hazard because of the large occupant loads and the concentration of people within confined spaces. The means of egress is an important factor in the design of such facilities. Theaters for the performing arts that require stages are considered particularly hazardous because of the amount of combustibles such as curtains, drops, scenery, construction materials and other accessories normally associated with stage operation. As such, special protection requirements applicable to stages and platforms are provided in Section 410 and Chapter 10.

A-2: Group A-2 includes occupancies in which people congregate in high densities for social entertainment, such as drinking and dancing (e.g., nightclubs, banquet halls, cabarets) and food and drink consumption (e.g., restaurants). The uniqueness of these occupancies is characterized by some or all of the following:

- Low lighting levels;
- Entertainment by a live band or recorded music generating above-normal sound levels;
- No theatrical stage accessories;
- Later-than-average operating hours;
- Tables and seating arranged or positioned so as to create ill-defined aisles;
- A specific area designated for dancing;
- Service facilities for alcoholic beverages and food; and
- High occupant load density.

The fire records are very clear in identifying that the characteristics listed above often cause a delayed awareness of a fire situation and confuse occupants regarding the appropriate response, resulting in an increased egress time and sometimes panic. Together, these factors may result in extensive life and property losses. These characteristics are only advisory in determining whether Group A-2 is the appropriate classification. Often there are additional characteristics that are unique to a project, which also must be taken into consideration when a classification is made.

Example: The Downtown Club, a popular local nightclub/dance hall, features a different band every weekend [see Figure 303.1(5)]. It is equipped with a bar and basic kitchen facilities so that beverages and appetizers can be served. There is a platform for a band to perform, a dance floor in front of the platform and numerous cocktail tables and chairs. The tables and chairs are not fixed, resulting in a hazardous arrangement because there are no distinct aisles. When the band performs, the house lights are dimmed and spotlights are keyed in on the performers. The club is equipped with a sound system that is used at loud levels. The club is open until 3:00 a.m.—the latest time the local jurisdiction will allow.

From this description of the Downtown Club, one can readily see that the appropriate classification is Group A-2. Sometimes, however, it is not this easy to determine the appropriate classification. In such cases, the building official must seek additional information regarding the functions of the building and each area within the building.

A-3: Structures in which people assemble for the purpose of social activities (such as entertainment, recre-

**Figure 303.1(3)
GROUP A-1**

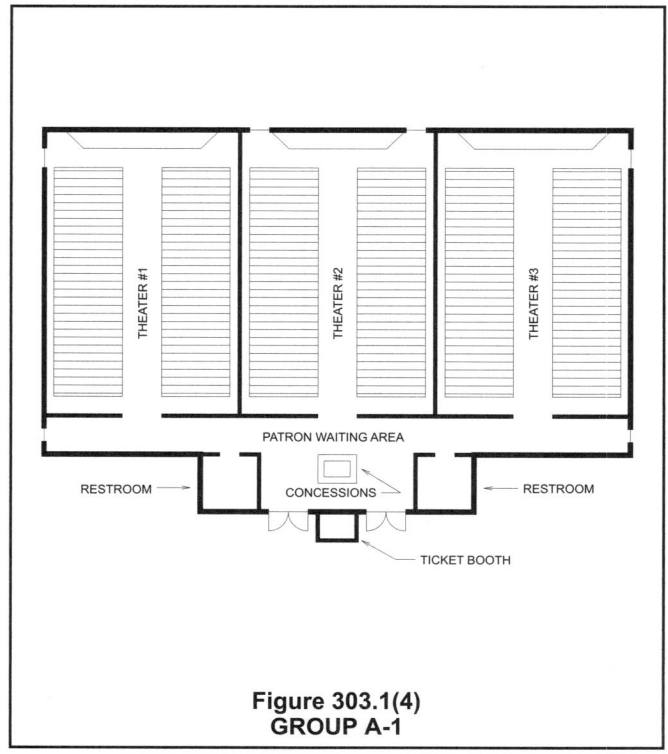

**Figure 303.1(4)
GROUP A-1**

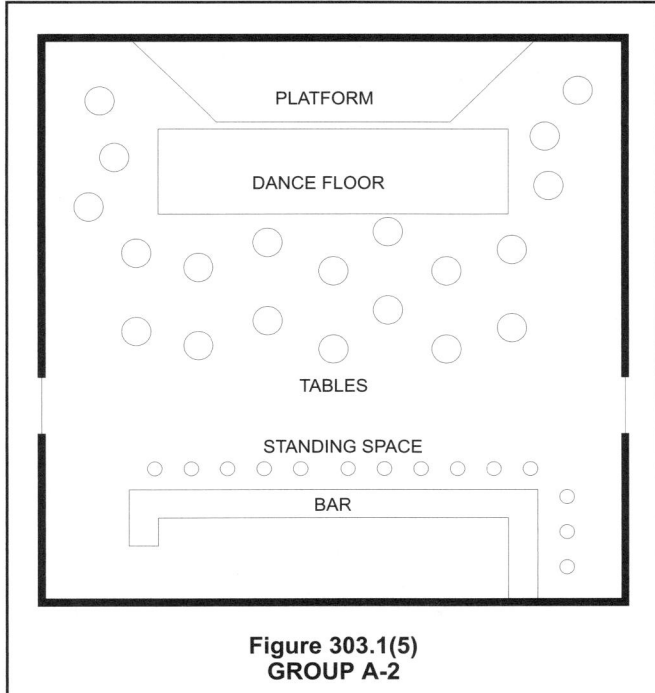

**Figure 303.1(5)
GROUP A-2**

ation and amusement) that are neither classified in Group A-1 or A-2 nor appropriately classified in Group A-4 or A-5 are to be classified in Group A-3. Exhibition halls, libraries, dance halls (not including food and drink), places of religious worship, museums, gymnasiums, recreation centers, health clubs, fellowship halls, indoor shooting galleries, bowling centers and billiard halls are among the facilities often classified in Group A-3. Also, since they most nearly resemble this occupancy classification, public and private spaces used for assembly are often classified in Group A-3. These include large courtrooms, meeting rooms and conference centers. Similarly, lecture rooms located in colleges, universities or in schools for students above the 12th grade that have an occupant load of 50 or more are also classified in Group A-3. Structures in which people gather exclusively for worship and other religious purposes are also classified as Group A-3. Although such worship and religious purposes are without restriction to any particular sect or creed, the intent of the code is to limit Group A-3 classification to occupancies that are specifically related to worship services, devotions and religious rituals.

Frequently, other occupancies are located within the same structure where religious services (Group A-3) are performed [e.g., classrooms (Group E), care for infants (Group I-4) and staff offices (Group B)]. When this occurs, and depending on their size, these occupancies must be considered as either accessory occupancies or other group occupancies. As indicated in Exception 5, religious educational classrooms with occupant loads less than 100 would be permitted to be classified as part of the Group A-3 occupancy. Any area that does not qualify as an acces-

sory occupancy must be classified in another group occupancy. Accordingly, the structure then contains multiple occupancies and is subject to the provisions of Section 508.3.

The fire hazard in terms of combustible contents (fuel load) in structures classified in Group A-3 is most often expected to be moderate to low. Since structures classified in Group A-3 vary widely as to the purpose for which they are used, the range of fuel load varies widely. For example, the fuel load in a library or an exhibition hall usually is considerably greater than that normally found in a gymnasium.

A-4: Structures provided with spectator seating in which people assemble to watch an indoor sporting event are to be classified as Group A-4. Arenas, skating rinks, swimming pools and tennis courts are among the facilities often classified as Group A-4. The distinguishing factor between Group A-4 and A-5 structures is whether the event is indoors or outdoors. Group A-4 facilities are limited to indoor structures only. The distinguishing factor between Group A-4 and Group A-3 facilities is the presence of a defined seating area. While Group A-3 facilities are indoors (i.e., tennis courts, swimming pools.), they typically do not have a defined seating area in which to view the event. Only facilities that are both indoors and have a defined seating area are to be classified as Group A-4.

A-5: Structures classified in Group A-5 are outdoor facilities where people assemble to view or participate in social and recreational activities (e.g., stadiums, grandstands, bleachers, coliseums). In order to qualify as an outdoor facility, the structure must be one where the products of combustion are freely and rapidly vented to the atmosphere (i.e., a structure without enclosures that would prevent the free movement of smoke from the occupied area to the outside). Any recreation facility that has exterior walls that enclose the facility and a roof that fully covers the area would not be classified in Group A-5, but rather in Group A-3 or A-4 depending on whether a seating area has been provided. In the case of a structure with a retractable roof, the more stringent occupancy classification (i.e., Group A-4) would be required.

Since occupancies classified in Group A-5 are primarily viewing and sports participation areas, the fuel load associated with them is very low (i.e., the structure itself and seats). Since the fuel load present is relatively low and the expectation is that smoke will be quickly evacuated from the structure, the relative fire hazard of occupancies classified in Group A-5 is expected to be low. The life safety hazard from panic that might occur in an emergency, however, is a serious concern; hence, the capability of large crowds to exit the structure quickly and orderly during emergencies is an important design consideration (see Section 1028).

Both Group A-4 and A-5 occupancies will include a variety of uses that support the viewing of sports and similar activities. There will likely be luxury seating suites, locker rooms, toilet facilities and press boxes, which are clearly part of the overall uses of the facility. There will also be offices, food concession stands and merchandise stands which by their use are different occupancies, but are probably within the accessory occupancy limits established in Section 508.2. Because of the multitiered design of most Group A-4 and A-5 occupancies, the limit for accessory occupancies to account for less than 10 percent of the story will need to be creatively applied. There may be full-fledged restaurants that are in the same building, but may be open to guests not limited to those attending an event. An Group A-2 occupancy designation is likely the most appropriate classification unless the restaurant is clearly accessory to the arena or stadium use.

SECTION 304
BUSINESS GROUP B

304.1 Business Group B. Business Group B occupancy includes, among others, the use of a building or structure, or a portion thereof, for office, professional or service-type transactions, including storage of records and accounts. Business occupancies shall include, but not be limited to, the following:

Airport traffic control towers
Ambulatory health care facilities
Animal hospitals, kennels and pounds
Banks
Barber and beauty shops
Car wash
Civic administration

Clinic—outpatient
Dry cleaning and laundries: pick-up and delivery stations and self-service
Educational occupancies for students above the 12th grade
Electronic data processing
Laboratories: testing and research
Motor vehicle showrooms
Post offices
Print shops
Professional services (architects, attorneys, dentists, physicians, engineers, etc.)
Radio and television stations
Telephone exchanges
Training and skill development not within a school or academic program

❖ The risks to life safety in the business occupancy classification are relatively low. Exposure to the potential effects of fire is limited because business-type facilities most often have low fuel loads, are normally occupied only during the daytime and, with some exceptions, are usually occupied for a set number of hours. The occupants, because of the nature of the use, are typically alert, ambulatory, conscious, aware of their surroundings and generally familiar with the building's features, particularly the means of egress. Historically, this occupancy has one of the better fire safety records for the protection of life and property.

This section identifies the general characteristics and lists examples of occupancies that are classified in Group B. Note that the description recognizes the need for limited storage spaces that are incidental to office occupancies. Classrooms and laboratories that are located in colleges, universities and academies for

Table 303.1
SMALL ASSEMBLY USES—OCCUPANCY ASSIGNMENT

TYPE OF BUILDING OR SPACE	OCCUPANT LOAD	
	< 50	50 or greater
Separate building Section 303.1, Exception 1	B	A
Independent tenant space Section 303.1, Exception 1	B	A
Accessory < 10% of story Section 508.2	A[a]	A[a]
Less than 750 square feet[b] Section 303.1, Exception 3	B or same as primary use[c]	B or same as primary use[c]
Less than 50 occupants[b] Section 303.1, Exception 2	B or same as primary use[c]	NA
Larger than 10% of story, larger than 750 square feet and more than 50 occupants	NA	A[d]

For SI: 1 square foot = 0.0929 m².
NA = Not applicable.
a. Separation from primary occupancy is not required.
b. Even where the spaces account for over 10 percent of the story on which they are located, Section 508.2 is not applicable.
c. Individual religious education rooms and auditoriums are designated Group A-3.
d. Assembly spaces, or any size accessory to a Group E occupancy, are designated as Group E regardless of occupant load (see Exception 4 to Section 303.1).

educating students above the 12th grade and that have an occupant load of less than 50 are classified in Group B. Classrooms with an occupant load of 50 or more are classified in Group A-3 (see Section 303.1). When lecture facilities for large groups (i.e., occupant load of 50 or more) are located within the same building where classrooms with an occupant load less than 50 are found, the building is a mixed occupancy (Groups A-3 and B) and is subject to the provisions of Section 508.

While civic administration covers a broad range of state and local government buildings, many such buildings will have a variety of uses and need to be considered under mixed occupancy provisions. Frequently police stations will include jails or holding cells. Fire stations will be a mix of offices, parking and maintenance facilities for the fire engines and living spaces for the fire fighters. Often a meeting room that is open to the public is also included. This type of facility is a mix of Group A, B, R and S occupancies.

Ambulatory health care facilities are those used to provide medical, or similar care, on less than a 24-hour basis to patients who are rendered incapable of self-preservation (see Section 202). Frequently called "day surgery centers" or "ambulatory surgical centers," ambulatory health care facilities perform procedures that render patients temporarily incapable of self-preservation due to the use of nerve blocks, sedation or anesthesia. Due to the condition of the patients, the need for medical staff to stabilize the patients before evacuation and the use of medical gases such as oxygen and nitrous oxide, these types of facilities pose greater fire and life safety hazards than other business occupancies. Accordingly, additional fire protection and means of egress requirements specific to ambulatory health care are provided in Section 422.

Facilities that provide medical services for inpatient care where the patients stay for more than 24 hours would be classified as Group I-2. Buildings used as sleep clinics would be classified as Group B since these spaces are not typical dwelling or sleeping units where people live, the occupants are assumed to be capable of self-preservation and the occupants are not living in a supervised environment. Although the patients in a sleep clinic may be sleeping, they can be easily awakened and alerted to an emergency as compared to the patients at an ambulatory health care facility.

304.1.1 Definition. The following word and term shall, for the purposes of this section and as used elsewhere in this code, have the meaning shown herein.

❖ This section contains a definition of a term that is associated with the subject matter of the section. It is important to emphasize that this term is not exclusively related to this chapter but are applicable everywhere the term is used in the code.

This term is also listed in Chapter 2 with a cross reference to this section. The use and application of all defined terms, including those defined herein, are set forth in Section 201.

CLINIC, OUTPATIENT. Buildings or portions thereof used to provide medical care on less than a 24-hour basis to individuals who are not rendered incapable of self-preservation by the services provided.

❖ Outpatient clinics generally consist of doctors' offices where various medical services can be provided. These clinics typically function during normal business hours (i.e., less than 24 hours) and, unlike ambulatory health care facilities, the patients are generally ambulatory and capable of self-preservation. This definition clarifies the difference between ambulatory surgery centers and the typical doctor's office. In many cities, outpatient clinics are open at all hours to be available to people who work a variety of shifts. The term "Urgent care" is often used to describe such facilities. An outpatient facility that is open 24/7 may still be classified as a Group B occupancy, provided all patients are outpatients and individual patients are not treated for periods in excess of 24 hours. The latter would describe a Group I-2 hospital.

SECTION 305
EDUCATIONAL GROUP E

305.1 Educational Group E. Educational Group E occupancy includes, among others, the use of a building or structure, or a portion thereof, by six or more persons at any one time for educational purposes through the 12th grade. Religious educational rooms and religious auditoriums, which are accessory to *places of religious worship* in accordance with Section 303.1 and have *occupant loads* of less than 100, shall be classified as A-3 occupancies.

❖ The risks to life safety in this occupancy vary with the composition of the facilities and also with the ages of the occupants. In general, children require more safeguards than do older, more mature persons.

This section identifies the criteria for classification of a building in Group E. The two fundamental characteristics of a Group E facility are as follows:

1. The facility is occupied by more than five persons (excluding the instructor); and

2. The purpose of the facility is for educating persons at the 12th-grade level and below, but not including more than five occupants $2^1/_2$ years of age or less.

Occupancies used for the education of persons above the 12th-grade level are not included in Group E. These facilities are occupied by adults who are not expected to require special supervision, direction or instruction in a fire or other emergency. By the same measure; however, they also are not closely supervised. Therefore, classrooms and laboratories located in colleges, universities and academies for students above the 12th grade are classified in Group B, because the occupancy characteristics and potential hazards to life safety present in these facilities more nearly resemble those of a business occupancy than educational occupancy. Please note, lecture halls for students above the 12th grade with an occupant load of 50 or more are classified in Group A-3 (see Section 303.1).

It is common for a school to also have gymnasiums (Group A-3), auditoriums (Group A-1), libraries (Group A-3), offices (Group B) and storage rooms (Group S-1). When this occurs, the building is a mixed occupancy and is subject to the provisions of Section 508. In accordance with Section 303.1, Exception 4, assembly spaces, such as the gymnasium, auditorium, library and cafeteria, do not have to be considered separate occupancies if used for school purposes (see commentary, Section 303.1). For such assembly functions to be considered part of the primary Group E occupancy, the assembly functions must be ancillary and supportive to the educational operation of the building.

Places of religious worship, religious educational rooms and religious auditoriums are often provided in the same building complex. These religious educational rooms and religious auditoriums are not to be considered separate occupancies (i.e., Group E) as long as the occupant load in these spaces is no greater than 99 people (see commentary, Section 303.1, Exception 5).

305.2 Day care. The use of a building or structure, or portion thereof, for educational, supervision or *personal care services* for more than five children older than $2^1/_2$ years of age, shall be classified as a Group E occupancy.

❖ Day care occupancies include facilities intended to be used for the care and supervision of more than five children where individual care is for a period of less than 24 hours per day and that do not contain more than five children who are $2^1/_2$ years of age or less. Facilities that provide care for more than five occupants greater than the 12th grade are to be classified as adult care facilities (Group I-4, Section 308.5.1) or Group R-3.

Day care centers are a special concern since they are generally occupied by preschool children who are less capable of responding to an emergency. The hazards found in a day care center are far greater than in normal educational facilities, not so much because of the occupant or fuel load, but because of the inability of the occupants to respond. See the more stringent requirement for spaces with one means of egress in Table 1015.1.

Children $2^1/_2$ years of age or less usually are not able to recognize an emergency situation, may not respond appropriately or simply may not be able to egress without assistance; thus, facilities that have more than five children $2^1/_2$ years of age or less are classified as child care facilities and considered to be Group I-4 (see Section 308.5.2). Table 308.3.1 summarizes the appropriate occupancy classifications for day care, adult care (more than five adults, less than 24 hours) and child care facilities.

SECTION 306
FACTORY GROUP F

306.1 Factory Industrial Group F. Factory Industrial Group F occupancy includes, among others, the use of a building or structure, or a portion thereof, for assembling, disassembling, fabricating, finishing, manufacturing, packaging, repair or processing operations that are not classified as a Group H hazardous or Group S storage occupancy.

❖ The purpose of this section is to identify the characteristics of occupancies that are classified in factory and industrial occupancies and to differentiate Groups F-1 and F-2.

Because of the vast number of diverse manufacturing and processing operations in the industrial community, it is more practical to classify such facilities by their level of hazard rather than their function. In industrial facilities, experience has shown that the loss of life or property is most directly related to fire hazards, particularly the fuel load contributed by the materials being fabricated, assembled or processed.

Statistics show that property losses are comparatively high in factory and industrial occupancies, but the record of fatalities and injuries from fire has been remarkably low. This excellent life safety record can, in part, be attributed to fire protection requirements of the code.

This section requires that all structures that are used for fabricating, finishing, manufacturing, packaging, assembling or processing products or materials are to be classified in either Group F-1 (moderate hazard) or F-2 (low hazard). These classifications are based on the relative level of hazard for the types of materials that are fabricated, assembled or processed. Where the products and materials in a factory present an extreme fire, explosion or health hazard, such facilities are classified in Group H (see Section 307). It should be noted that the term "Group F" is not a specific occupancy, but is a term that collectively applies to Groups F-1 and F-2.

306.2 Factory Industrial F-1 Moderate-hazard Occupancy. Factory industrial uses which are not classified as Factory Industrial F-2 Low Hazard shall be classified as F-1 Moderate Hazard and shall include, but not be limited to, the following:

Aircraft (manufacturing, not to include repair)
Appliances
Athletic equipment
Automobiles and other motor vehicles
Bakeries
Beverages: over 16-percent alcohol content
Bicycles
Boats
Brooms or brushes
Business machines
Cameras and photo equipment
Canvas or similar fabric
Carpets and rugs (includes cleaning)

Clothing
Construction and agricultural machinery
Disinfectants
Dry cleaning and dyeing
Electric generation plants
Electronics
Engines (including rebuilding)
Food processing
Furniture
Hemp products
Jute products
Laundries
Leather products
Machinery
Metals
Millwork (sash and door)
Motion pictures and television filming (without spectators)
Musical instruments
Optical goods
Paper mills or products
Photographic film
Plastic products
Printing or publishing
Recreational vehicles
Refuse incineration
Shoes
Soaps and detergents
Textiles
Tobacco
Trailers
Upholstering
Wood; distillation
Woodworking (cabinet)

❖ Structures classified in Group F-1 (moderate hazard) are occupied for the purpose of fabrication, finishing, manufacturing, packaging, assembly or processing of materials that are combustible or that use combustible products in the production process.

306.3 Factory Industrial F-2 Low-hazard Occupancy. Factory industrial uses that involve the fabrication or manufacturing of noncombustible materials which during finishing, packing or processing do not involve a significant fire hazard shall be classified as F-2 occupancies and shall include, but not be limited to, the following:

Beverages: up to and including 16-percent alcohol content
Brick and masonry
Ceramic products
Foundries
Glass products
Gypsum
Ice
Metal products (fabrication and assembly)

❖ Structures classified in Group F-2 (low hazard) are occupied for the purpose of fabrication, manufacturing or processing of noncombustible materials. It is acceptable for noncombustible products to be packaged in a combustible material, provided that the fuel load contributed by the packaging is negligible when compared to the amount of noncombustible product.

The use of a significant amount of combustible material to package or finish a noncombustible product, however, will result in a Group F-1 (moderate-hazard factory and industrial) classification.

To distinguish when the presence of combustible packaging constitutes a significant fuel load, possibly requiring the reclassification of the building or structure as Group F-1, a reasonable guideline to follow is the "single thickness" rule, which is when a noncombustible product is put in one layer of packaging material.

Examples of acceptable conditions in Group F-2 are:

* Vehicle engines placed on wood pallets for transportation after assembly;
* Washing machines in corrugated cardboard boxes; and
* Soft-drink glass bottles packaged in pressed paper boxes.

Occupancies involving noncombustible items packaged in more than one layer of combustible packaging material are most appropriately classified in Group F-1.

Typical examples of packaging that would result in a Group F-1 classification are:

* Chinaware wrapped in corrugated paper and placed in cardboard boxes;
* Glassware set in expanded foam forms and placed in cardboard boxes; and
* Fuel filters individually packed in pressed paper boxes, placed by the gross in a cardboard box and stacked on a pallet for transportation.

Factories and industrial facilities often have offices and areas where large quantities of materials are kept in the same building as manufacturing operations, fabrication processes and assembly processes. The stock areas are classified as either Group S-1 or S-2, depending on the combustibility of the materials stored. Areas used for offices that do not qualify as accessory occupancies (see Section 508.2) are classified in Group B. When these combinations of occupancies occur, as well as other combinations of occupancies, the building is subject to the mixed occupancy provisions in Section 508.

SECTION 307
HIGH-HAZARD GROUP H

[F] 307.1 High-hazard Group H. High-hazard Group H occupancy includes, among others, the use of a building or structure, or a portion thereof, that involves the manufacturing, processing, generation or storage of materials that constitute a physical or health hazard in quantities in excess of those allowed in *control areas* complying with Section 414, based on the maximum allowable quantity limits for control areas set forth in Tables 307.1(1) and 307.1(2). Hazardous occupancies are classified in Groups H-1, H-2, H-3, H-4 and H-5 and shall be in accordance with this section, the requirements of Section

415 and the *International Fire Code.* Hazardous materials stored, or used on top of roofs or canopies shall be classified as outdoor storage or use and shall comply with the *International Fire Code.*

Exceptions: The following shall not be classified as Group H, but shall be classified as the occupancy that they most nearly resemble.

1. Buildings and structures occupied for the application of flammable finishes, provided that such buildings or areas conform to the requirements of Section 416 and the *International Fire Code.*

2. Wholesale and retail sales and storage of flammable and combustible liquids in mercantile occupancies conforming to the *International Fire Code.*

3. Closed piping system containing flammable or combustible liquids or gases utilized for the operation of machinery or equipment.

4. Cleaning establishments that utilize combustible liquid solvents having a flash point of 140°F (60°C) or higher in closed systems employing equipment *listed* by an *approved* testing agency, provided that this occupancy is separated from all other areas of the building by 1-hour *fire barriers* constructed in accordance with Section 707 or 1-hour *horizontal assemblies* constructed in accordance with Section 712, or both.

5. Cleaning establishments that utilize a liquid solvent having a flash point at or above 200°F (93°C).

6. Liquor stores and distributors without bulk storage.

7. Refrigeration systems.

8. The storage or utilization of materials for agricultural purposes on the premises.

9. Stationary batteries utilized for facility emergency power, uninterrupted power supply or telecommunication facilities, provided that the batteries are provided with safety venting caps and ventilation is provided in accordance with the *International Mechanical Code.*

10. Corrosives shall not include personal or household products in their original packaging used in retail display or commonly used building materials.

11. Buildings and structures occupied for aerosol storage shall be classified as Group S-1, provided that such buildings conform to the requirements of the *International Fire Code.*

12. Display and storage of nonflammable solid and non-flammable or noncombustible liquid hazardous materials in quantities not exceeding the maximum allowable quantity per *control area* in Group M or S occupancies complying with Section 414.2.5.

13. The storage of black powder, smokeless propellant and small arms primers in Groups M and R-3 and special industrial explosive devices in Groups B, F, M and S, provided such storage conforms to the quantity limits and requirements prescribed in the *International Fire Code.*

❖ This section identifies the various types of facilities contained in the high-hazard occupancy. This occupancy classification relates to those facilities where the storage of materials or the operations are deemed to be extremely hazardous to life and property, especially when they involve the use of significant amounts of highly combustible, flammable or explosive materials, regardless of their composition (i.e., solids, liquids, gases or dust). Although they are not explosive or highly flammable, other hazardous materials, such as corrosive liquids, highly toxic materials and poisonous gases, still present an extreme hazard to life. Many materials possess multiple hazards, whether physical or health related.

There is a wide range of high-hazard operations in the industrial community; therefore, it is more practical to categorize such facilities in terms of the degree of hazard they present, rather than attempt to define a facility in terms of its function. This method is similar to that used to categorize factory (see Section 306) and storage (see Section 311) occupancies.

Group H is handled as a separate classification because it represents an unusually high degree of hazard that is not found in the other occupancies. It is important to isolate those industrial or storage operations that pose the greatest dangers to life and property and to reduce such hazards by providing systems or elements of protection through the regulatory provisions of building codes.

There are numerous provisions and exceptions throughout the code that cannot be used when one or more Group H occupancies are present.

Operations that, because of the materials utilized or stored, cause a building or portion of a building to be classified as a high-hazard occupancy are identified in this section. While buildings classified as Group H may not have a large occupant load, the unstable chemical properties of the materials contained on the premises constitute an above-average fuel load and serve as a potential danger to the surrounding area.

The dangers created by the high-hazard materials require special consideration for the abatement of the danger. The classification of a material as high hazard is based on information derived from National Fire Protection Association (NFPA) standards and the Code of Federal Regulations (DOL 29 CFR).

The wide range of materials utilized or stored in buildings creates an equally wide range of hazards to the occupants of the building, the building proper and the surrounding area. Since these hazards range from explosive to corrosive conditions, the high-hazard occupancy has been broken into four subclassifications: Groups H-1 through H-4. A fifth category, Group H-5, is used to represent structures that contain hazardous production material (HPM) facilities. Each of these subclassifications addresses materials that have similar characteristics and the protection requirements attempt to address the hazard involved. These subclassifications are defined by the properties of the materials involved with only occasional reference to specific materials. This performance-based criterion may involve additional research to identify a hazard, but it is the only way to remain current in a rapidly changing field. Material Safety Data Sheets (MSDS) will be a major source for information.

Additional information on hazardous materials can be found in Section 415 as well as the commentary to the *International Fire Code®* (IFC®).

Section 307.1 acknowledges that a building is not classified as a high-hazard occupancy unless the maximum allowable quantities per control area as prescribed in Tables 307.1(1) and 307.1(2) are exceeded, subject to the applicable control area provisions of Section 414.2. The maximum quantity limitations per control area prescribed in Tables 307.1(1) and 307.1(2) have been determined to be relatively safe when maintained in accordance with the IFC. Therefore, a building containing less than the maximum allowable quantities specified in Tables 307.1(1) and 307.1(2) would not be classified as a Group H occupancy but rather as the occupancy group it most nearly resembles.

Section 414.2 establishes the control area concept for regulating hazardous materials. This concept would allow the maximum allowable quantities of hazardous materials per control area in Tables 307.1(1) and 307.1(2) to be exceeded within a given building without classifying the building as a high-hazard occupancy by utilizing a multiple control area approach. The permitted number of control areas, maximum percentage of allowable quantities of hazardous materials per control area and degree of fire separation between control areas is regulated by Section 414.2 (see commentary, Section 414.2).

Section 307.1 also clarifies that hazardous materials outside of the building envelope should be classified as outdoor storage. As such, hazardous material quantities on roofs or canopies are not included in evaluating the occupancy classification of a building or structure. Canopies used to support gaseous hydrogen systems must comply with Section 406.5.3.1.

The exceptions list conditions that are exempt from a high-hazard classification because of the building's construction or use, the packaging of materials, the quantity of materials or the precautions taken to prevent fire. Even if a high-hazard material meets one of the exceptions, its storage and use must comply with the applicable provisions of Section 414 and the IFC.

Exception 1 exempts spray painting and similar operations within buildings from being classified as a high-hazard occupancy. This exception requires that all such operations, as well as the handling of flammable finishes, are in accordance with the provisions of Section 416 and the IFC; therefore, an adequately protected typical paint spray booth in a factory (Group F-1) would not result in a high-hazard occupancy classification for either the building or the paint spray area.

Exception 2 relies on the provisions of Section 3404.3.4.1 of the IFC to regulate the storage of flammable and combustible liquids for wholesale and retail sales and storage in mercantile occupancies. The overall permitted amount of flammable and combustible liquids is dependent on the class of liquid, storage arrangement, container size and level of sprinkler protection. For nonsprinklered buildings, the maximum allowable quantity per control area permitted by Table 3404.3.4.1 of the IFC is 1,600 gallons (6057 L) of Class IB, IC, II, and IIIA liquids with a maximum of 60 gallons (227 L) of Class IA liquids. Depending on storage and ceiling heights, buildings equipped with a sprinkler system with a minimum design density for an Ordinary Hazard Group 2 occupancy may have an aggregate total of 7,500 gallons (28 391 L) of Class IB, IC, II, and IIIA liquids with a maximum of 60 gallons (227 L) of Class IA liquids. The quantities of Class IB, IC, II and IIA liquids could be further increased depending on the potential storage conditions and enhanced degree of sprinkler protection. (See Section 3403.4.3.1 of the IFC for additional design information.) Again, it should be noted, that despite the increased quantities which far exceed the base quantity limitations of Table 307.1(1), compliance with this exception would result in the building not being classified as a Group H occupancy.

Exception 3 exempts closed systems that are used exclusively for the operation of machinery or equipment. The closed piping systems, which are essentially not open to the atmosphere, keep flammable or combustible liquids from direct exposure to external sources of ignition as well as prevent the users from coming in direct contact with liquids or harmful vapors. This exception would include systems such as oil-burning equipment, piping for diesel fuel generators and LP-gas cylinders for use in forklift trucks.

Exception 4 exempts cleaning establishments that utilize a closed system for all combustible liquid solvents with a flash point at or above 140°F (60°C). The reference to using equipment listed by an approved testing laboratory does not mean that the entire system needs to be approved, but rather the individual pieces of equipment. As with any mechanical equipment or appliance, it should bear the label of an approved agency and be installed in accordance with the manufacturer's installation instructions [see the *International Mechanical Code®* (IMC®)].

Exception 5 covers cleaning establishments that use solvents that have very high flash points [at least 200°F (93°C)] and that are exceedingly difficult to ignite. Such liquids can be used openly, but with due care.

Exception 6 exempts all retail liquor stores and liquor distribution facilities from the high-hazard occupancy classification, even though most of the contents are considered combustible liquids. The exception takes into account that alcoholic beverages are packaged in individual containers of limited size.

Exception 7 refers to refrigeration systems that utilize refrigerants that may be flammable or toxic.

Refrigeration systems do not alter the occupancy classification of the building, provided they are installed in accordance with the IMC. The IMC has specific limitations on the quantity and type of refrigerants that can be used, depending on the occupancy classification of the building.

Exception 8 exempts materials that are used for agricultural purposes, such as fertilizers, pesticides, fungicides, etc., when used on the premises. Agricultural materials stored for direct or immediate use are not usually of such quantities that would constitute a large fuel load or an exceptionally hazardous condition.

Exception 9 addresses battery storage rooms when used as part of an operating system, such as for providing standby power. The batteries used in installations of this type do not represent a significant health, safety or fire hazard. The electrolyte and battery casing contribute little fuel load to a fire. The release of hydrogen gas during the operation of battery systems is minimal. Ventilation in accordance with the IMC will disperse the small amounts of liberated hydrogen. This exception also assumes that rooms containing stationary storage battery systems are in compliance with Section 608 of the IFC.

Without Exception 10 certain products that technically are corrosive could cause grocery stores and other mercantile occupancies to be inappropriately classified as Group H-4. This exception allows the maximum allowable quantity per control area in Table 307.1(2) for corrosives to be exceeded in the retail display area. This would include such things as bleaches, detergents and other household cleaning supplies in normal-size containers. The exception also exempts the storage or manufacture of commonly used building materials, such as portland cement, from being inappropriately classified as Group H.

Exception 11 exempts buildings and structures used for the storage of aerosol products, provided they are protected in accordance with the provisions of NFPA 30B and the IFC. The aerosol storage requirements in the IFC, referred to in this exception, are based on the provisions of NFPA 30B. Compliance with the exception exempts buildings from complying with the code provisions for Group H, provided the storage of aerosol products comply with the applicable separation, storage limitations and sprinkler design requirements specified in the IFC and NFPA 30B.

Exception 12 permits certain products found in mercantile and storage occupancies, which may be comprised of hazardous materials, to exceed the maximum allowable quantity per control area of Tables 307.1(1) and 307.1(2). The products, however, must be comprised of nonflammable solids or liquids that are nonflammable or noncombustible. Materials could include swimming pool chemicals, which are typically

Class 2 or 3 oxidizers or industrial corrosive cleaning agents (see commentary, Section 414.2.5).

Exception 13 permits the base maximum allowable quantity per control area of black powder, smokeless propellant and small arms primers in Group M and R-3 occupancies to be exceeded, provided the material is stored in accordance with Chapter 33 of the IFC. The requirements are based on the provisions in NFPA 495. Similarly, special industrial explosive devices are found in a number of occupancies other than Group H (Groups B, F, M and S). Storage of these devices in accordance with the IFC is not required to have a high-hazard occupancy classification. Power drivers are commonly used in the construction industry, and there are stocks of these materials maintained for sale and use by the trade. The automotive airbag industry has evolved with the use of these devices, and they are located in automotive dealerships and personal use vehicles throughout society. The IFC currently exempts up to 50 pounds (23 kg) of these materials from regulation under Chapter 33 (explosives).

TABLE 307.1(1). See page 3-17.

❖ The maximum allowable quantities of high-hazard materials allowed in each control area before having to classify a part of the (or the entire) building as a high-hazard occupancy are given in the table. This table is referenced in Section 307.1. The materials listed in this table are classified according to their specific occupancy in Sections 307.3 through 307.5 and defined in Section 307.2. This table only contains materials applicable to Groups H-1, H-2 and H-3. The maximum allowable quantities per control area for Group H-4 materials are listed in Table 307.1(2).

The presence of any one or more of the materials listed in Table 307.1(1) in an amount greater than allowed requires that the building or area in which the material is contained be classified as a Group H, high-hazard occupancy.

If a building or area contains only the materials listed in either Table 307.1(1) or 307.1(2) in the maximum allowable quantity per control area or less, then that building or area would not be classified as a Group H, high-hazard occupancy. The possible increase in overall danger that might exist should this occur because of the storage and use of incompatible materials is an issue that the code does not specifically address. In such situations, the building official can seek the advice of chemical engineers, fire protection engineers, fire service personnel or other experts in the use of hazardous materials. Based on their advice, the building official can deem the building a high-hazard occupancy.

The maximum allowable quantity per control area listed in Table 307.1(1) is based on the concept of control areas as further regulated in Section 414. The quantities listed apply per control area. While every building area also represents a single control area, a given building may have multiple control areas, provided that the allowable amount within each control area is not exceeded and adequate fire-resistance-rated separation is provided between control areas. As indicated in Section 307.1, a building that utilizes multiple control areas and complies with the applicable provisions of Section 414 is not classified as Group H. The number, degree of separation and location of control areas are indicated in Section 414.

Table 307.7(1) is subdivided based on whether the material is in storage or in use in a closed or open system. Definitions of both closed and open systems are found in Section 307.2. Within these subdivisions, the appropriate maximum allowable quantity per control area is listed in accordance with the physical state (solid, liquid or gas) of the material. A column for gas in open systems is not indicated because hazardous gaseous materials should not be allowed in a system that is continuously open to the atmosphere. While hazardous materials within a closed or open system are considered to be in use, Note b clearly indicates that the aggregate quantity of hazardous materials in use and storage within a given control area should not exceed the quantity listed in Table 307.7(1) for storage. Without Note c, many common alcoholic beverages and household products containing a negligible amount of a hazardous material could result in a Group M occupancy being classified as a high hazard. Note c recognizes the reduced hazard of the materials based on their water miscibility and limited container size.

Notes d and e of Table 307.1(1) are significant in that, for certain materials, the maximum allowable amount may be increased due to the use of approved hazardous material storage cabinets, where the building is fully protected by an automatic sprinkler system, or both. The notes are intended to be cumulative in that up to four times the base maximum quantity may be allowed per control area, if both sprinklered and in cabinets, without classifying the building as Group H. While the use of cabinets is not always a feasible or practical method of storage, they do provide additional protection to warrant an increase if provided. Construction requirements for hazardous material storage cabinets are contained in the IFC. Note that the use of day boxes, gas cabinets, exhausted enclosures or listed safety cans would allow the same increase as for cabinets. Listed safety cans, which are primarily intended for flammable and combustible liquids, must be in compliance with UL 30 when used to increase the maximum allowable quantities permitted by Table 307.1(1).

While classified as a hazardous material, the code recognizes the relative hazard of Class IIIB liquids as compared to that of other flammable and combustible liquids by establishing a base maximum allowable quantity per control area of 13,200 gallons (49 962 L). As indicated in Note f, the quantity of Class I oxidizers and Class IIIB liquids would not be limited, provided the building is fully sprinklered in accordance with NFPA 13. Since any building that exceeds this maximum amount would be required to be classified as Group H and these buildings are required to be sprinklered, the maximum allowable amount would then be unlimited. As such, a Group H classification would not be warranted. The hazard presented by Class I oxidizers is that they slightly increase the burning rate of combustible materials that they may come into contact with during a fire. Class IIIB combustible liquids have flash points at or above 200°F (93°C). Motor oil is a typical example of a Class IIIB combustible liquid.

Note g recognizes that the hazard presented by certain materials is such that they may be stored or used only inside buildings that are fully sprinklered.

Note h clarifies for the user that while there is a combination maximum allowable quantity for flammable liquids, no individual class of liquid (Class IA, IB or IC) may exceed its own individual maximum allowable quantity.

Note i is a specific exception for inside storage tanks of combustible liquids that are connected to a fuel-oil piping system in accordance with Section 603.3.2 of the IFC. This exception applies to most oil-fired stationary equipment, whether in industrial, commercial or residential occupancies. NFPA 31 and NFPA 37 provide further guidance on the type of installations this exception is intending to permit. This exception would permit fuel-oil storage tanks containing a maximum of 660 gallons (2498 L) of combustible liquids within a building without being classified as a Group H-3 occupancy. This quantity limitation could be further increased to 3,000 gallons (11 356 L) for combustible liquids stored in protected above-ground tanks in rooms protected by an automatic sprinkler system complying with NFPA 13.

Note k permits a larger amount of Class 3 oxidizers in a building when used for maintenance and health purposes. The quantities proposed are reasonable for occupancies such as the health care industry where Class 3 oxidizers are used for maintenance purposes, sterilization and sanitation of equipment and operation sanitation. The method used to store the oxidizers is subject to the evaluation and approval of the building official. Note k also provides consistency with Note k of Table 2703.1.1(1) of the IFC.

Note l clarifies that the 125 pounds (57 kg) of storage permitted for consumer fireworks represents the net weight of the pyrotechnic composition of the fireworks in a nonsprinklered building. This amount represents approximately 12$\frac{1}{2}$ shipping cases (less than one and one-half pallet loads) of fireworks in a nonsprinklered storage condition. In cases where the net weight of the pyrotechnic composition of the fireworks is unknown, 25 percent of the gross weight of the fireworks is to be used. The gross weight is to include the weight of the packaging.

Note n provides an exception when the amount of hazardous material in storage and display in Group M and S occupancies meet the requirements of Section 414.2.5.

Note o clarifies that densely packed baled cotton is not considered a hazardous material when meeting the size and weight requirements of ISO 8115 and, as such, is not subject to the maximum allowable quantity per control area specified for combustible fibers.

Note p is added to clarify that vehicles with closed fuel systems should be treated no differently than machinery or equipment when considering the allowable quantities of materials within a building. This note also clarifies that the fuels contained within the fuel tanks of vehicles or motorized equipment are not to be considered when calculating the aggregate quantity of hazardous materials within a control area of a building. For example, when evaluating a parking garage with several hundred cars parked inside, the fuel tanks of vehicles are not counted. When motorized equipment, such as a floor buffer or forklift, is used, those fuels are not included as long as other code requirements are satisfied.

TABLE 307.1(2). See page 3-19.

❖ Table 307.1(2), similar to Table 307.1(1), specifies the maximum quantities of hazardous materials, liquids or chemicals allowed per control area before having to classify a part of the (or the entire) building as a high-hazard occupancy. Table 307.1(2), as referenced in Section 307.1, contains materials classified as Group H-4 in accordance with Section 307.6. While the materials listed in this table are considered health hazards, some materials may also possess physical hazard characteristics more indicative of materials classified as Group H-1, H-2 or H-3.

The maximum allowable quantities per control area listed in Table 307.1(2) are indicative of industry practice and assume the materials are properly stored and handled in accordance with the IFC. Group H-4 materials, while indeed hazardous, are primarily considered a handling problem and do not possess the same

fire, explosion or reactivity hazard associated with other hazardous materials. The base maximum allowable quantity per control area of 810 cubic feet (23 m³) for gases that are either corrosive or toxic is based on a standard-size chlorine cylinder. The use of 150 pounds (68 kg) as the baseline quantity for liquefied corrosive and toxic gases is intended to be consistent with the philosophical approach to the same maximum quantity permitted for liquefied oxidizing gases in Table 307.1(1). The 150-pound (68 kg) limitation allows a single cylinder of chlorine, which could be considered both a corrosive and oxidizing gas, to not result in either a Group H-3 or H-4 occupancy classification.

Without Note b, many common household products containing a negligible amount of a hazardous material could result in a Group M occupancy being classified as a high hazard. Note b recognizes the reduced hazard of the materials based on their water miscibility and limited container size.

Note c provides an exception when the amount of hazardous material in storage and display in Group M and S occupancies meets the requirements of Section 414.2.5.

Note d clearly indicates that the aggregate quantity of hazardous materials in use and storage, within a given control area, cannot exceed the quantity listed in the table for storage.

Notes e and f are identical to Notes d and e to Table 307.1(1) and allow up to four times the maximum allowed quantities. See the commentary for Notes d and e of Table 307.1(1).

Note g is significant in that, for certain materials, their hazard is so great that their maximum allowable quantity per control area may be stored in the building only when approved exhausted enclosures or gas cabinets are utilized.

307.1.1 Hazardous materials. Hazardous materials in any quantity shall conform to the requirements of this code, including Section 414, and the *International Fire Code*.

❖ The use of high-hazard materials must be regulated in accordance with Sections 414 and 415 as well as the applicable requirements of the IFC. While the building may be exempt from a high-hazard occupancy classification (i.e., Group H-1, H-2, H-3, H-4 or H-5), any potential hazard with regard to the use of storage of any hazardous material, regardless of quantity, must be abated.

[F] TABLE 307.1(1)
MAXIMUM ALLOWABLE QUANTITY PER CONTROL AREA OF HAZARDOUS MATERIALS POSING A PHYSICAL HAZARD[a, j, m, n, p]

MATERIAL	CLASS	GROUP WHEN THE MAXIMUM ALLOWABLE QUANTITY IS EXCEEDED	STORAGE[b] Solid pounds (cubic feet)	STORAGE[b] Liquid gallons (pounds)	STORAGE[b] Gas (cubic feet at NTP)	USE-CLOSED SYSTEMS[b] Solid pounds (cubic feet)	USE-CLOSED SYSTEMS[b] Liquid gallons (pounds)	USE-CLOSED SYSTEMS[b] Gas (cubic feet at NTP)	USE-OPEN SYSTEMS[b] Solid pounds (cubic feet)	USE-OPEN SYSTEMS[b] Liquid gallons (pounds)
Combustible liquid[c, i]	II	H-2 or H-3	N/A	120[d, e]	N/A	N/A	120[d]	N/A	N/A	30[d]
	IIIA	H-2 or H-3	N/A	330[d, e]	N/A	N/A	330[d]	N/A	N/A	80[d]
	IIIB	N/A	N/A	13,200[e, f]	N/A	N/A	13,200[f]	N/A	N/A	3,300[f]
Combustible fiber	Loose Baled[o]	H-3	(100) (1,000)	N/A	N/A	(100) (1,000)	N/A	N/A	(20) (200)	N/A
Consumer fireworks (Class C, Common)	1.4G	H-3	125[d, e, l]	N/A	N/A	N/A	N/A	N/A	N/A	N/A
Cryogenics, flammable	N/A	H-2	N/A	45[d]	N/A	N/A	45[d]	N/A	N/A	10[d]
Cryogenics, inert	N/A	N/A	N/A	N/A	NL	N/A	N/A	NL	N/A	N/A
Cryogenics, oxidizing	N/A	H-3	N/A	45[d]	N/A	N/A	45[d]	N/A	N/A	10[d]
Explosives	Division 1.1	H-1	1[e, g]	(1)[e, g]	N/A	0.25[g]	(0.25)[g]	N/A	0.25[g]	(0.25)[g]
	Division 1.2	H-1	1[e, g]	(1)[e, g]	N/A	0.25[g]	(0.25)[g]	N/A	0.25[g]	(0.25)[g]
	Division 1.3	H-1 or H-2	5[e, g]	(5)[e, g]	N/A	1[g]	(1)[g]	N/A	1[g]	(1)[g]
	Division 1.4	H-3	50[e, g]	(50)[e, g]	N/A	50[g]	(50)[g]	N/A	N/A	N/A
	Division 1.4G	H-3	125[d, e, l]	N/A	N/A	N/A	N/A	N/A	N/A	N/A
	Division 1.5	H-1	1[e, g]	(1)[e, g]	N/A	0.25[g]	(0.25)[g]	N/A	0.25[g]	(0.25)[g]
	Division 1.6	H-1	1[d, e, g]	N/A	N/A	N/A	N/A	N/A	N/A	N/A
Flammable gas	Gaseous	H-2	N/A	N/A	1,000[d, e]	N/A	N/A	1,000[d, e]	N/A	N/A
	Liquefied	H-2	N/A	(150)[d, e]	N/A	N/A	(150)[d, e]	N/A	N/A	N/A
Flammable liquid[c]	1A	H-2 or H-3	N/A	30[d, e]	N/A	N/A	30[d]	N/A	N/A	10[d]
	1B and 1C	H-2 or H-3	N/A	120[d, e]	N/A	N/A	120[d]	N/A	N/A	30[d]
Flammable liquid, combination (1A, 1B, 1C)	N/A	H-2 or H-3	N/A	120[d, e, h]	N/A	N/A	120[d, h]	N/A	N/A	30[d, h]
Flammable solid	N/A	H-3	125[d, e]	N/A	N/A	125[d]	N/A	N/A	25[d]	N/A
Inert gas	Gaseous	N/A	N/A	N/A	NL	N/A	N/A	NL	N/A	N/A
	Liquefied	N/A	N/A	N/A	NL	N/A	N/A	NL	N/A	N/A
Organic peroxide	UD	H-1	1[e, g]	(1)[e, g]	N/A	0.25[g]	(0.25)[g]	N/A	0.25[g]	(0.25)[g]
	I	H-2	5[d, e]	(5)[d, e]	N/A	1[d]	(1)	N/A	1[d]	(1)[d]
	II	H-3	50[d, e]	(50)[d, e]	N/A	50[d]	(50)[d]	N/A	10[d]	(10)[d]
	III	H-3	125[d, e]	(125)[d, e]	N/A	125[d]	(125)[d]	N/A	25[d]	(25)[d]
	IV	N/A	NL	NL	N/A	NL	NL	N/A	NL	NL
	V	N/A	NL	NL	N/A	NL	NL	N/A	NL	NL
Oxidizer	4	H-1	1[e, g]	(1)[e, g]	N/A	0.25[g]	(0.25)[g]	N/A	0.25[g]	(0.25)[g]
	3[k]	H-2 or H-3	10[d, e]	(10)[d, e]	N/A	2[d]	(2)[d]	N/A	2[d]	(2)[d]
	2	H-3	250[d, e]	(250)[d, e]	N/A	250[d]	(250)[d]	N/A	50[d]	(50)[d]
	1	N/A	4,000[e, f]	(4,000)[e, f]	N/A	4,000[f]	(4,000)[f]	N/A	1,000[f]	(1,000)[f]

(continued)

[F] TABLE 307.1(1)—continued
MAXIMUM ALLOWABLE QUANTITY PER CONTROL AREA OF HAZARDOUS MATERIALS POSING A PHYSICAL HAZARD[a, j, m, n, p]

MATERIAL	CLASS	GROUP WHEN THE MAXIMUM ALLOWABLE QUANTITY IS EXCEEDED	STORAGE[b]			USE-CLOSED SYSTEMS[b]			USE-OPEN SYSTEMS[b]	
			Solid pounds (cubic feet)	Liquid gallons (pounds)	Gas (cubic feet at NTP)	Solid pounds (cubic feet)	Liquid gallons (pounds)	Gas (cubic feet at NTP)	Solid pounds (cubic feet)	Liquid gallons (pounds)
Oxidizing gas	Gaseous	H-3	N/A	N/A	1,500[d, e]	N/A	N/A	1,500[d, e]	N/A	N/A
	Liquefied	H-3	N/A	(150)[d, e]	N/A	N/A	(150)[d, e]	N/A	N/A	N/A
Pyrophoric material	N/A	H-2	4[e, g]	(4)[e, g]	50[e, g]	1[g]	(1)[g]	10[g]	0	0
Unstable (reactive)	4	H-1	1[e, g]	(1)[e, g]	10[g]	0.25[g]	(0.25)[g]	2[e, g]	0.25[g]	(0.25)[g]
	3	H-1 or H-2	5[d, e]	(5)[d, e]	50[d, e]	1[d]	(1)[d]	10[d, e]	1[d]	(1)[d]
	2	H-3	50[d, e]	(50)[d, e]	250[d, e]	50[d]	(50)[d]	250[d, e]	10[d]	(10)[d]
	1	N/A	NL	NL	NL	NL	NL	NL	NL	NL
Water reactive	3	H-2	5[d, e]	(5)[d, e]	N/A	5[d]	(5)[d]	N/A	1[d]	(1)[d]
	2	H-3	50[d, e]	(50)[d, e]	N/A	50[d]	(50)[d]	N/A	10[d]	(10)[d]
	1	N/A	NL	NL	N/A	NL	NL	N/A	NL	NL

For SI: 1 cubic foot = 0.028 m³, 1 pound = 0.454 kg, 1 gallon = 3.785 L.

NL = Not Limited; N/A = Not Applicable; UD = Unclassified Detonable

a. For use of control areas, see Section 414.2.

b. The aggregate quantity in use and storage shall not exceed the quantity listed for storage.

c. The quantities of alcoholic beverages in retail and wholesale sales occupancies shall not be limited providing the liquids are packaged in individual containers not exceeding 1.3 gallons. In retail and wholesale sales occupancies, the quantities of medicines, foodstuffs, consumer or industrial products, and cosmetics containing not more than 50 percent by volume of water-miscible liquids with the remainder of the solutions not being flammable, shall not be limited, provided that such materials are packaged in individual containers not exceeding 1.3 gallons.

d. Maximum allowable quantities shall be increased 100 percent in buildings equipped throughout with an automatic sprinkler system in accordance with Section 903.3.1.1. Where Note e also applies, the increase for both notes shall be applied accumulatively.

e. Maximum allowable quantities shall be increased 100 percent when stored in approved storage cabinets, day boxes, gas cabinets or exhausted enclosures or in listed safety cans in accordance with Section 2703.9.10 of the International Fire Code. Where Note d also applies, the increase for both notes shall be applied accumulatively.

f. The permitted quantities shall not be limited in a building equipped throughout with an automatic sprinkler system in accordance with Section 903.3.1.1.

g. Permitted only in buildings equipped throughout with an automatic sprinkler system in accordance with Section 903.3.1.1.

h. Containing not more than the maximum allowable quantity per control area of Class IA, IB or IC flammable liquids.

i. The maximum allowable quantity shall not apply to fuel oil storage complying with Section 603.3.2 of the International Fire Code.

j. Quantities in parenthesis indicate quantity units in parenthesis at the head of each column.

k. A maximum quantity of 200 pounds of solid or 20 gallons of liquid Class 3 oxidizers is allowed when such materials are necessary for maintenance purposes, operation or sanitation of equipment. Storage containers and the manner of storage shall be approved.

l. Net weight of the pyrotechnic composition of the fireworks. Where the net weight of the pyrotechnic composition of the fireworks is not known, 25 percent of the gross weight of the fireworks, including packaging, shall be used.

m. For gallons of liquids, divide the amount in pounds by 10 in accordance with Section 2703.1.2 of the International Fire Code.

n. For storage and display quantities in Group M and storage quantities in Group S occupancies complying with Section 414.2.5, see Tables 414.2.5(1) and 414.2.5(2).

o. Densely packed baled cotton that complies with the packing requirements of ISO 8115 shall not be included in this material class.

p. The following shall not be included in determining the maximum allowable quantities:

1. Liquid or gaseous fuel in fuel tanks on vehicles.
2. Liquid or gaseous fuel in fuel tanks on motorized equipment operated in accordance with this code.
3. Gaseous fuels in piping systems and fixed appliances regulated by the International Fuel Gas Code.
4. Liquid fuels in piping systems and fixed appliances regulated by the International Mechanical Code.

[F] TABLE 307.1(2)
MAXIMUM ALLOWABLE QUANTITY PER CONTROL AREA OF HAZARDOUS MATERIAL POSING A HEALTH HAZARD[a, b, c, i]

MATERIAL	STORAGE[d]			USE-CLOSED SYSTEMS[d]			USE-OPEN SYSTEMS[d]	
	Solid pounds (cubic feet)	Liquid gallons (pounds)[e, f]	Gas (cubic feet at NTP)[e]	Solid pounds[e]	Liquid gallons (pounds)[e]	Gas (cubic feet at NTP)[e]	Solid pounds[e]	Liquid gallons (pounds)[e]
Corrosive	5,000	500	Gaseous 810[f] Liquefied (150)[h]	5,000	500	Gaseous 810[f] Liquefied (150)[h]	1,000	100
Highly toxic	10	(10)[h]	Gaseous 20[g] Liquefied (4)[g, h]	10	(10)[i]	Gaseous 20[g] Liquefied (4)[g, h]	3	(3)[i]
Toxic	500	(500)[h]	Gaseous 810[f] Liquefied (150)[f, h]	500	(500)[i]	Gaseous 810[f] Liquefied (150)[f, h]	125	(125)

For SI: 1 cubic foot = 0.028 m³, 1 pound = 0.454 kg, 1 gallon = 3.785 L.

a. For use of control areas, see Section 414.2.

b. In retail and wholesale sales occupancies, the quantities of medicines, foodstuffs, consumer or industrial products, and cosmetics, containing not more than 50 percent by volume of water-miscible liquids and with the remainder of the solutions not being flammable, shall not be limited, provided that such materials are packaged in individual containers not exceeding 1.3 gallons.

c. For storage and display quantities in Group M and storage quantities in Group S occupancies complying with Section 414.2.5, see Tables 414.2.5(1) and 414.2.5(2).

d. The aggregate quantity in use and storage shall not exceed the quantity listed for storage.

e. Maximum allowable quantities shall be increased 100 percent in buildings equipped throughout with an approved automatic sprinkler system in accordance with Section 903.3.1.1. Where Note f also applies, the increase for both notes shall be applied accumulatively.

f. Maximum allowable quantities shall be increased 100 percent when stored in approved storage cabinets, gas cabinets or exhausted enclosures as specified in the *International Fire Code*. Where Note e also applies, the increase for both notes shall be applied accumulatively.

g. Allowed only when stored in approved exhausted gas cabinets or exhausted enclosures as specified in the *International Fire Code*.

h. Quantities in parenthesis indicate quantity units in parenthesis at the head of each column.

i. For gallons of liquids, divide the amount in pounds by 10 in accordance with Section 2703.1.2 of the *International Fire Code*.

[F] 307.2 Definitions. The following words and terms shall, for the purposes of this section and as used elsewhere in this code, have the meanings shown herein.

❖ Definitions of terms that are associated with the content of this section are contained herein. These definitions can help in the understanding and application of the code requirements. It is important to emphasize that these terms are not exclusively related to this section but are applicable everywhere the term is used in the code. The purpose for including these definitions within this section is to provide more convenient access to them without having to refer back to Chapter 2. For convenience, these terms are also listed in Chapter 2 with a cross reference to this section. The use and application of all defined terms, including those defined herein, are set forth in Section 201.

AEROSOL. A product that is dispensed from an aerosol container by a propellant.

Aerosol products shall be classified by means of the calculation of their chemical heats of combustion and shall be designated Level 1, 2 or 3.

Level 1 aerosol products. Those with a total chemical heat of combustion that is less than or equal to 8,600 British thermal units per pound (Btu/lb) (20 kJ/g).

Level 2 aerosol products. Those with a total chemical heat of combustion that is greater than 8,600 Btu/lb (20 kJ/g), but less than or equal to 13,000 Btu/lb (30 kJ/g).

Level 3 aerosol products. Those with a total chemical heat combustion that is greater than 13,000 Btu/lb (30 kJ/g).

❖ The intent of the code is to regulate those aerosols that contain a flammable propellant, such as butane, isobutane or propane. An aerosol product such as whipped cream is a water-based material with a nonflammable propellant (nitrous oxide) and would, therefore, not be regulated as a hazardous material. The contents of the aerosol container may be dispensed in the form of a mist spray, foam, gel or aerated powder.

Because of the wide range of flammability of aerosol products, a classification system was established to determine the required level of fire protection. Categories are defined according to the aerosol's chemical heat of combustion expressed in Btus per pound (Btu/lb). Aerosol category classifications of Levels 1, 2 and 3 are used to avoid confusion with flammable liquid classifications.

Examples of Level 1 aerosol products are shaving gel, whipped cream and air fresheners. Level 1 aerosols are not regulated as a hazardous material and are essentially exempt from the requirements of this section. Examples of Level 2 aerosols include some hair sprays and insect repellents. Level 3 aerosols include carburetor cleaner and other petroleum-based aerosols.

While aerosols are defined as hazardous materials, note that they are not listed in Table 307.1(1) or 307.1(2) as having a maximum allowable quantity per control area. As stated in Exception 11 to Section

307.1, a building or structure used for aerosol storage is classified as Group S-1, provided the requirements of the IFC are satisfied; therefore, the Group H classification is not utilized since the design must satisfy the IFC in order to be in compliance.

AEROSOL CONTAINER. A metal can or a glass or plastic bottle designed to dispense an aerosol. Metal cans shall be limited to a maximum size of 33.8 fluid ounces (1000 ml). Glass or plastic bottles shall be limited to a maximum size of 4 fluid ounces (118 ml).

❖ All design criteria for the aerosol container, including the maximum size and minimum strength, are set by the U.S. Department of Transportation (DOTn 49 CFR). These container regulations are essential for safe transportation of aerosol products.

BALED COTTON. A natural seed fiber wrapped in and secured with industry accepted materials, usually consisting of burlap, woven polypropylene, polyethylene or cotton or sheet polyethylene, and secured with steel, synthetic or wire bands or wire; also includes linters (lint removed from the cottonseed) and motes (residual materials from the ginning process).

❖ This definition of standard "Baled cotton" is being included only to distinguish it from "Baled cotton, densely packed" (see the commentary to the definition of "Baled cotton, densely packed"). The Joint Cotton Industry Bale Packaging Committee (JCIBPC) represents all parts of the cotton industry and sets standards and specifications for packaging of cotton bales that include bale density. The JCIBPC specifications for baling of cotton require that all cotton bales be secured with fixed-length wire bands, polyester plastic strapping or cold-rolled, high-tensile steel strapping and then covered in fully coated woven polyolefin, polyethylene film or burlap.

BALED COTTON, DENSELY PACKED. Cotton made into banded bales with a packing density of at least 22 pounds per cubic foot (360 kg/m³), and dimensions complying with the following: a length of 55 inches (1397 ± 20 mm), a width of 21 inches (533.4 ± 20 mm) and a height of 27.6 to 35.4 inches (701 to 899 mm).

❖ Currently, over 99 percent of all U.S. cotton is pressed and stored as densely packed baled cotton, with bales meeting the weight and dimension requirements of ISO 8115. One reason that the cotton industry has chosen to use such bales is because they are very difficult to ignite, which allows the industry to transport them without being labeled as "flammable solids" or "dangerous goods" by the national or international transport authorities. It is intended that this definition be used to distinguish such bales from other combustible fibers.

In order to counteract some erroneous information regarding the combustibility characteristics of densely packed cotton bales, flammability research was conducted on baled cotton. The research demonstrated that densely packed baled cotton meeting the size and weight requirements of ISO 8115 is not a hazardous material. In view of that data, the U.S. Department of

Transportation (U.S. Coast Guard), the United Nations (U.N.) and the International Maritime Organization (IMO) have all removed the listing of baled cotton from the list of hazardous materials and from the list of flammable solids, provided the cotton bales are the densely packed type that meet the standard noted above. The research conclusions were:

1. Standard cotton fiber "passed" the Department of Transportation's spontaneous combustion test: the cotton did not exceed the oven temperature and was not classified as self-heating.

2. Cotton, as densely packed baled cotton, did not cause sustained smoldering propagation: an electric heater placed within the bales was unable to cause sustained smoldering propagation, because of the lack of oxygen inside the densely packed bale.

3. Cotton, as densely packed baled cotton, was exposed to ignition from a cigarette and a match and performed very well: no propagating combustion with either.

4. Cotton, as densely packed baled cotton, was exposed to ignition from the gas burner source in ASTM E 1590 (also known as California Technical Bulletin 129) of 12 L/min of propane gas for 180 seconds and passed all the criteria, including mass loss of less than 1.36 kg (3 pounds), heat release rate less than 100 kW and total heat release of less than 25 MJ in the first 10 minutes of the test.

BARRICADE. A structure that consists of a combination of walls, floor and roof, which is designed to withstand the rapid release of energy in an explosion and which is fully confined, partially vented or fully vented; or other effective method of shielding from explosive materials by a natural or artificial barrier.

Artificial barricade. An artificial mound or revetment a minimum thickness of 3 feet (914 mm).

Natural barricade. Natural features of the ground, such as hills, or timber of sufficient density that the surrounding exposures that require protection cannot be seen from the magazine or building containing explosives when the trees are bare of leaves.

❖ The use of barricades provides an alternative method of explosion control by minimizing the potential damage due to blast effects and flying debris in the event of an explosion (see Section 414.5.1). Depending on the detonable hazard involved, an effective barricade may be a blast-resistant structure or natural or artificial barrier as provided for in Chapter 34 of the IFC for the storage of explosives.

BOILING POINT. The temperature at which the vapor pressure of a liquid equals the atmospheric pressure of 14.7 pounds per square inch (psi) (101 kPa) gage or 760 mm of mercury. Where an accurate boiling point is unavailable for the material in question, or for mixtures which do not have a constant boiling point, for the purposes of this classification, the 20-percent evaporated point of a distillation performed in accordance with ASTM D 86 shall be used as the boiling point of the liquid.

❖ The boiling point of a liquid is significant in determining the appropriate division for Class I flammable liquids. Temperatures above the established boiling point for a given liquid would result in the atmospheric pressure no longer being able to keep the liquid in a liquid state. Liquids with low boiling points present a greater fire hazard because of the increased vapor pressure at normal ambient temperatures.

CLOSED SYSTEM. The use of a solid or liquid hazardous material involving a closed vessel or system that remains closed during normal operations where vapors emitted by the product are not liberated outside of the vessel or system and the product is not exposed to the atmosphere during normal operations; and all uses of compressed gases. Examples of closed systems for solids and liquids include product conveyed through a piping system into a closed vessel, system or piece of equipment.

❖ The difference between a closed system and an open system is whether the hazardous material involved in the process is exposed to the atmosphere. While not specific in the definition, certain gases are also allowed in closed systems, as indicated in Tables 307.7(1) and 307.7(2). Materials in closed or open systems are assumed to be "in use" as opposed to "in storage." Gases are always assumed to be in closed systems, since they would be immediately dispersed in an open system if exposed to the atmosphere without some means of containment (see the definition of "Open system").

COMBUSTIBLE DUST. Finely divided solid material that is 420 microns or less in diameter and which, when dispersed in air in the proper proportions, could be ignited by a flame, spark or other source of ignition. Combustible dust will pass through a U.S. No. 40 standard sieve.

❖ Combustible dusts are combustible solids in a finely divided state that are suspended in the air. An explosion hazard exists when the concentration of the combustible dust is within the explosive limits and exposed to an ignition source of sufficient energy and duration to initiate self-sustained combustion. A review of the occupancy classification for Group H-2 in Section 307.4 indicates that combustible dusts are classified in that occupancy group. The intent of that section is that when combustible dust is determined by an engineering analysis to meet the definition parameter that, in a given occupancy, it is dispersed in air in the proper proportions so as to be ignitable by an ignition source, then the deflagration hazard is sufficient to classify the occupancy in Group H-2. Combustible dust, as a material, that does not rise to that defined level of hazard in a particular building would not cause the building or portion thereof housing the hazard to be classified in Group H-2, but rather in the group that is most appropriate for the particular operation.

The tabular Maximum Allowable Quantity per Con-

trol Area (MAQ) (formerly called "exempt amounts") for combustible dust, previously included in the legacy building and fire codes, was deleted because of its questionable value given the complexities of dust explosion hazards. Determining a theoretical MAQ of combustible dust and the potential for a dust explosion requires a thorough evaluation and technical report based on the provisions of Section 104.7.2 of the IFC. Such determination is complex and requires evaluation far beyond the simple 1 pound per 1,000 cubic feet (16 g/m³) exempt amount/MAQ previously used by the legacy codes. Critical factors, such as particle size, material density, humidity and oxygen concentration, play a major role in the evaluation of the dust hazard and are much too complex to be simply addressed.

COMBUSTIBLE FIBERS. Readily ignitable and free-burning materials in a fibrous or shredded form, such as cocoa fiber, cloth, cotton, excelsior, hay, hemp, henequen, istle, jute, kapok, oakum, rags, sisal, Spanish moss, straw, tow, wastepaper, certain synthetic fibers or other like materials. This definition does not include densely packed baled cotton.

❖ Operations involving combustible fibers are typically associated with paper milling, recycling, cloth manufacturing, carpet and textile mills and agricultural operations, among others. The primary hazards associated with such operations involve the abundance of materials and their ready ignitability. Many organic fibers are prone to spontaneous ignition if improperly dried and kept in areas without sufficient ventilation. Densely packed baled cotton is a special type of combustible fiber that, based on its weight and dimension requirements, is not easily ignitable and is not a hazardous material.

COMBUSTIBLE LIQUID. A liquid having a closed cup flash point at or above 100°F (38°C). Combustible liquids shall be subdivided as follows:

Class II. Liquids having a closed cup flash point at or above 100°F (38°C) and below 140°F (60°C).

Class IIIA. Liquids having a closed cup flash point at or above 140°F (60°C) and below 200°F (93°C).

Class IIIB. Liquids having a closed cup flash point at or above 200°F (93°C).

The category of combustible liquids does not include compressed gases or cryogenic fluids.

❖ Combustible liquids differ from flammable liquids in that the closed cup flash point of all combustible liquids is at or above 100°F (38°C) (see the definition of "Flash point"). There are three categories of combustible liquids. The range of their closed-cup flash point dictates the class of combustible liquid. The flash point range of 100°F (38°C) to 140°F (60°C) for Class II liquids is based on a possible indoor ambient temperature exceeding 100°F (38°C). Only a moderate degree of heating would be required to bring the liquid to its flash point in this type of condition. Class III liquids, which have flash points higher than 140°F (38°C), would require a significant heat source besides ambient temperature conditions to reach their flash point (see the definition of "Flammable liquid"). Class IIIA has a closed-cup flash point range of 140°F (93°C). Class IIIB has a closed cup flash point at or above 200°F (93°C). Combustible liquids are primarily considered Group H-2 materials except for Class II and IIIA liquids that are considered Group H-3 when used or stored in normally closed containers or systems pressurized at less than 15 psig (103.4 kPa). Motor oil is a typical example of a Class IIIB combustible liquid. Note that Class IIIB liquids are not regulated to be classified as Group H per Table 307.1(1).While cryogenic fluids and compressed gases may be combustible, they are to be regulated separately from combustible liquids.

COMPRESSED GAS. A material, or mixture of materials, that:

1. Is a gas at 68°F (20°C) or less at 14.7 pounds per square inch atmosphere (psia) (101 kPa) of pressure; and

2. Has a boiling point of 68°F (20°C) or less at 14.7 psia (101 kPa) which is either liquefied, nonliquefied or in solution, except those gases which have no other health- or physical-hazard properties are not considered to be compressed until the pressure in the packaging exceeds 41 psia (282 kPa) at 68°F (20°C).

The states of a compressed gas are categorized as follows:

1. Nonliquefied compressed gases are gases, other than those in solution, which are in a packaging under the charged pressure and are entirely gaseous at a temperature of 68°F (20°C).

2. Liquefied compressed gases are gases that, in a packaging under the charged pressure, are partially liquid at a temperature of 68°F (20°C).

3. Compressed gases in solution are nonliquefied gases that are dissolved in a solvent.

4. Compressed gas mixtures consist of a mixture of two or more compressed gases contained in a packaging, the hazard properties of which are represented by the properties of the mixture as a whole.

❖ This term refers to all types of gases that are under pressure at normal room or outdoor temperatures inside their containers, including, but not limited to, flammable, nonflammable, highly toxic, toxic, cryogenic and liquefied gases. The vapor pressure limitations provide the distinction between a liquid and a gas. Gases are materials that boil at a temperature of 68°F (20°C) or less at a pressure of 14.7 psia (101.3 kPa). Liquefied and nonliquefied compressed gases are determined by the state of the gas at a temperature of 68°F (20°C). Nonliquefied gases are entirely gaseous, while liquefied gases are partially liquid.

CONTROL AREA. Spaces within a building where quantities of hazardous materials not exceeding the maximum allowable quantities per *control area* are stored, dispensed, used or handled. See also the definition of "Outdoor control area" in the *International Fire Code*.

❖ Control areas provide an alternative method for the use and storage of hazardous materials without classifying the building or structure as a high-hazard occupancy (Group H). This concept is based on regulating the allowable quantities of hazardous materials per control area, rather than per building area, by giving credit for further compart- mentation through the use of fire barriers and horizontal assemblies having a minimum fire-resistance rating of not less than 1 hour. The maximum quantities of hazardous materials within a given control area cannot exceed the amounts for a given material listed in either Table 307.1(1) or 307.1(2) (see commentary, Section 414.2). Control areas are not limited to within buildings. A storage area that is exposed to the elements (wind, rain, snow, etc.) also cannot exceed the maximum allowable quantity per control area.

CORROSIVE. A chemical that causes visible destruction of, or irreversible alterations in, living tissue by chemical action at the point of contact. A chemical shall be considered corrosive if, when tested on the intact skin of albino rabbits by the method described in DOTn 49 CFR, Part 173.137, such a chemical destroys or changes irreversibly the structure of the tissue at the point of contact following an exposure period of 4 hours. This term does not refer to action on inanimate surfaces.

❖ This definition is derived from DOL 29 CFR; Part 1910.1200. While corrosive materials do not present a fire, explosion or reactivity hazard, they do pose a handling and storage problem. Corrosive materials, therefore, are primarily considered a health hazard and are classified as Group H-4 material. Many corrosive chemicals are also strong oxidizing agents that require classification as a multiple hazard in accordance with Section 307.8.

CRYOGENIC FLUID. A liquid having a boiling point lower than -150°F (-101°C) at 14.7 pounds per square inch atmosphere (psia) (an absolute pressure of 101 kPa).

❖ Cryogenic fluids present a hazard because they are extremely cold. Should a spill occur, their extremely cold temperature affects other compounds exposed to the spilled cryogenic fluid. Cryogenic fluids may be flammable or nonflammable; however, nonflammable cryogenics may possess properties that cause them to support combustion or react severely with other materials. The code is only intended to classify flammable or oxidizing cryogenic fluids as a hazardous material.

DAY BOX. A portable magazine designed to hold explosive materials constructed in accordance with the requirements for a Type 3 magazine as defined and classified in Chapter 33 of the *International Fire Code*.

❖ A day box is an explosive magazine that is listed in Note e of Table 307.1(1).

DEFLAGRATION. An exothermic reaction, such as the extremely rapid oxidation of a flammable dust or vapor in air, in which the reaction progresses through the unburned material at a rate less than the velocity of sound. A deflagration can have an explosive effect.

❖ Materials that present a deflagration hazard usually burn very rapidly with the release of energy from a chemical reaction in the form of intense heat. Confined deflagration hazards under pressure can result in an explosion. Most hazardous materials that pose a severe deflagration hazard are classified as Group H-2 in accordance with Section 307.4 (see the definition of "Detonation").

DETONATION. An exothermic reaction characterized by the presence of a shock wave in the material which establishes and maintains the reaction. The reaction zone progresses through the material at a rate greater than the velocity of sound. The principal heating mechanism is one of shock compression. Detonations have an explosive effect.

❖ Detonations are distinguished from deflagrations (which are produced by explosive gases, dusts, vapors and mists) by the speed with which they propagate a blast effect. Detonations occur much faster than deflagrations, since they propagate a combustion zone at a velocity greater than the speed of sound. Deflagrations propagate a combustion zone at a velocity less than the speed of sound. The speed of sound is approximately 1,100 feet per second (336 m/s) at sea level. Both detonations and deflagrations may produce explosive results when they occur in a confined space. Materials that are considered a detonation hazard are classified as Group H-1 materials in accordance with Section 307.3.

DISPENSING. The pouring or transferring of any material from a container, tank or similar vessel, whereby vapors, dusts, fumes, mists or gases are liberated to the atmosphere.

❖ This term refers to a specific operation whereby the act of transferring a material occurs and that has a hazard associated with the liberation of the material in the forms listed in the definition. It is not "handling" and should not be confused with that term (see the definitions of "Closed system" and "Handling").

EXPLOSION. An effect produced by the sudden violent expansion of gases, which may be accompanied by a shock wave or disruption, or both, of enclosing materials or structures. An explosion could result from any of the following:

1. Chemical changes such as rapid oxidation, *deflagration* or *detonation*, decomposition of molecules and runaway polymerization (usually *detonation*s).

2. Physical changes such as pressure tank ruptures.

3. Atomic changes (nuclear fission or fusion).

❖ Materials that pose a threat of explosion are classified as Group H-1 when present in quantities exceeding the maximum allowable quantity per control area (MAQ) in Table 307.1(1) and are required to be kept in a detached storage building meeting the requirements of Section 415.4 of the code and Chapter 33 of the IFC.

EXPLOSIVE. A chemical compound, mixture or device, the primary or common purpose of which is to function by explosion. The term includes, but is not limited to, dynamite, black powder, pellet powder, initiating explosives, detonators, safety fuses, squibs, detonating cord, igniter cord, igniters and display fireworks, 1.3G (Class B, Special).

The term "explosive" includes any material determined to be within the scope of USC Title 18: Chapter 40 and also includes any material classified as an explosive other than consumer fireworks, 1.4G (Class C, Common) by the hazardous materials regulations of DOTn 49 CFR Parts 100-185.

High explosive. Explosive material, such as dynamite, which can be caused to detonate by means of a No. 8 test blasting cap when unconfined.

Low explosive. Explosive material that will burn or deflagrate when ignited. It is characterized by a rate of reaction that is less than the speed of sound. Examples of low explosives include, but are not limited to, black powder; safety fuse; igniters; igniter cord; fuse lighters; fireworks, 1.3G (Class B, Special) and propellants, 1.3C.

Mass-detonating explosives. Division 1.1, 1.2 and 1.5 explosives alone or in combination, or loaded into various types of ammunition or containers, most of which can be expected to explode virtually instantaneously when a small portion is subjected to fire, severe concussion, impact, the impulse of an initiating agent or the effect of a considerable discharge of energy from without. Materials that react in this manner represent a mass explosion hazard. Such an explosive will normally cause severe structural damage to adjacent objects. Explosive propagation could occur immediately to other items of ammunition and explosives stored sufficiently close to and not adequately protected from the initially exploding pile with a time interval short enough so that two or more quantities must be considered as one for quantity-distance purposes.

UN/DOTn Class 1 explosives. The former classification system used by DOTn included the terms "high" and "low" explosives as defined herein. The following terms further define explosives under the current system applied by DOTn for all explosive materials defined as hazard Class 1 materials. Compatibility group letters are used in concert with the division to specify further limitations on each division noted (i.e., the letter G identifies the material as a pyrotechnic substance or article containing a pyrotechnic substance and similar materials).

Division 1.1. Explosives that have a mass explosion hazard. A mass explosion is one which affects almost the entire load instantaneously.

Division 1.2. Explosives that have a projection hazard but not a mass explosion hazard.

Division 1.3. Explosives that have a fire hazard and either a minor blast hazard or a minor projection hazard or both, but not a mass explosion hazard.

Division 1.4. Explosives that pose a minor explosion hazard. The explosive effects are largely confined to the package and no projection of fragments of appreciable size or range is to be expected. An external fire must not cause virtually instantaneous explosion of almost the entire contents of the package.

Division 1.5. Very insensitive explosives. This division is comprised of substances that have a mass explosion hazard, but that are so insensitive there is very little probability of initiation or of transition from burning to *detonation* under normal conditions of transport.

Division 1.6. Extremely insensitive articles which do not have a mass explosion hazard. This division is comprised of articles that contain only extremely insensitive detonating substances and which demonstrate a negligible probability of accidental initiation or propagation.

❖ Explosives either detonate or deflagrate when initiated by either heat, shock or electric current. While these materials are normally designed and intended to be initiated by detonators under controlled conditions, heat, shock and electric current from uncontrolled sources may initiate these materials to produce an explosion. DOTn classifies explosives in six classes according to the degree of hazard posed by the material. The most dangerous of these materials is capable of almost simultaneous detonation of all of the material in a single load or store. The least-sensitive explosives produce blasts limited to the packages in which they are transported. This definition of explosives includes materials such as detonators, blasting agents and water gels; examples of these materials are listed in DOTy 27 CFR 55.23.

Explosive materials are subdivided into high, low, mass detonating and UN/DOTn Class 1 explosives. High explosives and mass detonating explosives are typically classified as Group H-1 and present a detonation hazard. Low explosives more commonly are classified as Group H-2, as they tend to deflagrate or burn upon ignition. Mass detonating devices present a greater threat to adjacent objects and structures. The IFC, therefore, contains provisions in the form of Table 3305.3 to deal with the separation distances for mass explosion hazards.

The definitions cited in this section are consistent with DOTn 49 CFR; Section 173.50. The hazards of this group of materials vary with the nature of the material, with some explosives being very sensitive and others less sensitive. Some explosives detonate, others deflagrate and the hazards of others are limited to intense burning. The classification system was designed to correlate with the system of classification developed under recommendations of the United Nations (UN), wherein all explosive materials are placed into a hazard class of Class 1. This class is further divided into six divisions: Divisions 1.1 through 1.6.

FIREWORKS. Any composition or device for the purpose of producing a visible or audible effect for entertainment purposes by combustion, deflagration or *detonation* that meets the definition of 1.4G fireworks or 1.3G fireworks as set forth herein.

Fireworks, 1.3G. (Formerly Class B, Special Fireworks.) Large fireworks devices, which are explosive materials, intended for use in fireworks displays and designed to produce audible or visible effects by combustion, deflagration or *detonation*. Such 1.3G fireworks include, but are not limited to, firecrackers containing more than 130 milligrams (2 grains) of explosive composition, aerial shells containing more than 40 grams of pyrotechnic composition, and other display pieces which exceed the limits for classification as 1.4G fireworks. Such 1.3G fireworks are also described as fireworks, UN0335 by the DOTn.

Fireworks, 1.4G. (Formerly Class C, Common Fireworks.) Small fireworks devices containing restricted amounts of pyrotechnic composition designed primarily to produce visible or audible effects by combustion. Such 1.4G fireworks which comply with the construction, chemical composition and labeling regulations of the DOTn for fireworks, UN0336, and the U.S. Consumer Product Safety Commission (CPSC) as set forth in CPSC 16 CFR: Parts 1500 and 1507, are not explosive materials for the purpose of this code.

❖ Any device containing an explosive material that produces an audible or visible effect through combustion, deflagration, detonation or explosion is considered a firework. Fireworks are divided into two categories, 1.4G and 1.3G, based on the amount of pyrotechnic composition present.

The definitions of "1.4G fireworks" and "1.3G fireworks" are derived from the U.S. Department of Transportation (DOTn 49 CFR) clarification system for transporting explosives and from NFPA 1124. The amount of pyrotechnic composition is the distinguishing factor between the two types of fireworks (see commentary to the definition of "Pyrotechnic composition"). Fireworks that contain a limited amount of pyrotechnic composition are classified as 1.4G fireworks. 1.4G fireworks represent a physical hazard (Group H-3), while display fireworks represent a detonation hazard (Group H-1).

The requirements for storage, display and labeling depend on the correct application of the definition of "1.4G fireworks." This definition reflects the construction, chemical composition and labeling requirements of the U.S. Consumer Product Safety Commission found in Title 16, Code of Federal Regulations, Parts 1500 and 1507. Also, 1.4G fireworks are not considered to be explosives in accordance with the provisions of Chapter 33 of the IFC.

FLAMMABLE GAS. A material that is a gas at 68°F (20°C) or less at 14.7 pounds per square inch atmosphere (psia) (101 kPa) of pressure [a material that has a boiling point of 68°F (20°C) or less at 14.7 psia (101 kPa)] which:

1. Is ignitable at 14.7 psia (101 kPa) when in a mixture of 13 percent or less by volume with air; or

2. Has a flammable range at 14.7 psia (101 kPa) with air of at least 12 percent, regardless of the lower limit.

The limits specified shall be determined at 14.7 psi (101 kPa) of pressure and a temperature of 68°F (20°C) in accordance with ASTM E 681.

❖ This term essentially refers to any type of compressed gas that burns in normal concentrations of oxygen in the air (see the definition of "Compressed gas"). The definition is consistent with the provisions of ASTM E 681.

FLAMMABLE LIQUEFIED GAS. A liquefied compressed gas which, under a charged pressure, is partially liquid at a temperature of 68°F (20°C) and which is flammable.

❖ This term essentially refers to any type of liquefied compressed gas that burns in normal concentrations of oxygen in the air (see the definition of "Compressed gas").

FLAMMABLE LIQUID. A liquid having a closed cup flash point below 100°F (38°C). Flammable liquids are further categorized into a group known as Class I liquids. The Class I category is subdivided as follows:

Class IA. Liquids having a flash point below 73°F (23°C) and a boiling point below 100°F (38°C).

Class IB. Liquids having a flash point below 73°F (23°C) and a boiling point at or above 100°F (38°C).

Class IC. Liquids having a flash point at or above 73°F (23°C) and below 100°F (38°C).

The category of flammable liquids does not include compressed gases or cryogenic fluids.

❖ While all flammable liquids have a closed cup flash point less than 100°F (38°C), the further classification of the Class I liquid is dependent on the boiling point (see the definition of "Boiling point"). The 100°F (38°C) flash point limitation for flammable liquids assumes possible indoor ambient temperature conditions of 100°F (38°C).

FLAMMABLE MATERIAL. A material capable of being readily ignited from common sources of heat or at a temperature of 600°F (316°C) or less.

❖ Many standardized tests, such as ASTM E 136 and NFPA 701a, have been developed to assess the flammability and fire hazards of materials. Both of these tests include objective criteria for evaluating the combustibility of different materials; however, great care must be taken in conducting and evaluating the results of such tests.

FLAMMABLE SOLID. A solid, other than a blasting agent or explosive, that is capable of causing fire through friction, absorption or moisture, spontaneous chemical change, or retained heat from manufacturing or processing, or which has an ignition temperature below 212°F (100°C) or which burns so vigorously and persistently when ignited as to create a serious hazard. A chemical shall be considered a flammable solid as determined in accordance with the test method of CPSC 16 CFR; Part 1500.44, if it ignites and burns with a self-sustained flame at a rate greater than 0.1 inch (2.5 mm) per second along its major axis.

❖ Flammable solids are combustible materials that ignite easily and burn rapidly. Solids that may cause a fire due to friction are considered flammable solids as well as metal powders that can be readily ignited. Examples of flammable solids include nitrocellulose and combustible metals, such as magnesium and titanium.

FLASH POINT. The minimum temperature in degrees Fahrenheit at which a liquid will give off sufficient vapors to form an ignitable mixture with air near the surface or in the container, but will not sustain combustion. The flash point of a liquid shall be determined by appropriate test procedure and apparatus as specified in ASTM D 56, ASTM D 93 or ASTM D 3278.

❖ The flash point is the characteristic used in the classification of flammable and combustible liquids. The Tag Closed Tester (ASTM D 56), the Pensky-Martens Closed Cup Tester (ASTM D 93) and the Small Scale Closed-Cup Apparatus (ASTM D 3278) are the referenced test procedures for determining the flash points of liquids. The applicability of the respective test method is dependent on the viscosity of the test liquid and the expected flash point.

HANDLING. The deliberate transport by any means to a point of storage or use.

❖ The term "handling" pertains to the transporting or movement of hazardous materials within a building. Handling presents a level of hazard that is of a lesser degree than that of use or dispensing operations but greater than storage. Material is handled only when it is transported from one point to another; it is the act of conveyance. The definition provides the means to determine proper controls necessary to provide safety in the transport mode. Specific handling requirements for various hazardous materials are contained in the IFC.

HAZARDOUS MATERIALS. Those chemicals or substances that are physical hazards or health hazards as defined and classified in this section and the *International Fire Code*, whether the materials are in usable or waste condition.

❖ The term "hazardous materials" refers to those materials that present either a physical or health hazard. A specific listing of hazardous materials is indicated in Sections 307.3, 307.4, 307.5 and 307.6. An occupancy containing greater than the maximum allowable quantity per control area of these materials as indicated in Table 307.1(1) or 307.1(2) is classified in one of the four high-hazard occupancy classifications.

HEALTH HAZARD. A classification of a chemical for which there is statistically significant evidence that acute or chronic health effects are capable of occurring in exposed persons. The term "health hazard" includes chemicals that are *toxic* or *highly toxic*, and corrosive.

❖ Materials that present risks to people from handling or exposure are considered health hazards. Examples of these types of materials are indicated in Section 307.6. Buildings and structures containing materials that present a health hazard in excess of the maximum allowable quantity per control area would be classified as Group H-4. Materials that present a health hazard may also present a physical hazard (see the definition of "Physical hazard") and must comply with the requirements of the code applicable to both hazards.

HIGHLY TOXIC. A material which produces a lethal dose or lethal concentration that falls within any of the following categories:

1. A chemical that has a median lethal dose (LD_{50}) of 50 milligrams or less per kilogram of body weight when administered orally to albino rats weighing between 200 and 300 grams each.

2. A chemical that has a median lethal dose (LD_{50}) of 200 milligrams or less per kilogram of body weight when administered by continuous contact for 24 hours (or less if death occurs within 24 hours) with the bare skin of albino rabbits weighing between 2 and 3 kilograms each.

3. A chemical that has a median lethal concentration (LC_{50}) in air of 200 parts per million by volume or less of gas or vapor, or 2 milligrams per liter or less of mist, fume or dust, when administered by continuous inhalation for 1 hour (or less if death occurs within 1 hour) to albino rats weighing between 200 and 300 grams each.

Mixtures of these materials with ordinary materials, such as water, might not warrant classification as *highly toxic*. While this system is basically simple in application, any hazard evaluation that is required for the precise categorization of this type of material shall be performed by experienced, technically competent persons.

❖ The definition is derived from DOL 29 CFR; Part 1910.1200. These materials are considered dangerous to humans when either inhaled, absorbed or injected through the skin or ingested orally. Highly toxic materials present a health hazard and are subsequently listed as Group H-4 in Section 307.6. Examples of highly toxic materials include gases such as arsine, fluorine and hydrogen cyanide, liquid acrylic acid and calcium cyanide in solid form.

Mixtures of these materials with ordinary materials, such as water, might not warrant a highly toxic classification. While this system is basically simple in application, any hazard evaluation that is required for the precise categorization of this type of material is to be performed by experienced, technically competent persons.

INCOMPATIBLE MATERIALS. Materials that, when mixed, have the potential to react in a manner that generates heat, fumes, gases or byproducts which are hazardous to life or property.

❖ These materials, whether in storage or in use, constitute a dangerous chemical combination. Determination of which chemicals in combination present a hazard is a difficult situation for the building official. MSDS alone may not provide all of the necessary information. When in doubt, the building official should seek additional information from the manufacturer of the chemicals involved, the building owner or from experts who are knowledgeable in industrial hygiene or chemistry.

INERT GAS. A gas that is capable of reacting with other materials only under abnormal conditions such as high temperatures, pressures and similar extrinsic physical forces. Within the context of the code, inert gases do not exhibit either physi-

cal or health properties as defined (other than acting as a simple asphyxiant) or hazard properties other than those of a compressed gas. Some of the more common inert gases include argon, helium, krypton, neon, nitrogen and xenon.

❖ Inert gases do not react readily with other materials under normal temperatures and pressures, but it is possible for a reaction to occur. For example, even nitrogen combines with some of the more active metals such as lithium and magnesium to form nitrides, and at high temperatures it will also combine with oxygen and other elements. The formation of compounds utilizing inert gases is also possible. Xenon combines with fluorine to form various fluorides, and with oxygen to form oxides. The compounds formed are crystalline solids. As indicated in Table 307.1(1), there are no maximum allowable quantity limitations specified in Table 307.1(1) for inert gases. As such, inert gases are not regulated by the code as hazardous materials with respect to a potential Group H occupancy classification.

OPEN SYSTEM. The use of a solid or liquid hazardous material involving a vessel or system that is continuously open to the atmosphere during normal operations and where vapors are liberated, or the product is exposed to the atmosphere during normal operations. Examples of open systems for solids and liquids include dispensing from or into open beakers or containers, dip tank and plating tank operations.

❖ See the commentary to the definition of "Closed system."

OPERATING BUILDING. A building occupied in conjunction with the manufacture, transportation or use of explosive materials. Operating buildings are separated from one another with the use of intraplant or intraline distances.

❖ Buildings used for storage of explosives are not magazines. This definition is included in this section to address that fact.

ORGANIC PEROXIDE. An organic compound that contains the bivalent -O-O- structure and which may be considered to be a structural derivative of hydrogen peroxide where one or both of the hydrogen atoms have been replaced by an organic radical. Organic peroxides can pose an explosion hazard (*detonation* or deflagration) or they can be shock sensitive. They can also decompose into various unstable compounds over an extended period of time.

Class I. Those formulations that are capable of deflagration but not *detonation*.

Class II. Those formulations that burn very rapidly and that pose a moderate reactivity hazard.

Class III. Those formulations that burn rapidly and that pose a moderate reactivity hazard.

Class IV. Those formulations that burn in the same manner as ordinary combustibles and that pose a minimal reactivity hazard.

Class V. Those formulations that burn with less intensity than ordinary combustibles or do not sustain combustion and that pose no reactivity hazard.

Unclassified detonable. Organic peroxides that are capable of *detonation*. These peroxides pose an extremely high explosion hazard through rapid explosive decomposition.

❖ The chemical structure of organic peroxides differs from that of hydrogen peroxide (an oxidizer) in that an organic radical replaces the hydrogen atoms. Organic chemicals are all carbon based. As a result, organic peroxides pose varying degrees of fire or explosion hazard in addition to their oxidizing properties. The classification of organic peroxides is based on the provisions of NFPA 43B. Proper material classification of organic peroxides is essential to determining the appropriate occupancy classification of the structure. Examples of organic peroxides include acetyl cyclohexane, sulfonyl peroxide and benzoyl peroxide. The actual class of these materials is dependent on the percentage of concentration by weight. Most organic peroxides are available as liquids, pastes or solids in a powder form.

OXIDIZER. A material that readily yields oxygen or other oxidizing gas, or that readily reacts to promote or initiate combustion of combustible materials and, if heated or contaminated, can result in vigorous self-sustained decomposition.

   Class 4. An oxidizer that can undergo an explosive reaction due to contamination or exposure to thermal or physical shock and that causes a severe increase in the burning rate of combustible materials with which it comes into contact. Additionally, the oxidizer causes a severe increase in the burning rate and can cause spontaneous ignition of combustibles.

   Class 3. An oxidizer that causes a severe increase in the burning rate of combustible materials with which it comes in contact.

   Class 2. An oxidizer that will cause a moderate increase in the burning rate of combustible materials with which it comes in contact.

   Class 1. An oxidizer that does not moderately increase the burning rate of combustible materials.

❖ The classification of oxidizers is based on the provisions of NFPA 430. Oxidizers, whether a solid, liquid or gas, yield oxygen or another oxidizing gas during a chemical reaction or readily react to oxidize combustibles. The rate of reaction varies with the class of oxidizer. Specific classification of oxidizers is important because of the varying degree of hazard. Example of oxidizers include liquid hydrogen peroxide, nitric acid, sulfuric acid and solids such as sodium chlorite, chromic acid and calcium hypochlorite. Many commercially available swimming pool chemicals are indicative of Class 2 or 3 oxidizers.

OXIDIZING GAS. A gas that can support and accelerate combustion of other materials.

❖ Oxidizers sometimes yield oxidizing gases during a chemical reaction. These gases are capable of supporting and accelerating the combustion of other materials. Examples of oxidizing gases include bromine, chlorine and fluorine.

PHYSICAL HAZARD. A chemical for which there is evidence that it is a combustible liquid, cryogenic fluid, explosive, flammable (solid, liquid or gas), organic peroxide (solid or liquid), oxidizer (solid or liquid), oxidizing gas, pyrophoric (solid, liquid or gas), unstable (reactive) material (solid, liquid or gas) or water-reactive material (solid or liquid).

❖ Those materials that present a detonation hazard, deflagration hazard or readily support combustion are considered physical hazards. Examples of the types of materials that present a physical hazard are included in the definition. Buildings and structures containing materials that present a physical hazard in excess of the maximum allowable quantity per control area would be classified in Group H-1, H-2 or H-3. Materials that present a physical hazard may also present a health hazard (see the definition of "Health hazard").

PYROPHORIC. A chemical with an autoignition temperature in air, at or below a temperature of 130°F (54.4°C).

❖ The definition is derived from DOL 29 CFR; Part 1910.1200. Pyrophoric materials, whether in a gas, liquid or solid form, are capable of spontaneous ignition at low temperatures. Examples of pyrophoric materials include silane and phosphine gas; liquid diethylaluminum chloride and inert solids, such as cesium, plutonium, potassium and robidium.

PYROTECHNIC COMPOSITION. A chemical mixture that produces visible light displays or sounds through a self-propagating, heat-releasing chemical reaction which is initiated by ignition.

❖ Pyrotechnic composition consists of those chemical components, including oxidizers, that cause fireworks to make noise or display light when ignited. The definition is derived from NFPA 1124. The amount of pyrotechnic composition is the determining factor in whether the storage area for consumer fireworks is classified as Group H-3. The pyrotechnic content of consumer fireworks is contained within a significant amount of packaging and nonexplosive materials used in their manufacture, which constitute the bulk of the weight of the fireworks devices.

TOXIC. A chemical falling within any of the following categories:

   1. A chemical that has a median lethal dose (LD_{50}) of more than 50 milligrams per kilogram, but not more than 500 milligrams per kilogram of body weight when administered orally to albino rats weighing between 200 and 300 grams each.

   2. A chemical that has a median lethal dose (LD_{50}) of more than 200 milligrams per kilogram, but not more than 1,000 milligrams per kilogram of body weight when administered by continuous contact for 24 hours (or less if death occurs within 24 hours) with the bare skin of albino rabbits weighing between 2 and 3 kilograms each.

   3. A chemical that has a median lethal concentration (LC_{50}) in air of more than 200 parts per million, but not more than 2,000 parts per million by volume of gas or vapor, or more than 2 milligrams per liter but not more than 20 mil-

ligrams per liter of mist, fume or dust, when administered by continuous inhalation for 1 hour (or less if death occurs within 1 hour) to albino rats weighing between 200 and 300 grams each.

❖ The definition is derived from DOL 29 CFR; Part 1910.1200. These materials are considered dangerous to humans when either inhaled, absorbed or injected through the skin or when orally ingested. Toxic materials differ from highly toxic materials with regard to the specified median lethal dose or concentration of a given chemical. Toxic materials present a health hazard and are subsequently listed as a Group H-4 material in Section 307.6. Examples of toxic materials include gases such as chlorine; phosgene and hydrogen fluoride; liquid alkyl alcohol; methyl isocyanide and phosphorous chloride and barium chloride, benzidine and sodium fluoride in solid form.

UNSTABLE (REACTIVE) MATERIAL. A material, other than an explosive, which in the pure state or as commercially produced, will vigorously polymerize, decompose, condense or become self-reactive and undergo other violent chemical changes, including explosion, when exposed to heat, friction or shock, or in the absence of an inhibitor, or in the presence of contaminants, or in contact with incompatible materials. Unstable (reactive) materials are subdivided as follows:

Class 4. Materials that in themselves are readily capable of *detonation* or explosive decomposition or explosive reaction at normal temperatures and pressures. This class includes materials that are sensitive to mechanical or localized thermal shock at normal temperatures and pressures.

Class 3. Materials that in themselves are capable of *detonation* or of explosive decomposition or explosive reaction but which require a strong initiating source or which must be heated under confinement before initiation. This class includes materials that are sensitive to thermal or mechanical shock at elevated temperatures and pressures.

Class 2. Materials that in themselves are normally unstable and readily undergo violent chemical change but do not detonate. This class includes materials that can undergo chemical change with rapid release of energy at normal temperatures and pressures, and that can undergo violent chemical change at elevated temperatures and pressures.

Class 1. Materials that in themselves are normally stable but which can become unstable at elevated temperatures and pressure.

❖ The classification of unstable (reactive) materials is based on provisions in NFPA 704. The different classes of unstable (reactive) materials reflect the degree of susceptibility of the materials to release energy. Unstable (reactive) materials polymerize, decompose or become self-reactive when exposed to heat, air, moisture, pressure or shock. Separation from incompatible materials is essential to minimizing the hazards. Examples of unstable (reactive) materials include nitromethane, perchloric acid, sodium perchlorate, vinyl acetate and acetic acid.

WATER-REACTIVE MATERIAL. A material that explodes; violently reacts; produces flammable, *toxic* or other hazardous gases; or evolves enough heat to cause autoignition or ignition of combustibles upon exposure to water or moisture. Water-reactive materials are subdivided as follows:

Class 3. Materials that react explosively with water without requiring heat or confinement.

Class 2. Materials that react violently with water or have the ability to boil water. Materials that produce flammable, *toxic* or other hazardous gases or evolve enough heat to cause autoignition or ignition of combustibles upon exposure to water or moisture.

Class 1. Materials that react with water with some release of energy, but not violently.

❖ These materials liberate significant quantities of heat when reacting with water. Combustible water-reactive materials are capable of self-ignition. Even noncombustible water-reactive materials present a hazard because of the heat liberated during their reaction with water, which is sufficient to ignite surrounding combustible materials. While a definition for Class 1 water-reactive materials is provided for informational purposes, the maximum allowable quantity per control area of these materials in accordance with Table 307.1(1) is not limited. The descriptions of each of the subdivisions is consistent with the approach used for the determination of water hazards in NFPA 704.

[F] 307.3 High-hazard Group H-1. Buildings and structures containing materials that pose a *detonation* hazard shall be classified as Group H-1. Such materials shall include, but not be limited to, the following:

Detonable pyrophoric materials

Explosives:

    Division 1.1
    Division 1.2
    Division 1.3

      Exception: Materials that are used and maintained in a form where either confinement or configuration will not elevate the hazard from a mass fire to mass explosion hazard shall be allowed in H-2 occupancies.

    Division 1.4

      Exception: Articles, including articles packaged for shipment, that are not regulated as an explosive under Bureau of Alcohol, Tobacco and Firearms regulations, or unpackaged articles used in process operations that do not propagate a *detonation* or deflagration between articles shall be allowed in H-3 occupancies.

    Division 1.5
    Division 1.6

Organic peroxides, unclassified detonable
Oxidizers, Class 4
Unstable (reactive) materials, Class 3 detonable and Class 4

❖ The contents of occupancies in Group H-1 present a detonation hazard. Examples of materials that create this hazard are listed in the section. The definitions for Group H-1 materials are contained in Section 307.2. Because of the explosion hazard potential associated with Group H-1 materials, occupancies in Group H-1, which exceed the maximum allowable quantity per control area indicated in Table 307.1(1), are required to be located in detached one-story buildings without basements (see commentary, Sections 415.4 and 508.3). H-1 occupancies cannot be located in a mixed occupancy building.

[F] 307.4 High-hazard Group H-2. Buildings and structures containing materials that pose a deflagration hazard or a hazard from accelerated burning shall be classified as Group H-2. Such materials shall include, but not be limited to, the following:

Class I, II or IIIA flammable or combustible liquids which are used or stored in normally open containers or systems, or in closed containers or systems pressurized at more than 15 psi (103.4 kPa) gage.
Combustible dusts
Cryogenic fluids, flammable
Flammable gases
Organic peroxides, Class I
Oxidizers, Class 3, that are used or stored in normally open containers or systems, or in closed containers or systems pressurized at more than 15 psi (103 kPa) gage
Pyrophoric liquids, solids and gases, nondetonable
Unstable (reactive) materials, Class 3, nondetonable
Water-reactive materials, Class 3

❖ The contents of occupancies in Group H-2 present a deflagration or accelerated burning hazard. Examples of materials that create this hazard are listed. The definitions for Group H-2 materials are contained in Section 307.2. Because of the severe fire or reactivity hazard associated with these types of materials, proper classification is essential in determining the applicable requirements with regard to the mitigation of these hazards.

[F] 307.5 High-hazard Group H-3. Buildings and structures containing materials that readily support combustion or that pose a physical hazard shall be classified as Group H-3. Such materials shall include, but not be limited to, the following:

Class I, II or IIIA flammable or combustible liquids that are used or stored in normally closed containers or systems pressurized at 15 pounds per square inch gauge (103.4 kPa) or less
Combustible fibers, other than densely packed baled cotton
Consumer fireworks, 1.4G (Class C, Common)
Cryogenic fluids, oxidizing
Flammable solids
Organic peroxides, Class II and III
Oxidizers, Class 2
Oxidizers, Class 3, that are used or stored in normally closed containers or systems pressurized at 15 pounds per square inch gauge (103 kPa) or less
Oxidizing gases

Unstable (reactive) materials, Class 2
Water-reactive materials, Class 2

❖ The contents of occupancies in Group H-3 present a hazard inasmuch as they contain materials that readily support combustion or that present a physical hazard. Examples of materials that create this hazard are listed. The definitions for Group H-3 materials are contained in Section 307.2. While Group H-3 materials are generally less of a fire or reactivity hazard than Group H-2 materials, they still present a greater physical hazard than materials not currently regulated as high hazard.

[F] 307.6 High-hazard Group H-4. Buildings and structures which contain materials that are health hazards shall be classified as Group H-4. Such materials shall include, but not be limited to, the following:

Corrosives
Highly toxic materials
Toxic materials

❖ The contents of occupancies in Group H-4 present a hazard inasmuch as they contain materials that are health hazards. Examples of these hazards are listed in this section. The definitions for Group H-4 materials are contained in Section 307.2. While reference is made to chemicals that cause these hazards, the data sheets for these chemicals, which are furnished by the applicant, will need considerable subjective evaluation. Some materials falling into the category of health hazard may also present a physical hazard and would, therefore, require the structure to be designed for multiple hazards in accordance with Section 307.8.

[F] 307.7 High-hazard Group H-5 structures. Semiconductor fabrication facilities and comparable research and development areas in which hazardous production materials (HPM) are used and the aggregate quantity of materials is in excess of those listed in Tables 307.1(1) and 307.1(2) shall be classified as Group H-5. Such facilities and areas shall be designed and constructed in accordance with Section 415.8.

❖ HPM includes flammable liquids and gases, corrosives, oxidizers and, in many instances, highly toxic materials (see the definition for "Hazardous production material" in Section 415.2). In determining the applicable requirements of other sections of the code, HPM facilities are considered to be Group H-5 occupancies. It is intended that the quantities of materials permitted in Table 415.8.2.1.1 will take precedence over Tables 307.1(1) and 307.1(2).

[F] 307.8 Multiple hazards. Buildings and structures containing a material or materials representing hazards that are classified in one or more of Groups H-1, H-2, H-3 and H-4 shall conform to the code requirements for each of the occupancies so classified.

❖ If materials are present that possess characteristics of more than one Group H, high-hazard occupancy, then the structure must be designed to protect against the hazards of each relevant high-hazard occupancy classification. For example, a material could be classified as both a Class 2 oxidizer (Group H-3) and a corrosive (Group H-4). If the given quantity exceeded the maximum allowable quantity per control area individually for both a Class 2 oxidizer and a corrosive, the structure is required to conform to the applicable requirements of both Groups H-3 and H-4.

SECTION 308
INSTITUTIONAL GROUP I

308.1 Institutional Group I. Institutional Group I occupancy includes, among others, the use of a building or structure, or a portion thereof, in which people are cared for or live in a supervised environment, having physical limitations because of health or age are harbored for medical treatment or other care or treatment, or in which people are detained for penal or correctional purposes or in which the liberty of the occupants is restricted. Institutional occupancies shall be classified as Group I-1, I-2, I-3 or I-4.

❖ Institutional occupancies are comprised of two basic types. The first relates to health care facilities that are intended to provide medical care or treatment for people who have physical or mental disabilities or diseases and other infirmities. This includes persons who are ambulatory and are capable of self-preservation as well as those who are restricted in their mobility or totally immobile and need assistance to escape a life-threatening situation, such as a fire (i.e., children $2^1/_2$ years of age or less and infirm persons. The second type of occupancy relates primarily to detention and correctional facilities. Since security is the major operational consideration in these kinds of facilities, the occupants (inmates) are under some form of restraint and may be rendered incapable of self-preservation without assistance in emergency situations.

The degree of hazards in each type of institutional facility identified in this section varies respective to each kind of occupancy. The code addresses each occupancy separately and the regulatory provisions throughout the code provide the proper means of protection so as to produce an acceptable level of safety to life and property.

This section identifies the occupancies that are included in the general term "Group I." Institutional occupancies are divided into four individual occupancy classifications: Groups I-1, I-2, I-3 and I-4. These classifications are based on the degree of detention and physical mobility of the occupants. The term "Group I" includes each of the individual institutional occupancy classifications.

Group I occupancies are distinguished from other occupancies and within the subgroups of the I occupancies based on whether the care is provided on a 24-hour basis. The intent is that this criteria is not specific to the hours of operation of the facility, but the length of time that care is provided for the patients, residents or those in day care. For example, an outpatient clinic that is open 24 hours a day is a Group B occupancy provided patients are treated as outpatients and there are no in-patients that would stay at the facility 24 hours or longer. Another example would be a "day care" facility that is open 24 hours to serve workers who work any shift and need to have children in "day care" while they work. Provided that individual children receive care for less than 24 hours, the occupancy is an I-4.

308.2 Group I-1. This occupancy shall include buildings, structures or parts thereof housing more than 16 persons, on a 24-hour basis, who because of age, mental disability or other reasons, live in a supervised residential environment that provides *personal care services*. The occupants are capable of responding to an emergency situation without physical assistance from staff. This group shall include, but not be limited to, the following:

Alcohol and drug centers
Assisted living facilities
Congregate care facilities
Convalescent facilities
Group homes
Halfway houses
Residential board and care facilities
Social rehabilitation facilities

A facility such as the above with five or fewer persons shall be classified as a Group R-3 or shall comply with the *International Residential Code* in accordance with Section 101.2. A facility such as above, housing at least six and not more than 16 persons, shall be classified as Group R-4.

❖ An occupancy classified in Group I-1 is characterized by four conditions: it is a health care facility, the number of occupants housed in such facilities is greater than 16 (in order to be consistent with the definition of "Residential care/assisted living facility"), there is 24-hour-a-day supervision and the occupants are capable of reaching safety in an emergency situation without the need of physical assistance by staff or others. The supervision for Group I-1 buildings is for counseling and assistance purposes, not for medical purposes.

Any building that has these characteristics but that contains an occupant load of more than five and not more than 16 is classified as Group R-4 (see Section 310.1). Any building that has these characteristics but contains an occupant load of five or less is classified as Group R-3 (see Section 310.1), or shall be constructed in accordance with the *International Residential Code* (IRC®). When the code allows the construction in accordance with the IRC, the only requirements that would apply would be those of the IRC. It is important to note that the IRC requires sprinklers in one- and two-family homes starting in the year 2011. Also, if the occupancy is classified as R-4 and built to the requirements of the IRC, it must be sprinklered in accordance with either NFPA 13 or NFPA 13R.

The occupant load for occupancy classification purposes refers to the number of residents only. The number of guests or staff is not included. Please note, however, that the number of guests and staff is included for means of egress purposes.

A Group I-1 occupancy is intended to contain care recipients capable of self-preservation. When such an occupancy begins to house care recipients who are not capable of self-preservation, then the occupancy classification would likely need to be changed to a Group I-2 due to the 24-hour nature of such facilities. This same situation would be true of an existing Group R-4 assisted living facility that began to house more than five occupants not capable of self-preservation. Such occupancies would also need to be reevaluated and reclassified as Group I-2. Such buildings could be dealt with as mixed occupancies. Generally, Groups I-1 and R-4 were not intended to contain occupants not capable of self-preservation but may have occupants that may no longer be capable of self-preservation that at one time would have been classified as capable.

For clarification purposes, a dormitory or apartment complex that houses only elderly people and has a nonmedically trained live-in manager is not classified as an institutional occupancy but rather as a residential occupancy (see Section 310.1). A critical phrase in the code to consider when evaluating this type of facility is "live in a supervised residential environment." Such dormitories or apartment complexes may contain features such as special emergency call switches monitored by health center staff that are located in each dwelling unit. These emergency call switches are a convenience and do not necessarily indicate infirmity of the occupants.

308.3 Group I-2. This occupancy shall include buildings and structures used for medical, surgical, psychiatric, nursing or custodial care for persons who are not capable of self-preservation. This group shall include, but not be limited to, the following:

Child care facilities
Detoxification facilities
Hospitals
Mental hospitals
Nursing homes

❖ An occupancy classified in Group I-2 is characterized by three conditions: it is a health care facility, there is 24-hour-a-day medical supervision for individual patients and some or all of the occupants require physical assistance by staff or others to reach safety in an emergency situation. Determining whether occupants are capable of self-preservation may be a difficult aspect of classifying an occupancy as Group I-1 or I-2. This assessment needs to be taken with caution and reliance on other state and federal guidelines and associated regulations may be necessary. Also, it is important to keep in mind that facilities that may be classified initially as Group I-1 (capable of self-preservation) or R-4 can very easily need to reclassified as Group I-2 if the abilities of occupants are not carefully evaluated and addressed.

The most common examples of facilities classified in Group I-2 are hospitals and nursing homes [see Figures 308.3(1) and 308.3(2)]. Other facilities included are detoxification facilities, child care facilities with more than five children 2¹/₂ years or less in age, both of

which provide care for inpatients who stay longer than 24 hours.

It is not uncommon to find dining rooms (Group A-2), staff offices (Group B), gift shops (Group M) and other nonmedically related areas in buildings classified as Group I-2. Unless the area of the nonmedical occupancy qualifies as an accessory occupancy (see Section 508.2), the building is considered as a mixed occupancy and subject to the provisions of Section 508. In addition to the general requirements contained in this section, Section 407 contains specific requirements for Group I-2.

Note that the terms "custodial care" and "care" as used throughout this section have the same meaning.

308.3.1 Definitions. The following words and terms shall, for the purposes of this section and as used elsewhere in this code, have the meanings shown herein.

❖ Definitions of terms that are associated with the content of this section are contained herein. These definitions can help in the understanding and application of the code requirements. It is important to emphasize that these terms are not exclusively related to this section but are applicable everywhere the term is used in the code. The purpose for including these definitions within this section is to provide more convenient access to them without having to refer back to Chapter 2. For convenience, these terms are also listed in Chapter 2 with a cross reference to this section. The use and application of all defined terms, including those defined herein, are set forth in Section 201.

CHILD CARE FACILITIES. Facilities that provide care on a 24-hour basis to more than five children, 2¹/₂ years of age or less.

DETOXIFICATION FACILITIES. Facilities that serve patients who are provided treatment for substance abuse on a 24-hour basis and who are incapable of self-preservation or who are harmful to themselves or others.

HOSPITALS AND MENTAL HOSPITALS. Buildings or portions thereof used on a 24-hour basis for the medical, psychiatric, obstetrical or surgical treatment of inpatients who are incapable of self-preservation.

NURSING HOMES. Nursing homes are long-term care facilities on a 24-hour basis, including both intermediate care facilities and skilled nursing facilities, serving more than five persons and any of the persons are incapable of self-preservation.

❖ This section defines more specifically the occupancies that fall within the classification of Group I-2. This includes the number of hours and other criteria such as age, number of patients and the type of treatment that occurs in such facilities.

Child care facilities housing more than five children 2¹/₂ years old and younger who stay for periods in excess of 24 hours are classified as Group I-2 because children that age are not generally capable of responding to an emergency and must be led or carried to safety. Under such circumstances, the occupants are considered nonambulatory. The distinguishing factor between Groups I-2 and I-4 is the amount of time the facility provides care; Group I-2 facilities provide care to inpatients on a 24-hour basis, while Group I-4 facilities must be less than 24 hours. It is also assumed that medical supervision is present in Group I-2 facilities. Figure 308.3.1 summarizes the different occupancy classifications for care facilities.

The general condition of patients being treated for some type of substance abuse can be such that the response of these individuals to notification of emergency can be varied and unpredictable. Detoxification facilities are any facility treating patients for periods exceeding 24 hours with any number of patients being treated for substance abuse who are not capable of self-preservation. Patients in such facilities may be under some level of lockdown and depending upon their stage in the detoxification process may be unable to evacuate unassisted.

Hospitals, like detoxification facilities, are not dependent upon the number of patients but instead on the types of activities that occur in such facilities and the fact that they provide care for patients for periods exceeding 24 hours.

Nursing homes are classified as Group I-2 when they contain more than five occupants (nonstaff) incapable of self-preservation. Residents of nursing homes tend to be long-term and care is provided for more than 24 hours. "Nursing home" occupancies with five or fewer occupants would be classified as Group R-3. Although the code does not specifically state this classification, it is essentially the classification such facilities would default to.

308.4 Group I-3. This occupancy shall include buildings and structures that are inhabited by more than five persons who are under restraint or security. An I-3 facility is occupied by persons who are generally incapable of self-preservation due to security measures not under the occupants' control. This group shall include, but not be limited to, the following:

> Correctional centers
> Detention centers
> Jails
> Prerelease centers
> Prisons
> Reformatories

Buildings of Group I-3 shall be classified as one of the occupancy conditions indicated in Sections 308.4.1 through 308.4.5 (see Section 408.1).

❖ An occupancy classified in Group I-3 is characterized by three conditions: it is a location where the occupants are under restraint or where security is closely supervised, there are more than five occupants and the occupants are not capable of self-preservation because the conditions of confinement are not under their control (i.e., they require assistance by the facilities' staff to reach safety in an emergency situation). The occupant load for occupancy classification purposes refers to the number of residents only. The number of guests or staff are not included. Please note, however, that the number of guests and staff are included for means of egress purposes.

Buildings that have these characteristics but that contain an occupant load of five or fewer are classified as a residential occupancy, but are still subject to special hardware requirements (see Chapter 10).

It is recognized that not all Group I-3 occupancies have the same level of restraint; thus, to distinguish these different levels, the code defines five different conditions of occupancy based on the degree of access to the exit discharge.

The five occupancy conditions are summarized in Figure 308.4.

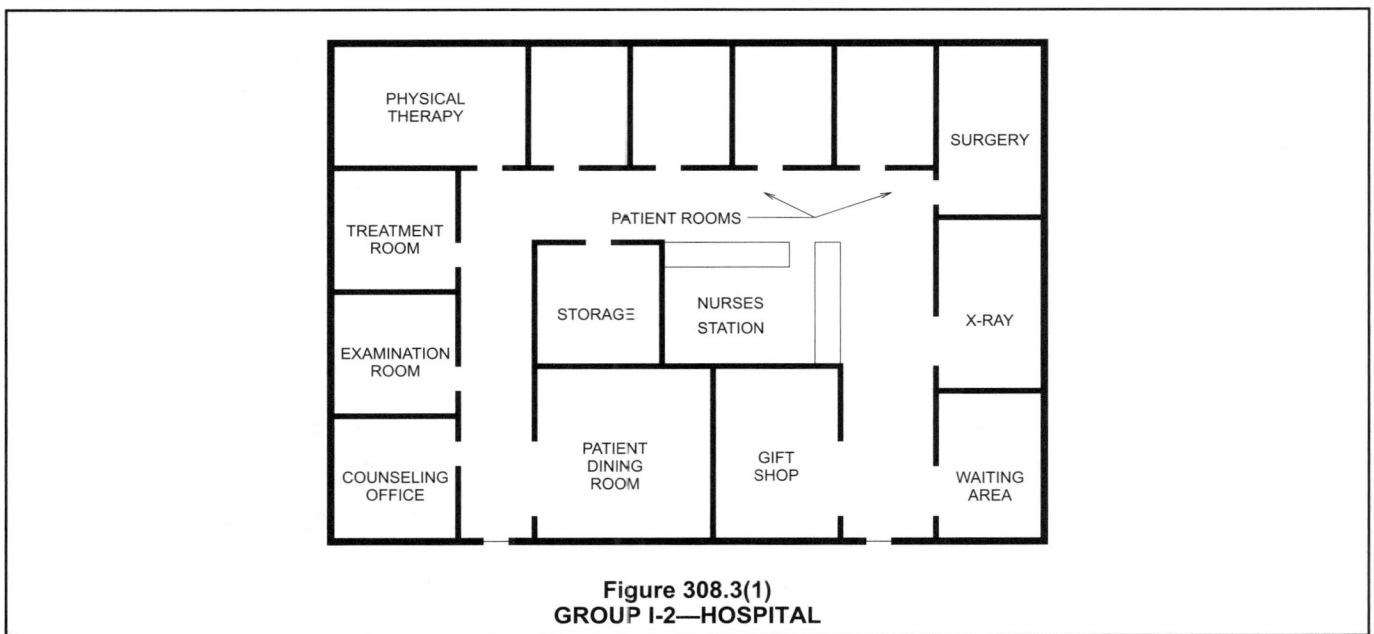

Figure 308.3(1)
GROUP I-2—HOSPITAL

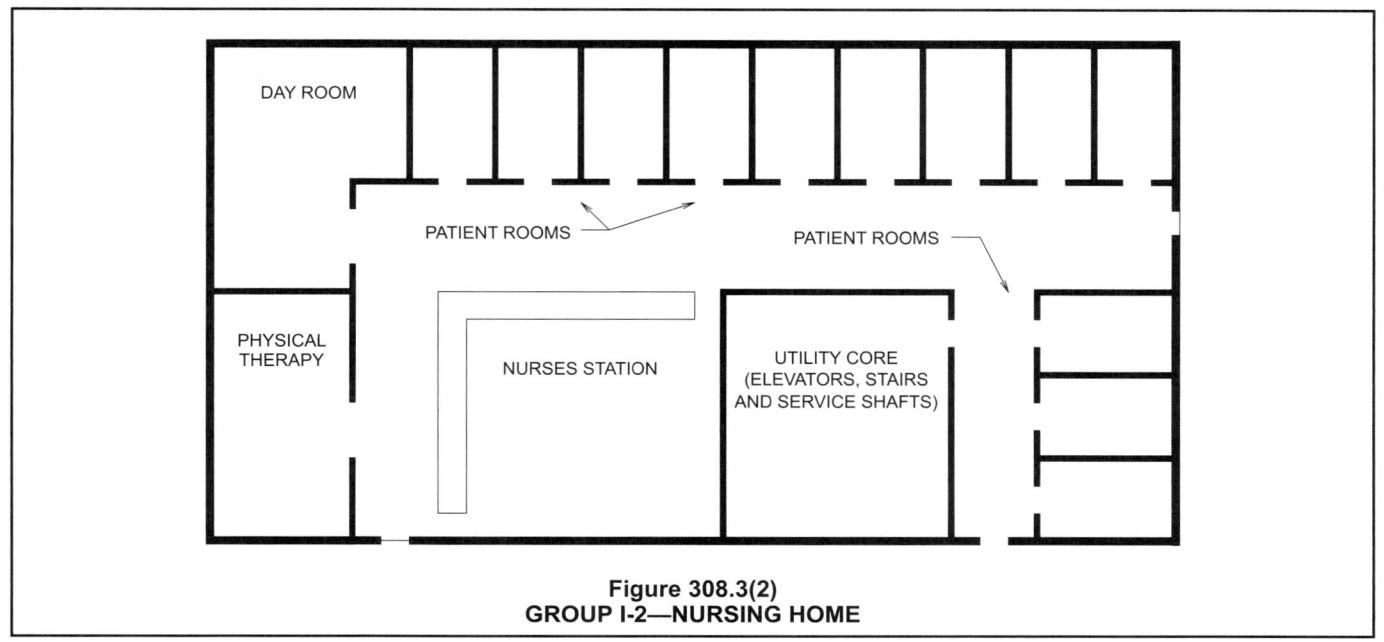

Figure 308.3(2)
GROUP I-2—NURSING HOME

24-HOUR CARE			
Age and capability of residents	1-5 occupants	6-16 occupants	Over 16 occupants
2¹/₂ years of age or less	R-3	I-2	I-2
Over 2¹/₂ years of age and capable of self-preservation	R-3	R-4	I-1
Over 2¹/₂ years of age and not capable of self-preservation and defined as child care facilities or nursing home	R-3	I-2	I-2
Over 2¹/₂ years of age and not capable of self-preservation and defined as hospital or detoxification facility	I-2	I-2	I-2

LESS THAN 24-HOUR CARE—DAY CARE		
Age and capability of residents	1-5 occupants	Over 5 occupants
2¹/₂ years of age or less	R-3	I-4 (Exception permits E)
Over 12th grade and capable of responding to emergency without physical assistance from staff	R-3	R-3
Over 12th grade and not capable of responding to emergency without physical assistance from staff	R-3	I-4
Over 2¹/₂ years of age and not capable of self-preservation	R-3	I-4

Figure 308.3.1
OCCUPANCY CLASSIFICATION OF CARE FACILITIES

308.4.1 Condition 1. This occupancy condition shall include buildings in which free movement is allowed from sleeping areas, and other spaces where access or occupancy is permitted, to the exterior via *means of egress* without restraint. A Condition 1 facility is permitted to be constructed as Group R.

❖ Condition 1 areas are those where the occupants have unrestrained access to the exterior of the building. As

such, a key or remote control release device is not needed for any occupant to reach the exterior of the building (exit discharge) at any time. These types of buildings are referred to as low-security facilities. A work-release center is a typical Condition 1 facility (see Figure 308.4.1).

Because of the lack of restraint associated with a Condition 1 building, it resembles a residential use more than a detention facility and, therefore, is permitted to be classified in Group R (see Section 310).

308.4.2 Condition 2. This occupancy condition shall include buildings in which free movement is allowed from sleeping areas and any other occupied smoke compartment to one or more other smoke compartments. Egress to the exterior is impeded by locked *exits*.

❖ Condition 2 areas are those in which the movement of occupants is not controlled within the exterior walls of the building (i.e., the occupants have unrestrained access within the building). As such, there is free movement by the occupants between smoke compartments (as created by smoke barriers); however, the occupants must rely on someone else to allow them to exit the building to the area of discharge. This is illustrated in Figure 308.4.2.

308.4.3 Condition 3. This occupancy condition shall include buildings in which free movement is allowed within individual smoke compartments, such as within a residential unit comprised of individual *sleeping units* and group activity spaces, where egress is impeded by remote-controlled release of *means of egress* from such a smoke compartment to another smoke compartment.

❖ Condition 3 areas are those in which free movement by the occupants is permitted within an individual smoke compartment; however, movement of occupants from one smoke compartment (as created by smoke barriers) to another smoke compartment and

from within the building to the exterior (exit discharge) is controlled by remote release locking devices. As such, the occupants in the facility are dependent on the staff for their release from each smoke compartment or to the exterior (exit discharge). This condition is illustrated in Figure 308.4.3.

308.4.4 Condition 4. This occupancy condition shall include buildings in which free movement is restricted from an occupied space. Remote-controlled release is provided to permit movement from *sleeping units*, activity spaces and other occupied areas within the smoke compartment to other smoke compartments.

❖ Condition 4 areas are those in which the movement of occupants from any room or space within a smoke compartment (as created by smoke barriers) to another smoke compartment or to the exterior (exit discharge) is controlled by remote release locking devices. Any movement within the facility requires activation of a remote control lock system to release the designated area (see Figure 308.4.4). The occupants within a Condition 4 area must rely on an activation system in the event of an emergency in order to evacuate the area.

Condition 4 facilities most often are penal facilities where the occupants are considered relatively safe to handle in large groups. As such, many occupants can be released simultaneously from their individual sleeping areas when they need to travel to dining areas or move to another area.

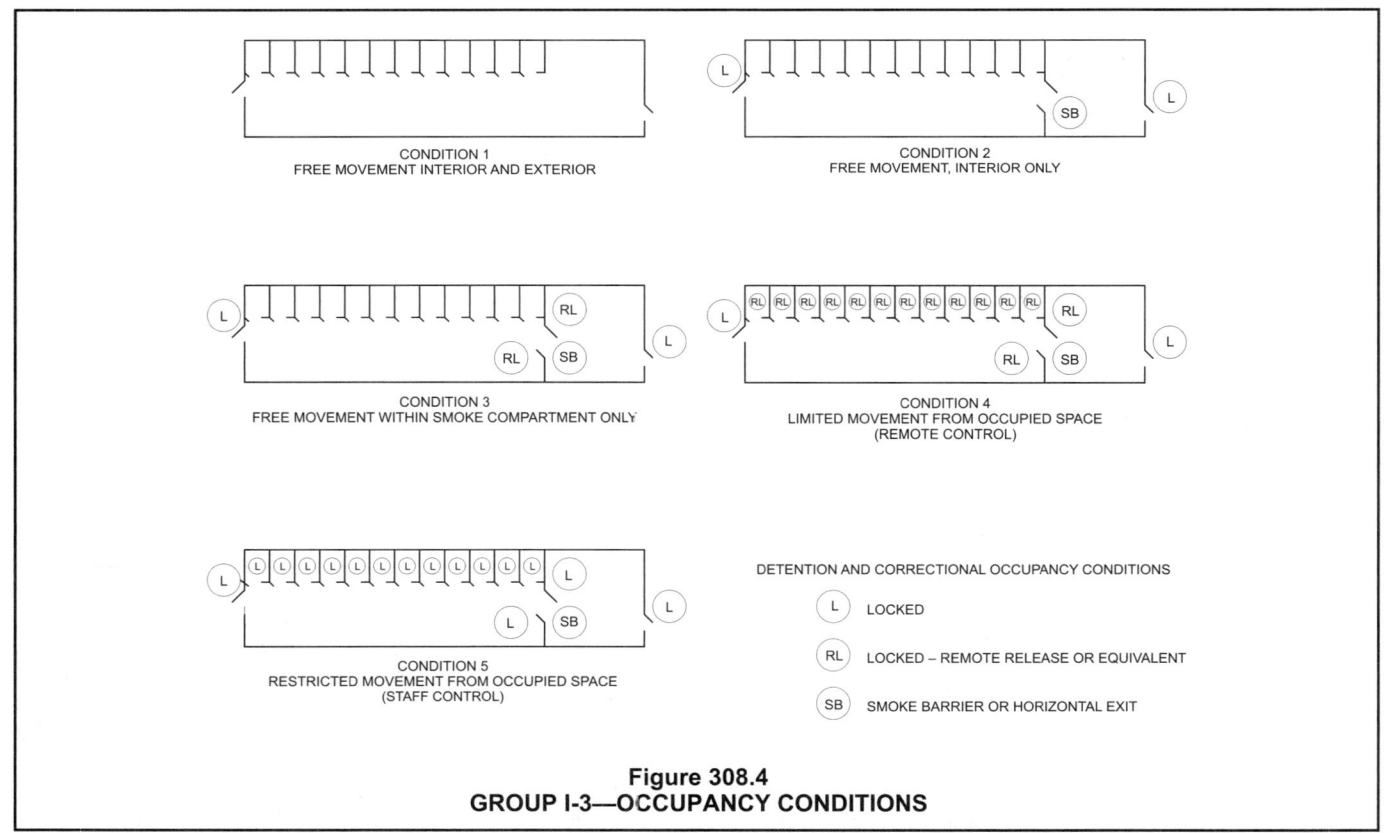

CONDITION 1
FREE MOVEMENT INTERIOR AND EXTERIOR

CONDITION 2
FREE MOVEMENT, INTERIOR ONLY

CONDITION 3
FREE MOVEMENT WITHIN SMOKE COMPARTMENT ONLY

CONDITION 4
LIMITED MOVEMENT FROM OCCUPIED SPACE
(REMOTE CONTROL)

CONDITION 5
RESTRICTED MOVEMENT FROM OCCUPIED SPACE
(STAFF CONTROL)

DETENTION AND CORRECTIONAL OCCUPANCY CONDITIONS

(L) LOCKED

(RL) LOCKED – REMOTE RELEASE OR EQUIVALENT

(SB) SMOKE BARRIER OR HORIZONTAL EXIT

Figure 308.4
GROUP I-3—OCCUPANCY CONDITIONS

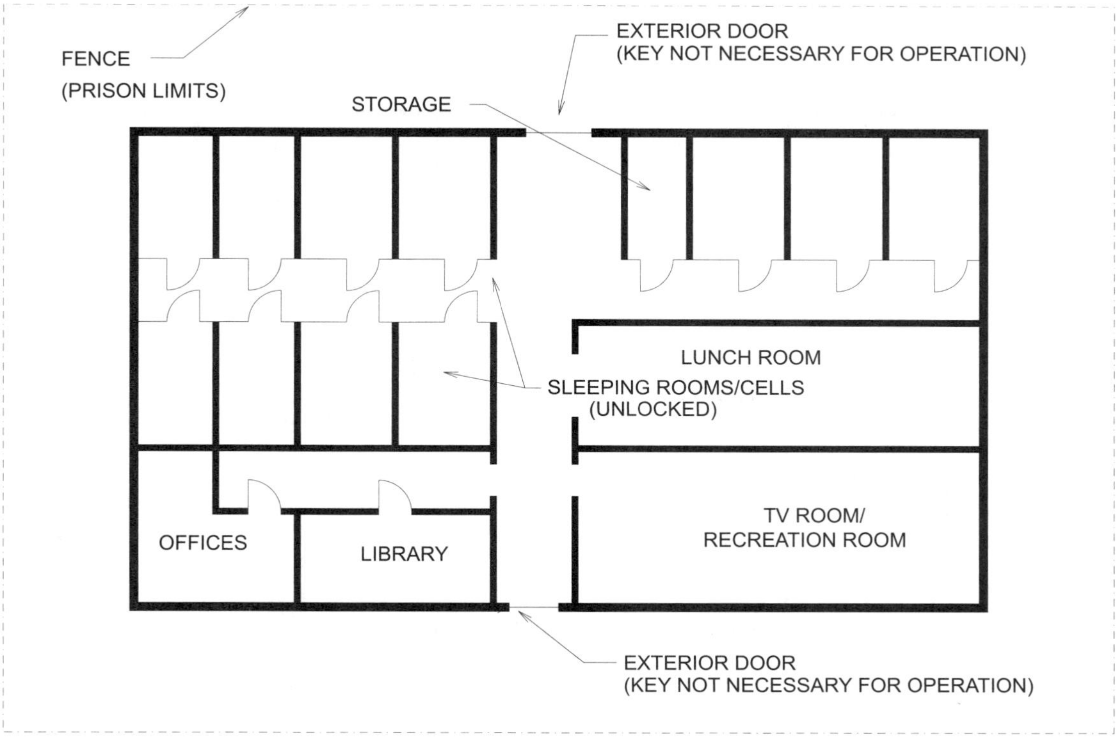

Figure 308.4.1
CONDITION 1

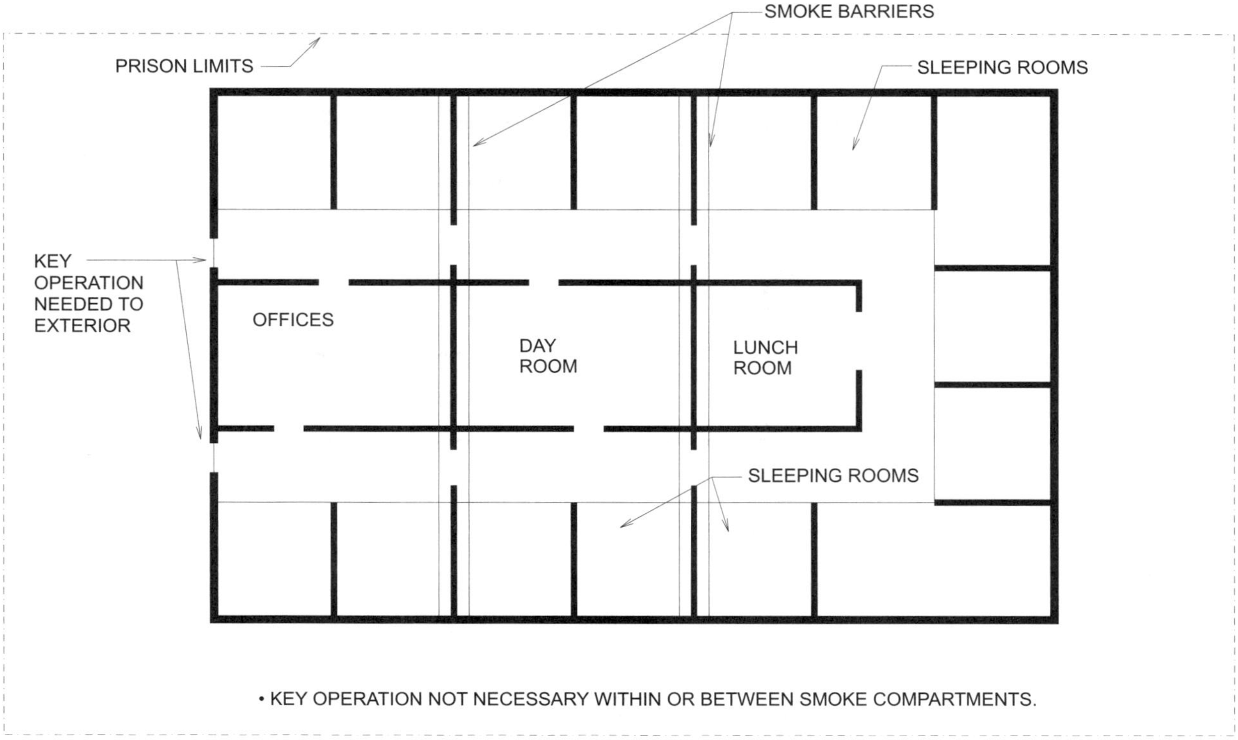

• KEY OPERATION NOT NECESSARY WITHIN OR BETWEEN SMOKE COMPARTMENTS.

Figure 308.4.2
CONDITION 2

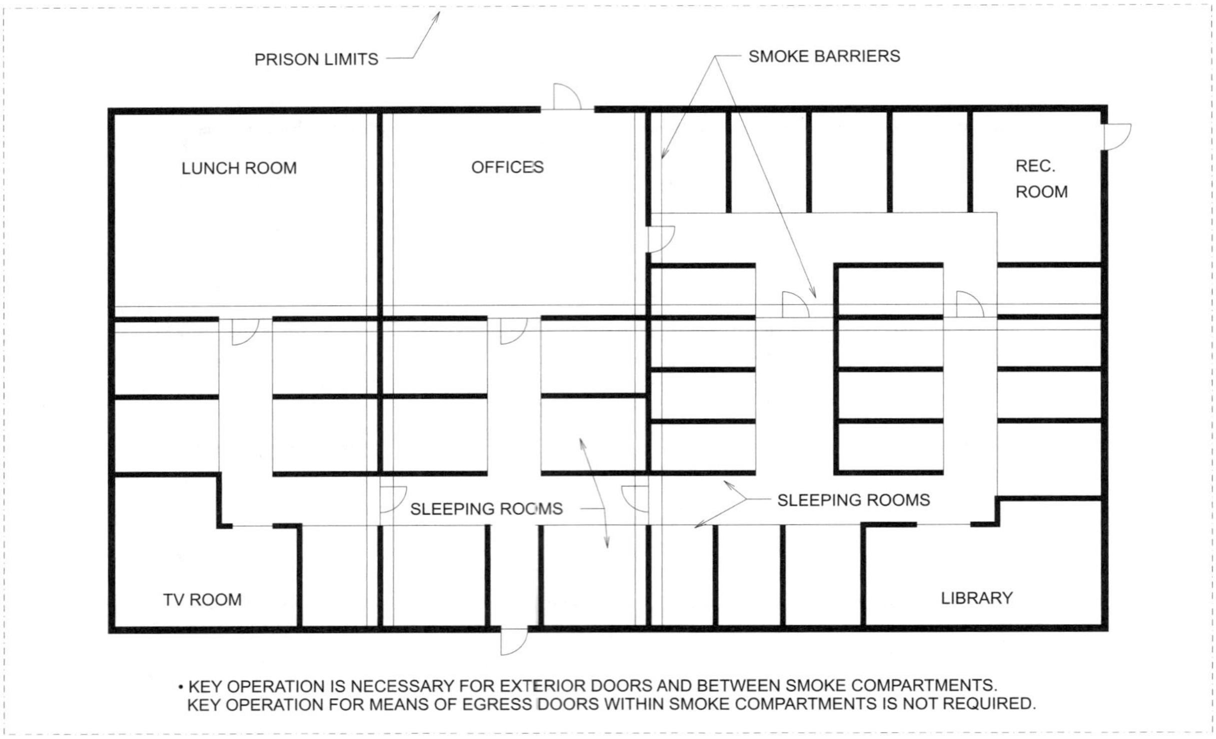
PRISON LIMITS
SMOKE BARRIERS
LUNCH ROOM
OFFICES
REC. ROOM
SLEEPING ROOMS
SLEEPING ROOMS
TV ROOM
LIBRARY
• KEY OPERATION IS NECESSARY FOR EXTERIOR DOORS AND BETWEEN SMOKE COMPARTMENTS.
KEY OPERATION FOR MEANS OF EGRESS DOORS WITHIN SMOKE COMPARTMENTS IS NOT REQUIRED.

Figure 308.4.3
CONDITION 3

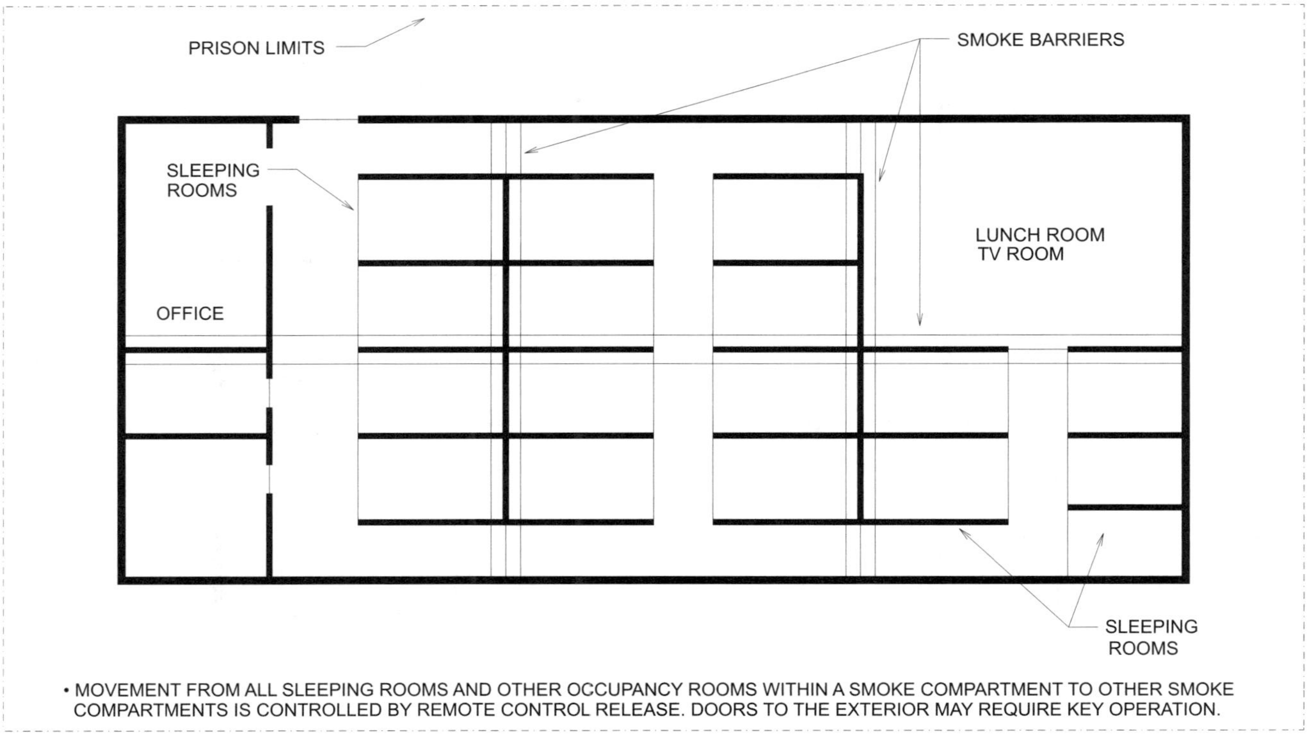
PRISON LIMITS
SMOKE BARRIERS
SLEEPING ROOMS
LUNCH ROOM
TV ROOM
OFFICE
SLEEPING ROOMS
• MOVEMENT FROM ALL SLEEPING ROOMS AND OTHER OCCUPANCY ROOMS WITHIN A SMOKE COMPARTMENT TO OTHER SMOKE COMPARTMENTS IS CONTROLLED BY REMOTE CONTROL RELEASE. DOORS TO THE EXTERIOR MAY REQUIRE KEY OPERATION.

Figure 308.4.4
CONDITION 4

308.4.5 Condition 5. This occupancy condition shall include buildings in which free movement is restricted from an occupied space. Staff-controlled manual release is provided to permit movement from *sleeping units*, activity spaces and other occupied areas within the smoke compartment to other smoke compartments.

❖ Condition 5 areas are those in which the occupants are not allowed free movement to any other room or space within a smoke compartment (as created by smoke barriers) to another smoke compartment or to the exterior (exit discharge) unless the locking device controlling their area of confinement is manually released by a staff member. Once released from an individual space, a staff member is responsible for unlocking all doors from that location to the next smoke compartment. This is the most restrictive occupancy condition, as each occupant must be released on an individual basis and escorted to other areas.

Condition 5 facilities are most often used for maximum security or solitary confinement areas where the occupants are considered to be dangerous to others, including staff members, and cannot safely be handled in large groups (see Figure 308.4.5).

308.5 Group I-4, day care facilities. This group shall include buildings and structures occupied by persons of any age who receive custodial care for less than 24 hours by individuals other than parents or guardians, relatives by blood, marriage or adoption, and in a place other than the home of the person cared for. A facility such as the above with five or fewer persons shall be classified as a Group R-3 or shall comply with the *International Residential Code* in accordance with Section 101.2. Places of worship during religious functions are not included.

❖ Facilities that contain provisions for the care of more than five adults (older than the 12th grade) or more than five children 2$\frac{1}{2}$ years of age or less are classified as Group I-4. Group I-4 facilities are less restrictive in some of the requirements (e.g., height and area) than the other Group I occupancies. Group I-4 facilities are intended to be used for less than 24 hours and are not intended to provide medical supervision. Day care facilities are not intended to be a residence for the people receiving care. The staff members are assumed not to be related to the individuals in the day care facilities. The premise of the provisions is that the numbers receiving care are exclusive of staff. Buildings that have five or fewer occupants receiving custodial care are to be classified as Group R-3, or shall be constructed in accordance with the IRC. The definition for Group I-4 facilities could be construed to include places of religious worship; therefore, the last sentence of Section 308.5 clarifies that Group I-4 provisions would not apply to places of religious worship simply providing child care services during worship and related religious functions. If the space is used at other times simply as a day care facility then it would be classified as Group I-4.

308.5.1 Adult care facility. A facility that provides accommodations for less than 24 hours for more than five unrelated

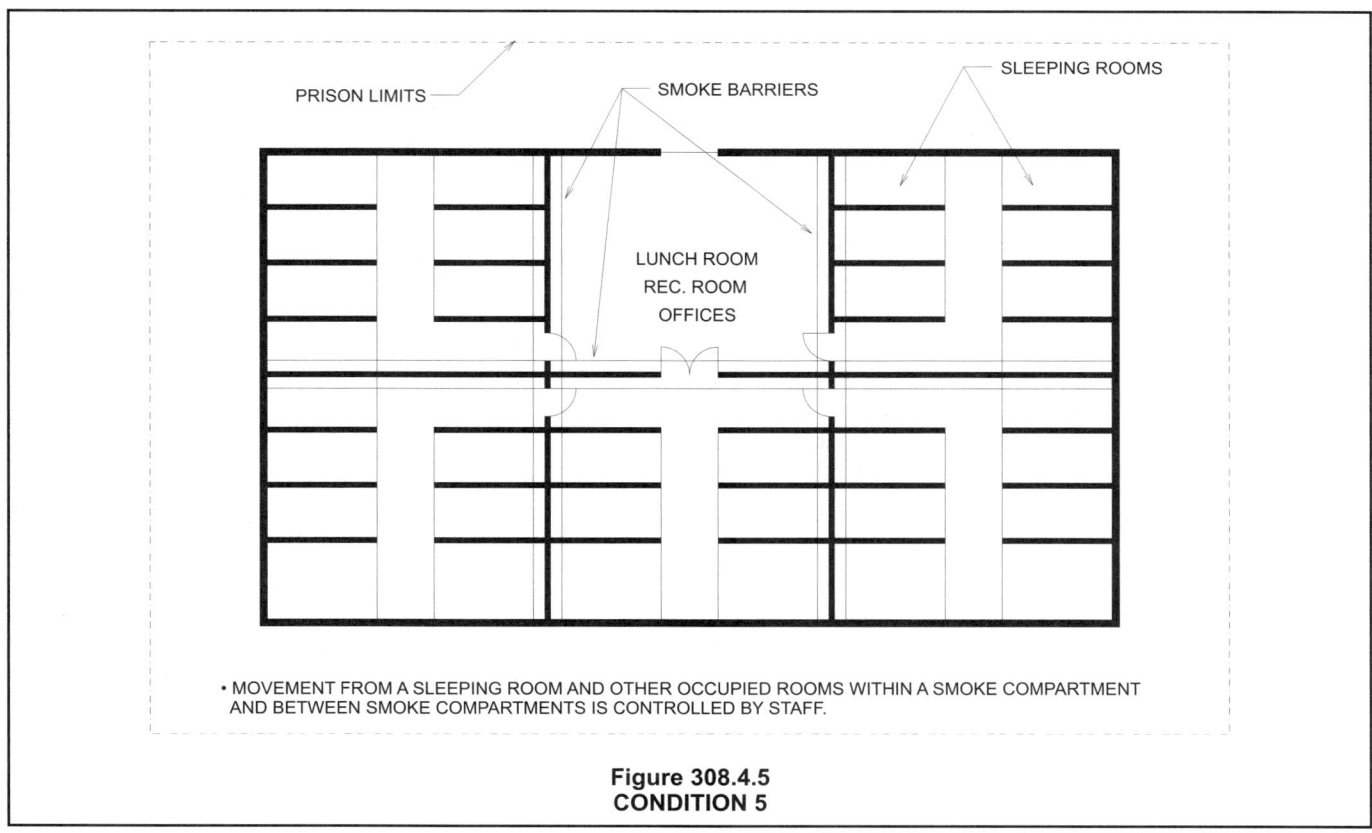

• MOVEMENT FROM A SLEEPING ROOM AND OTHER OCCUPIED ROOMS WITHIN A SMOKE COMPARTMENT AND BETWEEN SMOKE COMPARTMENTS IS CONTROLLED BY STAFF.

Figure 308.4.5
CONDITION 5

adults and provides supervision and *personal care services* shall be classified as Group I-4.

> **Exception:** A facility where occupants are capable of responding to an emergency situation without physical assistance from the staff shall be classified as Group R-3.

❖ Adult care facilities are assumed to be for people beyond the 12th grade that require some type of personal care (i.e., nonmedical). A facility where adults gather for social activities such as a community center or a YMCA is not an adult care facility (Group I) and would be regulated under other provisions of the code (Group A-3 or B). In addition, there must be more than five adults accommodated in the facility and they must not be related in any manner. Facilities in which the adults are related would be more appropriately classified in Group R. The exception clarifies that the classification of Group I-4 for an adult day care facility does not apply to facilities that provide services for adults who are capable of responding to an emergency unassisted. In that case, the facility is simply considered a Group R-3 occupancy. This classification correlates with Section 310.

Keep in mind that, while adult and child day care facilities with five or fewer occupants receiving care have the option of being designed as a Group R-3 occupancy or, under the IRC, adult day care facilities included in the exception to Section 308.5.1 are only allowed to be built to the requirements of the code for R-3 occupancies. The building official should make it clear that adults not capable of self-preservation are not to be present in the facility designed under the R-3 standards.

Remember that R-3 structures are required to be sprinkler protected, and that as a day care, the appropriate occupant load factor is 35 square feet (3.25 m²) per person and not that designated for dwelling units. Also note that, as a Group R-3 structure, only a single means of egress is permitted by Section 1021.2; however, the travel distance limitations of Section 1016 also apply, which might dictate the need for more than one exit. Despite the Group R-3 classification, the designer of such facilities might wish to add features such as additional exits, panic hardware and similar protective elements more reflective of an assembly-like use of a space by people unfamiliar with the structure. Section 2902.3 requires public toilets to be available to customers and visitors. Accessible toilet facilities as required for Group I-4 adult day care facilities would be required. In general an NFPA 13R sprinkler system would be required for this type of facility unless the adult day care is truly within a dwelling unit, in which case the sprinkler system is provided. Finally, if other activities or uses are occurring in the same building, the appropriate provisions for mixed occupancy (see Section 508) need to be applied.

308.5.2 Child care facility. A facility that provides supervision and personal care on less than a 24-hour basis for more than five children $2^{1}/_{2}$ years of age or less shall be classified as Group I-4.

> **Exception:** A child day care facility that provides care for more than five but no more than 100 children $2^{1}/_{2}$ years or less of age, where the rooms in which the children are cared for are located on a *level of exit discharge* serving such rooms and each of these child care rooms has an *exit* door directly to the exterior, shall be classified as Group E.

❖ As with Group I-2 child care facilities, the occupants of Group I-4 child care facilities are limited to $2^{1}/_{2}$ years of age or less. The distinguishing factor between the two occupancies is the amount of time the facility provides care for each individual; Group I-2 facilities provide care on a 24-hour basis while in Group I-4 facilities individual care must be less than 24 hours. It is also assumed that medical supervision is not present in Group I-4 facilities. Occupants $2^{1}/_{2}$ years of age or less are not typically capable of independently responding to an emergency and must be led or carried to safety. Under such circumstances, the occupants are considered nonambulatory.

A child care facility in which the number of occupants is greater than five but not more than 100 is permitted to be classified as Group E, provided the children are all located in rooms on the level of exit discharge that serve such rooms and all of the rooms have exit doors directly to the exterior. This exception is only applicable to rooms and spaces used for child care and is not intended to apply to accessory spaces such as restrooms, offices and kitchens. Many day care facilities primarily catering to those under primary school age tend to divide the children into three general categories based upon state laws and regulations. These include infant, toddler and preschool.

Some variations do occur in that larger day care facilities will have transition rooms for mobile infants or pre-K oriented rooms for those entering kindergarten. But basically there is a mixture of children $2^{1}/_{2}$ years or less and older children. The older children can automatically be in a facility classified as a Group E occupancy, but for the younger children the exception as discussed above would need to be applied to classify the entire occupancy as Group E. The total number of children can exceed 100 and the Group E classification is retained, provided that the number of children $2^{1}/_{2}$ years or less is limited to 100 or fewer. The infant and toddler rooms would need to have exits directly to the outside on the level of exit discharge. If the exception is not applied, the entire facility would need to be classified as Group I-4 or a mixed occupancy classification would be necessary.

By permitting the facility to be classified as Group E, the building would not be required to be sprinklered unless the fire area was greater than 12,000 square feet (115 m²). A Group I-4 facility would be required to be sprinklered regardless of the area. But as a Group E occupancy, panic hardware would be required in rooms and spaces exceeding 50 occupants.

SECTION 309
MERCANTILE GROUP M

309.1 Mercantile Group M. Mercantile Group M occupancy includes, among others, the use of a building or structure or a portion thereof, for the display and sale of merchandise and involves stocks of goods, wares or merchandise incidental to such purposes and accessible to the public. Mercantile occupancies shall include, but not be limited to, the following:

Department stores
Drug stores
Markets
Motor fuel-dispensing facilities
Retail or wholesale stores
Sales rooms

❖ The characteristics of occupancies classified in Group M are contained in this section. Because mercantile occupancies normally involve the display and sale of large quantities of combustible merchandise, the fuel load in such facilities can be relatively high, potentially exposing the occupants (customers and sales personnel) to a high degree of fire hazard. Mercantile operations often attract large crowds (particularly in large department stores and covered and open malls and especially during weekends and holidays). There are two factors that alleviate the risks to life safety: the occupant load normally has a low-to-moderate density and the occupants are alert, mobile and able to respond in an emergency situation. The degree of openness and the organization of the retail display found in most mercantile occupancies is generally orderly and does not present an unusual difficulty for occupant evacuation.

Listed here are general descriptions of the kinds of occupancies that are classified in Group M. Mercantile buildings most often have both a moderate occupant load and a high fuel load, which is in the form of fur-nishings and the goods being displayed, stored and sold [see Figure 309.1(1)].

The key characteristics that differentiate occupancies classified in Group M from those classified in Group B (see Section 304) are the larger quantity of goods or merchandise available for sale and the lack of familiarity of the occupants with the building, particularly its means of egress. To be classified in Group M, the goods that are on display must be accessible to the public. If a patron sees an item for sale, then that item is generally available for purchase at that time (i.e., there is a large stock of goods). If a store allows people to see the merchandise but it is not available on the premises, such as an automobile showroom, then the occupancy classification of business (Group B) should be considered. A mercantile building is open to the public, many of whom may not be regular visitors. A business building, however, is primarily occupied by regular employees who are familiar with the building arrangement and, most importantly, the exits. This awareness of the building and the exits can be an important factor in a fire emergency.

Automotive, fleet-vehicle, marine and self-service fuel-dispensing facilities, as defined in the IFC, are classified in the mercantile occupancy, as are the convenience stores often associated with such occupancies [see Figure 309.1(2)]. Quick-lube, tune-up, muffler and tire shops are not included in this classification. Those facilities that typically conduct automotive service and repair work are treated as a repair garage (Group S-1, also defined in the IFC).

Simply because a building containing a mercantile-type occupancy has a dense occupant load does not necessitate the need to classify the building as an assembly occupancy unless the activity includes an assembly-type area where purchasing of goods is a

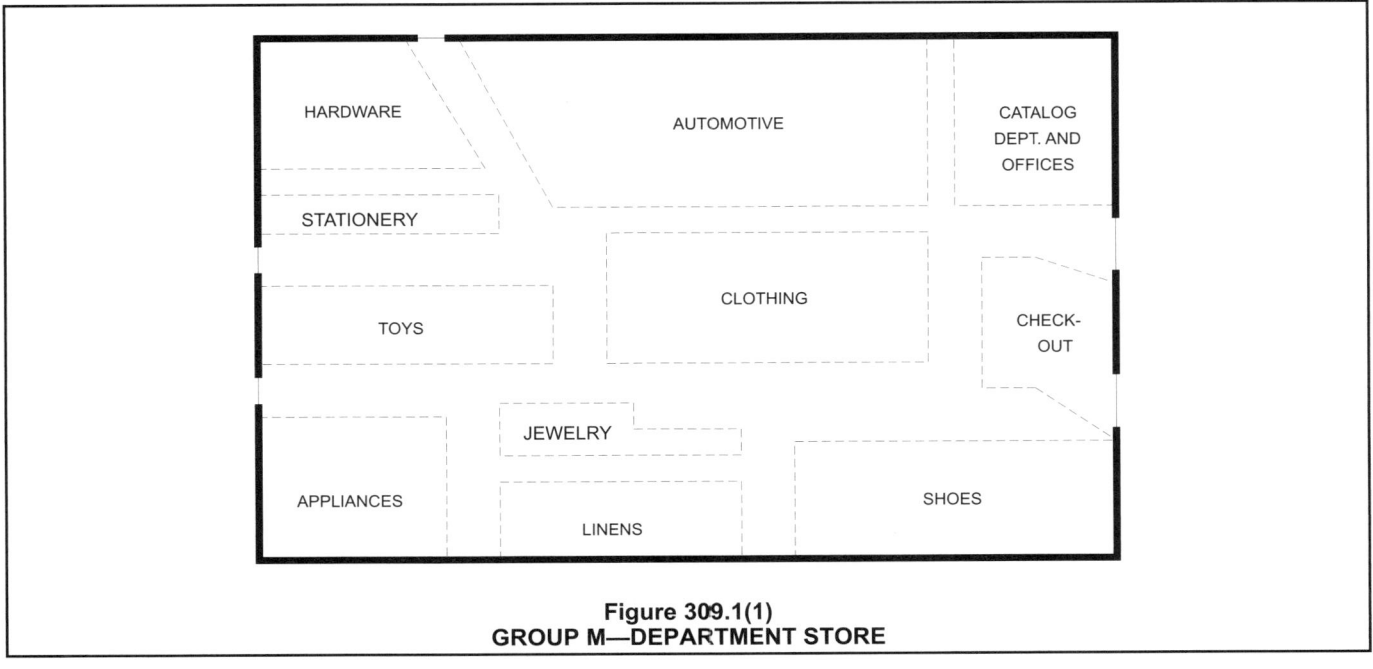

Figure 309.1(1)
GROUP M—DEPARTMENT STORE

group activity versus individual shoppers independently considering and purchasing merchandise. For example, a building in which auction sales occur may have a highly concentrated occupant load where the sales occur, but Section 309.1 describes mercantile occupancies as "the use of a building or structure or portion thereof for the display and sale of merchandise and involves stocks of goods, wares or merchandise incidental to such purposes and accessible to the public." However in an auction, the activity is dominated by an assembly use of the space as people gather to conduct and participate in the auction. As such, auction spaces need to be assigned a Group A-3 occupancy (IBC Interpretation No. 38-03). The presence of highly concentrated occupant loads does not in itself mandate an assembly use classification unless the activity is assembly in nature versus large numbers of people pursuing individual activities of acquisition.

309.2 Quantity of hazardous materials. The aggregate quantity of nonflammable solid and nonflammable or noncombustible liquid hazardous materials stored or displayed in a single *control area* of a Group M occupancy shall not exceed the quantities in Table 414.2.5(1).

❖ This section addresses an exception for control areas of mercantile occupancies containing certain nonflammable or noncombustible materials and health hazard gases that are stored in accordance with Table 414.2.5(1). This section allows Group H-4 materials, which present a health hazard rather than a physical hazard, as well as limited Group H-2 and H-3 materials, to be stored in both the retail display and stock areas of regulated mercantile occupancies in excess of the maximum allowable quantity per control area of Tables 307.1(1) and 307.1(2) without classifying the building as Group H. To correctly classify a building

where products that have hazardous properties are stored and sold, the code user must also be aware of the provisions contained in Section 307.1, Notes c and n of Table 307.1(1) and Notes b and c of Table 307.1(2). These provisions give the quantity limitations for specific high-hazard products in mercantile display areas, including medicines, foodstuffs, cosmetics and alcoholic beverages.

Without this option, many mercantile occupancies could technically be classified as Group H. The increased quantities of certain hazardous materials are based on the recognition that, while there is limited risk in mercantile occupancies, the packaging and storage arrangements can be controlled. For further information on the storage limitations required for these types of materials in mercantile occupancies, see Section 2703.11 of the IFC.

SECTION 310
RESIDENTIAL GROUP R

310.1 Residential Group R. Residential Group R includes, among others, the use of a building or structure, or a portion thereof, for sleeping purposes when not classified as an Institutional Group I or when not regulated by the *International Residential Code* in accordance with Section 101.2. Residential occupancies shall include the following:

R-1 Residential occupancies containing *sleeping units* where the occupants are primarily transient in nature, including:

   Boarding houses (transient)
   Hotels (transient)
   Motels (transient)

Congregate living facilities (transient) with 10 or fewer occupants are permitted to comply with the construction requirements for Group R-3.

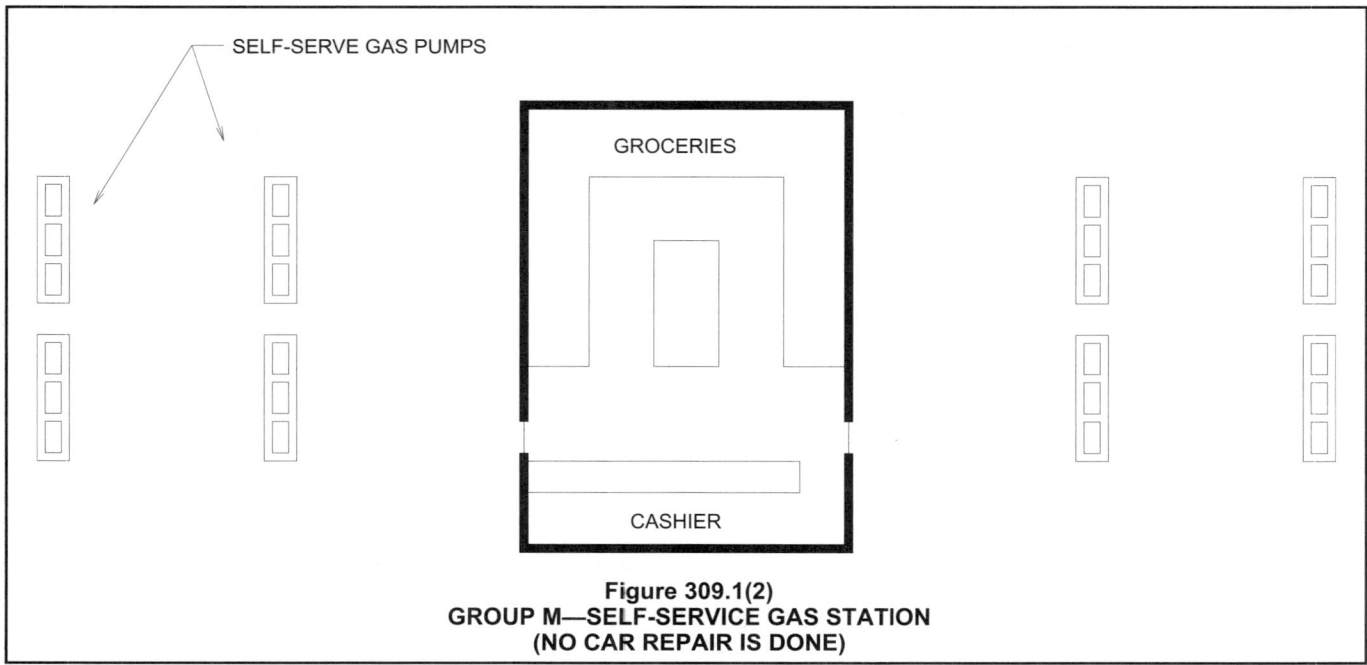

Figure 309.1(2)
GROUP M—SELF-SERVICE GAS STATION
(NO CAR REPAIR IS DONE)

R-2 Residential occupancies containing *sleeping units* or more than two *dwelling units* where the occupants are primarily permanent in nature, including:

Apartment houses
Boarding houses (nontransient)
Convents
Dormitories
Fraternities and sororities
Hotels (nontransient)
Live/work units
Monasteries
Motels (nontransient)
Vacation timeshare properties

Congregate living facilities with 16 or fewer occupants are permitted to comply with the construction requirements for Group R-3.

R-3 Residential occupancies where the occupants are primarily permanent in nature and not classified as Group R-1, R-2, R-4 or I, including:

Buildings that do not contain more than two *dwelling units.*
Adult care facilities that provide accommodations for five or fewer persons of any age for less than 24 hours.
Child care facilities that provide accommodations for five or fewer persons of any age for less than 24 hours.
Congregate living facilities with 16 or fewer persons.

Adult care and child care facilities that are within a single-family home are permitted to comply with the *International Residential Code.*

R-4 Residential occupancies shall include buildings arranged for occupancy as residential care/assisted living facilities including more than five but not more than 16 occupants, excluding staff.

Group R-4 occupancies shall meet the requirements for construction as defined for Group R-3, except as otherwise provided for in this code or shall comply with the *International Residential Code* provided the building is protected by an *automatic sprinkler system* installed in accordance with Section 903.2.8.

❖ Residential occupancies represent some of the highest fire safety risks of any of the occupancies listed in Chapter 3. There are several reasons for this condition:

Structures in the residential occupancy house the widest range of occupant types, i.e., infants to the aged, for the longest periods of time. As such, residential occupancies are more susceptible to the careless acts of the occupants; therefore, the consequences of exposure to the effects of fire are the most serious.

Most residential occupants are asleep approximately one-third of every 24-hour period. When sleeping, they are not likely to become immediately aware of a developing fire. Also, if awakened from sleep by the presence of fire, the residents often may not immediately react in a rational manner and delay their evacuation.

The fuel load in residential occupancies is often quite high, both in quantity and variety. Also, in the construction of residential buildings, it is common to use extensive amounts of combustible materials.

Another portion of the fire problem in residential occupancies relates to the occupants' lack of vigilance in the prevention of fire hazards. In their own domicile or residence, people tend to relax and are often prone to allow fire hazards to go unabated; thus, in residential occupancies, fire hazards tend to accrue over an extended period of time and go unnoticed or are ignored.

Most of the nation's fire problems occur in Group R buildings and, in particular, one- and two-family dwellings, which account for more than 80 percent of all deaths from fire in residential occupancies and about two-thirds of all fire fatalities in all occupancies. One- and two-family dwellings also account for more than 80 percent of residential property losses from fire and more than one-half of all property losses from fire.

Because of the relatively high fire risk and potential for loss of life in buildings classified in Groups R-1 (hotels and motels) and R-2 (apartments and dormitories), the code has stringent provisions for the protection of life in these occupancies. Group R-3 occupancies, however, are not generally considered to be in the same domain and, thus, are not subject to the same level of regulatory control as is provided in other occupancies.

Because of the growing trend to care for people in a residential environment, residential care/assisted living facilities are also classified as Group R. Specifically, these facilities are classified as Group R-4. Mainstreaming people who are recovering from alcohol or drug addiction and people who are developmentally disabled is reported to have therapeutic and social benefits. A residential environment often fosters this mainstreaming.

Buildings in Group R are described below. A building or part of a building is considered to be a residential occupancy if it is intended to be used for sleeping accommodations (including residential care/assisted living facilities) and is not an institutional occupancy. Institutional occupancies are similar to residential occupancies in many ways; however, they differ from each other in that institutional occupants are in a supervised environment, and, in the case of Groups I-2 and I-3 occupancies, are under some form of restraint or physical limitation that makes them incapable of complete self-preservation. The number of these occupants who are under supervision or are incapable of self-preservation is one distinguishing factor for being classified as an institutional or residential occupancy.

The term Group R refers collectively to the four individual residential occupancy classifications: Groups R-1, R-2, R-3 and R-4. These classifications are differentiated in the code based on the following criteria: (1) whether the occupants are transient or nontransient in nature; (2) the type and number of dwelling units or sleeping units contained in a single building and (3) the number of occupants in the facility.

R-1: The key characteristic of Group R-1 that differentiates it from other Group R occupancies is the number of transient occupants (i.e., those whose length of stay is not more than 30 days).

The most common building types classified in Group R-1 are hotels, motels and boarding houses. Group R-1 occupancies do not typically have cooking facilities in the unit. When a unit is not equipped with cooking facilities, it does not meet the definition of a "Dwelling unit" in Section 202. When this occurs, such units are treated as sleeping units for the application of code provisions [see Figure 310.1(1)]. Sleeping units are required to be separated from each other by fire partitions and horizontal assemblies (see Sections 420, 709.1 and 712.3). A recent trend in development is the construction of "extended-stay hotels." While these units may have all of the characteristics of a typical dwelling unit (i.e., cooking, living, sleeping, eating, sanitation), the length of stay is still typically not more than 30 days. As such, these buildings would still be classified as Group R-1. If the length of stay is more than 30 days, these buildings would be classified as Group R-2.

Other occupancies are often found in buildings classified in Group R-1. These occupancies include nightclubs (Group A-2), restaurants (Group A-2), gift shops (Group M), health clubs (Group A-3) and storage facilities (Group S-1). When this occurs, the building is a mixed occupancy and is subject to the provisions of Section 508.

Transient congregate living facilities with 10 or fewer occupants can be constructed to the standards of Group R-3 occupancies. The primary intent of this provision (new to the 2009 code) is to permit bed and breakfast type facilities to be established in existing single-family (one-family) structures. In comparison to the provision under Group R-2 which permits congregate living facility with fewer than 16 nontransient occupants to be built as an R-3, the R-3 "transient" facility is limited to 10 or fewer occupants in reflection of the fact that the occupants (other than a resident family) will lack familiarity with their surroundings.

R-2: The length of the occupants' stay plus the arrangement of the facilities provided are the basic factors that differentiate occupancies classified in Group R-2 from other occupancies in Group R. The occupants of facilities or areas classified in Group R-2 are primarily nontransient, capable of self-preservation and share their means of egress in whole or in part with other occupants outside of their sleeping area or dwelling unit. The separation between dwelling units must, at a minimum, meet the requirements contained in Sections 420, 709.1 and 712.3. Building types ordinarily classified in Group R-2 include apartments, boarding houses (when the occupants are not transient) and dormitories [see Figures 310.1(2) and 310.1(3)].

Individual dwelling units in Group R-2 are either rented by tenants or owned by the occupants. The

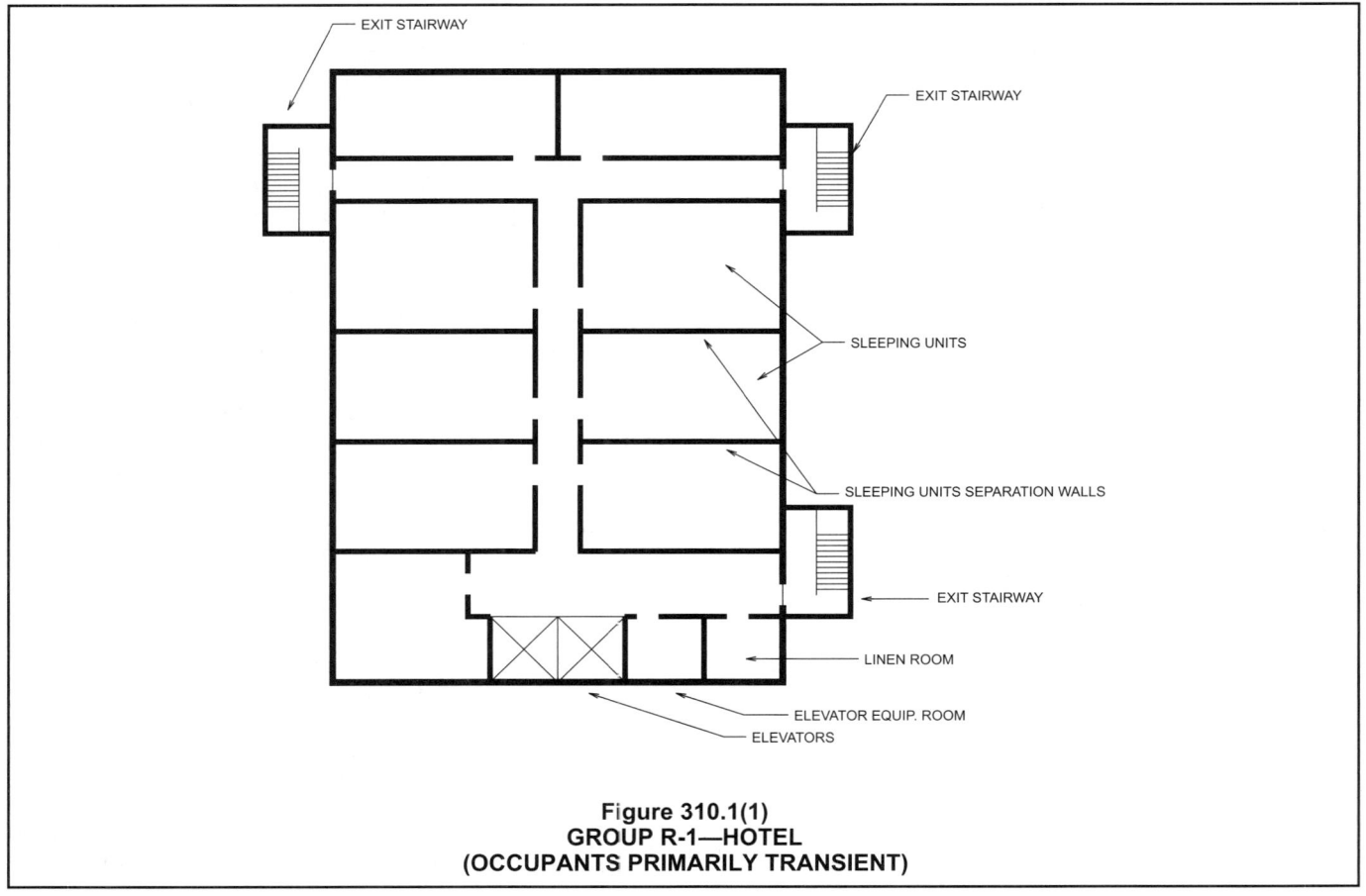

Figure 310.1(1)
GROUP R-1—HOTEL
(OCCUPANTS PRIMARILY TRANSIENT)

code does not make a distinction between either type of tenancy. Residential condominiums are treated in the code the same as Group R-2 apartments. Such condominiums are based on shared ownership of a building and related facilities. While an individual owner will have exclusive rights to a certain unit, the building, the lot the building sits upon, parking, common recreational facilities and similar features are owned in common by all the owners of individual dwelling units. In most cases condominiums do not establish separate lots and the walls between units are not setting on lot lines. Another type of shared ownership is referred to as a "co-op," short for co-operative. Occasionally a condominium will establish actual lots and lot lines distinguishing individual ownership. When the dwelling unit is located on a separate parcel of land, lot lines defining the parcel exist and the requirements for fire separation must be met.

Dormitories are generally associated with university or college campuses for use as student housing, but this is changing rapidly. Many dormitories are now being built as housing for elderly people who wish to live with other people their own age and who do not need 24-hour-a-day medical supervision. The only difference between the dormitory that has just been described and the dormitory found on a college campus is the age of its occupants. If the elderly people must have 24-hour-a-day medical supervision (i.e., a nurse or doctor on the premises), the building is no longer considered a residential occupancy but an institutional occupancy and would have to comply with the applicable provisions of the code for the appropriate Group I occupancy.

Similar to Group R-1, individual rooms in dormitories are sleeping units and are required to be separated from each other by fire partitions and horizontal assemblies in accordance with Sections 420, 709.1 and 712.3. When college classes are not in session, the rooms in dormitories are sometimes rented out for periods of less than 30 days to convention attendees and other visitors. When dormitories undergo this type of transient use, they more closely resemble Group R-1.

Buildings containing dormitories often contain other occupancies, such as cafeterias or dining rooms (Group A-2), recreation rooms (Group A-3) and office (Group B) or meeting rooms (Group A-3). When this occurs, the building is considered a mixed occupancy and is subject to the provisions of Section 508 [see Figure 310.1(4)].

Included in the listing of Group R-2 are live/work units. A live/work unit is a dwelling unit or sleeping unit in which a significant portion of the space includes a nonresidential use operated by the tenant. Reflecting

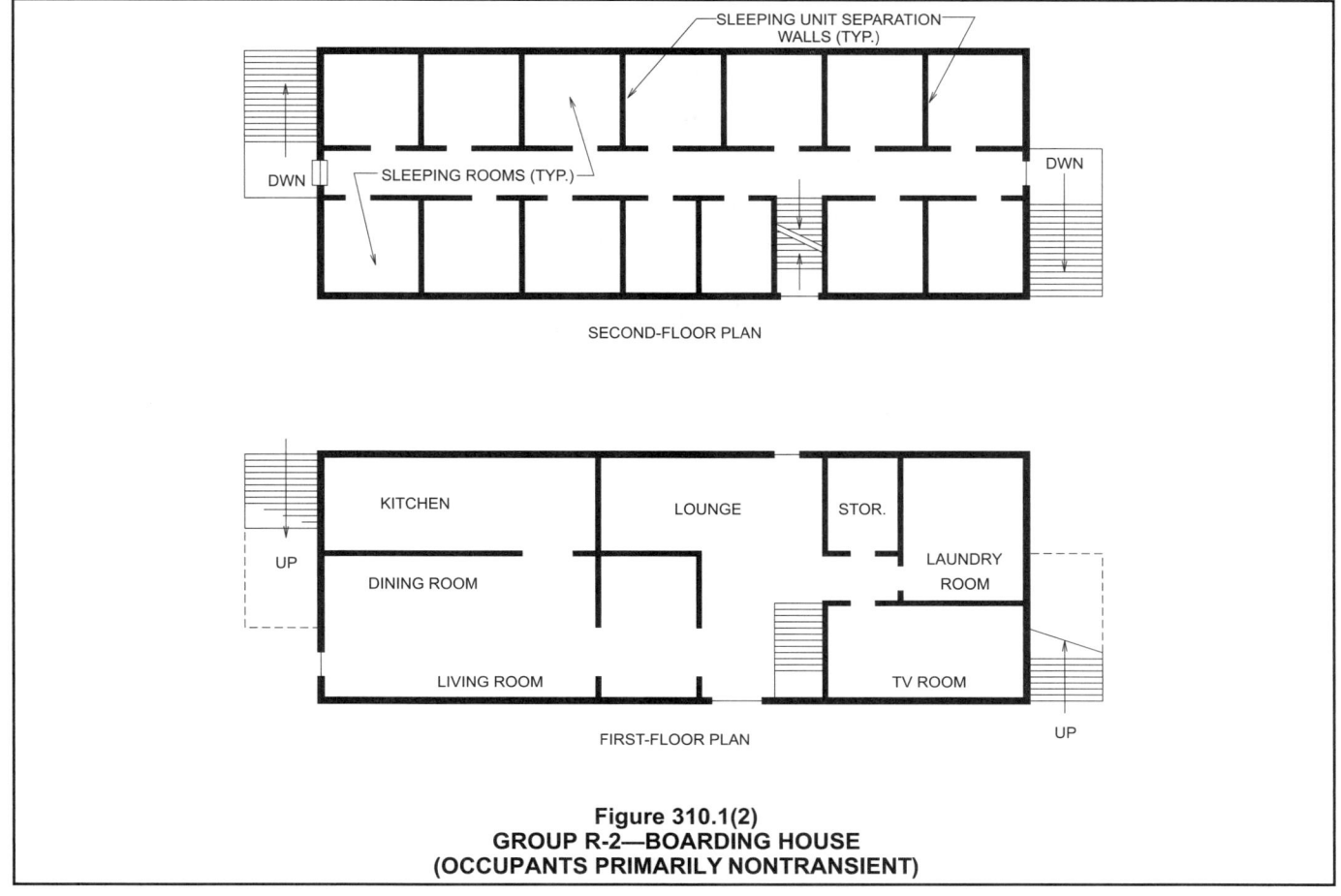

Figure 310.1(2)
GROUP R-2—BOARDING HOUSE
(OCCUPANTS PRIMARILY NONTRANSIENT)

a growing trend in urban neighborhoods and the reuse of existing buildings, live/work units must comply with the provisions of Section 419.

The intent of the congregate living facility reference is to better define when a congregate living facility is operating as a single-family home. Blended families are now commonplace and not necessarily defined strictly by blood or marriage. Small boarding houses, convents, dormitories, fraternities, sororities, monasteries and nontransient hotels and motels may be small enough to operate as a single-family unit and would be permitted to be constructed as Group R-3 occupancies as intended by the code. The threshold of 16 persons is consistent with the results of the most recent census, which has 98 percent of all homes in the U.S. containing less than 16 persons.

R-3: Group R-3 facilities include all detached one- and two-family dwellings and multiple (three or more) single-family dwellings (townhouses) more than three stories in height. Those buildings three or less stories in height are not classified as Group R-3 and are regulated by the IRC (see Section 101.2). Each pair of dwelling units in multiple single-family dwellings greater than three stories in height must be separated by fire walls (see Section 706) or by two exterior walls (see Table 602) in order to be classified as Group R-3. (Duplexes, buildings with two dwelling units, must be

detached from other structures in order to be regulated by the IRC.) A duplex attached to another duplex would be required to comply with the code and be classified as Group R-2 or R-3, depending on the presence of fire walls.

Buildings that are one- and two-family dwellings and multiple single-family dwellings less than three stories in height and that contain another occupancy (e.g., Groups B, M, I-4) must be regulated as a mixed occupancy in accordance with the code and are not required to comply with the provisions of the IRC [see Figures 310.1(5) and 310.1(6)]. Some mixed use dwelling units may qualify as live/work units under Section 419 and be classified as a Group R-2 occupancy.

In addition, institutional facilities other than hospitals, mental hospitals and detoxification facilities, that accommodate five or fewer people and all congregate living facilities with no more than 16 nontransient occupants or no more than 10 transient occupants, are to be classified as Group R-3. Where these small care facilities are provided as a portion of a private home, the intent of the code is that the requirements would be the same as for a single-family home. As such, the provisions of the IRC can still be utilized.

Note that a Group R-3 occupancy is permitted to accommodate a maximum of five occupants in resi-

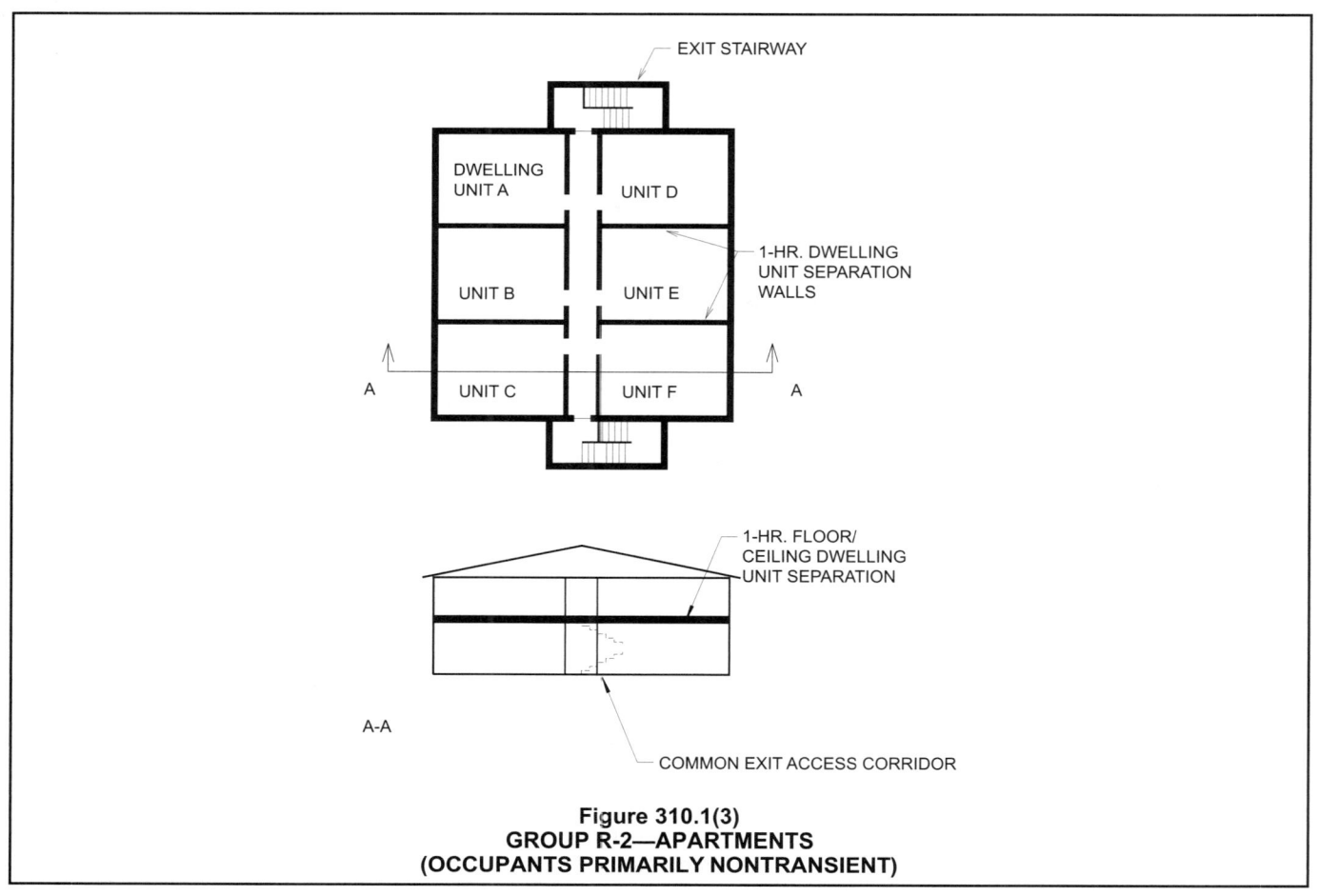

Figure 310.1(3)
GROUP R-2—APARTMENTS
(OCCUPANTS PRIMARILY NONTRANSIENT)

dential care or assisted living situations. Above that number of occupants, the proper classification is R-4. The occupants must be capable of responding to an emergency situation without physical assistance from staff as permitted in a Group I-1 occupancy or not capable of self-preservation as required for a Group I-2 occupancy. A facility that accommodates five persons who are "capable of responding to an emergency situation without physical assistance from staff" and five persons "who are not capable of self-preservation" cannot be classified as a Group R-3 because the total occupant load of 10 persons exceeds the permitted maximum of five occupants. The facility is a single occupancy; therefore, the entire facility must be assumed to be occupied by persons with the most restrictive capability when determining the occupancy classification. A facility that accommodates 16 or more occupants who are not capable of self-preservation is a Group I-2 occupancy.

Buildings that are classified as Group R-3, while limited in height, are not limited in the allowable area per floor as indicated in Table 503.

All dwelling units must be separated from each other by fire partitions and horizontal assemblies (in accordance with Sections 420, 709.1 and 712.3) unless required to be separated by fire walls.

See also the provisions and commentary for Section 308.5.1, which allows adult day care of any size to be a Group R-3 occupancy provided the occupants are capable of responding to an emergency situation without physical assistance from the staff.

R-4: When a limited number of people who require personal care live in a residential environment, a facility is no longer classified as Group I-1 but as a residential care/assisted living facility, Group R-4. Ninety-eight percent of households in the U.S. have less than 16 occupants; thus the limit of 16 would allow equal access for the disabled. The number of occupants includes those who receive care and is not intended to include staff. A Group R-4 occupancy is not permitted to include any number of occupants that "are not capable of self-preservation." If an existing Group R-4 facility has residents added who are not capable of self-preservation, then the occupancy has changed and a permit to change the occupancy must be obtained (IBC Interpretation No.16-03). Group R-4 facilities must satisfy the construction requirements of Group R-3 or shall comply with the requirements of the IRC. Note that the provisions of the code and IRC shall not be mixed. All of the requirements in the code for Group R-3 shall be followed, or all of the requirements in the IRC shall be followed. (IBC Interpretation No. 50-07).

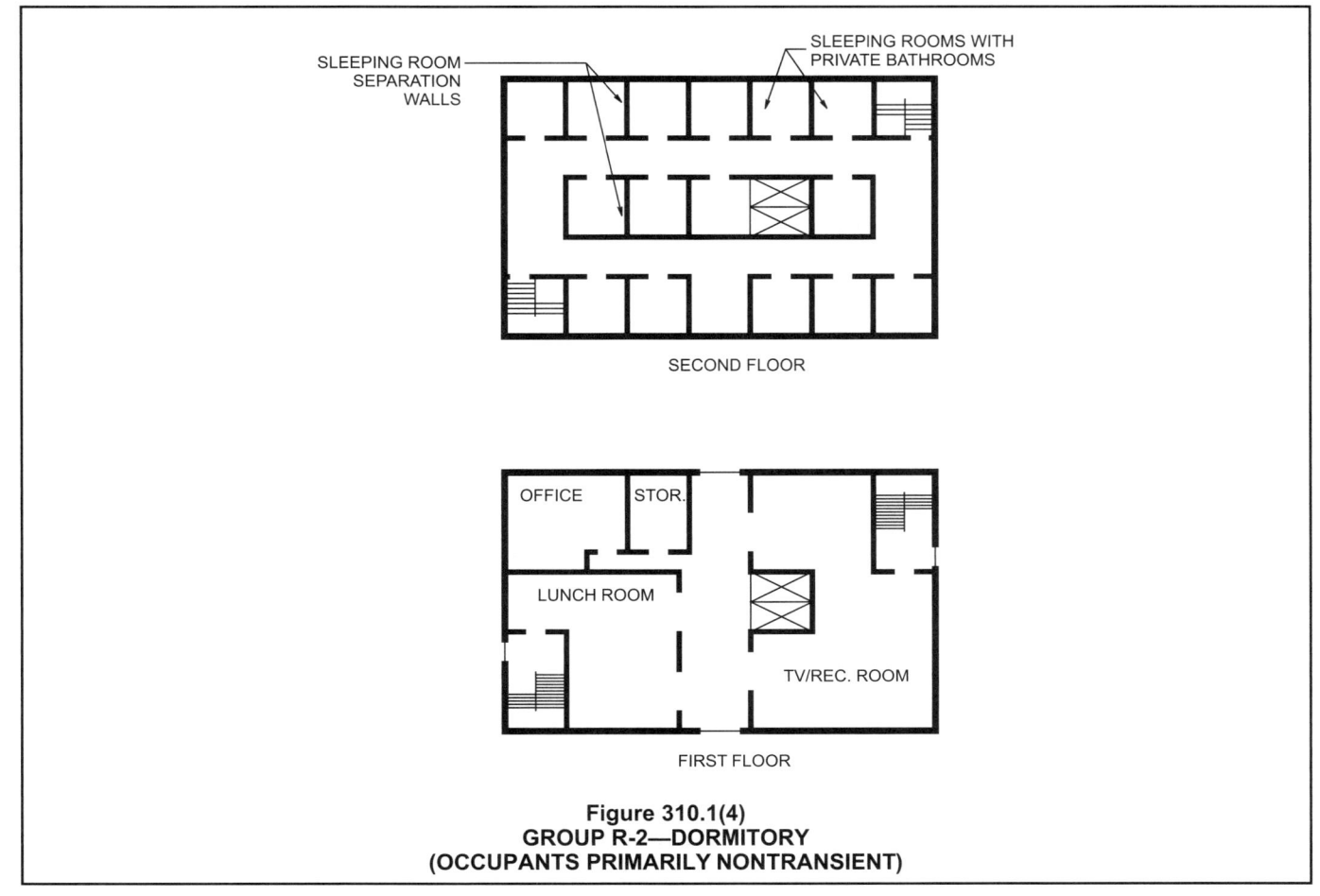

Figure 310.1(4)
GROUP R-2—DORMITORY
(OCCUPANTS PRIMARILY NONTRANSIENT)

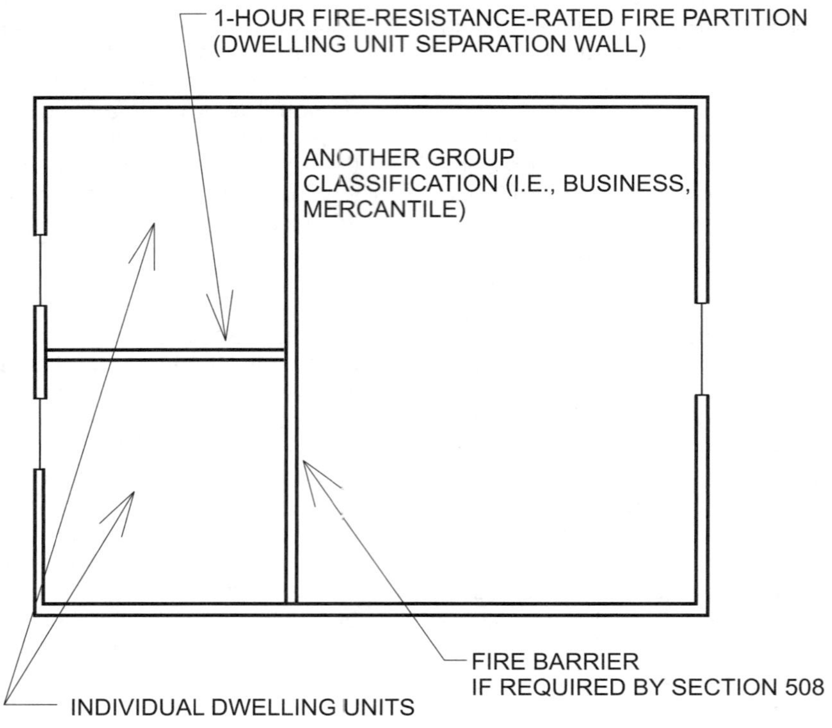

1-HOUR FIRE-RESISTANCE-RATED FIRE PARTITION
(DWELLING UNIT SEPARATION WALL)

ANOTHER GROUP
CLASSIFICATION (I.E., BUSINESS,
MERCANTILE)

FIRE BARRIER
IF REQUIRED BY SECTION 508

INDIVIDUAL DWELLING UNITS

Figure 310.1(5)
GROUP R-3—MIXED OCCUPANCY BUILDING

ONE- OR TWO-
FAMILY DWELLING
(GROUP R-3)

HORIZONTAL
ASSEMBLY IF REQUIRED
BY SECTION 508

GROUP BUSINESS (B)

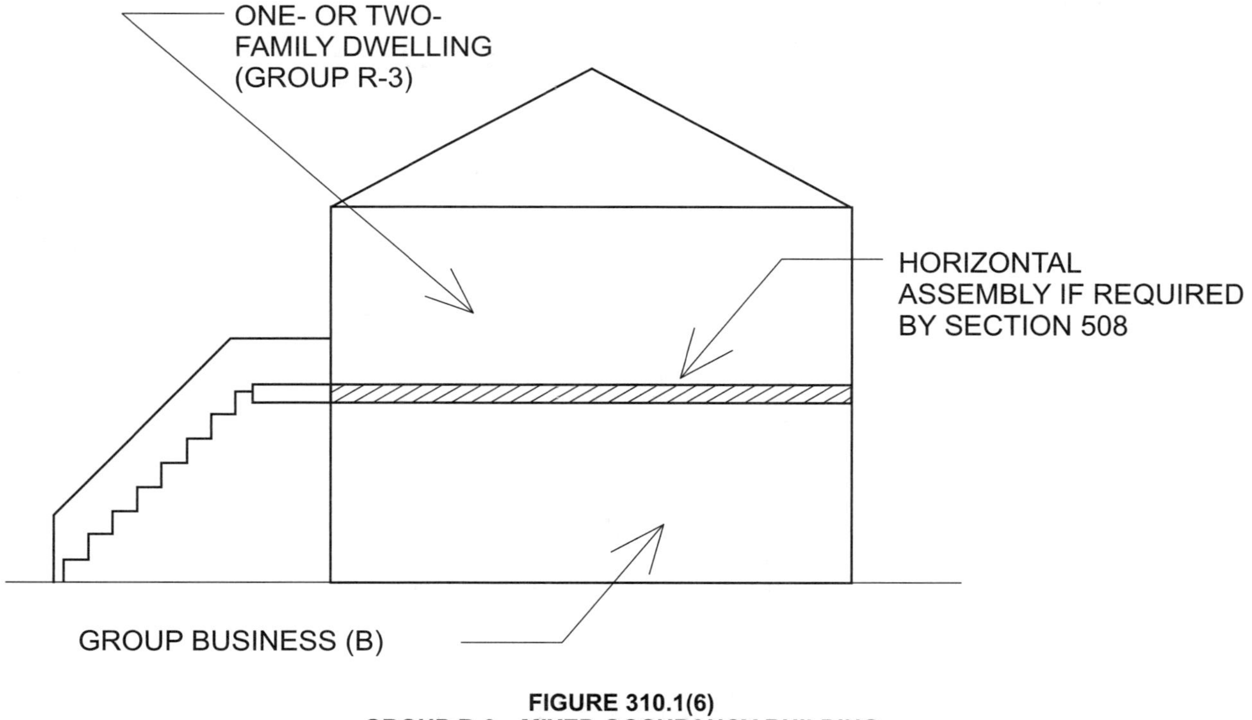

FIGURE 310.1(6)
GROUP R-3—MIXED OCCUPANCY BUILDING

310.2 Definitions. The following words and terms shall, for the purposes of this section and as used elsewhere in this code, have the meanings shown herein.

❖ Definitions of terms that are associated with the content of this section are contained herein. These definitions can help in the understanding and application of the code requirements. One should keep in mind, however, that in many cases, terms defined in the code may also be defined by ordinances and statutes of local and state governments. In such cases, code users must focus on the specific features that define the term relative to the code and not its generally held meaning. For convenience, these terms are also listed in Chapter 2 with a cross reference to this section. The use and application of all defined terms, including those defined herein, are set forth in Section 201.

BOARDING HOUSE. A building arranged or used for lodging for compensation, with or without meals, and not occupied as a single-family unit.

❖ A boarding house is a structure housing lodgers or boarders in which the occupants are provided lodging or meals and lodging for a fee. The individual rooms used for lodging usually do not contain all of the permanent living provisions of a dwelling unit (e.g., permanent cooking facilities). Most often, the term "boarding house" describes a facility that is primarily for transient occupants; however, these facilities might also be used for nontransient purposes. Depending on the extent of transiency, a boarding house could be classified as Group R-1 when an occupant typically stays for not more than 30 days or Group R-2 when the length of stay is greater than 30 days [see Section 310.1 and Figure 310.1(2)].

CONGREGATE LIVING FACILITIES. A building or part thereof that contains sleeping units where residents share bathroom and/or kitchen facilities.

❖ Congregate living facilities are those pertaining to group housing (i.e., dormitories, fraternities, convents) that combine individual sleeping quarters with communal facilities for food, care, sanitation and recreation. The number of occupants in the facility determines the appropriate occupancy classification. There are two thesholds: 10 and 16. A congregate living facility with 16 or fewer nontransient residents falls in the R-3 classification—for above 16 nontransient residents the classification is R-2. For transient residents, if there are 10 or fewer in the facility, it is also in the R-3 classification. If over 10 transient residents, it is an R-1 occupancy.

DORMITORY. A space in a building where group sleeping accommodations are provided in one room, or in a series of closely associated rooms, for persons not members of the same family group, under joint occupancy and single management, as in college dormitories or fraternity houses.

❖ Dormitories typically consist of a large room serving as a community sleeping room or many smaller rooms grouped together and serving as private or semiprivate sleeping rooms (sleeping units). A typical setting for dormitories is on college campuses; however, sleeping areas of a fire station and similar lodging facilities for occupants not of the same family group are also considered dormitories. Dormitories most often are not the permanent residence of the occupants. They are typically occupied only for a designated period of time, such as a school year. Though limited, the period of occupancy is usually more than 30 days, which provides the occupant with a familiarity of the structure such that the occupancy is not considered transient. A dormitory is classified as Group R-2 (see Section 310.1).

Structures containing a dormitory often have a cafeteria or central eating area and common recreational areas. When such conditions exist, the structure must comply with the mixed occupancy provisions of the code [see Section 508 and Figure 310.1(4)].

PERSONAL CARE SERVICE. The care of residents who do not require chronic or convalescent medical or nursing care. Personal care involves responsibility for the safety of the resident while inside the building.

❖ Used in conjunction with Group I-1 and R-4 facilities, this item refers to personal care for residents in a supervised environment who do not require medical supervision. The purpose of personal care service is to distinguish between residents who require medical supervision and those who require solely physical supervision (i.e., to provide safety for the residents while inside the building).

While personal care service is also provided in Group I-4 adult care and child care facilities, these care facilities are not intended to be a residence for the people receiving care.

RESIDENTIAL CARE/ASSISTED LIVING FACILITIES. A building or part thereof housing persons, on a 24-hour basis, who because of age, mental disability or other reasons, live in a supervised residential environment which provides *personal care services*. The occupants are capable of responding to an emergency situation without physical assistance from staff. This classification shall include, but not be limited to, the following: residential board and care facilities, assisted living facilities, halfway houses, group homes, congregate care facilities, social rehabilitation facilities, alcohol and drug abuse centers and convalescent facilities.

❖ Residential care/assisted living facilities are essentially Group I-1 facilities with a smaller number of occupants. The same provisions that are applicable to Group I-1 facilities are applicable to residential care/assisted living facilities (e.g., occupants are there on a 24-hour basis and are capable of responding to an emergency situation without physical assistance).

TRANSIENT. Occupancy of a *dwelling unit* or *sleeping unit* for not more than 30 days.

❖ The intent of this definition is to establish a time parameter to differentiate between transient and nontransient as listed under Groups R-1 and R-2. Real estate law often dictates that a lease must be created after 30 days and 30-day time periods are typically how extended-stay hotels and motels rent to people. Such a time period gives the occupant time to be familiar with the surroundings and, therefore, become more accustomed to any hazards of the built environment than an overnight guest would be or a guest who stays for just a few days. Since nontransient occupancies do not have the same level of protection in the code as transient occupancies, it is important to determine what makes an occupancy transient so as to provide consistency in enforcement.

Since the requirements for Type B units are tied to the facilities that are intended to be occupied as a residence under both the Group R-1 and Group R-2, this definition does not have a detrimental effect on matching the Fair Housing Act provisions.

SECTION 311
STORAGE GROUP S

311.1 Storage Group S. Storage Group S occupancy includes, among others, the use of a building or structure, or a portion thereof, for storage that is not classified as a hazardous occupancy.

❖ This section requires that all structures (or parts thereof) designed or occupied for the storage of moderate- and low-hazard materials are to be classified in either Group S-1 (moderate hazard) or S-2 (low hazard). Small storage areas are inherent in almost any activity or occupancy. Where these are less than 10 percent of the occupancy of the story, such storage might qualify as an accessory use under Section 508.2. Storage areas in excess of 10 percent of the floor in which they are located will need to be fully addressed under the mixed occupancy requirements of Section 508.

The life safety problems in structures used for storage of moderate- and low-hazard materials are minimal because the number of people involved in a storage operation is usually small and normal work patterns require the occupants to be dispersed throughout the facility.

The problems of fire safety, particularly as they relate to the protection of stored contents, are directly associated with the amount and combustibility of the materials (including packaging) that are housed on the premises.

Storage facilities typically contain significant amounts of combustible or noncombustible materials that are kept in a common area. Because of the combustion, flammability or explosive characteristics of certain materials (see Section 307), a structure (or portion thereof) that is used to store high-hazard materials, which exceeds the maximum allowable quantities or that does not meet one of the exceptions identified in Section 307.1, cannot be classified as Group S and is to be classified as Group H, high-hazard use, and is to comply with Section 307.

Storage occupancies consist of two basic types:

Groups S-1 and S-2, which are based on the properties of the materials being stored. The distinction between Groups S-1 and S-2 is similar to that between Groups F-1 and F-2, as outlined in Section 306.

311.2 Moderate-hazard storage, Group S-1. Buildings occupied for storage uses that are not classified as Group S-2, including, but not limited to, storage of the following:

Aerosols, Levels 2 and 3
Aircraft hangar (storage and repair)
Bags: cloth, burlap and paper
Bamboos and rattan
Baskets
Belting: canvas and leather
Books and paper in rolls or packs
Boots and shoes
Buttons, including cloth covered, pearl or bone
Cardboard and cardboard boxes
Clothing, woolen wearing apparel
Cordage
Dry boat storage (indoor)
Furniture
Furs
Glues, mucilage, pastes and size
Grains
Horns and combs, other than celluloid
Leather
Linoleum
Lumber
Motor vehicle repair garages complying with the maximum allowable quantities of hazardous materials listed in Table 307.1(1) (see Section 406.6)
Photo engravings
Resilient flooring
Silks
Soaps
Sugar
Tires, bulk storage of
Tobacco, cigars, cigarettes and snuff
Upholstery and mattresses
Wax candles

❖ Buildings in which combustible materials are stored and that burn with ease are classified in Group S-1, moderate-hazard storage occupancies. Examples of the kinds of materials that, when stored, are representative of occupancies classified in Group S-1 are also listed in this section.

As defined by the IFC, a repair garage is any structure used for servicing or repairing motor vehicles. Therefore, regardless of the extent of work done (e.g., quick lube, tune-up, muffler and tire shops, painting, body work, engine overhaul), repair garages are classified as Group S-1 (see Figure 311.2) and must be in compliance with Section 406.6. In addition, to avoid a Group H classification, the amounts of hazardous materials in the garage must be less than the maximum allowable quantity per control area permitted in Tables 307.1(1) and 307.1(2).

Aircraft hangars for storage, repair or both would be classified as Group S-1. This classification correlates with the actual use of such hangars which very frequently would include some level of repair work and also works with the requirements of NFPA 409. Previous editions allowed aircraft hangars simply for storage to be classified as Group S-2. Aircraft hangers accessory to one- and two-family structures remain a Group U occupancy.

311.3 Low-hazard storage, Group S-2. Includes, among others, buildings used for the storage of noncombustible materials such as products on wood pallets or in paper cartons with or without single thickness divisions; or in paper wrappings. Such products are permitted to have a negligible amount of plastic *trim*, such as knobs, handles or film wrapping. Group S-2 storage uses shall include, but not be limited to, storage of the following:

Asbestos
Beverages up to and including 16-percent alcohol in metal, glass or ceramic containers
Cement in bags
Chalk and crayons
Dairy products in nonwaxed coated paper containers
Dry cell batteries
Electrical coils
Electrical motors
Empty cans
Food products
Foods in noncombustible containers
Fresh fruits and vegetables in nonplastic trays or containers
Frozen foods
Glass
Glass bottles, empty or filled with noncombustible liquids
Gypsum board
Inert pigments

Ivory
Meats
Metal cabinets
Metal desks with plastic tops and *trim*
Metal parts
Metals
Mirrors
Oil-filled and other types of distribution transformers
Parking garages, open or enclosed
Porcelain and pottery
Stoves
Talc and soapstones
Washers and dryers

❖ Buildings in which noncombustible materials are stored are classified as Group S-2, low-hazard storage occupancies (see Figure 311.3). It is acceptable for stored noncombustible products to be packaged in combustible materials as long as the quantity of packaging is kept to an insignificant level.

As seen in Group F-1 and F-2 classifications, it is important to be able to distinguish when the presence of combustible packaging constitutes a significant fuel load. As such, a fuel load might require the building to be classified in Group S-1, moderate-hazard storage. A simple guideline to follow is the "single thickness" rule, which is when a noncombustible product is put in one layer of packaging material.

Examples of materials qualified for storage in Group S-2 storage facilities are as follows:

• Vehicle engines placed on wood pallets for transportation after assembly;

• Washing machines in corrugated cardboard boxes; and

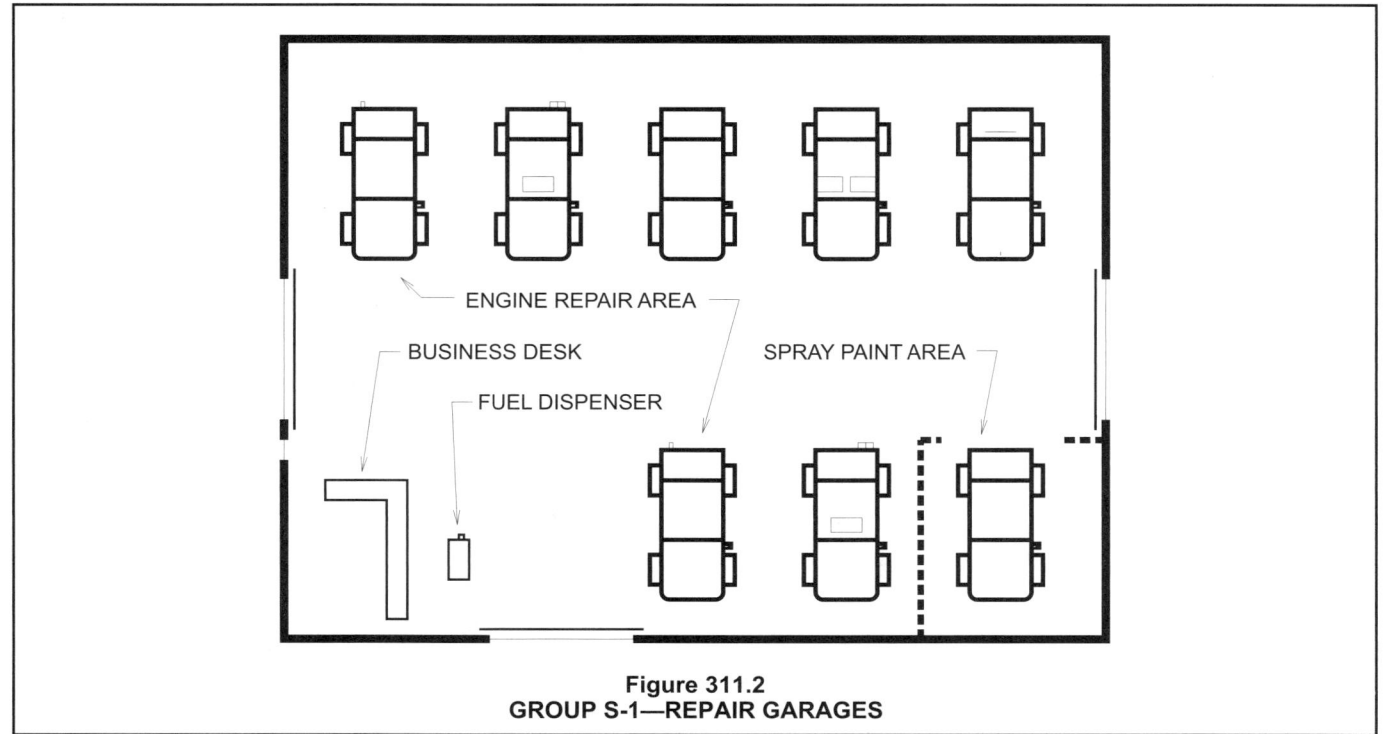

Figure 311.2
GROUP S-1—REPAIR GARAGES

- Soft-drink glass bottles packaged in pressed paper boxes.

Structures used to store noncombustible materials packaged in more than one layer of combustible packaging material are to be classified in Group S-1.

Examples of materials that, because of packaging, do not qualify for classification in Group S-2 are:

- Chinaware wrapped in corrugated paper and placed in cardboard boxes;
- Glassware set in expanded foam forms and placed in a cardboard box; and
- Fuel filters individually packed in pressed paper boxes, placed by the gross in a cardboard box and then stacked on a wood pallet for transportation.

An area of the IFC that is often related to Group S occupancies is Chapter 23, which regulates high- piled combustible storage [storage over 12 feet (3658 mm) in height or 6 feet (1829 mm) if the material is considered high hazard]. Chapter 23 of the IFC is focused not only on the type of materials being stored but also the height and configuration of such storage. It is important to note that not all Group S occupancies will contain high-piled storage and that high-piled storage is not limited to Group S occupancies. High-piled storage can be found in occupancies such as Group H or F.

Open and enclosed parking garages are classified as Group S-2 occupancies as long as no repair activities as discussed in the commentary to Section 311.2 occur in such buildings. A garage in a fire station, for example, that undertakes maintenance and repairs limited to cleaning, hose change, water fill, fire equipment upgrades or wheel removal for repair off premise would not constitute the same hazard associated with repair garages and would be appropriately classified as a Group S-2 classification.

SECTION 312
UTILITY AND MISCELLANEOUS GROUP U

312.1 General. Buildings and structures of an accessory character and miscellaneous structures not classified in any specific occupancy shall be constructed, equipped and maintained to conform to the requirements of this code commensurate with the fire and life hazard incidental to their occupancy. Group U shall include, but not be limited to, the following:

Agricultural buildings
Aircraft hangars, accessory to a one- or two-family residence (see Section 412.5)
Barns
Carports
Fences more than 6 feet (1829 mm) high
Grain silos, accessory to a residential occupancy
Greenhouses
Livestock shelters
Private garages
Retaining walls
Sheds
Stables
Tanks
Towers

❖ This section identifies the characteristics of occupancies classified in Group U. Structures that are classified in Group U are typically accessory to another building or structure and are not more appropriately classified in another occupancy. Miscellaneous storage buildings accessory to detached one- and two-family dwellings and multiple single-family dwellings (townhouses) not more than three stories in height, however, are intended to be designed and built in accordance with the IRC (see Section 101.2).

Structures classified as Group U, such as fences, equipment, foundations, retaining walls, etc., are somewhat outside the primary scope of the code (i.e.,

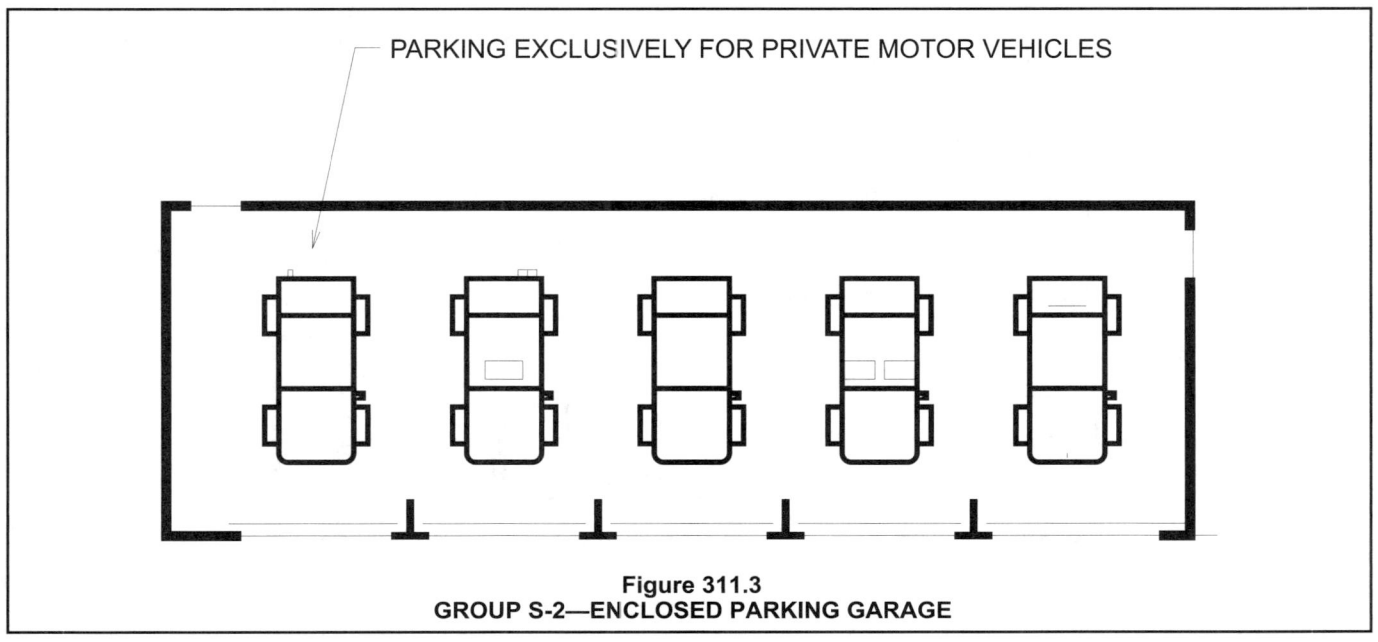

PARKING EXCLUSIVELY FOR PRIVATE MOTOR VEHICLES

Figure 311.3
GROUP S-2—ENCLOSED PARKING GARAGE

means of egress, fire resistance). They are not usually considered to be habitable or occupiable. Nevertheless, many code provisions do apply and need to be enforced (e.g., structural design and material performance).

Structures housing accessory equipment that is part of a utility or communications system are often classified as Group U occupancies when there is no intent that these structures be occupied except for servicing and maintaining the equipment within the structure. A pumphouse for a water or sewage system or an equipment building at the base of a telecommunications tower are examples of such buildings.

Group U occupancies are subject to the same structural loadings such as snow loads as other occupancies. Section 312.1 establishes that occupancies classified as utility and miscellaneous structures shall be constructed, equipped and maintained to conform to the code requirements that are commensurate with the fire and life hazards incidental to their occupancy. The structural design requirements for roofs are the minimum deemed necessary to withstand such elements. Allowing construction of a building with an accessory occupancy that could reasonably be expected to collapse under the snow loads known to prevail in a certain area is not in the best interest of public safety.

Bibliography

The following resource materials are referenced in this chapter or are relevant to the subject matter addressed in this chapter.

ASTM D 56-05, *Test Method for Flash Point by Tag Closed Tester.* West Conshohocken, PA: ASTM International, 2005.

ASTM D 93-07, *Test Methods for Flash Point by Pensky-Martens Closed Cup Tester.* West Conshohocken, PA: ASTM International, 2007.

ASTM D 3278-2004e01, *Test Methods for Flash Point of Liquids by Small Scale Closed Cup Apparatus.* West Conshohocken, PA: ASTM International, 2004.

ASTM E 136-04, *Test Method for Behavior of Materials in a Vertical Tube Furnace at 750°C.* West Conshohocken, PA: ASTM International, 2004.

ASTM E 681-04, *Test Method for Concentration Limits of Flammability of Chemicals (Vapors and Gases).* West Conshohocken, PA: ASTM International, 2004.

ASTM E 1590-07, Test Method for Fire Testing Mattresses. West Conshohocken, PA: ASTM International, 2007.

DOL 29 CFR; Part 1910-07, *Occupational Safety and Health Standards.* Washington, DC: United States Department of Labor, 2007.

DOTn 49 CFR; Part 100-199-05, *Specification for Transportation of Explosive and Other Dangerous Articles, Shipping Containers.* Washington, DC: United States Department of Transportation, 2005.

DOTy 27 CFR Chapter 1, Section 55.23-02, *Alcohol, Tobacco and Firearms.* Washington, DC: United States Department of Treasury, 2002.

IFC-09, *International Fire Code.* Washington, DC: International Code Council, 2009.

IMC-09, *International Mechanical Code.* Washington, DC: International Code Council, 2009.

IRC-09, *International Residential Code.* Washington, DC: International Code Council, 2009.

ISO 8115-86, *Cotton Bales–Dimensions and Density.* Geneva 20, Switzerland: International Organization for Standardization, 1986.

NFPA 13-07, *Installation of Sprinkler Systems.* Quincy, MA: National Fire Protection Association, 2007.

NFPA 13R-07, *Installation of Sprinkler Systems in Residential Occupancies Up to and Including Four Stories in Height.* Quincy, MA: National Fire Protection Association, 2007.

NFPA 30B-07, *Manufacture and Storage of Aerosol Products.* Quincy, MA: National Fire Protection Association, 2007.

NFPA 31-06, *Installation of Oil-burning Equipment.* Quincy, MA: National Fire Protection Association, 2006.

NFPA 37-06, *Installation and Use of Stationary Combustion Engines and Gas Turbines.* Quincy, MA: National Fire Protection Association, 2006.

NFPA 43B-93, *Safe Handling of Organic Peroxides.* Quincy, MA: National Fire Protection Association, 1993.

NFPA 409-04, *Aircraft Hangars,* Quincy, MA: National Fire Protection Association, 2004.

NFPA 430-04, *Storage of Liquid and Solid Oxidizers.* Quincy, MA: National Fire Protection Association, 2004.

NFPA 495-06, *Explosive Materials Code.* Quincy, MA: National Fire Protection Association, 2006.

NFPA 701a-04, *Methods of Fire Tests for Flame Propagation of Textiles and Films.* Quincy, MA: National Fire Protection Association, 2004.

NFPA 704-07, *Identification of the Hazards of Materials for Emergency Response.* Quincy, MA: National Fire Protection Association, 2007.

NFPA 1124-06, *Manufacture, Transportation, Storage and Retail Sales of Fireworks and Pyrotechnic Articles.* Quincy, MA: National Fire Protection Association, 2006.

UL 30-95, *Metal Safety Cans-with Revisions through December 2004*. Northbrook, IL: Underwriters Laboratories, Inc., 1995.

USC Title 16-04, *Code of Federal Regulations, Parts 1500 and 1507*. Washington, DC: United States Code, 2004.

USC Title 18, Chapter 40-07, *Importation, Manufacture, Distribution and Storage of Explosive Materials*. Washington, DC: United States Code, 2007.

Chapter 4:
Special Detailed Requirements Based On Use and Occupancy

General Comments

The provisions of Chapter 4 are supplemental to the remainder of the code. Chapter 4 contains provisions that may alter requirements found elsewhere in the code; however, the general requirements of the code still apply unless modified within the chapter. For example, the height and area limitations established in Chapter 5 apply to all special occupancies unless Chapter 4 contains height and area limitations. In this case, the limitations in Chapter 4 supersede those in other sections. An example of this is the height and area limitations given in Section 406.3.5, which supersede the limitations given in Table 503 and Section 503 for open parking garages.

The *International Fire Code*® (IFC®) contains provisions applicable to the storage, handling and use of hazardous substances, materials or devices and, therefore, must also be complied with when addressing such occupancies as those involving flammable and combustible liquids. Similarly, the *International Mechanical Code*® (IMC®) and the *International Plumbing Code*® (IPC®) include provisions for specific applications, such as hazardous exhaust systems and hazardous material piping.

In some instances, it may not be necessary to apply the provisions of Chapter 4. For example, if a covered mall building complies with the provisions of the code for Group M occupancies, then Section 402 does not apply. However, other sections that address a use, process or operation must be applied to that specific occupancy, such as Sections 410, 411 and 414.

Purpose

The purpose of Chapter 4 is to combine in one chapter the provisions of the code applicable to special uses and occupancies. Hazardous materials and operations may occur in more than one group; therefore, the applicable provisions for the specific hazardous occupancy or operation apply to multiple groups. Also, while the provisions for all structures are interrelated to form an overall protection system, by providing requirements for specific occupancies in Chapter 4 the package of protection features is more easily identified.

Chapter 4 contains the requirements for protecting special uses and occupancies. The provisions in this chapter reflect those occupancies and groups that require special consideration and are not addressed elsewhere in the code. The chapter includes requirements for buildings and conditions that apply to one or more groups, such as high-rise buildings or atriums. Special uses may also imply specific occupancies and operations, such as for Groups H-1, H-2, H-3, H-4 and H-5, application of flammable finishes and combustible storage; or for a specific occupancy within a much larger occupancy, such as covered mall buildings, motor-vehicle-related occupancies, special amusement buildings and aircraft-related occupancies. Finally, in order that the overall package of protection features can be easily understood, occupancies, such as Groups I-2 and I-3, live/work units and underground buildings, are addressed.

SECTION 401
SCOPE

401.1 Detailed use and occupancy requirements. In addition to the occupancy and construction requirements in this code, the provisions of this chapter apply to the special uses and occupancies described herein.

❖ This section provides guidance on how Chapter 4 is to be applied with respect to other sections of the code. Section 401.1 indicates that all other provisions of the code apply except as modified by Chapter 4.

The requirements contained in Chapter 4 are intended to apply to special uses and occupancies, as well as special construction features as defined in various sections in this chapter. These requirements are applicable in addition to other chapters of the code.

SECTION 402
COVERED MALL AND OPEN MALL BUILDINGS

402.1 Scope. The provisions of this section shall apply to buildings or structures defined herein as *covered mall buildings* not exceeding three floor levels at any point nor more than three *stories above grade plane*. Except as specifically required by this section, *covered mall buildings* shall meet applicable provisions of this code.

Exceptions:

1. Foyers and lobbies of Groups B, R-1 and R-2 are not required to comply with this section.

2. Buildings need not comply with the provisions of this section when they totally comply with other applicable provisions of this code.

❖ This section primarily addresses shopping centers with a maximum of three levels that consist of one or more sizeable anchor businesses, typically department stores, known as anchor buildings, and numerous smaller retail stores, all of which are interconnected by means of a mall or common pedestrian way. The mall may be covered, providing a climate-controlled environment, or open to the sky. The complex may include movie theaters, bowling lanes, ice arenas, offices, and dining and drinking establishments. Historically, anchor buildings were large department stores, and malls remain attached to department stores, but under this section anchor buildings can contain any occupancy with the exception of a Group H. The complex may also include single- or multiple-level buildings, with a vast majority of shopping centers being one or two levels, and with an occasional mezzanine. Many malls now contain large food courts with a variety of food vendors surrounding a common seating area.

This section addresses the special considerations associated with covered or open mall buildings, including construction, egress and fire protection systems. It does not, in general, apply to the anchor buildings around the perimeter of the covered or open mall building. To be considered an anchor building for purposes of applying the code, the building must have complete egress facilities, including the required number and capacity of exits, independent of the covered or open mall building.

Originally, the provisions of this section applied to the typical covered mall building: a one- to three-level structure consisting primarily of retail space and a covered pedestrian way. More recently, other types of covered mall buildings, such as airport passenger terminals and office centers, have also been constructed in accordance with this section.

Malls with an unroofed common pedestrian way, or open malls, are becoming common in the "sun belt" areas of the country and in similar climates around the world. For 2009 the code includes open malls within the broader definition of mall and allows the same benefits of the covered mall provisions, because an open-to-the-sky mall provides equivalent or better life safety and property protection. The key to the open mall concept is to have everything a covered mall building would have, except for the roof over the mall area. Without a roof over the mall area and with required openings from the grade level to the sky above, natural ventilation is provided and mechanical smoke control is no longer necessary in the mall area and adjoining tenant spaces. The term "covered mall building" includes open mall buildings as defined by this section. Unless noted otherwise, open mall buildings shall comply with all the provisions for covered mall buildings. Because many of the provisions of this section were written presuming a covered and enclosed building, the application of some provisions will need to be addressed carefully for an open mall so that the intent of the code is maintained.

This section is not intended to apply to the foyers and lobbies of business and residential occupancies

that may contain retail spaces, small restaurants and offices unrelated to the use of upper stories. If these lobbies have multiple levels open to each other, the provisions of Section 404, Atriums, may apply. In all such cases, the provisions for mixed occupancies found in Section 508 would apply. This section is also not intended to apply to large footprint retail buildings, which may include smaller retail spaces, such as banks, florists or coffee stands, around the perimeter of a central sales area. These buildings do not provide the intended concept of a central pedestrian mall that is distinct from the surrounding tenant spaces.

Covered or open mall buildings that comply with all other applicable provisions of the code need not comply with the provisions of this section; however, covered or open mall buildings that are designed and constructed to comply with the provisions of this section must also comply with the provisions of the code that are not otherwise modified in this section. For example, the egress provisions in Section 402.4 are applicable to covered or open mall buildings designed in accordance with this section. The provisions in Chapter 10 related to door swing and stairways are also applicable, however, since Section 402 contains no similar provisions.

402.2 Definitions. The following words and terms shall, for the purposes of this chapter and as used elsewhere in this code, have the meanings shown herein.

❖ Definitions of terms that are associated with the content of this section are contained herein. These definitions can help in the understanding and application of the code requirements. It is important to emphasize that these terms are not exclusively related to this section but are applicable everywhere the term is used in the code. The purpose for including these definitions within this section is to provide more convenient access to them without having to refer back to Chapter 2. For convenience, these terms are also listed in Chapter 2 with a cross reference to this section. The use and application of all defined terms, including those defined herein, are set forth in Section 201.

ANCHOR BUILDING. An exterior perimeter building of a group other than H having direct access to a *covered mall building* but having required *means of egress* independent of the mall.

❖ A key to understanding what distinguishes an anchor building from a tenant space is that anchor buildings are typically retail establishments (Group M), although this may not always be the case. The anchor building is typically some facility that, by its nature, draws a considerable number of people. The tenants in the adjoining covered or open mall building then seek to capitalize on this traffic generated by the anchor building. The scale or size of the building is not a primary factor in determining whether it is an anchor building; rather, its function is such that it draws people to the site in sufficient numbers so that other facilities can benefit from being located in the same facility (see Figure 402.2).

Generally, the anchor building will have its own identity and there is a high probability that it will have separate management and its own hours of operation. This will necessitate a means of egress that does not rely on the mall being available for its patrons to enter or leave. Similarly, the means of egress from the covered or open mall building cannot rely on the anchor building being open and available for patrons to exit through; hence, egress facilities for the anchor building must be independent of those for the covered or open mall building. An anchor building is a separate building from the covered or open mall building and must comply with the provisions of the code for its own identity, except as modified by this section (see Sections 402.6 and 402.7.3.1).

COVERED MALL BUILDING. A single building enclosing a number of tenants and occupants, such as retail stores, drinking and dining establishments, entertainment and amusement facilities, passenger transportation terminals, offices and other similar uses wherein two or more tenants have a main entrance into one or more malls. For the purpose of this chapter, *anchor buildings* shall not be considered as a part of the *covered mall building*. The term "*covered mall building*" shall include open mall buildings as defined below.

❖ The covered mall building is the entire area of the building (area of mall plus gross leasable area), excluding the anchor buildings. Passenger transportation terminals frequently are developed as wide concourses with small shops along the sides. For this reason, passenger transportation facilities are included. Transportation facilities used for freight or other purposes are not to be considered a covered mall building (see Figure 402.2). The term "covered mall building" also includes open mall buildings. Unless noted otherwise, open mall buildings shall comply with all the provisions for a covered mall building.

Mall. A roofed or covered common pedestrian area within a *covered mall building* that serves as access for two or more tenants and not to exceed three levels that are open to each other. The term "mall" shall include open malls as defined below.

❖ The mall is an interior, climate-controlled pedestrian way that is open to the tenant spaces within the mall building and typically connects to the anchor buildings. The term "mall" shall also include open mall. Unless noted otherwise, open malls must comply with all the provisions for malls.

Open mall. An unroofed common pedestrian way serving a number of tenants not exceeding three levels. Circulation at levels above grade shall be permitted to include open exterior balconies leading to *exits* discharging at grade.

❖ The open mall is an uncovered common pedestrian walk that is open to the sky above and to tenant spaces within the open mall building and typically connects to the anchor buildings. The size of the openings to the sky is in accordance with Section 402.5.2.

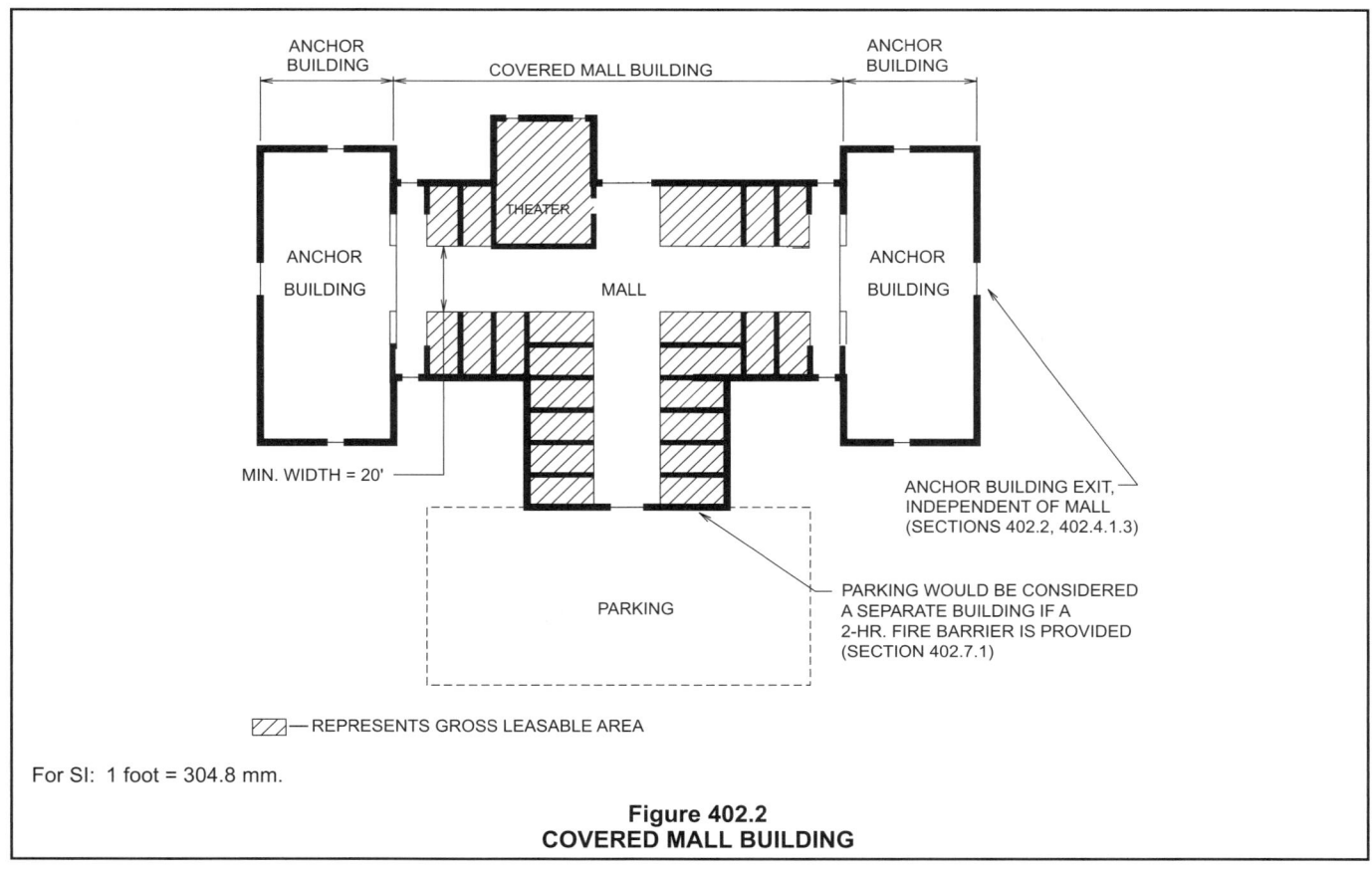

For SI: 1 foot = 304.8 mm.

Figure 402.2
COVERED MALL BUILDING

Unless noted otherwise, open malls must comply with all the provisions for malls.

Open mall building. Several structures housing a number of tenants, such as retail stores, drinking and dining establishments, entertainment and amusement facilities, offices, and other similar uses, wherein two or more tenants have a main entrance into one or more open malls. For the purpose of Chapter 4 of the *International Building Code*, *anchor buildings* are not considered as a part of the open mall building.

❖ The open mall building includes all of the buildings housing a number of tenants wherein two or more tenants have a main entrance into one or more open malls. Because the open mall is characterized by there not being a roof connecting one side of the pedestrian mall to the other, the covered mall "building" may actually be a collection of separate buildings which all rely on a shared pedestrian concourse for egress. Similar to the covered mall building, the open mall "building" does not include the anchor buildings. Unless noted otherwise, open mall buildings have to comply with all the provisions for covered mall building.

FOOD COURT. A public seating area located in the mall that serves adjacent food preparation tenant spaces.

❖ Typical mall building layouts include a central gathering area for food and drink consumption. These areas are usually located in the mall itself and chairs are provided for the public's use to consume the food and drink. This public area is usually surrounded by numerous tenant spaces where food is prepared and sold over the counter. A separate design occupant load is required to be calculated in accordance with Section 402.4.1.4.

GROSS LEASABLE AREA. The total floor area designed for tenant occupancy and exclusive use. The area of tenant occupancy is measured from the centerlines of joint partitions to the outside of the tenant walls. All tenant areas, including areas used for storage, shall be included in calculating gross leasable area.

❖ The gross leasable area represents the aggregate area available in the covered or open mall building for tenant occupancy. It does not include the area of the mall, unless portions of it are leased for the purposes of setting up separate tenant spaces (kiosks). The area is used to determine the design occupant load in accordance with Section 402.4.1.1.

402.3 Lease plan. Each *covered mall building* owner shall provide both the building and fire departments with a lease plan showing the location of each occupancy and its exits after the certificate of occupancy has been issued. No modifications or changes in occupancy or use shall be made from that shown on the lease plan without prior approval of the *building official*.

❖ The required lease plan is permitted to be submitted after the certificate of occupancy has been issued because many times the developer does not have the information at the time of construction. The location of tenant separations may not be known until leases are negotiated with prospective tenants. This may occur after the mall has opened.

During initial construction, it is anticipated that the building department will require tenant improvements to be submitted through the building permit process. After tenant spaces are prepared for occupancy and the lease plan is developed, subsequent modifications and changes must be submitted for approval prior to commencing construction or changing the use. It is important that the fire department receives copies of current lease plans, since not only does this help the fire department while performing fire prevention inspections, but also the lease plans assist in fire department response to an emergency.

402.4 Means of egress. Each tenant space and the *covered mall building* shall be provided with *means of egress* as required by this section and this code. Where there is a conflict between the requirements of this code and the requirements of this section, the requirements of this section shall apply.

❖ Each individual occupancy or tenant space within the covered or open mall building is required to have a means of egress that complies with other provisions of the code—especially Chapter 10. Keep in mind that the requirements of Section 402.4 will supersede some of the provisions of Chapter 10. In order to comply, travel through the mall area is considered as exit access and must comply with Section 402.4.4. Travel distance within a tenant space need only be measured to the entrance of the mall from the space (see Figure 402.4). As such, two distinct criteria must be met with regard to exit access travel distance: travel within the mall and travel within the tenant space.

The change that included open malls under Section 402 did not include provisions addressing means of egress in an open mall. As compared to a covered mall where the common mall is similar to exit access, an open mall could more easily be considered similar to exit discharge given that it is outside of the buildings. If access and egress is controlled by some sort of gates or barriers, it may be appropriate to treat those gates as "exits" from the open mall and measure distances from each tenant space to these gates. The intent of the proponent of the code change that added open malls to these provisions was simply to allow for an uncovered mall space wherein smoke control was not required. All other provisions were to be the same. Therefore, the intent was that the open mall portion be treated as if it were covered in order to determine compliance with the code on means of egress provisions. An approach that might make more sense than others would be to establish a line around the exterior side of all the buildings considered to be part of the open mall "building" and apply the means of egress provisions as if crossing that line is equivalent to exiting the building.

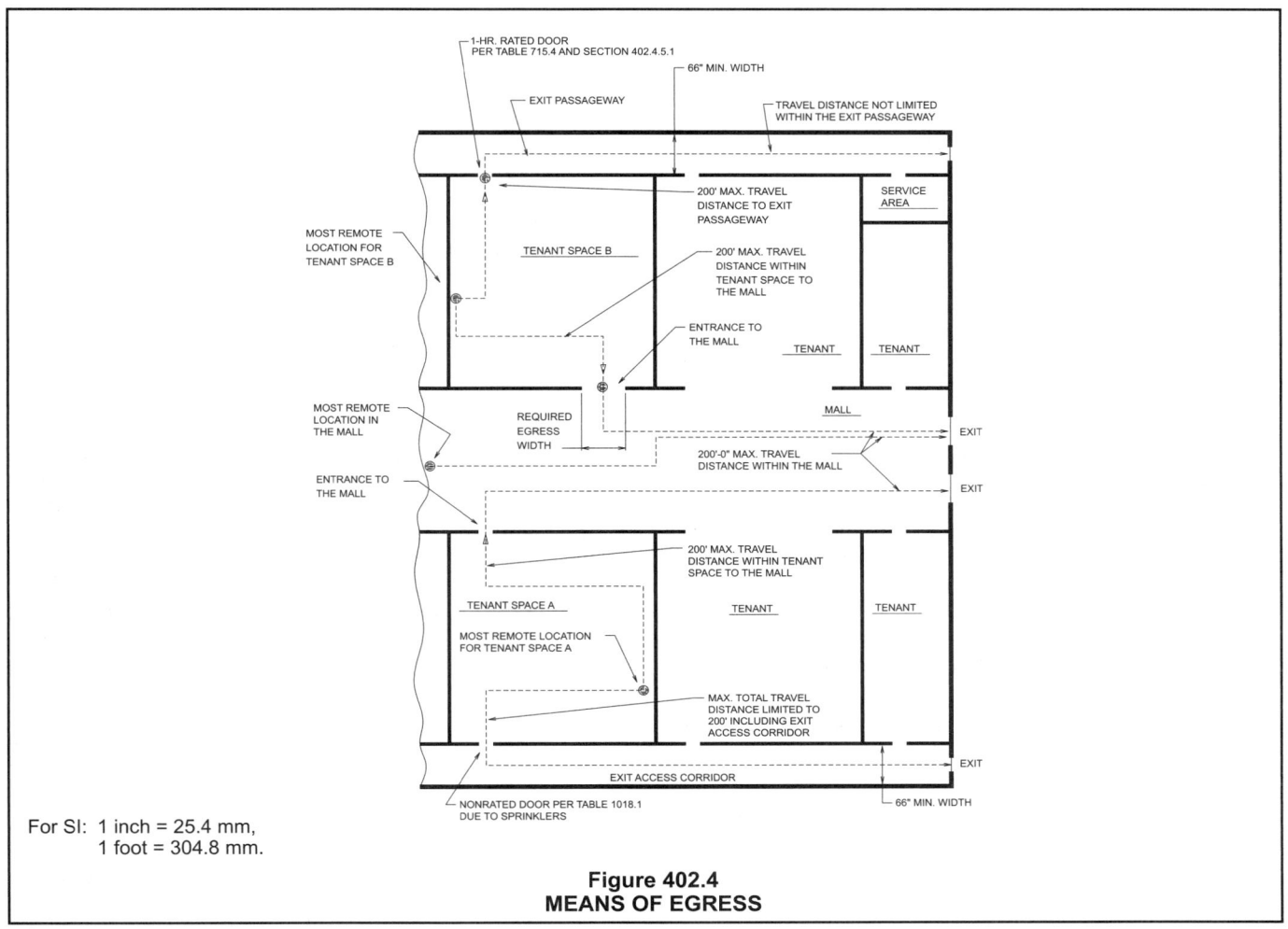

For SI: 1 inch = 25.4 mm,
1 foot = 304.8 mm.

Figure 402.4
MEANS OF EGRESS

402.4.1 Determination of occupant load. The *occupant load* permitted in any individual tenant space in a *covered mall building* shall be determined as required by this code. *Means of egress* requirements for individual tenant spaces shall be based on the *occupant load* thus determined.

❖ Since the tenant spaces of covered or open mall buildings can be used for varied occupancies, the design occupant loads will also vary. Although each tenant space contributes to the gross leasable area of Section 402.4.1.1, each must have its own occupant load calculated along with the applicable means of egress requirements. For example, a restaurant with its own dining area may occupy a tenant space. The design occupant load for this space is calculated in accordance with Table 1004.1.1 based on the assembly occupancy of the dining areas and on the commercial kitchen occupancy of the food preparation areas. Door swing direction and panic and fire exit hardware requirements would be based on this occupant load.

402.4.1.1 Occupant formula. In determining required *means of egress* of the mall, the number of occupants for whom *means of egress* are to be provided shall be based on gross leasable area of the *covered mall building* (excluding *anchor buildings*)

and the *occupant load* factor as determined by the following equation.

$$OLF = (0.00007)(GLA) + 25 \qquad \textbf{(Equation 4-1)}$$

where:

OLF = The *occupant load* factor (square feet per person).

GLA = The gross leasable area (square feet).

Exception: Tenant spaces attached to a *covered mall building* but with a *means of egress* system that is totally independent of the *covered mall building* shall not be considered as gross leasable area for determining the required *means of egress* for the *covered mall building*.

❖ The capacity of the exits serving the covered mall buildings must satisfy the calculated occupant load based on the gross leasable area. The occupant load factors (OLF) were determined empirically by surveying more than 270 covered mall shopping centers, studying mercantile occupancy parking requirements [5.0 cars per 1,000 square feet (93 m²) of gross leasable area] and observing the number of occupants per vehicle during peak seasons (4.0 per car).

The formula used for determining occupant load results in a smooth transition between the occupant load and the gross leasable area with a range of 30 to 50.

For example, if the gross leaseable area of the covered mall building is 400,000 square feet (37 160 m²), the calculated OLF is determined as follows:

$$OLF = (0.00007) (400,000) + 25$$

$$OLF = 53 \text{ square feet } (4.9 \text{ m}^2) \text{ per person}$$

Since the design occupant load factor cannot exceed 50 in accordance with Section 402.4.1.2, the required OLF would be 50 square feet (4.6 m²) per person.

If the gross leasable area of the mall building is 55,000 square feet (5109 m²), the calculated OLF is determined as follows:

$$OLF = (0.00007) (55,000) + 25$$

$$OLF = 29 \text{ square feet } (2.7 \text{ m}^2) \text{ per person}$$

Since the OLF need not be less than 30 in accordance with Section 402.4.1.2, the required occupant load factor would be 30 square feet (2.8 m²) per person. The code would permit the use of an occupant load factor of 29, since it would result in a higher design occupant load and consequently increased width of exits.

402.4.1.2 OLF range. The *occupant load* factor *(OLF)* is not required to be less than 30 and shall not exceed 50.

❖ Although not mandatory, an OLF of less than 30 may be used at the designer's discretion (see Section 1004.2). In no case should an OLF greater than 50 be used, since it would result in too small a design occupant load.

402.4.1.3 Anchor buildings. The *occupant load* of *anchor buildings* opening into the mall shall not be included in computing the total number of occupants for the mall.

❖ Although anchor buildings are allowed access to the mall portions of the covered or open mall building, the occupant load of the mall does not have to be increased for this access. People may certainly discharge from the anchor building into the mall on their own accord; however, the anchor building is required to provide sufficient exits for its occupant load at locations other than the mall.

402.4.1.4 Food courts. The *occupant load* of a food court shall be determined in accordance with Section 1004. For the purposes of determining the *means of egress* requirements for the mall, the food court *occupant load* shall be added to the *occupant load* of the *covered mall building* as calculated above.

❖ The occupant load for food courts must be determined in accordance with Section 1004. Typically, tables and chairs are provided for the public's use in a food court. As such, the design occupant load for a food court is usually calculated at a rate of one person for every 15 square feet (1.4 m²) of net floor space for an unconcentrated assembly occupancy. The occupant loads for all food courts are added to the mall occupant load calculated in accordance with Section 402.4.1.1

and Equation 4-1. Food courts are not considered part of the gross leasable area of the mall; therefore, the occupant loads of the food courts must be calculated separately and then added to the overall mall building occupant load.

402.4.2 Number of means of egress. Wherever the distance of travel to the mall from any location within a tenant space used by persons other than employees exceeds 75 feet (22 860 mm) or the tenant space has an *occupant load* of 50 or more, not less than two *means of egress* shall be provided.

❖ This section applies within each tenant space. It dictates the minimum number of paths of travel an occupant is to have available to avoid a fire incident in the occupied tenant space. While providing multiple egress doorways from every tenant space is unrealistic, a point does exist where alternative egress paths must be provided, based on the number of occupants at risk, the distance occupants must travel to reach the mall and the relative hazards associated with the occupancy of the space.

The 75-foot (22 860 mm) travel distance and the 49-person occupant load limitations represent an empirical judgment of the risks associated with a single means of egress from a tenant space based on the inherent risks associated with the occupancy (such as occupant mobility, occupant familiarity with the mall, occupant response and the fire growth rate). The requirement for two exits for 50 or more occupants applies to all occupants of the tenant space, including employees. The 75-foot (22 860 mm) travel distance does not apply to employees. Employee-only areas of the tenant space can require travel greater than 75 feet (22 860 mm) to reach the mall or other exit and a second means of egress is not required in spaces with an occupant load less than 50.

402.4.3 Arrangements of means of egress. Assembly occupancies with an *occupant load* of 500 or more shall be so located in the *covered mall building* that their entrance will be immediately adjacent to a principal entrance to the mall and shall have not less than one-half of their required *means of egress* opening directly to the exterior of the *covered mall building*.

❖ Tenant spaces which are assembly occupancies with an occupant load of 500 or more must be carefully located to minimize the hazards of egress. Movie theaters, nightclubs and large restaurants must be located on the perimeter of the covered or open mall building and adjacent to the mall's exits if their occupant loads exceed 499. A maximum of 50 percent of the means of egress from these assembly occupancies is permitted to discharge into the mall. Other means of egress from these tenant spaces must discharge directly to the exterior with a capacity of at least one-half the occupant load. For open malls, see the commentary to Section 402.4.

402.4.3.1 Anchor building means of egress. Required *means of egress* for *anchor buildings* shall be provided independently from the mall *means of egress* system. The *occupant load* of *anchor buildings* opening into the mall shall not be included in determining *means of egress* requirements for the mall. The path of egress travel of malls shall not exit through anchor buildings. Malls terminating at an *anchor building* where no other *means of egress* has been provided shall be considered as a dead-end mall.

❖ As indicated in the definition, the required means of egress for an anchor building must be provided independently of the mall and the means of egress from the mall. Since independent means of egress are provided, the occupant load of anchor buildings is not included in determining the egress requirements for the mall. If independent means of egress are not provided, the space cannot be considered an anchor building and is to be treated as any other tenant space of the mall. As such it would need to be included in the gross leasable area under Section 402.4.1.1.

402.4.4 Distance to exits. Within each individual tenant space in a *covered mall building*, the maximum distance of travel from any point to an *exit* or entrance to the mall shall not exceed 200 feet (60 960 mm).

The maximum distance of travel from any point within a mall to an *exit* shall not exceed 200 feet (60 960 mm).

❖ The maximum permissible travel distance from any point in the tenant space to the mall or from anywhere in the mall to an exit is 200 feet (60 960 mm). As such, the mall is treated like an aisle in a large store, the large store being the covered mall building. Although the distance is less than that permitted in Group M occupancies with an automatic sprinkler system, it must be recognized that this travel distance is measured from a point within the mall and not from the most remote point within the covered mall building (i.e., within a tenant space). Because of the uncertainty with respect to tenant improvements, it is more reliable to regulate travel distance within the mall instead of throughout the covered or open mall building.

For an open mall, see the commentary to Section 402.4. The application of this limit may depend on whether the open mall has a restricted entrance or exit points, or whether the mall can be treated as an exit discharge. Typically travel distance is not applicable outside of buildings, but in the case of an open mall where the path of the means of egress to a public way may be constrained, or not obvious, distance to the points where occupants transition from being "within the open mall building" to being "outside of the open mall building" should be addressed by the designer and building official. In a typical covered mall, the presence of a sprinkler system and other mall design features allow different treatments of distance to exits. In an open mall, the mall itself will not be sprinkler protected, but it is open to the air. This should also be considered in determining the applicability of this section.

402.4.5 Access to exits. Where more than one *exit* is required, they shall be so arranged that it is possible to travel in either direction from any point in a mall to separate *exits*. The mini-

mum width of an *exit passageway* or *corridor* from a mall shall be 66 inches (1676 mm).

Exception: Dead ends not exceeding a length equal to twice the width of the mall measured at the narrowest location within the dead-end portion of the mall.

❖ In order to accommodate the many occupants anticipated in a covered mall building, exits are to be distributed equally throughout the mall. If exits were congregated in certain areas, egress would be compounded by the convergence of a large number of people to one or more areas of the building. The corridors and exit access passageways from a mall must be at least 66 inches (1676 mm) wide.

The maximum dead end permitted in a mall is twice the width of the mall. For a mall with a minimum width of 20 feet (6096 mm), dead ends are permitted to be 40 feet (12 192 mm) or less in length. The allowance for dead ends is in recognition of the reduced hazard represented by a dead end in a mall and because the relatively large minimum width provides alternative paths of travel. Additionally, dead ends are not as critical in a mall, since the volume of the space and the automatic sprinkler system minimize the potential for the space to become untenable in a fire condition.

For open malls, see the commentary to Sections 402.4 and 402.4.4.

402.4.5.1 Exit passageways. Where *exit passageways* provide a secondary *means of egress* from a tenant space, doorways to the exit passageway shall be protected by 1-hour *fire door assemblies* that are self- or automatic-closing by smoke detection in accordance with Section 715.4.8.3.

❖ If exit passageways must be provided because of either travel distance limitations or the number of means of egress required from a tenant space, the passageways must be enclosed as required by the code. Additionally, the doors opening from the tenant spaces into the exit enclosures must be 1-hour rated. These doors must be self-closing or automatic-closing by smoke detection to provide the necessary protection as the occupants of a tenant space flee a fire situation.

402.4.6 Service areas fronting on exit passageways. Mechanical rooms, electrical rooms, building service areas and service elevators are permitted to open directly into *exit passageways*, provided the exit passageway is separated from such rooms with not less than 1-hour *fire barriers* constructed in accordance with Section 707 or *horizontal assemblies* constructed in accordance with Section 712, or both. The minimum *fire protection rating* of openings in the *fire barriers* shall be 1 hour.

❖ In a covered mall building, it is necessary to provide for services to the tenant spaces that are maintained by the mall management (e.g., water, electricity, telephone and fire protection). These services must be located in a common space controlled by the mall management and, therefore, cannot be located within the tenant spaces. Frequently, these services are logically located with direct access to exit passageways at the rear of the tenant spaces. For open mall buildings, the use of exit passageways will probably not occur as frequently, but where they do, they can also comply with this section.

Exit passageways are generally treated similar to exit stairways in that only openings from normally occupied spaces are permitted. This would prohibit doors or utility penetrations to such mechanical/electrical rooms. In the case of malls, the code allows an exception to that general rule, provided that the fire-resistance rating of the exit enclosure is maintained by appropriate opening protection, such as fire doors, fire dampers and through-penetration firestopping.

402.5 Mall width. For the purpose of providing required egress, malls are permitted to be considered as *corridors* but need not comply with the requirements of Section 1005.1 of this code where the width of the mall is as specified in this section.

❖ The pedestrian malls that serve as the exit access routes for the tenant space occupants are not required to comply with Section 1005.1. The design capacity of the mall width is already established by Section 402.5.1. No further egress width calculations are required by the code for mall corridors.

402.5.1 Minimum width. The minimum width of the mall shall be 20 feet (6096 mm). The mall width shall be sufficient to accommodate the *occupant load* served. There shall be a minimum of 10 feet (3048 mm) clear exit width to a height of 8 feet (2438 mm) between any projection of a tenant space bordering the mall and the nearest kiosk, vending machine, bench, display opening, food court or other obstruction to *means of egress* travel.

❖ The minimum width of a mall [20 feet (6096 mm)] is based on the need to provide adequate access to exits or, for open malls, the point where people transition

from within the open mall "building" to outside of the "building." Together with the automatic sprinkler system, the physical separation further reduces the need for a separation between tenant spaces and the mall.

So that an aggregate clear width of 20 feet (6096 mm) is always provided in the mall, a minimum 10-foot (3048 mm) clear and unobstructed space is to be maintained to a height of 8 feet (2438 mm) in front of, adjacent to and parallel to the storefronts. The requirement applies to kiosks, vending machines, benches, small stands, merchandise displays and any other potential obstruction to egress (see Figure 402.5.1).

402.5.2 Minimum width open mall. The minimum floor and roof opening width above grade shall be 20 feet (9096 mm) in open malls.

❖ An open mall requires a minimum width opening of 20 feet (9096 mm) from the lowest grade level to the sky above. This dimension aligns with Section 402.5.1, minimum width for egress. The width of the opening is measured perpendicular from the face of the tenant spaces, essentially across the pedestrian mall. This is similar to the mall width for covered malls. The opening, or the unroofed area, must extend from the lowest grade level of the common open mall to the sky above. This will provide natural ventilation to all levels. Balconies on the upper levels are permitted but may not project into the required 20-foot (9096 mm) width. Interior pedestrian bridges and overhangs at the end locations of the open mall are permitted (see Figure 402.5.2).

402.6 Types of construction. The area of any *covered mall building*, including *anchor buildings*, of Types I, II, III and IV construction, shall not be limited provided the *covered mall building* and attached *anchor buildings* and parking garages are surrounded on all sides by a permanent open space of not less than 60 feet (18 288 mm) and the *anchor buildings* do not

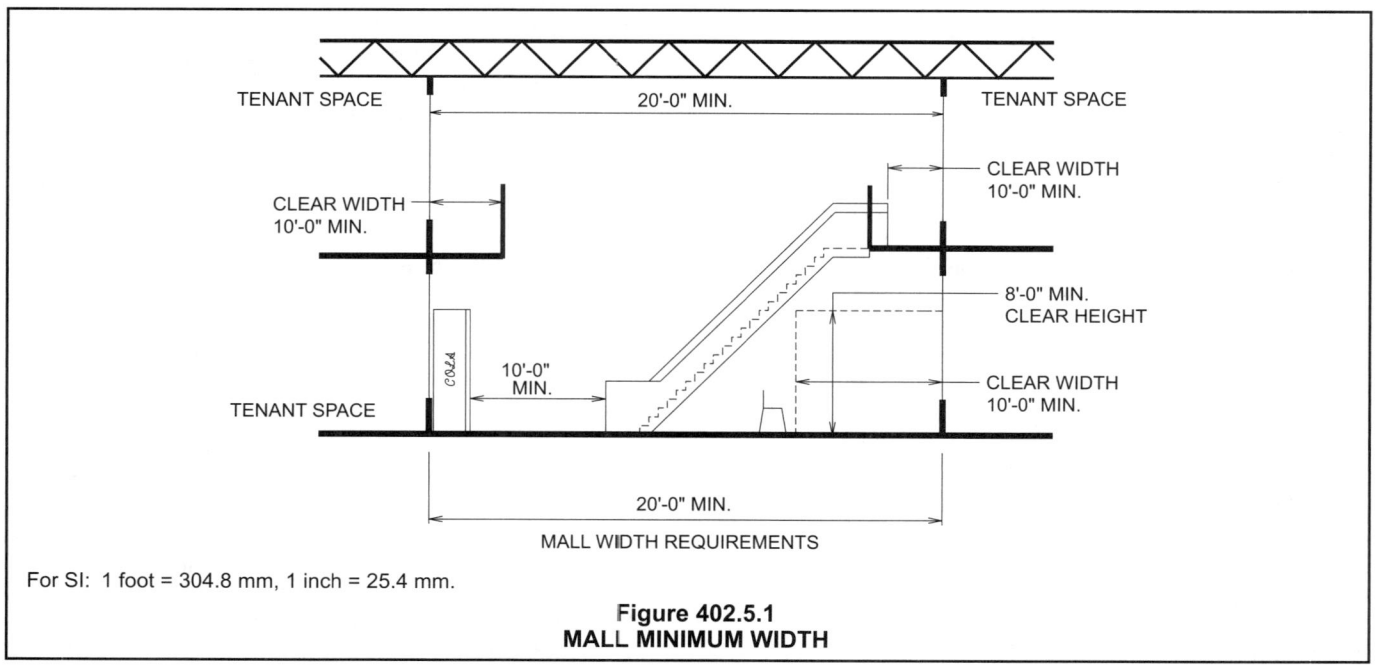

For SI: 1 foot = 304.8 mm, 1 inch = 25.4 mm.

**Figure 402.5.1
MALL MINIMUM WIDTH**

exceed three *stories above grade plane*. The allowable height and area of *anchor buildings* greater than three *stories above grade plane* shall comply with Section 503, as modified by Sections 504 and 506. The construction type of *open parking garages* and enclosed parking garages shall comply with Sections 406.3 and 406.4, respectively.

❖ Covered or open mall buildings are considered to be special types of unlimited area buildings and are exempt from the area limitations of Table 503 when they are limited to not more than three stories above grade plane. It should be noted that the height limitations in feet, specified in Table 503 based on the type of construction classification, are applicable to covered or open mall buildings. The allowance of an unlimited area anchor building is based on the restriction of construction types to noncombustible (Types I and II), noncombustible/combustible (Type III) and heavy timber (Type IV), and the effectiveness of the automatic sprinkler system. When the anchor building is over three stories above grade plane, it must comply with the general provisions of Chapter 5, such as Sections 503, 504 and 506. The last sentence of the code text serves as a reminder that the construction of parking garages is regulated by the general provisions found in Section 406 and that they are not regulated by the covered mall building's type of construction. It also

clarifies that the garage is not included in the unlimited area as stated in the first sentence of the code text.

402.6.1 Reduced open space. The permanent open space of 60 feet (18 288 mm) shall be permitted to be reduced to not less than 40 feet (12 192 mm), provided the following requirements are met:

1. The reduced open space shall not be allowed for more than 75 percent of the perimeter of the *covered mall building* and *anchor buildings*.

2. The *exterior wall* facing the reduced open space shall have a minimum *fire-resistance rating* of 3 hours.

3. Openings in the *exterior wall* facing the reduced open space shall have opening protectives with a minimum *fire protection rating* of 3 hours.

4. Group E, H, I or R occupancies are not within the *covered mall building* or *anchor stores*.

❖ This section allows covered or open mall buildings to reduce the required open space around the building as already permitted for similar unlimited area buildings regulated in Section 507. Where permitted in Section 507, unlimited area buildings require an open space around the building of 60 feet (18 288 mm) to prevent exposure fire spread. Section 507.5 permits five of the nine categories of unlimited area buildings to have a

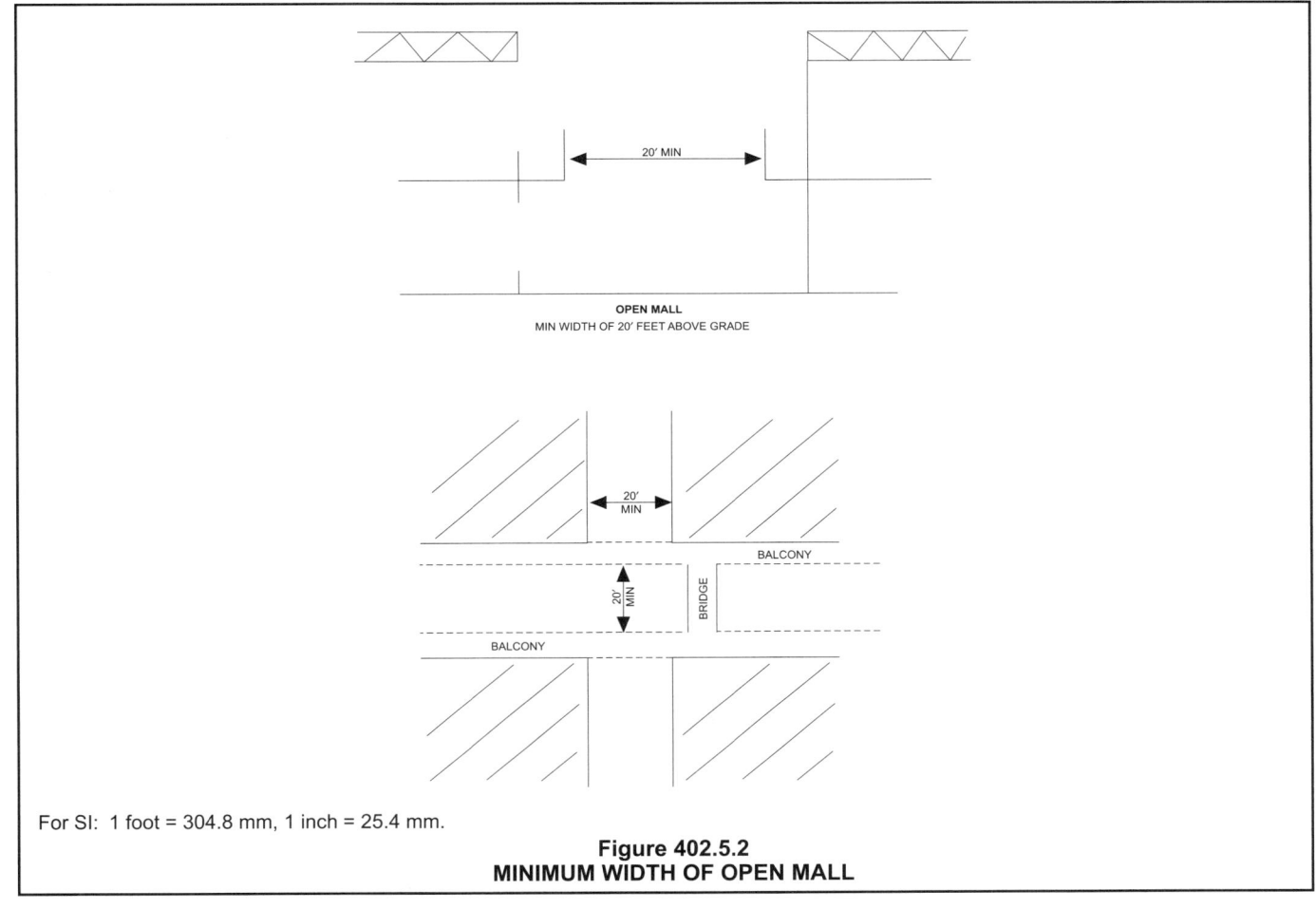

For SI: 1 foot = 304.8 mm, 1 inch = 25.4 mm.

Figure 402.5.2
MINIMUM WIDTH OF OPEN MALL

reduced open space of only 40 feet (12 192 mm). This reduction is permitted in these locations when the exterior wall in the required space area is rated for 3 hours and does not include more than 75 percent of the building perimeter. A significant majority of the uses described in Section 507 are found in the definition of a covered or open mall building. Therefore, it can be assumed that the fire load within a mall is similar to the fire load permitted in a code-compliant unlimited area building. Since the intent of the open space requirement is to prevent fire spread from one building to another and the fire loads are similar, allowing covered or open mall buildings to be subject to the same separation requirements as other unlimited area buildings does not reduce the level of fire protection afforded.

Exception 4 was added to keep this section aligned with the occupancies permitted to use the unlimited area provisions of Section 507.5. Group E, H, I and R occupancies are not prohibited from being within a covered or open mall building or an anchor store, but the exception would prohibit the use of reduced open space if one of these occupancies was a part of the covered or open mall building or anchor store.

402.7 Fire-resistance-rated separation. Fire-resistance-rated separation is not required between tenant spaces and the mall. Fire-resistance-rated separation is not required between a food court and adjacent tenant spaces or the mall.

❖ From an operational point of view, separating the tenant space from the mall is not practical in that customer flow would be restricted and, therefore, detract from the merchandising purpose of covered or open malls. Historical experience and reliability data on automatic sprinkler performance indicate that a separation between the tenant space and the mall is not warranted.

402.7.1 Attached garage. An attached garage for the storage of passenger vehicles having a capacity of not more than nine persons and *open parking garages* shall be considered as a separate building where it is separated from the *covered mall building* by not less than 2-hour *fire barriers* constructed in accordance with Section 707 or *horizontal assemblies* constructed in accordance with Section 712, or both.

> **Exception:** Where an *open parking garage* or enclosed parking garage is separated from the *covered mall building* or *anchor building* a distance greater than 10 feet (3048 mm), the provisions of Table 602 shall apply. Pedestrian walkways and tunnels that attach the *open parking garage* or enclosed parking garage to the *covered mall building* or *anchor building* shall be constructed in accordance with Section 3104.

❖ In accordance with Section 402.1, adjacent buildings may be considered separate buildings only if they are separated by exterior walls or fire walls. In recognition of the limited hazard presented by attached garages and open parking garages, this section permits such structures to be considered separate buildings, provided that a fire barrier or horizontal assemblies having a fire-resistance rating of at least 2 hours or both are provided. See the commentary to Section 707 relative to the construction of fire barriers and Section 712 relative to the construction of horizontal assemblies.

The 2-hour fire-resistance rating is consistent with the requirements of Table 508.4 for Groups M and S-2. If the 2-hour fire barrier or horizontal assembly is not provided, the parking structure can be considered part of the covered mall building and be limited. See the commentary to Section 3104 for the pedestrian walkways exception.

402.7.2 Tenant separations. Each tenant space shall be separated from other tenant spaces by a *fire partition* complying with Section 709. A tenant separation wall is not required between any tenant space and the mall.

❖ Covered mall buildings are essentially large unlimited area mixed-occupancy buildings. Rather than complying with Section 508.3 or 508.4, the building has a series of protections including the tenant separations specified in this section. In order to limit the spread of smoke, tenant separation walls are required to be fire partitions (see Section 709) having a fire-resistance rating of at least 1 hour and extending from the floor to the underside of the ceiling (see Figure 402.7.2). Extending tenant separations to the floor slab or roof deck above is not always practical or possible because of operation of the heating, ventilating and air-conditioning (HVAC) system. The effectiveness of the automatic sprinkler system is also a reason for not requiring tenant separations to extend above the ceiling, including attic spaces (see Section 709.4, Exception 4).

Again the open mall "building" provides a unique question regarding separation of tenant spaces. The open mall building may likely be a collection of separate structures treated as a one open mall "building." As such, the walls between the tenant spaces and the mall will be "interior walls" to the whole covered mall building complex, but they are also exterior walls as they separate interior from exterior environments. As part of a complex of buildings, the exterior walls of each individual tenant space facing the mall will not need to meet the wall and opening protection of Section 705, but need to comply with other exterior wall provisions of Chapter 14. The walls on the exterior face of the open mall "building" would be subject to the provisions of Section 705.

402.7.3 Anchor building separation. An *anchor building* shall be separated from the *covered mall building* by *fire walls* complying with Section 706.

> **Exception:** *Anchor buildings* of not more than three *stories above grade plane* that have an occupancy classification the same as that permitted for tenants of the *covered mall building* shall be separated by 2-hour fire-resistive *fire barriers* complying with Section 707.

❖ In general, anchor buildings are typically viewed as being separate from the covered or open mall building (see commentary, Sections 402.1 and 402.6). As separate buildings, the code requires that a fire wall complying with Section 706 be constructed. The exception to this section may be used in situations where the occupancy and size limitations of the anchor building are the same as what would be permitted for any other tenant within the covered mall building. In those situations, the code permits the separation to be constructed as a fire barrier instead of as a fire wall. The exception would not be applicable if the anchor building was over three stories above grade plane or if it contained an occupancy that was not normally permitted within the mall. An example would be a high-rise office building or a hotel that was connected with the covered mall building (see Figure 402.7.2 and commentary, Section 402.7.3.1).

For an open mall building where the anchor building is actually a separate building, the application of a fire wall may not be appropriate, but since it is considered a separate structure even under the covered mall scenario, it would be appropriate to have the anchor building provide exterior wall protection set in Table 602.

A growing trend in the development of covered mall and anchor buildings is the desire of the operators of the businesses in the anchor building to own the prop-

erty and the building of the anchor building. The original intent of the covered mall provisions was that this would apply to one property under the control of one owner. Therefore, the complex simply becomes a group of separate buildings, which eliminates the application of these provisions to the complex. Application of these mall provisions to a situation where the mall and anchor buildings are separate properties would be problematic, requiring special legal and technical considerations and treatment by the authority having jurisdiction.

402.7.3.1 Openings between anchor building and mall. Except for the separation between Group R-1 *sleeping units* and the mall, openings between *anchor buildings* of Type IA, IB, IIA and IIB construction and the mall need not be protected.

❖ As with tenant separations, openings between the anchor building and the pedestrian area of the mall need not be protected. Such separations would defeat the merchandising purpose of covered malls by restricting customer flow and visual access. While the openings may be unprotected, the separations between the anchor building and all tenant spaces in the covered mall building are to be constructed as required in Section 402.7.3 for a fire wall (see com-

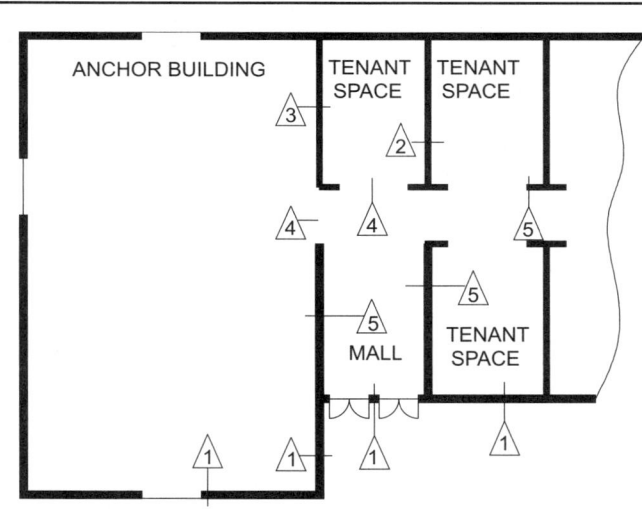

△1 WALL CONSTRUCTED AS REQUIRED FOR AN EXTERIOR WALL

△2 WALL CONSTRUCTED AS REQUIRED FOR A TENANT SEPARATION BY SECTION 402.7.2

△3 WALL CONSTRUCTED AS REQUIRED BY SECTION 402.7.3 (FIRE WALL OR FIRE BARRIER)

△4 UNPROTECTED OPENING

△5 WALL CONSTRUCTED AS REQUIRED BY SECTION 402.7

Figure 402.7.2
TENANT SPACE AND ANCHOR BUILDING SEPARATIONS

mentary, Section 706) or fire barrier (see commentary, Section 707) since the anchor building is considered a separate building (see Figure 402.7.2).

402.8 Interior finish. *Interior wall* and *ceiling finishes* within the mall and *exits* shall have a minimum *flame spread index* and smoke-developed index of Class B in accordance with Chapter 8. *Interior floor finishes* shall meet the requirements of Section 804.

❖ Previously, Sections 402 and 804 did not specifically state the requirements for interior finish within a covered mall building. Since a covered mall building is actually a mixed-use building, the most restrictive occupancy requirement should apply. The use of the mall has gone beyond its original intent of providing a wide-open walking space for people. Mall spaces have become locations to assemble to watch a dance recital, attend a fashion show, or for the local radio station to run a promotion. It has become clear that malls are assembly spaces themselves, not taking into account the tenant assembly spaces that open into them. Therefore, is seems logical to protect the mall spaces similar to Group A requirements.

[F] 402.9 Automatic sprinkler system. The *covered mall building* and buildings connected shall be equipped throughout with an *automatic sprinkler system* in accordance with Section 903.3.1.1, which shall comply with the following:

1. The *automatic sprinkler system* shall be complete and operative throughout occupied space in the *covered mall building* prior to occupancy of any of the tenant spaces. Unoccupied tenant spaces shall be similarly protected unless provided with *approved* alternative protection.

2. Sprinkler protection for the mall shall be independent from that provided for tenant spaces or anchors. Where tenant spaces are supplied by the same system, they shall be independently controlled.

Exception: An *automatic sprinkler system* shall not be required in spaces or areas of *open parking garages* constructed in accordance with Section 406.3.

❖ The covered or open mall building and connected buildings, such as anchor buildings, must be protected with an automatic sprinkler system to protect life and property effectively. As has been discussed throughout the section, numerous allowances (such as reduced tenant separations and elimination of area limitations) are based on the effectiveness of the automatic sprinkler system.

The sprinkler system is to be designed, installed, tested and maintained in accordance with Chapter 9, the IFC and NFPA 13. Additionally, the system must be installed such that any portion serving tenant spaces may be shut down independently without affecting the operation of the systems protecting the mall area. This special feature is in recognition of the frequent need to shut down the system so that changes can be made to it as a result of tenant improvements and modifications.

Section 909.12.3 requires operation of the sprinkler system to activate automatically the mechanical smoke control system (where an automatic control system is utilized). It is imperative that the zoning of the sprinkler system match the zoning of the smoke control system. This is necessary so that the area where water flow has occurred will also be the area from which smoke is removed.

[F] 402.9.1 Standpipe system. The *covered mall building* shall be equipped throughout with a standpipe system as required by Section 905.3.3.

❖ This section provides a reminder to the code user and a direct reference to the section that regulates the need for standpipes or hose connections for covered or open mall buildings (see commentary, Section 905.3.3).

402.10 Smoke control. Where a *covered mall building* contains an atrium, a smoke control system shall be provided in accordance with Section 404.5.

> **Exception:** A smoke control system is not required in *covered mall buildings* when an atrium connects only two stories.

❖ A smoke control system must be provided for all mall buildings that contain an atrium which connects more than two stories. An "Atrium" is defined as an opening connecting two or more stories other than enclosed stairways, elevators, hoistways, escalators, plumbing, electrical, air-conditioning or other equipment, which is closed at the top. Malls inherently contain unenclosed floor openings that may allow for the migration of smoke and hot gases resulting from a fire. See Sections 404.5 and 909 for a discussion of the requirements of the smoke control system, noting the exceptions listed for floor openings. Open malls and adjoining tenant spaces are not required to have a smoke control system because they are open at the top and would not meet the definition of an "Atrium." However, individual tenant spaces that contain an atrium and are part of an open mall "building" are not exempted from the smoke control requirement.

402.11 Kiosks. Kiosks and similar structures (temporary or permanent) shall meet the following requirements:

1. Combustible kiosks or other structures shall not be located within the mall unless constructed of any of the following materials:

   1.1. *Fire-retardant-treated wood* complying with Section 2303.2.

   1.2. Foam plastics having a maximum heat-release rate not greater than 100 kilowatts (105 Btu/h) when tested in accordance with the exhibit booth protocol in UL 1975.

   1.3. Aluminum composite material (ACM) having a *flame spread index* of not more than 25 and a smoke-developed index of not more than 450 when tested as an assembly in the maximum thickness intended for use in accordance with ASTM E 84 or UL 723.

2. Kiosks or similar structures located within the mall shall be provided with *approved* fire suppression detection devices.

3. The minimum horizontal separation between kiosks or groupings thereof and other structures within the mall shall be 20 feet (6096 mm).

4. Each kiosk or similar structure or groupings thereof shall have a maximum area of 300 square feet (28 m²).

❖ Other potential sources of combustibles within the mall area are kiosks and similar structures. As with the restrictions on plastic signs, in order to maintain the mall as a viable means of egress, the amount of combustibles within the mall must be minimized. The restriction on kiosk construction and location is also intended to minimize the potential for fire spread through the mall area. These restrictions apply to both permanent and temporary structures.

Kiosks and similar structures are permitted to be of noncombustible construction or may be of combustible construction where the provisions of Item 1 are followed. The code allows combustible kiosks to be made of fire-retardant-treated wood or certain foam plastics and aluminum composites that provide performance comparable to that of the fire-retardant-treated wood. As such, the amount of combustibles within the mall and the potential for fire spread through the mall are minimized.

If the kiosk or similar structure has a cover, the automatic sprinkler system within the covered mall building will not be able to control effectively a fire within the kiosk (or similar structure) in the early stages of development. Suppression and detection devices must, therefore, be installed within such structures when there is anything that shields the mall sprinkler system from areas within the kiosk.

In order to minimize a fire exposure hazard, kiosks and similar structures must be located at least 20 feet (6096 mm) from each other or be situated in groups that are appropriately separated from other kiosks. Although the structure itself is often noncombustible or of the limited types of acceptable combustible materials, it is recognized that combustibles may be displayed within the structure and, therefore, the separation minimizes the potential for fire spread from kiosk to kiosk. If a series of kiosks do not have 20 feet (6096 mm) separating them, they can be treated as a group and would be limited to 300 square feet (28 m²), as a group. The open space between individual kiosks in the group can be excluded from the aggregate area for determining the area of the group (see IBC Interpretation No. 11-07). Within a group of kiosks treated as one, there can be multiple "tenants" or one, and there are no specifics about the relative size of individual kiosks within the group (see IBC Interpretation Nos. 13-07 and 14-07). As required by Section 402.5.1, the kiosk must also be located at least 10 feet (3048 mm) from any projection of a tenant space.

To further control the use of kiosks and similar structures within the mall, their size is restricted to 300 square feet (28 m²) or less. This area restriction attempts to minimize the potential for a significant fire within the mall area. A kiosk larger than 300 square feet (28 m²) must be considered as a tenant space.

402.12 Children's playground structures. Structures intended as children's playgrounds that exceed 10 feet (3048 mm) in height and 150 square feet (14 m²) in area shall comply with Sections 402.12.1 through 402.12.4.

❖ One of the more recent additions to covered mall buildings provisions is dedicated play areas for children. Typically, the occasional "ride machine" has been replaced with jungle gyms, playhouses and activity areas. If these structures exceed 10 feet (3048 mm) in height and 150 square feet (14 m²) in area, then the provisions of Sections 402.12 through 402.12.4 must be followed. If a structure exceeds 10 feet (3048 mm), but has a footprint less than 300 square feet (28 m²); or if the structure is less than 10 feet (3048 mm) regardless of area covered, it is not covered by Section 402.12.

402.12.1 Materials. Children's playground structures shall be constructed of noncombustible materials or of combustible materials that comply with the following:

1. *Fire-retardant-treated wood.*

2. Light-transmitting plastics complying with Section 2606.

3. Foam plastics (including the pipe foam used in soft-contained play equipment structures) having a maximum heat-release rate not greater than 100 kilowatts when tested in accordance with UL 1975.

4. Aluminum composite material (ACM) meeting the requirements of Class A *interior finish* in accordance with Chapter 8 when tested as an assembly in the maximum thickness intended for use.

5. Textiles and films complying with the flame propagation performance criteria contained in NFPA 701.

6. Plastic materials used to construct rigid components of soft-contained play equipment structures (such as tubes, windows, panels, junction boxes, pipes, slides and decks) exhibiting a peak rate of heat release not exceeding 400 kW/m² when tested in accordance with ASTM E 1354 at an incident heat flux of 50 kW/m² in the horizontal orientation at a thickness of 6 mm.

7. Ball pool balls, used in soft-contained play equipment structures, having a maximum heat-release rate not greater than 100 kilowatts when tested in accordance with UL 1975. The minimum specimen test size shall be 36 inches by 36 inches (914 mm by 914 mm) by an average of 21 inches (533 mm) deep, and the balls shall be held in a box constructed of galvanized steel poultry netting wire mesh.

8. Foam plastics shall be covered by a fabric, coating or film meeting the flame propagation performance criteria of NFPA 701.

9. The floor covering placed under the children's playground structure shall exhibit a Class I *interior floor finish* classification, as described in Section 804, when tested in accordance with NFPA 253.

❖ As with other structures located within a covered mall building (such as kiosks), the amount of combustible materials must be rigidly limited. The requirements of Section 402.12.1 provide nine items of combustible materials that can be used to construct play equipment.

402.12.2 Fire protection. Children's playground structures located within the mall shall be provided with the same level of *approved* fire suppression and detection devices required for kiosks and similar structures.

❖ Playground structures within a mall must be provided with additional sprinkler heads to maintain the necessary protection.

402.12.3 Separation. Children's playground structures shall have a minimum horizontal separation from other structures within the mall of 20 feet (6090 mm).

❖ Combustible playground structures must be isolated from themselves, kiosks and the other mall construction by a distance of 20 feet (6090 mm). This allows for emergency personnel access and reduced fire spread.

402.12.4 Area limits. Children's playground structures shall not exceed 300 square feet (28 m^2) in area, unless a special investigation has demonstrated adequate fire safety.

❖ Playground structures are limited to the same 300-square-foot (28 m^2) area limitation as any other kiosk structure.

402.13 Security grilles and doors. Horizontal sliding or vertical security grilles or doors that are a part of a required *means of egress* shall conform to the following:

1. They shall remain in the full open position during the period of occupancy by the general public.

2. Doors or grilles shall not be brought to the closed position when there are 10 or more persons occupying spaces served by a single exit or 50 or more persons occupying spaces served by more than one exit.

3. The doors or grilles shall be openable from within without the use of any special knowledge or effort where the space is occupied.

4. Where two or more exits are required, not more than one-half of the exits shall be permitted to include either a horizontal sliding or vertical rolling grille or door.

❖ When improperly installed or when their use is not properly supervised, security grilles and doors can prohibit or delay egress beyond an acceptable time period. Such security devices are common in covered mall buildings. In every case, the grille must be openable to permit egress from the inside without the use of tools, keys or special knowledge or effort when the grille spans across a means of egress. For example, during business hours in a mercantile use, the grille must remain in its full, open position. The grille may be taken to a partially closed position when it provides the sole means of egress and no more than nine persons occupy the space. If a second means of egress is required, the grille may not be closed, as long as more than 49 persons occupy the space.

Because security grilles represent such a threat to prompt egress, the number of occupants exposed to the risk of a lack of supervision of the devices is limited. Except where a single means of egress is permitted, alternative means of egress must be available and not equipped with such devices. Security grilles must not be used on more than 50 percent of the exits.

[F] 402.14 Standby power. *Covered mall buildings* exceeding 50,000 square feet (4645 m^2) shall be provided with standby power systems that are capable of operating the emergency voice/alarm communication system.

❖ Covered mall buildings of a substantial size are required to have a standby power system for the emergency voice/alarm communication system. See Section 2702.2.14 of the commentary for further discussion of when a covered mall building exceeds 50,000 square feet (4645 m^2).

[F] 402.15 Emergency voice/alarm communication system. *Covered mall buildings* exceeding 50,000 square feet (4645 m^2) in total floor area shall be provided with an emergency voice/alarm communication system. Emergency voice/alarm communication systems serving a mall, required or otherwise, shall be accessible to the fire department. The system shall be provided in accordance with Section 907.5.2.2.

❖ An emergency voice/alarm communication system is required to allow the fire department full control during an emergency situation. Such a system is required when the covered mall building exceeds an aggregate area of 50,000 square feet (4645 m^2) (see commentary, Section 907.5.2.2).

402.16 Plastic signs. Plastic signs affixed to the storefront of any tenant space facing the mall shall be limited as specified in Sections 402.16.1 through 402.16.5.2.

❖ In order that the mall can be used as an exit access, combustible materials in the mall must be restricted; therefore, the size, area, location and amount of exposed plastic signs are restricted. This section restricts only plastic signs that are along the wall separating the tenant space from the mall (see Figure 402.16). If the sign housing and face panels are noncombustible or glass, there are no limitations. Light-transmitting plastic interior signs not located in the mall, such as those located within an individual tenant space, must comply with the requirements of Chapter 26.

402.16.1 Area. Plastic signs shall not exceed 20 percent of the wall area facing the mall.

❖ The signs located on the wall facing the mall must not exceed 20 percent of the wall area. The restrictions on size, coupled with the restrictions on location and height, are intended to prevent fire spread via the plastic signs.

402.16.2 Height and width. Plastic signs shall not exceed a height of 36 inches (914 mm), except that if the sign is vertical, the height shall not exceed 96 inches (2438 mm) and the width shall not exceed 36 inches (914 mm).

❖ Signs that are wider than they are high (horizontal panels) must not be more than 36 inches (914 mm) in height. The width of signs is controlled by the maximum area (see Section 402.16.1) and the location restrictions with respect to adjacent tenant spaces (see Section 402.16.3). Vertical signs must not be more than 36 inches (914 mm) in width and 96 inches (2438 mm) in height and are also subject to the area and location restrictions of other sections. The restrictions are intended to minimize horizontal and vertical fire spread through the mall area via the signs.

402.16.3 Location. Plastic signs shall be located a minimum distance of 18 inches (457 mm) from adjacent tenants.

❖ To minimize the potential for fire spread, plastic signs must be located at least 18 inches (457 mm) from adjacent tenant spaces.

402.16.4 Plastics other than foam plastics. Plastics other than foam plastics used in signs shall be light-transmitting plastics complying with Section 2606.4 or shall have a self-ignition temperature of 650°F (343°C) or greater when tested in accordance with ASTM D 1929, and a *flame spread index* not greater than 75 and smoke-developed index not greater than 450 when tested in the manner intended for use in accordance with ASTM E 84 or UL 723 or meet the acceptance criteria of Section 803.1.2.1 when tested in accordance with NFPA 286.

❖ To minimize the potential fire hazard, light-transmitting plastic signs located in a mall must comply with the specific material requirements of Section 2606.4 (see commentary, Section 2606.4) or must meet one of the alternative test methods that are permitted. Additionally, the backs and edges of the sign must be encased to reduce the likelihood of ignition and fire spread.

402.16.4.1 Encasement. Edges and backs of plastic signs in the mall shall be fully encased in metal.

❖ In order that the amount of exposed plastic materials in the mall is minimized, as well as to reduce the potential for ignition of a plastic panel, the edges and backs of plastic signs must be encased in metal. To conceptually understand this requirement, consider how easily a piece of paper ignites when a flame is held to its edge versus being held in the middle of the paper.

402.16.5 Foam plastics. Foam plastics used in signs shall have flame-retardant characteristics such that the sign has a maxi-

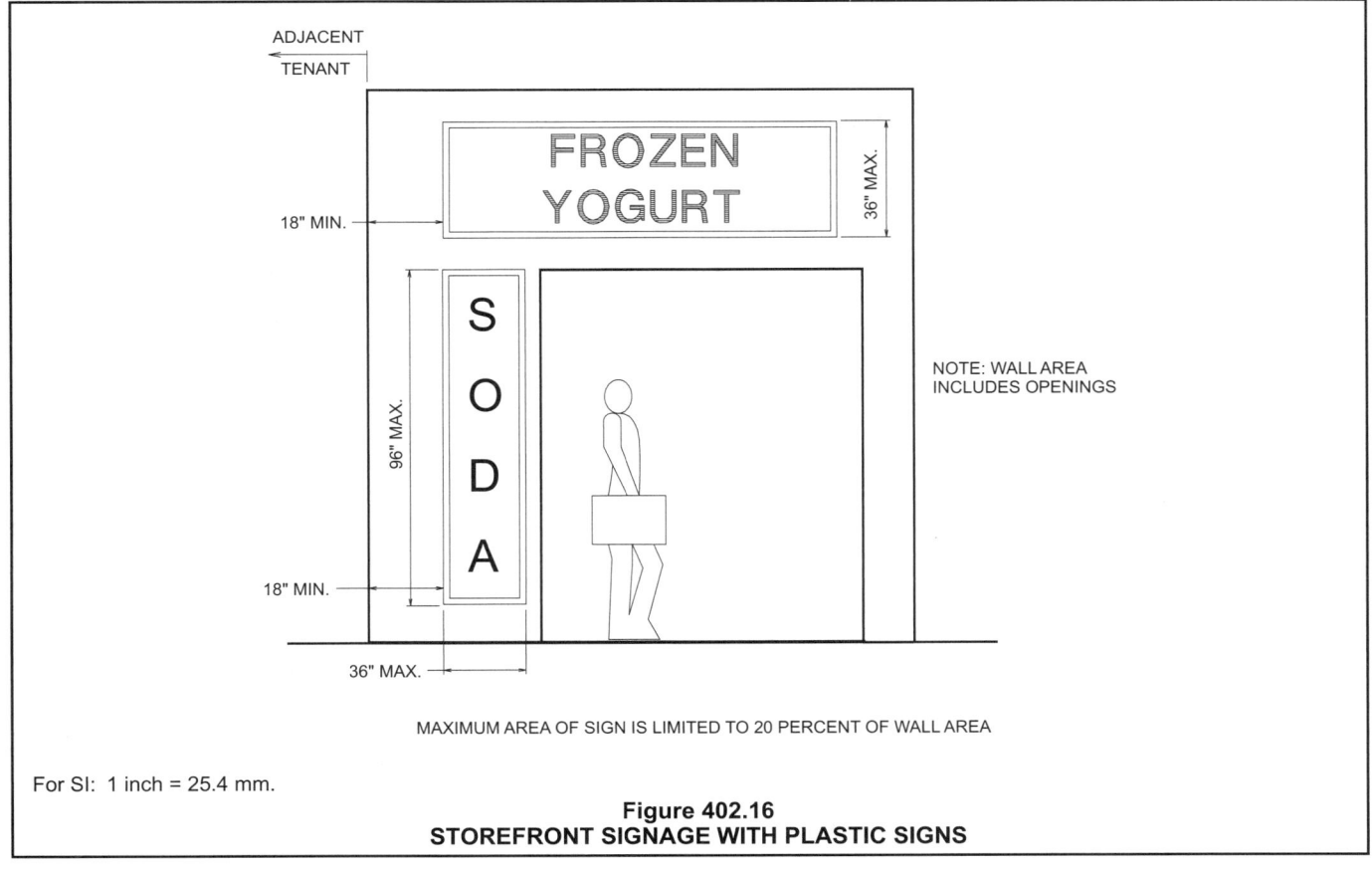

For SI: 1 inch = 25.4 mm.

Figure 402.16
STOREFRONT SIGNAGE WITH PLASTIC SIGNS

mum heat-release rate of 150 kilowatts when tested in accordance with UL 1975 and the foam plastics shall have the physical characteristics specified in this section. Foam plastics used in signs installed in accordance with Section 402.16 shall not be required to comply with the flame spread and smoke-developed indexes specified in Section 2603.3.

❖ Foam plastics that are used in signs located in a mall must comply with the provisions of Sections 402.16.5.1 and 402.16.5.2. The provisions of the code include specific material requirements for the foam plastics, such as the allowable heat-release rate, density and thickness permitted.

402.16.5.1 Density. The minimum density of foam plastics used in signs shall not be less than 20 pounds per cubic foot (pcf) (320 kg/m³).

❖ Foam plastics used in signs are limited to a minimum density of no less than 20 pounds per cubic foot (pcf) (320 kg/m³). This limitation lessens the likelihood that the sign will add a significant fuel load to the tenant space entrance.

402.16.5.2 Thickness. The thickness of foam plastic signs shall not be greater than $^1/_2$ inch (12.7 mm).

❖ The thickness of the foam plastic sign is also limited to lessen the likelihood that the sign will add a significant fuel load to the tenant space entrance.

[F] 402.17 Fire department access to equipment. Rooms or areas containing controls for air-conditioning systems, automatic fire-extinguishing systems or other detection, suppression or control elements shall be identified for use by the fire department.

❖ Consideration should be given to fire department access during an emergency. Fire department personnel may need to either determine that the fire protection systems are functioning properly or override the automatic features to manually activate or shut down a particular system. For this reason, the fire department must have access to controls for the air-conditioning and fire protection systems. The controls are to be clearly identified so that fire department personnel can properly operate them.

SECTION 403
HIGH-RISE BUILDINGS

403.1 Applicability. High-rise buildings shall comply with Sections 403.2 through 403.6.

Exception: The provisions of Sections 403.2 through 403.6 shall not apply to the following buildings and structures:

1. Airport traffic control towers in accordance with Section 412.3.

2. Open parking garages in accordance with Section 406.3.

3. Buildings with a Group A-5 occupancy in accordance with Section 303.1.

4. Special industrial occupancies in accordance with Section 503.1.1.

5. Buildings with a Group H-1, H-2 or H-3 occupancy in accordance with Section 415.

❖ "High-rise buildings" are defined in Chapter 2 as buildings with occupied floors located more than 75 feet (22 860 mm) above the lowest level of fire department vehicle access. Such buildings require special consideration relative to fire protection and occupant safety because of difficulties associated with smoke movement (stack effect), egress time, access by fire department personnel and perceived vulnerability to terrorist attack. This section contains provisions for high-rise buildings to address these special considerations.

While most of the provisions in Section 403 apply to all high-rise buildings, additional elevator requirements apply to high-rises in excess of 120 feet (36 576 mm). Further standards apply to high-rise buildings greater than 420 feet (128 m). These "super high-rises" must comply with enclosure structural integrity requirements (see Section 403.2.3), increased bond strength for sprayed fire-resistant materials (see Section 403.2.4), additional sprinkler risers (see Section 403.3.1) and have an additional exit stairway. However, these taller high-rises are not allowed to be designed with the reduced fire-resistance ratings permitted in Section 403.2.1.

The provisions of Section 403 apply to all high-rise buildings except those identified in the five exceptions. Exception 1 addresses airport control towers and is based on the limited fuel load and the limited number of persons occupying the tower (see Section 412.3). Places of outdoor assembly (Group A-5) and stand-alone open parking garages are exempted because of the free ventilation to the outside that exists in such structures. Low-hazard special industrial occupancies, which meet the criteria of Section 503.1.1, are exempt from the high-rise provisions. Finally, buildings with occupancies in Groups H-1, H-2 and H-3 are excluded from the provisions of this section because the fire hazard characteristics of such occupancies in a high-rise context have not yet been considered.

High-rise buildings must comply with all provisions of the code. Section 403 provides additional requirements for high-rise development.

- Section 403.2 specifies additional construction requirements, as well as some requirement reductions, applicable to constructing a high-rise building.

- Section 403.3 states the basic requirement for all high-rise buildings that an automatic sprinkler system be provided throughout the building. Specific standards unique to high-rise development are also provided.

- Section 403.4 specifies the various emergency detection and response systems that are required in a high-rise building.

- Section 403.5 provides additional means of egress system requirements for the occupants of a high-rise building.
- Section 403.6 provides elevator-related requirements for these structures.

The provisions are applicable to all buildings when the highest occupied floor is more than 75 feet (22 860 mm) above the lowest level of fire department vehicle access [see Figure 403.1(1)]. The lowest level of fire department vehicle access refers to the lowest ground level at which the fire department vehicle could be staged at the exterior of the building for carrying out fire-fighting operations. The definiton of a "High-rise building" comes from the International Symposium on Fire Safety in High-Rise Buildings sponsored by the General Services Administration (GSA). The definition developed at the symposium is as follows:

"A high-rise building is one in which emergency evacuation is not practical and in which fire must be fought internally because of height. The usual characteristics of such a building are:

A-1: Beyond the reach of fire department equipment;

A-2: Poses a potential for significant stack effect; and

A-3: Requires unreasonable evacuation time."

The 75-foot (22 860 mm) height threshold was determined from the effective reach of a 100-foot (30 480 mm) aerial apparatus based on the assumptions that the building will be set back from the curb, that access might be restricted by such things as parked cars and that obstructions, such as utility lines, would be present. The applicability of this section, however, is not based on the availability of such apparatus within the community.

Stack effect is illustrated in Figure 403.1(2), while typical evacuation times for buildings are shown in Figure 403.1(3).

403.2 Construction. The construction of high-rise buildings shall comply with the provisions of Sections 403.2.1 through 403.2.4.

❖ Section 403.2 includes subsections allowing the reduction of fire-resistance rating in some high-rise buildings, requiring minimum structural integrity for exit and elevator enclosures and setting bond strength for sprayed fire-resistant materials.

403.2.1 Reduction in fire-resistance rating. The *fire-resistance-rating* reductions listed in Sections 403.2.1.1 and 403.2.1.2 shall be allowed in buildings that have sprinkler control valves equipped with supervisory initiating devices and water-flow initiating devices for each floor.

❖ Since the overall reliability of the sprinkler system is greatly improved by the required control valves and water-flow devices, certain code modifications are allowed.

Sprinkler control valves with supervisory initiating devices and water-flow initiating devices must be provided for each floor to realize the reduction in rating. Annunciation by floor is intended to allow rapid identification of the fire location by the fire department. Sprinkler control valves on each floor are intended to permit

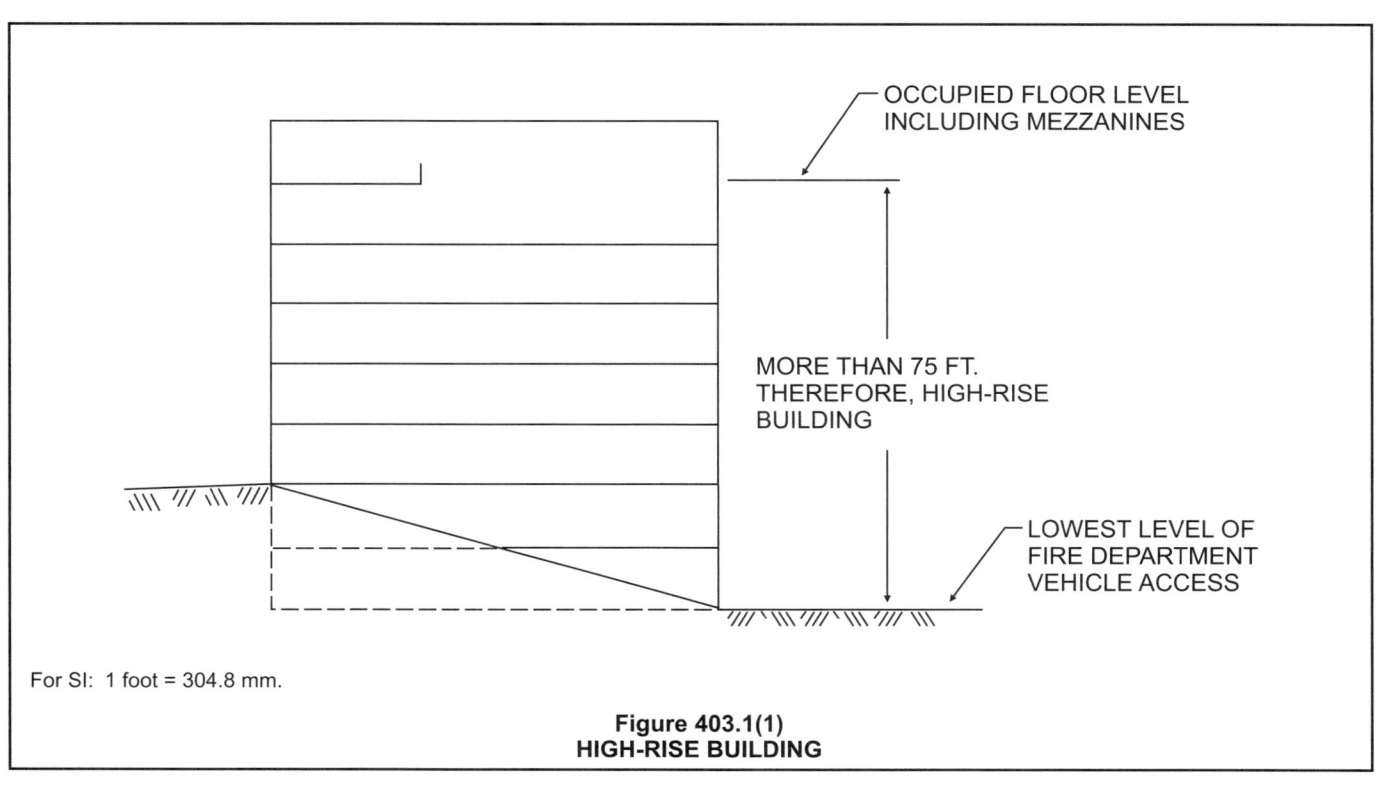

OCCUPIED FLOOR LEVEL
INCLUDING MEZZANINES

MORE THAN 75 FT.
THEREFORE, HIGH-RISE
BUILDING

LOWEST LEVEL OF
FIRE DEPARTMENT
VEHICLE ACCESS

For SI: 1 foot = 304.8 mm.

Figure 403.1(1)
HIGH-RISE BUILDING

servicing activated systems without impairing the water supply to large portions of the building. It is, therefore, necessary to equip the valves with supervisory initiating devices. NFPA 13 does not require annunciation per floor, since some systems may take on configurations that would not lend themselves to this type of zoning. In accordance with Section 102.4, the criteria for this section takes precedence over the permissive language of NFPA 13; therefore, in order to use the reductions allowed by this section, the sprinkler system must comply with these features. Without these sprinkler system enhancements, no reductions are permissible.

403.2.1.1 Type of construction. The following reductions in the minimum *fire-resistance rating* of the building elements in Table 601 shall be permitted as follows:

1. For buildings not greater than 420 feet (128 m) in *building height*, the *fire-resistance rating* of the building elements in Type IA construction shall be permitted to be reduced to the minimum *fire-resistance ratings* for the building elements in Type IB.

   Exception: The required *fire-resistance rating* of columns supporting floors shall not be permitted to be reduced.

2. In other than Group F-1, M and S-1 occupancies, the *fire-resistance rating* of the building elements in Type IB construction shall be permitted to be reduced to the *fire-resistance ratings* in Type IIA.

3. The *building height* and *building area* limitations of a building containing building elements with reduced *fire-resistance ratings* shall be permitted to be the same as the building without such reductions.

❖ This section specifies the allowed reductions of fire-resistance rating of building elements specified in

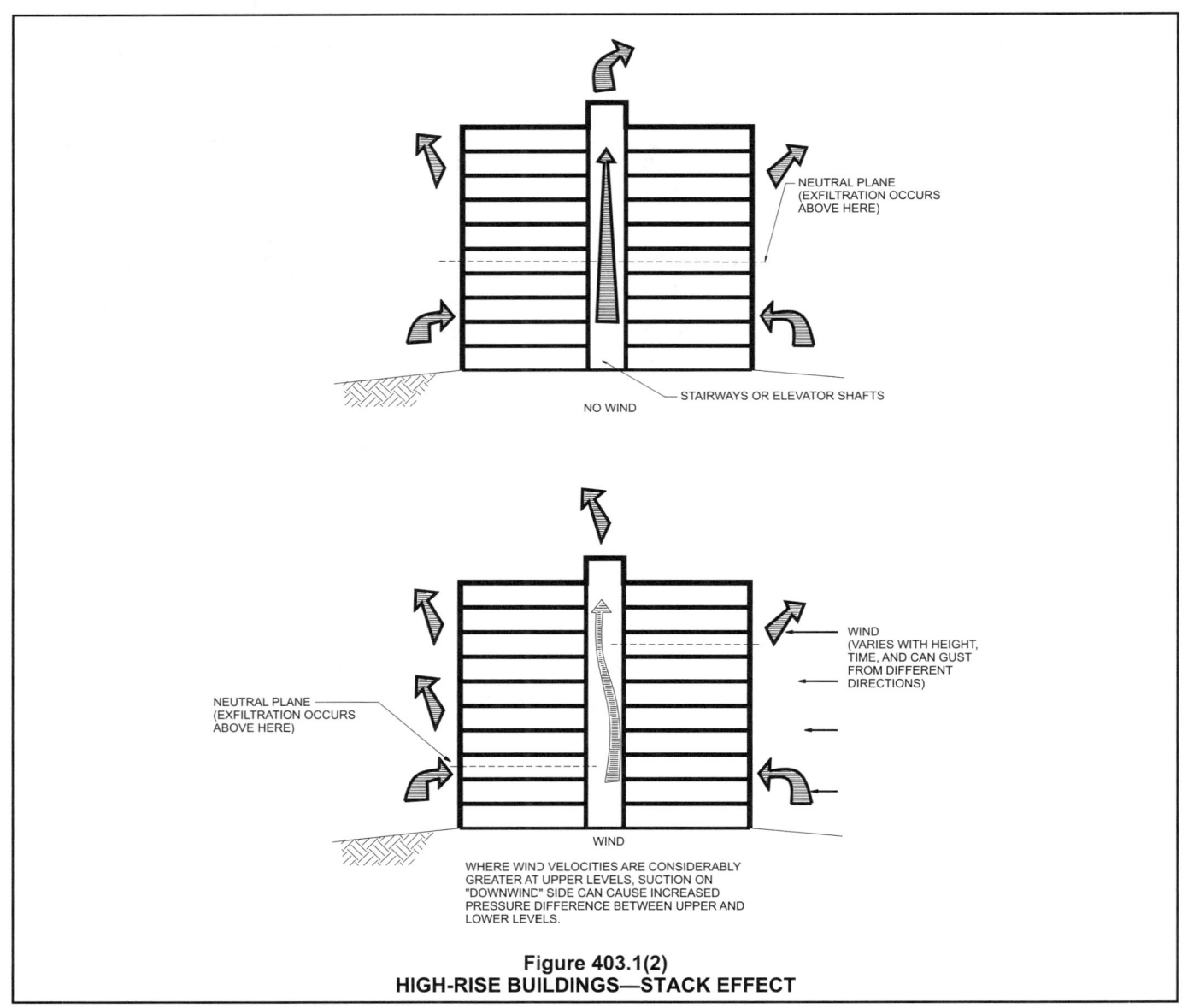

Figure 403.1(2)
HIGH-RISE BUILDINGS—STACK EFFECT

Table 601, without this being considered a reduction in the construction type.

- Item 1 allows high-rise buildings that must be of Type IA construction for building height and building area purposes to have building elements which comply with the fire-resistance rating of the Type IIB construction type as shown in Table 601. This reduction is not applicable to buildings with a height in excess of 420 feet (128 m) and does not apply to columns which are supporting floors in any building classified as a high-rise building.

- Item 2 allows high-rise buildings that must be of Type IB construction for building height and area purposes to have elements that comply with the fire-resistance rating of Type IIA construction as shown in Table 601. This reduction is not allowed if the building contains Group F-1, M or S-1 occupancies. This reduction is not allowed for these moderate-hazard buildings because of their customary higher fuel loads.

- Item 3 states that even though the fire-resistance ratings have been reduced to the equivalent of the next lower construction type, the construction type of the building is not considered reduced. A Type IB building with 1-hour fire-resistant protection of the structural frame is still a Type IB building.

In determining the minimum allowable type of construction for a high-rise building, Table 503 is applied in the usual fashion (see Section 503 and Table 503). Item 3 reminds the code user that even though reductions are taken, the allowable area for the building is that of the "original" type of construction and is not reduced because building elements have reduced fire protection. Section 403.2.1.1 is applied once the minimum type of construction has been determined.

For example, an eight-story high-rise office building is to be constructed and equipped with an automatic sprinkler system in accordance with Sections 403.2.1 and 403.3. The minimum allowable construction type is Type IB. In accordance with Section 403.2.1.1; however, the required fire-resistance ratings for the structural elements would be determined using the column for Type IIA construction in Tables 601 and 602. It should be noted that Table 503 does not permit the use of unprotected noncombustible construction in a high-rise building.

403.2.1.2 Shaft enclosures. For buildings not greater than 420 feet (128 m) in *building height*, the required *fire-resistance rating* of the *fire barriers* enclosing vertical shafts, other than *exit enclosures* and elevator hoistway enclosures, is permitted to be reduced to 1 hour where automatic sprinklers are installed within the shafts at the top and at alternate floor levels.

❖ As with previous sections, based on the effectiveness and reliability of a complete automatic sprinkler sys-

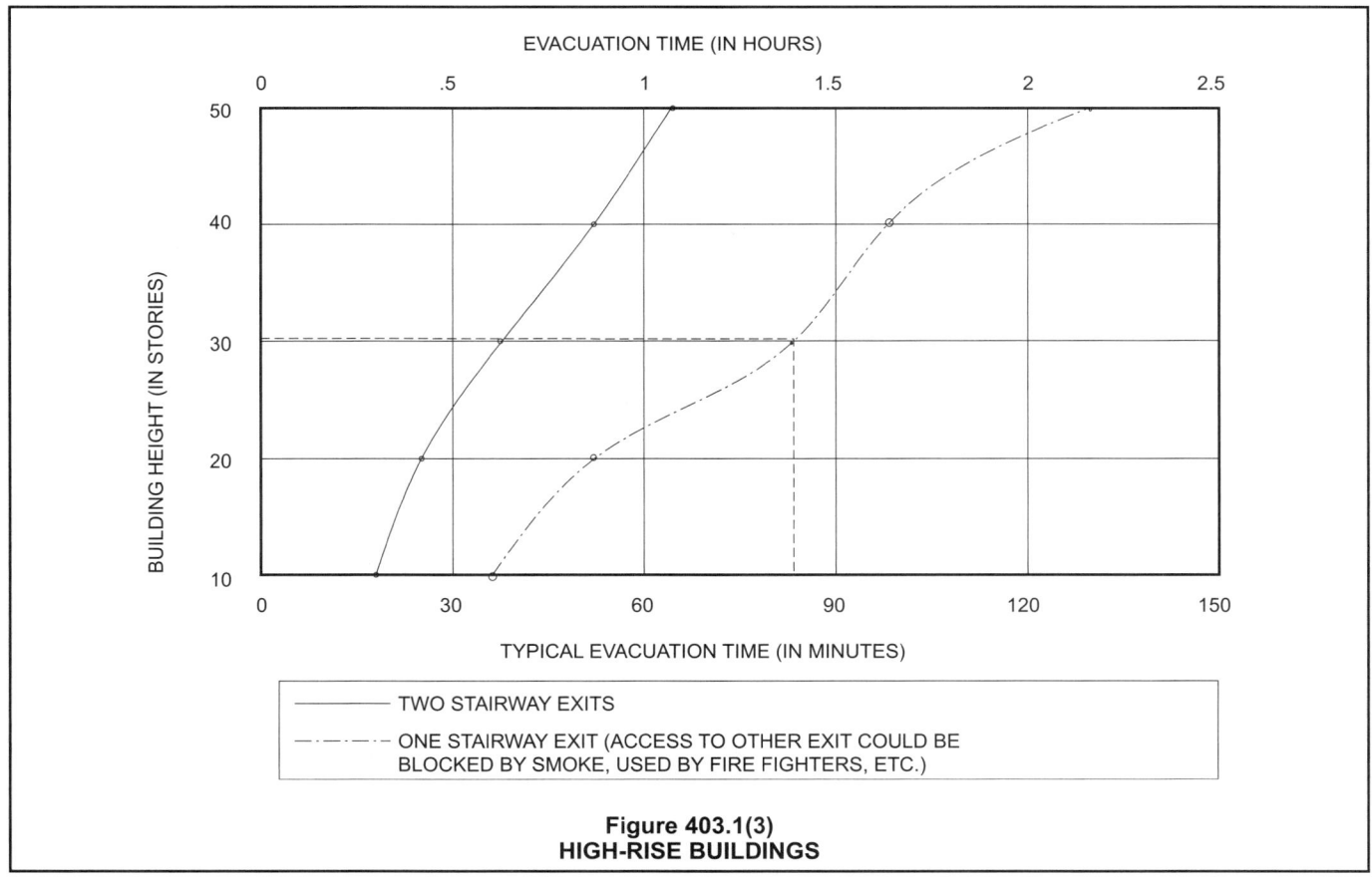

Figure 403.1(3)
HIGH-RISE BUILDINGS

tem in accordance with Sections 403.2.1 and 403.3, the fire-resistance rating of shaft enclosures may be reduced to 1 hour, provided the sprinklers are also installed within the shafts at the top and at alternate floor levels. This reduction is not applicable to shafts within buildings with a building height in excess of 420 feet (128 m).

The reduced fire-resistance rating also does not apply to shafts which are either exit enclosures or elevator hoistways (see also Section 403.2.3 for requirements applicable to these shafts.) The reduction does not apply to exit and elevator shafts since they may continue to be used for egress or for fire-fighter access during a fire; therefore, the integrity of the shaft must be maintained. If, for some reason, the fire penetrates the 1-hour shaft, it will affect conditions on other floors, hence, the need to maintain the protection for exit and elevator shafts. By the time the fire penetrates a 1-hour fire-resistance-rated shaft, the affected floors should be evacuated.

403.2.2 Seismic considerations. For seismic considerations, see Chapter 16.

❖ A reminder is provided in this section of the code that high-rise buildings are required to be designed for the effects of seismic activity. The reminder was primarily placed in Section 403.2.2, in addition to general design for seismic forces, to note that mechanical and electrical components are required to be designed for forces determined by Section 1613. This would also require that elevators be designed to resist the effects of the seismic forces. These elements are critical for the safety of high-rise buildings so that all of the required fire protection systems and the electrical components that secure them will be designed for the applicable seismic loads.

403.2.3 Structural integrity of exit enclosures and elevator hoistway enclosures. For high-rise buildings of occupancy category III or IV in accordance with Section 1604.5, and for all buildings that are more than 420 feet (128 m) in *building height*, *exit enclosures* and elevator hoistway enclosures shall comply with Sections 403.2.3.1 through 403.2.3.4.

❖ This section, new in the 2009 edition of the code, specifies a minimum standard for the integrity of the walls of exit enclosures and elevator hoistway enclosures. The requirement applies to all high-rises with a building height in excess of 420 feet (128 m). It also applies to any high-rise building, regardless of height, that is either a Class III or IV occupancy category as specified in Section 1604.5. Class IV occupancies are those that are considered to be essential in that their continuous use is needed, particularly in response to disasters. Occupancy Category III buildings include those occupancies that have relatively large numbers of occupants, where there is an elevated life safety hazard, or where the occupants' ability to respond to an emergency is limited. Similar standards were adopted by New York City in 2006 for application to all high-rise office buildings.

403.2.3.1 Wall assembly. The wall assemblies making up the *exit enclosures* and elevator hoistway enclosures shall meet or exceed Soft Body Impact Classification Level 2 as measured by the test method described in ASTM C 1629/C 1629M.

❖ The requirement applies to all wall assemblies of both vertical and horizontal exit enclosures, as well as elevator hoistway enclosures in the buildings specified in Section 403.2.3. The requirement applies regardless of the height or location of the particular enclosure. The standard applies to exit passageways that connect a vertical exit enclosure to the exterior of the building, as well as those that connect an "offset" in the vertical exit enclosures. The code relies on ASTM C 1629 and C 1629M for determining compliance with this requirement. The standard was developed specifically to test gypsum and fiber-reinforced cement panels; however, it can readily be used to test the impact resistance of other board and panel materials. For concrete and masonry walls, see Section 403.2.3.3. The code requires the entire wall assembly to withstand an impact of 195 pound force (867 N), as measured by the ASTM C 1629 Soft Body Impact Test. The test method used in ASTM C1629 is conducted in accordance with the ASTM E 695 test method, which covers the measurement of the relative resistance of wall, floor and roof construction to impact loading.

403.2.3.2 Wall assembly materials. The face of the wall assemblies making up the *exit enclosures* and elevator hoistway enclosures that are not exposed to the interior of the *exit enclosure* or elevator hoistway enclosure shall be constructed in accordance with one of the following methods:

1. The wall assembly shall incorporate not less than two layers of impact-resistant construction board each of which meets or exceeds Hard Body Impact Classification Level 2 as measured by the test method described in ASTM C 1629/C 1629M.

2. The wall assembly shall incorporate not less than one layer of impact-resistant construction material that meets or exceeds Hard Body Impact Classification Level 3 as measured by the test method described in ASTM C 1629/C 1629M.

3. The wall assembly incorporates multiple layers of any material, tested in tandem, that meet or exceed Hard Body Impact Classification Level 3 as measured by the test method described in ASTM C 1629/C 1629M.

❖ Section 403.2.3.2 requires that the face of the wall assembly that is not exposed to the shaft (or horizontal passageway)—the outside face—be protected by a material or materials that comply with a level of impact resistance. To comply with the requirement, at least two layers of Level 2 material or one layer of Level 3 material must be incorporated into the system. Level 2 material must withstand a Hard Body impact of 100 lbf (445 N) to comply with the standard. Level 3 material must withstand a Hard Body impact of 150 lbf (667 N). The code also permits the use of a system composed of multiple layers of different materials, provided the composite system can comply with a Level 3 Hard Body test.

403.2.3.3 Concrete and masonry walls. Concrete or masonry walls shall be deemed to satisfy the requirements of Sections 403.2.3.1 and 403.2.3.2.

❖ Concrete and masonry walls used for such enclosures are deemed to be adequate to meet the intent of this requirement without any testing under ASTM C 1629.

403.2.3.4 Other wall assemblies. Any other wall assembly that provides impact resistance equivalent to that required by Sections 403.2.3.1 and 403.2.3.2 for Hard Body Impact Classification Level 3, as measured by the test method described in ASTM C 1629/C 1629M, shall be permitted.

❖ This section allows other assemblies to be tested and approved for application in these walls. If a designer wished to include materials or a combination of materials not specified in Section 403.2.3.2 or 403.2.3.3, and wished to have the assembly tested, it would be acceptable as a wall for these enclosures if it complies with the Level 3 Hard Body test.

403.2.4 Sprayed fire-resistant materials (SFRM). The bond strength of the SFRM installed throughout the building shall be in accordance with Table 403.2.4.

❖ For buildings other than high-rise buildings, the minimum bond strength required for sprayed fire-resistant materials (SFRM) required by the code is 150 psf (667 N) when tested in accordance with ASTM E 736. Recommendation 6 of the National Institute of Standards and Technology (NIST) investigation of the World Trade Center (WTC) Report calls for improvement of the in-place performance of SFRM. For high-rises with a building height in the 75- to 420-foot (23 to 128 m) range, SFRM must have a bond strength of 430 psf (1913 N) wherever applied in the building. For high-rise buildings over 420 feet (128 m), the bond strength must be at least 1,000 psf (4448 N) wherever SFRM is applied in the building. While the NIST report was based on the specific events of the WTC fire and collapse, events far less dramatic can dislodge SFRM, such as elevator movement, building sway or maintenance activities. Using greater bond strengths will increase the probability that the protection will stay in place. These factors should provide for a longer time of safety.

TABLE 403.2.4
MINIMUM BOND STRENGTH

HEIGHT OF BUILDING[a]	SFRM MINIMUM BOND STRENGTH
Up to 420 feet	430 psf
Greater than 420 feet	1,000 psf

For SI: 1 foot = 304.8 mm, 1 pound per square foot (psf) = 0.0479 kW/m².
a. Above the lowest level of fire department vehicle access.

❖ See the commentary to Section 403.2.4.

[F] 403.3 Automatic sprinkler system. Buildings and structures shall be equipped throughout with an *automatic sprinkler*

system in accordance with Section 903.3.1.1 and a secondary water supply where required by Section 903.3.5.2.

Exception: An *automatic sprinkler system* shall not be required in spaces or areas of:

1. *Open parking garages* in accordance with Section 406.3.

2. Telecommunications equipment buildings used exclusively for telecommunications equipment, associated electrical power distribution equipment, batteries and standby engines, provided that those spaces or areas are equipped throughout with an automatic fire detection system in accordance with Section 907.2 and are separated from the remainder of the building by not less than 1-hour *fire barriers* constructed in accordance with Section 707 or not less than 2-hour *horizontal assemblies* constructed in accordance with Section 712, or both.

❖ Because of the difficulties associated with manual suppression of a fire in a high-rise building, all high-rise buildings regulated by this section are required to be protected throughout with an automatic sprinkler system. The sprinkler system is to be in accordance with Section 903.3.1.1, which requires compliance with NFPA 13. If a high-rise building contains an open parking garage along with Group B occupancies above, then the open parking garage portion of the building need not be protected with an automatic sprinkler system. See the commentary to Section 406.3 for a discussion of the fire hazards associated with open parking garages.

If a telecommunications equipment facility is part of a high-rise building, automatic sprinkler protection can be eliminated from certain areas if an automatic sprinkler fire detection system is provided and the telecommunication equipment areas are separated from other building areas with fire-resistance-rated construction of the specified ratings. This same exception is provided in Section 903.2, but is restated here to emphasize its application to all buildings, including high-rise buildings. This exception is based on the need to maintain uninterrupted operation of telecommunications equipment, which frequently includes emergency telephone and similar communication services. The presence of an automatic sprinkler system or other automatic fire suppression system may be detrimental to public safety should disruption of emergency services occur.

In order to increase the reliability of the sprinkler system should an earthquake disable the primary water supply, a secondary water supply is required for buildings at sites with the specified seismic design category listed in Section 903.3.5.2. The secondary water supply must be equal to the hydraulically calculated sprinkler demand and must have a duration of no less than 30 minutes.

[F] 403.3.1 Number of sprinkler risers and system design. Each sprinkler system zone in buildings that are more than 420 feet (128 m) in *building height* shall be supplied by a minimum of two risers. Each riser shall supply sprinklers on alternate floors. If more than two risers are provided for a zone, sprinklers on adjacent floors shall not be supplied from the same riser.

❖ For all high-rises with a building height in excess of 420 feet (128 m), this section requires the sprinkler system to be designed with additional risers for each sprinkler zone. This redundancy provides increased reliability of the automatic sprinkler system. At least two risers must be provided for each sprinkler zone. Recommendation 12 of the NIST WTC report called for the redundancy of active fire suppression systems to be increased to accommodate the greater risks associated with increasing building height and population. Providing two risers in each zone allows that if one riser is taken out of service, the other will be able to supply sprinklers on the floors above and below. This will impede any fire spread and allow the fire department time to respond and extinguish the fire.

[F] 403.3.1.1 Riser location. Sprinkler risers shall be placed in *exit enclosures* that are remotely located in accordance with Section 1015.2.

❖ This section requires the sprinkler risers to be located in the exit enclosures of the building which must be separated in accordance with the requirements of Section 1015.2. This separation reduces the possibility that one incident could incapacitate both risers and additionally provides the protection of the rated enclosures. While the section references the separation requirements of Chapter 10, Section 403.5.1 can result in an even greater separation of the exit enclosures in these taller high-rises. Also, the new Section 403.2.3 regarding the integrity of the exit enclosures is intended to make sure that the walls of the enclosures protecting the risers meet a minimum impact standard.

[F] 403.3.2 Water supply to required fire pumps. Required fire pumps shall be supplied by connections to a minimum of two water mains located in different streets. Separate supply piping shall be provided between each connection to the water main and the pumps. Each connection and the supply piping between the connection and the pumps shall be sized to supply the flow and pressure required for the pumps to operate.

> **Exception:** Two connections to the same main shall be permitted provided the main is valved such that an interruption can be isolated so that the water supply will continue without interruption through at least one of the connections.

❖ Fire pumps are installed in sprinkler and standpipe systems to pressurize the water supply for the minimum required sprinkler and standpipe operation. Fire pumps are only "required" to meet the system needs. Therefore, whether a particular high-rise building includes fire pumps will depend on interaction of the height of the building, the local water system and the designs of the sprinkler and standpipe systems in the building (see Section 913 of the IFC for more informa-

tion on fire pumps).

Where the systems require a fire pump, this section requires the fire pumps to be supplied by two water mains located in separate streets. Having two connections will greatly reduce the possibility of the loss of water due to a main break, given the valving which is a feature of a public water system. Each connection must be adequate to provide the flow and pressure needed for the fire pumps to operate. The requirement is not written to be limited in application; therefore, it is applicable to all high-rise buildings subject to Section 403. The exception is a performance-based provision that is not tied to any specific configuration.

403.4 Emergency systems. The detection, alarm and emergency systems of high-rise buildings shall comply with Sections 403.4.1 through 403.4.8.

❖ Section 403.4 lists the requirements for detection and emergency response systems applicable to high-rise buildings. These systems detect smoke and heat in these buildings and provide systems to notify occupants of the hazards and how to respond to the fire or other event. Systems and facilities are provided for emergency responders to assist the occupants and fight a fire. With the provisions of Section 403.6, high-rise buildings built under the 2009 edition of the code will provide greater safety for the occupants and emergency responders. Finally, the section provides requirements for standby and emergency power to keep vital systems operating in case of power loss, either during an event or power loss to the building.

[F] 403.4.1 Smoke detection. Smoke detection shall be provided in accordance with Section 907.2.13.1.

❖ Automatic smoke detectors are required in all high-rise buildings in locations so that a fire will be detected in its early stages of development. Smoke detectors must be installed in rooms that are not typically occupied, such as mechanical equipment rooms, elevator equipment rooms and similar spaces. Smoke detectors are also required in elevator lobbies. Finally, smoke detectors are required at various locations in the HVAC ducts. See Section 907.2.13 for the specific requirements for high-rise buildings.

[F] 403.4.2 Fire alarm systems. A fire alarm system shall be provided in accordance with Section 907.2.13.

❖ Fire alarm systems must be provided in high-rise buildings in accordance with the provisions of Section 907.2.13. Although a fire alarm system is not specifically listed as a requirement in Section 907.2.13, this section does require an automatic smoke detection system. Section 907.2.13.1.1 requires that the smoke detectors be connected to an automatic fire alarm system. Therefore, regardless of the occupancy category or categories in the high-rise building, a fire alarm system must comply with the code and NFPA 72, as applicable.

[F] 403.4.3 Emergency voice/alarm communication system. An emergency voice/alarm communication system shall be provided in accordance with Section 907.5.2.2.

❖ By definition, one characteristic of high-rise buildings is longer evacuation times. As such, the traditional fire alarm system, which usually results in simultaneous total building evacuation, is not practical; therefore, the alarm communication system should be able to:
- Direct the occupants of the fire zone to an area of refuge or exit;
- Notify the fire department of the existence of the fire; and
- Sound no alarms outside the fire zone until deemed desirable.

The alarm signal to the fire zone may include continuous sounding devices (bells, horns, chimes, etc.), as well as voice direction, which may momentarily silence the continuous sounding devices so as to be heard clearly. In accordance with Section 907.5.2.3, visible alarm notification appliances are required in the common and public areas of the high-rise buildings. Visible alarm appliances are also required in certain dwelling units and sleeping units, such as hotel guest rooms. The message of the voice/alarm is to be predetermined but need not be a recorded message. It would typically indicate that a fire has been reported at a specific location and that occupants should await further instructions or evacuate in accordance with the building's fire safety and evacuation plan. The voice/alarm signal will usually commence with a 3- to 10-second alert signal followed by the message. See the commentary to Section 907.5.2.2 for further requirements of the emergency voice/alarm communication system.

[F] 403.4.4 Emergency responder radio coverage. Emergency responder radio coverage shall be provided in accordance with Section 510 of the *International Fire Code*.

❖ High-rise buildings have posed a challenge to the traditional communication systems used by the fire service for fire-to-ground communications; therefore, emergency responder radio coverage complying with Section 510 of the IFC must be provided. The coverage is needed to coordinate with the fire department radio system to allow fireground officers to remain in communication with fire fighters working in various areas of a building.

As facilities grow larger and more complex, they become more challenging for effective fire response. Therefore, more fire protection features must be provided within the building. While modeling and other techniques may provide a good prediction as to whether a building will interfere with radio communications, the reality is that it is unknown if a building will need to have an enhanced radio system installed until after the building is constructed. The presumption is that high-rise buildings will more likely than not need to be provided with an emergency responder system. Section 510.1 of the IFC does offer two exceptions

which may be considered by the building official and fire official.

[F] 403.4.5 Fire command. A fire command center complying with Section 911 shall be provided in a location *approved* by the fire department.

❖ Fireground operations usually involve establishing an incident command post where the fire official can observe what is happening, control arriving personnel and equipment, and direct resources and fire-fighting operations effectively. Because of the difficulties in controlling a fire in a high-rise building, a separate room (enclosed in 1-hour-rated construction) within the building must be established as a fire command center. The room must be provided at a location that is acceptable to the fire department, usually along the front of the building or near the main entrance. The room must contain equipment necessary to monitor or control fire protection and other building service systems (see Section 911 for further information).

403.4.6 Smoke removal. To facilitate smoke removal in post-fire salvage and overhaul operations, buildings and structures shall be equipped with natural or mechanical ventilation for removal of products of combustion in accordance with one of the following:

1. Easily identifiable, manually operable windows or panels shall be distributed around the perimeter of each floor at not more than 50-foot (15 240 mm) intervals. The area of operable windows or panels shall not be less than 40 square feet (3.7 m^2) per 50 linear feet (15 240 mm) of perimeter.

   Exceptions:

   1. In Group R-1 occupancies, each *sleeping unit* or suite having an *exterior wall* shall be permitted to be provided with 2 square feet (0.19 m^2) of venting area in lieu of the area specified in Item 1.

   2. Windows shall be permitted to be fixed provided that glazing can be cleared by fire fighters.

2. Mechanical air-handling equipment providing one exhaust air change every 15 minutes for the area involved. Return and exhaust air shall be moved directly to the outside without recirculation to other portions of the building.

3. Any other *approved* design that will produce equivalent results.

❖ Section 403.4.6 requires new high-rise buildings constructed under the 2009 edition of the code to be provided with a smoke removal system to exhaust products of combustion following a fire incident. The intent of the requirement as clearly specified in the code is to facilitate the removal of smoke during the salvage and overhaul of a building after a fire has ended. "Overhaul and salvage operations" are generally understood to include searching the fire scene to detect and extinguish hidden fires or "hot spots." They also include controlling additional losses, stabilizing the incident scene by providing for fire-fighter safety and securing the structure. The smoke removal system is not a smoke control system similar to those required for atriums or underground buildings. The smoke removal system in a high-rise is not intended for use during a fire event. It is not intended to serve any health and life safety function, and, therefore, is only required to operate after, and not during, a fire event. In comparison, a smoke control system for an atrium is a life safety function and is designed to operate during a fire event to control smoke that is generated by the fire and keep it from migrating to other portions of the structure. This is not a smoke control system required elsewhere in the code; therefore, Section 909 is not referenced and is not applicable.

A smoke removal system can be a whole building system or a floor-by-floor localized system. It can also be provided by either natural means or a mechanical system. When a mechanical ventilation system is utilized, the controls for the smoke removal system are allowed to be independent of the building's HVAC system, or it may be an integral part of that or other ventilation systems present in the building. The technical requirements for smoke removal design for high-rise buildings are included in Section 403.4.6. Two primary options for accomplishing post-fire smoke removal are provided.

1. Natural Ventilation System. This design option relies on reasonably spaced and sized openings to allow air movement through a story. This is accomplished by providing operable windows or panels of a minimum size and spacing around the exterior of each story.

    There are two exceptions to this design option. Exception 1 applies to Group R-1 occupancy hotels. It permits each guestroom or suite to have a single 2-square-foot (0.19 m²) operable window or panel in lieu of the 40 square feet (3.7 m²). Exception 2 allows fixed windows to be used in lieu of operable windows or panels, if the glazing can be cleared by fire fighters. Traditional tempered glass may be an appropriate type of glazing for this exception provided it is not coated, or has an applied film that modifies its natural breaking characteristics. Designers should work with the building official and the fire department to determine what type of glazing might meet the intent of this exception.

2. Mechanical Ventilation System. This design option requires a mechanical air-handling system capable of exhausting the equivalent of one air change every 15 minutes. The code requires both the air being exhausted from the building and any return air that might be coming from the fire salvage areas to be exhausted outside of the building. When using the mechanical ventilation option, the design will most likely result in the use of dampers to zone floors and to introduce fresh air.

Since the smoke removal system is meant to be used after a fire for cleanup and recovery and not during a fire event, it is not necessary to connect this system to standby or emergency power.

In Item 3 of this section, a third design option is given. It simply states that any system that can provide results similar to the prescribed natural or mechanical systems in Item 1 or 2 can be used when such alternative has been approved by the building official. This provision restates the general code allowance for alternative methods permitted by Section 104.11.

[F] 403.4.7 Standby power. A standby power system complying with Chapter 27 shall be provided for standby power loads specified in Section 403.4.7.2.

❖ Standby power is required to increase the probability that the fire protection systems and elevators will continue to function in the event of failure of normal building service. By referencing Chapter 27, these provisions pick up the requirements of not only the *National Electrical Code* (NEC®), but also NFPA 110 and 111.

[F] 403.4.7.1 Special requirements for standby power systems. If the standby system is a generator set inside a building, the system shall be located in a separate room enclosed with 2-hour *fire barriers* constructed in accordance with Section 707 or *horizontal assemblies* constructed in accordance with Section 712, or both. System supervision with manual start and transfer features shall be provided at the fire command center.

❖ Standby power systems are intended to supply power automatically to selected loads in the event of failure of the normal power source. The system is to be capable of supplying the connected loads within 60 seconds of normal power failure. If the standby power system is a generator set in the building, the generator must be located in a room that is separated from the rest of the buildings by fire barriers or horizontal assemblies having a fire-resistance rating of at least 2 hours. The purpose of the enclosure is to decrease the probability that a fire can affect both the normal and standby power systems. It is important to note that this requirement is more restrictive than the provisions found in the NEC. The NEC would only require either the 2-hour fire protection or a sprinkler system, but not both. Because high-rise buildings are required to be sprinklered, these standby systems end up with both the sprinkler protection and the separation. Although not specifically stated, consideration should also be given to the routing of the conduit to minimize the impact of a single fire on both primary and standby electrical systems. The provisions of Chapter 27 are applicable to standby power systems.

[F] 403.4.7.2 Standby power loads. The following are classified as standby power loads:

1. Power and lighting for the fire command center required by Section 403.4.5;

2. Ventilation and automatic fire detection equipment for smokeproof enclosures; and

3. Standby power shall be provided for elevators in accordance with Sections 1007.4, 3003, 3007 and 3008.

❖ All fire protection and emergency systems required in Sections 403.4.5, 403.5.4 and 403.6 must be connected to the standby power system.

Elevators in a high-rise must be provided with emergency power. In addition, elevators provided for fire service access, for occupant self-evacuation or for accessible means of egress are also required to be provided with emergency power. Where fire service access elevators are required by Section 403.6.1, standby power is required for such elevators by Section 3007. Where occupant evacuation elevators are provided under Section 403.6.2, standby power is required for these elevators by Section 3008. Where elevators are being used as an accessible means of egress, they shall be provided with emergency power as specified in Section 1007 (see the commentaries of those respective sections).

When standby power is required for elevators, Section 3003 describes how it is to be provided. Under Section 3003, an arrangement must be made such that any an elevator may be connected to the standby power with elevator designated as the primary recipient of power once the standby system is activated. The primary elevator must be capable of serving all floors of a building. In extremely tall buildings, it may, therefore, be necessary to designate multiple elevators as being primary recipients in order to permit usage of both local and express elevators. Based on language in the NEC, the capacity of the standby system must be such that all equipment that has to be operational at the same time will be able to function. The system need not, however, be capable of supplying the load for all connected equipment simultaneously if automatic load shedding is provided. For example, if multiple elevators are connected to the system, the standby power system need only be capable of supplying one elevator such that the elevator provides access to all floors and that central controls restrict operation of other elevators at the same time. The intent of the provisions of Section 3003 is not only to keep one elevator in operation, but also to make sure that all elevators at least initially make it to the designated level and allow passengers to exit the elevators (see commentary, Section 3003).

Note that the requirements for elevators designated as fire service access elevators or elevators used for occupant evacuation have different and generally more restrictive requirements for standby power (see commentary, Sections 3007.7 and 3008.15).

See also the commentary to Section 403.5.3 regarding standby power for the release of stairway door locks.

[F] 403.4.8 Emergency power systems. An emergency power system complying with Chapter 27 shall be provided for emergency power loads specified in Section 403.4.8.1.

❖ Emergency power is required to ensure that fire protection systems and means of egress illumination systems will continue to function with only a very minor interruption in the event of a failure of normal building service. By referencing Chapter 27, these provisions pick up not only the requirements of the NEC, but also NFPA 110 and 111.

[F] 403.4.8.1 Emergency power loads. The following are classified as emergency power loads:

1. Exit signs and *means of egress* illumination required by Chapter 10;

2. Elevator car lighting;

3. Emergency voice/alarm communications systems;

4. Automatic fire detection systems;

5. Fire alarm systems; and

6. Electrically powered fire pumps.

❖ The illumination of exit signs and the means of egress as required by Sections 1006 and 1011, elevator car lighting, alarm and detection systems and electrically powered fire pumps must be considered emergency electrical systems. As such, the systems are to comply with the provisions of Chapter 27, the NEC and the referenced standards. Because of the critical nature of these systems and the potential for panic in the areas covered, the electrical load of these systems must be picked up within 10 seconds after failure of the normal power supply.

403.5 Means of egress and evacuation. The *means of egress* in high-rise buildings shall comply with Sections 403.5.1 through 403.5.6.

❖ Section 403.5 lists the requirements for means of egress systems applicable to high-rise buildings. Because of the size and height of high-rise buildings, they will contain large occupant loads which, when a building is evacuated, have longer evacuation times then low-rise buildings. Therefore, additional standards for egress systems are imposed for high-rise buildings. These include greater separation of the exit stairways in all high-rise buildings, requiring most stairways to be smokeproof enclosures, and for the tall high-rises over 420 feet (128 m) an extra stairway, in addition to those provided for the occupant load, is required. Luminous egress path markings, a new requirement in the 2009 edition of the code, are required for both vertical and horizontal exit enclosures. Section 403.5.3 addresses the practice of locking doors of an exit enclosure.

403.5.1 Remoteness of exit stairway enclosures. The required *exit stairway* enclosures shall be separated by a distance not less than 30 feet (9144 mm) or not less than one-fourth of the length of the maximum overall diagonal dimension of the building or area to be served, whichever is less. The distance shall be measured in a straight line between the nearest points of the *exit stairway* enclosures. In buildings with three or more *exit stairway* enclosures, at least two of the *exit stairway* enclosures shall comply with this section. Interlocking or *scissor stairs* shall be counted as one *exit stairway*.

❖ This section, new to the 2009 edition of the code, requires that two of the exit stairway enclosures be separated from each other by a minimum distance. Requiring a minimum separation, even as little as 30 feet (9144 mm), reduces the probability that a fire, which damages one enclosure, will also damage the second enclosure. This separation requirement must be met, as well as the separation standards of Section 1015.2. If the design of a building under Section 1015.2 results in two enclosures being separated by 30 feet (9144 mm) or more, then Section 403.5.1 is also satisfied. While Section 1015.2 requires that two means of egress be separated by at least one-half or one-third of the diagonal measurement of the area served, the distance is measured between the doors into those enclosures. Depending on the size and configuration of the building, two enclosures could directly adjoin each other and still comply with Section 1015. For high-rise buildings, Section 403.5.1 requires that the enclosures at all points be completely separated by the distance prescribed (see Figure 403.5.1).

Section 403.5.1 applies to all high-rise buildings subject to Section 403, regardless of the height of the building above 75 feet (22 860 mm). The requirement is either 30 feet (9144 mm), or one-fourth of the diagonal measurement of the area served, whichever is less. This is the same diagonal that is measured for compliance with Section 1015.2. For any building with a diagonal exceeding 120 feet (36 576 mm), the enclosures must be at least 30 feet (9144 mm) apart at their closest points. Only where the diagonal is less than 120 feet (36 576 mm) will the one-fourth of the diagonal come into play. For example, a building with a 100-foot (30 480 mm) maximum diagonal would only need 25 feet (7620 mm) between the two enclosures. In larger buildings, with an occupant load per story in excess of 500, a third stairway is required. Also in buildings in excess of 420 feet (128 m), an additional stairway is required. The separation required by Section 403.5.1 only applies to two of the stairway enclosures. Any required third or fourth stairway could be directly adjacent to one of the first two enclosures.

Interlocking stairway enclosures (two stairway enclosures "wrap" around each other as they descend through the building) or scissor stairways (two stairways in the same enclosure) have to be considered as one stairway for the purposes of this section. Clearly a scissor stairway is not two enclosures, and therefore would not meet the needs of this requirement. Simi-

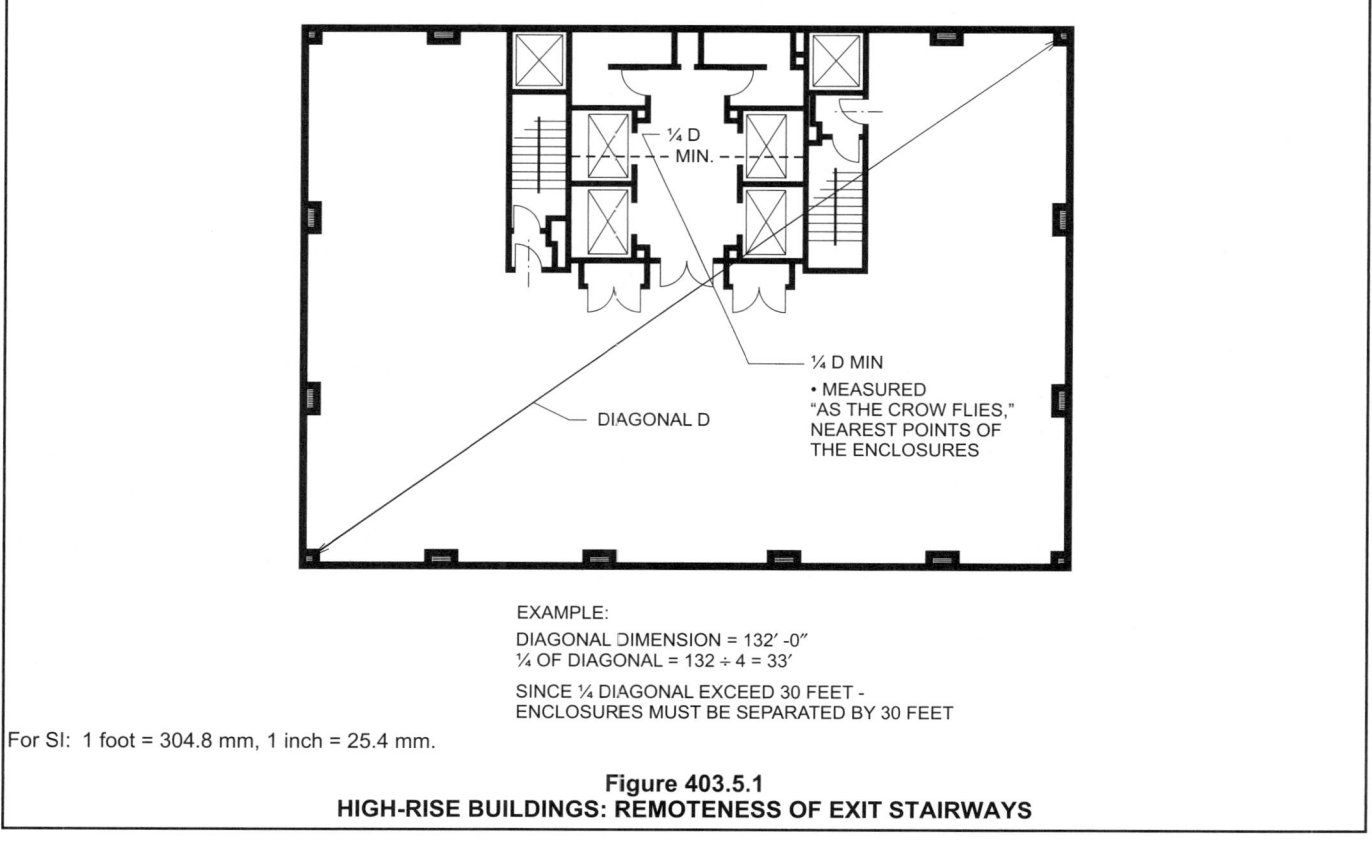

EXAMPLE:

DIAGONAL DIMENSION = 132′ -0″
¼ OF DIAGONAL = 132 ÷ 4 = 33′

SINCE ¼ DIAGONAL EXCEED 30 FEET -
ENCLOSURES MUST BE SEPARATED BY 30 FEET

For SI: 1 foot = 304.8 mm, 1 inch = 25.4 mm.

Figure 403.5.1
HIGH-RISE BUILDINGS: REMOTENESS OF EXIT STAIRWAYS

larly, interlocking enclosures often share a common wall with an enclosure on each side. Obviously, two stairway enclosures sharing a common enclosure wall would not meet the minimum separation requirement.

403.5.2 Additional exit stairway. For buildings other than Group R-2 that are more than 420 feet (128 m) in *building height*, one additional *exit stairway* meeting the requirements of Sections 1009 and 1022 shall be provided in addition to the minimum number of *exits* required by Section 1021.1. The total width of any combination of remaining *exit stairways* with one *exit stairway* removed shall not be less than the total width required by Section 1005.1. *Scissor stairs* shall not be considered the additional *exit stairway* required by this section.

> **Exception:** An additional *exit stairway* shall not be required to be installed in buildings having elevators used for occupant self-evacuation in accordance with Section 3008.

❖ For buildings in excess of 420 feet (128 m) in height, this provision requires an extra stairway to be provided in addition to the number of stairways (means of egress) required based on occupant load of the building as well as travel distance (see Sections 1015 and 1016). This requirement does not apply to high-rise buildings that are occupied as Group R-2 apartments or condominiums, but would apply to R-1 hotels. If a building is a mixed-occupancy structure, the building official should consider which occupancies are located above the 420-foot (128 m) level.

The intent of this provision is to accommodate a simultaneous occurance of evacuation of the occupants of a high-rise building and fire-fighting operations. The traditional approach to emergencies in a high-rise is to allow people on a floor affected by a fire to move up or down in the building to another story that is relatively safe from the effects of the fire. Fire fighters will usually set up a staging point one or two stories below the fire and commandeer one of the two stairways to move up to the fire floor. If for some reason it becomes necessary to evacuate the building during active fire fighting, the capacity of the means of egress system can be cut in half. This section implements, in part, Recommendation 17 of the NIST report. The report states that buildings should be designed to accommodate timely full-building evacuation of occupants when required by building-specific or large-scale emergencies, such as widespread power outages, major earthquakes, tornadoes, hurricanes without sufficient advance warning, fires, explosions or terrorist attacks.

The code requires that the extra stairway be sized so that the loss of any one stairway will still result in the remaining stairways being of sufficient size to accommodate the occupant load served by the stairways. For example, in a high-rise office building that has a calculated occupant load of 250 on each story, two stairways each designed to accommodate 125 occupants will be the minimum requirement for exit enclosures in accordance with Chapter 10. The minimum

stairway width of 44 inches (1118 mm) is sufficient to accommodate 146 people, or slightly more than needed for this building. In this building a third stairway would therefore need to be provided that was also at least 44 inches (1118 m) in width. The result is that if the fire fighters need to commandeer stairway 1, stairways 2 and 3, which each have a capacity to serve 146 people, will still be large enough to meet the occupant load requirements for the means of egress system. In larger buildings with occupant loads in excess of 500, a third stairway is required by Section 1015. In such buildings, Section 403.5.2 would require a fourth exit stairway.

The "additional" stairway has to comply with all the same requirements of any stairway in a high-rise building, including a design complying with smokeproof enclosures required by Section 403.5.4.

The additional stairway is not affected by the remoteness requirement of Section 403.5.1. That section applies to the first two exit stairway enclosures required in a building. Additional stairways, whether required for occupant load, travel distance or this section, also do not need to comply with the remoteness requirement of Section 403.5.1.

Finally, the exception to this section permits the installation of occupant evacuation elevators as an alternative to providing the additional exit stairway (see also the commentary to Sections 403.6.2 and 3008). The intent behind the exception is a recognition that the reason behind the requirement for an additional stairway is the loss of egress capacity under the unusual circumstance of simultaneous total building evacuation and fire fighting. Recognizing that occupant evacuation elevators provide an alternative way to meet a portion of the evacuation needs in these circumstances, the code allows the elevators as a substitute for the additional capacity provided by a third or fourth stairway. Where occupant evacuation elevators are proposed, the requirements of Section 3008 apply to all passenger elevators for general public use in the building (see commentary, Section 3008).

403.5.3 Stairway door operation. Stairway doors other than the *exit discharge* doors shall be permitted to be locked from the stairway side. Stairway doors that are locked from the stairway side shall be capable of being unlocked simultaneously without unlatching upon a signal from the fire command center.

❖ Section 1008.1.9.10 requires that all egress doors for interior stairways be readily openable from both sides. It is often desirable to control movement of people within a building and to provide additional security from external threats. This section permits locking of stairway doors from the stair side when all doors are capable of being simultaneously unlocked. Since high-rise buildings are difficult to evacuate and people are often relocated to another floor level during an emergency, access from the stairway to a floor could be essential in a fire or other emergency. Therefore, all stairway doors that are to be locked from the stairway side must have the capability of being unlocked on a signal from the fire command center. The unlocking of the door must not negate the latching feature, which is essential to the operation of the door as a fire door. Section 403.4.7.2, Item 1, by its reference to Section 403.4.5, requires the locking feature to be connected to the standby power system. When the door is unlocked during an emergency, it should not automatically relock on closure. Electrically powered locks should be designed such that when power to the locking device is interrupted, the lock is released. This is intended to enable doors to be operable from the inside of the stairway, and not locked, if power to the lock is interrupted. The building official should review the emergency release operation of stairway doors to determine that they remain unlocked.

403.5.3.1 Stairway communication system. A telephone or other two-way communications system connected to an *approved* constantly attended station shall be provided at not less than every fifth floor in each *stairway* where the doors to the *stairway* are locked.

❖ If the stairway doors are locked to restrict reentry as permitted in Section 403.5.3, a two-way communication system must be provided at no less than every fifth floor and must be connected to the standby power system. This system is required to be connected to a constantly attended location, which could be within the building, or to a central station that monitors fire alarms and is manned 24 hours a day, 7 days a week. Use of the fire command center is not recommended, since it may not be constantly attended. The system will permit occupants in the stairway to notify the attended location that the stairway doors need to be unlocked to access another floor or because conditions in the stairway prevent its continued use.

403.5.4 Smokeproof exit enclosures. Every required level *exit stairway* serving floors more than 75 feet (22 860 mm) above the lowest level of fire department vehicle access shall comply with Sections 909.20 and 1022.9.

❖ This section serves as a reminder that the smokeproof enclosure provisions of Section 1022.9 are applicable to high-rise buildings. Section 909.20 provides the standards on the construction of a smokeproof enclosure. It is important to understand that this requirement is only applicable in the portions of the building where the stack effect is more of a factor and the resulting spread of smoke will be the greatest. For example, if the high-rise building consists of a tower portion with a larger area on the bottom that is only a couple of stories in height, these requirements would only be applicable to stairways in the tower portion that serve floors that are more than 75 feet (22 860 mm) above the level where fire department vehicles would be located. The requirement would not apply to stairways in a low-rise portion of a building not exceeding 75 feet (22 860 mm), nor to stairways within the tower which do not serve floors more than 75 feet (22 860 mm) above the lowest level of fire department access.

403.5.5 Luminous egress path markings. Luminous egress path markings shall be provided in accordance with Section 1024.

❖ This section requires the provision of luminous egress path markings in all high-rise buildings containing any Group A, B, E, I, M or R-1 occupancy. The specific requirements are found in Section 1024. The requirements apply only to exit enclosures, vertical or horizontal. The markings are not required in the exit access or exit discharge portions of the means of egress system, nor at a horizontal exit.

403.5.6 Emergency escape and rescue. Emergency escape and rescue openings required by Section 1029 are not required.

❖ This section of the code reinforces other tradeoffs or allowances permitted in the code for buildings equipped with automatic sprinkler systems. Section 1029.1, Exception 1, does not require emergency escape and rescue openings for a building protected with an automatic sprinkler system. In addition, high-rise buildings have additional safeguards, alarm, monitoring and communication systems which balance the lack of rescue opening. Further exterior rescue is not feasible once the building contains occupied floors more than 75 feet (22 860 mm) above the lowest level of fire department access because of the limited height of most fire-fighting apparatus.

403.6 Elevators. Elevator installation and operation in high-rise buildings shall comply with Chapter 30 and Sections 403.6.1 and 403.6.2.

❖ This section explicitly states that elevator installation and operation are to be in accordance with Chapter 30, which contains provisions for the design, construction, installation, operation and maintenance of elevators. While a vital purpose of the elevator requirements in high-rise buildings have always been for the use of the fire department, this role has been enhanced in the 2009 code (see Section 403.6.1). Elevators have been one method of providing an accessible means of egress under the requirements of Section 1007, but have never been considered part of the means of egress system for the nondisabled occupants of a building or for those with a disability to self-evacuate. However in the 2009 code, elevators are allowed to augment the egress system for all occupants in buildings over 420 feet (128 m) in height. The allowance for self-evacuation is based upon occupants using the elevators prior to the elevators being recalled. This is much different from the requirements of Section 1007, under which fire department personnel use the elevators to assist in the rescue of occupants while those elevators are on Phase II emergency operation.

Chapter 30 provides important standards for the size of elevators, allowing ambulance stretchers to easily fit into at least one elevator serving high-rise buildings. Protection of the hoistway and machine room are also addressed.

403.6.1 Fire service access elevator. In buildings with an occupied floor more than 120 feet (36 576 mm) above the lowest level of fire department vehicle access, a minimum of one fire service access elevator shall be provided in accordance with Section 3007.

❖ For buildings with occupied floors in excess of 120 feet (36 576 mm) above the lowest level of fire department vehicle access, at least one of the elevators must comply with the fire service access elevator requirements, which are detailed in Section 3007 (see Figure 403.6.1). The standards in Section 3007 are another result of the NIST research following the WTC fire and collapse. The fire service access standards work with already existing emergency operation requirements for elevators to allow at least one elevator to remain in service for trained fire fighters to reach the upper levels of a building within a reasonable amount of time and to stage their fire-fighting operations at a level below the actual fire. These elevators must be provided with a lobby of a minimum size which is protected by at least 1-hour fire-resistance-rated smoke barriers. The lobby must be directly connected to one of the exit enclosures which must contain a standpipe as required by Section 905. These lobbies may also serve as the area of refuge required as part of an accessible means of egress. People waiting for assistance in egress will need to be helped to a safe location before one lobby or another is used as the staging location for fire fighting.

403.6.2 Occupant evacuation elevators. Where installed in accordance with Section 3008, passenger elevators for general public use shall be permitted to be used for occupant self-evacuation.

❖ New to the 2009 code is Section 3008, which provides the requirements for occupant evacuation elevators. When part of a fire safety and evacuation plan, occupant evacuation elevators are allowed in a building. A key element of applying Section 3008 is that all passenger elevators for general public use in the building must comply. Similar to the fire service access elevators, the occupant evacuation elevators must be provided with a lobby constructed of 1-hour fire-resistance-rated smoke barriers. The occupant evacuation elevators also require signs and other indicators, as well as a communication system connected to the fire command center to help occupants use the elevators. When provided, the requirements of the occupant evacuation elevators make for a substantially improved accessible means of egress over the requirements provided in Section 1007. Section 1007 is strictly for fire-fighter-assisted evacuation versus occupant self-evacuation as Section 3008 allows.

For any high-rise building, elevators may be designed for occupant self-evacuation. In fact, Section 3008 could be applied to any building as long as the design complies with all of the requirements.

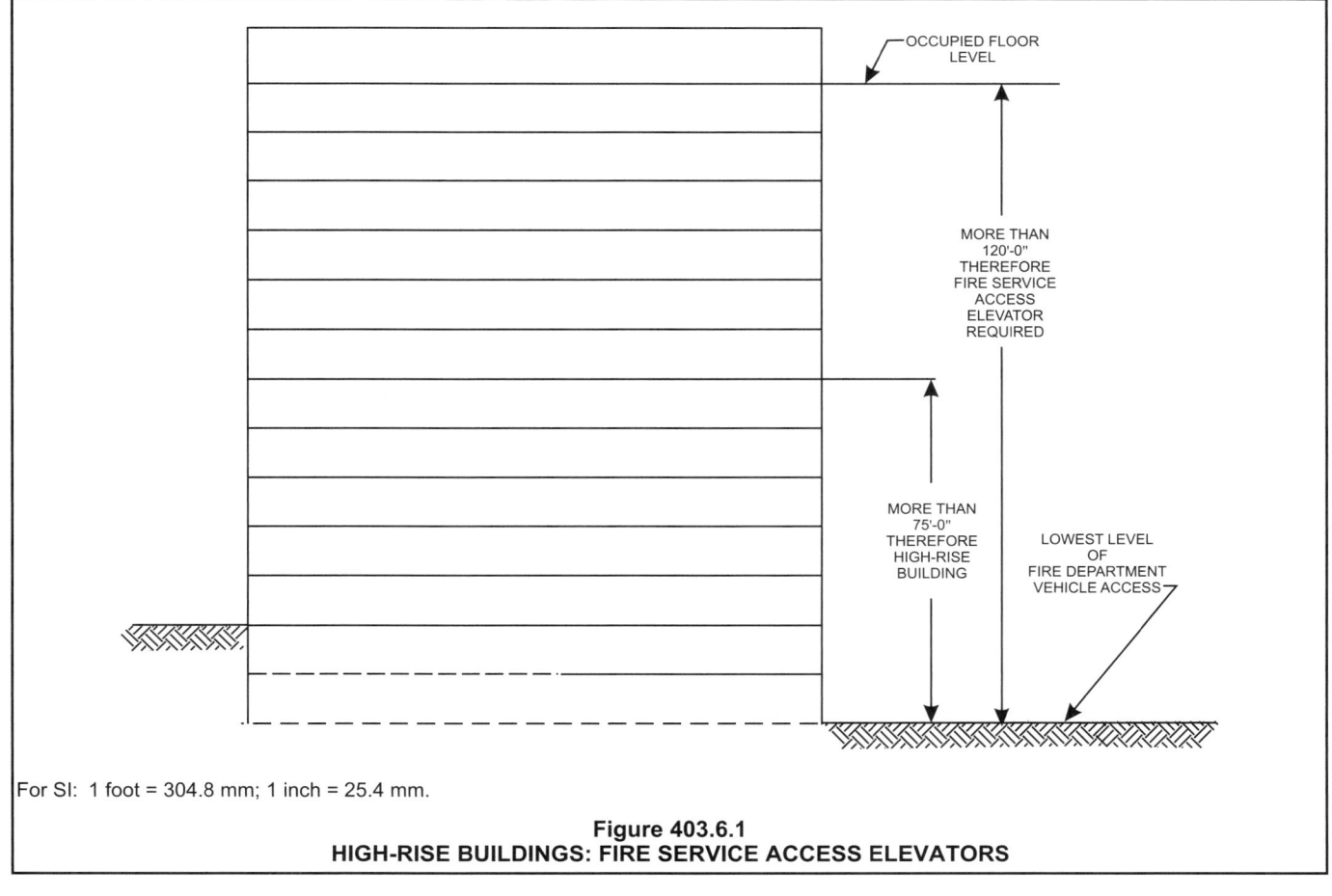

For SI: 1 foot = 304.8 mm; 1 inch = 25.4 mm.

Figure 403.6.1
HIGH-RISE BUILDINGS: FIRE SERVICE ACCESS ELEVATORS

Elevators are still prohibited by Section 1003.7 from serving as a means of egress for the general population of a building. However, in high-rise buildings in excess of 420 feet (128 m) in height, Section 403.5.2 requires there be an additional exit stairway provided in addition to those needed for egress capacity or travel distance purposes. This additional stairway is not required when the elevators in the building comply with Section 3008 and allow occupant self-evacuation.

SECTION 404
ATRIUMS

404.1 General. In other than Group H occupancies, and where permitted by Exception 5 in Section 708.2, the provisions of this section shall apply to buildings or structures containing vertical openings defined herein as "Atriums."

❖ Unprotected vertical openings are often identified as the factor responsible for fire spread in incidents involving fire fatalities or extensive property damage. Section 404 addresses one method for protection of these specific building features in lieu of providing a complete floor separation. As noted in this section, the atrium provisions are primarily an exception to the shaft requirements of Section 708. Atriums are regulated by this section and are considered acceptable when in compliance with the criteria of this section. Additionally, the code permits other types of vertical openings that are not addressed by this section. For example, Section 708.2 permits openings within residential dwelling units to be unprotected. Other unprotected vertical openings that are permitted by the code include covered malls (see Section 402); communicating spaces in buildings of Group I-3 (see Section 408.5); mezzanines (see Section 505); escalator or supplemental stairway openings (see Section 708.2); open parking garages (see Section 406.3); enclosed parking garages (see Section 406.4) or as otherwise noted in Section 708.2.

An atrium is a space within a building that extends vertically and connects two or more stories. Atriums are not to be considered unprotected vertical openings; rather, the vertical openings are protected by means other than enclosure of the shaft or a complete floor assembly. This section does not apply to spaces that comply with Section 708.2 through exceptions other than Exception 5. In other words, if compliance with Section 708.2 is achieved by meeting an exception other than that permitting an atrium, then compliance with Section 404 is not required. Likewise, if the provisions of Section 404 have been complied with, then the other exceptions to Section 708.2 do not need to be addressed. Spaces that are separated from the atrium by shaft enclosure assemblies complying with the provisions of Section 708 are not considered as part of the atrium. Section 404 applies only to the spaces that are contained within the atrium. Atriums are permitted in all buildings except Group H.

The exceptions in Section 708.2 are all considered legitimate alternatives to the shaft requirements, but were not developed with specific consideration of the application of multiple exceptions in the same building/space. In most cases this will be appropriate but can become complicated where, for example, an atrium is open to a space that is applying Exception 2 to Section 708.2. More discussion of this particular scenario is found in the commentary to Section 404.6.

404.1.1 Definition. The following word and term shall, for the purposes of this chapter and as used elsewhere in this code, have the meaning shown herein.

❖ Definitions of terms that are associated with the content of this section are contained herein. These definitions can help in the understanding and application of the code requirements. It is important to emphasize that these terms are not exclusively related to this section but are applicable everywhere the term is used in the code. The purpose for including this definition within this section is to provide more convenient access to it without having to refer back to Chapter 2. For convenience, this term is also listed in Chapter 2 with a cross reference to this section. The use and application of all defined terms, including that term defined herein, are set forth in Section 201.

ATRIUM. An opening connecting two or more *stories* other than enclosed *stairways*, elevators, hoistways, escalators, plumbing, electrical, air-conditioning or other equipment, which is closed at the top and not defined as a mall. Stories, as used in this definition, do not include balconies within assembly groups or *mezzanines* that comply with Section 505.

❖ The definition identifies that an atrium is a floor opening or a series of floor openings that connect the environments of adjacent stories. The definition of "Atrium" excludes enclosed stairways, elevators, hoistways and other similar openings in order to clarify that those elements would not fall under the purview as to what is considered an atrium, and therefore, the associated requirements found in Section 404 would not apply. What this does not preclude is the inclusion of elevators and open stairways within atriums. Such elements would need to be entirely within the atrium to meet the separation requirements found in Section 404.6. Building features, such as stairways, elevators, hoistways, escalators, plumbing, electrical, air conditioning or other equipment openings, are required to be enclosed in fire-resistance-rated shafts in accordance with Section 708.2 or must be protected with one of the methods specified in the exceptions to Section 708.2, The allowance of an atrium is one of the exceptions (5) found in Section 708.2. An atrium is not defined by size or use. A series of floor openings that are enclosed with exterior walls, yet open at the roof, would be considered a court and would be exempt from the requirements of Section 404. Balconies associated with assembly occupancies and mezzanines are not considered individual stories that would contribute to the classification of a space as an atrium.

404.2 Use. The floor of the atrium shall not be used for other than low fire hazard uses and only *approved* materials and decorations in accordance with the *International Fire Code* shall be used in the atrium space.

> **Exception:** The atrium floor area is permitted to be used for any *approved* use where the individual space is provided with an *automatic sprinkler system* in accordance with Section 903.3.1.1.

❖ Because an automatic sprinkler system at the ceiling of an atrium may not be effective for a fire on the floor of the atrium due to the ceiling height or obstructions to the sprinkler discharge, the use and activities of the floor level and the types of materials in the atrium space must be controlled. This section applies to all atriums regardless of their height or area. Low fire-hazard uses would limit the atrium floor to such functions as pedestrian walk-through areas, security desks and reception areas. Storage areas, fabrication areas and office areas would not be low fire-hazard uses. Chapter 8 of the IFC regulates the use of decorative materials and furnishings.

If the floor area is equipped with an automatic sprinkler system that can provide the required protection, then its use is not restricted. The exception stipulates that such areas must be equipped with an automatic sprinkler system as is required throughout the remainder of the atrium.

[F] 404.3 Automatic sprinkler protection. An *approved automatic sprinkler system* shall be installed throughout the entire building.

Exceptions:

1. That area of a building adjacent to or above the atrium need not be sprinklered provided that portion of the building is separated from the atrium portion by not less than 2-hour *fire barriers* constructed in accordance with Section 707 or *horizontal assemblies* constructed in accordance with Section 712, or both.

2. Where the ceiling of the atrium is more than 55 feet (16 764 mm) above the floor, sprinkler protection at the ceiling of the atrium is not required.

❖ One means of controlling the spread of fire and smoke through vertical openings is to control and extinguish the fire as early as possible; therefore, all floor areas that are connected by the atrium are to be protected with an approved sprinkler system, including the atrium space itself. The system is to be designed, installed, tested and maintained in accordance with the provisions of Section 903.3 and the IFC. Since Chapter 9 requires that the sprinkler system be supervised, the reliability of the sprinkler system is improved. Exception 1 clarifies that since the atrium protection is an alternative to the shaft enclosure requirements, the sprinkler system is not required in areas that are separated from the atrium by 2-hour fire barrier walls or horizontal assemblies that would otherwise be required for a shaft enclosure. Such fire barrier walls must conform to Section 707, while the hori-

zontal assemblies must comply with Section 712.

Exception 2 permits the required sprinkler system to be deleted from the ceiling areas of atriums where the vertical distance between the atrium floor and atrium ceiling is greater than 55 feet (16 764 mm). A ceiling height of more than 55 feet (16 764 mm) is the height at which the system is no longer effective and installing such systems provides little benefit. This exception does not alter the use limitations of the atrium as stated in Section 404.2, nor does it exempt any adjacent floor areas with lower ceiling heights that are included in the atrium boundary in accordance with Exception 1. It is important to note that if a smoke control system is required by Section 404.5, the smoke control design should not take sprinkler activation into account for design fires that are located in areas not protected with sprinklers or in areas where sprinklers are installed but, due to their location, may not be able to control a fire.

[F] 404.4 Fire alarm system. A fire alarm system shall be provided in accordance with Section 907.2.14.

❖ Section 907.2.14 of the code contains a requirement for a fire alarm system in atriums connecting more than two stories. In addition, when such atriums are located in Group A, E and M occupancies, an emergency voice communication system is required.

404.5 Smoke control. A smoke control system shall be installed in accordance with Section 909.

> **Exception:** Smoke control is not required for atriums that connect only two *stories.*

❖ In order to prevent the migration of smoke throughout interconnected levels of a building via the atrium, a mechanical smoke control system is to be installed in atriums connecting more than two stories in accordance with the provisions of Section 909. See Figure 404.6(1) for examples of when smoke control is and is not required. For spaces such as atriums, the primary method of smoke control in accordance with Section 909, is the exhaust method. Pressure differences in such a large space will be difficult to achieve and the area of fire origin when undertaking the pressurization method is not required to be tenable in accordance with Section 909.6. The airflow method is impractical with such large spaces. The airflow method is sometimes used in combination with the exhaust methods to protect openings into the atrium (see commentary, Section 909). The smoke control system for the atrium is required to be connected to a standby source of power in accordance with Sections 404.7 and 909.11.

Section 402.10 for covered mall buildings also references Section 404.5 for smoke control provisions when a mall contains an atrium. It should be noted that when a mall contains an atrium, all areas adjacent to the atrium without separation would need to be addressed when designing the smoke control system as required by Section 404.6, Exception 3.

As discussed in the commentary to Section 404.1, the atrium provisions are basically an exception to the shaft requirements in Section 708. Therefore, if another exception to Section 708.2 is chosen, such openings would not be considered atriums and smoke control would not be required.

404.6 Enclosure of atriums. Atrium spaces shall be separated from adjacent spaces by a 1-hour *fire barrier* constructed in accordance with Section 707 or a *horizontal assembly* constructed in accordance with Section 712, or both.

Exceptions:

1. A glass wall forming a smoke partition where automatic sprinklers are spaced 6 feet (1829 mm) or less along both sides of the separation wall, or on the room side only if there is not a walkway on the atrium side, and between 4 inches and 12 inches (102 mm and 305 mm) away from the glass and designed so that the entire surface of the glass is wet upon activation of the sprinkler system without obstruction. The glass shall be installed in a gasketed frame so that the framing system deflects without breaking (loading) the glass before the sprinkler system operates.

2. A glass-block wall assembly in accordance with Section 2110 and having a ³/₄-hour *fire protection rating*.

3. The adjacent spaces of any three *floors* of the atrium shall not be required to be separated from the atrium where such spaces are accounted for in the design of the smoke control system.

❖ One of the basic premises of atrium requirements is that an engineered smoke control system combined with an automatic fire sprinkler system that is properly supervised provide an adequate alternative to the fire-resistance rating of a shaft enclosure. It is also recognized that some form of a boundary is required to assist the smoke control system in containing smoke to just the atrium area. The basic requirement, therefore, is that the atrium space be separated from adjacent areas by fire barriers and horizontal assemblies having a fire-resistance rating of at least 1 hour.

Also, openings in the wall are required to be protected, in accordance with Section 707.6. In accordance with Section 708.4, shafts are required to have a fire-resistance rating of at least 2 hours if connecting more than three stories, and 1 hour when connecting two or three stories. The basis for the 1-hour requirement in Section 404.6 is that an automatic sprinkler system can be substituted for the 1 hour of fire resistance of a shaft enclosure. The allowance is consistent with the 1-hour fire-resistance-rating reduction permitted in high-rise buildings (see Section 403.2.1.2).

In lieu of a 1-hour fire-resistance-rated separation, Exception 1 allows adjacent spaces to be separated by glass walls where automatic sprinklers have been installed to protect the glass. The sprinklers are to be located so as to wet the entire surface of the glass wall.

If there is a floor surface on each side of the wall, both sides of the glass must be protected. The glass must be in a gasketed frame such that the framing system can deflect without breaking the glass.

Although this exception does not address obstructions or other window treatments, consideration must be given to locating such items to avoid interference with the required sprinkler heads.

Without specific test evidence, curtain rods, traverse rods, curtains and draperies must be located at least 12 inches (305 mm) from the window surface [see Figure 404.6(1)]. Any doors through the required 1-hour fire barrier wall must be ³/₄-hour rated in accordance with Table 715.4.

The sprinkler system required for Exception 1 is not intended to be a deluge system. Instead it is intended to protect the glazing material from breakage as a result of thermal shock. It is not necessary to activate all the sprinklers along the glazing material to provide such protection as long as the entire surface of the glazed panel is designed such that it can be wetted by the sprinkler system.

Exception 2 allows a glass-block wall assembly conforming to Section 2110. It is important to note that these glass-block assemblies do not require the sprinkler protection that is required by Exception 1.

Exception 3 recognizes the desire to have at least some floors open to the atrium, and permits a maximum of three. The three-floor restriction is consistent with the basic premise that the life safety hazard becomes significant when more than three floors are open and is also consistent with the allowances for covered mall buildings. It should be noted that the three floor levels may be at any height and need not be consecutive floor levels [see Figure 404.6(2)]. The exception also states that the smoke control design must account for these spaces. This particular reference to the smoke control design does not require that the 6-foot-high (1829 mm) layer required by Section 909.8.1 be maintained in these spaces. Instead it is saying that if a smoke control system is required by Section 404.5, such spaces must be accounted for in terms of the hazard they pose to the atrium and to smoke migrating to other adjacent spaces on other stories open to the atrium. Essentially these spaces have simply increased the possible design fires that may send smoke into the atrium, thus threatening to send smoke throughout the building and other adjoining spaces. If the atrium smoke control system is not designed to handle fires in these areas, the system may become overpowered. This exception can also permit a two-story atrium to have adjacent areas open to the atrium. In this case, since there is no required smoke control system, obviously there is no need to account for the added space (see IBC Interpretation No. 54-07).

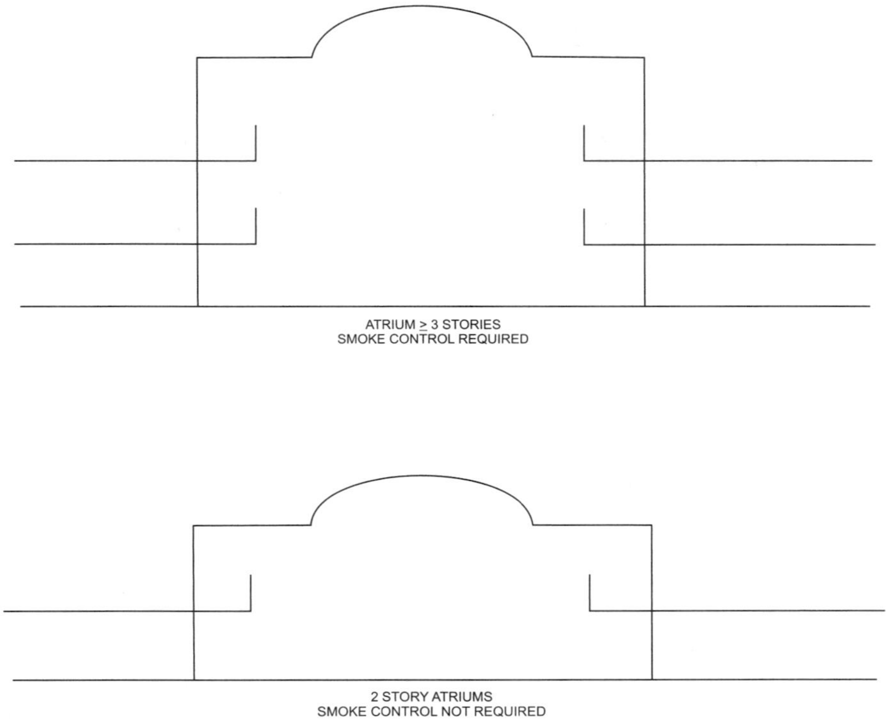

ATRIUM ≥ 3 STORIES
SMOKE CONTROL REQUIRED

2 STORY ATRIUMS
SMOKE CONTROL NOT REQUIRED

Figure 404.6(1)
ENCLOSURE OF ATRIUMS (EXCEPTION 1)

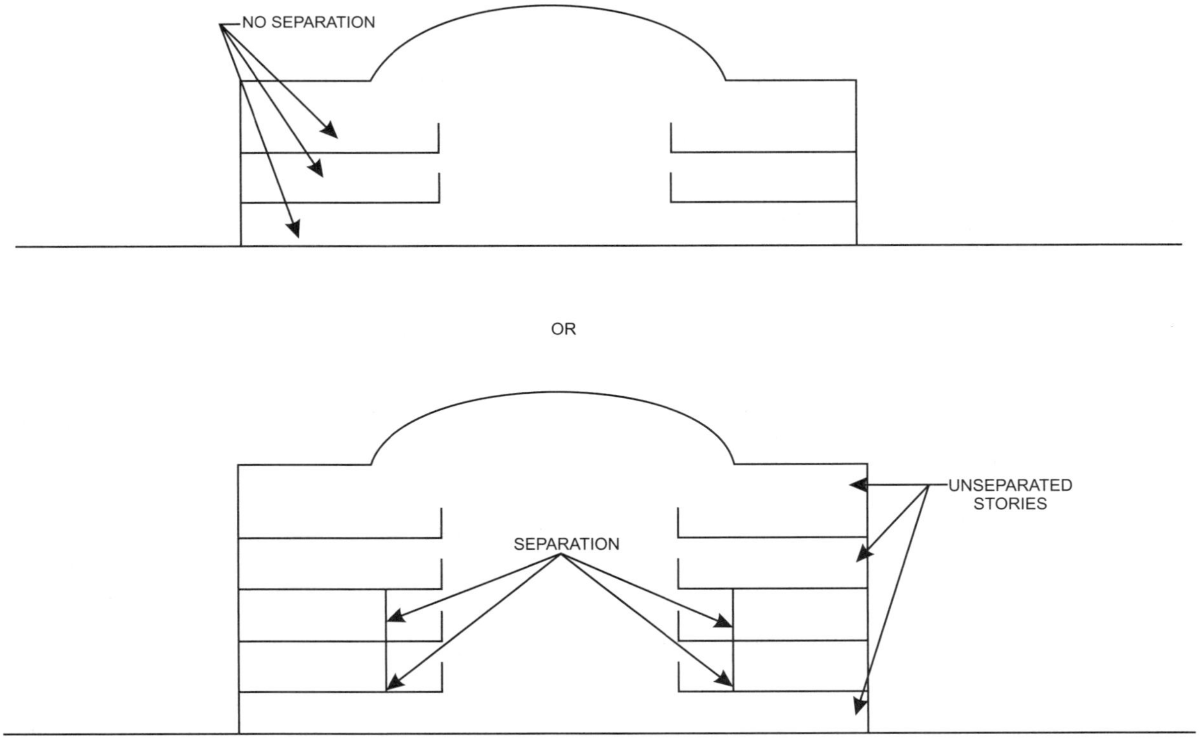

NO SEPARATION

OR

UNSEPARATED
STORIES

SEPARATION

Figure 404.6(2)
ENCLOSURE OF ATRIUMS (EXCEPTION 3)

As discussed in the commentary to Section 404.1, the atrium requirements are basically an exception to the shaft requirements in Section 708.2. More specifically, Section 708.2 has multiple exceptions that are all legitimate alternatives to the shaft requirements. If there are portions of the building applying other exceptions to Section 708.2 that are not separated from the atrium, they can be considered part of the atrium themselves and be subject to the smoke control layer height of 6 feet (1829 mm) or could be considered as adjacent spaces and need to be accounted for when designing the smoke control system, as discussed above. It should be noted that if the space is considered part of the atrium itself, the other exceptions in Section 708.2 need not apply as the atrium provisions are already being applied.

[F] 404.7 Standby power. Equipment required to provide smoke control shall be connected to a standby power system in accordance with Section 909.11.

❖ To enhance the reliability of the required smoke control system, this section of the code mandates that the mechanical exhaust must be connected to a standby power system. The reference to Section 909.11 allows the atrium's equipment to receive its primary power from the building's power system. A secondary standby power source is required should the building's primary system fail.

404.8 Interior finish. The *interior finish* of walls and ceilings of the atrium shall not be less than Class B with no reduction in class for sprinkler protection.

❖ Although Chapter 8 contains provisions governing the use of interior finishes, trim and decorative materials, specific limitations are provided for an atrium. Similar to the use limitations contained in Section 404.2, a minimum interior finish classification is specified to limit the fuel load within the atrium spaces. At least a Class B finish is mandated for the walls and ceilings of the atrium. Such a finish classification limits the material's flame spread index to a range of 26 through 75 and a smoke-developed index rating of 450 or less. The sprinkler reductions shown in Table 803.9 are not applicable to the interior finishes of an atrium space.

404.9 Travel distance. In other than the lowest level of the atrium, where the required *means of egress* is through the atrium space, the portion of exit access travel distance within the atrium space shall not exceed 200 feet (60 960 mm). The travel distance requirements for areas of buildings open to the atrium and where access to the *exits* is not through the atrium, shall comply with the requirements of Section 1016.

❖ On all floors with direct communication to the atrium for exit access (other than the floor level of the atrium), the total exit access travel distance through the atrium is limited to 200 feet (60 960 mm). Since smoke is being drawn to the atrium, the time allowed to reach an exit through the atrium is limited. Travel distance at the atrium floor level and the levels that do not communicate with the atrium for exit access in the building is regulated by Table 1016.1.

SECTION 405
UNDERGROUND BUILDINGS

405.1 General. The provisions of this section apply to building spaces having a floor level used for human occupancy more than 30 feet (9144 mm) below the finished floor of the lowest *level of exit discharge*.

Exceptions:

1. One- and two-family *dwellings*, sprinklered in accordance with Section 903.3.1.3.

2. Parking garages with automatic sprinkler systems in compliance with Section 405.3.

3. Fixed guideway transit systems.

4. Grandstands, *bleachers*, stadiums, arenas and similar facilities.

5. Where the lowest *story* is the only *story* that would qualify the building as an underground building and has an area not exceeding 1,500 square feet (139 m²) and has an *occupant load* less than 10.

6. Pumping stations and other similar mechanical spaces intended only for limited periodic use by service or maintenance personnel.

❖ An underground building presents a unique hazard to life safety. Due to its isolation and inaccessibility, occupants within the structure and fire fighters attempting to locate and suppress a fire are presented with a unique fire protection challenge.

Underground buildings that require the occupants of the lowest floor level to travel upwards for more than 30 feet (9144 mm) to reach the finished floor of the lowest level of exit discharge present a significant hazard to the occupants. As such, Section 405 is applicable to buildings with a floor level more than 30 feet (9144 mm) below the finished floor surface of the lowest level of exit discharge (see Figure 405.1).

To egress the structure, occupants must travel in an upward direction. The direction of occupant travel is the same as the direction that the products of combustion travel. As such, the occupants are potentially exposed to the products of combustion along the entire means of egress.

Fire fighters are also confronted by constant exposure to the products of combustion. Beginning their descent above the actual location of the fire source, fire fighters encounter an increasing amount of smoke, heat and flame as they attempt to locate and extinguish the fire source. These extreme conditions could significantly hinder the effectiveness of the fire department if not offset by appropriate fire protection requirements. The requirements for underground buildings are, in some ways, similar to those for high-rise structures. Both types of structures present an unusual hazard, since they are virtually inaccessible to exterior fire department suppression and rescue operations with the increased potential to trap occupants inside. To counteract these hazards, such structures are required by Section 405.2 to be built of noncombustible, fire-resistance-rated construction.

Additionally, they are required by Section 405.3 to be equipped with an automatic sprinkler system and a smoke control system in accordance with Section 405.5. Standby and emergency power systems are also required in these structures by Sections 405.8 and 405.9.

Structures regulated by Section 405 are also subject to all other applicable code provisions. Additionally, underground buildings to which this section does not apply are still subject to all other code provisions, including fire suppression (see Section 903); standpipe systems (see Section 905); fire alarm and detection (see Section 907) and emergency escape (see Section 1029).

There are six exceptions to the applicability of Section 405. These exceptions are in consideration of specific types of structures to which the requirements of this section are impractical, unnecessary or to which alternative provisions apply.

405.2 Construction requirements. The underground portion of the building shall be of Type I construction.

❖ Two of the key goals with underground construction are to limit the fuel load and to increase the fire resistance of underground buildings. To reasonably ensure that the building elements will remain structurally sound during exposure to fire, all structural elements of the underground portions are required to be of noncombustible,

protected construction; therefore, any portion of the structure located below grade level is required to be of Type I construction (see Figure 405.1).

Portions of the structure above grade, however, are permitted to be of any type of construction.

Any portion of the structure above ground level must conform to the height and area limitations of Table 503. This table does not regulate the maximum depth of an underground structure; therefore, the height, area and depth of most Type I structures are not limited.

[F] 405.3 Automatic sprinkler system. The highest level of *exit discharge* serving the underground portions of the building and all levels below shall be equipped with an *automatic sprinkler system* installed in accordance with Section 903.3.1.1. Water-flow switches and control valves shall be supervised in accordance with Section 903.4.

❖ One of the most effective preventive measures to fire growth is the installation of an automatic sprinkler system. Because of the unique conditions for occupant egress and fire department access in the underground portion of a building, automatic sprinkler protection is required. The level of exit discharge and all floor levels below are required to be sprinklered throughout in accordance with Section 903.3.1.1. This section does permit a portion of a building to extend above the level of exit discharge and not be equipped with an automatic sprinkler system. If, however, another code sec-

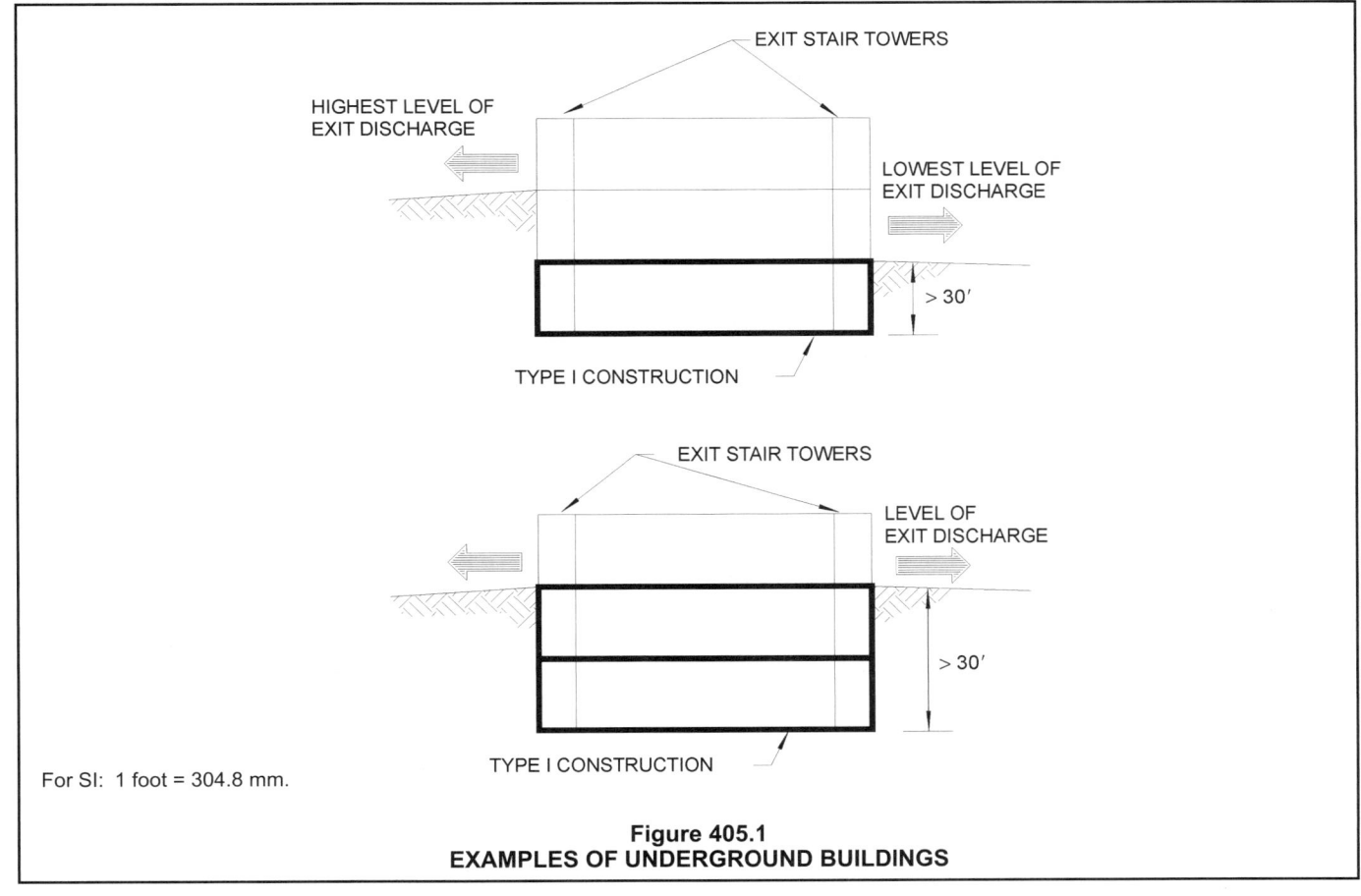

For SI: 1 foot = 304.8 mm.

Figure 405.1
EXAMPLES OF UNDERGROUND BUILDINGS

tion (see Section 403.3 or 903) requires an automatic sprinkler system in the above-ground portion, such a requirement would still be applicable. Note that a smoke control system is required in accordance with Section 405.5. Automatic sprinkler systems are essential elements of any smoke control system. Without suppression, the size of the fire or the resulting products of combustion will rapidly overwhelm most mechanical smoke control systems. The pressurization system requirements in Section 909 are based upon the use of sprinklers. Pressure differences needed to maintain the smoke to the area of origin would need to be higher in an unsprinklered building.

405.4 Compartmentation. Compartmentation shall be in accordance with Sections 405.4.1 through 405.4.3.

❖ Compartmentation is a key element in the egress and fire access plan for floor areas in an underground building. Subdivision into separate compartments through the use of smoke barriers (see Section 710) permits occupants to travel horizontally to escape the fire condition and provides a staging area for the fire service (see Figure 405.4).

405.4.1 Number of compartments. A building having a floor level more than 60 feet (18 288 mm) below the finished floor of the lowest *level of exit discharge* shall be divided into a minimum of two compartments of approximately equal size. Such compartmentation shall extend through the highest *level of exit discharge* serving the underground portions of the building and all levels below.

> **Exception:** The lowest *story* need not be compartmented where the area does not exceed 1,500 square feet (139 m²) and has an *occupant load* of less than 10.

❖ The 60-foot (18 288 mm) threshold is based on establishing a reasonable limitation on the required vertical travel distance for the occupants and fire service before the added protection of compartmentation is beneficial.

It is important to realize that the code requires, at a minimum, two compartments of approximately equal size. The maximum size of each compartment is not limited and is, therefore, a design decision subject to the approval of the building official.

An exception is permitted at the lowest level of the underground building as long as that level is relatively small in area [less than or equal to 1,500 square feet (139 m²)] and serves a low occupant load (less than 10), since an evacuation to the next higher level would not be expected to adversely affect that level. Note that this exception coordinates with Exception 5 to Section 405.1.

It should be noted that smoke control would be required for all underground buildings in accordance with Section 405.5 that fall under the scope of Section 405.1. Smoke control systems, which are required to be designed in accordance with Section 909, would generally require the use of smoke barriers to create compartments to achieve the design goals stated in Section 405.5.1. Therefore, even when compart-

mentation is not required for a building that has a floor level that is less than 60 feet (18 288 mm) below the finished floor of the level of exit discharge, compartmentation will probably still be utilized within a building to achieve the goals of Section 909. These compartments may be utilized in a different manner than as mandated in Section 405.4.1. In other words, such compartments may not span vertically from the finished floor of the level of exit discharge to the lowest level of the underground building. The compartments may be designated by story (see Figure 405.4.1).

The provisions for smoke control and compartmentation found in Section 405 were historically focused on smoke exhaust versus smoke control. More specifically, smoke control is for the safety of the occupants only, whereas smoke exhaust tends to focus on the needs of the fire department in managing smoke in underground conditions. The compartmentation requirements in Section 405.4 were meant to work with the smoke exhaust requirements and, at the very least, provide passive smoke management for the building for both the occupants and the fire department. As noted, the requirements of Section 909 as required by Section 405.5 have a different focus and will result in many compartments within the building.

405.4.2 Smoke barrier penetration. The compartments shall be separated from each other by a *smoke barrier* in accordance with Section 710. Penetrations between the two compartments shall be limited to plumbing and electrical piping and conduit that are firestopped in accordance with Section 713. Doorways shall be protected by *fire door assemblies* that are automatic-closing by smoke detection in accordance with Section 715.4.8.3 and are installed in accordance with NFPA 105 and Section 715.4.3. Where provided, each compartment shall have an air supply and an exhaust system independent of the other compartments.

❖ The smoke barrier walls and floor/ceiling assemblies that create the compartments must be fire-resistance rated for 1 hour and must meet the requirements of Section 710. Penetrations through the smoke barrier wall are limited to plumbing, piping and penetrations that are vital to the fire protection systems (i.e., sprinkler piping and electrical raceways). Special fire-resistance-rated doors are required to maintain the intended compartmentation, including gasketing and drop sills. Separate air distribution systems are required for each compartment. Components of the air distribution system are not permitted to penetrate the smoke barrier walls. These requirements are intended to help provide for the independence of the compartments separated by the smoke barrier walls by minimizing the penetrations therein. Note also that Section 712.9 addresses horizontal assemblies that are used as smoke barriers.

The separate air distribution systems required by this section are focused on the smoke compartments specifically required by this section and would not necessarily require separate systems for each compartment as created by the requirements of Section 405.5.

405.4.3 Elevators. Where elevators are provided, each compartment shall have direct access to an elevator. Where an elevator serves more than one compartment, an elevator lobby shall be provided and shall be separated from each compartment by a *smoke barrier* in accordance with Section 710. Doors shall be gasketed, have a drop sill and be automatic-closing by smoke detection in accordance with Section 715.4.8.3.

❖ Elevators are permitted to serve more than one compartment when properly protected. When different compartments utilize a common elevator, a smoke barrier consisting of a 1-hour fire-resistance-rated elevator lobby is required. This lobby provides additional separation between adjacent compartments, thus

helping to create their independence from the immediate effects of fire (see Figure 405.4). Essentially the elevator lobby as prescribed in this section creates another smoke compartment within the building.

[F] 405.5 Smoke control system. A smoke control system shall be provided in accordance with Sections 405.5.1 and 405.5.2.

❖ A smoke control system is required for all underground buildings. The smoke control system is an integral part of the required fire protection systems for underground buildings and is focused upon the safety of occupants while exiting the building during a fire. In the case where the building contains both underground and above-ground portions, such specific details and requirements are listed in Sections 405.5.1 and 405.5.2.

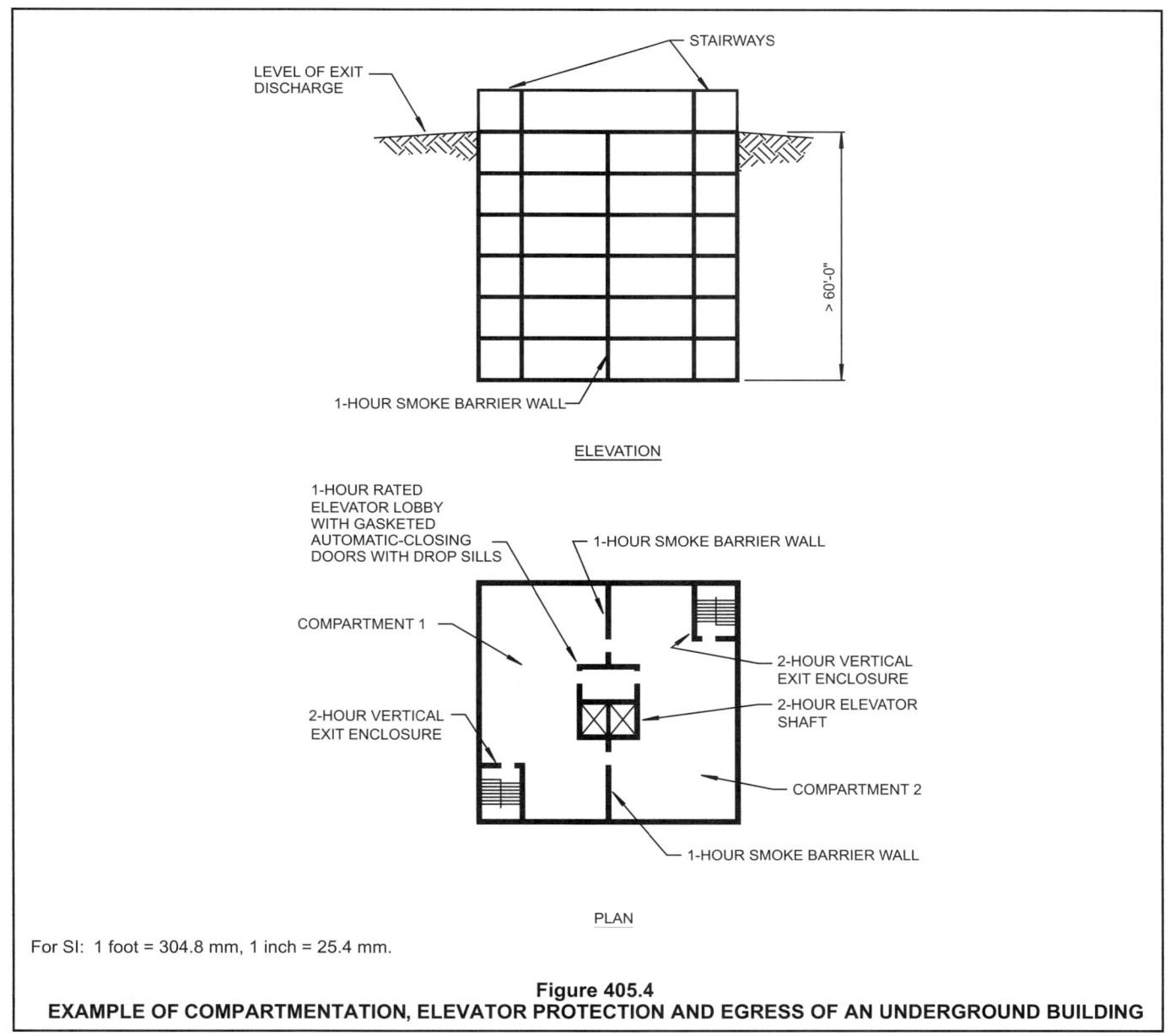

For SI: 1 foot = 304.8 mm, 1 inch = 25.4 mm.

Figure 405.4
EXAMPLE OF COMPARTMENTATION, ELEVATOR PROTECTION AND EGRESS OF AN UNDERGROUND BUILDING

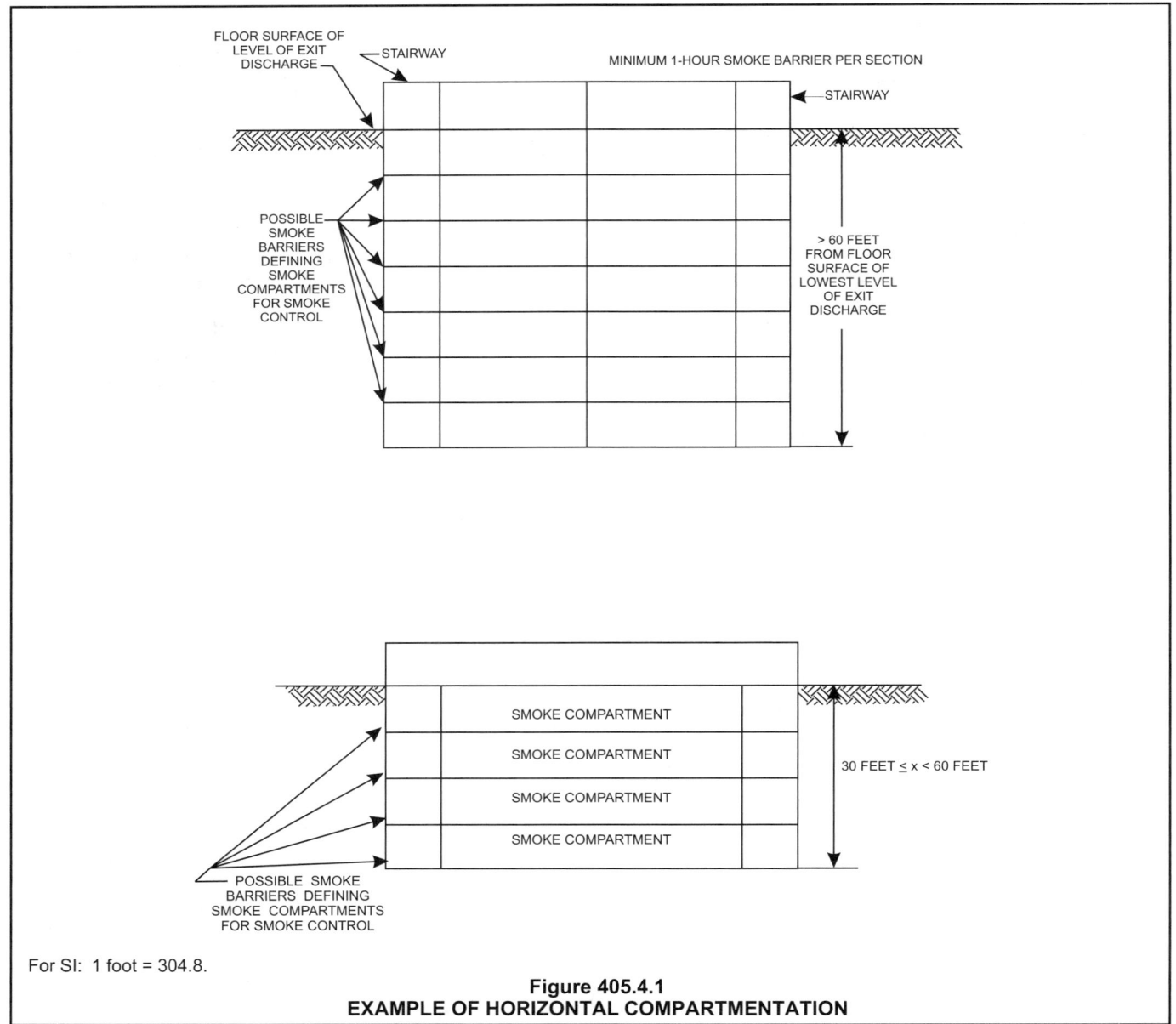

For SI: 1 foot = 304.8.

Figure 405.4.1
EXAMPLE OF HORIZONTAL COMPARTMENTATION

[F] 405.5.1 Control system. A smoke control system is required to control the migration of products of combustion in accordance with Section 909 and the provisions of this section. Smoke control shall restrict movement of smoke to the general area of fire origin and maintain *means of egress* in a usable condition.

❖ The general design requirements, special inspection and test requirements along with the applicable smoke control methods are contained within Section 909. The performance goal of the smoke control system is to contain the smoke and hot gases generated by a fire condition to the immediate area of origin. "Containing to the immediate area of origin" is usually taken to mean the compartment of origin. This allows the building occupants to access their required exits and evacuate the building before smoke movement traps them below grade. Generally the type of system will be a pressurization system where the smoke is managed by maintaining pressure differences across smoke barriers. It could be possible that a different type of system such as an exhaust system is required in larger more open spaces such as an atrium. It is important to note that the systems in Section 909 are intended for the protection of occupants and do not necessarily provide longer term capacity for the fire department. In addition, such systems are not intended as a method of smoke exhaust for salvage and overhaul after a fire. It is recognized that overhaul in underground buildings after a fire is far more complex due to the fact that the structure is entirely underground, but as Section 909 is currently written, such systems are not specifically designed with this role in mind.

As discussed in the commentary to Section 405.4.1, the smoke control system required by Section 405.5 will likely result in many smoke compartments in order to meet the design criteria of Section 909.

[F] 405.5.2 Compartment smoke control system. Where compartmentation is required, each compartment shall have an independent smoke control system. The system shall be automatically activated and capable of manual operation in accordance with Sections 907.2.18 and 907.2.19.

❖ The compartmentation referred to in this section is that required in Section 405.4.1 which requires the building be separated from the story of level of exit discharge to the lowest level in the building. It is not the intent that each smoke compartment (smoke zone) as created through the application of Section 405.5.1 would be required to have independent smoke control systems (see commentary, Section 405.4.2).

[F] 405.6 Fire alarm systems. A fire alarm system shall be provided where required by Sections 907.2.18 and 907.2.19.

❖ The ability to communicate and offer warning of a fire scenario can increase the time available for egress from the building. Underground buildings with a floor level greater than 60 feet (18 288 mm) below the finished floor of the level of exit discharge are, therefore, required to be provided with a manual fire alarm system. An emergency voice/alarm communication system is required as part of this system.

405.7 Means of egress. *Means of egress* shall be in accordance with Sections 405.7.1 and 405.7.2.

❖ This section is simply the introduction for means of egress requirements in underground buildings. Egress from underground buildings can be more challenging than buildings above grade and additional requirements are necessary.

405.7.1 Number of exits. Each floor level shall be provided with a minimum of two *exits*. Where compartmentation is required by Section 405.4, each compartment shall have a minimum of one *exit* and shall also have an *exit access* doorway into the adjoining compartment.

❖ The means of egress from underground buildings is an integral part of the safety precautions necessary to abate the hazards of people being located more than 30 feet (9144 mm) below the level of exit discharge. Sections 405.7.1 and 405.7.2 provide the minimum number of exits and requirements for smokeproof enclosures. These provisions create safe and usable elements that the building occupants can enter to flee the immediate effects of a fire located well below grade.

The arrangement of exits and exit access doors for an underground building is regulated by this section (see Figure 405.4).

Based on the requirements of this section and Section 405.4, an underground building having a floor level more than 30 feet (9144 mm) below the level of exit discharge requires a minimum of two exits, and cannot qualify as a single-exit building. Additional exits

may be required in accordance with Section 1021.1. Underground buildings requiring compartmentation in accordance with Section 405.4.1 must have an exit in each and every compartment plus a doorway into the adjoining compartment.

405.7.2 Smokeproof enclosure. Every required *stairway* serving floor levels more than 30 feet (9144 mm) below the finished floor of its *level of exit discharge* shall comply with the requirements for a smokeproof enclosure as provided in Section 1022.9.

❖ In order for exit stairways to provide the necessary means of egress, all stairways of an underground building must be constructed as smokeproof enclosures or pressurized stairways. The requirements for smokeproof enclosures are contained in Section 1022.9 and include a 2-hour enclosure along with limited access via a vestibule. The requirements for pressurized stairways are listed in Section 909.20. Pressurized stairways do not require a vestibule, but pressure differences as prescribed in Section 909.20.5 need to be provided and demonstrated through testing. These provisions for either a smokeproof enclosure or pressurized stairways are intended to maintain a protected path of travel for building occupants to egress the underground building and also allow the fire department to enter for rescue and fire-fighting operations.

[F] 405.8 Standby power. A standby power system complying with Chapter 27 shall be provided standby power loads specified in Section 405.8.1.

❖ All underground buildings regulated by this section are required to be provided with a standby power system to increase the probability that critical systems will be operational in the event of a loss of normal power supply. The purpose of the standby power system is to provide an alternative means of supplying power to selected building systems should the normal electrical source fail.

[F] 405.8.1 Standby power loads. The following loads are classified as standby power loads:

1. Smoke control system.

2. Ventilation and automatic fire detection equipment for smokeproof enclosures.

3. Fire pumps.

Standby power shall be provided for elevators in accordance with Section 3003.

❖ The standby power system is required to supply electrical power to equipment that is essential to emergency egress, fire-fighting and rescue operations.

[F] 405.8.2 Pick-up time. The standby power system shall pick up its connected loads within 60 seconds of failure of the normal power supply.

❖ This section of the code specifies how quickly the standby power system must be activated to provide the necessary power for the loads specified in Section 405.8.1. The standby power systems must be capable of providing full power to those items within 60 seconds of when the main electrical power source goes down. This maintains the level of safety necessary to evacuate the building and begin fire-fighting operations. The maximum time of 60 seconds is consistent with the requirements of the NEC.

[F] 405.9 Emergency power. An emergency power system complying with Chapter 27 shall be provided for emergency power loads specified in Section 405.9.1.

❖ In addition to the standby power system, underground buildings regulated by this section are also required to be provided with an emergency power system for certain building systems. These systems include fire warning devices, lighting and other equipment required in underground buildings.

[F] 405.9.1 Emergency power loads. The following loads are classified as emergency power loads:

1. Emergency voice/alarm communications systems.

2. Fire alarm systems.

3. Automatic fire detection systems.

4. Elevator car lighting.

5. *Means of egress* and exit sign illumination as required by Chapter 10.

❖ The emergency power system is required to supply electrical power to equipment that is essential to detecting and warning others of a fire condition and to provide for illuminated evacuation of the underground building. Because of the critical nature of the systems and the potential for panic in the areas covered, the loads must be picked up within 10 seconds after failure of the normal power supply. This maximum time is consistent with the requirements of NEC.

[F] 405.10 Standpipe system. The underground building shall be equipped throughout with a standpipe system in accordance with Section 905.

❖ Just as high-rise buildings or other large buildings with special features are required to be provided with a standpipe system, so are underground buildings. A Class I automatic wet or manual wet standpipe system must be provided. Standpipe systems allow the fire department or other trained personnel to fight the fires at basement levels where it is impractical to run supply lines from the fire department trucks or outside hydrants. A quick and convenient water source for fire department use is essential to containing an underground fire.

SECTION 406
MOTOR-VEHICLE-RELATED OCCUPANCIES

❖ Included in this section of the code are all the special use and occupancy requirements for those buildings and structures that house motor vehicles. Corresponding requirements for aircraft-related occupancies can be found in Section 412. By definition, all structures that provide services or are used for the storage or parking of motor vehicles are regulated by these provisions. These requirements are applicable regardless of the types of fuel source used to power the vehicles' motors, the number of vehicles present, whether the vehicles are privately or commercially owned and whether they are used strictly for passengers or for freight. Typically, fire hazards for motor-vehicle occupancies are low because of the limited amounts of combustibles present along with the steel frame and metal clad bodies. Motor vehicle structures normally have sufficient space around each vehicle for parking purposes, which serves to further limit the fuel load. The design occupant loads are usually low and pose little if any life safety risks.

This section contains requirements for five different motor-vehicle occupancies along with general requirements for parking garages. Section 406.1 addresses private garages and carports, which are classified as a Group U occupancy. All other parking garages are divided up into either open parking garages or enclosed parking garages. These Group S-2 structures represent a slightly higher hazard than private garages or carports. As such, the heights and areas of these structures along with their type of construction classification are rigorously controlled. Requirements are also provided for those buildings and structures that provide service, care and repair of the vehicles. Stations where motor fuels are sold and dispensed are classified as Group M occupancies. The requirements for such facilities are contained in Section 406.5. Garages that provide repair services for vehicles also represent a slightly higher hazard; therefore, Section 406.6 identifies those code requirements necessary to abate the hazards of repair garages and their Group S-1 occupancy classification.

Although not specifically referenced in the code, both Section 304 of the IMC and Section 305 of the *International Fuel Gas Code®* (IFGC®) address the location of mechanical equipment in both private and public garages.

406.1 Private garages and carports.

406.1.1 Classification. Buildings or parts of buildings classified as Group U occupancies because of the use or character of the occupancy shall not exceed 1,000 square feet (93 m²) in area or one *story* in height except as provided in Section 406.1.2. Any building or portion thereof that exceeds the limitations specified in this section shall be classified in the occupancy group other than Group U that it most nearly resembles.

❖ Whereas private garages are smaller than parking garages by definition, the impact of the fuel load represented by the vehicle is somewhat higher. Such garages are frequently attached to residential-type occupancies, with the potential for fire exposure to the residential area and the occupants therein. The code does not specifically prohibit the servicing of vehicles in private garages, since it would be impractical to enforce. For example, the occupant of a residential dwelling unit will occasionally work on a vehicle in an attached private garage. For this reason, there are special provisions for private garages.

Private garages and carports are usually considered accessory to the building they serve. Section 312 classifies these utility and miscellaneous uses as Group U. The code limits these structures to 1,000 square feet (93 m²) in area and one story in height. Those limitations correspond to a fuel load associated with the passenger motor vehicles that is matched by the specified separation requirements. Separation and construction details for private garages attached to dwelling units are provided in Section 406.1.4. This separation includes the ¹/₂-inch (12.7 mm) regular or ⁵/₈-inch (15.9 mm) Type X gypsum board applied to the garage side (see Section 406.1.4 for further details).

Any garage that exceeds the height and area limits of Section 406.1.1 or the area increase of Section 406.1.2 must be reclassified. Once the specified height and area limits have been exceeded, the garage must be classified for the increased hazards. This results in the garage being classified as Group S-2, and the provisions of Sections 406.2 and 406.3 or 406.4 must now be met.

406.1.2 Area increase. Group U occupancies used for the storage of private or pleasure-type motor vehicles where no repair work is completed or fuel is dispensed are permitted to be 3,000 square feet (279 m²) when the following provisions are met:

1. For a mixed occupancy building, the *exterior wall* and opening protection for the Group U portion of the building shall be as required for the major occupancy of the building. For such a mixed occupancy building, the allowable floor area of the building shall be as permitted for the major occupancy contained therein.

2. For a building containing only a Group U occupancy, the *exterior wall* shall not be required to have a *fire-resistance rating* and the area of openings shall not be limited when the *fire separation distance* is 5 feet (1524 mm) or more.

More than one 3,000-square-foot (279 m²) Group U occupancy shall be permitted to be in the same building, provided each 3,000-square-foot (279 m²) area is separated by *fire walls* complying with Section 706.

❖ The area of a private garage can be increased to 3,000 square feet (279 m²) when all of the conditions of this section are met. If the garage will only be used for the parking and storage of private or pleasure-type motor vehicles, then the area increase may be used. Care must be taken such that no repair work or fuel dispensing is done within these areas. The limitations to private- or pleasure-type motor vehicles prevent commercial motor vehicles and their increased hazards from being located in this larger private garage.

As part of the increased area provisions for private garages, the exterior walls and any openings therein must be properly protected. If the garage is part of another occupancy, then the exterior wall ratings for that occupancy apply to the garage. For example, an

automobile insurance office may have several parking bays attached to it. These bays will be used to bring damaged cars in for the purposes of evaluating the damage and estimating the cost of repairs for the vehicle owner. The insurance office spaces are usually classified as Group B. In this case, the allowable building area is based on the Group B occupancy, the garage area is limited to 3,000 square feet (279 m²) and the exterior walls and openings in the garages are rated as part of the Group B occupancy.

For other applications, where the buildings only will be used to park vehicles, the building is classified as Group U. The exterior walls and openings therein are not required to be rated or limited if the fire separation distance is 5 feet (1524 mm) or more.

Additional private garages can be added to a Group U building, but they must be separated with fire walls. The 3,000-square-foot (279 m²) limit is a maximum. To reduce the risk of a fire from involving other adjacent garage spaces, fire walls are required to compartmentalize and further isolate the hazards. If the garages are located below another occupancy, such as Group R-2 apartments, the fire walls will be required to continue through to the roof of the residential portion of the building.

406.1.3 Garages and carports. Carports shall be open on at least two sides. Carport floor surfaces shall be of *approved* noncombustible material. Carports not open on at least two sides shall be considered a garage and shall comply with the provisions of this section for garages.

Exception: Asphalt surfaces shall be permitted at ground level in carports.

The area of floor used for parking of automobiles or other vehicles shall be sloped to facilitate the movement of liquids to a drain or toward the main vehicle entry doorway.

❖ Carports are covered structures, either free-standing or attached to the side of a building, for the purposes of providing shelter for motor vehicles. Carports must be open on at least two sides. This allows a fire condition to vent quickly to the outside. This further prevents a fire from going undetected and usually limits the amount of other incidental storage that may be included in a fully enclosed garage. Attached carports that are not open on at least two sides are considered a garage and must then be separated from the residence or other occupancy (see Section 406.1.4 for further details).

The use of a combustible floor finish that could absorb flammable and combustible liquids commonly used and stored within garages and carports presents a potential safety hazard to the occupants of an attached building. The floors of private garages must be sloped towards the main vehicle doors. Although not specified, the floor must be positively sloped at some rate, such as 1 in 48 (2-percent slope), to prevent the accumulation of any spilled flammable and combustible liquids and their vapors. This minimizes the risk of the vapors building up to a point where a fire condition could result.

406.1.4 Separation. Separations shall comply with the following:

1. The private garage shall be separated from the *dwelling unit* and its *attic* area by means of a minimum $^1/_2$-inch (12.7 mm) gypsum board applied to the garage side. Garages beneath habitable rooms shall be separated from all habitable rooms above by not less than a $^5/_8$-inch (15.9 mm) Type X gypsum board or equivalent. Door openings between a private garage and the *dwelling unit* shall be equipped with either solid wood doors or solid or honeycomb core steel doors not less than $1^3/_8$ inches (34.9 mm) thick, or doors in compliance with Section 715.4.3. Openings from a private garage directly into a room used for sleeping purposes shall not be permitted. Doors shall be self-closing and self-latching.

2. Ducts in a private garage and ducts penetrating the walls or ceilings separating the *dwelling unit* from the garage shall be constructed of a minimum 0.019-inch (0.48 mm) sheet steel and shall have no openings into the garage.

3. A separation is not required between a Group R-3 and U carport, provided the carport is entirely open on two or more sides and there are not enclosed areas above.

❖ As indicated in Item 1, when a private garage is attached to a residence, the adjacent areas (including attics) are to be separated to provide a minimum level of protection. This separation is to be constructed of at least $^1/_2$-inch (12.7 mm) gypsum wallboard applied to the garage side. In a location where the garage is beneath any type of habitable space, the separation requirement is increased by requiring that $^5/_8$-inch (15.9 mm) Type X gypsum board or some other material that would provide an equivalent level of protection be used. Doors opening within the adjacent wall are required to be protected with a minimum $1^3/_8$-inch-thick (34 mm) solid core wood or $1^3/_8$-inch-thick (34 mm) solid or honeycomb steel door. Although the $1^3/_8$-inch-thick (34 mm) solid core wood and honeycomb steel doors are not listed as fire doors, they still provide adequate protection. Alternatively, doors with a 20-minute fire protection rating in accordance with Section 715.4.3 may be used. All door openings from the garage, whether solid core wood, honeycomb steel or 20-minute rated, are prohibited from opening directly into a bedroom or any other room used for sleeping purposes. It is important to note that this separation requirement, including the protection and the limitation on the openings, is only applicable to garages and not to carports.

With regards to Item 2, ducts in the walls and ceilings that do not penetrate into the garage are already protected by the gypsum wallboard; therefore, additional separation is not necessary. If the ducts do penetrate through the separation, then the specified protection must be provided.

In accordance with Item 3, a roofed structure open

on two or more sides without any enclosed uses above poses no special hazard to the occupants of the dwelling. If a fire were to start under the carport roof, the smoke, hot gases and flames would be able to escape out the open sides, providing the structure with an adequate amount of protection. See the commentary to Section 406.1.3 for additional related discussion.

Section 406.1.4 is silent regarding the separation of a private garage from other occupancies. Where a Group U private garage is attached to other occupancies, such mixed occupancy building needs to be designed in accordance with Section 508.3 for nonseparated mixed occupancies or Section 508.4 for separated occupancies. Under separated occupancies, Table 508.4 would require a 2-hour wall between a Group U and Group B occupancy in a building without sprinkler protection. It might also be possible to consider the private parking garage under the accessory use provisions of Section 508.2.

406.1.5 Automatic garage door openers. Automatic garage door openers, if provided, shall be *listed* in accordance with UL 325.

❖ This section provides a requirement that automatic door openers comply with UL 325. This provision has been in previous editions of the *International Residential Code*® (IRC®). The code now has the equivalent level of safety.

406.2 Parking garages.

406.2.1 Classification. Parking garages shall be classified as either open, as defined in Section 406.3, or enclosed and shall meet the appropriate criteria in Section 406.4. Also see Section 509 for special provisions for parking garages.

❖ Parking garages are considered to be storage occupancies (Group S-2). In addition to the provisions of Sections 406.2, 406.3 and 406.4, parking garages must also comply with the code provisions for Group S occupancies.

The overall building fire loading in parking garages is typically low because of the considerable amount of metal in vehicles, which absorbs heat and the average weight of combustibles per square foot being relatively low [approximately 2 pounds per square foot (9.8 kg/m² psf)]. Still, a vehicle fire may be quite extensive.

Parking garages are divided into one of two categories. Those parking garages that contain sufficient clear openings in their exterior walls that meet the requirements of Section 406.3.3.1 can be classified as open parking garages. All other parking garages must be considered as enclosed parking garages and must comply with Section 406.4. Special height and area provisions for parking garages located above or below other uses and occupancies are contained in Section 509.

406.2.2 Clear height. The clear height of each floor level in vehicle and pedestrian traffic areas shall not be less than 7 feet (2134 mm). Vehicle and pedestrian areas accommodating van-accessible parking required by Section 1106.5 shall conform to ICC A117.1.

❖ A clear height of 7 feet (2134 mm) is required for all parking garages. This is an exception to the 7 foot, 6 inch (2286 mm) minimum height required in Chapter 10 for the means of egress system. This minimum height permits free and unobstructed egress around the vehicles and exit access areas. Van-accessible parking areas and the vehicular route to them must comply with ICC A117.1, *Standard for Accessible and Usable Buildings and Facilities.*

406.2.3 Guards. *Guards* shall be provided in accordance with Section 1013. *Guards* serving as vehicle barrier systems shall comply with Sections 406.2.4 and 1013.

❖ For the same reasons as identified in the commentary to Section 1013, guards are required around all vertical openings in the floors and roofs of parking garages where the vertical distance between adjacent levels exceeds 30 inches (762 mm). Guards that serve a dual purpose of guard and vehicle barrier systems must also comply with Section 406.2.4.

406.2.4 Vehicle barrier systems. Vehicle barrier systems not less than 2 feet 9 inches (835 mm) high shall be placed at the end of drive lanes, and at the end of parking spaces where the vertical distance to the ground or surface directly below is greater than 1 foot (305 mm). Vehicle barrier systems shall comply with the loading requirements of Section 1607.7.3.

Exception: Vehicle storage compartments in a mechanical access parking garage.

❖ This section requires vehicle barrier systems at locations where a 1-foot (305 mm) dropoff exists at the end of parking spaces or drive lanes. Vehicle barrier systems are to be designed in accordance with Section 1607.7.3. The vehicle barrier systems are required to resist a single load of 6,000 pounds (2722 kg). The minimum height of the barriers is now 2 feet, 9 inches (835 mm), in recognition of the greater number of vehicles with higher profiles on the road today compared to when this requirement was first introduced into the code. Barriers are not required in areas where the employees of the garage park the vehicles, such as mechanical access garages.

406.2.5 Ramps. Vehicle ramps shall not be considered as required *exits* unless pedestrian facilities are provided. Vehicle ramps that are utilized for vertical circulation as well as for parking shall not exceed a slope of 1:15 (6.67 percent).

❖ Since the vehicular ramps of parking garages are open to all levels, they are directly exposed to the effects of smoke and hot gases. In addition, the vehicle ramps are often sloped at a rate greater than 1:12 (8-percent slope). These ramps, therefore, cannot be counted as part of the required exits from each level or tier. Certainly occupants can travel from their vehicles on sloped parking levels to access the required exit stairways; however, occupants cannot continuously travel down the vehicle ramps to reach the exit discharge unless the slope is 1:15 or less and sidewalks or other protected walking surfaces are provided.

406.2.6 Floor surface. Parking surfaces shall be of concrete or similar noncombustible and nonabsorbent materials.

The area of floor used for parking of automobiles or other vehicles shall be sloped to facilitate the movement of liquids to a drain or toward the main vehicle entry doorway.

Exceptions:

1. Asphalt parking surfaces shall be permitted at ground level.

2. Floors of Group S-2 parking garages shall not be required to have a sloped surface.

❖ To avoid acquiring a buildup of flammable liquids on the floor of parking garages, the floor finish is required to be noncombustible and nonabsorbent. Because of pollution concerns, however, garage floors are not automatically required to be drained into the building drainage system. This is consistent with requirements of the Environmental Protection Agency (EPA) for storm water drainage, as well as the philosophy of hazardous material handling (i.e., localize spills and treat them). If floor drains are provided, however, they must be installed in accordance with the IPC.

Exception 1 recognizes that many states have allowed the use of asphalt paving surfaces at grade levels of parking garages with no record of fire hazards. This exception does not allow asphalt paving to be used on stories above or below grade. Although not specified, the floor must be positively sloped at some rate, such as 1 in 48 (2-percent slope), to prevent the accumulation of any spilled flammable and combustible liquids and their vapors. This minimizes the risk of the vapors building up to a point where a fire condition could result.

Exception 2 was introduced into the code based on the recognition that many parking garages are constructed of prefabricated materials which prove difficult to achieve the required slope when assembled on site. The exception, however, waives the slope requirement for all Group S-2 parking facilities, not just those of prefabricated materials.

406.2.7 Mixed occupancy separation. Parking garages shall be separated from other occupancies in accordance with Section 508.1.

❖ If a building or structure consists not only of parking garages but also other uses and occupancies, then the building is treated as any other mixed occupancy. The building designer must choose one of the options available in Section 508 to address the mixed use and occupancy issues. Keep in mind that Sections 509.2, 509.3, 509.4, 509.7, 509.8 and 509.9 contain special provisions for mixed separations when the parking garages are located above or below other groups.

406.2.8 Special hazards. Connection of a parking garage with any room in which there is a fuel-fired appliance shall be by means of a vestibule providing a two-doorway separation.

> **Exception:** A single door shall be allowed provided the sources of ignition in the appliance are at least 18 inches (457 mm) above the floor.

❖ As part of the special use and occupancy requirements for parking garages, all possible ignition sources must be controlled and isolated. Specifically, all heating equipment must be located in rooms that are separated from the main areas where the vehicles are parked. Doors connecting the heating equipment rooms and the main parking area must be done with a vestibule or airlock arrangement such that one must pass through two doors prior to entering the other room. Again, this is done to minimize the possibility of any spilled flammable liquids and the resulting vapors from coming in contact with the ignition sources of the heating equipment.

An allowance is made if the heating equipment is located at least 18 inches (457 mm) above the floor of the separated room. In such a case, the vestibule/airlock arrangement with a double-door system is not required and access can be done with just a single door. If this exception is used, it must be understood by the building and fire officials that these special stipulations and conditions are part of the certificate of occupancy. Please note that there are more specific standards regarding the installation of mechanical equipment in public and private garages in Section 304 of the IMC and Section 305 of the IFGC.

406.2.9 Attached to rooms. Openings from a parking garage directly into a room used for sleeping purposes shall not be permitted.

❖ Just as a private garage cannot have openings into bedrooms (see Section 406.1.4, Item 1), other parking garages are likewise limited. The risks of a fire and smoke quickly spreading into adjacent areas where occupants may be sleeping are too great.

406.3 Open parking garages.

406.3.1 Scope. Except where specific provisions are made in Sections 406.3.2 through 406.3.13, other requirements of this code shall apply.

❖ The code defines four types of garages: private, enclosed parking, repair garages and open parking. With the exception of private garages, all such structures are classified as Group S. Unlike other storage occupancies in which the fuel load is evenly distributed, garages represent a different fire hazard.

Because of the generally fast-burning upholstery and the gasoline or diesel fuel content, fires in individual vehicles can be quite extensive. Because of both the overall combustible loading within a parking garage, which yields a low combustible fuel load [approximately 2 pounds per square foot (psf) (9.8 kg/m²)], and the considerable amount of metal in vehi-

cles, which absorbs heat, the overall fire hazards are low. Section 406.3 provides requirements that are unique to open parking garages, while Section 406.1 addresses private garages and Section 406.4 addresses enclosed parking garages. Because of the permanently open exterior walls of open parking garages, which permit the dissipation of heated gases, special provisions are made for heights and areas of such structures.

Open parking garages are classified as Group S-2 and all of the provisions for this group are applicable except as modified herein.

406.3.2 Definitions. The following words and terms shall, for the purposes of this chapter and as used elsewhere in this code, have the meanings shown herein.

❖ Definitions of terms that are associated with the content of this section are contained herein. These definitions can help in the understanding and application of the code requirements. It is important to emphasize that these terms are not exclusively related to this section but are applicable everywhere the term is used in the code. The purpose for including these definitions within this section is to provide more convenient access to them without having to refer back to Chapter 2. For convenience, these terms are also listed in Chapter 2 with a cross reference to this section. The use and application of all defined terms, including those defined herein, are set forth in Section 201.

MECHANICAL-ACCESS OPEN PARKING GARAGES. *Open parking garages* employing parking machines, lifts, elevators or other mechanical devices for vehicles moving from and to street level and in which public occupancy is prohibited above the street level.

❖ These types of garages are constructed with most of the same attributes as other garages. They have numerous parking levels that house the motor vehicles; however, public access to these upper levels is not permitted. Employees take control of the vehicles at the street level and drive into some sort of a vertical mechanical device, which conveys them to the upper parking levels. These structures are still provided with the required openings of Section 406.3.3.1.

OPEN PARKING GARAGE. A structure or portion of a structure with the openings as described in Section 406.3.3.1 on two or more sides that is used for the parking or storage of private motor vehicles as described in Section 406.3.4.

❖ Open parking garages are defined as having uniformly distributed openings on no less than two sides totaling no less than 40 percent of the building perimeter. The aggregate area of the openings is to be a minimum of 20 percent of the total wall area of all perimeter walls (see Figure 406.3.3.1).

RAMP-ACCESS OPEN PARKING GARAGES. *Open parking garages* employing a series of continuously rising floors or a series of interconnecting ramps between floors permitting the movement of vehicles under their own power from and to the street level.

❖ These types of garages employ vehicular ramps or sloped tiers that connect all of the levels of the structure. Whether the vehicles are self-parked or employee parked is immaterial.

406.3.3 Construction. *Open parking garages* shall be of Type I, II or IV construction. *Open parking garages* shall meet the design requirements of Chapter 16. For vehicle barrier systems, see Section 406.2.4.

❖ One of the basic premises of the provisions for open parking garages is that the overall fuel load is low and that the average fuel load per square foot is low. The construction of open parking garages, therefore, is restricted to Type I, II or IV, so that the combustible loading of cars is not exceeded by that of the structure.

406.3.3.1 Openings. For natural ventilation purposes, the exterior side of the structure shall have uniformly distributed openings on two or more sides. The area of such openings in *exterior walls* on a tier must be at least 20 percent of the total perimeter wall area of each tier. The aggregate length of the openings considered to be providing natural ventilation shall constitute a minimum of 40 percent of the perimeter of the tier. Interior walls shall be at least 20 percent open with uniformly distributed openings.

> **Exception:** Openings are not required to be distributed over 40 percent of the building perimeter where the required openings are uniformly distributed over two opposing sides of the building.

❖ Key attributes of an open parking garage, based on the fire records of such facilities, include the exterior wall openings and ventilation of the structure. This section requires that 40 percent of the building perimeter has openings that are uniformly distributed on no less than two sides of the structure. In addition to providing for adequate distribution, the section also requires that a minimum of 20 percent of the total perimeter wall area at each level must be open. The openings in the exterior wall must be free so that natural ventilation will occur without interior wall obstructions.

In instances where the open parking garage is provided with openings on two opposite sides of the structure, thereby providing cross ventilation, the 40-percent criterion does not apply; however, the openings must still meet the 20-percent criterion of the total perimeter wall area.

In every case, interior walls are to have distributed openings totaling 20 percent of the wall area to allow ventilation of all spaces (see Figure 406.3.3.1).

406.3.4 Uses. Mixed uses shall be allowed in the same building as an *open parking garage* subject to the provisions of Sections 402.7.1, 406.3.13, 508.1, 509.3, 509.4 and 509.7.

❖ In contrast to the limitations found in the legacy codes, under the code, open parking garages can be constructed as part of a mixed-occupancy building. However, if the open parking garage is in a mixed-occupancy building, it cannot be designed to take advantage of the height and area increases allowed by Sections 406.3.5 and 406.3.6. The code provides references to the placement of open parking garages as part of a covered mall building. It also provides specific references to three of the special building provisions in Section 509 where open parking garages may be located above or below other buildings. There is also a reference to the mixed-occupancy provisions of Section 508, wherein parking is no longer listed as an inci-

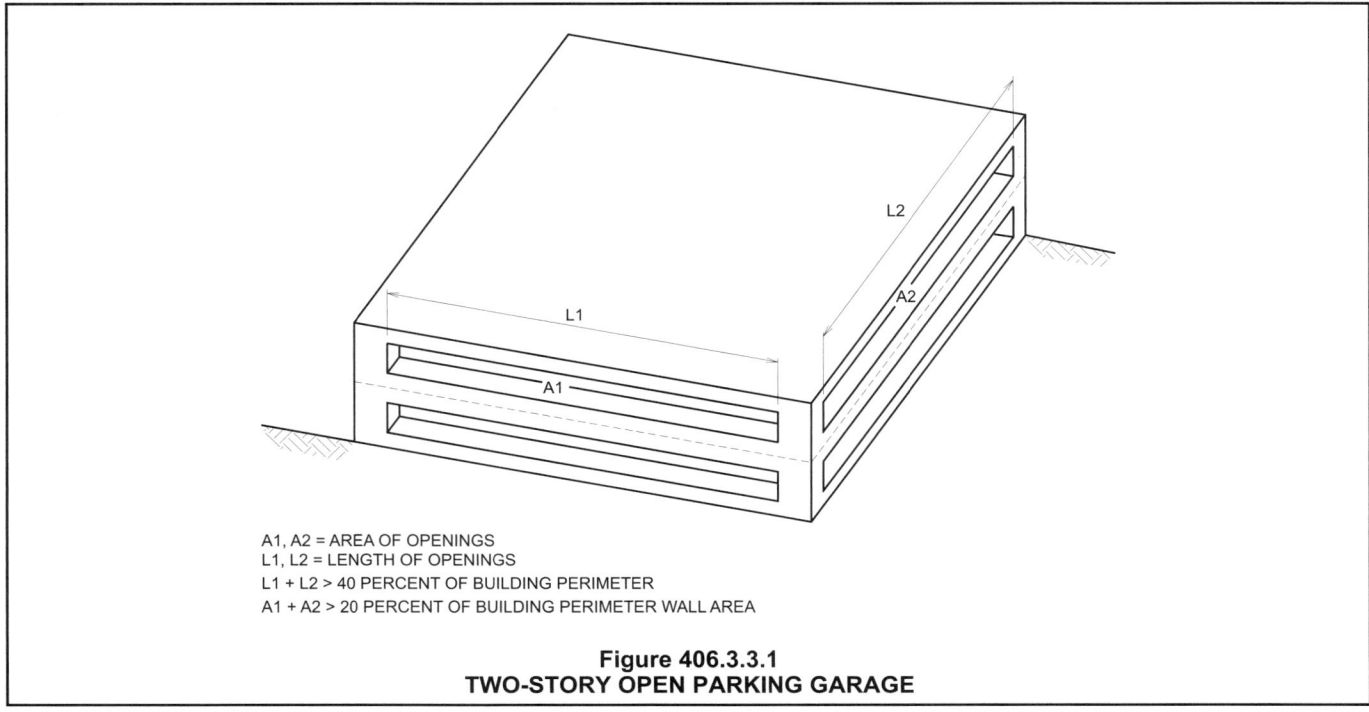

A1, A2 = AREA OF OPENINGS
L1, L2 = LENGTH OF OPENINGS
L1 + L2 > 40 PERCENT OF BUILDING PERIMETER
A1 + A2 > 20 PERCENT OF BUILDING PERIMETER WALL AREA

Figure 406.3.3.1
TWO-STORY OPEN PARKING GARAGE

dental accessory occupancy. Therefore, parking must be treated as an accessory use or an additional primary use in a mixed-occupancy building. There is also a reference to Section 406.3.13 where there are specific prohibitions for certain activities within an open parking garage.

406.3.5 Area and height. Area and height of *open parking garages* shall be limited as set forth in Chapter 5 for Group S-2 occupancies and as further provided for in Section 508.1.

❖ When an open parking garage is located within a building containing other occupancies, the code requires that the height and area limitations for the open parking garage be determined by using the normal provisions found in Chapter 5 for a Group S-2 occupancy. This essentially means that the garage would receive the same allowable areas of an enclosed parking garage or a storage building that was classified in the same occupancy.

406.3.5.1 Single use. When the *open parking garage* is used exclusively for the parking or storage of private motor vehicles, with no other uses in the building, the area and height shall be permitted to comply with Table 406.3.5, along with increases allowed by Section 406.3.6.

Exception: The grade-level tier is permitted to contain an office, waiting and toilet rooms having a total combined area of not more than 1,000 square feet (93 m²). Such area need not be separated from the *open parking garage*.

In *open parking garages* having a spiral or sloping floor, the horizontal projection of the structure at any cross section shall not exceed the allowable area per parking tier. In the case of an *open parking garage* having a continuous spiral floor, each 9 feet 6 inches (2896 mm) of height, or portion thereof, shall be considered a tier.

The clear height of a parking tier shall not be less than 7 feet (2134 mm), except that a lower clear height is permitted in mechanical-access *open parking garages* where *approved* by the *building official*.

❖ When an open parking garage is located in a building that is only used as a parking garage, the type of haz-

ard that is created by the occupancy is greatly reduced. Because of the openness requirements in Section 406.3.3.1 and the dissipation of heated gases, as well as the overall low fuel load, the height and area limitations for single-use open parking garages are not the same as for other buildings of Group S-2. The height requirements are based on actual fire experience and full-scale fire tests conducted in the United States, Great Britain, Japan and Switzerland. For these single-use buildings that comply with the other provisions of this section, Table 406.3.5 replaces Table 503 for the determination of heights and areas. Although these open parking garages are considered as being a single occupancy, the exception permits a limited amount of support-type spaces to be included in the structure.

TABLE 406.3.5. See below.

❖ Table 406.3.5 is used in the same manner as Table 503. For a given construction type, the height and area limitations can be determined from the table. For example, an open parking garage of Type IIB construction may be a maximum of eight tiers in height for a ramp access garage and the area would then be limited to 50,000 square feet (4545 m²) per tier.

The height restrictions in the table are measured as defined in Section 202. In this instance, parking may be permitted on the highest level of the structure, which also constitutes the roof of the structure. The areas provided are per tier.

406.3.6 Area and height increases. The allowable area and height of *open parking garages* shall be increased in accordance with the provisions of this section. Garages with sides open on three-fourths of the building's perimeter are permitted to be increased by 25 percent in area and one tier in height. Garages with sides open around the entire building's perimeter are permitted to be increased by 50 percent in area and one tier in height. For a side to be considered open under the above provisions, the total area of openings along the side shall not be less than 50 percent of the interior area of the side at each tier and such openings shall be equally distributed along the length of the tier.

TABLE 406.3.5
OPEN PARKING GARAGES AREA AND HEIGHT

TYPE OF CONSTRUCTION	AREA PER TIER (square feet)	HEIGHT (in tiers)		
		Ramp access	Mechanical access	
			Automatic sprinkler system	
			No	Yes
IA	Unlimited	Unlimited	Unlimited	Unlimited
IB	Unlimited	12 tiers	12 tiers	18 tiers
IIA	50,000	10 tiers	10 tiers	15 tiers
IIB	50,000	8 tiers	8 tiers	12 tiers
IV	50,000	4 tiers	4 tiers	4 tiers

For SI: 1 square foot = 0.0929 m².

Allowable tier areas in Table 406.3.5 shall be increased for *open parking garages* constructed to heights less than the table maximum. The gross tier area of the garage shall not exceed that permitted for the higher structure. At least three sides of each such larger tier shall have continuous horizontal openings not less than 30 inches (762 mm) in clear height extending for at least 80 percent of the length of the sides and no part of such larger tier shall be more than 200 feet (60 960 mm) horizontally from such an opening. In addition, each such opening shall face a street or *yard* accessible to a street with a width of at least 30 feet (9144 mm) for the full length of the opening, and standpipes shall be provided in each such tier.

Open parking garages of Type II construction, with all sides open, shall be unlimited in allowable area where the *building height* does not exceed 75 feet (22 860 mm). For a side to be considered open, the total area of openings along the side shall not be less than 50 percent of the interior area of the side at each tier and such openings shall be equally distributed along the length of the tier. All portions of tiers shall be within 200 feet (60 960 mm) horizontally from such openings or other natural ventilation openings as defined in Section 406.3.3.1. These openings shall be permitted to be provided in *courts* with a minimum dimension of 20 feet (6096 mm) for the full width of the openings.

❖ The area modifications permitted by Section 406.3.6 for open building perimeters that are similar to Section 506.2 are also applicable to the area restrictions in Table 406.3.5. The section essentially provides three different ways to qualify as an open parking garage and use the increased heights and areas.

- Open parking garages of Type IA construction are unlimited in area and height. This is consistent with Type I construction in Table 503.

- Open parking garages of Type IB construction are also unlimited in area, but are limited in height to 12 tiers (or 18 tiers for a mechanical-access garage which is sprinkler protected). If a Type IB garage meets either of the open perimeter criteria below, it can be one tier higher than allowed in the table.

- Where the building is of Type II or IV construction and 75 percent of the perimeter is open, it is permitted to be increased 25 percent in area per tier [or from 50,000 square feet (4645 m²) to 62,500 square feet (5806 m²)] and one tier in height above the limits for the type of construction in Table 406.3.5.

- Where the building is of Type II or IV construction and is open for the entire perimeter, a 50-percent increase in area is permitted, as well as one tier in height. This results in the same number of tiers as the 75-percent open perimeter, but in this case the area of each tier can be as large as 75,000 square feet (6968 m²).

In addition to complying with the opening criteria found in Section 406.3.3.1, the openings for an increased area and/or height open parking garage

must be at least equal to 50 percent of the interior area of each tier and must be at least 3 feet (914 mm) in vertical dimension. They must be found on 80 percent of the perimeter sides considered open and each such opening shall front on a yard or public way with a minimum dimension of 30 feet (9144 mm). Each part of enlarged tiers needs to be within 200 feet (60 960 mm) of such openings.

For Type II garages there is a third design option. If all sides are open, the area per tier is unlimited provided the building height does not exceed 75 feet (22 860 mm) above grade plane. As in the other options, the openings must be at least 50 percent of the interior of each tier, but in this case the openings must be evenly distributed around the perimeter. All portions of each tier must be within 200 feet (60 960 mm) of such openings, and, where necessary, some of the openings needed for this ventilation may be located in courts, provided the courts have a least dimension of 20 feet (6096 mm).

406.3.7 Fire separation distance. *Exterior walls* and openings in *exterior walls* shall comply with Tables 601 and 602. The distance to an adjacent *lot line* shall be determined in accordance with Table 602 and Section 705.

❖ The required fire separation is intended to provide sufficient open space adjacent to the exterior wall openings to allow for free ventilation and to reduce the probability for fire spread to adjacent structures. The exterior wall openings in an open parking garage are not subject to the opening limitations with a fire separation distance greater than 10 feet (3048 mm) (see Table 705.8, Note g).

406.3.8 Means of egress. Where persons other than parking attendants are permitted, *open parking garages* shall meet the *means of egress* requirements of Chapter 10. Where no persons other than parking attendants are permitted, there shall not be less than two 36-inch-wide (914 mm) *exit stairways*. Lifts shall be permitted to be installed for use of employees only, provided they are completely enclosed by noncombustible materials.

❖ Typical means of egress elements are required for open parking garages based on a design occupant load of one person per 200 square feet (19 m²) of gross floor area (see Table 1004.1.1). For mechanical-access open parking garages or for attendant-only ramp-access parking garages, the code would permit two exit stairs, each 36 inches (914 mm) wide.

406.3.9 Standpipes. Standpipes shall be installed where required by the provisions of Chapter 9.

❖ Depending on the height of an open parking garage, Section 905.3.1 may require a standpipe system. The standpipe system would aid the fire department in on-site fire-fighting operations, since Table 406.3.5 permits buildings that may not allow for conventional hoses and apparatus to be carried from ground operations to the fire floor.

406.3.10 Sprinkler systems. Where required by other provisions of this code, *automatic sprinkler systems* and standpipes shall be installed in accordance with the provisions of Chapter 9.

❖ Although sprinkler systems are not mandated by Section 903.2 for open parking garages, sprinkler systems may be required for other occupancies based on the special provisions of Section 509. In addition, where an open parking garage is in a mixed-occupancy building, the sprinkler requirements of the other occupancy in Section 903.2 could require the sprinkler system to be provided throughout the building. For example, an open parking garage on the lower levels of a Group R occupancy high-rise would need to be sprinkler protected.

406.3.11 Enclosure of vertical openings. Enclosure shall not be required for vertical openings except as specified in Section 406.3.8.

❖ Based on the use of open parking garages requiring convenient and frequent access between floor levels, enclosure of most vertical openings is not required or desired due to the opening requirements of Section 406.3.3.1. This would include the interior exit stairways as permitted by Section 1022.1, Exception 4. Where a designer chooses to provide an enclosure—for example, around an elevator hoistway—since the enclosure does not need to be rated, there is no requirement that a lobby be provided for the elevator (see IBC Interpretation No. 36-07).

406.3.12 Ventilation. Ventilation, other than the percentage of openings specified in Section 406.3.3.1, shall not be required.

❖ Mechanical ventilation systems are not required in a structure that is inherently open to the exterior atmosphere.

406.3.13 Prohibitions. The following uses and alterations are not permitted:

1. Vehicle repair work.

2. Parking of buses, trucks and similar vehicles.

3. Partial or complete closing of required openings in exterior walls by tarpaulins or any other means.

4. Dispensing of fuel.

❖ This section reinforces the concepts and requirements of Section 406.3.3.1 and the definition of "Open parking garage." Any other use or alteration of an open parking garage, which is designed to take advantage of the increased heights and areas, is specifically prohibited.

406.4 Enclosed parking garages.

406.4.1 Heights and areas. Enclosed vehicle parking garages and portions thereof that do not meet the definition of *open parking garages* shall be limited to the allowable heights and areas specified in Table 503 as modified by Sections 504, 506 and 507. Roof parking is permitted.

❖ Enclosed parking garages are considered to be storage occupancies. In addition to the provisions of Sec-

tions 406.2 and 406.4, enclosed parking garages must also comply with the code provisions for Group S-2 low-hazard storage occupancies.

The overall building fire load in enclosed parking garages is typically low because of the considerable amount of metal in vehicles that absorbs heat and the average weight of combustibles per square foot being relatively low [approximately 2 psf (9.8 kg/m²)]. Still, a vehicle fire may be quite extensive.

Special height and area limitations allowed for open parking garages in Table 406.3.5 are not provided for enclosed parking garages. The typical allowable building heights and areas of Table 503 are to be used for these Group S-2 structures. It should be noted that Section 509 contains special provisions for enclosed parking garages located beneath other groups and occupancies. See that section for further discussion of these special building arrangements.

406.4.2 Ventilation. A mechanical ventilation system shall be provided in accordance with the *International Mechanical Code.*

❖ Enclosed parking garages are required to be ventilated in accordance with the provisions of the IMC. If motor vehicles will be operated for a period of time exceeding 10 seconds, the ventilation return air for the enclosed parking garage ventilation system must be exhausted (see Section 404 of the IMC).

406.5 Motor fuel-dispensing facilities.

406.5.1 Construction. Motor fuel-dispensing facilities shall be constructed in accordance with the *International Fire Code* and Sections 406.5.1 through 406.5.3.

❖ Motor fuel-dispensing facilities are those areas where motor fuels are stored, sold or otherwise dispensed to vehicles. These buildings and structures are classified as Group M. Motor fuel-dispensing facilities must be constructed in accordance with the applicable requirements of Chapter 22 of the IFC. Specific details are provided for the special hazards relative to fuel-dispensing systems and areas.

406.5.2 Vehicle fueling pad. The vehicle shall be fueled on noncoated concrete or other *approved* paving material having a resistance not exceeding 1 megohm as determined by the methodology in EN 1081.

❖ The expansion of the use of hydrogen as a vehicle fuel necessitates that the code more clearly address the hazard of potential electrostatic discharge around fueling facilities. The preferred material for fueling pads is concrete, but the referenced standard provides alternative materials to achieve an antistatic performance of the pad.

406.5.3 Canopies. Canopies under which fuels are dispensed shall have a clear, unobstructed height of not less than 13 feet 6 inches (4115 mm) to the lowest projecting element in the vehicle drive-through area. Canopies and their supports over pumps shall be of noncombustible materials, *fire-retardant-treated wood* complying with Chapter 23, wood of Type IV sizes or of construction providing 1-hour *fire resistance.* Combustible materials used in or on a canopy shall comply with one of the following:

1. Shielded from the pumps by a noncombustible element of the canopy, or wood of Type IV sizes;

2. Plastics covered by aluminum facing having a minimum thickness of 0.010 inch (0.30 mm) or corrosion-resistant steel having a minimum base metal thickness of 0.016 inch (0.41 mm). The plastic shall have a *flame spread index* of 25 or less and a smoke-developed index of 450 or less when tested in the form intended for use in accordance with ASTM E 84 or UL 723 and a self-ignition temperature of 650°F (343°C) or greater when tested in accordance with ASTM D 1929; or

3. Panels constructed of light-transmitting plastic materials shall be permitted to be installed in canopies erected over motor vehicle fuel-dispensing station fuel dispensers, provided the panels are located at least 10 feet (3048 mm) from any building on the same lot and face yards or streets not less than 40 feet (12 192 mm) in width on the other sides. The aggregate areas of plastics shall not exceed 1,000 square feet (93 m²). The maximum area of any individual panel shall not exceed 100 square feet (9.3 m²).

❖ Almost all motor fuel-dispensing facilities are provided with a flat canopy-style roof structure that is column supported. Obviously these structures must be designed to meet the applicable wind and snow loads of Chapter 16 for life safety purposes. For fire safety purposes, the canopies must pose little fire risk and be located above any immediate fire hazards from the fuel-dispensing areas. Limitations are provided for all combustible materials used in the canopy construction.

406.5.3.1 Canopies used to support gaseous hydrogen systems.
Canopies that are used to shelter dispensing operations where flammable compressed gases are located on the roof of the canopy shall be in accordance with the following:

1. The canopy shall meet or exceed Type I construction requirements.

2. Operations located under canopies shall be limited to refueling only.

3. The canopy shall be constructed in a manner that prevents the accumulation of hydrogen gas.

❖ This section of Chapter 4 has been added as part of the overall package of new code requirements related to the gaseous hydrogen concept of fuel. These additional requirements are necessary to abate the inherent hazards while permitting the use of weather shelters. Not only must the canopies be of noncombustible materials but they must also have the fire-resistance ratings specified in Table 601 for either Type IA or IB construction.

406.6 Repair garages.

406.6.1 General.
Repair garages shall be constructed in accordance with the *International Fire Code* and Sections 406.6.1 through 406.6.6. This occupancy shall not include motor fuel-dispensing facilities, as regulated in Section 406.5.

❖ Repair garages are those in which provisions are made for the care, repair and painting of vehicles. Repair garages have an inherently higher fire hazard represented by numerous vehicles in some state of repair, with possible spray painting. As such, these buildings or structures are classified as Group S-1. Repair garages must also meet the applicable requirements of Chapter 22 of the IFC and Section 416 if there are paint-spraying operations.

All motor vehicle work that involves repairs or reconstruction of damaged or nonfunctioning vehicles is considered a higher hazard. These types of facilities provide a variety of services, including tune-ups, oil changes, engine or transmission overhauls and body work, and are all considered as repair garages. Additional protection and detection systems are required to match the increased hazards.

406.6.2 Mixed uses.
Mixed uses shall be allowed in the same building as a repair garage subject to the provisions of Section 508.1.

❖ Repair garages usually have integral office and administration areas or are part of an automobile dealership. To minimize the hazards and functions housed in the repair areas, the code reminds the user that these spaces are considered as different occupancies and must therefore comply with the general code requirements for mixed uses. The actual requirements will depend on the size of the spaces and how the designer chooses to deal with the different uses.

406.6.3 Ventilation.
Repair garages shall be mechanically ventilated in accordance with the *International Mechanical Code*. The ventilation system shall be controlled at the entrance to the garage.

❖ Repair garages are required to be ventilated in accordance with the provisions of the IMC. If motor vehicles will be operated for a period of time exceeding 10 seconds, the ventilation return air for the repair garage ventilation system must be exhausted (see Sections 403 and 404 of the IMC).

406.6.4 Floor surface.
Repair garage floors shall be of concrete or similar noncombustible and nonabsorbent materials.

Exception: Slip-resistant, nonabsorbent, *interior floor finishes* having a critical radiant flux not more than 0.45 W/cm², as determined by NFPA 253, shall be permitted.

❖ To avoid a buildup of flammable liquids on the floor of repair garages, the floor finish is in general required to be noncombustible and nonabsorbent. Because of pollution concerns, however, garage floors are not automatically required to be drained into the building drainage system. This is consistent with requirements of the EPA for storm water drainage, as well as the philosophy of hazardous material handling (i.e., localize spills and treat them). If floor drains are provided, however, they must be installed in accordance with the IPC.

406.6.5 Heating equipment. Heating equipment shall be installed in accordance with the *International Mechanical Code*.

❖ As part of the special use and occupancy requirements for repair garages, all possible ignition sources must be controlled and isolated. Specifically, all heating equipment must be installed in accordance with the IMC, which contains provisions that address items such as the elevation of the appliance above the floor and the use of overhead unit heaters.

[F] 406.6.6 Gas detection system. Repair garages used for repair of vehicles fueled by nonodorized gases, such as hydrogen and nonodorized LNG, shall be provided with a flammable gas detection system.

❖ Some gases contain additives, such as ethyl mercaptan, that produce pungent odors for easy sensory detection and recognition. If the repair garage repairs vehicles equipped with fuel systems that do not use these odorized gases, a gas detection system must be installed to replace the lost sensory detection capability.

[F] 406.6.6.1 System design. The flammable gas detection system shall be *listed* or *approved* and shall be calibrated to the types of fuels or gases used by vehicles to be repaired. Gas detectors or sensors shall be *listed* in accordance with UL 2075 and shall indicate the gases they are intended to detect. The gas detection system shall be designed to activate when the level of flammable gas exceeds 25 percent of the lower flammable limit (LFL). Gas detection shall also be provided in lubrication or chassis service pits of repair pits if garages used for repairing nonodorized LNG-fueled vehicles.

❖ The flammable gas detection system is designed to produce an alarm or signal when exposed to different concentrations of gases or vapor and the gas detector or gas sensor is a critical part of the system for the detection of the different gasses. This section requires gas detection equipment to be listed in accordance with UL 2075 and include an indication of the different gases it will detect. Under UL 2075, a set of flammable gases and concentrations (PPM) is developed for each detector or sensor and the manufacturer is required to provide information as to what gases and the concentrations the device is designed to detect. Tests under the standard then verify the performance of each detector or sensor for each gas it is designed to detect. The gases that the equipment will detect may be shown in the manufacturer's instructions rather than on the product. This section will require quick-lube-type facilities that change oil and lubricate vehicles to install gas detection systems in the pit area if they service vehicles that are equipped with LNG fuel systems using nonodorized LNG.

[F] 406.6.6.2 Operation. Activation of the gas detection system shall result in all of the following:

1. Initiation of distinct audible and visual alarm signals in the repair garage.

2. Deactivation of all heating systems located in the repair garage.

3. Activation of the mechanical ventilation system, where the system is interlocked with gas detection.

❖ Once the required flammable gas detection system identifies a gas level exceeding 25 percent of the lower explosive limit (LEL), it must do three things: alert the occupants of the emergency, cut power to all possible ignition sources and dilute the concentration level with additional ventilation air.

[F] 406.6.6.3 Failure of the gas detection system. Failure of the gas detection system shall result in the deactivation of the heating system, activation of the mechanical ventilation system when the system is interlocked with the gas detection system and cause a trouble signal to sound in an *approved* location.

❖ The required flammable gas detection system must include a default mode of operation. If the system malfunctions or loses power, it must in effect do the same operational requirements referenced in Section 406.6.6.2. This redundancy will lessen the likelihood of an explosive fire risk for the specialized motor fuels.

SECTION 407
GROUP I-2

407.1 General. Occupancies in Group I-2 shall comply with the provisions of Sections 407.1 through 407.9 and other applicable provisions of this code.

❖ The provisions contained in Section 407 reflect previous requirements of the other former model codes, full-scale fire tests and historical fire experience. Features unique to health care facilities, such as a low fuel load, the presence of staff personnel, operating practices and procedures and the regulatory process, were also included in developing the provisions. One of the concerns addressed by this section is the need to provide an acceptable level of protection without interfering with the operation of the facility.

Underlying the protection requirements for Group I-2 is a "defend-in-place" philosophy based on the difficulties associated with evacuating such facilities.

The first level of protection established is the individual room (typically a patient sleeping room). Horizontal evacuation to an adjacent smoke compartment provides the second level of protection. If necessary, the third level of protection involves evacuation of the floor or entire building.

This section has been created not only because of considerations unique to health care facilities, but also to facilitate the consideration of the entire package of protection features. The basic protection features include early detection, fire containment, horizontal evacuation and fire extinguishment. Additionally, Chapter 4 of the IFC contains criteria for emergency planning and preparedness.

As previously noted, the provisions of this section are based in part on a relatively low fuel load. Certain areas within Group I-2, however, will have a higher concentration of combustibles. These areas, as well as other areas that represent a specific hazard within Group I-2, are addressed in Section 508.2.5 as incidental accessory occupancy areas. Because of the nature of the activity, the incidental accessory occupancy area is required to be protected with an automatic fire-extinguishing system, separated with a fire barrier and horizontal assembly or both (see Table 508.2.5).

These provisions apply to all occupancies classified as Group I-2. The requirements of Section 407 are intended to be additional special use and occupancy provisions for facilities that provide care for more than five persons who are not capable of self-preservation. Other requirements for Group I-2 occupancies are provided throughout the code.

407.2 Corridors. *Corridors* in occupancies in Group I-2 shall be continuous to the *exits* and separated from other areas in accordance with Section 407.3 except spaces conforming to Sections 407.2.1 through 407.2.4.

❖ Since the occupants of areas classified as Group I-2 may be incapable of self-preservation, the means of egress serving such areas must facilitate the evacuation of such persons, as well as other people with mobility impairments. Accordingly, exit access corridors serving Group I-2 occupancies are required to be continuous to the exits and, except as provided for in Sections 407.2.1 through 407.2.4, separated from other use areas. As such, corridors in Group I-2 occupancies are intended to provide a protected path of travel that, once entered, leads directly and continuously to the exits (see also commentary, Sections 1014.2.2 and 1018.6). Note that a door through a horizontal exit constructed with a fire barrier or fire wall would constitute an exit (see Section 1025); however, moving through a smoke barrier into another smoke compartment is not a horizontal exit and therefore corridor continuity must be maintained on both sides of the smoke barrier.

To address the various operational requirements of occupancies classified as Group I-2, the code includes provisions that both permit various areas to be open to corridors and address suite arrangements of the exit access portions of the facilities. Standards for areas open to corridors are covered primarily in this chapter. The design of suites is regulated by provisions found in Sections 1014.2.3 though 1014.2.7. See the commentary to Section 1014.2 for information on suite arrangements.

For the ease of operations of hospitals, it is often desirable to have certain areas open to exit access corridors. Recognizing these needs and that such areas are equipped throughout with an automatic sprinkler system, the code permits specific areas to be open to corridors, provided they are properly arranged and equipped with compensating protection features (e.g., automatic fire detection). Sections 407.2.1 through 407.2.4 identify those areas and the protection features required to maintain the safety of the corridor.

In addition to practical reasons, such as in the case of nurses' stations, it is also desirable to designate certain areas being open to the corridor for treatment purposes. In many instances, the patient is encouraged to leave the patient room for both physical as well as social benefit. So that patients can be encouraged to socialize, the areas need to be open to the corridor. If the area were separated with only a door opening, many patients would not be willing to enter the room and benefit from group activities and social interaction.

Such waiting areas also serve a useful purpose during a fire emergency. If evacuation becomes necessary, patients usually will be relocated to an adjacent smoke compartment in which the areas open to the corridor will serve as a refuge area. Section 407.4.1 contains provisions for the net area that must be available for the relocation of patients. Waiting areas provide a space in which patients can be relocated such that treatment can continue during a fire emergency.

407.2.1 Waiting and similar areas. Waiting areas and similar spaces constructed as required for *corridors* shall be permitted to be open to a *corridor*, only where all of the following criteria are met:

1. The spaces are not occupied for patient *sleeping units*, treatment rooms, hazardous or incidental accessory occupancies in accordance with Section 508.2.

2. The open space is protected by an automatic fire detection system installed in accordance with Section 907.

3. The *corridors* onto which the spaces open, in the same smoke compartment, are protected by an automatic fire detection system installed in accordance with Section 907, or the smoke compartment in which the spaces are located is equipped throughout with quick-response sprinklers in accordance with Section 903.3.2.

4. The space is arranged so as not to obstruct access to the required *exits*.

❖ In the design of health care facilities, it is desirable to have small waiting areas and similar spaces open to the corridor. Such seating areas or gathering spaces provide residents and others a place to visit outside of the patient rooms. This provides physical and social benefits to the residents by encouraging them to leave their rooms and socialize with others. In consideration of the additional protection features specified, small waiting areas and similar spaces are permitted to be open to corridors in Group I-2 occupancies within each smoke compartment provided they are constructed to the corridor standards for Group I-2 occupancies, as specified in Section 407.3.

Patient rooms, as well as hazardous and incidental accessory occupancy areas, cannot be open to the corridors under this section. In addition, treatment areas, other than those for mental health treatment, cannot be open to corridors (see also the suite provisions in Section 1014.2). Areas similar to waiting rooms could include common dining areas, reception or information stations and patient check-in stations. If a dining area is open to a corridor, any associated food preparation area should probably remain separated from the corridor. General hospital office areas beyond those associated with waiting or reception areas also remain separated from the corridor (see Figure 407.2.1).

To reduce the likelihood that a fire within such an open space could develop beyond the incipient stage, thereby jeopardizing the integrity of the corridor, the area is to be equipped with an automatic fire detection system (see Section 907). The detectors are to be located so as to provide the appropriate coverage to the space.

There is no size limit set by the code for these spaces. Due to the unlimited size of the space and consequently the potential for a significant number of people in the area, an automatic fire detection system is also required in all corridors open to the spaces within the smoke compartment, or the smoke compartment in which such spaces are located must be protected throughout with quick-response sprinklers.

Waiting areas and similar spaces open to corridors must be arranged so as to not obstruct access to exits. There must be a clear path of travel of the minimum required width for the corridor, which should be maintained at all times. As such, furniture layouts must be arranged in a manner that obstruction of the exits does not occur under normal use.

407.2.2 Nurses' stations. Spaces for doctors' and nurses' charting, communications and related clerical areas shall be permitted to be open to the *corridor*, when such spaces are constructed as required for *corridors*.

❖ Nurses' stations need to be located to provide quick access to patients and to permit visual or audible monitoring of the patient rooms. For these and other practical reasons, nurses' stations are permitted to be open to the corridor, provided the walls surrounding the area are constructed as required for corridor walls (see Figure 407.2.2). Except as required in Section 407.7, the additional protection of automatic fire detection is not required because of the reduced risk to life safety and attendance by the staff (see Section 407.7). This section is not intended to apply to nurses' offices, administrative supply storage areas, drug distribution stations or similar areas with a higher fuel load and from which continual observation of the nursing unit is not essential.

407.2.3 Mental health treatment areas. Areas wherein mental health patients who are not capable of self-preservation are housed, or group meeting or multipurpose therapeutic spaces other than incidental accessory occupancies in accordance with Section 508.2.5, under continuous supervision by facility

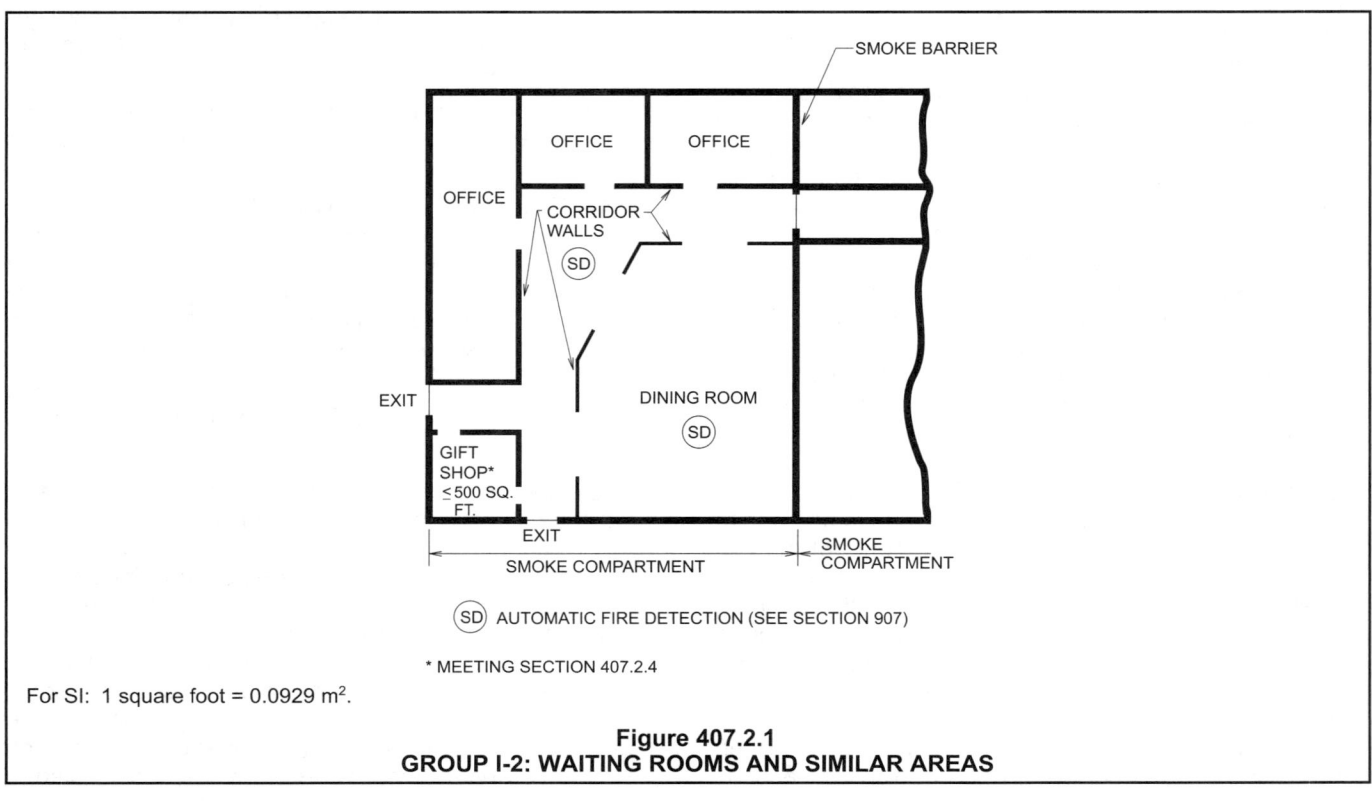

For SI: 1 square foot = 0.0929 m².

Figure 407.2.1
GROUP I-2: WAITING ROOMS AND SIMILAR AREAS

staff, shall be permitted to be open to the *corridor*, where the following criteria are met:

1. Each area does not exceed 1,500 square feet (140 m²).

2. The area is located to permit supervision by the facility staff.

3. The area is arranged so as not to obstruct any access to the required *exits*.

4. The area is equipped with an automatic fire detection system installed in accordance with Section 907.2.

5. Not more than one such space is permitted in any one smoke compartment.

6. The walls and ceilings of the space are constructed as required for *corridors*.

❖ To encourage residents of mental health facilities to participate in group meetings and therapeutic activities, treatment areas may be open to the corridor, provided that the area does not exceed 1,500 square feet (140 m²) and only one such area is provided within a smoke compartment. Other protection features that must be provided to permit the area to be open include: staff supervision; unobstructed access to exits; automatic fire detection in accordance with Section 907.2 within the area and the surrounding walls and ceilings constructed as required for corridors. These provisions are based on the increased mobility of the occupants and direct supervision by staff. This section applies to all areas of a building classified as Group I-2, including smoke compartments that contain patient sleeping rooms that house mental health patients who are not physically capable of self-preservation.

As mentioned in the commentary to Section 407.2, one of the primary reasons for permitting areas to be open to the corridor is to encourage and facilitate participation by the residents in group activities. If the space is not used for such treatment or therapeutic purposes, the need to have the area open to the corridor is reduced and the section is not applicable.

407.2.4 Gift shops. Gift shops less than 500 square feet (46.5 m²) in area shall be permitted to be open to the *corridor* provided the gift shop and storage areas are fully sprinklered and storage areas are protected in accordance with Section 508.2.5.

❖ Gift shops are not permitted to be open to a corridor in occupancies classified as Group I-2 unless they are less than 500 square feet (46 m²) in area and protected with the automatic sprinkler system required for the Group I-2 occupancy. The section requires any associated storage to be protected in accordance with Section 508.2.5. In the 2006 code, storage was listed as an incidental use in Section 508. In the 2009 code, Table 508.2.5 was amended and storage is no longer listed as an incidental accessory occupancy; however, no corresponding change was made to Section 407.2.4. At this time there is no specific requirement for storage areas referenced for protection of this section (see Figure 407.2.1).

407.3 Corridor walls. *Corridor* walls shall be constructed as smoke partitions in accordance with Section 711.

❖ Corridor walls are required to form a barrier to limit smoke transfer, thus providing protection for exit access routes, from separate areas containing combustible materials that could produce smoke and to

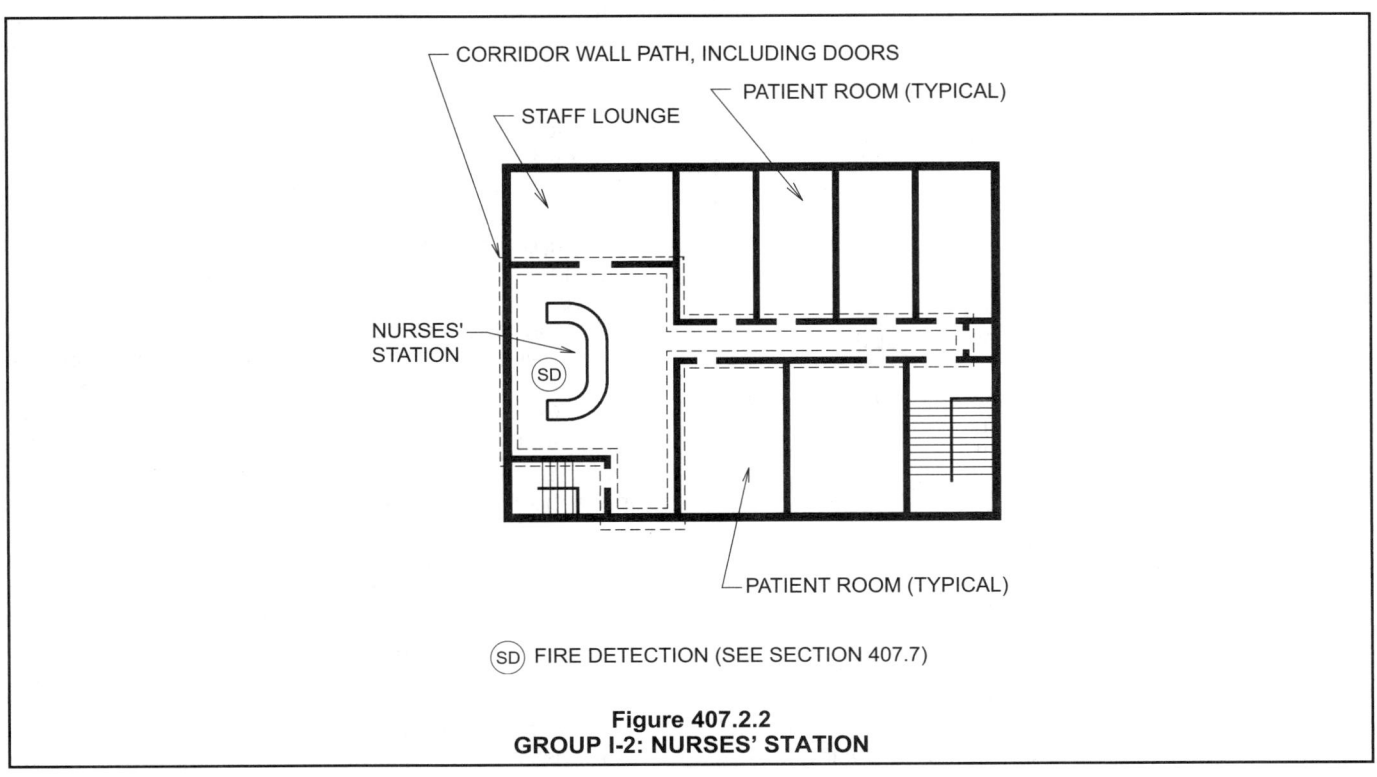

SD FIRE DETECTION (SEE SECTION 407.7)

Figure 407.2.2
GROUP I-2: NURSES' STATION

provide adequate separation of the patient sleeping rooms. This section essentially provides a reference to the requirements of Section 711 for smoke partitions. In a building protected throughout with an automatic sprinkler system, the probability that a fire will develop that is life threatening to persons outside the room of origin is reduced; therefore, corridor walls need only be able to resist the passage of smoke.

The corridor wall is not required to have a fire-resistance rating, but the wall is to terminate at the underside of the floor or roof deck above or to the underside of the ceiling membrane if it is capable of resisting the passage of smoke. The walls must be of materials consistent with the building's type of construction classification.

Note that the provisions of Section 508.2.5 must be considered in determining the fire-resistance rating of any corridor walls that also enclose incidental accessory occupancies (see Figure 407.3).

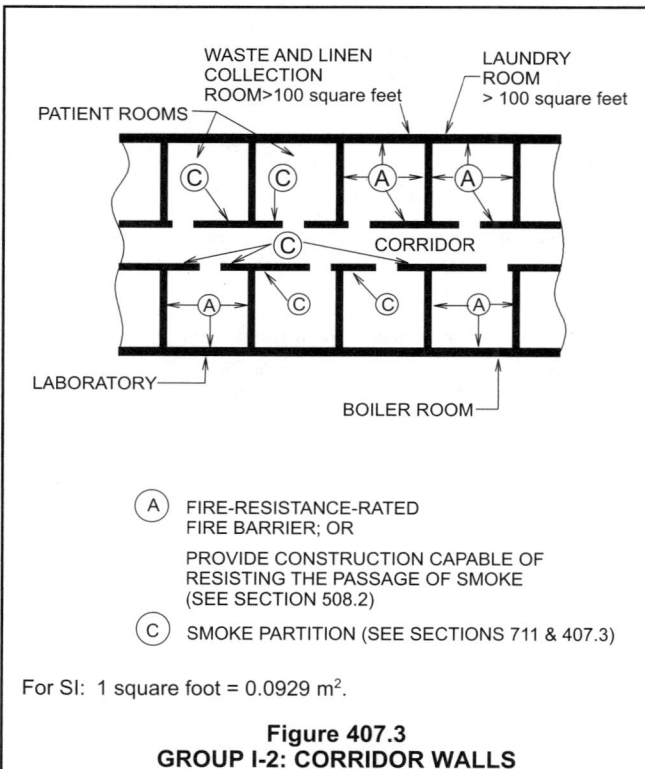

WASTE AND LINEN COLLECTION ROOM>100 square feet

LAUNDRY ROOM > 100 square feet

PATIENT ROOMS

CORRIDOR

LABORATORY

BOILER ROOM

(A) FIRE-RESISTANCE-RATED FIRE BARRIER; OR

PROVIDE CONSTRUCTION CAPABLE OF RESISTING THE PASSAGE OF SMOKE (SEE SECTION 508.2)

(C) SMOKE PARTITION (SEE SECTIONS 711 & 407.3)

For SI: 1 square foot = 0.0929 m².

**Figure 407.3
GROUP I-2: CORRIDOR WALLS**

407.3.1 Corridor doors. *Corridor* doors, other than those in a wall required to be rated by Section 508.2.5 or for the enclosure of a vertical opening or an *exit*, shall not have a required *fire protection rating* and shall not be required to be equipped with self-closing or automatic-closing devices, but shall provide an effective barrier to limit the transfer of smoke and shall be equipped with positive latching. Roller latches are not permitted. Other doors shall conform to Section 715.4.

❖ As with corridor wall construction (see commentary, Section 407.3), the door assemblies need only be capable of resisting the passage of smoke. Corridor doors in occupancies classified as Group I-2 are not required to be self-closing or automatic closing except

for doors serving incidental accessory occupancy areas as required in Section 508.2.5, doors in walls that enclose a vertical opening and exit doors. This provision is primarily intended to apply to patient room corridor doors. Doors in walls that separate incidental accessory occupancy areas in accordance with Section 508.2.5 are required to be self-closing to minimize the possibility that corridors will be affected by a fire originating in those adjacent areas. Self-closing or automatic-closing doors in enclosures of vertical openings are still required to avoid a breech in vertical opening protection that would enable rapid fire spread from story to story and for adequate protection of the exits. It is important to recognize that this section contains a specific set of requirements for the corridor doors in this occupancy and that the provisions of this section must be followed versus the provisions of Section 711.5 (see code and commentary, Section 102.1). Full-scale fire tests and historical experience indicate that the automatic sprinkler system will provide adequate protection to persons outside of the room of origin. It is also expected that staff will be adequately trained and will close the doors to patient sleeping rooms. The presence of the automatic sprinkler system, particularly one equipped with quick-response heads, increases the probability that staff will be able to close the door to the room of origin and thereby limit the size of the fire. Roller latches are not permitted on these doors, as they are not regarded as providing positive latching that will allow the door to act as a reliable smoke barrier.

407.3.2 Locking devices. Locking devices that restrict access to the patient room from the *corridor*, and that are operable only by staff from the *corridor* side, shall not restrict the *means of egress* from the patient room except for patient rooms in mental health facilities.

❖ Locking devices are permitted on patient room doors, provided egress from the patient room is not restricted. Such locking devices enable residents in long-term care facilities to secure their rooms and staff to secure unoccupied rooms. In such cases, it is expected that the staff still has access, via key or other opening device.

The need to lock corridor doors to restrain patients in mental health facilities, however, is recognized. When such locking is necessary, provisions must be made for continuous supervision and prompt release. Additional guidance on locking arrangements can be found in Section 408.4, which applies to occupancies in Group I-3. A decision needs to be made to determine if the building more closely represents an occupancy in Group I-2 or I-3. In occupancies in Group I-3, it is generally assumed that the occupants are capable of self-preservation once the doors are unlocked. This may not be the case in some mental health facilities and, therefore, the provision applicable to occupancies in Group I-2 may be more appropriate.

Similar protection may also be necessary in certain portions of other occupancies in Group I-2 where the movement of the residents must be controlled (i.e., patients with Alzheimer's disease). For special locking arrangements for suites or floors where elopement may be an issue, see Section 1008.1.9.6.

407.4 Smoke barriers. *Smoke barriers* shall be provided to subdivide every *story* used by patients for sleeping or treatment and to divide other *stories* with an *occupant load* of 50 or more persons, into at least two smoke compartments. Such *stories* shall be divided into smoke compartments with an area of not more than 22,500 square feet (2092 m²) and the travel distance from any point in a smoke compartment to a *smoke barrier* door shall not exceed 200 feet (60 960 mm). The *smoke barrier* shall be in accordance with Section 710.

❖ One of the basic premises of this section is that vertical evacuation of occupancies in Group I-2 is extremely difficult. It is essential, therefore, to provide a refuge area on every story used by patients for sleeping or treatment. The refuge area is intended to provide a protected area to which patients may be relocated if it becomes necessary to evacuate them. For this reason, each story used by patients for sleeping or treatment and all other stories with an occupant load of 50 or more must be divided into at least two smoke compartments. Each smoke compartment is limited to an area of no more than 22,500 square feet (2092 m²) (see the definition of "Smoke compartment" in Section 702). The travel distance from any point in a smoke compartment to a smoke barrier door is limited to 200 feet (60 960 mm). This is distinct from the travel distance to exits. This distance is limiting the distance that staff and patients must travel until reaching another smoke compartment and the refuge area it contains. Travel to an exit enclosure or exit door is not a substitute for travel to the smoke barrier and a separate compartment (see Figure 407.4).

Smoke barriers used to divide one smoke compartment from the next are to be constructed in accordance with Section 710 and with horizontal assemblies as specified in Section 407.4.3. The smoke barrier is different from that required for corridor walls in Section 407.3. Among other things, a smoke barrier is required to have a fire-resistance rating of 1 hour and, except as otherwise provided for in the exception to Section 710.4, provide a continuous separation through all concealed spaces to resist the passage of fire and smoke. The maximum area of the smoke compartment provides two protection features. First, the number of persons that are exposed to a fire will be restricted. Typical designs indicate that the number of persons exposed will usually be less than 100 and, more likely, 50 or less. Second, the overall dimensions of the smoke compartment must result in a maximum travel distance to the smoke barrier door of 200 feet (60 960 mm) for all design configurations.

407.4.1 Refuge area. At least 30 net square feet (2.8 m²) per patient shall be provided within the aggregate area of *corridors*, patient rooms, treatment rooms, lounge or dining areas and other low-hazard areas on each side of each *smoke barrier*. On floors not housing patients confined to a bed or litter, at least

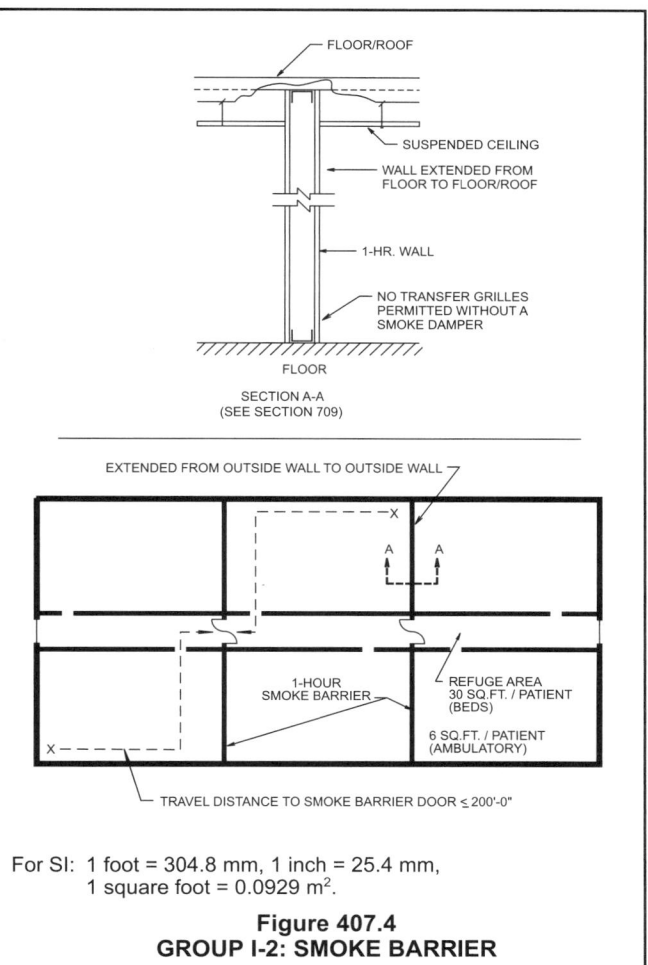

For SI: 1 foot = 304.8 mm, 1 inch = 25.4 mm,
1 square foot = 0.0929 m².

Figure 407.4
GROUP I-2: SMOKE BARRIER

6 net square feet (0.56 m²) per occupant shall be provided on each side of each *smoke barrier* for the total number of occupants in adjoining smoke compartments.

❖ To provide adequate space for occupants who might be relocated to an adjacent smoke compartment, minimum net areas per occupant are required. In determining the area available, it should be noted that the facility may still need to function as a medical care facility even while the patients are relocated. On floors housing nonambulatory patients (i.e., patient sleeping floors), a minimum of 30 net square feet (2.8 m²) per patient is required to be provided in low-hazard areas, such as waiting areas, corridors or patient rooms. This is to permit the relocation of the patient on a bed or litter to an area of safety on the same floor. The 30 net square feet (2.8 m²) is intended only to apply to the number of patients and not the calculated occupant load as determined using Table 1004.1.1.

For floors classified as Group I-2, which are not used for patient sleeping rooms (i.e., treatment floors), the net area required is 6 square feet (0.56 m²) per occupant. In this instance, the area required is based on the calculated occupant load as determined by Table 1004.1.1. The 6-square-foot (0.56 m²) minimum is based on an assumption that some occupants will be using walkers, wheelchairs and possibly a few litters for patients who are being transported through the area at the time of the fire.

Refuge areas required by this section are distinct from areas of refuge required by Section 1007 for accessible means of egress.

407.4.2 Independent egress. A *means of egress* shall be provided from each smoke compartment created by *smoke barriers* without having to return through the smoke compartment from which *means of egress* originated.

❖ To prevent creation of a dead-end smoke compartment, exits are to be arranged so as to permit access without returning through a smoke compartment from which egress originated. This section does not require an exit from within each smoke compartment. See Figures 407.4.2(1) and 407.4.2(2) for acceptable and unacceptable egress arrangements.

407.4.3 Horizontal assemblies. *Horizontal assemblies* supporting *smoke barriers* required by this section shall be designed to resist the movement of smoke and shall comply with Section 712.9.

❖ In order to maintain the integrity of a smoke compartment, the horizontal assemblies supporting the smoke barrier walls must also be designed to resist the passage of smoke. With the combination of these horizontal assemblies and the fire barriers, a complete smoke-protected compartment is achieved. Provisions of Section 712.9 address openings and penetrations of each horizontal assembly providing a barrier to smoke. Elevators and other large openings will have to be in a shaft enclosure. A lobby complying with Section 708.14.1 will need to be provided for such elevators (see commentary, Section 712.9).

[F] 407.5 Automatic sprinkler system. Smoke compartments containing patient *sleeping units* shall be equipped throughout with an *automatic sprinkler system* in accordance with Section 903.3.1.1. The smoke compartments shall be equipped with *approved* quick-response or residential sprinklers in accordance with Section 903.3.2.

❖ Recognizing the effectiveness of an automatic sprinkler system, the reference to Section 903.3.1.1 clarifies that the suppression system required by Section 903.2.5 must be a sprinkler system in accordance with the code and NFPA 13. Furthermore, in response to recent proven technology, smoke compartments containing patient rooms are required to be equipped throughout with quick-response or residential sprinklers.

The majority of fire injuries in hospitals begin with the ignition of clothing on a person, a mattress, a pillow, bedding or linen. The Building and Fire Research Laboratory of NIST conducted fire tests in patient rooms using quick-response sprinklers and smoke detectors. The test results were published in a July 1993 report entitled, "Measurement of Room Conditions and Response of Sprinklers and Smoke Detectors During a Simulated Two-bed Hospital Patient Room Fire." The report concluded that quick-response sprinklers actuated before the patient's life would be threatened by a fire in the his or her room.

Prior to the development of quick-response sprinkler technology, smoke detection in patient rooms classified as Group I-2 occupancies was typically required, due to the slower response time of standard sprinklers. While properly operating standard sprinklers are effective, the extent of fire growth and smoke production that can occur before sprinkler activation creates the need for early warning to enable faster response by staff and initiation of egress that is critical in occupancies containing persons incapable of self-preservation. With quick-response or residential sprinklers installed throughout the smoke compartment, and additional protection provided as indicated for spaces open to corridors, incidental accessory occupancy areas and certain specific health care facilities, smoke detectors in patient sleeping units are no longer required in Group I-2 occupancies.

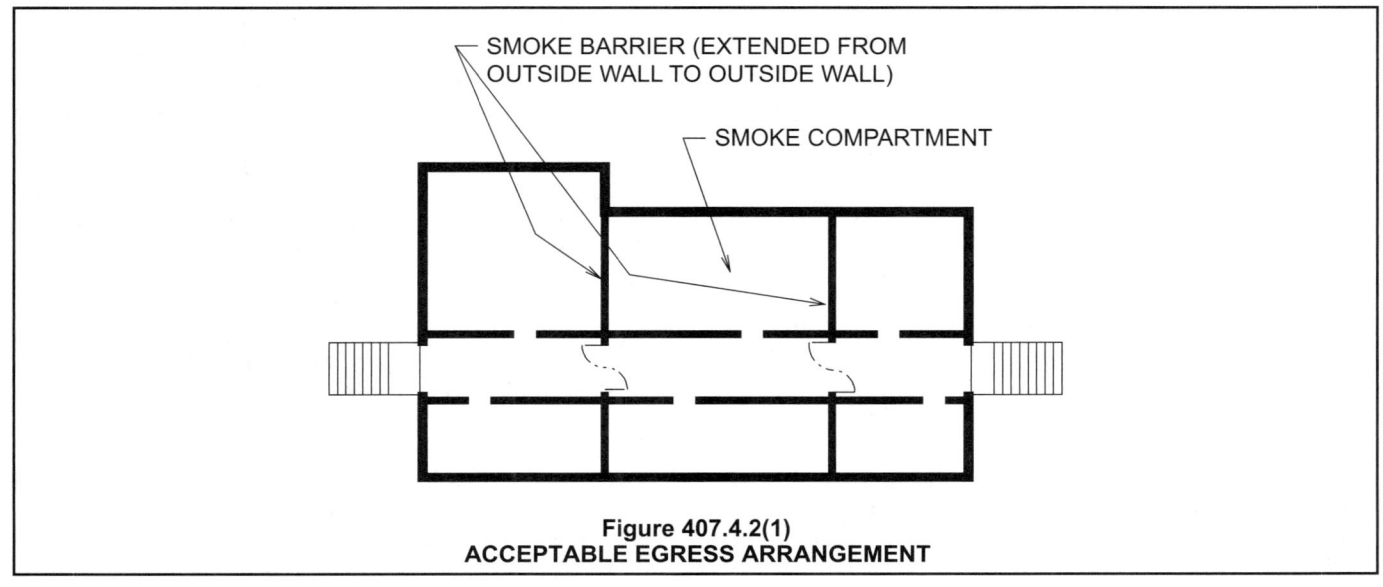

SMOKE BARRIER (EXTENDED FROM OUTSIDE WALL TO OUTSIDE WALL)

SMOKE COMPARTMENT

Figure 407.4.2(1)
ACCEPTABLE EGRESS ARRANGEMENT

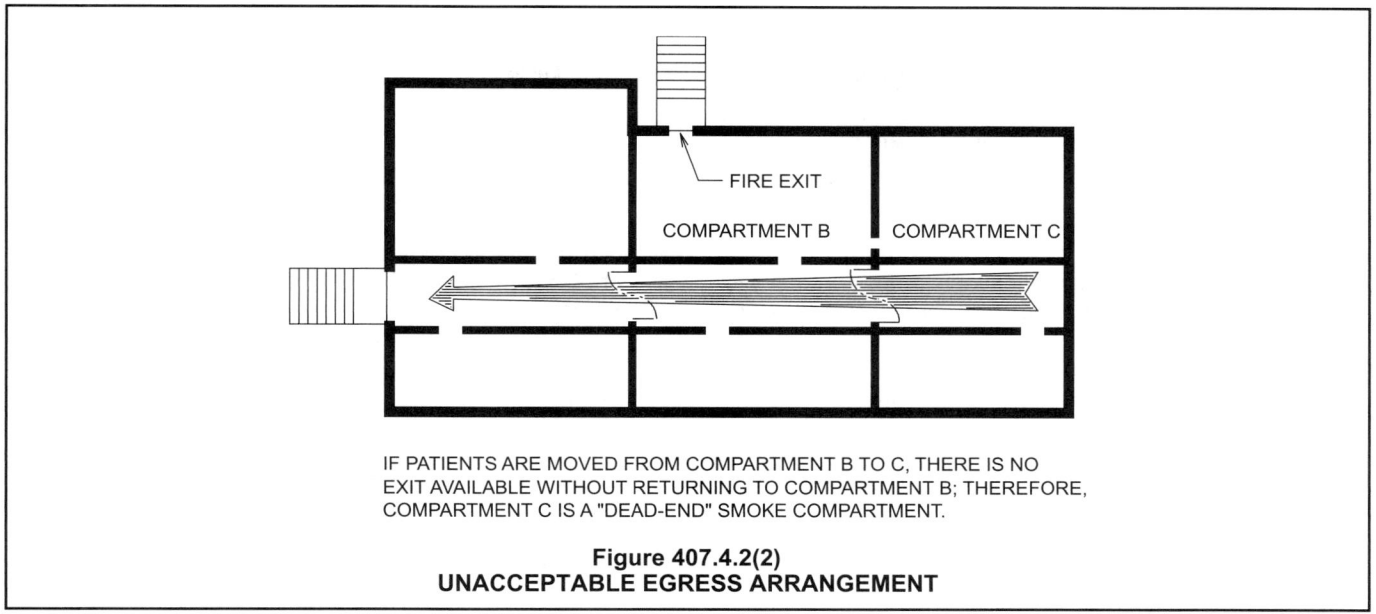

IF PATIENTS ARE MOVED FROM COMPARTMENT B TO C, THERE IS NO
EXIT AVAILABLE WITHOUT RETURNING TO COMPARTMENT B; THEREFORE,
COMPARTMENT C IS A "DEAD-END" SMOKE COMPARTMENT.

Figure 407.4.2(2)
UNACCEPTABLE EGRESS ARRANGEMENT

[F] 407.6 Fire alarm system. A fire alarm system shall be provided in accordance with Section 907.2.6.

❖ Fire alarms are essential to this occupancy where many of those receiving care and treatment cannot leave a fire incident area without the assistance of staff. The alarm system is a key part of the safeguards to protect patients. Standards for this alarm system are found in Sections 907.2.6 and 907.2.6.2.

[F] 407.7 Automatic fire detection. *Corridors* in nursing homes (both intermediate care and skilled nursing facilities), detoxification facilities and spaces permitted to be open to the *corridors* by Section 407.2 shall be equipped with an automatic fire detection system. Hospitals shall be equipped with smoke detection as required in Section 407.2.

Exceptions:

1. *Corridor* smoke detection is not required where patient *sleeping units* are provided with smoke detectors that comply with UL 268. Such detectors shall provide a visual display on the *corridor* side of each patient *sleeping unit* and an audible and visual alarm at the nursing station attending each unit.

2. *Corridor* smoke detection is not required where patient *sleeping unit* doors are equipped with automatic door-closing devices with integral smoke detectors on the unit sides installed in accordance with their listing, provided that the integral detectors perform the required alerting function.

❖ Automatic fire detection is required in areas open to corridors in occupancies classified as Group I-2 hospitals and corridors in nursing homes and detoxification facilities. In recognition of quick-response sprinkler technology and the fact that the sprinkler system is electronically supervised and doors to patient rooms are supervised by staff on a continual basis when in the open position, it is now believed that patient room smoke detectors are not required for adequate fire safety in patient sleeping units. In nursing homes and detoxification facilities, however, some redundancy is appropriate because such facilities typically have less control over furnishings and personal items and, thereby, result in a less predictable and usually higher fire hazard load than other Group I-2 occupancies. Also, there is generally less staff supervision in these facilities than in other health care facilities and, thus, less control over patient smoking and other fire causes. To provide additional protection against fires spreading from the room of origin, therefore, automatic fire detection is required in corridors of nursing homes and detoxification facilities. It should be noted that fire detection is not required in corridors of other Group I-2 occupancies except where otherwise specifically required in the code (see Section 907.2.6). Similarly, since areas open to the corridor very often are the room of fire origin and because such areas are no longer required by the code to be under visual supervision by staff, some redundancy to protection by the sprinkler system is requested. Accordingly, all areas open to corridors must be protected by an automatic fire detection system. This requirement provides an additional level of protection against sprinkler system failures or lapses in staff supervision. There are two exceptions to the requirement for an automatic fire detection system in corridors of nursing homes and detoxification facilities. In both exceptions, the alternative methods of protection specifically provide an equivalent level of safety to what is required in patient sleeping rooms.

Exception 1 requires smoke detectors to be located in the patient's room, which activate both a visual display on the corridor side of the patient room and a visual and audible alarm at the nurses' station serving or attending the room. Detectors complying with UL 268 are intended for open area protection and for connection to a normal power supply or as part of a fire alarm system.

This exception, however, is specifically designed so as not to require the detectors to activate the building's fire alarm system where approved patient room smoke detectors are installed and where visual and audible alarms are provided. This is in response to the concern over unwanted alarms. It should be noted that the required alarm signals will not necessarily indicate to staff that a fire emergency exists. The nursing call system may typically be used to identify numerous conditions within the room.

Exception 2 addresses the situation where smoke detectors are incorporated within automatic door-closing devices. The units are acceptable as long as the required alarm functions are still provided. Such units are usually listed as combination door closer and hold-open devices.

407.8 Secured yards. Grounds are permitted to be fenced and gates therein are permitted to be equipped with locks, provided that safe dispersal areas having 30 net square feet (2.8 m²) for bed and litter patients and 6 net square feet (0.56 m²) for ambulatory patients and other occupants are located between the building and the fence. Such provided safe dispersal areas shall not be located less than 50 feet (15 240 mm) from the building they serve.

❖ Group I-2 occupancies specializing in the treatment of mental disabilities, such as Alzheimer's, often provide a secured yard for patient use. During an emergency, it may be difficult to control residents if they are allowed to have direct access to a public way. This section essentially creates a safe area where people may move from the building without having unrestricted access to the public way.

407.9 Hyperbaric facilities. Hyperbaric facilities in Group I-2 occupancies shall meet the requirements contained in Chapter 20 of NFPA 99.

❖ For these unique treatment facilities, Chapter 20 of NFPA 99 provides appropriate design criteria. This standard is only referenced where such facilities are installed in a Group I-2 occupancy.

SECTION 408
GROUP I-3

408.1 General. Occupancies in Group I-3 shall comply with the provisions of Sections 408.1 through 408.10 and other applicable provisions of this code (see Section 308.4).

❖ The provisions of Section 408 address the unique features of detention and correctional occupancies in Group I-3. The provisions are based on full-scale fire tests, fire experience (in particular several multiple-death fires that occurred between 1974 and 1979) and the provisions for occupancies in Group I-2. With respect to evacuation, occupancies in Groups I-2 and I-3 are similar in that the occupants are typically not capable of self-preservation and, therefore, staff must assist in evacuation.

The general approach is a defend-in-place philosophy based on the difficulties associated with evacuation, especially in medium- and high-security facilities. The first level of protection is established as the room that is typically a sleeping room or cell. Horizontal evacuation to an adjacent smoke compartment provides the second level of protection, and vertical evacuation is the third level of protection. The evacuation difficulty associated with occupancies in Group I-3 is not occupant mobility; rather, it is the need to maintain security and, in some instances, to keep the residents segregated. There are also instances, such as protected witness facilities, in which the identity of the occupant must not be revealed to other occupants or the public.

It is recognized that the broad classification of Group I-3 includes a variety of locking arrangements based on the various levels of security required. As such, Section 308.4 contains definitions for five different occupancy conditions that are intended to represent the various locking arrangements and different levels of security (see commentary, Section 308.4). Regardless of the security level, fire protection and life safety features must be provided in order to achieve an acceptable level of protection without interfering with the operation of the facility and the need to maintain security.

This section has been developed out of considerations unique to detention and correctional facilities and to facilitate the consideration of the entire package of protection features. The basic protection features provided include early detection, fire containment, evacuation and fire extinguishment. Section 408.7 of the IFC contains provisions for emergency preparedness, including staff training, staffing requirements and the need to be able to obtain and identify emergency keys. Additional consideration should be given to restricting the fuel load within resident housing areas. It should also be noted that Section 508.2.5 contains provisions for incidental accessory occupancy areas within Group I-3. The incidental accessory occupancies represent different degrees of hazard than are associated with the hazards of the primary building area—in this case, the housing of occupants under physical restraint. The different hazard is typically related to the basic nature of the contents or the quantity of combustibles.

The provisions of Section 408 apply to all occupancies in Group I-3 and those portions of occupancies that are considered Group I-3, except Condition 1 facilities (see Figure 408.1). As defined in Section 308.4.1, Condition 1 facilities permit free egress even to the exterior and, therefore, such occupancies are subject to the same provisions applicable to Group R.

The elimination of Condition 1 from the provisions of Section 408 may result in some interesting applications of the code. For example, the corridor separation requirements in Section 1018.1 for occupancies in Group R are more restrictive than the corridor requirements for occupancies in Group I-3, Condition 2. This may result in a request to consider a facility as Condition 2 even though the building is really Condition 1. Such a request could be considered acceptable only if the additional requirements in the code and in the IFC for Condition 2 are met, including automatic fire suppression and staffing (see Section 408.7.2 of the IFC).

408.1.1 Definition. The following words and terms shall, for the purposes of this chapter and as used elsewhere in this code, have the meanings shown herein.

❖ Definitions of terms that are associated with the content of this section are contained herein. These definitions can help in the understanding and application of the code requirements. It is important to emphasize that these terms are not exclusively related to this section but are applicable everywhere the term is used in the code. The purpose for including these definitions within this section is to provide more convenient access to

them without having to refer back to Chapter 2. For convenience, these terms are also listed in Chapter 2 with a cross reference to this section. The use and application of all defined terms, including those defined herein, are set forth in Section 201.

CELL. A room within a housing unit in a detention or correctional facility used to confine inmates or prisoners.

❖ Cells are the smallest unit in the portion of a Group I-3 prison or jail for the housing of inmates. In contrast to other occupancies, many of the provisions in Section 408 refer to cells.

CELL TIER. Levels of cells vertically stacked above one another within a housing unit.

❖ A tier is one level of cells. There can be multiple tiers within a single housing unit as well as more than one cell tier within a story.

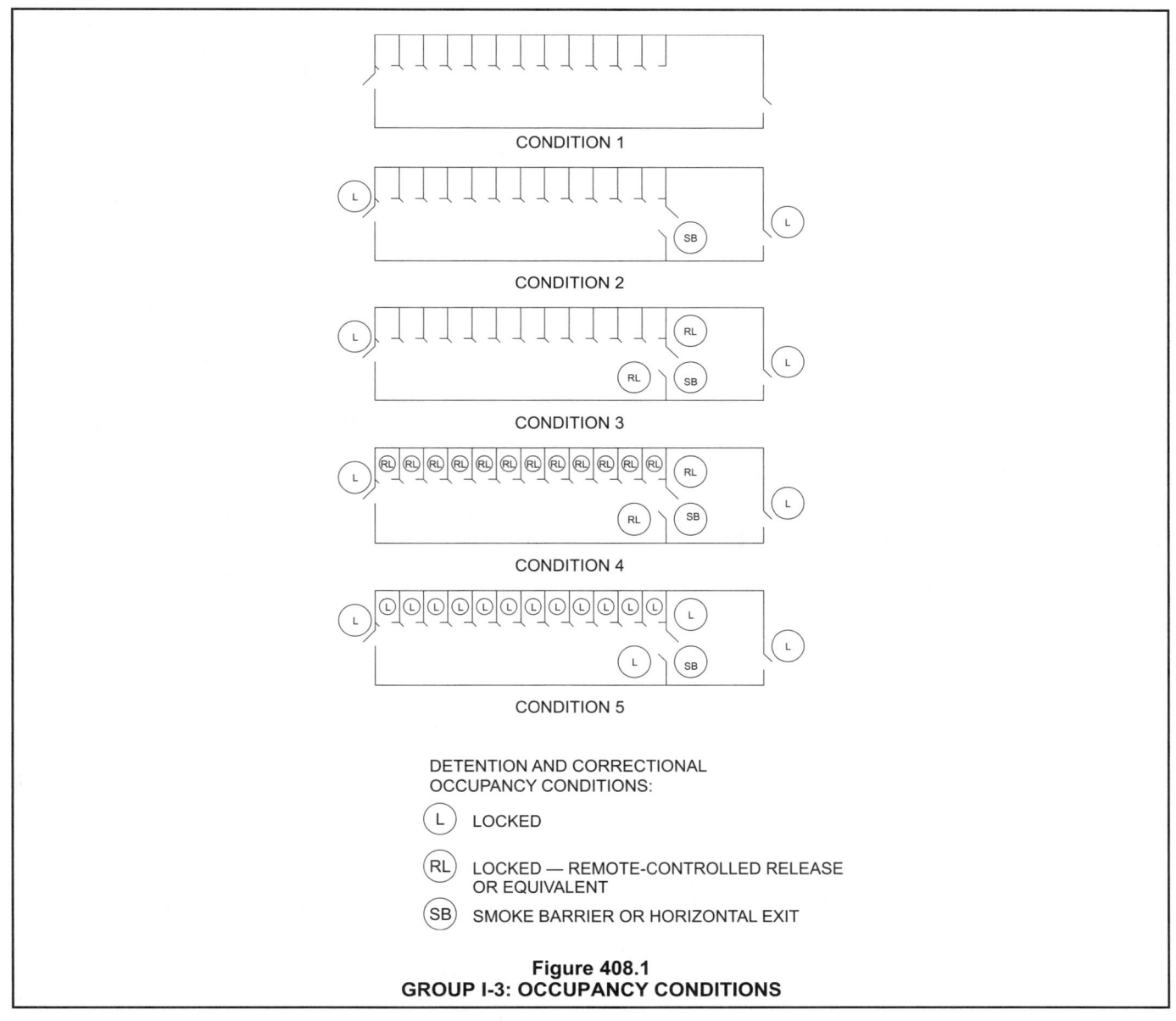

**Figure 408.1
GROUP I-3: OCCUPANCY CONDITIONS**

HOUSING UNIT. A dormitory or a group of cells with a common dayroom in Group I-3.

❖ A housing unit is a key delineation of space within a Group I-3 occupancy. Often inmates are restricted to their housing unit and not allowed to move freely outside of the unit. A housing unit can include more than one tier of cells. Because a housing unit is often designed as a multiple level area with a shared day room, it will typically share a common atmosphere (see Section 408.5 for the protection of vertical openings).

SALLYPORT. A security vestibule with two or more doors or gates where the intended purpose is to prevent continuous and unobstructed passage by allowing the release of only one door or gate at a time.

❖ While sallyports are not unique to Group I-3 occupancies, they are essential to the control and safe movement of inmates in a Group I-3 facility. Section 408.3.7 allows sallyports to be in the means of egress system. See the commentary for that section for an additional description of sallyport design and use.

408.2 Other occupancies. Buildings or portions of buildings in Group I-3 occupancies where security operations necessitate the locking of required *means of egress* shall be permitted to be classified as a different occupancy. Occupancies classified as other than Group I-3 shall meet the applicable requirements of this code for that occupancy provided provisions are made for the release of occupants at all times.

Means of egress from detention and correctional occupancies that traverse other use areas shall, as a minimum, conform to requirements for detention and correctional occupancies.

> **Exception:** It is permissible to exit through a *horizontal exit* into other contiguous occupancies that do not conform to detention and correctional occupancy egress provisions but that do comply with requirements set forth in the appropriate occupancy, as long as the occupancy is not a Group H use.

❖ In accordance with the provisions of Section 508, portions of an occupancy in Group I-3 may be classified as a separate occupancy and meet the provisions of the code for that occupancy. Since the area may be used by the residents, however, a need may exist for the other occupancy to also contain security provisions, such as the locking of egress doors. This section specifically permits such a condition as long as arrangements have been made for release of the occupants within these areas at any time they are occupied. Acceptable methods include having either staff operate the locks or remote release of the locks, similar to that which is provided in the housing areas.

Although it is indicated that the section applies to portions of occupancies in Group I-3, consideration should be given to allow the necessary security to be provided in separate buildings that are part of a detention or correctional facility. Applications of this section should be restricted to the buildings that must be secure because they are occupied by residents, and

arrangements for quick release of the locks must be provided at all times the building is occupied. If residents are to be permitted into an area or building without staff supervision, the residents should be able to initiate their own evacuation.

The means of egress provisions from this section and Chapter 10 that are applicable to Group I-3 are also applicable to the path through different occupancies. For example, if the means of egress from a cell block wing traverses through a Group A area, then the requirements of the code for the Group I-3 egress are also applicable to the path through the assembly occupancy.

The exception in the code permits egress through a different occupancy other than Group H, that is separated from the Group I-3 areas by either a fire wall or fire barrier. A horizontal exit complying with Section 1025 must be provided in this wall or barrier. This separation voids the requirements that Group I-3 egress provisions are also applicable to the different occupancies. The reason for this, of course, is that Group I-3 occupants have fled the immediate effects of the fire condition in the original detention or correctional areas and are now harbored in a refuge area.

408.3 Means of egress. Except as modified or as provided for in this section, the provisions of Chapter 10 shall apply.

❖ Because of the need to provide security in occupancies in Group I-3, many of the means of egress requirements of Chapter 10 have been modified for Group I-3. Except as modified in Section 408, however, all other provisions of Chapter 10 are applicable.

408.3.1 Door width. Doors to resident *sleeping units* shall have a clear width of not less than 28 inches (711 mm).

❖ Except for sleeping units that are provided to meet accessible unit provisions of Section 1107.5.5, sleeping unit doors may be a minimum of 28 inches (711 mm) in width. This section is not intended to negate the need to provide a 32-inch (813 mm) door if the room is intended to serve as part of an accessible route. In an occupancy in Group I-3, most sleeping units and even sections of the prison need not be accessible (see Sections 1103.2.14 and 1107.5.5).

408.3.2 Sliding doors. Where doors in a *means of egress* are of the horizontal-sliding type, the force to slide the door to its fully open position shall not exceed 50 pounds (220 N) with a perpendicular force against the door of 50 pounds (220 N).

❖ Swinging doors in occupancies in Group I-3 may not be acceptable for security reasons. If the door swings into a room, the door can easily be blocked and staff access can be prevented. If the door swings out, staff does not have control of the door and the residents can use the door as a means to overcome a staff member; therefore, horizontal sliding doors are used quite extensively and are permitted in the means of egress. The doors must be capable of being opened with a force of 50 pounds (220 N) even if a 50-pound (220 N) force is being applied perpendicular to the door. The maximum force permitted to open the door exceeds the restrictions in Section 1008.1.3 in recognition of the physical capabilities of staff members.

408.3.3 Guard tower doors. A hatch or trap door not less than 16 square feet (610 m²) in area through the floor and having minimum dimensions of not less than 2 feet (610 mm) in any direction shall be permitted to be used as a portion of the *means of egress* from guard towers.

❖ This provision allows the use of trap doors in the floor of an observation point as both the access and the means of egress from the tower. In order to provide the 360-degree visibility necessary for guard observation stations, the size of the base of each such elevated station must be kept to a minimum. These towers are usually limited to prison/jail staff personnel. The trap doors work in conjunction with the spiral stairway and ship ladder provisions of the next two sections.

408.3.4 Spiral stairways. *Spiral stairways* that conform to the requirements of Section 1009.9 are permitted for access to and between staff locations.

❖ Recognizing the physical capabilities of the staff, spiral stairways are permitted for access to and between staff locations. In multiple story facilities, spiral stairways are often used between staff control areas on different levels. As such, staff can move between the areas without entering adjacent housing areas and without losing the space that would be required for a traditional stairway. Spiral stairways provide an option for vertical circulation in guard towers and similar observation areas where the floor area is limited and the desire is to have the least amount of obstruction of floor area possible.

408.3.5 Ship ladders. Ship ladders shall be permitted for egress from control rooms or elevated facility observation rooms in accordance with Section 1009.11.

❖ Ship ladders are another alterative to spiral stairways for providing vertical paths of access and egress within a Group I-3 occupancy. These are intended for access to locations used by prison/jail staff. The design standard for ship ladders is found in Section 1009.11. In that section, the code limits any space accessed by a ship ladder to a maximum of 250 square feet (23 m²) and a maximum of three occupants (see commentary, Section 1009.11).

408.3.6 Exit discharge. *Exits* are permitted to discharge into a fenced or walled courtyard. Enclosed yards or *courts* shall be of a size to accommodate all occupants, a minimum of 50 feet (15 240 mm) from the building with a net area of 15 square feet (1.4 m²) per person.

❖ For security purposes, exits from areas in Group I-3, do not normally provide access to a public way. In some instances, the building is located in a complex that has perimeter walls preventing access to a public way but that permit occupants to move away from the building. In other instances or in complexes where the mixing of residents must be controlled, one or more yards or courts are provided for exit discharge. Such arrangements are acceptable, provided that the occu-

pants have an area that is at least 15 square feet (1.4 m²) per person and is located at least 50 feet (15 240 mm) from the building.

This 50-foot (15 240 mm) distance enables residents to move away from the building and should have adequate spatial separation—especially since the building is required to be protected with an automatic fire suppression system in accordance with Section 903.2.6. The 15-square-foot (1.4 m²) measurement is provided so that the residents have adequate space in which to stand and to move around since they may need to remain in the yard or court for some time.

408.3.7 Sallyports. A sallyport shall be permitted in a *means of egress* where there are provisions for continuous and unobstructed passage through the sallyport during an emergency egress condition.

❖ A sallyport is a compartment established for security purposes that restricts the movement of individuals. A sallyport consists of two or more security doors that do not normally operate simultaneously. One door will not normally open until the other doors are closed and locked. Sallyports are provided for security purposes to control movement and passage of residents from one area to another. The sallyport is to be arranged such that during an emergency, the doors may be opened simultaneously so as to permit free and unobstructed movement from the area. In many instances, the sallyport doors contain a manual release that can be operated by staff during an emergency to override the normal electric operation.

Although not addressed by this section, if the sallyport is in the path that the fire department or fire brigade will use to bring fire hoses, additional consideration may need to be given to permit the hose to go through the sallyport. An alternative would be to provide a standpipe connection on the housing-area side of the sallyport.

408.3.8 Exit enclosures. One of the required *exit enclosures* in each building shall be permitted to have glazing installed in doors and interior walls at each landing level providing access to the enclosure, provided that the following conditions are met:

1. The *exit enclosure* shall not serve more than four floor levels.

2. Exit doors shall not be less than ³/₄-hour *fire door assemblies* complying with Section 715.4.

3. The total area of glazing at each floor level shall not exceed 5,000 square inches (3 m²) and individual panels of glazing shall not exceed 1,296 square inches (0.84 m²).

4. The glazing shall be protected on both sides by an *automatic sprinkler system*. The sprinkler system shall be designed to wet completely the entire surface of any glazing affected by fire when actuated.

5. The glazing shall be in a gasketed frame and installed in such a manner that the framing system will deflect

without breaking (loading) the glass before the sprinkler system operates.

6. Obstructions, such as curtain rods, drapery traverse rods, curtains, drapes or similar materials shall not be installed between the automatic sprinklers and the glazing.

❖ In addressing the security needs in Group I-3, the limitation on openings in the enclosures for exits given in Sections 1022.1, 1022.3 and 1022.4 is modified to facilitate visibility into the exit.

A vertical exit that is open to view in a correctional facility is useful and necessary for two reasons. First, the movement of the inmates is observable, thus reducing the potential for concealment, physical attacks on other inmates and other undesirable activities that could otherwise take place in enclosed spaces that are not observable.

Second, the movement of inmates who are under physical restraint is observable from the exterior, thus increasing the level of safety for correctional officers. The alternative of closed-circuit television within the exits is not as desirable because the potential for malfunction of or intentional damage to the equipment makes this method a less reliable form of observation.

The conditions specified that permit glazing require full enclosure of the floor opening so that there is no direct communication between stories. These criteria provide a measure of fire integrity, if somewhat less than that otherwise required of a vertical exit enclosure.

408.4 Locks. Egress doors are permitted to be locked in accordance with the applicable use condition. Doors from a refuge area to the exterior are permitted to be locked with a key in lieu of locking methods described in Section 408.4.1. The keys to unlock the exterior doors shall be available at all times and the locks shall be operable from both sides of the door.

❖ The locking arrangements for egress doors depend on the occupancy condition (1, 2, 3, 4 or 5) given in Section 308.4. Except for Condition 1 buildings, the doors to the exterior from an occupancy in Group I-3 may be locked and need not be remotely controlled. The door locks are to be arranged so that they can be released from the exterior of the building, as well as the interior. This feature accomplishes two purposes. First, since the door can be unlocked from the outside, staff need not enter the housing area in which the fire may be located to release the locks. Second, many facilities do not permit staff members in the housing area to carry the keys for the exterior doors; therefore, there may be no benefit in requiring the door lock to be arranged to permit operation from the interior, although such an arrangement is not prohibited.

The provisions of Section 408.7.4 of the IFC concerning the marking of keys applies to both exterior and interior door locks. For this reason, it is usually desirable to minimize the number of keys required for an emergency; therefore, the exterior doors will most likely be keyed alike. Unfortunately, many facilities do not take this into consideration and the number of emergency keys becomes excessive to the extent that it is virtually impossible to identify each key by sight and touch.

408.4.1 Remote release. Remote release of locks on doors in a *means of egress* shall be provided with reliable means of operation, remote from the resident living areas, to release locks on all required doors. In Occupancy Conditions 3 or 4, the arrangement, accessibility and security of the release mechanism(s) required for egress shall be such that with the minimum available staff at any time, the lock mechanisms are capable of being released within 2 minutes.

Exception: Provisions for remote locking and unlocking of occupied rooms in Occupancy Condition 4 are not required provided that not more than 10 locks are necessary to be unlocked in order to move occupants from one smoke compartment to a refuge area within 3 minutes. The opening of necessary locks shall be accomplished with not more than two separate keys.

❖ The provisions of this section do not mandate remote release locks; rather, the facility determines the level and manner in which security is to be provided. Door-locking arrangements fall into one of the conditions defined in Section 308.4. If remote release locks are provided, the means of operating them must be external to the resident housing area so the staff is not required to enter the housing area to release the locks. Clearly, in Conditions 3 and 4, excessive delay in releasing locks puts the occupants at additional risk. Locks in Conditions 3 and 4 must be able to be released promptly (within 2 minutes) with the minimum staff that will be available. A control center is typically provided for two or more housing areas from which the locks are controlled. In evaluating the use and arrangement of the remote release, consideration must be given to the intended staffing levels at different times of the day, as well as the location(s) of the remote release.

The condition definitions of Section 308.4 are intended to relate to the time necessary to evacuate the residents to a refuge area, such as a smoke compartment; therefore, in Condition 4, manual release locks may be used only when no more than 10 locks have to be unlocked in order to move all occupants from one smoke compartment to another and provided that this can be accomplished within 3 minutes. The time period of 3 minutes is presumed to be a reasonable staff response time to a fire emergency. So that the time to release the locks is kept to a minimum, it cannot take more than two separate keys to release the 10 locks. Again, the intended staff levels of the facility need to be considered (see Figure 408.4.1).

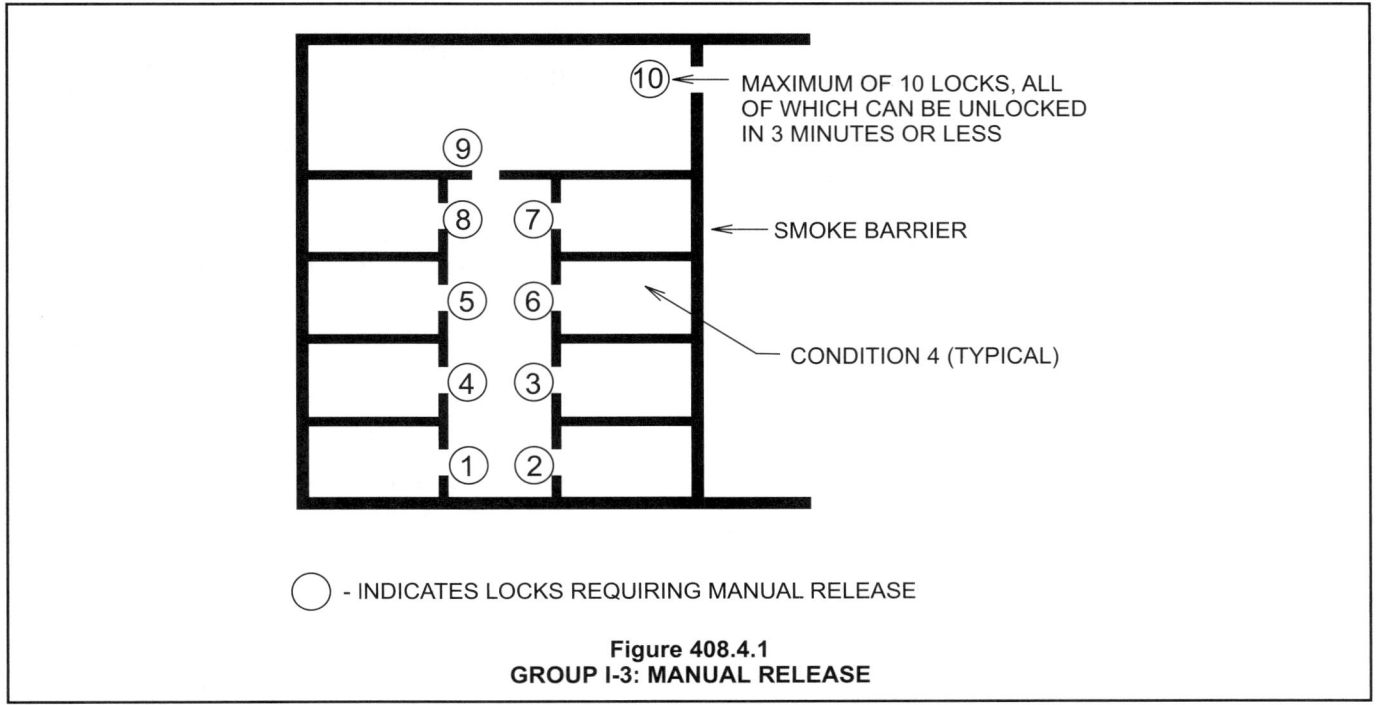

- INDICATES LOCKS REQUIRING MANUAL RELEASE

Figure 408.4.1
GROUP I-3: MANUAL RELEASE

408.4.2 Power-operated doors and locks. Power-operated sliding doors or power-operated locks for swinging doors shall be operable by a manual release mechanism at the door, and either emergency power or a remote mechanical operating release shall be provided.

> **Exception:** Emergency power is not required in facilities with 10 locks or less complying with the exception to Section 408.4.1.

❖ To increase the likelihood that the doors in the means of egress can be operated even during a power failure, two alternative locking arrangements are required whenever the doors or locks are electrically operated. First, a manual release mechanism is to be provided at the door. Also, a remote backup (second) mechanical operating door release system must be provided or the electrical door control system must be provided with an emergency power source. If provided, the mechanical operating door release system must be capable of being operated from outside the residential housing area.

This section does not apply to manually operated doors and locks that meet the requirements of the exception to Section 408.4.1.

408.4.3 Redundant operation. Remote release, mechanically operated sliding doors or remote release, mechanically operated locks shall be provided with a mechanically operated release mechanism at each door, or shall be provided with a redundant remote release control.

❖ If a mechanically operated remote release mechanism is provided as the remote release system, either a second, redundant system is required or each remote release door or lock is to be provided with a mechanically operated device at the door. The redundant oper-

ation is essential to increase the likelihood that the doors will continue to operate in the event of failure of the primary locking system. If provided, the redundant mechanical operating door release system must also be operable from outside the residential housing area.

This provision does not apply to power-operated doors and locks that are required to have alternative systems in accordance with Section 408.4.2.

408.4.4 Relock capability. Doors remotely unlocked under emergency conditions shall not automatically relock when closed unless specific action is taken at the remote location to enable doors to relock.

❖ So as to maintain means of egress doors in an open position during an emergency condition, once a door in a means of egress is unlocked via an emergency unlocking system, the door must not automatically relock when closed unless a specific action is taken at the remote location. The normal operation of such doors and locks is typically arranged so that the door will automatically relock upon closure.

Many electronic-locking systems operate such that the door automatically relocks upon closure for security reasons. To comply with the provisions of this section, however, a specific switch is usually provided that serves as an emergency control and does not permit the door to relock without operating it. This section does not require that doors automatically release upon activation of the fire alarm system.

408.5 Protection of vertical openings. Any vertical opening shall be protected by a shaft enclosure in accordance with Section 708, or shall be in accordance with Section 408.5.1.

❖ This section requires that vertical openings be protected in the manner established in Section 708. The section allows compliance with Section 408.5.1 as an alternative within a housing unit. If the provisions of Section 408.5.1 for floor openings are used, then Section 408.5.2 regarding plumbing chases is also available for use in the design.

408.5.1 Floor openings. Openings in floors within a housing unit are permitted without a shaft enclosure, provided all of the following conditions are met:

1. The entire normally occupied areas so interconnected are open and unobstructed so as to enable observation of the areas by supervisory personnel;

2. *Means of egress* capacity is sufficient for all occupants from all interconnected cell tiers and areas;

3. The height difference between the floor levels of the highest and lowest cell tiers shall not exceed 23 feet (7010 mm); and

4. Egress from any portion of the cell tier to an *exit* or *exit access* door shall not require travel on more than one additional floor level within the housing unit.

❖ This section is essentially an exception to the protection of vertical openings required by Section 708. It allows openings between the various floor levels and cell tiers within a single housing unit. This provision applies regardless of whether or not the communicating floor levels are stories or mezzanines as long as they are within the same housing unit [see Figures 408.5(1) and (2)].

The following is a discussion of the conditions that must be met:

1. The areas must be sufficiently open and unobstructed so that a fire in one part will be immediately obvious to the occupants and supervisory personnel in the area. The intent of the provision is not to require open cell fronts; rather, it is to enable the staff to observe events occurring on all levels from one location. This arrangement is usually desirable from a security standpoint [see Figure 408.5(1)].

2. The exit capacity required is determined based on the occupant load of all floor levels exiting simultaneously. It is not unusual for most of the residents to be on the main level in the common space during periods when they are not required to be in the sleeping units (cells); therefore, this level must be able to handle the potential occupant load of all cell tier levels.

3. The height differential between the floor levels of highest and lowest cell tiers must not exceed 23 feet (7010 mm). The limitation of 23 feet (7010 mm) expands the general code limitations on unprotected floor openings beyond two stories, because in Group I-3 it is not uncommon to have more than two interconnected levels in this open arrangement; however, the effective limitation in number of cell tiers is three. As with atriums (see Section 404), an automatic sprinkler system is required to control the spread of smoke and fire through vertical openings. Beyond the specified dimension of 23 feet (7010 mm), effective unobstructed sight becomes less attainable [see Figure 408.5(2)].

4. To minimize the potential that a fire in the day room would block access to all exits, the exits must be arranged such that it is possible to reach an exit on each level, or from the level next above or below. This would permit, for example, a three-tiered housing unit to have exits or exit access doors out of the housing unit, which are located on the lowest level and mid-level only; or on the lowest level and the top level only. In these two scenarios, the residents have access to an exit either from the level they are on or on the next level directly above or below.

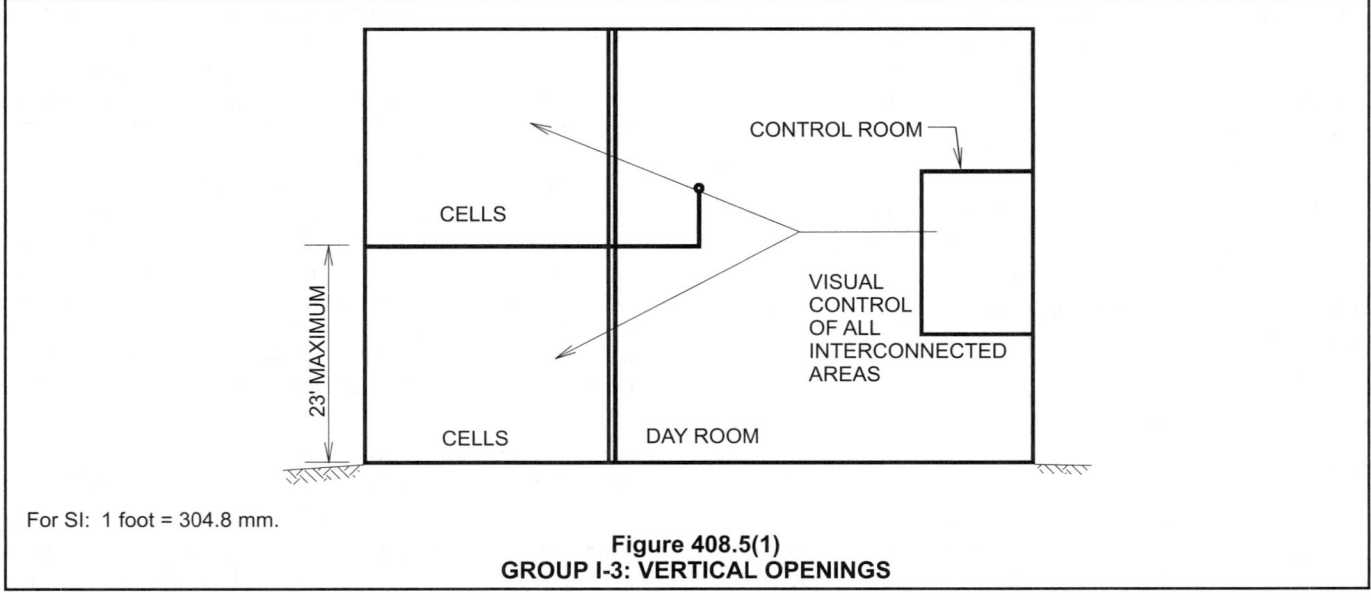

For SI: 1 foot = 304.8 mm.

Figure 408.5(1)
GROUP I-3: VERTICAL OPENINGS

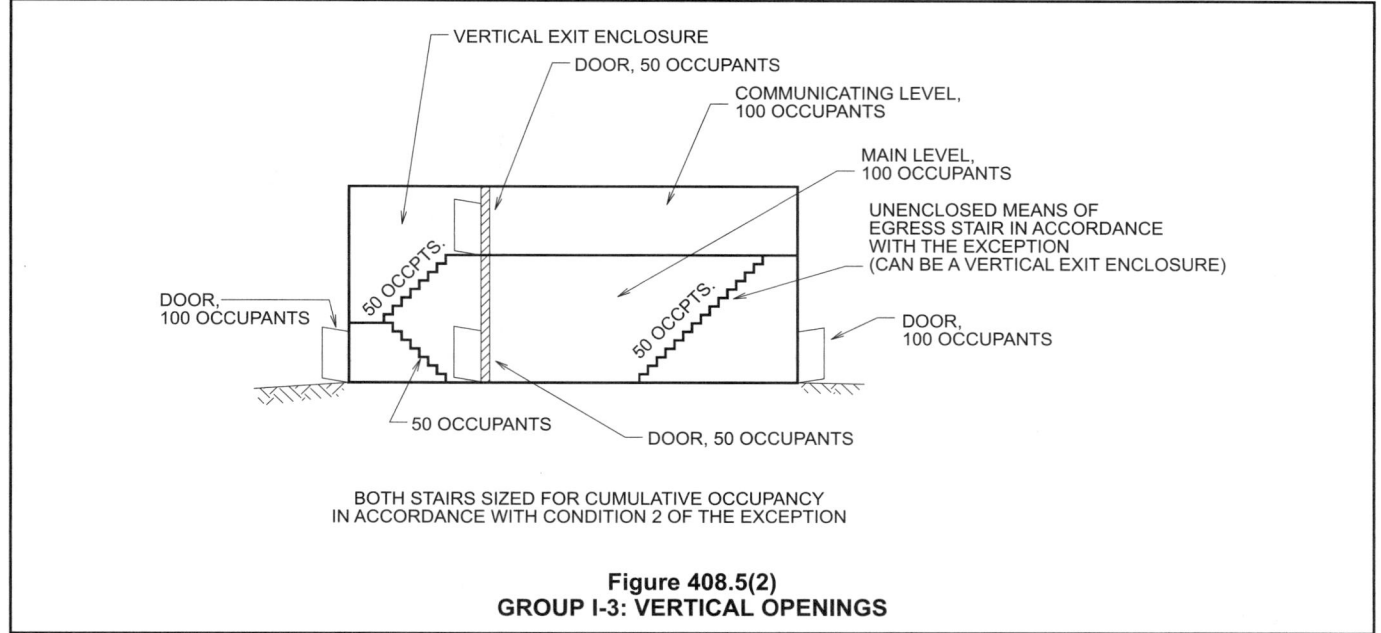

Figure 408.5(2)
GROUP I-3: VERTICAL OPENINGS

408.5.2 Shaft openings in communicating floor levels. Where a floor opening is permitted between communicating floor levels of a housing unit in accordance with Section 408.5.1, plumbing chases serving vertically staked individual cells contained with the housing unit shall be permitted without a shaft enclosure.

❖ Where a housing unit is designed in compliance with Section 408.5.1, this section allows plumbing chases within the same housing unit to also be unprotected where they penetrate through the floors separating cell tiers. If opening protection is provided within a housing unit under the provisions of Section 708, the provisions of this section cannot be used in a building design.

408.6 Smoke barrier. Occupancies in Group I-3 shall have *smoke barriers* complying with Sections 408.8 and 710 to divide every *story* occupied by residents for sleeping, or any other *story* having an *occupant load* of 50 or more persons, into at least two smoke compartments.

> **Exception:** Spaces having a direct *exit* to one of the following, provided that the locking arrangement of the doors involved complies with the requirements for doors at the *smoke barrier* for the use condition involved:
>
> 1. A *public way*.
> 2. A building separated from the resident housing area by a 2-hour fire-resistance-rated assembly or 50 feet (15 240 mm) of open space.
> 3. A secured *yard* or *court* having a holding space 50 feet (15 240 mm) from the housing area that provides 6 square feet (0.56 m²) or more of refuge area per occupant, including residents, staff and visitors.

❖ One of the basic premises of this section is that evacuation of a Group I-3 facility is impractical because of the need to maintain security. The security measures may also result in a delayed evacuation as compared to residential occupancies in which free egress is allowed; therefore, it is essential to provide a refuge area to which residents can be relocated when it becomes necessary to evacuate them from the housing units. For this reason, each cell tier and each floor with an occupant load of 50 or more is required to be divided into at least two smoke compartments using smoke barriers that comply with Sections 408.8 and 710 (see commentary, Section 710). It should be noted that Section 710 provides several exceptions to smoke barriers used in Group I-3 occupancies (e.g., construction, doors and opening protectives). A further exception allowing increased glazing is permitted by the provisions of Section 408.7.

The exception provides three alternatives that are common-sense options to movement into a smoke compartment. If the particular occupancy condition affords free, unobstructed access with no delay to the exterior, to a separate building or through a horizontal exit, then defend-in-place provisions are not needed.

408.6.1 Smoke compartments. The maximum number of residents in any smoke compartment shall be 200. The travel distance to a door in a *smoke barrier* from any room door required as *exit access* shall not exceed 150 feet (45 720 mm). The travel distance to a door in a *smoke barrier* from any point in a room shall not exceed 200 feet (60 960 mm).

❖ This section defines the maximum allowable size of smoke compartments by limiting the number of residents in any single smoke compartment to 200, the maximum door-to-door distance within the compartment to 150 feet (45 720 mm) and the maximum total travel distance to 200 feet (60 960 mm). These limitations are intended to enable evacuation of a smoke compartment for all of the occupants (see Figure 408.6.1).

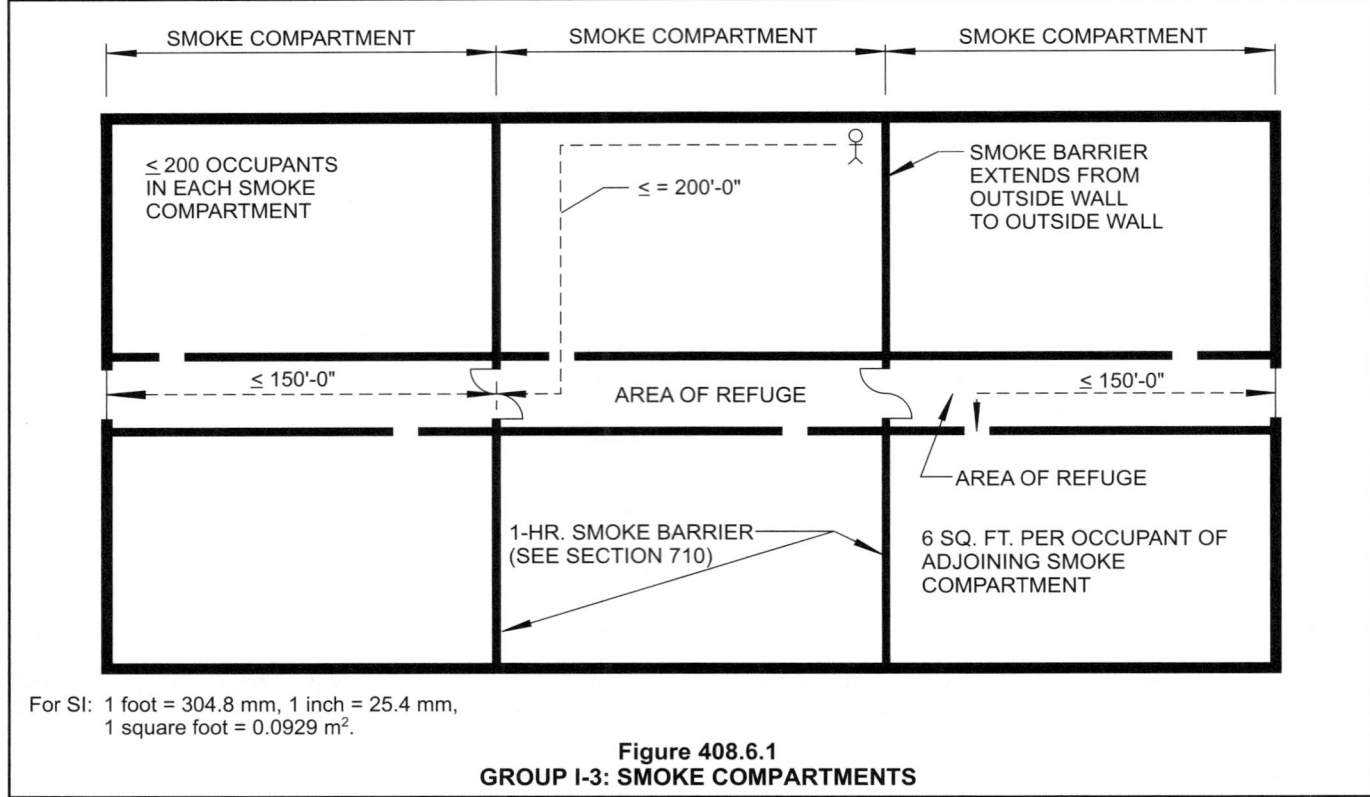

For SI: 1 foot = 304.8 mm, 1 inch = 25.4 mm, 1 square foot = 0.0929 m².

Figure 408.6.1
GROUP I-3: SMOKE COMPARTMENTS

408.6.2 Refuge area. At least 6 net square feet (0.56 m²) per occupant shall be provided on each side of each *smoke barrier* for the total number of occupants in adjoining smoke compartments. This space shall be readily available wherever the occupants are moved across the *smoke barrier* in a fire emergency.

❖ To provide adequate space for the residents evacuated to an adjacent smoke compartment, an additional 6 square feet (0.56 m²) per person must be readily available for the total number of occupants in the adjoining smoke compartment (see Figure 408.6.2). The 6-square-foot (0.56 m²) criterion enables the occupants to be evacuated to the adjacent smoke compartment without delay from crowding.

In a fire emergency, any action that must be taken to unlock the space will result in a delay in moving occupants into the refuge area. As such, the code requires the refuge area to be readily available in order to preclude consideration (as a refuge area) of a space that may be normally locked and, therefore, not immediately usable.

408.6.3 Independent egress. A *means of egress* shall be provided from each smoke compartment created by smoke barriers without having to return through the smoke compartment from which *means of egress* originates.

❖ To prevent creating a dead-end smoke compartment, exits are to be arranged so as to permit access without returning through a compartment from which egress originated. An exit is not required from within

each smoke compartment [see Figures 407.4.2(1) and (2) for acceptable and unacceptable exit arrangements].

408.7 Security glazing. In occupancies in Group I-3, windows and doors in 1-hour fire barriers constructed in accordance with Section 707, fire partitions constructed in accordance with Section 709 and smoke barriers constructed in accordance with Section 710 shall be permitted to have security glazing installed provided that the following conditions are met.

1. Individual panels of glazing shall not exceed 1,296 square inches (0.84 m²).

2. The glazing shall be protected on both sides by an *automatic sprinkler system*. The sprinkler system shall be designed to, when actuated, wet completely the entire surface of any glazing affected by fire.

3. The glazing shall be in a gasketed frame and installed in such a manner that the framing system will deflect without breaking (loading) the glass before the sprinkler system operates.

4. Obstructions, such as curtain rods, drapery traverse rods, curtains, drapes or similar materials shall not be installed between the automatic sprinklers and the glazing.

❖ This section allows sprinkler protected glazing in walls in a Group I-3 facility provided the wall is rated for 1 hour or less. As with many provisions in Section 408, this enhances the security of a facility by providing more opportunities for staff to monitor the activities of residents.

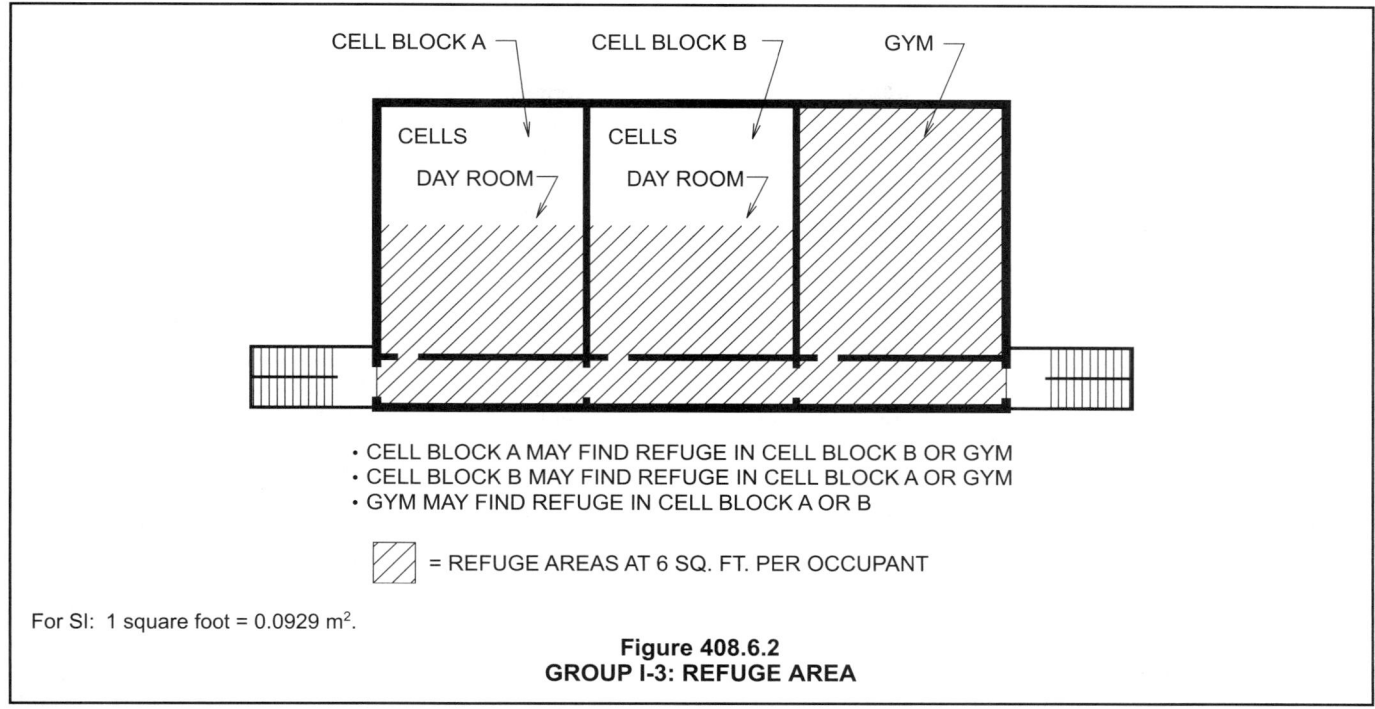

For SI: 1 square foot = 0.0929 m².

Figure 408.6.2
GROUP I-3: REFUGE AREA

408.8 Subdivision of resident housing areas. Sleeping areas and any contiguous day room, group activity space or other common spaces where residents are housed shall be separated from other spaces in accordance with Sections 408.8.1 through 408.8.4.

❖ Resident housing areas usually consist of sleeping areas and other contiguous common area spaces, such as day rooms or activity areas onto which sleeping areas open. The separation required by this section is intended to separate housing areas, where the occupants in those areas may be asleep, from other activity areas since people asleep are more vulnerable to a fire emergency. The separation requirements are based on the relative evacuation difficulty as determined by the occupancy condition. Sections 408.8.1 through 408.8.4 give the separation requirements for the various occupancy conditions for which separation is necessary.

408.8.1 Occupancy Conditions 3 and 4. Each sleeping area in Occupancy Conditions 3 and 4 shall be separated from the adjacent common spaces by a smoke-tight partition where the travel distance from the sleeping area through the common space to the *corridor* exceeds 50 feet (15 240 mm).

❖ A smoke-tight partition is required between the sleeping areas and common areas in Conditions 3 and 4 when travel distance from the sleeping area through the common area to an exit access corridor exceeds 50 feet (15 240 mm). This requirement recognizes the need to protect sleeping areas from nonsleeping areas due to the increased hazards associated with common areas, delayed recognition of fire hazards by sleeping persons and the restrictions on movement for these two occupancy conditions. Moreover, release from the sleeping area may be delayed and as such,

evacuation with extended travel through a common space may not be safe. In such cases, the sleeping area will be a refuge area until the fire is controlled and the occupants in the sleeping area can be safely evacuated.

It should be noted that the provisions contained in Section 1014.2 are applicable to this circumstance, and as such, sleeping areas, including those in Group I-3 occupancies, are precluded from egressing through another sleeping area. Additionally, egress is not allowed from sleeping areas through toilet rooms.

408.8.2 Occupancy Condition 5. Each sleeping area in Occupancy Condition 5 shall be separated from adjacent sleeping areas, *corridors* and common spaces by a smoke-tight partition. Additionally, common spaces shall be separated from the *corridor* by a smoke-tight partition.

❖ Given that movement from an occupied space in Condition 5 is even more restricted than for the other conditions, smoke-tight partitions are required to be installed between areas, sleeping areas, common spaces and the exit access corridor. Protection is provided because the staff-controlled release from the sleeping area is not as immediate as in other conditions.

408.8.3 Openings in room face. The aggregate area of openings in a solid sleeping room face in Occupancy Conditions 2, 3, 4 and 5 shall not exceed 120 square inches (77 419 mm²). The aggregate area shall include all openings including door undercuts, food passes and grilles. Openings shall be not more than 36 inches (914 mm) above the floor. In Occupancy Condition 5, the openings shall be closeable from the room side.

❖ Food pass-throughs, door undercuts and grilles are a functional necessity; therefore, openings are limited in size and close to the floor. The openings are required to be close to the floor to minimize the potential for the smoke to pass through the opening. By placing the opening close to the floor, the smoke will not pass through a smoke-tight barrier initially; rather, there will be some time delay as the smoke layer descends to 36 inches (914 mm) above the floor before smoke is transmitted through such openings.

408.8.4 Smoke-tight doors. Doors in openings in partitions required to be smoke tight by Section 408.8 shall be substantial doors, of construction that will resist the passage of smoke. Latches and door closures are not required on cell doors.

❖ If the partition is required to be capable of resisting the passage of smoke, door openings within the enclosure are also to be protected with an assembly that is capable of resisting the passage of smoke. As with other sections of the code that deal with doors and walls capable of resisting the passage of smoke, application of the section requires judgment by the building official. The intent is a solid door surface without louvered openings. Door closers and latches are not required. It is anticipated that the security lock will secure the door in the closed position. A door closer can interfere with the removal of the occupants not only in an emergency but also during normal operations.

408.9 Windowless buildings. For the purposes of this section, a windowless building or portion of a building is one with nonopenable windows, windows not readily breakable or without windows. Windowless buildings shall be provided with an engineered smoke control system to provide a tenable environment for exiting from the smoke compartment in the area of fire origin in accordance with Section 909 for each windowless smoke compartment.

❖ An engineered smoke control system is required for smoke compartments in which there are no openings through which the products of combustion can be vented. The intent of the system is to provide a tenable environment during the period it takes the occupants to egress from a smoke compartment that is the area of fire origin. As it often takes longer to evacuate a Group I-3 facility due to the locking and restricted egress paths, the maintenance of a tenable environment is essential. As defined for this section only, a windowless building is a building or portion thereof with only nonoperable windows, windows that are not readily breakable or without windows. The intent of this section is that staff must have some means to ventilate the products of combustion; therefore, if the window cannot be broken by items readily available to the staff, the area is considered windowless.

[F] 408.10 Fire alarm system. A fire alarm system shall be provided in accordance with Section 907.2.6.3.

❖ Fire alarm systems are required in Group I-3 facilities. See the commentary to Section 907.2.6.3 for detailed requirements for this system.

SECTION 409
MOTION PICTURE PROJECTION ROOMS

409.1 General. The provisions of Sections 409.1 through 409.5 shall apply to rooms in which ribbon-type cellulose acetate or other safety film is utilized in conjunction with electric arc, xenon or other light-source projection equipment that develops hazardous gases, dust or radiation. Where cellulose nitrate film is utilized or stored, such rooms shall comply with NFPA 40.

❖ Cellulose acetate or other safety film used in the presence of potential ignition sources, including projectors or related equipment, creates a potential hazard. The provisions of this section are intended to minimize the potential for exposure of the audience and other occupants to the hazards associated with the presence of ignition sources in close proximity to the fuel load.

Provisions are established for projection rooms in recognition of the hazards attendant to the projection of ribbon-type cellulose acetate or other safety film in such spaces.

Safety film is typically made of cellulose acetate, polyester or triacetate film. In the early 1950s, safety film began to replace cellulose nitrate film. Although the use and manufacture of cellulose nitrate film has virtually ceased in the United States, large quantities of such film still exist, primarily for archival purposes. Cellulose nitrate film presents a serious hazard because degradation under external heat below ignition temperatures causes a chemical reaction. This chemical reaction releases heat of sufficient quantity to raise the substance to ignition temperatures, resulting in spontaneous combustion. After ignition, rapid burning and the production of toxic and flammable gases takes place. If such film is used, the provisions of NPFA 40 are applicable as are the provisions of Section 306 of the IFC.

The provisions of this section apply only where the previously mentioned hazards exist. The use of ribbon-type cellulose acetate or other safety film in conjunction with electric arc, xenon or other light-source projection equipment that develops hazardous gases, dust or radiation invokes these provisions. The provisions of this section do not apply where nonprofessional projection equipment is used, except that statements specific to cellulose nitrate film apply if such film is present regardless of the nature of the equipment. The safety film used by the motion picture industry since 1951 has fire hazard characteristics similar to ordinary paper of the same thickness and form.

If cellulose nitrate film is used, the provisions of NFPA 40 apply. NFPA 40 contains minimum requirements to provide a reasonable level of protection for the storage and handling of cellulose nitrate film. The standard does not contain provisions for the manufacture of cellulose nitrate film, since it has not been manufactured in the United States since 1951.

409.1.1 Projection room required. Every motion picture machine projecting film as mentioned within the scope of this section shall be enclosed in a projection room. Appurtenant electrical equipment, such as rheostats, transformers and generators, shall be within the projection room or in an adjacent room of equivalent construction.

❖ The projection room is to be enclosed in accordance with Section 409.2 to isolate the spaces housing the potential hazard. To minimize the potential for arcs or sparks to come in contact with the film, electrical equipment, such as rheostats, transformers and generators, should be located in a separate room. This section permits such equipment to be in the same room, but it should be arranged to minimize the potential for arcs or sparks to come in contact with the film, reducing the potential for ignition. Additional requirements for electrical equipment permitted in the projection room and requirements for nonprofessional equipment are given in NFPA 70.

409.2 Construction of projection rooms. Every projection room shall be of permanent construction consistent with the construction requirements for the type of building in which the projection room is located. Openings are not required to be protected.

The room shall have a floor area of not less than 80 square feet (7.44 m²) for a single machine and at least 40 square feet (3.7 m²) for each additional machine. Each motion picture projector, floodlight, spotlight or similar piece of equipment shall have a clear working space of not less than 30 inches by 30 inches (762 mm by 762 mm) on each side and at the rear thereof, but only one such space shall be required between two adjacent projectors. The projection room and the rooms appurtenant thereto shall have a ceiling height of not less than 7 feet 6 inches (2286 mm). The aggregate of openings for projection equipment shall not exceed 25 percent of the area of the wall between the projection room and the auditorium. Openings shall be provided with glass or other *approved* material, so as to close completely the opening.

❖ The projection room is to be constructed as required by the construction type of the building. Since the room dimensions are small, ventilation is provided to reduce heat buildup and accumulation of dust particles. A minimum fire-resistance rating is not required unless by other provisions of the code. Limited openings are permitted and are not required to be protected so as to facilitate the intended function of projecting the image onto the auditorium screen. Openings to the auditorium must not exceed 25 percent of the area of the wall between the projection room and the auditorium. Glass or other approved material must be provided that will completely close the opening and prevent a sharing of atmospheres with the audience or seating area.

In order to minimize the buildup of heat in any area, adequate floor space must be available. The space is required to be at least 80 square feet (7.44 m²) in size for a single projector and an additional 40 square feet (3.7 m²) for each additional projector. The room must be of adequate size so that no less than 30 inches by 30 inches (762 mm by 762 mm) of clear working space

is available on each side and behind the projectors, floodlights, spotlights and similar equipment. For adjacent equipment, a clear space of 30 inches (762 mm) is adequate.

Whereas the projection room is considered an occupiable room, the ceiling height must not be less than 7 feet, 6 inches (2286 mm) (see commentary, Section 1208.2).

409.3 Projection room and equipment ventilation. Ventilation shall be provided in accordance with the *International Mechanical Code*.

❖ All projection rooms and the equipment located within the rooms must be properly ventilated to abate the hazards of any hot gases and smoke generated by the equipment. Further, any intense heat given off by the equipment must be directly exhausted to the exterior rather than be absorbed by the building's ventilation system. Sections 502.11.1 and 502.11.2 of the IMC contain requirements for projectors provided with an exhaust discharge and those without an exhaust connection.

409.3.1 Supply air. Each projection room shall be provided with adequate air supply inlets so arranged as to provide well-distributed air throughout the room. Air inlet ducts shall provide an amount of air equivalent to the amount of air being exhausted by projection equipment. Air is permitted to be taken from the outside; from adjacent spaces within the building, provided the volume and infiltration rate is sufficient; or from the building air-conditioning system, provided it is so arranged as to provide sufficient air when other systems are not in operation.

❖ Separate supply (see Section 409.3.1) and exhaust (see Section 409.3.2) requirements are provided for projection rooms. Exhaust rates are specified in Section 502.11.2 of the IMC for projectors without exhaust connection.

Sufficient air must be introduced into each projection room to allow the effects of hot gases, smoke and heat to be diluted and to provide a makeup air source for the exhausted air. Although the code does not prescriptively state where the air supply inlets are to be located, the intent is to create a well-ventilated room. Several air supply inlets should be provided so that the entire room has adequate air changes. The amount of air supplied to each projection room must be approximately equal to the volume of air that is exhausted according to the IMC. Makeup supply air can be obtained from any source as long as sufficient quantities are provided at all times that the exhaust system is operating.

409.3.2 Exhaust air. Projection rooms are permitted to be exhausted through the lamp exhaust system. The lamp exhaust system shall be positively interconnected with the lamp so that the lamp will not operate unless there is the required airflow. Exhaust air ducts shall terminate at the exterior of the building in such a location that the exhaust air cannot be readily recirculated into any air supply system. The projection room ventilation system is permitted to also serve appurtenant rooms, such as the generator and rewind rooms.

❖ Another method of providing exhaust is to exhaust the room itself. This method can only be used when the projectors do not have an integral lamp exhaust system. Exhaust rates for this type of system are specified in the IMC based on the type of projector. The exhaust ducts for the projection room must be terminated at the exterior of the building and cannot be combined with any other exhaust or return system present in the building. The intent of this requirement is so that the hot gases, smoke and heat will not be allowed to be disbursed elsewhere in the building.

409.3.3 Projection machines. Each projection machine shall be provided with an exhaust duct that will draw air from each lamp and exhaust it directly to the outside of the building. The lamp exhaust is permitted to serve to exhaust air from the projection room to provide room air circulation. Such ducts shall be of rigid materials, except for a flexible connector *approved* for the purpose. The projection lamp or projection room exhaust system, or both, is permitted to be combined but shall not be interconnected with any other exhaust or return system, or both, within the building.

❖ Projectors with an integral lamp exhaust system use air to cool the equipment during operation. The amount of air exhausted with these systems is part of the manufacturer's installation instructions. These integral exhaust systems must be directly connected to the mechanical exhaust systems. Further, the lamp exhaust system must be interconnected to the projector's power supply such that when the projector is operating, its exhaust system will be automatically activated.

409.4 Lighting control. Provisions shall be made for control of the auditorium lighting and the *means of egress* lighting systems of theaters from inside the projection room and from at least one other convenient point in the building.

❖ The projection room is usually an attended location, but it is also a likely area for a fire to originate. Lighting controls that provide adequate normal and emergency auditorium lighting must be provided in the projection room. Upon detection of a fire anywhere in the building, the operator can immediately provide adequate illumination for the occupants egressing the auditorium. A second location for the lighting controls is also required somewhere else in the building, so that entry into the projection room is not required to provide the necessary illumination, as it could be the location of fire origin. This second control is required to be conveniently located for the staff yet inaccessible to unauthorized persons.

409.5 Miscellaneous equipment. Each projection room shall be provided with rewind and film storage facilities.

❖ In order to keep all operations related to projection activities in the protected enclosure, rewind and film storage facilities are to be located in the projection room or in an adjacent room in accordance with Section 409.1.1.

SECTION 410
STAGES AND PLATFORMS

410.1 Applicability. The provisions of Sections 410.1 through 410.7 shall apply to all parts of buildings and structures that contain stages or platforms and similar appurtenances as herein defined.

❖ Historically, most significant theater fires originated on the stage. The 1903 Iroquois Theater fire in Chicago serves as a vivid example of a stage fire and its potentially tragic effects—602 people lost their lives. Hazards associated with stages include: combustible scenery and lighting suspended overhead; scenic elements, contents and acoustical treatment on the back and sides of the stage; workshops, scene docks and dressing rooms located around the stage perimeter and storage areas and property rooms located underneath the stage.

The protection requirements set forth in this section are intended to limit the threat from stage fires to an audience and reduce the likelihood of a large fire in the stage area. These provisions include construction restrictions, automatic sprinkler systems, ventilation, separation of the stage from the audience and compartmentalization of appurtenant rooms to the stage.

Based on historical events and the expertise of a broad range of professionals with experience in theater design, operations and fire events, a set of principles guides the requirements for theaters. These principles are summarized as follows:

1. The fire hazards of stages are not necessarily a function of type (legitimate, regular or thrust) but rather area and height. Accordingly, a stage area in excess of 1,000 square feet (93 m²) or a height in excess of 50 feet (15 240 mm) is the threshold that represents a significant potential for fuel load due to scenery and drops.

2. Stages that exceed 1,000 square feet (93 m²) or the 50-foot (15 240 mm) height threshold require a means of emergency ventilation.

3. Stages that exceed 1,000 square feet (93 m²) and the 50-foot (15 240 mm) height threshold require automatic sprinkler protection.

4. Stages and platforms are similar to floor construction; therefore, the type of construction should be consistent with that of the floor construction in the building.

5. Separation of the stage from the seating area and dressing rooms, property rooms and similar spaces is critical in order to provide a degree of fire containment.

Section 410.3 contains provisions for the construction of the stage and gallery. Stage openings, decorations, equipment and scenery are addressed in Sections 410.3.3 through 410.3.6. Stage ventilation is required in accordance with Section 410.3.7. Platforms must be constructed in accordance with Section 410.4. The construction of auxiliary stage spaces is addressed in Section 410.5. Sections 410.6 and 410.7 contain the requirements for automatic sprinkler systems and standpipes, respectively.

All parts of a building or structure that contain a stage, platform or similar appurtenance are required to comply with the provisions of this section.

410.2 Definitions. The following words and terms shall, for the purposes of this section and as used elsewhere in this code, have the meanings shown herein.

❖ Definitions of terms that are associated with the content of this section are contained herein. These definitions can help in the understanding and application of the code requirements. It is important to emphasize that these terms are not exclusively related to this section but are applicable everywhere the term is used in the code. The purpose for including these definitions within this section is to provide more convenient access to them without having to refer back to Chapter 2. For convenience, these terms are also listed in Chapter 2 with a cross reference to this section. The use and application of all defined terms, including those defined herein, are set forth in Section 201.

FLY GALLERY. A raised floor area above a stage from which the movement of scenery and operation of other stage effects are controlled.

❖ The fly gallery is an integral part of the stage construction. This building feature provides an elevated, limited-size working area where the stagehands can operate and control the movement of suspended scenery or other visual effects.

GRIDIRON. The structural framing over a stage supporting equipment for hanging or flying scenery and other stage effects.

❖ The gridiron is directly tied to the functions of the fly gallery. This structural element usually consists of lightweight trusses that are used to suspend not only scenery but also behind-the-scenes equipment.

PINRAIL. A rail on or above a stage through which belaying pins are inserted and to which lines are fastened.

❖ This structural beam provides support for the tying off of various ropes and cables that are used to raise and lower suspended scenery. Modern theaters typically do not use actual pins anymore, but there is a "lockrail" through which the cables run and are then locked to hold a fly or other suspended piece of scenery. These systems are generally motorized.

PLATFORM. A raised area within a building used for worship, the presentation of music, plays or other entertainment; the head table for special guests; the raised area for lecturers and speakers; boxing and wrestling rings; theater-in-the-round

stages; and similar purposes wherein there are no overhead hanging curtains, drops, scenery or stage effects other than lighting and sound. A temporary platform is one installed for not more than 30 days.

❖ Platforms are raised areas that are used for public performances and presentations that, except for lighting, do not incorporate any overhead hanging curtains, drops, scenery or stage effects. Additionally, while it is not specified, the fuel load on platforms (e.g., a podium for a lecturer or a boxing ring) is anticipated to be low.

Thus, since the fuel load on platforms is ordinarily low and there is no fuel load overhead in areas that would be difficult to access, the code requirements for platforms are less stringent than for stages.

Since many platforms are installed and used on a temporary basis, this definition specifically defines temporary use as 30 days or less. Temporary use is less likely to accumulate additional fuel loads, storage or waste materials on, beneath or above the platform.

PROSCENIUM WALL. The wall that separates the stage from the auditorium or assembly seating area.

❖ A proscenium wall separates the stage area from the audience. The performance on the stage is viewed through a large opening known as the main proscenium opening. Where the stage height is greater than 50 feet (15 240 mm), a proscenium wall of approved construction is required by the code (see Section 410.3.4).

STAGE. A space within a building utilized for entertainment or presentations, which includes overhead hanging curtains, drops, scenery or stage effects other than lighting and sound.

❖ The building feature known as a "stage" is defined as the area or space where the performers are located. The presentation area is usually elevated to provide clear sightlines for the audience. The stage area also includes the wing areas and backstage areas where curtains and scenery are placed.

410.3 Stages. Stage construction shall comply with Sections 410.3.1 through 410.3.7.

❖ This section indicates the scope of the code provisions that apply to stage construction.

410.3.1 Stage construction. Stages shall be constructed of materials as required for floors for the type of construction of the building in which such stages are located.

Exceptions:

1. Stages of Type IIB or IV construction with a nominal 2-inch (51 mm) wood deck, provided that the stage is separated from other areas in accordance with Section 410.3.4.

2. In buildings of Types IIA, IIIA and VA construction, a fire-resistance-rated floor is not required, provided the space below the stage is equipped with an automatic fire-extinguishing system in accordance with Section 903 or 904.

3. In all types of construction, the finished floor shall be constructed of wood or *approved* noncombustible materials. Openings through stage floors shall be equipped with tight-fitting, solid wood trap doors with *approved* safety locks.

❖ Stages, most simply, are floor construction, and as such, must conform to the requirements for all other floors in the building so that the building's designated construction classification will be consistent throughout.

Since wood is the material of choice for stage floors because it allows for nailing of sets and props and provides the best acoustics for dance and other performances, the code allows three exceptions to the requirement for stage construction.

Exception 1 permits stages in buildings of any construction classification to be constructed with a nominal 2-inch (51 mm) wood deck supported by either unprotected noncombustible construction (Type IIB) or heavy timber construction (Type IV) where the stage is separated from the audience by a proscenium wall built in accordance with Section 410.3.4.

Under Exception 2, stage floors are not required to be fire-resistance rated in buildings of Types IIA, IIIA and VA where the space below the stage is equipped with an automatic fire-extinguishing system installed in accordance with Section 903 or 904. In this case, the requirement for fire-resistance-rated floors is offset by the protection supported by the automatic fire-extinguishing system.

Exception 3 recognizes that stages are predominantly constructed with a finished floor of wood and identifies wood as an acceptable finish material for all construction classifications. It also permits openings in stage floors to be protected with tight-fitting solid wood doors with safety locks to secure them. The normal operation of stage trap doors does not typically allow them to have a fire-resistance rating.

410.3.1.1 Stage height and area. Stage areas shall be measured to include the entire performance area and adjacent backstage and support areas not separated from the performance area by fire-resistance-rated construction. Stage height shall be measured from the lowest point on the stage floor to the highest point of the roof or floor deck above the stage.

❖ As previously stated, many of the code's requirements for stages and platforms are triggered based on the volume of the stage or platform space. All of the wing areas and backstage areas are included as the stage area, unless they are separated with fire-resistance-rated construction. Stage height is not the height of the opening in the proscenium wall, but the height from the stage floor to the highest horizontal assembly that encloses the stage space. It will typically include the fly gallery and fly loft above the stage.

410.3.2 Galleries, gridirons, catwalks and pinrails. Beams designed only for the attachment of portable or fixed theater equipment, gridirons, galleries and catwalks shall be constructed of *approved* materials consistent with the requirements for the type of construction of the building; and a

fire-resistance rating shall not be required. These areas shall not be considered to be floors, stories, *mezzanines* or levels in applying this code.

Exception: Floors of fly galleries and catwalks shall be constructed of any *approved* material.

❖ This section identifies the components of the areas located above the stage that are used to conceal movable scenery from the audience and where the operation of such scenery and other stage effects is controlled.

The normal use and operation of a stage requires that equipment, normally rigging, be installed and rearranged as production requirements vary. This equipment is normally installed by clamping or welding to the structural framing over and around the stage, making protecting the framing by encasement or membranes infeasible; hence, a fire-resistance rating is not required for these elements. In order to control the fuel load in this area, however, the code requires that all of the elements must be of approved materials consistent with the requirements of the building's construction type. The exception to this requirement is that in all cases, the floors of fly galleries and catwalks may be constructed of any approved material regardless of the building's construction type. Such spaces are generally limited in area, access is restricted to authorized personnel and they require floor materials that will deaden sound so as not to disrupt the performance below. These auxiliary areas do not have to meet the code's requirements as another story, mezzanine level or equipment platforms.

410.3.3 Exterior stage doors. Where protection of openings is required, exterior *exit* doors shall be protected with *fire door assemblies* that comply with Section 715. Exterior openings that are located on the stage for *means of egress* or loading and unloading purposes, and that are likely to be open during occupancy of the theater, shall be constructed with vestibules to prevent air drafts into the auditorium.

❖ If exterior opening protectives are required, exit discharge doors directly from the stage to the outside must be fire door assemblies in accordance with Section 715.

Since one of the major concerns is containing a stage fire to the stage area, any exterior opening from the stage that is likely to be opened during a performance is to be constructed with a vestibule to prevent air drafts into the auditorium. This vestibule requirement applies to all exterior openings that are likely to be open regardless of their intended use.

410.3.4 Proscenium wall. Where the stage height is greater than 50 feet (15 240 mm), all portions of the stage shall be completely separated from the seating area by a proscenium wall with not less than a 2-hour *fire-resistance rating* extending continuously from the foundation to the roof.

❖ The protection afforded by a 2-hour fire separation and opening protection (see Section 410.3.5) is required where the stage height is greater than 50 feet (15 240 mm). Stages with a height greater than 50 feet (15 240 mm) permit multiple settings and large amounts of scenery and scenic elements in dense configurations. The height may reduce the effectiveness of the suppression system and the multiple settings hung over the stage may further obstruct the suppression system and impede access to a fire originating high above the stage. Stages with a height less than 50 feet (15 240 mm) do not require separation from the audience since they represent a limited fuel load potential caused by scenery, drops and curtains.

410.3.5 Proscenium curtain. Where a proscenium wall is required to have a *fire-resistance rating*, the stage opening shall be provided with a fire curtain complying with NFPA 80 or an *approved* water curtain complying with Section 903.3.1.1 or, in facilities not utilizing the provisions of smoke-protected assembly seating in accordance with Section 1028.6.2, a smoke control system complying with Section 909 or natural ventilation designed to maintain the smoke level at least 6 feet (1829 mm) above the floor of the *means of egress.*

❖ Fuel loads in a stage with a height greater than 50 feet (15 240 mm) most often are high and the resulting fire is severe. In such cases, to permit the audience to evacuate the seating area without being threatened directly or indirectly by a stage fire, this section requires the proscenium opening be protected with either a fire curtain (complying with NFPA 80) or an approved water curtain. Another available option is to provide a smoke control system or natural ventilation designed to prevent the accumulation of smoke within the first 6 feet (1829 mm) above the floor. This option is only permissible if a smoke-protected assembly seating design in accordance with Section 1028.6.2 is not utilized.

In accordance with NFPA 80, a fire curtain must be capable of preventing the passage of flame or smoke, thereby providing a certain level of protection for the audience when a stage fire occurs. It is notable that the fire curtain is not required to provide the same level of protection as is typically required for a fire door assembly. Also, since one of the goals is to protect against panic, flame and smoke must not be capable of passing through to the unexposed side of the fire curtain. The level of protection for the fire curtain is based upon evacuation time of the audience. Fire curtains are designed to provide 20 minutes of protection to allow the audience to evacuate.

NFPA 80 requirements for the fire curtain include, but are not limited to: testing of the fabric (NFPA 701, Test Method 2), a fire curtain sample assembly subjected to the standard fire test as specified in NFPA 251, activation (emergency and manual) and rate of closing speed. The speed of descent must be slowed for the last 8 feet (2438 mm) of travel in order to allow individuals on the stage to move away from the curtain and to avoid injury from being hit by the batten in the bottom pocket of the curtain. It should be noted that only stages with a height greater than 50 feet (15 240 mm) are required to have a proscenium wall and curtain.

Water curtains conforming with Section 903.3.1.1 are viewed as providing an equivalent degree of protection to that of a fire curtain. This, in part, recognizes that emergency ventilation is also required in accordance with Section 410.3.7, which will assist in controlling smoke movement. The duration of protection is, of course, continuous as long as the sprinkler system is in operation.

Lastly, today's assembly facilities, such as arenas and theaters in the round, make the use of a traditional fire curtain impractical. As such, a smoke control system or a natural ventilation system may be utilized in lieu of a fire curtain or approved water curtain, provided smoke-protected assembly seating is not being used.

410.3.6 Scenery. Combustible materials used in sets and scenery shall meet the fire propagation performance criteria of NFPA 701, in accordance with Section 806 and the *International Fire Code.* Foam plastics and materials containing foam plastics shall comply with Section 2603 and the *International Fire Code.*

❖ This section recognizes that scenery and sets are decorative materials. As such, combustible materials used in sets and scenery must meet the fire propagation performance criteria in accordance with the provisions of NFPA 701, Section 806 and the IFC. Section 806 references Section 2604 for specific requirements for foam plastics used as trim. It should also be noted that Section 2603 addresses the uses of foam plastics in insulation.

410.3.7 Stage ventilation. Emergency ventilation shall be provided for stages larger than 1,000 square feet (93 m²) in floor area, or with a stage height greater than 50 feet (15 240 mm). Such ventilation shall comply with Section 410.3.7.1 or 410.3.7.2.

❖ In addition to an automatic sprinkler system, emergency ventilation is one of the key fire protection components for stages that contain large fuel loads. Stages with large areas or heights permit multiple settings and large amounts of scenery and scenic elements in dense configurations. Increased height also permits multiple settings to be hung over the stage that may reduce the effectiveness of the suppression system or impede access to a fire originating high above the stage. Based on these factors and the potential hazard presented by them, stages larger than 1,000 square feet (93 m²) in area or with a height greater than 50 feet (15 240 mm) are required to be equipped with emergency ventilation to control smoke movement and minimize the potential for fire and smoke to spread to the seating area.

410.3.7.1 Roof vents. Two or more vents constructed to open automatically by *approved* heat-activated devices and with an aggregate clear opening area of not less than 5 percent of the area of the stage shall be located near the center and above the highest part of the stage area. Supplemental means shall be provided for manual operation of the ventilator. Curbs shall be provided as required for skylights in Section 2610.2. Vents shall be labeled.

❖ Where vents are used to ventilate a stage, the code requires that there be a minimum of two vents provided for redundancy. Those vents are to be located near the center and above the highest part of the stage, since this is the point at which smoke and hot gases are likely to accumulate. The minimum required aggregate clear opening area of the vents is required to be 5 percent of the floor area of the stage. The vents must open automatically by approved heat-activated devices, often fusible links, and must also be operable by a supplemental manual means. The manual means serves as a backup to the automatic detectors and permits the vent to be opened as a precaution prior to achieving the heat required to activate the heat device. The vents are to be provided with curbs as required for skylights in Section 2610.2. In addition, the vents must be labeled by an approved agency (see Section 1703.5 for further labeling requirements).

[F] 410.3.7.2 Smoke control. Smoke control in accordance with Section 909 shall be provided to maintain the smoke layer interface not less than 6 feet (1829 mm) above the highest level of the assembly seating or above the top of the proscenium opening where a proscenium wall is provided in compliance with Section 410.3.4.

❖ In addition to roof vents, another option for emergency ventilation of stages is provided in this section (i.e., smoke control). This section, in concert with the provisions contained in Section 909, provides the criteria for smoke control for stages larger than 1,000 square feet (93 m²) in floor area or with a stage height greater than 50 feet (15 240 mm). In such cases, the smoke layer interface must be maintained 6 feet (1829 mm) above the highest level of the assembly seating or above the top of the proscenium opening when one is provided. The smoke layer is maintained 6 feet (1829 mm) above the proscenium opening in order to prevent smoke from entering into the audience area.

410.4 Platform construction. Permanent platforms shall be constructed of materials as required for the type of construction of the building in which the permanent platform is located. Permanent platforms are permitted to be constructed of *fire-retardant-treated wood* for Types I, II and IV construction where the platforms are not more than 30 inches (762 mm) above the main floor, and not more than one-third of the room floor area and not more than 3,000 square feet (279 m²) in area. Where the space beneath the permanent platform is used for storage or any purpose other than equipment, wiring or plumbing, the floor assembly shall not be less than 1-hour fire-resistance-rated construction. Where the space beneath the permanent platform is used only for equipment, wiring or plumbing, the underside of the permanent platform need not be protected.

❖ This section establishes the scope of the code requirements for platform construction. A permanent platform is basically raised floor construction. As such, the construction must be consistent with that of all the floors in the building.

Permanent platforms can be of fire-retardant-treated wood in limited applications. In buildings of Type I, II or IV construction, platforms can be of fire-retardant-treated wood if they are no more than 30 inches (762 mm) in height, are no more than one-third of the room floor area and are 3,000 square feet (279 m²) or less in area. A platform that meets these three size limits poses a small fire risk relative to the normal fuel load in the room. As such, the platform deck and its supporting construction can be of fire-retardant-treated wood materials. If the platform exceeds any one of the three size limits, then it must be constructed of materials consistent with the building's type of construction classification. A basic premise of Section 410.4 is that the platform must not significantly increase the fire hazard of the space or building. As such, the space beneath platforms is regulated to appropriately abate the risk. Where the space beneath a permanent platform is utilized for any purpose other than electrical wiring or plumbing, the platform is to have a 1-hour fire-resistance rating. The purpose of the fire-resistance-rating requirement is to provide some structural integrity to the platform should a fire occur in the concealed space. In the case of permanent platforms where the space beneath the platform is only used for plumbing or electrical wiring, no further protection is required since the fire risk will be minimal.

410.4.1 Temporary platforms. Platforms installed for a period of not more than 30 days are permitted to be constructed of any materials permitted by the code. The space between the floor and the platform above shall only be used for plumbing and electrical wiring to platform equipment.

❖ Temporary platforms may be constructed of any material regardless of the building's construction classification due to their limited time of use and their normally limited size and fuel load. Because temporary platforms are permitted to be constructed of any approved material, the space beneath a temporary platform may never be used for any purpose other than electrical wiring or plumbing to the platform equipment. Such lines serving other areas may not be located beneath a temporary platform.

410.5 Dressing and appurtenant rooms. Dressing and appurtenant rooms shall comply with Sections 410.5.1 through 410.5.3.

❖ Because stages are open to the viewing audience and typically contain a substantial fuel load, it is essential to contain fires in rooms around the stage to the room of origin; therefore, this section contains provisions for the separation of such areas from the stage and from one another. Section 410.5.3 also contains provisions for adequate egress so that an occupant on the stage and in areas above and below it may be able to egress the area safely should a fire occur in the stage area.

410.5.1 Separation from stage. The stage shall be separated from dressing rooms, scene docks, property rooms, workshops, storerooms and compartments appurtenant to the stage and other parts of the building by *fire barriers* constructed in accordance with Section 707 or *horizontal assemblies* constructed in accordance with Section 712, or both. The minimum *fire-resistance rating* shall be 2 hours for stage heights greater than 50 feet (15 240 mm) and 1 hour for stage heights of 50 feet (15 240 mm) or less.

❖ Separation of the stage from appurtenant rooms provides a significant level of fire containment. Such containment is fundamental to limiting the spread of fire from adjacent spaces to the stage area, as well as the growth of fires in the stage area itself. The 2-hour fire-resistance rating for stages with a height greater than 50 feet (15 240 mm), such as large theatrical stages, acknowledges the significant potential for a large fuel load and is consistent with the 2-hour rating of Section 410.3.4. Although stages with a height of 50 feet (15 240 mm) or less represent a limited potential fuel load, a 1-hour fire-resistance rating will provide an additional level of compartmentation in the event a fire originates at the stage (which may not be sprinklered if the stage complies with Section 410.6, Exception 2). Such stages are likely to be found in educational occupancies or meeting halls.

410.5.2 Separation from each other. Dressing rooms, scene docks, property rooms, workshops, storerooms and compartments appurtenant to the stage shall be separated from each other by not less than 1-hour *fire barriers* constructed in accordance with Section 707 or *horizontal assemblies* constructed in accordance with Section 712, or both.

❖ As an additional level of protection, rooms and compartments appurtenant to the stage must be separated from one another by an approved 1-hour fire-resistance-rated fire barrier and horizontal assemblies. Rooms appurtenant to the stage often have very high fuel loads (i.e., property rooms and activities with a history of fire incidents, such as scenery workshops). This additional level of fire containment is intended to minimize fire growth and limit fires in these areas to the room of origin.

410.5.3 Stage exits. At least one *approved means of egress* shall be provided from each side of the stage and from each side of the space under the stage. At least one means of escape shall be provided from each fly gallery and from the gridiron. A steel ladder, *alternating tread device* or *spiral stairway* is permitted to be provided from the gridiron to a scuttle in the stage roof.

❖ Because of the relative fire hazards associated with stages, this section requires that a minimum of one approved means of egress be provided from each side of the stage and from each side of the occupied space under the stage. This is intended to provide the occupants of a stage ready access to evacuate the stage area in the event of a fire. It should be noted that in accordance with Section 1022.1, Exception 6, means of egress are not required to be enclosed if they are within the stage area. Fly galleries and gridirons are ordinarily occupied by a few persons at any one time who are likely to be very familiar with the stage operations. As such, these areas need only be provided with one approved means of egress. In these areas, lad-

ders, alternating tread devices or spiral stairways are permitted (see stairways Exception 5, Section 1015.6.1). In all cases, interior means of egress stairs serving these areas are not required to be enclosed (see Section 1022.1, Exception 6). A steel ladder, alternating tread stairways or spiral stairways from the gridiron to the roof are permitted. Additional means of egress provisions specific to stages are contained in Section 1015.6.

[F] 410.6 Automatic sprinkler system. Stages shall be equipped with an automatic fire-extinguishing system in accordance with Chapter 9. Sprinklers shall be installed under the roof and gridiron and under all catwalks and galleries over the stage. Sprinklers shall be installed in dressing rooms, performer lounges, shops and storerooms accessory to such stages.

Exceptions:

1. Sprinklers are not required under stage areas less than 4 feet (1219 mm) in clear height that are utilized exclusively for storage of tables and chairs, provided the concealed space is separated from the adjacent spaces by not less than ⁵/₈-inch (15.9 mm) Type X gypsum board.

2. Sprinklers are not required for stages 1,000 square feet (93 m²) or less in area and 50 feet (15 240 mm) or less in height where curtains, scenery or other combustible hangings are not retractable vertically. Combustible hangings shall be limited to a single main curtain, borders, legs and a single backdrop.

3. Sprinklers are not required within portable orchestra enclosures on stages.

❖ Stages contain significant quantities of combustible materials stored in, around and above the stage that are located in close proximity to large quantities of lighting equipment (i.e., scenery and lighting above the stage). There also is scenery on the sides and rear of the stage; shops located along the back and sides of the stage, and storage, props, trap doors and lifts under the stage floor. This combination of fuel load and ignition sources increases the potential for a fire. As such, stages and auxiliary areas, such as dressing rooms, workshops and storerooms, are required to be protected with an automatic fire-extinguishing system.

The references to Chapter 9 indicate that the fire-extinguishing system may be designed in accordance with NFPA 13 and Section 903.3.1.1 or, if not more than 20 sprinklers are required on any single connection, a limited area sprinkler system may be used (see Section 903.3.5.1.1).

Exception 1 applies to areas less than 4 feet (1219 mm) in clear height under stages that are used only for the storage of tables and chairs. Because of the limited fuel load present, such areas are not required to be protected with sprinklers, provided that the concealed space is separated from all adjacent spaces by no less than ⁵/₈-inch (15.9 mm) Type X gypsum board. This level of separation is intended to provide fire containment in the absence of sprinkler protection. This arrangement is often found in educational buildings where the room is used as a multipurpose room.

Exception 2 recognizes that stages 1,000 square feet (93 m²) or less in area and 50 feet (15 240 mm) or less in height represent a limited potential for combustibles that does not warrant the requirements for an automatic sprinkler system. The code further limits the amounts of combustible materials by not allowing any vertical retractable curtains, hangings, etc. It should be noted, however, that although sprinkler protection is not required, the requirements of Section 410.5 still apply.

Exception 3 acknowledges the limited hazards associated with portable orchestra enclosures. These elements are temporary in nature and are intended to improve the acoustics of the stage performances. These temporary enclosures do not lend themselves to temporary sprinkler heads; therefore, none are required.

[F] 410.7 Standpipes. Standpipe systems shall be provided in accordance with Section 905.

❖ Because of the potentially large fuel load and three-dimensional aspect of the fire hazard associated with stages greater than 1,000 square feet (93 m²) in area, a Class III wet standpipe system is required on each side of such stages. The standpipes are required to be equipped with both a 1¹/₂-inch (38 mm) and 2¹/₂-inch (64 mm) hose connection. The required design criteria for the standpipe system, including hoses and cabinets, is specified in Section 905.3.4.

SECTION 411
SPECIAL AMUSEMENT BUILDINGS

411.1 General. Special *amusement buildings* having an *occupant load* of 50 or more shall comply with the requirements for the appropriate Group A occupancy and Sections 411.1 through 411.8. Amusement buildings having an *occupant load* of less than 50 shall comply with the requirements for a Group B occupancy and Sections 411.1 through 411.8.

Exception: Amusement buildings or portions thereof that are without walls or a roof and constructed to prevent the accumulation of smoke.

For flammable *decorative materials*, see the *International Fire Code*.

❖ A special amusement building is one in which the egress is not readily apparent, is intentionally confounded or is not readily available (see Section 411.2). This section addresses the hazards associated with such use by requiring automatic fire detection (see Section 411.3), automatic sprinkler protection (see Section 411.4), alarm requirements (see Section 411.5), an emergency voice/alarm communication system (see Section 411.6) and specific means of egress lighting and marking (see Section 411.7). Additionally, only interior finish materials that meet the most stringent flame spread classification, Class A, are permitted in special amusement buildings.

In addition to the provisions of this section, other applicable requirements, such as means of egress (i.e., occupant load, travel distance, etc.) and building construction (i.e., type of construction, fire-resistance ratings, etc.), apply in accordance with the appropriate assembly group classification (see Section 303.1).

Special amusement buildings are considered assembly uses based on the definition in Section 303.1. This section further specifies that a Group A classification is only applicable when the building's design occupant load is 50 or more. The provisions of this section apply in addition to the other requirements for the appropriate assembly use, usually Group A-1 or A-3. Very small special amusement buildings are not required to be classified as Group A. If the design occupant load is less than 50, then the building is classified as Group B. This acknowledges the lesser hazards associated with smaller groups of people. Although smaller buildings are classified as Group B, they must still meet the requirements of this section as special amusement buildings.

Section 411 does not apply to a facility that is constructed to permit the free and immediate ventilation of the products of combustion to the outside. Such free and immediate ventilation addresses the hazard associated with many special amusement buildings relative to the rapid accumulation of smoke in a building in which the egress is not readily apparent, confounded or not readily available.

All flammable decorative materials used in special amusement buildings are required to follow the provisions of the IFC. The use of flammable and combustible materials in these types of buildings must be strictly regulated to limit the fire hazards to the building occupants (refer to Chapter 8 of the IFC for those restrictions).

411.2 Definition. The following word and term shall, for the purpose of this section and as used elsewhere in this code, have the meaning shown herein.

❖ The definition of a term that is associated with the content of this section is contained herein. This definition can help in the understanding and application of the code requirements. It is important to emphasize that this term is not exclusively related to this section but is applicable everywhere the term is used in the code. The purpose for including this definition within this section is to provide more convenient access to it without having to refer back to Chapter 2. For convenience, this term is also listed in Chapter 2 with a cross reference to this section. The use and application of all defined terms, including that defined herein, are set forth in Section 201.

SPECIAL AMUSEMENT BUILDING. A *special amusement building* is any temporary or permanent building or portion thereof that is occupied for amusement, entertainment or educational purposes and that contains a device or system that conveys passengers or provides a walkway along, around or over a course in any direction so arranged that the *means of egress* path is not readily apparent due to visual or audio distractions or is intentionally confounded or is not readily available because of the nature of the attraction or mode of conveyance through the building or structure.

❖ In general, a special amusement building is a building or portion thereof in which people gather (thus, an assembly occupancy) and in which egress is either not readily apparent due to distractions, is intentionally confounded (i.e., maze) or is not readily available. The definition includes all such facilities, including portable and temporary structures. The hazard associated with such buildings is not related to the permanence or length of use; therefore, seasonal uses (such as haunted houses at Halloween) and portable uses (carnival rides) are included if they meet the criteria in the definition.

[F] **411.3 Automatic fire detection.** *Special amusement buildings* shall be equipped with an automatic fire detection system in accordance with Section 907.

❖ The automatic fire detection system is required to provide early warning of fire and must comply with Section 907. The detection system is required regardless of the presence of staff in the building.

[F] **411.4 Automatic sprinkler system.** *Special amusement buildings* shall be equipped throughout with an *automatic sprinkler system* in accordance with Section 903.3.1.1. Where the *special amusement building* is temporary, the sprinkler water supply shall be of an *approved* temporary means.

Exception: Automatic sprinklers are not required where the total floor area of a temporary *special amusement building* is less than 1,000 square feet (93 m²) and the travel distance from any point to an *exit* is less than 50 feet (15 240 mm).

❖ One protection strategy to minimize the potential hazard to occupants is to control fire development. As such, special amusement buildings are required to be protected with an automatic sprinkler system. If the building is small [less than 1,000 square feet (93 m²)] and the travel distance to exits is short [less than 50 feet (15 240 mm)] and only used on a temporary basis (such as at Halloween), automatic sprinklers are not required. In such buildings, it is anticipated that automatic fire detection and the resulting alarm (see Section 411.5) will provide additional egress time for the limited number of occupants.

Since many special amusement buildings are portable or temporary, it is not practical to require a permanent, automatic sprinkler water supply as required in Chapter 9. Instead, the building official may allow a reliable, temporary water supply. There are other unique design considerations for the sprinkler system, such as drainage and pipe and sprinkler support, which may be necessary because of the movement of the structures from one location to another.

[F] **411.5 Alarm.** Actuation of a single smoke detector, the *automatic sprinkler system* or other automatic fire detection device shall immediately sound an alarm at the building at a *constantly attended location* from which emergency action can

be initiated including the capability of manual initiation of requirements in Section 907.2.12.2.

❖ Upon activation of either the automatic fire detection or the automatic sprinkler systems, an alarm is required to be sounded at a constantly attended location. The staff at the location is expected to be capable of then providing the required egress illumination, stopping the conflicting or confusing sounds and distractions and activating the exit marking required by Section 411.7. It is also anticipated that the staff would be capable of preventing additional people from entering the building.

[F] **411.6 Emergency voice/alarm communications system.** An emergency voice/alarm communications system shall be provided in accordance with Sections 907.2.12 and 907.5.2.2, which is also permitted to serve as a public address system and shall be audible throughout the entire *special amusement building*.

❖ An integral part of the fire protection systems required for special amusement buildings is an emergency voice/alarm communications system. This system can serve as a public address system to alert the building occupants of the fire emergency and provide them with the proper emergency instructions. The system must be installed in accordance with NFPA 72 and must be heard throughout the entire special amusement building. Upon activation, the system must sound an alert tone followed by the necessary voice instructions. These instructions can save valuable time in directing the building occupants to quickly and safely egress.

411.7 Exit marking. Exit signs shall be installed at the required *exit* or *exit access* doorways of amusement buildings in accordance with this section and Section 1011. *Approved* directional exit markings shall also be provided. Where mirrors, mazes or other designs are utilized that disguise the path of egress travel such that they are not apparent, *approved* and *listed* low-level exit signs that comply with Section 1011.4, and directional path markings *listed* in accordance with UL 1994, shall be provided and located not more than 8 inches (203 mm) above the walking surface and on or near the path of egress travel. Such markings shall become visible in an emergency. The directional exit marking shall be activated by the automatic fire detection system and the *automatic sprinkler system* in accordance with Section 907.2.12.2.

❖ During normal operation of a special amusement building, the illumination and marking of the egress path may not be adequate to allow for prompt egress from the building. This section clearly reminds the reader that exit signs must be provided at the required exit or exit access doors, and comply with the requirements of Section 1011. As special amusement spaces quite often consist of low-level lighting and/or spaces and features that can confuse a person's orientation, exit signs are required regardless of the number of required means of egress. Where obstructions and confusion may still exist because of the nature of the facility, low-level exit signs and directional path markings (listed in accordance with UL 1994) must be provided and located not higher than 8 inches (203 mm) above the walking surface in order to direct the occupants toward the exits. In an emergency, activation of the automatic fire detection system or automatic sprinkler system must activate the directional exit markings to become visible.

411.7.1 Photo luminescent exit signs. Where photo luminescent *exit* signs are installed, activating light source and viewing distance shall be in accordance with the listing and markings of the signs.

❖ Photo luminescent exit signs operate due to exposure to specific sources of light as indicated in their listing and labeling. In some situations, not all types of photo luminescent exit signs can be used and, as such, the normal lighting levels in the area where such exit signs are to be installed must be assessed to verify compatibilty.

411.8 Interior finish. The *interior finish* shall be Class A in accordance with Section 803.1.

❖ All interior finish materials must be tested for surface-burning performance in accordance with ASTM E 84. Due to the potential for fire to spread quickly in the relatively confined spaces in these structures, only Class A materials are permitted to be used as interior finish in a special amusement building. These special amusement buildings are not permitted the one classification reduction that would normally be allowed by Table 803.9 in a sprinklered building.

SECTION 412
AIRCRAFT-RELATED OCCUPANCIES

412.1 General. Aircraft-related occupancies shall comply with Sections 412.1 through 412.7 and the *International Fire Code.*

❖ Section 412 provides specific details for the construction of the full range of aircraft-related occupancies from residential aircraft hangars to those handling large commercial aircraft, as well as helistops and heliports. This section contains code requirements for some very specialized occupancies dealing with aircraft. Aircraft pose some of the same hazards associated with motor vehicles; therefore, some of the same requirements are applicable for life and fire safety issues. The unique issues surrounding aircraft traffic control towers are addressed in Section 412.3.

Sections 412.4 and 412.5 are composed of code provisions that regulate both commercial and residential aircraft hangars. The overall building fire load for hangars is typically low because of the considerable amounts of metal in the aircraft that can absorb heat and provide limited combustibility to sustain a fire. An aircraft fire, however, can be quite severe because of the flammability of the fuel contained in its tank(s). Provisions for regulating the exterior walls, basements and floor surfaces of commercial aircraft hangars are specified along with separation and fire suppression requirements.

Section 412.6 addresses specialized hangars that are used for the painting of aircraft with flammable materials. Just as Section 416 contains limitations and requirements for the application of flammable finishes, so does Section 412.6 for aircraft paint hangars. Use and occupancy requirements commensurate with the hazards are provided, including construction, fire protection systems and necessary ventilation.

Lastly, Section 412.7 identifies those code requirements applicable to helicopter landing and service ports. Although the code relies heavily on the requirements of NFPA 418 for heliports and helistops located on the roofs of buildings and structures, there are provisions applicable for all locations. This section contains requirements for size limitations, design and means of egress.

There are additional requirements applicable to these occupancies in Chapter 11, Aviation Facilities, as well as other chapters of the IFC.

412.2 Definitions. The following words and terms shall, for the purposes of this chapter and as used elsewhere in this code, have the meanings shown herein.

❖ The definition of terms that are associated with the content of this section is contained herein. These definitions can help in the understanding and application of the code requirements. It is important to emphasize that these terms are not exclusively related to this section but are applicable everywhere the terms are used in the code. The purpose for including these definitions within this section is to provide more convenient access to them without having to refer back to Chapter 2. For convenience, these terms are also listed in Chapter 2 with a cross reference to this section. The use and application of all defined terms, including those defined herein, are set forth in Section 201.

FIXED BASE OPERATOR (FBO). A commercial business granted the right by the airport sponsor to operate on an airport and provide aeronautical services, such as fueling, hangaring, tie-down and parking, aircraft rental, aircraft maintenance and flight instruction.

❖ Fixed base operator is a term of the aviation industry used to describe a firm that is permanently based at an airport and provides a variety of aircraft services. It is used in conjunction with Section 412.4.6 in the context of determining the appropriate level of fire suppression required in various aircraft hangars at an airport (see commentary, Section 412.4.6).

HELIPORT. An area of land or water or a structural surface that is used, or intended for the use, for the landing and taking off of helicopters, and any appurtenant areas that are used, or intended for use, for heliport buildings or other heliport facilities.

❖ A heliport includes not only the immediate landing and take-off pad, but also all other adjacent service areas. The fueling, maintenance, repairs or storage of helicopters may be done within or outside of a building or structure. These outside areas or enclosed spaces are considered as part of the heliport.

HELISTOP. The same as "heliport," except that no fueling, defueling, maintenance, repairs or storage of helicopters is permitted.

❖ A helistop, by definition, is limited only to the immediate landing and take-off pad. Examples of helistops would be the pad located on top of a hospital for the unloading of emergency room patients, a pad for discharging commuters outside of an office building or the pad used to load and unload tourists at a sight-seeing attraction.

RESIDENTIAL AIRCRAFT HANGAR. An accessory building less than 2,000 square feet (186 m²) and 20 feet (6096 mm) in *building height* constructed on a one- or two-family property where aircraft are stored. Such use will be considered as a residential accessory use incidental to the *dwelling*.

❖ A residential aircraft hangar is considered an accessory or auxiliary structure to a residential house similar to any other shed or detached garage. One- or two-family dwellings are required to be constructed in accordance with the IRC (see Section 101.2). Although that code is also applicable to accessory structures associated with one- or two-family dwellings, it is clear that this section of the code was intended to control aircraft hangars since the IRC is silent on the subject.

A residential aircraft hangar is limited to 2,000 square feet (186 m²) in area and 20 feet (6096 mm) in height. The hangar must be integral to a residential home on the same property. As such, the hangar would be considered the same as the private garage used to store the home's motor vehicles.

TRANSIENT AIRCRAFT. Aircraft based at another location and at the transient location for not more than 90 days.

❖ In the place of the undefined term "private aircraft" previously used in the code, a defined term "transient aircraft" has been added in the 2009 edition. Transient aircraft are those that are merely visiting an airport as compared to those that are based at that location. The definition is used in conjunction with Section 412.4.6 to establish the level of fire suppression needed in various aircraft hangars. Fixed-base operators, especially at larger airports, will have distinct hangars that are used for repair and maintenance of aircraft. The hangars used by transient aircraft are primarily a storage place for aircraft based at another location. This better identifies the intent of this type of aircraft hangar. Most frequently, the owner who wants to develop an aircraft hangar that fits the Group II category will do no "major maintenance" and will only store airplanes in his or her hangar (see commentary, Section 412.4.6).

412.3 Airport traffic control towers.

❖ Section 412.3 addresses airport traffic control towers. Although these structures do not house aircraft, they are included in this section for consistency. These structures pose unique hazards to occupants because of their extreme height and limited routes of escape. This section contains requirements governing the permitted types of construction and necessary egress along with the needed fire protection systems.

412.3.1 General. The provisions of Sections 412.3.1 through 412.3.6 shall apply to airport traffic control towers not exceeding 1,500 square feet (140 m²) per floor occupied only for the following uses:

1. Airport traffic control cab.

2. Electrical and mechanical equipment rooms.

3. Airport terminal radar and electronics rooms.

4. Office spaces incidental to the tower operation.

5. Lounges for employees, including sanitary facilities.

❖ Airport traffic control towers are structures designed for highly specific functions. These functions include the housing of vital electronic equipment, providing an elevated structure for electronic communication, such as radar, and providing an observation area that allows an unobstructed view of the ground and airspace for air traffic controllers. This functional configuration creates special hazards, including limited means of egress, limited fire-fighting accessibility to upper floors and vulnerability to exposure fires. The code provisions are based on providing adequate fire protection and life safety to a small number of occupants within a limited area [1,500 square feet (139 m²) per floor maximum], while allowing a structure configuration that accommodates the intended function. The intent of the code is to restrict the fuel load by limiting the structure construction to mostly noncombustible materials; minimizing combustible contents and potential ignition sources by limiting the use of the structure; providing automatic fire detection systems, adequate and reliable egress and fire department access; and providing standby power in towers over 65 feet (19 812 mm) in height. These measures result in prompt evacuation and fire department accessibility early in a fire event.

This section applies only to airport traffic control towers that do not exceed 1,500 square feet (139 m²) per floor and are limited to uses exclusively related to air traffic control purposes. Rooms and spaces related to the air traffic control function, such as equipment rooms, offices, lounges and restrooms, are permitted in the tower. The section does not apply to air traffic control towers containing other uses, including assembly areas, observation decks, restaurants and terminal operations. The 1,500-square-foot (139 m²) restriction is based, in part, on a survey of existing towers that would fall within the above restrictions. Regardless of the height of control towers, they are not considered high-rise buildings, and therefore they are exempt from Section 403.

412.3.2 Type of construction. Airport traffic control towers shall be constructed to comply with the height and area limitations of Table 412.3.2.

❖ The limited area per floor, low fuel load, adequate egress provisions and low occupant load present a relatively low potential fire hazard. The provisions of Table 503 are, therefore, modified by Table 412.3.2. The allowable height and area modifications permitted by Sections 504 and 506 do not apply to Table 412.3.2.

TABLE 412.3.2
HEIGHT AND AREA LIMITATIONS FOR
AIRPORT TRAFFIC CONTROL TOWERS

TYPE OF CONSTRUCTION	HEIGHT[a] (feet)	MAXIMUM AREA (square feet)
IA	Unlimited	1,500
IB	240	1,500
IIA	100	1,500
IIB	85	1,500
IIIA	65	1,500

For SI: 1 foot = 304.8 mm, 1 square foot = 0.0929 m².
a. Height to be measured from grade plane to cab floor.

❖ Table 412.3.2 functions the same as Table 503, except that the height restriction is based on feet only and not on stories, and the actual height is measured from grade plane to the finished floor of the cab or the highest occupied level. The number of stories is not a criterion, since many of the towers do not have occupied stories for the entire height between the ground and the occupiable level at the top. If the type of construction is known, the allowable height is given by the corresponding figure in the second column. If the height is known, the permitted construction types are determined from the first column. The third column restates the area-per-floor restriction given in Section 412.3.1.

412.3.3 Egress. A minimum of one *exit stairway* shall be permitted for airport traffic control towers of any height provided that the *occupant load* per floor does not exceed 15. The *stairway* shall conform to the requirements of Section 1009. The *stairway* shall be separated from elevators by a minimum distance of one-half of the diagonal of the area served measured in a straight line. The *exit stairway* and elevator hoistway are permitted to be located in the same shaft enclosure, provided they are separated from each other by a 4-hour *fire barrier* having no openings. Such *stairway* shall be pressurized to a minimum of 0.15 inch of water column (43 Pa) and a maximum of 0.35 inch of water column (101 Pa) in the shaft relative to the building with stairway doors closed. *Stairways* need not extend to the roof as specified in Section 1009.11. The provisions of Section 403 do not apply.

Exception: Smokeproof enclosures as set forth in Section 1022.9 are not required where required *stairways* are pressurized.

❖ The benefit of a second exit is greatly reduced when the two exits would be in such close proximity. With the low fuel load, limited area per floor and limited number of occupants, only one exit is required per floor provided that the occupant load of each floor is 15 or less. The occupant load restriction of 15 persons is based on the calculated occupant load for a business use of 1,500 square feet (139 m²) (see Table 1004.1.1).

Since only one exit stairway is provided, elevators must be remotely located from the stairway a distance of one-half of the diagonal of the area served, as if they were the second required exit. If conditions are such that the stairway is not usable but the elevators are, the elevators provide an alternative escape route from the tower and fire department access to each floor level. The elevators are to be considered an egress route only if the stairway is not usable.

The code does permit the required exit stairway and the normally provided elevator to be in the same shaft enclosure; however, to maintain their independence and usability, a 4-hour separation is required between the two building features. No intercommunicating openings are permitted between these two elements to increase the likelihood that one will remain operational during a fire situation. Additionally, an exit stairway located in the same shaft enclosure as the elevator hoistway must be pressurized in accordance with Section 909.20.5.

The exit stairway for an airport traffic control tower does not need to provide roof access. Because of the limited area of these buildings, the fire department most likely will not need access to the roof for fire-fighting operations. Roof access for repair and mechanical equipment maintenance is not mandated by the code. This section of the code also notes that the requirements for high-rise buildings in regards to stairway door operations and communication are also not applicable because of the limited occupant load.

Finally, an exception is provided for smokeproof enclosures. Assuming that the exit stairway is not in the same shaft enclosure as the elevator hoistway, the stairway must still be a smokeproof enclosure in accordance with Section 1022.9. Alternatively, the required exit stairway may be pressurized in accordance with Section 909.20.5. These provisions serve to maintain the required means of egress as the building occupants travel large vertical distances to the discharge doors and the public way.

[F] 412.3.4 Automatic fire detection systems. Airport traffic control towers shall be provided with an automatic fire detection system installed in accordance with Section 907.2.

❖ To ensure early fire detection and to alert the occupants to egress the building during the incipient stage of a fire event, an automatic smoke detection system is required in accordance with Section 907.2.22. The code does not require an automatic sprinkler system because of the limited fuel load, limited occupant load and protection afforded by early detection.

[F] 412.3.5 Standby power. A standby power system that conforms to Chapter 27 shall be provided in airport traffic control towers more than 65 feet (19 812 mm) in height. Power shall be provided to the following equipment:

1. Pressurization equipment, mechanical equipment and lighting.

2. Elevator operating equipment.

3. Fire alarm and smoke detection systems.

❖ To increase the reliability of the protection afforded by equipment provided for egress and fire department accessibility, a standby power system is required in airport traffic control towers that exceed 65 feet (19 812 mm) in height. The system must support the pressurization and mechanical equipment, egress illumination, elevator operation, fire alarm system and the automatic fire detection system. Height is measured in accordance with the definition found in Section 502. If the pressurized stairway alternatives in Section 412.3.3 are utilized, the mechanical equipment required to pressurize the stairway must be connected to the standby power system to provide the same degree of reliability required for the smokeproof enclosure. The reference to Chapter 27 incorporates some of the same standby power requirements contained under the high-rise provisions based on the premise that towers over 65 feet (19 812 mm) in height represent similar hazards to those of high-rise buildings.

412.3.6 Accessibility. Airport traffic control towers need not be *accessible* as specified in the provisions of Chapter 11.

❖ Airport traffic control towers are required to have an accessible route to the cab and the floor immediately below the cab (see Section 1104.4, Exception 3). Other areas of the control tower are typically employee work stations, which must be connected by an accessible route so that such stations can be approached, entered and exited (see Sections 1103.2.1 and 1104.3.1).

412.4 Aircraft hangars. Aircraft hangars shall be in accordance with Sections 412.4.1 through 412.4.6.

❖ The requirements of Section 412.4 address commercial (or nonresidential) aircraft hangars. It should be noted that most commercial aircraft hangars will not be limited in height in terms of feet (see Section 504.1) or in area (see Section 507) based on the presence of an automatic sprinkler system. All commercial aircraft hangars, however, must be regulated in regards to exterior wall fire-resistance ratings, basement limitations, floor surface construction requirements, heating equipment separation and finishing restrictions. All of these provisions serve to abate the hazards associated with large aircraft and their integral fuel tanks to acceptable fire safety levels.

412.4.1 Exterior walls. *Exterior walls* located less than 30 feet (9144 mm) from *lot lines* or a *public way* shall have a *fire-resistance rating* not less than 2 hours.

❖ To abate the hazards of a fire condition from spreading from a commercial aircraft hangar to adjacent buildings and structures, the code requires exterior walls that are located less than 30 feet (9144 mm) from lot lines or public ways to have fire-resistance-rated construction of not less than 2 hours. Fire-resistance-rated exterior walls permit the fire department additional time and protection as it attempts to take control of a fire situation in a hangar located less than 30 feet (9144 mm) from the lot lines or public ways.

412.4.2 Basements. Where hangars have basements, floors over basements shall be of Type IA construction and shall be made tight against seepage of water, oil or vapors. There shall be no opening or communication between basements and the hangar. Access to basements shall be from outside only.

❖ As part of the hangar requirements, the use and separation of any basement levels are rigidly controlled. A fire in the main level of a hangar could pose a severe hazard not only to the occupants of basements but also to the fire department itself. The floor/ceiling assembly located between the hangar and the first basement level below must be of Type IA construction. With a floor of 2-hour fire-resistance-rated construction along with the supporting construction, a fire on the main level should be contained. Not only is the floor construction required to be rated, but it must be properly sealed or otherwise made waterproof. This requirement prevents the chances of a fire or other liquids and vapors from seeping into the floor construction and leaking into the basement level. Further, the code prohibits any openings or access between the main level of the hangar and the basement level. This would include fire-resistance-rated shafts. Any opening into the basement has to be made from the exterior of the structure via areaway construction.

412.4.3 Floor surface. Floors shall be graded and drained to prevent water or fuel from remaining on the floor. Floor drains shall discharge through an oil separator to the sewer or to an outside vented sump.

> **Exception:** Aircraft hangars with individual lease spaces not exceeding 2,000 square feet (186 m²) each in which servicing, repairing or washing is not conducted and fuel is not dispensed shall have floors that are graded toward the door, but shall not require a separator.

❖ These provisions go hand in hand with the waterproof requirements of Section 412.4.2. The floors of all hangars must be positively sloped to prevent standing liquids. This requirement is not only for personnel safety but also to minimize the effects of any spilled flammable liquids. Floor drains, if provided, must discharge their contents into an oil separator or an outside sump. This prevents the flammable liquids from entering the jurisdiction's sewer system and causing additional hazards.

An exception has been included for those commercial aircraft hangars that are divided into individual tenant or lease spaces. Hangars with tenant or lease spaces of 2,000 square feet (186 m²) or less in area do not pose the same overall hazard. The likelihood of all the aircraft being used at the same time is remote; therefore, the hazards of ponding water or spilled fuel is remote. The floor surfaces of small aircraft hangar tenant or lease spaces only need to be sloped towards the main exterior wall openings. It is imperative that no other servicing, repair work or aircraft washing is done and that no fuel dispensing can occur in these small tenant lease spaces.

412.4.4 Heating equipment. Heating equipment shall be placed in another room separated by 2-hour *fire barriers* constructed in accordance with Section 707 or *horizontal assemblies* constructed in accordance with Section 712, or both. Entrance shall be from the outside or by means of a vestibule providing a two-doorway separation.

Exceptions:

1. Unit heaters and vented infrared radiant heating equipment suspended at least 10 feet (3048 mm) above the upper surface of wings or engine enclosures of the highest aircraft that are permitted to be housed in the hangar and at least 8 feet (2438 mm) above the floor in shops, offices and other sections of the hangar communicating with storage or service areas.

2. A single interior door shall be allowed, provided the sources of ignition in the appliances are at least 18 inches (457 mm) above the floor.

❖ As part of the special use and occupancy requirements for commercial aircraft hangars, all possible ignition sources must be controlled and isolated. Specifically, all heating equipment must be located in rooms that are separated from the main areas where the aircraft are parked. This separation (both fire barrier walls and horizontal assemblies) must be 2-hour fire-resistance-rated construction. Although not explicitly stated, all openings through the rated walls must be protected. Doors connecting the heating equipment rooms and the main hangar area must be done with a vestibule or airlock arrangement such that one must pass through two doors prior to entering the other room. Again, this is done to minimize the possibility of any spilled flammable liquids and the resulting vapors from coming in contact with the ignition sources of the heating equipment.

Two exceptions to the separation requirements are provided. Unit heaters or vented infrared radiant heating equipment that are carefully located high above not only the floor surfaces but also the fuel tanks and engine compartments of the aircraft pose little risk. An allowance is also made if the heating equipment is located at least 18 inches (457 mm) above the floor of the separated room. In such a case, the vestibule/airlock arrangement with a double door system is not required and can be done with just a single door. If either exception is used, care must be taken by the building and fire officials that these special stipulations and conditions are part of the certificate of occupancy.

412.4.5 Finishing. The process of "doping," involving use of a volatile flammable solvent, or of painting, shall be carried on in a separate detached building equipped with automatic fire-extinguishing equipment in accordance with Section 903.

❖ Any application of spraying or "doping" of flammable finishes or solvent treatments to aircraft is prohibited within the hangar. These types of operations must be done in a separate detached building that is provided with an automatic fire-extinguishing system. Although not specifically stated here, the intent of the code is to treat those types of buildings as an aircraft paint hangar in accordance with Section 412.6.

Doping is a type of lacquer used to protect, waterproof and make taut cloth surfaces of airplane wings. It is used on lighter-than-air, ultra-light and some light aircraft. It is essentially painting on fabric. Doping is not used on metallic surfaces; however, the use of flammable paints is also addressed in this section. When flammable finishes are applied, the process must occur in a separate building not attached to the hangar. Because the code text refers to Section 903, the intent is for an automatic sprinkler system to be installed, unless otherwise approved.

412.4.6 Fire suppression. Aircraft hangars shall be provided with a fire suppression system designed in accordance with NFPA 409, based upon the classification for the hangar given in Table 412.4.6.

Exception: When a fixed base operator has separate repair facilities on site, Group II hangars operated by a fixed base operator used for storage of transient aircraft only shall have a fire suppression system, but the system is exempt from foam requirements.

❖ To minimize the fire hazards associated with aircraft hangars, most hangars are required to be protected with a fire suppression system. Where required, the fire suppression system must be designed and installed in accordance with NFPA 409, which requires fire suppression based on the type and construction, and the activities in a given hangar. In the standard, the suppression requirements are broken down based on three categories: Group I, Group II and Group III. Table 412.4.6 designates which group designation applies to various sizes of fire areas within a hangar and the type of construction. For example a hangar that is 28,000 square feet (2601 m²) in Type IIB construction is a Group II hangar. Group I and II hangars are required to have fire suppression as specified in NFPA 409. In general, Group III hangars are exempt from providing fire suppression unless one or more of the hazardous operations listed in Section 412.4.6.1 occur within the hanger. In these situations fire suppression based on the appropriate portion of the standard for either Group I or II is required.

The exception would not require a foam system for Group II hangers if the hangar is essentially a parking garage for transient aircraft. The exception would likely only be applicable at a larger airport facility that has multiple hangars and separate hangars for repair operations.

TABLE 412.4.6. See below.

❖ As discussed Table 412.4.6 simply determines which hangar classification to which the fire suppression must be designed in accordance with NFPA 409. This is based upon the construction type of each building and its floor area. Table 412.4.6, Note a indicates that regardless of size or construction type, any hangar with a door opening greater than 28 feet high (8534 mm) requires that hangar to have fire suppression system required for a Group I. Note c provides a Group IV designation for any hangar located in a membrane structure.

412.4.6.1 Hazardous operations. Any Group III aircraft hangar according to Table 412.4.6 that contains hazardous operations including, but not limited to, the following shall be provided with a Group I or II fire suppression system in accordance with NFPA 409 as applicable:

1. Doping.

2. Hot work including, but not limited to, welding, torch cutting and torch soldering.

3. Fuel transfer.

4. Fuel tank repair or maintenance not including defueled tanks in accordance with NFPA 409, inerted tanks or tanks that have never been fueled.

5. Spray finishing operations.

6. Total fuel capacity of all aircraft within the unsprinklered single *fire area* in excess of 1,600 gallons (6057 L).

7. Total fuel capacity of all aircraft within the maximum single *fire area* in excess of 7,500 gallons (28 390 L) for a hangar with an *automatic sprinkler system* in accordance with Section 903.3.1.1.

❖ Any of the operations listed in Section 412.6.1 which are occurring in a Group III hangar will require, under NFPA 409, some level of fire suppression. The operations on the list are straight forward. Doping is clarified in Section 412.4.5.

412.4.6.2 Separation of maximum single fire areas. Maximum single *fire areas* established in accordance with hangar classification and construction type in Table 412.4.6 shall be separated by 2-hour *fire walls* constructed in accordance with Section 706.

❖ Table 412.4.6 places a maximum size limit on hangars based on type of construction. For a hangar structure to exceed these sizes requires the construction of fire walls to establish fire areas that stay within the limits. This section is more stringent that the definition of fire areas which allows fire barriers to establish a fire area. For hangars, the fire areas must be created by exterior walls of a building or a combination of exterior and fire walls.

412.5 Residential aircraft hangars. Residential aircraft hangars as defined in Section 412.2 shall comply with Sections 412.5.1 through 412.5.2.

❖ This section of the code contains provisions that account for small, limited-size aircraft hangars that are truly accessory and auxiliary to a dwelling unit. Housing developments located along or adjacent to small-scale airports are the most obvious application of these code requirements. As part of the scoping requirement of this section, the hangar must meet all of the criteria listed in its definition in Section 412.2. A hangar that exceeds the limitations must then meet the provisions of Section 412.4. Included in the code requirements are fire separation from the adjacent residence; adequate means of egress; smoke detection and alarms; independent mechanical and plumbing systems; and height and area limitations.

412.5.1 Fire separation. A hangar shall not be attached to a *dwelling* unless separated by a *fire barrier* having a *fire-resistance rating* of not less than 1 hour. Such separation shall be continuous from the foundation to the underside of the roof and

[F] TABLE 412.4.6
HANGAR FIRE SUPPRESSION REQUIREMENTS[a, b, c]

MAXIMUM SINGLE FIRE AREA, SQ. FT.	TYPE OF CONSTRUCTION								
	IA	IB	IIA	IIB	IIIA	IIIB	IV	VA	VB
≥ 40,001	Group I	Group I	Group I	Group I	Group I	Group I	Group I	Group I	Group I
40,000	Group II	Group II	Group II	Group II	Group II	Group II	Group II	Group II	Group II
30,000	Group III	Group II	Group II	Group II	Group II	Group II	Group II	Group II	Group II
20,000	Group III	Group III	Group II	Group II	Group II	Group II	Group II	Group II	Group II
15,000	Group III	Group III	Group III	Group II	Group III	Group II	Group III	Group II	Group II
12,000	Group III	Group III	Group III	Group III	Group III	Group III	Group III	Group II	Group II
8,000	Group III	Group III	Group III	Group III	Group III	Group III	Group III	Group III	Group II
5,000	Group III	Group III	Group III	Group III	Group III	Group III	Group III	Group III	Group III

For SI: 1 foot = 304.8 mm, 1 square foot = 0.0929 m².

a. Aircraft hangars with a door height greater than 28 feet shall be provided with fire suppression for a Group I hangar regardless of maximum fire area.

b. Groups shall be as classified in accordance with NFPA 409.

c. Membrane structures complying with Section 3102 shall be classified as a Group IV hangar.

unpierced except for doors leading to the *dwelling unit*. Doors into the *dwelling unit* must be equipped with self-closing devices and conform to the requirements of Section 715 with at least a 4-inch (102 mm) noncombustible raised sill. Openings from a hanger directly into a room used for sleeping purposes shall not be permitted.

❖ A residential aircraft hangar can either be a stand-alone detached structure or attached to the residential dwelling. If the hangar is in close proximity or attached to the dwelling, it must then be separated with ˄-hour fire-resistance-rated fire barriers. This fire separation must be continuous from the floor slab up to the roof sheathing. The only openings permitted in the fire separation are normal access doors. Windows or other vent openings are prohibited as are any openings between the hangar and the bedrooms of the dwelling. The doors must have a ³/₄-hour opening protective fire protection rating in accordance with Table 715.4. A 4-inch-high (102 mm) noncombustible step is required at the door along with a self-closing device. All of these fire separation requirements serve to isolate the fire hazards associated with the hangar from the occupants of the dwelling, similar to a private garage.

412.5.2 Egress. A hangar shall provide two *means of egress*. One of the doors into the *dwelling* shall be considered as meeting only one of the two *means of egress*.

❖ The hangar must be provided with two separate and remotely located means of egress. This redundancy provides persons with an alternative means of escaping a fire in the hangar. One of the means of egress can be the access door into the adjacent dwelling unit. The other means of egress could be the aircraft entrance door subject to the requirements of Sections 1008.1.1 and 1008.1.2 regarding door size and swing.

[F] 412.5.3 Smoke alarms. Smoke alarms shall be provided within the hangar in accordance with Section 907.2.21.

❖ A smoke alarm is required in the hangar space in accordance with Section 907.2.21. Similar to the smoke alarms required in each and every bedroom, in the immediate vicinity of the bedrooms and in each and every story of the dwelling, the hangar smoke alarm must be interconnected such that one alarm will activate all other alarms. An early detection and warning alert is essential to abate the hazards of an aircraft hangar located adjacent to a residence.

412.5.4 Independent systems. Electrical, mechanical and plumbing drain, waste and vent (DWV) systems installed within the hangar shall be independent of the systems installed within the *dwelling*. Building sewer lines shall be permitted to be connected outside the structures.

Exception: Smoke detector wiring and feed for electrical subpanels in the hangar.

❖ To maintain the fire separation requirements of Section 412.5.1, the only openings permitted are normal access doors. Likewise, the mechanical and plumbing

systems for the hangar must be independent of the systems within the residential house. The plumbing drain or waste line could discharge into its own building sewer line that then connects to the house's building sewer line. This connection must be done outside of both the house and the hangar. Electrical wiring serving as the feed for the subpanels in the hangar is permitted to penetrate the fire separation along with the wiring for the smoke alarm required by Section 412.5.3 and the necessary interconnection with the other interior alarms.

412.5.5 Height and area limits. Residential aircraft hangars shall not exceed 2,000 square feet (186 m²) in area and 20 feet (6096 mm) in *building height*.

❖ As previously stated in the definition in Section 412.2, the residential aircraft hangar is limited to 2,000 square feet (186 m²) in area and 20 feet (6096 mm) in building height. These limits control the fire hazard associated with such a use. The type and number of aircraft stored in the hangar are not limited. Since the hangar is considered part of the residential use, it can be constructed of any materials that are permitted for the house, including wood-frame construction.

[F] 412.6 Aircraft paint hangars. Aircraft painting operations where flammable liquids are used in excess of the maximum allowable quantities per *control area* listed in Table 307.1(1) shall be conducted in an aircraft paint hangar that complies with the provisions of Sections 412.6.1 through 412.6.6.

❖ This section provides requirements for aircraft-related structures that exceed normal hangar storage purposes. The painting or cleaning of all aircraft with flammable liquids must be carefully controlled in an aircraft paint hangar. To determine the applicability of these requirements, the building owner must provide a complete list of all flammable liquids intended to be used in the building along with their anticipated quantities. If the amounts exceed the maximum allowable quantities per control area, then the building must be classified as an aircraft paint hangar. See Table 307.1(1) and Section 414.2 for further discussion of the maximum allowable quantities per control area of hazardous materials.

[F] 412.6.1 Occupancy group. Aircraft paint hangars shall be classified as Group H-2. Aircraft paint hangars shall comply with the applicable requirements of this code and the *International Fire Code* for such occupancy.

❖ Similar to any other building containing hazardous materials in excess of the maximum allowable quantities per control area, the building or structure must be classified as Group H. Because of the flammable liquids present, the aircraft paint hangar is classified as Group H-2. Based on the equivalent risk theory, the requirements of Section 412.6 and other applicable portions of this code and the IFC must be followed for this Group H-2 occupancy.

412.6.2 Construction. The aircraft paint hangar shall be of Type I or II construction.

❖ All aircraft paint hangars must be constructed as a Type I or II building. Special height modifications are provided in Section 504.1 for aircraft paint hangars along with special area limitations in Section 507.9. If these modifications are not applicable, then the height and area limitations of Table 503 for a Group H-2 structure must be used. The requirement for noncombustible construction limits the fuel load that is added to the occupancy contents by the structure.

[F] 412.6.3 Operations. Only those flammable liquids necessary for painting operations shall be permitted in quantities less than the maximum allowable quantities per *control area* in Table 307.1(1). Spray equipment cleaning operations shall be conducted in a liquid use, dispensing and mixing room.

❖ To lessen the likelihood of a fire in the actual painting or aircraft cleaning areas, only flammable liquids necessary for those operations are permitted. The cleaning and maintenance of spray equipment is further limited to a liquid use, dispensing and mixing room. This requirement for separation and segregation of the different operations reduces the fire risks associated with the painting and cleaning services.

[F] 412.6.4 Storage. Storage of flammable liquids shall be in a liquid storage room.

❖ The storage of all flammable liquids must be limited to a liquid storage room, which is further defined in Section 415.2. The applicable requirements of Section 415 and the IFC must be followed for liquid storage rooms to reduce the fire risks to the rest of the hangar operation.

[F] 412.6.5 Fire suppression. Aircraft paint hangars shall be provided with fire suppression as required by NFPA 409.

❖ To minimize the fire hazards associated with aircraft paint hangars, all such buildings are required to be protected with a fire suppression system. This requirement is applicable regardless of the size of the hangar in terms of height or area or the types and quantities of aircraft that are being cleaned or painted. The fire suppression system must be designed and installed in accordance with referenced standard NFPA 409. This standard contains specific requirements for the suppression systems needed to properly protect paint hangars.

412.6.6 Ventilation. Aircraft paint hangars shall be provided with ventilation as required in the *International Mechanical Code*.

❖ Integral to the requirements for aircraft paint hangars are the ventilation provisions. Hazardous exhaust systems for controlling the overspray of painting and aircraft cleaning operations are required. These systems along with the necessary ventilation of the occupiable spaces must be in accordance with the IMC.

412.7 Heliports and helistops. Heliports and helistops shall be permitted to be erected on buildings or other locations where they are constructed in accordance with Sections 412.7.1 through 412.7.4.

❖ This section contains special use and occupancy requirements for a very unique and specialized aircraft occupancy—those related to helicopters. Because of the limited space requirements necessary for the landing and taking off of helicopters, these areas are more likely to be incorporated into other buildings and structures, such as hospitals or large office buildings than facilities for fixed-wing aircraft.

Included in these requirements are provisions for minimum clearance sizes, structural use and design, means of egress and a referenced standard for rooftop locations. Heliports and helistops pose less fire hazards than the storage of these aircraft in hangars, but increased life safety risks.

Heliports and helistops are permitted to be located anywhere as long as they meet the requirements of Section 412.7. Certainly federal, state and local governments may have restrictions on the locations of heliports and helistops for general aviation purposes; however, the code only addresses those fire and life safety hazards associated with their locations to other buildings and structures and the means of egress from the same.

412.7.1 Size. The landing area for helicopters less than 3,500 pounds (1588 kg) shall be a minimum of 20 feet (6096 mm) in length and width. The landing area shall be surrounded on all sides by a clear area having a minimum average width at roof level of 15 feet (4572 mm) but with no width less than 5 feet (1524 mm).

❖ The landing pad for small helicopters [less than 3,500 pounds (1588 kg) in weight] must be a minimum 20-foot-diameter (6096 mm) circle. In addition, a concentric circle must be provided around the landing pad, which provides a clear area with an average width of 15 feet (4572 mm) but with the least dimension of not less than 5 feet (1524 mm). This additional clear space provides an increased landing area during windy conditions when pinpoint landing is not possible. Further, this clear area maintains the necessary separation between the rotating blades and all adjacent construction.

412.7.2 Design. Helicopter landing areas and the supports thereof on the roof of a building shall be noncombustible construction. Landing areas shall be designed to confine any flammable liquid spillage to the landing area itself and provisions shall be made to drain such spillage away from any *exit* or *stairway* serving the helicopter landing area or from a structure housing such *exit* or *stairway*. For structural design requirements, see Section 1605.4.

❖ Landing areas (helistops) located on the roofs of buildings must be of noncombustible materials, including the supporting construction. This requirement is necessary to provide a structurally sound support for the additional weight of the helicopter and its loads. This section refers to Section 1605.4 for further structural design requirements. Section 1605.4 requires that the building designer must account for the increased roof loads, including impact loads in the structural design.

Rooftop landing areas must be sloped or diked to prevent any spillage of flammable fuel from the helicopter from entering the building. This is most important since penthouse doors or exit stairways could allow spilled hazardous materials and vapors to enter the building, which would pose an unacceptable fire hazard.

412.7.3 Means of egress. The *means of egress* from heliports and helistops shall comply with the provisions of Chapter 10. Landing areas located on buildings or structures shall have two or more *means of egress*. For landing areas less than 60 feet (18 288 mm) in length or less than 2,000 square feet (186 m²) in area, the second *means of egress* is permitted to be a fire escape, *alternating tread device* or ladder leading to the floor below.

❖ As with all means of egress, the required egress paths from heliports and helistops must be in accordance with Chapter 10. Rooftop landing areas must be provided with at least two remotely located means of egress so that the helicopter occupants have redundant means to leave the landing area and enter the building. For very small landing areas, the code would permit one exit back into the building while the other could take the form of a fire escape or ladder to the next lower floor level. These requirements are reprinted in Section 1021.1.3 for consistency purposes (see that section for further means of egress discussion).

412.7.4 Rooftop heliports and helistops. Rooftop heliports and helistops shall comply with NFPA 418.

❖ In addition to the specific requirements of Section 412.7, rooftop heliports and helistops must comply with reference standard NFPA 418. That standard provides further life safety and fire safety requirements associated with rooftop landing areas.

SECTION 413
COMBUSTIBLE STORAGE

413.1 General. High-piled stock or rack storage in any occupancy group shall comply with the *International Fire Code*.

❖ This section alerts the code user to the specific high-piled combustible storage requirements contained in Chapter 23 of the IFC. High-piled storage presents a hazard above that of normal combustible storage. By increasing the height of the storage, the ability for a fire to grow and thrive is increased dramatically. Therefore, such storage requires special consideration with regard to arrangement and fire protection design features.

Chapter 23 of the IFC provides requirements for the high-piled storage of combustible materials regardless

of the occupancy classification. High-piled storage can occur in many different occupancies, but is most typical in Group M, S and F occupancies. High-piled storage of combustible materials includes solid-piled, palletized, shelf or rack storage where the top of storage is in excess of 12 feet (3658 mm) in height or 6 feet (1829 mm) for high-hazard commodities. Commodity classifications for all types of products, as well as fire protection requirements, are provided in Chapter 23 of the IFC and the high-piled storage provisions of NFPA 13.

413.2 Attic, under-floor and concealed spaces. *Attic*, under-floor and concealed spaces used for storage of combustible materials shall be protected on the storage side as required for 1-hour fire-resistance-rated construction. Openings shall be protected by assemblies that are self-closing and are of noncombustible construction or solid wood core not less than 1³/₄ inch (45 mm) in thickness.

Exceptions:

1. Areas protected by *approved automatic sprinkler systems*.

2. Group R-3 and U occupancies.

❖ The severity of a potential fire hazard increases when combustibles are located within concealed spaces and similar areas that provide limited access to manual fire fighting. The areas typically have low supervision and, therefore, there is increased potential for a fire to develop and spread undetected through the building. This section regulates the minimum level of separation required between storage areas and the main occupiable area in nonsprinklered buildings. Since the intent is to protect against a fire in the storage area from endangering the other occupied areas of the building, the required 1-hour fire-resistance rating need only be achieved from the storage side. While any access openings in the 1-hour fire-resistant construction need not be rated, they must be self-closing and of either noncombustible construction or a minimum 1³/₄-inch (44 mm) thickness of solid wood core.

Exception 1 exempts the storage area from being separated by 1-hour fire-resistance-rated construction, provided the area is protected by an approved automatic sprinkler system. This exception only requires the sprinkler system in the attic, under-floor or concealed space. Complete sprinkler protection throughout the building is not required in order to be in compliance with the exception.

Exception 2 clarifies that storage in residential occupancies consisting of not more than two dwelling units (Group R-3) and utility structures are exempt from the separation requirement.

SECTION 414
HAZARDOUS MATERIALS

[F] 414.1 General. The provisions of Sections 414.1 through 414.7 shall apply to buildings and structures occupied for the manufacturing, processing, dispensing, use or storage of hazardous materials.

❖ This section, along with Sections 307 (High-hazard Group H) and 415 (Groups H-1, H-2, H-3, H-4 and H-5) and the IFC, are intended to be companion provisions for the treatment of occupancies that contain hazardous materials. Any building or structure utilizing hazardous materials, regardless of quantity, is to comply with all of the applicable provisions of both this code and the IFC. This section also contains design alternatives for the use and storage of hazardous materials without classifying the building as a high-hazard Group H occupancy through the use of control areas (Section 414.2) or the mercantile display option (Section 414.2.5). While Section 414 contains general construction-related requirements for high-hazard occupancies, they are not indicative of a specific Group H occupancy classification but are dictated by hazardous material requirements in the IFC. Construction-related provisions for specific Group H occupancies are contained in Section 415.

The provisions of Section 414 apply to the use and storage of hazardous materials whether or not the building is classified as Group H. Requirements for specific materials are contained in the IFC.

[F] 414.1.1 Other provisions. Buildings and structures with an occupancy in Group H shall also comply with the applicable provisions of Section 415 and the *International Fire Code*.

❖ Section 415 is referenced for specific provisions applicable to occupancies classified as Groups H-1, H-2, H-3, H-4 and H-5. Regardless of the actual quantity of hazardous materials present, the use and storage of all such materials are required to comply with the applicable provisions of the IFC.

[F] 414.1.2 Materials. The safe design of hazardous material occupancies is material dependent. Individual material requirements are also found in Sections 307 and 415, and in the *International Mechanical Code* and the *International Fire Code*.

❖ This section emphasizes that high-hazard occupancies are different than other occupancies in that they are material dependent. This section alerts the code user to companion provisions in both the IMC and the IFC. Section 307 contains specific parameters for when a high-hazard occupancy classification is warranted. Section 415 contains specific building requirements dependent on the actual Group H occupancy classification for the building or area.

[F] 414.1.2.1 Aerosols. Level 2 and 3 aerosol products shall be stored and displayed in accordance with the *International Fire Code*. See Section 311.2 and the *International Fire Code* for occupancy group requirements.

❖ When Level 2 and 3 aerosol products are stored or displayed in accordance with Chapter 28 of the IFC, they may be classified as a Group S-1 occupancy as stated in Section 311.2. Section 307.1, Exception 11 specifically exempts aerosol storage from being classified as a Group H occupancy when in compliance

with the IFC. The protection required by the IFC is important so that the hazards created by these aerosols are addressed. The reference to the IFC will also address those locations where limited quantities of aerosol products are allowed in other occupancies.

[F] 414.1.3 Information required. A report shall be submitted to the *building official* identifying the maximum expected quantities of hazardous materials to be stored, used in a closed system and used in an *open system*, and subdivided to separately address hazardous material classification categories based on Tables 307.1(1) and 307.1(2). The methods of protection from such hazards, including but not limited to *control areas*, fire protection systems and Group H occupancies shall be indicated in the report and on the *construction documents*. The opinion and report shall be prepared by a qualified person, firm or corporation *approved* by the *building official* and provided without charge to the enforcing agency.

For buildings and structures with an occupancy in Group H, separate floor plans shall be submitted identifying the locations of anticipated contents and processes so as to reflect the nature of each occupied portion of every building and structure.

❖ A detailed plan addressing storage of hazardous materials, as well as their use in both closed and open systems must be prepared and submitted to the building official. Such plan is essential for assisting fire department and other emergency response personnel in hazardous materials situations. A report, such as a Hazardous Materials Management Plan (HMMP), as indicated in the IFC, or other approved plan, should be submitted to aid fire department personnel in the building design preplanning phase.

[F] 414.2 Control areas. *Control areas* shall comply with Sections 414.2.1 through 414.2.5 and the *International Fire Code*.

❖ As defined in Section 307.2, control areas are spaces within a building where quantities of hazardous materials not exceeding the maximum allowable quantities per control area are stored, dispensed, used or handled.

This section, in conjunction with the maximum allowable quantity tables in Section 307, utilizes a limited density concept for hazardous materials through the use of control areas. The intent of the control area concept is to provide an alternative method for the handling of hazardous materials without classifying the occupancy as Group H. In order to not be considered Group H, the amount of hazardous materials within any single control area bounded by fire barriers, horizontal assemblies, fire walls and exterior walls cannot exceed the maximum allowable quantity for a specific material listed in Table 307.1(1) or 307.1(2) (see Figure 414.2). A control area may be an entire building or a portion thereof. Note that when an entire building is the control area, the entire maximum allowable quantity of material from Table 307.1(1) or 307.1(2) located on any story is subject to the limitations of Table 414.2.2 (see IFC Interpretation Nos 51-07 and 52-07).

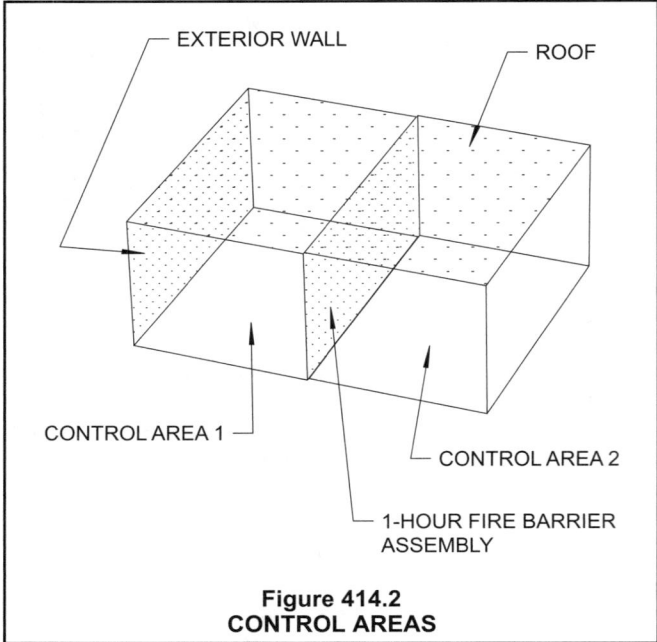

Figure 414.2
CONTROL AREAS

414.2.1 Construction requirements. *Control areas* shall be separated from each other by *fire barriers* constructed in accordance with Section 707 or *horizontal assemblies* constructed in accordance with Section 712, or both.

❖ Control areas are compartments of a building surrounded by fire barrier walls and fire-resistance-rated horizontal assemblies. If there are no fire barriers or fire-resistance-rated horizontal assemblies, the entire building is a single control area, for the purpose of applying these code provisions. Therefore, if more than the permitted maximum allowable quantities of Table 307.1(1) or 307.1(2) are anticipated in the building, additional control areas with minimum 1-hour fire barrier wall construction (2 hours where more than three stories) must be provided in order to not warrant a high-hazard occupancy classification. The provisions for required fire barriers also minimize the possibility of simultaneous involvement of multiple control areas due to a single fire condition. A fire in a single control area would involve only the amount of hazardous materials as limited by the maximum allowable quantities.

[F] 414.2.2 Percentage of maximum allowable quantities.
The percentage of maximum allowable quantities of hazardous materials per *control area* permitted at each floor level within a building shall be in accordance with Table 414.2.2.

❖ Table 414.2.2 specifies the percentage of maximum allowable quantities of hazardous materials per control area dependent on the location of a given floor level with respect to grade. The noted percentages are a percentage of the maximum allowable quantities of hazardous materials permitted per control area in accordance with Tables 307.1(1) and 307.1(2).
  For example, Table 307.1(1) would allow 240 gallons (908 L) of Class IB flammable liquid per control area in a fully sprinklered building [see Table 307.1(1),

Note d]. Table 414.2.2, in turn, would allow 75 percent of the maximum allowable quantity per control area for control areas located on the second floor level above grade. As such, 180 gallons (681 L) of Class IB flammable liquids per control area could be located on the second floor of a fully sprinklered building without classifying the building as a high-hazard occupancy.

TABLE 414.2.2. See page 4-89.

❖ The purpose of Table 414.2.2 is to establish the maximum allowable quantity of hazardous materials permitted in a building without classifying the building as a high-hazard occupancy based on the use of control areas. The number of control areas and permitted quantities of hazardous materials per control area are reduced when stored or used above the first floor level. This table also sets forth the minimum vertical fire barrier assemblies between adjacent control areas on the same floor. For floor levels above the third floor, a minimum 2-hour fire barrier is required from adjacent areas to aid fire department response due to the additional time necessary for them to access the hazardous material storage areas on upper floors. The required fire-resistance rating of the horizontal assembly above the control area is dictated by the required fire-resistance rating of the enclosure walls of the control area in order to maintain continuity and integrity. Special attention needs to be given to the rating of the floor of the control area, especially for levels below the fourth level. Section 414.2.4 requires that the floor of the control area and its supporting construction have a minimum 2-hour fire-resistance rating. In general, this will not affect floors above the third level since the continuity provisions of Section 707.5 when combined with the 2-hour separation requirement from Table 414.2.2 will ensure the floor that supports the fire barrier has an equivalent rating; however, in situations where Table 414.2.2 only requires the walls separating control areas to be of 1-hour fire-resistance-rated construction, Section 414.2.4 still would require a 2-hour fire-resistance-rated floor. This could greatly affect the design of a two-story building that was originally intended to be of Type IIB construction (see commentary, Section 414.2.4).
  The percentage of quantities of hazardous materials per control area per floor area is intended to be cumulative. A two-story building, therefore, could contain three control areas with each having 75 percent of the maximum allowable quantity on the second floor in addition to the four control areas containing 100 percent each of the maximum allowable quantity on the first floor. This condition would require that adequate fire-resistance-rated separation be provided.
  Note a clarifies that the maximum allowable quantity of hazardous materials per control area is based on Tables 307.1(1) and 307.1(2). The maximum permitted amount includes the increases allowed by either an automatic sprinkler system in accordance with NFPA 13, approved hazardous material storage cabinets or both where applicable.

Note b clarifies the fire barrier separation needed to establish the boundaries of the control area that include not only the vertical wall assemblies but also the floor/ceiling assemblies in order to be adequately separated from all adjacent interior spaces.

Example: Determine the maximum amount of Class IB flammable liquids that can be stored within a single-story, 10,000-square-foot (929 m²) nonsprinklered Group F-1 occupancy (see Figure 414.2.2) of Type IIB construction without classifying the storage area as Group H-2. Based on a maximum allowable quantity of 120 gallons (454 L) for Class IB flammable liquids from Table 307.1(1), a maximum of 120 gallons (454 L) can be stored in each of the four control areas; therefore, while the building may actually contain a total of 480 gallons (1817 L), a maximum of 120 gallons (454 L) is

permitted in each control area that is separated from all adjacent control areas by minimum 1-hour fire barriers in accordance with Section 707. The building, in this case, could still be classified as Group F-1. An automatic fire suppression system would not be required, since the 12,000-square-foot (1115 m²) threshold for suppression of Group F-1 fire areas is not exceeded (Section 903.2.3) and there are no control areas containing hazardous materials that exceed the maximum allowable quantities. Notes d and e of Table 307.1(1) would allow the base quantity of Class IB flammable liquids to be increased 100 percent in buildings protected with an automatic sprinkler system or when the material is stored in approved hazardous material storage cabinets. In this example, this would result in increasing the maximum allowable quantity of Class IB

[F] TABLE 414.2.2
DESIGN AND NUMBER OF CONTROL AREAS

FLOOR LEVEL		PERCENTAGE OF THE MAXIMUM ALLOWABLE QUANTITY PER CONTROL AREA[a]	NUMBER OF CONTROL AREAS PER FLOOR	FIRE-RESISTANCE RATING FOR FIRE BARRIERS IN HOURS[b]
Above grade plane	Higher than 9	5	1	2
	7-9	5	2	2
	6	12.5	2	2
	5	12.5	2	2
	4	12.5	2	2
	3	50	2	1
	2	75	3	1
	1	100	4	1
Below grade plane	1	75	3	1
	2	50	2	1
	Lower than 2	Not Allowed	Not Allowed	Not Allowed

a. Percentages shall be of the maximum allowable quantity per control area shown in Tables 307.1(1) and 307.1(2), with all increases allowed in the notes to those tables.

b. Fire barriers shall include walls and floors as necessary to provide separation from other portions of the building.

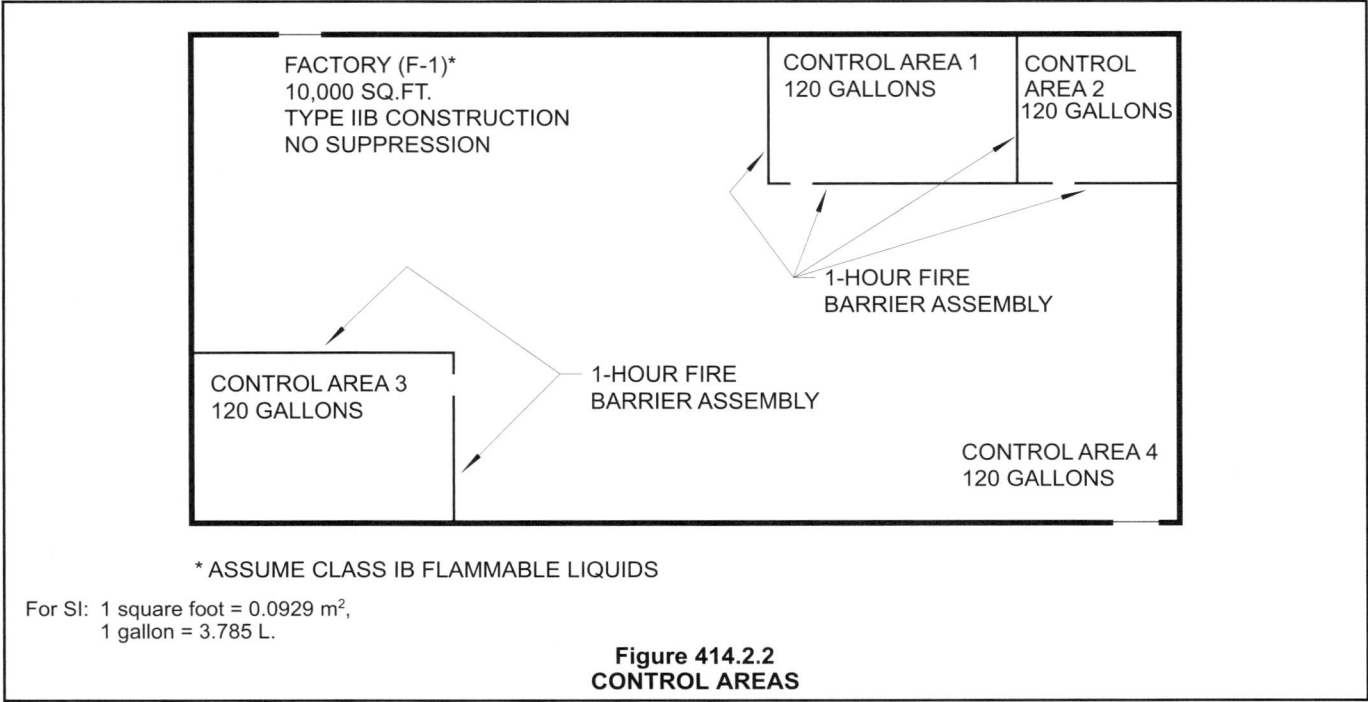

FACTORY (F-1)*
10,000 SQ.FT.
TYPE IIB CONSTRUCTION
NO SUPPRESSION

CONTROL AREA 1
120 GALLONS

CONTROL AREA 2
120 GALLONS

1-HOUR FIRE BARRIER ASSEMBLY

CONTROL AREA 3
120 GALLONS

1-HOUR FIRE BARRIER ASSEMBLY

CONTROL AREA 4
120 GALLONS

* ASSUME CLASS IB FLAMMABLE LIQUIDS

For SI: 1 square foot = 0.0929 m²,
1 gallon = 3.785 L.

Figure 414.2.2
CONTROL AREAS

flammable liquids by a factor of two; therefore, the building could now contain a total of 960 gallons (3634 L) with a maximum of 240 gallons (908 L) in each of the four control areas, separated as required by the code, and still maintain a Group F-1 classification. If both an automatic sprinkler system and hazardous material storage cabinets are used to protect Class IB flammable liquids, then the base quantity of Table 307.1(1) could be increased by a factor of four. The building in this example, therefore, with both sprinkler protection and approved cabinets, could contain a total of 1920 gallons (7267 L) with a maximum of 480 gallons (1817 L) in each of the four control areas, separated as required by the code, and still maintain a Group F-1 classification. The allowable increase in the maximum allowable quantities is offset by the additional level or levels of protection. The use of control areas provides a tradeoff based on building compartmentation. Fire protection (automatic sprinkler systems) and controlled storage through the use of approved hazardous material storage cabinets also adds a degree of protection, justifying the increased allowable quantities.

[F] 414.2.3 Number. The maximum number of *control areas* within a building shall be in accordance with Table 414.2.2.

❖ The maximum quantity of hazardous materials, therefore, which are permitted in a building without classifying it as a high-hazard occupancy, is regulated per control area and not per building area. The quantity limitation for the entire building would be established based on the number of permitted control areas on each floor of the building in accordance with Table 414.2.2. Based on the table, the first floor could contain four control areas with up to 100 percent of the maximum allowable quantity of hazardous materials per control area. For example, a single control area in a nonsprinklered building could contain up to 30 gallons (113 L) of Class 1A flammable liquids, 125 pounds (57 kg) of Class III organic peroxides, 250 pounds (113 kg) of Class 2 oxidizers and 500 gallons (1893 L) of corrosive liquids based on the maximum allowable quantities of Tables 307.1(1) and 307.1(2). Those quantities could be contained in each of four different control areas, provided that all control areas are separated from each other with minimum 1-hour fire barriers and horizontal assemblies. Please note that in order to have more control areas per floor than indicated in Table 414.2.2, a fire wall in accordance with Section 706 would be required in order to create separate additional building areas.

414.2.4 Fire-resistance-rating requirements. The required *fire-resistance rating* for *fire barriers* shall be in accordance with Table 414.2.2. The floor assembly of the *control area* and the construction supporting the floor of the *control area* shall have a minimum 2-hour *fire-resistance rating*.

Exception: The floor assembly of the *control area* and the construction supporting the floor of the *control area* are

allowed to be 1-hour fire-resistance rated in buildings of Types IIA, IIIA and VA construction, provided that both of the following conditions exist:

1. The building is equipped throughout with an *automatic sprinkler system* in accordance with Section 903.3.1.1; and

2. The building is three *stories* or less above *grade plane*.

❖ The fire separation requirements for control areas, both horizontal and vertical, is dependent on their location in a building in accordance with Table 414.2.2. The amount of hazardous materials per control area, as well as the number of control areas per floor, are reduced if stored or used above the first floor.

Where the control area is located above the first floor, the floor assembly and all supporting construction for the control area would require a minimum 2-hour fire-resistance rating. The required 2-hour fire-resistance rating of the floor construction only refers to the floor of the control area. The increased fire-resistance rating and reduced quantities are intended to aid fire department personnel. The use of control areas on upper floors provides an alternative method for multistory research and laboratory-type facilities that may need to use a limited amount of hazardous materials throughout various portions of the building. Without control areas, the maximum allowable quantity for a hazardous material would be limited to a single building area regardless of the overall size or height of the building. For example, if control areas are not utilized, a 50,000-square-foot (4645 m²) single-story building would be limited to the same quantity of hazardous materials as a two-story building with 5,000 square feet (464 m²) per floor.

Buildings of Type IIA, IIIA and VA construction are required to have floor construction with a minimum fire-resistance rating of 1-hour as indicated in Table 601. The exception recognizes the combination of a 1-hour horizontal assembly in conjunction with sprinkler protection as a reasonable alternative for the noted construction types. The three-story limitation is consistent with the fire-resistant-rating requirements for fire barriers in Table 414.2.2.

[F] 414.2.5 Hazardous material in Group M display and storage areas and in Group S storage areas. The aggregate quantity of nonflammable solid and nonflammable or noncombustible liquid hazardous materials permitted within a single *control area* of a Group M display and storage area, a Group S storage area or an outdoor *control area* is permitted to exceed the maximum allowable quantities per *control area* specified in Tables 307.1(1) and 307.1(2) without classifying the building or use as a Group H occupancy, provided that the materials are displayed and stored in accordance with the *International Fire Code* and quantities do not exceed the maximum allowable specified in Table 414.2.5(1).

In Group M occupancy wholesale and retail sales uses, indoor storage of flammable and combustible liquids shall not exceed the maximum allowable quantities per *control area* as indicated in Table 414.2.5(2), provided that the materials are displayed and stored in accordance with the *International Fire Code*.

The maximum quantity of aerosol products in Group M occupancy retail display areas, storage areas adjacent to retail display areas and retail storage areas shall be in accordance with the *International Fire Code*.

❖ This section addresses an option for control areas containing certain nonflammable or noncombustible hazardous materials that are stored in mercantile and storage occupancies, including outdoor control areas. This option would allow Group H-4 materials, which present a health hazard rather than a physical hazard, as well as limited Group H-2 and H-3 materials, such as oxidizers, to be stored in both retail display and stock areas of regulated mercantile occupancies and in storage-related occupancies in excess of the maximum allowable quantities of Tables 307.1(1) and 307.1(2) without classifying the building as Group H. Without this option, many mercantile and storage occupancies could be classified technically as Group H. The increased quantities of certain hazardous materials are based on the recognition that while there is limited risk in these occupancies, the packaging and storage arrangements can be controlled. For further information on the storage limitations required for these types of materials, see Section 2703.11.3 of the IFC.

This section, in conjunction with Table 414.2.5(1), establishes the maximum quantity of the indicated hazardous materials permitted within a single control area of a mercantile occupancy. As indicated in Table 414.2.5(1), this section only applies to certain nonflammable solids and nonflammable or noncombustible liquids. Please note that this option is not applicable to mercantile and storage occupancies containing hazardous materials other than those indicated in Table 414.2.5(1).

This section also addresses Group M occupancies utilized for the wholesale and retail sales of flammable and combustible liquids. Group M occupancies must be able to display flammable and combustible liquids for sale to the public. The maximum allowable quantities of flammable and combustible liquids per control area can exceed the limitations of Table 307.1(1), provided they are in compliance with the amounts in Table 414.2.5(2) and displayed and stored in accordance with Section 3404.3.6 of the IFC. Section 307.1, Exception 2 also addresses this design alternative to a Group H occupancy classification for flammable and combustible liquids in mercantile occupancies.

The retail sales of aerosol products requires compliance with Chapter 28 of the IFC and the applicable provisions of NFPA 30B.

TABLE 414.2.5(1). See page 4-92.

❖ Table 414.2.5(1) lists the hazardous materials eligible for the mercantile and storage occupancy option and the corresponding maximum permitted quantities per

control area depending on the extent of protection provided. The permitted quantities of each listed material are independent of each other, as well as the various classes or physical state of a specific material. For example, a given control area could contain up to the permitted maximum quantity of Class 2 solid oxidixers, Class 3 solid oxidizers and Class 2 liquid oxidizers, in addition to the permitted quantities of corrosive materials.

Notes b and c would allow the listed maximum quantity in Table 414.2.5(1) to be increased due to the use of sprinklers, approved hazardous materials storage cabinets or both. The notes are intended to be cumulative in that up to four times the listed amount may be allowed per control area, if the building is fully sprinklered and approved cabinets are utilized, without classifying the building as Group H.

Note d simply refers to Table 414.2.2 for the design and permitted number of control areas.

The 100-percent increase in maximum quantities for outdoor control areas permitted by Note f is based on the reduced exposure hazard to the building and its occupants. The increase encourages exterior storage applications without mandating sprinkler protection or approved hazardous material storage cabinets.

Notes g and h recognize that Class 2 and 3 solid oxidizers include several disinfectants that are commonly used in recreational, potable and wastewater treatment. Without these exceptions, the tabular maximum allowable quantities allowed in Group M and S occupancies would not be sufficient to sustain trade demand during times of peak usage. Because small containers of these materials have not been involved in losses, the exceptions permit additional containers of 10 pounds (5 kg) or less. Note that Section 2703.11.3.6 of the IFC limits the tabular quantities to individual containers of 100 pounds (45 kg) or less, whereas these exceptions give the retailer/wholesaler the option of increasing quantities on the shelves when the packaging sizes are reduced and limited to 10 pounds (5 kg) or less.

Note i recognizes the inherently higher level of protection and safety afforded by a sprinkler system and that, by definition, the only hazard presented by Class 1 oxidizers is that they slightly increase the burning rate of combustible materials with which they may come into contact during a fire. Materials with such properties present nowhere near the level of hazard of many ordinary commodities that might be found in a Group M or S occupancy, such as foam plastics. To put this matter into perspective, Class 1 oxidizers are materials with a degree of hazard similar to that of toilet bowl cleaner crystals. Note i also correlates with Table 307.1(1), Note f.

Note j further recognizes the lesser hazard of Class 1 oxidizers and the inherent safety of storing hazardous materials outdoors by allowing quantities to be unlimited in outdoor control areas. Note j also correlates with Table 2703.1.1(3) of the IFC.

[F] TABLE 414.2.5(1)
MAXIMUM ALLOWABLE QUANTITY PER INDOOR AND OUTDOOR CONTROL AREA IN GROUP M AND S OCCUPANCIES
NONFLAMMABLE SOLIDS AND NONFLAMMABLE AND NONCOMBUSTIBLE LIQUIDS [d, e, f]

CONDITION		MAXIMUM ALLOWABLE QUANTITY PER CONTROL AREA	
Material[a]	Class	Solids pounds	Liquids gallons
A. Health-hazard materials—nonflammable and noncombustible solids and liquids			
1. Corrosives[b, c]	Not Applicable	9,750	975
2. Highly toxics	Not Applicable	20[b, c]	2[b, c]
3. Toxics[b, c]	Not Applicable	1,000	100
B. Physical-hazard materials—nonflammable and noncombustible solids and liquids			
1. Oxidizers[b, c]	4	Not Allowed	Not Allowed
	3	1,150[g]	115
	2	2,250[h]	225
	1	18,000[i, j]	1,800[i, j]
2. Unstable (reactives)[b, c]	4	Not Allowed	Not Allowed
	3	550	55
	2	1,150	115
	1	Not Limited	Not Limited
3. Water (reactives)	3[b, c]	550	55
	2[b, c]	1,150	115
	1	Not Limited	Not Limited

For SI: 1 pound = 0.454 kg, 1 gallon = 3.785 L.

a. Hazard categories are as specified in the *International Fire Code*.

b. Maximum allowable quantities shall be increased 100 percent in buildings that are sprinklered in accordance with Section 903.3.1.1. When Note c also applies, the increase for both notes shall be applied accumulatively.

c. Maximum allowable quantities shall be increased 100 percent when stored in approved storage cabinets, in accordance with the *International Fire Code*. When Note b also applies, the increase for both notes shall be applied accumulatively.

d. See Table 414.2.2 for design and number of control areas.

e. Allowable quantities for other hazardous material categories shall be in accordance with Section 307.

f. Maximum quantities shall be increased 100 percent in outdoor control areas.

g. Maximum amounts are permitted to be increased to 2,250 pounds when individual packages are in the original sealed containers from the manufacturer or packager and do not exceed 10 pounds each.

h. Maximum amounts are permitted to be increased to 4,500 pounds when individual packages are in the original sealed containers from the manufacturer or packager and do not exceed 10 pounds each.

i. The permitted quantities shall not be limited in a building equipped throughout with an automatic sprinkler system in accordance with Section 903.3.1.1.

j. Quantities are unlimited in an outdoor control area.

TABLE 414.2.5(2). See page 4-93.

❖ Table 414.2.5(2) provides the maximum allowable quantity of flammable and combustible liquids per control area in mercantile occupancies. The limitations are based on the type of flammable and combustible liquid in the control area, the type of storage (rack, palletized, etc.) and the level of sprinkler protection.

The easier the flammable or combustible liquid is to ignite, the more restrictive is the quantity of liquid allowed. As such, the table severely limits the quantity of Class IA liquids. Class IIIB liquids, on the other hand, with a flash point above 200°F (93°C) are unlimited in a building with an automatic sprinkler system.

The automatic sprinkler system must be designed in accordance with NFPA 13. Flammable and combusti-ble liquids displayed on shelves of 6 feet (1829 mm) or less are treated as an Ordinary Hazard Group 2 occupancy in accordance with NFPA 13, which would require a minimum sprinkler density of 0.20 gallon per minute (0.72 L/min) per square foot over the most remote 1,500-square-foot (140 m²) area.

Because the flammable and combustible liquid is more exposed in individual packaging, the sprinkler system can provide better fire control. If the flammable and combustible liquids are displayed or stored in cartons, pallets or racks, the minimum sprinkler density is 0.21 gallons per minute (0.79 L/min) per square foot over the most remote 1,500-square-foot (140 m²) area. This type of display or storage is limited to a maximum height of 4 feet, 6 inches (1372 mm). This type of packaging is more difficult for the fire sprinkler system to handle so a greater density is required.

To allow a larger quantity and increased storage height of flammable and combustible liquids, the mercantile occupancy would require an enhanced sprinkler system in accordance with the applicable provisions of the IFC and NFPA 30. These provisions would require the sprinkler system to have a greater design density and operating area to adequately protect most storage conditions. For additional guidance, see Section 3404.3.6 of the IFC.

[F] 414.3 Ventilation. Rooms, areas or spaces of Group H in which explosive, corrosive, combustible, flammable or *highly toxic* dusts, mists, fumes, vapors or gases are or may be emitted due to the processing, use, handling or storage of materials shall be mechanically ventilated as required by the *International Fire Code* and the *International Mechanical Code.*

Ducts conveying explosives or flammable vapors, fumes or dusts shall extend directly to the exterior of the building without entering other spaces. Exhaust ducts shall not extend into or through ducts and plenums.

Exception: Ducts conveying vapor or fumes having flammable constituents less than 25 percent of their lower flammable limit (LFL) are permitted to pass through other spaces.

Emissions generated at workstations shall be confined to the area in which they are generated as specified in the *International Fire Code* and the *International Mechanical Code.*

The location of supply and exhaust openings shall be in accordance with the *International Mechanical Code.* Exhaust air contaminated by *highly toxic* material shall be treated in accordance with the *International Fire Code.*

A manual shutoff control for ventilation equipment required by this section shall be provided outside the room adjacent to the principal access door to the room. The switch shall be of the break-glass type and shall be labeled: VENTILATION SYSTEM EMERGENCY SHUTOFF.

❖ This section requires mechanical ventilation for occupancies utilizing hazardous materials when required by the IFC. The specific ventilation requirements are required to be in accordance with the applicable provisions in the IMC.

With regard to ducts conveying hazardous exhaust, the intent of this section is to minimize the potential for spreading hazardous exhaust to other parts of the building as a result of duct leakage or failure. In the event of a duct fire or explosion, other areas of the building could be jeopardized. Section 414.3 requires the exhaust flow to be maintained at concentrations below the lower flammability limit (LFL) of the contaminant. In all cases, ducts conveying hazardous exhaust must not extend into or through other ducts or plenum spaces, unless the flammable vapor-air mixtures are less than 25 percent of the LFL of the vapor being generated. The reference to work stations addresses spaces within a Group H-5 occupancy that utilize hazardous production materials within a fabrication area. The required control of emissions at work stations is intended to prevent exposure of the work station operator to hazardous fumes or vapors and to prevent hazardous concentrations of the materials used in the manufacturing process.

The exhaust system required by this section must be provided with an emergency manual shutoff control (kill switch) that will permit the exhaust system to be shut down without requiring personnel to enter the storage room. In the event of an emergency, it would be desirable to shut off an exhaust fan that is a source of ignition or is exhausting significant quantities of hazardous substances as a result of leaking storage containers. Under some circumstances, continued operation of an exhaust system could increase the level of hazard or cause the spread of fire.

TABLE [F] 414.2.5(2)
MAXIMUM ALLOWABLE QUANTITY OF FLAMMABLE AND COMBUSTIBLE LIQUIDS
IN WHOLESALE AND RETAIL SALES OCCUPANCIES PER CONTROL AREA[a]

TYPE OF LIQUID	MAXIMUM ALLOWABLE QUANTITY PER CONTROL AREA (gallons)		
	Sprinklered in accordance with note b densities and arrangements	Sprinklered in accordance with Tables 3404.3.6.3(4) through 3404.3.6.3(8) and Table 3404.3.7.5.1 of the *International Fire Code*	Nonsprinklered
Class IA	60	60	30
Class IB, IC, II and IIIA	7,500[c]	15,000[c]	1,600
Class IIIB	Unlimited	Unlimited	13,200

For SI: 1 foot = 304.8 mm, 1 square foot = 0.0929 m², 1 gallon = 3.785 L, 1 gallon per minute per square foot = 40.75 L/min/m².

a. Control areas shall be separated from each other by not less than a 1-hour fire barrier wall.

b. To be considered as sprinklered, a building shall be equipped throughout with an approved automatic sprinkler system with a design providing minimum densities as follows:

    1. For uncartoned commodities on shelves 6 feet or less in height where the ceiling height does not exceed 18 feet, quantities are those permitted with a minimum sprinkler design density of Ordinary Hazard Group 2.

    2. For cartoned, palletized or racked commodities where storage is 4 feet 6 inches or less in height and where the ceiling height does not exceed 18 feet, quantities are those permitted with a minimum sprinkler design density of 0.21 gallon per minute per square foot over the most remote 1,500-square-foot area.

c. Where wholesale and retail sales or storage areas exceed 50,000 square feet in area, the maximum allowable quantities are allowed to be increased by 2 percent for each 1,000 square feet of area in excess of 50,000 square feet, up to a maximum of 100 percent of the table amounts. A control area separation is not required. The cumulative amounts, including amounts attained by having an additional control area, shall not exceed 30,000 gallons.

To prevent tampering and unauthorized use of the shutoff control, such controls must be of the type that requires a seal to be broken before they can be actuated. The control must be clearly identified as to its purpose.

[F] 414.4 Hazardous material systems. Systems involving hazardous materials shall be suitable for the intended application. Controls shall be designed to prevent materials from entering or leaving process or reaction systems at other than the intended time, rate or path. Automatic controls, where provided, shall be designed to be fail safe.

❖ Process-type systems involving the use of hazardous materials generally involve many design variables. Many times, the building official may not be aware of the potentially dangerous chemical combination or use of incompatible materials that might be inherent in a given system. This section simply requires that all systems involving the use of hazardous materials should be designed to prevent an unwanted mixture of hazardous materials from occurring.

[F] 414.5 Inside storage, dispensing and use. The inside storage, dispensing and use of hazardous materials in excess of the maximum allowable quantities per *control area* of Tables 307.1(1) and 307.1(2) shall be in accordance with Sections 414.5.1 through 414.5.5 of this code and the *International Fire Code*.

❖ Sections 414.5.1 through 414.5.5 contain construction-related items for the inside storage and use of hazardous materials in excess of the maximum allowable quantities of Tables 307.1(1) and 307.1(2). The applicability of the provisions of Section 414.5 is dependent on the specific hazardous material requirements in the IFC.

[F] 414.5.1 Explosion control. Explosion control shall be provided in accordance with the *International Fire Code* as required by Table 414.5.1 where quantities of hazardous materials specified in that table exceed the maximum allowable quantities in Table 307.1(1) or where a structure, room or space is occupied for purposes involving explosion hazards as required by Section 415 or the *International Fire Code*.

❖ It is usually impractical to design a building to withstand the pressure created by an explosion; therefore, this section mandates an explosion relief system for all structures, rooms or spaces with occupancies involving explosion hazards. Explosions may result from the overpressurization of a containing structure, by physical/chemical means or by a chemical reaction. During an explosion, a sudden release of a high-pressure gas occurs and the energy is dissipated in the form of a shock wave.

All structures, rooms or spaces with occupancies involving explosion hazards must be equipped with some method of explosion control as required by Section 415 or the material-specific requirements in the IFC. Table 414.5.1 also specifies when explosion con-

trol is required based on certain materials or occupancies where the quantities of hazardous materials involved exceed the maximum allowable quantities in Table 307.1(1). Similarly, Section 911 of the IFC recognizes explosion (deflagration) venting and explosion (deflagration) prevention systems as acceptable methods of explosion control, where appropriate. The use of barricades or other explosion protective devices, such as magazines, may be permitted as the means of explosion control where indicated in the IFC as an acceptable alternative and approved by the building official.

TABLE 414.5.1. See page 4-95.

❖ This table designates when some methods of explosion control are required for specific material or special use conditions. The applicability of this table assumes the quantities of hazardous materials involved exceed the maximum allowable quantities in Table 307.1(1). Section 911 of the IFC provides design criteria for explosion (deflagration) venting. Explosion prevention (suppression) systems, where utilized, must comply with NFPA 69. Barricade construction must be designed and installed in accordance with NFPA 495. As indicated in Table 414.5.1, Note b, the IFC provides additional guidance as to the applicability and design criteria for explosion control methods.

[F] 414.5.2 Monitor control equipment. Monitor control equipment shall be provided where required by the *International Fire Code*.

❖ Monitor control equipment consists of limit controls such as liquid level controls for atmospheric tanks, temperature level controls, controls for hazardous materials required to be stored at other than ambient temperature and pressure relief devices for stationary tanks. The provisions for limit controls are located in Sections 2704.8 and 2705.1.4 of the IFC. This section provides a correlative cross reference to limit control requirements that may be dictated by a specific hazardous material in the IFC.

[F] 414.5.3 Automatic fire detection systems. Group H occupancies shall be provided with an automatic fire detection system in accordance with Section 907.2.

❖ This section requires an automatic fire (smoke) detection system for certain high-hazard occupancies. In accordance with Section 907.2.5, occupancies utilizing highly toxic gases, organic peroxides or oxidizers in excess of the maximum allowable quantities from Tables 307.1(1) and 307.1(2) would require an automatic smoke detection system installed in accordance with NFPA 72. Chapters 37, 39 and 40 of the IFC contain the specific design parameters as to when the automatic smoke detection system is required.

[F] 414.5.4 Standby or emergency power. Where mechanical ventilation, treatment systems, temperature control, alarm, detection or other electrically operated systems are required, such systems shall be provided with an emergency or standby power system in accordance with Chapter 27.

Exceptions:

1. Mechanical ventilation for storage of Class IB and Class IC flammable and combustible liquids in closed containers not exceeding 6.5 gallons (25 L) capacity.

2. Storage areas for Class 1 and 2 oxidizers.

3. Storage areas for Class II, III, IV and V organic peroxides.

4. Storage, use and handling areas for asphyxiant, irritant and radioactive gases.

5. For storage, use and handling areas for *highly toxic* or *toxic* materials, see Sections 3704.2.2.8 and 3704.3.4.2 of the *International Fire Code*.

6. Standby power for mechanical ventilation, treatment systems and temperature control systems shall not be required where an *approved* fail-safe engineered system is installed.

❖ A backup emergency power source is considered essential for required systems monitoring hazardous

[F] TABLE 414.5.1
EXPLOSION CONTROL REQUIREMENTS[a]

MATERIAL	CLASS	EXPLOSION CONTROL METHODS	
		Barricade construction	Explosion (deflagration) venting or explosion (deflagration) prevention systems[b]
HAZARD CATEGORY			
Combustible dusts[c]	—	Not Required	Required
Cryogenic flammables	—	Not Required	Required
Explosives	Division 1.1	Required	Not Required
	Division 1.2	Required	Not Required
	Division 1.3	Not Required	Required
	Division 1.4	Not Required	Required
	Division 1.5	Required	Not Required
	Division 1.6	Required	Not Required
Flammable gas	Gaseous	Not Required	Required
	Liquefied	Not Required	Required
Flammable liquid	IA[d]	Not Required	Required
	IB[e]	Not Required	Required
Organic peroxides	U	Required	Not Permitted
	I	Required	Not Permitted
Oxidizer liquids and solids	4	Required	Not Permitted
Pyrophoric gas	—	Not Required	Required
Unstable (reactive)	4	Required	Not Permitted
	3 Detonable	Required	Not Permitted
	3 Nondetonable	Not Required	Required
Water-reactive liquids and solids	3	Not Required	Required
	2[g]	Not Required	Required
SPECIAL USES			
Acetylene generator rooms	—	Not Required	Required
Grain processing	—	Not Required	Required
Liquefied petroleum gas-distribution facilities	—	Not Required	Required
Where explosion hazards exist[f]	Detonation	Required	Not Permitted
	Deflagration	Not Required	Required

a. See Section 414.1.3.
b. See the *International Fire Code*.
c. As generated during manufacturing or processing. See definition of "Combustible dust" in Chapter 3.
d. Storage or use.
e. In open use or dispensing.
f. Rooms containing dispensing and use of hazardous materials when an explosive environment can occur because of the characteristics or nature of the hazardous materials or as a result of the dispensing or use process.
g. A method of explosion control shall be provided when Class 2 water-reactive materials can form potentially explosive mixtures.

materials; therefore, when limit controls, detection systems or mechanical ventilation is required for a specific hazardous material, an emergency electrical system or standby power system is required.

Exception 1 correlates with industry treatment of portable container storage. Notably, storage of small, closed containers does not pose a risk that warrants ventilation for these materials. FM Data Sheet 7-29, *Flammable and Combustible Liquid Storage in Portable Containers*, does not require mechanical ventilation for flammable liquids in closed containers of not greater than 6.5 gallons (2290 L) individual capacity, with a flash point of not greater than 100°F (38°C) and a boiling point equal to or greater than 100°F (38°C). NFPA 30, *Flammable and Combustible Liquids Code*, also recognizes that closed container storage does not pose a risk that warrants ventilation (ventilation is required if there is open dispensing). These materials are in sealed containers in storage. Any loss of power would require an immediate cessation of operations, which would eliminate spill risk. By limiting the container size, the potential for accidental spills is significantly reduced.

Exceptions 2 and 3 address low-hazard oxidizers and organic peroxides that do not present a severe fire or reactivity hazard. Highly toxic and toxic materials (see Exception 5) must conform to applicable requirements of Chapter 37 of the IFC. For example, emergency power may be required for treatment systems utilized to process the accidental release of highly toxic or toxic compressed gases caused by a leak or rupture in storage cylinders or tanks. Without emergency power, all required monitoring systems, including the treatment system for neutralizing potential leaking gas, would be rendered inoperative if a power failure or other electrical system failure occurred.

Exception 4 exempts storage areas for asphyxiant, irritant or radioactive gases because, unlike the requirements for other hazard categories which use the Maximum Allowable Quantity (MAQ) per control area as a trigger threshold, the requirement for ventilation in storage areas containing asphyxiant, irritant and radioactive gases is not quantity based. The construction of compressed gas containers is robust compared to the containers used for other materials that may be of glass, plastic or paper. The integrity of the containers alone represents a major safeguard against likely failure. While leakage from containers is a consideration, the need for the reestablishment of power to the ventilation system within a 60-second period is not warranted given the fact that the requirement could be imposed for insignificant quantities of the gas, and given the fact that occupancy of a storage area during power outage is not the norm.

Exception 6 recognizes the use of an engineered system that is designed to always fail in the appropriate design mode without human intervention in lieu of the emergency power system. The intent of the exception is to permit alternative systems that are not subject to power interruptions. The exception, as noted, does not apply to detection and alarm systems, but addresses those systems essential to the removal of hazardous fumes and vapors from potentially occupied areas.

[F] 414.5.5 Spill control, drainage and containment. Rooms, buildings or areas occupied for the storage of solid and liquid hazardous materials shall be provided with a means to control spillage and to contain or drain off spillage and fire protection water discharged in the storage area where required in the *International Fire Code*. The methods of spill control shall be in accordance with the *International Fire Code*.

❖ This section references the IFC for material-specific occupancies containing materials in excess of the maximum allowable quantities, which would require some method of spill control, drainage and containment. The specific provisions for providing adequate spill control, drainage and containment, when required, are located in Section 2704.2 of the IFC.

[F] 414.6 Outdoor storage, dispensing and use. The outdoor storage, dispensing and use of hazardous materials shall be in accordance with the *International Fire Code*.

❖ This section requires the outdoor storage, dispensing and use of hazardous materials to be in accordance with the provisions of the IFC, regardless of the quantity. Certain provisions in the IFC, however, are only applicable when the maximum allowable quantities are exceeded. In general, the permitted quantity per outdoor control area exceeds that permitted for the inside storage or use of the same material. The outdoor storage or use of hazardous materials in excess of the maximum allowable quantities does not result in a high-hazard occupancy classification but simply dictates the need for additional requirements.

[F] 414.6.1 Weather protection. Where weather protection is provided for sheltering outdoor hazardous material storage or use areas, such areas shall be considered outdoor storage or use when the weather protection structure complies with Sections 414.6.1.1 through 414.6.1.3.

❖ This section provides the minimum construction requirements for outdoor storage areas of hazardous materials that require protection from the elements. The need for weather protection is dependent on the specific material requirements in the IFC. For ready access to materials, many such structures are commonly constructed as an attached canopy. Depending on the hazardous material involved, additional protection, such as fire suppression of the outside storage area, fire-resistance-rated exterior wall construction or a limitation on exterior wall openings, may be necessary.

Structures not in compliance with the provisions of this section would be regulated as inside storage, and as such, would be considered a high-hazard occupancy if the maximum allowable quantities of a specific hazardous material were exceeded.

[F] 414.6.1.1 Walls. Walls shall not obstruct more than one side of the structure.

Exception: Walls shall be permitted to obstruct portions of multiple sides of the structure, provided that the obstructed area does not exceed 25 percent of the structure's perimeter.

❖ The structure should be sufficiently open to allow for adequate cross ventilation. The intent of this section is to allow either one full side of a weather protection structure to be obstructed or, through the exception, to permit portions of multiple sides to be obstructed, provided the obstructed perimeter does not exceed 25 percent of the total perimeter.

[F] 414.6.1.2 Separation distance. The distance from the structure to buildings, *lot lines*, *public ways* or *means of egress* to a *public way* shall not be less than the distance required for an outside hazardous material storage or use area without weather protection.

❖ The structure is required to be located with respect to lot lines, public ways or the means of egress to the public way as required for the specific hazardous material provisions in the IFC.

[F] 414.6.1.3 Noncombustible construction. The overhead structure shall be of *approved* noncombustible construction with a maximum area of 1,500 square feet (140 m²).

Exception: The increases permitted by Section 506 apply.

❖ The overhead structure is required to be of noncombustible construction to eliminate the possibility of adding to the fuel load in a fire condition. The exception permits the area of the overhead structure to exceed the 1,500-square-foot (139 m²) area limitation if either excess frontage (see Section 506.2) or an automatic sprinkler system (see Section 506.3) is provided.

[F] 414.7 Emergency alarms. Emergency alarms for the detection and notification of an emergency condition in Group H occupancies shall be provided as set forth herein.

❖ An emergency alarm is required in all areas utilized for the storage, dispensing, use and handling of hazardous materials in accordance with Sections 414.7.1 through 414.7.3. This section assumes the area in question utilizes hazardous materials in excess of the maximum allowable quantities indicated in Tables 307.1(1) and 307.1(2).

[F] 414.7.1 Storage. An *approved* manual emergency alarm system shall be provided in buildings, rooms or areas used for storage of hazardous materials. Emergency alarm-initiating devices shall be installed outside of each interior *exit* or *exit access* door of storage buildings, rooms or areas. Activation of an emergency alarm-initiating device shall sound a local alarm to alert occupants of an emergency situation involving hazardous materials.

❖ A manual pull station or other emergency signal device approved by the building official must be provided outside of each egress door to hazardous material storage areas. Activation of the device is intended to warn the building occupants of a potential dangerous condi-

tion within the hazardous material storage area. The alarm signal required by this section is intended only to be a local alarm. A complete evacuation alarm system, excluding a local trouble alarm located within the immediate high-hazard area, is not required.

[F] 414.7.2 Dispensing, use and handling. Where hazardous materials having a hazard ranking of 3 or 4 in accordance with NFPA 704 are transported through *corridors* or *exit enclosures*, there shall be an emergency telephone system, a local manual alarm station or an *approved* alarm-initiating device at not more than 150-foot (45 720 mm) intervals and at each *exit* and *exit access* doorway throughout the transport route. The signal shall be relayed to an *approved* central, proprietary or remote station service or constantly attended on-site location and shall also initiate a local audible alarm.

❖ This section requires access to an approved supervised signaling device along the transport route when hazardous materials must be transported through corridors or exit enclosures. A spill or other incident involving hazardous materials within a corridor or exit may render it unusable for egress. A loud audible alarm, which relays a signal to a remote station, allows a quick response by emergency responders.

[F] 414.7.3 Supervision. Emergency alarm systems shall be supervised by an *approved* central, proprietary or remote station service or shall initiate an audible and visual signal at a constantly attended on-site location.

❖ This section requires an approved method of electrical supervision for the emergency alarm systems for the hazardous material storage and use areas required by Sections 414.7.1 and 414.7.2. The method of supervision should also be in compliance with the applicable provisions of NFPA 72.

SECTION 415
GROUPS H-1, H-2, H-3, H-4 AND H-5

[F] 415.1 Scope. The provisions of Sections 415.1 through 415.8 shall apply to the storage and use of hazardous materials in excess of the maximum allowable quantities per *control area* listed in Section 307.1. Buildings and structures with an occupancy in Group H shall also comply with the applicable provisions of Section 414 and the *International Fire Code*.

❖ This section establishes the application of Section 415 and references the IFC for additional specific hazardous material requirements. Section 415 is only applicable when the maximum allowable quantity of a hazardous material listed in either Table 307.1(1) or 307.1(2) is exceeded. The provisions of Section 414, however, are applicable wherever hazardous materials are stored or used, regardless of quantity.

[F] 415.2 Definitions. The following words and terms shall, for the purposes of this chapter and as used elsewhere in the code, have the meanings shown herein.

❖ This section contains definitions of terms that are associated with the subject matter of this chapter. It is important to emphasize that these terms are not exclusively related to this chapter but are applicable everywhere the term is used in the code. Definitions of terms can help in the understanding and application of the code requirements. The purpose for including these definitions within this chapter is to provide more convenient access to them without having to refer back to Chapter 2. For convenience, these terms are also listed in Chapter 2 with a cross reference to this section. The use and application of all defined terms, including those defined herein, are set forth in Section 202.

[F] CONTINUOUS GAS DETECTION SYSTEM. A gas detection system where the analytical instrument is maintained in continuous operation and sampling is performed without interruption. Analysis is allowed to be performed on a cyclical basis at intervals not to exceed 30 minutes.

❖ This term refers to a system that is capable of constantly monitoring the presence of highly toxic or toxic compressed gases at or below the permissible exposure limit (PEL) for the gas. A continuous gas detection system will provide notification of a leak or rupture in a compressed gas cylinder or tank in a storage or use condition.

[F] DETACHED BUILDING. A separate single-story building, without a basement or crawl space, used for the storage or use of hazardous materials and located an *approved* distance from all structures.

❖ The term is used to define the type of structure the code recognizes for the use and storage of hazardous materials in excess of the maximum allowed quantity per control area. While the definition addresses all hazardous materials, a detached storage building is only required for Group H-1, H-2 and H-3 structures as indicated in Sections 415.3.2 and 415.4, and Table 415.3.2. The location of the structure may be regulated by Section 415.3.1 and Table 415.3.1 based on the characteristics of the materials contained in the building.

[F] EMERGENCY CONTROL STATION. An *approved* location on the premises where signals from emergency equipment are received and which is staffed by trained personnel.

❖ This definition identifies the room or area located in the hazardous production materials (HPM) facility that is utilized for the purpose of receiving various alarms and signals. The smoke detectors located in the building's recirculation ventilation ducts, the gas-monitoring/detection system and the telephone/fire protective signaling systems located outside of HPM storage rooms are all required to be connected to the emergency control station. The location of the emergency control station must be approved by the building official. An approved location should be based on personnel being able to adequately monitor the necessary alarms and signals and on the fire department being able to gain access quickly when responding to emergency situations. Additionally, the room must be occu-

pied by persons who are trained to respond to the various alarms and signals in the appropriate fashion.

[F] EXHAUSTED ENCLOSURE. An appliance or piece of equipment that consists of a top, a back and two sides providing a means of local exhaust for capturing gases, fumes, vapors and mists. Such enclosures include laboratory hoods, exhaust fume hoods and similar appliances and equipment used to locally retain and exhaust the gases, fumes, vapors and mists that could be released. Rooms or areas provided with general ventilation, in themselves, are not exhausted enclosures.

❖ Exhausted enclosures, such as laboratory hoods or exhaust fume hoods, are utilized to contain hazardous fumes and vapors. The use of an approved exhausted enclosure may allow an increase in the maximum allowable quantity per control area of a given hazardous material. Exhausted enclosures are typically utilized when highly toxic or toxic compressed gases are involved. The exhausted enclosures are required to be of noncombustible construction, have specific ventilation criteria and be protected by an approved fire-extinguishing system.

[F] FABRICATION AREA. An area within a semiconductor fabrication facility and related research and development areas in which there are processes using hazardous production materials. Such areas are allowed to include ancillary rooms or areas such as dressing rooms and offices that are directly related to the fabrication area processes.

❖ This definition describes the basic component of an HPM facility. The code uses this definition to provide certain material limitations on both a quantity and density basis, and to require enclosure of the fabrication areas with fire barrier assemblies. The fabrication area of an HPM facility is the area where the hazardous materials are actively handled and processed. The fabrication area includes accessory rooms and spaces, such as work stations and employee dressing rooms.

[F] FLAMMABLE VAPORS OR FUMES. The concentration of flammable constituents in air that exceed 25 percent of their lower flammable limit (LFL).

❖ Vapors or fumes are considered to be flammable when they exceed 25 percent of their LFL. The LFL of a given vapor in air is the concentration at which flame propagation could occur in the presence of an ignition source (see the definition of "Lower flammable limit").

[F] GAS CABINET. A fully enclosed, noncombustible enclosure used to provide an isolated environment for compressed gas cylinders in storage or use. Doors and access ports for exchanging cylinders and accessing pressure-regulating controls are allowed to be included.

❖ Gas cabinets are used to provide adequate control for escaping gas in the event of a leaking cylinder of compressed gases. Gas cabinets are commonly used when dealing with highly toxic and toxic compressed gases. Sections 2703.8.6 and 3704.1.2 of the IFC provide additional construction and ventilation requirements for gas cabinets.

[F] GAS ROOM. A separately ventilated, fully enclosed room in which only compressed gases and associated equipment and supplies are stored or used.

❖ Gas rooms are used exclusively for the storage or use of hazardous gases in excess of the maximum allowable quantities per control area permitted by Tables 307.1(1) and 307.1(2). Gas rooms are commonly used as an alternative storage area for HPM gases in a Group H-5 facility.

[F] HAZARDOUS PRODUCTION MATERIAL (HPM). A solid, liquid or gas associated with semiconductor manufacturing that has a degree-of-hazard rating in health, flammability or instability of Class 3 or 4 as ranked by NFPA 704 and which is used directly in research, laboratory or production processes which have as their end product materials that are not hazardous.

❖ This definition identifies those specific materials that can be contained within an HPM facility. The restriction in the definition for only hazardous materials with a Class 3 or 4 rating is not intended to exclude materials that are less hazardous, but to clarify that materials of the indicated higher ranking are still permitted in an HPM facility without classifying the building as Group H. NFPA 704 is referenced in order to establish the degree of hazard ratings for all materials as related to health, flammability and instability risks.

[F] HPM FLAMMABLE LIQUID. An HPM liquid that is defined as either a Class I flammable liquid or a Class II or Class IIIA combustible liquid.

❖ This definition clarifies that an HPM liquid is essentially all classes of flammable or combustible liquids except Class IIIB combustible liquids, which have a flash point at or above 200°F (93°C). Class IIIB liquids, therefore, are not considered a hazardous production material.

[F] HPM ROOM. A room used in conjunction with or serving a Group H-5 occupancy, where HPM is stored or used and which is classified as a Group H-2, H-3 or H-4 occupancy.

❖ An HPM room in a Group H-5 facility is utilized for the storage and use of hazardous production materials in excess of the maximum allowable quantities permitted in Table 307.1(1) or 307.1(2). The rooms are, therefore, considered a Group H-2, H-3 or H-4 occupancy depending on the type of hazardous material.

[F] IMMEDIATELY DANGEROUS TO LIFE AND HEALTH (IDLH). The concentration of air-borne contaminants which poses a threat of death, immediate or delayed permanent adverse health effects, or effects that could prevent escape from such an environment. This contaminant concentration level is established by the National Institute of Occupational Safety and Health (NIOSH) based on both toxicity and flammability. It generally is expressed in parts per million by volume (ppm v/v) or milligrams per cubic meter (mg/m³). If adequate data do not exist for precise establishment of IDLH concentrations, an independent certified industrial hygienist, industrial toxicologist, appropriate regulatory agency or other

source *approved* by the *building official* shall make such determination.

❖ The definition of "Immediately dangerous to life and health (IDLH)" is the minimum concentration of an air-borne contaminant, such as a highly toxic compressed gas, that a person could be exposed to before the risk of permanent adverse side effects. The IDLH is important in determining the design of treatment systems for highly toxic or toxic compressed gases.

[F] LIQUID. A material that has a melting point that is equal to or less than 68°F (20°C) and a boiling point that is greater than 68°F (20°C) at 14.7 pounds per square inch absolute (psia) (101 kPa). When not otherwise identified, the term "liquid" includes both flammable and combustible liquids.

❖ This definition specifies the criteria to establish when material is considered a liquid based on its melting and boiling points. When the term "liquid" is referred to, it is intended to include both flammable and combustible liquids.

[F] LIQUID STORAGE ROOM. A room classified as a Group H-3 occupancy used for the storage of flammable or combustible liquids in a closed condition.

❖ Liquid storage rooms are utilized in Group H-5 facilities. These rooms are utilized exclusively for the storage of flammable and combustible liquids in closed containers in excess of the maximum allowable quantities per control area permitted by Tables 307.1(1) and 307.1(2). The storage room itself is considered a Group H-3 occupancy in accordance with Section 307.5.

[F] LIQUID USE, DISPENSING AND MIXING ROOM. A room in which Class I, II and IIIA flammable or combustible liquids are used, dispensed or mixed in open containers.

❖ This term refers to all nonstorage rooms utilized exclusively for flammable and combustible liquids other than Class IIIB liquids. Class IIIB liquids have a flash point in excess of 200°F (93°C) and are not considered hazardous.

[F] LOWER FLAMMABLE LIMIT (LFL). The minimum concentration of vapor in air at which propagation of flame will occur in the presence of an ignition source. The LFL is sometimes referred to as "LEL" or "lower explosive limit."

❖ When the vapor-to-air ratio is somewhere between the LFL and the upper flammable limit (UFL), fires and explosions can occur upon introduction of an ignition source. The UFL is the maximum vapor-to-air concentration above which propagation of flame will not occur. If a vapor-to-air mixture is below the LFL, it is described as being "too lean" to burn, and if it is above the UFL, it is "too rich" to burn.

[F] NORMAL TEMPERATURE AND PRESSURE (NTP). A temperature of 70°F (21°C) and a pressure of 1 atmosphere [14.7 psia (101 kPa)].

❖ This term refers to the standard room temperature at atmospheric pressure. Atmospheric pressure results from the weight of air elevated above the earth's surface. At sea level, the atmosphere exerts a pressure of 14.7 pounds per square inch (psi) (101 kPa). The properties of commercially available compressed gases are indicated at their normal temperature and pressure (NTP).

[F] PHYSIOLOGICAL WARNING THRESHOLD LEVEL. A concentration of air-borne contaminants, normally expressed in parts per million (ppm) or milligrams per cubic meter (mg/m³), that represents the concentration at which persons can sense the presence of the contaminant due to odor, irritation or other quick-acting physiological response. When used in conjunction with the permissible exposure limit (PEL) the physiological warning threshold levels are those consistent with the classification system used to establish the PEL. See the definition of "Permissible exposure limit (PEL)" in the *International Fire Code*.

❖ The term "physiological warning properties" is not defined. From a practical standpoint, the physiological warning properties are represented by a concentration of a contaminant that allows the average individual to sense its presence by a body warning signal including, but not limited to, odor, irritating effects such as stinging sensations, coughing, scratchy feeling in the throat, running of the eyes or nose and similar signals.

There may be a wide variability reported for some of the more common threshold levels including that of olfactory perception. Variations that may be encountered are due to a number of factors including the methods used in their determination, the population exposed and others. The requirements for gas detection established in the code are tied to the PEL, and there are several methods for determining the PEL inherent in the definition of that term. Including this definition intends to link the determination of the physiological warning threshold level to the data used to determine the PEL.

For example, the PEL as established by 29 CFR Part 1910.1000 is primarily based on data developed by the American Conference of Governmental Industrial Hygienists (ACGIH) called threshold limit values (TLVs) as referenced in the definition of PEL found in Section 3702.1 of the IFC. To substantiate the TLVs (PELs), the ACGIH publishes the *Documentation of the Threshold Limit Values* (TLVs®) and *Biological Exposure Indices* (BEIs®) where the user is provided with data used in their establishment. The significant commercially available toxic and highly toxic gases with published TLVs® are listed by ACGIH, and perception thresholds are provided.

These warning properties are considered, as evidenced by the documentation when the TLV and hence the PEL is established. It is appropriate that the data used in the base documents be used as the basis for determining the threshold level when such data is available. The use of data from other sources may be used in the absence of data within the system used for the establishment of the PEL, but where such data has been considered in determining the PEL such data should take precedent.

By providing a definition for physiological warning threshold level and guidance as to how it is to be applied, the code user is given guidance that carries out the intent of the provisions for gas detection that have been established in the code. See the commentary to Section 3704.2.2.10 of the IFC for further discussion of gas detection.

[F] SERVICE CORRIDOR. A fully enclosed passage used for transporting HPM and purposes other than required *means of egress*.

❖ Though HPM facility occupants may be exposed to limited HPM quantities during the course of their employment, their means of egress are protected from the HPM hazards by confining the HPM being transferred to its own passageway. A service corridor is only required when the HPM must be carried from a storage room or external area to a fabrication area through a passageway.

[F] SOLID. A material that has a melting point, decomposes or sublimes at a temperature greater than 68°F (20°C).

❖ The temperature at which a solid melts is the melting point. A material will begin to melt as heat is added to it and, thus, eventually change to a liquid state.

[F] STORAGE, HAZARDOUS MATERIALS.

1. The keeping, retention or leaving of hazardous materials in closed containers, tanks, cylinders or similar vessels, or

2. Vessels supplying operations through closed connections to the vessel.

❖ This term refers to all hazardous materials that are essentially being stored in a static condition. The material is considered in use once it is placed into action by either handling or transport or in a closed or open system.

[F] USE (MATERIAL). Placing a material into action, including solids, liquids and gases.

❖ This definition describes the active utilization mode of a material as opposed to the inactive nature of storage or the limited movement or transport involved in handling. This mode tends to be more hazardous in that the material is exposed to human or mechanical contact rather than being confined to a closed container, thus increasing the potential for spills or other releases of liquids, vapors, gases or solids.

[F] WORKSTATION. A defined space or an independent principal piece of equipment using HPM within a fabrication area where a specific function, laboratory procedure or research activity occurs. *Approved* or *listed* hazardous materials storage cabinets, flammable liquid storage cabinets or gas cabinets serving a workstation are included as part of the workstation. A workstation is allowed to contain ventilation equipment, fire protection devices, detection devices, electrical devices and other processing and scientific equipment.

❖ Workstations further subdivide a fabrication area and provide relatively self-contained, specialized areas where HPM processes are conducted. Workstation controls limit the quantity of materials and impose limitations on the design of these processes to include, but not be limited to, protection by local exhaust; sprinklers; automatic and emergency shutoffs; construction materials and HPM compatibility. Excess materials are prohibited and must be contained in storage rooms designed to accommodate such hazards.

[F] 415.3 Fire separation distance. Group H occupancies shall be located on property in accordance with the other provisions of this chapter. In Groups H-2 and H-3, not less than 25 percent of the perimeter wall of the occupancy shall be an *exterior wall*.

Exceptions:

1. Liquid use, dispensing and mixing rooms having a floor area of not more than 500 square feet (46.5 m²) need not be located on the outer perimeter of the building where they are in accordance with the *International Fire Code* and NFPA 30.

2. Liquid storage rooms having a floor area of not more than 1,000 square feet (93 m²) need not be located on the outer perimeter where they are in accordance with the *International Fire Code* and NFPA 30.

3. Spray paint booths that comply with the *International Fire Code* need not be located on the outer perimeter.

❖ This section specifies the location of Group H storage areas within a building. In order to provide adequate access for fire-fighting operations and venting of the products of combustion, Group H-2 and H-3 storage areas within a building must be located along an exterior wall.

Exception 1 recognizes the use of inside storage rooms that are utilized for operations involving flammable and combustible liquids. These types of rooms, which have no perimeter access, are permitted by the IFC and NFPA 30. The size of the room as well as the quantity of liquids, however, is restricted due to the lack of perimeter access and ventilation options.

Exception 2 is similar to Exception 1 except that a larger room area is permitted since the flammable and combustible liquids are in a static storage condition.

Spray paint booths are typically power-ventilated structures within a building that have their own enclosure separate from the exterior wall construction. The location and fire separation requirements for the spray paint booth must be in accordance with Section 416 and NFPA 33.

[F] 415.3.1 Group H occupancy minimum fire separation distance. Regardless of any other provisions, buildings containing Group H occupancies shall be set back to the minimum *fire separation distance* as set forth in Items 1 through 4 below. Distances shall be measured from the walls enclosing the occupancy to *lot lines*, including those on a *public way*. Distances to assumed *lot lines* established for the purpose of determining *exterior wall* and opening protection are not to be used to establish the minimum *fire separation distance* for buildings on sites where explosives are manufactured or used when separation is

provided in accordance with the quantity distance tables specified for explosive materials in the *International Fire Code*.

1. Group H-1. Not less than 75 feet (22 860 mm) and not less than required by the *International Fire Code*.

   Exceptions:

   1. Fireworks manufacturing buildings separated in accordance with NFPA 1124.

   2. Buildings containing the following materials when separated in accordance with Table 415.3.1:

      2.1. Organic peroxides, unclassified detonable.

      2.2. Unstable reactive materials, Class 4.

      2.3. Unstable reactive materials, Class 3 detonable.

      2.4. Detonable pyrophoric materials.

2. Group H-2. Not less than 30 feet (9144 mm) where the area of the occupancy exceeds 1,000 square feet (93 m²) and it is not required to be located in a detached building.

3. Groups H-2 and H-3. Not less than 50 feet (15 240 mm) where a detached building is required (see Table 415.3.2).

4. Groups H-2 and H-3. Occupancies containing materials with explosive characteristics shall be separated as required by the *International Fire Code*. Where separations are not specified, the distances required shall not be less than the distances required by Table 415.3.1.

❖ Due to the potentially volatile nature of hazardous materials, specific setback requirements are necessary for Group H occupancies. These provisions take precedence over provisions in the code that may specify a minimum fire separation distance based on building construction type and exposure (see Table 602). The listed conditions are dependent on the type of materials that are indicative of the specified Group H occupancies, the size of the hazardous material storage area and whether a detached building is required by Table 415.3.2. Buildings with explosive materials must comply with Table 415.3.1.

Exception 1 in Item 1 recognizes that fireworks manufacturing buildings present unique hazards due to the potential volume (net weight) of fireworks in any single building. NFPA 1124 specifies the minimum separation distances between all process buildings, public highways and other inhabited buildings.

Exception 2 addresses the fact that these specific materials have a different explosive hazard and, therefore, permits distances to be established using Table 415.3.1 and the modification that is found in Note b of the table.

When dealing with explosives, it is important to note that the base paragraph makes a distinction between the assumed lot lines (see Section 702) that are used for determining exterior wall and opening protection and those that are used for the separation of buildings, based on fire separation distance, where explosives are involved. Where explosives are involved, the separation distances are measured between structures and not to some imaginary property line that is assumed to be between them. It is reasonable to make this distinction since the separation distances for explosives far exceed the normal exterior wall and opening separation requirements.

TABLE 415.3.1. See page 4-103.

❖ Table 415.3.1 establishes minimum separation distances for the permanent storage of explosives from selected property classes. The intent of the table is to place explosive storage in a sufficiently remote location from occupied buildings, lot lines and other magazines in a manner to reduce exposure of such properties from damage if a detonation of a magazine occurs. The storage of explosives must also comply with the applicable provisions of Chapter 33 of the IFC and NFPA 495.

TABLE 415.3.2. See page 4-105.

❖ Table 415.3.2 establishes when hazardous materials must be stored in detached structures. The need for detached storage is a function of the type, physical state and quantity of material.

[F] 415.3.2 Detached buildings for Group H-1, H-2 or H-3 occupancy. The storage of hazardous materials in excess of those amounts listed in Table 415.3.2 shall be in accordance with the applicable provisions of Sections 415.4 and 415.5. Where a detached building is required by Table 415.3.2, there are no requirements for wall and opening protection based on *fire separation distance*.

❖ This section recognizes that detached buildings are required to be adequately separated from lot lines and other important buildings. As such, additional exposure protection for exterior walls is not needed. Table 415.3.2 requires detached buildings for some of the materials found in Groups H-2 and H-3 due to the large quantities involved. Even though these materials are less of a hazard when compared to those found in Group H-1, they do create a significant hazard in sufficiently large quantities.

[F] 415.4 Special provisions for Group H-1 occupancies. Group H-1 occupancies shall be in buildings used for no other purpose, shall not exceed one *story* in height and be without basements, crawl spaces or other under-floor spaces. Roofs shall be of lightweight construction with suitable thermal insulation to prevent sensitive material from reaching its decomposition temperature. Group H-1 occupancies containing materials that are in themselves both physical and health hazards in quantities exceeding the maximum allowable quantities per *control area* in Table 307.1.(2) shall comply with requirements for both Group H-1 and H-4 occupancies.

❖ Due to the explosion hazard potential associated with Group H-1 materials, Group H-1 occupancies are required to be in separate detached structures. The limitation of one story is based on the need to exit a building with a detonation hazard as soon as possible. Exiting from a second story or basement, even within a fire-resistance-rated stairway enclosure, may not offer sufficient protection where an explosion hazard exists. The one-story limitation is also intended to limit the volume of detonable materials that could be stored in any one structure.

Group H-1 occupancies that contain materials that present a health as well as a detonation hazard must also comply with the applicable requirements for a Group H-4 occupancy (see commentary, Section 307.8).

[F] 415.4.1 Floors in storage rooms. Floors in storage areas for organic peroxides, pyrophoric materials and unstable (reactive) materials shall be of liquid-tight, noncombustible construction.

❖ Noncombustible floors are required to prevent the structure from contributing to a fire scenario. The floors are also required to be liquid tight to prevent the spread of hazardous materials to areas outside the storage room.

[F] 415.5 Special provisions for Groups H-2 and H-3 occupancies. Groups H-2 and H-3 occupancies containing quantities of hazardous materials in excess of those set forth in Table 415.3.2 shall be in buildings used for no other purpose, shall not exceed one *story* in height and shall be without basements, crawl spaces or other under-floor spaces.

Groups H-2 and H-3 occupancies containing water-reactive materials shall be resistant to water penetration. Piping for conveying liquids shall not be over or through areas containing water reactives, unless isolated by *approved* liquid-tight construction.

Exception: Fire protection piping.

❖ This section in conjunction with Table 415.3.2 specifies when a Group H-2 or H-3 occupancy must be in a detached structure. Detached structures used for the storage of hazardous materials are not intended to be mixed-use occupancies.

The design and construction of buildings used for storing water-reactive materials must be such that water will not be permitted to come in contact with the stored materials. The building materials should be designed to resist the passage of flowing water. Piping conveying liquids, other than sprinkler system piping, is prohibited in any area containing water-reactive materials.

[F] 415.5.1 Floors in storage rooms. Floors in storage areas for organic peroxides, oxidizers, pyrophoric materials, unstable (reactive) materials and water-reactive solids and liquids shall be of liquid-tight, noncombustible construction.

❖ Noncombustible floors are required to prevent the structure from contributing to a fire scenario. The floors are also required to be liquid tight to prevent the spread of hazardous materials to areas outside the storage room.

[F] TABLE 415.3.1
MINIMUM SEPARATION DISTANCES FOR BUILDINGS CONTAINING EXPLOSIVE MATERIALS

QUANTITY OF EXPLOSIVE MATERIAL[a]		MINIMUM DISTANCE (feet)		
		Lot lines[b] and inhabited buildings[c]		
Pounds over	Pounds not over	Barricaded[d]	Unbarricaded	Separation of magazines[d, e, f]
2	5	70	140	12
5	10	90	180	16
10	20	110	220	20
20	30	125	250	22
30	40	140	280	24
40	50	150	300	28
50	75	170	340	30
75	100	190	380	32
100	125	200	400	36
125	150	215	430	38
150	200	235	470	42
200	250	255	510	46
250	300	270	540	48
300	400	295	590	54
400	500	320	640	58
500	600	340	680	62
600	700	355	710	64
700	800	375	750	66
800	900	390	780	70
900	1,000	400	800	72
1,000	1,200	425	850	78
1,200	1,400	450	900	82
1,400	1,600	470	940	86
1,600	1,800	490	980	88
1,800	2,000	505	1,010	90
2,000	2,500	545	1,090	98
2,500	3,000	580	1,160	104
3,000	4,000	635	1,270	116
4,000	5,000	685	1,370	122
5,000	6,000	730	1,460	130
6,000	7,000	770	1,540	136
7,000	8,000	800	1,600	144
8,000	9,000	835	1,670	150
9,000	10,000	865	1,730	156
10,000	12,000	875	1,750	164
12,000	14,000	885	1,770	174
14,000	16,000	900	1,800	180
16,000	18,000	940	1,880	188
18,000	20,000	975	1,950	196
20,000	25,000	1,055	2,000	210
25,000	30,000	1,130	2,000	224
30,000	35,000	1,205	2,000	238
35,000	40,000	1,275	2,000	248

(continued)

TABLE 415.3.1—continued
MINIMUM SEPARATION DISTANCES FOR BUILDINGS CONTAINING EXPLOSIVE MATERIALS

QUANTITY OF EXPLOSIVE MATERIAL[a]		MINIMUM DISTANCE (feet)		
		Lot lines[b] and inhabited buildings[c]		
Pounds over	Pounds not over	Barricaded[d]	Unbarricaded	Separation of magazines[d, e, f]
40,000	45,000	1,340	2,000	258
45,000	50,000	1,400	2,000	270
50,000	55,000	1,460	2,000	280
55,000	60,000	1,515	2,000	290
60,000	65,000	1,565	2,000	300
65,000	70,000	1,610	2,000	310
70,000	75,000	1,655	2,000	320
75,000	80,000	1,695	2,000	330
80,000	85,000	1,730	2,000	340
85,000	90,000	1,760	2,000	350
90,000	95,000	1,790	2,000	360
95,000	100,000	1,815	2,000	370
100,000	110,000	1,835	2,000	390
110,000	120,000	1,855	2,000	410
120,000	130,000	1,875	2,000	430
130,000	140,000	1,890	2,000	450
140,000	150,000	1,900	2,000	470
150,000	160,000	1,935	2,000	490
160,000	170,000	1,965	2,000	510
170,000	180,000	1,990	2,000	530
180,000	190,000	2,010	2,010	550
190,000	200,000	2,030	2,030	570
200,000	210,000	2,055	2,055	590
210,000	230,000	2,100	2,100	630
230,000	250,000	2,155	2,155	670
250,000	275,000	2,215	2,215	720
275,000	300,000	2,275	2,275	770

For SI: 1 pound = 0.454 kg, 1 foot = 304.8 mm, 1 square foot = 0.0929 m^2.

a. The number of pounds of explosives listed is the number of pounds of trinitrotoluene (TNT) or the equivalent pounds of other explosive.

b. The distance listed is the distance to lot line, including lot lines at public ways.

c. For the purpose of this table, an inhabited building is any building on the same lot that is regularly occupied by people. Where two or more buildings containing explosives or magazines are located on the same lot, each building or magazine shall comply with the minimum distances specified from inhabited buildings and, in addition, they shall be separated from each other by not less than the distance shown for "Separation of magazines," except that the quantity of explosive materials contained in detonator buildings or magazines shall govern in regard to the spacing of said detonator buildings or magazines from buildings or magazines containing other explosive materials. If any two or more buildings or magazines are separated from each other by less than the specified "Separation of Magazines" distances, then such two or more buildings or magazines, as a group, shall be considered as one building or magazine, and the total quantity of explosive materials stored in such group shall be treated as if the explosive were in a single building or magazine located on the site of any building or magazine of the group, and shall comply with the minimum distance specified from other magazines or inhabited buildings.

d. Barricades shall effectively screen the building containing explosives from other buildings, public ways or magazines. Where mounds or revetted walls of earth are used for barricades, they shall not be less than 3 feet in thickness. A straight line from the top of any side wall of the building containing explosive materials to the eave line of any other building, magazine or a point 12 feet above the centerline of a public way shall pass through the barricades.

e. Magazine is a building or structure, other than an operating building, approved for storage of explosive materials. Portable or mobile magazines not exceeding 120 square feet in area need not comply with the requirements of this code, however, all magazines shall comply with the International Fire Code.

f. The distance listed is permitted to be reduced by 50 percent where approved natural or artificial barriers are provided in accordance with the requirements in Note d.

[F] TABLE 415.3.2
DETACHED BUILDING REQUIRED

A DETACHED BUILDING IS REQUIRED WHEN THE QUANTITY OF MATERIAL EXCEEDS THAT LISTED HEREIN			
Material	Class	Solids and Liquids (tons)[a, b]	Gases (cubic feet)[a, b]
Explosives	Division 1.1	Maximum Allowable Quantity	Not Applicable
	Division 1.2	Maximum Allowable Quantity	
	Division 1.3	Maximum Allowable Quantity	
	Division 1.4	Maximum Allowable Quantity	
	Division 1.4[c]	1	
	Division 1.5	Maximum Allowable Quantity	
	Division 1.6	Maximum Allowable Quantity	
Oxidizers	Class 4	Maximum Allowable Quantity	Maximum Allowable Quantity
Unstable (reactives) detonable	Class 3 or 4	Maximum Allowable Quantity	Maximum Allowable Quantity
Oxidizer, liquids and solids	Class 3	1,200	Not Applicable
	Class 2	2,000	Not Applicable
Organic peroxides	Detonable	Maximum Allowable Quantity	Not Applicable
	Class I	Maximum Allowable Quantity	Not Applicable
	Class II	25	Not Applicable
	Class III	50	Not Applicable
Unstable (reactives) nondetonable	Class 3	1	2,000
	Class 2	25	10,000
Water reactives	Class 3	1	Not Applicable
	Class 2	25	Not Applicable
Pyrophoric gases	Not Applicable	Not Applicable	2,000

For SI: 1 ton = 906 kg, 1 cubic foot = 0.02832 m³, 1 pound = 0.454 kg.

a. For materials that are detonable, the distance to other buildings or lot lines shall be as specified in Table 415.3.1 based on trinitrotoluene (TNT) equivalence of the material. For materials classified as explosives, see Chapter 33 the *International Fire Code*. For all other materials, the distance shall be as indicated in Section 415.3.1.

b. "Maximum Allowable Quantity" means the maximum allowable quantity per control area set forth in Table 307.1(1).

c. Limited to Division 1.4 materials and articles, including articles packaged for shipment, that are not regulated as an explosive under Bureau of Alcohol, Tobacco and Firearms (BATF) regulations or unpackaged articles used in process operations that do not propagate a detonation or deflagration between articles, providing the net explosive weight of individual articles does not exceed 1 pound.

[F] 415.5.2 Waterproof room. Rooms or areas used for the storage of water-reactive solids and liquids shall be constructed in a manner that resists the penetration of water through the use of waterproof materials. Piping carrying water for other than *approved* automatic fire sprinkler systems shall not be within such rooms or areas.

❖ Similar to Section 415.5.1, storage rooms containing water-reactive materials must be waterproofed. Though water piping may not be run into or through such rooms, the code recognizes that automatic sprinkler systems are a more regulated type of water piping system and have a low leakage and failure rate when properly installed and maintained.

[F] 415.6 Group H-2. Occupancies in Group H-2 shall be constructed in accordance with Sections 415.6.1 through 415.6.4 and the *International Fire Code*.

❖ Sections 415.6.1 through 415.6.4 contain specific construction requirements for occupancies that contain Group H-2 materials in excess of the maximum allowable quantities. Such occupancies are also required to comply with all material-specific related provisions in the IFC.

[F] 415.6.1 Combustible dusts, grain processing and storage. The provisions of Sections 415.6.1.1 through 415.6.1.6 shall apply to buildings in which materials that produce com-bustible dusts are stored or handled. Buildings that store or handle combustible dusts shall comply with the applicable provisions of NFPA 61, NFPA 85, NFPA 120, NFPA 484, NFPA 654, NFPA 655 and NFPA 664, and the *International Fire Code*.

❖ Combustible dusts can generate high-pressure gas by combustion in air. The pressure created can destroy process equipment and structures. While there is a minimum concentration of dust required, there is no reliable upper concentration limitation beyond which combustion will not occur. The ignition source is often the process equipment itself. The provisions of this section primarily address the construction of a building that contains this hazard. The construction requirements are primarily intended to reduce the exposure hazard should a fire or explosion occur. To minimize the impact of the explosion on the structure, explosion control in accordance with Section 414.5.1 and the IFC is required. The provisions of this section apply to all buildings in which combustible dusts and particles are of sufficient quantity to be readily ignited and subject to explosion. The concentration that would be required to constitute such a hazard is dependent on the particle size. Additional guidance on the relative fire risk associated with various combustible dusts can be found in the referenced standards. In addition to the provisions of this section, compliance with the provisions of the IFC and the referenced standards is essential to minimize the potential ignition and the fire and explosion hazards. Chapter 13 of the IFC contains specific safety precautions for combustible dust-producing operations and references additional standards for explosion protection based on the type of process involved. Essentially, each of the referenced standards prescribes reasonable requirements for safety to life and property from fire and explosion. The standards also minimize the resulting damage should a fire or explosion occur. More specifically, they contain provisions for construction, ventilation, explosion venting, equipment, heating devices, dust control, fire protection and supplemental requirements related to electrical wiring and equipment, provisions concerning protection from sparks, cutting and welding and smoking and signage regulations.

[F] 415.6.1.1 Type of construction and height exceptions. Buildings shall be constructed in compliance with the height and area limitations of Table 503 for Group H-2; except that where erected of Type I or II construction, the heights and areas of grain elevators and similar structures shall be unlimited, and where of Type IV construction, the maximum height shall be 65 feet (19 812 mm) and except further that, in isolated areas, the maximum height of Type IV structures shall be increased to 85 feet (25 908 mm).

❖ The construction of buildings in which combustible dusts can be readily ignitable and subject to an explosion hazard is restricted to the height and area limits in Table 503 for Group H-2. The limitation on construction is intended to reduce the available fuel source that could ignite and serve as the source of a dust explosion. Grain elevators and similar structures that require a height in excess of that permitted in Table 503 may be of unlimited height if the structures are of Type I or II construction; or such structures may be 65 feet (19 812 mm) if they are of Type IV construction. If the structure is of Type IV construction and located so as not to constitute an explosion hazard, the allowable height may be increased to 85 feet (25 908 mm). Although the buildings are required to have an automatic sprinkler system in accordance with Section 903.2.4, the speed at which the combustion process or explosion could occur may reduce the effectiveness of the sprinkler system. Additional guidance on methods to prevent explosions can be found in Section 911 of the IFC and applicable referenced standards, such as NFPA 69, for explosion prevention systems.

[F] 415.6.1.2 Grinding rooms. Every room or space occupied for grinding or other operations that produce combustible dusts shall be enclosed with *fire barriers* constructed in accordance with Section 707 or *horizontal assemblies* constructed in accordance with Section 712, or both. The minimum *fire-resistance rating* shall be 2 hours where the area is not more than 3,000 square feet (279 m²), and 4 hours where the area is greater than 3,000 square feet (279 m²).

❖ Grinding rooms are to be separated from remaining parts of the building by construction having a fire-resis-

tance rating of at least 2 hours. If the area of the room is greater than 3,000 square feet (279 m²), the fire-resistance rating of the enclosure must be at least 4 hours. Not only is it the intent to protect the room from an explosion fire but also, assuming the explosion venting results in minimal structural damage, for the enclosure to contain the resulting fire, if any, to the room of origin. Without venting, it is doubtful whether the enclosure could withstand the forces exerted on the wall by an explosion.

[F] 415.6.1.3 Conveyors. Conveyors, chutes, piping and similar equipment passing through the enclosures of rooms or spaces shall be constructed dirt tight and vapor tight, and be of *approved* noncombustible materials complying with Chapter 30.

❖ The intent of this section is to restrict conveyors that pass through walls so that they do not serve as an ignition source or avenue of fire spread to adjacent areas. These conveyors and similar equipment must be constructed of noncombustible materials and be dirt and vapor tight.

[F] 415.6.1.4 Explosion control. Explosion control shall be provided as specified in the *International Fire Code*, or spaces shall be equipped with the equivalent mechanical ventilation complying with the *International Mechanical Code*.

❖ The pressure exerted by a combustible dust explosion typically ranges from 13 to 89 psi (89 to 614 kPa). It is impractical to construct a building that will withstand such pressures; therefore, either a means of explosion control must be provided in accordance with Section 911 of the IFC or an equivalent mechanical ventilation system must be provided. The mechanical ventilation system must be in accordance with the IMC and must be designed to control the dust concentration below hazardous levels. The mechanical ventilation is dependent on the concentration at which a dust explosion could occur. Additional guidance on the relative fire risk associated with various combustible dusts can be found in the referenced standards.

[F] 415.6.1.5 Grain elevators. Grain elevators, malt houses and buildings for similar occupancies shall not be located within 30 feet (9144 mm) of interior *lot lines* or structures on the same lot, except where erected along a railroad right-of-way.

❖ The 30-foot (9144 mm) separation between grain elevators (or similar structures with regard to lot lines) and other structures is intended to reduce the exposure hazard. It is intended not only to reduce the damage to the adjacent structure but also to minimize the potential that a fire in the adjacent structure would affect the grain elevator. Grain elevators are singled out because of the additional height permitted in Section 415.6.1.1 and the typical openness of the structure. The 30-foot (9144 mm) separation is not required if the grain elevator is located along a railroad right-of-way. The railroad right-of-way will offer some separation, except when a car is being loaded from the grain elevator, in which case the 30-foot (9144 mm) criterion may not be practical.

[F] 415.6.1.6 Coal pockets. Coal pockets located less than 30 feet (9144 mm) from interior *lot lines* or from structures on the same lot shall be constructed of not less than Type IB construction. Where more than 30 feet (9144 mm) from interior *lot lines*, or where erected along a railroad right-of-way, the minimum type of construction of such structures not more than 65 feet (19 812 mm) in *building height* shall be Type IV.

❖ Like grain elevators, coal pockets are typically open structures and there is a need to provide separation between the coal pocket and the adjacent structures to reduce the exposure fire risk. Type IB construction is the required minimum if a coal pocket is within 30 feet (9144 mm) of another structure or lot line. Like grain elevators, if the separation is in excess of 30 feet (9144 mm), Type IV construction is permitted to a maximum building height of 65 feet (19 812 mm).

[F] 415.6.2 Flammable and combustible liquids. The storage, handling, processing and transporting of flammable and combustible liquids in Groups H-2 and H-3 occupancies shall be in accordance with Sections 415.6.2.1 through 415.6.2.10, the *International Mechanical Code* and the *International Fire Code*.

❖ Although Section 415.6.2 and its subsections are part of Section 415.6, these sections apply to both Group H-2 and H-3 occupancies. The storage of flammable and combustible liquids constitutes a special hazard because of the potential fire severity and ease of ignition related to such liquids. The flash point is the most important criterion of the hazards associated with flammable and combustible liquids because it represents the lowest temperature at which sufficient vapor will be given off to form an ignitable or flammable mixture with air. Although it is the flammable vapors that burn or explode, flammable liquids (by definition) are normally stored above their flash point. Since flammable and combustible liquids are found in virtually every industrial plant and many other occupancies, minimum requirements for the storage of such liquids are set forth in the provisions of Chapter 34 of the IFC and NFPA 30.

Additionally, the provisions of Sections 415.6.2.1 through 415.6.2.10 permit the inside tank storage of flammable and combustible liquids. It is intended that all of these provisions, where applicable, be complied with in order to allow the installation of large tanks of flammable and combustible liquids within the building. These provisions were developed in part due to the cost of complying with environmental considerations regarding the installation of underground storage tanks. In downtown areas, above-ground inside storage tanks are commonly used in conjunction with emergency power generators. While not specifically defined, these provisions are intended for tanks proposed for fixed installation as opposed to containers and portable tanks, such as 55-gallon (208 L) drums.

[F] 415.6.2.1 Mixed occupancies. Where the storage tank area is located in a building of two or more occupancies and the quantity of liquid exceeds the maximum allowable quantity for one *control area*, the use shall be completely separated from adjacent occupancies in accordance with the requirements of Section 508.4.

❖ This section requires mandatory fire-resistance-rated separation of the Group H storage tank area from adjacent areas containing other groups. By referencing Section 508.4 for separated mixed occupancies, the storage tank area would require a minimum fire barrier construction, both horizontally and vertically, whenever a mixed-occupancy condition occurs.

[F] 415.6.2.1.1 Height exception. Where storage tanks are located within a building no more than one *story above grade plane*, the height limitation of Section 503 shall not apply for Group H.

❖ The intent of this exception is to allow Group H storage tank areas to be located on an upper floor in a multiple-story building. As long as the storage tank area is located within a building no more than one story, the height limitations of Table 503, with respect to both height in feet and number of stories, would not apply to the Group H fire area. The height limitation for the building would be dictated by the other occupancy groups in the building and the actual construction type. The height exception assumes the storage tank area complies with all of the applicable provisions of Section 415.6.2 and Chapter 34 of the IFC.

[F] 415.6.2.2 Tank protection. Storage tanks shall be noncombustible and protected from physical damage. *Fire barriers* or *horizontal assemblies* or both around the storage tank(s) shall be permitted as the method of protection from physical damage.

❖ The intent of this section is to reduce the potential for industrial accidents involving storage tanks by providing a physical barrier, such as a pipe bollard or barricade. This section would also allow the fire barrier or horizontal assembly that is enclosing the storage tank area to qualify as the means of protection from physical damage. Any method of physical protection, other than a fire barrier wall assembly, must be approved by the building official.

[F] 415.6.2.3 Tanks. Storage tanks shall be *approved* tanks conforming to the requirements of the *International Fire Code*.

❖ The design, construction and installation of all storage tanks for flammable and combustible liquids must comply with the applicable provisions of Chapter 34 of the IFC.

[F] 415.6.2.4 Suppression. Group H shall be equipped throughout with an *approved automatic sprinkler system*, installed in accordance with Section 903.

❖ As with any high-hazard fire area, the storage tank area is required to be protected with an automatic sprinkler system. The use of an alternative extinguishing agent in accordance with Section 904 would be subject to the approval of the building official. A foam-extinguishing system in lieu of an automatic sprinkler system may be a more viable option when protecting a localized flammable liquid storage tank.

[F] 415.6.2.5 Leakage containment. A liquid-tight containment area compatible with the stored liquid shall be provided. The method of spill control, drainage control and secondary containment shall be in accordance with the *International Fire Code.*

> **Exception:** Rooms where only double-wall storage tanks conforming to Section 415.6.2.3 are used to store Class I, II and IIIA flammable and combustible liquids shall not be required to have a leakage containment area.

❖ In order to prevent the spread of flammable and combustible liquids to adjacent nonstorage areas in the event of a leak in a storage tank, an adequately sized containment area is required. The sizing of the containment area should be designed to contain a spill from the largest vessel plus the fire protection water for the required duration. It is also important that the drainage system be sufficient to drain not only the flammable or combustible liquid but also the drainage water from the sprinkler system as well as hose streams. Since the outer wall of a double-walled storage tank acts as the secondary means of containment, the exception would allow the omission of the leakage containment area for all flammable and combustible liquids.

[F] 415.6.2.6 Leakage alarm. An *approved* automatic alarm shall be provided to indicate a leak in a storage tank and room. The alarm shall sound an audible signal, 15 dBa above the ambient sound level, at every point of entry into the room in which the leaking storage tank is located. An *approved* sign shall be posted on every entry door to the tank storage room indicating the potential hazard of the interior room environment, or the sign shall state: WARNING, WHEN ALARM SOUNDS, THE ENVIRONMENT WITHIN THE ROOM MAY BE HAZARDOUS. The leakage alarm shall also be supervised in accordance with Chapter 9 to transmit a trouble signal.

❖ This section requires an alarm system to indicate a leak in the storage tank. The alarm must be supervised, as well as sound a local signal. Automatic electrical supervision of the leakage alarm is required in accordance with Section 901.6.3.

[F] 415.6.2.7 Tank vent. Storage tank vents for Class I, II or IIIA liquids shall terminate to the outdoor air in accordance with the *International Fire Code.*

❖ Section 3404.2.7.3.3 of the IFC specifies the proper location of the tank vent to enable flammable vapors to be released to the outside air and dissipate without creating a hazard. It also requires the tank vent to terminate a minimum of 12 feet (3658 mm) above grade and 5 feet (1524 mm) from building openings and lot lines to reduce the possibility of ignitable concentrations of the vapors collecting near the discharge end of

the vent pipe, entering building openings or exposing adjoining property.

[F] 415.6.2.8 Room ventilation. Storage tank areas storing Class I, II or IIIA liquids shall be provided with mechanical ventilation. The mechanical ventilation system shall be in accordance with the *International Mechanical Code* and the *International Fire Code.*

❖ This section requires mechanical ventilation for storage tank areas in order to maintain any vapor buildup at safe levels until the hazard of a spill or leak can be abated. The ventilation requirements for flammable or combustible liquid storage tanks must be in accordance with the applicable provisions of the IMC and the IFC.

[F] 415.6.2.9 Explosion venting. Where Class I liquids are being stored, explosion venting shall be provided in accordance with the *International Fire Code.*

❖ Class I flammable liquids, when stored in sufficient quantities, can readily produce vapors within the explosive limitations. Explosion venting must be in accordance with the provisions of Sections 414.5.1 and 911 and Chapter 34 of the IFC.

[F] 415.6.2.10 Tank openings other than vents. Tank openings other than vents from tanks inside buildings shall be designed to ensure that liquids or vapor concentrations are not released inside the building.

❖ Additional provisions are necessary to minimize the possibility of an accidental release of flammable or combustible liquids or their vapors via any connection to the tank other than a vent. This includes requirements for liquid-tight openings below the liquid level in the tank, overflow protection and vapor recovery connections as indicated in Section 3404.2.7.5 of the IFC.

[F] 415.6.3 Liquefied petroleum gas facilities. The construction and installation of liquefied petroleum gas facilities shall be in accordance with the requirements of this code, the *International Fire Code,* the *International Mechanical Code,* the *International Fuel Gas Code* and NFPA 58.

❖ The key to this section is the reference to NFPA 58 for design and construction of liquefied petroleum gas facilities. NFPA 58 provides a comprehensive set of construction requirements for LP-gas distribution facilities, as well as bulk plants, and industrial plants in which LP-gas systems, storage systems, vaporizers, mixing systems and similar activities are involved.

[F] 415.6.4 Dry cleaning plants. The construction and installation of dry cleaning plants shall be in accordance with the requirements of this code, the *International Mechanical Code,* the *International Plumbing Code* and NFPA 32. Dry cleaning solvents and systems shall be classified in accordance with the *International Fire Code.*

❖ Dry cleaning plants that do not qualify for Exception 4 or 5 of Section 307.1 and in which the maximum allowable quantity per control area of dry cleaning solvents exceeds the values in Table 307.1(1) would be classified in Group H and would be required to comply with the applicable provisions of the IPC, the IMC and NFPA 32. It should be noted that Chapter 12 of the IFC also contains provisions for dry cleaning plants and references NFPA 32 for their maintenance. NFPA 32 contains provisions for the prevention and control of fire and explosion hazards associated with dry cleaning operations and for the protection of employees and the public. This standard addresses such issues as location, construction, building services, processes and equipment and fire control.

[F] 415.7 Groups H-3 and H-4. Groups H-3 and H-4 shall be constructed in accordance with the applicable provisions of this code and the *International Fire Code.*

❖ This section contains provisions that are applicable to all Group H-3 and H-4 occupancies in addition to material-specific requirements in the IFC. These requirements assume that the Group H-3 and H-4 occupancies contain more than the maximum allowable quantities of hazardous materials per control area permitted by Tables 307.1(1) and 307.1(2). Sections 307.5 and 307.6 contain a list of those materials that could be present in Group H-3 and H-4 occupancies.

[F] 415.7.1 Flammable and combustible liquids. The storage, handling, processing and transporting of flammable and combustible liquids in Group H-3 occupancies shall be in accordance with Section 415.6.2.

❖ This section refers back to Section 415.6.2, where the standards for the handling, storage and processing of these liquids are provided for both Group H-3 and H-2 occupancies.

[F] 415.7.2 Gas rooms. When gas rooms are provided, such rooms shall be separated from other areas by not less than 1-hour *fire barriers* constructed in accordance with Section 707 or *horizontal assemblies* constructed in accordance with Section 712, or both.

❖ In order to restrict the potential fire involvement of a storage room of hazardous gases within a Group H-3 or H-4 occupancy, a 1-hour fire-resistance-rated fire barrier or horizontal assemblies, or both, are required to separate gas rooms from other adjacent areas. The required fire-resistance-rated 1-hour barrier separation is not due to a mixed-occupancy condition but rather further compartmentation in a Group H-3 or H-4 occupancy.

[F] 415.7.3 Floors in storage rooms. Floors in storage areas for corrosive liquids and *highly toxic* or *toxic* materials shall be of liquid-tight, noncombustible construction.

❖ Storage rooms of corrosive liquids and highly toxic and toxic materials would be classified as a Group H-4 occupancy. Noncombustible floors are required to prevent the structure from contributing to a fire scenario. The floors are also required to be liquid tight to prevent the spread of liquids from a spill to areas outside the storage room.

[F] 415.7.4 Separation—highly toxic solids and liquids. *Highly toxic* solids and liquids not stored in *approved* hazardous materials storage cabinets shall be isolated from other hazardous materials storage by not less than 1-hour *fire barriers* constructed in accordance with Section 707 or *horizontal assemblies* constructed in accordance with Section 712, or both.

❖ The potential involvement of materials that may be incompatible in storage situations with highly toxic solids and liquids must be addressed. The required degree of separation, either by approved cabinets or 1-hour fire-resistance-rated fire barrier and horizontal assembly construction, is also intended to restrict the highly toxic materials from other combustibles.

[F] 415.8 Group H-5.

[F] 415.8.1 General. In addition to the requirements set forth elsewhere in this code, Group H-5 shall comply with the provisions of Sections 415.8.1 through 415.8.11 and the *International Fire Code.*

❖ This section contains the requirements for Group H-5 facilities that utilize HPM. Section 415.8 is intended to be a design option for buildings that utilize hazardous materials in excess of the maximum allowable quantities permitted by Tables 307.1(1) and 307.1(2). HPM facilities are considered unique high-hazard occupancies. The HPM occupancy classification assumes that while the manufacturing process involves the use of hazardous materials, the end product by itself is not hazardous.

The provisions of Section 415.8 resulted from nonuniform regulation of semiconductor fabrication facilities and are based on code changes submitted by groups associated with the semiconductor industry. These industries include electronic high-tech industries, such as semiconductor microchip fabrication, as well as the floppy disk and telecommunication industries. Additionally, there are other comparable activities involving similar technology that also use high-hazard materials. HPM include flammable liquids and gases, corrosives, oxidizers and, in many instances, highly toxic materials. The HPM may be a solid, liquid or gas with a degree of hazard rating in health, flammability or reactivity of Class 3 or 4 in accordance with NFPA 704, which is used directly in research, laboratory or production processes.

Structures that contain HPM facilities are required to meet the provisions of Section 415.8 as well as any other code provisions for a Group H-5 occupancy classification unless specifically modified in Section 415.8. Additionally, Chapter 18 of the IFC provides complementary fire safety controls that reduce the risk of hazard to an acceptable level in existing HPM buildings and fabrication areas.

[F] 415.8.2 Fabrication areas.

❖ The fabrication area of an HPM facility is where the hazardous materials are actively handled and processed. The fabrication area includes accessory rooms and spaces such as workstations and employee dressing rooms.

[F] 415.8.2.1 Hazardous materials in fabrication areas.

❖ Sections 415.8.2.1.1 and 415.8.2.1.2 regulate the quantities and densities of HPM for each fabrication area of a Group H-5 facility.

[F] 415.8.2.1.1 Aggregate quantities. The aggregate quantities of hazardous materials stored and used in a single fabrication area shall not exceed the quantities set forth in Table 415.8.2.1.1.

Exception: The quantity limitations for any hazard category in Table 415.8.2.1.1 shall not apply where the fabrication area contains quantities of hazardous materials not exceeding the maximum allowable quantities per *control area* established by Tables 307.1(1) and 307.1(2).

❖ This section regulates the total amount of hazardous materials, whether in use or storage, within a single fabrication area based on the density/quantity of material specified in Table 415.8.2.1.1. The exception permits a fabrication area to have a total quantity of HPM of either the quantity specified in Table 415.8.2.1.1 or the maximum allowable quantities per control area specified in Table 307.1(1) or 307.1(2), whichever is greater. For example, a small fabrication area may have a quantity in accordance with Tables 307.1(1) and 307.1(2) even though it may exceed the quantities specified in Table 415.8.2.1.1. Conversely, a large fabrication area may have an aggregate quantity that exceeds those specified in Tables 307.1(1) and 307.1(2) if it does not exceed the quantities specified in Table 415.8.2.1.1.

It should be reiterated that this section refers to an aggregate quantity of hazardous materials that are in use and storage within a single fabrication area. Section 415.8.2.1.2 further limits the amount of HPM that is being stored within a fabrication area.

TABLE 415.8.2.1.1. See page 4-111.

❖ Table 415.8.2.1.1 controls the permitted densities/quantities of hazardous materials in a single fabrication area. In accordance with Section 415.8.2.1.1, the quantities permitted in Tables 307.1(1) and 307.1(2) may be exceeded, provided the quantities in Table 415.8.2.1.1 are maintained subject to the storage limitations of Section 415.8.2.1.2. Conversely, the maximum quantities of Table 415.8.2.1.1 may be exceeded, provided the maximum allowable quantities of Tables 307.1(1) and 307.1(2) are not exceeded.

In accordance with Note a, hazardous materials in piping need not be included in evaluating the permitted quantity. Note b establishes that a specific quantity limitation, rather than a calculated density, is required for certain hazardous materials.

[F] 415.8.2.1.2 Hazardous production materials. The maximum quantities of hazardous production materials (HPM) stored in a single fabrication area shall not exceed the maximum allowable quantities per *control area* established by Tables 307.1(1) and 307.1(2).

❖ This section provides the overall limit of HPM that can be stored within a single fabrication area based on

Tables 307.1(1) and 307.1(2). Any storage in excess of the permitted maximum allowable quantities would be required to be stored in an HPM storage room, liquid storage room or gas room. The aggregate quantities of hazardous materials in both a storage and use condition within a single fabrication area are regulated by Section 415.8.2.1.1.

[F] 415.8.2.2 Separation. Fabrication areas, whose sizes are limited by the quantity of hazardous materials allowed by Table 415.8.2.1.1, shall be separated from each other, from *corridors* and from other parts of the building by not less than 1-hour *fire barriers* constructed in accordance with Section 707 or *horizontal assemblies* constructed in accordance with Section 712, or both.

Exceptions:

1. Doors within such *fire barrier* walls, including doors to *corridors*, shall be only self-closing *fire door assemblies* having a *fire protection rating* of not less than $^3/_4$ hour.

2. Windows between fabrication areas and corridors are permitted to be fixed glazing *listed* and labeled for a *fire protection rating* of at least $^3/_4$ hour in accordance with Section 715.

❖ Fabrication areas must be separated from other parts of the building, egress corridors and other fabrication areas by fire barriers and horizontal assemblies having a fire-resistance rating of at least 1 hour. While the potential fire exposure could warrant a higher fire-resistance rating, the additional protection features indicative of a Group H-5 facility result in a 1-hour fire-resistance rating being acceptable. The fire barriers must be constructed in accordance with Section 707. The horizontal assemblies must be constructed in accordance with Section 712. One reason for separating one fabrication area from another is the limitation of HPM quantities in a single fabrication area. If the need exists for quantities in excess of those permitted by Section 415.8.2.1.1, the area can be divided into two fabrication areas with 1-hour fire barriers and horizontal assemblies.

Exception 1 clarifies that all doors within the fire barrier walls that comprise the fabrication area must be self-closing and have a fire protection rating of $^3/_4$ hour; therefore, if a corridor wall was also part of the fabrication area, a $^3/_4$-hour opening protective would be required for the door opening. Openings in rated corridors that are not part of a fabrication area typically are required to only have a 20-minute fire protection rating. The doors are required to be self-closing since fabrication areas are "clean rooms" under positive pressure.

The $^3/_4$-hour requirement of Exception 2 will permit the use of fire protection-rated glazing or wired glass windows in accordance with Section 715.4. The use of windows is often desirable to facilitate the security of activities in the fabrication area as well as emergency response in the event of an accident.

[F] TABLE 415.8.2.1.1
QUANTITY LIMITS FOR HAZARDOUS MATERIALS IN A SINGLE FABRICATION AREA IN GROUP H-5[a]

HAZARD CATEGORY		SOLIDS (pounds per square feet)	LIQUIDS (gallons per square feet)	GAS (feet³ @ NTP/square feet)
PHYSICAL-HAZARD MATERIALS				
Combustible dust		Note b	Not Applicable	Not Applicable
Combustible fiber	Loose	Note b	Not Applicable	Not Applicable
	Baled	Notes b, c		
Combustible liquid	II		0.01	
	IIIA		0.02	
	IIIB	Not Applicable	Not Limited	Not Applicable
			0.04	
Combination Class I, II and IIIA				
Cryogenic gas	Flammable	Not Applicable	Not Applicable	Note d
	Oxidizing			1.25
Explosives		Note b	Note b	Note b
Flammable gas	Gaseous	Not Applicable	Not Applicable	Note d
	Liquefied			Note d
Flammable liquid	IA		0.0025	
	IB		0.025	
	IC	Not Applicable	0.025	Not Applicable
Combination Class IA, IB and IC			0.025	
Combination Class I, II and IIIA			0.04	
Flammable solid		0.001	Not Applicable	Not Applicable
Organic peroxide				
Unclassified detonable		Note b		
Class I		Note b		
Class II		0.025	Not Applicable	Not Applicable
Class III		0.1		
Class IV		Not Limited		
Class V		Not limited		
Oxidizing gas	Gaseous			1.25
	Liquefied	Not Applicable	Not Applicable	1.25
Combination of gaseous and liquefied				1.25
Oxidizer	Class 4	Note b	Note b	
	Class 3	0.003	0.03	
	Class 2	0.003	0.03	Not Applicable
	Class 1	0.003	0.03	
Combination	Class 1, 2, 3	0.003	0.03	
Pyrophoric material		Note b	0.00125	Notes d and e
Unstable reactive	Class 4	Note b	Note b	Note b
	Class 3	0.025	0.0025	Note b
	Class 2	0.1	0.01	Note b
	Class 1	Not Limited	Not Limited	Not Limited
Water reactive	Class 3	Note b	0.00125	
	Class 2	0.25	0.025	Not Applicable
	Class 1	Not Limited	Not Limited	
HEALTH-HAZARD MATERIALS				
Corrosives		Not Limited	Not Limited	Not Limited
Highly toxic		Not Limited	Not Limited	Note d
Toxics		Not Limited	Not Limited	Note d

For SI: 1 pound per square foot = 4.882 kg/m², 1 gallon per square foot = 40.7 L/m², 1 cubic foot @ NTP/square foot = 0.305 m³ @ NTP/m², 1 cubic foot = 0.02832 m³.

a. Hazardous materials within piping shall not be included in the calculated quantities.

b. Quantity of hazardous materials in a single fabrication shall not exceed the maximum allowable quantities per control area in Tables 307.1(1) and 307.1(2).

c. Densely packed baled cotton that complies with the packing requirements of ISO 8115 shall not be included in this material class.

d. The aggregate quantity of flammable, pyrophoric, toxic and highly toxic gases shall not exceed 9,000 cubic feet at NTP.

e. The aggregate quantity of pyrophoric gases in the building shall not exceed the amounts set forth in Table 415.3.2.

[F] 415.8.2.3 Location of occupied levels. Occupied levels of fabrication areas shall be located at or above the first *story above grade plane*.

❖ Due to the potential extensive use of hazardous materials, the occupiable levels of fabrication areas are not permitted in basements or other areas below grade. Maintaining fabrication areas at or above grade aids in the egress of the occupants and access for fire department operations.

[F] 415.8.2.4 Floors. Except for surfacing, floors within fabrication areas shall be of noncombustible construction.

Openings through floors of fabrication areas are permitted to be unprotected where the interconnected levels are used solely for mechanical equipment directly related to such fabrication areas (see also Section 415.8.2.5).

Floors forming a part of an occupancy separation shall be liquid tight.

❖ Fabrication area floors are required to be noncombustible. The intent of the requirement for noncombustible floors is to prevent the structure from being involved in the fire from a liquid that may seep into the floor system. The requirement is not intended to apply to floor-covering material.

This section permits the interconnection of the operation floor and a mechanical floor, provided that the mechanical floor is essentially unoccupied. The intent is to prevent leakage past a required separation and to allow multiple-level fabrication areas where the space above or below the operation level is used for mechanical equipment, ducts and similar service equipment. Where a rated separation is required, the floors must be liquid tight. Whereas the hazards often involve liquids, floors are required to be liquid tight to prevent the spread of liquids from a spill to areas outside the fabrication area. "Liquid tight" requires that floors be sealed where they intersect with walls. This can be accomplished by a cove base of the same material as the floor.

[F] 415.8.2.5 Shafts and openings through floors. Elevator shafts, vent shafts and other openings through floors shall be enclosed when required by Section 708. Mechanical, duct and piping penetrations within a fabrication area shall not extend through more than two floors. The *annular space* around penetrations for cables, cable trays, tubing, piping, conduit or ducts shall be sealed at the floor level to restrict the movement of air. The fabrication area, including the areas through which the ductwork and piping extend, shall be considered a single conditioned environment.

❖ Section 708.2 specifies the conditions when a shaft enclosure is required. This section provides the requirements for specific opening penetrations through a floor within a fabrication area. The maximum number of stories that are allowed to be interconnected is three, since penetrations are limited to two floors. The penetrations for duct and piping are to have the annular space protected. When such penetrations exist, the fire separation required by Section 415.8.2.2 is to separate the entire interconnected volume.

[F] 415.8.2.6 Ventilation. Mechanical exhaust ventilation at the rate of not less than 1 cubic foot per minute per square foot [0.0051 m³/(s · m²)] of floor area shall be provided throughout the portions of the fabrication area where HPM are used or stored. The exhaust air duct system of one fabrication area shall not connect to another duct system outside that fabrication area within the building.

A ventilation system shall be provided to capture and exhaust gases, fumes and vapors at workstations.

Two or more operations at a workstation shall not be connected to the same exhaust system where either one or the combination of the substances removed could constitute a fire, explosion or hazardous chemical reaction within the exhaust duct system.

Exhaust ducts penetrating occupancy separations shall be contained in a shaft of equivalent fire-resistance-rated construction. Exhaust ducts shall not penetrate *fire walls*.

Fire dampers shall not be installed in exhaust ducts.

❖ A major design factor is the required minimum ventilation. The minimum rate is 1 cubic foot per minute (cfm) per square foot (0.0051 m³/s · m²), or about eight air changes per hour for an 8-foot-high (2438 mm) fabrication area. Typically, at least 24 air changes per hour are provided for operational needs and where unclassified electrical systems are to be used. In order to prevent the spread of gases or fumes, the exhaust air duct system of any fabrication area must not be connected to any other ventilation system. With regard to workstations, the required ventilation system is intended to prevent exposure of the work station operator to hazardous fumes or vapors and prevent hazardous concentrations of materials used in the manufacturing process. Additionally, ventilation systems must be designed to reduce the possibility of a chemical reaction occurring between two or more HPM or a material's and exhaust system's components. Where two or more chemicals may react dangerously if exhausted through a common duct, they must be removed separately. NFPA 491, which is included in the *NFPA Fire Protection Guide on Hazardous Materials*, is a good first reference when trying to determine if any two agents are reactive.

Ventilation should not be interrupted by a fire damper when a fire or other emergency occurs involving a workstation. This helps reduce the likelihood that hazardous combustion byproducts or hazardous concentrations of HPM are not forced back into the workstation or clean room. Continuous ventilation through a duct enclosed in a fire-resistance-rated shaft is required. Fire walls define separations between buildings. Ducts must never penetrate a barrier common to another building or occupancy. This reduces the likelihood of tampering with or interrupting the duct integrity.

[F] 415.8.2.7 Transporting hazardous production materials to fabrication areas. HPM shall be transported to fabrication areas through enclosed piping or tubing systems that comply with Section 415.8.6.1, through service *corridors* complying with Section 415.8.4, or in *corridors* as permitted in the exception to Section 415.8.3. The handling or transporting of HPM within service *corridors* shall comply with the *International Fire Code*.

❖ Except as permitted in the exception to Section 415.8.3 for existing facilities, all HPM must be transported to the fabrication area by service corridors or piping (see Sections 415.8.4 and 415.8.6). This reduces the chance that an egress route will be affected by an accidental leak or spill. Chapter 18 of the IFC contains additional requirements for the transport of HPM through all egress components as well as service corridors.

[F] 415.8.2.8 Electrical.

❖ Sections 415.8.2.8.1 and 415.8.2.8.2 provide specific electrical requirements with respect to ventilation systems in fabrication areas and at workstations.

[F] 415.8.2.8.1 General. Electrical equipment and devices within the fabrication area shall comply with NFPA 70. The requirements for hazardous locations need not be applied where the average air change is at least four times that set forth in Section 415.8.2.6 and where the number of air changes at any location is not less than three times that required by Section 415.8.2.6. The use of recirculated air shall be permitted.

❖ Consistent with Chapter 27, electrical installations are required to comply with NFPA 70. This section specifically permits the dilution of concentrations of hazardous areas such that the hazardous location provisions of NFPA 70 need not apply. This is because some of the equipment used in HPM facilities cannot comply with the hazardous location provisions of NFPA 70. The nature of the fabrication process also dictates the use of the dilution concept. Typically, such facilities have a ventilation system designed to provide at least 24 air changes per hour.

At least 3 cfm per square foot (0.132 L/s/m²) or 24 air changes in an 8-foot-high (2438 mm) fabrication area is required at all locations in order not to apply the hazardous location provisions of NFPA 70. Also, an average rate of 4 cfm per square foot (0.176 L/s/m²) or 32 air changes in an 8-foot-high (2438 mm) fabrication area are required if the hazardous location provisions of NFPA 70 are not applied.

[F] 415.8.2.8.2 Workstations. Workstations shall not be energized without adequate exhaust ventilation. See Section 415.8.2.6 for workstation exhaust ventilation requirements.

❖ Accidental exposure to flammable fumes or vapors can occur at workstations where hazardous materials are used. Either a mechanical or electrical interlock must be provided to engage the required exhaust ventilation system before HPM enters the workstation. This reduces the likelihood of gas or vapor exposure.

[F] 415.8.3 Corridors. *Corridors* shall comply with Chapter 10 and shall be separated from fabrication areas as specified in Section 415.8.2.2. *Corridors* shall not contain HPM and shall not be used for transporting such materials, except through closed piping systems as provided in Section 415.8.6.3.

Exception: Where existing fabrication areas are altered or modified, HPM is allowed to be transported in existing *corridors,* subject to the following conditions:

1. Corridors. *Corridors* adjacent to the fabrication area where the *alteration* work is to be done shall comply with Section 1018 for a length determined as follows:

    1.1. The length of the common wall of the *corridor* and the fabrication area; and

    1.2. For the distance along the *corridor* to the point of entry of HPM into the *corridor* serving that fabrication area.

2. Emergency alarm system. There shall be an emergency telephone system, a local manual alarm station or other *approved* alarm-initiating device within *corridors* at not more than 150-foot (45 720 mm) intervals and at each *exit* and doorway. The signal shall be relayed to an *approved* central, proprietary or remote station service or the emergency control station and shall also initiate a local audible alarm.

3. Pass-throughs. Self-closing doors having a *fire protection rating* of not less than 1 hour shall separate pass-throughs from existing *corridors*. Pass-throughs shall be constructed as required for the *corridors* and protected by an *approved* automatic fire-extinguishing system.

❖ Corridors are required to comply with the requirements of Section 1018 and must be separated from fabrication areas in accordance with Section 415.8.2.2. Additionally, typical egress corridors in new buildings are not to be used for transporting HPM to the fabrication area. In accordance with Section 415.8.2.7, HPM must be transported by service corridors or piping (see Sections 415.8.4 and 415.8.6, and Figure 415.8.3).

The exception addresses HPM facilities that existed before the adoption and enforcement of this section. It permits the transport of HPM in egress corridors in existing buildings under the conditions specified in Items 1, 2 and 3. When alterations are made to a fabrication area, such corridors must be upgraded. Chapter 18 of the IFC provides additional requirements for the storage, handling and transporting of HPM materials in existing HPM facilities. Item 1 of the exception requires that corridors be upgraded to a 1-hour fire-resistance rating for the entire length of the wall common to the fabrication area. Additionally, the entire length of the corridor that is used for transporting HPM to the fabrication area being renovated must also be upgraded to have a 1-hour fire-resistance rating. Item 2 requires a means to notify trained personnel of an emergency condition in an egress corridor. A local alarm is required to alert the occupants of a potential hazardous condition. The intent of Item 3 is to permit a "pass-through" for providing HPM to a fabrication area. A pass-through, such as a storage cabinet, is used to store and receive HPM for the fabrication area. The pass-through must be separated from the egress corridor by 1-hour fire-resistance-rated construction, including a 1-hour-rated self-closing fire door, and be suppressed.

[F] 415.8.4 Service corridors.

❖ This section regulates the corridor that may be used for the transport of HPM in a Group H-5 facility of new construction. For the transport of HPM via piping systems, see Section 415.8.6.

[F] 415.8.4.1 Occupancy. Service corridors shall be classified as Group H-5.

❖ The intent of this section is to clarify that a potential separation due to the mixed occupancy provisions of Section 508.4 does not apply, since service corridors are considered part of the Group H-5 occupancy.

[F] 415.8.4.2 Use conditions. Service corridors shall be separated from *corridors* as required by Section 415.8.2.2. Service corridors shall not be used as a required *corridor*.

❖ Service corridors are used for transporting HPM in lieu of piping. The separation between a service corridor

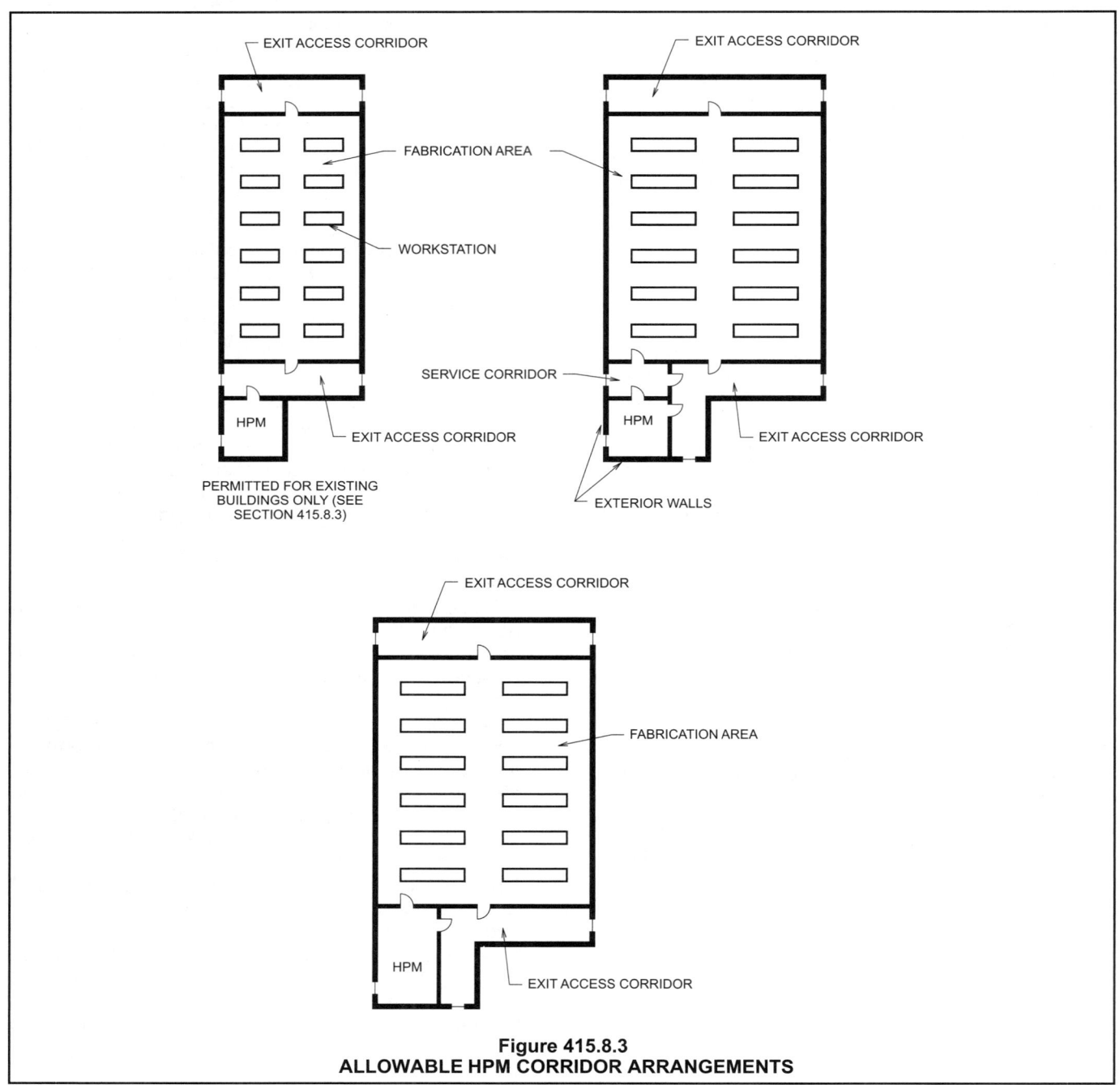

Figure 415.8.3
ALLOWABLE HPM CORRIDOR ARRANGEMENTS

and an egress corridor must be in accordance with Section 415.8.2.2. As such, a minimum 1-hour fire separation is required. Service corridors are also not intended to be used as a portion of the required means of egress.

[F] 415.8.4.3 Mechanical ventilation. Service corridors shall be mechanically ventilated as required by Section 415.8.2.6 or at not less than six air changes per hour, whichever is greater.

❖ Since the same HPM that may be present in the fabrication area may be transported in the service corridors, an adequate means of ventilation is required. This section requires a ventilation rate of either 1 cfm per square foot (0.044 L/s/m²) of floor area in accordance with Section 415.8.2.6 or six air changes per hour, whichever is greater depending on the size and volume of the service corridor.

[F] 415.8.4.4 Means of egress. The maximum distance of travel from any point in a service corridor to an *exit*, *exit access corridor* or door into a fabrication area shall not exceed 75 feet (22 860 mm). Dead ends shall not exceed 4 feet (1219 mm) in length. There shall be not less than two *exits*, and not more than one-half of the required *means of egress* shall require travel into a fabrication area. Doors from service corridors shall swing in the direction of egress travel and shall be self-closing.

❖ The exit access travel distance is limited to 75 feet (22 860 mm), except that the travel distance may be measured from a door to a fabrication area. In order to minimize the potential for an occupant to be trapped in a service corridor, dead ends must not exceed 4 feet (1219 mm). As such, the occupant would most likely be intimate with the source of the problem or located in an adjacent fabrication area from which egress is provided via exit access, not service, corridors.

An occupant at any location in a service corridor is required to have access to at least two means of egress. No more than 50 percent of the means of egress may be through a fabrication area that, in most cases, will be one of the means of egress. This assumes that, should an incident or problem occur, it will be in either the service corridor or the fabrication area—not both. Because of the potential for a hazardous condition, all doors to service corridors must be self-closing and swing in the direction of egress travel. Smoke-detector-actuated hold-open devices are not acceptable because smoke detectors may not respond to hazardous conditions that do not involve products of combustion.

[F] 415.8.4.5 Minimum width. The minimum clear width of a service corridor shall be 5 feet (1524 mm), or 33 inches (838 mm) wider than the widest cart or truck used in the corridor, whichever is greater.

❖ The minimum width requirement for the service corridor is to provide adequate clearance to get around the device used to transport the HPM in the event it is involved in an emergency incident in the service corridor. The size of the transport carts and trucks depends

on the type and quantity of materials utilized in the facility.

[F] 415.8.4.6 Emergency alarm system. Emergency alarm systems shall be provided in accordance with this section and Sections 414.7.1 and 414.7.2. The maximum allowable quantity per *control area* provisions shall not apply to emergency alarm systems required for HPM.

❖ This section requires an emergency alarm system in all areas where HPM is transported or stored. The applicability of either Section 414.7.1 or 414.7.2 depends on whether the HPM material is in a storage or use condition. This section also clarifies that the requirement for an emergency alarm system in a Group H-5 facility in the locations identified in Sections 415.8.4.6.1 through 415.8.6.4.3 is not dependent on whether the maximum allowable quantities per control area of Tables 307.1(1) or 307.1(2) are exceeded.

[F] 415.8.4.6.1 Service corridors. An emergency alarm system shall be provided in service corridors, with at least one alarm device in each service corridor.

❖ An emergency telephone system or manual alarm pull station is required in service corridors. Such devices must initiate an alarm at the emergency control station as well as activate a local audible device.

[F] 415.8.4.6.2 Exit access corridors and exit enclosures. Emergency alarms for *exit access corridors* and *exit enclosures* shall comply with Section 414.7.2.

❖ Since HPM materials would not be in exit access corridors or exit enclosures unless they were being transported to another approved area, the emergency alarm requirements of Section 414.7.2 for dispensing, use and handling must be complied with.

[F] 415.8.4.6.3 Liquid storage rooms, HPM rooms and gas rooms. Emergency alarms for liquid storage rooms, HPM rooms and gas rooms shall comply with Section 414.7.1.

❖ This section requires compliance with the emergency alarm requirements of Section 414.7.1 for hazardous materials in a storage condition. This section addresses storage areas, which by their designation contain HPM in quantities greater than those listed in Table 307.1(1) or 307.1(2).

[F] 415.8.4.6.4 Alarm-initiating devices. An *approved* emergency telephone system, local alarm manual pull stations, or other *approved* alarm-initiating devices are allowed to be used as emergency alarm-initiating devices.

❖ This section classifies what constitutes an approved alarm-initiating device (see commentary, Section 414.7).

[F] 415.8.4.6.5 Alarm signals. Activation of the emergency alarm system shall sound a local alarm and transmit a signal to the emergency control station.

❖ The alarm signal is required to be transmitted to the emergency control station in order to notify trained personnel of an emergency condition. A local alarm is required to alert the occupants of a potentially hazardous condition

[F] 415.8.5 Storage of hazardous production materials.

❖ Section 415.8.5 addresses general storage provisions for HPM materials in a Group H-5 facility.

[F] 415.8.5.1 General. Storage of HPM in fabrication areas shall be within *approved* or *listed* storage cabinets or gas cabinets or within a workstation. The storage of HPM in quantities greater than those listed in Section 1804.2 of the *International Fire Code* shall be in liquid storage rooms, HPM rooms or gas rooms as appropriate for the materials stored. The storage of other hazardous materials shall be in accordance with other applicable provisions of this code and the *International Fire Code*.

❖ Even though the amount of HPM in a fabrication area is controlled, it still must be within approved cabinets or a workstation. This requirement is intended to limit the exposure to occupants of the fabrication area to only the material in use within that area.

The larger amounts of HPM typically stored in separate areas present a hazard comparable to other Group H facilities; therefore, such storage rooms need to meet similar code requirements. Storage rooms containing HPM in quantities greater than permitted by Section 1804.2 of the IFC are required to comply with the applicable provisions of Section 415.8.5.2, depending on the state of the material.

It should be noted that the HPM quantity limitations per fabrication area are essentially identical to the limitations of Section 415.8.2.1.

[F] 415.8.5.2 Construction.

❖ This section deals with construction requirements for two types of storage rooms: those classified as HPM rooms or gas rooms and those utilized as liquid storage rooms. The size and separation of such rooms is dependent on the type of materials stored.

[F] 415.8.5.2.1 HPM rooms and gas rooms. HPM rooms and gas rooms shall be separated from other areas by *fire barriers* constructed in accordance with Section 707 or *horizontal assemblies* constructed in accordance with Section 712, or both. The minimum *fire-resistance rating* shall be 2 hours where the area is 300 square feet (27.9 m²) or more and 1 hour where the area is less than 300 square feet (27.9 m²).

❖ The amount of hazardous material in an HPM room or gas room is only limited by room size; therefore, since storage rooms in excess of 300 square feet (27.9 m²) are not limited by size or quantity of hazardous material, construction of fire barriers and horizontal assemblies of not less than 2-hour fire-resistance rating is required. It should be noted that this section assumes the storage areas have at least one exterior wall with a minimum 30-foot (9144 mm) fire separation distance (see Section 415.8.5.3). If the room is less than 300 square feet (27.9 m²), the construction can be rated for 1 hour.

[F] 415.8.5.2.2 Liquid storage rooms. Liquid storage rooms shall be constructed in accordance with the following requirements:

1. Rooms in excess of 500 square feet (46.5 m²) shall have at least one exterior door *approved* for fire department access.

2. Rooms shall be separated from other areas by *fire barriers* constructed in accordance with Section 707 or *horizontal assemblies* constructed in accordance with Section 712, or both. The *fire-resistance rating* shall be not less than 1 hour for rooms up to 150 square feet (13.9 m²) in area and not less than 2 hours where the room is more than 150 square feet (13.9 m²) in area.

3. Shelving, racks and wainscotting in such areas shall be of noncombustible construction or wood of not less than 1-inch (25 mm) nominal thickness.

4. Rooms used for the storage of Class I flammable liquids shall not be located in a basement.

❖ This section assumes storage rooms are utilized for the storage of flammable and combustible liquids in closed containers. The size of the rooms is severely restricted when only 1-hour fire barriers and 1-hour horizontal assemblies are provided to limit the size of a potential fire. For rooms greater than 150 square feet (14 m²) in area, a minimum 2-hour-rated construction is required because the size of the room or quantity of hazardous materials is not limited. Due to the need for readily available access in a fire scenario, all liquid storage rooms in excess of 500 square feet (46.5 m²) must have at least one exterior door that provides access to the room. In accordance with Section 415.8.5.3, all liquid storage rooms, regardless of size, must have at least one exterior wall.

The construction limitations on the shelving racks and wainscotting are intended to limit their potential involvement in a fire scenario. Due to the volatile nature of Class I flammable liquids and the need for quick fire department access, they must be located in liquid storage rooms at or above grade.

[F] 415.8.5.2.3 Floors. Except for surfacing, floors of HPM rooms and liquid storage rooms shall be of noncombustible liquid-tight construction. Raised grating over floors shall be of noncombustible materials.

❖ Similar to the floor construction of fabrication areas, the floors of HPM rooms and liquid storage rooms are required to be noncombustible and liquid tight (see commentary, Section 415.8.2.4).

[F] 415.8.5.3 Location. Where HPM rooms, liquid storage rooms and gas rooms are provided, they shall have at least one *exterior wall* and such wall shall be not less than 30 feet (9144 mm) from *lot lines*, including *lot lines* adjacent to *public ways*.

❖ Access from the exterior for fire-fighting operations is essential for storage rooms of hazardous materials. These rooms by their designation contain more than the maximum allowable quantities per control area listed in Tables 307.1(1) and 307.1(2). The 30-foot (9144 mm) limitation is based in part on the fact that the exterior wall need not have a fire-resistance rating depending on the construction type of the building (see Table 602). Such walls may also be used for explosion venting when such venting is required.

[F] 415.8.5.4 Explosion control. Explosion control shall be provided where required by Section 414.5.1.

❖ This section requires a method of explosion control only when an explosion hazard exists. Not all storage rooms are required to have explosion control. Section 414.5.1 requires a method of explosion control, such as explosion venting or barricade construction, depending on the specific hazardous material involved in accordance with the IFC. Table 414.5.1 provides guidance as to when explosion control is required for various hazardous materials.

[F] 415.8.5.5 Exits. Where two exits are required from HPM rooms, liquid storage rooms and gas rooms, one shall be directly to the outside of the building.

❖ Depending on the common path of egress travel available in accordance with Section 1014.3, two exits may be required from all HPM rooms, liquid storage rooms and gas rooms. For example, a liquid storage room containing Class IB flammable liquids in normally closed containers would be considered a Group H-3 occupancy. As such, a maximum common path of travel of 25 feet (7620 mm) would be permitted before a second exit, one of which was directly to the exterior, is required. Storage rooms are required to be located along an exterior wall. In addition to the requirements of this section, all liquid storage rooms in excess of 500 square feet (46 m²) are required to have at least one exterior door regardless of travel distance.

[F] 415.8.5.6 Doors. Doors in a *fire barrier* wall, including doors to *corridors*, shall be self-closing *fire door assemblies* having a *fire-protection rating* of not less than ³/₄ hour.

❖ Doors that are in an interior wall that comprise part of the storage room enclosure are required to have a minimum ³/₄-hour fire protection rating. The actual rating is dependent on the fire-resistance rating of the fire barrier separation required. While a 1-hour fire barrier would permit a door with a ³/₄-hour fire protection rating, a 2-hour fire barrier would require a door to have a 1¹/₂-hour fire protection rating in accordance with Table 715.4.

[F] 415.8.5.7 Ventilation. Mechanical exhaust ventilation shall be provided in liquid storage rooms, HPM rooms and gas rooms at the rate of not less than 1 cubic foot per minute per square foot (0.044 L/s/m²) of floor area or six air changes per hour, whichever is greater, for categories of material.

Exhaust ventilation for gas rooms shall be designed to operate at a negative pressure in relation to the surrounding areas and direct the exhaust ventilation to an exhaust system.

❖ Similar to fabrication areas and service corridors, mechanical exhaust ventilation is required (see commentary, Sections 415.8.2.6 and 415.8.4.3).

[F] 415.8.5.8 Emergency alarm system. An *approved* emergency alarm system shall be provided for HPM rooms, liquid storage rooms and gas rooms.

Emergency alarm-initiating devices shall be installed outside of each interior exit door of such rooms.

Activation of an emergency alarm-initiating device shall sound a local alarm and transmit a signal to the emergency control station.

An *approved* emergency telephone system, local alarm manual pull stations or other *approved* alarm-initiating devices are allowed to be used as emergency alarm-initiating devices.

❖ An emergency alarm system is required for all storage rooms containing hazardous materials in excess of the quantities permitted in Tables 307.1(1) and 307.1(2). See the commentary to Section 415.8.4.6 for similar emergency alarm requirements.

[F] 415.8.6 Piping and tubing.

❖ This section addresses the requirements for all piping and tubing utilized to transport HPM throughout a Group H-5 facility.

[F] 415.8.6.1 General. Hazardous production materials piping and tubing shall comply with this section and ASME B31.3.

❖ The nature of the HPM involved requires strict control of both the materials used in piping and the joint methods used to avoid leakage. ASME B 31.3 provides criteria for piping installation applicable to HPM facilities.

[F] 415.8.6.2 Supply piping and tubing.

❖ This section regulates the design requirements for all piping and tubing, including where it can be located within a Group H-5 facility.

[F] 415.8.6.2.1 HPM having a health-hazard ranking of 3 or 4. Systems supplying HPM liquids or gases having a health-hazard ranking of 3 or 4 shall be welded throughout, except for connections, to the systems that are within a ventilated enclosure if the material is a gas, or an *approved* method of drainage or containment is provided for the connections if the material is a liquid.

❖ The primary purpose of this section is to minimize the potential for a leak of HPM. The use of mechanical compression-type fittings or other nonwelded joints in areas of health hazard 3 or 4 must be limited to areas where any leaks will be vented and the materials exhausted. If the material is a liquid, then a method of drainage or containment is required to minimize the hazard.

[F] 415.8.6.2.2 Location in service corridors. Hazardous production materials supply piping or tubing in service corridors shall be exposed to view.

❖ Concealment of piping in service corridors must be avoided to provide a means of monitoring for leaks.

[F] 415.8.6.2.3 Excess flow control. Where HPM gases or liquids are carried in pressurized piping above 15 pounds per square inch gauge (psig) (103.4 kPa), excess flow control shall be provided. Where the piping originates from within a liquid storage room, HPM room or gas room, the excess flow control shall be located within the liquid storage room, HPM room or gas room. Where the piping originates from a bulk source, the excess flow control shall be located as close to the bulk source as practical.

❖ Excess flow control valves regulate the flow of hazardous materials within the piping system. The valves are designed to shut off in the event the predetermined flow is exceeded.

[F] 415.8.6.3 Installations in corridors and above other occupancies. The installation of HPM piping and tubing within the space defined by the walls of *corridors* and the floor or roof above, or in concealed spaces above other occupancies, shall be in accordance with Section 415.8.6.2 and the following conditions:

1. Automatic sprinklers shall be installed within the space unless the space is less than 6 inches (152 mm) in the least dimension.

2. Ventilation not less than six air changes per hour shall be provided. The space shall not be used to convey air from any other area.

3. Where the piping or tubing is used to transport HPM liquids, a receptor shall be installed below such piping or tubing. The receptor shall be designed to collect any discharge or leakage and drain it to an *approved* location. The 1-hour enclosure shall not be used as part of the receptor.

4. HPM supply piping and tubing and nonmetallic waste lines shall be separated from the *corridor* and from occupancies other than Group H-5 by *fire barriers* that have a *fire-resistance rating* of not less than 1 hour. Where gypsum wallboard is used, joints on the piping side of the enclosure are not required to be taped, provided the joints occur over framing members. Access openings into the enclosure shall be protected by *approved* fire protection-rated assemblies.

5. Readily accessible manual or automatic remotely activated fail-safe emergency shutoff valves shall be installed on piping and tubing other than waste lines at the following locations:

   5.1. At branch connections into the fabrication area.

   5.2. At entries into *corridors*.

   Exception: Transverse crossings of the *corridors* by supply piping that is enclosed within a ferrous pipe or tube for the width of the *corridor* need not comply with Items 1 through 5.

❖ The installation of HPM piping in the space above an egress corridor or another occupancy, as well as the cavity of the egress corridor wall, present a potential source of hazard to the building's occupants. The five requirements provided for such installations serve to mitigate the hazard by suppression, ventilation, containment and ignition control. In order to address the potential hazards, the containment must be a fire barrier with at least a 1-hour fire-resistance rating, drainage receptors, excess flow control and shutoff valves.

The elimination of the taping of the wallboard joints on the piping side of a rated assembly is in recognition of actual installation difficulties and the reduced likelihood of a fire on the interior of the wall cavity. To eliminate the taping of joints, however, the joints must occur over framing members.

When the piping traverses a corridor, the use of a coaxial enclosed pipe around the HPM piping is considered acceptable for providing the required separation and containment of a potential leak. The assumption is that the open ends of that pipe are in an HPM facility and, therefore, a leak into the outer casing can be monitored. If the adjacent areas that contain the open ends are not in an HPM facility, then the outerjacket method cannot be used.

[F] 415.8.6.4 Identification. Piping, tubing and HPM waste lines shall be identified in accordance with ANSI A13.1 to indicate the material being transported.

❖ To facilitate monitoring, detecting and controlling any possible leaks, all piping, tubing and HPM waste lines must be identified in accordance with ANSI A13.1. This standard contains criteria for properly identifying the piping system, including labeling and frequency or distribution of the signs or labels.

[F] 415.8.7 Continuous gas detection systems. A continuous gas detection system shall be provided for HPM gases when the physiological warning threshold level of the gas is at a higher level than the accepted PEL for the gas and for flammable gases in accordance with Sections 415.8.7.1 and 415.8.7.2.

❖ A gas detection system in the room or area utilized for the storage or use of HPM gases provides early notification of a leak that is occurring before the escaping gas reaches hazardous concentration levels.

[F] 415.8.7.1 Where required. A continuous gas detection system shall be provided in the areas identified in Sections 415.8.7.1.1 through 415.8.7.1.4.

❖ Sections 415.8.7.1.1 through 415.8.7.1.4 prescribe the locations in a Group H-5 facility when a gas detection system is required.

[F] 415.8.7.1.1 Fabrication areas. A continuous gas detection system shall be provided in fabrication areas when gas is used in the fabrication area.

❖ All fabrication areas that utilize HPM gases must have a gas detection system. It should be noted that gas detection is often installed within workstations as a means of early detection of leaks. Such detection is generally not acceptable as an alternative to gas detection for the fabrication area, since a leak may occur remote from the workstation.

[F] 415.8.7.1.2 HPM rooms. A continuous gas detection system shall be provided in HPM rooms when gas is used in the room.

❖ HPM rooms, which by definition contain more than the quantities of hazardous materials per control area permitted by Tables 307.1(1) and 307.1(2), are required to have a gas detection system.

[F] 415.8.7.1.3 Gas cabinets, exhausted enclosures and gas rooms. A continuous gas detection system shall be provided in gas cabinets and exhausted enclosures. A continuous gas detection system shall be provided in gas rooms when gases are not located in gas cabinets or exhausted enclosures.

❖ In the potential event of a leaking cylinder of a hazardous gas, gas cabinets, exhausted enclosures and gas rooms must have a gas detection system.

[F] 415.8.7.1.4 Corridors. When gases are transported in piping placed within the space defined by the walls of a *corridor* and the floor or roof above the *corridor*, a continuous gas detection system shall be provided where piping is located and in the *corridor*.

> **Exception:** A continuous gas detection system is not required for occasional transverse crossings of the corridors by supply piping that is enclosed in a ferrous pipe or tube for the width of the *corridor*.

❖ In addition to the requirements of Section 415.8.6.3 for HPM piping and tubing in an egress corridor, a gas detection system is required for early notification of a potential leak of an HPM gas.
The exception is similar to the exception in Section 415.8.6.3. In essence, the gas detection system required by this section as well as the additional provisions in Section 415.8.6.3 are not required if the exception is met (see commentary, Section 415.8.6.3).

[F] 415.8.7.2 Gas detection system operation. The continuous gas detection system shall be capable of monitoring the room, area or equipment in which the gas is located at or below all the following gas concentrations:

1. Immediately dangerous to life and health (IDLH) values when the monitoring point is within an exhausted enclosure, ventilated enclosure or gas cabinet.

2. Permissible exposure limit (PEL) levels when the monitoring point is in an area outside an exhausted enclosure, ventilated enclosure or gas cabinet.

3. For flammable gases, the monitoring detection threshold level shall be vapor concentrations in excess of 25 percent of the lower flammable limit (LFL) when the monitoring is within or outside an exhausted enclosure, ventilated enclosure or gas cabinet.

4. Except as noted in this section, monitoring for *highly toxic* and *toxic* gases shall also comply with Chapter 37 of the *International Fire Code*.

❖ This section requires gas detection systems to be capable of sensing a leak at or below the permissible exposure limit (PEL). This exposure limit regulated by OSHA to prevent adverse health effects is the breathing zone exposure limit for employees over an 8-hour time weighted average. In most cases, gas detection in the semiconductor industry is conducted in an exhausted enclosure, ventilated enclosure or gas cabinet and not in the breathing zone of the employee, and is designed to detect and alert employees of leaks inside exhausted enclosures, ventilated enclosures or gas cabinets and is not intended to estimate potential employee breathing zone exposures. The semiconductor industry addressed this by codifying NFPA 318 Section 10.9 to differentiate gas detection set points in exhausted enclosures (set at the IDLH) with gas detection when the monitoring point is in an area outside an exhausted enclosure, ventilated enclosure or gas cabinet. This section is consistent with the provisions of NFPA 318 Section 10.9 guidelines that are much more relevant to the type of monitoring performed in semiconductor manufacturing (inside exhausted enclosures, ventilated enclosures or gas cabinets). Monitoring in the semiconductor industry is designed to detect and alert employees of leaks inside exhausted enclosures, ventilated enclosures and gas cabinets, and is not intended to estimate potential employee breathing zone exposures. Therefore, set points are not required or recommended to be set at occupational exposure limits (e.g., TLVs or PELs). Additionally, the 25 percent LFL is consistent with both IMC Section 510.2 and NFPA 318 Section 10.9. Chapter 37 of the IFC contains additional requirements for the monitoring of highly toxic and toxic compressed gases.

[F] 415.8.7.2.1 Alarms. The gas detection system shall initiate a local alarm and transmit a signal to the emergency control station when a short-term hazard condition is detected. The alarm shall be both visual and audible and shall provide warning both inside and outside the area where the gas is detected. The audible alarm shall be distinct from all other alarms.

❖ The required local alarm is intended to alert occupants to a hazardous condition in the vicinity of where the HPM gases are being stored or used. The alarm is not intended to be an evacuation alarm; however, it is required to be monitored to hasten emergency personnel response.

[F] 415.8.7.2.2 Shutoff of gas supply. The gas detection system shall automatically close the shutoff valve at the source on gas supply piping and tubing related to the system being monitored for which gas is detected when a short-term hazard condition is detected. Automatic closure of shutoff valves shall comply with the following:

1. Where the gas detection sampling point initiating the gas detection system alarm is within a gas cabinet or exhausted enclosure, the shutoff valve in the gas cabinet or exhausted enclosure for the specific gas detected shall automatically close.

2. Where the gas detection sampling point initiating the gas detection system alarm is within a room and compressed gas containers are not in gas cabinets or an exhausted enclosure, the shutoff valves on all gas lines for the specific gas detected shall automatically close.

3. Where the gas detection sampling point initiating the gas detection system alarm is within a piping distribution manifold enclosure, the shutoff valve supplying the manifold for the compressed gas container of the specific gas detected shall automatically close.

> **Exception:** Where the gas detection sampling point initiating the gas detection system alarm is at the use location or

within a gas valve enclosure of a branch line downstream of a piping distribution manifold, the shutoff valve for the branch line located in the piping distribution manifold enclosure shall automatically close.

❖ Where gas detection systems are required, automatic emergency shutoff valves are required to stop the flow of hazardous material from possibly deteriorating further in an emergency.

[F] 415.8.8 Manual fire alarm system. An *approved* manual fire alarm system shall be provided throughout buildings containing Group H-5. Activation of the alarm system shall initiate a local alarm and transmit a signal to the emergency control station. The fire alarm system shall be designed and installed in accordance with Section 907.

❖ Due to the type and potential quantities of hazardous materials that could be present in a Group H-5 facility, a manual means of activating an evacuation alarm is essential for the safety of the occupants in an emergency situation. The alarm signal must be transmitted to the emergency control station. The fire alarm must also be in accordance with the applicable provisions of Section 907 and NFPA 72.

[F] 415.8.9 Emergency control station. An emergency control station shall be provided in accordance with Sections 415.8.9.1 through 415.8.9.3.

❖ Due to the extent of hazardous materials permitted in an HPM facility, an approved on-site location is needed where the alarm signals from emergency equipment can be received.

[F] 415.8.9.1 Location. The emergency control station shall be located on the premises at an *approved* location outside the fabrication area.

❖ The emergency control station must be located on-site at a location approved by the building official. The emergency control station must not be located in an area where hazardous materials are stored, used or transported, such as a fabrication area.

[F] 415.8.9.2 Staffing. Trained personnel shall continuously staff the emergency control station.

❖ Appropriate response by trained personnel is essential to any emergency event involving HPM materials.

[F] 415.8.9.3 Signals. The emergency control station shall receive signals from emergency equipment and alarm and detection systems. Such emergency equipment and alarm and detection systems shall include, but not be limited to, the following where such equipment or systems are required to be provided either in this chapter or elsewhere in this code:

1. *Automatic sprinkler system* alarm and monitoring systems.
2. Manual fire alarm systems.
3. Emergency alarm systems.
4. Continuous gas detection systems.
5. Smoke detection systems.
6. Emergency power system.

7. Automatic detection and alarm systems for pyrophoric liquids and Class 3 water-reactive liquids required in Section 1805.2.3.4 of the *International Fire Code.*
8. Exhaust ventilation flow alarm devices for pyrophoric liquids and Class 3 water-reactive liquids cabinet exhaust ventilation systems required in Section 1805.2.3.4 of the *International Fire Code.*

❖ This section specifies the types of systems that are to be monitored by the emergency control station.

[F] 415.8.10 Emergency power system. An emergency power system shall be provided in Group H-5 occupancies where required in Section 415.8.10.1. The emergency power system shall be designed to supply power automatically to required electrical systems when the normal electrical supply system is interrupted.

❖ A backup emergency power source is considered essential for systems that are monitoring and protecting hazardous materials in a Group H-5 occupancy. Without an emergency power system, all required electrical controls or equipment monitoring hazardous materials would be rendered inoperative if a power failure or other electrical system failure were to occur.

[F] 415.8.10.1 Required electrical systems. Emergency power shall be provided for electrically operated equipment and connected control circuits for the following systems:

1. HPM exhaust ventilation systems.
2. HPM gas cabinet ventilation systems.
3. HPM exhausted enclosure ventilation systems.
4. HPM gas room ventilation systems.
5. HPM gas detection systems.
6. Emergency alarm systems.
7. Manual fire alarm systems.
8. *Automatic sprinkler system* monitoring and alarm systems.
9. Automatic alarm and detection systems for pyrophoric liquids and Class 3 water-reactive liquids required in Section 1805.2.3.4 of the *International Fire Code.*
10. Flow alarm switches for pyrophoric liquids and Class 3 water-reactive liquids cabinet exhaust ventilation systems required in Section 1805.2.3.4 of the *International Fire Code.*
11. Electrically operated systems required elsewhere in this code or in the *International Fire Code* applicable to the use, storage or handling of HPM.

❖ This section specifies the types of systems within a Group H-5 occupancy that are required to be connected to an approved emergency power system. As indicated in Section 2702, all emergency power systems must be installed in accordance with the applicable requirements of NFPA 70, 110 and 111.

[F] 415.8.10.2 Exhaust ventilation systems. Exhaust ventilation systems are allowed to be designed to operate at not less than one-half the normal fan speed on the emergency power system where it is demonstrated that the level of exhaust will maintain a safe atmosphere.

❖ Emergency power for exhaust ventilation is required to prevent hazardous concentrations of HPM fumes or vapors in areas such as workstations or fabrication areas. Fans for exhaust ventilation draw a considerable amount of current when operating. Running exhaust fans at a reduced speed may be desirable when it will not endanger the operator or result in a hazardous condition; however, exhaust fans must not be run at a speed less than 50 percent of its rating, even if a slower speed will not produce a serious hazard.

[F] 415.8.11 Automatic sprinkler system protection in exhaust ducts for HPM.

❖ This section prescribes the sprinkler system requirements for exhaust ducts for HPM. The requirements depend on the construction materials of the exhaust duct.

[F] 415.8.11.1 Exhaust ducts for HPM. An *approved automatic sprinkler system* shall be provided in exhaust ducts conveying gases, vapors, fumes, mists or dusts generated from HPM in accordance with this section and the *International Mechanical Code.*

❖ An exhaust duct for HPM materials could convey flammable and combustible gases, fumes, vapors or ducts. To provide protection against the spread of fire within the exhaust system and to prevent a duct fire from involving the building, sprinkler protection is required in the exhaust duct. The use of an extinguishing agent other than water based on agent compatibility would be subject to local approval. Section 510 of the IMC contains additional requirements for protecting hazardous exhaust systems.

[F] 415.8.11.2 Metallic and noncombustible nonmetallic exhaust ducts. An *approved automatic sprinkler system* shall be provided in metallic and noncombustible nonmetallic exhaust ducts when all of the following conditions apply:

1. Where the largest cross-sectional diameter is equal to or greater than 10 inches (254 mm).

2. The ducts are within the building.

3. The ducts are conveying flammable gases, vapors or fumes.

❖ Sprinklers are required within each individual duct when all three of the following conditions exist:

1. Cross-sectional diameter at the widest point is equal to or exceeds 10 inches (254 mm).

2. Ducts are located within the building.

3. Ducts convey gases or vapors within the flammable range.

Figure 415.8.11.2 illustrates how to measure the cross-sectional diameter of various duct shapes. Provisions of this section require the square and rounded ducts to be protected by automatic sprinklers. The round or elliptical ducts depicted on the right side of the diagram are not required to be protected.

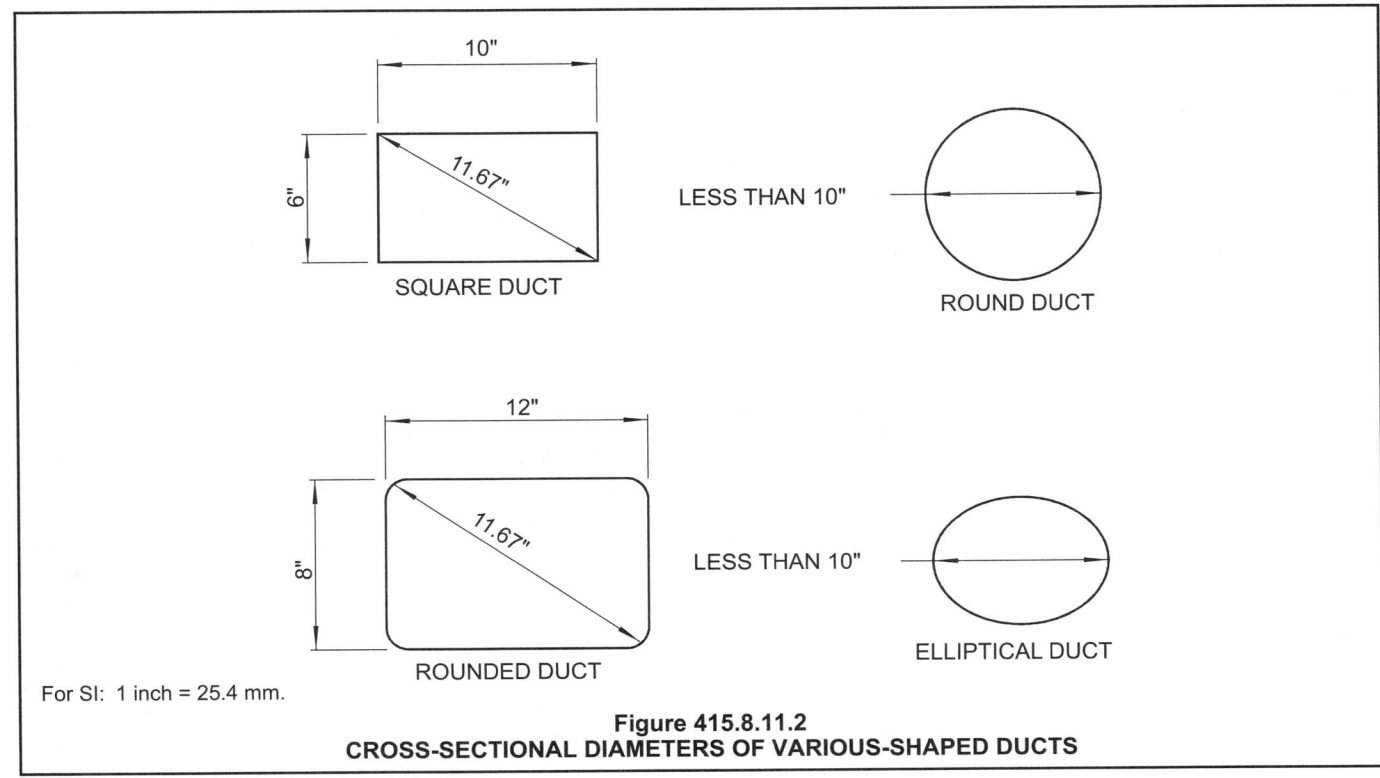

For SI: 1 inch = 25.4 mm.

Figure 415.8.11.2
CROSS-SECTIONAL DIAMETERS OF VARIOUS-SHAPED DUCTS

[F] 415.8.11.3 Combustible nonmetallic exhaust ducts. *Automatic sprinkler system* protection shall be provided in combustible nonmetallic exhaust ducts where the largest cross-sectional diameter of the duct is equal to or greater than 10 inches (254 mm).

Exceptions:

1. Ducts *listed* or *approved* for applications without automatic fire sprinkler system protection.

2. Ducts not more than 12 feet (3658 mm) in length installed below ceiling level.

❖ Galvanized steel is not an appropriate duct material for all substances handled in the broad category of hazardous exhaust systems. A duct material compatible with the exhaust must be selected, and factors such as corrosiveness, abrasion resistance, chemical resistance and operating temperatures must be taken into account. Section 510.8 of the IMC contains additional provisions for duct construction that is part of a hazardous exhaust system. As such, automatic sprinkler protection is also required for all combustible nonmetallic ducts with a cross-sectional diameter equal to or greater than 10 inches (254 mm).

Exception 1 states that automatic sprinklers are not required where the risk to people or property is limited, such as when nonmetallic ducts approved for installation without sprinklers are used. Exception 2 states that when ducts do not exceed 12 feet (3658 mm) in length and are installed exposed below ceiling level, sprinklers may be omitted. The limited duct length and exposure reduces the potential of a concealed fire hazard.

[F] 415.8.11.4 Automatic sprinkler locations. Sprinkler systems shall be installed at 12-foot (3658 mm) intervals in horizontal ducts and at changes in direction. In vertical ducts, sprinklers shall be installed at the top and at alternate floor levels.

❖ Adequate sprinkler coverage needs to be maintained to prevent the spread of fire within the exhaust system. All sprinkler system components must also be in compliance with the applicable provisions of NFPA 13.

SECTION 416
APPLICATION OF FLAMMABLE FINISHES

[F] 416.1 General. The provisions of this section shall apply to the construction, installation and use of buildings and structures, or parts thereof, for the spraying of flammable paints, varnishes and lacquers or other flammable materials or mixtures or compounds used for painting, varnishing, staining or similar purposes. Such construction and equipment shall comply with the *International Fire Code.*

❖ The principal hazards associated with paint spraying and spray booths originate from the presence of flammable liquids or powders and their vapors or mists. The purpose of this section is to provide requirements that address the hazards associated with spray appli-

cations and dipping or coating applications involving flammable paints, varnishes and lacquers. The requirements listed include such areas as ventilation, automatic sprinklers, control of ignition sources and proper operation of the equipment. In addition to the provisions of this section, the provisions of the IMC and Chapter 15 of the IFC also apply.

The provisions of this section apply to the indoor use of spray applications and dipping or coating applications involving flammable paints, varnishes and lacquers. Outdoor applications involving flammable paints, varnishes and lacquers are not covered, since overspray deposits are not likely to create hazardous conditions, and flammable vapor-air mixtures are minimized because of atmospheric dilution. Safeguards, such as ventilation, ignition control, material storage and waste disposal (see Chapter 15 of the IFC), should still apply. This reference to the IFC and its provisions regarding flammable finishes covers the application of flammable or combustible materials when applied as a spray by compressed air; "airless" or "hydraulic atomization" steam; electrostatically or by any other means in continuous or intermittent processes. They also cover the application of combustible powders when applied by powder spray guns, electrostatic powder spray guns, fluidized beds or electrostatic fluidized beds. The IFC and its referenced standards contain provisions relating to the location of spray areas; ignition sources; ventilation; liquid storage and handling; protection; operation; maintenance and training. The provisions found in Chapter 15 of the IFC also apply to processes in which articles or materials are passed through the contents of tanks, vats or containers of flammable or combustible liquids, including dipping; roll; flow and curtain coating; finishing; treating; cleaning and similar processes. Requirements for the location of dipping and coating processes; ventilation; equipment construction; liquid storage and handling; protection; operation; maintenance and training are also provided.

[F] 416.2 Spray rooms. Spray rooms shall be enclosed with not less than 1-hour *fire barriers* constructed in accordance with Section 707 or *horizontal assemblies* constructed in accordance with Section 712, or both. Floors shall be waterproofed and drained in an *approved* manner.

❖ A spray room is a power-ventilated, fully enclosed room used exclusively for open spraying of flammable and combustible materials. The entire spray room is considered the spray area. The primary difference between a spray room and a spray booth is that spray booths are partially open. A spray room is to be enclosed with fire barriers and horizontal assemblies having a fire-resistance rating of at least 1 hour. Waterproof floors are to be arranged to drain either to the outside of the building, to internal drains or to other suitable places. Properly designed and guarded drains or scuppers of sufficient number and size to dispose of all surplus water likely to be discharged by automatic sprinklers must be provided.

[F] 416.2.1 Surfaces. The interior surfaces of spray rooms shall be smooth and shall be so constructed to permit the free passage of exhaust air from all parts of the interior and to facilitate washing and cleaning, and shall be so designed to confine residues within the room. Aluminum shall not be used.

❖ Rough, corrugated or uneven surfaces are difficult to clean. Periodic cleaning of interior surfaces reduces the fire hazard posed by the accumulation of flammable or combustible coatings. Due to the physical properties of aluminum, it is unsuitable for cleaning and scraping of overspray residue.

[F] 416.3 Spraying spaces. Spraying spaces shall be ventilated with an exhaust system to prevent the accumulation of flammable mist or vapors in accordance with the *International Mechanical Code*. Where such spaces are not separately enclosed, noncombustible spray curtains shall be provided to restrict the spread of flammable vapors.

❖ The objective of ventilation is to remove flammable vapors and mists so as to minimize the potential for a flash fire or explosion. The spray area is required to be ventilated in accordance with the provisions of the IMC. When the spray process is not separated from the other operations or areas, a noncombustible spray curtain must be provided to restrict the spread of flammable vapors. The IFC requires ventilation systems and enclosures to be interlocked with the spraying equipment to ensure that fans operate and doors remain closed when in use.

[F] 416.3.1 Surfaces. The interior surfaces of spraying spaces shall be smooth and continuous without edges; shall be so constructed to permit the free passage of exhaust air from all parts of the interior and to facilitate washing and cleaning; and shall be so designed to confine residues within the spraying space. Aluminum shall not be used.

❖ See commentary, Section 416.2.1.

[F] 416.4 Spray booths. Spray booths shall be designed, constructed and operated in accordance with the *International Fire Code*.

❖ Detailed requirements for the design and construction of spray booths are give in Section 1504.3.2 of the IFC. The IFC also addresses requirements for ventilation of adjacent areas, fire protection, fire extinguishers, control of ignition sources, housekeeping during operations, lighting and operation control through interlocks.

[F] 416.5 Fire protection. An automatic fire-extinguishing system shall be provided in all spray, dip and immersing spaces and storage rooms and shall be installed in accordance with Chapter 9.

❖ Spray application operations are to be protected with an automatic fire suppression system. While an automatic sprinkler system is the most desirable type of suppression system, this section would allow the use of other effective extinguishing agents, such as dry chemical or foam. Because of the ease of ignition of flammable liquids, specifically spray paint applications of a flammable liquid, an automatic fire suppression system is required in rooms in which spray painting is conducted, as well as rooms in which flammable materials are used for painting, brushing, dipping or mixing on a regular basis. Dry sprinklers should not be used for the protection of spray operations, except possibly at the exhaust duct penetrations to the outside. Wet-pipe, preaction or deluge systems should be used so that water is placed on the fire in the shortest possible time. If the entire building is not protected with an automatic sprinkler system, then the system protecting the spray booth or room may be a limited area sprinkler system in accordance with Section 903.3.5.1.1. Locating sprinklers in paint spray booths presents an unusual problem, since the sprinkler may become clogged with paint. The most satisfactory solution is to locate the sprinklers in areas in a way that the paint spray will most likely not reach the sprinkler. Even in the most ideal locations, the sprinkler will need to be cleaned very frequently. A coating of grease with a low melting point, such as petroleum jelly, motor oil or a soft neutral soap, will facilitate cleaning of the sprinkler. Other alternatives that are commonly used include polyethylene or cellophane bags with a thickness of 0.003 inches (0.076 mm) or less or thin paper bags to protect the sprinklers.

Additional guidance on fire protection requirements for the application of flammable finishes can be found in Chapter 15 of the IFC.

SECTION 417
DRYING ROOMS

[F] 417.1 General. A drying room or dry kiln installed within a building shall be constructed entirely of *approved* noncombustible materials or assemblies of such materials regulated by the *approved* rules or as required in the general and specific sections of Chapter 4 for special occupancies and where applicable to the general requirements of Chapter 28.

❖ This section establishes specific safety requirements for drying rooms and dry kilns used in conjunction with the drying (accelerated seasoning) of lumber, and dryers or dehydrators used to reduce the moisture content of agricultural products. Included are drying operations for other combustible materials, such as certain building materials, textiles and fabrics. Drying operations associated with the application of flammable finishes are regulated by Section 416. The hazards associated with drying rooms and dry kilns relate to the volume of readily combustible materials being processed at temperatures conducive to their ignition in the event of a system malfunction.

The materials or assemblies used in the construction of drying rooms or dry kilns located within buildings must be noncombustible, regardless of the type of construction of the building itself (see commentary, Section 703.4 for a discussion of noncombustibility criteria) when required by the approved rules or Chapters 4 and 28. The use of noncombustible materials of construction for drying rooms and dry kilns provides for a measure of confinement of a fire occurring within and prevents the room or kiln from contributing to the fuel load exposed to an unwanted fire.

[F] 417.2 Piping clearance. Overhead heating pipes shall have a clearance of not less than 2 inches (51 mm) from combustible contents in the dryer.

❖ Piped steam and hot-water indirect heating systems used in drying rooms or dry kilns can operate at temperatures of up to several hundred degrees Fahrenheit and are considered a safe and appropriate method of providing heat to the drying processes. A minimum clearance of 2 inches (51 mm) between hydronic piping and adjacent combustible contents in the drying room or dry kiln is required to reduce the likelihood of contact ignition of the combustible contents. Though hydronic piping is frequently installed overhead, the clearance is required regardless of the pipes' location within the drying room or kiln.

[F] 417.3 Insulation. Where the operating temperature of the dryer is 175°F (79°C) or more, metal enclosures shall be insulated from adjacent combustible materials by not less than 12 inches (305 mm) of airspace, or the metal walls shall be lined with $^1/_4$-inch (6.35 mm) insulating mill board or other *approved* equivalent insulation.

❖ The insulation of metal drying room and kiln enclosures from adjacent combustible materials is required when the drying apparatus operates at or above 175°F (79°C). Insulation can be accomplished by the use of airspace or by lining the metal dryer walls with insulating material. The insulation will reduce the possibility that combustible materials in proximity to the drying room or dry kiln will be ignited by heat transfer from it. The temperature of 175°F (79°C) has been determined to be the highest allowable temperature without any clearance to combustibles.

[F] 417.4 Fire protection. Drying rooms designed for high-hazard materials and processes, including special occupancies as provided for in Chapter 4, shall be protected by an *approved* automatic fire-extinguishing system complying with the provisions of Chapter 9.

❖ The possibility of ignition increases in a drying room utilized in conjunction with hazardous materials, such as combustible fibers and other finely divided materials. The installation of an approved fire-extinguishing system in accordance with Chapter 9 will contain a fire within the drying room until the fire department (or industrial fire brigade) arrives. In sprinklered buildings, fire suppression can be easily and economically provided by installing sprinklers within the drying room or area. Wet-pipe, preaction or deluge systems should be used so that water is placed on the fire in the shortest possible time. Note that dryers used in conjunction with plywood, veneer and composite board mill operations are required by the IFC to be protected by an approved deluge water-spray system. See the commentary for Chapter 19 of the IFC for further information on lumber and woodworking facilities. In nonsprinklered or partially sprinklered buildings, the system protecting the drying room may be an

approved, limited area sprinkler system in accordance with Section 903.3.5.1.1. An alternative fire-extinguishing system consisting of dry chemical, carbon dioxide, clean agent or other approved nonwater-based fire-extinguishing system in accordance with Section 904 can provide an equivalent level of protection, depending on the nature of the drying room contents. See the commentary to Section 904 for a discussion of the various alternative fire-extinguishing systems available.

SECTION 418
ORGANIC COATINGS

❖ The manufacture of organic coatings encompasses operations that produce decorative and protective coatings for architectural uses, industrial products and other specialized purposes. Requirements of this chapter address the hazards associated with the manufacture of solvent-based organic coatings. Water-based materials are exempt from these requirements. These provisions are consistent with Chapter 20 of the IFC, which contains more detailed requirements for the manufacture of organic coatings. NFPA 35 also provides additional guidance on the maintenance of facilities utilized for the manufacture of organic coatings.

[F] 418.1 Building features. Manufacturing of organic coatings shall be done only in buildings that do not have pits or basements.

❖ Basements, pits or depressed first-floor construction are prohibited because of the tendency for hazardous material vapors to accumulate in low areas and the difficulty in fighting fires in basements and pits in such occupancies.

[F] 418.2 Location. Organic coating manufacturing operations and operations incidental to or connected therewith shall not be located in buildings having other occupancies.

❖ Incidental uses involve operations and activities closely related to the primary occupancy and are necessary for the efficient, continuous and safe performance of the manufacture of organic coatings. Administration, storage, shipping and receiving, as well as other related but not indispensable operations, should be located in separate buildings.

[F] 418.3 Process mills. Mills operating with close clearances and that process flammable and heat-sensitive materials, such as nitrocellulose, shall be located in a detached building or noncombustible structure.

❖ Milling of heat-sensitive materials, such as nitrocellulose, is an extraordinary hazard. Such operations must be located in separate, single-purpose buildings away from other uses and high-hazard operations. Pebble mills pose a special vapor ignition hazard caused by static electricity. Both the grinding material and inner lining of these mills are made of materials with good insulating characteristics. Because static electricity is produced during milling, it has nowhere to go. Generally, the inside of this type of mill is either partially or totally inerted using nitrogen or carbon dioxide gas to prevent ignition.

[F] 418.4 Tank storage. Storage areas for flammable and combustible liquid tanks inside of structures shall be located at or above grade and shall be separated from the processing area by not less than 2-hour *fire barriers* constructed in accordance with Section 707 or *horizontal assemblies* constructed in accordance with Section 712, or both.

❖ Tank storage located below grade is prohibited. Basements located under grade-level storage areas should be discouraged. Below-grade flammable liquid fires are extremely difficult to fight. Similarly, above-grade spills will flow to lower floors, possibly resulting in spill fires on more than one building level. Tank storage of raw materials must be confined to locations at or above grade level. Tank storage rooms must be separated from process and other storage areas by fire barriers and horizontal assemblies that are a minimum of 2-hour fire-resistance rated. If possible, these rooms should be able to be accessed on at least one side from the outside. Cutoff rooms with access from two or even three sides are considered ideal.

[F] 418.5 Nitrocellulose storage. Nitrocellulose storage shall be located on a detached pad or in a separate structure or a room enclosed with no less than 2-hour *fire barriers* constructed in accordance with Section 707 or *horizontal assemblies* constructed in accordance with Section 712, or both.

❖ Once ignited, nitrocellulose will continue to burn even in the absence of oxygen. Extra precautions must be used to prevent ignition and fire spread; therefore, nitrocellulose storage should be in either a separate detached structure, on a detached exterior pad or fully enclosed by a construction of minimum 2-hour fire-resistance rating complying with fire barrier and horizontal assembly requirements.

[F] 418.6 Finished products. Storage rooms for finished products that are flammable or combustible liquids shall be separated from the processing area by not less than 2-hour *fire barriers* constructed in accordance with Section 707 or *horizontal assemblies* constructed in accordance with Section 712, or both.

❖ Finished products, which are also classified as flammable or combustible liquids, must also comply with the applicable provisions of Chapter 34 of the IFC and NFPA 30. As a minimum, a 2-hour fire-resistance-rated construction consisting of horizontal assemblies and fire barriers with approved opening protectives is required between the processing area and the storage area to eliminate mutual involvement in a single fire scenario. A higher degree of fire separation may be required depending on the building's occupancy classification.

SECTION 419
LIVE/WORK UNITS

419.1 General. A live/work unit is a *dwelling unit* or *sleeping unit* in which a significant portion of the space includes a nonresidential use that is operated by the tenant and shall comply with Sections 419.1 through 419.8.

Exception: *Dwelling* or *sleeping units* that include an office that is less than 10 percent of the area of the *dwelling unit* shall not be classified as a live/work unit.

❖ These provisions allow a live/work unit that includes both living and working environments to be considered a single Group R-2 dwelling for application of the code. Several limitations and specific requirements are applied to both the living portion of the unit and the work portion of the unit. Prior to the adoption of these provisions, the code and the IRC did not allow residential live/work units in a form that is typically desirable for community development. This concept has become increasingly popular allowing design and construction of a public business, with employees working within a residence, allowing the public to enter the work area of the unit to acquire service. Examples of live/work commercial functions are artist's studios, beauty parlors, nail salons and chiropractor's offices. It is important to note that live/work is specifically not to apply to an in-home office (architect home office, consultant home office, etc.). The exception to Section 419.1 is intended to address these small home offices which involve less than 10 percent of the dwelling.

These concepts are throwbacks to an era of planning which created a community where residents could walk to all needed services such as the typical corner commercial store. Examples of this form of planning can be found in many older cities, as well as many "planned communities." Live/work units began to re-emerge in the 1990s through a development style known as "Traditional Neighborhood Design" (TND). More recently, adaptive reuse in many older urban structures in city centers incorporated the same live/work tools to provide a variety of business offerings combined with residential unit types.

Historically, building codes did not have to deal with many live/work issues because zoning codes generally precluded a mixing of uses within a neighborhood, much less within a building. However, recent planning trends have been adopted in many jurisdictions, encouraging mixing of commercial and residential uses, not just in neighborhoods, but also in buildings, and even within unit types, such as the live/work unit commonly found in TND projects.

The live/work approach is also driven by the desire to provide affordable housing. Many cities and towns also aggressively pursue affordable housing, with the IRC being a key tool in this effort. The IRC allows jurisdictions to produce a range of housing types at competitive market values, including the live/work unit.

However, there are no provisions for any use other than residential in the IRC. Since live/work units mix in a commercial use, they are driven out of the IRC into the code. In previous editions of the code, when this happens, the live/work units incur an increase in code-related construction requirements (use separation, construction type, egress, fire prevention) in excess of any risk present in the work function. The added requirements drive construction costs up, and inevitably drive the units out of the affordable housing range.

The provisions in the 2009 code for live/work units apply to the code criteria based on the Group R-2 provisions for construction. The occupant loads will be determined by the "function" of each space in accordance with Table 1004.1.1.

In the new IRC, there is an exception to Section R101.2 that allows live/work units complying with Section 419 of the code to be constructed in accordance with the IRC. The concept of live/work units being unseparated is dependent on how mixed use is applied.

Section 419 allows mixed use unseparated occupancies within the dwelling unit or sleeping unit as long as it meets the limits within this section. It is then to be classified as a Group R-2 occupancy. In previous codes, any combination of occupancies would have to be treated as a mixed use condition and would be considered either separated or nonseparated in accordance with the code. The building must then be classified in both occupancies and meet the more restrictive requirements for both. Under the new live/work requirements, the building is treated only as a Group R-2 occupancy despite including work environments. Special features that are common within a dwelling unit and are likely within the live/work unit are addressed in order to clearly delineate the means for designing a live/work unit.

419.1.1 Limitations. The following shall apply to all live/work areas:

1. The live/work unit is permitted to be a maximum of 3,000 square feet (279 m²);

2. The nonresidential area is permitted to be a maximum 50 percent of the area of each live/work unit;

3. The nonresidential area function shall be limited to the first or main floor only of the live/work unit; and

4. A maximum of five nonresidential workers or employees are allowed to occupy the nonresidential area at any one time.

❖ These provisions were meant to apply strictly to small businesses associated with dwelling and sleeping units. In fact the intent is that the main occupancy of the building is residential with some business activity within the building. The code limits the nonresidential aspect to a maximum of 50 percent of the area of each unit. In addition, the total area of the live/work unit is limited to 3,000 square feet (279 m²). Therefore, the total area of the work unit would be a maximum of 1,500 square feet (139 m²). Since a nonresidential use is being located in a dwelling unit or sleeping unit, the nonresidential area is limited to the first or main floor. Therefore, those coming to the place of business do not need to enter the residential portion of the building. Finally, in keeping with the intent that these are small occupancies and that such occupancies could not create unnecessary life safety concerns, the number of nonresidential workers (employees) is limited to five. This limit of five is not the limit on the number of occupants that can be located within the work area, but simply the number of workers from outside the household that can be there on a regular basis. The 1,500-square-foot (139 m²) limit on area would limit the number of occupants based upon the occupant load factors.

419.2 Occupancies. Live/work units shall be classified as a Group R-2 occupancy. Separation requirements found in Sections 420 and 508 shall not apply within the live/work unit when the live/work unit is in compliance with Section 419. High-hazard and storage occupancies shall not be permitted in a live/work unit. The aggregate area of storage in the nonresidential portion of the live/work unit shall be limited to 10 percent of the space dedicated to nonresidential activities.

❖ The entire live/work unit is to be classified as Group R-2 regardless of the types of business being conducted. This exempts such units from the requirements for separation in Sections 420 and 508 within the unit, but would still require the separation between each live/work unit.

The provisions that prohibit high-hazard and storage occupancies intend to avoid the accumulation of excessive and dangerous fire loads in residential related occupancies. This section would not prohibit the occupancy from containing the maximum allowable quantities of hazardous materials per control area within a building. Control areas are regulated in Section 414 (see the commentary for that section).

Storage is always a potential fire load with any business and to ensure that it does not become a large fire hazard it is limited to 10 percent of the nonresidential portion of the live/work unit. That would be a maximum of 150 square feet (14 m²)—about the size of a large closet.

419.3 Means of egress. Except as modified by this section, the provisions for Group R-2 occupancies in Chapter 10 shall apply to the entire live/work unit.

❖ This section requires compliance with Chapter 10 for means of egress unless the general requirements are modified by the following four subsections. This will address exiting requirements for both the Group R-2 occupancy and additional requirements for the nonresidential activity and the general public in such areas where they occur.

419.3.1 Egress capacity. The egress capacity for each element of the live/work unit shall be based on the *occupant load* for the function served in accordance with Table 1004.1.1.

❖ The egress capacity must be based upon the actual use of the space. Therefore if you had a mercantile type use in the live/work unit, the egress capacity must be based upon 30 square feet (2.8 m²) per person of the gross area of the mercantile space. If a 3,000-square-foot (279 m²) unit is equally divided for mercantile and residential use, in such a case the calculated occupant load would be determined by dividing the area by the square feet per person: 1,500 square feet/ 30 square feet (18.5 m²) per person = 50 occupants. In addition, the capacity for residential use is 200 square feet per person (gross area), which would be 1,500 square feet/200 square feet per person = 8 occupants.

419.3.2 Sliding doors. Where doors in a *means of egress* are of the horizontal-sliding type, the force to slide the door to its fully open position shall not exceed 50 pounds (220 N) with a perpendicular force against the door of 50 pounds (220 N).

❖ Section 1008.1.2, Exception 4, allows the use of sliding doors in Group R-2 occupancies. This section continues this allowance of such doors in live/work units, but puts an upper limit on door-opening force of 50 pounds (220 N). This is greater than the door-opening force indicated for sliding doors in Section 1008.1.3.

419.3.3 Spiral stairways. *Spiral stairways* that conform to the requirements of Section 1009.9 shall be permitted.

❖ Spiral staircases are allowed in dwelling units and this section simply emphasizes this allowance with a specific reference back to the section with the design criteria in Chapter 10.

419.3.4 Locks. Egress doors shall be permitted to be locked in accordance with Exception 4 of Section 1008.1.9.3.

❖ This section allows the use of dead bolts, night latches or security chains in live/work units with an occupant load of 10 or less. If the occupant load is greater than 10, other exceptions for locking devices may be applicable, as indicated in Section 1008.1.9.3, Exception 2.

419.4 Vertical openings. Floor openings between floor levels of a live/work unit are permitted without enclosure.

❖ One of the problems with the prior codes when designing and constructing a live/work unit was the requirement for enclosure of exit stairways. For ease of access, the live/work units can have open interior stairways for the residential and nonresidential portions of the dwelling unit, similar to a multiple-story dwelling unit (see Section 1022, Exception 3). If the units are separated and do have enclosed stairways, they would be treated as any other residential stairway enclosure within a dwelling unit; and no fire-resistant-rated construction is required.

419.5 Fire protection. The live/work unit shall be provided with a monitored fire alarm system where required by Section 907.2.9 and an *automatic sprinkler system* in accordance with Section 903.2.8.

❖ Since the unit is considered as Group R-2, the entire building would be required to be sprinklered in accordance with any of the three sprinkler standards, as appropriate. Note that the IRC Section 101.2 exception explicitly allows the use of NFPA 13D. The code requirements provide only a general reference to Group R sprinkler requirements. Live/work units are not specifically addressed; however, single-family homes, duplexes and townhouses are listed under Section 903.3.1.3.

This section requires the installation of a fire alarm system as required for a Group R-2 occupancy. Section 907.2.9 would only require a fire alarm system in certain cases. The requirements for a manual fire alarm are based upon the location of the Group R-2 dwelling unit (three or more stories above the lowest level of exit discharge or one story below the highest level of exit discharge) and the number of units (more than 16). If a building has a sprinkler system and would require a manual system, the manual aspect is no longer required. Instead the sprinkler system is required to be tied to the notification appliances and activate the system upon water flow.

419.6 Structural. Floor loading for the areas within a live/work unit shall be designed to conform to Table 1607.1 based on the function within the space.

❖ Since live/work units may have structural loads not normally anticipated by Group R-2 occupancies, the code specifically requires structural design of floor loadings to be addressed in accordance with Table 1607.1, based upon what is actually occurring in the space. For instance, if the nonresidential activity is a business, there may be equipment such as computers, files or a large copy machine which requires loading based on office loads of 100 psf (4788 Pa).

419.7 Accessibility. Accessibility shall be designed in accordance with Chapter 11.

❖ Accessibility to and within the live/work unit must be designed and constructed in compliance with Chapter 11. The business/work area on the first floor must be fully accessible. In accordance with Section 1103.2.13, the dwelling unit portion must be evaluated separately. If the structure has four or more units, or the site has more than 20 units, Type A unit or Type B unit requirements may be applicable. The exceptions under Section 1107.7 are applicable. The most common application will typically be the multistory dwelling unit (see Section 1107.7.2).

The live/work unit would also require an accessible toilet within the nonresidential part of the unit if the public will be entering the space. More specifically Section 2902.3 requires toilet facilities for tenant spaces intended for public utilization. Section 1107.3 requires rooms and spaces available to the public be accessible. Accessible spaces that may be utilized by the public in accordance with Section 1107.3 include other spaces such as kitchens, living and dining areas and any exterior spaces including patios, terraces and balconies that may be on the same level as the entrance to the unit. These may not be "available to the public" unless they are part of the function of the nonresidential activity. Such spaces that are strictly part of the residential function of the live/work unit are only required to be accessible to the degree that such units are required to be made accessible.

419.8 Ventilation. The applicable requirements of the *International Mechanical Code* shall apply to each area within the live/work unit for the function within that space.

❖ Similar to the egress and structural requirements the use of the space must be looked at individually to ensure ventilation being provided fits the use. For instance, IMC Table 403.3 would require a florist to have 15 cubic feet (0.42 m³) per minute of outdoor air per person. With a possible floor area of 1,500 square feet (139 m²) for the work area and a minimum of 8 persons per 1,000 square feet (92.9 m²) as required by Table 403.3 the cfm required would be as follows:

(8 occupants/1,000 square feet) x 1,500 square feet = 12 occupants

12 occupants x 15 cfm = 180 cfm of outside air ventilation

A two-bedroom dwelling unit would typically require 105 cfm (15 cfm per occupant x 7 occupants – 200 square feet per occupant in accordance with Table 1004.1.1). Although typical residential designed uses fresh air by the natural ventilation option. However, it is more typical that designs of residential dwelling units include the provision of fresh air by the natural ventilation option.

SECTION 420
GROUPS I-1, R-1, R-2, R-3

420.1 General. Occupancies in Groups I-1, R-1, R-2 and R-3 shall comply with the provisions of this section and other applicable provisions of this code.

❖ The nature of occupancies in Groups I-1, R-1, R-2 and R-3 is such that some level of protection against fire is needed for occupants in dwelling units and sleeping units. There remains a high frequency of fires where people live. These occupancies now need to also provide sprinkler protection. Requiring a minimum fire resistance to the construction separating both units and residential areas from nonresidential areas provides an extra level of protection in people's homes from unfortunate occurrences in their neighbor's homes. In accordance with Section 310.1, Group R-4 units are to be constructed to R-3 standards; therefore, these provisions also apply to sleeping units in a Group R-4. These separations are required between

live/work units as provided in Section 419, but separations within each live/work unit between residential and nonresidential uses are not required.

420.2 Separation walls. Walls separating *dwelling units* in the same building, walls separating *sleeping units* in the same building and walls separating *dwelling* or *sleeping units* from other occupancies contiguous to them in the same building shall be constructed as *fire partitions* in accordance with Section 709.

❖ The sleeping units or dwelling units that are in a single building are required to be separated by fire partitions complying with Section 709, or horizontal fire-resistance-rated assemblies complying with Section 712. Fire partitions are the least robust of all fire-resistance-rated walls called out in the code. These occupancies all require smoke alarms in the sleeping areas of the units, and occupancies in Groups I-1 and R-1 also require general fire alarms. In addition, these occupancies are all required to be sprinklered. The reason for the nominal fire-resistant separation is to account for the fact that individuals asleep in these units could respond more slowly to a fire; therefore, some amount of fire resistance is deemed necessary to protect these occupants.

Section 709.3 requires fire partitions to be not less than 1-hour fire-resistance rated. If the building's sprinkler protection is provided by a system complying with NFPA 13, the rating can be reduced to 30 minutes. Section 709.4 requires 1-hour-rated fire partitions to be supported by construction that has the same rating or better; however, this requirement is waived for these separation walls in buildings of Type IIB, IIIB and VB construction.

This requirement also applies to walls that separate these Group R and I occupancies from other occupancies in the building. Even if those other areas are considered accessory areas (see Section 508.2) to these occupancies or if the building is being developed under the nonseparated occupancies option of mixed occupancies contained in Section 508.3, these partitions and horizontal assemblies are still required.

420.3 Horizontal separation. Floor assemblies separating *dwelling units* in the same buildings, floor assemblies separating *sleeping units* in the same building and floor assemblies separating *dwelling* or *sleeping units* from other occupancies contiguous to them in the same building shall be constructed as *horizontal assemblies* in accordance with Section 712.

❖ See the commentary to Section 420.2. Section 712.3 requires floor assemblies providing this separation to be not less than 1-hour fire-resistance rated. If the building's sprinkler protection is provided by a system complying with NFPA 13, the rating can be reduced to 30 minutes. Section 712.4 requires 1-hour-rated fire partitions to be supported by construction that has the same rating or better; however, this requirement is waived for these floor assemblies in buildings of Type IIB, IIIB and VB construction.

SECTION 421
HYDROGEN CUTOFF ROOMS

[F] 421.1 General. When required by the *International Fire Code*, hydrogen cutoff rooms shall be designed and constructed in accordance with Sections 421.1 through 421.8

❖ This section is simply stating that all hydrogen cutoff rooms are to be constructed in accordance with the provisions contained in Section 421. Hydrogen cutoff rooms were created to address the increasing and emerging concepts of fuel cells that use hydrogen and actually generate hydrogen onsite to run the fuel cells. The IFGC requires in Section 2209.3.2.3 that generation, compression, storage and dispensing equipment related to hydrogen be located in one of three places inside buildings. The first is a hydrogen cutoff room in accordance with this section. The second is outside a cutoff room where the gaseous hydrogen system is listed and labeled for indoor installation and installed in accordance with the manufacturer's installation instructions. The third is in a dedicated hydrogen fuel dispensing area having an aggregate hydrogen delivery capacity no greater than 12 standard cubic feet (0.34 m³) per minute. The provisions of Section 421 address construction-related issues for hydrogen cutoff rooms such as location, fire-resistant separation, ventilation and safety features, such as gas detection and explosion control.

[F] 421.2 Definitions. The following words and terms shall, for the purposes of this chapter and as used elsewhere in this code, have the meanings shown herein.

❖ Definitions of terms that are associated with the content of this section are provided here. These definitions can help in the understanding and application of the code requirements. It is important to emphasize that these terms are not exclusively related to this section but are applicable everywhere the term is used in the code. The purpose for including these definitions within this section is to provide more convenient access to them without having to refer back to Chapter 2. For convenience, these terms are also listed in Chapter 2 with a cross reference to this section. The use and application of all defined terms, including those defined herein, are set forth in Section 201.

[F] GASEOUS HYDROGEN SYSTEM. An assembly of piping, devices and apparatus designed to generate, store, contain, distribute or transport a nontoxic, gaseous hydrogen-containing mixture having at least 95-percent hydrogen gas by volume and not more than 1-percent oxygen by volume. Gaseous hydrogen systems consist of items such as compressed gas containers, reactors and appurtenances, including pressure regulators, pressure relief devices, manifolds, pumps, compressors and interconnecting piping and tubing and controls.

❖ This term includes the source of hydrogen and all piping and devices between the source and the equipment being used. The gas in a hydrogen system is above the upper flammability limit (UFL) and is therefore "too rich" to burn. Any leakage, however, can quickly create conditions that will be explosive under ambient conditions.

[F] HYDROGEN CUTOFF ROOM. A room or space that is intended exclusively to house a gaseous hydrogen system.

❖ This term refers to an enclosed space used exclusively for a gaseous hydrogen system that requires construction and protection that are unique to the hazards associated with this use. The room itself may be considered as an incidental accessory occupancy or a Group H occupancy, depending upon the amount of hydrogen in such rooms. The definition itself should not be interpreted to prevent hydrogen piping systems from serving distributed hydrogen-using equipment and appliances located elsewhere on site or in the building. However the amount of hydrogen within such piping needs to be evaluated with respect to the maximum allowable quantities in Table 307.1(1).

[F] 421.3 Location. Hydrogen cutoff rooms shall not be located below grade.

❖ Restrictions against installation of cutoff rooms below grade are similar to those restricting the location of flammable and combustible liquids in basements. Explosion hazards are the primary concern, and placement of materials that have an ability to cause an explosion in below-grade spaces is not appropriate. Such spaces are more difficult to evacuate, create a fire explosion hazard to the structure above and are very difficult for the fire department to address.

[F] 421.4 Design and construction. Hydrogen cutoff rooms shall be classified with respect to occupancy in accordance with Section 302.1 and separated from other areas of the building by not less than 1-hour *fire barriers* constructed in accordance with Section 707 or *horizontal assemblies* constructed in accordance with Section 712, or both; or as required by Section 508.2, 508.3 or 508.4, as applicable.

❖ Hydrogen cutoff rooms are required to be separated by not less than a 1-hour fire-resistance-rated fire barriers or horizontal assemblies, which is consistent with the requirements in Table 508.2 for incidental accessory occupancies. In addition, the classification of the space will affect the separation requirements. Hydrogen cutoff rooms can contain any amount of hydrogen but if the maximum allowable quantities are exceeded, the cutoff room can no longer be an incidental accessory occupancy and instead must be classified as a Group H-2 occupancy. A Group H-2 could then be addressed as an accessory, nonseparated or separated occupancy. In all cases, separation in accordance with Table 508.4 would apply in addition to the applicable occupancy-specific requirements.

[F] 421.4.1 Opening protectives. Doors within the *fire barriers*, including doors to *corridors*, shall be self-closing in accordance with Section 715. Interior door openings shall be electronically interlocked to prevent operation of the hydrogen system when doors are opened or ajar or the room shall be provided with a mechanical exhaust ventilation system designed in accordance with Section 421.4.1.1.

❖ To allow for broader applications (such as emergency generators using fuel cell technology), interior wall openings are allowed as necessary for easy access to the systems. In order to allow these openings, Section 715 is referenced for the proper rating and closure of doors. This section requires that the doors be self-closing and interlocked with the hydrogen generating equipment. This is necessary because the rest of the building may not be properly ventilated to address the presence of hydrogen. If the doors are not interlocked with the hydrogen generation system, then continuous ventilation is required in accordance with Section 421.4.1.1.

[F] 421.4.1.1 Ventilation alternative. When an exhaust system is used in lieu of the interlock system required by Section 421.4.1, exhaust ventilation systems shall operate continuously and shall be designed to operate at a negative pressure in relation to the surrounding area. The average velocity of ventilation at the face of the door opening with the door in the fully open position shall not be less than 60 feet per minute (0.3048 m/s) with a minimum of 45 feet per minute (0.2287 m/s) at any point in the door opening.

❖ In applications where installation of interlocks is not practical, significant ventilation is required to ensure that a flammable mixture is not attained within the cutoff room and adjacent spaces. If exhaust ventilation is to be used as an alternative to an interlock, the exhaust system must be maintained in continuous operation so that a negative pressure with respect to spaces adjacent to the cutoff room is maintained.

[F] 421.4.2 Windows. Operable windows in interior walls shall not be permitted. Fixed windows shall be permitted when in accordance with Section 715.

❖ Operable windows are prohibited to further reduce the likelihood of allowing hydrogen from escaping into the room and entering other portions of the building that may not be properly ventilated. An operable window could inadvertently be left in the open or partially open position and go unnoticed. Fixed window openings must meet the requirements of Section 715 to ensure that the requirements for opening protectives are met. More specifically, the proper fire protection ratings for openings in fire barriers are required.

[F] 421.5 Ventilation. Cutoff rooms shall be provided with mechanical ventilation in accordance with the applicable provisions for repair garages in Chapter 5 of the *International Mechanical Code*.

❖ The purpose of this section is to prevent a dangerous accumulation of flammable gas in the room through the use of an exhaust ventilation system. The *Sourcebook for Hydrogen Applications* recommends ventilation at the rate of 1 cfm per square foot [0.00508 $m^3/(s \cdot m^2)$] of floor area, which is consistent with the requirements in Chapter 5 of the IMC.

[F] 421.6 Gas detection system. Hydrogen cutoff rooms shall be provided with an *approved* flammable gas detection system in accordance with Sections 421.6.1 through 421.6.3.

❖ Some gases contain additives that produce pungent odors for easy recognition. Systems using non-odorized gases, such as hydrogen and liquid natural gas, must utilize gas detection systems to detect leaks. This section specifically requires such detection due to the hazards associated with a build up of hydrogen at hazardous levels within a building.

[F] 421.6.1 System design. The flammable gas detection system shall be *listed* for use with hydrogen and any other flammable gases used in the room. The gas detection system shall be designed to activate when the level of flammable gas exceeds 25 percent of the lower flammability limit (LFL) for the gas or mixtures present at their anticipated temperature and pressure.

❖ The detection system must initiate the operations specified in Section 421.6.2 at any time that the flammable gas concentration exceeds one-fourth of the concentration necessary to support combustion. Early detection of the presence of a flammable gas will allow adequate mitigation procedures to be taken. Hydrogen fires are not normally extinguished until the supply of hydrogen has been shut off because of the danger of reignition or explosion. A gas detection system in the room or space housing a gaseous hydrogen system results in early notification of a leak that is occurring before the escaping gas reaches a hazardous concentration.

[F] 421.6.2 Operation. Activation of the gas detection system shall result in all of the following:

1. Initiation of distinct audible and visual alarm signals both inside and outside of the cutoff room.

2. Activation of the mechanical ventilation system.

❖ The detection system must activate the mechanical ventilation system that is required by Section 421.5, in addition to causing alarms to activate. Note that the mechanical ventilation alternative in Section 421.4.1.1 is continuously in operation.

The required local alarm is intended to alert the occupants to an emerging hazardous condition in the vicinity. The monitor control equipment must also initiate operation of the mechanical ventilation system in the event of a leak or rupture in the gaseous hydrogen system to prevent an accumulation of flammable gas.

[F] 421.6.3 Failure of the gas detection system. Failure of the gas detection system shall result in activation of the mechanical ventilation system, cessation of hydrogen generation and the sounding of a trouble signal in an *approved* location.

❖ Systems must be designed to be self-monitoring and fail-safe in that all safety systems are activated to alert any occupants that a problem exists and to prevent more hydrogen from being generated by any appliances in the room when hazardous conditions cannot be monitored.

[F] 421.7 Explosion control. Explosion control shall be provided in accordance with Chapter 9 of the *International Fire Code*.

❖ The requirements of this section are intended to address the circumstance resulting from a catastrophic failure of the cutoff room. These requirements are the final safeguard in case safety features such as interlocked doors, ventilation and gas detection systems should fail. An ignited hydrogen mixture produces large quantities of heat, causing a rapid expansion of the surrounding air. This can cause a pressure increase in a confined space and a catastrophic failure. Explosion control methods are identified in Section 911 of the IFC to prevent such a catastrophic failure. The explosion control requirements for hydrogen are consistent with the requirements in NFPA 50A and the *Sourcebook for Hydrogen Applications*.

[F] 421.8 Standby power. Mechanical ventilation and gas detection systems shall be connected to a standby power system in accordance with Chapter 27.

❖ The ventilation system and gas detection system are life safety systems and, therefore, must be dependable. Both safety systems must remain active in the event of a power failure of the primary power supply. Hydrogen is a colorless, odorless gas; a release might go undetected if detection systems are not functioning. The accumulation of hydrogen in an unventilated area can lead to mixtures in the flammable range if safety systems and mechanical ventilation systems are not in operation. Chapter 27 of the IFC addresses emergency and standby power requirements for emergency systems. It also allows an exception to the requirement for systems that are fail-safe (see IFC Section 2704.7, Exception 4). This exception may be used in cutoff rooms where hydrogen is generated, but not stored. Any storage of hydrogen within the cutoff room would not qualify for the exception because in the event of a power failure, there will be no way to detect or ventilate a release from a storage vessel.

SECTION 422
AMBULATORY HEALTH CARE FACILITIES

422.1 General. Occupancies classified as Group B ambulatory health care facilities shall comply with the provisions of Sections 422.1 through 422.6 and other applicable provisions of this code.

❖ Complex outpatient surgeries outside of the hospital are now commonplace. They are performed in facilities often called "day surgery centers" or "ambulatory surgical centers (ASCs)" because patients are able to walk in and walk out the same day. Procedures render patients temporarily incapable of self-preservation by application of nerve blocks, sedation or anesthesia. Patients in these facilities typically recover quickly. A definition of "Ambulatory health care facility" is provided in Section 202.

The code identifies health care Group I occupancies

as having 24-hour stay. Without a 24-hour stay, these surgery centers would have typically been classified as Group B. Strictly classifying such occupancies as a typical Group B occupancy would be inappropriate, as this would allow the rendering of an unlimited number of people incapable of self-preservation with no more protection than a business office. These types of facilities contain distinctly different hazards to life and safety than other business occupancies, such as:

1. Patients incapable of self-preservation require rescue by other occupants or fire personnel.

2. Medical staff must stabilize the patient prior to evacuation; therefore, staff may require evacuation as well.

3. Use of oxidizing medical gases, such as oxygen and nitrous oxide.

4. Prevalence of surgical fires.

In the past, there was a movement to classify ambulatory health care facilities as Group I-2 occupancies. This is a poor fit, because these are not hospitals. Federal and state jurisdictions have recognized that there is a middle ground somewhere between Groups B and I-2. These requirements provide a scaled approach to protection. The occupancy classification stays as Group B, but with some enhanced safety features focused on the concern with occupants being incapable of self-preservation on a temporary basis. The enhanced requirements are based on the concepts in the regulation of the Group I-2 occupancy requirements found in Section 407.

422.2 Smoke barriers. *Smoke barriers* shall be provided to subdivide every ambulatory care facility greater than 10,000 square feet (929 m²) into a minimum of two smoke compartments per *story*. The travel distance from any point in a smoke compartment to a *smoke barrier* door shall not exceed 200 feet (60 960 mm). The *smoke barrier* shall be installed in accordance with Section 710.

❖ In larger facilities, a smoke compartment is provided to allow a protect-in-place environment. These allow staff a safer environment to stabilize the patients before evacuation, and protection for fire personnel who may have to evacuate both patients and staff. The maximum size of a smoke compartment is limited to 10,000 square feet (929 m²), which is much lower than the 22,500 square feet (2090 m²) allowed for Group I-2 occupancies. The lower number is associated with the much smaller scale of such facilities. It is important to note that each floor must be divided into two smoke compartment if the floor area exceeds 10,000 square feet (929 m²) on each story. The travel distance of 200 feet (60 960 mm) to a door providing egress from the smoke compartment is the same requirement as for Group I-2 occupancies.

422.3 Refuge area. At least 30 net square feet (2.8 m²) per nonambulatory patient shall be provided within the aggregate area of *corridors*, patient rooms, treatment rooms, lounge or dining areas and other low-hazard areas on each side of each *smoke barrier*.

❖ This requirement is very similar to that found in Section 407.4.1 for Group I-2 occupancies. The purpose is to provide adequate space for patients on beds who might be relocated to an adjacent smoke compartment. Since the facility may still be in operation during the fire, space must be provided in the corridors, patient rooms, treatment rooms, lounge or dining areas and other low-hazard areas for both the patients being relocated and accommodate the patients being treated in that smoke compartment. Section 407.4.1 acknowledges that there may be floors that do not house patients who are confined to beds or litter.

422.4 Independent egress. A *means of egress* shall be provided from each smoke compartment created by smoke barriers without having to return through the smoke compartment from which *means of egress* originated.

❖ This section is identical to Section 407.4.2 and requires that occupants should be able to exit the building without having to re-enter a smoke compartment from where they started. Although the strategy is to protect in place, occupants should not have to travel back through a smoke compartment to ultimately exit the building. This prevents the creation of a dead-end smoke compartment [see Figures 407.4, 407.4.2(1) and 407.4.2(2)].

422.5 Automatic sprinkler systems. *Automatic sprinkler systems* shall be provided for ambulatory care facilities in accordance with Section 903.2.2.

❖ Section 422.5 is part of the package of enhanced Group B requirements to accommodate the risk to occupants that is higher than the typical business occupancy but does not warrant classification as a Group I-2 occupancy. The actual sprinkler requirements are located in Section 903.2.2. The specific requirements apply when there are four or more care recipients incapable of self-preservation or if one or more recipients are located on a story other than the level of exit discharge. The entire fire area meeting either one of the criteria must be sprinklered.

422.6 Fire alarm systems. A fire alarm system shall be provided in accordance with Section 907.2.2.1.

❖ As with the sprinkler requirements, the alarm requirements are actually located in Chapter 9. More specifically, Section 907.2.2.1 requires a manual fire alarm system in any fire area containing a Group B ambulatory health care facility. Furthermore, an automatic smoke detection system is required within ambulatory health care facility and public use spaces, such as lobbies and corridors, except where all areas of the building are protected by an automatic sprinkler system. In addition, the sprinkler system is required to activate the occupant notification appliances upon sprinkler water flow (see commentary, Section 907.2.2.1). The scope of the smoke detection was purposely narrowed to limit the amount of retrofit measures that a building

owner would be required to undertake if a new ambulatory health care tenant was established in an existing building.

SECTION 423
STORM SHELTERS

423.1 General. In addition to other applicable requirements in this code, storm shelters shall be constructed in accordance with ICC-500.

❖ Standard ICC-500 provides requirements for the design and construction of shelters to protect people from the violent winds of hurricanes and tornadoes. The standard includes special requirements for structural design, including wind loads that are considerably higher than the wind loads required by Chapter 16 for all structures.

Wind loads for storm shelters will be based upon wind speed contour maps developed specially for this standard. The wind load design requirements are relatively severe when compared to the wind speed maps in Chapter 16. Contour maps for wind speeds in hurricane prone regions were determined based upon a 10,000-year mean return period. The map shows 200 mph wind speeds on the coast of Florida and the Carolinas, and wind speeds higher than 200 mph in some locations. These are wind speeds associated with a Category 5 hurricane. Shelter design wind speeds in the central part of the United States (a region called "tornado alley") are as high as 250 mph.

Such high wind speeds, of course, produce flying debris, turning construction materials into deadly missiles. The standard contains specific test methods and pass-fail criteria for window and doors protection from flying debris.

ICC-500 addresses nonstructural issues, as well. Storm shelters for hurricanes will be required to house people for 24 hours. Tornado shelters will be required to house people for 2 hours. The standard addresses minimum requirements for ventilation air, sanitation facilities, potable water supply, lighting and other minimal power needs.

It should be noted that the entrances and exits to storm shelters will be required to be accessible. In addition, the occupant load requirements are such that some wheelchair space will be required.

423.1.1 Scope. This section applies to the construction of storm shelters constructed as separate detached buildings or constructed as safe rooms within buildings for the purpose of providing safe refuge from storms that produce high winds, such as tornados and hurricanes. Such structures shall be designated to be hurricane shelters, tornado shelters, or combined hurricane and tornado shelters.

❖ Most storm shelters are safe rooms within a bigger facility. A common type of shelter doubles as a gymnasium or classroom in a school. The purpose of the storm shelter is to provide refuge for people during a storm. The standard does not address the use of the shelter as a post-storm recovery facility, although it may well be used for that purpose.

423.2 Definitions. The following words and terms shall, for the purposes of this chapter and as used elsewhere in this code, have the meanings shown herein.

❖ Definitions of terms that are associated with the content of this section are contained herein. These definitions can help in the understanding and application of the code requirements. It is important to emphasize that these terms are not exclusively related to this section but are applicable everywhere the term is used in the code. The purpose for including these definitions within this section is to provide more convenient access to them without having to refer back to Chapter 2. For convenience, these terms are also listed in Chapter 2 with a cross reference to this section. The use and application of all defined terms, including those defined herein, are set forth in Section 201.

STORM SHELTER. A building, structure or portions(s) thereof, constructed in accordance with ICC 500 and designated for use during a severe wind storm event, such as a hurricane or tornado.

> **Community storm shelter.** A storm shelter not defined as a "Residential Storm Shelter."

> **Residential storm shelter.** A storm shelter serving occupants of *dwelling units* and having an *occupant load* not exceeding 16 persons.

❖ The significance of the definitions given in this section is to distinguish between residential shelters and community shelters. Residential shelters are somewhat more basic than community shelters. A residential shelter allows for less floor area per person than that required for a community shelter because the occupants are presumed to be more familiar with each other.

Bibliography

The following resource materials are referenced in this chapter or are relevant to the subject matter addressed in this chapter.

ANSI A13.1-96 Reaffirmed 2002, *Scheme for the Identification of Piping Systems*. New York: American National Standards Institute, 1996.

ASME B31.3-04, *Process Piping*. New York: American National Standards Institute, 2004.

ASTM C 1628-06, *Specification for Joints for Concrete Gravity Flow Sewer Pipe, Using Rubber Gaskets*. West Conshohocken, PA: ASTM International, 2006.

ASTM C 1629/C 1629M-06, *Standard Classification for Abuse-resistant Nondecorated Interior Gypsum Panel Products and Fiber-reinforced Cement Panels*. West Conshohocken, PA: ASTM International, 2006.

ASTM E 84-07, *Test Methods for Surface-burning Characteristics of Building Materials*. West Conshohocken, PA: ASTM International, 2007.

ASTM E 119-07, *Test Methods for Fire Tests of Building Construction and Materials*. West Conshohocken, PA: ASTM International, 2007.

ASTM E 695-03, *Method for Measuring Relative Resistance of Wall, Floor, and Roof Construction to Impact Loading*. West Conshohocken, PA: ASTM International, 2003.

ASTM E 736-00 (2006), *Test Method for Cohesion/Adhesion of Sprayed Fire-resistive Materials Applied to Structural Members*. Conshohocken, PA: ASTM International, 2006.

Bain, A., J. Barclay, T. Bose, F. Edeskuty, M. Farlie, J. Hansel, R. Hay and M. Swain (et al). *Sourcebook for Hydrogen Application*. Golden, CO: Hydrogen Research Institute and National Renewable Energy Laboratory, 1998.

Boring, Delbert F. and others. *Fire Protection Through Modern Building Codes*, 5th ed. Washington, DC: American Iron and Steel Institute, 1981.

Final Report of the National Construction Safety Team on the Collapses of the World Trade Center Towers. Washington, DC: National Institute of Standards and Technology, 2005.

Fire Protection Handbook, 20th ed. Quincy, MA: National Fire Protection Association, 2008.

Fothergill, John W. and John H. Klote. *Design of Smoke Control Systems for Buildings*. Atlanta: American Society for Heating Refrigerating and Air-Conditioning Engineers, Inc., 1983.

ICC/ANSI A117.1-03, *Accessible and Usable Buildings and Facilities*. Washington, DC: International Code Council, 2003.

ICC-500-08, *ICC/NSSA Standard on the Design and Construction of Storm Shelters*. Washington, DC: International Code Council, 2008.

IFC-09, *International Fire Code*. Washington, DC: International Code Council, 2009.

IFGC-09, *International Fuel Gas Code*, Washington, DC: International Code Council, 2009.

IMC-09, *International Mechanical Code*. Washington, DC: International Code Council, 2009.

IPC-09, *International Plumbing Code*. Washington, DC: International Code Council, 2009.

IRC-09, *International Residential Code*. Washington, DC: International Code Council, 2009.

NFPA 13-07, *Installation of Sprinkler Systems*. Quincy, MA: National Fire Protection Association, 2007.

NFPA 13-D-07, *Installation of Sprinkler Systems in One- and Two-family Dwellings and Manufactured Homes*. Quincy, MA: National Fire Protection Association, 2007.

NFPA 13R-07, *Installation of Sprinkler Systems in Residential Occupancies up to and Including Four Stories in Height.* Quincy, MA: National Fire Protection Association, 2007.

NFPA 30-08, *Flammable and Combustible Liquids Code.* Quincy, MA: National Fire Protection Association, 2008.

NFPA 30A-08, *Code for Motor Fuel-dispensing Facilities and Repair Garages.* Quincy, MA: National Fire Protection Association, 2008.

NFPA 30B-07, *Manufacture and Storage of Aerosol Products.* Quincy, MA: National Fire Protection Association, 2007.

NFPA 32-07, *Dry Cleaning Plants.* Quincy, MA: National Fire Protection Association, 2007.

NFPA 33-07, *Spray Application Using Flammable and Combustible Materials.* Quincy, MA: National Fire Protection Association, 2007.

NFPA 34-07, *Dipping and Coating Processes Using Flammable or Combustible Liquids.* Quincy, MA: National Fire Protection Association, 2007.

NFPA 35-05, *Manufacture of Organic Coatings.* Quincy, MA: National Fire Protection Association, 2005.

NFPA 40-07, *Storage and Handling of Cellulose Nitrate Film.* Quincy, MA: National Fire Protection Association, 2007.

NFPA 50A-99, *Gaseous Hydrogen Systems at Consumer Sites,* Quincy, MA: National Fire Protection Association, 1999.

NFPA 58-08, *Liquefied Petroleum Gas Code.* Quincy, MA: National Fire Protection Association, 2008.

NFPA 61-08, *Prevention of Fires and Dust Explosions in Agricultural Food Products Facilities.* Quincy, MA: National Fire Protection Association, 2008.

NFPA 68-07, *Explosion Protection by Deflagration Venting.* Quincy, MA: National Fire Protection Association, 2007.

NFPA 69-08, *Explosion Prevention Systems.* Quincy, MA: National Fire Protection Association, 2008.

NFPA 70-08, *National Electrical Code.* Quincy, MA: National Fire Protection Association, 2008.

NFPA 72-07, *National Fire Alarm Code.* Quincy, MA: National Fire Protection Association, 2007.

NFPA 80-07, *Fire Doors and Other Opening Protectives.* Quincy, MA: National Fire Protection Association, 2007.

NFPA 99-05, *Standard for Health Care Facilities.* Quincy, MA: National Fire Protection Association, 2005.

NFPA 110-05, *Emergency and Standby Power Systems.* Quincy, MA: National Fire Protection Association, 2005.

NFPA 111-05, *Stored Electrical Energy Emergency and Standby Power Systems.* Quincy, MA: National Fire Protection Association, 2005.

NFPA 120-04, *Coal Preparation Plants.* Quincy, MA: National Fire Protection Association, 2004.

NFPA 251-06, *Standard Methods of Tests of Building Endurance of Building Construction and Materials.* Quincy, MA: National Fire Protection Association, 2006.

NFPA 318-09, *Standard for the Protection of Semiconductor Fabrication Facilities.* Quincy, MA: National Fire Protection Association, 2009.

NFPA 409-04, *Aircraft Hangars.* Quincy, MA: National Fire Protection Association, 2004.

NFPA 418-06, *Standard for Heliports.* Quincy, MA: National Fire Protection Association, 2006.

NFPA 484-06, *Combustible Metals.* Quincy, MA: National Fire Protection Association, 2006.

NFPA 495-06, *Explosive Materials Code.* Quincy, MA: National Fire Protection Association, 2006.

NFPA 654-06, *Prevention of Fire and Dust Explosions from the Manufacturing, Processing and Handling of Combustible Particulate Solids.* Quincy, MA: National Fire Protection Association, 2006.

NFPA 655-07, *Prevention of Sulfur Fires and Explosions.* Quincy, MA: National Fire Protection Association, 2007.

NFPA 664-07, *Prevention of Fires and Explosions in Wood Processing and Woodworking Facilities.* Quincy, MA: National Fire Protection Association, 2007.

NFPA 701-04, *Standard Methods of Fire Tests for Flame-propagation of Textiles and Films.* Quincy, MA: National Fire Protection Association, 2007.

NFPA 704-07, *Flammable and Combustible Liquids Code.* Quincy, MA: National Fire Protection Association, 2007.

NFPA 1124-04, *Manufacture, Transportation and Storage of Fireworks and Pyrotechnic Articles.* Quincy, MA: National Fire Protection Association, 2004.

NFPA, *Fire Protection Guide on Hazardous Materials.* Quincy, MA: National Fire Protection Association, 2002.

NIST IR 5240-93, *Measurement of Room Conditions and Response of Sprinklers and Smoke Detectors During a Simulated Two-bed Hospital Patient Room Fire.* Gaithersburg, MD: National Institute of Standards and Technology, July 1993.

UL 217-06, *Single- and Multiple-station Smoke Detectors with Revisions through August 2005*. Northbrook, IL: Underwriters Laboratories Inc., 2006.

UL 268-06, *Smoke Detectors for Fire Protection Signaling*. Northbrook, IL: Underwriters Laboratories Inc., 2006.

UL 325-02, *Door, Drapery, Gate, Louver and Window Operations and Systems—with Revisions through February 2006*. Northbrook, IL: Underwriters Laboratories Inc., 2002.

UL 1994-04, *Standard for Luminous Egress Path Marking Systems, with Revisions through February of 2005*. Northbrook, IL: Underwriters Laboratories Inc., 2005.

UL 2075-07, *Standard for Gas and Vapor Detectors and Sensors*. Northbrook, IL: Underwriters Laboratories Inc., 2007.

Chapter 5:
General Building Heights and Areas

General Comments

Chapter 5 contains the provisions that regulate the minimum type of construction, area modifications, height modifications, mezzanines, unlimited area structures and special provisions for buildings, including parking garages. The occupancy or mix of occupancies in a building is of key importance to the application of these provisions.

Section 501.1 gives the scope of Chapter 5. It specifies that the provisions of Chapter 5 apply not only to new construction, but also to existing building additions.

Section 502 defines terms that are related to the use and application of Chapter 5 and other provisions of the code related to height and area.

Section 503 addresses the allowable height and area limitations of buildings. Table 503 establishes the limits for the height and area of a building as a function of the type of construction. It is a risk/safety table where the risk is based on the occupancy and the safety is related to the type of construction.

Section 504 describes the modifications to height limits in Table 503 that are allowed for buildings that have a sprinkler system installed in accordance with specified standards.

Section 505 provides the criteria and regulations for mezzanines and industrial equipment platforms.

Section 506 describes the allowable modifications to area limits given in Table 503 for buildings with either sprinkler systems or a larger separation distance from adjacent properties or public ways.

Section 507 includes requirements for buildings with no area limit. These provisions are commonly used for very large industrial warehouses or factories.

Section 508 provides requirements for buildings with mixed uses and occupancies.

Section 509 contains special options for buildings containing parking garages and other occupancies.

Chapter 5 is one of the most important chapters in the code because many other code requirements depend on the establishment of the minimum required type of construction for a building. A building has one minimum required construction type and must conform to those requirements, with certain exceptions for buildings incorporating parking garages (see Section 509). Without the correct determination of the minimum type of construction, misapplication of the code is probable.

Table 503 is the keystone in setting thresholds for building size based on the building's use and the materials with which it is constructed. It is aimed at providing an acceptable level of hazard for building occupants by limiting fire load and fire hazards relating to height, area and occupancy. In accordance with Table 503, buildings of higher construction types [noncombustible materials and a higher degree of fire-resistance rating of elements (see Chapter 6)] can be larger and higher than buildings of lower construction types (combustible and unprotected with fire-resistance ratings). The area and height thresholds set by Table 503 incorporate the limits that had been established by the three model codes that were "merged" when producing the code.

The limits of Table 503 are modified by the text of the chapter, particularly the height and area modifications of Sections 504 and 506. Sections 507 and 509 give additional conditions under which the height and area limitations of Table 503 are modified, such as in the case of an unlimited area building.

Purpose

The main purpose of Chapter 5 is to regulate the size of structures based on the specific hazards associated with their occupancy and the materials of which they are constructed. Chapter 5 also provides for adjustments to the allowable area and height, based on the presence of fire protection systems for building occupants and fire service personnel in the event of a fire.

The code review for the design of any building begins with Chapter 5, with the determination of the minimum required type of construction based on the use and size of the building. Misapplication of Chapter 5 can result in a multitude of related errors in subsequent code application, since many code requirements depend on the type of construction that is required for the building.

SECTION 501
GENERAL

501.1 Scope. The provisions of this chapter control the height and area of structures hereafter erected and additions to existing structures.

❖ Chapter 5 is applicable to all new and existing structures that are to be enlarged. Allowable height and area are evaluated on the basis of occupancy classification, type of construction, location on the property relative to lot lines and other structures and the presence of an automatic sprinkler system.

In the case of additions to existing buildings that are not separated by fire walls, the designer and plan reviewer must evaluate the entire building, including the addition, as if it were a new structure for purposes of determining allowable area and height. For instance, if an existing Type IIB building is to have an addition, the aggregate area of the new building (existing plus the addition) must be within the limits established by Table 503 for the allowable area and height of a Type IIB building, taking into account area and height modifications for open frontage and sprinklers.

If the aggregate area of the existing building and the addition exceeds the allowable area for Type IIB construction in Table 503, the addition is not permitted unless something is done to solve the allowable area problem. The solution could either be adding sprinklers to the building to get an allowable area increase in accordance with Section 506.3, providing a fire wall between the addition and the existing building or creating two buildings in accordance with Section 503.1. If a fire wall is used, the designer must check the allowable area of the existing building again because the perimeter conditions have changed. Specifically, if the allowable area of the existing building results from an increase due to open frontage (see Section 506.2), then the following question must be answered: With the reduced open perimeter (there is no longer an open perimeter where the fire wall and adjacent building will be located), does the existing building exceed the area limitations of Table 503 based on its use and construction type? If so, another solution must be found.

[F] 501.2 Address identification. New and existing buildings shall be provided with *approved* address numbers or letters. Each character shall be a minimum 4 inches (102 mm) high and a minimum of 0.5 inch (12.7 mm) wide. They shall be installed on a contrasting background and be plainly visible from the street or road fronting the property. Where access is by means of a private road and the building address cannot be viewed from the *public way*, a monument, pole or other *approved* sign or means shall be used to identify the structure.

❖ Identifying buildings, both new and existing, during an emergency (i.e., fire department, ambulances, medical, police) is greatly aided by the proper placement of address identification. In other than emergencies, the address identification serves as a convenience for people attempting to locate a building. The size and color criteria are intended to aid visibility from the street. When several structures are remotely located on a site or set back into a property, or at locations where multiple addresses are provided (e.g., strip malls), where the address is not readily visible from the public way, an approved method of identification will also be required, which will have characters posted in a location that will help in an emergency. The primary concern is for emergency personnel to locate the building without going through a lengthy search procedure. In the case of a strip mall, identification would be provided for the backs of buildings that face alleys or roads since the emergency response unit may often be directed to the back entrance.

SECTION 502
DEFINITIONS

502.1 Definitions. The following words and terms shall, for the purposes of this chapter and as used elsewhere in this code, have the meanings shown herein.

❖ Definitions of terms can help in the understanding and application of the code requirements. The purpose for including these definitions within this chapter is to provide more convenient access to them without having to refer back to Chapter 2. For convenience, these terms are also listed in Chapter 2 with a cross reference to this section.

AREA, BUILDING. The area included within surrounding *exterior walls* (or *exterior walls* and *fire walls*) exclusive of vent shafts and courts. Areas of the building not provided with surrounding walls shall be included in the *building area* if such areas are included within the horizontal projection of the roof or floor above.

❖ Allowable building areas (as established by the provisions of Chapter 5 and Table 503) are a function of the potential fire hazard and the level of fire endurance of the building's structural elements, as defined by the types of construction in Chapter 6. A building area is the "footprint" of the building; that is, the area measured within the perimeter formed by the inside surface of the exterior walls. This excludes spaces that are inside this perimeter and open to the outside atmosphere at the top, such as open shafts and courts (see Section 1206). When a portion of the building has no exterior walls, the area regulated by Chapter 5 is defined by the projection of the roof or floor above [see Figure 502.1(1)]. The roof overhang on portions of a building where there are exterior enclosure walls does not add to the building area because the area is defined by exterior walls.

BASEMENT. A *story* that is not a *story above grade plane* (see "*Story above grade plane*" in Section 202).

The definition of "Basement" does not apply to the provisions of Section 1612 for flood loads (see "Basement" in Section 1612.2).

❖ Unlike previous editions of the *International Codes®* (I-Codes®) where a story would be defined as a basement if any portion of the story was below grade, a basement is now defined as a story that has its floor surface below the adjoining ground level and that does not qualify as a story above grade plane (see the commentary to Chapter 2 for the definition of "Story above grade plane"). Figure 502.1(2) illustrates the application of the definition of "Story above grade plane." Since a basement is not a story above grade, it does not contribute to the height of the building for the purpose of applying the allowable building height in stories from Table 503. This definition of "Basement" applies to all sections of the code except for flood loads.

Basements in buildings that are located in flood hazard areas and subject to flood loads are defined differently than the general definition in this section. See Section 1612.2 for the definition of "Basement" with regard to flood loads.

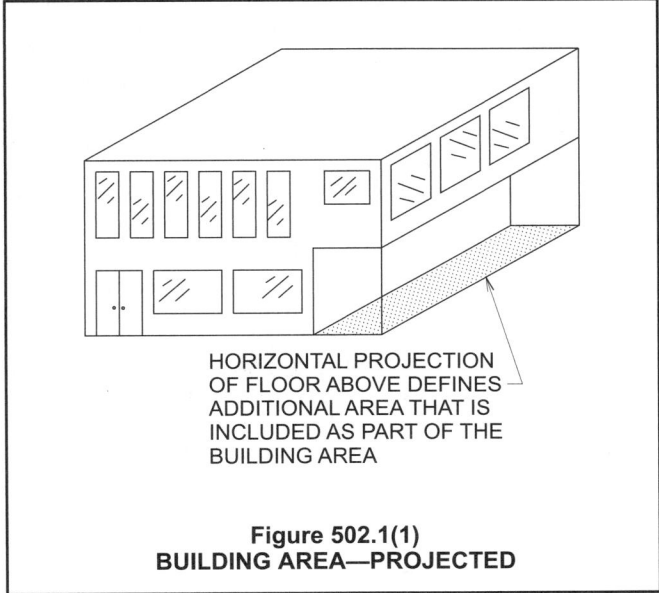

**Figure 502.1(1)
BUILDING AREA—PROJECTED**

EQUIPMENT PLATFORM. An unoccupied, elevated platform used exclusively for mechanical systems or industrial process equipment, including the associated elevated walkways, *stairs*, *alternating tread devices* and ladders necessary to access the platform (see Section 505.5).

❖ A distinction is made between equipment platforms and mezzanines by way of this definition. Equipment platforms, covered in Section 505.5, are unoccupied and used exclusively for housing equipment and providing access thereto, and are not subject to the requirements for mezzanines. Their purpose could also be to allow access for maintenance, repair or modification of elevated or very large equipment. Equipment platforms allow efficient use of high bay areas by locating infrequently accessed equipment or processes overhead without the occupant load or increasing the hazard to occupants in the room. Elevated floor areas that do not meet this definition would be subject to the requirements for mezzanines.

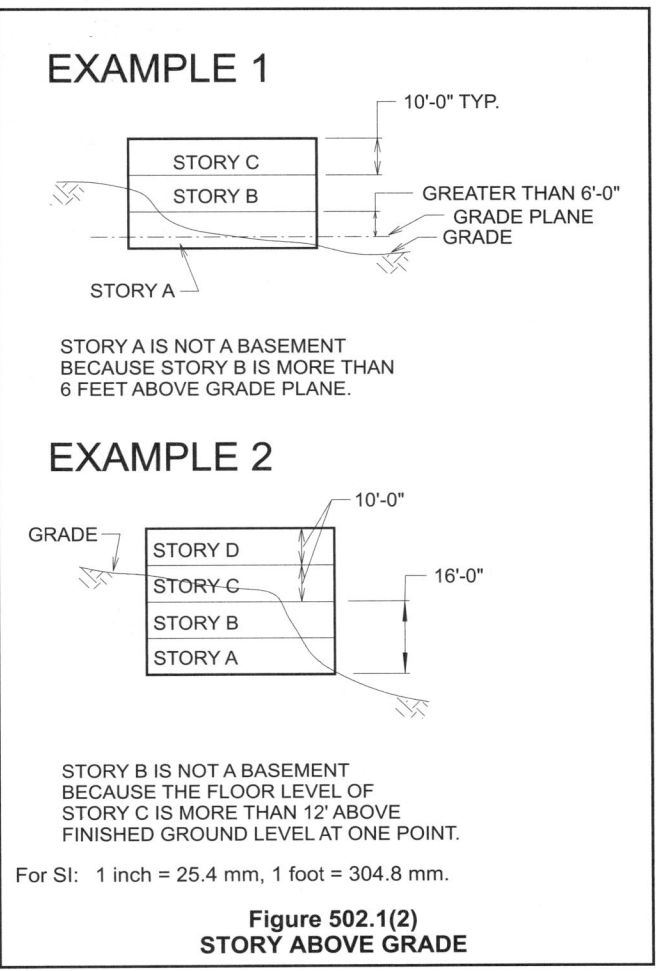

**Figure 502.1(2)
STORY ABOVE GRADE**

GRADE PLANE. A reference plane representing the average of finished ground level adjoining the building at *exterior walls*. Where the finished ground level slopes away from the *exterior walls*, the reference plane shall be established by the lowest points within the area between the building and the *lot line* or, where the *lot line* is more than 6 feet (1829 mm) from the building, between the building and a point 6 feet (1829 mm) from the building.

❖ This term is used in the definitions of "Basement" and "Story above grade plane." It is critical in determining the height of a building and the number of stories, which are regulated by this chapter. Since the finished ground surface adjacent to the building may vary (depending on site conditions), the mean average taken at various points around the building constitutes the grade plane. One method of determining the grade plane elevation is illustrated in Figure 502.1(3), where the ground slopes uniformly along the length of each exterior wall.

Where a site has a more complex slope, a more detailed calculation that takes into account the various segments of the perimeter walls must be taken. Figure

502.1(4) shows an example of a complex finished grade. A full calculation will show the grade plane to be at an elevation of 498.64 feet (151 986 mm). If a calculation is done based on just the four extreme corners, grade plane would be thought to be 495.5 feet (151 029 mm), an error of over 3 feet (914 mm).

Situations may arise where the ground adjacent to the building slopes away from the building because of site or landscaping considerations. In this case, the lowest finished ground level at any point between the building's exterior wall and a point 6 feet (1829 mm) from the building [or the lot line, if closer than 6 feet (1829 mm)] comes under consideration. These points are used to determine the elevation of the grade plane as illustrated in Figures 502.1(5) and 502.1(6).

In the context of the code, the term "grade" means the finished ground level at the exterior walls. While the grade plane is a hypothetical horizontal plane derived as indicated above, the grade is that which actually exists or is intended to exist at the completion of site work. The only situation where the grade plane and the grade are identical is when the site is perfectly level for a distance of 6 feet (1829 mm) from all exterior walls.

HEIGHT, BUILDING. The vertical distance from *grade plane* to the average height of the highest roof surface.

❖ This definition establishes the two points of measurement that determine the height of a building in feet. This measurement is used to determine compliance with the building height limitations of Section 503.1 and Table 503, as well as other sections of the code where the height of the building is a factor in the requirements (for example, see Section 1406.2.2).

The lower point of measurement is the grade plane (see the definition of "Grade plane"). The upper point of measurement is the roof surface of the building, with consideration given to sloped roofs (such as a hip or gable roof). In the case of sloped roofs, the average height would be used as the upper point of measurement, rather than the eave line or the ridge line. The average height of the roof is the mid-height between the roof eave and the roof ridge, regardless of the shape of the roof.

This definition also indicates that building height is measured to the highest roof surface. In the case of a building with multiple roof levels, the highest of the various roof levels must be used to determine the building height. If the highest of the various roof levels is a sloped roof, then the average height of that sloped roof must be used. The average height of multiple roof levels is not to be used to determine the building height.

A penthouse is not intended to affect the measurement of building height. By definition, a "Penthouse" is a structure that is built above the roof of a building (see Section 1502.1) and can be located above the maximum allowed roof height provided it complies with the limitations of Section 1509.

The distance that a building extends above ground also determines the relative hazards of that building. Simply stated, a taller building presents relatively greater safety hazards than a shorter building for several reasons, including fire service access and time for occupant egress. The code specifically defines how building height is measured to enable various code requirements, such as type of construction and fire suppression, to be consistent with those relative hazards [see Figure 502.1(7) for the computation of building height in terms of feet and stories].

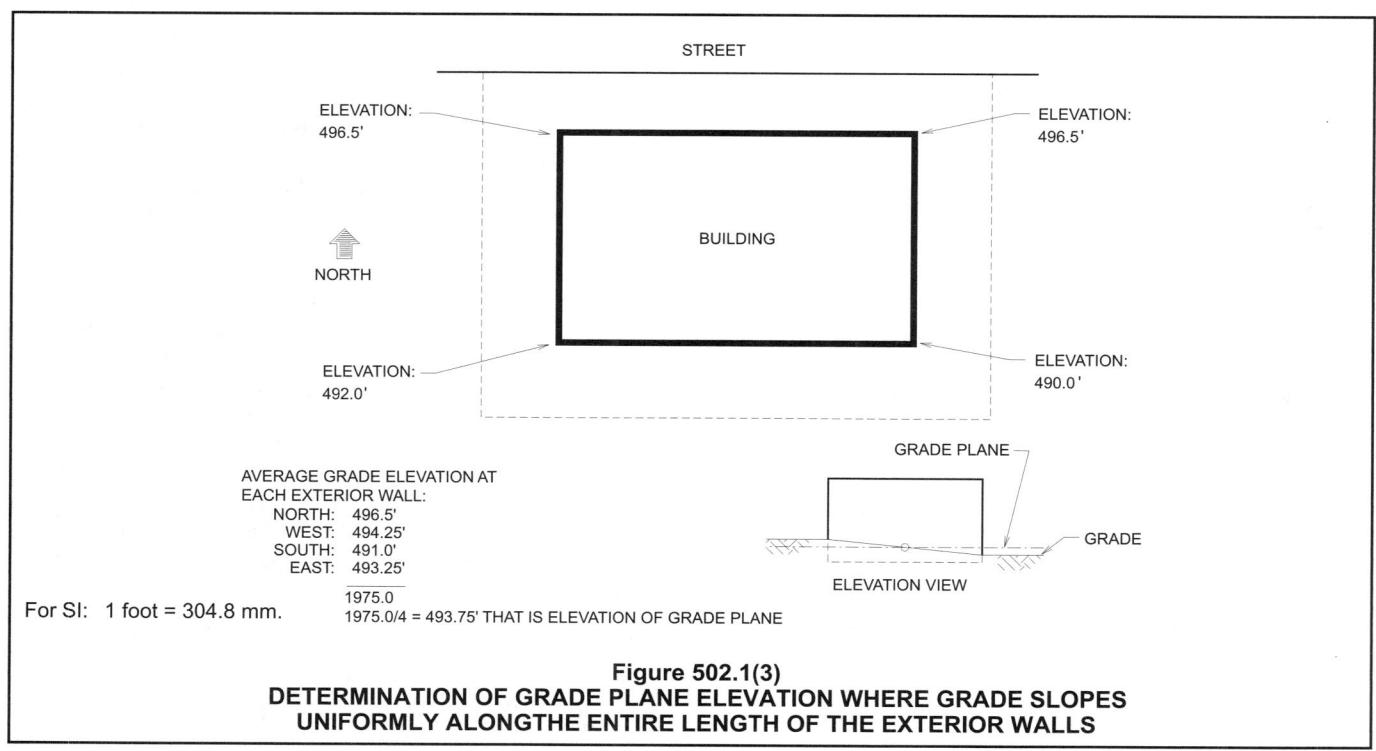

Figure 502.1(3)
DETERMINATION OF GRADE PLANE ELEVATION WHERE GRADE SLOPES UNIFORMLY ALONG THE ENTIRE LENGTH OF THE EXTERIOR WALLS

MEZZANINE. An intermediate level or levels between the floor and ceiling of any *story* and in accordance with Section 505.

❖ A common design feature in factories, warehouses and mercantile buildings is an intermediate loft, or platform, between the story levels of a building. This type of feature, or mezzanine, can be found in buildings of all occupancies. The code must deal with whether this intermediate level is another story of the building or whether it can simply be treated as part of the story in which it is contained. The basic rule is that the intermediate level must be less than one-third of the area of the story below (of the room in which it is located) in order to be considered a mezzanine. Requirements for mezzanines are found in Section 505.

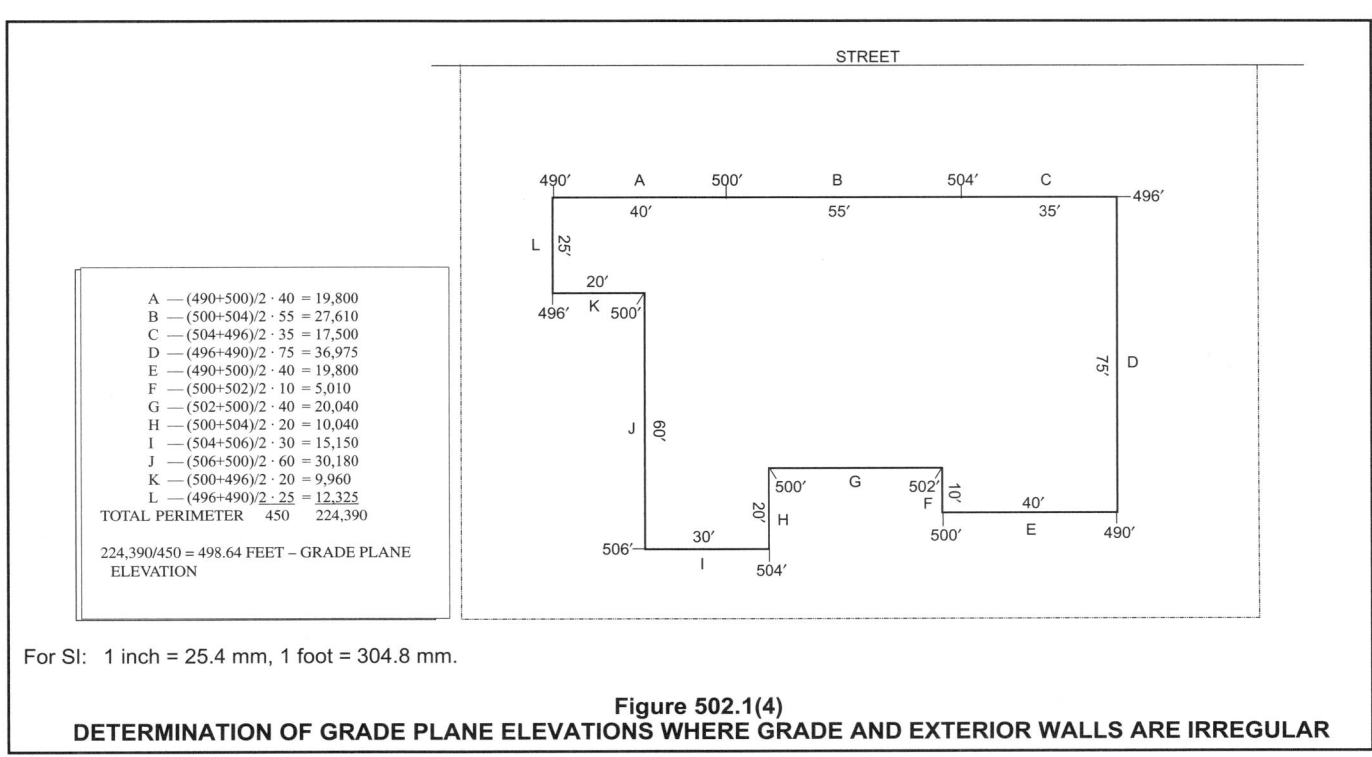

For SI: 1 inch = 25.4 mm, 1 foot = 304.8 mm.

Figure 502.1(4)
DETERMINATION OF GRADE PLANE ELEVATIONS WHERE GRADE AND EXTERIOR WALLS ARE IRREGULAR

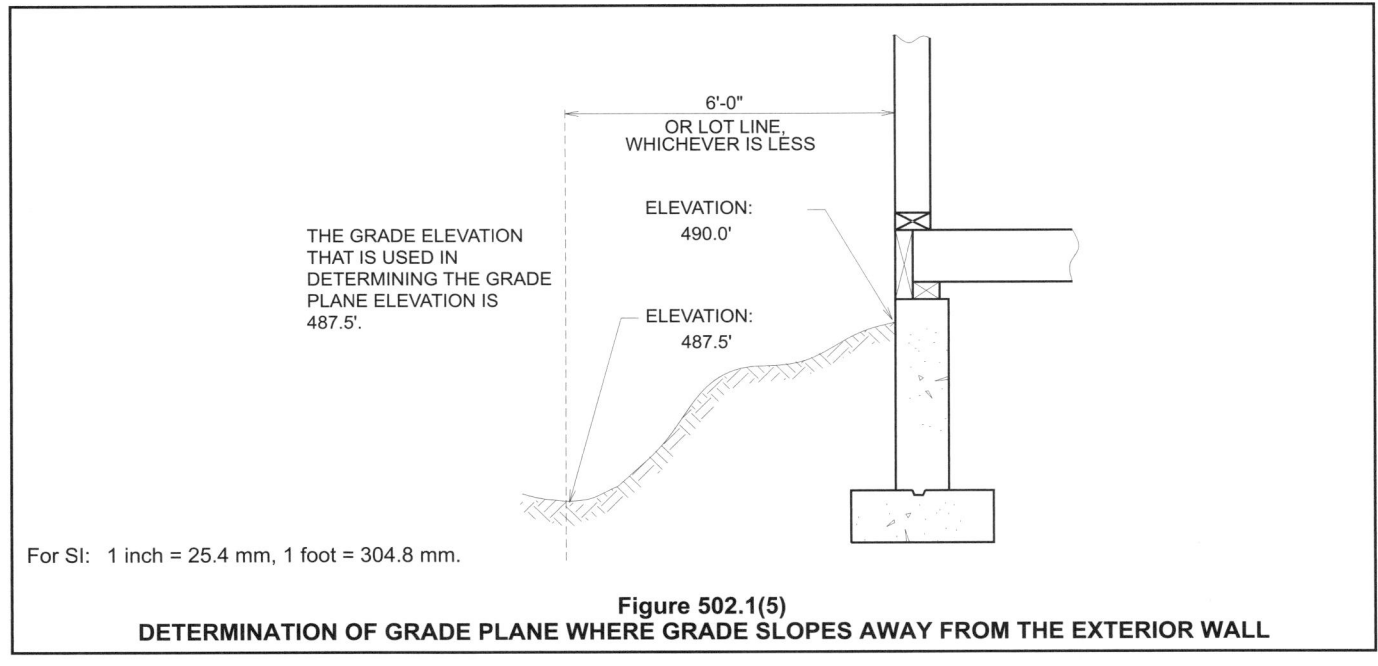

For SI: 1 inch = 25.4 mm, 1 foot = 304.8 mm.

Figure 502.1(5)
DETERMINATION OF GRADE PLANE WHERE GRADE SLOPES AWAY FROM THE EXTERIOR WALL

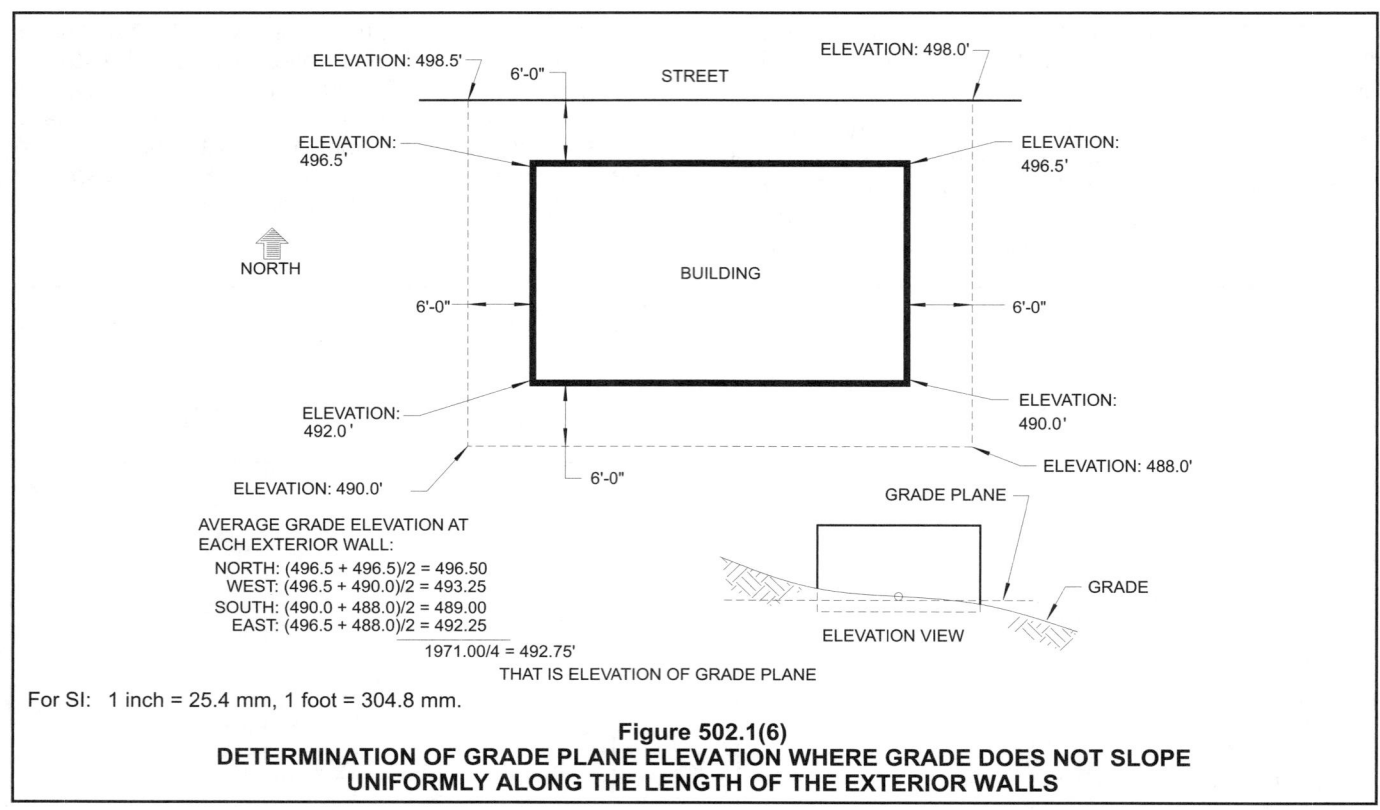

For SI: 1 inch = 25.4 mm, 1 foot = 304.8 mm.

Figure 502.1(6)
DETERMINATION OF GRADE PLANE ELEVATION WHERE GRADE DOES NOT SLOPE
UNIFORMLY ALONG THE LENGTH OF THE EXTERIOR WALLS

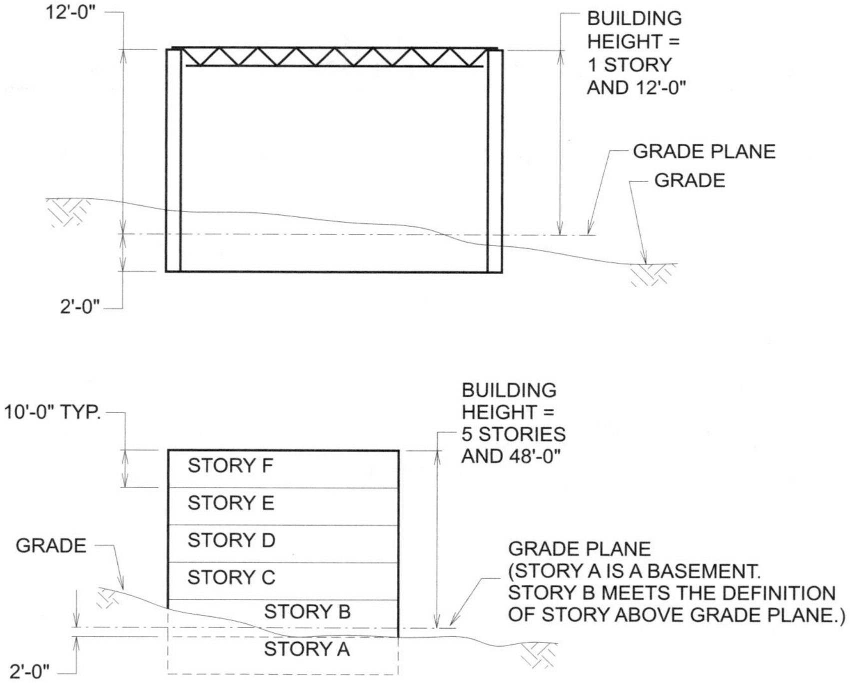

For SI: 1 inch = 25.4 mm, 1 foot = 304.8 mm.

Figure 502.1(7)
BUILDING HEIGHT

SECTION 503
GENERAL BUILDING HEIGHT AND
AREA LIMITATIONS

503.1 General. The *building height and area* shall not exceed the limits specified in Table 503 based on the type of construction as determined by Section 602 and the occupancies as determined by Section 302 except as modified hereafter. Each portion of a building separated by one or more *fire walls* complying with Section 706 shall be considered to be a separate building.

❖ The provisions for governing the height and area of buildings on the basis of occupancy classification and type of construction are established in this section. This section also establishes Table 503 as the primary tool for determining the minimum type of construction. All buildings are subject to these limitations unless more specific code requirements for a building type provide for different height or area limitations. For instance:

- Section 507 allows certain buildings to be unlimited in area due to lack of exposure, low hazard level, construction type, the presence of fire safety systems or a combination of these characteristics. Under these specific provisions, Table 503 does not apply unless referenced.
- Section 509 allows certain buildings with additional safeguards to adjust the heights and areas allowed by Table 503. Many of the provisions allow a mixture of construction types, and most involve a distinct portion of the structure being used for parking.
- Where more than one occupancy is present within one building, the provisions of Section 508 must be used in conjunction with Table 503 to determine the appropriate construction type for the occupancies involved.

Table 601 is used in conjunction with Table 503 to determine acceptable risk and fire safety levels for a building. Classification by occupancy, in accordance with the descriptions in Chapter 3, can be considered as establishing the level of "risk" associated with the use of a building. The various construction types, described in Chapter 6 and Table 601, can be thought of as various levels of safety in regard to fire resistance. Table 503 becomes a risk/safety matrix that sets a minimum level of safety (construction type) in accordance with the risk (the occupancy classification).

Fire walls are useful when a single building exceeds the allowable area limitation of Table 503. When fire walls (see Section 706) are used in a structure, multiple buildings are created. Each building created by the fire walls would be permitted to have its own occupancy classification and type of construction. Since multiple buildings would be created, each building would be evaluated separately in accordance with the height and area limitations of Table 503 (see commentary, Section 706).

The following is an example for the use of Table 503:

Example: Figure 503.1 shows a two-story office building of Type IIIB construction. Assuming the building is not sprinklered and does not qualify for any area or height increases, what is the maximum allowable area per story and height of the building? Given the size of the building, is Type IIIB construction adequate? What is the minimum required construction type?

Answer: From Table 503, the allowable area per story for a Group B occupancy of Type IIIB construction is 19,000 square feet (1765 m²). Since the actual area is 130 × 150 = 19,500 square feet (1812 m²), Type IIIB construction is not acceptable. The minimum required type of construction is Type I, IIA, IIB, IIIA or IV, depending on the desired materials for construction. All of these construction types for a Group B occupancy have allowable per-story areas greater than 19,500 square feet (1812 m²) in accordance with Table 503. The actual height in feet and stories is also within the table limits for these construction types.

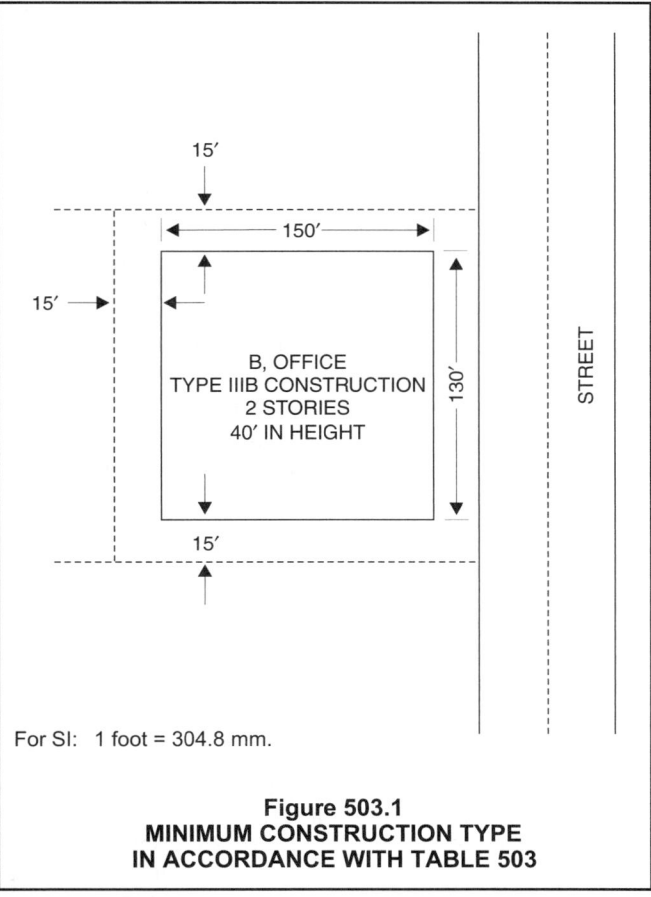

For SI: 1 foot = 304.8 mm.

**Figure 503.1
MINIMUM CONSTRUCTION TYPE
IN ACCORDANCE WITH TABLE 503**

503.1.1 Special industrial occupancies. Buildings and structures designed to house special industrial processes that require large areas and unusual *building heights* to accommodate craneways or special machinery and equipment, including, among others, rolling mills; structural metal fabrication shops and foundries; or the production and distribution of electric, gas or steam power, shall be exempt from the *building height and area* limitations of Table 503.

❖ This section provides an exemption from the limits of Table 503. The occupancies that may use this exemption are quite limited. The exemption is only applicable when large areas or unusual heights beyond that permitted by Table 503 are necessary to accommodate the specific manufacturing process.

It is the responsibility of the building official to determine when the application of this section is appropriate. The mere cost impact of the application of Table 503 does not dictate an exemption. The building official, in assessing the proposed construction, may wish to compare what protection features are being proposed to those features of other similar facilities.

TABLE 503. See page 5-9.

Table 503 is the foremost code table used in establishing equivalent risk (offsetting a building's inherent fire hazard—represented by occupancy—with materials and construction features). Sections 504 and 506 give height and area increases to the limits of Table 503 for buildings with certain features. Section 507 contains provisions by which buildings can in fact be unlimited in area; however, the height limitations of Table 503 would still apply.

Table 503 has three components:

1. The left column represents all of the occupancy classifications defined in Chapter 3.

2. The top row lists all of the types of construction subclassifications explained in Chapter 6. Each cell in this matrix contains the specific height (in stories above grade plane) and area (per story) limitations for the occupancy/type of construction combination in question.

3. Maximum building height in feet above grade plane is shown along a top row, just below the construction-type classifications. "Grade plane" is defined in Section 502.1.

In order to be considered in compliance with a type of construction, buildings must not exceed the maximum height based on number of stories shown in each cell of the table, nor exceed the maximum height in feet above grade plane shown along the upper row of the table.

Example 1: A proposed business building of Type IIIB construction is designed to be three stories above grade plane; however, the building height in feet above grade plane is 60 feet (18 288 mm). Since the height limitation for a Group B, Type IIIB building in Section 503 is 55 feet (16 764 mm), the proposed building could not be permitted as Type IIIB construction. If, however, the building is fully sprinklered in accordance with Section 504.2, an additional 20 feet (6096 mm) of height is allowed and Type IIIB would be acceptable.

Example 2: A mercantile building of Type IIB construction is proposed to be 50 feet (15 240 mm) and three stories above grade plane. It would not be permitted to be Type IIB construction unless fully sprinklered in accordance with Section 504.2, since it exceeds the number of stories permitted in Table 503 for a Group M building of Type IIB construction.

The area limitation appearing in each cell of the table represents the absolute maximum square footage per story for each occupancy/type of construction combination before increases are made in accordance with Section 506. The definition of "Area, building" in Section 502.1 indicates the application of the tabular areas set in Table 503.

Unprotected combustible construction (Type VB) is not permitted for buildings of Group H-1 or I-2. This type of construction—in particular, the fire hazard caused by the unprotected wood structure and the concealed spaces—does not provide the minimum protection required by the code to maintain equivalent risk in the presence of these occupancies.

For many groups in Type IA or IB construction, both the height above grade and the area per story are not limited by the code because of the required fire-resistance ratings of these construction types. In addition, the story height and building area of Group A-5 buildings (outdoor assembly for viewing) are unlimited in any construction type; however, the overall height of a Group A-5 building is limited to the number of feet above grade plane specified in the first row of Table 503. For example, an open stadium of Type IIIA construction can be of unlimited area and an unlimited number of stories, but the structure cannot exceed 65 feet (19 812 mm).

Simply put, the application of Table 503 is accomplished by identifying the occupancy classification of the building in question along the left column, as well as identifying the cell in that row that matches the building's height (in stories and feet) and area per story. The resulting type of construction (at the top of the resulting column) represents the minimum required type of construction for the building before the application of increases are made in accordance with Sections 504 and 506.

Example 3: Assuming no area or height increases for a two-story Group S-1 building 25 feet (7620 mm) above grade plane and 11,500 square feet (1115 m^2) per story, what is the minimum required construction type of the building in accordance with Table 503?

Answer: In accordance with the Group S-1 row of Table 503, the minimum required construction type is Type VA, which allows 14,000 square feet (1301 m^2) per story and limits the height to three stories and 50 feet (15 240 mm). The building cannot be Type VB construction, since it exceeds one story and 9,000 square feet (836 m^2) per story. Type IIB construction (noncombustible, unprotected) is also acceptable.

TABLE 503
ALLOWABLE BUILDING HEIGHTS AND AREAS[a]
Building height limitations shown in feet above grade plane. Story limitations shown as stories above grade plane.
Building area limitations shown in square feet, as determined by the definition of "Area, building," per story

GROUP		TYPE OF CONSTRUCTION								
		TYPE I		TYPE II		TYPE III		TYPE IV	TYPE V	
		A	B	A	B	A	B	HT	A	B
	HEIGHT(feet)	UL	160	65	55	65	55	65	50	40
		STORIES(S) AREA (A)								
A-1	S	UL	5	3	2	3	2	3	2	1
	A	UL	UL	15,500	8,500	14,000	8,500	15,000	11,500	5,500
A-2	S	UL	11	3	2	3	2	3	2	1
	A	UL	UL	15,500	9,500	14,000	9,500	15,000	11,500	6,000
A-3	S	UL	11	3	2	3	2	3	2	1
	A	UL	UL	15,500	9,500	14,000	9,500	15,000	11,500	6,000
A-4	S	UL	11	3	2	3	2	3	2	1
	A	UL	UL	15,500	9,500	14,000	9,500	15,000	11,500	6,000
A-5	S	UL	UL	UL	UL	UL	UL	UL	UL	UL
	A	UL	UL	UL	UL	UL	UL	UL	UL	UL
B	S	UL	11	5	3	5	3	5	3	2
	A	UL	UL	37,500	23,000	28,500	19,000	36,000	18,000	9,000
E	S	UL	5	3	2	3	2	3	1	1
	A	UL	UL	26,500	14,500	23,500	14,500	25,500	18,500	9,500
F-1	S	UL	11	4	2	3	2	4	2	1
	A	UL	UL	25,000	15,500	19,000	12,000	33,500	14,000	8,500
F-2	S	UL	11	5	3	4	3	5	3	2
	A	UL	UL	37,500	23,000	28,500	18,000	50,500	21,000	13,000
H-1	S	1	1	1	1	1	1	1	1	NP
	A	21,000	16,500	11,000	7,000	9,500	7,000	10,500	7,500	NP
H-2[d]	S	UL	3	2	1	2	1	2	1	1
	A	21,000	16,500	11,000	7,000	9,500	7,000	10,500	7,500	3,000
H-3[d]	S	UL	6	4	2	4	2	4	2	1
	A	UL	60,000	26,500	14,000	17,500	13,000	25,500	10,000	5,000
H-4	S	UL	7	5	3	5	3	5	3	2
	A	UL	UL	37,500	17,500	28,500	17,500	36,000	18,000	6,500
H-5	S	4	4	3	3	3	3	3	3	2
	A	UL	UL	37,500	23,000	28,500	19,000	36,000	18,000	9,000
I-1	S	UL	9	4	3	4	3	4	3	2
	A	UL	55,000	19,000	10,000	16,500	10,000	18,000	10,500	4,500
I-2	S	UL	4	2	1	1	NP	1	1	NP
	A	UL	UL	15,000	11,000	12,000	NP	12,000	9,500	NP
I-3	S	UL	4	2	1	2	1	2	2	1
	A	UL	UL	15,000	10,000	10,500	7,500	12,000	7,500	5,000
I-4	S	UL	5	3	2	3	2	3	1	1
	A	UL	60,500	26,500	13,000	23,500	13,000	25,500	18,500	9,000
M	S	UL	11	4	2	4	2	4	3	1
	A	UL	UL	21,500	12,500	18,500	12,500	20,500	14,000	9,000
R-1	S	UL	11	4	4	4	4	4	3	2
	A	UL	UL	24,000	16,000	24,000	16,000	20,500	12,000	7,000
R-2	S	UL	11	4	4	4	4	4	3	2
	A	UL	UL	24,000	16,000	24,000	16,000	20,500	12,000	7,000
R-3	S	UL	11	4	4	4	4	4	3	3
	A	UL	UL	UL	UL	UL	UL	UL	UL	UL
R-4	S	UL	11	4	4	4	4	4	3	2
	A	UL	UL	24,000	16,000	24,000	16,000	20,500	12,000	7,000
S-1	S	UL	11	4	2	3	2	4	3	1
	A	UL	48,000	26,000	17,500	26,000	17,500	25,500	14,000	9,000
S-2[b, c]	S	UL	11	5	3	4	3	5	4	2
	A	UL	79,000	39,000	26,000	39,000	26,000	38,500	21,000	13,500
U[c]	S	UL	5	4	2	3	2	4	2	1
	A	UL	35,500	19,000	8,500	14,000	8,500	18,000	9,000	5,500

For SI: 1 foot = 304.8 mm, 1 square foot = 0.0929 m^2.

A = building area per story, S = stories above grade plane, UL = Unlimited, NP = Not permitted.

a. See the following sections for general exceptions to Table 503:
    1. Section 504.2, Allowable building height and story increase due to automatic sprinkler system installation.
    2. Section 506.2, Allowable building area increase due to street frontage.
    3. Section 506.3, Allowable building area increase due to automatic sprinkler system installation.
    4. Section 507, Unlimited area buildings.
b. For open parking structures, see Section 406.3.
c. For private garages, see Section 406.1.
d. See Section 415.5 for limitations.

503.1.2 Buildings on same lot. Two or more buildings on the same lot shall be regulated as separate buildings or shall be considered as portions of one building if the *building height* of each building and the aggregate *building area* of the buildings are within the limitations of Table 503 as modified by Sections 504 and 506. The provisions of this code applicable to the aggregate building shall be applicable to each building.

❖ Ordinarily, two buildings on the same lot are considered independently for compliance with the requirements of the code. Section 705.3 requires that when two buildings are on the same lot an "imaginary" line be assumed between the buildings, which is used to determine the appropriate exterior wall fire-resistance ratings in Table 602 and opening protectives in Tables 705.8, 715.4 and 715.5 (also see the commentary for the definition of "Fire separation distance" in Section 702.1). The primary purpose of Section 503.1.2 is to eliminate the application of the provisions of Section 705.3 when two small buildings can be regulated as one larger building on the premise. In other words, if the two buildings under consideration were actually connected (making one building) and could meet the area and height limitations for one building based on construction type, the connecting interior wall would not be required to have been rated; therefore, it is inconsistent to require the protection of the facing exterior walls simply because there is not a physical connection between two portions of the same building.

For example, in Figure 503.1.2(1), Buildings A and B are being constructed on the same lot at the same time. If they are considered to be separate buildings, then the code requirements for exterior wall fire-resistance ratings and opening protectives (see Tables 602, 705.8, 715.4 and 715.5) would apply to each wall on the basis of the placement of an imaginary lot line between the two buildings.

Under this section, however, if they are considered to be one building (for minimum type of construction purposes), then there is no need for protection of the facing exterior walls and Tables 602, 705.8, 715.4 and 715.5 do not apply to the walls between the two buildings. These walls would have to meet exterior wall requirements for the type of construction specified in Table 601 for exterior bearing walls only. If they were nonbearing walls, then the required fire-resistance rating would be zero.

Although facing exterior walls between two buildings on the same lot are not subject to fire-resistance rating and opening size limitations by using this provision (except as previously indicated for bearing walls), the remainder of the exterior walls in both buildings must comply with Tables 601, 602, 705.8, 715.4 and 715.5. Three of the walls in Building A and three of the walls in Building B, therefore, would still be required to meet the requirements of Tables 601, 602, 705.8, 715.4 and 715.5, and the exemption would only apply to the two walls (one from each building) that face each other.

A special circumstance occurs when a large structure is divided by fire walls into two or more buildings. In such a case, the exception provided by Section 503.1.2 would not be applicable because it is likely that a fire wall has been established because the overall structure exceeds the allowable area for a single building. If the configuration of the buildings results in a court, an imaginary line should be established between the two buildings in order to determine exterior wall and opening protection requirements [see Figure 503.1.2(2)].

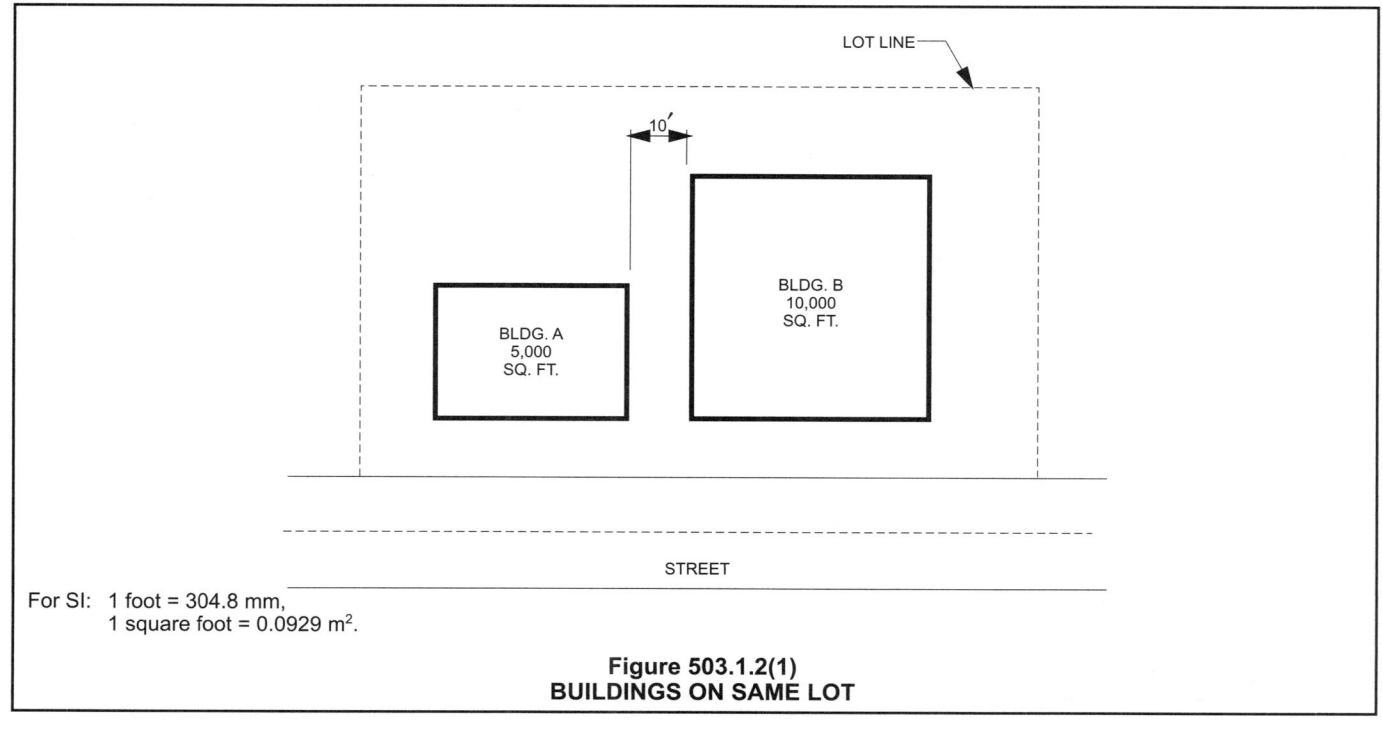

For SI: 1 foot = 304.8 mm,
1 square foot = 0.0929 m².

Figure 503.1.2(1)
BUILDINGS ON SAME LOT

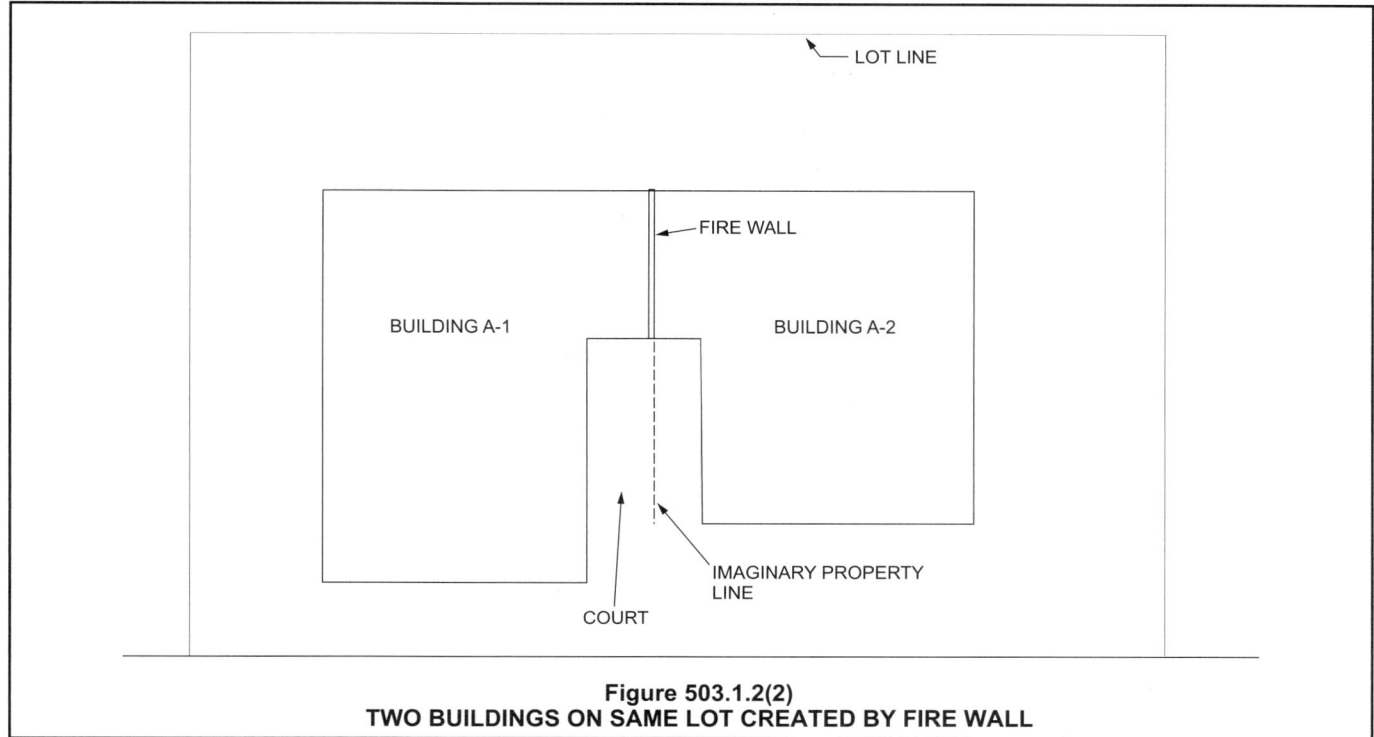

Figure 503.1.2(2)
TWO BUILDINGS ON SAME LOT CREATED BY FIRE WALL

503.1.3 Type I construction. Buildings of Type I construction permitted to be of unlimited tabular building heights and areas are not subject to the special requirements that allow unlimited area buildings in Section 507 or unlimited *building height* in Sections 503.1.1 and 504.3 or increased *building heights and areas* for other types of construction.

❖ Buildings permitted by Table 503 to be of unlimited height and area do not need to comply with the provisions of Section 504.2 or 506, which allow height and area increases. While most buildings of Type I construction are permitted to be unlimited in area, they are not required to comply with any of the provisions in Section 507 for unlimited area buildings. The requirements in Section 507 address special circumstances that permit a building to be unlimited in area. As there are no limitations to the size of these structures, the application of these sections would be superfluous. These buildings may be of unlimited size based on their type of construction alone. High-rise buildings are required to be sprinklered in accordance with Section 403.3, and this section should not be construed to be a release from that requirement if a building fits the definition of "High-rise building" in Section 202.

A special industrial occupancy building, in accordance with Section 503.1.1, is permitted to be unlimited in height. However, this does not require the type of construction for the building to be Type I. Any type of construction is permitted for special industrial occupancy buildings. Similarly, noncombustible roof structures are permitted to be unlimited in height on buildings of other than Type I construction.

SECTION 504
BUILDING HEIGHT

504.1 General. The *building height* permitted by Table 503 shall be increased in accordance with this section.

> **Exception:** The *building height* of one-story aircraft hangars, aircraft paint hangars and buildings used for the manufacturing of aircraft shall not be limited if the building is provided with an automatic fire-extinguishing system in accordance with Chapter 9 and is entirely surrounded by *public ways* or *yards* not less in width than one and one-half times the *building height*.

❖ Section 504 contains exceptions to the height limitations of Table 503 for certain buildings. These exceptions are made on the basis of the structure's proposed occupancy and of fire safety features included in the design of the structure. The principal exception given is the increase for buildings equipped throughout with an automatic sprinkler system (see Section 504.2). When a sprinkler system is installed either as an option or where required by other sections of the code, an increase in height may be applicable because of the increased protection provided by such a system. (See the exceptions to Section 504.2 that restrict this allowance.)

The exception permits fully suppressed aircraft hangars to exceed the limits of Table 503 as long as they have the specified open area surrounding the building. The exception is necessary to accommodate the size of large aircraft within the building, and the hazard is mitigated by the requirement for suppression and the very large open space provided by the yards and public ways. While the code refers to fire-extinguishing requirements in Chapter 9, the detailed suppression requirements for aircraft hangars are specified in Section 412.4.6. Aircraft paint hangars may be unlimited in area as well, under certain conditions (see commentary, Section 507.9).

504.2 Automatic sprinkler system increase. Where a building is equipped throughout with an *approved automatic sprinkler system* in accordance with Section 903.3.1.1, the value specified in Table 503 for maximum *building height* is increased by 20 feet (6096 mm) and the maximum number of *stories* is increased by one. These increases are permitted in addition to the *building area* increase in accordance with Sections 506.2 and 506.3. For Group R buildings equipped throughout with an *approved automatic sprinkler system* in accordance with Section 903.3.1.2, the value specified in Table 503 for maximum *building height* is increased by 20 feet (6096 mm) and the maximum number of *stories* is increased by one, but shall not exceed 60 feet (18 288 mm) or four *stories*, respectively.

Exceptions:

1. Buildings, or portions of buildings, classified as a Group I-2 occupancy of Type IIB, III, IV or V construction.

2. Buildings, or portions of buildings, classified as a Group H-1, H-2, H-3 or H-5 occupancy.

3. *Fire-resistance rating* substitution in accordance with Table 601, Note d.

❖ This section permits the building height limitations of Table 503 to be increased one story and 20 feet (6096 mm) when the building is protected throughout with an approved automatic sprinkler system. When used in this context, the phrase "equipped throughout" (see Section 903.3.1.1) means the entire structure is provided with sprinkler protection, and the only exceptions are the specific locations identified in Section 903.3.1.1.1 or in the applicable standard. If a building contains an area that meets the conditions of Section 903.3.1.1.1 or an exception in the applicable standard, then sprinklers may be omitted in that specific location of the building and the building is still considered protected throughout (see the commentary to Section 903.3.1.1.1 for application of the exceptions).

By referencing Section 903.3.1.1, the code requires a system in accordance with NFPA 13 (not NFPA 13R or NFPA 13D). Only buildings equipped throughout with systems installed in accordance with NFPA 13 qualify for the height increase. The systems and coverage criteria of NFPA 13R and 13D are specifically for residential occupancies and do not afford the same type of protection as an NFPA 13 system provides. More specifically, buildings of residential occupancies are permitted to receive an increase of one story and 20 feet (6096 mm) up to a maximum of four stories and 60 feet (18 288 mm) if equipped with a sprinkler system that conforms to Section 903.3.1.2 (an NFPA 13R system). This alternative is intended to address the dif-

ferences between application and protection provided by an NFPA 13 system versus an NFPA 13R system. Note that no increase is allowed for an NFPA 13D system, which is specifically scoped for one- and two-family residential buildings. If a Group R building is protected throughout with an NFPA 13 system in accordance with Section 903.3.1.1, then the four-story, 60-foot (18 288 mm) limitation would not apply. It should be noted that NFPA 13R systems are limited to buildings that are four stories or less in height above grade plane.

It should also be noted that the height increase is permitted even if the sprinkler system is required by Chapter 9 of the code, based on the use and size of the building. For instance, Section 903.2.8 requires all buildings containing Group R fire areas to be sprinklered. Even so, a Type VA, Group R-2 building protected throughout with an NFPA 13R system (in accordance with Section 903.3.1.2) is allowed up to four stories (one story more than the story limitation shown in Table 503 for Group R-2, Type VA construction) and up to 60 feet (18 288 mm) in height.

Please note that the height increase can be combined with an area increase in accordance with Section 506.3.

On the other hand, an NFPA 13R sprinkler system would not yield a height-in-stories increase for a Type IIB building of Group R-2, since Table 503 already permits four stories. The height-in-feet threshold, however, would be 60 feet (18 288 mm), a 5-foot (1524 mm) increase over what is allowed by Table 503 for Type IIB construction. If the Group R-2 building is protected throughout with a system in accordance with NFPA 13 (see Section 903.3.1.1), then the height thresholds would be five stories and 75 feet (22 860 mm) [the permitted one-story, 20-foot (6096 mm) increase over the Table 503 limits].

The three exceptions to this section each describe circumstances where the height increases allowed by this section would not be permitted.

Exception 1 indicates several types of construction are eliminated from the height increase for buildings, or portions of buildings, containing a Group I-2 occupancy. Each of these types of construction either lacks the fire-resistance rating deemed necessary or includes combustible construction materials in a use where the occupants must rely on the fire sprinkler system for defend-in-place protection.

Exception 2 indicates buildings, or portions of buildings, containing a Group H-1, H-2, H-3 or H-5 occupancy are not eligible for the general height increase because of the higher hazards associated with these occupancies.

Exception 3 states that when an automatic sprinkler system is installed to address the allowed trade-off of sprinklers for 1-hour fire-resistance-rated construction given in Note d of Table 601, the height increase allowed in this section cannot be taken. For example, in a building constructed using Type IIA construction, the required 1-hour fire-resistance rating of all structural members (with the exception of the exterior walls) could be reduced to zero if the building was equipped throughout with a sprinkler system in accordance with Section 903.3.1.1 (NFPA 13 system). However, the allowable height for the occupancy classification of, for example, Group B for Type IIA construction could not be increased by 20 feet (6096 mm) to 85 feet (25 908 mm), as would otherwise be allowed in this section.

504.3 Roof structures. Towers, spires, steeples and other roof structures shall be constructed of materials consistent with the required type of construction of the building except where other construction is permitted by Section 1509.2.4. Such structures shall not be used for habitation or storage. The structures shall be unlimited in height if of noncombustible materials and shall not extend more than 20 feet (6096 mm) above the allowable *building height* if of combustible materials (see Chapter 15 for additional requirements).

❖ Certain roof structures may exceed the height limitations of Table 503 in accordance with this section and Chapter 15. These roof structures on a noncombustible building (Type I or II) must be of noncombustible construction and, in some cases, must have ratings equivalent to those required for exterior walls (see Section 1509.5.2 for enclosed towers and spires). Section 1509.2.1 contains exceptions for penthouse-type structures, mechanical equipment enclosures and screens when not in proximity to a property line. Structures that are used for habitation or storage must be considered an additional story. The height limitation for noncombustible roof structures in accordance with this section is unlimited. Please note that Section 1509.5 and its subsections contain additional restrictions for structures of a certain size or use. Combustible roof structures are permitted to extend up to 20 feet (6096 mm) above the allowable height set in Table 503 or as increased in accordance with Section 504.2.

SECTION 505
MEZZANINES

505.1 General. A *mezzanine* or *mezzanines* in compliance with Section 505 shall be considered a portion of the *story* in which it is contained. Such *mezzanines* shall not contribute to either the *building area* or number of *stories* as regulated by Section 503.1. The area of the *mezzanine* shall be included in determining the *fire area* defined in Section 902. The clear height above and below the *mezzanine* floor construction shall not be less than 7 feet (2134 mm).

❖ Although mezzanines provide an additional or intermediate useable floor level in a building, they are not considered an additional story as long as they comply with the requirements of Section 505. Building height and area limitations are intended to offset the inherent fire hazard associated with specific occupancies and with materials and features of a specific construction type. Because of a mezzanine's restricted size and its required openness to the room or space in which it is located, a mezzanine does not contribute significantly to a building's inherent fire hazard; therefore, the area of a mezzanine is not considered when applying the provisions of Section 503.1 for building area limitations, and mezzanines are not considered in determining the height in stories of a building as regulated by Table 503. Please note that a mezzanine level is considered an occupied level that may affect the classification of a building as a high rise (see commentary, Section 403). The occupant and fuel load of the mezzanine should be taken into consideration, however, when determining the necessity for fire protection systems. As such, the area of the mezzanine is to be included in the calculation of the size of the fire area for sprinkler thresholds (see the commentary for the definition of "Fire area" in Section 902.1).

This section does not include any requirements for the construction of a mezzanine or for fire-resistance ratings; therefore, the mezzanine is to be constructed of materials consistent with the construction type of the building. Required fire-resistance ratings are determined on the basis of Table 601 for the appropriate construction type.

Mezzanines are required to have a ceiling height of not less than 7 feet (2134 mm), and the ceiling height below the mezzanine must also be 7 feet (2134 mm). Even habitable rooms located in mezzanines may have a ceiling height of 7 feet (2134 mm), in accordance with Exception 3 to Section 1208.2.

505.2 Area limitation. The aggregate area of a *mezzanine* or *mezzanines* within a room shall not exceed one-third of the floor area of that room or space in which they are located. The enclosed portion of a room shall not be included in a determination of the floor area of the room in which the *mezzanine* is located. In determining the allowable *mezzanine* area, the area of the *mezzanine* shall not be included in the floor area of the room.

Exceptions:

1. The aggregate area of *mezzanines* in buildings and structures of Type I or II construction for special industrial occupancies in accordance with Section 503.1.1 shall not exceed two-thirds of the floor area of the room.

2. The aggregate area of *mezzanines* in buildings and structures of Type I or II construction shall not exceed one-half of the floor area of the room in buildings and structures equipped throughout with an *approved automatic sprinkler system* in accordance with Section 903.3.1.1 and an *approved* emergency voice/alarm communication system in accordance with Section 907.5.2.2.

❖ So as not to contribute significantly to a building's inherent fire hazard, a mezzanine is restricted to a maximum of one-third of the area of the room with which it shares a common atmosphere. The area may consist of multiple mezzanines at the same or different floor levels, provided that the aggregate area does not exceed the one-third limitation. (This determination is based on the gross floor area of the mezzanines.) If the area limitation is exceeded, the provisions of this section do not apply and the level is considered an additional story.

In determining the allowable area of the mezzanine, the enclosed spaces of the room in which the mezzanine is contained are not to be included in calculating the room size. Although the mezzanine area is included in the calculation of fire area size, it is not included in the area of the room when computing the allowable mezzanine area. For example, a room contains 5,000 square feet (465 m²), 500 square feet (46 m²) of which are enclosed and not part of the common atmosphere with the mezzanine. A mezzanine may be provided in the room such that the area of the mezzanine is not more than 1,500 square feet (139 m²) [5,000 square feet − 500 square feet = 4,500 square feet × $^1/_3$ = 1,500 square feet (139 m²)]. The mezzanine level must be open, except that enclosed spaces are permitted for mezzanines in accordance with the exceptions to Section 505.4. Even if the mezzanine is enclosed, the allowable area of the mezzanine is always determined assuming the space is open to the room in which it is located.

By definition, in special industrial occupancies (see commentary, Section 503.1.1) the inherent fire hazard is very low; therefore, mezzanines in such occupancies located in buildings of Type I or II construction are permitted to constitute up to two-thirds of the area of the room in which they are located, in accordance with Exception 1. The limitation on construction types further reduces the fire hazard associated with such occupancies.

Exception 2 provides additional allowable area for mezzanines in Type I or II construction in any occupancy, when sprinklers, as well as an approved emergency voice/alarm communication system are installed. The trade-off for additional area for installation of a sprinkler system recognizes the additional time that sprinklers provide for occupant evacuation. The emergency voice/alarm communication system is also seen as a necessary feature of this trade-off so that occupants of a building become aware of a fire that could start in a remote part of the mezzanine, in order to allow evacuation before smoke spread could cause a problem for occupants. Experience has shown that occupants react more readily when given voice instructions than when given a general alarm.

Equipment platforms, as defined in Section 502.1, are not considered mezzanines (see Section 505.5 for requirements for equipment platforms).

505.3 Egress. Each occupant of a *mezzanine* shall have access to at least two independent *means of egress* where the *common path of egress travel* exceeds the limitations of Section 1014.3. Where a *stairway* provides a means of *exit access* from a *mezzanine*, the maximum travel distance includes the distance traveled on the *stairway* measured in the plane of the tread nosing.

Accessible means of egress shall be provided in accordance with Section 1007.

Exception: A single *means of egress* shall be permitted in accordance with Section 1015.1.

❖ A mezzanine can be likened to a single room when considering means of egress. As with rooms, if the occupant load of the mezzanine exceeds the limitation of Table 1015.1 for the specific use of the space, at least two independent means of egress must be provided for the mezzanine. For example, if a mezzanine containing office areas (Group B) has an occupant load exceeding 50, a second means of egress from the mezzanine is required. Additionally, if a mezzanine has one means of egress and it is by means of an open stair to the floor below, the travel distance from the most remote point on the mezzanine to the bottom of the stair may not exceed 75 feet (22 860 mm) in accordance with Section 1014.3 for common path of travel [see the exceptions to Section 1014.3 that allow 100 feet (30 480 mm) in some circumstances]. If the travel distance to the bottom of the stair exceeds the limits of Section 1014.3, then a second means of egress must be provided from the mezzanine.

Because mezzanines that comply with Section 505 are considered a portion of the story in which they are located in accordance with Section 505.1, the stair leading from it is not considered an exit stairway, but is part of exit access and is not required to be a vertical exit enclosure.

This section does not require that an exit be provided directly from the mezzanine level. Exception 2 to Section 505.4, however, requires mezzanines that are not open to the room in which they are located to have access to an exit directly from the mezzanine level, if two means of egress are provided. When two means of egress are required, however, they are to be located remote from each other in accordance with Section 1015.2, as for any other space, so that if one means of egress is blocked by fire or smoke, then the other will be presumed available.

When exits are not located on the same level as the mezzanine, the occupant load of the mezzanine is added to the room or space below and the required means of egress width for that room is determined accordingly (see Section 1005). For example, if a room (Group B) has an occupant load of 45 and a mezzanine (also Group B) has an occupant load of 15, the total occupant load for the space served is 60; therefore, the room must have two exit access doors in accordance with Table 1015.1. The mezzanine itself, however, needs only one means of egress.

The requirements for accessible means of egress are not in any way intended to be affected by these provisions. The requirements for accessibility relate only to the requirements for the occupancy and use of the space as provided in Chapters 10 and 11 of the code. If two means of egress are required from the mezzanine and an accessible route is required to the mezzanine, then two accessible means of egress must be provided from the mezzanine. If an accessible route is not required to the mezzanine or the mezzanine is not required to be accessible, then no accessible means of egress are required.

505.4 Openness. A *mezzanine* shall be open and unobstructed to the room in which such *mezzanine* is located except for walls not more than 42 inches (1067 mm) high, columns and posts.

Exceptions:

1. *Mezzanines* or portions thereof are not required to be open to the room in which the *mezzanines* are located, provided that the *occupant load* of the aggregate area of the enclosed space does not exceed 10.

2. A *mezzanine* having two or more *means of egress* is not required to be open to the room in which the *mezzanine* is located if at least one of the *means of egress* provides direct access to an *exit* from the *mezzanine* level.

3. *Mezzanines* or portions thereof are not required to be open to the room in which the *mezzanines* are located, provided that the aggregate floor area of the enclosed space does not exceed 10 percent of the *mezzanine* area.

4. In industrial facilities, *mezzanines* used for control equipment are permitted to be glazed on all sides.

5. In occupancies other than Groups H and I, that are no more than two *stories* above *grade plane* and equipped throughout with an *automatic sprinkler system* in accordance with Section 903.3.1.1, a *mezzanine* having two or more *means of egress* shall not be required to be open to the room in which the *mezzanine* is located.

❖ A mezzanine presents a unique fire threat to the occupant. If a mezzanine is closed off from the larger room, an undetected fire could develop such that it would jeopardize or eliminate the opportunity for occupant escape.

The exceptions address situations where the hazard is reduced. A low occupant load, typical of small mezzanines, would permit a mezzanine to be enclosed in accordance with Exception 1. Occupant load is calculated in accordance with Section 1004.1 for the use or function of the mezzanine space. Similarly, Exception 3 permits the enclosure of a limited portion of a mezzanine.

Exception 2 permits enclosure of the mezzanine based on the availability of an exit at the mezzanine level, and the mezzanine has at least two means of egress. The definition of "Exit" is important for this exception. "Exit" is defined in Section 1002.1 as an element that is separated by fire-resistance-rated construction and opening protectives and includes exterior doors; therefore, the mezzanine must be served by at least one vertical exit enclosure (see Section 1022), a horizontal exit (see Section 1025), an exit

passageway (see Section 1023), an exterior stair or ramp (see Section 1026) or an exterior door discharging directly at grade. In addition, another means of egress is required, which may be an open stairway to the room below.

Exception 4 addresses industrial facilities, where enclosure may be necessary for noise reduction or atmospheric control.

A fire in an enclosed mezzanine could easily go unrecognized for a longer period than in an open mezzanine. Therefore, Exception 5 permits enclosure of a mezzanine in a building that is fully protected by a sprinkler system and limited to two stories. Under this exception two means of egress must be provided for the mezzanine; however, a direct means of egress from the mezzanine to the room below is not necessary.

505.5 Equipment platforms. *Equipment platforms* in buildings shall not be considered as a portion of the floor below. Such *equipment platforms* shall not contribute to either the *building area* or the number of *stories* as regulated by Section 503.1. The area of the *equipment platform* shall not be included in determining the *fire area* in accordance with Section 903. *Equipment platforms* shall not be a part of any *mezzanine* and such platforms and the walkways, *stairs*, *alternating tread devices* and ladders providing access to an *equipment platform* shall not serve as a part of the *means of egress* from the building.

❖ "Equipment platform" is defined in Section 502.1 as an unoccupied, elevated platform used exclusively for supporting mechanical systems or industrial process equipment and providing access to them. If an elevated platform does not meet all the conditions of this definition, then it must be considered either a mezzanine or another story.

Equipment platforms are treated as part of the equipment they support (within the limitations of the subsections to this section), and do not contribute in any way to the area of the building, the number of stories, the area of any mezzanine or any fire area. If equipment platforms are located in the same room as a mezzanine, however, the aggregate area of the platforms and mezzanines is limited by Section 505.5.1.

The definition of "Equipment platform" includes the associated elevated walkways, stairs, alternating tread devices and ladders necessary to access the platform. Elements that serve an equipment platform are not permitted to serve as a means of egress for occupants of the building unless they meet all the requirements for means of egress in Chapter 10. Because they are not used for means of egress, the elements used to access these platforms could be something other than a stairway or ramp. For example, a permanent ladder could be used to access an equipment platform. Since the code does not address ladder construction for this circumstance, other appropriate standards, such as Occupational Safety and Health Administration (OSHA) standards, would need to be consulted for design.

Guards are required for elevated walking surfaces serving equipment platforms in accordance with Section 1013.1 (see Section 505.5.3).

505.5.1 Area limitations. The aggregate area of all *equipment platforms* within a room shall not exceed two-thirds of the area of the room in which they are located. Where an *equipment platform* is located in the same room as a *mezzanine*, the area of the *mezzanine* shall be determined by Section 505.2 and the combined aggregate area of the *equipment platforms* and *mezzanines* shall not exceed two-thirds of the room in which they are located.

❖ In determining the allowable area of an equipment platform, neither the enclosed spaces of the room in which the equipment platform is located nor the area of the equipment platform itself is to be included in calculating the room size. Whereas mezzanines are limited to one-third of the area of the room in which they are located, equipment platforms are permitted to be two-thirds of the area of the room. Whereas the area of mezzanines is included in the calculation of the fire area (see Section 505.1), the area of equipment platforms is not included as part of the fire area in accordance with Section 505.5. The area of mezzanines and equipment platforms, however, is summed when they are in the same room, and the aggregate area is limited to two-thirds of the area of the room. The total area of a mezzanine still cannot exceed one-third of the area of the room.

[F] 505.5.2 Fire suppression. Where located in a building that is required to be protected by an *automatic sprinkler system*, *equipment platforms* shall be fully protected by sprinklers above and below the platform, where required by the standards referenced in Section 903.3.

❖ In buildings or spaces that are required to be protected with an automatic fire suppression system, fire suppression above and below the equipment platform is needed so the equipment platform will not obstruct sprinkler coverage or delay sprinkler activation if a fire develops below the platform. As stated in Section 903.3.1.1, sprinkler installation is to be in accordance with NFPA 13. This section should not be construed to require sprinkler protection for equipment platforms where such a system is not otherwise required.

505.5.3 Guards. *Equipment platforms* shall have *guards* where required by Section 1013.1.

❖ Guards are required for equipment platforms in the same manner that they are required at open walking surfaces in other parts of the building, and are subject to the same requirements for height, design load and configuration. Exception 3 to Section 1013.3 allows spacing of guard balusters in Group F occupancies to only be close enough to prevent the passage of a 21-inch (533 mm) sphere.

SECTION 506
BUILDING AREA MODIFICATIONS

506.1 General. The *building areas* limited by Table 503 shall be permitted to be increased due to frontage (I_f) and *automatic sprinkler system* protection (I_s) in accordance with the following:

$$A_a = \left\{ A_t + \left[A_t \times I_f \right] + \left[A_t \times I_s \right] \right\}$$
(Equation 5-1)

where:

A_a = Allowable *building area* per *story* (square feet).

A_t = Tabular *building area* per *story* in accordance with Table 503 (square feet).

I_f = Area increase factor due to frontage as calculated in accordance with Section 506.2.

I_s = Area increase factor due to sprinkler protection as calculated in accordance with Section 506.3.

❖ Other than increasing the fire-resistance rating (using a higher construction type) of the building structure, in general there are two circumstances that would decrease a building's fire hazard. These are: (1) isolating the building from other structures and (2) equipping it with a fire suppression system. Equation 5-1 takes these circumstances into account. The equation determines the largest area for any single story of a building, taking into account the permissible increases due to open frontage and the installation of a fire suppression system. Equation 5-1 in conjunction with Sections 506.1 through 506.3 is used to determine the maximum allowable area on any individual story of a building. This information is then used in Sections 506.4 and 506.5 to establish the maximum allowed area in the total building.

The following examples demonstrate the determination of maximum area per story:

Example 1: A single-story, Group B building of Type IIB construction does not have a fire suppression system but is allowed a frontage increase of 50 percent in accordance with Section 506.2. In Equation 5-1, the following values would be given to the terms:

A_t = 23,000 square feet (1765 m²) (from Table 503)

I_f = 0.5

I_s = 0

The equation $A_a = \{A_t + [(A_t)(I_f)] + [(A_t)(I_s)]\}$
becomes A_a = 23,000 + [(23,000)(0.5)] + 0

= 23,000 + 11,500

= 34,500 square feet (3205 m²).

The allowable area per story for the building is 34,500 square feet (3205 m²).

Example 2: A five-story, Group R-2 building of Type IIA construction has a frontage increase of 75 percent in accordance with Section 506.2. The building is fully sprinklered with an NFPA 13 system and the increase yielded by Section 506.3 is 200 percent. In Equation 5-1, the following values would be given to the terms:

A_t = 24,000 square feet (2230 m²) (from Table 503)

I_f = 0.75

I_s = 2

Equation 5-1 becomes:

A_a = {24,000 + [(24,000)(0.75)] + [(24,000)(2)]}

= 24,000 + 18,000 + 48,000

= 90,000 square feet (8361 m²) per story

At this point, because the building is more than three stories in height, an additional calculation needs to be made in accordance with Section 506.4.1, which limits the aggregate area of all stories in the building. In accordance with that section, the total allowable area for the building may be no more than three times the maximum allowable area per story (A_a). Assuming all stories of the building have equal area, the maximum allowable area of each story is calculated as follows:

$(3)(A_a) = (3)(90,000) = 270,000$ square feet (25 083 m²) total allowable area in the building.

270,000/5 stories = 54,000 square feet (5017 m²) per story.

Therefore, if the total allowable area is divided equally among the five stories, the allowable area per story for the building is 54,000 square feet (5017 m²). The code does not require that the total allowable building area be equally divided on the five stories. For example, a building with a 90,000-square-foot (8361 m²) first story, a 60,000-square-foot (5574 m²) second story and three stories of 40,000 square feet (3716 m²) each would also be within the 270,000-square-foot (25 083 m²) maximum size determined in Example 2, and each story is 90,000 square feet (8361 m²) or less.

506.2 Frontage increase. Every building shall adjoin or have access to a *public way* to receive a *building area* increase for frontage. Where a building has more than 25 percent of its perimeter on a *public way* or open space having a minimum width of 20 feet (6096 mm), the frontage increase shall be determined in accordance with the following:

$$I_f = [F/P - 0.25]W/30 \qquad \textbf{(Equation 5-2)}$$

where:

I_f = Area increase due to frontage.

F = Building perimeter that fronts on a *public way* or open space having 20 feet (6096 mm) open minimum width (feet).

P = Perimeter of entire building (feet).

W = Width of *public way* or open space (feet) in accordance with Section 506.2.1.

❖ The allowable area of a building is permitted to be increased when it has a certain amount of frontage on streets or open spaces since this provides access to the structure by fire service personnel, a temporary refuge area for occupants as they leave the building in a fire emergency and a reduced exposure to and from adjacent structures.

This section indicates how the frontage increase, or the term "I_f" in Equation 5-1, is calculated. There is no requirement in the code that buildings must have at least 25 percent of their perimeter on a public way or open space; however, in order to qualify for an area increase, a building must have more than 25 percent of its perimeter on a public way or open space having a minimum width of at least 20 feet (6096 mm). When the calculations are done, the maximum percent increase for a fully open perimeter (full frontage—the entire perimeter fronts on a public way or open space) is 75 percent. Width is measured at right angles to the perimeter walls, but it is not the same as fire separation distance. The latter, as defined in Section 702.1, is measured to the centerline of public ways or to an imaginary line between buildings on the same lot. Open space can include the total width of the public way as well as the total open space between buildings on the same lot. "Public way" is defined in Section 1002.1.

Section 506.2.2 requires that an open space that is not a public way be on the same lot or dedicated for public use, and it must have access from a street or an approved fire lane in order to contribute to the frontage increase (see commentary, Section 506.2.2).

If a structure is divided into two or more buildings by fire walls complying with Section 706, the area modifications allowed by Section 506 must be determined based on each separate building within the structure. This especially comes into play in determining increases allowable based on frontage. The fire wall is essentially the perimeter wall for that side of the building and must be included in determining the P in Equation 5-2. Since there is another building on the other side of the fire wall, this portion of the perimeter is not considered fronting on a public way or yard and, therefore, is not included in F [see Figure 506.2(1)].

The term "W" in Equation 5-2 represents the weighted average of the width of any open space for the portion of the perimeter that fronts on a public way or open space. The minimum width required is 20 feet (6096 mm). The maximum width allowed in this equation is 30 feet (9144 mm). The following example demonstrates this point:

Example: Refer to Figure 506.2(2). In Equation 5-2, the terms would have the following values:

F = 500 feet (15 240 mm), since all sides of the building front on a public way or open space having 20 feet (6096 mm) minimum open width.

P = 500 feet (15 240 mm), the length of the entire perimeter.

W = The weighted average of the widths of the public way or open space =
[(20 ft.)(100 ft.) + (30 ft.)(150 ft.) + (30 ft.)(100 ft.) + (30 ft.)(150 ft.)]/500 = 28 feet (8534 mm).

Note in the above computation for W that, where the actual width exceeds 30 feet (9144 mm), a value of 30 feet (9144 mm) is used as required by Section 506.2.1.

Equation 5-2, $I_f = [(F/P) - 0.25](W/30)$, becomes:

I_f = [(500/500) - 0.25](28/30)

= (.75)(.93)

= 0.70

Therefore, the term "I_f" equals 0.70 when using Equation 5-1 for determining the allowable area per story (A_a).

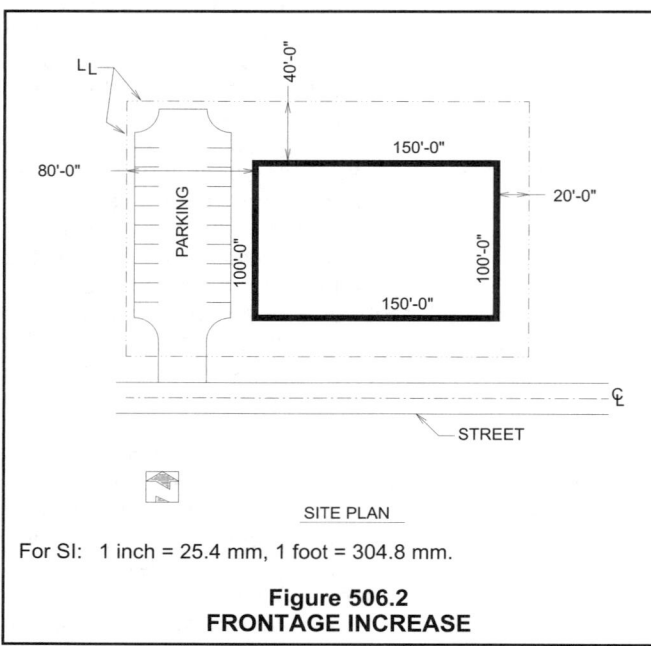

For SI: 1 inch = 25.4 mm, 1 foot = 304.8 mm.

Figure 506.2
FRONTAGE INCREASE

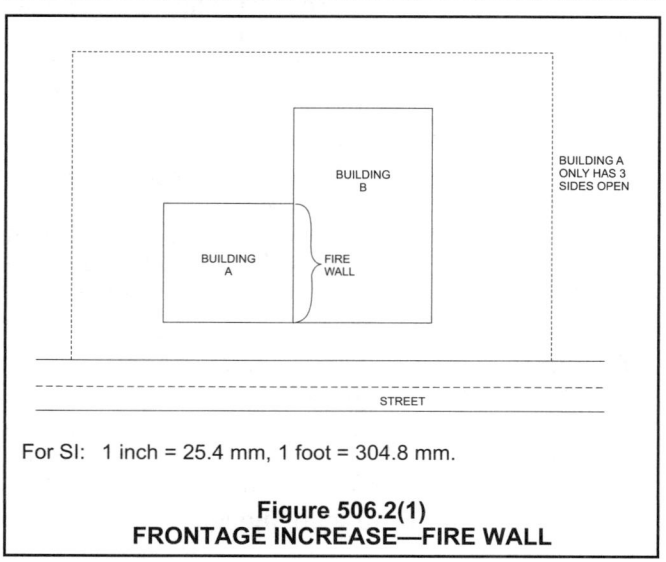

For SI: 1 inch = 25.4 mm, 1 foot = 304.8 mm.

Figure 506.2(1)
FRONTAGE INCREASE—FIRE WALL

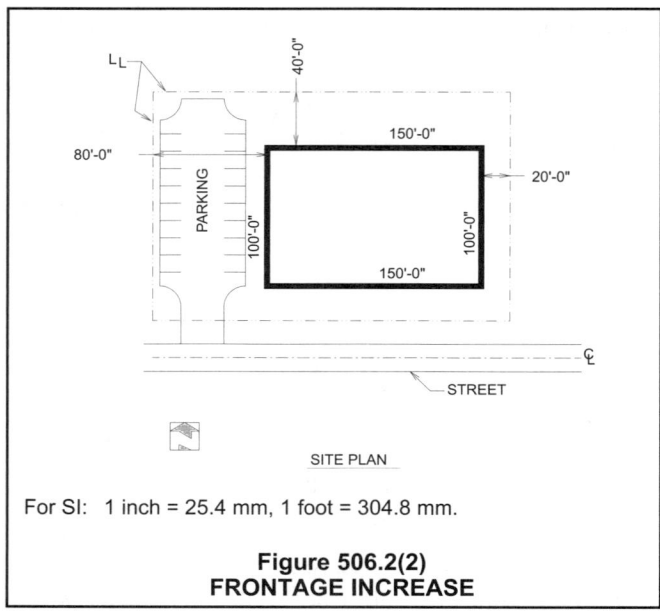

For SI: 1 inch = 25.4 mm, 1 foot = 304.8 mm.

Figure 506.2(2)
FRONTAGE INCREASE

506.2.1 Width limits. The value of W shall be at least 20 feet (6096 mm). Where the value of W varies along the perimeter of the building, the calculation performed in accordance with Equation 5-2 shall be based on the weighted average of each portion of *exterior wall* and open space where the value of W is greater than or equal to 20 feet (6096 mm). Where the value of W exceeds 30 feet (9144 mm), a value of 30 feet (9144 mm) shall be used in calculating the weighted average, regardless of the actual width of the open space. Where two or more buildings are on the same lot, W shall be measured from the exterior face of a building to the exterior face of an opposing building, as applicable.

Exception: The value of W divided by 30 shall be permitted to be a maximum of 2 when the building meets all requirements of Section 507 except for compliance with the 60-foot (18 288 mm) *public way* or *yard* requirement, as applicable.

❖ The amount of area increase (that is, the value of I_f) will vary depending on the minimum width of the open space used in the frontage increase calculation. The general requirement is that the value of W equal at least 20 feet (6096 mm) and that the value of W not exceed 30 feet (9144 mm). The value of W is the weighted average of the portions of the wall and each open space when the width of the open space is between 20 and 30 feet (6096 and 9144 mm). Except as provided in the exception, the width of open space used in this calculation cannot exceed 30 feet (9144 mm) even when part of the open space is wider than 30 feet (9144 mm). See the following two examples illustrating this requirement:

Example 1: In Figure 506.2.1(1), the value of W would be 26 feet (7925 mm), calculated as follows:

W = [(200 feet x 25 feet) + (100 feet x 20 feet) + (200 feet x 30 feet)]/500 feet = 26 feet (7925 mm).

Example 2: In Figure 506.2.1(2), the value of *W* would be 30 feet (9144 mm), since the minimum width of the open space that qualifies as open frontage is greater than 30 feet (9144 mm) on all sides of this building. The reason the value of *W* must be taken at 30 feet (9144 mm) and not 50 feet (15 240 mm) is that this section sets an upper limit on the value of the *W*/30 term in Equation 5-2.

Typically, when multiple buildings are located on the same lot, an imaginary lot line must be established somewhere between the buildings in order to deter-mine the fire separation distance. However, the value of *W* is defined as the width of the public way or open space, and is not dependent upon the fire separation distance. Therefore, for purposes of determining the value of *W* between buildings on the same lot, the entire distance between the buildings is permitted to be used, not solely the fire separation distance. Similarly, the entire width of a public way is used and not just the distance to the centerline of the public way.

The exception to Section 506.2.1 states that for certain buildings, the value of *W* divided by 30 is permitted to have a maximum value of 2.0. This exception applies to buildings that would be allowed to be unlimited in area in accordance with Section 507, save for the fact that the open area of 60 feet (18 288 mm) required by Section 507 is not met. Therefore, the weighted average of *W* would be calculated between 20 feet (6096 mm) and 60 feet (18 288 mm).

506.2.2 Open space limits. Such open space shall be either on the same lot or dedicated for public use and shall be accessed from a street or *approved fire lane*.

❖ The requirement that the open space be on the same lot is so that the owner or the jurisdiction can control the space that is assumed to be open for purposes of the area increase. One cannot encumber a neighbor's property with a requirement that the space will always

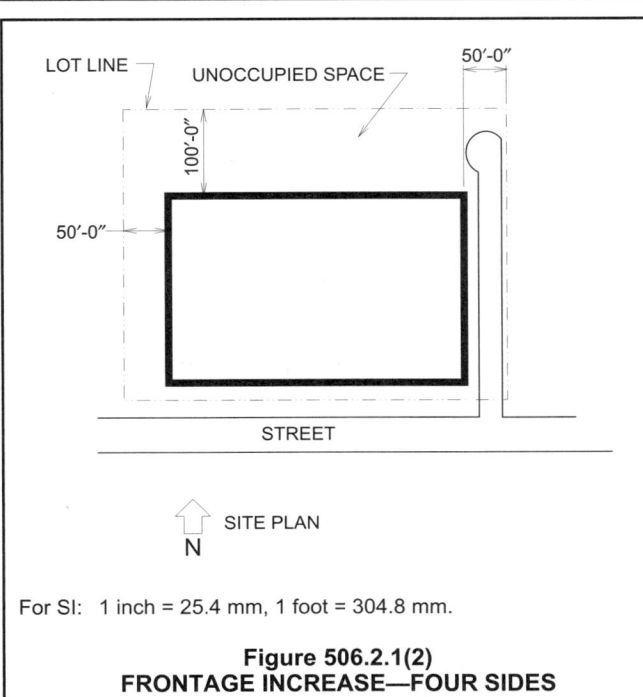

For SI: 1 inch = 25.4 mm, 1 foot = 304.8 mm.

Figure 506.2.1(1)
FRONTAGE INCREASE—THREE SIDES

For SI: 1 inch = 25.4 mm, 1 foot = 304.8 mm.

Figure 506.2.1(2)
FRONTAGE INCREASE—FOUR SIDES

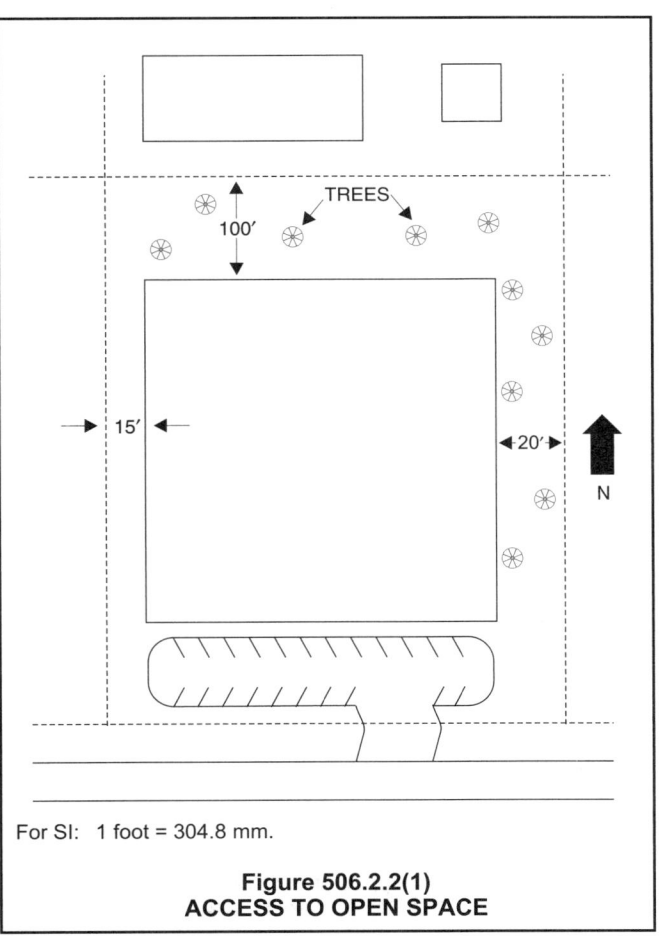

For SI: 1 foot = 304.8 mm.

Figure 506.2.2(1)
ACCESS TO OPEN SPACE

remain unoccupied.

Any part of the perimeter that is not accessible to the fire department by means of a street or fire lane cannot be considered open for the purposes of Section 506.2. For instance, if the back side of a building on a narrow lot cannot be reached by means of a fire lane on one side of the building (and there is no alley or street at the back), that portion of the perimeter is not considered open for purposes of frontage increase, even if there is actual open space exceeding 20 feet (6096 mm) in width. See Figure 506.2.2(1) as an illustration of this limitation.

This section does not require that a fire lane or street extend immediately adjacent to every portion of the perimeter that is considered open for purposes of the increase. Rather, access by a fire lane must be provided up to the open side such that fire department personnel can approach the side and pull hoses across the open area to fight a fire, and no corner of the building will impede the use of hoses and equipment on that side of the building. The following examples demonstrate this point.

Example 1: In Figure 506.2.2(1), the south and east side of the building facing the street can be considered open perimeter (frontage). The north side of the building cannot be considered open perimeter for pur-

poses of the increase, since it is not accessible from the street or a fire lane. Even though the 20-foot-wide yard can be included in open frontage because it does not provide a fire lane to the rear of the building, the 100-foot yard cannot be included.

Example 2: In Figure 506.2.2(2), all sides of the building are considered open perimeter (frontage) for purposes of the increase. Access up to each side of the building is provided by means of a fire lane or street.

Section 503 of the *International Fire Code®* (IFC®) specifies that access roads extend to within 150 feet (45 720 mm) of all portions of the exterior walls of a building; however, there are exceptions that would permit the omission of such roads under certain circumstances. One such exception is for buildings equipped throughout with an automatic sprinkler system.

The IFC also stipulates that the access roads must be at least 20 feet (6096 mm) in unobstructed width, although it also gives the building official authority to require greater widths if necessary for effective fire-fighting operations. The type of surface necessary for the approved fire lane is determined by the local building official with input from the fire department, but the road must be capable of supporting the imposed loads of fire apparatus and be surfaced so as to provide all-weather driving capabilities.

506.3 Automatic sprinkler system increase. Where a building is equipped throughout with an *approved automatic sprinkler system* in accordance with Section 903.3.1.1, the *building area* limitation in Table 503 is permitted to be increased by an additional 200 percent ($I_s = 2$) for buildings with more than one *story above grade plane* and an additional 300 percent ($I_s = 3$) for buildings with no more than one *story above grade plane*. These increases are permitted in addition to the height and *story* increases in accordance with Section 504.2.

> **Exception:** The *building area* limitation increases shall not be permitted for the following conditions:
>
> 1. The *automatic sprinkler system* increase shall not apply to *buildings* with an occupancy in Group H-1.
>
> 2. The *automatic sprinkler system* increase shall not apply to the *building area* of an occupancy in Group H-2 or H-3. For *buildings* containing such occupancies, the allowable *building area* shall be determined in accordance with Section 508.4.2, with the sprinkler system increase applicable only to the portions of the building not classified as Group H-2 or H-3.
>
> 3. *Fire-resistance rating* substitution in accordance with Table 601, Note d.

❖ This section permits an increase of the allowable building areas established in Table 503 for each type of construction if the building in question is equipped throughout with an automatic sprinkler system. This section provides for two levels of increase: 200 percent for multistory buildings and 300 percent for single-story buildings.

The scope of the phrase "protected throughout with

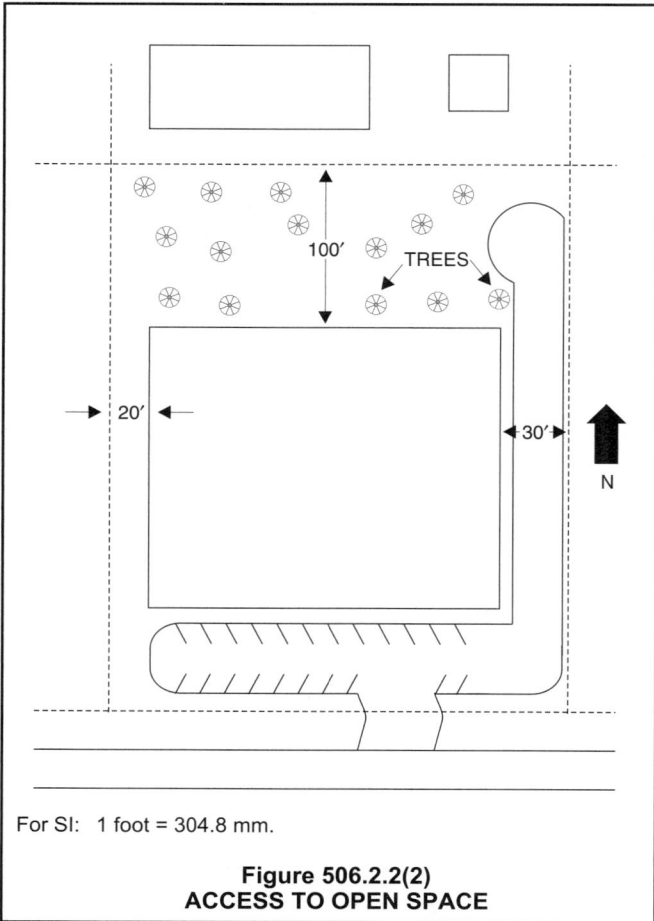

For SI: 1 foot = 304.8 mm.

Figure 506.2.2(2)
ACCESS TO OPEN SPACE

an automatic sprinkler system" means that the entire structure is to be provided with sprinkler protection designed and installed in accordance with NFPA 13, as stipulated in Section 903.3.1.1. It is intended that only buildings protected throughout the entire structure with a system designed in accordance with NFPA 13 be eligible for the sprinkler increase permitted by this section, except as specifically modified by the exceptions in Section 903.3.1.1.1. Those exceptions permit the omission of sprinklers in certain locations within buildings because of conditions that exist in those locations. Even if an exception is utilized in the sprinkler design, the building is still eligible for the area increase because the exempted locations either have a negligible impact on the fire load of the building or it is likely that other requirements will abate the hazard associated with these rooms. This section also clarifies that these area increases are cumulative with the height/story increases in Section 504. Please note that under Section 504.2 Group R-2 occupancy buildings are allowed a height increase when an NFPA 13R sprinkler system, as stipulated in Section 903.3.1.2, is provided. However, such a system will not allow an area increase under the requirements of this section. To achieve both height and area increases in a Group R occupancy building, an NFPA 13 system must be installed.

The three exceptions to this section address circumstances under which the area increases described in this section would not be allowed.

Exception 1 does not permit the area increase for buildings with Group H-1 occupancy areas because of the higher level of hazard that this group represents.

Exception 2 allows Group H-2 and H-3 occupancies to be placed in a building where the automatic increase would be taken, but with the limitation that the increase for automatic sprinklers would not apply to the Group H-2 or H-3 areas. These Group H-2 or H-3 areas would be required to be separated from the other uses, given that Sections 508.2.4, 508.3.3 and 508.4.4 specifically require separation of these occupancies from other occupancies. The required separation would be as given for separated occupancies in Section 508.4.4 and the allowable area would be based upon the sum of the ratios of the actual area of each occupancy to the allowable area per story (see commentary, Section 508.4.2). In the case of the Group H-2 or H-3 areas, the allowable area per story could not include the increase of 200 percent or 300 percent. This would have the effect of limiting the allowable areas of the entire building when a Group H-2 or H-3 area is being included.

Example: It is desired to construct a single-story Group F-1 factory using Type IIIB construction, with a 3,500-square-foot (325 m²) Group H-2 area. An automatic sprinkler system will be used throughout the building. The building does not qualify for any frontage increase.

Allowable area of the Group F-1 area with an automatic sprinkler system = 12,000 square feet (1115 m²)

plus 3[12,000 square feet (1115 m²)] = 48,000 square feet (4459 m²). So, without a Group H-2 area, the building could be 48,000 square feet (4459 m²) in area.

However, with the Group H-2 area present:

$$\frac{\text{Actual area of F-1}}{\text{Allowable area F-1}} + \frac{\text{Actual area of H-2}}{\text{Allowable area of H-2}}$$

$$\frac{\text{Actual area of F-1}}{36,000 \text{ square feet}} + \frac{3,500 \text{ square feet}}{7,000 \text{ square feet}} \leq 1$$

Solving this equation for the actual area of Group F-1, it is determined that the maximum area of Group F-1 that could be allowed would now be 18,000 square feet (1672 m²). Thus, while the high hazard is allowed, it does restrict the size of the host occupancy.

Exception 3 states that when an automatic sprinkler system is installed to address the allowed trade-off of sprinklers for 1-hour fire-resistance-rated construction given in Note d of Table 601, the area increases allowed in this section cannot be taken. For example, in a Group B building constructed using Type IIA construction, the required 1-hour fire-resistance rating of all structural members (with the exception of the exterior walls) could be reduced to zero if the building were fully equipped with a sprinkler system in accordance with Section 903.3.1.1 (NFPA 13); however, the allowable area will not be increased by the percentages that would otherwise be allowed in this section.

506.4 Single occupancy buildings with more than one story. The total allowable *building area* of a single occupancy building with more than one *story above grade plane* shall be determined in accordance with this section. The actual aggregate *building area* at all *stories* in the building shall not exceed the total allowable *building area*.

> **Exception:** A single basement need not be included in the total allowable *building area*, provided such basement does not exceed the area permitted for a building with no more than one *story above grade plane*.

❖ The total allowable area of a single occupancy building (the aggregate area of all stories) is determined in accordance with Section 506.4.1. This limit is based on determining the maximum allowable building area per story under the provisions of Sections 506.1 through 506.3. For buildings with two or more stories above grade plane, the total allowable area must be shared by all stories (see commentary, Section 506.4.1) and must be equal to or greater than the aggregate of the actual area of all stories. The area of a single basement, however, is not required to be counted as part of the total building area when evaluating total allowable area in accordance with Section 506.4.1. Therefore, if the total allowable area of a single occupancy building is 50,000 square feet (4645 m²) in accordance with Section 506.4.1, the total aggregate area of all stories above grade must not exceed 50,000 square feet (4645 m²). The presence of a single basement, however, is permitted to make the actual total area of the building larger than 50,000 square feet (4645 m²). If more than one basement is present, the extra basement levels must be counted as contributing to the total allowable area for the building.

Where a proposed building exceeds the total allowable building area determined under this section, the structure must either be built of a higher construction type that allows greater area, or it must be divided by fire walls into two or more buildings. The maximum allowable building area is determined for each building established by fire walls independently from the determinations for buildings on the other side of a fire wall. For fire walls, see Section 706.

506.4.1 Area determination. The total allowable *building area* of a single occupancy building with more than one *story above grade plane* shall be determined by multiplying the allowable *building area* per *story* (A_a), as determined in Section 506.1, by the number of *stories above grade plane* as listed below:

1. For buildings with two *stories above grade plane*, multiply by 2;

2. For buildings with three or more *stories above grade plane*, multiply by 3; and

3. No *story* shall exceed the allowable *building area* per *story* (A_a), as determined in Section 506.1, for the occupancies on that *story*.

Exceptions:

1. Unlimited area buildings in accordance with Section 507.

2. The maximum area of a building equipped throughout with an *automatic sprinkler system* in accordance with Section 903.3.1.2 shall be determined by multiplying the allowable area per *story* (A_a), as determined in Section 506.1, by the number of *stories above grade plane*.

❖ This section establishes a total allowable building area for a single occupancy building. The aggregate area of all stories, excluding a single basement, located within the building must not exceed the allowable building area established in this section. The determination of total allowable building area starts with the determination of the allowable building area per story. The per-story allowable area is established in Table 503 and may be increased under the provisions of Sections 506.1 through 506.3. The maximum allowable building area for the total building can be distributed on the various stories of the building provided any one story does not exceed the maximum allowable area per story.

The following examples demonstrate the determination of total building area as a function of allowable area per story:

Two-story building: A single occupancy building has an allowable area per story (A_a), determined in accordance with Section 506.1, of 10,000 square feet (929 m²). The total allowable area of the building is 10,000 square feet (929 m²) × 2 (number of stories) = 20,000

square feet (1858 m²). Each story above grade, therefore, may be 10,000 square feet (929 m²). In addition, the building may have one basement of no more than 10,000 square feet (929 m²) in accordance with the exception to Section 506.4.

Four-story building: A single occupancy building has an allowable area per story (A_a), determined in accordance with Section 506.1, of 10,000 square feet (929 m²). The total allowable area of the building is 10,000 square feet (929 m²) × 3 (the maximum multiplier) = 30,000 square feet (2787 m²). If each story is of uniform size, each story above grade may be 7,500 square feet (697 m²) [30,000 square feet (2787 m²)/4 stories]. In addition, the building may have one basement of no more than 10,000 square feet (929 m²) in accordance with the exception to Section 506.4.

If the allowable area per story is not exceeded for any single story, but the total allowable area is exceeded, a higher construction type must be used or the conditions of an area increase must be met. The following example demonstrates this point:

Building exceeds total allowable area: A four-story Group S-2 building of Type VA construction is proposed. The actual area per story is 34,000 square feet (3159 m²). The building has a fully open perimeter, but does not have a fire suppression system.

Section 506.2 yields an increase of 75 percent for full open frontage for this building, resulting in an allowable area per story (A_a) of 36,750 square feet (3414 m²) (21,000 square feet [1951 m²] × 1.75; see Section 506.2 for frontage increase calculation). The total allowable area of the building (aggregate of all stories above grade) is 36,750 square feet (3414 m²) × 3 (the maximum multiplier) = 110,250 square feet (10 242 m²). If each story is of uniform size, each story above grade may be 27,562 square feet (2560 m²) [110,250 square feet (10 242 m²)/4 stories]; therefore, the building does not comply. Two possible solutions for this problem are:

1. **Change construction type to Type IIIA:** Instead of Type VA construction, Type IIIA construction could be utilized. The allowable area per story (A_a) for Type IIIA construction is 39,000 square feet (3623 m²) × 1.75 = 68,250 square feet (6340 m²). The total allowable area of the building (aggregate area of all stories above grade) is 68,250 square feet (6340 m²) × 3 (the maximum multiplier) = 204,750 square feet (19 021 m²). If each story is of uniform size, each story above grade may be 51,187 square feet (4755 m²) [204,750 square feet (19 021 m²)/4 stories]. Since the actual area per story is 34,000 square feet (3159 m²), the building complies.

2. **Provide sprinklers:** Alternatively, if the building is equipped throughout with an automatic fire suppression system, the allowable area per story (A_a) for Type VA construction would be

21,000 square feet (1951 m²) × 3.75 = 78,750 square feet (7316 m²) [represents a 200-percent increase for sprinklers in accordance with Section 506.3, plus a 75-percent increase for frontage in accordance with Section 506.2 (see commentary, Sections 506.2 and 506.3)]. The total allowable area of the building (aggregate area of all stories above grade) is 78,750 square feet (7316 m²) × 3 (the maximum multiplier) = 236,250 square feet (21 948 m²). If each story is of uniform size, each story above grade may be 59,062 square feet (5487 m²) [236,250 square feet (21 948 m²)/4 stories]. Since the actual area per story is 34,000 square feet (3159 m²), the building complies.

This section does not require that the total allowable area of the building be equally distributed among the stories of a building, as long as no single story exceeds the allowable area per story (A_a) determined in accordance with Section 506.1. The following example demonstrates this point:

Building with stories of unequal area: Suppose the allowable area per story (A_a) for a single occupancy building is calculated in accordance with Section 506.1 to be 40,250 square feet (3739 m²) and the building has four stories. The total allowable area of the building (aggregate area of all stories above grade) is 40,250 square feet (3739 m²) × 3 (the maximum multiplier) = 120,750 square feet (11 218 m²). As long as no single story exceeds 40,250 square feet (3739 m²), the maximum allowable area could be divided among the stories unequally. For instance, the first story could have an area of 40,250 square feet (3739 m²) and the three upper stories could each have an area of {120,750 square feet (11 218 m²) - 40,250 square feet [3739 m²]}/3 = 26,833 square feet (2493 m²).

Exception 1 is necessary to establish that the use of the unlimited area provisions in Section 507 would obviate the need for compliance with Section 504. The unlimited area provisions apply to one- or two-story buildings, depending upon the occupancies and sprinkler provisions utilized.

Exception 2 is an increase allowed for residential sprinkler systems installed in accordance with NFPA 13R. The area increases allowed by Section 506.3 are only applicable when the building is to be equipped with a sprinkler system in accordance with NFPA 13—not NFPA 13R. The three-story multiplier limit or "three-story cap" was initially evaluated based upon use of area increases resulting from the installation of an NFPA 13 system. Since buildings equipped with an NFPA 13R system are not allowed an area increase, it was determined that the "three-story cap" had an unintentional detrimental impact on residential buildings. Specifically, for a four-story residential building with an NFPA 13R sprinkler system, the provi-

sions of Section 506.4.1 would reduce the area per story and the total building area. Given that the maximum height of buildings with NFPA 13R systems is four stories, this application of an increase in allowable area was determined to be warranted.

506.5 Mixed occupancy area determination. The total allowable *building area* for buildings containing mixed occupancies shall be determined in accordance with the applicable provisions of this section. A single basement need not be included in the total allowable *building area*, provided such basement does not exceed the area permitted for a building with no more than one *story above grade plane*.

❖ As with single occupancy buildings regulated by Section 506.4, mixed occupancy buildings are limited to a maximum area for an entire building. The maximum building area is a factor of the maximum area allowed per story as established under Sections 506.1 through 506.3. Mixed occupancies present a different mathematical problem with regard to the maximum total area allowed for the building. The total allowable area of a mixed occupancy building is determined in accordance with Section 506.5.1 or 506.5.2 depending on the number of stories above grade plane. All three of the mixed occupancy design options (accessory occupancies, nonseparated occupancies and separated occupancies) are acknowledged by this section and the rational sum of the ratio concept is maintained. For one-story mixed occupancy buildings, Section 506.5.1 defaults to Section 508.1. (In Section 508.1, the code user is directed to provisions for accessory occupancies, nonseparated occupancies and separated occupancies, each of which prescribes allowable area determination procedures for each mixed occupancy contingency.) For multistory buildings, Section 506.5.2 also defaults to Section 508.1, which limits the allowable area of a given story based on the applicable design method. Similar to single occupancy buildings, the area of a single basement in mixed occupancy buildings is not required to be counted as part of the total building area when evaluating total allowable area. If more than one basement is present, the extra basement levels must be counted as contributing to the total area for the building. Similar to single occupancy, multistory buildings, these mixed occupancy, multistory provisions only apply to buildings not permitted to be of unlimited area.

Where a proposed building exceeds the total allowable building area determined under this section and the referenced provisions of Section 508, the structure must either be built of a higher construction type that allows greater area, or it must be divided by fire walls into two or more buildings. The maximum allowable building area is determined for each building established by fire walls independently from the determination of the buildings on the other side of a fire wall. For fire walls, see Section 706.

506.5.1 No more than one story above grade plane. For buildings with no more than one *story above grade plane* and containing mixed occupancies, the total *building area* shall be determined in accordance with the applicable provisions of Section 508.1.

❖ Mixed occupancy, one story above grade buildings are required to comply with the applicable provisions of Section 508.1. The code provides three basic options for mixed occupancies: (1) accessory occupancies, (2) nonseparated occupancies and (3) separated occupancies. The availability of the options depends upon the area of each story desired for each different occupancy and the relationship of one occupancy to the other.

The accessory occupancy concept uses the same principle behind nonseparated occupancies, which is that the different occupancies within the building do not have to be separated by rated assemblies if the building complies throughout with the more restrictive code requirements. Unlike the nonseparated occupancies option, under the accessory occupancies option the main occupancy is used to determine the allowable area. Generally the code requirements for the accessory occupancy only need to be applied to the area occupied by the accessory occupancy unless the code requirement clearly indicates a larger area such as fire areas for determination of sprinkler system requirements.

Example: A single-story business office of 23,000 square feet (2137 m²) contains a lunch room that is 2,000 square feet (186 m²) in area. The main occupancy is Group B and the accessory occupancy is Group A-2. Since the Group A-2 occupancy has a floor area of less than 10 percent of the overall floor area of the single story, this is an accessory occupancy and there is no fire separation requirement between the Group A-2 and B occupancies. For purposes of evaluating the area limitation of Table 503, the building is evaluated solely as Group B.

In contrast to the accessory occupancy provisions illustrated above, the fundamental concept for nonseparated occupancies is that the allowable building areas are based upon the most restrictive requirements of each of the occupancies in the mixed occupancy building.

Example: A single-story Type IIB building contains nonseparated Group B and M occupancies. The Group B portion is 1,000 square feet (93 m²) in area while the Group M portion is 10,000 square feet (929 m²) in area. The building does not have a sprinkler system and does not qualify for any frontage increase. With the exception of Type I construction, the area limitation of Table 503 is more restrictive for Group M than for Group B. Therefore, for purposes of evaluating the area limitation of Table 503, the building is evaluated solely as Group M.

Actual area (Group M)/Allowable area
(Group M) ≤ 1.0

11,000/12,500 = 0.88 which is ≤ 1.0 ∴ OK

The separated occupancies concept requires the different occupancies to be separated in accordance with Table 508.4. The sum of the ratios of the actual floor area of each separated occupancy contained therein, as compared to the area allowed per story by Table 503 and as modified by Section 506 for each respective occupancy, must not exceed one.

Example: A single-story Type IIB building contains separated Group B and M occupancies. The Group B portion is 1,000 square feet (93 m²) in area while the Group M portion is 10,000 square feet (929 m²) in area. The building does not have a sprinkler system and does not qualify for any frontage increase.

$$\frac{\text{Actual area (B)}}{\text{Allowable area (B)}} + \frac{\text{Actual area of (M)}}{\text{Allowable area of (M)}} \leq 1.0$$

$$\frac{1,000}{23,000} + \frac{10,000}{12,500}$$

0.04 + 0.8 = 0.84 < 1.0 ∴ OK

506.5.2 More than one story above grade plane. For buildings with more than one *story above grade plane* and containing mixed occupancies, each *story* shall individually comply with the applicable requirements of Section 508.1. For buildings with more than three *stories above grade plane*, the total *building area* shall be such that the aggregate sum of the ratios of the actual area of each *story* divided by the allowable area of such *stories* based on the applicable provisions of Section 508.1 shall not exceed 3.

❖ When buildings containing mixed occupancies are two or more stories above grade plane, the same concept applied to Section 506.5.1 is applied to each story of the building. Once a mixed occupancy building exceeds three stories in height, the maximum allowed area for the total building is set at three times the sum of the ratios. The following examples illustrate the application of the total area limitations for each building as well as showing the variety of ways in which the provisions of Section 508, Mixed use and occupancy, can be used in the design of the building.

Example: Sprinklered, two-story Type IIA building.

First story: Group B occupancy—
3,000 square feet (278 m²)
Group S-2 occupancy—
23,000 square feet (2137 m²)

Second story: Group A-3 occupancy—
2,000 square feet (186 m²)
Group B occupancy—
24,000 square feet (2230 m²)

The first story qualifies for the nonseparated occupancy design method since the total square footage [26,000 square feet (2415 m²)] is less than that permitted for the most restrictive occupancy [Group

B—37,500 square feet (3484 m²)].

The second story is larger than that permitted for a Group A-3 occupancy; however, since the Group A-3 occupancy is less than 10 percent of the area of the second story, it qualifies as an accessory occupancy if subsidiary to the Group B occupancy.

In this instance, each story individually qualifies based on the applicable mixed-occupancy provision.

Example: Sprinklered, four-story Type IIB building, no frontage increase.

First story:	Group B occupancy— 20,000 square feet (1858 m²) Group M occupancy— 25,000 square feet (3159 m²)
Second story:	Group A-3 occupancy— 4,000 square feet (372 m²) Group B occupancy— 41,000 square feet (3809 m²)
Third story:	Group B occupancy— 15,000 square feet (1394 m²) Group S-1 occupancy— 15,000 square feet (1394 m²) Group F-1 occupancy— 15,000 square feet (1394 m²)
Fourth story:	Group B occupancy— 10,000 square feet (929 m²) Group S-2 occupancy— 35,000 square feet (3252 m²)

The first story does not qualify as an accessory or nonseparated occupancy. As separated occupancies, the ratio on the story is 0.96 [(20,000/69,000) + (25,000/37,500)].

The second story is larger than that permitted for a Group A-3 occupancy; however, since the Group A-3 occupancy is less than 10 percent of the area of the second story, it qualifies as an accessory occupancy if subsidiary to the Group B occupancy. The ratio on the story is 0.65 (45,000/69,000).

The third story qualifies as a nonseparated occupancy since the allowable area per story for the most restrictive occupancy (Group F-1; 46,500 square feet) is larger than the actual area on any individual story. The ratio of the floor is 0.97 (45,000/46,500).

The fourth floor qualifies as a nonseparated occupancy since the allowable area per floor for the most restrictive occupancy (Group B; 69,000 square feet) is larger than the actual area on any individual story. The ratio of the story is 0.65 (45,000/69,000).

The aggregate sum of the ratios is 3.23 (0.96 + 0.65 + 0.97 + 0.65). Since the aggregate sum of the ratios exceeds three, Type IIB construction would not be permitted without further modification.

SECTION 507
UNLIMITED AREA BUILDINGS

507.1 General. The area of buildings of the occupancies and configurations specified herein shall not be limited.

❖ This section addresses circumstances under which a building would be allowed to be constructed unlimited in building area. Depending upon the height (in stories) of the structure and the occupancy classification, the code presents different circumstances that would allow an unlimited area building. There is a common thread among the requirements for these facilities:

1. Except for the option in Section 507.2, the building is equipped throughout with an automatic sprinkler system.

2. The building is surrounded by increased open space, usually 60 feet (18 288 mm) in width.

3. The buildings are limited to one or two stories above grade plane, basements are not permitted.

In all cases the open space can occur either within public ways surrounding the site, or by yards provided on the lot between the building and the lot lines, or a combination of yards and public ways. Where a yard is used to achieve the open space, it must be on the same lot as the building receiving the benefit. With respect to the 60 feet (18 288 mm), the code does not specify how much of the public way can be included. (Fire separation distance, which is a different requirement, requires measurement to the centerline of the public way.) The open space must be provided in all directions around the perimeter of the building, not just measured at right angles to the building (see Figure 507.1). Two unlimited area buildings on the same property must be separated by 60 feet (18 288 mm) [or 40 feet (12 192 mm) if Section 507.5 is used] or they must be treated as the same building.

The open space located on the private property does not need to be dedicated to the public or publicly owned, but can be the location of parking, landscaping, roadways and other minor accessory features (tanks, generators, trash dumpster enclosures) (see IBC Interpretation No. 20-03). However, the yard cannot be occupied by any exterior use that is essentially a continuation of use of the building. For example, many big box retailers will have an adjoining lawn and garden merchandise area; or a lumber supply area that is only partially enclosed by walls, fencing and roof covering. This type of use would need to be considered part of the unlimited area building, and the 60 feet (18 288 mm) of open space provided beyond this area (see IBC Interpretation No. 03-05).

The open areas serve two key roles: separation of

these buildings from other buildings and ample space on all sides for fire-fighting operations. These buildings are limited to stories above grade plane because basements would not be directly accessible for fire fighting.

Please note that Section 507.5 permits a reduction in the open space in exchange for increased wall and opening protection.

507.2 Nonsprinklered, one story. The area of a Group F-2 or S-2 building no more than one *story* in height shall not be limited when the building is surrounded and adjoined by *public ways* or *yards* not less than 60 feet (18 288 mm) in width.

❖ By definition, occupancies of Groups F-2 and S-2 are not permitted to contain significant amounts of combustible materials (see Sections 306.3 and 311.3); therefore, because the fire load of the contents is lower, the hazard is lower. No other structures may be located within the 60-foot-wide (18 288 mm) open space required around the building, and the open space must be on the same lot and dedicated for public use to preclude any reduction of this isolation of the building. The type of construction is not restricted, and sprinklers are not required for unlimited area buildings containing Group F-2 and S-2 occupancies, provided the building is no more than one story above grade plane.

507.3 Sprinklered, one story. The area of a Group B, F, M or S building no more than one *story above grade plane*, or a Group A-4 building no more than one *story* above *grade plane* of other than Type V construction, shall not be limited when the building is provided with an *automatic sprinkler system* throughout in accordance with Section 903.3.1.1 and is surrounded and adjoined by *public ways* or *yards* not less than 60 feet (18 288 mm) in width.

Exceptions:

1. Buildings and structures of Types I and II construction for rack storage facilities that do not have access

by the public shall not be limited in height, provided that such buildings conform to the requirements of Sections 507.3, 903.3.1.1 and Chapter 23 of the *International Fire Code*.

2. The *automatic sprinkler system* shall not be required in areas occupied for indoor participant sports, such as tennis, skating, swimming and equestrian activities in occupancies in Group A-4, provided that:

   2.1. *Exit* doors directly to the outside are provided for occupants of the participant sports areas; and

   2.2. The building is equipped with a fire alarm system with manual fire alarm boxes installed in accordance with Section 907.

❖ Because of the excellent record in controlling and preventing fires, the installation of a sprinkler system throughout single-story buildings of the listed groups, individually or in combination, permits them to be unlimited in area, as long as they are surrounded by public ways or yards not less than 60 feet (18 288 mm) in width (IBC Interpretation No. 44-06). This also applies to two-story above-grade-plane buildings of these groups, with the exception of Group A-4 (see Section 507.4).

The life-safety hazards in buildings of these occupancies, because of the typical activities of the occupants and their level of awareness, are considered low enough that larger building areas and increased fire loads can be tolerated. For Groups B, F, M and S, this section may be applied to buildings of all construction types; it is not limited to noncombustible construction only. For Group A-4 indoor arenas, the unlimited area provision can be applied to buildings of Type I, II, III and IV construction only. This reflects the size of the occupant load in these types of facilities. The sprinkler system is required to be designed and installed in accordance with NFPA 13 (see Section 903.3.1.1). The

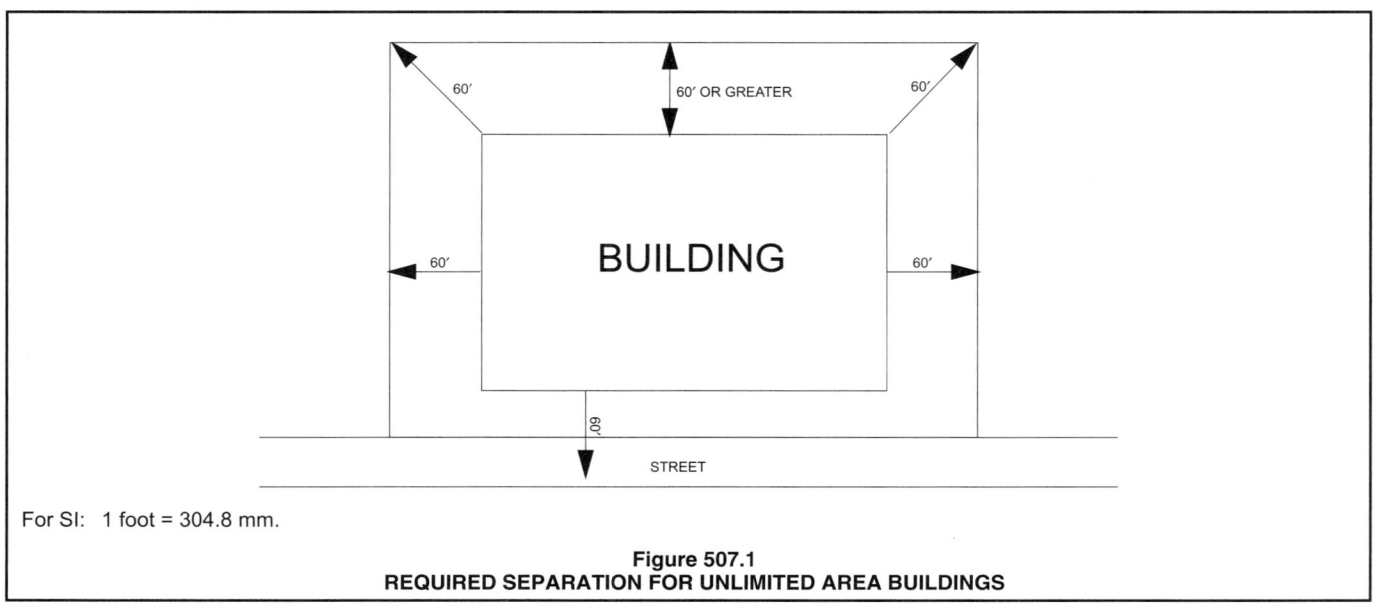

For SI: 1 foot = 304.8 mm.

Figure 507.1
REQUIRED SEPARATION FOR UNLIMITED AREA BUILDINGS

height of the unlimited area building cannot exceed the limits established in Table 503 and Section 504.2 for the occupancy and type of construction.

Exception 1 specifically permits Type I and II buildings with rack storage to be unlimited in height if they are surrounded by 60 feet (18 288 mm) of open space, do not permit access by the public and comply with NFPA 13 and Chapter 23 of the IFC. These standards contain provisions for storage configuration, in-rack sprinkler coverage and system requirements that offset the hazard introduced by increased height.

Exception 2 exempts providing a sprinkler system in certain areas of unlimited area Group A-4 occupancies where a lack of fuel loading and excessive ceiling heights would reduce the need for, and effectiveness of the system. Group A-4 occupancies frequently have indoor participant sports areas, such as tennis courts, skating rinks, swimming pools, baseball fields, basketball courts and equestrian activities, with spectator seating usually situated around the perimeter of the sports field or area. These types of indoor recreation areas often require very large, open areas with such high ceilings that the installation of an automatic sprinkler system in the immediate participant sport area would be ineffective. The potential for significant fire involvement in such an area is generally quite low because of the low fuel load; therefore, sprinkler coverage is unnecessary for the playing field in most of these buildings. These areas are, therefore, exempt from the suppression requirement of this section, provided the conditions regarding exiting and the required fire alarm system are met. When an indoor arena or sports facility is built without suppression in the participant sport area, occupants of the playing field or participant sport area must be able to exit the building directly from the playing area, without having to pass through other parts of the building. This eliminates the hazard of having to pass through higher fuel load areas such as locker rooms or concession areas. The building must also be equipped with a manual alarm system that complies with Section 907. This manual alarm system provides an additional and acceptable level of life safety in spite of the omission of sprinklers over the playing field area. This exception only applies to Group A-4 occupancies that are contained within an unlimited area building.

Omission of sprinkler coverage is permitted in the unlimited area building for the participant sport area only. All other areas are required to be equipped with an automatic sprinkler system. This includes all other rooms and spaces in the building, such as the spectator seating areas, locker rooms, restaurants, lounges, shops, arcades, skyboxes and storage areas.

507.3.1 Mixed occupancy buildings with Groups A-1 and A-2. Group A-1 and A-2 occupancies of other than Type V construction shall be permitted within mixed occupancy buildings of unlimited area complying with Section 507.3, provided:

1. Group A-1 and A-2 occupancies are separated from other occupancies as required for separated occupancies in Section 508.4.4 with no reduction allowed in the *fire-resistance rating* of the separation based upon the installation of an *automatic sprinkler system*;

2. Each area of the portions of the building used for Group A-1 or A-2 occupancies shall not exceed the maximum allowable area permitted for such occupancies in Section 503.1; and

3. All *exit* doors from Group A-1 and A-2 occupancies shall discharge directly to the exterior of the building.

❖ This section allows Group A-1 and A-2 occupancies in mixed occupancy, single-story unlimited area buildings under limited conditions. A typical example of a practical application of this would be the construction of a strip mall that is mainly for retail stores, but might contain a restaurant or a movie theater, or both. Group A-1 or A-2 buildings would not be permitted as stand-alone unlimited area buildings. Similar to the requirement in Section 507.3 for Group A-4 buildings, unlimited area buildings that contain a Group A-1 or A-2 occupancy are not permitted to be built of Type V construction.

The restrictions on the use of Group A-1 or A-2 occupancies in mixed occupancy, unlimited area buildings include: (1) required separation; (2) limited area of each Group A occupancy to the area allowed in Section 503.1; and (3) all required exits discharging to the exterior.

Item 1 states that the Group A-1 or A-2 occupancy is required to be separated from the rest of the unlimited area building by fire barriers in accordance with Section 508.4.4. For example, if a strip mall contains mercantile facilities next to a theater (Group A-1), the theater would need to be separated from the mercantile facilities by a 2-hour fire barrier, as determined from Table 508.4. Although the unlimited area building is required to be sprinklered, the required fire-resistance rating for the Group A-1 or A-2 occupancy separation must be based on the nonsprinklered entry in Table 508.4 (i.e., no reduction is permitted in the fire-resistance rating of the separation due to the presence of the sprinkler system).

Item 2 requires each Group A-1 or A-2 area to be limited to that allowed by Section 503.1. For example, if the strip mall mentioned above were Type IIB construction, the allowable area of the theater would be 34,000 square feet (3159 m²) [tabular value of 8,500 square feet (790 m²) × 4]. Note that the allowable area is increased by 300 percent as permitted by Section 506.3 for single-story buildings equipped with fire sprinklers. The allowable area could also be increased for frontage in accordance with Section 506.2, depending upon the amount of frontage that the Group A portion has, relative to its own perimeter. The important point is that the area limits for each Group A-1 or A-2 occupancy are based upon Section 503.1, not the tabular values of Table 503. Thus, because Section 503.1 states that the allowable area is limited to the values in Table 503.1 except as modified hereafter, the allowable increases given in Section 506 would be applicable. Also note that each Group A-1 or A-2 occupancy is evaluated on its own and not as an aggregate (i.e., the area of the Group A-1 and A-2 occupancies would not have to be added together).

Item 3 states that all exits from the Group A-1 or A-2 occupancy must discharge directly to the exterior. While other paths of travel available to the occupants are permitted to discharge back into the building, the required exits must discharge directly to the exterior such that an occupant in the Group A-1 or A-2 portion can exit the space without having to pass through other parts of the building. This eliminates the hazard of having to pass through higher fuel load areas.

507.4 Two story. The area of a Group B, F, M or S building no more than two *stories* above *grade plane* shall not be limited when the building is equipped throughout with an *automatic sprinkler system* in accordance with Section 903.3.1.1, and is surrounded and adjoined by *public ways* or *yards* not less than 60 feet (18 288 mm) in width.

❖ The rationale for allowing unlimited area two-story buildings in these four occupancy groups is the same as for one-story structures of these uses. The type of construction is not restricted. The number of stories are those above grade plane. Stories below grade plane (basements) are not permitted. Group A is not included in the occupancies that can be located within a two-story, unlimited area building because of the hazards associated with higher occupant loads. Even though limited Group A occupancies can be located in one-story unlimited area buildings, such occupancies cannot be located on either story of a two-story, unlimited area building. The provisions of Section 507.3.1 do not apply to two-story buildings.

507.5 Reduced open space. The *public ways* or *yards* of 60 feet (18 288 mm) in width required in Sections 507.2, 507.3, 507.4, 507.6 and 507.11 shall be permitted to be reduced to not less than 40 feet (12 192 mm) in width provided all of the following requirements are met:

1. The reduced width shall not be allowed for more than 75 percent of the perimeter of the building.

2. The *exterior walls* facing the reduced width shall have a minimum *fire-resistance rating* of 3 hours.

3. Openings in the *exterior walls* facing the reduced width shall have opening protectives with a minimum *fire protection rating* of 3 hours.

❖ The minimum width of the open space surrounding unlimited area buildings, with the exception of Group E buildings, may be reduced if the exterior walls are protected with 3-hour fire-resistance-rated construction and the openings are protected with 3-hour rated assemblies. All three conditions of this section must be met in order to reduce the open space to a minimum of 40 feet (12 192 mm). The criteria and standards for wall assemblies and opening protection are contained in Chapter 7. The rated walls and protected openings are required only for those portions of the walls that do not face at least 60 feet (18 288 mm) of unoccupied space or public way.

507.6 Group A-3 buildings of Type II construction. The area of a Group A-3 building no more than one *story* above *grade plane*, used as a *place of religious worship*, community hall, dance hall, exhibition hall, gymnasium, lecture hall, indoor swimming pool or tennis court of Type II construction, shall not be limited when all of the following criteria are met:

1. The building shall not have a stage other than a platform.

2. The building shall be equipped throughout with an *automatic sprinkler system* in accordance with Section 903.3.1.1.

3. The building shall be surrounded and adjoined by *public ways* or *yards* not less than 60 feet (18 288 mm) in width.

❖ Assembly buildings pose a greater risk in general, primarily because of their relatively large occupant loads and the density of such loads; therefore, the unlimited area provisions are more restrictive for assembly occupancies. Limited types of Group A-3 buildings are allowed as unlimited area buildings provided each is limited to one story above grade plane. In the case of the Group A-3 occupancies listed, the construction classifications allowed to be unlimited area include Type II, which are noncombustible structures, and Type III and IV (see Section 507.7), which are combustible structures. The types of Group A-3 occupancies listed—churches, dance halls, community halls, exhibition halls, gymnasiums, lecture halls, indoor swimming pools and tennis courts—are typically well-lit, open spaces, containing a low fuel load. Libraries and museums are excluded, as they generally have a higher fuel load, as are bowling alleys and pool halls that are often poorly lit and have higher levels of activity. The dance halls included in Group A-3 are not to be confused with nightclubs, which are Group A-2 occupancies.

As with the occupancies referenced in Sections 507.2, 507.3 and 507.4, the Group A-3 buildings listed in this section must have 60 feet (18 288 mm) of open space surrounding the structure. Similar to the occupancies referenced in Sections 507.3 and 507.4, a sprinkler system must be provided and is required to be designed and installed in accordance with NFPA 13. Group A-3 buildings are further restricted, however, in that the building must not have a stage (other than a platform). These requirements aid in the exiting of the building in an emergency.

507.7 Group A-3 buildings of Types III and IV construction. The area of a Group A-3 building no more than one *story above grade plane*, used as a *place of religious worship*, community hall, dance hall, exhibition hall, gymnasium, lecture hall, indoor swimming pool or tennis *court* of Type III or IV construction, shall not be limited when all of the following criteria are met:

1. The building shall not have a stage other than a platform.

2. The building shall be equipped throughout with an *automatic sprinkler system* in accordance with Section 903.3.1.1.

3. The assembly floor shall be located at or within 21 inches (533 mm) of street or grade level and all *exits* are provided with ramps complying with Section 1010.1 to the street or grade level.

4. The building shall be surrounded and adjoined by *public ways* or *yards* not less than 60 feet (18 288 mm) in width.

❖ The requirements for Group A-3 buildings of Type III or IV construction are similar to the provisions for buildings of Type II construction. Due to the permitted use of combustible materials in Group A-3 buildings of Type III and IV construction, these buildings are further restricted in that the assembly floor level of the building must not be higher than 21 inches (533 mm) above street or grade level. These requirements aid in the exiting of the building occupants in an emergency. In order to further assist in the evacuation of the building in an emergency, assembly floors which are not at grade level must be provided with ramps (i.e., no stairs) which lead to grade level. Note that the 21-inch (533 mm) height limitation for the assembly floor would prohibit the installation of a mezzanine, balcony or most raised floor surfaces in the building.

507.8 Group H occupancies. Group H-2, H-3 and H-4 occupancies shall be permitted in unlimited area buildings containing Group F and S occupancies, in accordance with Sections 507.3 and 507.4 and the limitations of this section. The aggregate floor area of the Group H occupancies located at the perimeter of the unlimited area building shall not exceed 10 percent of the area of the building nor the area limitations for the Group H occupancies as specified in Table 503 as modified by Section 506.2, based upon the percentage of the perimeter of each Group H floor area that fronts on a street or other unoccupied space. The aggregate floor area of Group H occupancies not located at the perimeter of the building shall not exceed 25 percent of the area limitations for the Group H occupancies as specified in Table 503. Group H occupancies shall be separated from the rest of the unlimited area building and from each other in accordance with Table 508.4. For two-story unlimited area buildings, the Group H occupancies shall not be located more than one *story above grade plane* unless permitted by the allowable height in stories and feet as set forth in Table 503 based on the type of construction of the unlimited area building.

❖ This section sets forth provisions for limited areas of Group H-2, H-3 and H-4 occupancies in unlimited area buildings. Group F facilities often inherently contain a certain amount of Group H-2, H-3 and H-4 occupancies when hazardous materials are needed for the industrial processes. Similarly, it is reasonable to provide for limited quantities of hazardous material storage in a storage (Group S) occupancy.

The allowable area of the Group H-2, H-3 or H-4 occupancy is dependent on the type of construction of the building and the location of the high-hazard occupancy in the building. If the high-hazard occupancy is surrounded on all sides by an unlimited area building, it is difficult for fire department personnel to locate and access that area; therefore, Group H-2, H-3 and H-4 occupancies that are not located at the perimeter of the building are limited to 25 percent of the area limitation specified in Table 503 for the building type of construction and occupancy classification, as shown in Figure 507.8(1).

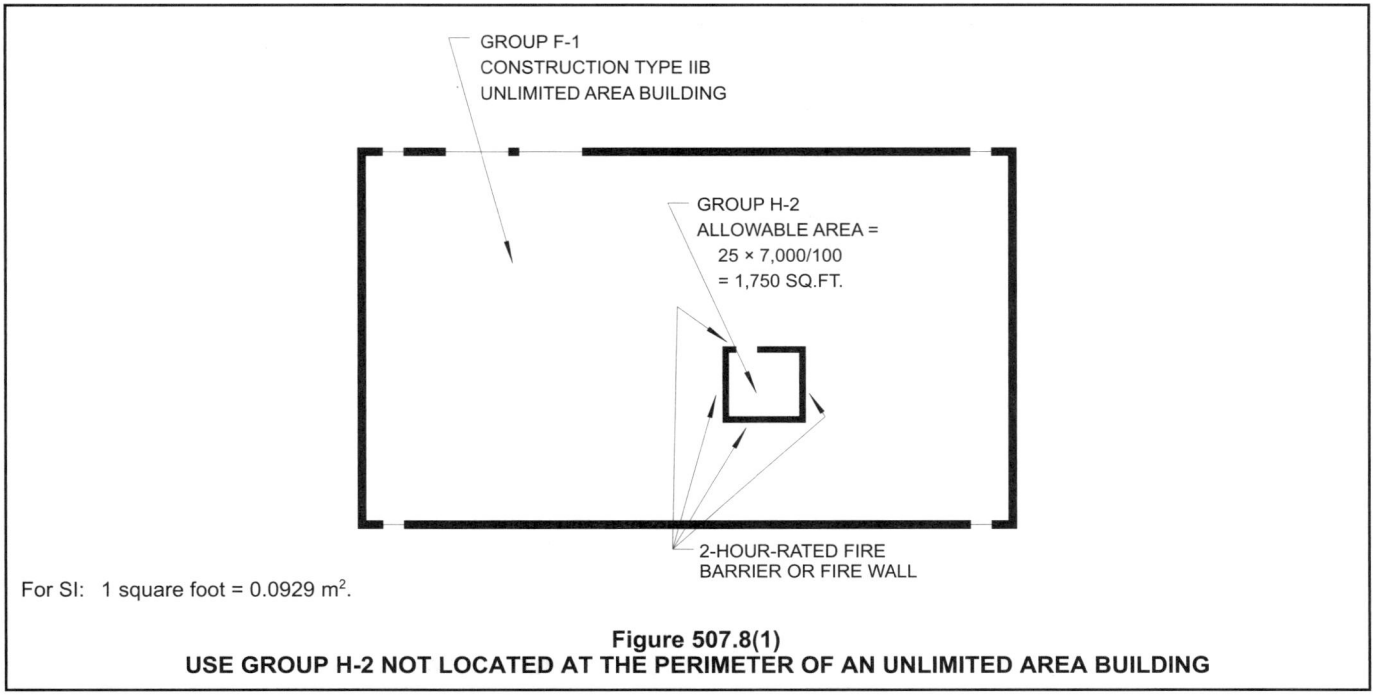

GROUP F-1
CONSTRUCTION TYPE IIB
UNLIMITED AREA BUILDING

GROUP H-2
ALLOWABLE AREA =
25 × 7,000/100
= 1,750 SQ.FT.

2-HOUR-RATED FIRE
BARRIER OR FIRE WALL

For SI: 1 square foot = 0.0929 m².

Figure 507.8(1)
USE GROUP H-2 NOT LOCATED AT THE PERIMETER OF AN UNLIMITED AREA BUILDING

More ready access to the high-hazard occupancy from the exterior of the building provides the fire department with an opportunity to respond more effectively to an incident. As such, Group H-2, H-3 and H-4 occupancies that are located on the perimeter of an unlimited area building are permitted to be a maximum of 10 percent of the total building area, or the maximum area permitted by Table 503 as modified by Section 506.2 for the Group H occupancy, whichever is less.

The increase permitted in Section 506.2 is to be based on the open frontage of Group H-2, H-3 and H-4 floor areas. In Figure 507.8(2), the Group H-2 floor area is located in the corner of the Group F-1 building, so that two walls of the perimeter of the Group H-2 floor area front an unoccupied open space. In accordance with Section 506.2, the area increase due to frontage (I_f) is calculated to be 25 percent, in accordance with the figure. Applying this increase to the area from Table 503, the allowable area for the Group H-2 occupancy is [7,000 square feet (650 m²)](1.25) = 8,750 square feet (813 m²). This is less than 10 percent of the total area per story [i.e., 15,000 square feet (1394 m²)], so 8,750 square feet (813 m²) is the maximum allowable area for the Group H-2 occupancy.

Locating the Group H-2 floor area further away from the perimeter of the Group F-1 building, as occurs in Figure 507.8(3), qualifies the Group H-2 floor area for an even greater allowable area increase [see Figure 507.8(3)].

This section also notes that the Group H occupancy must be on the first story above grade unless Table 503 allows for such occupancies on the second story above grade. This will depend on the type of construction. All the other size limitations still apply.

507.9 Aircraft paint hangar. The area of a Group H-2 aircraft paint hangar no more than one *story above grade plane* shall not be limited where such aircraft paint hangar complies with the provisions of Section 412.6 and is surrounded and adjoined by *public ways* or *yards* not less in width than one and one-half times the *building height*.

❖ Because of their specialized nature and the required size of these facilities, one-story Group H-2 aircraft paint hangars are not limited in area if they meet all the requirements of Sections 412.4 through 412.4.6. These sections require that the building be of Type I or II construction and have a fire suppression system in accordance with NFPA 409. This standard requires a foam-water suppression system in the paint areas and water sprinklers in all accessory areas. The criteria for the foam-water system depends on the size (and classification in accordance with NFPA 409) of the hangar. Other requirements of Section 412.4 include compartmentalized storage and use of flammable liquids, compliance with the IFC for spray application of flammable liquids and compliance with the *International Mechanical Code®* (IMC®) for ventilation of flammable-finish application areas (see commentary, Section 412.6).

In addition to compliance with Section 412.4, the hangar must be surrounded by open space. However, instead of the 60-foot (18 288 mm) width required for other buildings (see Sections 507.2, 507.3, 507.4,

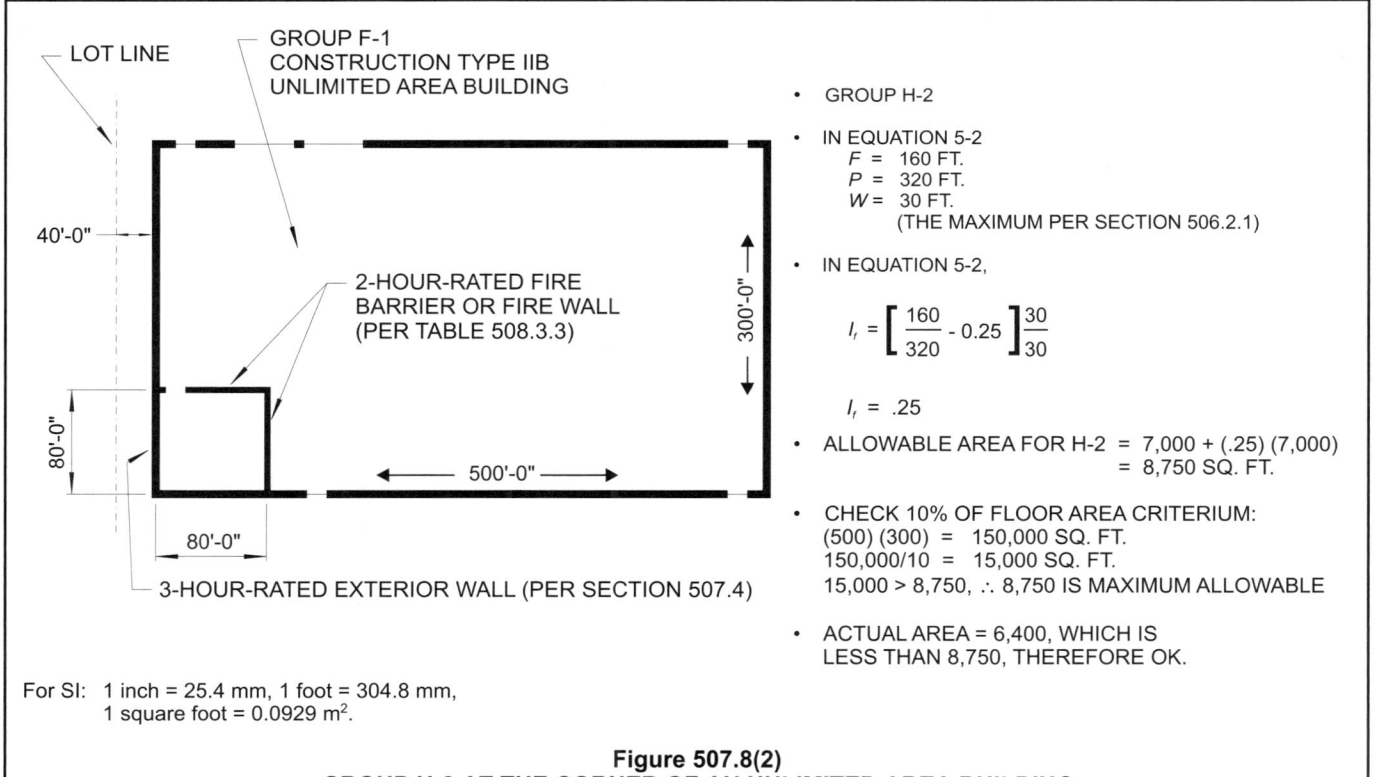

For SI: 1 inch = 25.4 mm, 1 foot = 304.8 mm,
 1 square foot = 0.0929 m².

Figure 507.8(2)
GROUP H-2 AT THE CORNER OF AN UNLIMITED AREA BUILDING

507.6, 507.7, 507.10 and 507.11), the open space surrounding the hangar must be at least one and one-half times the height of the building. Therefore, if the hangar was 30 feet (9144 mm) high, there would need to be at least 45 feet (13 716 mm) [30 feet (9144 mm) x 1.5] of open space surrounding the building.

Note that one-story paint hangars would also be permitted to exceed the height limit (in feet) of Table 503 if the exception to Section 504.1 is satisfied.

507.10 Group E buildings. The area of a Group E building no more than one *story above grade plane*, of Type II, IIIA or IV construction, shall not be limited when all of the following criteria are met:

1. Each classroom shall have not less than two *means of egress*, with one of the *means of egress* being a direct *exit* to the outside of the building complying with Section 1020.

2. The building is equipped throughout with an *automatic sprinkler system* in accordance with Section 903.3.1.1.

3. The building is surrounded and adjoined by *public ways* or *yards* not less than 60 feet (18 288 mm) in width.

❖ This section permits Group E structures to be built as an unlimited area building. A direct exit to the outside from each classroom must be provided in addition to another means of egress. It is clear in the wording that to fulfill this requirement, students must be able to egress the classroom directly to the outside without intervening corridors, exit passageways or exit enclosures. The unlimited area school building must also be fully sprinklered in accordance with NFPA 13, have open space on all sides of at least 60 feet (18 288 mm) and be no more than one story above grade plane.

Where accessory assembly spaces are included as part of the Group E occupancy classification in accordance with Exception 4 to Section 303.1, they would also be allowed to be unlimited in area under this section. The type of construction is limited to Type II, IIIA or IV.

507.11 Motion picture theaters. In buildings of Type II construction, the area of a motion picture theater located on the first *story above grade plane* shall not be limited when the building is provided with an *automatic sprinkler system* throughout in accordance with Section 903.3.1.1 and is surrounded and adjoined by *public ways* or *yards* not less than 60 feet (18 288 mm) in width.

❖ Type II motion picture theaters located in buildings of Type II construction that are no more than one story above grade plane may be unlimited in area by virtue of being of noncombustible construction, and having low fire loading and sprinkler protection in accordance with NFPA 13. The building must also be surrounded by at least 60 feet (18 288 mm) of open space.

507.12 Covered mall buildings and anchor stores. The area of *covered mall buildings* and *anchor stores* not exceeding three *stories* in height that comply with Section 402.6 shall not be limited.

❖ This section is simply included as a reference to Section 402.6 for covered malls. The definition of "Covered mall" has been expanded to include open malls as well. An open mall is very similar to a covered mall but the central pedestrian mall area is open to the sky. If an open mall complies with the provisions for covered malls under Section 402, it is allowed to be of unlimited area.

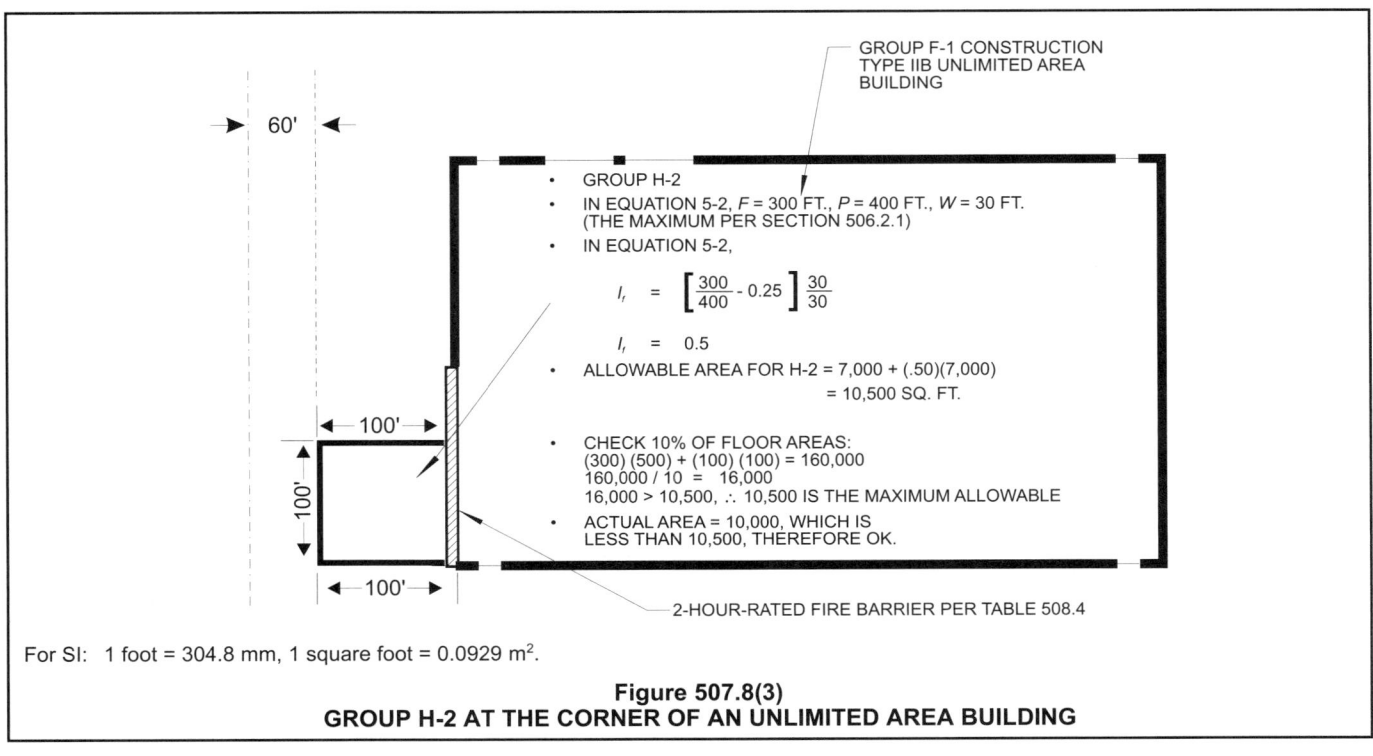

For SI: 1 foot = 304.8 mm, 1 square foot = 0.0929 m².

Figure 507.8(3)
GROUP H-2 AT THE CORNER OF AN UNLIMITED AREA BUILDING

SECTION 508
MIXED USE AND OCCUPANCY

508.1 General. Each portion of a building shall be individually classified in accordance with Section 302.1. Where a building contains more than one occupancy group, the building or portion thereof shall comply with the applicable provisions of Section 508.2, 508.3 or 508.4, or a combination of these sections.

Exceptions:

1. Occupancies separated in accordance with Section 509.

2. Where required by Table 415.3.2, areas of Group H-1, H-2 and H-3 occupancies shall be located in a separate and detached building or structure.

3. Uses within live/work units, complying with Section 419, are not considered separate occupancies.

❖ Very frequently, buildings are designed for multiple uses, and invariably that will result in the building having more than one occupancy classification. This section describes the provisions governing the condition when a building contains more than one use, or more than one of the occupancies identified in Sections 303 through 312. For example, a facility that appears to be used solely for storage often has a small office area for bookkeeping and other purposes. Because the office would be classified as a business occupancy, the building would be identified as a mixed occupancy, storage (Group S-1) and business (Group B).

The section contains three major parts:

1. Accessory occupancies including incidental accessory occupancies (see Section 508.2). This portion of the section addresses situations where a building contains a main occupancy with minor areas of a building containing occupancies that can be treated as part of the main occupancy or specific functions that need specific attention.

2. Mixed occupancies that are not separated from each other (see Section 508.3). This section addresses circumstances where the building contains more than one different occupancy classification in which certain requirements for each occupancy are addressed on a building-wide basis.

3. Mixed occupancies where occupancies of different hazard levels are separated (see Section 508.4). This section addresses circumstances where the building contains more than one occupancy classification but in which occupancies with different hazard levels are separated from each other by fire-resistance-rated construction.

The final sentence of this section reemphasizes the option for the designer to use just the provisions of one of the three parts, or options, and not to use the other options. But the code also allows a mixture of the options in different portions of a building. For example, a building could be designed to comply with

only Section 508.3 for nonseparated mixed occupancies and not comply with any of the provisions of either Section 508.2 or 508.4. In this example, the incidental accessory occupancy provisions of Section 508.2.5 would not apply. A different example could be to use the provisions of Section 508.3 on the first story of a building which perhaps had three or four different occupancies, but then separate the upper stories of the building from the first story and treat the upper stories under Section 508.4 for separated occupancies. A final example would be a building that was predominately one use but had small accessory areas on various stories. For this building the best option might be Section 508.2 regarding accessory occupancies. If this section is used in the design, then those items listed in Table 508.2.5 will have to be separated or protected as prescribed.

These options depend upon the limitations on their use, as stated in the respective sections of the code. Basically, the availability of the options depend upon the size of the areas desired for each different occupancy, and the relationship of one occupancy to the other.

Exception 1 provides a fourth set of options for mixed occupancies by referring to Section 509, which describes special circumstances of parking garages located above or below other uses.

Exception 2 eliminates the options of including Group H-1, H-2 or H-3 in the same building with other occupancies when quantities exceed certain amounts. This is required by Section 415.3.2 because of the nature of the hazard for high quantities of flammable, combustible and explosive materials.

Exception 3 excepts uses within a live/work unit from needing to comply with Section 508 provided the uses are within the limitations of Section 419. Live/work units provide the option for as much as 50 percent of nonresidential uses within what is classified as a Group R-2 dwelling unit.

508.2 Accessory occupancies. Accessory occupancies are those occupancies that are ancillary to the main occupancy of the building or portion thereof. Accessory occupancies shall comply with the provisions of Sections 508.2.1 through 508.2.5.3.

❖ Buildings often have rooms or spaces with an occupancy that is different from, but accessory to, the principal occupancy of the building. When such accessory areas are limited in size, they will not ordinarily represent a significantly different life safety hazard. This principle does not apply, however, to the incidental accessory occupancy areas indicated in Section 508.2.5 or where otherwise indicated in Sections 508.2.4 and 508.3.3 for areas classified as Group H.

The occupancy must be ancillary to the principal purpose for which the structure is occupied. This means that the purpose and function of the area is subordinate and secondary to the structure's primary function. As such, the activities that occur in accessory use areas are necessary for the principal occupancy to properly function and would not otherwise reasonably exist apart from the principal occupancy.

508.2.1 Area limitations. Aggregate accessory occupancies shall not occupy more than 10 percent of the *building area* of the *story* in which they are located and shall not exceed the tabular values in Table 503, without *building area* increases in accordance with Section 506 for such accessory occupancies.

❖ The aggregate area within a story devoted to the occupancies that are designated as accessory use areas must not be greater than 10 percent of the area of that story [see Figure 508.2.1(1)].

The area of the portion of the building devoted to an accessory occupancy must be less than the tabular building area permitted by Table 503, based on the group classification that most nearly resembles the accessory occupancy under consideration. Area increases for street frontage and automatic sprinkler protection based on the provisions of Sections 504 and 506 are not allowed [see Figure 508.2.1(2)].

508.2.2 Occupancy classification. Accessory occupancies shall be individually classified in accordance with Section 302.1. The requirements of this code shall apply to each portion of the building based on the occupancy classification of that space.

❖ Under Section 508.2, each accessory use is to be classified in accordance with Section 302 in the appropriate occupancy classification. Code requirements such as means of egress, the provision of sprinkler protection and structural load are to be determined for this occupancy as if it were a main occupancy of the building. For example, an accessory lunchroom located in a business office would need to be protected

with an automatic sprinkler system if the lunchroom's occupant load exceeds 100. The lunchroom would need to comply with all other code requirements applicable to a Group A-2 use.

Please note that Section 508.2.3 states that height and area of the building are based on the main occupancy; therefore, for determining the construction type for the total building, the building's floor area and height should be compared to the limits of Table 503 to determine the appropriate construction type. See Section 508.2.1 regarding the area of the accessory occupancy itself and see Section 508.2.3 regarding the height of the accessory occupancy itself.

When applying Chapter 9, many of the requirements for automatic sprinkler systems are based upon the size of the fire area in which the occupancy is located. For instance, a Group A-2 occupancy requires a sprinkler system when the fire area is greater than 5,000 square feet (465 m²) or when the fire area has an occupant load of 100 or more. By understanding the definition of a "Fire area," if there is no rated separation of the accessory occupancy from the main occupancy, the fire area will be everything between fire barriers, fire walls or exterior walls, and therefore could, and most likely will, contain both the accessory occupancy area as well as the main occupancy.

Continuing with this example, suppose that a single-story business office of 23,000 square feet (2137 m²) contains a lunchroom that is 2,000 square feet (186 m²) and an occupant load of 134. The main occupancy is a Group B, and the accessory occupancy is a Group A-2. Because the Group A-2 occupancy has a floor area of less than 10 percent of the overall floor area of

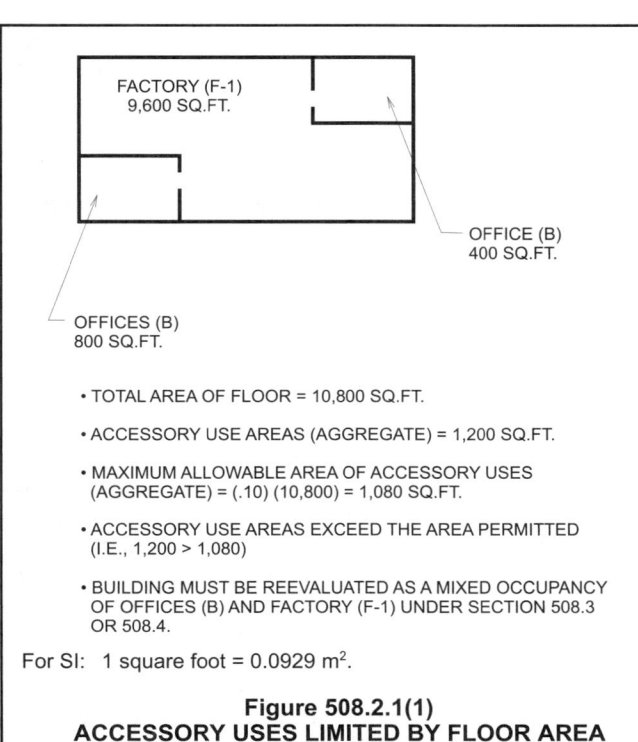

- TOTAL AREA OF FLOOR = 10,800 SQ.FT.

- ACCESSORY USE AREAS (AGGREGATE) = 1,200 SQ.FT.

- MAXIMUM ALLOWABLE AREA OF ACCESSORY USES (AGGREGATE) = (.10) (10,800) = 1,080 SQ.FT.

- ACCESSORY USE AREAS EXCEED THE AREA PERMITTED (I.E., 1,200 > 1,080)

- BUILDING MUST BE REEVALUATED AS A MIXED OCCUPANCY OF OFFICES (B) AND FACTORY (F-1) UNDER SECTION 508.3 OR 508.4.

For SI: 1 square foot = 0.0929 m².

Figure 508.2.1(1)
ACCESSORY USES LIMITED BY FLOOR AREA

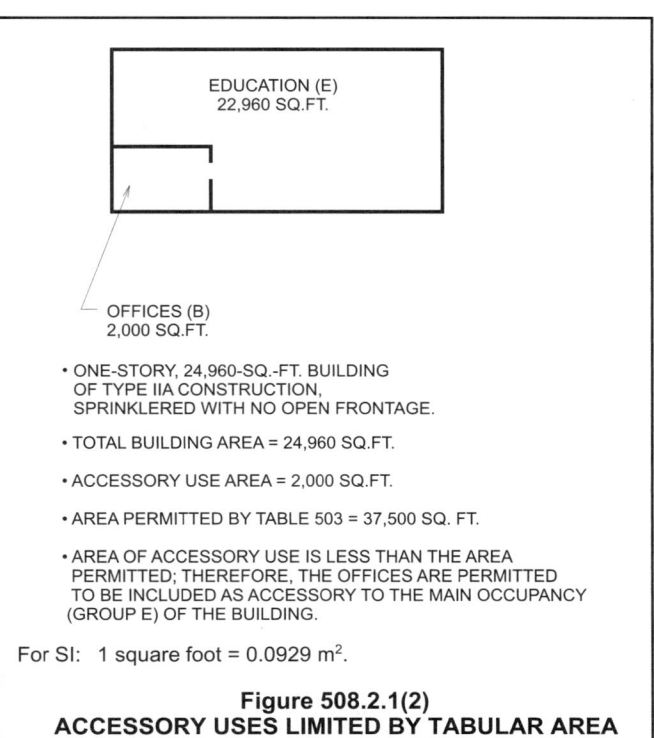

- ONE-STORY, 24,960-SQ.-FT. BUILDING OF TYPE IIA CONSTRUCTION, SPRINKLERED WITH NO OPEN FRONTAGE.

- TOTAL BUILDING AREA = 24,960 SQ.FT.

- ACCESSORY USE AREA = 2,000 SQ.FT.

- AREA PERMITTED BY TABLE 503 = 37,500 SQ. FT.

- AREA OF ACCESSORY USE IS LESS THAN THE AREA PERMITTED; THEREFORE, THE OFFICES ARE PERMITTED TO BE INCLUDED AS ACCESSORY TO THE MAIN OCCUPANCY (GROUP E) OF THE BUILDING.

For SI: 1 square foot = 0.0929 m².

Figure 508.2.1(2)
ACCESSORY USES LIMITED BY TABULAR AREA

the single story, this is an accessory use, and there is no fire separation requirement between the Group A-2 and Group B. A Group A-2 occupancy is required to be provided with an automatic sprinkler system when it is either in excess of 5,000 square feet (465 m²) or 100 occupants. Since this fire area is 23,000 square feet (2137 m²) and contains a Group A-2 occupancy with an occupant load of over 100 and is located within a fire area of greater than 5,000 square feet (465 m²) this fire area would need to be provided with an automatic sprinkler system throughout.

508.2.3 Allowable building area and height. The allowable *building area and height* of the building shall be based on the allowable *building area and height* for the main occupancy in accordance with Section 503.1. The height of each accessory occupancy shall not exceed the tabular values in Table 503, without increases in accordance with Section 504 for such accessory occupancies. The *building area* of the accessory occupancies shall be in accordance with Section 508.2.1.

❖ For accessory uses, the height and area of the building containing the accessory use is based upon the height and area of the main use because accessory uses are limited to 10 percent of the area of the story in which they are housed. The accessory occupancies' area and height are further limited in that neither can exceed the tabular area given by Table 503. For this determination, the height and area increases cannot be applied to the accessory use areas. In other words, the accessory use must be a minor part of the total area of the building. Therefore, the construction type for the total building will be based on the total area of the main and accessory occupancies, but the result may be to limit an accessory occupancy to certain stories of a building. Continuing with the Group A-2 accessory lunchroom in the Group B occupancy building: If the building is two stories of Type VB construction, the accessory Group A-2 lunchroom would be limited to a first-story location under the limits set in Table 503.

508.2.4 Separation of occupancies. No separation is required between accessory occupancies and the main occupancy.

Exceptions:

1. Group H-2, H-3, H-4 and H-5 occupancies shall be separated from all other occupancies in accordance with Section 508.4.

2. Incidental accessory occupancies required to be separated or protected by Section 508.2.5.

3. Group I-1, R-1, R-2 and R-3 *dwelling units* and *sleeping units* shall be separated from other *dwelling* or *sleeping units* and from accessory occupancies contiguous to them in accordance with the requirements of Section 420.

❖ When a designer is using Section 508.2, most accessory occupancies need not be separated from the main occupancy, unless that accessory use is a Group H-2, H-3, H-4 or H-5 high-hazard occupancy. In addition, those accessory occupancies listed in Table

508.2.5 as incidental accessory occupancies must be separated or protected as specified in the table. Please note that even where an accessory occupancy is not separated from the main occupancy under the provisions of Section 508, there may be other provisions of the code that might require a fire-resistance-rated separation. For instance, if it was desired not to provide a sprinkler system in the fire area discussed in the commentary example in Section 508.2.2, then a fire barrier would need to be installed to make the fire area containing the Group A-2 occupancy smaller than the threshold amount of 5,000 square feet (465 m²) and below the 100-occupant threshold. Or, another example is, in some cases there could be a wall between two different occupancies that is required by another section of the code to be fire-resistance rated (e.g., a separation wall is provided that also serves as a corridor wall that is required to be fire-resistance rated).

Exception 1 addresses the need to separate Group H-2, H-3, H-4 or H-5 occupancies because of the higher risk posed by these high-hazard occupancies. Please notice that Group H-1 occupancies are not mentioned because these occupancies are always required to be located in a separate building (see Section 415.4).

Exception 2 states that incidental accessory occupancies are required to be separated or protected in accordance with Sections 508.2.5 through 508.2.5.3. When using the accessory occupancy provisions of Section 508.2, the listed accessory incidental occupancies must be separated or protected in accordance with Table 508.2.5.

Exception 3 states that the separation between dwelling units and sleeping units and any accessory occupancies must still comply with Section 420. For example, a common lounge or manager's office that otherwise might qualify as an accessory occupancy within an apartment building must still be separated from adjoining dwelling units.

508.2.5 Separation of incidental accessory occupancies. The incidental accessory occupancies listed in Table 508.2.5 shall be separated from the remainder of the building or equipped with an automatic fire-extinguishing system, or both, in accordance with Table 508.2.5.

Exception: Incidental accessory occupancies within and serving a *dwelling unit* are not required to comply with this section.

❖ Incidental use areas are rooms or areas that constitute special hazards or risks to life safety. For most accessory occupancies no separation is required from the main occupancy because the limited area of an accessory occupancy will limit any exposure to the main occupancy that may be inherent in the accessory occupancy. However, for the listed accessory incidental occupancy, the code states that the separation between the building and its accessory area should not be eliminated. The protection for the rooms or areas identified in Table 508.2.5 is mandatory under the accessory occupancy design option. These protection requirements, however, are not applicable to incidental use areas that are located within and serve a dwelling unit. In conjunction with making the requirements of this section mandatory for the listed occupancies, two incidental areas were removed from Table 508.2.5 in the 2009 edition. Parking and storage are not regulated in the code as incidental accessory occupancies, but can be addressed either as a regular accessory occupancy without separation from the main occupancy or under one of the other mixed occupancy options.

The purpose of Table 508.2.5 is to identify the incidental accessory occupancy areas that require special protection and to indicate the required protection when the designer chooses to apply the accessory occupancy mixed use option. The protection requirements identified in Table 508.2.5 vary depending on the incidental accessory occupancy. In some cases, a specific type of protection is required, while in others there is an option. As indicated by Section 508.2.5.3, the requirement for an automatic fire-extinguishing system for the majority of those listed applies only to the incidental accessory occupancy room or area, not the entire building.

TABLE 508.2.5. See below.

❖ Table 508.2.5 identifies accessory incidental occupancies and the required separation or other protection to be provided. Where a fire-resistance-rated separation is required, the incidental accessory occupancy area must be separated from other portions of the building with fire barriers that comply with Section 707. Where Table 508.2.5 permits protection by an automatic fire-extinguishing system without fire barriers, the walls enclosing the incidental accessory occupancy area must comply with Section 508.2.5.2.

508.2.5.1 Fire-resistance-rated separation. Where Table 508.2.5 specifies a fire-resistance-rated separation, the incidental accessory occupancies shall be separated from the remainder of the *building* by a *fire barrier* constructed in accordance with Section 707 or a *horizontal assembly* constructed in accordance with Section 712, or both. Construction supporting 1-hour fire-resistance-rated *fire barriers* or *horizontal assemblies* used for incidental accessory occupancy separations in buildings of Type IIB, IIIB and VB construction are not required to be fire-resistance rated unless required by other sections of this code.

❖ Where a fire-resistance rating is required, the incidental accessory occupancy area must be separated from

TABLE 508.2.5
INCIDENTAL ACCESSORY OCCUPANCIES

ROOM OR AREA	SEPARATION AND/OR PROTECTION
Furnace room where any piece of equipment is over 400,000 Btu per hour input	1 hour or provide automatic fire-extinguishing system
Rooms with boilers where the largest piece of equipment is over 15 psi and 10 horsepower	1 hour or provide automatic fire-extinguishing system
Refrigerant machinery room	1 hour or provide automatic sprinkler system
Hydrogen cutoff rooms, not classified as Group H	1 hour in Group B, F, M, S and U occupancies; 2 hours in Group A, E, I and R occupancies.
Incinerator rooms	2 hours and automatic sprinkler system
Paint shops, not classified as Group H, located in occupancies other than Group F	2 hours; or 1 hour and provide automatic fire-extinguishing system
Laboratories and vocational shops, not classified as Group H, located in a Group E or I-2 occupancy	1 hour or provide automatic fire-extinguishing system
Laundry rooms over 100 square feet	1 hour or provide automatic fire-extinguishing system
Group I-3 cells equipped with padded surfaces	1 hour
Group I-2 waste and linen collection rooms	1 hour
Waste and linen collection rooms over 100 square feet	1 hour or provide automatic fire-extinguishing system
Stationary storage battery systems having a liquid electrolyte capacity of more than 50 gallons, or a lithium-ion capacity of 1,000 pounds used for facility standby power, emergency power or uninterrupted power supplies	1 hour in Group B, F, M, S and U occupancies; 2 hours in Group A, E, I and R occupancies.
Rooms containing fire pumps in nonhigh-rise buildings	2 hours; or 1 hour and provide automatic sprinkler system throughout the building
Rooms containing fire pumps in high-rise buildings	2 hours

For SI: 1 square foot = 0.0929 m², 1 pound per square inch (psi) = 6.9 kPa, 1 British thermal unit (Btu) per hour = 0.293 watts, 1 horsepower = 746 watts, 1 gallon = 3.785 L.

other occupancies with fire barriers that comply with Section 707 or horizontal assemblies complying with Section 712, or both when there are stories above or below the incidental accessory occupancy. Where Table 508.2.5 permits protection by an automatic fire-extinguishing system without fire barriers, the construction enclosing the incidental accessory occupancy area must resist the passage of smoke in accordance with Section 508.2.5.2. Where the construction surrounding an incidental accessory occupancy is only 1 hour, and the building is of nonrated construction (Type IIB, IIIB or VB), the rated construction does not need to be supported by 1-hour fire-resistance-rated construction. In all other instances, the construction supporting incidental occupancy separations must be supported by construction with at least the same rating as the separations.

508.2.5.2 Nonfire-resistance-rated separation and protection. Where Table 508.2.5 permits an automatic fire-extinguishing system without a *fire barrier*, the incidental accessory occupancies shall be separated from the remainder of the building by construction capable of resisting the passage of smoke. The walls shall extend from the top of the foundation or floor assembly below to the underside of the ceiling that is a component of a fire-resistance-rated floor assembly or roof assembly above or to the underside of the floor or roof sheathing, deck or slab above. Doors shall be self- or automatic-closing upon detection of smoke in accordance with Section 715.4.8.3. Doors shall not have air transfer openings and shall not be undercut in excess of the clearance permitted in accordance with NFPA 80. Walls surrounding the incidental accessory occupancy shall not have air transfer openings unless provided with smoke dampers in accordance with Section 711.7.

❖ Where Table 508.2.5 permits protection by an automatic fire-extinguishing system without fire barriers, the construction enclosing the incidental accessory occupancy area must resist the passage of smoke. While this section can be viewed as a performance standard, construction details for resisting the passage of smoke are provided in this section. Although the section specifically states that air transfer openings must be provided with smoke dampers, it is silent with respect to ducts. If ducts are penetrating this separation, the arrangement of the duct system should be analyzed to determine if it will allow smoke to pass through the wall and not restrict it to the incidental accessory occupancy.

The wall construction described here is required to be neither a smoke barrier conforming to Section 710 nor a smoke partition conforming to Section 711.

508.2.5.3 Protection. Except as specified in Table 508.2.5 for certain incidental accessory occupancies, where an automatic fire-extinguishing system or an *automatic sprinkler system* is provided in accordance with Table 508.2.5, only the space occupied by the incidental accessory occupancy need be equipped with such a system.

❖ The point of this section is that the fire protection system that is stipulated in Table 508.2.5 is required for the incidental accessory occupancy area only. In gen-

eral, the nature of these incidental occupancies is such that they are small areas that are not frequented by the building occupants very often in which a fire could get underway and go unnoticed for a longer time than a part of the building that is constantly occupied. One should also note that there are only certain instances where the table specifically requires a fire sprinkler system as opposed to a more general requirement for a fire-extinguishing system. There are two notable exceptions to the general provisions of this section. If fire pumps are located within a nonhigh-rise building, to reduce the fire barrier from 2 hours to 1 hour, the entire building must be sprinkler protected, not just the incidental accessory occupancy. All high-rise buildings subject to Section 403 are required to be sprinkler protected; therefore, fire pumps located in a high-rise building will be in an area that is both sprinkler protected and separated from the rest of the building by 2-hour fire-resistance-rated construction. The fire pump room requirements are consistent with NFPA 20.

508.3 Nonseparated occupancies. Buildings or portions of buildings that comply with the provisions of this section shall be considered as nonseparated occupancies.

❖ This section describes the second option a designer may choose to apply to a building that contains more than one occupancy classification. Except with respect to occupancies classified in Group H, this option is similar to the accessory use option (see Section 508.2) in that there is no requirement for various occupancies to be physically separated by any type of fire-resistance-rated assembly (see Figure 508.3).

The principle behind nonseparated uses is that the different occupancies within the same building do not have to be separated by fire-resistance-rated assemblies if the building complies throughout with the more restrictive code requirements for minimum construction type and fire protection systems (Chapter 9) applicable to the occupancies in the building. If the building is also a high-rise building, the most restrictive provisions of Section 403 applicable to the specific occupancies will also apply. Although each occupancy is separately classified as to its group, a fire-resistance-rated assembly is not required by the nonseparated uses option. A designer may choose to physically separate the occupancies; however, a fire-resistance rating of these separations would not be required by this section.

There are four basic steps to follow when applying the nonseparated uses concept:

Step 1: Determine which occupancy group classifications are present in the building.

Step 2: Determine the minimum type of construction based on the height and area of the building for each occupancy in accordance with this chapter and Table 503. Apply the requirement for the highest type of construction to the entire building (see Section 508.3.2).

Step 3: Apply the most restrictive provisions found in Chapter 9 throughout the building containing nonseparated occupancies. For a high-rise building, also apply the most restrictive provisions contained in Section 403 related to the occupancies present (see Section 508.3.1).

Step 4: Apply all other requirements of the code, except for Section 403 and Chapter 9, to each occupancy individually based on the specific occupancy of each space (e.g., means of egress) (see commentary, Chapter 10) (see Section 508.3.1).

508.3.1 Occupancy classification. Nonseparated occupancies shall be individually classified in accordance with Section 302.1. The requirements of this code shall apply to each portion of the building based on the occupancy classification of that space except that the most restrictive applicable provisions of Section 403 and Chapter 9 shall apply to the building or portion thereof in which the nonseparated occupancies are located.

❖ The nonseparated use option requires that each occupancy area be considered a different occupancy for purposes of application of code requirements and therefore requires that all code issues related to that occupancy be considered separately for the areas of the building where the occupancy is located. The exception to this is for the provisions of Chapter 9 and Section 403, which must be applied to the whole building containing the nonseparated occupancies.

The principle behind the nonseparated use concept is that the different occupancies within the same building do not have to be separated by fire-resistance-rated assemblies if the building complies throughout with the more restrictive code requirements for construction, fire protection systems and, where applicable, requirements for high-rise buildings. Although each occupancy group is separately classified as to its group, a fire-resistance-rated assembly is not required

under the nonseparated occupancies option. A designer may choose to physically separate the occupancies; however, a fire-resistance rating of these separations would not be required by this section, as stated in Section 508.3.3.

For example, if a building contains both Groups B and M, each portion of the building must comply with the code requirements for its respective group classification located in that portion. The occupant load for the area classified as Group B is based on the code requirements applicable to the business occupancy. Similarly, the occupant load for the Group M area is based on requirements applicable to the mercantile occupancy. Live loads for the Group B area are applied to the area of business occupancy. Group M live loads are applied to the area of mercantile occupancy. Exterior wall ratings based on fire separation distance (see Table 602) apply to exterior walls of the business occupancy for Group B buildings; the same is true for walls in Group M occupancies.

It is important to note the threshold requirements for fire protection systems are contained in Chapter 9. In some cases, they are simply based on the occupancy and in other cases they are based on height or area criteria. Also, the requirements to be met in some cases only apply to the fire area in which a given occupancy is contained, while in others they apply to the entire building.

For example, if the business occupancy as described in Figure 508.3 required a fire alarm system with manual fire alarm boxes, then one would be required for the entire building—even though Chapter 9 does not require a fire alarm system for a storage occupancy. Note that when determining if a fire alarm system is required and the threshold for the requirement is based upon occupant load, it is based on the individual occupancy (see commentary, Section 907.2).

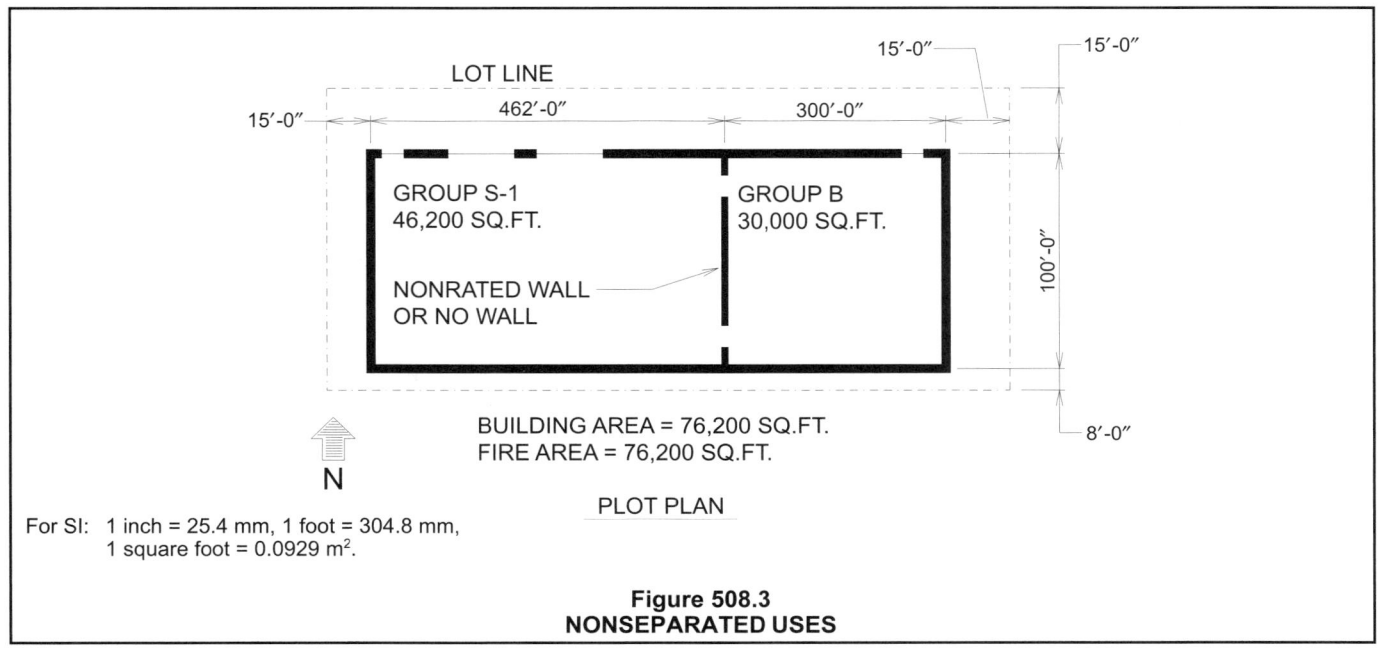

Figure 508.3
NONSEPARATED USES

508.3.2 Allowable building area and height. The allowable *building area and height* of the building or portion thereof shall be based on the most restrictive allowances for the occupancy groups under consideration for the type of construction of the building in accordance with Section 503.1.

❖ The fundamental concept underlying the nonseparated use option is that the allowable building heights and allowable building areas are based upon the most restrictive requirements of Table 503 applicable to each of the occupancy groups in the mixed use building.

For example, the tabular area for Group S-1 in Table 503 is less than the corresponding tabular area for Group B; therefore, Group S-1 results in a requirement for a higher type of construction for buildings of equal size and will, therefore, determine the minimum construction type of the building. If the building were, say, Type IIB construction, then the allowable area for the building would be 17,500 square feet (1626 m²), based upon the allowable area of Group S-1 occupancies for Type IIB construction in Table 503. Area increases for sprinklers and frontage could be applied.

508.3.3 Separation. No separation is required between nonseparated occupancies.

Exceptions:

1. Group H-2, H-3, H-4 and H-5 occupancies shall be separated from all other occupancies in accordance with Section 508.4.

2. Group I-1, R-1, R-2 and R-3 *dwelling units* and *sleeping units* shall be separated from other *dwelling* or *sleeping units* and from other occupancies contiguous to them in accordance with the requirements of Section 420.

❖ As is indicated by the name of this mixed use option, "nonseparated occupancies," the occupancy groups within the building need not be separated from each other by fire-resistance-rated construction. It must be noted that other portions of the code might require fire separation assemblies, and where the separations are required by other provisions, they must be provided even in a building designed under the nonseparated mixed use option. For instance, if it was desired not to provide a sprinkler system in the fire area discussed in the commentary example in Section 508.2.2, then a fire barrier would need to be installed to make the fire area containing the Group A-2 occupancy smaller than the threshold amount of 5,000 square feet (465 m²) and the occupant load less than 100. Another example is, in some cases it could be a wall between two different occupancies that is required by another section of the code to be fire-resistance rated (e.g., a separation wall is provided that also serves as a corridor wall that is required to be fire-resistance rated).

When one of the uses is a Group H-2, H-3, H-4 or H-5 high-hazard occupancy, the nonseparated use option could not be applied. Exception 2 states that the horizontal assemblies and walls required to separate dwelling units and sleeping units from each other and

from other occupancies need to comply with Section 420.

Example: A building contains a 46,200-square-foot (4292 m²) general warehouse and an accompanying 30,000-square-foot (2787 m²) regional dispatch office for a total area of 76,200 square feet (7079 m²). The building is a one-story, metal building. Assume that there is no allowable area increase for frontage. In applying the four-step procedure for nonseparated uses, the following holds true:

Step 1: The warehouse is classified in Group S-1 and the offices are classified in Group B.

Step 2: The minimum type of construction is based on the lesser allowable area of the building for each occupancy classification. First, determine if any increases can be obtained (no frontage increase—check for installation of a sprinkler system). In Step 3, the building is shown to be sprinklered throughout; therefore, a 300-percent increase can be applied to the tabular areas set in Table 503 (see Section 506.3). For the purpose of analysis, assume the building is of Type IIB construction. The Group B requirements would allow the building to be 92,000 square feet (8547 m²) (23,000 square feet + 300 percent of 23,000 square feet). The Group S-1 requirements, however, only allow a building of Type IIB construction to be 70,000 square feet (6503 m²) (17,500 square feet + 300 percent of 17,500 square feet). The most restrictive is the Group S-1 requirement; therefore, it governs. Hence, Type IIB construction would not allow adequate area for the S-1 occupancy. A higher construction type would be required. The minimum type of construction needs to be Type IIA.

Step 3: The fire area containing the occupancy classified in Group S-1 is more than 12,000 square feet (1115 m²); thus, in accordance with Section 903.2.10, this area must be sprinklered. The area classified as Group B ordinarily is not required to be equipped with sprinklers by Section 903; however, since the Group S-1 provision does require sprinkler protection, it is more restrictive than the requirement for Group B under Chapter 9. Therefore, the entire building must be sprinklered.

Step 4: The occupant load of each area should be calculated separately (based on the occupancy of that area) using Section 1004.1 and Table 1004.1.1.

Business: = 30,000 square feet (2787 m²) divided by 100 square feet (9 m²) per person = 300 people.

Storage: = 46,200 square feet (4292 m²) divided by 500 (warehouses) square feet (46 m²) per person = 93 people.

Total occupant load: 393 people

The evaluation of means of egress in accordance with Chapter 10 is then based on these occupant loads.

The load-bearing members in the north, east and west exterior walls of the building must provide a minimum 1-hour fire-resistance rating as required for Type IIA construction in Table 601. The nonload-bearing walls are also required to have a 1-hour fire-resistance rating because of the 15-foot (4572 mm) fire separation distance (see Table 602).

The fire separation distance on the south side of the mixed occupancy building is 8 feet (2438 mm). Table 602 requires the Group S-1 and B portions of this wall to have a 1-hour fire-resistance rating. The ratings required by this table apply to both load-bearing and nonload-bearing exterior walls. The substitution of an automatic sprinkler system in lieu of 1-hour fire-resistance-rated construction (as indicated in Note d of Table 602) does not apply to exterior walls; therefore, all exterior walls (whether load-bearing or nonload-bearing) in this building are required to have a 1-hour fire-resistance rating.

This analysis must be continued for other aspects of code compliance for each respective occupancy. Once all four steps have been satisfied, the building is deemed to have an equivalent level of safety as a building with only one occupancy. Any future changes in the occupancy, or the areas used by each occupancy, would require review and approval by the local jurisdiction in accordance with the provisions contained in Sections 105.1 and 3408.

508.4 Separated occupancies. Buildings or portions of buildings that comply with the provisions of this section shall be considered as separated occupancies.

❖ This section describes the third option, separated occupancies, that a designer may choose to apply when constructing a building that contains more than one occupancy classification. This third option differs from the first two alternatives (i.e., accessory uses and nonseparated uses) (see Sections 508.2 and 508.3) in four ways:

1. Occupancies in different uses that are to be evaluated as separated occupancies must be separated in accordance with Table 508.4. When the table requires a separation, the occupancies must be separated completely, both horizontally and vertically, by the provision of fire-resistance-rated fire barriers and horizontal assemblies (see Sections 707 and 712). It should be noted that Table 508.4 indicates "N" at the intersection of some occupancies. The "N" indicates that no rated separation is required. This is allowed where the adjacent occupancies pose similar levels of risk. Separations, where provided, will also create separate fire areas.

2. The determination of the minimum type of construction is based on both the height of each occupancy relative to the grade plane and the areas of each occupancy relative to the total area per story.

3. Allowable building area is determined based on the occupancies on each story through a comparison of the allowed area for each occupancy category and the actual floor area proposed for each occupancy.

4. In comparison to the nonseparated occupancy provisions in Section 508.3 where the most restrictive of the requirements of Chapter 9 apply throughout the building, under separated occupancies, requirements for such systems as sprinkler systems, alarms and standpipes are determined based on the areas containing each occupancy. It must be noted that the determination of sprinkler requirements in Section 903 is based on the occupancies present. The threshold for providing a sprinkler system is often based on the size of a fire area. In the cases where Table 508.4 does not require a separation, the result is that certain fire areas will contain more than one occupancy. Therefore when determining when sprinklers are required, the total size of the fire area needs to be considered, not just the portion containing each occupancy. For example, a proposed building is 16,000 square feet (1486 m²), there is 8,000 square feet (743 m²) of Group M occupancy and 8,000 square feet (743 m²) of Group B. According to Table 508.4, there does not need to be a separation between these two occupancies. Individually neither is large enough to require a sprinkler system. However, if no separation is provided, the fire area is 16,000 square feet (1486 m²), and in accordance with Section 903.2.7 a fire area containing a Group M occupancy and exceeding 12,000 square feet (1115 m²) must be sprinkler protected. (Fire areas are defined in Section 902 and are used in Section 903.2 to determine if sprinklers are required. Fire areas are a distinct requirement from separation of occupancies and the separation of occupancies by a fire-resistance-rated wall may not always create a separate fire area. Fire areas can be delineated by fire walls and fire barriers per Chapter 7.)

There are four basic steps to follow when using the separated occupancies option:

Step 1: Determine which occupancies are present in the building (see Section 508.4.1).

Step 2: Separate the occupancies in accordance with Table 508.4 with fire barrier walls and horizontal assemblies in accordance with Sections 707 and 712 (see Section 508.4.4).

Step 3: Apply all code requirements for each separated space individually based on the occupancy or occupancies present (i.e., design occupant load, means of egress elements, exterior wall requirements and fire protection). For the application of code provisions, each separated space is taken into consideration individually. Again, it must be remembered that Table 508.4 may not require a separation and, therefore, code requirements based on fire area must be addressed for the total fire area even though it contains more than one occupancy.

Step 4: Determine the minimum type of construction of a building based on the building height limitations of Sections 503 and 504 and the building area limitations of Sections 503 and 506 (see Sections 508.4.3 and 508.4.2, respectively).

> **Part A:** Determine the minimum type of construction required based on the height of each occupancy relative to the grade plane (see Section 508.4.3). This needs to be evaluated for both height in feet and stories above grade plane.

> **Part B:** Determine the minimum type of construction based on a weighted average of areas occupied by the various occupancies (see Section 508.4.2).

508.4.1 Occupancy classification. Separated occupancies shall be individually classified in accordance with Section 302.1. Each separated space shall comply with this code based on the occupancy classification of that portion of the building.

❖ In the separated uses option, occupancies in different uses that are to be evaluated as separated uses must be separated in accordance with Table 508.4. When a separation is required they must be separated completely, both horizontally and vertically, with fire barriers and horizontal assemblies (see Sections 707 and 712).

Except for certain occupancies, it is the designer's prerogative to use the accessory occupancies option, nonseparated occupancies option or the separated occupancies option when establishing a mixed occupancy building. It is also possible to apply both options within different portions or different stories of a building. Where a mixture of options is used, the design documentation needs to clearly show how the requirements of each option are applied in each portion of the building.

508.4.2 Allowable building area. In each *story*, the *building area* shall be such that the sum of the ratios of the actual *building area* of each separated occupancy divided by the allowable *building area* of each separated occupancy shall not exceed 1.

❖ For each story, it must be determined that the sum of the ratios of the actual floor area of each separated occupancy, respective to the most restrictive occupancy

contained therein as compared to the areas per story allowed by Table 503 and as modified by Section 506 for each respective occupancy, does not exceed one. In the evaluation of allowable area, intervening fire barriers between different fire areas containing the same occupancy are not a consideration. In determining the floor area per occupancy, all areas of the same occupancy are added together irrespective of the presence of multiple fire areas.

In determining the allowable areas for each occupancy, the tabular areas from Table 503 are permitted to be modified in accordance with the provisions of Section 506; thus, the allowable areas are intended to include the increases permitted for sprinklers and open perimeter. For determination of the allowable perimeter increase, use the entire building perimeter—not the occupancy perimeters. If considering the sprinkler increase, the entire building must be sprinklered, not just particular occupancies.

508.4.3 Allowable height. Each separated occupancy shall comply with the *building height* limitations based on the type of construction of the building in accordance with Section 503.1.

> **Exception:** Special provisions permitted by Section 509.

❖ The allowable height is occupancy dependant. As long as individual occupancies meet the height limitations based upon a measurement from grade plane then the building complies. For example, a building of Type IIB construction with no increases for sprinklers containing a Group B occupancy and a Group F-2 occupancy could be four stories. The only limitation in this case, in terms of building height, would be that the Group F-2 occupancy could not be located any higher than the third story.

508.4.4 Separation. Individual occupancies shall be separated from adjacent occupancies in accordance with Table 508.4.

❖ Table 508.4 provides the required separation between areas containing the separated occupancies. Separations must be of fire-resistance-rated construction for 1, 2 or 3 hours as specified in the cells of the table. Generally the required rating will be higher if a building is not fully protected by an automatic sprinkler system. The table states "NP" in certain cells. These occur where occupancies are not permitted to be in a mixed occupancy building, or where one or both occupancies cannot be in an unsprinklered building. Other cells contain the letter "N." For the adjoining occupancies represented by each such cell, no separation is required. Although certain adjacent occupancies are not required to be physically separated, they are still evaluated under the separated option, but note that, where code requirements are based on established fire areas, occupancies not separated must be considered as sharing the same fire area.

For example, a completely sprinklered building of Type VB construction contains areas devoted to business and assembly occupancies. The designer has chosen the separated occupancies option and has completely separated the areas containing the two different occupancies by fire barrier walls and horizontal assemblies having a minimum 1-hour fire-resistance rating in accordance with Table 508.4. This is found by consulting the box that intersects with "A, E" in the first column with "B, F-1, M, S-1" in the first row and, because the entire building is sprinklered, the fire-resistance rating of the fire barrier walls and horizontal assemblies is required to be 1 hour, as indicated in the column designated "S." Had the building not been sprinklered, the required rating would have been 2 hours, as indicated in the column designated "NS."

Please note that Table 508.4 contains groupings of some of the occupancies, including:

- A, E
- I-1, I-3, I-4
- R
- F-2, S-2, U
- B, F-1, M, S-1
- H-3, H-4, H-5

These are occupancies that share the same level of hazard with respect to fire safety. It is possible, therefore, to have two occupancies that comply with these separated use provisions that require no separation between them. For instance, a mixed occupancy of Groups B and M would not be required to have a separation between them, but the provisions for calculation of the sum of the ratios of actual areas to allowable areas would still be applied to this circumstance.

TABLE 508.4. See below.

❖ The purpose of Table 508.4 is to set forth the fire-resistance rating required for fire barrier walls and horizontal assemblies used to separate occupancies. The fire-resistance rating of the separation between different occupancies is based on the relative anticipated fire severity of the occupancies.

Note a: See the commentary for Section 903.2.5.2.

Note b: The fire-resistance rating of spaces used solely for private or pleasure vehicles may be reduced by 1 hour but never less than 1 hour.

Note c: See the commentary for Section 406.1.4.

Note d: Kitchens and the restaurants they serve need not be separated from each other by fire barrier walls.

Note d has often been applied to any A-2 occupancy such as not requiring separation between a school cafeteria and its kitchen.

Note e: Occupancies in the same classification do not need to be separated. For example, a Group A-2 restaurant is not required to have a rated separation from a Group A-3 museum.

Note f: See the commentary for Section 415.8.2.2 regarding separations within Group H-5 occupancy buildings.

TABLE 508.4
REQUIRED SEPARATION OF OCCUPANCIES (HOURS)

OCCUPANCY	A[d], E S	A[d], E NS	I-1, I-3, I-4 S	I-1, I-3, I-4 NS	I-2 S	I-2 NS	R S	R NS	F-2, S-2[b], U S	F-2, S-2[b], U NS	B, F-1, M, S-1 S	B, F-1, M, S-1 NS	H-1 S	H-1 NS	H-2 S	H-2 NS	H-3, H-4, H-5 S	H-3, H-4, H-5 NS
A[d], E	N	N	1	2	2	NP	1	2	N	1	1	2	NP	NP	3	4	2	3[a]
I-1, I-3, I-4	—	—	N	N	2	NP	1	NP	1	2	1	2	NP	NP	3	NP	2	NP
I-2	—	—	—	—	N	N	2	NP	2	NP	2	NP	NP	NP	3	NP	2	NP
R	—	—	—	—	—	—	N	N	1[c]	2[c]	1	2	NP	NP	3	NP	2	NP
F-2, S-2[b], U	—	—	—	—	—	—	—	—	N	N	1	2	NP	NP	3	4	2	3[a]
B, F-1, M, S-1	—	—	—	—	—	—	—	—	—	—	N	N	NP	NP	2	3	1	2[a]
H-1	—	—	—	—	—	—	—	—	—	—	—	—	N	NP	NP	NP	NP	NP
H-2	—	—	—	—	—	—	—	—	—	—	—	—	—	—	N	NP	1	NP
H-3, H-4, H-5	—	—	—	—	—	—	—	—	—	—	—	—	—	—	—	—	1[e, f]	NP

For SI: 1 square foot = 0.0929 m².
S   = Buildings equipped throughout with an automatic sprinkler system installed in accordance with Section 903.3.1.1.
NS = Buildings not equipped throughout with an automatic sprinkler system installed in accordance with Section 903.3.1.1.
N = No separation requirement.
NP = Not permitted.
a. For Group H-5 occupancies, see Section 903.2.5.2.
b. The required separation from areas used only for private or pleasure vehicles shall be reduced by 1 hour but to not less than 1 hour.
c. See Section 406.1.4.
d. Commercial kitchens need not be separated from the restaurant seating areas that they serve.
e. Separation is not required between occupancies of the same classification.
f. For H-5 occupancies, see Section 415.8.2.2.

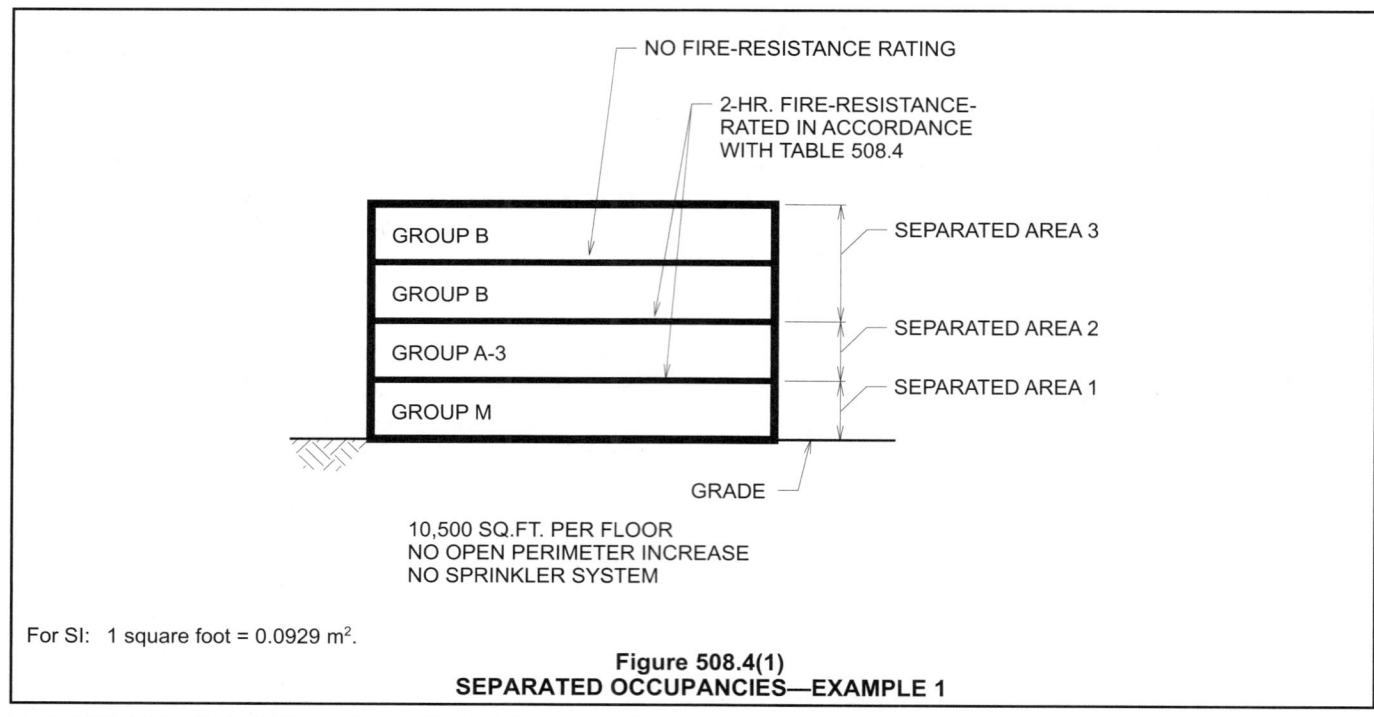

For SI: 1 square foot = 0.0929 m².

Figure 508.4(1)
SEPARATED OCCUPANCIES—EXAMPLE 1

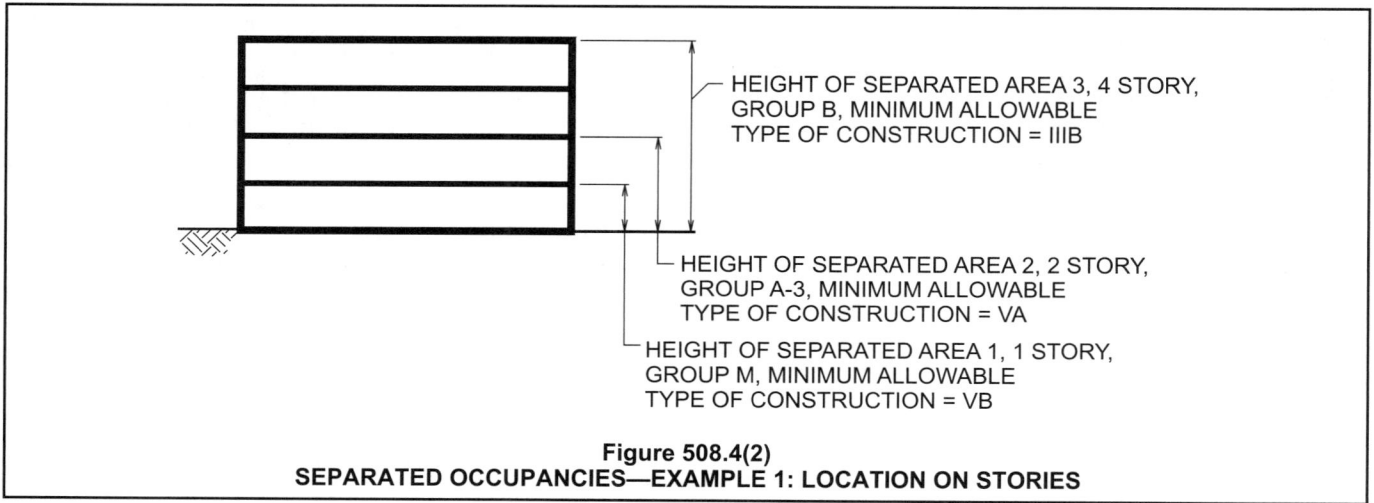

Figure 508.4(2)
SEPARATED OCCUPANCIES—EXAMPLE 1: LOCATION ON STORIES

508.4.4.1 Construction. Required separations shall be fire barriers constructed in accordance with Section 707 or horizontal assemblies constructed in accordance with Section 712, or both, so as to completely separate adjacent occupancies.

❖ This section is merely a reference to the appropriate provisions in Chapter 7 for fire-resistance-rated assemblies that are required to meet the provisions for adequate separation of occupancies.

The following three examples illustrate the application of the separated occupancies option.

Example 1: A four-story building contains a retail store on the first story, a lecture hall with fixed seats on the second story and offices on the top two stories. Each story is 10,500 square feet (975 m²). A sprinkler system is not provided throughout the building. As-

sume that there is no open perimeter increase [see Figure 508.4(1)]. In applying the four-step procedure for separated uses, the following holds true:

Step 1: First story is Group M for the entire story. Second story is Group A-3 for the entire story. Third and fourth stories are Group B for both stories.

Step 2: The separation requirements in accordance with Table 508.4 require a 2-hour fire-resistance rating for the horizontal assembly between the Group M occupancy and the Group A-3 occupancy. Likewise, a 2-hour fire-resistance-rated horizontal assembly is required between the Group A-3 occupancy and the Group B occupancy. Each occupancy is separated horizontally with an approved fire-resistance-rated assembly. There is no rating required for the horizontal assembly between the third and fourth stories because they are in the same occupancy.

Step 3: The occupancy-specific requirements need to be addressed. The building is essentially broken into the following three areas based upon the separations provided to achieve the separated occupancy requirements:

Group M area = 10,500 square feet (975 m²).

Group A-3 area = 10,500 square feet (975 m²).

Group B area = 21,000 square feet (1951 m²) [combined area for two stories, each 10,500 square feet (975 m²)].

Because of the construction required for each separation, these occupancy areas may also be separate fire areas.

Since the story containing the Group M occupancy is also a fire area that is less than 12,000 square feet (1115 m²) and is located below the fourth story, sprinklers are not required for the Group M occupancy.

The story containing the Group A-3 occupancy is also a fire area that is less than 12,000 square feet (1115 m²), but it is located on the second story and above the level of exit discharge. According to Section 903.2.1.3, a Group A-3 occupancy fire area in this location must be provided with a sprinkler system. In addition, this would mean that the Group M occupancy would be required to be sprinklered in accordance with Section 903.2.1 which requires that all stories between the Group A-3 occupancy and the level of exit discharge be sprinklered.

The third and fourth stories containing the Group B occupancy do not require sprinklers because neither the size nor height of these stories triggers any threshold in Section 903.

Additionally, utilize the provisions in other parts of the code for each occupancy area respective to means of egress, exterior wall requirements, etc., based on the occupancy located on each story. For example, the design occupant load of the area containing the Group M occupancy is based on the mercantile occupant load factor found in Table 1004.1.1 while the area containing the Group A-3 occupancy is based upon the number of fixed seats in the lecture hall in accordance with Section 1004.7.

Step 4:

Part A: Review for the minimum types of construction based on the height limitations for each of the three occupancies as follows [see Figure 508.4(2)]:

1. Group M, one story in height; the minimum type of construction is Type VB [one story and 40 feet (12 192 mm) in accordance with Table 503].

2. Group A-3, two stories in height; the minimum type of construction is Type VA [two stories and 50 feet (15 240 mm) in accordance with Table 503].

3. Group B, four stories in height; the minimum type of construction is Type IIIB [four stories and 55 feet (16 764 mm) in accordance with Table 503].

The highest type of construction for any one of the occupancies governs for the entire building; thus, the minimum allowable type of construction permitted for the building based on height is Type IIIB. The designer is permitted to choose any one of the construction types other than Types VA and VB for the design of the building. The separated occupancy option presumes that each occupancy is restricted to the story (or stories) for which it is designated. For this example, the first story is separated and designated for the Group M occupancy, and the Group M occupancy can only occur on that story. In comparison, the nonseparated option would assume that the Group M occupancy could occur on any of the four stories.

If the building were completely sprinklered, Section 504.2 would permit a height increase. Accordingly, under Table 503, the minimum type of construction allowed for the building would be Type VA. In addition, the minimum fire-resistance rating of each floor/ceiling assembly would need to be only 1 hour, in accordance with Table 508.4.

Part B: Since each story has only one occupancy, the maximum allowable area for a story of the building is derived directly from Table 503. Based on the allowable construction type determined, Type IIIB construction is acceptable for Groups B and M; however, it is not acceptable for Group A-3. Based on the given area of 10,500 square feet (975 m²) and without any area increases, the allowable types of construction permitted for Group A-3 based on area are Types IA, IB, IIA, IIIA, IV and VA. Although the Group A-3 story, as well as the story below, will be required to be sprinkler protected, because the whole building is not protected, area increases for sprinklers are not allowed in this example.

From Part A it was determined that Type VA construction is not permitted for the height of the building due to Group B; therefore, the allowable types of construction for the building based on height and area are Types IA, IB, IIA, IIIA and IV.

Example 2: Consider a 76,200-square-foot (7079 m²) building with a fire barrier wall separating a 46,200-square-foot (4292 m²) warehouse from a 30,000-square-foot (2787 m²) office space [see Figure 508.4(3)].

Step 1: The warehouse occupancy (which is also a fire area) is classified in Group S-2. The office occupancy (which is also a fire area) is classified in Group B.

Step 2: A 2-hour fire barrier is required in accordance with Table 508.4 for nonsprinklered buildings (1 hour for sprinklered buildings) between the two occupancies and extends continuously through all concealed spaces to the underside of the roof deck in accordance with Section 707.

Step 3: The separation required by Table 508.4 also creates two fire areas. The Group B area is one story and 30,000 square feet (2787 m²). A sprinkler system is not required by the code for a one-story Group B fire area. The Group S-2 fire area is 46,200 square feet (4292 m²), which is also not required to have a sprinkler system. However, the designer wants to be able to utilize Type IIB construction. Table 503 restricts Group S-2 occupancies within a Type IIB building to 26,000 feet (2416 m²); therefore, a sprinkler system will need to be provided under Section 506.3 to allow an increase in allowable area. For a single-story building, the increase is 300 percent; the allowable area for a Group S-2 building in Type IIB would then be 104,000 square feet (9662 m²). To allow this increase the entire building including both occupancies would need to be sprinkler protected.

Further, evaluate each occupany area in the same manner for compliance with all provisions of the code.

Step 4:

Part A: The proposed building is only one story in height. Based on height, any type of construction is acceptable for the occupancies under consideration.

Part B: Make a trial evaluation of the building assuming it to be Type IIB construction:

Actual area of Group S-2 = 46,200 square feet (4292 m²) and actual area of Group B = 30,000 square feet (2787 m²).

Group S-2 maximum allowable area, including allowable increases (allowable tabular area plus sprinkler increase) = 26,000 + (26,000)(3.0) = 104,000 square feet (9662 m²).

The maximum allowable area for Group B including allowable increases (allowable tabular area plus sprinkler increase) = 23,000 + (23,000)(3.0) = 92,000 square feet (8547 m²).

Therefore, Type IIB construction is an acceptable minimum type of construction. When the factor arrived at is substantially below 1.0, it indicates that the building might qualify as a lower type of construction or as nonseparated uses.

The building is deemed to comply with the code after each of the code requirements, as determined by

Steps 1 through 4, is satisfied.

In the case of multistory buildings, an evaluation of the allowable area must be performed for every story of the building that contains separated uses.

Example 3: Consider a 76,200-square-foot (7079 m²) building identical to that used in Example 2, except that the warehouse is classified in Group S-1. The Group S-1 warehouse is 46,000 square feet (4292 m²) and the Group B office space is 30,000-square-foot (2787 m²) [see Figure 508.4(4)].

Step 1: The warehouse is classified as Group S-1. The office is classified as Group B.

Step 2: Table 508.4 indicates that there is no separation requirement between Groups B and S-1, since they are in the same grouping of occupancies in the table.

Step 3: As there is no separation between the occupancies, there is only one fire area including both occupancies. The fire area for the occupancy of Groups B and S-1 is the entire area of the one-story building, or 76,200 square feet (7079 m²). In accordance with Section 903.2.9, since the fire area containing Group S-1 is greater than 12,000 square feet (1115 m²) the entire building is required to be sprinklered. Note that even if the uses were separated, Section 903.2.9 requires sprinklers throughout all buildings containing a Group S-1 fire area greater than 12,000 square feet (1115 m²).

Step 4:

Part A: Based on the one-story height, any type of construction is acceptable for the occupancies under consideration.

Part B: Make a trial evaluation of the building assuming it to be Type IIB construction.

Actual area of Group S-1 = 46,200 square feet (4292 m²) and actual area of Group B = 30,000 square feet (2787 m²).

Group S-1 maximum allowable area, including allowable increases (allowable tabular area plus sprinkler increase) = 17,500 + (17,500)(3.0) = 70,000 square feet (6503 m²).

The maximum allowable area for Group B including allowable increases (allowable tabular area plus sprinkler increase) = 23,000 + (23,000)(3.0) = 92,000 square feet (8547 m²).

Note that, in this case, the separated use option requires no separation between these particular uses, and the Type IIB classification is acceptable. Comparing this to the example for nonseparated uses, with the same dimensions of each occupancy, the Type IIB construction classification is not acceptable, and a Type IIA construction classification is required.

Since Section 508.3 (nonseparated occupancies) makes a reference to applying the most restrictive requirements of Section 403 and Chapter 9, there may be the assumption that under the separated occupancies option, requirements can be applied separately to each occupancy in all cases. This is not completely true. There are sections of Chapter 9, such as Section 903.2.6, 903.2.8 or 903.2.9, that require sprinklers throughout the entire building when a particular occupancy is present. Some sections in Chapter 9, however, do permit separate application of requirements, even though the term "fire area" is not used. For example, a warehouse building with an office would be categorized as Groups B (office) and S-1 (warehouse). If the separated uses option is selected and the office has an occupant load of at least 500 people, then a fire alarm system is only required in the Group B area and not the Group S-1 area (see Section 907.2). It should be noted that alarm requirements must address the path of egress for occupants and may require notification appliances along those paths.

With respect to the high-rise provisions of Section 403, the requirements generally apply throughout the building and are not specific to fire areas. There are limited provisions in Section 403 that are specific to one occupancy or another. Regardless of the option selection in a mixed occupancy building, requirements in Section 403 that are based on occupancy will need to be applied throughout the building.

Another approach available to the designer of structures that contain more than one occupancy requires that the occupancies are completely separated by fire walls (see Section 706), thus creating separate buildings for each use. This is the simplest option to apply

and analyze. When a fire wall is utilized, for code application purposes, two or more separate and independent buildings (see the definition of "Building" in Section 202) are created, and each building is reviewed individually.

SECTION 509
SPECIAL PROVISIONS

509.1 General. The provisions in this section shall permit the use of special conditions that are exempt from, or modify, the specific requirements of this chapter regarding the allowable heights and areas of buildings based on the occupancy classification and type of construction, provided the special condition complies with the provisions specified in this section for such condition and other applicable requirements of this code. The provisions of Sections 509.2 through 509.8 are to be considered independent and separate from each other.

❖ The subsections of Section 509 are exceptions to the general height and area limitations of Chapter 5. Most of the subsections deal with attached parking structures and contain conditions by which these can be attached to buildings without creating hardship in regard to allowable area and height for the building. These allowances are not always related to buildings containing parking below but can involve other uses and occupancies. Also many of these scenarios will be applied such that there may be multiple buildings above the parking garage. These allowances tend to involve a 3-hour fire-resistance-rated horizontal separation between the garage and other parts of the building or buildings.

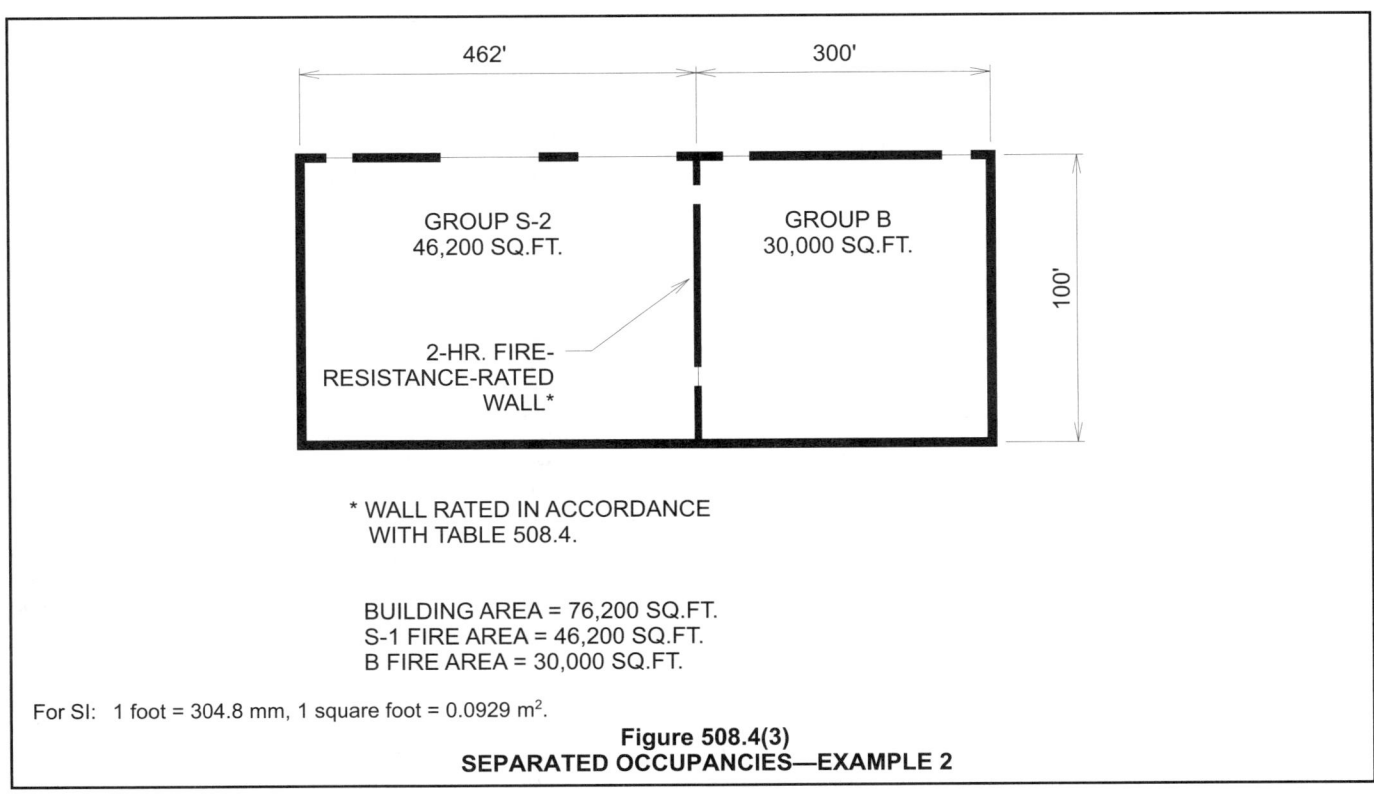

For SI: 1 foot = 304.8 mm, 1 square foot = 0.0929 m².

Figure 508.4(3)
SEPARATED OCCUPANCIES—EXAMPLE 2

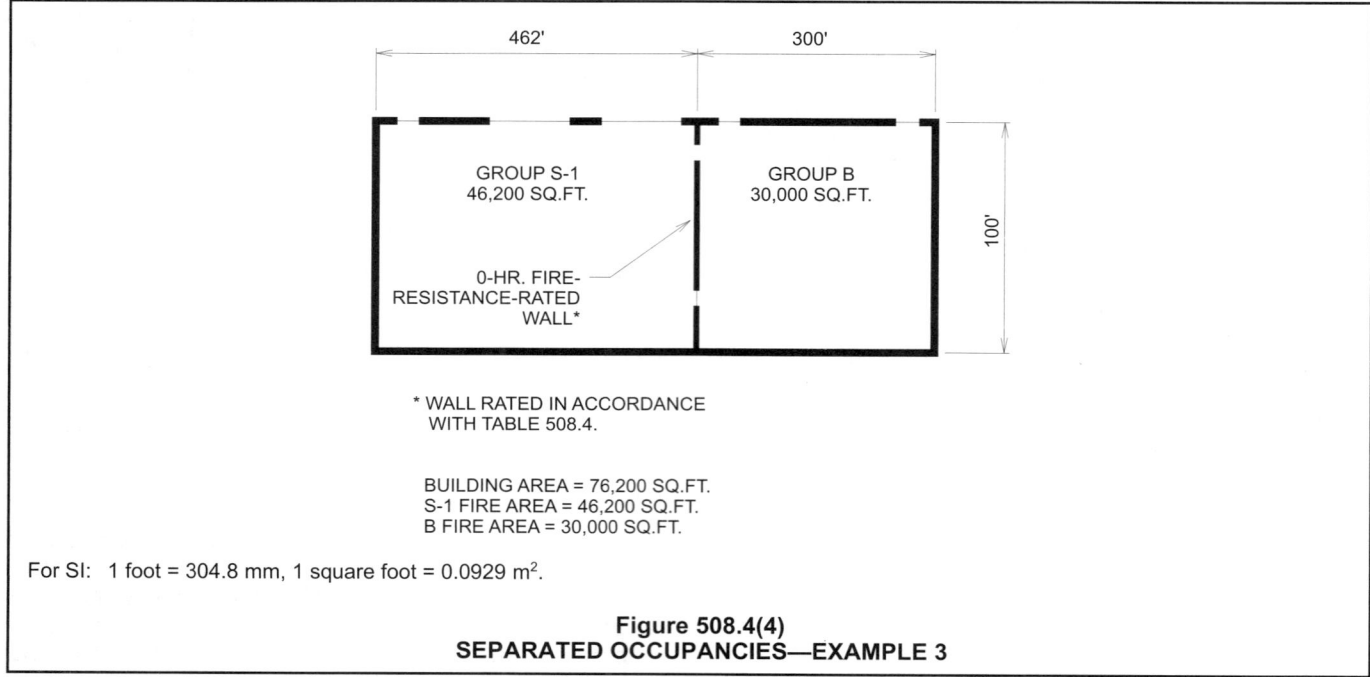

For SI: 1 foot = 304.8 mm, 1 square foot = 0.0929 m².

Figure 508.4(4)
SEPARATED OCCUPANCIES—EXAMPLE 3

509.2 Horizontal building separation allowance. A building shall be considered as separate and distinct buildings for the purpose of determining area limitations, continuity of *fire walls*, limitation of number of *stories* and type of construction where all of the following conditions are met:

1. The buildings are separated with a *horizontal assembly* having a minimum 3-hour *fire-resistance rating*.

2. The building below the *horizontal assembly* is no more than one *story above grade plane*.

3. The building below the *horizontal assembly* is of Type IA construction.

4. Shaft, *stairway*, ramp and escalator enclosures through the *horizontal assembly* shall have not less than a 2-hour *fire-resistance rating* with opening protectives in accordance with Section 715.4.

 Exception: Where the enclosure walls below the *horizontal assembly* have not less than a 3-hour *fire-resistance rating* with opening protectives in accordance with Section 715.4, the enclosure walls extending above the *horizontal assembly* shall be permitted to have a 1-hour *fire-resistance rating*, provided:

 1. The building above the *horizontal assembly* is not required to be of Type I construction;

 2. The enclosure connects less than four *stories*; and

 3. The enclosure opening protectives above the *horizontal assembly* have a minimum 1-hour *fire protection rating*.

5. The building or buildings above the *horizontal assembly* shall be permitted to have multiple Group A occupancy uses, each with an *occupant load* of less than 300, or Group B, M, R or S occupancies.

6. The building below the *horizontal assembly* shall be protected throughout by an *approved automatic sprinkler system* in accordance with Section 903.3.1.1, and shall be permitted to be any of the following occupancies:

 6.1. Group S-2 parking garage used for the parking and storage of private motor vehicles;

 6.2. Multiple Group A, each with an *occupant load* of less than 300;

 6.3. Group B;

 6.4. Group M;

 6.5. Group R; and

 6.6. Uses incidental to the operation of the building (including entry lobbies, mechanical rooms, storage areas and similar uses).

7. The maximum *building height* in feet (mm) shall not exceed the limits set forth in Section 503 for the building having the smaller allowable height as measured from the *grade plane*.

❖ Section 509.2 essentially allows a 3-hour fire-resistance-rated horizontal assembly to create separate buildings similar to the concept used for fire walls (see Figure 509.2). This allowance provides an extensive benefit for height and area in these structures. Buildings constructed under this section are frequently referred to as "pedestal" or "platform" buildings. It should be noted that multiple buildings may be located above the horizontal assembly. Structures built under this section are considered to be distinct buildings above and below the 3-hour fire-resistance-rated horizontal assembly. As distinct buildings, they are individually evaluated with respect to allowable building area, the number of stories and the type of construction. In addition, if a fire wall is needed to address building area issues in the upper building or buildings, the fire wall construction can stop at the 3-hour fire-resistance-rated horizontal assembly and does not need to extend to the foundation. However, other building systems and requirements must be evaluated using the total structure. For example, if the upper building is apartments and the lower building an open parking garage, both buildings will need to be protected by an automatic sprinkler system because of the requirement for buildings occupied by Group R occupancies (see Section 903.2.8).

There are seven conditions that set the limits of this exceptional design:

1. Separation of upper and lower buildings by a 3-hour fire-resistance-rated horizontal assembly.

2. The building area below the horizontal assembly is limited to having no more than one story above grade plane. There is no limit on the number of basement levels included below the single story above grade plane.

3. The building below the horizontal assembly is Type IA. As such it is allowed to be of unlimited area in accordance with Table 503.

4. All openings through the 3-hour fire-resistance-rated horizontal assembly are to be protected by a 2-hour shaft (the shaft openings having $1^1/_2$-hour protectives; see Table 715.4).

5. The uses in the upper building or buildings are limited to Group A uses where each Group A area has an occupant load of less than 300, or Group B, M, R or S occupancies.

6. The building below is allowed to be used exclusively for any of the occupancies listed in Items 6.1 through 6.6 or any combination of those occupancies. While historical versions of Section 509.2 limited the use of the lower building to parking, the provision has evolved to reflect a mixed occupancy setting of urban neighborhoods. The limit of the 300 occupants in Group A occupancies is also reflective of the history of these provisions from the legacy codes, and it reflects the intent of restricting the assembly spaces to smaller businesses so that the assembly use does not dominate the building.

7. The height of the combined buildings above and below the horizontal assembly is limited to the number of feet above grade plane allowed by Section 503 for the type of construction of the upper building. However, the charging language of Section 509.2 does not restrict the number of stories to the entire structure but only to that which is above the horizontal assembly. Thus, a Type VA building, protected by a NFPA 13R sprinkler system and containing a Group R-2 occupancy, can have four stories above the hori-

zontal assembly, provided the overall height of both buildings does not exceed 60 feet (18 288 mm) above grade plane.

This is one of the rare circumstances where there could be two different construction types in a single structure without being separated by a fire wall. It is possible that following the conventional provisions for mixed uses in Sections 508.3 and 508.4 would be less restrictive than the conditions of this section. Compliance with the general provisions of Section 508.3 or 508.4 is permissible, and this section should be viewed as an alternative means of compliance with Section 508.3 or 508.4.

There is one exception within the seven items: the exception to Item 4 indicates the conditions by which the shaft construction protecting openings in the horizontal separation may be 1-hour rated above the horizontal separation. This allowance is only for openings connecting a maximum of four stories. This would include the connection to the area below the horizontal assembly (see Figure 509.2).

A common example of a building constructed under the provisions of Section 509.2 is a Type IA building that contains parking at grade level and below, with up to four stories of Group R-2 apartments in a Type VA, wood frame constructed building above the separation (see Figure 509.2). In many city neighborhoods where zoning laws encourage a mixture of uses, the first story is often occupied with retail shops, service businesses and small restaurants. The height of this structure measured in feet is limited by the Type VA upper structure to maximum of 70 feet (21 336 mm) above grade plane. But the number of stories of the Type VA portion of the building is determined by starting at the horizontal separation between the construction types.

509.3 Group S-2 enclosed parking garage with Group S-2 open parking garage above. A Group S-2 enclosed parking garage with no more than one *story* above *grade plane* and located below a Group S-2 *open parking garage* shall be classified as a separate and distinct building for the purpose of determining the type of construction where all of the following conditions are met:

1. The allowable area of the building shall be such that the sum of the ratios of the actual area divided by the allowable area for each separate occupancy shall not exceed 1.

2. The Group S-2 enclosed parking garage is of Type I or II construction and is at least equal to the *fire-resistance* requirements of the Group S-2 *open parking garage*.

3. The height and the number of tiers of the Group S-2 *open parking garage* shall be limited as specified in Table 406.3.5.

4. The floor assembly separating the Group S-2 enclosed parking garage and Group S-2 *open parking garage* shall be protected as required for the floor assembly of the Group S-2 enclosed parking garage. Openings between the Group S-2 enclosed parking garage and Group S-2

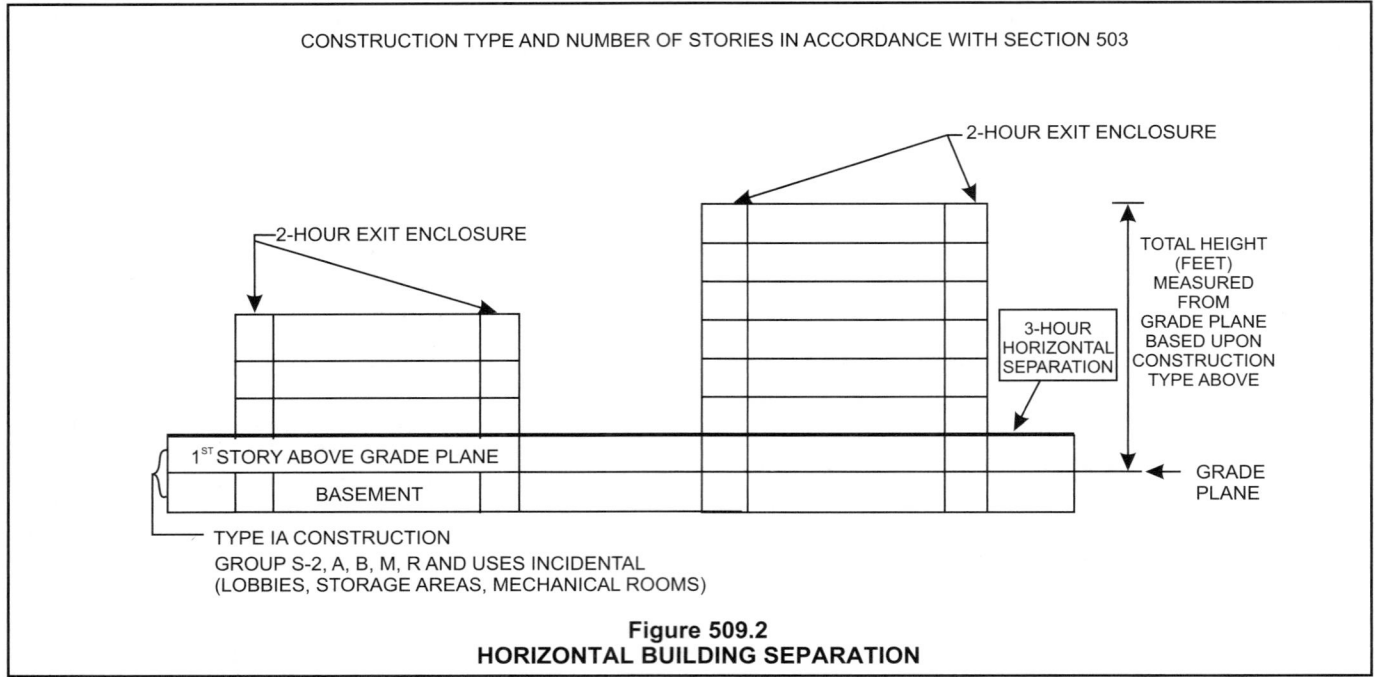

Figure 509.2
HORIZONTAL BUILDING SEPARATION

open parking garage, except *exit* openings, shall not be required to be protected.

5. The Group S-2 enclosed parking garage is used exclusively for the parking or storage of private motor vehicles, but shall be permitted to contain an office, waiting room and toilet room having a total area of not more than 1,000 square feet (93 m²), and mechanical equipment rooms incidental to the operation of the building.

❖ Parking garages of both types, enclosed and open, are addressed in Section 406. Special height and area allowances are given in Section 406.3.5 for open parking structures; however, these special height and area provisions are not applicable if any level of the parking garage does not meet the definition of "Open parking garage" by not having the requisite clear open area to the exterior. This would normally preclude having parking levels below grade in an open parking garage.

Section 509.3 contains provisions that would allow an open parking structure to take advantage of the special height and area limits for open parking structures in Section 406.3.5 while incorporating, in the same building, enclosed parking levels below grade. This is an alternative to treating the whole building as an enclosed parking garage.

There are five conditions listed that must be met in order to use this alternative:

1. Appropriate increases in accordance with Section 506 are to be considered for each portion, and for purposes of frontage increase the same measurement of open perimeter may apply to both the upper and lower garages.

2. The enclosed parking structure below must meet or exceed the fire resistance of the open

parking structure above. At a minimum, the enclosed parking structure must be of Type IIB construction (all noncombustible materials, without fire-resistance-rated protection); however, if the open parking structure above is Type VA, IIIA, IIA or IA, the enclosed parking structure would also have to meet or exceed the required ratings of the building elements from Table 601 for the construction type of the open parking structure.

3. Both the entire height of the building and the number of tiers are limited by Table 406.3.5; therefore, even if the first story above grade is part of the enclosed parking garage portion of the building, the height of the structure above grade plane could not exceed what would be permitted by Table 406.3.5 if the entire structure above grade plane were part of the open parking garage portion.

4. Protection of the floor assembly of the Group S-2 building depends on the construction type and Table 601 for the S-2 enclosed parking garage.

5. Certain accessory uses are permitted to be present when taking advantage of these alternative provisions, such as a small office and waiting area.

509.4 Parking beneath Group R. Where a maximum one *story above grade plane* Group S-2 parking garage, enclosed or open, or combination thereof, of Type I construction or open of Type IV construction, with grade entrance, is provided under a building of Group R, the number of stories to be used in determining the minimum type of construction shall be measured from the floor above such a parking area. The floor assembly between the parking garage and the Group R above shall comply with the type of construction required for the parking garage and shall also provide a *fire-resistance rating* not less than the mixed occupancy separation required in Section 508.4.

❖ This section permits an extra story (above the limits of Table 503), based on construction type, for Group R buildings with parking on the first story (see Figure 509.4). There are several conditions that must be met: the parking must be limited to one story above grade; must be Type IV (if open) or I (open or enclosed) construction; the entrance to the garage must be at grade; and a 1-hour horizontal assembly must be provided between the parking and the Group R occupancy in accordance with Table 508.4. The limitation of Table 503 for height above grade plane (in feet) is not changed under this circumstance. It should be noted that all buildings containing Group R occupancies are required to be sprinklered in the code. Therefore, the separation required would only be 1 hour, but the parking area must also be sprinklered.

For instance, a fully sprinklered Group R-2 building of Type VB construction would normally be permitted to be three stories above grade (Table 503 limit of two stories plus the additional story increase for sprinklers in Section 504.2). If it meets the conditions of this section for parking on the first story above grade, then it could actually be four stories above grade; however, the building height in feet cannot exceed 60 feet (18

288 mm) above grade plane in accordance with Table 503 for Type VB construction and Section 504.2, which allows a height increase for sprinklers. Figure 509.4 provides an example of a Group R-1 building over enclosed parking.

509.5 Group R-1 and R-2 buildings of Type IIIA construction. The height limitation for buildings of Type IIIA construction in Groups R-1 and R-2 shall be increased to six *stories* and 75 feet (22 860 mm) where the first floor assembly above the basement has a *fire-resistance rating* of not less than 3 hours and the floor area is subdivided by 2-hour fire-resistance-rated *fire walls* into areas of not more than 3,000 square feet (279 m²).

❖ This section contains special provisions for increasing the height of Type IIIA Group R-1 and R-2 buildings based upon increases in fire resistance and compartmentation. More specifically, the higher rating would apply to the floor structure of the first story above grade (3 hours), and the fire walls (2 hours) subdividing the building into floor areas of not more than 3,000 square feet (279 m²). The fire walls must extend from foundation to roof in accordance with the definition of "Fire wall" in Section 702.1. In addition, the fire walls must comply with the requirements in Section 706.

It should be noted that this section is independent of the fire area requirements of Section 707 but such separations would certainly be seen as fire areas or as separate buildings. In either case all buildings containing a Group R-1 or R-2 fire area would require sprin-

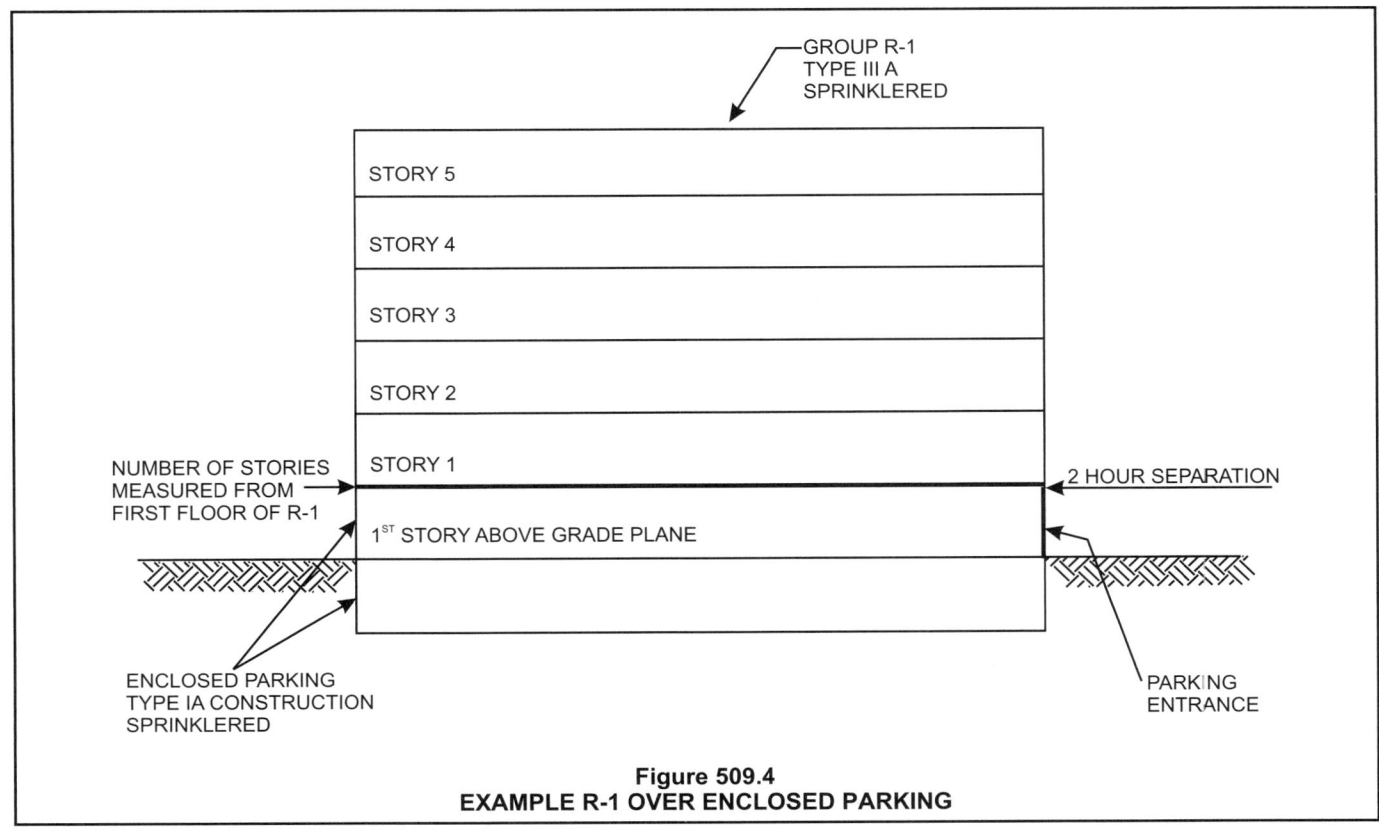

Figure 509.4
EXAMPLE R-1 OVER ENCLOSED PARKING

klers in accordance with Section 903.2.8.

This section constitutes a trade-off of building area for extra height, and depends on the materials and rating requirements for Type IIIA construction, as well as the extra fire resistance of the first story.

See Figure 509.5 for an example of a Type IIIA building height increase for a Group R-2 occupancy.

509.6 Group R-1 and R-2 buildings of Type IIA construction. The height limitation for buildings of Type IIA construction in Groups R-1 and R-2 shall be increased to nine *stories* and 100 feet (30 480 mm) where the building is separated by not less than 50 feet (15 240 mm) from any other building on the lot and from *lot lines*, the *exits* are segregated in an area enclosed by a

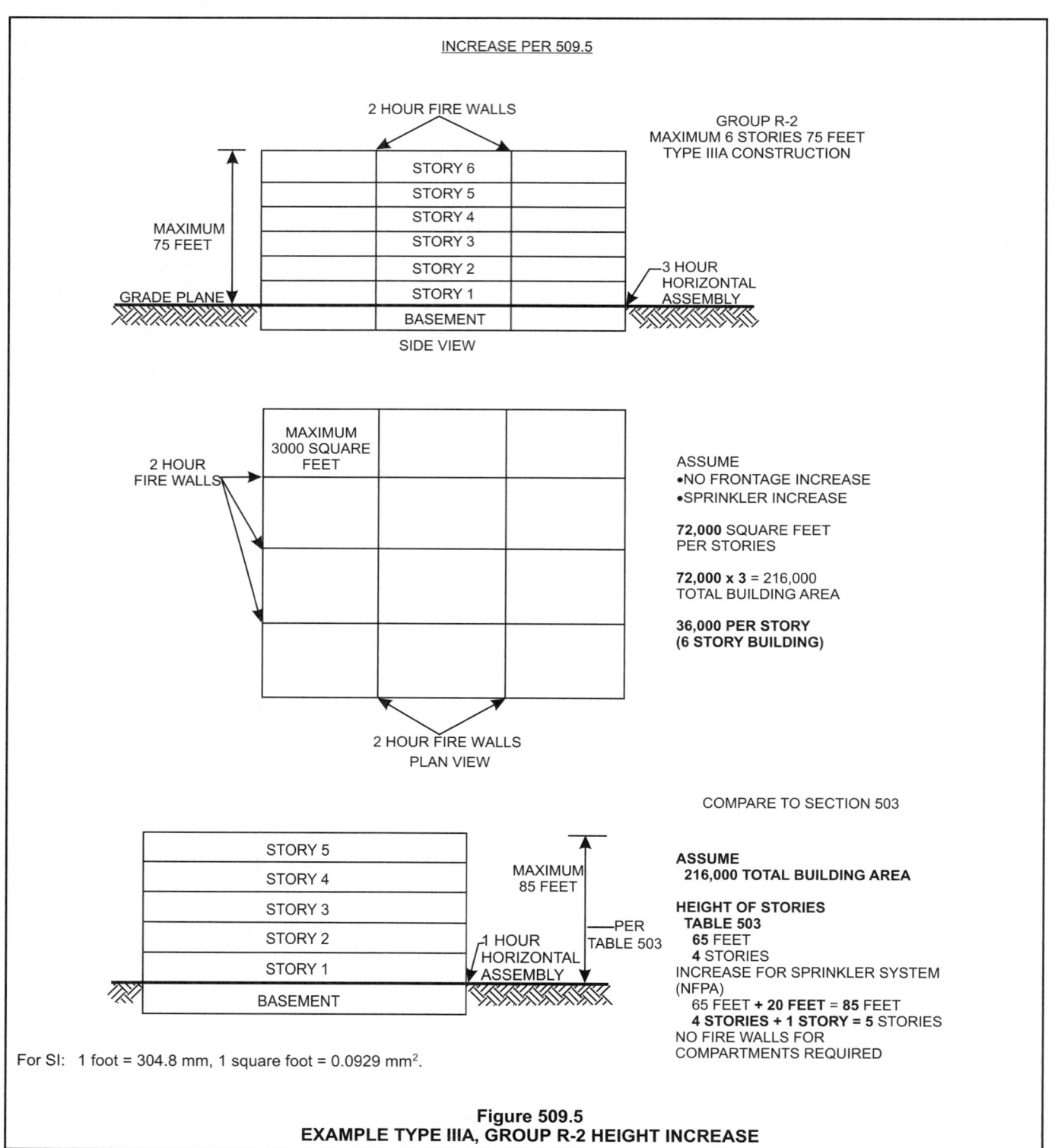

For SI: 1 foot = 304.8 mm, 1 square foot = 0.0929 mm².

Figure 509.5
EXAMPLE TYPE IIIA, GROUP R-2 HEIGHT INCREASE

2-hour fire-resistance-rated *fire wall* and the first floor assembly has a *fire-resistance rating* of not less than 1¹/₂ hours.

❖ This section contains special provisions for increasing the height of Type IIA, Group R-1 and R-2 buildings. The higher 1¹/₂-hour rating would apply to the floor assembly of the first story above grade, and the exits are required to be enclosed with fire walls extending from foundation to roof, otherwise meeting all the requirements of Section 706. Such fire walls would create separate fire areas or buildings for the purposes of the application of the rest of the code.

A separation distance of at least 50 feet (15 240 mm)

from other buildings on the same lot or property lines is also required to use this alternative. This section constitutes a trade-off of extra protection for exits for extra building height, and depends also on the materials and rating requirements for Type IIA construction, as well as the extra fire-resistance rating protecting the first story from the basement. The entire building must be sprinklered in accordance with Section 903.2.8. See Figure 509.6 for an example of a Type IIA building height increase for a Group R-2 occupancy.

Note that such buildings will likely be considered high-rise buildings and must comply with Section 403.

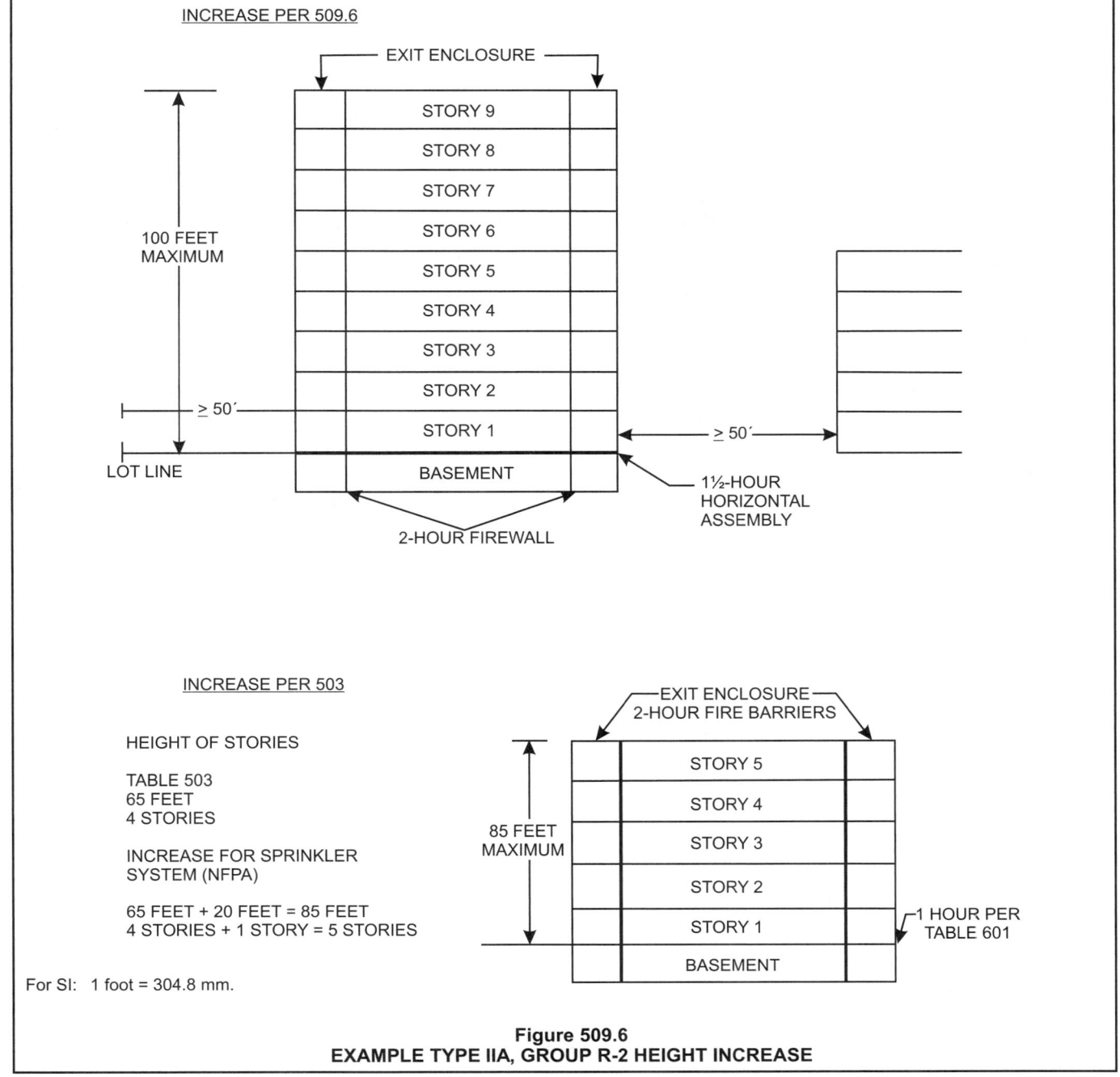

Figure 509.6
EXAMPLE TYPE IIA, GROUP R-2 HEIGHT INCREASE

509.7 Open parking garage beneath Groups A, I, B, M and R. *Open parking garages* constructed under Groups A, I, B, M and R shall not exceed the height and area limitations permitted under Section 406.3. The height and area of the portion of the building above the *open parking garage* shall not exceed the limitations in Section 503 for the upper occupancy. The height, in both feet and stories, of the portion of the building above the *open parking garage* shall be measured from *grade plane* and shall include both the *open parking garage* and the portion of the building above the parking garage.

❖ This section addresses a special mixed use condition, and is another circumstance wherein a building can be designated with two different types of construction for determining height and area. This provision is only to be applied when an open parking structure (Group S-2) is to be constructed below a Group A, I, B, M or R occupancy (see Figure 509.7). If an open parking structure is located below any other occupancy group, Section 508 must be applied, as for any other mixed use condition. In accordance with Section 509.1, this section is an alternative to the general mixed use provisions in Section 508 that can be applied where advantageous to the design of buildings with open parking on the lower levels.

In the application of Section 509.7, there are two criteria that must be met for code compliance (see Figure 509.7):

1. The height and area of the open parking structure comprising a part of a mixed-use group building must not exceed the limitations for open parking structures permitted in Section 406.3.5 and Table 406.3.5.

2. The allowable height of the occupancy located above the open parking structure is to be deter-

mined in accordance with Section 503 and Table 503. The height is the vertical distance (measured in feet and stories) from the grade plane to the top of the average height of the highest roof surface, in accordance with the definition of "Building height" in Section 502.1. Allowable heights and areas may be modified in accordance with Sections 504 and 506.

509.7.1 Fire separation. *Fire barriers* constructed in accordance with Section 707 or *horizontal assemblies* constructed in accordance with Section 712 between the parking occupancy and the upper occupancy shall correspond to the required *fire-resistance rating* prescribed in Table 508.4 for the uses involved. The type of construction shall apply to each occupancy individually, except that structural members, including main bracing within the *open* parking structure, which is necessary to support the upper occupancy, shall be protected with the more restrictive fire-resistance-rated assemblies of the groups involved as shown in Table 601. *Means of egress* for the upper occupancy shall conform to Chapter 10 and shall be separated from the parking occupancy by *fire barriers* having at least a 2-hour *fire-resistance rating* as required by Section 706 with self-closing doors complying with Section 715 or horizontal assemblies having at least a 2-hour *fire-resistance rating* as required by Section 712, with self-closing doors complying with Section 715. *Means of egress* from the *open parking garage* shall comply with Section 406.3.

❖ This section contains additional conditions for the use of Section 509.7 as an alternative to the general mixed use provisions in Section 508. It contains an additional five criteria:

1. The open parking structure and occupancy located above must be separated, both horizontally and vertically if necessary, by fire separa-

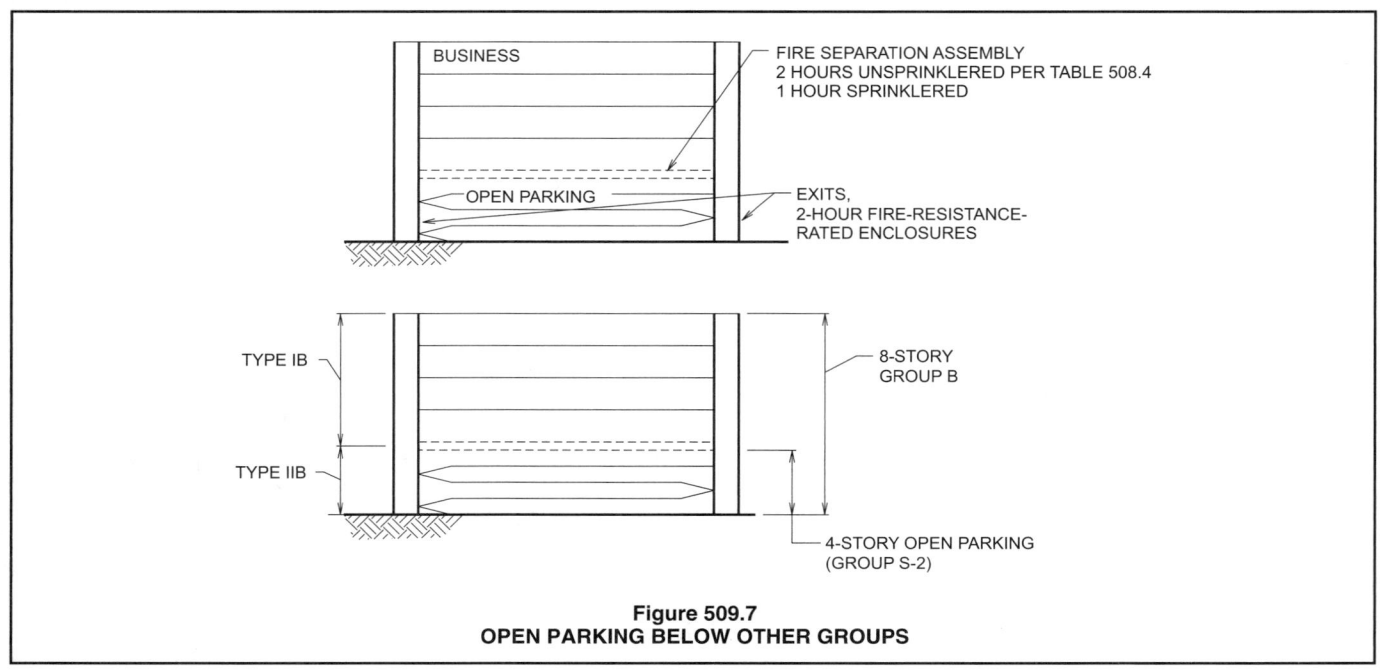

Figure 509.7
OPEN PARKING BELOW OTHER GROUPS

tion assemblies having a fire-resistance rating corresponding to that specified in Table 508.4.

> **Example:** An open parking structure, Group S-2, is located below an office (Group B); therefore, based on Table 508.4, all vertical and horizontal assemblies separating the two groups are required to have a minimum fire-resistance rating of 2 hours in an unsprinklered building and 1 hour in a sprinklered building.

2. The upper and lower portions of the building may be constructed of a separate type of construction (except as discussed in Item 5 below). The minimum type of construction for an open parking structure is Type IIB or IV, depending on the thresholds established in Table 406.3.5.

> **Example:** The open parking structure may be of Type IIB construction and the offices located above the open parking structure may be of Type IB construction.

3. Regardless of the construction types being used, all structural members, including main bracing in the open parking structure for the stability of the upper occupancy, must be rated in accordance with the most restrictive fire-resistance-rating requirement in accordance with Table 601, which would be that for Type IB construction.

> **Example:** Consider a building where the upper occupancy is of Type IB protected construction, and the open parking structure is of Type IIB unprotected construction. In accordance with Table 601, all load-bearing walls and structural frames, including columns and girders necessary to support the upper occupancy, must have at least a 2-hour fire-resistance rating, in accordance with the requirements of Type IB construction. This typically applies to the columns and bracing in the entire structure, beams supporting the floor separating the upper occupancy from the open parking structure and any transfer beams located in the open parking structure. The story of the open parking structure and any members supporting these stories, however, are permitted to have a zero fire-resistance rating in accordance with the requirements for Type IIB construction.

4. Means of egress facilities within and from the upper occupancy are to conform to Chapter 10. In addition, the egress facilities from the upper occupancy must be separated from the parking area by fire-resistance-rated wall assemblies of at least 2 hours, by fire barriers meeting the requirements of Section 707 or horizontal assemblies meeting Section 712. These egress facilities are required to maintain the 2-hour protection for their full height and must be continuous

to the level of exit discharge. The fire-resistance-rating reduction of vertical exit enclosures for structures less than four stories in height (see Section 1022.1) is not applicable to an exit passing through the open parking structure in accordance with this section. Door openings to the exit enclosure must have self-closing doors that comply with Section 715, with a fire protection rating of $1^1/_2$ hours in accordance with Table 715.4.

5. Means of egress facilities within and from the open parking structure are to conform to Section 406.3.8, which also references Chapter 10.

509.8 Group B or M with Group S-2 open parking garage. Group B or M occupancies located no higher than the first *story* above *grade plane* shall be considered as a separate and distinct building for the purpose of determining the type of construction where all of the following conditions are met:

1. The buildings are separated with a *horizontal assembly* having a minimum 2-hour *fire-resistance rating*.

2. The occupancies in the building below the *horizontal assembly* are limited to Groups B and M.

3. The occupancy above the *horizontal assembly* is limited to a Group S-2 *open parking garage*.

4. The building below the *horizontal assembly* is of Type I or II construction but not less than the type of construction required for the Group S-2 *open parking garage* above.

5. The height and area of the building below the *horizontal assembly* does not exceed the limits set forth in Section 503.

6. The height and area of the Group S-2 *open parking garage* does not exceed the limits set forth in Section 406.3. The height, in both feet and stories, of the Group S-2 *open parking garage* shall be measured from *grade plane* and shall include the building below the *horizontal assembly*.

7. Exits serving the Group S-2 *open parking garage* discharge directly to a street or *public way* and are separated from the building below the *horizontal assembly* by 2-hour *fire barriers* constructed in accordance with Section 707 or 2-hour *horizontal assemblies* constructed in accordance with Section 712, or both.

❖ This section deals with the inverse of the other circumstances in Section 509: the parking, which must be an open parking garage, is located above the other groups, in this case, Group B or M. This is a common type of construction in metropolitan areas. The parking garage is an open parking structure, with developers using the street level part of the building as an opportunity to provide retail space or other commercial space in a downtown application. The conditions under which this configuration of parking garage and Group B or M uses is similar to conditions for other circumstances is found in Section 509.

More specifically there are seven criteria that must be met.

1. Similar to Item 1 of Section 509.2, a horizontal separation is required to essentially divide the buildings into separate buildings. Again it is like having a horizontal fire wall. The separation is only required to be 2 hours, which is less than the 3 hours required by Item 1 of Section 509.2.

2. The occupancies below the separation are limited to Groups B and M, which relates to the likely application of such configurations.

3. The occupancy above the horizontal assembly is limited to a Group S-2 open parking garage.

4. The building below is required to be of Type I or II construction but must always be at least that of the Group S-2 above.

5. The Group B and M occupancies must comply with the height and area requirements of Section 503. As written this would allow increases in area and height as appropriate. Note that the height is generally restricted by the fact that such occupancies can be located no higher than the first story above grade plane.

6. The height and area of the Group S-2 occupancy shall not exceed that of Section 406.3. In addition the overall height (measured in both the number of stories and building height in feet) is further restricted by the fact that the occupancies below the horizontal assembly must be addressed.

7. Lastly, similar to Section 509.7.1, the exits from the Group S-2 must be protected within 2-hour fire barriers or horizontal assemblies and extend directly to a street or public way.

See Figure 509.8 for an application of this section.

509.9 Multiple buildings above Group S-2 parking garages. Where two or more buildings are provided above the *horizontal assembly* separating a Group S-2 *open* or closed *parking garage* from the buildings above in accordance with the special provisions in Sections 509.2, 509.3 or 509.8, the buildings above the *horizontal assembly* shall be regarded as separate and distinct buildings from each other and shall comply with all other provisions of this code as applicable to each separate and distinct building.

❖ A very common practice when applying the alternatives such as those found in Sections 509.2 and 509.3 is to have multiple buildings above the horizontal assemblies. Specifically, the concept in Section 509.2 is often called a "pedestal" building. In other words, the building below and including the horizontal assembly creates a "pedestal" on top of which many buildings, which are considered separate from one another, can be located. This section is simply clarifying that the multiple buildings located on top of the pedestal are considered separate from one another. This can be realized by there being multiple and distinct structures separated by yards or courts between different structures, or can be the result of the use of fire walls dividing a single structure into multiple buildings. Just as distinct buildings sitting on the ground, these multiple buildings above the horizontal assemblies can be of different construction types from the pedestal building and of different construction types from each other (see Figure 509.9).

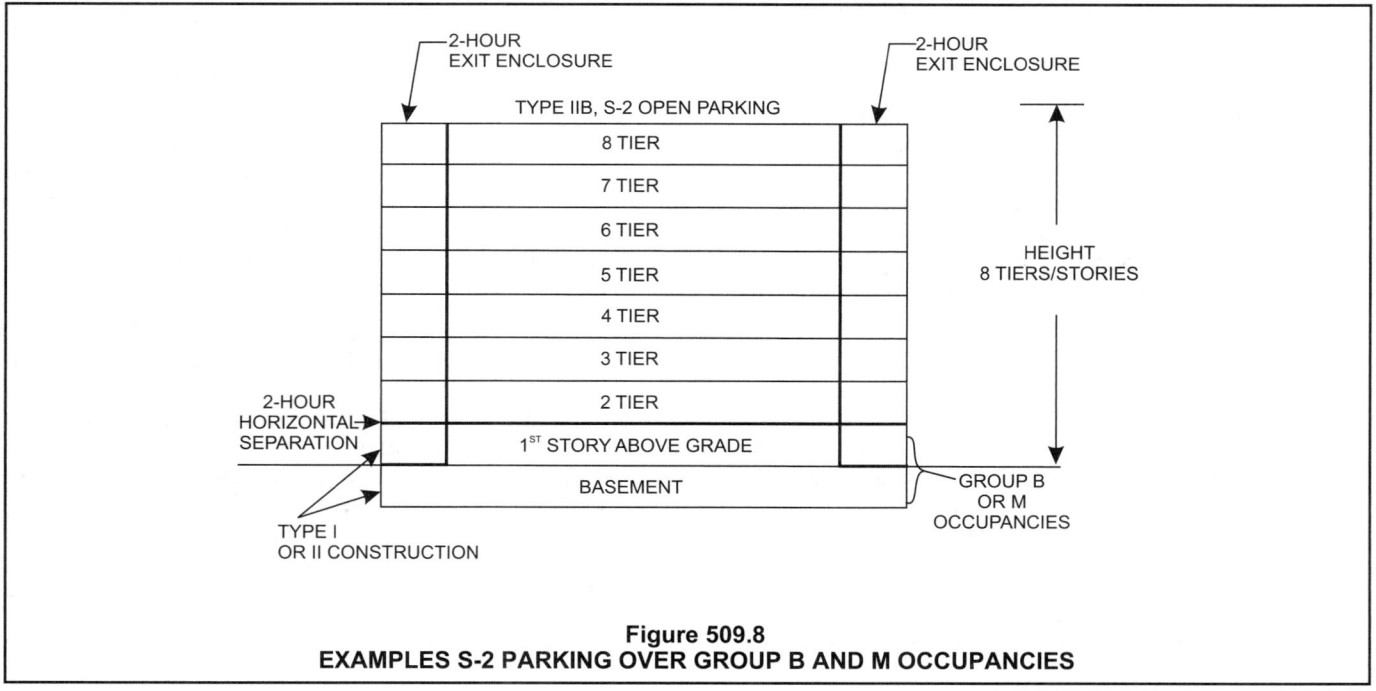

Figure 509.8
EXAMPLES S-2 PARKING OVER GROUP B AND M OCCUPANCIES

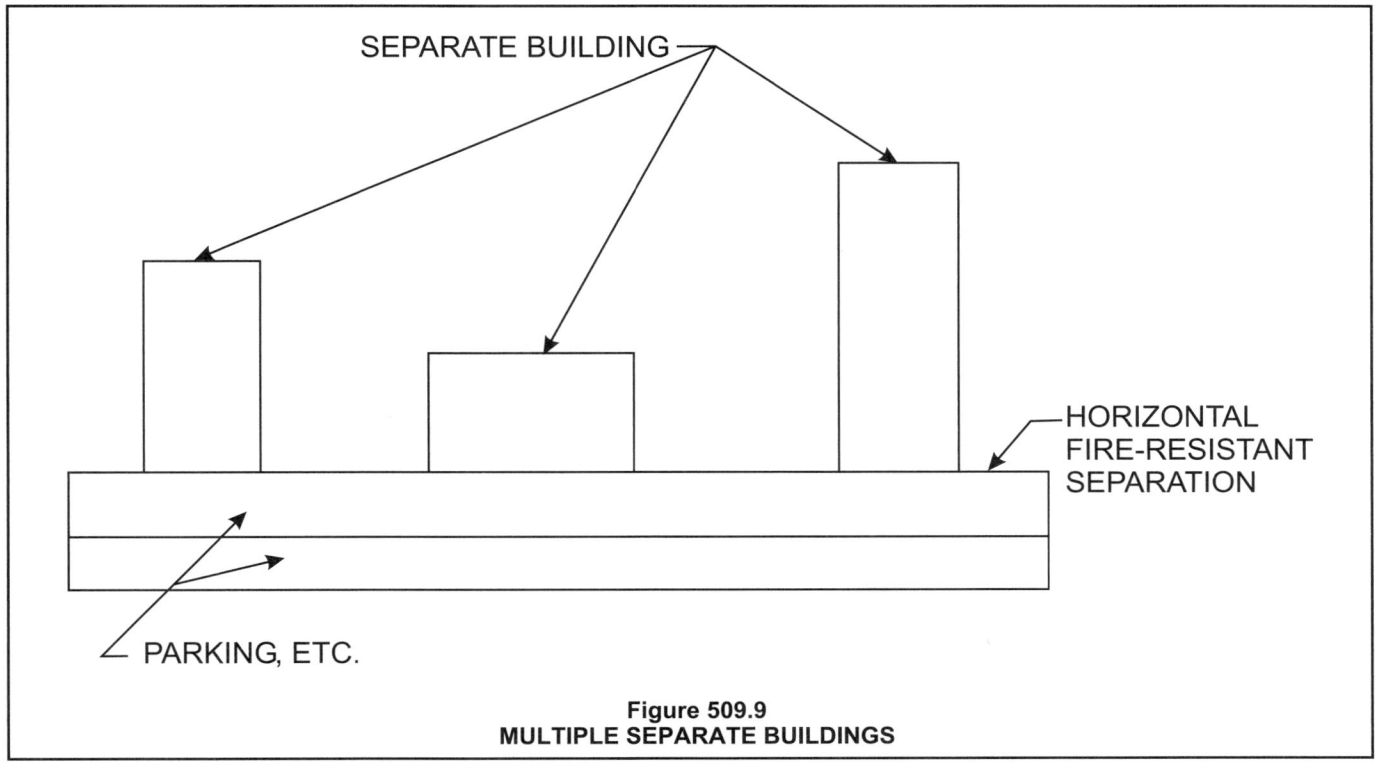

Figure 509.9
MULTIPLE SEPARATE BUILDINGS

Bibliography

The following resource materials are referenced in this chapter or are relevant to the subject matter addressed in this chapter.

29 CFR Part 1910, *Occupational Safety and Health Act.* Washington, DC: Occupational Safety and Health Administration.

IFC-09, *International Fire Code.* Washington DC: International Code Council, 2009.

IMC-09, *International Mechanical Code.* Washington DC: International Code Council, 2009.

NFPA 13-07, *Installation of Sprinkler Systems.* Quincy, MA: National Fire Protection Association, 2006.

NFPA 13D-07, *Installation of Sprinkler Systems in One- and Two-family Dwellings and Manufactured Homes.* Quincy, MA: National Fire Protection Association, 2006.

NFPA 13R-07, *Installation of Sprinkler Systems in Residential Occupancies Up to Four Stories in Height.* Quincy, MA: National Fire Protection Association, 2006.

NFPA 20-10, *Installation of Stationary Pumps for Fire Protection.* Quincy, MA: National Fire Protection Association, 2010.

NFPA 409-04, *Standard on Aircraft Hangars.* Quincy, MA: National Fire Protection Association, 2004.

Chapter 6:
Types of Construction

General Comments

Chapter 6 contains the requirements to classify buildings into one of five types of construction. Tables 601 and 602 provide the minimum hourly fire-resistance ratings for the structural elements based on the type of construction of the building and fire separation distance. Section 602 describes each construction type in detail. Section 603 describes the use of combustible materials in buildings of noncombustible construction.

Correct classification of a building by its type of construction is essential. Many code requirements applicable to a building, such as allowable height and area (see Chapter 5), are dependent on its type of construction. If a building is placed in an incorrect construction classification (for example, one that is overly restrictive), its owner may be penalized by increased construction costs. On the other hand, when a building is incorrectly classified in an overly lenient type of construction, it will not be constructed in a manner that takes into account the relative risks associated with its size or function. The provisions of this chapter, coupled with Chapter 3 and 5 and Tables 601 and 602, establish the basis for the "equivalent risk theory" on which the entire code is based.

Purpose

The purpose of classifying buildings or structures by their type of construction is to account for the response or participation that a building's structure will have in a fire condition originating within the building as a result of its occupancy or fuel load.

The code requires every building to be classified as one of five possible types of construction: Types I, II, III, IV and V. Each type of construction denotes the kinds of materials that are to be used [i.e., noncombustible steel, concrete, masonry, combustible (wood, plastic) or heavy timber (HT)], and the minimum fire-resistance ratings that are associated with the structural elements in a building having that classification, i.e., 0, 1, $1^{1}/_{2}$, 2 or 3 hours. Type I and II construction have building elements that are noncombustible. Type III construction has noncombustible exterior walls and combustible or noncombustible interior elements. Type IV construction has noncombustible exterior walls and HT interior elements. Type V construction has building elements that are combustible. Type I, II, III and V construction are further subdivided into two categories (IA and IB, IIA and IIB, IIIA and IIIB, VA and VB).

SECTION 601
GENERAL

601.1 Scope. The provisions of this chapter shall control the classification of buildings as to type of construction.

❖ This section requires that all buildings be assigned a type of construction classification as indicated in the "General Comments" above.

TABLE 601. See page 6-2.

❖ Table 601 has three components: the top rows list the types of construction and their subclassifications; the left column lists the various building elements that are regulated by the table and each cell in the table contains the minimum required fire-resistance rating, in hours, for the various elements based on the required type of construction of the building. "Building element" is defined in Section 702 as a fundamental component of building construction which needs to be constructed of the materials and having the fire-resistance rating specified in this chapter in order for a building to be classified in a type of construction. Notes a through g apply as specifically referenced in the table.

Type I, II, III and V construction are further subdivided into two categories (A and B). Type A and B con-

struction are not defined in the code. The designations simply refer to the hourly fire-resistance rating required for the structural elements. A Type A designation will have a higher fire-resistance rating for the structural element than a Type B designation. Sometimes, Type A and B are referred to as protected and unprotected construction, respectively. Please note this terminology does not refer to whether the building is sprinklered.

The following describes the items in the left column of the table, titled "Building Element."

Row 1: *Primary structural frame.* This category includes the structural (load-bearing) components of the building frame. Definitions of "Primary structural frame" and "Secondary members" are found in Section 202. Any structural item that provides direct connections to columns and bracing members that are designed to carry a gravity load is considered part of the structural frame. To delay vertical (i.e., gravity) load-carrying collapse of a building due to fire exposure for a theoretical amount of time, the components that make up the primary structural frame are required to maintain a minimum degree of fire resistance. The components defined as part of the primary structural frame, with the exception of Type IV construction, must also comply with Section 704. Secondary members (e.g., floor or roof panels without a connection to the column) are not considered part of the structural frame. (see Rows 5 and 6 of the table).

Row 2: *Bearing walls—exterior and interior.* Exterior bearing walls are the outermost walls that enclose the structure and support any structural load other than their own weight. Their required fire-resistance rating is established by the higher of two fire-resistance ratings. The first component of determining the fire-resistance rating is based on the type of construction of the building. The second component of determining the fire-resistance rating is based on the exterior wall's fire separation distance in Table 602. Whichever of the two requires the higher fire-resistance rating will dictate the minimum required fire-resistance rating of the exterior wall.

In addition to Tables 601 and 602, exterior walls must comply with Section 705 and Chapter 14. Also, Section 706.5.1 has fire-resistance-rating requirements for exterior walls on each side of the intersection of fire wall.

There are also several requirements related to exterior walls mentioned in Chapter 10. Section 1007.7 has fire-resistance-rating requirements for exterior walls adjacent to exterior areas for assisted rescue. Section 1022.6 has specific fire-resistance-rating requirements for exterior walls adjacent to an exit stairway. Section 1026.6 has fire-resistance-rating requirements for exterior walls adjacent to exterior exit stairways. Section 1027.5.2 has fire-resistance-rating requirements for exterior walls adjacent to an egress court.

Additionally, this category includes the structural (load-bearing) interior walls of a building. To delay vertical load-carrying collapse of a building due to fire exposure for a predetermined amount of time, the structural partitions are required to maintain a minimum degree of fire resistance. Primary structural frame elements supporting such walls must comply with Table 601, as well as have at least the same degree of fire resistance as the supported wall.

TABLE 601
FIRE-RESISTANCE RATING REQUIREMENTS FOR BUILDING ELEMENTS (hours)

BUILDING ELEMENT	TYPE I A	TYPE I B	TYPE II A[d]	TYPE II B	TYPE III A[d]	TYPE III B	TYPE IV HT	TYPE V A[d]	TYPE V B
Primary structural frame[g] (see Section 202)	3[a]	2[a]	1	0	1	0	HT	1	0
Bearing walls Exterior[f, g]	3	2	1	0	2	2	2	1	0
Bearing walls Interior	3[a]	2[a]	1	0	1	0	1/HT	1	0
Nonbearing walls and partitions Exterior	See Table 602								
Nonbearing walls and partitions Interior[e]	0	0	0	0	0	0	See Section 602.4.6	0	0
Floor construction and secondary members (see Section 202)	2	2	1	0	1	0	HT	1	0
Roof construction and secondary members (see Section 202)	1¹/₂[b]	1[b, c]	1[b, c]	0[c]	1[b, c]	0	HT	1[b, c]	0

For SI: 1 foot = 304.8 mm.

a. Roof supports: Fire-resistance ratings of primary structural frame and bearing walls are permitted to be reduced by 1 hour where supporting a roof only.

b. Except in Group F-1, H, M and S-1 occupancies, fire protection of structural members shall not be required, including protection of roof framing and decking where every part of the roof construction is 20 feet or more above any floor immediately below. Fire-retardant-treated wood members shall be allowed to be used for such unprotected members.

c. In all occupancies, heavy timber shall be allowed where a 1-hour or less fire-resistance rating is required.

d. An approved automatic sprinkler system in accordance with Section 903.3.1.1 shall be allowed to be substituted for 1-hour fire-resistance-rated construction, provided such system is not otherwise required by other provisions of the code or used for an allowable area increase in accordance with Section 506.3 or an allowable height increase in accordance with Section 504.2. The 1-hour substitution for the fire resistance of exterior walls shall not be permitted.

e. Not less than the fire-resistance rating required by other sections of this code.

f. Not less than the fire-resistance rating based on fire separation distance (see Table 602).

g. Not less than the fire-resistance rating as referenced in Section 704.10.

See the commentary for Section 602.1 regarding opening and penetration protection of bearing walls.

Row 3: *Nonload-bearing walls—exterior.* This category includes all exterior walls that only support their own weight. The fire-resistance rating, unlike for load-bearing walls, is based solely on the exterior wall's fire separation distance (Table 602). Where they occur in buildings of Type I or II construction, the walls can be constructed of fire-retardant-treated wood (FRTW) if no rating is required (see Section 603.1).

Row 4: *Nonload-bearing walls—interior.* This category includes all interior nonload-bearing walls and partitions (for example, the common wall separating two offices in the same suite). These walls must comply with all of the material requirements associated with their type of construction classification. Nonload-bearing interior walls and partitions are not required to have a fire-resistance rating. Where they occur in buildings of Type I or II construction and are to be constructed of wood, the wood must be fire-retardant treated as indicated in Section 603.1. Nonload-bearing interior walls may be required to be fire-resistance rated when they also serve another purpose. Some examples are where an interior wall also serves to separate mixed occupancies (see Section 508.4), dwelling units, tenant spaces in covered malls and sleeping units and where these are also corridor walls (see Sections 707 and 709).

Row 5: *Floor construction and secondary members.* Floor construction provides a natural fire compartment in a building by means of a horizontal barrier that retards the vertical passage of fire from floor to floor. In order to accomplish this, floor assemblies, including the beams and secondary structural members supporting the floor, must comply with Table 601 and Section 712. Ceilings are included if they are part of the tested assembly. A definition of "Secondary member" is provided in Section 202. Secondary members are not considered part of the primary structural frame.

Row 6: *Roof construction and secondary members.* Proper roof construction is necessary to prevent collapse from fire as well as potential impingement on adjacent buildings. Roof construction must comply with Table 601 and Section 712 (see the definition of "Secondary member" in Section 202). Additionally, when a portion of the roof construction is less than 20 feet (6096 mm) above the floor immediately below, the entire structural member must be protected to meet the minimum fire-resistance rating requirements. In Figure 601(2), the roof construction is assumed to be comprised of one structural member. While a majority of the roof construction is greater than 20 feet (6096 mm) above the floor below, a portion of the roof construction is only 15 feet (4572 mm) above the mezzanine. As such, the entire length of the structural member, or in this case, the entire roof must be rated for 1 hour. The fire-resis-

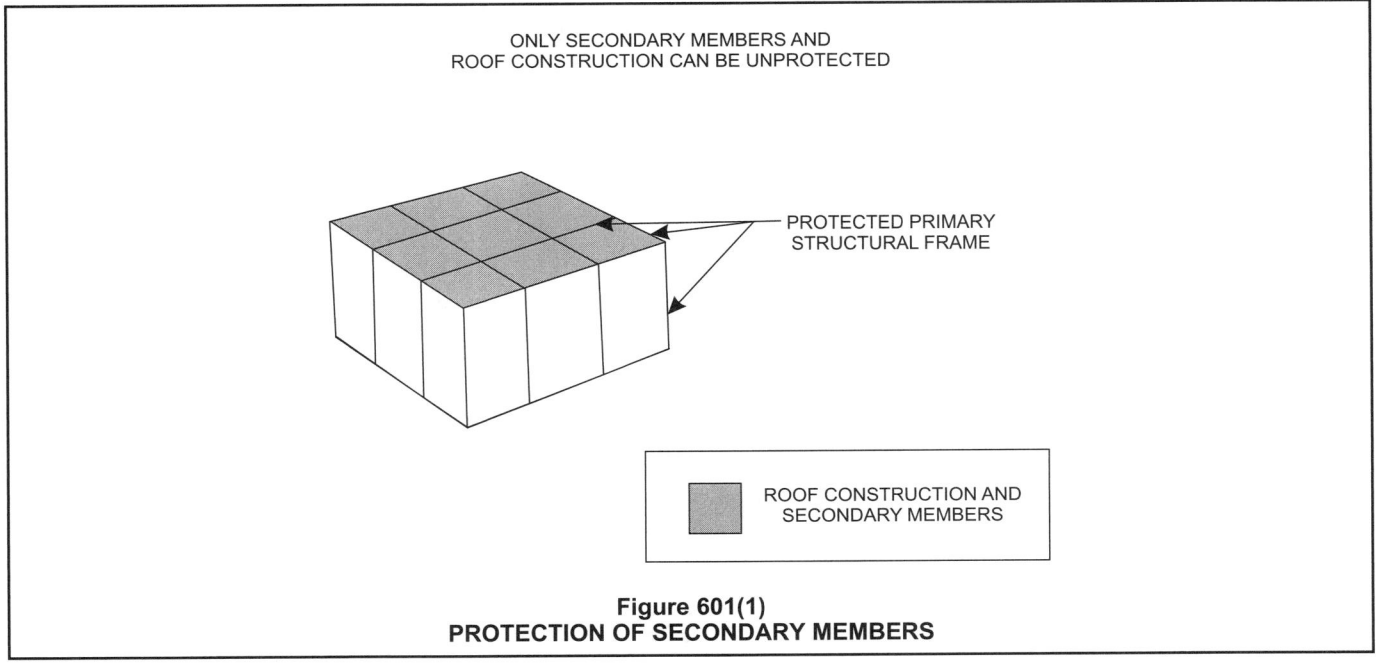

ONLY SECONDARY MEMBERS AND
ROOF CONSTRUCTION CAN BE UNPROTECTED

PROTECTED PRIMARY
STRUCTURAL FRAME

ROOF CONSTRUCTION AND
SECONDARY MEMBERS

Figure 601(1)
PROTECTION OF SECONDARY MEMBERS

tance rating is not permitted to terminate at the portion where the roof is 20 feet (6096 mm) above the floor below unless the structural member ends at that point. The fire-resistance rating of a column must also be continuous for the full height of the column and not reduced or eliminated at a height of 20 feet (6096 mm) and above (see Note b).

Note a permits the fire-resistance ratings of primary structural frame and interior load-bearing walls in buildings of Type IA and IB construction to be reduced by 1 hour if the members are supporting only the roof.

Note b applies to the construction of the roof and related secondary members in all types of construction. It allows these elements not to be of protected construction when all parts of the roof construction are more than 20 feet (6096 mm) above any floor below. This only applies to the secondary members of the structure and not to primary structural frame located within the roof or at this roof level [see Figure 601(1)]. This alternative is applicable for all occupancy classifications except Groups F-1, H, M and S-1. Figure 601(2) shows an example where a mezzanine reduces the clearance to the roof to less than 20 feet (6096 mm) for a portion of the total roof. Designs similar to Figure 601(2) do not comply with note b, and elimination of fire-resistance is not allowed for any of the roof.

In buildings of Type I and II construction, fire-retardant-treated wood (FRTW) may be utilized for unprotected roof members. Please note that FRTW is not required in the roof of buildings of Type IIIA or VA construction since combustible materials are already permitted by Sections 602.3 and 602.5, respectively.

Note c permits heavy timber (HT) construction to be utilized in the roof construction as an alternative to having a fire-resistance rating of 1 hour or less. Note that HT cannot be used in Type IA construction since the roof is required to have a rating greater than 1 hour (i.e., $1^1/_2$). The intent of the note is to allow the substitution of HT for 1-hour construction in roof construction; it is not intended to say that HT is 1-hour-rated construction.

Note d permits buildings of Type IIA, IIIA and VA construction to use an automatic sprinkler system in compliance with NFPA 13 as an alternative to 1-hour fire-resistance-rated construction. In order to utilize the substitution, the sprinkler system must not be required for any other reason, including for increases in allowable height and area or where it is required by Section 903 based on the occupancy of a building. The sprinkler system may not be used as an alternative to the fire-resistance rating for exterior walls. This means that this exception is applicable in very limited circumstances.

Note e is applicable only to interior nonload-bearing walls. While nonload-bearing interior walls are not required to have a rating in accordance with Table 601, other sections of the code may require a rating (e.g., corridor walls, dwelling unit separation, sleeping unit separation). If other sections of the code require a rating, this would override the requirements of Table 601 stating no rating is required. In addition, interior nonload-bearing walls in buildings of Type IV (Heavy Timber, HT) construction shall be of solid wood construction formed by not less than two layers of 1-inch (25 mm) matched boards or laminated construction 4 inches (102 mm) thick, or of 1-hour-rated construction (see Section 602.4.6).

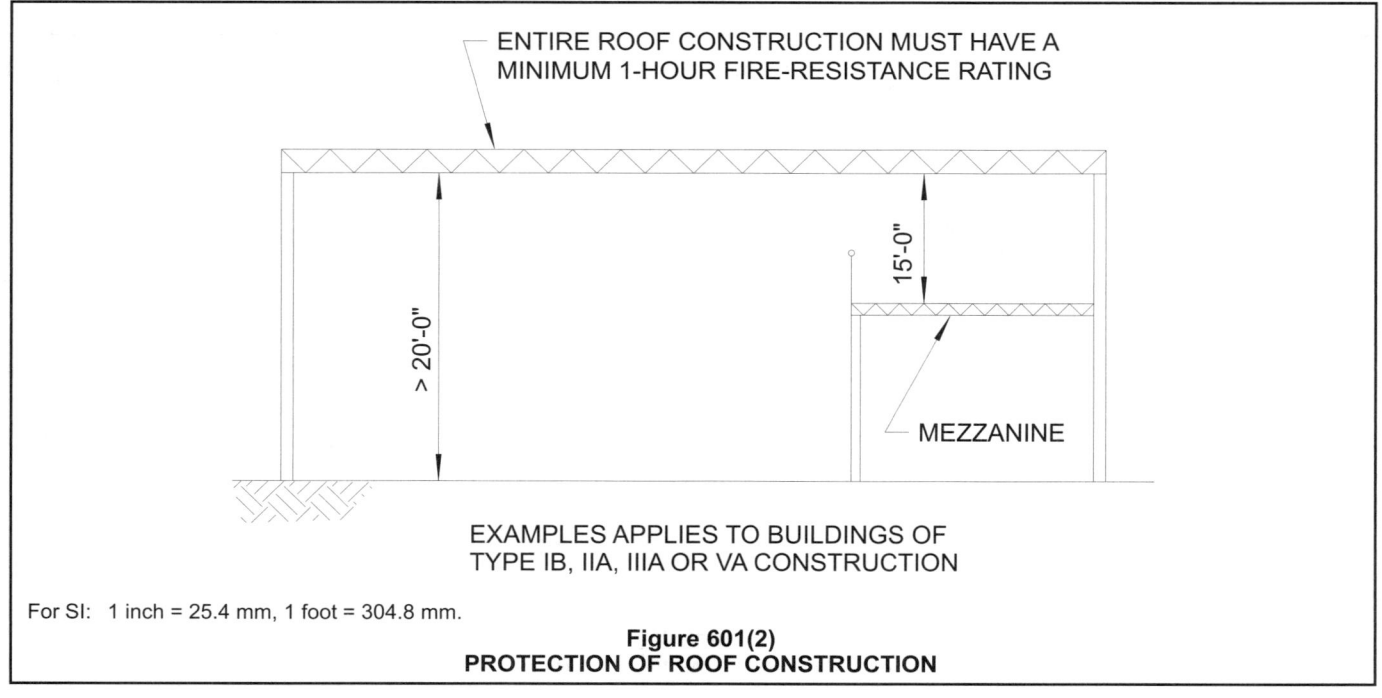

For SI: 1 inch = 25.4 mm, 1 foot = 304.8 mm.

Figure 601(2)
PROTECTION OF ROOF CONSTRUCTION

Note f is a reminder that exterior load-bearing walls must also comply with the fire-resistance rating listed in Table 602 based on the fire separation distance. Exterior load-bearing walls must satisfy the higher of the two fire-resistance ratings required by these two tables.

Note g is a reminder of the provisions of Section 704.10 that load-bearing structural members located within the exterior walls or "outside" of the structure need to comply with the highest of three fire-resistance ratings: the requirement for the primary structural frame; the requirement for exterior bearing walls; and the requirement from Table 602 based on the fire separation distance of the exterior wall. In Type III and IV construction, the rating for an exterior load-bearing wall is higher than that required for a primary structural frame alone.

TABLE 602. See below.

❖ Table 602 establishes the minimum fire-resistance ratings for all exterior walls. The required ratings are based on the fuel load, probable fire intensity of the various occupancy classifications and the fire separation distance [see Figure 602(1)]. In using the table, the occupancy classification of the building and the fire separation distance of the walls must be determined. Once determined, the required fire-resistance rating is obtained by referring to the appropriate row and column corresponding to these parameters. It should be noted that the fire-resistance-rating requirements of Table 601, which are based on construction type, also apply. Exterior load-bearing walls must conform to the higher of the fire-resistance ratings specified in Tables 601 and 602. Exterior nonload-bearing walls need only comply with Table 602.

Please note that where there is more than one building on a lot, an imaginary property line must be assumed between the buildings in accordance with Section 705.3. Figure 602(1) shows an assumed line halfway between the two buildings. The imaginary property line can be at any location between the structures. Wherever the line is established, fire separation distance for each wall must be measured from the line, and wall and opening protection based on the distance to the assumed location.

Note b refers to private garages and carports that are no greater than 3,000 square feet (279 m²) in area. As long as the fire separation distance is at least 5 feet (1524 mm), no fire-resistance rating would be required for the exterior walls. This footnote does not apply to other Group U occupancies not used for the storage of motor vehicles. Other Group U occupancies would have a minimum 1-hour fire-resistance rating for a fire separation distance of less than 10 feet (3048 mm).

Note c requires walls that are located on property lines between adjacent buildings (i.e., zero fire separation distance) to be constructed as fire walls and be rated in accordance with Table 706.4.

Note d permits the exterior walls of an open parking garage not to have a fire-resistance rating when the fire separation distance is at least 10 feet (3048 mm). The exception to Section 402.7.1 indicates that when a garage is more than 10 feet (3048 mm) from the covered mall, Table 602 is to be used rather than the 2-hour fire-resistance rating. At a distance of 10 feet (3048 mm), the fire separation distance would be between 5 feet (1524 mm) and 10 feet (3048 mm). Table 705.8 allows the openings to be unlimited for open parking garages at 10 feet (3048 mm) of fire separation distance. If the amount of openings is no longer limited, then the entire wall could be removed.

Note e clarifies that the fire-resistance rating of each wall in each story of a building must be determined separately. Should a multistory building be configured

TABLE 602
FIRE-RESISTANCE RATING REQUIREMENTS FOR EXTERIOR WALLS BASED ON FIRE SEPARATION DISTANCE[a, e]

FIRE SEPARATION DISTANCE = X (feet)	TYPE OF CONSTRUCTION	OCCUPANCY GROUP H[f]	OCCUPANCY GROUP F-1, M, S-1[g]	OCCUPANCY GROUP A, B, E, F-2, I, R, S-2[g], U[b]
X < 5[c]	All	3	2	1
5 ≤ X <10	IA	3	2	1
	Others	2	1	1
10 ≤ X < 30	IA, IB	2	1	1[d]
	IIB, VB	1	0	0
	Others	1	1	1[d]
X ≥ 30	All	0	0	0

For SI: 1 foot = 304.8 mm.
a. Load-bearing exterior walls shall also comply with the fire-resistance-rating requirements of Table 601.
b. For special requirements for Group U occupancies, see Section 406.1.2.
c. See Section 706.1.1 for party walls.
d. Open parking garages complying with Section 406 shall not be required to have a fire-resistance rating.
e. The fire-resistance rating of an exterior wall is determined based upon the fire separation distance of the exterior wall and the story in which the wall is located.
f. For special requirements for Group H occupancies, see Section 415.3.
g. For special requirements for Group S aircraft hangars, see Section 412.4.1.

such that the fire separation distance of each story is different, the required fire-resistance ratings associated with each of the exterior walls in each of those stories is established separately (see Figure 602B).

Note f provides a reference for Group H occupancies to Section 415.3 where there are specific stan-

dards for fire separation distances and wall and opening protection which supersede Table 602.

Note g provides a reference for aircraft hangars to Section 412.4.1 where there are specific standards for fire separation distances which supersede Table 602.

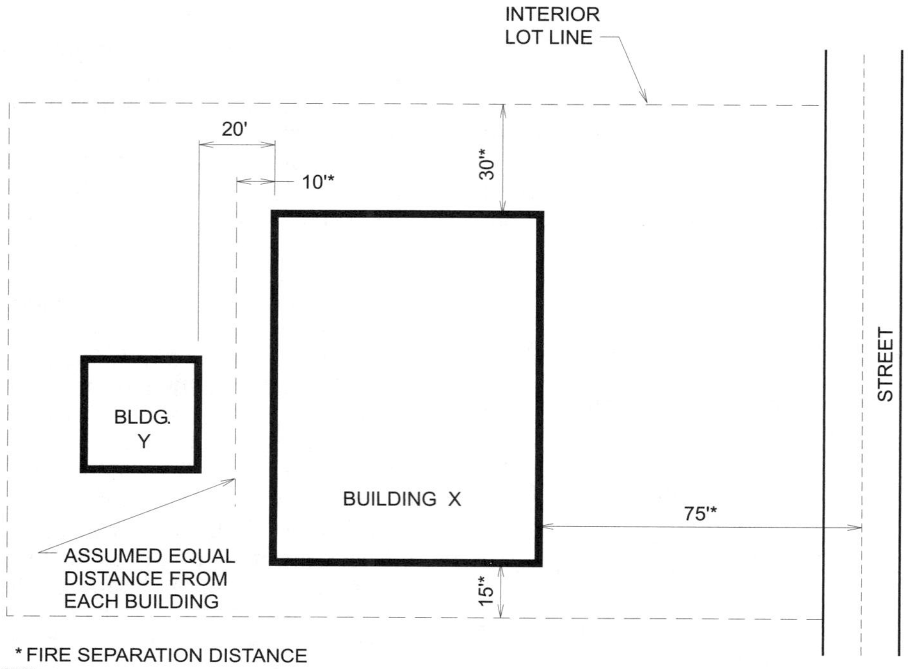

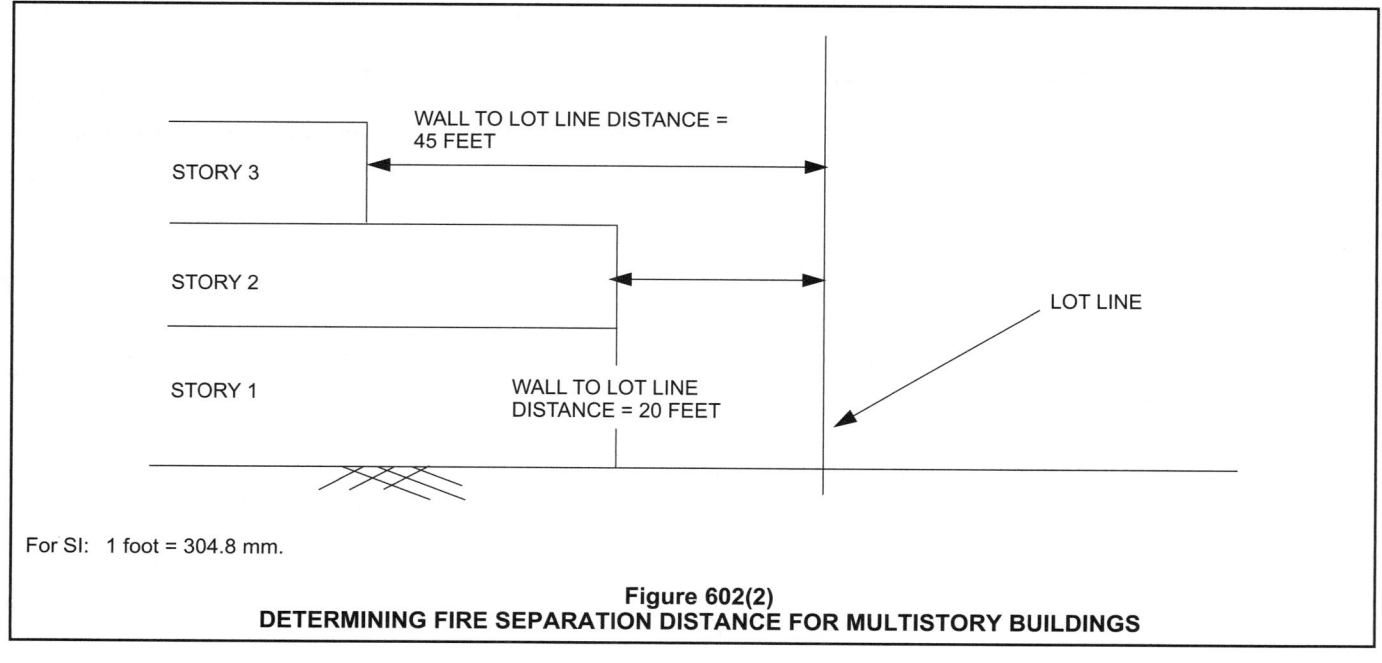

For SI: 1 foot = 304.8 mm.

Figure 602(1)
DETERMINING FIRE SEPARATION DISTANCE

For SI: 1 foot = 304.8 mm.

Figure 602(2)
DETERMINING FIRE SEPARATION DISTANCE FOR MULTISTORY BUILDINGS

SECTION 602
CONSTRUCTION CLASSIFICATION

602.1 General. Buildings and structures erected or to be erected, altered or extended in height or area shall be classified in one of the five construction types defined in Sections 602.2 through 602.5. The building elements shall have a *fire-resistance rating* not less than that specified in Table 601 and exterior walls shall have a *fire-resistance rating* not less than that specified in Table 602. Where required to have a *fire-resistance rating* by Table 601, building elements shall comply with the applicable provisions of Section 703.2. The protection of openings, ducts and air transfer openings in building elements shall not be required unless required by other provisions of this code.

❖ This section requires that each building or structure be put into one of five possible construction classifications: Type I, II, III, IV or V. All structural members are required to have a fire-resistance rating in accordance with Table 601. Additionally, the exterior walls of the structure must satisfy the requirements in Table 602, which bases the fire-resistance rating on the fire separation distance.

The use of multiple construction classifications in a single building is very limited and can only be done when specifically called out in the code. An example of combining types of construction is an office building of Type IIA construction located above an open parking structure of Type IIB construction, as described in Sections 509.7 and 509.7.1.

A more common example is where a single structure is divided into two compartments by using a fire wall, resulting in two separate buildings or structures—each of which may be of a different type of construction. While a structure may contain more than one building (for example, separation by a fire wall), each building is to be individually assigned a type of construction.

Also, a building may have elements that comply with the requirements of more than one type of construction, in which case the building as a whole must be assigned the less restrictive type of construction. The designer may have intended, however, to comply with a higher type of construction, in which case those elements not in compliance with the intended type of construction are to be brought into compliance. The selection of a construction type is the prerogative of the permit applicant. The applicant should have the designer indicate on the submitted construction documents which construction type permit has been selected. This will expedite the compliance review by the plans examiner.

Section 602.1 applies to both new construction and additions. The provisions in Chapter 34 on existing structures, Section 503 on general height and area limitations, Chapter 7 on fire and smoke protection features and construction and the applicable portions of the code depend on the requirements of this section. Where Table 601 requires a building element to meet a fire-resistance rating, that rating is determined based on the criteria in Section 703. Building elements required to have a fire-resistance rating by Table 601 do not necessarily need opening protectives or dampers for penetrations through these elements. Opening protectives or dampers are only required for building elements listed in Table 601 that are also a specific type of wall or horizontal assembly that is required by other criteria in the code to have protected openings.

Where Table 601 requires a building element to meet a fire-resistance rating, that rating is determined based on the criteria in Section 703. Because a building element is required to have fire-resistance by Table 601 does not require openings in these building elements or duct penetrations through these elements to be protected. However, a building element listed in Table 601 may also be a specific type of wall or horizontal assembly that is required by other criteria in the code to have protected openings. For example, an interior bearing wall inside a Type IIA building is required to have a fire-resistance rating of not less than 1 hour. However, the openings in that wall need not be protected unless that bearing wall is also serving another purpose. This wall could also be part of the walls establishing a control area on the second story. In accordance with Section 414.2.2, such wall would also have to be a 1-hour-rated fire barrier, and in accordance with Table 715.4, openings in this 1-hour-rated fire barrier would need to have 45-minute-rated opening protection. It is essential to check the provisions of Chapter 7 to determine where building elements are required, or not required, to have protected openings, transfer openings, ducts, joints and other penetrations.

602.1.1 Minimum requirements. A building or portion thereof shall not be required to conform to the details of a type of construction higher than that type which meets the minimum requirements based on occupancy even though certain features of such a building actually conform to a higher type of construction.

❖ These requirements permit design flexibility by allowing various building materials and components to be used. A building must, as a minimum, meet all of the requirements of a given type of construction to be classified as such, even though portions of that building meet the criteria of a higher construction type (i.e., greater fire-resistance ratings). This is consistent with the concept that the code is a minimum requirement. For example, a building classified as Type III construction is not prohibited from having construction that is superior, but it could not be reclassified into a higher type of construction unless it met all of the requirements for that construction type. In a normal situation, the design professional has identified the construction classification on the drawings. When this assignment has not been made, the building official is placed in a position of verifying the designer's intent and selecting the least-restrictive type that will meet all of the code requirements.

602.2 Types I and II. Types I and II construction are those types of construction in which the building elements listed in

Table 601 are of noncombustible materials, except as permitted in Section 603 and elsewhere in this code.

❖ Buildings of Type I and II construction are required to be constructed of noncombustible materials (see Section 703.4) and, therefore, are frequently referred to as "noncombustible construction." All Type I and II structural members have a fire-resistance rating as required by Tables 601 and 602. A typical example of a building of Type IA, IB or IIA construction would be a high-rise structure or a very large low-rise structure. These buildings are permitted to be relatively large in height and area due to the fire resistance afforded the structure's components. The structural members of a building of Type IIB construction do not have the same fire resistance as structural members in a building of Type IA, IB or IIA construction. As such, the height and area requirements are not as large for Type IIB construction (see Figure 602.2 for an example of Type I or II construction).

Type I and II construction are divided into four subclassifications: Types IA, IB, IIA and IIB. The difference among the four subclassifications is the degree of fire-resistance rating required for similar elements and assemblies. For example, the required rating for structural frame members in Type IA construction is 3 hours, for Type IB is 2 hours, for Type IIA is 1 hour and for Type IIB is 0 hours. Often, the fire-resistance ratings required by Tables 601 and 602 for structural elements are achieved by "fireproofing" structural members. Fireproofing is typically the process of creating a fire-resistance-rated assembly that incorporates the structural member by encapsulating it, either by boxing it in or by spraying on a coating to achieve the re-quired fire-resistance ratings. It should be noted that when a protective covering is used to provide the fire-resistance rating, it must be a noncombustible material, except as indicated in Section 603.1, Item 21.

Fire-retardant-treated wood (FRTW), although combustible, is permitted in limited uses in buildings of Type I and II construction (see Section 603 and Table 601, Note b). While FRTW is permitted in certain applications in buildings of Type I and II construction, it is not assumed to be fire-resistance rated, and generally does not afford any higher fire-resistance rating than untreated wood material. Other combustible items (as specified in Section 603.1) are also permitted in buildings of Type I or II construction.

602.3 Type III. Type III construction is that type of construction in which the exterior walls are of noncombustible materials and the interior building elements are of any material permitted by this code. *Fire-retardant-treated wood* framing complying with Section 2303.2 shall be permitted within *exterior wall* assemblies of a 2-hour rating or less.

❖ Buildings of Type III construction are made with both combustible and noncombustible materials. The exterior walls are required to be noncombustible with load-bearing exterior walls required to have a minimum 2-hour fire-resistance rating. Exterior nonload-bearing walls are not required by Table 601 to have a fire-resistance rating, but must comply with the provisions of Table 602. The elements within the perimeter established by the exterior walls (i.e., floors, roofs and walls) are permitted to be of combustible materials. An example of a typical building of Type III construction is a structure having its exterior walls constructed of concrete, masonry or other approved noncombustible ma-

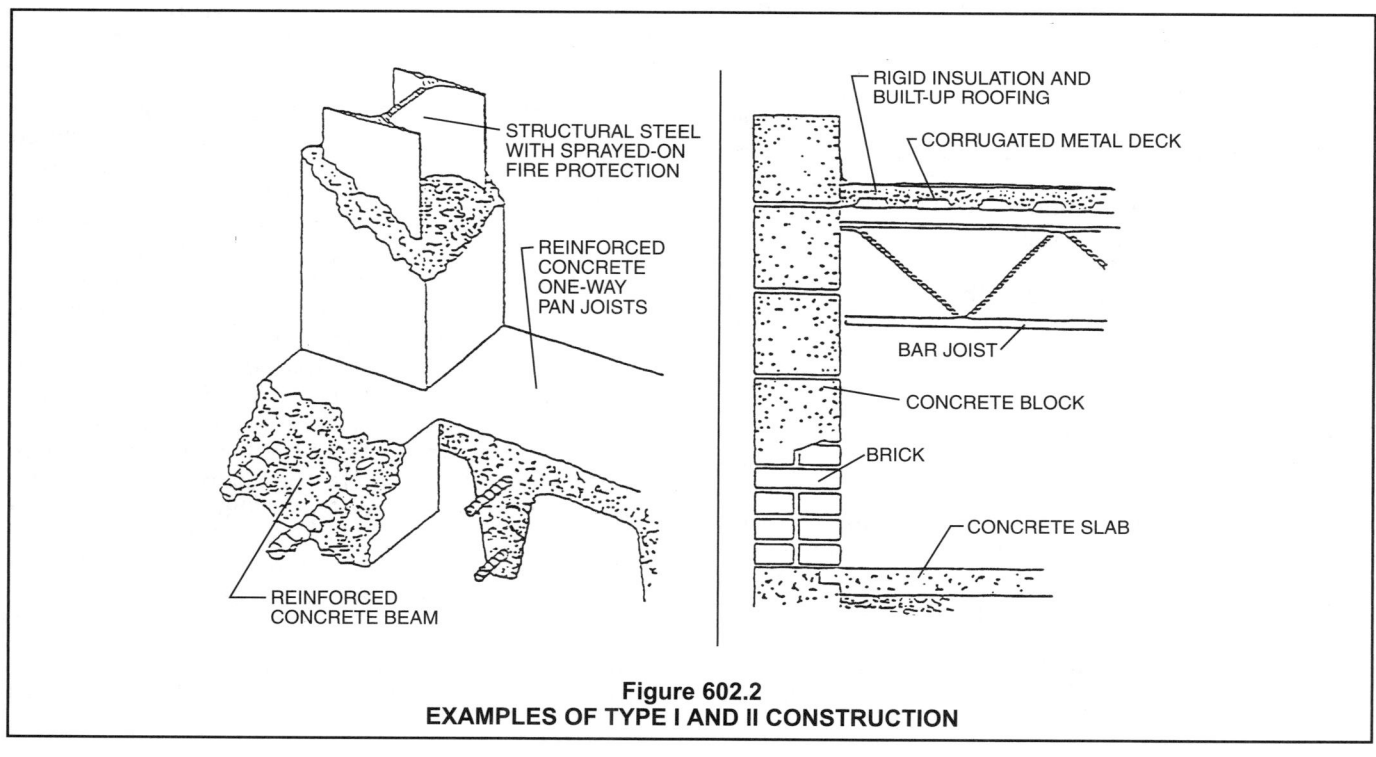

Figure 602.2
EXAMPLES OF TYPE I AND II CONSTRUCTION

terials, but with a wood frame floor, interior wall and roof construction. The structural members of a building of Type IIIB construction are not required to have a fire-resistance rating with the exception of the exterior load-bearing walls.

Although fire-retardant-treated wood (FRTW) does not meet the specifications of the code as a noncombustible material, it is permitted as a substitute for noncombustible materials for framing within exterior wall assemblies of Type III construction. The exterior surfaces of the walls must be of noncombustible materials. While the exterior walls are permitted to be either nonload-bearing or load-bearing, to apply the allowance for FRTW the required fire-resistance rating of the exterior wall must be no greater than 2 hours. FRTW is required to comply with the provisions in Section 2303.2.

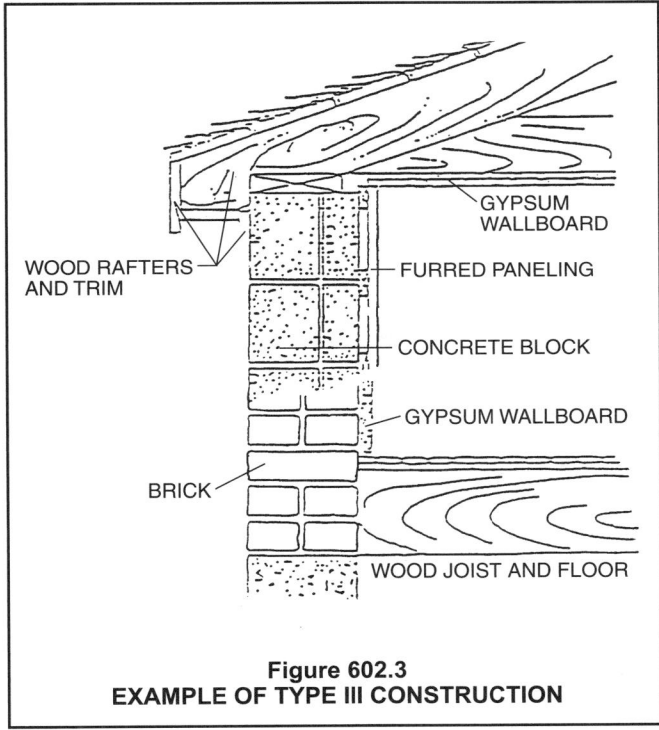

Figure 602.3
EXAMPLE OF TYPE III CONSTRUCTION

602.4 Type IV. Type IV construction (Heavy Timber, HT) is that type of construction in which the exterior walls are of noncombustible materials and the interior building elements are of solid or laminated wood without concealed spaces. The details of Type IV construction shall comply with the provisions of this section. *Fire-retardant-treated wood* framing complying with Section 2303.2 shall be permitted within exterior wall assemblies with a 2-hour rating or less. Minimum solid sawn nominal dimensions are required for structures built using Type IV construction (HT). For glued-laminated members the equivalent net finished width and depths corresponding to the minimum nominal width and depths of solid sawn lumber are required as specified in Table 602.4.

❖ This section provides the general regulations for Type IV (Heavy Timber, HT) construction. HT construction

requires the exterior walls to be constructed of noncombustible materials. The interior elements are required to be constructed of solid or laminated wood without any concealed spaces. All of the combustible structural elements are permitted to be unprotected because of the massive element sizes and the requirement that there not be any concealed spaces, such as soffits, plenums or suspended ceilings. Sections 602.4.1 through 602.4.7 provide specific requirements for the connection of structural members and minimum dimensions. An examination of Table 503 indicates that the allowable height and area for Type IV construction is greater than that permitted for buildings of Type IIB construction. This distinction is based on testing that demonstrated that HT structural members perform better structurally under fire conditions than comparable unprotected steel structural members because of charring, which insulates the wood mass.

As with Type III construction, fire-retardant-treated wood (FRTW) is permitted as a substitute for noncombustible materials within exterior wall assemblies of Type IV construction. Except as noted in Section 602.4.7, exterior structural members used externally must be noncombustible.

While the exterior walls are permitted to be either nonload-bearing or load-bearing, to apply the allowance for FRTW the required fire-resistance rating of the exterior wall must be no greater than 2 hours. FRTW is required to comply with the provisions of Section 2303.2.

TABLE 602.4. See page 6-10.

❖ Solid sawn wood members and glue-laminated timbers are manufactured using different methods and procedures and, therefore, do not have the same dimensions. However, they both have the same inherent fire-resistance capability. The dimensions noted in Sections 602.4.1 through 602.4.7 refer to the nominal dimensions of solid sawn lumber. These dimensions do not directly correlate to the actual dimensions of glue-laminated timbers. Table 602.4 provides a simple procedure to determine the dimensions that are required for glue-laminated timbers when designing to meet the requirements of Type IV construction. The table provides minimum dimensions for glue-laminated wood members for each set of dimensions specified in the specific provisions of Section 602.4. The requirements for glulam were developed to recognize that the reduced width of glulam members (compared to timbers with the same nominal dimension) shall be offset by increasing the depth to maintain similar cross-sectional areas. To use the table, compare the required minimum dimensions for sawn timber found in the two left-hand columns with the dimensions found in the two right-hand columns for glued-laminated wood members. For example, where the code requires a minimum sawn timber of 8 inches by 8 inches (203 mm by 203 mm), for a glulam wood member to be used, it would need to be a minimum of 6³/₄ inches wide by 8¹/₄ inches (171 mm by 210 mm) deep.

602.4.1 Columns. Wood columns shall be sawn or glued laminated and shall not be less than 8 inches (203 mm), nominal, in any dimension where supporting floor loads and not less than 6 inches (152 mm) nominal in width and not less than 8 inches (203 mm) nominal in depth where supporting roof and ceiling loads only. Columns shall be continuous or superimposed and connected in an *approved* manner.

❖ Minimum construction requirements and dimensions for timber columns are provided in this section. Columns are required to be a minimum of 8 inches (203 mm) nominal in any dimension if they support floor loads, or a minimum of 6 by 8 inches (152 by 203 mm) nominal if they support a roof and ceiling. Timber columns are required to be continuous or superimposed, positioned on or over each other, through floors for the entire height of the building. The design engineer or architect must provide details of all column connections. As with all structural members, each column must also be adequately fastened to other structural members in order to withstand the loads that will be placed upon the column. Some typical examples include reinforced concrete or metal caps, steel or iron column caps and timber splice plates [see Figures 602.4.1(1) and 602.4.1(2)].

602.4.2 Floor framing. Wood beams and girders shall be of sawn or glued-laminated timber and shall be not less than 6 inches (152 mm) nominal in width and not less than 10 inches (254 mm) nominal in depth. Framed sawn or glued-laminated timber arches, which spring from the floor line and support floor loads, shall be not less than 8 inches (203 mm) nominal in

TABLE 602.4
WOOD MEMBER SIZE

MINIMUM NOMINAL SOLID SAWN SIZE		MINIMUM GLUED-LAMINATED NET SIZE	
Width, inch	Depth, inch	Width, inch	Depth, inch
8	8	$6^{3}/_{4}$	$8^{1}/_{4}$
6	10	5	$10^{1}/_{2}$
6	8	5	$8^{1}/_{4}$
6	6	5	6
4	6	3	$6^{7}/_{8}$

For SI: 1 inch = 25.4 mm.

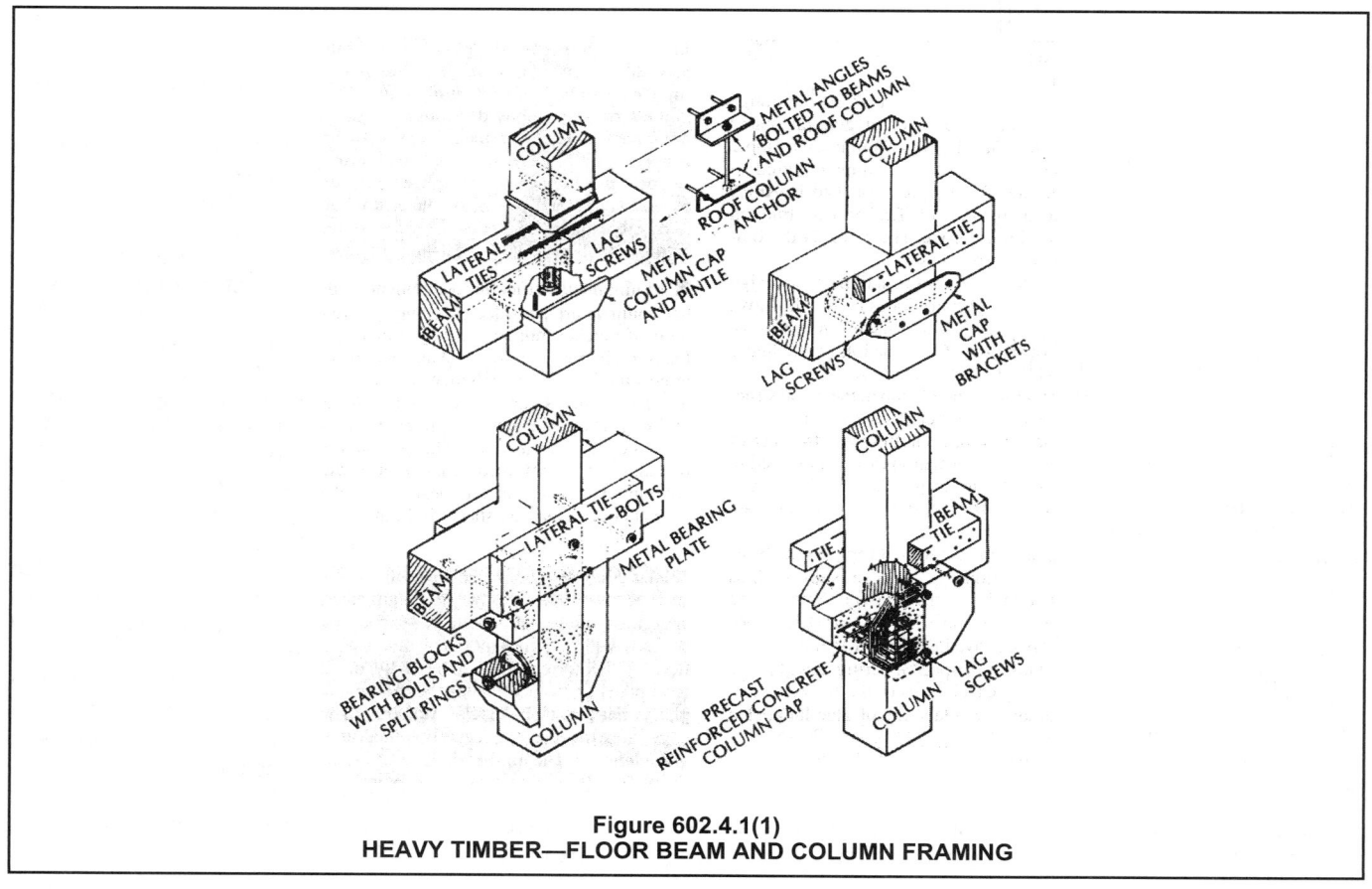

Figure 602.4.1(1)
HEAVY TIMBER—FLOOR BEAM AND COLUMN FRAMING

any dimension. Framed timber trusses supporting floor loads shall have members of not less than 8 inches (203 mm) nominal in any dimension.

❖ Minimum construction requirements and dimensions for floor framing are provided in this section. Girders are the principal horizontal structural members that support columns or beams. Beams are the structural members that support a floor or roof. Both girders and beams are required to be a minimum 6 inches (152 mm) wide and 10 inches (254 mm) deep. Both framed timber trusses supporting floor loads and framed sawn or glue-laminated timber arches that spring from the floor line and support floor loads are required to be at least 8 inches (203 mm) in any dimension.

602.4.3 Roof framing. Wood-frame or glued-laminated arches for roof construction, which spring from the floor line or from grade and do not support floor loads, shall have members not less than 6 inches (152 mm) nominal in width and have not less than 8 inches (203 mm) nominal in depth for the lower half of the height and not less than 6 inches (152 mm) nominal in depth for the upper half. Framed or glued-laminated arches for roof construction that spring from the top of walls or wall abutments, framed timber trusses and other roof framing, which do not support floor loads, shall have members not less than 4 inches (102 mm) nominal in width and not less than 6 inches (152 mm) nominal in depth. Spaced members shall be permitted to be composed of two or more pieces not less than 3 inches (76 mm) nominal in thickness where blocked solidly through-out their intervening spaces or where spaces are tightly closed by a continuous wood cover plate of not less than 2 inches (51 mm) nominal in thickness secured to the underside of the members. Splice plates shall be not less than 3 inches (76 mm) nominal in thickness. Where protected by *approved* automatic sprinklers under the roof deck, framing members shall be not less than 3 inches (76 mm) nominal in width.

❖ Minimum construction requirements and dimensions for arches and other types of roof framing are provided in this section. Other types of roof framing included in this section are heavy timber trusses with spaced members. When the members of a heavy timber truss are split and placed on either side of a main member, such as a web connecting a chord, each component of the web must be 3 inches (76 mm) or more in nominal thickness. The space between the two web members must be protected with a 2-inch-thick (51 mm) cover plate [see Figure 602.4.3(1)], or solidly filled with blocking [see Figure 602.4.3(2)]. The size of the roof framing members is dependent on the configuration used and is regulated by this section.

If a building of Type IV construction is equipped with approved automatic sprinklers under the roof deck, the minimum size of the roof framing members is reduced to 3 inches (76 mm). Roof framing members of a smaller size will have a lower resistance to fire than the 6-inch by 8-inch (152 mm by 203 mm) or 4-inch by 6-inch (102 mm by 152 mm) members required by this

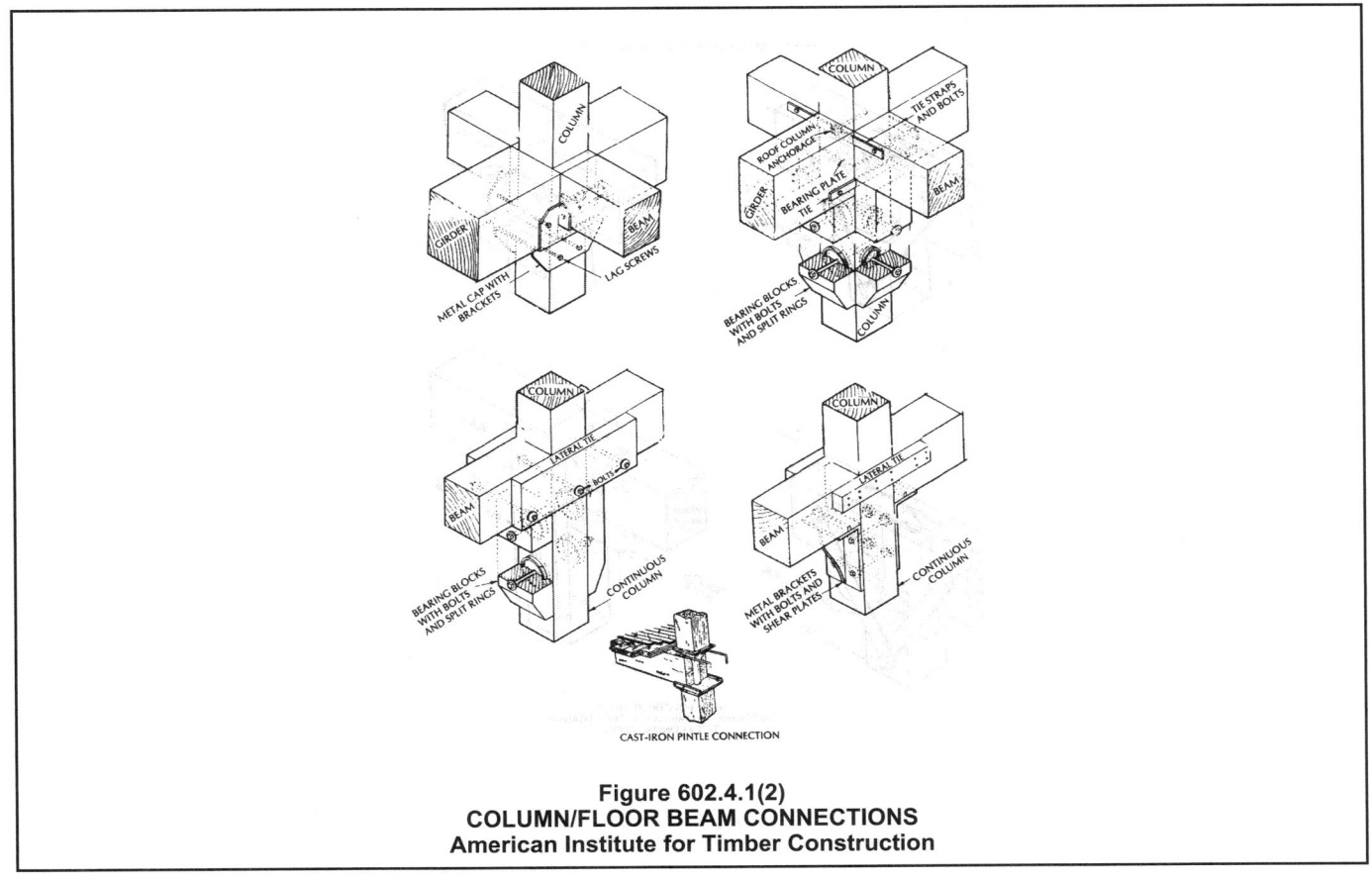

Figure 602.4.1(2)
COLUMN/FLOOR BEAM CONNECTIONS
American Institute for Timber Construction

section. The trade-off allowing smaller roof framing members when the building is equipped with an automatic sprinkler system is consistent with the concept of maintaining "equivalent risk" for the building.

602.4.4 Floors. Floors shall be without concealed spaces. Wood floors shall be of sawn or glued-laminated planks, splined or tongue-and-groove, of not less than 3 inches (76 mm) nominal in thickness covered with 1-inch (25 mm) nomi-

nal dimension tongue-and-groove flooring, laid crosswise or diagonally, or 0.5-inch (12.7 mm) particleboard or planks not less than 4 inches (102 mm) nominal in width set on edge close together and well spiked and covered with 1-inch (25 mm) nominal dimension flooring or $^{15}/_{32}$-inch (12 mm) wood structural panel or 0.5-inch (12.7 mm) particleboard. The lumber shall be laid so that no continuous line of joints will occur except at points of support. Floors shall not extend closer than

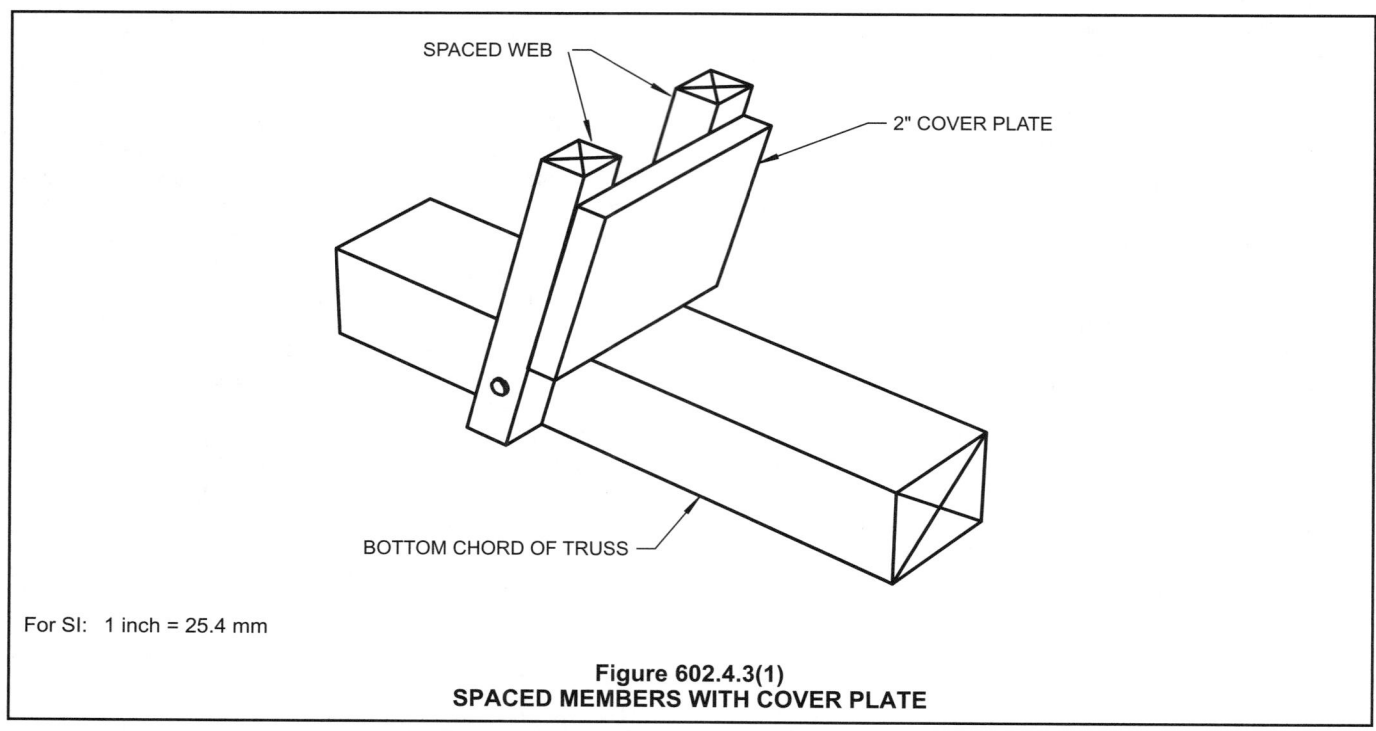

For SI: 1 inch = 25.4 mm

Figure 602.4.3(1)
SPACED MEMBERS WITH COVER PLATE

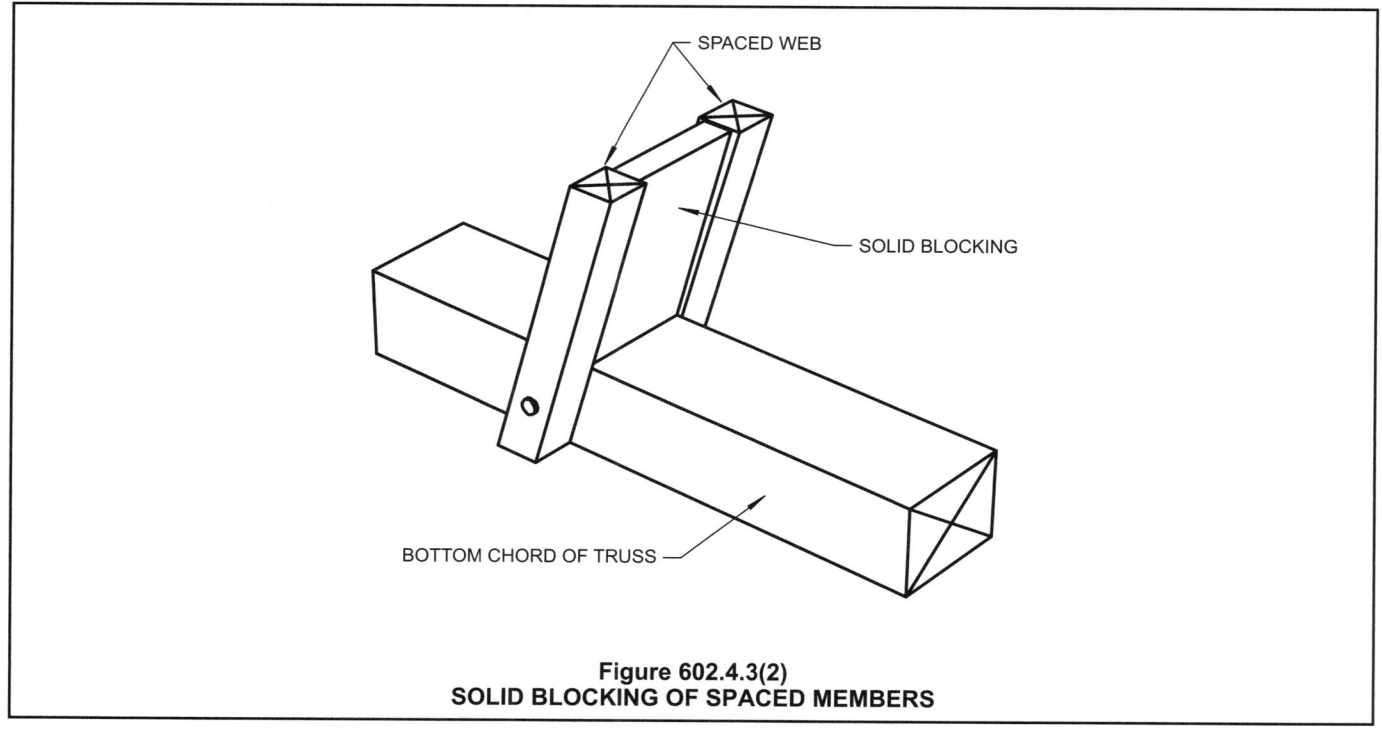

Figure 602.4.3(2)
SOLID BLOCKING OF SPACED MEMBERS

0.5 inch (12.7 mm) to walls. Such 0.5-inch (12.7 mm) space shall be covered by a molding fastened to the wall and so arranged that it will not obstruct the swelling or shrinkage movements of the floor. Corbeling of masonry walls under the floor shall be permitted to be used in place of molding.

❖ Heavy timber flooring is required to consist of minimum 3-inch-thick (76 mm) sawn or glue-laminated planks, splined floors or tongue-and-groove floors with an overlayment of 1-inch (25 mm) tongue-and-groove flooring, laid crosswise or diagonally. HT flooring may also consist of $^{1}/_{2}$-inch (12.7 mm) particleboard or planks at least 4 inches (102 mm) in width set on edge and secured together, with an appropriate over-layment, such as 1-inch (25 mm) hardwood flooring or a $^{15}/_{32}$-inch (12.7 mm) wood structural panel. Flooring in Type IV construction is not permitted to have concealed spaces because an undetected fire can spread quickly in combustible concealed floor spaces [see Figure 602.4.4(1)]. Because of the support afforded by adjacent members, continuous joints must only occur over supports.

Wood flooring must be fastened to supports that are perpendicular to the planking. Fastening must not be made to beams or girders that are parallel to the planks [see Figure 602.4.4(2)]. This precaution is intended to prevent separation of the planks because of differential movement of the beam relative to the girders and possible expansion/contraction due to differing moisture or humidity levels. This section requires a $^{1}/_{2}$-inch (12.7 mm) clearance between the wood flooring and exterior walls. This will prevent damage to the walls if the flooring expands due to rain during construction. This space also creates a potential flue for flames and hot gases. It should be emphasized that the integrity of the floor assembly must be maintained

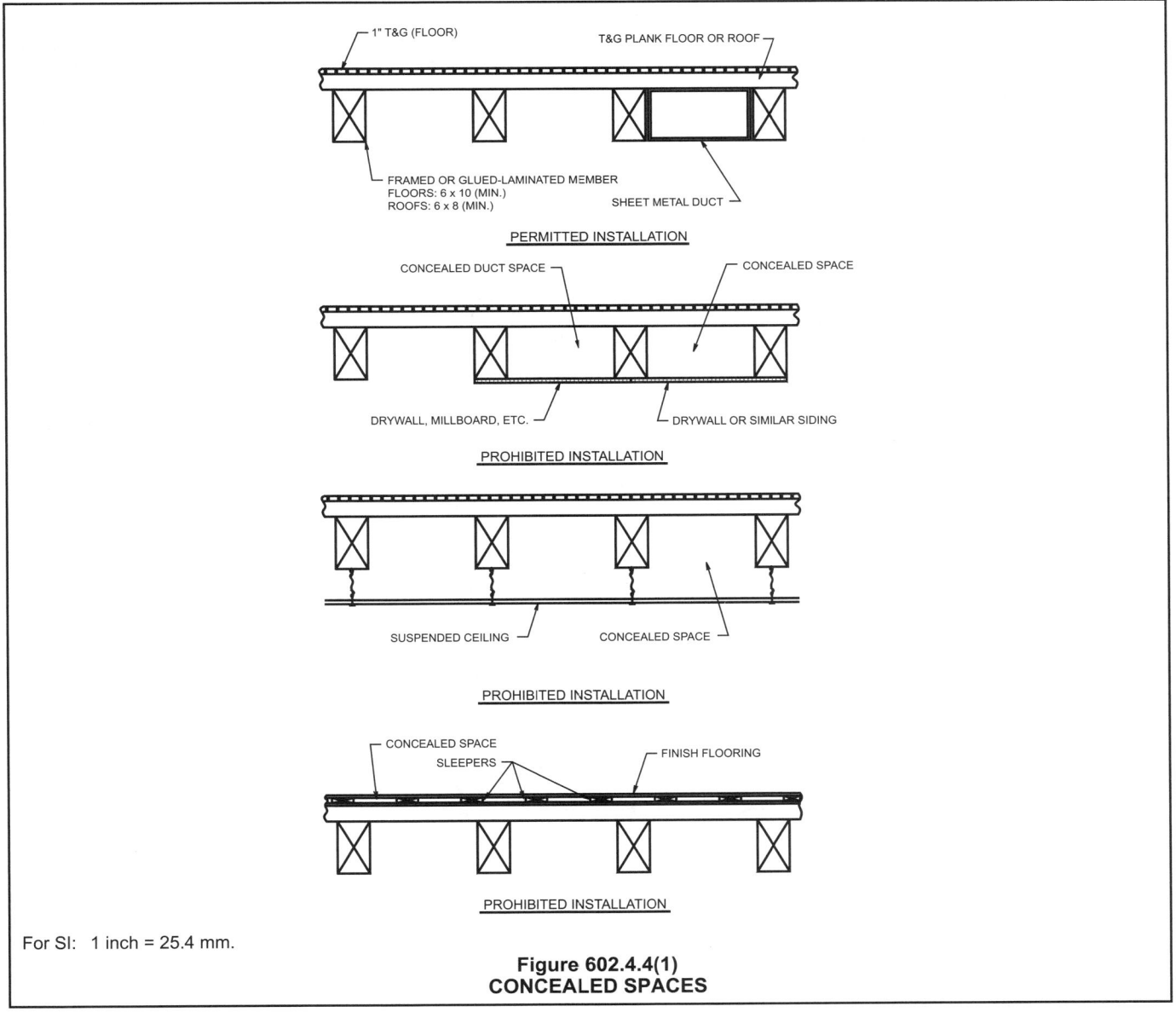

For SI: 1 inch = 25.4 mm.

Figure 602.4.4(1)
CONCEALED SPACES

to provide the equivalent of a 1-hour fire-resistance rating. In addition, the $^1/_2$-inch (12.7 mm) gap must be protected by a molding connected to the wall so that any possible contracting or expanding of the floor is not impeded. If masonry walls are utilized, corbeling of the masonry may be used as an alternate to the molding requirements.

602.4.5 Roofs. Roofs shall be without concealed spaces and wood roof decks shall be sawn or glued laminated, splined or tongue-and-groove plank, not less than 2 inches (51 mm) nominal in thickness, $1^1/_8$-inch-thick (32 mm) wood structural panel (exterior glue), or of planks not less than 3 inches (76 mm) nominal in width, set on edge close together and laid as required for floors. Other types of decking shall be permitted to be used if providing equivalent *fire resistance* and structural properties.

❖ Minimum construction requirements and dimensions for roof decks are provided in this section. As required for floors, roofs are not permitted to have concealed spaces [see Figure 602.4.4(1)]. If the materials used in roof construction are different from those described in this section, the roof must have a 1-hour fire-resistance rating and be of the same structural properties.

602.4.6 Partitions. Partitions shall be of solid wood construction formed by not less than two layers of 1-inch (25 mm) matched boards or laminated construction 4 inches (102 mm) thick, or of 1-hour fire-resistance-rated construction.

❖ Minimum construction requirements and dimensions for partitions are provided in this section. Partitions must either be formed by not less than two layers of 1-inch (25 mm) matched boards or laminated construction 4 inches (102 mm) thick if they are constructed of solid wood. Partitions are permitted to be constructed of materials other than solid wood if they have a 1-hour fire-resistance rating. An example of the use of alternative materials is when a fire-resistance

rating for an exit access corridor wall is required. It is common practice to utilize a 1-hour fire-resistance-rated stud and gypsum wallboard assembly between the exposed columns to form the walls of the exit access corridor.

602.4.7 Exterior structural members. Where a horizontal separation of 20 feet (6096 mm) or more is provided, wood columns and arches conforming to heavy timber sizes shall be permitted to be used externally.

❖ Heavy timber columns and arches that conform to minimum dimensional requirements may be used on the exterior if a fire separation distance of at least 20 feet (6096 mm) is maintained, although the exterior wall itself must be of noncombustible construction. If a fire separation distance of at least 20 feet (6096 mm) is maintained, the risk of exposure of the wood members to fire from an adjacent building is reduced, and the HT columns and arches are permitted to be exposed to the exterior.

If a building of Type IV construction has a fire separation distance of less than 20 feet (6096 mm), the wood columns and arches are to be located on the interior side of the exterior wall. The noncombustible construction of the exterior wall will provide some degree of protection to the interior timber members; therefore, placing the wood structural members inside an exterior wall is preferable to placing them within 20 feet (6096 mm) of a lot line or adjacent building with no exposure protection.

602.5 Type V. Type V construction is that type of construction in which the structural elements, *exterior walls* and interior walls are of any materials permitted by this code.

❖ Type V construction allows the use of all types of materials, both noncombustible and combustible, but is most commonly constructed of dimensional lumber (see Figure 602.5 for an example of Type V construc-

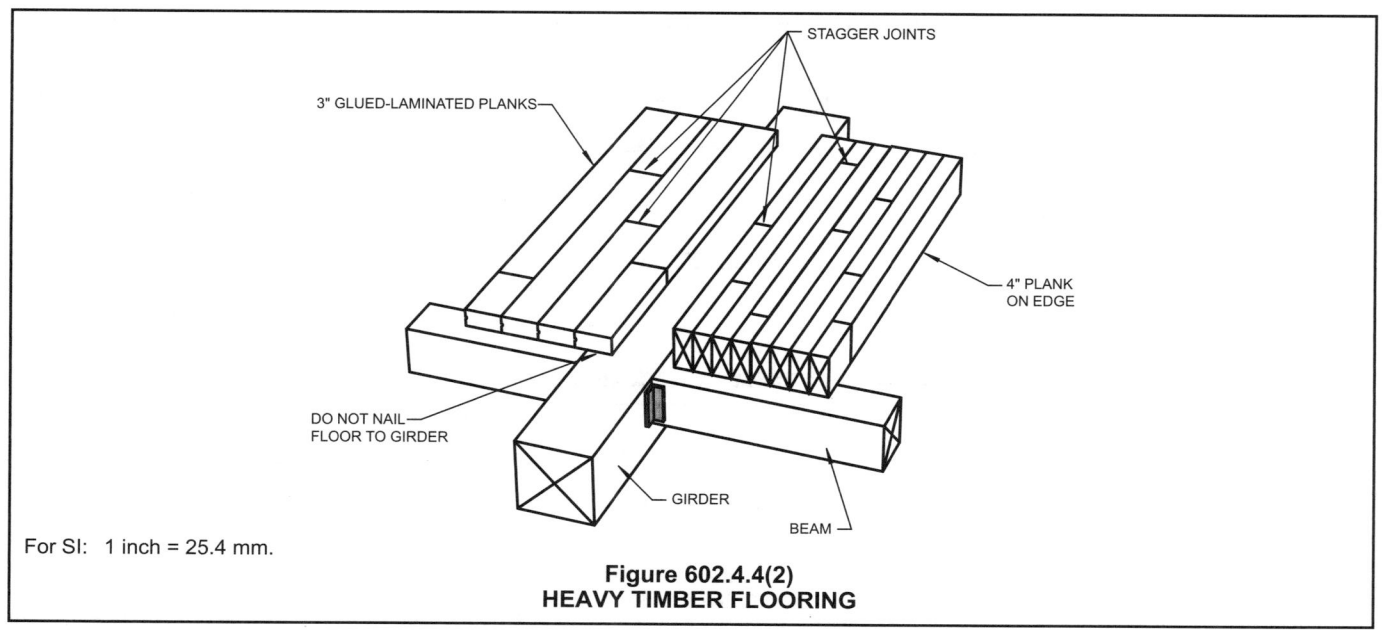

For SI: 1 inch = 25.4 mm.

STAGGER JOINTS

3" GLUED-LAMINATED PLANKS

4" PLANK ON EDGE

DO NOT NAIL FLOOR TO GIRDER

GIRDER

BEAM

Figure 602.4.4(2)
HEAVY TIMBER FLOORING

tion). It is divided into two subclassifications: Types VA and VB. An example of a typical building of Type VA construction is a wood frame building in which the interior and exterior load-bearing walls, floors, roofs [those members that are less than 20 feet (6096 mm) to the lowest member] and all structural members are protected to provide a minimum 1-hour fire-resistance rating. An example of a building of Type VB construction is found in the typical single-family home, where a fire-resistance rating is not required for the structural members. Type V construction is required to comply with Table 601 and Chapter 23.

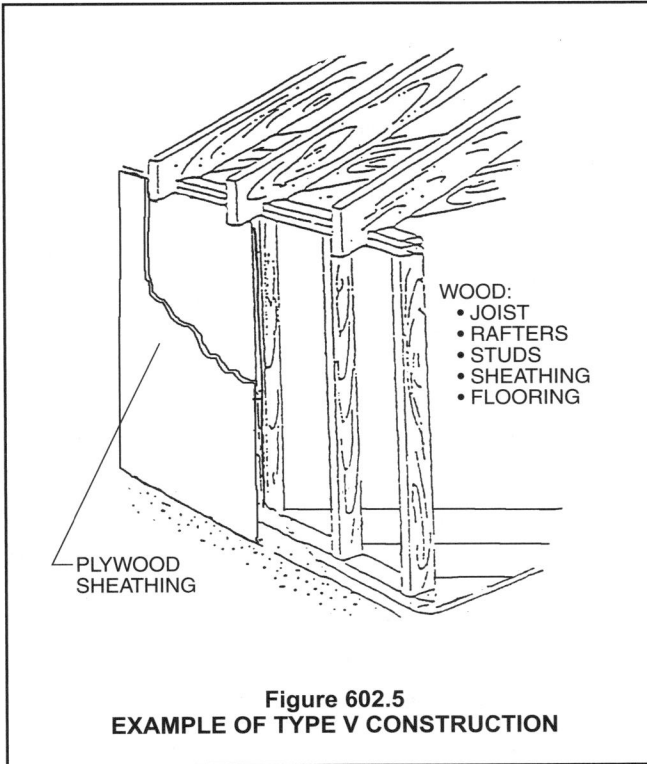

Figure 602.5
EXAMPLE OF TYPE V CONSTRUCTION

SECTION 603
COMBUSTIBLE MATERIAL IN TYPE I
AND II CONSTRUCTION

603.1 Allowable materials. Combustible materials shall be permitted in buildings of Type I or II construction in the following applications and in accordance with Sections 603.1.1 through 603.1.3:

1. *Fire-retardant-treated wood* shall be permitted in:

    1.1 Nonbearing partitions where the required *fire-resistance rating* is 2 hours or less.

    1.2. Nonbearing *exterior walls* where no fire rating is required.

    1.3. Roof construction, including girders, trusses, framing and decking.

    Exception: In buildings of Type IA construction exceeding two *stories above grade plane*, *fire-retardant-treated wood* is not permitted in

roof construction when the vertical distance from the upper floor to the roof is less than 20 feet (6096 mm).

2. Thermal and acoustical insulation, other than foam plastics, having a *flame spread index* of not more than 25.

    Exceptions:

    1. Insulation placed between two layers of noncombustible materials without an intervening airspace shall be allowed to have a *flame spread index* of not more than 100.

    2. Insulation installed between a finished floor and solid decking without intervening airspace shall be allowed to have a *flame spread index* of not more than 200.

3. Foam plastics in accordance with Chapter 26.

4. Roof coverings that have an A, B or C classification.

5. *Interior floor finish* and floor covering materials installed in accordance with Section 804.

6 Millwork such as doors, door frames, window sashes and frames.

7. *Interior wall and ceiling finishes* installed in accordance with Sections 801 and 803.

8. *Trim* installed in accordance with Section 806.

9. Where not installed over 15 feet (4572 mm) above grade, show windows, nailing or furring strips and wooden bulkheads below show windows, including their frames, aprons and show cases.

10 Finish flooring installed in accordance with Section 805.

11. Partitions dividing portions of stores, offices or similar places occupied by one tenant only and that do not establish a *corridor* serving an *occupant load* of 30 or more shall be permitted to be constructed of *fire-retardant-treated wood*, 1-hour fire-resistance-rated construction or of wood panels or similar light construction up to 6 feet (1829 mm) in height.

12. Stages and platforms constructed in accordance with Sections 410.3 and 410.4, respectively.

13. Combustible *exterior wall coverings*, balconies and similar projections and bay or oriel windows in accordance with Chapter 14.

14. Blocking such as for handrails, millwork, cabinets and window and door frames.

15. Light-transmitting plastics as permitted by Chapter 26.

16. Mastics and caulking materials applied to provide flexible seals between components of *exterior wall* construction.

17. Exterior plastic veneer installed in accordance with Section 2605.2.

18. Nailing or furring strips as permitted by Section 803.4.

19. Heavy timber as permitted by Note c to Table 601 and Sections 602.4.7 and 1406.3.

20. Aggregates, component materials and admixtures as permitted by Section 703.2.2.

21. Sprayed fire-resistant materials and intumescent and mastic fire-resistant coatings, determined on the basis of *fire-resistance* tests in accordance with Section 703.2 and installed in accordance with Sections 1704.12 and 1704.13, respectively.

22. Materials used to protect penetrations in fire-resistance-rated assemblies in accordance with Section 713.

23. Materials used to protect joints in fire-resistance-rated assemblies in accordance with Section 714.

24. Materials allowed in the concealed spaces of buildings of Types I and II construction in accordance with Section 717.5.

25. Materials exposed within plenums complying with Section 602 of the *International Mechanical Code*.

Note: The order of the 25 items of Section 603.1 was changed starting with the third printing of the code. For the first two printings, what appears above as Item 1 was shown as Item 25. Therefore, the balance of the items were all one number less.

❖ Section 603.1 provides a listing of circumstances where combustible materials can be used in buildings of Type I and II construction, which are otherwise required to be constructed of noncombustible materials. The list of 25 items frequently references the code user to other sections of the code where specific allowances are found. Most of these listed items, while of combustible materials, are either of minor quantities in the overall building or are of materials with fire-resistive properties.

Treated wood: Fire-retardant-treated wood (FRTW) does not meet the criteria in the code for a noncombustible material. It is, however, permitted as an alternative to noncombustible materials in specific locations in Type I and II construction (see Items 1, 11 and 13). For example, the use of FRTW in walls of Type I and II construction has been limited to nonload-bearing partitions with a fire-resistance rating of no greater than 2 hours and nonload-bearing exterior walls without a fire-resistance rating. Additionally, roofs in buildings of Type I and II construction are permitted to be constructed of FRTW. The exception to Item 1.3 does not permit FRTW in the roofs of buildings of Type I construction over two stories in height if the distance from the uppermost floor to the roof is less than 20 feet (6096 mm). If the distance is 20 feet (6096 mm) or greater, then FRTW is acceptable in the roof. FRTW is permitted in the roof of any Type I building two stories or less in height regardless of the distance from the floor to the roof. Similarly, FRTW is permitted in the roof of any Type II building (no height restrictions) regardless of the distance from the floor to the roof.

Untreated wood: Numerous items in the list of Section 603.1 permit the use of untreated wood in Type I and II construction. Blocking or nailers used to support fixtures, railings, cabinets or interior and exterior finishes are permitted within walls and partitions required to be of noncombustible construction in accordance with Item 14. Item 18 permits combustible nailers and blocking as stipulated in accordance with Section 803.11. Section 803.11.1 indicates that "furring strips not exceeding 1.75 inches (44 mm)" are permitted to be used in concrete or masonry construction for securing trim and finishes. Although locating these combustible elements within noncombustible frame partitions is not specifically identified in this section, the presence of combustible nailers within noncombustible construction types, other than concrete and masonry, represents an equivalent circumstance. Therefore, it is the intent of the code to permit the use of combustible nailers and blocking within Type I and II construction.

According to Item 11, partitions within a small store or office can include untreated wood in their construction (or FTRW), provided the partitions do not exceed 6 feet (1829 mm) in height. The limitations to 30 occupants and to not forming a corridor restrict this exception from introducing extensive amounts of combustible materials. Item 9 permits the use of untreated wood for show windows and related construction up to a height 15 feet (4572 mm) above grade. This permits a broader range of materials for storefronts without jeopardizing the overall integrity of these buildings.

Item 19 references three different code sections where heavy timber construction can occur within and on the outside of noncombustible buildings.

Plastics: Items 2, 3, 15 and 17 specifically address allowed plastics within or on the surfaces of Type I and II buildings. These items include allowances for thermal and acoustical insulation having a flame spread index of no greater than 25 when tested in accordance with ASTM E 84; foam and light-transmitting plastics complying with Chapter 26; and exterior plastic veneers complying with Section 2605.2. Plastic materials are often included in installations allowed for interior finishes and within the construction of the building, in accordance with Item 7.

Roof construction and roof coverings: While combustible roofs must be of FRTW if used in a Type I or II building, roof coverings, blocking, nailers and furring strips are also permitted to be combustible without the use of FRTW (see Items 4, 14 and 18).

"Roof covering" is defined as the membrane covering the roof that provides weather resistance, fire resistance and appearance. As long as a noncombustible roof deck (or FRTW) is provided as the structural element, foam plastic insulation, wood structural panels, nailing/furring strips and roof coverings may be applied.

Exterior wall coverings and projections: Item 13 provides a reference to Chapter 14 where provisions address combustible wall coverings and use of combustible materials in projections such as balconies and eaves. Combustible wall coverings need to be applied on top of a noncombustible wall meeting the required rating of Table 601. The coverings cannot be a substitute for the exterior face of a tested assembly. Plastic veneers (Item 17) are allowed in accordance with Section 2605.2.

Concealed spaces and construction: Many combustible materials are used, and allowed to be used within the elements of a building. Often these are used to help protect a fire-resistance-rated assembly where it is penetrated by pipes, ducts or conduit. Many of these materials are tested for the specific purposes and installations such as protection of joints or are used to actually resist, or react to, fire (see Items 16, 20, 21, 22, 23, 24 and 25).

Finish and interior materials: Finish materials for floors, walls and ceilings need to comply with various provisions of Chapter 8. Allowed materials do include combustible materials, but typically they are limited in flame spread index and smoke generation. Items 5, 7 and 8 address interior finishes. Stages and platforms can be combustible when in compliance with Section 410 (see Item 12). And, finally, doors and windows are not among the elements required to be noncombustible in accordance with Item 6.

603.1.1 Ducts. The use of nonmetallic ducts shall be permitted when installed in accordance with the limitations of the *International Mechanical Code.*

❖ Ducts are not addressed by construction-type requirements and the use of these must not be controlled by construction-type provisions. The *International Mechanical Code®* (IMC®) provides requirements for nonmetallic ducts that deal with the issue of flammability, flame spread, etc. This section clarifies that this chapter is not intended to override those requirements.

603.1.2 Piping. The use of combustible piping materials shall be permitted when installed in accordance with the limitations of the *International Mechanical Code* and the *International Plumbing Code.*

❖ Piping is not addressed by construction-type requirements and the use of piping must not be controlled by construction-type provisions. The *International Plumbing Code®* (IPC®) provides requirements for combustible piping materials, such as plastic, that deal with the issue of flammability, and flame spread. This section clarifies that this chapter is not intended to override those requirements.

603.1.3 Electrical. The use of electrical wiring methods with combustible insulation, tubing, raceways and related components shall be permitted when installed in accordance with the limitations of this code.

❖ Electrical wiring and equipment are not addressed by construction-type requirements and the use of these must not be controlled by construction-type provisions. NFPA 70, *The National Electrical Code,* provides requirements for combustible wiring materials that deal with the issue of flammability and flame spread. This section clarifies that this chapter is not intended to override those requirements.

Bibliography

The following resource material is referenced in this chapter or is relevant to the subject matter addressed in this chapter.

ASTM E 84-07, *Test Methods for Surface Burning Characteristics of Building Materials.* West Conshohocken, PA: ASTM International, 2007.

DOC PS 20-99, *American Softwood Lumber Standard.* Washington, DC: United States Department of Commerce, 1999.

Fire Protection through Modern Building Codes. Washington, DC: American Iron and Steel Institute, 1981.

IMC-09, *International Mechanical Code.* Washington, DC: International Code Council, 2009.

IPC-09, *International Plumbing Code.* Washington, DC: International Code Council, 2009.

NFPA 13-07, *Standard for the Installation of Sprinkler Systems.* Quincy, MA: National Fire Protection Association, 2007.

NFPA 70-08, *National Electrical Code.* Quincy, MA: National Fire Protection Association.

Technical Report No. 1, *Comparative Fire Test on Wood and Steel Joists.* Washington, DC: National Forest Products Association, 1961.

Timber Construction Manual, 2nd edition. American Institute of Timber Construction. New York: John Wiley and Sons Inc., 1974.

Wood Construction Data No. 5, *Heavy Timber Construction.* Washington, DC: American Forest and Paper Association, 2004.

Chapter 7:
Fire and Smoke Protection Features

General Comments

Chapter 7 provides detailed requirements for fire-resistance-rated construction, including structural members, walls, partitions and horizontal assemblies. Other portions of the code tell us when certain fire-resistance-rated elements are required. This chapter specifies how these elements are constructed, how openings in walls and partitions are protected, and how penetrations of such elements are protected.

Fire-resistance-rated construction is one form of fire protection in building design. It is often referred to as "passive protection." Fire-resistance-rated building elements provide resistance to the advance of fire, as opposed to active fire protection systems, such as automatic sprinkler systems, which actively attempt to suppress a fire.

At the core of fire-resistance-rated construction is test standard ASTM E 119. This test standard specifies how to test different building elements for fire resistance, and defines what the level of fire-resistance performance is, based upon this test.

Restriction of fire growth, or passive protection, has been a key part of building codes since their inception.

Other sections of the code require elements to have a fire-resistance rating or a degree of resistance to flame spread because of the relative hazard associated with the type of construction, the occupancy of the building, or the function of the element.

The construction features that are dealt with in this chapter that are required to have some degree of fire resistance include structural members (see Section 704), exterior walls (see Section 705), fire walls (see Section 706), fire barriers (see Section 707), shaft enclosures (see Section 708), fire partitions (see Section 709), smoke barriers (see Section 710), smoke partitions (see Section 711), and horizontal assemblies (see Section 712). Additionally, the method of protecting penetrations, joints, doors, windows, ducts, air transfer openings in the fire-resistant elements is covered in Sections 713, 714, 715 and 716. This chapter also covers draftstopping and fireblocking of concealed spaces in Section 717. The flame spread and smoke development of insulation are regulated by Section 719. As an alternative to tested assemblies, some prescriptive and calculated methods of determining fire-resistant assemblies are given in Sections 720 and 721.

SECTION 701
GENERAL

701.1 Scope. The provisions of this chapter shall govern the materials, systems and assemblies used for structural *fire resistance* and fire-resistance-rated construction separation of adjacent spaces to safeguard against the spread of fire and smoke within a building and the spread of fire to or from buildings.

❖ The provisions of Chapter 7 apply to the materials, assemblies and systems used to protect against the passage of fire and smoke. Each of the walls and partitions outlined herein provides various degrees of protection. The required fire-resistant rating varies with the potential fire hazard associated with type of construction, occupancy, height and area of the building and degree of protection for different elements of the means of egress. The potential fire hazard associated with various occupancies is reflected in Tables 508.3.3 and 707.3.9. Chapter 7 provides the details and the extent of the protection (horizontal and vertical continuity); however, the fire-resistance-rated construction is mandated by provisions in Chapters 4, 5, 6, 7 and 10.

Whenever the code mandates the use of materials or assemblies required to be noncombustible or have

a degree of fire or smoke resistance, the performance of the material or assembly is required to be evaluated in accordance with the provisions of this chapter and the referenced standards.

The important criteria for the terms "combustible" and "noncombustible" used in this chapter and elsewhere throughout the code are contained in Section 703.4.

SECTION 702
DEFINITIONS

702.1 Definitions. The following words and terms shall, for the purposes of this chapter, and as used elsewhere in this code, have the meanings shown herein.

❖ This section contains definitions of terms that are associated with the subject matter of this chapter. It is important to emphasize that these terms are not exclusively related to this chapter but are applicable everywhere the term is used in the code.

Definitions of terms help in the understanding and application of the code requirements. Locating definitions within this chapter provides ready access to terms used in defining requirements specific to fire-resistive requirements.

These terms are also listed in Chapter 2 with a cross-reference to this section. The use and application of all defined terms, including those defined herein, are set forth in Section 201.

ANNULAR SPACE. The opening around the penetrating item.

❖ The annular space is the space created between the outer surface of a penetrating item and the construction penetrated if left unfilled that provides a means of free passage of smoke, fire and products of combustion [see Figure 702.1(1) for an example of an annular space]. Sections 713.3.1, 713.4.1 and 713.4.2 contain the requirements for protection of annular spaces. Tested and listed through-penetration firestop systems (see Section 713.3.1.2), which are commonly used to seal these annular spaces, generally specify maximum and minimum allowable annular spaces, neither of which must be exceeded in order to ensure that the intended fire-resistance rating is achieved. The minimum and maximum annular space for a penetrating item is frequently not equal, as penetrating items are often located eccentrically in the hole (i.e., not perfectly centered).

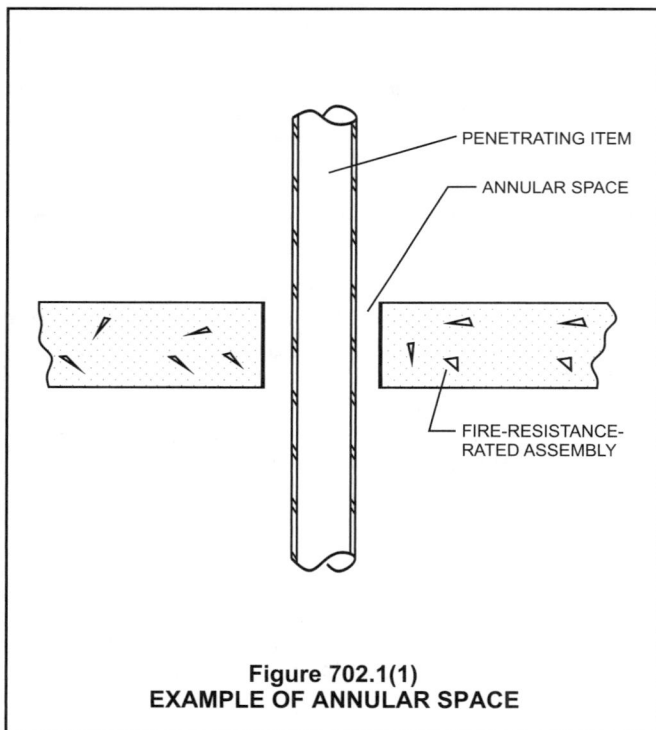

Figure 702.1(1)
EXAMPLE OF ANNULAR SPACE

BUILDING ELEMENT. A fundamental component of building construction, listed in Table 601, which may or may not be of fire-resistance-rated construction and is constructed of materials based on the building type of construction.

❖ Building elements include primary structural members, secondary structural members, exterior walls and interior partitions. The required fire-resistance ratings of the elements are found in Tables 601 and 602.

CEILING RADIATION DAMPER. A *listed* device installed in a ceiling membrane of a fire-resistance-rated floor/ceiling or roof/ceiling assembly to limit automatically the radiative heat transfer through an air inlet/outlet opening.

❖ See the definition of "Damper."

COMBINATION FIRE/SMOKE DAMPER. A *listed* device installed in ducts and air transfer openings designed to close automatically upon the detection of heat and resist the passage of flame and smoke. The device is installed to operate automatically, controlled by a smoke detection system, and where required, is capable of being positioned from a fire command center.

❖ A combination damper is used when the code requires not only a fire damper but also a smoke damper designed to limit the passage of smoke from one side of fire-resistance-rated construction to the other. Fire and smoke dampers are required at duct penetrations of shafts in accordance with Section 716.5.3. Both fire and smoke dampers are required at duct and air transfer openings in fire walls utilized as horizontal exits in accordance with Section 716.5.1. Both shall (subject to exceptions listed) also be required at duct penetrations of fire barrier walls utilized as horizontal exits in accordance with Section 716.5.2. Both shall (subject to exceptions listed) also be required at duct penetrations of fire partitions utilized as corridor partitions in accordance with Section 716.5.4.

The combination fire/smoke damper must be actuated automatically by a smoke detection system. Where the damper is part of a smoke control system in Section 909, it shall also be controlled from the fire command center.

DAMPER. See "*Ceiling radiation damper,*" "*Combination fire/smoke damper,*" "*Fire damper*" and "*Smoke damper.*"

❖ Fire dampers are used primarily in heating, ventilating and air-conditioning (HVAC) duct systems that pass through fire-resistance-rated walls or floors. Dampers may also be installed in rated walls independent of HVAC duct systems. Dampers are provided to maintain the fire-resistance rating of the penetrated assembly. Fire dampers are regulated by UL 555, ceiling dampers by UL 555C and smoke dampers by UL 555S.

DRAFTSTOP. A material, device or construction installed to restrict the movement of air within open spaces of concealed areas of building components such as crawl spaces, floor/ceiling assemblies, roof/ceiling assemblies and *attics.*

❖ Draftstopping is required in concealed combustible spaces to limit the movement of air, smoke and other products of combustion. Draftstopping materials are permitted to be combustible based on the rationale that a large and thick enough combustible material will act as a hindrance against the free movement of air, of flame/fire and of the products of combustion [see Figures 702.1(2) and 702.1(3) for typical draftstopping applications (also see Section 717)]. Although the term "draftstopping" would seem to imply that its primary purpose is to hinder the circulation of air within the space, its intended purpose is to stop the movement of fire and products of combustion, as evidenced by the fact that draftstopping can be omitted in some cases when appropriate automatic fire sprinkler pro-

tection is installed (see Sections 717.3.2, 717.3.3, 717.4.2 and 717.4.3).

F RATING. The time period that the *through-penetration firestop system* limits the spread of fire through the penetration when tested in accordance with ASTM E 814 or UL 1479.

❖ See the definition of "Through-penetration firestop system." The F rating is determined not only from a fire endurance test, but also includes a test for mechanical

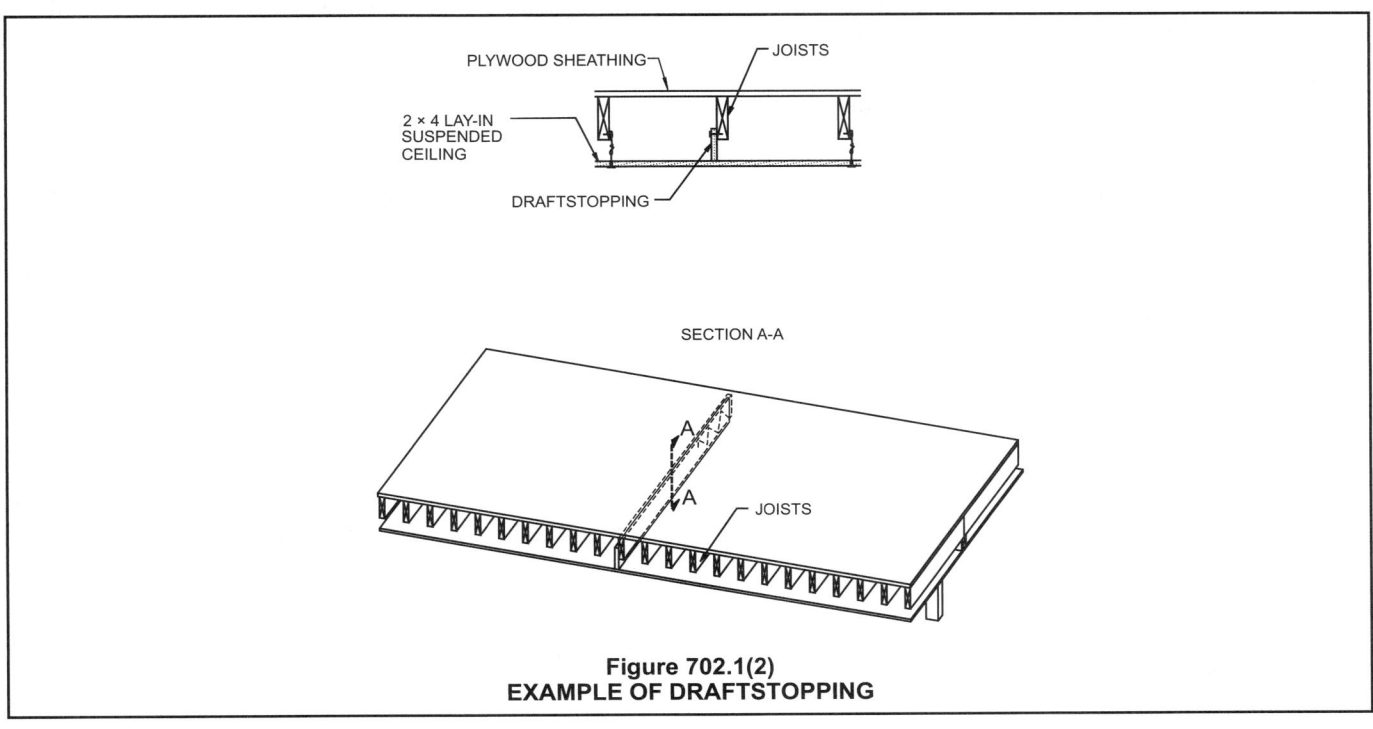

Figure 702.1(2)
EXAMPLE OF DRAFTSTOPPING

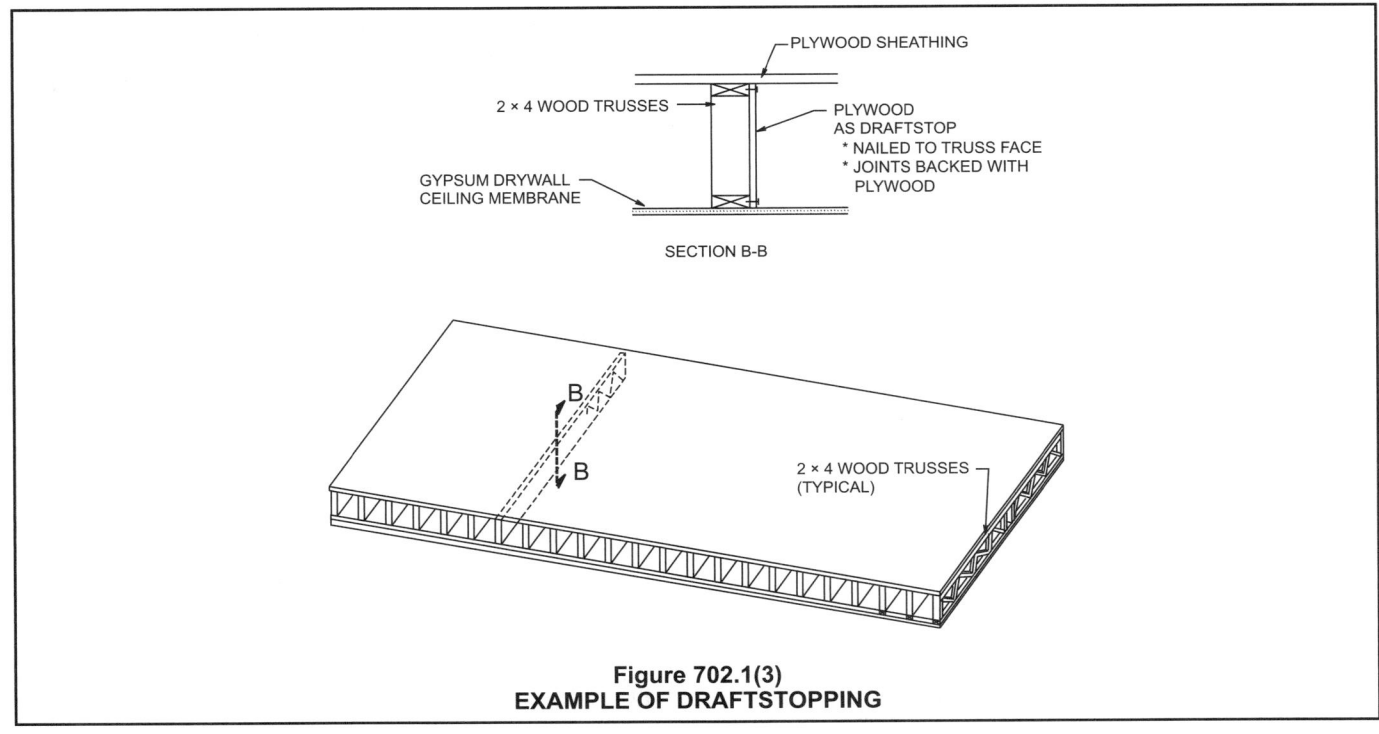

Figure 702.1(3)
EXAMPLE OF DRAFTSTOPPING

integrity of the through-penetration seal after fire exposure, known as the "hose stream test." The F rating indicates that the firestop system is capable of stopping the fire, flame and hot gases from passing through the assembly at the penetration.

FIRE BARRIER. A fire-resistance-rated wall assembly of materials designed to restrict the spread of fire in which continuity is maintained.

❖ The term represents wall assemblies with a fire-resistance rating that are constructed in accordance with Section 707. Even though the definition applies to walls, horizontal assemblies can be fire barriers, also. See the definition of "Horizontal assembly" and the requirements in Section 712 that apply to floor and roof assemblies designed to restrict the spread of fire. See Figure 702.1(4) for an example of a fire barrier.

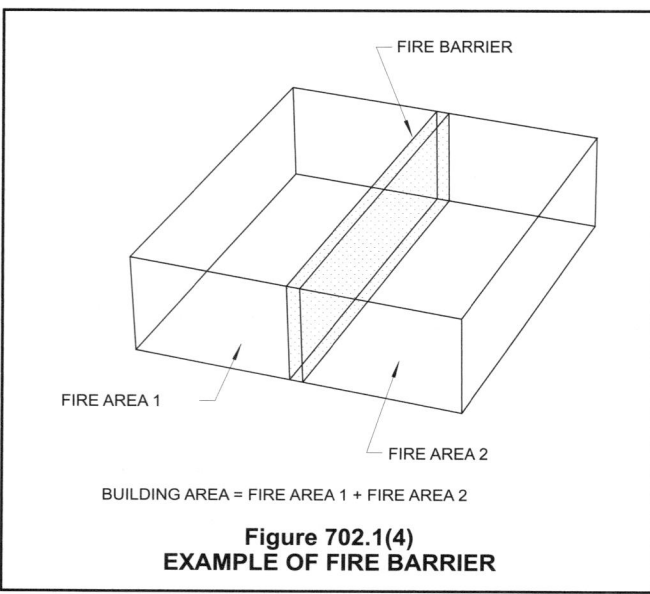

BUILDING AREA = FIRE AREA 1 + FIRE AREA 2

Figure 702.1(4)
EXAMPLE OF FIRE BARRIER

FIRE DAMPER. A *listed* device installed in ducts and air transfer openings designed to close automatically upon detection of heat and resist the passage of flame. *Fire dampers* are classified for use in either static systems that will automatically shut down in the event of a fire, or in dynamic systems that continue to operate during a fire. A dynamic *fire damper* is tested and rated for closure under elevated temperature airflow.

❖ See the definition of "Damper."

FIRE DOOR. The door component of a *fire door* assembly.

❖ A fire door is the primary component of a fire door assembly.
 The fire protection rating assigned to a tested fire door is only valid if the door is installed in a labeled frame with appropriate hardware. Installation requirements within the code reference NFPA 80, *Standard for Fire Doors and Fire Windows*. Door ratings are expressed in minutes or hours. Field modification of doors are primarily limited to the mounting of listed hardware.

FIRE DOOR ASSEMBLY. Any combination of a *fire door*, frame, hardware and other accessories that together provide a specific degree of fire protection to the opening.

❖ Fire door assemblies, (door, frame and hardware) are required to be tested using the appropriate standard and then installed in accordance with NFPA 80. Side hinged doors, hardware and frames are often manufactured separately with manufacturers' and listing agencies defining acceptable combinations of assembly components that have been tested together.

FIRE PARTITION. A vertical assembly of materials designed to restrict the spread of fire in which openings are protected.

❖ Fire partitions are used as wall assemblies to separate adjacent tenant spaces in covered mall buildings, dwelling units, and sleeping rooms, and to enclose

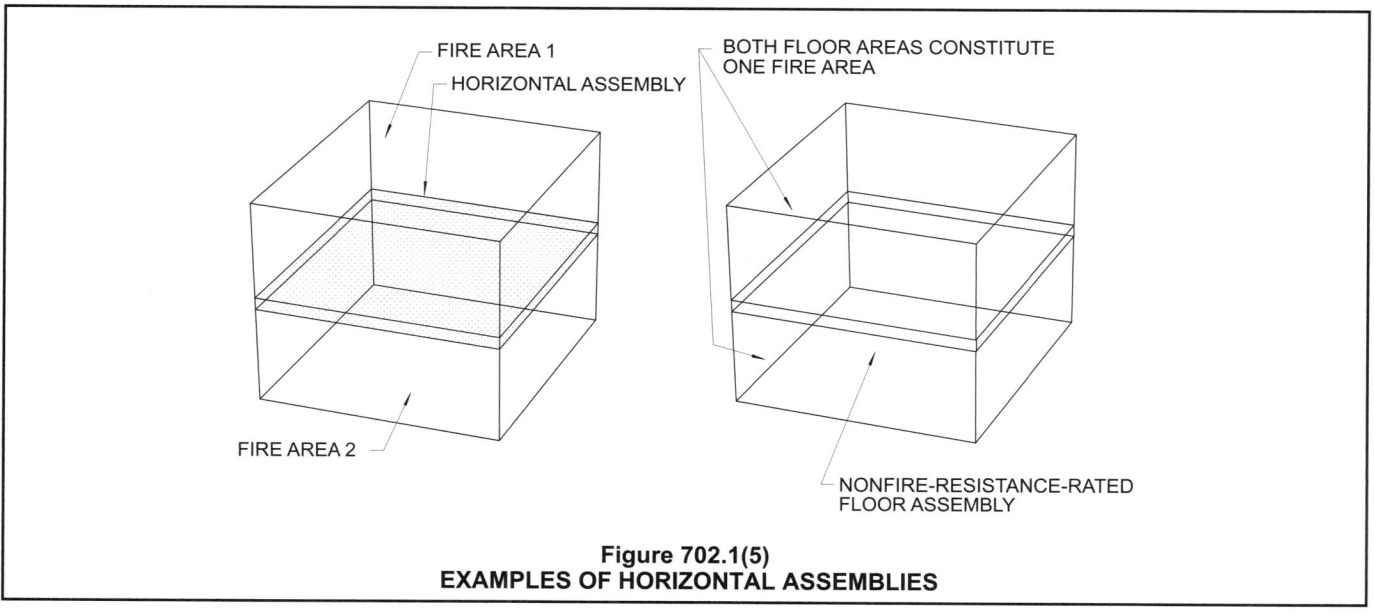

Figure 702.1(5)
EXAMPLES OF HORIZONTAL ASSEMBLIES

corridors and elevator lobbies. Section 709 establishes the construction requirements for fire partitions. The fire-resistance ratings, continuity requirements and opening protective requirements for fire partitions are usually less restrictive than those for fire barriers.

FIRE PROTECTION RATING. The period of time that an opening protective will maintain the ability to confine a fire as determined by tests prescribed in Section 715. Ratings are stated in hours or minutes.

❖ The term "fire protection rating" applies to the fire performance of an opening protective, such as a fire door, which is determined through tests performed in accordance with NFPA 252 or UL 10C.

FIRE RESISTANCE. That property of materials or their assemblies that prevents or retards the passage of excessive heat, hot gases or flames under conditions of use.

❖ All materials offer some degree of fire resistance. A sheet of plywood has a low level of fire resistance as compared to a concrete block, which has a higher level of fire resistance. The fire resistance of a material or an assembly is evaluated by testing performed in accordance with ASTM E 119. Tested materials will be assigned a fire-resistance rating consistent with the demonstrated performance.

FIRE-RESISTANCE RATING. The period of time a building element, component or assembly maintains the ability to confine a fire, continues to perform a given structural function, or both, as determined by the tests, or the methods based on tests, prescribed in Section 703.

❖ This refers to the period of time a building element, component or assembly maintains the ability to confine a fire, continues to perform a given structural function, or both, as determined by tests or the methods based on tests, prescribed in Section 703.

The fire-resistance rating is developed using standardized test methods (i.e., ASTM E 119, etc.). Assemblies rated under these tests are deemed to be able to perform their function for a specified period of time under specific fire conditions (standard time-temperature curve).

The fire-resistance rating is not intended to be a prediction of the actual length of time that an assembly will perform its intended function under actual fire conditions. Although the time-temperature curves of standardized fire test methods are usually selected to approximate at least some real-life fire conditions, the very wide range of actual fire conditions makes the listed fire-resistance rating more of a nominal, comparative index than a predictor of fire-resistance time in any given fire incident.

FIRE-RESISTANT JOINT SYSTEM. An assemblage of specific materials or products that are designed, tested and fire-resistance rated in accordance with either ASTM E 1966 or UL 2079 to resist for a prescribed period of time the passage of fire through joints made in or between fire-resistance-rated assemblies.

❖ In order to maintain the fire-resistant integrity of fire-resistance-rated assemblies, joints that occur within an assembly or between adjacent assemblies must be protected through an installation that has been tested in accordance with ASTM E 1966 or UL 2079. Some common examples of applications where a fire-resistant joint system would be required are expansion joints in fire-resistance-rated floors or walls and the junction between fire-resistance-rated floors and walls [see Figure 702.1(9) for examples]. The regular joints that occur within a uniform assembly are most often tested as part of fire testing (e.g., in accordance with ASTM E 119) for that entire assembly. The required details for these joints are specified in the listings for the underlying assembly. These joints do not need additional testing in accordance with ASTM E 1966 or UL 2079. An example of such joints is the joints between individual sheets of gypsum board in a gypsum-sheathed stud wall. Consequently, other than the joints covering or filling the gaps within an assembly, the need for ASTM E 1966 or UL 2079 tested joint systems is usually for the joints between dissimilar assemblies or adjacent assemblies.

FIRE SEPARATION DISTANCE. The distance measured from the building face to one of the following:

1. The closest interior *lot line*;

2. To the centerline of a street, an alley or public way; or

3. To an imaginary line between two buildings on the property.

The distance shall be measured at right angles from the face of the wall.

❖ Fire separation distance is the distance from the exterior wall of the building to one of the three following property locations, measured perpendicular to the exterior wall face: an interior lot line [see Figure 702.1(6)], the centerline of a street or public way [see Figure 702.1(7)] or an imaginary line between two buildings on the same property [see Figure 702.1(8)]. The imaginary line can be located anywhere between the two buildings; it is the designer's choice, but once established, the location of the line applies to both buildings and cannot be revised.

The distance can vary with irregular-shaped lots and buildings, as shown in Figures 702.1(6) and 702.1(7). When applying the exterior wall requirements of Table 602, the required exterior wall fire-resistance rating might vary along a building side; for example, where the lot line is not parallel to the exterior wall.

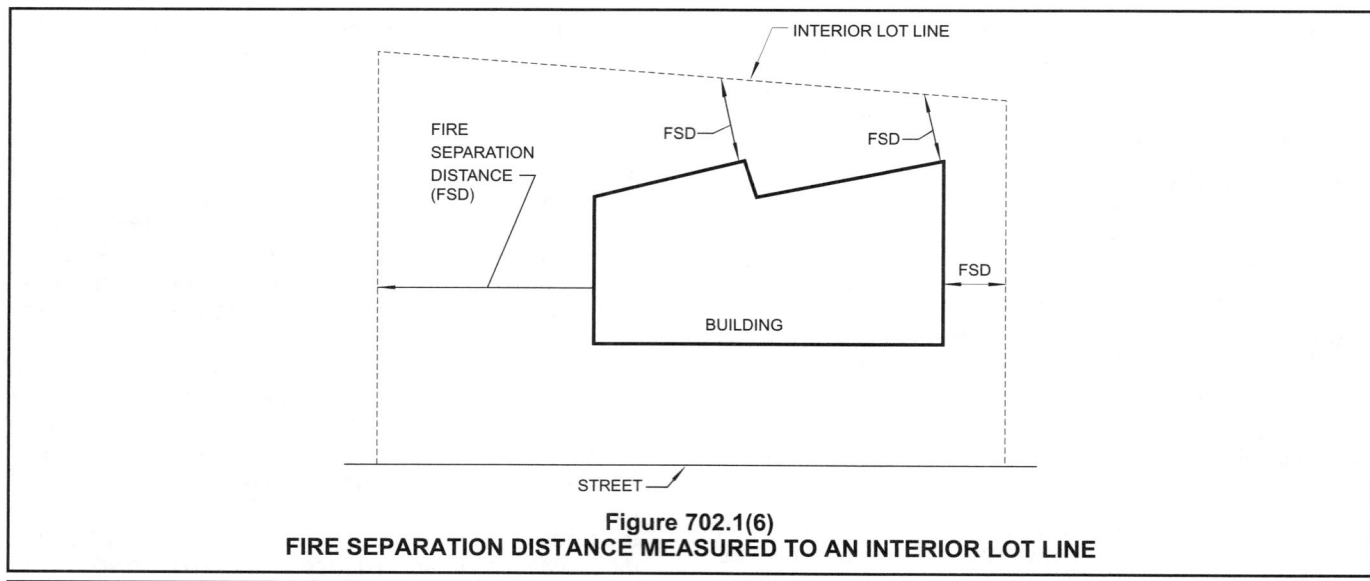

Figure 702.1(6)
FIRE SEPARATION DISTANCE MEASURED TO AN INTERIOR LOT LINE

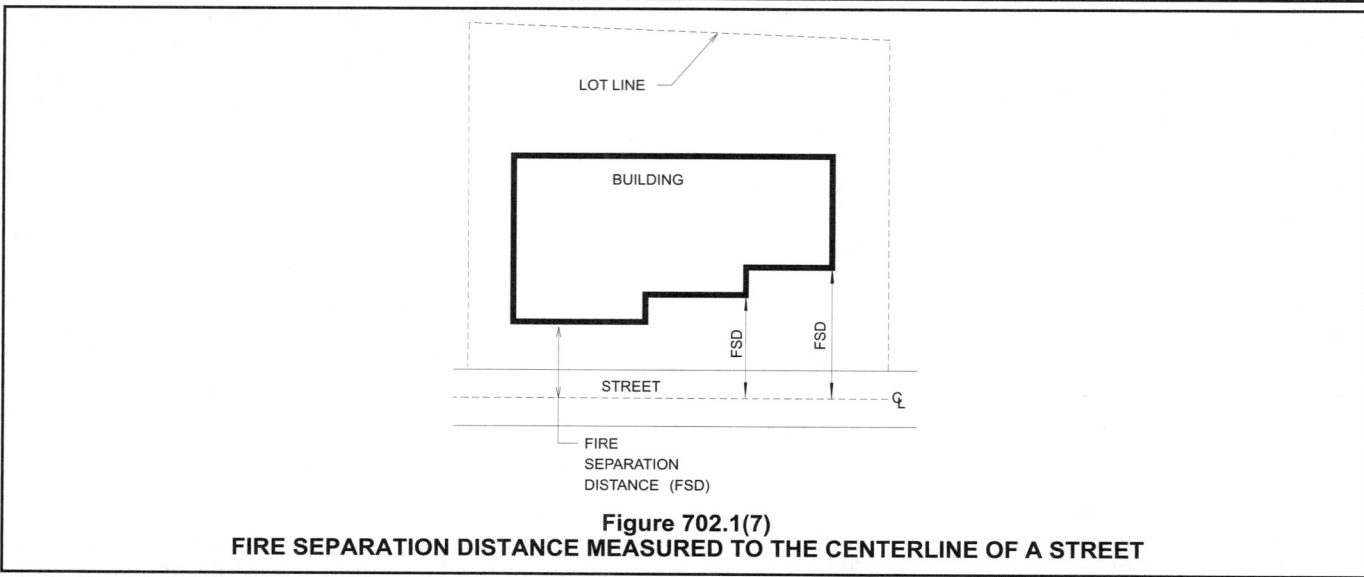

Figure 702.1(7)
FIRE SEPARATION DISTANCE MEASURED TO THE CENTERLINE OF A STREET

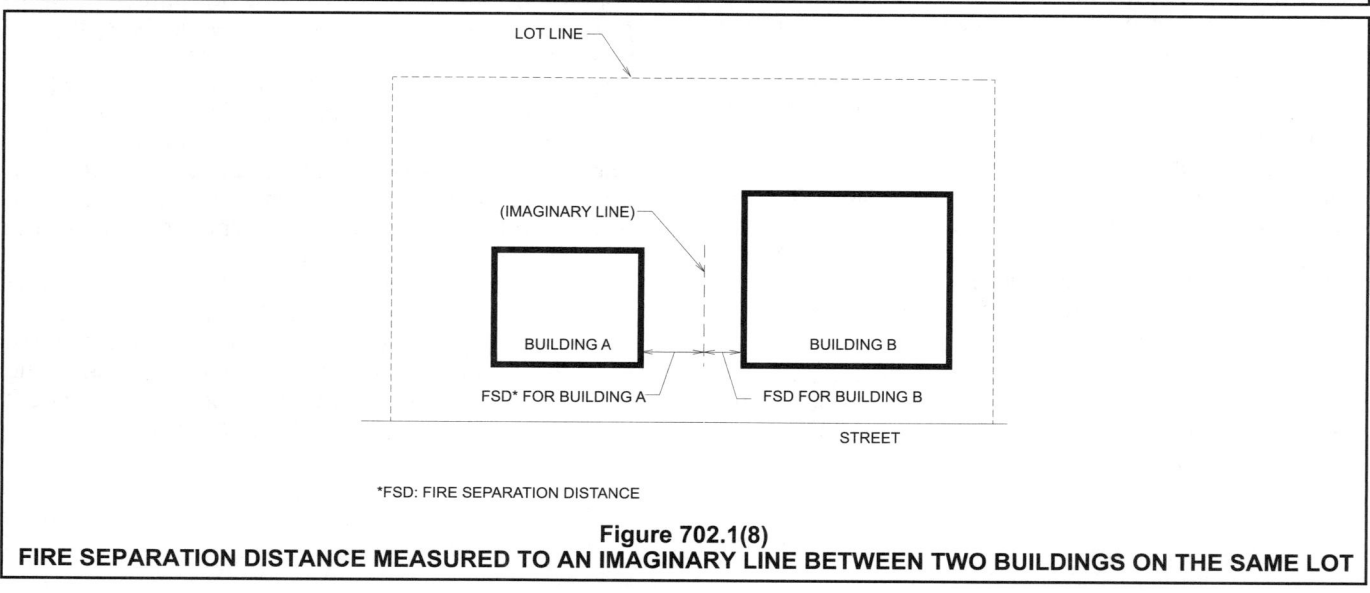

*FSD: FIRE SEPARATION DISTANCE

Figure 702.1(8)
FIRE SEPARATION DISTANCE MEASURED TO AN IMAGINARY LINE BETWEEN TWO BUILDINGS ON THE SAME LOT

FIRE WALL. A fire-resistance-rated wall having protected openings, which restricts the spread of fire and extends continuously from the foundation to or through the roof, with sufficient structural stability under fire conditions to allow collapse of construction on either side without collapse of the wall.

❖ Fire walls must meet the construction requirements in Section 706. The requirements for fire walls are much more restrictive than for fire barriers or fire partitions. The material constituting the fire wall must be noncombustible in all construction types except Type V. The vertical and horizontal continuity requirements are much more restrictive and opening protective. A fire wall, unlike the fire barrier and fire partition, must be constructed so it will remain in place if the construction on either side of it collapses. However, the fire wall is not required to remain in place if construction on both sides of it collapses (i.e., the fire wall is not required to be a free-standing cantilever wall). Fire walls are used to divide a structure into separate buildings (see the definition of "Areas, building" in Section 502.1). To be considered separate buildings, the division must be vertical. The code applies the term "fire wall" to vertically constructed assemblies only and not to horizontal assemblies.

FIRE WINDOW ASSEMBLY. A window constructed and glazed to give protection against the passage of fire.

❖ Fire windows are "opening protectives" and contain glazing (see Section 715.5). They are required to be tested in accordance with NFPA 257 or UL 9 and are then to be installed in accordance with NFPA 80.

FIREBLOCKING. Building materials or materials for use as fireblocking, installed to resist the free passage of flame to other areas of the building through concealed spaces.

❖ Fireblocking is required to hinder the concealed spread of flame, heat and other products of combustion within hollow spaces inside of walls or floor/ceiling assemblies. This is done by periodically subdividing that space, as indicated in Section 717.2, using construction materials that have some resistance to fire and by sealing the openings around penetrations through those materials.
    Some fireblocking materials are permitted to be combustible based on the rationale that a substantial combustible material will provide a barrier adequate to perform the intended function (also see Section 717).

FLOOR FIRE DOOR ASSEMBLY. A combination of a *fire door*, a frame, hardware and other accessories installed in a horizontal plane, which together provide a specific degree of fire protection to a through-opening in a fire-resistance-rated floor (see Section 712.8).

❖ Floor fire door assemblies are required to be tested in accordance with ASTM E 119 and are used to protect openings in fire-resistance-rated floors. They are an alternative for protecting a floor opening, such as an access opening to mechanical equipment. See the commentary to Section 712.8 for additional information on floor fire door assemblies.

HORIZONTAL ASSEMBLY. A fire-resistance-rated floor or roof assembly of materials designed to restrict the spread of fire in which continuity is maintained.

❖ A horizontal assembly is a component for completing compartmentation. Horizontal assemblies have all openings and penetrations protected equal to the rating for the fire-resistance-rated floor or roof assembly [see Figure 702.1(5)].

JOINT. The linear opening in or between adjacent fire-resistance-rated assemblies that is designed to allow independent movement of the building in any plane caused by thermal, seismic, wind or any other loading.

❖ This term defines the opening created when two fire-resistance-rated assemblies meet and an open space occurs between the assemblies [see Figure 702.1(9) for examples of building joints].

MEMBRANE PENETRATION. An opening made through one side (wall, floor or ceiling membrane) of an assembly.

❖ This term refers to the situation where a penetration is made of a single layer of a fire-resistance-rated assembly. Sections 713.3.2 and 713.4.1.2 establish criteria where the penetration of a single membrane of an assembly is permitted while still considering the assembly to have the required fire-resistance rating [see Figure 702.1(10) for an example of a single membrane penetration].
    Where a penetration is made completely through a fire-resistance-rated wall or floor/ceiling assembly, it must be protected in accordance with Section 713.3.1, 713.4.1.1 or 713.4.2 as applicable for the assembly penetrated.

MEMBRANE-PENETRATION FIRESTOP. A material, device or construction installed to resist for a prescribed time period the passage of flame and heat through openings in a protective membrane in order to accommodate cables, cable trays, conduit, tubing, pipes or similar items.

❖ Sections 713.3.2 and 713.4.1.2 both refer to protection methods for membrane penetrations of fire-resistance-rated walls, floor/ceiling assemblies and roof/ceiling assemblies.

MINERAL FIBER. Insulation composed principally of fibers manufactured from rock, slag or glass, with or without binders.

❖ This term provides a definition for a type of insulation that is commonly used or accepted within rated assemblies [see Table 720.1(2), Item 15-1.11] or as a fireblocking material (see Section 717.2.1).

MINERAL WOOL. Synthetic vitreous fiber insulation made by melting predominately igneous rock or furnace slag, and other inorganic materials, and then physically forming the melt into fibers.

❖ This term provides a definition for a type of insulation that is commonly used or accepted within rated assemblies [see Table 720.1(2), Item 15-1.15] or as a fireblocking material (see Section 717.2.1).

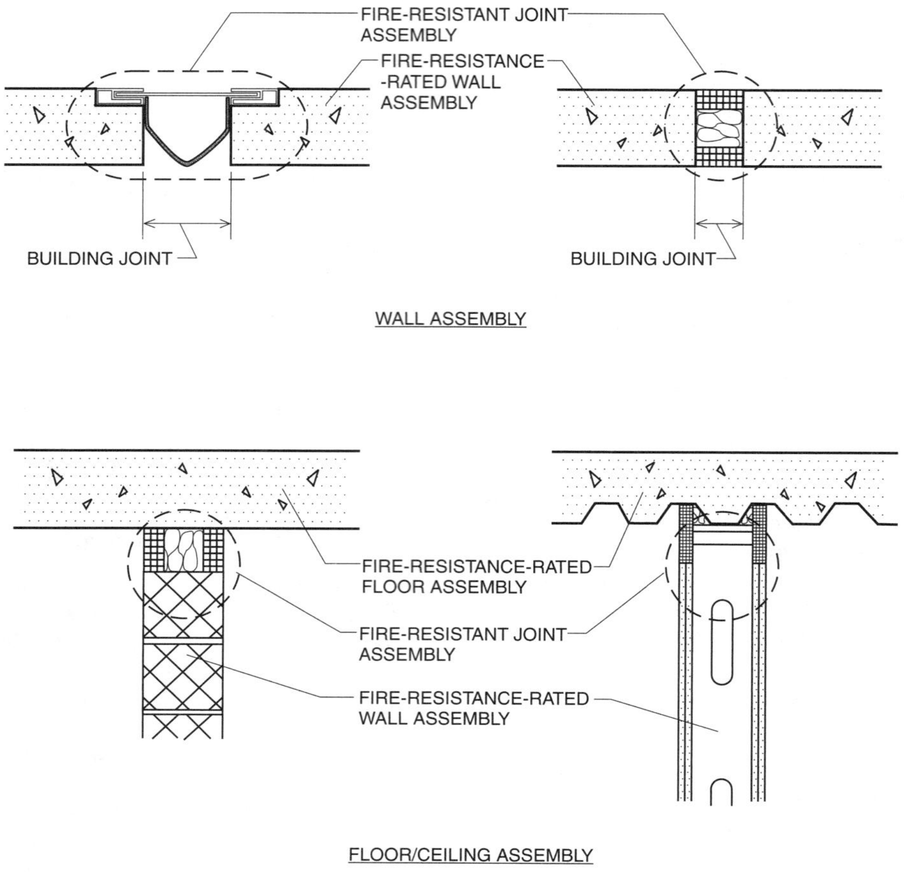

FIRE-RESISTANT JOINT ASSEMBLY

FIRE-RESISTANCE-RATED WALL ASSEMBLY

BUILDING JOINT

BUILDING JOINT

WALL ASSEMBLY

FIRE-RESISTANCE-RATED FLOOR ASSEMBLY

FIRE-RESISTANT JOINT ASSEMBLY

FIRE-RESISTANCE-RATED WALL ASSEMBLY

FLOOR/CEILING ASSEMBLY

Figure 702.1(9)
EXAMPLES OF FIRE-RESISTANT JOINT SYSTEMS

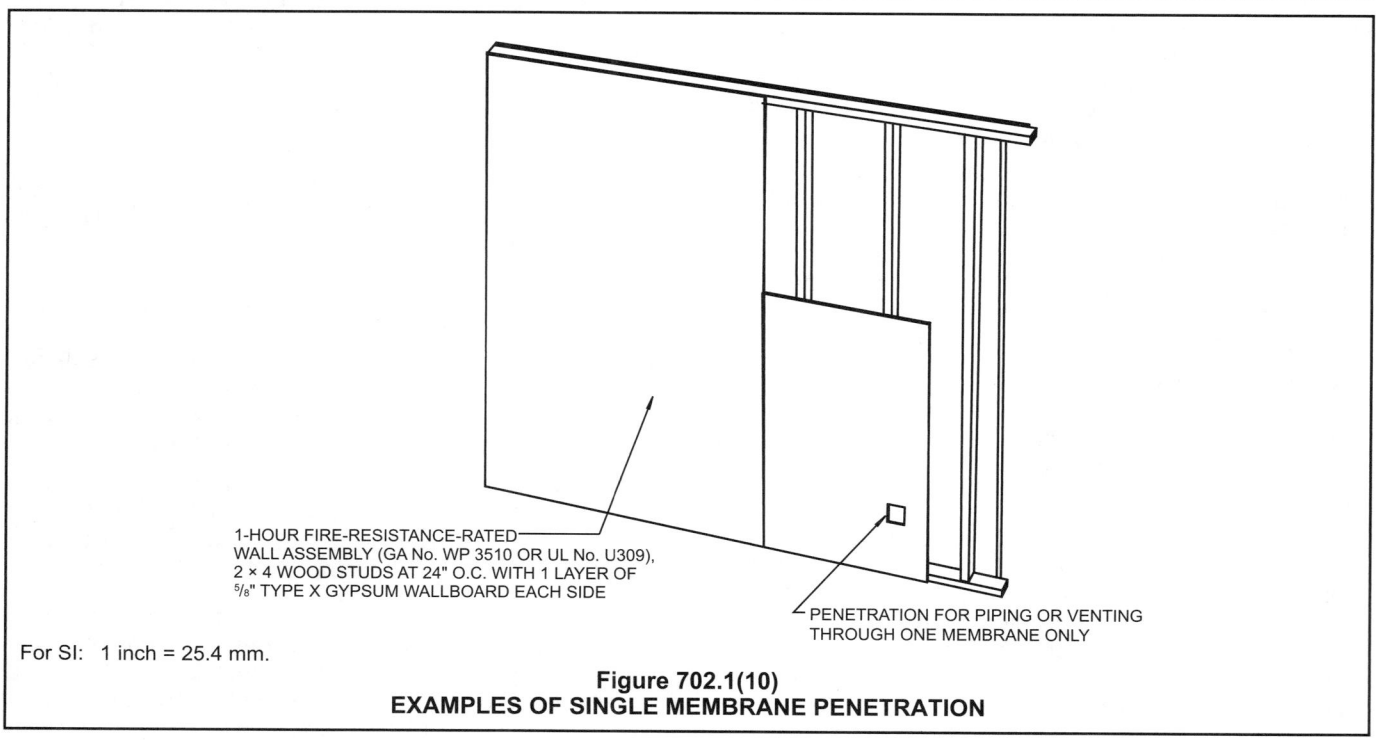

1-HOUR FIRE-RESISTANCE-RATED WALL ASSEMBLY (GA No. WP 3510 OR UL No. U309), 2 × 4 WOOD STUDS AT 24" O.C. WITH 1 LAYER OF ⁵/₈" TYPE X GYPSUM WALLBOARD EACH SIDE

PENETRATION FOR PIPING OR VENTING THROUGH ONE MEMBRANE ONLY

For SI: 1 inch = 25.4 mm.

Figure 702.1(10)
EXAMPLES OF SINGLE MEMBRANE PENETRATION

PENETRATION FIRESTOP. A through-penetration firestop or a *membrane-penetration firestop*.

❖ See the definitions of "Membrane-penetration firestop" and "Through-penetration firestop system."

SELF-CLOSING. As applied to a *fire door* or other opening protective, means equipped with an device that will ensure closing after having been opened.

❖ A self-closing opening protective refers to a fire or smoke door assembly equipped with a listed closer for doors that must be maintained in the normally closed position. When the door is opened and released, the self-closing feature returns the door to the closed position. It is important to distinguish between the terms "self-closing" and "automatic closing" because they are not interchangeable. "Automatic closing" refers to an opening protective that is normally in the open position (see Section 715.4.8.2). Opening protectives with automatic closers are often held open and then returned to the closed position upon activation of fire detectors or smoke detectors or loss of power, which automatically releases the hold-open device allowing the door to close.

SHAFT. An enclosed space extending through one or more *stories* of a building, connecting vertical openings in successive floors, or floors and roof.

❖ Shafts are successive openings in the floors. A shaft is required to be enclosed with fire-resistance-rated assemblies to help prevent the vertical spread of fire and resist the spread of products of combustion from story to story. Stairway and elevator floor openings are examples of shafts. Provisions for vertical shafts are found in Section 708. These provisions are applicable to floor/ceiling openings required to be enclosed as specified in Section 708.2.

SHAFT ENCLOSURE. The walls or construction forming the boundaries of a shaft.

❖ Fire-resistance-rated walls forming shaft enclosures must be constructed as fire barrier walls in accordance with Section 707. Usually, vertical ducts penetrating multiple floors or buildings are required to be in shaft enclosures. Unlike a common chase wall, a shaft wall must be fire-resistance rated from the room side and from the shaft side.

SMOKE BARRIER. A continuous membrane, either vertical or horizontal, such as a wall, floor or ceiling assembly, that is designed and constructed to restrict the movement of smoke.

❖ A smoke barrier is a fire-resistance-rated assembly that is different from a fire partition, fire barrier or fire wall. Smoke barriers include walls and floor/ceiling assemblies that are constructed with a 1-hour fire-resistance rating and are one of the components in a smoke compartment. In Group I-2 and I-3 occupancies, smoke barriers are intended to create adjacent smoke compartments to which building occupants can be safely and promptly relocated during a fire, thus preventing the need to have complete and immediate egress from the building. For these occupancies, com-

plete egress from the building would not be practical in most cases, due to restrictions on the mobility of the occupants. To maintain tenability in the adjacent smoke compartment, the smoke barrier is therefore intended to resist the spread of fire and hinder the movement of smoke. Smoke barriers are also used to compartment a building into separate smoke control zones when using the provisions of Section 909. The construction requirements for a smoke barrier provide resistance to the transmission of smoke [see Figure 702.1(11) and Section 710].

SMOKE COMPARTMENT. A space within a building enclosed by *smoke barriers* on all sides, including the top and bottom.

❖ Smoke compartments create spaces that protect occupants from the products of combustion produced by a fire in an adjacent smoke compartment and to restrict smoke to the compartment of fire origin.

SMOKE DAMPER. A *listed* device installed in ducts and air transfer openings designed to resist the passage of smoke. The device is installed to operate automatically, controlled by a smoke detection system, and where required, is capable of being positioned from a fire command center.

❖ See the definition of "Damper."

SPLICE. The result of a factory and/or field method of joining or connecting two or more lengths of a *fire-resistant joint system* into a continuous entity.

❖ In many instances, the actual length of a joint required to be protected by a fire-resistant joint system exceeds the length of the joint system, especially when a prefabricated system is used. When two or more lengths of a joint system are necessary to protect a joint, the ends of each section must be joined by a splice to maintain the fire-resistive integrity of the assembly. Splices may be either factory or field installed, but in either case, the splice must be tested in accordance with UL 2079.

T RATING. The time period that the penetration firestop system, including the penetrating item, limits the maximum temperature rise to 325°F (163°C) above its initial temperature through the penetration on the nonfire side when tested in accordance with ASTM E 814 or UL 1479.

❖ See the definition of "Through-penetration firestop system." Some through penetrations having a high thermal conductivity (e.g., metal pipes, cables) will not be able to achieve a T rating without some method to prevent the heat from being transmitted from the fire side to the nonfire side. For metallic pipes, insulation must be applied to the through penetrations for some distance on one or, more often, both sides of the fire-resistance-rated assembly. Thus, when a T rating is desired, the seemingly unusual situation can arise of needing to insulate items that would normally not need insulation, such as drain pipes or conduit. Some plastic pipes are also unable to achieve a T rating without some insulation, particularly those used and, therefore, tested as "open" systems, where hot gases from the test furnace are able to flow within the plastic pipe once the pipe burns through. The type and thickness of insulation required, as well as the length over which it must be applied, are all indicated in the through-penetration firestop system listings.

Some highly intumescent firestop products can close (choke down) a plastic pipe early enough during a fire test to prevent hot gases from heating up the pipe from inside, thus allowing plastic pipe through-penetration firestop systems to be developed that have a T rating without the need for any insulation.

In the case of cable through penetrations, the need for heat dissipation would usually make the application of additional insulating material unacceptable. The few cable through-penetration firestop systems that do have a T rating achieve this by enclosing the cables within some protective enclosure for a distance from the rated assembly, thus allowing any heat transmitted through the cables during fire exposure to be dissipated, resulting in the exposed section of cabling to be cool enough for the duration of the fire test exposure to meet the T rating requirement [see Figure 702.1(12)].

Overall, due to the unavoidable difficulty in having a conductive or semiconductive through penetrant obtain a T rating, less than 5 percent of all tested and listed through-penetration firestop systems have a T rating. This is because the T rating indicates that the firestop system is not only capable of stopping the fire, flame and hot gases from passing through the assembly at the penetration as an F rating does, but it also must limit the temperature transfer to the unexposed side of the assembly.

THROUGH PENETRATION. An opening that passes through an entire assembly.

❖ This term is different from a single "membrane penetration." It defines the situation where an opening is created to accommodate an object that passes completely through an assembly from one side to the other [see Figure 702.1(12) for an example of a through penetration]. It is important to realize that this term applies only to an item that passes all the way through the assembly; back-to-back membrane penetrations, such as outlet boxes on each side of a wall assembly, should not be considered a through penetration.

THROUGH-PENETRATION FIRESTOP SYSTEM. An assemblage of specific materials or products that are designed, tested and fire-resistance rated to resist for a prescribed period of time the spread of fire *through penetrations*. The F and T rating criteria for penetration firestop systems shall be in accordance with ASTM E 814 or UL 1479. See definitions of "F rating" and "T rating."

❖ One method of protection for penetrations through fire walls, fire barriers, fire partitions and fire-resistance-rated floor/ceiling assemblies is to provide a through-penetration firestop system (see Sections 712.3.1 and 712.4.1.1). Through-penetration firestop systems maintain the required protection from the spread of fire, passage of hot gases and transfer of heat. The protection is often provided by an intumescent material. Upon exposure to high temperatures, this material expands as much as eight to 10 times its original volume, forming a high-strength char.

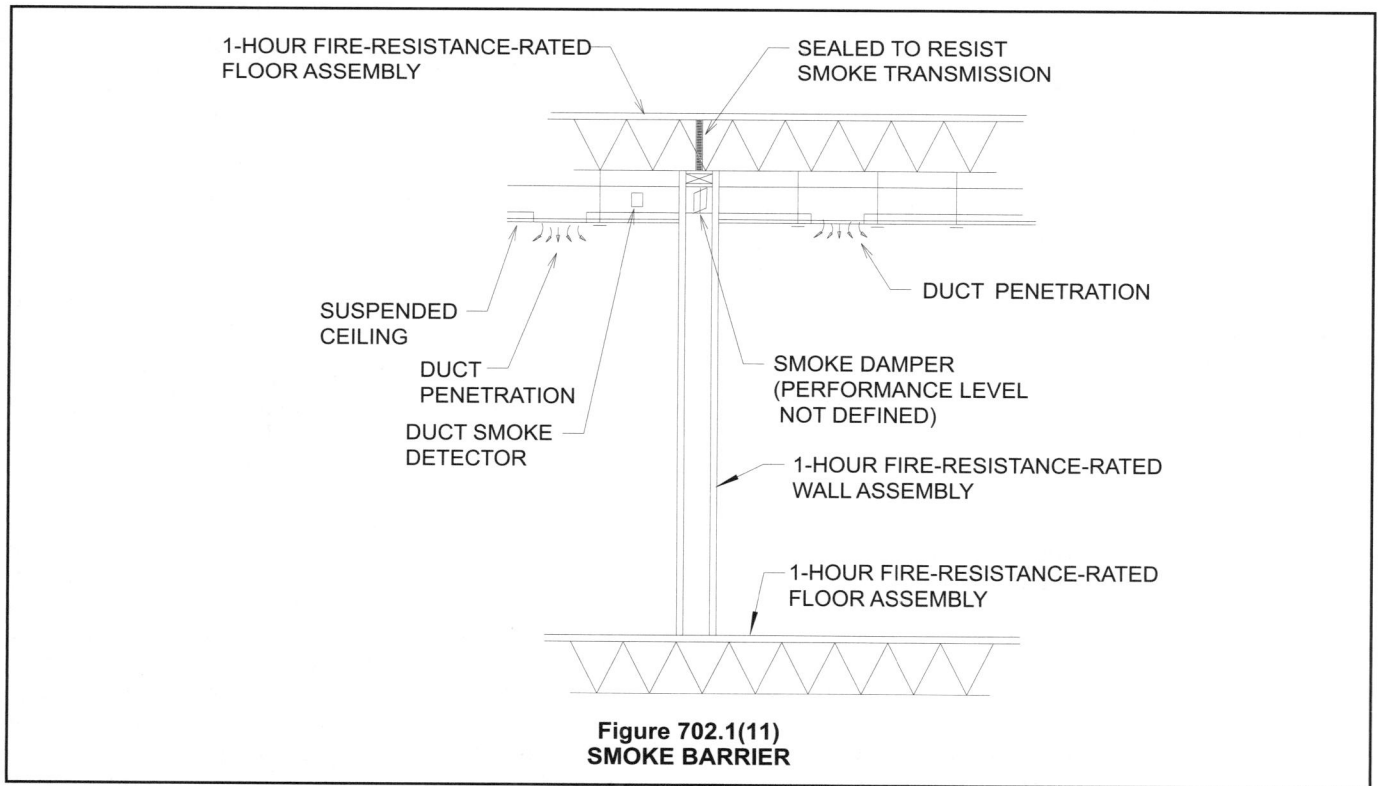

**Figure 702.1(11)
SMOKE BARRIER**

This is one of several types of through-penetration firestop systems available. This definition is based on information from three sources: Section 3.1.1 of ASTM E 814; a compilation of definitions from ASTM standards and the *Fire Resistance Directory* by Underwriters Laboratories Inc.

SECTION 703
FIRE-RESISTANCE RATINGS AND FIRE TESTS

703.1 Scope. Materials prescribed herein for *fire resistance* shall conform to the requirements of this chapter.

❖ Standard fire test methods for determining fire-resistance ratings and combustibility of materials are covered in this section. Section 703.2 addresses the acceptable testing method to be used in determining the fire-resistance ratings of building assemblies and structural elements, while Section 703.3 provides alternative methods. Section 703.4 defines methods to determine if a material is combustible or noncombustible. The flame spread rating provisions in Section 703.4.2 are used in evaluating the acceptability of composite materials. The minimum fire-resistance ratings for structural elements are specified in Table 601, based on the type of construction classification. These fire-resistance ratings are one of the elements taken into consideration when determining the allowable height and area of buildings. By using materials and construction methods that result in the required fire-resistance ratings (see commentary, Table 503), the inherent fire hazards associated with different uses are, in part, addressed. Other sections of the code also contain provisions for required fire-resistance ratings in specific applications based on the need to separate uses, processes, fire areas or building areas.

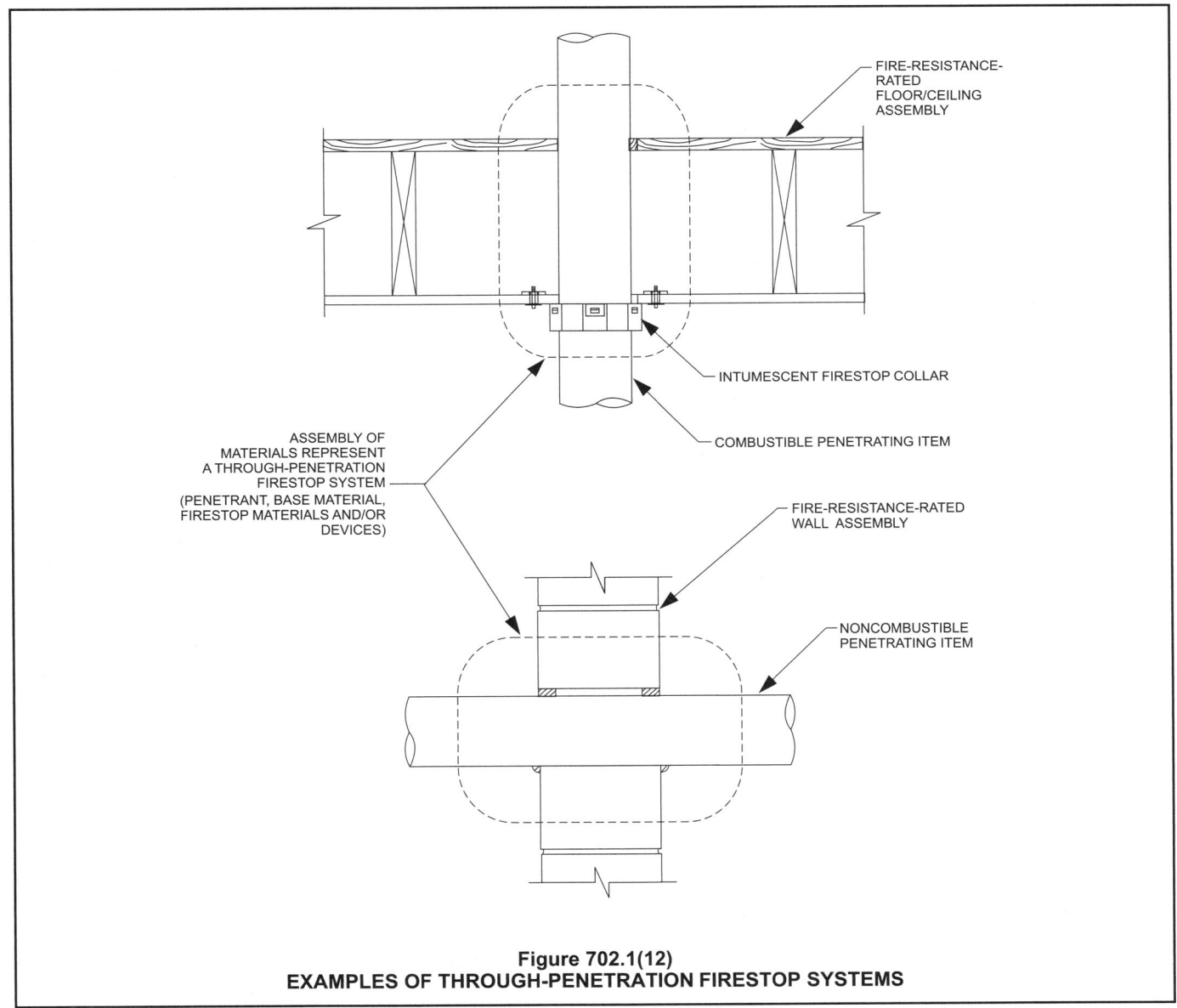

Figure 702.1(12)
EXAMPLES OF THROUGH-PENETRATION FIRESTOP SYSTEMS

703.2 Fire-resistance ratings. The *fire-resistance rating* of building elements, components or assemblies shall be determined in accordance with the test procedures set forth in ASTM E 119 or UL 263 or in accordance with Section 703.3. Where materials, systems or devices that have not been tested as part of a fire-resistance-rated assembly are incorporated into the building element, component or assembly, sufficient data shall be made available to the *building official* to show that the required *fire-resistance rating* is not reduced. Materials and methods of construction used to protect joints and penetrations in fire-resistance-rated building elements, components or assemblies shall not reduce the required *fire-resistance rating*.

Exception: In determining the *fire-resistance rating* of exterior bearing walls, compliance with the ASTM E 119 or UL 263 criteria for unexposed surface temperature rise and ignition of cotton waste due to passage of flame or gases is required only for a period of time corresponding to the required *fire-resistance rating* of an exterior nonbearing wall with the same *fire separation distance*, and in a building of the same group. When the *fire-resistance rating* determined in accordance with this exception exceeds the *fire-resistance rating* determined in accordance with ASTM E 119 or UL 263, the fire exposure time period, water pressure and application duration criteria for the hose stream test of ASTM E 119 or UL 263 shall be based upon the *fire-resistance rating* determined in accordance with this exception.

❖ The fire-resistance properties of materials and assemblies must be measured and specified in accordance with a common test standard. For this reason, the fire-resistance ratings of building assemblies and structural elements are required to be determined in accordance with ASTM E 119 or UL 263. The ASTM E 119 and UL 263 test methods evaluate the ability of an assembly to contain a fire, to retain its structural integrity, or both, in terms of endurance time during the test conditions. The test standard also contains conditions for measuring heat transfer through membrane elements protecting framing or surfaces. The fire exposure is based on the standard time-temperature curve (see Figure 703.2).

The extent to which the test specimen is representative of in-place conditions is determined by the test standard and what is actually tested. For example, the test standard does not provide for the effect on fire endurance of conventional openings in the assembly, such as electrical outlets, plumbing pipes, etc., unless specifically included in the construction assembly tested. The test reports will, however, typically indicate an allowance for openings for steel electrical outlet boxes that do not exceed 16 square inches (0.0103 m²) in area, provided that the area of such openings does not exceed 100 square inches (0.0645 m²) for any 100 square feet (9 m²) of wall area. Since, in actuality, assemblies are often penetrated by cables, conduits, ducts, pipes, vents and similar materials, such penetrations that are not part of the tested assembly must be adequately protected and sealed for the assembly to perform as required in resisting the spread of fire. This section, therefore, also mandates that adequate documentation be submitted in order for the building official to verify that the fire-resistance rating of the assembly is not compromised by the penetration. In addition, the materials and methods used to protect joints in or between fire-resistance-rated assemblies must be submitted to determine that the required fire-resistance ratings are maintained. The documentation submitted as evidence that the required fire-resistance ratings are not reduced can be in the form of specific product, assembly or system listings (see the definition of "Listed" in Section 902.1), engineering judgments provided by accredited testing laboratories, product manufacturers that are based on similar tested conditions or, less frequently, data reports for specific fire tests conducted to evaluate a specific penetration method. However, care must be taken to determine that the materials and methods of the proposed penetration or joint protection correspond to the details of the actual penetration or joint protection assembly that was tested. In the case of engineering judgments submitted for through-penetration firestops or fire-resistive joint systems, the publication, "Recommended IFC Guidelines for Evaluating Firestop Systems in Engineering Judgments (EJ's)" can be used as a tool in evaluating the submitted documentation. It is available as a free download from the International Firestop Council (www.firestop.org).

ASTM E 119 or UL 263 states that it is not intended to determine the following:

• Contribution of the tested assembly to the fire hazard;

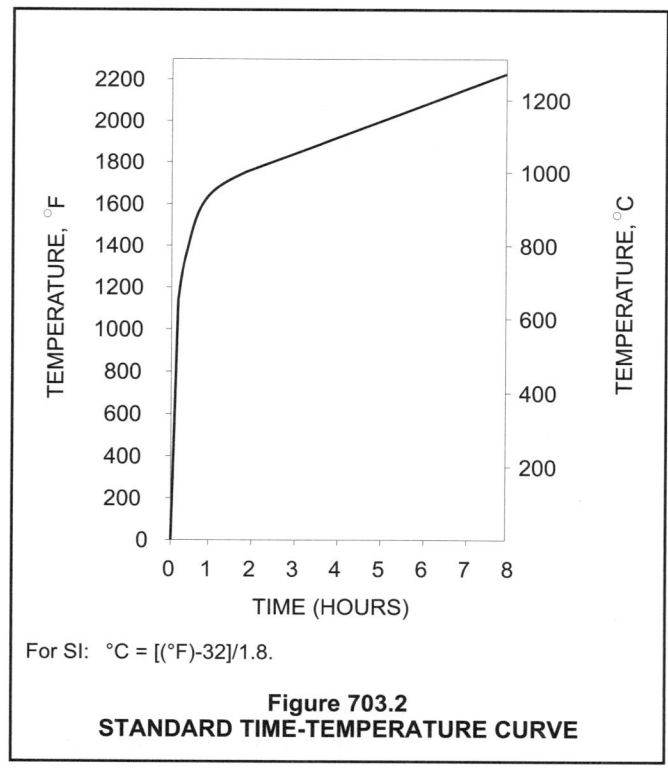

For SI: °C = [(°F)-32]/1.8.

Figure 703.2
STANDARD TIME-TEMPERATURE CURVE

- Measurement of the degree of control or limitation of the passage of smoke or products of combustion;

- Simulation of the fire behavior at joints between building elements, such as floor-to-wall or wall-to-wall connections; or

- Measurement of the flame spread over the surface of the tested element.

Documentation that a component or assembly has been tested to ASTM E 119 or UL 263 is typically in the form of a report issued by an approved, qualified testing agency. The test report contains certain information, including the time periods of fire resistance (e.g., 1 hour, 1$^1/_2$ hours, 2 hours, etc.) and other significant details of the material or assembly. In lieu of an actual copy of a test report, the building official may accept the use of a design number found in a compendium of ASTM E 119 or UL 263 test reports such as, but not limited to, the *Fire Resistance Design Manual* published by the Gypsum Association, the *Fire Resistance Directory* published by Underwriters Laboratories Inc. or *Certification Listings* published by Warnock Hersey. Each publication includes a detailed description of the assembly, its hourly fire-resistance rating and other pertinent details, such as specifications of materials and alternative assembly details.

These publications, detail the design of columns, beams, walls, partitions, and floor/ceiling and roof/ceiling assemblies that have been tested or analyzed in accordance with ASTM E 119 or UL 263. The designs are classified in accordance with their use and fire-resistance rating. Often, wall, partition, and floor/ceiling sound transmission class (STC) ratings. Many floor/ceiling assemblies also include impact insulation class (IIC) ratings. The designs indicate whether an assembly was tested under load-bearing or nonload-bearing conditions, and what structural height limitations there are for nonload-bearing assemblies. Many of the publications have introductory material that provides details on the uses and limitations of the tested assemblies, such as the amount of penetrations permitted in an assembly for electrical outlet boxes.

The exception to the section applies to exterior load-bearing walls that are required to have a fire-resistance rating for structural integrity and not as a fire barrier (e.g., due to fire separation distance in accordance with Table 704). The exception is necessary because ASTM E 119 or UL 263 does not distinguish between structural and fire-containment features of a wall. For example, assume that a concrete double-tee wall assembly was tested in accordance with ASTM E 119 or UL 263. The flanges of the double-tees were 1$^1/_2$ inches (38 mm) thick and the load applied to the wall was 9,600 pounds per lineal foot (14 150 kg/m). Because the temperature of the unexposed surface

rose an average of 250°F (121°C) in 21 minutes, the fire-resistance rating assigned to the assembly was 21 minutes. The fire exposure was continued after the unexposed surface temperature criterion was exceeded and the wall assembly continued to support the applied load for 2 hours, at which time the test was terminated. Thus, the wall should be acceptable where structural performance for 2 hours is required and prevention of fire spread from building to building is not under consideration because of the fire separation distance. The exception, therefore, allows exterior load-bearing walls to be tested after the first failure criterion of ASTM E 119 or UL 263 is reached, as long as the element continued to sustain the applied load, and a higher structural fire-resistance rating is then determined. It should be noted that the hose stream test procedure is to be based on the fire exposure time period related to the structural fire-resistance rating.

703.2.1 Nonsymmetrical wall construction. Interior walls and partitions of nonsymmetrical construction shall be tested with both faces exposed to the furnace, and the assigned *fire-resistance rating* shall be the shortest duration obtained from the two tests conducted in compliance with ASTM E 119 or UL 263. When evidence is furnished to show that the wall was tested with the least fire-resistant side exposed to the furnace, subject to acceptance of the *building official*, the wall need not be subjected to tests from the opposite side (see Section 705.5 for *exterior walls*).

❖ In applying Sections 703.2 and 703.2.1, it is the intent of the code that the required fire-resistance rating of interior walls be based on a fire exposure from either side. For example, it is appropriate to expect the required fire performance of an exit access corridor wall from the corridor side, as well as from the room side. If the corridor becomes involved in a fire, the corridor wall must restrict the spread of fire into adjacent spaces; however, Section 704.5 does allow exterior walls with a fire separation distance of more than 5 feet (1524 mm) to be fire-resistance rated from the interior only.

Section 703.2.1 permits testing from only one side of nonsymmetrical assemblies, provided it can be demonstrated that the tested side results in the lower value (i.e., the nontested side, if tested, would result in a higher value). For example, a 2-inch by 4-inch (51 mm by 102 mm) wood stud wall with two layers of $^5/_8$-inch (15.9 mm) Type X gypsum board on one side and only one layer on the opposite side is an illustration of a "nonsymmetrical" assembly. This assembly would be permitted to be tested only from the side with the single layer of gypsum board. This is because testing from the side with the two layers would result in a higher rating.

703.2.2 Combustible components. Combustible aggregates are permitted in gypsum and portland cement concrete mixtures for fire-resistance-rated construction. Any component material or admixture is permitted in assemblies if the resulting tested assembly meets the fire-resistance test requirements of this code.

❖ The combustibility of materials is regulated by the code when it has been determined that the material or assembly could significantly contribute to a fire hazard; however, the limited use of combustible materials, such as combustible aggregates, is permitted as long as the desired fire-resistance rating can still be obtained. The use of a combustible aggregate in the material or assembly will not significantly contribute to fire growth or severity.

703.2.3 Restrained classification. Fire-resistance-rated assemblies tested under ASTM E 119 or UL 263 shall not be considered to be restrained unless evidence satisfactory to the *building official* is furnished by the *registered design professional* showing that the construction qualifies for a restrained classification in accordance with ASTM E 119 or UL 263. Restrained construction shall be identified on the plans.

❖ The effect on the fire-resistance rating of a structural element that is part of an assembly is addressed, to an extent, through the distinction in ASTM E 119 or UL 263 between "restrained" and "unrestrained" conditions. The standard defines this distinction as follows: "A restrained condition in fire tests, as used in this test method, is one in which expansion of the supports of a load-carrying element resulting from the effects of the fire is resisted by forces external to the element. An unrestrained condition is one in which the load-carrying element is free to expand and rotate at its supports." In evaluating fire-resistance ratings of structural elements, the conditions of support must be consistent with the tested condition as described in the test report.

A restrained classification yields higher fire-resistance ratings than unrestrained; therefore, the code takes the conservative approach, and thus the lesser rating by assuming the in-place conditions to be unrestrained unless structural documentation is provided that supports a restrained condition.

703.3 Alternative methods for determining fire resistance. The application of any of the alternative methods listed in this section shall be based on the fire exposure and acceptance criteria specified in ASTM E 119 or UL 263. The required *fire resistance* of a building element, component or assembly shall be permitted to be established by any of the following methods or procedures:

1. Fire-resistance designs documented in sources.

2. Prescriptive designs of fire-resistance-rated building elements, components or assemblies as prescribed in Section 720.

3. Calculations in accordance with Section 721.

4. Engineering analysis based on a comparison of building element, component or assemblies designs having *fire-resistance ratings* as determined by the test procedures set forth in ASTM E 119 or UL 263.

5. Alternative protection methods as allowed by Section 104.11.

❖ When based on the fire exposure and acceptance criteria of ASTM E 119 or UL 263, an analytical method can be used in lieu of an actual ASTM E 119 or UL 263 fire test to establish the fire-resistance rating of a single component or assembly.

1. The commentary to Section 703.2 also identifies additional resource materials relative to test reports.

2 & 3. Sections 720 and 721 provide further methods of determining fire-resistance ratings (see the commentary to these sections).

For example, methods for determining fire-resistance ratings of concrete or masonry assemblies have been standardized in ACI 216.1/TMS 0216.1 (see commentary, Section 721.1), as well as in the following:

- PCI MNL 124, Precast/Prestressed Concrete Institute.

- *Reinforced Concrete Fire Resistance*, Concrete Reinforcing Steel Institute.

- CMIFC SR267.01B, Concrete and Masonry Industry Fire Safety Committee.

- *A Compilation of Fire Tests on Concrete Masonry Assemblies*, National Concrete Masonry Association.

- NCMA TEK 6A, National Concrete Masonry Association.

- NCMA 35D, National Concrete Masonry Association.

4. An approved engineering analysis allows the fire protection properties of a component or assembly to be determined using information generated from previous fire tests through empirical calculations. There are several analytical methods available to designers for many types of construction materials. An engineering analysis available for calculating fire-resistance ratings of heavy timber construction uses, as a basis of procedure, an empirical mathematical model that generates a conservative design for the minimum dimensions of a wood beam or column. These dimensions account for the changes that occur in the member due to charring of the wood after any given period of fire exposure. AISI's *Designing Fire Protection for Steel Columns* presents the factors that influence the fire-resistance ratings of steel columns in frequently used sizes and shapes. The publication also contains a discussion of the fire protection materials most often used with steel columns and provides accepted methods for calculating the fire-resistance rating of steel columns predicated on the ASTM E119 or UL

263 fire exposure standard. Similarly, AISI's *Designing Fire Protection for Steel Beams* presents the factors that influence the fire-resistance ratings of steel beams in frequently used sizes. As typical test reports specify minimum beam sizes only, three different ways of evaluating beam substitutions are provided. Since the ASTM E 119 or UL 263 test furnace is incapable of testing assemblies that incorporate large structural members, AISI's *Designing Fire Protection for Steel Trusses* proposes concepts for determining fire protection designs by applying existing test data. The application of these methods, as well as information on the fire resistance of archaic construction, are summarized in the *International Code Council® (ICC®) publication, 2001 Guidelines for Determining Fire Resistance of Building Elements*. It is important to emphasize that all analytical methods of calculating fire-resistance ratings must be substantiated as being based on the fire exposure and acceptance criteria of ASTM E 119 or UL 263. Upon submission of such documentation to the building official, the analytical methods mentioned herein, or any other such analytical methods, can be approved as the basis for showing compliance with the fire-resistance ratings required by the code.

5. As indicated in Section 104.11, alternative protection may be provided if approved by the building official. An example of an alternative protection method is the water-filled steel column system. The system consists of hollow, liquid-filled columns that are interconnected with pipe loops to allow the water to circulate freely. Engineering studies confirmed by tests have shown that critical temperatures are not reached in the steel columns as long as they remain filled with liquid.

703.4 Noncombustibility tests. The tests indicated in Sections 703.4.1 and 703.4.2 shall serve as criteria for acceptance of building materials as set forth in Sections 602.2, 602.3 and 602.4 in Type I, II, III and IV construction. The term "noncombustible" does not apply to the flame spread characteristics of *interior finish* or *trim* materials. A material shall not be classified as a noncombustible building construction material if it is subject to an increase in combustibility or flame spread beyond the limitations herein established through the effects of age, moisture or other atmospheric conditions.

❖ The restrictions on the use of combustible materials are primarily found in Chapter 6; however, certain sections of Chapter 7 (such as use of combustibles in Section 703.2.2 and fire walls in Section 706.3) contain restrictions on the use of combustible materials or separate provisions, depending on whether the material is combustible or noncombustible. Sections 703.4.1 and 703.4.2 contain the appropriate test criteria by which a material is to be evaluated to ascertain whether it is combustible or noncombustible.

Materials that are considered noncombustible must be capable of maintaining the required performance characteristics (noncombustibility and flame spread ratings) regardless of age, moisture or other atmospheric conditions. If exposure to atmospheric conditions results in an increase in combustibility or flame spread rating beyond the limitations specified, the material is considered combustible.

The criteria established by this section are primarily based on the National Board of Fire Underwriters' (NBFU) *National Building Code*, 1955 edition, which permitted a noncombustible material to contain a limited amount of combustible material, provided that it did not contribute to fire propagation. The requirement for noncombustibility is considered to apply to certain materials used for walls, roofs and other structural elements, but not to surface finish materials. The determination of whether a material is noncombustible from the standpoint of minimum required clearances to heating appliances, flues or other sources of high temperature is based on the ASTM E 136 definition of "Noncombustibility," not the prescriptive definition in Section 703.4.2. Note that the *International Mechanical Code® (IMC®)* contains a definition of "Noncombustible materials," which does not contain provisions for the use of composite materials (defined in Section 703.4.2) as acceptable noncombustible materials.

703.4.1 Elementary materials. Materials required to be noncombustible shall be tested in accordance with ASTM E 136.

❖ Materials intended to be classified as noncombustible are to be tested in accordance with ASTM E 136. In accordance with Section 8 of ASTM E 136, such materials are acceptable as noncombustible when at least three of four specimens tested conform to all of the following criteria:

1. The recorded temperature of the surface and interior thermocouples shall not at any time during the test rise more than 54°F (30°C) above the furnace temperature at the beginning of the test.

2. There shall not be flaming from the specimen after the first 30 seconds.

3. If the weight loss of the specimen during testing exceeds 50 percent, the recorded temperature of the surface and interior thermocouples shall not at any time during the test rise above the furnace air temperature at the beginning of the test, and there shall not be flaming of the specimen.

The use of the test standard is limited to elementary materials and excludes laminated and coated materials because of the uncertainties associated with more complex materials and with products that cannot be tested in a realistic configuration. The test standard is also limited to solid materials and does not measure the self-heating tendencies of large masses of materials, such as resin-impregnated mineral fiber insulation.

The defined furnace temperature in the standard, 1,382°F (750°C), is representative of temperatures that are known to exist during building fires, although temperatures between 1,800°F (982°C) and 2,200°F (1204°C) are frequently achieved during intense fires. For most building materials, however, complete burning of the combustible fraction will occur as readily at 1,382°F (750°C) as compared to higher temperatures. The criterion requiring four specimens to be tested recognizes the variable nature of the measurements and the fact that there are difficulties in observing the presence and duration of flaming.

The need to measure and limit the duration of flaming and the rise in temperature recognizes that a brief period of flaming and a small amount of heating are not considered serious limitations on the use of building materials. Test results have shown that such criteria limit the combustible portion of noncombustible materials to a maximum of 3 percent. The 50-percent weight-loss limitation precludes the possibility that combustion of low-density materials will occur so rapidly that the recorded temperature rise and the measured flaming duration will be less than the prescribed limitations. The 50-percent limitation is considered appropriate for materials that contain appreciable quantities of combined water or gaseous components.

703.4.2 Composite materials. Materials having a structural base of noncombustible material as determined in accordance with Section 703.4.1 with a surfacing not more than 0.125 inch (3.18 mm) thick that has a *flame spread index* not greater than 50 when tested in accordance with ASTM E 84 or UL 723 shall be acceptable as noncombustible materials.

❖ In recognition that an essentially noncombustible material with a thin combustible coating will not contribute appreciably to an ambient fire, this section provides criteria by which a composite material may be determined acceptable as a noncombustible material. The structural base of the composite material must meet the criteria for elementary materials (see Section 703.4.1). The material may have a surface not more than $^1/_8$-inch (3.2 mm) thick applied to the noncombustible base. This surface is required to have a flame spread rating no greater than 50 when tested in accordance with ASTM E 84 or UL 723. For a discussion of both the ASTM E 84 and UL 723 test methods, refer to the commentary to Section 803.1.

In accordance with this section, material such as gypsum board—which consists of a noncombustible base and a combustible (paper) surface—is considered to be a composite material. As noted in the commentary to Section 703.4, a composite material is considered a noncombustible material in the context of the code. The IMC, however, intentionally includes only limited applications of the criteria of this section in the definition of a noncombustible material. Therefore, composite materials are not considered acceptable noncombustible materials from the standpoint of clearances to heating appliances, flues or other sources of high temperature because of their potentially poor response to radiant heat exposure.

703.5 Fire-resistance-rated glazing. Fire-resistance-rated glazing, when tested in accordance with ASTM E 119 or UL 263 and complying with the requirements of Section 707, shall be permitted. Fire-resistance-rated glazing shall bear a *label* or other identification showing the name of the manufacturer, the test standard and the identifier "W-XXX," where the "XXX" is the *fire-resistance rating* in minutes. Such *label* or identification shall be issued by an agency and shall be permanently affixed to the glazing.

❖ Fire-resistance-rated glazing is allowed in three separate locations, depending on which tests the glazing has passed. The label is required to provide the location and the fire-resistance rating in minutes. There are door labels, window labels and wall labels. This section indicates that fire-resistance-rated glazing bearing the "W" label has passed ASTM E 119 or UL 263 and is considered a wall and, therefore, is not subject to the limitations on doors and windows. This section also indicates that fire-resistance-rated glazing with the "W" designation on the label is good for fire barriers, as well as shaft walls and stair enclosures, and, of course, it would be used for a fire partition, smoke partition or smoke barrier also. It could not be used for a fire wall.

Fire-resistance-rated glazing is also mentioned in Section 715.2 for doors and windows, Section 715.4.7.3.1 for identification of doors, and Section 715.5.9.1 for identification of windows.

703.6 Marking and identification. *Fire walls*, *fire barriers*, *fire partitions*, *smoke barriers* and smoke partitions or any other wall required to have protected openings or penetrations shall be effectively and permanently identified with signs or stenciling. Such identification shall:

1. Be located in accessible concealed floor, floor-ceiling or *attic* spaces;

2. Be repeated at intervals not exceeding 30 feet (914 mm) measured horizontally along the wall or partition; and

3. Include lettering not less than 0.5 inch (12.7 mm) in height, incorporating the suggested wording: "FIRE AND/OR SMOKE BARRIER—PROTECT ALL OPENINGS," or other wording.

> **Exception:** Walls in Group R-2 occupancies that do not have a removable decorative ceiling allowing access to the concealed space.

❖ The concern addressed by this code requirement is the need for installed fire-resistance-rated assemblies to maintain their fire resistance over the life of the building. This identification will allow tradespeople, craftsmen, installers, maintenance workers or inspectors to know that the wall is fire-resistance rated and openings or penetrations of it must be protected.

SECTION 704
FIRE-RESISTANCE RATING OF
STRUCTURAL MEMBERS

704.1 Requirements. The *fire-resistance ratings* of structural members and assemblies shall comply with this section and the requirements for the type of construction as specified in Table

601. The *fire-resistance ratings* shall not be less than the ratings required for the fire-resistance-rated assemblies supported by the structural members.

> **Exception:** *Fire barriers*, *fire partitions*, *smoke barriers* and *horizontal assemblies* as provided in Sections 707.5, 709.4, 710.4 and 712.4, respectively.

❖ This section contains provisions that apply to structural members that are required to have a fire-resistance rating. The required rating of structural members is usually based on the type of construction and Table 601. Other sections of the code, such as the continuity provisions of Sections 707.5, 709.4, 710.4 and 711.4, may require structural members to have a fire-resistance rating because they support fire-resistance-rated construction. The minimum required fire-resistance rating for the structural member is, therefore, the more restrictive of the above two criteria.

There are instances in which the code does not require the supporting structural members to have a fire-resistance rating. Examples of these include the supporting construction for tenant and sleeping unit separations, exit access corridors and smoke partitions in Type IIB, IIIB, and VB construction (see Sections 709.4 and 710.4).

Protection of structural members differs for various members. Required column fire-resistance protection must be afforded in accordance with Section 704.2. Certain primary structural frames (see definition in Section 202) must be protected by individual encasement in accordance with Section 704.3. Secondary structural members (see definition in Section 202) may be protected by membrane protection in accordance with Section 704.4. Special requirements apply to fire-resistance protection of trusses (see Section 704.5).

Section 704 addresses issues such as protection of attachments to structural members (see Section 704.6), embedment of service facilities (see Section 704.8), impact protection (see Section 704.9), exterior structural members (see Section 704.10), bottom flange protection (see Section 704.11), seismic isolation systems (see Section 704.12) and sprayed fire-resistant materials (SFRM) (see Section 704.13).

704.2 Column protection. Where columns are required to be fire-resistance rated, the entire column shall be provided individual encasement protection by protecting it on all sides for the full column length, including connections to other structural members, with materials having the required fire-resistance rating. Where the column extends through a ceiling, the encasement protection shall be continuous from the top of the foundation or floor/ceiling assembly below through the ceiling space to the top of the column.

❖ Columns required to be fire-resistance rated must be encased on all four sides. This is true even if the column is located on the exterior of the structure. The ceiling membrane or ceiling protection of a floor/ceiling or roof/ceiling fire-resistance-rated assembly is prohibited from being considered as the protection for columns required to have a fire-resistance rating; therefore, the materials that encase or provide individual protection must continue through the concealed space above a ceiling, even if the ceiling membrane is part of a fire-resistance-rated assembly (see Figure 704.2) Connections of structural members to columns must also be individually protected (see Section 704.6).

704.3 Protection of the primary structural frame other than columns. Members of the primary structural frame other than columns that are required to have a *fire-resistance rating* and support more than two floors or one floor and roof, or support a *load-bearing wall* or a nonload-bearing wall more than

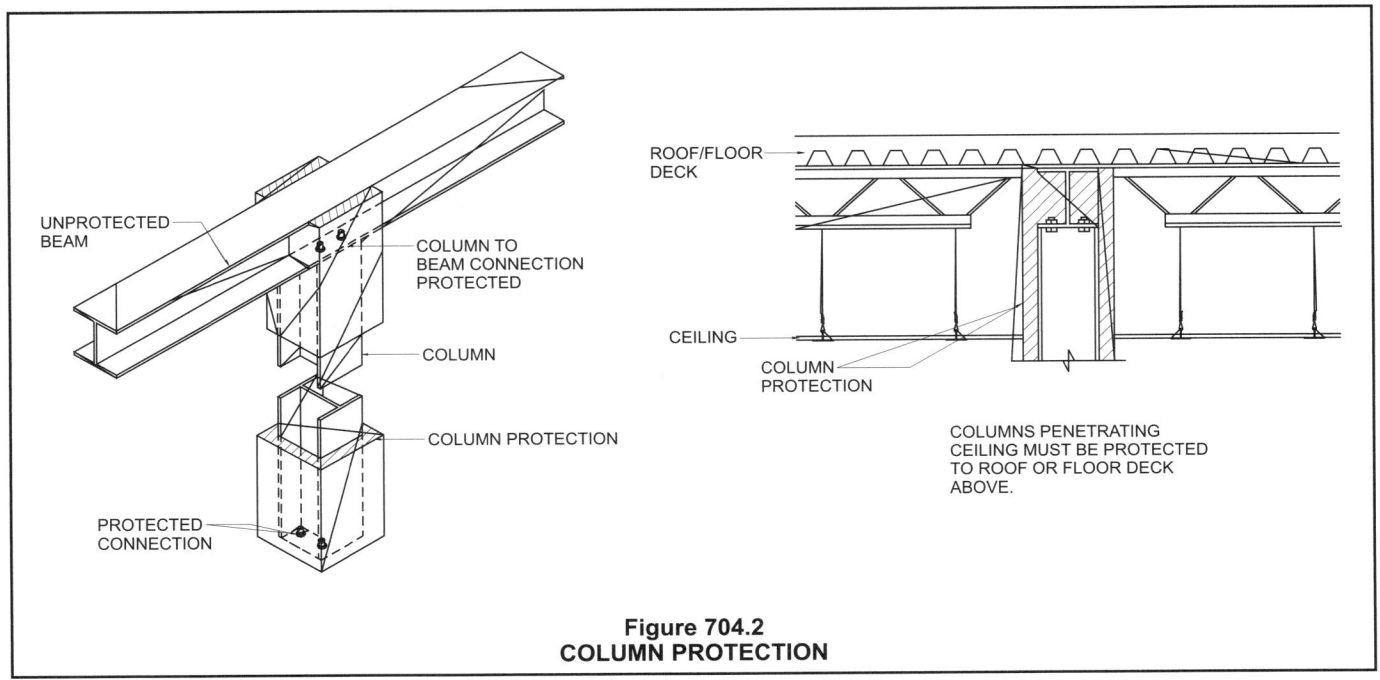

Figure 704.2
COLUMN PROTECTION

two *stories* high, shall be provided individual encasement protection by protecting them on all sides for their full length, including connections to other structural members, with materials having the required *fire-resistance* rating.

> **Exception:** Individual encasement protection on all sides shall be permitted on all exposed sides provided the extent of protection is in accordance with the required *fire-resistance rating*, as determined in Section 703.

❖ All primary structural frames do not need to comply with this section. Only those primary structural frame members that are required to have a fire-resistance rating and support more than two floors, more than one floor and a roof, a bearing wall of any height or a nonload-bearing wall more than two stories high are required to be protected by individual encasement. Individual encasement, though not defined, is attained by using tested design assemblies that provide gypsum board applied directly to the member or to studs that are directly attached to all sides of the member, mastic and intumescent coatings, and sprayed fire-resistant materials.

Individual encasement is required for the primary structural frame members receiving tributary loads from multiple levels in order to reduce the risk associated with catastrophic failure. The risk represented by structural collapse during a fire is significantly increased in multiple-story buildings, since the occupants require additional time to egress and a single structural element that supports multiple elements requires more conservative protection methods.

Though the text states that the encasement must be on all four sides of the structural member, the exception will allow tested assemblies that have the encasement on only the exposed sides of a member. Figure 704.3 is an example of individual encasement protection.

704.4 Protection of secondary members. Secondary members that are required to have a *fire-resistance rating* shall be protected by individual encasement protection, by the membrane or ceiling of a *horizontal assembly* in accordance with Section 712, or by a combination of both.

❖ Where fire-resistance-rated structural members are required, membrane or ceiling fire-resistance protection is only permitted for structural members (primary or secondary) that support two floors or less, only one floor and a roof, or a nonload-bearing wall two stories or less in height. Please note that ceiling membrane protection is never allowed for column protection (see Section 704.2). Figure 704.4 is an example of membrane protection.

Remember that the fire-resistance requirements are established based upon a fire test in accordance with ASTM E 119 or UL 263. Although the fire temperature in the furnace is the same, the test protocol in both standards has different temperature rise limits for the elements and different temperature couple locations between a structural member test and a wall or horizontal assembly test. Therefore, although a wall may obtain a 1-hour fire-resistance rating, it does not mean that a column placed within that wall would also obtain a 1-hour fire-resistance rating. Also a floor/ceiling assembly rated for 2 hours cannot be assumed to protect a beam that was not in the tested assembly. Additionally, if the beam was in the tested assembly, the fire-resistance rating of the floor/ceiling may be different than the beam. It is important that the structural members located within the assemblies were also in the tested floor/ceiling and roof/ceiling assemblies.

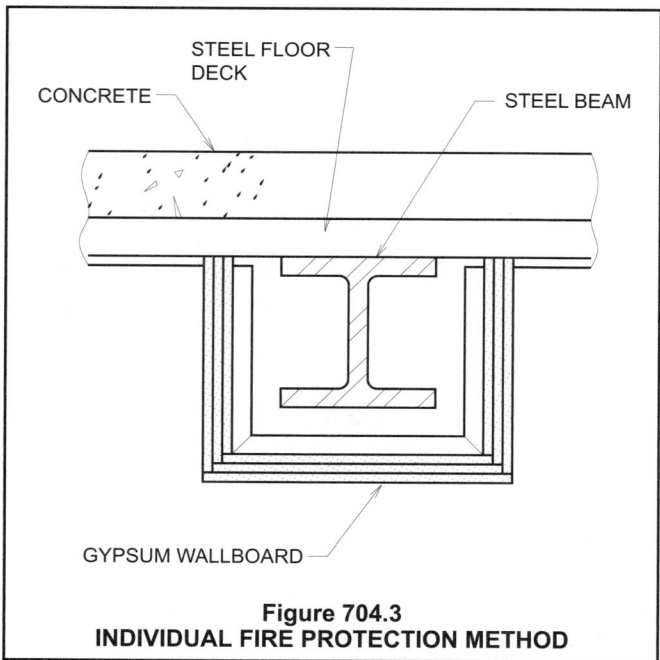

Figure 704.3
INDIVIDUAL FIRE PROTECTION METHOD

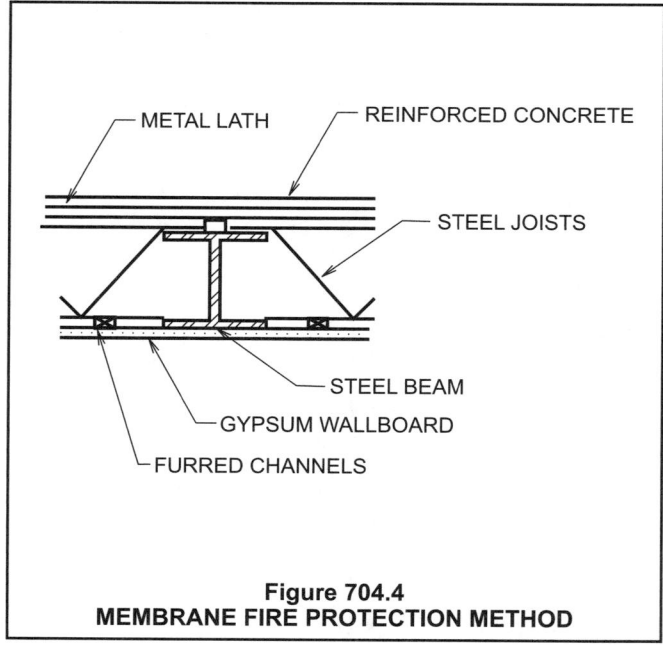

Figure 704.4
MEMBRANE FIRE PROTECTION METHOD

704.4.1 Light-frame construction. King studs and boundary elements that are integral elements in *load-bearing walls* of light-frame construction shall be permitted to have required *fire-resistance ratings* provided by the membrane protection provided for the *load-bearing wall.*

❖ Historically, codes have considered king and jack studs in light-frame construction as standard parts of the wall assembly. King studs have essentially the same function, load ratio and thermal properties as the other studs in the load-bearing wall, and there is no need for them to be considered separate distinct column elements. In seismic design, boundary elements for light-frame steel construction are required to be designed for amplified seismic forces. Rather than placing multiple studs, designers frequently specify larger members, which still serve the function of studs. These boundary elements are not always at openings.

704.5 Truss protection. The required thickness and construction of fire-resistance-rated assemblies enclosing trusses shall be based on the results of full-scale tests or combinations of tests on truss components or on *approved* calculations based on such tests that satisfactorily demonstrate that the assembly has the required *fire resistance.*

❖ Trusses can be of an almost infinite number of configurations and sizes. By nature, they are typically designed and constructed of different structural components. This leads to the impracticality of such trusses being tested. The code acknowledges this by allowing full-scale tests, tests on the individual components or some form of calculations that demonstrate that the structural elements that form the truss are provided with the requisite protection to attain the fire-resistance rating.

704.6 Attachments to structural members. The edges of lugs, brackets, rivets and bolt heads attached to structural members shall be permitted to extend to within 1 inch (25 mm) of the surface of the fire protection.

❖ This section is intended to limit the transfer of heat to the structural element from the connection element by requiring the connection element to be protected with at least 1 inch (25 mm) of some form of fire-resistance protection. Attaching elements are not allowed to transfer heat through the protection and to the structural element. At least 1 inch (25 mm) of cover is required on all the attaching elements.

704.7 Reinforcing. Thickness of protection for concrete or masonry reinforcement shall be measured to the outside of the reinforcement except that stirrups and spiral reinforcement ties are permitted to project not more than 0.5-inch (12.7 mm) into the protection.

❖ Since load-bearing assemblies must sustain the superimposed load while being subjected to standard time-temperature fire conditions, concrete, masonry and reinforcing steel must be able to provide the required strength at elevated temperatures. This involves providing adequate cover over the steel reinforcement so that the stress induced in the reinforcement is less than its yield stress, which is commonly referred to as "yield strength." Tests of steel reinforcing bars show that at a temperature of approximately 1,100°F (593°C), the yield strength of steel is reduced to approximately 50 percent of that at ambient temperature conditions. Similar tests of prestressing tendons show that at approximately 800°F (427°C) the tensile strength is approximately 50 percent of that at ambient conditions. Cover requirements, therefore, are typically based on limiting the reinforcing and prestressing steel temperatures to 1,100 and 800°F (593 and 427°C), respectively. This section allows for stirrups and trusses to project into the required cover by a minimal distance, acknowledging the practicality of construction tolerances while providing adequate cover to the main reinforcing steel, which is located inside of the stirrups and truss. This also acknowledges that the stirrups are not continuous for the length of the member and are spaced at intermittent intervals. Therefore, they only reduce the actual cover to the main reinforcing steel at a percentage of its length. The 1/2-inch (12.7 mm) dimension reflects a No. 4 reinforcing bar.

704.8 Embedments and enclosures. Pipes, wires, conduits, ducts or other service facilities shall not be embedded in the required fire protective covering of a structural member that is required to be individually encased.

❖ Piping, wires, conduits, ducts and other service facilities are not to be embedded in the fire protective covering of a structural member that is required to be individually encased in accordance with Section 704.3. The fire protection performance of encasement materials is critical to achieve the required fire-resistance rating, and would be impaired if the continuity of the encasement is interrupted. This does not, however, prevent the installation if these items are installed in hollow spaces that may be created by the encasement method. When items are installed in such hollow spaces, penetrations of the protective assembly must be properly protected so that the fire-resistance rating is maintained.

704.9 Impact protection. Where the fire protective covering of a structural member is subject to impact damage from moving vehicles, the handling of merchandise or other activity, the fire protective covering shall be protected by corner guards or by a substantial jacket of metal or other noncombustible material to a height adequate to provide full protection, but not less than 5 feet (1524 mm) from the finished floor.

Exception: Corner protection is not required on concrete columns in open or enclosed parking garages.

❖ To aid in the reliability of the fire protective covering, protection must be provided if the covering is subject to impact from moving vehicles, handling of merchandise or other activities. The impact protection is to be provided to at least 5 feet (1524 mm) above the finished floor or as needed to prevent damage.

The type of protection required is not specified in the code, but should be determined based upon the type of fire protection used as well as the impact that the protection may be expected to encounter. For example, if the fire-resistance protection of a column is provided by either spray-on fireproofing or gypsum board, that would be more likely to be damaged than a concrete-encased column. Additionally, if the column was in a grocery store and the only expected item to impact the column was a shopping cart, then it may be determined to be acceptable to simply cover the gypsum board with a piece of thin sheet metal. On the other hand, if the column is located in a warehouse or parking garage where a vehicle, such as a forklift or automobile, could strike it, then the protection may require the installation of steel plates or bollards.

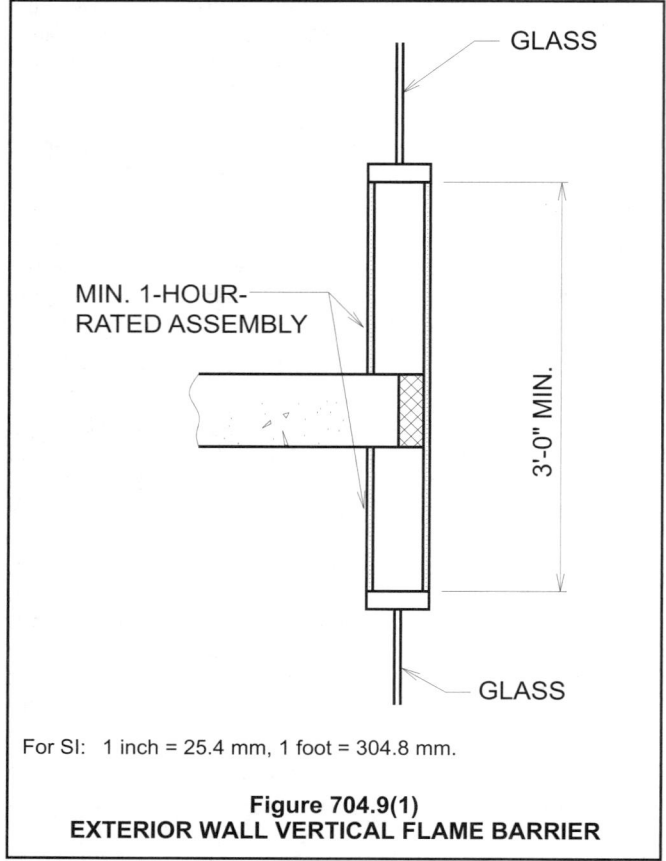

For SI: 1 inch = 25.4 mm, 1 foot = 304.8 mm.

Figure 704.9(1)
EXTERIOR WALL VERTICAL FLAME BARRIER

704.10 Exterior structural members. Load-bearing structural members located within the *exterior walls* or on the outside of a building or structure shall be provided with the highest *fire-resistance rating* as determined in accordance with the following:

1. As required by Table 601 for the type of building element based on the type of construction of the building;

2. As required by Table 601 for exterior bearing walls based on the type of construction; and

3. As required by Table 602 for *exterior walls* based on the *fire separation distance*.

❖ Exterior load-bearing structural members, such as columns or girders, must have the same fire-resistance rating as required for exterior load-bearing walls. As such, the required fire-resistance rating is the higher rating of that found in Table 601 for type of construction for structural elements or bearing walls, or as required in Table 602 based upon the separation distance.

704.11 Bottom flange protection. Fire protection is not required at the bottom flange of lintels, shelf angles and plates, spanning not more than 6 feet (1829 mm) whether part of the primary structural frame or not, and from the bottom flange of lintels, shelf angles and plates not part of the primary structural frame, regardless of span.

❖ Structural frame elements located over an opening in a wall required to be fire-resistance rated, particularly a masonry wall with an exposed steel lintel or angle over the wall opening, are covered by this section. The bottom flanges of such lintels or angles if less than 6 feet (1829 mm) in length are not required to be protected even if part of the primary structural frame, regardless of the span.

704.12 Seismic isolation systems. Fire-resistance ratings for the isolation system shall meet the *fire-resistance rating* required for the columns, walls or other structural elements in which the isolation system is installed in accordance with Table 601. Isolation systems required to have a *fire-resistance rating* shall be protected with *approved* materials or construction assemblies designed to provide the same degree of *fire resistance* as the structural element in which it is installed when tested in accordance with ASTM E 119 or UL 263 (see Section 703.2).

Such isolation system protection applied to isolator units shall be capable of retarding the transfer of heat to the isolator unit in such a manner that the required gravity load-carrying capacity of the isolator unit will not be impaired after exposure to the standard time-temperature curve fire test prescribed in ASTM E 119 or UL 263 for a duration not less than that required for the *fire-resistance rating* of the structure element in which it is installed.

Such isolation system protection applied to isolator units shall be suitably designed and securely installed so as not to dislodge, loosen, sustain damage or otherwise impair its ability to accommodate the seismic movements for which the isolator unit is designed and to maintain its integrity for the purpose of providing the required fire-resistance protection.

❖ This section states that the fire-resistance requirements for seismic isolation systems are the same as other structural elements required in Table 601. Additionally, the fire-resistance method used must be able to withstand the anticipated movement of the system without the fire-resistance application methods and systems affecting its functionality during a seismic event. Elements of the system can either be individually protected or contained within a tested assembly. Another key element of this section is that, when tested under exposure to the standard time-temperature curve test, the structural elements are still able to carry the gravity load as designed.

704.13 Sprayed fire-resistant materials (SFRM). Sprayed fire-resistant materials (SFRM) shall comply with Sections 704.13.1 through 704.13.5.

❖ The intent of this section is to increase the in-place durability of spray-applied fire-resistant materials (SFRM). The National Institute of Standards and Technology's (NIST) investigation on the World Trade Center (WTC) tragedy documented that the proximate cause of the actual collapse was the action of a building contents fire on light steel members in the absence of spray-applied fire-resistant material, which had been dislodged. Events far less dramatic than an airplane attack have been known to dislodge SFRM including elevator movement, building sway or maintenance activities, if is not adhered properly. These code requirements, as well as the special testing requirements in Section 1704.12, are necessary to increase the in-place capability of SRFM assemblies. Section 403.2.4 gives the bond strength requirements for SFRM.

704.13.1 Fire-resistance rating. The application of SFRM shall be consistent with the *fire-resistance rating* and the listing, including, but not limited to, minimum thickness and dry density of the applied SFRM, method of application, substrate surface conditions and the use of bonding adhesives, sealants, reinforcing or other materials.

❖ In order that the fire-resistance rating that is actually realized by sprayed fire-resistant materials equals or exceeds the required fire-resistance rating, it is necessary to install the materials in accordance with their listing. In this section, minimum thicknesses, proper conditions of the substrate, method of application and correct bonding materials are all items that are necessary to ensure the proper performance of these materials.

704.13.2 Manufacturer's installation instructions. The application of SFRM shall be in accordance with the manufacturer's installation instructions. The instructions shall include, but are not limited to, substrate temperatures and surface conditions and SFRM handling, storage, mixing, conveyance, method of application, curing and ventilation.

❖ Sections 704.13.1 and 704.13.2 require that the application of SFRM be in accordance with all terms and conditions of the listing and the manufacturer's instructions.

704.13.3 Substrate condition. The SFRM shall be applied to a substrate in compliance with Sections 704.13.3.1 through 704.13.3.2.

❖ The in-place adhesion of SFRM can be reduced by a factor of 10 when applied over certain primers when compared to the adhesion obtained by the rated material applied on bare clean steel. This section gives the requirements for the surface condition of the substrate receiving the SFRM.

704.13.3.1 Surface conditions. Substrates to receive SFRM shall be free of dirt, oil, grease, release agents, loose scale and any other condition that prevents adhesion. The substrates shall also be free of primers, paints and encapsulants other than those fire tested and *listed* by a nationally recognized testing agency. Primed, painted or encapsulated steel shall be allowed, provided that testing has demonstrated that required adhesion is maintained.

❖ The condition of the substrate that SFRM is applied to is critical in attaining proper adhesion of the SFRM. Events as simple as elevator movement, building sway or maintenance activities can dislodge SFRM if it is not adhered properly.

As mentioned above, studies have shown that the in-place adhesion of SFRM can be reduced by a factor of 10 when applied over certain primers when compared to the adhesion obtained by the rated material applied on bare clean steel.

704.13.3.2 Primers, paints and encapsulants. Where the SFRM is to be applied over primers, paints or encapsulants other than those specified in the listing, the material shall be field tested in accordance with ASTM E 736. Where testing of the SFRM with primers, paints or encapsulants demonstrates that required adhesion is maintained, SFRM shall be permitted to be applied to primed, painted or encapsulated wide flange steel shapes in accordance with the following conditions:

1. The beam flange width does not exceed 12 inches (305 mm); or

2. The column flange width does not exceed 16 inches (400 mm); or

3. The beam or column web depth does not exceed 16 inches (400 mm).

4. The average and minimum bond strength values shall be determined based on a minimum of five bond tests conducted in accordance with ASTM E 736. Bond tests conducted in accordance with ASTM E 736 shall indicate a minimum average bond strength of 80 percent and a minimum individual bond strength of 50 percent, when compared to the bond strength of the SFRM as applied to clean uncoated $^1/_8$-inch-thick (3-mm) steel plate.

❖ When there is some sort of material, such as paint or primer, on the steel structural members where it is desired to install SFRM, it is still possible to apply the SFRM to the steel without removing the material. However, this section specifies a field test in accordance with ASTM E 736 be done on the SFRM so applied. This ensures that a bond strength is achieved and that the material will remain in place. Note that there are limitations on the sizes of structural members that can utilize this field test (Conditions 1-3).

704.13.4 Temperature. A minimum ambient and substrate temperature of 40°F (4.44°C) shall be maintained during and for a minimum of 24 hours after the application of the SFRM, unless the manufacturer's installation instructions allow otherwise.

❖ This section gives the minimum temperature for the air and the material itself, which must be maintained during application and for 24 hours after, unless the manufacturer's instructions allow otherwise.

704.13.5 Finished condition. The finished condition of SFRM applied to structural members or assemblies shall not, upon complete drying or curing, exhibit cracks, voids, spalls, delamination or any exposure of the substrate. Surface irregularities of SFRM shall be deemed acceptable.

❖ This section and the special inspections required by Section 1704.12 establish the acceptance criteria for the finished condition of the SFRM.

SECTION 705
EXTERIOR WALLS

705.1 General. *Exterior walls* shall comply with this section.

❖ In order to accomplish these functions, this section regulates the fire protection capability of the exterior wall, including limitations on openings in the walls. The requirements are based on the fire separation distance, the occupancy that represents a relative fuel loading and the percentage of openings in the wall. Section 705.8.5 also regulates the location of openings in order to prevent ignition of materials in adjacent stories.

All other applicable provisions of the code concerning exterior walls still apply. The provisions of this section refer to the fire-resistance requirements contained in Tables 601 and 602 (see Section 704.5). Table 601 establishes minimum fire-resistance ratings for exterior load-bearing walls based on the type of construction that addresses the structural integrity of the building under fire conditions. Table 602 establishes ratings based on fire separation distance in an effort to address conflagration concerns.

705.2 Projections. Cornices, eave overhangs, exterior balconies and similar projections extending beyond the *exterior wall* shall conform to the requirements of this section and Section 1406. Exterior egress balconies and *exterior exit stairways* shall also comply with Sections 1019 and 1026, respectively. Projections shall not extend beyond the distance determined by the following three methods, whichever results in the lesser projection:

1. A point one-third the distance from the exterior face of the wall to the *lot line* where protected openings or a combination of protected and unprotected openings are required in the *exterior wall*.

2. A point one-half the distance from the exterior face of the wall to the *lot line* where all openings in the *exterior wall* are permitted to be unprotected or the building is equipped throughout with an *automatic sprinkler system* installed under the provisions of Section 705.8.2.

3. More than 12 inches (305 mm) into areas where openings are prohibited.

Buildings on the same lot and considered as portions of one building in accordance with Section 705.3 are not required to comply with this section.

❖ Horizontal projections, similar to walls and openings, are a potential source of the spread of fire from build-

ing to building. Some projections from the building exterior wall may even shelter combustible materials, thus limits on the allowable projections are necessary.

The projection is measured from the actual exterior wall and is limited to the lesser distance determined by one of three methods. The combining of these three methods is a result of the legacy codes having different methods of determining the allowed projections. Fortunately, only one of the three methods can result in the least projection and, therefore, the other two will not govern.

- All methods. No projections are allowed from exterior walls less than 3 feet (914 mm to 9144 mm) from the property line.

- Method 1. For nonsprinklered buildings with exterior walls from 3 feet to 30 feet (914 mm) from the lot line (actual or assumed in accordance with Section 705.3), Method 1 always governs. For sprinklered buildings with exterior walls from 3 feet to 20 feet (914 mm to 6096 mm) from the lot line, Method 1 applies again, since it results in the lesser projection.

- Method 2. For nonsprinklered buildings with exterior walls over 30 feet (9144 mm) from the lot line, Method 2 applies, since Method 1 is not applicable because protected or unprotected openings are not required. Unprotected openings are unlimited. For sprinklered buildings with exterior walls over 20 feet (6096 mm) from the lot line, Method 2 applies, since unprotected openings are unlimited.

- Method 3 never governs.

See Figure 705.2 for a graphical representation of the above points.

For buildings on the same property that can meet the code as one building in accordance with the exception to Section 705.2, the projections are not limited.

705.2.1 Type I and II construction. Projections from walls of Type I or II construction shall be of noncombustible materials or combustible materials as allowed by Sections 1406.3 and 1406.4.

❖ The code requires projections from walls of Type I and II construction to be noncombustible. An exception for combustible materials is for balconies (see Section 1406.3) and bay windows (see Section 1406.4). Buildings of Type I through IV construction require the exterior walls to be noncombustible; however, the code places a higher restriction on projections on the two highest types of construction, while Section 705.2.2 allows combustible projections on buildings of Type III and IV construction. This approach reflects the general requirements that all elements of Type I and II buildings be noncombustible, while those of Type III and IV are permitted to be constructed of combustible interior materials.

705.2.2 Type III, IV or V construction. Projections from walls of Type III, IV or V construction shall be of any *approved* material.

❖ When combustible materials are used, the provisions of Section 705.2.3 must be followed (see also Section 1406).

705.2.3 Combustible projections. Combustible projections located where openings are not permitted or where protection of openings is required shall be of at least 1-hour fire-resistance-rated construction, Type IV construction, *fire-retardant-treated wood* or as required by Section 1406.3.

Exception: Type V construction shall be allowed for R-3 occupancies.

❖ This section states the requirements for protecting combustible projections and does not state when combustible projections are allowed. That is covered in Sections 705.2.1 and 705.2.2. There are two locations where protection is required for combustible projections.

- One is a combustible projection that extends closer than 3 feet (914 mm) from the lot line. That is the point at which Table 705.8 prohibits

openings. The only scenario that would require this would be an exterior wall between 3 feet and 4.5 feet (914 mm and 1372 mm) from the lot line, with a projection as close as 2 feet (610 mm) to the property line. That projection, if allowed to be combustible, would have to be protected in accordance with this section.

- The other is a combustible projection that extends closer to the lot line than the plane at which protected openings are required. That plane is at 5 feet (1524 mm) for nonsprinklered buildings. Table 705.8 requires protected openings for nonsprinklered buildings when the exterior wall is less than 5 feet (1524 mm) from the property line. Table 705.8 never requires protected openings for a sprinklered building.

The bottom line is that for nonsprinklered buildings combustible projections that extend to within 5 feet (1524 mm) of the lot line must be protected as required in this section. For sprinklered buildings, only projections that extend past a plane 3 feet (914 mm) from the lot line need to be protected in accordance with this section.

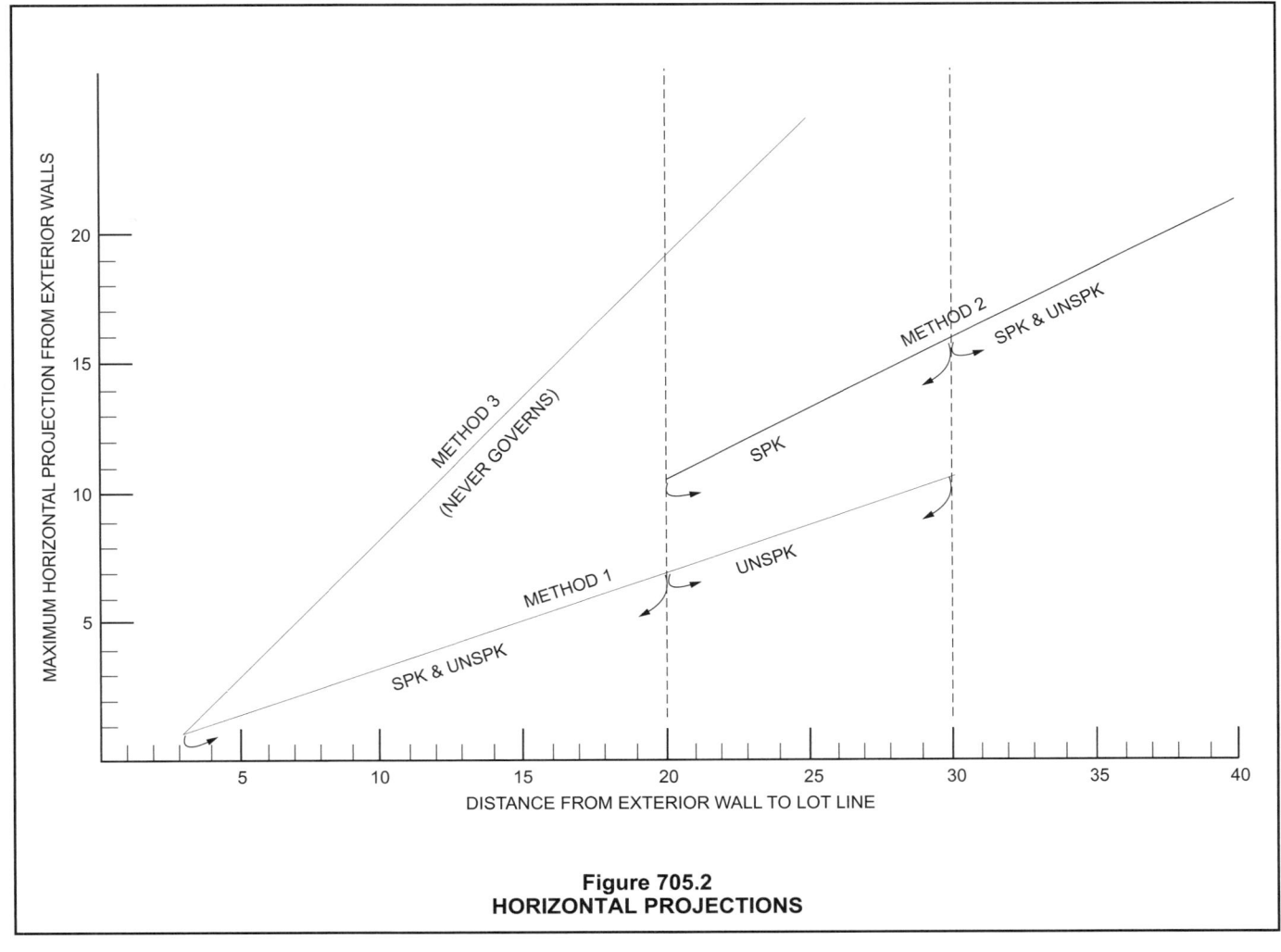

Figure 705.2
HORIZONTAL PROJECTIONS

Since the maximum horizontal projection from the exterior wall is one-third the distance from the exterior face of the exterior wall to the lot line (see commentary, Section 705.2), an exterior wall over 7.5 feet (2286 mm) from the lot line can never have a projection that extends to within 5 feet (1524 mm) of the lot line. This answers the obvious question should the entire projection be protected or just the portion of the projection within the 5 feet (1524 mm). From a practical standpoint, the entire projection must be protected.

705.3 Buildings on the same lot. For the purposes of determining the required wall and opening protection and roof-covering requirements, buildings on the same lot shall be assumed to have an imaginary line between them.

Where a new building is to be erected on the same lot as an existing building, the location of the assumed imaginary line with relation to the existing building shall be such that the *exterior wall* and opening protection of the existing building meet the criteria as set forth in Sections 705.5 and 705.8.

Exception: Two or more buildings on the same lot shall either be regulated as separate buildings or shall be considered as portions of one building if the aggregate area of such buildings is within the limits specified in Chapter 5 for a single building. Where the buildings contain different occupancy groups or are of different types of construction, the area shall be that allowed for the most restrictive occupancy or construction.

❖ This section addresses buildings on the same lot and requires that an imaginary lot line be established between buildings in order to determine exterior wall fire ratings and opening protectives (see the definition of "Fire separation distance"). This section takes the approach of limiting the conflagration hazard between buildings on the same property.

The exception permits two buildings on the same lot to be exempt from Sections 705.5 and 705.8 when considered as one building in accordance with Section 503.1.2. The provisions of Sections 705.8.5 and 705.8.6 would still apply, since a need exists to restrict fire spread between stories within a building. Although not specifically identified in the exception, Section 705.11 would not apply, since it is dependent on the exterior wall being required to have a fire-resistance rating in accordance with Section 705.5, which relates to measurement of fire separation distance. The last sentence of the exception reminds the user that the normal code requirements would still be applicable once the two buildings are considered one. Therefore, if two types of construction are involved, then the lowest type of construction would be assumed for the entire building because there is no fire wall between them. In applying this last sentence, it is probably easier if the user imagines what the code requirement would be if the buildings were pushed together. In such a case, it would be easier to see that a fire wall would be needed between portions of a building that are of two different construction types or the entire building would have to be viewed as the lowest possi-

ble type of construction. In regards to occupancies and associated allowable areas, they would be separated in accordance with Section 508.4 or considered as nonseparated occupancies in accordance with Section 508.3 and, therefore, use the most restrictive allowable area.

In order to help make sense of this last sentence of the exception, the idea of imagining the two buildings being pushed together, as mentioned above, is helpful, then realizing that simply pulling the building apart should not increase or lessen the code requirements for that "single" building.

705.4 Materials. *Exterior walls* shall be of materials permitted by the building type of construction.

❖ The material (combustible or noncombustible) requirements for exterior walls are found in Sections 602.1 through 602.5. Only Type V construction allows exterior walls to be combustible construction. Other types of construction require exterior walls, or at least the framing members, to be noncombustible. Type I and II construction allows limited use of fire-retardant wood for exterior nonload-bearing wall. All types of construction allow insulation, exterior wall coverings and interior finish to be combustible within limits. See Sections 603, 719, 1405.5, 1717.5 and for materials allowed within and on framed exterior walls.

705.5 Fire-resistance ratings. *Exterior walls* shall be fire-resistance rated in accordance with Tables 601 and 602 and this section. The required *fire-resistance rating* of *exterior walls* with a *fire separation distance* of greater than 10 feet (3048 mm) shall be rated for exposure to fire from the inside. The required *fire-resistance rating* of *exterior walls* with a *fire separation distance* of less than or equal to 10 feet (3048 mm) shall be rated for exposure to fire from both sides.

❖ Table 601 states the requirements for the fire-resistance ratings for load-bearing exterior walls. All exterior load-bearing walls except Types IIB and VB require some degree of fire-resistance rating. Table 602 states the requirements for the fire-resistance ratings for both load-bearing exterior walls and nonload-bearing exterior walls. Table 602 is based on the fire separation distance, type of construction and occupancy group; whereas Table 601 requirements are only a function of the type of construction. Whichever of the two tables results in the highest fire-resistance rating would be used.

Fire exposure from the outside of the building is thought to be from other buildings either on the same property or another property, therefore, the spread of fire from building to building is lessened with increased fire separation distance. For years the code required the fire-resistance rating to be from both sides of the exterior wall when the fire separation distance was 5 feet (1524 mm) or less. During the development of the code, it was decided that exterior walls that had no required fire-resistance rating according to Tables 601 and 602 could have unlimited unprotected openings (see Section 705.8.1, Exception 2). This would allow an exterior wall in a Type IIB or VB building of nearly any occupancy group to have unlimited openings and a nonfire-resistance-rated wall at a fire separation distance of 10 feet (3048 mm). There could be another building 15 feet (4572 mm) from the first building with an exterior wall without fire-resistance rating from the exterior exposure because that building would have a fire separation distance of 5 feet (1524 mm). The consensus now is that a fire separation distance of 10 feet (3048 mm) is more practical for allowing the fire-resistance ratings of the exterior wall to be based on interior surface exposure only. Now, the fire-resistance rating of the exterior wall is from both sides when the fire separation distance is 10 feet (3048 mm) or less.

705.6 Structural stability. The wall shall extend to the height required by Section 705.11 and shall have sufficient structural stability such that it will remain in place for the duration of time indicated by the required *fire-resistance rating*.

❖ Structural stability of fire-resistance-rated construction is an important concern. Section 705.6 requires that the fire-resistant-rated exterior wall must be constructed to remain in place for the required duration of time equal to the required fire-resistance-rated construction. Similar sections describe stability, support and continuity for fire walls (see Section 706.2), fire barriers (see Section 707.5.1), fire partitions (see Section 709.4), smoke barriers (see Section 710.4) and horizontal assemblies (see Section 712.4). This section, on structural integrity for exterior walls, does not require that the wall remains in place when the structure collapses. That language is only used for fire wall structural integrity.

The ability of the exterior wall to remain in place for the required duration of time will require that the supporting elements also be fire-resistance rated for the same duration of time as the wall. In light-frame platform construction, this will require that the band joist or beam supporting the floor and wall above must also be fire-resistant construction. Although the floor construction itself may not be required to be fire-resistance-rated construction in Type IIB and VB construction, some effort must be made to ensure that the floor joists, at least at the exterior wall, are fire-resistance-rated construction. Although the floor framing acts as a lateral support for the exterior wall, this section does not require that the entire floor system be fire-resistance-rated construction. To state otherwise would prohibit Type IIB and VB buildings with a fire separation distance of less than 10 feet (3048 mm). Only the structural element within the floor system that supports the vertical load of the wall must be fire-resistance-rated construction.

Fire-resistance-rated exterior walls are required to extend to the roof construction or to the top of the parapet if one is required (see Section 705.11). This begs the question—in conventional light-frame platform construction, is the floor system supported by the lower exterior wall and supporting the exterior wall above part of the exterior wall? If so, how and to what limits do you go to provide a fire-resistance rating?

This is a valid concern in Type IIB and VB construction with a fire separation distance of less than 10 feet (3048 mm) because the floor system is not required to be fire-resistance rated, but the exterior wall is required to have a fire-resistance rating. Both the continuity and the structural integrity issues are illustrated in Figure 705.6.

When parapet walls are not required, the exterior wall for fire-resistance rating purposes stops at the roof/ceiling construction.

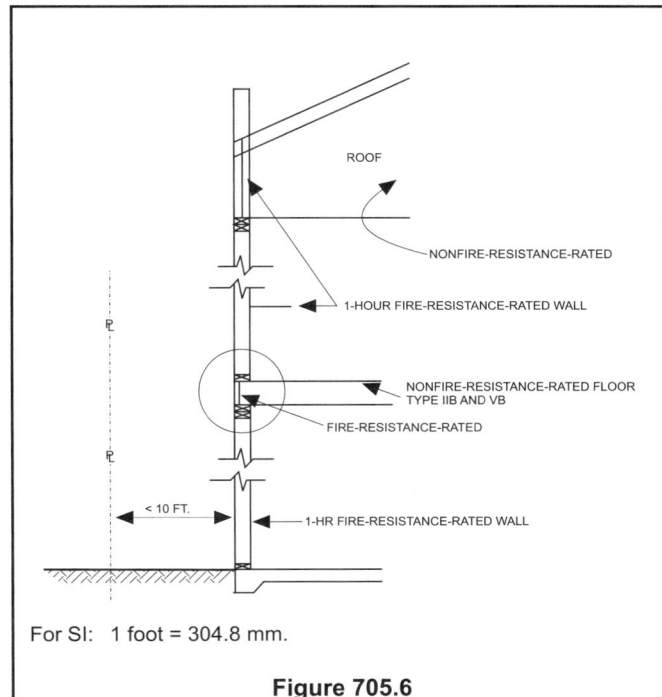

For SI: 1 foot = 304.8 mm.

Figure 705.6
TYPE IIB AND VB EXTERIOR
FIRE-RESISTANCE-RATED WALL CONTINUITY AND
STRUCTURAL STABILITY

705.7 Unexposed surface temperature. Where protected openings are not limited by Section 705.8, the limitation on the rise of temperature on the unexposed surface of *exterior walls* as required by ASTM E 119 or UL 263 shall not apply. Where protected openings are limited by Section 705.8, the limitation on the rise of temperature on the unexposed surface of *exterior walls* as required by ASTM E 119 or UL 263 shall not apply provided that a correction is made for radiation from the unexposed *exterior wall* surface in accordance with the following formula:

$$A_e = A + (A_f \times F_{eo}) \qquad \textbf{(Equation 7-1)}$$

where:

A_e = Equivalent area of protected openings.

A = Actual area of protected openings.

A_f = Area of *exterior wall* surface in the *story* under consideration exclusive of openings, on which the temperature limitations of ASTM E 119 or UL 263 for walls are exceeded.

F_{eo} = An "equivalent opening factor" derived from Figure 705.7 based on the average temperature of the unexposed wall surface and the *fire-resistance rating* of the wall.

❖ The purpose of this provision is to allow the use of an exterior fire-resistance-rated wall assembly that does not meet all of the conditions of acceptance of ASTM E 119 or UL 263. The only condition that may not apply is the limitation on the rise of temperature on the unexposed surface of the wall. To pass the standard, the rise of temperature cannot exceed 250°F (121°C) above the initial temperature. All listed fire-resistance-rated exterior walls and partitions meet all the conditions of acceptance. If a particular required exterior fire-resistance-rated wall was not listed by an approved testing agency for the required duration of time, then this section could be applicable.

For exterior walls with fire separation distances of 20 feet (6096 mm) or greater, the limitation on the rise of temperature on the unexposed surface of the exterior wall does not apply. This is practical because at 20 feet (6096 mm) an unlimited area of protected openings can be installed, and protected openings are treated as windows or doors, and the standard fire test for windows and doors does not have a limit on the temperature rise. To accept a nonlisted exterior wall assembly, the failed test data would have to be provided showing that the assembly did indeed pass all the conditions for acceptance, except rise of temperature on the unexposed surface.

For required exterior walls with fire separation distances of less than 20 feet (6096 mm), the limitation on the rise of temperature on the unexposed surface of the exterior wall may not apply. To accept a nonlisted exterior wall assembly, the failed test data would have to be provided showing that the assembly did indeed pass all the conditions for acceptance, except rise of temperature on the unexposed surface. The test data must also state the average temperature of the unexposed surface at the time duration for which the exterior wall is required to be rated. Therefore, the nonlisted wall assembly can still be used if a correction is made for the radiation from the unexposed exterior wall surface. This is done by using Figure 705.7 to convert the wall surface, which has too much temperature rise to an equivalent amount of protected openings. Compliance with the opening provisions of Section 705.8.1 would then be required using the converted value. See Example 705.7:

Example 705.7:

The exterior bearing wall of a Type VA nonsprinklered building with a Group B occupancy is 16 feet (4877 mm) from an interior lot line. The designer chose to use an exterior wall assembly, which failed the ASTM E 119 fire test because of the rise of temperature of

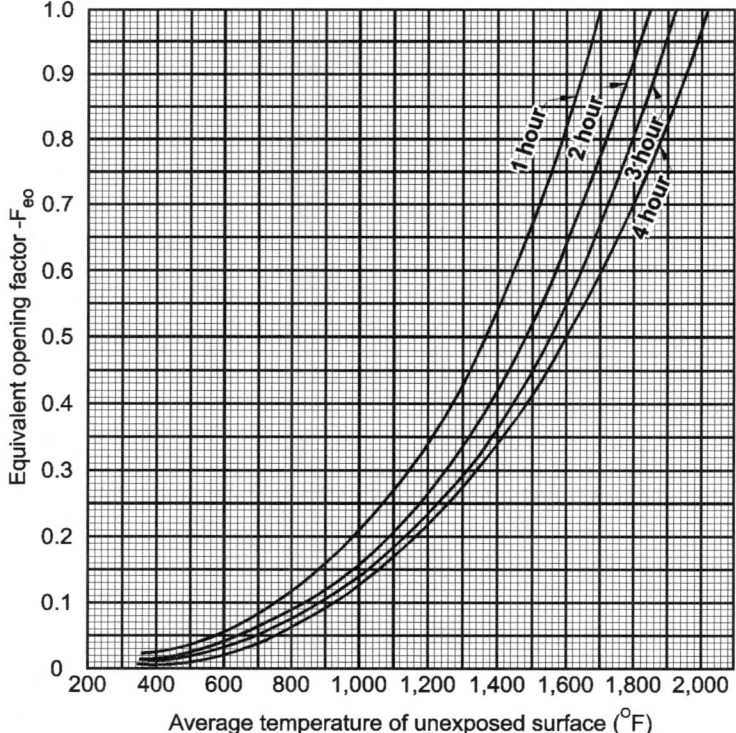

FIGURE 705.7
EQUIVALENT OPENING FACTOR

For SI: °C = [(°F) - 32] / 1.8.

the unexposed surface. The designer supplies the failed test data that clearly shows that the wall met all the conditions of acceptance for a 1-hour fire-resistance rating, except the rise of temperature on the unexposed side, which was an average of 1200°F (649°C) at 1-hour fire duration. Since the temperature rise exceeds the allowed, one must treat the wall surface as though it had an equivalent area of protected openings.

Solution:

- From Table 602: the exterior wall must be fire-resistance rated for 1 hour.

- From Table 705.8: the wall may have 75 percent protected wall openings or 25 percent unprotected wall openings or a combination of the two in accordance with Section 705.8.4.

- From drawing of exterior wall elevation (example 705.7):

     Total area of wall = 700 sq. ft.
     Total area of protected openings = 0
     Total area of unprotected openings = 120 sq. ft.
     Total area of exterior wall surface minus the openings = 580 sq. ft.

- Compute the equivalent area of protected openings using Equation 7-1 from page 7-25:

- Calculate the area from the drawings of the actual area of protected openings which is A. $A = 0$

- Calculate A_f from the drawings. It is the total area of the wall surface minus the openings. 700 sq. ft.-120 sq. ft. = 580 sq. ft.

- Obtain F_{eo} from Figure 705.7. Using the given average temperature rise on the unexposed side, which was 1200°F, and the curve on the figure for 1 hour, and reading to the left legend we find that F_{eo} for this assembly is 0.34.

- Equation 7-1 yields: $A_e = 0 + (580 \times 0.34)$ or $A_e = 197$ sq. ft.

     For SI: 1 square foot = 0.0929 m²,
     1°C = (°F-32)/1.8.

Having solved for the equivalent area of protected openings, one can proceed as normal by checking compliance with Section 705.8.4, with a mix of protected and unprotected openings in the exterior fire-resistance-rated wall.

- From Table 705.8: 75 percent of the total wall area may be protected openings or 0.75 x 700 = 525 sq. ft.

- From Table 705.8: 25 percent of the total wall area may be unprotected openings or 0.25 x 700 = 175 sq. ft.

- From Equation 7-2: $(A_p/a_p) + (A_u/a_u) \leq 1$ or have too many openings in the wall.

- From Equation 7-2 and using A_e computed above for A_p, have $(197 \div 525) + (120 \div 175) \leq 1$ or 0.375 + 0.685 = 1.06, which is not less than or equal to 1. The wall does not meet the code because there are too many openings.

705.8 Openings. Openings in *exterior walls* shall comply with Sections 705.8.1 through 705.8.6.

❖ The requirements of this section limit the allowable area of openings in exterior walls and are applicable to buildings with walls with fire separation distances less than 30 feet (9144 mm). The limitations on openings in exterior walls is a function of fire separation distance (see definition in this chapter) and the degree of protection provided for the opening. The degree of protection of the openings is either unprotected in a nonsprinklered building (UP, NS), unprotected in a sprinklered building (UP, S) or protected openings (P). Protected openings are openings with fire doors, fire shutters or fire window assemblies that comply with Sections 715.4 and 715.5. Sprinklered buildings are buildings with a NFPA 13 sprinkler system complying with standard NFPA 13. Buildings with only a system complying with standard NFPA 13 shall be considered nonsprinklered for the purpose of opening limitations in exterior walls. The percentage of openings allowed is in Table 705.8, subject to the exceptions in Section 705.8.1.

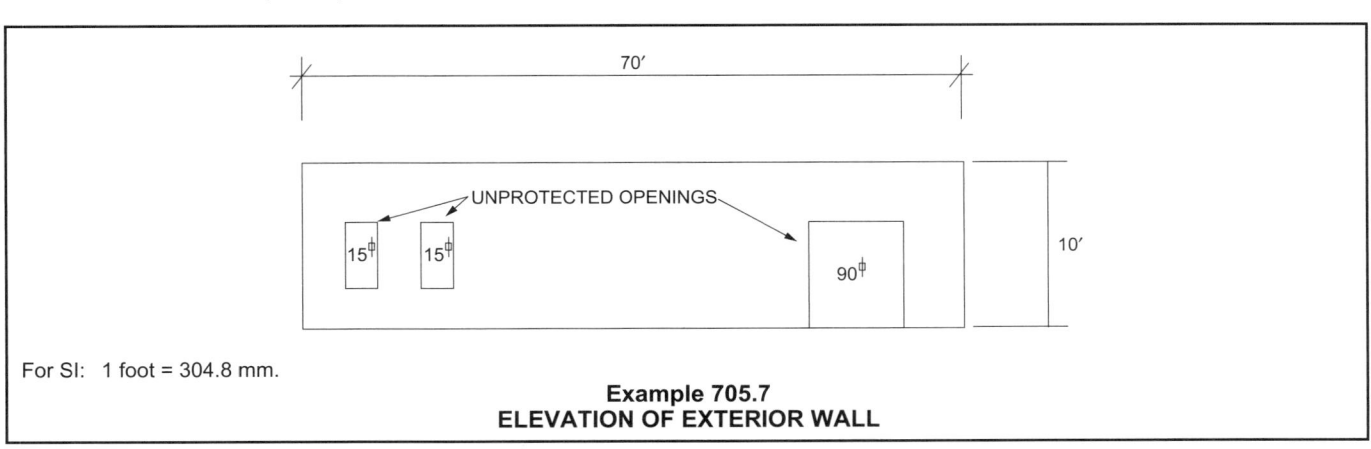

For SI: 1 foot = 304.8 mm.

Example 705.7
ELEVATION OF EXTERIOR WALL

TABLE 705.8
MAXIMUM AREA OF EXTERIOR WALL OPENINGS BASED ON FIRE SEPARATION DISTANCE AND DEGREE OF OPENING PROTECTION

FIRE SEPARATION DISTANCE (feet)	DEGREE OF OPENING PROTECTION	ALLOWABLE AREA[a]
0 to less than 3[b, c]	Unprotected, Nonsprinklered (UP, NS)	Not Permitted
	Unprotected, Sprinklered (UP, S)[i]	Not Permitted
	Protected (P)	Not Permitted
3 to less than 5[d, e]	Unprotected, Nonsprinklered (UP, NS)	Not Permitted
	Unprotected, Sprinklered (UP, S)[i]	15%
	Protected (P)	15%
5 to less than 10[e, f]	Unprotected, Nonsprinklered (UP, NS)	10%[h]
	Unprotected, Sprinklered (UP, S)[i]	25%
	Protected (P)	25%
10 to less than 15[e, f, g]	Unprotected, Nonsprinklered (UP, NS)	15%[h]
	Unprotected, Sprinklered (UP, S)[i]	45%
	Protected (P)	45%
15 to less than 20[f, g]	Unprotected, Nonsprinklered (UP, NS)	25%
	Unprotected, Sprinklered (UP, S)[i]	75%
	Protected (P)	75%
20 to less than 25[f, g]	Unprotected, Nonsprinklered (UP, NS)	45%
	Unprotected, Sprinklered (UP, S)[i]	No Limit
	Protected (P)	No Limit
25 to less than 30[f, g]	Unprotected, Nonsprinklered (UP, NS)	70%
	Unprotected, Sprinklered (UP, S)[i]	No Limit
	Protected (P)	No Limit
30 or greater	Unprotected, Nonsprinklered (UP, NS)	No Limit
	Unprotected, Sprinklered (UP, S)[i]	Not Required
	Protected (P)	Not Required

For SI: 1 foot = 304.8 mm.
UP, NS = Unprotected openings in buildings not equipped throughout with an automatic sprinkler system in accordance with Section 903.3.1.1.
UP, S   = Unprotected openings in buildings equipped throughout with an automatic sprinkler system in accordance with Section 903.3.1.1.
P   = Openings protected with an opening protective assembly in accordance with Section 705.8.2.
a. Values indicated are the percentage of the area of the exterior wall, per story.
b. For the requirements for fire walls of buildings with differing heights, see Section 706.6.1.
c. For openings in a fire wall for buildings on the same lot, see Section 706.8.
d. The maximum percentage of unprotected and protected openings shall be 25 percent for Group R-3 occupancies.
e. Unprotected openings shall not be permitted for openings with a fire separation distance of less than 15 feet for Group H-2 and H-3 occupancies.
f. The area of unprotected and protected openings shall not be limited for Group R-3 occupancies, with a fire separation distance of 5 feet or greater.
g. The area of openings in an open parking structure with a fire separation distance of 10 feet or greater shall not be limited.
h. Includes buildings accessory to Group R-3.
i. Not applicable to Group H-1, H-2 and H-3 occupancies.

705.8.1 Allowable area of openings. The maximum area of unprotected and protected openings permitted in an *exterior wall* in any *story* of a building shall not exceed the percentages specified in Table 705.8.

Exceptions:

1. In other than Group H occupancies, unlimited unprotected openings are permitted in the first *story* above grade either:

   1.1. Where the wall faces a street and has a *fire separation distance* of more than 15 feet (4572 mm); or

   1.2. Where the wall faces an unoccupied space. The unoccupied space shall be on the same lot or dedicated for public use, shall not be less than 30 feet (9144 mm) in width and shall have access from a street by a posted fire lane in accordance with the *International Fire Code.*

2. Buildings whose exterior bearing walls, exterior nonbearing walls and exterior primary structural frame are not required to be fire-resistance rated shall be permitted to have unlimited unprotected openings.

❖ The allowable area is given as a percentage of the total wall area at any story. Each story must comply independently.

Exception 1.1 only applies to the exterior walls on the first story above grade. For all occupancies except Group H, the first-story exterior wall that faces a street, with at least 15 feet (4572 mm) from the wall to the centerline of the street, shall be allowed unlimited unprotected openings.

Exception 1.2 also applies to the exterior walls on only the first story above grade. For all occupancies, except Group H, the first-story exterior wall that faces an unoccupied space on the same lot or an unoccupied space on property dedicated for public use, at least 30 feet (9144 mm) wide with access from a street by a fire lane, shall be allowed unlimited unprotected openings.

Since the first story above grade is generally readily available for fire department access, unprotected openings are not limited in the above exception. See Figure 705.8.1 for examples of first story openings.

Exception 2 is for openings in all exterior walls in all stories where Tables 601 and 602 do not require any exterior wall (bearing or nonbearing) or primary structural member to be fire-resistance rated. Therefore, this exception is only applicable to Type IIB and VB construction with a fire separation distance of 10 feet (3048 mm) or greater. This allows unlimited unprotected openings in Type IIB and VB construction for all exterior walls facing a fire separation distance of 10 feet (3048 mm) or more.

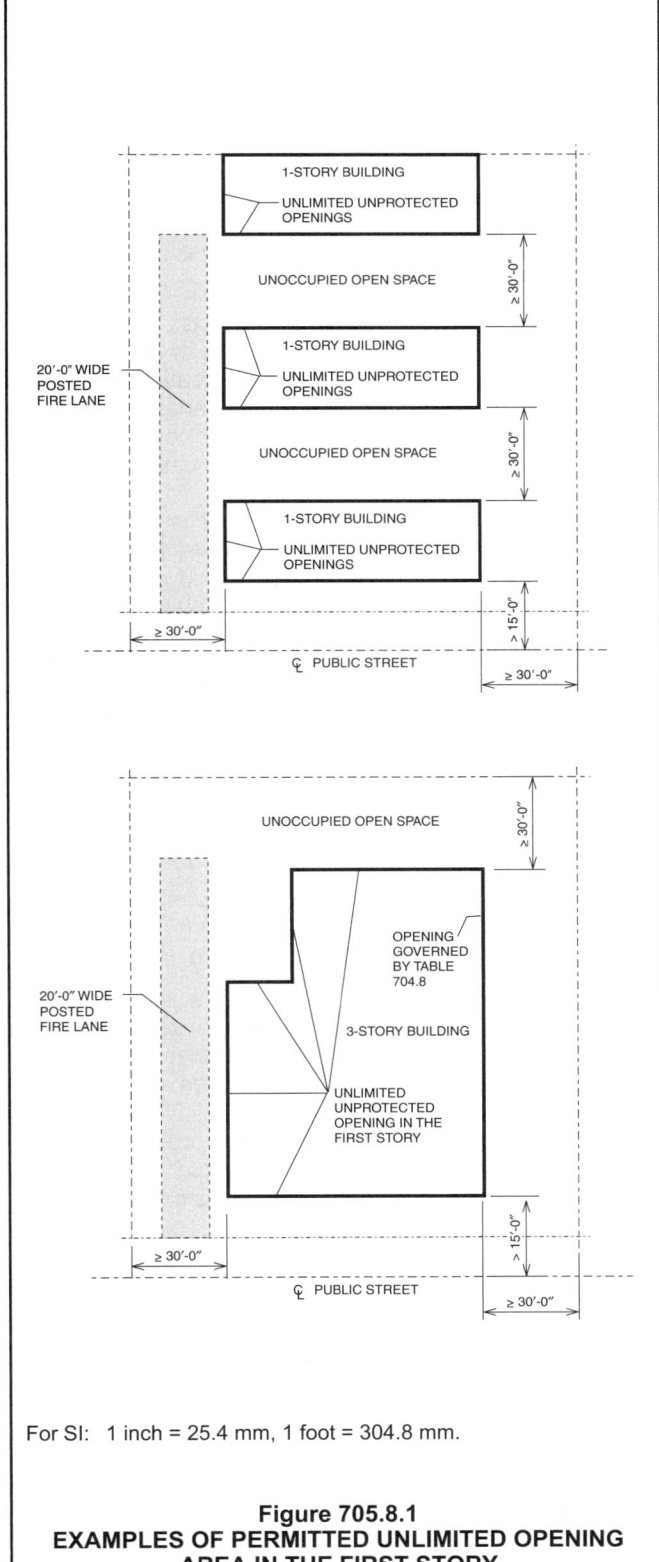

For SI: 1 inch = 25.4 mm, 1 foot = 304.8 mm.

Figure 705.8.1
EXAMPLES OF PERMITTED UNLIMITED OPENING AREA IN THE FIRST STORY

705.8.2 Protected openings. Where openings are required to be protected, *fire doors* and fire shutters shall comply with Section 715.4 and *fire window assemblies* shall comply with Section 715.5.

> **Exception:** Opening protectives are not required where the building is equipped throughout with an *automatic sprinkler system* in accordance with Section 903.3.1.1 and the exterior openings are protected by a water curtain using automatic sprinklers *approved* for that use.

❖ To be considered a protected opening, the opening must have a listed and labeled fire door, fire shutter or fire window meeting the requirements of Section 714.5 or 715.5. The required fire protection rating of doors in protected openings of exterior walls shall be in accordance with Table 715.4. Protected window assemblies shall have the fire protection rating as indicated in Table 715.5.

The exception to Section 705.8.2 is irrelevant for most occupancies. Table 705.8 allows the same percentage of opening area for protected openings and unprotected openings in sprinklered buildings and Table 705.8 does not have the water curtain requirement as the exception does. In a building sprinklered throughout with a full NFPA 13 system, there can be the same percentage of openings, whether the openings are protected or unprotected.

The exception to Section 705.8.2 would be relevant for Group H occupancies since Note i to Table 705.8 makes the unprotected sprinklered category not applicable to Group H-1, H-2 and H-3 occupancies.

705.8.3 Unprotected openings. Where unprotected openings are permitted, windows and doors shall be constructed of any *approved* materials. Glazing shall conform to the requirements of Chapters 24 and 26.

❖ Unprotected openings can be windows and doors of any approved materials. Glazing must meet the structural requirements and safety requirements as applicable in Chapters 24 and 26. All openings in wind-borne debris regions must be protected by impact-resistant material in accordance with Section 1609.1.2. Sheds without exterior walls must also meet Section 705.8 and Table 705.8.

705.8.4 Mixed openings. Where both unprotected and protected openings are located in the *exterior wall* in any *story* of a building, the total area of openings shall be determined in accordance with the following:

$(A_p/a_p)+(A_u/a_u)\leq 1$ **(Equation 7-2)**

where:

A_p = Actual area of protected openings, or the equivalent area of protected openings, A_e (see Section 705.7).

a_p = Allowable area of protected openings.

A_u = Actual area of unprotected openings.

a_u = Allowable area of unprotected openings.

❖ The opening limitations in Table 705.8 vary according to the degrees of opening protection. The table estab-

lishes the acceptable percentage for openings if all the openings were of one degree of protection. For instance, at 12 feet (3658 mm) if all the openings are unprotected in a nonsprinklered building, only 15 percent of the wall area is allowed to be unprotected openings. In that same exterior wall if all the openings were protected, 45 percent of the wall area is allowed to be protected openings. If any exterior wall area in any story has both protected and unprotected openings, then the actual areas of both types of openings must meet Equation 7-2.

Example 705.8.4:

In Type IIA buildings, the exterior wall is required to be 1-hour fire-resistance-rated construction by Table 602. The wall has a fire separation distance of 12 feet (3658 mm). The building is nonsprinklered. The wall in the first story has a total area of 1,000 sq. ft. (93 m²). The designer has 150 sq. ft. (14 m²) of protected openings in the wall and 100 sq. ft. (9 m²) of unprotected openings. Does this comply with the code, particularly Section 705.8.4?

Solution:

- Use Equation 7-2.
- From the building in the example, the protected openings are 150 sq. ft. This value is A_p.
- From Table 705.8, the allowable area if only protected openings are used is 45 percent. Forty-five percent of 1,000 sq. ft. is 450 sq. ft. (1,000 x 0.45 = 450). This value is A_p in the equation.
- From the building in the example, the unprotected openings are 100 sq. ft. This value is A_u in the equation.
- From Table 705.8, the allowable area if only unprotected openings are used is 15 percent. Fifteen percent of 1,000 sq. ft. is 150 sq. ft. (1,000 x 0.15 = 150). This value is a_u in the equation.
- Equation 7-2 $(A_p/a_p) + (A_u/a_u) = 1$ or (150 ÷ 450) + (100 ÷ 150) = 1 or 0.333 + 0.666 = 1
- This amount of openings exactly complies with the code. Had the sum of the ratios exceeded 1, the opening mixture would not be allowed.

For SI: 1 square foot = 0.0929 m².

705.8.5 Vertical separation of openings. Openings in *exterior walls* in adjacent *stories* shall be separated vertically to protect against fire spread on the exterior of the buildings where the openings are within 5 feet (1524 mm) of each other horizontally and the opening in the lower *story* is not a protected opening with a *fire protection rating* of not less than $^3/_4$ hour. Such openings shall be separated vertically at least 3 feet (914 mm) by spandrel girders, *exterior walls* or other similar assemblies that have a *fire-resistance rating* of at least 1 hour or by flame barriers that extend horizontally at least 30 inches (762 mm) beyond the *exterior wall*. Flame barriers shall also have a *fire-resistance rating* of at least 1 hour. The unexposed surface temperature limitations specified in ASTM E 119 or UL 263 shall not apply to the flame barriers or vertical separation unless otherwise required by the provisions of this code.

Exceptions:

1. This section shall not apply to buildings that are three *stories* or less above *grade plane*.

2. This section shall not apply to buildings equipped throughout with an *automatic sprinkler system* in accordance with Section 903.3.1.1 or 903.3.1.2.

3. Open parking garages.

❖ Where unprotected openings occur in adjacent stories, a fire that breaks out of an opening in a lower story can spread vertically to upper stories of the building. Experience has shown that a fire can break out at a lower level and then quickly spread upwards to a number of other stories simply due to the flame plume from the lower level impinging on the windows above and causing the combustible materials in those upper levels to ignite. Therefore, the code prescribes a vertical panel or horizontal flame barrier to minimize the possibility of vertical flame spread. The vertical panel is to be at least 3 feet (914 mm) in height and have a fire-resistance rating of at least 1 hour [see Figure 705.8.5(1)]. Full-scale fire tests have shown that the vertical panel should extend at least 30 inches (762 mm) above the finished floor level. This is to prevent the transfer of radiant heat to furnishings and other combustibles located in the lower portion of the room in the story above the assumed fire exposure. Horizontal flame barriers are required to extend at least 30 inches (762 mm) beyond the face of the exterior wall

and have a fire-resistance rating of at least 1 hour [see Figure 705.8.5(2)].

This section requires such protection when the opening on the lower level is not fire protection rated for $^3/_4$ hour or more. The use of glazing is one example of why this section is so specific. Wired glass, for example, has been shown to retain its integrity with limited fire exposure when exposed to a fire condition for a certain period of time. For this reason, wired glass has historically been recognized as a suitable exterior opening protective within size and mounting methods defined by its listing. However, wired glass permits a significant percentage, typically recognized as high as 50 percent, of radiant heat energy to pass through. Since the initial concern is heat radiated from the flame plume, wired glass does little to prevent ignition of combustibles in the room on the upper floor.

This section permits the elimination of the unexposed surface temperature limitations prescribed by ASTM E 119 for flame barriers and vertical shields. Fire spread between floors is primarily the result of radiant heat, not conductive heat transfer. Full-scale tests have shown that only a small degree of fire resistance is required for adequate protection since radiant energy is the primary concern and glazing in the unprotected openings above the story of fire origin may break within 1 minute.

Protection is not required for buildings three stories or less in height. This exception is consistent with Table 503, which permits buildings of three stories in height to be of Type IIB, IIIB and VB construction when these buildings do not require a fire-resistance rating for the floor construction.

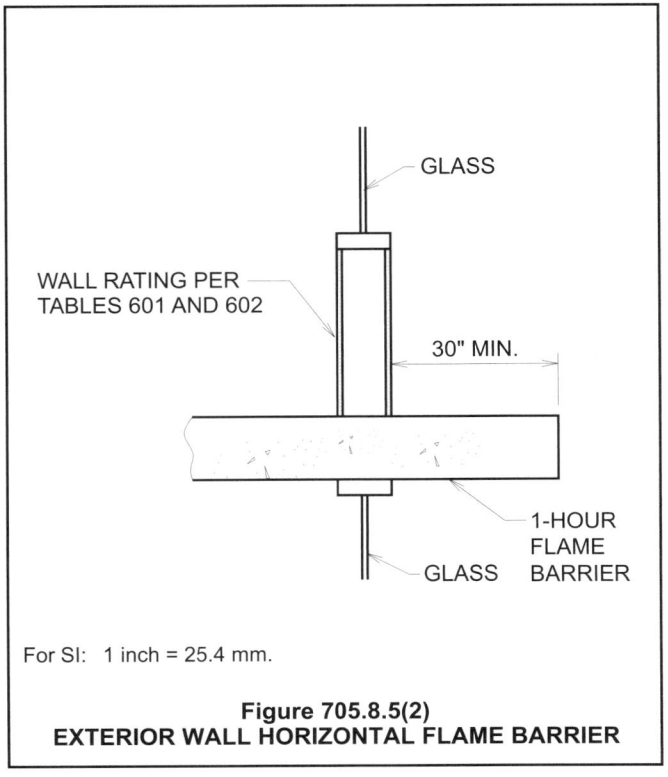

For SI: 1 inch = 25.4 mm, 1 foot = 304.8 mm.

Figure 705.8.5(1)
EXTERIOR WALL VERTICAL FLAME BARRIER

For SI: 1 inch = 25.4 mm.

Figure 705.8.5(2)
EXTERIOR WALL HORIZONTAL FLAME BARRIER

The exception for buildings protected with an automatic sprinkler system is consistent with the code's approach to limiting fire size and, thus, limiting spread of fire beyond the area or room of origin. The exception is based on the effectiveness and reliability of automatic sprinkler systems in decreasing fire size. The reference to Section 903.3.1.1 or 903.3.1.2 reinforces that the system must be a code-compliant automatic sprinkler system in accordance with NFPA 13 as modified by Section 903.3.1.1 or NFPA 13R as modified by Section 903.3.1.2. Exception 3 acknowledges the practicality of constructing open parking structures and the need to have a significant amount of openings for natural ventilation.

705.8.6 Vertical exposure. For buildings on the same lot, opening protectives having a *fire protection rating* of not less than $^3/_4$ hour shall be provided in every opening that is less than 15 feet (4572 mm) vertically above the roof of an adjacent building or structure based on assuming an imaginary line between them. The opening protectives are required where the *fire separation distance* between the imaginary line and the adjacent building or structure is less than 15 feet (4572 mm).

Exceptions:

1. Opening protectives are not required where the roof assembly of the adjacent building or structure has a

fire-resistance rating of not less than 1 hour for a minimum distance of 10 feet (3048 mm) from the *exterior wall* facing the imaginary line and the entire length and span of the supporting elements for the fire-resistance-rated roof assembly has a *fire-resistance rating* of not less than 1 hour.

2. Buildings on the same lot and considered as portions of one building in accordance with Section 705.3 are not required to comply with Section 705.8.6.

❖ A fire in a building that is adjacent to a taller building can be the source of fire exposure to openings in the taller building. Although the height of a fire plume is dependent on several factors, this section requires $^3/_4$-hour or greater opening protectives in the wall where the openings are within 15 feet (457 mm) vertically above the roof of a building that is within a horizontal fire separation distance of 15 feet (457 mm) (see Figure 705.8.6). Full-scale tests have indicated that exterior flame plumes may extend higher than 16 feet (4877 mm) above a window using a fuel load of 8 pounds per square foot (psf) (39 kg/m²). Since this provision is based on a fire exposure from the roof of the lower building, it does not apply where the roof construction is 1-hour fire-resistance rated, which reduces the potential for fire exposure.

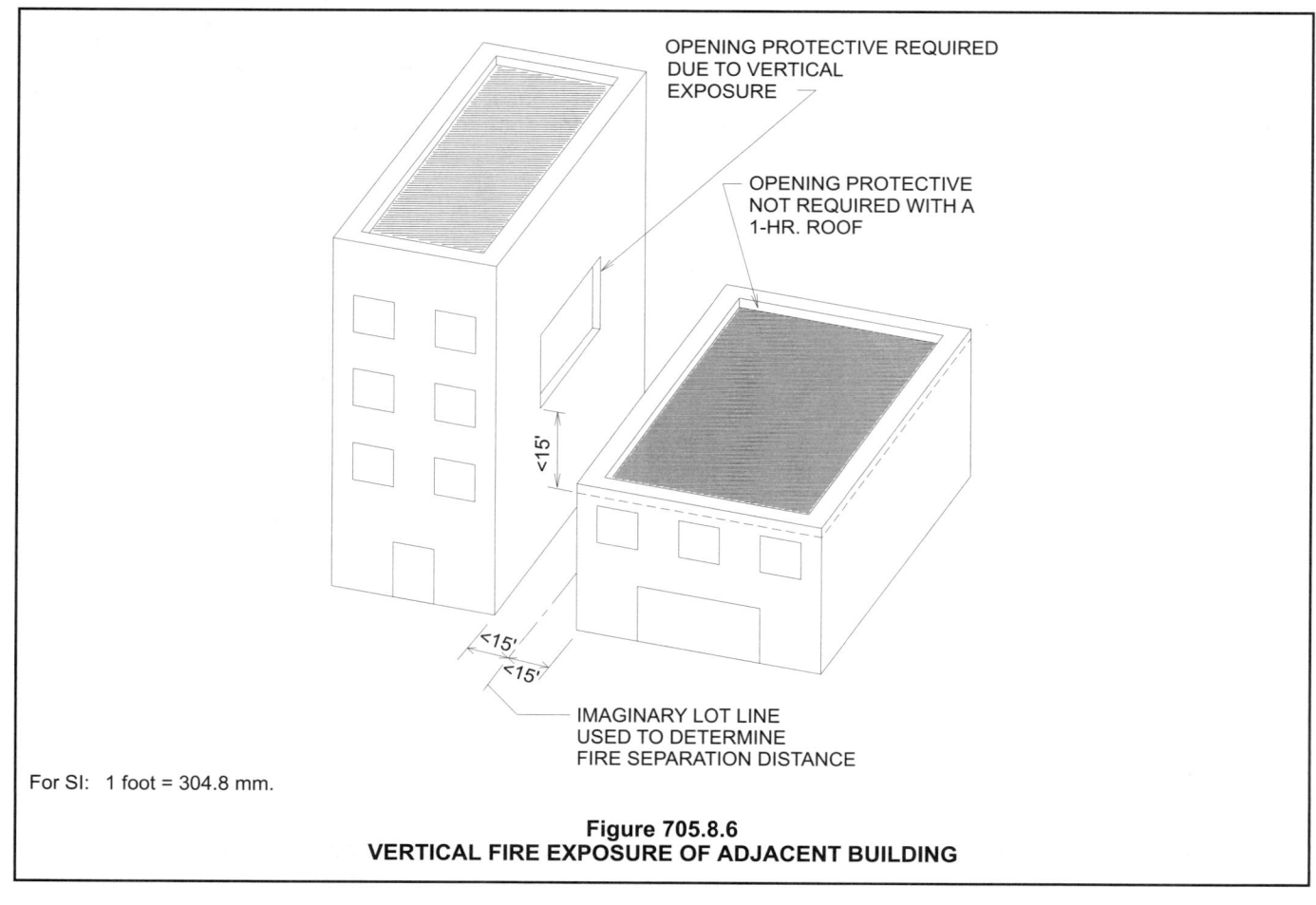

For SI: 1 foot = 304.8 mm.

Figure 705.8.6
VERTICAL FIRE EXPOSURE OF ADJACENT BUILDING

705.9 Joints. Joints made in or between *exterior walls* required by this section to have a *fire-resistance rating* shall comply with Section 714.

> **Exception:** Joints in *exterior walls* that are permitted to have unprotected openings.

❖ Joints, such as expansion or seismic, are another form of openings in exterior walls and, therefore, must be considered with regard to maintaining the fire-resistance ratings of the exterior walls. This section requires that all joints located in exterior walls required to be fire-resistance rated are to be protected by a joint system that has a fire-resistance rating and complies with the requirements of Section 714 (see commentary, Section 714). The exception to joint protection here and in Section 714 is for exterior walls that are permitted by Table 705.8 to have unprotected openings. Therefore, this section requires fire-resistant joint systems to protect joints in exterior walls of nonsprinklered buildings with a fire separation distance of 5 feet (1524 mm) or less; in sprinklered buildings with a fire separation distance of 3 feet (914 mm) or less; and in Group H-2 or H-3 buildings with a fire separation distance of 15 feet (4572 mm) or less (see note e to Table 705.8, Note e). This implies that there may be joints in fire-resistance-rated walls with a fire separation distance of less than 3 feet (914 mm). In this sense, joints are not treated like openings that are not permitted in walls with a fire separation distance of less than 3 feet (914 mm).

705.9.1 Voids. The void created at the intersection of a floor/ceiling assembly and an exterior curtain wall assembly shall be protected in accordance with Section 714.4.

❖ See the commentary to Section 714.4.

705.10 Ducts and air transfer openings. Penetrations by air ducts and air transfer openings in fire-resistance-rated *exterior walls* required to have protected openings shall comply with Section 716.

> **Exception:** Foundation vents installed in accordance with this code are permitted.

❖ This section requires air outlets in exterior walls, including air intake and air exhaust outlets, to be protected by fire dampers where Table 705.8 requires protected openings. Therefore, fire dampers would be required in openings based upon the percentage of the wall required by Table 705.8 to have protected openings, and the actual amount of noted openings desired in the wall. Note that there can be no openings in a wall with a fire separation distance of less than 3 feet (914 mm). Fire damper assembly standards are in accordance with Section 716. Exhaust outlets that cannot have fire dampers, such as commercial kitchen hood exhausts, cannot be located in exterior walls in locations where openings are required to be protected in accordance with Table 705.8. This is also stated in Section 506.3.12.2 of the IMC.

705.11 Parapets. Parapets shall be provided on *exterior walls* of buildings.

> **Exceptions:** A parapet need not be provided on an *exterior wall* where any of the following conditions exist:
>
> 1. The wall is not required to be fire-resistance rated in accordance with Table 602 because of *fire separation distance*.
>
> 2. The building has an area of not more than 1,000 square feet (93 m²) on any floor.
>
> 3. Walls that terminate at roofs of not less than 2-hour fire-resistance-rated construction or where the roof, including the deck or slab and supporting construction, is constructed entirely of noncombustible materials.
>
> 4. One-hour fire-resistance-rated *exterior walls* that terminate at the underside of the roof sheathing, deck or slab, provided:
>
>   4.1. Where the roof/ceiling framing elements are parallel to the walls, such framing and elements supporting such framing shall not be of less than 1-hour fire-resistance-rated construction for a width of 4 feet (1220 mm) for Groups R and U and 10 feet (3048 mm) for other occupancies, measured from the interior side of the wall.
>
>   4.2. Where roof/ceiling framing elements are not parallel to the wall, the entire span of such framing and elements supporting such framing shall not be of less than 1-hour fire-resistance-rated construction.
>
>   4.3. Openings in the roof shall not be located within 5 feet (1524 mm) of the 1-hour fire-resistance-rated *exterior wall* for Groups R and U and 10 feet (3048 mm) for other occupancies, measured from the interior side of the wall.
>
>   4.4. The entire building shall be provided with not less than a Class B roof covering.
>
> 5. In Groups R-2 and R-3 where the entire building is provided with a Class C roof covering, the *exterior wall* shall be permitted to terminate at the underside of the roof sheathing or deck in Type III, IV and V construction, provided:
>
>   5.1. The roof sheathing or deck is constructed of *approved* noncombustible materials or of *fire-retardant-treated wood* for a distance of 4 feet (1220 mm); or
>
>   5.2. The roof is protected with 0.625-inch (16 mm) Type X gypsum board directly beneath the underside of the roof sheathing or deck, supported by a minimum of nominal 2-inch (51 mm) ledgers attached to the sides of the roof framing members for a minimum distance of 4 feet (1220 mm).

6. Where the wall is permitted to have at least 25 percent of the *exterior wall* areas containing unprotected openings based on *fire separation distance* as determined in accordance with Section 705.8.

❖ Parapets are required to restrict the spread of fire from building to building and are, therefore, a function of the fire separation distance. Parapets are not required to extend above the roofs on all buildings. Exceptions 1 and 6 apply to all buildings.

- Exception 1 exempts exterior walls with 30 feet (9144 mm) or greater fire separation distance for all types of construction except Types IIB and VB. Exterior walls of Types IIB and VB (except for Group H occupancy) with greater than 10 feet (3048 mm) of fire separation distance are exempt from parapets.

- Exception 6 exempts all sprinklered buildings from parapet walls when the exterior walls have a fire separation distance of 5 feet (1524 mm) or greater. Exception 6 exempts all nonsprinklered buildings from parapets with a fire separation distance of 15 feet (4572 mm) or greater. Exception 6 is the greater exemption for all but Type IIB and VB nonsprinklered buildings, when Exception 1 is better. These two exceptions yield the following:
  ○ Parapet walls are not required on sprinklered buildings with a fire separation distance of 5 feet (1524 mm) or greater.
  ○ Parapet walls are not required for nonsprinklered Type IIB and VB construction with a fire separation distance of 10 feet (3048 mm) or more.
  ○ Parapet walls are not required for nonsprinklered buildings of all types of construction other than Types IIB and VB with a fire separation distance of 15 feet (4572 mm) or more.

- Exception 2 exempts buildings with 1,000 square feet (93 m²) or less on any floor.

- Exceptions 3, 4 and 5. While Exceptions 1, 2 and 6 completely exempt the exterior walls from parapets, Exceptions 3, 4 and 5 deal with special credit for the roof construction.

Exceptions 3, 4 and 5 require the exterior walls to extend to the roof deck or underside of roof sheathing and require special attention to the roof construction to eliminate the need for the parapet extension, thus they are not subject to any fire separation distance and can have a fire separation distance of zero. Exceptions 1, 2 and 6 simply exempt the walls from extending above the roof without any special roof construction and do not say that the walls must extend to the roof deck.

Exception 1 applies to bearing and nonload-bearing walls and would also exempt bearing walls from parapets if the bearing wall was not required to have a fire-resistance rating by Table 602, even if said bearing wall was required to have a fire-resistance rating by Table 601.

705.11.1 Parapet construction. Parapets shall have the same *fire-resistance rating* as that required for the supporting wall, and on any side adjacent to a roof surface, shall have noncombustible faces for the uppermost 18 inches (457 mm), including counterflashing and coping materials. The height of the parapet shall not be less than 30 inches (762 mm) above the point where the roof surface and the wall intersect. Where the roof slopes toward a parapet at a slope greater than two units vertical in 12 units horizontal (16.7-percent slope), the parapet shall extend to the same height as any portion of the roof within a *fire separation distance* where protection of wall openings is required, but in no case shall the height be less than 30 inches (762 mm).

❖ Parapet wall construction shall be combustible or noncombustible material depending on the exterior wall requirements of the type of construction and shall be fire-resistance-rated construction as required for the exterior wall. The interior wall covering facing the roof including the flashing shall be noncombustible to a height of 18 inches (457 mm) above the roof. The required height of the parapet shall be 30 inches (762 mm) above the roof surface unless the roof slopes upward away from the wall on a pitch of 2 in 12 or greater. In that case, the last part of this section requires a higher parapet in some cases, depending on the fire separation distance. When the slope of the roof is over 2 in 12, the parapet shall extend to a height equal to the height of the roof at the point determined as follows:

- In a nonsprinklered building, Table 705.8 requires protected openings for any fire separation distance between 3 and 5 feet (914 and 1524 mm), therefore, 5 feet (1524 mm) is the height to which the parapet wall extension must extend, but must always extend at least 30 inches (762 mm) above the roof at the exterior wall. For a nonsprinklerd building, the height of the roof at the plane where protected openings are required according to Table 705.8 is the height at 5 feet (1524 mm) from the common lot line or assumed property line [see Figure 705.11.1(1)].

- For a sprinklered building, Table 705.8 never requires protected openings. Therefore, for a sprinklered building the parapet need only extend 30 inches (762 mm) above the roof, regardless of the slope of the roof or the fire separation distance [see Figure 705.11.1(2)].

SECTION 706
FIRE WALLS

❖ Fire walls serve to create separate buildings for the purpose of allowable area and type of construction requirements (see the definition of "Area, building" in Section 502.1). All the code provisions that apply to buildings would be applied individually to the "building" on each side of the fire wall. Because of the special allowances for allowable areas, a fire wall must provide a higher level of fire safety, continuity and structural integrity than other types of fire-resistance-rated walls. When the designer utilizes a fire wall, the design of the fire wall must be in accordance with this section.

It is not intended that fire walls can be provided in the horizontal plane to create separate buildings. They cannot be utilized to increase the allowable building height. However, offsetting two vertical sections of fire walls is permissible as long as the required fire-resistance rating and structural stability are maintained.

706.1 General. Each portion of a building separated by one or more *fire walls* that comply with the provisions of this section shall be considered a separate building. The extent and location of such *fire walls* shall provide a complete separation. Where a *fire wall* also separates occupancies that are required to be separated by a *fire barrier* wall, the most restrictive requirements of each separation shall apply.

❖ The provisions of this section apply to assemblies that are required to have a fire-resistance rating and are required to be constructed as "fire barriers." As addressed in Section 706.3, fire barriers are used for separating exits, incidental use areas, shafts, hazardous materials control areas and fire areas. Fire barriers provide a higher degree of protection than fire par-

titions (see Section 708), but lack the inherent structural integrity of fire walls. Fire barriers limit the number of openings. Fire barrier wall assemblies must be continuous from the top of a fire-resistance-rated floor/ceiling assembly to the bottom of the floor or roof slab/deck above. Unlike fire partitions, addressed in Section 708, there are no circumstances under which a fire barrier wall is permitted to terminate at a ceiling.

Fire barriers are used for a variety of purposes, such as mixed occupancies, areas of refuge separations, shafts, and exit and floor opening enclosures. Fire barriers also include interior walls that serve to subdivide a space by separating one fire area from an adjacent fire area and for separating incidental use areas (see Section 508.2.2.1). Fire-resistance-rated assemblies used to separate exit access corridors in many applications, as well as tenant, dwelling unit and guestroom separations, are fire partitions (see Section 708). The provisions of this section provide minimum requirements for the fire-resistance rating, continuity, combustibility and protection of openings and penetrations in order to help maintain the reliability of the fire separation assembly. As with any fire-resistance-rated assembly, consideration must be given to the openings and penetrations that are provided within the assembly. The intent is to maintain the fire-resistance rating of the assembly. These sections recognize that fire spread beyond a fire-resistance-rated compartment is

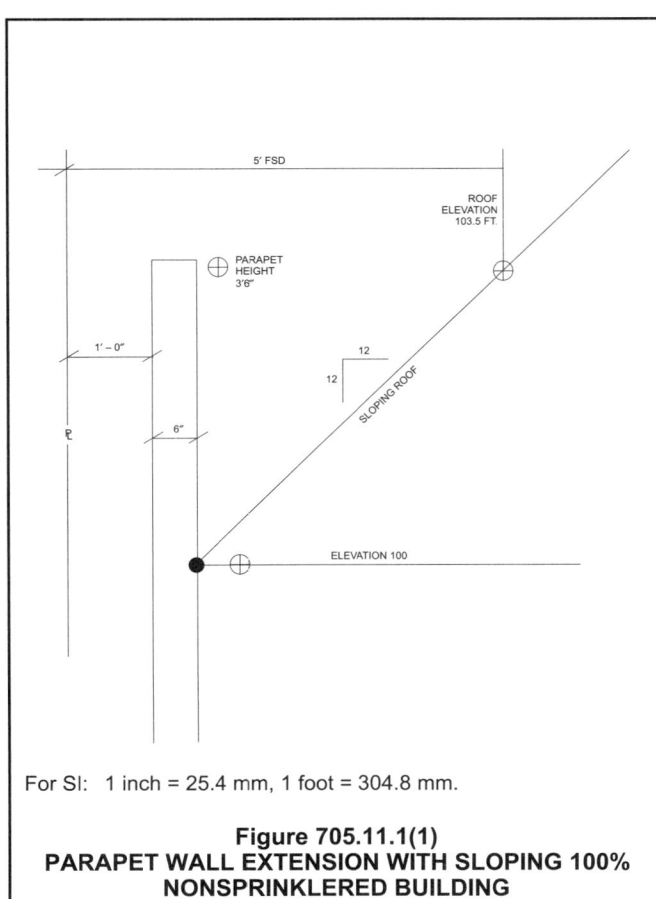

For SI: 1 inch = 25.4 mm, 1 foot = 304.8 mm.

Figure 705.11.1(1)
PARAPET WALL EXTENSION WITH SLOPING 100% NONSPRINKLERED BUILDING

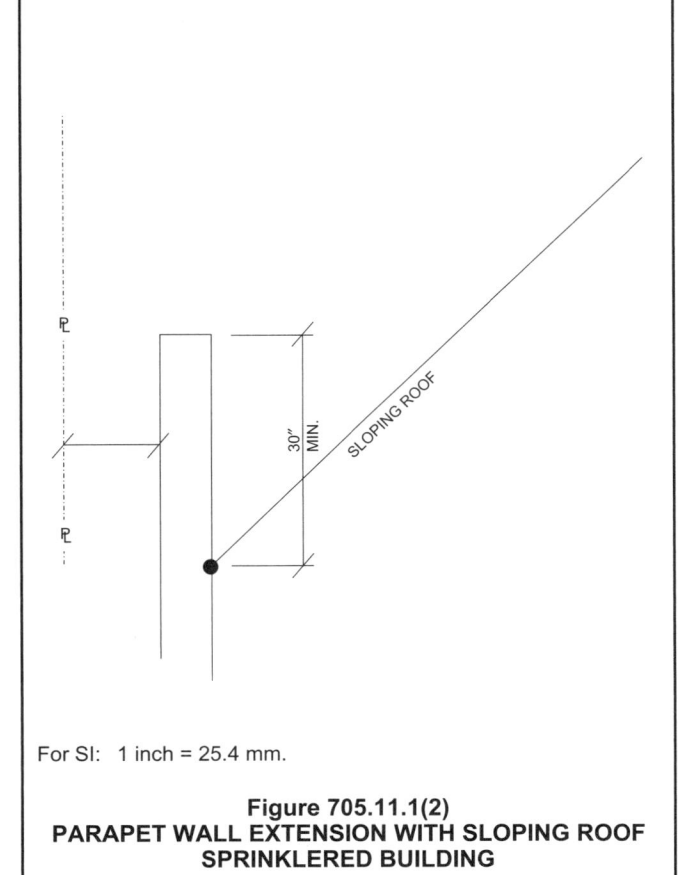

For SI: 1 inch = 25.4 mm.

Figure 705.11.1(2)
PARAPET WALL EXTENSION WITH SLOPING ROOF SPRINKLERED BUILDING

often attributed to the protection given to any opening or penetration in the fire barrier, or the lack thereof.

Since the fire barrier is intended to provide a reliable subdivision of areas, the construction that structurally supports the assembly is required to provide at least the same hourly fire-resistance rating as the fire barrier. This is applicable regardless of the type of construction of the building. Structural stability is regulated by Section 706.5.

706.1.1 Party walls. Any wall located on a *lot line* between adjacent buildings, which is used or adapted for joint service between the two buildings, shall be constructed as a *fire wall* in accordance with Section 706. Party walls shall be constructed without openings and shall create separate buildings.

> **Exception:** Openings in a party wall separating an *anchor building* and a mall shall be in accordance with Section 402.7.3.1.

❖ A party wall is a fire wall on an interior lot line, adapted for joint use by both buildings. It is distinguished from other fire walls in that it is on the property line and serves to separate buildings usually owned by two separate parties. When two separate structures are built up to the property line, the designer has the option of using two separate exterior walls with zero fire separation distance or a single party wall. Since there is a real property line involved, the prohibition for openings between the two buildings is important and even utilities cannot penetrate the party wall. Unlike fire walls in other buildings, which can have openings in them, party walls cannot have any openings in them (see Section 706.8 for opening limits on other fire walls).

The exception to allow openings is important since many anchor stores are actually owned by the major department store, while the mall is owned by a separate entity. The fact that there is a real property line at the separation walls between an anchor and a mall means that technically there is a party wall, but openings are normally present and a necessary function of the mall and anchor store.

706.2 Structural stability. *Fire walls* shall have sufficient structural stability under fire conditions to allow collapse of construction on either side without collapse of the wall for the duration of time indicated by the required *fire-resistance rating*.

❖ Since the collapse of one building from fire conditions should not cause the collapse of an adjacent building, a fire wall is required to be capable of withstanding the collapse of the construction on either side of the wall. An exception to the provision is found in Exception 5 to Section 706.6. In this situation, the issue of structural stability would still apply if the fire was in the buildings located above the horizontal separation required in Section 509.2, Item 1. However, it would clearly be impossible for the buildings above the separation or for

that matter the fire wall to be able to withstand the collapse of the building beneath it. Since the building below is of Type IA construction and the separation has a 3-hour fire-resistance rating, this should not be a problem. See 706.6, Exception 5, regarding fire walls in this situation.

There are various methods of designing and constructing fire walls for structural stability during a fire. Among the systems used are cantilevered or freestanding walls, laterally supported and tied walls, and double wall construction. For masonry walls, the NCMA TEK Bulletin 5-8B contains helpful information. For an example with a masonry fire wall, see Figure 706.2(1). The Gypsum Association *Fire Resistance Design Manual* contains construction details for area separation walls (party wall/fire walls) which have been accepted as fire walls. For an example of a gypsum board fire wall, see Figure 706.2 (2).

The collapse of structural members at some distance from the wall can generally be ignored, but the structural members that are supported by the wall or that are attached directly to the wall for lateral support do require special attention.

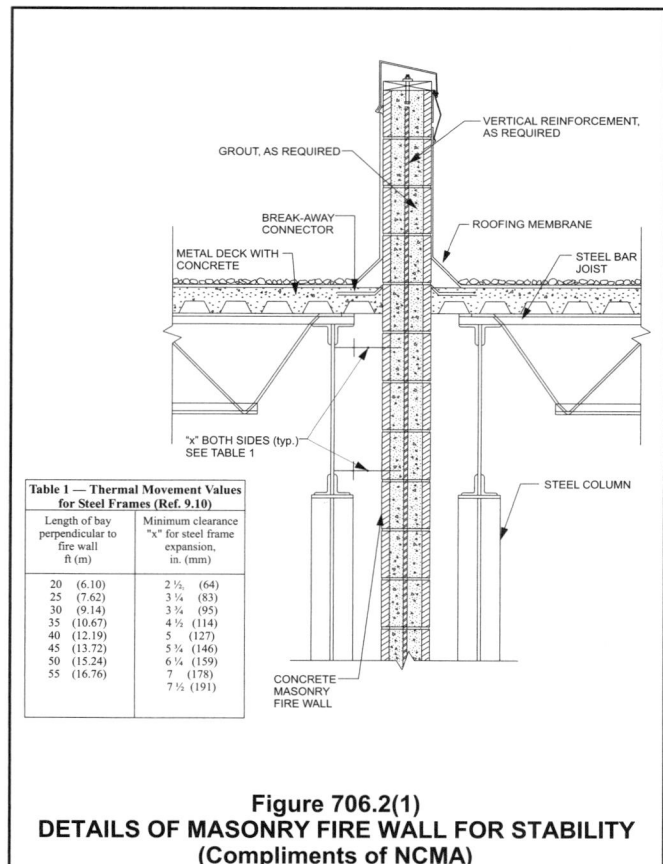

Table 1 — Thermal Movement Values for Steel Frames (Ref. 9.10)	
Length of bay perpendicular to fire wall ft (m)	Minimum clearance "x" for steel frame expansion, in. (mm)
20 (6.10)	2 ½ (64)
25 (7.62)	3 ¼ (83)
30 (9.14)	3 ¾ (95)
35 (10.67)	4 ½ (114)
40 (12.19)	5 (127)
45 (13.72)	5 ¾ (146)
50 (15.24)	6 ¼ (159)
55 (16.76)	7 (178)
	7 ½ (191)

Figure 706.2(1)
DETAILS OF MASONRY FIRE WALL FOR STABILITY
(Compliments of NCMA)

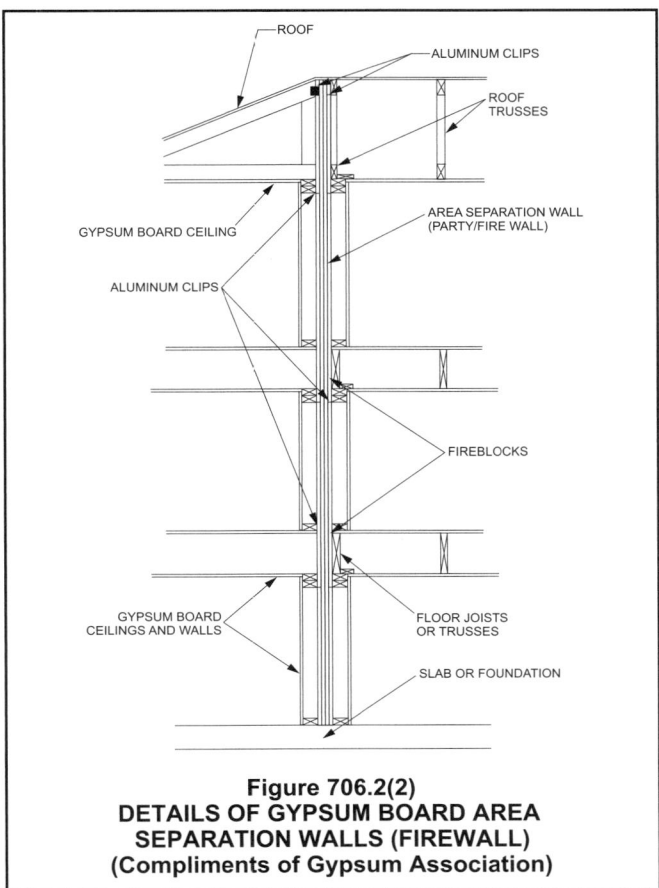

Figure 706.2(2)
DETAILS OF GYPSUM BOARD AREA
SEPARATION WALLS (FIREWALL)
(Compliments of Gypsum Association)

706.3 Materials. *Fire walls* shall be of any *approved* noncombustible materials.

Exception: Buildings of Type V construction.

❖ This section requires that fire walls be constructed of noncombustible materials unless both buildings are of Type V (combustible) construction. This is consistent with the provisions of Section 602, which require exterior walls of buildings of Type I, II, III and IV to be built of noncombustible construction, while buildings of Type V construction are permitted to be built of combustible exterior walls.

706.4 Fire-resistance rating. *Fire walls* shall have a *fire-resistance rating* of not less than that required by Table 706.4.

❖ The required fire-resistance rating must comply with the more restrictive occupancy of Table 706.4. For example, if two buildings of Type II construction with occupancies in Groups S-1 and B are built of any construction type and separated by a fire wall, the fire wall is required to have a fire-resistance rating of 3 hours (in accordance with Table 706.4 based on Group S-1). The provisions of Section 705.1 should also be reviewed when determining the required fire-resistance rating. If the occupancies involved require a separation, or are not permitted in a mixed-occupancy building, then the most restrictive requirement between that and the fire wall provisions will apply.

TABLE 706.4
FIRE WALL FIRE-RESISTANCE RATINGS

GROUP	FIRE-RESISTANCE RATING (hours)
A, B, E, H-4, I, R-1, R-2, U	3[a]
F-1, H-3[b], H-5, M, S-1	3
H-1, H-2	4[b]
F-2, S-2, R-3, R-4	2

a. In Type II or V construction, walls shall be permitted to have a 2-hour fire-resistance rating.
b. For Group H-1, H-2 or H-3 buildings, also see Sections 415.4 and 415.5.

❖ The fire-resistance ratings represent a relationship between the fuel load and an exposure to fire severity that is equivalent to the standard time-temperature curve. The relationship is based on a number of full-scale fire tests by the U.S. Department of Commerce's National Bureau of Standards (NBS) conducted in the 1920s to determine how actual building fires compared with the temperatures represented by the standard time-temperature curve (see commentary, Section 703.2).

An analysis of tests documented by S.H. Ingberg indicates that the weight per square foot (psf) of ordinary combustibles, such as wood and paper, with a heat of combustion of 7,000 to 8,000 British thermal units (Btu) per pound (16 282 to 18 608 kJ/kg), is related to hourly fire severity as described in Figure 706.4(1). By comparing Table 706.4 to Figure 706.4(2), it shows that the table is not based solely on the average fuel load of the occupancy. Not only are other factors, such as occupant density and evacuation capability, incorporated into the required fire-resistance rating, but it must also be recognized that fuel loads in a building are not evenly distributed. For example, the average fuel load for an occupancy in Group B may be 5 to 10 psf (24 to 49 kg/m²). The fuel load in file and storage rooms, however, may be 10 to 20 psf (49 to 98 kg/m²). While other sections of the code may address these areas by mandating an automatic fire suppression system with a partition capable of resisting the passage of smoke or a fire-resistance-rated enclosure of the incidental use area, the potential for higher fuel loads

Average fuel load psf	kg/m²	Equivalent fire endurance (hours)
5	24.4	$^1/_2$
7$^1/_2$	36.6	$^3/_4$
10	48.8	1
15	73.2	1$^1/_2$
20	97.6	2
30	146.5	3
40	195.3	4$^1/_2$
50	244.1	6
60	292.9	7$^1/_2$

For SI: 1 pound per square foot = 4.882 kg/m².

Figure 706.4(1)
RELATIONSHIP BETWEEN
FUEL LOAD AND FIRE ENDURANCE

is also factored into determining the required fire-resistance rating.

The minimum fire-resistance ratings required in Table 706.4 generally correlate with the fire-resistance ratings required by Table 508.3.3; however, Table 706.4 contains provisions that acknowledge the need for more conservative ratings for fire walls due to their unique role in creating separate buildings.

Due to the unique nature of the hazards presented by occupancies of Groups H-1, H-2 and H-3, Table 706.4 includes Note b, which references Sections 415.4 and 415.5 for specific requirements relative to the construction of such buildings (see commentary, Sections 415.4 and 415.5).

Occupancy	Combustibles in occupancy (psf)	Fire severity (hours)
Assembly	5 to 10	$^1/_2$ to 1
Business	5 to 10	$^1/_2$ to 1
Educational	5 to 10	$^1/_2$ to 1
Factory—Industrial		
Low hazard	0 to 10	0 to 1
Moderate hazard	10 to 25	1 to 3
Hazardous	Variable	Variable
Institutional	5 to 10	$^1/_2$ to 1
Mercantile	10 to 20	1 to 2
Residential	5 to 10	$^1/_2$ to 1
Storage		
Low hazard	0 to 10	0 to 1
Moderate hazard	10 to 30	1 to 1

For SI: 1 pound per square foot = 4.882 kg/m².

Figure 706.4(2)
OCCUPANCIES—FUEL LOAD—FIRE SEVERITY

706.5 Horizontal continuity. *Fire walls* shall be continuous from *exterior wall* to *exterior wall* and shall extend at least 18 inches (457 mm) beyond the exterior surface of *exterior walls*.

Exceptions:

1. *Fire walls* shall be permitted to terminate at the interior surface of combustible exterior sheathing or siding provided the *exterior wall* has a *fire-resistance rating* of at least 1 hour for a horizontal distance of at least 4 feet (1220 mm) on both sides of the *fire wall*. Openings within such *exterior walls* shall be protected by opening protectives having a *fire protection rating* of not less than $^3/_4$ hour.

2. *Fire walls* shall be permitted to terminate at the interior surface of noncombustible exterior sheathing, exterior siding or other noncombustible exterior finishes provided the sheathing, siding, or other exterior noncombustible finish extends a horizontal distance of at least 4 feet (1220 mm) on both sides of the *fire wall*.

3. *Fire walls* shall be permitted to terminate at the interior surface of noncombustible exterior sheathing where the building on each side of the *fire wall* is protected by an *automatic sprinkler system* installed in accordance with Section 903.3.1.1 or 903.3.1.2.

❖ Historically, the codes have addressed the hazards of fire exposure at the fire wall from only a vertical perspective; namely, at the roof (see Section 705.6). Section 705.5 addresses a similar fire hazard concern from the horizontal perspective; namely, at the intersection of the fire wall and the exterior wall. The horizontal exposure is similarly addressed in Section 1020.1.4 relative to stair enclosure exposure.

The 18-inch (457 mm) extension is intended to abate the potential for fire to travel from one building to the other around the fire wall. The 18-inch (457 mm) extension is required to extend the full height of the fire wall. The three exceptions acknowledge the effect certain types of exterior wall construction will have on fire breaching the exterior wall and exposing the adjacent building; namely, fire-resistance-rated construction (see Exception 1), noncombustible finish materials (see Exception 2) and noncombustible sheathing materials coupled with sprinkler protection (see Exception 3) provide the necessary barrier to limit fire spread across the exterior surface. The difference between Exceptions 2 and 3 is that Exception 2 would not permit a combustible exterior finish to be placed over the noncombustible exterior wall construction, while Exception 3 would allow a combustible exterior finish, provided the building is sprinklered with either an NFPA 13 or 13R system.

706.5.1 Exterior walls. Where the *fire wall* intersects *exterior walls*, the *fire-resistance rating* and opening protection of the *exterior walls* shall comply with one of the following:

1. The *exterior walls* on both sides of the *fire wall* shall have a 1-hour *fire-resistance rating* with $^3/_4$-hour protection where opening protection is required by Section 705.8. The *fire-resistance rating* of the *exterior wall* shall extend a minimum of 4 feet (1220 mm) on each side of the intersection of the *fire wall* to *exterior wall*. Exterior wall intersections at *fire walls* that form an angle equal to or greater than 180 degrees (3.14 rad) do not need *exterior wall* protection.

2. Buildings or spaces on both sides of the intersecting *fire wall* shall assume to have an imaginary *lot line* at the *fire wall* and extending beyond the exterior of the *fire wall*. The location of the assumed line in relation to the *exterior walls* and the *fire wall* shall be such that the *exterior wall* and opening protection meet the requirements set forth in Sections 705.5 and 705.8. Such protection is not required for *exterior walls* terminating at *fire walls* that

form an angle equal to or greater than 180 degrees (3.14 rad).

❖ This section deals only with fire walls that terminate at an exterior wall that does not continue in the same horizontal plane, but the exterior walls on each side of the fire wall form an angle of less than 180 degrees (3.14 rad). For example, the required fire wall may end at an

inside corner of an L-shaped building. Should this occur, not only does Section 706.5 apply, but this section has extra requirements to mitigate the exterior wall fire exposure from one building to another. This section states two methods of complying [see Figures 706.5.1(1), 706.5.1(2) and 706.5.1(3)].

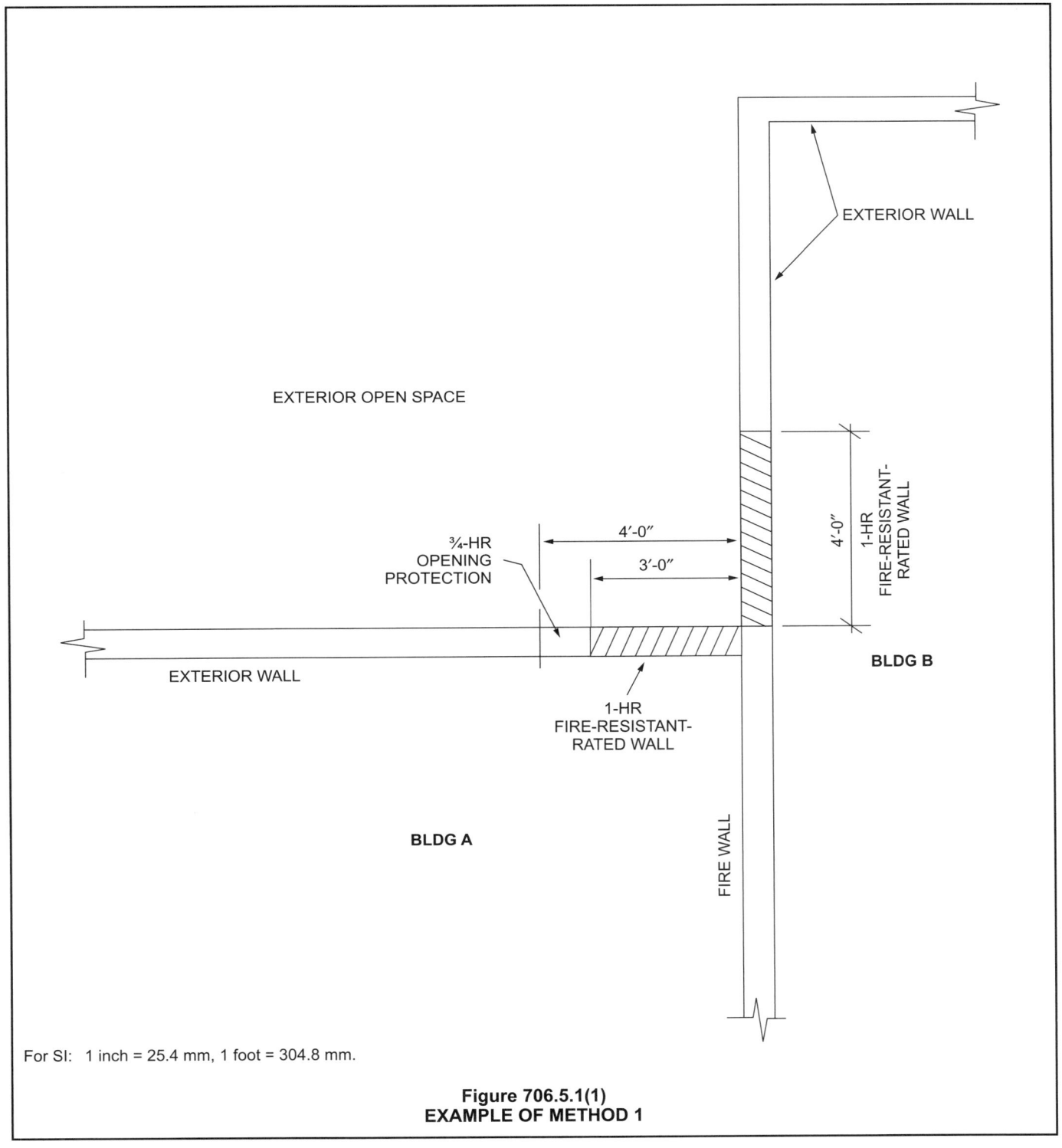

For SI: 1 inch = 25.4 mm, 1 foot = 304.8 mm.

Figure 706.5.1(1)
EXAMPLE OF METHOD 1

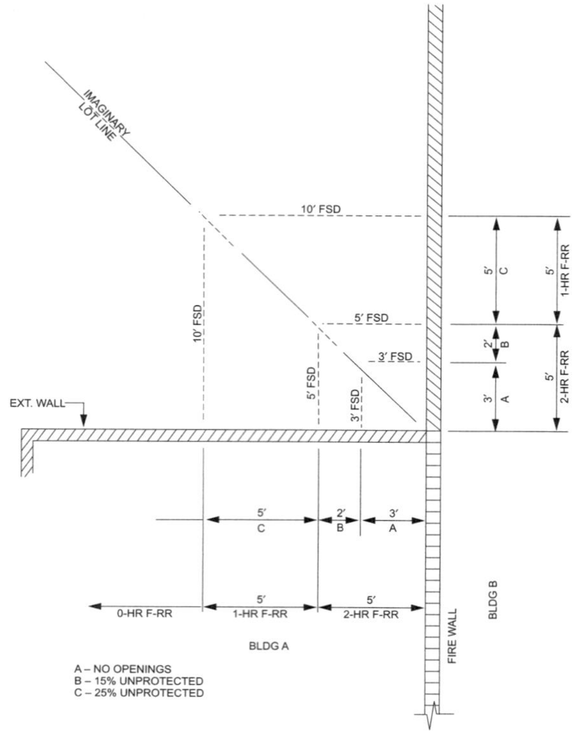

Figure 706.5.1(2)
EXAMPLE OF METHOD 2

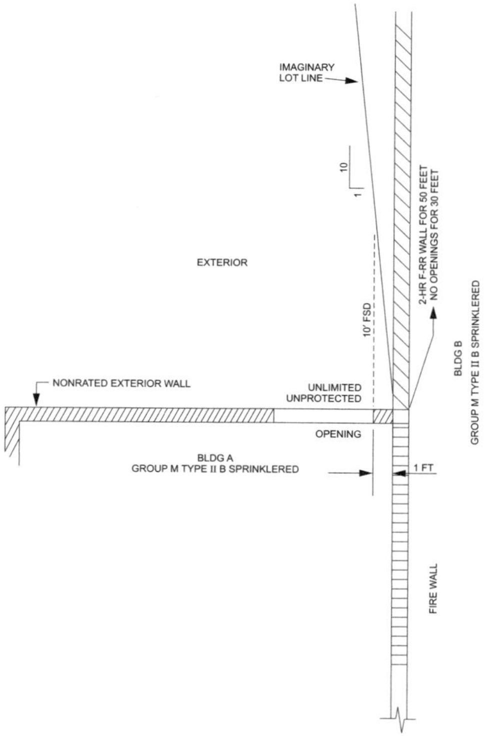

Figure 706.5.1(3)
EXAMPLE OF METHOD 3

706.5.2 Horizontal projecting elements. *Fire walls* shall extend to the outer edge of horizontal projecting elements such as balconies, roof overhangs, canopies, marquees and similar projections that are within 4 feet (1220 mm) of the *fire wall*.

Exceptions:

1. Horizontal projecting elements without concealed spaces, provided the *exterior wall* behind and below the projecting element has not less than 1-hour fire-resistance-rated construction for a distance not less than the depth of the projecting element on both sides of the *fire wall*. Openings within such *exterior walls* shall be protected by opening protectives having a *fire protection rating* of not less than $^3/_4$ hour.

2. Noncombustible horizontal projecting elements with concealed spaces, provided a minimum 1-hour fire-resistance-rated wall extends through the concealed space. The projecting element shall be separated from the building by a minimum of 1-hour fire-resistance-rated construction for a distance on each side of the *fire wall* equal to the depth of the projecting element. The wall is not required to extend under the projecting element where the building *exterior wall* is not less than 1-hour fire-resistance rated for a distance on each side of the *fire wall* equal to the depth of the projecting element. Openings within such *exterior walls* shall be protected by opening protectives having a *fire protection rating* of not less than $^3/_4$ hour.

3. For combustible horizontal projecting elements with concealed spaces, the *fire wall* need only extend through the concealed space to the outer edges of the projecting elements. The *exterior wall* behind and below the projecting element shall be of not less than 1-hour fire-resistance-rated construction for a distance not less than the depth of the projecting elements on both sides of the *fire wall*. Openings within such *exterior walls* shall be protected by opening protectives having a fire-protection rating of not less than $^3/_4$ hour.

❖ Fire walls are typically used to reduce the building area for the purpose of code compliance; however, many structures divided by fire walls are still detailed as a contiguous structure with projecting elements located on the exterior wall that extend across the fire wall. These projecting elements represent a potential conduit for fire to be transferred from one side of the building to the other side of the building. Even if the projecting elements are terminated at a distance in close proximity to the fire wall, the potential for fire spread to the adjacent projecting element still exists. In this section, 4 feet (1219 mm) (measured from the fire wall) is considered the appropriate threshold.

This section requires the fire wall to extend to the outer edge of the projecting element when such an element is located within 4 feet (1219 mm) of the fire wall. For example, in a contiguous residential structure with interior lot lines and fire walls or party walls between each dwelling unit, and where balconies are located within 4 feet (1219 mm) of the party or fire wall, then the fire/party wall must be extended to the outer edge of the balcony or to a point in line with the outer edge of the balcony.

Exception 1 acknowledges the reduction in hazard due to the lack of a concealed space crossing the fire wall; however, since the projecting element provides a connection between the buildings, there is still a potential for fire to transfer across the wall, thus the need for a rated exterior wall separation between the inside of the building and the projection. Exceptions 2 and 3 address projecting elements with concealed spaces of noncombustible and combustible construction, respectively.

706.6 Vertical continuity. *Fire walls* shall extend from the foundation to a termination point at least 30 inches (762 mm) above both adjacent roofs.

Exceptions:

1. Stepped buildings in accordance with Section 706.6.1.

2. Two-hour fire-resistance-rated walls shall be permitted to terminate at the underside of the roof sheathing, deck or slab, provided:

    2.1. The lower roof assembly within 4 feet (1220 mm) of the wall has not less than a 1-hour *fire-resistance rating* and the entire length and span of supporting elements for the rated roof assembly has a *fire-resistance rating* of not less than 1 hour.

    2.2. Openings in the roof shall not be located within 4 feet (1220 mm) of the *fire wall*.

    2.3. Each building shall be provided with not less than a Class B roof covering.

3. Walls shall be permitted to terminate at the underside of noncombustible roof sheathing, deck or slabs where both buildings are provided with not less than a Class B roof covering. Openings in the roof shall not be located within 4 feet (1220 mm) of the *fire wall*.

4. In buildings of Type III, IV and V construction, walls shall be permitted to terminate at the underside of combustible roof sheathing or decks, provided:

    4.1. There are no openings in the roof within 4 feet (1220 mm) of the *fire wall*,

    4.2. The roof is covered with a minimum Class B roof covering, and

    4.3. The roof sheathing or deck is constructed of *fire-retardant-treated wood* for a distance of 4 feet (1220 mm) on both sides of the wall or the roof is protected with $^5/_8$-inch (15.9 mm) Type X gypsum board directly beneath the underside of the roof sheathing or deck, supported by a minimum of 2-inch (51 mm) nominal ledgers attached to the sides of the roof framing members for a minimum distance of 4 feet (1220 mm) on both sides of the *fire wall*.

5. In buildings designed in accordance with Section 509.2, *fire walls* located above the 3-hour *horizontal assembly* required by Section 509.2, Item 1 shall be permitted to extend from the top of this *horizontal assembly*.

❖ One of the primary purposes of a fire wall is to provide protection from an exposure fire. A fire wall must be continuous from the foundation to or through the roof. The fire wall is required to extend to 30 inches (762 mm) above the roof surface, except where other provisions are met that will restrict the spread of fire (see Exceptions 1 through 5). If the two buildings have different roof levels, the fire wall must comply with the provisions of Section 706.6.1.

Exception 2 to this section's requirement for a parapet applies to buildings where: (a) the maximum required fire-resistance rating of the fire wall is 2 hours; (b) the roof assembly located within 4 feet (1219 mm) of the fire wall has a minimum fire-resistance rating of 1 hour (including all supporting elements and structural members); (c) no roof openings are located within 4 feet (1219 mm) of the fire wall; and (d) the buildings on either side of the fire wall have a minimum Class B roof covering. The provisions of Exception 2 are not applicable to buildings with combustible roof construction where a fire wall is required to have a fire-resistance rating of 3 hours or greater. Buildings where the fire wall is required to have a 3-hour or greater fire-resistance rating must comply with the general provisions of Section 706.6, or with Exceptions 3 or 4 in order to qualify for the parapet extension exception.

Exception 3 states that, since the intent of the parapet is to prevent fire spread over the roof, it is not required if the entire roof is of noncombustible construction; the fire wall is continuous to the underside of

the roof deck, roof slab or roof sheathing; the roof-covering material has a minimum Class B classification in accordance with the provisions of Section 1505.3 (see commentary, Section 1505.3) and there are no roof openings located within 4 feet (1219 mm) on either side of the fire wall.

The intent of the requirement for a Class B roof covering is due to the fact that roof coverings are not considered part of the roof, and therefore, are not subject to the noncombustibility requirements of this section. This section requires that the roof covering be effective against moderate fire test exposure to reduce the potential for fire spread across the roof. The restriction on the location of roof openings is intended to minimize the potential of the spread of fire from one building to an adjacent building via roof openings, such as skylights, pentrations or roof windows. This fire spread could occur from something such as a fan that breaches the roof opening in the involved building and exposes roof openings in the adjacent building to direct fire or to burning brands, which could land on the roof, burn through the roof opening and ignite combustibles in the interior of the adjacent building. The 4-foot (1219 mm) threshold provides a reasonable approximation of a fire plume, beyond which the exposure to fire of the adjacent building is reduced. Exception 4 to the requirement in Section 706.6 for a parapet is the condition in buildings of Type III, IV and V construction where the sheathing or deck is constructed of noncombustible materials or approved fire-retardant-treated wood for a distance of 4 feet (1219 mm) on both sides of the wall (see commentary, Section 2303.2 for additional information on fire-retardant-treated wood) or where the prescribed application of $^5/_8$-inch (16 mm) Type X gypsum board is applied to the

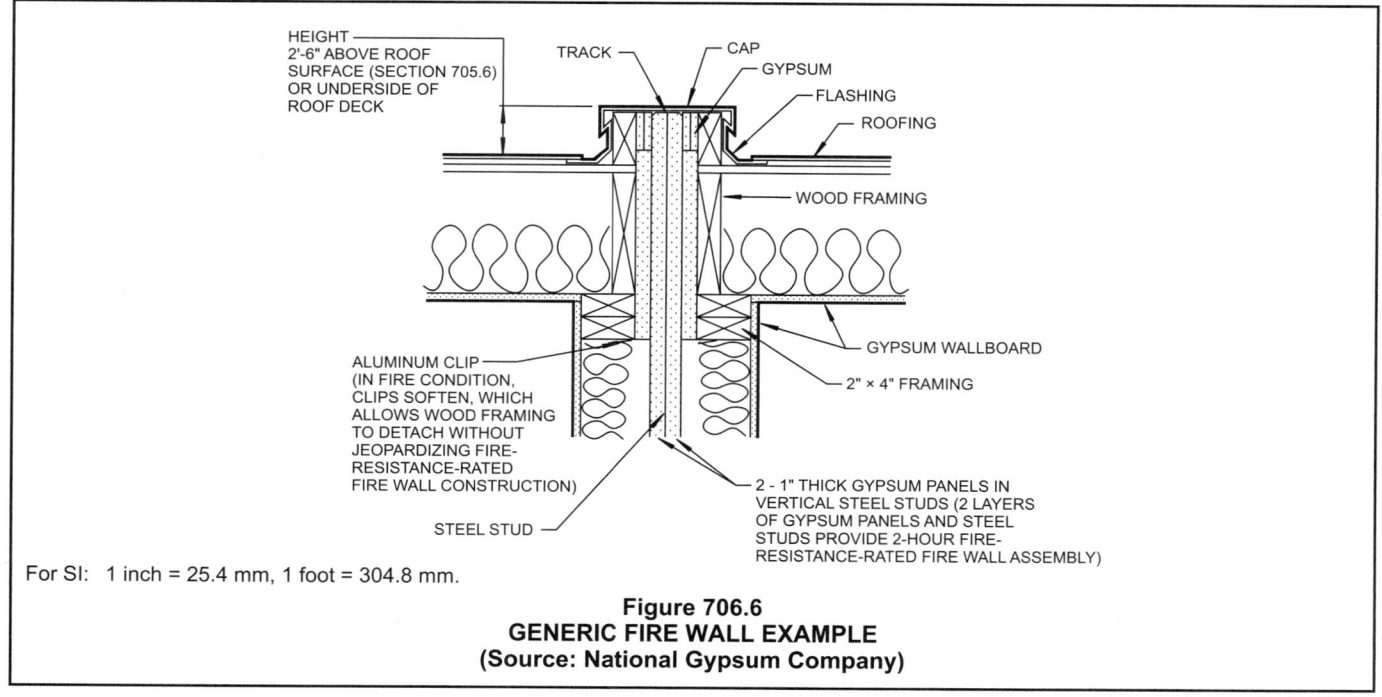

For SI: 1 inch = 25.4 mm, 1 foot = 304.8 mm.

Figure 706.6
GENERIC FIRE WALL EXAMPLE
(Source: National Gypsum Company)

underside of the deck for a distance of 4 feet (1219 mm) on both sides of the wall. The roof coverings on both buildings must have a minimum Class B rating and no roof openings are to be located within 4 feet (1219 mm) of the fire wall (see commentary, Section 508.2). The intent is the same: to resist the passage of flame beyond the fire wall.

706.6.1 Stepped buildings. Where a *fire wall* serves as an *exterior wall* for a building and separates buildings having different roof levels, such wall shall terminate at a point not less than 30 inches (762 mm) above the lower roof level, provided the *exterior wall* for a height of 15 feet (4572 mm) above the lower roof is not less than 1-hour fire-resistance-rated construction from both sides with openings protected by fire assemblies having a *fire protection rating* of not less than $^3/_4$ hour.

> **Exception:** Where the *fire wall* terminates at the underside of the roof sheathing, deck or slab of the lower roof, provided:
>
> 1. The lower roof assembly within 10 feet (3048 mm) of the wall has not less than a 1-hour *fire-resistance rating* and the entire length and span of supporting elements for the rated roof assembly has a fire-resistance rating of not less than 1 hour.
>
> 2. Openings in the lower roof shall not be located within 10 feet (3048 mm) of the *fire wall.*

❖ The basic provisions of Section 706.6 require that a parapet extend a minimum of 30 inches (762 mm) above the roof surfaces on both sides of the fire wall, which would then require fire walls separating buildings with different roof heights to extend 30 inches (762 mm) above the highest roof surface. The provisions of Section 706.6.1 address situations where a

fire wall separates adjacent buildings with a difference in roof height. These provisions acknowledge that fire exposure to the exterior wall of an adjacent building from the roof of a lower building represents, to a certain extent, a reduced hazard.

This section retains the fire wall extension above the lower roof surface and places a 15-foot (4572 mm) limit on rated wall construction and opening protectives, while the exception allows the fire wall extension to be eliminated as long as a fire barrier, in the form of 1-hour fire-rated roof construction, that extends a minimum of 10 feet (3048 mm) from the fire wall, is provided. The option provided by this section permits openings in an exterior wall that extend above the fire wall, where the fire wall extends 30 inches (762 mm) above the lower roof surface and the exterior wall extending above the fire wall is constructed as a 1-hour fire-resistance-rated wall, rated for exposure from both sides, for a distance of 15 feet (4572 mm) above the lower roof surface. Where the fire wall and exterior wall comply with these requirements, openings are permitted in the exterior wall extension, provided that all openings located within 15 feet (4572 mm) of the lower roof surface are protected with $^3/_4$-hour fire-rated opening protectives. Openings located more than 15 feet (4572 mm) above the lower roof surface are not required to have opening protectives [see Figure 706.6.1(1)]. The provisions of this section are similar to those contained in Section 704.10 for vertical exposure of exterior openings. They require that when the difference in height between the roof surfaces of the adjacent buildings is less than 15 feet (4572 mm), the exterior wall extension is required to extend to the underside of the upper roof deck [see Figure 706.6.1(2)].The exception permits openings in

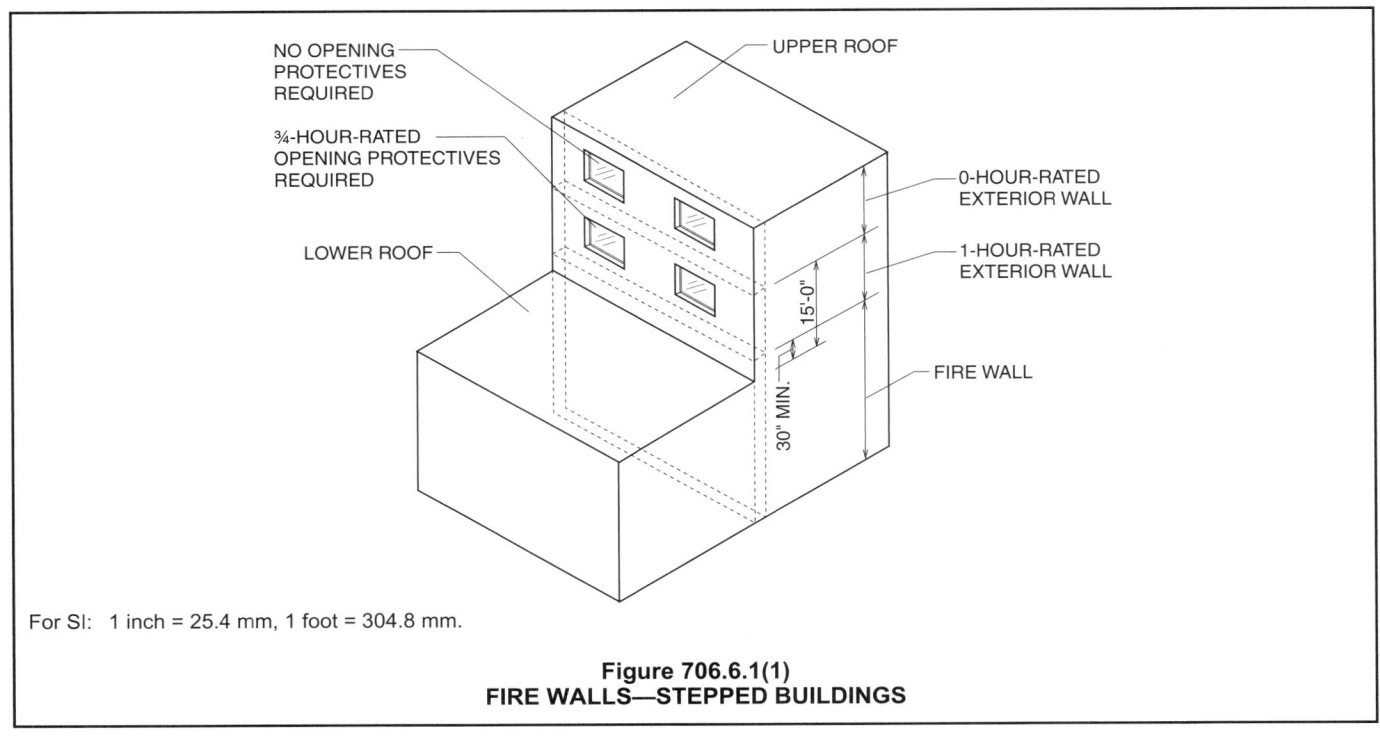

NO OPENING PROTECTIVES REQUIRED

¾-HOUR-RATED OPENING PROTECTIVES REQUIRED

LOWER ROOF

UPPER ROOF

0-HOUR-RATED EXTERIOR WALL

1-HOUR-RATED EXTERIOR WALL

FIRE WALL

15'-0"

30" MIN.

For SI: 1 inch = 25.4 mm, 1 foot = 304.8 mm.

Figure 706.6.1(1)
FIRE WALLS—STEPPED BUILDINGS

the exterior wall that extends above the fire wall, where: (a) the fire wall terminates at the bottom of the roof deck of the lower roof; (b) the lower roof assembly has a minimum 1-hour fire-resistance rating for a minimum distance of 10 feet (3048 mm) from the fire wall and (c) there are no openings located in the lower roof within 10 feet (3048 mm) of the fire wall. Openings located in the exterior wall above the lower roof surface are not required to have opening protectives [see Figure 706.6.1(3)]. It must be noted that all structural elements, including beams, columns and bearing walls that provide support for the fire-resistance-rated roof assembly, must also have a 1-hour fire-resistance rating for their entire length or span in order to maintain the effectiveness of the rated roof assembly. The provisions of this section are similar to those contained in the section for the protection of the exterior walls of exit stairways. The provisions for stepped buildings apply only where a fire wall is required in between the buildings, and do not apply to party walls on real property lines.

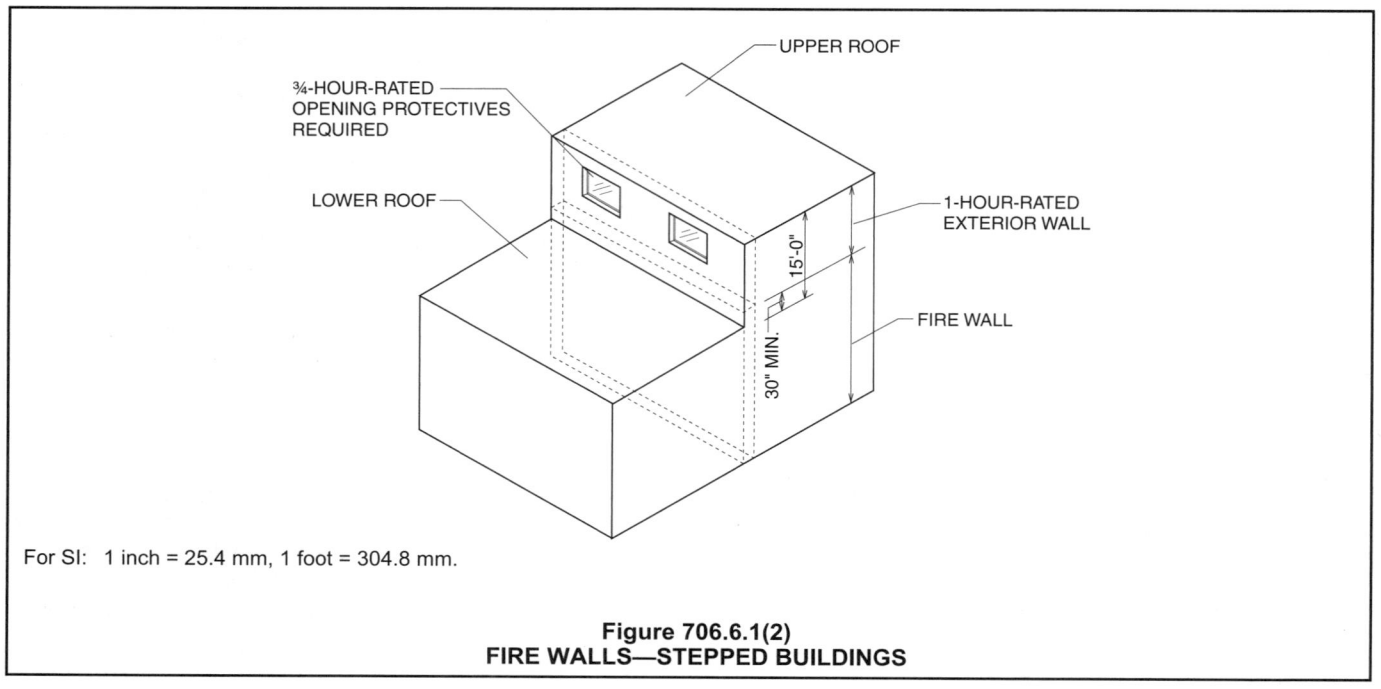

For SI: 1 inch = 25.4 mm, 1 foot = 304.8 mm.

Figure 706.6.1(2)
FIRE WALLS—STEPPED BUILDINGS

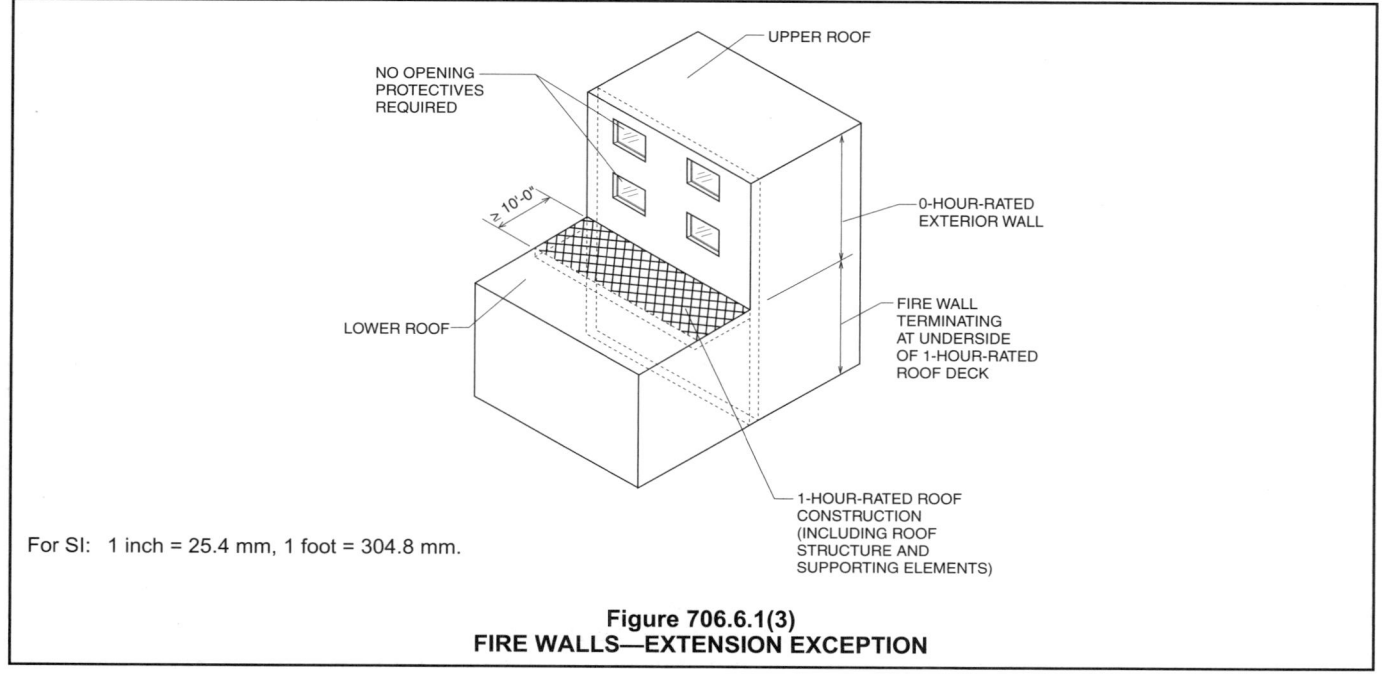

For SI: 1 inch = 25.4 mm, 1 foot = 304.8 mm.

Figure 706.6.1(3)
FIRE WALLS—EXTENSION EXCEPTION

706.7 Combustible framing in fire walls. Adjacent combustible members entering into a concrete or masonry *fire wall* from opposite sides shall not have less than a 4-inch (102 mm) distance between embedded ends. Where combustible members frame into hollow walls or walls of hollow units, hollow spaces shall be solidly filled for the full thickness of the wall and for a distance not less than 4 inches (102 mm) above, below and between the structural members, with noncombustible materials *approved* for fireblocking.

❖ In order to retain the fire-resistance capability of the wall where combustible members will frame into it, hollow walls or walls of hollow units must be solidly filled for the thickness of the wall and for a distance of not less than 4 inches (102 mm) above, below and between the structural members. Consistent with the construction of the walls, the fireblocking materials are to be noncombustible and approved for fireblocking in accordance with Section 717.2. If combustible members enter both sides of a fire wall, there must be at least 4 inches (102 mm) of masonry between the embedded ends.

In order to maintain structural integrity where wood beams or joists are supported in and on masonry fire walls, the joist and beams should be shaped so that they can rotate out of the pocket without exerting undue force on the wall.

706.8 Openings. Each opening through a *fire wall* shall be protected in accordance with Section 715.4 and shall not exceed 156 square feet (15 m²). The aggregate width of openings at any floor level shall not exceed 25 percent of the length of the wall.

Exceptions:

1. Openings are not permitted in party walls constructed in accordance with Section 706.1.1.

2. Openings shall not be limited to 156 square feet (15 m²) where both buildings are equipped throughout with an *automatic sprinkler system* installed in accordance with Section 903.3.1.1.

❖ In order to maintain the integrity of the fire wall, the maximum area and percent of openings in the wall are restricted. When provided, the openings must be properly protected so that the fire-resistance rating of the wall is maintained. This section prescribes the maximum area and the percent of openings that may be permitted in a fire wall at any one floor level. The provisions must be used in concert with Section 706.1.1, which limits openings for party walls.

Fire wall openings have restrictive limitations in their size and total area because of the critical function that a fire wall serves. To maintain the required fire performance of the fire wall, each opening through a fire wall is restricted in area to 156 square feet (11 m²), and the aggregate width of all openings at any one floor level may not constitute more than 25 percent of the length of the wall. The 156-square-foot (11 m²) limitation provides a reasonable size through which industrial machinery may pass and corresponds with the maximum area limitations of many tested fire doors.

Recognizing the effectiveness of automatic sprinklers, the 156-square-foot (11 m²) opening limitation does not apply where the buildings on both sides of the fire wall are fully sprinklered (see Exception 2); however, the aggregate width of all openings in a fire wall at any one floor level is still limited to 25 percent of the length of the wall.

706.9 Penetrations. Penetrations of *fire walls* shall comply with Section 713.

❖ In order to maintain the integrity of the required fire-resistance rating, penetrations through the fire wall must be properly protected. Acceptable protection methods for various penetrations of fire walls are identified in Sections 713.2 and 713.3.

706.10 Joints. Joints made in or between *fire walls* shall comply with Section 714.

❖ Joints, such as expansion or seismic, are another form of openings in fire walls and, therefore, must be considered with regard to maintaining the fire-resistance ratings of fire walls. This section requires all joints that are located in fire walls to be protected by a joint system with a fire-resistance rating and to comply with the requirements of Section 714.

706.11 Ducts and air transfer openings. Ducts and air transfer openings shall not penetrate *fire walls*.

Exception: Penetrations by ducts and air transfer openings of *fire walls* that are not on a *lot line* shall be allowed provided the penetrations comply with Section 716. The size and aggregate width of all openings shall not exceed the limitations of Section 706.8.

❖ The general provisions of this section mirror those of Section 706.1.1 for party walls. The exception permits duct and transfer openings for fire walls not located on a lot line, provided the maximum aggregate area provisions of Section 706.8 are met and that the openings are protected in accordance with Section 716.

SECTION 707
FIRE BARRIERS

707.1 General. *Fire barriers* installed as required elsewhere in this code or the *International Fire Code* shall comply with this section.

❖ The provisions of this section apply to assemblies that are required to have a fire-resistance rating and to be constructed as "fire barriers." As addressed in Section 707.3, fire barriers are used for separating exits, incidental use areas, shafts, hazardous materials control areas and fire areas. Fire barriers provide a higher degree of protection than fire partitions (see Section 708), but lack the inherent structural integrity of fire walls. Fire barriers limit the number of openings. Fire barrier wall assemblies must be continuous from the top of a fire-resistance-rated floor/ceiling assembly to the bottom of the floor or roof slab/deck above. Unlike fire partitions addressed in Section 709, there are no circumstances under which a fire barrier wall is permitted to terminate at a ceiling.

Fire barriers are used for a variety of purposes, such as mixed occupancies, areas of refuge separations, shafts, and exit and floor opening enclosures. Fire barriers also include interior walls that serve to subdivide a space by separating one fire area from an adjacent fire area and for separating incidental use areas (see Section 508.2.5). Fire-resistance-rated assemblies used to separate exit access corridors in many applications, as well as tenant, dwelling unit and guestroom separations, are fire partitions (see Section 709). The provisions of this section provide minimum requirements for the fire-resistance rating, continuity, combustibility and protection of openings and penetrations in order to help maintain the reliability of the fire separation assembly. As with any fire-resistance-rated assembly, consideration must be given to the openings and penetrations that are provided within the assembly. The intent is to maintain the fire-resistance rating of the assembly. These sections recognize that fire spread beyond a fire-resistance-rated compartment is often attributed to the protection given to any opening or penetration in the fire barrier, or the lack thereof.

Since the fire barrier is intended to provide a reliable subdivision of areas, the construction that structurally supports the assembly is required to provide at least the same hourly fire-resistance rating as the fire barrier. This is applicable regardless of the type of construction of the building. Structural stability is regulated by Section 707.5.

707.2 Materials. *Fire barriers* shall be of materials permitted by the building type of construction.

❖ The types of materials used in fire barriers are to be consistent with Sections 602 through 602.5 for the type of construction classification of the building. The fire-resistance ratings of fire barriers used to separate mixed occupancies are determined in accordance with Section 508.4 (see commentary, Section 508.4). Fire barriers are permitted to be of combustible materials in Type III, IV and V construction, and are required to be of noncombustible materials in Type I and II construction.

707.3 Fire-resistance rating. The *fire-resistance rating* of *fire barriers* shall comply with this section.

❖ The elements that are identified in Section 707 must be constructed as required for fire barriers and must be fire-resistance rated as required by the code sections referenced in Section 707.

707.3.1 Shaft enclosures. The *fire-resistance rating* of the *fire barrier* separating building areas from a shaft shall comply with Section 708.4.

❖ See Section 708 for the fire-resistance rating for shaft enclosures and for the locations that require shaft enclosures and the exceptions to shaft enclosures.

707.3.2 Exit enclosures. The *fire-resistance rating* of the *fire barrier* separating building areas from an *exit* shall comply with Section 1022.1.

❖ See Section 1022.1 for the fire-resistance rating for exit enclosures and for the locations that require exit enclosures and the exceptions to exit enclosures. Exit enclosures may be fire barriers complying with this section and horizontal assemblies complying with Section 712.

707.3.3 Exit passageway. The *fire-resistance rating* of the *fire barrier* separating building areas from an *exit* passageway shall comply with Section 1023.3.

❖ See Section 1023.3 for the fire-resistance rating for fire barriers forming exit passageways. Exit passageway enclosures may be fire barriers complying with this section or horizontal assemblies complying with Section 712.

707.3.4 Horizontal exit. The *fire-resistance rating* of the separation between building areas connected by a horizontal *exit* shall comply with Section 1025.1.

❖ See Section 1025.1 for the fire-resistance rating of the fire barriers forming horizontal exits. Horizontal exits may be formed by fire walls meeting Section 706, fire barriers complying with this section or horizontal assemblies complying with Sections 712 and 1025.2.

707.3.5 Atriums. The *fire-resistance rating* of the *fire barrier* separating atriums shall comply with Section 404.6

❖ See Section 404.6 for the fire-resistance rating of the fire barrier walls separating the atrium area from other building use areas. Horizontal assemblies meeting Section 712 can also provide the separation of the atrium area.

707.3.6 Incidental accessory occupancies. The *fire barrier* separating incidental accessory occupancies from other spaces in the building shall have a *fire-resistance rating* of not less than that indicated in Table 508.2.5.

❖ Table 508.2.5 states the fire-resistance requirements for fire barrier walls separating incidental use areas. For some incidental uses, there is a sprinkler option in lieu of the fire-resistance-rated fire barrier wall. With the sprinkler option, the wall is not required to be fire-resistance rated but must be capable of resisting the passage of smoke. Those nonrated wall assemblies are not required to be constructed as fire barriers.

707.3.7 Control areas. *Fire barriers* separating *control areas* shall have a *fire-resistance rating* of not less than that required in Section 414.2.4.

❖ Control areas refer to areas within a building where limited quantities of hazardous materials are stored or used. Section 414.2.4 references Table 414.2.4 for the design and number of control areas. The required fire-resistance rating for the fire barriers separating the control areas from the rest of the area is also found within the table.

707.3.8 Separated occupancies. Where the provisions of Section 508.4 are applicable, the *fire barrier* separating mixed occupancies shall have a *fire-resistance rating* of not less than that indicated in Table 508.4 based on the occupancies being separated.

❖ Where separation of mixed occupancies is the chosen option for mixed uses in a building, the required fire separation is either a fire barrier, meeting the requirements of this section, or a horizontal fire separation assembly, meeting the requirements of Section 712 (see Section 508.4).

707.3.9 Fire areas. The *fire barriers* or *horizontal assemblies*, or both, separating a single occupancy into different *fire areas* shall have a *fire-resistance rating* of not less than that indicated in Table 707.3.9. The *fire barriers* or *horizontal assemblies*, or both, separating *fire areas* of mixed occupancies shall have a *fire-resistance rating* of not less than the highest value indicated in Table 707.3.9 for the occupancies under consideration.

❖ One of the alternatives available in addressing fire protection systems in many buildings is to divide the building into separate fire areas (see the definition of "Fire areas" in Section 709). Since many of the fire suppression system thresholds (see Section 903.2) are based upon the size of the fire area, separation of a single occupancy into small fire areas can be an acceptable method for avoiding the use of sprinklers. This is a classic type of design decision: sprinklers versus compartmentation. If the separation is provided, each fire area may be evaluated separately for purposes of determining the applicable provisions of the code. Table 707.3.9 provides the minimum required fire-resistance ratings of the fire barrier wall or horizontal assembly separating two fire areas of the same occupancy groups.

Areas separated with fire barriers are not considered separate buildings; they are considered separate fire areas. Two areas must be separated by a fire wall or exterior walls to be considered separate buildings. Two areas separated with fire barriers are still considered part of a single building. This distinction is critical in determining compliance with allowable height and area, and other code provisions.

TABLE 707.3.9
FIRE-RESISTANCE RATING REQUIREMENTS FOR FIRE BARRIER ASSEMBLIES OR HORIZONTAL ASSEMBLIES BETWEEN FIRE AREAS

OCCUPANCY GROUP	FIRE-RESISTANCE RATING (hours)
H-1, H-2	4
F-1, H-3, S-1	3
A, B, E, F-2, H-4, H-5, I, M, R, S-2	2
U	1

707.4 Exterior walls. Where *exterior walls* serve as a part of a required fire-resistance-rated shaft or *exit* enclosure, or separa-

tion, such walls shall comply with the requirements of Section 705 for *exterior walls* and the fire-resistance-rated enclosure or separation requirements shall not apply.

Exception: *Exterior walls* required to be fire-resistance rated in accordance with Section 1019 for exterior egress balconies, Section 1022.6 for *exit* enclosures and Section 1026.6 for exterior *exit* ramps and *stairways*.

❖ If an area is required to be enclosed by fire barriers and an exterior wall constitutes part of the enclosure, the exterior wall is only required to comply with the fire-resistance-rating requirements in Section 705, unless the exterior wall is protecting part of an exterior egress balcony, an exit enclosure or an exterior stairway or ramp (see commentary, Sections 1019, 1022.6 and 1026.6). The intent of the fire barrier requirements is to subdivide or enclose areas to protect them from a fire in the building. In general, the exterior wall need only have a fire-resistance rating if required for structural stability (see Table 601) or because of exterior exposure potential (see Table 602 and Section 705.5). The exception, however, points to three sections of the code where the exterior wall must be rated for reasons other than the fire separation distance.

707.5 Continuity. Fire barriers shall extend from the top of the floor/ceiling assembly below to the underside of the floor or roof sheathing, slab or deck above and shall be securely attached thereto. Such *fire barriers* shall be continuous through concealed spaces, such as the space above a suspended ceiling.

❖ To minimize the potential for fire spread from one area to another over a fire barrier wall, such fire barrier assemblies must be continuous from the fire-resistance-rated floor/ceiling assembly below to the underside of the floor slab or roof deck above (see Figure 707.5). To maintain the efficiency of the fire barrier, it must be continuous through all concealed spaces (such as a space above a suspended ceiling), be constructed tight and securely attached to the underside of the floor slab or roof deck.

707.5.1 Supporting construction. The supporting construction for a *fire barrier* shall be protected to afford the required *fire-resistance rating* of the *fire barrier* supported. Hollow vertical spaces within a *fire barrier* shall be fireblocked in accordance with Section 717.2 at every floor level.

Exceptions:

1. The maximum required *fire-resistance rating* for assemblies supporting *fire barriers* separating tank storage as provided for in Section 415.6.2.1 shall be 2 hours, but not less than required by Table 601 for the building construction type.

2. Shaft enclosures shall be permitted to terminate at a top enclosure complying with Section 708.12.

3. Supporting construction for 1-hour *fire barriers* required by Table 508.2.5 in buildings of Type IIB, IIIB and VB construction is not required to be

fire-resistance rated unless required by other sections of this code.

❖ In general, fire barriers must be supported by construction having an equivalent fire-resistance rating. If the supporting structure is a primary structural frame (see definition) and supports a fire barrier wall more than two stories in height, the fire-resistance rating for the supporting structure must be protected by the individual encasement method in Section 704.3. If the supporting members are a secondary structural member, then the supporting structure can be protected by membrane protection as in Section 712 for horizontal assemblies. The intent of this requirement is to prevent the effectiveness of the assembly from being circumvented by a fire that threatens the supporting elements. The requirement for the supporting construction to be fire-resistance rated applies to buildings of all types of construction, even to buildings of Type IIB, IIIB and VB construction for all fire barrier walls except those separating incidental use areas.

Exception 1 is not an exception at all, but a requirement that supporting structures for fire barriers separating flammable or combustible tank storage be 2-hour fire-resistance rated.

Exception 2 is an exception to the continuity requirement for shaft enclosure walls.

Exception 3 allows only incidental use area separation walls in Type IIB, IIIB and VB construction to be supported on nonfire-resistance-rated construction if no other code section requires the supporting elements to be fire-resistance rated.

Fire barrier walls will usually be built on top of a floor and will terminate at the floor above. Should a fire barrier wall, as in the case of some shaft walls, be constructed through a floor, any hollow space within that wall could provide a passage for fire or smoke and, therefore, must be fireblocked as specified in the last sentence of Section 707.5.1. Any hollow vertical spaces within fire barrier walls must be fireblocked in accordance with Section 717.2.2.

707.6 Openings. Openings in a *fire barrier* shall be protected in accordance with Section 715. Openings shall be limited to a maximum aggregate width of 25 percent of the length of the wall, and the maximum area of any single opening shall not exceed 156 square feet (15 m²). Openings in *exit* enclosures and *exit* passageways shall also comply with Sections 1022.3 and 1023.5, respectively.

Exceptions:

1. Openings shall not be limited to 156 square feet (15 m²) where adjoining floor areas are equipped throughout with an *automatic sprinkler system* in accordance with Section 903.3.1.1.

2. Openings shall not be limited to 156 square feet (15 m²) or an aggregate width of 25 percent of the length of the wall where the opening protective is a *fire door* serving an *exit* enclosure.

3. Openings shall not be limited to 156 square feet (15 m²) or an aggregate width of 25 percent of the length of the wall where the opening protective has been tested in accordance with ASTM E 119 or UL 263 and has a minimum *fire-resistance rating* not less than the *fire-resistance rating* of the wall.

4. Fire window assemblies permitted in atrium separation walls shall not be limited to a maximum aggregate width of 25 percent of the length of the wall.

5. Openings shall not be limited to 156 square feet (15 m²) or an aggregate width of 25 percent of the length of the wall where the opening protective is a *fire door* assembly in a *fire barrier* separating an *exit* enclosure from an *exit* passageway in accordance with Section 1022.2.1.

❖ Buildings, in order to provide utility, must provide access into different areas that include the need to provide openings. Section 707.7 defines what openings are permitted and how they have to be protected to maintain the integrity of the fire barrier. To maintain the viability of the fire barrier, the aggregate width of open-

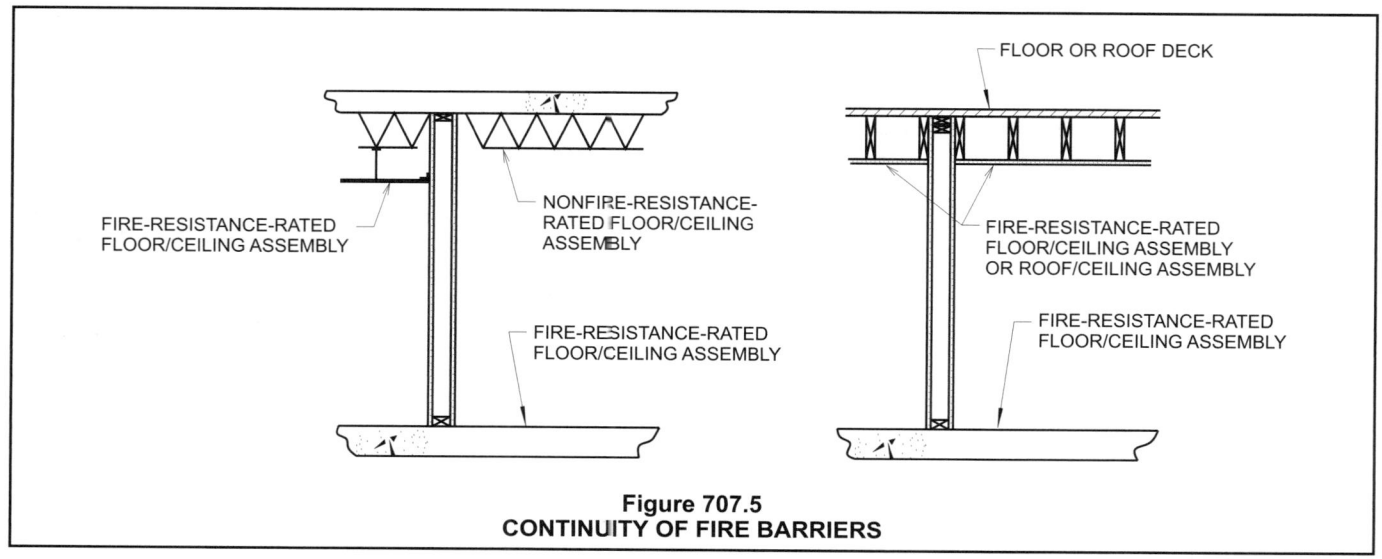

Figure 707.5
CONTINUITY OF FIRE BARRIERS

FLOOR OR ROOF DECK

FIRE-RESISTANCE-RATED FLOOR/CEILING ASSEMBLY

NONFIRE-RESISTANCE-RATED FLOOR/CEILING ASSEMBLY

FIRE-RESISTANCE-RATED FLOOR/CEILING ASSEMBLY OR ROOF/CEILING ASSEMBLY

FIRE-RESISTANCE-RATED FLOOR/CEILING ASSEMBLY

FIRE-RESISTANCE-RATED FLOOR/CEILING ASSEMBLY

ings is restricted to a maximum of 25 percent of the length of the wall. This limitation is based on the fact that the criteria for opening protectives do not include limitations on unexposed surface temperature or radiant heat transfer. Consistent with typical listing limitations, a single opening protective is limited to a maximum of 156 square feet (15 m²). It should be noted, however, that certain opening protectives, such as fire windows, are often limited to much smaller areas per opening.

Traditionally, the limiting size on openings in fire-resistance-rated walls has been based on the maximum sizes identified in the fire door listings. The previous size limitation was 120 square feet (11 m²) due to the limitations of such listings. In 2006, the code increased the maximum permitted size of an opening protective to 156 square feet (15 m²) in a fire barrier, based upon the current listing limitations of steel fire doors. The maximum length of any single opening is limited to 13 feet, 6 inches (4114 mm) for the same reason during the testing process.

The reference to Sections 1022.3 and 1023.5 specifies that in exit enclosures and exit passageways, only openings for the purpose of exiting from normally occupied spaces are permitted. Spaces that are not normally occupied, such as janitor closets or mechanical and electrical rooms, are not permitted to open directly into these protected exit systems, since fire in those areas may produce large volumes of heat and smoke that could readily enter the exit enclosure and delay or prevent egress. In order to maintain the required fire-resistance rating of the assembly, opening protectives must have a fire protection rating in accordance with Section 715. The reference to Section 715 is intended to identify the required fire protection rating for the opening protective, as indicated in Table 715.4, installation requirements and the applicable test standards.

Openings in fire barriers are not limited to 156 square feet (15 m²) when all fire areas separated by the assembly are equipped throughout with automatic sprinkler protection (see Exception 1). This exception is similar to the one made for fire walls (see Section 706.8), based on the historical fire record of automatic sprinkler systems. Although the openings in the fire barrier are not limited in size under this exception, they are still required to be protected by opening protectives that meet the requirements of Section 715.

Exception 2 acknowledges the practicality of the 25-percent limitation for walls of an exit enclosure. Most exit enclosures are of such limited size that the placement of the fire door in the wall of the enclosure often exceeds 25 percent of the wall.

Exception 3 addresses new opening protective products that have been tested to the more rigorous provisions of ASTM E 119 or UL 263 rather than, or in addition to, the opening protective standard NFPA 252 or UL 10C. Since the opening protective has been tested to the same standard as the wall itself, it is then logical to allow such an opening protective without restrictions (see commentary, Section 703.5).

Exception 4 acknowledges the inclusive requirements for openings into atriums addressed in Section 404 that include glazing material allowances not found elsewhere in Chapter 7.

Exception 5 is similar to Exception 2 in that the wall area between the exit enclosure and the exit passageway is usually so very small that the door opening will nearly always exceed the 25-percent limitation.

707.7 Penetrations. Penetrations of *fire barriers* shall comply with Section 713.

❖ In order to maintain the integrity of the fire barrier, penetrations into and through the fire-resistance-rated wall must be properly protected. Acceptable methods for various penetrations of fire barriers are identified in Sections 713.3 through 713.3.1.2. The provisions of Section 707.7.1 must be used when the penetration is into an exit enclosure or exit passageway.

707.7.1 Prohibited penetrations. Penetrations into an *exit* enclosure or an *exit* passageway shall be allowed only when permitted by Section 1022.4 or 1023.6, respectively.

❖ This section reminds the code user that although the penetration firestop systems do provide protection for the penetration, the code prohibits most penetrations through exit enclosures. Only penetrations of items, such as sprinkler piping, necessary ductwork for stair pressurization and electrical conduit that serve the exit enclosure, are allowed. There can never be a penetration through a fire barrier that separates an adjacent exit enclosure (see Sections 1022.4 and 1023.6).

707.8 Joints. Joints made in or between *fire barriers*, and joints made at the intersection of *fire barriers* with underside of the floor or roof sheathing, slab or deck above, shall comply with Section 714.

❖ This section regulates joints or linear openings created between building assemblies, which are sometimes referred to as construction, expansion or seismic joints. These joints are most often created where the structural design of a building necessitates a separation between building components in order to accommodate anticipated structural displacements caused by thermal expansion and contraction, seismic activity, wind or other loads. Figure 714.1 illustrates some of the most common locations of these joints.

These linear openings create a "weak link" in fire-resistance-rated assemblies, which can compromise the integrity of the tested assembly by allowing an avenue for the passage of fire and the products of combustion through the assembly. In order to maintain the efficacy of the fire-resistance-rated assembly, these openings must be protected by a joint system with a fire-resistance rating equal to the adjacent assembly. It is not the intent of this section to regulate joints installed in assemblies that are provided to control shrinkage cracking, such as a saw-cut control joint in concrete (see Section 714).

707.9 Ducts and air transfer openings. Penetrations in a *fire barrier* by ducts and air transfer openings shall comply with Section 716.

❖ Section 716 details the protection of ducts and air transfer openings at the point where they penetrate a fire-resistance-rated assembly. Section 716.5 indicates which situations will require the installation of a damper. As stated in Section 716.1.1, if a duct does not require a damper, the penetration of that duct through a fire-resistance-rated assembly must be protected as a through penetration in accordance with Section 713.

SECTION 708
SHAFT ENCLOSURES

708.1 General. The provisions of this section shall apply to shafts required to protect openings and penetrations through floor/ceiling and roof/ceiling assemblies. Shaft enclosures shall be constructed as *fire barriers* in accordance with Section 707 or *horizontal assemblies* in accordance with Section 712, or both.

❖ This section applies to vertical shafts, including those referenced by other sections of the code, namely: interior exit stairways (see Section 1022.1); refuse and linen-handling chutes (see Section 708.13); and elevator and dumbwaiter hoistways (see Section 708.14). All openings or penetrations in floor/ceiling or roof/ceiling assemblies are required to be protected with a vertical shaft enclosure, unless one of the exceptions provided for in Section 708.2 applies. Some common types of vertical openings besides those just mentioned are vertical exhaust ducts, gas flues, metal chimneys, and vertical supply and return ducts. Shafts are required to be enclosed in fire-resistance-rated fire barriers (see Section 707), or a combination of fire barriers and horizontal assemblies (see Section 712).

708.2 Shaft enclosure required. Openings through a floor/ceiling assembly shall be protected by a shaft enclosure complying with this section.

Exceptions:

1. A shaft enclosure is not required for openings totally within an individual residential *dwelling unit* and connecting four *stories* or less.

2. A shaft enclosure is not required in a building equipped throughout with an *automatic sprinkler system* in accordance with Section 903.3.1.1 for an escalator opening or *stairway* that is not a portion of the *means of egress* protected according to Item 2.1 or 2.2.

    2.1. Where the area of the floor opening between *stories* does not exceed twice the horizontal projected area of the escalator or *stairway* and the opening is protected by a draft curtain and closely spaced sprinklers in accordance with NFPA 13. In other than Groups B and M, this

application is limited to openings that do not connect more than four *stories*.

    2.2. Where the opening is protected by *approved* power-operated automatic shutters at every penetrated floor. The shutters shall be of noncombustible construction and have a *fire-resistance rating* of not less than 1.5 hours. The shutter shall be so constructed as to close immediately upon the actuation of a smoke detector installed in accordance with Section 907.3 and shall completely shut off the well opening. Escalators shall cease operation when the shutter begins to close. The shutter shall operate at a speed of not more than 30 feet per minute (152.4 mm/s) and shall be equipped with a sensitive leading edge to arrest its progress where in contact with any obstacle, and to continue its progress on release therefrom.

3. A shaft enclosure is not required for penetrations by pipe, tube, conduit, wire, cable and vents protected in accordance with Section 713.4.

4. A shaft enclosure is not required for penetrations by ducts protected in accordance with Section 716.6. Grease ducts shall be protected in accordance with the *International Mechanical Code*.

5. In other than Group H occupancies, a shaft enclosure is not required for floor openings complying with the provisions for atriums in Section 404.

6. A shaft enclosure is not required for *approved* masonry chimneys where *annular space* is fireblocked at each floor level in accordance with Section 717.2.5.

7. In other than Groups I-2 and I-3, a shaft enclosure is not required for a floor opening or an air transfer opening that complies with the following:

    7.1. Does not connect more than two *stories*.

    7.2. Is not part of the required *means of egress* system.

    7.3. Is not concealed within the construction of a wall or a floor/ceiling assembly.

    7.4. Is not open to a *corridor* in Group I and R occupancies.

    7.5. Is not open to a *corridor* on nonsprinklered floors in any occupancy.

    7.6. Is separated from floor openings and air transfer openings serving other floors by construction conforming to required shaft enclosures.

    7.7. Is limited to the same smoke compartment.

8. A shaft enclosure is not required for automobile ramps in open and enclosed parking garages constructed in accordance with Sections 406.3 and 406.4, respectively.

9. A shaft enclosure is not required for floor openings between a *mezzanine* and the floor below.

10. A shaft enclosure is not required for joints protected by a *fire-resistant joint system* in accordance with Section 714.

11. A shaft enclosure shall not be required for floor openings created by unenclosed *stairs* or ramps in accordance with Exception 3 or 4 in Section 1016.1.

12. Floor openings protected by floor *fire doors* in accordance with Section 712.8.

13. In Group I-3 occupancies, a shaft enclosure is not required for floor openings in accordance with Section 408.5.

14. A shaft enclosure is not required for elevator hoistways in open or enclosed parking garages that serve only the parking garage.

15. In open or enclosed parking garages a shaft enclosure is not required to enclose mechanical exhaust or supply duct systems when such duct system is contained within and serves only the parking garage.

16. Where permitted by other sections of this code.

❖ In the code, floor openings that connect two or more stories are to be enclosed in shafts that are constructed in accordance with this section. The purpose is to confine a fire to the floor of origin and to prevent the fire or the products of the fire (smoke, heat and hot gases) from spreading to other levels. The exceptions identify floor openings that are not required to be protected by a shaft enclosure.

Exception 1 identifies that a floor opening connecting no more than four stories within a dwelling unit is permitted without a shaft enclosure. This is generally consistent with the permitted omission of stairway enclosures in such uses in accordance with Exception 3 to Section 1122.1.

Exception 2 addresses stairways and escalators in fully sprinklered buildings that are not part of the required means of egress. Item 2.1 requires a draft curtain to protect the opening, the size of which is limited based on the configuration of the stairway or escalator [see Figure 708.2(1)]. This option allows such floor openings to occur for the full height of the building, re-

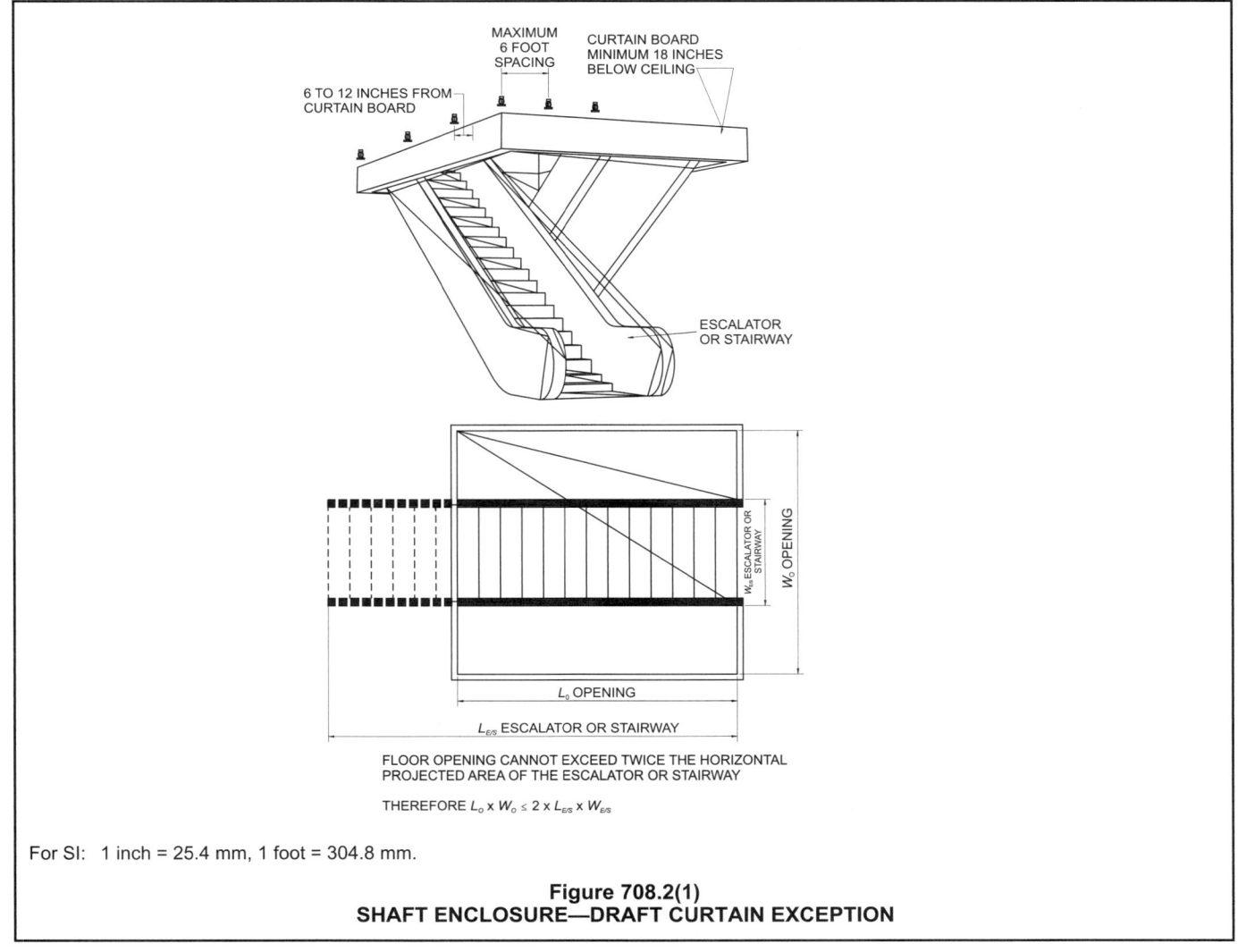

MAXIMUM
6 FOOT
SPACING

CURTAIN BOARD
MINIMUM 18 INCHES
BELOW CEILING

6 TO 12 INCHES FROM
CURTAIN BOARD

ESCALATOR
OR STAIRWAY

$W_{E/S}$ ESCALATOR OR STAIRWAY

W_o OPENING

L_o OPENING

$L_{E/S}$ ESCALATOR OR STAIRWAY

FLOOR OPENING CANNOT EXCEED TWICE THE HORIZONTAL
PROJECTED AREA OF THE ESCALATOR OR STAIRWAY

THEREFORE $L_o \times W_o \leq 2 \times L_{E/S} \times W_{E/S}$

For SI: 1 inch = 25.4 mm, 1 foot = 304.8 mm.

Figure 708.2(1)
SHAFT ENCLOSURE—DRAFT CURTAIN EXCEPTION

gardless of the number of stories for Groups B and M. For all other occupancies, there is a four-story limitation (three floors permitted to contain openings). Item 2.2 requires automatic shutters at each floor opening and is not limited as to height or occupancy [see Figure 708.2(2)].

Exceptions 3 through 6 recognize other floor openings that are permitted without a shaft enclosure in accordance with the various referenced sections.

The requirements Exception 5 need to be understood because of their effect on other sections and also since there is often confusion regarding the application of the atrium requirements. First of all, it is important to understand that the atrium provisions of Section 404 are only one option for protecting an opening through a floor. While floor openings are in general required to be protected by a shaft, the 13 exceptions in Section 708.2 provide a variety of options that may be used in lieu of a shaft. Therefore, an atrium is only one option for protecting the floor opening. It is for this reason that Section 404.1 provides a reference back to Section 708.2. Just because two stories or more are open to each other does not automatically mean that the atrium provisions are applicable.

It should also be noted that the first edition of the code included malls within Exception 5. The removal of covered mall buildings from the scope of Exception 5 has resulted in those buildings being regulated for shaft enclosure provisions in a manner similar to most other buildings. Removing malls from Exception 5 will only have a minimal influence on covered mall buildings that are two stories in height. In general, most such malls will utilize the provisions of Exception 7 to create an unprotected opening between the two stories; however, three-story covered mall buildings with floor openings that connect all three stories must comply with the provisions of Section 404 for atriums, be protected with a shaft or meet one of the other more limited exceptions found in Section 708.2.

Exception 7 addresses the issue of floor openings and air transfer openings that are not a part of the required means of egress. For example, convenience stairways are often provided for building occupants. Such stairways serve as a means of communication between two adjacent floors and are not used as a required means of egress component. To maintain the integrity of the exit access corridor, such stairways may not connect with an exit access corridor in Groups I and R [see Figure 708.2(3)] because occupants can be sleeping and the integrity of the corridor system is especially important under such conditions.

Supplemental stairways are also not permitted to be connected to any other floor opening that connects to an additional floor level. Such stairways must be separated from such floor openings by construction that complies with this section for shaft enclosures. This requirement limits the use of this provision so that an

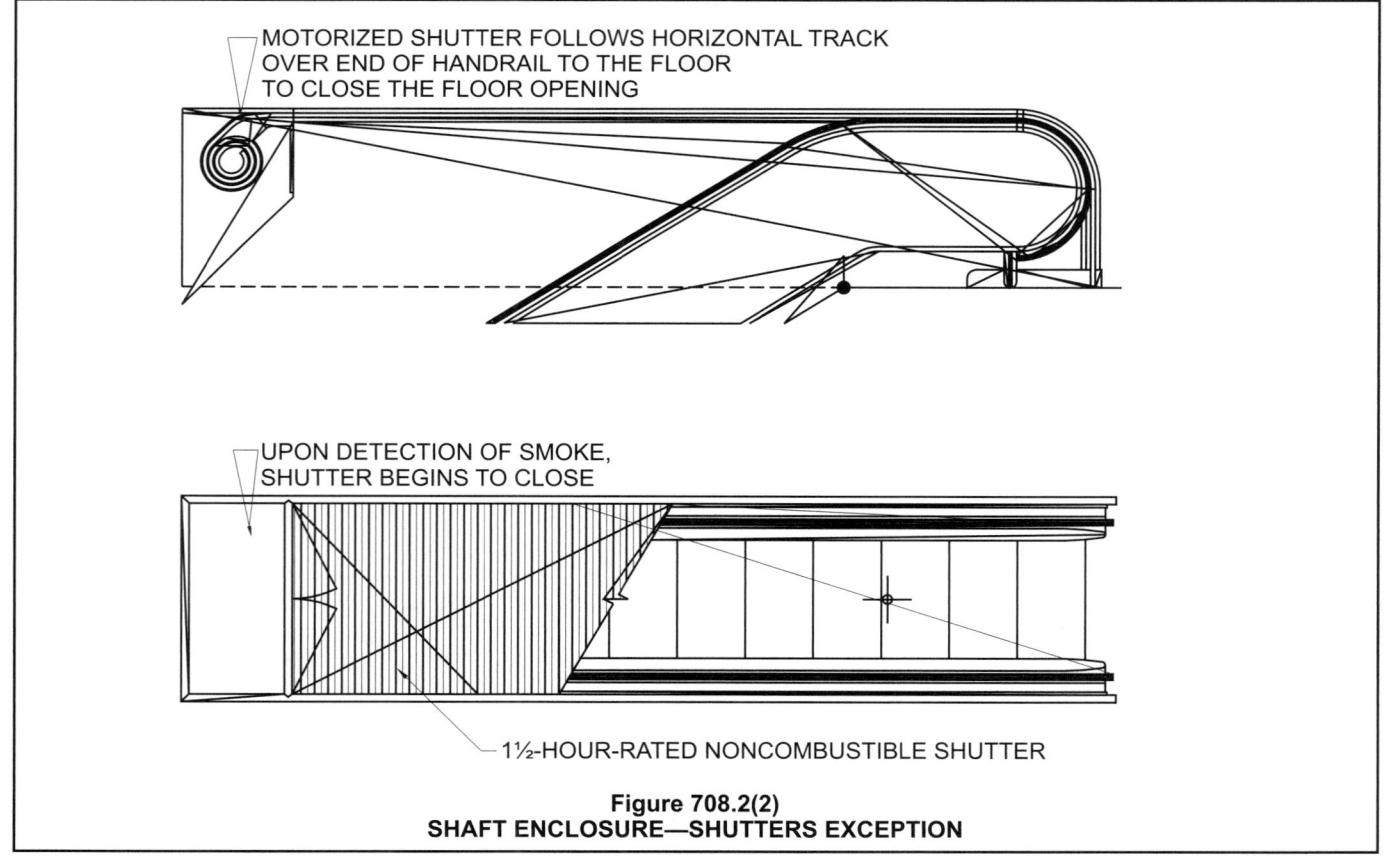

MOTORIZED SHUTTER FOLLOWS HORIZONTAL TRACK OVER END OF HANDRAIL TO THE FLOOR TO CLOSE THE FLOOR OPENING

UPON DETECTION OF SMOKE, SHUTTER BEGINS TO CLOSE

1½-HOUR-RATED NONCOMBUSTIBLE SHUTTER

Figure 708.2(2)
SHAFT ENCLOSURE—SHUTTERS EXCEPTION

opening between two stories does not openly communicate with another opening to an additional story and, therefore, interconnect three or more different levels. Stairways must also comply with other requirements of Chapter 10, such as headroom, handrails, guards, and tread and riser relationships.

Exception 8 addresses the practicality of a shaft enclosure for automobile ramps in both open and enclosed parking garages. A mezzanine is defined by the code as an additional level within a story with size limitations; therefore, Exception 9 does not require it to be enclosed as a separate story (see Section 505.1).

A joint protected in accordance with Section 714 is considered a viable method of protection between floor levels, thus Exception 10 does not require the additional protection offered by a shaft enclosure. The purpose of the joint is to restore or complete the assembly that has a gap provided for a purpose, such as expansion or seismic movement.

Exception 11 provides a direct reference to Section 1022.1 provisions so that open stairways that serve as a means of egress are permitted without enclosure protection.

Exception 12 is a floor fire door requirement that is also detailed in Section 712.8. These floor fire doors can be used to close a floor opening because they have a fire-resistance rating and, therefore, meet requirements which are similar to those of the floor itself.

Exception 13 recognizes that Section 408.5 has special requirements for cell blocks in a jail or prison where tiered cells are utilized.

Exception 14 recognizes that elevator hoistways that only serve a parking structure (open or enclosed) need not be enclosed in fire-resistance-rated shaft construction. The tiered levels of parking are open anyway, therefore, to provide fire-resistance-rated construction for the hoistway enclosures serves no useful purpose. Please note that if the elevator hoistway serves floors that are not part of the parking, then the fire-resistance-rated shafts would be required throughout the entire length of the elevator hoistway.

Exception 15 is similar to Exception 14. Fire-resistance-rated shafts are not required for vertical ductwork that is contained within and serves only the parking structure.

Exception 16 states that vertical openings between stories may be regulated by other sections of the code.

708.3 Materials. The shaft enclosure shall be of materials permitted by the building type of construction.

❖ Material used for the construction of shaft walls must comply with the requirements for partitions based on the building construction type in Section 602.

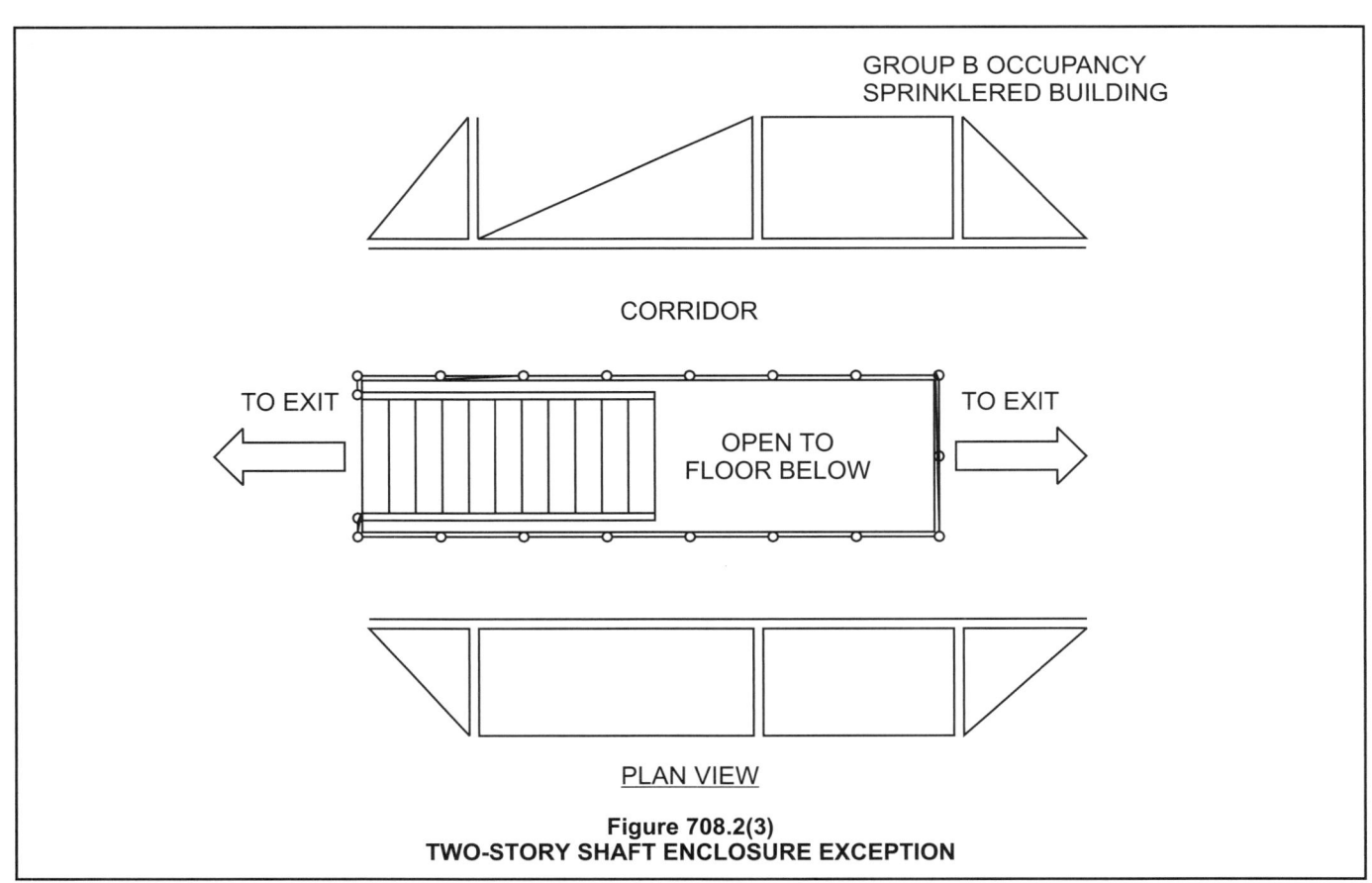

Figure 708.2(3)
TWO-STORY SHAFT ENCLOSURE EXCEPTION

708.4 Fire-resistance rating. Shaft enclosures shall have a *fire-resistance rating* of not less than 2 hours where connecting four *stories* or more, and not less than 1 hour where connecting less than four *stories*. The number of *stories* connected by the shaft enclosure shall include any basements but not any *mezzanines*. Shaft enclosures shall have a *fire-resistance rating* not less than the floor assembly penetrated, but need not exceed 2 hours. Shaft enclosures shall meet the requirements of Section 703.2.1.

❖ The required fire-resistance rating for a shaft enclosure is related to the number of stories being connected. Please note that the shaft enclosure is required to be fire-resistance rated even though the floor may not be required to be fire-resistance rated, such as in Type IIB, IIIB or VB construction. The fire-resistance rating of shaft enclosures must be 2 hours in buildings four or more stories in height. The fire-resistance rating of shaft enclosures that connect less than four stories is required to be only 1 hour; however, in the case of a Type IA or IB, the shaft enclosure for a building height of three stories or less would have to be 2 hours. That is because the floors are 2-hour fire-resistance rated in those types of construction. The fire-resistance integrity of the floor construction must be maintained. The shaft wall construction shall never be allowed to be less than the fire-resistance rating of the floors being penetrated, but needs not to exceed 2 hours (see Figure 708.4). Another case where a two- or three-story building will require a 2-hour shaft enclo-

sure is a mixed-occupancy building using the separated use option with a floor separating the occupancies that is required to be rated for 2 hours.

There is one exception to the 2-hour fire-resistance rating for shaft enclosures in buildings meeting the high-rise provisions of Section 403. Section 403.2.1.2 allows a reduction to 1 hour for shaft enclosures other than exits and elevator hoistways.

The reference to Section 703.2.1 is a 2009 revision to this section. The intent is that fire-resistance-rated shaft-enclosure walls must be rated from fire exposure from both sides; that is, they must be symmetrical assemblies or assume the rating of the least rated side. This has always been the case, but this makes it abundantly clear.

708.5 Continuity. Shaft enclosures shall be constructed as *fire barriers* in accordance with Section 707 or *horizontal assemblies* constructed in accordance with Section 712, or both, and shall have continuity in accordance with Section 707.5 for *fire barriers* or Section 712.4 for *horizontal assemblies* as applicable.

❖ This section references the fire barrier and horizontal assembly provisions for details on the continuity requirements. For fire barriers, the continuity section is Section 707.5, which also contains the supporting structure provisions. For horizontal assemblies, Section 712.4 describes the continuity provisions. See the commentary for both of those sections.

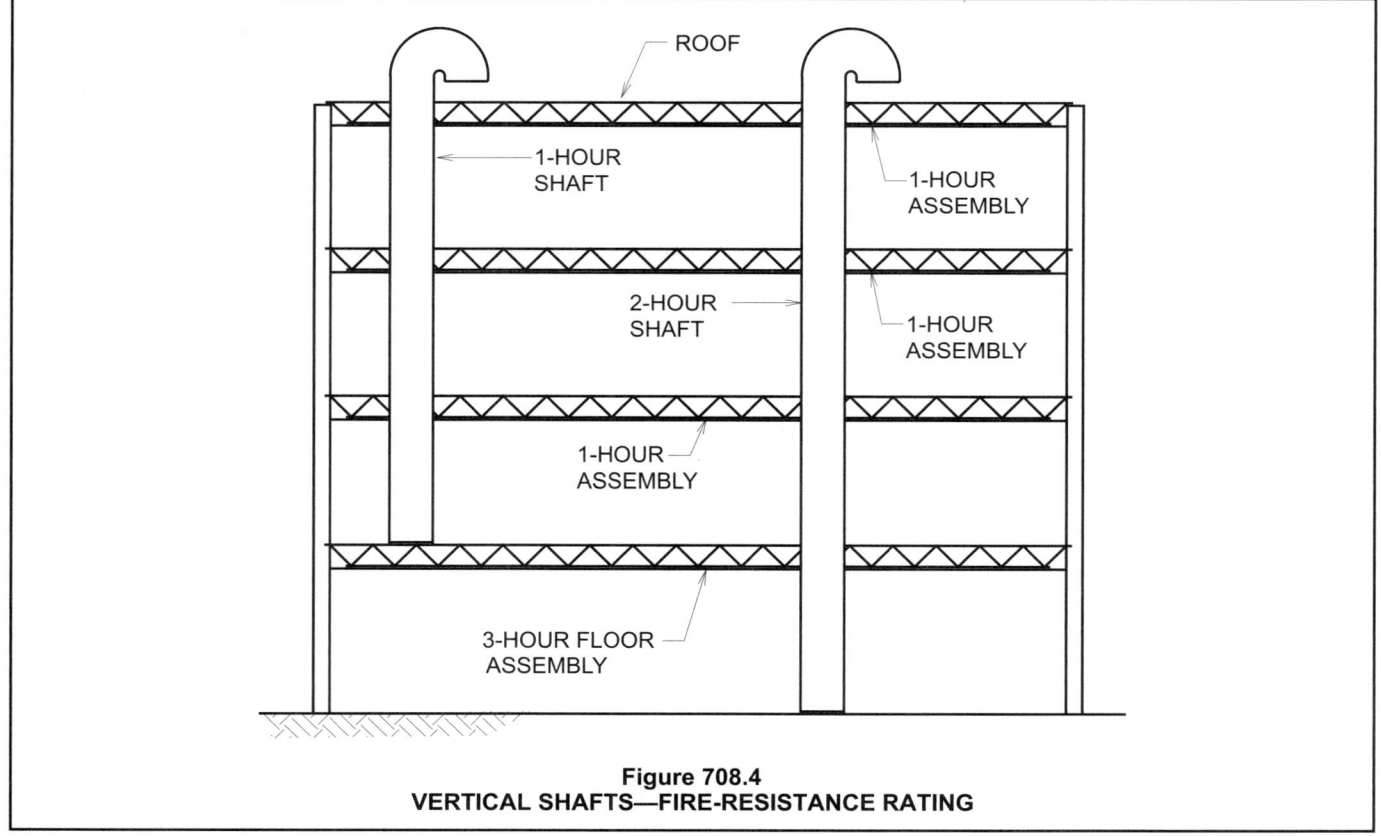

Figure 708.4
VERTICAL SHAFTS—FIRE-RESISTANCE RATING

708.6 Exterior walls. Where *exterior walls* serve as a part of a required shaft enclosure, such walls shall comply with the requirements of Section 705 for *exterior walls* and the fire-resistance-rated enclosure requirements shall not apply.

> **Exception:** *Exterior walls* required to be fire-resistance rated in accordance with Section 1019.2 for exterior egress balconies, Section 1022.6 for *exit* enclosures and Section 1026.6 for exterior *exit* ramps and *stairways*.

❖ If an exterior wall constitutes part of a shaft enclosure, the exterior wall is only required to comply with the fire-resistance-rating requirements in Section 705. The intent of the fire barrier requirements is to subdivide or enclose areas to protect them from a fire within the building. The exterior wall need only have a fire-resistance rating if required for structural stability (see Table 601) or because of an exterior exposure potential. Exterior balcony, exit enclosure, and ramp and stair requirements are referenced by the exception as areas required to be protected for reasons other than structural or fire separation distances (see commentary, Section 707.6).

708.7 Openings. Openings in a shaft enclosure shall be protected in accordance with Section 715 as required for *fire barriers*. Doors shall be self- or automatic-closing by smoke detection in accordance with Section 715.4.8.3.

❖ The integrity of the shaft enclosures must be maintained with approved opening protectives (see Section 715). An example of a protected opening is illustrated in Figure 708.7. Pipe, tube and conduit penetrations must be protected in accordance with Section 708.8, and all joints must be protected in accordance with

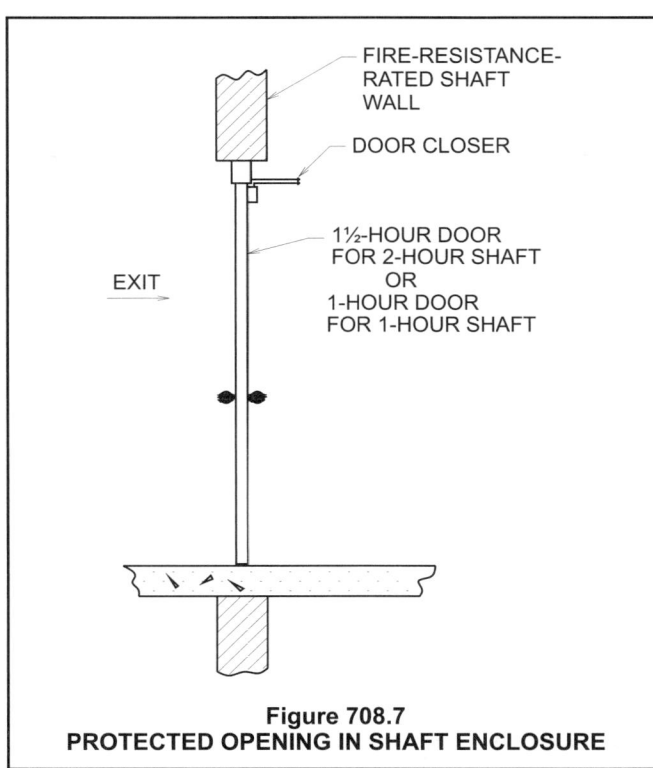

EXIT →

FIRE-RESISTANCE-RATED SHAFT WALL

DOOR CLOSER

1½-HOUR DOOR FOR 2-HOUR SHAFT
OR
1-HOUR DOOR FOR 1-HOUR SHAFT

Figure 708.7
PROTECTED OPENING IN SHAFT ENCLOSURE

Section 708.9. The provisions in Section 715 are not specific to shaft protection regarding openings. The provisions in Section 715 specific to fire barriers are applicable to shaft enclosures.

708.7.1 Prohibited openings. Openings other than those necessary for the purpose of the shaft shall not be permitted in shaft enclosures.

❖ In fire barrier walls forming shaft enclosures, openings are limited to only those necessary for the shaft to serve its intended purpose.

708.8 Penetrations. Penetrations in a shaft enclosure shall be protected in accordance with Section 713 as required for *fire barriers*.

❖ See the commentary to Section 713.

708.8.1 Prohibited penetrations. Penetrations other than those necessary for the purpose of the shaft shall not be permitted in shaft enclosures.

❖ The provisions of Section 708.8.1 mirror those of Section 708.7.1 for openings. Because a shaft provides a potential method for a fire or the products of combustion to spread throughout a building, especially when driven due to the stack effect, the code restricts penetrations so that the effectiveness of the shaft is maintained. For instance, horizontal plumbing piping cannot penetrate the shaft walls of a shaft serving vertical air ducts.

708.9 Joints. Joints in a shaft enclosure shall comply with Section 714.

❖ See the commentary to Section 714.

708.10 Ducts and air transfer openings. Penetrations of a shaft enclosure by ducts and air transfer openings shall comply with Section 716.

❖ These provisions reference the applicable protection features of Section 716 for duct openings. The limitations of Sections 708.7.1 and 708.8.1 must be remembered. For exit enclosures, it is important to note that duct openings are limited to those in Section 1022.4. An exit enclosure may not have any type of interior duct opening for general ventilation purposes even if it is protected by a damper. Section 716.5.3 indicates which situations will require the installation of a damper and what type of damper is needed. As stated in Section 716.1.1, if a duct is exempted from the need for a damper, the penetration of that duct through a fire-resistance-rated assembly must be protected as a through penetration in accordance with Section 713.

708.11 Enclosure at the bottom. Shafts that do not extend to the bottom of the building or structure shall comply with one of the following:

1. They shall be enclosed at the lowest level with construction of the same *fire-resistance rating* as the lowest floor through which the shaft passes, but not less than the rating required for the shaft enclosure.

2. They shall terminate in a room having a use related to the purpose of the shaft. The room shall be separated from

the remainder of the building by *fire barriers* constructed in accordance with Section 707 or *horizontal assemblies* constructed in accordance with Section 712, or both. The *fire-resistance rating* and opening protectives shall be at least equal to the protection required for the shaft enclosure.

3. They shall be protected by *approved fire dampers* installed in accordance with their listing at the lowest floor level within the shaft enclosure.

❖ In order to limit the spread of fire from story to story through vertical shafts, an enclosure is required throughout the length of the shaft including the bottom, if the shaft does not extend to the bottom floor of the building. This section gives three separate methods of providing protection for the bottom of shafts, which are not at the bottom of the building.

Method 1 is illustrated in Figure 708.11(1). Please note that the bottom of the shaft is supported by a horizontal assembly meeting Section 712. However, if the horizontal structural members are primary structural frames and not secondary structural members and the structural floor member supports shaft walls three stories or more in height, then individual encasement is required (see Section 704.3).

Method 2 is illustrated in Figure 708.11(2). This method recognizes that the purpose of some shafts cannot be accomplished if the bottom must be en-

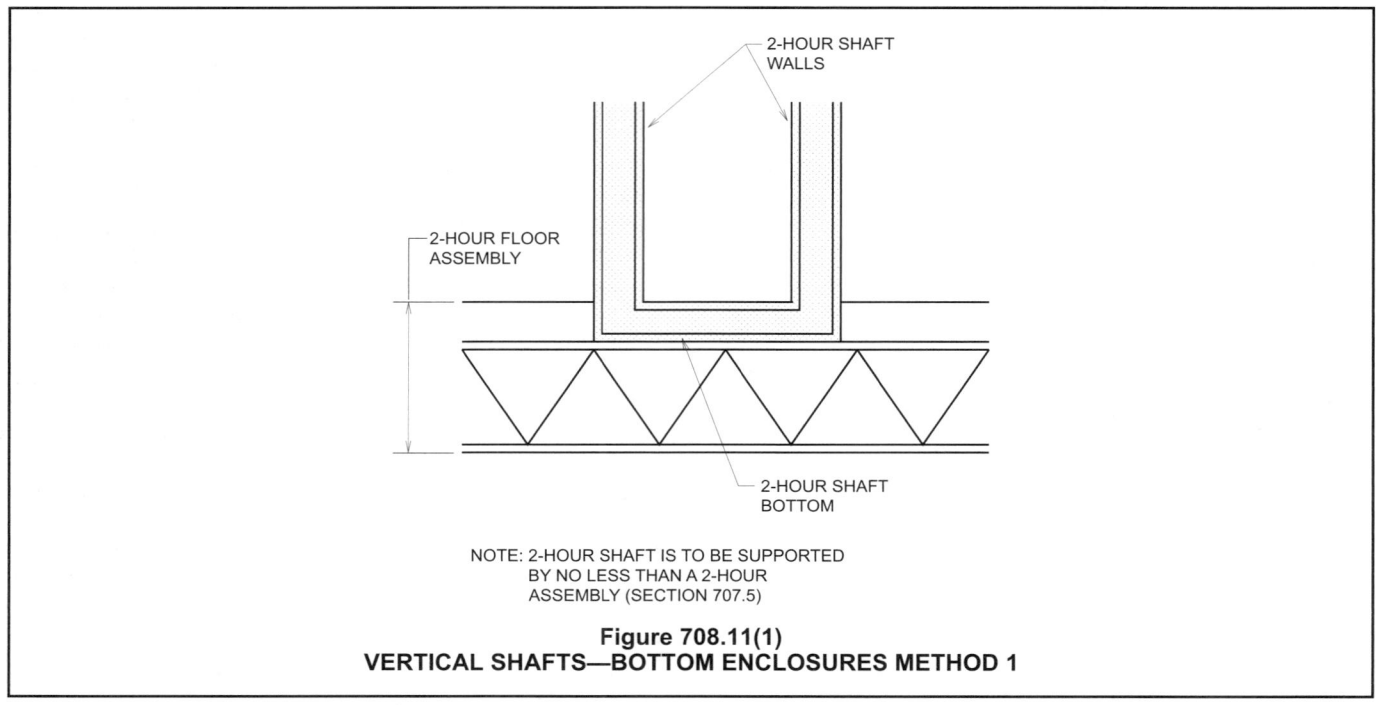

NOTE: 2-HOUR SHAFT IS TO BE SUPPORTED
BY NO LESS THAN A 2-HOUR
ASSEMBLY (SECTION 707.5)

Figure 708.11(1)
VERTICAL SHAFTS—BOTTOM ENCLOSURES METHOD 1

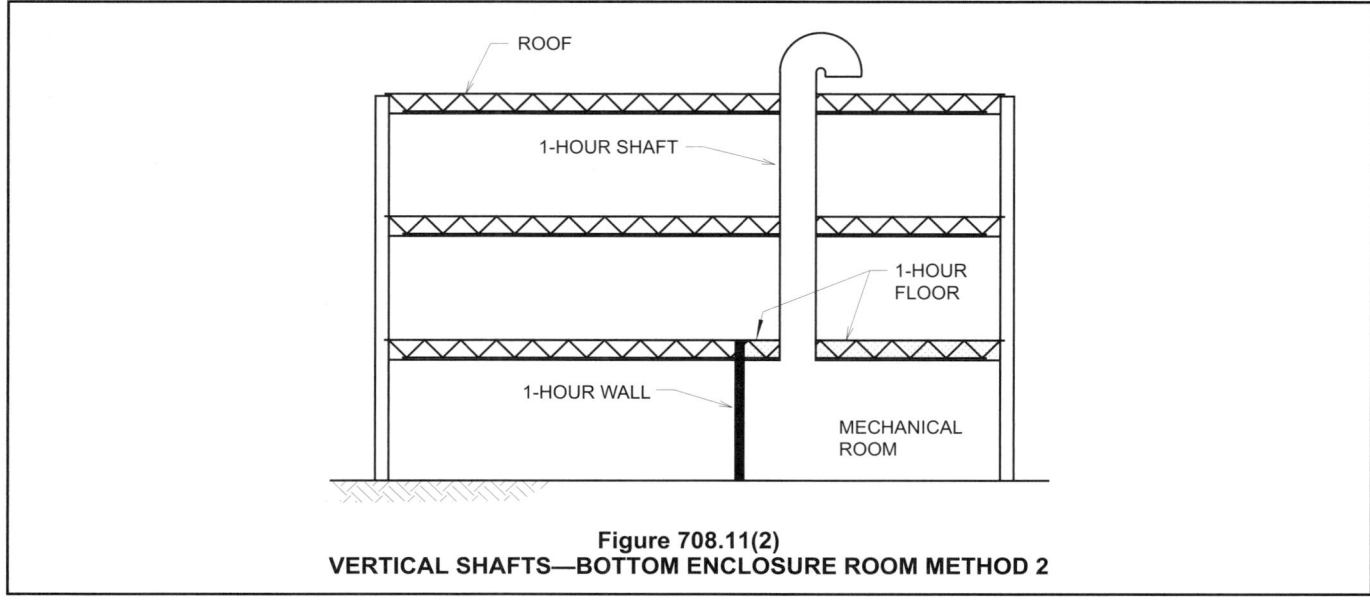

Figure 708.11(2)
VERTICAL SHAFTS—BOTTOM ENCLOSURE ROOM METHOD 2

closed as in Method 1; therefore, a room is permitted at the bottom of the shaft, but the function of the room must be related to the purpose of the shaft. Two common examples of this situation are either mechanical rooms, which have vents from the room to the roof or refuse chutes, which terminate at a room at the bottom for collection. Exceptions 1 and 2 to this section relate to this method.

Method 3 has no illustration. This method is similar to Method 1, but allows a fire damper to be provided at the bottom of the enclosure. Please note that this method predates the requirement for both smoke and fire dampers in ducts penetrating shaft walls (see Section 716.5.3). There must be compliance with both code sections, therefore, both fire and smoke dampers must be provided with this method unless the smoke damper is exempted by Section 716.5.3.

Exceptions:

1. The fire-resistance-rated room separation is not required, provided there are no openings in or penetrations of the shaft enclosure to the interior of the building except at the bottom. The bottom of the shaft shall be closed off around the penetrating items with materials permitted by Section 717.3.1 for draftstopping, or the room shall be provided with an *approved* automatic fire suppression system.

2. A shaft enclosure containing a refuse chute or laundry chute shall not be used for any other purpose and shall terminate in a room protected in accordance with Section 708.13.4.

3. The fire-resistance-rated room separation and the protection at the bottom of the shaft

are not required provided there are no combustibles in the shaft and there are no openings or other penetrations through the shaft enclosure to the interior of the building.

Exception 1 is illustrated by both Figures 708.11(3) and 708.11(4). This is an exception to Method 2 and eliminates the need for the room to be separated. This is rarely used. With this exception, there cannot be any openings or penetrations into the shaft enclosures anywhere except at the bottom and top. A gas-fired appliance flue is a good example.

Exception 2 is also an exception to Method 2. This exception will allow the refuse and linen chute termination rooms to be enclosed in only 1-hour fire-resistance-rated construction, regardless of the fact that the shaft enclosure may have to be 2-hour fire-resistance rated and, thus Method 2 would require a 2-hour room separation. This exception is specific to laundry and refuse chute termination rooms.

Exception 3 is illustrated by Figure 708.11(5). This is an exception to the entire section and an exception to all three methods. While the figure illustrates a very small light well in a three-story building, this exception is not limited to three stories and the size of the shaft is not limited. The area and volume of the shaft can be considered as part of the story in which the bottom of the shaft terminates. The conditions of the exception are based on the fact that the shaft contents will not contribute to the fuel load and that the shaft serves and connects to only one story—the story in which it terminates. This is similar to an atrium with no openings except in the first story. Unlike an atrium, which is exempted from shaft enclosures, this actually is a shaft with shaft wall construction on all stories except the first.

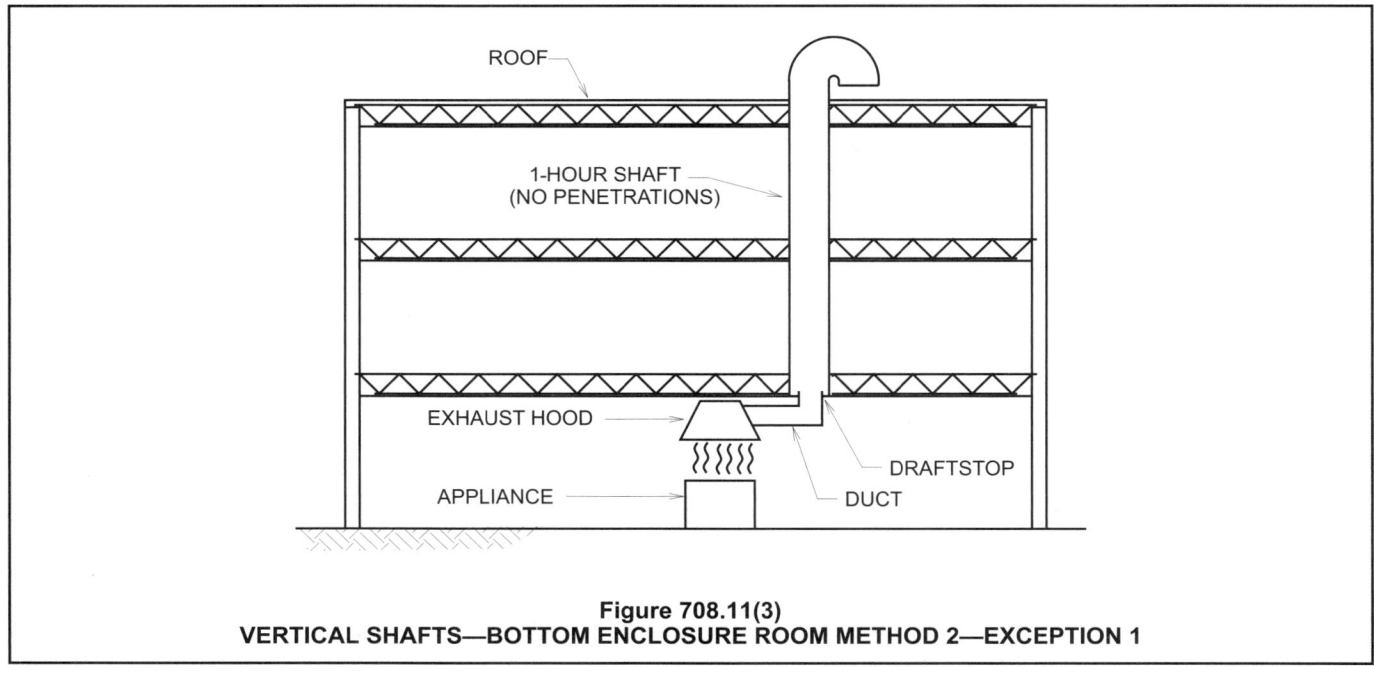

Figure 708.11(3)
VERTICAL SHAFTS—BOTTOM ENCLOSURE ROOM METHOD 2—EXCEPTION 1

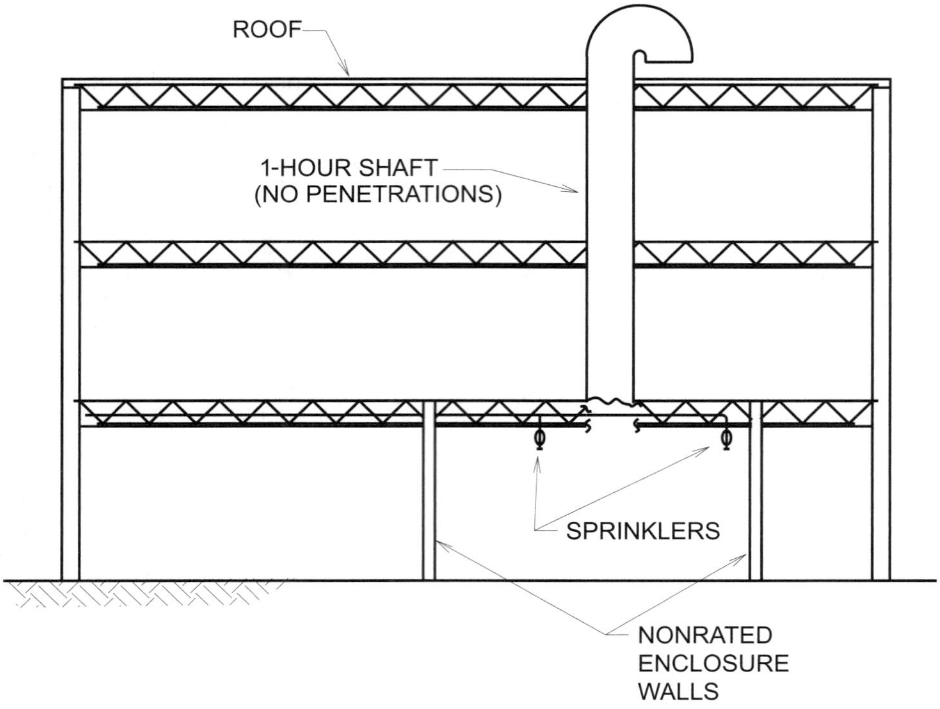

ROOF

1-HOUR SHAFT
(NO PENETRATIONS)

SPRINKLERS

NONRATED
ENCLOSURE
WALLS

Figure 708.11(4)
VERTICAL SHAFTS—BOTTOM ENCLOSURE WITH SPRINKLERS—METHOD 2—EXCEPTION 1

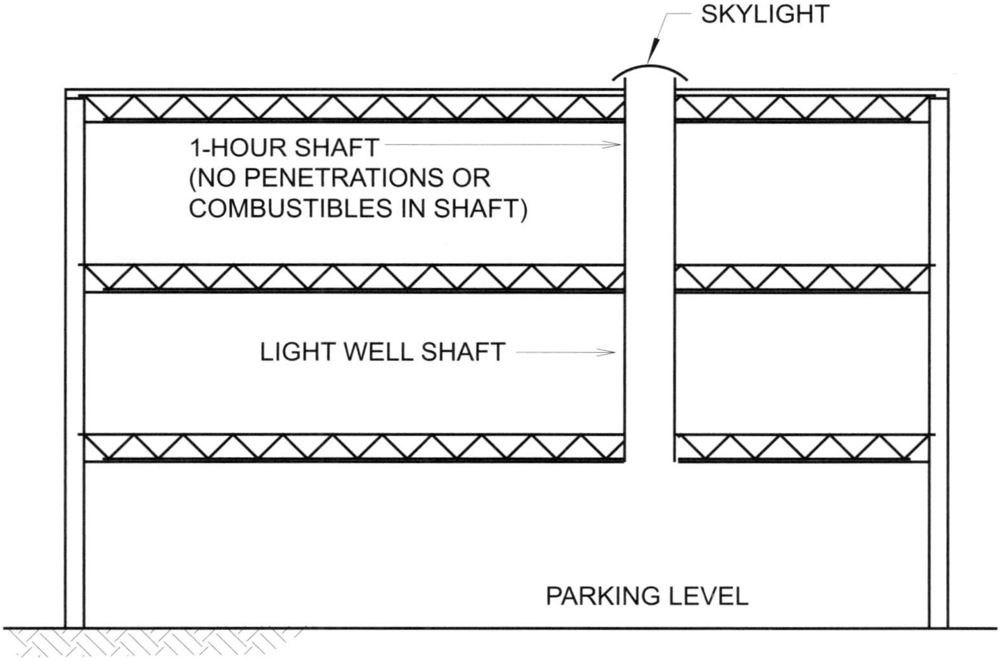

SKYLIGHT

1-HOUR SHAFT
(NO PENETRATIONS OR
COMBUSTIBLES IN SHAFT)

LIGHT WELL SHAFT

PARKING LEVEL

Figure 708.11(5)
VERTICAL SHAFTS—BOTTOM ENCLOSURE EXCEPTION 3

708.12 Enclosure at the top. A shaft enclosure that does not extend to the underside of the roof sheathing, deck or slab of the building shall be enclosed at the top with construction of the same *fire-resistance rating* as the topmost floor penetrated by the shaft, but not less than the *fire-resistance rating* required for the shaft enclosure.

❖ Proper shaft enclosures must include all sides and the top unless the top of the shaft is also the roof of the building. Because the purpose of the shaft is to limit the spread of fire within the building, when the top of the shaft extends to the underside of the roof sheathing, deck or slab, or the shaft extends above the roof line, then the code does not require any type of fire-resistance rating or protection of openings at the top. The fire-resistance rating for the top of the shaft must not be less than the required fire resistance of the shaft enclosure from the slab upwards (see Figure 708.12). The required rating for the top of the shaft that extends to the sheathing or roof deck must be consistent with the requirements of Table 601 and Section 712 for roof construction. The top of the shaft must be constructed using a horizontal assembly (see Section 712) with the proper fire-resistance rating. It is not permissible to simply take a fire barrier, such as the assembly used for the shaft wall, and turn it horizontally.

The code is silent on whether or not a fire damper is appropriate for use to enclose the top of a shaft. Fire dampers are expressly permitted to be used at the bottom of a fire-resistance-rated shaft in Section 708.11. The damper listing should be reviewed before any damper is installed at this location.

708.13 Refuse and laundry chutes. Refuse and laundry chutes, access and termination rooms and incinerator rooms shall meet the requirements of Sections 708.13.1 through 708.13.6.

Exception: Chutes serving and contained within a single *dwelling unit*.

❖ Refuse- and laundry-handling systems represent a location that provides a rapid fire spread potential. The emphasis on the vertical chute systems in this section is because such systems interconnect multiple stories and can contain combustible material. Additionally, these systems all too often receive an ignition source, such as discarded chemicals and materials capable of producing spontaneous combustion, smoking materials that have not been extinguished and hot ashes, embers and charcoal. Such systems are also subject to acts of vandalism. The construction, fire suppression and termination requirements of refuse- and laundry-handling systems are interrelated in other code sections. The exception recognizes the small size, limited use and occupant control of refuse and laundry chutes commonly installed within dwelling units.

708.13.1 Refuse and laundry chute enclosures. A shaft enclosure containing a refuse or laundry chute shall not be used for any other purpose and shall be enclosed in accordance with Section 708.4. Openings into the shaft, including those from access rooms and termination rooms, shall be protected in accordance with this section and Section 715. Openings into chutes shall not be located in *corridors*. Doors shall be self- or automatic-closing upon the actuation of a smoke detector in accordance with Section 715.4.8.3, except that heat-activated closing devices shall be permitted between the shaft and the termination room.

❖ This section requires that the vertical systems or chutes be enclosed in a fire-resistance-rated shaft in

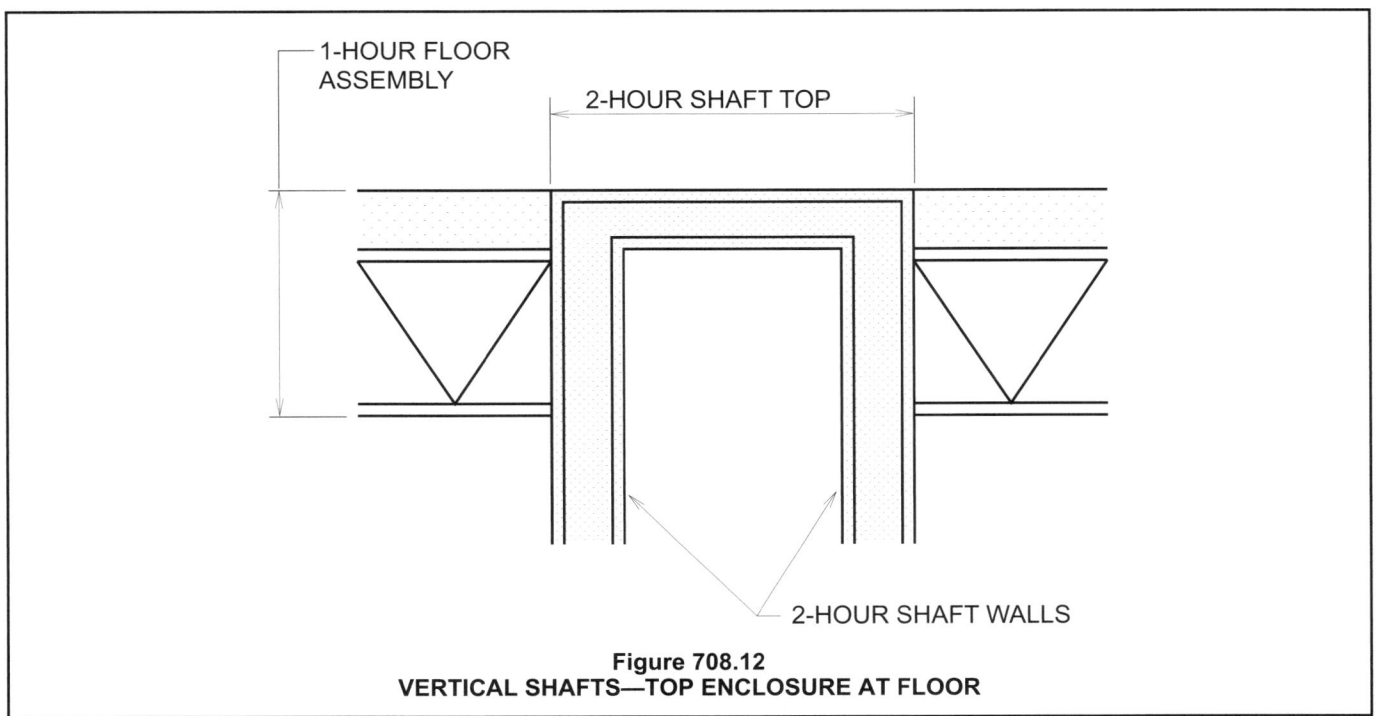

Figure 708.12
VERTICAL SHAFTS—TOP ENCLOSURE AT FLOOR

accordance with Section 708.4. As with all shafts, the fire-resistance integrity of the shaft must be protected at all penetrations of the shaft enclosure. All access openings and termination points must be treated as shaft enclosure penetrations requiring opening protectives. Refuse and laundry access openings are prohibited from being located in corridors or exits. While a corridor is a very convenient location for access to these chutes, in order to access the chute from a corridor, an intervening room (see Section 708.13.3) with an additional fire-rated opening protective provides redundancy that is justified by fire experience.

708.13.2 Materials. A shaft enclosure containing a refuse or laundry chute shall be constructed of materials as permitted by the building type of construction.

❖ The types of materials used in shaft enclosures are to be consistent with Sections 602 through 602.5 for the type of construction of the building. Fire barriers or horizontal assemblies that form shaft walls are permitted to be of combustible construction in Type III, IV and V construction and are required to be of noncombustible materials in Type I and II construction.

708.13.3 Refuse and laundry chute access rooms. Access openings for refuse and laundry chutes shall be located in rooms or compartments enclosed by not less than 1-hour *fire barriers* constructed in accordance with Section 707 or *horizontal assemblies* constructed in accordance with Section 712, or both. Openings into the access rooms shall be protected by opening protectives having a *fire protection rating* of not less than $^3/_4$ hour. Doors shall be self- or automatic-closing upon the detection of smoke in accordance with Section 715.4.8.3.

❖ The chutes themselves must have a rated fire door with closers in accordance with Section 708.13.1; therefore, the fire door in the shaft wall must be rated

for 60 or 90 minutes, depending on the height of the shaft enclosure. In addition to this protection, such openings into the chute shaft must be located in a room or compartment. Said room or compartment must be enclosed by fire barrier walls (see Section 707) and horizontal assemblies (see Section 712) with fire-resistance ratings of 1 hour. The door openings into the access room shall be fire-resistance rated for 45 minutes (see Figure 708.13.3). This requirement is necessary to provide an added measure of protection should the access door to the shaft fail to close due to refuse being stuck in the door and because these rooms are a catchall for combustible material.

708.13.4 Termination room. Refuse and laundry chutes shall discharge into an enclosed room separated from the remainder of the building by not less than 1-hour *fire barriers* constructed in accordance with Section 707 or *horizontal assemblies* constructed in accordance with Section 712, or both. Openings into the termination room shall be protected by opening protectives having a *fire protection rating* of not less than $^3/_4$ hour. Doors shall be self- or automatic-closing upon the detection of smoke in accordance with Section 715.4.8.3. Refuse chutes shall not terminate in an incinerator room. Refuse and laundry rooms that are not provided with chutes need only comply with Table 508.2.5.

❖ The chute termination rooms, like the access rooms, are a source of collection of combustible material, and thus must be separated from all other parts of the building by fire barriers (see Section 707) and horizontal assemblies (see Section 712). The fire-resistance ratings are the same for access rooms. See also Exception 2 of Section 708.11 which allows the shaft to terminate into such room. The incidental use separation requirements for waste and linen separation rooms in Table 508.2.5 are only applicable to collec-

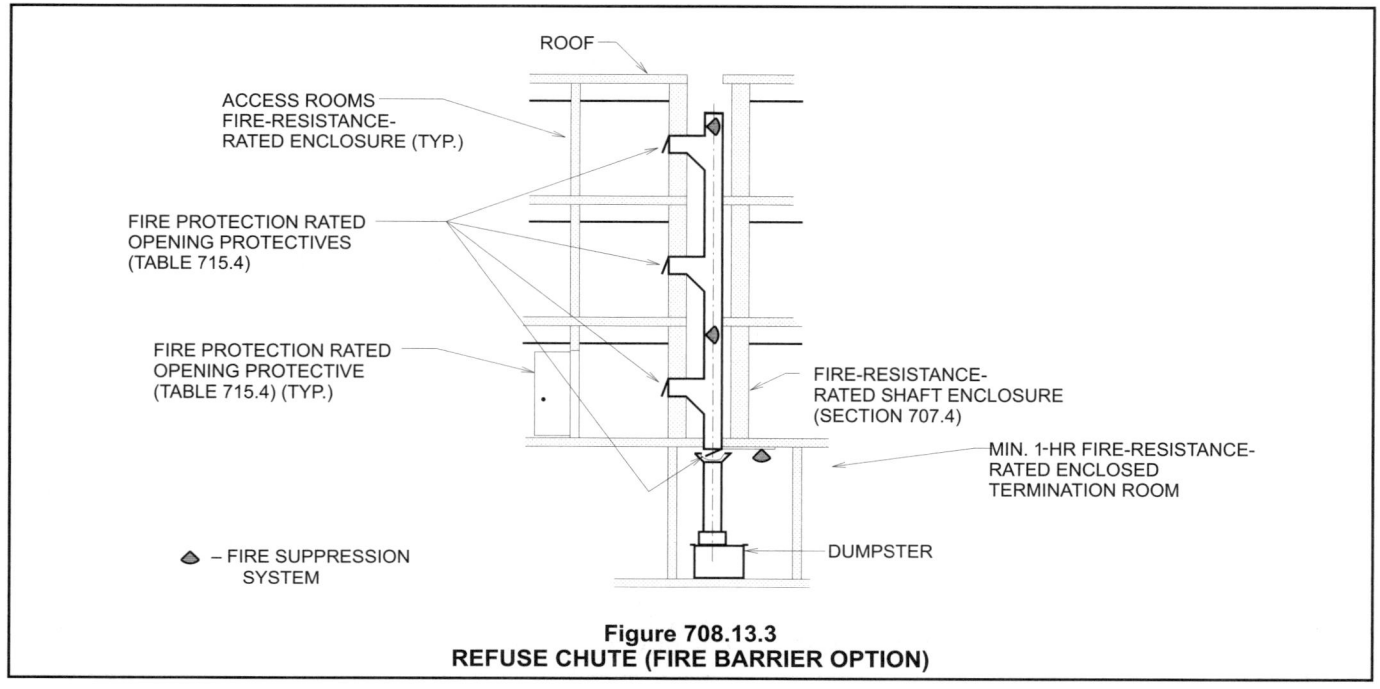

Figure 708.13.3
REFUSE CHUTE (FIRE BARRIER OPTION)

tion rooms not connected to chutes. The termination room must never be connected to an incinerator room.

708.13.5 Incinerator room. Incinerator rooms shall comply with Table 508.2.5.

❖ This section refers to the incidental use separation table. Table 508.2.5 requires a 2-hour fire-resistance enclosure for such rooms and sprinkler protection.

708.13.6 Automatic sprinkler system. An *approved automatic sprinkler system* shall be installed in accordance with Section 903.2.11.2.

❖ Laundry and refuse chutes, in addition to meeting the requirements for shaft enclosures, must be provided with automatic sprinkler coverage. Most buildings with chutes will need to be fully sprinklered by other sections of the code. The chutes must have additional heads at the top and in the termination room and in the chute at every other floor level. Section 13.15.2.1 of NFPA 13 gives more information on the location of the sprinkler heads in the shafts. The need for sprinkler protection addresses a potential fire in the chute or in the container below the chute.

708.14 Elevator, dumbwaiter and other hoistways. Elevator, dumbwaiter and other hoistway enclosures shall be constructed in accordance with Section 708 and Chapter 30.

❖ The hoistway enclosure is the fixed structure consisting of vertical walls or partitions that isolates the enclosure from all other building areas or from an adjacent enclosure in which the hoistway doors and door assemblies are installed. With the exception of observation elevators, the hoistway is normally enclosed with fire barriers (see Section 708.4). A hoistway enclosure for fire spread purposes may not be required if suitable protection measures are provided. Section 708.2 lists exceptions for shaft enclosures around floor openings and is applicable to all hoistways. In addition, shaft enclosures are not required for elevators located in an atrium, since there is no penetration of floor assemblies. Elevator hoistways are enclosed to ensure that flame, smoke and hot gases from a fire do not have an avenue of travel from one floor to another through a concealed space (see the discussion of stack effect in the commentary to Section 708.14.1). Enclosures are also provided to restrict contact with moving equipment and to protect people from falling.

708.14.1 Elevator lobby. An enclosed elevator lobby shall be provided at each floor where an elevator shaft enclosure connects more than three *stories*. The lobby enclosure shall separate the elevator shaft enclosure doors from each floor by *fire partitions*. In addition to the requirements in Section 709 for *fire partitions*, doors protecting openings in the elevator lobby enclosure walls shall also comply with Section 715.4.3 as required for *corridor* walls and penetrations of the elevator lobby enclosure by ducts and air transfer openings shall be protected as required for *corridors* in accordance with Section 716.5.4.1. Elevator lobbies shall have at least one *means of*

egress complying with Chapter 10 and other provisions within this code.

Exceptions:

1. Enclosed elevator lobbies are not required at the street floor, provided the entire street floor is equipped with an *automatic sprinkler system* in accordance with Section 903.3.1.1.

2. Elevators not required to be located in a shaft in accordance with Section 708.2 are not required to have enclosed elevator lobbies.

3. Enclosed elevator lobbies are not required where additional doors are provided at the hoistway opening in accordance with Section 3002.6. Such doors shall be tested in accordance with UL 1784 without an artificial bottom seal.

4. Enclosed elevator lobbies are not required where the building is protected by an *automatic sprinkler system* installed in accordance with Section 903.3.1.1 or 903.3.1.2. This exception shall not apply to the following:

   4.1. Group I-2 occupancies;

   4.2. Group I-3 occupancies; and

   4.3. High-rise buildings.

5. Smoke partitions shall be permitted in lieu of *fire partitions* to separate the elevator lobby at each floor where the building is equipped throughout with an *automatic sprinkler system* installed in accordance with Section 903.3.1.1 or 903.3.1.2. In addition to the requirements in Section 711 for smoke partitions, doors protecting openings in the smoke partitions shall also comply with Sections 711.5.2, 711.5.3, and 715.4.8 and duct penetrations of the smoke partitions shall be protected as required for *corridors* in accordance with Section 716.5.4.1.

6. Enclosed elevator lobbies are not required where the elevator hoistway is pressurized in accordance with Section 708.14.2.

7. Enclosed elevator lobbies are not required where the elevator serves only *open parking garages* in accordance with Section 406.3.

❖ An elevator lobby or one of its alternatives found in the exceptions to Section 708.14.1 is required whenever an elevator shaft connects more than three stories. Elevator shafts often constitute the largest vertical shaft in multiple-story buildings. Hoistways have the potential for accumulating and spreading hot smoke and gases from a fire to other stories in a building. The chimney or stack effect helps with the upward spread of the products of combustion. It is recognized that smoke is a major factor in fire deaths and smoke often migrates to areas remote from the source of the fire. Due to the typical operation and movement of the hoistway doors, they generally cannot provide the level of reduced air leakage that is required to reduce the spread of smoke.

Chapter 7 provisions in general address the isolation or compartmentation of fire areas, rooms and stories, which prohibits the growth of fires. Analyses of fires in multiple-story buildings have documented the movement of smoke to upper levels. In the 1980 fire at the MGM Grand Hotel in Las Vegas, 70 of the 84 deaths occupied the 14th to 24th floors, although the fire was on the first level. The Johnson City Retirement fire in 1989 also had a fire originating on the first level with all but two of the 16 fatalities occurring on the upper floors due to smoke movement via vertical shafts that included the elevator hoistway.

The elevator lobby requirement further isolates the fire-resistance-rated elevator shaft enclosures.

The lobby enclosure walls must separate the elevator hoistway doors from the rest of the building by 1-hour fire-resistance-rated partitions constructed for fire partitions. As a point of difference, the chute access rooms required fire barrier construction, whereas, the elevator lobby requirements are for fire partition construction. The elevator lobby separation does not require a horizontal assembly for the floor/ceiling, either, as does a chute access room. But on a positive note, this section references Sections 715.4.3 and 716.5.4.1 for door openings and duct openings. Therefore, the elevator lobby separation requires the 20-minute doors to be smoke and draft control doors (see Section 715.4.3) and also requires duct penetrations to have smoke and fire dampers (see Section 716.5.4.1) unless exempted by those sections. The purpose of the special opening protectives in the elevator lobby enclosures is to delay or prevent the vertical spread of smoke to other floors through the hoistway doors and the shaft.

This section also requires at least one means of egress from the elevator lobby enclosure. Multiple-story buildings often have security concerns that are solved by controlled access from elevator lobbies to the remainder of the floor. Section 708.14.1 clearly establishes that every elevator lobby shall have at least one means of egress ensuring that no occupant in a lobby is left isolated from escape. While this section requires one door from the elevator lobby, every room or space on the floor including the lobby must have access to the required number of exits (usually two) on that story without traveling through tenant spaces. In other words, once the occupant leaves the elevator lobby, he or she must have access to two exits. Egress through elevator lobbies from corridors on both sides is also allowed.

Two questions often arise. One, can a space have its only exit access path through an elevator lobby? The answer is yes, if it meets all the other egress requirements. Second, can an exit enclosure open into an elevator lobby? The answer is yes. An elevator lobby is a normally occupied space in the same manner that a corridor is a normally occupied space. Of course, the elevator hoistway doors can never open into an exit enclosure.

There are seven exceptions that eliminate the requirement for enclosed elevator lobbies or modify the enclosure.

Exception 1 removes the requirement for an enclosed elevator lobby on the street floor of a building when the entire street floor is provided with automatic sprinkler protection. The exception is notable in that it is not conditioned upon the entire building being sprinklered but just the first floor. All fully sprinklered buildings would be relieved of the elevator lobby requirements on the first story.

Exception 2 eliminates the requirement for enclosed elevator lobbies if the elevator is not required to be enclosed in a fire-resistance-rated shaft. Exception 14 to Section 708.2 exempts elevator hoistways in buildings exclusively used for parking. Although Exception 5 is for floor openings in an atrium, it is often considered to allow elevator hoistways located entirely within the atrium to be unenclosed. If an elevator is not required to have a fire-resistance-rated enclosure, it serves no purpose to require a fire-resistance-rated elevator lobby enclosure.

Exception 3 eliminates the need for a lobby area, but still requires the additional layer of fire resistance and smoke control. Doors, in addition to the normal hoistway door, are provided in front of the hoistway opening. These doors or door must meet the following two requirements:

1. They must be easily opened in accordance with Section 3002.6. This ensures that if the door or doors do close that they can be opened by someone who arrives at that level on the elevator, such as fire department personnel. This is particularly important if the doors are horizontal power-operated doors, such as those in Section 1008.1.4.3. Section 3002.6 states that the doors shall be readily openable from the car side without a key, tool, special knowledge or effort.

2. Additional door or doors must be tested in accordance with UL 1784. This test is titled "Air Leakage Tests of Door Assemblies." This test does not provide a fire-resistance rating for the door. This test has no failure criteria as it only measures the leakage rate. The leakage rate for smoke and draft control doors in Section 715.4.3.1 is a maximum of 3 cubic feet per minute per square foot (0.0014 m^3/s) of door opening at 0.10 inch of water column pressure (0.02 kPa). It is assumed that they must meet the leakage rate requirements in Section 715.4.3.1, and they must be identified as smoke and draft control doors in accordance with Section 715.4.6.3.

Exception 4 applies to other than Group I-2, Group I-3 and high-rise buildings. An elevator lobby is not required if the building is protected by an NFPA 13 or NFPA 13R sprinkler system. This recognizes the effectiveness of sprinkler systems in controlling the spread of smoke. This exception does not apply to buildings with a floor level over 75 feet (22 860 mm) above the lowest level of fire department vehicle access, nor does this exception apply to Group I-2 and I-3 regardless of height.

Exception 5 still requires an elevator lobby. This exception requires either an NFPA 13 or 13R sprinkler system. This exception allows nonfire-resistance-rated smoke partitions to enclose the lobby in lieu of the normal 1-hour fire-resistance-rated fire partitions. Door openings in the lobby enclosure must still meet the air leakage rate in Section 711.5.2 and must be self-closing or automatic closing. Duct penetrations through the smoke partitions must also be provided with smoke dampers in accordance with the referenced Section 716.5.4.1. This is not much of an exception. Use Exception 5 for Group I-2, I-3 and high-rise buildings; otherwise use Exception 4.

In Exception 6, elevator hoistway pressurization can be used in lieu of elevator lobbies in all buildings regardless of height. This is similar to pressurization of stair enclosures, which is used in lieu of smokeproof enclosures for stairs in Section 909.20.5. The hoistway pressurization requirements are in Section 708.14.2.

Exception 7 is similar to Exception 14 in Section 708.2. parking tiers are open between levels by necessity anyway. Elevators used exclusively to serve only parking levels do not need enclosures, nor do they need lobby separation.

708.14.1.1 Areas of refuge. Areas of refuge shall be provided as required in Section 1007.

❖ This is simply a cross reference to the accessible egress requirements. Where a required accessible floor is four or more stories above the level of exit discharge, at least one elevator must comply with Section 1007.4. Section 1007.4 requires the elevator that is used as an accessible means of egress to be accessed from an area of refuge or a horizontal exit. Of course, elevators in fully sprinklered buildings are exempt from the area of refuge requirement.

708.14.2 Enclosed elevator lobby. Where elevator hoist-way pressurization is provided in lieu of required enclosed elevator lobbies, the pressurization system shall comply with this section.

❖ Where Exception 6 to Section 708.14.1 is used, the design and operation of the pressurization system shall comply with this section. These requirements in the code are now very similar to the pressurization requirements found in Section 909.20.5, the stair pressurization alternative to smokeproof enclosures.

708.14.2.1 Pressurization requirements. Elevator hoistways shall be pressurized to maintain a minimum positive pressure of 0.10 inches of water (25 Pa) and a maximum positive pressure of 0.25 inches of water (67 Pa) with respect to adjacent occupied space on all floors. This pressure shall be measured at the midpoint of each hoistway door, with all elevator cars at the floor of recall and all hoistway doors on the floor of recall open and all other hoistway doors closed. The opening and closing of hoistway doors at each level must be demonstrated during

this test. The supply air intake shall be from an outside, uncontaminated source located a minimum distance of 20 feet (6096 mm) from any air exhaust system or outlet.

❖ This section states the minimum and the maximum positive pressure that must be achieved by the smoke control mechanical pressurization system. The minimum positive pressure is 0.10 inch of water column (0.02 kPa), the same as required for stairway pressurization in Section 909.20.5. The maximum pressure is 0.25 inch of water column (0.06 kPa) which is a little less than the maximum allowed for stairway pressurization. The minimum pressure is to ensure that the stack effect is overcome and the maximum pressure is an upper limit to ensure that the doors will operate properly. This section requires a test when the system is complete. The pressures are measured at the midpoint of each hoistway door with all elevator cars at the recall floor and all the hoistway doors open on that level. This simulates the Phase I recall requirements in Section 3003.2. Hoistway doors are then tested on each level to ensure proper operation.

The supply air intake for the pressurization system must be located at least 20 feet (6096 mm) away from any source of contamination to ensure that the hoistway remains tenable through the fire event or well into it before the elevators can no longer be used. Also, if smoke is drawn into the supply air, the system will only spread smoke and not prevent its spread.

708.14.2.2 Rational analysis. A rational analysis complying with Section 909.4 shall be submitted with the *construction documents*.

❖ Section 909.4 recognizes that there are many factors involved in a smoke control system, including stack effect due to height, temperature effect of fire, wind effect, interaction of the HVAC system, the weather and the egress time, all of which must be evaluated. The report must be submitted with the permit documents. Most importantly, the duration of operation of the smoke control system is a function of 1.5 times the egress time or 20 minutes, whichever is less.

708.14.2.3 Ducts for system. Any duct system that is part of the pressurization system shall be protected with the same *fire-resistance rating* as required for the elevator shaft enclosure.

❖ All ductwork necessary for hoistway pressurization must be protected from the effects of fire by enclosure in fire-resistance-rated construction equivalent to that required for the elevator hoistway shaft enclosure.

708.14.2.4 Fan system. The fan system provided for the pressurization system shall be as required by this section.

❖ This section details the requirements for the mechanical system used for pressurization of the hoistway enclosure.

708.14.2.4.1 Fire resistance. When located within the building, the fan system that provides the pressurization shall be protected with the same *fire-resistance rating* required for the elevator shaft enclosure.

❖ The only way to ensure that the mechanical pressurization system can operate during a fire is to locate it in a safe place. If located within the building it must be in an enclosed room protected with the same fire-resistance-rated construction required for the hoistway enclosure.

708.14.2.4.2 Smoke detection. The fan system shall be equipped with a smoke detector that will automatically shut down the fan system when smoke is detected within the system.

❖ The airflow must be free of smoke or it will only increase the likelihood of smoke spreading throughout the building. The smoke detector required by this section should be located on the intake side of the blower fan.

708.14.2.4.3 Separate systems. A separate fan system shall be used for each elevator hoistway.

❖ This section requires that each hoistway enclosure have its own mechanical system.

708.14.2.4.4 Fan capacity. The supply fan shall either be adjustable with a capacity of at least 1,000 cfm (.4719 m³/s) per door, or that specified by a *registered design professional* to meet the requirements of a designed pressurization system.

❖ The fan capacity should be as specified by the registered design professional to meet the operational ranges of pressure at each door or be adjustable with a capacity of at least 1,000 cfm (0.4719 m³/s) per hoistway door. In either case, it is subject to field testing and adjustments to meet the pressure ranges.

708.14.2.5 Standby power. The pressurization system shall be provided with standby power from the same source as other required emergency systems for the building.

❖ The elevator hoistway pressurization system is an emergency system and must have provisions for standby power like other emergency systems. Section 2702 states the requirements that standby power systems must meet.

708.14.2.6 Activation of pressurization system. The elevator pressurization system shall be activated upon activation of the building fire alarm system or upon activation of the elevator lobby smoke detectors. Where both a building fire alarm system and elevator lobby smoke detectors are present, each shall be independently capable of activating the pressurization system.

❖ This section requires that the pressurization system will be activated when the general building fire alarm system or either an elevator lobby smoke detector is activated. All buildings using this pressurization option will more than likely be required to have both. High-rise buildings require elevator lobby smoke detectors, but other buildings may not. Section 909.12 in the smoke control section of the code will require smoke detectors to activate the pressurization system if the design requires it to operate to remove the smoke.

708.14.2.7 Special inspection. *Special inspection* for performance shall be required in accordance with Section 909.18.8. System acceptance shall be in accordance with Section 909.19.

❖ A field inspection will be required to evaluate the performance of the completed system (see commentary, Sections 909.18 and 909.19).

708.14.2.8 Marking and identification. Detection and control systems shall be marked in accordance with Section 909.14.

❖ See the commentary to Section 909.14.

708.14.2.9 Control diagrams. Control diagrams shall be provided in accordance with Section 909.15.

❖ See the commentary to Section 909.15.

708.14.2.10 Control panel. A control panel complying with Section 909.16 shall be provided.

❖ See the commentary to Section 909.16.

708.14.2.11 System response time. Hoistway pressurization systems shall comply with the requirements for smoke control system response time in Section 909.17.

❖ See the commentary to Section 909.17.

SECTION 709
FIRE PARTITIONS

709.1 General. The following wall assemblies shall comply with this section.

1. Walls separating *dwelling units* in the same building as required by Section 420.2.

2. Walls separating *sleeping units* in the same building as required by Section 420.2.

3. Walls separating tenant spaces in *covered mall buildings* as required by Section 402.7.2.

4. Corridor walls as required by Section 1018.1.

5. Elevator lobby separation as required by Section 708.14.1.

❖ Fire partitions are wall assemblies that enclose an exit corridor, separate dwelling units, separate sleeping units and separate tenants in a covered mall building. There are some exceptions to the requirement that corridor walls and elevator lobby walls be fire partitions. Those exceptions are found in Sections 708.14.1 and 1018.1.

Corridor walls not required to be fire-resistance rated by Table 1018.1 are not required to meet this section. Elevator lobby walls not required by the exceptions to Section 708.14.1 to be fire-resistance rated are not required to meet this section.

The term "fire partition" is defined in Section 702. This section contains fire-resistance-rating requirements, continuity requirements, opening requirements, penetration requirements, joint requirements, ducts and air transfer-opening requirements for fire partitions.

For horizontal assemblies separating dwelling and sleeping units, see Section 712.

709.2 Materials. The walls shall be of materials permitted by the building type of construction.

❖ The types of materials used in fire partitions are to be consistent with Sections 602.2 through 602.5 for the type of construction classification of the building. Fire partitions are permitted to be of combustible materials in Type II, IV and V construction, and are required to be of noncombustible materials in Type I and II construction, except as allowed by Exception 25 to Section 603.1.

709.3 Fire-resistance rating. Fire partitions shall have a *fire-resistance rating* of not less than 1 hour.

Exceptions:

1. Corridor walls permitted to have a $^1/_2$ hour *fire-resistance rating* by Table 1018.1.

2. *Dwelling unit* and *sleeping unit* separations in buildings of Type IIB, IIIB and VB construction shall have *fire-resistance ratings* of not less than $^1/_2$ hour in buildings equipped throughout with an *automatic sprinkler system* in accordance with Section 903.3.1.1.

❖ The requirement is for fire partitions to be 1-hour fire-resistance rated, with two exceptions.

In Exception 1, corridor walls not required by Table 1018.1 to be fire-resistance rated need not comply with this section, since in accordance with Section 1018.1 they are not fire partitions. Table 1018.1, also includes a reduction of the 1-hour fire-resistance rating down to 30 minutes for sprinklered Group R occupancies.

In Exception 2, a reduction in the fire-resistance rating for dwelling and sleeping unit separation is allowed for Type IIB, IIIB and VB construction which is sprinklered with an NFPA 13 sprinkler system. This is not allowed with an NFPA 13R system. The reduction is from 1 hour to only 30 minutes. There is a similar reduction for the horizontal assemblies separating dwelling and sleeping units in Section 712.3.

709.4 Continuity. Fire partitions shall extend from the top of the foundation or floor/ceiling assembly below to the underside of the floor or roof sheathing, slab or deck above or to the fire-resistance-rated floor/ceiling or roof/ceiling assembly above, and shall be securely attached thereto. If the partitions are not continuous to the sheathing, deck or slab, and where constructed of combustible construction, the space between the ceiling and the sheathing, deck or slab above shall be fireblocked or draftstopped in accordance with Sections 717.2 and 717.3 at the partition line. The supporting construction shall be protected to afford the required *fire-resistance rating* of the wall supported, except for walls separating tenant spaces in *covered mall buildings*, walls separating *dwelling units*, walls separating *sleeping units* and *corridor* walls in buildings of Type IIB, IIIB and VB construction.

Exceptions:

1. The wall need not be extended into the crawl space below where the floor above the crawl space has a minimum 1-hour *fire-resistance rating*.

2. Where the room-side fire-resistance-rated membrane of the *corridor* is carried through to the underside of the floor or roof sheathing, deck or slab of a fire-resistance-rated floor or roof above, the ceiling of the *corridor* shall be permitted to be protected by the use of ceiling materials as required for a 1-hour fire-resistance-rated floor or roof system.

3. Where the *corridor* ceiling is constructed as required for the *corridor* walls, the walls shall be permitted to terminate at the upper membrane of such ceiling assembly.

4. The fire partitions separating tenant spaces in a *covered mall building*, complying with Section 402.7.2, are not required to extend beyond the underside of a ceiling that is not part of a fire-resistance-rated assembly. A wall is not required in *attic* or ceiling spaces above tenant separation walls.

5. Fireblocking or draftstopping is not required at the partition line in Group R-2 buildings that do not exceed four *stories above grade plane*, provided the *attic* space is subdivided by draftstopping into areas not exceeding 3,000 square feet (279 m²) or above every two *dwelling units*, whichever is smaller.

6. Fireblocking or draftstopping is not required at the partition line in buildings equipped with an *automatic sprinkler system* installed throughout in accordance with Section 903.3.1.1 or 903.3.1.2, provided that automatic sprinklers are installed in combustible floor/ceiling and roof/ceiling spaces.

❖ To minimize the potential for fire spread from the exposed side of the fire partition to the unexposed side, such partitions must be continuous from the floor assembly to the underside of a fire-resistance-rated floor/ceiling assembly or roof/ceiling assembly. In the absence of a rated floor/ceiling or roof/ceiling assembly, the fire partition is to be continuous to the floor slab or roof deck above [see Figures 709.4(1) and 709.4(2)(A)]. All hollow vertical spaces in the fire partition must be fireblocked at the ceiling and floor or roof levels in accordance with Section 717.2, and draftstopped in line with the wall with materials in accordance with Section 717.3. Fire partitions are not required to be continuous through concealed spaces, whereas fire barriers are to be continuous (see Section 707.4). This is the primary difference between fire barriers and fire partitions.

Fire partitions serving as exit access corridor walls and tenant and guestroom separation walls in buildings of Type IIB, IIIB and VB construction are not required to be supported by structural elements having the same fire-resistance rating. The primary purpose of rated corridor walls is to prevent fire or smoke spread from a room to the corridor in order to maintain the protection of the means of egress. The secondary purpose is that the rated corridor wall also prevents fire or smoke spread from a corridor to adjacent rooms should the fire involve the corridor; therefore, the fire-resistance rating of the floor construction supporting the wall is not critical to the performance of the corridor wall.

The construction supporting tenants and guestroom separation walls is exempted from the requirements for a fire-resistance rating, since the compartmentalization provided by these walls is similar to that provided by corridor walls. Additionally, if the supporting construction were required to be fire-resistance rated, this section would effectively result in all buildings containing guestrooms (occupancies in Groups R-1, R-2 and I-1) being built of protected construction only. The remaining construction types are not included, since the supporting structural elements (floors and columns) are required to have at least a 1-hour rating (in accordance with Table 601), which is the maximum fire-resistance rating required for corridor walls. Similarly, tenant separations in covered mall buildings of Type IIB construction are not required to be supported by construction having an equivalent fire-resistance rating.

While supporting members in Types IIB, IIIB and VB are not required to be protected by the same fire resistance as the separation walls between a dwelling and sleeping unit, the horizontal floor assembly between dwelling and sleeping units is required to be fire-resistance-rated construction by Section 712.3.

Exception 1 modifies the continuity requirement for where a crawl space is found below. This exception provides a blanket exclusion for extending the partition into the crawl space, it may be prudent to consider each building on an individual basis. If the crawl space is arranged or provided so that each dwelling unit has access to it or it may be intended for storage or equipment, then it may be wise to consider the crawl space as a part of the dwelling unit. Therefore, it may be reasonable to apply the exception to circumstances where the crawl space is not considered to be a part of the dwelling unit.

Although the exception requires that the floor above the crawl space have a 1-hour fire-resistive rating, it would still be permissible to use the provisions of Section 712.3.3 and eliminate the ceiling membrane if the crawl space is considered as being an unusable space.

Exception 2 allows for only one side (the room side) of a fire-resistance-rated wall assembly to be continuous to the rated assembly above. The membrane on the corridor side of the wall is carried up to the underside of the corridor ceilings, provided the corridor ceiling is constructed of materials that have been tested as part of a 1-hour assembly [see Figure 709.4(2)(B)].

Exception 3 allows for a corridor to be constructed in a "tunnel" fashion—with rated walls and a rated top. In these cases, the walls are not required to extend to the underside of a rated assembly. This is a unique provision in the code and the fact that it permits a wall assembly to be turned into a horizontal position should not be applied to other sections of the code or other

rated assemblies. In this one situation, the code will permit an assembly that has been tested as a wall to be turned and used horizontally. Because of the differences in the fire test and the fact that a horizontal test is generally a more severe condition, most assemblies will not obtain the same fire-resistance rating when installed in a horizontal position. In this case, because one of the primary purposes of a corridor is to stop the spread of smoke versus necessarily stopping a fire, this level of protection and construction has been accepted [see Figure 709.4(2)(C)].

In order to limit the spread of smoke, tenant separation walls in a mall are required to be fire partitions (see Sections 402.7.2 and 708.1) having a fire-resistance rating of a least 1 hour and extending from the floor to the underside of the ceiling. Exception 4 acknowledges that extending tenant separations to the floor slab or roof deck above is not always practical or possible because of operation of the HVAC system and in some cases due to particularly tall story heights. The effectiveness of the automatic sprinkler system, as well as the location of the fuel loading, is also a reason for not requiring tenant separations to extend above the ceiling, including attic spaces.

Within combustible construction in the situations where the fire partition is stopped at the ceiling line instead of being continuous up to the roof deck, Exception 5 permits draftstopping or fireblocking above every other dwelling unit or at a maximum spacing of 3,000 square feet (279 m²). This design requirement allows for proper ventilation of the attic space above dwelling units. In typical multiple-family construction, with dwelling units along the front and back of the building, trusses run front to rear with soffit vents at each end. If draftstopping or fireblocking were required at each dwelling unit, it would block cross ventilation, eliminating the use of ridge vents. Soffit and ridge venting allow natural air circulation that, in turn, lowers the roof sheathing temperature in the winter, relieving many of the problems associated with ice dams. This exception is the same as permitted for draftstopping in Section 717.4.2, Exception 3.

Exception 6 recognizes the added protection afforded a building that is equipped throughout with an automatic sprinkler system in accordance with Section 903.3.1.1 or 903.3.1.2 and NFPA 13 and 13R, respectively, and permits those fire partitions used for dwelling unit and guestroom separations to terminate at the underside of the ceiling membrane without the need for fireblocking or draftstopping above the partition. For buildings equipped with a sprinkler system that conforms to NFPA 13, the attic and other concealed areas are required to be sprinklered, thus providing protection that offsets the fact that the dwelling unit separations, fireblocking or draftstopping do not extend to the deck above. In buildings equipped throughout with an NFPA 13R system, the attic and other concealed areas may not be sprinklered. NFPA 13R allows the omission of sprinkler protection in combustible attics and concealed spaces, provided the space is not used for living purposes or storage. However, when using this exception, sprinkler protection is necessary and required, since the protection is considered available to control fires in the incipient stage and keep unoccupied concealed spaces and attic areas from becoming involved. For these buildings, the exception requires sprinkler protection and the fire partitions do not need to extend to the deck above.

Although this exception provides a blanket exclusion for fireblocking or draftstopping within floor/ceiling and roof/ceiling spaces, this may need to be reviewed on a case-by-case basis. The intent of this section is to address concealed spaces that are not used for any purpose. When an attic space is provided, it would be necessary to decide if the attic is simply a concealed space or if it is an occupied portion of the dwelling unit. For example, if each dwelling unit had access to the attic above its unit by means of a pull-down ladder and an area of the attic was provided with flooring so that storage could be placed there, it would be prudent to consider that attic as being a part of the dwelling unit and, therefore, requiring a fire partition, instead of draftstopping, to be provided between the adjacent attics.

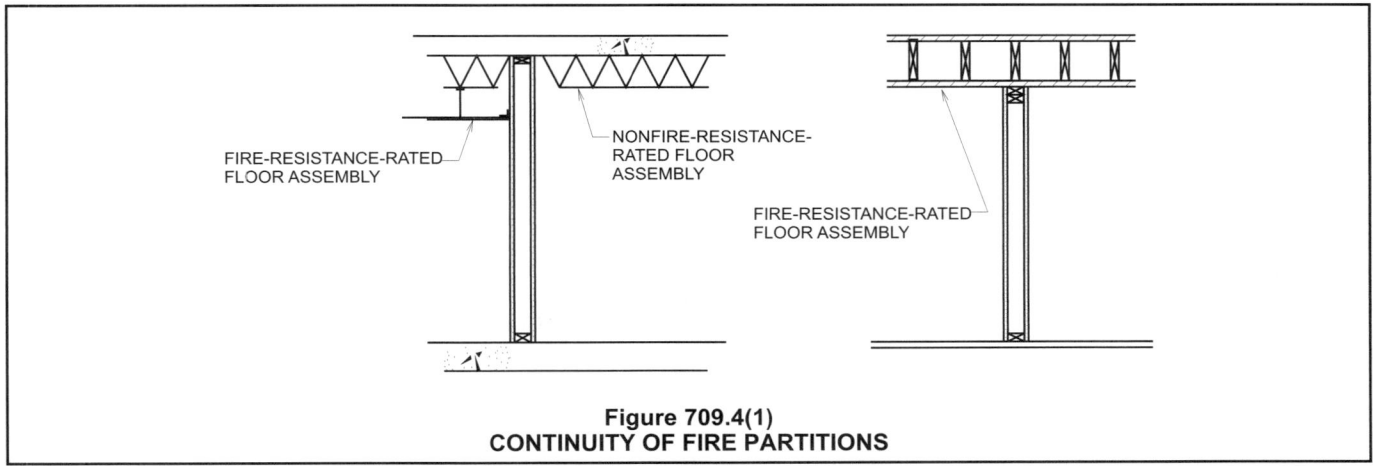

Figure 709.4(1)
CONTINUITY OF FIRE PARTITIONS

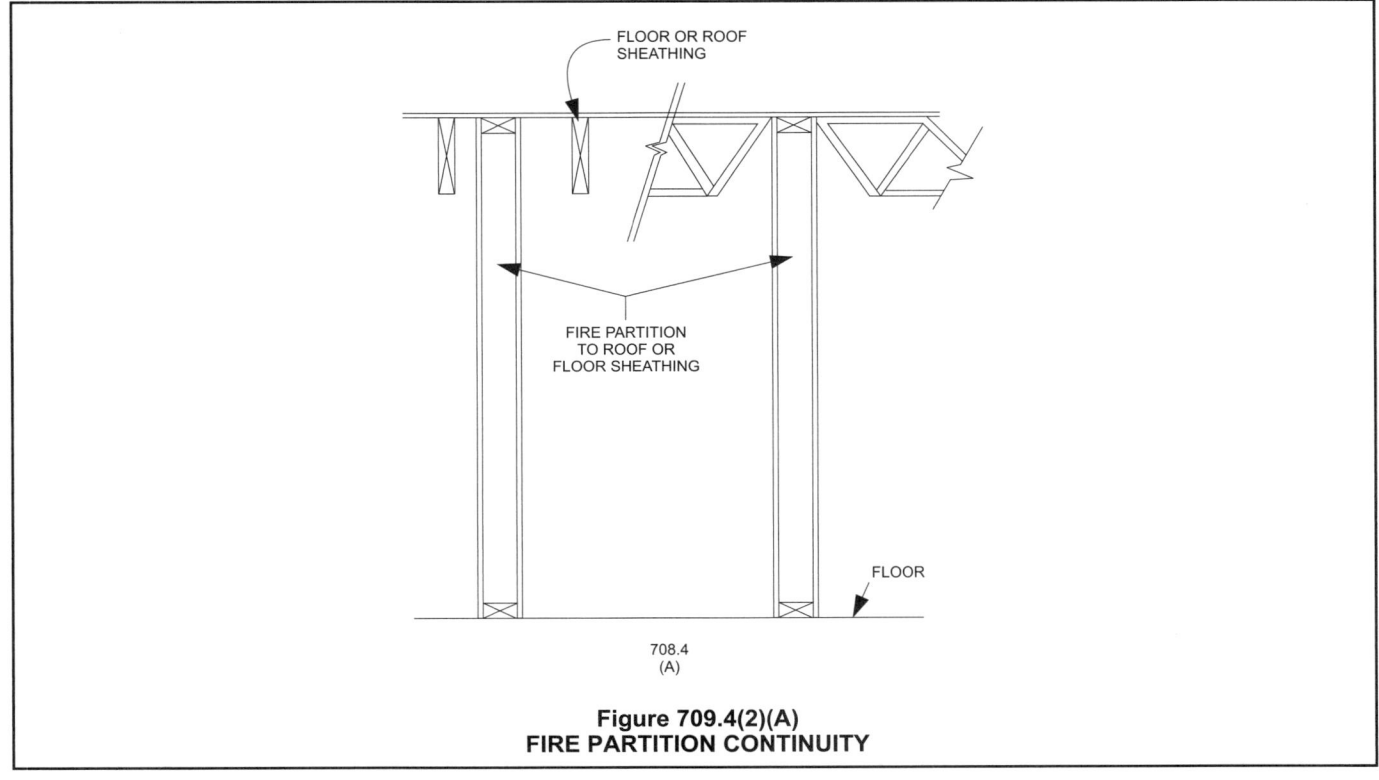

Figure 709.4(2)(A)
FIRE PARTITION CONTINUITY

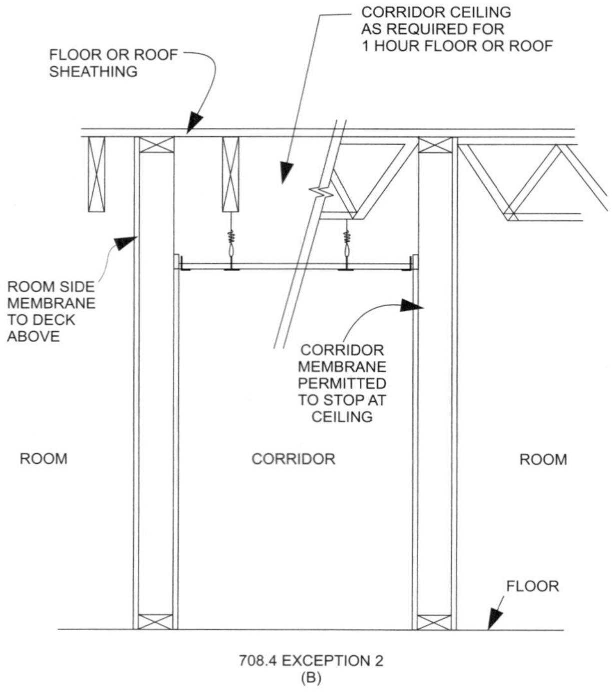

CORRIDOR CEILING
AS REQUIRED FOR
1 HOUR FLOOR OR ROOF

FLOOR OR ROOF
SHEATHING

ROOM SIDE
MEMBRANE
TO DECK
ABOVE

CORRIDOR
MEMBRANE
PERMITTED
TO STOP AT
CEILING

ROOM   CORRIDOR   ROOM

FLOOR

708.4 EXCEPTION 2
(B)

Figure 709.4(2)(B)
FIRE PARTITION CONTINUITY

FLOOR OR ROOF

CORRIDOR CEILING AS
REQUIRED FOR CORRIDOR
WALL

WALL MEMBRANE
CONTINUES TO
UPPER MEMBRANE
OF CEILING

FIRE
PARTITION

ROOM   CORRIDOR   ROOM

FLOOR

708.4 EXCEPTION 3
(C)

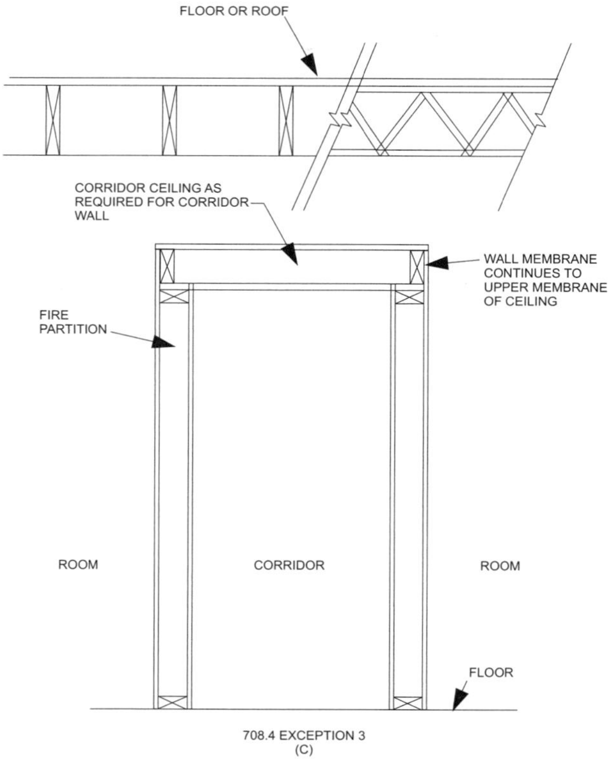

Figure 709.4(2)(C)
FIRE PARTITION CONTINUITY

709.5 Exterior walls. Where *exterior walls* serve as a part of a required fire-resistance-rated separation, such walls shall comply with the requirements of Section 705 for *exterior walls*, and the fire-resistance-rated separation requirements shall not apply.

> **Exception:** *Exterior walls* required to be fire-resistance rated in accordance with Section 1019.2 for exterior egress balconies, Section 1022.6 for *exit* enclosures and Section 1026.6 for exterior *exit* ramps and *stairways*.

❖ This section pertains to exterior walls that are a part of an enclosure, such as corridors on the exterior face of a building or elevator lobbies that have exterior walls. Such exterior walls do not need to be fire-resistance-rated construction unless required by Table 601 or 602. The exception noted in this section is identical to Sections 707.4 and 708.6 for fire barriers and shaft enclosures. See the commentary to both of those sections.

709.6 Openings. Openings in a *fire partition* shall be protected in accordance with Section 715.

❖ Section 715.4 includes the requirements for fire doors in fire partitions. Generally, $^1/_3$-hour doors are required for corridors and $^3/_4$-hour doors for tenant dwelling unit and sleeping unit separations in accordance with Table 715.4. Where $^1/_2$-hour fire-resistive assemblies are permitted, the doors are required to be $^1/_3$-hour doors. It is important to note that Section 715.4.3 requires smoke and draft control doors for corridors. Section 715.5 includes the requirements for windows located in a corridor.

709.7 Penetrations. Penetrations of *fire partitions* shall comply with Section 713.

❖ This section simply states that penetrations through fire partitions are required to conform to the requirements of Sections 713.2 and 713.3. This is the same as the penetration requirements for fire barriers.

709.8 Joints. Joints made in or between *fire partitions* shall comply with Section 714.

❖ Joints, such as expansion or seismic, are another form of opening in fire partitions, and, therefore, must be considered with regard to maintaining the fire-resistance ratings of these walls. This section requires all joints that are located in fire partitions to be protected by a joint system that has a fire-resistance rating and complies with the requirements of Section 714 (see commentary, Section 714).

709.9 Ducts and air transfer openings. Penetrations in a *fire partition* by ducts and air transfer openings shall comply with Section 716.

❖ Section 716 details the protection of ducts and air transfer openings at the point where they penetrate a fire-resistance-rated assembly. Sections 716.5 and 716.5.4 indicate which situations will require the installation of a damper. As stated in Section 716.1.1, if a duct does not require a damper, the penetration of that duct through a fire-resistance-rated assembly must be protected as a through penetration in accordance with Section 713.

SECTION 710
SMOKE BARRIERS

710.1 General. *Smoke barriers* shall comply with this section.

❖ Smoke barriers divide areas of a building into separate smoke compartments. A smoke barrier is designed to resist fire and smoke spread so that occupants can be evacuated or relocated to adjacent smoke compartments. This concept has proven effective in Group I-2 and I-3 occupancies. Sections 407.4 and 408.6 identify where smoke barriers are required. Also, while not cross referenced in this section, smoke barriers may be utilized in other applications, such as part of a smoke control system (see Section 909.5) and accessible means of egress (see Section 1007.6.2), accessible areas of refuge (see Section 1007.6), compartmentation of underground buildings (see Section 405.4.2) and elevator lobbies in underground buildings (see Section 405.4.3).

Other than the wall itself, all of the elements in the smoke barrier that can potentially allow smoke travel through the smoke barrier are required to have a quantified resistance to leakage. This includes doors, joints, through penetrations and dampers. The maximum leakage limits are as established in the individual code sections referenced for each element.

710.2 Materials. *Smoke barriers* shall be of materials permitted by the building type of construction.

❖ The types of materials used in smoke barriers are to be consistent with Sections 602.2 through 602.5 for the type of construction classification of the building. Smoke barriers are permitted to be of combustible construction in Type III, IV and V construction, and are required to be of noncombustible materials in Type I and II construction, except as permitted in the exceptions to Section 603.1.

710.3 Fire-resistance rating. A 1-hour *fire-resistance rating* is required for *smoke barriers*.

> **Exception:** *Smoke barriers* constructed of minimum 0.10-inch-thick (2.5 mm) steel in Group I-3 buildings.

❖ Smoke barriers are intended to create an area of refuge; therefore, they are to be capable of resisting the passage of smoke (see Section 709.4). Smoke barriers are also required to have a fire-resistance rating of at least 1 hour. The smoke barrier is not intended or expected to be exposed to fire for extended periods and is, therefore, not required to have a fire-resistance rating exceeding 1 hour. The occupancies in which smoke barriers are required are also generally required to be sprinklered (see Section 903.2). The exception of this section allows smoke barriers in occupancies in Group I-3 to be constructed of nominal 0.10-inch (2.5 mm) steel plate. This exception to the 1-hour fire-resistance-rated assembly recognizes the security needs of such facilities and at the same time provides the requisite smoke barrier performance.

710.4 Continuity. *Smoke barriers* shall form an effective membrane continuous from outside wall to outside wall and from the top of the foundation or floor/ceiling assembly below to the underside of the floor or roof sheathing, deck or slab above, including continuity through concealed spaces, such as those found above suspended ceilings, and interstitial structural and mechanical spaces. The supporting construction shall be protected to afford the required *fire-resistance rating* of the wall or floor supported in buildings of other than Type IIB, IIIB or VB construction.

> **Exception:** Smoke-barrier walls are not required in interstitial spaces where such spaces are designed and constructed with ceilings that provide resistance to the passage of fire and smoke equivalent to that provided by the smoke-barrier walls.

❖ Smoke barriers are to be continuous from outside wall to outside wall and from the top of the foundation or floor/ceiling assembly below to the underside of the floor or roof sheathing, deck or slab. The provisions require the barrier to be continuous through all concealed and interstitial spaces, including suspended ceilings and the space between the ceiling and the floor or roof sheathing, deck or slab above. Smoke barriers are not required to extend through interstitial spaces if the space is designed and constructed such that fire and smoke will not spread from one smoke compartment to another; therefore, the construction assembly forming the bottom of the interstitial space must provide the required fire-resistance rating and be capable of resisting the passage of smoke from the spaces below.

As mentioned in the commentary for the general provisions of Section 710.1, the air leakage performance of the smoke barrier itself is not typically regulated. This is because of the general assumption that a barrier which provides a fire-resistance rating will be capable of limiting the spread of smoke through it. When a smoke barrier is being used with a smoke control system the barrier does have limitations on the air leakage (see Section 909.5). Therefore, under those circumstances and when using the exception in Section 710.4, it may be necessary to provide a "hard ceiling" instead of a lay-in suspended ceiling in order to prevent the spread of smoke.

Since the primary performance of smoke barriers is to achieve protection on the fire floor, the supporting construction is not required to provide the same degree of fire resistance for buildings of Type IIB, IIIB and VB construction.

As with fire partitions serving as exit access corridor walls (see Section 709.4), these three construction types are identified because the floor construction is not otherwise required to have a fire-resistance rating, and it is not considered essential to only require fire-resistance-rated floor construction because the floor is supporting a smoke barrier.

When designing for occupancies in Groups I-2 and I-3, it is often desirable to incorporate a horizontal exit and a smoke barrier. The code does not prevent such a combination, but in such cases, the wall construction must meet the provisions of this section as well as

Section 706 for fire barrier assemblies and Section 1025 for horizontal exits. The most restrictive provisions of each section would apply. For example, the building would be required to have a fire barrier with a fire-resistance rating of at least 2 hours (see Section 1025.2) and duct penetrations would need to be protected with a combination fire and smoke damper (see Sections 716.5.2 and 716.5.5) or with separate fire and smoke dampers.

Furthermore, the supporting construction would then be required to provide at least the same fire-resistance rating as the wall in all types of construction (see Section 707.5.1).

710.5 Openings. Openings in a *smoke barrier* shall be protected in accordance with Section 715.

Exceptions:

1. In Group I-2, where doors are installed across *corridors*, a pair of opposite-swinging doors without a center mullion shall be installed having vision panels with fire-protection-rated glazing materials in fire-protection-rated frames, the area of which shall not exceed that tested. The doors shall be close fitting within operational tolerances, and shall not have undercuts in excess of $^3/_4$-inch, louvers or grilles. The doors shall have head and jamb stops, astragals or rabbets at meeting edges and shall be automatic-closing by smoke detection in accordance with Section 715.4.8.3. Where permitted by the door manufacturer's listing, positive-latching devices are not required.

2. In Group I-2, horizontal sliding doors installed in accordance with Section 1008.1.4.3 and protected in accordance with Section 715.

❖ In order to maintain the integrity of the smoke barrier, door assemblies and double means of egress corridor doors are required to have a fire protection rating of at least 20 minutes when tested in accordance with Section 715. In occupancies in Group I-2, however, double means of egress cross-corridor doors are not required to have positive latching hardware, since center mullions are prohibited, as they could serve as a possible obstruction in moving beds through the door opening. Without a center mullion, the latching hardware would utilize the floor below and the door frame above for latching purposes. The required stops and door-closing devices will assist in keeping the doors closed. Doors, other than cross-corridor doors, are required to be provided with positive latches as required for the fire protection rating of the door.

When double egress corridor doors are provided, they are required to have a vision panel consisting of protection-rated glazing, in accordance with Section 715.4.7. Vision panels are required to reduce the likelihood of an injury caused by opening the door into a person standing at or approaching the door from the opposite side. Vision panels also enable an individual to be alerted to a fire or smoke condition on the other side before opening the door. Since the primary purpose of a smoke barrier is to resist smoke spread, the doors are not to have undercuts, louvers or grilles. The only openings permitted in the door assembly are those clearances that are necessary for the proper operation of the door. The opening that occurs at the meeting edges of the door is to be protected with rabbets or astragals (see Figure 710.5). In this manner, the doors will close completely, thus preventing the spread of smoke. Stops are to be provided on the head and jambs of all doors in smoke barriers. Such doors must also comply with UL 1784 in accordance with Section 715.4.3.1, which requires these doors to have a leakage rating of 3 cfm per square foot [0.02 $m^3/(s \cdot m^2)$] or less as tested at a pressure of 0.10 inch of water (0.02 kPa).

710.6 Penetrations. Penetrations of *smoke barriers* shall comply with Section 713.

❖ The provisions for penetrations of smoke barriers are found in Sections 713.2, 713.3 and 713.5. To prevent through penetrations from becoming a source of significant smoke spread across the smoke barrier, the firestop system provided for each through penetration must have not only the appropriate F rating, but must also have a leakage rating (L rating) of 5 cfm per square foot [2.78 $m^3/(s \cdot m^2)$] or less as tested at a pressure of 0.30 inch of water (0.07 kPa) (see commentary, Section 713.5).

710.7 Joints. Joints made in or between *smoke barriers* shall comply with Section 714.

❖ Joints, such as expansion or seismic, are another form of openings in smoke barriers and, therefore, must be considered with regard to maintaining the fire-resistance ratings of these walls. This section requires all joints that are located in smoke barriers to be protected by a joint system that has a fire-resistance rating and complies with the requirements of Section 714 (see commentary, Section 714). To prevent joints from becoming a source of significant smoke spread across the smoke barrier, fire-resistive joint systems used in a smoke barrier must have a leakage rating (L rating) of 5 cfm/linear foot (0.01 m^3/g-in), in accordance with

Section 714.6 or less (see commentary, Section 714.6).

710.8 Ducts and air transfer openings. Penetrations in a *smoke barrier* by ducts and air transfer openings shall comply with Section 716.

❖ To prevent ducts and air transfer openings from becoming a source of significant smoke spread across the smoke barrier, a listed smoke damper must be provided for duct or air transfer openings penetrating the smoke barrier (see Section 716.5.5). The smoke damper leakage rating, as tested in accordance with UL 555S, shall not be less than Class II (see Section 716.3.2).

Section 716.5 indicates which situations will require the installation of a damper. For smoke barriers, there is one exception to that requirement. As stated in Section 716.1.1, if a duct does not require a damper, the penetration of that duct through a fire-resistance-rated assembly must be protected as a through penetration in accordance with Section 713.

SECTION 711
SMOKE PARTITIONS

711.1 General. Smoke partitions installed as required elsewhere in the code shall comply with this section.

❖ Unlike a 1-hour fire-resistance-rated smoke barrier, smoke partitions are nonrated walls that serve to resist the spread of fire and the unmitigated movement of smoke for an unspecified period of time. A smoke partition is intended to provide less protection than a smoke barrier, and is not required to be continuous through the concealed spaces and through the ceiling. The construction of a smoke partition is described in this section; however, the level of performance or a method of testing them is not provided. Some elements of this section (see Sections 711.5 and 711.7) will provide a performance level and test method to verify the protection provided by doors or dampers. In addition, some specific features of a smoke partition may be triggered by other sections of the code.

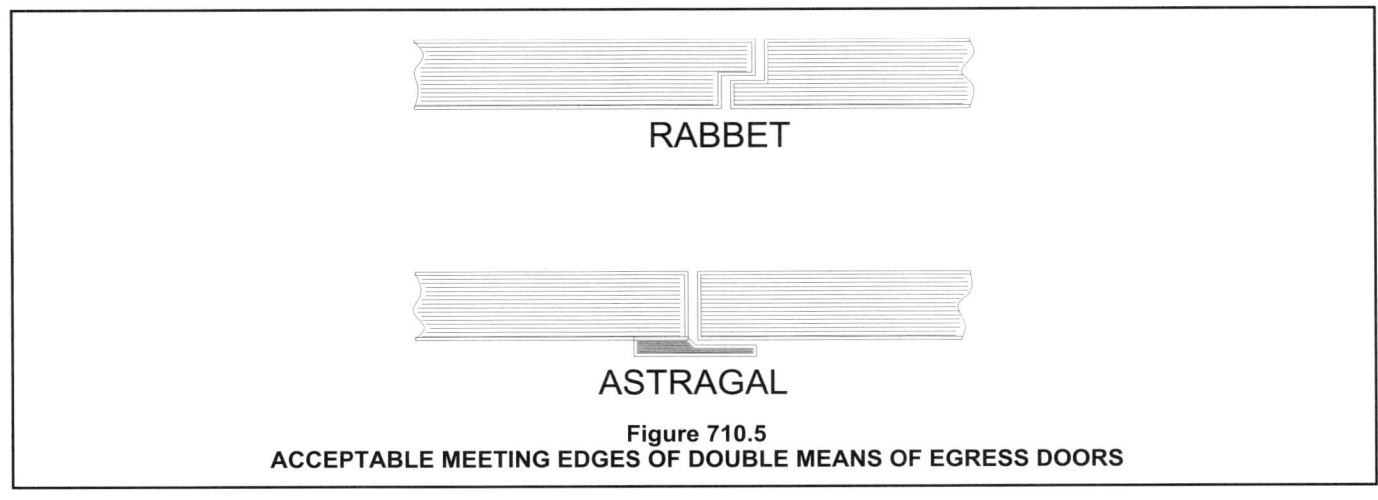

Figure 710.5
ACCEPTABLE MEETING EDGES OF DOUBLE MEANS OF EGRESS DOORS

The application of smoke partitions is still fairly limited. The smoke partition requirements apply to Section 407.3, which permits corridor walls in Group I-2 to have no fire-resistance rating, but it does require that they be constructed as smoke partitions. A smoke partition can also be used under the provisions of Section 708.14.1, Exception 5, to form the elevator lobby in a sprinklered building.

It is important to realize that like all other elements found in Chapter 7, a smoke partition is not required unless the specific text of the code states that it is required. For example, the two sections listed in the previous paragraph do require the use of a smoke partition. On the other hand, Section 508.2.5.2, when dealing with incidental uses, will permit the elimination of a fire barrier and only require the wall constructed around the incidental use area to be "capable of resisting the passage of smoke." In this circumstance, the wall is not required to comply with the smoke partition requirements of Section 710 and is only required to comply with the details listed in Section 508.2.5.2.

711.2 Materials. The walls shall be of materials permitted by the building type of construction.

❖ As with most wall or partition assemblies, except for fire walls, there are no additional requirements or restrictions on the materials used, except that they meet the requirements for the type of construction.

711.3 Fire-resistance rating. Unless required elsewhere in the code, smoke partitions are not required to have a *fire-resistance rating*.

❖ The primary purpose of smoke partitions is to prevent the ready and quick passage of smoke into corridors in Group I-2 or for elevator lobby protection in a sprinklered building. The automatic sprinkler system installed in these instances eliminates the need for a fire-resistance-rated assembly; however, the issue of smoke propagation must be addressed with the use of a smoke partition. Because a performance level or a method of testing the partition is not established, any rated or nonrated assembly would be permitted, provided it is approved by the building official.

711.4 Continuity. Smoke partitions shall extend from the top of the foundation or floor below to the underside of the floor or roof sheathing, deck or slab above or to the underside of the ceiling above where the ceiling membrane is constructed to limit the transfer of smoke.

❖ The continuity provisions for smoke partitions are similar to those for fire partitions (see Section 709.4), except that the issue is smoke spread, not fire. Therefore, the allowance for termination at the underside of the ceiling membrane (as opposed to the underside of the floor or roof deck above) relates to the ability of the ceiling membrane to limit the spread of smoke. Typical "lay in" ceiling tiles, for instance, would probably not serve this function; however, a drywall ceiling that is taped and finished would be an example of construction that could resist the passage of smoke.

711.5 Openings. Windows shall be sealed to resist the free passage of smoke or be automatic-closing upon detection of smoke. Doors in smoke partitions shall comply with this section.

❖ Limiting smoke movement in buildings with smoke partitions includes opening protectives that resist the passage of smoke.

711.5.1 Louvers. Doors in smoke partitions shall not include louvers.

❖ Louvers in doors are a ready opening to allow smoke to move from one area to the remainder of the building. This would defeat the sole purpose of a smoke partition.

711.5.2 Smoke and draft control doors. Where required elsewhere in the code, doors in smoke partitions shall meet the requirements for a smoke and draft control door assembly tested in accordance with UL 1784. The air leakage rate of the door assembly shall not exceed 3.0 cubic feet per minute per square foot [0.015424 $m^3/(s \cdot m^2)$] of door opening at 0.10 inch (24.9 Pa) of water for both the ambient temperature test and the elevated temperature exposure test. Installation of smoke doors shall be in accordance with NFPA 105.

❖ Only doors in smoke partitions that are required elsewhere in the code to be smoke and draft control doors must comply with section. Section 407.3.1 requires corridor doors in Group I-2 to "limit the transfer of smoke"; therefore, those doors must meet this section. Section 708.14.1, Exception 5, requires smoke and draft control doors for the smoke partition exception to elevator lobby enclosures (see commentary, Section 715.4.3.1).

711.5.3 Self- or automatic-closing doors. Where required elsewhere in the code, doors in smoke partitions shall be self- or automatic-closing by smoke detection in accordance with Section 715.4.8.3.

❖ The requirement for doors to be self-closing or automatic closing does not automatically apply to all walls that must be built as smoke partitions. That requirement must be mandated elsewhere in the code for this provision to apply. In the case of corridor walls for Group I-2 occupancies, there is no requirement for self-closing or automatic-closing doors so as to accommodate the normal operational requirements of that occupancy. In that situation, it is anticipated that doors would be manually closed by personnel in the event of a fire emergency based upon a defend-in-place strategy. However, having doors arranged for automatic closing or self-closing is not prohibited, as it would potentially increase the reliability of smoke partitions. Exception 5 that allows elevator lobby smoke partition enclosures in lieu of fire partitions does require closers on the doors to the lobby as a part of the conditions for the exception.

711.6 Penetrations and joints. The space around penetrating items and in joints shall be filled with an *approved* material to limit the free passage of smoke.

❖ There is no requirement for a material used to seal penetrations or joints to be applied to both sides of the wall, as would usually be the case when the penetrations and joints are in a fire-resistance-rated wall. The intent of this section will have been met as long as the space around penetrating items and in joints is filled in a way that would prevent a continuous channel from one side of the wall to the other. In selecting a material to seal these gaps, consideration should be given to the fact that smoke may have a temperature that is above the normal ambient temperature. The code does not mandate any specific temperature resistance; however, as one possible reference point, Underwriters Laboratories use a gas temperature of 325°F (163°C) above ambient to represent "warm smoke." With a common ambient temperature of 75°F (24°C), the "warm smoke" temperature would therefore be 400°F (204°C). Sealing materials that would not maintain their integrity to at least this temperature would be inferior choices for the purpose of limiting the free passage of smoke. Porous materials (e.g., fibrous insulation) that would not provide any significant resistance to the passage of smoke through them would also be inferior choices for sealing of joints and through penetrations for this purpose. Again, because the level of performance is not clearly established, the word "approved" and the building official's decision will determine what materials are required or accepted.

711.7 Ducts and air transfer openings. The space around a duct penetrating a smoke partition shall be filled with an *approved* material to limit the free passage of smoke. Air transfer openings in smoke partitions shall be provided with a *smoke damper* complying with Section 716.3.2.2.

> **Exception:** Where the installation of a *smoke damper* will interfere with the operation of a required smoke control system in accordance with Section 909, *approved* alternative protection shall be utilized.

❖ Smoke dampers are required to be provided to maintain the integrity of the smoke partition as a means to prevent the spread of smoke when an "air transfer opening" exists. If a ducted system is used, then the code specifies that the annular space around the duct must be protected. Because the code does not define what an "air transfer opening" is, the provision may not be consistently enforced. The general intent is that a ducted system does not require a damper, but an opening such as an air transfer grille or louvered opening would be dampered. Since Section 716.1.1 requires ducts without dampers to be protected as a penetration, the code requires a ducted system to be protected in accordance with Section 711.6 and just the first sentence of Section 711.7. The exception relating to required operation of a smoke control system is consistent with the exceptions given elsewhere regarding air transfer openings and ducts with fire dampers or smoke dampers (see Section 716). An effective smoke control system will protect the building against the spread of smoke, and it is important to ensure such installation of a smoke damper will not lessen the effectiveness of a smoke control system.

SECTION 712
HORIZONTAL ASSEMBLIES

712.1 General. Floor and roof assemblies required to have a *fire-resistance rating* shall comply with this section. Nonfire-resistance-rated floor and roof assemblies shall comply with Section 713.4.2.

❖ Horizontal assemblies may be either a floor or roof assembly and may rely on the ceiling as a part of a floor/ceiling or roof/ceiling assembly. The ceiling assembly is often an integral part of a fire-resistance-rated floor/ceiling or roof/ceiling assembly; therefore, the integrity of the ceiling assembly must be maintained in order to reduce the potential for premature failure of the floor or roof of a building.

Much of the information contained in the section is obtained from and based on conditions of testing various assemblies. One focus of the section is on maintaining the integrity of ceiling membranes, which are an integral component of the fire-resistance rating of floor/ceiling and roof/ceiling assemblies.

712.2 Materials. The floor and roof assemblies shall be of materials permitted by the building type of construction.

❖ The types of materials used in floor and roof assemblies are to be consistent with Sections 602.2 through 602.5 for the type of construction classification of the building. Floors and roofs are permitted to be of combustible construction in Type III, IV and V construction, and are required to be of noncombustible materials in Type I and II construction, except as permitted in the exceptions to Section 603.1 or in Notes c and d from Table 601.

712.3 Fire-resistance rating. The *fire-resistance rating* of floor and roof assemblies shall not be less than that required by the building type of construction. Where the floor assembly separates mixed occupancies, the assembly shall have a *fire-resistance rating* of not less than that required by Section 508.4 based on the occupancies being separated. Where the floor assembly separates a single occupancy into different *fire areas*, the assembly shall have a *fire-resistance rating* of not less than that required by Section 707.3.9. *Horizontal assemblies* separating *dwelling units* in the same building and *horizontal assemblies* separating *sleeping units* in the same building shall be a minimum of 1-hour fire-resistance-rated construction.

> **Exception:** *Dwelling unit* and *sleeping unit* separations in buildings of Type IIB, IIIB and VB construction shall have *fire-resistance ratings* of not less than $^1/_2$ hour in buildings equipped throughout with an *automatic sprinkler system* in accordance with Section 903.3.1.1.

❖ Floor and roof assemblies could have a fire-resistance rating for several reasons:

- Table 601 requires for the type of construction.

- Section 508.4 requires for separation of mixed uses.

- Separate fire areas of like or different occupancies in order to reduce the size of fire areas to below threshold values for sprinkler requirements in Section 903.

- Separation of dwelling units in the same building.

- Separation of sleeping units in Group R-1 hotels, R-2 and I-1.

As an example, consider a nonsprinklered two-story building of Type IIA construction and a Group B occupancy, with the basement having a storage occupancy in Group S-2. Assume the designer has chosen Section 508.3.3 for separated occupancies as the option for compliance with the mixed-occupancy condition. In this case, even though Table 601 requires a 1-hour fire-resistance-rated floor throughout the building, the assembly separating the basement from the first floor of the building is also regulated by Section 508.3 for mixed occupancies. Ultimately, Section 508.3.3 references Table 508.3.3, which requires that the floor/ceiling construction separating Group S-2 from Group B provide a 2-hour fire-resistance-rated separation.

Chapter 6 also contains provisions governing materials that may be used in the construction of such assemblies based on the type of construction of the building whether combustible or noncombustible materials are used.

The requirement for 1-hour floor assemblies separating dwelling and sleeping units mirrors that found in Section 708 for fire partitions. There are other sections of the code that require a fire-resistance rating for the floor or roof assembly or where other provisions are dependent on the fire-resistance rating or construction of the assembly (examples of such sections include Sections 704.10, 705.6 and the exceptions to Section 1024.1).

The exception for dwelling and sleeping unit separations in unprotected types of construction is based on the protection provided by a sprinkler system installed in accordance with Section 903.3.1.1 (NFPA 13). This presumably will reduce the potential fire exposure of the unit separation to that which makes a minimum $^1/_2$-hour fire-resistance rating adequate. The reason that only unprotected types of construction are permitted to take advantage of this trade-off is because Table 601 would not otherwise require the floor construction of such structures to be fire-resistance rated, whereas all other types of construction require the floor construction to provide at least a 1-hour fire-resistance rating.

712.3.1 Ceiling panels. Where the weight of lay-in ceiling panels, used as part of fire-resistance-rated floor/ceiling or roof/ceiling assemblies, is not adequate to resist an upward force of 1 pound per square foot (48 Pa), wire or other *approved* devices shall be installed above the panels to prevent vertical displacement under such upward force.

❖ Where a ceiling membrane constitutes part of a fire-resistance-rated floor/ceiling or roof/ceiling assembly, the ability of the ceiling membrane to remain in place and not be displaced by the upward pressure of a fire condition is necessary to maintain the viability of the fire-resistance rating. Figure 712.3.1 shows a floor/ceiling assembly that uses hold-down clips to keep the ceiling panels in place. When the weight of the ceiling panel is less than 1 pound per square foot (psf) (48 Pa), wire or other approved means must be provided to prevent uplift of the ceiling panels. Manufacturers' literature can be used to determine the weight of the panel.

712.3.2 Access doors. Access doors shall be permitted in ceilings of fire-resistance-rated floor/ceiling and roof/ceiling assemblies provided such doors are tested in accordance with ASTM E 119 or UL 263 as horizontal assemblies and labeled by an *approved agency* for such purpose.

❖ Access doors are often necessary in order to service mechanical and plumbing systems above the ceiling. This section states that if such doors are used where the ceiling provides part of the protection, they must be

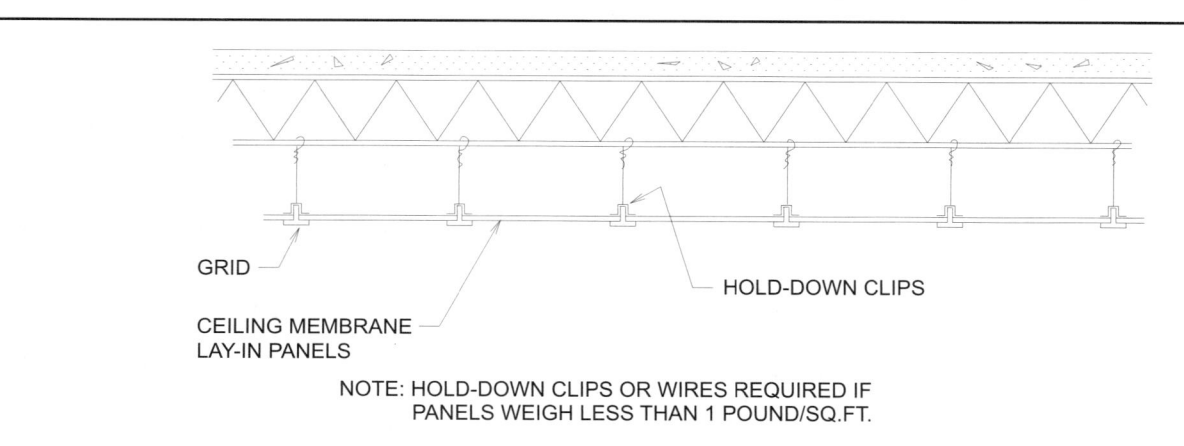

GRID

CEILING MEMBRANE LAY-IN PANELS

HOLD-DOWN CLIPS

NOTE: HOLD-DOWN CLIPS OR WIRES REQUIRED IF PANELS WEIGH LESS THAN 1 POUND/SQ.FT.

For SI: 1 pound per square foot = 4.882 kg/m².

Figure 712.3.1
CEILING PANELS USED IN FIRE-RESISTANCE-RATED ASSEMBLIES

tested in accordance with ASTM E 119 or UL 263 as a horizontal assembly. This makes it clear that the standard fire test for doors (NFPA 80 or 257) is not acceptable. This ensures that the thermal transmission through the access door and its affect on the assembly is considered. The provisions of this section are not applicable if the ceiling membrane does not provide any portion of the fire-resistive protection. Therefore, in a nonrated ceiling, this access door requirement would not apply.

712.3.3 Unusable space. In 1-hour fire-resistance-rated floor assemblies, the ceiling membrane is not required to be installed over unusable crawl spaces. In 1-hour fire-resistance-rated roof assemblies, the floor membrane is not required to be installed where unusable *attic* space occurs above.

❖ The section is a special exception for specific assemblies to allow the deletion of floor or roof decking or the ceiling membrane from a required fire-resistance-rated floor/ceiling or roof/ceiling assembly used in certain applications. In an attic application, the floor sheathing may be deleted from a fire-resistance-rated assembly as long as the joist and ceiling remain identical to the tested assembly and the attic space is not usable space where potential combustible materials or ignition sources may be located [see Figure 712.3.3(1)]. Over a crawl space, the ceiling membrane may be deleted subject to the same conditions [see Figure 712.3.3(2)]. In determining whether a space is "unusable," the building official must verify that combustible materials other than construction elements will not be located therein and ignition sources will be minimal. As such, pipes, conduit and ducts may be permitted in an unusable space.

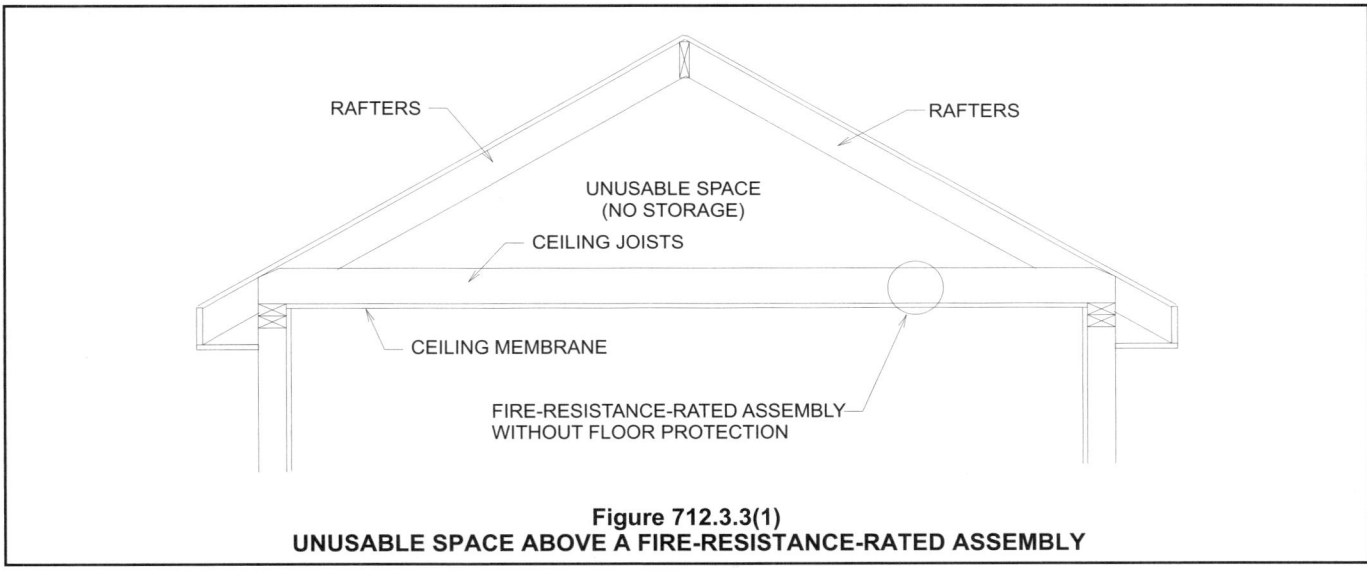

Figure 712.3.3(1)
UNUSABLE SPACE ABOVE A FIRE-RESISTANCE-RATED ASSEMBLY

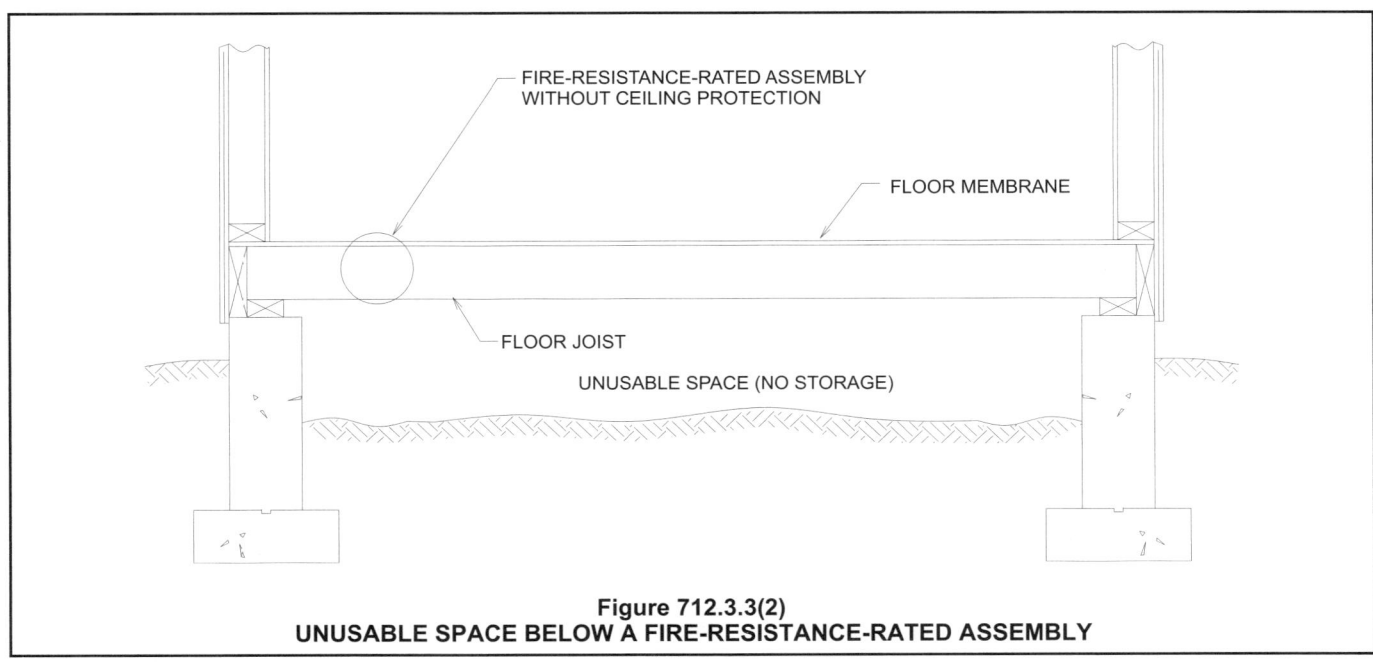

Figure 712.3.3(2)
UNUSABLE SPACE BELOW A FIRE-RESISTANCE-RATED ASSEMBLY

712.4 Continuity. Assemblies shall be continuous without openings, penetrations or joints except as permitted by this section and Sections 708.2, 713.4, 714 and 1022.1. Skylights and other penetrations through a fire-resistance-rated roof deck or slab are permitted to be unprotected, provided that the structural integrity of the fire-resistance-rated roof assembly is maintained. Unprotected skylights shall not be permitted in roof assemblies required to be fire-resistance rated in accordance with Section 704.10. The supporting construction shall be protected to afford the required *fire-resistance rating* of the *horizontal assembly* supported.

> **Exception:** In buildings of Type IIB, IIIB or VB construction, the construction supporting the *horizontal assembly* is not required to be fire-resistance-rated at the following:
>
> 1. Horizontal assemblies at the separations of incidental uses as specified by Table 508.2.5, provided the required *fire-resistance rating* does not exceed 1 hour.
>
> 2. Horizontal assemblies at the separations of *dwelling units* and *sleeping units* as required by Section 420.3.
>
> 3. Horizontal assemblies at *smoke barriers* constructed in accordance with Section 710.

❖ All floors, roofs and ceilings of horizontal assemblies are to be continuous without openings or penetrations, except as permitted by this section. The continuity of the assembly is critical to its ability to limit fire and smoke spread. The continuity provision applies regardless of whether a fire-resistance rating is required, since floor/ceiling assemblies are also intended to restrict vertical smoke movement [see Figure 712.4(1)]. Penetrations or openings of the assembly are permitted in accordance with Section 708.2, 713.4 or 714, provided that the fire-resistance rating, if required, is maintained [see Figure 712.4(2)]. The fire-resistance rating required by Table 601 for roof construction is intended to minimize the threat of premature structural failure of the roof construction under fire conditions. These provisions, with the exception of the requirements of Section 705.10 and

706.6, are not intended to create a barrier in order to contain the fire within the building. Nonfire-resistance-rated penetrations, and skylight and roof window assemblies are, therefore, permitted to be installed in fire-resistance-rated roof assemblies, provided that the structural integrity of the roof assembly is not reduced and the provisions of Section 705.10 for protection of vertical exposure do not apply (see commentary, Section 705.10). The issue of structural integrity refers to the effect the collapse of a skylight assembly, under fire conditions, would have on the roof structure. Section 708.12, regarding the extension of a shaft to the roof level, would also permit the same unprotected openings to the exterior. Code users should also review the fireblocking and draftstopping requirements that are found in Section 717. Fireblocking and draftstopping requirements apply to combustible concealed locations and it is a separate issue from fire-resistance ratings and may impose additional requirements for the assembly.

The exception deals with three specific applications of horizontal assembly where it is unnecessary to provide fire-resistance rating of the supporting construction of horizontal assemblies in buildings of Type IIB, IIIB or VB construction, which are types of construction where Table 601 would never require the horizontal assembly or the supporting structural members to have a fire-resistance rating. This exception exempts the supporting construction of horizontal assemblies in the same manner as the code currently exempts the supporting construction of fire barriers, and fire partitions and smoke barriers, but only in those circumstances where the horizontal assembly is a component of the same fire containment assembly as the fire barrier or fire partition or smoke barrier. It is not reasonable to exempt construction supporting a fire containment assembly for some components of the assembly but not for other components. If the exemptions for buildings of Type IIB, IIIB and VB construction are valid, they should be applied to the entire fire containment assembly, not just a portion of it.

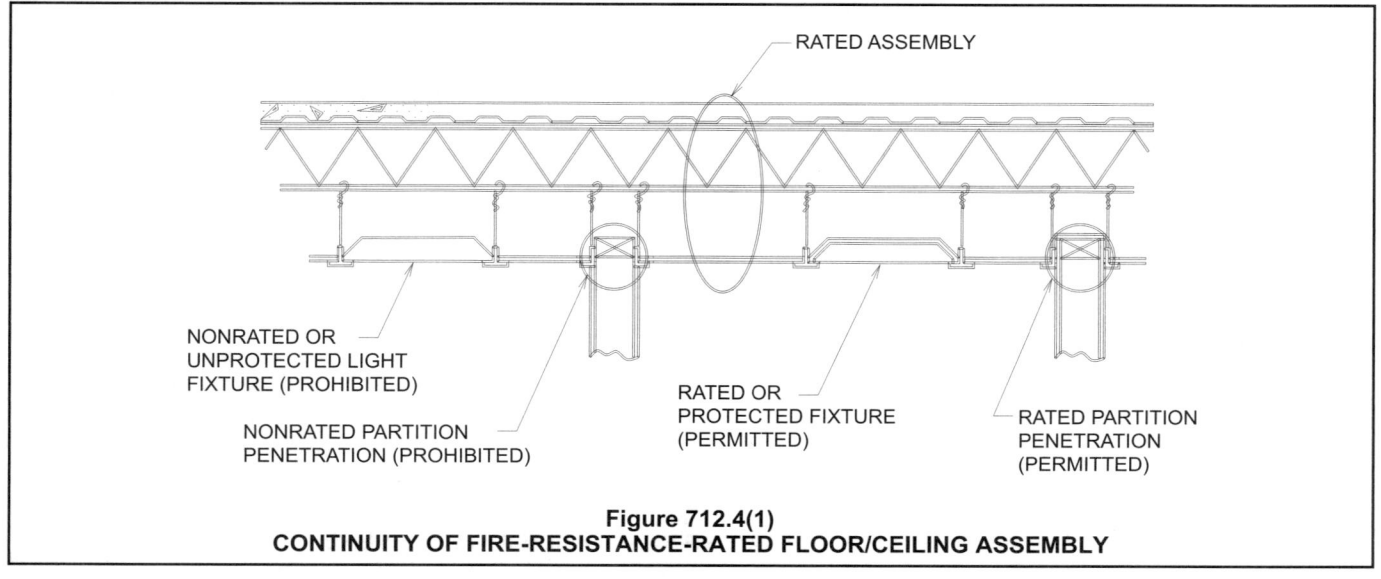

Figure 712.4(1)
CONTINUITY OF FIRE-RESISTANCE-RATED FLOOR/CEILING ASSEMBLY

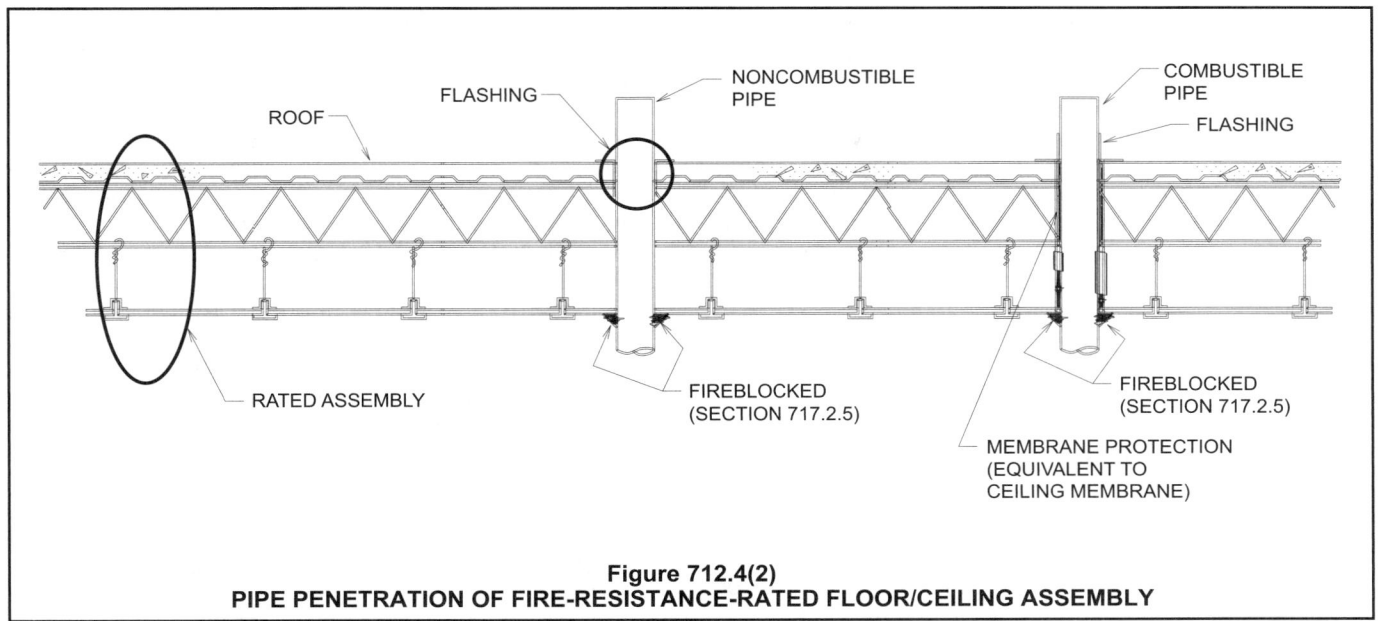

Figure 712.4(2)
PIPE PENETRATION OF FIRE-RESISTANCE-RATED FLOOR/CEILING ASSEMBLY

712.5 Penetrations. Penetrations of *horizontal assemblies* shall comply with Section 713.

❖ Penetration protection for horizontal assemblies is found in Sections 713.2 and 713.4. It should be remembered that as mentioned in the commentary for Section 713.4, the horizontal assembly requirements also apply to nonrated assemblies. The penetration of nonfire-resistance-rated assemblies is regulated in Section 713.4.2, unless dealt with in accordance with Section 708.

712.6 Joints. Joints made in or between *horizontal assemblies* shall comply with Section 714. The void created at the intersection of a floor/ceiling assembly and an exterior curtain wall assembly shall be protected in accordance with Section 714.4.

❖ Joints, such as expansion or seismic, are another form of openings in horizontal assemblies and, therefore, must be considered with regard to maintaining the fire-resistance ratings of these floors or roofs. This section requires all joints that are located in rated horizontal assemblies to be protected by a joint system that has a fire-resistance rating and complies with the requirements of Section 713 (see commentary, Section 713). The provisions of this section also include the protection of joints/voids that may occur at the intersection of the horizontal assembly and an exterior wall.

712.7 Ducts and air transfer openings. Penetrations in *horizontal assemblies* by ducts and air transfer openings shall comply with Section 716.

❖ See the commentary to Section 716.

712.8 Floor fire door assemblies. Floor *fire door* assemblies used to protect openings in fire-resistance-rated floors shall be tested in accordance with NFPA 288, and shall achieve a *fire-resistance rating* not less than the assembly being penetrated. Floor *fire door* assemblies shall be labeled by an

approved agency. The *label* shall be permanently affixed and shall specify the manufacturer, the test standard and the *fire-resistance rating*.

❖ Floor fire doors are specifically addressed as an opening protective in horizontal floor assemblies. The code requires that these doors have a fire-resistance rating instead of just a fire protection rating. By requiring a fire-resistance rating, it will restore the opening in the horizontal assembly to the same level of protection that was originally established during the fire test by limiting the temperature transmission to the unexposed side of the assembly.

712.9 Smoke barrier. Where *horizontal assemblies* are required to resist the movement of smoke by other sections of this code in accordance with the definition of *smoke barrier*, penetrations and joints in such *horizontal assemblies* shall be protected as required for *smoke barriers* in accordance with Sections 713.5 and 714.6. Regardless of the number of *stories* connected by elevator shaft enclosures, doors located in elevator shaft enclosures that penetrate the *horizontal assembly* shall be protected by enclosed elevator lobbies complying with Section 708.14.1. Openings through *horizontal assemblies* shall be protected by shaft enclosures complying with Section 708. *Horizontal assemblies* shall not be allowed to have unprotected vertical openings.

❖ This section is intended to clarify the requirements for horizontal assemblies that are used to support smoke barrier walls, such as in Group I-2 occupancies, where smoke barriers are required by Section 407.4 to subdivide floors. It is clear from the definition for "Smoke barrier" that it can be a horizontal assembly. Furthermore, in order to provide for the continuity of the smoke protection for smoke compartments created by vertical smoke barriers to provide for relative safe areas for horizontal movement of patients in a fire emergency, it follows that the floors supporting those smoke barrier walls should also be able to resist the passage or movement of smoke through the assembly to maintain the appropriate level of protection for the occupants. Generally, Group I-2 occupants occupancies are moved into a smoke barrier that is away from the area where the fire occurred so that they can remain until further moved as necessary or until the fire has been extinguished by the responding fire department. Therefore, the integrity of these horizontal assemblies is essential for the protection of the occupants of the building.

SECTION 713
PENETRATIONS

713.1 Scope. The provisions of this section shall govern the materials and methods of construction used to protect *through penetrations* and *membrane penetrations* of *horizontal assemblies* and fire-resistance-rated wall assemblies.

❖ This section addresses the specific requirements for maintaining the integrity of fire-resistance-rated assemblies at penetrations. The provisions of this section apply to penetrations of fire-resistance-rated walls (see Section 713.3), fire-resistance-rated horizontal assemblies (see Sections 713.4.1 and 713.4.1.4) and nonfire-resistance-rated assemblies (see Section 713.4.2). Penetrations of fire-resistance-rated assemblies range from combustible pipe and tubing to noncombustible wiring with combustible covering to noncombustible items, such as pipe, tube, conduit and ductwork.

Each type of penetration requires a specific method of protection, which is based on the type of fire-resistance-rated assembly that is penetrated and the type of penetrating item. To determine the type of penetration protection required, the first step is to identify whether the penetrated assembly is a fire-resistance-rated wall; a fire-resistance-rated horizontal assembly or a nonfire-resistance-rated horizontal assembly. Next, identify the type of penetrating item in the applicable section and determine the applicable method of penetration protection necessary.

713.1.1 Ducts and air transfer openings. Penetrations of fire-resistance-rated walls by ducts that are not protected with *dampers* shall comply with Sections 713.2 through 713.3.3. Penetrations of *horizontal assemblies* not protected with a shaft as permitted by Exception 4 of Section 708.2, and not required to be protected with fire *dampers* by other sections of this code, shall comply with Sections 713.4 through 713.4.2.2. Ducts and air transfer openings that are protected with *dampers* shall comply with Section 716.

❖ Duct penetrations are typically protected with fire dampers in accordance with Section 716.5; however, if the code does not require and the designer does not provide such dampers, as they may interfere with the system design, this section allows for such removal, provided the penetration is protected as part of a tested assembly (see Section 713.3.1.1) or is protected with some type of through-penetration firestop

system (see Section 713.3.1.2).

This section coordinates with Section 716.1.1, which states the appropriate section to go to in order to determine the proper protection for ducts when they penetrate a fire-resistance-rated wall assembly. Where ducts or air transfer openings penetrate a fire-resistance-rated wall assembly, fire dampers can be installed to maintain the rating of the assembly as required by Section 716. However, in addition to maintaining the fire-resistive integrity of the floor, wall or ceiling or roof assembly, the damper also serves an additional function as an active component of the fire management design. Unlike through-penetration firestop systems, fire dampers can restrict the spread of fire through the air duct system within a building to areas remote from the fire, or into a building from the outside. In addition, the air duct system can be used for the purpose of emergency smoke control, either manually or automatically. Although fire dampers and through-penetration firestop systems both are intended to maintain the rating of the assembly at the point where an opening occurs for the ventilation system, they are not optional or equivalent alternatives for one another.

The placement (or the elimination) of fire dampers within the ventilation system is based on additional factors other than the point at which the duct penetrates a fire-resistance-rated assembly.

Other sections of the code may permit fire dampers to be eliminated due to the installation of an automatic sprinkler system, or the design may not include a fire damper to interface with the ventilation system and fire management design at specific locations. This section addresses the protection of the fire rating of the assembly when such conditions occur and provides that the duct penetration is protected as part of a tested assembly (see Section 713.3.1.1) or is protected with some type of through-penetration firestop system (see Section 713.3.1.2), tested and listed for use on ducts without fire dampers.

713.2 Installation details. Where sleeves are used, they shall be securely fastened to the assembly penetrated. The space between the item contained in the sleeve and the sleeve itself and any space between the sleeve and the assembly penetrated shall be protected in accordance with this section. Insulation and coverings on or in the penetrating item shall not penetrate the assembly unless the specific material used has been tested as part of the assembly in accordance with this section.

❖ Sleeves are typically installed when the opening penetrates an assembly with interior voids, such as a steel stud-framed wall. The sleeve must be securely fastened to the penetrated assembly to prevent the sleeve from becoming dislodged and adversely affecting the performance of the annular space protection. The space between the sleeve and the penetrating item, as well as between the sleeve and the rated assembly, must be protected in accordance with Section 713.3.1. Without attention to these conditions, a sleeved penetration may not perform as required by the code. A sleeve may also exist when a segment of plastic or metal pipe or tube is cast into a concrete floor, thus creating the needed opening for an anticipated penetration. In this case, the need for secure fastening to the underlying assembly is essentially guaranteed, and there would be no space needing to be sealed between the sleeve and the assembly being penetrated.

Section 713.3.1 requires that combustible penetrations be protected with assemblies that have been tested in accordance with ASTM E 119 or UL 263 (see Section 713.3.1) or ASTM E 814 (see Section 713.3.1.2). If a sleeve is used in conjunction with the penetrating items, then the annular space within and around the sleeve must also be protected with materials that meet the conditions of the test standards. The purpose of this section is to provide the building official with the necessary text to enforce the conditions of the penetration protection standards.

If combustible insulation material or coverings on the penetrating item could provide a path for the line of fire to travel through the assembly, the method of protection must have also been fire tested using that specific type and thickness of insulation. This section clarifies that either the insulation or covering must be part of the tested assembly, whether the assembly was tested under ASTM E 119 or UL 263 or is a through-penetration firestop system tested in accordance with ASTM E 814, or else the insulation or covering is not allowed to pass through the penetration. In the latter case, the insulation or covering should be removed from the penetrated item at the point of penetration through the fire-resistance-rated assembly, assuming that the removal would not cause other problems and that the removal would be compliant with other applicable codes [e.g., *International Plumbing Code®* (IPC®), IMC].

713.3 Fire-resistance-rated walls. Penetrations into or through *fire walls*, *fire barriers*, *smoke barrier* walls and *fire partitions* shall comply with Sections 713.3.1 through 713.3.3. Penetrations in *smoke barrier* walls shall also comply with Section 713.5.

❖ In order to maintain the integrity of the fire-resistance-rated wall assembly (fire walls, fire barriers, smoke barriers and fire partitions), penetrations into and through the rated assembly must be properly protected. Acceptable protection methods for various penetrations of wall assemblies are identified in Sections 713.3.1 through 713.3.4. Additionally, penetrations of smoke barriers must meet the test requirements for air leakage given in Section 713.5.

713.3.1 Through penetrations. Through penetrations of fire-resistance-rated walls shall comply with Section 713.3.1.1 or 713.3.1.2.

Exception: Where the penetrating items are steel, ferrous or copper pipes, tubes or conduits, the *annular space* between

the penetrating item and the fire-resistance-rated wall is permitted to be protected as follows:

1. In concrete or masonry walls where the penetrating item is a maximum 6-inch (152 mm) nominal diameter and the area of the opening through the wall does not exceed 144 square inches (0.0929 m²), concrete, grout or mortar is permitted where it is installed the full thickness of the wall or the thickness required to maintain the *fire-resistance rating*; or

2. The material used to fill the *annular space* shall prevent the passage of flame and hot gases sufficient to ignite cotton waste when subjected to ASTM E 119 or UL 263 time-temperature fire conditions under a minimum positive pressure differential of 0.01 inch (2.49 Pa) of water at the location of the penetration for the time period equivalent to the *fire-resistance rating* of the construction penetrated.

❖ Combustible cables, wires, pipes, tubes and conduits that penetrate fire-resistance-rated walls are required to be properly protected [see Figures 713.3.1(1) and 713.3.1(2)]. These penetrations are required to be tested in accordance with ASTM E 119 or UL 263 (see Section 713.3.1.1) or ASTM E 814 (see Section 713.3.1.2).

Because combustible materials have a greater propensity to spread fire through a penetration, the requirements for combustible penetrating items are considerably more restrictive than for noncombustible penetrating items, which can use the exception. Penetration of fire-resistance-rated walls by cables, wires, pipes, tubes, conduits and vents represent a weak link in the continuity of the required fire-resistance rating and are required to be properly protected. These penetrations are to be protected with materials that will maintain the integrity of the wall for the duration of the required fire-resistance rating. In lieu of a system tested in accordance with ASTM E 119 or UL 263 (see Section 713.3.1.1), ASTM E 814 or UL 1479 (see Section 713.3.1.2), annular space protection in accordance with the exception is acceptable. Annular space protection is permitted for the protection of penetrations of wall assemblies by noncombustible items. The annular space is defined in Section 702.1 as the perimeter space between the penetrating item and the rated assembly. If sleeves are used, the annular space includes the space between the penetrating item and the sleeve, as well as the space between the sleeve and the rated assembly [see Figure 713.3.1(3)]. Item 1 of the exception permits the use of concrete, grout or mortar to provide annular space protection for certain penetrations of concrete and masonry wall assemblies. The concrete, grout or mortar must be provided for the full thickness of the wall, unless evidence can be provided that demonstrates that the required fire- resistance rating can be achieved with a lesser depth.

Concrete, grout and mortar have traditionally been used as protection for the annular space in penetrations of concrete and masonry walls. The presumption has been that experience has shown this form of protection to be viable.

Under the exception, because the penetrating items are limited and because the annular space protection is not based on specific performance testing of each penetration arrangement and detail, the performance of the protection is dependent on the annular space protection material alone. As such, Item 2

requires that the material be prequalified as to its ability to prevent the passage of flame and hot gases sufficient to ignite cotton waste and when subjected to the time-temperature criteria of ASTM E 119 or UL 263. This is consistent with the criteria required for through-penetration protection systems (ASTM E 814) in which the T rating is not required. Since it is very likely that the penetration in an actual fire will be exposed to a positive pressure, this section specifies that the test fire exposure include a positive pressure of 0.01 inch (0.25 mm) of water column as a further means to verify the performance of this protection method as required by the code.

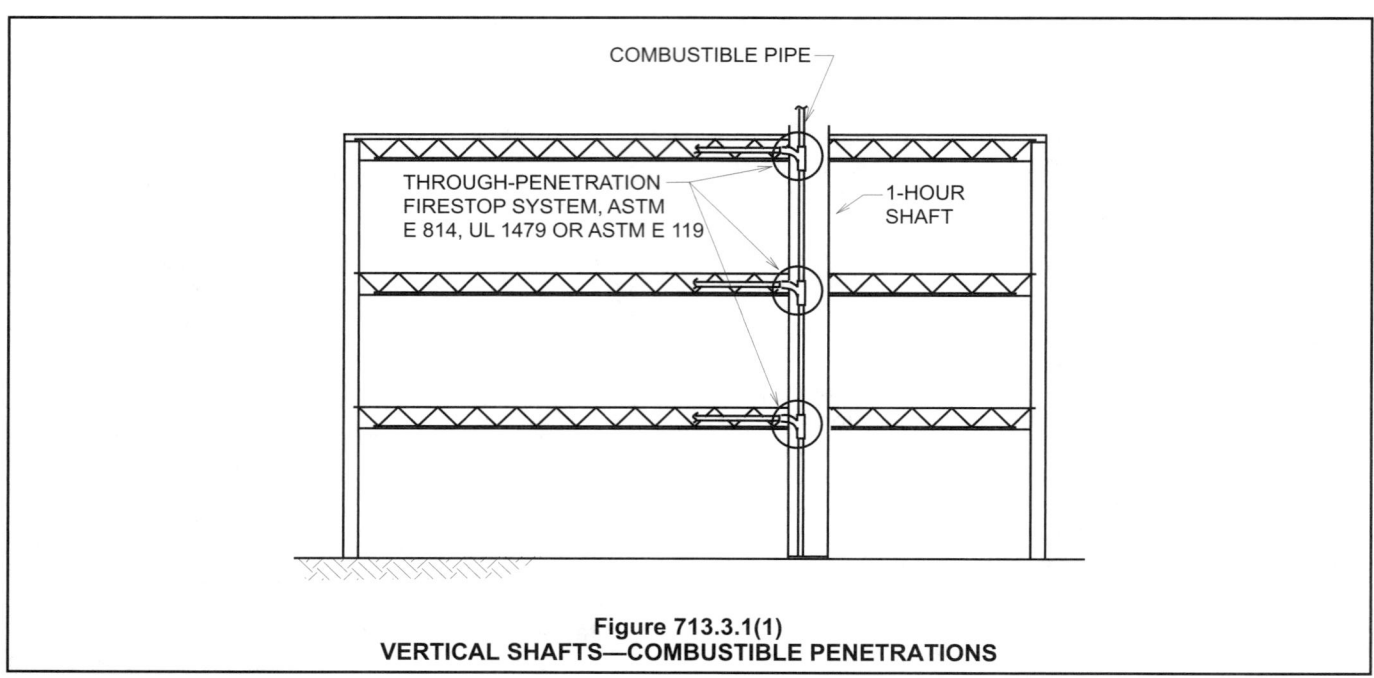

Figure 713.3.1(1)
VERTICAL SHAFTS—COMBUSTIBLE PENETRATIONS

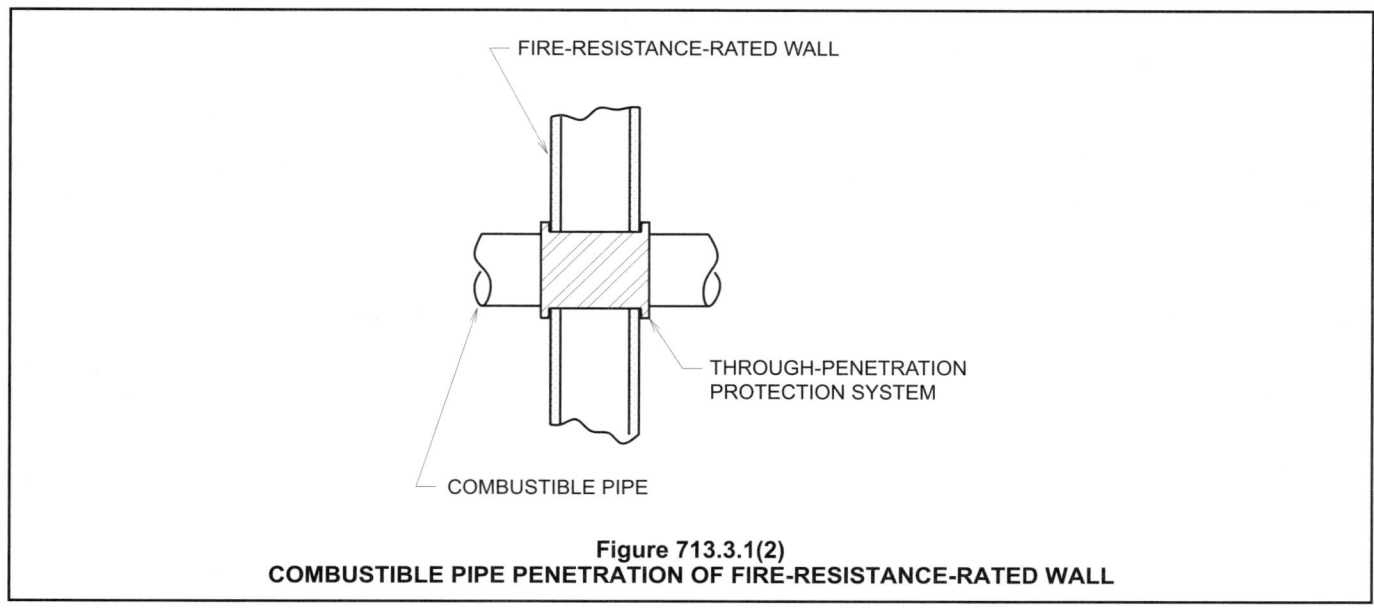

Figure 713.3.1(2)
COMBUSTIBLE PIPE PENETRATION OF FIRE-RESISTANCE-RATED WALL

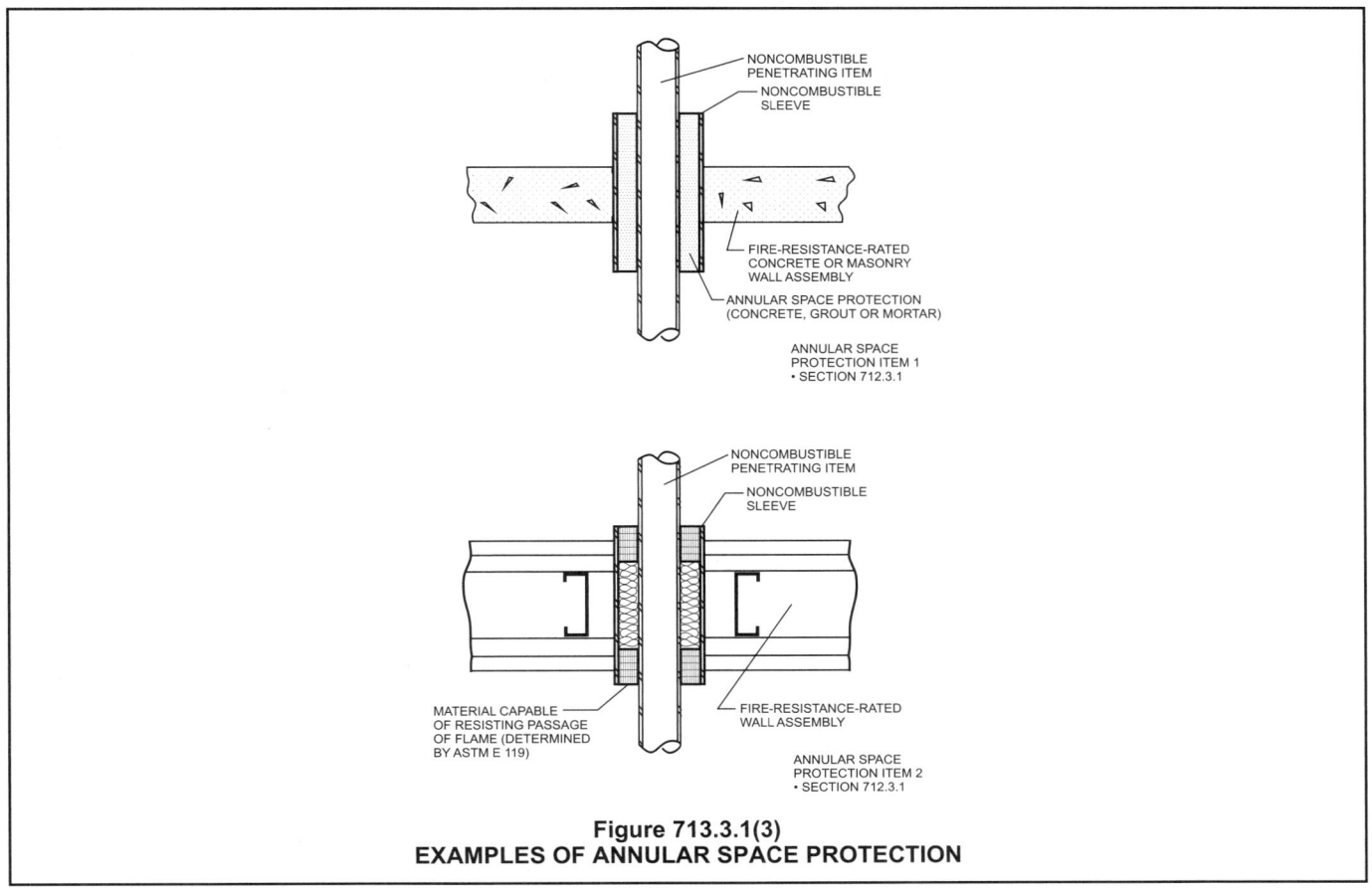

Figure 713.3.1(3)
EXAMPLES OF ANNULAR SPACE PROTECTION

713.3.1.1 Fire-resistance-rated assemblies. Penetrations shall be installed as tested in an *approved* fire-resistance-rated assembly.

❖ This section requires, as an option, that the tested assembly in accordance with ASTM E 119 or UL 263 include the penetrations along with the proposed type of protection—the entire assembly is then tested. As an option, this section is not used frequently due to the limitations placed on the tested assembly relative to its application. Penetration protection is most often provided in accordance with the exception to Section 713.3.1 and the provisions of Section 713.3.1.2.

713.3.1.2 Through-penetration firestop system. *Through penetrations* shall be protected by an *approved* penetration firestop system installed as tested in accordance with ASTM E 814 or UL 1479, with a minimum positive pressure differential of 0.01 inch (2.49 Pa) of water and shall have an F rating of not less than the required *fire-resistance rating* of the wall penetrated.

❖ In order to maintain effective compartmentation of a building to restrict the spread of fire, all penetrations through fire-resistance-rated assemblies must be protected. In recent history, there have been several examples of buildings where extensive fire damage and loss of life was attributed, at least in part, to lack of or improper installation of penetration protection. In the absence of a through-penetration firestop system to protect penetrations of fire-resistance-rated assemblies, the potential exists for fire to spread beyond the initial area of fire origin. One report on the source of origin of fires in buildings indicated that 23 percent of all building fires originate from electrical systems. This fact, coupled with the number of penetrations through walls and floors created by electrical distribution piping, helps to underscore the necessity for the protection of all penetrations through rated assemblies. It must be noted that penetrations are not limited to electrical systems only, but that openings to accommodate plumbing and mechanical systems also contribute to the number of penetrations through any given assembly.

Through-penetration firestop system consist of specific materials or an assembly of materials that are designed to restrict the passage of fire and hot gases for a prescribed period of time through openings made in fire-resistance-rated walls or horizontal assemblies. In certain instances, the through-penetration firestop system is also required to limit the transfer of heat from the fire side to the unexposed side. In order to determine the effectiveness of a through-penetration firestop system in restricting the passage of fire and the transfer of heat, firestop systems are required to be subjected to fire testing. ASTM E 814 or UL 1479 are the test methods developed specifically for the evaluation of a firestop system's ability to resist the passage of flame and hot gases, withstand thermal stresses and restrict transfer of heat through the penetrated assembly.

The basic provisions of ASTM E 814 and UL 1479 require that a test assembly consisting of a specific wall or floor construction, containing through penetrations of various types and sizes, be constructed. The through-penetration firestop system to be tested is installed in accordance with the manufacturer's instructions around the penetrations. The test assembly is then exposed to a test fire that corresponds to the time-temperature curve established by ASTM E 119 or UL 263 (see the commentary to Section 703.2 for more discussion on ASTM E 119 or UL 263). After the fire test exposure, the through-penetration firestop system is subjected to a hose stream, which evaluates the ability of the through-penetration firestop system to resist the effects of erosion and thermal shock. After completion of the ASTM E 814 or UL 1479 procedure, two ratings for the test subject are established, which indicate how the test specimen withstands exposure to the test fire for a specified period of time.

Ratings for the through-penetration firestop system are generated based on the results of the testing, and are reported as an F (flame) rating and a T (temperature) rating. The F rating indicates the period of time, in hours, the tested through-penetration firestop system remained in place without allowing the passage of fire during exposure or water during the hose stream. The T rating indicates the time, in hours, it took for the temperature, as recorded by thermocouples placed at specified locations on the unexposed side of the test assembly, to reach 325°F (162°C) above ambient. It must be noted that in order to obtain a T rating, the system must first obtain an F rating. F ratings are required for all through-penetration firestop system and must be equal to the fire-resistance rating of the assembly penetrated. A T rating is not required for wall penetrations, but a minimum of 1 hour is required where the through-penetration firestop system is installed in a floor assembly where the penetrating item is a pipe, tube or conduit that is in direct contact with a combustible material. This requirement is intended to minimize the potential for ignition of the combustible material on the unexposed side of the assembly due to elevated temperatures transmitted via the pipe, tube or conduit.

An ASTM standard practice is available that can be used in cases where a standardized methodology is desired for the inspection of installed through-penetration firestop systems. Such a standardized procedure can be particularly useful in cases where a third-party inspection firm is used. ASTM E 2174, *Standard Practice for On-site Inspection of Installed Firestops*, includes items such as preconstruction meetings to review submittals, details, variances, building mock-ups for destructive testing and scheduling witnessed inspection of a specified percentage of installations or conducting destructive testing on a specified percentage of installations. This standard practice also identifies what actions should be taken based on the percentage of noncompliant installations identified, as well as providing a standardized reporting form.

Two of the most common types of materials used in through-penetration firestop system are intumescent and endothermic materials. Intumescent materials expand to approximately eight to 10 times their original volume when exposed to temperatures exceeding 250°F (121°C). The expansion of the material fills the voids or openings within the penetration to resist the passage of flame, while the outer layer of the expanded intumescent material forms an insulating charred layer that assists in limiting the transfer of heat. The expansion properties of intumescent materials allow them to seal openings left by combustible penetrating items that burn away during a fire, but do not retard heat as well as endothermic materials. Intumescent materials are typically used with combustible penetrating items or where a higher T rating is not required.

Endothermic materials provide protection through chemically bound water released in the form of steam when exposed to temperatures exceeding 600°F (316°C). This released water provides a cooling of the penetration and retards heat transfer through the penetration. Endothermic materials tend to be superlatively resistant to heat transfer and have higher T ratings, but do not expand to fill voids left by combustible penetrating items that burn away during a fire. Endothermic materials are, therefore, typically used with noncombustible penetrating items and where a higher T rating is required.

713.3.2 Membrane penetrations. Membrane penetrations shall comply with Section 713.3.1. Where walls or partitions are required to have a *fire-resistance rating*, recessed fixtures shall be installed such that the required fire-resistance will not be reduced.

Exceptions:

1. Membrane penetrations of maximum 2-hour fire-resistance-rated walls and partitions by steel electrical boxes that do not exceed 16 square inches (0.0103 m²) in area, provided the aggregate area of the openings through the membrane does not exceed 100 square inches (0.0645 m²) in any 100 square feet (9.29 m²) of wall area. The *annular space* between the wall membrane and the box shall not exceed $^1/_8$ inch (3.1 mm). Such boxes on opposite sides of the wall or partition shall be separated by one of the following:

   1.1. By a horizontal distance of not less than 24 inches (610 mm) where the wall or partition is constructed with individual noncommunicating stud cavities;

   1.2. By a horizontal distance of not less than the depth of the wall cavity where the wall cavity is filled with cellulose loose-fill, rockwool or slag mineral wool insulation;

   1.3. By solid fireblocking in accordance with Section 717.2.1;

1.4. By protecting both outlet boxes with *listed* putty pads; or

1.5. By other *listed* materials and methods.

2. Membrane penetrations by *listed* electrical boxes of any material, provided such boxes have been tested for use in fire-resistance-rated assemblies and are installed in accordance with the instructions included in the listing. The *annular space* between the wall membrane and the box shall not exceed $^1/_8$ inch (3.1 mm) unless *listed* otherwise. Such boxes on opposite sides of the wall or partition shall be separated by one of the following:

2.1. By the horizontal distance specified in the listing of the electrical boxes;

2.2. By solid fireblocking in accordance with Section 717.2.1;

2.3. By protecting both boxes with *listed* putty pads; or

2.4. By other *listed* materials and methods.

3. Membrane penetrations by electrical boxes of any size or type, which have been *listed* as part of a wall opening protective material system for use in fire-resistance-rated assemblies and are installed in accordance with the instructions included in the listing.

4. Membrane penetrations by boxes other than electrical boxes, provided such penetrating items and the *annular space* between the wall membrane and

the box, are protected by an *approved membrane penetration* firestop system installed as tested in accordance with ASTM E 814 or UL 1479, with a minimum positive pressure differential of 0.01 inch (2.49 Pa) of water, and shall have an F and T rating of not less than the required *fire-resistance rating* of the wall penetrated and be installed in accordance with their listing.

5. The *annular space* created by the penetration of an automatic sprinkler, provided it is covered by a metal escutcheon plate.

❖ Membrane penetrations are defined in Section 702 as "an opening made through one side (wall, floor or ceiling membrane) of an assembly." Therefore, they are different from a through penetration in that they do not pass through the entire assembly. Membrane penetrations are treated similar to through penetrations; namely, noncombustible conduits, pipes and tubes that penetrate only one membrane of a fire-resistance-rated wall assembly are required to have the annular space protected in accordance with Section 713.3.1 or comply with the exception of that section. Combustible penetrations through only one membrane of a wall assembly would require either a tested assembly (see Section 713.3.1.1) or a through-penetration firestop system that complies with Section 713.3.1.2 for the penetrated membrane.

Penetrations, such as electrical outlet boxes, can affect the fire-resistance rating of an assembly. The criteria in Exception 1 [see Figure 713.3.2(1)] limits the

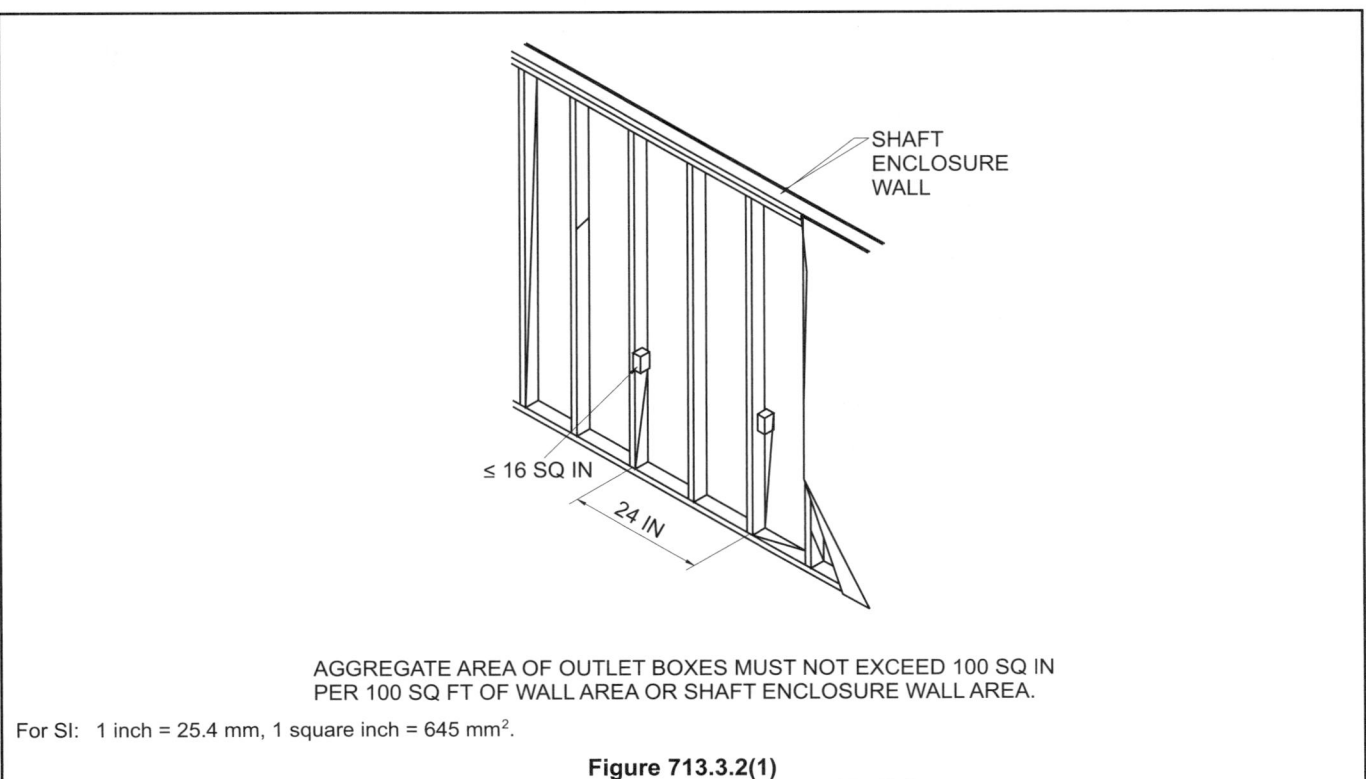

AGGREGATE AREA OF OUTLET BOXES MUST NOT EXCEED 100 SQ IN PER 100 SQ FT OF WALL AREA OR SHAFT ENCLOSURE WALL AREA.

For SI: 1 inch = 25.4 mm, 1 square inch = 645 mm².

Figure 713.3.2(1)
OUTLET BOXES IN RATED ASSEMBLIES

size of steel electrical outlet boxes [16 square inches (0.0103 m²)] and the amount allowed in a 100-square-foot (9.3 m²) area [100 square inches (0.0645 m²)]. Although not directly stated in the code, based on the supporting text in one of the legacy codes, when membrane penetrations of electrical outlet boxes were first regulated in 1979, it would appear that the original intent was to permit 100 square inches (0.0645 m²) of openings on each side of a 100 square foot (9.3 m²) wall section. These criteria are consistent with the criteria determined from fire tests that have generally shown that, within these limitations, these penetrations will not adversely affect the fire-resistance rating of the wall. The Design Information Section of the UL Fire Resistance Directory states that the opening in a wallboard facing is to be cut such that the clearance between the box and the wallboard does not exceed ¹/₈ inch (3.2 mm).

Electrical outlet boxes of steel or materials other than steel (see Exception 2) can be utilized in the fire-resistance-rated wall if they have been specifically tested as part of a fire-resistance-rated assembly. Steel outlet boxes that exceed the size limitations of Exception 1 would also require the same prequalification by testing. Exceptions 1 and 2 also allow electrical boxes on opposite sides of the wall if they meet one of several criteria, such as separation by a horizontal distance of 24 inches (610 mm).

Exception 3 relates to electrical boxes listed as part of an assembly by a testing agency. Certification and listing agencies have published listings covering proprietary compositions that are used to maintain the hourly ratings of fire-resistive walls and partitions incorporating flush-mounted devices such as outlet boxes, electrical cabinets and mechanical cabinets penetrating membranes of fire-resistance-rated assemblies. The individual systems indicate the specific applications and the method of installation for which the materials have been evaluated. The basic standards used to investigate these products are 263 and ASTM E 119. For example, UL classifies these materials and systems as "Wall Opening Protective Materials." This category includes classifications for both generic steel electrical boxes, as well as specific types and models of outlet and switch boxes composed of other materials, all listed for specific usage in fire-resistive-rated wall assemblies. The UL listings for wall opening protective materials indicate that, depending upon the testing conducted for the individual listing, their use can allow for any combination of (1) reducing the spacing between boxes contained on opposite sides of the wall, (2) increasing the size of the boxes, (3) increasing the density of boxes, and/or (4) allowing the use of boxes on each side of staggered stud walls. Because these systems are tested for the specific end-use applications, the individual and aggregate restrictions on maximum sizes and quantities [(i.e., 16 square inches (10 323 mm²) for an individual box, and the aggregate maximum of 100 square inches per 100 square feet.) (0.0645 m² per 9.3 m²)] are not required

for these systems to maintain the fire-resistance ratings of the assemblies penetrated.

Exception 4 creates a direct parallel between the requirements for electrical outlet boxes and other membrane penetration boxes, such as dryer exhaust boxes, washing machine hose connection boxes, fire or police alarm boxes, manual fire alarm boxes, switch boxes, valve boxes, special purpose boxes, electrical panels and hose cabinets. The protection systems are to be tested for use in fire-resistance-rated assemblies and installed in accordance with the instructions included in the listings. However, because these utility boxes can exceed 100 square inches (0.0645 m²) aggregate area, both an F and T rating are required in order to be directly equivalent to the fire-resistance rating of the assemblies penetrated. Given that these are membrane penetrations, there is a greater likelihood that someone could unknowingly place or store combustible materials, potentially even furniture and bedding, directly in contact with the unpenetrated membrane on the opposite side of the wall. This could significantly increase the threat of fire spread. The information provided for each classification would include the model numbers for the products, a description of the rated assemblies, the spacing limitations for the boxes and the installation details.

Exception 5 provides an alternative to the annular space protection provisions of Section 713.3.1 for fire sprinkler piping that penetrates a single membrane, provided that the annular space around the pipe is completely covered by an escutcheon plate of noncombustible material [see Figure 713.3.2(2)]. The nature of the hazard posed by single membrane penetrations of sprinkler piping is limited, due to the size of

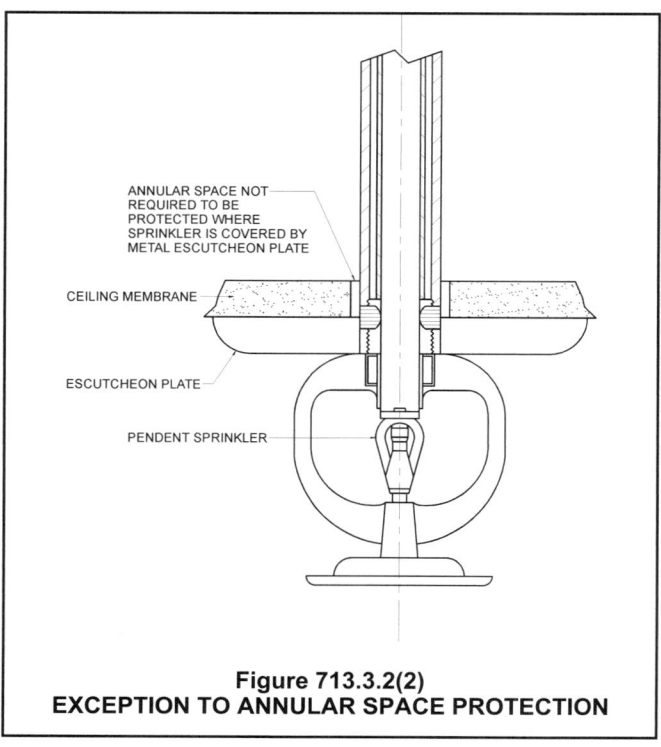

Figure 713.3.2(2)
EXCEPTION TO ANNULAR SPACE PROTECTION

the opening, the potential amount of openings and the presence of a sprinkler system. The installation of a noncombustible escutcheon provides protection against free passage of fire through the annular space, as well as allowing for the movement of sprinkler piping without breaking during a seismic event. These provisions correlate with the requirements of the National Earthquake Hazard Reduction Program (NEHRP) provisions recommended by the Building Seismic Safety Council (BSSC) for installation of sprinkler systems to resist seismic forces.

713.3.3 Dissimilar materials. Noncombustible penetrating items shall not connect to combustible items beyond the point of firestopping unless it can be demonstrated that the fire-resistance integrity of the wall is maintained.

❖ This section limits the common practice of using a short metal nipple to penetrate a rated assembly, firestopping for the metal penetration (which is substantially less expensive than firestopping for plastic) and then connecting to plastic pipe or conduit on the room side of the wall. Arguably, there is a distance at which such connection is safe; however, this distance is variable and cannot be specified in the body of the code, hence the requirement for demonstration of fire-resistance integrity. An identical provision is found in Section 713.4.1.4 regarding penetrations of horizontal assemblies.

713.4 Horizontal assemblies. Penetrations of a floor, floor/ceiling assembly or the ceiling membrane of a roof/ceiling assembly not required to be enclosed in a shaft by Section 708.2 shall be protected in accordance with Sections 713.4.1 through 713.4.2.2.

❖ Penetrations of horizontal assemblies are required to be protected by a shaft enclosure unless otherwise exempted by Section 708.2 (also see Section 708.2, Exceptions 3 and 4). The acceptable methods by which a floor or roof assembly can be interrupted by a penetration are identified in Sections 713.4 through 713.4.2.2.

As noted in the commentary to Section 712, roof construction that is required to have a fire-resistance rating is intended to minimize the threat of premature structural failure of the roof construction under fire conditions and not to create a barrier to contain fire within the building. Section 712.4 permits unprotected penetrations in fire-resistance-rated roof assemblies, provided that the structural integrity of the roof assembly is not reduced.

713.4.1 Fire-resistance-rated assemblies. Penetrations of the fire-resistance-rated floor, floor/ceiling assembly or the ceiling membrane of a roof/ceiling assembly shall comply with Sections 713.4.1.1 through 713.4.1.4. Penetrations in horizontal *smoke barriers* shall also comply with 713.5.

❖ This section is similar to the requirement found in Section 713.3, but it applies to horizontal assemblies. The application of this section is limited to floors, floor/ceil-

ing assemblies or the ceiling membrane of a roof/ceiling assembly. This section, therefore, does not apply to the roof of a roof/ceiling assembly or a roof, which by itself, provides the required rating.

713.4.1.1 Through penetrations. Through penetrations of fire-resistance-rated *horizontal assemblies* shall comply with Section 713.4.1.1.1 or 713.4.1.1.2.

Exceptions:

1. Penetrations by steel, ferrous or copper conduits, pipes, tubes or vents or concrete or masonry items through a single fire-resistance- rated floor assembly where the *annular space* is protected with materials that prevent the passage of flame and hot gases sufficient to ignite cotton waste when subjected to ASTM E 119 or UL 263 time-temperature fire conditions under a minimum positive pressure differential of 0.01 inch (2.49 Pa) of water at the location of the penetration for the time period equivalent to the *fire-resistance rating* of the construction penetrated. Penetrating items with a maximum 6-inch (152 mm) nominal diameter shall not be limited to the penetration of a single fire-resistance-rated floor assembly, provided the aggregate area of the openings through the assembly does not exceed 144 square inches (92 900 mm^2) in any 100 square feet (9.3 m^2) of floor area.

2. Penetrations in a single concrete floor by steel, ferrous or copper conduits, pipes, tubes or vents with a maximum 6-inch (152 mm) nominal diameter, provided the concrete, grout or mortar is installed the full thickness of the floor or the thickness required to maintain the *fire-resistance rating*. The penetrating items shall not be limited to the penetration of a single concrete floor, provided the area of the opening through each floor does not exceed 144 square inches (92 900 mm^2).

3. Penetrations by *listed* electrical boxes of any material, provided such boxes have been tested for use in fire-resistance-rated assemblies and installed in accordance with the instructions included in the listing.

❖ The code addresses the penetration of fire-resistance-rated horizontal separations in much the same way as penetrations through vertical assemblies; namely, by requiring the penetrations to either be part of a total assembly or be protected with a through-penetration firestop system (see commentary, Section 713.3.1).

Exception 1 is similar to Exception 2 of Section 713.3.1. This exception allows the noncombustible penetrating item to connect two stories (penetrate one floor), regardless of the size of the penetrating item [see Figure 713.4.1(1)]. Where the size of the individual penetration [6-inch diameter (152 mm)] and the aggregate area [144 square inches (0.095 m^2)] is limited in any 100 square feet (9.29 m^2), the code allows such penetrations of noncombustible construction to be through an unlimited number of floors.

Exception 2 allows a 6-inch-diameter (152 mm) penetration through a concrete floor similar to Exception 1 of Section 713.3.1. Where the area of the penetration is limited to 144 square inches (0.095 mm²), such a penetration is permitted to be through an unlimited number of floors [see Figure 713.4.1(2) for methods of annular space protection for Exceptions 1 and 2].

Exception 3 mirrors Exception 2 to Section 713.3.2

for membrane penetrations, but is applicable to through penetrations of a horizontal assembly, provided the outlet box has been tested.

713.4.1.1.1 Installation. *Through penetrations* shall be installed as tested in the *approved* fire-resistance-rated assembly.

❖ See the commentary to Section 713.3.1.1. This section addresses membrane penetrations which were tested as a part of the normal fire test of the assembly.

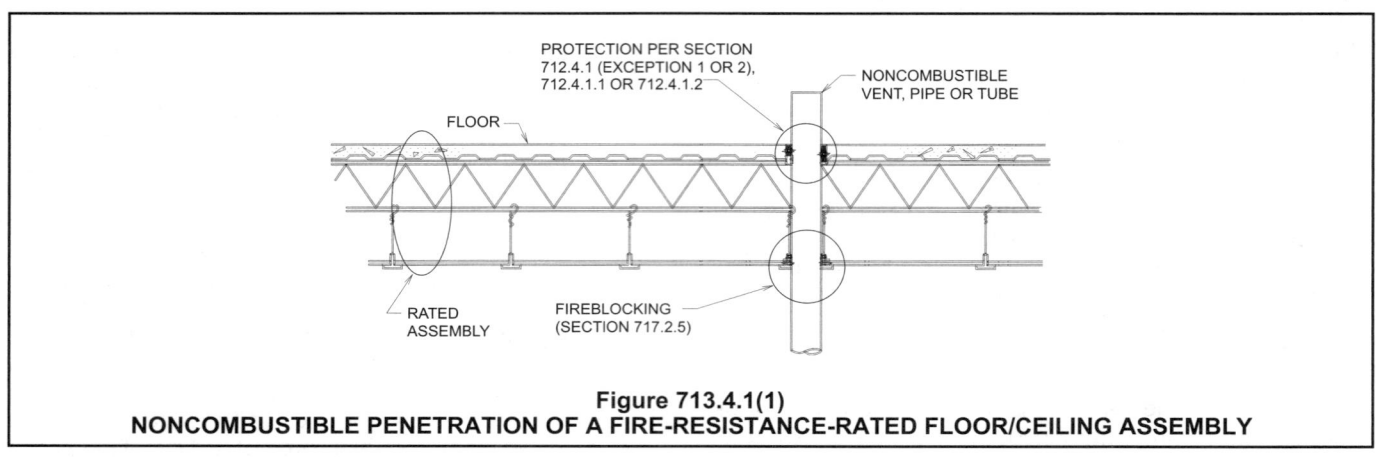

Figure 713.4.1(1)
NONCOMBUSTIBLE PENETRATION OF A FIRE-RESISTANCE-RATED FLOOR/CEILING ASSEMBLY

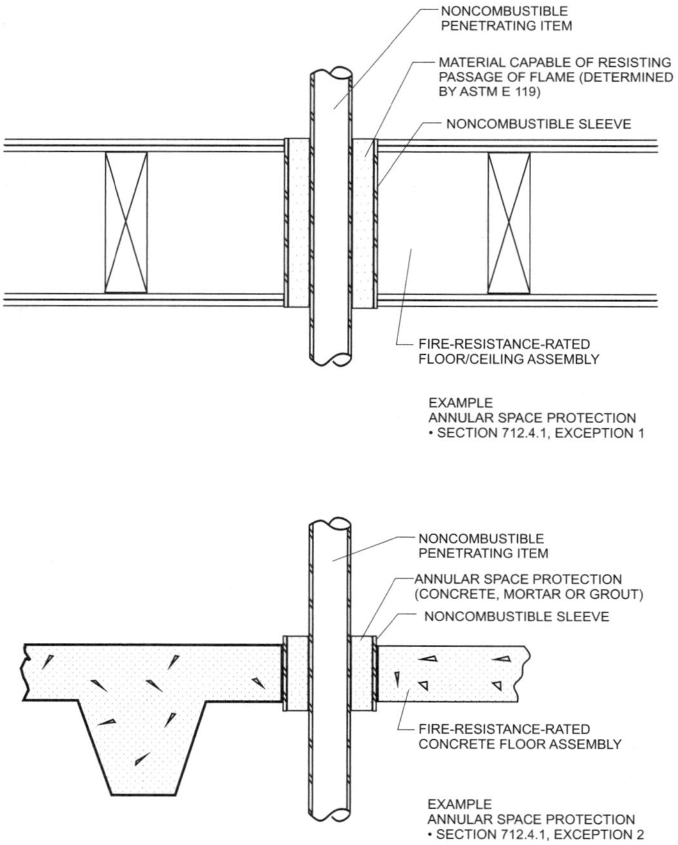

Figure 713.4.1(2)
EXAMPLES OF ANNULAR SPACE PROTECTION WITH SLEEVES

713.4.1.1.2 Through-penetration firestop system. *Through penetrations* shall be protected by an *approved through-penetration firestop system* installed and tested in accordance with ASTM E 814 or UL 1479, with a minimum positive pressure differential of 0.01 inch of water (2.49 Pa). The system shall have an F rating/T rating of not less than 1 hour but not less than the required rating of the floor penetrated.

> **Exception:** Floor penetrations contained and located within the cavity of a wall above the floor or below the floor do not require a T rating.

❖ This section differs slightly from Section 713.3.1.2 for wall penetrations in that a T rating is required for penetrations of horizontal assemblies, but is not required for penetrations of wall assemblies (see commentary, Section 713.3.1.2).

When permitted as an alternative to a shaft enclosure, according to Exception 3 of Section 708.2, cables, cable trays, conduits, tubes and pipes may penetrate a floor assembly when an approved through-penetration firestop system is used. An approved through-penetration firestop system is one that has been tested in accordance with ASTM E 814 or UL 1479. The test method determines the performance of the protection system with respect to exposure to a standard time-temperature fire and hose stream test. The performance of the protection system is dependent on the specific assembly of materials tested, including the number, type and size of penetrations and the types of floors or walls in which it is installed. It should also be noted that tests have been conducted at various pressure differentials; however, the current criterion used is 0.01 inch (2.49 Pa) of water gauge, and only tests with such minimum pressure throughout the test period are to be accepted. In evaluating test reports, the building official must determine that the tested assembly is truly representative of the manner in which the system is to be installed. It should be noted that the ASTM E 814 test establishes two ratings: the F rating, which identifies the ability of the material to resist the passage of flame, and the T rating, which identifies the thermal transmission characteristics of the material or assembly. The exception is intended to set aside the T rating requirement where the penetrating item is located in a cavity of a wall and is separated from contact with adjacent materials in the occupiable floor area.

Meeting the T-rating requirement can be very challenging for penetrants that are metallic and, therefore, conduct heat. A through-penetration firestop system for such penetrants would not typically meet the T-rating requirement simply by sealing the penetrant within the floor. Additional insulation or covering of the through penetrant is typically required above and below the floor to prevent heat transmission via the penetrant. Listings for through-penetration firestop systems will indicate the T rating obtained for any specific system. Figure 713.4.1.1.2 shows an example of typical firestop details that would be needed for a cable bundle to meet the T rating requirement of Section

713.4.1.1.2. For other types of metallic penetrants, listed systems with suitable T ratings would typically have insulation wrapped around the penetrant for some distance above the floor and below the floor.

713.4.1.2 Membrane penetrations. Penetrations of membranes that are part of a *horizontal assembly* shall comply with Section 713.4.1.1.1 or 713.4.1.1.2. Where floor/ceiling assemblies are required to have a *fire-resistance rating*, recessed fixtures shall be installed such that the required *fire resistance* will not be reduced.

Exceptions:

1. *Membrane penetrations* by steel, ferrous or copper conduits, pipes, tubes or vents, or concrete or masonry items where the *annular space* is protected either in accordance with Section 713.4.1.1 or to prevent the free passage of flame and the products of combustion. The aggregate area of the openings through the membrane shall not exceed 100 square inches (64 500 mm²) in any 100 square feet (9.3 m²) of ceiling area in assemblies tested without penetrations.

2. Ceiling membrane penetrations of maximum 2-hour *horizontal assemblies* by steel electrical boxes that do not exceed 16 square inches (10 323 mm²) in area, provided the aggregate area of such penetrations does not exceed 100 square inches (44 500 mm²) in any 100 square feet (9.29 m²) of ceiling area, and the annular space between the ceiling membrane and the box does not exceed 1/8 inch (3.2 mm).

3. Membrane penetrations by electrical boxes of any size or type, which have been *listed* as part of an opening protective material system for use in *horizontal assemblies* and are installed in accordance with the instructions included in the listing.

4. *Membrane penetrations* by *listed* electrical boxes of any material, provided such boxes have been tested for use in fire-resistance-rated assemblies and are installed in accordance with the instructions included in the listing. The *annular space* between the ceiling membrane and the box shall not exceed 1/8 inch (3.2 mm) unless *listed* otherwise.

5. The *annular space* created by the penetration of a fire sprinkler, provided it is covered by a metal eschutcheon plate.

❖ Penetrations of fire-resistance-rated floor/ceiling and roof/ceiling assemblies that have a ceiling membrane as a component of the tested assembly, such as steel or wood joist assemblies using acoustical tiles or gypsum wallboard, are limited to those penetrations that are listed in the tested assembly (see Section 713.4.1.1.1) or those protected in accordance with ASTM E 814 [see Section 713.4.1.1.2 and Figure 713.4.1.2(1)]. Penetrations of a ceiling membrane by items that are not part of a tested assembly create a point for possible fire penetration into the space above the ceiling membrane, which in turn could jeopardize the fire-resistance rating of the assembly. The exceptions contain protection requirements for specific

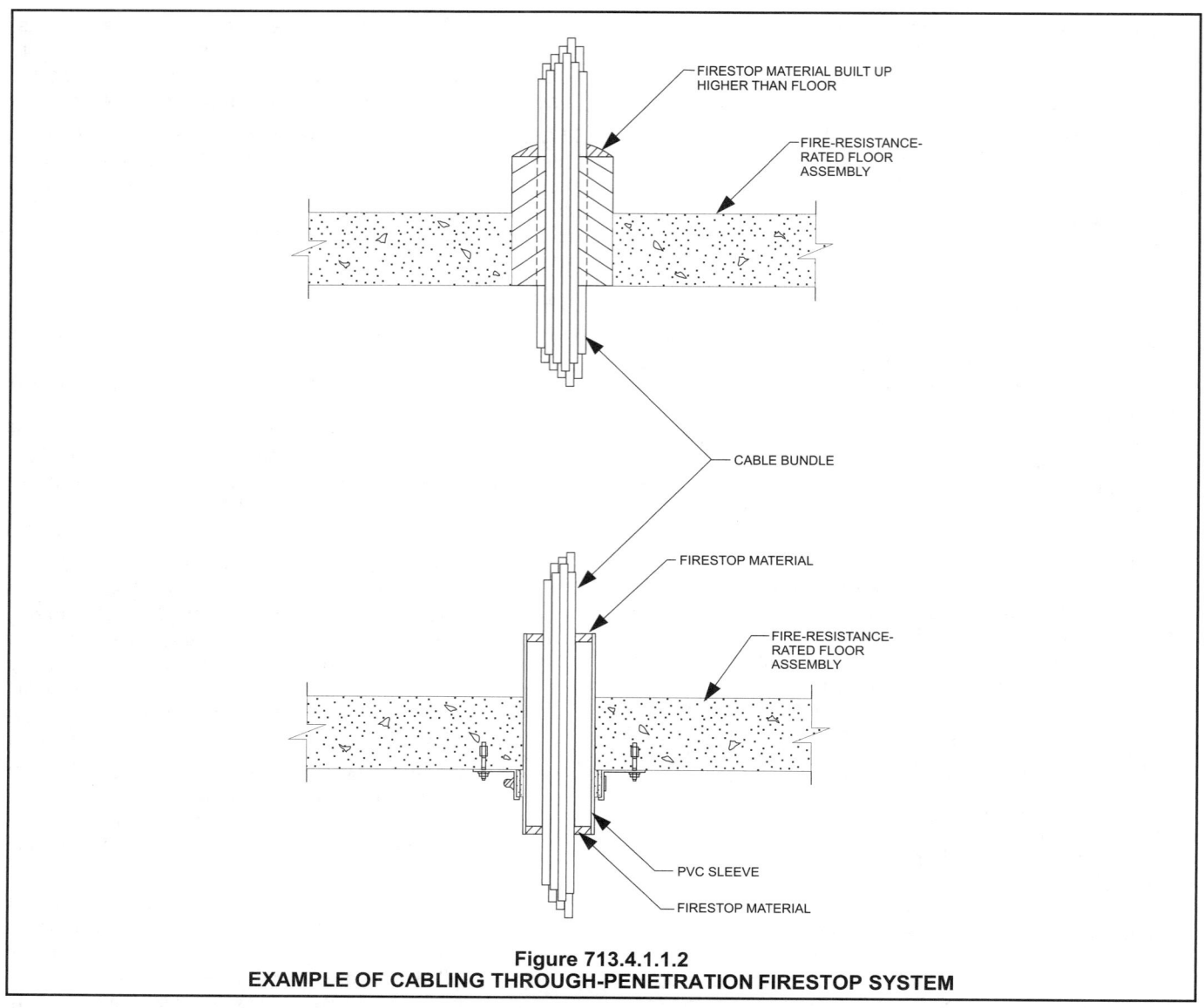

Figure 713.4.1.1.2
EXAMPLE OF CABLING THROUGH-PENETRATION FIRESTOP SYSTEM

types of penetrations that are not required to be part of the tested assembly.

Exception 1 allows for a maximum aggregate amount of noncombustible pipe, tube, vents and conduit penetrations of a floor or a ceiling that is part of a fire-resistance-rated assembly to be 100 square inches (0.0645 m²) per 100 square feet (9.3 m²) of ceiling area. The annular space of the penetration can be protected either by fireblocking (see Section 717.2.5), annular space protection (see Section 713.4.1.1, Exception 1) or a through-penetration firestop system tested to ASTM E 814 [see Section 713.4.1.1.2 and Figure 713.4.1.2(2)].

Annular space protection is not an option for combustible items because it does not have the ability to protect the resulting void created when the combustible penetrating item burns away.

Exception 2 allows penetrations of electrical outlet boxes that can affect the fire-resistance rating of an assembly. The criteria of this section limit the size of noncombustible electrical outlet boxes that penetrate a ceiling membrane to an aggregate area that shall not exceed 100 square inches (0.0645 m²) in 100 square feet (9.3 m²) of ceiling area. All of the openings at the penetrations of the outlets are required to be protected as mentioned above. The area limitations are consistent with the criteria determined from fire tests, which have shown that within these limitations, these penetrations will not adversely affect the fire-resistance rating of the floor/ceiling assembly.

The exception permits electrical outlet boxes and fittings, which have been evaluated and found suitable for this purpose, to be located in the floor or ceiling of the fire-resistance-rated assembly. Reference sources for fire-resistance-rated assemblies, such as the UL *Fire Resistance Directory*, state that certain tested floor assemblies will permit the installation of electrical and communication connection inserts that appear in the UL listing category "Outlet Boxes and Fittings Classified for Fire Resistance" without reducing the fire-resistance rating of the assembly. However, classified outlet boxes and fittings installed in floor/ceiling assemblies that do not specify their use will jeopardize the rating unless compensating protection is provided.

Exception 3 is related to membrane penetrations of fire-resistance-rated assemblies in Section 713.4.1.2 for membrane penetrations.

This new exception will permit additional tested and listed systems to be used for membrane penetrations in fire-resistance-rated assemblies. Many of these systems already exist and are being used in the marketplace. The code recognizes current common practice of a proven, regulated technology.

Section 713.4.1.2 already permits several exceptions to the basic requirement for membrane penetrations to be installed so that the required fire-resistance rating will not be reduced by the membrane penetrations. In the same way, certification and listing agencies have published listings covering proprietary compositions that are used to maintain the hourly ratings of fire-resistant walls and partitions incorporating flush-mounted devices, such as outlet boxes, electrical cabinets and mechanical cabinets, penetrating membranes of fire-resistance-rated assemblies. The individual systems indicate the specific applications and the method of installation for which the materials have been evaluated. The basic standards used to investigate these products are UL 263 and ASTM E 119.

For example, UL classifies nonmetallic outlet boxes for installation in floors, walls and partitions, or ceilings in accordance with the provisions of NFPA 70. These systems are required to provide a degree of fire resistance when installed in the particular floors, walls or ceiling assemblies. The systems listed for this application include nonmetallic outlet and switch boxes for use in fire-resistance-rated wall assemblies. Listing information includes the model numbers for the products, a description of the rated assemblies in which they can be used, the spacing limitations for the boxes and the installation details.

Product listings specify the conditions under which listed metallic outlet and switch boxes may be installed within fire-resistance-rated wall assemblies constructed with wood or steel studs and gypsum board facings. Listings also exist for nonmetallic outlet boxes along with the conditions under which such outlet and switch boxes may be installed within fire-resistant wall assemblies. With either type of outlet or switch box, it may be possible to install the boxes under less stringent conditions when such boxes are used in conjunction with wall-opening protective materials. Use of wall-opening protective materials may allow for any combination of (1) reducing the spacing between boxes contained on opposite sides of the wall, (2) increasing the size of the boxes, (3) increasing the density of boxes installed or (4) allowing the use of boxes on each side of staggered stud walls. The individual systems tested for compliance in these categories indicate the specific applications and the method of installation for which the materials have been evaluated.

Exceptions 4 and 5 are identical to those of Section 713.3.2, Exception 3 (see commentary, Section 713.3.2) and Section 713.4.1.2, Exception 3.

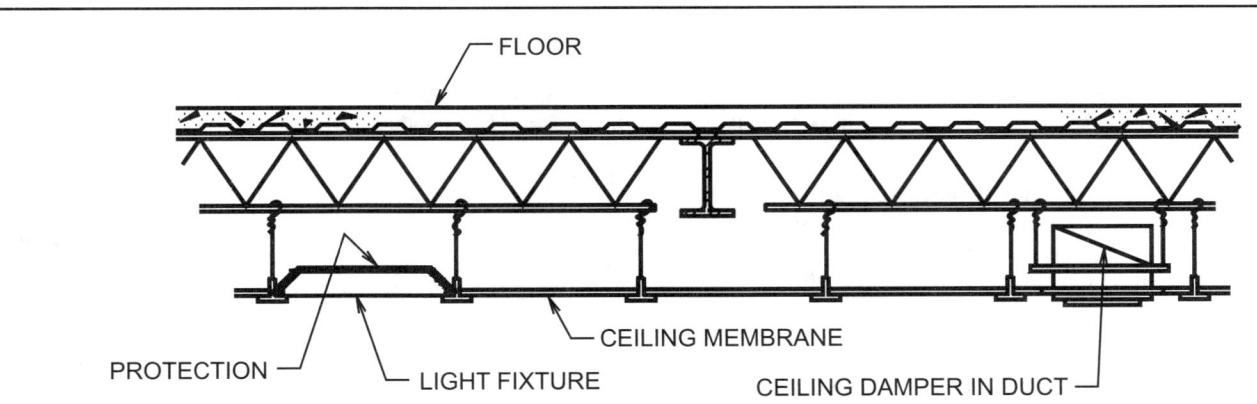

NOTE: 100 SQUARE INCHES OF OPENING PERMITTED IN EACH 100 SQUARE FEET OF CEILING AREA

For SI: 1 square inch = 645 mm², 1 square foot = 0.0929 m².

Figure 713.4.1.2(1)
PROTECTION OF PENETRATIONS THROUGH CEILING MEMBRANE OF FIRE-RESISTANCE-RATED ASSEMBLY

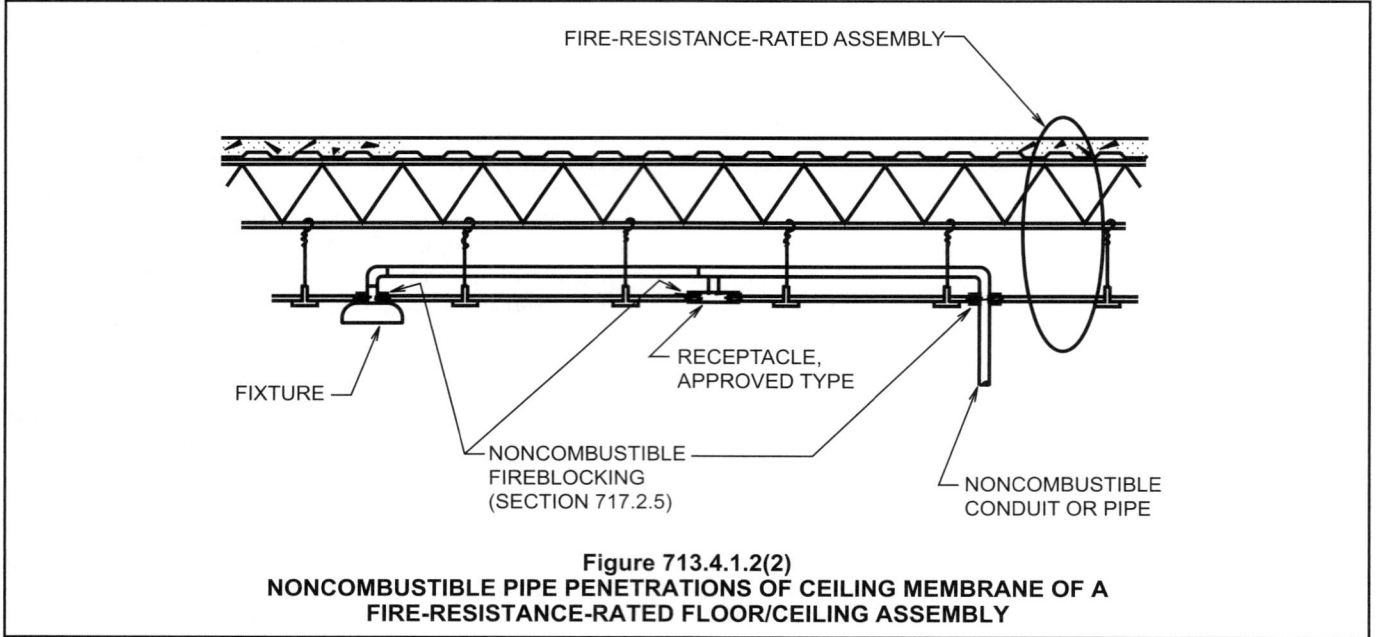

FIRE-RESISTANCE-RATED ASSEMBLY

FIXTURE

RECEPTACLE,
APPROVED TYPE

NONCOMBUSTIBLE
FIREBLOCKING
(SECTION 717.2.5)

NONCOMBUSTIBLE
CONDUIT OR PIPE

Figure 713.4.1.2(2)
NONCOMBUSTIBLE PIPE PENETRATIONS OF CEILING MEMBRANE OF A
FIRE-RESISTANCE-RATED FLOOR/CEILING ASSEMBLY

713.4.1.3 Ducts and air transfer openings. Penetrations of *horizontal assemblies* by ducts and air transfer openings shall comply with Section 716.

❖ When a membrane of a horizontal assembly is penetrated by a duct or air transfer opening, this section references the requirements of Section 716. Horizontal assemblies are addressed by Section 716.6, but the provisions of Sections 716.1 through 716.5 provide details and requirements.

713.4.1.4 Dissimilar materials. Noncombustible penetrating items shall not connect to combustible materials beyond the point of firestopping unless it can be demonstrated that the fire-resistance integrity of the *horizontal assembly* is maintained.

❖ This section mirrors that of Section 713.3.3 for direct penetrations through walls (see commentary, Section 713.3.3).

713.4.2 Nonfire-resistance-rated assemblies. Penetrations of nonfire-resistance-rated floor or floor/ceiling assemblies or the ceiling membrane of a nonfire-resistance-rated roof/ceiling assembly shall meet the requirements of Section 708 or shall comply with Section 713.4.2.1 or 713.4.2.2.

❖ This section limits the penetrations of nonfire-resistance-rated floor assemblies to prevent the migration of smoke through a building, despite being of an unprotected type of construction. Permitted protection methods are outlined in Sections 713.4.2.1 and 713.4.2.2.

Where penetrations of the ceiling membrane of a nonfire-resistance-rated roof/ceiling assembly occur, the penetrations are required to be protected by fireblocking in accordance with the requirements of Section 717.2.5 to restrict the spread of fire through the concealed roof space above the ceiling membrane (see commentary, Section 717.2.5).

713.4.2.1 Noncombustible penetrating items. Noncombustible penetrating items that connect not more than three *stories* are permitted, provided that the *annular space* is filled to resist the free passage of flame and the products of combustion with an *approved* noncombustible material or with a fill, void or cavity material that is tested and classified for use in through-penetration firestop systems.

❖ Noncombustible penetrations connecting three stories or less (penetrating two or less floors) are permitted when the annular space of the penetrating item is fireblocked at each floor line in accordance with Section 717.2.5 (see Figure 713.4.2.1).

713.4.2.2 Penetrating items. Penetrating items that connect not more than two *stories* are permitted, provided that the *annular space* is filled with an *approved* material to resist the free passage of flame and the products of combustion.

❖ This section is not limited as to the combustibility of the penetrating item. This section allows both combustible and noncombustible penetrations of a single nonfire-resistance-rated floor (thus connecting two stories), provided the annular space is fireblocked in accordance with Section 717.2.5. This section does not require a noncombustible firestop to protect the annular space, while Section 713.4.2.1 does require a noncombustible firestop material.

713.5 Penetrations in smoke barriers. Penetrations in *smoke barriers* shall be tested in accordance with the requirements of UL 1479 for air leakage. The air leakage rate of the penetration assemblies measured at 0.30 inch (7.47 Pa) of water in both the ambient temperature and elevated temperature tests, shall not exceed:

1. 5.0 cfm per square foot ($0.025m^3 / s \cdot m^2$) of penetration opening for each *through-penetration firestop system*; or

2. A total cumulative leakage of 50 cfm (0.024m³/s) for any 100 square feet (9.3 m²) of wall area, or floor area.

❖ Smoke barriers are utilized to create compartments within a building where, under fire conditions, they are intended to provide protection from both fire and smoke. As such compartments are critical in providing the necessary level of fire and smoke protection, enforceable language is deemed necessary to help ensure that the intention of the code is met. Previous language in the code included terms such as "limit." "restrict" and "resist" to indicate the degree of smoke protection required, creating inconsistent application of the provisions.

The L rating provides a quantitative indication of the through-penetration system's ability to resist the passage of smoke. Although the test is performed using air, the flow properties of air and smoke are sufficiently close for engineering purposes as to provide a reason-able quantification of smoke leakage. The L-rating test is an optional test within the UL 1479 test standard; therefore, not all through-penetration firestop systems that are tested and listed will have this information. In the 2005 UL *Fire Resistance Directory* (Vol. 2), there are approximately 550 through-penetration firestop systems that are listed that also have an L rating of 5 or less. This should provide a sufficient selection to allow for code compliance in all situations.

Smoke barriers, in general, have limited areas where smoke will leak through, such as at doors, dampers, penetrations and joints. Fire door assemblies and smoke dampers have previously had allowable leakage rates established by the code and referenced standards. When the 2006 edition of the code added this text for penetrations and also joint systems (see Section 713.6), all of the typical areas where leakage in smoke barriers can occur have been addressed.

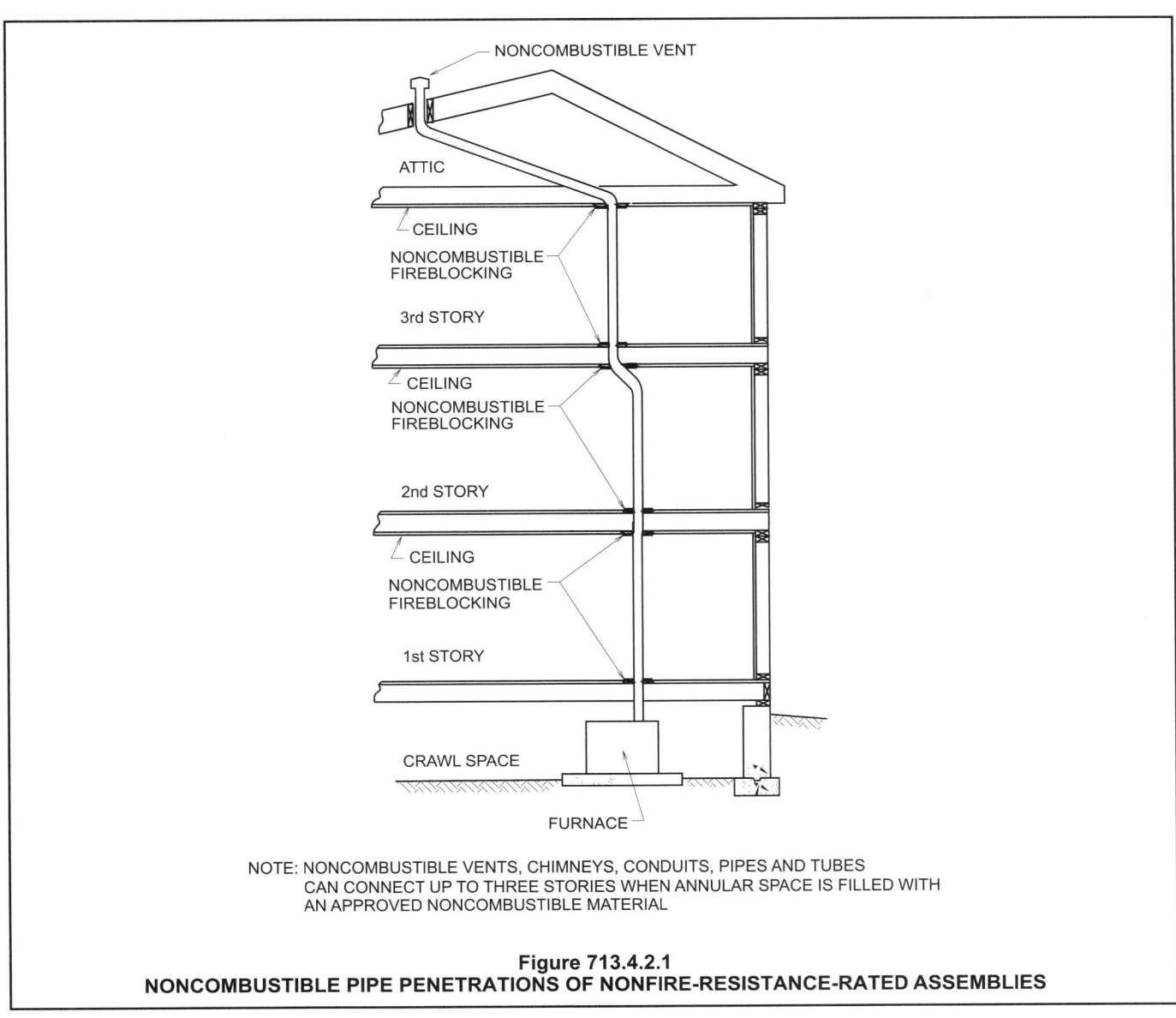

NOTE: NONCOMBUSTIBLE VENTS, CHIMNEYS, CONDUITS, PIPES AND TUBES CAN CONNECT UP TO THREE STORIES WHEN ANNULAR SPACE IS FILLED WITH AN APPROVED NONCOMBUSTIBLE MATERIAL

Figure 713.4.2.1
NONCOMBUSTIBLE PIPE PENETRATIONS OF NONFIRE-RESISTANCE-RATED ASSEMBLIES

SECTION 714
FIRE-RESISTANT JOINT SYSTEMS

714.1 General. Joints installed in or between fire-resistance-rated walls, floor or floor/ceiling assemblies and roofs or roof/ceiling assemblies shall be protected by an *approved fire-resistant joint system* designed to resist the passage of fire for a time period not less than the required *fire-resistance rating* of the wall, floor or roof in or between which it is installed. *Fire-resistant joint systems* shall be tested in accordance with Section 714.3. The void created at the intersection of a floor/ceiling assembly and an exterior curtain wall assembly shall be protected in accordance with Section 714.4.

Exception: *Fire-resistant joint systems* shall not be required for joints in all of the following locations:

1. Floors within a single *dwelling unit*.

2. Floors where the joint is protected by a shaft enclosure in accordance with Section 708.

3. Floors within atriums where the space adjacent to the atrium is included in the volume of the atrium for smoke control purposes.

4. Floors within malls.

5. Floors and ramps within open and enclosed parking garages or structures constructed in accordance with Sections 406.3 and 406.4, respectively.

6. *Mezzanine* floors.

7. Walls that are permitted to have unprotected openings.

8. Roofs where openings are permitted.

9. Control joints not exceeding a maximum width of 0.625 inch (15.9 mm) and tested in accordance with ASTM E 119 or UL 263.

❖ This section regulates joints or linear openings created between building assemblies, which are sometimes referred to as head-of-wall, expansion or seismic joints. These joints are most often created where the structural design of a building necessitates a separation between building components in order to accommodate anticipated structural displacements caused by thermal expansion and contraction, seismic activity, wind or other loads. Figure 714.1 illustrates some of the most common locations of these joints.

Seismic joints in multiple-story buildings are intended to allow differential lateral displacement of separate portions of a building during a seismic event. Expansion joints permit separate portions of the structural frame to contract and expand due to temperature change or wind sway without adversely affecting the building functions or structural integrity.

Joints can also occur in fire barriers at the intersection of the top of a wall and the underside of the floor or roof above (head of wall); within a wall or floor at any specific point to accommodate the structural design or at the edge of the floor at the intersection of the floor and exterior wall. All of these linear openings create a "weak link" in fire-resistance-rated assemblies that can compromise the integrity of the tested assembly by allowing an avenue for the passage of fire and the products of combustion through the assembly. In order to maintain the efficacy of the fire-resistance-rated as-

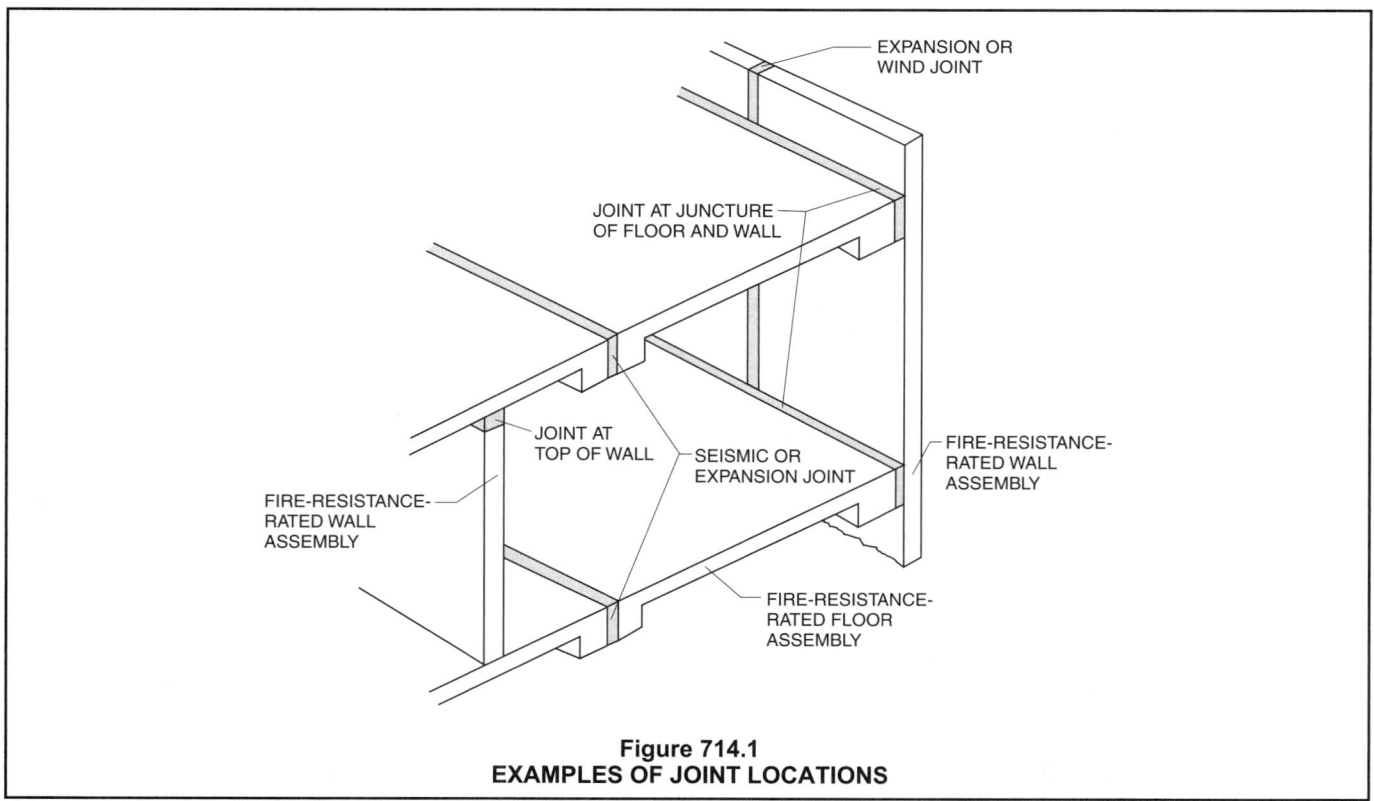

Figure 714.1
EXAMPLES OF JOINT LOCATIONS

sembly, these openings must be protected by a fire-re-sistance-rated joint system with a rating equal to the assembly in the same plane. Where two assemblies intersect, the fire rating of the joint must be the same as the fire rating of the assembly (or assemblies) of the same plane as the assembly where the joint occurs. It is not the intent of this section to regulate joints installed in assemblies that are provided to control shrinkage cracking, such as a saw-cut control joint in concrete. This section contains nine locations where a fire-resistant joint system is not required to be installed to protect the joint. These generally are locations where a separation or protected openings are not required or where the joint occurs within an area that is bounded by other means of protection. Exception 1 states that a fire-resistant joint is not required for joints contained within a single dwelling unit. This exception is similar to Section 708.2, Exception 1. Exception 2 exempts fire-resistant joints completely enclosed in a shaft that complies with Section 708. Exception 3 exempts the requirement for fire-resistant joints that are located in floors in an atrium that complies with Section 404. This exception is similar to Section 708.2, Exception 5. Exception 4 exempts joints located in malls that comply with Section 402. Although this appears to be a general exception that would affect all floors within a mall, the limitations of Section 708.2, Exception 5, should still be considered. The application of this section will really only exempt malls that are two stories in height. If the mall is over two stories, then Section 708.2 would either require shaft protection for any openings, or the mall could be protected as an atrium. If the atrium provisions are used, then Exception 3 of this section would be applicable. Exception 5 exempts joints located in open parking structures that comply with Section 406.3 and is similar to Section 708.2, Exception 8. Exception 6 exempts joints in mezzanines that comply with Section 505 and is similar to Section 708.2, Exception 9. As a type of opening/penetration, a joint represents a similar hazard to exterior wall fire exposures as other openings. Exception 7 acknowledges that if an exterior wall is permitted to have unprotected openings (see Table 704.8), then the joint does not require protection as well. A joint is a much smaller opening than an unprotected opening. Exception 8 exempts joints located in roof decks that are not required to have protected openings and is similar to that contained in Section 712.4, which allows skylights and other penetrations in rated roof decks, provided the structural integrity is not affected and the limitations found in Section 714.4.1, which exclude the roof deck are met (see commentary, Section 714.4.1). As discussed earlier, it is not the intent of this section to regulate construction joints, the stopping and starting points for two successive concrete pours or control joints, which are intended to prevent shrinkage cracking (such as saw-cut control joints) in concrete or masonry. Therefore, Exception 9 acknowledges UL 2079, as referenced in Section 714.3, which permits control joints with a maximum joint width of $^5/_8$ inch (15.9 mm)

to be tested in accordance with ASTM E 1966 or UL 2079. This allowance is in recognition of fire test data that has existed since the 1960s regarding control joints and perimeter relief joints in fire-rated systems, the performance of these devices and systems in the field and a rational approach regarding minor movement [defined here as a maximum of $^5/_8$ inch (15.9 mm)] where no appreciable damage or fatigue is incurred by the joint system that would have a profound impact on the joint's fire performance. Since all buildings move slightly, it is not prudent to have to cycle and test joint systems that only move a few millimeters. Although their movement is limited, control joint systems must still pass the rigors of ASTM E 119 or UL 263 in accordance with this exception.

714.2 Installation. *Fire-resistant joint systems* shall be securely installed in or on the joint for its entire length so as not to dislodge, loosen or otherwise impair its ability to accommodate expected building movements and to resist the passage of fire and hot gases.

❖ This section requires that joint systems be installed for its full length or height, due to the fact that the openings protected by these types of joints are most often continuous.

714.3 Fire test criteria. *Fire-resistant joint systems* shall be tested in accordance with the requirements of either ASTM E 1966 or UL 2079. Nonsymmetrical wall joint systems shall be tested with both faces exposed to the furnace, and the assigned *fire-resistance rating* shall be the shortest duration obtained from the two tests. When evidence is furnished to show that the wall was tested with the least fire-resistant side exposed to the furnace, subject to acceptance of the *building official*, the wall need not be subjected to tests from the opposite side.

Exception: For *exterior walls* with a horizontal *fire separation distance* greater than 5 feet (1524 mm), the joint system shall be required to be tested for interior fire exposure only.

❖ In order to determine the anticipated protection provided by a given assembly, the hourly fire-resistance rating must be determined by testing the joint system in accordance with UL 2079 or ASTM E 1966. This standard includes specific criteria for test specimen preparation, placement, configuration, size and testing conditions of joint systems. Also included in the text protocol is the minimum positive pressure differential to be used for the test, which is intended to assist in evaluating whether or not the joint will remain in place during a fire condition. A joint splice must be tested, since the presence and orientation of a splice in a joint system can affect the fire performance of the joint. These splices may occur where the length of the joint to be protected exceeds the length of a prefabricated joint or where a cold joint occurs in a field-installed system. The maximum joint width must be tested when in a fully expanded or extended condition in order to evaluate the fire-resistance rating. Joints that are designed to transfer structural building loads are required to have a superimposed load during the test, which is consistent with the requirements for the testing of load-bearing fire-resistance-rated assemblies. Finally, the test addresses requirements for joints that are intended to accommodate building movement, such as expansion, seismic and wind sway joints, including preconditioning cycling, which is intended to allow for evaluation of the joint's ability to withstand cyclical movement over its anticipated life.

Test data has indicated that the orientation of nonsymmetrical joints can have an effect on the performance of the joint. As a result, in accordance with this section, all joint systems must be tested for fire exposure from both sides so that the required protection will be provided regardless of which side of the joint is exposed to fire. The exception for joints in exterior walls correlates with Section 704.5 for the fire exposure rating of exterior walls.

Though not required by the code, an ASTM standard practice is available that can be used in cases where a standardized methodology is desired for the inspection of installed fire-resistant joint systems and perimeter fire barrier systems. Such a standardized procedure can be particularly useful in cases where a third-party inspection firm is used. ASTM E 2393, *Standard Practice for On-site Inspection of Installed Fire Resistive Joint Systems and Perimeter Fire Barriers*, includes items such as preconstruction meetings to review submittals, details, variances, building mock-ups for destructive testing and scheduling a witnessed inspection of a specified percentage of installations or conducting destructive testing on some types of installed joint systems. This standard practice also identifies what actions should be taken based on the percentage of noncompliant installations identified, as well as providing a standardized reporting form.

714.4 Exterior curtain wall/floor intersection. Where fire resistance-rated floor or floor/ceiling assemblies are required, voids created at the intersection of the exterior curtain wall assemblies and such floor assemblies shall be sealed with an *approved* system to prevent the interior spread of fire. Such systems shall be securely installed and tested in accordance with ASTM E 2307 to prevent the passage of flame for the time period at least equal to the *fire-resistance rating* of the floor assembly and prevent the passage of heat and hot gases sufficient to ignite cotton waste. Height and fire-resistance requirements for curtain wall spandrels shall comply with Section 705.8.5.

❖ The void created between a floor and a curtain wall can range anywhere between 1 to 12 inches (25 to 305 mm) or more, which clearly requires sealing to prevent the spread of flames and products of combustion between adjacent stories.

A fire test standard was specifically developed to evaluate the interface between a fire-resistance-rated horizontal assembly and an exterior curtain wall. This particular standard is ASTM E 2307, *Standard Test Method for Determining Fire Resistance of Perimeter Fire Barrier Systems Using Intermediate-scale, Multistory Test Apparatus*. The "perimeter fire barrier

system" is the assembly of materials that prevents the passage of flame and hot gases at the void space between the interior surface of the wall assembly and the adjacent edge of the floor. For the purposes of the ASTM E 2307 test standard, the interior face is at the interior surface of the wall's framework. The width of the joint, which has maximum allowable dimensions specified in the perimeter fire barrier system listings, is therefore the distance between the edge of the framing nearest the floor and the adjacent floor edge. The void space or cavity between framing members is not considered joint space. Figure 714.4 shows the typical components that are specified in listings of tested perimeter fire barrier systems.

Tested and listed perimeter fire barrier systems do not include the interior finished wall (e.g., "knee wall") details. This makes the systems applicable to any and all finished wall configurations. The existence of the interior wall, even if made of fire-resistant materials, such as fire-resistance-rated gypsum board, does not eliminate the need to have an appropriately tested material or system to protect the curtain wall from interior fire spread at the perimeter gap, unless that interior wall detail has been specifically tested and shown to meet the requirements of this section of the code.

Although not required by the code, an ASTM standard practice is available that can be used in cases where a standardized methodology is desired for the inspection of installed perimeter fire barrier systems. Such a standardized procedure can be particularly useful in cases where a third-party inspection firm is used. ASTM E 2393 includes items such as preconstruction meetings to review submittals, details, variances, building mock-ups for destructive testing and scheduling a witnessed inspection of a specified percentage of installations or conducting destructive testing on a specified percentage of installations. This standard practice also identifies what actions should be taken based on the percentage of noncompliant installations identified, as well as providing a standardized reporting form.

714.4.1 Exterior curtain wall/nonfire-resistance-rated floor assembly intersections. Voids created at the intersection of exterior curtain wall assemblies and nonfire-resistance-rated floor or floor/ceiling assemblies shall be sealed with an *approved* material or system to retard the interior spread of fire and hot gases between *stories*.

❖ Where the joint between walls involves a nonfire-resistance-rated floor and an exterior curtain wall, there is no reason to try to maintain fire-resistance rating with a rated joint system. However, spread of smoke is a concern, and, therefore, the code calls for a tight joint to protect the rapid spread of smoke from a floor of fire origin to other floors of the building.

714.5 Spandrel wall. Height and fire-resistance requirements for curtain wall spandrels shall comply with Section 705.8.5. Where Section 705.8.5 does not require a fire-resistance-rated

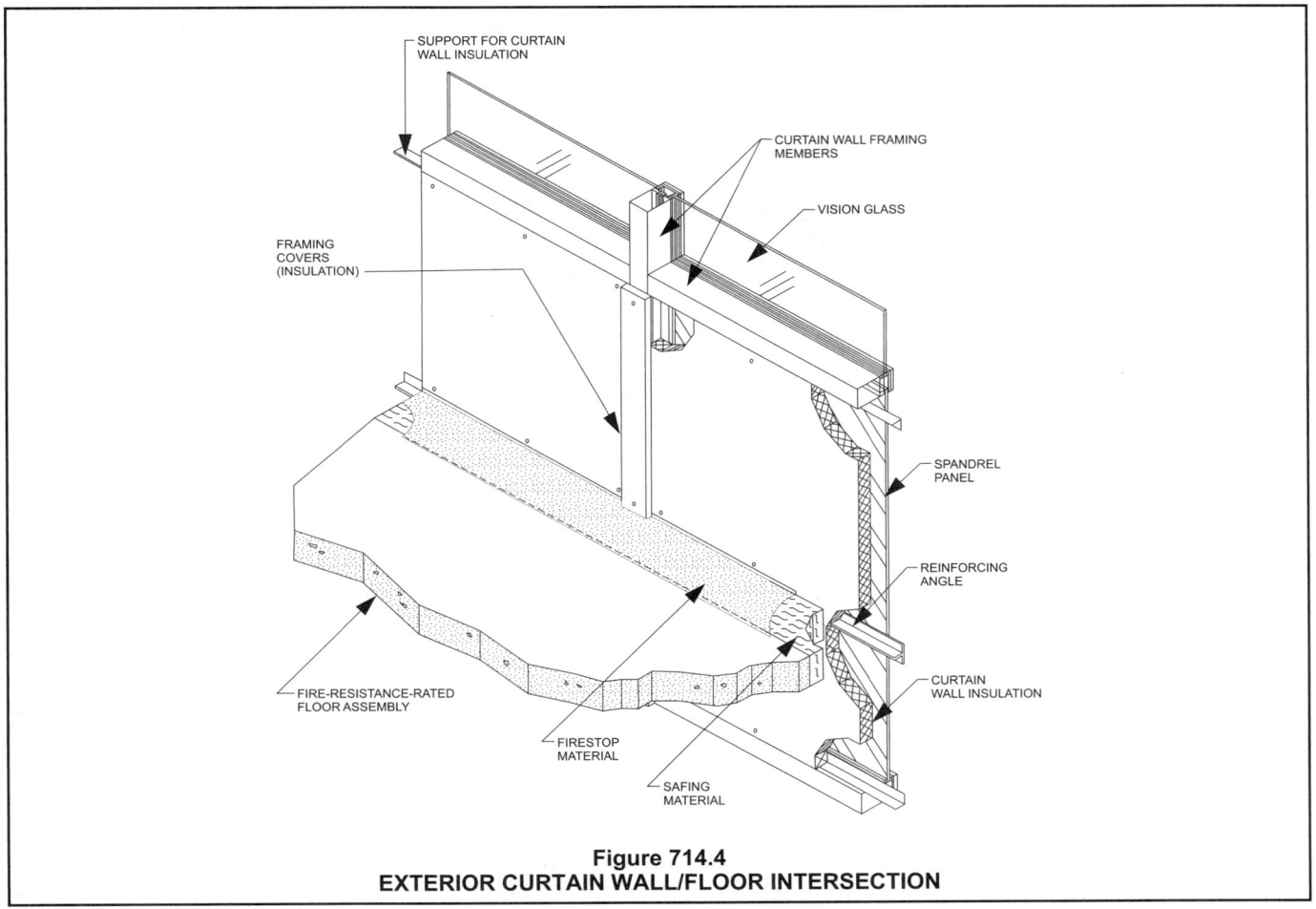

Figure 714.4
EXTERIOR CURTAIN WALL/FLOOR INTERSECTION

spandrel wall, the requirements of Section 714.4 shall still apply to the intersection between the spandrel wall and the floor.

❖ This provision serves as a cross reference to the vertical separation of opening requirements found within Section 704.9. This section still requires that where a fire-resistance-rated floor intersects with a nonrated spandrel wall that the void space must still be protected by an approved joint system.

714.6 Fire-resistant joint systems in smoke barriers. Fire-resistant joint systems in smoke barriers, and joints at the intersection of a horizontal *smoke barrier* and an exterior curtain wall, shall be tested in accordance with the requirements of UL 2079 for air leakage. The air leakage rate of the joint shall not exceed 5 cfm per lineal foot (0.00775 m³/s · m) of joint at 0.30 inch (7.47 Pa) of water for both the ambient temperature and elevated temperature tests.

❖ The leakage, or L rating, provides a quantitative indication of a fire-resistant joint system's ability to resist the passage of smoke. Although the test is performed using air, the flow properties of air and of smoke are sufficiently close for engineering purposes as to provide a reasonable quantification of smoke leakage. The L-rating test is an optional test within the UL 2079

test standard; therefore, not all fire-resistant joint systems that are tested and listed will have this information. In the 2005 UL *Fire Resistance Directory* (Vol. 2), there are 335 fire-resistant joint systems that are listed that also have an L rating of 5 or less. This should provide a sufficient selection to allow for code compliance in all situations.

SECTION 715
OPENING PROTECTIVES

715.1 General. Opening protectives required by other sections of this code shall comply with the provisions of this section.

❖ This section regulates two types of opening protectives: fire doors (see Section 715.4) and fire protection-rated glazing (see Section 715.5). As covered in Section 715.2, fire-resistance-rated glazing is not addressed by the balance of the requirements in Section 715. Fire doors are a type of opening protective and are installed in openings in fire-resistance-rated assemblies, including fire walls, fire barriers, fire partitions and exterior walls where the openings are required to be protected. Where an exterior load-bearing wall requires a fire-resistance rating in accordance with Table 601, it is for structural integrity purposes.

The openings are not required to be protected unless an opening protective is required by another section of the code, including Section 704.8.

In addition to fire doors, another form of an opening protective is fire protection-rated glazing (commonly referred to as "fire windows"). Fire windows refer to the entire assembly, which may consist of a frame, and the glazing material and mounting components. A glass-block assembly is also considered a window assembly.

The fire protection ratings in this section are based on the acceptance criteria of NFPA 252, 257 and UL 10A, 14B or 14C. The fire protection rating acceptance criteria are not the same as that required for a fire-resistance rating for building structural elements. Fire-resistance ratings of building construction are determined by ASTM E 119 or UL 263. The fire protection rating required for an opening protective is generally less than the required fire resistance of the wall (see Tables 715.4 and 715.5). This is based upon the ability of a wall to have material or a fuel package directly against the assembly while fire doors and windows are assumed to have the fuel package remote from the surface of the assembly. Sections 715.4 and 715.5 (and subsections) reference the appropriate test standards and require that all opening protectives be labeled (see Sections 715.4.6 and 715.5.9), with the exception being oversized doors that require a certificate of inspection by an approved agency (see Section 715.4.6.2).

715.2 Fire-resistance-rated glazing. Fire-resistance-rated glazing tested as part of a fire-resistance-rated wall assembly in accordance with ASTM E 119 or UL 263 and labeled in accordance with Section 703.5 shall be permitted in *fire doors* and *fire window assemblies* in accordance with their listings and shall not otherwise be required to comply with this section.

❖ There are glazing materials used and tested as a wall assembly under ASTM E 119 or UL 263 and NFPA 251, *Standard Methods of Tests of Fire Endurance of Building Construction and Materials*, that are not covered by Section 715. Because these materials can meet the same fire-resistance requirements of the wall, they are not required to be regulated as an opening that would require the lower level of protection that a fire protection rating provides.

715.3 Alternative methods for determining fire protection ratings. The application of any of the alternative methods *listed* in this section shall be based on the fire exposure and acceptance criteria specified in NFPA 252, NFPA 257 or UL 9. The required *fire resistance* of an opening protective shall be permitted to be established by any of the following methods or procedures:

1. Designs documented in *approved* sources.

2. Calculations performed in an *approved* manner.

3. Engineering analysis based on a comparison of opening protective designs having *fire protection ratings* as determined by the test procedures set forth in NFPA 252, NFPA 257 or UL 9.

4. Alternative protection methods as allowed by Section 104.11.

❖ Section 715.3 contains language similar to Section 703.3 regarding fire-resistance ratings. The intent of this section is to recognize approved calculations, an engineering analysis or alternative protection methods as permitted in Section 104.11 as acceptable alternative means to determine a fire protection rating. The section is intended to avoid some of the common practices of providing closely spaced sprinklers near glazing in lieu of using fire protection or fire-resistance-rated glazing products without any testing or analysis.

715.4 Fire door and shutter assemblies. Approved *fire door* and fire shutter assemblies shall be constructed of any material or assembly of component materials that conforms to the test requirements of Section 715.4.1, 715.4.2 or 715.4.3 and the *fire protection rating* indicated in Table 715.4. *Fire door* frames with transom lights, sidelights or both shall be permitted in accordance with Section 715.4.5. *Fire door* assemblies and shutters shall be installed in accordance with the provisions of this section and NFPA 80.

Exceptions:

1. Labeled protective assemblies that conform to the requirements of this section or UL 10A, UL 14B and UL 14C for tin-clad *fire door* assemblies.

2. Floor *fire door* assemblies in accordance with Section 712.8.

❖ When the code refers to a fire door assembly, the intent is that the term "fire door" (see the definitions in Section 702) applies to the door itself, while the term "fire door assembly" includes the fire door, the frame, related hardware and accessories, unless otherwise noted (see the definition of "Fire door assembly"). Therefore, it is important to realize that a fire door by itself is not generally accepted. An assembly that includes the door is needed. A door by itself cannot provide the protection; it needs to be supported by the hinges, held into a specific frame and generally have latching and closing hardware. By requiring an assembly tested to the specific standards, all of the components needed for protection will be provided. For example, although there is not a direct mention in the section relative to mandating latching hardware, the referenced test standard, NFPA 80, addresses the requirements relative to positive latch devices.

Note that reference to Section 715.4.5 is made for fire door assemblies that include transoms and a sidelight. These assemblies are outside the scope of NFPA 80. The required fire protection rating exceeds $^3/_4$ hour.

Labels that indicate compliance with UL 10A, 14B or 14C are acceptable (see Exception 1). Exception 2 exempts floor fire door assemblies from the requirements of this section, since Section 712.8 requires floor fire doors to comply with a different test standard (NFPA 288) and their rating must not be less than the rating of the horizontal assembly being penetrated (see the commentary to Section 712.8 for a discussion of floor fire doors).

TABLE 715.4. See below.

❖ This table lists the minimum fire protection ratings for fire doors relative to the nature and fire-resistance rating of the wall. Once the purpose and fire-resistance rating of the wall are identified, the minimum required fire protection rating of the door can be determined using the table. Typically, the minimum permitted fire door fire protection rating is less than the required fire-resistance rating of the wall.

The hourly designation in the table indicates the duration of fire test exposure and is actually called the "fire protection rating." Although no longer used in the code, fire doors are sometimes improperly classified by an alphabetical letter that once designated the fire ratings of doors. The Annex of NFPA 80 still lists these classifications for older existing doors and could be used to evaluate an existing door in an existing building.

In addition to the fire protection ratings in Table 715.4, other sections of the code prescribe additional performance criteria for fire door assemblies. For example, Section 715.4.4 requires doors in exit enclosures to have a maximum temperature increase on the unexposed surface of 450°F (232°C) after 30 minutes. Although temperature rise is not a condition of acceptance, test standards require that unexposed surface temperatures be measured and documented in the test report.

One item that could also be linked to the exterior wall requirements in this table is the possibility of using the exception that is found in Section 705.8.2. That provision would eliminate the requirement for a fire protected rated opening, provided the building is equipped with an automatic sprinkler system and the openings are protected by an approved water curtain (see Section 705.8.2 for the complete requirements).

715.4.1 Side-hinged or pivoted swinging doors. *Fire door* assemblies with side-hinged and pivoted swinging doors shall be tested in accordance with NFPA 252 or UL 10C. After 5 minutes into the NFPA 252 test, the neutral pressure level in the furnace shall be established at 40 inches (1016 mm) or less above the sill.

❖ Fire door assemblies that contain side-hinged or pivoted swinging doors are to be tested in accordance with the referenced test standards.

When first introduced in one of the early editions of the code, the requirement for the door to be tested with a pressure differential (positive pressure at the top and negative pressure at the bottom) and a specific neutral pressure point was a major revision in the way that doors were tested. The intent of this provision is to better approximate the conditions found in an actual building fire where smoke, heat and hot gasses rise to the top of the room and increase the pressure in the upper portion of the room. Test method UL 10C contains the requirements for positive pressure testing.

715.4.2 Other types of assemblies. *Fire door* assemblies with other types of doors, including swinging elevator doors and fire shutter assemblies, shall be tested in accordance with NFPA 252 or UL 10B. The pressure in the furnace shall be maintained as nearly equal to the atmospheric pressure as possible. Once established, the pressure shall be maintained during the entire test period.

❖ This section also references NFPA 252 for fire door assemblies containing any door other than typical side-hinged-type doors. NFPA 252 is a general test standard that is not specific to any type of door. UL 10B

TABLE 715.4
FIRE DOOR AND FIRE SHUTTER FIRE PROTECTION RATINGS

TYPE OF ASSEMBLY	REQUIRED ASSEMBLY RATING (hours)	MINIMUM FIRE DOOR AND FIRE SHUTTER ASSEMBLY RATING (hours)
Fire walls and fire barriers having a required fire-resistance rating greater than 1 hour	4	3
	3	3[a]
	2	$1^1/_2$
	$1^1/_2$	$1^1/_2$
Fire barriers having a required fire-resistance rating of 1 hour:		
Shaft, exit enclosure and exit passageway walls	1	1
Other fire barriers	1	$^3/_4$
Fire partitions:		
Corridor walls	1	$^1/_3$ [b]
	0.5	$^1/_3$ [b]
Other fire partitions	1	$^3/_4$
	0.5	$^1/_3$
Exterior walls	3	$1^1/_2$
	2	$1^1/_2$
	1	$^3/_4$
Smoke barriers	1	$^1/_3$ [b]

a. Two doors, each with a fire protection rating of $1^1/_2$ hours, installed on opposite sides of the same opening in a fire wall, shall be deemed equivalent in fire protection rating to one 3-hour fire door.
b. For testing requirements, see Section 715.4.3.

was developed for negative pressure testing of fire door and fire shutter assemblies and is viewed as a viable test method, provided that the pressure is maintained nearly or at atmospheric pressure. It is important to distinguish that the pressure testing that is required for side-hinged or pivoted swinging doors by Section 715.4.1 is not applicable to this section and other door types.

715.4.3 Door assemblies in corridors and smoke barriers. *Fire door* assemblies required to have a minimum *fire protection rating* of 20 minutes where located in *corridor* walls or *smoke barrier* walls having a *fire-resistance rating* in accordance with Table 715.4 shall be tested in accordance with NFPA 252 or UL 10C without the hose stream test.

Exceptions:

1. Viewports that require a hole not larger than 1 inch (25 mm) in diameter through the door, have at least a 0.25-inch-thick (6.4 mm) glass disc and the holder is of metal that will not melt out where subject to temperatures of 1,700°F (927°C).

2. *Corridor* door assemblies in occupancies of Group I-2 shall be in accordance with Section 407.3.1.

3. Unprotected openings shall be permitted for *corridors* in multitheater complexes where each motion picture auditorium has at least one-half of its required *exit* or *exit access doorways* opening directly to the exterior or into an *exit* passageway.

4. Horizontal sliding doors in *smoke barriers* that comply with Sections 408.3 and 408.8.4 in occupancies in Group I-3.

❖ NFPA 252 and UL 10C standards require a hose stream test on all fire door assemblies. The hose stream test is intended to provide a measurement of structural integrity by evaluating the ability of the assembly to withstand impact. It has come to be accepted that a hose stream is not justified for 20-minute doors.

Exception 1 accepts certain viewports that are not required to be tested as a part of a door assembly. The use of viewports is fairly common in corridors of a Group R-1 occupancy or other similar locations. Because of the limited size and materials, the overall performance of the door will not be greatly affected by including the viewport. Exception 2 serves as a cross reference to the requirements for a Group I-2 occupancy, which does not require the door to have a fire protection rating. Exception 3 has little application now due to the requirements of Sections 1017.1 and 903.2.1.1, Item 4. This exception was originally intended to allow the removal of latching hardware on the corridor door and, therefore, reduce the level of noise that was made when it was opened during a movie. Since the code now requires all multitheater complexes to be sprinklered and corridors in sprinklered Group A occupancies do not require a fire-resistive rating, the other provisions of the code would eliminate the need for the door to have a fire protection rating and the latch that was considered the

problem is then generally not required.

Exception 4 serves as a cross reference to Section 408 for Group I-3 occupancies that contains requirements specific to smoke-tight door assemblies for specific locations within resident housing areas.

715.4.3.1 Smoke and draft control. *Fire door* assemblies shall also meet the requirements for a smoke and draft control door assembly tested in accordance with UL 1784. The air leakage rate of the door assembly shall not exceed 3.0 cubic feet per minute per square foot (0.01524 m^3/s · m^2) of door opening at 0.10 inch (24.9 Pa) of water for both the ambient temperature and elevated temperature tests. Louvers shall be prohibited. Installation of smoke doors shall be in accordance with NFPA 105.

❖ Fire doors in corridors and smoke barriers are also required to meet the criteria of UL 1784. This standard measures the movement of smoke through a door assembly. The installation of the doors is to be done according to NFPA 105 requirements. This standard, titled *Standard for the Installation of Smoke Door Assemblies*, is a companion to the previously referenced NFPA 80. The criteria for air leakage are also provided in this section.

715.4.3.2 Glazing in door assemblies. In a 20-minute *fire door* assembly, the glazing material in the door itself shall have a minimum fire-protection-rated glazing of 20 minutes and shall be exempt from the hose stream test. Glazing material in any other part of the door assembly, including transom lights and sidelights, shall be tested in accordance with NFPA 257 or UL 9, including the hose stream test, in accordance with Section 715.5.

❖ Glazing in 20-minute-rated doors must also have a 20-minute fire protection rating and is not required to pass the hose stream test. Sidelights and transoms, which are not part of the fire door but instead are part of the fire door assembly (see definitions in Section 702), are required to pass the hose stream test as detailed in NFPA 257 or UL 9.

715.4.4 Doors in exit enclosures and exit passageways. *Fire door* assemblies in *exit* enclosures and *exit* passageways shall have a maximum transmitted temperature end point of not more than 450°F (250°C) above ambient at the end of 30 minutes of standard fire test exposure.

Exception: The maximum transmitted temperature rise is not required in buildings equipped throughout with an *automatic sprinkler system* installed in accordance with Section 903.3.1.1 or 903.3.1.2.

❖ This section requires that door assemblies utilized in vertical exit enclosures and exit passageways as defined in Chapter 10 comply with the requirements of Section 715 and be tested in accordance with either NFPA 252 or UL 10C. This section adds a requirement that door construction is limited to a temperature rise on the unexposed side of 450°F (232°C) during the first 30 minutes of the standard fire test (NFPA 252). The labels for these doors are required to indicate that the temperature on the unexposed side is 450°F

(232°C) or less above ambient temperature.

It should be noted that the temperature rise criterion is not otherwise a limitation with respect to a door receiving a fire protection rating. Therefore, simply specifying a 1¹/₂-hour fire protection rating on the door does not ensure the additional requirements of Section 715.4.4 are met. The basis for limiting the temperature rise of the unexposed surface of exit doors to 450°F (232°C) is that a higher allowable temperature would provide enough radiant heat to discourage or even prevent building occupants from closely approaching or passing by the door assembly during a fire emergency. The intent is to protect the occupants in the vertical exit enclosure and exit passageway from excessive radiant heat. The exception provides relief from providing doors meeting temperature rise ratings when buildings are equipped throughout with either an NFPA 13 or 13R automatic sprinkler system.

715.4.4.1 Glazing in doors. Fire-protection-rated glazing in excess of 100 square inches (0.065 m²) shall be permitted in *fire door* assemblies when tested as components of the door assemblies and not as glass lights, and shall have a maximum transmitted temperature rise of 450°F (250°C) in accordance with Section 715.4.4.

> **Exception:** The maximum transmitted temperature rise is not required in buildings equipped throughout with an *automatic sprinkler system* installed in accordance with Section 903.3.1.1 or 903.3.1.2.

❖ This section permits larger glazing panels in fire doors, which are to have a temperature rise of 450°F (232°C). This increased glazing is to be tested as a component of the door. The test methods to be utilized are listed in Section 715.4, and include NFPA 252, UL 10B and UL 10C. To provide an equivalent level of safety glazing in fire doors that are over 100 square inches (0.065 m²) in area is required to limit temperature rise just as required for the door itself. The 100-square-inch (0.065 m²) threshold is consistent with the glazing size limitations contained in exceptions listed in Section 715.4.7.1. The exception reduces transmitted temperature requirements when automatic sprinklers are installed throughout the building.

715.4.5 Fire door frames with transom lights and sidelights. Door frames with transom lights, sidelights, or both, shall be permitted where a ³/₄-hour *fire protection rating* or less is required in accordance with Table 715.4. Where a *fire protection rating* exceeding ³/₄-hour is required in accordance with Table 715.4, *fire door* frames with transom lights, sidelights, or both, shall be permitted where installed with fire-resistance-rated glazing tested as an assembly in accordance with ASTM E119 or UL 263.

❖ The purpose of this section is to address the use of fire-resistance-rated glazing in fire door frames with transforms or sidelights where the fire protection rating exceeds ³/₄ hour. For instances where the required rating is ³/₄ hour, the test requirements of Section 715.4.3.2 apply. For instances where the required rating exceeds ³/₄ hour, this section applies and requires

testing of the assembly in accordance with ASTM E 119 or UL 263. Testing the assemblies to these test criteria exposes the glazing to the appropriate temperature rise criteria to substantiate the higher fire protection rating.

715.4.6 Labeled protective assemblies. *Fire door* assemblies shall be labeled by an *approved agency*. The *labels* shall comply with NFPA 80, and shall be permanently affixed to the door or frame.

❖ The requirement that fire doors be labeled leads to the provisions of Section 1703.5 being applied. A label indicates that the door assembly has not only passed the required fire test but that a follow-up inspection was also performed during production (see Section 1703.5.2). The building official should, therefore, verify that fire door assemblies are properly labeled, and not rely solely on a test report. To provide specific guidance, compliance with NFPA 80 is required.

715.4.6.1 Fire door labeling requirements. *Fire doors* shall be labeled showing the name of the manufacturer or other identification readily traceable back to the manufacturer, the name or trademark of the third-party inspection agency, the *fire protection rating* and, where required for *fire doors* in *exit* enclosures and *exit* passageways by Section 715.4.4, the maximum transmitted temperature end point. Smoke and draft control doors complying with UL 1784 shall be labeled as such and shall also comply with Section 715.4.6.3. Labels shall be *approved* and permanently affixed. The *label* shall be applied at the factory or location where fabrication and assembly are performed.

❖ Labels on fire doors apply to the door only. The building official should verify that the remaining portions of the assembly (door frame, hardware and accessories) are also labeled for use with a labeled fire door. In addition to the appropriate fire protection rating, the label for exit doors is required to indicate the temperature rise on the unexposed surface after 30 minutes (see commentary, Section 715.4.4). A label is also to serve as evidence of both a required fire protection rating and third-party inspection—not solely as a manufacturer's identification (see commentary, Section 1703.5). To ensure appropriate labeling, it is required to occur at the actual location where fabrication and assembly occur.

715.4.6.2 Oversized doors. Oversized *fire doors* shall bear an oversized *fire door label* by an *approved agency* or shall be provided with a certificate of inspection furnished by an *approved* testing agency. When a certificate of inspection is furnished by an *approved* testing agency, the certificate shall state that the door conforms to the requirements of design, materials and construction, but has not been subjected to the fire test.

❖ Recognizing that doors may exceed the required tested size, the code indicates that oversized doors that have a certificate of inspection from an approved testing agency are acceptable. For example, rolling steel-type fire doors are normally listed for 120 square feet (11 m²) with no dimension in excess of 12 feet (3658 mm). Rolling steel-type doors that exceed these dimensions are permitted as long as they have certification indicating that they are constructed of materials of the same grade, thickness, shape, etc., as labeled doors. This is normally indicated with a certificate that reads "Oversized Fire Door Certificate."

The code does not specify what is considered to be an "oversized door." This information would be dependent on the test standard and the dimensions listed within it. Note that the provisions of Sections 706.8 and 707.6 allow openings up to 156 square feet (14 m²) in area.

715.4.6.3 Smoke and draft control door labeling requirements. Smoke and draft control doors complying with UL 1784 shall be labeled in accordance with Section 715.4.6.1 and shall show the letter "S" on the fire rating *label* of the door. This marking shall indicate that the door and frame assembly are in compliance when *listed* or labeled gasketing is also installed.

❖ When doors are required to be approved for smoke and draft control, they require a special label containing the letter "S." Without this label, it is very difficult to assess in the field that the door meets UL 1784. This also reminds the installer that listed or labeled gasketing is required to be installed with the door and frame.

715.4.6.4 Fire door frame labeling requirements. *Fire door* frames shall be labeled showing the names of the manufacturer and the third-party inspection agency.

❖ It is required that the fire door frame be labeled with the manufacturer's name and third-party testing agency to assist in the assessment of the door in the field. The actual rating is not required to be included.

715.4.7 Glazing material. Fire-protection-rated glazing conforming to the opening protection requirements in Section 715.4 shall be permitted in *fire door* assemblies.

❖ This section of the code addresses all types of glazing material in fire protection-rated doors, no matter if tested glazing material or wired glass is used.

715.4.7.1 Size limitations. Fire-protection-rated glazing used in *fire doors* shall comply with the size limitations of NFPA 80.

Exceptions:

1. Fire-protection-rated glazing in *fire doors* located in *fire walls* shall be prohibited except where serving in a *fire door* in a horizontal *exit*, a self-closing swinging door shall be permitted to have a vision panel of not more than 100 square inches (0.065 m²) without a dimension exceeding 10 inches (254 mm).

2. Fire-protection-rated glazing shall not be installed in *fire doors* having a $1^1/_2$-hour *fire protection rating* intended for installation in *fire barriers*, unless the glazing is not more than 100 square inches (0.065 m²) in area.

❖ NFPA 80 permits fire protection-rated glazing in fire doors with a fire protection rating up to 3 hours. For $^1/_2$-, $^1/_3$- and $^3/_4$-hour doors, the allowable area is a function of the glazed area tested. Exception 1 allows for

fire protection-rated glazing to be used as vision panels with a maximum area of 100 square inches (0.065 m²) in fire doors that are installed in fire walls and serve as horizontal exits. This is to recognize the life-safety benefit of the visual observance that these panels allow individuals exiting the building in an emergency situation.

Exception 2 allows fire protection-rated glazing to be used in openings with a maximum area of 100 square inches (0.065 m²) within $1^1/_2$-hour fire-resistance-treated fire doors that are installed in fire barriers.

715.4.7.2 Exit and elevator protectives. *Approved* fire-protection-rated glazing used in *fire door* assemblies in elevator and *exit* enclosures shall be so located as to furnish clear vision of the passageway or approach to the elevator, ramp or *stairway*.

❖ The purpose of vision panels in exit doors and elevator doors is to permit observation by occupants who may be on the opposite side of the door before opening it. The limitations on the size of panels in Section 715.4.6.1 still apply.

715.4.7.3 Labeling. Fire-protection-rated glazing shall bear a *label* or other identification showing the name of the manufacturer, the test standard and information required in Section 715.5.9.1 that shall be issued by an *approved agency* and shall be permanently affixed to the glazing.

❖ This provision of the code maintains proper labeling for each piece of fire protection-rated glazing. The label must specify the name or fully identifying logo of the manufacturer, the test standard that was used to evaluate the glass and the rating established by the test.

715.4.7.3.1 Identification. For fire protection-rated glazing, the *label* shall bear the following four-part identification: "D – H or NH – T or NT – XXX." "D" indicates that the glazing shall be used in *fire door* assemblies and that the glazing meets the fire protection requirements of NFPA 252. "H" shall indicate that the glazing meets the hose stream requirements of NFPA 252. "NH" shall indicate that the glazing does not meet the hose stream requirements of the test. "T" shall indicate that the glazing meets the temperature requirements of Section 715.4.4.1. "NT" shall indicate that the glazing does not meet the temperature requirements of Section 715.4.4.1. The placeholder "XXX" shall specify the fire-protection-rating period, in minutes.

❖ By following the provisions of this section, the glazing utilized in fire door assemblies can be easily identified for verification of its appropriate application. The "D" designation indicates the glazing can be used in a fire door assembly, with the remaining identifiers providing specific information as to the glazing's capability to meet the hose stream test and temperature limits. Similar identification methods for glazing are found in Section 715.5.9.1 for fire protection-rated glazing used in fire window assemblies.

715.4.7.4 Safety glazing. Fire-protection-rated glazing installed in *fire doors* in areas subject to human impact in hazardous locations shall comply with Chapter 24.

❖ This section provides a cross reference to the safety glazing requirements of Chapter 24, specifically Section 2406.3. Not only do glass or glazed panels need to be fire-resistance rated, but the glass also must pass the test requirements of CPSC 16 CFR, Part 1201, if they installed in locations subject to human impact loads. Locations where glazing presents a hazard to occupants upon impact are identified in Section 2406.3. This requirement reduces the hazards of someone falling into a fire-resistance-rated glass panel.

Earlier editions of codes permitted wired glass to meet a lower level of impact resistance. This was done because at the time of the original exclusion, wired glass was the only product that could effectively provide the fire protection ratings that were needed in rated assemblies. Therefore, the code accepted a lower level of impact resistance when testing the safety glazing aspect of wired glass. Because many newer products such as the ceramic-based glazing are now available on the market, the exclusion for wired glass was eliminated. Therefore, if wired glass or any other type of glazing material that needs a fire protection rating is installed where safety glazing is required (see Section 2406.3), the glazing must also meet the impact loads of Section 2406.1.

715.4.8 Door closing. *Fire doors* shall be self- or automatic-closing in accordance with this section.

Exceptions:

1. *Fire doors* located in common walls separating *sleeping units* in Group R-1 shall be permitted without automatic- or self-closing devices.

2. The elevator car doors and the associated hoistway enclosure doors at the floor level designated for recall in accordance with Section 3003.2 shall be permitted to remain open during Phase I emergency recall operation.

❖ In order for fire doors to be effective, they must be in the closed position; therefore, the preferred arrangement is to install self-closing doors. Recognizing that operational practices often require doors to be open for an extended period of time, automatic-closing doors are permitted as long as this opening will not pose a threat to occupant safety and the doors will self-latch. Automatic-closing devices enable the opening to be protected during a fire condition. The basic requirement for closing devices and specific requirements for automatic-closing and self-closing devices are given in NFPA 80 (see Section 715.4.7.2).

Exception 1 discusses doors in the separation walls of side-by-side sleeping units in Group R-1 occupancies. This allows two or more sleeping units to be opened to each other so that a suite can be established, if desired. If the adjacent rooms are rented by separate people, the doors will generally be closed. If the rooms are rented together, this permits a larger room to be created simply by opening the door. Because most rooms only open to one adjacent room, most configurations would still result in a solid fire partition without any door openings between this suite of rooms and any other sleeping units.

Exception 2 coordinates with the requirements related to elevator recall that are found in Chapter 30. Since the hoistway doors generally provide the shaft opening protection, this exception is needed for this situation.

715.4.8.1 Latch required. Unless otherwise specifically permitted, single *fire doors* and both leaves of pairs of side-hinged swinging *fire doors* shall be provided with an active latch bolt that will secure the door when it is closed.

❖ This section merely reinforces the acceptance criteria of fire test standards for fire doors. A door that does not latch would generally be an ineffective barrier against the spread of a fire and would be unable to withstand the pressures of a fire in the adjacent space.

715.4.8.2 Automatic-closing fire door assemblies. Automatic-closing *fire door* assemblies shall be self-closing in accordance with NFPA 80.

❖ NFPA 80 requires doors to be automatic closing with a closing device and a separate hold/release device or hold-open mechanism that closes upon activation of an automatic fire detector acceptable to the building official. This provision simply ensures that when the door is released that it will move to the closed position.

715.4.8.3 Smoke-activated doors. Automatic-closing doors installed in the following locations shall be automatic-closing by the actuation of smoke detectors installed in accordance with Section 907.3 or by loss of power to the smoke detector or hold-open device. Doors that are automatic-closing by smoke detection shall not have more than a 10-second delay before the door starts to close after the smoke detector is actuated:

1. Doors installed across a *corridor*.

2. Doors that protect openings in *exits* or *corridors* required to be of fire-resistance-rated construction.

3. Doors that protect openings in walls that are capable of resisting the passage of smoke in accordance with Section 508.2.5.2.

4. Doors installed in *smoke barriers* in accordance with Section 710.5.

5. Doors installed in *fire partitions* in accordance with Section 709.6.

6. Doors installed in a *fire wall* in accordance with Section 706.8.

7. Doors installed in shaft enclosures in accordance with Section 708.7.

8. Doors installed in refuse and laundry chutes and access and termination rooms in accordance with Section 708.13.

9. Doors installed in the walls for compartmentation of underground buildings in accordance with Section 405.4.2.

10. Doors installed in the elevator lobby walls of underground buildings in accordance with Section 405.4.3.

11. Doors installed in smoke partitions in accordance with Section 711.5.3.

❖ Since the integrity of the means of egress or general occupant safety can be compromised by smoke, doors at certain locations are required to be automatic closing upon the detection of smoke if the door is not self-closing. The automatic closer is also to activate upon loss of power to the smoke detector and to the hold-open device. NFPA 72, as referenced in Chapter 9, contains criteria relative to the number and location of detectors necessary for door release service. NFPA 80 also provides criteria for closing devices required for a fire door.

715.4.8.4 Doors in pedestrian ways. Vertical sliding or vertical rolling steel *fire doors* in openings through which pedestrians travel shall be heat activated or activated by smoke detectors with alarm verification.

❖ Where vertical sliding or rolling doors are provided as an opening protective through which pedestrian travel is intended, automatic-closing actuation is not permitted via a smoke detector unless alarm verification is provided. Sudden closing of these doors and the potential of false actuation of smoke detectors create a hazard to occupants moving through the opening. Although pedestrian travel is intended in openings regulated by this section, a common oversight is that the vertical rolling or sliding door in the closed position is not permitted to be considered an egress component, as stated in Section 1008.1.2.

715.4.9 Swinging fire shutters. Where fire shutters of the swinging type are installed in exterior openings, not less than one row in every three vertical rows shall be arranged to be readily opened from the outside, and shall be identified by distinguishing marks or letters not less than 6 inches (152 mm) high.

❖ The fire department needs access to openings for ventilation purposes and entry; therefore, if an extensive amount of swinging shutters are used, at least one in every three vertical rows must be readily openable from the outside. The operable shutter must also be easily recognized for use by the fire department. The identification is on the exterior.

715.4.10 Rolling fire shutters. Where fire shutters of the rolling type are installed, such shutters shall include *approved* automatic-closing devices.

❖ Rolling fire shutters require a detecting device to initiate the closing sequence.

715.5 Fire-protection-rated glazing. Glazing in *fire window assemblies* shall be fire-protection rated in accordance with this section and Table 715.5. Glazing in *fire door* assemblies shall comply with Section 715.4.7. Fire-protection-rated glazing shall be tested in accordance with and shall meet the acceptance criteria of NFPA 257 or UL 9. Fire-protection-rated glazing shall also comply with NFPA 80. Openings in nonfire-resistance-rated *exterior wall* assemblies that require protection in accordance with Section 705.3, 705.8, 705.8.5 or 705.8.6 shall have a fire-protection rating of not less than $^3/_4$ hour.

Exceptions:

1. Wired glass in accordance with Section 715.5.4.

2. Fire protection-rated glazing in 0.5-hour fire-resistance-rated partitions is permitted to have an 0.33-hour fire-protection rating.

❖ The appropriate test standards referenced for fire windows (fire protection-rated glazing) are NFPA 257 or UL 9, which are pass/fail tests with a specified time of exposure. It is important to realize that this section only applies to materials that are tested in accordance with the two identified standards and that are intended as an opening protective in a wall that requires openings with a fire protection rating. The distinction between a fire protection rating and a fire-resistance rating must be understood. Glazing that has a fire-resistance rating is exempted from the requirements of Section 715 (see Section 715.2). In general, the location where fire protection-rated openings may be used includes exterior walls, certain fire barriers, fire partitions and smoke barriers. Other specific sections of the code, such as Section 706.5.2, Exception 1, and Section 1022.6, may also require fire window assemblies to comply with this section, while fire windows are prohibited by Sections 1022.3 and 1023.5 even though they are a rated assembly. The basic premise of the referenced standard is that it evaluates the effectiveness of windows when used as opening protectives to remain in place during the specified time of exposure. In addition to being subjected to a predetermined fire condition, the assembly is also subjected to a hose stream impact test. The test procedure does not measure or evaluate heat transmission or radiation through the assembly.

The first exception not only provides a cross reference to the wired glass provisions of Section 715.5.4, but it also establishes that the wired glass assemblies are deemed to be acceptable where a fire window requires a $^3/_4$-hour rating (see Section 715.5.4). There is also an exception that would allow a $^1/_3$-hour rating for openings in $^1/_2$-hour-rated fire partitions. Fire partitions are only allowed to be $^1/_2$-hour rated in accordance with the exceptions to Section 709.3 and Table 1018.1.

TABLE 715.5
FIRE WINDOW ASSEMBLY FIRE PROTECTION RATINGS

TYPE OF ASSEMBLY		REQUIRED ASSEMBLY RATING (hours)	MINIMUM FIRE WINDOW ASSEMBLY RATING (hours)
Interior walls:	Fire walls	All	NP^a
	Fire barriers	>1	NP^a
		1	$^3/_4$
	Smoke barriers	1	$^3/_4$
	Fire partitions	1	$^3/_4$
		$^1/_2$	$^1/_3$
Exterior walls		>1	$1^1/_2$
		1	$^3/_4$
Party wall		All	NP

NP = Not Permitted.

a. Not permitted except as specified in Section 715.2.

❖ Table 715.5 prescribes the minimum fire protection ratings for fire windows, relative to the type and fire-resistance rating of the wall. Once the purpose and fire-resistance rating of the wall are identified, the minimum required fire protection rating of windows is determined by using the table. The requirements in this table are not to be confused with the ratings for fire doors provided in Table 715.4 since the requirements do differ.

715.5.1 Testing under positive pressure. NFPA 257 or UL 9 shall evaluate fire-protection-rated glazing under positive pressure. Within the first 10 minutes of a test, the pressure in the furnace shall be adjusted so at least two-thirds of the test specimen is above the neutral pressure plane, and the neutral pressure plane shall be maintained at that height for the balance of the test.

❖ Under fire conditions, there is likely to be positive pressure acting on the window; therefore, this section requires that tested window assemblies be subjected to positive pressure conditions during the test. This requirement is consistent with standards used for testing window assemblies in the U.S. and other countries.

715.5.2 Nonsymmetrical glazing systems. Nonsymmetrical fire-protection-rated glazing systems in *fire partitions, fire barriers* or in *exterior walls* with a *fire separation distance* of 5 feet (1524 mm) or less pursuant to Section 705 shall be tested with both faces exposed to the furnace, and the assigned *fire protection rating* shall be the shortest duration obtained from the two tests conducted in compliance with NFPA 257 or UL 9.

❖ This section requires that glazing systems that by design are not the same on either side be tested on both sides. The rating resulting from these tests will be determined from the lower performing side. These requirements are for fire-type barriers (versus smoke) within buildings and when used on an exterior wall with a small separation distance [less than 5 feet (1524 mm)]. The concern here is the same as for nonsymmetrical wall assemblies in Section 703.2.1.

715.5.3 Safety glazing. Fire-protection-rated glazing installed in *fire window assemblies* in areas subject to human impact in hazardous locations shall comply with Chapter 24.

❖ This section provides a cross reference to the safety glazing requirements of Chapter 24, specifically Section 2406.3. Not only do glass or glazed panels need to be fire-resistance rated, but the glass also must pass the test requirements of CPSC 16 CFR, Part 1201, if installed in locations subject to human impact loads. Locations where glazing presents a hazard to occupants upon impact are identified in Section 2406.3. This requirement reduces the hazards of someone falling into a fire-resistance-rated glass panel.

Earlier editions of codes permitted wired glass to meet a lower level of impact resistance. This was done because at the time of the original exclusion, wired glass was the only product that could effectively provide the fire protection ratings that were needed in rated assemblies. Therefore, the code accepted a lower level of impact resistance when testing the safety glazing aspect of wired glass. Because many newer products such as the ceramic-based glazing are now available on the market, the exclusion for wired glass was eliminated. Therefore, if wired glass or any other type of glazing material that needs a fire protection rating is installed where safety glazing is required (see Section 2406.3), the glazing must also meet the impact loads of Section 2406.1.

715.5.4 Wired glass. Steel window frame assemblies of 0.125-inch (3.2 mm) minimum solid section or of not less than nominal 0.048-inch-thick (1.2 mm) formed sheet steel members fabricated by pressing, mitering, riveting, interlocking or welding and having provision for glazing with $^1/_4$-inch (6.4 mm) wired glass where securely installed in the building construction and glazed with $^1/_4$-inch (6.4 mm) labeled wired glass shall be deemed to meet the requirements for a $^3/_4$-hour *fire window assembly*. Wired glass panels shall conform to the size limitations set forth in Table 715.5.4.

❖ An approved fire window assembly normally includes a tested fire window frame and glazing materials. This section permits the use of tested $^1/_4$-inch (6.4 mm) wired glass with a specific steel frame to be considered as a $^3/_4$-hour fire window assembly without the need to pass the NFPA 257 or UL 9 test standard (see the exception to Section 715.5). So although the assembly is not tested, it is "deemed to meet the requirements for a $^3/_4$-hour fire window assembly."

This section has a significant impact on wall assemblies that are not of masonry construction. Most fire window frames are tested for use in masonry construction only. The individual evaluation of the frame will indicate whether it may be used in drywall construction. Fire window frames evaluated for drywall construction are further subdivided into those intended to be supported by a noncombustible floor and those intended to be installed above the floor.

Wired glass for fire windows must be installed in either a tested fire window assembly (see Section 715.5) or a steel frame that conforms to this section.

The maximum size limitations in Table in 715.5.4 are based on actual tests of fire windows and fire doors with vision panels of $^1/_4$-inch-thick (6.4 mm) wired glass.

TABLE 715.5.4
LIMITING SIZES OF WIRED GLASS PANELS

OPENING FIRE PROTECTION RATING	MAXIMUM AREA (square inches)	MAXIMUM HEIGHT (inches)	MAXIMUM WIDTH (inches)
3 hours	0	0	0
$1^1/_2$-hour doors in exterior walls	0	0	0
1 and $1^1/_2$ hours	100	33	10
$^3/_4$ hour	1,296	54	54
20 minutes	Not Limited	Not Limited	Not Limited
Fire window assemblies	1,296	54	54

For SI: 1 inch = 25.4 mm, 1 square inch = 645.2 mm^2

❖ Based on the rating of the opening, the maximum area, height and width of wired glass panels are to be in accordance with the table. The limitations are based on maximum sizes of wired glass panels that have been tested in accordance with ASTM E 163, which was referenced in earlier editions of the legacy codes. Although Section 715.5.4 states that the wired glass assemblies are deemed to meet the $^3/_4$-hour fire protection rating, Table 715.5.4 includes higher rated assemblies. This is because Section 715.4.7.1 references this table for wired glass used in fire doors.

715.5.5 Nonwired glass. Glazing other than wired glass in *fire window assemblies* shall be fire-protection-rated glazing installed in accordance with and complying with the size limitations set forth in NFPA 80.

❖ This section merely reinforces Section 715.5.

715.5.6 Installation. Fire-protection-rated glazing shall be in the fixed position or be automatic-closing and shall be installed in *approved* frames.

❖ In order to adequately protect the opening, protection must be in place during a fire event. This is achieved by either having the glazing in a permanently fixed position or having the glazing protective actuated by a detection device.

715.5.7 Window mullions. Metal mullions that exceed a nominal height of 12 feet (3658 mm) shall be protected with materials to afford the same *fire-resistance rating* as required for the wall construction in which the protective is located.

❖ Fire windows are normally tested in panels not exceeding 1,296 square inches (0.85 m²) and 54 inches (1372 mm) in width or length, respectively. The code does, however, permit panels to be installed adjacent to one another. This section requires that mullions higher than 12 feet (3658 mm) meet the criteria for fire

resistance (ASTM E 119 or UL 263), not fire protection (NFPA 257).

715.5.8 Interior fire window assemblies. Fire-protection-rated glazing used in *fire window assemblies* located in *fire partitions* and *fire barriers* shall be limited to use in assemblies with a maximum *fire-resistance rating* of 1 hour in accordance with this section.

❖ Since fire protection-rated glazing, including wired glass, has a $^3/_4$-hour fire-resistance rating, fire windows may only be used in fire partitions (1-hour rated in accordance with Section 709.3) and fire barriers having a fire-resistance rating of 1 hour.

715.5.8.1 Where $^3/_4$-hour fire protection window assemblies permitted. Fire-protection-rated glazing requiring 45-minute opening protection in accordance with Table 715.5 shall be limited to *fire partitions* designed in accordance with Section 709 and *fire barriers* utilized in the applications set forth in Sections 707.3.6 and 707.3.8 where the *fire-resistance rating* does not exceed 1 hour.

❖ This section identifies the specific 1-hour-rated assemblies that are referred to in Section 715.5.8. Fire partitions are 1-hour fire-resistance rated in accordance with Section 709.3. Fire barriers, on the other hand, can have ratings as high as 4 hours (see Table 508.4). This section limits the use of fire windows to fire barriers that are 1-hour rated, 1-hour incidental use separations and 1-hour mixed-occupancy separations. It must be noted that fire barriers used to separate exit elements (see Sections 707.3.2, 707.3.3 and 707.3.4) are not permitted to have fire windows, even where the separation is 1-hour fire-resistance rated. Since fire walls (see Section 706) are not included in the list of permitted locations and because they require a minimum rating of 2 hours, fire windows are not allowed in such walls.

715.5.8.2 Area limitations. The total area of windows shall not exceed 25 percent of the area of a common wall with any room.

❖ The area limitations for interior fire windows varies from that found in Section 708.7 for fire barriers. Fire barriers are limited to 25 percent of the length of the wall while this section bases the limitation on the area of the wall. Therefore, when dealing with a fire barrier, both the 25 percent length and area would be applicable. When dealing with a fire partition, only the area limitation would apply.

The phrase "common wall" will limit the size to the actual visible wall that is shared and can be seen in the two rooms. The area of the wall that continues through the interstitial space above a ceiling would not be included when determining the wall size. In addition, if the ceiling height differs on the two sides of the wall, the lowest ceiling height would be used to determine what is considered the "common wall."

715.5.9 Labeling requirements. Fire-protection-rated glazing shall bear a *label* or other identification showing the name of the manufacturer, the test standard and information required in Section 715.5.9.1 that shall be issued by an *approved agency* and shall be permanently affixed to the glazing.

❖ This section requires fire protection-rated (see Section 715.5) glazing to be labeled by an approved testing agency. This requirement enables the code user to verify that a fire window has been tested to the correct standard and that the test was properly performed by experienced people with correct equipment. This section details the necessary information required on the fire window label and how the label is to be applied. These descriptive requirements are used to verify that fire window assemblies have been correctly tested. The label on each and every fire window assembly must contain the manufacturer's name, test standard used and fire protection rating. The label must be applied at the manufacturer's plant by the third-party inspection agency. As a reminder, fire-resistance-rated glazing is not regulated by this section (see Section 715.2). Labeling requirements for fire-resistance-rated glazing can be found in Section 715.4.7.3.1.

715.5.9.1 Identification. For fire-protection-rated glazing, the *label* shall bear the following two-part identification: "OH – XXX." "OH" indicates that the glazing meets both the fire protection and the hose-stream requirements of NFPA 257 or UL 9 and is permitted to be used in openings. "XXX" represents the fire-protection rating period, in minutes, that was tested.

❖ By following the provisions of this section, the glazing utilized in fire window assemblies can be easily identified for its appropriate application. As noted, the identification includes "OH" to denote that the glazing meets both the fire protection test and the hose stream test. These designations are required, along with the fire protection rating, to be given in minutes. Similar identification methods for glazing are found in Section 715.4.7.3.1 for fire protection-rated glazing used in fire door assemblies.

SECTION 716
DUCTS AND AIR TRANSFER OPENINGS

716.1 General. The provisions of this section shall govern the protection of duct penetrations and air transfer openings in assemblies required to be protected.

❖ Fire dampers, smoke dampers and combination fire/smoke dampers protect openings created by duct penetrations and transfer openings in assemblies required to be protected. Ceiling radiation dampers protect duct penetrations, which only penetrate the ceiling membrane of a fire-resistance-rated assembly. This section includes installation, testing rating and actuation requirements in Sections 716.2, 716.3.1, 716.3.2, and 716.3.3, respectively. Section 716.4 indicates that dampers are to bear the label of an approved agency. When dampers are provided, they must be properly maintained and, therefore, must be accessible (see

Section 716.4). Section 716.5 indicates the conditions in which dampers are required at penetrations through vertical assemblies. Duct penetrations and transfer openings through horizontal assemblies, including requirements for ceiling dampers, are regulated in Section 716.6. The provisions of Section 716 are duplicated in Section 607 of the IMC. All the requirements for protection of duct penetrations and transfer openings are found in this section. This section is directly referenced by sections that address vertical and horizontal fire-resistance-rated assemblies, as follows:

Exterior walls:	Section 705.10
Fire walls:	Section 706.11
Fire barriers:	Section 707.9
Shafts:	Section 708.10
Fire partitions:	Section 709.9
Smoke barriers:	Section 710.8
Smoke partitions :	Section 711.7
Horizontal assemblies:	Section 712.7

716.1.1 Ducts that penetrate fire-resistance-rated assemblies without dampers. Ducts that penetrate fire-resistance-rated assemblies and are not required by this section to have *dampers* shall comply with the requirements of Sections 713.2 through 713.3.3. Ducts that penetrate *horizontal assemblies* not required to be contained within a shaft and not required by this section to have *dampers* shall comply with the requirements of Sections 713.4 through 713.4.2.2.

❖ This section identifies how duct penetrations are to be properly protected, either in accordance with the requirements of Section 716 using fire or fire/smoke dampers where required or in accordance with Section 713 with materials that have been tested to either ASTM E 814, ASTM E 119 or UL 263 to maintain the fire-resistance rating of the fire assembly. This section is coordinated with Sections 713.3 (fire-resistance-rated walls) and 713.4 (horizontal assemblies). If a duct does not require a damper, the penetration of that duct through a fire-resistance-rated assembly must be protected as a through penetration in accordance with Section 713. If a duct does require a fire damper, smoke damper or combination fire/smoke damper, the through-penetration protection requirements of Section 713 will not apply. The reason is that these dampers are fire tested as a complete assembly installed within a fire-resistance-rated wall assembly or horizontal assembly. The test and resulting listing will include not only the damper but also all required mounting hardware for the duct at the point where it penetrates the wall or floor. As long as the complete installation details in the listing are complied with, there is no need for any additional protection (e.g., in accordance with Section 713) to ensure that the fire-resistance rating of the wall assembly or horizontal assembly is maintained. However, it is possible to pass the fire test and hence obtain a listing for a fire damper or smoke damper without completely preventing smoke migration outside the duct, such as through the gap between the outside surface of the duct and the inside surface of the hole made to accommodate the duct. This can be quite apparent in some cases, as daylight might be visible from one side of the wall assembly or floor assembly to the other side. This is because the tested dampers generally require a minimum gap be provided between the damper and the wall or floor it is installed in. Although not required by code, this gap is sometimes closed with a sealant, in order to provide more complete smoke resistance for the wall assembly or floor assembly. It is imperative that an intumescent firestop sealant not be used on the outside of the duct to seal this gap. The expansion pressure of the intumescent sealant can buckle the duct, thus hindering or preventing the proper operation of the damper under fire conditions.

716.1.1.1 Ducts that penetrate nonfire-resistance-rated assemblies. The space around a duct penetrating a nonfire-resistance-rated floor assembly shall comply with Section 716.6.3.

❖ The purpose of this section is to point the user to the annular space protection requirements for ducts that penetrate nonfire-resistance-rated assemblies, when applicable. The referenced section, Section 716.6.3, provides options for protecting these types of penetrations, two of which contain requirements for annular space protection.

716.2 Installation. *Fire dampers, smoke dampers, combination fire/smoke dampers* and *ceiling radiation dampers* located within air distribution and smoke control systems shall be installed in accordance with the requirements of this section, the manufacturer's installation instructions and the *dampers'* listing.

❖ This section performs two regulating functions. First, it requires that dampers be installed in accordance with the manufacturer's installation instructions and listing. Such instructions will result in an installation that not only protects the penetration when the damper is actuated, but that also results in a protected opening should the duct collapse. Second, this section introduces Sections 716.2.1 and 716.2.2, which address the relationship of dampers with smoke control systems and hazardous exhaust ducts.

716.2.1 Smoke control system. Where the installation of a fire *damper* will interfere with the operation of a required smoke control system in accordance with Section 909, *approved* alternative protection shall be utilized. Where mechanical systems including ducts and *dampers* utilized for normal building ventilation serve as part of the smoke control system, the expected performance of these systems in smoke control mode shall be addressed in the rational analysis required by Section 909.4.

❖ Fire dampers are not permitted where they will obstruct or inhibit the proper operation of a required smoke control system. In a smoke control system, fire dampers are replaced with an alternative means of protection, subject to the approval of the building official. Alternative protection may be in the form of a

fire-resistance-rated shaft enclosure, steel subducts extending at least 22 inches (559 mm) vertically in an exhaust shaft having continuous upward airflow or a fire-resistance-rated horizontal duct enclosure. If a smoke control system is installed that is not required by the code, it is still subject to the same fire damper requirements as a required smoke control system; however, such nonmandated systems would be permitted to employ fire dampers instead of the mandatory alternative protection prescribed in this section for required smoke control systems. A required smoke control system is an important life-safety system, and the code-mandated provisions here combine with Section 909 design requirements to ensure operability. This section does more than simply allow fire dampers to be omitted in required smoke control systems; it mandates the omission of fire dampers where such system operation could be jeopardized. Lastly, this section allows the normal building ventilation system (HVAC) to be employed to serve as the smoke control system, as long as the performance is substantiated.

716.2.2 Hazardous exhaust ducts. *Fire dampers* for hazardous exhaust duct systems shall comply with the *International Mechanical Code.*

❖ This section refers the user to the IMC. While not stated in the IMC, this type of duct penetrations is typically not provided with fire dampers, as this may interfere with the exhausting of hazardous materials. Instead of fire damper protection, horizontal ducts are typically placed in a fire-resistance-rated horizontal shaft, which in turn, penetrates the wall.

716.3 Damper testing, ratings and actuation. *Damper* testing, ratings and actuation shall be in accordance with Sections 716.3.1 through 716.3.3.

❖ The purpose of this section is to introduce the requirements for the testing (see Section 716.3.1), rating (see Sections 716.3.2.1 through 716.3.2.3) and actuation (see Sections 716.3.3.1 through 716.3.3.4) of fire dampers, smoke dampers, combination fire/smoke dampers and ceiling radiation dampers.

716.3.1 Damper testing. *Dampers* shall be *listed* and bear the *label* of an *approved* testing agency indicating compliance with the standards in this section. *Fire dampers* shall comply with the requirements of UL 555. Only *fire dampers* labeled for use in dynamic systems shall be installed in heating, ventilation and air-conditioning systems designed to operate with fans on during a fire. *Smoke dampers* shall comply with the requirements of UL 555S. *Combination fire/smoke dampers* shall comply with the requirements of both UL 555 and UL 555S. *Ceiling radiation dampers* shall comply with the requirements of UL 555C.

❖ The purpose of this section is to provide the physical testing requirements for fire dampers, smoke dampers, combination fire/smoke dampers and ceiling radiation dampers. A fire damper is a device designed to close automatically upon detection of heat, to interrupt migratory airflow and to restrict the passage of flame through duct penetrations of rated assemblies. Fire dampers are required to be tested in accordance with UL 555. The criteria for acceptance of fire dampers require that they close and latch automatically, limit movement (warping) to prescribed limitations and remain in the opening for the fire exposure period for which they are rated. In addition to being tested, fire dampers must be labeled and, in accordance with Section 1705.3.3, require a follow-up service.

Fire dampers are classified by UL 555 for use in static and dynamic airflow conditions. Fire dampers installed in air distribution systems that remain in operation after smoke or heat from a fire is detected (a dynamic airflow condition) must be labeled for such use. Static fire dampers may not operate properly under dynamic conditions; therefore, fire dampers used in systems designed with dynamic airflow must be tested and labeled for closure under the anticipated airflow and pressure conditions.

The manufacturer's installation instructions must be followed. Fire dampers are to be installed in the assembly such that an opening created by a duct collapse is properly protected. Dampers are to be installed in accordance with their tested application, either vertically or horizontally. Although the test criteria of UL 555 are not a function of a horizontal versus vertical application, there are specific tests required for spring-operated dampers that are not required for gravity-operated types; therefore, the type of damper used requires the building official to determine if it is acceptable for the given application.

Fire dampers are required to have the minimum fire protection rating established in Table 716.3.2.1. A smoke damper is a device installed in a duct, typically in smoke control systems or to protect smoke barrier penetrations, which operates to seal the duct against smoke leakage. Fire dampers tested to UL 555 are not rated for smoke leakage. Fire dampers are activated using a fusible link or other heat-responsive device that activates the damper at a predetermined temperature. Smoke dampers, on the other hand, are tested for leakage in accordance with the requirements of UL 555S and are activated by a smoke detector in the duct or area to be served.

Ceiling radiation dampers, formerly know as ceiling dampers, are designed to limit the radiative heat transfer through an air inlet/outlet opening in the ceiling membrane of a fire-resistance-rated floor/ceiling or roof/ceiling assembly. Ceiling radiation dampers are evaluated either in accordance with UL 555C or as part of the floor/ceiling or roof/ceiling assembly tested to ASTM E 119 or UL 263. Ceiling radiation dampers evaluated in accordance with UL 555C are investigated for use in lieu of hinged-door-type dampers commonly specified in the listing of fire-resistance-rated floor/ceiling or roof/ceiling assemblies. UL 555C also provides criteria for the construction of the ceiling dampers. Ceiling radiation dampers investigated as part of the fire-resistance-rated floor/ceiling or roof/ceiling assembly are specified in the listing of

the floor/ceiling or roof/ceiling assembly. The description of the tested assembly will include a description of the ceiling damper and its installation. A fire damper is a device designed to close automatically upon detection of heat, to interrupt migratory airflow and to restrict the passage of flame through duct penetrations of rated assemblies. Since ceiling and fire dampers have different design criteria, a fire damper must not be used in place of a ceiling damper.

716.3.2 Damper rating. *Damper* ratings shall be in accordance with Sections 716.3.2.1 through 716.3.2.3.

❖ The purpose of this section is to provide direction to the requirements for the ratings of fire dampers (see Section 716.3.2.1), smoke dampers (see Section 716.3.2.2) and combination fire/smoke dampers (see Section 716.3.2.3).

716.3.2.1 Fire damper ratings. *Fire dampers* shall have the minimum *fire protection rating* specified in Table 716.3.2.1 for the type of penetration.

❖ This section primarily serves the purpose of requiring fire damper ratings to be determined in accordance with Table 716.3.2.1, which 716.3.2.1 sets the minimum damper ratings based upon the type of assemblies being penetrated.

TABLE 716.3.2.1
FIRE DAMPER RATING

TYPE OF PENETRATION	MINIMUM DAMPER RATING (hours)
Less than 3-hour fire-resistance-rated assemblies	1.5
3-hour or greater fire-resistance-rated assemblies	3

❖ The table summarizes the required hourly ratings for fire dampers based on the fire-resistance-rated assembly that is being penetrated by the air distribution system. The left-hand column lists the associated hourly ratings of either wall or horizontal assemblies. The code user enters the line item in the left-hand column based on the applicable fire-resistance rating of the element penetrated and reads across to the right column for the appropriate fire damper rating. These fire damper ratings are established based on tests conducted in accordance with UL 555 and represent the ratings necessary to maintain the integrity of the rated wall or horizontal assembly.

716.3.2.2 Smoke damper ratings. *Smoke damper* leakage ratings shall not be less than Class II. Elevated temperature ratings shall not be less than 250°F (121°C).

❖ In accordance with UL 555S, smoke dampers are marked with a temperature rating starting at 250°F (121°C) and rising in increments of 100°F (38°C). Leakage ratings range from Classes I to III in accordance with the standard and are further identified for tested pressure ranging from 4 to 12 inches (102 to 305 mm) of water. Classes I and II leakage ratings meet the code requirement for the damper to be not less than Class II.

716.3.2.3 Combination fire/smoke damper ratings. *Combination fire/smoke dampers* shall have the minimum *fire protection rating* specified for *fire dampers* in Table 716.3.2.1 for the type of penetration and shall also have a minimum Class II leakage rating and a minimum elevated temperature rating of 250°F (121°C).

❖ This section primarily serves the purpose of requiring the ratings of combination fire/smoke dampers to comply with the requirements for both fire and smoke dampers, including the fire protection rating specified in Table 716.3.2.1 and the leakage rating specified in Section 716.3.2.2.

716.3.3 Damper actuation. *Damper* actuation shall be in accordance with Sections 716.3.3.1 through 716.3.3.4 as applicable.

❖ The purpose of this section is to provide direction to the requirements for the actuation of fire dampers (see Section 716.3.3.1), smoke dampers (see Section 716.3.3.2), combination fire/smoke dampers (see Section 716.3.3.3) and ceiling radiation dampers (see Section 716.3.3.4).

716.3.3.1 Fire damper actuation device. The *fire damper* actuation device shall meet one of the following requirements:

1. The operating temperature shall be approximately 50°F (10°C) above the normal temperature within the duct system, but not less than 160°F (71°C).

2. The operating temperature shall be not more than 350°F (177°C) where located in a smoke control system complying with Section 909.

❖ The purpose of this section is to prescribe the actuation (operating temperature) requirements for fire dampers. The thresholds for actuation are based on UL 555. For static systems, UL 555 identifies 160°F (71°C) as the minimum and 215°F (102°C) as the maximum temperature (see Item 1). For dynamic systems, the standard includes a minimum value of 160°F (71°C) and a maximum value of 350°F (177°C) (see Item 2).

716.3.3.2 Smoke damper actuation. The *smoke damper* shall close upon actuation of a *listed* smoke detector or detectors installed in accordance with Section 907.3 and one of the following methods, as applicable:

1. Where a *smoke damper* is installed within a duct, a smoke detector shall be installed in the duct within 5 feet (1524 mm) of the *damper* with no air outlets or inlets between the detector and the *damper*. The detector shall be *listed* for the air velocity, temperature and humidity anticipated at the point where it is installed. Other than in mechanical smoke control systems, *dampers* shall be closed upon fan shutdown where local smoke detectors require a minimum velocity to operate.

2. Where a *smoke damper* is installed above *smoke barrier* doors in a *smoke barrier*, a spot-type detector *listed* for releasing service shall be installed on either side of the *smoke barrier* door opening.

3. Where a *smoke damper* is installed within an air transfer opening in a wall, a spot-type detector *listed* for releasing service shall be installed within 5 feet (1524 mm) horizontally of the *damper*.

4. Where a *smoke damper* is installed in a *corridor* wall or ceiling, the *damper* shall be permitted to be controlled by a smoke detection system installed in the *corridor*.

5. Where a total-coverage smoke detector system is provided within areas served by a heating, ventilation and air-conditioning (HVAC) system, *smoke dampers* shall be permitted to be controlled by the smoke detection system.

❖ This section provides specific requirements to ensure that an installed detector will detect smoke (see Item 1). Additionally, for passive systems (not part of a smoke control system), the location of detectors is critical to their performance. Smoke dampers are installed in not only duct penetrations, but also transfer openings. A method for properly controlling dampers in such arrangements is provided in Item 3. Where smoke detection is provided throughout a corridor or an entire area, such is more effectively than a duct-type detector. Such detection is necessary for controlling dampers (see Items 4 and 5). Smoke dampers may also be operated remotely, especially when used as part of a smoke control system. Combination fire and smoke dampers can be activated by either heat or smoke sensors. Once the primary heat-sensing device has been activated, a remote control system can be used to open the damper to permit its use as a smoke damper in the smoke removal system. Then, if the temperature rises to the damper's maximum degradation test temperature, the secondary heat sensor closes the damper again.

716.3.3.3 Combination fire/smoke damper actuation. *Combination fire/smoke damper* actuation shall be in accordance with Sections 716.3.3.1 and 716.3.3.2. *Combination fire/smoke dampers* installed in smoke control system shaft penetrations shall not be activated by local area smoke detection unless it is secondary to the smoke management system controls.

❖ This section primarily serves the purpose of requiring the actuation of combination fire/smoke dampers to comply with the requirements for both fire and smoke dampers, including the requirements in Sections 716.3.3.1 and 716.3.3.2. Further, when these dampers are used as part of a smoke control system, this section prohibits their activation by local area smoke detectors because this could render the smoke control system inoperable.

716.3.3.4 Ceiling radiation damper actuation. The operating temperature of a *ceiling radiation damper* actuation device shall be 50°F (27.8°C) above the normal temperature within the duct system, but not less than 160°F (71°C).

❖ The purpose of this section is to prescribe the actuation (operating temperature) requirements for ceiling radiation dampers. The thresholds for actuation are similar to those required for fire dampers used in static systems.

716.4 Access and identification. Fire and smoke *dampers* shall be provided with an *approved* means of access, which is large enough to *permit* inspection and maintenance of the *damper* and its operating parts. The access shall not affect the integrity of fire-resistance-rated assemblies. The access openings shall not reduce the *fire-resistance rating* of the assembly. Access points shall be permanently identified on the exterior by a *label* having letters not less than $^1/_2$ inch (12.7 mm) in height reading: FIRE/SMOKE DAMPER, SMOKE DAMPER or FIRE DAMPER. Access doors in ducts shall be tight fitting and suitable for the required duct construction.

❖ Fire and smoke dampers and combination dampers must be properly maintained so that they will operate as intended. The need to maintain dampers, as well as reset them and replace fusible links after operation, requires that they be accessible as a part of the initial design in the code. The access doors may need to be fire doors in accordance with Section 715, depending on the location of the door, in order to maintain a fire-resistance rating. Access doors in the duct itself need not have a fire protection rating.

716.5 Where required. *Fire dampers, smoke dampers* and *combination fire/smoke dampers* shall be provided at the locations prescribed in Sections 716.5.1 through 716.5.7 and

716.6. Where an assembly is required to have both *fire dampers* and *smoke dampers, combination fire/smoke dampers* or a *fire damper* and a *smoke damper* shall be required.

❖ In order to maintain the fire-resistance rating of fire and smoke vertical walls, barriers and partitions, code-required dampers must be provided for all duct penetrations. Fire dampers are to be installed in the assembly such that an opening created by a duct collapse is properly protected. The manufacturer's installation instructions, which should indicate how the damper is to be installed, must be followed so that the operation and functioning of the installed assembly will be consistent with that required by the code.

Figure 716.5 shows the difference between where a fire damper is required and where one is not. In this example, the designer has chosen to separate mixed occupancies in accordance with the option of Section 508.3.3. The occupancy separation wall (fire barrier) that is required to be fire-resistance rated is also required to have the duct penetration protected by a fire damper, while the nonrated wall requires no such protection. Occasionally the visible gap between the outside of the dampered duct and the inside surface of the hole in a wall or floor that accommodates the duct is closed with a sealant. Although not required by code, this is occasionally done to provide more complete smoke resistance for the wall assembly or floor assembly. The installation of the damper assembly in accordance with the listing and manufacturer's instructions will provide the required fire resistance, even without sealing any apparent gaps between the duct and the hole. It is imperative that an intumescent firestop sealant not be used on the outside of the duct to seal this gap. The expansion pressure of the intu-

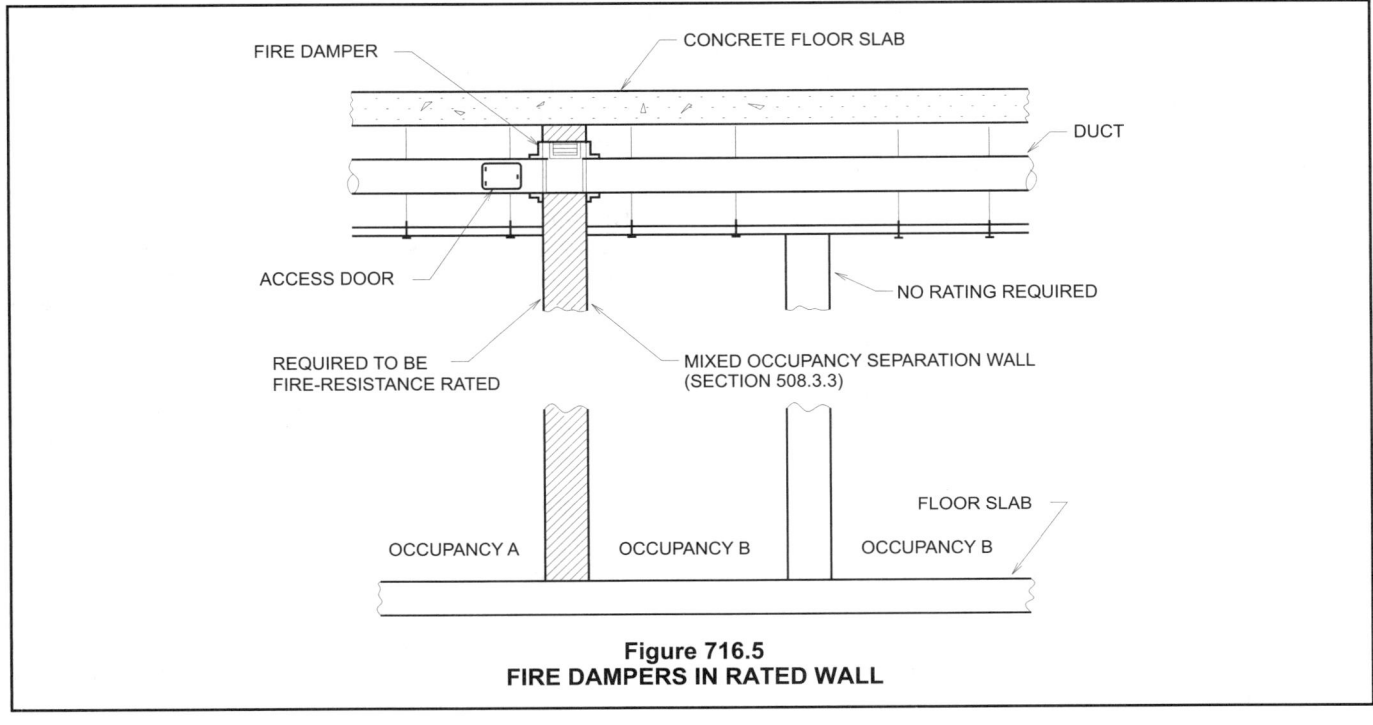

Figure 716.5
FIRE DAMPERS IN RATED WALL

mescent sealant can buckle the duct, thus hindering or preventing the proper operation of the damper under fire conditions.

The code requires that a fire damper not be used as a smoke damper unless the location of the installation is appropriate for the dual purpose. The most likely location for this to occur would be where the air distribution system also serves as a smoke control system and a duct penetrates a fire-resistance-rated assembly. Section 716.2.1 requires alternative protection to be provided where fire dampers will interfere with the operation of a smoke control system. A combination fire and smoke damper, such as that described above, may be approved for use as alternative protection.

716.5.1 Fire walls. Ducts and air transfer openings permitted in *fire walls* in accordance with Section 706.11 shall be protected with *listed fire dampers* installed in accordance with their listing.

❖ Fire walls can create separate buildings within a structure. These multiple buildings may be on a single lot or, in the case of zero lot line construction, the fire wall may be located on the lot line. In the case where the fire wall separates different buildings on different lots, the code does not permit openings (see Sections 705.1.1 and 705.11). In such instances, a duct penetration is not permitted. Where the fire wall is not located on a lot line, the wall is permitted to be penetrated by a duct or transfer opening, provided the opening is protected with a listed fire damper.

716.5.1.1 Horizontal exits. A *listed smoke damper* designed to resist the passage of smoke shall be provided at each point a duct or air transfer opening penetrates a *fire wall* that serves as a horizontal *exit*.

❖ The purpose of this section is to recognize that horizontal exits are required to resist the passage of smoke, as well as fire. This section, therefore, requires that smoke dampers be installed at ducts or transfer openings that penetrate a fire wall.

716.5.2 Fire barriers. Ducts and air transfer openings of *fire barriers* shall be protected with *approved fire dampers* installed in accordance with their listing. Ducts and air transfer openings shall not penetrate *exit* enclosures and *exit* passageways except as permitted by Sections 1022.4 and 1023.6, respectively.

Exception: *Fire dampers* are not required at penetrations of *fire barriers* where any of the following apply:

1. Penetrations are tested in accordance with ASTM E 119 or UL 263 as part of the fire-resistance-rated assembly.

2. Ducts are used as part of an *approved* smoke control system in accordance with Section 909 and where the use of a *fire damper* would interfere with the operation of a smoke control system.

3. Such walls are penetrated by ducted HVAC systems, have a required *fire-resistance rating* of 1 hour or less, are in areas of other than Group H and are in buildings equipped throughout with an *automatic*

sprinkler system in accordance with Section 903.3.1.1 or 903.3.1.2. For the purposes of this exception, a ducted HVAC system shall be a duct system for conveying supply, return or exhaust air as part of the structure's HVAC system. Such a duct system shall be constructed of sheet steel not less than No. 26 gage thickness and shall be continuous from the air-handling appliance or equipment to the air outlet and inlet terminals.

❖ This section includes three exceptions to the general requirement on how open duct and air transfer openings in rated construction are to be protected. It is important that the fire dampers installed when a duct or air transfer opening penetrates a fire barrier are listed for the fire rating of the barrier wall. These provisions emphasize that ducts and air transfer openings not penetrate exit enclosures and exit passageways unless permitted by Sections 1022 and 1023. Penetration of an exit has the potential to allow smoke or flame to be introduced into the enclosure from other parts of the building, effectively blocking the exit path. Exception 1 states that fire dampers are not required for duct penetrations of fire barriers when the assembly has been tested in accordance with ASTM E 119 or UL 263 without fire dampers and the required fire-resistance rating is obtained. In this case, the penetration assembly has been demonstrated as preserving the fire-resistance rating of the wall assembly.

Exception 2 reinforces the provisions of Section 716.2.1 (see commentary). A fire damper may interfere with the operation of the smoke control system; however, some form of alternative protection must be provided due to the duct penetrating the fire barrier.

Exception 3 states that, in fully sprinklered buildings containing occupancies other than Group H, fire dampers may be omitted in duct penetrations of walls having a fire-resistance rating of 1 hour or less.

716.5.2.1 Horizontal exits. A *listed smoke damper* designed to resist the passage of smoke shall be provided at each point a duct or air transfer opening penetrates a *fire barrier* that serves as a horizontal *exit*.

❖ The purpose of this section is to recognize that horizontal exits are required to resist the passage of smoke, as well as fire. This section, therefore, requires that smoke dampers be installed at ducts or transfer openings that penetrate a fire barrier.

716.5.3 Shaft enclosures. Shaft enclosures that are permitted to be penetrated by ducts and air transfer openings shall be protected with *approved* fire and smoke *dampers* installed in accordance with their listing.

Exceptions:

1. *Fire dampers* are not required at penetrations of shafts where:

1.1. Steel exhaust subducts are extended at least 22 inches (559 mm) vertically in exhaust shafts, provided there is a continuous airflow upward to the outside; or

1.2. Penetrations are tested in accordance with ASTM E 119 or UL 263 as part of the fire-resistance-rated assembly; or

1.3. Ducts are used as part of an *approved* smoke control system designed and installed in accordance with Section 909 and where the *fire damper* will interfere with the operation of the smoke control system; or

1.4. The penetrations are in parking garage exhaust or supply shafts that are separated from other building shafts by not less than 2-hour fire-resistance-rated construction.

2. In Group B and R occupancies equipped throughout with an *automatic sprinkler system* in accordance with Section 903.3.1.1, *smoke dampers* are not required at penetrations of shafts where:

2.1. Kitchen, clothes dryer, bathroom and toilet room exhaust openings are installed with steel exhaust subducts, having a minimum wall thickness of 0.187-inch (0.4712 mm) (No. 26 gage);

2.2. The subducts extend at least 22 inches (559 mm) vertically; and

2.3. An exhaust fan is installed at the upper terminus of the shaft that is powered continuously in accordance with the provisions of Section 909.11, so as to maintain a continuous upward airflow to the outside.

3. *Smoke dampers* are not required at penetration of exhaust or supply shafts in parking garages that are separated from other building shafts by not less than 2-hour fire-resistance-rated construction.

4. *Smoke dampers* are not required at penetrations of shafts where ducts are used as part of an *approved* mechanical smoke control system designed in accordance with Section 909 and where the *smoke damper* will interfere with the operation of the smoke control system.

5. *Fire dampers* and *combination fire/smoke dampers* are not required in kitchen and clothes dryer exhaust systems when installed in accordance with the *International Mechanical Code*.

❖ This section requires both fire and smoke dampers at duct and air transfer openings in the shaft wall. The fire damper is required due to the penetration of a fire-resistance-rated wall. The smoke damper is required in order to limit the migration of smoke to other parts of the building via the shaft and the chimney effect (see commentary, Section 708.14.1). The exceptions allow the omission of fire dampers and smoke dampers under certain conditions. Exception 1 addresses fire dampers, and Exceptions 2, 3 and 4 address smoke dampers. Exception 4 addresses the interface with smoke control systems as designed in accordance

with Section 909 and coordinates with Section 716.2.1.

Exception 1.1 recognizes that the presence of a vertical subduct in a shaft will also offer some degree of fire resistance should the duct outside of the shaft collapse. Steel exhaust ducts that consist of a 22-inch (559 mm) vertical upturn in the shaft need not be protected with a fire damper if there is a continuous upward airflow to the outside. The continuous airflow will create a negative pressure in the shaft as compared to adjacent spaces, thereby minimizing the spread of hot gases from the shaft [see Figure 716.5.3.1(1)].

Exceptions 1.2 and 1.3 are identical to Exceptions 1 and 2 to Section 716.5.2 (see commentary, Section 716.5.2).

Exception 1.4 states that fire dampers may also be omitted in exhaust and supply shafts that serve a garage and are separated from all other shafts in the building by a 2-hour fire separation assembly [see Figure 716.5.3.1(2)]. Requiring fire dampers in garage exhaust and supply shafts would not significantly prevent the spread of smoke and fire within the garage, since the vehicle ramp from floor to floor is a much greater conduit of smoke and fire in a garage.

Lastly, Exception 5 exempts fire dampers and combination fire/smoke dampers from kitchen and clothes dryer exhaust systems that are installed in accordance with the IMC. This is to recognize the hazards associated with obstructing the airflow for these installations.

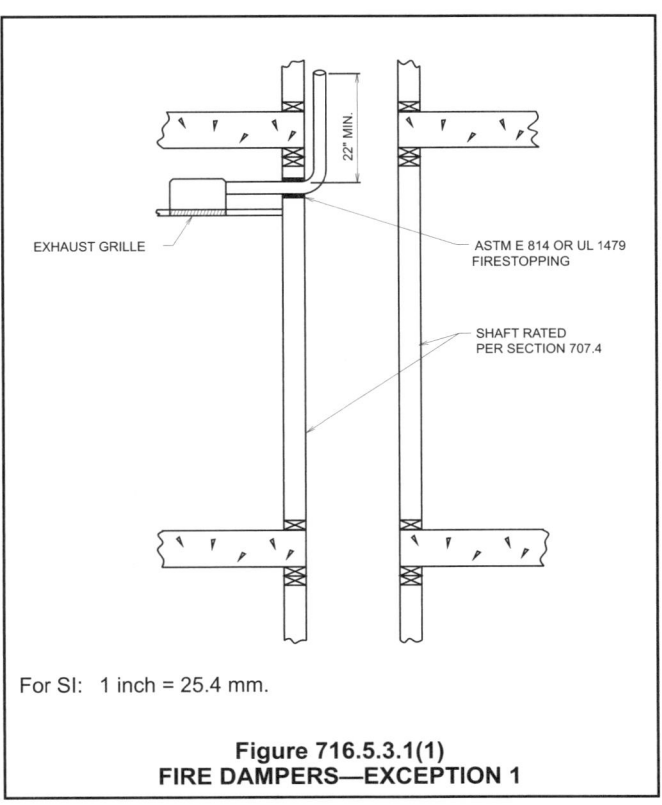

For SI: 1 inch = 25.4 mm.

Figure 716.5.3.1(1)
FIRE DAMPERS—EXCEPTION 1

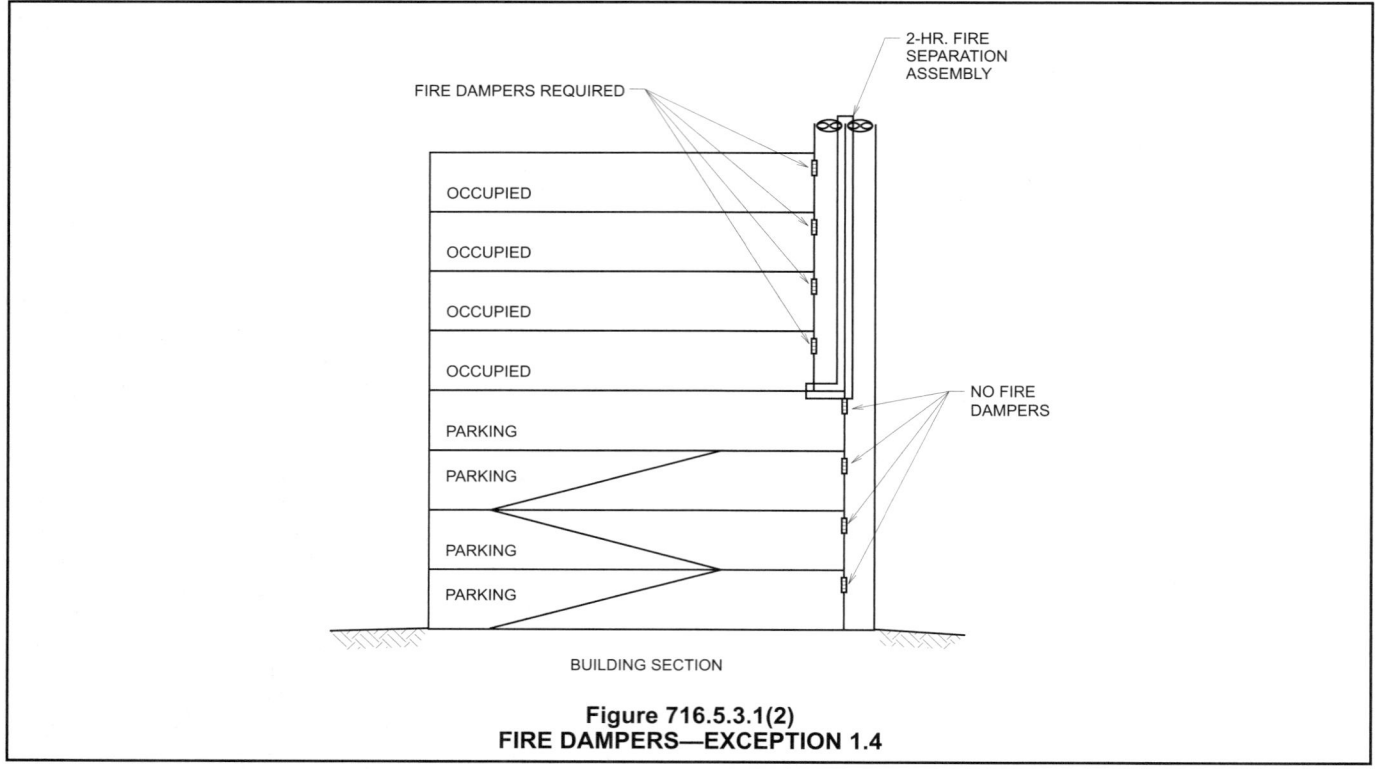

Figure 716.5.3.1(2)
FIRE DAMPERS—EXCEPTION 1.4

716.5.4 Fire partitions. Ducts and air transfer openings that penetrate *fire partitions* shall be protected with *listed fire dampers* installed in accordance with their listing.

Exceptions: In occupancies other than Group H, *fire dampers* are not required where any of the following apply:

1. Corridor walls in buildings equipped throughout with an *automatic sprinkler system* in accordance with Section 903.3.1.1 or 903.3.1.2 and the duct is protected as a *through penetration* in accordance with Section 713.

2. Tenant partitions in *covered mall buildings* where the walls are not required by provisions elsewhere in the code to extend to the underside of the floor or roof sheathing, slab or deck above.

3. The duct system is constructed of *approved* materials in accordance with the *International Mechanical Code* and the duct penetrating the wall complies with all of the following requirements:

    3.1. The duct shall not exceed 100 square inches (0.06 m²).

    3.2. The duct shall be constructed of steel a minimum of 0.0217 inch (0.55 mm) in thickness.

    3.3. The duct shall not have openings that communicate the *corridor* with adjacent spaces or rooms.

    3.4. The duct shall be installed above a ceiling.

    3.5. The duct shall not terminate at a wall register in the fire-resistance-rated wall.

    3.6. A minimum 12-inch-long (305 mm) by 0.060-inch-thick (1.52 mm) steel sleeve shall be centered in each duct opening. The sleeve shall be secured to both sides of the wall and all four sides of the sleeve with minimum 1¹/₂-inch by 1¹/₂-inch by 0.060-inch (38 mm by 38 mm by 1.52 mm) steel retaining angles. The retaining angles shall be secured to the sleeve and the wall with No. 10 (M5) screws. The *annular space* between the steel sleeve and the wall opening shall be filled with mineral wool batting on all sides.

❖ Fire dampers installed when a duct or air transfer opening penetrates a fire partition or corridor wall must be listed for the fire rating of the partition and installed in accordance with the listing. This section includes three exceptions to the general requirement that duct and air transfer openings in rated construction be protected.

Exceptions 1 and 3 specifically address circumstances wherein fire dampers are not required in ducts penetrating fire partitions, in areas other than Group H.

Exception 2 coordinates with Section 709.4, Exception 4, and addresses standard construction designs for mall buildings as they relate to the openness of ceiling spaces. The exception removes an implied requirement for dampers where logic would dictate none are required. This provision recognizes that walls separating tenants in a covered mall are required to be rated but are allowed to stop at unrated ceilings and unrated storefronts, providing no true separation. Therefore, it does not make sense to require a damper through the wall. If the tenant separation wall is required to be continuous to a roof deck or rated floor assembly for another purpose, the fire dampers would then be required.

Exception 1 states that dampers are not required to protect penetrations of tenant separation or corridor walls in buildings protected throughout with an approved automatic sprinkler system and protected as a through penetration [see Figure 716.5.4(1)]. The reference to Sections 903.3.1.1 and 903.3.1.2 establishes that the exception only applies to buildings equipped throughout with an automatic sprinkler system designed and installed in accordance with either NFPA 13 or 13R and does not apply to buildings with an NFPA 13D system.

Exception 3 states that fire dampers are not required where a steel duct penetrates a wall but does not have openings serving adjacent rooms or spaces [see Figure 716.5.4(2)]. This exception has several criteria, one of which is to utilize a sleeve with mineral wool batting on all sides. Steel ducts have been shown to have the ability to remain in place during severe fire exposures. With no openings into the corridor to allow for smoke and heat transfer in the protected corridor enclosure, a fire damper is not necessary to attain the required fire performance of the penetration.

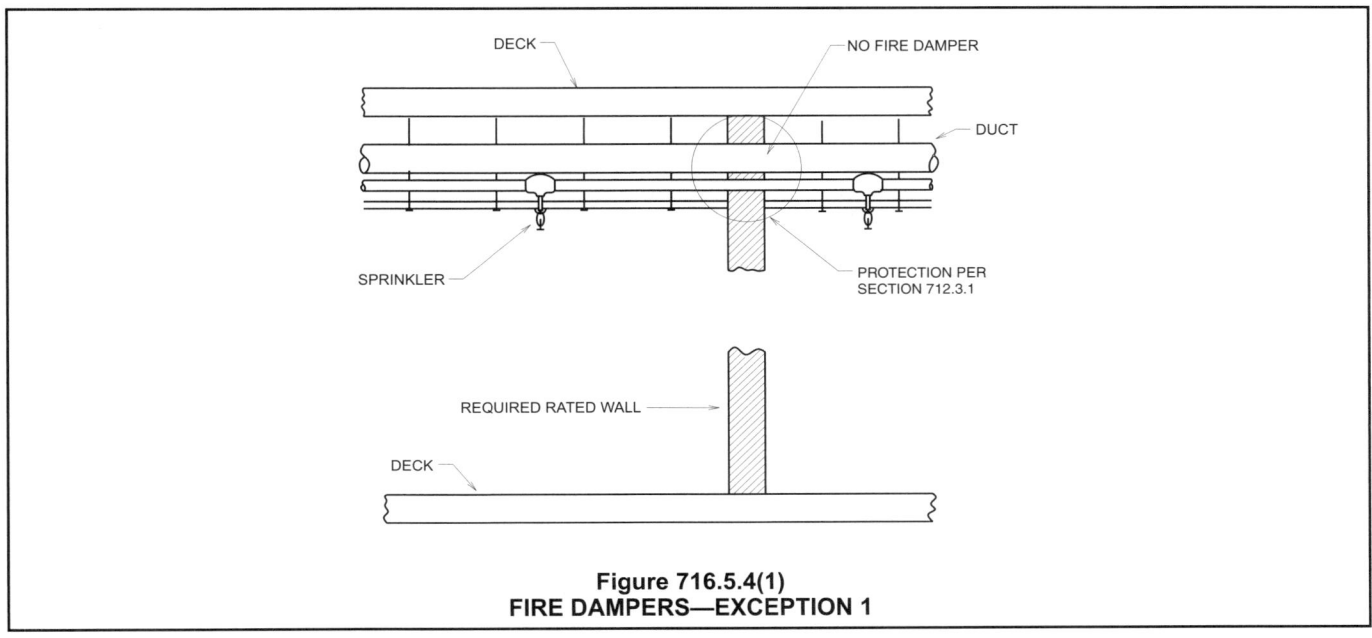

Figure 716.5.4(1)
FIRE DAMPERS—EXCEPTION 1

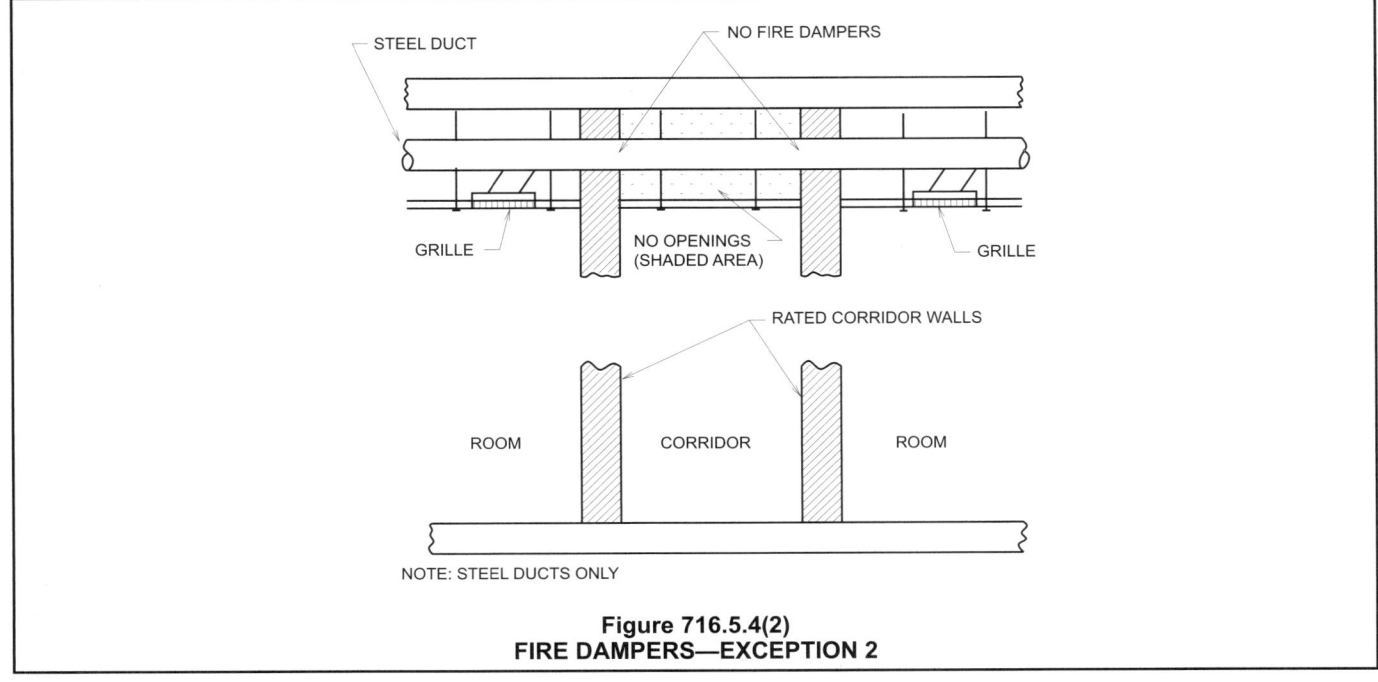

Figure 716.5.4(2)
FIRE DAMPERS—EXCEPTION 2

716.5.4.1 Corridors. A *listed smoke damper* designed to resist the passage of smoke shall be provided at each point a duct or air transfer opening penetrates a *corridor* enclosure required to have smoke and draft control doors in accordance with Section 715.4.3.

Exceptions:

1. *Smoke dampers* are not required where the building is equipped throughout with an *approved* smoke control system in accordance with Section 909, and *smoke dampers* are not necessary for the operation and control of the system.

2. *Smoke dampers* are not required in *corridor* penetrations where the duct is constructed of steel not less than 0.019 inch (0.48 mm) in thickness and there are no openings serving the *corridor*.

❖ This section is coordinated with the provisions of Section 715.4.3 regarding opening protection for rated corridor walls. Similar to Section 715.4.3, which requires that the fire doors also be tested for smoke and draft control, this section requires that duct and air transfer openings be protected with smoke dampers.

Exception 1 cites the concern of the damper interfering with the smoke control system (see commentary, Section 716.2.1).

Exception 2 is similar to Exceptions 3.2 and 3.3 of Section 716.5.4 for fire dampers. Also see the commentary to Section 716.5.5, which addresses smoke dampers in smoke barrier walls.

716.5.5 Smoke barriers. A *listed smoke damper* designed to resist the passage of smoke shall be provided at each point a duct or air transfer opening penetrates a *smoke barrier*. *Smoke dampers* and *smoke damper* actuation methods shall comply with Section 716.3.3.2.

Exception: *Smoke dampers* are not required where the openings in ducts are limited to a single smoke compartment and the ducts are constructed of steel.

❖ Smoke barriers are required for both Group I-2 (see Section 407.4) and I-3 (see Section 408.6) occupancies. The concept of required smoke barriers is to limit the migration of smoke. In fact, Section 710.4 requires that smoke barriers form an effective continuous membrane from outside wall to outside wall and from floor slab to floor or roof deck above. This would include continuity through concealed spaces, such as those found above suspended ceilings and interstitial structural and mechanical spaces.

To prevent smoke migration across the smoke barrier where duct penetrations occur, an approved damper designed to resist the passage of smoke is to be provided at each point where a duct penetrates a smoke barrier. The damper closes upon detection of smoke by an approved smoke detector within the duct. If the duct penetration is above a smoke barrier door assembly, smoke detectors for adjacent doors may also be used to operate the damper and the duct detector may be eliminated. A reference to Section 716.3.3.2 is provided for activation mechanisms.

Additionally, smoke dampers are not required in a fully ducted system that does not contain any vents or openings that would spread smoke across the smoke barrier; in smoke control systems or in duct penetrations of smoke barriers where all supply and return diffusers and registers in the duct are contained in a single smoke compartment.

The designer must evaluate the specific conditions of the installation to determine which class of damper should be installed. If a fire damper is required (i.e., the smoke barrier is also a mixed-occupancy separation or corridor wall), a combination fire and smoke damper or separate fire and smoke dampers may be installed.

716.5.6 Exterior walls. Ducts and air transfer openings in fire-resistance-rated *exterior walls* required to have protected openings in accordance with Section 705.10 shall be protected with *listed fire dampers* installed in accordance with their listing.

❖ The purpose of this section is to recognize that under certain conditions exterior walls are required fire-resistance rated. This section by reference to be Section 705.10 requires that fire dampers be installed at ducts or transfer openings that penetrate an exterior wall that is required to be fire-resistance rated.

716.5.7 Smoke partitions. A *listed smoke damper* designed to resist the passage of smoke shall be provided at each point that an air transfer opening penetrates a smoke partition. *Smoke dampers* and *smoke damper* actuation methods shall comply with Section 716.3.3.2.

Exception: Where the installation of a *smoke damper* will interfere with the operation of a required smoke control system in accordance with Section 909, *approved* alternative protection shall be utilized.

❖ The purpose of this section is to recognize that penetrations of smoke partitions by air transfer openings need to resist the transfer of smoke and, therefore, are required to be provided with a smoke damper. This is also consistent with the requirements in Section 711.7. This section also requires the actuation of the damper to comply with Section 716.3.3.2. The exception cites the concern of the damper interfering with the smoke control system (see commentary, Section 716.2.1).

716.6 Horizontal assemblies. Penetrations by ducts and air transfer openings of a floor, floor/ceiling assembly or the ceiling membrane of a roof/ceiling assembly shall be protected by a shaft enclosure that complies with Section 708 or shall comply with Sections 716.6.1 through 716.6.3.

❖ In general, all floor openings that connect two or more stories are to be enclosed in shafts that are constructed in accordance with Section 708. Exception 4 to Section 708.2 references Section 716.6 for duct openings (penetration). Sections 716.6.1 through 716.6.3 identify conditions where a shaft is not required and either a fire damper (see Sections 716.6.1 and 716.6.3) or ceiling radiation damper (see Section 716.6.2) is required. There is an exception within Section 716.6.1 that allows the elimination of the fire damper. However, ceiling radiation dampers would be required in some cases (see Section 716.6.2.1).

716.6.1 Through penetrations. In occupancies other than Groups I-2 and I-3, a duct constructed of *approved* materials in accordance with the *International Mechanical Code* that penetrates a fire-resistance-rated floor/ceiling assembly that connects not more than two *stories* is permitted without shaft enclosure protection, provided a *listed fire damper* is installed at the floor line or the duct is protected in accordance with Section 713.4. For air transfer openings, see Exception 7 to Section 708.2.

Exception: A duct is permitted to penetrate three floors or less without a *fire damper* at each floor, provided such duct meets all of the following requirements:

1. The duct shall be contained and located within the cavity of a wall and shall be constructed of steel having a minimum wall thickness of 0.187 inches (0.4712 mm) (No. 26 gage).

2. The duct shall open into only one *dwelling or sleeping unit* and the duct system shall be continuous from the unit to the exterior of the building.

3. The duct shall not exceed 4-inch (102 mm) nominal diameter and the total area of such ducts shall not exceed 100 square inches (0.065 m²) in any 100 square feet (9.3 m²) of floor area.

4. The annular space around the duct is protected with materials that prevent the passage of flame and hot gases sufficient to ignite cotton waste where subjected to ASTM E 119 or UL 263 time-temperature conditions under a minimum positive pressure differential of 0.01 inch (2.49 Pa) of water at the location of the penetration for the time period equivalent to the *fire-resistance rating* of the construction penetrated.

5. Grille openings located in a ceiling of a fire-resistance-rated floor/ceiling or roof/ceiling assembly shall be protected with a *listed ceiling radiation damper* installed in accordance with Section 716.6.2.1.

❖ Penetrations of a fire-resistance-rated horizontal assembly by air ducts are permitted where no more than two stories are connected (i.e., the duct penetrates a single floor) and a fire damper is provided at the floor line. The fire damper must be tested in accordance with UL 555 [see Figure 716.6.1(1)]. A roof assembly penetrated by a duct that is open to the atmosphere is not required to have a fire damper installed at the penetrations because it is considered acceptable for the fire to vent to the outside atmosphere and heat or fire extension will do no harm [see Figure 716.6.1(2)]. However, a ceiling radiation damper would still be needed to protect a roof/ceiling assembly in accordance with Section 716.6.2, Item 2. As noted, there is an exception to this section that allows the elimination of fire dampers at each floor when the ducts extend through a maximum of three floors. There are several criteria that must be followed to allow this, including:

- Minimum duct thickness (increases time for heat to penetrate and provides more durability) and enclosure within wall cavity.

- Limited to a single sleeping unit or dwelling unit (to reduce the likelihood that one dwelling unit or sleeping unit will be able to endanger another).

- Size of the duct opening (smaller opening means smaller amounts of fire effluents).

- Annular protection (reduces the chance of duct leakage from penetrated walls/barriers).

- Ceiling radiation dampers at grilles in fire-resistant ceilings (maintains ceiling fire resistance).

All criteria center on limiting the impact of hot gases being present and spreading from one area to another. As noted, Exception 7 of Section 708.2 addresses air transfer openings.

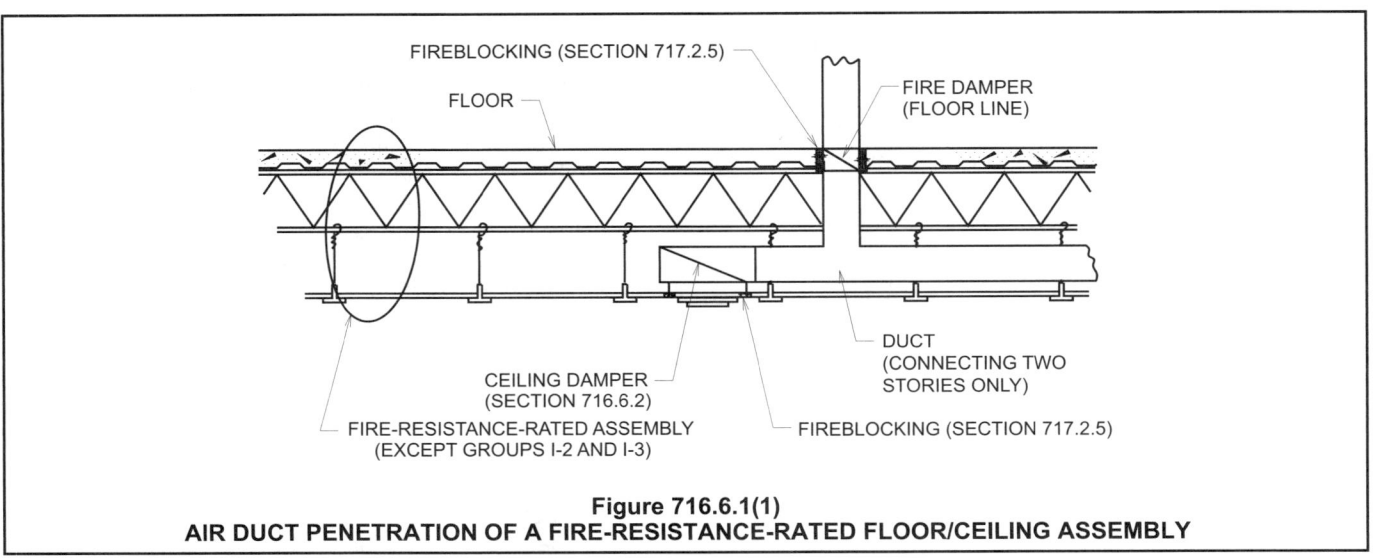

Figure 716.6.1(1)
AIR DUCT PENETRATION OF A FIRE-RESISTANCE-RATED FLOOR/CEILING ASSEMBLY

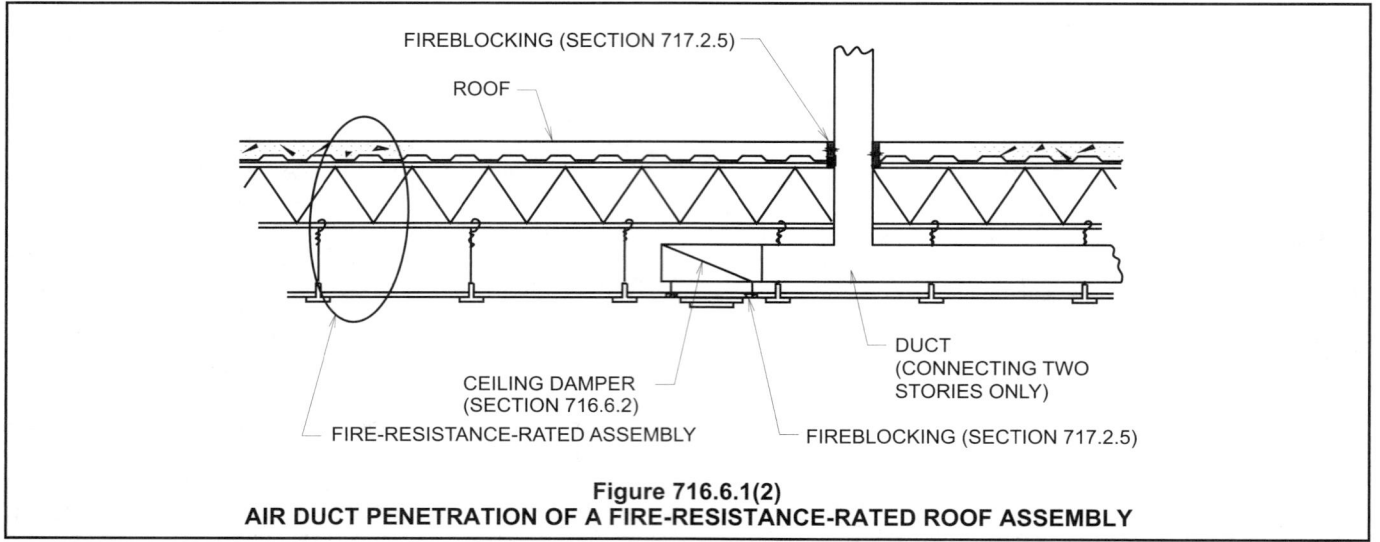

Figure 716.6.1(2)
AIR DUCT PENETRATION OF A FIRE-RESISTANCE-RATED ROOF ASSEMBLY

716.6.2 Membrane penetrations. Ducts and air transfer openings constructed of *approved* materials in accordance with the *International Mechanical Code* that penetrate the ceiling membrane of a fire-resistance-rated floor/ceiling or roof/ceiling assembly shall be protected with one of the following:

1. A shaft enclosure in accordance with Section 708.

2. A *listed ceiling radiation damper* installed at the ceiling line where a duct penetrates the ceiling of a fire-resistance-rated floor/ceiling or roof/ceiling assembly.

3. A *listed ceiling radiation damper* installed at the ceiling line where a diffuser with no duct attached penetrates the ceiling of a fire-resistance-rated floor/ceiling or roof/ceiling assembly.

❖ Unless the duct system is protected with a shaft enclosure in accordance with Sections 708 and 713.4, a ceiling radiation damper, as described in Section 716.6.2.1, is to be installed at the line where the ceiling is both penetrated by a noncombustible air duct and is an integral component of the fire-resistance rating.

The requirement for a ceiling radiation damper applies to ductwork that penetrates a ceiling membrane from above and continues downwards, or where a diffuser with no ductwork attached below penetrates a ceiling membrane. Section 716.6.1 contains the requirements for fire dampers where the duct penetrates through the entire assembly [see Figures 716.6.1(1) and 716.6.1(2) for ceiling radiation damper and fire damper locations].

Ceiling radiation dampers are designed to limit the radiative heat transfer through an air inlet or outlet opening in the ceiling membrane of a fire-resistance-rated floor/ceiling or roof/ceiling assembly.

716.6.2.1 Ceiling radiation dampers. *Ceiling radiation dampers* shall be tested as part of a fire-resistance-rated floor/ceiling or roof/ceiling assembly in accordance with ASTM E 119 or UL 263. *Ceiling radiation dampers* shall be installed in accordance with the details *listed* in the fire-resistance-rated assembly and the manufacturer's installation

instructions and the listing. *Ceiling radiation dampers* are not required where either of the following applies:

1. Tests in accordance with ASTM E 119 or UL 263 have shown that *ceiling radiation dampers* are not necessary in order to maintain the *fire-resistance rating* of the assembly.

2. Where exhaust duct penetrations are protected in accordance with Section 713.4.1.2, are located within the cavity of a wall and do not pass through another *dwelling unit* or tenant space.

❖ Ceiling radiation dampers, formerly known as ceiling dampers, are evaluated either in accordance with UL 555C or as part of the floor/ceiling or roof/ceiling assembly tested to ASTM E 119 or UL 263. Ceiling radiation dampers evaluated in accordance with UL 555C are investigated for use in lieu of hinged-door-type dampers commonly specified in the listing of fire-resistance-rated floor/ceiling or roof/ceiling assemblies. UL 555C also provides criteria for the construction of the ceiling radiation dampers. If the fire-resistance-rated floor/ceiling or roof/ceiling assembly does not incorporate a hinged-door-type damper, a ceiling radiation damper may not be utilized in the assembly. Ceiling radiation dampers investigated as part of the fire-resistance-rated floor/ceiling or roof/ceiling assembly are specified in the listing of the floor/ceiling or roof/ceiling assembly. The description of the tested assembly will include a description of the ceiling radiation damper and its installation. Both types of ceiling radiation dampers shall be installed in accordance with the details listed in the fire-resistance-rated assembly and the manufacturer's installation instructions and listing.

The "Design Information Section" of the UL *Fire Resistance Directory* also includes two duct outlet protection systems. Duct outlet protection system A may be used where specified in the fire-resistance-rated floor/ceiling or roof/ceiling assembly. Duct outlet protection system B may be used in lieu of hinged-door-type dampers commonly specified in the listing of fire-resistance-rated floor/ceiling or roof/ceiling assemblies.

Ceiling radiation dampers are not required to be installed if they comply with Section 713.4.1.2, are limited to a single dwelling unit or tenant spaces and are contained within a wall. Where this criteria is met, Section 713.4.1.2 would allow either assemblies tested to ASTM E 119 or UL 263 or the use of a through-penetration firestop system tested to ASTM E 814 or UL1479 by way of the references to Section 713.4.1.1.1 or 713.4.1.1.2.

716.6.3 Nonfire-resistance-rated floor assemblies. Duct systems constructed of *approved* materials in accordance with the *International Mechanical Code* that penetrate nonfire-resistance-rated floor assemblies shall be protected by any of the following methods:

1. A shaft enclosure in accordance with Section 708.

2. The duct connects not more than two *stories*, and the *annular space* around the penetrating duct is protected with an *approved* noncombustible material that resists the free passage of flame and the products of combustion.

3. The duct connects not more than three *stories*, and the *annular space* around the penetrating duct is protected with an *approved* noncombustible material that resists the free passage of flame and the products of combustion and a *fire damper* is installed at each floor line.

Exception: *Fire dampers* are not required in ducts within individual residential *dwelling units*.

❖ Floor assemblies that are not rated still require isolation in some form based upon the requirements of Section 716.6.3 when duct systems penetrate the assemblies. Three conditions or requirements are listed.

First, the ducts can be part of a shaft enclosure in accordance with Section 708. This coordinates with the requirements for membrane penetrations in Section 716.6.2 and no rated damper is required. The shaft construction required in Section 708.4 would be at least 1 hour.

Second, duct systems that connect not more than two stories (penetrate one floor) are permitted, provided the annular space around the duct is sealed with a noncombustible material that will not allow free passage of flame or smoke. The two-story arrangement does not require a rated damper [see Figure 716.6.3(1)].

Third, when air ducts connect three levels or less and thus penetrate two floors or less, fire dampers must be provided at each floor line. A $1^1/_2$-hour-rated fire damper (see Table 716.3.2.1) is required even though the floor is not required to have a fire-resistance rating. The damper can be viewed as an alternative to the 1-hour shaft enclosure requirement in Section 716.6.3 [see Figure 716.6.3(2)]. Fire dampers, unlike door opening protectives with multiple ratings, are only rated for $1^1/_2$ or 3 hours.

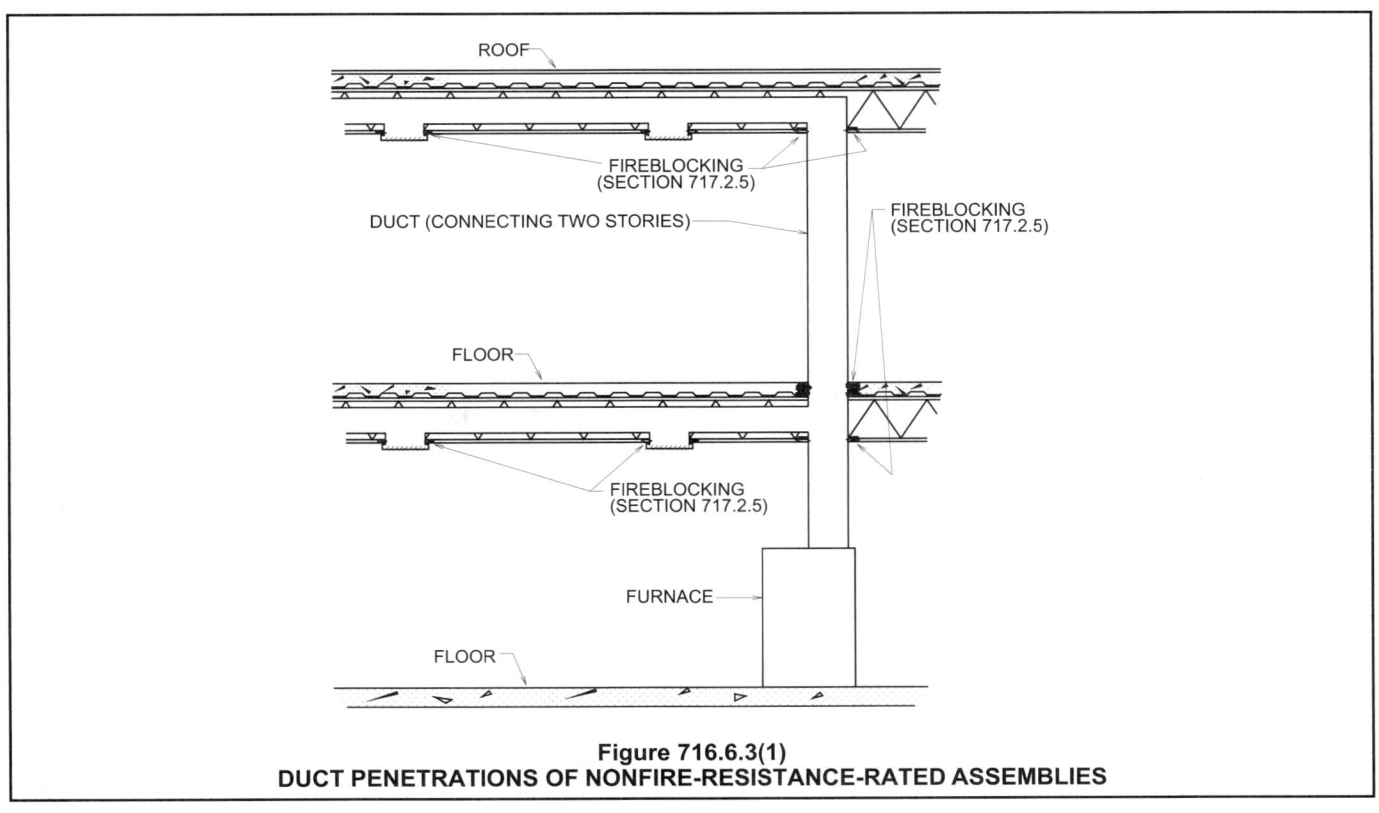

Figure 716.6.3(1)
DUCT PENETRATIONS OF NONFIRE-RESISTANCE-RATED ASSEMBLIES

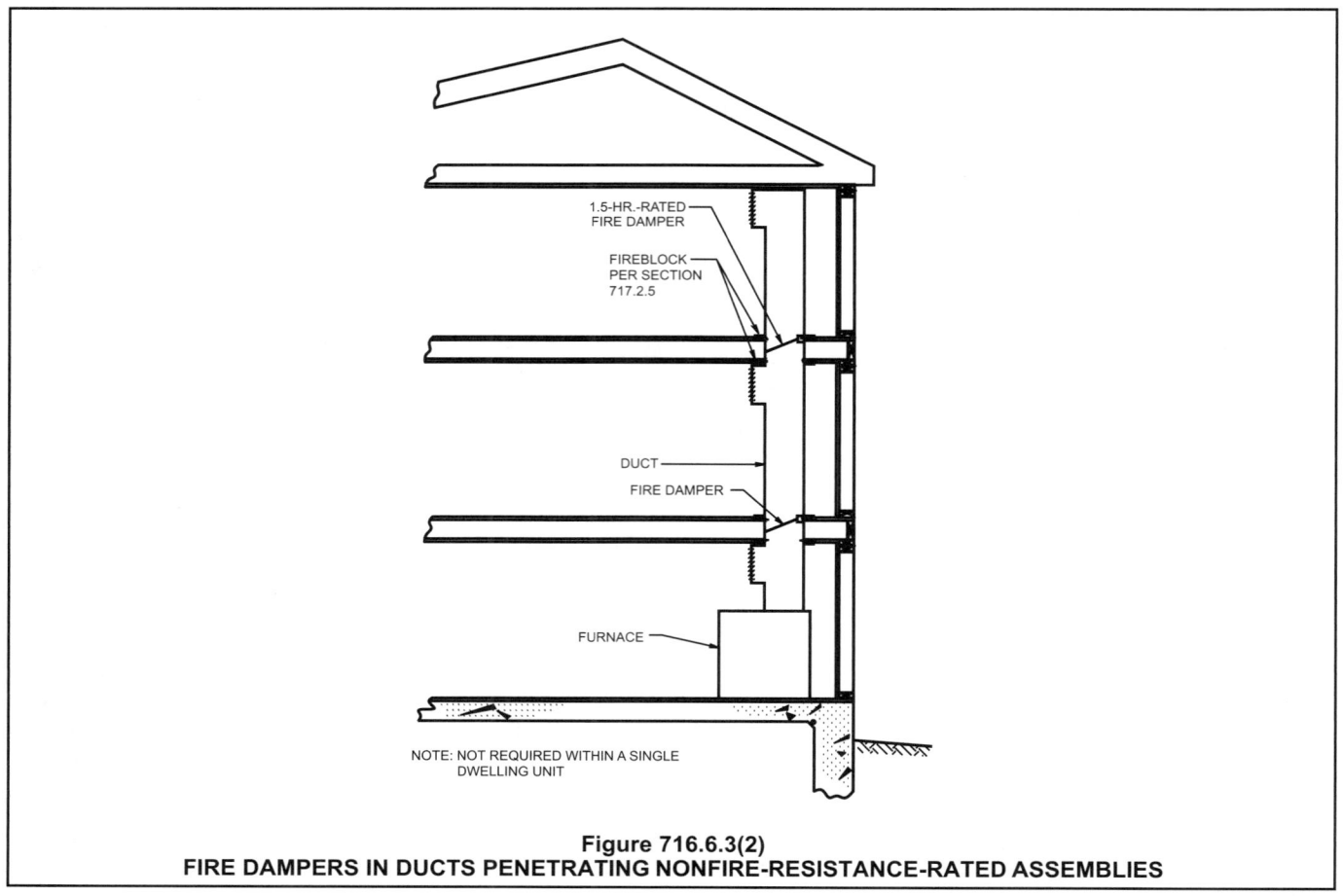

1.5-HR.-RATED
FIRE DAMPER

FIREBLOCK
PER SECTION
717.2.5

DUCT

FIRE DAMPER

FURNACE

NOTE: NOT REQUIRED WITHIN A SINGLE
DWELLING UNIT

Figure 716.6.3(2)
FIRE DAMPERS IN DUCTS PENETRATING NONFIRE-RESISTANCE-RATED ASSEMBLIES

716.7 Flexible ducts and air connectors. Flexible ducts and air connectors shall not pass through any fire-resistance-rated assembly. Flexible air connectors shall not pass through any wall, floor or ceiling.

❖ Flexible air ducts are prohibited from penetrating fire-resistance-rated assemblies. Flexible air connectors are not permitted to pass through any wall, floor or ceiling (fire-resistance rated or not). An inadequate seal at the assembly penetration could allow smoke or flame to penetrate the assembly. Flexible air ducts and connectors can be constructed of both combustible and noncombustible components; therefore, the duct's resistance to the passage of fire could be less than the resistance of the penetrated assembly. All construction assemblies, whether fire-resistance rated or not, contain some inherent resistance to the spread of fire.

SECTION 717
CONCEALED SPACES

717.1 General. Fireblocking and draftstopping shall be installed in combustible concealed locations in accordance with this section. Fireblocking shall comply with Section 717.2. Draftstopping in floor/ceiling spaces and *attic* spaces shall comply with Sections 717.3 and 717.4, respectively. The permitted use of combustible materials in concealed spaces of buildings of Type I or II construction shall be limited to the applications indicated in Section 717.5.

❖ During a fire, flame, smoke and gases will spread via the paths of least resistance. Certain assemblies create void spaces, which will not only affect the spread of fire but, since the voids are concealed, will also make access for suppression more difficult.

The code has established two means by which fire spread within void spaces can be controlled: fireblocking and draftstopping. Fireblocking involves the use of building materials to prevent the movement of flame and gases to other areas through concealed spaces. Draftstopping involves the use of building materials to prevent the movement of air, smoke, gases and flames to other areas through large concealed spaces. For example, the protection in an attic space is draftstopping, while the protection in the cavity of a wall assembly is fireblocking.

Section 717.2.1 identifies the materials that are acceptable for use as fireblocks and Section 717.3.1 specifies the acceptable materials for draftstops. Sections 717.2, 717.3 and 717.4 address where fireblocking and draftstopping are required.

Fireblocks and draftstops are to be installed in combustible concealed spaces to prevent the spread of flame, smoke and gases. If a fire condition exists in a concealed space, the fireblocks and draftstops will help contain the fire until it can be suppressed. This section also includes provisions for permitted combustibles in concealed spaces in buildings of Type I and II construction (see Section 717.5 and Section 603.1, Exception 21).

717.2 Fireblocking. In combustible construction, fireblocking shall be installed to cut off concealed draft openings (both vertical and horizontal) and shall form an effective barrier between floors, between a top *story* and a roof or *attic* space. Fireblocking shall be installed in the locations specified in Sections 717.2.2 through 717.2.7.

❖ This section merely states the performance requirements associated with fireblocking. The intent of fireblocking is to reduce the ability of fire, smoke and gases from moving to different parts of the building through combustible concealed spaces.

717.2.1 Fireblocking materials. Fireblocking shall consist of the following materials:

1. Two-inch (51 mm) nominal lumber.

2. Two thicknesses of 1-inch (25 mm) nominal lumber with broken lap joints.

3. One thickness of 0.719-inch (18.3 mm) wood structural panels with joints backed by 0.719-inch (18.3 mm) wood structural panels.

4. One thickness of 0.75-inch (19.1 mm) particleboard with joints backed by 0.75-inch (19 mm) particleboard.

5. One-half-inch (12.7 mm) gypsum board.

6. One-fourth-inch (6.4 mm) cement-based millboard.

7. Batts or blankets of mineral wool, mineral fiber or other *approved* materials installed in such a manner as to be securely retained in place.

❖ This approved list of fireblocking materials has not been tested for fire resistance, but is simply deemed as acceptable materials for slowing the spread of flame and products of combustion. Their effectiveness depends on their correct installation, something which must be inspected at the time of construction.

Various insulation materials have been evaluated for use as fireblocking and have been shown to perform well if installed in accordance with their specifications. For insulating material to function well as fireblocking, it must completely fill the area where fireblocking is required so that no unblocked passages exist.

Certain insulating materials have been evaluated for use as fireblocks. The evaluation reports that specify the configuration and attachment for the insulating materials in the construction assemblies are based on tests performed on the assemblies. Although the code does not specify a test standard for fireblocking, these materials have been tested using the ASTM E 119 or UL 263 standard and their performance during the test is compared to the performance of other conventional fireblocking material. The use of insulation for fireblocking is only for the 10-foot (3048 mm) horizon-

tal fireblocking in walls as required by Item 2 in Section 717.2.2. This is likely related to the fact that Section 717.2.2 only requires fireblocking at floors and ceilings and that fires burn more vigorously vertically than they do horizontally; therefore, more durable materials, such as wood, would be required.

Loose-fill insulation will not remain in place under fire conditions and, therefore, the code requires that it be tested if it is to be used as fireblocking. In the process of performing the test procedure, it is necessary to install the insulation so that it is securely attached and will be retained during the test; thus, specifications for its use as a fireblock are created.

Certain construction designs employing materials for sound transmission control require wider-than-usual concealed wall spaces. Fireblocking of these spaces at locations required by the code can be challenging. Simply because a space is filled with insulation does not guarantee that fireblocking will be accomplished. This section allows the use of batts or blankets of mineral or glass fiber insulation to serve as fireblocking in these spaces, but the installation details must be approved and inspected in the field to show that all concealed spaces are effectively fireblocked at the required locations.

717.2.1.1 Batts or blankets of mineral wool or mineral fiber. Batts or blankets of mineral wool or mineral fiber or other *approved* nonrigid materials shall be permitted for compliance with the 10-foot (3048 mm) horizontal fireblocking in walls constructed using parallel rows of studs or staggered studs.

❖ Batts or blankets of fiberglass insulation are sufficient to serve as fireblocking when installed in the manner described in the section (see commentary, Section 717.2.1).

717.2.1.2 Unfaced fiberglass. Unfaced fiberglass batt insulation used as fireblocking shall fill the entire cross section of the wall cavity to a minimum height of 16 inches (406 mm) measured vertically. When piping, conduit or similar obstructions are encountered, the insulation shall be packed tightly around the obstruction.

❖ Unfaced fiberglass is a batt insulation that can be used as fireblocking when installed as described in this section (see commentary, Section 717.2.1.)

717.2.1.3 Loose-fill insulation material. Loose-fill insulation material, insulating foam sealants and caulk materials shall not be used as a fireblock unless specifically tested in the form and manner intended for use to demonstrate its ability to remain in place and to retard the spread of fire and hot gases.

❖ Loose-fill insulation materials are generally not suitable for use as fireblocking because, by their very nature, they are not intended to span an open gap and serve the purpose intended for fireblocking. However, the code does allow for the possibility that some types of materials with some cohesiveness might work. Therefore, the code would require testing. This should be full-scale testing using a fire generated in accordance with ASTM E 119.

717.2.1.4 Fireblocking integrity. The integrity of fireblocks shall be maintained.

❖ The continued integrity of the fireblocking is necessary to ensure that it is in place at the time a fire occurs. Some of these materials, over time, with repair work or remodeling work being done, will be removed, or have holes put in the material, or be subject to other activity that will make the material ineffective as a fireblock.

717.2.1.5 Double stud walls. Batts or blankets of mineral or glass fiber or other *approved* nonrigid materials shall be allowed as fireblocking in walls constructed using parallel rows of studs or staggered studs.

❖ See the commentary to Section 717.2.1.

717.2.2 Concealed wall spaces. Fireblocking shall be provided in concealed spaces of stud walls and partitions, including furred spaces, and parallel rows of studs or staggered studs, as follows:

1. Vertically at the ceiling and floor levels.

2. Horizontally at intervals not exceeding 10 feet (3048 mm).

❖ The intent of this section is to prevent the spread of fires within walls. Losses in this area have been seen in existing structures of balloon construction. This section requires both vertical and horizontal blocking. Vertical blocking simply requires fireblocking at floors and ceilings; therefore, a tall warehouse-type building may not need any vertical fireblocking. Horizontal blocking, however, would be required every 10 feet (3048 mm). Section 717.2.1 provides the various methods for fireblocking. One method addressed is the use of mineral or glass fiber insulation, which is only allowed for horizontal fireblocking (see commentary, Section 717.2.1).

717.2.3 Connections between horizontal and vertical spaces. Fireblocking shall be provided at interconnections between concealed vertical stud wall or partition spaces and concealed horizontal spaces created by an assembly of floor joists or trusses, and between concealed vertical and horizontal spaces such as occur at soffits, drop ceilings, cove ceilings and similar locations.

❖ To prevent fire and smoke from spreading between vertical and horizontal spaces, fireblocks are required [see Figures 717.2.3(1-3)]. If not provided, a concealed fire could spread throughout the floor level because of the interconnection of the concealed combustible wall and ceiling spaces.

   In platform framing where the gypsum wallboard is continued to the top plate of the wall, then typically either the gypsum board or the top plate will serve to cut off the connection between the vertical and horizontal construction. Therefore, with many typical details, no additional fireblocking is needed at this location.

717.2.4 Stairways. Fireblocking shall be provided in concealed spaces between *stair* stringers at the top and bottom of the run. Enclosed spaces under *stairs* shall also comply with Section 1009.6.3.

❖ Fireblocks are required at the top and bottom of concealed spaces between stairway stringers (see Figure 717.2.4). Similar to fireblocks between horizontal and vertical spaces, fireblocks at the top and bottom of a run will provide a barrier between the floor below the stairway and the floor that the stairway serves. Additionally, a reference is made to Section 1009.5.3, which requires the interior of the space under stairways to contain 1-hour fire-resistant-rated construction or the stairway to be fire resistant (whichever is greater).

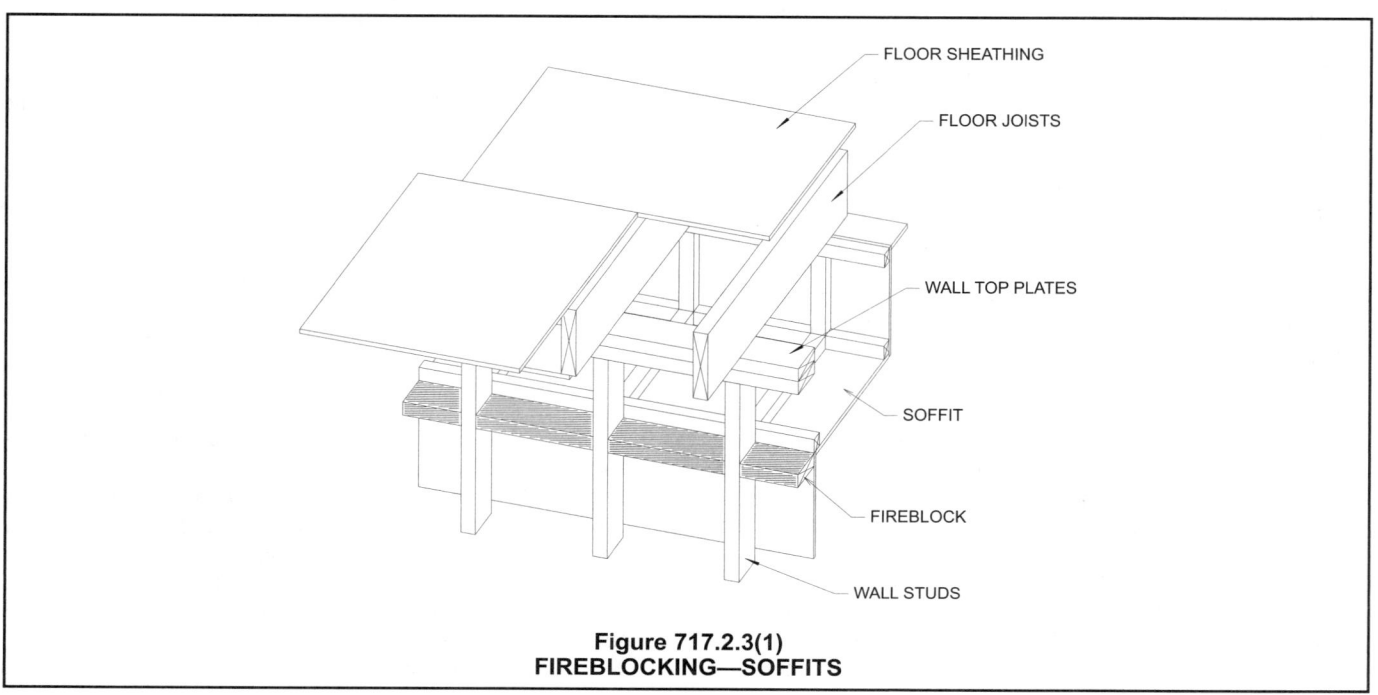

Figure 717.2.3(1)
FIREBLOCKING—SOFFITS

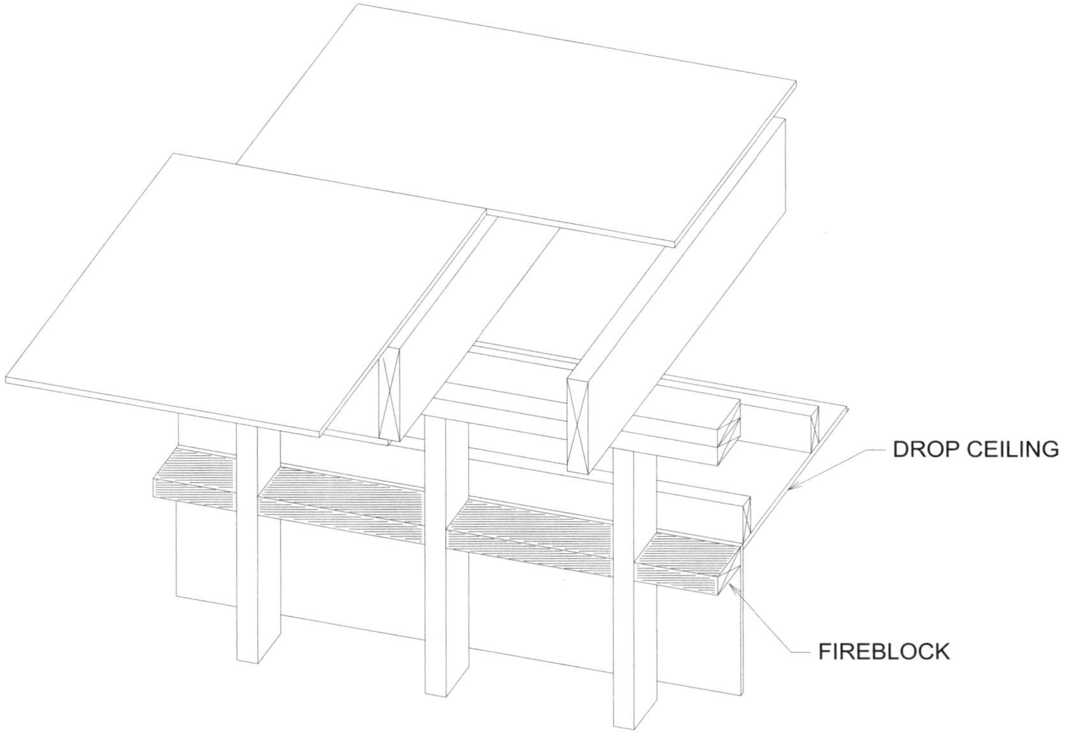

Figure 717.2.3(2)
FIREBLOCKING—DROP CEILINGS

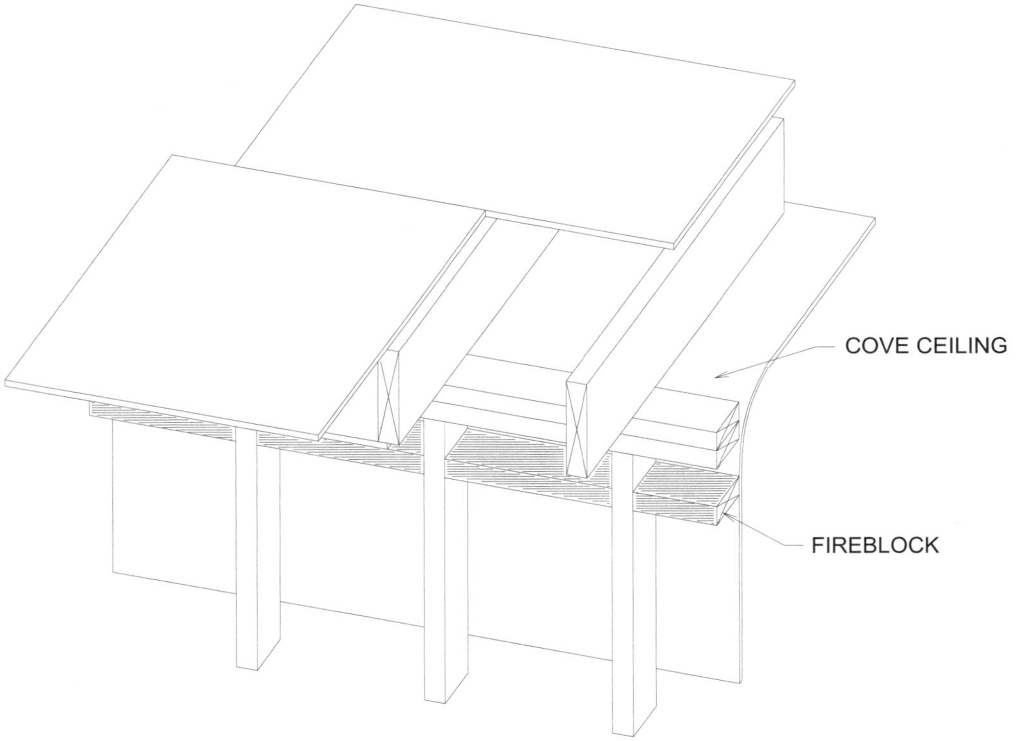

DROP CEILING

FIREBLOCK

Figure 717.2.3(3)
FIREBLOCKING—COVE CEILING

COVE CEILING

FIREBLOCK

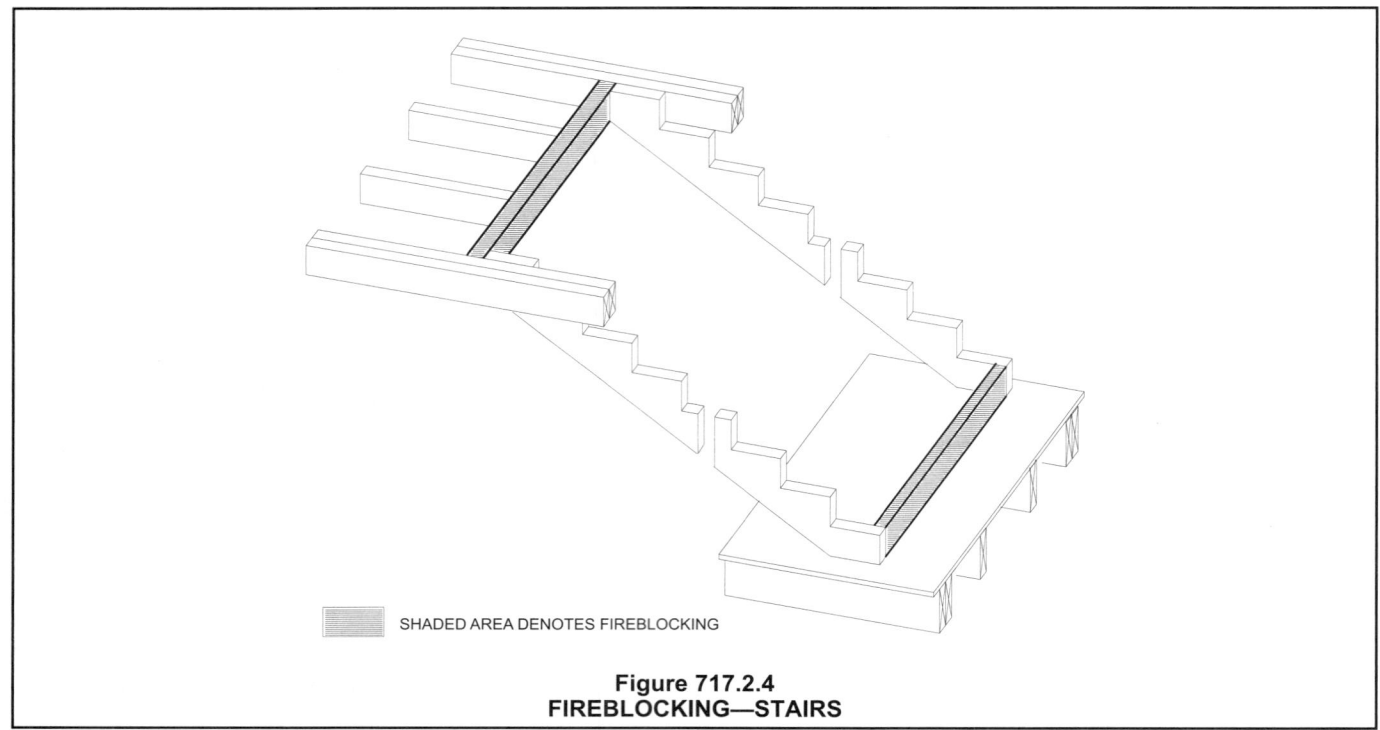

SHADED AREA DENOTES FIREBLOCKING

Figure 717.2.4
FIREBLOCKING—STAIRS

717.2.5 Ceiling and floor openings. Where required by Exception 6 of Section 708.2, Exception 1 of Section 713.4.1.2 or Section 713.4.2, fireblocking of the *annular space* around vents, pipes, ducts, chimneys and fireplaces at ceilings and floor levels shall be installed with a material specifically tested in the form and manner intended for use to demonstrate its ability to remain in place and resist the free passage of flame and the products of combustion.

❖ Sections 707 and 712 allow penetrations of ceilings and floors to be fireblocked [see Figures 717.2.3(1-4)]. It should be noted that depending on the type of penetration and the number of floors penetrated (or stories connected), a shaft may be required (see Section 708.2). Additional requirements for penetrations through rated floor/ceiling and roof/ceiling assemblies and shaft requirements can be found in Section 712.4.

UL 103 and 127 specify that only metal fireblocks and insulation shields are to be used with factory-built chimneys and fireplaces. The IMC contains a reference similar to that made in the UL standards. When fire-resistance-rated floor/ceiling assemblies are penetrated, the higher degree of protection afforded by tested through-penetration protection systems is required.

717.2.5.1 Factory-built chimneys and fireplaces. Factory-built chimneys and fireplaces shall be fireblocked in accordance with UL 103 and UL 127.

❖ UL 103 and 127 specify that only metal fireblocks and insulation shields are to be used with factory-built chimneys and fireplaces. The IMC contains a reference similar to that made in the UL standards. When fire-resistance-rated floor/ceiling assemblies are pen-

etrated, the higher degree of protection afforded by tested through-penetration protection systems is required.

717.2.6 Architectural trim. Fireblocking shall be installed within concealed spaces of *exterior wall* finish and other exterior architectural elements where permitted to be of combustible construction as specified in Section 1406 or where erected with combustible frames, at maximum intervals of 20 feet (6096 mm), so that there will be no open space exceeding 100 square feet (9.3 m^3). Where wood furring strips are used, they shall be of *approved* wood of natural decay resistance or *preservative-treated* wood. If noncontinuous, such elements shall have closed ends, with at least 4 inches (102 mm) of separation between sections.

Exceptions:

1. Fireblocking of cornices is not required in single-family *dwellings*. Fireblocking of cornices of a two-family *dwelling* is required only at the line of *dwelling unit* separation.

2. Fireblocking shall not be required where installed on noncombustible framing and the face of the *exterior wall* finish exposed to the concealed space is covered by one of the following materials:

   2.1. Aluminum having a minimum thickness of 0.019 inch (0.5 mm).

   2.2. Corrosion-resistant steel having a base metal thickness not less than 0.016 inch (0.4 mm) at any point.

   2.3. Other *approved* noncombustible materials.

❖ Combustible exterior wall finish and exterior architectural elements are required to be fireblocked at a maximum of 20-foot (6096 mm) intervals and a 100-square-foot (9.3 m²) maximum area to prevent the spread of fire or smoke through concealed spaces (see Figure 717.2.7). Adjacent sections of exterior trim need not be considered as contributing to the 20-foot (6096 mm) limitation if there is a minimum 4-inch (102 mm) separation between sections of exterior trim and the ends are closed. Trim on the building exterior presents the same concern as interior combustible concealed wall spaces and, therefore, requires fireblocking. To increase the long-term durability of wood used on the exterior, a certain level of decay resistance is required.

717.2.7 Concealed sleeper spaces. Where wood sleepers are used for laying wood flooring on masonry or concrete fire-resistance-rated floors, the space between the floor slab and the underside of the wood flooring shall be filled with an *approved* material to resist the free passage of flame and products of combustion or fireblocked in such a manner that there will be no open spaces under the flooring that will exceed 100 square feet (9.3 m²) in area and such space shall be filled solidly under permanent partitions so that there is no communication under the flooring between adjoining rooms.

Exceptions:

1. Fireblocking is not required for slab-on-grade floors in gymnasiums.

2. Fireblocking is required only at the juncture of each alternate lane and at the ends of each lane in a bowling facility.

❖ Concealed spaces created by floor sleepers must be fireblocked into areas no greater than 100 square feet (9.3 m²) or such spaces must be filled with materials capable of resisting flame, smoke or gases (see Figure 717.2.7).

Since floor sleepers may cover large areas, this fireblock or material will restrict a fire from spreading within the concealed space to other areas of the floor. This section is commonly applied when a raised,

wood-finish floor surface is installed in Type I or II construction.

717.3 Draftstopping in floors. In combustible construction, draftstopping shall be installed to subdivide floor/ceiling assemblies in the locations prescribed in Sections 717.3.2 through 717.3.3.

❖ Concealed spaces created by floor sleepers must be fireblocked into areas no greater than 100 square feet (9.3 m²) or such spaces must be filled with materials capable of resisting flame, smoke or gases (see Figure 717.2.7).

Since floor sleepers may cover large areas, this fireblock or material will restrict a fire from spreading within the concealed space to other areas of the floor. This section is commonly applied when a raised, wood-finish floor surface is installed in Type I or II construction.

717.3.1 Draftstopping materials. Draftstopping materials shall not be less than $^1/_2$-inch (12.7 mm) gypsum board, $^3/_8$-inch (9.5 mm) wood structural panel, $^3/_8$-inch (9.5 mm) particleboard, 1-inch (25-mm) nominal lumber, cement fiberboard, batts or blankets of mineral wool or glass fiber, or other *approved* materials adequately supported. The integrity of draftstops shall be maintained.

❖ Similar to fireblocks in small concealed spaces, draftstops also act as a barrier to smoke and gases. Because of the large areas that require this barrier, however, the code permits certain types of sheathing material, such as $^1/_2$-inch (12.7 mm) gypsum board, $^3/_8$-inch (9.5 mm) plywood and particleboard, 1-inch nominal lumber, cement fiberboard, mineral wool and glass fiber batts or blankets, or other approved materials. The draftstop material must be properly supported and capable of remaining in place when subjected to initial fire exposure. It is reasonable to assume that materials that are acceptable for fireblocking purposes (see Section 717.2.1) would perform as well for draftstopping situations, so many items are duplicated

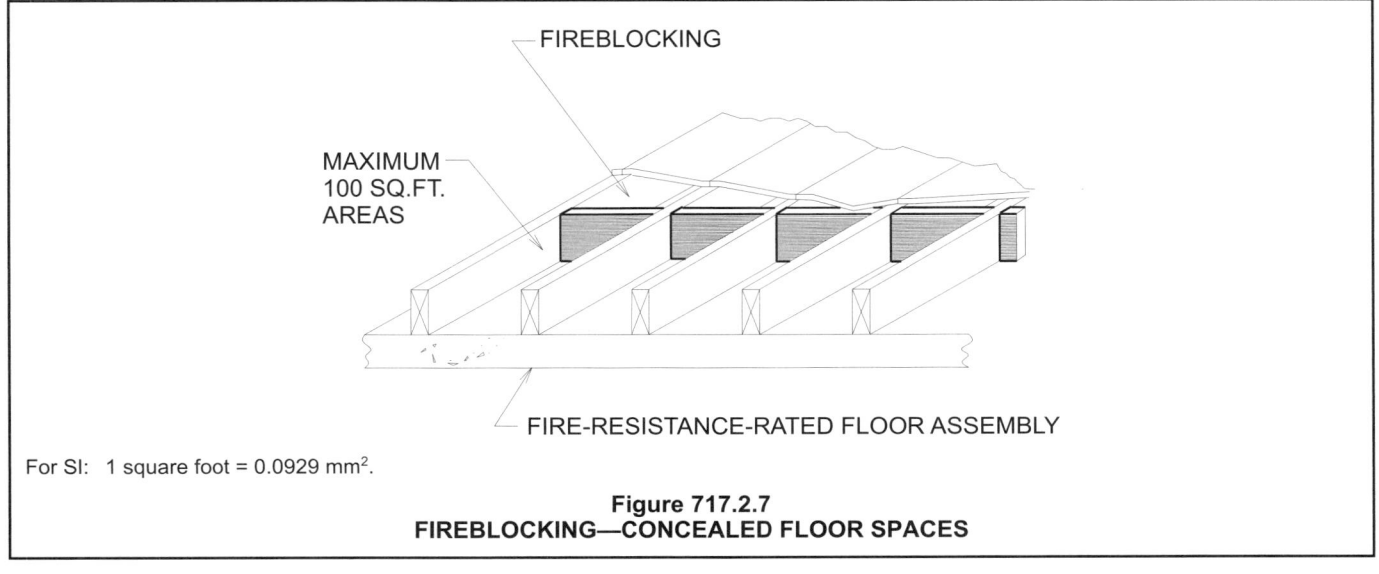

FIREBLOCKING

MAXIMUM
100 SQ.FT.
AREAS

FIRE-RESISTANCE-RATED FLOOR ASSEMBLY

For SI: 1 square foot = 0.0929 mm².

Figure 717.2.7
FIREBLOCKING—CONCEALED FLOOR SPACES

here. Although the code does provide an extensive laundry list of accepted materials, it is important to note that the code does allow the use of any other approved material when it is adequately supported.

717.3.2 Groups R-1, R-2, R-3 and R-4. Draftstopping shall be provided in floor/ceiling spaces in Group R-1 buildings, in Group R-2 buildings with three or more *dwelling units*, in Group R-3 buildings with two *dwelling units* and in Group R-4 buildings. Draftstopping shall be located above and in line with the *dwelling unit* and *sleeping unit* separations.

Exceptions:

1. Draftstopping is not required in buildings equipped throughout with an *automatic sprinkler system* in accordance with Section 903.3.1.1.

2. Draftstopping is not required in buildings equipped throughout with an *automatic sprinkler system* in accordance with Section 903.3.1.2, provided that automatic sprinklers are also installed in the combustible concealed spaces.

❖ To maintain the integrity of dwelling or sleeping unit separation walls in buildings of Groups R-1, R-2, R-3 and R-4, draftstopping is to be provided when the dwelling or sleeping unit separation wall is not continuous to the floor sheathing above. The draftstopping must be installed directly above the dwelling or sleeping unit separation wall (see Figure 717.3.2). The dwelling or sleeping unit separation wall (see Section 708), plus the draftstopping above are considered a barrier to the spread of fire and smoke. As such, the draftstop offers some level of protection to the occupants of one dwelling or sleeping unit from a fire occurring in another dwelling or sleeping unit.

The exception indicates that the draftstopping need not be provided if sprinklers are installed above and below the ceiling, since sprinkler activation will control the spread of fire. The sprinkler system must be installed in accordance with Section 903.3.1.1 or 903.3.1.2 and NFPA 13 or 13R. The NFPA 13 sprinkler system (see Section 903.3.1.1) will generally require the sprinklers to be installed within combustible floor spaces unless the space meets the exceptions found within the standard. The NFPA 13R sprinkler system (see Section 903.3.1.2) would generally not require the sprinkler system within the concealed floor space. However, when using the provisions of Exception 2, the NFPA 13R system must be extended into the floor space despite the normal exclusion within the standard. See Section 102.4, which supports that the code requirement will control in this situation.

717.3.3 Other groups. In other groups, draftstopping shall be installed so that horizontal floor areas do not exceed 1,000 square feet (93 m²).

> **Exception:** Draftstopping is not required in buildings equipped throughout with an *automatic sprinkler system* in accordance with Section 903.3.1.1.

❖ Unless the spaces above and below the ceiling are sprinklered (NFPA 13 system), draftstopping is to be provided in all groups except Groups R-1, R-2, R-3 and R-4 such that the open space does not exceed 1,000 square feet (93 m²) (see Figure 717.3.3).

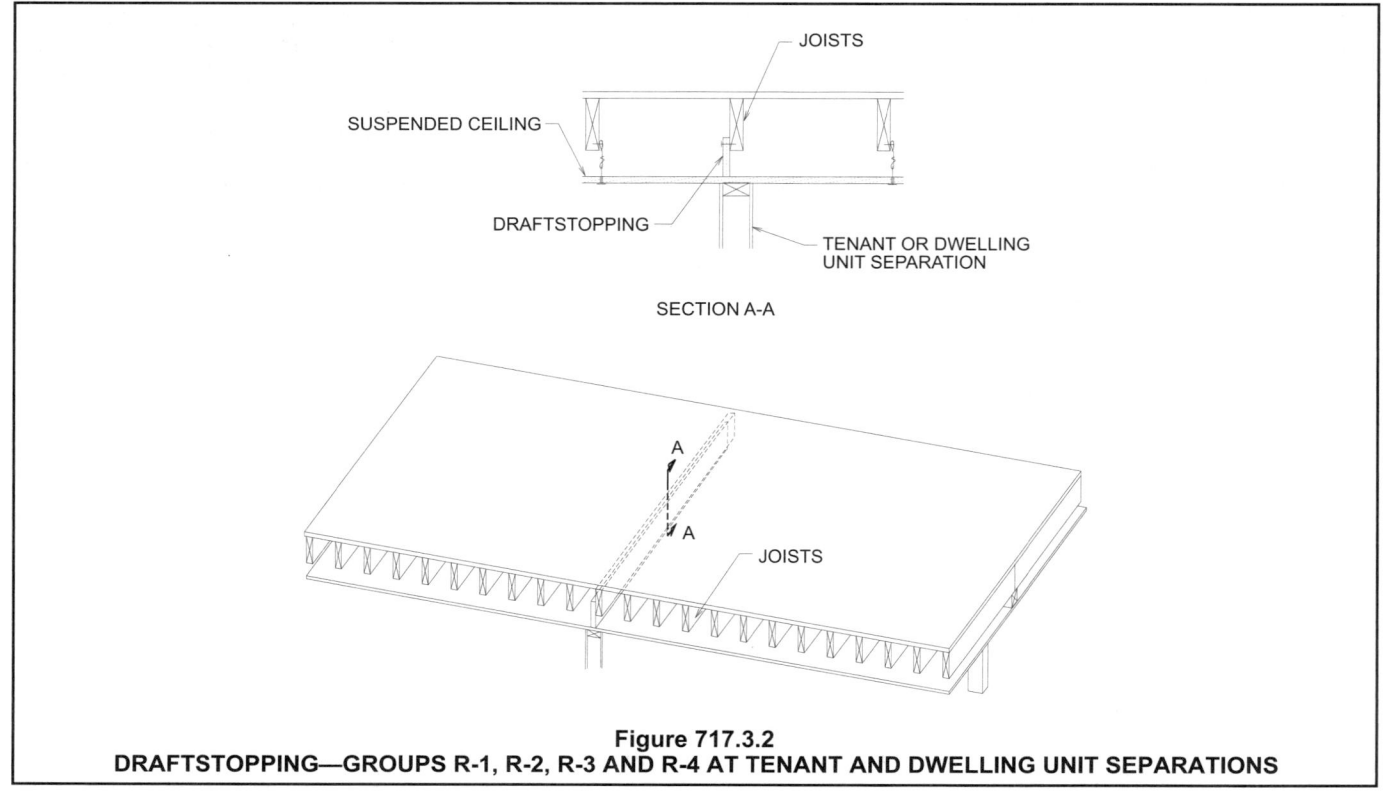

Figure 717.3.2
DRAFTSTOPPING—GROUPS R-1, R-2, R-3 AND R-4 AT TENANT AND DWELLING UNIT SEPARATIONS

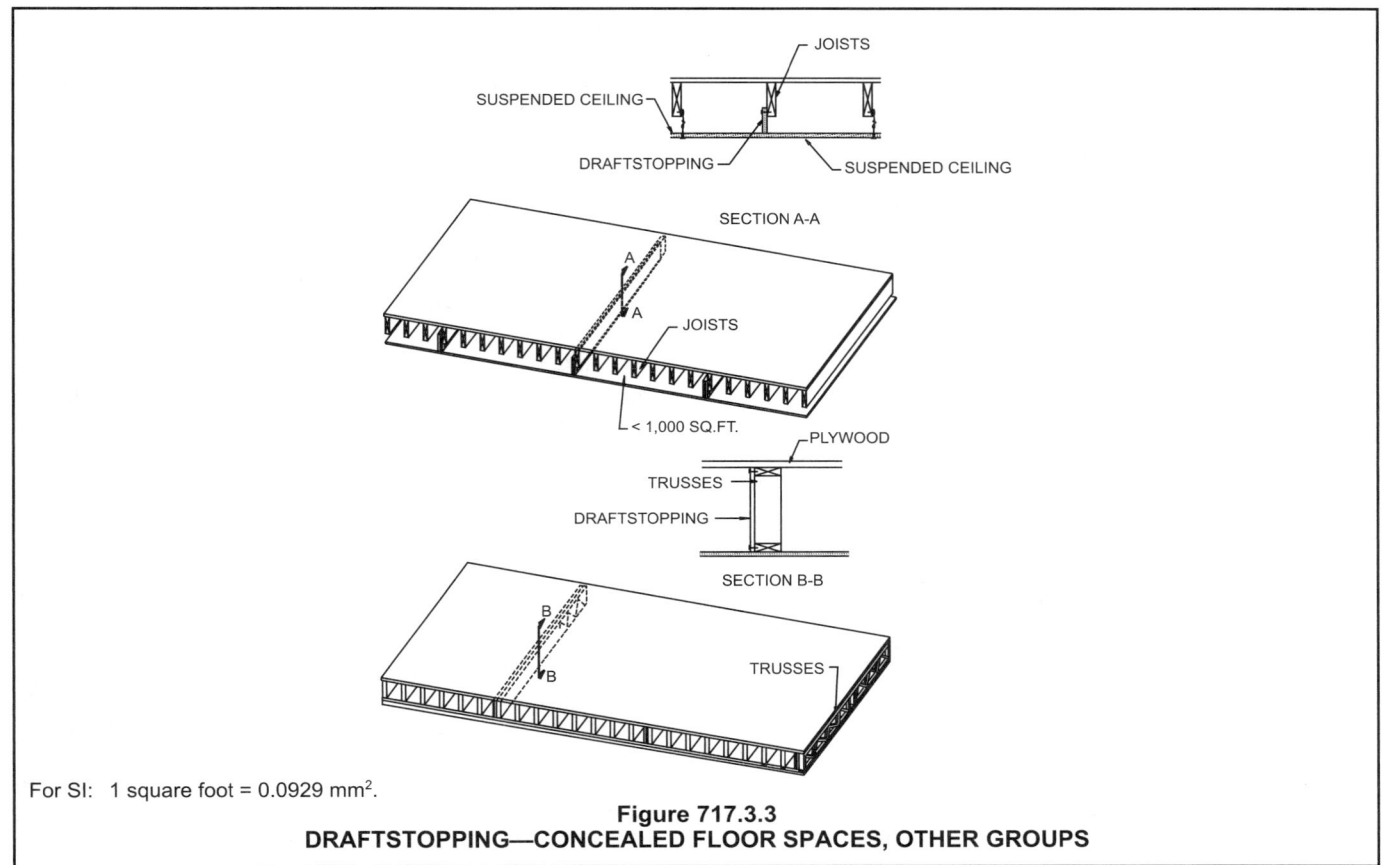

For SI: 1 square foot = 0.0929 mm².

Figure 717.3.3
DRAFTSTOPPING—CONCEALED FLOOR SPACES, OTHER GROUPS

717.4 Draftstopping in attics. In combustible construction, draftstopping shall be installed to subdivide *attic* spaces and concealed roof spaces in the locations prescribed in Sections 717.4.2 and 717.4.3. Ventilation of concealed roof spaces shall be maintained in accordance with Section 1203.2.

❖ Fires that spread to attics that are not properly draftstopped often cause considerable damage. For this reason, draftstopping is required in attic and concealed roof spaces in accordance with Sections 717.4.2 and 717.4.3. The text also reminds the code user that, although the space must be compartmented, it still must be ventilated in accordance with the provisions of Chapter 12 in order to eliminate moisture or condensation or to cool the attic.

717.4.1 Draftstopping materials. Materials utilized for draftstopping of *attic* spaces shall comply with Section 717.3.1.

❖ See the commentary to Section 717.3.1.

717.4.1.1 Openings. Openings in the partitions shall be protected by self-closing doors with automatic latches constructed as required for the partitions.

❖ Section 1209.2 requires attic access. The placement of draftstopping in the attic may interfere with the ability of the fire department to gain access to all attic spaces. This section requires that if draftstopping is provided with access openings to adjacent draftstopped portions of the attic, the openings must be

constructed of draftstopping materials and be equipped with self-closing mechanisms in order to ensure that the draftstop will perform its intended function.

717.4.2 Groups R-1 and R-2. Draftstopping shall be provided in *attics*, mansards, overhangs or other concealed roof spaces of Group R-2 buildings with three or more *dwelling units* and in all Group R-1 buildings. Draftstopping shall be installed above, and in line with, *sleeping unit* and *dwelling unit* separation walls that do not extend to the underside of the roof sheathing above.

Exceptions:

1. Where *corridor* walls provide a *sleeping unit* or *dwelling unit* separation, draftstopping shall only be required above one of the *corridor* walls.

2. Draftstopping is not required in buildings equipped throughout with an *automatic sprinkler system* in accordance with Section 903.3.1.1.

3. In occupancies in Group R-2 that do not exceed four *stories above grade plane*, the *attic* space shall be subdivided by draftstops into areas not exceeding 3,000 square feet (279 m²) or above every two *dwelling units*, whichever is smaller.

4. Draftstopping is not required in buildings equipped throughout with an *automatic sprinkler system* in accordance with Section 903.3.1.2, pro-

vided that automatic sprinklers are also installed in the combustible concealed spaces.

❖ To maintain the integrity of dwelling and sleeping unit separation walls (see commentary, Section 717.3.2), draftstopping is to be provided above dwelling or sleeping unit separation walls that do not extend to the roof sheathing [see Figures 717.4.2(1) and 717.4.2(2)]. This provision is limited to Group R-1 and R-2 occupancies. Groups R-3 and R-4 are regulated by Section 717.4.3.

Exception 1 clarifies that for corridor walls that serve as dwelling or sleeping unit separations, draftstopping is only required above one of the two corridor walls. It should be noted that this option only applies to roof/attic spaces and not to floors (see Sections 717.3.2 and 708.4).

Exceptions 2 and 4 omit the requirement for draftstopping attics that are sprinklered in accordance with Section 903.3.1.1 (see Exception 2) or 903.3.1.2 (see Exception 4). This exception would require the attic to be sprinklered even though it may not be required by the standard. NFPA 13R typically would not require an attic to be sprinklered. This is consistent with the exceptions to Section 717.3.2 for draftstopping of floor spaces (see commentary, Section 717.3.2).

The draftstopping requirements can be reduced in Group R-2 occupancies of four stories or less by Exception 3. This exception allows the area between draftstopping to be increased to 3,000 square feet (279 m²) or above every two dwelling units, whichever

is smaller. This provides the opportunity for a simpler draftstopping installation when the draftstops are not required to be in line with the dwelling or sleeping unit separation walls. Group R-2 buildings are required by Section 903.2.7 to be sprinklered; sprinklers reduce the need for attic draftstopping. The four-story restriction will match with the limitations of Section 903.3.1.2 for NFPA 13R sprinkler systems. Therefore by providing this level of draftstopping, buildings that use the NFPA 13R systems will not need to comply with Exception 4, which is discussed above. If Section 903.3.1.1 (an NFPA 13 sprinkler system) is used, then sprinklers will generally be required within the attic space and Exception 2 (discussed above) would be applicable.

717.4.3 Other groups. Draftstopping shall be installed in *attics* and concealed roof spaces, such that any horizontal area does not exceed 3,000 square feet (279 m²).

> **Exception:** Draftstopping is not required in buildings equipped throughout with an *automatic sprinkler system* in accordance with Section 903.3.1.1.

❖ Since Group R-3 and R-4 and nonresidential buildings are not always subdivided into small tenant spaces, concealed roof spaces and attics are to be subdivided into areas not exceeding 3,000 square feet (279 m²). The exception indicates that draftstopping is not required when the spaces above and below the ceiling are provided coverage by an automatic sprinkler system, in accordance with NFPA 13.

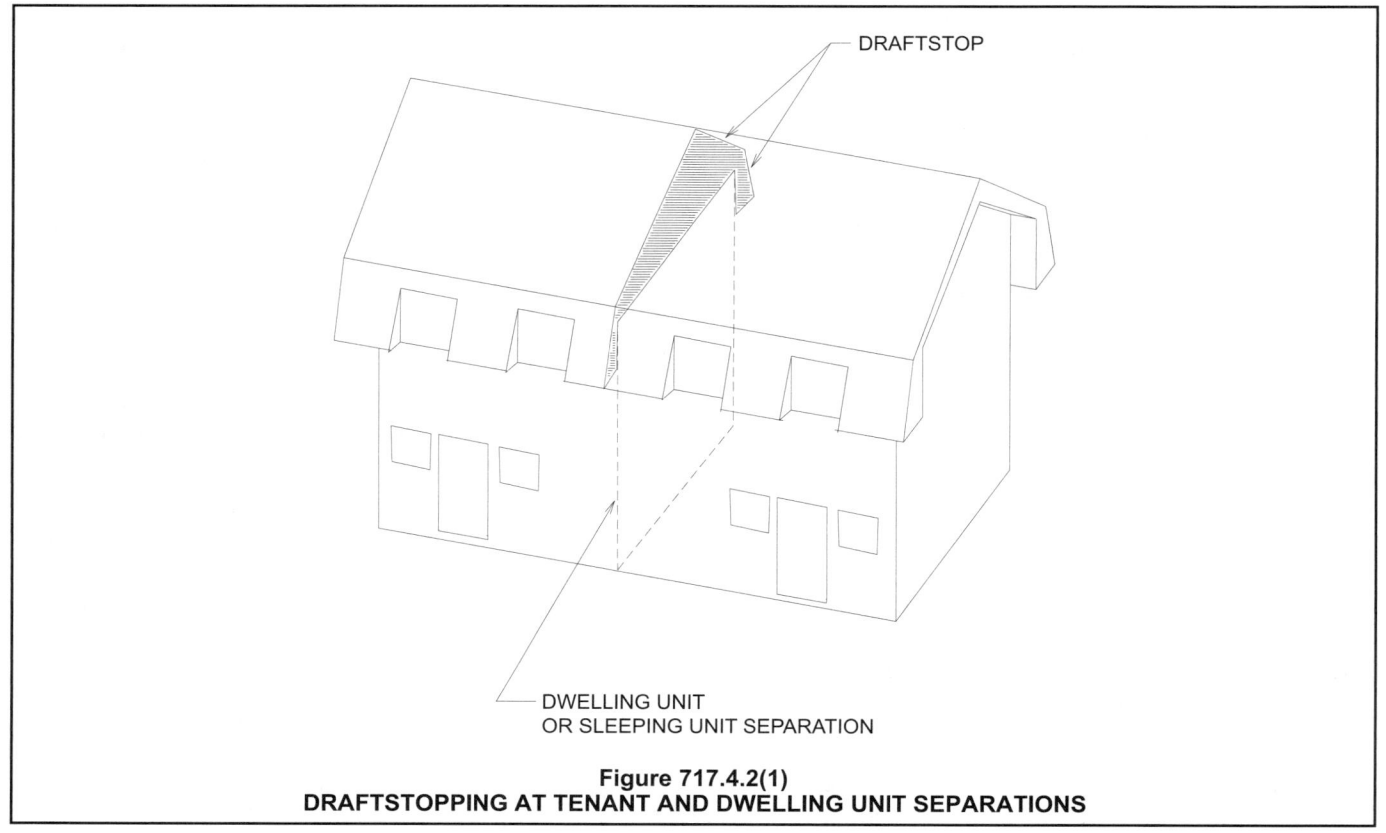

Figure 717.4.2(1)
DRAFTSTOPPING AT TENANT AND DWELLING UNIT SEPARATIONS

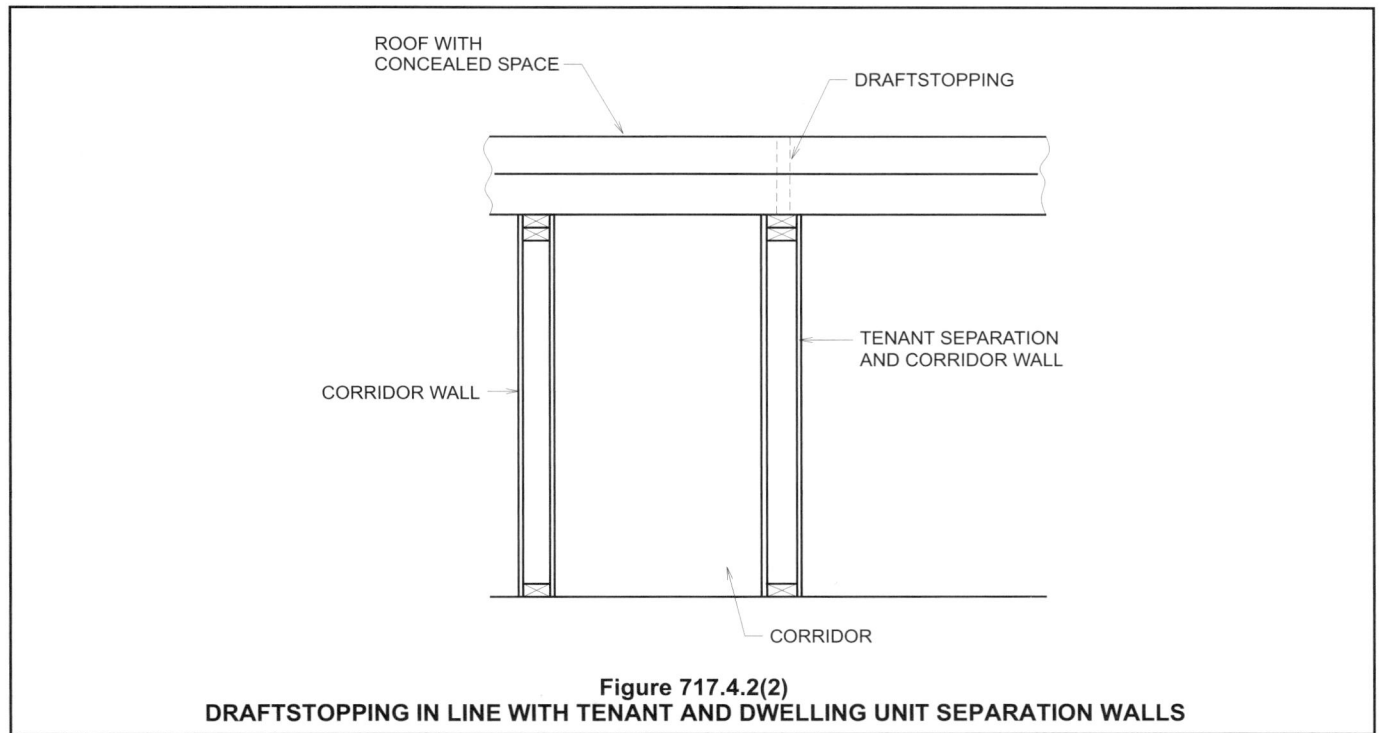

Figure 717.4.2(2)
DRAFTSTOPPING IN LINE WITH TENANT AND DWELLING UNIT SEPARATION WALLS

717.5 Combustible materials in concealed spaces in Type I or II construction. Combustible materials shall not be permitted in concealed spaces of buildings of Type I or II construction.

Exceptions:

1. Combustible materials in accordance with Section 603.

2. Combustible materials exposed within plenums complying with Section 602 of the *International Mechanical Code*.

3. Class A *interior finish* materials classified in accordance with Section 803.

4. Combustible piping within partitions or shaft enclosures installed in accordance with the provisions of this code.

5. Combustible piping within concealed ceiling spaces installed in accordance with the *International Mechanical Code* and the *International Plumbing Code*.

6. Combustible insulation and covering on pipe and tubing, installed in concealed spaces other than plenums, complying with Section 719.7.

❖ This section is necessary because Section 717 would generally exclude Type I and II buildings from the concealed space draftstopping and fireblocking provisions. The use of combustibles in nonstructural applications is limited to the listed exceptions to prevent fire spread within building elements. The severity of a potential problem increases when combustibles are located within concealed spaces that are inaccessible to manual fire fighting. This section regulates the use of combustibles within concealed spaces of noncombustible buildings. This section is referenced from Exception 21 of Section 603.1.

SECTION 718
FIRE-RESISTANCE
REQUIREMENTS FOR PLASTER

718.1 Thickness of plaster. The minimum thickness of gypsum plaster or portland cement plaster used in a fire-resistance-rated system shall be determined by the prescribed fire tests. The plaster thickness shall be measured from the face of the lath where applied to gypsum lath or metal lath.

❖ The fire resistance of plaster is dependent on material variations, such as the aggregate mixture (sand content), as well as bonding to the plaster base. The use of wire or wire fabric reinforces the plaster mixes applied to plaster bases and, therefore, ensures increased fire resistance. As such, the makeup of the plastered element is required when evaluating the fire-resistance rating of plaster. Because the type of plaster used affects the fire-resistance rating, it is essential that the tests conducted utilize a plaster mix similar to that which is to be used in actual construction Depending on the type of plaster, additional tests may not be required. If the plaster is a type that is judged to be equivalent to a tested type (see Section 718.2), additional testing is not required.

Similar to the requirements of Section 704.7, it is the thickness of the protective covering that is important. Therefore, the plaster is to be measured from the face of the lath, which reinforces the plaster.

Code users should also be aware that Sections 720 and 721 contain a number of items related to the use of plaster in fire-resistance ratings. Section 720 contains a number of prescriptive assemblies that can be used to obtain a fire-resistance rating. Section 721 contains methods of calculating fire-resistive assemblies and the contribution of plaster to other assemblies. For more information, see Sections 721.2.1.4, 721.3.2 and 721.4.1. Although many people are not familiar with these provisions, they are permitted under the provisions of Section 703.3 and provide another option for establishing rated assemblies.

718.2 Plaster equivalents. For fire-resistance purposes, $^1/_2$ inch (12.7 mm) of unsanded gypsum plaster shall be deemed equivalent to $^3/_4$ inch (19.1 mm) of one-to-three gypsum sand plaster or 1 inch (25 mm) of portland cement sand plaster.

❖ This section indicates that prescribed thicknesses of unsanded gypsum plaster, one-to-three sanded gypsum plaster, as well as portland cement sand plaster are considered equivalent in terms of their fire-resistance-rating characteristics. Once tests are conducted to determine the fire resistance of an assembly using one of these types, equivalency to the other prescribed types can be determined without subsequent tests.

718.3 Noncombustible furring. In buildings of Type I and II construction, plaster shall be applied directly on concrete or masonry or on *approved* noncombustible plastering base and furring.

❖ Buildings of Type I and II construction are required to be predominantly of noncombustible construction. This section requires the plastering base and furring to be noncombustible. The intent of this provision is to not permit a combustible cavity behind the finished plaster, which may be an area for a potentially concealed fire buildup.

718.4 Double reinforcement. Plaster protection more than 1 inch (25 mm) in thickness shall be reinforced with an additional layer of *approved* lath embedded at least $^3/_4$ inch (19.1 mm) from the outer surface and fixed securely in place.

Exception: Solid plaster partitions or where otherwise determined by fire tests.

❖ The additional reinforcement is intended to decrease the likelihood of the plaster loosening when subjected to high temperatures. The plaster relies on the lath to provide a finish that is structurally sound. As the thickness of the plaster increases, additional reinforcement is required to hold the plaster in place.

718.5 Plaster alternatives for concrete. In reinforced concrete construction, gypsum plaster or portland cement plaster is permitted to be substituted for $^1/_2$ inch (12.7 mm) of the required poured concrete protection, except that a minimum thickness of $^3/_8$ inch (9.5 mm) of poured concrete shall be provided in reinforced concrete floors and 1 inch (25 mm) in reinforced concrete columns in addition to the plaster finish. The

concrete base shall be prepared in accordance with Section 2510.7.

❖ Concrete elements, such as reinforced slabs and columns, rely in part on the thickness of the element to provide a fire-resistance rating. In order to reduce the concrete thickness, this section permits up to $^1/_2$ inch (12.7 mm) of plaster to be substituted for $^1/_2$ inch (12.7 mm) of concrete without affecting the fire-resistance rating of the element. While plaster can be substituted for concrete for fire-resistance purposes, a minimum amount of concrete is still required to be provided for structural integrity.

SECTION 719
THERMAL- AND SOUND-INSULATING MATERIALS

719.1 General. Insulating materials, including facings such as vapor retarders and *vapor-permeable membranes*, similar coverings and all layers of single and multilayer reflective foil insulations, shall comply with the requirements of this section. Where a flame spread index or a smoke-developed index is specified in this section, such index shall be determined in accordance with ASTM E 84 or UL 723. Any material that is subject to an increase in flame spread index or smoke-developed index beyond the limits herein established through the effects of age, moisture or other atmospheric conditions shall not be permitted.

Exceptions:

1. Fiberboard insulation shall comply with Chapter 23.

2. Foam plastic insulation shall comply with Chapter 26.

3. Duct and pipe insulation and duct and pipe coverings and linings in plenums shall comply with the *International Mechanical Code*.

4. All layers of single and multilayer reflective plastic core insulation shall comply with Section 2613.

❖ This section addresses the potential fire or flame spread hazards of insulating materials installed in building spaces. Other provisions of this code or other I-Codes regulate insulating materials used for duct and plenum insulation [see the IMC and the *International Energy Conservation Code®* (IECC®)] and foam plastic insulation (see Section 2603). Insulating materials can affect fire development and fire spread and are, therefore, regulated accordingly.

This section addresses the various insulating materials that may be installed in building spaces, including insulating batts, blankets, fills (including vapor barriers and vapor-permeable membranes) or other coverings. Fiberboard insulation is regulated by Chapter 23. Cellulose loose-fill insulation is regulated by Section 719.6. Foam plastic insulation is regulated by Chapter 26. The applicable test method is ASTM E 84 or UL 723 (see commentary, Section 803.1).

The requirements of Section 703.2, regarding the potential that adding items into a fire-resistance-rated assembly may reduce the rating of the tested assembly, must be considered. The addition of insulation has the potential to either add to or reduce the fire-resistance rating of an assembly. The reduction of ratings is more of a problem when insulation is added into a horizontal assembly.

719.2 Concealed installation. Insulating materials, where concealed as installed in buildings of any type of construction, shall have a flame spread index of not more than 25 and a smoke-developed index of not more than 450.

> **Exception:** Cellulose loose-fill insulation that is not spray applied, complying with the requirements of Section 719.6, shall only be required to meet the smoke-developed index of not more than 450.

❖ Concealed insulation no longer acts as an interior finish unless the covering material is removed or the fire is in the concealed building space. To limit the contribution of the insulation to a fire condition, the material must have a flame spread index of no more than 25 and a smoke-developed index of no more than 450. This is illustrated in Figure 719.2. Cellulose loose-fill insulation is regulated by the U.S. Consumer Product Safety Commission (CPSC) in accordance with Section 719.6. Spray-applied cellulose is regulated by Section 719.3.1 (see commentary, Section 719.4).

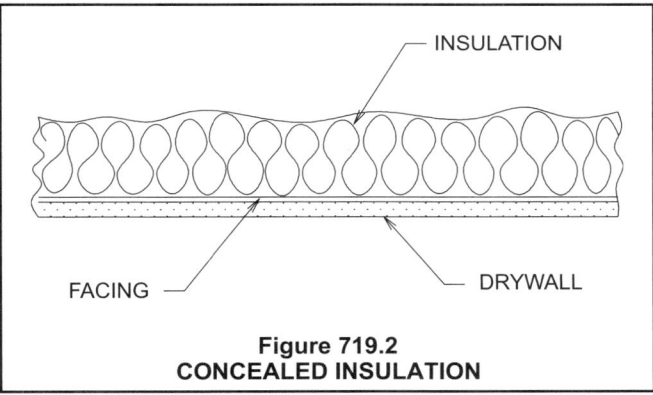

Figure 719.2
CONCEALED INSULATION

719.2.1 Facings. Where such materials are installed in concealed spaces in buildings of Type III, IV or V construction, the flame spread and smoke-developed limitations do not apply to facings, coverings, and layers of reflective foil insulation that are installed behind and in substantial contact with the unexposed surface of the ceiling, wall or floor finish.

> **Exception:** All layers of single and multilayer reflective plastic core insulation shall comply with Section 2613.

❖ In buildings of Type III, IV and V construction, the flame spread indices do not apply to facing material, provided it is installed behind (and in substantial contact with) the unexposed surface of the ceiling, floor or wall finish. For example, when paper-backed insulation is placed directly on top of a ceiling, the paper facing is not required to meet the flame spread index; however, if the same material is applied to the underside of a roof deck and the paper facing is exposed to the attic space, the paper facing must meet the flame spread criteria. The poten-

tial for flame spread is greatly diminished when the facings are installed in direct contact with the finish material because of the lack of airspace to support a fire if it were to be exposed to a source of ignition.

The purpose of the vapor retarder being placed on the warm-in-winter side of the building element is to prevent moisture vapor in the warm interior from reaching the dew-point temperature on a cold surface in the roof, wall or floor system. The generally accepted rating of a vapor retarder is 1 perm [57 mg/(s · m² · Pa)]. It is assumed that the permeance of an adequate vapor barrier will not exceed 1 perm [57 mg/(s · m² · P · a)] (see commentary, Section 1203.2).

719.3 Exposed installation. Insulating materials, where exposed as installed in buildings of any type of construction, shall have a flame spread index of not more than 25 and a smoke-developed index of not more than 450.

> **Exception:** Cellulose loose-fill insulation that is not spray applied complying with the requirements of Section 719.6 shall only be required to meet the smoke-developed index of not more than 450.

❖ Exposed insulating materials represent the same fire exposure hazard as any other exposed material, such as interior finish. The flame spread index for all applications of insulation materials is limited to 25, compared to 200 for interior finishes installed in rooms or spaces of certain occupancies (see commentary, Chapter 8). As with all interior finishes, the smoke-developed index is limited to 450. Cellulose loose-fill insulation is regulated by the CPSC in accordance with Section 719.6. Spray-applied cellulose is regulated by Section 719.3.1 (see commentary, Section 719.4).

It is important to remember that insulation materials that are open and exposed within an attic or crawl space are still considered as being "exposed" and, therefore, regulated by this section. The general requirements of this section do not differ from the provisions of Section 719.2. The main difference is that the facing materials within concealed installations may be exempt from the flame spread and smoke-development rating (see Section 719.2.1) while the exposed facings would be limited as stated in this section.

719.3.1 Attic floors. Exposed insulation materials installed on *attic* floors shall have a critical radiant flux of not less than 0.12 watt per square centimeter when tested in accordance with ASTM E 970.

❖ ASTM E 970 is a test method that was developed by the insulation industry to evaluate the fire hazard of exposed attic insulation, and is referenced in the material standards for insulation. Cellulose loose-fill insulation is required to comply with CPSC 16 CFR, Part 1209 (see Section 719.6), which requires testing by this standard. Spray-applied cellulose insulation that is not subject to the CPSC standard is also subject to the ASTM E 970 testing by virtue of this section.

719.4 Loose-fill insulation. Loose-fill insulation materials that cannot be mounted in the ASTM E 84 or UL 723 apparatus without a screen or artificial supports shall comply with the flame spread and smoke-developed limits of Sections 719.2 and 719.3 when tested in accordance with CAN/ULC S102.2.

> **Exception:** Cellulose loose-fill insulation shall not be required to be tested in accordance with CAN/ULC S102.2, provided such insulation complies with the requirements of Section 719.2 or 719.3, as applicable, and Section 719.6.

❖ The exception serves to make a distinction between cellulose insulation, which is spray applied using a water-mist applicator, and cellulose or other loose-fill insulation, which is poured or blown in place. Spray-applied cellulose insulation can be exposed on vertical and horizontal ceiling-type surfaces so it is treated as any other insulating material in regard to testing. Cellulose loose-fill insulation, which is poured or blown in, is regulated by CPSC requirements (see Section 719.6) and is exempt from the test procedure described in this section.

719.5 Roof insulation. The use of combustible roof insulation not complying with Sections 719.2 and 719.3 shall be permitted in any type of construction provided it is covered with *approved* roof coverings directly applied thereto.

❖ Foam plastic roof insulation is required to comply with Chapter 26. This section allows combustible insulation other than foam plastic insulation to be incorporated in the roof assembly without declassifying the type of construction. This provision coordinates with Section 603.1, Item 4.

719.6 Cellulose loose-fill insulation. Cellulose loose-fill insulation shall comply with CPSC 16 CFR, Part 1209 and CPSC 16 CFR, Part 1404. Each package of such insulating material shall be clearly labeled in accordance with CPSC 16 CFR, Part 1209 and CPSC 16 CFR, Part 1404.

❖ Cellulose loose-fill insulation is federally regulated by the CPSC. Parts 1209 and 1404 of CPSC 16 CFR contain various requirements that regulate the product to avoid excessive flammability or significant fire hazards. The smoke-developed index for cellulose loose-fill insulation must be determined by the ASTM E 84 test and must be 450 or less. The intent of this section is that the procedure for the smoke-developed index should be done by an ASTM E 84 test and not by the test procedures specified in Section 719.4.

719.7 Insulation and covering on pipe and tubing. Insulation and covering on pipe and tubing shall have a flame spread index of not more than 25 and a smoke-developed index of not more than 450.

> **Exception:** Insulation and covering on pipe and tubing installed in plenums shall comply with the *International Mechanical Code.*

❖ This section maintains the general provision for insulation by requiring a maximum flame spread index of 25 and smoke-developed index of 450, as stated in Sections 719.2 and 719.3; however, if exposed in a plenum, the IMC limits the smoke-developed index to 50.

SECTION 720
PRESCRIPTIVE FIRE RESISTANCE

720.1 General. The provisions of this section contain prescriptive details of fire-resistance-rated building elements, components or assemblies. The materials of construction listed in Tables 720.1(1), 720.1(2), and 720.1(3) shall be assumed to have the *fire-resistance ratings* prescribed therein. Where materials that change the capacity for heat dissipation are incorporated into a fire-resistance-rated assembly, fire test results or other substantiating data shall be made available to the *building official* to show that the required fire-resistance-rating time period is not reduced.

❖ Section 703.3 permits the fire-resistance ratings to be determined in a number of ways, including those found in Section 720 (see Section 703.3, Item 2). In this section, there are many prescriptive details for fire-resistance-rated construction, particularly those materials and assemblies listed in Table 720.1(1) for structural parts; Table 720.1(2) for walls and partitions and Table 720.1(3) for floor and roof systems. For the most part, the listed items have been tested in accordance with the fire-resistance ratings indicated. In addition, a similar footnote to all of the tables allows the acceptance of generic assemblies that are listed in the Gypsum Association's *Fire-resistance Design Manual,* GA 600. It is important to review all of the applicable footnotes when using a material or assembly from one of the tables.

As stated above, the fire-resistance ratings for the walls and partitions outlined in Table 720.1(2) are based on actual tests. For reinforced concrete walls, it is important to note the type of aggregate used. The difference in aggregates is quite significant for a 4-hour fire-resistance-rated wall, as it amounts to a difference in thickness of almost 2 inches (51 mm). For hollow-unit masonry walls, the thickness required for a particular fire-endurance rating is the equivalent thickness as defined in Section 721.3.1 for concrete masonry and Section 721.4.1.1 for clay masonry.

Table 720.1(3) provides fire-resistance ratings for floor/ceiling and roof/ceiling assemblies. Note n, which exempts unusable space from the flooring and ceiling requirements is especially important. This exemption is consistent with Section 711.3.3.

Often, materials, such as insulation, are added to fire-resistance-rated assemblies. The code requires substantiating fire test data to show that when the materials are added, they do not reduce the required fire-endurance time period. As an example, adding insulation to a floor/ceiling assembly may change its capacity to dissipate heat and, particularly for noncombustible assemblies, the fire-resistance rating may be changed.

Although the primary intent of the provision is to cover those cases where thermal insulation is added, the language is intentionally broad so that it applies to any material that might be added to the assembly.

720.1.1 Thickness of protective coverings. The thickness of fire-resistant materials required for protection of structural members shall be not less than set forth in Table 720.1(1), except as modified in this section. The figures shown shall be the net thickness of the protecting materials and shall not include any hollow space in back of the protection.

❖ In accordance with this section, the required thickness of insulating material used to provide fire resistance to a structural member cannot be less than the dimension established by Table 720.1(1), except for permitted modifications. An example of the minimum thickness of concrete required for a structural steel column is shown in Figure 720.1.1(1). Figure 720.1.1(2) illustrates the minimum concrete thickness requirements for protecting reinforcing steel in concrete columns, beams, girders and trusses. Refer to Section 714 for additional provisions regarding structural members.

720.1.2 Unit masonry protection. Where required, metal ties shall be embedded in bed joints of unit masonry for protection of steel columns. Such ties shall be as set forth in Table 720.1(1) or be equivalent thereto.

❖ Items 1-3.1 through 1-3.4 of Table 720.1(1) require horizontal joint reinforcement. This section stipulates that if ties are required to connect the masonry to the steel column, such ties are to be located in the bed joints, which is where the horizontal reinforcement is to be located.

720.1.3 Reinforcement for cast-in-place concrete column protection. Cast-in-place concrete protection for steel columns shall be reinforced at the edges of such members with wire ties of not less than 0.18 inch (4.6 mm) in diameter wound spirally around the columns on a pitch of not more than 8 inches (203 mm) or by equivalent reinforcement.

❖ In order for a concrete-encased steel column to maintain a fire-resistance rating, it is imperative that the concrete remain in place. Similar to a reinforced concrete column with spiral reinforcement (see Section 721.2.4.2), the ties required by this section are necessary for the concrete to remain in place around the column.

720.1.4 Plaster application. The finish coat is not required for plaster protective coatings where they comply with the design

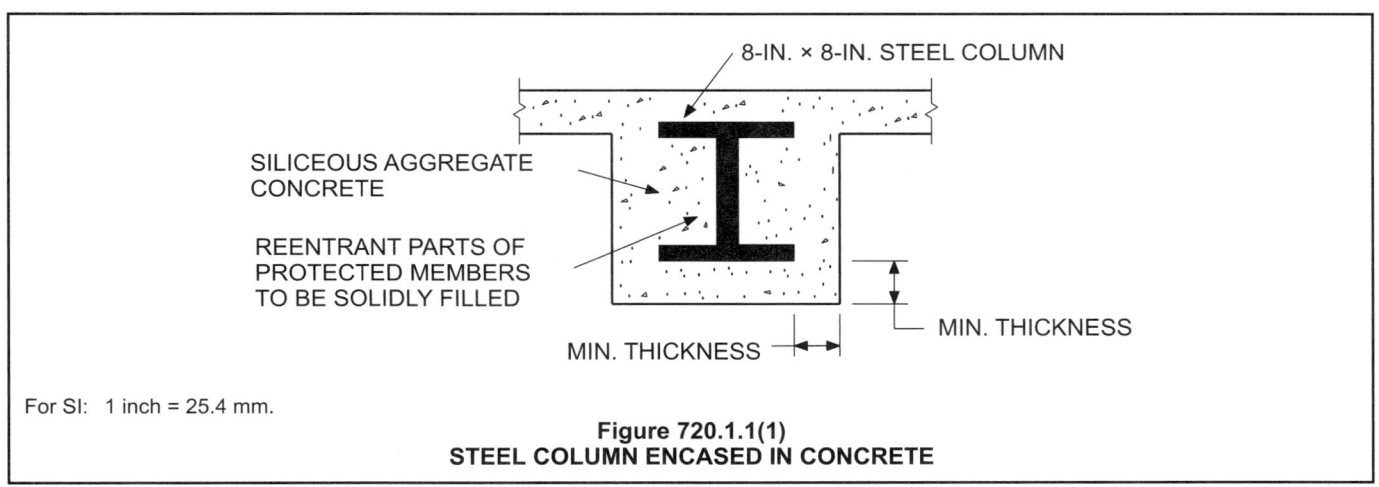

For SI: 1 inch = 25.4 mm.

Figure 720.1.1(1)
STEEL COLUMN ENCASED IN CONCRETE

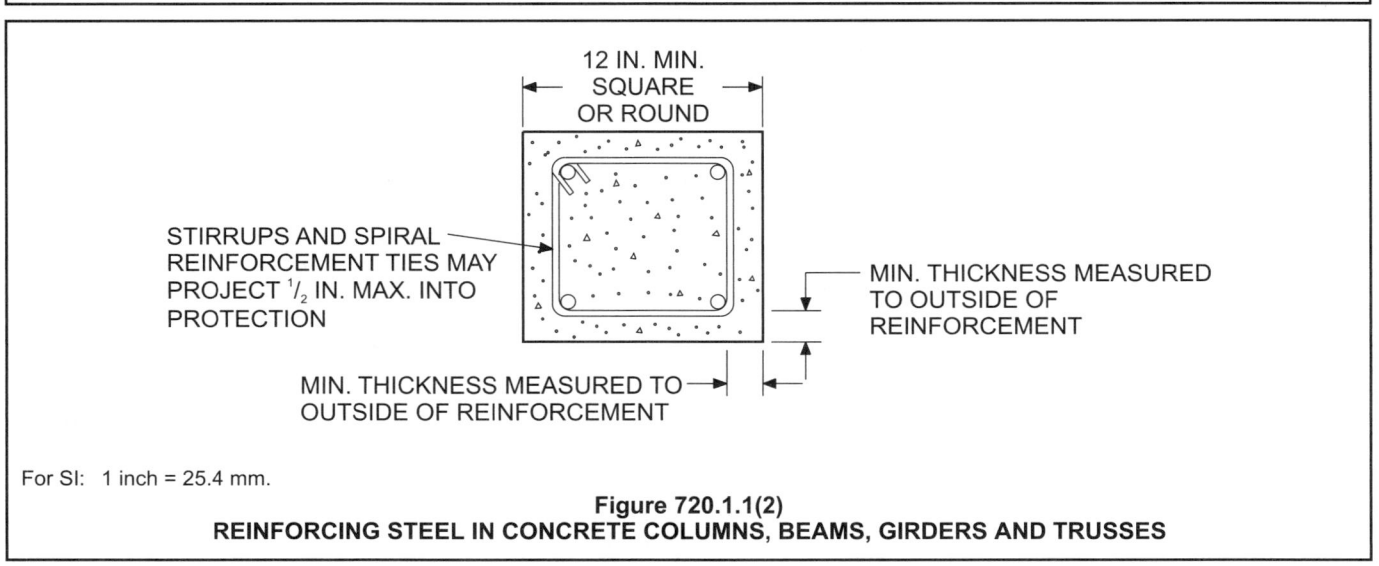

For SI: 1 inch = 25.4 mm.

Figure 720.1.1(2)
REINFORCING STEEL IN CONCRETE COLUMNS, BEAMS, GIRDERS AND TRUSSES

mix and thickness requirements of Tables 720.1(1), 720.1(2) and 720.1(3).

❖ The thickness of plaster for prescriptive assemblies, such as Item 1-4.1 in Table 720.1(1), Item 12-1.1 in Table 720.1(2) and Item 11-1.1 in Table 720.1(3), does not include the finish coat. Only the scratch coat and brown coat are needed to achieve the indicated fire-resistance ratings in the tables.

720.1.5 Bonded prestressed concrete tendons. For members having a single tendon or more than one tendon installed with equal concrete cover measured from the nearest surface, the cover shall not be less than that set forth in Table 720.1(1). For members having multiple tendons installed with variable concrete cover, the average tendon cover shall not be less than that set forth in Table 720.1(1), provided:

1. The clearance from each tendon to the nearest exposed surface is used to determine the average cover.

2. In no case can the clear cover for individual tendons be less than one-half of that set forth in Table 720.1(1). A minimum cover of $^3/_4$ inch (19.1 mm) for slabs and 1 inch (25 mm) for beams is required for any aggregate concrete.

3. For the purpose of establishing a *fire-resistance rating*, tendons having a clear covering less than that set forth in Table 720.1(1) shall not contribute more than 50 percent of the required ultimate moment capacity for members less than 350 square inches (0.226 m²) in cross-sectional area and 65 percent for larger members. For structural design purposes, however, tendons having a reduced cover are assumed to be fully effective.

❖ As the ultimate-moment capacity of the member is critical to its behavior under fire conditions, the code requires the reduction for those tendons having cover to be less than that specified by the code. Behavior at service loads, however, is less affected by the heat of a fire; therefore, the code permits those tendons with reduced cover to be fully effective (see Figure 720.1.5).

TABLE 720.1(1). See page 7-134.

❖ An example of the minimum thickness of concrete required for a structural steel column is shown in Figure 720.1.1(1). As shown, this figure depicts Item 1-1.5 of Table 720.1(1). Figure 720.1.1(2) depicts a reinforced concrete column based on Items 5-1.1 (see commentary, Section 720.1).

TABLE 720.1(2). See page 7-138.

❖ An example of the minimum thickness of concrete required for a wall is shown in Figure 720.1(2)(a). As shown, this figure depicts Item 4-1.1 of Table 720.1(2). Figure 720.1(2)(b) depicts a noncombustible stud exterior wall based on Item 15-1.4 (see commentary, Section 720.1).

TABLE 720.1(3). See page 7-146.

❖ An example of a 1-hour-rated wood floor or roof truss is shown in Figure 720.1(3). As shown, this figure depicts Item 21-1.1 (see commentary, Section 720.1).

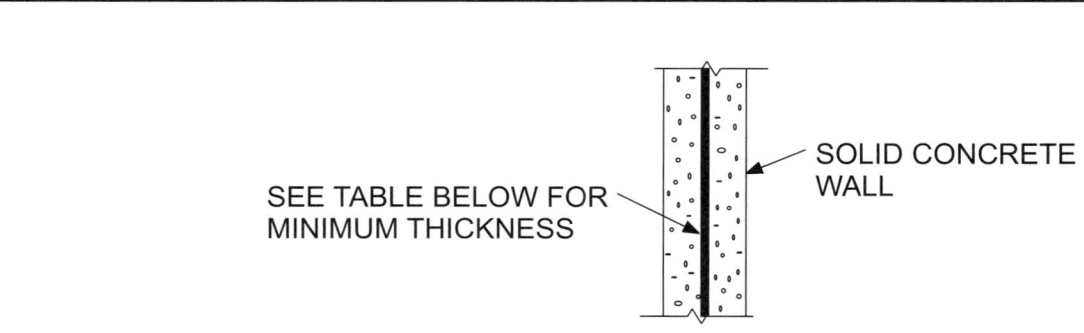

MATERIAL	ITEM NUMBER	CONSTRUCTION	MINIMUM FINISHED THICKNESS FACE-TO-FACE (inches)			
			4 hour	3 hour	2 hour	1 hour
Solid concrete	4-1.1	Siliceous aggregate concrete	7.0	6.2	5.0	3.5
		Carbondate aggregate concrete	6.6	5.7	4.6	3.2
		Sand-lightweight concrete	5.4	4.6	3.8	2.7
		Lightweight concrete	5.1	4.4	3.6	2.5

For SI: 1 inch = 25.4 mm.

Figure 720.1(2)(a)
CONCRETE WALL

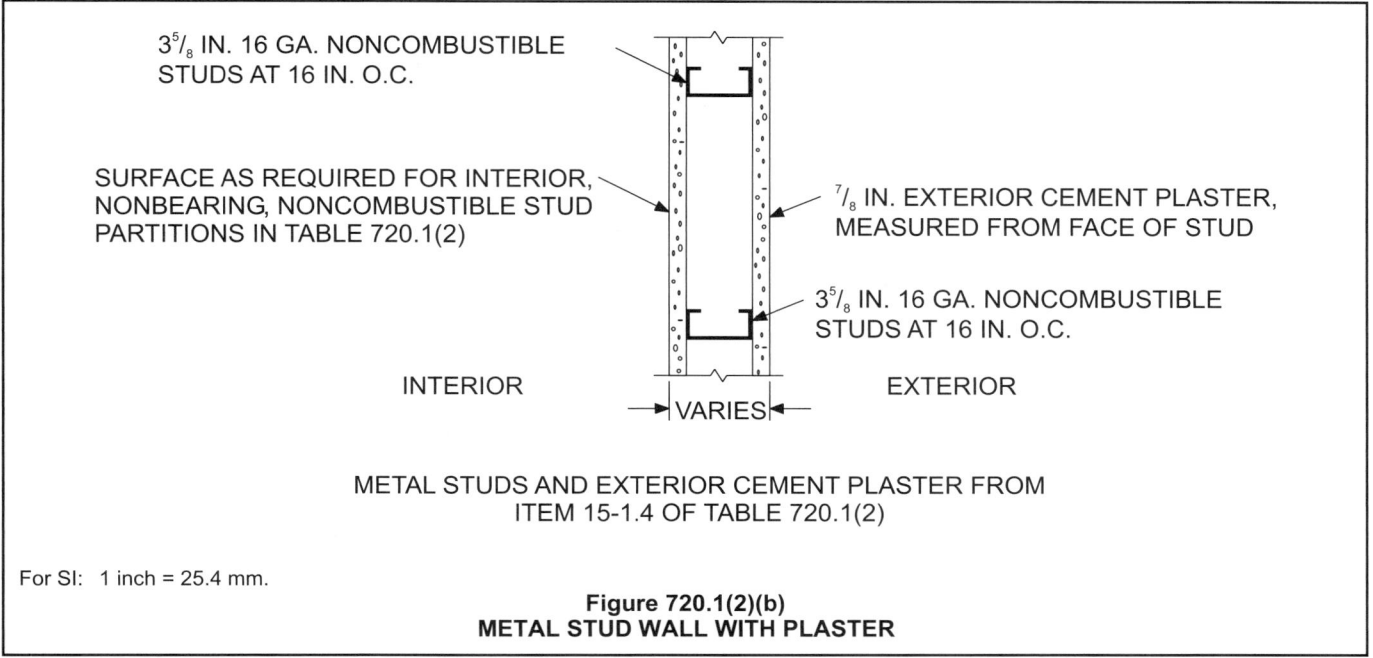

3⁵/₈ IN. 16 GA. NONCOMBUSTIBLE STUDS AT 16 IN. O.C.

SURFACE AS REQUIRED FOR INTERIOR, NONBEARING, NONCOMBUSTIBLE STUD PARTITIONS IN TABLE 720.1(2)

⁷/₈ IN. EXTERIOR CEMENT PLASTER, MEASURED FROM FACE OF STUD

3⁵/₈ IN. 16 GA. NONCOMBUSTIBLE STUDS AT 16 IN. O.C.

INTERIOR

VARIES

EXTERIOR

METAL STUDS AND EXTERIOR CEMENT PLASTER FROM ITEM 15-1.4 OF TABLE 720.1(2)

For SI: 1 inch = 25.4 mm.

Figure 720.1(2)(b)
METAL STUD WALL WITH PLASTER

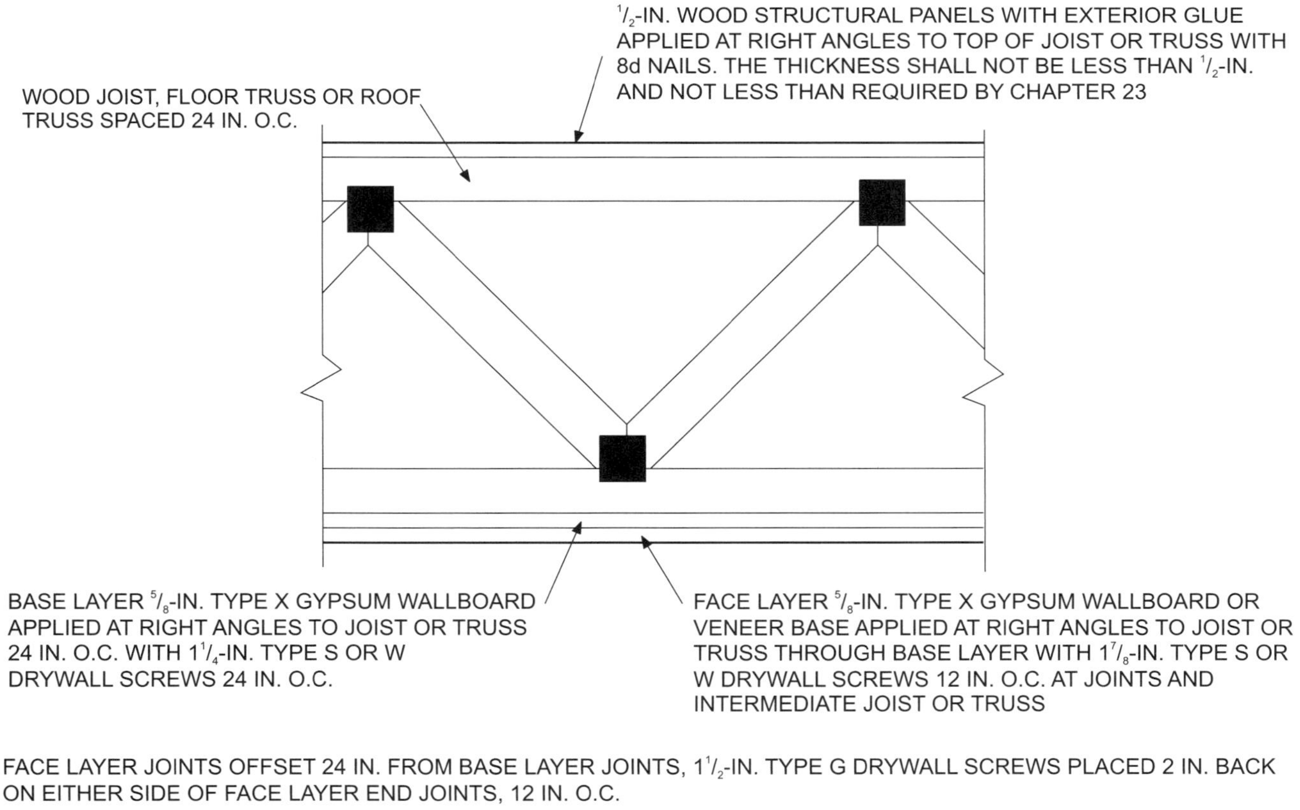

¹/₂-IN. WOOD STRUCTURAL PANELS WITH EXTERIOR GLUE APPLIED AT RIGHT ANGLES TO TOP OF JOIST OR TRUSS WITH 8d NAILS. THE THICKNESS SHALL NOT BE LESS THAN ¹/₂-IN. AND NOT LESS THAN REQUIRED BY CHAPTER 23

WOOD JOIST, FLOOR TRUSS OR ROOF TRUSS SPACED 24 IN. O.C.

BASE LAYER ⁵/₈-IN. TYPE X GYPSUM WALLBOARD APPLIED AT RIGHT ANGLES TO JOIST OR TRUSS 24 IN. O.C. WITH 1¹/₄-IN. TYPE S OR W DRYWALL SCREWS 24 IN. O.C.

FACE LAYER ⁵/₈-IN. TYPE X GYPSUM WALLBOARD OR VENEER BASE APPLIED AT RIGHT ANGLES TO JOIST OR TRUSS THROUGH BASE LAYER WITH 1⁷/₈-IN. TYPE S OR W DRYWALL SCREWS 12 IN. O.C. AT JOINTS AND INTERMEDIATE JOIST OR TRUSS

FACE LAYER JOINTS OFFSET 24 IN. FROM BASE LAYER JOINTS, 1¹/₂-IN. TYPE G DRYWALL SCREWS PLACED 2 IN. BACK ON EITHER SIDE OF FACE LAYER END JOINTS, 12 IN. O.C.

For SI: 1 inch = 25.4 mm.

Figure 720.1(3)
WOOD FLOOR TRUSS

TABLE 720.1(1)
MINIMUM PROTECTION OF STRUCTURAL PARTS BASED ON TIME PERIODS
FOR VARIOUS NONCOMBUSTIBLE INSULATING MATERIALS[m]

STRUCTURAL PARTS TO BE PROTECTED	ITEM NUMBER	INSULATING MATERIAL USED	MINIMUM THICKNESS OF INSULATING MATERIAL FOR THE FOLLOWING FIRE-RESISTANCE PERIODS (inches)			
			4 hour	3 hour	2 hour	1 hour
1. Steel columns and all of primary trusses	1-1.1	Carbonate, lightweight and sand-lightweight aggregate concrete, members 6″ × 6″ or greater (not including sandstone, granite and siliceous gravel).[a]	$2^1/_2$	2	$1^1/_2$	1
	1-1.2	Carbonate, lightweight and sand-lightweight aggregate concrete, members 8″ × 8″ or greater (not including sandstone, granite and siliceous gravel).[a]	2	$1^1/_2$	1	1
	1-1.3	Carbonate, lightweight and sand-lightweight aggregate concrete, members 12″ × 12″ or greater (not including sandstone, granite and siliceous gravel).[a]	$1^1/_2$	1	1	1
	1-1.4	Siliceous aggregate concrete and concrete excluded in Item 1-1.1, members 6″ × 6″ or greater.[a]	3	2	$1^1/_2$	1
	1-1.5	Siliceous aggregate concrete and concrete excluded in Item 1-1.1, members 8″ × 8″ or greater.[a]	$2^1/_2$	2	1	1
	1-1.6	Siliceous aggregate concrete and concrete excluded in Item 1-1.1, members 12″ × 12″ or greater.[a]	2	1	1	1
	1-2.1	Clay or shale brick with brick and mortar fill.[a]	$3^3/_4$	—	—	$2^1/_4$
	1-3.1	4″ hollow clay tile in two 2″ layers; $^1/_2$″ mortar between tile and column; $^3/_8$″ metal mesh 0.046″ wire diameter in horizontal joints; tile fill.[a]	4	—	—	—
	1-3.2	2″ hollow clay tile; $^3/_4$″ mortar between tile and column; $^3/_8$″ metal mesh 0.046″ wire diameter in horizontal joints; limestone concrete fill;[a] plastered with $^3/_4$″ gypsum plaster.	3	—	—	—
	1-3.3	2″ hollow clay tile with outside wire ties 0.08″ diameter at each course of tile or $^3/_8$″ metal mesh 0.046″ diameter wire in horizontal joints; limestone or trap-rock concrete fill[a] extending 1″ outside column on all sides.	—	—	3	—
	1-3.4	2″ hollow clay tile with outside wire ties 0.08″ diameter at each course of tile with or without concrete fill; $^3/_4$″ mortar between tile and column.	—	—	—	2
	1-4.1	Cement plaster over metal lath wire tied to $^3/_4$″ cold-rolled vertical channels with 0.049″ (No. 18 B.W. gage) wire ties spaced 3″ to 6″ on center. Plaster mixed 1:2 $^1/_2$ by volume, cement to sand.	—	—	$2^1/_2$[b]	$^7/_8$
	1-5.1	Vermiculite concrete, 1:4 mix by volume over paperbacked wire fabric lath wrapped directly around column with additional 2″ × 2″ 0.065″/0.065″ (No. 16/16 B.W. gage) wire fabric placed $^3/_4$″ from outer concrete surface. Wire fabric tied with 0.049″ (No. 18 B.W. gage) wire spaced 6″ on center for inner layer and 2″ on center for outer layer.	2	—	—	—
	1-6.1	Perlite or vermiculite gypsum plaster over metal lath wrapped around column and furred $1^1/_4$″ from column flanges. Sheets lapped at ends and tied at 6″ intervals with 0.049″ (No. 18 B.W. gage) tie wire. Plaster pushed through to flanges.	$1^1/_2$	1	—	—
	1-6.2	Perlite or vermiculite gypsum plaster over self-furring metal lath wrapped directly around column, lapped 1″ and tied at 6″ intervals with 0.049″ (No. 18 B.W. gage) wire.	$1^3/_4$	$1^3/_8$	1	—
	1-6.3	Perlite or vermiculite gypsum plaster on metal lath applied to $^3/_4$″ cold-rolled channels spaced 24″ apart vertically and wrapped flatwise around column.	$1^1/_2$	—	—	—
	1-6.4	Perlite or vermiculite gypsum plaster over two layers of $^1/_2$″ plain full-length gypsum lath applied tight to column flanges. Lath wrapped with 1″ hexagonal mesh of No. 20 gage wire and tied with doubled 0.035″ diameter (No. 18 B.W. gage) wire ties spaced 23″ on center. For three-coat work, the plaster mix for the second coat shall not exceed 100 pounds of gypsum to $2^1/_2$ cubic feet of aggregate for the 3-hour system.	$2^1/_2$	2	—	—

(continued)

TABLE 720.1(1)—continued
**MINIMUM PROTECTION OF STRUCTURAL PARTS BASED ON TIME PERIODS
FOR VARIOUS NONCOMBUSTIBLE INSULATING MATERIALS[m]**

STRUCTURAL PARTS TO BE PROTECTED	ITEM NUMBER	INSULATING MATERIAL USED	MINIMUM THICKNESS OF INSULATING MATERIAL FOR THE FOLLOWING FIRE-RESISTANCE PERIODS (inches)			
			4 hour	3 hour	2 hour	1 hour
1. Steel columns and all of primary trusses (continued)	1-6.5	Perlite or vermiculate gypsum plaster over one layer of $^1/_2$″ plain full-length gypsum lath applied tight to column flanges. Lath tied with doubled 0.049″ (No. 18 B.W. gage) wire ties spaced 23″ on center and scratch coat wrapped with 1″ hexagonal mesh 0.035″ (No. 20 B.W. gage) wire fabric. For three-coat work, the plaster mix for the second coat shall not exceed 100 pounds of gypsum to $2^1/_2$ cubic feet of aggregate.	—	2	—	—
	1-7.1	Multiple layers of $^1/_2$″ gypsum wallboard[c] adhesively[d] secured to column flanges and successive layers. Wallboard applied without horizontal joints. Corner edges of each layer staggered. Wallboard layer below outer layer secured to column with doubled 0.049″ (No. 18 B.W. gage) steel wire ties spaced 15″ on center. Exposed corners taped and treated.	—	—	2	1
	1-7.2	Three layers of $^5/_8$″ Type X gypsum wallboard.[c] First and second layer held in place by $^1/_8$″ diameter by $1^3/_8$″ long ring shank nails with $^5/_{16}$″ diameter heads spaced 24″ on center at corners. Middle layer also secured with metal straps at mid-height and 18″ from each end, and by metal corner bead at each corner held by the metal straps. Third layer attached to corner bead with 1″ long gypsum wallboard screws spaced 12″ on center.	—	—	$1^7/_8$	—
	1-7.3	Three layers of $^5/_8$″ Type X gypsum wallboard,[c] each layer screw attached to $1^5/_8$″ steel studs 0.018″ thick (No. 25 carbon sheet steel gage) at each corner of column. Middle layer also secured with 0.049″ (No. 18 B.W. gage) double-strand steel wire ties, 24″ on center. Screws are No. 6 by 1″ spaced 24″ on center for inner layer, No. 6 by $1^5/_8$″ spaced 12″ on center for middle layer and No. 8 by $2^1/_4$″ spaced 12″ on center for outer layer.	—	$1^7/_8$	—	—
	1-8.1	Wood-fibered gypsum plaster mixed 1:1 by weight gypsum-to-sand aggregate applied over metal lath. Lath lapped 1″ and tied 6″ on center at all end, edges and spacers with 0.049″ (No. 18 B.W. gage) steel tie wires. Lath applied over $^1/_2$″ spacers made of $^3/_4$″ furring channel with 2″ legs bent around each corner. Spacers located 1″ from top and bottom of member and a maximum of 40″ on center and wire tied with a single strand of 0.049″ (No. 18 B.W. gage) steel tie wires. Corner bead tied to the lath at 6″ on center along each corner to provide plaster thickness.	—	—	$1^5/_8$	—
	1-9.1	Minimum W8x35 wide flange steel column (w/d ≥ 0.75) with each web cavity filled even with the flange tip with normal weight carbonate or siliceous aggregate concrete (3,000 psi minimum compressive strength with 145 pcf ± 3 pcf unit weight). Reinforce the concrete in each web cavity with a minimum No. 4 deformed reinforcing bar installed vertically and centered in the cavity, and secured to the column web with a minimum No. 2 horizontal deformed reinforcing bar welded to the web every 18″ on center vertically. As an alternative to the No. 4 rebar, $^3/_4$″ diameter by 3″ long headed studs, spaced at 12″ on center vertically, shall be welded on each side of the web midway between the column flanges.	—	—	—	See Note n
2. Webs or flanges of steel beams and girders	2-1.1	Carbonate, lightweight and sand-lightweight aggregate concrete (not including sandstone, granite and siliceous gravel) with 3″ or finer metal mesh placed 1″ from the finished surface anchored to the top flange and providing not less than 0.025 square inch of steel area per foot in each direction.	2	$1^1/_2$	1	1
	2-1.2	Siliceous aggregate concrete and concrete excluded in Item 2-1.1 with 3″ or finer metal mesh placed 1″ from the finished surface anchored to the top flange and providing not less than 0.025 square inch of steel area per foot in each direction.	$2^1/_2$	2	$1^1/_2$	1
	2-2.1	Cement plaster on metal lath attached to $^3/_4$″ cold-rolled channels with 0.04″ (No. 18 B.W. gage) wire ties spaced 3″ to 6″ on center. Plaster mixed $1:2\ ^1/_2$ by volume, cement to sand.	—	—	$2^1/_2$[b]	$^7/_8$

(continued)

TABLE 720.1(1)—continued
MINIMUM PROTECTION OF STRUCTURAL PARTS BASED ON TIME PERIODS
FOR VARIOUS NONCOMBUSTIBLE INSULATING MATERIALS^m

STRUCTURAL PARTS TO BE PROTECTED	ITEM NUMBER	INSULATING MATERIAL USED	MINIMUM THICKNESS OF INSULATING MATERIAL FOR THE FOLLOWING FIRE-RESISTANCE PERIODS (inches)			
			4 hour	3 hour	2 hour	1 hour
2. Webs or flanges of steel beams and girders (continued)	2-3.1	Vermiculite gypsum plaster on a metal lath cage, wire tied to 0.165″ diameter (No. 8 B.W. gage) steel wire hangers wrapped around beam and spaced 16″ on center. Metal lath ties spaced approximately 5″ on center at cage sides and bottom.	—	$^7/_8$	—	—
	2-4.1	Two layers of $^5/_8$″ Type X gypsum wallboard^c are attached to U-shaped brackets spaced 24″ on center. 0.018″ thick (No. 25 carbon sheet steel gage) $1^5/_8$″ deep by 1″ galvanized steel runner channels are first installed parallel to and on each side of the top beam flange to provide a $^1/_2$″ clearance to the flange. The channel runners are attached to steel deck or concrete floor construction with approved fasteners spaced 12″ on center. U-shaped brackets are formed from members identical to the channel runners. At the bent portion of the U-shaped bracket, the flanges of the channel are cut out so that $1^5/_8$″ deep corner channels can be inserted without attachment parallel to each side of the lower flange. As an alternate, 0.021″ thick (No. 24 carbon sheet steel gage) 1″× 2″ runner and corner angles may be used in lieu of channels, and the web cutouts in the U-shaped brackets may be omitted. Each angle is attached to the bracket with $^1/_2$″-long No. 8 self-drilling screws. The vertical legs of the U-shaped bracket are attached to the runners with one $^1/_2$″ long No. 8 self-drilling screw. The completed steel framing provides a $2^1/_8$″ and $1^1/_2$″ space between the inner layer of wallboard and the sides and bottom of the steel beam, respectively. The inner layer of wallboard is attached to the top runners and bottom corner channels or corner angles with $1^1/_4$″-long No. 6 self-drilling screws spaced 16″ on center. The outer layer of wallboard is applied with $1^3/_4$″-long No. 6 self-drilling screws spaced 8″ on center. The bottom corners are reinforced with metal corner beads.	—	—	$1^1/_4$	—
	2-4.2	Three layers of $^5/_8$″ Type X gypsum wallboard^c attached to a steel suspension system as described immediately above utilizing the 0.018″ thick (No. 25 carbon sheet steel gage) 1″ × 2″ lower corner angles. The framing is located so that a $2^1/_8$″ and 2″ space is provided between the inner layer of wallboard and the sides and bottom of the beam, respectively. The first two layers of wallboard are attached as described immediately above. A layer of 0.035″ thick (No. 20 B.W. gage) 1″ hexagonal galvanized wire mesh is applied under the soffit of the middle layer and up the sides approximately 2″. The mesh is held in position with the No. 6 $1^5/_8$″-long screws installed in the vertical leg of the bottom corner angles. The outer layer of wallboard is attached with No. 6 $2^1/_4$″-long screws spaced 8″ on center. One screw is also installed at the mid-depth of the bracket in each layer. Bottom corners are finished as described above.	—	$1^7/_8$	—	—

3. Bonded pretensioned reinforcement in prestressed concrete^e	3-1.1	Carbonate, lightweight, sand-lightweight and siliceous^f aggregate concrete
		    Beams or girders
		
		    Solid slabs^h
			4^g	3^g
				
				2
					$2^1/_2$
					
					$1^1/_2$
						$1^1/_2$
						
						1

4. Bonded or unbonded post-tensioned tendons in prestressed concrete^e, i	4-1.1	Carbonate, lightweight, sand-lightweight and siliceous^f aggregate concrete Unrestrained members:
		    Solid slabs^h
		    Beams and girders^j
		        8″ wide
		        greater than 12″ wide
			—
			
			3
				2
				
				$4^1/_2$
				$2^1/_2$
					$1^1/_2$
					
					$2^1/_2$
					2
						—
						
						$1^3/_4$
						$1^1/_2$

	4-1.2	Carbonate, lightweight, sand-lightweight and siliceous aggregate Restrained members:^k
		    Solid slabs^h
		    Beams and girders^j
		        8″ wide
		        greater than 12″ wide
			$1^1/_4$
			
			$2^1/_2$
			2
				1
				
				2
				$1^3/_4$
					$^3/_4$
					
					$1^3/_4$
					$1^1/_2$
						—
						
						—
						—

(continued)

TABLE 720.1(1)—continued
MINIMUM PROTECTION OF STRUCTURAL PARTS BASED ON TIME PERIODS
FOR VARIOUS NONCOMBUSTIBLE INSULATING MATERIALS[m]

STRUCTURAL PARTS TO BE PROTECTED	ITEM NUMBER	INSULATING MATERIAL USED	MINIMUM THICKNESS OF INSULATING MATERIAL FOR THE FOLLOWING FIRE-RESISTANCE PERIODS (inches)			
			4 hour	3 hour	2 hour	1 hour
5. Reinforcing steel in reinforced concrete columns, beams girders and trusses	5-1.1	Carbonate, lightweight and sand-lightweight aggregate concrete, members 12″ or larger, square or round. (Size limit does not apply to beams and girders monolithic with floors.)	$1^1/_2$	$1^1/_2$	$1^1/_2$	$1^1/_2$
		Siliceous aggregate concrete, members 12″ or larger, square or round. (Size limit does not apply to beams and girders monolithic with floors.)	2	$1^1/_2$	$1^1/_2$	$1^1/_2$
6. Reinforcing steel in reinforced concrete joists[l]	6-1.1	Carbonate, lightweight and sand-lightweight aggregate concrete.	$1^1/_4$	$1^1/_4$	1	$^3/_4$
	6-1.2	Siliceous aggregate concrete.	$1^3/_4$	$1^1/_2$	1	$^3/_4$
7. Reinforcing and tie rods in floor and roof slabs[l]	7-1.1	Carbonate, lightweight and sand-lightweight aggregate concrete.	1	1	$^3/_4$	$^3/_4$
	7-1.2	Siliceous aggregate concrete.	$1^1/_4$	1	1	$^3/_4$

For SI: 1 inch = 25.4 mm, 1 square inch = 645.2 mm², 1 cubic foot = 0.0283 m³, 1 pound per cubic foot = 16.02 kg/m³.

a. Reentrant parts of protected members to be filled solidly.

b. Two layers of equal thickness with a $^3/_4$-inch airspace between.

c. For all of the construction with gypsum wallboard described in Table 720.1(1), gypsum base for veneer plaster of the same size, thickness and core type shall be permitted to be substituted for gypsum wallboard, provided attachment is identical to that specified for the wallboard and the joints on the face layer are reinforced, and the entire surface is covered with a minimum of $^1/_{16}$-inch gypsum veneer plaster.

d. An approved adhesive qualified under ASTM E 119 or UL 263.

e. Where lightweight or sand-lightweight concrete having an oven-dry weight of 110 pounds per cubic foot or less is used, the tabulated minimum cover shall be permitted to be reduced 25 percent, except that in no case shall the cover be less than $^3/_4$ inch in slabs or $1^1/_2$ inches in beams or girders.

f. For solid slabs of siliceous aggregate concrete, increase tendon cover 20 percent.

g. Adequate provisions against spalling shall be provided by U-shaped or hooped stirrups spaced not to exceed the depth of the member with a clear cover of 1 inch.

h. Prestressed slabs shall have a thickness not less than that required in Table 720.1(3) for the respective fire-resistance time period.

i. Fire coverage and end anchorages shall be as follows: Cover to the prestressing steel at the anchor shall be $^1/_2$ inch greater than that required away from the anchor. Minimum cover to steel-bearing plate shall be 1 inch in beams and $^3/_4$ inch in slabs.

j. For beam widths between 8 inches and 12 inches, cover thickness shall be permitted to be determined by interpolation.

k. Interior spans of continuous slabs, beams and girders shall be permitted to be considered restrained.

l. For use with concrete slabs having a comparable fire endurance where members are framed into the structure in such a manner as to provide equivalent performance to that of monolithic concrete construction.

m. Generic fire-resistance ratings (those not designated as PROPRIETARY* in the listing) in GA 600 shall be accepted as if herein listed.

n. No additional insulating material is required on the exposed outside face of the column flange to achieve a 1-hour fire-resistance rating.

TABLE 720.1(2)
RATED FIRE-RESISTANCE PERIODS FOR VARIOUS WALLS AND PARTITIONS [a, o, p]

MATERIAL	ITEM NUMBER	CONSTRUCTION	MINIMUM FINISHED THICKNESS FACE-TO-FACE[b] (inches)			
			4 hour	3 hour	2 hour	1 hour
1. Brick of clay or shale	1-1.1	Solid brick of clay or shale[c].	6	4.9	3.8	2.7
	1-1.2	Hollow brick, not filled.	5.0	4.3	3.4	2.3
	1-1.3	Hollow brick unit wall, grout or filled with perlite vermiculite or expanded shale aggregate.	6.6	5.5	4.4	3.0
	1-2.1	4″ nominal thick units at least 75 percent solid backed with a hat-shaped metal furring channel $^3/_4$″ thick formed from 0.021″ sheet metal attached to the brick wall on 24″ centers with approved fasteners, and $^1/_2$″ Type X gypsum wallboard attached to the metal furring strips with 1″-long Type S screws spaced 8″ on center.	—	—	5[d]	—
2. Combination of clay brick and load-bearing hollow clay tile	2-1.1	4″ solid brick and 4″ tile (at least 40 percent solid).	—	8	—	—
	2-1.2	4″ solid brick and 8″ tile (at least 40 percent solid).	12	—	—	—
3. Concrete masonry units	3-1.1[f, g]	Expanded slag or pumice.	4.7	4.0	3.2	2.1
	3-1.2[f, g]	Expanded clay, shale or slate.	5.1	4.4	3.6	2.6
	3-1.3[f]	Limestone, cinders or air-cooled slag.	5.9	5.0	4.0	2.7
	3-1.4[f, g]	Calcareous or siliceous gravel.	6.2	5.3	4.2	2.8
4. Solid concrete[h, i]	4-1.1	Siliceous aggregate concrete.	7.0	6.2	5.0	3.5
		Carbonate aggregate concrete.	6.6	5.7	4.6	3.2
		Sand-lightweight concrete.	5.4	4.6	3.8	2.7
		Lightweight concrete.	5.1	4.4	3.6	2.5
5. Glazed or unglazed facing tile, nonload-bearing	5-1.1	One 2″ unit cored 15 percent maximum and one 4″ unit cored 25 percent maximum with $^3/_4$″ mortar-filled collar joint. Unit positions reversed in alternate courses.	—	$6^3/_8$	—	—
	5-1.2	One 2″ unit cored 15 percent maximum and one 4″ unit cored 40 percent maximum with $^3/_4$″ mortar-filled collar joint. Unit positions side with $^3/_4$″ gypsum plaster. Two wythes tied together every fourth course with No. 22 gage corrugated metal ties.	—	$6^3/_4$	—	—
	5-1.3	One unit with three cells in wall thickness, cored 29 percent maximum.	—	—	6	—
	5-1.4	One 2″ unit cored 22 percent maximum and one 4″ unit cored 41 percent maximum with $^1/_4$″ mortar-filled collar joint. Two wythes tied together every third course with 0.030″ (No. 22 galvanized sheet steel gage) corrugated metal ties.	—	—	6	—
	5-1.5	One 4″ unit cored 25 percent maximum with $^3/_4$″ gypsum plaster on one side.	—	—	$4^3/_4$	—
	5-1.6	One 4″ unit with two cells in wall thickness, cored 22 percent maximum.	—	—	—	4
	5-1.7	One 4″ unit cored 30 percent maximum with $^3/_4$″ vermiculite gypsum plaster on one side.	—	—	$4^1/_2$	—
	5-1.8	One 4″ unit cored 39 percent maximum with $^3/_4$″ gypsum plaster on one side.	—	—	—	$4^1/_2$

(continued)

TABLE 720.1(2)—continued
RATED FIRE-RESISTANCE PERIODS FOR VARIOUS WALLS AND PARTITIONS [a, o, p]

MATERIAL	ITEM NUMBER	CONSTRUCTION	MINIMUM FINISHED THICKNESS FACE-TO-FACE[b] (inches)			
			4 hour	3 hour	2 hour	1 hour
6. Solid gypsum plaster	6-1.1	$^3/_4$″ by 0.055″ (No. 16 carbon sheet steel gage) vertical cold-rolled channels, 16″ on center with 2.6-pound flat metal lath applied to one face and tied with 0.049″ (No. 18 B.W. Gage) wire at 6″ spacing. Gypsum plaster each side mixed 1:2 by weight, gypsum to sand aggregate.	—	—	—	2^d
	6-1.2	$^3/_4$″ by 0.05″ (No. 16 carbon sheet steel gage) cold-rolled channels 16″ on center with metal lath applied to one face and tied with 0.049″ (No. 18 B.W. gage) wire at 6″ spacing. Perlite or vermiculite gypsum plaster each side. For three-coat work, the plaster mix for the second coat shall not exceed 100 pounds of gypsum to $2^1/_2$ cubic feet of aggregate for the 1-hour system.	—	—	$2^1/_2{}^d$	2^d
	6-1.3	$^3/_4$″ by 0.055″ (No. 16 carbon sheet steel gage) vertical cold-rolled channels, 16″ on center with $^3/_8$″ gypsum lath applied to one face and attached with sheet metal clips. Gypsum plaster each side mixed 1:2 by weight, gypsum to sand aggregate.	—	—	—	2^d
	6-2.1	Studless with $^1/_2$″ full-length plain gypsum lath and gypsum plaster each side. Plaster mixed 1:1 for scratch coat and 1:2 for brown coat, by weight, gypsum to sand aggregate.	—	—	—	2^d
	6-2.2	Studless with $^1/_2$″ full-length plain gypsum lath and perlite or vermiculite gypsum plaster each side.	—	—	$2^1/_2{}^d$	2^d
	6-2.3	Studless partition with $^3/_8$″ rib metal lath installed vertically adjacent edges tied 6″ on center with No. 18 gage wire ties, gypsum plaster each side mixed 1:2 by weight, gypsum to sand aggregate.	—	—	—	2^d
7. Solid perlite and portland cement	7-1.1	Perlite mixed in the ratio of 3 cubic feet to 100 pounds of portland cement and machine applied to stud side of $1^1/_2$″ mesh by 0.058-inch (No. 17 B.W. gage) paper-backed woven wire fabric lath wire-tied to 4″-deep steel trussed wire[j] studs 16″ on center. Wire ties of 0.049″ (No. 18 B.W. gage) galvanized steel wire 6″ on center vertically.	—	—	$3^1/_8{}^d$	—
8. Solid neat wood fibered gypsum plaster	8-1.1	$^3/_4$″ by 0.055-inch (No. 16 carbon sheet steel gage) cold-rolled channels, 12″ on center with 2.5-pound flat metal lath applied to one face and tied with 0.049″ (No. 18 B.W. gage) wire at 6″ spacing. Neat gypsum plaster applied each side.	—	—	2^d	—
9. Solid wallboard partition	9-1.1	One full-length layer $^1/_2$″ Type X gypsum wallboard[e] laminated to each side of 1″ full-length V-edge gypsum coreboard with approved laminating compound. Vertical joints of face layer and coreboard staggered at least 3″.	—	—	2^d	—
10. Hollow (studless) gypsum wallboard partition	10-1.1	One full-length layer of $^5/_8$″ Type X gypsum wallboard[e] attached to both sides of wood or metal top and bottom runners laminated to each side of 1″ × 6″ full-length gypsum coreboard ribs spaced 2″ on center with approved laminating compound. Ribs centered at vertical joints of face plies and joints staggered 24″ in opposing faces. Ribs may be recessed 6″ from the top and bottom.	—	—	—	$2^1/_4{}^d$
	10-1.2	1″ regular gypsum V-edge full-length backing board attached to both sides of wood or metal top and bottom runners with nails or $1^5/_8$″ drywall screws at 24″ on center. Minimum width of rumors $1^5/_8$″. Face layer of $^1/_2$″ regular full-length gypsum wallboard laminated to outer faces of backing board with approved laminating compound.	—	—	$4^5/_8{}^d$	—

(continued)

TABLE 720.1(2)—continued
RATED FIRE-RESISTANCE PERIODS FOR VARIOUS WALLS AND PARTITIONS [a, o, p]

MATERIAL	ITEM NUMBER	CONSTRUCTION	MINIMUM FINISHED THICKNESS FACE-TO-FACE[b] (inches)			
			4 hour	3 hour	2 hour	1 hour
11. Noncombustible studs—interior partition with plaster each side	11-1.1	$3^1/_4'' \times 0.044''$ (No. 18 carbon sheet steel gage) steel studs spaced 24″ on center. $^5/_8''$ gypsum plaster on metal lath each side mixed 1:2 by weight, gypsum to sand aggregate.	—	—	—	$4^3/_4''^d$
	11-1.2	$3^3/_8'' \times 0.055''$ (No. 16 carbon sheet steel gage) approved nailable[k] studs spaced 24″ on center. $^5/_8''$ neat gypsum wood-fibered plaster each side over $^3/_8''$ rib metal lath nailed to studs with 6d common nails, 8″ on center. Nails driven $1^1/_4''$ and bent over.	—	—	$5^5/_8$	—
	11-1.3	$4'' \times 0.044''$ (No. 18 carbon sheet steel gage) channel-shaped steel studs at 16″ on center. On each side approved resilient clips pressed onto stud flange at 16″ vertical spacing, $^1/_4''$ pencil rods snapped into or wire tied onto outer loop of clips, metal lath wire-tied to pencil rods at 6″ intervals, 1″ perlite gypsum plaster, each side.	—	$7^5/_8''^d$	—	—
	11-1.4	$2^1/_2'' \times 0.044''$ (No. 18 carbon sheet steel gage) steel studs spaced 16″ on center. Wood fibered gypsum plaster mixed 1:1 by weight gypsum to sand aggregate applied on $^3/_4$-pound metal lath wire tied to studs, each side. $^3/_4''$ plaster applied over each face, including finish coat.	—	—	$4^1/_4''^d$	—
12. Wood studs interior partition with plaster each side	12-1.1[l, m]	$2'' \times 4''$ wood studs 16″ on center with $^5/_8''$ gypsum plaster on metal lath. Lath attached by 4d common nails bent over or No. 14 gage by $1^1/_4''$ by $^3/_4''$ crown width staples spaced 6″ on center. Plaster mixed $1:1^1/_2$ for scratch coat and 1:3 for brown coat, by weight, gypsum to sand aggregate.	—	—	—	$5^1/_8$
	12-1.2[l]	$2'' \times 4''$ wood studs 16″ on center with metal lath and $^7/_8''$ neat wood-fibered gypsum plaster each side. Lath attached by 6d common nails, 7″ on center. Nails driven $1^1/_4''$ and bent over.	—	—	$5^1/_2''^d$	—
	12-1.3[l]	$2'' \times 4''$ wood studs 16″ on center with $^3/_8''$ perforated or plain gypsum lath and $^1/_2''$ gypsum plaster each side. Lath nailed with $1^1/_8''$ by No. 13 gage by $^{19}/_{64}''$ head plasterboard blued nails, 4″ on center. Plaster mixed 1:2 by weight, gypsum to sand aggregate.	—	—	—	$5^1/_4$
	12-1.4[l]	$2'' \times 4''$ wood studs 16″ on center with $^3/_8''$ Type X gypsum lath and $^1/_2''$ gypsum plaster each side. Lath nailed with $1^1/_8''$ by No. 13 gage by $^{19}/_{64}''$ head plasterboard blued nails, 5″ on center. Plaster mixed 1:2 by weight, gypsum to sand aggregate.	—	—	—	$5^1/_4$
13. Noncombustible studs—interior partition with gypsum wallboard each side	13-1.1	0.018″ (No. 25 carbon sheet steel gage) channel-shaped studs 24″ on center with one full-length layer of $^5/_8''$ Type X gypsum wallboard[e] applied vertically attached with 1″ long No. 6 drywall screws to each stud. Screws are 8″ on center around the perimeter and 12″ on center on the intermediate stud. The wallboard may be applied horizontally when attached to $3^5/_8''$ studs and the horizontal joints are staggered with those on the opposite side. Screws for the horizontal application shall be 8″ on center at vertical edges and 12″ on center at intermediate studs.	—	—	—	$2^7/_8''^d$
	13-1.2	0.018″ (No. 25 carbon sheet steel gage) channel-shaped studs 25″ on center with two full-length layers of $^1/_2''$ Type X gypsum wallboard[e] applied vertically each side. First layer attached with 1″-long, No. 6 drywall screws, 8″ on center around the perimeter and 12″ on center on the intermediate stud. Second layer applied with vertical joints offset one stud space from first layer using $1^5/_8''$ long, No. 6 drywall screws spaced 9″ on center along vertical joints, 12″ on center at intermediate studs and 24″ on center along top and bottom runners.	—	—	$3^5/_8''^d$	—
	13-1.3	0.055″ (No. 16 carbon sheet steel gage) approved nailable metal studs[e] 24″ on center with full-length $^5/_8''$ Type X gypsum wallboard[e] applied vertically and nailed 7″ on center with 6d cement-coated common nails. Approved metal fastener grips used with nails at vertical butt joints along studs.	—	—	—	$4^7/_8$

(continued)

TABLE 720.1(2)—continued
RATED FIRE-RESISTANCE PERIODS FOR VARIOUS WALLS AND PARTITIONS [a, o, p]

MATERIAL	ITEM NUMBER	CONSTRUCTION	MINIMUM FINISHED THICKNESS FACE-TO-FACE[b] (inches)			
			4 hour	3 hour	2 hour	1 hour
14. Wood studs—interior partition with gypsum wallboard each side	14-1.1[h, m]	$2'' \times 4''$ wood studs $16''$ on center with two layers of $3/8''$ regular gypsum wallboard[e] each side, 4d cooler[n] or wallboard[n] nails at $8''$ on center first layer, 5d cooler[n] or wallboard[n] nails at $8''$ on center second layer with laminating compound between layers, joints staggered. First layer applied full length vertically, second layer applied horizontally or vertically.	—	—	—	5
	14-1.2[l, m]	$2'' \times 4''$ wood studs $16''$ on center with two layers $1/2''$ regular gypsum wallboard[e] applied vertically or horizontally each side[k], joints staggered. Nail base layer with 5d cooler[n] or wallboard[n] nails at $8''$ on center face layer with 8d cooler[n] or wallboard[n] nails at $8''$ on center.	—	—	—	$5^1/_2$
	14-1.3[l, m]	$2'' \times 4''$ wood studs $24''$ on center with $5/8''$ Type X gypsum wallboard[e] applied vertically or horizontally nailed with 6d cooler[n] or wallboard[n] nails at $7''$ on center with end joints on nailing members. Stagger joints each side.	—	—	—	$4^3/_4$
	14-1.4[l]	$2'' \times 4''$ fire-retardant-treated wood studs spaced $24''$ on center with one layer of $5/8''$ Type X gypsum wallboard[e] applied with face paper grain (long dimension) parallel to studs. Wallboard attached with 6d cooler[n] or wallboard[n] nails at $7''$ on center.	—	—	—	$4^3/_4{}^d$
	14-1.5[l, m]	$2'' \times 4''$ wood studs $16''$ on center with two layers $5/8''$ Type X gypsum wallboard[e] each side. Base layers applied vertically and nailed with 6d cooler[n] or wallboard[n] nails at $9''$ on center. Face layer applied vertically or horizontally and nailed with 8d cooler[n] or wallboard[n] nails at $7''$ on center. For nail-adhesive application, base layers are nailed $6''$ on center. Face layers applied with coating of approved wallboard adhesive and nailed $12''$ on center.	—	—	6	—
	14-1.6[l]	$2'' \times 3''$ fire-retardant-treated wood studs spaced $24''$ on center with one layer of $5/8''$ Type X gypsum wallboard[e] applied with face paper grain (long dimension) at right angles to studs. Wallboard attached with 6d cement-coated box nails spaced $7''$ on center.	—	—	—	$3^5/_8{}^d$
15. Exterior or interior walls	15-1.1[l, m]	Exterior surface with $3/4''$ drop siding over $1/2''$ gypsum sheathing on $2'' \times 4''$ wood studs at $16''$ on center, interior surface treatment as required for 1-hour-rated exterior or interior $2'' \times 4''$ wood stud partitions. Gypsum sheathing nailed with $1^3/_4''$ by No. 11 gage by $7/16''$ head galvanized nails at $8''$ on center. Siding nailed with 7d galvanized smooth box nails.	—	—	—	Varies
	15-1.2[l, m]	$2'' \times 4''$ wood studs $16''$ on center with metal lath and $3/4''$ cement plaster on each side. Lath attached with 6d common nails $7''$ on center driven to $1''$ minimum penetration and bent over. Plaster mix 1:4 for scratch coat and 1:5 for brown coat, by volume, cement to sand.	—	—	—	$5^3/_8$
	15-1.3[l, m]	$2'' \times 4''$ wood studs $16''$ on center with $7/8''$ cement plaster (measured from the face of studs) on the exterior surface with interior surface treatment as required for interior wood stud partitions in this table. Plaster mix 1:4 for scratch coat and 1:5 for brown coat, by volume, cement to sand.	—	—	—	Varies
	15-1.4	$3^5/_8''$ No. 16 gage noncombustible studs $16''$ on center with $7/8''$ cement plaster (measured from the face of the studs) on the exterior surface with interior surface treatment as required for interior, nonbearing, noncombustible stud partitions in this table. Plaster mix 1:4 for scratch coat and 1:5 for brown coat, by volume, cement to sand.	—	—	—	Varies[d]

(continued)

TABLE 720.1(2)—continued
RATED FIRE-RESISTANCE PERIODS FOR VARIOUS WALLS AND PARTITIONS [a, o, p]

MATERIAL	ITEM NUMBER	CONSTRUCTION	MINIMUM FINISHED THICKNESS FACE-TO-FACE[b] (inches)			
			4 hour	3 hour	2 hour	1 hour
15. Exterior or interior walls (continued)	15-1.5[m]	$2^1/_4'' \times 3^3/_4''$ clay face brick with cored holes over $^1/_2''$ gypsum sheathing on exterior surface of $2'' \times 4''$ wood studs at 16'' on center and two layers $^5/_8''$ Type X gypsum wallboard[e] on interior surface. Sheathing placed horizontally or vertically with vertical joints over studs nailed 6'' on center with $1^3/_4'' \times$ No. 11 gage by $^7/_{16}''$ head galvanized nails. Inner layer of wallboard placed horizontally or vertically and nailed 8'' on center with 6d cooler[n] or wallboard[n] nails. Outer layer of wallboard placed horizontally or vertically and nailed 8'' on center with 8d cooler[n] or wallboard[n] nails. All joints staggered with vertical joints over studs. Outer layer joints taped and finished with compound. Nail heads covered with joint compound. 0.035 inch (No. 20 galvanized sheet gage) corrugated galvanized steel wall ties $^3/_4''$ by $6^5/_8''$ attached to each stud with two 8d cooler[n] or wallboard[n] nails every sixth course of bricks.	—	—	10	—
	15-1.6[l, m]	$2'' \times 6''$ fire-retardant-treated wood studs 16'' on center. Interior face has two layers of $^5/_8''$ Type X gypsum with the base layer placed vertically and attached with 6d box nails 12'' on center. The face layer is placed horizontally and attached with 8d box nails 8'' on center at joints and 12'' on center elsewhere. The exterior face has a base layer of $^5/_8''$ Type X gypsum sheathing placed vertically with 6d box nails 8'' on center at joints and 12'' on center elsewhere. An approved building paper is next applied, followed by self-furred exterior lath attached with $2^1/_2''$, No. 12 gage galvanized roofing nails with a $^3/_8''$ diameter head and spaced 6'' on center along each stud. Cement plaster consisting of a $^1/_2''$ brown coat is then applied. The scratch coat is mixed in the proportion of 1:3 by weight, cement to sand with 10 pounds of hydrated lime and 3 pounds of approved additives or admixtures per sack of cement. The brown coat is mixed in the proportion of 1:4 by weight, cement to sand with the same amounts of hydrated lime and approved additives or admixtures used in the scratch coat.	—	—	$8^1/_4$	—
	15-1.7[l, m]	$2'' \times 6''$ wood studs 16'' on center. The exterior face has a layer of $^5/_8''$ Type X gypsum sheathing placed vertically with 6d box nails 8'' on center at joints and 12'' on center elsewhere. An approved building paper is next applied, followed by 1'' by No. 18 gage self-furred exterior lath attached with 8d by $2^1/_2''$ long galvanized roofing nails spaced 6'' on center along each stud. Cement plaster consisting of a $^1/_2''$ scratch coat, a bonding agent and a $^1/_2''$ brown coat and a finish coat is then applied. The scratch coat is mixed in the proportion of 1:3 by weight, cement to sand with 10 pounds of hydrated lime and 3 pounds of approved additives or admixtures per sack of cement. The brown coat is mixed in the proportion of 1:4 by weight, cement to sand with the same amounts of hydrated lime and approved additives or admixtures used in the scratch coat. The interior is covered with $^3/_8''$ gypsum lath with 1'' hexagonal mesh of 0.035 inch (No. 20 B.W. gage) woven wire lath furred out $^5/_{16}''$ and 1'' perlite or vermiculite gypsum plaster. Lath nailed with $1^1/_8''$ by No. 13 gage by $^{19}/_{64}''$ head plasterboard glued nails spaced 5'' on center. Mesh attached by $1^3/_4''$ by No. 12 gage by $^3/_8''$ head nails with $^3/_8''$ furrings, spaced 8'' on center. The plaster mix shall not exceed 100 pounds of gypsum to $2^1/_2$ cubic feet of aggregate.	—	—	$8^3/_8$	—
	15-1.8[l, m]	$2'' \times 6''$ wood studs 16'' on center. The exterior face has a layer of $^5/_8''$ Type X gypsum sheathing placed vertically with 6d box nails 8'' on center at joints and 12'' on center elsewhere. An approved building paper is next applied, followed by $1^1/_2''$ by No. 17 gage self-furred exterior lath attached with 8d by $2^1/_2''$ long galvanized roofing nails spaced 6'' on center along each stud. Cement plaster consisting of a $^1/_2''$ scratch coat, and a $^1/_2''$ brown coat is then applied. The plaster may be placed by machine. The scratch coat is mixed in the proportion of 1:4 by weight, plastic cement to sand. The brown coat is mixed in the proportion of 1:5 by weight, plastic cement to sand. The interior is covered with $^3/_8''$ gypsum lath with 1'' hexagonal mesh of No. 20 gage woven wire lath furred out $^5/_{16}''$ and 1'' perlite or vermiculite gypsum plaster. Lath nailed with $1^1/_8''$ by No. 13 gage by $^{19}/_{64}''$ head plasterboard glued nails spaced 5'' on center. Mesh attached by $1^3/_4''$ by No. 12 gage by $^3/_8''$ head nails with $^3/_8''$ furrings, spaced 8'' on center. The plaster mix shall not exceed 100 pounds of gypsum to $2^1/_2$ cubic feet of aggregate.	—	—	$8^3/_8$	—

(continued)

|

TABLE 720.1(2)—continued
RATED FIRE-RESISTANCE PERIODS FOR VARIOUS WALLS AND PARTITIONS [a, o, p]

MATERIAL	ITEM NUMBER	CONSTRUCTION	MINIMUM FINISHED THICKNESS FACE-TO-FACE[b] (inches)			
			4 hour	3 hour	2 hour	1 hour
15. Exterior or interior walls (continued)	15-1.9	4″ No. 18 gage, nonload-bearing metal studs, 16″ on center, with 1″ portland cement lime plaster [measured from the back side of the $^3/_4$-pound expanded metal lath] on the exterior surface. Interior surface to be covered with 1″ of gypsum plaster on $^3/_4$-pound expanded metal lath proportioned by weight—1:2 for scratch coat, 1:3 for brown, gypsum to sand. Lath on one side of the partition fastened to $^1/_4$″ diameter pencil rods supported by No. 20 gage metal clips, located 16″ on center vertically, on each stud. 3″ thick mineral fiber insulating batts friction fitted between the studs.	—	—	$6^1/_2$[d]	—
	15-1.10	Steel studs 0.060″ thick, 4″ deep or 6″ at 16″ or 24″ centers, with $^1/_2$″ Glass Fiber Reinforced Concrete (GFRC) on the exterior surface. GFRC is attached with flex anchors at 24″ on center, with 5″ leg welded to studs with two $^1/_2$″-long flare-bevel welds, and 4″ foot attached to the GFRC skin with $^5/_8$″ thick GFRC bonding pads that extend $2^1/_2$″ beyond the flex anchor foot on both sides. Interior surface to have two layers of $^1/_2$″ Type X gypsum wallboard.[e] The first layer of wallboard to be attached with 1″-long Type S buglehead screws spaced 24″ on center and the second layer is attached with $1^5/_8$″-long Type S screws spaced at 12″ on center. Cavity is to be filled with 5″ of 4 pcf (nominal) mineral fiber batts. GFRC has $1^1/_2$″ returns packed with mineral fiber and caulked on the exterior.	—	—	$6^1/_2$	—
	15-1.11	Steel studs 0.060″ thick, 4″ deep or 6″ at 16″ or 24″ centers, respectively, with $^1/_2$″ Glass Fiber Reinforced Concrete (GFRC) on the exterior surface. GFRC is attached with flex anchors at 24″ on center, with 5″ leg welded to studs with two $^1/_2$″-long flare-bevel welds, and 4″ foot attached to the GFRC skin with $^5/_8$″-thick GFRC bonding pads that extend $2^1/_2$″ beyond the flex anchor foot on both sides. Interior surface to have one layer of $^5/_8$″ Type X gypsum wallboard[e], attached with $1^1/_4$″-long Type S buglehead screws spaced 12″ on center. Cavity is to be filled with 5″ of 4 pcf (nominal) mineral fiber batts. GFRC has $1^1/_2$″ returns packed with mineral fiber and caulked on the exterior.	—	—	—	$6^1/_8$
	15-1.12[q]	2″ × 6″ wood studs at 16″ with double top plates, single bottom plate; interior and exterior sides covered with $^5/_8$″ Type X gypsum wallboard, 4′ wide, applied horizontally or vertically with vertical joints over studs, and fastened with $2^1/_4$″ Type S drywall screws, spaced 12″ on center. Cavity to be filled with $5^1/_2$″ mineral wool insulation.	—	—	—	$6^3/_4$
	15-1.13[q]	2″ × 6″ wood studs at 16″ with double top plates, single bottom plate; interior and exterior sides covered with $^5/_8$″ Type X gypsum wallboard, 4′ wide, applied vertically with all joints over framing or blocking and fastened with $2^1/_4$″ Type S drywall screws, spaced 12″ on center. R-19 mineral fiber insulation installed in stud cavity.	—	—	—	$6^3/_4$
	15-1.14[q]	2″ × 6″ wood studs at 16″ with double top plates, single bottom plate; interior and exterior sides covered with $^5/_8$″ Type X gypsum wallboard, 4′ wide, applied horizontally or vertically with vertical joints over studs, and fastened with $2^1/_4$″ Type S drywall screws, spaced 7″ on center.	—	—	—	$6^3/_4$
	15-1.15[q]	2″ × 4″ wood studs at 16″ with double top plates, single bottom plate; interior and exterior sides covered with $^5/_8$″ Type X gypsum wallboard and sheathing, respectively, 4′ wide, applied horizontally or vertically with vertical joints over studs, and fastened with $2^1/_4$″ Type S drywall screws, spaced 12″ on center. Cavity to be filled with $3^1/_2$″ mineral wool insulation.	—	—	—	$4^3/_4$
	15-1.16[q]	2″ × 6″ wood studs at 24″ centers with double top plates, single bottom plate; interior and exterior side covered with two layers of $^5/_8$″ Type X gypsum wallboard, 4′ wide, applied horizontally with vertical joints over studs. Base layer fastened with $2^1/_4$″ Type S drywall screws, spaced 24″ on center and face layer fastened with Type S drywall screws, spaced 8″ on center, wallboard joints covered with paper tape and joint compound, fastener heads covered with joint compound. Cavity to be filled with $5^1/_2$″ mineral wool insulation.	—	—	$7^3/_4$	—

TABLE 720.1(2)—continued
RATED FIRE-RESISTANCE PERIODS FOR VARIOUS WALLS AND PARTITIONS [a, o, p]

MATERIAL	ITEM NUMBER	CONSTRUCTION	MINIMUM FINISHED THICKNESS FACE-TO-FACE[b] (inches)			
			4 hour	3 hour	2 hour	1 hour
15. Exterior or interior walls (continued)	15-2.1[d]	$3^5/_8''$ No. 16 gage steel studs at 24″ on center or 2″ × 4″ wood studs at 24″ on center. Metal lath attached to the exterior side of studs with minimum 1″ long No. 6 drywall screws at 6″ on center and covered with minimum $3/_4''$ thick portland cement plaster. Thin veneer brick units of clay or shale complying with ASTM C 1088, Grade TBS or better, installed in running bond in accordance with Section 1405.10. Combined total thickness of the portland cement plaster, mortar and thin veneer brick units shall be not less than $1^3/_4''$. Interior side covered with one layer of $5/_8''$ thick Type X gypsum wallboard attached to studs with 1″ long No. 6 drywall screws at 12″ on center.				6
	15-2.2[d]	$3^5/_8''$ No. 16 gage steel studs at 24″ on center or 2″ × 4″ wood studs at 24″ on center. Metal lath attached to the exterior side of studs with minimum 1″ long No. 6 drywall screws at 6″ on center and covered with minimum $3/_4''$ thick portland cement plaster. Thin veneer brick units of clay or shale complying with ASTM C 1088, Grade TBS or better, installed in running bond in accordance with Section 1405.10. Combined total thickness of the portland cement plaster, mortar and thin veneer brick units shall be not less than 2″. Interior side covered with two layers of $5/_8''$ thick Type X gypsum wallboard. Bottom layer attached to studs with 1″ long No. 6 drywall screws at 24″ on center. Top layer attached to studs with $1^5/_8''$ long No. 6 drywall screws at 12″ on center.			$6^7/_8$	
	15-2.3[d]	$3^5/_8''$ No. 16 gage steel studs at 16″ on center or 2″ × 4″ wood studs at 16″ on center. Where metal lath is used, attach to the exterior side of studs with minimum 1″ long No. 6 drywall screws at 6″ on center. Brick units of clay or shale not less than $2^5/_8''$ thick complying with ASTM C 216 installed in accordance with Section 1405.6 with a minimum 1″ air space. Interior side covered with one layer of $5/_8''$ thick Type X gypsum wallboard attached to studs with 1″ long No. 6 drywall screws at 12″ on center.				$7^7/_8$
	15-2.4[d]	$3^5/_8''$ No. 16 gage steel studs at 16″ on center or 2″ × 4″ wood studs at 16″ on center. Where metal lath is used, attach to the exterior side of studs with minimum 1″ long No. 6 drywall screws at 6″ on center. Brick units of clay or shale not less than $2^5/_8''$ thick complying with ASTM C 216 installed in accordance with Section 1405.6 with a minimum 1″ air space. Interior side covered with two layers of $5/_8''$ thick Type X gypsum wallboard. Bottom layer attached to studs with 1″ long No. 6 drywall screws at 24″ on center. Top layer attached to studs with $1^5/_8''$ long No. 6 drywall screws at 12″ on center.			$8^1/_2$	
16. Exterior walls rated for fire resistance from the inside only in accordance with Section 705.5.	16-1.1[q]	2″ × 4″ wood studs at 16″ centers with double top plates, single bottom plate; interior side covered with $5/_8''$ Type X gypsum wallboard, 4′ wide, applied horizontally unblocked, and fastened with $2^1/_4''$ Type S drywall screws, spaced 12″ on center, wallboard joints covered with paper tape and joint compound, fastener heads covered with joint compound. Exterior covered with $3/_8''$ wood structural panels, applied vertically, horizontal joints blocked and fastened with 6d common nails (bright) — 12″ on center in the field, and 6″ on center panel edges. Cavity to be filled with $3^1/_2''$ mineral wool insulation. Rating established for exposure from interior side only.	—	—	—	$4^1/_2$

(continued)

TABLE 720.1(2)—continued
RATED FIRE-RESISTANCE PERIODS FOR VARIOUS WALLS AND PARTITIONS [a, o, p]

MATERIAL	ITEM NUMBER	CONSTRUCTION	MINIMUM FINISHED THICKNESS FACE-TO-FACE [b] (inches)			
			4 hour	3 hour	2 hour	1 hour
16. Exterior walls rated for fire resistance from the inside only in accordance with Section 705.5. (continued)	16-1.2[q]	$2'' \times 6''$ (51mm x 152 mm) wood studs at 16″ centers with double top plates, single bottom plate; interior side covered with $^5/_8''$ Type X gypsum wallboard, 4′ wide, applied horizontally or vertically with vertical joints over studs and fastened with $2^1/_4''$ Type S drywall screws, spaced 12″ on center, wallboard joints covered with paper tape and joint compound, fastener heads covered with joint compound, exterior side covered with $^7/_{16}''$ wood structural panels fastened with 6d common nails (bright) spaced 12″ on center in the field and 6″ on center along the panel edges. Cavity to be filled with $5^1/_2''$ mineral wool insulation. Rating established from the gypsum-covered side only.	—	—	—	$6^9/_{16}$
	16-1.3	$2'' \times 6''$ wood studs at 16″ centers with double top plates, single bottom plates; interior side covered with $^5/_8''$ Type X gypsum wallboard, 4′ wide, applied vertically with all joints over framing or blocking and fastened with $2^1/_4''$ Type S drywall screws spaced 7″ on center. Joints to be covered with tape and joint compound. Exterior covered with $^3/_8''$ wood structural panels, applied vertically with edges over framing or blocking and fastened with 6d common nails (bright) at 12″ on center in the field and 6″ on center on panel edges. R-19 mineral fiber insulation installed in stud cavity. Rating established from the gypsum-covered side only.	—	—	—	$6^1/_2$

For SI: 1 inch = 25.4 mm, 1 square inch = 645.2 mm², 1 cubic foot = 0.0283 m³.

a. Staples with equivalent holding power and penetration shall be permitted to be used as alternate fasteners to nails for attachment to wood framing.

b. Thickness shown for brick and clay tile is nominal thicknesses unless plastered, in which case thicknesses are net. Thickness shown for concrete masonry and clay masonry is equivalent thickness defined in Section 721.3.1 for concrete masonry and Section 721.4.1.1 for clay masonry. Where all cells are solid grouted or filled with silicone-treated perlite loose-fill insulation; vermiculite loose-fill insulation; or expanded clay, shale or slate lightweight aggregate, the equivalent thickness shall be the thickness of the block or brick using specified dimensions as defined in Chapter 21. Equivalent thickness may also include the thickness of applied plaster and lath or gypsum wallboard, where specified.

c. For units in which the net cross-sectional area of cored brick in any plane parallel to the surface containing the cores is at least 75 percent of the gross cross-sectional area measured in the same plane.

d. Shall be used for nonbearing purposes only.

e. For all of the construction with gypsum wallboard described in this table, gypsum base for veneer plaster of the same size, thickness and core type shall be permitted to be substituted for gypsum wallboard, provided attachment is identical to that specified for the wallboard, and the joints on the face layer are reinforced and the entire surface is covered with a minimum of $^1/_{16}$-inch gypsum veneer plaster.

f. The fire-resistance time period for concrete masonry units meeting the equivalent thicknesses required for a 2-hour fire-resistance rating in Item 3, and having a thickness of not less than $7^5/_8$ inches is 4 hours when cores which are not grouted are filled with silicone-treated perlite loose-fill insulation; vermiculite loose-fill insulation; or expanded clay, shale or slate lightweight aggregate, sand or slag having a maximum particle size of $^3/_8$ inch.

g. The fire-resistance rating of concrete masonry units composed of a combination of aggregate types or where plaster is applied directly to the concrete masonry shall be determined in accordance with ACI 216.1/TMS 0216. Lightweight aggregates shall have a maximum combined density of 65 pounds per cubic foot.

h. See also Note b. The equivalent thickness shall be permitted to include the thickness of cement plaster or 1.5 times the thickness of gypsum plaster applied in accordance with the requirements of Chapter 25.

i. Concrete walls shall be reinforced with horizontal and vertical temperature reinforcement as required by Chapter 19.

j. Studs are welded truss wire studs with 0.18 inch (No. 7 B.W. gage) flange wire and 0.18 inch (No. 7 B.W. gage) truss wires.

k. Nailable metal studs consist of two channel studs spot welded back to back with a crimped web forming a nailing groove.

l. Wood structural panels shall be permitted to be installed between the fire protection and the wood studs on either the interior or exterior side of the wood frame assemblies in this table, provided the length of the fasteners used to attach the fire protection is increased by an amount at least equal to the thickness of the wood structural panel.

m. The design stress of studs shall be reduced to 78 percent of allowable F'_c with the maximum not greater than 78 percent of the calculated stress with studs having a slenderness ratio l_e/d of 33.

n. For properties of cooler or wallboard nails, see ASTM C 514, ASTM C 547 or ASTM F 1667.

o. Generic fire-resistance ratings (those not designated as PROPRIETARY* in the listing) in the GA 600 shall be accepted as if herein listed.

p. NCMA TEK 5-8A shall be permitted for the design of fire walls.

q. The design stress of studs shall be equal to a maximum of 100 percent of the allowable F'_c calculated in accordance with Section 2306.

TABLE 720.1(3)
MINIMUM PROTECTION FOR FLOOR AND ROOF SYSTEMS[a, q]

FLOOR OR ROOF CONSTRUCTION	ITEM NUMBER	CEILING CONSTRUCTION	THICKNESS OF FLOOR OR ROOF SLAB (inches)				MINIMUM THICKNESS OF CEILING (inches)			
			4 hour	3 hour	2 hour	1 hour	4 hour	3 hour	2 hour	1 hour
1. Siliceous aggregate concrete	1-1.1	Slab (no ceiling required). Minimum cover over nonprestressed reinforcement shall not be less than $^3/_4$″ [b].	7.0	6.2	5.0	3.5	—	—	—	—
2. Carbonate aggregate concrete	2-1.1		6.6	5.7	4.6	3.2	—	—	—	—
3. Sand-lightweight concrete	3-1.1		5.4	4.6	3.8	2.7	—	—	—	—
4. Lightweight concrete	4-1.1		5.1	4.4	3.6	2.5	—	—	—	—
5. Reinforced concrete	5-1.1	Slab with suspended ceiling of vermiculite gypsum plaster over metal lath attached to $^3/_4$″ cold-rolled channels spaced 12″ on center. Ceiling located 6″ minimum below joists.	3	2	—	—	1	$^3/_4$	—	—
	5-2.1	$^3/_8$″ Type X gypsum wallboard[c] attached to 0.018 inch (No. 25 carbon sheet steel gage) by $^7/_8$″ deep by $2^5/_8$″ hat-shaped galvanized steel channels with 1″-long No. 6 screws. The channels are spaced 24″ on center, span 35″ and are supported along their length at 35″ intervals by 0.033″ (No. 21 galvanized sheet gage) galvanized steel flat strap hangers having formed edges that engage the lips of the channel. The strap hangers are attached to the side of the concrete joists with $^5/_{32}$″ by $1^1/_4$″ long power-driven fasteners. The wallboard is installed with the long dimension perpendicular to the channels. All end joints occur on channels and supplementary channels are installed parallel to the main channels, 12″ each side, at end joint occurrences. The finished ceiling is located approximately 12″ below the soffit of the floor slab.	—	—	$2^1/_2$	—	—	—	$^5/_8$	—
6. Steel joists constructed with a poured reinforced concrete slab on metal lath forms or steel form units[d, e]	6-1.1	Gypsum plaster on metal lath attached to the bottom cord with single No. 16 gage or doubled No. 18 gage wire ties spaced 6″ on center. Plaster mixed 1:2 for scratch coat, 1:3 for brown coat, by weight, gypsum-to-sand aggregate for 2-hour system. For 3-hour system plaster is neat.	—	—	$2^1/_2$	$2^1/_4$	—	—	$^3/_4$	$^5/_8$
	6-2.1	Vermiculite gypsum plaster on metal lath attached to the bottom chord with single No.16 gage or doubled 0.049-inch (No. 18 B.W. gage) wire ties 6″ on center.	—	2	—	—	—	$^5/_8$	—	—
	6-3.1	Cement plaster over metal lath attached to the bottom chord of joists with single No. 16 gage or doubled 0.049″ (No. 18 B.W. gage) wire ties spaced 6″ on center. Plaster mixed 1:2 for scratch coat, 1:3 for brown coat for 1-hour system and 1:1 for scratch coat, 1:1 $^1/_2$ for brown coat for 2-hour system, by weight, cement to sand.	—	—	—	2	—	—	—	$^5/_8$[f]
	6-4.1	Ceiling of $^5/_8$″ Type X wallboard[c] attached to $^7/_8$″ deep by $2^5/_8$″ by 0.021 inch (No. 25 carbon sheet steel gage) hat-shaped furring channels 12″ on center with 1″ long No. 6 wallboard screws at 8″ on center. Channels wire tied to bottom chord of joists with doubled 0.049 inch (No. 18 B.W. gage) wire or suspended below joists on wire hangers.[g]	—	—	$2^1/_2$	—	—	—	$^5/_8$	—
	6-5.1	Wood-fibered gypsum plaster mixed 1:1 by weight gypsum to sand aggregate applied over metal lath. Lath tied 6″ on center to $^3/_4$″ channels spaced $13^1/_2$″ on center. Channels secured to joists at each intersection with two strands of 0.049 inch (No. 18 B.W. gage) galvanized wire.	—	—	$2^1/_2$	—	—	—	$^3/_4$	—

(continued)

2009 INTERNATIONAL BUILDING CODE® COMMENTARY

TABLE 720.1(3)—continued
MINIMUM PROTECTION FOR FLOOR AND ROOF SYSTEMS[a, q]

FLOOR OR ROOF CONSTRUCTION	ITEM NUMBER	CEILING CONSTRUCTION	THICKNESS OF FLOOR OR ROOF SLAB (inches)				MINIMUM THICKNESS OF CEILING (inches)			
			4 hour	3 hour	2 hour	1 hour	4 hour	3 hour	2 hour	1 hour
7. Reinforced concrete slabs and joists with hollow clay tile fillers laid end to end in rows $2^1/_2''$ or more apart; reinforcement placed between rows and concrete cast around and over tile.	7-1.1	$^5/_8''$ gypsum plaster on bottom of floor or roof construction.	—	—	8^h	—	—	—	$^5/_8$	—
	7-1.2	None	—	—	—	$5^1/_2{}^i$	—	—	—	—
8. Steel joists constructed with a reinforced concrete slab on top poured on a $^1/_2''$ deep steel deck.[e]	8-1.1	Vermiculite gypsum plaster on metal lath attached to $^3/_4''$ cold-rolled channels with 0.049'' (No. 18 B.W. gage) wire ties spaced 6'' on center.	$2^1/_2{}^j$	—	—	—	$^3/_4$	—	—	—
9. 3'' deep cellular steel deck with concrete slab on top. Slab thickness measured to top.	9-1.1	Suspended ceiling of vermiculite gypsum plaster base coat and vermiculite acoustical plaster on metal lath attached at 6'' intervals to $^3/_4''$ cold-rolled channels spaced 12'' on center and secured to $1^1/_2''$ cold-rolled channels spaced 36'' on center with 0.065'' (No. 16 B.W. gage) wire. $1^1/_2''$ channels supported by No. 8 gage wire hangers at 36'' on center. Beams within envelope and with a $2^1/_2''$ airspace between beam soffit and lath have a 4-hour rating.	$2^1/_2$	—	—	—	$1^1/_8{}^k$	—	—	—
10. $1^1/_2''$-deep steel roof deck on steel framing. Insulation board, 30 pcf density, composed of wood fibers with cement binders of thickness shown bonded to deck with unified asphalt adhesive. Covered with a Class A or B roof covering.	10-1.1	Ceiling of gypsum plaster on metal lath. Lath attached to $^3/_4''$ furring channels with 0.049'' (No. 18 B.W. gage) wire ties spaced 6'' on center. $^3/_4''$ channel saddle tied to 2'' channels with doubled 0.065'' (No. 16 B.W. gage) wire ties. 2'' channels spaced 36'' on center suspended 2'' below steel framing and saddle-tied with 0.165'' (No. 8 B.W. gage) wire. Plaster mixed 1:2 by weight, gypsum-to-sand aggregate.	—	—	$1^7/_8$	1	—	—	$^3/_4{}^l$	$^3/_4{}^l$
11. $1^1/_2''$-deep steel roof deck on steel-framing wood fiber insulation board, 17.5 pcf density on top applied over a 15-lb asphalt-saturated felt. Class A or B roof covering.	11-1.1	Ceiling of gypsum plaster on metal lath. Lath attached to $^3/_4''$ furring channels with 0.049'' (No. 18 B.W. gage) wire ties spaced 6'' on center. $^3/_4''$ channels saddle tied to 2'' channels with doubled 0.065'' (No. 16 B.W. gage) wire ties. 2'' channels spaced 36'' on center suspended 2'' below steel framing and saddle tied with 0.165'' (No. 8 B.W. gage) wire. Plaster mixed 1:2 for scratch coat and 1:3 for brown coat, by weight, gypsum-to-sand aggregate for 1-hour system. For 2-hour system, plaster mix is 1:2 by weight, gypsum-to-sand aggregate.	—	—	$1^1/_2$	1	—	—	$^7/_8{}^g$	$^3/_4{}^l$

(continued)

TABLE 720.1(3)—continued
MINIMUM PROTECTION FOR FLOOR AND ROOF SYSTEMS[a, q]

FLOOR OR ROOF CONSTRUCTION	ITEM NUMBER	CEILING CONSTRUCTION	THICKNESS OF FLOOR OR ROOF SLAB (inches)				MINIMUM THICKNESS OF CEILING (inches)			
			4 hour	3 hour	2 hour	1 hour	4 hour	3 hour	2 hour	1 hour
12. $1^1/_2$″ deep steel roof deck on steel-framing insulation of rigid board consisting of expanded perlite and fibers impregnated with integral asphalt waterproofing; density 9 to 12 pcf secured to metal roof deck by $1/_2$″ wide ribbons of waterproof, cold-process liquid adhesive spaced 6″ apart. Steel joist or light steel construction with metal roof deck, insulation, and Class A or B built-up roof covering.[e]	12-1.1	Gypsum-vermiculite plaster on metal lath wire tied at 6″ intervals to $3/_4$″ furring channels spaced 12″ on center and wire tied to 2″ runner channels spaced 32″ on center. Runners wire tied to bottom chord of steel joists.	—	—	1	—	—	—	$7/_8$	—
13. Double wood floor over wood joists spaced 16″ on center.[m,n]	13-1.1	Gypsum plaster over $3/_8$″ Type X gypsum lath. Lath initially applied with not less than four $1^1/_8$″ by No. 13 gage by $19/_{64}$″ head plasterboard blued nails per bearing. Continuous stripping over lath along all joist lines. Stripping consists of 3″ wide strips of metal lath attached by $1^1/_2$″ by No. 11 gage by $1/_2$″ head roofing nails spaced 6″ on center. Alternate stripping consists of 3″ wide 0.049″ diameter wire stripping weighing 1 pound per square yard and attached by No.16 gage by $1^1/_2$″ by $3/_4$″ crown width staples, spaced 4″ on center. Where alternate stripping is used, the lath nailing may consist of two nails at each end and one nail at each intermediate bearing. Plaster mixed 1:2 by weight, gypsum-to-sand aggregate.	—	—	—	—	—	—	—	$7/_8$
	13-1.2	Cement or gypsum plaster on metal lath. Lath fastened with $1^1/_2$″ by No. 11 gage by $7/_{16}$″ head barbed shank roofing nails spaced 5″ on center. Plaster mixed 1:2 for scratch coat and 1:3 for brown coat, by weight, cement to sand aggregate.	—	—	—	—	—	—	—	$5/_8$
	13-1.3	Perlite or vermiculite gypsum plaster on metal lath secured to joists with $1^1/_2$″ by No. 11 gage by $7/_{16}$″ head barbed shank roofing nails spaced 5″ on center.	—	—	—	—	—	—	—	$5/_8$
	13-1.4	$1/_2$″ Type X gypsum wallboard[c] nailed to joists with 5d cooler[o] or wallboard[o] nails at 6″ on center. End joints of wallboard centered on joists.	—	—	—	—	—	—	—	$1/_2$
14. Plywood stressed skin panels consisting of $5/_8$″-thick interior C-D (exterior glue) top stressed skin on 2″ × 6″ nominal (minimum) stringers. Adjacent panel edges joined with 8d common wire nails spaced 6″ on center. Stringers spaced 12″ maximum on center.	14-1.1	$1/_2$″-thick wood fiberboard weighing 15 to 18 pounds per cubic foot installed with long dimension parallel to stringers or $3/_8$″ C-D (exterior glue) plywood glued and/or nailed to stringers. Nailing to be with 5d cooler[o] or wallboard[o] nails at 12″ on center. Second layer of $1/_2$″ Type X gypsum wallboard[c] applied with long dimension perpendicular to joists and attached with 8d cooler[o] or wallboard[o] nails at 6″ on center at end joints and 8″ on center elsewhere. Wallboard joints staggered with respect to fiberboard joints.	—	—	—	—	—	—	—	1

(continued)

TABLE 720.1(3)—continued
MINIMUM PROTECTION FOR FLOOR AND ROOF SYSTEMS[a, q]

FLOOR OR ROOF CONSTRUCTION	ITEM NUMBER	CEILING CONSTRUCTION	THICKNESS OF FLOOR OR ROOF SLAB (inches)				MINIMUM THICKNESS OF CEILING (inches)			
			4 hour	3 hour	2 hour	1 hour	4 hour	3 hour	2 hour	1 hour
15. Vermiculite concrete slab proportioned 1:4 (portland cement to vermiculite aggregate) on a $1^1/_2$″-deep steel deck supported on individually protected steel framing. Maximum span of deck 6′-10″ where deck is less than 0.019 inch (No. 26 carbon steel sheet gage) or greater. Slab reinforced with 4″ × 8″ 0.109/0.083″ (No. $^{12}/_{14}$ B.W. gage) welded wire mesh.	15-1.1	None	—	—	—	3^j	—	—	—	—
16. Perlite concrete slab proportioned 1:6 (portland cement to perlite aggregate) on a $1^1/_4$″-deep steel deck supported on individually protected steel framing. Slab reinforced with 4″ × 8″ 0.109/0.083″ (No. $^{12}/_{14}$ B.W. gage) welded wire mesh.	16-1.1	None	—	—	—	$3^1/_2{}^j$	—	—	—	—
17. Perlite concrete slab proportioned 1:6 (portland cement to perlite aggregate) on a $9/_{16}$″-deep steel deck supported by steel joists 4′ on center. Class A or B roof covering on top.	17-1.1	Perlite gypsum plaster on metal lath wire tied to $3/_4$″ furring channels attached with 0.065″ (No. 16 B.W. gage) wire ties to lower chord of joists.	—	2^p	2^p	—	—	$7/_8$	$3/_4$	—
18. Perlite concrete slab proportioned 1:6 (portland cement to perlite aggregate) on $1^1/_4$″-deep steel deck supported on individually protected steel framing. Maximum span of deck 6′-10″ where deck is less than 0.019″ (No. 26 carbon sheet steel gage) and 8′-0″ where deck is 0.019″ (No. 26 carbon sheet steel gage) or greater. Slab reinforced with 0.042″ (No. 19 B.W. gage) hexagonal wire mesh. Class A or B roof covering on top.	18-1.1	None	—	$2^1/_4{}^p$	$2^1/_4{}^p$	—	—	—	—	—

(continued)

TABLE 720.1(3)—continued
MINIMUM PROTECTION FOR FLOOR AND ROOF SYSTEMS[a, q]

FLOOR OR ROOF CONSTRUCTION	ITEM NUMBER	CEILING CONSTRUCTION	THICKNESS OF FLOOR OR ROOF SLAB (inches)				MINIMUM THICKNESS OF CEILING (inches)			
			4 hour	3 hour	2 hour	1 hour	4 hour	3 hour	2 hour	1 hour
19. Floor and beam construction consisting of 3″-deep cellular steel floor unit mounted on steel members with 1:4 (proportion of portland cement to perlite aggregate) perlite-concrete floor slab on top.	19-1.1	Suspended envelope ceiling of perlite gypsum plaster on metal lath attached to $^3/_4$″ cold-rolled channels, secured to $1^1/_2$″ cold-rolled channels spaced 42″ on center supported by 0.203 inch (No. 6 B.W. gage) wire 36″ on center. Beams in envelope with 3″ minimum airspace between beam soffit and lath have a 4-hour rating.	2[p]	—	—	—	1[l]	—	—	—
20. Perlite concrete proportioned 1:6 (portland cement to perlite aggregate) poured to $^1/_8$″ thickness above top of corrugations of $1^5/_{16}$″-deep galvanized steel deck maximum span 8′-0″ for 0.024″ (No. 24 galvanized sheet gage) or 6′ 0″ for 0.019″ (No. 26 galvanized sheet gage) with deck supported by individually protected steel framing. Approved polystyrene foam plastic insulation board having a flame spread not exceeding 75 (1″ to 4″ thickness) with vent holes that approximate 3 percent of the board surface area placed on top of perlite slurry. A 2′ by 4′ insulation board contains six $2^3/_4$″ diameter holes. Board covered with $2^1/_4$″ minimum perlite concrete slab.	20-1.1	None	—	—	Varies	—	—	—	—	—

(continued)

TABLE 720.1(3)—continued
MINIMUM PROTECTION FOR FLOOR AND ROOF SYSTEMS[a, q]

FLOOR OR ROOF CONSTRUCTION	ITEM NUMBER	CEILING CONSTRUCTION	THICKNESS OF FLOOR OR ROOF SLAB (inches)				MINIMUM THICKNESS OF CEILING (inches)			
			4 hour	3 hour	2 hour	1 hour	4 hour	3 hour	2 hour	1 hour
(continued) 20. Slab reinforced with mesh consisting of 0.042″ (No. 19 B.W. gage) galvanized steel wire twisted together to form 2″ hexagons with straight 0.065″ (No. 16 B.W. gage) galvanized steel wire woven into mesh and spaced 3″. Alternate slab reinforcement shall be permitted to consist of 4″ × 8″, 0.109/0.238″ (No. 12/4 B.W. gage), or 2″ × 2″, 0.083/0.083″ (No. 14/14 B.W. gage) welded wire fabric. Class A or B roof covering on top.	20-1.1	None	—	—	Varies	—	—	—	—	—
21. Wood joists, wood I-joists, floor trusses and flat or pitched roof trusses spaced a maximum 24″ o.c. with $^1/_2$″ wood structural panels with exterior glue applied at right angles to top of joist or top chord of trusses with 8d nails. The wood structural panel thickness shall not be less than nominal $^1/_2$″ nor less than required by Chapter 23.	21-1.1	Base layer $^5/_8$″ Type X gypsum wallboard applied at right angles to joist or truss 24″ o.c. with $1^1/_4$″ Type S or Type W drywall screws 24″ o.c. Face layer $^5/_8$″ Type X gypsum wallboard or veneer base applied at right angles to joist or truss through base layer with $1^7/_8$″ Type S or Type W drywall screws 12″ o.c. at joints and intermediate joist or truss. Face layer Type G drywall screws placed 2″ back on either side of face layer end joints, 12″ o.c.	—	—	—	Varies	—	—	—	$1^1/_4$
22. Steel joists, floor trusses and flat or pitched roof trusses spaced a maximum 24″ o.c. with $^1/_2$″ wood structural panels with exterior glue applied at right angles to top of joist or top chord of trusses with No. 8 screws. The wood structural panel thickness shall not be less than nominal $^1/_2$″ nor less than required by Chapter 23.	22-1.1	Base layer $^5/_8$″ Type X gypsum board applied at right angles to steel framing 24″ on center with 1″ Type S drywall screws spaced 24″ on center. Face layer $^5/_8$″ Type X gypsum board applied at right angles to steel framing attached through base layer with $1^5/_8$″ Type S drywall screws 12″ on center at end joints and intermediate joints and $1^1/_2$″ Type G drywall screws 12 inches on center placed 2″ back on either side of face layer end joints. Joints of the face layer are offset 24″ from the joints of the base layer.	—	—	—	Varies	—	—	—	$1^1/_4$
23. Wood I-joist (minimum joist depth $9^1/_4$″ with a minimum flange depth of $^{15}/_{16}$″ and a minimum flange cross-sectional area of 2.3 square inches) at 24″ o.c. spacing with 1 inch by 4 inch (nominal) wood furring strip spacer applied parallel to and covering the bottom of the bottom flange of each member, tacked in place. 2″ mineral wool insulation, 3.5 pcf (nominal) installed adjacent to the bottom flange of the I-joist and supported by the 1″ × 4″ furring strip spacer.	23-1.1	$^1/_2$″ deep single leg resilient channel 16″ on center (channels doubled at wallboard end joints), placed perpendicular to the furring strip and joist and attached to each joist by $1^7/_8$″ Type S drywall screws. $^5/_8$″ Type C gypsum wallboard applied perpendicular to the channel with end joints staggered at least 4′ and fastened with $1^1/_8$″ Type S drywall screws spaced 7″ on center. Wallboard joints to be taped and covered with joint compound.	—	—	—	Varies	—	—	—	$^5/_8$

(continued)

TABLE 720.1(3)—continued
MINIMUM PROTECTION FOR FLOOR AND ROOF SYSTEMS[a, q]

FLOOR OR ROOF CONSTRUCTION	ITEM NUMBER	CEILING CONSTRUCTION	THICKNESS OF FLOOR OR ROOF SLAB (inches)				MINIMUM THICKNESS OF CEILING (inches)			
			4 hour	3 hour	2 hour	1 hour	4 hour	3 hour	2 hour	1 hour
24. Wood I-joist (minimum I-joist depth $9^1/_4''$ with a minimum flange depth of $1^1/_2''$ and a minimum flange cross-sectional area of 5.25 square inches; minimum web thickness of $3/_8''$) @ 24″ o.c., $1^1/_2''$ mineral wool insulation (2.5 pcf—nominal) resting on hat-shaped furring channels.	24-1.1	Minimum 0.026″ thick hat-shaped channel 16″ o.c. (channels doubled at wallboard end joints), placed perpendicular to the joist and attached to each joist by $1^5/_8''$ Type S drywall screws. $5/_8''$ Type C gypsum wallboard applied perpendicular to the channel with end joints staggered and fastened with $1^1/_8''$ Type S drywall screws spaced 12″ o.c. in the field and 8″ o.c. at the wallboard ends. Wallboard joints to be taped and covered with joint compound.	—	—	—	Varies	—	—	—	$5/_8$
25. Wood I-joist (minimum I-joist depth $9^1/_4''$ with a minimum flange depth of $1^1/_2''$ and a minimum flange cross-sectional area of 5.25 square inches; minimum web thickness of $7/_{16}''$) @ 24″ o.c., $1^1/_2''$ mineral wool insulation (2.5 pcf—nominal) resting on resilient channels.	25-1.1	Minimum 0.019″ thick resilient channel 16″ o.c. (channels doubled at wallboard end joints), placed perpendicular to the joist and attached to each joist by $1^5/_8''$ Type S drywall screws. $5/_8''$ Type C gypsum wallboard applied perpendicular to the channel with end joints staggered and fastened with 1″ Type S drywall screws spaced 12″ o.c. in the field and 8″ o.c. at the wallboard ends. Wallboard joints to be taped and covered with joint compound.	—	—	—	Varies	—	—	—	$5/_8$
26. Wood I-joist (minimum I-joist depth $9^1/_4''$ with a minimum flange thickness of $1^1/_2''$ and a minimum flange cross-sectional area of 2.25 square inches; minimum web thickness of $3/_8''$) @ 24″ o.c.	26-1.1	Two layers of $1/_2''$ Type X gypsum wallboard applied with the long dimension perpendicular to the I-joists with end joints staggered. The base layer is fastened with $1^5/_8''$ Type S drywall screws spaced 12″ o.c. and the face layer is fastened with 2″ Type S drywall screws spaced 12″ o.c. in the field and 8″ o.c. on the edges. Face layer end joints shall not occur on the same I-joist as base layer end joints and edge joints shall be offset 24″ from base layer joints. Face layer to also be attached to base layer with $1^1/_2''$ Type G drywall screws spaced 8″ o.c. placed 6″ from face layer end joints. Face layer wallboard joints to be taped and covered with joint compound.	—	—	—	Varies	—	—	—	1
27. Wood I-joist (minimum I-joist depth $9^1/_2''$ with a minimum flange depth of $1^{15}/_{16}''$ and a minimum flange cross-sectional area of 1.95 square inches; minimum web thickness of $3/_8''$) @ 24″ o.c.	27-1.1	Minimum 0.019″ thick resilient channel 16″ o.c. (channels doubled at wallboard end joints), placed perpendicular to the joist and attached to each joist by $1^5/_8''$ Type S drywall screws. Two layers of $1/_2''$ Type X gypsum wallboard applied with the long dimension perpendicular to the I-joists with end joints staggered. The base layer is fastened with $1^1/_4''$ Type S drywall screws spaced 12″ o.c. and the face layer is fastened with $1^5/_8''$ Type S drywall screws spaced 12″ o.c. Face layer end joints shall not occur on the same I-joist as base layer end joints and edge joints shall be offset 24″ from base layer joints. Face layer to also be attached to base layer with $1^1/_2''$ Type G drywall screws spaced 8″ o.c. placed 6″ from face layer end joints. Face layer wallboard joints to be taped and covered with joint compound.	—	—	—	Varies	—	—	—	1

(continued)

TABLE 720.1(3)—continued
MINIMUM PROTECTION FOR FLOOR AND ROOF SYSTEMS[a, q]

FLOOR OR ROOF CONSTRUCTION	ITEM NUMBER	CEILING CONSTRUCTION	THICKNESS OF FLOOR OR ROOF SLAB (inches)				MINIMUM THICKNESS OF CEILING (inches)			
			4 hour	3 hour	2 hour	1 hour	4 hour	3 hour	2 hour	1 hour
28. Wood I-joist (minimum I-joist depth $9^1/_4$″ with a minimum flange depth of $1^1/_2$″ and a minimum flange cross-sectional area of 2.25 square inches; minimum web thickness of $^3/_8$″) @ 24″ o.c. Unfaced fiberglass insulation is installed between the I-joists supported on the upper surface of the flange by stay wires spaced 12″ o.c.	28-1.1	Base layer of $^5/_8$″ Type C gypsum wallboard attached directly to I-joists with $1^5/_8$″ Type S drywall screws spaced 12″ o.c. with ends staggered. Minimum 0.0179″ thick hat-shaped $^7/_8$-inch furring channel 16″ o.c. (channels doubled at wallboard end joints), placed perpendicular to the joist and attached to each joist by $1^5/_8$″ Type S drywall screws after the base layer of gypsum wallboard has been applied. The middle and face layers of $^5/_8$″ Type C gypsum wallboard applied perpendicular to the channel with end joints staggered. The middle layer is fastened with 1″ Type S drywall screws spaced 12″ o.c. The face layer is applied parallel to the middle layer but with the edge joints offset 24″ from those of the middle layer and fastened with $1^5/_8$″ Type S drywall screws 8″ o.c. The joints shall be taped and covered with joint compound.	—	—	—	Varies	—	—	$2^3/_4$	—
29. Channel-shaped 18 gage steel joists (minimum depth 8″) spaced a maximum 24″ o.c. supporting tongue-and-groove wood structural panels (nominal minimum $^3/_4$″ thick) applied perpendicular to framing members. Structural panels attached with 1-$^5/_8$″ Type S-12 screws spaced 12″ o.c.	29-1.1	Base layer $^5/_8$″ Type X gypsum board applied perpendicular to bottom of framing members with $1^1/_8$″ Type S-12 screws spaced 12″ o.c. Second layer $^5/_8$″ Type X gypsum board attached perpendicular to framing members with $1^5/_8$″ Type S-12 screws spaced 12″ o.c. Second layer joints offset 24″ from base layer. Third layer $^5/_8$″ Type X gypsum board attached perpendicular to framing members with $2^3/_8$″ Type S-12 screws spaced 12″ o.c. Third layer joints offset 12″ from second layer joints. Hat-shaped $^7/_8$-inch rigid furring channels applied at right angles to framing members over third layer with two $2^3/_8$″ Type S-12 screws at each framing member. Face layer $^5/_8$″ Type X gypsum board applied at right angles to furring channels with $1^1/_8$″ Type S screws spaced 12″ o.c.	—	—	Varies	—	—	—	$3^3/_8$	—

Table 720.1(3) Notes.

For SI: 1 inch = 25.4 mm, 1 foot = 304.8 mm, 1 pound = 0.454 kg, 1 cubic foot = 0.0283m³,
1 pound per square inch = 6.895 kPa, 1 pound per lineal foot = 1.4882 kg/m.

a. Staples with equivalent holding power and penetration shall be permitted to be used as alternate fasteners to nails for attachment to wood framing.

b. When the slab is in an unrestrained condition, minimum reinforcement cover shall not be less than $1^5/_8$ inches for 4-hour (siliceous aggregate only); $1^1/_4$ inches for 4- and 3-hour; 1 inch for 2-hour (siliceous aggregate only); and $^3/_4$ inch for all other restrained and unrestrained conditions.

c. For all of the construction with gypsum wallboard described in this table, gypsum base for veneer plaster of the same size, thickness and core type shall be permitted to be substituted for gypsum wallboard, provided attachment is identical to that specified for the wallboard, and the joints on the face layer are reinforced and the entire surface is covered with a minimum of $^1/_{16}$-inch gypsum veneer plaster.

d. Slab thickness over steel joists measured at the joists for metal lath form and at the top of the form for steel form units.

e. (a) The maximum allowable stress level for H-Series joists shall not exceed 22,000 psi.
   (b) The allowable stress for K-Series joists shall not exceed 26,000 psi, the nominal depth of such joist shall not be less than 10 inches and the nominal joist weight shall not be less than 5 pounds per lineal foot.

f. Cement plaster with 15 pounds of hydrated lime and 3 pounds of approved additives or admixtures per bag of cement.

g. Gypsum wallboard ceilings attached to steel framing shall be permitted to be suspended with $1^1/_2$-inch cold-formed carrying channels spaced 48 inches on center, which are suspended with No. 8 SWG galvanized wire hangers spaced 48 inches on center. Cross-furring channels are tied to the carrying channels with No. 18 SWG galvanized wire hangers spaced 48 inches on center. Cross-furring channels are tied to the carrying channels with No. 18 SWG galvanized wire (double strand) and spaced as required for direct attachment to the framing. This alternative is also applicable to those steel framing assemblies recognized under Note q.

h. Six-inch hollow clay tile with 2-inch concrete slab above.

i. Four-inch hollow clay tile with $1^1/_2$-inch concrete slab above.

j. Thickness measured to bottom of steel form units.

k. Five-eighths inch of vermiculite gypsum plaster plus $^1/_2$ inch of approved vermiculite acoustical plastic.

l. Furring channels spaced 12 inches on center.

m. Double wood floor shall be permitted to be either of the following:
   (a) Subfloor of 1-inch nominal boarding, a layer of asbestos paper weighing not less than 14 pounds per 100 square feet and a layer of 1-inch nominal tongue-and-groove finished flooring; or
   (b) Subfloor of 1-inch nominal tongue-and-groove boarding or $^{15}/_{32}$-inch wood structural panels with exterior glue and a layer of 1-inch nominal tongue-and-groove finished flooring or $^{19}/_{32}$-inch wood structural panel finish flooring or a layer of Type I Grade M-1 particleboard not less than $^5/_8$-inch thick.

n. The ceiling shall be permitted to be omitted over unusable space, and flooring shall be permitted to be omitted where unusable space occurs above.

o. For properties of cooler or wallboard nails, see ASTM C 514, ASTM C 547 or ASTM F 1667.

p. Thickness measured on top of steel deck unit.

q. Generic fire-resistance ratings (those not designated as PROPRIETARY* in the listing) in the GA 600 shall be accepted as if herein listed.

SECTION 721
CALCULATED FIRE RESISTANCE

❖ Due to the prescriptive nature of Section 721, this section contains explanatory material, examples and reference materials related to calculating fire-resistance ratings of materials and assemblies. As such, commentary is provided where necessary to provide the basis for a provision or to illustrate the application of the requirement.

The provisions of this section provide a number of useful alternatives for establishing compliance with the code. The provisions in this section can be used not only to calculate the fire-resistive rating for an assembly (concrete, concrete masonry, clay masonry, steel and wood) but it also contains provisions that would permit modifications or changes to a tested assembly. For example, steel assemblies (see Section 721.5) will permit the selection of different sized or shaped members depending on the weight-to-heated-perimeter ratio (see Sections 721.5.2.1.2 and 721.5.2.2). In addition, the amount of spray-applied fire-resistant materials can be modified from a tested assembly to either increase or decrease the coverage dependent on the member and the spray-applied material.

The section also addresses items that are much less exotic and may be helpful in a number of situations. Section 721.6 will provide methods for calculating the fire resistance of wood assemblies, including light-framed walls and floors or heavy timber exposed

members. Section 721.4.1 can be used for masonry walls (including existing conditions) where additional protection, such as filled cores, or an additional wythe, such as a brick veneer, is added.

While this section is not often used by many people, those that have used it will find that it is often helpful in solving difficult situations related to the protection of certain assemblies. Therefore, taking the time to become familiar with this section can be very beneficial. The provisions of this section include methods for calculating fire resistance as follows:

Concrete Assemblies:
    Walls–721.2.1
    Floor and roof slabs–721.2.2
    Beams–721.2.3
    Columns–721.2.3

Concrete Masonry:
    Walls–721.3.2
    Lintels–721.3.4
    Columns–721.3.5

Clay Brick and Tile Masonry:
    Walls–721.4.1
    Lintels–721.4.3
    Columns–721.4.4

Steel Assemblies:
    Columns using:
        Gypsum wallboard–721.5.1.2

Spray-applied materials–721.5.1.3
Concrete or masonry–721.5.1.4

Beams using:
Spray-applied materials–721.5.2.2

Trusses using:
Spray-applied materials–721.5.2.3

Wood Assemblies:
Walls, floors and roofs–721.6.2

Exposed Wood Members:
Beams and columns–721.6.3

721.1 General. The provisions of this section contain procedures by which the *fire resistance* of specific materials or combinations of materials is established by calculations. These procedures apply only to the information contained in this section and shall not be otherwise used. The calculated *fire resistance* of concrete, concrete masonry and clay masonry assemblies shall be permitted in accordance with ACI 216.1/TMS 0216. The calculated *fire resistance* of steel assemblies shall be permitted in accordance with Chapter 5 of ASCE 29. The calculated *fire resistance* of exposed wood members and wood decking shall be permitted in accordance with Chapter 16 of ANSI/AF&PA *National Design Specification for Wood Construction (NDS)*.

❖ Section 703.3 permits the fire-resistance ratings to be determined in a number of ways, including those found in Section 721 (see Section 703.3, Item 3). The provisions of this section contain procedures by which the fire resistance of specific materials or combinations of materials is established by calculations. These procedures apply only to the information contained in this section and are not to be otherwise used. The calculated fire resistance of concrete, concrete masonry and clay masonry assemblies are also permitted in accordance with ACI 216.1/TMS 0216.1 while steel assemblies can use ASCE 29 (see commentary, Section 703.3).

721.1.1 Definitions. The following words and terms shall, for the purposes of this chapter and as used elsewhere in this code, have the meanings shown herein.

CERAMIC FIBER BLANKET. A mineral wool insulation material made of alumina-silica fibers and weighing 4 to 10 pounds per cubic foot (pcf) (64 to 160 kg/m³).

❖ This form of insulation is used in conjunction with the provisions of Section 721.2.1.3.1, where joints in precast walls must be insulated (protected) in order to maintain the fire-resistance integrity of the precast wall panel (see Code Figure 721.2.1.3.1).

CONCRETE, CARBONATE AGGREGATE. Concrete made with aggregates consisting mainly of calcium or magnesium carbonate, such as limestone or dolomite, and containing 40 percent or less quartz, chert or flint.

❖ Concrete made with this type of aggregate is listed in Tables 720.1(1) (Items 1, 2, 3, 4, 5, 6 and 7) and 720.1(2) (Item 4) with prescriptive details regarding concrete fire-resistance ratings. This type of aggregate is also indicated for concrete assemblies that re-

quire calculations to determine the fire-resistance rating in accordance with Section 721.2 (i.e., Table 721.2.1.1) (also see Section 702.1).

CONCRETE, CELLULAR. A lightweight insulating concrete made by mixing a preformed foam with portland cement slurry and having a dry unit weight of approximately 30 pcf (480 kg/m³).

❖ Concrete made with this type of aggregate is listed in Table 720.1(3) (Items 9 and 19) with prescriptive details regarding concrete fire-resistance ratings. This type of aggregate is also indicated for concrete assemblies that require calculations to determine the fire-resistance rating in accordance with Section 721.2.1 [i.e., Table 721.2.1.2(1), Note a].

CONCRETE, LIGHTWEIGHT AGGREGATE. Concrete made with aggregates of expanded clay, shale, slag or slate or sintered fly ash or any natural lightweight aggregate meeting ASTM C 330 and possessing equivalent fire-resistance properties and weighing 85 to 115 pcf (1360 to 1840 kg/m³).

❖ Concrete made with this type of aggregate is listed in Tables 720.1(1) (Items 1, 2, 3, 4, 5, 6 and 7), 720.1(2) (Item 4) and 720.1(3) (Item 4) with prescriptive details regarding concrete fire-resistance ratings. This type of aggregate is also indicated for concrete assemblies that require calculations to determine the fire-resistance rating in accordance with Section 721.2.1 (i.e., Table 721.2.1.1) (also see Section 702.1).

CONCRETE, PERLITE. A lightweight insulating concrete having a dry unit weight of approximately 30 pcf (480 kg/m³) made with perlite concrete aggregate. Perlite aggregate is produced from a volcanic rock which, when heated, expands to form a glass-like material of cellular structure.

❖ Concrete made with this type of aggregate is listed in Tables 720.1(1) (Item 1), 720.1(2) (Item 1) and 720.1(3) (Items 7, 16, 17, 18, 19 and 20) with prescriptive details regarding concrete fire-resistance ratings. This type of aggregate is also indicated for concrete assemblies that require calculations to determine the fire-resistance rating in accordance with Section 721.2.1 (i.e., Section 721.2.1.1.2).

CONCRETE, SAND-LIGHTWEIGHT. Concrete made with a combination of expanded clay, shale, slag, slate, sintered fly ash, or any natural lightweight aggregate meeting ASTM C 330 and possessing equivalent fire-resistance properties and natural sand. Its unit weight is generally between 105 and 120 pcf (1680 and 1920 kg/m³).

❖ Concrete made with this type of aggregate is listed in Tables 720.1(1) (Items 1, 2, 3, 4, 5, 6 and 7), 720.1(2) (Item 4) and 720.1(3) (Item 3) with prescriptive details regarding concrete fire-resistance ratings. This type of aggregate is also indicated for concrete assemblies that require calculations to determine the fire-resistance rating in accordance with Section 721.2.1 (i.e., Table 721.2.1.1) (also see Section 702.1).

CONCRETE, SILICEOUS AGGREGATE. Concrete made with normal-weight aggregates consisting mainly of silica or compounds other than calcium or magnesium carbonate, which contains more than 40-percent quartz, chert or flint.

❖ Concrete made with this type of aggregate is listed in Tables 720.1(1) (Items 1, 2, 4, 5, 6 and 7), 720.1(2) (Items 3 and 4) and 720.1(3) (Item 1) with prescriptive details regarding concrete fire-resistance ratings. This type of aggregate is also indicated for concrete assemblies that require calculations to determine the fire-resistance rating in accordance with Section 721.2.1 (i.e., Table 721.2.1.1) (also see Section 702.1).

CONCRETE, VERMICULITE. A lightweight insulating concrete made with vermiculite concrete aggregate which is laminated micaceous material produced by expanding the ore at high temperatures. When added to a portland cement slurry the resulting concrete has a dry unit weight of approximately 30 pcf (480 kg/m³).

❖ Concrete made with this type of aggregate is listed in Tables 720.1(1) (Item 1), 720.1(2) (Item 1) and 720.1(3) (Item 15) with prescriptive details regarding concrete fire-resistance ratings. This type of aggregate is also indicated for concrete assemblies that require calculations to determine the fire-resistance rating in accordance with Section 721.2.1 [i.e., Table 721.2.1.2(1), Note a].

GLASS FIBERBOARD. Fibrous glass roof insulation consisting of inorganic glass fibers formed into rigid boards using a binder. The board has a top surface faced with asphalt and kraft reinforced with glass fiber.

❖ Depending on the type and location of insulation in walls, floors and roofs, the insulation may impact the fire-resistance rating (see Sections 703.2 and 720.1). Glass fiber insulation is specifically listed in Table 721.6.2(5).

MINERAL BOARD. A rigid felted thermal insulation board consisting of either felted mineral fiber or cellular beads of expanded aggregate formed into flat rectangular units.

❖ This form of insulation is listed in Table 720.1(2), Items 15-1.9, 15-1.10 and 15-1.11.

721.2 Concrete assemblies. The provisions of this section contain procedures by which the *fire-resistance ratings* of concrete assemblies are established by calculations.

721.2.1 Concrete walls. Cast-in-place and precast concrete walls shall comply with Section 721.2.1.1. Multiwythe concrete walls shall comply with Section 721.2.1.2. Joints between precast panels shall comply with Section 721.2.1.3. Concrete walls with gypsum wallboard or plaster finish shall comply with Section 721.2.1.4.

721.2.1.1 Cast-in-place or precast walls. The minimum equivalent thicknesses of cast-in-place or precast concrete walls for *fire-resistance ratings* of 1 hour to 4 hours are shown in Table 721.2.1.1. For solid walls with flat vertical surfaces, the equivalent thickness is the same as the actual thickness. The values in Table 721.2.1.1 apply to plain, reinforced or prestressed concrete walls.

❖ Although there have been few fire tests of concrete walls (other than concrete masonry), there have been

many fire tests of concrete slabs tested as floors or roofs. Fire tests of floors or roofs are considered to be more severe than those of walls because of the loading conditions (tension cracks associated with bending moment). In addition, most ASTM E 119 or UL 263 fire tests of floor assemblies have been conducted while the assembly was supported within restraining frames. As concrete assemblies are heated and tend to expand, the expansion is resisted by the restraining frame. These restraining forces are usually much greater than the superimposed loads supported by load-bearing walls; thus, floor or roof assemblies are subjected to both vertical superimposed loads and horizontal restraining loads during fire tests. By contrast, load-bearing walls are subjected only to superimposed loads.

The fire endurance of masonry or concrete walls is nearly always governed by the ASTM E 119 or UL 263 criteria for temperature rise of the unexposed surface (i.e., the "heat transmission" end point). For flat concrete slabs or panels, the heat transmission fire endurance depends primarily on the aggregate type and thickness and is essentially the same for floors as it is for walls.

TABLE 721.2.1.1
MINIMUM EQUIVALENT THICKNESS OF
CAST-IN-PLACE OR PRECAST CONCRETE WALLS,
LOAD-BEARING OR NONLOAD-BEARING

CONCRETE TYPE	MINIMUM SLAB THICKNESS (inches) FOR FIRE-RESISTANCE RATING OF				
	1-hour	1¹/₂-hour	2-hour	3-hour	4-hour
Siliceous	3.5	4.3	5.0	6.2	7.0
Carbonate	3.2	4.0	4.6	5.7	6.6
Sand-Lightweight	2.7	3.3	3.8	4.6	5.4
Lightweight	2.5	3.1	3.6	4.4	5.1

For SI: 1 inch = 25.4 mm.

❖ The data in Table 721.2.1.1 was derived from Portland Cement Association (PCA) Bulletin 223, "Fire Endurances of Concrete Slabs as Influenced by Thickness, Aggregate Type and Moisture," and PCA Publication T-140, "Fire Resistance of Reinforced Concrete Floors."

721.2.1.1.1 Hollow-core precast wall panels. For hollow-core precast concrete wall panels in which the cores are of constant cross section throughout the length, calculation of the equivalent thickness by dividing the net cross-sectional area (the gross cross section minus the area of the cores) of the panel by its width shall be permitted.

❖ The method for determining equivalent thickness of masonry units was developed because the cores in masonry units taper. The method is, of course, also applicable to hollow-core precast concrete panels; however, because the cores in hollow-core panels do not taper, the equivalent thickness can be calculated by dividing the net cross-sectional area of the panel by its width.

721.2.1.1.2 Core spaces filled. Where all of the core spaces of hollow-core wall panels are filled with loose-fill material, such as expanded shale, clay, or slag, or vermiculite or perlite, the *fire-resistance rating* of the wall is the same as that of a solid wall of the same concrete type and of the same overall thickness.

❖ The PCA report "Tests of Fire Resistance and Strength of Walls of Concrete Masonry Units" shows that filling cores of concrete masonry units with loose lightweight aggregates increases the fire endurance to a duration significantly longer than that of solid masonry units of the same total thickness. It is reasonable to assume that the same relationship exists for walls made of hollow-core panels.

721.2.1.1.3 Tapered cross sections. The thickness of panels with tapered cross sections shall be that determined at a distance $2t$ or 6 inches (152 mm), whichever is less, from the point of minimum thickness, where t is the minimum thickness.

❖ Some precast concrete wall panel sections (e.g., certain single-tee units) have tapered surfaces so the thickness varies, as shown in Figure 721.2.2.1.2. In fire tests, it has been customary to monitor the unexposed surface temperature at the location shown in the figure.

721.2.1.1.4 Ribbed or undulating surfaces. The equivalent thickness of panels with ribbed or undulating surfaces shall be determined by one of the following expressions:

For $s \geq 4t$, the thickness to be used shall be t

For $s \leq 2t$, the thickness to be used shall be t_e

For $4t > s > 2t$, the thickness to be used shall be

$$t + \left(\frac{4t}{s} - 1\right)\left(t_e - t\right)$$ (Equation 7-3)

where:

s = Spacing of ribs or undulations.

t = Minimum thickness.

t_e = Equivalent thickness of the panel calculated as the net cross-sectional area of the panel divided by the width, in which the maximum thickness used in the calculation shall not exceed $2t$.

❖ The portion of a ribbed panel that can be used in calculating equivalent thickness, t_e, is shown in Figure 721.2.2.1.3. Note that the procedure outlined gives no credit for stems of double tees or of similar ribbed panels and clearly indicates that the minimum thickness must be used for such sections.

721.2.1.2 Multiwythe walls. For walls that consist of two wythes of different types of concrete, the *fire-resistance ratings* shall be permitted to be determined from Figure 721.2.1.2.

❖ The graphs in the Figure 721.2.1.2 were derived from a report entitled "Fire Endurance of Two-course Floors and Roofs" in the *ACI Journal*.

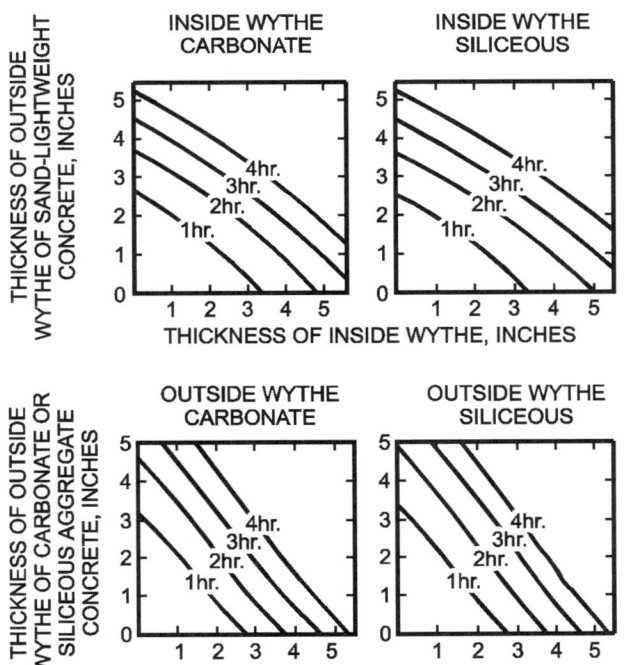

For SI: 1 inch = 25.4 mm.

FIGURE 721.2.1.2
FIRE-RESISTANCE RATINGS OF TWO-WYTHE CONCRETE WALLS

721.2.1.2.1 Two or more wythes. The *fire-resistance rating* for wall panels consisting of two or more wythes shall be permitted to be determined by the formula:

$$R = (R_1^{0.59} + R_2^{0.59} + ... + R_n^{0.59})^{1.7} \qquad \text{(Equation 7-4)}$$

where:

R = The fire endurance of the assembly, minutes.

R_1, R_2, and R_n = The fire endurances of the individual wythes, minutes. Values of $R_n^{0.59}$ for use in Equation 7-4 are given in Table 721.2.1.2(1). Calculated *fire-resistance ratings* are shown in Table 721.2.1.2(2).

❖ Equation 7-4 was developed by the U.S. National Bureau of Standards (NBS) in the early 1940s and first appeared in Appendix B of BMS 92. Verification of the accuracy of the equation is given in "Fire Endurance of Two-course Floors and Roofs" in the *ACI Journal.*

This section and the associated tables allow both the determination of the fire endurance of a particular assembly [see Table 721.2.1.2(1)] and what thickness is required to achieve a certain fire endurance [see Table 721.2.1.2(2)].

Table 721.2.1.2(1). See below.

❖ See the commentary to Section 721.2.1.2.1.

TABLE 721.2.1.2(2)
FIRE-RESISTANCE RATINGS BASED ON $R^{0.59}$

R [a], MINUTES	$R^{0.59}$
60	11.20
120	16.85
180	21.41
240	25.37

a. Based on Equation 7-4.

❖ See the commentary to Section 721.2.1.2.1.

721.2.1.2.2 Foam plastic insulation. The *fire-resistance ratings* of precast concrete wall panels consisting of a layer of foam plastic insulation sandwiched between two wythes of concrete shall be permitted to be determined by use of Equation 7-4. Foam plastic insulation with a total thickness of less than 1 inch (25 mm) shall be disregarded. The R_n value for thickness of foam plastic insulation of 1 inch (25 mm) or greater, for use in the calculation, is 5 minutes; therefore $R_n^{0.59} = 2.5$.

❖ This value was for foam plastic insulation 1-inch (25 mm) thick and greater, determined from a fire test conducted on a panel that consisted of a 2-inch (51 mm) base slab of carbonate aggregate concrete, a 1-inch (25 mm) layer of cellular polystyrene insulation and a 2-inch (51 mm) face slab of carbonate aggregate concrete. The resulting fire endurance was 2 hours. From Equation 7-4, the contribution of the 1-inch (25 mm) layer of foam polystyrene was calculated to be 5 minutes.

Presumably, a comparable *R*-value for a 1-inch (25 mm) layer of foam polyurethane would be somewhat greater than for a 1-inch (25 mm) layer of foam polystyrene, but test values are not available. The above value for polystyrene is conservative for foam polyurethane.

Equation 7-4 can be rewritten as:

$$R^{0.59} = (R_1^{0.59} + R_2^{0.59} + ... R_n^{0.59})$$

This form of the equation is useful in making a quick determination of whether or not an assembly qualifies for a particular classification when used in conjunction with Tables 721.2.1.2(1) and 721.2.1.2(2).

EXAMPLE:

GIVEN: A sandwich wall panel consists of two $2^1/_2$-inch (63 mm) wythes of normal-weight concrete

TABLE 721.2.1.2(1)
VALUES OF $R_n^{0.59}$ FOR USE IN EQUATION 7-4

TYPE OF MATERIAL	THICKNESS OF MATERIAL (inches)											
	$1^1/_2$	2	$2^1/_2$	3	$3^1/_2$	4	$4^1/_2$	5	$5^1/_2$	6	$6^1/_2$	7
Siliceous aggregate concrete	5.3	6.5	8.1	9.5	11.3	13.0	14.9	16.9	18.8	20.7	22.8	25.1
Carbonate aggregate concrete	5.5	7.1	8.9	10.4	12.0	14.0	16.2	18.1	20.3	21.9	24.7	27.2[c]
Sand-lightweight concrete	6.5	8.2	10.5	12.8	15.5	18.1	20.7	23.3	26.0[c]	Note c	Note c	Note c
Lightweight concrete	6.6	8.8	11.2	13.7	16.5	19.1	21.9	24.7	27.8[c]	Note c	Note c	Note c
Insulating concrete[a]	9.3	13.3	16.6	18.3	23.1	26.5[c]	Note c	Note c	Note c	Note c	Note c	Note c
Airspace[b]	—	—	—	—	—	—	—	—	—	—	—	—

For SI: 1 inch = 25.4 mm, 1 pound per cubic foot = 16.02 kg/m³.

a. Dry unit weight of 35 pcf or less and consisting of cellular, perlite or vermiculite concrete.

b. The $R_n^{0.59}$ value for one $^1/_2''$ to 3 $^1/_2''$ airspace is 3.3. The $R_n^{0.59}$ value for two $^1/_2''$ to 3 $^1/_2''$ airspaces is 6.7.

c. The fire-resistance rating for this thickness exceeds 4 hours.

with a 2-inch (51 mm) layer of foam polystyrene between them.

FIND: Does the panel qualify for a 3-hour fire-resistance rating?

SOLUTION: From Table 721.2.1.2(1) in the code, the value of $R_{n0.59}$ for a carbonate aggregate concrete is higher than for siliceous aggregate concrete, but because the type of concrete was not given, the value for siliceous should be used. From Section 721.2.1.2.2, the value of $R_n^{0.59}$ for the 2-inch (51 mm) layer of foam polystyrene is 2.5.

$R^{0.59} = 8.1 + 2.5 + 8.1 = 18.7$

From Table 721.2.1.2(2), a 3-hour rating is required to have an R-0.59 value of 21.41.

18.7 < 21.41, thus the wall does not qualify for a 3-hour rating.

721.2.1.3 Joints between precast wall panels. Joints between precast concrete wall panels which are not insulated as required by this section shall be considered as openings in walls. Uninsulated joints shall be included in determining the percentage of openings permitted by Table 705.8. Where openings are not permitted or are required by this code to be protected, the provisions of this section shall be used to determine the amount of joint insulation required. Insulated joints shall not be considered openings for purposes of determining compliance with the allowable percentage of openings in Table 705.8.

721.2.1.3.1 Ceramic fiber joint protection. Figure 721.2.1.3.1 shows thicknesses of ceramic fiber blankets to be used to insulate joints between precast concrete wall panels for various panel thicknesses and for joint widths of $^3/_8$ inch (9.5 mm) and 1 inch (25 mm) for *fire-resistance ratings* of 1 hour to 4 hours. For joint widths between $^3/_8$ inch (9.5 mm) and 1 inch

(25 mm), the thickness of ceramic fiber blanket is allowed to be determined by direct interpolation. Other tested and labeled materials are acceptable in place of ceramic fiber blankets.

❖ Figure 721.2.1.3.1 was derived from data in the report entitled "Fire Tests of Joints Between Precast Concrete Wall Units: Effect of Various Joint Treatments" in the *PCI Journal*.

EXAMPLE:

FIND: Determine the thickness of a ceramic fiber blanket needed for a 2-hour fire-resistance rating for joints between 5-inch-thick (127 mm) precast concrete wall panels made of siliceous aggregate concrete if the maximum joint width is $^7/_8$-inch (22.2 mm).

SOLUTION: Figure 721.2.1.3.1 indicates a minimum 0.7-inch (18 mm) thickness of ceramic fiber blanket for 5-inch (127 mm) panels for a 2-hour rating of a $^3/_4$-inch (19.5 mm) wide joint and 2.1 inches (53 mm) for a 1-inch (25 mm) wide joint. By interpolation, the thickness is computed as follows:

$t = 2.1 - (2.1 - 0.7)(1 - ^7/_8) = 1.93$ inches

Therefore, the required thickness for a 2-hour rating is 1.93 inches (49 mm).

721.2.1.4 Walls with gypsum wallboard or plaster finishes. The *fire-resistance rating* of cast-in-place or precast concrete walls with finishes of gypsum wallboard or plaster applied to one or both sides shall be permitted to be calculated in accordance with the provisions of this section.

❖ The information contained in this section is based on fire endurance tests on unit masonry walls with gypsum wallboard and *The Supplement to the National Building Code of Canada 1980* (NRCC 17724) by the

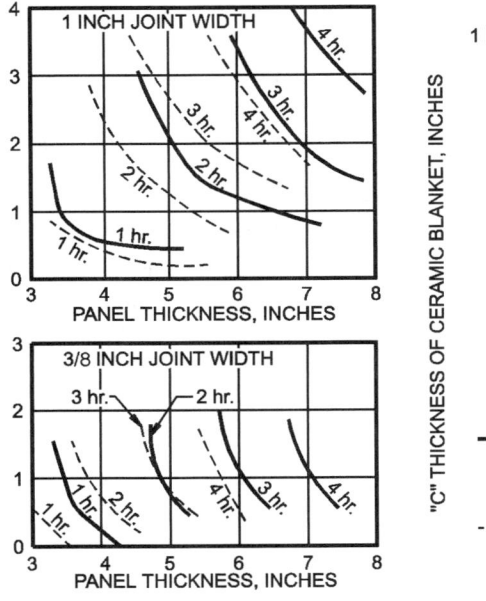

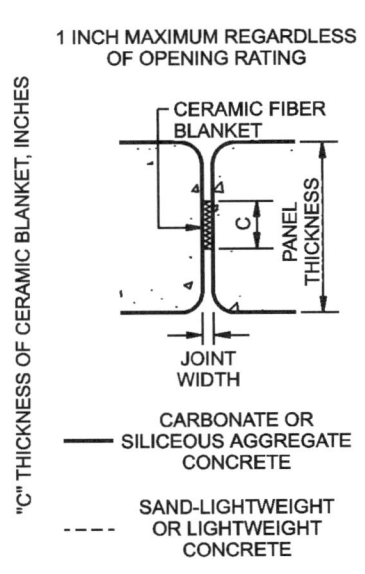

For SI: 1 inch = 25.4 mm.

FIGURE 721.2.1.3.1
CERAMIC FIBER JOINT PROTECTION

National Research Council of Canada.

For instance, a wall with a rating of 1 hour that has a 1-inch layer of Portland cement-sand plaster on metal lath applied would add 30 minutes to the rating. Therefore, the fire-resistance rating would now be $1^{1}/_{2}$ hours.

721.2.1.4.1 Nonfire-exposed side. Where the finish of gypsum wallboard or plaster is applied to the side of the wall not exposed to fire, the contribution of the finish to the total *fire-resistance rating* shall be determined as follows: The thickness of the finish shall first be corrected by multiplying the actual thickness of the finish by the applicable factor determined from Table 721.2.1.4(1) based on the type of aggregate in the concrete. The corrected thickness of finish shall then be added to the actual or equivalent thickness of concrete and *fire-resistance rating* of the concrete and finish determined from Table 721.2.1.1, Figure 721.2.1.2 or Table 721.2.1.2(1).

❖ The fire resistance of concrete walls is generally determined by temperature rise on the unexposed surface (i.e., the heat transmission end point) (see commentary, Section 721.2.1.1). The time required to reach the heat transmission end point (fire-resistance rating) is primarily dependent upon the thickness of the concrete and the type of aggregate used to make the concrete. When additional finishes are applied to the unexposed side of the wall, the time required to reach the heat transmission end point is delayed and the fire-resistance rating of the wall is thus increased. The increase in the rating contributed by the finish can be determined by considering the finish as adding to the thickness of concrete; however, since the finish material and concrete may have different insulating properties, the actual thickness of the finish may need to be corrected to be compatible with the type of aggregate used in the concrete. The correction is made by multiplying the actual finish thickness by the factor determined from Table 721.2.1.4(1), and then adding the corrected thickness to the thickness of the concrete. This equivalent thickness is used to determine the fire-resistance rating from Table 721.2.1.1, Figure 721.2.1.2 or Table 721.2.1.2(1).

TABLE 721.2.1.4(1). See below.

❖ See the commentary to Section 721.2.1.4.1.

721.2.1.4.2 Fire-exposed side. Where gypsum wallboard or plaster is applied to the fire-exposed side of the wall, the contribution of the finish to the total *fire-resistance rating* shall be determined as follows: The time assigned to the finish as established by Table 721.2.1.4(2) shall be added to the *fire-resistance rating* determined from Table 721.2.1.1 or Figure 721.2.1.2, or Table 721.2.1.2(1) for the concrete alone, or to the rating determined in Section 721.2.1.4.1 for the concrete and finish on the nonfire-exposed side.

❖ Where finishes are added to the fire-exposed side of a concrete wall, their contribution to the total fire-resistance rating is based primarily upon the ability of the finish to remain in place, thus affording protection to the concrete wall. Table 721.2.1.4(2) lists the times that have been assigned to finishes on the fire-exposed side of the wall. These time-assigned values are based upon actual fire tests. The time-assigned values are added to the fire-resistance rating of the wall alone or to the rating determined for the wall and any finish on the unexposed side.

TABLE 721.2.1.4(2). See page 7-161.

❖ See the commentary to Section 721.2.1.4.2.

721.2.1.4.3 Nonsymmetrical assemblies. For a wall having no finish on one side or different types or thicknesses of finish on each side, the calculation procedures of Sections 721.2.1.4.1 and 721.2.1.4.2 shall be performed twice, assuming either side of the wall to be the fire-exposed side. The *fire-resistance rating* of the wall shall not exceed the lower of the two values.

Exception: For an *exterior wall* with a *fire separation distance* greater than 5 feet (1524 mm) the fire shall be assumed to occur on the interior side only.

❖ Except for exterior walls having more than 5 feet (1524 mm) of horizontal separation, Section 704.5 requires that walls be rated for exposure to fire from both sides;

TABLE 721.2.1.4(1)
MULTIPLYING FACTOR FOR FINISHES ON NONFIRE-EXPOSED SIDE OF WALL

TYPE OF FINISH APPLIED TO CONCRETE OR CONCRETE MASONRY WALL	TYPE OF AGGREGATE USED IN CONCRETE OR CONCRETE MASONRY			
	Concrete: siliceous or carbonate Concrete Masonry: siliceous or carbonate; solid clay brick	Concrete: sand-lightweight Concrete Masonry: clay tile; hollow clay brick; concrete masonry units of expanded shale and <20% sand	Concrete: lightweight Concrete Masonry: concrete masonry units of expanded shale, expanded clay, expanded slag, or pumice < 20% sand	Concrete Masonry: concrete masonry units of expanded slag, expanded clay, or pumice
Portland cement-sand plaster	1.00	0.75[a]	0.75[a]	0.50[a]
Gypsum-sand plaster	1.25	1.00	1.00	1.00
Gypsum-vermiculite or perlite plaster	1.75	1.50	1.25	1.25
Gypsum wallboard	3.00	2.25	2.25	2.25

For SI: 1 inch = 25.4 mm.

a. For portland cement-sand plaster $^{5}/_{8}$ inch or less in thickness and applied directly to the concrete or concrete masonry on the nonfire-exposed side of the wall, the multiplying factor shall be 1.00.

therefore, two calculations must be performed, which assumes each side of the wall to be the fire-exposed side. Two calculations are not necessary for exterior walls with more than 5 feet (1524 mm) of horizontal separation or for other walls that are symmetrical (i.e., walls having the same type and thickness of finish on each side). The calculated fire-resistance rating of the wall is the lower of the two ratings determined, assuming that each side is the fire-exposed side.

721.2.1.4.4 Minimum concrete fire-resistance rating. Where finishes applied to one or both sides of a concrete wall contribute to the *fire-resistance rating*, the concrete alone shall provide not less than one-half of the total required *fire-resistance rating*. Additionally, the contribution to the *fire resistance* of the finish on the nonfire-exposed side of a *load-bearing wall* shall not exceed one-half the contribution of the concrete alone.

❖ Where gypsum wallboard or plaster finishes are applied to a concrete wall, the calculated fire-resistance rating for the concrete alone should not be less than one-half the required fire-resistance rating. This limitation is necessary to ensure that the concrete wall is of sufficient thickness to withstand fire exposure.

EXAMPLE:

GIVEN: An exterior bearing wall of a building of Type IB construction with 4 feet (1216 mm) of horizontal separation is required to have a 2-hour fire-resistance rating. The wall will be cast in place with siliceous aggregate concrete. The interior will be finished with a $^1/_2$-inch (12.7 mm) thickness of gypsum wallboard applied to steel furring members.

FIND: What is the minimum thickness of concrete required?

SOLUTION:

First calculation: Assume the interior to be the fire-exposed side.

1. From Table 721.2.1.4(2), the $^1/_2$-inch (12.7 mm) gypsum wallboard has a time-assigned value of 15 minutes; therefore, the fire-resistance rating that must be developed by the concrete must not be less than $1^3/_4$ hours (2 hours - 15 minutes).

2. Since Table 721.2.1.1 does not include a minimum thickness requirement corresponding to $1^3/_4$ hours, direct interpolation between the values for $1^1/_2$ and 2 hours is acceptable. The interpolation results in a required thickness of 4.65 inches (118 mm) of concrete.

Second calculation: Assume the exterior to be the fire-exposed side.

1. From Table 721.2.1.4(1), the multiplying factor for gypsum wallboard and siliceous aggregate concrete is 1.25; therefore, the corrected thickness for $^1/_2$ inch (12.7 mm) of gypsum wallboard

TABLE 721.2.1.4(2)
TIME ASSIGNED TO FINISH MATERIALS ON FIRE-EXPOSED SIDE OF WALL

FINISH DESCRIPTION	TIME (minute)
Gypsum wallboard	
$^3/_8$ inch	10
$^1/_2$ inch	15
$^5/_8$ inch	20
2 layers of $^3/_8$ inch	25
1 layer $^3/_8$ inch, 1 layer $^1/_2$ inch	35
2 layers $^1/_2$ inch	40
Type X gypsum wallboard	
$^1/_2$-inch	25
$^5/_8$ inch	40
Portland cement-sand plaster applied directly to concrete masonry	See Note a
Portland cement-sand plaster on metal lath	
$^3/_4$ inch	20
$^7/_8$ inch	25
1 inch	30
Gypsum sand plaster on $^3/_8$-inch gypsum lath	
$^1/_2$ inch	35
$^5/_8$ inch	40
$^3/_4$ inch	50
Gypsum sand plaster on metal lath	
$^3/_4$ inch	50
$^7/_8$ inch	60
1 inch	80

For SI: 1 inch = 25.4 mm.

a. The actual thickness of portland cement-sand plaster, provided it is $^5/_8$ inch or less in thickness, shall be permitted to be included in determining the equivalent thickness of the masonry for use in Table 721.3.2.

is 0.63 inch (16.02 mm) [1.25 inches (32 mm) × $^1/_2$ inch (12.7 mm)].

2. Table 721.2.1.1 requires 5 inches (127 mm) of siliceous aggregate concrete for a 2-hour fire-resistance rating; therefore, the actual thickness of concrete required is 4.37 inches (111 mm) [5 inches (127 mm)–0.63 inches (16 mm)].

3. Since the thickness of concrete required when assuming the interior side to be the fire-exposed side is greater, the minimum concrete thickness required to achieve a 2-hour fire-resistance rating is 4.65 inches (118 mm).

4. Section 721.2.1.4.4 requires that the concrete alone provides no less than one-half the total required rating; thus, the concrete must provide at least a 1-hour rating. From Table 721.2.1.1, it can be seen that only 3.5 inches (89 mm) of siliceous aggregate concrete is required for 1 hour, whereas 4.65 inches (118 mm) will be provided.

For a similar example problem, see the commentary to Section 721.3.2.4.

721.2.1.4.5 Concrete finishes. Finishes on concrete walls that are assumed to contribute to the total *fire-resistance rating* of the wall shall comply with the installation requirements of Section 721.3.2.5.

❖ See the commentary to Section 721.3.2.5.

721.2.2 Concrete floor and roof slabs. Reinforced and prestressed floors and roofs shall comply with Section 721.2.2.1. Multicourse floors and roofs shall comply with Sections 721.2.2.2 and 721.2.2.3, respectively.

❖ The fire test criteria for temperature rise of the unexposed surface and the ability to resist superimposed loads (heat transmission and structural criteria, respectively) must both be considered in determining the fire resistance of floors and roofs. Section 721.2.2 deals with heat transmission and Section 721.2.3 deals with structural criteria.

721.2.2.1 Reinforced and prestressed floors and roofs. The minimum thicknesses of reinforced and prestressed concrete floor or roof slabs for *fire-resistance ratings* of 1 hour to 4 hours are shown in Table 721.2.2.1.

❖ The criterion limiting the average temperature rise to 250°F (392°C) and the maximum rise at one point to 325°F (527°C) is often referred to as the "heat transmission end point." For solid concrete slabs, the heat transmission end point is primarily a function of slab thickness and aggregate type. Other factors that affect heat transmission to a lesser degree are moisture content of the concrete, aggregate size, mortar content and air content. Factors that have very little effect on heat transmission are cement content and strength; type; amount and location of reinforcement, provided these items are within the normal range of usage. The values in Table 721.2.2.1 apply to concrete slabs reinforced with bars or welded wire fabric, as well as to prestressed slabs.

TABLE 721.2.2.1
MINIMUM SLAB THICKNESS (inches)

CONCRETE TYPE	FIRE-RESISTANCE RATING (hour)				
	1	1$^1/_2$	2	3	4
Siliceous	3.5	4.3	5.0	6.2	7.0
Carbonate	3.2	4.0	4.6	5.7	6.6
Sand-lightweight	2.7	3.3	3.8	4.6	5.4
Lightweight	2.5	3.1	3.6	4.4	5.1

For SI: 1 inch = 25.4 mm.

❖ See the commentary to Section 721.2.2.1.

721.2.2.1.1 Hollow-core prestressed slabs. For hollow-core prestressed concrete slabs in which the cores are of constant cross section throughout the length, the equivalent thickness shall be permitted to be obtained by dividing the net cross-sectional area of the slab including grout in the joints, by its width.

❖ The method for determining equivalent thickness of masonry units was developed because the cores in masonry units taper. The method is, of course, also applicable to hollow-core precast concrete panels; however, because the cores in hollow-core panels do not taper, the equivalent thickness can be calculated by dividing the net cross-sectional area of the panel by its width.

721.2.2.1.2 Slabs with sloping soffits. The thickness of slabs with sloping soffits (see Figure 721.2.2.1.2) shall be determined at a distance 2*t* or 6 inches (152 mm), whichever is less, from the point of minimum thickness, where *t* is the minimum thickness.

❖ Some precast concrete wall panel sections (e.g., certain single-tee units) have tapered surfaces, so the thickness varies, as shown in Figure 721.2.2.1.2. In fire tests, it is customary to monitor the unexposed surface temperature at the location shown in the figure.

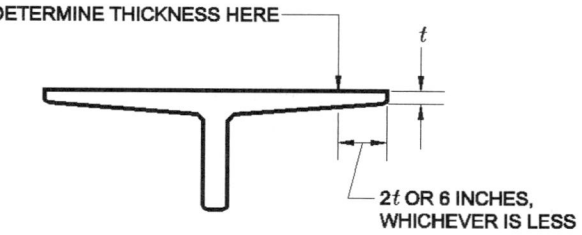

DETERMINE THICKNESS HERE

t

2*t* OR 6 INCHES, WHICHEVER IS LESS

For SI: 1 inch = 25.4 mm.

FIGURE 721.2.2.1.2
DETERMINATION OF SLAB THICKNESS
FOR SLOPING SOFFITS

❖ See the commentary to Section 721.2.2.1.2.

721.2.2.1.3 Slabs with ribbed soffits. The thickness of slabs with ribbed or undulating soffits (see Figure 721.2.2.1.3) shall be determined by one of the following expressions, whichever is applicable:

For $s > 4t$, the thickness to be used shall be t

For $s \leq 2t$, the thickness to be used shall be t_e

For $4t > s > 2t$, the thickness to be used shall be

$$t + \left(\frac{4t}{s} - 1\right)\left(t_e - t\right) \qquad \text{(Equation 7-5)}$$

where:

s = Spacing of ribs or undulations.

t = Minimum thickness.

t_e = Equivalent thickness of the slab calculated as the net area of the slab divided by the width, in which the maximum thickness used in the calculation shall not exceed $2t$.

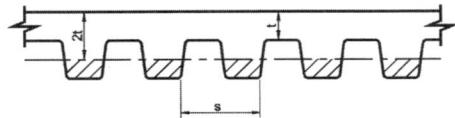

NEGLECT SHADED AREA IN CALCULATION OF EQUIVALENT THICKNESS

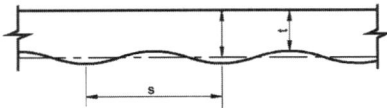

For SI: 1 inch = 25.4 mm.

FIGURE 721.2.2.1.3
SLABS WITH RIBBED OR UNDULATING SOFFITS

❖ The portion of a ribbed slab that can be used in calculating equivalent thickness, t_e, is shown in Figure 721.2.2.1.3. Note that the procedure does not give credit for joists in one-way or two-way joisted floors or in double tees. For such sections, the minimum thickness of the deck slab must be used.

EXAMPLE:

FIND: Determine the fire-resistance rating of the floor section shown in Figure 721.2.2.1.3 (1) if the units were made of siliceous aggregate concrete.

SOLUTION:

s = 12 inches.

t = 4 inches.

Therefore:

$4t > s > 2t$ (16 inches > 12 inches > 8 inches)

$$t_e = \frac{(4")(12") + (5")(1.6") + (\frac{1}{2}")(1")(1.6")}{12"}$$

t_e = 4.8 inches < $2t$.

Therefore the thickness to be used, t_s:

$$t_s = 4" + \left(\frac{(4)(4")}{12} - 1\right)(4.8" - 4")$$

t_s = 4.27 inches.

❖ From Table 721.2.2.1, using interpolation, the fire-resistance rating of the floor section shown is 1.48 hours or 1 hour, 29 minutes.

721.2.2.2 Multicourse floors. The *fire-resistance ratings* of floors that consist of a base slab of concrete with a topping (overlay) of a different type of concrete shall comply with Figure 721.2.2.2.

❖ The information in this section is based on the report entitled "Fire Endurance of Two-course Floors and Roofs" in the *ACI Journal*.

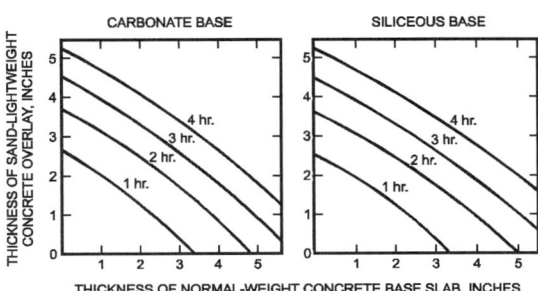

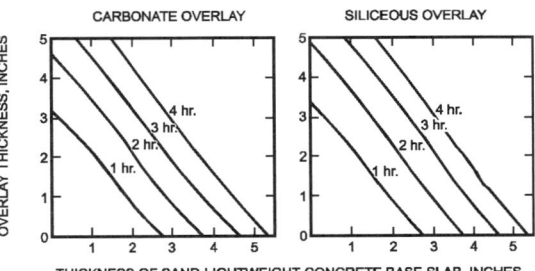

For SI: 1 inch = 25.4 mm.

FIGURE 721.2.2.2
FIRE-RESISTANCE RATINGS FOR TWO-COURSE
CONCRETE FLOORS

721.2.2.3 Multicourse roofs. The *fire-resistance ratings* of roofs which consist of a base slab of concrete with a topping (overlay) of an insulating concrete or with an insulating board and built-up roofing shall comply with Figures 721.2.2.3(1) and 721.2.2.3(2).

❖ The information in this section is based on the report entitled "Fire Endurance of Two-course Floors and Roofs" in the *ACI Journal*.

721.2.2.3.1 Heat transfer. For the transfer of heat, three-ply built-up roofing contributes 10 minutes to the *fire-resistance rating*. The *fire-resistance rating* for concrete assemblies such as those shown in Figure 721.2.2.3(1) shall be increased by 10 minutes. This increase is not applicable to those shown in Figure 721.2.2.3(2).

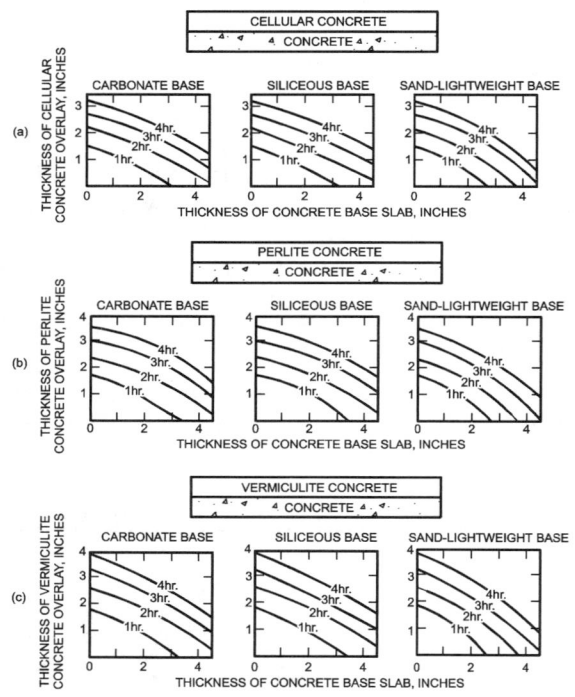

For SI: 1 inch = 25.4 mm.

FIGURE 721.2.2.3(1)
FIRE-RESISTANCE RATINGS FOR CONCRETE
ROOF ASSEMBLIES

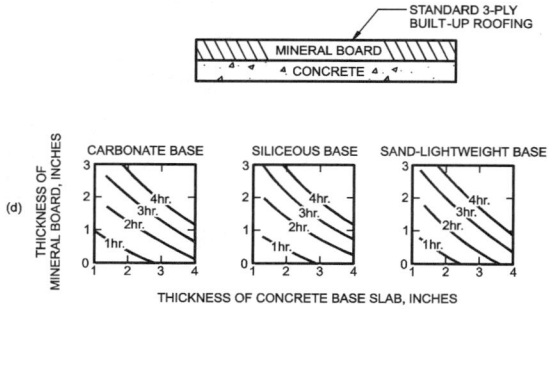

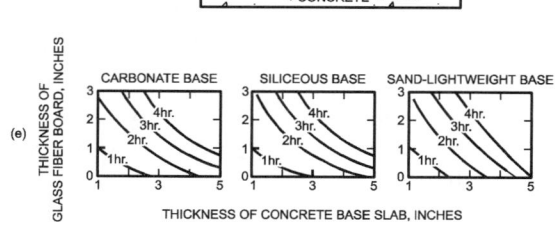

For SI: 1 inch = 25.4 mm.

FIGURE 721.2.2.3(2)
FIRE-RESISTANCE RATINGS FOR CONCRETE
ROOF ASSEMBLIES

721.2.2.4 Joints in precast slabs. Joints between adjacent precast concrete slabs need not be considered in calculating the slab thickness provided that a concrete topping at least 1 inch (25 mm) thick is used. Where no concrete topping is used, joints must be grouted to a depth of at least one-third the slab thickness at the joint, but not less than 1 inch (25 mm), or the joints must be made fire resistant by other *approved* methods.

❖ Based on data developed by Underwriters Laboratories, where a concrete topping is not used over precast concrete floors, joints must be grouted. If a concrete topping at least 1 inch thick (25 mm) is used, the joints need not be grouted.

721.2.3 Concrete cover over reinforcement. The minimum thickness of concrete cover over reinforcement in concrete slabs, reinforced beams and prestressed beams shall comply with this section.

❖ See the commentary to Sections 721.2.3.1 and 721.2.3.2.

721.2.3.1 Slab cover. The minimum thickness of concrete cover to the positive moment reinforcement shall comply with Table 721.2.3(1) for reinforced concrete and Table 721.2.3(2) for prestressed concrete. These tables are applicable for solid or hollow-core one-way or two-way slabs with flat undersurfaces. These tables are applicable to slabs that are either cast in place or precast. For precast prestressed concrete not covered elsewhere, the procedures contained in PCI MNL 124 shall be acceptable.

❖ The temperature of the tensile reinforcement depends on the thickness of cover and aggregate type. For unrestrained slabs, the values shown in Tables 721.2.3(1) and 721.2.3(2) are the cover thickness needed to keep the tensile reinforcement below 1,100°F (1922°C) and 800°F (1382°C), respectively. For restrained slabs, the temperature of the tensile reinforcement is not as critical, and thus, a minimum cover of $^3/_4$ inch (19.5 mm) is specified in PCA Bulletin 223, "Fire Endurance of Concrete Slabs as Influenced by Thickness, Aggregate Type and Moisture."

TABLE 721.2.3(1). See page 7-165.

❖ See the commentary to Section 721.2.3.1.

TABLE 721.2.3(2). See page 7-165.

❖ See the commentary to Section 721.2.3.1.

TABLE 721.2.3(3). See page 7-165.

❖ See the commentary to Section 721.2.3.2.

TABLE 721.2.3(4). See page 7-166.

❖ See the commentary to Section 721.2.3.2.

TABLE 721.2.3(5). See page 7-166.

❖ See the commentary to Section 721.2.3.2.

721.2.3.2 Reinforced beam cover. The minimum thickness of concrete cover to the positive moment reinforcement (bottom steel) for reinforced concrete beams is shown in Table 721.2.3(3) for *fire-resistance ratings* of 1 hour to 4 hours.

❖ For reinforced concrete beams, the critical steel temperature is 1,100°F (1922°C), so the effect of aggregate type is minimal. For prestressed concrete, the comparable temperature is 800°F (1382°C), so aggregate type must be considered. The data in Tables 721.2.3(3) and 721.2.3(4) was derived from the fire tests of a series of beam specimens that ranged in width between 2 and 24 inches (51 and 607 mm). Other variables in the series included aggregate type and amount of reinforcement. Tests were conducted at the Portland Cement Association. The results of the fire tests of beams conducted at Underwriters Laboratories Inc., were also analyzed. Charts graphically showing the resulting data are shown in Figure A4 of PCI Manual 124, "Design for Fire Resistance of Precast Prestressed Concrete."

721.2.3.3 Prestressed beam cover. The minimum thickness of concrete cover to the positive moment prestressing tendons (bottom steel) for restrained and unrestrained prestressed concrete beams and stemmed units shall comply with the values shown in Tables 721.2.3(4) and 721.2.3(5) for *fire-resistance ratings* of 1 hour to 4 hours. Values in Table 721.2.3(4) apply to beams 8 inches (203 mm) or greater in width. Values in Table 721.2.3(5) apply to beams or stems of any width, provided the cross-section area is not less than 40 square inches (25 806 mm²). In case of differences between the values determined from Table 721.2.3(4) or 721.2.3(5), it is permitted to use the smaller value. The concrete cover shall be calculated in accordance with Section 721.2.3.3.1. The minimum concrete cover for nonprestressed reinforcement in prestressed concrete beams shall comply with Section 721.2.3.2.

❖ See the commentary to Section 721.2.3.2.

TABLE 721.2.3(1)
COVER THICKNESS FOR REINFORCED CONCRETE FLOOR OR ROOF SLABS (inches)

CONCRETE AGGREGATE TYPE	FIRE-RESISTANCE RATING (hours)									
	Restrained					Unrestrained				
	1	$1^1/_2$	2	3	4	1	$1^1/_2$	2	3	4
Siliceous	$^3/_4$	$^3/_4$	$^3/_4$	$^3/_4$	$^3/_4$	$^3/_4$	$^3/_4$	1	$1^1/_4$	$1^5/_8$
Carbonate	$^3/_4$	$^3/_4$	$^3/_4$	$^3/_4$	$^3/_4$	$^3/_4$	$^3/_4$	$^3/_4$	$1^1/_4$	$1^1/_4$
Sand-lightweight or lightweight	$^3/_4$	$^3/_4$	$^3/_4$	$^3/_4$	$^3/_4$	$^3/_4$	$^3/_4$	$^3/_4$	$1^1/_4$	$1^1/_4$

For SI: 1 inch = 25.4 mm.

TABLE 721.2.3(2)
COVER THICKNESS FOR PRESTRESSED CONCRETE FLOOR OR ROOF SLABS (inches)

CONCRETE AGGREGATE TYPE	FIRE-RESISTANCE RATING (hours)									
	Restrained					Unrestrained				
	1	$1^1/_2$	2	3	4	1	$1^1/_2$	2	3	4
Siliceous	$^3/_4$	$^3/_4$	$^3/_4$	$^3/_4$	$^3/_4$	$1^1/_8$	$1^1/_2$	$1^3/_4$	$2^3/_8$	$2^3/_4$
Carbonate	$^3/_4$	$^3/_4$	$^3/_4$	$^3/_4$	$^3/_4$	1	$1^3/_8$	$1^5/_8$	$2^1/_8$	$2^1/_4$
Sand-lightweight or lightweight	$^3/_4$	$^3/_4$	$^3/_4$	$^3/_4$	$^3/_4$	1	$1^3/_8$	$1^1/_2$	2	$2^1/_4$

For SI: 1 inch = 25.4 mm.

TABLE 721.2.3(3)
MINIMUM COVER FOR MAIN REINFORCING BARS OF REINFORCED CONCRETE BEAMS[c]
(APPLICABLE TO ALL TYPES OF STRUCTURAL CONCRETE)

RESTRAINED OR UNRESTRAINED[a]	BEAM WIDTH[b] (inches)	FIRE-RESISTANCE RATING (hours)				
		1	$1^1/_2$	2	3	4
Restrained	5	$^3/_4$	$^3/_4$	$^3/_4$	1[a]	$1^1/_4$[a]
	7	$^3/_4$	$^3/_4$	$^3/_4$	$^3/_4$	$^3/_4$
	≥ 10	$^3/_4$	$^3/_4$	$^3/_4$	$^3/_4$	$^3/_4$
Unrestrained	5	$^3/_4$	1	$1^1/_4$	—	—
	7	$^3/_4$	$^3/_4$	$^3/_4$	$1^3/_4$	3
	≥ 10	$^3/_4$	$^3/_4$	$^3/_4$	1	$1^3/_4$

For SI: 1 inch = 25.4 mm, 1 foot = 304.8 mm.

a. Tabulated values for restrained assemblies apply to beams spaced more than 4 feet on center. For restrained beams spaced 4 feet or less on center, minimum cover of $^3/_4$ inch is adequate for ratings of 4 hours or less.

b. For beam widths between the tabulated values, the minimum cover thickness can be determined by direct interpolation.

c. The cover for an individual reinforcing bar is the minimum thickness of concrete between the surface of the bar and the fire-exposed surface of the beam. For beams in which several bars are used, the cover for corner bars used in the calculation shall be reduced to one-half of the actual value. The cover for an individual bar must be not less than one-half of the value given in Table 721.2.3(3) nor less than $^3/_4$ inch.

721.2.3.3.1 Calculating concrete cover. The concrete cover for an individual tendon is the minimum thickness of concrete between the surface of the tendon and the fire-exposed surface of the beam, except that for ungrouted ducts, the assumed cover thickness is the minimum thickness of concrete between the surface of the duct and the fire-exposed surface of the beam. For beams in which two or more tendons are used, the cover is assumed to be the average of the minimum cover of the individual tendons. For corner tendons (tendons equal distance from the bottom and side), the minimum cover used in the calculation shall be one-half the actual value. For stemmed members with two or more prestressing tendons located along the vertical centerline of the stem, the average cover shall be the distance from the bottom of the member to the centroid of the tendons. The actual cover for any individual tendon shall not be less than one-half the smaller value shown in Tables 721.2.3(4) and 721.2.3(5), or 1 inch (25 mm), whichever is greater.

721.2.4 Concrete columns. Concrete columns shall comply with this section.

❖ Most building code provisions for reinforced concrete columns are based on two reports: "Fire Tests of Building Columns" of the Associated Factory Mutual Insurance Companies and "Fire Resistance of Concrete Columns" by the NBS. Sizes of the tested columns were 12 inches (305 mm), 16 inches (406 mm) and 18 inches (457 mm). Nearly all of the columns withstood 4-hour fire tests that were essentially conducted in accordance with ASTM E 119 or UL 263. Most of the tests were stopped after 4 hours, but some were continued to 8 hours. The shortest duration was 3 hours. At the time of the tests few, if any, concrete columns were smaller than 12 inches (305 mm), but smaller columns have been in use for many years since then.

Fire tests conducted in Europe on smaller columns

TABLE 721.2.3(4)
MINIMUM COVER FOR PRESTRESSED CONCRETE BEAMS 8 INCHES OR GREATER IN WIDTH

RESTRAINED OR UNRESTRAINED[a]	CONCRETE AGGREGATE TYPE	BEAM WIDTH[b] (inches)	FIRE-RESISTANCE RATING (hours)				
			1	$1^1/_2$	2	3	4
Restrained	Carbonate or siliceous	8	$1^1/_2$	$1^1/_2$	$1^1/_2$	$1^3/_4$[a]	$2^1/_2$[a]
	Carbonate or siliceous	≥ 12	$1^1/_2$	$1^1/_2$	$1^1/_2$	$1^1/_2$	$1^7/_8$[a]
	Sand lightweight	8	$1^1/_2$	$1^1/_2$	$1^1/_2$	$1^1/_2$	2[a]
	Sand lightweight	≥ 12	$1^1/_2$	$1^1/_2$	$1^1/_2$	$1^1/_2$	$1^5/_8$[a]
Unrestrained	Carbonate or siliceous	8	$1^1/_2$	$1^3/_4$	$2^1/_2$	5[c]	—
	Carbonate or siliceous	≥ 12	$1^1/_2$	$1^1/_2$	$1^7/_8$[a]	$2^1/_2$	3
	Sand lightweight	8	$1^1/_2$	$1^1/_2$	2	$3^1/_4$	—
	Sand lightweight	≥ 12	$1^1/_2$	$1^1/_2$	$1^5/_8$	2	$2^1/_2$

For SI: 1 inch = 25.4 mm, 1 foot = 304.8 mm.

a. Tabulated values for restrained assemblies apply to beams spaced more than 4 feet on center. For restrained beams spaced 4 feet or less on center, minimum cover of $^3/_4$ inch is adequate for 4-hour ratings or less.
b. For beam widths between 8 inches and 12 inches, minimum cover thickness can be determined by direct interpolation.
c. Not practical for 8-inch-wide beam but shown for purposes of interpolation.

TABLE 721.2.3(5)
MINIMUM COVER FOR PRESTRESSED CONCRETE BEAMS OF ALL WIDTHS

RESTRAINED OR UNRESTRAINED[a]	CONCRETE AGGREGATE TYPE	BEAM AREA[b] A (square inches)	FIRE-RESISTANCE RATING (hours)				
			1	$1^1/_2$	2	3	4
Restrained	All	$40 \leq A \leq 150$	$1^1/_2$	$1^1/_2$	2	$2^1/_2$	—
	Carbonate or	$150 < A \leq 300$	$1^1/_2$	$1^1/_2$	$1^1/_2$	$1^3/_4$	$2^1/_2$
	siliceous	$300 < A$	$1^1/_2$	$1^1/_2$	$1^1/_2$	$1^1/_2$	2
	Sand lightweight	$150 < A$	$1^1/_2$	$1^1/_2$	$1^1/_2$	$1^1/_2$	2
Unrestrained	All	$40 \leq A \leq 150$	2	$2^1/_2$	—	—	—
	Carbonate or	$150 < A \leq 300$	$1^1/_2$	$1^3/_4$	$2^1/_2$	—	—
	siliceous	$300 < A$	$1^1/_2$	$1^1/_2$	2	3[c]	4[c]
	Sand lightweight	$150 < A$	$1^1/_2$	$1^1/_2$	2	3[c]	4[c]

For SI: 1 inch = 25.4 mm, 1 foot = 304.8 mm.

a. Tabulated values for restrained assemblies apply to beams spaced more than 4 feet on center. For restrained beams spaced 4 feet or less on center, minimum cover of $^3/_4$ inch is adequate for 4-hour ratings or less.
b. The cross-sectional area of a stem is permitted to include a portion of the area in the flange, provided the width of the flange used in the calculation does not exceed three times the average width of the stem.
c. U-shaped or hooped stirrups spaced not to exceed the depth of the member and having a minimum cover of 1 inch shall be provided.

indicate that fire endurance is greater for larger columns.

TABLE 721.2.4
MINIMUM DIMENSION OF CONCRETE COLUMNS (inches)

TYPES OF CONCRETE	FIRE-RESISTANCE RATING (hours)				
	1	1¹/₂	2[a]	3[a]	4[b]
Siliceous	8	9	10	12	14
Carbonate	8	9	10	11	12
Sand-lightweight	8	8¹/₂	9	10¹/₂	12

For SI: 1 inch = 25 mm.

a. The minimum dimension is permitted to be reduced to 8 inches for rectangular columns with two parallel sides at least 36 inches in length.

b. The minimum dimension is permitted to be reduced to 10 inches for rectangular columns with two parallel sides at least 36 inches in length.

❖ Table 721.2.4 reflects the information cited in the reports listed in the commentary to Section 721.2.4 and in the report "Investigations on Building Fires; Part VI; Fire Resistance of Reinforced Concrete Columns" in the National Building Studies Research Paper No. 18. This table is for columns that may be exposed to fire on all four sides. Notes a and b to Table 721.2.4 permit a smaller minimum dimension for columns than required in the table, provided the longer dimension of the column is at least 36 inches (914 mm).

721.2.4.1 Minimum size. The minimum overall dimensions of reinforced concrete columns for *fire-resistance ratings* of 1 hour to 4 hours for exposure to fire on all sides shall comply with this section.

❖ Column requirements are based on provisions in ACI 216.1-07/TMS 0216.1-07, *Code Requirements for Determining Fire Resistance of Concrete and Masonry Construction Assemblies*, the successor to ACI 216.1-97/TMS 0216.1-97.

721.2.4.1.1 Concrete strength less than or equal to 12,000 psi. For columns made with concrete having a specified compressive strength, f'_c, of less than or equal to 12,000 psi (82.7 MPa), the minimum dimension shall comply with Table 721.2.4.

❖ For concrete with compressive strength less than 12,000 psi (1827 MPa), the minimum dimensions necessary to achieve certain fire-resistance ratings are provided by Table 721.2.4. Concrete compressive strengths used in building construction are normally in the range of 4,000 to 6,000 psi (27 580 to 41 370 kPA), so this section and Table 721.2.4 are operative most of the time in determining fire-resistance ratings for concrete.

721.2.4.1.2 Concrete strength greater than 12,000 psi. For columns made with concrete having a specified compressive strength, f'_c, greater than 12,000 psi (82.7 MPa), for fire-resistance ratings of 1 hour to 4 hours the minimum dimension shall be 24 inches (610 mm).

❖ Concrete with a compressive strength greater than 12,000 psi (82.7 MPa) is very rarely used in building construction. The concrete is very stiff, with a very low water to cement ratio, and therefore is very difficult to work. Because it contains very small air pockets and voids, the material conducts heat relatively quickly when compared to more common concrete with compressive strengths in the 4,000 to 6,000 psi (27 580 to 41 370 kPA) range. Therefore, the minimum thickness necessary to achieve a fire-resistance rating of 1 to 4 hours is considerably higher.

721.2.4.2 Minimum cover for R/C columns. The minimum thickness of concrete cover to the main longitudinal reinforcement in columns, regardless of the type of aggregate used in the concrete and the specified compressive strength of concrete, f'_c, shall not be less than 1 inch (25 mm) times the number of hours of required *fire resistance* or 2 inches (51 mm), whichever is less.

❖ The minimum concrete cover refers to the main longitudinal reinforcement, not ties or spirals.

721.2.4.3 Tie and spiral reinforcement. For concrete columns made with concrete having a specified compressive strength, f'_c, greater than 12,000 psi (82.7 MPa), tie and spiral reinforcement shall comply with the following:

1. The free ends of rectangular ties shall terminate with a 135-degree (2.4 rad) standard tie hook.

2. The free ends of circular ties shall terminate with a 90-degree (1.6 rad) standard tie hook.

3. The free ends of spirals, including at lap splices, shall terminate with a 90-degree (1.6 rad) standard tie hook.

The hook extension at the free end of ties and spirals shall be the larger of six bar diameters and the extension required by Section 7.1.3 of ACI 318. Hooks shall project into the core of the column.

❖ The intent of the provisions in this section is to prevent ties or spirals from disengaging from the longitudinal reinforcement should the concrete cover over the ties or spirals be lost during a fire. The section expands on the provisions found in ACI 216.1/TMS 0216.1 by addressing spiral reinforcement, which is typically used for lateral reinforcement in round columns.

721.2.4.4 Columns built into walls. The minimum dimensions of Table 721.2.4 do not apply to a reinforced concrete column that is built into a concrete or masonry wall provided all of the following are met:

1. The *fire-resistance rating* for the wall is equal to or greater than the required rating of the column;

2. The main longitudinal reinforcing in the column has cover not less than that required by Section 721.2.4.2; and

3. Openings in the wall are protected in accordance with Table 715.4.

Where openings in the wall are not protected as required by Section 715.4, the minimum dimension of columns required to have a *fire-resistance rating* of 3 hours or less shall be 8 inches (203 mm), and 10 inches (254 mm) for columns required to have a *fire-resistance rating* of 4 hours, regardless of the type of aggregate used in the concrete.

❖ Table 721.2.4 does not apply to columns built into concrete or masonry walls that meet all three of the conditions in this section.

721.2.4.5 Precast cover units for steel columns. See Section 721.5.1.4.

721.3 Concrete masonry. The provisions of this section contain procedures by which the *fire-resistance ratings* of concrete masonry are established by calculations.

721.3.1 Equivalent thickness. The equivalent thickness of concrete masonry construction shall be determined in accordance with the provisions of this section.

721.3.1.1 Concrete masonry unit plus finishes. The equivalent thickness of concrete masonry assemblies, T_{ea}, shall be computed as the sum of the equivalent thickness of the concrete masonry unit, T_e, as determined by Section 721.3.1.2, 721.3.1.3 or 721.3.1.4, plus the equivalent thickness of finishes, T_{ef}, determined in accordance with Section 721.3.2:

$$T_{ea} = T_e + T_{ef} \qquad \text{(Equation 7-6)}$$

721.3.1.2 Ungrouted or partially grouted construction. T_e shall be the value obtained for the concrete masonry unit determined in accordance with ASTM C 140.

❖ See the commentary to Section 721.2.3.2.

721.3.1.3 Solid grouted construction. The equivalent thickness, T_e, of solid grouted concrete masonry units is the actual thickness of the unit.

721.3.1.4 Airspaces and cells filled with loose-fill material. The equivalent thickness of completely filled hollow concrete masonry is the actual thickness of the unit when loose-fill materials are: sand, pea gravel, crushed stone, or slag that meet ASTM C 33 requirements; pumice, scoria, expanded shale, expanded clay, expanded slate, expanded slag, expanded fly ash, or cinders that comply with ASTM C 331; or perlite or vermiculite meeting the requirements of ASTM C 549 and ASTM C 516, respectively.

721.3.2 Concrete masonry walls. The *fire-resistance rating* of walls and partitions constructed of concrete masonry units shall be determined from Table 721.3.2. The rating shall be based on the equivalent thickness of the masonry and type of aggregate used.

❖ It has been accepted practice to determine the fire-resistance rating of concrete masonry walls based on the type of aggregate used to manufacture them and the equivalent thickness of solid material in the wall. Equivalent thicknesses shown in Table 721.3.2 have been developed and refined through actual fire testing.

To determine the minimum equivalent thickness, Section 721.3.1.2 references ASTM C 140.

EXAMPLE:

FIND: Determine the equivalent thickness of a standard 8-inch by 8-inch by 16-inch (203 mm by 203 mm by 406 mm) concrete masonry unit as shown in Figure

721.3.2. The unit is normal weight with sand and gravel aggregate.

SOLUTION: The equivalent thickness is actually the average thickness of the solid material in the unit. The equivalent thickness is determined by the following:

$$T_e = 1728 \, A/(L \times H)$$

where:

T_e = Equivalent thickness, in.

A = Net volume of unit, ft^3.

L = Length of unit, in.

H = Height of unit, in.

Net volume, A, is determined by the following equation:

$$A = C/D$$

where:

C = Dry weight of unit, pounds.

D = Density of unit, pounds per cubic foot.

From data furnished by the manufacturer, the dry unit weight is 44 pounds (20 kg) and the density is 135 pounds per cubic foot (pcf) (2115 kg/m^2).

$$T_e = \frac{\left(1,728 \ \text{in.}^3/\text{ft.}^3\right)\left(0.326 \ \text{ft.}^3\right)}{(15.625 \ \text{in.})(7.625 \ \text{in.})}$$

T_e = 4.73 inches.

The fire-resistance rating for this type of unit with siliceous gravel aggregate can now be determined from Table 721.3.2. In this case, the fire-resistance rating will fall between 2.25 and 2.5 hours.

TABLE 721.3.2. See page 7-169.

❖ See the commentary to Section 721.3.2.

721.3.2.1 Finish on nonfire-exposed side. Where plaster or gypsum wallboard is applied to the side of the wall not exposed to fire, the contribution of the finish to the total *fire-resistance rating* shall be determined as follows: The thickness of gypsum wallboard or plaster shall be corrected by multiplying the actual thickness of the finish by applicable factor determined from Table 721.2.1.4(1). This corrected thickness of finish shall be added to the equivalent thickness of masonry and the *fire-resistance rating* of the masonry and finish determined from Table 721.3.2.

❖ The information contained in this section is based on *Fire Endurance Tests on Unit Masonry Walls with Gypsum Wallboard* and the *Supplement to the National Building Code of Canada 1980* by the National Research Council of Canada.

The fire resistance of concrete masonry walls is generally determined by temperature rise on the unexposed surface (i.e., the "heat transmission" end point). The time required to reach the heat transmission end point (fire-resistance rating) is primarily dependent upon the equivalent thickness of the masonry and the type of aggregate used to make the concrete. When additional finishes are applied to the unexposed side of the wall, the time required to reach the heat transmission end point is delayed; thus, the fire-resistance rating of the wall is increased. The increase in the rating contributed by the finish can be determined by considering the finish as an addition to the equivalent thickness of masonry; however, since the finish material and masonry may have different insulating properties, the actual thickness of the finish must be multiplied by a correction factor to be compatible with the type of aggregate used in the concrete. The correction is made by multiplying the actual finish thickness by the factor determined from Table 721.2.1.4(1), then adding the modified thickness to the equivalent thickness of masonry. This combined equivalent thickness is used to determine the fire-resistance rating from Table 721.3.2.

721.3.2.2 Finish on fire-exposed side. Where plaster or gypsum wallboard is applied to the fire-exposed side of the wall, the contribution of the finish to the total *fire-resistance rating* shall be determined as follows: The time assigned to the finish as established by Table 721.2.1.4(2) shall be added to the *fire-resistance rating* determined in Section 721.3.2 for the masonry alone, or in Section 721.3.2.1 for the masonry and finish on the nonfire-exposed side.

❖ Where finishes are added to the fire-exposed side of a concrete masonry wall, their contribution to the total fire-resistance rating is based primarily upon their ability to remain in place, thus affording protection to the masonry wall. Table 721.2.1.4(2) lists the times that have been assigned to finishes on the fire-exposed side of the wall. These time-assigned values are based upon actual fire tests. The time-assigned values are added to the fire-resistance rating of the wall alone or to the rating determined for the wall and any finish on the unexposed side.

721.3.2.3 Nonsymmetrical assemblies. For a wall having no finish on one side or having different types or thicknesses of finish on each side, the calculation procedures of this section shall be performed twice, assuming either side of the wall to be

the fire-exposed side. The *fire-resistance rating* of the wall shall not exceed the lower of the two values calculated.

Exception: For *exterior walls* with a *fire separation distance* greater than 5 feet (1524 mm) the fire shall be assumed to occur on the interior side only.

❖ Except for exterior walls having more than 5 feet (1524 mm) of fire separation distance, Section 704.5 requires that walls be rated for exposure to fire from both sides; therefore, two calculations must be performed, which assumes each side to be the fire-exposed side. Two calculations are not necessary for exterior walls with a fire separation distance of more than 5 feet (1524 mm) or for other walls that are symmetrical (i.e., walls having the same type and thickness of finish on each side). The calculated fire-resistance rating must not exceed the lower of the two ratings determined, assuming that each side is the fire-exposed side.

721.3.2.4 Minimum concrete masonry fire-resistance rating. Where the finish applied to a concrete masonry wall contributes to its *fire-resistance rating*, the masonry alone shall provide not less than one-half the total required *fire-resistance rating*.

❖ Where gypsum wallboard or plaster finishes are applied to a concrete masonry wall, the calculated fire-resistance rating for the concrete alone should be not less than one-half the required fire-resistance rating. This limitation is necessary to ensure that the masonry wall is of sufficient thickness to withstand fire exposure.

EXAMPLE:

GIVEN: A wall required to have a 4-hour fire-resistance rating will be constructed with concrete masonry units of expanded shale aggregate. The wall will be finished on each side with a layer of $^1/_2$-inch (12.7 mm) gypsum wallboard.

FIND: What is the minimum equivalent thickness of concrete masonry required?

TABLE 721.3.2
MINIMUM EQUIVALENT THICKNESS (inches) OF BEARING OR NONBEARING CONCRETE MASONRY WALLS[a,b,c,d]

TYPE OF AGGREGATE	FIRE-RESISTANCE RATING (hours)														
	$^1/_2$	$^3/_4$	1	$1^1/_4$	$1^1/_2$	$1^3/_4$	2	$2^1/_4$	$2^1/_2$	$2^3/_4$	3	$3^1/_4$	$3^1/_2$	$3^3/_4$	4
Pumice or expanded slag	1.5	1.9	2.1	2.5	2.7	3.0	3.2	3.4	3.6	3.8	4.0	4.2	4.4	4.5	4.7
Expanded shale, clay or slate	1.8	2.2	2.6	2.9	3.3	3.4	3.6	3.8	4.0	4.2	4.4	4.6	4.8	4.9	5.1
Limestone, cinders or unexpanded slag	1.9	2.3	2.7	3.1	3.4	3.7	4.0	4.3	4.5	4.8	5.0	5.2	5.5	5.7	5.9
Calcareous or siliceous gravel	2.0	2.4	2.8	3.2	3.6	3.9	4.2	4.5	4.8	5.0	5.3	5.5	5.8	6.0	6.2

For SI: 1 inch = 25.4 mm.

a. Values between those shown in the table can be determined by direct interpolation.

b. Where combustible members are framed into the wall, the thickness of solid material between the end of each member and the opposite face of the wall, or between members set in from opposite sides, shall not be less than 93 percent of the thickness shown in the table.

c. Requirements of ASTM C 55, ASTM C 73, ASTM C 90 or ASTM C 744 shall apply.

d. Minimum required equivalent thickness corresponding to the hourly fire-resistance rating for units with a combination of aggregate shall be determined by linear interpolation based on the percent by volume of each aggregate used in manufacture.

SOLUTION:

Since the wall has the same type and thickness of finish on each side, only one calculation is required.

1. The $\frac{1}{2}$-inch (12.7 mm) gypsum wallboard on the fire-exposed side has a time-assigned value of 15 minutes in accordance with Table 721.2.1.4(2).

2. Therefore, the fire resistance required to be provided by the masonry and gypsum wallboard on the unexposed side is 3 hours and 45 minutes (4 hours minus 15 minutes).

3. From Table 721.2.1.4(1), the corrected thickness of gypsum wallboard on the unexposed side is $\frac{1}{2}$ inch (12.7 mm) ($1.00 \times \frac{1}{2}$ inch).

4. From Table 721.3.2, the minimum equivalent thickness of masonry, including the corrected thickness of gypsum wallboard, required for a rating of 3 hours and 45 minutes is 4.9 inches (124 mm).

5. Therefore, the equivalent thickness of masonry required is 4.4 inches (112 mm) [4.9 inches (124 mm) minus 0.5 inches (12.7 mm)].

6. From Table 721.3.2, it can be determined that 4.4 inches (112 mm) of expanded shale aggregate concrete masonry will provide fire resistance of 3 hours; therefore, the requirement that the masonry alone provide at least one-half of the total required rating is satisfied.

For a similar example problem, see the commentary to Section 721.2.1.4.4.

721.3.2.5 Attachment of finishes. Installation of finishes shall be as follows:

1. Gypsum wallboard and gypsum lath applied to concrete masonry or concrete walls shall be secured to wood or steel furring members spaced not more than 16 inches (406 mm) on center (o.c.).

2. Gypsum wallboard shall be installed with the long dimension parallel to the furring members and shall have all joints finished.

3. Other aspects of the installation of finishes shall comply with the applicable provisions of Chapters 7 and 25.

721.3.3 Multiwythe masonry walls. The *fire-resistance rating* of wall assemblies constructed of multiple wythes of masonry materials shall be permitted to be based on the *fire-resistance rating* period of each wythe and the continuous airspace between each wythe in accordance with the following formula:

$$R_A = (R_1^{0.59} + R_2^{0.59} + ... + R_n^{0.59} + A_1 + A_2 + ... + A_n)^{1.7}$$

(Equation 7-7)

where:

R_A = *Fire-resistance rating* of the assembly (hours).

$R_1, R_2, ..., R_n$ = *Fire-resistance rating* of wythes for 1, 2, *n* (hours), respectively.

$A_1, A_2, ..., A_n$ = 0.30, factor for each continuous airspace for 1, 2, ...*n*, respectively, having a depth of $\frac{1}{2}$ inch (12.7 mm) or more between wythes.

❖ This calculation method was first published in Report BMS 92 of the National Bureau of Standards (NBS).

721.3.4 Concrete masonry lintels. *Fire-resistance ratings* for concrete masonry lintels shall be determined based upon the nominal thickness of the lintel and the minimum thickness of concrete masonry or concrete, or any combination thereof, covering the main reinforcing bars, as determined according to Table 721.3.4, or by *approved* alternate methods.

TABLE 721.3.4
MINIMUM COVER OF LONGITUDINAL REINFORCEMENT IN FIRE-RESISTANCE-RATED REINFORCED CONCRETE MASONRY LINTELS (inches)

NOMINAL WIDTH OF LINTEL (inches)	FIRE-RESISTANCE RATING (hours)			
	1	2	3	4
6	$1\frac{1}{2}$	2	—	—
8	$1\frac{1}{2}$	$1\frac{1}{2}$	$1\frac{3}{4}$	3
10 or greater	$1\frac{1}{2}$	$1\frac{1}{2}$	$1\frac{1}{2}$	$1\frac{3}{4}$

For SI: 1 inch = 25.4 mm.

721.3.5 Concrete masonry columns. The *fire-resistance rating* of concrete masonry columns shall be determined based upon the least plan dimension of the column in accordance with Table 721.3.5 or by *approved* alternate methods.

TABLE 721.3.5
MINIMUM DIMENSION OF CONCRETE MASONRY COLUMNS (inches)

FIRE-RESISTANCE RATING (hours)			
1	2	3	4
8 inches	10 inches	12 inches	14 inches

For SI: 1 inch = 25.4 mm.

721.4 Clay brick and tile masonry. The provisions of this section contain procedures by which the *fire-resistance ratings* of clay brick and tile masonry are established by calculations.

721.4.1 Masonry walls. The *fire-resistance rating* of masonry walls shall be based upon the equivalent thickness as calculated in accordance with this section. The calculation shall take into account finishes applied to the wall and airspaces between wythes in multiwythe construction.

721.4.1.1 Equivalent thickness. The *fire-resistance ratings* of walls or partitions constructed of solid or hollow clay masonry units shall be determined from Table 721.4.1(1) or 721.4.1(2). The equivalent thickness of the clay masonry unit shall be determined by Equation 7-8 when using Table 721.4.1(1). The *fire-resistance rating* determined from Table 721.4.1(1) shall be permitted to be used in the calculated *fire-resistance rating* procedure in Section 721.4.2.

$$T_e = V_n/LH$$

(Equation 7-8)

where:

T_e = The equivalent thickness of the clay masonry unit (inches).

V_n = The net volume of the clay masonry unit (inch³).

L = The specified length of the clay masonry unit (inches).

H = The specified height of the clay masonry unit (inches).

❖ It has been accepted practice to determine the fire-resistance ratings of clay masonry walls by the ASTM E 119 or UL 263 test method. In fact, Table 721.4.1(1) is based on tests performed by various laboratories in conformance with ASTM E 119 or UL 263. Most notable are the NBS, Ohio State University Experiment Station and the University of California at Berkeley. New methods, however, have been developed to evaluate the fire resistance of clay masonry by the equivalent thickness method, similar to concrete masonry. This approach was used in conjunction with extensive testing previously performed on clay masonry to develop Table 721.4.1(1).

721.4.1.1.1 Hollow clay units. The equivalent thickness, T_e, shall be the value obtained for hollow clay units as determined in accordance with Equation 7-8. The net volume, V_n, of the units shall be determined using the gross volume and percentage of void area determined in accordance with ASTM C 67.

721.4.1.1.2 Solid grouted clay units. The equivalent thickness of solid grouted clay masonry units shall be taken as the actual thickness of the units.

721.4.1.1.3 Units with filled cores. The equivalent thickness of the hollow clay masonry units is the actual thickness of the unit when completely filled with loose-fill materials of: sand, pea gravel, crushed stone, or slag that meet ASTM C 33 requirements; pumice, scoria, expanded shale, expanded clay, expanded slate, expanded slag, expanded fly ash, or cinders in compliance with ASTM C 331; or perlite or vermiculite meeting the requirements of ASTM C 549 and ASTM C 516, respectively.

TABLE 721.4.1(1)
FIRE-RESISTANCE PERIODS OF CLAY MASONRY WALLS

MATERIAL TYPE	MINIMUM REQUIRED EQUIVALENT THICKNESS FOR FIRE RESISTANCE[a, b, c] (inches)			
	1 hour	2 hour	3 hour	4 hour
Solid brick of clay or shale[d]	2.7	3.8	4.9	6.0
Hollow brick or tile of clay or shale, unfilled	2.3	3.4	4.3	5.0
Hollow brick or tile of clay or shale, grouted or filled with materials specified in Section 721.4.1.1.3	3.0	4.4	5.5	6.6

For SI: 1 inch = 25.4 mm.

a. Equivalent thickness as determined from Section 721.4.1.1.

b. Calculated fire resistance between the hourly increments listed shall be determined by linear interpolation.

c. Where combustible members are framed in the wall, the thickness of solid material between the end of each member and the opposite face of the wall, or between members set in from opposite sides, shall be not less than 93 percent of the thickness shown.

d. For units in which the net cross-sectional area of cored brick in any plane parallel to the surface containing the cores is at least 75 percent of the gross cross-sectional area measured in the same plane.

TABLE 721.4.1(2)
FIRE-RESISTANCE RATINGS FOR BEARING STEEL FRAME
BRICK VENEER WALLS OR PARTITIONS

WALL OR PARTITION ASSEMBLY	PLASTER SIDE EXPOSED (hours)	BRICK FACED SIDE EXPOSED (hours)
Outside facing of steel studs: $1/2''$ wood fiberboard sheathing next to studs, $3/4''$ airspace formed with $3/4'' \times 1\,5/8''$ wood strips placed over the fiberboard and secured to the studs; metal or wire lath nailed to such strips, $3\,3/4''$ brick veneer held in place by filling $3/4''$ airspace between the brick and lath with mortar. Inside facing of studs: $3/4''$ unsanded gypsum plaster on metal or wire lath attached to $5/16''$ wood strips secured to edges of the studs.	1.5	4
Outside facing of steel studs: $1''$ insulation board sheathing attached to studs, $1''$ airspace, and $3\,3/4''$ brick veneer attached to steel frame with metal ties every 5th course. Inside facing of studs: $7/8''$ sanded gypsum plaster (1:2 mix) applied on metal or wire lath attached directly to the studs.	1.5	4
Same as above except use $7/8''$ vermiculite—gypsum plaster or $1''$ sanded gypsum plaster (1:2 mix) applied to metal or wire.	2	4
Outside facing of steel studs: $1/2''$ gypsum sheathing board, attached to studs, and $3\,3/4''$ brick veneer attached to steel frame with metal ties every 5th course. Inside facing of studs: $1/2''$ sanded gypsum plaster (1:2 mix) applied to $1/2''$ perforated gypsum lath securely attached to studs and having strips of metal lath 3 inches wide applied to all horizontal joints of gypsum lath.	2	4

For SI: 1 inch = 25.4 mm.

721.4.1.2 Plaster finishes. Where plaster is applied to the wall, the total *fire-resistance rating* shall be determined by the formula:

$$R = (R_n^{0.59} + pl)^{1.7} \quad \text{(Equation 7-9)}$$

where:

R = The *fire-resistance rating* of the assembly (hours).

R_n = The *fire-resistance rating* of the individual wall (hours).

pl = Coefficient for thickness of plaster.

Values for $R_n^{0.59}$ for use in Equation 7-9 are given in Table 721.4.1(3). Coefficients for thickness of plaster shall be selected from Table 721.4.1(4) based on the actual thickness of plaster applied to the wall or partition and whether one or two sides of the wall are plastered.

❖ The variables for use in Equation 7-9 for determining the fire-resistance rating of plastered clay masonry walls are based on "Fire Resistance Classifications of Building Construction," BMS 92 of the NBS. These values were derived from fire test results. The average thickness of plaster applied in the series of tests ranged from $\frac{1}{2}$ inch (12.7 mm) to $\frac{3}{4}$ inch (19.1 mm). The thickness for which the coefficients of plaster in Table 721.4.1(4) are given are those most likely to be applied to a clay masonry wall. Ratings for other thicknesses provided can be obtained by substituting the appropriate coefficients in the formula.

A test of four hollow concrete unit walls shows the effect of one coat of plaster on the fire-exposed side to be about the same as for one coat of plaster on the unexposed side. Tests have not been made with plaster on the unexposed side of only clay hollow walls; however, the ratings resulting from Equation 7-9 for clay or tile masonry walls for plaster on one side are believed to have a sufficient margin of safety to be applicable to either the exposed or unexposed side of the wall.

EXAMPLE:

GIVEN: A 4-inch (102 mm) solid brick wall is required to have a 3-hour fire-resistance rating with clay masonry units.

FIND: The thickness of one side of plaster required to attain a 3-hour rating.

SOLUTION:

1. From Table 721.4.1(1), a 2-hour solid brick wall is 3.8 inches (99 mm) of equivalent thickness and a 3-hour wall is 4.9 inches (124 mm). Through interpolation (Note a), a 4-inch (102 mm) wall is approximately a 131-minute (2.18 hours) fire rating.

2. From Table 721.4.1(4), plaster has a coefficient of 0.3 for $\frac{1}{2}$-inch-thick (12.7 mm) plaster, 0.37 for $\frac{5}{8}$-inch-thick (15.9 mm) plaster or 0.45 for $\frac{3}{4}$-inch-thick (19.1 mm) plaster.

3. From Equation 7-9:
   $$R = (R_n^{0.59} + pl)^{1.7}$$

where:

R = 3

R_n = 2.18

3 = $[(2.18)^{0.59} + pl]^{1.7}$

$$\sqrt[1.7]{3} = \sqrt[1.7]{[158 + pl)^{1.7}]}$$

1.91 = $1.58 + pl$

pl = 0.326

4. Therefore, one coat of $\frac{5}{8}$-inch-thick (15.9 mm) sanded gypsum plaster on a 4-inch (102 mm) solid brick masonry wall would result in a 3-hour fire-resistance rating.

721.4.1.3 Multiwythe walls with airspace. Where a continuous airspace separates multiple wythes of the wall or partition, the total *fire-resistance rating* shall be determined by the formula:

$$R = (R_1^{0.59} + R_2^{0.59} + ... + R_n^{0.59} + as)^{1.7} \quad \text{(Equation 7-10)}$$

where:

R = The *fire-resistance rating* of the assembly (hours).

R_1, R_2 and R_n = The *fire-resistance rating* of the individual wythes (hours).

as = Coefficient for continuous airspace.

Values for $R_n^{0.59}$ for use in Equation 7-10 are given in Table 721.4.1(3). The coefficient for each continuous airspace of $\frac{1}{2}$ inch to $3\frac{1}{2}$ inches (12.7 to 89 mm) separating two individual wythes shall be 0.3.

❖ Tests have shown that a continuous airspace separating two wythes of masonry can also increase the fire-resistance rating of a clay masonry wall. According to BMS 92 of the NBS, the coefficient for continuous airspace (*as*) between $\frac{1}{2}$ inch (12.7 mm) and $3\frac{1}{2}$ inches (89 mm) is 0.3. Equation 7-10 provides the formula for calculating the fire-resistance ratings involving an airspace if the fire-resistance rating of the wythes separated by the airspace is known. The fire-resistance ratings for various wythes can be obtained from Tables 721.4.1(1) and 721.4.1(2). The procedure for using Equation 7-10 is the same as Equation 7-9, except that a coefficient for airspace (*as*) is substituted for the coefficient of plaster (*pl*).

TABLE 721.4.1(3)
VALUES OF $R_n^{0.59}$

$R_n^{0.59}$	R (hours)
1	1.0
2	1.50
3	1.91
4	2.27

TABLE 721.4.1(4)
COEFFICIENTS FOR PLASTER, pl [a]

THICKNESS OF PLASTER (inch)	ONE SIDE	TWO SIDE
$^1/_2$	0.3	0.6
$^5/_8$	0.37	0.75
$^3/_4$	0.45	0.90

For SI: 1 inch = 25.4 mm.
a. Values listed in table are for 1:3 sanded gypsum plaster.

TABLE 721.4.1(5)
REINFORCED MASONRY LINTELS

NOMINAL LINTEL WIDTH (inches)	MINIMUM LONGITUDINAL REINFORCEMENT COVER FOR FIRE RESISTANCE (inch)			
	1 hour	2 hour	3 hour	4 hour
6	$1^1/_2$	2	NP	NP
8	$1^1/_2$	$1^1/_2$	$1^3/_4$	3
10 or more	$1^1/_2$	$1^1/_2$	$1^1/_2$	$1^3/_4$

For SI: 1 inch = 25.4 mm.
NP = Not permitted.

TABLE 721.4.1(6)
REINFORCED CLAY MASONRY COLUMNS

COLUMN SIZE	FIRE-RESISTANCE RATING (hour)			
	1	2	3	4
Minimum column dimension (inches)	8	10	12	14

For SI: 1 inch = 25.4 mm.

721.4.1.4 Nonsymmetrical assemblies. For a wall having no finish on one side or having different types or thicknesses of finish on each side, the calculation procedures of this section shall be performed twice, assuming either side to be the fire-exposed side of the wall. The *fire resistance* of the wall shall not exceed the lower of the two values determined.

> **Exception:** For *exterior walls* with a *fire separation distance* greater than 5 feet (1524 mm), the fire shall be assumed to occur on the interior side only.

❖ Except for exterior walls having a fire separation distance of more than 5 feet (1524 mm), Section 704.5 requires that walls be rated for exposure to fire from both sides; therefore, two calculation procedures must be performed, which assumes each side to be the fire-exposed side. Two calculations are not necessary for exterior walls with a fire separation distance of more than 5 feet (1524 mm) or for other walls that are symmetrical (i.e., having the same type and thickness of finish on each side). The calculated fire-resistance rating must not exceed the lower of the two values determined, assuming that each side is the fire-exposed side.

721.4.2 Multiwythe walls. The *fire-resistance rating* for walls or partitions consisting of two or more dissimilar wythes shall be permitted to be determined by the formula:

$$R = (R_1^{0.59} + R_2^{0.59} + ... + R_n^{0.59})^{1.7}$$ **(Equation 7-11)**

where:

R = The *fire-resistance rating* of the assembly (hours).

R_1, R_2 and R_n = The *fire-resistance rating* of the individual wythes (hours).

Values for $R_n^{0.59}$ for use in Equation 7-11 are given in Table 721.4.1(3).

❖ Typically, the fire-resistance-rating period for clay masonry walls is determined by the temperature rise on the unexposed side of the wall. Equation 7-11 is based on this criterion. According to the general theory of heat transmission, if walls of the same material are exposed to a heat source that maintains a constant surface temperature on the exposed side and the unexposed side is protected against heat loss, the time in which a given temperature will be attained on the unexposed side will vary as the square of the wall thickness does.

In the standard ASTM E 119 or UL 263 test, which involves specified conditions of temperature measurement and a fire that increases the temperature at the exposed surface of the wall as the test proceeds, the time required to attain a given temperature rise on the unexposed side will differ when the temperature on the exposed side remains constant at the initial exposure temperature for any period. A degree of correlation between test results and calculations can be obtained by assuming the variation to be according to a lower power of n than the second. The fire resistance of the wall can then be expressed by the formula:

$$R = (CV)^n$$

where

R = Fire-resistance period;

C = Coefficient depending on the material, design of the wall and the units of measurement of R and V;

V = Volume of solid material per unit area of wall surface; and

n = Exponent depending on the rate of increase of temperature at the exposed face of the wall.

For walls of a given material and design, it was found that an increase of 50 percent in volume of solid material per unit area of wall surface resulted in a 100-percent increase in the fire-resistance period. This correlates to a value of 1.7 for n. A value of n less than 2 is expected, since an increasing temperature on the exposed surface would tend to shorten the fire-resistance rating of walls that would qualify for a higher rating.

The fire-resistance rating of a wall may be expressed in terms of the fire-resistance rating of the conjoined wythes or laminae of the wall as follows:

$$R_1 = (C_1 V_1)^n$$
$$R_2 = (C_2 V_2)^n$$
$$R_n = (C_n V_n)^n$$

where R_1, R_2 and R_n are the fire-resistance ratings of each conjoined wythe.

The fire-resistance period of the composite wall will be:

$$R = \left(\sum_{i=1}^{n} C_i V_i \right)^{n_0}$$

$$R = \left(C_1 V_1 + C_2 V_2 + \ldots C_n V_n \right)^{n_0}$$

Substituting for $C_1 V_1$, $C_2 V_2$ and $C_n V_n$ from the above relations yields:

$$R = \left(R_1^{1/n_0} + R_2^{1/n_0} + R_n^{1/n_0} \right)^{n_0}$$

Substituting 1.7 for n_0 and 0.59 for $1/n_0$, the general formula becomes:

$$R = \left(R_1^{0.59} + R_2^{0.59} \ldots + R_n^{0.59} \right)^{1.7}$$

Equation 7-11 was developed by the NBS in the early 1940s and first appeared in Appendix B of BMS 92 titled "Fire Resistance Classifications of Building Construction." It is noted that the fire-resistance rating has been expressed in terms of the fire-resistance rating of the component laminae of the wall, which need not be of the same material and design.

721.4.2.1 Multiwythe walls of different material. For walls that consist of two or more wythes of different materials (concrete or concrete masonry units) in combination with clay masonry units, the *fire-resistance rating* of the different materials shall be permitted to be determined from Table 721.2.1.1 for concrete; Table 721.3.2 for concrete masonry units or Table 721.4.1(1) or 721.4.1(2) for clay and tile masonry units.

❖ A multiwythe wall (i.e., a wall consisting of two or more dissimilar materials) has a greater fire-resistance rating than a simple summation of the fire-endurance rating of the various layers. Equation 7-11 permits a calculated fire-resistance rating if the fire-resistance endurance rating is known for each dissimilar material. This section lists the applicable tables from Section 721 that can be used in conjunction with clay masonry walls to determine the total fire-resistance rating of a multiwythe wall composed of a combination of concrete, concrete masonry or clay masonry units.

721.4.3 Reinforced clay masonry lintels. *Fire-resistance ratings* for clay masonry lintels shall be determined based on the nominal width of the lintel and the minimum covering for the longitudinal reinforcement in accordance with Table 721.4.1(5).

721.4.4 Reinforced clay masonry columns. The *fire-resistance ratings* shall be determined based on the last plan dimension of the column in accordance with Table 721.4.1(6). The minimum cover for longitudinal reinforcement shall be 2 inches (51 mm).

721.5 Steel assemblies. The provisions of this section contain procedures by which the *fire-resistance ratings* of steel assemblies are established by calculations.

721.5.1 Structural steel columns. The *fire-resistance ratings* of steel columns shall be based on the size of the element and the type of protection provided in accordance with this section.

721.5.1.1 General. These procedures establish a basis for determining the *fire resistance* of column assemblies as a function of the thickness of fire-resistant material and, the weight, W, and heated perimeter, D, of steel columns. As used in these sections, W is the average weight of a structural steel column in pounds per linear foot. The heated perimeter, D, is the inside perimeter of the fire-resistant material in inches as illustrated in Figure 721.5.1(1).

721.5.1.1.1 Nonload-bearing protection. The application of these procedures shall be limited to column assemblies in which the fire-resistant material is not designed to carry any of the load acting on the column.

721.5.1.1.2 Embedments. In the absence of substantiating fire-endurance test results, ducts, conduit, piping, and similar mechanical, electrical, and plumbing installations shall not be embedded in any required fire-resistant materials.

721.5.1.1.3 Weight-to-perimeter ratio. Table 721.5.1(1) contains weight-to-heated-perimeter ratios (W/D) for both contour and box fire-resistant profiles, for the wide flange shapes most often used as columns. For different fire-resistant protection profiles or column cross sections, the weight-to-heated-perimeter ratios (W/D) shall be determined in accordance with the definitions given in this section.

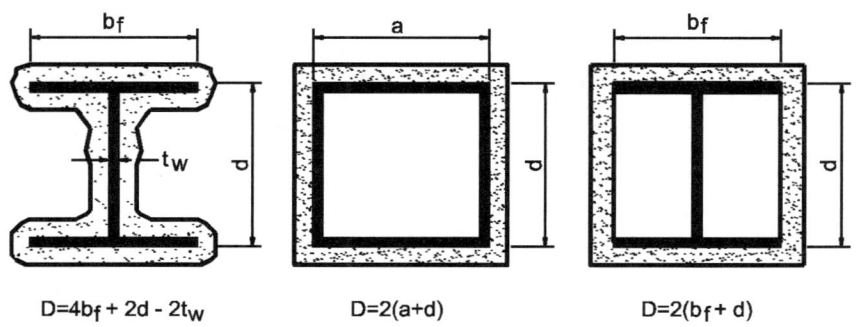

FIGURE 721.5.1(1)
DETERMINATION OF THE HEATED PERIMETER OF STRUCTURAL STEEL COLUMNS

TABLE 721.5.1(1)
_W/D_ RATIOS FOR STEEL COLUMNS

STRUCTURAL SHAPE	CONTOUR PROFILE	BOX PROFILE	STRUCTURAL SHAPE	CONTOUR PROFILE	BOX PROFILE
W14 × 233	2.49	3.65	W10 × 112	1.78	2.57
× 211	2.28	3.35	× 100	1.61	2.33
× 193	2.10	3.09	× 88	1.43	2.08
× 176	1.93	2.85	× 77	1.26	1.85
× 159	1.75	2.60	× 68	1.13	1.66
× 145	1.61	2.39	× 60	1.00	1.48
× 132	1.52	2.25	× 54	0.91	1.34
× 120	1.39	2.06	× 49	0.83	1.23
× 109	1.27	1.88	× 45	0.87	1.24
× 99	1.16	1.72	× 39	0.76	1.09
× 90	1.06	1.58	× 33	0.65	0.93
× 82	1.20	1.68			
× 74	1.09	1.53	W8 × 67	1.34	1.94
× 68	1.01	1.41	× 58	1.18	1.71
× 61	0.91	1.28	× 48	0.99S	1.44
× 53	0.89	1.21	× 40	0.83	1.23
× 48	0.81	1.10	× 35	0.73	1.08
× 43	0.73	0.99	× 31	0.65	0.97
			× 28	0.67	0.96
W12 × 190	2.46	3.51	× 24	0.58	0.83
× 170	2.22	3.20	× 21	0.57	0.77
× 152	2.01	2.90	× 18	0.49	0.67
× 136	1.82	2.63			
× 120	1.62	2.36	W6 × 25	0.69	1.00
× 106	1.44	2.11	× 20	0.56	0.82
× 96	1.32	1.93	× 16	0.57	0.78
× 87	1.20	1.76	× 15	0.42	0.63
× 79	1.10	1.61	× 12	0.43	0.60
× 72	1.00	1.48	× 9	0.33	0.46
× 65	0.91	1.35			
× 58	0.91	1.31	W5 × 19	0.64	0.93
× 53	0.84	1.20	× 16	0.54	0.80
× 50	0.89	1.23			
× 45	0.81	1.12	W4 × 13	0.54	0.79
× 40	0.72	1.00			

For SI: 1 pound per linear foot per inch = 0.059 kg/m/mm.

TABLE 721.5.1(2)
PROPERTIES OF CONCRETE

PROPERTY	NORMAL-WEIGHT CONCRETE	STRUCTURAL LIGHTWEIGHT CONCRETE
Thermal conductivity (k_c)	0.95 Btu/hr · ft · °F	0.35 Btu/hr · ft · °F
Specific heat (c_c)	0.20 Btu/lb °F	0.20 Btu/lb °F
Density (P_c)	145 lb/ft^3	110 lb/ft^3
Equilibrium (free) moisture content (m) by volume	4%	5%

For SI: 1 inch = 25.4 mm, 1 foot = 304.8 mm, 1 lb/ft^3 = 16.0185 kg/m^3, Btu/hr · ft · °F = 1.731 W/(m · K).

TABLE 721.5.1(3)
THERMAL CONDUCTIVITY OF CONCRETE OR CLAY
MASONRY UNITS

DENSITY (d_m) OF UNITS (lb/ft^3)	THERMAL CONDUCTIVITY (K) OF UNITS (Btu/hr · ft · °F)
Concrete Masonry Units	
80	0.207
85	0.228
90	0.252
95	0.278
100	0.308
105	0.340
110	0.376
115	0.416
120	0.459
125	0.508
130	0.561
135	0.620
140	0.685
145	0.758
150	0.837
Clay Masonry Units	
120	1.25
130	2.25

For SI: 1 pound per cubic foot = 16.0185 kg/m^3, Btu/hr · ft · °F = 1.731 W/(m · K).

TABLE 721.5.1(4)
WEIGHT-TO-HEATED-PERIMETER RATIOS (*W/D*)
FOR TYPICAL WIDE FLANGE BEAM AND GIRDER SHAPES

STRUCTURAL SHAPE	CONTOUR PROFILE	BOX PROFILE	STRUCTURAL SHAPE	CONTOUR PROFILE	BOX PROFILE
W36 × 300	2.47	3.33	W24 × 68	0.92	1.21
× 280	2.31	3.12	× 62	0.92	1.14
× 260	2.16	2.92	× 55	0.82	1.02
× 245	2.04	2.76			
× 230	1.92	2.61	W21 × 147	1.83	2.60
× 210	1.94	2.45	× 132	1.66	2.35
× 194	1.80	2.28	× 122	1.54	2.19
× 182	1.69	2.15	× 111	1.41	2.01
× 170	1.59	2.01	× 101	1.29	1.84
× 160	1.50	1.90	× 93	1.38	1.80
× 150	1.41	1.79	× 83	1.24	1.62
× 135	1.28	1.63	× 73	1.10	1.44
			× 68	1.03	1.35
W33 × 241	2.11	2.86	× 62	0.94	1.23
× 221	1.94	2.64	× 57	0.93	1.17
× 201	1.78	2.42	× 50	0.83	1.04
× 152	1.51	1.94	× 44	0.73	0.92
× 141	1.41	1.80			
× 130	1.31	1.67	W18 × 119	1.69	2.42
× 118	1.19	1.53	× 106	1.52	2.18
			× 97	1.39	2.01
W30 × 211	2.00	2.74	× 86	1.24	1.80
× 191	1.82	2.50	× 76	1.11	1.60
× 173	1.66	2.28	× 71	1.21	1.59
× 132	1.45	1.85	× 65	1.11	1.47
× 124	1.37	1.75	× 60	1.03	1.36
× 116	1.28	1.65	× 55	0.95	1.26
× 108	1.20	1.54	× 50	0.87	1.15
× 99	1.10	1.42	× 46	0.86	1.09
			× 40	0.75	0.96
W27 × 178	1.85	2.55	× 35	0.66	0.85
× 161	1.68	2.33			
× 146	1.53	2.12	W16 × 100	1.56	2.25
× 114	1.36	1.76	× 89	1.40	2.03
× 102	1.23	1.59	× 77	1.22	1.78
× 94	1.13	1.47	× 67	1.07	1.56
× 84	1.02	1.33	× 57	1.07	1.43
			× 50	0.94	1.26
			× 45	0.85	1.15
W24 × 162	1.85	2.57	× 40	0.76	1.03
× 146	1.68	2.34	× 36	0.69	0.93
× 131	1.52	2.12	× 31	0.65	0.83
× 117	1.36	1.91	× 26	0.55	0.70
× 104	1.22	1.71			
× 94	1.26	1.63	W14 × 132	1.83	3.00
× 84	1.13	1.47	× 120	1.67	2.75
× 76	1.03	1.34	× 109	1.53	2.52

(continued)

TABLE 721.5.1(4)—continued
WEIGHT-TO-HEATED-PERIMETER RATIOS (*W/D*)
FOR TYPICAL WIDE FLANGE BEAM AND GIRDER SHAPES

STRUCTURAL SHAPE	CONTOUR PROFILE	BOX PROFILE	STRUCTURAL SHAPE	CONTOUR PROFILE	BOX PROFILE
W14 × 99	1.39	2.31	W10 × 30	0.79	1.12
× 90	1.27	2.11	× 26	0.69	0.98
× 82	1.41	2.12	× 22	0.59	0.84
× 74	1.28	1.93	× 19	0.59	0.78
× 68	1.19	1.78	× 17	0.54	0.70
× 61	1.07	1.61	× 15	0.48	0.63
× 53	1.03	1.48	× 12	0.38	0.51
× 48	0.94	1.35			
× 43	0.85	1.22	W8 × 67	1.61	2.55
× 38	0.79	1.09	× 58	1.41	2.26
× 34	0.71	0.98	× 48	1.18	1.91
× 30	0.63	0.87	× 40	1.00	1.63
× 26	0.61	0.79	× 35	0.88	1.44
× 22	0.52	0.68	× 31	0.79	1.29
			× 28	0.80	1.24
W12 × 87	1.44	2.34	× 24	0.69	1.07
× 79	1.32	2.14	× 21	0.66	0.96
× 72	1.20	1.97	× 18	0.57	0.84
× 65	1.09	1.79	× 15	0.54	0.74
× 58	1.08	1.69	× 13	0.47	0.65
× 53	0.99	1.55	× 10	0.37	0.51
× 50	1.04	1.54			
× 45	0.95	1.40	W6 × 25	0.82	1.33
× 40	0.85	1.25	× 20	0.67	1.09
× 35	0.79	1.11	× 16	0.66	0.96
× 30	0.69	0.96	× 15	0.51	0.83
× 26	0.60	0.84	× 12	0.51	0.75
× 22	0.61	0.77	× 9	0.39	0.57
× 19	0.53	0.67			
× 16	0.45	0.57	W5 × 19	0.76	1.24
× 14	0.40	0.50	× 16	0.65	1.07
W10 × 112	2.14	3.38	W4 × 13	0.65	1.05
× 100	1.93	3.07			
× 88	1.70	2.75			
× 77	1.52	2.45			
× 68	1.35	2.20			
× 60	1.20	1.97			
× 54	1.09	1.79			
× 49	0.99	1.64			
× 45	1.03	1.59			
× 39	0.94	1.40			
× 33	0.77	1.20			

For SI: Pounds per linear foot per inch = 0.059 kg/m/mm.

TABLE 721.5.1(5)
FIRE RESISTANCE OF CONCRETE MASONRY PROTECTED STEEL COLUMNS

COLUMN SIZE	CONCRETE MASONRY DENSITY POUNDS PER CUBIC FOOT	MINIMUM REQUIRED EQUIVALENT THICKNESS FOR FIRE-RESISTANCE RATING OF CONCRETE MASONRY PROTECTION ASSEMBLY, T_e (inches)				COLUMN SIZE	CONCRETE MASONRY DENSITY POUNDS PER CUBIC FOOT	MINIMUM REQUIRED EQUIVALENT THICKNESS FOR FIRE-RESISTANCE RATING OF CONCRETE MASONRY PROTECTION ASSEMBLY, T_e (inches)			
		1-hour	2-hour	3-hour	4-hour			1-hour	2-hour	3-hour	4-hour
W14 × 82	80	0.74	1.61	2.36	3.04	W10 × 68	80	0.72	1.58	2.33	3.01
	100	0.89	1.85	2.67	3.40		100	0.87	1.83	2.65	3.38
	110	0.96	1.97	2.81	3.57		110	0.94	1.95	2.79	3.55
	120	1.03	2.08	2.95	3.73		120	1.01	2.06	2.94	3.72
W14 × 68	80	0.83	1.70	2.45	3.13	W10 × 54	80	0.88	1.76	2.53	3.21
	100	0.99	1.95	2.76	3.49		100	1.04	2.01	2.83	3.57
	110	1.06	2.06	2.91	3.66		110	1.11	2.12	2.98	3.73
	120	1.14	2.18	3.05	3.82		120	1.19	2.24	3.12	3.90
W14 × 53	80	0.91	1.81	2.58	3.27	W10 × 45	80	0.92	1.83	2.60	3.30
	100	1.07	2.05	2.88	3.62		100	1.08	2.07	2.90	3.64
	110	1.15	2.17	3.02	3.78		110	1.16	2.18	3.04	3.80
	120	1.22	2.28	3.16	3.94		120	1.23	2.29	3.18	3.96
W14 × 43	80	1.01	1.93	2.71	3.41	W10 × 33	80	1.06	2.00	2.79	3.49
	100	1.17	2.17	3.00	3.74		100	1.22	2.23	3.07	3.81
	110	1.25	2.28	3.14	3.90		110	1.30	2.34	3.20	3.96
	120	1.32	2.38	3.27	4.05		120	1.37	2.44	3.33	4.12
W12 × 72	80	0.81	1.66	2.41	3.09	W8 × 40	80	0.94	1.85	2.63	3.33
	100	0.91	1.88	2.70	3.43		100	1.10	2.10	2.93	3.67
	110	0.99	1.99	2.84	3.60		110	1.18	2.21	3.07	3.83
	120	1.06	2.10	2.98	3.76		120	1.25	2.32	3.20	3.99
W12 × 58	80	0.88	1.76	2.52	3.21	W8 × 31	80	1.06	2.00	2.78	3.49
	100	1.04	2.01	2.83	3.56		100	1.22	2.23	3.07	3.81
	110	1.11	2.12	2.97	3.73		110	1.29	2.33	3.20	3.97
	120	1.19	2.23	3.11	3.89		120	1.36	2.44	3.33	4.12
W12 × 50	80	0.91	1.81	2.58	3.27	W8 × 24	80	1.14	2.09	2.89	3.59
	100	1.07	2.05	2.88	3.62		100	1.29	2.31	3.16	3.90
	110	1.15	2.17	3.02	3.78		110	1.36	2.42	3.28	4.05
	120	1.22	2.28	3.16	3.94		120	1.43	2.52	3.41	4.20
W12 × 40	80	1.01	1.94	2.72	3.41	W8 × 18	80	1.22	2.20	3.01	3.72
	100	1.17	2.17	3.01	3.75		100	1.36	2.40	3.25	4.01
	110	1.25	2.28	3.14	3.90		110	1.42	2.50	3.37	4.14
	120	1.32	2.39	3.27	4.06		120	1.48	2.59	3.49	4.28

(continued)

TABLE 721.5.1(5)—continued
FIRE RESISTANCE OF CONCRETE MASONRY PROTECTED STEEL COLUMNS

NOMINAL TUBE SIZE (inches)	CONCRETE MASONRY DENSITY, POUNDS PER CUBIC FOOT	MINIMUM REQUIRED EQUIVALENT THICKNESS FOR FIRE-RESISTANCE RATING OF CONCRETE MASONRY PROTECTION ASSEMBLY, T_e (inches)				NOMINAL PIPE SIZE (inches)	CONCRETE MASONRY DENSITY, POUNDS PER CUBIC FOOT	MINIMUM REQUIRED EQUIVALENT THICKNESS FOR FIRE-RESISTANCE RATING OF CONCRETE MASONRY PROTECTION ASSEMBLY, T_e (inches)			
		1-hour	2-hour	3-hour	4-hour			1-hour	2-hour	3-hour	4-hour
$4 \times 4 \times {}^1/_2$ wall thickness	80	0.93	1.90	2.71	3.43	4 double extra strong 0.674 wall thickness	80	0.80	1.75	2.56	3.28
	100	1.08	2.13	2.99	3.76		100	0.95	1.99	2.85	3.62
	110	1.16	2.24	3.13	3.91		110	1.02	2.10	2.99	3.78
	120	1.22	2.34	3.26	4.06		120	1.09	2.20	3.12	3.93
$4 \times 4 \times {}^3/_8$ wall thickness	80	1.05	2.03	2.84	3.57	4 extra strong 0.337 wall thickness	80	1.12	2.11	2.93	3.65
	100	1.20	2.25	3.11	3.88		100	1.26	2.32	3.19	3.95
	110	1.27	2.35	3.24	4.02		110	1.33	2.42	3.31	4.09
	120	1.34	2.45	3.37	4.17		120	1.40	2.52	3.43	4.23
$4 \times 4 \times {}^1/_4$ wall thickness	80	1.21	2.20	3.01	3.73	4 standard 0.237 wall thickness	80	1.26	2.25	3.07	3.79
	100	1.35	2.40	3.26	4.02		100	1.40	2.45	3.31	4.07
	110	1.41	2.50	3.38	4.16		110	1.46	2.55	3.43	4.21
	120	1.48	2.59	3.50	4.30		120	1.53	2.64	3.54	4.34
$6 \times 6 \times {}^1/_2$ wall thickness	80	0.82	1.75	2.54	3.25	5 double extra strong 0.750 wall thickness	80	0.70	1.61	2.40	3.12
	100	0.98	1.99	2.84	3.59		100	0.85	1.86	2.71	3.47
	110	1.05	2.10	2.98	3.75		110	0.91	1.97	2.85	3.63
	120	1.12	2.21	3.11	3.91		120	0.98	2.02	2.99	3.79
$6 \times 6 \times {}^3/_8$ wall thickness	80	0.96	1.91	2.71	3.42	5 extra strong 0.375 wall thickness	80	1.04	2.01	2.83	3.54
	100	1.12	2.14	3.00	3.75		100	1.19	2.23	3.09	3.85
	110	1.19	2.25	3.13	3.90		110	1.26	2.34	3.22	4.00
	120	1.26	2.35	3.26	4.05		120	1.32	2.44	3.34	4.14
$6 \times 6 \times {}^1/_4$ wall thickness	80	1.14	2.11	2.92	3.63	5 standard 0.258 wall thickness	80	1.20	2.19	3.00	3.72
	100	1.29	2.32	3.18	3.93		100	1.34	2.39	3.25	4.00
	110	1.36	2.43	3.30	4.08		110	1.41	2.49	3.37	4.14
	120	1.42	2.52	3.43	4.22		120	1.47	2.58	3.49	4.28
$8 \times 8 \times {}^1/_2$ wall thickness	80	0.77	1.66	2.44	3.13	6 double extra strong 0.864 wall thickness	80	0.59	1.46	2.23	2.92
	100	0.92	1.91	2.75	3.49		100	0.73	1.71	2.54	3.29
	110	1.00	2.02	2.89	3.66		110	0.80	1.82	2.69	3.47
	120	1.07	2.14	3.03	3.82		120	0.86	1.93	2.83	3.63
$8 \times 8 \times {}^3/_8$ wall thickness	80	0.91	1.84	2.63	3.33	6 extra strong 0.432 wall thickness	80	0.94	1.90	2.70	3.42
	100	1.07	2.08	2.92	3.67		100	1.10	2.13	2.98	3.74
	110	1.14	2.19	3.06	3.83		110	1.17	2.23	3.11	3.89
	120	1.21	2.29	3.19	3.98		120	1.24	2.34	3.24	4.04
$8 \times 8 \times {}^1/_4$ wall thickness	80	1.10	2.06	2.86	3.57	6 standard 0.280 wall thickness	80	1.14	2.12	2.93	3.64
	100	1.25	2.28	3.13	3.87		100	1.29	2.33	3.19	3.94
	110	1.32	2.38	3.25	4.02		110	1.36	2.43	3.31	4.08
	120	1.39	2.48	3.38	4.17		120	1.42	2.53	3.43	4.22

For SI: 1 inch = 25.4 mm, 1 pound per cubic feet = 16.02 kg/m³.
Note: Tabulated values assume 1-inch air gap between masonry and steel section.

TABLE 721.5.1(6)
FIRE RESISTANCE OF CLAY MASONRY PROTECTED STEEL COLUMNS

COLUMN SIZE	CLAY MASONRY DENSITY, POUNDS PER CUBIC FOOT	MINIMUM REQUIRED EQUIVALENT THICKNESS FOR FIRE-RESISTANCE RATING OF CLAY MASONRY PROTECTION ASSEMBLY, T_e (inches)				COLUMN SIZE	CLAY MASONRY DENSITY, POUNDS PER CUBIC FOOT	MINIMUM REQUIRED EQUIVALENT THICKNESS FOR FIRE-RESISTANCE RATING OF CLAY MASONRY PROTECTION ASSEMBLY, T_e (inches)			
		1-hour	2-hour	3-hour	4-hour			1-hour	2-hour	3-hour	4-hour
W14 × 82	120	1.23	2.42	3.41	4.29	W10 × 68	120	1.27	2.46	3.26	4.35
	130	1.40	2.70	3.78	4.74		130	1.44	2.75	3.83	4.80
W14 × 68	120	1.34	2.54	3.54	4.43	W10 × 54	120	1.40	2.61	3.62	4.51
	130	1.51	2.82	3.91	4.87		130	1.58	2.89	3.98	4.95
W14 × 53	120	1.43	2.65	3.65	4.54	W10 × 45	120	1.44	2.66	3.67	4.57
	130	1.61	2.93	4.02	4.98		130	1.62	2.95	4.04	5.01
W14 × 43	120	1.54	2.76	3.77	4.66	W10 × 33	120	1.59	2.82	3.84	4.73
	130	1.72	3.04	4.13	5.09		130	1.77	3.10	4.20	5.13
W12 × 72	120	1.32	2.52	3.51	4.40	W8 × 40	120	1.47	2.70	3.71	4.61
	130	1.50	2.80	3.88	4.84		130	1.65	2.98	4.08	5.04
W12 × 58	120	1.40	2.61	3.61	4.50	W8 × 31	120	1.59	2.82	3.84	4.73
	130	1.57	2.89	3.98	4.94		130	1.77	3.10	4.20	5.17
W12 × 50	120	1.43	2.65	3.66	4.55	W8 × 24	120	1.66	2.90	3.92	4.82
	130	1.61	2.93	4.02	4.99		130	1.84	3.18	4.28	5.25
W12 × 40	120	1.54	2.77	3.78	4.67	W8 × 18	120	1.75	3.00	4.01	4.91
	130	1.72	3.05	4.14	5.10		130	1.93	3.27	4.37	5.34

STEEL TUBING						STEEL PIPE					
NOMINAL TUBE SIZE (inches)	CLAY MASONRY DENSITY, POUNDS PER CUBIC FOOT	MINIMUM REQUIRED EQUIVALENT THICKNESS FOR FIRE-RESISTANCE RATING OF CLAY MASONRY PROTECTION ASSEMBLY, T_e (inches)				NOMINAL PIPE SIZE (inches)	CLAY MASONRY DENSITY, POUNDS PER CUBIC FOOT	MINIMUM REQUIRED EQUIVALENT THICKNESS FOR FIRE-RESISTANCE RATING OF CLAY MASONRY PROTECTION ASSEMBLY, T_e (inches)			
		1-hour	2-hour	3-hour	4-hour			1-hour	2-hour	3-hour	4-hour
4 × 4 × 1/2 wall thickness	120	1.44	2.72	3.76	4.68	4 double extra strong 0.674 wall thickness	120	1.26	2.55	3.60	4.52
	130	1.62	3.00	4.12	5.11		130	1.42	2.82	3.96	4.95
4 × 4 × 3/8 wall thickness	120	1.56	2.84	3.88	4.78	4 extra strong 0.337 wall thickness	120	1.60	2.89	3.92	4.83
	130	1.74	3.12	4.23	5.21		130	1.77	3.16	4.28	5.25
4 × 4 × 1/4 wall thickness	120	1.72	2.99	4.02	4.92	4 standard 0.237 wall thickness	120	1.74	3.02	4.05	4.95
	130	1.89	3.26	4.37	5.34		130	1.92	3.29	4.40	5.37
6 × 6 × 1/2 wall thickness	120	1.33	2.58	3.62	4.52	5 double extra strong 0.750 wall thickness	120	1.17	2.44	3.48	4.40
	130	1.50	2.86	3.98	4.96		130	1.33	2.72	3.84	4.83
6 × 6 × 3/8 wall thickness	120	1.48	2.74	3.76	4.67	5 extra strong 0.375 wall thickness	120	1.55	2.82	3.85	4.76
	130	1.65	3.01	4.13	5.10		130	1.72	3.09	4.21	5.18
6 × 6 × 1/4 wall thickness	120	1.66	2.91	3.94	4.84	5 standard 0.258 wall thickness	120	1.71	2.97	4.00	4.90
	130	1.83	3.19	4.30	5.27		130	1.88	3.24	4.35	5.32
8 × 8 × 1/2 wall thickness	120	1.27	2.50	3.52	4.42	6 double extra strong 0.864 wall thickness	120	1.04	2.28	3.32	4.23
	130	1.44	2.78	3.89	4.86		130	1.19	2.60	3.68	4.67
8 × 8 × 3/8 wall thickness	120	1.43	2.67	3.69	4.59	6 extra strong 0.432 wall thickness	120	1.45	2.71	3.75	4.65
	130	1.60	2.95	4.05	5.02		130	1.62	2.99	4.10	5.08
8 × 8 × 1/4 wall thickness	120	1.62	2.87	3.89	4.78	6 standard 0.280 wall thickness	120	1.65	2.91	3.94	4.84
	130	1.79	3.14	4.24	5.21		130	1.82	3.19	4.30	5.27

TABLE 721.5.1(7)
MINIMUM COVER (inch) FOR STEEL COLUMNS
ENCASED IN NORMAL-WEIGHT CONCRETE[a]
[FIGURE 721.5.1(6)(c)]

STRUCTURAL SHAPE	FIRE-RESISTANCE RATING (hours)				
	1	1½	2	3	4
W14 × 233				1½	2
× 176			1		
× 132		1			2½
× 90	1			2	
× 61			1½		
× 48					3
× 43		1½		2½	
W12 × 152			1		2½
× 96		1		2	
× 65	1				
× 50			1½		3
× 40		1½		2½	
W10 × 88	1			2	
× 49					3
× 45	1	1½	1½		
× 39				2½	3½
× 33			2		
W8 × 67		1			3
× 58			1½		
× 48	1			2½	
× 31		1½			3½
× 21			2		
× 18				3	4
W6 × 25		1½	2		3½
× 20				3	
× 16	1	2			4
× 15					
× 9	1½		2½	3½	

For SI: 1 inch = 25.4 mm.
a. The tabulated thicknesses are based upon the assumed properties of normal-weight concrete given in Table 721.5.1(2).

TABLE 721.5.1(8)
MINIMUM COVER (inch) FOR STEEL COLUMNS
ENCASED IN STRUCTURAL LIGHTWEIGHT CONCRETE[a]
[FIGURE 721.5.1(6)(c)]

STRUCTURAL SHAPE	FIRE-RESISTANCE RATING (HOURS)				
	1	1½	2	3	4
W14 × 233				1	1½
× 193					
× 74	1	1	1	1½	2
× 61					
× 43			1½	2	2½
W12 × 65			1½		2
× 53	1	1	1		
× 40			1½	2	2½
W10 × 112					2
× 88	1			1½	
× 60		1			
× 33			1½	2	2½
W8 × 35					2½
× 28	1	1		2	
× 24			1½		3
× 18		1½	2½		

For SI: 1 inch = 25.4 mm.
a. The tabulated thicknesses are based upon the assumed properties of structural lightweight concrete given in Table 721.5.1(2).

TABLE 721.5.1(9)
MINIMUM COVER (inch) FOR STEEL COLUMNS IN NORMAL-WEIGHT PRECAST COVERS[a]
[FIGURE 721.5.1(6)(a)]

STRUCTURAL SHAPE	FIRE-RESISTANCE RATING (hours)				
	1	1 1/2	2	3	4
W14 × 233			1 1/2		3
× 211				2 1/2	
× 176					3 1/2
× 145		1 1/2	2		
× 109	1 1/2			3	
× 99					
× 61					4
× 43		2	2 1/2	3 1/2	4 1/2
W12 × 190			1 1/2	2 1/2	3 1/2
× 152					
× 120		1 1/2	2		
× 96				3	
× 87	1 1/2				4
× 58					
× 40		2	2 1/2	3 1/2	4 1/2
W10 × 112					3 1/2
× 88		1 1/2	2	3	
× 77	1 1/2				4
× 54		2	2 1/2	3 1/2	
× 33					4 1/2
W8 × 67		1 1/2	2	3	
× 58					4
× 48	1 1/2	2	2 1/2	3 1/2	
× 28					
× 21					4 1/2
× 18		2 1/2	3	4	
W6 × 25		2	2 1/2	3 1/2	
× 20	1 1/2				4 1/2
× 16			3		
× 12	2	2 1/2		4	
× 9					5

For SI: 1 inch = 25.4 mm.

a. The tabulated thicknesses are based upon the assumed properties of normal-weight concrete given in Table 721.5.1(2).

TABLE 721.5.1(10)
MINIMUM COVER (inch) FOR STEEL COLUMNS IN STRUCTURAL LIGHTWEIGHT PRECAST COVERS[a]
[FIGURE 721.5.1(6)(a)]

STRUCTURAL SHAPE	FIRE-RESISTANCE RATING (hours)				
	1	1 1/2	2	3	4
W14 × 233					2 1/2
× 176				2	
× 145			1 1/2		
× 132	1 1/2	1 1/2			3
× 109					
× 99				2 1/2	
× 68			2		
× 43				3	3 1/2
W12 × 190					2 1/2
× 152				2	
× 136			1 1/2		3
× 106					
× 96	1 1/2	1 1/2		2 1/2	
× 87					
× 65			2		
× 40				3	3 1/2
W10 × 112				2	
× 100			1 1/2		3
× 88					
× 77	1 1/2	1 1/2		2 1/2	
× 60			2		
× 39				3	3 1/2
× 33		2			
W8 × 67			1 1/2	2 1/2	3
× 48		1 1/2			
× 35	1 1/2		2		3 1/2
× 28				3	
× 18		2	2 1/2		4
W6 × 25			2	3	3 1/2
× 15	1 1/2	2			4
× 9			2 1/2	3 1/2	

For SI: 1 inch = 25.4 mm.

a. The tabulated thicknesses are based upon the assumed properties of structural lightweight concrete given in Table 721.5.1(2).

721.5.1.2 Gypsum wallboard protection. The *fire resistance* of structural steel columns with weight- to-heated-perimeter ratios (*W/D*) less than or equal to 3.65 and which are protected with Type X gypsum wallboard shall be permitted to be determined from the following expression:

$$R = 130 \left[\frac{h\left(W'/D\right)}{2} \right]^{0.75}$$ (Equation 7-12)

where:

R = *Fire resistance* (minutes).

h = Total thickness of gypsum wallboard (inches).

D = Heated perimeter of the structural steel column (inches).

W' = Total weight of the structural steel column and gypsum wallboard protection (pounds per linear foot).

W' = *W* + 50*hD*/144.

❖ See the example given after Section 721.5.1.2.1

721.5.1.2.1 Attachment. The gypsum wallboard shall be supported as illustrated in either Figure 721.5.1(2) for *fire-resistance ratings* of 4 hours or less, or Figure 721.5.1(3) for *fire-resistance ratings* of 3 hours or less.

❖ EXAMPLE:

GIVEN: A W12 × 136 column protected by $^1/_2$-inch (12.7 mm) Type X gypsum wallboard as shown in Figure 721.5.1.2.1.

FIND: The fire-resistance rating of the column.

SOLUTION:

1. The heated perimeter *D* = 2(12.4 inches + 13.4-inches) = 51.6 inches (1311 mm).

2. *W'* = 136 + (50)(0.5)(51.6)/144 = 144.96.

3. *W'*/Δ= 144.96/51.62 = 2.81.

4. *R* = 130[(0.5)(2.81)/2]$^{0.75}$ = 99.7 minutes (1.66 hours).

5. Alternatively, from Table 721.5.1(1), the *W/D* ratio for a W12 × 136 is tabulated as 2.63. From Figure 721.5.1(4) using a *W/D* ratio of 2.63 and $^1/_2$-inch (12.7 mm) gypsum wallboard, an *R*-value of approximately 1.6 hours is obtained.

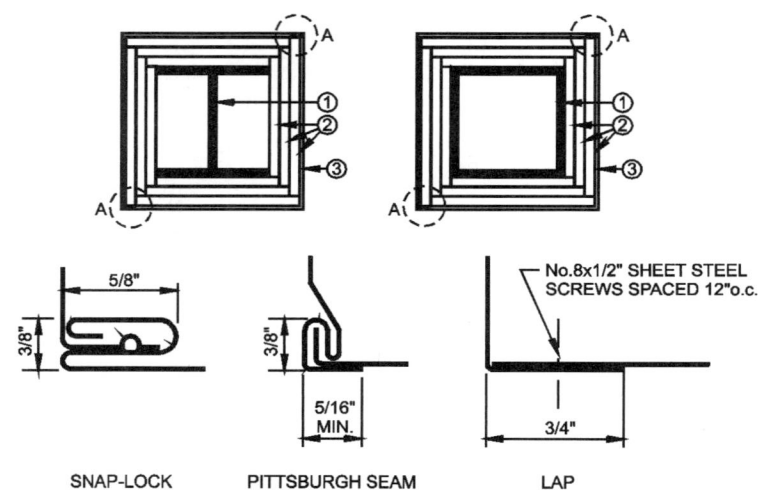

CORNER JOINT DETAILS (A)

FIGURE 721.5.1(2)
GYPSUM WALLBOARD PROTECTED STRUCTURAL STEEL
COLUMNS WITH SHEET STEEL COLUMN COVERS

For SI: 1 inch = 25.4 mm, 1 foot = 305 mm.

1. Structural steel column, either wide flange or tubular shapes.

2. Type X gypsum wallboard in accordance with ASTM C 36. For single-layer applications, the wallboard shall be applied vertically with no horizontal joints. For multiple-layer applications, horizontal joints are permitted at a minimum spacing of 8 feet, provided that the joints in successive layers are staggered at least 12 inches. The total required thickness of wallboard shall be determined on the basis of the specified fire-resistance rating and the weight-to-heated-perimeter ratio (*W/D*) of the column. For fire-resistance ratings of 2 hours or less, one of the required layers of gypsum wallboard may be applied to the exterior of the sheet steel column covers with 1-inch- long Type S screws spaced 1 inch from the wallboard edge and 8 inches on center. For such installations, 0.0149-inch minimum thickness galvanized steel corner beads with 1$^1/_2$-inch legs shall be attached to the wallboard with Type S screws spaced 12 inches on center.

3. For fire-resistance ratings of 3 hours or less, the column covers shall be fabricated from 0.0239-inch minimum thickness galvanized or stainless steel. For 4-hour fire-resistance ratings, the column covers shall be fabricated from 0.0239-inch minimum thickness stainless steel. The column covers shall be erected with the Snap Lock or Pittsburgh joint details.

For fire-resistance ratings of 2 hours or less, column covers fabricated from 0.0269-inch minimum thickness galvanized or stainless steel shall be permitted to be erected with lap joints. The lap joints shall be permitted to be located anywhere around the perimeter of the column cover. The lap joints shall be secured with $^1/_2$-inch-long No. 8 sheet metal screws spaced 12 inches on center.

The column covers shall be provided with a minimum expansion clearance of $^1/_8$ inch per linear foot between the ends of the cover and any restraining construction.

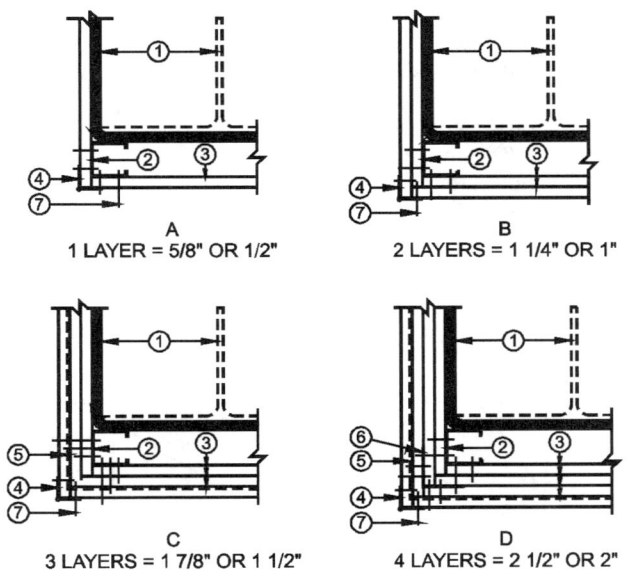

FIGURE 721.5.1(3)
GYPSUM WALLBOARD PROTECTED STRUCTURAL STEEL
COLUMNS WITH STEEL STUD/SCREW ATTACHMENT SYSTEM

For SI: 1 inch = 25.4 mm, 1 foot = -305 mm.

1. Structural steel column, either wide flange or tubular shapes.
2. $1^5/_8$-inch deep studs fabricated from 0.0179-inch minimum thickness galvanized steel with $1^5/_{16}$ or $1^7/_{16}$-inch legs. The length of the steel studs shall be $^1/_2$ inch less than the height of the assembly.
3. Type X gypsum wallboard in accordance with ASTM C 36. For single-layer applications, the wallboard shall be applied vertically with no horizontal joints. For multiple-layer applications, horizontal joints are permitted at a minimum spacing of 8 feet, provided that the joints in successive layers are staggered at least 12 inches. The total required thickness of wallboard shall be determined on the basis of the specified fire-resistance rating and the weight-to-heated-perimeter ratio (*W/D*) of the column.
4. Galvanized 0.0149-inch minimum thickness steel corner beads with $1^1/_2$-inch legs attached to the wallboard with 1-inch-long Type S screws spaced 12 inches on center.
5. No. 18 SWG steel tie wires spaced 24 inches on center.
6. Sheet metal angles with 2-inch legs fabricated from 0.0221-inch minimum thickness galvanized steel.
7. Type S screws, 1 inch long, shall be used for attaching the first layer of wallboard to the steel studs and the third layer to the sheet metal angles at 24 inches on center. Type S screws $1^3/_4$-inch long shall be used for attaching the second layer of wallboard to the steel studs and the fourth layer to the sheet metal angles at 12 inches on center. Type S screws $2^1/_4$ inches long shall be used for attaching the third layer of wallboard to the steel studs at 12 inches on center.

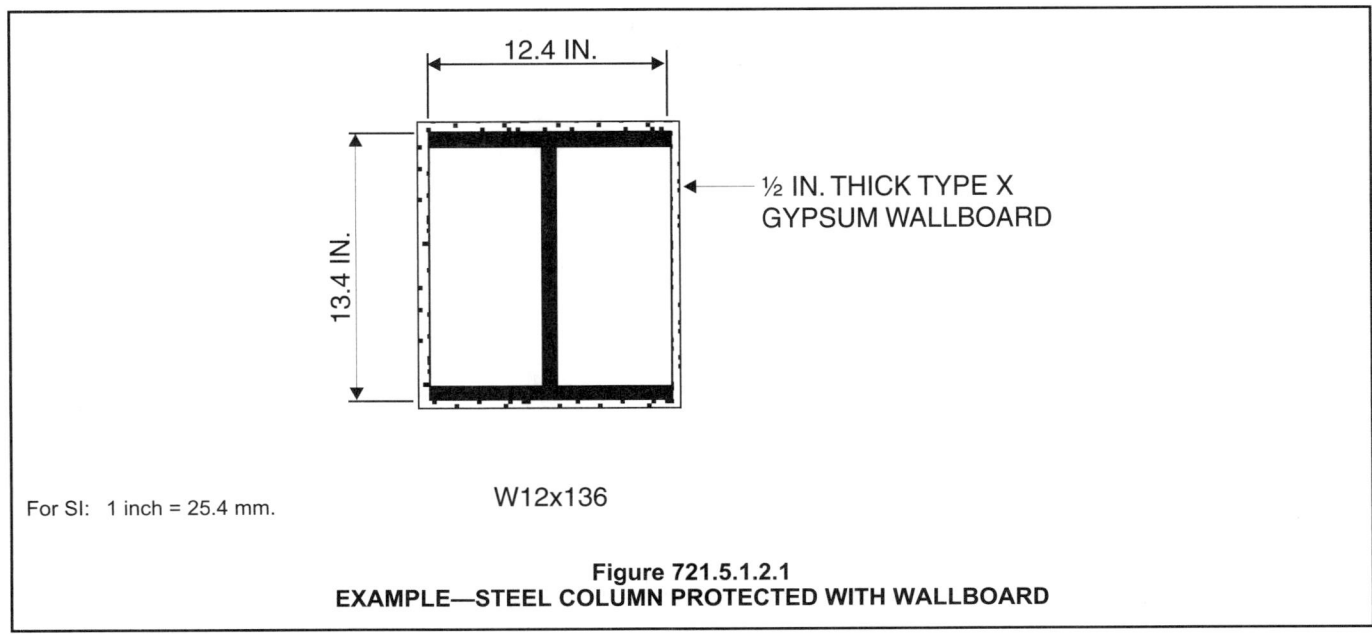

For SI: 1 inch = 25.4 mm.

Figure 721.5.1.2.1
EXAMPLE—STEEL COLUMN PROTECTED WITH WALLBOARD

721.5.1.2.2 Gypsum wallboard equivalent to concrete. The determination of the *fire resistance* of structural steel columns from Figure 721.5.1(4) is permitted for various thicknesses of gypsum wallboard as a function of the weight-to-heated-perimeter ratio (*W/D*) of the column. For structural steel columns with weight-to-heated-perimeter ratios (*W/D*) greater than 3.65, the thickness of gypsum wallboard required for specified *fire-resistance ratings* shall be the same as the thickness determined for a *W*14 × 233 wide flange shape.

721.5.1.3 Sprayed fire-resistant materials. The *fire resistance* of wide-flange structural steel columns protected with sprayed fire-resistant materials, as illustrated in Figure

721.5.1(5), shall be permitted to be determined from the following expression:

$$R=\left[C_1\left(W/D\right)+C_2\right]h \qquad \textbf{(Equation 7-13)}$$

where:

R = *Fire resistance* (minutes).

h = Thickness of sprayed fire-resistant material (inches).

D = Heated perimeter of the structural steel column (inches).

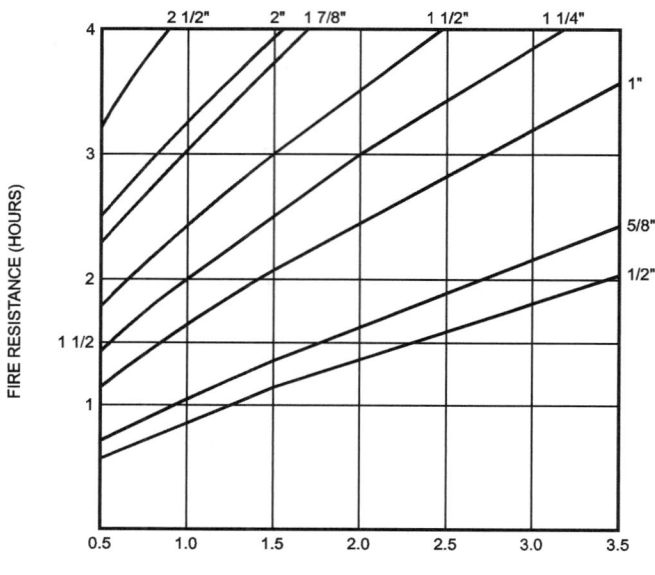

For SI: 1 inch = 25.4 mm, 1 pound per linear foot/inch = 0.059 kg/m/mm.

FIGURE 721.5.1(4)
FIRE RESISTANCE OF STRUCTURAL STEEL COLUMNS PROTECTED WITH VARIOUS THICKNESSES OF
TYPE X GYPSUM WALLBOARD

a. The *W/D* ratios for typical wide flange columns are listed in Table 721.5.1(1). For other column shapes, the *W/D* ratios shall be determined in accordance with Section 720.5.1.1.

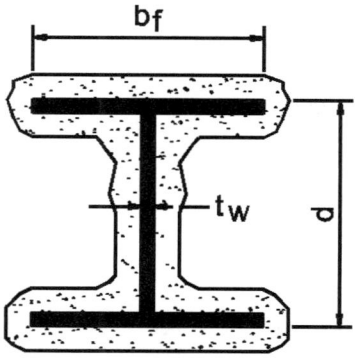

FIGURE 721.5.1(5)
WIDE FLANGE STRUCTURAL STEEL COLUMNS WITH SPRAYED FIRE-RESISTANT MATERIALS

C_1 and C_2 = Material-dependent constants.

W = Weight of structural steel columns (pounds per linear foot).

The *fire resistance* of structural steel columns protected with intumescent or mastic fire-resistant coatings shall be determined on the basis of fire-resistance tests in accordance with Section 703.2.

❖ This section sets forth procedures for determining the fire resistance of structural steel columns protected with spray-applied cementitious or mineral fiber fire protection materials. These procedures do not apply to intumescent or mastic fire-resistant coatings that must be tested in accordance with the requirements in Section 703.2 . These procedures are based upon an empirical equation, which includes two material-dependent constants. As a result, in order to apply this equation, the values of these two constants must be determined for specific fire protection materials. The purpose of this section is to provide guidance for the determination of these constants so that the resulting equation will be reasonably accurate, and yet slightly conservative, over the range of column shapes for the test data that is available.

Two different techniques are available for determining the two constants. The first requires a knowledge of thermal conductivity and specific heat of the fire protection material at elevated temperatures. Data of this nature is both difficult and expensive to obtain with any reasonable degree of accuracy. As a result, this technique will probably not be widely used, and accordingly, it will not be described in this section.

The second technique involves the use of Equation 7-13 as a means for interpolating between ASTM E 119 or UL 263 fire endurance test results on different structural steel columns. Since this technique will undoubtedly be the most widely used, it is described in detail in this section. It is, however, important to recognize that a wide variety of both large- and small-scale tests can be used to accurately determine the required constants. Inherent in Equation 7-13 is the fact that the ratio of fire endurance time to the thickness of fire protection material (R/h) varies linearly as a function of the increasing W/D ratio of the protected steel column. These concepts are graphically illustrated in the example that follows. It has been found that this assumption is reasonably accurate for lightweight [density less than 50 pcf (801 kg/m³)] spray-applied materials.

If ASTM E 119 or UL 263 fire endurance test results are available for a specific fire protection material on two different structural steel column shapes, the constants C_1 and C_2 can be determined directly. The resulting equation can then be used to determine the thickness of fire protection material required for any specified fire endurance rating when applied to structural steel columns with W/D ratios between the largest and smallest column for the actual test results that are available.

To determine the constants, at least four ASTM E 119 or UL 263 fire endurance tests or a combination of two ASTM E 119 or UL 263 fire endurance tests and six small-scale tests [3-foot-long (914 mm) specimens] are conducted. If the results of the small-scale tests are used, at least two of the test assemblies are necessary to duplicate the ASTM E 119 or UL 263 test assemblies for the purpose of establishing correlation. Regardless of the combination of small- and large-scale tests selected, at least two tests are conducted on the largest and two on the smallest columns to establish the limits of applicability to the resulting equation. The constants C_1 and C_2 are to be determined on the basis of the lowest ratios of fire endurance time to fire protection thickness (R/h) for these columns.

In addition, the test data need to be evaluated with respect to the assumption that the ratio of fire endurance to fire protection thickness (R/h) is reasonably constant for a given column shape [(W/D) ratio]. The tests conducted on columns of the same shape are designed so that the resulting fire endurance times are approximately $1^1/_2$ hours and $3^1/_2$ hours. In evaluating the R/h ratios resulting from tests on the same column shape, differences in the range of 10 percent are typical. Differences greater than 20 percent may, however, suggest that Equation 7-13 is not applicable to the specific fire protection material under consideration, and further examination of the test data is warranted.

EXAMPLE:

GIVEN: The fire endurance test data given in Figure 721.5.1.3.

FIND: The thickness of a spray-applied cementitious material required for a 3-hour rating on a W12 × 136.

SOLUTION:

1. Determine C_1 and C_2. From the end points of the graph:

   R/h = 75 for a W/D ratio of 0.6 and

   R/h = 200 for a W/D ratio of 2.5.

   From this, two equations with two unknowns are created:

   75 = C_1 (0.6) + C_2.

   200 = C_1 (2.5) + C_2.

   Solving these results in: C_1 = 65.79; C_2 = 35.53 and Equation 7-13, with "h" in the denominator, becomes:

   R/h = 65.79 (W/D) + 35.53.

2. From Table 721.5.1(1), a W12 × 136 has a W/D ratio of 1.82.

3. The required fire resistance is 180 minutes. Calculating for h yields:

   180 = [(65.79)(1.82) + 35.53]h

   h = 1.16 inches (29 mm).

Therefore, the required thickness of the spray-applied material is 1.16 inches (29 mm).

721.5.1.3.1 Material-dependent constants. The material-dependent constants, C_1 and C_2, shall be determined for specific fire-resistant materials on the basis of standard fire endurance tests in accordance with Section 703.2. Unless evidence is submitted to the *building official* substantiating a broader application, this expression shall be limited to determining the *fire resistance* of structural steel columns with weight-to-heated-perimeter ratios (*W/D*) between the largest and smallest columns for which standard fire-resistance test results are available.

721.5.1.3.2 Identification. Sprayed fire-resistant materials shall be identified by density and thickness required for a given *fire-resistance rating*.

721.5.1.4 Concrete-protected columns. The *fire resistance* of structural steel columns protected with concrete, as illustrated in Figure 721.5.1(6) (a) and (b), shall be permitted to be determined from the following expression:

$$R = R_o(1 + 0.03_m) \qquad \text{(Equation 7-14)}$$

where:

$$R_o = 10\,(W/D)^{0.7} + 17\,(h^{1.6}/k_c^{0.2}) \times [1 + 26\,\{H/p_c c_c h\,(L + h)\}^{0.8}]$$

As used in these expressions:

R = Fire endurance at equilibrium moisture conditions (minutes).

R_o = Fire endurance at zero moisture content (minutes).

m = Equilibrium moisture content of the concrete by volume (percent).

W = Average weight of the steel column (pounds per linear foot).

D = Heated perimeter of the steel column (inches).

h = Thickness of the concrete cover (inches).

k_c = Ambient temperature thermal conductivity of the concrete (Btu/hr ft °F).

H = Ambient temperature thermal capacity of the steel column = 0.11W (Btu/ ft °F).

p_c = Concrete density (pounds per cubic foot).

c_c = Ambient temperature specific heat of concrete (Btu/lb °F).

L = Interior dimension of one side of a square concrete box protection (inches).

❖ The values were determined by the procedures indicated in the report "Fire Endurance of Concrete-protected Columns" in the *ACI Journal*.

EXAMPLE:

GIVEN: A W8 × 28 steel column encased in lightweight concrete [density = 110 pcf (1762 kg/m³)] shown in Figure 721.5.1.4, with all reentrant spaces filled with concrete cover 1.25 inches (32 mm) and a moisture content of 5 percent. The web thickness is 0.285 inch.

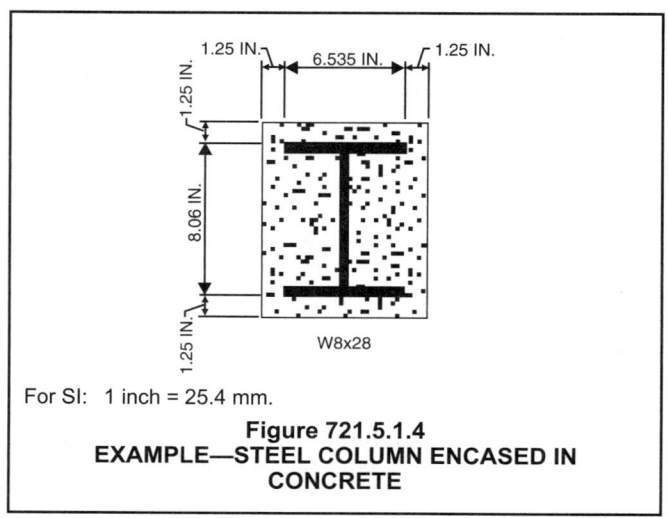

For SI: 1 inch = 25.4 mm.

Figure 721.5.1.4
EXAMPLE—STEEL COLUMN ENCASED IN CONCRETE

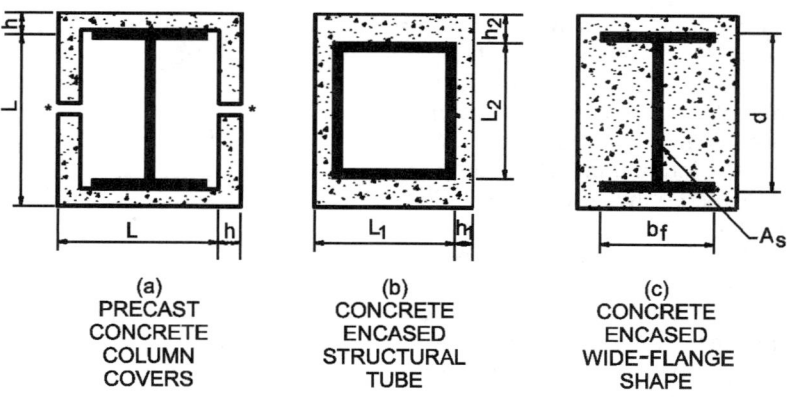

(a)
PRECAST
CONCRETE
COLUMN
COVERS

(b)
CONCRETE
ENCASED
STRUCTURAL
TUBE

(c)
CONCRETE
ENCASED
WIDE-FLANGE
SHAPE

FIGURE 721.5.1(6)
CONCRETE PROTECTED STRUCTURAL STEEL COLUMNS[a,b]

a. When the inside perimeter of the concrete protection is not square, *L* shall be taken as the average of L_1 and L_2. When the thickness of concrete cover is not constant, *h* shall be taken as the average of h_1 and h_2.

b. Joints shall be protected with a minimum 1 inch thickness of ceramic fiber blanket but in no case less than one-half the thickness of the column cover (see Section 720.2.1.3).

FIND: The fire-resistance rating.

SOLUTION:

D = 4(6.535) + 2(8.06) - 2(0.285) = 41.69 inches (1059 mm).

W/D = 28 lb/ft/41.69² = 0.67 lb/ft-in.

h = 1.25 inches (32 mm).

k_c = 0.35 Btu/hr · ft · °F.

c_c = 0.20 Btu/lb · °F.

r_c = 110 pcf (1762 kg/m³).

L = (6.535 + 8.06)/2 = 7.30 inches (185 mm).

A_s = 8.25 square inches (0.005 m²).

H = 11(28) + [(110)(0.20)/144][(6.535)(8.06) - 8.25] = 9.87

$$R_o = 10(0.67)^{0.7} + 17\left(\frac{(125)^{1.6}}{(0.35)^{0.2}}\right) \times$$

$$\left(1 + 26\left[\frac{9.87}{(110)(0.2)(125)(7.3 + 125)}\right]^{0.8}\right)$$

R_o = 99 minutes.

Therefore,

R = 99[1 + 0.33(5)] = 114 minutes.

For comparison purposes, the minimum cover requirement for a W8 × 28 steel column from Table 721.5.1(8) is 1 inch (25 mm) for a 1¹/₂-hour rating. The column does not quite meet the required fire rating for a 2-hour column since 1.5 inches (38 mm) is required; therefore, the fire resistance of the column is 1¹/₂ hours.

721.5.1.4.1 Reentrant space filled. For wide-flange steel columns completely encased in concrete with all reentrant spaces filled [Figure 721.5.1(6)(c)], the thermal capacity of the concrete within the reentrant spaces shall be permitted to be added to the thermal capacity of the steel column, as follows:

$$H = 0.11W + (p_c c_c/144)(b_f d - A_s) \qquad \textbf{(Equation 7-15)}$$

where:

b_f = Flange width of the steel column (inches).

d = Depth of the steel column (inches).

A_s = Cross-sectional area of the steel column (square inches).

721.5.1.4.2 Concrete properties unknown. If specific data on the properties of concrete are not available, the values given in Table 721.5.1(2) are permitted.

721.5.1.4.3 Minimum concrete cover. For structural steel column encased in concrete with all reentrant spaces filled, Figure 721.5.1(6)(c) and Tables 721.5.1(7) and 721.5.1(8) indicate the thickness of concrete cover required for various *fire-resistance ratings* for typical wide-flange sections. The thicknesses of concrete indicated in these tables also apply to structural steel columns larger than those listed.

721.5.1.4.4 Minimum precast concrete cover. For structural steel columns protected with precast concrete column covers as shown in Figure 721.5.1(6)(a), Tables 721.5.1(9) and 721.5.1(10) indicate the thickness of the column covers required for various *fire-resistance ratings* for typical wide-flange shapes. The thicknesses of concrete given in these tables also apply to structural steel columns larger than those listed.

721.5.1.4.5 Masonry protection. The *fire resistance* of structural steel columns protected with concrete masonry units or clay masonry units as illustrated in Figure 721.5.1(7), shall be permitted to be determined from the following expression:

$$R = 0.17 (W/D)^{0.7} + [0.285 (T_e^{1.6}/K^{0.2})]$$
$$[1.0 + 42.7 \{(A_s/d_m T_e)/(0.25p + T_e)\}^{0.8}]$$

$$\textbf{(Equation 7-16)}$$

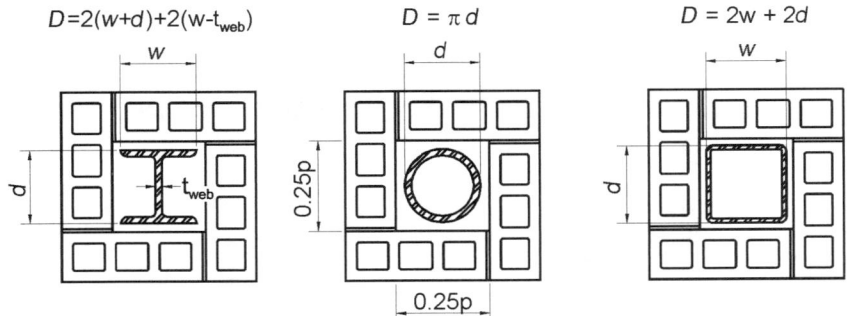

For SI: 1 inch = 25.4 mm.

FIGURE 721.5.1(7)
CONCRETE OR CLAY MASONRY PROTECTED STRUCTURAL STEEL COLUMNS

d = Depth of a wide flange column, outside diameter of pipe column, or out-side dimension of structural tubing column (inches).

t_{web} = Thickness of web of wide flange column (inches).

w = Width of flange of wide flange column (inches).

where:

R = *Fire-resistance rating* of column assembly (hours).

W = Average weight of steel column (pounds per foot).

D = Heated perimeter of steel column (inches) [see Figure 721.5.1(7)].

T_e = Equivalent thickness of concrete or clay masonry unit (inches) (see Table 721.3.2 Note a or Section 721.4.1).

K = Thermal conductivity of concrete or clay masonry unit (Btu/hr · ft · °F) [see Table 721.5.1(3)].

A_s = Cross-sectional area of steel column (square inches).

d_m = Density of the concrete or clay masonry unit (pounds per cubic foot).

p = Inner perimeter of concrete or clay masonry protection (inches) [see Figure 721.5.1(7)].

721.5.1.4.6 Equivalent concrete masonry thickness. For structural steel columns protected with concrete masonry, Table 721.5.1(5) gives the equivalent thickness of concrete masonry required for various *fire-resistance ratings* for typical column shapes. For structural steel columns protected with clay masonry, Table 721.5.1(6) gives the equivalent thickness of concrete masonry required for various *fire-resistance ratings* for typical column shapes.

721.5.2 Structural steel beams and girders. The *fire-resistance ratings* of steel beams and girders shall be based upon the size of the element and the type of protection provided in accordance with this section.

721.5.2.1 Determination of *fire resistance*. These procedures establish a basis for determining resistance of structural steel beams and girders which differ in size from that specified in *approved* fire-resistance-rated assemblies as a function of the thickness of fire-resistant material and the weight (W) and heated perimeter (D) of the beam or girder. As used in these sections, W is the average weight of a structural steel member in pounds per linear foot (plf). The heated perimeter, D, is the inside perimeter of the fire-resistant material in inches as illustrated in Figure 721.5.2.

721.5.2.1.1 Weight-to-heated perimeter. The weight-to-heated-perimeter ratios (W/D), for both contour and box fire-resistant protection profiles, for the wide flange shapes most often used as beams or girders are given in Table 721.5.1(4). For different shapes, the weight-to-heated-perimeter ratios (W/D) shall be determined in accordance with the definitions given in this section.

721.5.2.1.2 Beam and girder substitutions. Except as provided for in Section 721.5.2.2, structural steel beams in *approved* fire-resistance-rated assemblies shall be considered the minimum permissible size. Other beam or girder shapes shall be permitted to be substituted provided that the weight-to-heated-perimeter ratio (W/D) of the substitute beam is equal to or greater than that of the beam specified in the *approved* assembly.

❖ This section defines a general rule for the substitution of different steel beam and girder shapes in all fire-resistant assemblies. In the past, the substitution of larger beams has been permitted based upon the thickness of web and flange elements. Extensive research at Underwriters Laboratories has proven that the heat transfer to a protected steel beam or girder is a direct function of the W/D ratio. As a result, beam substitutions should be based upon W/D ratios, as defined in this section. The significance of the thickness of web and flange elements is inherently included in the determination of W/D ratios. Figure 721.5.2 illustrates the procedure for determining heated perimeters (D), and Table 721.5.1(4) provides W/D ratios for the most commonly used wide flange beam and girder shapes.

721.5.2.2 Sprayed fire-resistant materials. The provisions in this section apply to structural steel beams and girders protected with sprayed fire-resistant materials. Larger or smaller beam and girder shapes shall be permitted to be substituted for beams specified in *approved* unrestrained or restrained fire-resistance-rated assemblies, provided that the thickness of

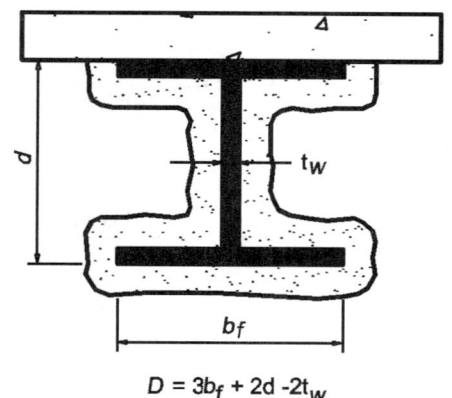

$D = 3b_f + 2d - 2t_w$

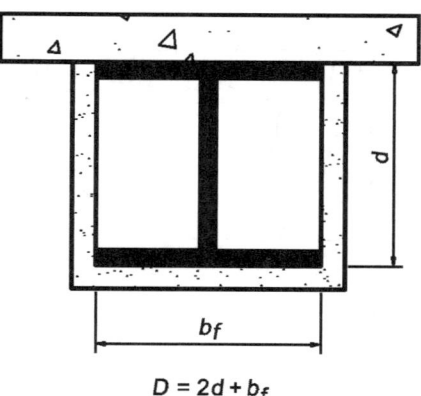

$D = 2d + b_f$

FIGURE 721.5.2
DETERMINATION OF THE HEATED PERIMETER OF STRUCTURAL STEEL BEAMS AND GIRDERS

the fire-resistant material is adjusted in accordance with the following expression:

$$h_2 = h_1 \left[(W_1 / D_1) + 0.60 \right] / \left[(W_2 / D_2) + 0.60 \right]$$

(Equation 7-17)

where:

h = Thickness of sprayed fire-resistant material in inches.

W = Weight of the structural steel beam or girder in pounds per linear foot.

D = Heated perimeter of the structural steel beam in inches.

Subscript 1 refers to the beam and fire-resistant material thickness in the *approved* assembly.

Subscript 2 refers to the substitute beam or girder and the required thickness of fire-resistant material.

The *fire resistance* of structural steel beams and girders protected with intumescent or mastic fire-resistant coatings shall be determined on the basis of fire-resistance tests in accordance with Section 703.2.

❖ This section defines an equation for adjusting the thickness of spray-applied cementitious and mineral fiber materials as a function of *W/D* ratios. This equation was developed by Underwriters Laboratories, and appropriate limitations have been included. The minimum *W/D* ratio of 0.37 in Section 721.5.2.2.1 will prevent the use of this equation for determining the fire resistance of very small shapes that have not been tested. The $^3/_8$-inch (9.5 mm) minimum thickness of protection is a practical limit based upon the most commonly used spray-applied fire protection materials.

EXAMPLE:

GIVEN: Determine the thickness of spray-applied fire protection required to provide a 2-hour fire-resistance rating for a W12 × 16 beam to be substituted for a W8 × 15 beam requiring 1.44 inches (37 mm) of protection for the same rating.

SOLUTION:

From Table 721.5.1(4):

W_1/D_1 = 0.54 for W8 × 15

W_2/D_2 = 0.45 for W12 × 16

h_1 = 1.44 inches.

h_2 = $\left[\dfrac{0.54 + 0.60}{0.45 + 0.60} \right] \times 1.44$

= 1.56 inches (39 mm).

It is noted in this section that intumescent or mastic fire-resistant coatings are required to be tested in accordance with the requirements of Section 703.2.

721.5.2.2.1 Minimum thickness. The use of Equation 7-17 is subject to the following conditions:

1. The weight-to-heated-perimeter ratio for the substitute beam or girder (W_2/D_2) shall not be less than 0.37.

2. The thickness of fire protection materials calculated for the substitute beam or girder (T_1) shall not be less than $^3/_8$ inch (9.5 mm).

3. The unrestrained or restrained beam rating shall not be less than 1 hour.

4. When used to adjust the material thickness for a restrained beam, the use of this procedure is limited to steel sections classified as compact in accordance with the AISC *Specification for Structural Steel Buildings*, (AISC 360-05).

❖ See the commentary to Section 721.5.2.2.

721.5.2.3 Structural steel trusses. The *fire resistance* of structural steel trusses protected with fire-resistant materials sprayed to each of the individual truss elements shall be permitted to be determined in accordance with this section. The thickness of the fire-resistant material shall be determined in accordance with Section 721.5.1.3. The weight-to-heated-perimeter ratio (*W/D*) of truss elements that can be simultaneously exposed to fire on all sides shall be determined on the same basis as columns, as specified in Section 721.5.1.1. The weight-to-heated-perimeter ratio (*W/D*) of truss elements that directly support floor or roof assembly shall be determined on the same basis as beams and girders, as specified in Section 721.5.2.1.

The *fire resistance* of structural steel trusses protected with intumescent or mastic fire-resistant coatings shall be determined on the basis of fire-resistance tests in accordance with Section 703.2.

❖ This section describes the application of spray-applied fire protection to structural steel trusses when each truss element is individually protected with spray-applied materials. The thickness of protection is determined using the column equation specified in Section 721.5.1.3. For trusses, the column equation is more technically correct than the beam equation, since it requires greater thicknesses of protection than the beam equation. Most truss elements can be exposed to fire on all four sides simultaneously. As a result, the heated perimeter of truss elements should be determined in the same manner as for columns. An exception is, however, included for top chord elements that directly support floor or roof construction. The heated perimeter of such elements should be determined in the same manner as for beams and girders.

This section notes that intumescent and mastic fire-resistant materials must be tested in accordance with the requirements in Section 703.2.

721.6 Wood assemblies. The provisions of this section contain procedures by which the *fire-resistance ratings* of wood assemblies are established by calculations.

❖ The information contained in this section is based on Technical Paper No. 222 of the Division of Building Research Council of Canada titled "Fire Endurance of Light-framed and Miscellaneous Assemblies" and NBS Report BMS 92. The fire-resistance rating is equal to the sum of the time assigned to the membranes [see Table 721.6.2(1)], the time assigned to the framing members [see Table 721.6.2(2)] and the time assigned for additional contribution by other protective measures, such as insulation [see Table 721.6.2(5)]. The membrane on the unexposed side is not included in the calculations. It is assumed that once the structural members fail, the entire assembly fails. The time assigned to the individual membranes is individual contributions to the overall fire-resistance rating of the complete assembly. The assigned time is not to be confused with the fire-resistance rating of the membranes. Fire-resistance rating takes into account the rise in temperature on the unexposed side of the membrane. Times that have been assigned to membranes on the fire-exposed side of the assembly are based on their ability to remain in place during fire tests.

The fire-resistance rating of wood-frame assemblies is equal to the sum of the time assigned to the various components (membranes) on the fire-exposed side and the structural members. Interior walls and partitions, exterior walls within 5 feet (1524 mm) of a lot line or assumed lot line and some shaft walls must be designed for fire resistance by assuming that either side of the wall is exposed to fire.

Mineral fiber insulation provides additional protection to wood studs by shielding them from exposure to the furnace, thus delaying the time of collapse. The use of reinforcement in the membrane exposed to fire also adds to the fire resistance by extending the time to failure. Special care must be taken to ensure that all insulation materials used in conjunction with this method satisfy the weight criteria of Table 721.6.2(5).

The following are examples of how to use Section 721.6:

EXAMPLE 1:

GIVEN: The exterior wall assembly shown in Figure 721.6 having a layer of $^{15}/_{32}$-inch (12 mm) wood structural panel bonded with exterior glue covered with a layer of $^{5}/_{8}$-inch (15.8 mm) gypsum wallboard and attached to studs spaced at 16 inches (406 mm) on center (o.c.).

FIND: Does the wall assembly qualify as a 1-hour fire-resistant wall assembly?

SOLUTION: From Tables 721.6.2(1) and 721.6.2(2):

Wood structural panel	= 10 minutes.
Gypsum wallboard	= 30 minutes.
Wood studs	= 20 minutes.
Total	= 60 minutes.

Therefore, the wall does qualify as a 1-hour wall. If the wall is an interior wall, both sides would be required to be fire protected with at least 40 minutes of membrane coverings (60 minutes - 20 minutes for wood frame).

It should be noted that Section 721.6.2.3 requires the exterior side to be protected in accordance with Table 721.6.2(3) or any membrane that is assigned a time of at least 15 minutes as listed in Table 721.6.2(1) and as illustrated in Figure 721.6.

It should also be noted that if the wall cavities between the studs are filled with mineral fiber batts weighing no less than that specified in Table 721.6.2(5), the $^{15}/_{32}$-inch (12 mm) plywood membrane layer could be eliminated because the insulation adds 15 minutes of fire resistance, as indicated in Section 721.6.2.5 and Table 721.6.2(5). Thus, adding the contribution times for the $^{5}/_{8}$-inch (15.9 mm) gypsum board, the wood framing and the insulation (30 minutes + 20 minutes + 15 minutes), the resultant rating for the wall would be 65 minutes and meet a 1-hour fire-resistance rating.

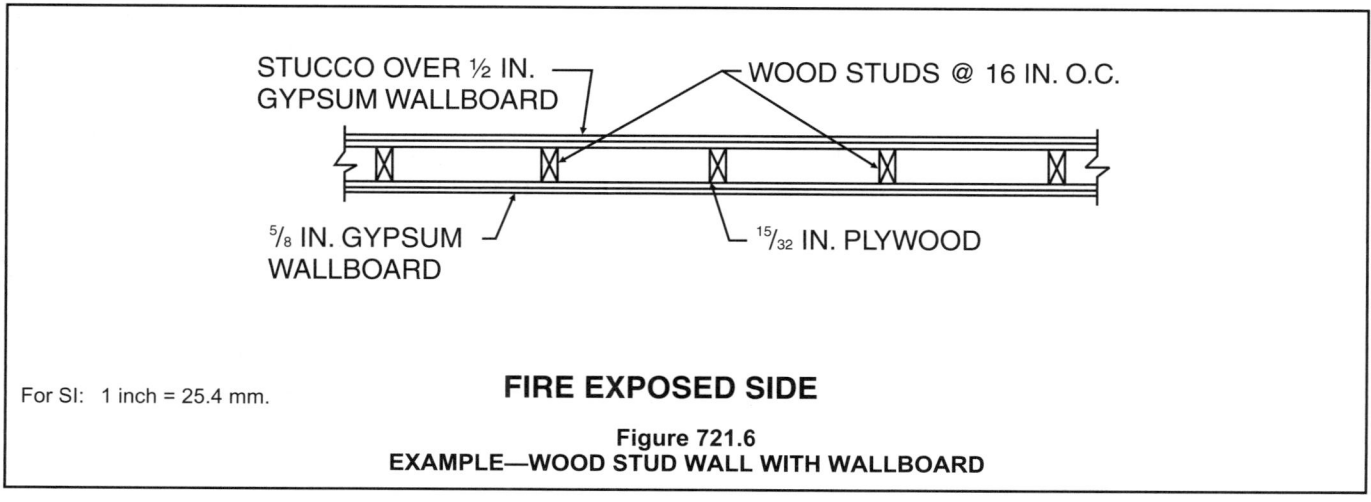

STUCCO OVER ½ IN. GYPSUM WALLBOARD

WOOD STUDS @ 16 IN. O.C.

$^{5}/_{8}$ IN. GYPSUM WALLBOARD

$^{15}/_{32}$ IN. PLYWOOD

For SI: 1 inch = 25.4 mm.

FIRE EXPOSED SIDE

Figure 721.6
EXAMPLE—WOOD STUD WALL WITH WALLBOARD

EXAMPLE 2:

GIVEN: A floor/ceiling assembly using wood joists spaced at 16 inches (406 mm) o.c., protected on the bottom side (ceiling side) with two layers of $^1/_2$-inch (12.7 mm) Type X gypsum wall board and protected on the upper side (floor side) with a $^{15}/_{32}$-inch (12 mm) plywood subfloor, a $^3/_8$-inch (15.9 mm) panel-type underlayment and carpet.

FIND: Does the floor/ceiling assembly meet the requirements of a 1-hour fire-resistance-rated assembly?

SOLUTION: Referring to Sections 721.6.2.1 and 721.6.2.4, Table 721.6.2(1) indicates that the time contribution for each layer of $^1/_2$-inch (12.7 mm) Type X gypsum wallboard is 25 minutes. The time of contribution for wood joists 16 inches (406 mm) o.c. is listed in Table 721.6.2(2) as 10 minutes. By adding the two layers of gypsum board (2 × 25 minutes) to the wood frame (10 minutes), a fire-resistance rating of 60 minutes, or 1 hour, can be obtained. It should be noted that Section 721.6.2.4 requires the upper membrane to be specified as in Table 721.6.2(4) or any membrane that has a time of contribution of at least 15 minutes as listed in Table 721.6.2(1).

If the above example had been a roof/ceiling assembly, the upper membrane would have been treated the same. If the proposed assembly is a ceiling with an attic above, Section 721.6.2.4 notes the exception to Section 711.3.3, which allows the elimination of the upper membrane.

The fastening requirements for assemblies developed by Section 721.6 should be in accordance with Chapter 23 as stated in Section 721.6.2.6.

TABLE 721.6.2(1)
TIME ASSIGNED TO WALLBOARD MEMBRANES[a, b, c, d]

DESCRIPTION OF FINISH	TIME[e] (minutes)
$^3/_8$-inch wood structural panel bonded with exterior glue	5
$^{15}/_{32}$-inch wood structural panel bonded with exterior glue	10
$^{19}/_{32}$-inch wood structural panel bonded with exterior glue	15
$^3/_8$-inch gypsum wallboard	10
$^1/_2$-inch gypsum wallboard	15
$^5/_8$-inch gypsum wallboard	30
$^1/_2$-inch Type X gypsum wallboard	25
$^5/_8$-inch Type X gypsum wallboard	40
Double $^3/_8$-inch gypsum wallboard	25
$^1/_2$-inch + $^3/_8$-inch gypsum wallboard	35
Double $^1/_2$-inch gypsum wallboard	40

For SI: 1 inch = 25.4 mm.

a. These values apply only when membranes are installed on framing members which are spaced 16 inches o.c.

b. Gypsum wallboard installed over framing or furring shall be installed so that all edges are supported, except $^5/_8$-inch Type X gypsum wallboard shall be permitted to be installed horizontally with the horizontal joints staggered 24 inches each side and unsupported but finished.

c. On wood frame floor/ceiling or roof/ceiling assemblies, gypsum board shall be installed with the long dimension perpendicular to framing members and shall have all joints finished.

d. The membrane on the unexposed side shall not be included in determining the fire resistance of the assembly. When dissimilar membranes are used on a wall assembly, the calculation shall be made from the least fire-resistant (weaker) side.

e. The time assigned is not a finished rating.

TABLE 721.6.2(2)
TIME ASSIGNED FOR CONTRIBUTION OF WOOD FRAME[a, b, c]

DESCRIPTION	TIME ASSIGNED TO FRAME (minutes)
Wood studs 16 inches o.c.	20
Wood floor and roof joists 16 inches o.c.	10

For SI: 1 inch = 25.4 mm.

a. This table does not apply to studs or joists spaced more than 16 inches o.c.

b. All studs shall be nominal 2 × 4 and all joists shall have a nominal thickness of at least 2 inches.

c. Allowable spans for joists shall be determined in accordance with Sections 2308.8, 2308.10.2 and 2308.10.3.

721.6.1 General. This section contains procedures for calculating the fire-resistance ratings of walls, floor/ceiling and roof/ceiling assemblies based in part on the standard method of testing referenced in Section 703.2.

721.6.1.1 Maximum fire-resistance rating. Fire resistance ratings calculated for assemblies using the methods in Section 721.6 shall be limited to a maximum of 1 hour.

721.6.1.2 Dissimilar membranes. Where dissimilar membranes are used on a wall assembly, the calculation shall be made from the least fire-resistant (weaker) side.

❖ All wood assemblies by which the fire-resistance rating is calculated must have a membrane on each side; however, the calculation shall be made from the least fire-resistant side.

721.6.2 Walls, floors and roofs. These procedures apply to both load-bearing and nonload-bearing assemblies.

721.6.2.1 Fire-resistance rating of wood frame assemblies. The fire-resistance rating of a wood frame assembly is equal to the sum of the time assigned to the membrane on the fire-exposed side, the time assigned to the framing members and the time assigned for additional contribution by other protective measures such as insulation. The membrane on the unexposed side shall not be included in determining the *fire resistance* of the assembly.

❖ See the commentary to Section 721.6.

721.6.2.2 Time assigned to membranes. Table 721.6.2(1) indicates the time assigned to membranes on the fire-exposed side.

721.6.2.3 Exterior walls. For an *exterior wall* with a *fire separation distance* greater than 5 feet (1524 mm), the wall is assigned a rating dependent on the interior membrane and the framing as described in Tables 721.6.2(1) and 721.6.2(2). The membrane on the outside of the nonfire-exposed side of *exterior walls* with a *fire separation distance* greater than 5 feet (1524 mm) may consist of sheathing, sheathing paper and siding as described in Table 721.6.2(3).

721.6.2.4 Floors and roofs. In the case of a floor or roof, the standard test provides only for testing for fire exposure from below. Except as noted in Section 703.3, Item 5, floor or roof assemblies of wood framing shall have an upper membrane consisting of a subfloor and finished floor conforming to Table 721.6.2(4) or any other membrane that has a contribution to *fire resistance* of at least 15 minutes in Table 721.6.2(1).

721.6.2.5 Additional protection. Table 721.6.2(5) indicates the time increments to be added to the *fire resistance* where glass fiber, rockwool, slag mineral wool or cellulose insulation is incorporated in the assembly.

721.6.2.6 Fastening. Fastening of wood frame assemblies and the fastening of membranes to the wood framing members shall be done in accordance with Chapter 23.

721.6.3 Design of fire-resistant exposed wood members. The *fire-resistance rating*, in minutes, of timber beams and columns with a minimum nominal dimension of 6 inches (152 mm) is equal to:

Beams: $2.54Zb\ [4 - 2(b/d)]$ for beams which may be exposed to fire on four sides.

(Equation 7-18)

$2.54Zb\ [4 - (b/d)]$ for beams which may be exposed to fire on three sides.

(Equation 7-19)

Columns: $2.54Zd\ [3 - (d/b)]$ for columns which may be exposed to fire on four sides

(Equation 7-20)

TABLE 721.6.2(3)
MEMBRANE[a] ON EXTERIOR FACE OF WOOD STUD WALLS

SHEATHING	PAPER	EXTERIOR FINISH
$5/8$-inch T & G lumber		Lumber siding
$5/16$-inch exterior glue wood structural panel		Wood shingles and shakes
$1/2$-inch gypsum wallboard	Sheathing paper	$1/4$-inch wood structural panels—exterior type
$5/8$-inch gypsum wallboard		$1/4$-inch hardboard
$1/2$-inch fiberboard		Metal siding
		Stucco on metal lath
		Masonry veneer
		Vinyl siding
None	—	$3/8$-inch exterior-grade wood structural panels

For SI: 1 pound/cubic foot = 16.0185 kg/m^2.
a. Any combination of sheathing, paper and exterior finish is permitted.

TABLE 721.6.2(4)
FLOORING OR ROOFING OVER WOOD FRAMING[a]

ASSEMBLY	STRUCTURAL MEMBERS	SUBFLOOR OR ROOF DECK	FINISHED FLOORING OR ROOFING
Floor	Wood	$15/32$-inch wood structural panels or $11/16$ inch T & G softwood	Hardwood or softwood flooring on building paper resilient flooring, parquet floor felted-synthetic fiber floor coverings, carpeting, or ceramic tile on $3/8$-inch-thick panel-type underlay
			Ceramic tile on $1 1/4$-inch mortar bed
Roof	Wood	$15/32$-inch wood structural panels or $11/16$ inch T & G softwood	Finished roofing material with or without insulation

For SI: 1 inch = 25.4 mm.
a. This table applies only to wood joist construction. It is not applicable to wood truss construction.

TABLE 721.6.2(5)
TIME ASSIGNED FOR ADDITIONAL PROTECTION

DESCRIPTION OF ADDITIONAL PROTECTION	FIRE RESISTANCE (minutes)
Add to the fire-resistance rating of wood stud walls if the spaces between the studs are completely filled with glass fiber mineral wool batts weighing not less than 2 pounds per cubic foot (0.6 pound per square foot of wall surface) or rockwool or slag material wool batts weighing not less than 3.3 pounds per cubic foot (1 pound per square foot of wall surface), or cellulose insulation having a nominal density not less than 2.6 pounds per cubic foot.	15

For SI: 1 pound/cubic foot = 16.0185 kg/m^3.

2.54Zd [3 -($d/2b$)] for columns which may be exposed to fire on three sides.

(Equation 7-21)

where:

b = The breadth (width) of a beam or larger side of a column before exposure to fire (inches).

d = The depth of a beam or smaller side of a column before exposure to fire (inches).

Z = Load factor, based on Figure 721.6.3(1).

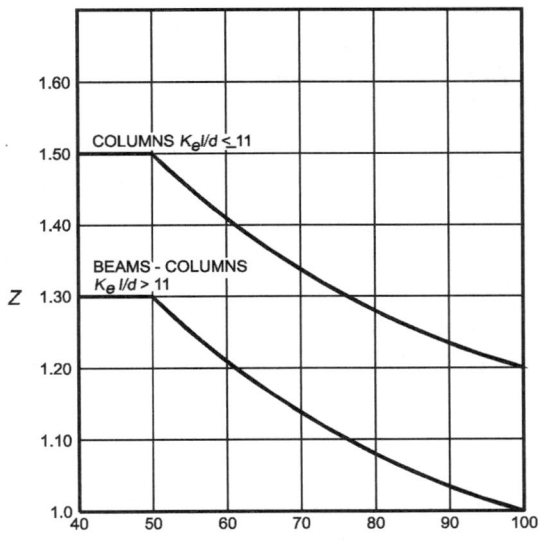

FIGURE 721.6.3(1)
LOAD FIGURE

K_e = The effective length factor as noted in Figure 721.6.3(2).
l = The unsupported length of columns (inches).

721.6.3.1 Equation 7-21. Equation 7-21 applies only where the unexposed face represents the smaller side of the column. If a column is recessed into a wall, its full dimension shall be used for the purpose of these calculations.

721.6.3.2 Allowable loads. Allowable loads on beams and columns are determined using design values given in AF&PA NDS.

721.6.3.3 Fastener protection. Where minimum 1-hour *fire resistance* is required, connectors and fasteners shall be protected from fire exposure by $1^1/_2$ inches (38 mm) of wood, or other *approved* covering or coating for a 1-hour rating. Typical details for commonly used fasteners and connectors are shown in AITC Technical Note 7.

721.6.3.4 Minimum size. Wood members are limited to dimensions of 6 inches (152 mm) nominal or greater. Glued-laminated timber beams utilize standard laminating combinations except that a core lamination is removed. The tension zone is moved inward and the equivalent of an extra nominal 2-inch-thick (51 mm) outer tension lamination is added.

Bibliography

The following resource materials are referenced in this chapter or are relevant to the subject matter addressed in this chapter.

Abrams and Gustaferro. "Fire Endurance of Two-course Floors and Roofs." *Journal of the American Concrete Institute.* American Concrete Institute, February 1969.

ACI 216.1-07/TMS 0216.1-07, *Code Requirements for Determining Fire Resistance of Concrete and Masonry Construction Assemblies.* Farmington Hills, MI: American Concrete Institute, 2007 .

ACI 530/ASCE 5/TMS 402-02, *Building Code Requirements for Masonry Structures.* Detroit, MI: American

BUCKLING MODES						
THEORETICAL K_e VALUE	0.5	0.7	1.0	1.0	2.0	2.0
RECOMMENDED DESIGN K_e WHEN IDEAL CONDITIONS APPROXIMATED	0.65	0.80	1.2	1.0	2.10	2.4
END CONDITION CODE	ROTATION FIXED, TRANSLATION FIXED					
	ROTATION FREE, TRANSLATION FIXED					
	ROTATION FIXED, TRANSLATION FREE					
	ROTATION FREE, TRANSLATION FREE					

FIGURE 721.6.3(2)
EFFECTIVE LENGTH FACTORS

Concrete Institute; New York: American Society of Civil Engineers; Boulder, CO: The Masonry Council, 2002.

ACI 530.1/ASCE 6/TMS 602-02, *Specifications for Masonry Structures*. Detroit, MI: American Concrete Institute; New York: American Society of Civil Engineers; Boulder, CO: The Masonry Council, 2002.

Allen L.W., M. Galbreath and W.W. Stanzak. *Fire Endurance Tests on Unit Masonry Walls with Gypsum Wallboard* (NRCC 13901). Division of Building Research, National Research Council of Canada.

Analytical Methods of Determining Fire Endurance of Concrete and Masonry Members—Model Code Approved Procedures (SR267.01B). Skokie, IL: Concrete and Masonry Industry Fire Safety Committee, 1987.

ASCE 29-05, *Standard Calculation Methods for Structural Fire Protection*, Reston, VA: American Society of Civil Engineers Structural Engineering Institute, 2005.

ASME A17.1/CSA B44-2007, *Safety Code for Elevators and Escalators,* New York, NY: American Society of Mechanical Engineers, 2007.

ASTM C 140-01, *Standard Test Method for Sampling and Testing Concrete Masonry Units*. West Conshohocken, PA: ASTM International, 2001.

ASTM E 84-04, *Test Method for Surface Burning Characteristics of Building Materials*. West Conshohocken, PA: ASTM International, 2004.

ASTM E 970-00, *Test Method for Critical Radiant Flux of Exposed Attic Floor Insulation Using a Radiant Heat Energy Source*. West Conshohocken, PA: ASTM International, 2000.

ASTM E 119-00, *Test Methods for Fire Tests of Building Construction and Materials*. West Conshohocken, PA: ASTM International, 2000.

ASTM E 136-99, *Test Method for Behavior of Materials in a Vertical Tube Furnace at 750°C*. West Conshohocken, PA: ASTM International, 1999.

ASTM E 152-81a, *Methods of Fire Tests of Door Assemblies*. West Conshohocken, PA: ASTM International, 1981.

ASTM E 163-84, *Methods of Fire Tests of Window Assemblies*. West Conshohocken, PA: ASTM International, 1984.

ASTM E 736-00 (2006), *Test Method for Cohesion/Adhesions of Sprayed Fire-resistive Materials Applied to Structural Members*. West Conshohocken, PA: ASTM International, 2006.

ASTM E 814-02, *Test Method for Fire Tests of Through-penetration Firestops*. West Conshohocken, PA: ASTM International, 2002.

ASTM E 1966-01, *Test Method for Fire-resistant Joint Systems*. West Conshohocken, PA: ASTM International, 2001.

ASTM E 2174, *Standard Practice for On-site Inspection of Installed Firestops*. West Conshohocken, PA: ASTM International, 2009.

ASTM E 2307-04e0, *Standard Test Method for Determining Fire Resistance of Perimeter Fire Barrier Systems Using Intermediate-scale, Multistory Test Apparatus*. West Conshohocken, PA: ASTM International, 2004.

ASTM E 2393, *Standard Practice for On-site Inspection of Installed Fire-resistive Joint Systems and Perimeter Fire Barriers*. West Conshohocken, PA: ASTM International, 2004. 2009

BMS 92, *Fire-resistance Classifications of Construction Building*. National Bureau of Standards, 1942.

Boring, D.F., J.C. Spence and W.G. Wells. *Fire Protection Through Modern Building Codes*. Washington, DC: American Iron and Steel Institute, 1981.

Building Materials Directory. Northbrook, IL: Underwriters Laboratories Inc., 1996.

Christian, W.J. and J.E. Waterman. "Flame Spread in Corridors: Effects of Location and Area of Wall Finish." *Fire Journal*, July 1971.

CPSC 16 CFR, Part 1201-77, *Safety Standard for Architectural Glazing Material*, Washington, DC: Consumer Product Safety Commission, 1977.

CPSC 16 CFR, Part 1209-98, *Interim Safety Standard for Cellulose Insulation*. Washington, DC: Consumer Product Safety Commission, 1998.

CPSC 16 CFR, Part 1404-98, *Cellulose Insulation*. Washington, DC: Consumer Product Safety Commission, 1998.

Designing Fire Protection for Steel Beams. Washington, DC: American Iron and Steel Institute, 1984.

Designing Fire Protection for Steel Columns. Washington, DC: American Iron and Steel Institute, 1980.

Designing Fire Protection for Steel Trusses. Washington, DC: American Iron and Steel Institute, 1981.

"Fire Endurance of Concrete Slabs as Influenced by Thickness, Aggregate Type and Moisture." *PCA Bulletin 223*. Skokie, IL: Portland Cement Association.

Fire-resistance Classifications of Building Construction. Gaithersburg, MD: National Bureau of Standards, 1942.

Fire-resistance Directory. Northbrook, IL: Underwriters Laboratories Inc., 1996.

Fire Tests of Building Columns. Johnston, RI: Associated Factory Mutual Insurance Companies, 1921.

"Fire Tests of Joints Between Precast Concrete Wall Units: Effect of Various Joint Treatments." *PCI Journal*. September–October, 1975.

Fisher, F., B. MacCracken and R.B. Williamson. "Room Fire Tests of Textile Wallcoverings." *Service to Industry Report No. 85-4*. University of California Fire Research Laboratory, 1989.

GA 600-03, *Fire-resistance Design Manual*. Washington, DC: Gypsum Association, 2003.

Galbreath, Murdock. "Fire Resistance of Light-framed and Miscellaneous Assemblies." Technical Paper No. 222 of the Division of Building Research, National Research Council of Canada.

Harmanthy, T.Z. and T.T. Lie. "Fire Endurance of Concrete-protected Columns." *ACI Journal*, Proceedings, Vol. 71, No. 2, 1974.

Hull and Inburg. "Fire Resistance of Concrete Columns." Technological Papers of the Bureau of Standards, No. 271, February 24, 1925.

ICC 2001 *Guidelines for Determining Fire Resistance of Building Elements*, Falls Church, VA: International Code Council, 2001.

IECC-09, *International Energy Conservation Code*, Washington, DC: International Code Council, 2009.

IMC-09, *International Mechanical Code*. Washington, DC: International Code Council, 2009.

Ingberg, S.H. "Tests of Severity of Building Fires." *NFPA Quarterly*, Vol. 22, No. 1, pp. 43-61. 2006.

"Investigations on Building Fires, Part VI: Fire Resistance of Reinforced Concrete Columns." National Building Studies Research Paper No. 18. London, England: Her Majesty's Stationary Office, 1953.

IPC-09, *International Plumbing Code*. Washington, DC: International Code Council, 2009.

Menzel, Carl A. *Tests of Fire Resistance and Strength of Walls of Concrete Masonry Units*. Skokie, IL: Portland Cement Association, 1934.

National Building Code. New York: National Board of Fire Underwriters, 1995.

NCMA TEK Bulletin 5-8B, *Details for Concrete Masonry Fire Walls*. Herndon, VA: National Concrete Masonry Association.

NCMA TEK 6A-91, *Fire-resistance Rating of Concrete Masonry Assembly*. Herndon, VA: National Concrete Masonry Association, 1991.

NCMA TEK 35D-91, *Fire Safety with Concrete Masonry*. Herndon, VA: National Concrete Masonry Association, 1991.

NFPA 13-02, *Installation of Sprinkler Systems*. Quincy, MA: National Fire Protection Association, 2002.

NFPA 13D-02, *Installation of Sprinkler Systems in One- and Two-family Dwellings and Manufactured Homes*. Quincy, MA: National Fire Protection Association, 2002.

NFPA 13R-02, *Installation of Sprinkler Systems in Residential Occupancies Up to Four Stories in Height*. Quincy, MA: National Fire Protection Association, 2002.

NFPA 72-07, *National Fire Alarm Code*. Quincy, MA: National Fire Protection Association, 2007.

NFPA 72-02, *National Fire Alarm Code*. Quincy, MA: National Fire Protection Association, 2002.

NFPA 80-99, *Standard Fire Doors and Windows*. Quincy, MA: National Fire Protection Association, 1999.

NFPA 105-03, *Standard for the Installation of Smoke Door Assemblies*. Quincy, MA: National Fire Protection Association, 2003.

NFPA 252-03, *Methods of Fire Tests of Door Assemblies*. Quincy, MA: National Fire Protection Association, 2003.

NFPA 257-00, *Methods of Fire Tests of Window Assemblies*. Quincy, MA: National Fire Protection Association, 2000.

NFPA 288-07, *Standard Method of Fire Tests of Floor Fire Door Assemblies Installed Horizontally in Fire-resistance-rated Floor Systems*. Quincy, MA: National Fire Protection Association, 2007.

NRCC 17724, *Supplement to the National Building Code of Canada 1980*. Associate Committee on the National Building Code. National Research Council of Canada, 1980.

PCA Publication T-140, *Fire Resistance of Referenced Concrete Floors*. Skokie, IL: Portland Cement Association.

PCI, "Fire Tests of Joints Between Precast Concrete Wall Panels: Effect of Various Joint Treatments." *PCI Journal*, Gusta-ferro, AH and Abrams, M.S. V20, n5, pages 44-64. September–October 1975.

PCI MNL 124-89, *Design for Fire Resistance of Precast Prestressed Concrete*. Chicago: Precast/Prestressed Concrete Institute, 1989.

Reinforced Concrete Fire Resistance. Schaumburg, IL: Concrete Reinforcing Steel Institute, 1980.

SFPE Handbook of Fire Protection Engineering, 2nd. Quincy, MA: National Fire Protection Association, 1995.

The Moroney Report, Vol. 5, Issue 2, Second Quarter 1993. Bedford Park, IL: John J. Moroney & Company, 1993.

UL 9-2000, *Fire Tests of Window Assemblies—with Revisions through April 2005*. Northbrook, IL: Underwriters Laboratories Inc., 2000.

UL 10A-98, *Standard for Safety Tin-clad Fire Doors—with Revisions through July 1998*. Northbrook, IL: Underwriters Laboratories Inc., 1998.

UL 10B-97, *Fire Tests of Door Assemblies—with Revisions through October 2001*. Northbrook, IL: Underwriters Laboratories Inc., 1997.

UL 14B-98, *Sliding Hardware for Standard, Horizontally Mounted Tin-clad Fire Doors—with Revisions through October 1996*. Northbrook, IL: Underwriters Laboratories Inc., 1998.

UL 10C-98, *Positive Pressure Fire Tests of Door Assemblies—with Revisions through November 2001*. Northbrook, IL: Underwriters Laboratories Inc., 2001.

UL 14C-99, *Swinging Hardware for Standard Tin-clad Fire Doors Mounted Singly and in Pairs—with Revisions through October 1996*. Northbrook, IL: Underwriters Laboratories Inc., 1999.

UL 103-01, *Chimneys, Factory-built, Residential-type and Building Heating Appliance—with Revisions through February 1996*. Northbrook, IL: Underwriters Laboratories Inc., 2001.

UL 127-96, *Factory-built Fireplaces*. Northbrook, IL: Underwriters Laboratories Inc., 1996.

UL 263-03, *Standard for Fire Test of Building Construction and Materials*. Northbrook, IL: Underwriters Laboratories Inc., 2003.

UL 555C-96, *Ceiling Dampers*. Northbrook, IL: Underwriters Laboratories Inc., 1996.

UL 555-99, *Fire Dampers*. Northbrook, IL: Underwriters Laboratories Inc., 1999.

UL 555S-99, *Leakage Rated Dampers for Use in Smoke Control Systems*. Northbrook, IL: Underwriters Laboratories Inc. 1999.

UL 723-03, *Standard for Test for Surface Burning Characteristics of Building Materials—with Revisions through May 2005*. Northbrook, IL: Underwriters Laboratories Inc., 2003.

UL 1479-03, *Fire Tests of Through-penetration Firestops—with Revisions through April 2007*. Northbrook, IL: Underwriters Laboratories Inc., 2003.

UL 1784-01, *Air Leakage Tests of Door Assemblies—with Revisions through December 2004*. Northbrook, IL: Underwriters Laboratories Inc., 2001.

UL 2079-04, *Tests for Fire Resistance for Building Joint Systems—with Revisions through March 2006*. Northbrook, IL: Underwriters Laboratories Inc., 2004.

UL Fire Resistance Directory (Vol. 2), 2005, Northbrook, IL: Underwriters Laboratories Inc., 2005.

Chapter 8:
Interior Finishes

General Comments

Past fire experience has shown that interior finish and decorative materials are key elements in the development and spread of fire. In some cases, as with Boston's devastating Cocoanut Grove fire in 1942, interior decorations were the first materials ignited. In many other cases, the interior finish materials became involved in the early stages of fire and contributed to its early growth and spread.

The provisions of Chapter 8 require materials used as interior finishes and decorations to have a flame spread index or meet certain flame propagation criteria based on the relative fire hazard associated with the occupancy.

The performance of the material is evaluated based on test standards. The design professional or permit applicant is responsible for determining and providing documentation on the flame spread potential and ability to propagate flames of a material in order to support the permit application. The building official then evaluates the information supplied to ascertain that compliance with the applicable code provisions have been achieved. It is also critical that a field inspection verifies that the material is installed in accordance with the approved documents and the code.

Purpose

This chapter contains the performance requirements for controlling fire growth within buildings by restricting interior finish materials and decorative materials.

SECTION 801
GENERAL

801.1 Scope. Provisions of this chapter shall govern the use of materials used as *interior finishes*, *trim* and *decorative materials*.

❖ This chapter contains requirements for materials used as interior finish, trim or decorative materials. These materials must conform to the flame spread index limitations, the heat release and flashover limitations or flame propagation limitations established by this chapter. This chapter is similar to Chapter 8 of the *International Fire Code®* (IFC®) in intent but is focused only on new construction and only on interior finish, trim and a limited amount of decorative materials. The IFC addresses both new and existing buildings and also some key building contents.

801.2 Interior wall and ceiling finish. The provisions of Section 803 shall limit the allowable fire performance and smoke development of *interior wall and ceiling finish* materials based on occupancy classification.

❖ The provisions of this chapter require materials used as interior finishes to have limited flame spread and smoke-developed indexes (or a limited heat and smoke release), based on the relative fire hazard and occupant vulnerability associated with the occupancy classification. Section 803 provides the necessary performance of wall and ceiling finishes required in test standards based upon the occupancy classifica-

tion. The primary test standard is ASTM E 84 or UL 723).

801.3 Interior floor finish. The provisions of Section 804 shall limit the allowable fire performance of *interior floor finish* materials based on occupancy classification.

❖ Just as for wall and ceiling finish materials, Chapter 8 provides limitations on flame propagation for materials used as floor finish. Section 804 provides the test requirements and criteria for floor finish materials.

[F] 801.4 Decorative materials and trim. *Decorative materials* and *trim* shall be restricted by combustibility and the flame propagation performance criteria of NFPA 701, in accordance with Section 806.

❖ This section is specific to decorative materials and trim, and their specific hazards versus the interior finish within a space. Section 806 addresses the tests for these materials, such as NFPA 701, which provides an indication of the potential for flame propagation of materials such as draperies. Additionally, materials, such as foam plastic used as interior trim, are limited in their combustibility through size, density and coverage limits (see commentary, Section 806).

801.5 Applicability. For buildings in flood hazard areas as established in Section 1612.3, *interior finishes*, *trim* and *decorative materials* below the design flood elevation shall be flood-damage-resistant materials.

❖ All building materials located below the design flood elevation (DFE), including interior finishes, trim and decorative materials, must be resistant to flood damage. Many elevated buildings located in designated flood hazard areas have enclosed lower areas. "Flood-damage-resistant materials" are defined in Section 1612.2 as "any construction material capable of withstanding direct and prolonged contact with floodwaters without sustaining any damage that requires more than cosmetic repair." The term "prolonged contact" means at least 72 hours, and the term "cosmetic repair" means cleaning the affected surfaces with commercially available cleaners and repainting as necessary. For further guidance, refer to FEMA FIA-TB #2, *Flood-resistant Material Requirements for Buildings Located in Special Flood Hazard Areas*; FEMA FIA-TB #7, *Wet Floodproofing Requirements for Structures Located in Special Flood Hazard Areas*; and FEMA FIA-TB #8, *Corrosion Protection for Metal Connectors in Coastal Areas for Structures Located in Special Flood Hazard Areas*.

801.6 Application. Combustible materials shall be permitted to be used as finish for walls, ceilings, floors and other interior surfaces of buildings.

❖ This section simply points out that this chapter allows combustible materials as interior finish. This would include the finish of walls that are required to be "noncombustible" due to the building's type of construction or other code requirements. This is supported by the provisions of Section 603.1, which allows limited combustible materials in Type I and II construction. Item 5 of Section 603.1 lists interior floor finish and interior finish, trim and millwork, such as doors, door frames, window sashes and frames, as allowable combustible materials. The majority of materials that are used as finish materials are combustible in nature; however, all materials in regulated environments are required to exhibit maximum flame spread and smoke-developed indexes or an equivalent performance on heat and smoke release in an appropriate room-corner fire test. Section 803 clarifies the limits on such materials by limiting the flame spread index and smoke-developed index of such materials in certain portions of the building based upon the occupancy.

801.7 Windows. Show windows in the exterior walls of the first *story* above grade shall be permitted to be of wood or of unprotected metal framing.

❖ First-story windows used as display spaces are exempt from finish requirements. These areas are subject to frequent redecoration; however, the enforcement of finish requirements would be impractical. The exemption is limited to the first story above grade plane, which is considered to provide ready access for the fire department from the exterior for effective manual fire suppression efforts.

801.8 Foam plastics. Foam plastics shall not be used as *interior finish* except as provided in Section 803.4. Foam plastics shall not be used as interior *trim* except as provided in Section 806.3 or 2604.2. This section shall apply both to exposed foam plastics and to foam plastics used in conjunction with a textile or vinyl facing or cover.

❖ This section establishes that foam plastic materials are not permitted to be used as interior finish materials unless the foam plastic complies with Section 2603.4, 2603.8 or 2604. Sections 2603.4 and 2603.4.1 generally require that foam plastic materials be covered by an approved thermal barrier, such as 0.5-inch (12.7 mm) gypsum wallboard. When covered in this manner, the foam plastic is no longer the interior finish material. Section 2603.9 provides a means by which foam plastics can be installed without a thermal barrier. A thermal barrier is not required when foam plastic insulation has met the following two criteria: (1) it has been tested in accordance with FM Procedure 4880, UL Subject 1040 or UL 1715, or with NFPA 286 (with the pass/fail criteria of Section 803.1.2.1) and (2) the foam plastic meets the flame spread requirements of this chapter, Section 803 (see commentary, Section 2603.8).

The reference to Section 2604 is for the use of foam plastic as trim.

Sections 806.3 and 2604.2 do allow the use of a limited amount of foam plastic as interior trim when certain criteria is met. The IFC contains the same allowance for interior trim (see commentary, Section 806.3).

It should be noted that the IFC further regulates some building contents with regard to foam plastic. More specifically, Section 807.4.1.1 of the IFC places maximum heat release rate criteria on foam plastic used for decorative purposes, stage scenery or exhibit booths when tested in accordance with UL 1975. This code typically does not regulate building contents.

SECTION 802
DEFINITIONS

802.1 General. The following words and terms shall, for the purposes of this chapter and as used elsewhere in this code, have the meanings shown herein.

❖ This section contains definitions of terms that are associated with the subject matter of this chapter. It is important to emphasize that these terms are not exclusively related to this chapter, but are applicable everywhere the term is used in the code.

EXPANDED VINYL WALL COVERING. Wall covering consisting of a woven textile backing, an expanded vinyl base coat layer and a nonexpanded vinyl skin coat. The expanded base coat layer is a homogeneous vinyl layer that contains a blowing agent. During processing, the blowing agent decomposes, causing this layer to expand by forming closed cells. The total thickness of the wall covering is approximately 0.055 inch to 0.070 inch (1.4 mm to 1.78 mm).

❖ Expanded vinyl wall coverings are manufactured with a textile backing, an expanded vinyl base coat layer consisting of closed cells and a vinyl skin coat. Expanded vinyl wall coverings are subject to the requirements of Section 803.7.

FLAME SPREAD. The propagation of flame over a surface.

❖ The rate at which flames travel along the surface of a combustible finish material directly impacts the speed with which a fire spreads within a room or space, and is, therefore, regulated by this chapter.

FLAME SPREAD INDEX. A comparative measure, expressed as a dimensionless number, derived from visual measurements of the spread of flame versus time for a material tested in accordance with ASTM E 84 or UL 723.

❖ The ASTM E 84 (or UL 723) test method renders measurements of surface flame spread (and smoke density) in comparison with test results obtained by using select red oak as a control material. Red oak is used as a control material for furnace calibration because it is a fairly uniform grade of lumber that is readily available nationally, is uniform in thickness and moisture content, and generally gives consistent and reproducible results. The results of this test simply provide a relative understanding of flame spread potential. The flame spread index is sometimes abbreviated as FSI.

INTERIOR FINISH. *Interior finish* includes *interior wall and ceiling finish* and *interior floor finish*.

❖ This is a more general term that addresses all exposed surfaces, which includes walls, ceilings and floors. Interior finish material is exposed to the interior space enclosed by these building elements.

INTERIOR FLOOR FINISH. The exposed floor surfaces of buildings including coverings applied over a finished floor or *stair*, including risers.

❖ This definition clarifies which part of the interior finish is considered the floor. Floors are treated differently than walls and ceilings as they pose a minimal fire hazard in comparison to that potentially created by the interior finish on walls and ceilings (see commentary, Section 804). Materials that are installed above the structural floor element and exposed to the room or space are considered floor finish materials. Such materials are, therefore, subject to the requirements of Section 804. This section restricts the combustibility and ability to propagate a fire based upon radiant exposure levels.

[F] INTERIOR FLOOR-WALL BASE. Interior floor finish trim used to provide a functional and/or decorative border at the intersection of walls and floors.

❖ This definition, which addresses interior floor-wall base trim materials, provides an understanding and clarification of these types of products versus other interior trim materials. In many cases, floor covering material is just seamlessly turned up or used at the intersection of the floor and the wall, thus becoming the floor-wall base trim. Because of their location at the floor line, floor-wall base materials are not likely to be involved in a fire until the floor covering is also involved, usually at room flashover (see also commentary, Section 806.6).

INTERIOR WALL AND CEILING FINISH. The exposed interior surfaces of buildings, including but not limited to: fixed or movable walls and partitions; toilet room privacy partitions; columns; ceilings; and interior wainscoting, paneling or other finish applied structurally or for decoration, acoustical correction, surface insulation, structural fire resistance or similar purposes, but not including trim.

❖ A material that is applied to ceilings as well as walls, columns, partitions (including the privacy partitions in bathrooms that could pose a significant threat in larger bathrooms if unrated) and other vertical interior surfaces whether fixed or movable. The application of this material may be for structural, decorative, acoustical, structural fire resistance and other similar reasons. Trim, such as baseboard, door or window casing, is not considered interior wall and ceiling finish. Interior wall and ceiling finish is regulated by Section 803.

SITE-FABRICATED STRETCH SYSTEM. A system, fabricated on site and intended for acoustical, tackable or aesthetic purposes, that is comprised of three elements: (a) a frame (constructed of plastic, wood, metal or other material) used to hold fabric in place, (b) a core material (infill, with the correct properties for the application), and (c) an outside layer, comprised of a textile, fabric or vinyl, that is stretched taunt and held in place by tension or mechanical fasteners via the frame.

❖ Site-fabricated stretch systems are interior finish materials that are stretched taut across walls or ceilings with a frame that holds a fabric and core. These systems are now being used extensively because they can stretch to cover decorative walls and ceilings with unusual looks and shapes. The systems consist of three parts: a fabric (or vinyl), a frame and an infill core material. This type of product is not exclusive to any particular manufacturer. It is important to point out that these materials are not curtains or drapes because they are not free hanging (see commentary, Section 803.9).

SMOKE-DEVELOPED INDEX. A comparative measure, expressed as a dimensionless number, derived from measurements of smoke obscuration versus time for a material tested in accordance with ASTM E 84.

❖ The ASTM E 84 test method of measuring the density of smoke emitted from combustible materials determines the smoke-developed index. This value is only comparative and provides only a relative understanding of smoke generation potential. The smoke-developed index is sometimes abbreviated as SDI.

TRIM. Picture molds, chair rails, baseboards, handrails, door and window frames and similar decorative or protective materials used in fixed applications.

❖ As interior trim, this material is usually combustible and permanently affixed. Trim is primarily located around door and window openings (frames), around walls at floors (baseboard) and on walls (chair rail). Interior trim material should only constitute 10 percent of the wall or ceiling area (see Section 806.5).

SECTION 803
WALL AND CEILING FINISHES

803.1 General. *Interior wall and ceiling finish* materials shall be classified for fire performance and smoke development in accordance with Section 803.1.1 or 803.1.2, except as shown in Sections 803.2 through 803.13. Materials tested in accordance with Section 803.1.2 shall not be required to be tested in accordance with Section 803.1.1.

❖ Wall and ceiling interior finish and trim materials are required to have limits on flame spread and smoke-developed indexes as prescribed in Section 803.9 based upon occupancy. ASTM E 84 or UL 723) is one of the tests available to demonstrate compliance with the requirements of Section 803. This test method has been around since 1944 [see Figure 803.1(1)]. This is the primary method used, but NFPA 286 is an alternate optional test that is discussed in the commentary for Section 803.1.2. ASTM E 84 is intended to determine the relative burning behavior of materials on exposed surfaces, such as walls and ceilings, by visually observing the flame spread along the test specimen [see Figure 803.1(2)]. Flame spread and smoke-developed indexes are then reported. The test method is not appropriate for materials that are not capable of supporting themselves or of being supported in the test tunnel. There may also be concerns with materials that drip, melt or delaminate and with very thin materials. A distinction is made, therefore, for textiles, expanded vinyl wall or ceiling coverings and foam plastic insulation materials (see Sections 803.6, 803.7 and 806.3).

ASTM E 84 establishes a flame spread index based on the area under a curve when the actual flame spread distance is plotted as a function of time. The code has divided the acceptable range of flame spread indexes (0-200) into three classes: Class A (0-25), Class B (26-75) and Class C (76-200). For all three classes the code has established a common acceptable range of smoke-developed index: 0-450. An indication of relative fire performance is as follows: an inorganic reinforced-cement board has a flame spread index of zero, while select grade red oak wood flooring has a flame spread index of 100. Not to preclude more detailed information resulting from an ASTM E 84 test report, Figure 803.1(3) identifies the typical flame spread properties of certain building materials.

803.1.1 Interior wall and ceiling finish materials. Interior wall and ceiling finish materials shall be classified in accordance with ASTM E 84 or UL 723. Such *interior finish* materials shall be grouped in the following classes in accordance with their flame spread and *smoke-developed indexes.*

Class A: Flame spread index 0-25; smoke-developed index 0-450.

Class B: Flame spread index 26-75; smoke-developed index 0-450.

Class C: Flame spread index 76-200; smoke-developed index 0-450.

Exception: Materials tested in accordance with Section 803.1.2.

❖ Wall and ceiling interior finish and trim materials are required to have limits on flame spread and smoke-developed indexes as prescribed in Section 803.9, based upon occupancy. ASTM E 84 (or UL 723) is one of the tests available to demonstrate compliance with the requirements of Section 803. This test method has been around since 1944 [see Figure 803.1(1)]. This is the primary method used; but, as noted earlier, NFPA 286 is an alternative test, which is discussed in the commentary for Section 803.1.2. ASTM E 84 is intended to determine the relative burning behavior of materials on exposed surfaces, such as ceilings and

Figure 803.1(1)
ASTM E 84 TUNNEL TEST

Figure 803.1(2)
FLAME IN TUNNEL TEST

walls, by visually observing the flame spread along the test specimen [see Figure 803.1(2)]. Flame spread and smoke developed indexes are then reported. The test method is not appropriate for materials that are not capable of supporting themselves or of being supported in the test tunnel. There may also be concerns with materials that drip, melt or delaminate and that are very thin. A distinction is made, therefore, for textile wall or ceiling covering, expanded vinyl wall or ceiling coverings and foam plastic insulation materials (see Sections 803.5, 803.6, 803.7 and 803.8).

ASTM E 84 establishes a flame spread index based on the area under a curve when the actual flame spread distance is plotted as a function of time. The code has divided the acceptable range of flame spread indexes (0-200) into three classes: Class A (0-25), Class B (26-75) and Class C (76-200). For all three classes the code has established a common acceptable range of smoke-developed index of 0-450. An indication of relative fire performance is as follows: an inorganic reinforced-cement board has a flame spread and smoke-developed index of zero, while select grade red oak wood flooring has a flame spread and smoke-developed index of 100. Not to preclude more detailed information resulting from an ASTM E 84 test report, Figure 803.1(3) identifies the typical flame spread properties of certain building materials.

Material	Flame spread
Glass-fiber sound-absorbing planks	15 to 30
Mineral-fiber sound-absorbing panels	10 to 25
Shredded wood fiberboard (treated)	20 to 25
Sprayed cellulose fibers (treated)	20
Aluminum (with baked enamel finish on one side)	5 to 10
Asbestos-cement board	0
Brick or concrete block	0
Cork	175
Gypsum board (with paper surface on both sides)	10 to 25
Northern pine (treated)	20
Southern pine (untreated)	130 to 190
Plywood paneling (untreated)	75 to 275
Plywood paneling (treated)	100
Carpeting	10 to 600
Concrete	0

Figure 803.1(3)
TYPICAL FLAME SPREAD OF COMMON MATERIALS

803.1.2 Room corner test for interior wall or ceiling finish materials. *Interior wall or ceiling finish* materials shall be permitted to be tested in accordance with NFPA 286. Interior wall or ceiling finish materials tested in accordance with NFPA 286 shall comply with Section 803.1.2.1.

❖ The alternative test method for determining compliance for interior wall and ceiling finish and trim, other than textiles, is found in test standard NFPA 286. This

test is known as a "room-corner" fire test and is similar to that referenced for textile wall coverings in Section 803.1.3.1 (NFPA 265) [see Figures 803.1.2(1) and 803.1.2(2)]. In this test, materials are mounted covering three walls of the compartment (excluding the wall containing the door) and the ceiling. In the case where testing is only for ceiling finish properties, the sample only needs to be mounted on the ceiling. Then a fire source consisting of a gas burner is placed in one corner, flush against both walls (furthest from the door-

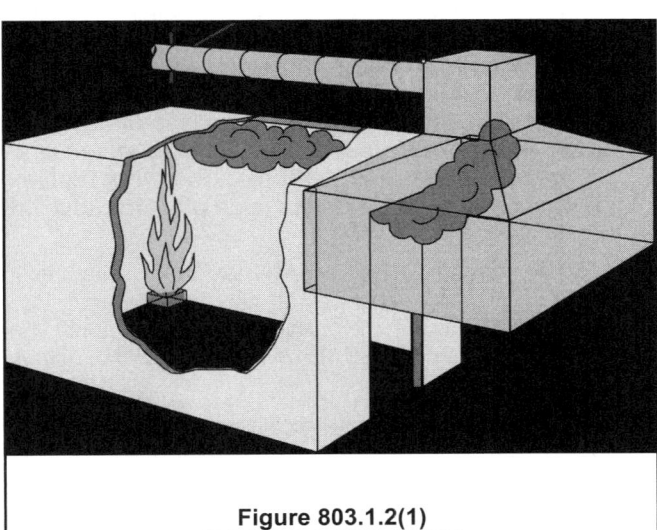

Figure 803.1.2(1)
ROOM CORNER TEST

Figure 803.1.2(2)
FLAME IN ROOM CORNER TEST

way) of the compartment with the following exposure conditions:

- 40 kilowatts (kW) for 5 minutes, then
- 160 kW for 10 minutes.

The test then measures heat release and smoke release through the collection of the fire effluents and measurement of oxygen concentrations in the exhaust duct. Heat release is calculated by the oxygen consumption principle, which has shown that heat release is a function of the decrease in oxygen concentration in the fire effluents. Thus, exhaust duct measurements include temperatures, pressures and smoke values for use in the calculations. Temperatures and heat fluxes are also measured in the room. This generally provides a more realistic understanding of the fire hazard associated with the materials.

The NFPA 286 test method does not contain pass/fail criteria; however, the code provides such criteria in Section 803.1.2.1.

803.1.2.1 Acceptance criteria for NFPA 286. During the 40 kW exposure, the *interior finish* shall comply with Item 1. During the 160 kW exposure, the *interior finish* shall comply with Item 2. During the entire test, the *interior finish* shall comply with Items 3 and 4.

1. During the 40 kW exposure, flames shall not spread to the ceiling.

2. During the 160 kW exposure, the *interior finish* shall comply with the following:

   2.1. Flame shall not spread to the outer extremity of the sample on any wall or ceiling.

   2.2. Flashover, as defined in NFPA 286, shall not occur.

3. The peak rate of heat release throughout the NFPA 286 test shall not exceed 800 kW.

4. The total smoke released throughout the NFPA 286 test shall not exceed 1,000 m².

❖ As noted in Section 803.1.2, there are two levels of exposure during an NFPA 286 fire test in order to better represent a growing fire: 40 kW fire size for 5 minutes and 160 kW for 10 minutes. The 40 kW exposure represents the beginning of a fire where the initial spread is critical; therefore, the stated criteria is that the fire cannot spread to the ceiling. The 160 kW exposure is obviously a more intense fire situation and the criteria relates to preventing flashover (as defined by NFPA 286) and the extent of flame spread throughout the entire test assembly. There is also a total smoke production criterion of 1,000 m². It should be noted that the criteria used in NFPA 265 and NFPA 286 to determine if flashover has occurred would include any two of the following:

- Heat release exceeds 1 megawatt (MW).
- Heat flux at the floor exceeds 20 kW/m².
- Average upper layer temperature exceeds 1112°F (600°C).
- Flames exit the doorway.

- Auto ignition of paper target on the floor occurs.

It should be noted that the code has additional criteria when applying NFPA 286 to new buildings, which is a maximum peak heat release rate of 800 kW. The reasoning for this criterion relates to the fact that some poorer performing materials can achieve compliance with the flashover criteria but could have a higher peak heat release rate. Generally, the current criteria without the 800-kW limit will catch poorer performing materials. The 800-kW peak heat release rate limit is applicable to new buildings only, in accordance with the criteria in Section 803.2.1.

803.1.3 Room corner test for textile wall coverings and expanded vinyl wall coverings. Textile wall coverings and expanded vinyl wall coverings shall meet the criteria of Section 803.1.3.1 when tested in the manner intended for use in accordance with the Method B protocol of NFPA 265 using the product-mounting system, including adhesive.

❖ This particular section allows textile materials and expanded vinyl wall coverings when tested in accordance with Method B protocol of NFPA 265. NFPA 265 is known as a full-scale "room-corner" fire test. Past research conducted with this kind of configuration has shown that flame spread indexes produced by ASTM E 84 may not reliably predict the fire behavior of textile wall coverings. A more reliable test procedure was developed at the University of California and involved the use of a room-corner fire test with a gas diffusion burner(s). The research findings are described in a report from the University of California Fire Research Laboratory titled, "Room Fire Experiments of Textile Wall Coverings." The NFPA 265 test method is only slightly different from NFPA 286. NFPA 286 is more severe in three ways:

1. The gas diffusion burner is used to expose the material on the wall in the room. The fire test starts with a heat release rate exposure of 40 kW for the first 5 minutes in both tests, but is then followed by 150 kW for 10 minutes in NFPA 265 and by 160 kW for 10 minutes in NFPA 286.

2. The gas burner is placed 2 inches (51 mm) away from each of the walls in NFPA 265, whereas it is placed flush against both walls in NFPA 286.

3. The test sample is mounted on the walls only in accordance with NFPA 265, while it is mounted both on the walls and ceiling in accordance with NFPA 286.

A key result of the difference in intensity and location of the gas burner is that the burner flame does not reach the ceiling during the 150 kW exposure (while it does reach the ceiling during the 160 kW exposure in NFPA 286). Therefore, NFPA 265 is not considered to be suitable for testing ceiling coverings.

Based on the use of test Method B from NFPA 265, the sample must meet the criteria of Section 803.1.3.1. This test helps to determine the contribution of textile wall coverings to overall fire growth and spread in a compartment fire. This test also exposes the textile to an ignition source while mounted on the wall. It should be noted that the IFC allows the use of test Method A or B in NFPA 265. The testing requirements in the IFC apply specifically to existing buildings. In the Method A test protocol, 2-foot-wide (610 mm) strips of the material are mounted on the two walls closest to the corner with the burner, whereas, in the Method B test protocol, the sample is mounted completely covering three walls (except for the wall containing the door); therefore, Method B test protocol is more severe. As noted, these textiles are exposed to a prescribed heat release rate of 40 kW for 5 minutes, which is then followed by a heat release rate of 150 kW for 10 minutes.

803.1.3.1 Acceptance criteria for NFPA 265. During the 40 kW exposure the *interior finish* shall comply with Item 1. During the 150 kW exposure, the *interior finish* shall comply with Item 2. During the entire test, the *interior finish* shall comply with Item 3.

1. During the 40 kW exposure, flames shall not spread to the ceiling.

2. During the 150 kW exposure, the *interior finish* shall comply with the following:

    2.1. Flame shall not spread to the outer extremities of the samples on the 8-foot by 12-foot (203 mm by 305 mm) walls.

    2.2. Flashover, as described in NFPA 265, shall not occur.

3. The total smoke released throughout the NFPA 265 test shall not exceed 1,000 m².

❖ To pass Method B test protocol, there are several criteria, including avoiding flashover, as defined in NFPA 265. See the commentary to Section 803.1.3, which discusses limitations on the flame spread of the sample in both the 40 kW and 160 kW exposure and limits the total smoke released to 1000 m², just as in NFPA 286.

803.1.4 Acceptance criteria for textile and expanded vinyl wall or ceiling coverings tested to ASTM E 84 or UL 723. Textile wall and ceiling coverings and expanded vinyl wall and ceiling coverings shall have a Class A flame spread index in accordance with ASTM E 84 or UL 723 and be protected by an *automatic sprinkler system* installed in accordance with Section 903.3.1.1 or 903.3.1.2. Test specimen preparation and mounting shall be in accordance with ASTM E 2404.

❖ Textile wall coverings and expanded vinyl wall coverings dramatically increase the fuel load in a room. As such, additional protection is required as a safety factor for rooms finished with these types of material. ASTM E 84 alone is not considered a sufficient test of the flame spread properties of textile wall coverings because they are too thin when tested, and thus sprinklers would be required in addition to a Class A flame spread rating as a safety factor.

803.2 Thickness exemption. Materials having a thickness less than 0.036 inch (0.9 mm) applied directly to the surface of walls or ceilings shall not be required to be tested.

❖ Thin materials used as interior finish, such as wallpaper, are exempt from the testing requirements of this section. These materials represent a very low fuel load and minimal risk of significant flame spread.

803.3 Heavy timber exemption. Exposed portions of structural members complying with the requirements for buildings of Type IV construction in Section 602.4 shall not be subject to *interior finish* requirements.

❖ This section is simply intended to clarify the code regarding heavy timber, which is commonly exposed to the interior without any interior finish. The concerns over the contribution of heavy timber members of a structure are dealt with in the general requirements of Chapter 6 and other provisions of the code that allow heavy timber.

803.4 Foam plastics. Foam plastics shall not be used as *interior finish* except as provided in Section 2603.9. This section shall apply both to exposed foam plastics and to foam plastics used in conjunction with a textile or vinyl facing or cover.

❖ This section is simply intended to serve as a reminder that the provisions of Chapter 26 apply when the material is foam plastic. The special approval provisions of Section 2603.9 would be required to allow a foam plastic to be used in an exposed interior application. The special approval would involve testing in accordance with NFPA 286, which is also used for acceptance of other materials in this section.

803.5 Textile wall coverings. Where used as interior wall finish materials, textile wall coverings, including materials having woven or nonwoven, napped, tufted, looped or similar surface and carpet and similar textile materials, shall be tested in the manner intended for use, using the product mounting system, including adhesive, and shall comply with the requirements of Section 803.1.2, 803.1.3 or 803.1.4.

❖ The concern that this section addresses for mounting or attaching any interior finish is that the materials or method of attachment will not cause a significant increase in the flame spread characteristics of the interior finish material. For textiles and expanded vinyl wall coverings, the variety of adhesives and mounting methods are such that the entire system needs to be tested in the Steiner tunnel tests or the room-corner test specified in Section 803.1.3 or 803.1.4.

803.6 Textile ceiling coverings. Where used as interior ceiling finish materials, textile ceiling coverings, including materials having woven or nonwoven, napped, tufted, looped or similar surface and carpet and similar textile materials, shall be tested in the manner intended for use, using the product mounting system, including adhesive, and shall comply with the requirements of Section 803.1.2 or 803.1.4.

❖ The concern that this section addresses for mounting or attaching any interior finish is that the materials or method of attachment will not cause a significant increase in the flame spread characteristics of the interior finish material. For textiles and expanded vinyl wall coverings, the variety of adhesives and mounting methods are such that the entire system needs to be tested in the Steiner tunnel tests or the room-corner test specified in Section 803.1.3 or 803.1.4.

803.7 Expanded vinyl wall coverings. Where used as interior wall finish materials, expanded vinyl wall coverings shall be tested in the manner intended for use, using the product mounting system, including adhesive, and shall comply with the requirements of Section 803.1.2, 803.1.3 or 803.1.4.

❖ The concern that this section addresses for mounting or attaching any interior finish is that the materials or method of attachment will not cause a significant increase in the flame spread characteristics of the interior finish material. For textiles and expanded vinyl wall coverings, the variety of adhesives and mounting methods are such that the entire system needs to be tested in the Steiner tunnel tests or the room-corner test specified in Section 803.1.3 or 803.1.4.

803.8 Expanded vinyl ceiling coverings. Where used as interior ceiling finish materials, expanded vinyl ceiling coverings shall be tested in the manner intended for use, using the product mounting system, including adhesive, and shall comply with the requirements of Section 803.1.2 or 803.1.4.

❖ The concern that this section addresses for mounting or attaching any interior finish is that the materials or method of attachment will not cause a significant increase in the flame spread characteristics of the interior finish material. For textiles and expanded vinyl wall coverings, the variety of adhesives and mounting methods are such that the entire system needs to be tested in the Steiner tunnel tests or the room-corner test specified in Section 803.1.3 or 803.1.4.

803.9 Interior finish requirements based on group. *Interior wall and ceiling finish* shall have a flame spread index not greater than that specified in Table 803.9 for the group and location designated. *Interior wall and ceiling finish* materials tested in accordance with NFPA 286 and meeting the acceptance criteria of Section 803.1.2.1, shall be permitted to be used where a Class A classification in accordance with ASTM E 84 or UL 723 is required.

❖ The requirements for flame spread indexes for interior finish materials applied to walls and ceilings are contained in Table 803.9 (see Sections 804 and 805 for floor finish requirements). The referenced test for determining flame spread indexes is ASTM E 84 or UL 723, which establishes a relative measurement of flame spread across the surface of the material. The classifications used in Table 803.9 are defined in Section 803.1. See the commentary to Section 803.1 for additional information on the uses and limitations of the test procedure. This section also allows the use of NFPA 286 as noted in the exception to Section 803.1 and Section 803.2 (without the exception). Section 803.5, in particular, states that materials that pass NFPA 286 in accordance with the pass/fail criteria in Section 803.2.1 can be used as an alternative to

ASTM E 84. Passing NFPA 286 means that the material would be considered equivalent to Class A. In other words, it can be used as a replacement for all categories, as Class A is the most restrictive (see commentary, Section 803.1.2 for more detail on the test).

When evaluating test reports, the method of application and substrate material are considered. For example, adhesives that soften at moderate temperatures will allow wall or ceiling finishes to drop or peel. This not only increases the susceptibility of the material to ignition but also exposes the substrate material. The substrate material can also affect the interior finish rating of the surface material. It is not uncommon for a thin, combustible finish applied to a noncombustible material to obtain a low flame spread index, while the same material applied to a combustible substrate or applied without a substrate could exhibit a significantly high tendency to spread flame.

TABLE 803.9. See page 8-9.

❖ This table prescribes the minimum requirements for interior finishes applied to walls and ceilings; therefore, the use of a Class A material in an area that requires a minimum Class B material is always allowed. Likewise, when the table requires Class C materials, Classes A and B can also be used.

The requirements are based on the use of the space. To determine the applicable criteria, first determine whether the space is an exit passageway or exit enclosure (stairway), a corridor or a room or enclosed space. Interior finishes in spaces that are not separated from a corridor (for example, waiting areas in business or health care facilities) must comply with the requirements for a corridor space. As shown in the table, the code places more emphasis on the allowable flame spread index for exits than for enclosed rooms because of the critical nature and relative importance that is placed on maintaining the integrity of exits to evacuate the building.

Numerous notes amend the basic requirements of Table 803.9. Notes a through l apply only where they are specifically referenced in the table.

803.10 Stability. *Interior finish* materials regulated by this chapter shall be applied or otherwise fastened in such a manner that such materials will not readily become detached where subjected to room temperatures of 200°F (93°C) for not less than 30 minutes.

❖ Interior finishes are not to become detached under exposure to elevated temperatures [200°F (93°C)] for 30 minutes. There is not a standard test method yet developed to evaluate this requirement. Other sections of the code, however, offer some additional guidance. For example, the performance of the method of attachment of finish materials during a fire-resistance test will usually be an adequate indication of performance. Section 2504 contains requirements for gypsum wallboard that is not part of a fire-resistance-rated assembly. Interior finishes that become detached may add to the spread of fire, as well as create a hazard for fire fighters.

803.11 Application of interior finish materials to fire-resistance-rated structural elements. Where *interior finish* materials are applied on walls, ceilings or structural elements required to have a *fire-resistance rating* or to be of noncombustible construction, they shall comply with the provisions of this section.

❖ Where an assembly is required to have a fire-resistance rating or be of noncombustible materials, interior finish materials are required to be applied in accordance with Sections 803.4.1 through 803.4.4. These methods are to prevent other types of fire spread onto the exposed surface through the use of fireblocking and similar methods.

803.11.1 Direct attachment and furred construction. Where walls and ceilings are required by any provision in this code to be of fire-resistance-rated or noncombustible construction, the *interior finish* material shall be applied directly against such

construction or to furring strips not exceeding $1^3/_4$ inches (44 mm) applied directly against such surfaces. The intervening spaces between such furring strips shall comply with one of the following:

1. Be filled with material that is inorganic or non-combustible;

2. Be filled with material that meets the requirements of a Class A material in accordance with Section 803.1.1 or 803.1.2; or

3. Be fireblocked at a maximum of 8 feet (2438 mm) in any direction in accordance with Section 717.

❖ Interior finish materials are required to be applied directly to the exposed surface of structural elements or to the furring attached to such surfaces. All concealed spaces created by furring are to be fireblocked at a maximum of 8-foot (2438 mm) intervals in any direction. By applying the finish directly to the surface, the

TABLE 803.9
INTERIOR WALL AND CEILING FINISH REQUIREMENTS BY OCCUPANCY[k]

GROUP	SPRINKLERED[l]			NONSPRINKLERED		
	Exit enclosures and exit passageways[a, b]	Corridors	Rooms and enclosed spaces[c]	Exit enclosures and exit passageways[a, b]	Corridors	Rooms and enclosed spaces[c]
A-1 & A-2	B	B	C	A	A[d]	B[e]
A-3[f], A-4, A-5	B	B	C	A	A[d]	C
B, E, M, R-1	B	C	C	A	B	C
R-4	B	C	C	A	B	B
F	C	C	C	B	C	C
H	B	B	C[g]	A	A	B
I-1	B	C	C	A	B	B
I-2	B	B	B[h, i]	A	A	B
I-3	A	A[j]	C	A	A	B
I-4	B	B	B[h, i]	A	A	B
R-2	C	C	C	B	B	C
R-3	C	C	C	C	C	C
S	C	C	C	B	B	C
U	No restrictions			No restrictions		

For SI: 1 inch = 25.4 mm, 1 square foot = 0.0929m².

a. Class C interior finish materials shall be permitted for wainscotting or paneling of not more than 1,000 square feet of applied surface area in the grade lobby where applied directly to a noncombustible base or over furring strips applied to a noncombustible base and fireblocked as required by Section 803.11.1.

b. In exit enclosures of buildings less than three stories above grade plane of other than Group I-3, Class B interior finish for nonsprinklered buildings and Class C interior finish for sprinklered buildings shall be permitted.

c. Requirements for rooms and enclosed spaces shall be based upon spaces enclosed by partitions. Where a fire-resistance rating is required for structural elements, the enclosing partitions shall extend from the floor to the ceiling. Partitions that do not comply with this shall be considered enclosing spaces and the rooms or spaces on both sides shall be considered one. In determining the applicable requirements for rooms and enclosed spaces, the specific occupancy thereof shall be the governing factor regardless of the group classification of the building or structure.

d. Lobby areas in Group A-1, A-2 and A-3 occupancies shall not be less than Class B materials.

e. Class C interior finish materials shall be permitted in places of assembly with an occupant load of 300 persons or less.

f. For places of religious worship, wood used for ornamental purposes, trusses, paneling or chancel furnishing shall be permitted.

g. Class B material is required where the building exceeds two stories.

h. Class C interior finish materials shall be permitted in administrative spaces.

i. Class C interior finish materials shall be permitted in rooms with a capacity of four persons or less.

j. Class B materials shall be permitted as wainscotting extending not more than 48 inches above the finished floor in corridors.

k. Finish materials as provided for in other sections of this code.

l. Applies when the exit enclosures, exit passageways, corridors or rooms and enclosed spaces are protected by an automatic sprinkler system installed in accordance with Section 903.3.1.1 or 903.3.1.2.

potential for the back side to contribute to a fire is diminished and, as a result, the effect on the fire-resistance rating and the creation of combustible cavities in noncombustible walls are minimized.

Likewise, fireblocks in furred construction also diminish the likelihood of fire involvement between the finish and the interior wall surface. The use of wood nailing strips is intended to be limited to that which is necessary for the installation of finish and trim materials. As such, the amount of combustible material is restricted so that the performance of the element to which it is attached will not be adversely affected. It should be noted that the type of construction classification of the building is not intended to regulate the combustibility of the interior finish materials.The fire performance characteristics of the interior finish materials are regulated by Section 803.1.

803.11.2 Set-out construction. Where walls and ceilings are required to be of fire-resistance-rated or noncombustible construction and walls are set out or ceilings are dropped distances greater than specified in Section 803.11.1, Class A finish materials, in accordance with Section 803.1.1 or 803.1.2, shall be used except where *interior finish* materials are protected on both sides by an *automatic sprinkler system* in accordance with Section 903.3.1.1 or 903.3.1.2, or attached to noncombustible backing or furring strips installed as specified in Section 803.11.1. The hangers and assembly members of such dropped ceilings that are below the main ceiling line shall be of noncombustible materials, except that in Types III and V construction, *fire-retardant-treated wood* shall be permitted. The construction of each set-out wall shall be of fire-resistance-rated construction as required elsewhere in this code.

❖ Where walls and ceilings are required to be of fire-resistance-rated or noncombustible construction and walls are set out or ceilings are dropped distances greater than specified in Section 803.11.1, Class A finish materials shall be used except where interior finish materials are protected on both sides by an automatic sprinkler system or attached to noncombustible backing or furring strips installed as specified in Section 803.11.1. The hangers and assembly members of such dropped ceilings that are below the main ceiling line shall be of noncombustible materials, except that in Type III and V construction, fire-retardant-treated wood (FRTW) shall be permitted. The construction of each set-out wall shall be of fire-resistance-rated construction as required elsewhere in this code.

Where wide spaces are created, one of the following methods may be used to protect the concealed spaces:

- Use of Class A finish material;
- Sprinkler protection within the concealed space and outside;
- Intimately backing the finish material with noncombustible material or FRTW in Type III

and V construction, separating the finish material from the concealed space; or

- Filling the intervening space between furring strips (see Figure 803.11.2) with inorganic or noncombustible material; or
- Class A material or firestopping such spaces a maximum of 8 feet (2438 mm) in any direction.

Class B and C materials may be used on Methods 2, 3 and 4.

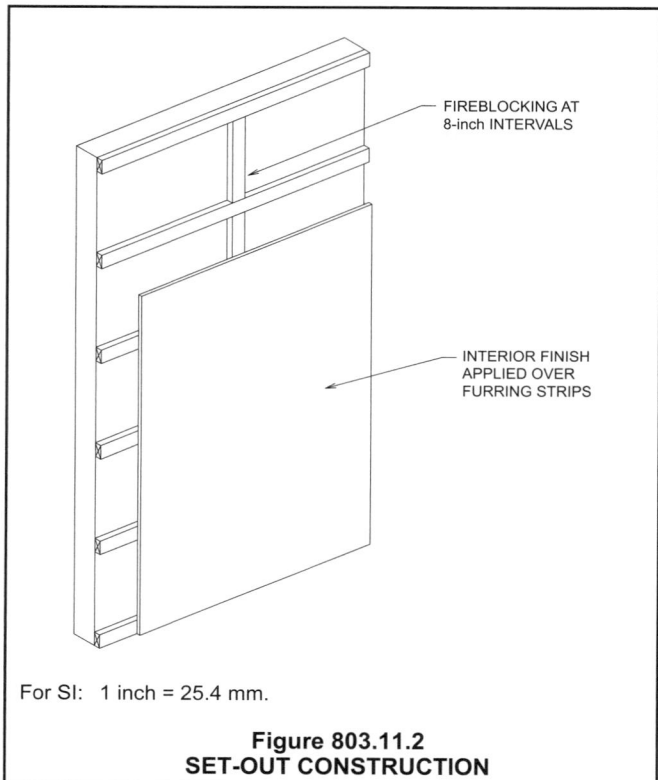

FIREBLOCKING AT 8-inch INTERVALS

INTERIOR FINISH APPLIED OVER FURRING STRIPS

For SI: 1 inch = 25.4 mm.

Figure 803.11.2
SET-OUT CONSTRUCTION

803.11.3 Heavy timber construction. Wall and ceiling finishes of all classes as permitted in this chapter that are installed directly against the wood decking or planking of Type IV construction or to wood furring strips applied directly to the wood decking or planking shall be fireblocked as specified in Section 803.11.1.

❖ Attachment directly to heavy timber decking or planking eliminates the concealed spaces (see Section 803.3). Furring strips of any depth are allowed when concealed spaces are treated according to Section 803.11.1. Again, the intent is to reduce the likelihood of fire spread between these spaces.

803.11.4 Materials. An interior wall or ceiling finish that is not more than $^1/_4$-inch (6.4 mm) thick shall be applied directly against a noncombustible backing.

Exceptions:

1. Noncombustible materials.

2. Materials where the qualifying tests were made with the material suspended or furred out from the noncombustible backing.

❖ Combustible materials that have a maximum thickness of 0.25 inch (6.4 mm) should be applied directly against a noncombustible backing and not to an added substrate or furred away from the backing, which may increase significantly the spread of fire (see Figure 803.11.4). Exception 1 allows Class A materials, because of their low flame spread index, to be furred away from the noncombustible backing. Exception 2 also allows materials that perform in tests similar to directly attached materials when furred or suspended from a noncombustible backing.

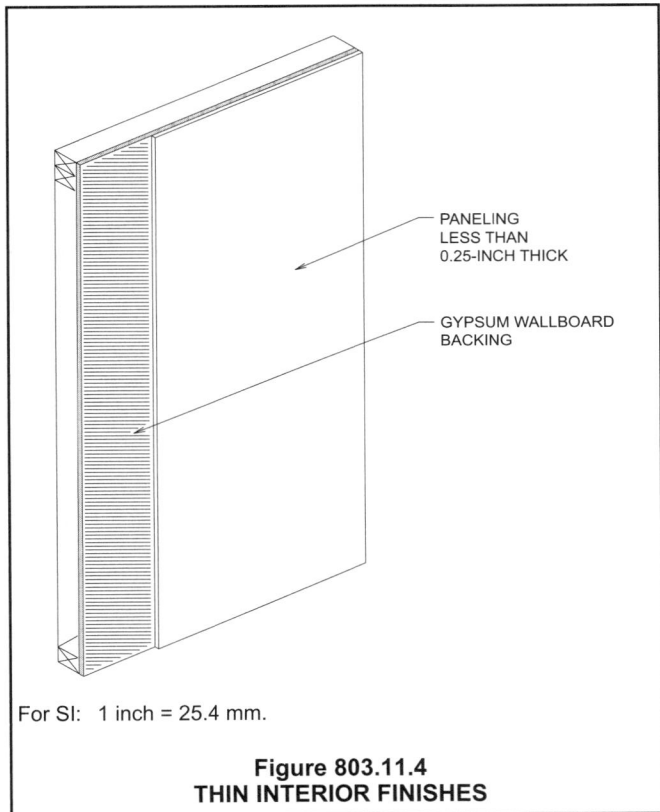

PANELING
LESS THAN
0.25-INCH THICK

GYPSUM WALLBOARD
BACKING

For SI: 1 inch = 25.4 mm.

Figure 803.11.4
THIN INTERIOR FINISHES

803.12 High-density polyethylene (HDPE). Where high-density polyethylene is used as an *interior finish*, it shall comply with the requirements of Section 803.1.2.

❖ High-density polyethylene (HDPE) is a thermoplastic that when it burns, it gives off considerable energy and produces a flammable liquids fire. Recent full-scale room-corner tests using NFPA 286 have demonstrated a significant hazard with some of these (HDPE) materials. These tests had to be terminated prior to the standard 15-minute duration due to flashover occurring, yet there was still much of the product left to burn. Extensive flammable liquid pool fires occurred during the tests. However, the Steiner tunnel test is not a suitable measure of the performance of this material because the calculation of

flame spread index does not take into account the unique hazards known for this material. Therefore, HDPE is required to be tested in a full-scale room-corner test in accordance with Section 803.1.2.

803.13 Site-fabricated stretch systems. Where used as interior wall or interior ceiling finish materials, site-fabricated stretch systems shall be tested in the manner intended for use, and shall comply with the requirements of Section 803.1.1 or 803.1.2. If the materials are tested in accordance with ASTM E 84 or UL 723, specimen preparation and mounting shall be in accordance with ASTM E 2573.

❖ The ASTM E 05 committee on fire standards has issued a standard practice, ASTM E 2573, for specimen preparation and mounting of site-fabricated stretch systems. Until now, there was no correct mandatory way to test these systems. These systems are now being used extensively because they can stretch to cover decorative walls and ceilings with unusual looks and shapes. The systems consist of three parts: a fabric (or vinyl), a frame and an infill core material. The testing has often been done on each component separately, which is inappropriate and unsafe, instead of testing the composite system.

SECTION 804
INTERIOR FLOOR FINISH

804.1 General. *Interior floor finish* and floor covering materials shall comply with Sections 804.2 through 804.4.1.

Exception: Floor finishes and coverings of a traditional type, such as wood, vinyl, linoleum or terrazzo, and resilient floor covering materials that are not comprised of fibers.

❖ This section regulates the design and installation of floor finish and floor covering materials. Traditional floor coverings; such as wood, vinyl, terrazzo and other resilient floor covering material, must be exempt from this section. Smooth surface floor coverings generally contribute minimally to a fire. The focus is more upon textile floor coverings, such as carpets.

804.2 Classification. *Interior floor finish* and floor covering materials required by Section 804.4.1 to be of Class I or II materials shall be classified in accordance with NFPA 253. The classification referred to herein corresponds to the classifications determined by NFPA 253 as follows: Class I, 0.45 watts/cm^2 or greater; Class II, 0.22 watts/cm^2 or greater.

❖ The use of a classification system eliminates the need to state the actual critical radiant flux value for a product to meet the identification requirements of Section 804. Over the years, a classification system has been found to be much easier for the industry to follow and still provides the building official with the information required to verify compliance. The test required to measure the combustibility of floor coverings is NFPA 253. This standard is a radiant floor panel test, which basically simulates materials subjected to heat from a fire above. The primary concern with flooring is related to the spread of a fire that has already been ignited within a space or room to a different room. Figure

804.2 shows the test apparatus. The critical heat flux indicates the threshold value above which flame spread occurs in the testing environment.

Figure 804.2
RADIANT FLOOR PANEL TEST (NFPA 253)

804.3 Testing and identification. *Interior floor finish* and floor covering materials shall be tested by an agency in accordance with NFPA 253 and identified by a hang tag or other suitable method so as to identify the manufacturer or supplier and style, and shall indicate the *interior floor finish* or floor covering classification according to Section 804.2. Carpet-type floor coverings shall be tested as proposed for use, including underlayment. Test reports confirming the information provided in the manufacturer's product identification shall be furnished to the building official upon request.

❖ The only method to ascertain that a floor meets the criteria of this section is to request a copy of the test report for the specific material being installed; therefore, it is critical that the carpeting be properly identified in order to verify that acceptable materials are being provided in the appropriate locations. The identification is to be provided on the material itself since a manufacturer's designation is required.

804.4 Interior floor finish requirements. In all occupancies, *interior floor finish* and floor covering materials in *exit* enclo-

sures, *exit* passageways, corridors and rooms or spaces not separated from corridors by full-height partitions extending from the floor to the underside of the ceiling shall withstand a minimum critical radiant flux as specified in Section 804.4.1.

❖ The finish flooring requirements for all means of egress components, such as stairs and corridors, as well as floors to rooms that open to these elements, are regarded as more critical than other building floor surfaces. As such, the minimum critical radiant flux must be as prescribed in Section 804.4.1.

804.4.1 Minimum critical radiant flux. *Interior floor finish* and floor covering materials in exit enclosures, *exit* passageways and corridors shall not be less than Class I in Groups I-1, I-2 and I-3 and not less than Class II in Groups A, B, E, H, I-4, M, R-1, R-2 and S. In all areas, floor covering materials shall comply with the DOC FF-1 "pill test" (CPSC 16 CFR, Part 1630).

Exception: Where a building is equipped throughout with an *automatic sprinkler system* in accordance with Section 903.3.1.1 or 903.3.1.2, Class II materials are permitted in any area where Class I materials are required, and materials complying with the DOC FF-1 "pill test" (CPSC 16 CFR, Part 1630) are permitted in any area where Class II materials are required.

❖ This section prescribes the minimum requirements for interior floor finish materials that are used in exit enclosures, exit passageways and exit access corridors. The criteria are based on the occupancy classification and the relationship of the space to the egress system. Similar to Table 803.5, the occupancy classification designation is meant to apply to the actual occupancy of the space and not necessarily the overall building classification.

Classifications I and II as used in this section are defined in Section 804.2, and are based on the results of the NFPA 253 test procedure. DOC FF-1, also referred to as the "Methenamine Pill Test," was developed as a means of preventing the distribution of highly flammable soft floor coverings within the United States. The test essentially evaluates the performance of the floor covering when subject to a cigarette-type ignition by using a small methenamine tablet. All carpeting greater than 24 square feet (2.2 m²) in area sold in the United States is required by federal law to pass this test procedure as a minimum.

Recognizing the ability of automatic sprinkler systems to control a fire and the minimal contribution of interior floor finishes to the early stages of fire growth, the exception allows the required interior floor finish ratings to be reduced when an automatic sprinkler system is provided throughout the building. The reference to Section 903.3.1.1 or 903.3.1.2 clarifies that the system is to be installed in accordance with NFPA 13 or 13R. In cases where Class II materials are required and an automatic sprinkler system is provided, the minimum requirement is that the material meet the DOC FF-1 test criteria.

SECTION 805
COMBUSTIBLE MATERIALS IN
TYPES I AND II CONSTRUCTION

805.1 Application. Combustible materials installed on or embedded in floors of buildings of Type I or II construction shall comply with Sections 805.1.1 through 805.1.3.

Exception: Stages and platforms constructed in accordance with Sections 410.3 and 410.4, respectively.

❖ This section provides several methods of allowing combustible materials in the floors of Type I or II construction. Generally, the focus is on fire spread in concealed spaces. If not properly addressed, combustible floor surfaces may eventually contribute to a fire within a concealed space. Sections 410.3 and 410.4 permit wood flooring in stages.

805.1.1 Subfloor construction. Floor sleepers, bucks and nailing blocks shall not be constructed of combustible materials, unless the space between the fire-resistance-rated floor assembly and the flooring is either solidly filled with noncombustible materials or fireblocked in accordance with Section 717, and provided that such open spaces shall not extend under or through permanent partitions or walls.

❖ Sleepers, bucks and nailing blocks are permitted to be of combustible materials only where the void space is filled with a noncombustible material or is fireblocked in accordance with Section 717. Section 717 permits a maximum concealed space area of 100 square feet (9.3 m²). In either case, the open spaces cannot extend under or through permanent partitions or walls. The purpose of the fill or fireblocking is to reduce the impact of a fire in a concealed combustible space in the floor. Likewise, fire spread around partitions or walls through the concealed space in the floor is also intended to be prevented.

805.1.2 Wood finish flooring. Wood finish flooring is permitted to be attached directly to the embedded or fireblocked wood sleepers and shall be permitted where cemented directly to the top surface of fire-resistance-rated floor assemblies or directly to a wood subfloor attached to sleepers as provided for in Section 805.1.1.

❖ Wood floor finish materials may be applied directly to the wood sleepers, provided that the space is protected in accordance with Section 805.1.1. Wood floor finish materials may also be applied directly to the top surface of a fire-resistance-rated assembly or directly to a wood subfloor that is attached to sleepers that are either fireblocked or provide noncombustible fill between the sleepers in accordance with Section 805.1.1.

805.1.3 Insulating boards. Combustible insulating boards not more than $^1/_2$ inch (12.7 mm) thick and covered with finish flooring are permitted where attached directly to a noncombustible floor assembly or to wood subflooring attached to sleepers as provided for in Section 805.1.1.

❖ The addition of combustible insulating boards with a maximum thickness of 0.5 inch (12.7 mm) will not significantly increase the fuel load when applied as prescribed by this section or attached to sleepers as provided for in Section 805.1.1.

In all occupancies, interior floor finish in exit enclosures, exit passageways, exit access corridors and rooms or spaces not separated from exit access corridors by full-height partitions extending from the floor to the underside of the ceiling must withstand a minimum critical radiant flux as specified in Section 804.4.1. The focus with the critical radiant flux is combustibility and potential for flame spread of the exposed materials. Section 805 focuses upon fire spread through concealed spaces created by floors.

[F] SECTION 806
DECORATIVE MATERIALS AND TRIM

[F] 806.1 General requirements. In occupancies in Groups A, E, I and R-1 and dormitories in Group R-2, curtains, draperies, hangings and other *decorative materials* suspended from walls or ceilings shall meet the flame propagation performance criteria of NFPA 701 in accordance with Section 806.2 or be noncombustible.

In Groups I-1 and I-2, combustible *decorative materials* shall meet the flame propagation criteria of NFPA 701 unless the *decorative materials*, including, but not limited to, photographs and paintings, are of such limited quantities that a hazard of fire development or spread is not present. In Group I-3, combustible decorative materials are prohibited.

Fixed or movable walls and partitions, paneling, wall pads and crash pads applied structurally or for decoration, acoustical correction, surface insulation or other purposes shall be considered *interior finish* if they cover 10 percent or more of the wall or of the ceiling area, and shall not be considered *decorative materials* or furnishings.

In Group B and M occupancies, fabric partitions suspended from the ceiling and not supported by the floor shall meet the flame propagation performance criteria in accordance with Section 806.2 and NFPA 701 or shall be noncombustible.

❖ The requirements in this section apply to decorative materials installed in Group A, E, I and R-1 occupancies and dormitories in Group R-2 occupancies. The list of groups is based on the number of persons and the condition or capabilities of the occupants to evacuate quickly in an emergency. Decorative materials must be noncombustible or meet the flame propagation performance criteria of Section 806.2 and NFPA 701.

The requirements for Groups I-1 and I-2 become more stringent because of the nature of the occupants and activities in these types of facilities. Combustible materials in these facilities must meet the flame propagation performance criteria of NFPA 701, but a limited amount of decorative materials, such as photographs and paintings, need not be flame retardant. No amount of combustible decorative materials are permitted in Group I-3 facilities.

In any occupancy classification when a movable wall, partition, paneling, wall pads or crash pads cover larger areas, they need to be dealt with as interior finishes instead of as decorative materials. Such areas would not be considered decorative materials if the area covered is greater than 10 percent of the wall or ceiling area. This is clarified in the third paragraph of this section.

This section also clarifies that in Group B and M occupancies fabric partitions suspended from the ceiling, but not physically contacting the floor; should be treated as decorative materials similar to curtains or draperies and comply with the flame propagation performance criteria of NFPA 701.

[F] 806.1.1 Noncombustible materials. The permissible amount of noncombustible decorative material shall not be limited.

❖ Where decorative materials are classified as noncombustible, it is presumed that they will contribute little, if at all, to the growth and spread of fire; therefore, the quantity of noncombustible decorative materials is not limited.

[F] 806.1.2 Combustible decorative materials. The permissible amount of *decorative materials* meeting the flame propagation performance criteria of NFPA 701 shall not exceed 10 percent of the specific wall or ceiling area to which it is attached.

Exceptions:

1. In auditoriums in Group A, the permissible amount of decorative material meeting the flame propagation performance criteria of NFPA 701 shall not exceed 75 percent of the aggregate wall area where the building is equipped throughout with an *automatic sprinkler system* in accordance with Section 903.3.1.1 and where the material is installed in accordance with Section 803.11.

2. The amount of fabric partitions suspended from the ceiling and not supported by the floor in Group B and M occupancies shall not be limited.

❖ Meeting the flame propagation performance criteria of NFPA 701 does not mean that materials will not burn, only that they are going to spread flame relatively slowly. These materials are, therefore, limited to a maximum of 10 percent of the total wall and ceiling area of the space under consideration. Unlike incidental trim, decorative materials are not necessarily distributed evenly throughout the room. Additionally, consideration of the long-term maintenance of the materials, including possible periodic retreatment, should be taken into account.

There are two exceptions to the 10-percent limitation in Section 806.1.2. Exception 1 is for Group A auditoriums that would allow 75-percent coverage of the walls and ceilings (instead of the limit of 10 percent) if the space is sprinklered in accordance with NFPA 13 and the material is applied in accordance with Section 803.4. Exception 2 correlates with the general requirements in Section 806.1 and further emphasizes that an unlimited amount of fabric ceiling partitions are al-

lowed as long as they meet the flame propagation performance of NFPA 701.

[F] 806.2 Acceptance criteria and reports. Where required by Section 806.1, *decorative materials* shall be tested by an agency and meet the flame propagation performance criteria of NFPA 701 or such materials shall be noncombustible. Reports of test results shall be prepared in accordance with NFPA 701 and furnished to the *building official* upon request.

❖ The standard test method to be used to evaluate the ability of a material to propagate flame is NFPA 701, which contains two test methods. Test 1 is a less severe test than Test 2 and uses smaller test specimens. Test 1 is intended for lighter weight materials and single layer fabrics: the limit is a linear density of 700 g/m^2 (21 oz/yd^2). Test 2 is a more severe test than Test 1 and uses larger test specimens. It is intended for higher density materials and multilayered fabrics. It also applies to vinyl-coated fabric blackout linings and lined draperies using any density, because they have been shown to give misleading results using Test 1. NFPA 701 sets out the types of materials, including fabrics, that should be tested using each method.

Essentially, NFPA 701 provides a mechanism to distinguish between materials that allow flames to spread quickly and those that do not when using a small fire exposure.

Materials tested only to NFPA 701 are not permitted for use as interior finish materials, but instead are generally used as shades, swags, curtains and other similar materials. These tests are used to determine whether materials propagate flame beyond the area exposed to the ignition source. They are not intended to indicate whether the material tested will resist the propagation of flame under fire exposures more extreme than the test conditions.

[F] 806.3 Foam plastic. Foam plastic used as *trim* in any occupancy shall comply with Section 2604.2.

❖ This section establishes that some dense foam plastic materials may be used as interior trim if the thickness, width and area of coverage is specifically limited as required in Section 2604.2.

Section 2604.2 provides the following maximum values: 0.5-inch (12.7 mm) thickness, 8-inch (203 mm) width and 10 percent of the aggregate wall and ceiling area. Section 2604.2 also establishes a minimum density for foam permitted for use as trim: 20 pounds per cubic foot (320 kg/m^3). Minimum instead of maximum density is specified because the denser the foam plastic material, the less likely it is that it will generate misleading ASTM E 84 test results due to having insufficient material for fire testing of the foam [see Figures 806.5(1) and 806.5(2)].

This section specifically calls out a numerical flame spread index limitation of 75 for foam plastic used as trim, which is a Class B flame spread index. No thermal barrier is required for the use of foam plastic as trim and the smoke-developed index is not limited. It should be noted that Section 803.7.3 of the IFC, which regulates the use of foam plastic trim for existing construction, has essentially identical requirements to those found in this section. Neither this section nor Section 2603.4.1.11 requires a thermal barrier for interior trim.

[F] 806.4 Pyroxylin plastic. Imitation leather or other material consisting of or coated with a pyroxylin or similarly hazardous base shall not be used in Group A occupancies.

❖ Use of pyroxylin plastics, also known as cellulose nitrate plastics, as imitation leather (or coated to imitation leather) is strictly prohibited in Group A occupancies because of the normally high occupant loads of such occupancies. This type of material is very hazardous because it has the potential to develop explosive atmospheres with high heat emission and it will begin decomposition at temperatures starting at 300°F (149°C). In actual fact, cellulose nitrate use has generally declined considerably because it tends to become somewhat unstable and is easily ignitable. Specifically, use of cellulose nitrate for motion picture film was discontinued in 1951.

[F] 806.5 Interior trim. Material, other than foam plastic used as interior *trim*, shall have a minimum Class C flame spread and smoke-developed index when tested in accordance with ASTM E 84 or UL 723, as described in Section 803.1.1. Combustible *trim*, excluding handrails and guardrails, shall not exceed 10 percent of the specific wall or ceiling area in which it is attached.

❖ In occupancies of any group, unless otherwise noted in the code, the minimum classification of all trim must be at least Class C. Additionally, combustible trim may not exceed 10 percent of the area of the aggregate wall or ceiling area. The 10-percent calculation does not need to include handrails and guardrails. Although

a Class C rating may be lower than the rating required for a particular building or facility, this quantity of combustible material will not significantly increase the fuel load and flame spread hazards [see Figures 806.5(1) and 806.5(2)].

[F] 806.6 Interior floor-wall base. *Interior floor-wall base* that is 6 inches (152 mm) or less in height shall be tested in accordance with Section 804.2 and shall not be less than Class II. Where a Class I floor finish is required, the floor-wall base shall be Class I.

Exception: Interior *trim* materials that comply with Section 806.5.

❖ Section 806.6 addresses the issue of testing and regulation of interior floor-wall base trim materials. In many cases, the floor covering material is just seamlessly turned up or used at the intersection of the floor and the wall and thus it becomes the floor-wall base trim.

In the 2006 code, these materials could have been considered interior trim in accordance with Sections 804.1 and 806.5 and would have been required to be tested in accordance with ASTM E 84 or UL 723, even though the floor covering may have been required to be tested in accordance with NFPA 253. Based on the small amount of material used, it is very difficult to test these materials in a reliable manner, upside down in the ASTM E 84 test method. Because of their location, at the floor line, floor-wall base materials are not likely to be involved in a fire until the floor covering is also involved, usually at room flashover. Thus, it is reasonable that floor-wall base materials meet the same criteria as floor coverings. The section specifies that floor-wall base materials 6 inches (152 mm) or less in height be tested in accordance with NFPA 253, and the proposal provides requirements for this application.

Wall area: 2 walls (20 × 12) + 2 walls (10 × 12) = 720 sq. ft.
   10% of walls = 720 × 0.10 = 72 sq. ft.
   Decorative trim coverage = 72 sq. ft.
Ceiling 20 × 10 = 200 sq. ft.
   10% of ceiling = 200 × 0.10 = 20
   Decorative trim coverage = 20 sq. ft.

For SI: 1 foot = 304.8 mm,
       1 square foot = 0.0929 m².

Figure 806.5(1)
TRIM COVERAGE AREA

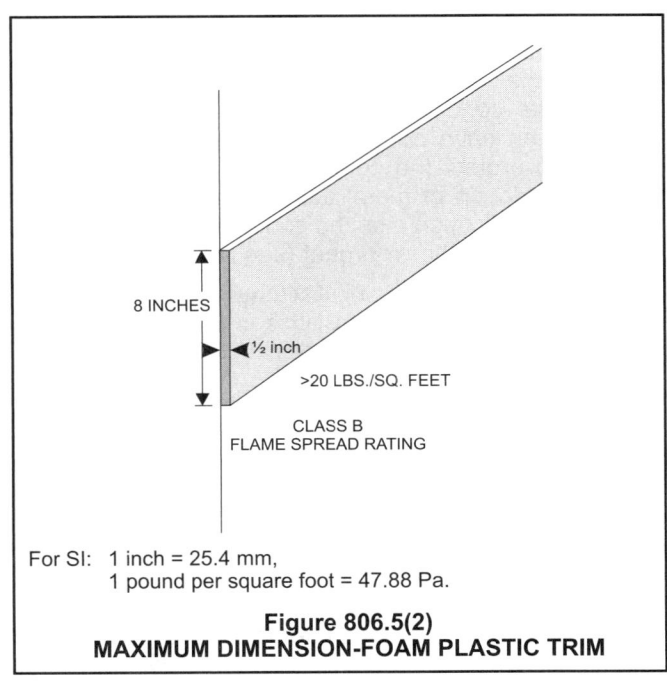

8 INCHES

½ inch

>20 LBS./SQ. FEET

CLASS B
FLAME SPREAD RATING

For SI: 1 inch = 25.4 mm,
       1 pound per square foot = 47.88 Pa.

Figure 806.5(2)
MAXIMUM DIMENSION-FOAM PLASTIC TRIM

SECTION 807
INSULATION

807.1 Insulation. Thermal and acoustical insulation shall comply with Section 719.

❖ This section simply references Section 719 for requirements for insulation materials to clarify that these materials can be used when they meet the requirements of that section.

SECTION 808
ACOUSTICAL CEILING SYSTEMS

808.1 Acoustical ceiling systems. The quality, design, fabrication and erection of metal suspension systems for acoustical tile and lay-in panel ceilings in buildings or structures shall conform with generally accepted engineering practice, the provisions of this chapter and other applicable requirements of this code.

❖ This section regulates the design and installation of metal suspension systems for acoustical tile and acoustical lay-in panel ceilings. Suspended ceiling systems basically consist of a grid of metal channels or T-bars suspended by steel hangers (wires, rods or flats) from the structure above. Light fixtures, air diffusers and acoustical panels may be placed between the metal framing members.

808.1.1 Materials and installation. Acoustical materials complying with the *interior finish* requirements of Section 803 shall be installed in accordance with the manufacturer's recommendations and applicable provisions for applying *interior finish*.

❖ Acoustical materials used in metal suspension systems are regulated for the purpose of limiting flame spread and smoke development in the occupancies shown in Table 803.9.

Metal suspension system components for acoustical ceilings that might be subjected to the severe environmental conditions of high humidity (fog) or salt spray may have coatings ranked according to their ability to protect the components from deterioration. The installation of metal suspension systems in any exterior application must be considered as an installation in a severe environment (see ASTM C 635).

808.1.1.1 Suspended acoustical ceilings. Suspended acoustical ceiling systems shall be installed in accordance with the provisions of ASTM C 635 and ASTM C 636.

❖ ASTM C 635 covers metal ceiling suspension systems used primarily to support acoustical tile or acoustical lay-in panels. Suspension systems have three structural classifications, as described below:

- Light-duty systems—Used primarily for residential and light commercial structures where ceiling loads other than acoustical tile or lay-in panels are not anticipated.
- Intermediate-duty systems—Used primarily for ordinary commercial structures where some

ceiling loads, due to light fixtures and air diffusers, are anticipated.
- Heavy-duty systems—Used primarily for commercial structures in which the quantities and weights of ceiling fixtures (lights, air diffusers, etc.) are greater than those for an ordinary commercial structure.
- ASTM C 636 covers the installation of suspension systems for acoustical tile and lay-in panels.

While the practices described in this standard have equal application to fire-resistance-rated suspension systems, additional requirements may be imposed to obtain the fire endurance classification of particular floor/ceiling or roof/ceiling assemblies (see Section 808.1.1.2).

808.1.1.2 Fire-resistance-rated construction. Acoustical ceiling systems that are part of fire-resistance-rated construction shall be installed in the same manner used in the assembly tested and shall comply with the provisions of Chapter 7.

❖ A suspended acoustical ceiling system by itself does not possess a fire-resistance rating. Acoustical ceiling systems are part of an assembly that has a specified fire-resistance rating. Acoustical lay-in panels must remain in place for the tested assembly to be effective.

Bibliography

The following resource materials are referenced in this chapter or are relevant to the subject matter addressed in this chapter.

ASTM E 84-07, *Test Method for Surface-burning Characteristics of Building Materials*. West Conshohocken, PA: ASTM International, 2007.

Building Materials Directory. Northbrook, IL: Underwriters Laboratories Inc., 2009.

Christian, W.J., and T.E. Waterman. "Flame Spread in Corridors: Effects of Location and Area of Wall Finish." *NFPA Fire Journal*, July 1971.

Cohn, B.M., "Plastics and Rubber," Section 3, Chapter 10, *Fire Protection Handbook*. Quincy, MA: National Fire Protection Association, 1991.

DOC FF-1 (CPSC 16 CFR, Part 1630-07), *Standard for the Surface Flammability of Carpets and Rugs*. Bethesda, MD: Consumer Product Safety Commission, 2007.

Evaluation Plan for Determination of Flood Resistance of Building Elements. ICC Evaluation Service, Inc., 2000.

Evans, D. "Unique Interiors on the Las Vegas Strip." *Fire Protection Engineering*, pages 6-14. Society of Fire Protection Engineers, Fall 2004.

FEMA FIA-TB #2, *Flood-resistant Material Requirements for Buildings Located in Special Flood Hazard*

Areas. Washington, DC: Federal Emergency Management Agency, 1993.

FEMA FIA-TB #7, *Wet Floodproofing Requirements for Structures Located in Special Flood Hazard Areas*. Washington, DC: Federal Emergency Management Agency, 1993.

FEMA FIA-TB #8, *Corrosion Protection for Metal Connectors in Coastal Areas for Structures Located in Special Flood Hazard Areas*. Washington, DC: Federal Emergency Management Agency, 1996.

Fisher, F.L., B. McCracken and R.B. Williamson. "Room Fire Experiments of Textile Wall Coverings, Final Report of All Materials Tested Between March 1985 and January 1986." *Service to Industry Report*. ES-7853, Report 86-2, p.142. Berkley, CA: California University, March 1986.

"Flammable Decorations, Lack of Exits Create Tragedy at Cocoanut Grove." *Fire Engineering*, August 1977.

FM 4880 (2005), *American National Standard for Evaluating Insulated Wall or Wall and Roof/Ceiling Assemblies, Plastic Interior Finish Materials, Plastic Exterior Building Panels, Wall/Ceiling Coating Systems, Interior and Exterior Finish Systems*. Johnston, RI: Factory Mutual Global Research, 2005.

Hirschler, M.M. "Fire Testing of Interior Finish," *Fire Protection Engineering*, pages 16-22. Society of Fire Protection Engineers, Fall 2004.

IFC-09, *International Fire Code*. Washington, DC:, International Code Council, 2009.

"Looking Back at the Cocoanut Grove." *Fire Journal*. November, 1982.

NFIP Technical Bulletin Series. Washington, DC: National Flood Insurance Program. [Online]. Available: www.fema.gov//library.

NFPA 13-07, *Installation of Sprinkler Systems*. Quincy, MA: National Fire Protection Association, 2007.

NFPA 13R-07, *Installation of Sprinkler Systems in Residential Occupancies Up to and Including Four Stories in Height*. Quincy, MA: National Fire Protection Association, 2007.

NFPA 253-06, *Test for Critical Radiant Flux of Floor Covering Systems Using a Radiant Heat Energy Source*. Quincy, MA: National Fire Protection Association, 2006.

NFPA 265-07, *Standard Fire Tests for Evaluating Room Fire Growth Contribution of Textile Wall Coverings*. Quincy, MA: National Fire Protection Association, 2007.

NFPA 286-06, *Standard Methods of Fire Tests for Evaluating Contribution of Wall and Ceiling Interior Finish to Room Fire Growth*. Quincy, MA: National Fire Protection Association, 2006.

NFPA 701-2004, *Standard Methods of Fire Tests for Flame Propagation of Textiles and Films*. Quincy, MA: National Fire Protection Association, 2004.

Quintiere, J. "Surface Flame Spread," Section 2, Chapter 14. *SFPE Handbook of Fire Protection Engineering*. Boston, MA: Society of Fire Protection Engineers, 1995.

UL 1040-96, *Fire Test of Insulated Wall Construction—with Revisions through June 2001*. Northbrook, IL: Underwriters Laboratories Inc., 1996.

UL 723-03, *Standard for Test for Surface Burning Characteristics of Building Materials—with Revisions through May 2005*. Northbrook, IL: Underwriters Laboratories Inc., 2003.

UL 1715-97, *Fire Test of Interior Finish Material—with Revisions through March 2004*. Northbrook, IL: Underwriters Laboratories Inc., 1997.

UL 1975-06, *Fire Test of Foamed Plastics Used for Decorative Purposes*. Northbrook, IL: Underwriters Laboratories Inc., 2006.

Chapter 9:
Fire Protection Systems

General Comments

The provisions required by Chapter 9 are just one aspect of the overall fire protection system of a building or structure. All fire protection requirements contained in the code must be considered as a package or overall system. Noncompliance with any part of the overall system may cause other parts of the system to fail. Failure to install the systems in accordance with code provisions may result in an increased loss of life and property due to a reduction in the level of protection provided.

It is important that every effort be made to verify the proper design and installation of a given fire protection system, especially those that result in construction alternatives and other code trade-offs.

The requirements found in Chapter 9 can be considered active fire safety provisions. They are provisions directed at containing and abating the fire once it has erupted. This chapter is almost a direct copy of Chapter 9 in the *International Fire Code®* (IFC®). The IFC, however, contains specific provisions that are only applicable to existing buildings. The IFC also contains periodic testing criteria that are not duplicated in this chapter. Proper testing, inspection and maintenance of the vari-ous systems, however, are critical to establish the reliability of the system. Additionally, Chapter 9 references and adopts numerous National Fire Protection Association (NFPA) standards, including the acceptance testing criteria within the standard. The referenced standards will also contain more specific design and installation criteria than are found in this chapter. As noted in Section 102.4, where differences occur between code provisions and the referenced standard, the code provisions apply.

Purpose

Fire protection systems may serve one or more purposes in providing adequate protection from fire and hazardous material exposure. The purpose of Chapter 9 is to prescribe the minimum requirements for an active system or systems of fire protection to perform the following functions: to detect a fire; to alert the occupants or fire department of a fire emergency; to control smoke and to control or extinguish the fire. Generally, the requirements are based on the occupancy and height and area of the building, as these are the factors that most affect fire-fighting capabilities and the relative hazard of a specific space or area.

SECTION 901
GENERAL

901.1 Scope. The provisions of this chapter shall specify where *fire protection systems* are required and shall apply to the design, installation and operation of *fire protection systems*.

❖ Chapter 9 contains requirements for fire protection systems that may be provided in a building. These include automatic suppression systems; standpipe systems; fire alarm and detection systems; smoke control systems; smoke and heat vents and portable fire extinguishers and emergency alarm systems. Besides indicating the conditions under which respective systems are required, this chapter contains the design, installation, maintenance and operational criteria for fire protection systems. While the chapter requires proper maintenance for the reliability of the systems, the actual maintenance provisions (periodic testing, inspections and maintenance) are found in the IFC.

This chapter also addresses the requirements for fire command centers, fire department connections, fire pumps and emergency radio systems. These features all directly relate to the proper function of fire protection systems.

901.2 Fire protection systems. *Fire protection systems* shall be installed, repaired, operated and maintained in accordance with this code and the *International Fire Code*.

Any *fire protection system* for which an exception or reduction to the provisions of this code has been granted shall be considered to be a required system.

Exception: Any *fire protection system* or portion thereof not required by this code shall be permitted to be installed for partial or complete protection provided that such system meets the requirements of this code.

❖ The fire protection system is an integral component of the protection features of the building and is required to be properly installed, repaired, operated and maintained in accordance with the code and the IFC. Improperly installed or maintained systems can negate any anticipated protection and, in fact, create a hazard in itself.

While the code may not require a protection system for a specific building or portion thereof, due to its occupancy, the fire protection system would still be considered a required system if some other code trade-off, exception or reduction was taken based on the installation of the fire protection system. For example, a typical small office building may not require an automatic sprinkler system solely due to its Group B occupancy classification; however, if an exit access corridor fire-resistance-rating reduction is taken in accordance with Table 1018.1 for buildings equipped throughout with an NFPA 13 sprinkler system, that sprinkler system would be considered a required system.

The exception acknowledges that a building owner or designer may elect to install a fire protection system that is not required by the code. Even though such a system is not required, it must comply with the applicable provisions of this chapter and the IFC. This requirement is predicated on the concept that any fire protection system not installed in accordance with the code is intrinsically lacking because it could give a false impression of properly installed protection.

For example, if a building owner chooses to provide sprinkler protection in a certain area and such protection is not required by the code, the system must be installed in accordance with the applicable NFPA sprinkler standard (13,13R or 13D) and other applicable requirements of the code, such as water supply and supervision. The extent of the protection thus provided would not be regulated.

901.3 Modifications. No person shall remove or modify any *fire protection system* installed or maintained under the provisions of this code or the *International Fire Code* without approval by the *building official*.

❖ This section emphasizes the principle that systems installed and maintained in compliance with the codes and standards in effect at the time they were placed in service must remain in an operative condition at all times. Protection must not be diminished in any existing building except for the purpose of conducting tests, maintenance or repairs. The length of service interruptions should be kept to a minimum. The building official must be notified of any service interruptions and should carefully evaluate the continued operation or occupancy of buildings and structures where protection is interrupted.

901.4 Threads. Threads provided for fire department connections to sprinkler systems, standpipes, yard hydrants or any other fire hose connection shall be compatible with the connections used by the local fire department.

❖ Incompatible fire service threads have been a major problem throughout the history of modern fire suppression. Efforts to standardize fire service threads in the United States began as early as 1898. Following the Great Baltimore Fire of 1904, NFPA adopted a national standard for fire hose thread with diameters $2^1/_2$ inches (64 mm) and larger; however, to this day many jurisdictions continue to use their own (historic) thread standards. A number of jurisdictions have pioneered the use of nonthreaded, quarter-turn or "Storz" fire hose couplings in place of the national and local traditional thread standards.

901.5 Acceptance tests. *Fire protection systems* shall be tested in accordance with the requirements of this code and the *International Fire Code*. When required, the tests shall be conducted in the presence of the *building official*. Tests required by this code, the *International Fire Code* and the standards listed in this code shall be conducted at the expense of the owner or the owner's representative. It shall be unlawful to occupy portions of a structure until the required *fire protection systems* within that portion of the structure have been tested and *approved*.

❖ All fire protection systems are to be subjected to an acceptance test to determine that the system will operate in the manner required by the code. Acceptance tests are usually part of the final inspection procedures required by Section 110.3.10. Specific acceptance test procedures are provided in other sections as well as the referenced standards. In most instances, the acceptance test procedures require 100-percent operation of the testable system components to determine that they are functioning as required. Often, the design professional may require additional testing to verify that the system operates as designed, which may be beyond the code requirements.

The inclusion of the requirement for acceptance tests in the code is not intended to assign responsibility for witnessing the tests. The responsibility to witness the acceptance test is an administrative issue that each municipality must address. Because the acceptance test is critical during design and construction and is a requirement of occupancy, the requirement is located in the code. The section also clarifies that it is the owner's responsibility to conduct the test and the role of the building official to witness—not conduct—the test. Typically, the owner will assign the responsibility to the installing contractor.

Access to all flow test connections, valves and points of fluid discharge required by other sections is to be provided at locations acceptable to the building official. Consideration should be given to the location of sprinkler system inspector test connections and the main drain valves, as well as to the valves necessary to test standpipe and fire pump installations.

Consideration should also be given to the discharge of water from the various test connections for aqueous systems that should be arranged to discharge to the outside or to a drain with sufficient capacity to handle the anticipated water flow. If the test connection is piped to the outside, then protection should be furnished to prevent unnecessary damage to the property by providing splash blocks or similar protection. When the building drainage system is used, the indirect waste provisions of Chapter 7 of the *International Plumbing Code®* (IPC®) are to apply.

Partial occupancy of any structure should not be permitted unless all fire protection systems for the occupied areas have been tested and approved. Even so, the code assumes that full protection for all areas will be provided as expeditiously as possible. The installation of many fire protection systems and the associated code trade-offs permitted for a given occupancy assume complete building protection, not just in the occupied areas. Final approval of all partial occupancy conditions is subject to the building official.

901.6 Supervisory service. Where required, *fire protection systems* shall be monitored by an supervising station in accordance with NFPA 72.

❖ All required fire protection systems are to be supervised as a means of determining at any time that the system is operational. Acceptable methods of supervision are provided in NFPA 72.

901.6.1 Automatic sprinkler systems. Automatic sprinkler systems shall be monitored by an *approved* supervising station.

Exceptions:

1. A supervising station is not required for *automatic sprinkler systems* protecting one- and two-family dwellings.

2. Limited area systems serving fewer than 20 sprinklers.

❖ This section highlights that automatic sprinkler systems must be monitored by an approved supervising station. All water supply control valves and waterflow switches are required to be electrically supervised as indicated in Section 903.4. A central station, remote supervising station or proprietary supervising station are approved services recognized in NFPA 72 (see commentary, Section 903.4.1).

Automatic sprinkler systems in one- and two-family dwellings are typically designed in accordance with NFPA 13D, which does not require electrical supervision (see Exception 1). Limited area sprinkler systems are generally supervised by their connection to the domestic water service (see Exception 2). As such, limited area sprinkler systems with fewer than 20 sprinkler heads are not required to have local alarms or a fire department connection.

901.6.2 Fire alarm systems. Fire alarm systems required by the provisions of Section 907.2 of this code and Sections 907.2 and 907.3 of the *International Fire Code* shall be monitored by an *approved* supervising station in accordance with Section 907.6.5.

Exceptions:

1. Single- and multiple-station smoke alarms required by Section 907.2.11.

2. Smoke detectors in Group I-3 occupancies.

3. Supervisory service is not required for *automatic sprinkler systems* in one- and two-family dwellings.

❖ Fire alarm systems are required to automatically transmit an alarm and any trouble signals to the fire department through one of the approved methods in NFPA 72. This includes systems in new buildings where required by Section 907.2 of both the code and the IFC and in existing buildings where required by Section 907.3 of the IFC.

Exception 1 exempts single- and multiple-station smoke alarms from being supervised due to the poten-

tial for unwanted false alarms. Similarly, due to the concern over unwanted alarms, smoke detectors in Group I-3 occupancies need only sound an approved alarm signal, which automatically notifies staff (see Section 907.2.6.3.1). Smoke detectors in Group I-3 occupancies are typically subject to misuse and abuse. Frequent unwanted alarms would negate the effectiveness of the system (see Exception 2).

Exception 3 clarifies that sprinkler systems in one- and two-family dwellings are not part of a dedicated fire alarm system and are typically designed in accordance with NFPA 13D, which does not require electrical supervision.

901.6.3 Group H. Manual fire alarm, automatic fire-extinguishing and emergency alarm systems in Group H occupancies shall be monitored by an *approved* supervising station.

Exception: When *approved* by the *building official*, on-site monitoring at a *constantly attended location* shall be permitted provided that notifications to the fire department will be equal to those provided by an *approved* supervising station.

❖ Given the varied nature and quantity of hazardous materials that could be present, all fire protection systems in Group H occupancies are required to be monitored by an approved supervising station. The exception, however, permits the building official to approve an on-site monitoring system. Many companies that routinely deal with hazardous materials employ their own fire brigades and have emergency response procedures in place.

901.7 Fire areas. Where buildings, or portions thereof, are divided into *fire areas* so as not to exceed the limits established for requiring a *fire protection system* in accordance with this chapter, such *fire areas* shall be separated by *fire barriers* constructed in accordance with Section 707 or *horizontal assemblies* constructed in accordance with Section 712, or both, having a *fire-resistance rating* of not less than that determined in accordance with Section 707.3.9.

❖ This section provides specific guidance on how a building needs to be divided into fire areas in order to avoid the required installation of a fire protection system. A single occupancy group would require fire barriers or horizontal assemblies in order to create multiple fire areas (see Table 707.3.9), each having an area below the threshold for fire protection system installation.

SECTION 902
DEFINITIONS

902.1 Definitions. The following words and terms shall, for the purposes of this chapter, and as used elsewhere in this code, have the meanings shown herein.

❖ Definitions of terms can help in the understanding and application of the code requirements. These definitions are included within this chapter to provide more convenient access to them without having to refer back to Chapter 2. For convenience, these terms are also listed in Chapter 2 with a cross reference to this section. The use and application of all defined terms, including those defined here, are set forth in Section 201.

Certain requirements in the code are based on code provisions in the IFC, *International Mechanical Code®* (IMC®) and IPC. A review of definitions included in those codes will aid in the understanding of many of the requirements contained in the code.

[F] ALARM NOTIFICATION APPLIANCE. A fire alarm system component such as a bell, horn, speaker, light or text display that provides audible, tactile or visible outputs, or any combination thereof.

❖ The code requires that fire alarm systems be equipped with approved alarm notification appliances so that in an emergency, the fire alarm system will notify the occupants of the need for evacuation or implementation of the fire emergency plan. Alarm notification devices required by the code are of two general types: visible and audible. Except for emergency voice/alarm communication systems, once the fire alarm system has been activated, all visible and audible communication alarms are required to activate. Emergency voice/ alarm communication systems are special signaling systems that are activated selectively in response to specific emergency conditions but have the capability to be activated throughout the building if necessary.

[F] ALARM SIGNAL. A signal indicating an emergency requiring immediate action, such as a signal indicative of fire.

❖ This is a general term for all types of supervisory and trouble signals. An example would be a supervisory (tamper) switch on a sprinkler control valve. It could also be the response to a specific device that is not part of the alarm notification system but that causes a specific function such as a smoke detector for elevator recall. The activation of the device does not necessarily indicate that there is a fire; however, the level of protection may have been compromised (see the definition of "Fire alarm signal").

[F] ALARM VERIFICATION FEATURE. A feature of automatic fire detection and alarm systems to reduce unwanted alarms wherein smoke detectors report alarm conditions for a minimum period of time, or confirm alarm conditions within a given time period, after being automatically reset, in order to be accepted as a valid alarm-initiation signal.

❖ False fire (evacuation) alarms are a nuisance. For this reason the code specifies that alarms activated by smoke detectors are not to be sounded until the alarm signal is verified by cross-zoned detectors in a single protected area or by system features that will retard the alarm until the signal is determined to be valid. Valid alarm initiation signals can be determined by detectors that report alarm conditions for a minimum period of time or that, after being reset, continue to report an alarm condition. The alarm verification feature may not retard signal activation for a period of more than 60 seconds and must not apply to alarm-initi-

tiating devices other than smoke detectors (which may be connected to the same circuit). Alarm verification is not the same as presignal features that delay an alarm signal for more than 1 minute and that are allowed only where specifically permitted by the authority having jurisdiction.

[F] ANNUNCIATOR. A unit containing one or more indicator lamps, alphanumeric displays or other equivalent means in which each indication provides status information about a circuit, condition or location.

❖ This refers to the panel that displays the status of the monitored fire protection systems and devices. It is not the fire alarm control unit; however, the control panel may include an annunciator.

[F] AUDIBLE ALARM NOTIFICATION APPLIANCE. A notification appliance that alerts by the sense of hearing.

❖ Audible alarms that are part of a fire alarm system must be loud enough to be heard in every occupied space of a building. Section 907.5.2.1.1 prescribes the minimum sound pressure level for all audible alarm notification appliances depending on the occupancy of the building and the function of the space.

[F] AUTOMATIC. As applied to fire protection devices, a device or system providing an emergency function without the necessity for human intervention and activated as a result of a predetermined temperature rise, rate of temperature rise or combustion products.

❖ This term, when used in conjunction with fire protection systems or devices, means that the system or device will perform its intended function without a person being present or performing any task in its control or operation. The device or system has the inherent capability to detect a developing fire condition and perform some predetermined function. Automatic devices and systems operate completely without human presence or intervention.

[F] AUTOMATIC FIRE-EXTINGUISHING SYSTEM. An *approved* system of devices and equipment which automatically detects a fire and discharges an *approved* fire-extinguishing agent onto or in the area of a fire.

❖ This term is the generic name for all types of automatic fire-extinguishing systems, including the most common type—the automatic sprinkler system. See Section 904 for requirements for particular alternative automatic fire-extinguishing systems, such as wet-chemical, dry-chemical, foam, carbon dioxide, halon and clean-agent systems.

[F] AUTOMATIC SMOKE DETECTION SYSTEM. A fire alarm system that has initiation devices that utilize smoke detectors for protection of an area such as a room or space with detectors to provide early warning of fire.

❖ This definition, new in the 2009 edition of the code, is part of a complete Section 907 reorganization and clarification effort that has taken place since the previous edition. Part of that effort was to improve the correlation of terminology used in Section 907 with respect to the terms "automatic fire detection system," "automatic fire alarm system" and "automatic smoke detection system." All of those terms were used previously in the section and caused confusion as to exactly what the code intended to require. Especially problematic was the fact that an "automatic fire detection system" was sometimes being interpreted as including an automatic sprinkler system, the sprinkler heads being "automatic heat detectors," which was never the intent of the section.

[F] AUTOMATIC SPRINKLER SYSTEM. An automatic sprinkler system, for fire protection purposes, is an integrated system of underground and overhead piping designed in accordance with fire protection engineering standards. The system includes a suitable water supply. The portion of the system above the ground is a network of specially sized or hydraulically designed piping installed in a structure or area, generally overhead, and to which automatic sprinklers are connected in a systematic pattern. The system is usually activated by heat from a fire and discharges water over the fire area.

❖ An automatic sprinkler system is one type of automatic fire-extinguishing system. Automatic sprinkler systems are the most common, and their life safety attributes are widely recognized. The code specifies three types of automatic sprinkler systems: one installed in accordance with NFPA 13, one in accordance with NFPA 13R and the other in accordance with NFPA 13D. To be considered for most code design alternatives, a building automatic sprinkler system must be installed throughout in accordance with NFPA 13 (see Section 903.3.1.1).

In a fire, sprinklers automatically open and discharge water onto the fire in a spray pattern that is designed to contain or extinguish the fire. Originally, automatic sprinkler systems were developed just for the protection of buildings and their contents. Because of the development and improvements in sprinkler head response time and water distribution, however, automatic sprinkler systems are now also considered a life safety system. Proper operation of an automatic sprinkler system requires careful selection of the sprinkler heads so that water in sufficient quantity at adequate pressure and properly distributed will be available to suppress the fire. Note that the context of the use of the term "fire area" in the last sentence of the definition is to refer to the area in which the fire is occurring, not in the context of the defined term "fire area."

There are many different types of automatic sprinkler systems—wet pipe, dry pipe, preaction, antifreeze and various combinations. Sprinklers can be pendant, upright or sidewall and can be designed for standard or extended coverage. Additional information can be found in NFPA 13.

[F] AVERAGE AMBIENT SOUND LEVEL. The root mean square, A-weighted sound pressure level measured over a 24-hour period, or the time any person is present, whichever time period is less.

❖ The ambient noise that can be expected depends on the occupancy of the building. To attract the attention of the occupants, the audible alarm devices must be heard above the ambient noise in the space. For this reason, the alarm devices must have minimum sound pressure levels above the average ambient sound level. Section 907.5.2.1.1 prescribes the minimum sound pressure levels for the audible alarm notification appliances for all occupancy conditions.

Although it is possible to measure the ambient sound within an occupied space, the alarm notification devices are usually designed and installed before buildings are occupied, thus it is typically a careful analysis of the types of uses within a space that will determine the average ambient sound level. If, after the building is occupied, the alarm notification devices are below expected audibility, a field measurement may be necessary to determine whether or not the design assumptions are correct.

[F] CARBON DIOXIDE EXTINGUISHING SYSTEMS. A system supplying carbon dioxide (CO_2) from a pressurized vessel through fixed pipes and nozzles. The system includes a manual- or automatic-actuating mechanism.

❖ Carbon dioxide (CO_2) extinguishing systems are useful in extinguishing fires in specific hazards or equipment in occupancies where an inert electrically nonconductive medium is essential or desirable and where cleanup of other extinguishing agents, such as dry-chemical residue, presents a problem. The system works by displacing the oxygen in an enclosed area by flooding the space with carbon dioxide. To effectively flood the enclosure, automatic door and window closers and control dampers for the mechanical ventilation system must be installed.

These types of gaseous extinguishing systems have some inherent disadvantages that should be considered before selection. Because the oxygen is being displaced, occupants should not be in the space for a period after discharge, depending on the concentration of CO_2 to be achieved. Additionally, the discharge rate can result in a rapid increase in pressure within the space where the system is discharged. However, where water is not a desired means of suppression carbon dioxide and other gaseous suppression systems can be very effective. NFPA 12 contains minimum requirements for the design, installation, testing, inspection, approval, operation and maintenance of carbon dioxide extinguishing systems.

[F] CEILING LIMIT. The maximum concentration of an air-borne contaminant to which one may be exposed, as published in DOL 29 CFR Part 1910.1000.

❖ This term is used in Section 908.3 and indicates the threshold at which a gas detection system is required for highly toxic and toxic materials. It represents the maximum level of exposure for employees or occupants to hazardous air contaminants during any part of a normal workday. DOL 29 CFR Part 1910.1000 provides acceptable ceiling limits of contamination for various substances.

[F] CLEAN AGENT. Electrically nonconducting, volatile or gaseous fire extinguishant that does not leave a residue upon evaporation.

❖ The two categories of clean agents are halocarbon compounds and inert gas agents. Halocarbon compounds include bromine, carbon, chloride, fluorine, hydrogen and iodine. Halocarbon compounds suppress the fire through a combination of breaking the chemical chain reaction of the fire, reducing the ambient oxygen supporting the fire and reducing the ambient temperature of the fire origin to reduce the propagation of fire. The clean agents that are inert gas agents contain primary components consisting of helium, neon or argon, or a combination of all three. Inert gases work by reducing the oxygen concentration around the fire origin to a level that does not support combustion (see commentary, Section 904.10).

[F] CONSTANTLY ATTENDED LOCATION. A designated location at a facility staffed by trained personnel on a continuous basis where alarm or supervisory signals are monitored and facilities are provided for notification of the fire department or other emergency services.

❖ These locations are intended to receive trouble, supervisory and fire alarm signals transmitted by the fire protection equipment installed within a protected facility. It is the intent of this section to have both an approved location and personnel who are acceptable to the authority having jurisdiction responsible for actions taken when the fire protection system requires attention. The term "constantly attended" implies 24-hour surveillance of the system, at the designated location.

[F] DELUGE SYSTEM. A sprinkler system employing open sprinklers attached to a piping system connected to a water supply through a valve that is opened by the operation of a detection system installed in the same areas as the sprinklers. When this valve opens, water flows into the piping system and discharges from all sprinklers attached thereto.

❖ A deluge system applies large quantities of water or foam throughout the protected area by means of a system of open sprinklers. In a fire, the system is activated by a fire detection system that makes it possible to apply water to a fire more quickly and to cover a larger area than with a conventional automatic sprinkler system, which depends on sprinklers being activated individually as the fire spreads. As the definition indicates, the sprinklers are open. There is no fusible link so when water is admitted into the system by the fire detection system, it flows through the piping and is immediately discharged through the sprinkler heads.

Deluge systems are particularly beneficial in hazardous areas where the fuel loads (combustible contents) are of such a nature that fire may grow with exceptional rapidity and possibly flash ahead of the operations of conventional automatic sprinklers.

[F] DETECTOR, HEAT. A fire detector that senses heat—either abnormally high temperature or rate of rise, or both.

❖ In a fire, heat is released that causes the temperature in a room or space to increase. Automatic fire detectors that sense abnormally high temperature or rate of temperature rise are known as heat detectors. These include fixed temperature detectors, rate compensation detectors and rate-of-rise detectors. The code requires all automatic fire detectors to be smoke detectors, except that heat detectors tested and approved in accordance with NFPA 72 may be used as an alternative to smoke detectors in rooms and spaces where, during normal operation, products of combustion are present in sufficient quantity to actuate a smoke detector.

[F] DRY-CHEMICAL EXTINGUISHING AGENT. A powder composed of small particles, usually of sodium bicarbonate, potassium bicarbonate, urea-potassium-based bicarbonate, potassium chloride or monoammonium phosphate, with added particulate material supplemented by special treatment to provide resistance to packing, resistance to moisture absorption (caking) and the proper flow capabilities.

❖ A dry-chemical system extinguishes a fire by placing a chemical barrier between the fire and oxygen, which acts to smother a fire. This system is best known for protection for commercial ranges, commercial fryers and exhaust hoods. Wet-chemical extinguishing systems, however, are more commonly used for new installations in commercial cooking equipment.

The type of dry chemical to be used in the extinguishing system is a function of the hazard expected. The type of dry chemical used in a system must not be changed, unless it has been proven changeable by a testing laboratory; is recommended by the manufacturer of the equipment and is acceptable to the fire code official for the hazard expected. Additional guidance on the use of various dry-chemical agents can be found in NFPA 17, which gives minimum requirements for the design, installation, testing, inspection, approval, operation and maintenance of dry-chemical extinguishing systems.

[F] ELEVATOR GROUP. A grouping of elevators in a building located adjacent or directly across from one another that responds to a common hall call button(s).

❖ This definition, new in the 2009 edition of the code, is part of a complete Section 907 reorganization and clarification effort. Part of that effort was to improve the correlation of terminology used in Section 907 to make it clear to the user the correct intent of the code. This definition clarifies the application of the emergency voice/alarm communication system requirements in Section 907.5.2.2 as to the locations, called "paging zones," where system speakers are required (see commentary, Section 907.5.2.2).

[F] EMERGENCY ALARM SYSTEM. A system to provide indication and warning of emergency situations involving hazardous materials.

❖ Because of the potentially volatile nature of hazardous materials, an emergency alarm system is required outside of interior building rooms or areas containing hazardous materials in excess of the maximum allowable quantities permitted in Tables 307.1(1) and 307.1.(1). The intent of the emergency alarm, upon actuation by an alarm-initiating device, is to alert the occupants to an emergency condition involving hazardous materials. The initiation of the emergency alarm can be by manual or automatic means depending on the hazard and the specific requirements for the type of hazard. See Sections 908.

[F] EMERGENCY VOICE/ALARM COMMUNICATIONS. Dedicated manual or automatic facilities for originating and distributing voice instructions, as well as alert and evacuation signals pertaining to a fire emergency, to the occupants of a building.

❖ An emergency voice/alarm communication system is a special feature of fire alarm systems in buildings with special evacuation considerations, such as a high-rise building or a large assembly space. Emergency voice/alarm communication systems automatically communicate a fire emergency message to all occupants of a building on a general or selective basis. Such systems also enable the fire service to manually transmit voice instructions to the building occupants about a fire emergency condition and the action to be taken for evacuation or movement to another area of the building. Although most systems use prerecorded messages, some now use computer synthesized voices to communicate messages that allow customized messages unique to the facility.

[F] FIRE ALARM BOX, MANUAL. This is primarily known as a "manual fire alarm box" and is addressed in the commentary under that term.

[F] FIRE ALARM CONTROL UNIT. A system component that receives inputs from automatic and manual fire alarm devices and may be capable of supplying power to detection devices and transponder(s) or off-premises transmitter(s). The control unit may be capable of providing a transfer of power to the notification appliances and transfer of condition to relays or devices.

❖ The fire alarm control unit (panel) acts as a point where all signals initiated within the protected building are received before the signal is transmitted to a constantly attended location. As the name implies, it also contains controls to test and manually activate or silence systems.

[F] FIRE ALARM SIGNAL. A signal initiated by a fire alarm-initiating device such as a manual fire alarm box, automatic fire detector, waterflow switch or other device whose activation is indicative of the presence of a fire or fire signature.

❖ This signal is transmitted to a fire alarm control unit as a warning that requires immediate action. The person-

nel at the constantly attended location are trained to immediately respond to a fire alarm signal, which indicates the presence of a fire. A fire alarm signal assumes an actual fire has been detected (see the definition of "Alarm signal"). The fire alarm signal is not the signal used to notify the occupants of an emergency condition. Such an action would involve the audible alarm, visual alarm or emergency voice/alarm notification appliances.

[F] FIRE ALARM SYSTEM. A system or portion of a combination system consisting of components and circuits arranged to monitor and annunciate the status of fire alarm or supervisory signal-initiating devices and to initiate the appropriate response to those signals.

❖ Fire alarm systems are installed in buildings to limit fire casualties and property losses by notifying the occupants of the building, the local fire department or both of an emergency condition. The alarm notification appliances associated with fire alarm systems are intended to be evacuation alarms. All fire alarm systems must be designed and installed to comply with NFPA 72. The term is among the most generic terms used in the code. It does not necessarily imply an automatic or manual system nor does it identify what type of notification, if any should be provided. The definition only indicates that an appropriate response must be provided but does not indicate what that response must be. The appropriate responses are identified within the respective sections of Section 907.

[F] FIRE AREA. The aggregate floor area enclosed and bounded by fire walls, *fire barriers*, *exterior walls* or *horizontal assemblies* of a building. Areas of the building not provided with surrounding walls shall be included in the fire area if such areas are included within the horizontal projection of the roof or floor next above.

❖ This term is used to describe a specific and controlled area within a building that may consist of a portion of the floor area within a single story, one entire story or the combined floor area of several stories, depending on how these areas are enclosed and separated from other floor areas. Where a fire barrier with a fire-resistance rating in accordance with Section 707.3.9 divides the floor area of a one-story building, the floor area on each side of the wall would constitute a separate fire area. If a horizontal assembly separating the two stories in a two-story building is fire-resistance rated in accordance with Section 712.3, each story would be a separate fire area. In cases where mezzanines are present, the floor area of the mezzanine is included in the fire area calculations, even though the area of the mezzanine does not contribute to the building area calculations. See the commentary to Sections 707.3.9 and 712.3 for further information.

Note that fire walls are one way of creating fire areas but are typically used to create separate buildings.

[F] FIRE COMMAND CENTER. The principal attended or unattended location where the status of detection, alarm communications and control systems is displayed, and from which the system(s) can be manually controlled.

❖ Fire command centers are communication centers where dedicated manual and automatic facilities are located for the origination, control and transmission of information and instruction pertaining to a fire emergency to the occupants (including fire department personnel) of the building. Fire command centers must provide facilities for the control and display of the status of all fire protection (detection, signaling, etc.) systems. These stations must be located in secure areas as approved by the authority having jurisdiction. Often this is a location near the primary building entrance. Fire command centers also may be combined with other building operations and security facilities when permitted by the authority having jurisdiction; however, operating controls for use by the fire department must be clearly marked.

[F] FIRE DETECTOR, AUTOMATIC. A device designed to detect the presence of a fire signature and to initiate action.

❖ Automatic fire detectors include all approved devices designed to detect the presence of a fire and automatically initiate emergency action. These include smoke-sensing fire detectors, heat-sensing fire detectors, flame-sensing fire detectors, gas-sensing fire detectors and other fire detectors that operate on other principles as approved by the fire code official. Automatic fire detectors must be selected based on the type and size of fire to be detected and the response required. The automatic fire detector sends a signal to a processing unit to initiate some predetermined action. The processing unit may be internal to the device as is the case with single-station smoke detectors or it may be an external unit as in the case of a fire alarm control unit. Automatic fire detectors must be approved, installed and tested to comply with the code and NFPA 72.

[F] FIRE PROTECTION SYSTEM. *Approved* devices, equipment and systems or combinations of systems used to detect a fire, activate an alarm, extinguish or control a fire, control or manage smoke and products of a fire or any combination thereof.

❖ A fire protection system is any approved device or equipment used singly or in combination, either manually or automatically, and that is intended to detect a fire, notify the building occupants of a fire or suppress the fire. Fire protection systems include fire suppression systems, standpipe systems, fire alarm systems, fire detection systems, smoke control systems and smoke vents. All fire protection systems must be approved by the fire code official and tested in accordance with the referenced standards and Section 901.6.

[F] FIRE SAFETY FUNCTIONS. Building and fire control functions that are intended to increase the level of life safety for occupants or to control the spread of harmful effects of fire.

❖ In many cases automatic fire detectors are installed even in buildings not required to have a fire alarm sys-

tem. These fire detectors perform specific functions such as releasing door hold-open devices, activating elevator recall, smoke damper activation or air distribution system shutdown (see Section 907.3).

[F] FOAM-EXTINGUISHING SYSTEM. A special system discharging a foam made from concentrates, either mechanically or chemically, over the area to be protected.

❖ Foam-extinguishing systems must be of an approved type and installed and tested to comply with NFPA 11, 11A and 16. All foams are intended to exclude oxygen from the fire, cool the area of the fire and insulate adjoining surfaces from heat caused by fires. Foam systems are commonly used to extinguish flammable or combustible liquid fires (see commentary, Section 904.7). While water applied by an automatic sprinkler system can only act horizontally upon the surface that it reaches, foam-extinguishing agents have the ability to act vertically in addition to horizontally; and, unlike gaseous extinguishing agents, foam does not dissipate rapidly where there is no confined space. Thus, foam systems are also used where there is a need to fill a nonconfined space with extinguishing material as in the case of certain industrial applications.

[F] HALOGENATED EXTINGUISHING SYSTEM. A fire-extinguishing system using one or more atoms of an element from the halogen chemical series: fluorine, chlorine, bromine and iodine.

❖ Halon is a colorless, odorless gas that inhibits the chemical reaction of fire. Halon extinguishing systems are useful in occupancies such as computer rooms where an electrically nonconductive medium is essential or desirable and where cleanup of other extinguishing agents presents a problem. The halon extinguishing system must to be of an approved type and installed and tested to comply with NFPA 12A.

Halon extinguishing agents have been identified as a source of emissions resulting in the depletion of the stratospheric ozone layer. For this reason, production of new supplies of halon has been phased out. Alternative gaseous extinguishing agents, such as clean agents, have been developed as alternatives to halon.

[F] INITIATING DEVICE. A system component that originates transmission of a change-of-state condition, such as in a smoke detector, manual fire alarm box or supervisory switch.

❖ All fire protection systems consist of devices, which upon use or actuation, will initiate the intended operation. A manual fire alarm box, for example, upon actuation will transmit a fire alarm signal. In the case of a single-station device, the initiating device and the notification appliance are one in the same.

[F] MANUAL FIRE ALARM BOX. A manually operated device used to initiate an alarm signal.

❖ Manual fire alarm boxes are commonly known as pull stations. Manual fire alarm boxes include all manual devices used to activate a manual fire alarm system and have many configurations, depending on the manufacturer. All manual fire alarm devices, however, must be approved and installed in accordance with NFPA 72 for the particular application. Manual fire alarm boxes may be combined in guard tour boxes.

[F] MULTIPLE-STATION ALARM DEVICE. Two or more single-station alarm devices that are capable of interconnection such that actuation of one causes all integral or separate audible alarms to operate. It also can consist of one single-station alarm device having connections to other detectors or to a manual fire alarm box.

❖ This definition refers to a combination of similar or different types of alarm devices that could be interconnected. The actuation of any two devices, whether a smoke detector or manual fire alarm box, will activate the required audible alarms at all interconnected devices.

[F] MULTIPLE-STATION SMOKE ALARM. Two or more single-station alarm devices that are capable of interconnection such that actuation of one causes the appropriate alarm signal to operate in all interconnected alarms.

❖ In occupancies with sleeping areas, occupants must be notified in a fire so that they can promptly evacuate the premises. In accordance with the requirements of NFPA 72, multiple-station smoke alarms are self-contained, smoke-activated alarm devices built in accordance with UL 217 that can be interconnected with other devices so that all integral or separate alarms will operate when any one device is activated.

[F] NOTIFICATION ZONE. See "Zone, notification."

❖ This term is more commonly known as "Zone, notification." See also commentary for "Zone, notification."

[F] NUISANCE ALARM. An alarm caused by mechanical failure, malfunction, improper installation or lack of proper maintenance, or an alarm activated by a cause that cannot be determined.

❖ A nuisance alarm is essentially any alarm that occurs as a result of a condition that does not arise during the normal operation of the equipment. A nuisance alarm is not the same as a false alarm. A person who intentionally initiates an alarm by using a manual pull station or a person who accidentally initiates a smoke detector is not initiating a nuisance alarm. A nuisance alarm is, by nature, a factor of the system itself.

[F] RECORD DRAWINGS. Drawings ("as builts") that document the location of all devices, appliances, wiring sequences, wiring methods and connections of the components of a fire alarm system as installed.

❖ To verify that the system has been installed to comply with the code and applicable referenced standards, complete as-built drawings of the fire alarm system must be available on site for review.

[F] SINGLE-STATION SMOKE ALARM. An assembly incorporating the detector, the control equipment and the alarm-sounding device in one unit, operated from a power supply either in the unit or obtained at the point of installation.

❖ A single-station smoke alarm is a self-contained alarm device that detects visible or invisible particles of com-

bustion. Its function is to detect a fire in the immediate area of the detector location. Single-station smoke alarms are individual units with the capability to stand alone. Where the single-station smoke alarms are interconnected with other single-station devices, they would be considered a multiple-station smoke alarm system. Single-station smoke alarms are not capable of notifying or controlling any other fire protection equipment or systems. They may be battery powered, directly connected to the building power supply or a combination of both. Single-station smoke alarms must be built to comply with UL 217 and are to be installed as required by Section 907.2.11.

[F] SMOKE ALARM. A single- or multiple-station alarm responsive to smoke.

❖ This is a general term that applies to both single- and multiple-station smoke alarms that are not part of an automatic fire detection system. It is the generic term for any device that both detects the products of combustion and initiates an alarm signal for occupant notification.

[F] SMOKE DETECTOR. A *listed* device that senses visible or invisible particles of combustion.

❖ These devices are considered early warning devices and have saved many people from smoke inhalation and burns. Smoke detectors have a wide range of uses, from sophisticated fire detection systems for industrial and commercial uses to residential. A smoke detector is a device, typically listed in accordance with UL 268, that activates a fire alarm system. These system smoke detectors contain only the components required to detect the products of combustion and activate a fire alarm system and are, therefore, different from single- and multiple-station smoke alarms.

Smoke detectors typically consist of two types: ionization and photoelectric. An ionization detector contains a small amount of radioactive material that ionizes the air in a sensing chamber and causes a current to flow through the air between two charged electrodes. When smoke enters the chamber, the particles cause a reduction in the current. When the level of conductance decreases to a preset level, the detector responds with an alarm.

A photoelectric smoke detector consists primarily of a light source, a light beam and a photosensitive device. When smoke particles enter the light beam, they reduce the light intensity in the photosensitive device. When obscuration reaches a preset level, the detector initiates an alarm.

SMOKEPROOF ENCLOSURE. An exit stairway designed and constructed so that the movement of the products of combustion produced by a fire occurring in any part of the building into the enclosure is limited.

❖ A smokeproof enclosure is intended to provide an effective barrier to the entry of smoke into an exit stairway, thereby offering an additional level of protection for occupants of high-rise and underground structures.

[F] STANDPIPE SYSTEM, CLASSES OF. Standpipe classes are as follows:

❖ A standpipe system is typically an arrangement of vertical piping located in exit stairways that allows fire-fighting personnel to connect hand-carried hoses at each level to manually extinguish fires. Section 905 and NFPA 14 recognize three different classes of standpipe systems. For a further discussion of standpipe classes and types, see the commentary to Section 905.

Class I system. A system providing 2^1/$_2$-inch (64 mm) hose connections to supply water for use by fire departments and those trained in handling heavy fire streams.

❖ A Class I standpipe system is intended for use by trained fire service personnel as a readily available water source for manual fire-fighting operations. A Class I standpipe system is equipped with only 2^1/$_2$-inch (64 mm) hose connections to allow the fire service to attach the appropriate hose and nozzles. A Class I standpipe system is not equipped with hose stations, which include a cabinet, hose and nozzle.

Class II system. A system providing 1^1/$_2$-inch (38 mm) hose stations to supply water for use primarily by the building occupants or by the fire department during initial response.

❖ A Class II standpipe system is intended for use by building occupants or by the fire department for manual suppression. The hose stations defined in NFPA 14 as part of the Class II standpipe system include a hose rack, hose nozzle, hose and hose connection. The intent of providing the hose is for use by properly trained personnel. Occupant-use hose stations should only be provided where they can be used by people who have been properly trained in the use of the hose and nozzle.

Class III system. A system providing 1^1/$_2$-inch (38 mm) hose stations to supply water for use by building occupants and 2^1/$_2$-inch (64 mm) hose connections to supply a larger volume of water for use by fire departments and those trained in handling heavy fire streams.

❖ A Class III standpipe system is intended for use by building occupants as well as trained fire service personnel. The 1^1/$_2$-inch (38 mm) hose station is for use by the building occupants or fire department for manual fire suppression and the 2^1/$_2$-inch (64 mm) hose connection is intended for use primarily by fire service personnel or those who have received training in the use of the larger hoses. Class III systems allow the fire department to select the types of hose necessary based on the fire hazard present. If the fire is effectively controlled by an automatic sprinkler system, the smaller hose size may be all that is necessary for fire department mop up operations.

[F] STANDPIPE, TYPES OF. Standpipe types are as follows:

❖ Section 905 recognizes five types of standpipe systems. The use of each type of system depends on specific occupancy conditions and the presence of an automatic sprinkler system. For a further discussion of standpipe classes and types, see the commentary to Section 905.3.1.

Automatic dry. A dry standpipe system, normally filled with pressurized air, that is arranged through the use of a device, such as dry pipe valve, to admit water into the system piping automatically upon the opening of a hose valve. The water supply for an automatic dry standpipe system shall be capable of supplying the system demand.

❖ A typical automatic dry standpipe system has an automatic water supply retained by a dry pipe valve. The dry pipe valve clapper is kept in place by air placed in the standpipe system under pressure. Once a standpipe hose valve is opened, the air is released from the system, allowing water to fill the system through the dry pipe valve. This system is traditionally used in areas where the temperature falls below 40°F (4°C); where a wet system could freeze and possibly burst the pipe or simply not be available when needed.

Automatic wet. A wet standpipe system that has a water supply that is capable of supplying the system demand automatically.

❖ An automatic wet standpipe system is used in locations where the entire system would remain above 40°F (4°C). Because the system is pressurized with water, an immediate release of water occurs when a hose connection valve is opened. This is the most generally preferred type of standpipe but it is not necessarily the required type unless so stipulated.

Manual dry. A dry standpipe system that does not have a permanent water supply attached to the system. Manual dry standpipe systems require water from a fire department pumper to be pumped into the system through the fire department connection in order to meet the system demand.

❖ A manual dry standpipe system is filled with water only when the fire service is present. Typically, the fire service connects the discharge from a water source, such as a pumper truck, to the fire department connection of a manual dry standpipe system. When the fire service has suppressed the fire and is preparing to leave, the system is drained of the remaining water. Manual dry standpipe systems are commonly installed in open parking structures.

Manual wet. A wet standpipe system connected to a water supply for the purpose of maintaining water within the system but does not have a water supply capable of delivering the system demand attached to the system. Manual-wet standpipe systems require water from a fire department pumper (or the like) to be pumped into the system in order to meet the system demand.

❖ A manual-wet standpipe system is connected to an automatic water supply, but the supply is not capable of providing the system demand. The manual wet system could be one that is connected with the sprinkler system such that it is capable of supplying the demand for the sprinkler system but not for the standpipe. The standpipe system demand is met when the fire service provides additional water through the fire department connection from the discharge of a water source, such as a pumper truck.

Semiautomatic dry. A dry standpipe system that is arranged through the use of a device, such as a deluge valve, to admit water into the system piping upon activation of a remote control device located at a hose connection. A remote control activation device shall be provided at each hose connection. The water supply for a semiautomatic dry standpipe system shall be capable of supplying the system demand.

❖ This type of dry standpipe is a special design that uses a solenoid-activated valve to retain the automatic water supply. Once the standpipe hose valve is opened, a signal is sent to the deluge valve retaining the automatic water supply to allow water to fill the system. This kind of system is used in areas where the temperature falls below 40°F (4°C), where a wet system would otherwise freeze. As such, there is no semi-automatic wet system type.

[F] SUPERVISING STATION. A facility that receives signals and at which personnel are in attendance at all times to respond to these signals.

❖ The supervising station is the location where all fire protection-system-related signals are sent and where trained personnel are present to respond to an emergency. The supervising station may be an approved central station, a remote supervising station, a proprietary supervising station or other constantly attended location approved by the fire code official. Each type of supervising station must comply with the applicable specific provisions described in NFPA 72.

[F] SUPERVISORY SERVICE. The service required to monitor performance of guard tours and the operative condition of fixed suppression systems or other systems for the protection of life and property.

❖ The supervisory service is responsible for maintaining the integrity of the fire protection system by notifying the supervising station of a change in protection system status.

Guard tours are recognized as a nonrequired (voluntary) system. If a guard tour is provided, the signals from that system can be transmitted through the supervisory service to the supervision station. Guard tours are not a required part of a fire alarm system.

[F] SUPERVISORY SIGNAL. A signal indicating the need of action in connection with the supervision of guard tours, the fire suppression systems or equipment or the maintenance features of related systems.

❖ Activation of a supervisory signal-initiating device transmits a signal indicating that a change in the status of the fire protection system has occurred and that action must be taken. These signals are the basis for the actions taken by the attendant at the supervising station. These signals do not indicate an emergency condition but indicate that a portion of the system is not functioning in the manner in which it should and that if the condition is not corrected it could impair the ability of the fire protection system to perform properly. A supervisory signal is also a part of the nonrequired guard tour system.

[F] SUPERVISORY SIGNAL-INITIATING DEVICE. An initiation device, such as a valve supervisory switch, water-level indicator or low-air pressure switch on a dry-pipe sprinkler system, whose change of state signals an off-normal condition and its restoration to normal of a fire protection or life safety system, or a need for action in connection with guard tours, fire suppression systems or equipment or maintenance features of related systems.

❖ The supervisory signal-initiating device detects a change in protection system status. Examples of a supervisory signal-initiating device include a flow switch to detect movement of water through the system and a tamper switch to detect when someone shuts off a water control valve.

[F] TIRES, BULK STORAGE OF. Storage of tires where the area available for storage exceeds 20,000 cubic feet (566 m³).

❖ This definition describes a storage space that is larger than what would be found in most typical mercantile and storage occupancies. Because of its size and the volume of combustible material it would house, it poses an extraordinary hazard for fire protection.

The volume is based on the legacy code definition, which was based on 10,000 passenger vehicle tires weighing an average of 25 pounds (11 kg) each, rather than the volume of the stored tires. Assuming a 24-inch by 24-inch space (610 mm by 610 mm) for an average passenger vehicle tire and a 6-inch (152 mm) thickness, the result is 20,000 cubic feet (566 m³):

10,000 tires x 2 ft x 2 ft x 0.5 ft = 20,000 ft³

The 20,000 cubic feet (566 m³) represents the actual volume of stored materials based on an equivalent height and area for passenger vehicle tires as shown in the calculation above and does not include circulation area or other portions of the building. Rather, it focuses on how much of the material is present. Although the definition uses the term "area" rather than "volume," it is the volume that becomes the threshold consideration. Still, the area where the tires are stored implies the footprint used for storage. It is not the intent to apply this to areas outside of those used for bulk tire storage.

Buildings used for the bulk storage of tires are classified as Group S-1 occupancies in accordance with Section 311.2. All Group S-1 occupancies, regardless of square footage, must be equipped with an NFPA 13 automatic sprinkler system if used for the bulk storage of tires as required by Section 903.2.9.2. Chapter 25 of the IFC also requires that bulk tire storage buildings be further designed to comply with NFPA 13, and Chapter 23 of the IFC includes additional requirements for high-piled rubber tire storage as a high-hazard commodity (see commentary, Chapters 23 and 25 of the IFC).

[F] TROUBLE SIGNAL. A signal initiated by the fire alarm system or device indicative of a fault in a monitored circuit or component.

❖ This type of signal indicates that there has been an abnormal change in the normal status of the fire detection system or devices and that a response is required to determine the nature of the fault condition. The trouble signal is only associated with electronic portions of a fire protection system. Physical conditions such as a closed valve are monitored electronically and would report as a supervisory signal rather than a trouble signal. A valve supervisory switch, or "tamper switch," for example, would perform such a function.

[F] VISIBLE ALARM NOTIFICATION APPLIANCE. A notification appliance that alerts by the sense of sight.

❖ Visible alarm notification appliances are located anywhere an occupant notification system is required, where occupants may be hearing impaired and in sleeping accommodations of Group I-1 and R-1 occupancies. These alarm notification devices must be located and oriented so that they will display alarm signals throughout the required space. Visible alarms, when provided, are typically installed in the public and common areas of buildings (see commentary, Section 907.5.2.3).

[F] WET-CHEMICAL EXTINGUISHING SYSTEM. A solution of water and potassium-carbonate-based chemical, potassium-acetate-based chemical or a combination thereof, forming an extinguishing agent.

❖ This extinguishing agent is a suitable alternative to the use of a dry chemical, especially when protecting commercial kitchen range hoods. There is less cleanup time after system discharge. Wet chemical solutions are considered to be relatively harmless and normally have no lasting effect on the skin or respiratory system. These solutions may produce temporary irritation, which is usually mild and disappears when contact is eliminated. These systems must be preengineered and labeled. NFPA 17A applies to the design, installation, operation, testing and maintenance of wet-chemical extinguishing systems.

[F] WIRELESS PROTECTION SYSTEM. A system or a part of a system that can transmit and receive signals without the aid of wire.

❖ These systems use radio frequency transmitting devices that comply with the special requirements for supervision of low-power wireless systems in NFPA 72. Wireless devices have the advantage of flexibility in positioning. Consequently, portable wireless notification devices are frequently used in existing facilities where visual devices are not present throughout.

[F] ZONE. A defined area within the protected premises. A zone can define an area from which a signal can be received, an area to which a signal can be sent or an area in which a form of control can be executed.

❖ Zoning a system is important to emergency personnel in locating a fire. When an alarm is designated to a specific zone, it allows the fire service to immediately respond to the area where the fire is in progress instead of searching the entire building for the origin of an alarm.

[F] ZONE, NOTIFICATION. An area within a building or facility covered by notification appliances which are activated simultaneously.

❖ This definition is provided to clarify the code by making a clear distinction between fire alarm system initiation device zones required by Section 907.6.3 and the zones that may be designed into occupant notification device systems in a building. The term is used primarily in the exceptions for sprinkler systems found in the manual fire alarm system requirements in Section 907.2 and its subsections. Note that the code does not require audible and visible occupant notification device systems to be zoned; if such zones are provided, it is a matter of the system design engineer's judgement. The voice paging component in high-rise building emergency voice/alarm communication systems is, however, required to be zoned in accordance with Section 907.5.2.2.

SECTION 903
AUTOMATIC SPRINKLER SYSTEMS

[F] 903.1 General. *Automatic sprinkler systems* shall comply with this section.

❖ This section identifies the conditions requiring an automatic sprinkler system for all occupancies. The need for an automatic sprinkler system may depend on not only the occupancy but also the occupant load, fuel load, height and area of the building as well as firefighting capabilities. Section 903.2 addresses all occupancy conditions requiring an automatic sprinkler system. Section 903.3 contains the installation requirements for all sprinkler systems in addition to the requirements of NFPA 13, NFPA 13R and NFPA 13D. The supervision and alarm requirements for sprinkler systems are contained in Section 903.4, whereas Section 903.5 refers to the IFC, which references testing and maintenance requirements for sprinkler systems found in Section 901 and NFPA 25.

Unless specifically allowed by the code, residential sprinkler systems installed in accordance with NFPA 13R or NFPA 13D are not recognized for reductions or exceptions permitted by other sections of the code. NFPA 13 systems provide the level of protection associated with adequate fire suppression for all occupancies. NFPA 13R and NFPA 13D systems are intended more to provide adequate time for egress but not necessarily for complete suppression of the fire. Figure 903.2 lists examples of where the various sprinkler standards differ in application.

The area values contained in this section are intended to apply to fire areas, which are comprised of all floor areas bounded by fire barriers, fire walls or exterior walls. The minimum required fire-resistance rating of fire barrier assemblies that define a fire area is specified in Table 707.3.9. Because the areas are defined as fire areas, fire barriers, horizontal assemblies, fire walls or exterior walls are the only acceptable means of subdividing a building into smaller areas instead of installing an automatic sprinkler system. Where fire barrier and exterior walls define multiple fire areas within a single building, a fire wall defines separate buildings within one structure. Also note that some of the threshold limitations result in a requirement to install an automatic sprinkler system throughout the building while others may require only specific fire areas to be sprinklered.

Another important point is that one fire area may include floor areas in more than one story of a building (see the commentary to the definition of "Fire area" in Section 902.1).

The application of mixed occupancies and fire areas must be carefully researched. Often the required separation between occupancies for the purposes of applying the separated mixed-use option in Section 508.4 will result in a separation that is less than what is required to define the boundaries of a fire area. It is possible to have two different occupancies within a given fire area, treated as separated uses but with code requirements applicable to both, since they are not separated by the rating required for fire areas.

[F] 903.1.1 Alternative protection. Alternative automatic fire-extinguishing systems complying with Section 904 shall be permitted in lieu of automatic sprinkler protection where recognized by the applicable standard and *approved* by the fire code official.

❖ This section permits the use of an alternative automatic fire-extinguishing system when approved by the fire code official as a means of compliance with the occupancy requirements of Section 903. Although the use of an alternative extinguishing system allowed by Section 904, such as a carbon dioxide system or clean-agent system, would satisfy the requirements of Section 903.2, it would not be considered an acceptable alternative for the purposes of exceptions, reductions or other code alternatives that would be applicable if an automatic sprinkler system were installed.

[F] 903.2 Where required. Approved *automatic sprinkler systems* in new buildings and structures shall be provided in the locations described in Sections 903.2.1 through 903.2.12.

Exception: Spaces or areas in telecommunications buildings used exclusively for telecommunications equipment, associated electrical power distribution equipment, batteries and standby engines, provided those spaces or areas are equipped throughout with an automatic smoke detection system in accordance with Section 907.2 and are separated from the remainder of the building by not less than 1-hour *fire barriers* constructed in accordance with Section 707 or

not less than 2-hour *horizontal assemblies* constructed in accordance with Section 712, or both.

❖ Sections 903.2.1 through 903.2.12 identify the conditions requiring an automatic sprinkler system (see Figure 903.2). The type of sprinkler system must be one that is permitted for the specific occupancy condition. An NFPA 13R sprinkler system, for example, may not be installed to satisfy the sprinkler threshold requirements for a mercantile occupancy (see Section 903.2.7). As indicated in Section 903.3.1.2, the use of an NFPA 13R sprinkler system is limited to Group R occupancies not exceeding four stories in height.

Where the thresholds for sprinkler protection include the number of occupants and the fire area, it is important to remember that the proper application is the determination of the hazard present. If the actual fire area is less than the fire area threshold but the total occupant load exceeds the occupant load threshold, it is still necessary to determine whether or not the occupant load of that occupancy is present. For example, if the occupant load threshold is 300 for a given occupancy, it is necessary to determine that there are 300 occupants of that occupancy classification. Just because the fire area has an occupant load exceeding the occupant threshold does not in itself indicate that sprinkler protection is required. The code requires automatic sprinkler protection when the occupant load for the specific occupancy exceeds the established threshold. In applying the occupant load thresholds, it is important to note that they are to be evaluated and applied per occupancy and not as an aggregate of all occupancies that may be present in the building.

There is one exception for those spaces or areas used exclusively for telecommunications equipment. The telecommunications industry has continually stressed the need for the continuity of telephone service, and the ability to maintain this service is of prime importance. This service is a vital link between the community and the various life safety services, including fire, police and emergency medical services. The integrity of this communications service can be jeopardized not only by fire, but also by water, from whatever the source.

It must be emphasized that the exception applies only to those spaces or areas that are used exclusively for telecommunications equipment. Historically, those spaces have a low incidence of fire events. Fires in telecommunications equipment are difficult to start and, if started, grow slowly, thus permitting early detection. Such fires are typically of the smoldering type, do not spread beyond the immediate area and generally self-extinguish.

Note, however, that this exception requires a fire-resistance-rated separation from other portions of the building.

Occupancy	Threshold	Exception
All occupancies	Buildings with floor level ≥ 55 feet above fire department vehicle access and occupant load ≥ 30.	Airport control towers, open parking structures. F-2, U
Assembly (A-1, A-3, A-4)	Fire area > 12,000 sq. ft. or fire area occupant load > 300 or fire area above/below level of exit discharge. Multitheater complex (A-1 only)	Participant sport arenas at level of exit discharge. A-3, A-4
Assembly (A-2)	Fire area > 5,000 sq. ft. or fire area occupant load > 100 or fire area above/below level of exit discharge.	None
Assembly (A-5)	Accessory areas > 1,000 sq. ft.	None
Ambulatory health care facility (B)	> Three care recipients incapable of self-preservation or any care recipients incapable of self-preservation above or below level of exit discharge.	None
Educational (E)	Fire area > 12,000 sq. ft. or below level of exit discharge.	Each classroom has exterior door at grade.
Factory (F-1) Mercantile (M) Storage (S-1)	Fire area > 12,000 sq. ft. or building > three stories or combined fire area > 24,000 sq. ft. Woodworking > 2,500 sq. ft. (F-1 only). Display and sale of upholstered furniture (M only) Bulk storage of tires > 20,000 cu. ft. (S-1 only).	None
High hazard (H-1, H-2, H-3, H-4, H-5)	Sprinklers required.	None
Institutional (I-1, I-2, I-3, I-4)	Sprinklers required.	None
Residential (R)	Sprinklers required.	None
Repair garage (S-1)	Fire area > 12,000 sq. ft. or ≥ two stories (including basement) with fire area > 10,000 sq. ft. or repair garage servicing vehicles in basement or servicing commercial trucks/buses in fire area > 5,000 sq. ft.	None
Parking garage (S-1)	Enclosed automobile parking—sprinklers required. Commercial trucks/buses parking area > 5,000 sq. ft.	None
Parking garage (S-2)	Fire area > 12,000 sq. ft. or fire area > 5,000 sq. ft. for storage of commercial trucks/buses; or beneath other groups. (enclosed parking)	Not if beneath Group R-3
Covered malls (402.8)	Sprinklers required.	Attached open parking structures.
High rises (403.2, 403.3)	> 75 feet above fire department vehicle access.	Airport traffic control towers, open garages. A-5
Unlimited area buildings (507)	A-3, A-4, B, E, F, M, S: one story. B, F, M, S: two story.	One story F-2 or S-2.

Note: Thresholds located in Section 903.2 unless noted. See also Table 903.2.11.6 for additional required suppression systems.
For SI: 1 foot = 304.8 mm, 1 square foot = 0.0929 m^2.

Figure 903.2
SUMMARY OF OCCUPANCY-RELATED AUTOMATIC SPRINKLER THRESHOLDS

[F] 903.2.1 Group A. An *automatic sprinkler system* shall be provided throughout buildings and portions thereof used as Group A occupancies as provided in this section. For Group A-1, A-2, A-3 and A-4 occupancies, the *automatic sprinkler system* shall be provided throughout the floor area where the Group A-1, A-2, A-3 or A-4 occupancy is located, and in all floors from the Group A occupancy to, and including, the nearest *level of exit discharge* serving the Group A occupancy. For Group A-5 occupancies, the *automatic sprinkler system* shall be provided in the spaces indicated in Section 903.2.1.5.

❖ Group A occupancies are characterized by a significant number of people who are not familiar with their surroundings. The requirement for a suppression system reflects the additional time needed for egress. The protection is also intended to extend to the occupants of the assembly group from unobserved fires in other building areas located between the floor level containing the assembly occupancy and the level of exit discharge serving such occupancies. The only exception to the coverage is for Group A-5 occupancies that are open to the atmosphere. Such occupancies require only certain aspects to be sprinklered, such as certain concession stands and other enclosed use areas (see commentary, Section 903.2.1.5).

The requirement for sprinklers is based on the location and function of the space. It is not dependant on whether or not the area is provided with exterior walls. IFC Committee Interpretation No. 25-05 takes up this issue and states, in part, that "where no surrounding exterior walls are provided along the perimeter of the building, the building area is used to identify and determine applicable fire area." Outdoor areas such as pavilions and patios may have no walls but will have an occupant load and other factors that identify the assembly occupancy as such. If any of the thresholds are reached requiring sprinkler protection, then sprinkler protection must be provided whether there are exterior walls or not.

[F] 903.2.1.1 Group A-1. An *automatic sprinkler system* shall be provided for Group A-1 occupancies where one of the following conditions exists:

1. The *fire area* exceeds 12,000 square feet (1115 m²);

2. The *fire area* has an *occupant load* of 300 or more;

3. The *fire area* is located on a floor other than a *level of exit discharge* serving such occupancies; or

4. The *fire area* contains a multitheater complex.

❖ Group A-1 occupancies are identified as assembly occupancies with fixed seating, such as theaters. In addition to the high occupant load associated with these types of facilities, egress is further complicated by the possibility of low lighting levels customary during performances. The fuel load in these buildings is usually of a type and quantity that would support fairly rapid fire development and sustained duration.

Theaters with stages pose a greater hazard. Sections 410.6 and 410.7 require stages to be equipped with an automatic sprinkler system and standpipe system, respectively. The proscenium opening must also be protected. These features compensate for the additional hazards associated with stages in Group A-1 occupancies.

This section lists four conditions that require installing a suppression system in a Group A-1 occupancy. Condition 1 requires that, if any one fire area of Group A-1 exceeds 12,000 square feet (1115 m²), the automatic fire suppression system is to be installed throughout the entire story or floor level where a Group A-1 occupancy is located, regardless of whether the building is divided into more than one fire area. However, if all the fire areas are less than 12,000 square feet (1115 m²) (and less than the other thresholds), then sprinklers would not be required. Compartmentalization into multiple fire areas in compliance with Chapter 7 is deemed an adequate alternative to sprinkler protection.

Condition 2 establishes the minimum number of occupants for which a suppression system is considered necessary. The determination of the actual occupant load must be based on Section 1004.

Condition 3 accounts for occupant egress delay when traversing a stairway requiring a sprinkler system, regardless of the size of occupant load. In such cases alternative emergency escape elements such as windows may not be available, making the suppression needs all the greater. It is not necessary for the occupant load to exceed 300 on a level other than the level of exit discharge serving such occupancy. Any number of Group A-1 occupants on the alternative level would be cause to apply the requirement for sprinklers. The text does not make reference to "story" but uses the term "floor," which could include mezzanines and basements.

Condition 4 states that a sprinkler system is required for multiplex theater complexes to account for the delay associated with the notification of adjacent compartmentalized spaces where the occupants may not be immediately aware of an emergency.

[F] 903.2.1.2 Group A-2. An *automatic sprinkler system* shall be provided for Group A-2 occupancies where one of the following conditions exists:

1. The *fire area* exceeds 5,000 square feet (464.5 m²);

2. The *fire area* has an *occupant load* of 100 or more; or

3. The *fire area* is located on a floor other than a *level of exit discharge* serving such occupancies.

❖ Group A-2 assembly occupancies are intended for food or drink consumption, such as banquet halls, nightclubs and restaurants. Occupancies in Group A-2 involve life safety factors such as a high occupant density, flexible fuel loading, movable furnishings and limited lighting; therefore, they must be protected with an automatic sprinkler system under any of the listed conditions.

In the case of an assembly use, the purpose of the automatic sprinkler system is to provide life safety from fire as well as preserving property. By requiring fire suppression in areas through which the occupants may egress, including the level of exit discharge serving such occupancies, the possibility of unobserved fire development affecting occupant egress is minimized.

The 5,000-square-foot threshold for the automatic sprinkler system reflects the higher degree of life safety hazard associated with Group A-2 occupancies. As alluded to earlier, Group A-2 occupancies could have low lighting levels, loud music, late hours of operation, dense seating with ill-defined aisles and alcoholic beverage service. These factors in combination could delay fire recognition, confuse occupant response and increase egress time.

Although the calculated occupant load for a 5,000 square-foot (465 m²) space at 15 square feet (1.4 m²) per occupant would be over 100, the occupant load threshold in Condition 2 is meant to reflect the concern for safety in these higher density occupancies. Although the major reason for establishing the occupant threshold at 100 was due to several recent nightclub incidents, the requirement is not limited to nightclubs or banquet facilities but to all Group A-2 occupancies. Any restaurant with an occupant load greater than 100 would require sprinkler protection as well. This includes fast food facilities with no low lighting or alcohol sales. The similar intent of Condition 3 is addressed in the commentary to Section 903.2.1.1.

These conditions require sprinklers throughout the fire area containing the Group A-2 occupancy, regardless of the number of fire areas present.

[F] 903.2.1.3 Group A-3. An *automatic sprinkler system* shall be provided for Group A-3 occupancies where one of the following conditions exists:

1. The *fire area* exceeds 12,000 square feet (1115 m²);

2. The *fire area* has an *occupant load* of 300 or more; or

3. The *fire area* is located on a floor other than a *level of exit discharge* serving such occupancies.

❖ Group A-3 occupancies are assembly occupancies intended for worship, recreation or amusement and other assembly uses not classified elsewhere in Group A, such as churches, museums and libraries. While Group A-3 occupancies could potentially have a high occupant load, they normally do not have the same potential combination of life safety hazards associated with Group A-2 occupancies. As with most assembly occupancies, however, most of the occupants are typically not completely familiar with their surroundings. When any of the three listed conditions are applicable, an automatic sprinkler system is required throughout the fire area containing the Group A-3 occupancy and in all floors between the Group A occupancy and exit discharge that serves that occupancy (see commentary, Sections 903.2.1 and 903.2.1.1).

The exception exempts the participant sport area of Group A-3 occupancies from automatic sprinkler system requirements because these areas are typically

large open spaces with relatively low fuel loads. The exception includes only the participant sport area, such as an indoor swimming pool or the court area of an indoor tennis court.

Note that if the exception is claimed and sprinklers are omitted from the sport area, the building would not be considered completely sprinklered in accordance with Section 903.3.1.1 for purposes of allowing construction alternatives, such as height and area increases, corridor rating reduction and other code alternatives. Care must also be exercised in allowing the exception to be taken in buildings that have the potential for the sports participant area to be used for other than participant sports purposes, including but not limited to banquets, exhibits, rummage sales and similar activities or events which would have a much higher fuel load than was ever contemplated by the exception.

[F] 903.2.1.4 Group A-4. An *automatic sprinkler system* shall be provided for Group A-4 occupancies where one of the following conditions exists:

1. The *fire area* exceeds 12,000 square feet (1115 m²);

2. The *fire area* has an *occupant load* of 300 or more; or

3. The *fire area* is located on a floor other than a *level of exit discharge* serving such occupancies.

❖ Group A-4 occupancies are assembly uses intended for viewing of indoor sporting events and activities such as arenas, skating rinks and swimming pools. The occupant load density may be high depending on the extent and style of seating, such as bleachers or fixed seats, and the potential for standing-room viewing.

When any of the three listed conditions are applicable, an automatic sprinkler system is required throughout the fire area containing the Group A-4 occupancy and in all floors between the Group A occupancy and exit discharge (see commentary, Sections 903.2.1 and 903.2.1.1). Similar to Group A-3 occupancies, the participant sport areas on the main floor of Group A-4 occupancies are exempt from the sprinkler system requirement (see commentary, Section 903.2.1.3).

[F] 903.2.1.5 Group A-5. An *automatic sprinkler system* shall be provided for Group A-5 occupancies in the following areas: concession stands, retail areas, press boxes and other accessory use areas in excess of 1,000 square feet (93 m²).

❖ Group A-5 occupancies are assembly uses intended for viewing of outdoor activities. This occupancy classification could include amusement park structures, grandstands and open stadiums. A sprinkler system is not required in the open area of Group A-5 occupancies because the buildings would not accumulate smoke and hot gases. A fire in open areas would also be obvious to all spectators.

Enclosed areas such as retail areas, press boxes and concession stands require sprinklers if they are in excess of 1,000 square feet (93 m²). The 1,000-square-foot (93 m²) accessory use area is not intended to be an aggregate condition but rather per space. Thus, a press box that is 2,500 square feet (232 m²) in area would need to be subdivided into areas less than 1,000 square feet (93 m²) each in order to be below the threshold for sprinklers. There is no specific requirement for the separation of these spaces. It is assumed, however, that the separation would be a solid barrier of some type but without a required fire-resistance rating.

The provision is meant to mirror that in Section 1028.6.2.3, which exempts press boxes and storage facilities less than 1,000 square feet (93 m²) in area from sprinkler requirements in smoke-protected assembly seating areas.

[F] 903.2.2 Group B ambulatory health care facilities. An *automatic sprinkler system* shall be installed throughout all fire areas containing a Group B ambulatory health care facility occupancy when either of the following conditions exists at any time:

1. Four or more care recipients are incapable of self-preservation.

2. One or more care recipients who are incapable of self-preservation are located at other than the *level of exit discharge* serving such an occupancy.

❖ Group B ambulatory health care facilities are Group B occupancies, with an enhanced set of requirements to account for the fact that patients may be incapable of self-preservation and require rescue by other occupants or fire personnel. There are several aspects to the enhanced features, including smoke compartments, sprinklers and fire alarms. More specifically, the requirements for sprinklers are based on four or more patients at any given time being incapable of self-preservation or any number of patients who are incapable of self-preservation located on a floor other than the level of exit discharge that serves the Group B Ambulatory health care facility. The sprinkler requirement is limited to the fire area that contains the Group B ambulatory health care facility (see commentary, Section 422).

[F] 903.2.3 Group E. An *automatic sprinkler system* shall be provided for Group E occupancies as follows:

1. Throughout all Group E *fire areas* greater than 12,000 square feet (1115 m²) in area.

2. Throughout every portion of educational buildings below the lowest *level of exit discharge* serving that portion of the building.

> **Exception:** An *automatic sprinkler system* is not required in any area below the lowest *level of exit discharge* serving that area where every classroom throughout the building has at least one exterior *exit* door at ground level.

❖ Group E occupancies are limited to educational purposes through the 12th grade and day care centers serving children older than 2½ years of age. The 12,000-square-foot (1115 m²) fire area threshold for the sprinkler system was established to allow smaller

schools and day care centers to be nonsprinklered to minimize the economic impact on these facilities. The 12,000-square-foot (1115 m²) threshold is similar to that used for several other occupancies, such as Group M occupancies.

Sprinklers would also be required in portions of the building located below the level of exit discharge serving that occupancy. However, there is an exception that would allow the omission of the automatic sprinkler system for the Group E fire area if there is a direct exit to the exterior from each classroom at ground level. The occupants must be able to go from the classroom directly to the outside without passing through intervening corridors, passageways or exit enclosures.

[F] 903.2.4 Group F-1. An *automatic sprinkler system* shall be provided throughout all buildings containing a Group F-1 occupancy where one of the following conditions exists:

1. A Group F-1 *fire area* exceeds 12,000 square feet (1115 m²).

2. A Group F-1 *fire area* is located more than three stories above *grade plane*.

3. The combined area of all Group F-1 *fire areas* on all floors, including any mezzanines, exceeds 24,000 square feet (2230 m²).

❖ Because of the difficulty in manually suppressing a fire involving a large area, occupancies of Group F-1 must be protected throughout with an automatic sprinkler system if the fire area is in excess of 12,000 square feet (1115 m²), if the total of all fire areas of Group F-1 in the building is in excess of 24,000 square feet (2230 m²) or if the Group F-1 fire area is located more than three stories above grade plane. This is one of the few locations in the code where the total floor area of the building is aggregated for application of a code requirement. The stipulated conditions for when an automatic sprinkler system is required also apply to Group M (see Section 903.2.7) and S-1 (see Section 903.2.9) occupancies.

The following examples illustrate how the criteria are intended to be applied. If a building contains a single fire area of Group F-1 and the fire area is 13,000 square feet (1208 m²), an automatic sprinkler system is required throughout the entire building. However, if this fire area is separated into two fire areas and neither is in excess of 12,000 square feet (1115 m²), an automatic fire sprinkler system is not required. To be considered separate fire areas, the areas must be separated by fire barriers or horizontal assemblies having a fire-resistance rating as required in Table 707.3.9.

If a 30,000-square-foot (2787 m²) Group F-1 building was equally divided into separate fire areas of 10,000 square feet (929 m²) each, an automatic sprinkler system would still be required throughout the entire building. Because the aggregate area of all fire areas exceeds 24,000 square feet (2230 m²), additional compartmentation will not eliminate the need for an automatic sprinkler system. However, the use of a fire wall to separate the structure into two buildings would reduce the aggregate area of each building to less than 24,000 square feet (2230 m²) and each fire area to less than 12,000 square feet (1115 m²), which would offset the need for an automatic sprinkler system.

[F] 903.2.4.1 Woodworking operations. An *automatic sprinkler system* shall be provided throughout all Group F-1 occupancy *fire areas* that contain woodworking operations in excess of 2,500 square feet (232 m²) in area which generate finely divided combustible waste or use finely divided combustible materials.

❖ Because of the potential amount of combustible dust that could be generated during woodworking operations, an automatic sprinkler system is required throughout a fire area when it contains a woodworking operation that exceeds 2,500 square feet (232 m²) in area. Facilities where woodworking operations take place, such as cabinet making, are considered Group F-1 occupancies. The intent of the phrase "finely divided combustible waste" is to describe particle concentrations that are in the explosive range (see Chapter 13 of the IFC for discussion of dust-producing operations).

The extent of sprinkler coverage is only intended to be for the Group F-1 occupancy involved in the woodworking activity. If the fire area is larger than 2,500 square feet (232 m²) but the woodworking area is 2,500 square feet (232 m²) or less, sprinkers are not required. It is not the intent to require the installation of sprinkers throughout the builing but rather in the fire area where the hazard may be present.

[F] 903.2.5 Group H. *Automatic sprinkler systems* shall be provided in high-hazard occupancies as required in Sections 903.2.5.1 through 903.2.5.3.

❖ Group H occupancies are those intended for the manufacturing, processing or storage of hazardous materials that constitute a physical or health hazard. To be considered a Group H occupancy, the amount of hazardous materials is assumed to be in excess of the maximum allowable quantities permitted by Tables 307.1(1) and 307.1(2).

[F] 903.2.5.1 General. An *automatic sprinkler system* shall be installed in Group H occupancies.

❖ This section requires an automatic sprinkler system in all Group H occupancies. Even though in some instances the hazard associated with the occupancy may be one that is not a fire hazard, an automatic sprinkler system is still required to minimize the potential for fire spreading to the high-hazard use; that is, the sprinklers protect the high-hazard area from fire outside the area. This section does not prohibit the use of an alternative automatic fire-extinguishing system in accordance with Section 904. When a water-based system is not compatible with the hazardous materials involved and thus creates a dangerous condition, an alternative fire-extinguishing system should be used. For example, combustible metals, such as magnesium and titanium, have a serious record of involvement with fire and are typically not compatible with water (see commentary, Chapter 36 of the IFC).

Where control areas are used to regulate the quantity of hazardous material within a building, the building is not considered a Group H occupancy. Unless a building would be required by some other code provision to be protected with sprinklers, control areas can be used to control the allowable quantities of hazardous materials in a building so as to not warrant a Group H classification and its mandatory sprinkler requirements.

[F] 903.2.5.2 Group H-5. An *automatic sprinkler system* shall be installed throughout buildings containing Group H-5 occupancies. The design of the sprinkler system shall not be less than that required by this code for the occupancy hazard classifications in accordance with Table 903.2.5.2. Where the design area of the sprinkler system consists of a *corridor* protected by one row of sprinklers, the maximum number of sprinklers required to be calculated is 13.

❖ Group H-5 occupancies are structures that are typically used as semiconductor fabrication facilities and comparable research laboratory facilities that use hazardous production materials (HPM). Many of the materials used in semiconductor fabrication present unique hazards. Many of the materials are toxic, while some are corrosive, water reactive or pyrophoric. Fire protection for these facilities is aimed at preventing incidents from escalating and producing secondary threats beyond a fire, such as the release of corrosive or toxic materials. Because of the nature of Group H-5 facilities, the overall amount of hazardous materials can far exceed the maximum allowable quantities given in Tables 307.1(1) and 307.1(2). Although the amount of HPM material is restricted in fabrication areas, the quantities of HPM in storage rooms normally will be in excess of those allowed by the tables. Additional requirements for Group H-5 facilities are located in Chapter 18 of the IFC and Section 415.8.

This section also specifies the sprinkler design criteria, based on NFPA 13, for various areas in a Group H-5 occupancy (see commentary, Table 903.2.5.2). When the corridor design area sprinkler option is used, a maximum of 13 sprinklers must be calculated. This exceeds the requirements of NFPA 13 for typical egress corridors, which are that a maximum of either five or seven calculated sprinklers, depending on the extent of protected openings in the corridor. The increased number of calculated corridor sprinklers is based on the additional hazard associated with the movement of hazardous materials in corridors of Group H-5 facilities.

[F] TABLE 903.2.5.2
GROUP H-5 SPRINKLER DESIGN CRITERIA

LOCATION	OCCUPANCY HAZARD CLASSIFICATION
Fabrication areas	Ordinary Hazard Group 2
Service corridors	Ordinary Hazard Group 2
Storage rooms without dispensing	Ordinary Hazard Group 2
Storage rooms with dispensing	Extra Hazard Group 2
Corridors	Ordinary Hazard Group 2

❖ Table 903.2.5.2 designates the appropriate occupancy hazard classification for the various areas within a Group H-5 facility. The listed occupancy hazard classifications correspond to specific sprinkler system design criteria in NFPA 13. Ordinary Hazard Group 2 occupancies, for example, require a minimum design density of 0.20 gpm/ft^2 (8.1 L/min/m^2) with a minimum design area of 1,500 square feet (139 m^2). An Extra Hazard Group 2 occupancy, in turn, requires a minimum design density of 0.40 gpm/ft^2 (16.3 L/min/m^2) with a minimum operating area of 2,500 square feet (232 m^2). The increased overall sprinkler demand for Extra Hazard Group 2 occupancies is based on the potential use and handling of substantial amounts of hazardous materials, such as flammable or combustible liquids.

[F] 903.2.5.3 Pyroxylin plastics. An *automatic sprinkler system* shall be provided in buildings, or portions thereof, where cellulose nitrate film or pyroxylin plastics are manufactured, stored or handled in quantities exceeding 100 pounds (45 kg).

❖ Cellulose nitrate (pyroxylin) plastics pose unusual and substantial fire risks. Pyroxylin plastics are the most dangerous and unstable of all plastic compounds. The chemically bound oxygen in their structure permits them to burn vigorously in the absence of atmospheric oxygen. Although these compounds produce approximately the same amount of energy as paper when they burn, pyroxylin plastics burn at a rate as much as 15 times greater than comparable common combustibles. When burning, these materials release highly flammable and toxic combustion byproducts. Consequently, cellulose nitrate fires are very difficult to control. Although this section specifies a sprinkler threshold quantity of 100 pounds, the need for additional fire protection should be considered for pyroxylin plastics in any amount.

Although the code includes cellulose nitrate "film" in its requirements, cellulose nitrate motion picture film has not been used in the United States since the 1950s. All motion picture film produced since that time is what is typically called "safety film." Consequently, the only application for this section relative to motion picture film is where it may be used in laboratories or storage vaults that are dedicated to film restoration and archives. The protection of these facilities is addressed in Sections 306.2 and 4204.2, both in the IFC.

[F] 903.2.6 Group I. An *automatic sprinkler system* shall be provided throughout buildings with a Group I *fire area*.

Exception: An *automatic sprinkler system* installed in accordance with Section 903.3.1.2 or 903.3.1.3 shall be allowed in Group I-1 facilities.

❖ The Group I occupancy is divided into four individual occupancy classifications based on the degree of detention, supervision and physical mobility of the occupants. The evacuation difficulties associated with the building occupants creates the need to incorporate a defend-in-place philosophy of fire protection in occupancies of Group I. For this reason, all such occupancies are to be protected with an automatic sprinkler system.

Of particular note, this section encompasses all Group I-3 occupancies where more than five persons are detained. There has been considerable controversy concerning the use of automatic sprinklers in detention and correctional occupancies. Special design considerations can be taken into account to alleviate the perceived problems with sprinklers in sleeping units. Sprinklers that reduce the likelihood of vandalism as well as the potential to hang oneself are commercially available. Knowledgeable designers can incorporate certain design features to increase reliability and decrease the likelihood of damage to the system.

Group I-4 occupancies would include either adult-only care facilities or occupancies that provide personal care for more than five children 2$^1/_2$ years of age or less on a less than 24-hour basis. Because the degree of assistance and the time needed for egress cannot be gauged, an automatic sprinkler system is required. As defined in Section 202, a Group I-4 child care facility located at the level of exit discharge and accommodating no more than 100 children, with each child care room having an exit directly to the exterior would be classified as a Group E occupancy, and an automatic sprinkler system would not be required unless dictated by the requirements in Section 903.2.2.

The exception permits Group I-1 occupancies to be protected throughout with either an NFPA 13R or 13D sprinkler system instead of a standard NFPA 13 sprinkler system. The exception recognizes the perceived mobility of the occupants in a Group I-1 facility as well as the basic life safety intent to protect the main occupiable areas. However, use of this exception would result in the building not qualifying as a fully sprinklered building in accordance with NFPA 13 for any applicable code alternatives.

[F] 903.2.7 Group M. An *automatic sprinkler system* shall be provided throughout buildings containing a Group M occupancy where one of the following conditions exists:

1. A Group M *fire area* exceeds 12,000 square feet (1115 m^2).

2. A Group M *fire area* is located more than three stories above *grade plane*.

3. The combined area of all Group M *fire areas* on all floors, including any mezzanines, exceeds 24,000 square feet (2230 m²).

4. A Group M occupancy is used for the display and sale of upholstered furniture.

❖ The sprinkler threshold requirements for Group M occupancies are identical to those of Group F-1 and S-1 occupancies (see commentary, Section 903.2.4). Additionally, Group M occupancies used for the display and sale of upholstered furniture must be sprinklered. Upholstered furniture has the potential for rapid growing and high heat release fires. This hazard is increased substanially when there are numerous upholstered furniture items on display. Such fires put the occupants and emergency responders at risk. This requirement is regardless of whether the upholstered furniture has passed any fire-retardant tests.

The code does not specifically address what constitutes upholstered furniture but by simple dictionary definition, upholstered furniture is furniture, expecially seats covered with padding, springs, webbing and fabric or leather covers. The code does not make any distinction between levels of padding and upholstery provided on furniture, which was intentional. The proponent's reason statement for code change F135-07/08 stated, in part, that "the American Home Furnishings Alliance (AHFA) and the National Home Furnishings Association (NHFA) have examined proposals for exempting vendors of certain constructions of furniture and concluded that such exemptions would be impractical for local code officials to enforce. This is the case because the internal construction of furniture cannot be established reliably without deconstructing it."

In addition, the code currently does not make any size or fire load distinctions within Group M occupancies storing and displaying upholstered furniture. The original code change read "A Group M occupancy is used primarily for the display and sale of upholstered furniture." In approving the code change, the committee included (and the ICC membership approved) a modification to delete the word "primarily" and reasoned as follows (emphasis added): "The committee indicated its sense that future efforts on the topic need to address Group F and S upholstered furniture occupancies as well, and that a reasonable sprinkler threshold needs to be added to provide some relief to the small businesses that will now be affected. The modification removes a subjective term ("primarily") that the committee felt could create serious enforcement inconsistencies."

Automatic sprinkler systems for mercantile occupancies are typically designed for an Ordinary Hazard Group 2 classification in accordance with NFPA 13. If high-piled storage (see Section 903.2.7.1) is anticipated; however, additional levels of fire protection may be required. Also, some merchandise in mercantile occupancies, such as aerosols, rubber tires, paints and certain plastic commodities, even at limited storage heights, are considered beyond the standard Class I through IV commodity classification assumed for mercantile occupancies in NFPA 13 and may warrant additional fire protection.

[F] 903.2.7.1 High-piled storage. An *automatic sprinkler system* shall be provided in accordance with the *International Fire Code* in all buildings of Group M where storage of merchandise is in high-piled or rack storage arrays.

❖ Regardless of the size of the Group M fire area, an automatic sprinkler system may be required in a high-piled storage area. High-piled storage includes piled, palletized, bin box, shelf or rack storage of Class I through IV commodities to a height greater than 12 feet (3658 mm) and certain high-hazard commodities greater than 6 feet (1829 mm). Chapter 23 of the IFC provides a package of requirements that may include sprinkler protection depending upon the size of the high-piled storage area. The design standard for the sprinkler protection of high-piled storage is NFPA 13, which addresses the many different types and configurations of high-piled storage.

[F] 903.2.8 Group R. An *automatic sprinkler system* installed in accordance with Section 903.3 shall be provided throughout all buildings with a Group R *fire area*.

❖ This section requires sprinklers in any building that contains a Group R fire area. This includes uses such as hotels, apartment buildings, group homes and dormitories. There are no minimum criteria and no exceptions.

It should be noted that buildings constructed under the *International Residential Code*® (IRC®) are not included in Group R and would not, therefore, be subject to these particular requirements. However, the 2009 IRC requires sprinklers in all new townhouses and, beginning January 1, 2011, in all new one- and two-family dwellings. The IRC is a stand-alone code for the construction of detached one- and two-family dwellings and multiple single-family dwellings (townhouses) no more than three stories in height with a separate means of egress and addresses the requirements for sprinklers in a different way. That is, all of the provisions for new construction that affect those buildings are to be covered exclusively by the IRC and are not to be covered by another *International Code*®. Buildings that do not fall within the scope of the IRC would be classified in Group R and be subject to these provisions. This is stated clearly in IFC Committee Interpretation No. 29-03.

With respect to life safety, the need for a sprinkler system is dependent on the occupants' proximity to the fire and the ability to respond to a fire emergency. Group R occupancies could contain occupants who may require assistance to evacuate, such as infants, those with a disability or who may simply be asleep. While the presence of a sprinkler system cannot always protect occupants in residential buildings who are aware of the ignition and either do not respond or respond inappropriately, it can prevent fatalities outside of the area of fire origin regardless of the occupants' response. Section 903.3.2 requires quick-response or residential sprinklers in all Group R occupancies. Full-scale fire tests have demonstrated the ability of quick-response and residential sprinklers to maintain tenability from flaming fires in the room of fire origin.

Where a different occupancy is located in a building with a residential occupancy, the provisions of this section still apply and the entire building is required to be provided with an automatic sprinkler system regardless of the type of mixed-use condition considered. This is consistent with the mixed-use provisions in Chapter 5.

[F] 903.2.9 Group S-1. An *automatic sprinkler system* shall be provided throughout all buildings containing a Group S-1 occupancy where one of the following conditions exists:

1. A Group S-1 *fire area* exceeds 12,000 square feet (1115 m²).

2. A Group S-1 *fire area* is located more than three stories above *grade plane*.

3. The combined area of all Group S-1 *fire areas* on all floors, including any mezzanines, exceeds 24,000 square feet (2230 m²).

4. A Group S-1 *fire area* used for the storage of commercial trucks or buses where the *fire area* exceeds 5,000 square feet (464 m²).

❖ An automatic sprinkler system must be provided throughout all buildings where the fire area containing a Group S-1 occupancy exceeds 12,000 square feet (1115 m²), where more than three stories in height, where the combined fire area on all floors, including mezzanines, exceeds 24,000 square feet (2230 m²) or where the Group S-1 fire area used for the storage of commercial trucks or buses exceeds 5,000 square feet (464 m²).

The first three sprinkler threshold requirements for Group S-1 occupancies are identical to those of Group F-1 and M (see commentary, Sections 903.2.4 and 903.2.7). Group S-1 occupancies, such as warehouses and self-storage buildings, are assumed to be used for the storage of combustible materials. While high-piled storage does not change the Group S-1 occupancy classification, sprinkler protection, if required, may have to comply with the additional requirements of Chapter 23 of the IFC. High-piled stock or rack storage in any occupancy must comply with the code.

[F] 903.2.9.1 Repair garages. An *automatic sprinkler system* shall be provided throughout all buildings used as repair garages in accordance with Section 406, as shown:

1. Buildings having two or more *stories above grade plane*, including basements, with a *fire area* containing a repair garage exceeding 10,000 square feet (929 m²).

2. Buildings no more than one *story above grade plane*, with a *fire area* containing a repair garage exceeding 12,000 square feet (1115 m²).

3. Buildings with repair garages servicing vehicles parked in basements.

4. A Group S-1 *fire area* used for the repair of commercial trucks or buses where the *fire area* exceeds 5,000 square feet (464 m²).

❖ Automatic sprinklers may be required in repair garages, depending on the quantity of combustibles present, their location and floor area. In addition any Group S-1 fire area intended for the repair of commercial buses or trucks that exceeds 5,000 square feet (464 m²) would require sprinklers. This is the same criteria as Group S-1 occupancies and Group S-2 enclosed parking garages storing commercial buses and trucks. Repair garages may contain significant quantities of flammable liquids and other combustible materials. These occupancies are typically considered Ordinary Hazard Group 2 occupancies as defined in NFPA 13. Portions of repair garages used for parts cleaning using flammable or combustible liquids may require automatic sprinkler protection. If quantities of hazardous materials exceed the limitations in Section 307 for maximum allowable quantities per control area, the repair garage would be reclassified as a Group H occupancy.

[F] 903.2.9.2 Bulk storage of tires. Buildings and structures where the area for the storage of tires exceeds 20,000 cubic feet (566 m³) shall be equipped throughout with an *automatic sprinkler system* in accordance with Section 903.3.1.1.

❖ This section specifies when an automatic sprinkler system is required for the bulk storage of tires based on the volume of the storage area as opposed to a specific number of tires. Even in fully sprinklered buildings, tire fires pose significant problems to fire departments. Tire fires produce thick smoke and are difficult to extinguish by sprinklers alone. NFPA 13 contains specific fire protection requirements for the storage of rubber tires.

Whether the volume of tires is divided into different fire areas or not is irrelevant to the application of this section. If the total for all areas where tires are stored is great enough that the resultant storage volume exceeds 20,000 cubic feet (566 m³), the building must be sprinklered throughout. See the commentary to Section 902.1 definition of "Tires, bulk storage of" for further information.

[F] 903.2.10 Group S-2 enclosed parking garages. An *automatic sprinkler system* shall be provided throughout buildings classified as enclosed parking garages in accordance with Section 406.4 as follows:

1. Where the *fire area* of the enclosed parking garage exceeds 12,000 square feet (1115 m²); or

2. Where the enclosed parking garage is located beneath other groups.

> **Exception:** Enclosed parking garages located beneath Group R-3 occupancies.

❖ Fire records have shown that fires in parking structures typically fully involve only a single automobile with minor damage to adjacent vehicles. An enclosed parking garage, however, does not allow the dissipation of smoke and hot gases as readily as an open parking structure, which is also considered a Group S-2 occupancy. If the enclosed parking garage has a fire area greater than 12,000 square feet (1115 m²) or is located beneath another occupancy group, the enclosed parking garage must be protected with an automatic sprinkler system. This requirement that the enclosed parking garage located beneath other occupancy groups is required to be sprinklered is based on the potential for a fire to develop undetected, which would endanger the occupants of the other occupancy. The 12,000-square-foot (1115 m²) threshold is similar to other occupancies such as Groups M and S-1.

It should be noted that while open parking garages are considered a Group S-2 occupancy, they are not required by the provisions of this section to be equipped with an automatic sprinkler system. Section 406.3.10 indicates that sprinklers in open parking garages are only required when so specified in other sections of the code.

The exception exempts enclosed garages in buildings where the garages are located below a Group R-3 occupancy. The exception is essentially moot since the code requires all buildings with a Group R occupancy to be sprinklered throughout. Because the entire building with the residential occupancy is required to be sprinklered according to Section 903.2.8, the garage would be sprinklered as well. It should be noted that if the Group R-3 occupancy was protected with a 13D system, the enclosed parking garage would not require sprinklers.

[F] 903.2.10.1 Commercial parking garages. An *automatic sprinkler system* shall be provided throughout buildings used for storage of commercial trucks or buses where the *fire area* exceeds 5,000 square feet (464 m²).

❖ Because of the larger-sized vehicles involved in commercial parking structures, such as trucks or buses, a more stringent sprinkler threshold is required. Bus garages may also be located adjacent to passenger terminals (Group A-3) that have a substantial occupant load. Commercial parking requires only a single vehicle in order to be classified as commercial parking.

The criterion for sprinkler protection is based on the size of the fire area and not the size of the commercial parking. If the commercial parking involves only 1,000 square feet (93 m²) but the fire area exceeds 5,000 square feet (464 m²), sprinkler protection is required.

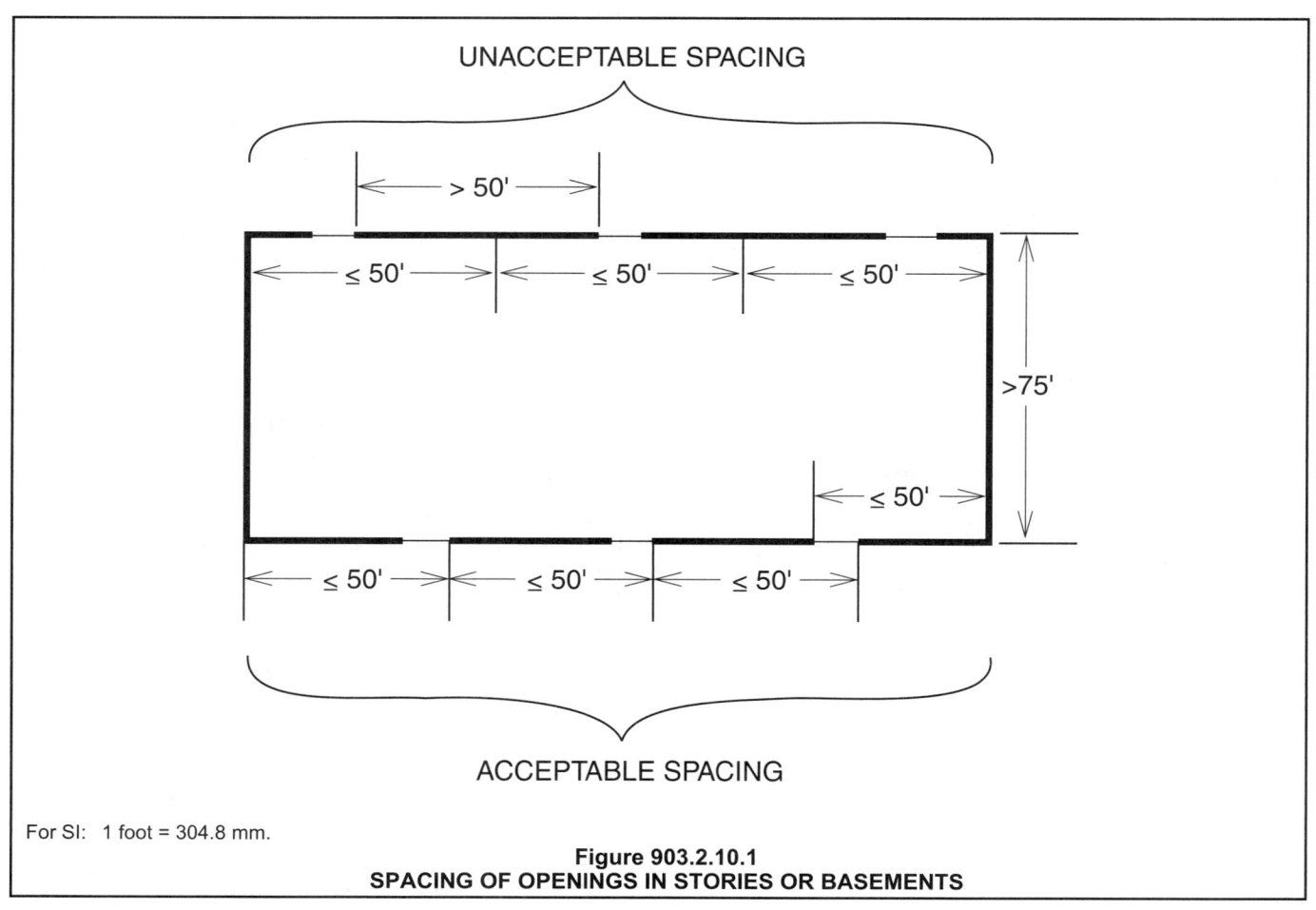

For SI: 1 foot = 304.8 mm.

Figure 903.2.10.1
SPACING OF OPENINGS IN STORIES OR BASEMENTS

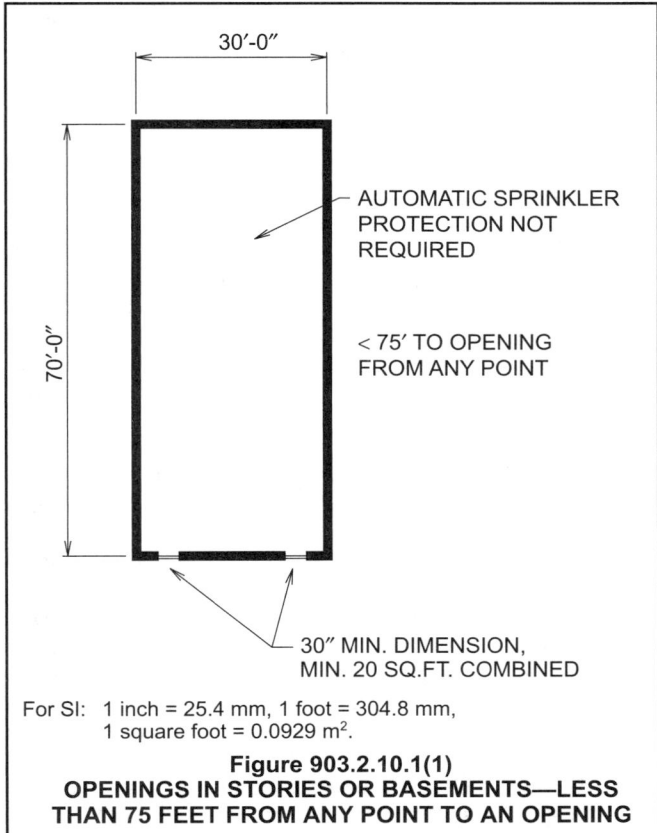

For SI: 1 inch = 25.4 mm, 1 foot = 304.8 mm,
1 square foot = 0.0929 m².

Figure 903.2.10.1(1)
**OPENINGS IN STORIES OR BASEMENTS—LESS
THAN 75 FEET FROM ANY POINT TO AN OPENING**

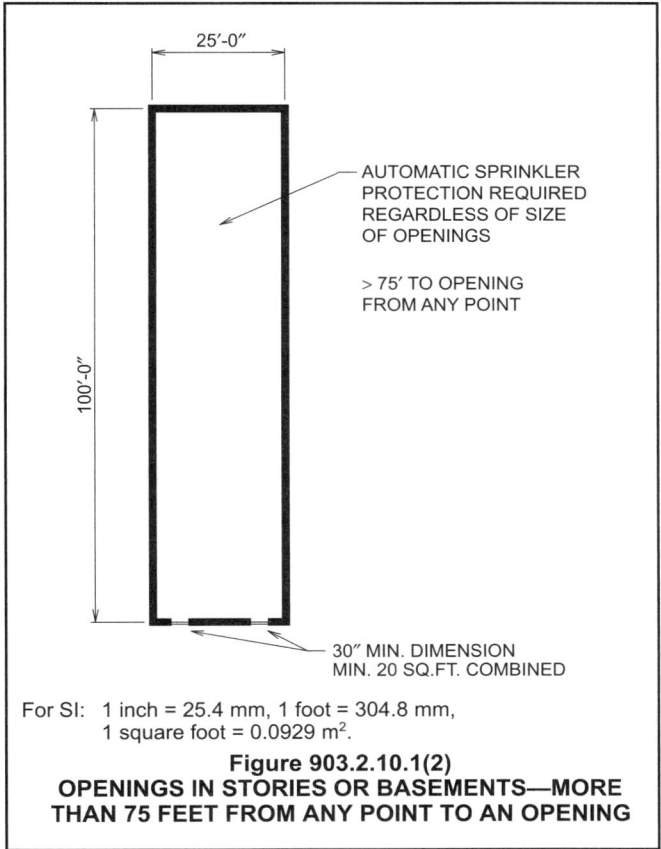

For SI: 1 inch = 25.4 mm, 1 foot = 304.8 mm,
1 square foot = 0.0929 m².

Figure 903.2.10.1(2)
**OPENINGS IN STORIES OR BASEMENTS—MORE
THAN 75 FEET FROM ANY POINT TO AN OPENING**

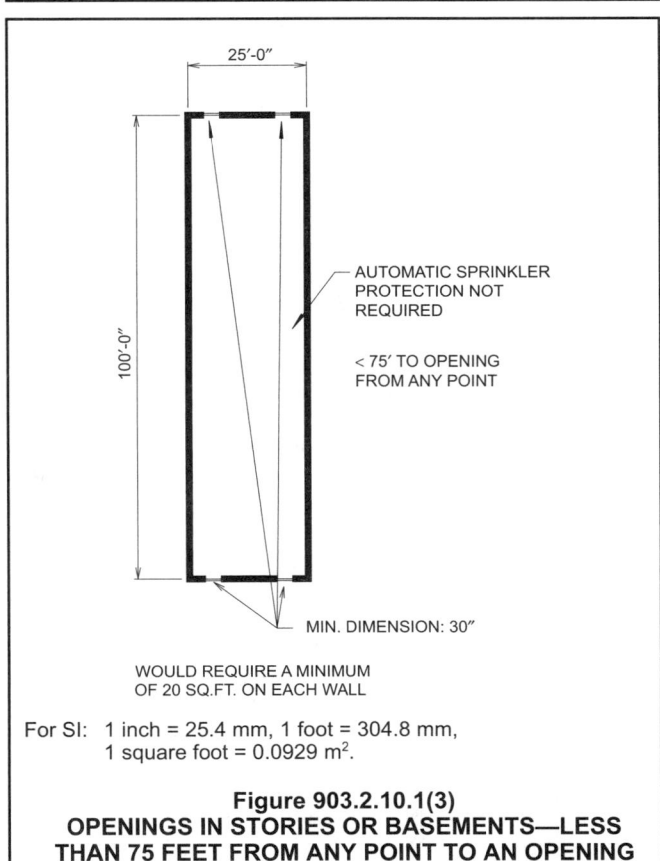

For SI: 1 inch = 25.4 mm, 1 foot = 304.8 mm,
1 square foot = 0.0929 m².

Figure 903.2.10.1(3)
**OPENINGS IN STORIES OR BASEMENTS—LESS
THAN 75 FEET FROM ANY POINT TO AN OPENING**

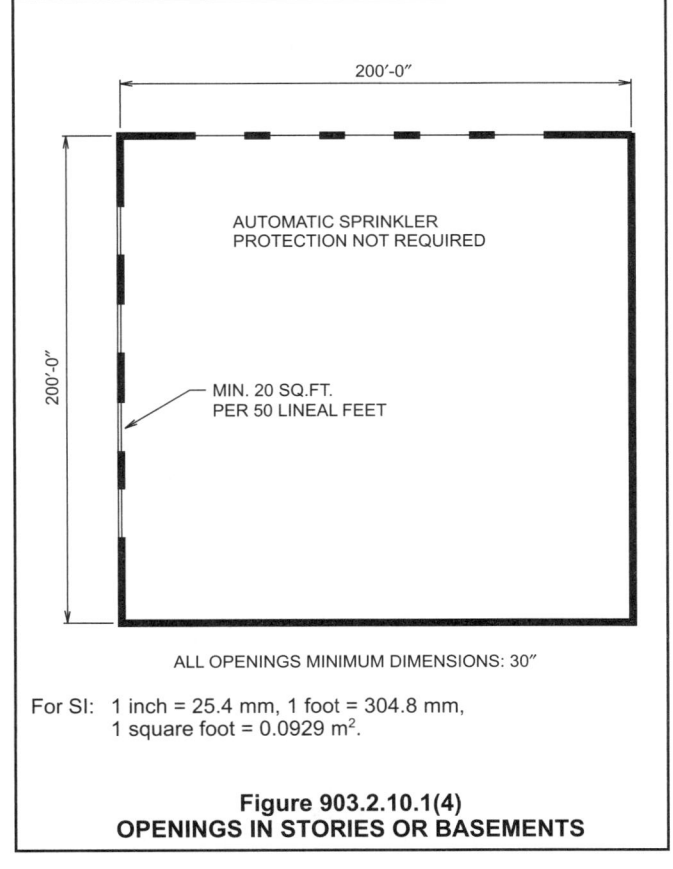

For SI: 1 inch = 25.4 mm, 1 foot = 304.8 mm,
1 square foot = 0.0929 m².

Figure 903.2.10.1(4)
OPENINGS IN STORIES OR BASEMENTS

[F] 903.2.11 Specific building areas and hazards. In all occupancies an *automatic sprinkler system* shall be installed for building design or hazards in the locations set forth in Sections 903.2.11.1 through 903.2.11.6.

Exception: Groups R-3 and U.

❖ Sections 903.2.11.1 through 903.2.11.2 specify certain conditions under which an automatic sprinkler system is required, even in otherwise nonsprinklered buildings. As indicated in the exception, the listed conditions in the noted sections are applicable to all occupancies except Groups R-3 and U. Most structures that qualify as Group U do not typically have the type of conditions stipulated in Sections 903.2.11.1 through 903.2.11.1.3.

The exception for Group R-3 occupancies is consistent with other noted sprinkler exceptions for Group R-3 occupancies, such as Section 903.2.10 for enclosed garages. Similar to Section 903.2.10, the exception does not really apply to Group R-3 since all Group R occupancies regardless of size are required to be sprinklered in accordance with Section 903.2.8.

[F] 903.2.11.1 Stories without openings. An *automatic sprinkler system* shall be installed throughout all *stories*, including basements, of all buildings where the floor area exceeds 1,500 square feet (139.4 m²) and where there is not provided at least one of the following types of *exterior wall* openings:

1. Openings below grade that lead directly to ground level by an exterior *stairway* complying with Section 1009 or an outside ramp complying with Section 1010. Openings shall be located in each 50 linear feet (15 240 mm), or fraction thereof, of *exterior wall* in the *story* on at least one side. The required openings shall be distributed such that the lineal distance between adjacent openings does not exceed 50 feet (15 240 mm).

2. Openings entirely above the adjoining ground level totaling at least 20 square feet (1.86 m²) in each 50 linear feet (15 240 mm), or fraction thereof, of *exterior wall* in the story on at least one side. The required openings shall be distributed such that the lineal distance between adjacent openings does not exceed 50 feet (15 240 mm).

❖ Because of both the lack of openings in exterior walls for access by the fire department for fire fighting and rescue and the problems associated with venting the products of combustion during fire suppression operations, all stories, including any basements of buildings that do not have adequate openings as defined in this section, must be equipped with an automatic sprinkler system. This section applies to stories without sufficient exterior openings where the floor area exceeds 1,500 square feet (139 m²) and where the building is not otherwise required to be fully sprinklered. The requirement for an automatic sprinkler system in this section applies only to the affected area and does not mandate sprinkler protection throughout the entire building.

Stories without openings, as defined in this section, are stories that do not have at least 20 square feet (1.9 m²) of opening leading directly to ground level in each

50 lineal feet (15 240 mm) or fraction thereof on at least one side. Since exterior doors will provide openings of 20 square feet (1.9 m²), or slightly less in some occupancies, exterior stairways and ramps in each 50 lineal feet (15 240 mm) are considered acceptable.

This section specifically states that the required openings be distributed such that the lineal distance between adjacent openings does not exceed 50 feet (15 240 mm). If the openings in the exterior wall are located without regard to the location of the adjacent openings, it is possible that segments of the exterior wall will not have the required access to the interior of the building for fire-fighting purposes. Any arrangement of required stairways, ramps or openings that results in a portion of the wall 50 feet (15 240 mm) or more in length with no openings to the exterior does not meet the intent of the code that access be provided in each 50 lineal feet (15 240 mm) (see Figure 903.2.11.1).

One application of this section has been addressed in the 2009 edition of the *International Code Interpretations* book and deals with automotive service shops that have below-grade service areas where employees perform oil changes and other minor maintenance services. The below-grade areas are typically open to the grade-level service bays via openings providing access to the underside of the vehicles without requiring the vehicle to be lifted into the air. Inasmuch as the below-grade space has no openings directly to the exterior, the question was asked if it would be regulated as a windowless story and thus be required to be equipped with an automatic fire suppression system in accordance with Section 903.2.11.1.

The answer to that question is no. Due to the openness between the adjacent service levels, the below-grade area would be more appropriately regulated similar to a mezzanine rather than a story. A mezzanine is not regulated as a separate story but rather as part of the same story that it serves. Therefore, if the below-grade service level is in compliance with the applicable provisions of Section 505, the windowless story provisions of Section 903.2.11.1 would be evaluated based on the exterior wall openings of the main level and not the service mezzanine below. The direct interconnections between the two adjacent floor levels by multiple service openings provide access to the lower service area for fire fighting and rescue operations. As such, it would not be regulated as a windowless story.

The requirement to sprinkler the basement is independent of mixed-use conditions. Whether the basement is separated or nonseparated is irrelevant to the need for sprinkler protection. Nor does the requirement to provide sprinklers in the basement imply that sprinklers must be provided elsewhere. This requirement is applicable to the basement or any story without openings irrespective of other code provisions.

These provisions are also not based upon the size of a fire area but rather upon the size of the basement. Thus, subdividing the basement into multiple fire areas would have no effect on the requirement. However, one benefit of the multiple fire areas could be that each fire area could have a separate limited area sprinkler system with less than 20 sprinklers.

[F] 903.2.11.1.1 Opening dimensions and access. Openings shall have a minimum dimension of not less than 30 inches (762 mm). Such openings shall be accessible to the fire department from the exterior and shall not be obstructed in a manner that fire fighting or rescue cannot be accomplished from the exterior.

❖ To qualify, an opening must not be less than 30 inches (762 mm) in least dimension and must be accessible to the fire department from the exterior. The minimum opening dimension gives fire department personnel access to the interior of the story or basement for fire-fighting and rescue operations and provides openings that are large enough to vent the products of combustion.

[F] 903.2.11.1.2 Openings on one side only. Where openings in a *story* are provided on only one side and the opposite wall of such *story* is more than 75 feet (22 860 mm) from such openings, the *story* shall be equipped throughout with an *approved automatic sprinkler system*, or openings as specified above shall be provided on at least two sides of the *story*.

❖ If openings are provided on only one side, an automatic sprinkler system would still be required if the opposite wall of the story is more than 75 feet (22 860 mm) from existing openings. An alternative to providing the automatic sprinkler system would be to design openings on at least two sides of the exterior of the building. As long as the story being considered is not a basement, the openings on two sides can be greater than 75 feet (22 860 mm) from any portion of the floor. In basements, if any portion is more than 75 feet (22 860 mm) from the openings, the entire basement must be equipped with an automatic sprinkler system, as indicated in Section 903.2.11.1.3. Providing openings on more than one wall allows cross ventilation to vent the products of combustion [see Figures 903.2.11.1(1-4)].

[F] 903.2.11.1.3 Basements. Where any portion of a basement is located more than 75 feet (22 860 mm) from openings required by Section 903.2.11.1, the basement shall be equipped throughout with an *approved automatic sprinkler system*.

❖ The 75-foot (22 860 mm) distance is intended to be measured in the line of travel—not in a straight line perpendicular to the wall. Where obstructions, such as walls or other partitions, are present in a basement, the walls and partitions enclosing any room or space must have openings that provide an equivalent degree of fire department access to that provided by the openings prescribed in Section 903.2.11.1 for exterior walls. If an equivalent degree of fire department access to all portions of the floor area is not provided, the basement would require an automatic sprinkler system.

[F] 903.2.11.2 Rubbish and linen chutes. An *automatic sprinkler system* shall be installed at the top of rubbish and linen chutes and in their terminal rooms. Chutes extending through three or more floors shall have additional sprinkler heads installed within such chutes at alternate floors. Chute sprinklers shall be accessible for servicing.

❖ The requirement for suppression is within the chute itself and in the chute terminal room but not within the required shaft that encloses the chute. Section 21.16.2.1 of NFPA 13 indicates that the sprinkler coverage is intended to be at or above the top service opening and then at alternate floors inside the chute. The need for sprinkler protection addresses a potential fire in the chute or in the container below it.

[F] 903.2.11.3 Buildings 55 feet or more in height. An *automatic sprinkler system* shall be installed throughout buildings with a floor level having an *occupant load* of 30 or more that is located 55 feet (16 764 mm) or more above the lowest level of fire department vehicle access.

Exceptions:

1. Airport control towers.

2. Open parking structures.

3. Occupancies in Group F-2.

❖ Because of the difficulties associated with manual suppression of a fire in buildings in excess of 55 feet (16 764 mm) in height, an automatic sprinkler system is required throughout the building regardless of occupancy. Buildings that qualify for a sprinkler system under Section 903.2.11.3 are not necessarily high-rise buildings as defined in Section 202.

The listed exceptions are occupancies that, based on height only, do not require an automatic sprinkler system. Airport control towers and open parking structures are also exempt from the high-rise provisions of Section 403. Although an automatic sprinkler system is not required in open parking structures, a sprinkler system may still be needed, depending on the building construction type and the area and number of parking tiers (see Table 406.3.5).

[F] 903.2.11.4 Ducts conveying hazardous exhausts. Where required by the *International Mechanical Code*, automatic sprinklers shall be provided in ducts conveying hazardous exhaust, or flammable or combustible materials.

Exception: Ducts in which the largest cross-sectional diameter of the duct is less than 10 inches (254 mm).

❖ Section 510 of the IMC addresses the requirements for hazardous exhaust systems. To protect against the spread of fire within a hazardous exhaust system and to prevent a duct fire from involving the building, an automatic sprinkler system must be installed to protect the exhaust duct system. Where materials conveyed in the ducts are not compatible with water, alternative extinguishing agents should be used. The fire suppression requirement is intended to apply to exhaust systems having an actual fire hazard. An automatic sprinkler system in the duct would be of little value for an exhaust system that conveys only nonflammable or noncombustible materials, fumes, vapors or gases.

The exception recognizes the reduced hazard associated with smaller ducts and the impracticality of installing sprinkler protection. Another exception in the IMC indicates that laboratory hoods that meet specific provisions of the IMC are not required to be suppressed. Because the IMC is more specific in this regard, it should be consulted for the proper application of the exception.

[F] 903.2.11.5 Commercial cooking operations. An *automatic sprinkler system* shall be installed in commercial kitchen exhaust hood and duct system where an *automatic sprinkler system* is used to comply with Section 904.

❖ An automatic suppression system is required for commercial kitchen exhaust hood and duct systems where required by Section 609 of the IFC or by the IMC to have a Type I hood. Type I hoods are required for commercial cooking equipment that produces grease-laden vapors or smoke. Section 904.11 recognizes that alternative extinguishing systems other than an automatic sprinkler system may be used. Where an automatic sprinkler system is used for commercial cooking operations, it must comply with the requirements identified in Section 904.11.4.

[F] 903.2.11.6 Other required suppression systems. In addition to the requirements of Section 903.2, the provisions indicated in Table 903.2.11.6 also require the installation of a fire suppression system for certain buildings and areas.

❖ In addition to Section 903.2, requirements for automatic fire suppression systems are also found elsewhere in the code as indicated in Table 903.2.11.6.

TABLE 903.2.11.6. See next column.

❖ Table 903.2.11.6 identifies other sections of the code that require an automatic fire suppression system based on the specific occupancy or use. The table does not identify the various sections of the code that contain design alternatives based on the use of an automatic fire suppression system, typically an automatic sprinkler system.

[F] 903.2.12 During construction. *Automatic sprinkler systems* required during construction, *alteration* and demolition operations shall be provided in accordance with Chapter 14 of the *International Fire Code*.

❖ Chapter 33 of the code and Chapter 14 of the IFC address fire safety requirements during construction, alteration or demolition work. Working sprinkler systems should remain operative at all times unless it is absolutely necessary to shut down the system because of the proposed work. All sprinkler system impairments should be rectified as quickly as possible unless specific prior approval has been obtained from the fire code official. Buildings with a required sprinkler system should not be occupied unless the sprinkler

[F] TABLE 903.2.11.6
ADDITIONAL REQUIRED SUPPRESSION SYSTEMS

SECTION	SUBJECT
402.9	Covered malls
403.2, 403.3	High-rise buildings
404.3	Atriums
405.3	Underground structures
407.5	Group I-2
410.6	Stages
411.4	Special amusement buildings
412.4.6, 412.4.6.1, 412.6.5	Aircraft hangars
415.6.2.4	Group H-2
416.4	Flammable finishes
417.4	Drying rooms
507	Unlimited area buildings
508.2.5	Incidental accessory occupancies
1028.6.2.3	Smoke-protected assembly seating
IFC	Sprinkler system requirements as set forth in Section 903.2.11.6 of the *International Fire Code*

system has been installed and tested consistent with Section 901.5. If the system must be placed out of service, the requirements of Section 901.7 of the IFC are necessary to address the temporary impairment to the fire protection system.

[F] 903.3 Installation requirements. *Automatic sprinkler systems* shall be designed and installed in accordance with Sections 903.3.1 through 903.3.6.

❖ Specific design, installation and testing criteria are given for automatic sprinkler systems in the sections and subsections that follow, as well as an indication of the applicability of a nationally recognized standard in the area. The information required to complete a thorough review of an automatic sprinkler system is listed in Figure 903.3.

[F] 903.3.1 Standards. Sprinkler systems shall be designed and installed in accordance with Sections 903.3.1.1, unless otherwise permitted by Sections 903.3.1.2 and 903.3.1.3.

❖ Automatic sprinkler systems are to be installed to comply with the code and NFPA 13, 13R or 13D. As provided for in Section 102.4, where differences occur between the code and NFPA 13, 13R or 13D, the code applies. The fire code official also has the authority to approve the type of sprinkler system to be installed. See Figure 903.3.1 for typical design parameters for each type of sprinkler system.

1. Information required on shop drawings includes:

— Name of owner and occupant
— Location, including street address
— Point of compass
— Graphic indication of scale
— Ceiling construction
— Full-height cross section
— Location of fire walls
— Location of partitions
— Occupancy of each area or room
— Location and size of blind spaces and closets
— Any questionable small enclosures in which no sprinklers are to be installed
— Size of city main in street, pressure and whether dead end or circulation and, if dead end, direction and distance to nearest circulating main, city main test results
— Other source of water supply, with pressure or elevation
— Make, type and orifice size of sprinkler
— Temperature rating and location of high-temperature sprinklers
— Limitations on extended coverage sprinklers or other special sprinkler types
— Number of sprinklers on each riser and on each system by floors and total area by each system on each floor
— Make, type, model and size of alarm or dry pipe valve
— Make, type, model and size of preaction or deluge valve
— Type and location of alarm bells
— Backflow prevention method and details
— Total number of sprinklers on each dry pipe system or preaction deluge system
— Approximate capacity in gallons or each dry pipe system
— Setting for pressure-reducing valves
— Pipe size, type, and schedule of wall thickness
— Cutting lengths of pipe (or center-to-center dimensions)
— Type of fittings, riser nipples and size, and all welds and bends
— Type and location of hangers, inserts and sleeves
— Calculations of loads and details for sway bracing
— All control valves, checks, drain pipes, flushing, and test pipes
— Size and location of standpipe risers and hose outlets
— Small hand-hose equipment
— Underground pipe size, length, location, weight, material, point of connection to city main; the type of valves, meters and valve pits; and the depth that top of the pipe is laid below grade
— Size and location of hydrants along with hose-houses
— Size and location of fire department connections
— When the equipment is to be installed as an addition to an old group of sprinklers without additional feed from the yard system, enough of the old system shall be indicated on the plans to show the total number of sprinklers to be supplied and to make all connections clear
— Information to be provided on the hydraulic nameplate
— Name, address and phone number of contractor and sprinkler designer
— Hydraulic reference points shall be shown by a number and/or letter designation and shall correspond with comparable reference points shown on the hydraulic calculation sheets
— System design criteria showing the minimum rate of water application (density), the design area of water application and the water required for hose streams both inside and outside
— Actual calculated requirements showing the total quantity of water and the pressure required at a common reference point for each system
— Elevation data showing elevations of sprinklers, junction points and supply or reference points
— Protected wall openings if room design method is used

2. Information required on calculations includes:

— Location
— Name of owner and occupant
— Building identification
— Description of hazard
— Name and address of contractor and designer
— Name of approving agency

3. System design requirements include:

— Design area of water application
— Minimum rate of water application (density)
— Area of sprinkler coverage
— Hazard or commodity classification
— Building height
— Storage height
— Storage method
— Total water requirements, as calculated, including allowance for hose demand water supply information and allowance for in-rack sprinklers
— Location and elevation static and residual test gauge with relation to the riser reference point
— Size and location of hydrants used for flow test data
— Flow location
— Static pressure, psi
— Residual pressure, psi
— Flow, gpm
— Date
— Time
— Test conducted by whom
— Sketch to accompany gridded system calculations to indicate flow quantities and directions for lines with sprinklers operated in the remote area

4. Additional information necessary for complete review includes:

— Sprinkler description and discharge constant (K value)
— Hydraulic reference points
— Flow, gpm
— Pipe diameter (actual internal diameter)
— Pipe length
— Equivalent pipe length for fittings and components
— Friction loss in psi per foot of pipe
— Total friction loss between reference points
— Elevation difference between reference points
— Required pressure in psi at each reference point
— Velocity pressures and normal pressure if included in calculations
— Notes to indicate starting points, reference to other sheets or classification of date
— Information on antifreeze solution (type and quantity)
— Water treatment system information including reason for treatment and program details

5. Included with the submittal must be a graph sheet showing water supply curves and system requirements including:

— Hose demand plotted on semilogarithmic graph paper so as to present a graphic summary of the complete hydraulic calculations
— Sprinkler system demand including in-rack sprinklers (if applicable)

Figure 903.3
SAMPLE SPRINKLER SYSTEM DRAWING AND DATA SUBMITTALS

	NFPA 13	NFPA 13R	NFPA 13D
Extent of protection	Equip throughout (Section 903.3.1.1)	Occupied spaces (Section 903.3.1.2)	Occupied spaces (Section 903.3.1.3)
Scope	All occupancies	Low-rise residential	One- and two-family dwellings
Sprinkler design	Density/area concept	4-head design	2-head design
Sprinklers	All types	Residential only	Residential only
Duration	30 minutes (minimum)	30 minutes	10 minutes
Advantages	Property and life protection	Life safety/tenability	Life safety/tenability

Figure 903.3.1
NFPA 13, NFPA 13R, NFPA 13D SYSTEMS

[F] 903.3.1.1 NFPA 13 sprinkler systems. Where the provisions of this code require that a building or portion thereof be equipped throughout with an *automatic sprinkler system* in accordance with this section, sprinklers shall be installed throughout in accordance with NFPA 13 except as provided in Section 903.3.1.1.1.

❖ NFPA 13 contains the minimum requirements for the design and installation of automatic water sprinkler systems and exposure protection sprinkler systems. The requirements contained in the standard include the character and adequacy of the water supply and the selection of sprinklers, piping, valves and all of the materials and accessories. The standard does not include requirements for installation of private fire service mains and their appurtenances; installation of fire pumps or construction and installation of gravity and pressure tanks and towers.

NFPA 13 defines seven classifications or types of water sprinkler systems: wet pipe (see Figure 903.3.1.1, dry pipe; preaction or deluge; combined dry pipe and preaction; antifreeze systems; sprinkler systems that are designed for a special purpose and outside sprinklers for exposure protection. While numerous variables must be considered in selecting the proper type of sprinkler system, the wet-pipe sprinkler system is recognized as the most effective and efficient. The wet-pipe system is also the most reliable type of sprinkler system, because water under pressure is available at the sprinkler. Therefore, wet-pipe sprinkler systems are recommended wherever possible.

The extent of coverage and distribution of sprinklers is based on the NFPA 13 standard. Numerous conditions exist in the standard where sprinklers are specifically required and also where they may or may not be located. Once it is determined that the sprinkler system is to be in accordance with NFPA 13, that standard must be reviewed for installation details. For example, exterior spaces such as combustible canopies are required to be equipped with sprinklers according to Section 8.15.7 of NFPA 13 where the canopy extends for a distance of 4 feet (1219 mm) or more. A 3-foot (914 mm) combustible canopy would not require sprinklers nor would a 6-foot (1829 mm) canopy constructed of noncombustible materials, provided there is no combustible storage under the canopy.

Because installation is required to be in accordance with NFPA 13, if the standard allows for the omission of sprinklers in any location, then the building is still considered as sprinklered throughout. For example, Section 8.15.8.1.1 of NFPA 13 allows sprinklers to be omitted from bathrooms in certain circumstances. If sprinklers are not provided in the bathrooms due to the conditions stipulated in NFPA 13, the building would still be considered as sprinklered throughout in accordance with the code, NFPA 13 and the IFC.

Exceptions for the use of NFPA 13R and 13D systems are addressed throughout the code when exceptions based upon the use of sprinklers are provided. More specifically, if the use of these other standards is appropriate it will be noted within the exception. For a building to be considered "equipped throughout" with an NFPA 13 sprinkler system, complete protection must be provided in accordance with the referenced standard, subject to the exempt locations indicated in Section 903.3.1.1.1.

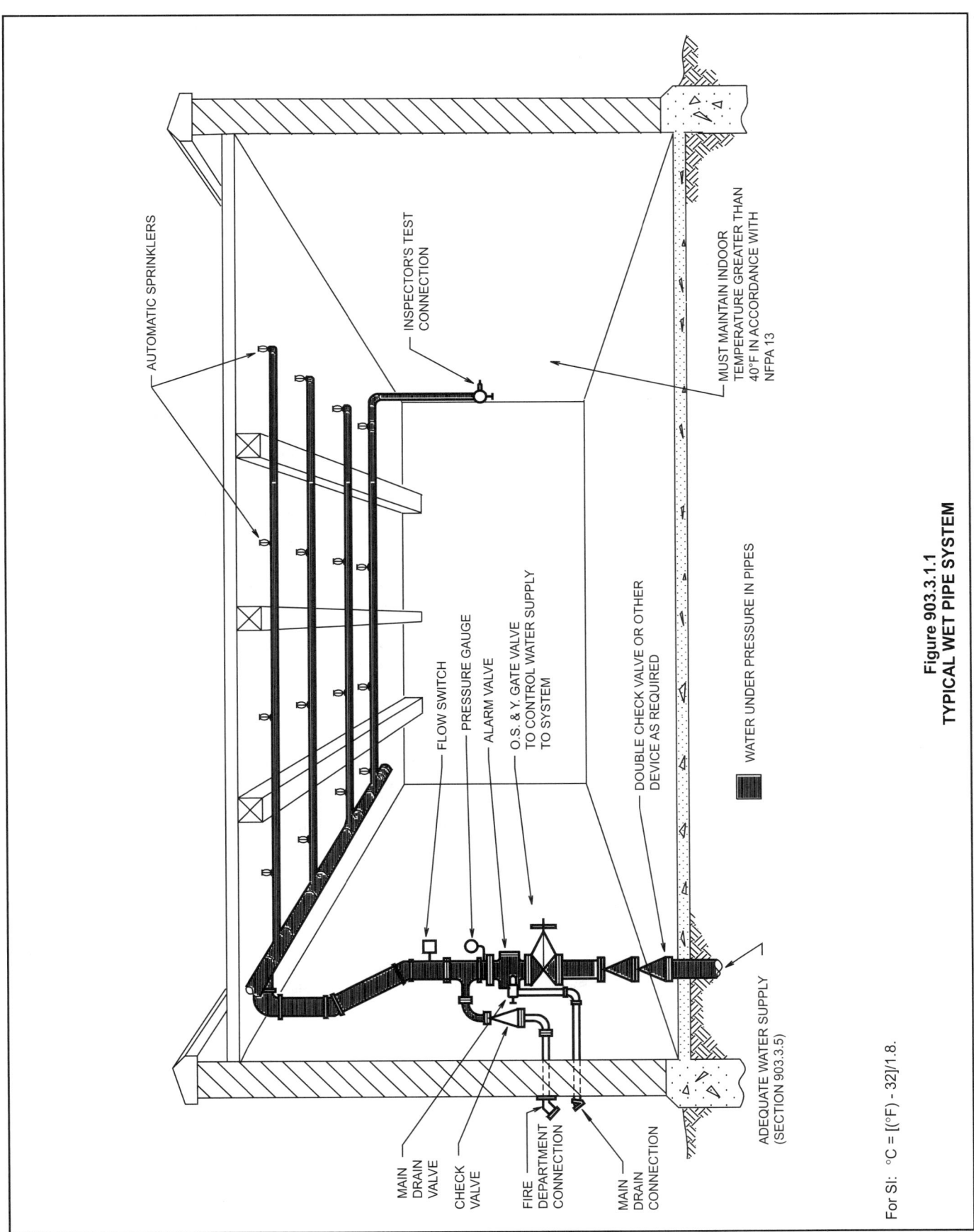

AUTOMATIC SPRINKLERS

INSPECTOR'S TEST CONNECTION

MUST MAINTAIN INDOOR TEMPERATURE GREATER THAN 40°F IN ACCORDANCE WITH NFPA 13

FLOW SWITCH

PRESSURE GAUGE

ALARM VALVE

O.S. & Y. GATE VALVE TO CONTROL WATER SUPPLY TO SYSTEM

DOUBLE CHECK VALVE OR OTHER DEVICE AS REQUIRED

WATER UNDER PRESSURE IN PIPES

MAIN DRAIN VALVE

CHECK VALVE

FIRE DEPARTMENT CONNECTION

MAIN DRAIN CONNECTION

ADEQUATE WATER SUPPLY (SECTION 903.3.5)

Figure 903.3.1.1
TYPICAL WET PIPE SYSTEM

For SI: °C = [(°F) - 32]/1.8.

[F] 903.3.1.1.1 Exempt locations. Automatic sprinklers shall not be required in the following rooms or areas where such rooms or areas are protected with an *approved* automatic fire detection system in accordance with Section 907.2 that will respond to visible or invisible particles of combustion. Sprinklers shall not be omitted from any room merely because it is damp, of fire-resistance-rated construction or contains electrical equipment.

1. Any room where the application of water, or flame and water, constitutes a serious life or fire hazard.

2. Any room or space where sprinklers are considered undesirable because of the nature of the contents, when *approved* by the fire code official.

3. Generator and transformer rooms separated from the remainder of the building by walls and floor/ceiling or roof/ceiling assemblies having a *fire-resistance rating* of not less than 2 hours.

4. Rooms or areas that are of noncombustible construction with wholly noncombustible contents.

5. Fire service access elevator machine rooms and machinery spaces.

❖ This section allows the omission of sprinkler protection in certain locations if an approved automatic fire detection system is installed. Buildings in compliance with one of the four listed conditions would still be considered fully sprinklered throughout in compliance with the code and NFPA 13 and thus are eligible for all applicable code alternatives, exceptions or reductions. Elimination of the sprinkler system in a sensitive area is subject to the approval of the fire code official.

Condition 1 addresses restrictions where the application of water could create a hazardous condition. For example, sprinkler protection is to be avoided where it is not compatible with certain stored materials (i.e., some water-reactive hazardous materials). Combustible metals, such as magnesium and aluminum, may burn so intensely that the use of water to attempt fire control will only intensify the reaction.

It is not the intent of Condition 2 to omit sprinklers solely because of a potential for water damage. A desire to not sprinkler a certain area (such as a computer room or operating room) does not fall within the limitations of the exception unless there is something unique about the space that would result in water being incompatible. A computer room can be adequately protected using a preaction sprinkler system in combination with an alternative gaseous suppression agent. The intent of Condition 2 is to consider whether or not the contents would react adversely to the application of water. It is important to note that the fire code official must approve the use of this item.

Condition 3 recognizes the low fuel load and low occupancy hazards associated with generator and transformer rooms and, therefore, allows the omission of sprinkler protection if the rooms are separated from adjacent areas by 2-hour fire-resistance-rated construction. This condition assumes the room is not used for any combustible storage. This condition is similar to Section 8.15.10.3 of NFPA 13, which exempts electrical equipment rooms from sprinkler protection, provided the room is dedicated to the use of dry-type electrical equipment, is constructed as a 2-hour fire-resistance-rated enclosure and is not used for combustible storage.

Condition 4 requires the construction of the room or area, as well as the contents, to be noncombustible. An example would be an area in an unprotected steel frame building (Type IIB construction) used for steel or concrete block storage. Neither involves any significant combustible packaging or sources of ignition, and few combustibles are present (see Figure 903.3.1).

Condition 5 addresses the concern for elevator machine rooms and machinery spaces associated with fire service access elevators as required for buildings with occupied floors greater than 120 feet (36.58 m) from the lowest level of fire department access by Sections 403.6.1 and 3007. These elevators need to work during fire situations and their operation cannot be threatened by the activation of a sprinkler in a machine room or space that may affect the operation of the elevator. Fire service access elevators are required to be continuously monitored at the fire command center in accordance with Section 3007.6.

[F] 903.3.1.2 NFPA 13R sprinkler systems. Automatic sprinkler systems in Group R occupancies up to and including four stories in height shall be permitted to be installed throughout in accordance with NFPA 13R.

❖ NFPA 13R contains design and installation requirements for a sprinkler system to aid in the detection and control of fires in low-rise (four stories or less) residential occupancies.

Sprinkler systems designed in accordance with NFPA 13R are intended to prevent flashover (total involvement) in the room of fire origin and to improve the chance for occupants to escape or be evacuated. The design criteria in NFPA 13R are similar to those in NFPA 13 except that sprinklers may be omitted from areas in which fatal fires in residential occupancies do not typically originate (bathrooms, closets, attics, porches, garages and concealed spaces).

A common question is whether a mixed occupancy building which contains a Group R occupancy could still use NFPA 13R for the design. If one of the mixed use occupancies would require a sprinkler system throughout the building in accordance with NFPA 13, then a 13R system would not be allowed. If, however, the only reason a sprinkler system is being installed is because there is a Group R fire area within the building, then an NFPA 13R system would be an appropriate design choice. The areas that are not classified as Group R would require protection in accordance with NFPA 13.

It must be noted that although the building would be considered sprinklered throughout in accordance with NFPA 13R, not all of the code sprinkler alternatives could be applied. Any alternative that requires the installation of an NFPA 13 system would not be applicable if a portion of the building utilizes an NFPA 13R system.

The code provisions that allow for an increase in building height according to Section 504.2 do not compound this section. NFPA 13R is applicable to buildings that are up to four stories in height. If the design of a residential building intends to take advantage of the sprinkler height increase so that the building is five stories or more, the sprinkler system must be an NFPA 13 system. Because this section limits the height to four stories, that is the maximum height for a building that can utilize an NFPA 13R system. This is consistent with the scoping provisions in the NFPA 13R standard.

The limitation of four stories in height is to be measured with respect to the established grade plane, which is consistent with IFC Interpretation No. 43-03. As such, a basement would not be considered a story above grade for purposes of determining the applicability of this section.

[F] 903.3.1.2.1 Balconies and decks. Sprinkler protection shall be provided for exterior balconies, decks and ground floor patios of *dwelling units* where the building is of Type V construction, provided there is a roof or deck above. Sidewall sprinklers that are used to protect such areas shall be permitted to be located such that their deflectors are within 1 inch (25 mm) to 6 inches (152 mm) below the structural members and a maximum distance of 14 inches (356 mm) below the deck of the exterior balconies and decks that are constructed of open wood joist construction.

❖ Balconies, decks and patios in buildings of Type V construction and used for Group R occupancies are required to have sprinkler protection when there is a roof or deck above. This is in addition to the requirements of NFPA 13R, which primarily addresses the life safety of occupants and not property protection. The intent is to address hazards such as grilling and similar activities. Since NFPA 13R does not require such coverage, there is potential that a fire on a balcony could grow much too large for the system within the building to handle. The concern is that a potential exterior balcony fire could spread to unprotected floor/ceiling assemblies and attic spaces and result in major property damage. Section 308.1.4 of the IFC specifically addresses restrictions on open flame cooking devices used on combustible balconies. Note that sprinklers are not intended to be provided in closets found on such balconies.

Regardless of whether the exterior walking surface is attached to the building and called a balcony or is a freestanding structure such as a deck or patio the concern for fire ignition in the area adjacent to the exterior wall is the same. Sidewall sprinklers should be selected based on the area of coverage and climate. If the potential for freezing exists, a dry sidewall sprinkler should be used. Where the overhanging deck or balcony is extensive, an extended coverage sprinkler should be selected.

[F] 903.3.1.3 NFPA 13D sprinkler systems. Automatic sprinkler systems installed in one- and two-family dwellings and townhouses shall be permitted to be installed throughout in accordance with NFPA 13D.

❖ NFPA 13D contains design and installation requirements for a sprinkler system to aid in the detection and control of fires in one- and two-family dwellings, mobile homes and townhouses. Similar to NFPA 13R, sprinkler systems designed in accordance with NFPA 13D are intended to prevent flashover (total involvement) in the room of fire origin and to improve the chance for occupants to escape or be evacuated. Although the allowable omission of sprinklers in certain areas of the dwelling unit in NFPA 13D is similar to that in NFPA 13R, the water supply requirements are less restrictive. NFPA 13D uses a two-head sprinkler design with a 10-minute duration requirement, while NFPA 13R uses a four-head sprinkler design with a 30-minute duration requirement. The decreased water supply requirement emphasizes the main intent of NFPA 13D to control the fire and maintain tenability during evacuation of the residence.

Since the fire code official has the authority to approve the type of sprinkler system, this section may be used to prevent the use of a specific type of sprinkler system that may be inappropriate for a particular type of occupancy.

[F] 903.3.2 Quick-response and residential sprinklers. Where automatic sprinkler systems are required by this code, quick-response or residential automatic sprinklers shall be installed in the following areas in accordance with Section 903.3.1 and their listings:

1. Throughout all spaces within a smoke compartment containing patient sleeping units in Group I-2 in accordance with this code.

2. *Dwelling units*, and *sleeping units* in Group R and I-1 occupancies.

3. Light-hazard occupancies as defined in NFPA 13.

❖ This section requires the use of either listed quick-response or residential automatic sprinklers depending on the type of sprinkler system required to achieve faster and more effective suppression in certain areas. Residential sprinklers are required in all types of residential buildings that would permit the use of an NFPA 13R or 13D sprinkler system.

Quick-response and residential sprinklers are similar in nature. They use a lighter material for the operating mechanism, thus reducing the heat lag in the element. The faster the heat can be absorbed, the sooner the sprinkler will begin to discharge water. Quick-response sprinklers have shown that they operate up to 25 percent faster than traditional sprinklers and create conditions in the room of origin that significantly increase the tenability of the environment. In tests performed by Factory Mutual (FM) for the Federal Emergency Management Agency (FEMA), the gas temperature in the room of origin was 550°F (288°C) with quick-response sprinklers, while it was 1,470°F (799°C) or conventional sprinklers at the time of sprinkler activation. More importantly, while the carbon monoxide (CO) level was 1,860 ppm for conventional sprinklers, the CO level when tested with quick-response sprinklers was only around 350 ppm. Comparatively, the National Institute of Occupational safety and Health (NIOSH) considers the IDLH (immediately dangerous to life and health) level of CO to be 1,200 ppm. Thus, quick-response sprinklers have been shown to add significantly to the life safety effects of standard sprinkler systems.

Condition 1 reiterates the requirements of Section 407.5 to use approved quick-response or residential sprinklers in smoke compartments containing patient sleeping units in Group I-2 occupancies. Even though properly operating standard sprinklers are effective, the extent of fire growth and smoke production that can occur before sprinkler activation creates the need for early warning to enable faster response by staff and initiation of egress that is critical in occupancies containing persons incapable of self-preservation. The faster response time associated with quick-response or residential sprinklers increases the probability that the sprinklers will actuate before the patient's life would be threatened by a fire in his or her room.

Because of the kind of occupants sleeping in Group R and I-1 occupancies, as indicated in Condition 2, a faster responding type of sprinkler is desirable. Similar to the first condition, because occupants will be sleeping, the use of quick-response sprinklers creates additional safety by reducing sprinkler response time, thereby increasing the time available for egress and allowing for the time necessary for occupants to wake up and recognize the emergency event.

Condition 3 recognizes light-hazard occupancies in accordance with NFPA 13. These could include restaurants, schools, office buildings, churches and similar occupancies where the fire load and potential heat release of combustible contents are low.

[F] 903.3.3 Obstructed locations. Automatic sprinklers shall be installed with due regard to obstructions that will delay activation or obstruct the water distribution pattern. Automatic sprinklers shall be installed in or under covered kiosks, displays, booths, concession stands, or equipment that exceeds 4 feet (1219 mm) in width. Not less than a 3-foot (914 mm) clearance shall be maintained between automatic sprinklers and the top of piles of combustible fibers.

> **Exception:** Kitchen equipment under exhaust hoods protected with a fire-extinguishing system in accordance with Section 904.

❖ To provide adequate sprinkler coverage, sprinkler protection must be extended under any obstruction that exceeds 4 feet (1219 mm) in width. Large air ducts are another common obstruction where sprinklers are routinely extended beneath the duct. The 3-foot (914 mm) storage clearance requirement for combustible fibers is caused by their potential high heat release. Most storage conditions require only a minimum 18-inch (457 mm) storage clearance to combustibles, depending on the type of sprinklers used and their actual storage conditions.

The exception recognizes that an alternative extinguishing system is permitted for commercial cooking systems in place of sprinkler protection for exhaust hoods that may be more than 4 feet (1219 mm) wide.

The application of this section is more critical to the ongoing use of the space. The obstruction conditions, therefore, should have already been addressed during plan review and installation inspection. This section gives the fire official and building owner adequate information to avoid the most typical obstruction-related issues in terms of proper sprinkler coverage.

[F] 903.3.4 Actuation. *Automatic sprinkler systems* shall be automatically actuated unless specifically provided for in this code.

❖ The intent of this section is to eliminate the need for occupant intervention during a fire. As such, it is assumed that it will not be necessary for a person to manually open a valve or perform some other physical activity in order to allow the sprinkler system to activate.

Wet-pipe and dry-pipe sprinkler systems, for example, are essentially fail-safe systems in the sense that, if the system is in proper operating condition, it will operate once a sprinkler fuses. Dry systems have an inherent time lag for water to reach the sprinkler; therefore, the response is not as fast as for a wet-pipe system. Other types of sprinkler systems, such as preaction and deluge, rely on the actuation of a detection system to operate the sprinkler valve.

[F] 903.3.5 Water supplies. Water supplies for *automatic sprinkler systems* shall comply with this section and the standards referenced in Section 903.3.1. The potable water supply shall be protected against backflow in accordance with the requirements of this section and the *International Plumbing Code*.

❖ To be effective, all sprinkler systems must have an adequate supply of water. The criteria for an acceptable water supply are contained in the standards referenced in Section 903.3.1. For example, NFPA 13 contains criteria for different types of water supplies as well as the methods to determine the pressure, flow capabilities and capacity necessary to get the intended performance from a sprinkler system. An acceptable water supply could consist of a reliable municipal supply, a gravity tank or a fire pump with a pressure tank or a combination of these.

This section also establishes the requirements for protecting the potable water system against a nonpotable source, such as stagnant water retained within the sprinkler piping. As stated in Section 608.16.4 of the IPC, an approved double check valve device or reduced pressure principle backflow preventer is required.

[F] 903.3.5.1 Domestic services. Where the domestic service provides the water supply for the *automatic sprinkler system*, the supply shall be in accordance with this section.

❖ This section establishes the scope of domestic services for limited area sprinkler systems and residential combination services.

[F] 903.3.5.1.1 Limited area sprinkler systems. Limited area sprinkler systems serving fewer than 20 sprinklers on any single connection are permitted to be connected to the domestic service where a wet automatic standpipe is not available. Limited area sprinkler systems connected to domestic water supplies shall comply with each of the following requirements:

1. Valves shall not be installed between the domestic water riser control valve and the sprinklers.

   Exception: An *approved* indicating control valve supervised in the open position in accordance with Section 903.4.

2. The domestic service shall be capable of supplying the simultaneous domestic demand and the sprinkler demand required to be hydraulically calculated by NFPA 13, NFPA 13R or NFPA 13D.

❖ Use of limited area sprinkler systems is primarily limited to fire areas or other areas where the number of sprinklers does not exceed 19.

The use of limited area sprinkler systems is restricted to cases in which the code requires a limited number of sprinklers and not a complete automatic sprinkler system. For example, limited area sprinkler systems may be used to protect stages; storage and workshop areas; stories, including basements, without openings; painting rooms; trash rooms and chutes; furnace rooms; kitchens and hazardous exhaust systems and incidental accessory occupancies as regulated in Section 508.2.5. When a wet automatic standpipe is not available, limited-area sprinkler systems may be connected to the domestic water supply.

The water supply to the sprinkler system is to be controlled only by the same valve that controls the domestic water supply to the building; no shutoff valves are permitted in the sprinkler system piping. These restrictions increase the likelihood that the sprinkler system will be operational should a fire occur. Likewise, if the sprinkler system needs restoration after having operated in response to a fire or needs repairs requiring that the water supply be shut off, this section increases the probability that the system will be restored quickly, because having the domestic water supply to the entire building shut off is an inconvenience to occupants that they will not tolerate for very long.

The exception recognizes the value of standard sprinkler system valve supervision complying with Section 903.4 as providing the level of system reliability contemplated by this section.

Documentation, usually in the form of hydraulic calculations, must be submitted demonstrating that the domestic water system is adequate to supply the sprinkler demand in addition to the peak domestic demand. The domestic demand would normally be determined as stated in the IPC.

[F] 903.3.5.1.2 Residential combination services. A single combination water supply shall be allowed provided that the

domestic demand is added to the sprinkler demand as required by NFPA 13R.

❖ NFPA 13R permits a common supply main to a building to serve both the sprinkler system and domestic services if the domestic demand is added to the sprinkler demand. NFPA 13R systems do not provide the same level of property protection as NFPA 13 systems.

[F] 903.3.5.2 Secondary water supply. A secondary on-site water supply equal to the hydraulically calculated sprinkler demand, including the hose stream requirement, shall be provided for high-rise buildings assigned to Seismic Design Category C, D, E or F as determined by this code. The secondary water supply shall have a duration of not less than 30 minutes as determined by the occupancy hazard classification in accordance with NFPA 13.

Exception: Existing buildings.

❖ The intent of this section is that a secondary water supply be provided on the high-rise building site in order to provide a high level of functional reliability for the fire protection systems in the event a seismic event disables the primary water supply for high-rise buildings assigned to Seismic Design Category C, D, B or F. The categories are described in Section 1613.5.

The text's specific wording that the secondary supply be "on-site" rather than "to the site" would preclude the use, for example, of a second connection to the municipal supply remote from the primary connection to the municipal supply to achieve compliance with this requirement. It should be noted, however, that the fire code official has the authority to modify the provisions of the code in accordance with Section 104.10 should there be practical difficulties preventing precise compliance with the text.

The required amount of water is equal to the hydraulically calculated sprinkler demand plus hose stream demand for a minimum 30-minute period dependent upon the appropriate occupancy hazard classification in NFPA 13.

The exception recognizes the infeasibility of requiring a secondary water supply in existing high-rise buildings.

[F] 903.3.6 Hose threads. Fire hose threads and fittings used in connection with *automatic sprinkler systems* shall be as prescribed by the fire code official.

❖ The threads on connections and fittings that the fire department will use to connect a hose must be compatible with the fire department threads.

Design documents must specify the type of thread to be used in order to be compatible with the local fire department equipment after consultation and coordination with the fire code official. The criteria typically apply to fire department connections for sprinkler and standpipe systems, standpipe hose connections, yard hydrants and wall hydrants.

The majority of fire departments in the United States use the American National Fire Hose Connection Screw Thread (NH) also commonly known as NST and NS. NFPA 1963 gives the screw thread dimensions and the thread size of threaded connections, with nominal sizes ranging from $^3/_4$ inch (19 mm) to 6 inches (152 mm) for the NH thread. Although efforts to standardize fire hose threads began after the Boston conflagration in 1872, there are still many different screw threads, some of which give the appearance of compatibility with the NH thread. While NFPA 1963 may be used as a guide, the code does not require that any particular standard be used. Rather, it is important that the fire code official be consulted for the appropriate thread selection. The intent is that the threads match those of the local department identically so that adapters are not required within the fire department's own district.

[F] 903.4 Sprinkler system supervision and alarms. All valves controlling the water supply for *automatic sprinkler systems*, pumps, tanks, water levels and temperatures, critical air pressures and waterflow switches on all sprinkler systems shall be electrically supervised by a *listed* fire alarm control unit.

Exceptions:

1. *Automatic sprinkler systems* protecting one- and two-family *dwellings*.

2. Limited area systems serving fewer than 20 sprinklers.

3. *Automatic sprinkler systems* installed in accordance with NFPA 13R where a common supply main is used to supply both domestic water and the *automatic sprinkler system*, and a separate shutoff valve for the *automatic sprinkler system* is not provided.

4. Jockey pump control valves that are sealed or locked in the open position.

5. Control valves to commercial kitchen hoods, paint spray booths or dip tanks that are sealed or locked in the open position.

6. Valves controlling the fuel supply to fire pump engines that are sealed or locked in the open position.

7. Trim valves to pressure switches in dry, preaction and deluge sprinkler systems that are sealed or locked in the open position.

❖ The reliability data on automatic sprinkler systems clearly indicate that a closed valve is the leading cause of sprinkler system failure. There are also a number of other critical elements that contribute to successful sprinkler system operation, including, but not limited to, pumps, water tanks and air pressure maintenance devices; therefore, this section requires that the various critical elements that contribute to an available water supply and to the function of the sprinkler system be electrically supervised.

Automatic sprinkler systems in one- and two-family dwellings are typically designed to comply with NFPA 13D, which does not require electrical supervision (see Exception 1).

Limited area sprinkler systems are generally supervised by their connection to the domestic water service (see Exception 2). Compliance with the exception

means that none of the following alarm provisions are applicable to limited area systems. Consequently, limited area sprinkler systems do not require local alarms or supervision. Electrical supervision is required only if a control valve is installed between the riser control valve and the sprinkler system piping.

Similar to limited area sprinkler systems, electrical supervision is not required for NFPA 13R residential combination services when a shutoff valve is not installed (see Exception 3). NFPA 13R sprinkler systems are supervised in that the only way to shut off the sprinkler system is to also shut off the domestic water supply.

The valves discussed in Exceptions 4 through 7 can be sealed or locked in the open position because they do not control the sprinkler system water supply.

[F] 903.4.1 Monitoring. Alarm, supervisory and trouble signals shall be distinctly different and shall be automatically transmitted to an *approved* supervising station or, when *approved* by the fire code official, shall sound an audible signal at a *constantly attended location*.

Exceptions:

1. Underground key or hub valves in roadway boxes provided by the municipality or public utility are not required to be monitored.

2. Backflow prevention device test valves located in limited area sprinkler system supply piping shall be locked in the open position. In occupancies required to be equipped with a fire alarm system, the backflow preventer valves shall be electrically supervised by a tamper switch installed in accordance with NFPA 72 and separately annunciated.

❖ Automatic sprinkler systems must be supervised as a means of determining that the system is operational. A valve supervisory switch operating as a normally open or normally closed switch is usually used. NFPA 72 does not permit valve supervisory switches to be connected to the same zone circuit as the waterflow switch unless it is specifically arranged to actuate a signal that is distinctive from the circuit trouble condition signal.

Required sprinkler systems are to be monitored by an approved supervising service to comply with NFPA 72. Types of supervising stations recognized in NFPA 72 include central stations, remote supervising stations or proprietary supervising stations.

A central station is an independent off-site facility operated and maintained by personnel whose primary business is to furnish, maintain, record and supervise a signaling system. A proprietary system is similar to a central station system; however, a proprietary system is typically an on-site facility monitoring a number of buildings on the same site for the same owner. A remote station system has an alarm signal that is transmitted to a remote location acceptable to the authority having jurisdiction and that is attended 24 hours a day. The receiving equipment is usually located at a fire station, police station, regional emergency communications center or telephone answering service. An alternative use to the three previous supervising methods is an audible signal that can be transmitted to a constantly attended location approved by the fire code official.

Exception 1 recognizes that underground key or hub valves in roadway boxes are not normally supervised or required to be supervised by this section or NFPA 13.

Exception 2 acknowledges that local water utilities and environmental authorities in many instances require, by local ordinances, that backflow prevention devices be installed in limited-area sprinkler system piping. To make the testing and maintenance of backflow prevention devices easier, test valves are installed on each side of the device. These valves are typically indicating-type valves and can function as shutoff valves for the sprinkler system, and therefore require some level of supervision.

Because these infrequently used valves may be the only feature of protection requiring supervision in occupancies not otherwise required to be equipped with a fire alarm system, Exception 2 permits these valves to be locked in the open position; however, if the occupancy is protected by a fire alarm system, these valves must be equipped with approved valve supervisory devices connected to the fire alarm control panel on a separate (supervisory) zone so that the supervisory signal is transmitted to the designated receiving station. Installation and testing of backflow preventers in sprinkler systems are regulated in Sections 312.10 (testing) and 608.16.4 (devices) of the IPC.

[F] 903.4.2 Alarms. *Approved* audible devices shall be connected to every *automatic sprinkler system*. Such sprinkler waterflow alarm devices shall be activated by waterflow equivalent to the flow of a single sprinkler of the smallest orifice size installed in the system. Alarm devices shall be provided on the exterior of the building in an *approved* location. Where a fire alarm system is installed, actuation of the *automatic sprinkler system* shall actuate the building fire alarm system.

❖ The audible alarm, sometimes referred to as the "outside ringer" or "water-motor gong," sounds when the sprinkler system has activated. The alarm device may be electrically operated or it may be a true water-motor gong operated by a paddle-wheel-type attachment to the sprinkler system riser that responds to the flow of water in the piping. Though no longer the alarm device of choice, water-motor gongs do have the advantage of not being subject to power failures within or outside the protected building (see Sections 6.9 and 8.17 of NFPA 13 for further information on these devices). The alarm must be installed on the exterior of the building in a location approved by the fire code official. This location is often in close proximity to the fire department connection (FDC), serving a collateral function of helping the responding fire apparatus engineer more promptly locate the FDC.

The alarm is not intended to be an evacuation alarm. The requirement is also not intended to be an indirect requirement for a fire alarm system. Unless a fire alarm system is required by some other code provision, only the exterior alarm device is required. However, when a fire alarm system is installed, the sprinkler system must also be interconnected with the fire alarm system so that when the sprinkler system actuates, it sounds the evacuation alarms required for the fire alarm system.

The primary purpose of the exterior alarm is to notify people outside the building that the sprinkler system is in operation. Originally, it was to act as a supplemental alert so that passersby could notify the fire department of the condition. However, because the code now requires electronic supervision of sprinkler systems, that function is mostly moot. The exterior notification now primarily serves the function of alerting the arriving fire department of which building or sprinkler system is in operation before staging fire-fighting activities for the building.

[F] 903.4.3 Floor control valves. *Approved* supervised indicating control valves shall be provided at the point of connection to the riser on each floor in high-rise buildings.

❖ In high-rise buildings, sprinkler control valves with supervisory initiating devices must be installed at the point of connection to the riser on each floor. Sprinkler control valves on each floor are intended to permit servicing activated systems without impairing the water supply to large portions of the building.

[F] 903.5 Testing and maintenance. Sprinkler systems shall be tested and maintained in accordance with the *International Fire Code*.

❖ Section 901 contains requirements for the testing and maintenance of sprinkler systems. Acceptance tests are necessary to verify that the system performs as intended by design and by the code. Periodic testing and maintenance are essential to verify that the level of protection designed into the building will be operational whenever a fire occurs. Water-based extinguishing systems must be tested and maintained as required by NFPA 25.

SECTION 904
ALTERNATIVE AUTOMATIC
FIRE-EXTINGUISHING SYSTEMS

[F] 904.1 General. Automatic fire-extinguishing systems, other than *automatic sprinkler systems*, shall be designed, installed, inspected, tested and maintained in accordance with the provisions of this section and the applicable referenced standards.

❖ Section 904 covers alternative fire-extinguishing systems that use extinguishing agents other than water. Alternative automatic fire-extinguishing systems include wet-chemical, dry-chemical, foam, carbon dioxide, halon and clean-agent suppression systems. In addition to the provisions of Section 904, the indicated referenced standards include specific installation, maintenance and testing requirements for all systems.

[F] 904.2 Where required. Automatic fire-extinguishing systems installed as an alternative to the required *automatic sprinkler systems* of Section 903 shall be *approved* by the fire code official. Automatic fire-extinguishing systems shall not be considered alternatives for the purposes of exceptions or reductions allowed by other requirements of this code.

❖ One of the main considerations in selecting an extinguishing agent should be the compatibility of the agent with the hazard. The fire code official is responsible for approving an alternative extinguishing agent. The approval should be based on the compatibility of the agent with the hazard and the potential effectiveness of the agent to suppress a fire involving the hazards present.

 The code places limitations on alternative systems in that they may not be credited toward a building being equipped throughout with an automatic sprinkler system where the sprinkler system is an alternative to a code requirement.

[F] 904.2.1 Commercial hood and duct systems. Each required commercial kitchen exhaust hood and duct system required by Section 609 of the *International Fire Code* or Chapter 5 of the *International Mechanical Code* to have a Type I hood shall be protected with an approved automatic fire-extinguishing system installed in accordance with this code.

❖ This section requires an effective suppression system to combat fire on the cooking surfaces of grease-producing appliances and within the hood and exhaust system of a commercial kitchen installation. Type I hoods, including the duct system, must be protected with an approved automatic fire-extinguishing system because they are used for handling grease-laden vapors or smoke, whereas Type II hoods handle fumes, steam, heat and odors. Type I hoods are typically required for commercial food heat-processing equipment, such as deep fryers, griddles, charbroilers, broilers and open burner stoves and ranges. For additional guidance on the requirements for Type I and II hoods, see the commentary to Section 507 of the IMC.

[F] 904.3 Installation. Automatic fire-extinguishing systems shall be installed in accordance with this section.

❖ The installation of automatic fire-extinguishing systems must comply with the requirements of Sections 904.3.1 through 904.3.5 in addition to the installation criteria contained in the referenced standard for the proposed type of alternative extinguishing system.

[F] 904.3.1 Electrical wiring. Electrical wiring shall be in accordance with NFPA 70.

❖ NFPA 70 regulates the design and installation of electrical systems and equipment. All electrical work must also be in compliance with any specific electrical classifications and conditions contained in the referenced standards for each type of system.

 Chapter 27 of the code and Section 605 of the IFC

contain provisions that also reference NFPA 70. Those sections also contain additional information that must be applied when addressing electrical issues.

[F] 904.3.2 Actuation. Automatic fire-extinguishing systems shall be automatically actuated and provided with a manual means of actuation in accordance with Section 904.11.1.

❖ To increase the reliability of the system and to provide the opportunity to initiate the system as a preventive measure, a manual means to activate the system is required (see commentary, Section 904.11.1).

[F] 904.3.3 System interlocking. Automatic equipment interlocks with fuel shutoffs, ventilation controls, door closers, window shutters, conveyor openings, smoke and heat vents and other features necessary for proper operation of the fire-extinguishing system shall be provided as required by the design and installation standard utilized for the hazard.

❖ Shutting off fuel supplies will eliminate potential ignition sources in the protected area. Automatic door and window closers and dampers for forced-air ventilation systems are intended to maintain the desired concentration level of the extinguishing agent in the protected area. See the commentary for Section 904.11.2 for information on system interconnections in commercial cooking fire-extinguishing systems.

[F] 904.3.4 Alarms and warning signs. Where alarms are required to indicate the operation of automatic fire-extinguishing systems, distinctive audible and visible alarms and warning signs shall be provided to warn of pending agent discharge. Where exposure to automatic-extinguishing agents poses a hazard to persons and a delay is required to ensure the evacuation of occupants before agent discharge, a separate warning signal shall be provided to alert occupants once agent discharge has begun. Audible signals shall be in accordance with Section 907.6.2.

❖ Safeguards are necessary to prevent injury or death to personnel in areas where the atmosphere will be made hazardous by oxygen depletion due to agent discharge in a confined space. The "where alarms are required" phrase is referring to requirements that will be found in the referenced installation standards indicated in Sections 904.5 through 904.11, as applicable. Predischarge alarms that will operate on fire detection system activation must be installed within and at entrances to the affected areas.

 Where required by the appropriate installation standard, an extinguishing agent discharge delay feature shall also be provided to allow evacuation of personnel prior to agent discharge. Warning and instructional signs are also to be posted, preferably at the entrances to and within the protected area. See Section 4.5.6.1 of NFPA 12 for additional information on carbon dioxide system alarms, Section 4.3.5 of NFPA 12A for additional information on Halon system alarms and Section 4.3.5 of NFPA 2001 for additional information on clean agent system alarms.

[F] 904.3.5 Monitoring. Where a building fire alarm system is installed, automatic fire-extinguishing systems shall be monitored by the building fire alarm system in accordance with NFPA 72.

❖ Automatic fire-extinguishing systems need not be electrically supervised unless the building is equipped with a fire alarm system. This section recognizes the fact that a fire alarm system is not required in all buildings. However, because most alternative fire-extinguishing systems require the space to be evacuated before the system is discharged, they are equipped with evacuation alarms. Interconnection of the fire-extinguishing system evacuation alarm with the building evacuation alarm results in an increased level of hazard notification for the occupants in addition to the electrical supervision of the fire-extinguishing system.

[F] 904.4 Inspection and testing. Automatic fire-extinguishing systems shall be inspected and tested in accordance with the provisions of this section prior to acceptance.

❖ The completed installation must be tested and inspected to determine that the system has been installed in compliance with the code and will function as required. Full-scale acceptance tests must be conducted as required by the applicable referenced standard.

[F] 904.4.1 Inspection. Prior to conducting final acceptance tests, the following items shall be inspected:

1. Hazard specification for consistency with design hazard.
2. Type, location and spacing of automatic- and manual-initiating devices.
3. Size, placement and position of nozzles or discharge orifices.
4. Location and identification of audible and visible alarm devices.
5. Identification of devices with proper designations.
6. Operating instructions.

❖ This section identifies those items that need to be verified or visually inspected prior to the final acceptance tests. All equipment should be listed, approved and installed in accordance with the manufacturer's recommendations.

[F] 904.4.2 Alarm testing. Notification appliances, connections to fire alarm systems and connections to *approved* supervising stations shall be tested in accordance with this section and Section 907 to verify proper operation.

❖ Components of fire-extinguishing systems related to alarm devices and their supervision must be tested before the system is approved. Alarm devices must be tested to satisfy the requirements of NFPA 72.

[F] 904.4.2.1 Audible and visible signals. The audibility and visibility of notification appliances signaling agent discharge or system operation, where required, shall be verified.

❖ This section requires verification upon completion of the system installation of the audibility and visibility of notification appliances in the area affected by the extinguishing agent discharge of the alternative automatic fire extinguishing system.

[F] 904.4.3 Monitor testing. Connections to protected premises and supervising station fire alarm systems shall be tested to verify proper identification and retransmission of alarms from automatic fire-extinguishing systems.

❖ Where monitoring of fire-extinguishing systems is required, such as by Section 904.3.5, all connections related to the supervision of the system must be tested to verify they are in proper working order.

[F] 904.5 Wet-chemical systems. Wet-chemical extinguishing systems shall be installed, maintained, periodically inspected and tested in accordance with NFPA 17A and their listing.

❖ NFPA 17A contains minimum requirements for the design, installation, operation, testing and maintenance of wet-chemical preengineered extinguishing systems. Equipment that is typically protected with wet-chemical extinguishing systems includes restaurant, commercial and institutional hoods; plenums; ducts and associated cooking equipment. Strict compliance with the manufacturer's installation instructions is vital for a viable installation.

Wet-chemical solutions used in extinguishing systems are relatively harmless and there is usually no lasting significant effect on a person's skin, respiratory system or clothing. These solutions may produce a mild, temporary irritation but the symptoms will usually disappear when contact is eliminated.

[F] 904.6 Dry-chemical systems. Dry-chemical extinguishing systems shall be installed, maintained, periodically inspected and tested in accordance with NFPA 17 and their listing.

❖ NFPA 17 contains the minimum requirements for the design, installation, testing, inspection, approval, operation and maintenance of dry-chemical extinguishing systems.

The fire code official has the authority to approve the type of dry-chemical extinguishing system to be used. NFPA 17 identifies three types of dry-chemical extinguishing systems: total flooding, local application and hand hose-line systems. Only total flooding and local application systems are considered automatic extinguishing systems.

The types of hazards and equipment that can be protected with dry-chemical extinguishing systems include: flammable and combustible liquids and combustible gases; combustible solids, which melt when involved in a fire; electrical hazards, such as transformers or oil circuit breakers; textile operations subject to flash surface fires; ordinary combustibles such as wood, paper or cloth and restaurant and commercial hoods, ducts and associated cooking appliance hazards, such as deep fat fryers and some plastics, depending on the type of material and configuration.

Total flooding dry-chemical extinguishing systems are used only where there is a permanent enclosure surrounding the hazard that is adequate to enable the required concentration to be built up. The total area of unclosable openings must not exceed 15 percent of the total area of the sides, top and bottom of the enclosure. Consideration must be given to eliminating the probable sources of reignition within the enclosure because the extinguishing action of dry-chemical systems is transient.

Local application of dry-chemical extinguishing systems is to be used for extinguishing fires where the hazard is not enclosed or where the enclosure does not conform to the requirements for total flooding systems. Local application systems have successfully protected hazards involving flammable or combustible liquids, gases and shallow solids, such as paint deposits.

NFPA 17 also discusses pre-engineered dry-chemical systems consisting of components designed to be installed in accordance with pretested limitations as tested and labeled by a testing agency. Pre-engineered systems must be installed within the limitations that have been established by the testing agency and may include total flooding, local application or a combination of both types of systems.

The type of dry chemical used in the extinguishing system is a function of the hazard to be protected. The type of dry chemical used in a system should not be changed unless it has been proven changeable by a testing laboratory, is recommended by the manufacturer of the equipment and is acceptable to the fire code official for the hazard being protected. Additional guidance on the use of various dry-chemical agents can be found in NFPA 17.

[F] 904.7 Foam systems. Foam-extinguishing systems shall be installed, maintained, periodically inspected and tested in accordance with NFPA 11 and NFPA 16 and their listing.

❖ NFPA 11 covers the characteristics of foam-producing materials used for fire protection and the requirements for design, installation, operation, testing and maintenance of equipment and systems, including those used in combination with other fire-extinguishing agents. The minimum requirements are covered for flammable and combustible liquid hazards in local areas within buildings, storage tanks and indoor and outdoor processing areas.

Low-expansion foam is defined as an aggregation of air-filled bubbles resulting from the mechanical expansion of a foam solution by air with a foam-to-solution volume ratio of less than 20:1. It is most often used to protect flammable and combustible liquid hazards.

Also, low-expansion foam may be used for heat radiation protection. Combined-agent systems involve the application of low-expansion foam to a hazard simultaneously or sequentially with dry-chemical powder.

NFPA 11 gives minimum requirements for the installation, design, operation, testing and maintenance of medium- and high-expansion foam systems. Medium-expansion foam is defined as an aggregation of air-filled bubbles resulting from the mechanical expansion of a foam solution by air or other gases with a foam-to-solution volume ratio of 20:1 to 200:1. High-expansion foam has a foam-to-solution volume ratio of 200:1 to approximately 1,000:1.

Medium-expansion foam may be used on solid fuel and liquid fuel fires where some degree of in-depth coverage is necessary (for example, for the total flooding of small, enclosed or partially enclosed volumes, such as engine test cells, transformer rooms, etc.). High-expansion foam is most suitable for filling volumes in which fires exit at various levels. For example, high-expansion foam can be used effectively against high-rack storage fires in enclosures such as in underground passages, where it may be dangerous to send personnel to control fires involving liquefied natural gas (LNG) and liquefied petroleum gas (LP-gas), and to provide vapor dispersion control for LNG and ammonia spills. High-expansion foam is particularly suited for indoor fires in confined spaces, since it is highly susceptible to wind and lack-of-confinement effects.

NFPA 16 contains the minimum requirements for open-head deluge-type foam-water sprinkler systems and foam-water spray systems. The systems are especially applicable to the protection of most flammable liquid hazards and have been used successfully to protect aircraft hangars and truck loading racks.

[F] 904.8 Carbon dioxide systems. Carbon dioxide extinguishing systems shall be installed, maintained, periodically inspected and tested in accordance with NFPA 12 and their listing.

❖ NFPA 12 provides minimum requirements for the design, installation, testing, inspection, approval, operation and maintenance of carbon dioxide extinguishing systems.

Carbon dioxide extinguishing systems are useful in extinguishing fires in specific hazards or equipment in occupancies where an inert electrically nonconductive medium is essential or desirable and where cleanup of other extinguishing agents, such as dry-chemical residue, presents a problem. Carbon dioxide systems have satisfactorily protected the following: flammable liquids; electrical hazards, such as transformers, oil switches, rotating equipment and electronic equipment; engines using gasoline and other flammable liquid fuels; ordinary combustibles, such as paper, wood and textiles and hazardous solids.

The fire code official has the authority to approve the type of carbon dioxide system to be installed. NFPA 12 defines four types of carbon dioxide systems: total flooding, local application, hand hose lines and standpipe and mobile supply systems. Only total flooding and local application systems are automatic suppression systems.

Total-flooding systems may be used where there is a permanent enclosure around the hazard that is adequate to allow the required concentration to be built up and maintained for the required period of time, which varies for different hazards. Examples of hazards that have been successfully protected by total flooding systems include rooms, vaults, enclosed machines, ducts, ovens and containers and their contents.

Local application systems may be used for extinguishing surface fires in flammable liquids, gases and shallow solids where the hazard is not enclosed or the enclosure does not conform to the requirements for a total-flooding system. Examples of hazards that have been successfully protected by local application systems include dip tanks, quench tanks, spray booths, oil-filled electric transformers and vapor vents.

[F] 904.9 Halon systems. Halogenated extinguishing systems shall be installed, maintained, periodically inspected and tested in accordance with NFPA 12A and their listing.

❖ NFPA 12A contains minimum requirements for the design, installation, testing, inspection, approval, operation and maintenance of Halon 1301 extinguishing systems. Halon 1301 fire-extinguishing systems are useful in specific hazards, equipment or occupancies where an electrically nonconductive medium is essential or desirable and where cleanup of other extinguishing agents presents a problem.

Halon 1301 systems have satisfactorily protected gaseous and liquid flammable materials; electrical hazards, such as transformers, oil switches and rotating equipment; engines using gasoline and other flammable fuels; ordinary combustibles, such as paper, wood and textiles and hazardous solids. Halon 1301 systems have also satisfactorily protected electronic computers, data processing equipment and control rooms.

The fire code official has the authority to approve the type of halogenated extinguishing system to be installed. NFPA 12A defines two types of halogenated extinguishing systems: total flooding and local application. Total-flooding systems may be used where there is a fixed enclosure around the hazard that is adequate to enable the required halon concentration to be built up and maintained for the required period of time to enable the effective extinguishing of the fire. Total-flooding systems may provide fire protection for rooms, vaults, enclosed machines, ovens, containers, storage tanks and bins.

Local application systems are used where there is not a fixed enclosure around the hazard or where the fixed enclosure around the hazard is not adequate to enable an extinguishing concentration to be built up and maintained in the space. Hazards that may be successfully protected by local application systems include dip tanks, quench tanks, spray booths, oil-filled electric transformers and vapor vents.

Two other considerations in selecting the proper extinguishing system are ambient temperature and the personnel hazards associated with the agent. The ambient temperature of the enclosure for a total-flooding system must be above 70°F (21°C) for halon 1301 systems. Special consideration must also be given to the use of halon systems when the temperatures are in excess of 900°F (482°C) because halon will readily decompose at such temperatures and the products of decomposition can be extremely irritating if inhaled, even in small amounts.

Halon 1301 total-flooding systems must not be used in concentrations greater than 10 percent in normally occupied areas. Where personnel cannot vacate the area within 1 minute, Halon 1301 total-flooding systems must not be used in normally occupied areas with concentrations greater than 7 percent. Halon 1301 total-flooding systems may be used with concentrations of up to 15 percent if the area is not normally occupied and the area can be evacuated within 30 seconds.

The use of halogenated extinguishing systems has become a concern with respect to the potential environmental effects of halon. Halongenated fire-extinguishing agents have been identified as a source of emissions, resulting in the depletion of the stratospheric ozone layer and, in accordance with the Montreal Protocol, the ceasing of its production in January 1994. Therefore, the supply of halon is limited and new supplies of halogenated extinguishing agents will not be available in the future. Existing supplies of halon can, however, continue to be used in existing, undischarged systems or to recharge discharged systems. This newfound need for halon supplies has given rise to new industries geared to the ranking, recycling and reclamation of existing halon supplies. Alternative "clean agent" extinguishing agents have been developed to replace halogenated agents (see Section 904.10).

[F] 904.10 Clean-agent systems. Clean-agent fire-extinguishing systems shall be installed, maintained, periodically inspected and tested in accordance with NFPA 2001 and their listing.

❖ NFPA 2001 contains minimum requirements for the design, installation, testing, inspection and operation of clean-agent fire-extinguishing systems. A clean agent is an electrically nonconducting suppression agent that is volatile or gaseous at discharge and does not leave a residue on evaporation. Clean-agent fire-extinguishing systems are installed in locations that are enclosed and have openings in the protected area that can be sealed on activation of the alarm to provide effective clean-agent concentrations. A clean-agent fire-extinguishing system should not be installed in locations that cannot be sealed unless testing has shown that adequate concentrations can be developed and maintained.

The two categories of clean agents are halocarbon compounds and inert gas agents. Halocarbon compounds include bromine, carbon, chlorine, fluorine, hydrogen and iodine. Halocarbon compounds suppress fire by a combination of breaking the chemical chain reaction of the fire, reducing the oxygen supporting the fire and reducing the ambient temperature of the fire origin to reduce the propagation of the fire. Inert gas agents contain primary components consisting of helium, neon, argon or a combination of these. Inert gases work by reducing the oxygen concentration around the fire origin to a level that does not support combustion.

Clean-agent fire-extinguishing systems were developed in response to the demise of halon as an acceptable fire-extinguishing agent because of its harmful effect on the environment. Although the original hope for a halon substitute was that these new clean agents could be directly and proportionally substituted for halon agents in existing systems (drop in replacements), research has shown that clean agents are less efficient in extinguishing fires than are the halons they were intended to replace and require approximately 60 percent more agent by weight and volume in storage to do the same job. Additionally, the physical and chemical characteristics of clean agents differ sufficiently from halon to require different nozzles in addition to the need for larger storage vessels. Existing piping systems should be salvaged for use with clean agents only if they are carefully evaluated and determined to be hydraulically compatible with the flow characteristics of the new agent.

This section also relies on strict adherence to the system manufacturer's design and installation instructions for code compliance. As with many of the alternative fire suppression systems covered in this chapter, clean-agent systems are, for the most part, subjected by their manufacturers to a testing and listing program conducted by an approved testing agency. In such testing and listing programs, the clean agent is listed for use with specific equipment and equipment is listed for use with specific clean agents. The resultant listings include reference to the manufacturer's installation manuals, thereby giving the fire code official another valuable resource for reviewing and approving clean-agent systems.

Although clean agents have found a limited market for local application uses, such as a replacement for Halon 1211 in portable fire extinguishers, their primary application is in total-flooding systems and they are available in both engineered and preengineered configurations.

Engineered clean-agent systems are specifically designed for protection of a particular hazard, whereas preengineered systems are designed to operate within predetermined limitations up to the noted maximums, thus allowing broader applicability to a variety of hazard applications.

Total flooding systems are used where there is a fixed enclosure around the hazard that is adequate to enable the required clean-agent concentration to build up and be maintained within the space long enough to extinguish the fire. Such applications can include vaults, ovens, containers, tanks, computer rooms, paint lockers or enclosed machinery. In selecting the clean agent to be used in a given application, careful consideration must be given to whether the protected area is a normally occupied space, because different agents have different levels of concentration at which they may be a health hazard to occupants of the area.

The fire code official has the authority to approve the type of clean-agent system to be installed and should become familiar with the unique characteristics and hazards of clean-agent extinguishing systems using all available resources on the subject.

[F] 904.11 Commercial cooking systems. The automatic fire-extinguishing system for commercial cooking systems shall be of a type recognized for protection of commercial cooking equipment and exhaust systems of the type and arrangement protected. Preengineered automatic dry- and wet-chemical extinguishing systems shall be tested in accordance with UL 300 and *listed* and *labeled* for the intended application. Other types of automatic fire-extinguishing systems shall be *listed* and *labeled* for specific use as protection for commercial cooking operations. The system shall be installed in accordance with this code, its listing and the manufacturer's installation instructions. Automatic fire-extinguishing systems of the following types shall be installed in accordance with the referenced standard indicated, as follows:

1. Carbon dioxide extinguishing systems, NFPA 12.

2. *Automatic sprinkler systems*, NFPA 13.

3. Foam-water sprinkler system or foam-water spray systems, NFPA 16.

4. Dry-chemical extinguishing systems, NFPA 17.

5. Wet-chemical extinguishing systems, NFPA 17A.

Exception: Factory-built commercial cooking recirculating systems that are tested in accordance with UL 710B and *listed*, *labeled* and installed in accordance with Section 304.1 of the *International Mechanical Code*.

❖ The history of commercial kitchen exhaust systems shows that the mixture of flammable grease and effluents carried by such systems and the potential for the cooking equipment to act as an ignition source contribute to a higher level of hazard for kitchen exhaust systems than is normally found in many other exhaust systems. Furthermore, fire in a grease exhaust duct can produce temperatures of 2,000°F (1093°C) or more and heat radiating from the duct can ignite nearby combustibles. As a result, the code requires exhaust systems serving grease-producing equipment to include fire suppression to protect the cooking surfaces, hood, filters and exhaust duct to confine a fire to the hood and duct system, thus reducing the likelihood of it spreading to the structure.

In addition to the general requirements of this section, five industry standards are referenced for the installation of fire-extinguishing systems protecting commercial food heat-processing equipment and kitchen exhaust systems. Design professionals should specify and design fire-extinguishing systems to comply with these referenced standards. Only the installation of fire-extinguishing systems is regulated by these references. Where preengineered automatic dry- and wet-chemical extinguishing systems are installed, they must be listed and labeled for the specific cooking operation and tested in accordance with UL 300. Design and construction requirements for the specific types of fire-extinguishing systems are found in the respective sections of the referenced standards.

Regulatory requirements for the approval and installation of fire-extinguishing systems are the same as the approval required for all mechanical equipment and appliances. This section, therefore, requires extinguishing systems to be listed and labeled by an approved agency and installed in accordance with their listing and the manufacturer's installation instructions.

The exception allows factory-built commercial cooking recirculating systems to be installed if they have been tested and listed in accordance with UL 710B. It is important that they be installed in accordance with the manufacturer's installation instructions so that the listing requirements are met. An improper installation could result in hazardous vapors being discharged back into the kitchen.

Commercial cooking recirculating systems consist of an electric cooking appliance and an integral or matched packaged hood assembly. The hood assembly consists of a fan, collection hood, grease filter, fire damper, fire-extinguishing system and air filter, such as an electrostatic precipitator. These systems are tested for fire safety and emissions. The grease vapor (condensible particulate matter) in the effluent at the system discharge is not allowed to exceed a concentration of 5.0 mg/m^3. Recirculating systems are not used with fuel-fired appliances because the filtering systems do not remove combustion products. Kitchens require ventilation in accordance with Chapter 4 of the IMC.

Although the provisions in Section 904.11 address many of the specifics for commercial kitchens, additional information regarding commercial cooking suppression systems is located in Sections 904.2 and 904.3. This information is supplemental to that and should be considered together in developing the design for commercial cooking suppression systems.

[F] 904.11.1 Manual system operation. A manual actuation device shall be located at or near a *means of egress* from the cooking area a minimum of 10 feet (3048 mm) and a maximum of 20 feet (6096 mm) from the kitchen exhaust system. The manual actuation device shall be installed not more than 48 inches (1200 mm) or less than 42 inches (1067 mm) above the floor and shall clearly identify the hazard protected. The manual actuation shall require a maximum force of 40 pounds (178 N) and a maximum movement of 14 inches (356 mm) to actuate the fire suppression system.

Exception: *Automatic sprinkler systems* shall not be required to be equipped with manual actuation means.

❖ The manual device, usually a pull station, mechanically activates the suppression system. The typical

system uses a mechanical circuit of cables under tension to hold the system in the armed (cocked) mode. Melting of a fusible link or actuation of a manual pull station causes the cable to lose tension, which, in turn, starts the discharge of the suppression agent. The manual actuation device must be readily and easily usable by the building occupants; therefore, the device must not require excessive force or range of movement to cause actuation.

In order to allow the actuation device to be used most effectively, the specified mounting height is intended to be consistent with the NFPA 17A standards and be handicapped accessible. This includes the requirement to identify the actuation device with the hazard protected. Where multiple kitchen appliances are provided, properly identifying which device relates to which appliance is very important. Required signage should be readily visible in the hazard area and capable of conveying information quickly and concisely.

Manual actuation is not required for automatic sprinkler systems because the typical system design will employ closed heads and wet system piping. A manual actuation valve would serve no purpose because sprinkler heads are already supplied with pressurized water and will discharge water only when the individual fusible elements open the heads.

[F] 904.11.2 System interconnection. The actuation of the fire suppression system shall automatically shut down the fuel or electrical power supply to the cooking equipment. The fuel and electrical supply reset shall be manual.

❖ The actuation of any fire suppression system must automatically shut off all sources of fuel or power to all cooking equipment located beneath the exhaust hood and protected by the suppression system. This requirement is intended to shut off all heat sources that could reignite or intensify a fire. Shutting off a fuel and power supply to cooking appliances will eliminate an ignition source and allow the cooking surfaces to cool down. This shutdown is accomplished with mechanical or electrical interconnections between the suppression system and a shutoff valve or switch located on the fuel or electrical supply.

Common fuel shutoff valves include mechanical-type gas valves and electrical solenoid-type gas valves. Contactor-type switches or shunt-trip circuit breakers can be used for electrically heated appliances. The fuel or electric source must not be automatically restored after the suppression system has been actuated.

Chemical-type fire-extinguishing systems discharge for only a limited time and can discharge only once before recharge and reset; therefore, precautions must be taken to prevent a fire from reigniting. After a fire is detected and the initial suppressant discharge begins, the fuel and power supply will be locked out, thereby preventing the operation of the appliances until all systems are again ready for operation. Fuel and power supply shutoff must be manually restored by resetting a mechanical linkage or holding (latching)-type circuit.

[F] 904.11.3 Carbon dioxide systems. When carbon dioxide systems are used, there shall be a nozzle at the top of the ventilating duct. Additional nozzles that are symmetrically arranged to give uniform distribution shall be installed within vertical ducts exceeding 20 feet (6096 mm) and horizontal ducts exceeding 50 feet (15 240 mm). *Dampers* shall be installed at either the top or the bottom of the duct and shall be arranged to operate automatically upon activation of the fire-extinguishing system. Where the *damper* is installed at the top of the duct, the top nozzle shall be immediately below the *damper*. Automatic carbon dioxide fire-extinguishing systems shall be sufficiently sized to protect against all hazards venting through a common duct simultaneously.

❖ This section states specific design requirements for nozzle locations, dampers and ducts for carbon dioxide extinguishing systems that may be used to protect commercial cooking systems. These requirements are intended to supersede similar, more general provisions in NFPA 12. Because carbon dioxide (CO_2) is a gaseous suppressant, dampers are required in the ductwork to define the atmosphere where the fire event would be. A specific concentration of CO_2 is necessary and dampers are required to define and contain the suppressant. The discharge cools exposed surfaces in addition to depriving the fire of oxygen. Although not mentioned specifically in this section, the applicable provisions of NFPA 12 should also be applied because the system is a CO_2 system as referenced in Section 904.8.

[F] 904.11.3.1 Ventilation system. Commercial-type cooking equipment protected by an automatic carbon dioxide-extinguishing system shall be arranged to shut off the ventilation system upon activation.

❖ Shutting down the ventilation system upon activation of the CO_2 extinguishing system maintains the desired concentration of carbon dioxide to suppress the fire. Leakage of gas from the protected area should be kept to a minimum. Where leakage is anticipated, additional quantities of carbon dioxide must be provided to compensate for any losses.

[F] 904.11.4 Special provisions for automatic sprinkler systems. *Automatic sprinkler systems* protecting commercial-type cooking equipment shall be supplied from a separate, readily accessible, indicating-type control valve that is identified.

❖ This section requires a separate control valve in the water line to the sprinklers protecting the cooking and ventilating system. The additional valve allows the flexibility to shut off the system for repairs or for cleanups after sprinkler discharge without taking the entire system out of service.

[F] 904.11.4.1 Listed sprinklers. Sprinklers used for the protection of fryers shall be tested in accordance with UL 199E,

listed for that application and installed in accordance with their listing.

❖ Sprinklers specifically listed for such use must be used when protecting deep-fat fryers. These specially listed sprinklers use finer water droplets than standard spray sprinklers. The water spray lowers the temperature below a point where the fire can sustain itself and reduces the possibility of expanding the fire. UL 199E addresses these special sprinklers and includes performance tests for deep-fat fryer extinguishment and also deep-fat fryer cooking temperature splash. The selection of inappropriate sprinklers for deep-fat fryer protection can increase the hazards during water application rather than suppressing the fire.

SECTION 905
STANDPIPE SYSTEMS

[F] 905.1 General. Standpipe systems shall be provided in new buildings and structures in accordance with this section. Fire hose threads used in connection with standpipe systems shall be *approved* and shall be compatible with fire department hose threads. The location of fire department hose connections shall be *approved*. In buildings used for high-piled combustible storage, fire protection shall be in accordance with the *International Fire Code*.

❖ Standpipe systems are required in buildings to provide a quick, convenient water source for fire department use where hose lines would otherwise be impractical, such as in high-rise buildings. Standpipe systems can also be used prior to deployment of hose lines from fire department apparatus. The requirements for standpipes are based on practical requirements of typical fire-fighting operations and the nationally recognized standard NFPA 14.

The threads on connections to which the fire department may connect a hose must be compatible with the fire department hose threads (see commentary, Section 903.3.6). Chapter 23 of the IFC requires a Class I standpipe system in exit passageways of buildings used for high-piled storage. Note that if a building containing high-piled storage does not contain an exit passageway, then standpipes would not be required. High-piled storage involves the solid-piled, bin box, palletized or rack storage of Class I through IV commodities over 12 feet (3658 mm) high. High-hazard commodities stored higher than 6 feet (1829 mm) are also considered high piled.

[F] 905.2 Installation standard. Standpipe systems shall be installed in accordance with this section and NFPA 14.

❖ This section requires the installation of standpipe systems to comply with the applicable provisions of NFPA 14 in addition to Section 905. NFPA 14 contains the minimum requirements for the installation of standpipe and hose systems for buildings and structures. The standard addresses additional requirements not addressed in the code, such as pressure limitations, minimum flow rates, piping specifications, hose connection details, valves, fittings, hangers and the testing and inspection of standpipes. The periodic inspection, testing and maintenance of standpipe systems must comply with NFPA 25.

Section 905 and NFPA 14 recognize three classes of standpipe systems: Class I, II or III. The type of system required depends on building height, building area, type of occupancy and the extent of automatic sprinkler protection. Section 905 also recognizes five types of standpipe systems: automatic dry, automatic wet, manual dry, manual wet and semiautomatic dry. The use of each type of system is limited to the building conditions and locations identified in Section 905.3. The classes and types of standpipe systems are defined in Section 902.1.

[F] 905.3 Required installations. Standpipe systems shall be installed where required by Sections 905.3.1 through 905.3.7 and in the locations indicated in Sections 905.4, 905.5 and 905.6. Standpipe systems are allowed to be combined with *automatic sprinkler systems*.

> **Exception:** Standpipe systems are not required in Group R-3 occupancies.

❖ Standpipe systems are installed in buildings based on the occupancy, fire department accessibility and special conditions that may require manual fire suppression exceeding the capacity of a fire extinguisher. Standpipe systems are most commonly required for buildings that exceed the height threshold requirement in Section 905.3.1 or the area threshold requirement in Section 905.3.2. Specific occupancies such as covered mall buildings, stages and underground buildings, because of their use or occupancy, also require a standpipe system.

This section also states that a standpipe system does not have to be separate from an installed sprinkler system. It is common practice in multistory buildings for the standpipe system risers to also serve as risers for the automatic sprinkler systems.

In these instances, precautions need to be taken so that the operation of one system will not interfere with the operation of the other system. Therefore, control valves for the sprinkler system must be installed where the sprinklers are connected to the standpipe riser at each floor level. This allows the standpipe system to remain operational, even if the sprinkler system is shut off at the floor control valve.

The exception recognizes that standpipe systems in Group R-3 occupancies would be of minimal value to the fire department and would send the wrong message to the occupants of a dwelling unit. In the case of multiple single-family dwellings, each dwelling unit has a separate entrance and is separated from the other units by 1-hour fire partitions. These conditions permit ready access to fires and also provide for a degree of fire containment through compartmentation, which is not always present in other occupancies.

[F] 905.3.1 Height. Class III standpipe systems shall be installed throughout buildings where the floor level of the highest *story* is located more than 30 feet (9144 mm) above the lowest level of fire department vehicle access, or where the floor level of the lowest *story* is located more than 30 feet (9144 mm) below the highest level of fire department vehicle access.

Exceptions:

1. Class I standpipes are allowed in buildings equipped throughout with an *automatic sprinkler system* in accordance with Section 903.3.1.1 or 903.3.1.2.

2. Class I manual standpipes are allowed in *open parking garages* where the highest floor is located not more than 150 feet (45 720 mm) above the lowest level of fire department vehicle access.

3. Class I manual dry standpipes are allowed in *open parking garages* that are subject to freezing temperatures, provided that the hose connections are located as required for Class II standpipes in accordance with Section 905.5.

4. Class I standpipes are allowed in basements equipped throughout with an *automatic sprinkler system*.

5. In determining the lowest level of fire department vehicle access, it shall not be required to consider:

   5.1. Recessed loading docks for four vehicles or less; and

   5.2. Conditions where topography makes access from the fire department vehicle to the building impractical or impossible.

❖ Given the available manpower on the fire department vehicle, standard fire-fighting operations and standard hose sizes, a 30-foot (9144 mm) vertical distance is generally considered the maximum height to which a typical fire department engine company can practically and readily extend its hose lines. Thus, the maximum vertical travel (height) threshold is based on the time it would take a typical fire department engine (pumper) company to manually suppress a fire. The standpipe connection reduces the time needed for the fire department to extend hose lines up or down stairways to advance and apply water to the fire. For this use, a minimum Class III standpipe system is required.

With respect to the height of the building, the threshold is measured from the level at which the fire department can gain access to the building directly from its vehicle and begin vertical movement. Floor levels above grade are measured from the lowest level of fire department vehicle access to the highest floor level above [see Figure 905.3.1(1)]. If a building contains floor levels below the level of fire department vehicle access, the measurement is made from the highest level of fire department vehicle access to the lowest floor level. In cases where a building has more than one level of fire department vehicle access, the most restrictive measurement is used because it is not known at which level the fire department will access the building. In other words, the vertical distance is to be measured from the more restrictive level of fire department vehicle access to the level of the highest (or lowest, if below) floor [see Figure 905.3.1(2)].

The threshold based on the height of the building is independent of the occupancy of the building, the area of the building or the presence of an automatic sprinkler system. This is based on the universal need to be able to provide a water supply for fire suppression in any building and on the limitations of the physical effort necessary to extend hose lines vertically.

Before discussing the exceptions it is important to understand the differences between the different classes and operational characteristics of standpipes. More detailed information is included in Section 902.1 for the definitions of the different classes and types of standpipes.

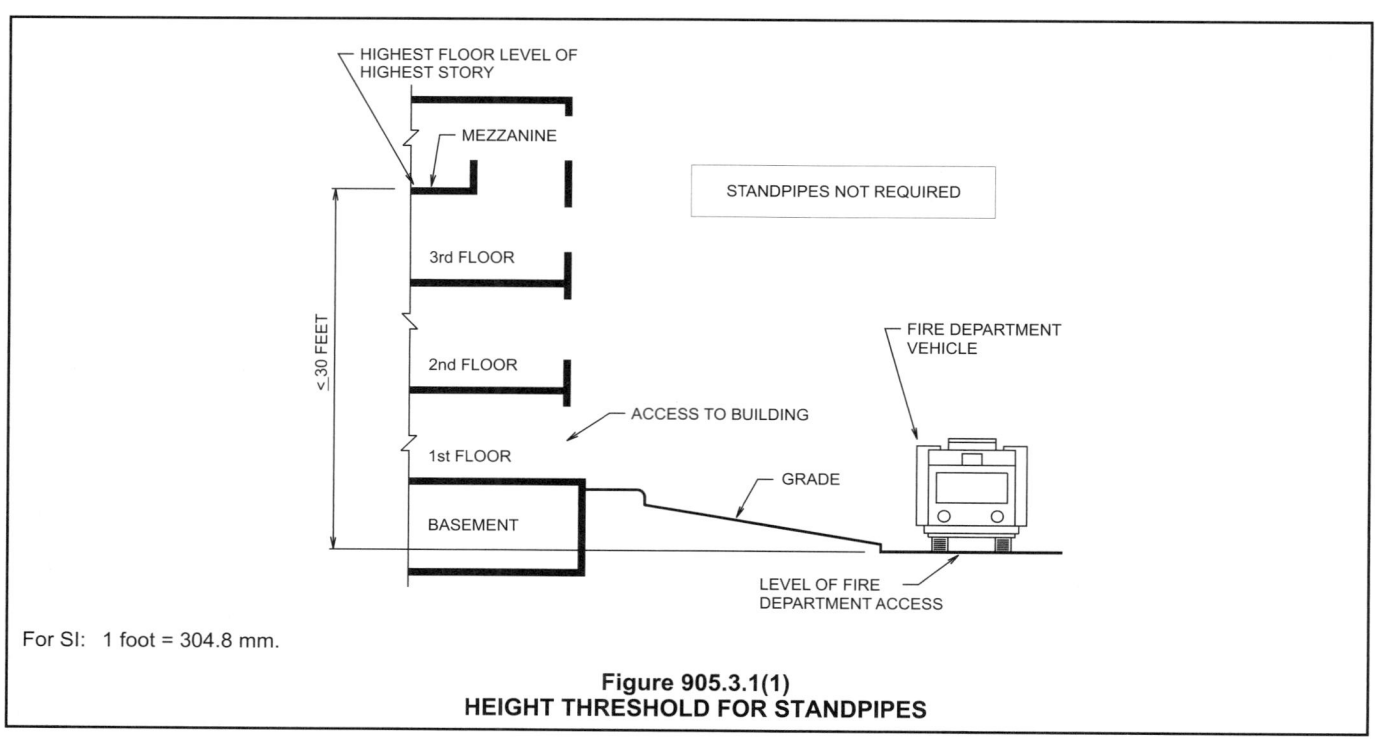

For SI: 1 foot = 304.8 mm.

Figure 905.3.1(1)
HEIGHT THRESHOLD FOR STANDPIPES

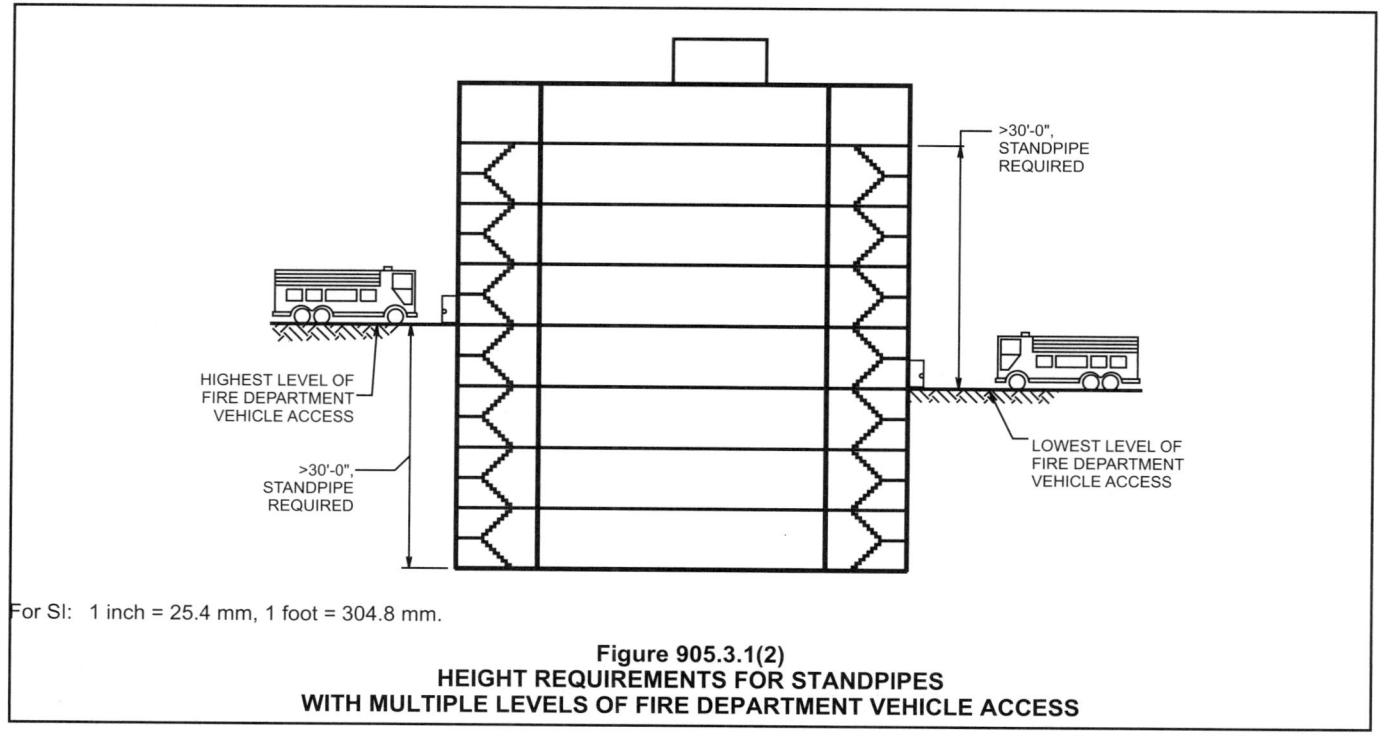

For SI: 1 inch = 25.4 mm, 1 foot = 304.8 mm.

Figure 905.3.1(2)
HEIGHT REQUIREMENTS FOR STANDPIPES
WITH MULTIPLE LEVELS OF FIRE DEPARTMENT VEHICLE ACCESS

Standpipes can be dry or wet, manual, automatic or semiautomatic. Automatic systems can be either wet or dry. Manual systems can be either wet or dry. A semiautomatic system is always in association with a dry system.

The code is written such that it could be assumed the default is an automatic wet system. This is, however, not the case. The requirement is left to the design standard, NFPA 14. Section 5.4.1.1 of NFPA 14 indicates that Class I standpipes can be manual if the building is not high-rise. Section 5.4.1.4 of the standard indicates that a Class I standpipe must be wet except where the pipe is subject to freezing. Thus, where a Class I standpipe is installed, possibly as a part of Exception 1, the system can be manual wet if the building is not a high rise. This is consistent with IFC Committee Interpretation No. 33-03. As long as the building is not high rise, it can be provided with a Class I standpipe system that is manual wet.

Class II and III standpipes are required to be automatic-wet or semiautomatic wet except where the piping is subject to freezing according to Section 5.4.3 of NFPA 14. They cannot be manual. Only Class I standpipes can be manual and only be used under the conditions noted in this code. Note that other sections of the code may specify whether the system must be automatic or not. If the requirement is not noted elsewhere in the code, then the decision to use an automatic or manual system is left to the designer.

Exception 1 recognizes the fact that with a fully operational automatic sprinkler system, the time that the fire department has to extend hoses within the building is substantially increased and that the amount of effort required is greatly reduced. Consequently a single Class I connection can be provided. The second, 1$^1/_2$-inch (38 mm) connection is allowed to be omitted. NFPA 14 also has a similar provision but is more restrictive as it only eliminates the hose station and additionally requires a 2$^1/_2$ inch by 1 $^1/_2$ inch (65 mm by 40 mm) reducer and a cap attached with a chain (see Section 7.3.4.1). In accordance with section 102.7, the IFC would take precedence and the reducer and cap would not be required.

Exception 2 identifies one of the issues relative to open parking garages. This exception allows for the garage, when not more than 150 feet (45 720 mm) in height above the lowest level of fire department access, to have a wet standpipe but without additional operating pressure until the fire department connects and begins pumping into the system. This makes sense since normal operations typically do not begin until after the fire department is on the scene and has made its initial assessments. This is generally considered to be the maximum safe height for pumpers to overcome the hydrostatic head presented by 150 feet (45 720 mm) of water. Careful considerations should be made since not all fire departments have equipment capable of this type of pumping capacity.

Exception 3 is similar to the prior exception but with the added provision that the standpipe can be dry if subject to freezing, regardless of height. Because the standpipe will be without water and dependant upon the fire department to provide both water and pressure, standpipe outlets must be spaced more frequently, as noted in Section 905.5 for Class II standpipes so that fire fighters can connect and begin operations quicker. The exception does not require Class II outlets; only that the spacing be consistent with the requirement for Class II.

Exception 4 is similar to Exception 1 but only addresses sprinklers in the basement. Thus it is possible to use this exception if only the basement is protected by automatic sprinklers. However, Class I connections can only be provided in the basements—not on the upper floors. The exception cannot be used for stories above grade unless the entire building is sprinklered and, therefore, compliant with Exception 1.

Exception 5 provides additional information about what must be considered when determining building height with respect to the level of fire department vehicle access. The first item is a practical one that excludes loading docks of a limited size. The second item notes that although it may be possible to have a fire department vehicle arrive adjacent to the building at a low level, it may not be possible for the fire department to access the building from that level. An example of this condition would be where a road surface is located below a building built on a bluff. Although the fire department vehicles can approach from the lower road, fire department personnel cannot access the building from that lower level. Thus, the standpipe requirement would not be based on the road below the bluff.

[F] 905.3.2 Group A. Class I automatic wet standpipes shall be provided in nonsprinklered Group A buildings having an *occupant load* exceeding 1,000 persons.

Exceptions:

1. Open-air-seating spaces without enclosed spaces.

2. Class I automatic dry and semiautomatic dry standpipes or manual wet standpipes are allowed in buildings where the highest floor surface used for human occupancy is 75 feet (22 860 mm) or less above the lowest level of fire department vehicle access.

❖ The main concern in assembly occupancies with a high occupant load is evacuation. Many occupants may not be familiar with either their surroundings or the egress arrangement in the building. This section also assumes the building is not sprinklered; therefore, control and suppression of the fire is left to the fire department.

Exception 1 exempts open-air seating without enclosed spaces, such as grandstands and bleachers. In such occupancies, a buildup of smoke and hot gases is not possible because these structures are open to the atmosphere.

Exception 2 states that in lieu of a Class I automatic wet standpipe, automatic dry and semiautomatic dry

Class I standpipes are permitted in buildings that are not considered to be a high rise.

[F] 905.3.3 Covered mall buildings. A *covered mall building* shall be equipped throughout with a standpipe system where required by Section 905.3.1. *Covered mall buildings* not required to be equipped with a standpipe system by Section 905.3.1 shall be equipped with Class I hose connections connected to the *automatic sprinkler system* sized to deliver water at 250 gallons per minute (946.4 L/min) at the most hydraulically remote hose connection while concurrently supplying the *automatic sprinkler system* demand. The standpipe system shall be designed not to exceed a 50 pounds per square inch (psi) (345 kPa) residual pressure loss with a flow of 250 gallons per minute (946.4 L/min) from the fire department connection to the hydraulically most remote hose connection. Hose connections shall be provided at each of the following locations:

1. Within the mall at the entrance to each *exit* passageway or *corridor*.

2. At each floor-level landing within enclosed stairways opening directly on the mall.

3. At exterior public entrances to the mall.

4. At other locations as necessary so that the distance to reach all portions of a tenant space does not exceed 200 feet (60 960 mm) from a hose connection.

❖ Covered mall buildings are only required to have a standpipe system if Section 905.3.1 requires such features. If standpipes are not required due to building height, Class I hose connections are still required that are connected to the automatic sprinkler system. Also to ensure that both the sprinkler system and hose connections will function at an acceptable level, the system must be sized for both the sprinkler demand and the hose connection demand. This section specifies a minimum flow rate and a maximum pressure loss to the most remote hose connection so that the fire department can gain full use of the hose connection during a fire. The hose connections required when a standpipe system is not are at key locations, such as entrances to exit passageways and at entrances to the mall. Note that these locations are essentially the same locations required for Class I hose connections in Section 905.4, except that this section also requires that all portions of tenant spaces do not exceed 200 feet (60 960 mm) from a hose connection.

[F] 905.3.4 Stages. Stages greater than 1,000 square feet in area (93 m²) shall be equipped with a Class III wet standpipe system with $1^1/_2$-inch and $2^1/_2$-inch (38 mm and 64 mm) hose connections on each side of the stage.

Exception: Where the building or area is equipped throughout with an *automatic sprinkler system*, a $1^1/_2$-inch (38 mm) hose connection shall be installed in accordance with NFPA 13 or in accordance with NFPA 14 for Class II or III standpipes.

❖ Because of the potentially large fuel load and three-dimensional aspect of the fire hazard associated with stages greater than 1,000 square feet (93 m²) in area, Class III standpipes are required on each side of

these large stages. The standpipes must be equipped with a $1^1/_2$-inch (38 mm) hose connection and a $2^1/_2$-inch (64 mm) hose connection. The $1^1/_2$-inch (38 m²) connection is for the hose requirement in Section 905.3.4.1. The $2^1/_2$-inch (64 mm) connection is to provide greater flexibility for the fire department in its fire-fighting operations.

Stages, as used in this section, are those stages defined in Section 410.2, which include overhead hanging curtains, drops, scenery or stage effects other than lighting and sound. These were traditionally referred to as "legitimate stages." It is not an appropriate application of this section to require standpipes for elevated areas in banquet rooms, or theatrical platforms where the higher fuel loads associated with a legitimate stage do not exist.

The exception recognizes the benefit of the building or area being sprinklered. If so, then only a single $1^1/_2$-inch (38 mm) connection is required. This hose connection is intended to be used by the fire department and apply less water from the hose due to the suppression activity of the sprinkler system. Hose threads must be compatible with those of the fire department as required in Section 903.3.6.

In a fully sprinklered building it is acceptable to supply the hose connections through the same standpipe as the sprinklers. This is reflected in the reference to both NFPA 13, which acknowledges this concept, and NFPA 14, which contains similar provisions. If the provisions of NFPA 14 are used, although the standpipe must be wet and Class II in its installation, the design of the water supply and interconnection of systems can be in accordance with the requirements for Class II as well as for Class III standpipes.

[F] 905.3.4.1 Hose and cabinet. The $1^1/_2$-inch (38 mm) hose connections shall be equipped with sufficient lengths of $1^1/_2$-inch (38 mm) hose to provide fire protection for the stage area. Hose connections shall be equipped with an *approved* adjustable fog nozzle and be mounted in a cabinet or on a rack.

❖ The $1^1/_2$-inch (38 mm) standpipe hose installed for stages greater than 1,000 square feet (93 m²) in area is intended for use by stage personnel who have been trained to use it. The length of hose provided is a function of the size and configuration of the stage. This includes by definition the entire performance area and adjacent backstage and support areas not fire separated from the performance area. The effective reach of the fire stream from the fog nozzle is a function of the available water supply, and in particular, the pressure. Fog nozzles typically require 100 pounds per square inch (psi) (690 kPa) for optimum performance.

[F] 905.3.5 Underground buildings. Underground buildings shall be equipped throughout with a Class I automatic wet or manual wet standpipe system.

❖ Underground buildings present unique hazards to life safety because of their isolation and inaccessibility. Additional fire protection and fire-fighting measures for the fire department are required to compensate for the lack of exterior access for fire suppression and rescue operations (see Section 405).

[F] 905.3.6 Helistops and heliports. Buildings with a helistop or heliport that are equipped with a standpipe shall extend the standpipe to the roof level on which the helistop or heliport is located in accordance with Section 1107.5 of the *International Fire Code*.

❖ If a building already has a standpipe (required or not) and also contains a helistop or heliport, the standpipe is required to extend to the roof level to make use of the fact that a protection feature is available. Section 1107.5 of the IFC requires a 2¹/₂-inch (64 mm) standpipe outlet to be within 150 feet (45 675 mm) of all portions of the heliport or helistop area and be either Class I or III.

[F] 905.3.7 Marinas and boatyards. Standpipes in marinas and boatyards shall comply with Chapter 45 of the *International Fire Code*.

❖ Section 4504.2 of the IFC contains the specifics as to when standpipes are required at marinas. Marinas and boatyards have unique challenges for fire fighting. Although there is water readily available, it is not easily or effectively capable of being applied to a fire at such a facility. A fire in such facilities can spread from structure to structure and from vessel to vessel with no effective way to attack and control it. Section 4504.2 of the IFC references NFPA 303 for the standpipe requirements and additionally requires that no point on the marina pier or float system exceed 150 feet (45 720 mm) from a standpipe hose connection (see commentary, Section 4504.2 of the IFC).

[F] 905.4 Location of Class I standpipe hose connections. Class I standpipe hose connections shall be provided in all of the following locations:

1. In every required *stairway*, a hose connection shall be provided for each floor level above or below grade. Hose connections shall be located at an intermediate floor level landing between floors, unless otherwise *approved* by the fire code official.

2. On each side of the wall adjacent to the *exit* opening of a *horizontal exit*.

   Exception: Where floor areas adjacent to a *horizontal exit* are reachable from *exit stairway* hose connections by a 30-foot (9144 mm) hose stream from a nozzle attached to 100 feet (30 480 mm) of hose, a hose connection shall not be required at the *horizontal exit*.

3. In every *exit* passageway, at the entrance from the *exit* passageway to other areas of a building.

   Exception: Where floor areas adjacent to an *exit* passageway are reachable from *exit stairway* hose connections by a 30-foot (9144 mm) hose stream from a nozzle attached to 100 feet (30 480 mm) of hose, a hose connection shall not be required at the entrance from the *exit* passageway to other areas of the building.

4. In covered mall buildings, adjacent to each exterior public entrance to the mall and adjacent to each entrance from an *exit* passageway or *exit corridor* to the mall.

5. Where the roof has a slope less than four units vertical in 12 units horizontal (33.3-percent slope), each standpipe shall be provided with a hose connection located either on the roof or at the highest landing of a *stairway* with *stair* access to the roof. An additional hose connection shall be provided at the top of the most hydraulically remote standpipe for testing purposes.

6. Where the most remote portion of a nonsprinklered floor or *story* is more than 150 feet (45 720 mm) from a hose connection or the most remote portion of a sprinklered floor or *story* is more than 200 feet (60 960 mm) from a hose connection, the fire code official is authorized to require that additional hose connections be provided in *approved* locations.

❖ Hose connections are required for the fire department to make use of the standpipe system. Since the fire department will typically access the building using the stairways, and most fire departments do not permit entry to the fire floor without an operating hose line, a hose connection must be installed for each floor level of each required enclosed stairway.

Item 1 also specifies that the hose connections are to be located at intermediate landings between floors. This reduces congestion at the stairway door and may reduce the hose lay distance. The hose connections, however, are still permitted at each floor level of the exit stair instead of at the intermediate landing if this arrangement is approved by the fire code official.

Because horizontal exits are also primary entrances to the fire floor, Item 2 states that hose connections must also be provided at each horizontal exit. The construction of the fire separation assembly used as the horizontal exit will protect the fire fighters while they are connecting to the standpipe system. The hose connections are to be located on each side of the horizontal exit to enable fire fighters to be in a protected area, regardless of the location of the fire. The exception acknowledges that there may already be a hose connection in close proximity to the horizontal exit if there is a stairway adjacent to the horizontal exit. The intent is to allow fewer standpipe outlets if the area can be adequately covered by the standpipes in stairways since those are the standpipes typically used by the fire department.

Item 3 states that an exit passageway in a building required to have a standpipe system is typically used as an extension of a required exit stairway. This allows use of the exit passageway for fire-fighting staging operations in the same way as an exit stair. The exception acknowledges that there may already be a hose connection in close proximity to the exit passageway. If there is a stairway containing a hose connection in close proximity to the exit passageway that can meet the 30-foot (9144 mm) hose stream from a nozzle attached to 100 feet (30 480 mm) of hose, then an additional standpipe is not required. The intent is to allow fewer standpipe outlets if the area can be adequately covered by the standpipes in stairways since those are the standpipes typically used by the fire department.

In covered mall buildings, Item 4 requires hose connections at the entrance to exits and the exterior entrances to allow fire personnel to have a support line as soon as they enter the building.

Depending on the slope of the roof, Item 5 requires a hose connection to aid in the suppression of roof fires, either because of the nature of the construction of the roof or the equipment on the roof, as well as for exposure protection.

Hose connections in each exit stairway result in hose connections being located based on the travel distances permitted in Table 1016.1, which recognizes that most fire departments carry standpipe hose packs with 150 feet (45 720 mm) of hose or possibly with 100 feet (30 480 mm) of hose and an additional 50-foot (15 240 mm) section that could be easily connected.

With the typical travel distance permitted in nonsprinklered buildings of 200 feet (60 960 mm), reasonable coverage is provided when the effective reach of a fire stream is considered. Depending on the arrangement of the floor, however, all areas may not be effectively protected. Although this situation could easily be corrected by locating additional hose connections on the floor, such connections may rarely be used because of the difficulty in identifying their location during a fire and the fact that most fire departments require an operational hose line before they enter the fire floor. Because longer travel distances are allowed in sprinklered buildings, the problem is increased, but the need for prompt manual suppression is reduced by the presence of the sprinkler system. Item 6 gives the fire code official the authority to require additional hose connections if needed.

[F] 905.4.1 Protection. Risers and laterals of Class I standpipe systems not located within an enclosed *stairway* or pressurized enclosure shall be protected by a degree of *fire resistance* equal to that required for vertical enclosures in the building in which they are located.

> **Exception:** In buildings equipped throughout with an *approved automatic sprinkler system*, laterals that are not located within an enclosed *stairway* or pressurized enclosure are not required to be enclosed within fire-resistance-rated construction.

❖ To minimize the potential for damage to the standpipe systems from a fire, the risers and laterals must be located in an enclosure having the same fire-resistance rating as required for a vertical or shaft enclosure within the building. The required fire-resistance rating for the enclosure can be determined as detailed in Section 708.4.

The enclosure is not required if the building is equipped throughout with an approved automatic sprinkler system. The potential for damage to the standpipe system is minimized by the protection pro-

vided by the sprinkler system. The automatic sprinkler system may be either an NFPA 13 or 13R system, depending on what was permitted for the building occupancy.

If the stair enclosure is not required to have a rated enclosure, such as in an open parking garage, the laterals are similarly not required to be in an enclosure. The protection afforded the vertical riser in the stairway must be the same as that afforded the laterals. If the stairway is not required by other sections of the code to be located in a rated enclosure then the laterals are not required to be in rated protection either.

[F] 905.4.2 Interconnection. In buildings where more than one standpipe is provided, the standpipes shall be interconnected in accordance with NFPA 14.

❖ In cases where there are multiple Class I standpipe risers, the risers must be supplied from and interconnected to a common supply line. The required fire department connection must serve all of the sprinklers or standpipes in the building.

[F] 905.5 Location of Class II standpipe hose connections. Class II standpipe hose connections shall be accessible and located so that all portions of the building are within 30 feet (9144 mm) of a nozzle attached to 100 feet (30 480 mm) of hose.

❖ Sections 905.5.1 through 905.5.3 specify the requirements for Class II standpipe hose connections. Class II standpipe systems are primarily intended for use by the building occupants.

This section for Class II standpipes does not specifically require the hose station and uses the term "hose connection" with a location based upon 100 feet (30 480 mm) of hose. However, the definition of Class II and III standpipes and Section 7.3.3.1 of NFPA 14 specifically require hose stations. Section 905.2 specifically references NFPA 14.

Although NFPA 14 requires a hose station, the decision as to whether a hose station is required may be one that is affected by the policies and procedures of the local fire department. It should be remembered that Class II hose connections and hose stations are intended for occupant use and not necessarily for fire department use. The fire department typically uses the Class I connection that is compatible with $2^1/_2$ inch (64 mm) hose.

[F] 905.5.1 Groups A-1 and A-2. In Group A-1 and A-2 occupancies with *occupant loads* of more than 1,000, hose connections shall be located on each side of any stage, on each side of the rear of the auditorium, on each side of the balcony and on each tier of dressing rooms.

❖ Because of the high occupant load density in Group A-1 and A-2 occupancies, providing additional means for controlling fires in their initial stage is important to enable prompt evacuation of the building. This section is independent of the Class I standpipe requirement for stages based on square footage as indicated in Section 905.3.4.

[F] 905.5.2 Protection. Fire-resistance-rated protection of risers and laterals of Class II standpipe systems is not required.

❖ Class II standpipe systems are normally not located in exit stairways; standpipe hose connections are located near the protected area to allow quick access. Therefore, it is likely that neither the risers nor the laterals would be located in any enclosure.

[F] 905.5.3 Class II system 1-inch hose. A minimum 1-inch (25 mm) hose shall be permitted to be used for hose stations in light-hazard occupancies where investigated and *listed* for this service and where *approved* by the fire code official.

❖ This section permits the use of 1-inch (25 mm) listed noncollapsible hose as an alternative to 1¹/₂-inch (38 mm) hose, subject to the approval of the fire code official. This alternative is limited to light-hazard occupancies, such as office buildings and certain assembly occupancies that tend to have lower fuel loads, since a smaller hose can discharge less water.

[F] 905.6 Location of Class III standpipe hose connections. Class III standpipe systems shall have hose connections located as required for Class I standpipes in Section 905.4 and shall have Class II hose connections as required in Section 905.5.

❖ Class III standpipe systems that have both a 2¹/₂-inch (64 mm) hose connection and a 1¹/₂-inch (38 mm) hose connection must comply with the applicable requirements of Sections 905.4, 905.5 and 905.6. Thus, it is necessary to review and comply with all applicable provisions.

[F] 905.6.1 Protection. Risers and laterals of Class III standpipe systems shall be protected as required for Class I systems in accordance with Section 905.4.1.

❖ Because Class III standpipe systems are intended for use by fire-suppression personnel, they must be located in construction that has a fire-resistance rating equivalent to that of the vertical or shaft enclosure requirements of the building (see commentary, Section 905.4.1).

[F] 905.6.2 Interconnection. In buildings where more than one Class III standpipe is provided, the standpipes shall be interconnected in accordance with NFPA 14.

❖ As indicated in Section 905.4.2 for Class I standpipe systems, multiple standpipe risers must be interconnected with a common supply line. An indicating valve is typically installed at the base of each riser so that individual risers can be taken out of service without affecting the water supply or the operation of other standpipe risers.

[F] 905.7 Cabinets. Cabinets containing fire-fighting equipment such as standpipes, fire hoses, fire extinguishers or fire department valves shall not be blocked from use or obscured from view.

❖ This section does not require that cabinets be provided to contain fire protection equipment. However, if they are provided, cabinets must be readily visible and accessible at all times. Sections 905.7.1 and 905.7.2

contain additional criteria for the construction and identification of the cabinets. Where cabinets are located in fire-resistance-rated assemblies, the integrity of the assembly must be maintained. Cabinet design for hose connections, control valves or other devices that require manual operation should be such that there is sufficient clearance between the cabinet body and the device to allow grasping of the device (quite likely with a gloved hand) and prompt operation of it.

[F] 905.7.1 Cabinet equipment identification. Cabinets shall be identified in an *approved* manner by a permanently attached sign with letters not less than 2 inches (51 mm) high in a color that contrasts with the background color, indicating the equipment contained therein.

Exceptions:

1. Doors not large enough to accommodate a written sign shall be marked with a permanently attached pictogram of the equipment contained therein.

2. Doors that have either an *approved* visual identification clear glass panel or a complete glass door panel are not required to be marked.

❖ This section specifies the minimum criteria to make the signs readily visible. Different color combinations may be approved by the fire code official if the color contrast between the letters and the background is vivid enough to make the sign visible at an approved distance. The exceptions address alternatives to letter signage if the cabinet is still conspicuously identified or the contents are readily visible.

[F] 905.7.2 Locking cabinet doors. Cabinets shall be unlocked.

Exceptions:

1. Visual identification panels of glass or other *approved* transparent frangible material that is easily broken and allows access.

2. *Approved* locking arrangements.

3. Group I-3.

❖ Ready access to all fire-fighting equipment in the cabinet is essential. The exceptions, however, recognize the need to lock cabinets for security reasons and to prevent theft or vandalism. See also the commentary to Section 906.8.

[F] 905.8 Dry standpipes. Dry standpipes shall not be installed.

Exception: Where subject to freezing and in accordance with NFPA 14.

❖ Wet standpipe systems are preferred because they tend to be the most reliable type of standpipe system; therefore, dry standpipes are prohibited unless subject to freezing. For example, Class I manual standpipe systems, which do not have a permanent water supply, are permitted in open parking structures. This recognizes that open parking structures are not heated and that most fires are limited to the vehicle of origin.

The use of any dry standpipe system instead of a wet standpipe should take into consideration the added response time and its effect on the occupancy characteristics of the building.

[F] 905.9 Valve supervision. Valves controlling water supplies shall be supervised in the open position so that a change in the normal position of the valve will generate a supervisory signal at the supervising station required by Section 903.4. Where a fire alarm system is provided, a signal shall also be transmitted to the control unit.

Exceptions:

1. Valves to underground key or hub valves in roadway boxes provided by the municipality or public utility do not require supervision.

2. Valves locked in the normal position and inspected as provided in this code in buildings not equipped with a fire alarm system.

❖ As with sprinkler systems, water control valves for standpipe systems must be electrically supervised as a means of determining that the system is operational (see commentary, Section 903.4).

Exception 1 recognizes that underground key or hub valves in roadway boxes are not normally supervised or need to be supervised whether the building contains a standpipe system or an automatic sprinkler system.

Exception 2 does not require the control valves for the standpipes to be electrically monitored if they are locked in the normal position and a fire alarm system is not installed in the building. When a fire alarm system is installed, the control valves for the standpipes must be electrically monitored and tied into the supervision required for the fire alarm system.

[F] 905.10 During construction. Standpipe systems required during construction and demolition operations shall be provided in accordance with Section 3311.

❖ As stated in Section 3311, at least one standpipe is required during construction of buildings four stories or more in height or during demolition of standpipe-equipped buildings. Standpipe systems must be accessible and operable during construction and demolition operations to assist in any potential fire (see commentary, Sections 3311.1 and 3311.2).

SECTION 906
PORTABLE FIRE EXTINGUISHERS

[F] 906.1 Where required. Portable fire extinguishers shall be installed in the following locations.

1. In new and existing Group A, B, E, F, H, I, M, R-1, R-2, R-4 and S occupancies.

    Exception: In new and existing Group A, B and E occupancies equipped throughout with quick response sprinklers, portable fire extinguishers shall be required only in locations specified in Items 2 through 6.

2. Within 30 feet (9144 mm) of commercial cooking equipment.

3. In areas where flammable or combustible liquids are stored, used or dispensed.

4. On each floor of structures under construction, except Group R-3 occupancies, in accordance with Section 1415.1 of the *International Fire Code*.

5. Where required by the *International Fire Code* sections indicated in Table 906.1.

6. Special-hazard areas, including but not limited to laboratories, computer rooms and generator rooms, where required by the fire code official.

❖ Portable fire extinguishers are required in certain instances to give the occupants the means to suppress a fire in its incipient stage. The capability for manual fire suppression can contribute to the protection of the occupants, especially if there are evacuation difficulties associated with the occupancy or the specific hazard in the area. To be effective, personnel must be properly trained in the use of portable fire extinguishers.

Because of the high-hazard nature of building contents, portable fire extinguishers are required in occupancies in Group H.

Portable fire extinguishers are required in occupancies in Groups A, B, E, F, I, M, R-1, R-2, R-4 and S because of the need to control the fire in its early stages and because evacuation can be slowed by the density of the occupant load, the capability of the occupants to evacuate or the overall fuel load in the building. Because the code typically focuses upon new buildings, this section is specific in identifying this section's applicability for both new and existing buildings.

Portable fire extinguishers are required in areas containing special hazards such as commercial cooking equipment and specific hazardous operations, as indicated in Table 906.1. Because of the potential extreme fire hazard associated with such areas or occupancy conditions, prompt extinguishment of the fire is critical.

Portable fire extinguishers are required in all buildings under construction, except in occupancies in Group R-3. The extinguishers are intended for use by construction personnel to suppress a fire in its incipient stages.

Portable fire extinguishers are also required in laboratories, computer rooms and other work spaces in which fire hazards may exist based on the use of the space. Many of these will be addressed by the required occupancy group criteria or by the specific hazard provisions of Table 906.1. Laboratories, for example, may not be considered Group H, but still use limited amounts of hazardous materials that would make manual means of fire extinguishment desirable.

The exception acknowledges the reliable advantages of an automatic sprinkler system designed to comply with NFPA 13. Group A, B and E occupancies are considered light hazard occupancies in NFPA 13. Light-hazard occupancies must be protected with quick-response sprinklers (see Section 903.3.2). The faster-acting sprinklers and lower fuel load associated with Group A, B and E occupancies counter the need for portable fire extinguishers. The desire is to have occupants evacuate the building whenever possible rather than fight the fire.

[F] TABLE 906.1
ADDITIONAL REQUIRED PORTABLE FIRE
EXTINGUISHERS IN THE INTERNATIONAL FIRE CODE

IFC SECTION	SUBJECT
303.5	Asphalt kettles
307.5	Open burning
308.1.3	Open flames—torches
309.4	Powered industrial trucks
1105.2	Aircraft towing vehicles
1105.3	Aircraft welding apparatus
1105.4	Aircraft fuel-servicing tank vehicles
1105.5	Aircraft hydrant fuel-servicing vehicles
1105.6	Aircraft fuel-dispensing stations
1107.7	Heliports and helistops
1208.4	Dry cleaning plants
1415.1	Buildings under construction or demolition
1417.3	Roofing operations
1504.4.1	Spray-finishing operations
1505.4.2	Dip-tank operations
1506.4.2	Powder-coating areas
1904.2	Lumberyards/woodworking facilities
1908.8	Recycling facilities
1909.5	Exterior lumber storage
2003.5	Organic-coating areas
2106.3	Industrial ovens
2205.5	Motor fuel-dispensing facilities
2210.6.4	Marine motor fuel-dispensing facilities
2211.6	Repair garages
2306.1	Rack storage
2404.12	Tents and membrane structures
2508.2	Tire rebuilding/storage
2604.2.6	Welding and other hot work
2903.6	Combustible fibers
3403.2.1	Flammable and combustible liquids, general
3404.3.3.1	Indoor storage of flammable and combustible liquids
3404.3.7.5.2	Liquid storage rooms for flammable and combustible liquids
3405.4.9	Solvent distillation units

(continued)

[F] TABLE 906.1—continued
ADDITIONAL REQUIRED PORTABLE FIRE
EXTINGUISHERS IN THE INTERNATIONAL FIRE CODE

IFC SECTION	SUBJECT
3406.2.7	Farms and construction sites—flammable and combustible liquids storage
3406.4.10.1	Bulk plants and terminals for flammable and combustible liquids
3406.5.4.5	Commercial, industrial, governmental or manufacturing establishments—fuel dispensing
3406.6.4	Tank vehicles for flammable and combustible liquids
3606.5.7	Flammable solids
3808.2	LP-gas
4504.4	Marinas

❖ Table 906.1 lists those sections of the IFC that represent specific occupancy conditions requiring portable fire extinguishers for incipient fire control. Wherever the code requires a fire extinguisher because of one of the listed occupancy conditions, it may identify the required rating of the extinguisher that is compatible with the hazard involved in addition to referencing Section 906.

[F] 906.2 General requirements. Portable fire extinguishers shall be selected, installed and maintained in accordance with this section and NFPA 10.

Exceptions:

1. The travel distance to reach an extinguisher shall not apply to the spectator seating portions of Group A-5 occupancies.

2. Thirty-day inspections shall not be required and maintenance shall be allowed to be once every three years for dry-chemical or halogenated agent portable fire extinguishers that are supervised by a *listed* and *approved* electronic monitoring device, provided that all of the following conditions are met:

   2.1. Electronic monitoring shall confirm that extinguishers are properly positioned, properly charged and unobstructed.

   2.2. Loss of power or circuit continuity to the electronic monitoring device shall initiate a trouble signal.

   2.3. The extinguishers shall be installed inside of a building or cabinet in a noncorrosive environment.

   2.4. Electronic monitoring devices and supervisory circuits shall be tested every three years when extinguisher maintenance is performed.

   2.5. A written log of required hydrostatic test dates for extinguishers shall be maintained by the owner to verify that hydrostatic tests are conducted at the frequency required by NFPA 10.

3. In Group I-3, portable fire extinguishers shall be permitted to be located at staff locations.

❖ NFPA 10 contains minimum requirements for the selection, installation and maintenance of portable fire extinguishers. Portable fire extinguishers are investigated and rated in conformance to NFPA 10 and listed under a variety of standards. Portable fire extinguishers must be labeled and rated for use on fires of the type, severity and hazard class protected.

NFPA 10 notes that more frequent inspections may be necessary where conditions warrant. For existing installations, a history of recent fires, vandalism, physical abuse and theft should be considered in determining if more frequent inspections are needed. For both existing and new facilities, determining the frequency of inspections should consider the environmental conditions in which the extinguisher will be located, including corrosiveness and temperature variations; and, the possibility of obstructions that may place the extinguisher out of reach in case of an emergency.

Exception 1 recognizes the openness to the atmosphere associated with Group A-5 occupancies. A fire in open areas is more obvious to all spectators. Group A-5 occupancies also do not accumulate smoke and hot gases because they are not enclosed spaces. Group A-5 occupancies also tend to be more subject to the corrosive conditions of an outdoor environment, and may include freeze/thaw cycles that are detrimental to fire extinguishers.

Exception 2 acknowledges a 30-day inspection interval similar to NFPA 10. An electronic monitoring device can determine whether or not the fire extinguisher is still present and whether or not its contents are still at the proper charge. The use of such devices, being relatively new, is allowed if it is limited to dry-chemical and halogenated agents with the additional safeguards noted in the list. Where inspection intervals may be more frequent, as discussed above, the use of electronic monitoring may have even greater benefit and is acknowledged as such in NFPA 10. The log, noted in the exception, can be a written log or a print-out of the electronic log maintained by the electronic monitoring device. This exception provides the building owner with an alternative to the contract inspections popularly used.

Exception 3 recognizes that portable fire extinguishers located throughout the facility are at times tampered with, removed and/or used for weapons by inmates in a detention or correctional setting. This exception would protect the extinguishers from damage or removal by inmates while still making them available to staff and employees for use in an emergency situation.

[F] 906.3 Size and distribution. The size and distribution of portable fire extinguishers shall be in accordance with Sections 906.3.1 through 906.3.4.

❖ Proper selection and distribution of portable fire extinguishers are essential to having adequate protection for the building structure and the occupancy conditions within. This section introduces the sections that

provide those requirements. Determination of the desired type of portable fire extinguisher depends on the character of the fire anticipated, building occupancy, specific hazards and ambient temperature conditions [see commentary, Tables 906.3(1) and 906.3(2)].

[F] 906.3.1 Class A fire hazards. The minimum sizes and distribution of portable fire extinguishers for occupancies that involve primarily Class A fire hazards shall comply with Table 906.3(1).

❖ Class A fires generally involve materials considered to be "ordinary combustibles," such as wood, cloth, paper, rubber and most plastics [see commentary, Table 906.3(1)].

[F] TABLE 906.3(1)
FIRE EXTINGUISHERS FOR CLASS A FIRE HAZARDS

	LIGHT (Low) HAZARD OCCUPANCY	ORDINARY (Moderate) HAZARD OCCUPANCY	EXTRA (High) HAZARD OCCUPANCY
Minimum Rated Single Extinguisher	2-A[c]	2-A	4-A[a]
Maximum Floor Area Per Unit of A	3,000 square feet	1,500 square feet	1,000 square feet
Maximum Floor Area for Extinguisher[b]	11,250 square feet	11,250 square feet	11,250 square feet
Maximum Travel Distance to Extinguisher	75 feet	75 feet	75 feet

For SI: 1 foot = 304.8 mm, 1 square foot = 0.0929m², 1 gallon = 3.785 L.

a. Two 2$^1/_2$-gallon water-type extinguishers shall be deemed the equivalent of one 4-A rated extinguisher.

b. Annex E.3.3 of NFPA 10 provides more details concerning application of the maximum floor area criteria.

c. Two water-type extinguishers each with a 1-A rating shall be deemed the equivalent of one 2-A rated extinguisher for Light (Low) Hazard Occupancies.

❖ Table 906.3(1), which parallels Table 6.2.1.1 of NFPA 10, establishes the minimum number and rating of fire extinguishers for Class A fires in any particular occupancy. The occupancy classifications are further defined in NFPA 10. The maximum area that a single fire extinguisher can protect is determined based on the rating of the fire extinguisher. The travel distance limitation of 75 feet (22 860 mm) is intended to be the actual walking distance along a normal path of travel to the extinguisher. For this reason, it is necessary to select fire extinguishers that comply with both the distribution criteria and travel distance limitation for a specific occupancy classification.

[F] 906.3.2 Class B fire hazards. Portable fire extinguishers for occupancies involving flammable or combustible liquids with depths less than or equal to 0.25-inch (6.35 mm) shall be selected and placed in accordance with Table 906.3(2).

Portable fire extinguishers for occupancies involving flammable or combustible liquids with a depth of greater than 0.25-inch (6.35 mm) shall be selected and placed in accordance with NFPA 10.

❖ Class B fires involve flammable and combustible liquids, oil-based paints, alcohols, solvents, flammable gases and similar materials. Selection of these extinguishers is made based on the depth of the liquid that could become involved in a fire. If the depth is $1/_4$-inch (6.35 mm) or less, selection is made using Table 906.3(2). Class B extinguishers for greater liquid depth, characterized in NFPA 10 as "appreciable depth," must be selected and installed in accordance with Section 6.3.2 of NFPA 10 [see commentary, Table 906.3(2)].

[F] TABLE 906.3(2)
FLAMMABLE OR COMBUSTIBLE LIQUIDS WITH
DEPTHS LESS THAN OR EQUAL TO 0.25 INCH

TYPE OF HAZARD	BASIC MINIMUM EXTINGUISHER RATING	MAXIMUM TRAVEL DISTANCE TO EXTINGUISHERS (feet)
Light (Low)	5-B	30
	10-B	50
Ordinary (Moderate)	10-B	30
	20-B	50
Extra (High)	40-B	30
	80-B	50

For SI: 1 inch = 25.4 mm, 1 foot = 304.8 mm.

Note: For requirements on water-soluble flammable liquids and alternative sizing criteria, see Section 5.5 of NFPA 10.

❖ Fires involving flammable or combustible liquids present a severe hazard challenge regardless of occupancy. Table 906.3(2), which parallels Table 6.3.1.1 of NFPA 10, prescribes the minimum portable fire extinguisher requirements where flammable or combustible liquids are limited in depth [0.25 inch (6 mm) or less]. As can be seen in the table, the size of the extinguisher is directly related to the travel distance to the extinguisher for each given occupancy classification. These fire extinguisher provisions are independent of whether other fixed automatic fire-extinguishing systems are installed. For occupancy conditions involving flammable or combustible liquids in potential depths greater than 0.25 inch (6 mm), the selection and spacing criteria of NFPA 10 must be used in addition to any applicable requirements in Chapter 34 of the IFC and NFPA 30.

[F] 906.3.3 Class C fire hazards. Portable fire extinguishers for Class C fire hazards shall be selected and placed on the basis of the anticipated Class A or B hazard.

❖ Class C fires involve energized electrical equipment where the electrical nonconductivity of the extinguishing agent is critical. The need for this class of extinguisher is simply based on the presence of the hazard in an occupancy and no numerical rating is required.

[F] 906.3.4 Class D fire hazards. Portable fire extinguishers for occupancies involving combustible metals shall be selected and placed in accordance with NFPA 10.

❖ Class D fires involve flammable solids, the bulk of which are combustible metals including, but not limited to, magnesium, potassium, sodium and titanium. Most Class D extinguishers will have a special low-velocity nozzle or discharge wand to gently apply the agent in large volumes to avoid disrupting any finely divided

burning materials. Extinguishing agents are also available in bulk and can be applied with a scoop or shovel. While Class D extinguishers are often referred to as "dry chemical" fire extinguishers, they are more properly called "dry powder" fire extinguishers because their mechanism of extinguishment is by a smothering action rather than by chemical reaction with the combustion process.

There are several Class D fire extinguisher agents available—some will handle multiple types of metal fires, others will not. Sodium carbonate-based extinguishers are used to control sodium, potassium, and sodium-potassium alloy fires but have limited use on other metals. This material smothers and forms a crust. Sodium chloride-based extinguishers contain sodium chloride salt and a thermoplastic additive. The plastic melts to form an oxygen-excluding crust over the metal, and the salt dissipates heat. This powder is useful on most alkali metals, including magnesium, titanium, aluminum, sodium, potassium and zirconium. Graphite-based extinguishers contain dry graphite powder that smothers burning metals. Unlike sodium chloride powder extinguishers, the graphite powder fire extinguishers can be used on very hot burning metal fires such as lithium, but the powder will not stick to and extinguish flowing or vertical lithium fires. The graphite powder acts as a heat sink as well as smothering the metal fire. See the commentary to Section 3606.5.7 of the IFC for a discussion of extinguishing flammable solid fires.

[F] 906.4 Cooking grease fires. Fire extinguishers provided for the protection of cooking grease fires shall be of an *approved* type compatible with the automatic fire-extinguishing system agent and in accordance with Section 904.11.5 of the *International Fire Code*.

❖ The combination of high-efficiency cooking appliances and hotter burning cooking media creates a potentially severe fire hazard. Although commercial cooking systems must have an approved exhaust hood and be protected by an approved automatic fire-extinguishing system, a manual means of extinguishment is desirable to attack a fire in its incipient stage.

As indicated in Section 904.11.5, a Class K-rated portable fire extinguisher must be located within 30 feet (9144 mm) of travel distance of commercial-type cooking equipment. Class K-rated extinguishers have been specifically tested on commercial cooking appliances using vegetable or animal oils or fats. These portable fire extinguishers are usually of sodium bicarbonate or potassium bicarbonate dry-chemical type.

[F] 906.5 Conspicuous location. Portable fire extinguishers shall be located in conspicuous locations where they will be readily accessible and immediately available for use. These locations shall be along normal paths of travel, unless the fire code official determines that the hazard posed indicates the need for placement away from normal paths of travel.

❖ Fire extinguishers must be located in readily accessible locations along normal egress paths. This increases the occupants, familiarity with the location of the fire extinguishers. When considering location, the most frequent occupants should be considered. These are the occupants who would become most familiar with the fire-extinguisher placement. For most buildings, it is the employees who are most familiar with their surroundings; therefore, a good understanding of employee operations is important for proper extinguisher placement.

[F] 906.6 Unobstructed and unobscured. Portable fire extinguishers shall not be obstructed or obscured from view. In rooms or areas in which visual obstruction cannot be completely avoided, means shall be provided to indicate the locations of extinguishers.

❖ Portable fire extinguishers must be located where they are readily visible at all times. If visual obstruction cannot be avoided, the location of the extinguishers must be marked by an approved means of identification. This could include additional signage, lights, arrows or other means approved by the fire code official. Unobstructed does not necessarily mean visible from all angles within the space. Often, columns or furnishings may obscure the extinguisher from one direction or another. These are not by themselves obstructions. The intent is that the extinguisher is not hidden but rather can be readily found. If the extinguisher is placed in the wall behind a door, it is clearly obstructed since it cannot be easily viewed. An extinguisher on a wall that is visible from most of the space would be considered unobstructed.

[F] 906.7 Hangers and brackets. Hand-held portable fire extinguishers, not housed in cabinets, shall be installed on the hangers or brackets supplied. Hangers or brackets shall be securely anchored to the mounting surface in accordance with the manufacturer's installation instructions.

❖ Portable fire extinguishers not housed in cabinets are usually mounted on walls or columns using securely fastened hangers. Brackets must be used where the fire extinguishers need to be protected from impact or other potential physical damage.

[F] 906.8 Cabinets. Cabinets used to house portable fire extinguishers shall not be locked.

Exceptions:

1. Where portable fire extinguishers subject to malicious use or damage are provided with a means of ready access.

2. In Group I-3 occupancies and in mental health areas in Group I-2 occupancies, access to portable fire extinguishers shall be permitted to be locked or to be located in staff locations provided the staff has keys.

❖ Cabinets housing fire extinguishers must not be locked in order to provide quick access in an emergency. Exception 1, however, allows the cabinets to be locked in occupancies where vandalism, theft or other malicious behavior is possible. Exception 2 also permits cabinets housing fire extinguishers to be locked or to be located in staff locations in Group I-3 occupancies and mental health areas in Group I-2 occupan-

cies. Occupants in Group I-3 areas of jails, prisons or similar restrained occupancies should not have access to fire extinguishers because they could possibly be used as a weapon or be subject to vandalism. Staff adequately trained in the use of fire extinguishers are assumed to have ready access to the keys for the cabinets at all times.

[F] 906.9 Extinguisher installation. The installation of portable fire extinguishers shall be in accordance with Sections 906.9.1 through 906.9.3.

❖ This section introduces the installation criteria for portable fire extinguishers based on the weight of the unit.

[F] 906.9.1 Extinguishers weighing 40 pounds or less. Portable fire extinguishers having a gross weight not exceeding 40 pounds (18 kg) shall be installed so that their tops are not more than 5 feet (1524 mm) above the floor.

❖ Due to the varying height and physical strength levels of persons who might be called upon to operate a portable fire extinguisher, the mounting height of the extinguisher must be commensurate with its weight so that it may be easily retrieved by anyone from its mounting location and placed into use.

[F] 906.9.2 Extinguishers weighing more than 40 pounds. Hand-held portable fire extinguishers having a gross weight exceeding 40 pounds (18 kg) shall be installed so that their tops are not more than 3.5 feet (1067 mm) above the floor.

❖ See the commentary to Section 906.9.1.

[F] 906.9.3 Floor clearance. The clearance between the floor and the bottom of installed hand-held portable fire extinguishers shall not be less than 4 inches (102 mm).

❖ The clearance between the floor and the bottom of installed hand-held extinguishers must not be less than 4 inches (102 mm) to facilitate cleaning beneath the unit and reduce the likelihood of the extinguisher becoming dislodged during cleaning operations (floor mopping, sweeping, etc.).

[F] 906.10 Wheeled units. Wheeled fire extinguishers shall be conspicuously located in a designated location.

❖ Wheeled fire extinguishers consist of a large-capacity (up to several hundred pounds of agent) fire extinguisher assembly (either stored-pressure or pressure transfer type) equipped with a carriage and wheels and discharge hose. They are constructed so that one able-bodied person could move the unit to the fire area and begin extinguishment unassisted. Wheeled fire extinguishers are capable of delivering greater flow rates and stream range for various extinguishing agents than hand-held portable fire extinguishers. Wheeled fire extinguishers are generally more effective in high-hazard areas and, as with any extinguisher, must be readily available and stored in an approved location. The wheeled fire extinguisher should be located a safe distance from the hazard area so that it will not become involved in the fire or access to it compromised by a fire. These units are typically found at airport fueling ramps, refineries, bulk plants and similar locations where high-challenge fires may be encountered. The extinguishing agents available in wheeled units include carbon dioxide, dry chemical, dry powder and foam.

SECTION 907
FIRE ALARM AND DETECTION SYSTEMS

[F] 907.1 General. This section covers the application, installation, performance and maintenance of fire alarm systems and their components.

❖ Fire alarm systems in new buildings, which typically include manual fire alarm systems and automatic smoke detection systems, must be installed in accordance with Section 907 and NFPA 72. Fire alarm systems in existing buildings are regulated by Section 907.3 and Chapter 46 of the IFC.

Manual fire alarm systems are installed in buildings to limit fire casualties and property losses. Fire alarm systems do this by promptly notifying the occupants of the building of an emergency, which increases the time available for evacuation. Similarly, when fire alarm systems are supervised, the fire department will be promptly notified and its response time relative to the onset of the fire will be reduced.

Automatic smoke detection systems are required under certain conditions to increase the likelihood that a fire is detected and occupants are given an early warning. The detection system is a system of devices and associated hardware that activates the alarm system. The automatic detecting devices are to be smoke detectors, unless a condition exists that calls for the use of a different type of detector.

[F] 907.1.1 Construction documents. *Construction documents* for fire alarm systems shall be of sufficient clarity to indicate the location, nature and extent of the work proposed and show in detail that it will conform to the provisions of this code, the *International Fire Code*, and relevant laws, ordinances, rules and regulations, as determined by the fire code official.

❖ Construction documents for fire alarm systems must be submitted for review to determine compliance with the code, the IFC and NFPA 72. All of the information required by this section may not be available during the design stage and initial permit process. Later submission of more detailed shop drawings may be required in accordance with Section 907.1.2. These provisions are intended to reflect the minimum scope of information needed to determine code compliance. When the work can be briefly described on the application form, the fire code official may utilize judgement in determining the need for more detailed documents.

[F] 907.1.2 Fire alarm shop drawings. Shop drawings for fire alarm systems shall be submitted for review and approval prior to system installation, and shall include, but not be limited to, all of the following:

1. A floor plan that indicates the use of all rooms.
2. Locations of alarm-initiating devices.
3. Locations of alarm notification appliances, including candela ratings for visible alarm notification appliances.
4. Location of fire alarm control unit, transponders and notification power supplies.
5. Annunciators.
6. Power connection.
7. Battery calculations.
8. Conductor type and sizes.
9. Voltage drop calculations.
10. Manufacturers' data sheets indicating model numbers and listing information for equipment, devices and materials.
11. Details of ceiling height and construction.
12. The interface of fire safety control functions.
13. Classification of the supervising station.

❖ Since the fire protection contractor(s) may not have been selected at the time a permit is issued for construction of a building, detailed shop drawings for fire alarm systems may not be available. Because they provide the information necessary to determine code compliance, as specified in this section, they must be submitted and approved by the fire code official before the contractor can begin installing the system.

[F] 907.1.3 Equipment. Systems and components shall be *listed* and *approved* for the purpose for which they are installed.

❖ The components of the fire alarm system must be approved for use in the planned system. NFPA 72 requires all devices, combinations of devices, appliances and equipment to be labeled for their proposed use. The testing agency will test the components for use in various types of systems and stipulate the use of the component on the label. Evidence of listing and labeling of the system components must be submitted with the shop drawings. In some instances, the entire system may be labeled.

At least one major testing agency, Underwriters Laboratories, Inc. (UL), has a program in which alarm installation and service companies are issued a certificate and become listed by the agency as being qualified to design, install and maintain local, auxiliary, remote station or proprietary fire alarm systems. The listed companies may then issue a certificate showing that the system is in compliance with Section 907. Terms of the company certification by UL include the company being responsible for keeping accurate system documentation, including as-built record drawings, acceptance test records and complete maintenance records on a given system. The company is also responsible for the required periodic inspection and testing of the system under contract with the owner. A similar program has been available for many years for central station alarm service, whereas the UL program is relatively new to the industry. Even though this company and system listing program is not required by the code or NFPA 72, it can be a valuable tool for the fire code official in determining compliance with the referenced standard.

Another issue that must be considered is the compatibility of the system components as required by NFPA 72. The labeling of system components discussed above should include any compatibility restrictions for components. Compatibility is primarily an issue of the ability of smoke detectors and fire alarm control panels (FACPs) to function properly when interconnected and affects the two-wire type of smoke detectors, which obtain their operating power over the same pair of wires used to transmit signals to the FACP (the control unit initiating device circuits). Laboratories will test for component compatibility either by actual testing or by reviewing the circuit parameters of both the detector and the FACP. Generally, if both the two-wire detector and the FACP are of the same brand, there should not be a compatibility problem. Nevertheless, the fire code official must be satisfied that the components are listed as being compatible. Failure to comply with the compatibility requirements of NFPA 72 can lead to system malfunction or failure at a critical time.

[F] 907.2 Where required—new buildings and structures. An *approved* fire alarm system installed in accordance with the provisions of this code and NFPA 72 shall be provided in new buildings and structures in accordance with Sections 907.2.1 through 907.2.23 and provide occupant notification in accordance with Section 907.5, unless other requirements are provided by another section of this code.

A minimum of one manual fire alarm box shall be provided in an *approved* location to initiate a fire alarm signal for fire alarm systems employing automatic fire detectors or waterflow detection devices. Where other sections of this code allow elimination of fire alarm boxes due to sprinklers, a single fire alarm box shall be installed.

Exceptions:

1. The manual fire alarm box is not required for fire alarm systems dedicated to elevator recall control and supervisory service.

2. The manual fire alarm box is not required for Group R-2 occupancies unless required by the fire code official to provide a means for fire watch personnel to initiate an alarm during a sprinkler system impairment event. Where provided, the manual fire alarm box shall not be located in an area that is accessible to the public.

❖ This section specifies the occupancies or conditions in new buildings or structures that require some form of fire alarm system which is either a manual fire alarm system (manual fire alarm boxes) or an automatic smoke detection system. These systems must, upon activation, provide occupant notification throughout the area protected by the system unless other alternative provisions are allowed by this section.

Manual fire alarm systems must be installed in cer-

tain occupancies depending on the number of occupants, capabilities of the occupants and height of the building. An automatic smoke detection system must be installed in those occupancies and conditions where the need to detect the fire is essential to evacuation or protection of the occupants. The requirements for automatic smoke detection are generally based on the evacuation needs of the occupants and whether the occupancy includes sleeping accommodations.

Fire alarm systems must be installed in accordance with the code and NFPA 72. NFPA 72 identifies the minimum performance, location, mounting, testing and maintenance requirements for fire alarm systems. Smoke detectors must be used, except when ambient conditions would prohibit their use. In that case other detection methods may be used. The manufacturer's literature will identify the limitations on the use of smoke detectors, including environmental conditions such as humidity, temperature and airflow.

Only certain occupancies are required to have either a manual fire alarm or automatic fire detection system installed (see Figure 907.2). The need for either system is generally determined by the number of occupants, the height of the building or the ability of the occupants for self-preservation. Note that generally the fire alarm requirements are based upon occupancy not on fire area. Figure 907.2 contains the conditions that require either system must be installed in a building. The extent that an alarm system must be installed in a building once it has been etermined that such a system is required is based on a couple of factors. One if it is the only occupancy in the building then it would be required throughout the building. If the building is a mixed occupancy, it can wither be separated or nonseparated. If the occupancy is separated in accordance with Section 508.4, then the alarm system is only required whihin that portion of the building. If the building is considered a nonseparated mixed occupancy building, then Section 508.3.1 states that "the code apply to each portion of the building based upon the occupancy classification of that space and that the most restrictive applicable provisions of Section 403 and Chapter 9 shall apply to the building or portion thereof in which the nonseparated occupancies are located." Therefore, if you had a Group A occupancy in a nonseparated mixed occupancy (containing other occupancies such as Group B and M) where the Group A occupancy exceeds an occupant load of 300, then the entire nonseparated mixed occupancy would require the alarm system. Note that Section 508.3.1 focuses on each space to determine occupancy and requirements. Once the occupant load is determined, then any requirements such as fire alarms would be required throughout. The code does not address whether or not a nonseparated mixed occupancy has a completely independent means of egress such as in a strip mall. Additionally, take a building containing primarily Group A occupancies the code does not clearly address whether such occupancies within a building should be looked at as an aggregate or whether it is each individual space. Multiplex theaters, for example, could have many theaters all under 300, but since as an aggregate contain more than 300 and the fact that they are required to share at least 50 percent of their egress it may be viewed as a single occupancy and require an alarm system. Again, the code does not address this issue but the key to addressing these situations appears to be separation and egress.

Figure 907.2 contains the threshold requirements for when a manual fire alarm system or an automatic fire detection system is required based on the occupancy group. It is important to remember that although the requirement for manual pull stations may not apply (e.g., sprinklered buildings), alarm and occupant notification may still be required. Sections 907.2.11 through

907.2.23 contain additional requirements for fire alarm systems depending on special occupancy conditions such as atriums, high-rise buildings or covered mall buildings.

The single manual fire alarm box required by this section is needed to provide a means of manually activating a fire alarm system that only contains automatic devices such as sprinkler waterflow switches or smoke detectors. Its primary use is for alarm system maintenance technicians to be able to manually activate the fire alarm system in the event of a fire during the time the system or portions of the system is down for maintenance. Note that this requirement is not subject to any of the exceptions in Sections 907.2.1 through 907.2.23 that might waive the need for manual fire alarm boxes in certain buildings. Exception 1 rec-

MANUAL FIRE ALARM SYSTEM	
Occupancy Group(s)	Threshold
Assembly (A-1, A-2, A-3, A-4, A-5)	All with an occupant load of > 300 (907.2.1)
Business (B)	Total occupant load of > 500; or, > 100 above/below level of exit discharge; or, in Group B fire areas containing an ambulatory health care facility (AHCF). (907.2.2)
Educational (E)	> 50 occupants (several exceptions for manual fire alarm box placement) (907.2.3)
Factory (F-1, F-2)	> 2 stories with occupant load of > 500 above/below lowest level of exit discharge (exception for sprinklers) (907.2.4)
High hazard (H)	Group H-5 and in occupancies for manufacture of organic coatings. (907.2.5)
Institutional (I-1, I-2, I-3, I-4)	All (exceptions for I-1 and I-2 manual fire alarm box placement and private mode signaling) (907.2.6)
Mercantile (M)	Total occupant load of > 500; or, occupant load of >100 above/below level of exit discharge (907.2.7)
Hotels (R-1)	All (exceptions for < 2 stories with sleeping units having exit directly to exterior; sprinklers) (907.2.8.1)
Apartments (R-2)	If units > 3 stories above lowest level of exit discharge; or, > 1 story below highest level of exit discharge; or, >16 units (exceptions for < 2 stories with sleeping units having exit directly to exterior; sprinklers) (907.2.9.1)
Residential care/assisted living (R-4)	All (exceptions for sprinklers, manual fire alarm boxes at staff locations and direct exit to exterior) (907.2.10.1)
AUTOMATIC SMOKE DETECTION SYSTEM	
Business (B) Ambulatory healthcare facilities (AHCF)	AHCF plus public use areas outside of it including public corridors and elevator lobbies (exception for sprinklers) (907.2.2.1)
High hazard (H)	Highly toxic gases, organic peroxides, oxidizers (907.2.5)
Institutional (I-1, I-2, I-3)	All, in specific areas (exceptions for corridors, waiting areas and habitable spaces in I-1 and I-2; occupant load and sprinklers in I-3) (907.2.6.1, 907.2.6.2, 907.2.6.3.3)
Hotels (R-1)	All, in interior corridors (exception for buildings without interior corridors and with sleeping units having exit directly to exterior) (907.2.8.2)
Residential care/assisted living (R-4)	All, in corridors, waiting areas open to corridors, non-sleeping area habitable spaces and kitchens (exceptions for sprinklers and sleeping units having exit directly to exterior) (907.2.10.2)

Figure 907.2
SUMMARY OF MANUAL FIRE ALARM AND AUTOMATIC SMOKE DETECTION SYSTEM THRESHOLDS

ognizes the specialized nature of fire alarm systems installed only for emergency elevator control and supervision. Exception 2 waives the single manual fire alarm box but gives the fire code official authority to require it in sprinklered buildings for use by fire watch personnel or sprinkler maintenance personnel to be able to manually activate the fire alarm system in the event of a fire during the time the sprinkler system is down for maintenance.

[F] 907.2.1 Group A. A manual fire alarm system that activates the occupant notification system in accordance with Section 907.5 shall be installed in Group A occupancies having an *occupant load* of 300 or more. Portions of Group E occupancies occupied for assembly purposes shall be provided with a fire alarm system as required for the Group E occupancy.

> **Exception:** Manual fire alarm boxes are not required where the building is equipped throughout with an *automatic sprinkler system* installed in accordance with Section 903.3.1.1 and the occupant notification appliances will activate throughout the notification zones upon sprinkler waterflow.

❖ Group A occupancies are typically occupied by a significant number of people who are not completely familiar with their surroundings. For this reason, a manual fire alarm system is required in Group A occupancies with an occupant load of 300 or more to aid in the prompt evacuation of the occupants.

The exception allows the omission of manual fire alarm boxes in buildings equipped throughout with an automatic sprinkler system if activation of the sprinkler system will activate the building evacuation alarms associated with the manual fire alarm system.

This section also permits assembly-type areas in Group E occupancies to comply with Section 907.2.3 instead of the requirements of this section. A typical high school, for example, contains many areas used for assembly purposes such as a gymnasium, cafeteria, auditorium or library; however, they all exist to serve as an educational facility as their main function. The exception does not eliminate the fire alarm system and occupant notification system, but rather permits them to be initiated automatically by the sprinkler waterflow switch(es) instead of by the manual fire alarm boxes. It also reduces the possibility of mischievous or malicious false alarms being turned in by manual fire alarm boxes in venues where large numbers of people congregate.

[F] 907.2.1.1 System initiation in Group A occupancies with an occupant load of 1,000 or more. Activation of the fire alarm in Group A occupancies with an *occupant load* of 1,000 or more shall initiate a signal using an emergency voice/alarm communications system in accordance with Section 907.5.2.2.

> **Exception:** Where *approved*, the prerecorded announcement is allowed to be manually deactivated for a period of time, not to exceed 3 minutes, for the sole purpose of allow-

ing a live voice announcement from an *approved, constantly attended location*.

❖ In order to afford authorized personnel the ability to selectively evacuate or manage occupant relocation in large assembly venues, this section requires the fire alarm system to operate through an emergency voice/alarm communications system. The exception allows the automatic alarm signals to be overridden for live voice instructions, if the live voice instructions do not exceed 3 minutes. The location from which the live voice announcement originates must be constantly attended and approved by the fire code official. See also the commentary to Section 907.5.2.2.

[F] 907.2.2 Group B. A manual fire alarm system shall be installed in Group B occupancies where one of the following conditions exists:

1. The combined Group B *occupant load* of all floors is 500 or more.

2. The Group B *occupant load* is more than 100 persons above or below the lowest *level of exit discharge*.

3. The Group B *fire area* contains a Group B ambulatory health care facility.

> **Exception:** Manual fire alarm boxes are not required where the building is equipped throughout with an *automatic sprinkler system* installed in accordance with Section 903.3.1.1 and the occupant notification appliances will activate throughout the notification zones upon sprinkler waterflow.

❖ Group B occupancies generally involve individuals or groups of people in separate office areas. As a result, the occupants are not necessarily aware of what is going on in other parts of the building. Group B buildings with large occupant loads, even in single-story buildings, or where a substantial number of occupants are above or below the level of exit discharge, increase the difficulty of alerting the occupants of a fire. This is especially true in nonsprinklered buildings with given occupant load thresholds. In the 2009 edition of the code, Occupancy Group B includes ambulatory health care facilities which present a higher level of life hazard than the typical Group B occupancy. The fact that the occupants of such facilities may be rendered incapable of self-preservation for limited periods of time makes the need for a fire alarm system critical. See the commentary to Section 202, definition of "Ambulatory health care facility" and Section 907.2.2.1.

The exception does not eliminate the fire alarm system, but rather permits it to be initiated automatically by the sprinkler waterflow switch(es) instead of by the manual fire alarm boxes.

[F] 907.2.2.1 Group B ambulatory health care facilities. Fire areas containing Group B ambulatory health care facilities shall be provided with an electronically supervised automatic smoke detection system installed within the ambulatory health care facility and in public use areas outside of tenant spaces, including public *corridors* and elevator lobbies.

Exception: Buildings equipped throughout with an *automatic sprinkler system* in accordance with Section 903.3.1.1, provided the occupant notification appliances will activate throughout the notification zones upon sprinkler waterflow.

❖ Years ago, few surgical procedures were performed outside of a hospital. Today, complex outpatient surgeries conducted outside of a hospital are commonplace. They are performed in facilities often called "day surgery centers" or "ambulatory surgical centers" because patients are able to walk in and walk out the same day. Procedures render patients temporarily incapable of self-preservation by application of nerve blocks, sedation or anesthesia; however, they do typically recover quickly.

The IBC identifies health care Group I occupancies as including a 24-hour stay. Without a 24-hour stay, these surgery centers have been classified as Group B which allowed the staff to render an unlimited number of people incapable of self-preservation with no more protection than a business office. Since these types of facilities contain distinctly different hazards to life safety than other Group B occupancies, they are now required to have a higher level of life safety and fire protection as evidenced by the requirements of this section as well as Section 903.2.2 and the construction provisions of the code.

The exception does not eliminate the fire alarm system, but rather permits it to be initiated automatically by the sprinkler waterflow switch(es) instead of by the smoke detection system.

[F] 907.2.3 Group E. A manual fire alarm system that activates the occupant notification system in accordance with Section 907.5 shall be installed in Group E occupancies. When *automatic sprinkler systems* or smoke detectors are installed, such systems or detectors shall be connected to the building fire alarm system.

Exceptions:

1. A manual fire alarm system is not required in Group E occupancies with an *occupant load* of less than 50.

2. Manual fire alarm boxes are not required in Group E occupancies where all of the following apply:

   2.1. Interior *corridors* are protected by smoke detectors.

   2.2. Auditoriums, cafeterias, gymnasiums and similar areas are protected by *heat detectors* or other *approved* detection devices.

   2.3. Shops and laboratories involving dusts or vapors are protected by *heat detectors* or other *approved* detection devices.

   2.4. The capability to activate the evacuation signal from a central point is provided.

   2.5. In buildings where normally occupied spaces are provided with a two-way communication system between such spaces and a constantly attended receiving station from where a general evacuation alarm can be sounded, except in locations specifically designated by the fire code official.

3. Manual fire alarm boxes shall not be required in Group E occupancies where the building is equipped throughout with an *approved automatic sprinkler system* installed in accordance with Section 903.3.1.1, the notification appliances will activate on sprinkler waterflow and manual activation is provided from a normally occupied location.

❖ Group E occupancies involve groups of people distributed throughout a number of classrooms or small rooms. Occupants in one area of the building would not necessarily be aware of an emergency in another part of the building unless a system is provided to alert them. Because of the age and maturity of the occupants, more time may be needed to safely evacuate the building. The requirement for a manual fire alarm system in Group E occupancies is not dependent on the location of the level of exit discharge.

Exception 1 exempts Group E occupancies from requiring a fire alarm system when the occupant load is less than 50. This would exempt small day care centers that serve children older than $2^1/_2$ years of age or a small Sunday school classroom at a church.

Exception 2 exempts manual fire alarm boxes in interior corridors, laboratories, auditoriums, cafeterias, gymnasiums and similar spaces based on the installation of heat/smoke detectors and the extent of supervision. The applicability of Exception 2 is independent of whether an automatic sprinkler system is installed. If an automatic smoke detection system is installed, it must be connected to the building fire alarm system.

Exception 3 allows the omission of the manual fire alarm boxes in Group E occupancies equipped throughout with an automatic sprinkler system if the actuation of the sprinkler system will activate the occupant notification system associated with the manual fire alarm system. See Section 903.2.3 for sprinkler requirements in Group E buildings.

[F] 907.2.4 Group F. A manual fire alarm system that activates the occupant notification system in accordance with Section 907.5 shall be installed in Group F occupancies where both of the following conditions exist:

1. The Group F occupancy is two or more *stories* in height; and

2. The Group F occupancy has a combined *occupant load* of 500 or more above or below the lowest *level of exit discharge*.

Exception: Manual fire alarm boxes are not required where the building is equipped throughout with an *automatic sprinkler system* installed in accordance with Section 903.3.1.1 and the occupant notification appliances will activate throughout the notification zones upon sprinkler waterflow.

❖ This section is intended to apply to large multistory manufacturing facilities. For this reason, a manual fire alarm system would be required only if the building were at least two stories in height and had 500 or more occupants above or below the level of exit discharge. An unlimited area, two-story Group F occupancy complying with Section 507.4 of the code would be indicative of an occupancy requiring a manual fire alarm system.

Buildings in compliance with Section 507.4 of the code, and large manufacturing facilities in general, however, must be fully sprinklered and would thus be eligible for the exception. The exception does not eliminate the fire alarm system but rather permits it to be initiated automatically by the sprinkler system waterflow switch(es) instead of by the manual fire alarm boxes.

[F] 907.2.5 Group H. A manual fire alarm system that activates the occupant notification system shall be installed in Group H-5 occupancies and in occupancies used for the manufacture of organic coatings. An automatic smoke detection system that activates the occupant notification system shall be installed for *highly toxic* gases, organic peroxides and oxidizers in accordance with Chapters 37, 39 and 40, respectively, of the *International Fire Code*.

❖ Because of the nature and potential quantity of hazardous materials in Group H-5 occupancies, a manual means of activating an occupant notification system is essential for the safety of the occupants. In accordance with Section 415.8.8, the activation of the alarm system must initiate a local alarm and transmit a signal to the emergency control station. The manual fire alarm system requirement for the building is in addition to the emergency alarm requirements in Section 415.8.4.6 (see also Section 908.2).

Occupancies involved in the manufacture of organic coatings present special hazardous conditions because of the unstable character of the materials, such as nitrocellulose. Good housekeeping and control of ignition sources is critical. Section 418 of the code and Chapter 20 of the IFC contain additional requirements for organic coating manufacturing processes.

This section also requires an automatic smoke detection system in certain occupancy conditions involving either highly toxic gases or organic peroxides and oxidizers. The need for the automatic smoke detection system may depend on the class of materials and additional levels of fire protection provided. This requirement also assumes the quantity of materials is in excess of the maximum allowable quantities shown in Tables 307.1(1) and 307.1(2).

[F] 907.2.6 Group I. A manual fire alarm system that activates the occupant notification system shall be installed in Group I occupancies. An automatic smoke detection system that acti-

vates the occupant notification system shall be provided in accordance with Sections 907.2.6.1, 907.2.6.2 and 907.2.6.3.3.

Exceptions:

1. Manual fire alarm boxes in resident or patient sleeping areas of Group I-1 and I-2 occupancies shall not be required at *exits* if located at all nurses' control stations or other constantly attended staff locations, provided such stations are visible and continuously accessible and that travel distances required in Section 907.4.2 are not exceeded.

2. Occupant notification systems are not required to be activated where private mode signaling installed in accordance with NFPA 72 is *approved* by the fire code official.

❖ Because the protection and possible evacuation of the occupants in Group I occupancies are most often dependent on the response by staff, occupancies in Group I must be protected with a manual fire alarm system and in certain instances, as described in Sections 907.2.6.1, 907.2.6.2 and 907.2.6.3, an automatic smoke detection system. In Group I-1, smoke alarms are also required in accordance with Section 907.2.6.1.1.

It is not the intent of this section to require a smoke detection system throughout all Group I occupancies. Smoke detectors are only generally required in the corridors and in waiting rooms that are open to corridors, unless noted otherwise. IFC Committee Interpretation No. 36-03 makes it clear that the Group I provisions only require a manual fire alarm system with smoke detectors in selected areas. To reduce the potential for unwanted alarms, manual fire alarm boxes may be located at the nurses' control stations or another constantly attended location.

Exception 1 reduces the likelihood of accidental or malicious false alarm system activations by manual means by allowing the pull stations to be located in a more controlled area. It assumes the approved location is always accessible by staff and within 200 feet (60 960 mm) of travel distance. Exception 2 allows the common practice in Group I occupancies of only notifying the staff instead of all building occupants in the event of a fire, subject to the approval of the fire code official.

[F] 907.2.6.1 Group I-1. An automatic smoke detection system shall be installed in *corridors*, waiting areas open to *corridors* and *habitable spaces* other than *sleeping units* and kitchens. The system shall be activated in accordance with Section 907.5.

Exceptions:

1. Smoke detection in *habitable spaces* is not required where the facility is equipped throughout with an *automatic sprinkler system* installed in accordance with Section 903.3.1.1.

2. Smoke detection is not required for exterior balconies.

❖ Occupancies in Group I-1 tend to be compartmentalized into small rooms so that a fire in one area of the building would not easily be noticed by occupants in another part of the building. Therefore, smoke detection is required in areas other than sleeping units and kitchens. Sleeping units are required by Section 907.2.6.1.1 to be equipped with single- and multiple-station smoke alarms in accordance with Section 907.2.11.

Since Group I-1 occupancies may not be supervised by staff and to reduce the likelihood that a fire within a waiting area open to the corridor or the corridor itself could develop beyond the incipient stage, thereby jeopardizing the building egress, these areas must be equipped with automatic smoke detection.

Exception 1 allows smoke detectors to be eliminated from habitable spaces if the building is equipped throughout with an NFPA 13 automatic sprinkler system. The sprinkler system should control any fire and perform occupant notification through actuation of the waterflow switch and subsequent activation of the building alarm notification appliances. A sprinkler system is required for all Group I occupancies in accordance with Section 903.2.6.

Exception 2 allows for omitting smoke detectors from exterior balconies for environmental reasons and does not require the installation of an alternative type of detector. The exterior balconies are assumed to be sufficiently open to the atmosphere to readily allow the dissipation of smoke and hot gases.

[F] 907.2.6.1.1 Smoke alarms. Single- and multiple-station smoke alarms shall be installed in accordance with Section 907.2.11.

❖ As with dwelling units or sleeping units in any occupancy, this section requires that single- and multiple-station smoke alarms be installed in accordance with Section 907.2.11. Section 907.2.11.2 deals specifically with the requirements for Group I-1.

[F] 907.2.6.2 Group I-2. An automatic smoke detection system shall be installed in *corridors* in nursing homes (both intermediate care and skilled nursing facilities), detoxification facilities and spaces permitted to be open to the *corridors* by Section 407.2. The system shall be activated in accordance with Section 907.5. Hospitals shall be equipped with smoke detection as required in Section 407.

Exceptions:

1. *Corridor* smoke detection is not required in smoke compartments that contain patient sleeping units where such units are provided with smoke detectors that comply with UL 268. Such detectors shall provide a visual display on the *corridor* side of each patient *sleeping unit* and shall provide an audible and visual alarm at the nursing station attending each unit.

2. *Corridor* smoke detection is not required in smoke compartments that contain patient *sleeping units* where patient *sleeping unit* doors are equipped with automatic door-closing devices with integral smoke detectors on the unit sides installed in accordance with their listing, provided that the integral detectors perform the required alerting function.

❖ Automatic smoke detection is required in areas permitted to be open to corridors in occupancies classified as Group I-2 and corridors in nursing homes and detoxification facilities. In recognition of quick-response sprinkler technology and the fact that the sprinkler system is electronically supervised, and because the doors to patient sleeping units are continuously supervised by staff when in the open position, it is now believed that smoke detectors are not required for adequate fire safety in patient sleeping units.

In nursing homes and detoxification facilities, however, some redundance is appropriate because such facilities typically have less control over furnishings and personal items, thereby resulting in a less predictable and usually higher fire hazard load than other Group I-2 occupancies. Also, there is generally less staff supervision in these facilities than in other health care facilities and thus less control over patient smoking and other fire causes. Therefore, to provide additional protection against fires spreading from the room of origin, smoke detection is required in corridors of nursing homes and detoxification facilities.

Smoke detection is not required in corridors of other Group I-2 occupancies except where otherwise specifically required in the code. Similarly, because areas open to the corridor very often are the room of fire origin, and such areas are no longer required by the code to be under visual supervision by staff, some redundance to protection by the sprinkler system is requested. Accordingly, all areas open to corridors must be protected by an automatic smoke detection system. This requirement provides an additional level of protection against sprinkler system failures or lapses in staff supervision.

These requirements are not applicable to hospitals. The scope of this section clearly indicates that its provisions are only applicable to detoxification facilities and nursing homes. Hospitals are noted as being subject to the provisions in Section 407.2. IFC Committee Interpretation No. 37-03 addresses this issue. Section 407.2 of the IBC notes that smoke detection is only required for spaces open to corridors, such as waiting areas and mental health treatment areas where patients are not capable of self-preservation (see commentary, Section 407.2).

There are two exceptions to the requirement for an automatic fire detection system in corridors of nursing homes and detoxification facilities. Both exceptions provide an alternative method for redundant protection in patient sleeping units. For this reason, they provide either a backup to the notification of a fire or containment of fire in the room of origin.

Exception 1 requires smoke detectors in patient sleeping units that activate both a visual display on the corridor side of the patient sleeping unit and a visual and audible alarm at the nurses' station serving or attending the room. Detectors complying with UL 268 are intended for open area protection and for connection to a normal power supply or as part of a fire alarm system. This exception, however, is specifically designed not to require the detectors to activate the building fire alarm system where approved patient sleeping unit smoke detectors are installed and where visual and audible alarms are provided. This is in response to the concern over unwanted alarms. The required alarm signals will not necessarily indicate to staff that a fire emergency exists because the nursing call system may typically be used to identify numerous conditions within the room.

Exception 2 addresses the situation where smoke detectors are incorporated within automatic door-closing devices. The units are acceptable as long as the required alarm functions are still provided. Such units are usually listed as combination door closer and hold-open devices.

[F] 907.2.6.3 Group I-3 occupancies. Group I-3 occupancies shall be equipped with a manual fire alarm system and automatic smoke detection system installed for alerting staff.

❖ Because of the evacuation difficulties associated with Group I-3 occupancies and the dependence on adequate staff response, a manual fire alarm system and an automatic smoke detection system are required subject to the special occupancy conditions in Sections 907.2.6.3.1 through 907.2.6.3.3. This section recognizes that the evacuation of Group I-3 occupancies depends on an effective staff response. Section 408.7 of the IFC contains the requirements for an emergency plan, including employee training, staff availability, the need for occupants to notify staff and the need for the proper keys for unlocking doors for staff in Group I-3 occupancies.

[F] 907.2.6.3.1 System initiation. Actuation of an automatic fire-extinguishing system, a manual fire alarm box or a fire detector shall initiate an *approved* fire alarm signal which automatically notifies staff.

❖ This section specifies the systems that, upon activation, must initiate the required alarm signal immediately and automatically to the staff so that staff will respond in a timely manner.

[F] 907.2.6.3.2 Manual fire alarm boxes. Manual fire alarm boxes are not required to be located in accordance with Section 907.4.2 where the fire alarm boxes are provided at staff-attended locations having direct supervision over areas where manual fire alarm boxes have been omitted.

❖ Because of the potential for intentional false alarms and the resulting disruption to the facility, manual fire alarm boxes in Group I-3 occupancies may be either locked or made inaccessible to the occupants.

907.2.6.3.2.1 Manual fire alarm boxes in detainee areas. Manual fire alarm boxes are allowed to be locked in areas occupied by detainees, provided that staff members are present

within the subject area and have keys readily available to operate the manual fire alarm boxes.

❖ The locking of manual fire alarm boxes is permitted only in areas where staff members are present and keys are readily available to them to unlock the boxes, or where the alarm boxes are located in a manned staff location that has direct supervision of the Group I-3 area.

[F] 907.2.6.3.3 Automatic smoke detection system. An automatic smoke detection system shall be installed throughout resident housing areas, including *sleeping units* and contiguous day rooms, group activity spaces and other common spaces normally accessible to residents.

Exceptions:

1. Other *approved* smoke detection arrangements providing equivalent protection, including, but not limited to, placing detectors in exhaust ducts from cells or behind protective guards *listed* for the purpose, are allowed when necessary to prevent damage or tampering.

2. *Sleeping units* in Use Conditions 2 and 3 as described in Section 308.

3. Smoke detectors are not required in *sleeping units* with four or fewer occupants in smoke compartments that are equipped throughout with an *automatic sprinkler system* installed in accordance with Section 903.3.1.1.

❖ Evacuation of Group I-3 facilities is impractical because of the need to maintain security. An automatic smoke detection system is therefore required to provide early warning of a fire.

As indicated in Exception 1, the installation of automatic smoke detectors must take into account the need to protect the detector from vandalism by residents. As a result, detectors may have to be located in return air ducts or be protected by a substantial physical barrier.

Since occupants in Use Condition 2 or 3 are not locked in their sleeping units, Exception 2 reduces the need for smoke detection.

Exception 3 allows smoke detectors to be omitted in sleeping units housing no more than four occupants on the basis that in a building that is protected throughout with an approved automatic sprinkler system, the system will provide both detection and suppression functions. Group I facilities are assumed to be fully sprinklered throughout in accordance with NFPA 13 as required by Section 903.2.6. The limitation of four occupants reduces the potential fuel load (mattresses, clothes, etc.) and the likelihood of involvement over an extended area.

[F] 907.2.7 Group M. A manual fire alarm system that activates the occupant notification system in accordance with Section 907.5 shall be installed in Group M occupancies where one of the following conditions exists:

1. The combined Group M *occupant load* of all floors is 500 or more persons.

2. The Group M *occupant load* is more than 100 persons above or below the lowest *level of exit discharge*.

Exceptions:

1. A manual fire alarm system is not required in *covered mall buildings* complying with Section 402.

2. Manual fire alarm boxes are not required where the building is equipped throughout with an *automatic sprinkler system* installed in accordance with Section 903.3.1.1 and the occupant notification appliances will automatically activate throughout the notification zones upon sprinkler waterflow.

❖ Group M occupancies have the potential for large numbers of occupants who may not be familiar with their surroundings. The installation of a fire alarm system increases the ability to alert the occupants of a fire. Note that the occupant thresholds must be considered independently. If the total occupant load is 500 or more persons, a manual fire alarm system is required. If more than 100 persons are above or below the level of exit discharge, a manual fire alarm system is required.

This section also specifies that the manual fire alarm boxes must, upon activation, provide occupant notification throughout the Group M occupancy.

The extent of fire alarm application is based upon the area in which the Group M occupancy is located. If the building is considered as a separated mixed occupancy, then the fire alarm system is only required in the individual occupancy in which the occupant load exceeds the threshold quantity. The rest of the building would not require a fire alarm system. This approach is noted in Section 508.4.1, which states that each separated space must comply with the code based upon the occupancy classification of that portion of the building. If the Group M occupancy was part of a nonseparated mixed use building then the alarm system would be required in the entire building in accordance with Section 508.3.1. The determination as to when such a system is required would be based solely upon the Group M occupant load.

Exception 1 recognizes the increased level of fixed automatic protection inherently required in covered mall buildings including an automatic sprinkler system, and, possibly, a smoke control system. Covered mall buildings are also required to contain an emergency voice communication system (see Section 907.2.20).

Exception 2 does not eliminate the fire alarm system, but rather allows it to be initiated automatically by sprinkler system waterflow switch(es) instead of by manual fire alarm boxes. Buildings with a fire area containing a Group M occupancy in excess of 12,000 square feet (1115 m²) must be equipped with an auto-matic sprinkler system complying with Section 903.2.7.

[F] 907.2.7.1 Occupant notification. During times that the building is occupied, the initiation of a signal from a manual fire alarm box or from a waterflow switch shall not be required to activate the alarm notification appliances when an alarm signal is activated at a *constantly attended location* from which evacuation instructions shall be initiated over an emergency voice/alarm communication system installed in accordance with Section 907.5.2.2.

❖ Occupants in a mercantile occupancy may assume the alarm is a false alarm or act inappropriately and thus delay evacuation of the building. To prevent such a dangerous situation, the manual fire alarm system may be part of an emergency voice/alarm communication system. The signal is to be sent to a constantly attended location on site from which evacuation instructions can be given.

It should be noted that, although the alarm notification alternative allows for the manual use of an emergency voice/alarm communication system, the alternative does not remove the requirement for audible and visual notification devices.

[F] 907.2.8 Group R-1. Fire alarm systems and smoke alarms shall be installed in Group R-1 occupancies as required in Sections 907.2.8.1 through 907.2.8.3.

❖ Because residents of Group R-1 occupancies may be asleep and are usually transients who are unfamiliar with the building, and because such buildings contain numerous small rooms so that the occupants may not notice a fire in another part of the building, occupancies in Group R-1 must have a manual fire alarm system and an automatic smoke detection system installed throughout. Requirements for single- or multiple-station smoke alarms in sleeping units are contained in Section 907.2.11.1.

[F] 907.2.8.1 Manual fire alarm system. A manual fire alarm system that activates the occupant notification system in accordance with Section 907.5 shall be installed in Group R-1 occupancies.

Exceptions:

1. A manual fire alarm system is not required in buildings not more than two *stories* in height where all individual *sleeping units* and contiguous *attic* and crawl spaces to those units are separated from each other and public or common areas by at least 1-hour *fire partitions* and each individual *sleeping unit* has an *exit* directly to a *public way*, *exit court* or *yard*.

2. Manual fire alarm boxes are not required throughout the building when all of the following conditions are met:

2.1. The building is equipped throughout with an *automatic sprinkler system* installed in accordance with Section 903.3.1.1 or 903.3.1.2;

2.2. The notification appliances will activate upon sprinkler waterflow; and

2.3. At least one manual fire alarm box is installed at an *approved* location.

❖ This section is specific to manual fire alarm systems and requires such systems in all Group R-1 occupancies, with two exceptions.

Exception 1 eliminates the requirement for a manual fire alarm system if the sleeping units have an exit discharging directly to a public way, exit court or yard. Even though the building may be two stories in height, the sleeping units on each floor must have access directly to an approved exit at grade level. The use of an exterior exit access balcony with exterior stairs serving the second floor does not constitute an exit directly at grade. The minimum 1-hour fire-resistance rating required for adequate separation of the sleeping units must be maintained.

Exception 2 does not omit the fire alarm system but rather permits it to be initiated automatically by sprinkler system waterflow switch(es) in lieu of manual fire alarm boxes. The sprinkler system must activate the occupant notification system and at least one manual fire alarm box shall be installed at an approved location. See the commentary to Section 907.2 for a discussion of the single manual fire alarm box.

The exceptions do not affect the independent provision in Section 907.2.11 for single- or multiple-station smoke alarms.

[F] 907.2.8.2 Automatic smoke detection system. An automatic smoke detection system that activates the occupant notification system in accordance with Section 907.5 shall be installed throughout all interior *corridors* serving *sleeping units*.

Exception: An automatic smoke detection system is not required in buildings that do not have interior *corridors* serving *sleeping units* and where each *sleeping unit* has a *means of egress* door opening directly to an *exit* or to an exterior *exit access* that leads directly to an *exit*.

❖ This section requires an automatic smoke detection system within interior corridors. Such systems make use of smoke detectors for alarm initiation in accordance with Section 907.2, with one exception.

The exception provides that automatic fire detectors are not required in motels and hotels that do not have interior corridors and in which sleeping units have a door opening directly to an exterior exit access that leads directly to the exits. The intent of the exception is that the exit access from the sleeping unit door be exterior and not require reentering the building prior to entering the exit. Since the exit access is outside, the need for detectors other than the smoke alarms required by Section 907.2.8.3 in sleeping units is greatly reduced.

[F] 907.2.8.3 Smoke alarms. Single- and multiple-station smoke alarms shall be installed in accordance with Section 907.2.11.

❖ The actual requirements for single- and multiple-station smoke alarms are located in Section 907.2.11. That section requires that the single- and multiple-station smoke alarms within sleeping units be connected

to the emergency electrical system. Automatic activation of the fire alarm system is avoided to reduce unnecessary alarms within such buildings.

[F] 907.2.9 Group R-2. Fire alarm systems and smoke alarms shall be installed in Group R-2 occupancies as required in Sections 907.2.9.1 and 907.2.9.2.

❖ This section introduces the fire alarm system and smoke alarm requirements for Group R-2 occupancies.

[F] 907.2.9.1 Manual fire alarm system. A manual fire alarm system that activates the occupant notification system in accordance with Section 907.5 shall be installed in Group R-2 occupancies where:

1. Any *dwelling unit* or *sleeping unit* is located three or more *stories* above the lowest *level of exit discharge*;

2. Any *dwelling unit* or *sleeping unit* is located more than one *story* below the highest *level of exit discharge* of *exits* serving the *dwelling unit* or *sleeping unit*; or

3. The building contains more than 16 *dwelling units* or *sleeping units*.

Exceptions:

1. A fire alarm system is not required in buildings not more than two *stories* in height where all *dwelling units* or *sleeping units* and contiguous *attic* and crawl spaces are separated from each other and public or common areas by at least 1-hour *fire partitions* and each *dwelling unit* or *sleeping unit* has an *exit* directly to a *public way*, *exit court* or *yard*.

2. Manual fire alarm boxes are not required where the building is equipped throughout with an *automatic sprinkler system* installed in accordance with Section 903.3.1.1 or 903.3.1.2 and the occupant notification appliances will automatically activate throughout the notification zones upon a sprinkler waterflow.

3. A fire alarm system is not required in buildings that do not have interior *corridors* serving *dwelling units* and are protected by an *approved automatic sprinkler system* installed in accordance with Section 903.3.1.1 or 903.3.1.2, provided that *dwelling units* either have a *means of egress* door opening directly to an exterior *exit access* that leads directly to the *exits* or are served by open-ended *corridors* designed in accordance with Section 1026.6, Exception 4.

❖ The occupants of Group R-2 occupancies are not considered to be as transient as those of Group R-1, which increases the probability that residents can more readily notify each other of a fire. Therefore, Group R-1 occupancies must have a manual fire alarm system with audible and visual notification appliances subject to the exceptions in Section 907.2.8.1, whereas Group R-2 occupancies are required to have only a manual fire alarm system as stipulated in one of the three listed conditions. The threshold conditions are meant to be applied independently of each other.

Exception 1 eliminates the requirement for a manual fire alarm system if the sleeping units have an exit discharging directly to a public way, exit court or yard. Even though the building may be two stories in height, the sleeping units on each floor must have access directly to an approved exit at grade level. The use of an exterior exit access balcony with exterior stairs serving the second floor does not constitute an exit directly at grade. The minimum 1-hour fire-resistance rating required for adequate separation of the sleeping units must be maintained.

Exception 2 does not omit the fire alarm system but rather permits it to be initiated automatically by sprinkler system waterflow switch(es) in lieu of manual fire alarm boxes. The sprinkler system must activate the occupant notification system. This exception does not affect the independent provisions of Section 907.2.11.

Exception 3 allows the omission of a fire alarm system in fully sprinklered buildings (NFPA 13 or 13R) with no interior corridors and that exit directly to an exterior exit access or have open-ended corridors. The important thing to note is that the sprinkler system is not required to activate alarm notification appliances since a fire alarm system would not be required. Only the sprinkler alarms required by Section 903.4 would be required.

[F] 907.2.9.2 Smoke alarms. Single- and multiple-station smoke alarms shall be installed in accordance with Section 907.2.11.

❖ The actual requirements for single- and multiple-station smoke alarms are located in Section 907.2.11. That section requires that the single- and multiple-station smoke alarms within sleeping units be connected to the emergency electrical system. Automatic activation of the fire alarm system is avoided to reduce unnecessary alarms within such buildings.

[F] 907.2.10 Group R-4. Fire alarm systems and smoke alarms shall be installed in Group R-4 occupancies as required in Sections 907.2.10.1 through 907.2.10.3.

❖ This section, based on the Group R-2 requirements for manual fire alarm systems and Group I-1 requirements for automatic smoke detection systems, is new in the 2009 edition of the code and adds manual fire alarm and automatic smoke detection system requirements for new Group R-4 occupancies. Reviewing the occupancy categories in Chapter 3 of the code, a Group R-4 could be considered either a small Group I-1 or a Group R-2 with occupants that have special needs or limitations. A further review found that both Group I-1 and R-2 occupancies had fire alarm requirements for new buildings, but Group R-4 did not, even though the IFC required a fire alarm system retroactively in existing Group R-4 occupancies (see Section 4603.6.7 of the IFC).

[F] 907.2.10.1 Manual fire alarm system. A manual fire alarm system that activates the occupant notification system in

accordance with Section 907.5 shall be installed in Group R-4 occupancies.

Exceptions:

1. A manual fire alarm system is not required in buildings not more than two *stories* in height where all individual *sleeping units* and contiguous *attic* and crawl spaces to those units are separated from each other and public or common areas by at least 1-hour *fire partitions* and each individual *sleeping unit* has an *exit* directly to a *public way*, *exit court* or *yard*.

2. Manual fire alarm boxes are not required throughout the building when the following conditions are met:

   2.1. The building is equipped throughout with an *automatic sprinkler system* installed in accordance with Section 903.3.1.1 or 903.3.1.2;

   2.2. The notification appliances will activate upon sprinkler waterflow; and

   2.3. At least one manual fire alarm box is installed at an *approved* location.

3. Manual fire alarm boxes in resident or patient sleeping areas shall not be required at *exits* where located at all nurses' control stations or other constantly attended staff locations, provided such stations are visible and continuously accessible and that travel distances required in Section 907.4.2.1 are not exceeded.

❖ This section is specific to manual fire alarm systems and requires such systems in all Group R-4 occupancies, with three exceptions. Exception 1 eliminates the requirement for a manual fire alarm system if the sleeping units have an exit discharging directly to a public way, exit court or yard. Even though the building may be two stories in height, the sleeping units on each floor must have access directly to an approved exit at grade level. The use of an exterior exit access balcony with exterior stairs serving the second floor does not constitute an exit directly at grade. The minimum 1-hour fire-resistance rating required for adequate separation of the sleeping units must be maintained.

Exception 2 does not omit the fire alarm system but rather permits it to be initiated automatically by sprinkler system waterflow switch(es) in lieu of manual fire alarm boxes. The sprinkler system must activate the occupant notification system and at least one manual fire alarm box shall be installed at an approved location. See the commentary to Section 907.2 for a discussion of the single manual fire alarm box.

Exception 3 reduces the likelihood of accidental or malicious false alarm system activations by manual means by allowing the pull stations to be located in a more controlled area. It assumes the approved location is always accessible by staff and within 200 feet (60 960 mm) of travel distance.

[F] 907.2.10.2 Automatic smoke detection system. An automatic smoke detection system that activates the occupant notification system in accordance with Section 907.5 shall be installed in *corridors*, waiting areas open to *corridors* and *habitable spaces* other than *sleeping units* and kitchens.

Exceptions:

1. Smoke detection in *habitable spaces* is not required where the facility is equipped throughout with an *automatic sprinkler system* installed in accordance with Section 903.3.1.1.

2. An automatic smoke detection system is not required in buildings that do not have interior *corridors* serving *sleeping units* and where each *sleeping unit* has a *means of egress* door opening directly to an *exit* or to an exterior *exit access* that leads directly to an *exit*.

❖ Occupancies in Group R-4 can be compartmentalized into small rooms so that a fire in one area of the building would not easily be noticed by occupants in another part of the building. Therefore, smoke detection is required in areas other than sleeping units and kitchens. Sleeping units are required by Section 907.2.10.3 to be equipped with single- and multiple-station smoke alarms in accordance with Section 907.2.11.

Since Group R-4 occupancies may not be supervised by staff and to reduce the likelihood that a fire within a waiting area open to the corridor or the corridor itself could develop beyond the incipient stage, thereby jeopardizing the building egress, these areas must be equipped with automatic smoke detection.

Exception 1 allows smoke detectors to be eliminated from habitable spaces if the building is equipped throughout with an NFPA 13 automatic sprinkler system. The sprinkler system should control any fire and perform occupant notification through actuation of the waterflow switch and subsequent activation of the building alarm notification appliances. A sprinkler system is required for all Group R occupancies in accordance with Section 903.2.8.

The exception provides that automatic fire detectors are not required in buildings that do not have interior corridors and in which sleeping units have a door opening to an exterior exit access that leads directly to the exits. The intent of the exception is that the exit access from the sleeping unit door be exterior and not require reentering the building prior to entering the exit. Since the exit access is outside, the need for detectors other than the smoke alarms required by Section 907.2.10.3 in sleeping units is greatly reduced.

[F] 907.2.10.3 Smoke alarms. Single- and multiple-station smoke alarms shall be installed in accordance with Section 907.2.11.

❖ The actual requirements for single- and multiple-station smoke alarms are located in Section 907.2.11. That section requires that the single- and multiple-station smoke alarms within sleeping units be connected to the emergency electrical system. Automatic activation of the fire alarm system is avoided to reduce unnecessary alarms within such buildings.

[F] 907.2.11 Single- and multiple-station smoke alarms. *Listed* single- and multiple-station smoke alarms complying with UL 217 shall be installed in accordance with Sections 907.2.11.1 through 907.2.11.4 and NFPA 72.

❖ Single- and multiple-station smoke alarms have evolved as one of the most important fire safety features in residential and similar occupancies having sleeping occupants. The value of early fire warning in these occupancies has been repeatedly demonstrated in fires involving both successful and unsuccessful smoke alarm performance.

For successful smoke alarm operation and performance, single- and multiple-station smoke alarms must be listed in accordance with UL 217 and installed to comply with the code and Chapter 11 of NFPA 72, which contains the minimum requirements for the selection, installation, operation and maintenance of fire warning equipment for use in family living units. These devices are called "smoke alarms" rather than "smoke detectors" because they are independent of a fire alarm system and include an integral alarm notification device.

[F] 907.2.11.1 Group R-1. Single- or multiple-station smoke alarms shall be installed in all of the following locations in Group R-1:

1. In sleeping areas.

2. In every room in the path of the *means of egress* from the sleeping area to the door leading from the *sleeping unit*.

3. In each *story* within the *sleeping unit*, including basements. For *sleeping units* with split levels and without an intervening door between the adjacent levels, a smoke alarm installed on the upper level shall suffice for the adjacent lower level provided that the lower level is less than one full *story* below the upper level.

❖ Because the occupants of a sleeping unit or suite may be asleep and unaware of a fire developing in the room or in the egress path, single- or multiple-station smoke alarms must be provided in the sleeping unit and in any intervening room between the sleeping unit and the exit access door from the room. If the sleeping unit or suite involves more than one level, a smoke alarm must also be installed on every level. See the commentary to Section 202 definition of "Sleeping unit."

Smoke alarms are required in split-level arrangements, except those that meet the conditions described in Item 3. In accordance with Section 907.2.11.3, all smoke alarms within a sleeping unit or suite must be interconnected so that actuation of one alarm will actuate all smoke alarms within the sleeping unit or suite.

[F] 907.2.11.2 Groups R-2, R-3, R-4 and I-1. Single- or multiple-station smoke alarms shall be installed and maintained in Groups R-2, R-3, R-4 and I-1 regardless of *occupant load* at all of the following locations:

1. On the ceiling or wall outside of each separate sleeping area in the immediate vicinity of bedrooms.

2. In each room used for sleeping purposes.

> **Exception:** Single- or multiple-station smoke alarms in Group I-1 shall not be required where smoke detectors are provided in the sleeping rooms as part of an automatic smoke detection system.

3. In each *story* within a *dwelling unit*, including basements but not including crawl spaces and uninhabitable *attics*. In *dwellings* or *dwelling units* with split levels and without an intervening door between the adjacent levels, a smoke alarm installed on the upper level shall suffice for the adjacent lower level provided that the lower level is less than one full *story* below the upper level.

❖ Because the occupants of a dwelling unit may be asleep and unaware of a fire developing in the room or in an area within the dwelling unit that will affect their ability to escape, single- or multiple-station smoke alarms must be installed in every bedroom, in the vicinity of all bedrooms (e.g., hallways leading to the bedrooms) and on each story of the dwelling unit (see Figure 907.2.11.2 and the commentary to Section 202 definition of "dwelling unit").

If a sprinkler system was installed throughout the building in accordance with NFPA 13, 13R or 13D, if applicable, smoke alarms would still be required in the bedrooms even if residential sprinklers were used.

Smoke alarms are required in split-level arrangements. As required by Section 907.2.11.3, all smoke alarms within a dwelling unit must be interconnected so that actuation of one alarm will actuate the alarms in all detectors within the dwelling unit.

These provisions do not apply to one- and two-family dwellings and multiple single-family dwellings (townhouses) not more than three stories in height with a separate means of egress that are regulated by the IRC. The IRC is intended to be a stand-alone document but if the residential units do not fall within the scope of the IRC or for other reasons are intended to be subject to this code, then the requirements of this section would apply. IFC Committee Interpretation No. 42-03 addresses this condition and contains additional explanatory information about the IRC and its relationship to the other *International Codes*® (I-Codes®).

Although the occupants of a sleeping unit in a Group I-1 occupancy may be asleep, they are still considered capable of self-preservation. Regardless, smoke alarms are required in sleeping units. The exception allows single- or multiple-station smoke alarms to be eliminated in the room if an automatic fire detection system that includes in-room system smoke detectors is installed as required by Section 907.2.6.

[F] 907.2.11.3 Interconnection. Where more than one smoke alarm is required to be installed within an individual *dwelling unit* or *sleeping unit* in Group R-1, R-2, R-3 or R-4, the smoke alarms shall be interconnected in such a manner that the activation of one alarm will activate all of the alarms in the individual unit. The alarm shall be clearly audible in all bedrooms over background noise levels with all intervening doors closed.

❖ The installation of smoke alarms in areas remote from the sleeping area will be of minimal value if the alarm is not heard by the occupants. Interconnection of multiple smoke alarms within an individual dwelling unit or sleeping unit is required in order to alert a sleeping occupant of a remote fire within the unit before the combustion products reach the smoke alarm in the sleeping area and thus provide additional time for evacuation.

The term "interconnection" is intended to allow the use of not only hard-wired systems, but also those that use radio signals (wireless systems) (see Section 907.7.1). UL has listed smoke detectors that use this technology. It is presumed that on safely evacuating the unit or room of fire origin, an occupant will notify

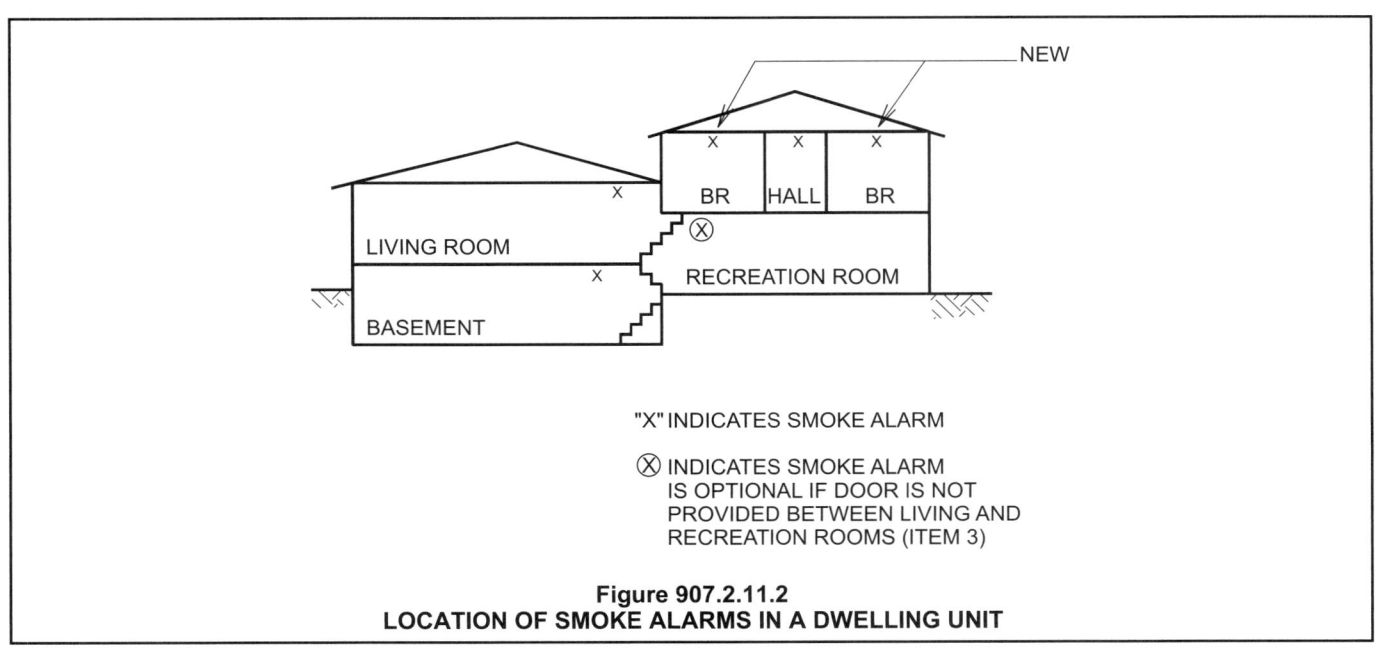

"X" INDICATES SMOKE ALARM

Ⓧ INDICATES SMOKE ALARM IS OPTIONAL IF DOOR IS NOT PROVIDED BETWEEN LIVING AND RECREATION ROOMS (ITEM 3)

Figure 907.2.11.2
LOCATION OF SMOKE ALARMS IN A DWELLING UNIT

other occupants by actuating the manual fire alarm system or using other available means.

[F] 907.2.11.4 Power source. In new construction, required smoke alarms shall receive their primary power from the building wiring where such wiring is served from a commercial source and shall be equipped with a battery backup. Smoke alarms with integral strobes that are not equipped with battery backup shall be connected to an emergency electrical system. Smoke alarms shall emit a signal when the batteries are low. Wiring shall be permanent and without a disconnecting switch other than as required for overcurrent protection.

> **Exception:** Smoke alarms are not required to be equipped with battery backup where they are connected to an emergency electrical system.

❖ Smoke alarms are required to use AC as a primary power source and battery power as a secondary source to improve their reliability. For example, during a power outage, the probability of fire is increased because of the use of candles or lanterns for temporary light. Required backup battery power is intended to provide continued functioning of the smoke alarms. Smoke alarms are commonly designed to emit a recurring signal when batteries are low and need to be replaced.

Certain occupancies may already have an emergency electrical system in the building to monitor other building system conditions. The emergency electrical system provides a level of reliability equivalent to battery backup.

[F] 907.2.12 Special amusement buildings. An automatic smoke detection system shall be provided in *special amusement buildings* in accordance with Sections 907.2.12.1 through 907.2.12.3.

❖ Special amusement buildings are buildings in which the means of egress is not readily apparent, is intentionally confounded or is not readily available. Special amusement buildings must also comply with the provisions of Section 411.

The approved automatic smoke detection system is required to provide early warning of a fire. The detection system is required regardless of the presence of staff in the building. The exception recognizes that the ambient conditions in some special amusement buildings may preclude the use of automatic smoke detectors. In those instances, an alternative detection device must be used for early detection of a fire.

[F] 907.2.12.1 Alarm. Activation of any single smoke detector, the *automatic sprinkler system* or any other automatic fire detection device shall immediately sound an alarm at the building at a *constantly attended location* from which emergency action can be initiated, including the capability of manual initiation of requirements in Section 907.2.12.2.

❖ Upon activation of either a smoke detector or other automatic fire detection device or the automatic sprinkler system, an alarm must sound at a constantly attended location. The staff at the location is expected

to be capable of then providing the required egress illumination, stopping the conflicting or confusing sounds and distractions and activating the exit marking required by Section 907.2.12.2. The staff is also expected to be capable of preventing additional people from entering the building.

[F] 907.2.12.2 System response. The activation of two or more smoke detectors, a single smoke detector equipped with an alarm verification feature, the *automatic sprinkler system* or other *approved* fire detection device shall automatically:

1. Cause illumination of the *means of egress* with light of not less than 1 foot-candle (11 lux) at the walking surface level;

2. Stop any conflicting or confusing sounds and visual distractions;

3. Activate an *approved* directional *exit* marking that will become apparent in an emergency; and

4. Activate a prerecorded message, audible throughout the *special amusement building*, instructing patrons to proceed to the nearest *exit*. Alarm signals used in conjunction with the prerecorded message shall produce a sound which is distinctive from other sounds used during normal operation.

❖ Once a fire has been detected, measures must be taken to stop the confusion or distractions. Additionally, the egress path must be illuminated and marked. These measures must occur automatically upon detection of the fire or sprinkler waterflow. A prerecorded message that can be heard throughout the building instructing the occupants to proceed to the nearest exit must be automatically activated. The message and alarm signals should be designed to prevent panic. The prerecorded message capability is in addition to the emergency voice/alarm communication system requirement of Section 907.2.12.3. The wiring of all devices must comply with NFPA 72.

[F] 907.2.12.3 Emergency voice/alarm communication system. An emergency voice/alarm communication system, which is also allowed to serve as a public address system, shall be installed in accordance with Section 907.5.2.2 and be audible throughout the entire *special amusement building*.

❖ Because of the problem associated with evacuating special amusement buildings, an emergency voice/alarm communication system is required (see also Section 907.5.2.2). This section allows the system to also serve as a public address system to have the capability to alert the occupants of a fire and give them evacuation instructions. The system must be designed so that once the voice alarm is activated, the typical public address function is superseded by the voice alarm. Because a manual override must be provided, it is possible that the same microphone used for the public address can be used for the override. However, a separate action would be necessary so that the override function can be used once the voice alarm is active.

[F] 907.2.13 High-rise buildings. Buildings with a floor used for human occupancy located more than 75 feet (22 860 mm) above the lowest level of fire department vehicle access shall be provided with an automatic smoke detection system in accordance with Section 907.2.13.1, a fire department communication system in accordance with Section 907.2.13.2 and an emergency voice/alarm communication system in accordance with Section 907.5.2.2.

Exceptions:

1. Airport traffic control towers in accordance with Sections 907.2.22 and 412.

2. *Open parking garages* in accordance with Section 406.3.

3. Buildings with an occupancy in Group A-5 in accordance with Section 303.1.

4. Low-hazard special occupancies in accordance with Section 503.1.1.

5. Buildings with an occupancy in Group H-1, H-2 or H-3 in accordance with Section 415.

6. In Group I-1 and I-2 occupancies, the alarm shall sound at a *constantly attended location* and general occupant notification shall be broadcast by the emergency voice/alarm communication system.

❖ High-rise buildings require additional fire protection systems because of the difficulties with smoke movement, egress time and fire department access. As a result, this section requires both an automatic fire alarm system and an emergency voice/alarm communication system (see commentary, Section 907.5.2.2). Exceptions 1 through 5 are the same as those in Section 403.1 regarding the applicability of the high-rise provisions.

Exception 1 addresses airport traffic control towers and is based on the limited fuel load and the limited number of persons occupying the tower. Open parking garages and places of outdoor assembly (Group A-5) are exempted by Exceptions 2 and 3, respectively, because of the free ventilation to the outside that exists in such structures. In Exception 4, low-hazard special industrial occupancies may be exempted when approved by the fire code official. Such buildings should be evaluated based on the occupant load and the hazards of the occupancy and its contents to determine whether the protection features required by Section 403 of the code are necessary. Buildings with occupancies in Groups H-1, H-2 and H-3 are excluded from the requirements of this section by Exception 5 because the fire hazard characteristics of these occupancies have not yet been considered in high-rise buildings. Exception 6 recognizes the supervised environment typical of institutional uses and the reliance placed on staff to act appropriately in an emergency. As is the case for most voice alarms, the key is in being able to deliver specific information to the people who can affect a safe egress—whether it be the public, employees, or both.

[F] 907.2.13.1 Automatic smoke detection. Automatic smoke detection in high-rise buildings shall be in accordance with Sections 907.2.13.1.1 and 907.2.13.1.2.

❖ This section simply introduces the fire alarm and detection system requirements for high-rise buildings.

[F] 907.2.13.1.1 Area smoke detection. Area smoke detectors shall be provided in accordance with this section. Smoke detectors shall be connected to an automatic fire alarm system. The activation of any detector required by this section shall operate the emergency voice/alarm communication system in accordance with Section 907.5.2.2. Smoke detectors shall be located as follows:

1. In each mechanical equipment, electrical, transformer, telephone equipment or similar room which is not provided with sprinkler protection.

2. In each elevator machine room and in elevator lobbies.

❖ Automatic smoke detectors are required in all high-rise buildings in certain locations so that a fire will be detected in its early stages of development. The detectors must be connected to the automatic fire alarm system and be capable of initiating operation of the emergency voice/alarm communication system.

This section divides the automatic smoke detection requirement into two categories. Smoke detectors must be installed in rooms that are not typically occupied. This includes rooms used for: mechanical equipment, electrical equipment, transformer equipment and telephone equipment where such rooms do not have automatic sprinkler protection. In most cases, these rooms will have sprinkler protection by virtue of being in a high-rise building and will therefore not require smoke detectors. However, in elevator machine rooms and elevator lobbies, smoke detectors are required regardless of sprinkler protection.

Note that smoke detection and smoke alarms may be required based upon occupancy related requirements elsewhere in Section 907.2.

[F] 907.2.13.1.2 Duct smoke detection. Duct smoke detectors complying with Section 907.3.1 shall be located as follows:

1. In the main return air and exhaust air plenum of each air-conditioning system having a capacity greater than 2,000 cubic feet per minute (cfm) (0.94 m³/s). Such detectors shall be located in a serviceable area downstream of the last duct inlet.

2. At each connection to a vertical duct or riser serving two or more stories from a return air duct or plenum of an air-conditioning system. In Group R-1 and R-2 occupancies, a smoke detector is allowed to be used in each return air riser carrying not more than 5,000 cfm (2.4 m³/s) and serving not more than 10 air-inlet openings.

❖ Smoke detectors must be installed in the main return air and exhaust air plenum of each air-conditioning system having a design capacity exceeding 2,000 cubic feet per minute (cfm) (0.94 m³/s). Systems with design capacities equal to or less than 2,000 cfm (0.94 m³/s) are exempt from this requirement because their small size limits their capacity for spreading smoke to parts of the building not already involved with fire.

The area that could be served by a 2,000-cfm (0.94 m³) system (approximately 5 tons of cooling capacity) is comparatively small; therefore, the distribution of smoke in a system of that size would be minimal. Smoke detectors must be located so that they monitor the total airflow within the system. If a single detector is unable to sample the total airflow at all times, then multiple detectors are required. The smoke detectors must be made accessible for maintenance and inspection. Many failures and false alarms are caused by a lack of maintenance and cleaning of the smoke detectors.

Consistent with Section 606.2.3 of the IMC, return air risers serving two or more stories must have smoke detectors installed at each story. Item 2 allows the use of a single listed smoke detector in each return air riser in a Group R-1 or R-2 occupancy if the capacity of each riser does not exceed 5,000 cfm (2.4 m³/s) and does not serve more than 10 air-inlet openings. This alternative recognizes that it is not as necessary in buildings dedicated to residential occupancies only to monitor the return air from each story prior to intermixing the return air in the common riser.

[F] 907.2.13.2 Fire department communication system. Where a wired communication system is *approved* in lieu of a radio coverage system in accordance with Section 510 of the *International Fire Code*, the wired fire department communication system shall be designed and installed in accordance with NFPA 72 and shall operate between a fire command center complying with Section 911, elevators, elevator lobbies, emergency and standby power rooms, fire pump rooms, *areas of refuge* and inside enclosed *exit stairways*. The fire department communication device shall be provided at each floor level within the enclosed *exit stairway*.

❖ High-rise buildings have posed a challenge to the traditional communication systems used by the fire service for fire-to-ground communications to assist fire ground officers in communicating with the fire fighters working in various areas of the building. Where testing of the building's public safety communications system required by Section 403.4.4 of the code shows that the signal strengths are not satisfactory, Section 510.1, Exception 1, of the IFC allows for the alternative of installation of a wired communication system designed in accordance with this section. The system must be capable of operating between the fire command center and every elevator, elevator lobby, emergency/standby power room, fire pump room, area of refuge and exit stairway. Note that this section does not offer specific criteria as to what constitutes an acceptable wired communication system or its components. It could be a component of an emergency voice/alarm communication system that complies with Section 907.5.2.2 or a building's telephone system. In any event, when applying Section 510.1, Exception 1, of the IFC and this section, the concurrent approval of the fire and building code officials is required.

[F] 907.2.14 Atriums connecting more than two stories. A fire alarm system shall be installed in occupancies with an atrium that connects more than two *stories*, with smoke detection installed throughout the atrium. The system shall be activated in accordance with Section 907.5. Such occupancies in Group A, E or M shall be provided with an emergency voice/alarm communication system complying with the requirements of Section 907.5.2.2.

❖ Buildings containing an atrium that connects more than two stories are to be equipped with a fire alarm system that can be used to notify building occupants to begin evacuating in case of a fire. The alarm system must be initiated in accordance with Section 907.4, which requires that in buildings containing an atrium, the alarm system is to be initiated by the sprinkler system and any automatic or manual fire alarm-initiating devices found in the atrium as well as elsewhere in the building. It does not intend to require certain features to be installed within the atrium but rather is simply requiring that any such features present initiate the occupant notification system. It would not necessarily be appropriate to also initiate the smoke control system upon activation of the alarm system within a building containing an atrium (see Section 909.12.2).

Groups A, E and M must have an emergency voice/alarm communication system that complies with Section 907.5.2.2 because of the number of persons to be evacuated and the lack of familiarity with the location of exits that is typical of occupants in Groups A and M. The alarm system is intended to warn occupants entering the atrium because smoke is being drawn to the atrium.

[F] 907.2.15 High-piled combustible storage areas. An automatic smoke detection system shall be installed throughout high-piled combustible storage areas where required by Section 2306.5 of the *International Fire Code*.

❖ Section 2306.5 of the IFC requires an automatic fire detection system in high-piled combustible storage areas depending on the commodity class, the size of the high-piled storage area and the presence of an automatic sprinkler system. High-piled storage is the storage of Class I through IV commodities in piles, bin boxes, on pallets or in racks more than 12 feet (3658 mm) high or for high-hazard commodities stored higher than 6 feet (1829 mm). Chapter 23 of the IFC and NFPA 13 contain additional requirements for all high-piled storage conditions.

[F] 907.2.16 Aerosol storage uses. Aerosol storage rooms and general-purpose warehouses containing aerosols shall be provided with an *approved* manual fire alarm system where required by the *International Fire Code*.

❖ Chapter 28 of the IFC and NFPA 30B contain additional guidance on the storage of and fire protection requirements for aerosol products. The requirements for storing the various levels of aerosol products are dependent on the level of sprinkler protection, the type of storage and the quantity of aerosol products. Although aerosol product fires generally involve property loss as opposed to loss of life, installation of a manual fire alarm system could aid in the prompt evacuation of the occupants. Fires involving aerosol products can spread rapidly through a building that is not properly protected and controlled.

[F] 907.2.17 Lumber, wood structural panel and veneer mills. Lumber, wood structural panel and veneer mills shall be provided with a manual fire alarm system.

❖ Any facility using mechanical methods to process wood into finished products produces debris and the potential for combustible dust. Such facilities include mills that produce solid wood lumber and wood veneers as well as those that manufacture structural wood panels such as waferboard, oriented strandboard, composite wood panels or plywood. Good housekeeping and control of ignition sources are therefore essential. To aid in the quick evacuation of occupants in an emergency, Section 1904.2.1of the IFC requires a manual fire alarm system in lumber, wood structural panel and veneer mills that contain product dryers because of their potential as a source of ignition. A manual fire alarm system is not required, however, if the dryers and all other potential sources of ignition are protected by a supervised automatic sprinkler system.

[F] 907.2.18 Underground buildings with smoke control systems. Where a smoke control system is installed in an underground building in accordance with this code, automatic smoke detectors shall be provided in accordance with Section 907.2.18.1.

❖ As indicated in Section 405.5.2, each compartment of an underground building must have a smoke control/exhaust system that can be activated both automatically and manually. Floor levels more than 60 feet (18 288 mm) below the lowest level of exit discharge must be compartmented. Compartmentation is a key element in the egress and fire access plan for floor areas in an underground building. The smoke control system must not only facilitate egress during a fire, but also improve fire department access to the fire source by maintaining visibility that is otherwise impossible given the inability of the fire service to manually ventilate the underground portion of the building (see commentary, Section 405.4.1).

[F] 907.2.18.1 Smoke detectors. A minimum of one smoke detector *listed* for the intended purpose shall be installed in the following areas:

1. Mechanical equipment, electrical, transformer, telephone equipment, elevator machine or similar rooms.

2. Elevator lobbies.

3. The main return and exhaust air plenum of each air-conditioning system serving more than one *story* and located in a serviceable area downstream of the last duct inlet.

4. Each connection to a vertical duct or riser serving two or more floors from return air ducts or plenums of heating,

ventilating and air-conditioning systems, except that in Group R occupancies, a *listed* smoke detector is allowed to be used in each return air riser carrying not more than 5,000 cfm (2.4 m³/s) and serving not more than 10 air-inlet openings.

❖ Automatic smoke detectors are required in certain locations in all underground buildings so that a fire will be detected in its early stages of development. Underground buildings are similar to high-rise buildings in that they present an unusual hazard by being virtually inaccessible to exterior fire department suppression and rescue operations with the increased potential to trap occupants inside the structure. For this reason, the smoke detector location requirements for underground buildings are similar to those in Section 907.2.13.1 for high-rise buildings (see commentary, Section 907.2.13.1).

The requirement for a smoke detector in the main return and exhaust air plenum of an air-conditioning system in an underground building, however, differs from that of a high-rise building in that it is not a function of capacity [2,000 cfm (0.94 m³/s)] but rather a function of whether the system serves more than one floor level. There is more concern over the threat of smoke movement from floor to floor because the products of combustion cannot be vented directly to the atmosphere.

[F] 907.2.18.2 Alarm required. Activation of the smoke control system shall activate an audible alarm at a *constantly attended location.*

❖ The audible alarm is required to notify qualified personnel immediately that the smoke control system has been activated and to put emergency procedures into action quickly.

[F] 907.2.19 Deep underground buildings. Where the lowest level of a structure is more than 60 feet (18 288 mm) below the finished floor of the lowest *level of exit discharge*, the structure shall be equipped throughout with a manual fire alarm system, including an emergency voice/alarm communication system installed in accordance with Section 907.5.2.2.

❖ The ability to communicate and offer warning of a fire can increase the time available for egress from the building. Underground structures located more than 60 feet (18 288 mm) below the level of exit discharge must therefore have a manual fire alarm system. An emergency voice/alarm communication system is also required as part of this system (see commentary, Section 907.5.2.2).

[F] 907.2.20 Covered mall buildings. *Covered mall buildings* exceeding 50,000 square feet (4645 m²) in total floor area shall be provided with an emergency voice/alarm communication system. An emergency voice/alarm communication system serving a mall, required or otherwise, shall be accessible to the fire department. The system shall be provided in accordance with Section 907.5.2.2.

❖ Because of the potentially large number of occupants and their unfamiliarity with their surroundings, an emergency voice/alarm communication system, accessible by the fire department, is required to aid in evacuation of covered mall buildings exceeding 50,000 square feet (4645 m²) in total floor area. Anchor stores are not included as part of the covered mall building (see commentary, Section 402 definition of "Covered mall building").

[F] 907.2.21 Residential aircraft hangars. A minimum of one single-station smoke alarm shall be installed within a residential aircraft hangar as defined in Section 412.3.1 and shall be interconnected into the residential smoke alarm or other sounding device to provide an alarm which will be audible in all sleeping areas of the *dwelling*.

❖ Residential aircraft hangars are assumed to be on the same property as a one- or two-family dwelling. Section 412.3 contains additional requirements for the construction of residential aircraft hangars. The hangar could be located immediately adjacent to the dwelling unit if it is separated by 1-hour fire-resistance-rated construction. Because of the potentially close proximity of the aircraft and its flammability and fuel source, at least one smoke alarm is required in the hangar that is interconnected to the residential smoke alarms. It should be noted, however, that the requirement for a smoke alarm is also applicable to residential aircraft hangars that are detached from the dwelling unit. Because a minimum separation distance is not specified, a fire in the hangar could still present a serious fire hazard to the dwelling unit.

[F] 907.2.22 Airport traffic control towers. An automatic smoke detection system that activates the occupant notification system in accordance with Section 907.5 shall be provided in airport control towers in all occupiable and equipment spaces.

> **Exception:** Audible appliances shall not be installed within the control tower cab.

❖ Airport traffic control towers must be designed to comply with Section 412. These structures are unique in that they can be built to excessive heights, depending upon construction type, are permitted to have one exit stairway and are typically nonsprinklered. Section 903.2.11.3 specifically exempts airport control towers from the requirements of an automatic sprinkler system. An automatic fire detection system is required, however, for early warning notification of the occupants in an emergency. Equipment spaces are included because they may be areas within an airport traffic control tower where a fire may begin, but may not be occupied. Early warning of a fire in these areas is required so as to alert the occupants and emergency forces. The exception recognizes the sensitive nature of the operations that take place in the cab located at the top of the tower and prohibits the installation of audible alarm notification devices there. Notification of occupants within the cab is to be by visual notification appliances only.

[F] 907.2.23 Battery rooms. An automatic smoke detection system shall be installed in areas containing stationary storage battery systems with a liquid capacity of more than 50 gallons (189 L).

❖ Stationary lead-acid battery systems are commonly used for standby power, emergency power or uninterrupted power supplies. The release of hydrogen gas during battery system operation is usually minimal. Adequate ventilation will disperse the small amounts of liberated hydrogen. Because standby power and emergency power systems control many important building emergency systems and functions, a supervised automatic smoke detection system is required for early warning notification of a hazardous condition. Section 608 of the IFC contains additional requirements, including the need for safety venting; room enclosure requirements; spill control and neutralization provisions; ventilation criteria; signage and seismic protection. Section 508.2.5 also requires that such rooms in certain occupancies be separated by 1-hour construction.

[F] 907.3 Fire safety functions. Automatic fire detectors utilized for the purpose of performing fire safety functions shall be connected to the building's fire alarm control unit where a fire alarm system is required by Section 907.2. Detectors shall, upon actuation, perform the intended function and activate the alarm notification appliances or activate a visible and audible supervisory signal at a *constantly attended location*. In buildings not equipped with a fire alarm system, the automatic fire detector shall be powered by normal electrical service and, upon actuation, perform the intended function. The detectors shall be located in accordance with NFPA 72.

❖ When the code requires installation of automatic fire detectors to perform a specific function, such as elevator recall or smokeproof enclosure ventilation, or when detectors are installed to comply with a permitted alternative, such as door-closing devices, these detectors must be connected to the building's automatic fire alarm system if the building is required by the code to have such a system.

In addition to performing its intended function (for example, closing a door), if a detector is activated, it must also activate either the building alarm devices (if one is present) or a supervisory signal at a constantly attended location. This requirement recognizes that these detectors and the devices they control are part of the building fire protection system and are expected to perform as designed. If they are connected to a fire alarm system, they will have the supervision necessary for operational reliability. If they are not connected to and supervised by a fire alarm system, they still must be supervised through the constantly attended location.

An exception is provided for fire safety function detectors in buildings not required to have a fire alarm system. The fire safety function detectors must be powered by the building electrical system and be located as required by NFPA 72. Without this exception, these detectors could not be expected to perform as intended because there would be no power supply.

Note that in the IFC, this section is entitled "Where required in existing buildings and structures." but since the IBC does not apply retroactively, such requirements do not appear here. That is why all sections of Section 907 following Section 907.2.23 do not exactly correlate with the content of Section 907 of the IFC.

[F] 907.3.1 Duct smoke detectors. Smoke detectors installed in ducts shall be *listed* for the air velocity, temperature and humidity present in the duct. Duct smoke detectors shall be connected to the building's fire alarm control unit when a fire alarm system is required by Section 907.2. Activation of a duct smoke detector shall initiate a visible and audible supervisory signal at a *constantly attended location* and shall perform the intended fire safety function in accordance with this code and the *International Mechanical Code.* Duct smoke detectors shall not be used as a substitute for required open area detection.

Exceptions:

1. The supervisory signal at a *constantly attended location* is not required where duct smoke detectors activate the building's alarm notification appliances.

2. In occupancies not required to be equipped with a fire alarm system, actuation of a smoke detector shall activate a visible and an audible signal in an *approved* location. Smoke detector trouble conditions shall activate a visible or audible signal in an *approved* location and shall be identified as air duct detector trouble.

❖ It is not the intent of this section to send a signal to the fire department or to activate the alarm notification devices within a building. Instead, this section requires that a supervisory signal be sent to a constantly attended location. Smoke detectors must be connected to a fire alarm system where such systems are installed. Connection to the fire alarm system will activate a visible and audible supervisory signal at a constantly attended location, which will alert building supervisory personnel that a smoke alarm has activated and will also provide electronic supervision of the duct detectors, thereby indicating any problems that may develop in the detector system circuitry or power supply.

Exception 1 allows activation of the building alarm notification appliances in place of a supervisory signal. Causing the occupant notification system to sound would alert the occupants of the building that an alarm condition exists within the air distribution system, thereby performing the same function as a supervisory signal sent to a constantly attended location.

Exception 2 recognizes the fact that not all buildings are required to have a fire alarm system. A visible and audible signal must be activated at an approved location that will alert building supervisory personnel to take action. Additionally, the duct smoke detectors must be electronically supervised to indicate trouble (system fault) in the detector system circuitry or power supply. A trouble condition must activate a distinct visible or audible signal at a location that will alert the responsible personnel.

[F] 907.3.2 Delayed egress locks. Where delayed egress locks are installed on *means of egress* doors in accordance with Section 1008.1.9.6, an automatic smoke or heat detection system shall be installed as required by that section.

❖ This section alerts the code user to additional requirements in Section 1008.1.9.7 that tie the operation of egress doors into the activation of an automatic fire detection system. A smoke or heat detection system is required to unlock delayed egress locks upon activation. The heat detection system can be the sprinkler system. For example, a similar requirement is found in Section 1008.1.4.3 that requires horizontal sliding doors used as a component of the means of egress, where required to be rated, to be self-closing or automatic-closing upon smoke detection. Also, access-controlled egress doors in occupancies as required by Section 1008.1.4.4 must be capable of being automatically unlocked by activation of an automatic fire detection system, if one is installed.

[F] 907.3.3 Elevator emergency operation. Automatic fire detectors installed for elevator emergency operation shall be installed in accordance with the provisions of ASME A17.1 and NFPA 72.

❖ This new section in the 2009 edition supplements and provides correlation with Section 607.1 of the IFC by making it clear that automatic fire detection devices used to initiate Phase I emergency recall of elevators are to be installed in accordance with both ASME A17.1 and NFPA 72.

[F] 907.3.4 Wiring. The wiring to the auxiliary devices and equipment used to accomplish the above fire safety functions shall be monitored for integrity in accordance with NFPA 72.

❖ In order to provide a reasonable level of integrity and reliability to the installation of automatic fire detection devices and related equipment installed to perform various fire safety functions in accordance with Section 907.3, this section requires that all wiring interconnecting such devices and equipment be monitored for integrity in accordance with NFPA 72.

[F] 907.4 Initiating devices. Where manual or automatic alarm initiation is required as part of a fire alarm system, the initiating devices shall be installed in accordance with Sections 907.4.1 through 907.4.3.

❖ This section introduces Sections 907.4.1 through 907.4.3, which contain requirements for the various types of manual or automatic fire alarm-initiating devices

[F] 907.4.1 Protection of fire alarm control unit. In areas that are not continuously occupied, a single smoke detector shall be provided at the location of each fire alarm control unit, notification appliance circuit power extenders, and supervising station transmitting equipment.

Exceptions:

1. Where ambient conditions prohibit installation of a smoke detector, a *heat detector* shall be permitted.

2. The smoke detector shall not be required where the building is equipped throughout with an *automatic sprinkler system* in accordance with Section 903.3.1.1 or 903.3.1.2.

❖ This is a new section in the 2009 edition to address the smoke detector that was formerly required by NFPA 72 to ensure the fire alarm system is capable of performing its function in the event of a fire in the vicinity of the fire alarm control unit and other critical components of the system. This smoke detector will activate the fire alarm control unit and allow it to either notify occupants or transmit a signal to a remote monitoring location before the fire impairs the fire alarm control unit. Exception 1 parallels Section 907.4.3 by allowing a heat detector to be installed in lieu of a smoke detector in areas where the ambient environment is hostile to smoke detectors and could lead to unwanted alarm activations. Exception 2 waives the requirement for a smoke detector in favor of the higher level of protection afforded by an automatic sprinkler system.

[F] 907.4.2 Manual fire alarm boxes. Where a manual fire alarm system is required by another section of this code, it shall be activated by fire alarm boxes installed in accordance with Sections 907.4.2.1 through 907.4.2.5.

❖ This section specifies the requirements for manual fire alarm boxes that are part of a required manual fire alarm system.

[F] 907.4.2.1 Location. Manual fire alarm boxes shall be located not more than 5 feet (1524 mm) from the entrance to each *exit*. Additional manual fire alarm boxes shall be located so that travel distance to the nearest box does not exceed 200 feet (60 960 mm).

❖ Manual fire alarm boxes must be located in the path of egress and be readily accessible to the occupants. They must be located within 5 feet (1524 mm) of the entrance to each exit on every story of the building. This would include the need to locate manual fire alarm boxes near each horizontal exit, as well as entrances to stairs and exit doors to the exterior.

Manual fire alarm boxes are located near exits so that an adequate number of devices are available in the path of egress to transmit an alarm in a timely manner. These locations also encourage the actuation of a manual fire alarm box on the fire floor prior to entering the stair, resulting in the alarm being received from the actual fire floor and not another floor along the path of egress.

The location also presumes that individuals will be evacuating the area where the fire originated. When evacuation of the fire area is unlikely, consideration

could be given to putting manual fire alarm boxes in more convenient places. Examples of such instances would be officer stations in Group I-3 occupancies and nurses' stations in Group I-2 occupancies.

The 200-foot (60 960 mm) travel distance limitation is consistent with the exit access travel distance permitted for most nonsprinklered occupancies. If the 200-foot (60 960 mm) travel distance to a manual fire alarm box is exceeded, even in a fully sprinklered building, additional manual fire alarm boxes would be required.

[F] 907.4.2.2 Height. The height of the manual fire alarm boxes shall be a minimum of 42 inches (1067 mm) and a maximum of 48 inches (1372 mm) measured vertically, from the floor level to the activating handle or lever of the box.

❖ Manual fire alarm boxes must be reachable by the occupants of the building. They must also be mounted high enough to reduce the likelihood of damage or false alarms from something accidentally striking the device. Therefore, manual fire alarm boxes must be mounted a minimum of 42 inches (1067 mm) and a maximum of 48 inches (1372 mm) above the floor level. The 48-inch (1372 mm) measurement corresponds to the maximum unobstructed side-reach height by a person in a wheelchair.

[F] 907.4.2.3 Color. Manual fire alarm boxes shall be red in color.

❖ Manual fire alarm boxes are to be painted or manufactured in a distinctive and traditional red color to provide a visual cue to help building occupants identify the device.

[F] 907.4.2.4 Signs. Where fire alarm systems are not monitored by a supervising station, an *approved* permanent sign shall be installed adjacent to each manual fire alarm box that reads: WHEN ALARM SOUNDS CALL FIRE DEPARTMENT.

Exception: Where the manufacturer has permanently provided this information on the manual fire alarm box.

❖ This section has limited application because, as indicated in Section 907.6.5, fire alarm systems generally must be monitored by an approved supervising station. When a system is not monitored, such as possibly a fire alarm system that is not required by code, adequate signage must be displayed to tell occupants what response actions must be taken. Most building occupants assume that when an alarm device is activated, the fire department will automatically be notified as well. The sign must be conspicuously located next to the manual fire alarm box unless it is mounted on the manual fire alarm box itself by the manufacturer.

[F] 907.4.2.5 Protective covers. The fire code official is authorized to require the installation of *listed* manual fire alarm box protective covers to prevent malicious false alarms or to provide the manual fire alarm box with protection from physical damage. The protective cover shall be transparent or red in color with a transparent face to permit visibility of the manual fire alarm box. Each cover shall include proper operating instructions. A protective cover that emits a local alarm signal shall not be installed unless *approved*. Protective covers shall not project more than that permitted by Section 1003.3.3.

❖ Although manual fire alarm boxes must be readily available to all occupants in buildings required to have a manual fire alarm system, this section allows the use of protective covers if they are approved by the fire code official. Protective covers are commonly used to reduce either the potential for intentional false alarms or vandalism. They also provide protection in locations where the manual fire alarm boxes may be exposed to physical damage, such as in gymnasiums, indoor tennis courts and the like.

[F] 907.4.3 Automatic smoke detection. Where an automatic smoke detection system is required it shall utilize smoke detectors unless ambient conditions prohibit such an installation. In spaces where smoke detectors cannot be utilized due to ambient conditions, *approved* automatic *heat detectors* shall be permitted.

❖ Smoke detectors must be used, except when ambient conditions would prohibit their use. This section would allow a heat detector to be installed in lieu of a smoke detector in areas where the ambient environment is hostile to smoke detectors and could lead to unwanted alarm activations. The smoke detector manufacturer's literature will identify the limitations on the use of smoke detectors, including environmental conditions such as humidity, temperature and airflow.

907.4.3.1 Automatic sprinkler system. For conditions other than specific fire safety functions noted in Section 907.3, in areas where ambient conditions prohibit the installation of smoke detectors, an *automatic sprinkler system* installed in such areas in accordance with Section 903.3.1.1 or 903.3.1.2 and that is connected to the fire alarm system shall be *approved* as automatic heat detection.

❖ This section states that automatic heat detection is not required when buildings are fully sprinklered in accordance with NFPA 13 or 13R. The presence of a sprinkler system exempts areas where a heat detector can be installed in place of a smoke detector, such as in storage or furnace rooms. The sprinkler head in this case essentially acts as a heat detection device. Note that this provision does not apply to the fire safety functions indicated in Section 907.3.

[F] 907.5 Occupant notification systems. A fire alarm system shall annunciate at the panel and shall initiate occupant notification upon activation, in accordance with Sections 907.5.1 through 907.5.2.3.4. Where a fire alarm system is required by another section of this code, it shall be activated by:

1. Automatic fire detectors.
2. Sprinkler waterflow devices.
3. Manual fire alarm boxes.

4. Automatic fire-extinguishing systems.

> **Exception:** Where notification systems are allowed elsewhere in Section 907 to annunciate at a *constantly attended location.*

❖ This section makes it clear that fire alarm system activation begins first by activating the fire alarm control panel and then by notifying the occupants of an alarm condition and goes on to introduce all of the components of an occupant notification system contained in Sections 907.5.1 through 907.5.2.3.4.

It also lists the system components that are to act as alarm initiation devices. The exception is a recognition that there are places in the code where an alternative to occupant notification is an alarm notification at a constantly attended location. The exception is intended to clarify the code so that there is no question as to whether this general provision for alarm activation is superseded by the other sections addressing the alarm notification at a constantly attended location.

[F] 907.5.1 Presignal feature. A presignal feature shall not be installed unless *approved* by the fire code official and the fire department. Where a presignal feature is provided, a signal shall be annunciated at a *constantly attended location approved* by the fire department, in order that occupant notification can be activated in the event of fire or other emergency.

❖ A presignal feature on a fire alarm system allows the occupant notification devices to activate in selected, constantly attended locations only and from which human intervention is required to activate a general occupant notification signal. Alternatively, this feature can be programmed to delay the general alarm notification for more than 1 minute before it will automatically be activated by the control panel. In either presignal scenario, remote transmission of the alarm signal to the fire department is immediate. See NFPA 72 for additional information on the presignal feature.

Improper use of the presignal feature has been a contributing factor in several multiple-death fire incidents. In most instances, the staff failed to activate the general alarm quickly and the occupants of the building were unaware of the fire. Therefore, the use of a presignal feature is discouraged by the code. A presignal feature may be used only if it is approved by the fire code official and the fire department.

[F] 907.5.2 Alarm notification appliances. Alarm notification appliances shall be provided and shall be *listed* for their purpose.

❖ The code requires that fire alarm systems be equipped with approved alarm notification appliances so that in an emergency, the fire alarm system will notify the occupants of the need for evacuation or implementation of the fire emergency plan. Alarm notification devices required by the code are of two general types: visible and audible. Except for voice/alarm signaling systems, once the system has been activated, all visible and audible alarms are required to activate. Voice/alarm signaling systems are special signaling systems that are activated selectively in response to specific emergency conditions.

[F] 907.5.2.1 Audible alarms. Audible alarm notification appliances shall be provided and emit a distinctive sound that is not to be used for any purpose other than that of a fire alarm.

Exception: Visible alarm notification appliances shall be allowed in lieu of audible alarm notification appliances in critical care areas of Group I-2 occupancies.

❖ To attract the attention of building occupants, audible alarms must be distinctive, using a sound that is unique to the fire alarm system and used for no other purpose than alerting occupants to a fire emergency. Other emergencies, such as tornados, etc., must be signaled by another sound different from the fire signal.

The exception recognizes that the occupants in critical care areas of Group I-2 occupancies are usually incapacitated. The audible alarms may have the effect of unnecessarily disrupting the patients who are most likely not capable of self-preservation. Likewise, audible alarms in operating theaters of hospitals could be hazardous because an alarm activation could startle a surgeon during a delicate procedure. Critical care areas are also assumed to be adequately staffed at all times and ready to respond upon activation of a visible alarm device.

907.5.2.1.1 Average sound pressure. The audible alarm notification appliances shall provide a sound pressure level of 15 decibels (dBA) above the average ambient sound level or 5 dBA above the maximum sound level having a duration of at least 60 seconds, whichever is greater, in every *occupiable space* within the building. The minimum sound pressure levels shall be: 75 dBA in occupancies in Groups R and I-1; 90 dBA in mechanical equipment rooms and 60 dBA in other occupancies.

❖ To attract the attention of building occupants, this section requires that the distinctive audible alarms must be capable of being heard above the ambient noise level in a space. The indicated levels are considered the minimum pressure differential that will be perceivable by most people. The minimum sound levels indicated in this section must be provided in all occupied spaces within the building, including bathrooms, mechanical spaces and other areas where people are likely to be. It does not include such building cavity spaces as interstitial areas above a suspended ceiling, vent shafts or crawl spaces.

907.5.2.1.2 Maximum sound pressure. The maximum sound pressure level for audible alarm notification appliances shall be 110 dBA at the minimum hearing distance from the audible appliance. Where the average ambient noise is greater than 95 dBA, visible alarm notification appliances shall be provided in accordance with NFPA 72 and audible alarm notification appliances shall not be required.

❖ In no case may the sound pressure level exceed 110 dBA at the minimum hearing distance from the audible

appliance. This is consistent with ADA requirements. Sound pressures above that level can cause pain or even permanent hearing loss. In such cases, audible alarms are not required to be installed but visual alarms would be necessary to compensate for the lack of audibility.

It should also be noted that in certain work areas, Occupational Safety and Health Administration (OSHA) requires employees to wear hearing protection, possibly preventing them from hearing an audible alarm. Additionally, the noise factor in these areas is high enough that an audible alarm may not be discernible. In these areas, as well as in others, the primary method of indicating a fire can be by a visible signal. Employees must be capable of identifying such a signal as indicating a fire.

907.5.2.2 Emergency voice/alarm communication systems. Emergency voice/alarm communication systems required by this code shall be designed and installed in accordance with NFPA 72. The operation of any automatic fire detector, sprinkler waterflow device or manual fire alarm box shall automatically sound an alert tone followed by voice instructions giving *approved* information and directions for a general or staged evacuation in accordance with the building's fire safety and evacuation plans required by Section 404. In high-rise buildings, the system shall operate on a minimum of the alarming floor, the floor above and the floor below. Speakers shall be provided throughout the building by paging zones. At a minimum, paging zones shall be provided as follows:

1. Elevator groups.

2. *Exit stairways.*

3. Each floor.

4. *Areas of refuge* as defined in Section 1002.1.

Exception: In Group I-1 and I-2 occupancies, the alarm shall sound in a constantly attended area and a general occupant notification shall be broadcast over the overhead page.

❖ The primary purpose of an emergency voice/alarm communication system is to provide dedicated manual and automatic facilities for the origination, control and transmission of information and instructions pertaining to a fire alarm emergency to the occupants of a building. This section identifies that notification speakers are required throughout the building with a minimum of one speaker in each paging zone when an emergency voice/alarm communication system is required. The system may sound a general alarm or be a selective system in which only certain areas of the building receive the alarm indication for staged evacuation. See Chapter 4 for evacuation plan requirements. The intent is to provide the capability to send out selective messages to individual areas; however, it does not prohibit the same message to be sent to all areas. In high-rise buildings, a minimum area of notification must include the alarming floor and one floor above and one floor below it.

This section also identifies the minimum paging

zone arrangement. This does not preclude further zone divisions for logical staged evacuation in accordance with an approved evacuation plan.

This section also indicates that the emergency voice/alarm system is to be initiated as all other fire alarm systems are initiated. The functional operation of the system begins with an alert tone (usually 3 to 10 seconds in duration) followed by the evacuation signal (message). It is important to remember that the voice alarm system is not an "audible alarm." It has its own specific criteria for installation and approval according to NFPA 72. Consequently the sound pressure requirements for audible alarms do not apply to voice alarm systems. For voice alarm systems, the intent is communication and an understanding of what is being said, not volume.

The exception is similar to the one to Section 907.5.2.1 and recognizes the supervised environment typical of institutional uses and the reliance placed on staff to act appropriately in an emergency. As is the case for most voice alarms, the key is in being able to deliver specific information to the people who can affect a safe egress—whether this is the public, employees, or both.

[F] 907.5.2.2.1 Manual override. A manual override for emergency voice communication shall be provided on a selective and all-call basis for all paging zones.

❖ The intent of this section is to provide the ability to transmit live voice instructions over any previously initiated signals or prerecorded messages for all zones. This would include the ability to override the voice message at once throughout the building or to be able to select individual paging zones for the message override.

[F] 907.5.2.2.2 Live voice messages. The emergency voice/alarm communication system shall also have the capability to broadcast live voice messages by paging zones on a selective and all-call basis.

❖ This would include the ability to provide the live voice message at once throughout the building or to be able to select individual paging zones to receive the message. Speakers used for background music must not be used unless specifically listed for fire alarm system use. NFPA 72 has additional requirements for the placement, location and audibility of speakers used as part of an emergency voice/alarm communication system.

[F] 907.5.2.2.3 Alternate uses. The emergency voice/alarm communication system shall be allowed to be used for other announcements, provided the manual fire alarm use takes precedence over any other use.

❖ In certain circumstances which should be approved by the fire code officials, the emergency voice/alarm communications system could be used to convey information other than fire alarm-related items. This could include severe weather warnings that might require evacuation or relocation, lockdown instructions (see commentary to Section 404.3.3 of the IFC) and similar

approved messages. In the event of such usage, the system must respond immediately to manual fire alarm box activations.

[F] 907.5.2.2.4 Emergency power. Emergency voice/alarm communications systems shall be provided with an *approved* emergency power source.

❖ Because the emergency voice/alarm communication system is a critical aid in evacuating the building, the system must be connected to an approved emergency power source complying with Section 2702.

[F] 907.5.2.3 Visible alarms. Visible alarm notification appliances shall be provided in accordance with Sections 907.5.2.3.1 through 907.5.2.3.4.

Exceptions:

1. Visible alarm notification appliances are not required in *alterations*, except where an existing fire alarm system is upgraded or replaced, or a new fire alarm system is installed.

2. Visible alarm notification appliances shall not be required in *exits* as defined in Section 1002.1.

3. Visible alarm notification appliances shall not be required in elevator cars.

❖ This section contains alarm system requirements for occupants who are hearing impaired. Visible alarm notification appliances are to be installed in conjunction with the audible devices and located and oriented so that they will display alarm signals throughout a space. It is not the intent of the code to offer visible alarm signals as an option to audible alarm signals. Both are required. However, the code acknowledges conditions when audible alarms may be of little or no value, such as when the ambient sound level exceeds 105 dBA. In such cases, Section 907.5.2.1, similar to NFPA 72, allows for visible alarm notification appliances in the area.

Exception 1 states that visible alarm devices are not required in previously approved existing fire alarm systems or as part of minor alterations to existing fire alarm systems. Extensive modifications to an existing fire alarm system such as an upgrade or replacement would require the installation of visible alarm devices even if the previous existing system neither had them nor required them. The main reason is a combination of simple economics and practical application. Many existing systems that do not have visible signal devices do not have the wiring capability to include such devices. To make the necessary changes to the existing system, a total replacement of the existing system may need to take place. In many cases this is cost prohibitive. Thus, if the alteration is small, the system can be left as is, without the visual devices. The second consideration is scope. If the alternation involves only a limited area, it could be confusing to have part of the area equipped with visual devices and part without. This is not good practice, as the alarm could be confusing. If an entire floor is being altered, then it becomes subject to consideration for an upgrade to an alarm system with visual devices. If only an office is being remodeled, then the implication is that the upgrade to visual devices may not be warranted. This determination will be subjective in many cases and should be applied based on the life safety benefit and financial expenses involved and whether adequate audible devices are present for full coverage.

In Exception 2, visible alarm devices are not required in exit elements because of the potential distraction during evacuation. Exits, as defined in Section 1002.1, could include exit enclosures or exit passageways but not exit access corridors. In tall buildings, exiting may be phased based on alarm zone. If the alarm floor and adjacent floors are notified of the emergency but the remainder of the building is not, then a visual device in the stairway would be confusing to those people who may not be coming from the alarm floor.

Previously, some jurisdictions were requiring visible alarm notification appliances to be installed in elevator cars since there was no exception in the code or NFPA 72 to allow omission of this type of notification appliance in elevator cars. Exception 3 will eliminate any confusion regarding the need to install visible notification appliances in elevator cars. The rationale for not installing visible notification appliances in elevator cars is the same as for exit enclosures; high light intensity from these notification appliances may cause confusion and disorientation. Also, elevator passengers are "captive" in that they cannot respond to such devices until the elevator arrives at its destination or is recalled by the Phase I emergency recall feature, which could lead to passenger panic.

[F] 907.5.2.3.1 Public and common areas. Visible alarm notification appliances shall be provided in public areas and common areas.

❖ Visible alarm notification appliances must provide coverage in all areas open to the public as well as all shared or common areas (e.g., corridors, public restrooms, shared offices, classrooms, medical exam rooms, etc.). Areas where visible alarm notification appliances are not required include private offices, mechanical rooms or similar spaces. The intent with this section is to replicate the provisions included in the Americans with Disabilities Act Accessibility Guidelines for Buildings and Facilities (ADAAG). See definition in Chapter 11 for "Common use."

[F] 907.5.2.3.2 Employee work areas. Where employee work areas have audible alarm coverage, the notification appliance circuits serving the employee work areas shall be initially designed with a minimum of 20-percent spare capacity to account for the potential of adding visible notification appliances in the future to accommodate hearing impaired employee(s).

❖ This section provides for spare capacity on notification circuits to allow for those with hearing impairments to be accommodated as necessary. This spare capacity is intended to eliminate the potential for overloading notification circuits when a hearing-impaired person is

hired and needs to be accommodated, but reduces the initial construction cost as such alarms may not be necessary in every situation. This section is intended to apply to employee work areas that are not common areas.

[F] 907.5.2.3.3 Groups I-1 and R-1. Group I-1 and R-1 *dwelling units* or *sleeping units* in accordance with Table 907.5.2.3.3 shall be provided with a visible alarm notification appliance, activated by both the in-room smoke alarm and the building fire alarm system.

❖ Fire alarm systems in Group I-1 and R-1 sleeping accommodations must be equipped with visible alarms to the extent stated in Table 907.5.2.3.3. The visible alarm notification devices in these rooms are to be activated by both the required in-room smoke alarm and the building fire alarm system. All visible alarm notification appliances in a building, however, need not be activated by individual room smoke alarms. It is not a requirement that the accessible sleeping units be provided with visible alarm notification appliances even though some elderly patients or residents may be both mobility and hearing impaired.

**[F] TABLE 907.5.2.3.3
VISIBLE ALARMS**

NUMBER OF SLEEP UNITS	SLEEPING ACCOMMODATIONS WITH VISIBLE ALARMS
6 to 25	2
26 to 50	4
51 to 75	7
76 to 100	9
101 to 150	12
151 to 200	14
201 to 300	17
301 to 400	20
401 to 500	22
501 to 1,000	5% of total
1,001 and over	50 plus 3 for each 100 over 1,000

❖ This table specifies the minimum number of sleeping units that are to be equipped with visible and audible alarms. The numbers are based on the total number of sleeping accommodations in the facility. The requirements in this table are intended to be consistent with the ADAAG.

[F] 907.5.2.3.4 Group R-2. In Group R-2 occupancies required by Section 907 to have a fire alarm system, all *dwelling units* and *sleeping units* shall be provided with the capability to support visible alarm notification appliances in accordance with ICC A117.1.

❖ Group R-2 occupancies with a fire alarm system are required to have all dwelling units wired to support visible alarm notification appliances. This includes all dwelling and sleeping units, not just those classified as either Type A or B. In accordance with Sections 1005.2

through 1005.4.4 of ICC A117.1, the building alarm system wiring must be extended to the unit smoke detectors so that audible/visible alarm notification appliances may be connected to the building fire alarm system to notify residents with hearing impairments of an emergency situation. Chapter 11 of the code contains additional information on the classification criteria and requirements for accessible dwelling units.

[F] 907.6 Installation. A fire alarm system shall be installed in accordance with this section and NFPA 72.

❖ This section specifies the requirements for fire alarm system installation and also references the installation requirements of NFPA 72.

[F] 907.6.1 Wiring. Wiring shall comply with the requirements of NFPA 70 and NFPA 72. Wireless protection systems utilizing radio-frequency transmitting devices shall comply with the special requirements for supervision of low-power wireless systems in NFPA 72.

❖ Wiring for fire alarm systems must be installed so that it is secure and will function reliably in an emergency. The code requires that the wiring for fire alarm systems meet the requirements of NFPA 70 and NFPA 72. This requirement is in addition to the general requirements for electrical installations set forth in Chapter 27 of the IBC. For reliability, systems that use radio-frequency transmitting devices for signal transmission are required to have supervised transmitting and receiving equipment that conforms to the special requirements contained in NFPA 72. This requirement is in addition to the general requirements for supervision in Section 907.6.5.

[F] 907.6.2 Power supply. The primary and secondary power supply for the fire alarm system shall be provided in accordance with NFPA 72.

Exception: Back-up power for single-station and multiple-station smoke alarms as required in Section 907.2.11.4.

❖ The operation of fire alarm systems is essential to life safety in buildings and must be reliable in the event the normal power supply fails. For proper operation of fire alarm systems, this section requires that the primary and secondary power supplies comply with NFPA 72. This is in addition to the general requirements for electrical installations in Chapter 27. NFPA 72 offers three alternatives for secondary supply: a 24-hour storage battery; storage batteries with a 4-hour capacity and a generator or multiple generators.

NFPA 72 requires that the primary and secondary power supplies for remotely located control equipment essential to the system operation must conform to the requirements for primary and secondary power supplies for the main system. Also, NFPA 72 contains requirements for monitoring the integrity of primary power supplies and requires a backup power supply.

[F] 907.6.3 Zones. Each floor shall be zoned separately and a zone shall not exceed 22,500 square feet (2090 m²). The length of any zone shall not exceed 300 feet (91 440 mm) in any direction.

Exception: *Automatic sprinkler system* zones shall not exceed the area permitted by NFPA 13.

❖ Since the fire alarm system also aids emergency personnel in locating the fire, the system must be zoned to shorten response time to the fire location. Zoning is also critical if the fire alarm system initiates certain other fire protection systems or control features, such as smoke control systems.

At a minimum, each floor of a building must constitute one zone of the system. If the floor area exceeds 22,500 square feet (2090 m²), additional zones per floor are required. The maximum length of a zone is 300 feet (91 440 mm).

The exception states that NFPA 13 defines the maximum areas to be protected by one sprinkler system and that the sprinkler system need not be designed to meet the 22,500-square-foot (2090 m²) area limitations for a fire alarm system zone. For example, NFPA 13 permits a sprinkler system riser in a light-hazard occupancy to protect an area of 52,000 square feet (4831 m²) per floor. In accordance with the exception, a single waterflow switch, and consequently a single fire alarm system zone, would be acceptable. If other alarm-initiating devices are present on the floor, they would need to be zoned separately to meet the 22,500-square-foot (2098 m²) limitation.

It is not intended that this section apply to sprinkler systems. This section only applies where a fire alarm system is required in accordance with Section 907. Unless the building is categorized as a high rise and must comply with Section 907.6.3.2, the code does not mandate the zoning of sprinkler systems per floor. With today's fully addressable fire alarm systems, each detector effectively becomes its own zone. The intent with zoning is to identify and limit the search area for fire alarm systems. Addressable devices will indicate the precise location of the alarm condition, thereby eliminating the need for the zoning contemplated by this section when approved by the fire code official in accordance with Section 104.11.

[F] 907.6.3.1 Zoning indicator panel. A zoning indicator panel and the associated controls shall be provided in an *approved* location. The visual zone indication shall lock in until the system is reset and shall not be canceled by the operation of an audible-alarm silencing switch.

❖ The zoning indicator panel, which can be the fire alarm control unit or a separate fire alarm annunciator panel (FAAP), must be installed in a location approved by the fire code official. One of the key considerations in determining panel placement is whether or not the panel is located to permit ready access by emergency responders. Once an alarm-initiating device within a zone has been activated, the annunciation of the zone must lock in until the system is reset.

[F] 907.6.3.2 High-rise buildings. In high-rise buildings, a separate zone by floor shall be provided for each of the following types of alarm-initiating devices where provided:

1. Smoke detectors.

2. Sprinkler waterflow devices.

3. Manual fire alarm boxes.

4. Other *approved* types of automatic fire detection devices or suppression systems.

❖ High-rise buildings must have a separate zone by floor for each indicated type of alarm-initiating device. Although this feature may be desirable in all buildings, the incremental cost difference is substantially higher in low-rise buildings in which basic fire alarm systems are installed. State-of-the-art fire alarm systems installed in high-rise buildings are addressable and by their nature automatically provide this minimum zoning.

[F] 907.6.4 Access. Access shall be provided to each fire alarm device and notification appliance for periodic inspection, maintenance and testing.

❖ Automatic fire detectors, especially smoke detectors, require periodic cleaning to reduce the likelihood of malfunction. Section 907.8 and NFPA 72 require inspection and testing at regular intervals. Access to perform the required inspections, necessary maintenance and testing is a particularly important consideration for those detectors that are installed within a concealed space, such as an air duct.

[F] 907.6.5 Monitoring. Fire alarm systems required by this chapter or by the *International Fire Code* shall be monitored by an *approved* supervising station in accordance with NFPA 72.

Exception: Monitoring by a supervising station is not required for:

1. Single- and multiple-station smoke alarms required by Section 907.2.11.

2. Smoke detectors in Group I-3 occupancies.

3. *Automatic sprinkler systems* in one- and two-family dwellings.

❖ Fire alarm systems required by Section 907 are required to be electrically supervised by one of the methods prescribed in NFPA 72.
   Exception 1 exempts single- and multiple-station smoke alarms from being supervised due to the potential for unwanted false alarms.
   Exception 2 recognizes a similar problem in Group I-3 occupancies. Accordingly, due to the concern over unwanted alarms, smoke detectors in Group I-3 occupancies need only sound an approved alarm signal that automatically notifies staff (see Section 907.2.6.3.1). Smoke detectors in such occupancies are typically subject to misuse and abuse, and frequent unwanted alarms would negate the effectiveness of the system.
   Exception 3 clarifies that sprinkler systems in one-

and two-family dwellings are not part of a dedicated fire alarm system and are typically designed in accordance with NFPA 13D, which does not require electrical supervision.

[F] 907.6.5.1 Automatic telephone-dialing devices. Automatic telephone-dialing devices used to transmit an emergency alarm shall not be connected to any fire department telephone number unless *approved* by the fire chief.

❖ Upon initiation of an alarm, supervisory or trouble signal, an automatic telephone-dialing device takes control of the telephone line for the reliability of transmission of all signals. The device, however, must not be connected to the fire department telephone number unless specifically approved by the fire department because that could disrupt any potential emergency (911) calls. NFPA 72 contains additional guidance on such devices, including digital alarm-communicator systems.

[F] 907.7 Acceptance tests and completion. Upon completion of the installation, the fire alarm system and all fire alarm components shall be tested in accordance with NFPA 72.

❖ A complete performance test of the fire alarm system must be conducted to determine that the system is operating as required by the code. The acceptance test must include a test of each circuit, alarm-initiating device, alarm notification appliance and any supplementary functions, such as activation of closers and dampers. The operation of the primary and secondary (emergency) power supplies must also be tested, as well as the supervisory function of the control panel. Section 901.5 assigns responsibility for conducting the acceptance tests to the owner or the owner's representative.
   NFPA 72 contains specific acceptance test procedures. Additional guidance on periodic testing and inspection can also be obtained from Section 907.8 and NFPA 72.

[F] 907.7.1 Single- and multiple-station alarm devices. When the installation of the alarm devices is complete, each device and interconnecting wiring for multiple-station alarm devices shall be tested in accordance with the smoke alarm provisions of NFPA 72.

❖ To determine that smoke alarms have been properly installed and are ready to function as intended, they must be actuated during an acceptance test. The test also confirms that interconnected detectors will operate simultaneously as required. The responsibility for conducting the acceptance tests rests with the owner or the owner representative as stated in Section 901.5.

[F] 907.7.2 Record of completion. A record of completion in accordance with NFPA 72 verifying that the system has been installed and tested in accordance with the *approved* plans and specifications shall be provided.

❖ In accordance with NFPA 72, this section requires a written statement from the installing contractor that the fire alarm system has been tested and installed in compliance with the approved plans and the manufacturer's specifications. Any deviations from the approved plans or the applicable provisions of NFPA 72 are to be noted in the record of completion.

[F] 907.7.3 Instructions. Operating, testing and maintenance instructions and record drawings ("as-builts") and equipment specifications shall be provided at an *approved* location.

❖ To permit adequate testing, maintenance and troubleshooting of the installed fire alarm system, an owner's manual with complete installation instructions must be kept on site or in another approved location. The instructions include a description of the system, operating procedures and testing and maintenance requirements.

[F] 907.8 Inspection, testing and maintenance. The maintenance and testing schedules and procedures for fire alarm and fire detection systems shall be in accordance with Section 907.9 of the *International Fire Code*.

❖ Fire alarms and smoke detection systems are to be inspected, tested and maintained in accordance with Sections 907.9.1 through 907.9.5 of the IFC and the applicable requirements of NFPA 72. It is the building owner's responsibility to keep these systems operable at all times.

SECTION 908
EMERGENCY ALARM SYSTEMS

[F] 908.1 Group H occupancies. Emergency alarms for the detection and notification of an emergency condition in Group H occupancies shall be provided in accordance with Section 414.7.

❖ Emergency alarm systems provide indication and warning of emergency situations involving hazardous materials. An emergency alarm system is required in all Group H occupancies, as indicated in Sections 2704.9 and 2705.4.4 of the IFC, as well as Group H-5 HPM facilities as indicated in Section 908.2. The Group H occupancy classification assumes the storage or use of hazardous materials exceeds the maximum allowable quantities specified in Tables 307.1(1) and 307.1(2).

An emergency alarm system should include an emergency alarm-initiating device outside each interior door of hazardous material storage areas, a local alarm device and adequate supervision.

Even though ozone gas-generator rooms (see Section 908.4), repair garages (see Section 908.5) and refrigeration systems (see Section 908.6) are not typically classified as Group H occupancies, the potential hazards associated with these occupancy conditions are great enough to require additional means of early warning detection.

[F] 908.2 Group H-5 occupancy. Emergency alarms for notification of an emergency condition in an HPM facility shall be provided as required in Section 415.8.4.6. A continuous gas-detection system shall be provided for HPM gases in accordance with Section 415.8.7.

❖ In addition to hazardous material storage areas as regulated by Section 2704.9 of the IFC, Section 1803.12.1 of the IFC also requires emergency alarms for service corridors, exit access corridors and exit enclosures because of the potential transport of hazardous materials through these areas. Section 1803.13 of the IFC requires a continuous gas detection system for early detection of leaks in areas where HPM gas is used. Gas detection systems are required to initiate a local alarm and transmit a signal to the emergency control station upon detection (see commentary, Sections 1803.12 and 1803.13 of the IFC.

[F] 908.3 Highly toxic and toxic materials. A gas detection system shall be provided to detect the presence of *highly toxic* or *toxic* gas at or below the permissible exposure limit (PEL) or ceiling limit of the gas for which detection is provided. The system shall be capable of monitoring the discharge from the treatment system at or below one-half the immediately dangerous to life and health (IDLH) limit.

> **Exception:** A gas-detection system is not required for *toxic* gases when the physiological warning threshold level for the gas is at a level below the accepted PEL for the gas.

❖ A gas detection system in the room or area utilized for indoor storage or the use of highly toxic or toxic gases provides early notification of a leak that is occurring before the escaping gas reaches hazardous exposure concentration levels. The exception recognizes that certain toxic compressed gases do not pose a severe exposure hazard. Those toxic gases whose properties under standard conditions are still below the 8-hour weighted average concentration for the permitted exposure limit (PEL) are exempt from the requirement for a gas detection system.

This section also specifies the discharge requirements for treatment system performance to establish a maximum allowable concentration of highly toxic or toxic gases at the point of discharge to the atmosphere. The concentration level of one-half IDLH represents a minimum acceptable level of dilution at the point of discharge where the location of discharge is away from the general public. Where the treatment system processes more than one type of compressed gas, the maximum allowable concentration must be based on the release rate, quantity and IDLH for the gas that poses the worst-case release scenario.

[F] 908.3.1 Alarms. The gas detection system shall initiate a local alarm and transmit a signal to a constantly attended control station when a short-term hazard condition is detected. The alarm shall be both visible and audible and shall provide warning both inside and outside the area where gas is detected. The audible alarm shall be distinct from all other alarms.

Exception: Signal transmission to a constantly attended control station is not required when not more than one cylinder of *highly toxic* or *toxic* gas is stored.

❖ The required local alarm is intended to alert occupants to a hazardous condition in the vicinity of the inside storage room or area. The alarm is not intended to be an evacuation alarm; however, it is required to be monitored to hasten emergency personnel response. The exception allows the omission of supervision for the gas detection system where a single cylinder of highly toxic or toxic gas is stored. It should be noted that this section is only intended to apply when the maximum allowable quantity of a specific gas per control area is exceeded. A single cylinder would not require a gas detection system if it contained less than the maximum allowable quantities indicated in Tables 307.1(1) and 307.1(2).

[F] 908.3.2 Shutoff of gas supply. The gas detection system shall automatically close the shutoff valve at the source on gas supply piping and tubing related to the system being monitored for whichever gas is detected.

Exception: Automatic shutdown is not required for reactors utilized for the production of *highly toxic* or *toxic* compressed gases where such reactors are:

1. Operated at pressures less than 15 pounds per square inch gauge (psig) (103.4 kPa).

2. Constantly attended.

3. Provided with readily accessible emergency shutoff valves.

❖ Actuation of the gas detection system is required to close automatically all valves controlling highly toxic or toxic gases. This degree of protection is deemed necessary to provide a life safety measure in areas where highly toxic and toxic gases are being utilized for filling, dispensing or other operations.

In most situations, dispensing and use operations are assumed to be constantly attended because of the potential hazard and need to monitor operations involving highly toxic and toxic compressed gases.

The exception recognizes the decreased level of hazard for constantly attended operations involving reactors in manufacturing processes with low operating pressures. With low operating pressures, the rate at which the volume of the affected space would be filled by the released gas is minimized. Readily accessible manual shutoff valves would thus be permitted in lieu of an automatic-closing shutoff valve.

[F] 908.3.3 Valve closure. The automatic closure of shutoff valves shall be in accordance with the following:

1. When the gas-detection sampling point initiating the gas detection system alarm is within a gas cabinet or exhausted enclosure, the shutoff valve in the gas cabinet or exhausted enclosure for the specific gas detected shall automatically close.

2. Where the gas-detection sampling point initiating the gas detection system alarm is within a gas room and compressed gas containers are not in gas cabinets or exhausted enclosures, the shutoff valves on all gas lines for the specific gas detected shall automatically close.

3. Where the gas-detection sampling point initiating the gas detection system alarm is within a piping distribution manifold enclosure, the shutoff valve for the compressed container of specific gas detected supplying the manifold shall automatically close.

Exception: When the gas-detection sampling point initiating the gas-detection system alarm is at a use location or within a gas valve enclosure of a branch line downstream of a piping distribution manifold, the shutoff valve in the gas valve enclosure for the branch line located in the piping distribution manifold enclosure shall automatically close.

❖ This section specifies the conditions where the shutoff valve for the gas supply is required to automatically close. The requirement depends on the location of the gas detection sampling point, which would initiate the actuation of the gas detection system, and the use of gas cabinets, exhaust enclosures or a piping distribution manifold enclosure.

[F] 908.4 Ozone gas-generator rooms. Ozone gas-generator rooms shall be equipped with a continuous gas-detection system that will shut off the generator and sound a local alarm when concentrations above the PEL occur.

❖ To monitor the potential buildup of dangerous levels of ozone, a gas detection system is required to, upon actuation, shut off the generator and sound a local alarm. Ozone gas generators are commonly used in water treatment applications. The ozone gas-generator room should not be a normally occupied area or be used for the storage of combustibles or other hazardous materials. Section 3705 of the IFC contains additional requirements for ozone gas generators.

[F] 908.5 Repair garages. A flammable-gas detection system shall be provided in repair garages for vehicles fueled by nonodorized gases in accordance with Section 406.6.6.

❖ As indicated in Section 406.6, an approved flammable-gas detection system is required for garages used for repair of vehicles fueled by nonodorized gases, such as hydrogen and nonodorized LNG. To prevent a hazardous potential buildup of flammable gas caused by normal leakage and use conditions, the flammable-gas detection system is required to activate when the level of flammable gas exceeds 25 percent of the lower explosive limit (LEL) (see commentary, Section 406.6).

[F] 908.6 Refrigerant detector. Machinery rooms shall contain a refrigerant detector with an audible and visual alarm. The detector, or a sampling tube that draws air to the detector, shall be located in an area where refrigerant from a leak will concentrate. The alarm shall be actuated at a value not greater than the corresponding TLV-TWA values for the refrigerant classification indicated in the *International Mechanical Code*. Detectors and alarms shall be placed in *approved* locations.

❖ A refrigerant-specific detector is required for the purpose of leak detection, early warning and actuation of emergency exhaust systems. Depending on the density of the refrigerants, leakage may collect near the floor, near the ceiling or disperse equally throughout. Refrigerant detector locations must be carefully considered. Most refrigerants are heavier than air; thus, floor depressions and pits are natural areas for accumulation. The code is silent on the location of sensors due to the endless variety of equipment room designs. The key to properly locating a detector in the machinery room is to remember that occupant safety is the primary objective, and the danger is in breathing refrigerant. Placing the sensor below the common breathing height of 5 feet (1525 mm) provides an additional safety margin because all commonly used halocarbon refrigerants are three to five times heavier than air. When undistributed by airflow, such escaping refrigerant will fall to the floor, seeking the lowest levels, and fill the room from the bottom up. Since pits, stairwells or trenches are likely to fill with refrigerant first, detectors should be also be placed in any of these areas that may be occupied. The alarm actuation threshold is dictated by the TLV-TWA values for the refrigerant classification in accordance with Table 1103.1 of the IMC.

Manufacturer's instructions for detectors provide installation guidance for the location and required number of detectors for any given room size.

The exception recognizes that ammonia refrigerant is "self-alarming" because of its strong odor and that ammonia machinery rooms are required to be continuously ventilated in accordance with the IMC.

Most general machinery rooms are unoccupied for long periods of time. Because of this, a refrigeration leak may go undetected, allowing a buildup of refrigerant that can pose a threat to the building occupants and the maintenance personnel who will be required to enter the machinery room. Also, the refrigerants may or may not be detectable by the sense of smell, depending upon the chemical nature and concentration in air of the refrigerant. This can be especially critical when a toxic refrigerant is used in the refrigeration system.

There are three levels of refrigerant exposure that are defined by the American Conference of Government Industrial Hygienists (ACGIH). Level 1 is the allowable exposure limit (AEL), which is the level at which a person can be exposed for 8 hours per day for 40 hours per week without having an adverse effect on health. At this level, a person exposed to the refrigerant should not suffer any adverse health effects. Level 2 is the short-term exposure limit (STEL), which is defined as three times the AEL. At this level, a person should not be exposed for more than 30 minutes at a time. Persons working in a machinery room having this concentration of refrigerant should be equipped with respiratory protection. Level 3 is the emergency exposure limit

(EEL). At this level, persons should not be in the room without a self-contained breathing apparatus.

Early detection of leaking refrigerant is dependent upon the location of the refrigerant detectors. If improperly located, a refrigerant leak could go undetected for an undesirable period of time, thus allowing a significant amount of refrigerant to escape. It is therefore advantageous to locate the detectors in positions that will provide early detection and minimize the amount of refrigerant leakage.

Factors to be considered when choosing locations for detectors are the airflow patterns of the room, the particular refrigerant density and the fact that the primary hazard to the occupants is through inhalation. Specifically, halocarbon refrigerants are oxygen displacers; that is, they are heavier than air and will occupy the lowest areas of the room first. This suggests that the location of detectors should be below the breathing zone height. Additionally, detectors should be located so as to prevent the normal ventilation system from interfering with detection. Placing detectors between the refrigeration system and exhaust fan inlets should help to ensure that the presence of refrigerant will be detected. Depending on the size of the machinery room and the number and type of refrigeration systems, more than one detector may be necessary. Manufacturer's installation instructions for the refrigeration detection system should be followed when choosing the location and number of sensors for a particular machinery room application.

SECTION 909
SMOKE CONTROL SYSTEMS

[F] 909.1 Scope and purpose. This section applies to mechanical or passive smoke control systems when they are required by other provisions of this code. The purpose of this section is to establish minimum requirements for the design, installation and acceptance testing of smoke control systems that are intended to provide a tenable environment for the evacuation or relocation of occupants. These provisions are not intended for the preservation of contents, the timely restoration of operations or for assistance in fire suppression or overhaul activities. Smoke control systems regulated by this section serve a different purpose than the smoke- and heat-venting provisions found in Section 910. Mechanical smoke control systems shall not be considered exhaust systems under Chapter 5 of the *International Mechanical Code*.

❖ This section is clarifying the intent of smoke control provisions, which is to provide a tenable environment to occupants during evacuation and relocation and not to protect the contents, enable timely restoration of operations or facilitate fire suppression and overhaul activities. There are provisions for high-rise buildings in Section 403.4.6 that are focused upon the removal of smoke for post fire and overhaul operations, which is very different than the smoke control provisions in Section 909. Another element addressed in this section is that smoke control systems serve a different purpose than smoke and heat vents (see Section 910). This eliminates any confusion that smoke and heat vents can be used as a substitution for smoke control. Additionally, a clarification is provided to note that smoke control systems are not considered an exhaust system in accordance with Chapter 5 of the IMC. This is due to the fact that such systems are unique in their operation and are not necessarily designed to exhaust smoke but are focused upon tenability for occupants during egress. It should be noted that the smoke control provisions are duplicated in Chapter 5 of the IMC.

It is important to note that these provisions only apply when smoke control is required by other sections of the code. The code requires smoke management within atrium spaces (see Section 404.5) and underground buildings (see Section 405.5). High-rise facilities require smokeproof exit enclosures in accordance with Sections 909.20 and 1019.1.8 (see Section 403.5.4). Also, covered mall buildings that contain atriums that connect more than two stories require smoke control (see Section 402.10).

Section 909 focuses primarily on mechanical smoke control systems, but there are many instances within the code where smoke is required to be managed in a passive way through the use of concepts such as smoke compartments. Smoke compartments are formed through the use of smoke barriers in accordance with Section 710. Smoke barriers can be used simply as a passive smoke management system or can be a design component of a mechanical smoke control system in accordance with Section 909. Some examples of occupancies requiring passive systems include hospitals, nursing homes and similar facilities (Group I-2 occupancies) and detention facilities (Group I-3 occupancies) (see Sections 407.4 and 408.6).

In some cases, mechanical smoke control in accordance with Section 909 is allowed as an option for compliance. More specifically if a Group I-3 contains windowless areas of the facility, natural or mechanical smoke management is required (see Section 408.9).

In the last several years, smoke control provisions have become more complex. The reason is related to the fact that smoke is a complex problem, while a generic solution of six air changes has repeatedly and scientifically been shown to be inadequate. Six air changes per hour does not take into account factors such as buoyancy; expansion of gases; wind; the geometry of the space and of communicating spaces; the dynamics of the fire, including heat release rate; the production and distribution of smoke and the interaction of the building systems.

Smoke control systems can be either passive or active. Active systems are sometimes referred to as mechanical. Passive smoke control systems take advantage of smoke barriers surrounding the zone in which the fire event occurs or high bay areas that act as reservoirs to control the movement of smoke to other areas of the building. Active systems utilize pres-

sure differences to contain smoke within the event zone or exhaust flow rates sufficient to slow the descent of the upper-level smoke accumulation to some predetermined position above necessary exit paths through the event zone. On rare occasions, there is also a possibility of controlling the movement of smoke horizontally by opposed airflow, but this method requires a specific architectural geometry to function properly that does not create an even greater hazard.

Essentially, there are three methods of mechanical or active smoke control that can be used separately or in combination within a design: pressurization, exhaust and, in rare and very special circumstances, opposed airflow.

Of course, all of these active approaches can be used in combination with the passive method.

Typically, the mechanical pressurization method is used in high-rise buildings when pressurizing stairways and for zoned smoke control. Pressurization is not practical in large open spaces such as atriums or malls, since it is difficult to develop the required pressure differences due to the large volume of the space.

The exhaust method is typically used in large open spaces such as atriums and malls. As noted, the pressurization method would not be practical within large spaces. The opposed airflow method, which basically uses a velocity of air horizontally to slow the movement of smoke, is typically applied in combination with either a pressurization method or exhaust method within hallways or openings into atriums and malls.

The application of each of these methods will be dependent on the specifics of the building design. Smoke control within a building is fundamentally an architecturally driven problem. Different architectural geometries first dictate the need or lack thereof for smoke control, and then define the bounds of available solutions to the problem.

[F] 909.2 General design requirements. Buildings, structures or parts thereof required by this code to have a smoke control system or systems shall have such systems designed in accordance with the applicable requirements of Section 909 and the generally accepted and well-established principles of engineering relevant to the design. The *construction documents* shall include sufficient information and detail to adequately describe the elements of the design necessary for the proper implementation of the smoke control systems. These documents shall be accompanied by sufficient information and analysis to demonstrate compliance with these provisions.

❖ This section simply states that when smoke control systems are required by the code, the design is required to be in accordance with the provisions of this section. As noted in the commentary to Section 909.1, there are instances within the code that have smoke management systems that are purely passive in nature and do not reference Section 909.

This section stresses that designs in accordance with this section need to follow "generally accepted and well-established principles of engineering relevant to the design," essentially requiring a certain level of qualifications in the applicable areas of engineering to prepare such designs. The primary engineering disciplines tend to be fire engineering and mechanical engineering. It should be noted that each state in the U.S. typically requires minimum qualifications to undertake engineering design. Two important resources when designing smoke control systems are ICC's *A Guide to Smoke Control in the 2006 IBC*® and American Society of Heating, Refrigerating and Air-Conditioning Engineers' (ASHRAE) *Design of Smoke Management Systems*. Additionally, Section 909.8 requires the use of NFPA 92B for the design of smoke control systems using the exhaust method. This standard has many relevant aspects beyond the design that are beneficial. In particular, Annex B provides resources in terms of determination of fire size for design. ICC's *A Guide to Smoke Control in the 2006 IBC*® also provides guidance on design fires.

A key element covered in this section is the need for detailed and clear construction documents so that the system is installed correctly. In most complex designs, the key to success is appropriate communication to the contractors as to what needs to be installed. The more complex a design becomes, the more likely there is to be construction errors. Most smoke control systems are complex, which is why special inspections in accordance with Section 909.3 and Chapter 17 are critical for smoke control systems. Additionally, in order for the design to be accepted, analyses and justifications need to be provided in enough detail to evaluate for compliance. Adequate documentation is critical to the commissioning, inspection, testing and maintenance of smoke control systems and significantly contributes to the overall reliability and effectiveness of such systems.

[F] 909.3 Special inspection and test requirements. In addition to the ordinary inspection and test requirements which buildings, structures and parts thereof are required to undergo, smoke control systems subject to the provisions of Section 909 shall undergo *special inspections* and tests sufficient to verify the proper commissioning of the smoke control design in its final installed condition. The design submission accompanying the *construction documents* shall clearly detail procedures and methods to be used and the items subject to such inspections and tests. Such commissioning shall be in accordance with generally accepted engineering practice and, where possible, based on published standards for the particular testing involved. The special inspections and tests required by this section shall be conducted under the same terms in Section 1704.

❖ Due to the complexity and uniqueness of each design, special inspection and testing must be conducted. The designer needs to provide specific recommendations for special inspection and testing within his or her documentation. In fact, the code specifies in Chapter 17 that special inspection agencies for smoke control have expertise in fire protection engineering, mechanical engineering and certification as air balancers.

Since the designs are unique to each building, there probably will not be a generic approach available to inspect and test such systems. The designer can and should, however, use any available published standards or guides when developing the special inspection and testing requirements for that particular design. ICC's *A Guide to Smoke Control in the 2006 IBC*® provides some background on such inspections, Also, ASHRAE Guideline 5 is a good starting place, but only as a general outline. In addition, NFPA 92A and NFPA 92B also have extensive testing, documentation and maintenance requirements that may be a good resource. NFPA 92B is referenced in Section 909.8 for the design of smoke control systems using the exhaust method. Each system will require a unique commissioning plan that can be developed only after careful and thoughtful examination of the final design and all of its components and interrelationships. Generally, these provisions may be included in design standards or engineering guides.

[F] 909.4 Analysis. A rational analysis supporting the types of smoke control systems to be employed, their methods of operation, the systems supporting them and the methods of construction to be utilized shall accompany the submitted *construction documents* and shall include, but not be limited to, the items indicated in Sections 909.4.1 through 909.4.6.

❖ This section indicates that simply determining airflow, exhaust rates and pressures to maintain tenable conditions is not adequate. There are many factors that could alter the effectiveness of a smoke control system, including stack effect, temperature effect of fire, wind effect, heating, ventilating and air-conditioning (HVAC) system interaction and climate, as well as the placement, quantity of inlets/outlets and velocity of supply and exhaust air. These factors are addressed in the sections that follow. Additionally, the duration of operation of any smoke control system is mandated at a minimum of 20 minutes or 1.5 times the egress time, whichever is less. The code cannot reasonably anticipate every conceivable building arrangement and condition the building may be subject to over its life and must depend on such factors being addressed through a rational analysis.

[F] 909.4.1 Stack effect. The system shall be designed such that the maximum probable normal or reverse stack effect will not adversely interfere with the system's capabilities. In determining the maximum probable stack effect, altitude, elevation, weather history and interior temperatures shall be used.

❖ Stack effect is the tendency for air to rise within a heated building when the temperature is colder on the exterior of the building. Reverse stack effect is the tendency for air to flow downward within a building when the interior is cooler than the exterior of the building. This air movement can affect the intended operation of a smoke control system. If stack effect is great enough, it may overcome the pressures determined during the design analyses and allow smoke to enter areas outside the zone of origin (see Figure 909.4.1).

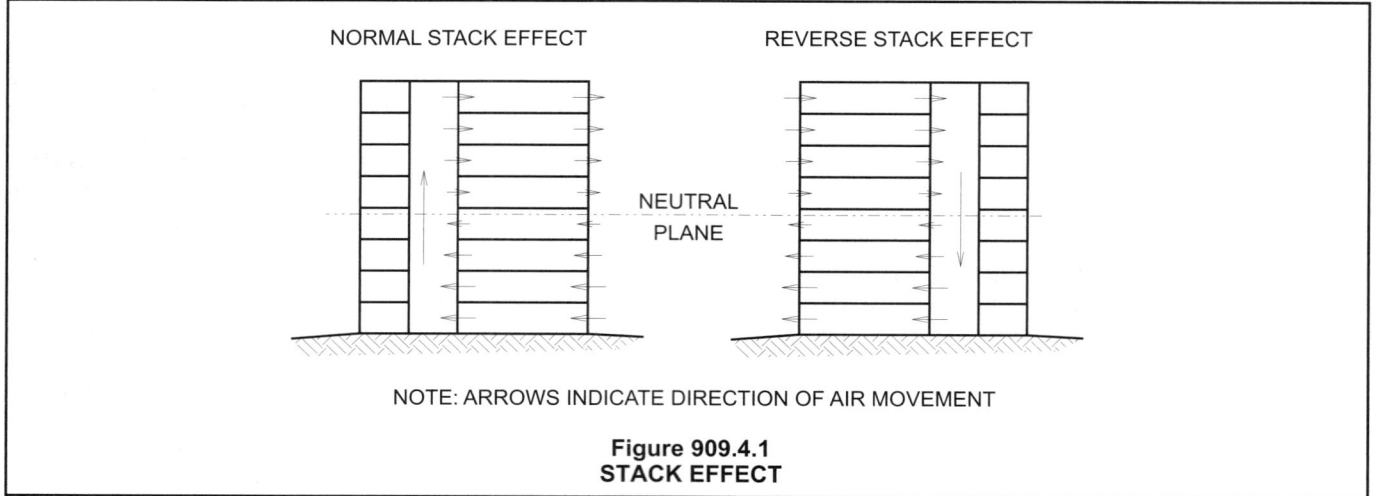

Figure 909.4.1
STACK EFFECT

[F] 909.4.2 Temperature effect of fire. Buoyancy and expansion caused by the design fire in accordance with Section 909.9 shall be analyzed. The system shall be designed such that these effects do not adversely interfere with the system's capabilities.

❖ This section requires that the design account for the effect temperature may have on the success of the system. When air or any gases are heated they will expand. This expansion makes the gases lighter and therefore more buoyant. The buoyancy of hot gases is important when the design is to exhaust such gases from a location in or close to the ceiling; therefore, if sprinklers are part of the design, as required by Section 909, the gases may be significantly cooler than an unsprinklered fire, making it more difficult to remove the smoke and alter the plume dynamics. The fact that air expands when heated needs to be accounted for in the design.

When using the pressurization method, the expansion of hot gases needs to be accounted for, since it will take a larger volume of air to create the necessary pressure differences to maintain the area of fire origin in negative pressure. The expansion of the gases has the effect of pushing the hot gases out of the area of fire origin. Since sprinklers will tend to cool the gases, the effect of expansion is lower. The pressure differences required in Section 909.6.1 are specifically based on a sprinklered building. If the building is nonsprinklered, higher pressure differences may be required. The minimum pressure difference for certain unsprinklered ceiling height buildings is as follows:

Ceiling height (feet)	Minimum pressure difference (inch water gage)
9	.10
15	.14
21	.18

This is a very complex issue that needs to be part of the design analysis. It needs to address the type and reaction of the fire protection systems, ceiling heights and the size of the design fire.

[F] 909.4.3 Wind effect. The design shall consider the adverse effects of wind. Such consideration shall be consistent with the wind-loading provisions of Chapter 16.

❖ The effect of wind on a smoke control system within a building is very complex. It is generally known that wind exerts a load upon a building. The loads are looked at as windward (positive pressure) and leeward (negative pressure). The velocity of winds will vary based on the terrain and the height above grade; therefore, the height of the building and surrounding obstructions will have an effect on these velocities. These pressures alter the operation of fans, especially propeller fans, thus altering the pressure differences and airflow direction in the building. There is not an easy solution to dealing with these effects. In fact, little research has been done in this area.

It should be noted that in larger buildings a wind study is normally undertaken for the structural design. The data from those studies can be used in the analysis of the effects on the pressures and airflow within the building with regard to the performance of the smoke control system.

[F] 909.4.4 HVAC systems. The design shall consider the effects of the heating, ventilating and air-conditioning (HVAC) systems on both smoke and fire transport. The analysis shall include all permutations of systems status. The design shall consider the effects of the fire on the HVAC systems.

❖ If not properly configured to shut down or included as part of the design, the HVAC system can alter the smoke control design. More specifically, if dampers are not provided between smoke zones within the HVAC system ducts, smoke could be transported from one zone to another. Additionally, if the HVAC system places more supply air than assumed for the smoke control system design, the velocity of the air may adversely affect the fire plume or a positive pressure may be created. Generally, an analysis of the smoke control design and the HVAC system in all potential modes should occur and be noted within the design documentation as well as incorporated into inspection, testing and maintenance procedures. This is critical as these systems need to be maintained and tested to help ensure that they operate and shut down systems as required.

[F] 909.4.5 Climate. The design shall consider the effects of low temperatures on systems, property and occupants. Air inlets and exhausts shall be located so as to prevent snow or ice blockage.

❖ This section is focused on properly protecting equipment from weather conditions that may affect the reliability of the design. For instance, extremely cold or hot air may damage critical equipment within the system when pulled directly from the outside. Some listings of duct smoke detectors are for specific temperature ranges; therefore, placing such detectors within areas exposed to extreme temperatures may void the listing. Also, the equipment and air inlets and outlets should be designed and located so as to not collect snow and ice that could block air from entering or exiting the building.

[F] 909.4.6 Duration of operation. All portions of active or passive smoke control systems shall be capable of continued operation after detection of the fire event for a period of not less than either 20 minutes or 1.5 times the calculated egress time, whichever is less.

❖ The intent of the smoke control provisions is to provide a tenable environment for occupants to either evacuate or relocate to a safe place. Evacuation and relocation activities include notifying occupants, possible investigation time for occupants, decision time and the actual travel time. In order to achieve this goal, the code has established 20 minutes or 1.5 times the calculated egress time, whichever is less, as a minimum time for evacuation or relocation. Basically this allows a designer to undertake an analysis to more closely determine the necessary time required. The code provides a safety factor of 1.5 times the egress time to account for uncertainty related to human behavior. It is stressed that the 20-minute duration as well as the calculated egress time, whichever approach is chosen, begins after the detection of the fire event and notification to the building occupants to evacuate has occurred, since occupants need to be alerted before evacuation can occur. The calculation of evacuation time needs to include delays with notification and the start of evacuation (i.e., premovement time, etc.). It is stressed that the code states 20 minutes or 1.5 times the egress time, whichever is less (i.e., 20 minutes is a maximum). Egress of occupants can be addressed through hand calculations or through the use of computerized egress models. Some of the more advanced models can address a variety of factors, including the building layout, different sizes of people, different movement speeds and different egress paths available. With these types of programs, the actual time can be even more precisely calculated. Of course it is cautioned that in many cases these models provide the optimal time for egress. The safety factor of 1.5 within the code is

intended to address many of these uncertainties.

Note that this section applies to all types of smoke control designed in accordance with Section 909. Also, most smoke control systems will typically have the ability to run for longer than the 20 minute maximum as they are on standby power and may be able to continue to achieve the tenability goals. In some cases even if the system runs longer than 20 minutes, the tenability may not be able to continue. It simply depends on the system design and the fire hazards within the building.

System response as required in Section 909.17 needs to be accounted for when determining the ability of the smoke control system to keep the smoke layer interface at the appropriate level (see commentary, Section 909.17).

[F] 909.5 Smoke barrier construction. *Smoke barriers* shall comply with Section 710, and shall be constructed and sealed to limit leakage areas exclusive of protected openings. The maximum allowable leakage area shall be the aggregate area calculated using the following leakage area ratios:

1. Walls: $A/A_w = 0.00100$

2. *Exit* enclosures: $A/A_w = 0.00035$

3. All other shafts: $A/A_w = 0.00150$

4. Floors and roofs: $A/A_F = 0.00050$

where:

A = Total leakage area, square feet (m^2).

A_F = Unit floor or roof area of barrier, square feet (m^2).

A_w = Unit wall area of barrier, square feet (m^2).

The leakage area ratios shown do not include openings due to doors, operable windows or similar gaps. These shall be included in calculating the total leakage area.

❖ Part of the strategy of smoke control systems, particularly smoke control systems using the pressurization method (often termed zoned smoke control) is the use of smoke barriers to divide a building into separate smoke zones (or compartments). This strategy is used in both passive and mechanical systems. It should be noted that not all walls, ceilings or floors would be considered smoke barriers. Only walls that designate separate smoke zones within a building need to be constructed as smoke barriers. This section is simply providing requirements for walls, floors and ceilings that are used as smoke barriers. It should be noted that it is possible that a smoke control system utilizing the exhaust method may not need to utilize a smoke barrier to divide the building into separate smoke zones; therefore, the evaluation of barrier construction and leakage area may not be necessary and, as noted, is primarily focused upon designs using the pressurization method.

In order for smoke to not travel from one smoke zone to another, specific construction requirements are necessary in accordance with the code. It should be noted that openings such as doors and windows are dealt with separately within Section 909.5.2 from openings such as cracks or penetrations.

[F] 909.5.1 Leakage area. The total leakage area of the barrier is the product of the *smoke barrier* gross area multiplied by the allowable leakage area ratio, plus the area of other openings such as gaps and operable windows. Compliance shall be determined by achieving the minimum air pressure difference across the barrier with the system in the smoke control mode for mechanical smoke control systems. Passive smoke control systems tested using other *approved* means such as door fan testing shall be as *approved* by the fire code official.

❖ It is impossible for walls and floors to be constructed that are completely free from openings that may allow the migration of smoke; therefore, leakage needs to be compensated for within the design by calculating the leakage area of walls, ceilings and floors. The factors provided in Section 909.5, which originate from ASHRAE's provisions on leaky buildings, are used to calculate the total leakage area. The total leakage area is then used in the design process to determine the proper amount of air to create the required pressure differences across these surfaces that form smoke zones. These pressure differences then need to be verified when the system is in smoke control mode.

Additionally, Section 909.5 provides ratios to determine the maximum allowable leakage in walls, exit enclosures, shafts, floors and roofs. These leakage areas are critical in determining whether the proper pressure differences are provided when utilizing the pressurization method of smoke control. Pressure differences will decrease as the openings get larger.

[F] 909.5.2 Opening protection. Openings in *smoke barriers* shall be protected by automatic-closing devices actuated by the required controls for the mechanical smoke control system. Door openings shall be protected by *fire door assemblies* complying with Section 715.4.3.

Exceptions:

1. Passive smoke control systems with automatic-closing devices actuated by spot-type smoke detectors *listed* for releasing service installed in accordance with Section 907.3.

2. Fixed openings between smoke zones that are protected utilizing the airflow method.

3. In Group I-2, where such doors are installed across corridors, a pair of opposite-swinging doors without a center mullion shall be installed having vision panels with fire protection-rated glazing materials in fire protection-rated frames, the area of which shall not exceed that tested. The doors shall be close-fitting within operational tolerances and shall not have undercuts, louvers or grilles. The doors shall have head and jamb stops, astragals or rabbets at meeting edges and shall be automatic-closing by smoke detection in accordance with Section 715.4.8.3. Positive-latching devices are not required.

4. Group I-3.

5. Openings between smoke zones with clear ceiling heights of 14 feet (4267 mm) or greater and bank-down capacity of greater than 20 minutes as determined by the design fire size.

❖ Similar to concerns of smoke leakage between smoke zones, openings may compromise the necessary pressure differences between smoke zones. Openings in smoke barriers, such as doors and windows, must be either constantly or automatically closed when the smoke control system is operating. This section requires that doors be automatically closed through the activation of an automatic closing device linked to the smoke control system. Essentially, when the smoke control system is activated, all openings are automatically closed. This most likely would mean that the mechanism that activates the smoke control system would also automatically close all openings. The smoke control system will be activated by a specifically zoned smoke detection or sprinkler system as required by Sections 909.12.2 and 909.12.3.

In terms of actual opening protection, Section 909.5.2 is simply referring the user to Section 715.4.3 for specific construction requirements for doors located in smoke barriers. Note that smoke barriers are different from fire barriers, since the intended measure of performance is different. One is focused on fire spread from the perspective of heat, the other from the perspective of smoke passage. Smoke barriers do require a 1-hour fire-resistance rating.

There are several exceptions to this particular section. Exception 1 is specifically for passive systems. Passive systems, as noted, are systems in which there is no use of mechanical systems. Instead, the system operates primarily upon the configuration of barriers and layout of the building to provide smoke control. Passive systems can use spot-type detectors to close doors that constitute portions of a smoke barrier. Essentially, this means a full fire alarm system would not be required. Instead, single station detectors would be allowed to close the doors. Such doors would need to fail in the closed position if power is lost. The specifics as to approved devices would be found in NFPA 72.

Exception 2 is based on the fact that some systems take advantage of the opposed airflow method such that smoke is prevented from migrating past the doors. Therefore, since the design already accounts for potential smoke migration at these openings through the use of air movement, it is unnecessary to require the barrier to be closed.

Exception 3 is specifically related to the unique requirements for Group I-2 occupancies. Essentially, a very specific alternative, which meets the functional needs of Group I-2 occupancies, is provided. One aspect of the alternative approach is that doors have vision panels with approved fire protection-rated glazing in fire protection-rated frames of a size that does not exceed the type tested.

Exception 4 allows an exemption from the automatic-closing requirements for all Group I-3 occupancies. This is related to the fact that facilities that have occupants under restraint or with specific security restrictions have unique requirements in accordance with Section 408 of the code. These requirements accomplish the intent of providing reliable barriers between each smoke zone since, for the most part, such facilities will have a majority of doors closed and in a locked position due to the nature of the facility. The staff very closely controls these types of facilities.

Exception 5 relates to the behavior of smoke. The assumption is that smoke rises due to the buoyancy of hot gases, and if the ceiling is sufficiently high, the smoke layer will be contained for a longer period of time before it begins to move into the next smoke zone. Therefore, it is not as critical that the doors automatically close. This allowance is dependent on the specific design fire for a building. See Section 909.9 for more information on design fire determination. Different size design fires create different amounts of smoke that, depending on the layout of the building, may migrate in different ways throughout the building. This section mandates that smoke cannot begin to migrate into the next smoke zone for at least 20 minutes. This is consistent with the 20-minute maximum duration of operation of smoke control systems required in Section 909.4.6. It should be noted that a minimum of 14-foot (4267 mm) ceilings are required to take advantage of this exception. This exception would require an engineering analysis.

[F] 909.5.2.1 Ducts and air transfer openings. Ducts and air transfer openings are required to be protected with a minimum Class II, 250°F (121°C) *smoke damper* complying with Section 716.

❖ Another factor that adds to the reliability of smoke barriers is the protection of ducts and air transfer openings within smoke barriers. Left open, these openings may allow the transfer of smoke between smoke zones. These ducts and air transfer openings most often are part of the HVAC system. Damper operation and the reaction with the smoke control system will be evaluated during acceptance testing. It should be noted that there are duct systems used within a smoke control design that are controlled by the smoke control system and should not automatically close upon detection of smoke via a smoke damper.

It should be noted that a smoke damper works differently than a fire damper. Fire dampers react to heat via a fusible link, while smoke dampers activate upon the detection of smoke. The smoke dampers used should be rated as Class II, 250°F (121°C). The class of the smoke damper refers to its level of performance relative to leakage. The temperature rating is related to its ability to withstand the heat of smoke resulting from a fire. It should be noted that although smoke barriers are only required to utilize smoke dampers, there may be many instances where a fire damper is also required. For instance, the smoke barrier may also be used as a fire barrier. Also, Section 716.5.3 would

require penetration of shafts to contain both a smoke and fire damper. Therefore, in some cases both a smoke damper and fire damper would be required. There are listings specific to combination smoke and fire dampers. Note that the exceptions to Section 716.5.3 recognize that smoke and fire dampers may interfere with a smoke control design.

More specific requirements about dampers can be found in Chapter 7 of this code and Chapter 6 of the IMC.

[F] 909.6 Pressurization method. The primary mechanical means of controlling smoke shall be by pressure differences across smoke barriers. Maintenance of a tenable environment is not required in the smoke control zone of fire origin.

❖ There are several methods or strategies that may be used to control smoke movement. One of these methods is pressurization, wherein the system primarily utilizes pressure differences across smoke barriers to control the movement of smoke. Basically, if the area of fire origin maintains a negative pressure, then the smoke will be contained to that smoke zone. A typical approach used to obtain a negative pressure is to exhaust the fire floor. This is a fairly common practice in high-rise buildings. Stairway enclosures also utilize the concept of pressurization by keeping the stairway enclosure under positive pressure. The pressurization method in large open spaces, such as malls and atria, is impractical since it would take a large quantity of supply air to create the necessary pressure differences. It should be noted that pressurization is mandated as the primary method for mechanical smoke control design-but this is related to the primary methods historically used for smoke control in high-rise buildings. Currently high-rise buildings do not require smoke control. Airflow and exhaust methods are only allowed when appropriate. The exhaust method is the most commonly applied method due to the use of the atrium provisions in Section 404.5.

The pressurization method does not require that tenable conditions be maintained in the smoke zone where the fire originates. Maintaining this area tenable would be impossible, based on the fact that pressures from the surrounding smoke zones would be placing a negative pressure within the zone of origin to keep the smoke from migrating.

Pressurization is used often with exit stair enclosures. This method provides a positive pressure within the stair enclosure to resist the passage of smoke. Stair pressurization is one method of compliance for stairways in high-rise or underground buildings where the floor surface is located more than 75 feet (22 860 mm) above the lowest level of fire department vehicle access or more than 30 feet (9144 mm) below the floor surface of the lowest level of exit discharge. It should be noted that there are two methods found in the code that address smoke movement—smokeproof enclosures or pressurized stairs. A smokeproof enclosure requires a certain fire-resistance rating along with access through a ventilated vestibule or an exterior balcony. The vestibule can be ventilated in two ways: using natural ventilation or mechanical ventilation as outlined in Sections 909.20.3 and 909.20.4. The pressurization method requires a sprinklered building and a minimum pressure difference of 0.15 inch (37 Pa) of water and a maximum of 0.35 inch (87 Pa) of water. These pressure differences are to be available with all doors closed under maximum stack pressures (see Sections 909.20 and 1022.9 for more details).

As noted, the pressurization method utilizes pressure differences across smoke barriers to achieve control of smoke. Sections 909.6.1 and 909.6.2 provide the criteria for smoke control design in terms of minimum and maximum pressure differences.

In summary, the pressurization method is used in two ways. The first is through the use of smoke zones where the zone of origin is exhausted, creating a negative pressure. The second is stair pressurization that creates a positive pressure within the stair to avoid the penetration of smoke. Note that the code allows the use of a smokeproof enclosure instead of pressurization.

[F] 909.6.1 Minimum pressure difference. The minimum pressure difference across a *smoke barrier* shall be 0.05-inch water gage (0.0124 kPa) in fully sprinklered buildings.

In buildings permitted to be other than fully sprinklered, the smoke control system shall be designed to achieve pressure differences at least two times the maximum calculated pressure difference produced by the design fire.

❖ The minimum pressure difference is established as $^1/_2$-inch water gage (12 Pa) in fully sprinklered buildings. This particular criterion is related to the pressures needed to overcome buoyancy and the pressures generated by the fire, which include expansion. This particular criterion is based upon a sprinklered building. The pressure difference would need to be higher in a building that is not sprinklered. Additionally, the pressure difference needs to be provided based upon the possible stack and wind effects present.

[F] 909.6.2 Maximum pressure difference. The maximum air pressure difference across a *smoke barrier* shall be determined by required door-opening or closing forces. The actual force required to open *exit* doors when the system is in the smoke control mode shall be in accordance with Section 1008.1.2. Opening and closing forces for other doors shall be determined by standard engineering methods for the resolution of forces and reactions. The calculated force to set a side-hinged, swinging door in motion shall be determined by:

$$F = F_{dc} + K(WA\Delta P)/2(W - d) \qquad \textbf{(Equation 9-1)}$$

where:

A = Door area, square feet (m²).

d = Distance from door handle to latch edge of door, feet (m).

F = Total door opening force, pounds (N).

F_{dc} = Force required to overcome closing device, pounds (N).

K = Coefficient 5.2 (1.0).

W = Door width, feet (m).

ΔP = Design pressure difference, inches of water (Pa).

❖ The maximum pressure difference is based primarily upon the force needed to open and close doors. The code establishes maximum opening forces for doors. This maximum opening force cannot be exceeded, taking into account the pressure differences across a doorway in a pressurized environment. Essentially, based on the opening force requirements of Section 1008.1.3, the maximum pressure difference can be calculated in accordance with Equation 9-1. In accordance with Chapter 10, the maximum opening force of a door has three components, including:

> Door latch release:
> > Maximum of 15 pounds (67 N)
> Set door in motion:
> > Maximum of 30 pounds (134 N)
> Swing to full open position:
> > Maximum of 15 pounds (67 N)

Equation 9-1 is used to calculate the total force to set the door into motion when in the smoke control mode; therefore, the limiting criteria would be 30 pounds (134 N). It should be noted that although the accessibility requirements related to door opening force are more restrictive in Section 404.2.8 of ICC A117.1, fire doors do not require compliance with these requirements.

[F] 909.7 Airflow design method. When *approved* by the fire code official, smoke migration through openings fixed in a permanently open position, which are located between smoke control zones by the use of the airflow method, shall be permitted. The design airflow shall be in accordance with this section. Airflow shall be directed to limit smoke migration from the fire zone. The geometry of openings shall be considered to prevent flow reversal from turbulent effects.

❖ This method is only allowed when approved by the building official. As the title states, this method utilizes airflow to avoid the migration of smoke across smoke barriers. This has been referred to as opposed airflow. Specifically, this method is suited for the protection of smoke migration through doors and related openings fixed in a permanently open position. This method consists of providing a particular velocity of air based upon the temperature of the smoke and the height of the opening. The temperature of the smoke will depend on the design fire that is established for the particular building. The higher the temperature of the smoke and the larger the opening, the higher the velocity necessary to maintain the smoke from migrating into the smoke zone. It should be noted that the airflow method seldom works for large openings, since the velocity to oppose the smoke becomes too high. This method tends to work better for smaller openings, such as pass-through windows. Equation 9-2 provides the method to calculate the necessary velocity.

[F] 909.7.1 Velocity. The minimum average velocity through a fixed opening shall not be less than:

$$v = 217.2 \, [h \, (T_f - T_o)/(T_f + 460)]^{1/2} \quad \text{(Equation 9-2)}$$

For SI: $v = 119.9 \, [h \, (T_f - T_o)/T_f]^{1/2}$

where:

h = Height of opening, feet (m).

T_f = Temperature of smoke, °F (K).

T_o = Temperature of ambient air, °F (K).

v = Air velocity, feet per minute (m/minute).

❖ This section provides the formula for the minimum average velocity through a fixed opening. The minimum velocity is based on the velocity needed to prevent the smoke from migrating into the smoke zone. Consideration needs to be given to the eventual exhaust of the air introduced for this approach. See the commentary to Section 909.7 for further discussion.

[F] 909.7.2 Prohibited conditions. This method shall not be employed where either the quantity of air or the velocity of the airflow will adversely affect other portions of the smoke control system, unduly intensify the fire, disrupt plume dynamics or interfere with exiting. In no case shall airflow toward the fire exceed 200 feet per minute (1.02 m/s). Where the formula in Section 909.7.1 requires airflow to exceed this limit, the airflow method shall not be used.

❖ The airflow method has a limitation on maximum velocity. This limitation is based upon the fact that air may distort the flame and cause additional entrainment and turbulence; therefore, having a high velocity of air entering the zone of fire origin has the potential of increasing the amount of smoke produced. The velocity may also interact with other portions of the smoke control design. For instance, the pressure differences in other areas of the building may be altered, which may exceed the limitations of Sections 909.6.1 and 909.6.2. This section requires that when a velocity of over 200 feet per minute (1.02 m/sec) is calculated, the airflow method is not allowed. The solution may result in requiring a barrier such as a wall or door.

If the airflow design method is chosen to protect areas communicating with an atrium, the air added to the smoke layer needs to be accounted for in the exhaust rate.

[F] 909.8 Exhaust method. When *approved* by the fire code official, mechanical smoke control for large enclosed volumes, such as in atriums or malls, shall be permitted to utilize the exhaust method. Smoke control systems using the exhaust method shall be designed in accordance with NFPA 92B.

❖ This method is only allowed when approved by the building official. The primary application of the exhaust method is in large spaces, such as atriums and malls and is the most widely used method in the IBC. The strategy of this method is to keep the smoke layer at a certain level within the space. This is primarily accomplished through exhausting smoke. The amount of exhaust depends upon the design fire [see Figure 909.8(1)]. Essentially, fires produce different amounts and properties of smoke based on the material being burned, and size and placement of the fire; therefore, NFPA 92B is referenced for the design of such systems. NFPA 92B presents several ways to address the control of smoke, which includes the use of the following tools:

- Scale Modeling (Small scale testing)—Utilizes the concept of scaling to allow small scale tests to be conducted to understand the smoke movement within a space.
  - Benefits—More realistic understanding of smoke movement in spaces with unusual configurations or projections than algebraic calculations.
  - Disadvantages—Expensive and the application of results is limited to the uniqueness of the space being analyzed.
- Algebraic (Calculations—similar to 2003 IBC)—Empirically derived (based upon testing) modeling in its simplest form.
  - Benefits—Simple, cost-effective analysis.
  - Disadvantages—Limited applicability due to the range of values they were derived from, only appropriate with certain types of design fires, typically over conservative outputs that increase equipment needs, equipment costs and can impact aesthetics and architectural design.
- Computer Modeling [Computational fluid dynamics (CFD) or zone models]—Combination of theory and empirical values to determine the smoke movement and fire induced conditions within a space and effectiveness of the smoke control system.
  - Benefits—More realistic understanding of smoke movement in spaces with unusual configurations or projections and less expensive than scale modeling. Helps significantly in designing smoke control systems tailored to spaces and achieving cost-effective designs, and can help limit the impact to architectural design.
  - Disadvantages—Computing time and cost can be longer than algebraic calculations but benefits typically outweigh this disadvantage. Early planning is important and can limit these adverse impacts.

In terms of computer modeling, as noted, there are essentially two methods that include zone models and CFD models. Zone models are based upon the unifying assumption that in any room or space where the effects of the fire are present there are distinct layers (hot upper layer, cool lower layer). In real life such distinct layers do not exist. Some examples of zone models used in such applications include Consolidated Model of Fire Growth and Smoke Transport (C-FAST) and Available Safe Egress Time (ASET). See Section 3-7 of the *SFPE Handbook of Fire Protection Engineering* for further information. CFD models take this much further and actually divide the space into thousands or millions of interconnected "cells" or "fields." The model then evaluates the fire dynamics and heat and mass in each individual cell and how it interacts with those adjacent to it. The use of such models becomes more accurate with more numerous and smaller cells but the computing power and expertise required is much higher than for zone models. As noted the use of either types of models can be advantageous but such use must be undertaken by someone qualified. Proper review and verification of the input and output is critical. The most popular model in the area of CFD with regard to fire is the Fire Dynamics Simulator (FDS) developed by National Institute for Standards and Technology (NIST). Other models such as Fluent are sometimes used (Fluent Inc.).

Depending upon the space being evaluated, some design strategies may provide a better approach than others. Past editions of the code smoke control provisions for the exhaust method mandated the use of the algebraic methods with a steady fire. This of course also mandated a mechanical system be used, whereas NFPA 92B allows an overall review of smoke layer movement and whether the design goals, which in this case are mandated by the code, can be met. Therefore, if it can be shown that the smoke layer interface can be held at 6 feet (1829 mm) as mandated in Section 909.8.1 for the design operation time required by Section 909.4.6 without mechanical ventilation, then the space would comply with Section 909. NFPA 92B presents several design approaches. This allows more flexibility in design than that found in previous editions of the code.

NFPA 92B as a standard does not set the minimum smoke layer interface height or duration for system operation. Such criteria is found within Sections 909.8.1 and 909.4.6, respectively. See the commentary for those sections.

If the algebraic approach is used, consideration of three types of fire plumes may be required to determine which one is the most demanding in terms of smoke removal needs based upon the space being assessed. They include:

Axisymmetric plumes—Smoke rises unimpeded by walls, balconies or similar projections [see Figure 909.8(2)].

Balcony spill plumes—Smoke flows under and around edges of a horizontal projection [see Figure 909.8(3)].

Window plumes—Smoke flows through an opening into a large-volume space [see Figure 909.8(4)].

It should be noted that prior to the reference to NFPA 92B in the code, the balcony spill and window plume calculations had been eliminated from the smoke control requirements of the code due to concerns with the applicability of those calculations. The major difference is that NFPA 92B does not mandate the use of such equations as did previous editions of the IBC. The use of such equations will depend upon the design fires agreed upon for the particular design and whether an algebraic approach is chosen. These equations are used to determine a mass flow rate of smoke to ultimately determine the required exhaust volume for that space. If the potential for a balcony or window spill plumes is known to exist within the space, then appropriate measures need to be taken to address these, as they typically result in more onerous exhaust and supply requirements. Part of the reason for the initial deletion of these equations was the fact that such scenarios are not as likely or their impact is significantly reduced in sprinklered buildings. There is also some concern with the applicability of the balcony spill plume equation in a variety of applications. These potential fire scenarios and resulting plumes may further the need to undertake a CFD analysis to address such hazards more appropriately and effectively.

Another key aspect that NFPA 92B included within the algebraic methods is equations to determine that a minimum number of exhaust inlets are available to prevent plugholing. Plugholing occurs when air from below the smoke layer is pulled through the smoke layer into the smoke exhaust inlets. As such, if plugholing occurs, some of the fan capacity is used to exhaust air rather than smoke and thus can affect the ability to maintain the smoke layer at or above the design height. Scale modeling and computer fire modeling would demonstrate these potential problems during the testing and analysis, respectively [see Figure 909.8(5)].

It should be noted that this section specifically references NFPA 92B for the design of smoke control using the exhaust method. Therefore the requirements in NFPA 92B related to testing, documentation and maintenance would not be applicable though they may be a good resource. Equipment and controls would be part of the design; therefore, related provisions of NFPA 92B would apply. Generally, the code addresses equipment and controls in a similar fashion.

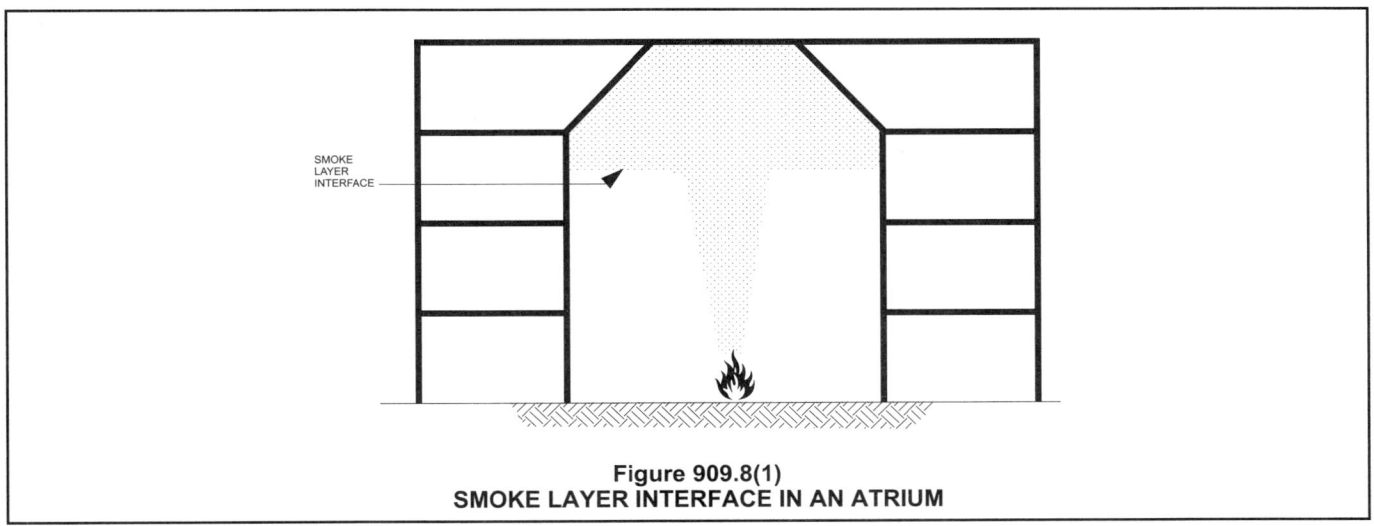

Figure 909.8(1)
SMOKE LAYER INTERFACE IN AN ATRIUM

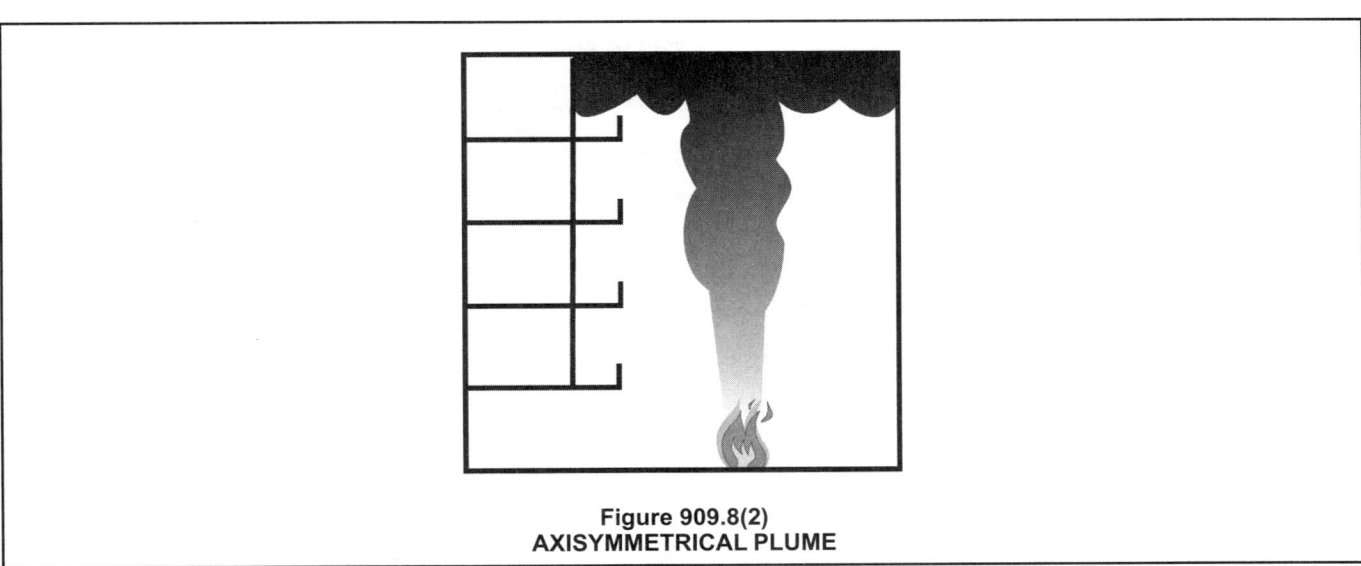

Figure 909.8(2)
AXISYMMETRICAL PLUME

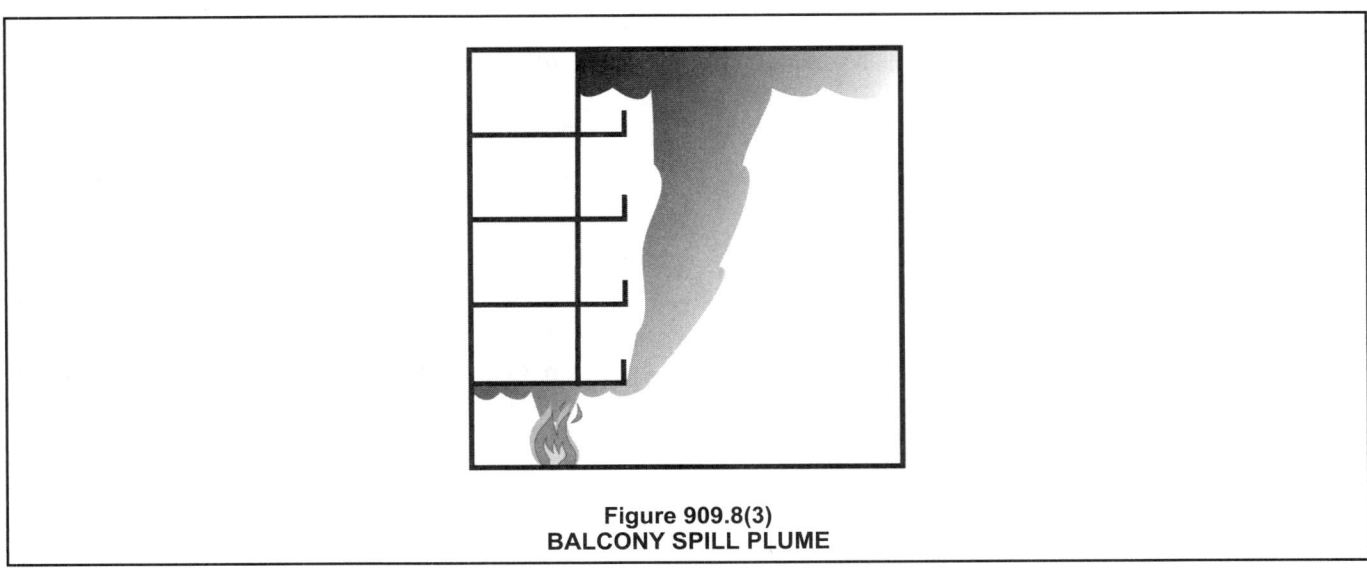

Figure 909.8(3)
BALCONY SPILL PLUME

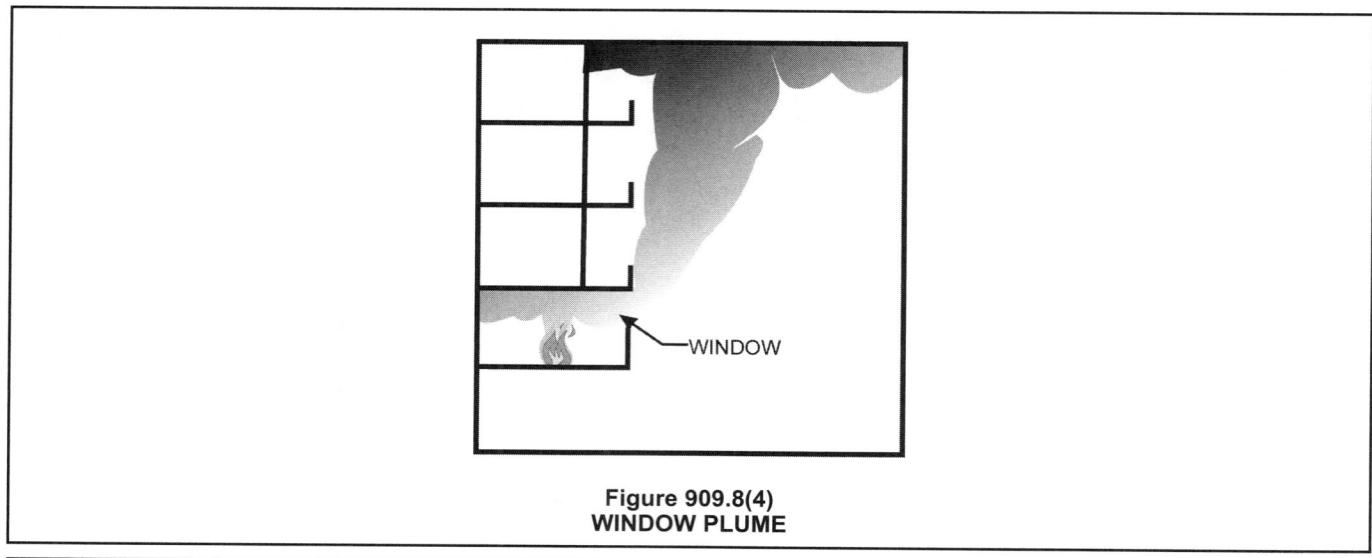

Figure 909.8(4)
WINDOW PLUME

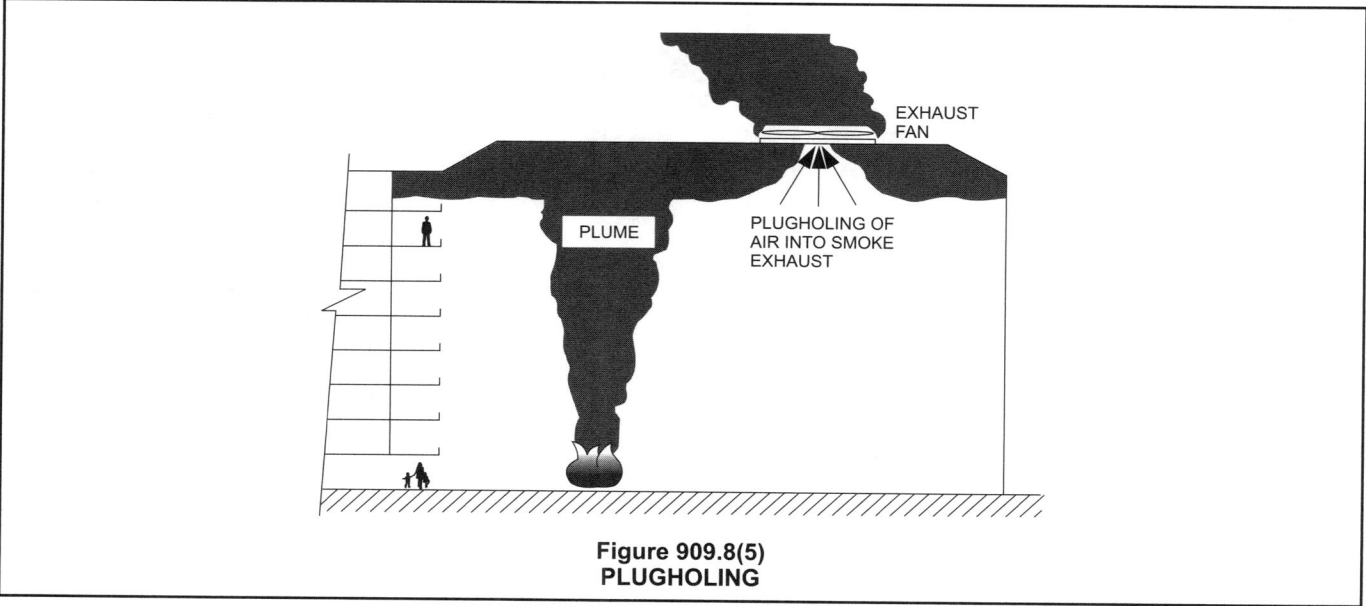

Figure 909.8(5)
PLUGHOLING

[F] 909.8.1 Smoke layer. The height of the lowest horizontal surface of the smoke layer interface shall be maintained at least 6 feet (1829 mm) above any walking surface that forms a portion of a required egress system within the smoke zone.

❖ The design criteria to be used when applying NFPA 92B is to maintain the smoke layer interface at least 6 feet (1829 mm) above any walking surface that is considered part of the required egress within the particular smoke zone, such as an atrium, for 20 minutes or 1.5 times the calculated egress time (see Section 909.4.6). Chapter 10 considers the majority of occupiable space as part of the means of egress system. Also keep in mind that the criteria of 6 feet (1829 mm) does not apply just to the main floor surface of the mall or atrium but to any level where occupants may be exposed (for example, balconies) [see Figure 909.8.1(1)].

The code uses the terminology "lowest horizontal surface of the accumulating smoke layer interface."

NFPA 92B has several definitions related to smoke layer, which include the following:

Smoke layer. The accumulated thickness of smoke below a physical barrier.

Smoke layer interface. The theoretical boundary between a smoke layer and the smoke-free air. (Note: This boundary is at the beginning of the transition zone.)

First indication of smoke. The boundary between the transition zone and the smoke-free air.

Transition zone. The layer between the smoke layer interface and the first indication of smoke in which the smoke layer temperature decreases to ambient.

The transition zone may be several feet thick (large open space) or may barely exist (small area with intense fire).

See also Figure 909.8.1(2).

NFPA 92B provides algebraic equations to determine the first indication of smoke but is limited to very specific conditions such as a uniform cross section, specific aspect ratios, steady or unsteady fires and no smoke exhaust operating. When using algebraic equations for smoke layer interface looking at different types of plumes, the smoke layer interface terminology is used and the user enters the desired smoke layer interface height. Zone models use simplifying assumptions so the layers are distinct from one another. In contrast, when CFD or scale modeling is used, the data must be analyzed to verify that the smoke layer interface is located at or above the 6 feet (1829 mm) during the event. This is not a simple analysis, as CFD and scale modeling provide more detail on actual smoke behavior; therefore, the location of the smoke layer interface may not be initially clear without some level of analysis. Again it depends on the depth of the transition layer. This may require reviewing tenability within the transition zone. Tenability limits need to be agreed upon by the stakeholders involved. Using CFD or scale modeling would likely need to occur through the alternative methods and materials section (see Section 104.11) due to the need to review tenability limits. It should be noted that NFPA 92B Annex A suggests that there are methods to determine where the smoke layer interface and first indication of smoke are located when undertaking CFD and scale modeling using a limited number of point measurements.

Also, Section 909.8.1 specifies a minimum distance for the smoke layer interface from any walking surface whereas Section 4.5.3 of NFPA 92B has provisions that simply allow the analysis to demonstrate tenability regardless of where the layer height is located above the floor. Defining tenability can be more difficult as there is not a standard definition. Any design using that approach would need to be addressed through Section 104.11.

Note that the response time of the system components (detection, activation, ramp up time, shutting down HVAC, opening/closing doors and dampers, etc.) needs to be accounted for when analyzing the location of the smoke layer interface in relation to the duration of operations stated in Section 909.4.6 (see commentary, Section 909.17).

[F] 909.9 Design fire. The design fire shall be based on a rational analysis performed by the *registered design professional* and *approved* by the fire code official. The design fire shall be based on the analysis in accordance with Section 909.4 and this section.

❖ The design fire is the most critical element in the smoke control system design. The fire is what produces the smoke to be controlled by the system; thus, the size of the fire directly impacts the quantity of smoke being produced. This section ensures that the design fire be determined through a rational analysis by a registered design professional with knowledge in this area. Such professionals should have experience in the area of fire dynamics, fire engineering and general building design, including mechanical systems. When determining the design fire the designer should work with various stakeholders to determine the types of hazards and combustible materials (fire scenarios) on a permanent as well as temporary basis (i.e.,

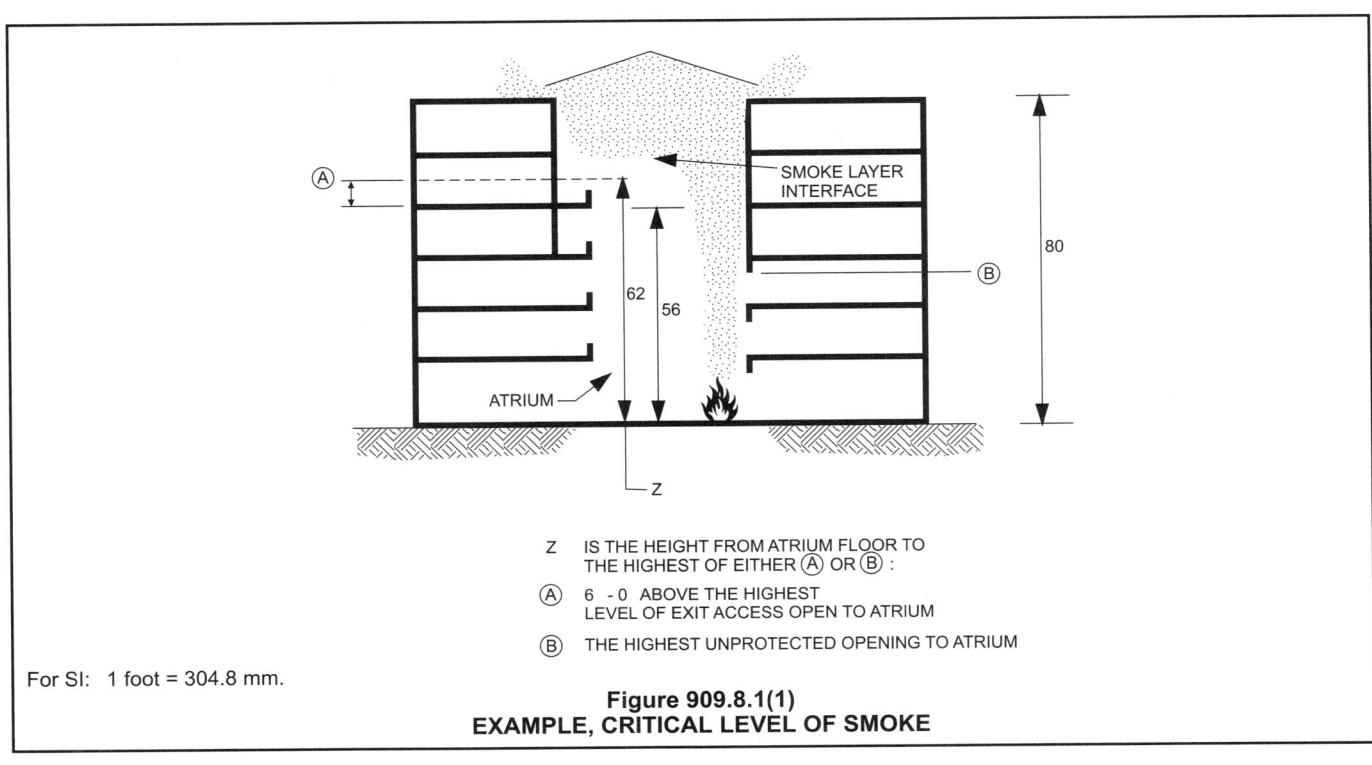

Z   IS THE HEIGHT FROM ATRIUM FLOOR TO
    THE HIGHEST OF EITHER Ⓐ OR Ⓑ :

Ⓐ   6 - 0   ABOVE THE HIGHEST
    LEVEL OF EXIT ACCESS OPEN TO ATRIUM

Ⓑ   THE HIGHEST UNPROTECTED OPENING TO ATRIUM

For SI: 1 foot = 304.8 mm.

Figure 909.8.1(1)
EXAMPLE, CRITICAL LEVEL OF SMOKE

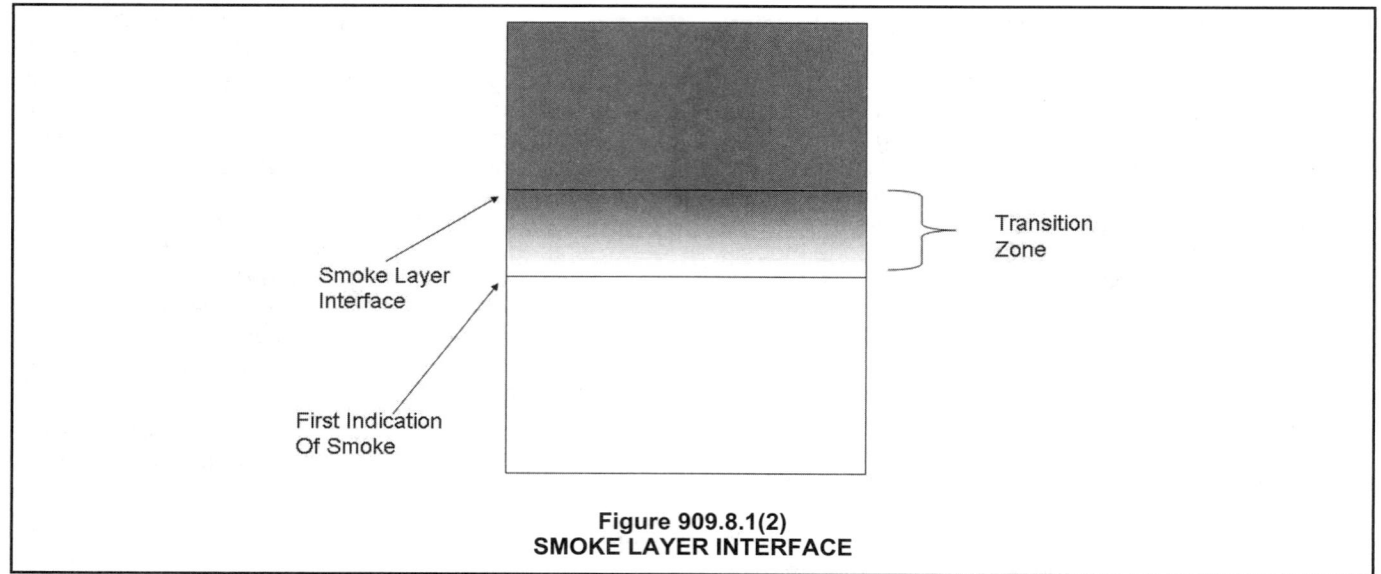

Figure 909.8.1(2)
SMOKE LAYER INTERFACE

Christmas/holiday decorative materials or scenery, temporary art exhibits) that may be present throughout the use of the building once occupied. Those hazards then need to be translated to potential design fires to be used when determining the smoke layer interface height for the duration as determined by Section 909.4.6. See the commentary for Section 909.9.3 for potential sources when determining design fires.

This section also does not mandate the type of fire (i.e., steady versus unsteady). A steady fire assumes a constant heat release rate over a period of time, where unsteady fires do not. An unsteady fire includes the growth and decay phases of the fire, as well as the peak heat release rate. An unsteady fire will hit a peak heat release rate when burning in the open, like an axisymmetric fire. An unsteady fire is a more realistic view of how fires actually burn. It should be noted that fires can be a combination of unsteady and steady fires when the sufficient fuel is available. In other words, the fire initially grows (unsteady) then reaches a steady state and burns for sometime at a particular heat release rate before decay occurs.

Design fire information should therefore typically include growth rate, peak heat release rate, duration and decay as well as information related to fire locations and products of combustion yield (CO, smoke, etc.) that are produced by the various design fires that are deemed credible for the space.

To provide an order of magnitude of fire sizes obtained from various combustibles, the following data from fire tests is provided. The following heat release rates, found in Section 3, Chapter 3-1 of the 2nd edition of the *SFPE Handbook of Fire Protection Engineering*, are peak heat release rates:

Plastic trash bags/paper trash:
114-332 Btu/sec (120-350 kw)

Latex foam pillow:
114 Btu/sec approximately (120 kw)

Dry Christmas tree:
475-618 Btu/sec (500-650 kw)

Sofa:
2,852 Btu/sec approximately (3,000 kw)

Plywood wardrobe:
2,947-6,084 Btu/sec (3,100-6,400 kw)

[F] 909.9.1 Factors considered. The engineering analysis shall include the characteristics of the fuel, fuel load, effects included by the fire and whether the fire is likely to be steady or unsteady.

❖ This section simply provides more detail on the factors that should be taken under consideration when determining the design fire size. To determine the appropriate fire size, an engineering analysis is necessary that takes into account the following elements: fuel (potential burning rates), fuel load (how much), effects included by the fire (smoke particulate size and density), steady or unsteady (burn steadily or simply peak and dissipate) and likelihood of sprinkler activation (based on height and distance from the fire).

[F] 909.9.2 Separation distance. Determination of the design fire shall include consideration of the type of fuel, fuel spacing and configuration.

❖ The design fire size may also be affected by surrounding combustibles, which may have the effect of increasing the fire size. More specifically, there is concern that if sufficient separation is not maintained between combustibles, then a larger design fire is likely. The code does not provide extensive detail on this as such determination is left to the rational analysis undertaken by the design professional. NFPA 92B provides one method in which you determine the critical separation distance, R. This is based upon fire size and the critical radiant heat flux for nonpiloted ignition. Nonpiloted ignition means the radiated heat from the fire without direct flame contact will ignite adjacent combustibles.

[F] 909.9.3 Heat-release assumptions. The analysis shall make use of best available data from *approved* sources and shall not be based on excessively stringent limitations of combustible material.

❖ This section is merely stressing the fact that data obtained for use in a rational analysis needs to come from relevant and appropriate sources. Data can be obtained from groups such as NIST or from Annex B of NFPA 92B. Data from fire tests is available and is a good resource for such analysis. As noted earlier, such data is not prevalent (see also Chapter 8, Analysis of Design Fires of *A Guide to Smoke Control in the 2006 IBC®* and Section 3, Chapter 3-1, of the *SFPE Handbook of Fire Protection Engineering*).

[F] 909.9.4 Sprinkler effectiveness assumptions. A documented engineering analysis shall be provided for conditions that assume fire growth is halted at the time of sprinkler activation.

❖ This section raises a few questions regarding activation of sprinklers and their impact on the fire both in terms of their ability to "control" as well as "extinguish" a fire. The first is concerning an assumption that sprinklers will immediately control the fire as soon as they are activated (i.e., control results in limiting further growth and maintaining the heat release rate at approximately the same fire size as when the sprinklers activated). This assumption may be true in some cases, but for high ceilings the sprinkler may not activate or may be ineffective. Sprinklers may be ineffective in high spaces, since by the time they are activated the fire is too large to control. Essentially, the fire plume may push away and evaporate the water before it actually reaches the seat of the fire. In addition the fire may be shielded from sprinkler spray so that insufficient quantities of water reach the fuel. These are common problems with high-piled storage as well as other fires including retail and has been shown in actual tests. Also, if the fire becomes too large before the sprinklers are activated, the available water supply and pressure for the system may be compromised. Additionally, based on the layout of the room and the movement of the fire effluents, the wrong sprinklers could be activated, which leads to a larger fire size and depletion of the available water supply and pressure.

Another issue is whether the sprinklers "control" or "extinguish" the fire. Typical sprinklers are assumed only to control fires as opposed to extinguishing them. Sprinklers may be able to extinguish the fire but it should not automatically be assumed. A fire that is controlled will achieve steady state and maintain a certain fire size, which is very different from a fire that is actually extinguished.

Based upon these concerns, each scenario needs to be looked at individually to determine whether sprinklers would be effective in halting the growth or extinguishing the fire. More specifically, the evaluation

should include droplet size, density and area of coverage and should also be based on actual test results.

[F] 909.10 Equipment. Equipment including, but not limited to, fans, ducts, automatic *dampers* and balance *dampers*, shall be suitable for its intended use, suitable for the probable exposure temperatures that the rational analysis indicates and as *approved* by the fire code official.

❖ Section 909.10 and subsequent sections are primarily related to the reliability of the system components to provide a smoke control system that works according to the design. One of the largest concerns when using smoke control provisions is the overall reliability of the system. Such systems have many different components, such as smoke and fire dampers; fans; ducts and controls associated with such components. The more components a system has, the less reliable it becomes. In fact, one approach in providing a higher level of reliability is utilizing the normal building systems such as the HVAC to provide the smoke control system. Basically, systems used every day are more likely to be working appropriately, since they are essentially being tested daily; however, there are many components that are specific to the smoke control system, such as exhaust fans in an atrium or the smoke control panel.

Also, there is not a generic prescriptive set of requirements as to how all smoke control system elements should operate, since each design may be fairly unique. The specifics on operation of such a system need to be included within the design and construction documents. Most components used in smoke control systems are elements used in many other applications such as HVAC systems; therefore, the basic mechanisms of a fan used in a smoke control system may not be different, although they may be applied differently.

[F] 909.10.1 Exhaust fans. Components of exhaust fans shall be rated and certified by the manufacturer for the probable temperature rise to which the components will be exposed. This temperature rise shall be computed by:

$$T_s = (Q_c/mc) + (T_a) \qquad \textbf{(Equation 9-3)}$$

where:

c = Specific heat of smoke at smoke layer temperature, Btu/lb°F (kJ/kg · K).

m = Exhaust rate, pounds per second (kg/s).

Q_c = Convective heat output of fire, Btu/s (kW).

T_a = Ambient temperature, °F (K).

T_s = Smoke temperature, °F (K).

Exception: Reduced T_s as calculated based on the assurance of adequate dilution air.

❖ Fans used for smoke control systems must be able to tolerate the possible elevated temperatures to which they will be exposed. Again, like many other factors this depends upon the specifics of the design fire. Essentially, Equation 9-3 requires the calculation of the potential temperature rise. The exhaust fans must be specifically rated and certified by the manufacturer to be able to handle these rises in temperature. There is an exception that allows reduction of the temperature if it can be shown that adequate temperature reduction will occur. In many cases if the exhaust fans are near the ceiling, the smoke will be much cooler than the value resulting from Equation 9-3 since the smoke may cool considerably by the time it reaches the ceiling. Also, sprinkler activation will assist in cooling the smoke further.

[F] 909.10.2 Ducts. Duct materials and joints shall be capable of withstanding the probable temperatures and pressures to which they are exposed as determined in accordance with Section 909.10.1. Ducts shall be constructed and supported in accordance with the *International Mechanical Code*. Ducts shall be leak tested to 1.5 times the maximum design pressure in accordance with nationally accepted practices. Measured leakage shall not exceed 5 percent of design flow. Results of such testing shall be a part of the documentation procedure. Ducts shall be supported directly from fire-resistance-rated structural elements of the building by substantial, noncombustible supports.

> **Exception:** Flexible connections (for the purpose of vibration isolation) complying with the *International Mechanical Code*, that are constructed of *approved* fire-resistance-rated materials.

❖ The next essential component of a smoke control system is the integrity of the ducts to transport supply and exhaust air. The integrity of ducts is also important for an HVAC system, but is more critical in this case since it is not simply a comfort issue but one of life safety. The key concern with ducts in smoke control systems is that they can withstand elevated temperatures and that there will be minimal leakage. The concern with leakage is the potential of leaking smoke into another smoke zone or not providing the proper amount of supply air to support the system.

More specifically, all ducts need to be leak tested to 1.5 times the maximum static design pressure. The leakage resulting should be no more than 5 percent of the design flow. For example, a duct that has a design flow of 300 cubic feet per minute (cfm) (0.141 m³/s) would be allowed 15 cfm (0.007 m³/s) of leakage when exposed to a pressure equal to 1.5 times the design pressure for that duct. The tests should be in accordance with nationally accepted practices. This criterion will often limit ductwork for smoke control systems to lined systems, since the amount of leakage in such systems is much less.

As part of the concern for possible exposure to fire and fire products, the ducts are required to be supported by way of substantial noncombustible supports connected to the fire-resistance-rated structural elements of the building. As noted, the system needs to able to run for 20 minutes starting from the detection of the fire. The exception to this section is really more of an acknowledgement that flexible connections for vibration isolation are acceptable when constructed of approved fire-resistance-rated materials. More specif-

ically, it is often necessary to use such connections for connecting the duct to the fan. These connections cannot necessarily meet the requirements of the main section, but are a minimal part of the ductwork and as long as they perform adequately with regard to fire resistance they are permitted. Note that the term "approved" is used to determine the required fire resistance, therefore, flexibility is provided. The code does not specifically address this determination but perhaps a relationship to the duration or operation and these flexible connections could be made to determine the necessary performance.

[F] 909.10.3 Equipment, inlets and outlets. Equipment shall be located so as to not expose uninvolved portions of the building to an additional fire hazard. Outside air inlets shall be located so as to minimize the potential for introducing smoke or flame into the building. Exhaust outlets shall be so located as to minimize reintroduction of smoke into the building and to limit exposure of the building or adjacent buildings to an additional fire hazard.

❖ The intent of this section is to minimize the likelihood of smoke being reintroduced into the building due to poorly placed outdoor air inlets and exhaust air outlets; therefore, placing one right next to another on the exterior of the building would be inappropriate. In addition, wind and other adverse conditions should be considered when choosing locations for these inlets and outlets. Particular attention should be paid to introducing exhausted smoke into another smoke zone. Also, smoke should be exhausted in a direction that will not introduce it into surrounding buildings or facilities. Within the building itself, the supply air and exhaust outlets should also be strategically located. The exhaust inlets and supply air should be evenly distributed to reduce the likelihood of a high velocity of air that may disrupt the fire plume and also push smoke back into occupied areas. See the commentary to Section 909.8 for discussion on avoiding plugholing.

[F] 909.10.4 Automatic dampers. Automatic *dampers*, regardless of the purpose for which they are installed within the smoke control system, shall be *listed* and conform to the requirements of *approved*, recognized standards.

❖ This section addresses the reliability of any dampers used within a smoke control system. This particular provision requires that the dampers be listed and conform to the appropriate recognized standards. More specifically, Section 716 contains more detailed information on the specific requirements for smoke and fire dampers. Smoke and fire dampers should be listed in accordance with UL 555S and 555, respectively. Also, remember that each smoke control design is unique and the sequence and methods used to activate the dampers may vary from design to design. This information needs to be addressed in the construction documents.

Another factor to take into account, with regard to timing of the system, is the fact that some dampers react more quickly than others, simply due to the particular smoke damper characteristics. Additionally, during the commissioning of the system, the damper is going to be exposed to many repetitions. These repetitions need to be accounted for in the overall reliability of the system.

[F] 909.10.5 Fans. In addition to other requirements, belt-driven fans shall have 1.5 times the number of belts required for the design duty, with the minimum number of belts being two. Fans shall be selected for stable performance based on normal temperature and, where applicable, elevated temperature. Calculations and manufacturer's fan curves shall be part of the documentation procedures. Fans shall be supported and restrained by noncombustible devices in accordance with the requirements of Chapter 16. Motors driving fans shall not be operated beyond their nameplate horsepower (kilowatts), as determined from measurement of actual current draw, and shall have a minimum service factor of 1.15.

❖ Part of the overall reliability requires that fans used to provide supply air and exhaust capacity will be functioning when necessary; therefore, a safety factor of 1.5 is placed upon the required belts for fans. All fans used as part of a smoke control system must provide 1.5 times the number of required belts with a minimum of two belts for all fans.

This section also points out that the fan chosen should fit the specific application. It should be able to withstand the temperature rise as calculated in Section 909.10.1 and generally be able to handle typical exposure conditions, such as location and wind. For instance, propeller fans are highly sensitive to the effects of wind. When located on the windward side of a building, wall-mounted, nonhooded propeller fans are not able to compensate for wind effects. Additionally, even hooded propeller fans located on the leeward side of the building may not adequately compensate for the decrease in pressure caused by wind effects. In general, when designing a system, it should be remembered that field conditions might vary from the calculations; therefore, flexibility should be built into the design that would account for things such as variations in wind conditions.

Finally, this section stresses that fan motors not be operated beyond their rated horsepower.

[F] 909.11 Power systems. The smoke control system shall be supplied with two sources of power. Primary power shall be from the normal building power systems. Secondary power shall be from an *approved* standby source complying with Chapter 27 of this code. The standby power source and its transfer switches shall be in a room separate from the normal power transformers and switch gears and ventilated directly to and from the exterior. The room shall be enclosed with not less than 1-hour *fire barriers* constructed in accordance with Section 707 or horizontal assemblies constructed in accordance with Section 712, or both.

❖ As with any life safety system, a level of redundancy with regard to power supply is required to enable the functioning of the system during a fire. The primary source is the building's normal power system. The secondary power system is by means of standby power. One of the key elements is that standby power

systems are intended to operate within 60 seconds of loss of primary power. It should be noted that the primary difference between standby power and emergency power is that emergency power must operate within 10 seconds of loss of primary power versus 60 seconds. This section also requires isolation from normal building power systems via a 1-hour fire barrier, 1-hour horizontal assembly, or both, depending upon the location within the building. This increases the reliability and reduces the likelihood that a single event could remove both power supplies.

[F] 909.11.1 Power sources and power surges. Elements of the smoke management system relying on volatile memories or the like shall be supplied with uninterruptable power sources of sufficient duration to span a 15-minute primary power interruption. Elements of the smoke management system susceptible to power surges shall be suitably protected by conditioners, suppressors or other *approved* means.

❖ Smoke management systems have many components, sometimes highly sensitive electronics, that are adversely affected by any interruption in or sudden surges of power. Therefore, Section 909.11.1 requires that any components of a smoke control system, such as volatile memories, be supplied with an uninterruptible power system for the first 15 minutes of loss of primary power. Volatile memory components will lose memory upon any loss of power no matter how short the time period. Once the 15 minutes elapses, these elements can be transitioned to the already operating standby power supply.

With regard to components sensitive to power surges, they need to be provided with surge protection in the form of conditioners, suppressors or other approved means.

[F] 909.12 Detection and control systems. Fire detection systems providing control input or output signals to mechanical smoke control systems or elements thereof shall comply with the requirements of Section 907. Such systems shall be equipped with a control unit complying with UL 864 and *listed* as smoke control equipment.

Control systems for mechanical smoke control systems shall include provisions for verification. Verification shall include positive confirmation of actuation, testing, manual override, the presence of power downstream of all disconnects and, through a preprogrammed weekly test sequence, report abnormal conditions audibly, visually and by printed report.

❖ This section is focused upon two main elements. The first is the proper operation and monitoring of the fire detection system that activates the smoke control system through compliance with Section 907 and UL 864. This requires a specific listing as smoke control equipment. UL 864 has a subcategory (UUKL) specific to fire alarm control panels for smoke control system applications.

The second aspect is related to the mechanical elements of the smoke control system once the system is activated. In particular, there is a focus upon verification of activities. Verification would include the following two aspects according to the second paragraph of this section:

1. The system is able to verify actuations, testing, manual overrides and the presence of power downstream. This would require information reported back to the smoke control panel, which can be accomplished via the weekly test sequence or through full electronic monitoring of the system.

2. Conduct a preprogrammed weekly test that simulates an actual (smoke) event to test the components of the system. These components would include elements such as smoke dampers, fans and doors. Abnormal conditions need to be reported in three ways:

   a. Audibly;

   b. Visually; and

   c. Printed report.

It should be noted that electrical monitoring of the control components is not required (supervision). Such supervision verifies integrity of the conductors from a fire alarm control unit to the control system input. The weekly test is considered sufficient verification of system performance and is often termed end-to-end verification. In other words, the control system input provides the expected results. Verification can be accomplished through any sensor that is calibrated to distinguish between the difference between proper operation and a fault condition. For fans, proper operation means that the fan is moving air within the intent of its design. Fault conditions include power failure, broken fan belts, adverse wind effects, a locked rotor condition and/or filters or large ducts that are blocked causing significantly reduced airflow. In addition to differential pressure transmitters and sail switches, this can be accomplished by the present state of the art current sensors. More discussion on verification for elements such as ducts and fire doors is discussed in Chapter 9 of *A Guide to Smoke Control in the 2006 IBC®*.

Also, the fact that a smoke control system is nondedicated (integrated with an HVAC system) does not mean that it is automatically being tested on a daily basis. It is cautioned that simply depending upon occupant discomfort, for example, is sometimes an insufficient indicator of a fully functioning smoke control system. There may be various modes in which the HVAC system could operate that may not exercise the smoke control features and the sequence in which the system should operate. An example is an air-conditioning system operating only in full recirculating mode versus exhaust mode. This failure will likely not affect occupants and will not exercise the exhaust function. Plus, doors, which may be part of the smoke barrier, may not need to be closed in normal building operations but would need to be closed during smoke control system operation. This is why this section does not necessarily differentiate between dedicated

and nondedicated smoke control systems and requires the system components to be tested.

It is important to note that this weekly test sequence is not an actual smoke event and is only intended to activate the system to ensure that the components are working correctly.

Although NFPA 92B is only referenced for design Sections 7.3.1 and 7.3.8 of that standard coordinate with this section. More specifically, Section 7.3.8 also requires the weekly test but as provided for by the UL 864UUKL-listed smoke control panel. NFPA 92B Section 7.3.6.2 requires off-normal indication at the smoke control panel within 200 seconds when a positive confirmation is failed to be achieved. Section 909.12 only requires that abnormal conditions be reported weekly.

[F] 909.12.1 Wiring. In addition to meeting requirements of NFPA 70, all wiring, regardless of voltage, shall be fully enclosed within continuous raceways.

❖ Wiring is required to be placed within continuous raceways, which provides an additional level of reliability for the system. The definition of the term "raceway" in NFPA 70 lists several acceptable types of complying raceway that can be used; however, manufactured cable assemblies such as metal-clad cable (Type MC) or armored cable (Type AC) are not included.

[F] 909.12.2 Activation. Smoke control systems shall be activated in accordance with this section.

❖ The activation of a smoke control system is dependent on when such a system is required. Mechanical smoke control systems, which could include pressurization, airflow or exhaust methods, require an automatic activation mechanism. When using a passive system, which depends upon compartmentation, spot-type detectors are acceptable for the release of door closers and similar openings. Whereas with more complex mechanical systems, such activation needs to go beyond single station detectors and be part of an automatic coordinated system.

[F] 909.12.2.1 Pressurization, airflow or exhaust method. Mechanical smoke control systems using the pressurization, airflow or exhaust method shall have completely automatic control.

❖ Automatic activation of such systems is especially critical as tenability is much more difficult to achieve if a delay occurred waiting for manual activation of the system. See Sections 909.6 for the pressurization method, 909.7 for the airflow design method and 909.8 for the exhaust method.

[F] 909.12.2.2 Passive method. Passive smoke control systems actuated by *approved* spot-type detectors *listed* for releasing service shall be permitted.

❖ This section recognizes that a passive system does not address smoke containment through mechanical means; therefore, it does not need to be "automatically activated" except in cases where smoke barriers have openings. These openings would be required to have smoke detectors to close openings where required by the design. Although spot-type detectors are technically automatic, they are not part of a more coordinated system of activation as needed for mechanical smoke control systems. Such detectors are simply standalone devices that fail in the fail safe position. In other words, if the power were lost a door on a magnetic hold would simply close.

[F] 909.12.3 Automatic control. Where completely automatic control is required or used, the automatic-control sequences shall be initiated from an appropriately zoned *automatic sprinkler system* complying with Section 903.3.1.1, manual controls that are readily accessible to the fire department and any smoke detectors required by engineering analysis.

❖ When automatic activation is required, it must be accomplished by a properly zoned automatic sprinkler system and, if the engineering analysis requires them, smoke detectors. Manual control for the fire department needs to be provided. An important point with this particular requirement is that smoke control systems are engineered systems and a prescribed smoke detection system may not fit the needs of the specific design. Other types of detectors, such as beam detectors (within an atrium), may be used and could be more useful and more practical from a maintenance standpoint. Also, it may not be practical or appropriate for the building's fire alarm system to activate such systems, as it may alter the effectiveness of the system by pulling smoke through the building versus removing or containing the smoke. For example, a building with an atrium may have several floors below the space. If a fire occurs in one of the floors not associated with the atrium, the atrium smoke control system could possibly pull smoke throughout the building if the detection is zoned incorrectly.

[F] 909.13 Control air tubing. Control air tubing shall be of sufficient size to meet the required response times. Tubing shall be flushed clean and dry prior to final connections and shall be adequately supported and protected from damage. Tubing passing through concrete or masonry shall be sleeved and protected from abrasion and electrolytic action.

❖ Control tubing is a method that uses pneumatics to operate components such as the opening and closing of dampers. Due to the sophistication of electronic systems today, control tubing is becoming less common.

These particular requirements provide the criteria for properly designing and installing control tubing. Essentially, it is up to the design professional to determine the size requirements and to properly design appropriate supports. This information needs to be detailed within the construction documents. Additionally, due to the effect of moisture and other contaminants on control tubing, it must be flushed clean then dried before installation.

[F] 909.13.1 Materials. Control-air tubing shall be hard-drawn copper, Type L, ACR in accordance with ASTM B 42,

ASTM B 43, ASTM B 68, ASTM B 88, ASTM B 251 and ASTM B 280. Fittings shall be wrought copper or brass, solder type in accordance with ASME B 16.18 or ASME B 16.22. Changes in direction shall be made with appropriate tool bends. Brass compression-type fittings shall be used at final connection to devices; other joints shall be brazed using a BCuP5 brazing alloy with solidus above 1,100°F (593°C) and liquids below 1,500°F (816°C). Brazing flux shall be used on copper-to-brass joints only.

Exception: Nonmetallic tubing used within control panels and at the final connection to devices provided all of the following conditions are met:

1. Tubing shall be *listed* by an *approved* agency for flame and smoke characteristics.

2. Tubing and connected devices shall be completely enclosed within a galvanized or paint-grade steel enclosure having a minimum thickness of 0.0296 inch (0.7534 mm) (No. 22 gage). Entry to the enclosure shall be by copper tubing with a protective grommet of neoprene or teflon or by suitable brass compression to male barbed adapter.

3. Tubing shall be identified by appropriately documented coding.

4. Tubing shall be neatly tied and supported within the enclosure. Tubing bridging cabinets and doors or moveable devices shall be of sufficient length to avoid tension and excessive stress. Tubing shall be protected against abrasion. Tubing serving devices on doors shall be fastened along hinges.

❖ This section addresses the materials allowed for control air tubing along with approved methods of connection. All of this information needs to be documented, as it will be subject to review by the special inspector.

[F] 909.13.2 Isolation from other functions. Control tubing serving other than smoke control functions shall be isolated by automatic isolation valves or shall be an independent system.

❖ This section requires separation of control tubing used for other functions through the use of isolation valves or a completely separate system. This is due to the difference in requirements for control tubing used in a smoke control system versus other building systems. The isolation of the control air tubing for a smoke control system needs to be specifically noted on the construction documents.

[F] 909.13.3 Testing. Control air tubing shall be tested at three times the operating pressure for not less than 30 minutes without any noticeable loss in gauge pressure prior to final connection to devices.

❖ As part of the acceptance testing of the smoke control system, the control air tubing will be pressure tested three times the operating pressure for 30 minutes or more. The performance criteria as to whether the control tubing is considered a failure is when there is any noticeable loss in gauge pressure prior to final connection of devices during the 30-minute duration test.

[F] 909.14 Marking and identification. The detection and control systems shall be clearly marked at all junctions, accesses and terminations.

❖ This section requires that all portions of the fire detection system that activate the smoke control system be marked and identified appropriately. This includes all applicable fire alarm-initiating devices, the respective junction boxes, all data-gathering panels and fire alarm control panels. Additionally, all components of the smoke control system, which are not considered a fire detection system, are required to be properly identified and marked. This would include all applicable junction boxes, control tubing, temperature control modules, relays, damper sensors, automatic door sensors and air movement sensors.

[F] 909.15 Control diagrams. Identical control diagrams showing all devices in the system and identifying their location and function shall be maintained current and kept on file with the fire code official, the fire department and in the fire command center in a format and manner *approved* by the fire chief.

❖ The purpose of control diagrams is to provide consistent information on the system in several key locations, including the building department, the fire department and the fire command center. If a fire command center is not required or provided, the diagrams need to be located such that they can be readily accessed during an emergency. Some possible locations may be the security office, the building manager's office or, if possible, within the smoke control panel. This information is intended to assist in the use and operation of the smoke control system. The format of the control diagram is as approved by the fire chief. This is necessary since the fire department is the agency that will be using such a system during a fire and when the system is tested in the future. The more clearly the information is communicated, the more effective the smoke control system will be.

It should be noted that the fire department may want all smoke control systems within a jurisdiction to follow a particular protocol for control diagrams. Generally, the control diagrams should indicate the required reaction of the system in all scenarios. The status or position of every fan and damper in every scenario must be clearly identified.

[F] 909.16 Fire-fighter's smoke control panel. A fire-fighter's smoke control panel for fire department emergency response purposes only shall be provided and shall include manual control or override of automatic control for mechanical smoke control systems. The panel shall be located in a fire command center complying with Section 911 in high- rise buildings or buildings with smoke-protected assembly seating. In all other buildings, the fire-fighter's smoke control panel shall be installed in an *approved* location adjacent to the fire alarm control panel. The fire-fighter's smoke control panel shall comply with Sections 909.16.1 through 909.16.3.

❖ One of the elements that makes a smoke control system effective is that its activity is successfully communicated to the fire department and the fire department is able to manually operate the system. The following sections provide requirements for a control panel specifically for smoke control systems. This panel is required to be located within a fire command center when it is located in a high-rise building or there is smoke-protected seating. Section 403.4.5 would require a fire command center for high-rise buildings. Smoke-protected seating does not require a fire command center in Chapter 10 but this provision would ensure that one exists and contains the smoke control panel. Facilities with smoke-protected seating tend to be larger facilities that, at the very least, would already have a central security center if not a fire command center as required by the jurisdiction. All other locations would only need to provide the panel in an approved location as long as it is located with the fire alarm panel. The specific location will depend on the needs of the fire department in that jurisdiction. The reason not all fire-fighter smoke control panels need to be located in a fire command center is that many smoke control systems are located in a building containing an atrium that may only be three stories in height. A 200 square foot fire command center would be excessive for such buildings. There are two components that include the requirements for the display and for the controls. This control panel will provide an ability to override any other controls whether manual or automatic within the building as they relate to the smoke control system.

Note that the publication *Guide to Smoke Control in the 2006 IBC* (Klote and Evans 2007) goes into more detail about the fire fighter smoke control panel requirements.

[F] 909.16.1 Smoke control systems. Fans within the building shall be shown on the fire-fighter's control panel. A clear indication of the direction of airflow and the relationship of components shall be displayed. Status indicators shall be provided for all smoke control equipment, annunciated by fan and zone, and by pilot-lamp-type indicators as follows:

1. Fans, *dampers* and other operating equipment in their normal status—WHITE.

2. Fans, *dampers* and other operating equipment in their off or closed status—RED.

3. Fans, *dampers* and other operating equipment in their on or open status—GREEN.

4. Fans, *dampers* and other operating equipment in a fault status—YELLOW/AMBER.

❖ This section denotes what should be displayed on the control panel. The display is required to include all fans, an indication of the direction of airflow and the relationship of the components. Also, status lights are required, and this section sets out specific standardized colors to indicate normal status, closed status, open status and fault status. A standardized approach increases the likelihood that the fire department will be able to quickly become familiar with a system. Since the fire department has the ability to override the automatic functions of the system, this information is critical.

[F] 909.16.2 Smoke control panel. The fire-fighter's control panel shall provide control capability over the complete smoke-control system equipment within the building as follows:

1. ON-AUTO-OFF control over each individual piece of operating smoke control equipment that can also be controlled from other sources within the building. This includes *stairway* pressurization fans; smoke exhaust fans; supply, return and exhaust fans; elevator shaft fans and other operating equipment used or intended for smoke control purposes.

2. OPEN-AUTO-CLOSE control over individual *dampers* relating to smoke control and that are also controlled from other sources within the building.

3. ON-OFF or OPEN-CLOSE control over smoke control and other critical equipment associated with a fire or smoke emergency and that can only be controlled from the fire-fighter's control panel.

> **Exceptions:**
>
> 1. Complex systems, where *approved*, where the controls and indicators are combined to control and indicate all elements of a single smoke zone as a unit.
>
> 2. Complex systems, where *approved*, where the control is accomplished by computer interface using *approved*, plain English commands.

❖ This section sets the requirements as to which controls need to be provided for the fire department on the control panel.

There are two aspects to the controls. The controls will include on-auto-off and open-auto-close settings or will be strictly on-off or open-close. If the system or component can be set on automatic (auto), it can be controlled from other locations beyond the fire command center. This would include an automatic smoke detection system or by manual activation. If a control only contains on-off or open-close settings, the only way the system component can be controlled is in the fire command center.

It should be noted that components such as fans are usually associated with on-off-type controls, whereas components such as dampers are associated with open-close-type controls.

[F] 909.16.3 Control action and priorities. The fire-fighter's control panel actions shall be as follows:

1. ON-OFF and OPEN-CLOSE control actions shall have the highest priority of any control point within the building. Once issued from the fire-fighter's control panel, no automatic or manual control from any other control point within the building shall contradict the control action. Where automatic means are provided to interrupt normal, nonemergency equipment operation or produce a specific result to safeguard the building or equipment (i.e., duct freezestats, duct smoke detectors, high-temperature cutouts, temperature-actuated linkage and similar devices), such means shall be capable of being overridden by the fire-fighter's control panel. The last

control action as indicated by each fire-fighter's control panel switch position shall prevail. In no case shall control actions require the smoke control system to assume more than one configuration at any one time.

> **Exception:** Power disconnects required by NFPA 70.

2. Only the AUTO position of each three-position fire-fighter's control panel switch shall allow automatic or manual control action from other control points within the building. The AUTO position shall be the NORMAL, nonemergency, building control position. Where a fire-fighter's control panel is in the AUTO position, the actual status of the device (on, off, open, closed) shall continue to be indicated by the status indicator described above. When directed by an automatic signal to assume an emergency condition, the NORMAL position shall become the emergency condition for that device or group of devices within the zone. In no case shall control actions require the smoke control system to assume more than one configuration at any one time.

❖ This section clarifies that when a component of the system is placed in an on-off or open-close configuration, no other control point in the building, whether automatic or manual, can override the action established in the fire command center. If a system component is configured in auto mode, it can be controlled from locations within the building beyond the fire command center. Some controls are specifically designed to only allow an action from the fire command center.

[F] 909.17 System response time. Smoke-control system activation shall be initiated immediately after receipt of an appropriate automatic or manual activation command. Smoke control systems shall activate individual components (such as *dampers* and fans) in the sequence necessary to prevent physical damage to the fans, *dampers*, ducts and other equipment. For purposes of smoke control, the fire-fighter's control panel response time shall be the same for automatic or manual smoke control action initiated from any other building control point. The total response time, including that necessary for detection, shutdown of operating equipment and smoke control system startup, shall allow for full operational mode to be achieved before the conditions in the space exceed the design smoke condition. The system response time for each component and their sequential relationships shall be detailed in the required rational analysis and verification of their installed condition reported in the required final report.

❖ This particular section provides the criteria as to when the smoke control system is required to begin operation. Whether or not the activation is manual or automatic, this criteria clarifies that the system be initiated immediately. Also, it requires that components activate in a sequence that will not potentially damage the fans, dampers, ducts and other equipment. Unrealistic timing of the system has the potential of creating an unsuccessful system. Delays in the system can be seen in slow dampers, fans that ramp up or down, systems that poll slowly and intentional built-in delays. These factors can add significantly to the reaction time of the system and may hamper achieving the design goals.

The key element is that the system be fully operational before the smoke conditions exceed the design parameters. The design should include these possible delays when analyzing the smoke layer interface location. The sequence of events need to be justified within the design analysis and described clearly in the construction documents.

[F] 909.18 Acceptance testing. Devices, equipment, components and sequences shall be individually tested. These tests, in addition to those required by other provisions of this code, shall consist of determination of function, sequence and, where applicable, capacity of their installed condition.

❖ In order to achieve a certain level of performance, the smoke control system needs to be thoroughly tested. Section 909.18 requires that all devices, equipment components and sequences be individually tested.

[F] 909.18.1 Detection devices. Smoke or fire detectors that are a part of a smoke control system shall be tested in accordance with Chapter 9 in their installed condition. When applicable, this testing shall include verification of airflow in both minimum and maximum conditions.

❖ Detection devices are required to be tested in accordance with the fire protection requirements found in Chapter 9. Also, since such detectors may be subject to higher air velocities than typical detectors, their operation needs to be verified in the minimum and maximum anticipated airflow conditions.

[F] 909.18.2 Ducts. Ducts that are part of a smoke control system shall be traversed using generally accepted practices to determine actual air quantities.

❖ This section requires ducts that are part of the smoke control system to be tested to show that the proper amount of air is flowing. It should be noted that Section 909.10.2 requires that the ducts be leak tested to 1.5 times the maximum design pressure. Such leakage is not allowed to exceed 5 percent of the design flow.

[F] 909.18.3 Dampers. *Dampers* shall be tested for function in their installed condition.

❖ This section notes that all dampers need to be inspected to meet the function for which they are installed. For instance, a damper that is to be open when the system is in smoke control mode should be verified to be open when testing the system. Also, a damper may have a specific timing associated with its operation that would need to be verified though testing.

[F] 909.18.4 Inlets and outlets. Inlets and outlets shall be read using generally accepted practices to determine air quantities.

❖ Similar to ducts, the appropriate amount of air that is entering or exiting the inlets and outlets, respectively, must be checked.

[F] 909.18.5 Fans. Fans shall be examined for correct rotation. Measurements of voltage, amperage, revolutions per minute (rpm) and belt tension shall be made.

❖ This section requires the testing of fans for the following: correct rotation, voltage, amperage, revolutions per minute and belt tension. These features are key in

having the system run as designed.

A common problem with fans is that they are often installed in the reverse direction. Also, to verify the reliability of the fans, elements such as the appropriate voltage and belt tension need to be tested.

[F] 909.18.6 Smoke barriers. Measurements using inclined manometers or other *approved* calibrated measuring devices shall be made of the pressure differences across *smoke barriers*. Such measurements shall be conducted for each possible smoke control condition.

❖ As discussed in Section 909.5.1, the testing of pressure differences across smoke barriers needs to be measured in the smoke control mode. As noted in Section 909.18.6, such testing is to be performed for every possible smoke control condition, and the measurements will be taken using an inclined manometer or other approved methods. Electronic devices are also available. Qualified individuals must calibrate these types of devices. Additionally, before using an alternate method of testing, the building official needs to approve it.

[F] 909.18.7 Controls. Each smoke zone equipped with an automatic-initiation device shall be put into operation by the actuation of one such device. Each additional device within the zone shall be verified to cause the same sequence without requiring the operation of fan motors in order to prevent damage. Control sequences shall be verified throughout the system, including verification of override from the fire-fighter's control panel and simulation of standby power conditions.

❖ This section requires the overall testing of the system. More specifically, each zone needs to individually initiate the smoke control system by the activation of an automatic initiation device. Once that has occurred, all other devices within each zone need to be verified that they will activate the system, but to avoid damage, the fans do not need to be activated.

In addition to determining that all the appropriate devices initiate the system, it must also be verified that all of the controls on the fire-fighter control panel initiate the appropriate aspects of the smoke control system, including the override capability.

Finally, the initiation and availability of the standby power system need to be verified.

[F] 909.18.8 Special inspections for smoke control. Smoke control systems shall be tested by a special inspector.

❖ Smoke control systems require special inspection since they are unique and complex life safety systems. Section 1704.16 provides the same requirements for special inspection as presented in Sections 909.18.8.1 and 909.18.8.2.

[F] 909.18.8.1 Scope of testing. *Special inspections* shall be conducted in accordance with the following:

1. During erection of ductwork and prior to concealment for the purposes of leakage testing and recording of device location.

2. Prior to occupancy and after sufficient completion for the purposes of pressure-difference testing, flow measurements, and detection and control verification.

❖ Special inspections need to occur at two different stages during construction to facilitate the necessary inspections. The first round of special inspections occurs before concealment of the ductwork or fire protection elements. The special inspector needs to verify the leakage as noted in Section 909.10.2. Additionally, the location of all fire protection devices needs to be verified and documented at this time.

The second round of special inspections occurs just prior to occupancy. The inspections include the verification of pressure differences across smoke barriers, as required in Sections 909.5.1 and 909.18.6, the verification of appropriate volumes of airflow as noted in the design and finally the verification of the appropriate operation of the detection and control mechanisms as required in Sections 909.18.1 and 909.18.7. These tests need to occur just prior to occupancy, since the test result will more clearly represent actual conditions. This also makes a strong design and quality assurance during construction critical, as it is very costly and difficult in most cases to make changes at this stage. Note that the test does not actually place smoke into the space and demonstrate the smoke layer interface location. Instead, the testing is focused on all the elements of the design such as airflow and duct closure as prescribed by the specific design.

[F] 909.18.8.2 Qualifications. *Special inspection* agencies for smoke control shall have expertise in fire protection engineering, mechanical engineering and certification as air balancers.

❖ As noted in Section 909.3, special inspections are required for smoke control systems. This means a certain level of qualification that would include the need for expertise in fire protection engineering, mechanical engineering and certification as air balancers.

[F] 909.18.8.3 Reports. A complete report of testing shall be prepared by the special inspector or *special inspection* agency. The report shall include identification of all devices by manufacturer, nameplate data, design values, measured values and identification tag or *mark*. The report shall be reviewed by the responsible *registered design professional* and, when satisfied that the design intent has been achieved, the responsible *registered design professional* shall seal, sign and date the report.

❖ Once the special inspections are complete, documentation of the activity is required. This documentation is to be prepared in the form of a report that identifies all devices by manufacturer, nameplate data, design values, measured values and identification or mark.

[F] 909.18.8.3.1 Report filing. A copy of the final report shall be filed with the fire code official and an identical copy shall be maintained in an *approved* location at the building.

❖ The report needs to be reviewed, approved and then signed, sealed and dated. This report is to be provided to the building official and a copy is also to remain in the building in an approved location. When a fire command

center is required this is the best location for such documents. Otherwise, a location such as the security office or building manager's office might be appropriate.

[F] 909.18.9 Identification and documentation. Charts, drawings and other documents identifying and locating each component of the smoke control system, and describing its proper function and maintenance requirements, shall be maintained on file at the building as an attachment to the report required by Section 909.18.8.3. Devices shall have an *approved* identifying tag or *mark* on them consistent with the other required documentation and shall be dated indicating the last time they were successfully tested and by whom.

❖ Additional documentation that needs to be maintained includes charts, drawings and other related documentation that assists in the identification of each aspect of the smoke control system. This documentation is where information, such as the last time a device or component was successfully tested and by whom, is recorded. This will serve as the main documentation for the system. Again, the fire command center, if required, is the most appropriate location for such information (see commentary, Section 909.18.8.3.1).

[F] 909.19 System acceptance. Buildings, or portions thereof, required by this code to comply with this section shall not be issued a certificate of occupancy until such time that the fire code official determines that the provisions of this section have been fully complied with and that the fire department has received satisfactory instruction on the operation, both automatic and manual, of the system.

Exception: In buildings of phased construction, a temporary certificate of occupancy, as *approved* by the fire code official, shall be allowed provided that those portions of the building to be occupied meet the requirements of this section and that the remainder does not pose a significant hazard to the safety of the proposed occupants or adjacent buildings.

❖ This section stipulates that the certificate of occupancy cannot be issued unless the smoke control system has been accepted. It is essential that the system be inspected and approved since it is a life safety system. There is an exception for buildings that are constructed in phases where a temporary certificate of occupancy is allowed. For example, a building where the portion requiring smoke control is not yet occupied so egress concerns through that space are not relevant. This space needs to be separated by smoke barriers (different smoke zone).

909.20 Smokeproof enclosures. Where required by Section 1022.9, a smokeproof enclosure shall be constructed in accordance with this section. A smokeproof enclosure shall consist of an enclosed interior *exit stairway* that conforms to Section 1022.1 and an open exterior balcony or ventilated vestibule meeting the requirements of this section. Where access to the roof is required by the *International Fire Code*, such access shall be from the smokeproof enclosure where a smokeproof enclosure is required.

❖ In a building that serves stories where the floor surface is located more than 75 feet (22 860 mm) above the level of fire department access or more than 30 feet (9144 mm) below the level of exit discharge stairways, either a smokeproof enclosure or a pressurized stairway is required.

A smokeproof enclosure essentially takes advantage of fire-resistance-rated construction surrounding the stair shaft and a buffer zone created by an outside balcony or ventilated vestibule to avoid the accumulation of smoke within the exit stairway. The premise is that if access to the smokeproof enclosure provides a sufficient buffer to the effects of smoke via ventilation or simply being located outside, the smoke will not migrate into the enclosure.

The vestibule as noted is required to be ventilated. There are two alternatives provided in Section 909.20. They include natural ventilation, covered in Section 909.20.3, and mechanical ventilation, covered in Section 909.20.4. It should be noted that a smokeproof enclosure is not considered a pressurized stairway. The method of pressurizing stairways is recognized as an alternative to the smokeproof enclosures in Section 909.20.5, but is not required.

909.20.1 Access. Access to the *stair* shall be by way of a vestibule or an open exterior balcony. The minimum dimension of the vestibule shall not be less than the required width of the *corridor* leading to the vestibule but shall not have a width of less than 44 inches (1118 mm) and shall not have a length of less than 72 inches (1829 mm) in the direction of egress travel.

❖ As noted, access to the stair in a smokeproof enclosure is via an exterior balcony or a ventilated vestibule. This section provides the minimum dimensions for access to the smokeproof enclosure through a vestibule and then travel space to the entrance into the stairway. If a vestibule is chosen, the minimum width is established as the larger of 44 inches (1118 mm) or the width of the corridor. Additionally, when entering a vestibule there needs to be at least 72 inches (1829 mm) in the direction of travel (see Figure 909.20.1). It

should be noted that the pressurized stairway alternative would not require a vestibule.

909.20.2 Construction. The smokeproof enclosure shall be separated from the remainder of the building by not less than 2-hour *fire barriers* constructed in accordance with Section 707 or *horizontal assemblies* constructed in accordance with Section 712, or both. Openings are not permitted other than the required *means of egress* doors. The vestibule shall be separated from the *stairway* by not less than 2-hour *fire barriers* constructed in accordance with Section 707 or *horizontal assemblies* constructed in accordance with Section 712, or both. The open exterior balcony shall be constructed in accordance with the *fire-resistance rating* requirements for floor assemblies.

❖ This section sets the basic construction requirements for the stairway and the associated vestibule or exterior balcony. Essentially, a 2-hour fire-resistance-rated fire barrier or horizontal assembly is required between the enclosure and the rest of the building.

Additionally, vestibules need to be separated from the stairway by a 2-hour fire barrier or horizontal assembly. The exterior balcony, if used, would only be required to comply with the fire-rating requirements for floor construction. The opening protection for doors into the stairway from the vestibule are only required to be fire protection-rated for 20 minutes (see Sections 909.20.3.2 and 909.20.4.1).

909.20.2.1 Door closers. Doors in a smokeproof enclosure shall be self- or automatic closing by actuation of a smoke detector in accordance with Section 715.4 and shall be installed at the floor-side entrance to the smokeproof enclosure. The actuation of the smoke detector on any door shall activate the closing devices on all doors in the smokeproof enclosure at all levels. Smoke detectors shall be installed in accordance with Section 907.3.

❖ In order to maintain the separation between the smokeproof enclosure and the rest of the building, the doors

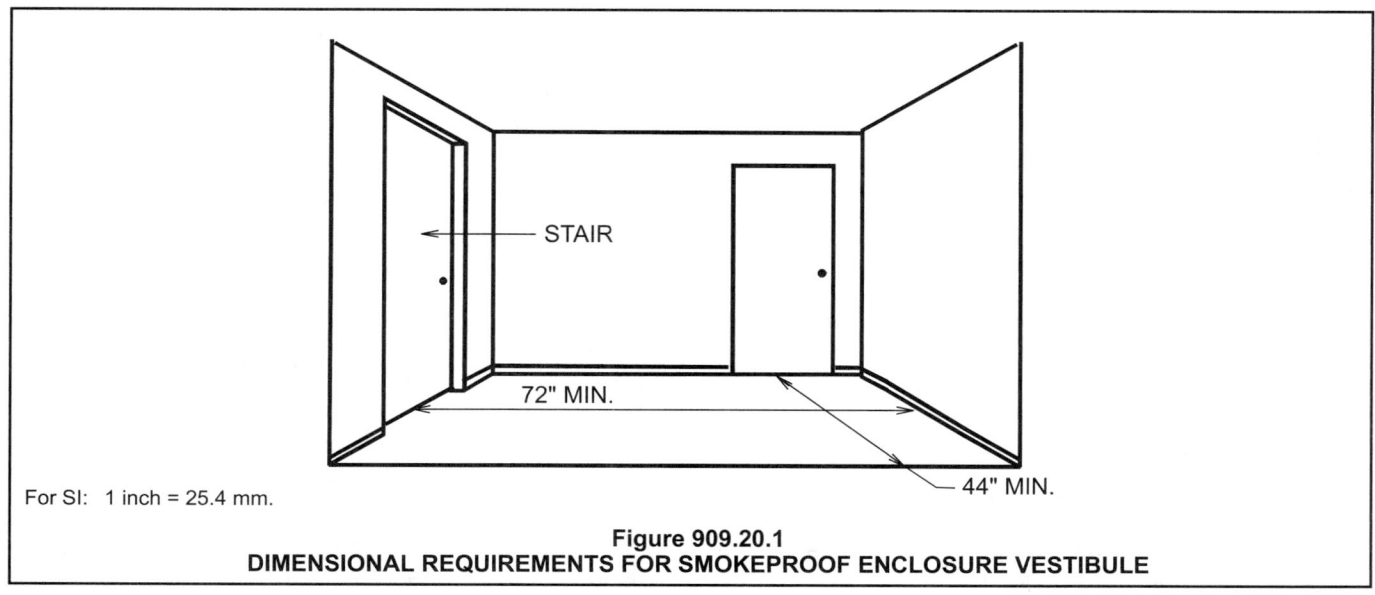

For SI: 1 inch = 25.4 mm.

Figure 909.20.1
DIMENSIONAL REQUIREMENTS FOR SMOKEPROOF ENCLOSURE VESTIBULE

need to be self-closing or automatic closing through the actuation of a smoke detector. Also, when a smoke detector is activated on any floor, doors at all levels are required to close for that particular smokeproof enclosure.

909.20.3 Natural ventilation alternative. The provisions of Sections 909.20.3.1 through 909.20.3.3 shall apply to ventilation of smokeproof enclosures by natural means.

❖ Smoke and hot gases generated by fire may enter smokeproof enclosures through the doorways. This section provides for the diffusion of smoke and gases to the outside of the building by means of unenclosed openings to the outside.

909.20.3.1 Balcony doors. Where access to the *stairway* is by way of an open exterior balcony, the door assembly into the enclosure shall be a *fire door assembly* in accordance with Section 715.4.

❖ Where an open exterior balcony is used for entry into the stairway, the doorways to the stairway are required to have a $1^1/_2$-hour fire protection rating as required for 2-hour fire-resistance-rated wall assemblies in accordance with the requirements of Section 715.4. If a vestibule is used the rating of the door from the vestibule into the stairway would only be required to be fire protection rated for 20 minutes. The door into the vestibule would require a fire protection rating of $1^1/_2$ hours.

909.20.3.2 Vestibule doors. Where access to the *stairway* is by way of a vestibule, the door assembly into the vestibule shall be a *fire door assembly* complying with Section 715.4. The door assembly from the vestibule to the *stairway* shall have not less than a 20-minute *fire protection rating* complying with Section 715.4.

❖ When a smokeproof enclosure is of the type that employs a vestibule as the means of access to an enclosed exit stairway, the entry into the vestibule from the exit access must consist of a fire door assembly having a $1^1/_2$-hour fire protection rating. This requirement is commensurate with the 2-hour fire-resistance-rated smokeproof enclosure construction requirement and the associated fire protection ratings specified in Table 715.4. Essentially, the fire door assemblies and the enclosure walls serve as the primary exit enclosure protection from fires occurring in adjacent spaces.

The entry from a vestibule into the stairway requires a 20-minute fire-protection-rated door assembly. These doorways are not normally subject to direct fire exposure and, therefore, only a nominal fire protection rating is required as a standard of construction that would minimize air leakage and thus reduce the possibilities of smoke and hot gases penetrating the stairway in amounts that could become a threat to life safety.

Door assemblies must comply with the requirements of Section 715.4 (see Figure 909.20.3.2).

909.20.3.3 Vestibule ventilation. Each vestibule shall have a minimum net area of 16 square feet (1.5 m²) of opening in a wall facing an outer *court*, *yard* or *public way* that is at least 20 feet (6096 mm) in width.

❖ Natural ventilation is allowed for vestibules that have a minimum net opening of 16 square feet (1.49 m²) and open into an outer court, yard or public way that is at least 20 feet (6096 mm) in width. Basically, this option is intended to take advantage of the fact that smoke may disperse into an open area before entering the stair enclosure.

909.20.4 Mechanical ventilation alternative. The provisions of Sections 909.20.4.1 through 909.20.4.4 shall apply to ventilation of smokeproof enclosures by mechanical means.

❖ When an exterior balcony is not possible or when the clear open area cannot be achieved, then the mechanical ventilation alternative is necessary.

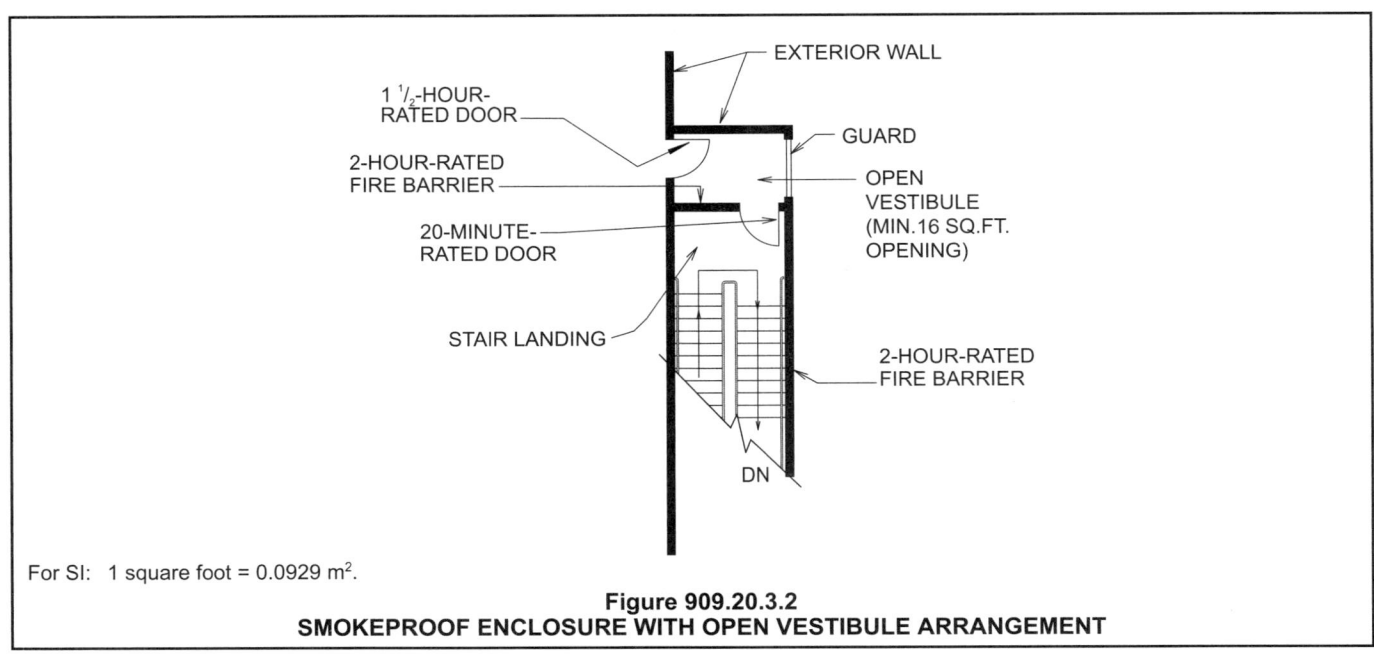

For SI: 1 square foot = 0.0929 m².

Figure 909.20.3.2
SMOKEPROOF ENCLOSURE WITH OPEN VESTIBULE ARRANGEMENT

909.20.4.1 Vestibule doors. The door assembly from the building into the vestibule shall be a *fire door assembly* complying with Section 715.4.3. The door assembly from the vestibule to the *stairway* shall not have less than a 20-minute *fire protection rating* and meet the requirements for a smoke door assembly in accordance with Section 715.4.3. The door shall be installed in accordance with NFPA 105.

❖ The entry into the vestibule from the exit access must consist of a fire door assembly having a 1½-hour fire protection rating. This requirement is commensurate with the 2-hour fire-resistance-rated smokeproof enclosure construction requirement and the associated fire protection ratings specified in Table 715.4. Essentially, the fire door assemblies and the enclosure walls serve as the primary exit enclosure protection from fires occurring in adjacent spaces.

The entry from a vestibule into the stairway requires a 20-minute fire protection-rated door assembly that also meets the requirements for a smoke door assembly. Vestibule doors into stairways for the natural ventilation option (see Section 909.20.3.2) would not require the smoke door assembly rating in accordance with Section 715.4.3. The natural ventilation option requires an opening to the outside that is intended to compensate for this difference (see Figure 909.20.4.1).

909.20.4.2 Vestibule ventilation. The vestibule shall be supplied with not less than one air change per minute and the exhaust shall not be less than 150 percent of supply. Supply air shall enter and exhaust air shall discharge from the vestibule through separate, tightly constructed ducts used only for that purpose. Supply air shall enter the vestibule within 6 inches (152 mm) of the floor level. The top of the exhaust register shall be located at the top of the smoke trap but not more than 6 inches (152 mm) down from the top of the trap, and shall be entirely within the smoke trap area. Doors in the open position shall not obstruct duct openings. Duct openings with controlling *dampers* are permitted where necessary to meet the design requirements, but *dampers* are not otherwise required.

❖ This section provides the basic criteria for the mechanical ventilation of the vestibule during a fire situation. Such ventilation is required to be activated via a smoke detector in accordance with Section 909.20.6. The basic requirements are for one air change per minute within the vestibule. Also the exhaust is required to be 150 percent of the supply. The exhaust is required to exceed the supply so that air will not be pushed into the stairway due to a positive pressure being created within the vestibule. It should be noted that there is not a minimum pressure difference requirement. Instead, the code simply prescribes a particular air change rate (exhaust).

This section also prescribes criteria as to where supply air and exhaust air are to enter and exit, respectively, and requires that those paths are clear (see Figure 909.20.4.2).

It should be noted that this section does not specify when and how such systems should activate. Activation is addressed by Section 909.20.6.

909.20.4.2.1 Engineered ventilation system. Where a specially engineered system is used, the system shall exhaust a quantity of air equal to not less than 90 air changes per hour from any vestibule in the emergency operation mode and shall be sized to handle three vestibules simultaneously. Smoke detectors shall be located at the floor-side entrance to each vestibule and shall activate the system for the affected vestibule.

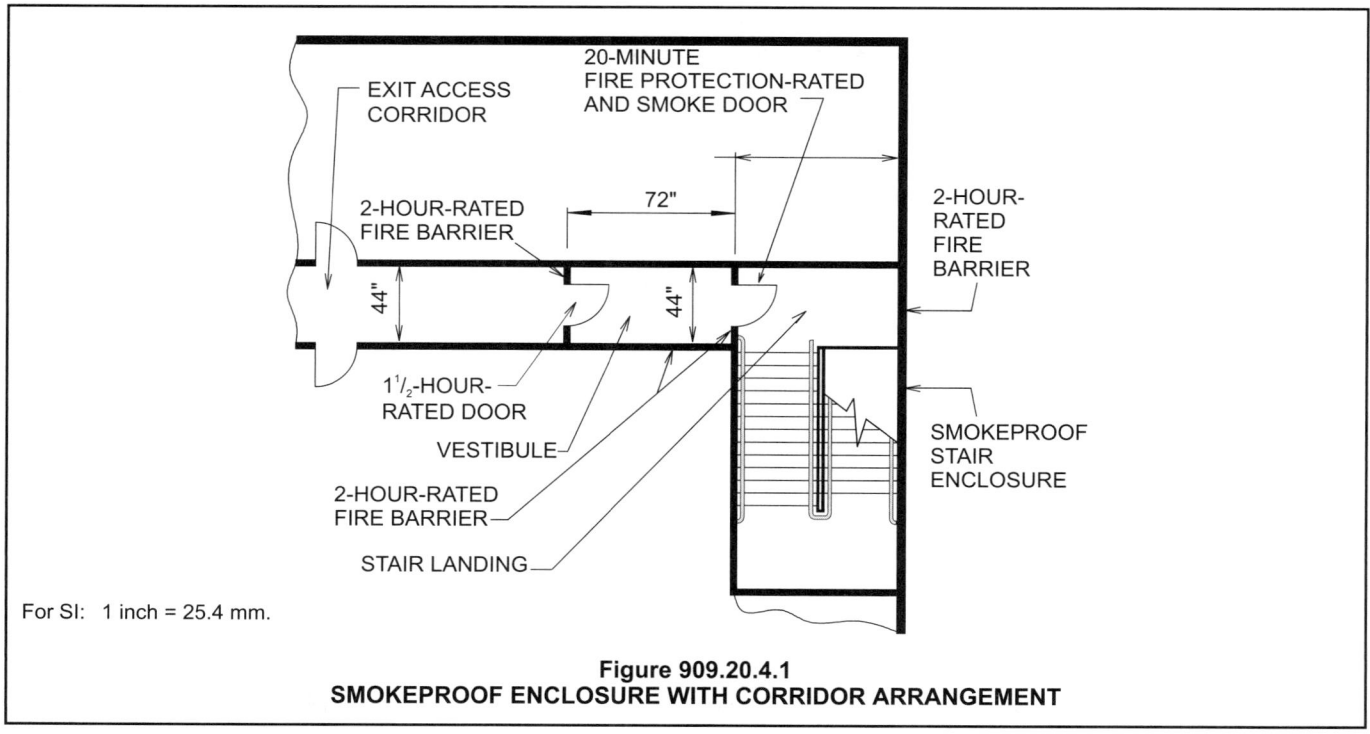

For SI: 1 inch = 25.4 mm.

Figure 909.20.4.1
SMOKEPROOF ENCLOSURE WITH CORRIDOR ARRANGEMENT

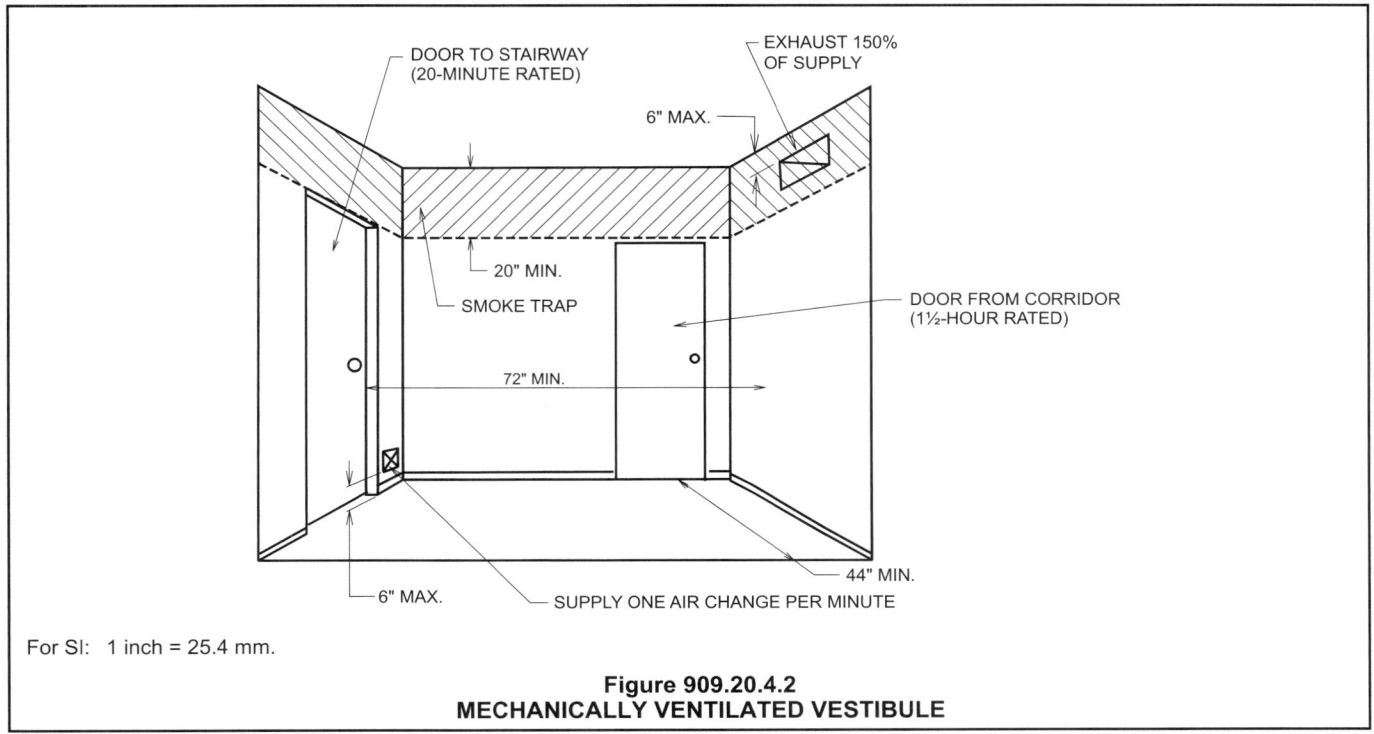

For SI: 1 inch = 25.4 mm.

Figure 909.20.4.2
MECHANICALLY VENTILATED VESTIBULE

Smoke detectors shall be installed in accordance with Section 907.3.

❖ If the method prescribed in Section 909.20.4.2 for mechanical ventilation of vestibules is not chosen, then minimum criteria is provided for engineered systems. Essentially, the designer can create a design that meets the flexibility needs of the building and at the same time meets basic criteria. Once designed and installed, the vestibules must be ventilated at 90 air changes per hour, which is 30 more air changes per hour than the prescriptive solution would require. The exhaust rate is not mandated. It is simply left to the designer to achieve the 90 air changes per hour within the space (vestibule). Additionally, this system needs to be sized to handle at least three vestibules simultaneously. Finally, the engineered systems are required to operate in accordance with the initiation of a smoke detector located on the floor-side entrance of the affected vestibule; therefore, a smoke detector would be required on each floor at the entrance to each vestibule. The system would only need to be activated on the floors where smoke is detected.

909.20.4.3 Smoke trap. The vestibule ceiling shall be at least 20 inches (508 mm) higher than the door opening into the vestibule to serve as a smoke and heat trap and to provide an upward-moving air column. The height shall not be decreased unless *approved* and justified by design and test.

❖ This section requires a 20-inch (508 mm) difference in height between the door openings into the vestibule and the ceiling. This is intended to allow an additional safeguard to assist in the containment of smoke inside the vestibule and outside the stair shaft. Essentially, as

the title states, it provides a smoke trap. This arrangement also has the effect of keeping the air movement upward.

909.20.4.4 Stair shaft air movement system. The *stair* shaft shall be provided with a dampered relief opening and supplied with sufficient air to maintain a minimum positive pressure of 0.10 inch of water (25 Pa) in the shaft relative to the vestibule with all doors closed.

❖ In order to prevent smoke from migrating into the stairway from the vestibule, a positive pressure of 0.10 inch (25 Pa) of water in the stair shaft relative to the vestibule is required. This pressure difference is to be available when all doors are closed. This pressure difference is lower than that required by the stair pressurization alternative. This would not be considered a pressurized stair. This pressure difference would need to be tested to obtain approval once constructed.

909.20.5 Stair pressurization alternative. Where the building is equipped throughout with an *automatic sprinkler system* in accordance with Section 903.3.1.1, the vestibule is not required, provided that interior *exit stairways* are pressurized to a minimum of 0.10 inches of water (25 Pa) and a maximum of 0.35 inches of water (87 Pa) in the shaft relative to the building measured with all *stairway* doors closed under maximum anticipated conditions of stack effect and wind effect.

❖ This method is allowed only when the building is fully sprinklered. This is partially related to the fact that these pressure differences were developed based upon a sprinklered fire. It should be noted that smokeproof enclosures are not required to be in fully sprinklered buildings, but the areas where smokeproof enclosures are required are often sprinklered buildings (i.e., high-rise buildings). This alternative would not require vestibules or an exterior exit balcony. The criteria for smoke control design is provided in terms of minimum and maximum pressure differences of 0.10 inch (37 Pa) of water and 0.35 inch (87 Pa) of water, respectively, between the shaft and the building. This pressure difference is to be achieved when all doors are closed and maximum conditions of wind and stack effect have been taken into account. It should be noted that additional limitations may be placed on the maximum pressure differences for pressurized stairs due to the lower opening forces required in order to comply with Section 1008.1.3. If the maximum pressure difference of .35 would exceed the requirements of Section 1008.1.3, the maximum pressure difference would need to be lowered. Also note that Section 404.2.8 of ICC A117.1 would not require opening forces to be lowered for accessibility purposes if the door is a fire door. Finally, as with all other smoke control systems addressed in Section 909, such systems need to be designed through a rational analysis, tested and documented as such.

909.20.6 Ventilating equipment. The activation of ventilating equipment required by the alternatives in Sections 909.20.4 and 909.20.5 shall be by smoke detectors installed at each floor level at an *approved* location at the entrance to the smokeproof enclosure. When the closing device for the *stair* shaft and vestibule doors is activated by smoke detection or power failure, the mechanical equipment shall activate and operate at the required performance levels. Smoke detectors shall be installed in accordance with Section 907.3.

❖ This section clarifies that the activation mechanism for both mechanical means of smoke management for exit enclosures in Sections 909.20.4 and 909.20.5 should be via a smoke detector located at each level outside the door leading into the vestibule and stairway, respectively. For systems that use automatic-closing devices on the doors, whether for vestibules in smokeproof enclosures or for pressurized stairs, if the door closes, the system must activate. This includes normal activation of the smoke detector or in the event of a power failure. Essentially, if there is a power failure it will operate in a fail-safe manner even if there is not a fire.

909.20.6.1 Ventilation systems. Smokeproof enclosure ventilation systems shall be independent of other building ventilation systems. The equipment, control wiring, power wiring and ductwork shall comply with one of the following:

1. Equipment, control wiring, power wiring and ductwork shall be located exterior to the building and directly connected to the smokeproof enclosure or connected to the smokeproof enclosure by ductwork enclosed by not less than 2-hour *fire barriers* constructed in accordance with Section 707 or *horizontal assemblies* constructed in accordance with Section 712, or both.

2. Equipment, control wiring, power wiring and ductwork shall be located within the smokeproof enclosure with intake or exhaust directly from and to the outside or through ductwork enclosed by not less than 2-hour *fire barriers* constructed in accordance with Section 707 or *horizontal assemblies* constructed in accordance with Section 712, or both.

3. Equipment, control wiring, power wiring and ductwork shall be located within the building if separated from the remainder of the building, including other mechanical equipment, by not less than 2-hour *fire barriers* constructed in accordance with Section 707 or *horizontal assemblies* constructed in accordance with Section 712, or both.

   Exceptions:

   1. Control wiring and power wiring utilizing a 2-hour rated cable or cable system.

   2. Where encased with not less than 2 inches (51 mm) of concrete.

❖ Smokeproof enclosures and pressurized stair shaft ventilation systems must be independent of other building ventilating systems. This section provides three options for the location of the ductwork associated with smokeproof enclosures and stair pressurization mechanical equipment. The three options include the following:

   • Located on the exterior of the building and directly connected to the smokeproof enclosure;

   • Within the smokeproof enclosure; or

   • Within the building but separated from the remainder of the building by 2-hour fire- barriers or horizontal barriers, or both.

909.20.6.2 Standby power. Mechanical vestibule and *stair* shaft ventilation systems and automatic fire detection systems shall be powered by an *approved* standby power system conforming to Section 403.4.7 and Chapter 27.

❖ This section requires standby power for mechanical systems for pressurized stair shafts, mechanically ventilated vestibules, stair shafts for smokeproof enclosures and any automatic fire detection systems related to the activation of such systems.

909.20.6.3 Acceptance and testing. Before the mechanical equipment is *approved*, the system shall be tested in the presence of the *building official* to confirm that the system is operating in compliance with these requirements.

❖ This section requires verification of successful testing of the system to the building official. The requirements of Sections 909.18 and 909.19 should be addressed when testing smokeproof enclosures or pressurized stairs. It should be noted that the IFC does not contain the provisions for smokeproof enclosures and pressurized stairways, but it may be in the best interest to also notify the fire department of such testing and acceptance.

SECTION 910
SMOKE AND HEAT VENTS

[F] 910.1 General. Where required by this code or otherwise installed, smoke and heat vents, or mechanical smoke exhaust systems, and draft curtains shall conform to the requirements of this section.

Exceptions:

1. Frozen food warehouses used solely for storage of Class I and II commodities where protected by an *approved automatic sprinkler system.*

2. Where areas of buildings are equipped with early suppression fast-response (ESFR) sprinklers, automatic smoke and heat vents shall not be required within these areas.

❖ Smoke and heat vents must be provided in buildings, structures or portions thereof where required by Section 910.2. It should be noted that Chapter 23 of the IFC would also be applicable (see commentary, Section 910.2.2). The systems must be designed, installed, maintained and operated in accordance with the provisions of this section.

The purpose of smoke and heat vents has historically been related to the needs of fire fighters. More specifically, smoke and heat vents, when activated, have the potential effect of lifting the height of the smoke layer and providing more tenable conditions to undertake fire-fighting activities. Other potential benefits include a decrease in property damage and the creation of more tenable conditions for occupants. The purpose of draft curtains, as addressed in Section 910.3.5, is both to contain the smoke in certain areas and potentially increase the speed in the activation of the smoke and heat vents.

Exception 1 recognizes the "building-within-a-building" nature of typical frozen food warehouses. As such, smoke from a fire within a freezer would be contained within the freezer, thus negating the usefulness of smoke and heat vents at the roof level.

Exception 2 recognizes the negative effect that smoke and heat vents can have on the operation of early suppression fast response (ESFR) sprinklers. Those negative effects include diverting heat away from the sprinklers, which could delay their activation or result in the activation of more sprinklers in areas away from the source of the fire, which may overwhelm the system. This section coordinates with the ESFR exception for draft curtains in Section 910.3.5. Both smoke and heat vents and draft curtains have a negative effect on ESFR sprinkler systems.

The intent of the code change that added this exception was to not require smoke and heat vents when ESFR sprinklers were used. The term "required" was used versus "prohibited" in an attempt to allow the installation of manual smoke and heat venting in some cases. Note j in Table 2306.2 of the IFC correlates with this exception but has a slightly different applicability. First, Note j only applies to high-piled storage. Second, the footnote does not differentiate between automatic

or manual smoke and heat vents; it simply does not require them when an ESFR system is used.

[F] 910.2 Where required. Smoke and heat vents shall be installed in the roofs of one-story buildings or portions thereof occupied for the uses set forth in Sections 910.2.1 and 910.2.2.

❖ Smoke and heat vents are only required in single-story buildings and then only as required by the provisions of Sections 910.2.1 and 910.2.2.

Section 910.2.1 addresses Group F-1 or S-1 occupancies over 50,000 square feet (4656 m²) of undivided area (regardless of high-piled storage).

Section 910.2.2 addresses high-piled storage areas as required by Section 2306.7 of the IFC.

[F] 910.2.1 Group F-1 or S-1. Buildings and portions thereof used as a Group F-1 or S-1 occupancy having more than 50,000 square feet (4645 m²) in undivided area.

Exception: Group S-1 aircraft repair hangars.

❖ Large-area buildings with moderate to heavy fire loads present special challenges to the fire department in disposing of the smoke generated in a fire. In order to provide the fire department with the ability to rapidly and efficiently dispose of smoke in large-area Groups F-1 and S-1 buildings exceeding 50,000 square feet (4645 m²) in undivided area without the exposure of personnel to the dangers associated with cutting ventilation holes in the roof, smoke and heat vents (or, alternatively, mechanical smoke removal facilities) must be provided.

In order to subdivide a more than 50,000-square-foot (4645 m²) undivided area as one method of avoiding the use of smoke and heat vents, the dividing element would only need to be a partition constructed of materials equivalent to the construction of a draft curtain but that would extend from floor to ceiling in the space being separated. A fire barrier, smoke barrier, fire partition or smoke partition would be more than what is required and would therefore be an acceptable method of dividing the area.

This requirement is independent of the requirements related to high-piled storage in Section 910.2.2. Smoke and heat vent area requirements in Table 910.3 tend to be more restrictive for high-piled storage. High-piled storage is not occupancy specific.

[F] 910.2.2 High-piled combustible storage. Buildings and portions thereof containing high-piled combustible stock or rack storage in any occupancy group in accordance with Section 413 and the *International Fire Code.*

❖ This section alerts the code user to the specific high-piled combustible storage requirements contained in Chapter 23 of the IFC. High-piled storage, whether solid piled, palletized or in racks, in excess of 12 feet (3658 mm) in height [6 feet (1829 mm) for high-hazard commodities] requires specific consideration, including fire protection design features and smoke and heat vents in order to be adequately protected. Not all high-piled storage will require the use of smoke and heat vents and draft curtains. In fact, if the high-piled storage is properly sprinklered (in accordance with Chapter 23 of the IFC and NFPA 13), draft curtains are not required (see commentary, Chapter 23 and Table 2306.2 of the IFC). In addition, where ESFR sprinklers are used, smoke and heat venting is not required. See Section 910.1, Exception 2 and Note j to Table 2306.2 of the IFC.

[F] 910.3 Design and installation. The design and installation of smoke and heat vents and draft curtains shall be as specified in Sections 910.3.1 through 910.3.5.2 and Table 910.3.

❖ Careful design and installation of smoke and heat vents is vital to their efficient operation in case of fire. The design criteria for these fire protection tools are organized for convenience and ready reference in Table 910.3, which is referenced by this section.

TABLE 910.3. See page 9-113.

❖ When smoke and heat vents and draft curtains are required, Table 910.3 identifies the required vent area in terms of ratio of vent area to floor area and draft curtain area and depth requirements. The table is essentially divided into two parts. The first part is for Group F-1 and S-2 occupancies, while the second portion is for high-piled combustible storage (organized by commodity type).

In applying the provisions of the table, note that the term "high hazard" is only referring to the high-piled storage commodity type, not the occupancy group classification (see Section 2303.6 of the IFC). Smoke and heat venting requirements for high-piled storage are not occupancy specific and originate from Chapter 23 of the IFC. The focus is upon the commodity classification, configuration and the size of the high-piled storage area. Chapter 23 of the IFC only requires smoke and heat venting and draft curtains for larger storage areas and areas that do not utilize ESFR sprinklers. The required vent areas vary based upon the commodity classification (I through IV or high hazard) and height of storage. The higher the storage, the higher the potential for a larger fire. Two options ("Option 1" and "Option 2") are given for commodity classifications I through IV and high hazard. The only significance to these options is that one option allows a lower vent-to-floor area ratio if a deeper draft curtain is chosen, simply providing some credit for the fact that the deeper draft curtains are more likely to contain more smoke than a shorter draft curtain; thus, the area contained by the draft curtains also varies. If draft curtains are not required, than these options are not necessary and the lower vent/area ratio can be used. Note c specifically allows the use of the lower vent-to-floor ratio (Option 1) where smoke and heat vents are required and draft curtains are not. Note d simply explains what "H" stands for as used in Column 3, Row 1. The last column of the table provides the maximum distance from walls or draft curtains that a smoke and heat vent can be located when the vent is adjacent to a wall or draft curtain. This would not apply to vents located in the middle of a curtained area. Footnote b addresses how this is to be measured.

[F] 910.3.1 Design. Smoke and heat vents shall be *listed* and labeled to indicate compliance with UL 793.

❖ This section specifically requires that all smoke and heat vents be both listed and labeled in accordance with UL 793. This provides consistency and a level of quality when smoke and heat vents are required. The standard addresses smoke and heat vents that automatically operate during fires via nonelectrical means. Automatic vents listed and labeled to this standard can be operated both automatically and manually. There are main mechanisms for activation that include a heat responsive device or on the action of a plastic cover shrinking and falling from place due to fire exposure.

[F] 910.3.2 Vent operation. Smoke and heat vents shall be capable of being operated by *approved* automatic and manual means. Automatic operation of smoke and heat vents shall conform to the provisions of Sections 910.3.2.1 through 910.3.2.3.

❖ Since vents are used as a component of an active venting system, the releasing device is required to be automatic, such as a fusible link. The next several subsections of this section provide requirements for automatic activation of smoke and heat vents. The fusible link ratings are prescribed for nonsprinklered buildings but the strategy will vary in sprinklered buildings.

In addition to automatic operation of the vents, a manual means of operating them by the fire department during fire suppression operations must also be provided. It should be remembered that one of the main reasons smoke and heat vents were initially introduced was to reduce the need for fire fighters to have to ventilate the fire by getting on the roof of the burning building, often having to traverse large expanses of roof in order to get to the area over the fire and breaching the roof manually. Accordingly, the mechanisms for release and the needs of the fire department should be carefully considered.

[F] 910.3.2.1 Gravity-operated drop-out vents. Automatic smoke and heat vents containing heat-sensitive glazing designed to shrink and drop out of the vent opening when exposed to fire shall fully open within 5 minutes after the vent cavity is exposed to a simulated fire, represented by a time-temperature gradient that reaches an air temperature of 500°F (260°C) within 5 minutes.

❖ This section establishes minimum performance criteria for drop-out vents, which include a nonmetallic, clear or opaque glazing element designed to shrink from its frame and fall away when exposed to heat from a fire. Such vent design must be capable of completely opening the roof vent within 5 minutes of exposure to a simulated fire represented by a time-temperature gradient that reaches an air temperature of 500°F (260°C) within 5 minutes. Drop-out vents tested in accordance with UL 793 must begin to operate at a maximum temperature of 286°F (141°C) in order to be labeled.

[F] TABLE 910.3
REQUIREMENTS FOR DRAFT CURTAINS AND SMOKE AND HEAT VENTS[a]

OCCUPANCY GROUP AND COMMODITY CLASSIFICATION	DESIGNATED STORAGE HEIGHT (feet)	MINIMUM DRAFT CURTAIN DEPTH (feet)	MAXIMUM AREA FORMED BY DRAFT CURTAINS (square feet)	VENT-AREA-TO-FLOOR-AREA RATIO[c]	MAXIMUM SPACING OF VENT CENTERS (feet)	MAXIMUM DISTANCE FROM VENTS TO WALL OR DRAFT CURTAIN[b] (feet)
Group F-1 and S-1	—	$0.2 \times H^d$ but ≥ 4	50,000	1:100	120	60
High-piled Storage (see Section 910.2.2) Class I-IV commodities (Option 1)	≤ 20	6	10,000	1:100	100	60
	$> 20 \leq 40$	6	8,000	1:75	100	55
High-piled Storage (see Section 910.2.2) Class I-IV commodities (Option 2)	≤ 20	4	3,000	1:75	100	55
	$> 20 \leq 40$	4	3,000	1:50	100	50
High-piled Storage (see Section 910.2.2) High-hazard commodities (Option 1)	≤ 20	6	6,000	1:50	100	50
	$> 20 \leq 30$	6	6,000	1:40	90	45
High-piled Storage (see Section 910.2.2) High-hazard commodities (Option 2)	≤ 20	4	4,000	1:50	100	50
	$> 20 \leq 30$	4	2,000	1:30	75	40

For SI: 1 foot = 304.8 mm, 1 square foot = 0.0929 m².

a. Additional requirements for rack storage heights in excess of those indicated shall be in accordance with Chapter 23. For solid-piled storage heights in excess of those indicated, an approved engineered design shall be used.

b. Vents adjacent to walls or draft curtains shall be located within a horizontal distance not greater than the maximum distance specified in this column as measured perpendicular to the wall or draft curtain that forms the perimeter of the draft curtained area.

c. Where draft curtains are not required, the vent area to floor area ratio shall be calculated based on a minimum draft curtain depth of 6 feet (Option 1).

d. "H" is the height of the vent, in feet, above the floor.

[F] 910.3.2.2 Sprinklered buildings. Where installed in buildings provided with an *approved automatic sprinkler system*, smoke and heat vents shall be designed to operate automatically.

❖ Where smoke and heat vents are installed in sprinklered buildings, their operation must be automatic and coordinated with the operation of the sprinkler system. Caution should be exercised in the design of smoke and heat vents and the required draft curtains so that the draft curtains do not interfere with the operation of the automatic sprinklers, since locating a draft curtain too close to a sprinkler head could prevent proper water distribution over the fire. In addition, draft curtains will contain smoke and hot gases and can direct them away from the area where the fire is actually burning, thus activating sprinklers in the wrong area. This has the potential of overwhelming the sprinkler system.

More specifically, the fusible link operating temperatures should be coordinated with sprinkler head operating temperatures. The premature operation of a vent-opening mechanism could retard the operation of higher temperature-rated sprinkler heads by dissipating the level of heat needed to make the fusible link of the sprinkler(s) operate.

Delaying the operation of sprinklers can have the negative effect of causing an excessive number of sprinklers to operate, including some located outside the immediate area of fire danger. Concern over this issue has increased with the introduction of new sprinkler technology, such as the use of ESFR sprinklers, which are designed to act quickly to apply larger volumes of water to extinguish rather than simply control the fire. For that reason, ESFR sprinklers are specifically exempted in Sections 910.1 from the smoke and heat vent requirements.

[F] 910.3.2.3 Nonsprinklered buildings. Where installed in buildings not provided with an *approved automatic sprinkler system*, smoke and heat vents shall operate automatically by actuation of a heat-responsive device rated at between 100°F (38°C) and 220°F (104°C) above ambient.

Exception: Gravity-operated drop-out vents complying with Section 910.3.2.1.

❖ Where smoke and heat vents are installed in buildings that are not equipped with an automatic sprinkler system, their operation must be automatic, with their operating elements set at between 100 and 220°F (38 and 104°C) above the ambient temperature of the area in which they are installed. Smoke and heat vents in nonsprinklered buildings do not have concerns with sprinkler interaction; therefore, the operation is specifically prescribed. The exception indicates that gravity-operated drop-out vents are not subject to this requirement because of their higher required operating temperatures and unique design.

[F] 910.3.3 Vent dimensions. The effective venting area shall not be less than 16 square feet (1.5 m²) with no dimension less than 4 feet (1219 mm), excluding ribs or gutters having a total width not exceeding 6 inches (152 mm).

❖ This section prescribes the minimum clear area required for each individual smoke and heat vent, exclusive of any obstructions (see Figure 910.3.4). The design of the aggregate vent area actually needed is based, in part, on the area defined by the draft curtains and the depth of the curtained area, the objective of the design being to prevent either smoke from spilling out of the curtained area or the smoke interface from interfering with egress visibility. It has also been argued that draft curtains are intended to speed the operation of the smoke and heat vents by keeping the smoke in a smaller area.

[F] 910.3.4 Vent locations. Smoke and heat vents shall be located 20 feet (6096 mm) or more from adjacent *lot lines* and *fire walls* and 10 feet (3048 mm) or more from *fire barriers*.

Vents shall be uniformly located within the roof in the areas of the building where the vents are required to be installed by Section 910.2 with consideration given to roof pitch, draft curtain location, sprinkler location and structural members.

❖ This section has two functions, the first being a focus on hazards to adjacent buildings and the second being proper function of smoke and heat vents through proper placement.

In terms of adjacent properties, this section requires a minimum distance to lot lines and fire walls and then a minimum distance to fire barriers. The first set of distances focuses upon separate buildings and exposures, whereas the distance to fire barriers is less restrictive since it focuses upon different uses and occupancies within the same building (see Figure 910.3.4).

To enhance vent performance within the area containing the smoke and heat vents, such vents need to be uniformly spaced. Consideration of issues such as sprinkler location and roof pitch are also essential to proper vent location.

[F] 910.3.5 Draft curtains. Where required by Table 910.3, draft curtains shall be installed on the underside of the roof in accordance with this section.

Exception: Where areas of buildings are equipped with ESFR sprinklers, draft curtains shall not be provided within

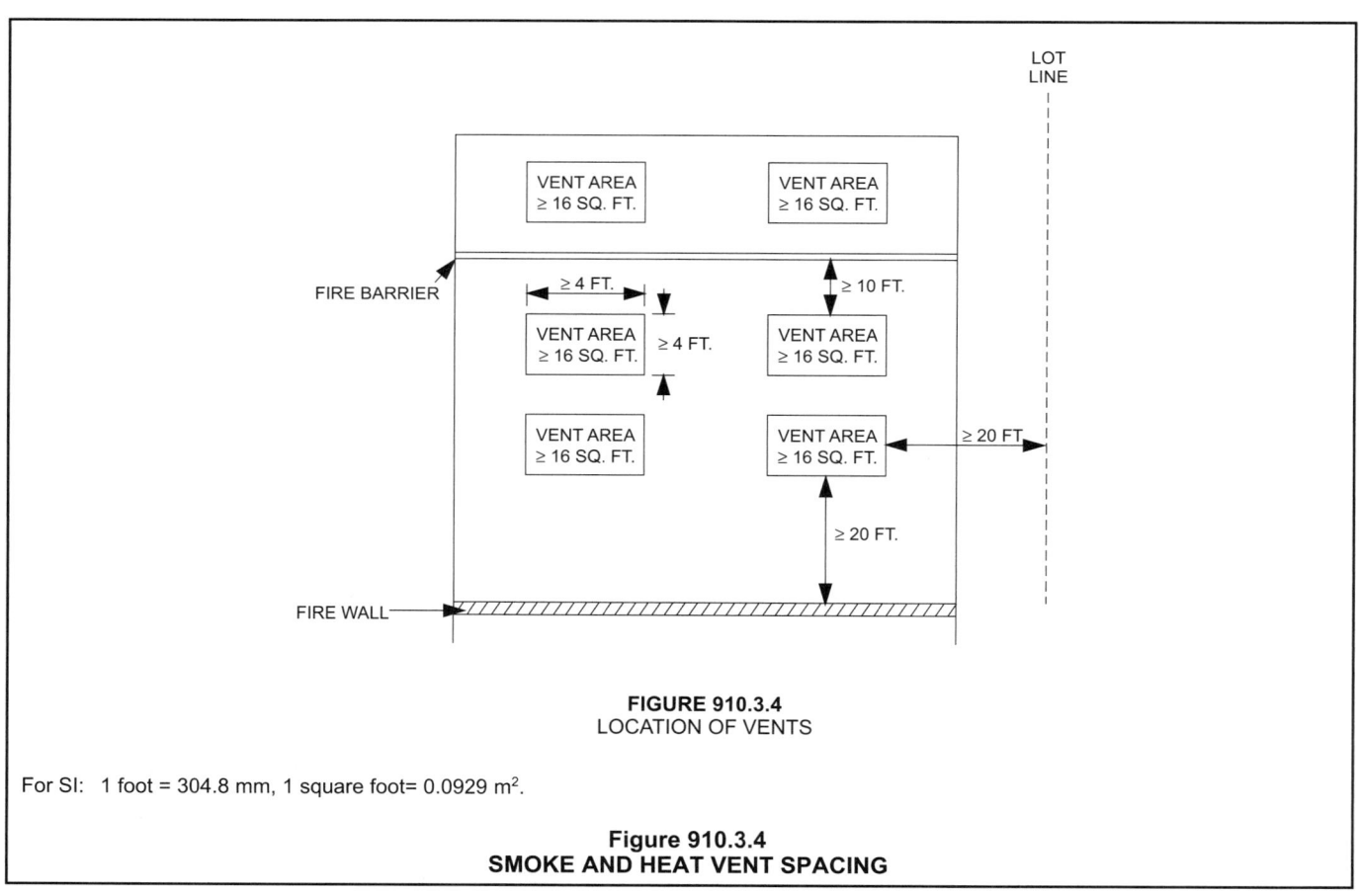

FIGURE 910.3.4
LOCATION OF VENTS

For SI: 1 foot = 304.8 mm, 1 square foot= 0.0929 m².

Figure 910.3.4
SMOKE AND HEAT VENT SPACING

these areas. Draft curtains shall only be provided at the separation between the ESFR sprinklers and the non-ESFR sprinklers.

❖ Draft curtains (sometimes termed "curtain boards") are required to be installed in conjunction with smoke and heat vents in accordance with Table 910.3 and as required by Section 413 of the code and Chapter 23 of the IFC. They are installed within and at the perimeter of a protected area to restrict smoke and heat movement beyond the area of fire origin or the protected area and enhance smoke and heat removal through the roof vents. Table 2306.2 of the IFC does not require draft curtains in sprinklered buildings. Instead, only smoke and heat vents are required in certain cases (larger areas of high-piled storage). The extent of the protection is addressed in Chapter 23 of the IFC. If draft curtains are required by Chapter 23 of the IFC, they need only extend 15 feet (4572 mm) beyond the high-piled storage area. This section also contains an exception for draft curtains when ESFR sprinklers are used. This exception would only apply to Group S-1 and F-1 occupancies as required in Section 910.2.1 because as noted, Chapter 23 of the IFC only requires draft curtains in unsprinklered buildings (see commentary, Table 2306.2 of the IFC). It should be noted that draft curtains are required between areas containing ESFR sprinklers and areas containing standard response sprinklers.

[F] 910.3.5.1 Construction. Draft curtains shall be constructed of sheet metal, lath and plaster, gypsum board or other *approved* materials which provide equivalent performance to resist the passage of smoke. Joints and connections shall be smoke tight.

❖ In order not to contribute to the fire load of a building and to increase the likelihood that draft curtains will remain intact under fire conditions, they must be constructed of noncombustible materials or an approved equivalent (see the commentary to Section 703.4 for further information on noncombustibility), but are not required to possess a fire-resistance rating. Draft curtains need only be capable of resisting the passage of smoke.

[F] 910.3.5.2 Location and depth. The location and minimum depth of draft curtains shall be in accordance with Table 910.3.

❖ The requirements for depth and location of draft curtains are provided in Table 910.3 based on the occupancy being Group S-1 or F-1 or, in the case of high-piled storage, the commodity classification of the stored materials and the height of the storage. High-piled storage areas would only be subject to these requirements when a building exceeds a certain minimum high-piled storage area threshold and not sprinklered. See the commentary to Table 910.3 for an explanation of how the options presented in the table should be applied.

[F] 910.4 Mechanical smoke exhaust. Where *approved* by the fire code official, engineered mechanical smoke exhaust shall be an acceptable alternate to smoke and heat vents.

❖ This section recognizes that providing a mechanical smoke exhaust system may, under certain circumstances, be more desirable, practical or efficient than installing automatic smoke and heat roof vents. The intent of Sections 910.4.1 through 910.4.6 is to create a mechanical system that performs at least as efficiently as smoke and heat vents designed in accordance with Section 910.3. Installation of an alternative mechanical smoke exhaust system is subject to the specific approval of the building official and fire code official so that the design can be reviewed and the operational sequence and control information can be shared with the fire department. Note that this smoke exhaust system is not considered a smoke control system. As discussed earlier, Section 910 is focused upon the needs of fire fighters in fighting a fire and the overhaul after the fire is extinguished. Section 909 addresses smoke control systems that focus upon tenable conditions for evacuation. Smoke control more specifically looks at the fire hazard and provides a system focused upon achieving certain life safety goals. The smoke exhaust system in this section is simply exhausting smoke at a rate that is not linked to a particular fire size.

[F] 910.4.1 Location. Exhaust fans shall be uniformly spaced within each draft-curtained area and the maximum distance between fans shall not be greater than 100 feet (30 480 mm).

❖ One or more smoke exhaust fans must be provided in each area defined by draft curtains, and when more than one fan is provided in a curtained area the fans must be spaced uniformly within that area, no more that 100 feet (30 480 mm) apart. Locating fans in this manner will enhance the uniform removal of smoke from curtained areas and reduce the likelihood of smoke spillage under the draft curtains. If draft curtains are not required, the fans simply need to meet the maximum separation distances and be uniformly distributed.

[F] 910.4.2 Size. Fans shall have a maximum individual capacity of 30,000 cfm (14.2 m³/s). The aggregate capacity of smoke exhaust fans shall be determined by the equation:

$$C = A \times 300 \qquad \textbf{(Equation 9-4)}$$

where:

C = Capacity of mechanical ventilation required, in cubic feet per minute (m³/s).

A = Area of roof vents provided in square feet (m²) in accordance with Table 910.3.

❖ The intent of the sizing requirements of this section is to provide a smoke exhaust rate at least equivalent to the venting capacity provided by roof vents. The exhaust rate required by this section, based on Equation 9-10, is equivalent to 300 cubic feet per minute per square foot (153 m³/s · m²) of the roof vent area required by Table 910.3, with no single fan exceeding a 30,000 cfm (14.2 m³/s) rate.

For example, a Group F-1 factory with maximum-sized draft-curtained areas of 50,000 square feet (4545 m²) would be required to have a total vent area of 500 square feet (46 m²) in each curtained area in accordance with Table 910.3. The mechanical exhaust rate required based on Equation 9-4 would then be 500 x 300 = 150,000 cfm (70.8 m³/s), which could be supplied by five 30,000 cfm (14.2 m³/s) fans spaced in accordance with Section 910.4.1 in each curtained area.

[F] 910.4.3 Operation. Mechanical smoke exhaust fans shall be automatically activated by the *automatic sprinkler system* or by *heat detectors* having operating characteristics equivalent to those described in Section 910.3.2. Individual manual controls of each fan unit shall also be provided.

❖ The activation of the mechanical smoke exhaust system must be capable of being accomplished by actuation of the automatic sprinkler system or, in nonsprinklered buildings, by heat detectors with a temperature rating of between 100 and 220°F (38 and 104°C) as required for smoke vents in Section 910.3.2.3 and manual controls. The manual control is for fire department use to increase the reliability of the system and to allow the fire department to activate the exhaust system to assist in the removal of smoke during or after a fire. Since manual control of the system is primarily for fire department use, the location of the controls should be subject to approval by the fire department. While not specifically stated in this section, a manual fire alarm system, if provided, can be more prone to intentional false activations; therefore, activation of the smoke exhaust system should not be allowed by this means. In addition, such systems should not be operated by the activation of smoke detection because the system may be activated before the sprinklers operate, which could be detrimental to the success of the sprinklers in suppressing the fire.

[F] 910.4.4 Wiring and control. Wiring for operation and control of smoke exhaust fans shall be connected ahead of the main disconnect and protected against exposure to temperatures in excess of 1,000°F (538°C) for a period of not less than 15 minutes. Controls shall be located so as to be immediately accessible to the fire service from the exterior of the building and protected against interior fire exposure by not less than 1-hour *fire barriers* constructed in accordance with Section 707 or *horizontal assemblies* constructed in accordance with Section 712, or both.

❖ Unless the mechanical smoke exhaust system also functions as a component of a smoke control system, standby power is not specifically required (see commentary, Sections 909.11 and 2702). In order to provide an enhanced level of operational reliability, this section requires that the power supply to smoke exhaust fans must be provided from a circuit connected on the supply side (i.e., ahead of) the building's main electrical service disconnecting means. Note that this is one of the sources of standby power recognized by the *National Electric Code®*, NFPA 70, Section 701.11(E). Such a circuit connected "ahead of the main" must still have its own approved overcurrent protection.

This section also requires that the wiring for smoke exhaust fans be thermally protected in a manner approved by the building official that will protect the wiring from heat damage in the event of an interior fire. This protection could be provided by an approved wiring material listed for the temperature application, by physical protection with approved materials or assemblies or by installation outside of the building.

Since smoke exhaust systems are a vital fire-fighting tool, their operating controls are also required to be protected from interior fire exposure by 1-hour fire barriers constructed in accordance with Section 707, horizontal assemblies constructed in accordance with Section 712, or both. Exterior access to the controls allows fire department personnel to promptly operate the system from a protected area without entering the building. Controls should also be clearly identified in an approved, permanent manner.

[F] 910.4.5 Supply air. Supply air for exhaust fans shall be provided at or near the floor level and shall be sized to provide a minimum of 50 percent of required exhaust. Openings for supply air shall be uniformly distributed around the periphery of the area served.

❖ The introduction of makeup air is critical to the proper operation of all exhaust systems. Too little makeup air will cause a negative pressure to develop in the area being exhausted, thereby reducing the exhaust flow.

This section requires that makeup air be introduced to the area equipped with a mechanical smoke exhaust system in order to maintain the required exhaust flow. Since the system can only exhaust as much air as is introduced into the area, and this section allows mechanical or gravity makeup air openings to provide only 50 percent of the required makeup air, this section allows the designer to rely upon infiltration air to provide up to the remaining 50 percent of the design makeup air required to allow the system to perform. Although not specifically stated in this section, where a mechanical makeup air source is utilized, it should be electrically interlocked and controlled by a single start switch, such that makeup air is always being supplied when the smoke exhaust system is in operation.

The even distribution of makeup air is important because if too much air is coming from one particular direction, it has the potential to vary the dynamics of the fire and the ability of the system to capture the smoke.

[F] 910.4.6 Interlocks. In combination comfort air-handling/smoke removal systems or independent comfort air-handling systems, fans shall be controlled to shut down in accordance with the *approved* smoke control sequence.

❖ This section was created to reduce the likelihood that the HVAC system will interfere with the proper function of the smoke exhaust system. It is important to emphasize that the system described in Section 910.4 is a smoke exhaust system and not a smoke control system; therefore, the actual fire performance is not as clearly understood. The concern is that HVAC systems should not work against the intended operation of the smoke exhaust system. In some case, the system may be a combination system where shutdown is not necessary or appropriate. It really depends on how the smoke exhaust system has been designed.

SECTION 911
FIRE COMMAND CENTER

[F] 911.1 General. Where required by other sections of this code and in all buildings classified as high-rise buildings by this code, a fire command center for fire department operations shall be provided and shall comply with Sections 911.1.1 through 911.1.5.

❖ Fireground operations usually involve establishing an incident command post where the incident command officer can observe what is happening; control arriving personnel and equipment and direct the resources and fire-fighting operations effectively. Because of the difficulties in controlling a fire in a high-rise building, a protected, readily accessible, separate room for this purpose within the building must be established to assist the incident command officer (see commentary, Section 902.1 for the definition of "Fire command center").

A fire command center is also required in buildings containing smoke-protected assembly seating to house the fire-fighter's smoke control panel. Facilities with smoke-protected seating tend to be larger facilities that, at the very least, would already have a central security center which could also function as a fire command center where approved by the jurisdiction (see commentary, Section 909.16).

[F] 911.1.1 Location and access. The location and accessibility of the fire command center shall be *approved* by the fire chief.

❖ Due to its importance to fire suppression and rescue operations, the fire command center must be provided at a location that is acceptable to the fire department, usually near the front of the building near the main entrance, so that the first arriving command officer can access it quickly and undertake operations. Since fireground operations are based on local operational procedures, it is only reasonable that the fire chief of the jurisdiction have approval authority over the location of and access to the fire command center.

[F] 911.1.2 Separation. The fire command center shall be separated from the remainder of the building by not less than a 1-hour *fire barrier* constructed in accordance with Section 707 or *horizontal assembly* constructed in accordance with Section 712, or both.

❖ Again, due to its importance to fire suppression and rescue operations, the fire command center must be separated from the remainder of the building by 1-hour fire barriers and horizontal assemblies, including opening protectives, to protect the room, its contents and the occupants from an incident in adjacent areas of the building and to limit noise and distractions during command operations within the room.

[F] 911.1.3 Size. The room shall be a minimum of 200 square feet (19 m²) with a minimum dimension of 10 feet (3048 mm).

❖ This section is intended to provide a minimum size and configuration of the fire command center that allows sufficient space for the necessary command personnel to effectively perform the required tasks associated with a fire command center without interfering with each other. Fire command centers need to be designed to accommodate several emergency response commanders wearing full protective equipment and also provide space to review building emergency plans during incidents, colocate decision makers within the Incident Command System (ICS) and interpret fire protection system and building system information generated by the features required by Section 911.1.5. Given the multiple uses of the fire command center, a room any smaller would compromise the effectiveness of incident management.

[F] 911.1.4 Layout approval. A layout of the fire command center and all features required by this section to be contained therein shall be submitted for approval prior to installation.

❖ The flow of critical tactical information into, within and out of a fire command center is, by its very nature, both high in volume and intense in nature and has a direct bearing on the safety of building occupants and the emergency response forces at work at an incident. For that reason, the layout and arrangement of the fire command center must comport with the operational procedures of the local fire department to optimize the receipt, processing and dissemination of operational information and orders. Accordingly, the fire code official must review and approve the arrangement of the fire command center prior to the installation of any of the controls and features required by Section 911.1.5. Consistent with Section 911.1.1, given the operational importance of the fire command center, the fire code official should work closely with the jurisdiction's fire chief to make sure that all operational needs are identified and met during the design stages.

[F] 911.1.5 Required features. The fire command center shall comply with NFPA 72 and shall contain the following features:

1. The emergency voice/alarm communication system control unit.

2. The fire department communications system.

3. Fire detection and alarm system annunciator.

4. Annunciator unit visually indicating the location of the elevators and whether they are operational.

5. Status indicators and controls for air distribution systems.

6. The fire-fighter's control panel required by Section 909.16 for smoke control systems installed in the building.

7. Controls for unlocking *stairway* doors simultaneously.

8. Sprinkler valve and waterflow detector display panels.

9. Emergency and standby power status indicators.

10. A telephone for fire department use with controlled access to the public telephone system.

11. Fire pump status indicators.

12. Schematic building plans indicating the typical floor plan and detailing the building core, *means of egress*, fire protection systems, fire-fighting equipment and fire department access and the location of *fire walls, fire barriers, fire partitions, smoke barriers* and smoke partitions.

13. Work table.

14. Generator supervision devices, manual start and transfer features.

15. Public address system, where specifically required by other sections of this code.

16. Elevator fire recall switch in accordance with ASME A17.1.

17. Elevator emergency or standby power selector switch(es), where emergency or standby power is provided.

❖ The fire command center must contain all equipment necessary to enable the incident commander to monitor or control fire protection and other building service systems as listed in this section (see also commentary, Section 909.16). The number and types of features required by this section can create a large volume of data, thus reinforcing the need for an approved layout as required by Section 911.1.4.

SECTION 912
FIRE DEPARTMENT CONNECTIONS

[F] 912.1 Installation. Fire department connections shall be installed in accordance with the NFPA standard applicable to the system design and shall comply with Sections 912.2 through 912.5.

❖ A fire department connection (FDC) is required as part of a water-based suppression system as the auxiliary water supply. These connections give the fire department the capability of supplying the necessary water to the automatic sprinkler or standpipe system at a sufficient pressure. The FDC also serves as an alternative source of water should a valve in the primary water supply be closed. A fire department connection does not, however, constitute an automatic water source. See Figure 903.3.1.1 for a typical FDC arrangement on a wet pipe sprinkler system.

The requirements for the FDC depend on the type of sprinkler system installed and whether a standpipe system is installed. NFPA 13 and 13R, for example, include design considerations for FDCs that are an auxiliary water supply source for automatic sprinkler systems; NFPA 14 is the design standard to use for FDCs serving standpipe systems. Threads for FDCs to sprinkler systems, standpipes, yard hydrants or any other fire hose connection must be approved (NFPA 1963 may be utitlized as part of the approval or as otherwise approved) and be compatible with the connections used by the local fire department (see commentary, Sections 903.3.6 and 905.1).

[F] 912.2 Location. With respect to hydrants, driveways, buildings and landscaping, fire department connections shall be so located that fire apparatus and hose connected to supply the system will not obstruct access to the buildings for other fire apparatus. The location of fire department connections shall be *approved* by the fire chief.

❖ This section specifies that the FDC must be located so that vehicles and hose lines will not interfere with access to the building for the use of other fire department apparatus. The location of potential connected hose lines to the FDC and hydrants must be preplanned with the fire department. Many fire departments have a policy restricting the distance that a FDC may be from a fire hydrant. Some also have policies that indicate the maximum distance from the nearest point of fire department vehicle access (often, the curb). Since fireground operations are based on local operational procedures, it is only reasonable that the fire chief of the jurisdiction have approval authority over the location of and access to the FDC.

Landscaping can also be a hindrance to fire department operations. Even where the FDC is visible, the extensive use of landscaping may make access difficult. Landscaping also changes over time. What may not have been an obstruction when it was planted can sometimes grow into an obstruction over time.

[F] 912.2.1 Visible location. Fire department connections shall be located on the street side of buildings, fully visible and recognizable from the street or nearest point of fire department vehicle access or as otherwise *approved* by the fire chief.

❖ FDCs must be readily visible and easily accessed. A local policy constituting what is readily visible and accessible needs to be established. While the intent is clearly understandable, its application can vary widely. A precise policy is the best way to avoid ambiguous directives that result in inconsistent and arbitrary enforcement. Usually, the policy will address issues such as location on the outside of the building and proximity to fire hydrants.

Landscaping is often used to hide the FDCs from the public. This can greatly hamper the efforts of the fire department in staging operations and supplying water to the fire protection systems. Landscaping must be designed so that it does not obstruct the visibility of the FDC. Since fireground operations are based on local operational procedures, it is only reasonable that the fire chief of the jurisdiction have final approval authority over the visibility of and access to the FDC.

[F] 912.2.2 Existing buildings. On existing buildings, wherever the fire department connection is not visible to approaching fire apparatus, the fire department connection shall be indicated by an *approved* sign mounted on the street front or on the side of the building. Such sign shall have the letters "FDC" at least 6 inches (152 mm) high and words in letters at least 2 inches (51 mm) high or an arrow to indicate the location. All such signs shall be subject to the approval of the fire code official.

❖ The section acknowledges that FDCs on existing buildings may not always be readily visible from the street or nearest point of fire department vehicle access. In those instances, the location of the connection must be clearly marked with signage. The FDC may be located on the side of the building or in an alley, not visible to arriving fire-fighting forces. A sign is necessary so that those driving the arriving apparatus know where to maneuver the vehicle to get close to the FDC.

[F] 912.3 Access. Immediate access to fire department connections shall be maintained at all times and without obstruction by fences, bushes, trees, walls or any other fixed or moveable object. Access to fire department connections shall be *approved* by the fire chief.

> **Exception:** Fences, where provided with an access gate equipped with a sign complying with the legend requirements of Section 912.4 and a means of emergency operation. The gate and the means of emergency operation shall be *approved* by the fire chief and maintained operational at all times.

❖ The FDC must be readily accessible to fire fighters and allow fire-fighting personnel an adequate area to maneuver a hose for the connection. Landscaping design must not block a clear view of the FDC from arriving fire department vehicles. Depending on the type of landscaping materials, an active maintenance program may be necessary to maintain ready access over time. This section also recognizes that the obstructing objects regulated here can be either fixed or moveable (such as outdoor furnishings, shopping cart queue areas, etc.). Note that no specific dimension is given as was the case in previous editions of the code. This performance language avoids previous misinterpretations that the code intended to allow obstructions to FDC access as long as they were kept 3 feet (914 mm) away. Since fireground operations are based on local operational procedures, it is only reasonable that the fire chief of the jurisdiction have final approval authority over the access to the FDC.

The exception recognizes the practical fact that sometimes, security or other considerations make installation of a fence around a building necessary as long as the fence meets the stated criteria. The sign requirement intends to provide a visual location cue to

approaching fire apparatus where the height of the fence may obscure the visibility of the FDC.

[F] 912.3.1 Locking fire department connection caps. The fire code official is authorized to require locking caps on fire department connections for water-based *fire protection systems* where the responding fire department carries appropriate key wrenches for removal.

❖ This section allows for the FDC caps to be equipped with locks as long as the fire departments that respond to that building or facility have the appropriate key wrenches. This avoids vandalism and affords a more functional FDC when needed. Locking caps, even more so than regular FDC caps, need proper maintenance so that they can be removed when required. Any time that an additional mechanical function is added to something that is exposed to the elements, it must be done with the understanding that the corrosive nature of the elements can place the FDC out of commission if the cap cannot be removed (see Figure 912.3.1).

Figure 912.3.1
LOCKING FDC CAPS
(Photo courtesy of Knox Company)

[F] 912.3.2 Clear space around connections. A working space of not less than 36 inches (762 mm) in width, 36 inches (914 mm) in depth and 78 inches (1981 mm) in height shall be provided and maintained in front of and to the sides of wall-mounted fire department connections and around the circumference of free-standing fire department connections, except as otherwise required or *approved* by the fire chief.

❖ Care must be taken so that fences, utility poles, barricades and other obstructions do not prevent access to and use of FDCs. A clear space of 3 feet (914 mm) must be maintained in front of and to either side of wall-mounted FDCs and around free-standing FDCs to allow easy hose connections to the fitting and efficient use of spanner wrenches and other tools needed by the apparatus engineer.

Though not specifically mentioned in this section, it is also important that FDCs be installed with the hose connections well above adjoining grade to accommodate the free turning of a spanner wrench when connecting hoses to the FDC.

[F] 912.3.3 Physical protection. Where fire department connections are subject to impact by a motor vehicle, vehicle impact protection shall be provided in accordance with Section 312 of the *International Fire Code.*

❖ Section 312 of the IFC requires vehicle impact protection by placing steel posts filled with concrete around the FDC. Section 312 of the IFC gives the specifications for the posts.

[F] 912.4 Signs. A metal sign with raised letters at least 1 inch (25 mm) in size shall be mounted on all fire department connections serving automatic sprinklers, standpipes or fire pump connections. Such signs shall read: AUTOMATIC SPRINKLERS or STANDPIPES or TEST CONNECTION or a combination thereof as applicable. Where the fire department connection does not serve the entire building, a sign shall be provided indicating the portions of the building served.

❖ The purpose of the sign is to provide the responding fire fighters with the correct information on which portions of a building are served by the fire department connection. They identify the type of system or zone served by a given FDC. Many buildings include multiple sets of fire department connections which are not interconnected, such as separate connections for the building sprinkler system and the dry standpipe system in open parking structures. Some buildings may have only a partial sprinkler system, such as rehabilitated buildings where a sprinkler system is only installed on certain floors or a building that only has basement sprinklers in accordance with Section 903.2.11.

Signs may also distinguish FDCs from fire pump test headers. Usually, FDCs may be distinguished from fire pump test headers by the types of couplings provided. FDCs are customarily equipped with female couplings, while fire pump test headers usually have separately valved male couplings. Furthermore, fire pump test headers are equipped with one 2^1/$_2$-inch (64 mm) outlet for each 250 gallon per minute (gpm) (16 L/s) of rated capacity.

Raised letters are required so that any repainting or fading of the colors on the sign will not affect its ability to be read. Each letter must be at least 1 inch (25 mm) in height so that the wording is clear. Often the wording may be abbreviated such that "AUTOMATIC SPRINKLERS" reads as "AUTO. SPKR." Existing signs may use language slightly different than that noted in the code. As long as the information is adequately communicated, there should be no reason to require new signage to replace existing ones (see Figure 912.4).

Figure 912.4
FIRE DEPARTMENT CONNECTION WITH SIGN

[P] 912.5 Backflow protection. The potable water supply to automatic sprinkler and standpipe systems shall be protected against backflow as required by the *International Plumbing Code.*

❖ Section 608.16.4 of the IPC requires all connections to automatic sprinkler systems and standpipe systems to be equipped with a means to protect the potable water supply. The means of backflow protection can be either a double check-valve assembly or a reduced-pressure-principle backflow preventer. This, in general, assumes a FDC is required. For example, a limited-area sprinkler system off the domestic supply does not necessarily require a FDC and would not require backflow protection.

SECTION 913
FIRE PUMPS

[F] 913.1 General. Where provided, fire pumps shall be installed in accordance with this section and NFPA 20.

❖ This section contains specific installation requirements for fire pumps supplying water to fire protection systems. Inspection, testing and maintenance requirements comply with NFPA 20 unless noted otherwise. Applicable maintenance standards are also identified.

Fire pumps are installed in sprinkler and standpipe systems to pressurize the water supply for the minimum required sprinkler and standpipe operation. They are considered a design feature or component of the system. Fire pumps can improve only the pressure of the incoming water supply, not the volume of water available.

When the volume from a water supply is not adequate to supply sprinkler or standpipe demand, water tanks for private fire protection, improvements in the size and capacity of fire mains or water distribution systems or all of these for the installation of a fire pump are needed.

When fire pumps are required to meet the pressure requirements of sprinkler and standpipe systems, they must be installed and tested in accordance with NFPA 20.

[F] 913.2 Protection against interruption of service. The fire pump, driver and controller shall be protected in accordance with NFPA 20 against possible interruption of service through damage caused by explosion, fire, flood, earthquake, rodents, insects, windstorm, freezing, vandalism and other adverse conditions.

❖ This section lists hazards that must be taken into account when determining the extent of protection required for the fire pump and its auxiliary equipment. A pump room in a building that is protected against the listed hazards in compliance with the code would be considered in compliance. Because fire pumps are also typically located in separate detached structures, geographical and security issues must also be considered.

913.2.1 Protection of fire pump rooms. Fire pumps shall be located in rooms that are separated from all other areas of the building by 2-hour *fire barriers* constructed in accordance with Section 707 or 2-hour *horizontal assemblies* constructed in accordance with Section 712, or both.

Exceptions:

1. In other than high-rise buildings, separation by 1-hour *fire barriers* constructed in accordance with Section 707 or 1-hour *horizontal assemblies* constructed in accordance with Section 712, or both, shall be permitted in buildings equipped throughout with an *automatic sprinkler system* in accordance with Section 903.3.1.1 or 903.3.1.2.

2. Separation is not required for fire pumps physically separated in accordance with NFPA 20.

❖ This section correlates the NFPA 20 requirements for separation of fire pumps with the IFC and specifies the required degree of fire-resistance as a requirement in the code so that designers and building officials are made aware of it. It requires fire pumps to be separated using either fire-resistive barriers or horizontal assemblies when the pump is located inside of a building or by using spatial separation (physical distance) when the fire pump is located outside of the building it serves.

The requirements are based on the criteria found in NFPA 20. Table 508.2.5, when used as a means for dealing with mixed uses in a building, and this section require fire pump units located inside of high-rise buildings to have a minimum 2-hour separation from the remainder of the building. The 2-hour separation requirement is consistent with the requirements in NFPA 20. Separation must be in the form of fire barriers used in conjunction with a horizontal assembly, as applicable, with both being of 2-hour fire-resistance-rated construction.

In all other buildings, consistent with similar exceptions elsewhere in the code, Exception 1 and Table 508.2.5, if applicable, recognize the protection afforded by complete automatic sprinkler protection in accordance with NFPA 13 or 13R as being equivalent to 1 hour of fire resistance.

A fire pump unit also may be located outside of a building. This is a common practice because it limits the exposure of a fire pump unit from a fire inside of the building it is serving. When a fire pump unit is located outdoors, Exception 2 recognizes spatial separation of at least 50 feet (15 240 mm) in accordance with Section 5.12 of NFPA 20 as an acceptable alternative to fire-resistance-rated protection.

[F] 913.3 Temperature of pump room. Suitable means shall be provided for maintaining the temperature of a pump room or pump house, where required, above 40°F (5°C).

❖ As previously noted for sprinkler systems, standpipe systems and other water-based fire protection systems, pump rooms or pump houses must be maintained at a temperature of 40°F (4°C) or above to prevent the system from freezing. This is consistent with Section 5.12.2.1 of NFPA 20.

[F] 913.3.1 Engine manufacturer's recommendation. Temperature of the pump room, pump house or area where engines are installed shall never be less than the minimum recommended by the engine manufacturer. The engine manufacturer's recommendations for oil heaters shall be followed.

❖ If the engine manufacturer's recommended minimum temperature is higher than the minimum established in Section 913.3, that recommendation must be complied with. Maintaining the desired engine temperature enhances the startability of the engine. Maintaining water heaters and oil heaters as required for diesel engines, for example, will improve the starting capabilities of the fire pump and reduce engine wear and the drain on batteries.

[F] 913.4 Valve supervision. Where provided, the fire pump suction, discharge and bypass valves, and isolation valves on the backflow prevention device or assembly shall be supervised open by one of the following methods:

1. Central-station, proprietary or remote-station signaling service.

2. Local signaling service that will cause the sounding of an audible signal at a *constantly attended location*.

3. Locking valves open.

4. Sealing of valves and *approved* weekly recorded inspection where valves are located within fenced enclosures under the control of the owner.

❖ As was the case with sprinkler systems, water control valves that are a part of the fire pump installation must be supervised in the open position so that the system is operational when needed and also to reduce the chance of a system failure (see commentary, Section 903.4). In most cases the required water-based extinguishing system, which the fire pump is an integral component of, will be electrically supervised. Locking or sealing valves open as the only means of supervision may not be permitted, depending on the type of valve. Section 903.4, for example, specifically exempts jockey pump control valves from being electrically supervised if they are sealed or locked in the open position.

[F] 913.4.1 Test outlet valve supervision. Fire pump test outlet valves shall be supervised in the closed position.

❖ Fire pump test outlet valves are for performance testing of the fire pump and do not control the available water supply to either a sprinkler system or a standpipe system. These valves are normally in a closed position and are supervised accordingly.

[F] 913.5 Acceptance test. Acceptance testing shall be done in accordance with the requirements of NFPA 20.

❖ Chapter 14 of NFPA 20 details the procedure for conducting a fire pump acceptance test. This test is run to determine that the installation matches the sprinkler or standpipe system design criteria, the approved shop drawings and the pump manufacturer's performance specifications. The test is to be conducted in the presence of the building official in accordance with Section 901.5 by the installing contractor and representatives of the pump manufacturer and the controller manufacturer. Where the pump engine and/or transfer switch are separately supplied components, their manufacturer representative must also be present.

SECTION 914
EMERGENCY RESPONDER SAFETY FEATURES

[F] 914.1 Shaftway markings. Vertical shafts shall be identified as required by Sections 914.1.1 and 914.1.2.

❖ This section was developed to prevent fire fighters from falling through shafts when entering buildings off ladders placed on the exterior of the building or from passing through an interior opening that leads to a shaft.

[F] 914.1.1 Exterior access to shaftways. Outside openings accessible to the fire department and that open directly on a hoistway or shaftway communicating between two or more floors in a building shall be plainly marked with the word "SHAFTWAY" in red letters at least 6 inches (152 mm) high on a white background. Such warning signs shall be placed so as to be readily discernible from the outside of the building.

❖ All exterior wall openings that are accessible to fire fighters by way of ladders and aerial equipment and open directly into shafts or hoistways communicating between two or more floors must be clearly marked (see Figure 914.1.1). The markings serve to warn emergency responders that the opening is unsafe for laddering to gain access to upper floors and could result in a fall to the bottom of the shaft if such a mistake were made.

[F] 914.1.2 Interior access to shaftways. Door or window openings to a hoistway or shaftway from the interior of the building shall be plainly marked with the word "SHAFTWAY"

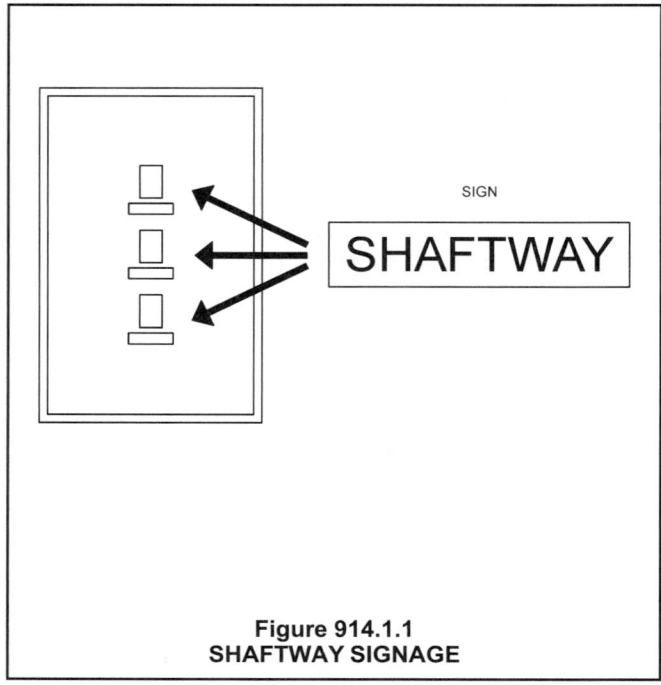

Figure 914.1.1
SHAFTWAY SIGNAGE

in red letters at least 6 inches (152 mm) high on a white background. Such warning signs shall be placed so as to be readily discernible.

Exception: Markings shall not be required on shaftway openings that are readily discernible as openings onto a shaftway by the construction or arrangement.

❖ Openings into shaftways from the interior of the building pose a threat to fire fighters when visibility is poor. Interior shaft openings must be marked so that they are plainly visible from the interior of the building. If fire fighters can readily identify an opening into a shaft by the way the opening is constructed, the shaft opening need not be marked, keeping in mind that the fire fighter may be feeling his or her way in heavy smoke or darkness.

[F] 914.2 Equipment room identification. Fire protection equipment shall be identified in an *approved* manner. Rooms containing controls for air-conditioning systems, sprinkler risers and valves or other fire detection, suppression or control elements shall be identified for the use of the fire department. *Approved* signs required to identify fire protection equipment and equipment location shall be constructed of durable materials, permanently installed and readily visible.

❖ In an emergency, it is vitally important that the fire department and other emergency responders be able to quickly locate and access critical controls for fire protection systems. Obstructed or poorly marked equipment can cause delays in fire-fighting operations while fire fighters locate other hose stations and stretch additional hose, for example. Valves and other controls are often located in rooms or other enclosures and their location must be clearly identified with written or pictographic signs, which must be clearly visible and legible. Signs using the NFPA 170 symbols for fire protection equipment can provide standardized markings throughout a jurisdiction. White reflective symbols on a red reflective background are effective. For exterior signs, heavy-gage, sign-grade aluminum is recommended. Interior signs may be constructed of plastic, light-gage aluminum or other approved, durable, water-resistant material. As a general rule, fire protection piping, cabinets, enclosures, wiring, equipment and accessories are red or are identified by red or red/white markings. The manner of identification is subject to the approval of the building official.

SECTION 915
EMERGENCY RESPONDER RADIO COVERAGE

[F] 915.1 General. Emergency responder radio coverage shall be provided in all new buildings in accordance with Section 510 of the *International Fire Code*.

❖ The provisions of Section 915 are new to the 2009 edition and are concerned with the reliability of portable radios used by emergency responders inside of buildings. This is in keeping with the philosophy inherent in the I-Codes® that, when a facility grows too large or complex for effective fire response, fire protection features must be provided within the building. See the commentary to Section 510 of the IFC (and its companion Appendix J) for complete information on this topic.

Bibliography

The following resource materials are referenced in this chapter or are relevant to the subject matter addressed in this chapter.

Americans with Disabilities Act Accessibility Guidelines for Buildings and Facilities (ADAAG). Washington, DC: US. Architectural and Transportation Barriers Compliance Board, 1994.

Americans With Disabilities Act (ADA) Accessibility Guidelines for Buildings and Facilities; Architectural Barriers Act (ABA) Accessibility Guidelines; (ADAAG). Washington, DC: US. Architectural and Transportation Barriers Compliance Board, 2004.

ASME A 17.1/CSA B 44-07, *Safety Code for Elevators and Escalators*. New York, NY: American Society of Mechanical Engineers, 2007.

Automatic Sprinkler Systems Handbook. Quincy, MA: National Fire Protection Association, 2007.

Bryan, John L. *Automatic Sprinkler and Standpipe Systems*, 4th edition. Quincy, MA: National Fire Protection Association, 2006.

DOJ 28 CFR, Part 36-91, *Americans With Disabilities Act*. Washington, DC: U.S. Department of Justice, 1991.

DOJ 28 CFR, Part 36 (Appendix A)-91, *ADA Guidelines for Buildings and Facilities*. Washington, DC: U.S. Department of Justice, 1991.

DOL 29 CFR, Part 1910.1000, *Air Contaminants*. Washington, DC: U.S. Department of Labor, 1974.

DOTn 49 CFR, Part 100-185-05, *Hazardous Materials Regulations*. Washington, DC: United States Department of Transportation, 2005.

Fire Protection Handbook, 19th edition. Quincy, MA: National Fire Protection Association, 2003.

ICC/ANSI A117.1-03, *Accessible and Usable Buildings and Facilities*. Washington, DC: International Code Council, 2003.

IFC-09, *International Fire Code*. Washington, DC: International Code Council, 2009.

IMC-09, *International Mechanical Code*. Washington, DC: International Code Council, 2009.

International Code Interpretations. Washington, DC: International Code Council, 2009.

International Fire Engineering Guidelines. Australian Building Codes Board, 2005.

IPC-09, *International Plumbing Code*. Washington, DC: International Code Council, 2009.

Klote, J. and D. Evans. *A Guide to Smoke Control in the 2006 IBC*. Washington, DC: International Code Council, 2007.

Klote, J. and J. Milke. *Principles of Smoke Management*. Atlanta, GA: American Society of Heating, Refrigerating and Air-Conditioning Engineers, 2002.

Morgan, H.P. and N.R. Marshall. "*Smoke Control Measures in Covered Two-story Shopping Malls Having Balconies as Pedestrian Walkways*." BRE CP 11/79, Borehamwood, 1979.

National Fire Alarm Code Handbook. Quincy, MA: National Fire Protection Association, 2007.

National Institute of Standards and Technology. *Final Report of the National Construction Safety Team on the Collapses of the World Trade Center Towers*. United States Government Printing Office: Washington, DC, September 2005.

NFPA 10-07, *Portable Fire Extinguishers*. Quincy, MA: National Fire Protection Association, 2007.

NFPA 11-05, *Low Expansion Foam and Combined Agent Systems*. Quincy, MA: National Fire Protection Association, 2005.

NFPA 11A-99, *Standard for Medium and High-expansion Foam Systems*. Quincy, MA: National Fire Protection Association, 1999.

NFPA 12-05, *Carbon Dioxide Extinguishing Systems*. Quincy, MA: National Fire Protection Association, 2005.

NFPA 12A-04, *Halon 1301 Fire Extinguishing Systems*. Quincy, MA: National Fire Protection Association, 2004.

NFPA 13-07, *Installation of Sprinkler Systems*. Quincy, MA: National Fire Protection Association, 2007.

NFPA 13D-07, *Installation of Sprinkler Systems in One- and Two-family Dwellings and Manufactured Homes*. Quincy, MA: National Fire Protection Association, 2007.

NFPA 13R-07, *Installation of Sprinkler Systems in Residential Occupancies Up to and Including Four Stories in Height*. Quincy, MA: National Fire Protection Association, 2007.

NFPA 14-07, *Standpipe and Hose Systems*. Quincy, MA: National Fire Protection Association, 2007.

NFPA 16-07, *Installation of Deluge Foam-water Sprinkler and Foam-water Spray Systems*. Quincy, MA: National Fire Protection Association, 2007.

NFPA 17-02, *Dry-chemical Extinguishing Systems*. Quincy, MA: National Fire Protection Association, 2002.

NFPA 17A-02, *Wet-chemical Extinguishing Systems*. Quincy, MA: National Fire Protection Association, 2002.

NFPA 20-07, *Installation of Centrifugal Fire Pumps*. Quincy, MA: National Fire Protection Association, 2007.

NFPA 24-07, *Installation of Private Fire Service Mains*. Quincy, MA: National Fire Protection Association, 2007.

NFPA 25-08, *Inspection, Testing and Maintenance of Water-based Fire Protection Systems*. Quincy, MA: National Fire Protection Association, 2008.

NFPA 30-08, *Flammable and Combustible Liquids Code*. Quincy, MA: National Fire Protection Association, 2008.

NFPA 30B-07, *Manufacture and Storage of Aerosol Products*. Quincy, MA: National Fire Protection Association, 2007.

NFPA 70-08, *National Electrical Code*. Quincy, MA: National Fire Protection Association, 2008.

NFPA 72-07, *National Fire Alarm Code*. Quincy, MA: National Fire Protection Association, 2007.

NFPA 92A-09, *Smoke Control Systems Utilizing Barriers and Pressure Differences*. Quincy, MA: National Fire Protection Association, 2009.

NFPA 92B-05, *Smoke Management Systems in Malls, Atria and Large Areas*. Quincy, MA: National Fire Protection Association, 2005.

NFPA 170-06, *Standard for Fire Safety and Emergency Symbols*. Quincy, MA: National Fire Protection Association, 2006.

NFPA 204-07, *Standard for Smoke and Heat Venting*. Quincy, MA: National Fire Protection Association, 2007.

NFPA 303-06, Fire *Protection Standard for Marinas and Boatyards*. Quincy, MA: National Fire Protection Association, 2006.

NFPA 1963-09, *Fire Hose Connections*. Quincy, MA: National Fire Protection Association, 2009.

NFPA 2001-08, *Clean Agent Fire Extinguishing Systems*. Quincy, MA: National Fire Protection Association, 2008.

National Institute of Standards and Technology. *Final Report of the National Construction Safety Team on the Collapses of the World Trade Center Towers*. Washington, DC: United States Government Printing Office, September 2005.

SFPE Engineering Guide to Performance-based Fire Protection Analysis and Design of Buildings. Quincy, MA: National Fire Protection Association, 2004.

Smoke Control in Fire Safety Design. London: E. & F.N. Spon Ltd., 1979.

The SFPE Handbook of Fire Protection Engineering 4th edition, Quincy, MA: National Fire Protection Association, 2008.

UL 33-03, *Standard for Heat-responsive Links for Fire Protection Service*. Northbrook, IL: Underwriters Laboratories Inc., 2003.

UL 217-06, *Single- and Multiple-station Smoke Detectors*. Northbrook, IL: Underwriters Laboratories Inc., 2006.

UL 268-06, *Smoke Detectors for Fire Alarm Signaling Systems*. Northbrook, IL: Underwriters Laboratories Inc., 2006.

UL 300-05, *Standard for Fire Testing of Fire Extinguishing Systems for Protection of Restaurant Cooking Areas*. Northbrook, IL: Underwriters Laboratories Inc., 2005.

UL 555-06, *Fire Dampers*. Northbrook, IL: Underwriters Laboratories Inc., 2006.

UL 555S-99, *Smoke Dampers*. Northbrook, IL: Underwriters Laboratories Inc.,1999.

UL 710B-04, *Recirculating Systems*. Northbrook, IL: Underwriters Laboratories Inc., 2004.

UL 793-03, *Standard for Automatically Operated Roof Vents for Smoke and Heat*. Northbrook, IL: Underwriters Laboratories Inc., 2003.

UL 864-03, *Standard for Control Units and Accessories for Fire Alarm Systems—with Revisions through October 2003*. Northbrook, IL: Underwriters Laboratories Inc., 2003.

Chapter 10:
Means of Egress

General Comments

The general criteria set forth in Chapter 10 regulating the design of the means of egress are established as the primary method for protection of people in buildings. Chapter 10 provides the minimum requirements for means of egress in all buildings and structures. Both prescriptive and performance language is utilized in this chapter to provide for a basic approach in the determination of a safe exiting system for all occupancies. It addresses all portions of the egress system and includes design requirements as well as provisions regulating individual components. The requirements detail the size, arrangement, number and protection of means of egress components. Functional and operational characteristics also are specified for the components that will permit their safe use without special knowledge or effort.

A zonal approach to egress provides a general basis for the chapter's format through regulation of the exit access, exit and exit discharge portions of the means of egress. Section 1001 includes the administrative provisions. Section 1002 shows the definitions of terms that are primarily associated with Chapter 10. Sections 1003 through 1013 include general provisions that apply to all three components of a means of egress system: exit access, exit and exit discharge. The exit access requirements are in Sections 1014 through 1019, the exit requirements are in Sections 1020 through 1026 and the exit discharge requirements are in Section 1027. Section 1028 includes those means of egress requirements that are unique to an assembly occupancy. Emergency escape and rescue opening requirements are in Section 1029. Chapter 10 requirements are repeated in Chapter 10 of the *International Fire Code*® (IFC®). The IFC has one additional section at the end of the chapter dealing with maintenance of the means of egress (see commentary, Section 1001.3). For means of egress requirements in existing buildings, refer to Chapter 34 of the code or IFC Chapter 46.

The evolution of means of egress requirements has been influenced by lessons learned from real fire incidents. While contemporary fires may reinforce some of these lessons, one must view each incident as an opportunity to assess critically the safety and reasonability of current regulations.

Cooperation among the developers of model codes and standards has resulted in agreement on many basic terms and concepts. The text of the code, including this chapter, is consistent with these national uniformity efforts.

National uniformity in an area such as means of egress has many benefits for the building official and other code users. At the top of the list are the lessons to be learned from experiences throughout the nation and the world, which can be reported in commonly used terminology and conditions that we can all relate to and clearly understand.

Purpose

A primary purpose of codes in general and building codes in particular is to safeguard life in the presence of a fire. Integral to this purpose is the path of egress travel for occupants to escape and avoid a fire. Means of egress can be considered the lifeline of a building. The principles on which means of egress are based and that form the fundamental criteria for requirements are to provide a system:

1. That will give occupants alternative paths of travel to a place of safety to avoid fire.

2. That will shelter occupants from fire and the products of combustion.

3. That will accommodate all occupants of a structure.

4. That is clear, unobstructed, well marked and illuminated and in which all components are under control of the user without requiring any tools, keys or special knowledge or effort.

History is marked with the severe loss of life from fire. Early as well as contemporary multiple fire fatalities can be traced to a compromise of one or more of the above principles.

Life safety from fire is a matter of successfully evacuating or relocating the occupants of a building to a place of safety. As a result, life safety is a function of time: time for detection, time for notification and time for safe egress. The fire growth rate over a period of time is also a critical factor in addressing life safety. Other sections of the code, such as protection of vertical openings (see Chapter 7), interior finish (see Chapter 8), fire suppression and detection systems (see Chapter 9) and numerous others, also have an impact on life safety. This chapter addresses the issues related to the means available to relocate or evacuate building occupants.

SECTION 1001
ADMINISTRATION

1001.1 General. Buildings or portions thereof shall be provided with a *means of egress* system as required by this chapter. The provisions of this chapter shall control the design, construction and arrangement of *means of egress* components required to provide an *approved means of egress* from structures and portions thereof.

❖ The minimum requirements for emergency evacuation, or means of egress, are to be incorporated in all new structures as specified in this chapter. The system shall include exit access, exit and exit discharge and address the needs for all occupants of the facility. Such application would be effective on the date the code is adopted and placed into effect.

1001.2 Minimum requirements. It shall be unlawful to alter a building or structure in a manner that will reduce the number of *exits* or the capacity of the *means of egress* to less than required by this code.

❖ A fundamental concept in life safety design is that the means of egress system is to be constantly available throughout the life of a building. Any change in the building or its contents, either by physical reconstruction, alteration or by a change of occupancy, is cause to review the resulting egress system. At a minimum, a building's means of egress is to be continued as initially approved. If a building or portion thereof has a change of occupancy, the complete egress system is to be evaluated and approved for compliance with the current code requirements for new occupancies (see Chapter 34).

The means of egress in an existing building that experiences a change of occupancy, such as from Group S-2 (storage) to A-3 (assembly), would require reevaluation for code compliance based on the new occupancy. Similarly, the means of egress in an existing Group A-3 occupancy in which additional seating is to be provided, thereby increasing the occupant load, would require reevaluation for code compliance based on the increased occupant load.

The temptation is to temporarily remove egress components or other fire protection features from service during an alteration, repair to or temporary occupancy of a building. During such times, a building is frequently more vulnerable to fire and the rapid spread of products of combustion. Either the occupants should not occupy those spaces where the means of egress has been compromised by the construction or compensating fire safety features should be considered that will provide equivalent safety for the occupants. Occupants in adjacent areas may also require access to the egress facilities in the area under construction.

[F] 1001.3 Maintenance. *Means of egress* shall be maintained in accordance with the *International Fire Code*.

❖ This section provides a cross reference to the code requirements that address the maintenance of the means of egress in an existing building. The means of egress must be maintained so that occupants are not prevented from exiting the building quickly in case of an emergency.

Sections 1002 through 1029 in the code are repeated in the IFC. These sections are maintained by the building code committees so that there will be consistency between the two documents. Note the [B] in front of the main section headings in the IFC. Note the [F] in front of this section. This means that this section is maintained by the International Fire Code Development Committee. Additionally, the IFC includes Section 1030, which applies to maintenance of the means of egress. For means of egress in existing buildings, refer to IFC Chapter 46 and Chapter 34 of the code.

SECTION 1002
DEFINITIONS

1002.1 Definitions. The following words and terms shall, for the purposes of this chapter and as used elsewhere in this code, have the meanings shown herein.

❖ Definitions of terms can help in the understanding and application of the code requirements. The purpose for including these definitions in this chapter is to provide more convenient access to them without having to refer back to Chapter 2.

These terms are also listed in Chapter 2 with a cross reference to this section. The use and application of all defined terms, including those defined herein, are set forth in Section 201.

ACCESSIBLE MEANS OF EGRESS. A continuous and unobstructed way of egress travel from any *accessible* point in a building or facility to a *public way*.

❖ Accessible means of egress requirements are needed to provide those persons with physical disabilities or mobility impairments a means of egress to exit the building. Because of physical limitations, some occupants may need assistance to exit a building. See Section 1007 for requirements establishing areas where people can safely wait for assisted rescue. Chapter 4 of the IFC also includes requirements in the fire safety and evacuation plans for specific planning to address occupants who may need assistance in evacuation during emergencies. In addition, Chapter 9 of the code includes requirements for emergency evacuation notification for persons with hearing and vision disabilities.

The accessible means of egress requirements may not be the same route as that required for ingress into the building (see Sections 1104 and 1105). For example, a two-story building requires one accessible route to connect all accessible spaces within the building. The accessible route to the second level is typically by an elevator. During a fire emergency, persons with mobility impairments on the second level would be moving to the exit stairways for assisted rescue, not back the way they came in, via the elevator.

AISLE. An unenclosed *exit access* component that defines and provides a path of egress travel.

❖ Aisles and aisle accessways are both utilized as part of the means of egress in facilities where tables, seats, displays or other furniture may limit the path of travel. The aisle accessways lead to the main aisles that lead to the exits from the space and building [see Figure 1002.1(1)]. While both may result in a confined path of travel, an aisle is an unenclosed component, while a corridor would be an enclosed component of the means of egress. See Sections 1017 and 1028 for requirements for aisles.

AISLE ACCESSWAY. That portion of an *exit access* that leads to an *aisle*.

❖ As illustrated in Figure 1002.1(1), an aisle accessway is intended for one-way travel or limited two-way travel. The space between tables, seats, displays or other furniture (i.e., aisle accessway) utilized for means of egress will lead to a main aisle. See Sections 1017 and 1028 for requirements for aisle accessways.

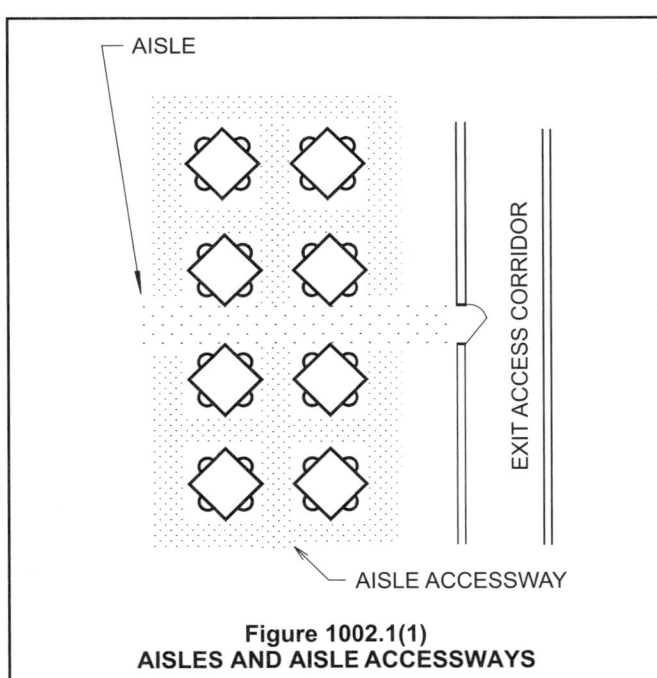

Figure 1002.1(1)
AISLES AND AISLE ACCESSWAYS

ALTERNATING TREAD DEVICE. A device that has a series of steps between 50 and 70 degrees (0.87 and 1.22 rad) from horizontal, usually attached to a center support rail in an alternating manner so that the user does not have both feet on the same level at the same time.

❖ An alternating tread device is commonly used in areas that would otherwise be provided with a ladder where there is not adequate space for a full stairway. Where these devices are permitted is specifically listed (i.e., Section 1015.3). The device is used extensively in industrial facilities for worker access to platforms or equipment. Requirements are found with stairways in Section 1009.

AREA OF REFUGE. An area where persons unable to use *stairways* can remain temporarily to await instructions or assistance during emergency evacuation.

❖ The area of refuge is a temporary waiting area used during emergency evacuations for persons who are unable to exit the building using the stairways. The fire safety plans (in accordance with IFC Section 404) include the locations of areas of refuge so that the fire department will know where people may be waiting for rescue assistance. See Section 1007 for where areas of refuge are required at stairways and elevators. Areas of refuge have requirements for separation, size, signage, instructional information and two-way communication systems.

BLEACHERS. Tiered seating supported on a dedicated structural system and two or more rows high and is not a building element (see "*Grandstands*").

❖ Bleachers, folding and telescopic seating and grandstands are essentially unique forms of tiered seating that are supported on a dedicated structural system. All types are addressed in ICC 300, *Standard on Bleachers, Folding and Telescopic Seating and Grandstands*, the safety standard for these types of seating arrangements. Bleachers often do not have backrests. The travel path across the bleachers is not restricted to designated rows, aisles and aisle accessways. Without backrests, occupants can traverse from row to row without traveling to the designated egress aisles (see Section 1028.1.1). The term "Building element" is defined in Section 702.1. An individual bench seat directly attached to a floor system is not a bleacher. The terms "bleacher" and "grandstand" are basically interchangeable. There is no cut-off in size or number of seats that separates bleachers and grandstands.

COMMON PATH OF EGRESS TRAVEL. That portion of *exit access* which the occupants are required to traverse before two separate and distinct paths of egress travel to two *exits* are available. Paths that merge are common paths of travel. Common paths of egress travel shall be included within the permitted travel distance.

❖ The common path of egress travel is a concept used to refine travel distance criteria. A common path of travel is the route an occupant will travel where the only way in is also the only way out, similar to a dead-end corridor. The length of a common path of egress travel is limited so that the means of egress path of travel provides a choice before the occupant has traveled an excessive distance (see Section 1014.3). This reduces the possibility that, although the exits are remote from one another, a single fire condition will render both paths unavailable.

CORRIDOR. An enclosed *exit access* component that defines and provides a path of egress travel to an *exit*.

❖ Corridors are regulated in the code because they serve as principal elements of travel in many means of egress systems within buildings. Typically, corridors have walls that extend from the floor to the ceiling. They need not extend above the ceiling or have doors in their openings unless a fire-resistance rating is required (see Section 1018).

While both aisles and corridors may result in a confined path of travel, an aisle is an unenclosed component, while a corridor would be an enclosed component of the means of egress. The enclosed character of the corridor restricts the sensory perception of the user. A fire located on the other side of the corridor wall, for example, may not be as readily seen, heard or smelled by the occupants traveling through the egress corridor. The code does not specifically state what is considered "enclosed" when corridors are not fire-resistance rated. When an egress path is bounded by partial-height walls, such as work-station partitions in an office, issues would be if the walls provided a confined path of travel and limited fire recognition in adjacent spaces by restricting line of sight, hearing and smell.

DOOR, BALANCED. A door equipped with double-pivoted hardware so designed as to cause a semicounter balanced swing action when opening.

❖ Balanced doors are commonly used to decrease the force necessary to open the door or to reduce the length of the door swing. Balanced doors typically reduce the clear opening width more than normally hinged doors [see Figure 1002.1(2) and Section 1008.1.10.2].

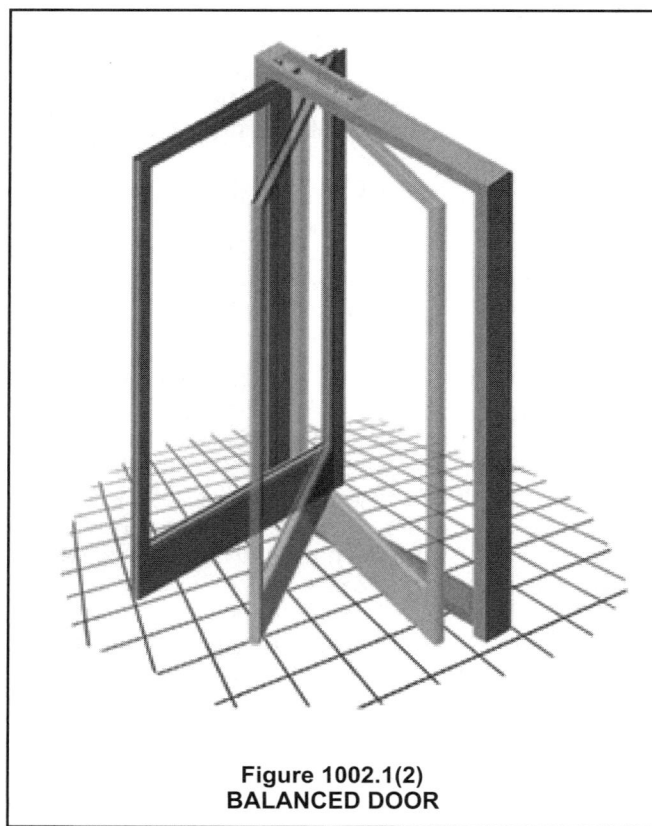

Figure 1002.1(2)
BALANCED DOOR

EGRESS COURT. A court or *yard* which provides access to a *public way* for one or more *exits*.

❖ The egress court requirements address situations where the exit discharge portion of the means of egress passes through confined areas near the building and therefore faces a hazard not normally found in the exit discharge (see Section 1027.5).

EMERGENCY ESCAPE AND RESCUE OPENING. An operable window, door or other similar device that provides for a means of escape and access for rescue in the event of an emergency.

❖ These are commonly windows that are sized and located such that they can be used to exit a building directly from a basement or bedroom during an emergency condition. The openings are also used by emergency personnel to rescue the occupants in a building (see Section 1029). They are never considered to be exit or exit access components for purposes of meeting minimum number of exit requirements.

EXIT. That portion of a *means of egress* system which is separated from other interior spaces of a building or structure by fire-resistance-rated construction and opening protectives as required to provide a protected path of egress travel between the *exit access* and the *exit discharge*. Exits include exterior exit doors at the *level of exit discharge*, vertical *exit enclosures*, *exit passageways*, *exterior exit stairways*, exterior *exit ramps* and *horizontal exits*.

❖ Exits are the critical element of the means of egress system that the building occupants travel through to reach the exterior at the level of exit discharge. Exit stairways and ramps from upper and lower stories must be separated from adjacent areas with fire-resistance-rated construction. The fire-resistance-rated construction serves as a barrier between the fire and the means of egress and protects the occupants while they travel through the exit. Separation by fire-resistance-rated construction is not required, however, where the exit leads directly to the exterior at the level of exit discharge (e.g., exterior door at grade). Figure 1002.1(3) illustrates three different types of exits: interior exit stairway, exterior exit stairway and exterior exit door.

A horizontal exit, while not discharging to the outside, does discharge to another building or refuge area. The door to the refuge area is through a fire wall or fire barrier (see the definition for "Exit, horizontal" and Section 1025).

EXIT ACCESS. That portion of a *means of egress* system that leads from any occupied portion of a building or structure to an *exit*.

❖ The exit access portion of the means of egress consists of all floor areas that lead from usable spaces within the building to the exit or exits serving that floor area. Crawl spaces and concealed attic and roof spaces are not considered to be part of the exit access. As shown in Figure 1002.1(5), the exit access begins at the furthest points within each room or space and ends at the entrance to the exit.

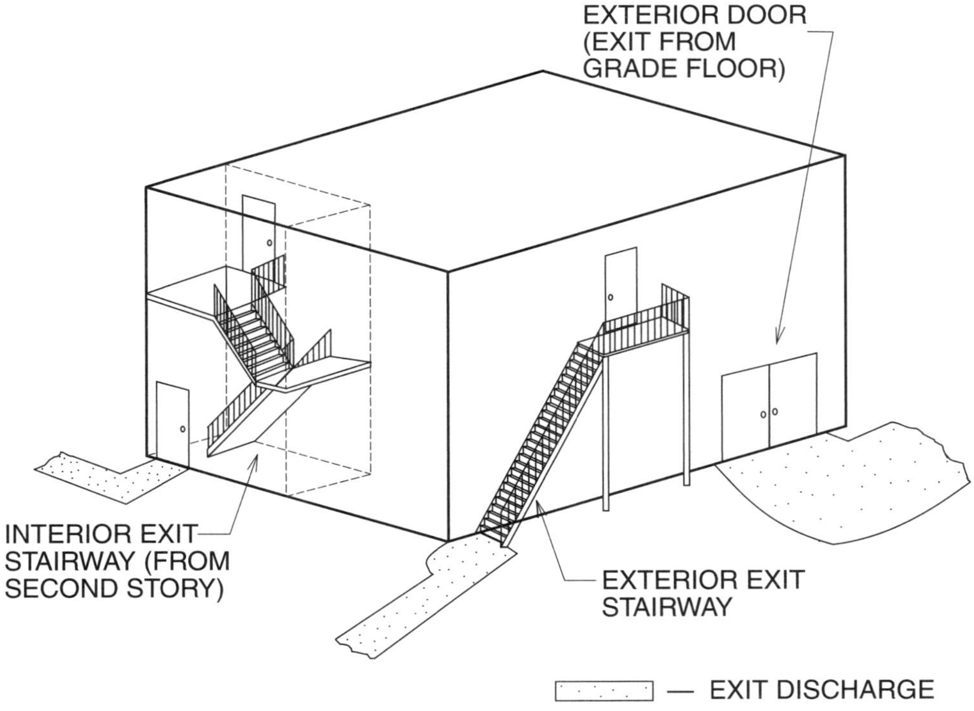

EXTERIOR DOOR
(EXIT FROM
GRADE FLOOR)

INTERIOR EXIT
STAIRWAY (FROM
SECOND STORY)

EXTERIOR EXIT
STAIRWAY

⬚⋯⬚ — EXIT DISCHARGE

Figure 1002.1(3)
EXIT

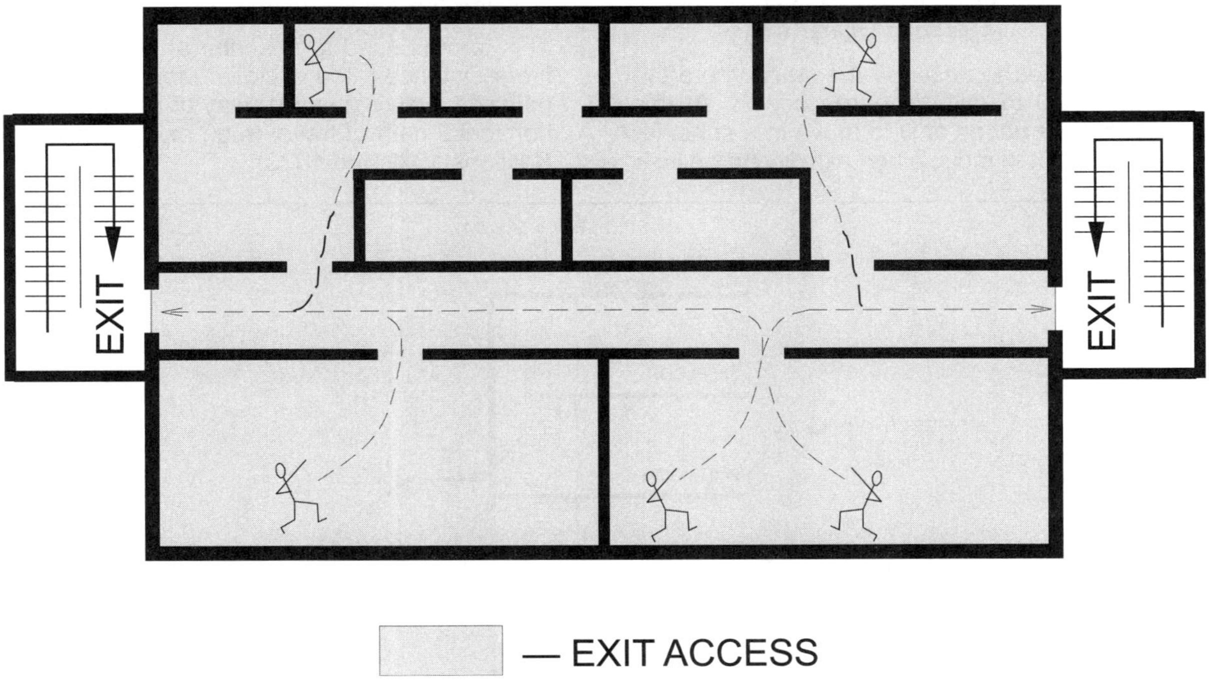

EXIT

EXIT

⬚ — EXIT ACCESS

Figure 1002.1(5)
EXIT ACCESS

EXIT ACCESS DOORWAY. A door or access point along the path of egress travel from an occupied room, area or space where the path of egress enters an intervening room, corridor, unenclosed *exit access stair* or unenclosed *exit access ramp*.

❖ Exit access doorways are used to design many critical aspects of the means of egress including arrangement, number, separation, opening protection and exit sign placement. The term "doorway" has traditionally been limited to those situations where an actual opening, either with or without a door, is present. With "access point" the term "exit access doorway" is inclusive of specific points in the means of egress which may not include a "door" such as when an unenclosed exit access stairway is used in the egress path (see Section 1016.1, Exceptions 3 and 4).

EXIT DISCHARGE. That portion of a *means of egress* system between the termination of an *exit* and a *public way*.

❖ The exit discharge will typically begin when the building occupants reach the exterior at or very near grade level. It provides occupants with a path of travel away from the building. All components between the building and the public way are considered to be the exit discharge, regardless of the distance. In areas of sloping terrain, it is possible to have steps or stairs in the exit discharge leading to the public way. The exit discharge is part of the means of egress and, therefore, its components are subject to the requirements of the code [see Figures 1002.1(3) and 1002.1(6) and Section 1027].

EXIT DISCHARGE, LEVEL OF. The *story* at the point at which an *exit* terminates and an *exit discharge* begins.

❖ The term is intended to describe the story where the transition from exit to exit discharge occurs. At this level, the occupant needs only to move in a substantially horizontal path to move along exit discharge [see

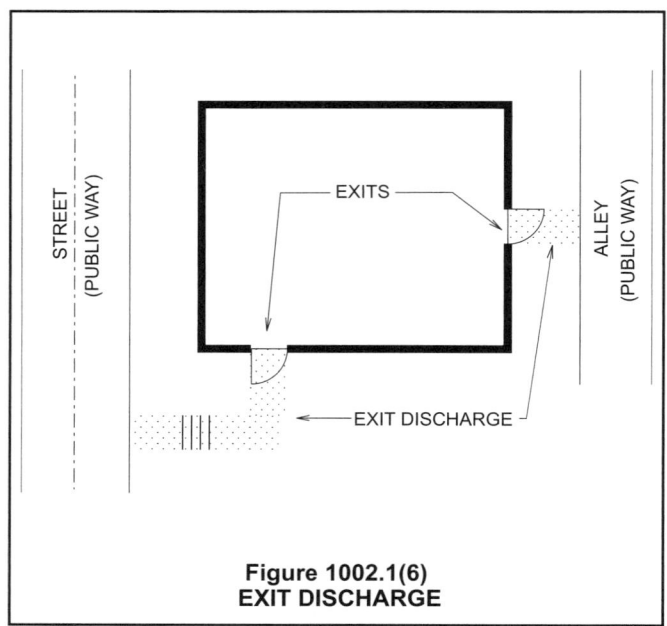

Figure 1002.1(6)
EXIT DISCHARGE

Figure 1002.1(7)]. Since the level is a volume rather than a horizontal plane, exterior exit steps may be part of the exit discharge when they provide access to the level that is closest to grade.

EXIT ENCLOSURE. An *exit* component that is separated from other interior spaces of a building or structure by fire-resistance-rated construction and opening protectives, and provides for a protected path of egress travel in a vertical or horizontal direction to the *exit discharge* or the *public way*.

❖ This term is used to describe an exit that is within a fire-resistance-rated enclosure for a generally vertical path of travel (e.g., a stairway or ramp) or a generally horizontal path of travel (e.g., exit passageway) (see Sections 1022 and 1023).

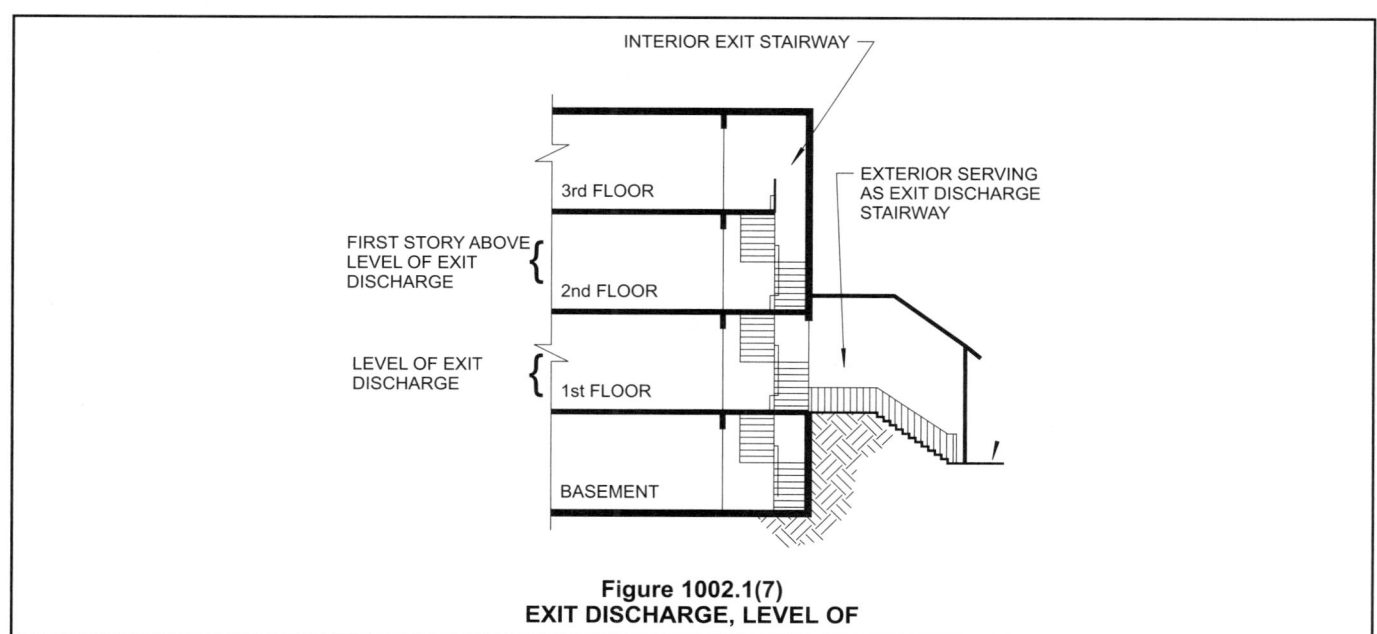

Figure 1002.1(7)
EXIT DISCHARGE, LEVEL OF

EXIT, HORIZONTAL. A path of egress travel from one building to an area in another building on approximately the same level, or a path of egress travel through or around a wall or partition to an area on approximately the same level in the same building, which affords safety from fire and smoke from the area of incidence and areas communicating therewith.

❖ This term refers to a fire-resistance-rated wall that subdivides a structure into multiple compartments and provides an effective barrier to protect occupants from a fire condition within one of the compartments. After occupants pass through a horizontal exit, they must be provided not only with sufficient space to gather but also with another exit, such as an exterior door or exit stairway, through which they can exit the building. Figure 1002.1(4) depicts the exits serving a single building that is subdivided with a fire-resistance-rated wall (see Section 1025).

EXIT PASSAGEWAY. An *exit* component that is separated from other interior spaces of a building or structure by fire-resistance-rated construction and opening protectives, and provides for a protected path of egress travel in a horizontal direction to the *exit discharge* or the *public way*.

❖ This term refers to a horizontal portion of the means of egress that serves as an exit element. Since an exit passageway is considered an exit element, it must be protected and separated as required by the code for exits (see Section 1023). Exit passageways between a vertical exit enclosure and an exterior exit door are typically found on the level of exit discharge to provide a protected path from a centrally located exit stairway to the exit discharge. In taller buildings that reduce floor sizes as they move up (sometimes called a wedding cake building), exit passageways may be utilized at "transfer floors" as stairway locations shift to move the vertical shafts in as the floor size decreases. Exit passageways that lead to an exterior exit door are commonly used in malls to satisfy the travel distance in buildings having a large floor area.

FIRE EXIT HARDWARE. Panic hardware that is *listed* for use on *fire door assemblies*.

❖ Where a door that is required to be of fire-resistance-rated construction has panic hardware, the hardware is required to be listed for use on the fire door. Thus, fire door hardware has been tested to function properly when exposed to the effects of a fire (see the definition for "Panic hardware" and Section 1008.1.10).

FLIGHT. A continuous run of rectangular treads, *winders* or combination thereof from one landing to another.

❖ Two points of clarification for stairways have been addressed by the definition of "Flight." First, a flight is made up of the treads and risers that occur between landings. Therefore, a stairway connecting two stories that includes an intermediate landing consists of two flights. Secondly, the inclusion of winders within a stairway does not create multiple flights. Winders are simply treads within a flight and are often combined with rectangular treads within the same flight.

FLOOR AREA, GROSS. The floor area within the inside perimeter of the *exterior walls* of the building under consideration, exclusive of vent shafts and courts, without deduction for corridors, stairways, closets, the thickness of interior walls, columns or other features. The floor area of a building, or portion thereof, not provided with surrounding *exterior walls* shall be the usable area under the horizontal projection of the roof or floor above. The gross floor area shall not include shafts with no openings or interior courts.

❖ Gross floor area is that area measured within the perimeter formed by the inside surface of the exterior walls. The area of all occupiable and nonoccupiable spaces, including mechanical and elevator shafts, toilets, closets, mechanical equipment rooms, etc., is included in the gross floor area. This area could also include any covered porches, carports or other exterior space intended to be used as part of the building's occupiable space. This gross and net floor areas are primarily used for the determination of occupant load in accordance with Table 1004.1.1.

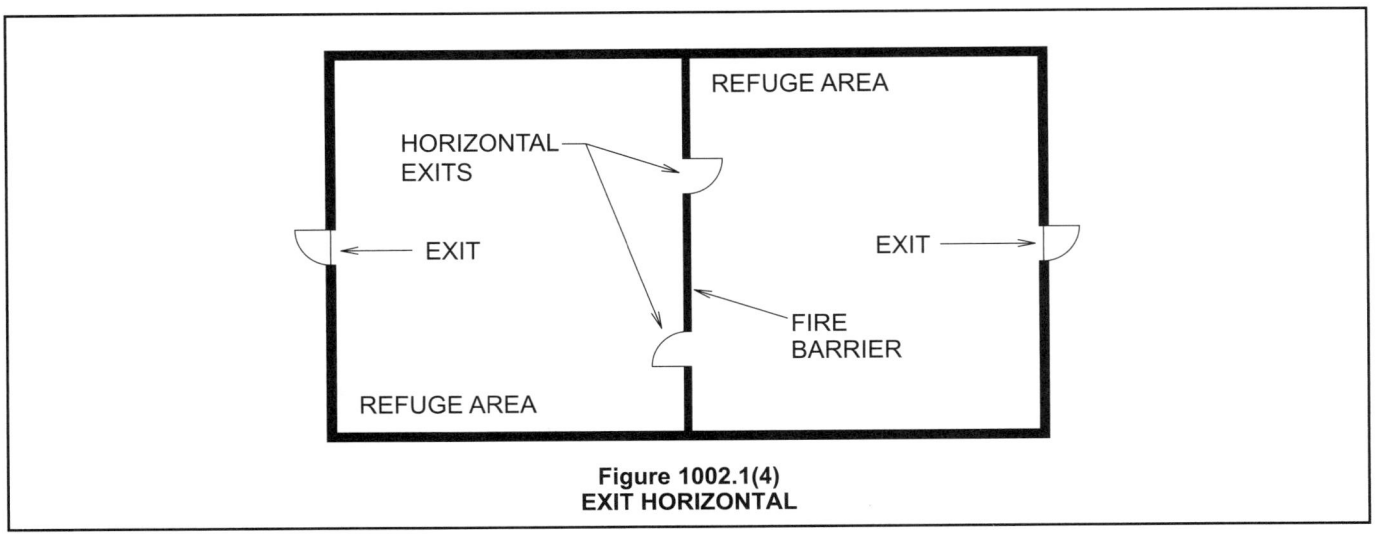

Figure 1002.1(4)
EXIT HORIZONTAL

FLOOR AREA, NET. The actual occupied area not including unoccupied accessory areas such as corridors, stairways, toilet rooms, mechanical rooms and closets.

❖ This area is intended to be only the room areas that are used for specific occupancy purposes and does not include circulation areas, such as corridors or stairways, and service and utility spaces, such as toilet rooms and mechanical and electrical equipment rooms. Floor area, net and gross, is utilized in Table 1004.1.1 to determine occupant load for a space.

FOLDING AND TELESCOPIC SEATING. Tiered seating having an overall shape and size that is capable of being reduced for purposes of moving or storing and is not a building element.

❖ Bleachers, folding and telescopic seating and grandstands are essentially unique forms of tiered seating that are supported on a dedicated structural system. All types are addressed in ICC 300, the safety standard for these types of seating arrangements. Folding and telescopic seating are commonly used in gymnasiums and sports arenas where the seating can be configured in a variety of ways for various types of events. While telescopic seating may be attached to a wall, the system when pulled out or folded includes its main support system (see Section 1028.1.1).

GRANDSTAND. Tiered seating supported on a dedicated structural system and two or more rows high and is not a building element (see "*Bleachers*").

❖ Bleachers, folding and telescopic seating and grandstands are essentially unique forms of tiered seating that are supported on a dedicated structural system. All types are addressed in the safety standard for these types of seating arrangements, ICC 300. Grandstands can be found at a county fair ground, along a parade route or within indoor facilities. Examples are sports arenas and public auditoriums, as well as churches and gallery-type lecture halls. The term "building element" is defined in Section 702.1. Individual bench seats directly attached to a floor system are not a grandstand. The terms "bleacher" and "grandstand" are basically interchangeable. There is no cut-off in size or number of seats that separates bleachers and grandstands (see Section 1028.1.1).

GUARD. A building component or a system of building components located at or near the open sides of elevated walking surfaces that minimizes the possibility of a fall from the walking surface to a lower level.

❖ Guards are sometimes mistakenly referred to as "guardrails." In actuality, the guard consists of the entire vertical portion of the barrier, not just the top rail (see "Handrail" and Section 1013). The purpose of guards is to prevent falls at dropoffs adjacent to walking surfaces.

HANDRAIL. A horizontal or sloping rail intended for grasping by the hand for guidance or support.

❖ Handrails are provided along walking surfaces that lead from one elevation to another, such as ramps, and stairways and their associated landings. Handrails are generally circular in shape. Noncircular shapes could also be acceptable provided that they can be gripped by hand for support and guidance and for checking possible falls on the adjacent walking surface. In addition to being necessary in normal day-to-day use, handrails are especially needed in times of emergency when the pace of egress travel is hurried and the probability for occupant instability while traveling along the sloped or stepped walking surface is greater. Handrails, by themselves, are not intended to be used in place of guards to prevent people from falling over the edge. Where guards and handrails are used together, the handrail is a separate element typically attached to the inside surface of the guard. The top guard cannot be used as a required handrail, except within dwelling units (see Section 1012).

MEANS OF EGRESS. A continuous and unobstructed path of vertical and horizontal egress travel from any occupied portion of a building or structure to a *public way*. A means of egress consists of three separate and distinct parts: the *exit access*, the *exit* and the *exit discharge*.

❖ The means of egress is the path traveled by building occupants to leave the building and the site on which it is located. It includes all interior and exterior elements that the occupants must utilize as they make their way from every room and usable space within the building to a public way such as a street or alley. The elements that make up the means of egress create the lifeline that occupants utilize to travel out of the structure and to a safe distance from the structure. The means of egress provisions of this chapter strive to provide a reasonable level of life safety in every structure. The means of egress provisions are subdivided into three distinct portions (see the definitions of "Exit access," "Exit" and "Exit discharge").

MERCHANDISE PAD. A merchandise pad is an area for display of merchandise surrounded by *aisles*, permanent fixtures or walls. Merchandise pads contain elements such as nonfixed and moveable fixtures, cases, racks, counters and partitions as indicated in Section 105.2 from which customers browse or shop.

❖ Merchandise pads would most likely be found in large stores with changing displays of clothes or furniture. This is not a raised display-only area. These areas allow customers to move between displays or racks. In regards to means of egress, merchandise pads could be considered analogous to areas of fixed seating or groups of tables. The aisle accessways are within the merchandise pad and lead to the aisles on the outside edges of the merchandise pads. Not all stores will contain merchandise pads (e.g. a typical grocery store with fixed shelves and aisles).

NOSING. The leading edge of treads of *stairs* and of landings at the top of *stairway flights*.

❖ Limiting the extent of the tread nosings results in a stair that is easy to use. If too large, they are a tripping hazard when walking up a stairway, and reduce the effective tread depth when walking down the stairway [see Figures 1009.4.2 and 1009.4.5(1)].

OCCUPANT LOAD. The number of persons for which the *means of egress* of a building or portion thereof is designed.

❖ In addition to the limitation on the maximum occupant load for a space, the code also requires the determination of the occupant load that is to be utilized for the design of the means of egress system. This occupant load is also utilized to determine the required number of plumbing fixtures (see Chapter 29) and when automatic sprinkler systems or fire alarm and detection systems are required (see Chapter 9).

PANIC HARDWARE. A door-latching assembly incorporating a device that releases the latch upon the application of a force in the direction of egress travel.

❖ Panic hardware is commonly used in educational and assembly-type spaces where the number of occupants who would use a doorway during a short time frame in an emergency is high in relation to an occupancy with a less dense occupant load, such as an office building. The hardware is required so that the door can be easily opened during an emergency when pressure on a door from a crush of people could tender normal hardware inoperable. Not all types of panic hardware are permitted on fire-rated doors (see the definition for "Fire exit hardware" and Section 1008.1.10).

PHOTOLUMINESCENT. Having the property of emitting light that continues for a length of time after excitation by visible or invisible light has been removed.

❖ An example of photoluminescent material is paint or tape that is charged by exposure to light. When the lights are turned off, the product will "glow" in the dark. Products utilized to meet the requirements for luminous egress path markings in high-rise buildings (see Sections 411.7 and 1024) or exit signs (see Section 1011.4) may be photoluminescent or self-luminous. A variety of materials can comply with the referenced standards for egress path markings—UL 1994, *Standard for Safety of Low Level Path Marking and Lighting Systems* and ASTM E 2072, *Standard Specification for Photoluminescent (Phosphorescent) Safety Markings*—and for signs—UL 924, *Standard for Safety Emergency Lighting and Power Equipment.*

PUBLIC WAY. A street, alley or other parcel of land open to the outside air leading to a street, that has been deeded, dedicated or otherwise permanently appropriated to the public for public use and which has a clear width and height of not less than 10 feet (3048 mm).

❖ The public way marks the termination of the exit discharge portion of the means of egress system. It is the final destination for occupants, and is presumed to be safe from the emergency occurring in the structure or that it will directly connect to other routes so that occupants can move a distance away from the danger. The 10-foot (3048 mm) width is consistent with the exit discharge requirements in Section 1027.

RAMP. A walking surface that has a running slope steeper than one unit vertical in 20 units horizontal (5-percent slope).

❖ This definition is needed to determine the threshold at which the ramp requirements apply to a walking surface. Walking surfaces steeper than specified in the definition are subject to the ramp requirements in Sections 1010 and 1028.

SCISSOR STAIR. Two interlocking *stairways* providing two separate paths of egress located within one stairwell enclosure.

❖ A scissor or interlocking stairway is sometimes used in high-rise buildings or to increase exit capacity of a stairwell enclosure. In this configuration, two independent stairway paths are located within the same exit enclosure and may or may not be visually open to one another. When interlocking stairways are separated from each other with compliant fire barriers and horizontal assemblies, they are not considered scissor stairways (see Section 1015.2.1).

SELF-LUMINOUS. Illuminated by a self-contained power source, other than batteries, and operated independently of external power sources.

❖ Self-luminous products do not need an outside light source to charge them like photoluminescent materials do. Products utilized to meet the requirements for luminous egress path markings in high-rise buildings (see Sections 411.7 and 1024) or exit signs (see Section 1011.4) may be photoluminescent or self-luminous. A variety of materials can comply with the referenced standards, for egress path markings—UL 1994, and ASTM E 2072—and for signs—UL 924.

SMOKE-PROTECTED ASSEMBLY SEATING. Seating served by *means of egress* that is not subject to smoke accumulation within or under a structure.

❖ An example of smoke-protected assembly seating is an open outdoor grandstand or an indoor arena with a smoke control system. The code has less stringent requirements for certain aspects of smoke-protected assembly seating than for seating that is not smoke protected, since occupants are subject to less hazard from the accumulation of smoke and fumes during a fire event. For example, an assembly dead-end aisle is permitted to be longer for a smoke-protected assembly area. For system requirements, see Section 909.

STAIR. A change in elevation, consisting of one or more risers.

❖ All steps, even a single step, are defined as a stair. This makes the stair requirements applicable to all steps unless specifically exempt in the code.

STAIRWAY. One or more *flights* of *stairs*, either exterior or interior, with the necessary landings and platforms connecting them, to form a continuous and uninterrupted passage from one level to another.

❖ It is important to note that this definition characterizes a stairway as connecting one level to another. The term "level" is not to be confused with "story." Steps that connect two levels, one of which is not considered a "story" of the structure, would be considered a stairway. For example, a set of steps between the basement level in an areaway and the outside ground level would be considered a stairway. A series of steps between the floor of a story and a mezzanine within that story would also be considered a stairway (see definitions for "Flight," "Stairway, exterior" and "Stairway, interior" and Sections 1009, 1021 and 1026).

STAIRWAY, EXTERIOR. A *stairway* that is open on at least one side, except for required structural columns, beams, *handrails* and *guards*. The adjoining open areas shall be either *yards*, *courts* or *public ways*. The other sides of the exterior stairway need not be open.

❖ This definition is needed since the code requirements for an exterior stairway are different than for an interior stairway. Exterior stairways are typically exits when providing egress from upper and lower stories, or exit discharge when outside the exit door at the level of exit discharge. For general stairway requirements, see Section 1009. For specific openness requirements for exterior stairways, see Section 1026.

STAIRWAY, INTERIOR. A *stairway* not meeting the definition of an *exterior stairway*.

❖ This definition is needed since the requirements for an interior stairway are more stringent than those for an exterior stairway (see the definition for "Stairway, exterior" and Sections 1016 and 1021). Interior stairways can be exit access elements or exit elements.

STAIRWAY, SPIRAL. A *stairway* having a closed circular form in its plan view with uniform section-shaped treads attached to and radiating from a minimum-diameter supporting column.

❖ Spiral stairways are permitted as part of a means of egress in limited circumstances given in Section 1009.9. Spiral staircases could be used for supplemental/convenience stairways in other locations. Spiral stairways are commonly used where a small number of occupants use the stairway and the floor space for the stair is very limited. Spiral stairways are typically supported by a center pole. Requirements are found with stairways in Section 1009.

SUITE. A group of patient treatment rooms or patient sleeping rooms within Group I-2 occupancies where staff are in attendance within the *suite*, for supervision of all patients within the suite and the suite is in compliance with the requirements of Sections 1014.2.2 through 1014.2.7.

❖ The concept for suites to function within the code without corridor width or rating requirements were accepted to allow staff to have clear and unobstructed supervision of patients in specific treatment and sleeping rooms. It is not intended for day rooms or business sections of the hospital. This term is only applicable to suites of patient rooms in Group I-2 occupancies, and should not be confused with similar layouts in other parts of the hospital or within other occupancies that may be referred to as a "suite."

WINDER. A tread with nonparallel edges.

❖ Winders are used as components of stairs that change direction, just as "fliers" (straight treads) are components in straight stairs. A winder performs the same function as a tread, but its shape allows the additional function of a gradual turning of the stairway direction. The tread depth of a winder at the walkline and the minimum tread depth at the narrow end control the turn made by each winder. Winders are not landings. Winder treads are limited to curved or spiral stairways with all groups or in all stairways within dwelling units (see Section 1009.4.3).

SECTION 1003
GENERAL MEANS OF EGRESS

1003.1 Applicability. The general requirements specified in Sections 1003 through 1013 shall apply to all three elements of the *means of egress* system, in addition to those specific requirements for the *exit access*, the *exit* and the *exit discharge* detailed elsewhere in this chapter.

❖ The text of Chapter 10 is subdivided into 29 sections. The requirements in the chapter address the three parts of a means of egress system: the exit access, the exit and the exit discharge. This section specifies that the requirements of Sections 1003 through 1013 apply to the components of all three parts of the system. For example, the stair tread and riser dimensions in Section 1009 apply to exit access stairways such as those leading from a small mezzanine and also apply to enclosed exit stairways according to the vertical exit enclosure requirements in Section 1022.

1003.2 Ceiling height. The *means of egress* shall have a ceiling height of not less than 7 feet 6 inches (2286 mm).

Exceptions:

1. Sloped ceilings in accordance with Section 1208.2.

2. Ceilings of dwelling units and sleeping units within residential occupancies in accordance with Section 1208.2.

3. Allowable projections in accordance with Section 1003.3.

4. Stair headroom in accordance with Section 1009.2.

5. Door height in accordance with Section 1008.1.1.

6. Ramp headroom in accordance with Section 1010.5.2.

7. The clear height of floor levels in vehicular and pedestrian traffic areas in parking garages in accordance with Section 406.2.2.

8. Areas above and below *mezzanine* floors in accordance with Section 505.1.

❖ Generally, the specified ceiling height is the minimum allowed in any part of the egress path. The exceptions are intended to address conditions where the code allows the ceiling height to be lower than specified in this section.

This section is consistent with the minimum ceiling height for other areas as specified in Section 1208.

1003.3 Protruding objects. Protruding objects shall comply with the requirements of Sections 1003.3.1 through 1003.3.4.

❖ This section identifies the sections that apply to protruding objects and helps to improve awareness of these safety and accessibility-related provisions.

1003.3.1 Headroom. Protruding objects are permitted to extend below the minimum ceiling height required by Section 1003.2 provided a minimum headroom of 80 inches (2032 mm) shall be provided for any walking surface, including walks, *corridors*, *aisles* and passageways. Not more than 50 percent of the ceiling area of a *means of egress* shall be reduced in height by protruding objects.

> **Exception:** Door closers and stops shall not reduce headroom to less than 78 inches (1981 mm).

A barrier shall be provided where the vertical clearance is less than 80 inches (2032 mm) high. The leading edge of such a barrier shall be located 27 inches (686 mm) maximum above the floor.

❖ This provision is applicable to all components of the means of egress. Specifically, the limitations in this section and those in Sections 1003.3.2 and 1003.3.3 provide a reasonable level of safety for those who are preoccupied or not paying attention while walking, as well as for people with impaired vision.

Minimum dimensions for headroom clearance are specified in this section. The minimum headroom clearance over all walking surfaces is required to be maintained at 80 inches (2032 mm). This minimum headroom clearance is consistent with the requirements in Section 1009.2 for stairs and Section 1010.5.2 for ramps. Allowance must be made for door closers and stops, since their design and function necessitates placement within the door opening. The minimum headroom clearance for door closers and stops is allowed to be 78 inches (1981 mm) [see Figure 1003.3.1(1)]. The 2-inch (51 mm) projection into the doorway height is reasonable since these devices are normally mounted away from the center of the door opening, thus minimizing the potential for contact with

a person moving through the opening. This is consistent with the exception in Section 1008.1.1.1.

The limitation on overhangs is of primary importance to those individuals with visual impairments. When vertical clearance along a walking surface is less than 80 inches (2032 mm), such as underneath the stairway on the ground floor, some sort of barrier that is detectable by a person using a cane must be provided. This can be a full-height wall, a rail at or below 27 inches (686 mm), a planter, fixed seating, etc. A low curb is not effective as a barrier. A person with visual impairments might mistake it for a stair tread, step up onto it and strike their head. A rail at handrail height would not be detectable by a person using a cane, and he or she could possibly walk into the rail before detecting it. Also, when making decisions on the choice of type of barrier, keep in mind that persons of shorter stature and children have a detectable range that may be below 27 inches (686 mm) [see Figure 1003.3.1(2)].

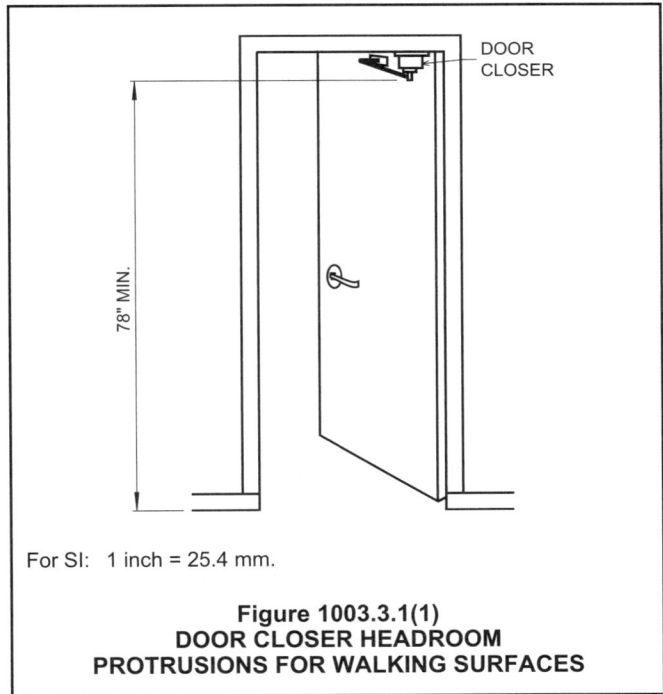

For SI: 1 inch = 25.4 mm.

Figure 1003.3.1(1)
DOOR CLOSER HEADROOM
PROTRUSIONS FOR WALKING SURFACES

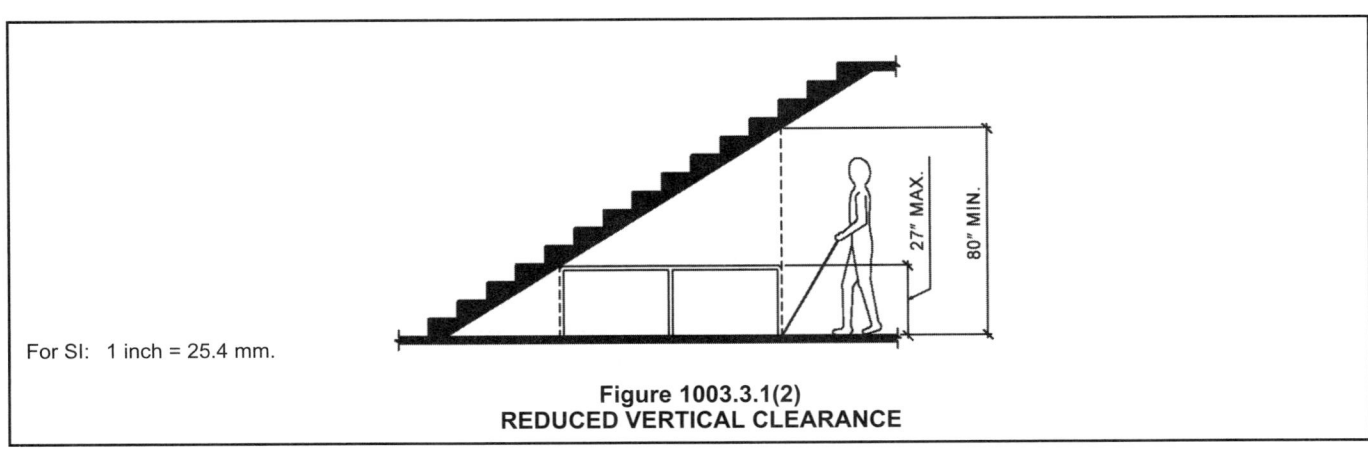

For SI: 1 inch = 25.4 mm.

Figure 1003.3.1(2)
REDUCED VERTICAL CLEARANCE

1003.3.2 Post-mounted objects. A free-standing object mounted on a post or pylon shall not overhang that post or pylon more than 4 inches (102 mm) where the lowest point of the leading edge is more than 27 inches (686 mm) and less than 80 inches (2032 mm) above the walking surface. Where a sign or other obstruction is mounted between posts or pylons and the clear distance between the posts or pylons is greater than 12 inches (305 mm), the lowest edge of such sign or obstruction shall be 27 inches (686 mm) maximum or 80 inches (2032 mm) minimum above the finished floor or ground.

> **Exception:** These requirements shall not apply to sloping portions of *handrails* between the top and bottom riser of *stairs* and above the *ramp* run.

❖ Post-mounted objects, such as signs or some types of drinking fountains or phone boxes, are not permitted to overhang more than 4 inches (102 mm) past the post where the bottom edge is located higher than 27 inches (686 mm) above the walking surface [see Figure 1003.3.2(1)]. Since the minimum required height of doorways, stairways and ramps in the means of egress is 80 inches (2032 mm), protruding objects located higher than 80 inches (2032 mm) above the walking surface are not regulated. Protrusions that are located lower than 27 inches (686 mm) above the walking surface are also permitted since they are more readily detected by a person using a long cane, provided that the minimum required width of the egress element is maintained. This is consistent with the post-mounted objects requirements in Section 307.3 of ICC A117.1, *Accessible and Usable Buildings and Facilities.* The intent is to reduce the potential for accidental impact for a person who is visually impaired.

When signs are provided on multiple posts, the posts must be located closer than 12 inches (305 mm) apart, or the bottom edge of the sign must be lower than 27 inches (686 mm) so it is within detectable cane range or above 80 inches (2032 mm) so that it is above headroom clearances [see Figure 1003.3.2(2)].

The exception is intended for handrails that are located along a stairway or ramp. The extensions at the top and bottom of stairways and ramps must meet the requirements for protruding objects.

1003.3.3 Horizontal projections. Structural elements, fixtures or furnishings shall not project horizontally from either side more than 4 inches (102 mm) over any walking surface between the heights of 27 inches (686 mm) and 80 inches (2032 mm) above the walking surface.

> **Exception:** *Handrails* are permitted to protrude $4^1/_2$ inches (114 mm) from the wall.

❖ Protruding objects could slow down the egress flow through a passageway and injure someone hurriedly passing by or someone with a visual impairment. Persons with a visual impairment, who use a long cane for guidance, must have sufficient warning of a protruding object. Where protrusions are located higher than 27 inches (686 mm) above the walking surface, the cane will most likely not encounter the protrusion before the person collides with the object.

Additionally, people with poor visual acuity or poor depth perception may have difficulty identifying protruding objects higher than 27 inches (686 mm). Therefore, objects such as lights, signs and door hardware, located between 27 inches (686 mm) and 80 inches (2032 mm) above the walking surface, are not permitted to extend more than 4 inches (102 mm) from each wall (see Figure 1003.3.3). The requirement for protrusions into the door clear width in Section 1008.1.1.1 is different because it deals with allowances for panic hardware on a door. It is not the intent of this section to prohibit column, pilasters or wing walls to project into a corridor as long as adequate egress width is maintained. These types of structural elements are detectable by persons using a long cane.

The exception is an allowance for handrails when they are provided along a wall, such as in some hospitals or nursing homes. The $4^1/_2$ inches (114 mm) is intended to be consistent with projections by handrails into the required width of stairways and ramps in Section 1012.8. There are additional requirements when talking about the required width (see Section 1005.1).

1003.3.4 Clear width. Protruding objects shall not reduce the minimum clear width of *accessible routes*.

❖ The intent of this section is to limit the projections into an accessible route so that a minimum clear width of 36 inches (914 mm) is maintained along the route. ICC A117.1 is referenced by Chapter 11 for technical requirements for accessibility. ICC A117.1, Section 403.5, allows the accessible route to be reduced in width to 32 inches (914 mm) for segments not to exceed 24 inches (635 mm) in length and spaced a minimum of 48 inches (1219 mm) apart.

1003.4 Floor surface. Walking surfaces of the *means of egress* shall have a slip-resistant surface and be securely attached.

❖ As the pace of exit travel becomes hurried during emergency situations, the probability of slipping on smooth or slick floor surfaces increases. To minimize the hazard, all floor surfaces in the means of egress are required to be slip resistant. The use of hard floor materials with highly polished, glazed, glossy or finely finished surfaces should be avoided.

Field testing and uniform enforcement of the concept of slip resistance are not practical. One method used to establish slip resistance is that the static coefficient of friction between leather [Type 1 (Vegetable Tanned) of Federal Specification KK-L-165C] and the floor surface is greater than 0.5. Laboratory test procedures, such as ASTM D 2047, can determine the static coefficient of resistance. Bulletin No. 4 titled "Surfaces" issued by the U.S. Architectural and Transportation Barriers Compliance Board (ATBCB or Access Board) contains further information regarding slip resistance.

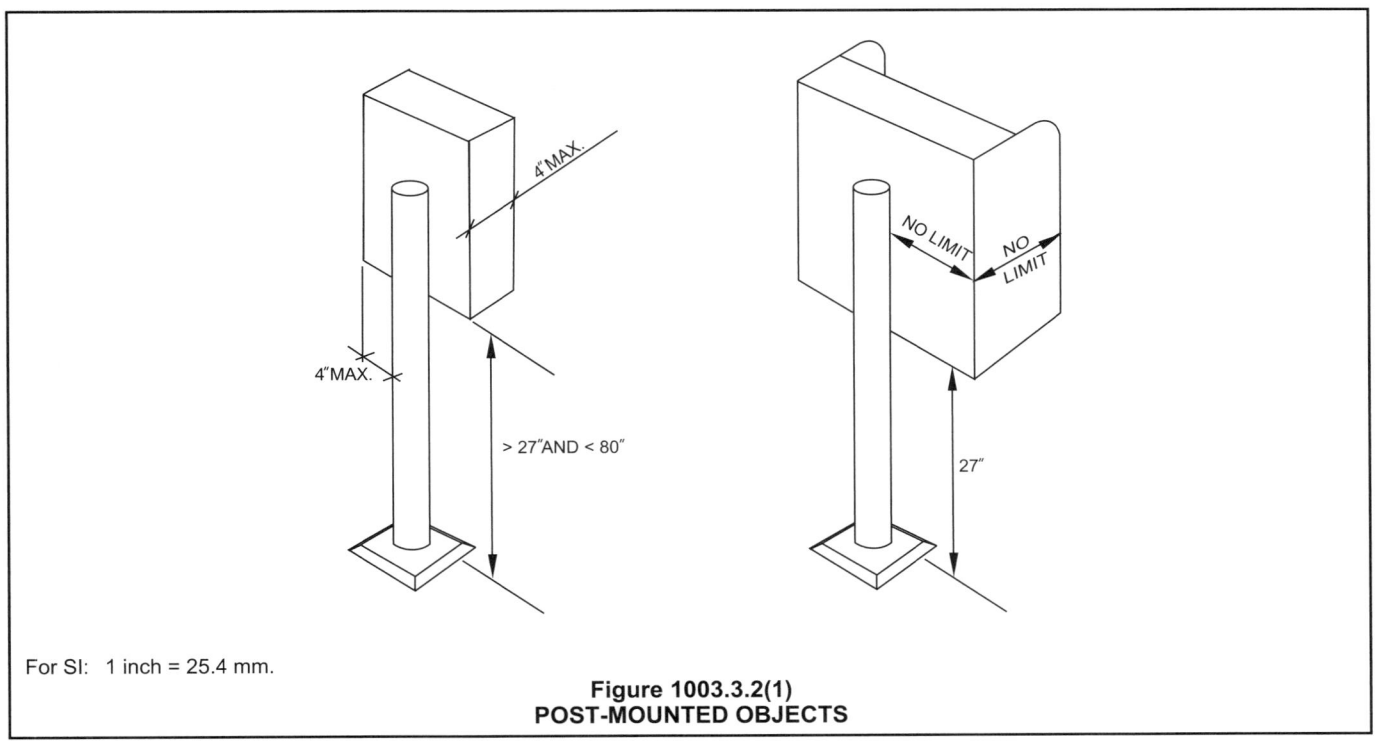

For SI: 1 inch = 25.4 mm.

Figure 1003.3.2(1)
POST-MOUNTED OBJECTS

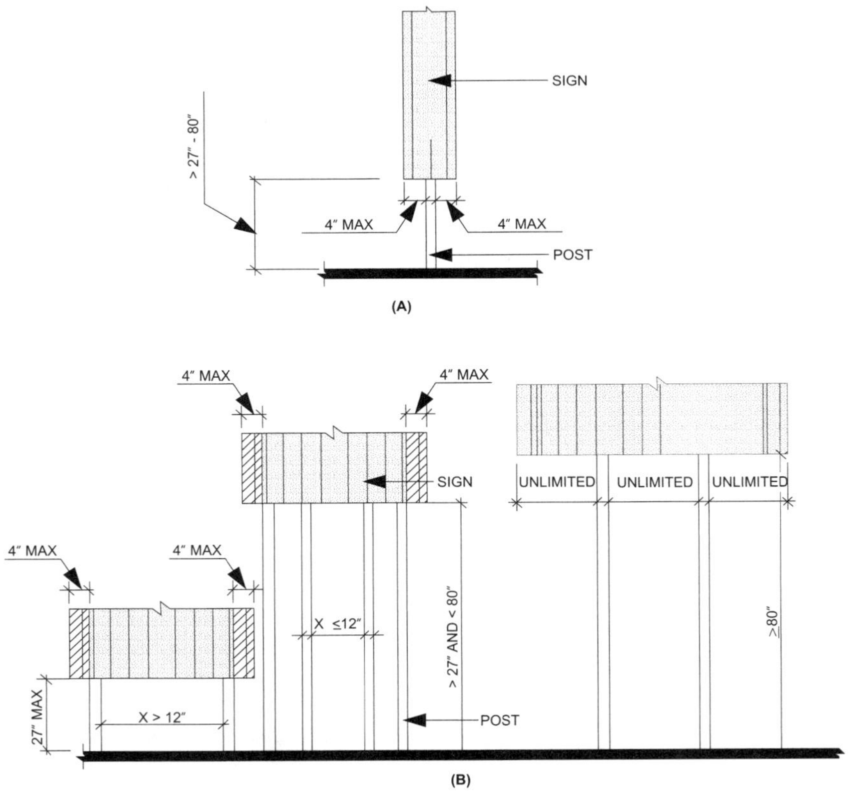

For SI: 1 inch = 25.4 mm.

Figure 1003.3.2(2)
POST-MOUNTED PROTRUDING OBJECTS

For SI: 1 inch = 25.4 mm.

Figure 1003.3.3
EXAMPLES OF HORIZONTAL PROJECTIONS

1003.5 Elevation change. Where changes in elevation of less than 12 inches (305 mm) exist in the *means of egress*, sloped surfaces shall be used. Where the slope is greater than one unit vertical in 20 units horizontal (5-percent slope), *ramps* complying with Section 1010 shall be used. Where the difference in elevation is 6 inches (152 mm) or less, the *ramp* shall be equipped with either handrails or floor finish materials that contrast with adjacent floor finish materials.

Exceptions:

1. A single step with a maximum riser height of 7 inches (178 mm) is permitted for buildings with occupancies in Groups F, H, R-2, R-3, S and U at exterior doors not required to be *accessible* by Chapter 11.

2. A *stair* with a single riser or with two risers and a tread is permitted at locations not required to be *accessible* by Chapter 11, provided that the risers and treads comply with Section 1009.4, the minimum depth of the tread is 13 inches (330 mm) and at least one *handrail* complying with Section 1012 is provided within 30 inches (762 mm) of the centerline of the normal path of egress travel on the *stair*.

3. A step is permitted in *aisles* serving seating that has a difference in elevation less than 12 inches (305 mm) at locations not required to be *accessible* by Chapter

11, provided that the risers and treads comply with Section 1028.11 and the *aisle* is provided with a *handrail* complying with Section 1028.13.

Throughout a story in a Group I-2 occupancy, any change in elevation in portions of the *exit access* that serve nonambulatory persons shall be by means of a *ramp* or sloped walkway.

❖ Minor changes in elevation, such as a single step that is located in any portion of the means of egress (i.e., exit access, exit or exit discharge) may not be readily apparent during normal use or emergency egress and are considered to present a potential tripping hazard. Where the elevation change is less than 12 inches (305 mm), a ramp or sloped surface is specified to make the transition from higher to lower levels. This is intended to reduce accidental falls associated with the tripping hazard of an unseen step. Ramps must be constructed in accordance with Section 1010.1. The presence of the ramp must be readily apparent from the directions from which it is approached. Handrails are one method of identifying the change in elevation. In lieu of handrails, the surface of the ramp must be finished with materials that visually contrast with the surrounding floor surfaces. The walking surface of the ramp should contrast both visually and physically.

None of the exceptions are permitted along an accessible route required for either entry or egress from a space or building.

Exception 1 allows up to a 7-inch (178 mm) step at exterior doors to avoid blocking the outward swing of the door by a buildup of snow or ice in locations that are not used by the public on a regular basis (see Figure 1003.5). This exception is coordinated with Exception 2 of Section 1008.1.5, and is only applicable in occupancies that have relatively low occupant densities, such as factory and industrial structures. This exception is not applicable to exterior doors that are required to serve as an accessible entrance or that are part of a required accessible route. If this exception is utilized at a Group R-2 or R-3 occupancy, the designer may want to consider the issues of potential tripping hazards if this is a common entrance for a large number of occupants.

Exception 2 allows the transition from higher to lower elevations to be accomplished through the construction of stairs with one or two risers. The pitch of the stairway, however, must be shallower than that required for typical stairways (see Section 1009.3). Since the total elevation change is limited to 12 inches (305 mm), each riser must be approximately 6 inches (152 mm) in height. The elevation change must be readily apparent from the directions from which it is approached. At least one handrail is required. It must be constructed in accordance with Section 1012 and located so as to provide a graspable surface from the normal walking path.

Exception 3 is basically a cross reference to the assembly provisions for stepped aisles in Section 1025.

None of the exceptions are permitted in a Group I-2 occupancy (e.g., nursing home, hospital) in areas where nonambulatory persons may need access. The mobility impairments of these individuals require additional consideration.

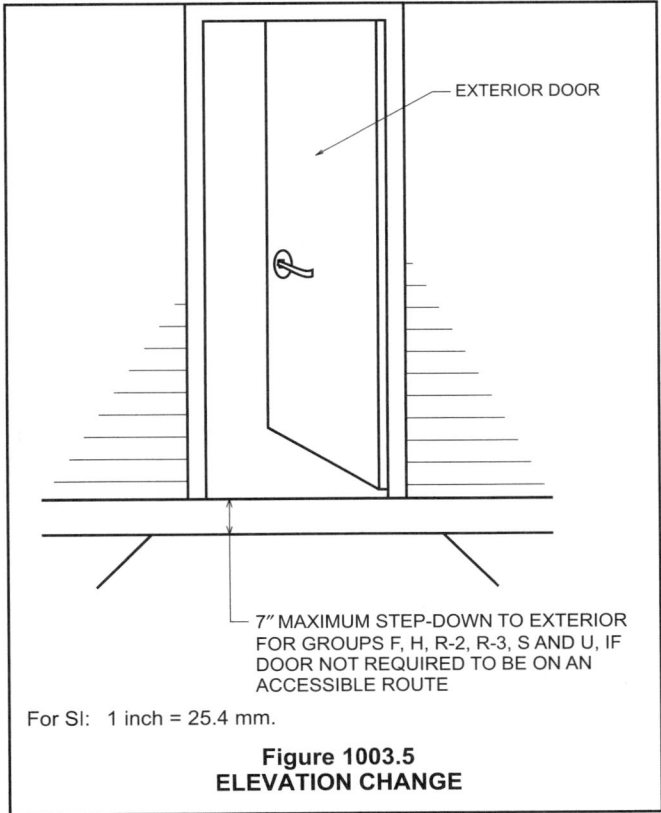

EXTERIOR DOOR

7" MAXIMUM STEP-DOWN TO EXTERIOR FOR GROUPS F, H, R-2, R-3, S AND U, IF DOOR NOT REQUIRED TO BE ON AN ACCESSIBLE ROUTE

For SI: 1 inch = 25.4 mm.

Figure 1003.5
ELEVATION CHANGE

1003.6 Means of egress continuity. The path of egress travel along a *means of egress* shall not be interrupted by any building element other than a *means of egress* component as specified in this chapter. Obstructions shall not be placed in the required width of a *means of egress* except projections permitted by this chapter. The required capacity of a *means of egress* system shall not be diminished along the path of egress travel.

❖ The purpose of this section requires that the entire means of egress path is clear of obstructions that could reduce the egress capacity at any point. The egress path is also not allowed to be reduced in width such that the design occupant load would not be served. Note, however, that the egress path could be reduced in width in situations where it is wider than required by the code based on the occupant load. For example, if the required width of a corridor was 52 inches (1321 mm) based on the number of occupants using the corridor and the corridor provided was 96 inches (2438 mm) in width, the corridor would be allowed to be reduced to the minimum required width of 52 inches (1321 mm) since that width would still serve the number of occupants required by the code. In the context of this section, a "means of egress component" would most likely be a door or doorway.

1003.7 Elevators, escalators and moving walks. Elevators, escalators and moving walks shall not be used as a component of a required *means of egress* from any other part of the building.

Exception: Elevators used as an *accessible means of egress* in accordance with Section 1007.4.

❖ Generally, the code does not allow elevators, escalators and moving sidewalks to be used as a required means of egress. The concern is that due to possible power outages, escalators and moving sidewalks may not provide a safe and reliable means of egress that is available for use at all times.

Elevators are not typically used for unassisted evacuation during fire emergencies. However, in taller buildings fire fighters use the elevators for both staging to fight the fire and assisted evacuation. They can verify that the shaft is not full of smoke, that the elevators will remain operational and, since they know the fire location, what floors the elevator is safe to access. In accordance with the exception, elevators are allowed to be part of an accessible means of egress (i.e., assisted evacuation), provided they comply with the requirements of Section 1007.4. Where elevators are required to serve as part of the accessible means of egress is addressed in Section 1007.2.1. There are new provisions for fire service elevators and occupant evacuation elevators for high-rises in Sections 403.6, 3007 and 3008. These specific provisions will provide a level of safety that would meet the intent of the means of egress provisions in Chapter 10.

SECTION 1004
OCCUPANT LOAD

1004.1 Design occupant load. In determining *means of egress* requirements, the number of occupants for whom *means of egress* facilities shall be provided shall be determined in accordance with this section. Where occupants from accessory areas egress through a primary space, the calculated *occupant load* for the primary space shall include the total *occupant load* of the primary space plus the number of occupants egressing through it from the accessory area.

❖ The design occupant load is the number of people that are intended to occupy a building or portion thereof at any one time; essentially the number for which the means of egress is to be designed. It is the largest number derived by the application of Sections 1004.1 through 1004.9. There is a limit to the density of occupants permitted in an area to enable a reasonable amount of freedom of movement (see Section 1004.2). The design occupant load is also utilized to determine the required plumbing fixture count (see commentary, Chapter 29) and other building requirements, such as automatic sprinkler systems and fire alarm and detection systems (see Chapter 9).

The intent of this section is to indicate the procedure by which design occupant loads are determined. This is particularly important because accurate determination of design occupant load is fundamental to the proper design of any means of egress system. For determining the means of egress capacity, when occupants from an accessory area move through another area to exit, the combined number must be utilized for egress capacity. It is not the intent of this section to "double count" occupants. For example, the means of egress from a lobby must be sized for the cumulative occupant load of the adjacent office spaces if the occupants must travel through the lobby to reach an exit. Likewise, if an adjacent room has an egress route independent of the lobby, the occupant load of that room would not be combined with the occupant loads of the other rooms that pass through that lobby. If a portion of the adjacent room's occupant load is to travel through the lobby, only that portion would be combined with the lobby occupant load for determining lobby egress (see Figure 1004.1). This is particularly important in determining the capacity and the number of means of egress.

1004.1.1 Areas without fixed seating. The number of occupants shall be computed at the rate of one occupant per unit of area as prescribed in Table 1004.1.1. For areas without fixed seating, the *occupant load* shall not be less than that number determined by dividing the floor area under consideration by the occupant per unit of area factor assigned to the occupancy as set forth in Table 1004.1.1. Where an intended use is not listed in Table 1004.1.1, the *building official* shall establish a use based on a listed use that most nearly resembles the intended use.

> **Exception:** Where *approved* by the *building official*, the actual number of occupants for whom each occupied space, floor or building is designed, although less than those deter-

mined by calculation, shall be permitted to be used in the determination of the design *occupant load*.

❖ The numbers for floor area per occupant in the table reflect common and traditional occupant density based on empirical data for the density of similar spaces. The number determined, using the occupant load rates in Table 1004.1.1, generally establishes the minimum occupant load for which the egress facilities of the rooms, spaces and building must be designed. The design occupant load is also utilized for other code requirements, such as determining the required plumbing fixture count (see commentary, Chapter 29) and other building requirements, including automatic sprinkler systems and fire alarm and detection systems (see Chapter 9).

It is difficult to predict the many conditions by which a space within a building will be occupied over time. An assembly banquet room in a hotel, for example, could be arranged with rows of chairs to host a business seminar one day and with mixed tables and chairs to host a dinner reception the next day. In some instances, the room will be arranged with no tables and very few chairs to accommodate primarily standing occupants. In such a situation, the egress facilities must safely accommodate the maximum number of persons permitted to occupy the space. When determining the occupant load of this type of occupancy, the various arrangements (e.g., tables and chairs, chairs only, standing space) should be recognized. The worst-case scenario should be utilized to determine the requirements for the means of egress elements.

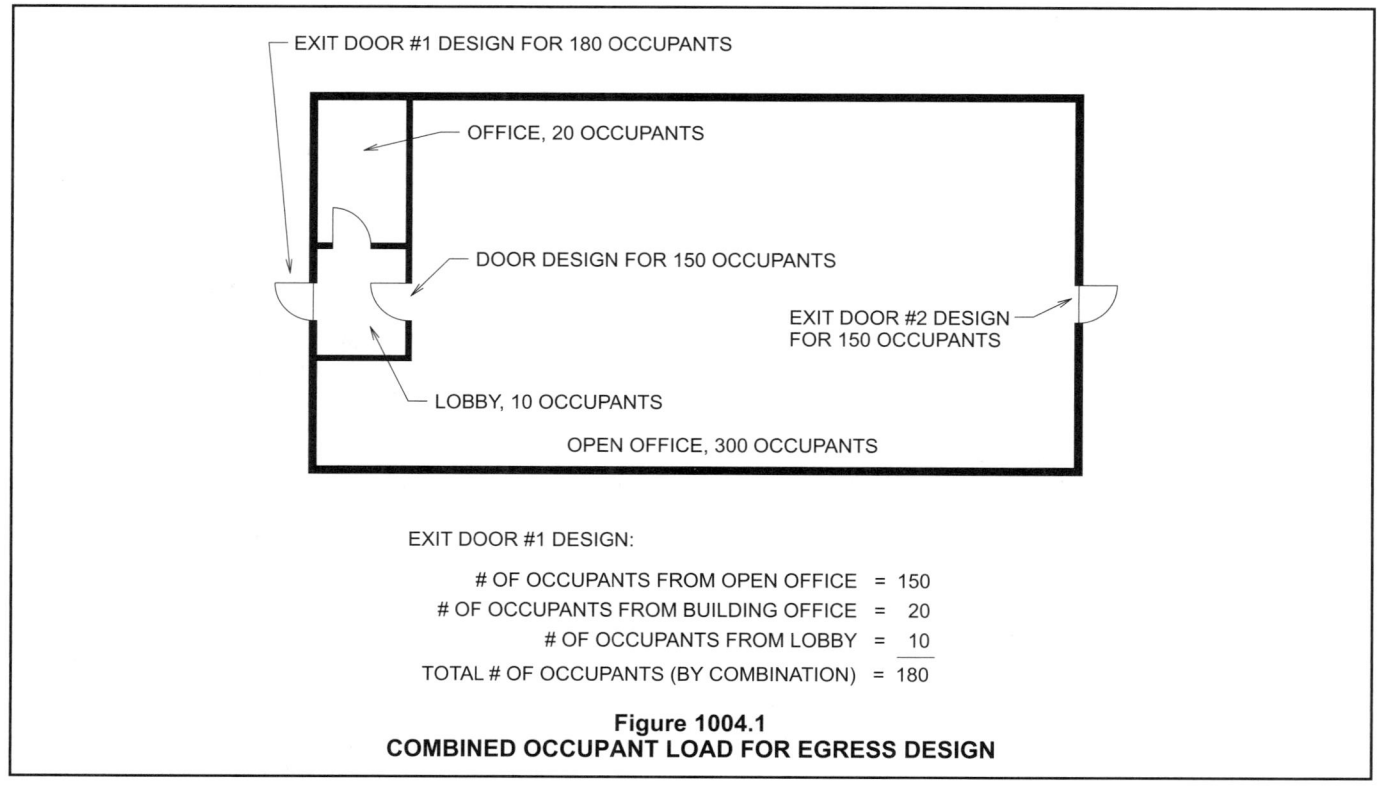

EXIT DOOR #1 DESIGN FOR 180 OCCUPANTS

OFFICE, 20 OCCUPANTS

DOOR DESIGN FOR 150 OCCUPANTS

EXIT DOOR #2 DESIGN FOR 150 OCCUPANTS

LOBBY, 10 OCCUPANTS

OPEN OFFICE, 300 OCCUPANTS

EXIT DOOR #1 DESIGN:

# OF OCCUPANTS FROM OPEN OFFICE	= 150
# OF OCCUPANTS FROM BUILDING OFFICE	= 20
# OF OCCUPANTS FROM LOBBY	= 10
TOTAL # OF OCCUPANTS (BY COMBINATION)	= 180

Figure 1004.1
COMBINED OCCUPANT LOAD FOR EGRESS DESIGN

This is consistent with the requirements for multiple use spaces addressed in Section 302.1.

While some of the values in the table utilize the net floor area, most utilize the gross floor area. See the commentary to Table 1004.1.1 and the definitions for "Floor area, gross" and "Floor area, net" for additional discussion and examples.

The occupant load determined in accordance with this section is typically the minimum occupant load upon which means of egress requirements are to be based. Some occupancies may not typically contain an occupant load totally consistent with the occupant load density factors of Table 1004.1.2. The exception is intended to address the limited circumstances where the actual occupant load is less than the calculated occupant load. Previously, designing for a reduced occupant load was permitted only through the variance process. With this exception, the building official can make a determination if a design that would use the actual occupant load was permissible. The building official may want to create specific conditions for approval. For example, the building official could choose to permit the actual occupant load to be utilized to determine the plumbing fixture count, but not the means of egress or sprinkler design; the determination could be that the reduced occupant load may be utilized in a specific area, such as in the storage warehouse, but not in the factory or office areas. Another point to consider would be the potential of the space being utilized for different purposes at different times, or the potential of a future change of tenancy without knowledge of the building department. Any special considerations for such unique uses must be documented and justified. Additionally, the owner must be aware that such special considerations will impact the future use of the building with respect to the means of egress and other protection features.

TABLE 1004.1.1. See next column.

❖ Table 1004.1.1 establishes minimum occupant densities based on the function or actual use of the space (not group classification). The table presents the maximum floor area allowance per occupant based on studies and counts of the number of occupants in typical buildings. The use of this table, then, results in the minimum occupant load for which rooms, spaces and the building must be designed. While an assumed normal occupancy may be viewed as somewhat less than that determined by the use of the table factors, such a normal occupant load is not necessarily an appropriate design criterion. The greatest hazard to the occupants occurs when an unusually large crowd is present. The code does not limit the occupant load density of an area, except as provided for in Section 1004.2, but once the occupant load is established, the means of egress must be designed for at least that capacity. If it is intended that the occupant load will exceed that calculated in accordance with the table, then the occupant load is to be based on the estimated actual number of people, but not to exceed the maximum allow-

TABLE 1004.1.1
MAXIMUM FLOOR AREA ALLOWANCES PER OCCUPANT

FUNCTION OF SPACE	FLOOR AREA IN SQ. FT. PER OCCUPANT
Accessory storage areas, mechanical equipment room	300 gross
Agricultural building	300 gross
Aircraft hangars	500 gross

Airport terminal
  Baggage claim
  Baggage handling
  Concourse
  Waiting areas
	
	20 gross
	300 gross
	100 gross
	15 gross

Assembly Gaming floors (keno, slots, etc.)	 11 gross
Assembly with fixed seats	See Section 1004.7

Assembly without fixed seats
  Concentrated (chairs only—not fixed)
  Standing space
  Unconcentrated (tables and chairs)
	
	7 net
	5 net
	15 net

Bowling centers, allow 5 persons for each lane including 15 feet of runway, and for additional areas	7 net
Business areas	100 gross
Courtrooms—other than fixed seating areas	40 net
Day care	35 net
Dormitories	50 gross

Educational
  Classroom area
  Shops and other vocational room areas
	
	20 net
	50 net

Exercise rooms	50 gross
H-5 Fabrication and manufacturing areas	200 gross
Industrial areas	100 gross

Institutional areas
  Inpatient treatment areas
  Outpatient areas
  Sleeping areas
	
	240 gross
	100 gross
	120 gross

Kitchens, commercial	200 gross

Library
  Reading rooms
  Stack area
	
	50 net
	100 gross

Locker rooms	50 gross

Mercantile
  Areas on other floors
  Basement and grade floor areas
  Storage, stock, shipping areas
	
	60 gross
	30 gross
	300 gross

Parking garages	200 gross
Residential	200 gross

Skating rinks, swimming pools
  Rink and pool
  Decks
	
	50 gross
	15 gross

Stages and platforms	15 net
Warehouses	500 gross

For SI: 1 square foot = 0.0929 m².

ance in accordance with Section 1004.2. Therefore, the occupant load of the office or business areas in a storage warehouse or nightclub is to be determined using the occupant load factor most appropriate to that space—one person for each 100 square feet (9 m²) of gross floor area.

The use of net and gross floor areas as defined in Section 1002.1 is intended to provide a refinement in the occupant load determination. The gross floor area technique applied to a building only allows the deduction of the plan area of the exterior walls, vent shafts and interior courts from the plan area of the building.

The net floor area permits the exclusion of certain spaces that would be included in the gross floor area. The net floor area is intended to apply to the actual occupied floor areas. The area used for permanent building components, such as shafts, fixed equipment, thicknesses of walls, corridors, stairways, toilet rooms, mechanical rooms and closets, is not included in net floor area. For example, consider a restaurant dining area with dimensions measured from the inside of the enclosing walls of 80 feet by 60 feet (24 384 mm by 18 288 mm) (see Figure 1004.1.1). Within the restaurant area is a 6-inch (152 mm) privacy wall running the length of the room [80 feet by 0.5 feet = 40 square feet (3.7 m²)], a fireplace [40 square feet (3.7 m²)] and a cloak room [60 square feet (5.6 m²)]. Each of these areas is deducted from the restaurant area, resulting in a net floor area of 4,660 square feet (433 m²). Since the restaurant intends to have unconcentrated seating that involves loose tables and chairs, the resulting occupant load is 311 persons (4,660 divided by 15). As the definition of "Floor area, net" indicates, certain spaces are to be excluded from the gross floor area to derive the net floor area. The key point in this definition is that the net floor area is to include the actual occupied area and does not include spaces uncharacteristic of that occupancy.

In determining the occupant load of a building with mixed groups, each floor area of a single occupancy must be separately analyzed, such as required by Section 1004.9. The occupant load of the business portion of an office/warehouse building is determined at a rate of one person for each 100 square feet (9 m²) of office space, whereas the occupant load of the warehouse portion is determined at the rate of one person for each 300 square feet (28 m²). There may even be different uses within the same room. For ex-

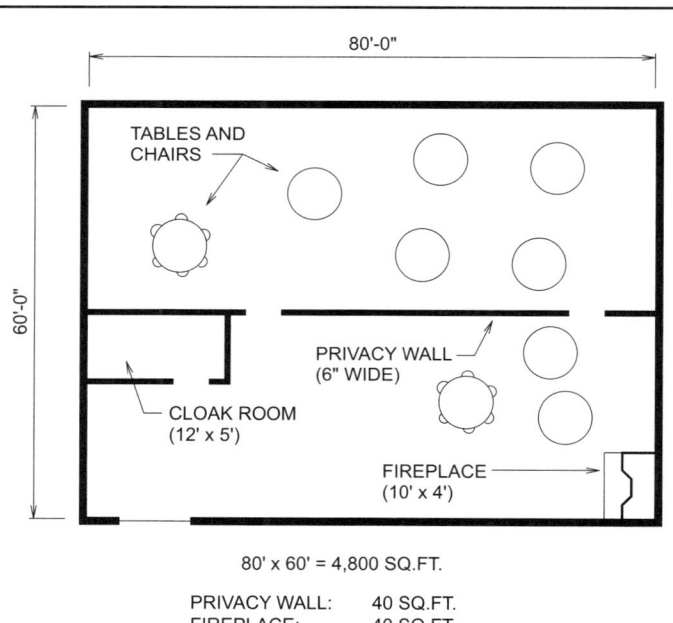

80' x 60' = 4,800 SQ.FT.

PRIVACY WALL:	40 SQ.FT.
FIREPLACE:	40 SQ.FT.
CLOAK ROOM:	60 SQ.FT.
TOTAL:	140 SQ.FT.

(TOTAL AREA WITHIN WALLS) - (EXCLUDED ITEMS) = (NET FLOOR AREA)
4,800 SQ.FT. - 140 SQ.FT. = 4,660 SQ.FT.

(NET FLOOR AREA)/(TABLE 1004.1.2 VALUE) = (OCCUPANT LOAD)
4,660 SQ.FT./15 SQ.FT. PER OCCUPANT = 311 OCCUPANTS

For SI: 1 inch = 25.4 mm, 1 foot = 304.8 mm,
1 square foot = 0.0929 m².

Figure 1004.1.1
TYPICAL NET FLOOR AREA OCCUPANT LOAD CALCULATION

ample, a restaurant dining room would have seating but may also have a waiting area with standing room, a take-out window with a queue line or employee areas behind a bar or reception desk.

If a specific type of facility is not found in the table, the occupancy it most closely resembles should be utilized. For example, a training room in a business office may utilize the 20-square-feet (1.86 m²) net established for educational classroom areas, or a dance or karate studio may use the occupant load for rinks and pools for the studio areas.

Table 1004.1.1, in accordance with Section 1004.1.1, presents a method of determining the absolute base minimum occupant load of a space that the means of egress is to accommodate.

The table occupant loads are based on the stereotypical configuration of spaces. For example, the dorm requirements were written based on dormitories with sleeping rooms with two to four students, a gang bathroom and a meeting/study lounge on each floor. Dormitory buildings that operate like army barracks may have a heavier occupant load, while facilities with groups of rooms with private bathrooms, living and even kitchenette areas may have a lower occupant load. Industrial facilities are based on typical fabricating plants. Warehouses are based on consistent movement in and out of product by employees. Factories with largely mechanized operations or warehouses that contain long-term storage are other examples where discussion with the building official and the application of the exception in Section 1004.1.1 might be considered.

In addition to the table, Section 402 contains the basis for calculating the occupant load of a covered mall building; however, Table 1004.1.1 should be used for determining the occupant load of each anchor store.

1004.2 Increased occupant load. The *occupant load* permitted in any building, or portion thereof, is permitted to be increased from that number established for the occupancies in Table 1004.1.1, provided that all other requirements of the code are also met based on such modified number and the *occupant load* does not exceed one occupant per 7 square feet (0.65 m²) of occupiable floor space. Where required by the *building official*, an *approved aisle*, seating or fixed equipment diagram substantiating any increase in *occupant load* shall be submitted. Where required by the *building official*, such diagram shall be posted.

❖ An increased occupant load is permitted above that developed by using Table 1004.1.1; for example, utilizing the actual occupant load. However, if the occupant load exceeds that which is determined in accordance with Section 1004.1.1, the building official has the authority to require aisle, seating and equipment diagrams to confirm that: all occupants have access to an exit, the exits provide sufficient capacity for all occupants and compliance with this section is attained.

The maximum area of 7 square feet (0.65 m²) per

occupant should allow for sufficient occupant movement in actual fire situations. This is not a conflict with the standing space provisions of 5 square feet (0.46 m²) net in accordance with Table 1004.1.1. Standing space is typically limited to a portion of a larger area, such as the area immediately in front of the bar or the waiting area in a restaurant, while the rest of the dining area would use 15 square feet (1.4 m²) net per occupant.

1004.3 Posting of occupant load. Every room or space that is an assembly occupancy shall have the *occupant load* of the room or space posted in a conspicuous place, near the main *exit* or *exit access doorway* from the room or space. Posted signs shall be of an *approved* legible permanent design and shall be maintained by the owner or authorized agent.

❖ Each room or space used for an assembly occupancy is required to display the approved occupant load. The placard must be posted in a visible location (near the main entrance) (see Figure 1004.3 for an example of an occupant load limit sign).

The posting is required to provide a means by which to determine that the maximum approved occupant load is not exceeded. This permanent and readily visible sign provides a constant reminder to building personnel and is a reference for building officials during periodic inspections.

While the composition and organization of information in the sign are not specified, information must be recorded in a permanent manner. This means that a sign with changeable numbers would not be acceptable.

NOTICE

FOR YOUR SAFETY

OCCUPANCY

IS LIMITED TO:

428

PERSONS

BY ORDER OF
THE CODE OFFICIAL
Keep Posted Under Penalty Of Law

**Figure 1004.3
EXAMPLE OF OCCUPANT LOAD LIMIT SIGN**

1004.4 Exiting from multiple levels. Where *exits* serve more than one floor, only the *occupant load* of each floor considered individually shall be used in computing the required capacity of the *exits* at that floor, provided that the *exit* capacity shall not decrease in the direction of egress travel.

❖ The sum total capacity of the exits that serve a floor is not to be less than the occupant load of the floor as determined by Section 1004.1. If an exit, such as a stairway, also serves a second floor, and the required capacity of the exit serving the occupants of the second floor is greater than the first floor, the greater capacity would govern the egress components that the occupants of the floors share. For example, if an exit stairway serves two floors, with occupant loads of 300 on the lower floor and 500 on the upper floor, assuming that two stairways serve each floor, the two stairways would be designed for a capacity of 250 people each, using the upper-floor occupant load of 500 as the basis of determination. Note that the doors to the stairways on the lower floor would be designed for a capacity of 150, and the doors to the stairways on the upper floor would be designed for a capacity of 250. Reversing these two floors would result in the portion of the stairways that serves the upper floor to be designed for a capacity of 150 and the stairways that serve the lower floor to be designed for 250. Requiring the egress component to be designed for the largest tributary occupant load accommodates the worst-case situation.

Also note that the capacity of the exits is based on the occupant load of one floor. The occupant loads are not combined with other floors for the exit design. It is assumed that the peak demand or flow of occupants from more than one floor level at a common point in the means of egress will not occur simultaneously, except as provided for in Sections 1004.5 (Egress convergence) and 1004.6 (Mezzanine levels).

1004.5 Egress convergence. Where *means of egress* from floors above and below converge at an intermediate level, the capacity of the *means of egress* from the point of convergence shall not be less than the sum of the two floors.

❖ Convergence of occupants can occur whenever the occupants of one floor travel down and occupants of a lower floor travel up and meet at a common, intermediate egress component. The intermediate component may or may not be another occupiable floor and, most often, is an exit door [see Figures 1004.5(1) and 1004.5(2)].

The entire premise of egress convergence is based on the assumption of simultaneous notification (i.e., all occupants of all floors begin moving toward the exits at the same time). As illustrated in Figure 1004.5(3), the occupants of the first floor will have exited the building by the time the occupants of the second floor have reached the exit discharge door. However, as illustrated in Figure 1004.5(1), the occupants of a basement will reach the discharge door simultaneously with the second-floor occupants, thereby creating the need for sizing the components for a larger combined occupant load.

An egress convergence situation can also be created when an intermediate floor level is not present, as illustrated in Figure 1004.5(2). Again, under the assumption of simultaneous notification, occupants of both floors would reach the exit discharge door at approximately the same time, invoking the requirements for a larger egress capacity.

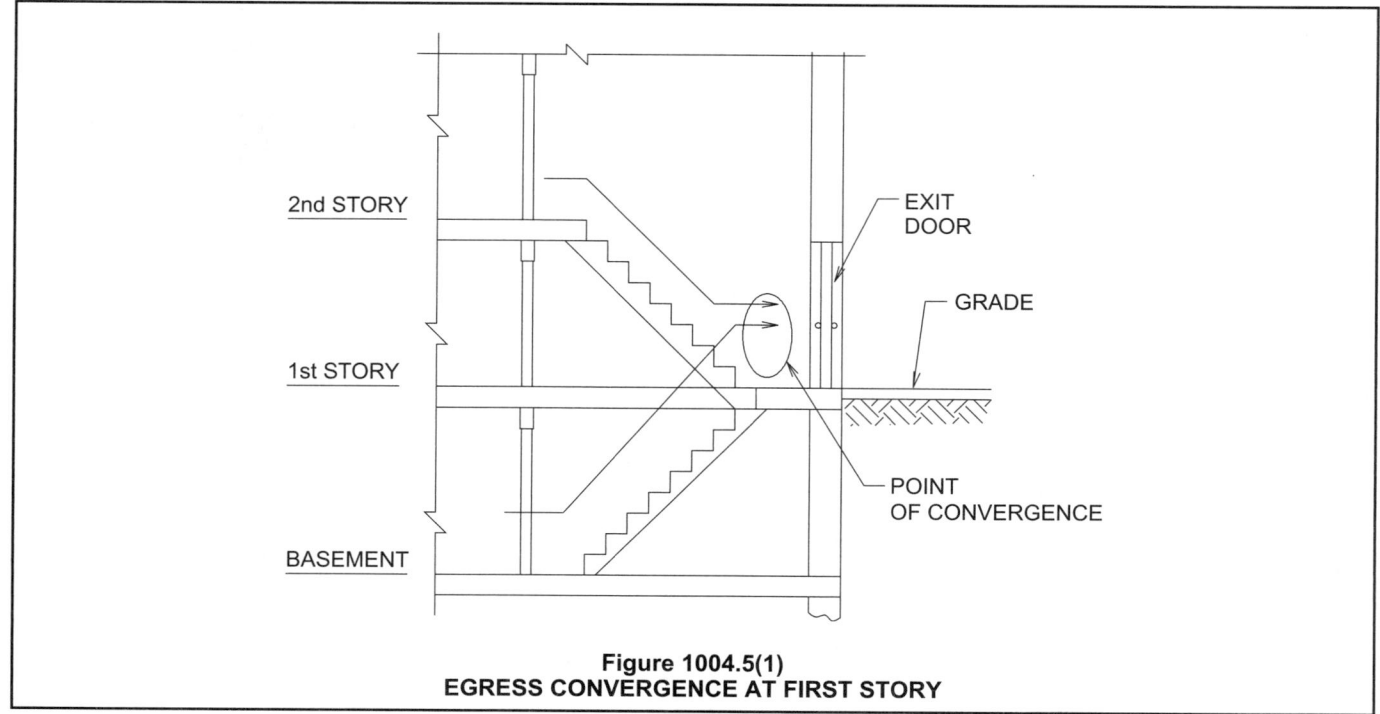

Figure 1004.5(1)
EGRESS CONVERGENCE AT FIRST STORY

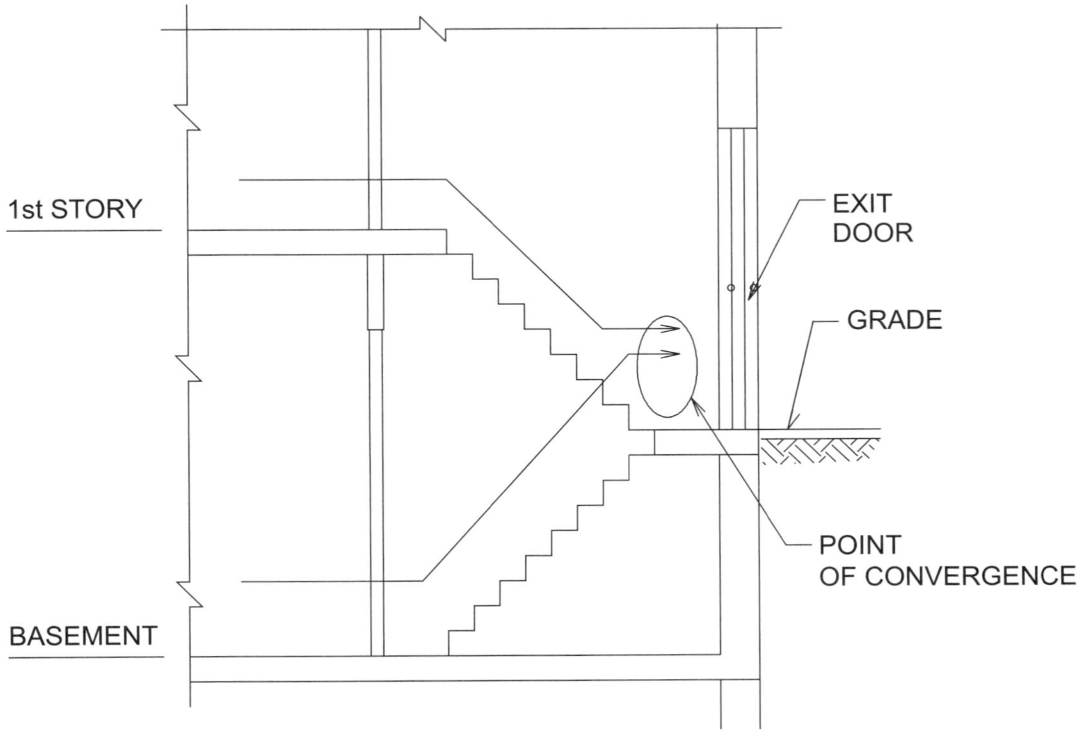

1st STORY

EXIT
DOOR

GRADE

POINT
OF CONVERGENCE

BASEMENT

Figure 1004.5(2)
EGRESS CONVERGENCE AT INTERMEDIATE LEVEL

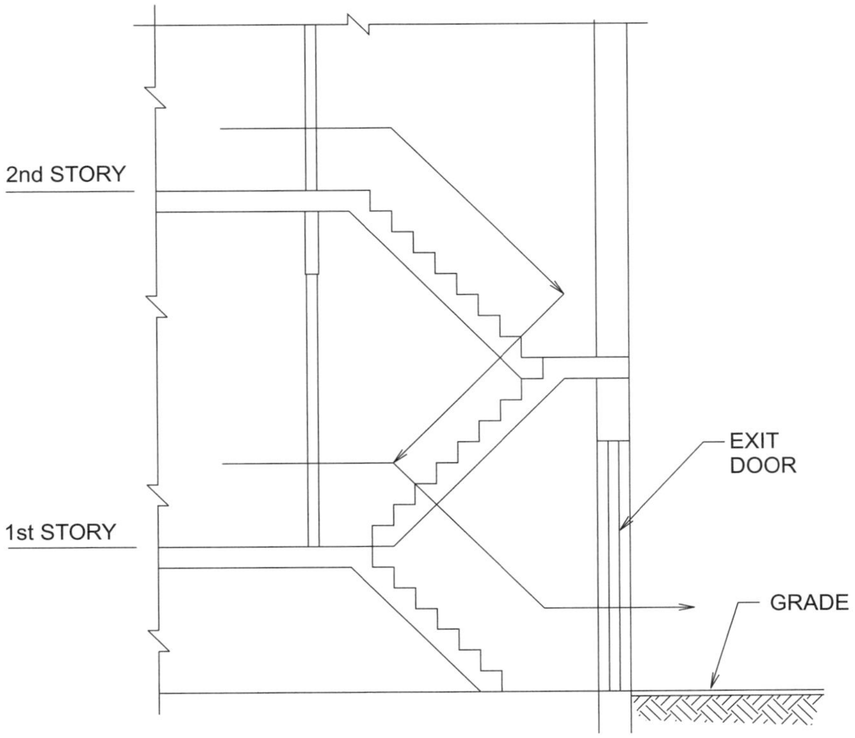

2nd STORY

1st STORY

EXIT
DOOR

GRADE

Figure 1004.5(3)
NO EGRESS CONVERGENCE

1004.6 Mezzanine levels. The *occupant load* of a *mezzanine* level with egress onto a room or area below shall be added to that room or area's *occupant load*, and the capacity of the exits shall be designed for the total *occupant load* thus established.

❖ The egress requirements for mezzanines are handled similar to those addressed in Section 1004.1 with accessory areas versus the requirements for exiting from multiple levels in Section 1004.4. That is, that portion of the mezzanine occupant load that discharges to the floor below is to be added to the occupant load of the space on the floor below. The sizing and number of the egress components must reflect this combined occupant load. This does not apply to the means of egress from a mezzanine that does not require travel through another level (i.e., an exit stairway serving the mezzanine). Section 505 contains additional criteria for the means of egress from mezzanines.

1004.7 Fixed seating. For areas having fixed seats and *aisles*, the *occupant load* shall be determined by the number of fixed seats installed therein. The *occupant load* for areas in which fixed seating is not installed, such as waiting spaces and *wheelchair spaces*, shall be determined in accordance with Section 1004.1.1 and added to the number of fixed seats.

For areas having fixed seating without dividing arms, the *occupant load* shall not be less than the number of seats based on one person for each 18 inches (457 mm) of seating length.

The *occupant load* of seating booths shall be based on one person for each 24 inches (610 mm) of booth seat length measured at the backrest of the seating booth.

❖ The occupant load in an area with fixed seats is readily determined. In spaces with a combination of fixed and loose seating, the occupant load is determined by a combination of the occupant density number from Table 1004.1.1 and a count of the fixed seats.

For bleachers, booths and other seating facilities without dividing arms, the occupant load is simply based on the number of people that can be accommodated in the length of the seat. Measured at the hips, an average person occupies about 18 inches (457 mm) on a bench. In a booth, additional space is necessary for "elbow room" while eating. In a circular or curved booth or bench, the measurement should be taken just a few inches from the back of the seat, which is where a person's hips would be located (see Figure 1004.7).

Some assembly spaces may have areas for standing or waiting. For example, some large sports stadiums have "standing room only" areas that they use for sell-out games. The Globe Theater in England has standing room in an area at the front of the theater. This section is not intended to assign an occupant load to the typical circulation aisles in an assembly space. Occupant load for wheelchair spaces should be based on the number of wheelchairs and companion seats that the space was designed for. As specified in Section 1004.9, if the wheelchair spaces may also be utilized for standing space, the occupant load must be determined by the worst-case scenario.

1004.8 Outdoor areas. Yards, patios, courts and similar outdoor areas accessible to and usable by the building occupants shall be provided with *means of egress* as required by this chapter. The *occupant load* of such outdoor areas shall be assigned by the *building official* in accordance with the anticipated use. Where outdoor areas are to be used by persons in addition to the occupants of the building, and the path of egress travel from the outdoor areas passes through the building, *means of egress*

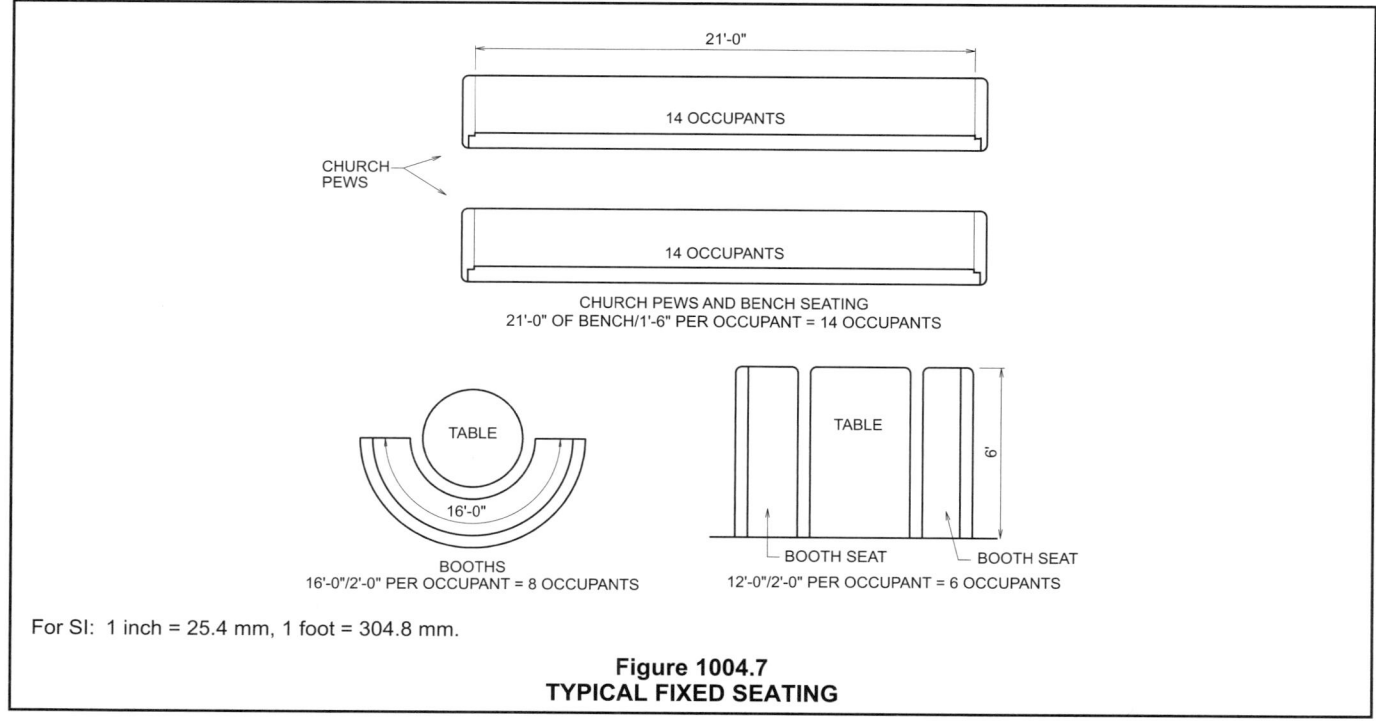

CHURCH PEWS

21'-0"

14 OCCUPANTS

14 OCCUPANTS

CHURCH PEWS AND BENCH SEATING
21'-0" OF BENCH/1'-6" PER OCCUPANT = 14 OCCUPANTS

TABLE

16'-0"

BOOTHS
16'-0"/2'-0" PER OCCUPANT = 8 OCCUPANTS

TABLE

BOOTH SEAT   BOOTH SEAT

12'-0"/2'-0" PER OCCUPANT = 6 OCCUPANTS

For SI: 1 inch = 25.4 mm, 1 foot = 304.8 mm.

Figure 1004.7
TYPICAL FIXED SEATING

requirements for the building shall be based on the sum of the *occupant loads* of the building plus the outdoor areas.

Exceptions:

1. Outdoor areas used exclusively for service of the building need only have one *means of egress*.

2. Both outdoor areas associated with Group R-3 and individual dwelling units of Group R-2.

❖ This section addresses the means of egress of outdoor areas such as yards, patios and courts. The primary concern is for those outdoor areas used for functions that may include occupants other than the building occupants or solely by the building occupants where egress from the outdoor area is back through the building to reach the exit discharge. An example is an interior court of an office building where assembly functions are held during normal business hours for persons other than the building occupants. When court occupants must egress from the interior court back through the building, the building's egress system is to be designed for the building occupants, plus the assembly occupants from the interior court. Another example would be an outdoor dining area that exited back through the restaurant.

The occupant load is to be assigned by the building official based on use. It is suggested that the design occupant load be determined in accordance with Section 1004.1.1.

Exception 1 describes conditions where the occupant load is very limited, such as areas where an interior courtyard had strictly plants or mechanical equipment. If the courtyard was open for building occupants, other than maintenance personnel, to use the space, the space must be designed with the occupant loads in Table 1004.1.1. Balconies or patios associated with individual dwelling units, in Exception 2, would typically be used by the occupants of the unit. Means of egress can be back through the building in accordance with Section 1014.2.

1004.9 Multiple occupancies. Where a building contains two or more occupancies, the *means of egress* requirements shall apply to each portion of the building based on the occupancy of that space. Where two or more occupancies utilize portions of the same *means of egress* system, those egress components shall meet the more stringent requirements of all occupancies that are served.

❖ Since the means of egress systems are designed for the specific occupancy of a space, the provisions of this chapter are to be applied based on the actual occupancy conditions of the space served.

For example, a hospital is classified as Group I-2 and normally includes the associated administrative or business functions found in the same building. Chapter 3 would permit the entire building to be constructed to the more restrictive provisions for Group I-2; however, each area of the building need only have the means of egress designed in accordance with the actual occupancy conditions, such as Groups I-2 and B. If the corridor serves only the occupants in the busi-

ness use (i.e., administrative staff), and is not intended to serve as a required means of egress for patients, the corridor need only be 36 or 44 inches (914 or 1118 mm) in width, depending on the occupant load.

Where the corridor is used by both Group I-2 and B occupancies, it must meet the most stringent requirement. For example, if a corridor in the business area is also used for the movement of beds (i.e., exit access from a patient care area), it would need to be a minimum of 96 inches (2438 mm) in clear width.

SECTION 1005
EGRESS WIDTH

1005.1 Minimum required egress width. The *means of egress* width shall not be less than required by this section. The total width of *means of egress* in inches (mm) shall not be less than the total *occupant load* served by the *means of egress* multiplied by 0.3 inches (7.62 mm) per occupant for stairways and by 0.2 inches (5.08 mm) per occupant for other egress components. The width shall not be less than specified elsewhere in this code. Multiple *means of egress* shall be sized such that the loss of any one *means of egress* shall not reduce the available capacity to less than 50 percent of the required capacity. The maximum capacity required from any *story* of a building shall be maintained to the termination of the *means of egress*.

Exception: *Means of egress* complying with Section 1028.

❖ The code requires the utilization of two methods to determine the minimum width of egress components. While this section provides a methodology for determining required widths based on the design occupant load, calculated in accordance with Section 1004.1, other sections provide minimum widths of various components. The actual width that is provided is to be the larger of the two widths. For this section, the sum of the capacities of the means of egress components that serve each space must equal or exceed the occupant load of that space. For example, the combined width of all the exit stairways from a floor would be considered to determine if the stairways had adequate capacity for everyone to evacuate the building. All elements must meet the minimum width requirements specified in other sections (e.g., Section 1008.1.1 for doors; Sections 1007.3 and 1009.1 for stairs.).

This section establishes the necessary width of each egress component on a "per occupant" basis. Means of egress components are separated between "stairs" and "other"; "other" being doors, doorways, corridors, ramps, aisles, etc. The requirements for stairways are wider due to the slowdown of travel to negotiate the steps. When the required occupant capacity of an egress component is determined, multiplication by the appropriate factor results in the required clear width of the component in inches, based on capacity. Similarly, if the clear width of a component is known, division by the appropriate factor results in the permitted capacity of that component. For example, two exit access doorways from a room with an occupant load of 300 would each have a required capacity of not less than 150.

Based on the minimum required clear door width (32 inches clear width per door divided by 0.2 inch per occupant = 160 occupants), two 32-inch clear width doors would meet both the minimum clear width (Section 1008.1.1) and the capacity requirements. Two exits from a space with an occupant load of 450 would each have a required capacity of not less than 225, necessitating more doors or larger door leaves.

The following typical calculations illustrate calculations for stairways from a typical two-story, two-exit office building:

1. Determine the minimum required stairway width with a second-floor occupant load of 350:
   - 350 occupants by 0.3 inches = 105 inches (2667 mm) minimum;
   - 105 inches divided by 2 stairways is $52^1/_2$ inches (1334 mm) minimum per stairway; or
   - Section 1009.1 prescribes that the width of an interior stairway cannot be less than 44 inches (1118 mm).

   The capacity criteria are more restrictive and, therefore, the minimum required width for each stairway is $52^1/_2$ inches (1334 mm).

2. Determine the minimum required stairways width with a second-floor occupant load of 90:
   - 90 occupants by 0.3 inches = 27 inches (686 mm) minimum;
   - 27 inches divided by two stairways is $13^1/_2$ inches (343 mm); or
   - Section 1009.1 prescribes that the width of an interior stairway cannot be less than 44 inches (1118 mm). Note that the stair width reduction in Section 1009.1, Exception 1, is applicable only when the entire occupant load of a story is less than 50.

   The minimum clear width requirements are more restrictive and, therefore, the minimum required width for each stairway is 44 inches (1118 mm).

   The maximum capacity of a 44-inch stairway is 44 inches divided by 0.3 inches per occupant = 146 occupants. Therefore, a floor level with two exit stairways could have 292 occupants before the capacity would control the stairway egress width.

   The traditional unit of measurement of egress capacity was based on a "unit exit width" that was to simulate the body ellipse with a basic dimensional width of 22 inches (559 mm)— approximately the shoulder width of an average adult male. This unit exit width was combined with assumed egress movement (such as single file or staggered file) to result in an egress capacity per unit exit width for various occupancies. This assumption simplifies the dynamic egress process since contemporary studies have indicated that people do not egress in such precise and predictable movements. As traditionally used in the codes, the method of determining capacity per unit of clear width implies a higher level of accuracy than can realistically be achieved. The resulting factors preserve the features of the past practices that can be documented, while providing a more straightforward method of determining egress capacity.

   Previous editions of the code also included allowances for the presence of an automatic sprinkler system. In all occupancies other than Groups H-1, H-2, H-3, H-4 and I-2, a factor of 0.2 inches (5.08 mm) of width per person was required for stairway travel and 0.15 inches (3.81 mm) per person was utilized for all other egress components. For sprinklered buildings, other than Groups H-1, H-2, H-3, H-4 and I-2, the allowances for the capacity for a typical door (32 inches by 0.15 = 213 occupants versus 32 inches by 0.2 = 160) has been reduced by 25 percent and the capacity for stairways (44 inches by 0.2 = 220 versus 44 inches by 0.3 = 146) has been reduced by over 33 percent. These allowances continue to be permitted in existing buildings [see Table 3412.6.11(1) or IFC Section 4604.7)].

   The code does require that when multiple means of egress are required, the loss of any one path would not reduce the available capacity to less than 50 percent. The 50-percent minimum of the required egress capacity results in a fairly uniform distribution of egress paths. This requirement does not, however, require that the capacities be equally distributed when more than two means of egress are provided. An egress design with a dramatic imbalance of egress component capacities relative to occupant load distribution should be reviewed closely to avoid a needless delay in egressing a floor or area. The balancing of the means of egress components, in accordance with the distribution of the occupant load, is reasonable and, in some cases, necessary for facilities having mixed occupancies with dramatically different occupant loads.

   The requirement that the maximum capacity from any floor is to be provided all the way along the exit to the termination, typically down the stairway to the exterior exit door at the level of exit discharge, results in an egress width that is adequate for the exit discharge.

   The purpose of the exception is to require assembly seating to comply with the more specific provisions in Section 1028.

1005.2 Door encroachment. Doors, when fully opened, and handrails shall not reduce the required *means of egress* width by more than 7 inches (178 mm). Doors in any position shall not reduce the required width by more than one-half. Other nonstructural projections such as trim and similar decorative features shall be permitted to project into the required width a maximum of 1¹/₂ inches (38 mm) on each side.

Exception: The restrictions on a door swing shall not apply to doors within individual dwelling units and sleeping units of Group R-2 and dwelling units of Group R-3.

❖ Projections or restrictions in the required width can impede and restrict occupant travel, causing egress to occur less efficiently than contemplated. The swing of a door, such as from a room into a corridor, and any handrails along the route are permitted projections.

Historically this section has looked at doors on one wall at a time. Doors located across the hall from one another are not considered additive when considering protrusion limits. Doors would not typically be opened to the full extent at exactly the same moment, nor would they remain open at 90 degrees and totally blocking the hall because of the maximum limitation of 7 inches (178 mm) when fully open. A cross-reference back to this section from the exceptions for width in corridors (Section 1018.3), aisles (Section 1017.1), exit passageways (Section 1023.2) and exit courts (Section 1027.5.1) reinforces the fact that this provision is generally applicable for these types of confined routes.

Handrails are not required along corridors, level aisles, exit passageways and exit corridors; however, if provided, both Section 1003.3.3 and 1005.2 would be applicable. The current text is not clear on if handrails are evaluated separately where located across from each other. However, continuous handrails along both sides would be more of continuous horizontal obstruction than doors. Handrails are sometimes provided along the hallways in hospitals or nursing homes to aid the residents. Bumper guards along the walls are not handrails. Assuming a corridor width that is at the required width, there are the following two scenarios.

• If handrails are provided on one wall only, Section 1003.3.3 with a limitation of 4¹/₂ inches (144 mm) for handrail project would be more restric-

tive, and would not require an increase in corridor width.

• Where handrails are provided on both walls, Section 1005.2 would limit the projection to 3¹/₂ inches (89 mm) from each side, or with a 4¹/₂ inch (144 mm) projection would result in an additional 2 inches (51 mm) in corridor width due to the 7-inch (178 mm) limitation for projection into the required corridor width.

In regard to the handrail projection limitations of 4½ inches (144 mm) on each side of ramps and stairways as set forth in Section 1012.7 or stepped and sloped aisles in Section 1028, they are still applicable since they are specific provisions as opposed to the general provisions of Section 1005.2. Therefore, the 7-inch (178 mm) limitation on handrails into the required width does not apply to stairways and ramps.

Regarding door encroachment there are two tests. The arc created by the door's outside edge cannot project into more than one-half of the required corridor width. When opened to its fullest extent, the door cannot project more than 7 inches (178 mm) into the required width, which is the dimension of a door leaf thickness excluding the hardware as shown in Figure 1005.2. Door hardware encroachment is addressed separately in Section 1005.3. These projections are permitted because they are considered to be temporary and do not significantly impede the flow. Occupants will compensate for the projection by a reduction in the natural cushion they retain between themselves and a boundary, known as the edge effect.

The door swing restrictions do not apply within dwelling units since the occupant load is very low. Based on the intent of this section, other situations that could be approved by the official having jurisdiction would be situations where the opening door would not block the egress, such as the door at the end of a corridor, or the room was not typically occupied, such as a janitor's closet.

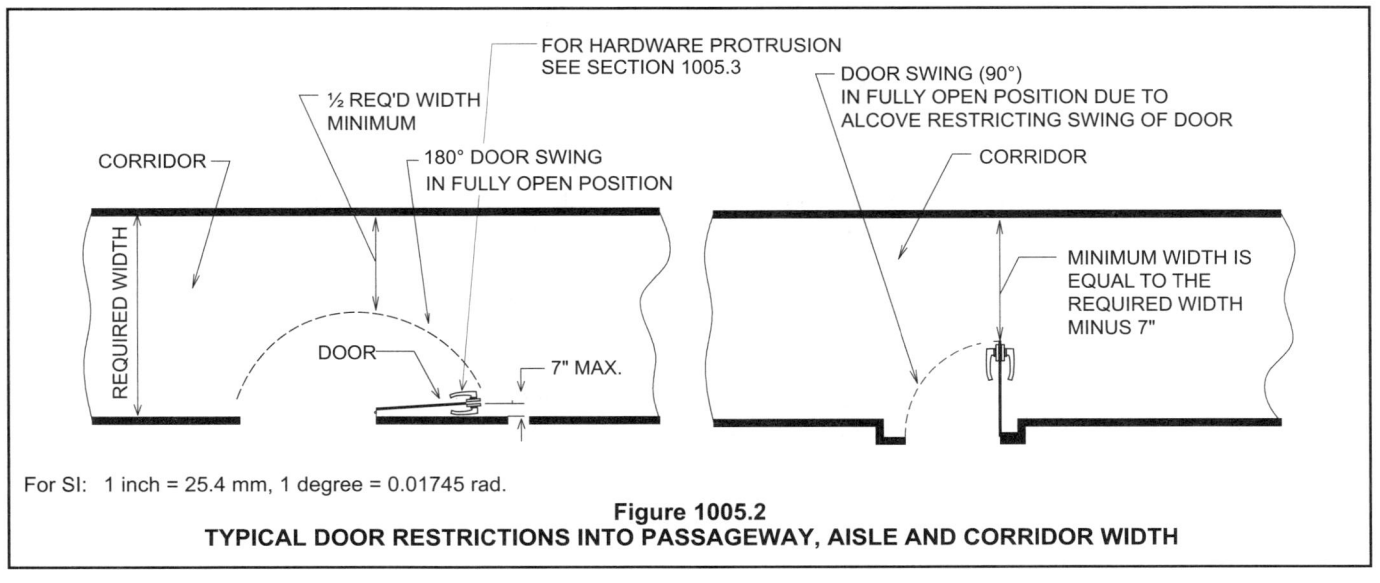

For SI: 1 inch = 25.4 mm, 1 degree = 0.01745 rad.

Figure 1005.2
TYPICAL DOOR RESTRICTIONS INTO PASSAGEWAY, AISLE AND CORRIDOR WIDTH

Item such as baseboards, chair rails, pilasters, etc., are limited to protruding over the required width of the corridor a maximum of $1^1/_2$ inches (38 mm). However, once again, Section 1003.3.3 would be applicable when the corridor was wider than required.

1005.3 Door hardware encroachment. Surface-mounted latch release hardware shall be exempt from inclusion in the 7-inch (178 mm) maximum projection requirement of Section 1005.2 when:

1. The hardware is mounted to the side of the door facing the corridor width when the door is in the open position; and

2. The hardware is mounted not less than 34 inches (865 mm) or more than 48 inches (1220 mm) above the finished floor.

❖ The provision indicates that hardware facing the corridor when the door is fully open need not be considered when determining the allowable door encroachment into a corridor of 7 inches (178 mm) maximum in Section 1005.2. The allowance is applicable only for hardware mounted within height range of 34 inches to 48 inches (865 to 1220 mm) which is consistent with the range for means of egress door hardware height as established in Section 1008.1.9.2. Where hardware extended across the door, such as panic hardware, the 4-inch (102 mm) projection in the door opening is addressed in Section 1008.1.1.1.

SECTION 1006
MEANS OF EGRESS ILLUMINATION

1006.1 Illumination required. The *means of egress*, including the *exit discharge*, shall be illuminated at all times the building space served by the *means of egress* is occupied.

Exceptions:

1. Occupancies in Group U.

2. *Aisle accessways* in Group A.

3. Dwelling units and sleeping units in Groups R-1, R-2 and R-3.

4. Sleeping units of Group I occupancies.

❖ All means of egress must be illuminated by artificial lighting during the entire time a building is occupied, so that the paths of exit travel are always visible and available for evacuation of the occupants during emergencies. The code makes a special point of noting that the exit discharge must also be provided with adequate illumination so that occupants can safely find the public way should the emergency occur at night.

Three of the exceptions are for occupancies where the constant illumination of the means of egress would interfere with the use of space, such as sleeping areas or theater aisles during a performance.

Bear in mind that means of egress lighting is not

emergency lighting. For emergency lighting requirements, see Sections 1006.3 and 1006.4.

1006.2 Illumination level. The *means of egress* illumination level shall not be less than 1 foot-candle (11 lux) at the walking surface.

Exception: For auditoriums, theaters, concert or opera halls and similar assembly occupancies, the illumination at the walking surface is permitted to be reduced during performances to not less than 0.2 foot-candle (2.15 lux), provided that the required illumination is automatically restored upon activation of a premises' fire alarm system where such system is provided.

❖ The intensity of lighting illuminating the entire means of egress, including open plan spaces, aisles, corridors and passageways, exit stairways, exit doors and places of exit discharge at the walking surface or floor level must not be less than 1 foot-candle (11 lux). It has been found that this low level of lighting renders enough visibility for the occupants to evacuate the building safely.

It is important to note that this lighting level is measured at the floor in order to make the floor surface visible. Levels of illumination above the floor may be higher or lower, thus allowing lights on the steps to be used rather than general area lights.

The exception addresses occupancies where low light level is needed for the function of the space. The level of intensity of aisle lighting in such spaces may be reduced to 0.2 foot-candle (2.15 lux), but only during the time of a performance. This intensity of illumination is sufficient to distinguish the aisles and stairs leading to the egress doors and is not a source of distraction during a performance. It is not the intent of the exception to require a fire alarm system but to require a connection to the egress lighting where a fire alarm system is provided.

1006.3 Illumination emergency power. The power supply for *means of egress* illumination shall normally be provided by the premises' electrical supply.

In the event of power supply failure, an emergency electrical system shall automatically illuminate all of the following areas:

1. *Aisles* and unenclosed egress *stairways* in rooms and spaces that require two or more *means of egress*.

2. *Corridors*, *exit enclosures* and *exit passageways* in buildings required to have two or more *exits*.

3. Exterior egress components at other than their *levels of exit discharge* until *exit discharge* is accomplished for buildings required to have two or more *exits*.

4. Interior *exit discharge* elements, as permitted in Section 1027.1, in buildings required to have two or more *exits*.

5. Exterior landings as required by Section 1008.1.6 for *exit discharge* doorways in buildings required to have two or more *exits*.

The emergency power system shall provide power for a duration of not less than 90 minutes and shall consist of storage batteries, unit equipment or an on-site generator. The installation of the emergency power system shall be in accordance with Chapter 27.

❖ The means of egress must be illuminated, especially in times of emergency when the occupants must have a lighted path of exit travel in order to evacuate the building safely. The code is very specific in the description of the areas that are required to be illuminated by the emergency power system.

The locations include:

1. Dedicated egress routes in larger rooms or spaces. For example, an aisle in an open office plan but not within individual offices that egress through the open area. Exit access stairways from a mezzanine or from a second floor (as permitted by Section 1016.1, Exceptions 3 and 4) must also be illuminated.

2. Essential portions of the interior egress system, such as exit stairways and corridors in larger buildings, must be illuminated. Exit passageways are used in facilities with long travel distances, such as malls, or are an extension of the exit stairway enclosure. This would include exit stairways that were not required to be enclosed in accordance with the exceptions in Section 1022.1.

3. Means of egress systems that use exit balconies (see Section 1019) or open exterior exit stairways or ramps (see Section 1027) need emergency exit illumination along these components until egress at grade is "reached" or "achieved."

4. Interior exit discharge elements, such as lobbies and vestibules (see Section 1027.1), where stairways discharge into these elements instead of directly to the exterior of a building.

5. Exterior portions of the exit discharge. Note that only the portion of the exterior discharge that is immediately adjacent to the building exit discharge door is required to have emergency illumination and not the entire exterior discharge path to the public way.

So that there will be a continuing source of electrical energy for maintaining the illumination of the means of egress when there is a loss of the main power supply, the means of egress lighting system must be connected to an emergency electrical system that consists of storage batteries, unit equipment or an on-site generator. This emergency power-generating facility must be capable of supplying electricity for at least 90 minutes, thereby giving the occupants sufficient time to leave the premises. In most cases, where the loss of the main electrical supply is attributed to a malfunction in the distribution system of the electric power company, experience has shown that such power outages do not usually last as long as 90 minutes.

1006.4 Performance of system. Emergency lighting facilities shall be arranged to provide initial illumination that is at least an average of 1 foot-candle (11 lux) and a minimum at any point of 0.1 foot-candle (1 lux) measured along the path of egress at floor level. Illumination levels shall be permitted to decline to 0.6 foot-candle (6 lux) average and a minimum at any point of 0.06 foot-candle (0.6 lux) at the end of the emergency lighting time duration. A maximum-to-minimum illumination uniformity ratio of 40 to 1 shall not be exceeded.

❖ This section provides the criteria of the illumination levels of the emergency lighting system. The initial average level for the main egress paths is the same as for the overall means of egress illumination in Section 1006.2. The reduction of illumination recognizes the performance characteristics over time of some types of power supplies, such as batteries. The minimum levels are sufficient for the occupants to egress from the building. In addition, the emergency lighting system is a secondary system that is spaced along the main egress routes. It will not provide the same level of general lighting over the route as what can be provided by the building lighting system.

The maximum illumination uniformity ratio of 40 means that the variation in the illumination levels is not to exceed that number. For example, a minimum of 0.06 foot-candle (0.6 lux) would establish a maximum illumination of 2.4 foot-candle (24 lux) in an adjacent area. This is to establish a variation limit such that the means of egress can be seen as a person walks from bright to darker areas along the egress path.

SECTION 1007
ACCESSIBLE MEANS OF EGRESS

1007.1 Accessible means of egress required. *Accessible means of egress* shall comply with this section. *Accessible* spaces shall be provided with not less than one *accessible means of egress*. Where more than one *means of egress* are required by Section 1015.1 or 1021.1 from any *accessible* space, each *accessible* portion of the space shall be served by not less than two *accessible means of egress*.

Exceptions:

1. *Accessible means of egress* are not required in alterations to existing buildings.

2. One *accessible means of egress* is required from an *accessible mezzanine* level in accordance with Section 1007.3, 1007.4 or 1007.5.

3. In assembly areas with sloped or stepped *aisles*, one *accessible means of egress* is permitted where the common path of travel is *accessible* and meets the requirements in Section 1028.8.

❖ The Architectural and Transportation Barriers Compliance Board (Access Board) is revising and updating its accessibility guidelines for buildings and facilities covered by the Americans with Disabilities Act of 1990 (ADA) and the Architectural Barriers Act of 1968 (ABA). The final ADA/ABA Guidelines, published by the Access Board in July 2004, serve as the basis for the minimum standards when adopted by other federal agencies responsible for issuing enforceable standards. The plan is to eventually use this new document in place of the Uniform Federal Accessibility Standard (UFAS) and the Americans with Disabilities Act Accessibility Guidelines (ADAAG). The ADA/ABA Guidelines, Section 207/F207, references the 2000 edition of the code with 2001 Supplement, as well as 2003 edition of the code, for accessible means of egress requirements. The ICC is very proud to be recognized for its work regarding accessible means of egress in this manner. Refer to the Access Board web site (www.access-board.gov) for more specific information and the current status of this adoption process.

The accessible means of egress requirements may not be the same route as that required for ingress into the building (see Sections 1104 and 1105). For example, a two-story building requires one accessible route to connect all accessible spaces within the building. The accessible route to the second level is typically by an elevator. During a fire emergency, persons with mobility impairments on the second level would be moving to the exit stairways for assisted rescue, not back the way they came in via the elevator.

This section establishes the minimum requirements for means of egress facilities serving all spaces that are required to be accessible to people with physical disabilities. Previously, attention had been focused on response to the civil-rights-based issue of providing adequate access for people with physical disabilities into and throughout buildings. Concerns about life safety and evacuation of people with mobility impairments were frequently cited as reasons for not embracing widespread building accessibility, in the best interest of the disabled community.

The provisions for accessible means of egress are predominantly, though not exclusively, intended to address the safety of persons with a mobility impairment. These requirements reflect the balanced philosophy that accessible means of egress are to be provided for occupants who have gained access into the building but are incapable of independently utilizing the typical means of egress facilities, such as the exit stairways. By making such provisions, the code now addresses means of egress for all building occupants, with and without physical disabilities.

Any space that is not required by the code to be accessible in accordance with Chapter 11 is not required to be provided with accessible means of egress. This may include an entire story, a portion of a story or an individual room.

In new construction and additions, accessible means of egress are required in the same number as the general means of egress, up to a maximum of two. For example, in buildings, stories or spaces required by Section 1015.1 or 1021 to have three or more exits or exit access doors, a minimum of two accessible means of egress is required. The number of exits or exit access doors is based on occupant load; therefore, no matter how large the total occupant load of the space, two fully complying accessible means of egress are considered to provide sufficient capacity for those building occupants with a mobility impairment.

The accessibility requirements are based on the required means of egress from both individual spaces and the building as a whole. Therefore, the facilities with multiple large assembly rooms, such as banquet halls or multiplex theaters where the second exit from the space is often a door directly to the outside, may require additional accessible means of egress from the building due to space requirements.

While there are no dispersement requirements specific to accessible means of egress or travel distance limitations where there is no area of refuge requirement (see Sections 1007.3, 1007.4 and 1007.6), the code requires exits to be distinct and independent (Section 1021.1). Therefore, if a stairway and elevator being used for accessible means of egress were adjacent to each other, this would only count as one accessible means of egress.

An accessible means of egress is required to provide a continuous path of travel to a public way. This principle is consistent with the general requirements for all means of egress, as reflected in Section 1003.1 and in the definition of "Means of egress" in Section 1002. This section also emphasizes the intent that accessible means of egress must be available to a person with a mobility impairment, such as a person in a wheelchair. Some mobility impairments do not allow for self-evacuation along a stairway; therefore, utilization of the exit and exit discharge may require assistance. The safety and fire evacuation plans (see IFC Section 404) require planning for all occupants of a building. This assistance is typically with the fire department or other trained personnel, either along the exit stairways or in buildings five stories or taller, with the elevator system or a combination of both (see commentary, Section 1007.2.1). It is required that accessible routes, areas of refuge and exterior areas of rescue assistance are indicated on these plans. These plans must be approved by the local fire official and reviewed annually.

The exceptions address special situations where accessible means of egress requirements need special consideration. Note that these are exceptions for accessible means of egress; not an exception for accessible entrance requirements (see Section 1105).

Exception 1 indicates that existing buildings that are undergoing alterations are not required to be provided with accessible means of egress as part of that alteration. In many cases, meeting the requirements for accessible means of egress, especially the 48-inch (1219 mm) clear stair width required in nonsprinklered buildings, would be considered technically infeasible. However, if an accessible means of egress was part of the original construction, it must be maintained in accordance with Section 3411.2.

Exception 2 is a special consideration for mezzanines. The size of mezzanines is limited to a portion of the space below. Most are open to the space below; thus, with same atmosphere and line of sight, fire recognition is quicker than in a two-story situation. If the elevator used for ingress has not gone into fire department recall, that system could be used for self-evacuation. There are three different scenarios:

1. If the mezzanine is exempted from accessibility, such as a mechanical mezzanine (see Section 1103.2.9), or small enough not to be required to be accessible (see Section 1104.4, Exception 1), no accessible means of egress is required.

2. If a mezzanine is required to be accessible (see Section 1104.4) but does meet the provisions for spaces with one means of egress (see Section 1015.1), the exit access stairway must meet the provisions of Section 1007.3.

3. If a mezzanine is required to be accessible (see Section 1104.4) and is required to have two means of egress, at least one of the exit access stairways must meet the provisions of Section 1007.3.

While under Scenarios 2 and 3 it is optional to have the elevator meet the requirements of Section 1007.4, the provisions for standby power at elevators are based on fire department assisted rescue. This is an expensive option that would likely never be used during a fire event. Where platform lifts can be used in new construction (see Section 1109.7), they are so limited that it is not likely that they will provide the accessible route to a mezzanine. If they do, Section 1007.5 does allow for platform lifts to serve as part of the accessible route for accessible means of egress when they have standby power.

Exception 3 is in consideration of the practical difficulties of providing accessible routes in assembly areas with sloped floors and stepped aisles. Rooms with more than 50 persons are required to have two means of egress; therefore, each accessible seating location is required to have access to two accessible means of egress. Depending on the slope of the seating arrangement, this can be difficult to achieve, especially in small theaters. A maximum travel distance of 30 feet (9144 mm) for ambulatory persons moving from the last seat in dead-end aisles or from box-type seating arrangements to where they have access to a choice of means of egress routes has been established in Section 1028.8. In accordance with Exception 3, persons using wheelchair seating spaces have the same maximum 30-foot (9144 mm) travel distance from the accessible seating locations to a cross aisle or out of the room to an adjacent corridor or space where two choices for accessible means of egress are provided. Note that there are increases in travel distance for smoke-protected seating and small spaces, such as boxes, galleries or balconies. For additional information, see Section 1028.8.

1007.2 Continuity and components. Each required *accessible means of egress* shall be continuous to a *public way* and shall consist of one or more of the following components:

1. *Accessible routes* complying with Section 1104.

2. *Interior exit stairways* complying with Sections 1007.3 and 1022.

3. *Exterior exit stairways* complying with Sections 1007.3 and 1026.

4. Elevators complying with Section 1007.4.

5. Platform lifts complying with Section 1007.5.

6. *Horizontal exits* complying with Section 1025.

7. *Ramps* complying with Section 1010.

8. *Areas of refuge* complying with Section 1007.6.

Exceptions:

1. Where the *exit discharge* is not *accessible*, an exterior area for assisted rescue must be provided in accordance with Section 1007.7.

2. Where the *exit stairway* is open to the exterior, the *accessible means of egress* shall include either an *area of refuge* in accordance with Section 1007.6 or an exterior area for assisted rescue in accordance with Section 1007.7.

❖ This section identifies the various building features that can serve as elements of an accessible means of egress. Accessible routes are readily recognizable as to how they can provide accessible means of egress; however, some nontraditional principles have been established for the total concept of accessible means of egress. This is evident in that stairways and elevators are also identified as elements that can comprise part of an accessible means of egress. For example, elevators are generally not available for egress during a fire, while stairways are not independently usable by a person in a wheelchair. The concept of accessible means of egress includes the idea that evacuating people with a mobility impairment may require the assistance of others. In some situations, provisions are also included for creating an area of refuge wherein people can safely await either further instructions or evacuation assistance. Refuge areas can also be established by utilizing horizontal exits. All of these elements can be arranged in the manner prescribed in this section to provide accessible means of egress.

Typically, the accessible way into a single-story facility is also one of the accessible means of egress. Exceptions 1 and 2 are intended to address problems that most typically arise for the second accessible means of egress. Exception 1 addressed situations where site constraints or configurations may result in difficulty providing an accessible exterior route for exit discharge. The alternative for exterior areas for assisted rescue is offered for these types of situations. A change in elevation across a site may result in exit doors leading to exterior steps in the exit discharge, or the path may not be accessible because it is too steep or not of materials that would provide for a stable and firm surface (see the commentary to Section 1007.7). Exception 2 is an option for exterior exit stairways. Since exceptions are alternatives, an alternative would be for the exterior exit stairway to comply with Section 1007.3. This exception would allow for an exterior exit stairway to include either an area of refuge inside the facility or an exterior area of rescue assistance. If the exterior area for assisted rescue is chosen, the separation requirements would override the exterior exit stairway separation exceptions in Section 1026.6.

1007.2.1 Elevators required. In buildings where a required *accessible* floor is four or more stories above or below a *level of exit discharge*, at least one required *accessible means of egress* shall be an elevator complying with Section 1007.4.

Exceptions:

1. In buildings equipped throughout with an *automatic sprinkler system* installed in accordance with Section 903.3.1.1 or 903.3.1.2, the elevator shall not be required on floors provided with a *horizontal exit* and located at or above the *levels of exit discharge*.

2. In buildings equipped throughout with an *automatic sprinkler system* installed in accordance with Section 903.3.1.1 or 903.3.1.2, the elevator shall not be required on floors provided with a ramp conforming to the provisions of Section 1010.

❖ Elevators are the most common and convenient means of providing access to the upper floors in multistory buildings. As such, elevators represent a prime candidate for accessible means of egress from such buildings, especially in light of the difficulties involved in carrying a person up or down a stairway for multiple levels. The primary consideration for elevators as an accessible means of egress is that the elevator will be available and protected during a fire event to allow for fire department assisted rescue. Typically it is not the intent that people use the elevator for self-evacuation due to the hazards associated with smoke in the elevator shaft or the elevator taking people to the floor with a direct fire hazard. There are some new technological advances for "fire service access elevators" and "occupant evacuation elevators" that are discussed in Sections 403, 3007 and 3008.

This section addresses where an elevator must serve as part of an accessible means of egress. See Section 1104 for when elevators are required for the accessible route into a building. By a reference to Section 1007.4, both an area of refuge and a standby source of power for the elevator is required. The standby power requirement establishes a higher degree of reliability that the elevator will be available and usable by reducing the likelihood of power loss caused by fire or other conditions of power failure.

In buildings having four or more stories above or below the level of exit discharge, it is unreasonable to rely solely on exit stairways for all of the required accessible means of egress. This is the point at which complete reliance on assisted evacuation down the stairs will not be effective or adequate because of the limited availability of either experienced personnel who are trained to carry people safely (i.e., fire fighters) or the availability of special devices (i.e., self-braking stairway descent equipment or evacuation

chairs). In this case, the code requires that at least one elevator, serving all floors of the building, is to serve as one of the required accessible means of egress. This should not represent a hardship, since elevators are typically provided in such buildings for the convenience of the occupants.

On a flat site, "buildings with four or more stories above the level of exit discharge" would typically be a five-story building. The level of exit discharge is the first floor level, and the fifth floor is the fourth story above that. A story four stories below the level of exit discharge would be the fourth basement level. The verbiage is such that a building built on a sloped site can take into consideration that people may be exiting the building from different levels on different sides of the building (see Figure 1007.2.1).

Exception 1 establishes that accessible egress elevator service to floor levels at or above the level of exit discharge is not necessary under specified conditions. The conditions are that the building is equipped throughout with an automatic sprinkler system in accordance with NFPA 13 or NFPA 13R (see Section 903.3.1.1 or 903.3.1.2) and the floors not serviced by an accessible egress elevator are provided with a horizontal exit. The presence of an automatic sprinkler system significantly reduces the potential fire hazard and provides for increased evacuation time. The combination of automatic sprinklers and a horizontal exit provides adequate protection for the occupants despite their distance to the level of exit discharge. This exception does not apply to floor levels below the level of exit discharge, since such levels are typically below grade and do not have the added advantage of exterior openings that are available for fire-fighting or rescue purposes. This option is most often utilized when a defend-in-place approach to occupant protection is utilized, such as in a hospital, nursing home or jail. Keep in mind that the horizontal exit (see Section 1025) creates large refuge areas that have separation requirements and capacity requirements that exceed area of refuge requirements.

Exception 2 specifies that a building sprinklered throughout in accordance with NFPA 13 or NFPA 13R (see Section 903.3.1.1 or 903.3.1.2), with ramp access to each level, such as in a sports stadium, is not required to also have an elevator for accessible means of egress. The reasoning behind this is that the issue of carrying people down stairways does not occur because the ramps may be utilized instead.

1007.3 Stairways. In order to be considered part of an *accessible means of egress*, an *exit access stairway* as permitted by Section 1016.1 or *exit stairway* shall have a clear width of 48 inches (1219 mm) minimum between handrails and shall either incorporate an *area of refuge* within an enlarged floor-level landing or shall be accessed from either an *area of refuge* complying with Section 1007.6 or a *horizontal exit*.

Exceptions:

1. The *area of refuge* is not required at open *exit access* or *exit stairways* as permitted by Sections 1016.1 and

1022.1 in buildings that are equipped throughout with an *automatic sprinkler system* installed in accordance with Section 903.3.1.1 or 903.3.1.2.

2. The clear width of 48 inches (1219 mm) between *handrails* is not required at *exit access stairway* as permitted by Section 1016.1 or *exit stairways* in buildings equipped throughout with an *automatic sprinkler system* installed in accordance with Section 903.3.1.1 or 903.3.1.2.

3. *Areas of refuge* are not required at *exit stairways* in buildings equipped throughout with an *automatic sprinkler system* installed in accordance with Section 903.3.1.1 or 903.3.1.2.

4. The clear width of 48 inches (1219 mm) between *handrails* is not required for *exit stairways* accessed from a *horizontal exit*.

5. *Areas of refuge* are not required at *exit stairways* serving *open parking garages*.

6. *Areas of refuge* are not required for smoke protected seating areas complying with Section 1028.6.2.

7. The *areas of refuge* are not required in Group R-2 occupancies.

❖ Stairways, while not part of an accessible route, can serve as part of the accessible means of egress when they are used as part of an assisted evacuation route. The starting point for these requirements is that the stairways must be 48 inches (1219 mm) clear width between handrails; and either include or be accessed directly by a location where people can wait for assisted evacuation. This place to wait can be either an "area of refuge" (see Section 1007.6) or a "refuge area" created by a horizontal exit (see Section 1025).

There are many mobility impairments that can limit or negate a person's ability to walk up and down the stairs. The taller the building, the higher the percentage of the population that will be affected. For example, an elderly person or a person with a broken foot may be able to get down a couple of flights, but not

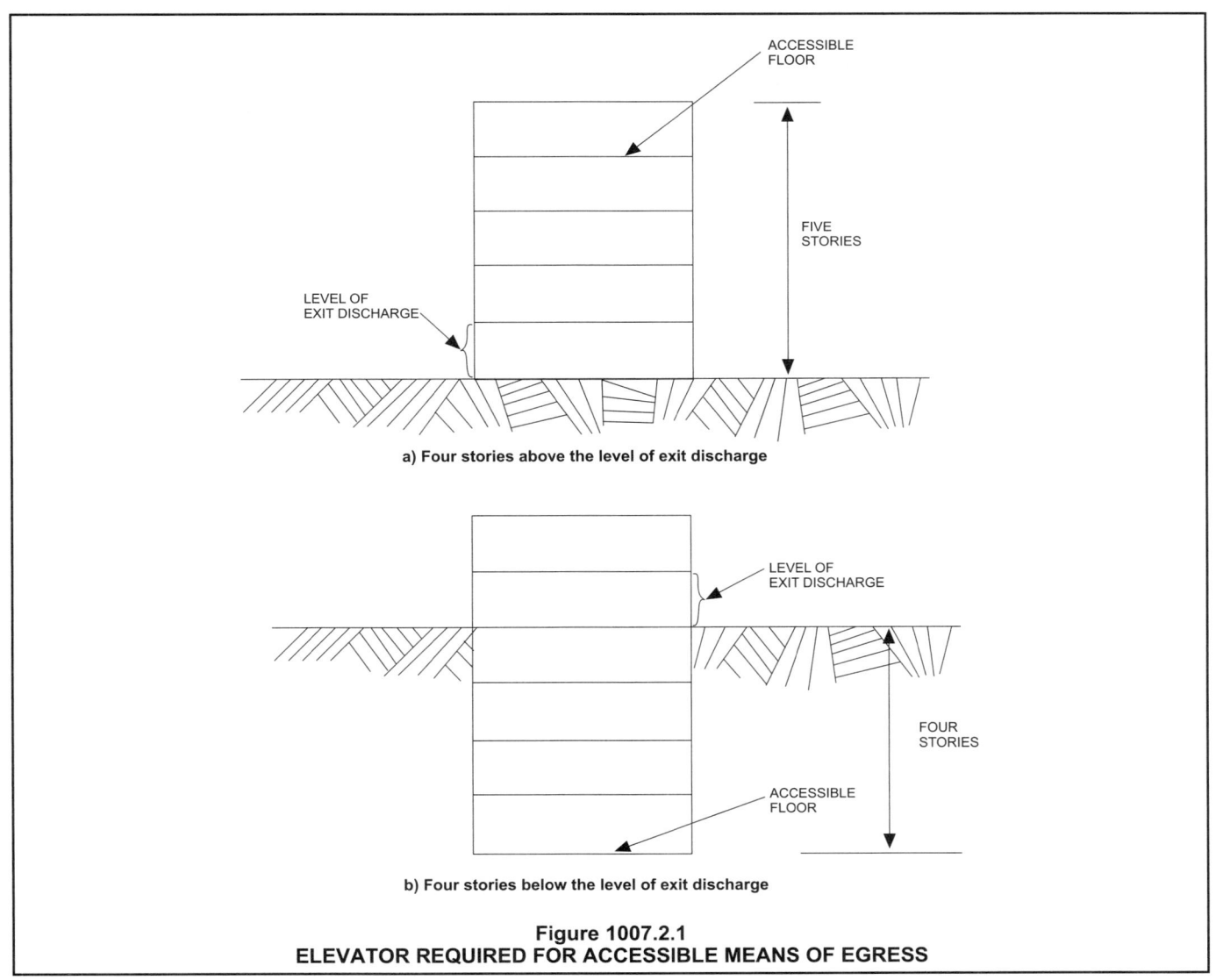

a) Four stories above the level of exit discharge

b) Four stories below the level of exit discharge

Figure 1007.2.1
ELEVATOR REQUIRED FOR ACCESSIBLE MEANS OF EGRESS

from an upper floor in a high-rise.

Note that this section is for exit stairways as addressed in Sections 1022 and 1026 and exit access stairways as addressed in Section 1016.1. Therefore, exit access stairways between stories could be considered part of an accessible means of egress, but not exit access steps within the same level, such as steps in a corridor or room leading to an exit or exit access doorway. Most often, exit stairways are enclosed. However, there are allowances in Sections 1016.1, 1022 and 1026 for exit access and exit stairways to be permitted to be open in certain circumstances. Open stairways that can be part of an accessible means of egress include the following. Section 1016.1, Exceptions 1 and 2, deal with open exit stairways in open parking garages and open outdoor stadiums (which are addressed in this section by Exceptions 5 and 6). Section 1016.1, Exception 3, deals with open exit access stairways between two levels. Section 1016.1, Exception 4, deals with open exit access stairways between the first and second floor of a building sprinklered with an NFPA 13 system. Section 1022 allows for open exit stairways in facilities such as very small second floors or basements (i.e., less than 10 occupants); outdoor sports facilities; within individual dwelling units; open parking garages; jails; stages; and balconies, press boxes and galleries in large assembly facilities. Exterior stairways are required to be open in accordance with Section 1026.3.

The dimension of 48 inches (1219 mm) clear width between handrails is sufficient to enable two or three persons to carry a person up or down to the level of exit discharge where access to a public way is afforded.

The enclosed exit stairway, in combination with an area of refuge, can provide for safety from fire in one of two ways. One approach is for the fire-resistance-rated stairway enclosure to afford the necessary safety. To accomplish this, the landing within the stairway enclosure must be able to contain the wheelchair. The concept is that the person in the wheelchair will remain on the stairway landing for a period of time awaiting further instructions or evacuation assistance; therefore, the stairway landing must be able to accommodate the wheelchair without obstructing the use of the stairway by other egressing occupants. An enlarged, story-level landing is required within the stairway enclosure and must be of sufficient size to accommodate the number of wheelchairs [see Figure 1007.3(1)].

The other approach is to utilize an enclosed exit stairway that is accessed from an area of refuge complying with Section 1007.6. Under this approach, the stairway is made safe by virtue of its access being in an area that is separated and protected from the point of fire origin. An area of refuge can be created by constructing a vestibule adjacent and with direct access to the stair enclosure [see Section 1007.6 and Figure

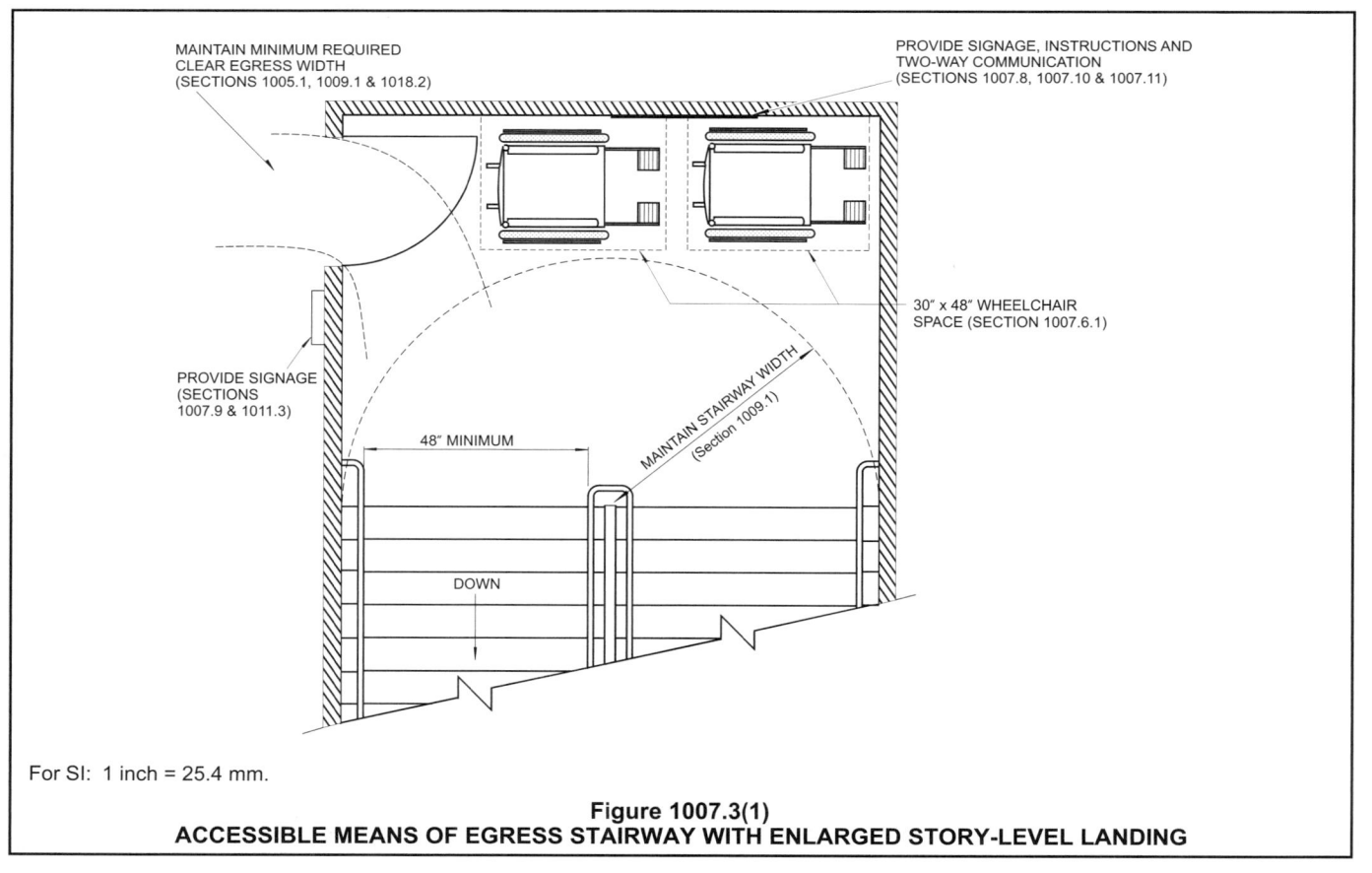

For SI: 1 inch = 25.4 mm.

Figure 1007.3(1)
ACCESSIBLE MEANS OF EGRESS STAIRWAY WITH ENLARGED STORY-LEVEL LANDING

1007.3(2)]. This is similar in theory to the approach of an enlarged landing within the stairway enclosure. Again, the general mean of egress path must be available past the wheelchair spaces.

In the case of a horizontal exit [see Figure 1007.3(3)], each floor area on either side of the exit is considered a refuge area (see commentary, Section 1025.1) by virtue of the construction and separation requirements for horizontal exits. The discharge area is always assumed to be the nonfire side and, therefore, is protected from fire. Therefore, any stairs within this refuge area are not required to have areas of refuge.

Exceptions may be combined. They can be for either the area of refuge or the 48-inch (1219 mm) stairway width requirement. It is very important to note that an exception for the area of refuge in Section 1007.3 or 1007.4 is not an exception for the accessible means of egress. The accessible route must be available to the stairway or elevator so that people with mobility impairments and emergency responders can meet up as soon as possible.

For clarity in discussion of the exceptions, the intent is that the open exit access stairways in Section 1016.1 should be treated the same as exit stairways for purposes of accessible means of egress. This reference would not include any other exit access stairways. While these stairs are considered exit access stairways for purposes of travel distance measurements, they do provide means of egress between floor levels; therefore, the same stairway that serves the

ambulatory population for between floor levels can also serve as part of the accessible means of egress. Open exit access stairway requirements for mezzanines and in assembly seating arrangements are addressed in Exceptions 2 and 3 to Section 1007.1.

Exceptions 1, 2 and 3 are in recognition of the increased level of safety and evacuation time that is afforded in a sprinklered occupancy. The expectation is that a supervised system will reduce the threat of fire by reliably controlling and confining the fire to the immediate area of origin. There is also the additional safety of the sprinkler system requirements for automatic notification when the system is activated. This has been substantiated by a study of accessible means of egress conducted for the General Services Administration (GSA). A report issued by the National Institute for Standards and Technology (NIST), NISTIR 4770, titled "*Staging Areas for Persons with Mobility Limitations,*" concluded that the operation of a properly designed sprinkler system eliminates the life threat to all building occupants, regardless of their individual physical abilities and is a superior form of protection as compared to areas of refuge. It was deemed that the ability of a properly designed and operational automatic sprinkler system to control a fire at its point of origin and to limit production of toxic products to a level that is not life-threatening to all occupants of the building, including persons with disabilities, eliminates the need for areas of refuge.

Exceptions 2 and 4 deal with stairway width. Exceptions 1, 3, 5, 6 and 7 are exceptions for the area of ref-

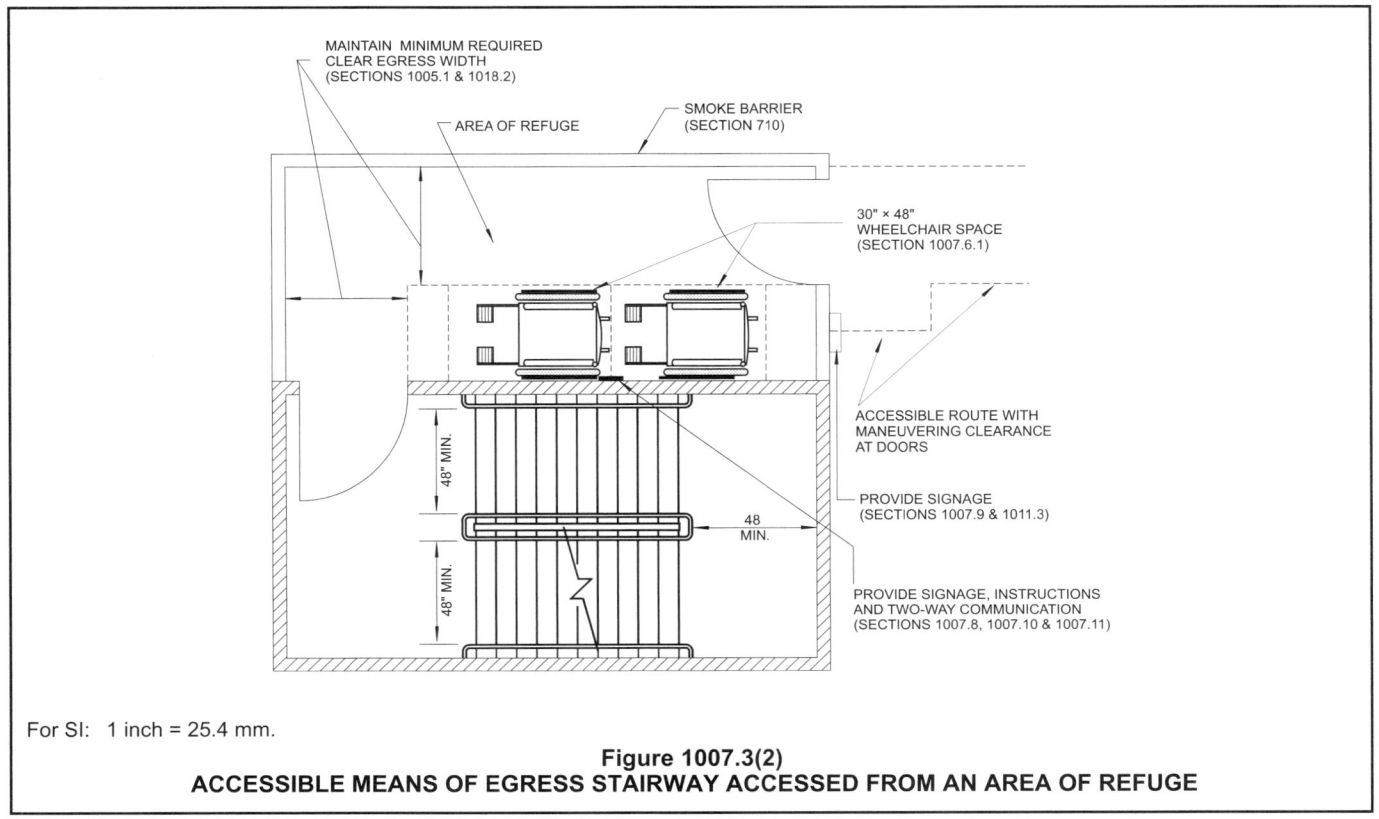

For SI: 1 inch = 25.4 mm.

Figure 1007.3(2)
ACCESSIBLE MEANS OF EGRESS STAIRWAY ACCESSED FROM AN AREA OF REFUGE

uge. This is not an exception for the accessible route to the exit, just the area of refuge at the exit.

Exception 1 is in recognition of the unenclosed exit stairways (see Section 1022.1) and limited open exit access stairways (see Section 1016.1, Exceptions 3 and 4). If the stairway provisions allow for the stair to be unenclosed and the building is sprinklered with an NFPA 13 or 13R system, then the area of refuge is not required. If the building is not sprinklered with an NFPA 13 or 13R system, an area of refuge and the 48-inch (1219 mm) clear width between handrails on the stairway would still be required; however, the stairway is still permitted to be unenclosed. Since there is no stairway enclosure to attach the area of refuge to, the location of enclosure is open to interpretation. To meet the intent, the enclosure should be in the immediate proximity of the top of the open stairway.

Exception 2 allows the stairway width to go back to the base requirements in Section 1009.1 in buildings sprinklered in accordance with NFPA 13 or NFPA 13R for both unenclosed and enclosed exit stairways (see Sections 1020 and 1026) and the limited open exit access stairways (see Section 1016.1, Exceptions 3 and 4). Exception 2 is often used in conjunction with Exceptions 1 and 3. With the sprinkler system in place, there is more opportunity for the fire department to bring in evacuation chairs or possibly bring people to the elevator for evacuation, thus the extra width for carrying someone down the stairway is not needed. This is safer for both emergency responders and the evacuees.

Exception 3 is for the area of refuge for all exit stairways, interior and exterior (see Sections 1022 and 1026) where the building is sprinklered throughout with an NFPA 13 or 13R system. Exception 3 is an alternative to Section 1007.2, Exception 2, for exterior exit stairways. Remember, exterior exit stairways are

between levels, not the steps leading from the level of exit discharge. Again, this is not an exception for the accessible route to the exit, just the area of refuge at the exit.

Exception 4 allows the stairway width to go back to the base requirements in Section 1009.1 when the stairway is within the refuge area created by a horizontal exit. This exception considers that the extra exiting time will permit the egress down the stairway to be more deliberate. Horizontal exits are often used in hospitals or jails when the defense scenario is defend in place rather than evacuation.

Exceptions 5 and 6 are for structures where the natural ventilation of the products of combustion that will be afforded by the exterior openings or smoke protection required of such structures (see Sections 406.3, 909 and 1028.6.2). The most immediate hazard for occupants in a fire incident is exposure to smoke and fumes. Floor areas in open parking structures communicate sufficiently with the outdoors such that the need for protection from smoke is not necessary; therefore, open parking garages are exempted from the requirements for an area of refuge (see also the exception to Section 1007.6.2). It is because of this level of natural ventilation that parking garage exit stairways are not required to be enclosed (see Section 1022.1, Exception 4). The logic for exterior sports facilities and smoke-protected seating is the same: if there is no accumulation of smoke, there is no need for areas of refuge, even when a sprinkler system is not included.

Exception 7 is in recognition of the dwelling unit separation and fire-resistant-rated corridors in Group R-2 facilities (see Sections 420 and 1018). Effectively, each dwelling unit can serve as a protected area. Since the current text requires all Group R structures to be sprinklered (see Section 903.2.8), Exceptions 1, 2 and 3 could also be utilized.

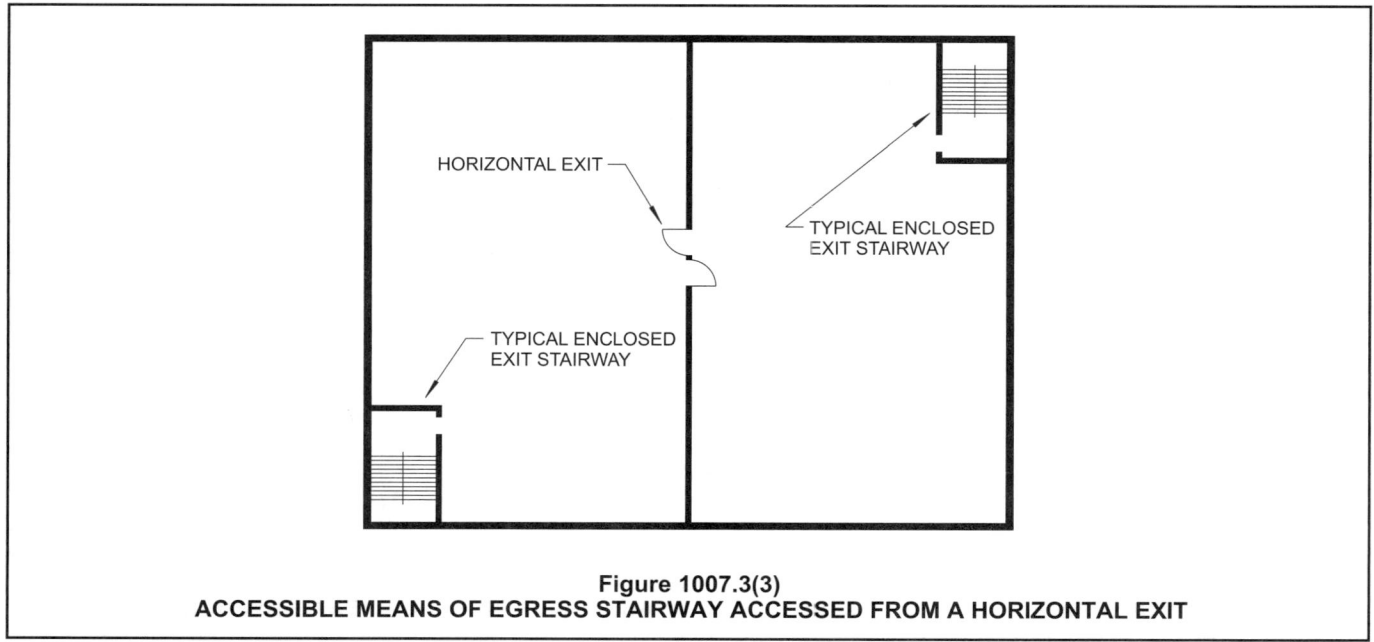

Figure 1007.3(3)
ACCESSIBLE MEANS OF EGRESS STAIRWAY ACCESSED FROM A HORIZONTAL EXIT

1007.4 Elevators. In order to be considered part of an *accessible means of egress*, an elevator shall comply with the emergency operation and signaling device requirements of Section 2.27 of ASME A17.1. Standby power shall be provided in accordance with Chapter 27 and Section 3003. The elevator shall be accessed from either an *area of refuge* complying with Section 1007.6 or a *horizontal exit*.

Exceptions:

1. Elevators are not required to be accessed from an *area of refuge* or *horizontal exit* in *open parking garages*.

2. Elevators are not required to be accessed from an *area of refuge* or *horizontal exit* in buildings and facilities equipped throughout with an *automatic sprinkler system* installed in accordance with Section 903.3.1.1 or 903.3.1.2.

3. Elevators not required to be located in a shaft in accordance with Section 708.2 are not required to be accessed from an *area of refuge* or *horizontal exit*.

4. Elevators are not required to be accessed from an *area of refuge* or *horizontal exit* for smoke protected seating areas complying with Section 1028.6.2.

❖ Elevators are the most common and convenient means of providing access to upper and lower floors in multistory buildings. As such, elevators represent a prime candidate for accessible means of egress from such buildings, especially in light of the difficulties involved in carrying a person in a wheelchair up or down a stairway. The primary consideration for elevators as an accessible means of egress is that the elevator will be available and protected during a fire event. See Sections 403, 3007 and 3008 for new provisions in high-rise buildings for "fire service access elevators" and "occupant evacuation elevators."

This section addresses the use of an elevator as part of an accessible means of egress by requiring both a backup source of power for the elevator and access to the elevator from an area of refuge or a horizontal exit. For situations where elevators are required to be part of one of the accessible means of egress, see Section 1007.2.1. Note that an elevator lobby that is off a fire-resistance-rated corridor must also comply with Section 708.14.1. The backup power requirement establishes a higher degree of reliability that the elevator will be available and usable by reducing the likelihood of power loss caused by fire or other conditions. Requiring access from an area of refuge or a horizontal exit affords the same degree of fire safety as described for stairways (see commentary, Section 1007.3). Additionally, the reference to Chapter 27 and Section 3003 clarifies that the elevator will comply with the emergency operation features that relate to operating an elevator under fire conditions (see commentary, Sections 2702.2.5, 2702.2.19 and 3003). Elevators on an accessible route are also required to meet the accessibility provisions of ICC A117.1 (see commentary, Sections 1109.6 and 3001.3).

Exception 2 is for the area of refuge for all elevators where the building is sprinklered throughout with an NFPA 13 or 13R system. Again, this is not an exception for the accessible route to the exit, just the area of refuge at the elevator with standby power. Exception 2 is in recognition of the increased level of safety and evacuation time that is afforded in a sprinklered occupancy. The expectation is that a supervised system will reduce the threat of fire by reliably controlling and confining the fire to the immediate area of origin. There is also the additional safety of the sprinkler system requirements for automatic notification when the system is activated. This has been substantiated by a study of accessible means of egress conducted for the GSA. A report issued by NIST, NIST IR 4770, titled "*Staging Areas for Persons with Mobility Impairments,*" concluded that the operation of a properly designed sprinkler system eliminates the life threat to all building occupants, regardless of their individual physical abilities and is a superior form of protection as compared to areas of refuge. It was deemed that the ability of a properly designed and operational automatic sprinkler system to control a fire at its point of origin and to limit production of toxic products to a level that is not life threatening to all occupants of the building, including persons with disabilities, eliminates the need for areas of refuge.

If a level in an open parking garage contains accessible parking spaces or is part of the route to and from those spaces, that level is required to have accessible means of egress. Exception 1, for open parking structures, is in recognition of the natural ventilation of the products of combustion that will be afforded by the exterior openings required of such structures (see Section 406.3.3.1). The most immediate hazard for occupants in a fire incident is exposure to smoke and fumes. Floor areas in open parking structures are sufficiently exposed to the outdoors; thus, the need for protection from smoke is not necessary. Therefore, open parking garages are exempt from the requirements for an area of refuge or horizontal exit to access an elevator that is utilized as part of the accessible means of egress. The same idea holds true for smoke-protected seating areas, in accordance with Exception 4. The protection offered by the smoke control system allows for adequate evacuation time before there is danger from smoke and fume accumulation.

Exception 2 allows for elevators that are not required to be enclosed by Section 708.2 to not to have an area of refuge or be accessed by a horizontal exit. If there is no shaft enclosure around the elevator, construction of a smoke-tight compartment immediately in front of the elevator doors would be very difficult. While there are many items listed under Section 708.2, combined with the height requirements in Section 1007.2.1, typically this would be elevators in atriums or in open and enclosed parking garages. Again, the nature of the location adjacent to an atrium or with the open ramps in parking garages would minimize the chances of smoke accumulation at the elevators.

1007.5 Platform lifts. Platform (wheelchair) lifts shall not serve as part of an *accessible means of egress*, except where allowed as part of a required *accessible route* in Section 1109.7, Items 1 through 9. Standby power shall be provided in accordance with Chapter 27 for platform lifts permitted to serve as part of a *means of egress*.

❖ Previously, there have been concerns about whether a platform lift will be reliably available at all times. However, ASME A 18.1, the standard for platform lifts, no longer requires key operation. It is important to note that platform lifts are not prohibited by the code. They simply cannot be counted as a required accessible means of egress in other than locations where they are allowed as part of the accessible route into a space (see commentary, Section 1109.7). When platform lifts are utilized as part of an accessible means of egress, they must come equipped with standby power. Note that platform lifts cannot be used to meet accessible means of egress requirements for a situation that utilizes Section 1109.7, Item 10. Accessible means of egress must be provided at other locations.

In existing buildings undergoing alterations, platform lifts are allowed as part of an accessible route (see commentary, Section 3411.8.3) at any location as long as they are compliant with ASME A 18.1. Note that accessible means of egress are not required in existing buildings undergoing an alteration (see Sections 1007.1 and 3411.6).

1007.5.1 Openness. Platform lifts on an *accessible means of egress* shall not be installed in a fully enclosed hoistway.

❖ Platform lifts on an accessible means of egress must be open to permit others in the building to view the user and allow communication between the rider and others in the building. Lifts are now permitted to penetrate floors by ASME A 18.1, and thus may be in fire-rated enclosures to maintain the separation between different floors. Even when fire separation is not an issue, nothing prevents the designer from installing a lift in a fully enclosed shaft. The openness requirement is in recognition that platform lifts do not have the fire-fighter service or two-way communication required in elevators. Current text is silent on whether the platform lift being within the same enclosure as the exit stairway would provide enough visibility to allow this platform lift to be considered open.

1007.6 Areas of refuge. Every required *area of refuge* shall be *accessible* from the space it serves by an *accessible means of egress*. The maximum travel distance from any *accessible* space to an *area of refuge* shall not exceed the travel distance permitted for the occupancy in accordance with Section 1016.1. Every required *area of refuge* shall have direct access to a *stairway* within an *exit enclosure* complying with Sections 1007.3 and 1022 or an elevator complying with Section 1007.4. Where an elevator lobby is used as an *area of refuge*, the shaft and lobby shall comply with Section 1022.9 for

smokeproof enclosures except where the elevators are in an *area of refuge* formed by a *horizontal exit* or *smoke barrier*.

Exceptions:

1. A *stairway* serving an *area of refuge* is not required to be enclosed where permitted in Sections 1016.1 and 1022.1.

2. A *smokeproof enclosure* is not required for an elevator lobby used as an *area of refuge* where the elevator is not required to be enclosed.

❖ Areas of refuge, when provided, are an important component of the fire and safety evacuation plans for the buildings. These areas must be included in the plans required by IFC Section 404.

An area of refuge is of no value as part of an accessible means of egress if it is not accessible. The code states an obvious but essential requirement: the path that leads to an area of refuge must qualify as an accessible means of egress. This provision is required so that there will be an accessible route leading from every accessible space to each required area of refuge. For consistency in principle with the general means of egress design concepts, the code also limits the travel distance to the area of refuge. The limitation is the same distance as specified in Section 1016.1 for maximum exit access travel distance. This equates the maximum travel distance required to reach an exit with the maximum distance required to reach an area of refuge. It should be noted that an area of refuge is not necessarily an exit in the classic sense. For example, when the area of refuge is an enlarged, story-level landing within an exit stairway, the area of refuge is within the exit and the maximum travel distance for both the conventional exit and the accessible area of refuge is measured to the same point (the entrance to the exit stairway). If the area of refuge is a vestibule immediately adjacent to an enclosed exit stairway, the maximum travel distance for the required accessible means of egress is measured to the entrance of the area of refuge and, for the travel distance to the exit, to the entrance of the exit stairway (see Figure 1007.6). In the case of accessible means of egress with an elevator, the maximum travel distance may end up being measured along two different paths, with the only consistency between the conventional means of egress and the accessible means of egress being the maximum travel distance (see Figure 1007.6). The travel distance within an area of refuge is not directly regulated, but will be limited by the general provisions for maximum exit access travel distance, which are always applicable.

In summary, the code takes a reasonably consistent approach for both conventional and accessible means of egress by limiting the distance one must travel to reach a safe area from which further egress to a public way is available.

To ensure that there is continuity in an accessible means of egress, the code requires that every area of refuge have direct access to either an exit stairway (see Sections 1007.3 and 1022.1) or an elevator (see Section 1007.4). This, again, may be viewed as stating the obvious, but it is necessary so that the egress layout does not involve entering an area of refuge and then having to leave that protected area before gaining access to a stairway or elevator. Once an occupant reaches the safety of an area of refuge, that level of protection must be continuous until the vertical transportation element (the stairway or elevator) is reached.

If one chooses to comply with the accessible means of egress requirements by providing an accessible elevator with an area of refuge in the form of an elevator lobby, the elevator shaft and the lobby are required to be constructed in accordance with Sections 708.14 and 1022.9. The elevator lobby must be protected from smoke moving up through the shaft by the elevator lobbies having an additional vestibule for access or pressurization of the shafts and lobbies (see Section 909.20). Elevator hoistway door assemblies are, by nature, not substantially air tight. Significant quantities of air can move throughout the shaft and adjacent lobby because of stack effect, especially in tall buildings. The requirements provide additional assurance that the elevator will not be rendered unavailable because of smoke movement into the elevator shaft. If the elevator is in a refuge area that is formed by the use of a horizontal exit (i.e., fire walls or fire barriers in accordance with Section 1025.2) or smoke compart-ments formed by smoke barriers (see Sections 407.4 and 408.6), it is presumed that the refuge area is relatively free from smoke; therefore, the extra protection of Sections 708.14 and 1022.9 would not be needed.

Exception 1 is in recognition of a situation unique to low-rise buildings with open stairways when the building is not fully sprinklered; therefore, the exceptions in Section 1007.3 are not applicable. Sections 1016.1 and 1022.1 allow limited situations where the stairway between floor levels may be permitted to be unenclosed. For example, a two-story office building would be permitted to be unsprinklered with 50 percent of its means of egress stairways open between floors. An area of refuge would still be required at the top of that stairway. Exception 1 clarifies that the area of refuge being there does not require the stairway to be enclosed.

Exception 2 is in recognition of elevators that are permitted to be unenclosed in exterior locations, parking garages or atriums but the building may not be sprinklered. If an area of refuge is provided in front of these elevators, the smokeproof enclosure is not required for the elevator shaft. Enclosing an exterior elevator shaft and lobby might offer more potential for smoke collection compared to an unenclosed situation.

1007.6.1 Size. Each *area of refuge* shall be sized to accommodate one *wheelchair space* of 30 inches by 48 inches (762 mm by 1219 mm) for each 200 occupants or portion thereof, based on the *occupant load* of the *area of refuge* and areas served by the *area of refuge*. Such *wheelchair spaces* shall not reduce the

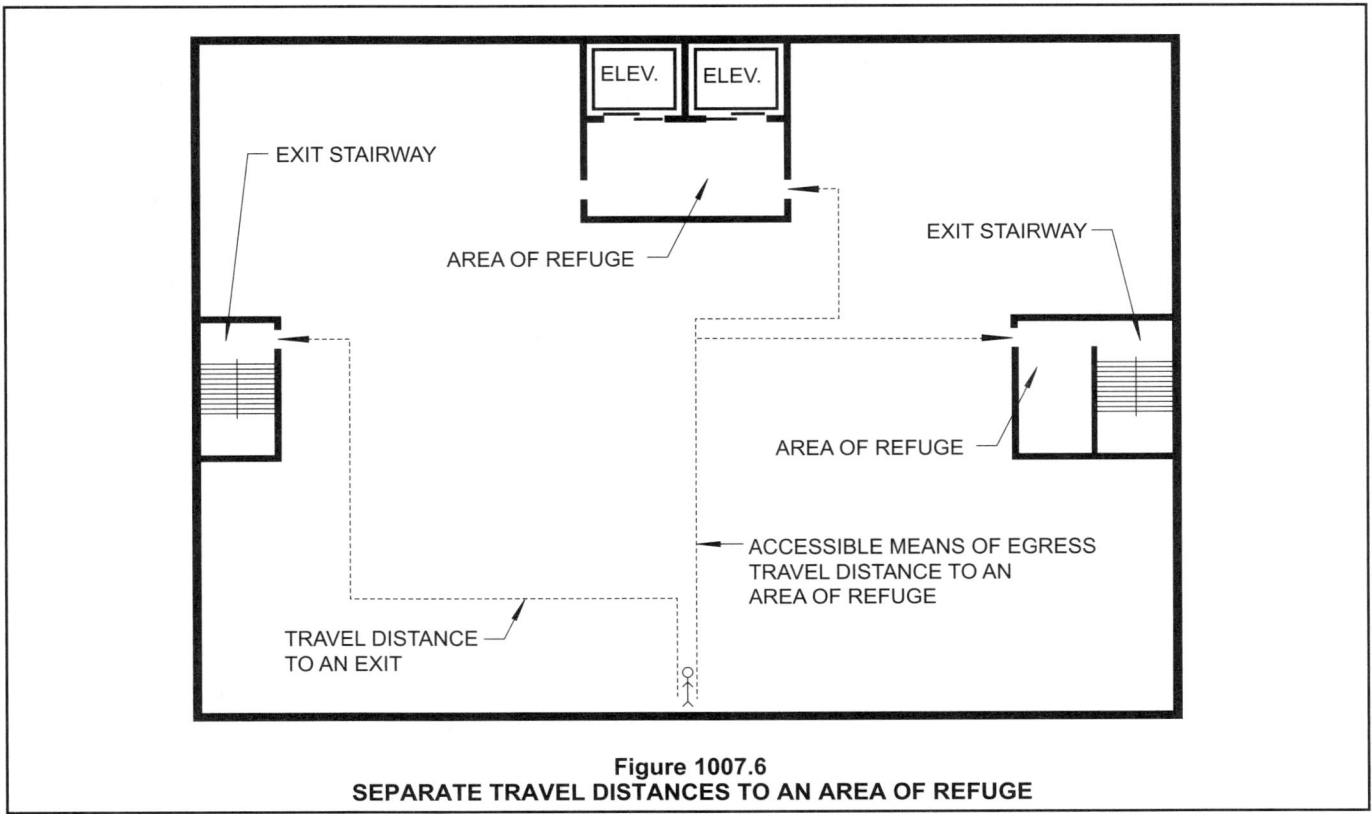

Figure 1007.6
SEPARATE TRAVEL DISTANCES TO AN AREA OF REFUGE

required *means of egress* width. Access to any of the required *wheelchair spaces* in an *area of refuge* shall not be obstructed by more than one adjoining *wheelchair space*.

❖ The number of wheelchair spaces that are required to be provided in an area of refuge is intended to represent broadly the expected population of the average building. As one point of measurement, a 1977 survey conducted by the National Center for Health indicated that one in 333 civilian, noninstitutionalized persons uses a wheelchair. Originally, a ratio of one space for space served by the area of refuge. The 1990 ADA currently utilizes the criterion of one space for each 200 occupants, based on the space served by the area of refuge. Given the variations and difficulties involved in accurately predicting a representative ratio for application to all occupancies, it was concluded that a requirement for one space for each 200 occupants based on the area of refuge itself, plus the areas served by the area of refuge, represents a reasonable criterion. Very few buildings would ever require more than four wheelchair spaces on a floor, since nearly all buildings with an occupant load greater than 400 per floor would be sprinklered and using Exception 3 in Section 1007.3.

Arrangement of the required wheelchair spaces is critical so as not to interfere with the means of egress for ambulatory occupants (see Section 1009.5). Since the design concept is that wheelchair occupants will move to the area of refuge and await further instructions or evacuation assistance, the spaces must be located so as not to reduce the required means of egress width of the stairway, door, corridor or other egress path through the area of refuge.

In order to provide for orderly maneuvering of wheelchairs, this section states that access to any of the required wheelchair spaces cannot be obstructed by more than one adjoining wheelchair space. For example, this precludes an arrangement that three or more wheelchairs could be stacked down a dead-end corridor. This also effectively limits the difficulty any given wheelchair occupant would have in reaching or leaving a given wheelchair space, as well as providing easier access to all wheelchair spaces by persons providing evacuation assistance.

1007.6.2 Separation. Each area of refuge shall be separated from the remainder of the story by a smoke barrier complying with Section 710 or a horizontal exit complying with Section 1025. Each area of refuge shall be designed to minimize the intrusion of smoke.

Exception: Areas of refuge located within an exit enclosure.

❖ The minimum standard for construction of an area of refuge is a smoke barrier, in accordance with Section 710. This establishes a minimum degree of performance by means of a 1-hour fire-resistance rating, including opening protectives and a minimum degree of performance against the intrusion of smoke into an enclosed area of refuge, as specified in Sections 710.4

and 710.5. By nature of the connection to the stair enclosure and/or elevator shaft, the smoke barrier requirement for extension from exterior wall to exterior wall is replaced by connection to the shaft enclosure.

An alternative is to provide a refuge area created by a horizontal exit complying with Section 1025. Horizontal exits are formed by fire walls or fire barriers with a minimum fire-resistance rating of 2 hours. The horizontal exit must extend vertically through all levels of the building, or on a horizontal assembly with a fire-resistance rating of 2 hours with no unprotected openings (see Section 1025.2). The other provisions for horizontal exits for additional egress elements, opening protection and capacity must also be complied with.

This section does not require an area of refuge within an exit stairway to be designed to prevent the intrusion of smoke. This was based on a study of areas of refuge conducted by NIST for the GSA, which concluded that a story-level landing within a fire-resistance-rated exit stairway would provide a satisfactory staging area for evacuation assistance.

1007.6.3 Two-way communication. *Areas of refuge* shall be provided with a two-way communication system complying with Sections 1007.8.1 and 1007.8.2.

❖ If a building includes areas of refuge at the stairway or elevators, each area of refuge must include a two-way communication system. If the building uses one of the exceptions for areas of refuge, Section 1007.8 would still require a two-way communication system at the elevator. This way anyone needing assistance can communicate where they are and that they need help evacuating. This system is an important part of the fire and safety evacuation plans required by IFC Section 404.

1007.7 Exterior area for assisted rescue. The exterior area for assisted rescue must be open to the outside air and meet the requirements of Section 1007.6.1. Separation walls shall comply with the requirements of Section 705 for *exterior walls*. Where walls or openings are between the area for assisted rescue and the interior of the building, the building *exterior walls* within 10 feet (3048 mm) horizontally of a nonrated wall or unprotected opening shall have a *fire-resistance rating* of not less than 1 hour. Openings within such *exterior walls* shall be protected by opening protectives having a *fire protection rating* of not less than $^3/_4$ hour. This construction shall extend vertically from the ground to a point 10 feet (3048 mm) above the floor level of the area for assisted rescue or to the roof line, whichever is lower.

❖ The exterior area of assisted rescue is an alternative for situations where either the exit is via an unenclosed exterior stairway or the exit discharge is not accessible because of steps or a steep site. Note that exterior exit stairways between stories could also comply with Section 1007.3. The protection provided by an exterior area for assisted rescue would be equivalent to that required for an area of refuge. Note that there is no exception for the exterior area of assisted rescue for buildings that contain sprinkler systems. The separation requirements are similar to exterior exit stairways (see Section 1023.6). The exceptions for exterior exit stairway protection in Section 1026.6 would not be applicable where they would also include an exterior area for assisted rescue. Providing a location that was 10 feet (3048 mm) away from the exterior wall would not serve as a viable alternative for having a fire-resistant-rated exterior wall. Persons waiting for assistance to move away from the building must have a minimum level of protection from a fire in the building.

An example where the exterior area for assisted rescue would be a good alternative is along the rear of a strip mall. The front of the facility may be accessible, but if two accessible means of egress are required, then an accessible route must be provided out the rear of each of the tenant spaces. Often in these types of malls, the rear entrance doubles as the loading dock or service entrance. Because of the associated change in elevation, the ramps required to allow accessible exit discharge could be extensive. In addition, the ramps could become damaged over time by trucks maneuvering into the loading dock areas. In this situation, interior areas of refuge are not always a positive alternative. Tenants may tend to use them as convenient storage areas. If persons with a mobility impairment wait for assisted rescue outside of the building, they are already protected from smoke and fumes—the deadliest of the fire hazards. Being visible should also result in a shorter period of time before assisted rescue is affected [see Figures 1007.8(1) and 1007.8(2)].

Exterior areas for assisted rescue, when provided, are an important component of the fire and safety evacuation plans for the buildings. These areas must be included in the plans required by IFC Section 404.

1007.7.1 Openness. The exterior area for assisted rescue shall be at least 50 percent open, and the open area above the guards shall be so distributed as to minimize the accumulation of smoke or toxic gases.

❖ The openness criteria for exterior areas of assisted rescue are similar to the requirements for exterior balconies. The purpose is to ensure that a person at an exterior area of rescue assistance is not in danger from smoke and fumes. The criteria are to address the situation where the area is open to outside air, but a combination of roof overhangs and perimeter walls or guards could still trap enough smoke that the safety of the occupants would be jeopardized.

1007.7.2 Exterior exit stairway. *Exterior exit stairways* that are part of the *means of egress* for the exterior area for assisted rescue shall provide a clear width of 48 inches (1219 mm) between handrails.

❖ Any steps that lead from an exterior area for assisted rescue to grade must have a clear width of 48 inches (1219 mm) between handrails. The additional width is to permit adequate room to assist a mobility impaired person down the steps and to a safe location.

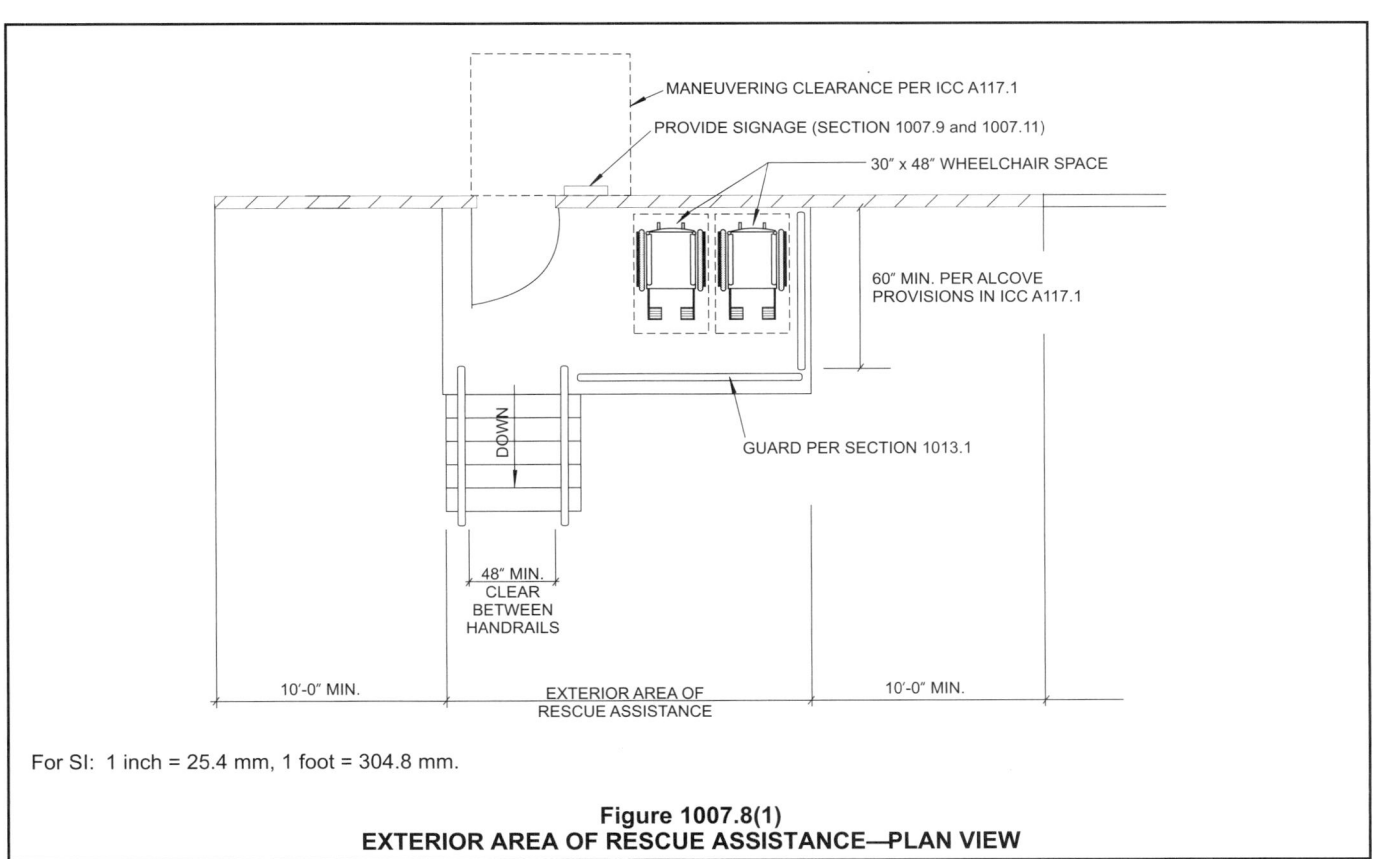

For SI: 1 inch = 25.4 mm, 1 foot = 304.8 mm.

Figure 1007.8(1)
EXTERIOR AREA OF RESCUE ASSISTANCE—PLAN VIEW

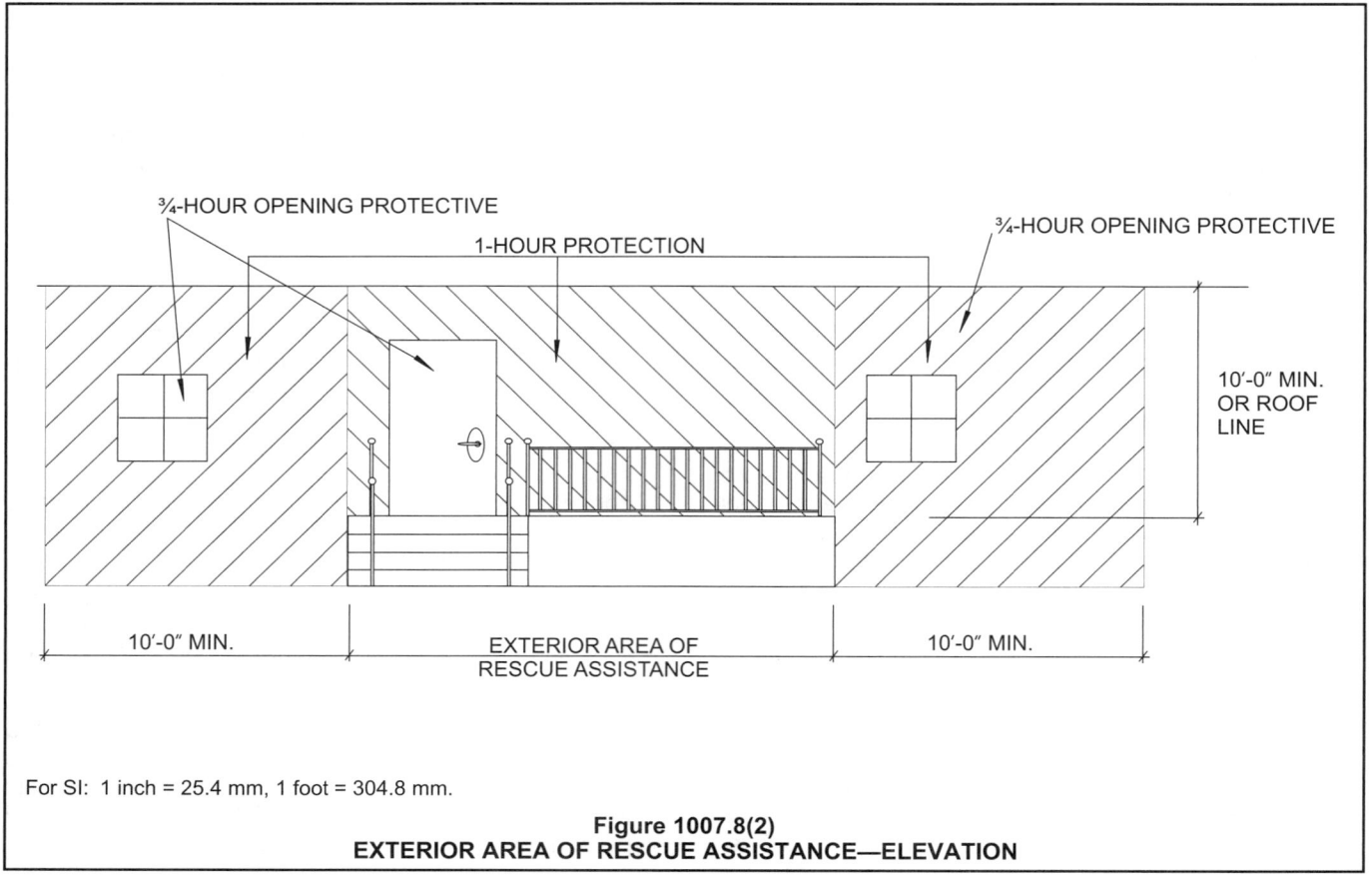

For SI: 1 inch = 25.4 mm, 1 foot = 304.8 mm.

Figure 1007.8(2)
EXTERIOR AREA OF RESCUE ASSISTANCE—ELEVATION

1007.8 Two-way communication. A two-way communication system shall be provided at the elevator landing on each *accessible* floor that is one or more stories above or below the *story* of *exit discharge* complying with Sections 1007.8.1 and 1007.8.2.

Exceptions:

1. Two-way communication systems are not required at the elevator landing where the two-way communication system is provided within *areas of refuge* in accordance with Section 1007.6.3.

2. Two-way communication systems are not required on floors provided with *exit ramps* conforming to the provisions of Section 1010.

❖ Unless provided in areas of refuge, in multistory buildings a two-way communication system must be located at the elevator landing of each accessible floor level other than the level of exit discharge. The system is intended to offer a means of communication to individuals with mobility impairment, either permanent or temporary, who need assistance during an emergency situation. Such a system can be useful not only in the event of a fire but also in the case of a natural or technological disaster by providing emergency responders with the location of individuals who will require assistance in being safely evacuated from floor levels above or below the discharge level.

The first exception exempts the requirement for lo-

cating the communication systems at the elevator landings where the building is provided with complying areas of refuge. Since areas of refuge are required by Section 1007.6.3 to be equipped with two-way communication systems, there is limited need to provide such additional systems at the elevator landings. However, where multistory buildings are not provided with areas of refuge, such as is the case with most sprinklered buildings, the installation of communications systems at the elevator landings is important to those individuals unable to negotiate egress stairways during an emergency. As a result, both sprinklered and nonsprinklered multistory buildings will be provided with the means for two-way communication at all accessible floor levels other than the level of exit discharge.

A second exception applies to floors levels that utilize exit ramps as vertical accessible means of egress elements. Where complying ramps are available for independent evacuation, such as occurs in a sports stadium, the two-way communication system is not required at the elevator landings.

Because of immediate visibility of a person at an exterior area of rescue assistance and the fact that they are typically located at the level of exit discharge, two-way communication systems are not required.

If the option of horizontal exits is utilized, the code does not currently address if a two-way communication system should be provided within a refuge area without an elevator. Since the horizontal exit is not typically recognizable by a person not familiar with the building plan, the most logical location for the two-way communication would seem to be adjacent to the exit stairway that was located within the refuge area.

1007.8.1 System requirements. Two-way communication systems shall provide communication between each required location and the fire command center or a central control point location *approved* by the fire department. Where the central control point is not constantly attended, a two-way communication system shall have a timed automatic telephone dial-out capability to a monitoring location or 911. The two-way communication system shall include both audible and visible signals.

❖ Use of an elevator, stair enclosure or other area of refuge as part of an accessible means of egress requires a person to wait for evacuation assistance or relevant instructions. The two-way communication system allows this person to inform emergency personnel of his or her location and to receive additional instructions or assistance as needed.

The arrangement and design of the two-way communication system is specified in Section 1007.8.1. In addition to the required locations specified in Section 1007.6.3 for areas of refuge or Section 1007.8 for elevator landings, a communication device is also required to be located in a high-rise building's fire command center or at a central control point whose location is approved by the fire department. "Central control point" is not a defined term. However, given the intent and function of the two-way communication system, a central control point is a location where an individual answers the call for assistance and either provides or requests aid for a person who needs help. A suitable central control point is often not available in low-rise buildings or in a high-rise building where the central control point may not be manned on a 24-hour basis. In order that a caller may reach an appropriate emergency location, such as the fire department, the fire department must approve the configuration of the system. A central control point could be the lobby of a building constantly staffed by a security officer, a public safety answering point such as a 9-1-1 center or a central supervising station in a Group I occupancy. There could be a combination solution—such as a system configured to automatically call 9-1-1 when the central control point within the building is not manned. The communication system provides visual signals for the hearing impaired and audible signals to assist the vision impaired.

1007.8.2 Directions. Directions for the use of the two-way communication system, instructions for summoning assistance via the two-way communication system and written identification of the location shall be posted adjacent to the two-way communication system.

❖ Guidance to the users of the two-way communication system is also specified. Operating instructions for the two-way communication system must be posted and the instructions are to include a means of identifying the physical location of the communication device. If a signal from a two-way communication system terminates to a public safety answering point, such as a fire department communication center, current 9-1-1 telephony technology only reports the address of the location of the emergency—it does not report a floor or area from the address reporting the emergency. The "identification of the location" posted adjacent to the communication system should ensure that most discrete location information can be provided to the central control point. This will aid emergency responders, especially in high-rise buildings or corporate campuses with multiple multistory structures.

1007.9 Signage. Signage indicating special accessibility provisions shall be provided as shown:

1. Each door providing access to an *area of refuge* from an adjacent floor area shall be identified by a sign stating: AREA OF REFUGE.

2. Each door providing access to an exterior area for assisted rescue shall be identified by a sign stating: EXTERIOR AREA FOR ASSISTED RESCUE.

Signage shall comply with the ICC A117.1 requirements for visual characters and include the International Symbol of Accessibility. Where exit sign illumination is required by Section 1011.2, the signs shall be illuminated. Additionally, tactile signage complying with ICC A117.1 shall be located at each door to an *area of refuge* and exterior area for assisted rescue in accordance with Section 1011.3.

❖ Signage enables an occupant to become aware of an area of refuge and/or the exterior area for rescue assistance. The assistance areas must have both visual and tactile signage that states either AREA OF REFUGE or EXTERIOR AREA FOR ASSISTED RESCUE and includes the International Symbol of Accessibility. The approach that the code takes for identification of the area of refuge is comparable to the general provisions for identification of exits, including the requirement for lighted signage. Tactile EXIT signage, including raised letters and Braille, is also required for the benefit of persons with a visual impairment.

The current text does not clearly indicate how to identify a refuge area formed by a horizontal exit. In hospitals and jails, where this option is typically utilized, the location of the horizontal exits is part of the staff training for the fire safety and evacuation plans.

1007.10 Directional signage. Direction signage indicating the location of the other *means of egress* and which are *accessible means of egress* shall be provided at the following:

1. At *exits* serving a required *accessible* space but not providing an *approved accessible means of egress*.

2. At elevator landings.

3. Within *areas of refuge*.

❖ The additional signage required by this section is intended to advise persons of the location of all means of egress and which also serve as accessible means of egress. Since not all of the exits will necessarily be accessible means of egress, it is appropriate to provide this information at exit stairways and, particularly, at all elevators, whether they are or are not part of an accessible means of egress. Directional signage is not required to meet tactile signage requirements.

1007.11 Instructions. In *areas of refuge* and exterior areas for assisted rescue, instructions on the use of the area under emergency conditions shall be posted. The instructions shall include all of the following:

1. Persons able to use the exit stairway do so as soon as possible, unless they are assisting others.

2. Information on planned availability of assistance in the use of stairs or supervised operation of elevators and how to summon such assistance.

3. Directions for use of the two-way communications system where provided.

❖ The instructions provided at the exterior area of rescue assistance and the areas of refuge will differ. The required instructions on the proper use of the area of refuge and the communication system provide a higher degree of assurance that the communication system will accomplish the intended function and occupants will behave as expected. A two-way communication system will not be of much value if a person in that area does not know how to operate it. Also, since the area of refuge is required by Section 1007.9 to be identified as such, ambulatory occupants may mistakenly conclude that they should remain in that area. The instructions remind ambulatory occupants that they should continue to egress as soon as possible.

The same concern pertains to the instructions for the two-way communication system in horizontal exits.

For an exterior area of assisted rescue, a two-way communication system is not provided, so this portion of the instructions is not needed. However, instructions for any ambulatory persons to move to the exit discharge is still required, as well as information on how assistance will be provided at this location.

Since each building's means of egress and fire and safety evacuation plans are unique, specific requirements for verbiage are not indicated, but will depend on the situation.

SECTION 1008
DOORS, GATES AND TURNSTILES

1008.1 Doors. *Means of egress* doors shall meet the requirements of this section. Doors serving a *means of egress* system shall meet the requirements of this section and Section 1020.2. Doors provided for egress purposes in numbers greater than required by this code shall meet the requirements of this section.

Means of egress doors shall be readily distinguishable from the adjacent construction and finishes such that the doors are easily recognizable as doors. Mirrors or similar reflecting materials shall not be used on *means of egress* doors. *Means of egress* doors shall not be concealed by curtains, drapes, decorations or similar materials.

❖ The general requirements for doors are in this section and the following subsections. The reference to Section 1020.2 is intended to emphasize that exterior exit doors must lead to a route that will allow a path to a public street or alley (see definition for "Public way"). Doors need to be easily recognizable for immediate use in an emergency condition. Thus, the code specifies that doors are not to be hidden in such a manner that a person would have trouble seeing where to egress.

1008.1.1 Size of doors. The minimum width of each door opening shall be sufficient for the *occupant load* thereof and shall provide a clear width of 32 inches (813 mm). Clear openings of doorways with swinging doors shall be measured between the face of the door and the stop, with the door open 90 degrees (1.57 rad). Where this section requires a minimum clear width of 32 inches (813 mm) and a door opening includes two door leaves without a mullion, one leaf shall provide a clear opening width of 32 inches (813 mm). The maximum width of a swinging door leaf shall be 48 inches (1219 mm) nominal. *Means of egress* doors in a Group I-2 occupancy used for the movement of beds shall provide a clear width not less than $41^1/_2$ inches (1054 mm). The height of door openings shall not be less than 80 inches (2032 mm).

Exceptions:

1. The minimum and maximum width shall not apply to door openings that are not part of the required *means of egress* in Group R-2 and R-3 occupancies.

2. Door openings to resident sleeping units in Group I-3 occupancies shall have a clear width of not less than 28 inches (711 mm).

3. Door openings to storage closets less than 10 square feet (0.93 m²) in area shall not be limited by the minimum width.

4. Width of door leaves in revolving doors that comply with Section 1008.1.4.1 shall not be limited.

5. Door openings within a dwelling unit or sleeping unit shall not be less than 78 inches (1981 mm) in height.

6. Exterior door openings in dwelling units and sleeping units, other than the required *exit* door, shall not be less than 76 inches (1930 mm) in height.

7. In other than Group R-1 occupancies, the minimum widths shall not apply to interior egress doors within a dwelling unit or sleeping unit that is not required to be an *Accessible unit*, *Type A unit* or *Type B unit*.

8. Door openings required to be accessible within Type B units shall have a minimum clear width of 31.75 inches (806 mm).

❖ The size of a door opening determines its capacity as a component of egress and its ability to fulfill its function in normal use. A door opening must meet certain minimum criteria as to its width and height in order to be used safely and to provide accessibility to people with physical disabilities. Doorways that are not in the means of egress are not limited in size by this section. However, doors that are used for egress purposes, including additional doors over and above the number of means of egress required by the code, are required to meet the requirements of this section unless one of the exceptions applies.

The minimum clear width of an egress doorway for occupant capacity is based on the portion of the occupant load (see Section 1004.1) intended to utilize the doorway for egress purposes multiplied by the egress width per occupant from Section 1005.1. The capacity of a 32-inch (813 mm) clear width door is 32/0.2 = 160 occupants. The 0.15-inch (3.81 mm) allowance for capacity in sprinklered buildings is permitted only in ex-

isting buildings [see Sections 1005.1 and 3412.6.11(1)]. The clear width of a swinging door opening is the horizontal dimension measured between the face of the door and the door stops when the door is in the 90-degree (1.57 rad) position [see Figure 1008.1.1(1)].

Using the face of the door as the measurement point is consistent with the provisions of ICC A117.1 and the ADAAG Review Advisory Committee. Further, this measurement is not intended to prohibit other projections into the required clear width, such as latching or panic hardware [see the commentary to Section 1008.1.1.1 and Figure 1008.1.1(2) for further discussion on the specific projections allowed in the required

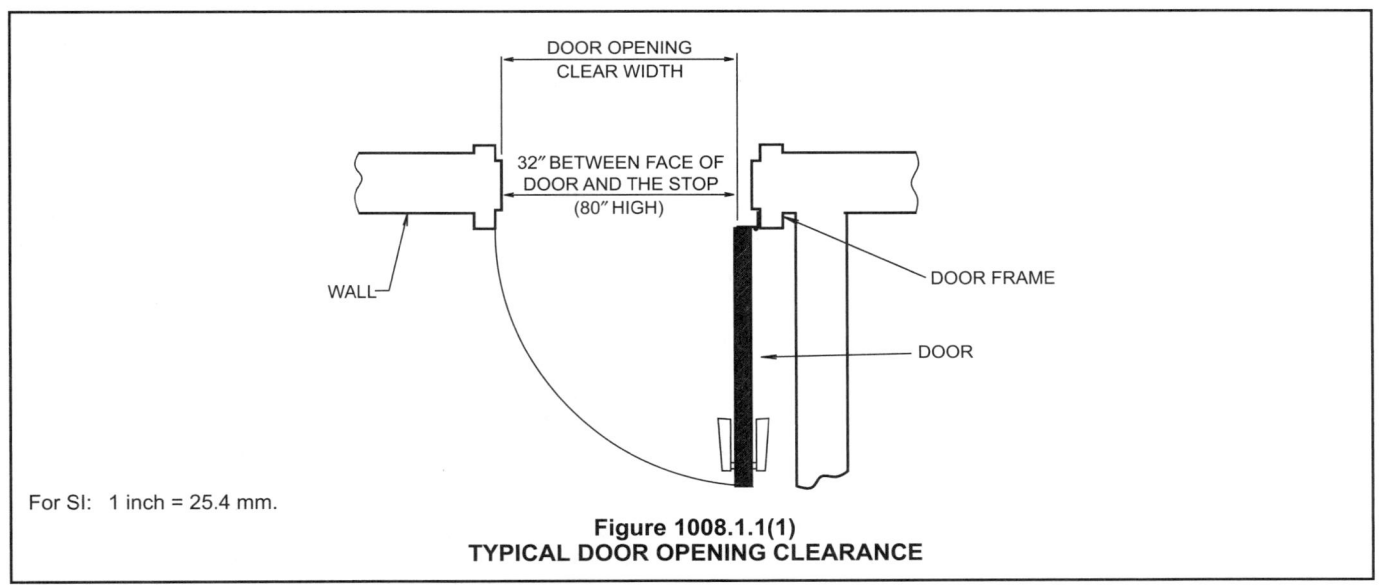

For SI: 1 inch = 25.4 mm.

Figure 1008.1.1(1)
TYPICAL DOOR OPENING CLEARANCE

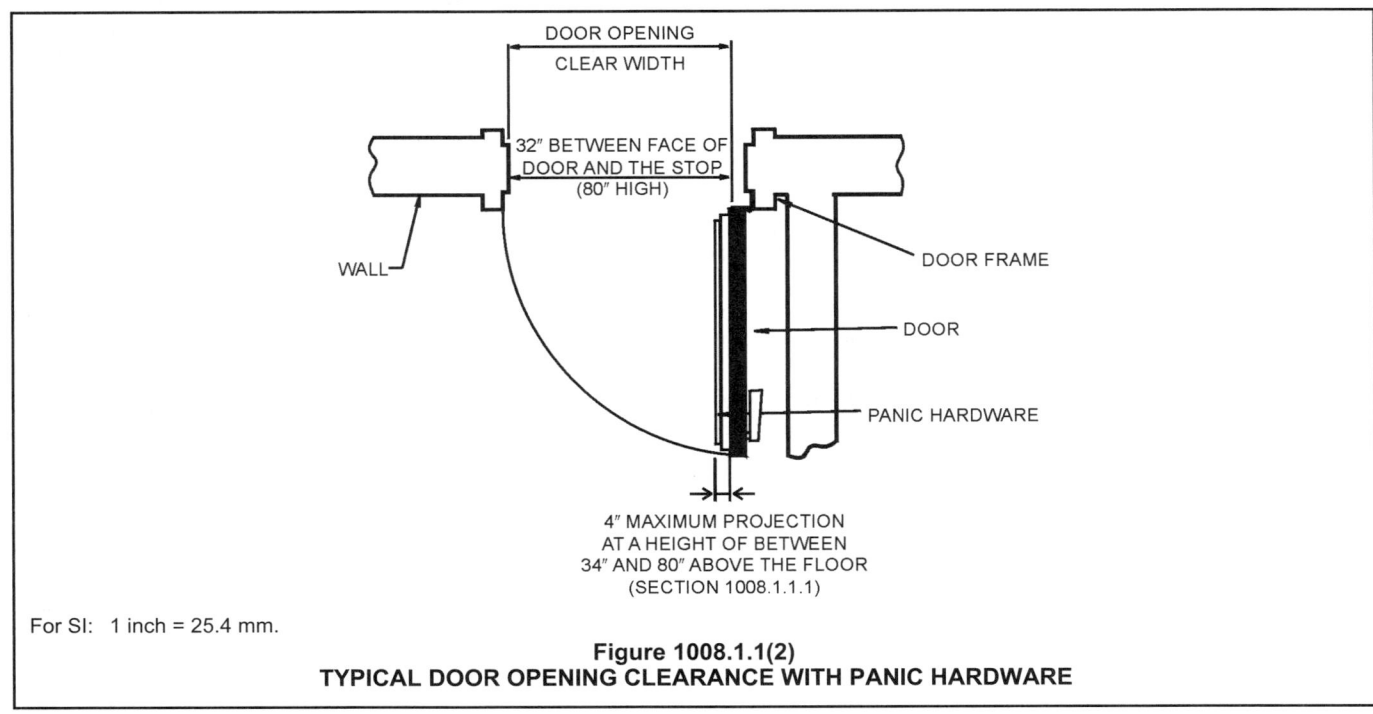

For SI: 1 inch = 25.4 mm.

Figure 1008.1.1(2)
TYPICAL DOOR OPENING CLEARANCE WITH PANIC HARDWARE

clear width]. For nonswinging means of egress doors, such as a sliding door, the clear width is to be measured from the face of the door jambs.

The minimum clear width in a doorway of 32 inches (813 mm) is to allow passage of a wheelchair as well as persons utilizing walking devices or other support apparatus. Similarly, because of the difficulties that a person with physical disabilities would have in opening a pair of doors simultaneously, the 32-inch (813 mm) minimum must be provided by a single door leaf.

Note that in some cases, with standard door construction and hardware, a 36-inch-wide (914 mm) door is the narrowest door that can be used while still providing the minimum clear width of 32 inches (813 mm). A standard 34-inch-wide (864 mm) door has less than a 32-inch (813 mm) clear opening depending on the thickness of the opposing doorstop, the door thickness and the type of hinge. The building designer must verify that the swinging door specified will in fact provide the required clear width. A minimum clear width of $41^1/_2$ inches (1054 mm) is required for doors in any portion of Group I-2 where bed movement is needed to evacuate patients from the area in the event of a fire.

The maximum width for a means of egress door leaf in a swinging door is 48 inches (1219 mm) because larger doors are difficult to handle and are of sizes that typically are not fire tested. The maximum width only applies to swinging doors and not to horizontal sliding doors.

Minimum door heights are required to provide clear headroom for the users. A minimum height of 80 inches (2032 mm) has been empirically derived as sufficient for most users. Note that although the clear height of a doorway is not specified, typical door frame dimensions will render an opening very close to 80 inches (2032 mm) in clear height. The exception in Sections 1003.3.1 and 1008.1.1.1 allows for door closers and doorstops to be as low as 78 inches (1981 mm).

Exception 1 is very limited in scope and is primarily intended to permit decorative-type doors, e.g., café doors, in dwelling units. This exception addresses spaces that are provided with two or more means of egress when only one is required. These nonrequired egress elements are exempted from the minimum and maximum dimensions.

Exception 2 permits the continued use of doors to resident sleeping rooms (cells) in occupancies in Group I-3 according to current practices.

Exception 3 permits doors to storage closets in any occupancy classification less than 10 square feet (0.9 m²) in area to be less than 32 inches (813 mm). This provision is intended to include those closets that can be reached in an arm's length and thus do not require full passage into the closet to be functional.

Exception 4 permits the door leaves in a revolving door assembly to be of any width when the revolving door passage width complies with Section 1008.1.4.1, which provides for adequate egress width when collapsed into a bookfold position.

Exception 5 permits the doorway within a dwelling or sleeping unit to be a minimum of 78 inches (1981 mm) in clear height. This is deemed acceptable because of the familiarity persons in a dwelling or sleeping unit usually have with the egress system and the lack of adverse injury statistics relating to such doors. Note that this exception does not apply to exterior doors of a townhouse or the main entrance doors leading to the hallway in hotels or apartment buildings. However, exterior doors could use the limited exception for doorstops and closers in Sections 1003.3.1 and 1008.1.1.1.

Exception 6 permits exterior doorways to a dwelling or sleeping unit, except for the required exit door, to be a minimum of 76 inches (1930 mm) in clear height. Accordingly, the required exterior exit door to a dwelling or sleeping unit must be 80 inches (2032 mm) in height (exterior doors are not within the scope of Exception 5), but other exterior doors are allowed to be a height of only 76 inches (1930 mm). This provision allows for the continued use of 76-inch-high (1930 mm) sliding patio doors and swinging doors sized to replace such doors.

Exception 7 allows interior means of egress doors in dwelling or sleeping units to have a clear width less than 32 inches (813 mm). Since the doors specified do not serve a dwelling or sleeping unit that is required to be Accessible, Type A or B units, smaller doors are allowed. This exception is not applicable to Group R-1. The requirement for all doorways within a Group R-1 unit to be sized to provide access to persons with physical disabilities is applicable to both entrance doors and passage doors. Because of the social interaction and visitation that often occur in lodging facilities, a door opening sized for accessibility (e.g., wheelchairs, walkers, canes, crutches) is deemed necessary to allow people with disabilities to visit a friend's, colleague's or relative's unit. In addition, wider doors provide an additional benefit to all persons handling luggage and bulky items, or for the situation when an Accessible unit is not available.

Exception 8 addresses the clear width of doors within a Type B dwelling or sleeping unit. The $31^3/_4$ inch (806 mm) dimension effectively allows for 2 foot, 10 inch (864 mm) doors to be used inside the unit. This is consistent with the correlative text in ICC A117.1 for Type B units. Again, note that the exterior door to the garden-style apartments or the main door to the hallway from units in an apartment building are not covered by this exception. Refer to Chapter 11 for additional information related to Type B dwelling and sleeping units.

1008.1.1.1 Projections into clear width. There shall not be projections into the required clear width lower than 34 inches (864 mm) above the floor or ground. Projections into the clear opening width between 34 inches (864 mm) and 80 inches (2032 mm) above the floor or ground shall not exceed 4 inches (102 mm).

Exception: Door closers and door stops shall be permitted to be 78 inches (1980 mm) minimum above the floor.

❖ This section of the code provides specific allowances for projection into the required clear widths of means of egress doors. These allowances directly correspond with the method of measuring the required clear width of the door as specified in Section 1008.1.1. A reasonable range of projections for door hardware and trim has been established by these requirements. The use of the means of egress door by a wheelchair occupant will not be significantly impacted by small projections located in inconspicuous areas. The key to these allowances is their location. Projections are allowed at a height between 34 inches (864 mm) and 80 inches (2032 mm). Below the 34-inch (864 mm) height, the code does not permit any projections since they would decrease the available width for wheelchair operation. The full 32-inch (813 mm) width must be provided at this location. At 34 inches (864 mm) and higher, projections of up to and including 4 inches (102 mm) are permitted. The 4-inch (102 mm) projection is consistent with the allowances of Section 1003.3.3. This section permits door hardware, such as panic hardware, to extend into the clear width, yet maintain accessibility for persons with physical disabilities [see Figure 1008.1.1(2)].

Allowance must be made for door closers and stops, since their design and function necessitates placement within the door opening. The minimum headroom clearance for door closers and stops is allowed to be 78 inches (1981 mm) [see Figure 1003.3.1(1)]. The 2-inch (51 mm) projection into the doorway height is reasonable since these devices are normally mounted away from the center of the door opening, thus minimizing the potential for contact with a person moving through the opening. This is consistent with the exception in Section 1003.3.1.

1008.1.2 Door swing. Egress doors shall be of the pivoted or side-hinged swinging type.

Exceptions:

1. Private garages, office areas, factory and storage areas with an *occupant load* of 10 or less.

2. Group I-3 occupancies used as a place of detention.

3. Critical or intensive care patient rooms within suites of health care facilities.

4. Doors within or serving a single dwelling unit in Groups R-2 and R-3.

5. In other than Group H occupancies, revolving doors complying with Section 1008.1.4.1.

6. In other than Group H occupancies, horizontal sliding doors complying with Section 1008.1.4.3 are permitted in a *means of egress*.

7. Power-operated doors in accordance with Section 1008.1.4.2.

8. Doors serving a bathroom within an individual sleeping unit in Group R-1.

9. In other than Group H occupancies, manually operated horizontal sliding doors are permitted in a *means of egress* from spaces with an *occupant load* of 10 or less.

Doors shall swing in the direction of egress travel where serving an *occupant load* of 50 or more persons or a Group H occupancy.

❖ Generally, egress doors are required to be the side swinging type. The hardware can be either a hinge or a pivot [see Figure 1002.1(2)]. Side swinging doors are familiar to all occupants in the method of operation. Door designs with pivots at the top and bottom are permitted by this section since the door action itself has little difference between the side-hinged-type door.

The code has several conditions where it allows doors that are not side-hinged swinging type.

Examples of the doors permitted in Exception 1 are an overhead garage door and sliding door. Exception 1 allows doors other than the swinging type for the listed occupancies where the number of occupants is very low.

Exception 2 allows for sliding-type doors that are commonly used in prisons and jails.

Exception 3 allows for sliding doors between the nursing areas and patient rooms in critical care and intensive care suites. Patients are not typically moving around on their own in these areas, visitors are extremely limited, the glass doors allow a better view for nurse supervision and the sliding option allows for equipment locations without worrying about the door swing. See also Sections 407, 1008.1.9.6 and 1014.2.2 through 1014.2.7 for these types of areas.

Exception 4 allows for sliding-type doors or pocket-type doors within or serving individual units in a residential occupancy. Residents are typically familiar with the door operation. The use of sliding doors on the interior of dwelling units is permitted by the Department of Housing and Urban Development (HUD) accessibility requirements and ICC A117.1 for Accessible, Type A and Type B units.

Exception 5 allows for revolving doors that meet the requirements of Section 1008.1.4.1.

Exceptions 6 and 7 allow for power-operated swinging or power-operated horizontal sliding doors that meet the requirements of Section 1008.1.4.2 or 1008.1.4.3. These exceptions are intended to allow wide span openings to be used in a means of egress. This is to enhance the movement of the general population as well as people with mobility impairments to areas of safety without obstructions, since the specified doors afford simple operation by persons for both typical and emergency operation.

Exception 8 allows for pocket doors between the bathrooms and living or sleeping space within hotel rooms. Since the bathroom is most commonly placed immediately inside the entrance to the room, a side-swinging door could sometimes be an obstruction for a person entering carrying suitcases. Familiarity with these types of doors and minimal occupant loads makes this situation acceptable.

Exception 9 partially overlaps the allowances in Exception 1 for horizontal sliding doors by matching the 10 or less occupant load, but extends the use to all other groups except for high hazard. For example, some emergency rooms or clinics use patio-type horizontal sliding doors to separate patient care rooms while still allowing visual supervision, or pocket doors may be used for access to a bathroom within a private office. The allowance for such a door will provide greater design flexibility and efficiency, while at the same time maintaining an acceptable level of safety.

A side-hinged door must swing in the direction of egress travel where the required occupant capacity of the room is 50 or more. As such, a room with two doors and an occupant load of 99 would require both doors to swing in the direction of egress travel, even though each door has a calculated occupant usage of less than 50. At this level of occupant load, the possibility exists that, in an emergency situation, a compact line of people could form at a closed door that swings in a direction opposite the egress flow. This could delay or eliminate the first person's ability to open the door inward with the rest of the queue behind the person.

In a Group H occupancy, the threat of rapid fire buildup, or worse, is such that any delay in egress caused by door swing may jeopardize the opportunity for all occupants to evacuate the premises. For this reason, all egress doors in Group H occupancies are to swing in the direction of egress.

1008.1.3 Door opening force. The force for pushing or pulling open interior swinging egress doors, other than *fire doors*, shall not exceed 5 pounds (22 N). For other swinging doors, as well as sliding and folding doors, the door latch shall release when subjected to a 15-pound (67 N) force. The door shall be set in motion when subjected to a 30-pound (133 N) force. The door shall swing to a full-open position when subjected to a 15-pound (67 N) force.

❖ The ability of all potential users to be physically capable of opening an egress door is a function of the forces required to open the door. The 5-pound (22 N) maximum force for pushing and pulling interior swinging doors without closers that are part of the means of egress inside a building is based on that which has been deemed appropriate for people with a physical limitation due to size, age or disability. The operating force is permitted to be higher for all exterior doors, interior swinging doors that are not part of the means of egress, doors that are part of the means of egress but also serve as opening protectives in fire-resistance-rated walls (i.e., fire doors), sliding doors and folding doors. This recognizes that doors with closers, particularly fire doors, require greater operating forces

in order to close fully in an emergency where combustion gases may be exerting pressure on the door assembly. Similarly, exterior doors are exempted because air pressure differentials and strong winds may prevent doors from being automatically closed. A maximum force of 15 pounds (67 N) is required for operating the latching mechanism. Once unlatched, a maximum force of 30 pounds (133 N) is applied to the latch side of the leaf to start the door in motion by overcoming its stationary inertia. Once in motion, it must not take more than 15 pounds (67 N) of force to keep the door in motion until it reaches its full open position and the required clear width is available. To conform to this requirement on a continual basis, door closers must be adjusted periodically and door fits must also be checked and adjusted when necessary.

1008.1.3.1 Location of applied forces. Forces shall be applied to the latch side of the door.

❖ See the commentary for door opening forces in Section 1008.1.3.

1008.1.4 Special doors. Special doors and security grilles shall comply with the requirements of Sections 1008.1.4.1 through 1008.1.4.5.

❖ This section simply defines the scope of the code requirements for special doors such as revolving doors, power-operated horizontal sliding doors, access controlled egress doors and security grilles.

1008.1.4.1 Revolving doors. Revolving doors shall comply with the following:

1. Each revolving door shall be capable of collapsing into a bookfold position with parallel egress paths providing an aggregate width of 36 inches (914 mm).

2. A revolving door shall not be located within 10 feet (3048 mm) of the foot of or top of *stairs* or escalators. A dispersal area shall be provided between the *stairs* or escalators and the revolving doors.

3. The revolutions per minute (rpm) for a revolving door shall not exceed those shown in Table 1008.1.4.1.

4. Each revolving door shall have a side-hinged swinging door which complies with Section 1008.1 in the same wall and within 10 feet (3048 mm) of the revolving door.

5. Revolving doors shall not be part of an *accessible route* required by Section 1007 and Chapter 11.

❖ Revolving doors must comply with all five provisions.
    Item 1: One of the causes contributing to the loss of lives in the 1942 Cocoanut Grove fire in Boston was that the revolving doors at the club's entrance could not collapse for emergency egress and there was not an alternative means of egress adjacent to the revolving doors. Thus, in the panic of the fire, the door became jammed and the club's occupants were trapped.
    As a result of this fire experience, all revolving doors, including those for air structures, now are required to be equipped with a collapse feature. A

bookfold operation is where all leaves collapse parallel to each other and to the direction of egress [see Figure 1008.1.4.1(1)]. A bookfold operation creates two openings of approximately equal width. The sum of the widths is not to be less than 36 inches (914 mm) so that a stream of pedestrians may use each side of the opening.

Item 2: If a stairway or escalator delivers users to a landing in front of a revolving door at a greater rate than the capacity of the door, a compact line of people will develop. Lines of people formed on a stairway or escalator create an unsafe situation, since stairways

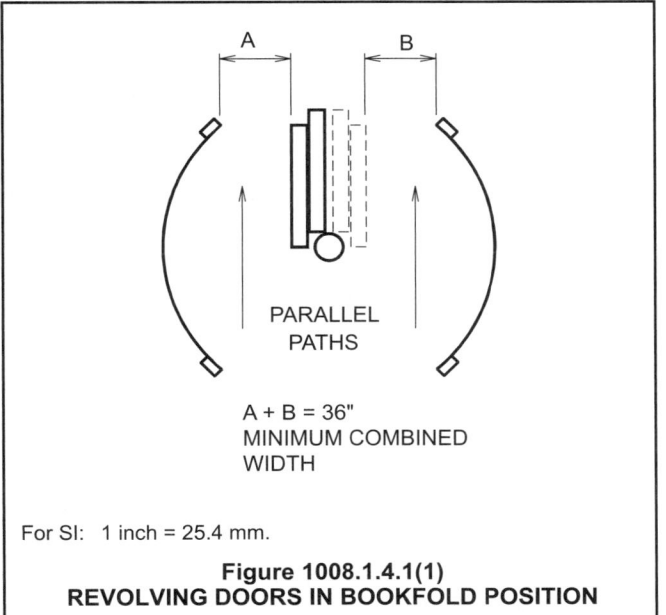

For SI: 1 inch = 25.4 mm.

Figure 1008.1.4.1(1)
REVOLVING DOORS IN BOOKFOLD POSITION

and escalators are not intended to be used as standing space for persons who may be waiting to use the revolving doors. Therefore, to avoid congestion at a revolving door that under normal operation has a maximum delivery capacity of users, a dispersal area is required between the stairways or escalators and the revolving doors to allow for the queuing of people as they enter the door. Accordingly, to create a dispersal area for users of a revolving door, the door is not to be placed closer than 10 feet (3048 mm) from the foot or top of a stairway or escalator.

Item 3: Door speeds also directly relate to the capacity of a revolving door, which is calculated by multiplying the number of leaves (wings) by the revolutions per minute (rpm). For example, if you have a four-leaf door moving at 10 rpm, the door will allow 40 people to move in either direction in 1 minute.

Item 4: In case a revolving door malfunctions or becomes obstructed, the adjacent area is to be equipped with a conventional side-hinged door to provide users with an immediate alternative way to exit a building. The side-hinged door is intended to be used as a relief device for people lined up to use the revolving door or who desire to avoid it because of a physical disability or other reason. It also can be used when the revolving door is obstructed or out of service. The swinging door is to be immediately adjacent to the revolving door so that its availability is obvious [see Figure 1008.1.4.1(2)]. A single swinging door can be located between two revolving doors in order to comply with this provision.

Item 5: While some revolving doors may be considered part of a means of egress, they cannot be considered part of a required accessible route for either ingress of egress. This requirement is consistent with ICC A117.1, which also prohibits revolving gates and

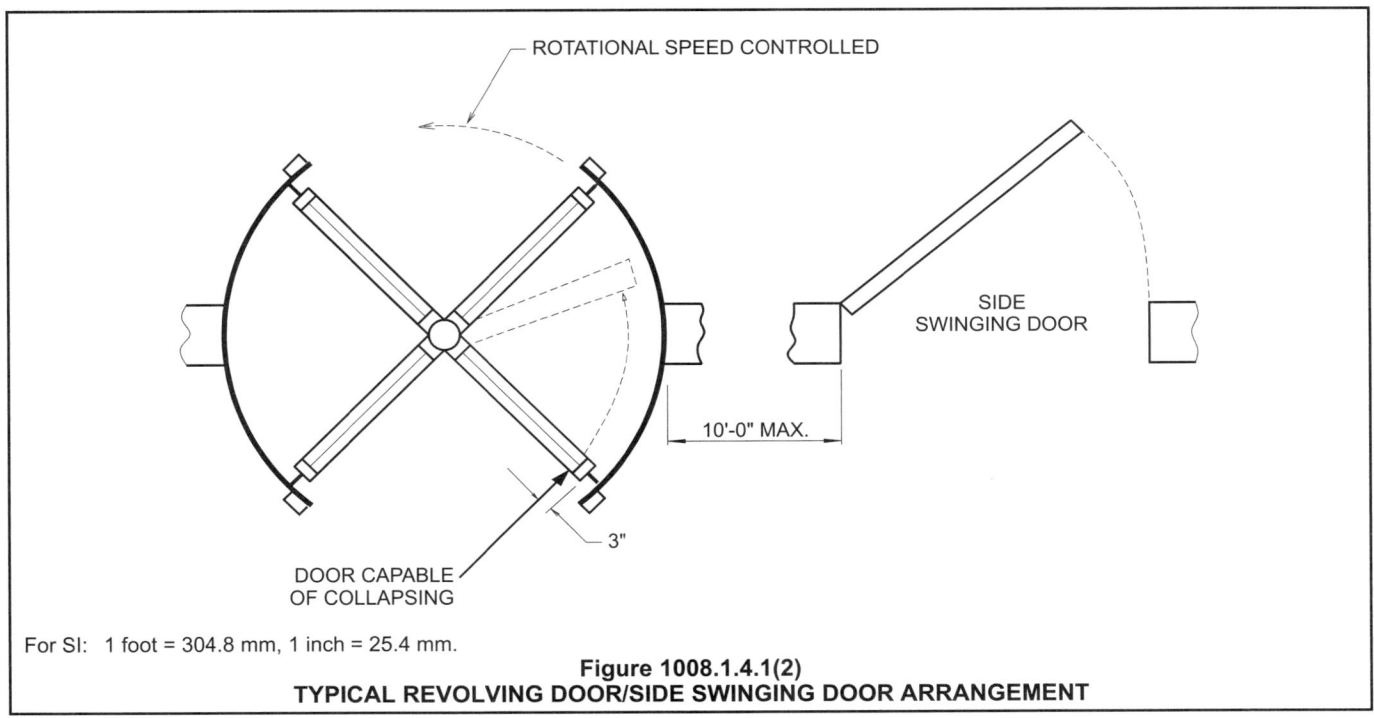

For SI: 1 foot = 304.8 mm, 1 inch = 25.4 mm.

Figure 1008.1.4.1(2)
TYPICAL REVOLVING DOOR/SIDE SWINGING DOOR ARRANGEMENT

turnstiles along the only accessible route.

A route through a hinged or sliding door differs remarkably from that provided through a revolving door. For a revolving door, the route includes a turn into the doorway, an arcing path of travel as the door revolves, followed by a change of direction when leaving the door. Items that may cause difficulty for anyone with mobility impairments could involve the overall doorway diameter, the number of leaves and their relative angle, and the configuration of the return walls surrounding the revolving door. Additionally, the speed of the door movement if motorized, or the force required for movement if not motorized, would be a concern for anyone who needed to keep both hands on their device to move forward (e.g., walker or wheelchair).

Automatic revolving doors, if large enough, may be usable by many people who use wheelchairs. However, the intent of this section is that these types of doors not be the only means of passage at an entrance or exit. An alternative door in full compliance with this section is considered necessary because some people with disabilities may be uncertain of the usability, or may not have enough strength or speed to use them. Although manufacturers have developed safety criteria, certain questions remain, such as the appropriate maximum and minimum speeds.

TABLE 1008.1.4.1
REVOLVING DOOR SPEEDS

INSIDE DIAMETER (feet-inches)	POWER-DRIVEN-TYPE SPEED CONTROL (rpm)	MANUAL-TYPE SPEED CONTROL (rpm)
6-6	11	12
7-0	10	11
7-6	9	11
8-0	9	10
8-6	8	9
9-0	8	9
9-6	7	8
10-0	7	8

For SI: 1 inch = 25.4 mm, 1 foot = 304.8 mm.

❖ Door speeds also directly relate to the capacity of a revolving door, which are calculated by multiplying the number of leaves (wings) by the revolutions per minute (rpm). For example, if you have a four-leaf door moving at 10 rpm, the door will allow 40 people to move in either direction in 1 minute.

1008.1.4.1.1 Egress component. A revolving door used as a component of a *means of egress* shall comply with Section 1008.1.4.1 and the following three conditions:

1. Revolving doors shall not be given credit for more than 50 percent of the required egress capacity.

2. Each revolving door shall be credited with no more than a 50-person capacity.

3. Each revolving door shall be capable of being collapsed when a force of not more than 130 pounds (578 N) is applied within 3 inches (76 mm) of the outer edge of a wing.

❖ A revolving door can be incorporated, to a very limited extent, in a means of egress. Compliance with these three additional conditions is required.

Condition 1 limits the exit capacity that revolving doors can provide in a building. This is so that 50 percent of the capacity has conventional egress components and is not dependent on mechanical devices or fail-safe mechanisms.

Condition 2 limits the capacity of any one revolving door for the same reasons as stated in Condition 1. Each revolving door is therefore limited to a 50-person capacity.

Condition 3 limits the collapse force to 130 pounds (578 N), as opposed to the 180-pound (792 N) value listed in Section 1008.1.4.1.2. Revolving doors used as means of egress are not permitted to have the collapse force exceed 130 pounds (578 N) under any circumstances.

1008.1.4.1.2 Other than egress component. A revolving door used as other than a component of a *means of egress* shall comply with Section 1008.1.4.1. The collapsing force of a revolving door not used as a component of a *means of egress* shall not be more than 180 pounds (801 N).

Exception: A collapsing force in excess of 180 pounds (801 N) is permitted if the collapsing force is reduced to not more than 130 pounds (578 N) when at least one of the following conditions is satisfied:

1. There is a power failure or power is removed to the device holding the door wings in position.

2. There is an actuation of the *automatic sprinkler system* where such system is provided.

3. There is an actuation of a smoke detection system which is installed in accordance with Section 907 to provide coverage in areas within the building which are within 75 feet (22 860 mm) of the revolving doors.

4. There is an actuation of a manual control switch, in an *approved* location and clearly defined, which reduces the holding force to below the 130-pound (578 N) force level.

❖ This section addresses revolving doors that are not used to serve any portion of the occupant egress capacity. For example, where adjacent side-hinged doors have more than the required egress capacity, the revolving door would not be part of the required means of egress.

The maximum collapse force of 180 pounds (792 N), applied within 3 inches (76 mm) of the outer edge of a wing, is based on industry standards to accommodate normal use conditions and other forces that may act on the leaves, such as that caused by wind or air pressure. An exception for revolving doors that are not a component of a required means of egress allows the collapse force to exceed 180 pounds (792 N) in normal operating conditions provided that a force of not more than 130 pounds (578 N) is required whenever any one of the listed conditions is satisfied.

1008.1.4.2 Power-operated doors. Where *means of egress* doors are operated by power, such as doors with a photoelectric-actuated mechanism to open the door upon the approach of a person, or doors with power-assisted manual operation, the design shall be such that in the event of power failure, the door is capable of being opened manually to permit *means of egress* travel or closed where necessary to safeguard *means of egress*. The forces required to open these doors manually shall not exceed those specified in Section 1008.1.3, except that the force to set the door in motion shall not exceed 50 pounds (220 N). The door shall be capable of swinging from any position to the full width of the opening in which such door is installed when a force is applied to the door on the side from which egress is made. Full-power-operated doors shall comply with BHMA A156.10. Power-assisted and low-energy doors shall comply with BHMA A156.19.

Exceptions:

1. Occupancies in Group I-3.

2. Horizontal sliding doors complying with Section 1008.1.4.3.

3. For a biparting door in the emergency breakout mode, a door leaf located within a multiple-leaf opening shall be exempt from the minimum 32-inch (813 mm) single-leaf requirement of Section 1008.1.1, provided a minimum 32-inch (813 mm) clear opening is provided when the two biparting leaves meeting in the center are broken out.

❖ For convenience purposes, power-operated doors are intended to facilitate the normal nonemergency flow of persons through a doorway. Where a power-operated or assisted door is also required to be an egress door, the door must conform to the requirements of this section. The essential characteristic is that the door is to be manually openable from any position to its full open position at any time, with or without a power failure or a failure of a door mechanism. Hence, both swinging and horizontal sliding doors, complying with this section, may be used, provided the door can be operated manually from any position as a swinging door and that the minimum required clear width for egress capacity is not less than 32 inches (813 mm). Note that the opening forces of Section 1008.1.3 are applicable, except that the 30-pound (133 N) force needed to set the door in motion is increased to 50 pounds (220 N) as an operational tolerance in the design of the power-operated door.

In accordance with Exception 1, power-operated doors in detention and correctional occupancies (Group I-3) are not required to be manually operable by the occupants (inmates) for security reasons, but otherwise are required to conform to Section 408. Section 1008.1.4.2 does not apply to horizontal sliding doors that do not meet the provisions of this section (i.e., breakout panels for horizontal power-operated doors) but do comply with Section 1008.1.4.3.

Exception 2 states that power-operated doors that meet the requirements of this section are not required to meet the requirements of Section 1008.1.4.3 for horizontal sliding doors that are not capable of swing operation in the event of power failure.

Exception 3 allows an individual leaf of a four-panel biparting door to be less than 32 inches (813 mm) wide, provided 32 inches (813 mm) of clear space is available when the two center biparting leaves are broken out as part of the emergency breakaway feature.

1008.1.4.3 Horizontal sliding doors. In other than Group H occupancies, horizontal sliding doors permitted to be a component of a *means of egress* in accordance with Exception 6 to Section 1008.1.2 shall comply with all of the following criteria:

1. The doors shall be power operated and shall be capable of being operated manually in the event of power failure.

2. The doors shall be openable by a simple method from both sides without special knowledge or effort.

3. The force required to operate the door shall not exceed 30 pounds (133 N) to set the door in motion and 15 pounds (67 N) to close the door or open it to the minimum required width.

4. The door shall be openable with a force not to exceed 15 pounds (67 N) when a force of 250 pounds (1100 N) is applied perpendicular to the door adjacent to the operating device.

5. The door assembly shall comply with the applicable *fire protection rating* and, where rated, shall be self-closing or automatic closing by smoke detection in accordance with Section 715.4.8.3, shall be installed in accordance with NFPA 80 and shall comply with Section 715.

6. The door assembly shall have an integrated standby power supply.

7. The door assembly power supply shall be electrically supervised.

8. The door shall open to the minimum required width within 10 seconds after activation of the operating device.

❖ Horizontal sliding doors are permitted in the means of egress, in other than rooms or areas of Group H, under the conditions set forth in this section. Horizontal sliding doors are not permitted to be used in Group H occupancies because of the potential for delaying or impeding egress from those areas and the additional risk to occupants in hazardous occupancies. Note that this section regulates egress doors that do not meet all of the requirements of Section 1008.1.4.2 (e.g., a power-operated horizontal sliding door that does not have "breakout" capabilities to allow the door panels to swing if power is lost).

Such doors permitted in other groups are typically in the open position and close either because of a fire or to provide some degree of separation, often for security purposes to restrict movement to certain parts of a building. When in the closed position, the doors must be easily operated with forces similar to the limitations for swinging doors.

All eight of the criteria listed in this section must be met for a horizontal sliding door since there is a concern that it must be able to be easily opened under all conditions.

Additionally, the door must be openable even if a force of 250 pounds (1100 N) is being applied perpendicular to it, as may occur if a group of people were pushing on it.

Since the doors are manually operable, they need not automatically open or close during a loss of power; however, a standby power supply must be provided. The primary power supply must be supervised so that an alarm is received at a constantly attended location (such as a security desk) on loss of the primary power. If the doors are also serving as fire doors, they must be automatic or self-closing in accordance with Section 715.

Since the maximum swinging door leaf width limitations of Section 1008.1.1 do not apply, a maximum opening time of 10 seconds is permitted. It should be noted, however, that the door need not open fully within the 10 seconds; rather, it must open to the required width. For example, if the door is protecting an opening that is 10 feet (3048 mm) wide, but the minimum required width of the opening is 32 inches (813 mm) (as determined by Section 1008.1.1), the door need only open 32 inches (813 mm) within the 10-second criterion. In fact, the door may have controls such that the automatic opening feature only opens the door to a width of 32 inches (813 mm). If additional width is required, it can be accomplished by manual means and, possibly, by an additional activation of the operating device.

1008.1.4.4 Access-controlled egress doors. The entrance doors in a *means of egress* in buildings with an occupancy in Group A, B, E, I-2, M, R-1 or R-2 and entrance doors to tenant spaces in occupancies in Groups A, B, E, I-2, M, R-1 and R-2 are permitted to be equipped with an *approved* entrance and egress access control system which shall be installed in accordance with all of the following criteria:

1. A sensor shall be provided on the egress side arranged to detect an occupant approaching the doors. The doors shall be arranged to unlock by a signal from or loss of power to the sensor.

2. Loss of power to that part of the access control system which locks the doors shall automatically unlock the doors.

3. The doors shall be arranged to unlock from a manual unlocking device located 40 inches to 48 inches (1016 mm to 1219 mm) vertically above the floor and within 5 feet (1524 mm) of the secured doors. Ready access shall be provided to the manual unlocking device and the device shall be clearly identified by a sign that reads "PUSH TO EXIT." When operated, the manual unlocking device shall result in direct interruption of power to the lock—independent of the access control system electronics—and the doors shall remain unlocked for a minimum of 30 seconds.

4. Activation of the building fire alarm system, if provided, shall automatically unlock the doors, and the doors shall remain unlocked until the fire alarm system has been reset.

5. Activation of the building automatic sprinkler or fire detection system, if provided, shall automatically unlock the doors. The doors shall remain unlocked until the fire alarm system has been reset.

6. Entrance doors in buildings with an occupancy in Group A, B, E or M shall not be secured from the egress side during periods that the building is open to the general public.

❖ Doors can be protected by controlled egress or free egress. A controlled egress door requires permission from the access control system to allow someone to enter or exit through the door. This type is used where entry and exit must be logged or where areas on both sides of the passage must be controlled. An entrance door that was locked or controlled from the exterior, but allowed free egress at any time, such as with a panic bar or other standard hardware, would not be an access controlled egress door. This locking arrangement is for situations where both the ingress and egress out of the door is controlled by some type of entry system, such as a key pad or card swipe.

Security in buildings is a major concern to owners and occupants from the perspective of property preservation and personal physical safety. Since many occupancies are partially occupied around the clock, after normal business hours or on weekends, it is necessary that an adequate level of security be provided without jeopardizing the egress capabilities of the occupants.

This section permits the building entrance doors in a means of egress and entrance doors to tenant spaces in occupancies of Groups A, B, E, I-2, M, R-1 and R-2 to be secured while maintaining them as a means of egress. Items 1 through 6 provide additional life safety measures to permit easier egress during normal and emergency situations. Occupancies in Groups F, S and H are not included here because of their increased hazard due to an increase in fuel load and other potentially life-threatening activities. This potential increase in life-threatening circumstances requires an immediate egress capacity without the necessary "waiting period" afforded by the access control of this section.

Item 1 requires that such doors be provided with an automatic exit sensor typically operating on an infrared, microwave or sonic principle. This sensor is required to automatically release the lock upon an occupant approaching the door from the egress side or when there is a loss of power to the sensor. This provision is written as "performance-based," where any means of sensor design can be utilized to allow the doors to unlock in an emergency situation. It is important to note that, during loss of power, the doors would be required to be "fail-safe" (prioritizing safety over security). The current text does not indicate to what distance the sensor should be set. Since Item 3 sets the exit button at within 5 feet (1524 mm) from the door, it would seem that the sensor would be set for a distance less than the location of the manual exit device.

Item 2 requires that if there is a loss of power to the access control system itself, the doors must unlock. Access controlled egress doors are held secure with fail-safe devices (such as magnetic locks and electric door-strikes) so that these doors will automatically unlock when power to the locking device is interrupted. In some instances, the access controller may be powered from a different source than the locking device itself. In these cases, a loss of power to the access controller (while power remains applied to the locking device) must also cause the egress door to automatically unlock.

Item 3 requires that there be a manual exit device, such as a push button, within 5 feet (1524 mm) of the door, mounted 40 to 48 inches (1016 to 1219 mm) above the floor, unobstructed and with a clearly identifiable sign that says "PUSH TO EXIT." When operated, the manual exit device is to interrupt (independent of the access control electronics) the power to the lock directly and cause the doors to remain unlocked for a minimum of 30 seconds.

Items 4 and 5 require the building fire alarm system, automatic fire detection system or sprinkler system, if provided, to be interfaced with the access control system to unlock automatically the doors on activation. The doors are to remain unlocked until the fire alarm system is reset. This is so that the building entrance doors with controlled access will remain unlocked and open until fire fighters responding to the alarm have entered.

Item 6 requires that during the hours the building is open to the general public, doors equipped with an access control system will not require the use of the system from the egress side in Group A, B, E and M occupancies. Thus, the building entrance doors in these occupancies are allowed to be secured from both directions only during off hours when the building occupant load will generally be reduced. In Groups I-2, R-1 and R-2, there is no restriction on the time period that access-controlled egress doors may be used, because the staff and residents of the building are expected to be familiar with the building entrance and locking systems. Hence, with a building access control system, the number of individuals who are not familiar with the building is limited, and they are usually accompanied by staff or a resident.

To summarize, it is important to keep in mind that an egress door equipped with an access control system must always allow egress whether power is present or not. The egress door must be "fail-safe" and must assume this "fail-safe" condition when power is removed from any part of the access control system. In other words, if the access control system loses power, the egress door must be capable of being opened. People must be kept from being involuntarily locked inside buildings.

1008.1.4.5 Security grilles. In Groups B, F, M and S, horizontal sliding or vertical security grilles are permitted at the main exit and shall be openable from the inside without the use of a key or special knowledge or effort during periods that the space is occupied. The grilles shall remain secured in the full-open position during the period of occupancy by the general public. Where two or more *means of egress* are required, not more than one-half of the *exits* or *exit access doorways* shall be equipped with horizontal sliding or vertical security grilles.

❖ This section really functions as an exception to several sections, including Sections 1008.1.2 (Door swing) and 1008.1.9.3 (Locks and latches) and permits the use of these security grilles under conditions that are similar to those found in Section 402 for covered mall buildings. These security grilles will be open when the space is occupied and will, therefore, not obstruct any egress path.

1008.1.5 Floor elevation. There shall be a floor or landing on each side of a door. Such floor or landing shall be at the same elevation on each side of the door. Landings shall be level except for exterior landings, which are permitted to have a slope not to exceed 0.25 unit vertical in 12 units horizontal (2-percent slope).

Exceptions:

1. Doors serving individual dwelling units in Groups R-2 and R-3 where the following apply:

   1.1. A door is permitted to open at the top step of an interior *flight* of *stairs*, provided the door does not swing over the top step.

   1.2. Screen doors and storm doors are permitted to swing over *stairs* or landings.

2. Exterior doors as provided for in Section 1003.5, Exception 1, and Section 1020.2, which are not on an *accessible route*.

3. In Group R-3 occupancies not required to be *Accessible units*, *Type A units* or *Type B units*, the landing at an exterior doorway shall not be more than $7^3/_4$ inches (197 mm) below the top of the threshold, provided the door, other than an exterior storm or screen door, does not swing over the landing.

4. Variations in elevation due to differences in finish materials, but not more than $^1/_2$ inch (12.7 mm).

5. Exterior decks, patios or balconies that are part of *Type B* dwelling units, have impervious surfaces and that are not more than 4 inches (102 mm) below the finished floor level of the adjacent interior space of the dwelling unit.

❖ Changes in floor surface elevation at a door, however small, often are slip or trip hazards. This is because persons passing through a door, including those who may have some mobility impairments, usually do not expect changes in floor surface elevation or are not able to recognize them because of the intervening door leaf. Under emergency conditions, a fall in a doorway could result not only in injury to the falling occupant but also interruption of orderly egress by other occupants. The exterior landing is allowed to slope to drain.

The size of this landing is set by Section 1008.1.6. In accordance with Exception 4, the floor surface elevation of the landing is to be at the same elevation plus or minus ¹/₂ inch (12.7 mm) (see Figure 1008.1.5).

Note that some of the exceptions indicate which direction the door swings to allow the exceptions while others do not limit the door swing direction.

Exception 1, which applies to nontransient residential occupancies, recognizes that occupants are familiar with the stair and landing arrangements. Note that an interior or exterior door (other than screen or storm doors) is not allowed to swing over a stair.

Exception 2 references two other locations. Section 1003.5, Exception 1, permits a 7-inch (178 mm) change in elevation at exterior doors in Groups F, H, R-2, R-3, S and U if they are not on an accessible route. The door could swing in either direction for this exception and may actually be required to swing out in accordance with Section 1008.1.2. A reference to Section 1020.2 does not address a change in elevation but does address exterior exit doors.

In accordance with Exception 3, for a residential unit, the step-down is limited to 7³/₄ inches (197 mm) and the exterior door cannot swing over the exterior landing. A screen door or storm door could swing over the exterior landing. This is consistent with the exception to Section 1008.1.7.

Exception 4 addresses a change in floor finish material.

In accordance with Exception 5, certain exterior doors of Type B dwelling or sleeping units are also exempt from the floor surface requirements of Section 1008.1.5. Please note that this exception is not applicable for the primary entrance door (see Section 1105.1.6). Exterior doors that open out onto an exterior deck, patio or balcony are allowed a 4-inch (102 mm) step-down. Type B units are established by Chapter 11 for residential occupancies containing four or more dwelling or sleeping units. In order to use this exception, the exterior decks, patios or balconies must be of solid and impervious construction, such as concrete or wood. A 4-inch (102 mm) step from inside the unit down to the exterior surfaces is allowed for weather purposes. This allowance is consistent with the provisions of ICC A117.1 and the Federal Fair Housing Accessibility Guidelines (FHAG).

1008.1.6 Landings at doors. Landings shall have a width not less than the width of the *stairway* or the door, whichever is greater. Doors in the fully open position shall not reduce a required dimension by more than 7 inches (178 mm). When a landing serves an *occupant load* of 50 or more, doors in any position shall not reduce the landing to less than one-half its required width. Landings shall have a length measured in the direction of travel of not less than 44 inches (1118 mm).

Exception: Landing length in the direction of travel in Groups R-3 and U and within individual units of Group R-2 need not exceed 36 inches (914 mm).

❖ Door landings are at either side of the door. Landings can overlap floor surfaces within a room or corridor, overlap an exterior porch or balcony, or share the landings for stairways. The floor surface elevation of the area, described by the doorway width and length measured from the face of the door on both sides of the threshold at a distance equal to the door width, is to be at the same elevation plus or minus ¹/₂ inch (12.7 mm) (see Figure 1008.1.5). The 7-inch (178 mm) encroachment and one-half required width limitations are consistent with Section 1005.2 for door encroachment. Section 1005.2 deals with egress width and is referenced from aisles (see Section 1017.1), corridors (see Section 1018.3), exit passageways (see Section 1023.2) and egress courts (see Section 1027.5.1).

This section also is intended to address landings at

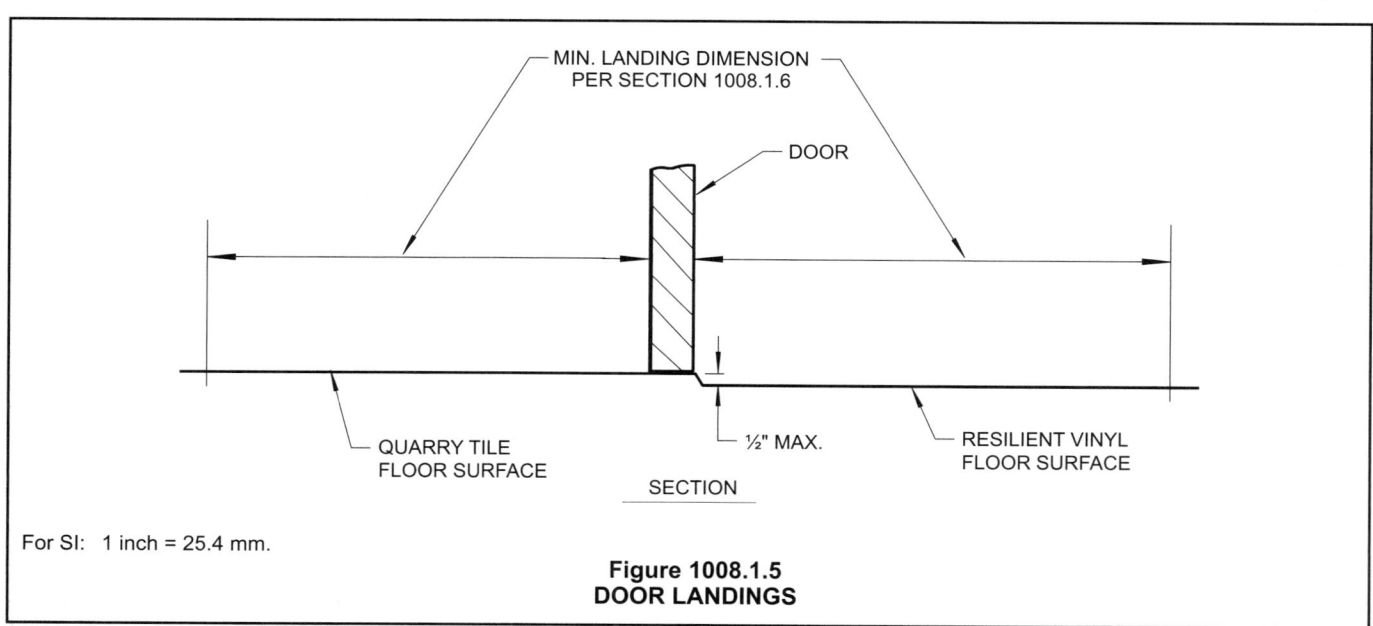

For SI: 1 inch = 25.4 mm.

**Figure 1008.1.5
DOOR LANDINGS**

the entrance door to enclosed stairways (also see Section 1009.5 for stairway landings). The width of a landing at a door in a stairway is to be not less than the width of the stairway or the door, whichever is greater [see Figure 1009.5(4) for an example of these provisions].

No matter what size the door or stair landing is, door landings are to have the floor elevation requirements of Section 1008.1.5 extending at least 44 inches (1118 mm) in the direction of egress travel.

The reduction in landing length from 44 inches minimum to 36 inches minimum for certain residential occupancies is in recognition of their low occupant load.

1008.1.7 Thresholds. Thresholds at doorways shall not exceed $^3/_4$ inch (19.1 mm) in height for sliding doors serving dwelling units or $^1/_2$ inch (12.7 mm) for other doors. Raised thresholds and floor level changes greater than $^1/_4$ inch (6.4 mm) at doorways shall be beveled with a slope not greater than one unit vertical in two units horizontal (50-percent slope).

Exception: The threshold height shall be limited to $7^3/_4$ inches (197 mm) where the occupancy is Group R-2 or R-3; the door is an exterior door that is not a component of the required *means of egress*; the door, other than an exterior storm or screen door, does not swing over the landing or step; and the doorway is not on an *accessible route* as required by Chapter 11 and is not part of an *Accessible unit, Type A unit* or *Type B unit.*

❖ A threshold is a potential tripping hazard and a barrier to accessibility by people with mobility impairments. For these reasons, thresholds for all doorways, except exterior sliding doors serving dwelling units, are to be a maximum of $^1/_2$ inch (12.7 mm) high. Exterior sliding doors serving dwelling units, however, are permitted to be $^3/_4$ inch (19.1 mm) high because of practical design considerations, concern for deterioration of the doorway because of snow and ice buildup and lack of adequate drainage in severe climates. Raised threshold and floor level changes at doorways without edge treatment [see Figure 1008.1.7(1)] are permitted to be $^1/_4$ inch (6.3 mm) high vertically.

Raised threshold and floor level changes with edges beveled with a slope not greater than one unit vertical in two units horizontal (1:2) (50-percent slope) [see Figure 1008.1.7(2)] are permitted to be $^1/_2$ inch (12.7 mm) high, with its parts no more than $^1/_4$ inch (6.3 mm) high and the remainder with beveled edges no more than 1:2 [see Figures 1008.1.7(3) and 1008.1.7(4)]. This kind of threshold treatment provides for minimum

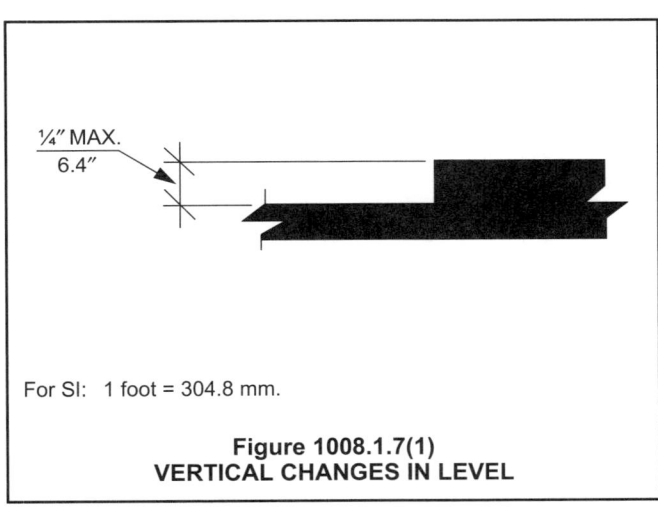

For SI: 1 foot = 304.8 mm.

Figure 1008.1.7(1)
VERTICAL CHANGES IN LEVEL

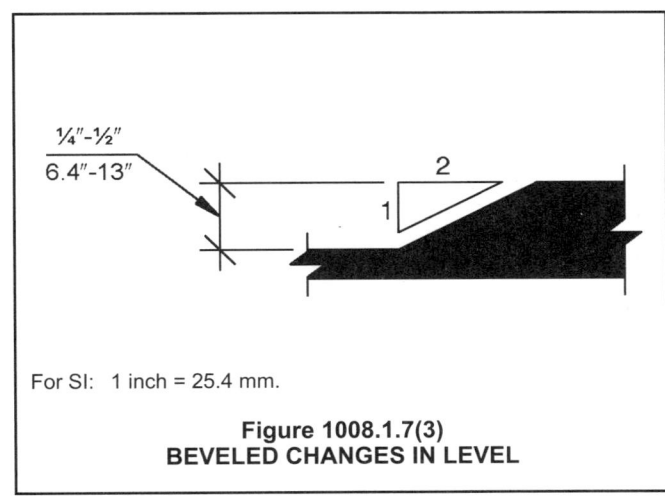

For SI: 1 inch = 25.4 mm.

Figure 1008.1.7(3)
BEVELED CHANGES IN LEVEL

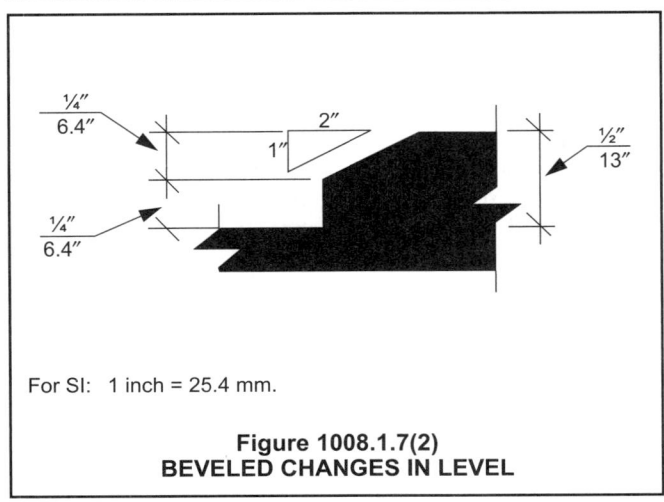

For SI: 1 inch = 25.4 mm.

Figure 1008.1.7(2)
BEVELED CHANGES IN LEVEL

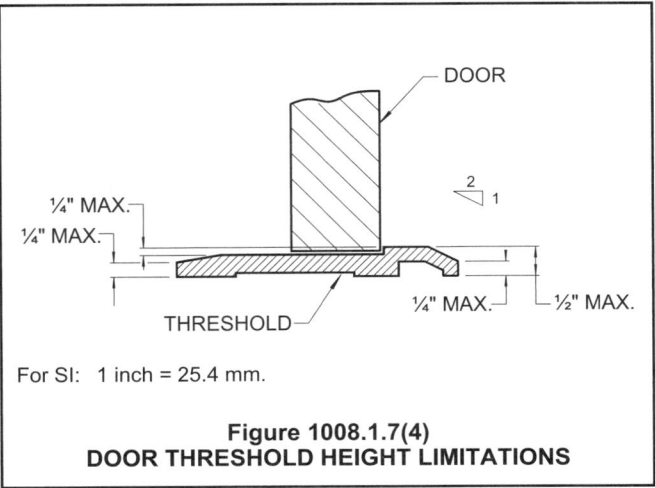

For SI: 1 inch = 25.4 mm.

Figure 1008.1.7(4)
DOOR THRESHOLD HEIGHT LIMITATIONS

obstructions for wheelchair users and limits the trip hazard for those with other mobility disabilities.

The exception permits a $7^3/_4$-inch (197 mm) threshold at exterior doors for dwelling or sleeping units not required to be Accessible, Type A or Type B units. The terminology is consistent with Section 1008.1.5, Exception 3; however, this exception is to the threshold requirements, not the landing elevations.

1008.1.8 Door arrangement. Space between two doors in a series shall be 48 inches (1219 mm) minimum plus the width of a door swinging into the space. Doors in a series shall swing either in the same direction or away from the space between the doors.

Exceptions:

1. The minimum distance between horizontal sliding power-operated doors in a series shall be 48 inches (1219 mm).

2. Storm and screen doors serving individual dwelling units in Groups R-2 and R-3 need not be spaced 48 inches (1219 mm) from the other door.

3. Doors within individual dwelling units in Groups R-2 and R-3 other than within *Type A* dwelling units.

❖ Door arrangement is required to be such that an occupant's use of a means of egress doorway is not hampered by the operation of a preceding door located in the same line of travel so that the occupant flow can be smooth through the openings. Successive doors in a single egress path (i.e., in a series) can cause such interference. The 4-foot (1219 mm) clear distance between doors when the first door is open allows an occupant, including a person using a wheelchair, to move past one door and its swing before beginning the operation of the next door [see Figure 1008.1.8(1)]. Note that where doors in a series are not arranged in a straight line, the intent of the code is to provide sufficient space to enable occupants to negotiate the second door without being encumbered by the first door's swing arc. To facilitate accessibility, the space between doors should provide sufficient clear space for a wheelchair [30 inches by 48 inches (762 mm by 1219 mm)] beyond the arc of the door swing [see Figure 1008.1.8(2)]. Additionally, the approach and access provisions of ICC A117.1 should be considered for any doors along an accessible route.

The exception is to permit horizontal sliding power-operated doors (see Sections 1008.1.4.2 and 1008.1.4.3) to be designed with a lesser distance between them in a series arrangement because they are customarily designed to open simultaneously or in sequence such that movement through them is unhampered. Storm and screen doors on residential dwelling units need not be spaced at 48 inches (1219 mm) since it would be impractical, and they do not operate the same as doors in a series. Doors within dwelling units of Group R-2 or R-3 that are not Type A dwelling units (see Section 1107) are also permitted to have a lesser distance between doors, because the accessibility provisions do not apply. There are requirements in Chapter 10 of the ICC A117.1 for door arrangements within Accessible and Type A dwelling and sleeping units.

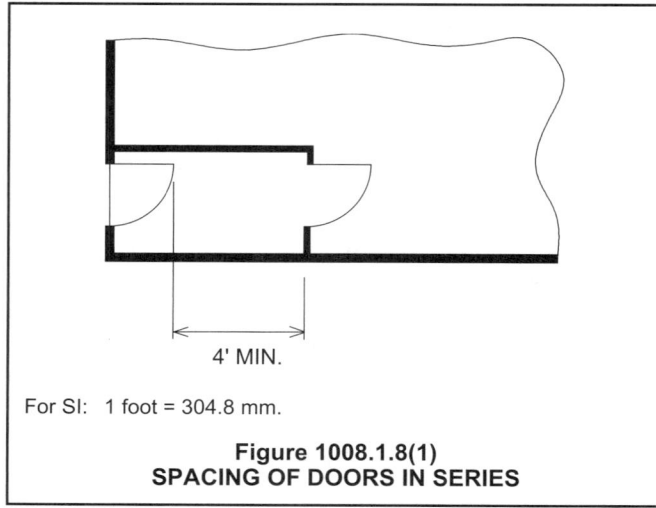

For SI: 1 foot = 304.8 mm.

Figure 1008.1.8(1)
SPACING OF DOORS IN SERIES

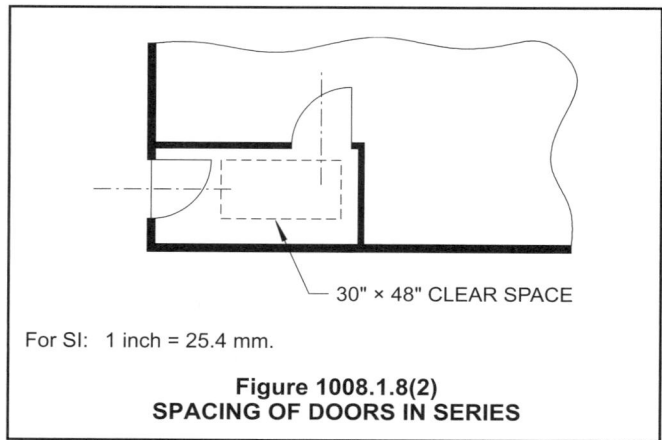

For SI: 1 inch = 25.4 mm.

Figure 1008.1.8(2)
SPACING OF DOORS IN SERIES

1008.1.9 Door operations. Except as specifically permitted by this section egress doors shall be readily openable from the egress side without the use of a key or special knowledge or effort.

❖ When installed for security purposes, locks and latches can intentionally prohibit the use of an egress door and thus interfere with or prevent the egress of occupants at the time of a fire. While the security of property is important for many, the life safety of occupants is essential for everyone. Where security and life safety objectives conflict, alternative measures, such as those permitted by each of the exceptions in Section 1008.1.9.3, may be applicable.

Egress doors are permitted to be locked, but must be capable of being unlocked and readily openable from the side from which egress is to be made. The outside of a door can be key locked as long as the inside—the side from which egress is to be made—can be unlocked without the use of tools, keys or special knowledge or effort. For example, an unlocking operation that is integral with an unlatching operation is acceptable.

Examples of special knowledge would be a combination lock or an unlocking device or deadbolt in an unknown, unexpected or hidden location. Special effort would dictate the need for unusual and unexpected physical ability to unlock or make the door fully available for egress.

Where a pair of egress door leaves is installed, with or without a center mullion, the general requirement is that each leaf must be provided with its own releasing or unlatching device so as to be readily openable. Door arrangements or devices that depend on the release of one door before the other can be opened are not to be used except as permitted by Section 1008.1.9.4.

1008.1.9.1 Hardware. Door handles, pulls, latches, locks and other operating devices on doors required to be *accessible* by Chapter 11 shall not require tight grasping, tight pinching or twisting of the wrist to operate.

❖ Any doors that are located along an accessible route for ingress or egress must have door hardware that is easy to operate by a person with limited mobility. This would include all elements of the door hardware used in typical door operation, such as door levers, locks, security changes, etc. This requirement is also an advantage for persons with arthritis in their hands. Items such as small, full-twist thumb turns or smooth circular knobs are examples of hardware that is not acceptable.

Some people with disabilities are unable to grasp objects with their hands or twist their wrists. Such people are unable to operate, or have great difficulty in operating, door hardware other than lever-operated mechanisms, push-type mechanisms and U-shaped handles. Door hardware that can be operated with a closed fist or a loose grip accommodates the greatest range of users. Hardware operated by simultaneous hand and finger movement requires greater dexterity and coordination and should be avoided for doors along an accessible route (see Figure 1008.1.9.1).

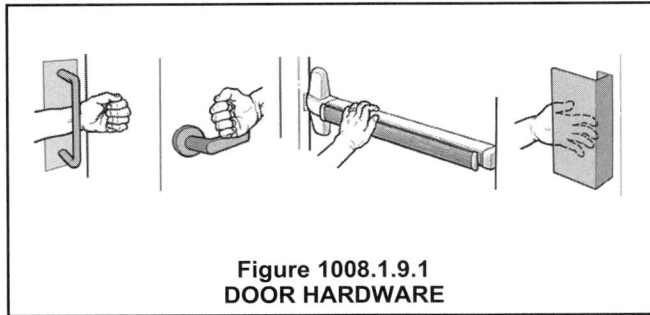

Figure 1008.1.9.1
DOOR HARDWARE

1008.1.9.2 Hardware height. Door handles, pulls, latches, locks and other operating devices shall be installed 34 inches (864 mm) minimum and 48 inches (1219 mm) maximum above the finished floor. Locks used only for security purposes and not used for normal operation are permitted at any height.

Exception: Access doors or gates in barrier walls and fences protecting pools, spas and hot tubs shall be permitted to have operable parts of the release of latch on self-latching devices at 54 inches (1370 mm) maximum above the finished floor or ground, provided the self-latching devices are not also self-locking devices operated by means of a key, electronic opener or integral combination lock.

❖ The requirements in this section place the door hardware at a level that is usable by most people, including a person using a wheelchair. The exception allows security locks to be placed at any height. An example would be an unframed glass door at the front door of a tenant space in a mall that has the lock near the floor level. The lock is only used when the store is not open for business. Such locks are not required for the normal operation of the door.

The exception permits a special allowance for security latches at pools, spas and hot tubs. The concern is that the 48-inch (1219 mm) maximum height would place the security latch within reach of children. The 54-inch (1372 mm) maximum height is intended to override the maximum 48-inch (1219 mm) reach range in ICC A117.1. This compromise addresses concerns for children's safety and still maintains accessibility to a reasonable level. This same exception is found in Section 1109.12, Exception 7, for the accessibility requirements for operable parts. This is consistent with the ADA/ABA Accessibility Guidelines and ANSI/NSPI-8 1996, *Model Barrier Code for Residential Swimming Pools, Spas, and Hot Tubs*.

1008.1.9.3 Locks and latches. Locks and latches shall be permitted to prevent operation of doors where any of the following exists:

1. Places of detention or restraint.

2. In buildings in occupancy Group A having an *occupant load* of 300 or less, Groups B, F, M and S, and in *places of religious worship*, the main exterior door or doors are permitted to be equipped with key-operated locking devices from the egress side provided:

    2.1. The locking device is readily distinguishable as locked;

    2.2. A readily visible durable sign is posted on the egress side on or adjacent to the door stating: THIS DOOR TO REMAIN UNLOCKED WHEN BUILDING IS OCCUPIED. The sign shall be in letters 1 inch (25 mm) high on a contrasting background; and

    2.3. The use of the key-operated locking device is revokable by the *building official* for due cause.

3. Where egress doors are used in pairs, *approved* automatic flush bolts shall be permitted to be used, provided that the door leaf having the automatic flush bolts has no doorknob or surface-mounted hardware.

4. Doors from individual dwelling or sleeping units of Group R occupancies having an *occupant load* of 10 or less are permitted to be equipped with a night latch, dead bolt or security chain, provided such devices are openable from the inside without the use of a key or tool.

5. *Fire doors* after the minimum elevated temperature has disabled the unlatching mechanism in accordance with listed fire door test procedures.

❖ Where security and life safety objectives conflict, alternative measures, such as those permitted by each of the exceptions, may be applicable.

Exception 1 is needed for jails and prisons or locations where someone must be kept inside for their own safety (i.e., dementia wards, psychiatric wards).

Exception 2 permits a locking device, such as a double-cylinder dead bolt, on the main entrance door. It must be immediately apparent that these doors are locked. For example, such locking devices may have an integral indicator that automatically reflects the "locked" or "unlocked" status of the device. In addition, a sign must be provided that clearly states that the door is to be unlocked when the building is occupied. The sign on the door not only reminds employees to unlock the door, but also advises the public that an unacceptable arrangement exists if one finds the door locked. Ideally, the individual who encounters the locked door will notify management and possibly the building official. Note that the use of the key-locking device is revocable by the building official. The locking arrangement is not permitted on any door other than the main exit and, therefore, the employees, security and cleaning crews will have access to other exits without requiring the use of a key. This allowance is not limited just to multiple-exit buildings but also to small buildings with one exit. This option is an alternative to the panic hardware required by Section 1008.1.10.

In Exception 3, an automatic flush bolt device is one that is internal to the inactive leaf of a pair of doors. The device has a small "knuckle" that extends from the in-

active leaf into an opening in the active leaf. When the active leaf is opened, the bolt is automatically retracted. When the active leaf is closed, the knuckle is pressed into the inactive leaf by the active leaf, extending the flush bolt(s), in the head or sill of the inactive leaf (see Figure 1008.1.9.3).

Automatic flush bolts on one leaf of a pair of egress doors are acceptable, provided the leaf with the automatic flush bolts is not equipped with a doorknob or other hardware that would imply to the user that the door leaf is unlatched independently of the companion leaf.

Exception 4 addresses the need for security in residential dwelling and sleeping units such as hotel rooms, apartments, dormitory rooms or townhouses. The occupants are familiar with the operation of the indicated devices, which are intended to be relatively simple to operate without the use of a key or tool. Note that this exception only applies to the door from the dwelling unit.

Exception 5 is in recognition of listed procedures for fire doors, which include the disabling of the locking mechanism.

1008.1.9.4 Bolt locks. Manually operated flush bolts or surface bolts are not permitted.

Exceptions:

1. On doors not required for egress in individual dwelling units or sleeping units.

2. Where a pair of doors serves a storage or equipment room, manually operated edge- or surface-mounted bolts are permitted on the inactive leaf.

3. Where a pair of doors serves an *occupant load* of less than 50 persons in a Group B, F or S occupancy, man-

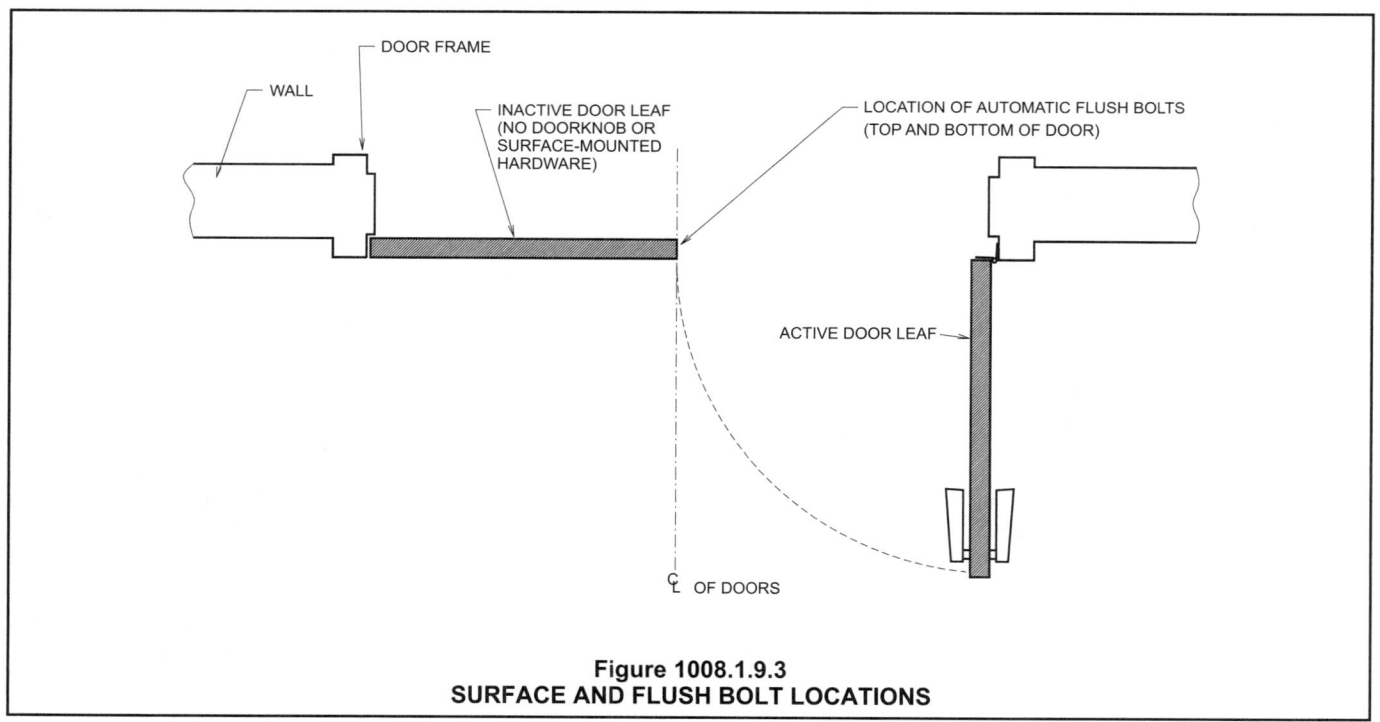

Figure 1008.1.9.3
SURFACE AND FLUSH BOLT LOCATIONS

ually operated edge- or surface-mounted bolts are permitted on the inactive leaf. The inactive leaf shall contain no doorknobs, panic bars or similar operating hardware.

4. Where a pair of doors serves a Group B, F or S occupancy, manually operated edge- or surface-mounted bolts are permitted on the inactive leaf provided such inactive leaf is not needed to meet egress width requirements and the building is equipped throughout with an *automatic sprinkler system* in accordance with Section 903.3.1.1. The inactive leaf shall contain no doorknobs, panic bars or similar operating hardware.

5. Where a pair of doors serves patient care rooms in Group I-2 occupancies, self-latching edge- or surface-mounted bolts are permitted on the inactive leaf provided that the inactive leaf is not needed to meet egress width requirements and the inactive leaf contains no doorknobs, panic bars or similar operating hardware.

❖ This section is applicable to doors that are intended and required to be for means of egress purposes or are identified as a means of egress, such as by an "Exit" sign or other device. Doors, as well as a second leaf in a doorway that is provided for a purpose other than means of egress, such as for convenience or building operations, should be arranged or identified so as not to be mistaken as a means of egress. The use of manually operated flush bolts or surface bolts on means of egress doors have traditionally been prohibited due to the inability of users to quickly identify and operate such devices under emergency conditions.

This section prohibits installation of manually operated flush and surface bolts except in limited situations. The exceptions allowing the use of such hardware are intended to expand the use of manually operated edge- or surface-mounted bolts under specified conditions while maintaining an appropriate degree of safety for the building occupants.

Exception 1 allows bolt locks at some doors within an individual dwelling or sleeping unit. Even then, such bolts may only be used on doors not required for egress (see Section 1008.1.9.3, Exception 4, for security of doors from individual dwelling and sleeping units). Flush and surface bolts represent locking devices that are difficult to operate because of their location and operation (see Figure 1008.1.9.4).

Exception 2 provides for edge-mounted or surface-mounted bolts on the inactive leaf of a pair of doors from storage or equipment areas. Double doors are often provided to allow for the easy removal or replacement of large pieces of equipment or bulk movement of goods.

Exceptions 3 and 4 offer two options for limited doors in Group B, F and S occupancies. Again, the wider door is sometimes needed for the movement of equipment. Automatic flush bolts and removable center posts can be easily damaged and difficult to main-

tain in areas of frequent door usage. Revisions to the requirements for door hardware on such pairs of doors will increase building functionality while maintaining a very high degree of occupant safety.

In Exception 3, the number of occupants within the space must be less than 50, the active leaf must meet means of egress requirements and the inactive leaf must not have any operating hardware so that it could be mistaken for an egress door.

In accordance with Exception 4, if the Group B, F or S building is sprinklered throughout with an NFPA 13 system, the room served by the double door can have any occupant load if the inactive leaf is not needed for egress and has no operating hardware.

Exception 5 is in recognition of the clinical needs for movement of equipment into some patient sleeping and treatment rooms in hospital and nursing home environments. Again, the inactive leaf must not be needed for means of egress or have any operating hardware. This is consistent with Section 407.3.1, which allows for staff to operate patient sleeping and treatment room doors during emergency events. The doors would still have to meet smoke barrier opening protective requirements. The clear width of $41^1/_2$ inches (1054 mm) requirements in Section 1008.1.1 would still have to be met with the active door leaf.

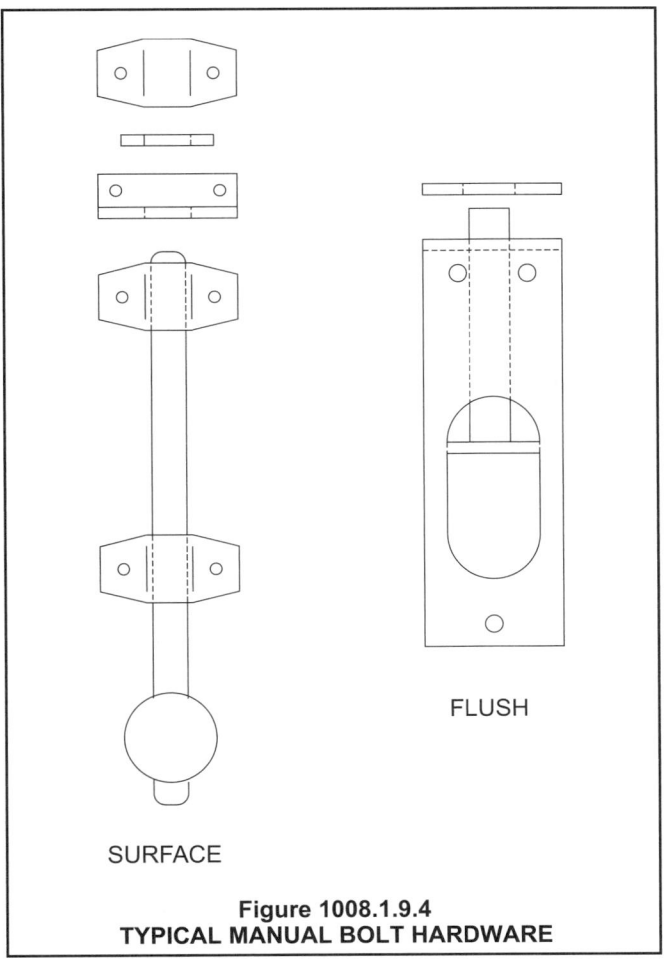

FLUSH

SURFACE

Figure 1008.1.9.4
TYPICAL MANUAL BOLT HARDWARE

1008.1.9.5 Unlatching. The unlatching of any door or leaf shall not require more than one operation.

Exceptions:

1. Places of detention or restraint.

2. Where manually operated bolt locks are permitted by Section 1008.1.9.4.

3. Doors with automatic flush bolts as permitted by Section 1008.1.9.3, Exception 3.

4. Doors from individual dwelling units and sleeping units of Group R occupancies as permitted by Section 1008.1.9.3, Exception 4.

❖ The code prohibits the use of multiple locks or latching devices on a door, which would be a safety hazard in an emergency situation. The exceptions address locations where multiple locks or latching devices are acceptable.

1008.1.9.5.1 Closet and bathroom doors in Group R-4 occupancies. In Group R-4 occupancies, closet doors that latch in the closed position shall be openable from inside the closet, and bathroom doors that latch in the closed position shall be capable of being unlocked from the ingress side.

❖ The intent of this provision is to address possible entrapment concerns in group homes. If a closet door has a door latch, the closet door must be openable from both inside and outside. This will ensure that someone cannot get stuck inside a closet by accident. If a closet does not latch, no interior hardware is required. In case a resident needs assistance in a bathroom, the bathroom door must have a type of hardware that would allow the door to be unlocked from the outside by staff. This requirement is unique to Group R-4. There are not similar requirements for Group I-1 assisted living facilities or Group I-2 nursing homes; however, some facilities install such devices to increase patient safety.

1008.1.9.6 Special locking arrangements in Group I-2. *Approved* delayed egress locks shall be permitted in a Group I-2 occupancy where the clinical needs of persons receiving care require such locking. Delayed egress locks shall be permitted in such occupancies where the building is equipped throughout with an *automatic sprinkler system* in accordance with Section 903.3.1.1 or an *approved* automatic smoke or heat detection system installed in accordance with Section 907, provided that the doors unlock in accordance with Items 1 through 6 below. A building occupant shall not be required to pass through more than one door equipped with a delayed egress lock before entering an *exit*.

1. The doors unlock upon actuation of the *automatic sprinkler system* or automatic fire detection system.

2. The doors unlock upon loss of power controlling the lock or lock mechanism.

3. The door locks shall have the capability of being unlocked by a signal from the fire command center, a nursing station or other *approved* location.

4. The procedures for the operation(s) of the unlocking system shall be described and *approved* as part of the emergency planning and preparedness required by Chapter 4 of the *International Fire Code*.

5. All clinical staff shall have the keys, codes or other means necessary to operate the locking devices.

6. Emergency lighting shall be provided at the door.

Exception: Items 1 through 3 shall not apply to doors to areas where persons, because of clinical needs, require restraint or containment as part of the function of a mental hospital.

❖ The intent of these provisions is to address the special safety needs for wards that may include dementia or Alzheimer's patients/residents. Due to concerns over possible elopement, there must be a balance between maintaining a safe and secure environment for patients/residents and emergency evacuation requirements. Items 1 through 3 deal with when the delayed egress locks would be automatically unlocked. Items 4 and 5 deal with staff and fire department awareness of the issues. Item 6 requires emergency lighting at the delayed egress door to ensure visability for unlocking during a possible power outage. The exception allows for the automatic unlocking to not be included as a requirement in mental hospitals due to additional safety concerns for the public.

1008.1.9.7 Delayed egress locks. *Approved, listed,* delayed egress locks shall be permitted to be installed on doors serving any occupancy except Group A, E and H occupancies in buildings that are equipped throughout with an *automatic sprinkler system* in accordance with Section 903.3.1.1 or an *approved* automatic smoke or heat detection system installed in accordance with Section 907, provided that the doors unlock in accordance with Items 1 through 6 below. A building occupant shall not be required to pass through more than one door equipped with a delayed egress lock before entering an *exit*.

1. The doors unlock upon actuation of the *automatic sprinkler system* or automatic fire detection system.

2. The doors unlock upon loss of power controlling the lock or lock mechanism.

3. The door locks shall have the capability of being unlocked by a signal from the fire command center.

4. The initiation of an irreversible process which will release the latch in not more than 15 seconds when a force of not more than 15 pounds (67 N) is applied for 1 second to the release device. Initiation of the irreversible process shall activate an audible signal in the vicinity of the door. Once the door lock has been released by the application of force to the releasing device, relocking shall be by manual means only.

   Exception: Where approved, a delay of not more than 30 seconds is permitted.

5. A sign shall be provided on the door located above and within 12 inches (305 mm) of the release device reading: PUSH UNTIL ALARM SOUNDS. DOOR CAN BE OPENED IN 15 [30] SECONDS.

6. Emergency lighting shall be provided at the door.

❖ This locking system is called delayed egress due to Item 4, which allows for an alarm to sound before someone is able to egress. These types of locks are permitted for situations where there are concerns about security, and one such delay is not considered detrimental to occupant evacuation. For security reasons, special locking arrangements are permitted for doors in a means of egress serving occupancies other than those in Groups A, E and H. The arrangements are not permitted in assembly or educational occupancies because the resulting delay in egress is not acceptable given the greater number of occupants who may be unfamiliar with the space or of a young age. Such a delay would also be inconsistent with Section 1008.1.10, which requires the installation of panic hardware on doors in such uses. Also, the delay from Group H would be unreasonable given the potential for rapid fire buildup in such areas.

Because of the possible delay caused by the controlled locking device in the egress door, the building must be provided throughout with compensating fire protection features to promptly warn occupants of a fire condition. All of the listed conditions must be met in order to permit use of such a locking device.

Item 1 interconnects the lock with an automatic sprinkler system in accordance with NFPA 13 or, alternatively, an automatic fire detection system in accordance with Section 907, which is required to be installed throughout the building. Such systems are to provide occupants with an early warning of a fire event, and thus additional time for egress. Note that the provision for an automatic fire detection system does not include the use of single- or multiple-station detectors. Also note that actuation of the automatic sprinkler or fire detection system is to unlock the control device so as to permit the egress door to be readily and immediately openable.

Item 2 is so that the control device is "fail safe." Since the operation of the device is dependent on electrical power, in the event of electrical power loss to the lock or locking mechanism, the egress doors must be readily and immediately openable from the side from which egress is to be made.

Item 3 specifies that the door must be capable of being manually unlocked by a signal sent from a fire command center. Personnel at that location are intended to be the first alerted to an emergency event and are expected to take appropriate action to unlock all egress doors equipped with special locking arrangements. This will permit the locks to be deactivated prior to the sprinklers or smoke detection system detecting the problem or in case of other nonfire emergencies such as an earthquake.

Item 4 specifies the operational characteristics of the locking control device that is similar to a panic device. A user must apply a minimum 15-pound (67 N) force to the release device for at least 1 second, at which time an audible alarm will sound and the device will automatically start to unlock the door. The 1-second duration is to prevent initiation of the unlocking process because of an inadvertent bump or accidental contact against the device. The unlocking cycle is irreversible; once it is started, it does not stop. Once the cycle starts, the door is required to be unlocked in no more than 15 seconds. When the door is unlocked at the end of the 15-second delay, it stays unlocked until someone comes to the door and manually relocks it. A method of automatically relocking the door from a remote location such as a central control station or security office is not permitted; therefore, the first users to the door may face a delay, but after that other users would be able to exit immediately.

The exception will permit the building official to allow the time delay prior to opening to be increased beyond the basic 15 seconds, but never to the point where the delay would exceed 30 seconds.

If a user continues to exert the 15-pound (67 N) force for more than 1 second, the door is required to be openable after 15 seconds from the start of the force application.

The sign required by Item 5 informs the user of the type of unlocking device and that the door will become available for egress. An undesirable consequence of the door not unlocking immediately is if the user assumes it will never be available and then proceeds to another exit door.

Item 6 provides emergency lighting at the door so that the user can read the sign required by Item 5.

1008.1.9.8 Electromagnetically locked egress doors. Doors in the *means of egress* that are not otherwise required to have panic hardware in buildings with an occupancy in Group A, B, E, M, R-1 or R-2 and doors to tenant spaces in Group A, B, E, M, R-1 or R-2 shall be permitted to be electromagnetically locked if equipped with *listed* hardware that incorporates a built-in switch and meet the requirements below:

1. The *listed* hardware that is affixed to the door leaf has an obvious method of operation that is readily operated under all lighting conditions.

2. The *listed* hardware is capable of being operated with one hand.

3. Operation of the *listed* hardware releases to the electromagnetic lock and unlocks the door immediately.

4. Loss of power to the *listed* hardware automatically unlocks the door.

❖ In limited occupancy groups, doors that are electromagnetically locked during building occupancy are now permitted to be utilized in the means of egress if equipped with listed hardware that incorporates a built-in switch that meets specified conditions.

As a general rule, means of egress door hardware shall be operable by manual operation to provide for occupant control of the egress system. Locking devices are typically prohibited as they can interfere or prevent efficient egress through the door during an emergency situation. However, owner concerns that must be considered sometimes require a greater degree of security. In specific occupancies, doors in the means of egress are now permitted to be electromagnetically locked if equipped with listed hardware that incorporates a built-in switch that interrupts the power supply to the electromagnetic lock and unlocks the door. The use of this type of locking system provides for a greater degree of security than that offered by other methods addressed in the code, including delayed egress locking systems and egress access control systems.

The allowance for electronically locked egress doors is limited to low- and moderate-hazard occupancies where security can be a major concern. The listed hardware that incorporates a built-in switch has been tested by Underwriters Laboratories (UL). When the occupant prepares to use the door hardware, the method of operating the hardware must be obvious, even under poor lighting conditions. The operation shall be accomplished through the use of a single hand. This is consistent with the general requirement that the door be readily openable without the use of special knowledge or effort. The unlocking of the door must occur immediately on the operation of the hardware by interrupting the power supply to the electromagnetic lock. As an additional safeguard, the loss of power to the hardware shall automatically unlock the door.

Where these special provisions are utilized, the requirements of IBC Section 1008.1.10 regarding panic hardware remain applicable. In Group A and E occupancies having occupant loads of 50 or more, the door hardware must also comply with the requirements for panic hardware.

1008.1.9.9 Locking arrangements in correctional facilities. In occupancies in Groups A-2, A-3, A-4, B, E, F, I-2, I-3, M and S within correctional and detention facilities, doors in *means of egress* serving rooms or spaces occupied by persons whose movements are controlled for security reasons shall be permitted to be locked when equipped with egress control devices which shall unlock manually and by at least one of the following means:

1. Activation of an *automatic sprinkler system* installed in accordance with Section 903.3.1.1;

2. Activation of an *approved* manual alarm box; or

3. A signal from a *constantly attended location*.

❖ Correctional facilities can include a variety of uses where detainees may be gathered for eating, recreational activities, education, technical training, job training, etc. Correctional facilities can also contain types of support services, such as a store, storage areas or hospital area. Security is still a concern within these areas. This provision will allow the correctional facility to maintain security on all areas. Most commonly the doors would be opened by staff from a central control point under Item 3, but Items 1 and 2 allow

for other alternatives in lower security facilities. This provision is not intended to apply to these groups when located outside of a detention or correctional facility.

1008.1.9.10 Stairway doors. *Interior stairway means of egress* doors shall be openable from both sides without the use of a key or special knowledge or effort.

Exceptions:

1. *Stairway* discharge doors shall be openable from the egress side and shall only be locked from the opposite side.

2. This section shall not apply to doors arranged in accordance with Section 403.5.3.

3. In *stairways* serving not more than four stories, doors are permitted to be locked from the side opposite the egress side, provided they are openable from the egress side and capable of being unlocked simultaneously without unlatching upon a signal from the fire command center, if present, or a signal by emergency personnel from a single location inside the main entrance to the building.

❖ Based on adverse fire experience where occupants have become trapped in smoke-filled stairway enclosures, stairway doors generally must be arranged to permit reentry into the building without the use of any tools, keys or special knowledge or effort. For security reasons, this restriction does not apply to the discharge door from the stairway enclosure which is often to the outside. Section 403 for high-rise buildings permits the locking of the doors from the stairway side, provided the doors are capable of being unlocked from a fire command station and there is a communication system within the stairway enclosure that allows contact with the fire command station. It would be reasonable to permit this arrangement in buildings other than high-rise buildings.

Exception 3 addresses the need for security. The exception is limited to four-story buildings to provide a short travel distance to the stairway discharge door for the building occupants. In addition, to allow quick entrance for fire fighters and emergency responders, a means of simultaneously unlocking all of the doors by emergency personnel must be provided. This provision further requires that the stairway doors be unlocked without unlatching. Stairway doors will typically be fire door assemblies, and their continued latching is necessary to maintain the integrity of the fire-resistive separation for the exit enclosure. The remote unlocking signal shall be initiated from the fire command station, if provided, or a single point of signal initiation at an approved location inside the building's main entrance.

1008.1.10 Panic and fire exit hardware. Doors serving a Group H occupancy and doors serving rooms or spaces with an *occupant load* of 50 or more in a Group A or E occupancy shall not be provided with a latch or lock unless it is panic hardware or *fire exit hardware*.

Exception: A main *exit* of a Group A occupancy in compliance with Section 1008.1.9.3, Item 2.

Electrical rooms with equipment rated 1,200 amperes or more and over 6 feet (1829 mm) wide that contain overcurrent devices, switching devices or control devices with *exit* or *exit access* doors shall be equipped with panic hardware or *fire exit hardware*. The doors shall swing in the direction of egress travel.

❖ Doors that are part of a means of egress from the locations listed in this section shall not be provided with a latch or lock unless it is panic hardware or fire exit hardware. (See the commentary for the definitions for "Fire exit hardware" or "Panic hardware" and Sections 1008.1.10.1 and 1008.1.10.2).

Panic hardware is also required for all Group H occupancies because of the physical hazards of these spaces.

Panic hardware is required on all doors that provide means of egress to rooms and spaces of assembly and educational (Groups A and E) occupancies with an occupant load of 50 or more. These uses are characterized by higher occupant load densities. Whereas doors from an assembly or educational room with an occupant load of less than 50 do not require panic hardware, a door that provides means of egress for two or more such rooms would require panic hardware when the combination of spaces has a total occupant load of 50 or more.

The exception is to clarify that the provisions for key-locking hardware at the main exit in Group A occupancies are permitted instead of panic hardware at those specific locations. (For the Group A exception, see the commentary to Section 1008.1.9.3, Item 2.)

Certain electrical rooms are required to have panic hardware. This provision is consistent with the requirement for panic hardware in the NFPA 70: *National Electrical Code®* (NEC). This requirement is applicable only where multiple conditions are present. The type of room regulated creates a potentially hazardous environment. In the event of an electrical accident, the more immediate egress provided by the panic hardware is desirable.

1008.1.10.1 Installation. Where panic or *fire exit hardware* is installed, it shall comply with the following:

1. Panic hardware shall be *listed* in accordance with UL 305;

2. *Fire exit hardware* shall be *listed* in accordance with UL 10C and UL 305;

3. The actuating portion of the releasing device shall extend at least one-half of the door leaf width; and

4. The maximum unlatching force shall not exceed 15 pounds (67 N).

❖ As its name implies, panic hardware is special unlatching and unlocking hardware that is intended to simplify the unlatching and unlocking operation to a single 15-pound (67 N) force applied in the direction of egress (see Figure 1008.1.10.1). In a panic situation

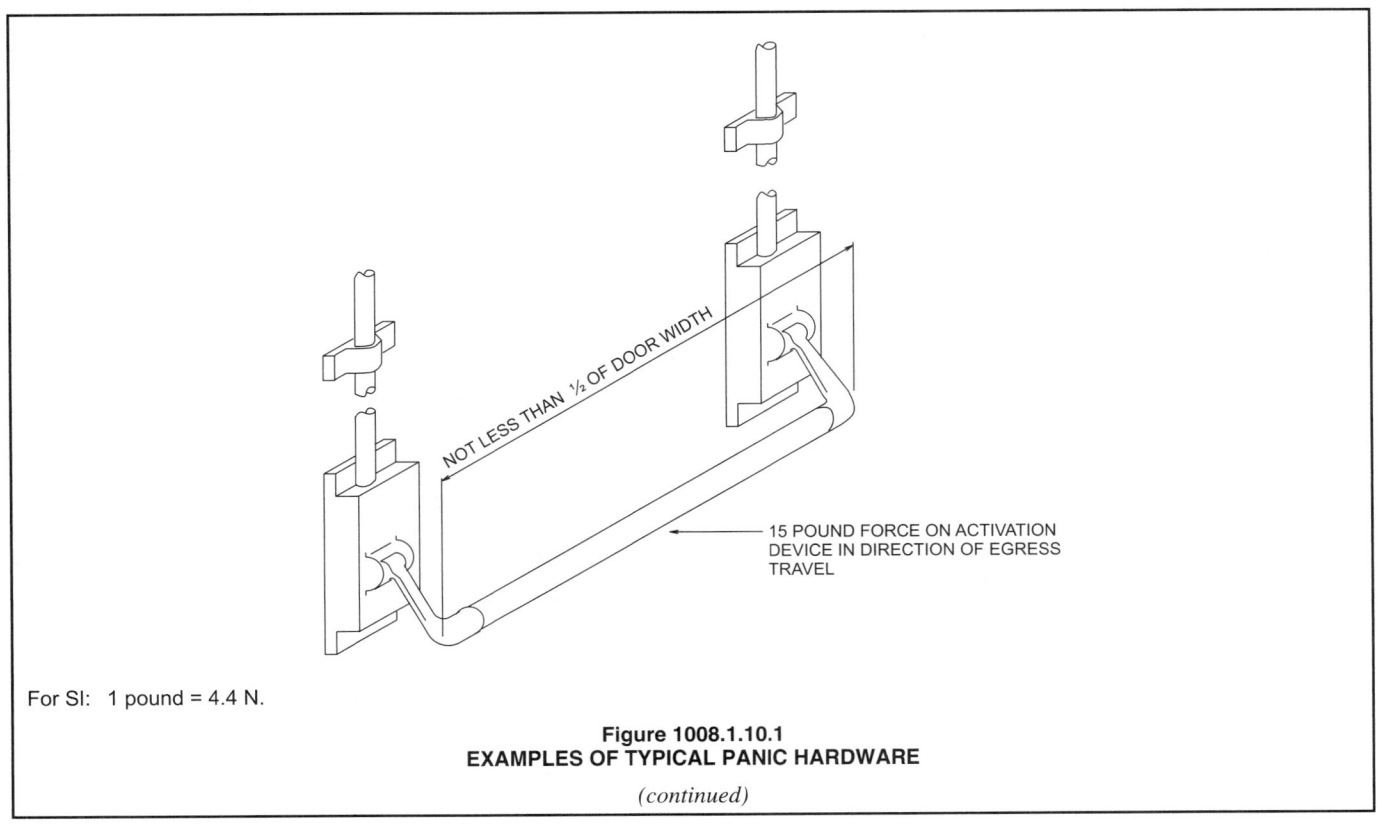

For SI: 1 pound = 4.4 N.

Figure 1008.1.10.1
EXAMPLES OF TYPICAL PANIC HARDWARE

(continued)

with a rush of persons trying to utilize a door, the conventional devices, such as doorknobs or thumb turns, may cause sufficient delay so as to create a crush at the door and prevent or slow the opening operation.

The locational specifications for the activating panel or bar are based on ready availability and access to the unlatching device. Note that the section requires the activating portion to extend at least one-half the width of the door leaf. Panic and fire exit hardware must be listed. UL 305, *Standard for Safety Panic Hardware*, includes construction and performance requirements dealing with endurance, emergency operation, elevated ambient exposure and low-temperature impact tests to ensure that the panic device operates properly. (For panic hardware on a balanced door, see Section 1008.1.10.2.) The activation device must be mounted between 34 inches and 48 inches (864 mm and 1219 mm) high in accordance with Section 1008.1.9.2. Section 1008.1.1.1 was specifically added to allow the panic hardware to extend the full width of the door as long as it does not protrude more than 4 inches (102 mm) into the door clear width.

Standard panic hardware or "Listed Panic Hardware" is not approved for use on fire door assemblies. Panic hardware and fire exit hardware can be similar in appearance.

Where a fire door, such as to an exit stairway, is required to be equipped with panic hardware, the hardware must accomplish the dual objectives of panic hardware and continuity of the enclosure in which it is located—thus the reference to UL 10C, *Standard for Safety Positive Pressure Fire Tests of Door Assem-*

blies. In this case, fire exit hardware is to be provided that meets both objectives and requirements, since panic hardware is not tested for use on fire doors. There are standard test procedures designed to evaluate the performance of panic and fire exit hardware from the panic standpoint as well as from a fire protection standpoint. "Fire door assemblies" are defined in Section 715.4 as a combination of doors, frame, hardware and other accessories required to provide a specific degree of fire and smoke barrier protection to the opening in a fire wall, fire barrier, fire partition, smoke barriers or exterior wall required to have a fire-resistance rating.

Fire doors must close and positively latch in order to protect exit stairways, corridors and other areas of the building from the spread of smoke and fire. Additionally, fire doors are required to self-close and automatically latch after each use. Positive latching of fire doors is not related to the locking of the door and should never be confused with locking or security issues.

The requirement for positive latching means that dogging devices are not permitted on fire exit hardware. A dogging device is an option on the hardware that allows for the panic hardware to be locked in the fully depressed position. A dogging device mechanically defeats the latching feature of panic hardware preventing the door from positively latching when in the closed position. The dogging device is typically manually activated with a small wrench or tool and is activated through a hole adjacent to the activation bar.

Fire exit hardware must be labeled. Typical locations are on either end of the hardware. Information on

Figure 1008.1.10.1—continued
EXAMPLES OF TYPICAL PANIC HARDWARE

the label must include the words "listed" and "fire exit hardware" and indicate a control or serial number. The label on the fire door itself should indicate that it is a fire door suitable for use with fire exit hardware.

1008.1.10.2 Balanced doors. If *balanced doors* are used and panic hardware is required, the panic hardware shall be the push-pad type and the pad shall not extend more than one-half the width of the door measured from the latch side.

❖ The provisions for balanced doors ensure that the occupants push only on the latch side of the door since the hinge side of a balanced door pivots "against" the direction of egress (see the commentary for the definition of "Door, balanced").

1008.2 Gates. Gates serving the means of egress system shall comply with the requirements of this section. Gates used as a component in a *means of egress* shall conform to the applicable requirements for doors.

> **Exception:** Horizontal sliding or swinging gates exceeding the 4-foot (1219 mm) maximum leaf width limitation are permitted in fences and walls surrounding a stadium.

❖ This section specifies that all of the requirements for doors also apply to gates, except gates that surround a stadium are allowed to exceed 4 feet (1219 mm) in width. Usually a large gate is required to adequately serve a stadium crowd for egress purposes.

1008.2.1 Stadiums. Panic hardware is not required on gates surrounding stadiums where such gates are under constant immediate supervision while the public is present, and where safe dispersal areas based on 3 square feet (0.28 m²) per occupant are located between the fence and enclosed space. Such required safe dispersal areas shall not be located less than 50 feet (15 240 mm) from the enclosed space. See Section 1027.6 for *means of egress* from safe dispersal areas.

❖ Panic hardware is impractical for large gates that surround stadiums. Normally, these gates are opened and closed by the stadium's ground crew that is constantly in attendance during the use of such gates. The safe dispersal area requirement provides for the safety of the crowd if for some reason the gate is not open. The safe dispersal area is to be between the stadium enclosure and the surrounding fence and the area to be occupied is not to be closer than 50 feet (15 240 mm) to the stadium enclosure.

1008.3 Turnstiles. Turnstiles or similar devices that restrict travel to one direction shall not be placed so as to obstruct any required *means of egress*.

> **Exception:** Each turnstile or similar device shall be credited with no more than a 50-person capacity where all of the following provisions are met:
>
> 1. Each device shall turn free in the direction of egress travel when primary power is lost, and upon the manual release by an employee in the area.
>
> 2. Such devices are not given credit for more than 50 percent of the required egress capacity.

3. Each device is not more than 39 inches (991 mm) high.

4. Each device has at least $16^1/_2$ inches (419 mm) clear width at and below a height of 39 inches (991 mm) and at least 22 inches (559 mm) clear width at heights above 39 inches (991 mm).

Where located as part of an *accessible route*, turnstiles shall have at least 36 inches (914 mm) clear at and below a height of 34 inches (864 mm), at least 32 inches (813 mm) clear width between 34 inches (864 mm) and 80 inches (2032 mm) and shall consist of a mechanism other than a revolving device.

❖ This section provides for a limited use of turnstiles to serve as a means of egress component. The exception to this section limits each turnstile to a maximum of 50 persons of egress capacity. The turnstile must comply with all four listed items to be considered to serve any part of the occupant load for means of egress. The turnstiles must rotate freely both when there is a loss of power and when they are manually released. Note that the 50-person limit applies to each individual turnstile. These provisions are similar to the revolving door provisions in Section 1008.1.4.1.

If turnstiles are located along an accessible route, the route for persons using mobility devices must be something other than a revolving device, such as a swinging gate. A common example would be the turnstiles for automatic ticket taking, such as at the entrance to a mass transit platform.

1008.3.1 High turnstile. Turnstiles more than 39 inches (991 mm) high shall meet the requirements for revolving doors.

❖ Where a turnstile is higher than 39 inches (991 mm), the restriction to egress is much like a revolving door. Thus, the egress limitations for revolving doors in Section 1008.1.4.1 apply to this type of turnstile. If a high turnstile does not meet the revolving door requirements for doors that are an egress component, it is not to be included as serving a portion of the means of egress. It would be necessary to provide doors in these areas for egress. High turnstiles may not be part of an accessible route for ingress or egress.

1008.3.2 Additional door. Where serving an *occupant load* greater than 300, each turnstile that is not portable shall have a side-hinged swinging door which conforms to Section 1008.1 within 50 feet (15 240 mm).

❖ This section addresses a common egress condition for sports arenas where a number of turnstiles are installed for ticket taking. Portable turnstiles are moved from the egress path for proper exiting capacity. Permanent turnstiles are not considered as providing any of the required egress capacity when serving an occupant load greater than 300, no matter how many turnstiles are installed. Doors are required to provide occupants with a path of egress other than through the turnstiles. The doors are to be located within 50 feet (15 240 mm) of the turnstiles.

SECTION 1009
STAIRWAYS

1009.1 Stairway width. The width of *stairways* shall be determined as specified in Section 1005.1, but such width shall not be less than 44 inches (1118 mm). See Section 1007.3 for *accessible means of egress stairways*.

Exceptions:

1. *Stairways* serving an *occupant load* of less than 50 shall have a width of not less than 36 inches (914 mm).

2. *Spiral stairways* as provided for in Section 1009.9.

3. *Aisle stairs* complying with Section 1028.

4. Where an incline platform lift or stairway chairlift is installed on *stairways* serving occupancies in Group R-3, or within dwelling units in occupancies in Group R-2, a clear passage width not less than 20 inches (508 mm) shall be provided. If the seat and platform can be folded when not in use, the distance shall be measured from the folded position.

❖ To provide adequate space for occupants traveling in opposite directions and to permit the intended full egress capacity to be developed, minimum dimensions are dictated for means of egress stairways. A minimum width of 44 inches (1118 mm) is required for stairway construction to permit two columns of users to travel in the same or opposite directions. The reference to Section 1005.1 is for the determination of stairway width based on occupant load. The larger of the two widths is to be used.

Exception 1 recognizes the relatively small occupant loads of less than 50 that permit a staggered file of users when traveling in the same direction. When traveling in opposite directions, one column of users must stop their ascent (or descent) to permit the opposite column to continue. Again, considering the relatively small occupant loads, any disruption of orderly flow will be infrequent. The use of this exception is limited to buildings where the occupant load of each upper story and/or basement is less than 50.

Exception 2 permits a spiral stairway to have a minimum width of 26 inches (660 mm) when it conforms to Section 1009.9, on the basis that the configuration of a spiral stairway will allow nothing other than single-file travel.

Exception 3 provides for the aisle stair widths that are specified in Section 1028.

Exception 4 addresses the use of inclined platform lifts or stairway chairlifts for individual dwelling units. For clarification on the types of lifts, see the commentary to Section 1109.7. Both types of lifts may be installed to aid persons with mobility impairments in their homes. The code and ASME A18.1 allow for a reduction in the width of the stair to a minimum of 20 inches (508 mm) of clear passageway to be maintained on a stairway where a lift is located. If a portion of the lift, such as a platform or seat, can be folded, the minimum clear dimension is to be measured from the folded po-

sition. If the lift cannot be folded, then the 20 inches (508 mm) is measured from the fixed position. The track for these lifts typically extends 9 to 12 inches (229 to 305 mm) from the wall, making the 20-inch (508 mm) clear measurement actually 24 to 27 inches (610 to 686 mm) from the edge of the track.

The code does not have any specific provisions for where incline platform lifts are utilized along stairways in locations other than within dwelling units. Section 1109.7 limits the use of platform lifts in new construction to mainly areas with minimal occupant loads or where elevators and ramps are impractical. Section 3411.8.3 allows for platform lifts anywhere in existing buildings in order to gain accessibility for persons with mobility impairments. When in the closed and off position, the platform lifts should not block the clear width required for the stairway, or use of the handrails. The industry is currently working on different options to address the concern that the lift may be in operation during an event that requires evacuation.

1009.2 Headroom. *Stairways* shall have a minimum headroom clearance of 80 inches (2032 mm) measured vertically from a line connecting the edge of the *nosings*. Such headroom shall be continuous above the *stairway* to the point where the line intersects the landing below, one tread depth beyond the bottom riser. The minimum clearance shall be maintained the full width of the *stairway* and landing.

Exceptions:

1. *Spiral stairways* complying with Section 1009.9 are permitted a 78-inch (1981 mm) headroom clearance.

2. In Group R-3 occupancies; within dwelling units in Group R-2 occupancies; and in Group U occupancies that are accessory to a Group R-3 occupancy or accessory to individual dwelling units in Group R-2 occupancies; where the *nosings* of treads at the side of a *flight* extend under the edge of a floor opening through which the *stair* passes, the floor opening shall be allowed to project horizontally into the required headroom a maximum of $4^3/_4$ inches (121 mm).

❖ This headroom requirement is necessary to avoid an obstruction to orderly flow and to provide visibility to the users so that the desired path of travel can be planned and negotiated. Height is a vertical measurement above every point along the stairway stepping and walking surfaces, with minimum height measured vertically from the tread nosing or from the surface of a landing or platform up to the ceiling [see Figure 1009.2(1)].

Sections 1003.2 and 1208.2 require a minimum ceiling height within a room of 7 feet, 6 inches (2307 mm). A bulkhead or doorway at the bottom of the stairway would be allowed to meet the minimum headroom height of 80 inches (2032 mm), as permitted in Section 1003.3.

Exception 1, allowing for a clear headroom of 6 feet, 6 inches (1981 mm) for spiral stairs, correlates with the provisions of Section 1009.9.

Exception 2 recognizes a common method of stairwell construction in which the stringer on the open side of a stair is supported by the same floor joists or wall that supports the edge of the opening through which the stairway passes to the floor above; thus resulting in the stairway being wider at the lower portion than at the top portion. In this case headroom is not required for a distance of up to 4³/₄ inches (121 mm) measured horizontally from the edge of the opening above to the handrail or guard system on the lower portion of the stairway; effectively not measuring for headroom at the sides of the lower portion of the open stairway. The 4³/₄ inches (121 mm) maximum is derived from the finished width of a typical 2 by 4 supporting wall and is not critical to obstructing orderly flow or visibility in the desired path of travel [see Figure 1009.2(2)].

1009.3 Walkline. The walkline across *winder* treads shall be concentric to the direction of travel through the turn and located 12 inches (305 mm) from the side where the *winders* are narrower. The 12-inch (305 mm) dimension shall be measured from the widest point of the clear *stair* width at the walking surface of the *winder*. If *winders* are adjacent within the *flight*, the point of the widest clear *stair* width of the adjacent *winders* shall be used.

❖ This requirement is essential for smooth, consistent travel on stairs that turn with winder treads. It provides a standard location for the regulation of the uniform tread depth of winders. Due to the wide range of anthropometrics of stairway users, there is no one line that all persons will travel on stairs; however, the code recognizes a standard location of a walkline is essential to design and enforcement. Each footfall of the user through the turn can be associated with an arc to describe the path traveled. As a user ascends or descends the flight, the turning at each step should be consistent through the turn. The walkline is established concentric, or having the same center (approximately parallel) as the arc of travel of the user. The tread depth dimension at the walkline is one of two tread depths across the width of the stair at which winder tread depth is regulated, cited in Section 1009.4.2. The second is the minimum tread depth. Regulation at these two points controls the angularity of the turn and the configuration of the flight. In order to establish consistently shaped winders, tread depths must always be measured concentric to the arc of travel. The walkline is unique as the only line or path of travel where winder tread depth is controlled by the same minimum tread depth as rectangular treads. However, Exception 2 of Section 1009.4.4 recognizes winder tread depth need not be compared to rectangular tread depths for dimensional uniformity in the same flight because the location of the walkline is chosen for the purpose of providing a standard and cannot be specific to the variety of actual paths followed by all users. This specific line location is determined by measuring along each nosing edge 12 inches (305 mm) from the extreme of the clear width of the stair at the surface of the winder tread or the limit of where the foot might be placed in use of the stair. If adjacent winders are present, the point of the widest clear stair width at the surface of the tread in the group of adjacent consecutive winders is used to provide the reference from which the 12-inch (305 mm) dimension will be measured along each nosing. The tread depth may be determined by measuring between adjacent nosings at these determined intersections of the nosings with the walkline. It is important to note that the clear stair width is only that portion of the stair width that is clear for passage. Portions of the stair beyond the clear width are not consequential to use of the stair, consistent travel or location of the walkline.

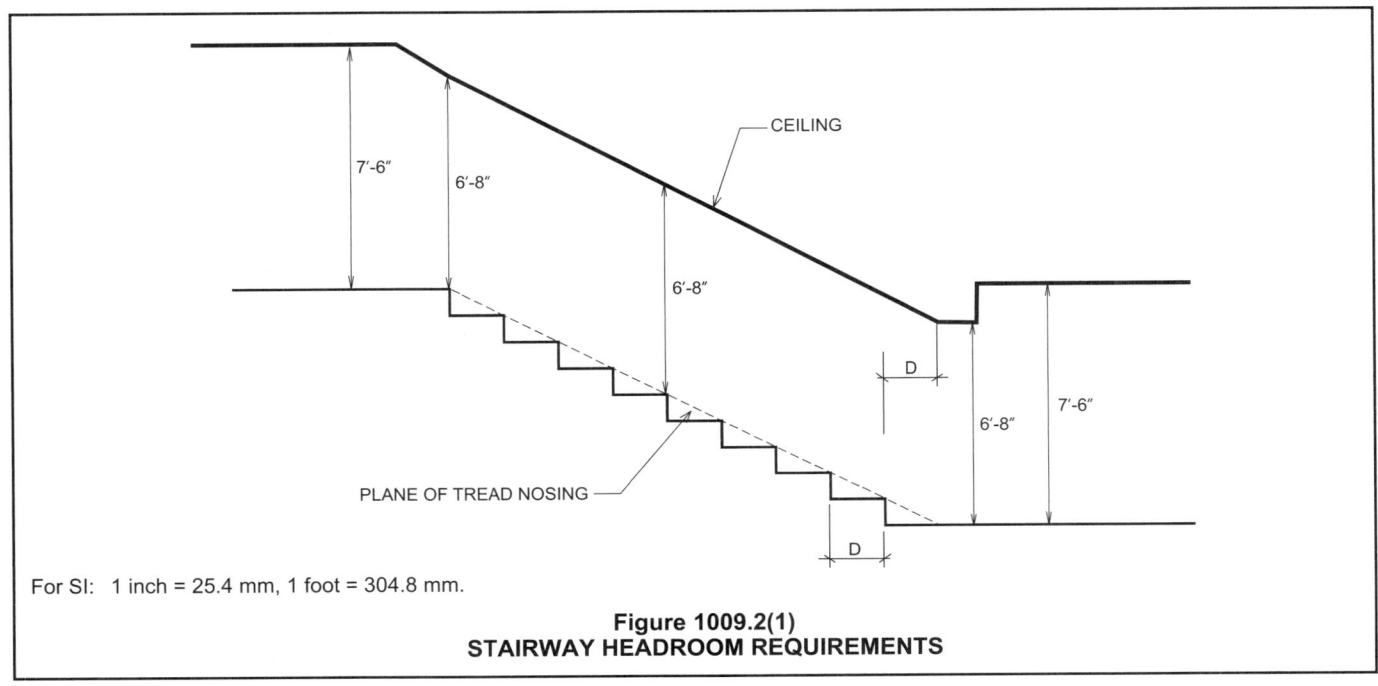

For SI: 1 inch = 25.4 mm, 1 foot = 304.8 mm.

Figure 1009.2(1)
STAIRWAY HEADROOM REQUIREMENTS

Figure 1009.2(2)
EXAMPLE OF SECTION 1009.2, EXCEPTION 2

1009.4 Stair treads and risers. *Stair* treads and risers shall comply with Sections 1009.4.1 through 1009.4.5.

❖ The provisions for treads and risers contribute to the efficient use of the stairway, facilitating smooth and consistent travel. This section provides dimensional ranges and tolerances for the component elements to allow the flexibility required to design and construct a stair or a flight of stairs which are elements of a stairway. The allowed proportion of maximum riser height and minimum tread depth provide for a maximum angle of ascent but there is no maximum tread depth to consider with the minimum riser height that would define a minimum angle for a stairway. Nor is the proportion of riser height to tread depth compared with the limitations of the length of the user's stride on stairways that is significantly foreshortened from the user's stride on the level. For this reason, care should be taken when incorporating larger tread depths and controlling the point at which a tread might be wide enough to require more than one step to cross, which can vary significantly when considering ascent and descent movement patterns. Especially in

areas where all segments of the public might use the stairs, those persons requiring two smaller sequential steps to cross the tread would progress at significantly different rates than those that might be able to stretch or jump and lead to dangerous complications, especially in egress. Of equal significance is the use of shorter risers without increasing tread depth resulting in a proportion that could cause overstepping. With these same limitations for proportion in mind however, by controlling the minimum depth of rectangular treads and the minimum depth and angularity of winder treads these components can control the configuration of the plan of a flight of stairs to provide for smooth and consistent travel.

1009.4.1 Dimension reference surfaces. For the purpose of this section, all dimensions are exclusive of carpets, rugs or runners.

❖ Carpets, rugs and runners, like furniture, are frequently changed by the occupants and are not regulated by the code. For this reason it is essential that the riser height and tread depth be regulated exclusive of these transitory surfaces to provide an enforceable standard. This practice minimizes the possible variation due to the removal of nonpermanent carpeting throughout the life of a structure and provides a standard enforcement methodology that will provide consistency across the build environment for all users. When owners or occupants add carpeting, rugs or runners, they need to add it to all tread and landing surfaces in the stairway. It is important that the tread and landing surfaces are consistent and comply with the code prior to the addition of carpet. This methodology of enforcement makes it unnecessary to reconstruct floor and stair elevations in the stairway when nonpermanent carpet surfaces are changed that do not require a building permit and eliminates the resulting variations in the built environment that will not comply with the tolerance in Section 1009.4.4 (see Figure 1009.4.1).

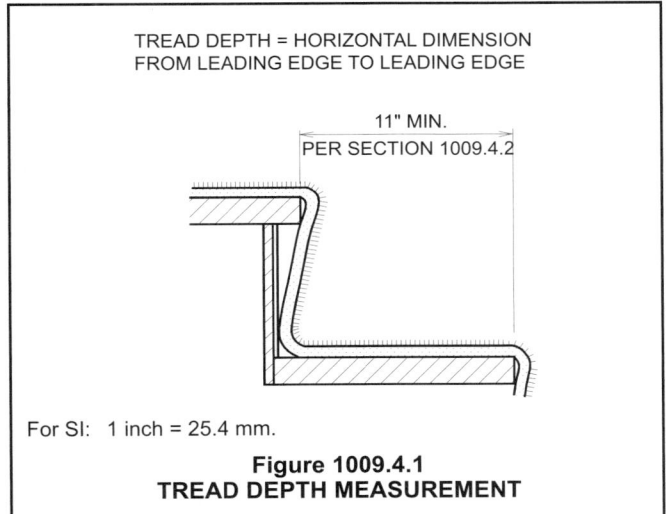

For SI: 1 inch = 25.4 mm.

**Figure 1009.4.1
TREAD DEPTH MEASUREMENT**

1009.4.2 Riser height and tread depth. *Stair* riser heights shall be 7 inches (178 mm) maximum and 4 inches (102 mm) minimum. The riser height shall be measured vertically between the leading edges of adjacent treads. Rectangular tread depths shall be 11 inches (279 mm) minimum measured horizontally between the vertical planes of the foremost projection of adjacent treads and at a right angle to the tread's leading edge. *Winder* treads shall have a minimum tread depth of 11 inches (279 mm) measured between the vertical planes of the foremost projection of adjacent treads at the intersections with the walkline and a minimum tread depth of 10 inches (254 mm) within the clear width of the *stair*.

Exceptions:

1. *Alternating tread devices* in accordance with Section 1009.10.

2. Ship ladders in accordance with Section 1009.11.

3. *Spiral stairways* in accordance with Section 1009.9.

4. *Aisle stairs* in assembly seating areas where the *stair* pitch or slope is set, for sightline reasons, by the slope of the adjacent seating area in accordance with Section 1028.11.2.

5. In Group R-3 occupancies; within dwelling units in Group R-2 occupancies; and in Group U occupancies that are accessory to a Group R-3 occupancy or accessory to individual dwelling units in Group R-2 occupancies; the maximum riser height shall be 7³/₄ inches (197 mm); the minimum tread depth shall be 10 inches (254 mm); the minimum *winder* tread depth at the walkline shall be 10 inches (254 mm); and the minimum *winder* tread depth shall be 6 inches (152 mm). A *nosing* not less than ³/₄ inch (19.1 mm) but not more than 1¹/₄ inches (32 mm) shall be provided on *stairways* with solid risers where the tread depth is less than 11 inches (279 mm).

6. See Section 3404.1 for the replacement of existing *stairways*.

7. In Group I-3 facilities, *stairways* providing access to guard towers, observation stations and control rooms, not more than 250 square feet (23 m²) in area, shall be permitted to have a maximum riser height of 8 inches (203 mm) and a minimum tread depth of 9 inches (229 mm).

❖ The riser height—the vertical dimension from tread surface to tread surface or tread surface to landing surface—is typically limited to no more than 7 inches (178 mm) nor less than 4 inches (102 mm). The minimum tread depth—the horizontal distance from the leading edge (nosing) of one tread to the leading edge (nosing) of the next adjacent tread or landing—is typically limited to no less than 11 inches (279 mm) [see Figure 1009.4.2]. The minimum tread depth of 11 inches (279 mm) is intended to accommodate the largest shoe size found in 95 percent of the adult population, allowing for an appropriate overhang of the foot beyond the tread nosing while descending a stairway. Tread depths under 11 inches (279 mm) could cause a larger overhang (depending on the size of the foot) and could force users with larger feet to descend a stairway increasing the angle of their foot to the line of travel. Based on the probability of adequate foot placement, the rate of misstep with various step sizes and consideration for the user's comfort and energy expenditure, it was agreed that the 11-inch (279 mm) minimum tread depth and maximum 7-inch (178 mm) riser height resulted in the reasonable proportion of riser height and tread depth for stairway construction. A minimum riser height of 4 inches (102 mm) is considered to allow the visual identification of the presence of the riser in ascent or descent.

The precise location of rectangular tread depth and riser measurements is to be perpendicular to the tread's nosing or leading edge. This is to duplicate the user's anticipated foot placement in traveling the stairway.

The size for a winder tread is also considered for proper foot placement along the walkline [see Figure 1009.8 and the commentary for Section 1009.3]. The dimensional requirements are consistent with the straight tread.

The exceptions apply only to the extent of the text of each exception. For example, the entire text of Section

1009.4.2 is set aside for spiral stairways conforming to Section 1009.9 (see Exception 3). However, Exception 5 allows a different maximum riser and minimum tread under limited conditions, but retains the minimum riser height and measurement method of Section 1009.4.2.

The requirements for dimensional uniformity are found in Section 1009.4.4.

Exceptions 1 and 2 are for unique elements of vertical egress. Section 1009.4.2 is not applicable to these elements because of their construction issues and limited application. For a discussion on alternating tread devices, see Section 1009.10. For ships ladders, see Section 1009.11.

Exception 3 is for spiral staircases, a unique type of stairway. Section 1009.4.2 is not applicable to this stair type, again because of construction issues and limited applications. For a discussion on spiral staircases, see Section 1009.9.

Exception 4 provides a practical exception where assembly facilities are designed for viewing. See Sections 1028.11 through 1028.11.3 for assembly aisle stair-limiting dimensions.

Exception 5 allows revisions to the 7 inches/11 inches (178 mm/279 mm) riser/tread requirements for Group R-3 and any associated utility (such as barns, connected garages or detached garages) and within individual units of Group R-2 and their associated utility areas (such as attached garages). This change is allowed because of the low occupant load and the high degree of occupant familiarity with the stairways. When this exception is taken for stairways that have solid risers, each tread is required to have a nosing projection with a minimum dimension of $^3/_4$ inch (19.1 mm) and maximum dimension of $1^1/_4$ inches (32 mm) where the tread depth is less than 11 inches (279 mm). Nosing projections are created where the nosing of the tread above extends beyond the trailing edge of the tread below or when a solid riser is angled under the tread above and connected to the trailing edge of the

tread below. Nosing projections are not required for residential stairs with open risers and 10-inch (254 mm) treads. A nosing projection provides a greater stepping surface for those ascending the stairway. For users descending the stairway, the nosing projection allows the toe of the foot to be placed further away from the riser above, providing the necessary clearance for the heel of the foot as it swings down in an arc to its position on the tread [see Figure 1009.4.3].

Exception 6 allows for the replacement of an existing stair. Where a change of occupancy would require compliance with current standards, this exception allows a stairway that may be steeper than that permitted, provided it does not constitute a hazard (see Section 3404.1). This language is consistent with that found for existing stairways in the *International Existing Building Code®* (IEBC®).

Exception 7 allows steeper stairs in spaces of not more than 250 square feet (23 m²) in correctional facilities (Group I-3) with a maximum riser height of 8 inches (203 mm) and a minimum tread depth of 9 inches (229 mm) because of the minimal occupant load and the familiarity of the users with the stairway.

1009.4.3 Winder treads. *Winder* treads are not permitted in *means of egress stairways* except within a dwelling unit.

Exceptions:

1. Curved *stairways* in accordance with Section 1009.8.

2. *Spiral stairways* in accordance with Section 1009.9.

❖ The intent of this section is to coordinate the general provisions for stairway tread and riser dimensions in Section 1009.4.2 with the provisions for winder treads permitted in curved and spiral stairways (see Sections 1009.8 and 1009.9). Winders are permitted in means of egress stairways within dwelling units where occupant loads are smaller and do not block visual clues that prompt the autonomic adjustments in the gait of the user as well as increased familiarity. This is typically fan-shaped treads at the turning of an L-shaped

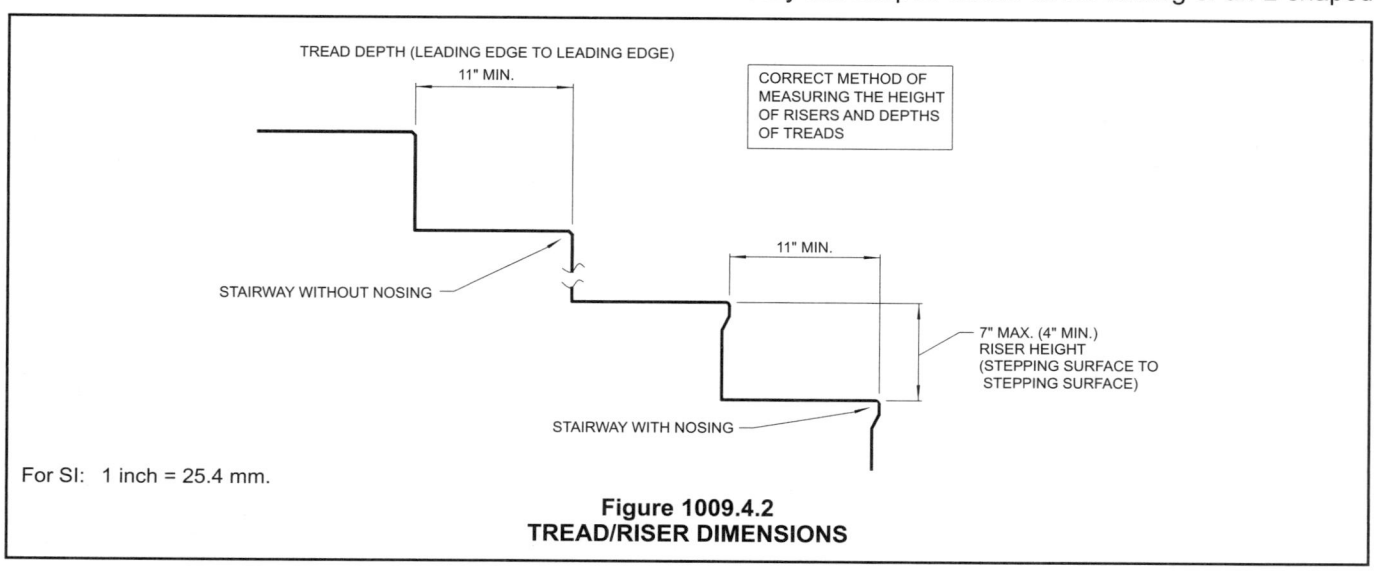

**Figure 1009.4.2
TREAD/RISER DIMENSIONS**

For SI: 1 inch = 25.4 mm.

stairway (see Figure 1009.4.3 and Section 1009.4.2, Exception 5). This is consistent with provisions in the *International Residential Code*® (IRC®).

Winders are used to change the direction of a flight by introducing a consistent incremental turn associated with each tread. The risk of injury in the use of stairways constructed with winders is considered to be greater than for stairways constructed as straight runs where users may be restricted by the presence of other users limiting visual clues or influencing the rate of travel. Additional user attention in the turn and the aid of the turn in arresting falls similar to turns at landings is also understood to negate this.

The employment of winders in stairway construction may necessitate the change of the user's gait in both ascent and descent where the tread depth of the winder is not equal to the tread depth of any rectangular treads in the same flight. For example, a person descending a straight flight of stairs will develop a particular gait conforming to the proportion of the riser height and tread depth that will be consistent throughout the flight. However, in a flight that includes winders and rectangular treads the user must accommodate a change in the proportion of the riser height and tread depth if the winder tread depth increases or decreases as determined by the path of travel chosen. Visual clues are important to the users' autonomic responses to alter the path of travel, the length of the stride, or a combination of both that may result in nonconcentric movement. To assure users of the visual clues necessary to alter their gait and limit the need to alter the path of travel in conditions of higher occupant loading, flights with winders must meet the specific safety provisions listed for curved or spiral stairways unless they are within a dwelling unit.

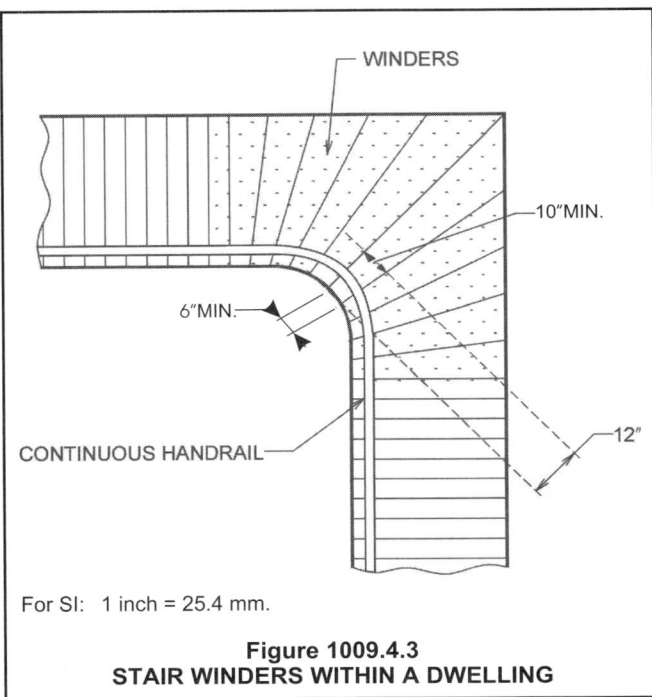

For SI: 1 inch = 25.4 mm.

Figure 1009.4.3
STAIR WINDERS WITHIN A DWELLING

1009.4.4 Dimensional uniformity. *Stair* treads and risers shall be of uniform size and shape. The tolerance between the largest and smallest riser height or between the largest and smallest tread depth shall not exceed $3/_8$ inch (9.5 mm) in any *flight* of stairs. The greatest *winder* tread depth at the walkline within any *flight* of stairs shall not exceed the smallest by more than $3/_8$ inch (9.5 mm).

Exceptions:

1. Nonuniform riser dimensions of *aisle stairs* complying with Section 1028.11.2.

2. Consistently shaped *winders*, complying with Section 1009.4.2, differing from rectangular treads in the same *stairway flight*.

Where the bottom or top riser adjoins a sloping *public way*, walkway or driveway having an established grade and serving as a landing, the bottom or top riser is permitted to be reduced along the slope to less than 4 inches (102 mm) in height, with the variation in height of the bottom or top riser not to exceed one unit vertical in 12 units horizontal (8-percent slope) of *stairway* width. The *nosings* or leading edges of treads at such nonuniform height risers shall have a distinctive marking stripe, different from any other *nosing* marking provided on the *stair flight*. The distinctive marking stripe shall be visible in descent of the *stair* and shall have a slip-resistant surface. Marking stripes shall have a width of at least 1 inch (25 mm) but not more than 2 inches (51 mm).

❖ Dimensional uniformity in the design and construction of stairways contributes to safe stairway use. When ascending or descending a stair, users establish a gait based on the autonomic expectation or "feel" that each step taken will be at the same height and will land in approximately the same position on the tread as the previous steps in the pattern. A substantial change in tread or riser dimensions in a stairway flight in excess of the allowed dimensional tolerance can break the rhythm and cause a misstep, stumbling or physical strain that may result in a fall or serious injury. Therefore, this section limits the dimensional variations to a tolerance of $3/_8$ inch (9.5 mm) between the largest and smallest riser or tread dimension in a flight of stairs. A "flight" of stairs is defined as a run of stairs between landings.

For special conditions of construction and as a practical matter, this section allows some greater variations in stairway tread and riser dimensions than the general limitations specified above. Exception 1 addresses conditions where the seating in assembly facilities is on a sloping gradient (for sightline purposes), and the aisle stairs become an integral part of the arrangement. Exception 2 addresses winder treads, which must be consistent along the walkline [see Figure 1009.4.4(1)] when compared to other winder treads in the same flight but are not required to meet the tolerance when compared to the uniform dimension of rectangular treads in the same flight.

This section also addresses the situation where the bottom riser of a flight of stairways meets a sloped landing, such as a public way, walk or driveway [see

Figure 1009.4.4(2)]. Because the sidewalk landing is sloped perpendicular to the stairway run, stepping off the bottom tread on one side will result in a higher riser than stepping off the bottom tread on the other side. This is permitted provided the bottom riser is marked so that someone using the stairs will be aware of the hazard of a nonuniform riser.

1009.4.5 Profile. The radius of curvature at the leading edge of the tread shall be not greater than $^9/_{16}$ inch (14.3 mm). Beveling of *nosings* shall not exceed $^9/_{16}$ inch (14.3 mm). Risers shall be solid and vertical or sloped under the tread above from the underside of the *nosing* above at an angle not more than 30 degrees (0.52 rad) from the vertical. The leading edge (*nosings*) of treads shall project not more than $1^1/_4$ inches (32 mm) beyond the tread below and all projections of the leading edges shall be of uniform size, including the leading edge of the floor at the top of a *flight*.

Exceptions:

1. Solid risers are not required for *stairways* that are not required to comply with Section 1007.3, provided that the opening between treads does not permit the passage of a sphere with a diameter of 4 inches (102 mm).

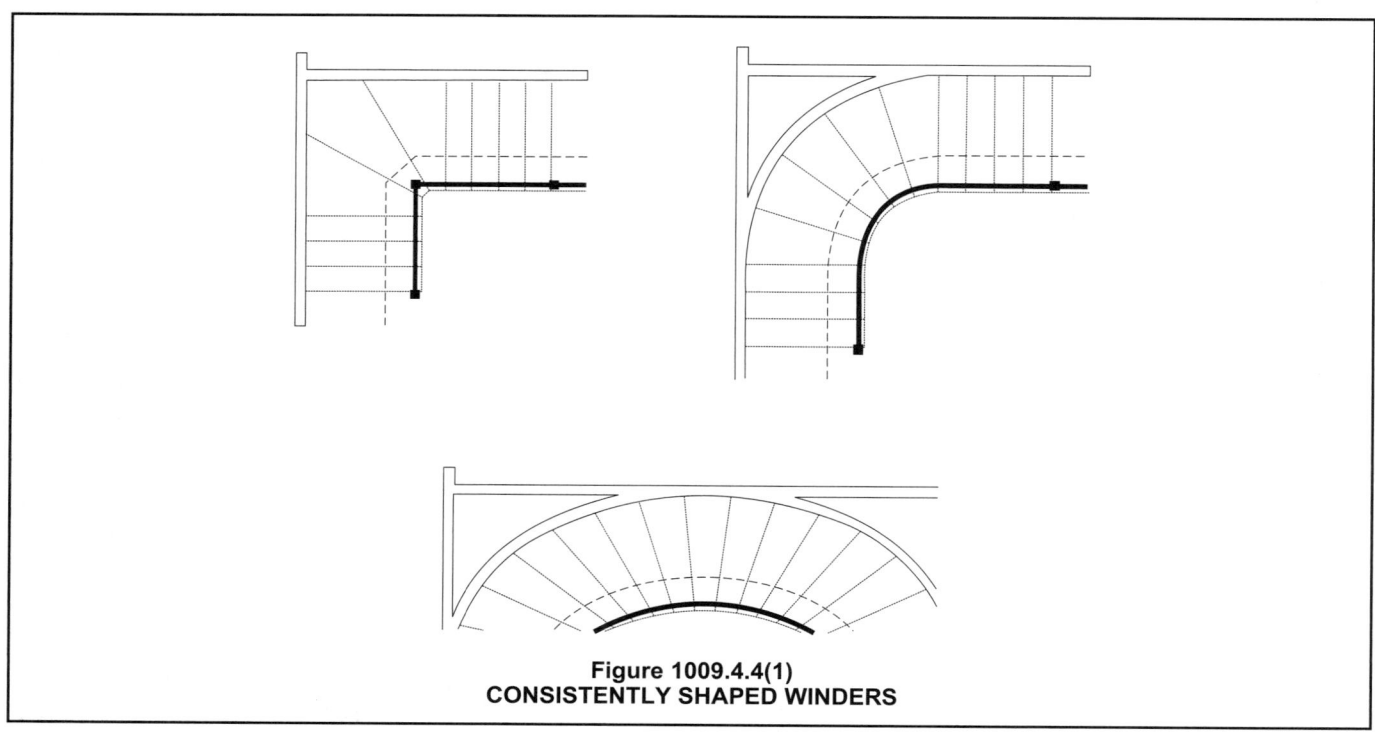

Figure 1009.4.4(1)
CONSISTENTLY SHAPED WINDERS

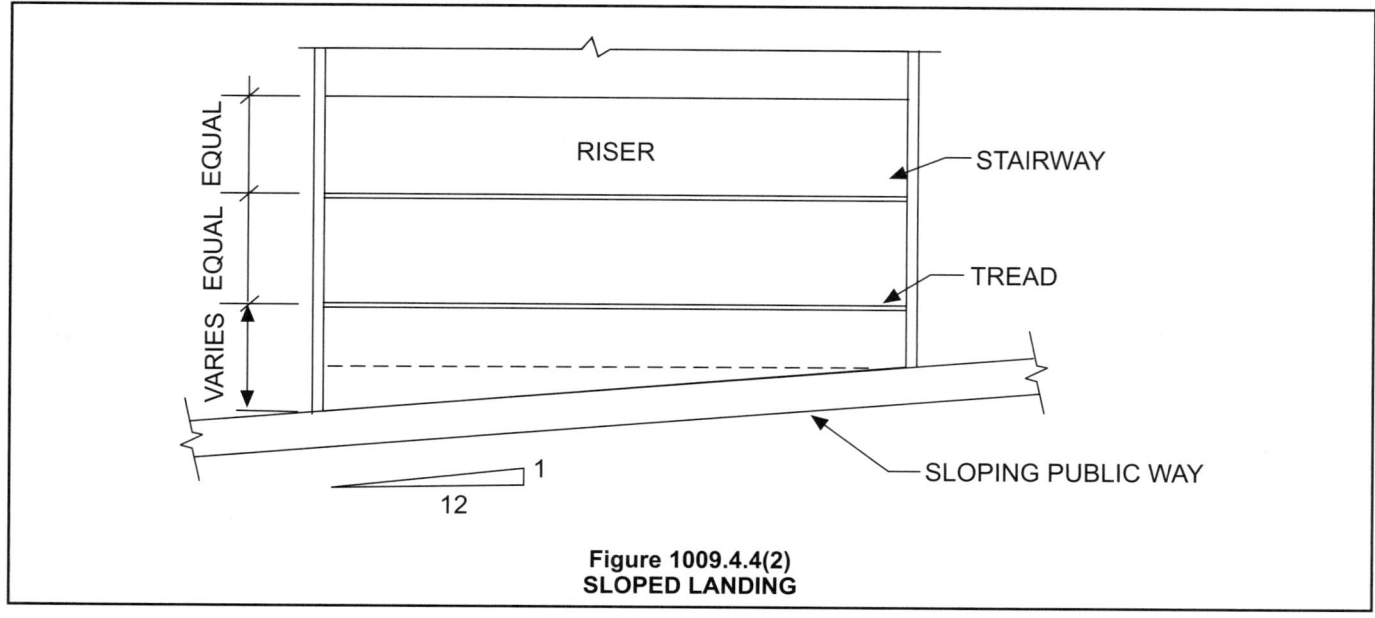

Figure 1009.4.4(2)
SLOPED LANDING

2. Solid risers are not required for occupancies in Group I-3 or in Group F, H and S occupancies other than areas accessible to the public. There are no restrictions on the size of the opening in the riser.

3. Solid risers are not required for *spiral stairways* constructed in accordance with Section 1009.9.

4. Solid risers are not required for *alternating tread devices* constructed in accordance with Section 1009.10.

❖ The profiles of treads and risers contribute to stairway safety. The radius or bevel of the nosing eases the otherwise square edge of the tread and prevents irregular chipping that can become a maintenance issue seriously affecting the safe use of the stair. In addition, it eliminates a sharp square edge that will cause greater injury in falls and allows light modeling reflecting light at various angles, providing a certain contrast from the other surfaces of the stair allowing easier visual location of the start of the tread surface. The maximum radius of curvature at the leading edge of the tread is intended to allow descending foot placement on a surface that does not pitch the foot forward or allow the ball of the foot to slide off the treads and ascending foot placement to slide on to the tread without catching on a square edge. If a stairway design uses a beveled nosing configuration, the bevel is limited to a depth of $^1/_2$ inch (12.7 mm). A nosing projection allows the descending foot to be placed further forward on the tread and the heel to then clear the nosing of the tread above as it swings down in an arc landing on a tread that is effectively deeper than if no nosing projection is used. Nosing projections are so common in stair design that they are noticed by users when absent as affecting their gait. Treads with vertical risers are allowed with or without a nosing projection. A nosing projection may also be accommodated by slanting the riser under the tread above. The nosing projection is limited to 1$^1/_4$ inch (32 mm) maximum. Treads designed with rounding or bevel on the underside would reduce the chance that a user's foot might catch while ascending the stairway [see Figure 1009.4.5(1)].

The code does not address when a riser could contain openings and still be considered "solid." Exception 1 allows the use of open risers on all stairs that are not part of an accessible means of egress. Where the riser is allowed to be open, the opening is limited to be consistent with the requirements for guards [see Figure 1009.4.5(2)]. The maximum radius for the leading edge, however, is still required. The code does not reference ICC A117.1 for stairways, because stairways are not part of an accessible route; however, the code and standard provide opening limitations in tread surfaces. Section 1009.6.1 does allow for treads to have a maximum opening that allows for a 1$^1/_4$-inch (32 mm) sphere.

Exceptions 2, 3 and 4 recognize that open risers are commonly used for stairs in occupancies such as detention facilities, storage, industrial and high-hazard areas for practical reasons. In detention facilities, open risers provide a greater degree of security and supervision because of the fact that people cannot effectively conceal themselves behind the stair. Factories, high-hazard buildings and storage facilities have areas where workers may need the open risers to decrease the chance of spillage, water or snow accumulating on the stairs. Open risers are necessary for adequate foot placement in spiral stairways and alternating tread devices. The 4-inch (102 mm) opening limitations of Exception 1 are not applicable to these stairs.

1009.5 Stairway landings. There shall be a floor or landing at the top and bottom of each *stairway*. The width of landings shall not be less than the width of *stairways* they serve. Every

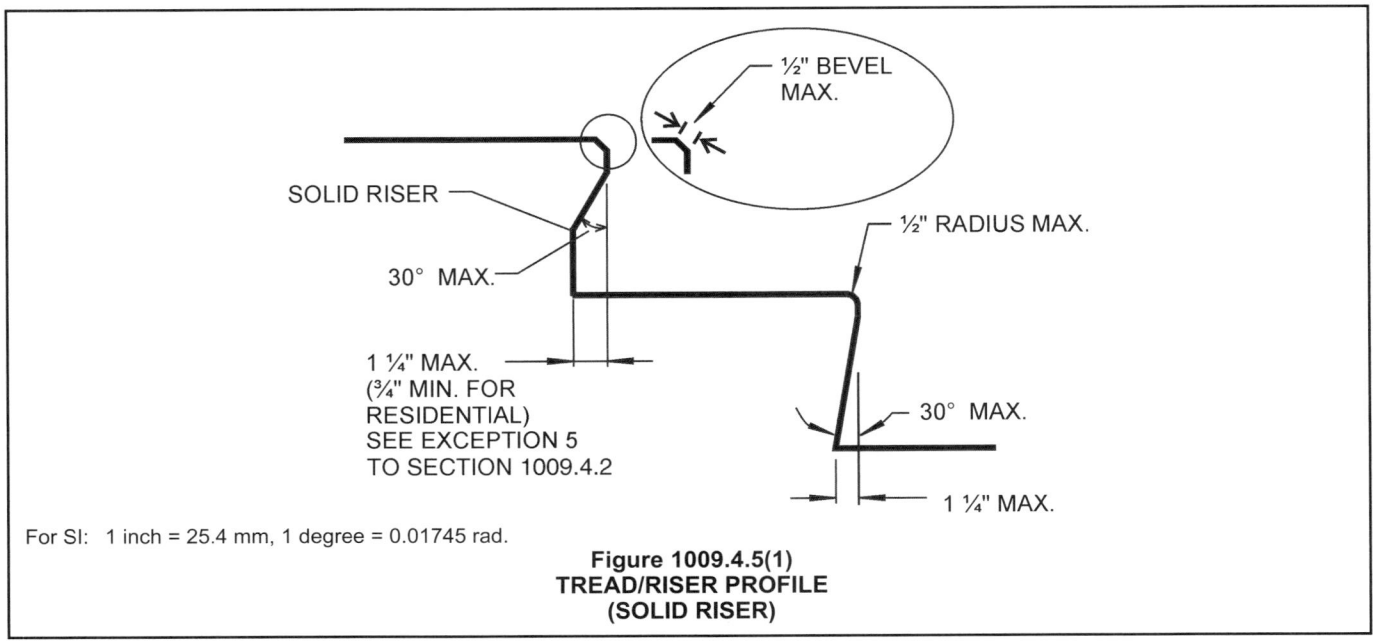

For SI: 1 inch = 25.4 mm, 1 degree = 0.01745 rad.

Figure 1009.4.5(1)
TREAD/RISER PROFILE
(SOLID RISER)

landing shall have a minimum dimension measured in the direction of travel equal to the width of the *stairway*. Such dimension need not exceed 48 inches (1219 mm) where the *stairway* has a straight run. Doors opening onto a landing shall not reduce the landing to less than one-half the required width. When fully open, the door shall not project more than 7 inches (178 mm) into a landing. When *wheelchair spaces* are required on the *stairway* landing in accordance with Section 1007.6.1, the *wheelchair space* shall not be located in the required width of the landing and doors shall not swing over the *wheelchair spaces*.

Exception: *Aisle stairs* complying with Section 1028.

❖ A level portion of a stairway provides users with a place to rest in their ascent or descent, to enter a stairway and to adjust their gait before continuing. Land-

ings also break up the run of a stairway especially at a turn to aid in the arrest of falls that may occur (see Section 1009.7).

The minimum size (width and depth) of all landings in a stairway is determined by the actual width of the stairway. If Section 1009.1 requires a stairway to have a width of at least 44 inches (1118 mm) and the stairway is constructed with that minimum width, then all landings serving that stairway must be at least 44 inches (1118 mm) wide and 44 inches (1118 mm) deep [see Figure 1009.5(1)]. If a stairway is constructed wider than required, landings must increase accordingly so as to not create a bottleneck situation in the egress travel. However, when a stairway is configured so that it has a straight run, the minimum dimension of the landing between flights in the direction of egress

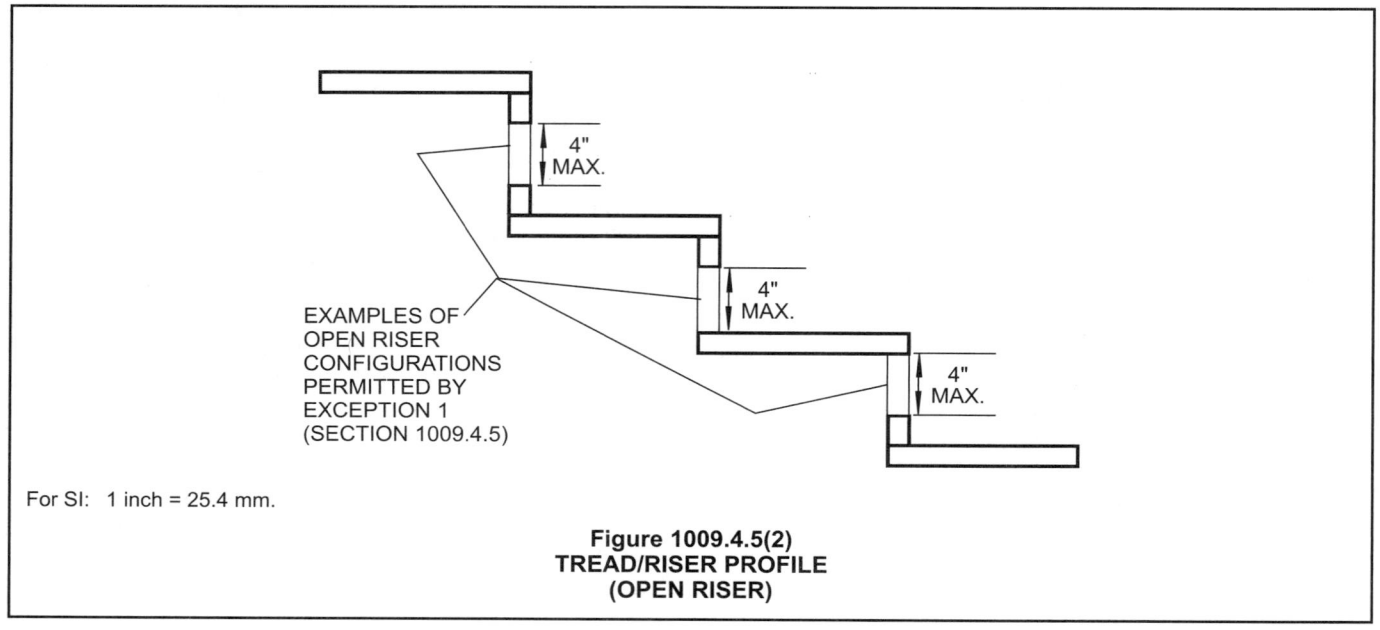

EXAMPLES OF OPEN RISER CONFIGURATIONS PERMITTED BY EXCEPTION 1 (SECTION 1009.4.5)

4" MAX.
4" MAX.
4" MAX.

For SI: 1 inch = 25.4 mm.

Figure 1009.4.5(2)
TREAD/RISER PROFILE
(OPEN RISER)

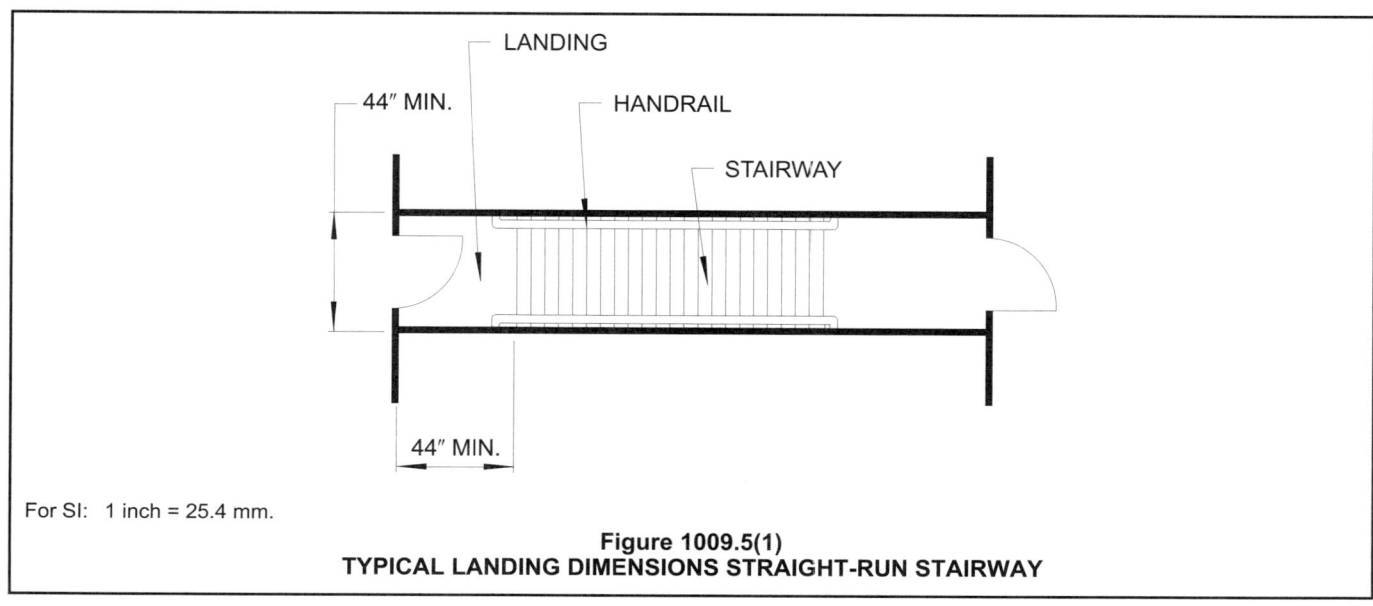

LANDING
44″ MIN.
HANDRAIL
STAIRWAY
44″ MIN.

For SI: 1 inch = 25.4 mm.

Figure 1009.5(1)
TYPICAL LANDING DIMENSIONS STRAIGHT-RUN STAIRWAY

travel is not required to exceed 48 inches (1219 mm), even though the actual width of the stair may exceed 48 inches (1219 mm) [see Figure 1009.5(2)].

It is not the intent of this section to require that a stairway landing be shaped as a square or rectangle. A landing turning the stairway 90 degrees (1.57 rad) or more with a curved or segmented outside periphery would be permitted, as long as the landing provides an area described by an arc with a radius equal to the actual stairway width [see Figure 1009.5(3)]. In this case, the space necessary for means of egress will be available.

The last portion of the requirement limits the extent to which doors that swing on to landings may interfere or encroach upon the required landing space. This limits the arc of the door swing on a landing, so that the effect on the means of egress is minimized [see Figure 1009.5(4)]. This is consistent with a door opening into an exit access corridor in Section 1005.2. For safety reasons and to ensure the means of egress is continually available for everyone, where an area of refuge/wheelchair space must be located on a landing, the wheelchair spaces must not be within the required landing width and the entrance door to the stair enclosure may not swing over the wheelchair spaces [see Figure 1007.3(1)].

Exception 1 provides for aisle stairs where the requirements are in Section 1028.

1009.6 Stairway construction. All *stairways* shall be built of materials consistent with the types permitted for the type of construction of the building, except that wood handrails shall be permitted for all types of construction.

❖ In keeping with the different levels of fire protection provided by each of the five basic types of construction designated in Chapter 6, the materials used for stairway construction must meet the appropriate combustibility/noncombustibility requirements indicated in Section 602 for the particular type of construction of the

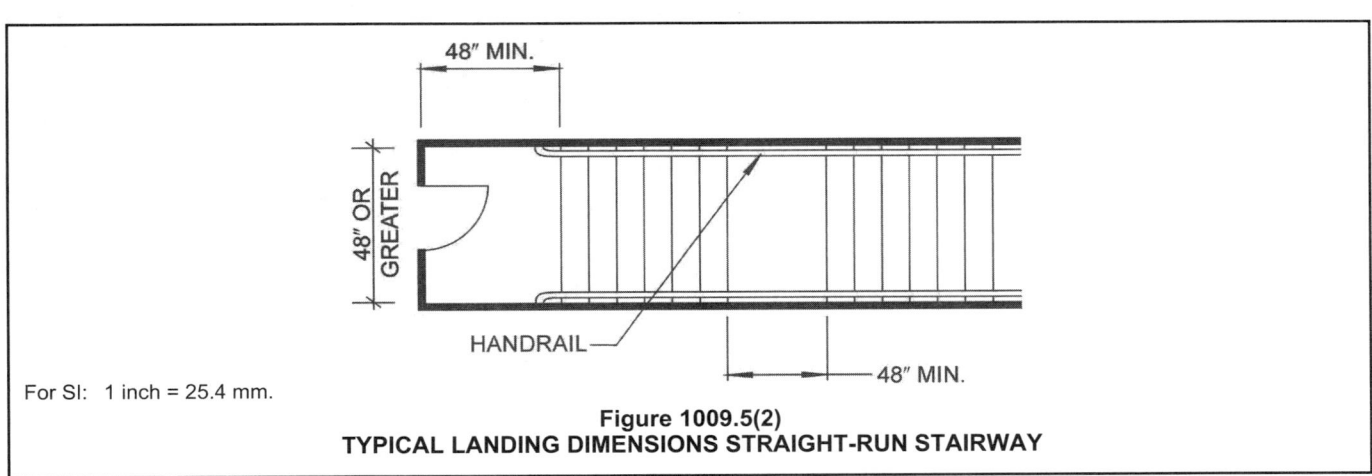

For SI: 1 inch = 25.4 mm.

Figure 1009.5(2)
TYPICAL LANDING DIMENSIONS STRAIGHT-RUN STAIRWAY

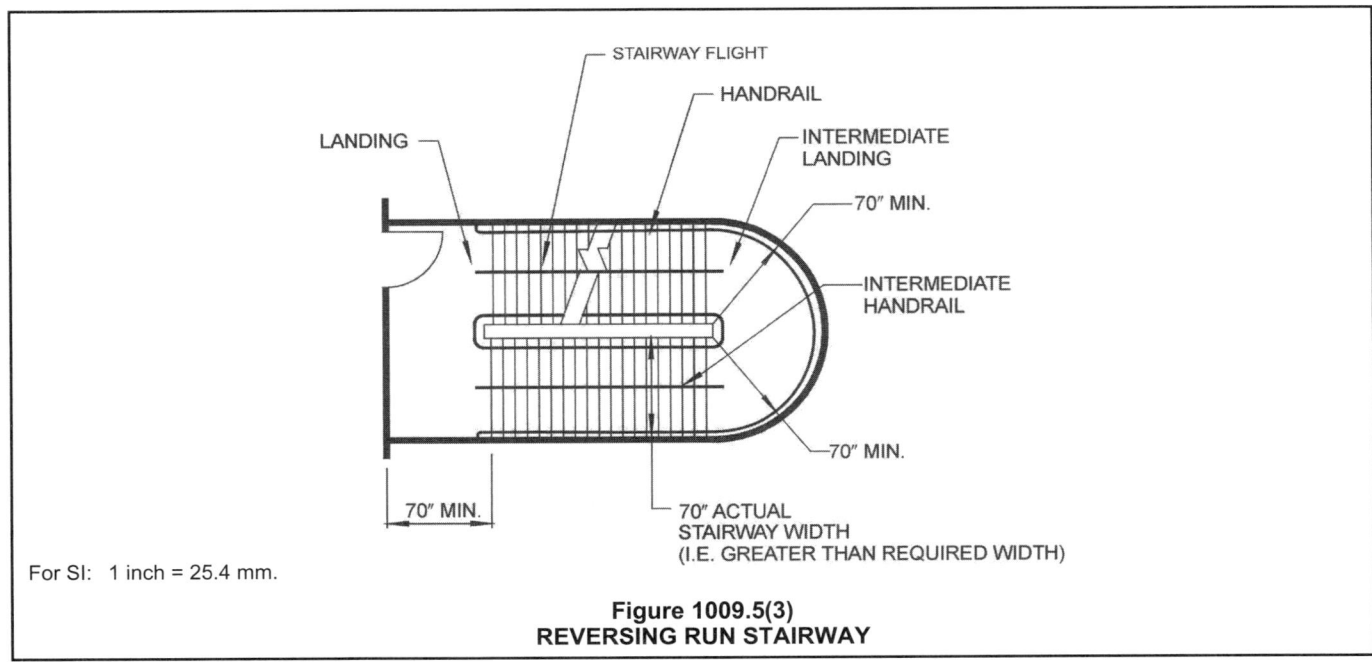

For SI: 1 inch = 25.4 mm.

Figure 1009.5(3)
REVERSING RUN STAIRWAY

building in which the stairway is located. This is required whether or not the stair is part of the required means of egress. Any structure supporting the stairway and the stairway enclosure must be fire-resistance rated consistent with the construction type; however, the stairway components inside the enclosure need only comply with the material limits for the type of construction.

If desired, wood handrails may be used on the basis that the fuel load contributed by this combustible component of stairway construction is insignificant and will not pose a fire hazard.

1009.6.1 Stairway walking surface. The walking surface of treads and landings of a *stairway* shall not be sloped steeper than one unit vertical in 48 units horizontal (2-percent slope) in any direction. *Stairway* treads and landings shall have a solid surface. Finish floor surfaces shall be securely attached.

Exceptions:

1. Openings in stair walking surfaces shall be a size that does not permit the passage of $^1/_2$-inch-diameter (12.7 mm) sphere. Elongated openings shall be placed so that the long dimension is perpendicular to the direction of travel.

2. In Group F, H and S occupancies, other than areas of parking structures accessible to the public, openings in treads and landings shall not be prohibited provided a sphere with a diameter of $1^1/_8$ inches (29 mm) cannot pass through the opening.

❖ It is the intent of this section that both landing and stair treads be solid and level with firmly attached surface materials; however, the 1:48 slope should be adequate to allow for drainage to limit the chance for an

accumulation of water where someone might slip.

The exceptions permit the use of open grate-type material or slotted grill for stairway treads and landings in two different situations.

Exception 1 allows for a maximum $^1/_2$-inch (12.7 mm) opening on stairway treads in public areas and serving any use (see Figure 1009.6.1). This is very beneficial on exterior stairways where snow, ice or water may accumulate. The $^1/_2$-inch (12.7 mm) limitation is consistent with ICC A117.1 and federal accessibility requirements. The opening limitation is small enough that most shoe heels will not get stuck. If a slotted grill pattern is used, the slots must run side to side on the

Figure 1009.6.1
OPEN TREAD IN ACCORDANCE WITH EXCEPTION 1

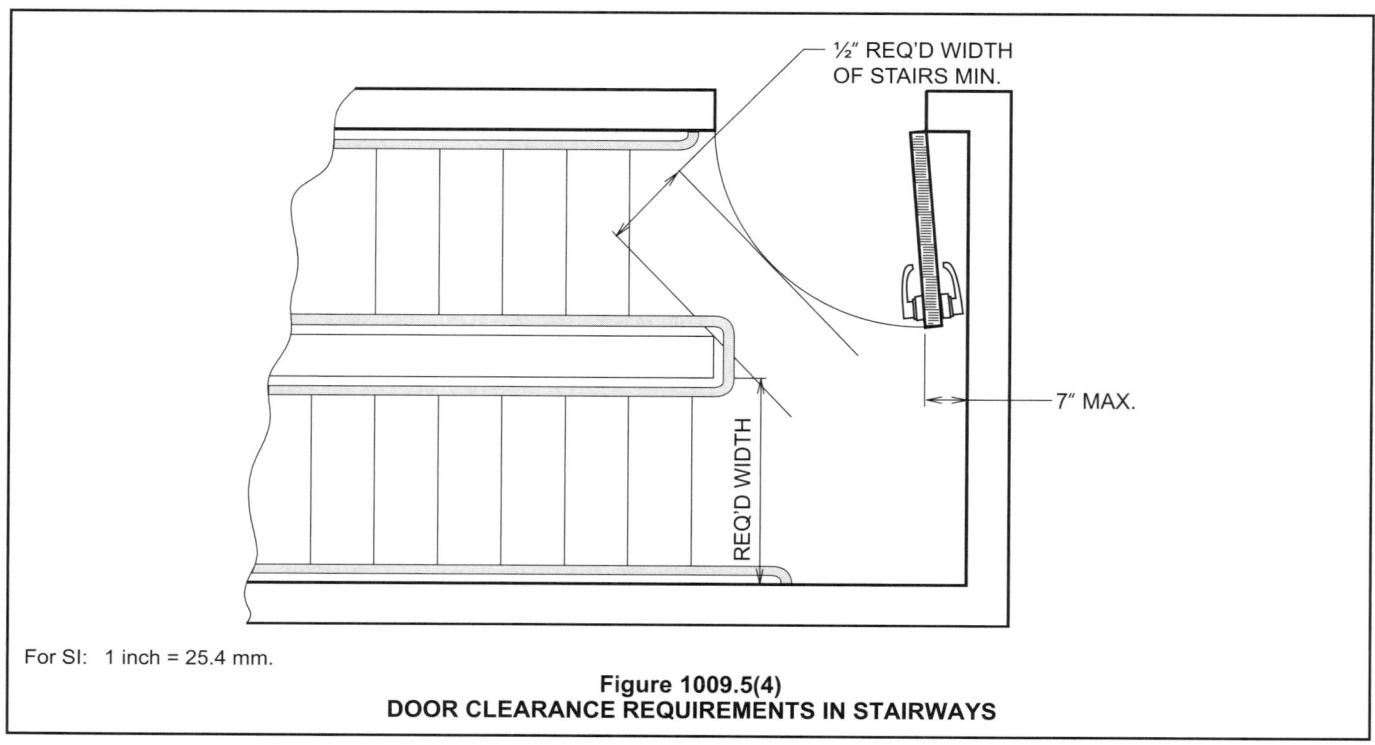

For SI: 1 inch = 25.4 mm.

Figure 1009.5(4)
DOOR CLEARANCE REQUIREMENTS IN STAIRWAYS

stairway tread, not front to back.

Exception 2 is applicable in factory, industrial, storage and high-hazard occupancies. This provision is intended to apply primarily to stairs that provide access to areas not required to be accessible, such as pits, catwalks, tanks, equipment platforms, roofs or mezzanines. Walking surfaces with limited-size openings are typically used because open grate-type material is less susceptible to accumulation of dirt, debris or moisture, as well as being more resistant to corrosion. Most commercially available grate material is manufactured with a maximum nominal 1-inch (25 mm) opening; therefore, the limitation that the openings not allow the passage of a sphere of $1^{1}/_{8}$ inch (29 mm) diameter allows the use of most material as well as accounting for manufacturing tolerances.

1009.6.2 Outdoor conditions. Outdoor *stairways* and outdoor approaches to *stairways* shall be designed so that water will not accumulate on walking surfaces.

❖ Outdoor stairways and approaches to stairways are to be constructed with a slope that complies with Section 1009.6.1 or are required to be protected such that walking surfaces do not accumulate water. While not specifically stated, any interior locations, such as near a pool, should also have the stair designed to limit the accumulation of water in order to maintain slip resistance (see Section 1003.4).

Where exterior stairways are used in moderate or severe climates, there may also be a concern to protect the stairway from accumulations of snow and ice to provide a safe path of egress travel at all times, including winter. Maintenance of the means of egress in the IFC requires an unobstructed path to allow for full instant use in case of a fire or emergency (see IFC Section 1030.2). Typical methods for protecting these egress elements include roof overhangs or canopies; heated slabs; grated treads and landings as permitted in Section 1005.1; or when approved by the building official, a reliable snow removal maintenance program.

1009.6.3 Enclosures under stairways. The walls and soffits within enclosed usable spaces under enclosed and unenclosed *stairways* shall be protected by 1-hour fire-resistance-rated construction or the *fire-resistance rating* of the stairway enclosure, whichever is greater. Access to the enclosed space shall not be directly from within the stair enclosure.

Exception: Spaces under *stairways* serving and contained within a single residential dwelling unit in Group R-2 or R-3 shall be permitted to be protected on the enclosed side with $^{1}/_{2}$-inch (12.7 mm) gypsum board.

There shall be no enclosed usable space under *exterior exit stairways* unless the space is completely enclosed in 1-hour fire-resistance-rated construction. The open space under *exterior stairways* shall not be used for any purpose.

❖ This section addresses the fire hazard of storage under a stairway, both inside and outside a structure. The stairway must be protected from a storage area under it, even if the stair is not required to be enclosed. The

section also requires that the storage area not open into a stair enclosure. This limits the potential of a fire that starts in the storage area from affecting the stair enclosure. The exception provides specific criteria for separation for storage areas under an interior stairway for the indicated residential occupancies.

1009.7 Vertical rise. A *flight* of *stairs* shall not have a vertical rise greater than 12 feet (3658 mm) between floor levels or landings.

Exceptions:

1. *Aisle stairs* complying with Section 1028.

2. *Alternating tread devices* used as a *means of egress* shall not have a rise greater than 20 feet (6096 mm) between floor levels or landings.

❖ Between landings and platforms, the vertical rise is to be measured from one landing walking surface to another (see Figure 1009.7). The limited height provides a reasonable interval for users with physical limitations to rest on a level surface and also serves to alleviate potential negative psychological effects of long and uninterrupted stairway flights.

Exception 1 provides for aisle stairs in assembly occupancies that are regulated by Section 1028 and not by this section.

Exception 2 allows for 20 feet (6096 mm) between landings for alternating tread devices given their limited application and low occupant loads. In addition, Exception 2 recognizes that additional vertical rise is needed for steeper devices used where space is often too restrictive for a landing.

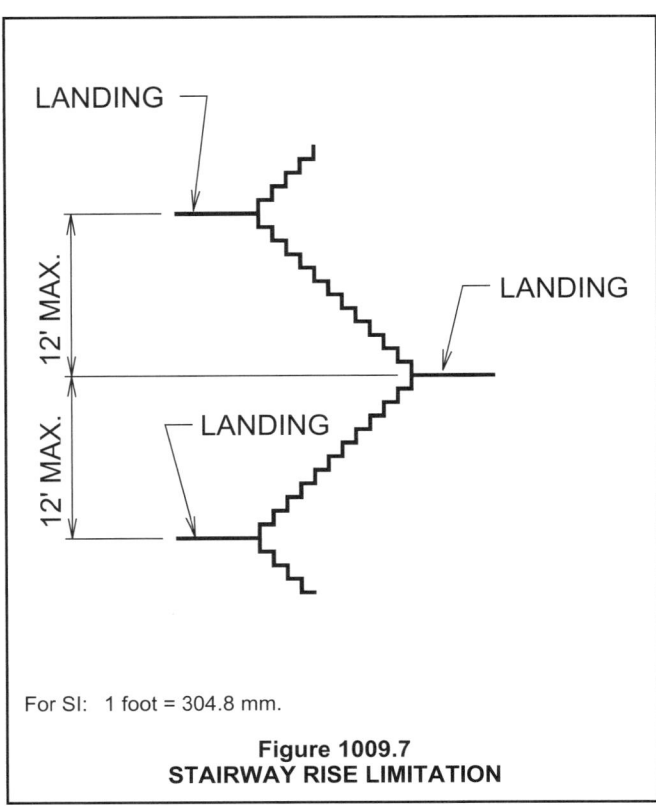

For SI: 1 foot = 304.8 mm.

Figure 1009.7
STAIRWAY RISE LIMITATION

1009.8 Curved stairways. Curved *stairways* with *winder* treads shall have treads and risers in accordance with Section 1009.4 and the smallest radius shall not be less than twice the required width of the *stairway*.

> **Exception:** The radius restriction shall not apply to curved *stairways* for occupancies in Group R-3 and within individual dwelling units in occupancies in Group R-2.

❖ Curved stairway construction consists of a series of winder treads that form a stairway configuration. Options are many, including circular, S-shaped, oval, elliptical, hourglass, etc. The commentary to Section 1009.4.3 regarding the possible event of nonconcentric movement on stairways with winders also applies to curved stairways. This type of stairway is allowed to be used as a component of a means of egress when tread and riser dimensions meet the requirements or exceptions of Section 1009.4. This section also requires that the shorter radius must be equal to or greater than twice the required width of the stairway (see Figure 1009.8).

This exception for residential units eliminates the minimum radius requirement where the occupants are familiar with the extent of the turning of the stair through the curve.

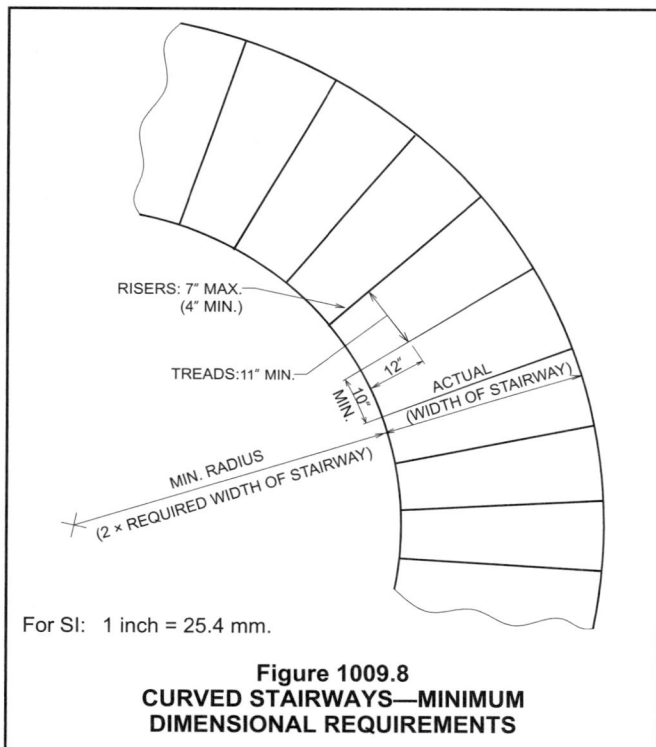

For SI: 1 inch = 25.4 mm.

Figure 1009.8
CURVED STAIRWAYS—MINIMUM
DIMENSIONAL REQUIREMENTS

1009.9 Spiral stairways. *Spiral stairways* are permitted to be used as a component in the *means of egress* only within dwelling units or from a space not more than 250 square feet (23 m²) in area and serving not more than five occupants, or from galleries, catwalks and *gridirons* in accordance with Section 1015.6.

A *spiral stairway* shall have a 7¹/₂-inch (191 mm) minimum clear tread depth at a point 12 inches (305 mm) from the narrow edge. The risers shall be sufficient to provide a headroom of 78 inches (1981 mm) minimum, but riser height shall be not more than 9¹/₂ inches (241 mm). The minimum *stairway* clear width at and below the *handrail* shall be 26 inches (660 mm).

❖ Spiral stairways are generally constructed with a fixed center pole that serves as either the primary or the only means of support from which pie-shaped treads radiate to form a winding stairway.

The commentary to Section 1009.4.3 regarding the possible event of nonconcentric movement on stairways with winders also applies to spiral stairways. The nature of stairway construction is such that it does not serve well when used in emergencies that require immediate evacuation, nor does a spiral stairway configuration permit the handling of a large occupant load in an efficient and safe manner. Furthermore, it is impossible for fire service personnel to use a spiral stairway at the same time and in a direction opposite that being used by occupants to exit the premises, possibly causing a serious delay in fire-fighting operations. Therefore, this section allows only very limited use of spiral stairways when part of a required means of egress. Spiral stairways may be used in any occupancy as long as such stairways are not a component of a required means of egress.

Spiral stairways are required to have dimensional uniformity. The stairway must have a clear width of at least 26 inches (660 mm) at and below the handrail. The depth of the treads must not be less than 7¹/₂ inches (191 mm) measured at a point that is 12 inches (305 mm) out from the narrow edge (see Figure 1009.9). Riser heights are required to be the same

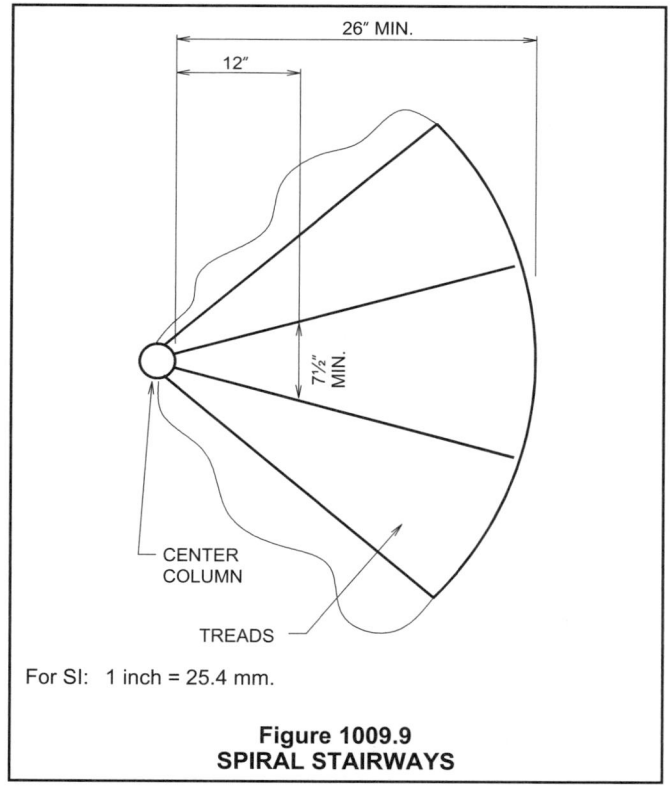

For SI: 1 inch = 25.4 mm.

Figure 1009.9
SPIRAL STAIRWAYS

throughout the stairway, but are not to exceed 9¹/₂ inches (241 mm). Minimum headroom of 6 feet, 6 inches (1981 mm) is required.

1009.10 Alternating tread devices. *Alternating tread devices* are limited to an element of a *means of egress* in buildings of Groups F, H and S from a mezzanine not more than 250 square feet (23 m²) in area and which serves not more than five occupants; in buildings of Group I-3 from a guard tower, observation station or control room not more than 250 square feet (23 m²) in area and for access to unoccupied roofs.

❖ This type of device is constructed in such a way that each tread alternates with each adjacent tread so that the device consists of a system of right-footed and left-footed treads (see Figure 1009.10).

The use of center stringer construction, half-treads and an incline that is considerably steeper than allowed for ordinary stairway construction makes the alternating tread device unique. However, because of its structural feature, only single-file use of the device (between handrails) is possible, thus preventing the occupants from passing one another. The pace of occupant travel is set by the slowest user, a condition that could become critical in an emergency situation. Furthermore, it is impossible for fire service personnel to use an alternating tread device at the same time and in a direction opposite that being used by occupants to exit the premises, possibly causing a serious delay in fire-fighting operations. For these reasons, this section greatly restricts the use of alternating tread devices as a means of egress. Alternating tread devices may be used in any occupancy as long as such stairways are not a component of a required means of egress.

Alternating tread devices are considered a modest improvement to ladder construction and, therefore, can be used as an unoccupied roof access in accordance with the requirements of Section 1009.13.

1009.10.1 Handrails of alternating tread devices. *Handrails* shall be provided on both sides of *alternating tread devices* and shall comply with Section 1012.

❖ For the safety of occupants, this section references the dimensional requirements for handrail locations to be used in conjunction with the special construction features of alternating tread devices provided in Section 1009.10. Because of the steepness of these devices, additional clearances are required so that hand movement will not be encumbered by obstructions.

1009.10.2 Treads of alternating tread devices. *Alternating tread devices* shall have a minimum projected tread of 5 inches (127 mm), a minimum tread depth of 8¹/₂ inches (216 mm), a minimum tread width of 7 inches (178 mm) and a maximum riser height of 9¹/₂ inches (241 mm). The projected tread depth shall be measured horizontally between the vertical planes of the foremost projections of adjacent treads. The riser height shall be measured vertically between the leading edges of adjacent treads. The combination of riser height and projected tread depth provided shall result in an alternating tread device angle that complies with Section 1002. The initial tread of the device

shall begin at the same elevation as the platform, landing or floor surface.

Exception: *Alternating tread devices* used as an element of a *means of egress* in buildings from a mezzanine area not more than 250 square feet (23 m²) in area which serves not more than five occupants shall have a minimum projected tread of 8¹/₂ inches (216 mm) with a minimum tread depth of 10¹/₂ inches (267 mm). The rise to the next alternating tread surface should not be more than 8 inches (203 mm).

❖ Alternating tread stairways (see Section 1009.10) are required to have tread depths of at least 8¹/₂ inches (216 mm) and widths of 7 inches (178 mm) or more. With the center support, the total width will be more than 15 inches (381 mm).

The risers are to be not more than 9¹/₂ inches (241 mm) when measured from tread to alternating tread (next adjacent tread to the left or right). The rise between treads on the same side would be 19 inches (482 mm) maximum. Tread projections are not to be less than 5 inches (127 mm) when measured from tread nosing to the next tread nosing (the tread above or below, see Figure 1009.10). Applying the limiting dimensions stated above would result in a device with a very steep incline (approximately 4:1). Just using the dimensions could result in an alternating tread device with an angle of over 75 degrees (1.3 rad). In any case, the overall angle of the device must be between 50 and 70 degrees (0.87 and 1.22 rad) as indicated in the definition for "Alternating tread devices" in Section 1002.1.

For alternating tread devices used as a means of egress from small-area mezzanines as prescribed in the exception, the treads must project at least 8¹/₂ inches (216 mm) as compared to the 5 inches (127 mm) stated above; treads are to be at least 10¹/₂ inches (267 mm) in depth [compared to 8¹/₂ inches (216 mm)] and risers are not to exceed 8 inches (203 mm) in height [compared to 9¹/₂ inches (341 mm)]. Applying these latter limiting dimensions would result in a device of lesser incline [approximate slope of 2:1 or 62 degrees (1.08 rad)] and a more comfortable and safer device to use for egress travel.

1009.11 Ship ladders. Ship ladders are permitted to be used in Group I-3 as a component of a *means of egress* to and from control rooms or elevated facility observation stations not more than 250 square feet (23 m²) with not more than three occupants and for access to unoccupied roofs.

Ship ladders shall have a minimum tread depth of 5 inches (127 mm). The tread shall be projected such that the total of the tread depth plus the *nosing* projection is no less than 8¹/₂ inches (216 mm). The maximum riser height shall be 9¹/₂ inches (241 mm).

Handrails shall be provided on both sides of ship ladders. The minimum clear width at and below the *handrails* shall be 20 inches (508 mm).

❖ Ship ladders look similar to alternating tread devices; however, the treads form more of a ladder shape rather than being staggered to either side of a center support (see Figure 1009.11). Ship ladders can be used in correctional facilities for access to small control rooms, observation stations and unoccupied roofs.

Where approved by the code official, ship ladders could be used for access to unoccupied roofs in other occupancies (see Sections 1009.13 and 1009.13.1).

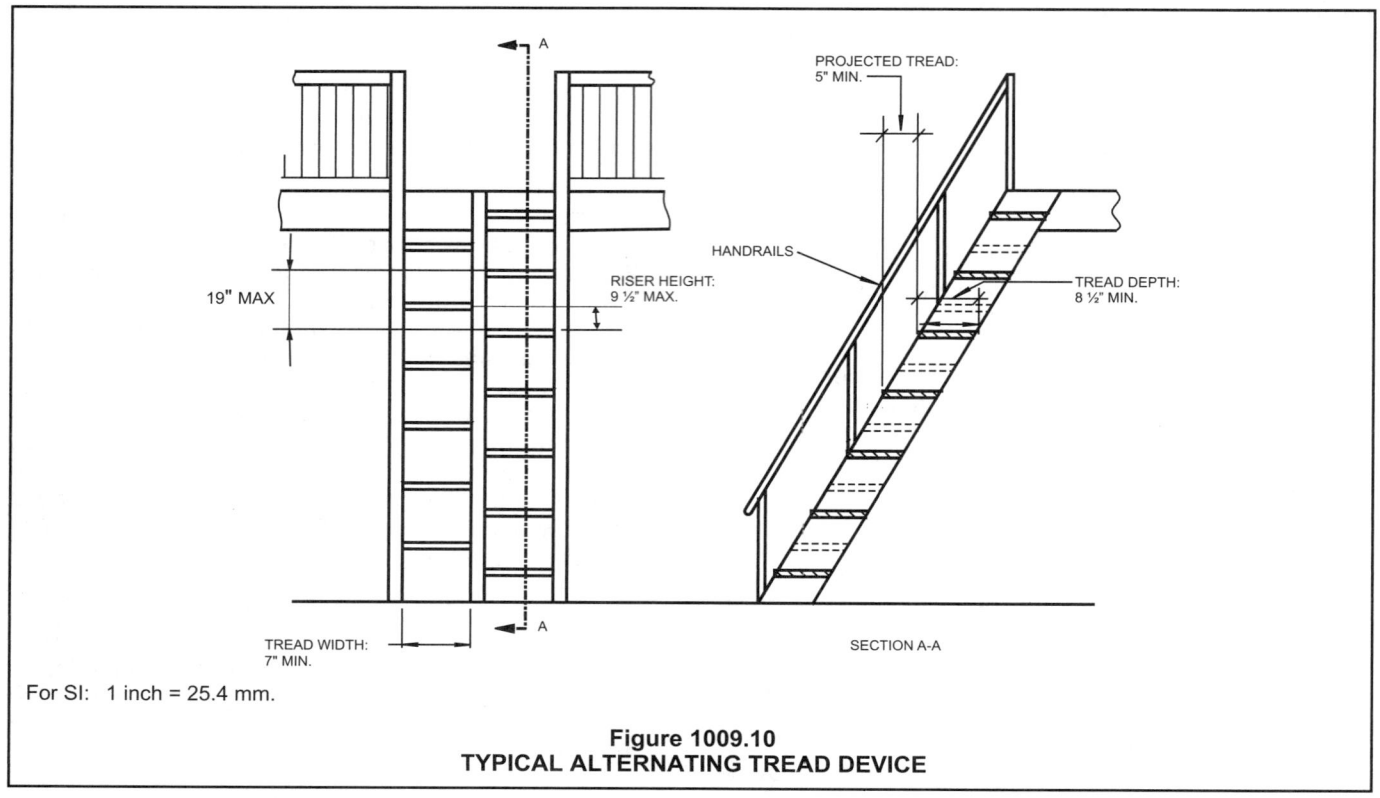

For SI: 1 inch = 25.4 mm.

Figure 1009.10
TYPICAL ALTERNATING TREAD DEVICE

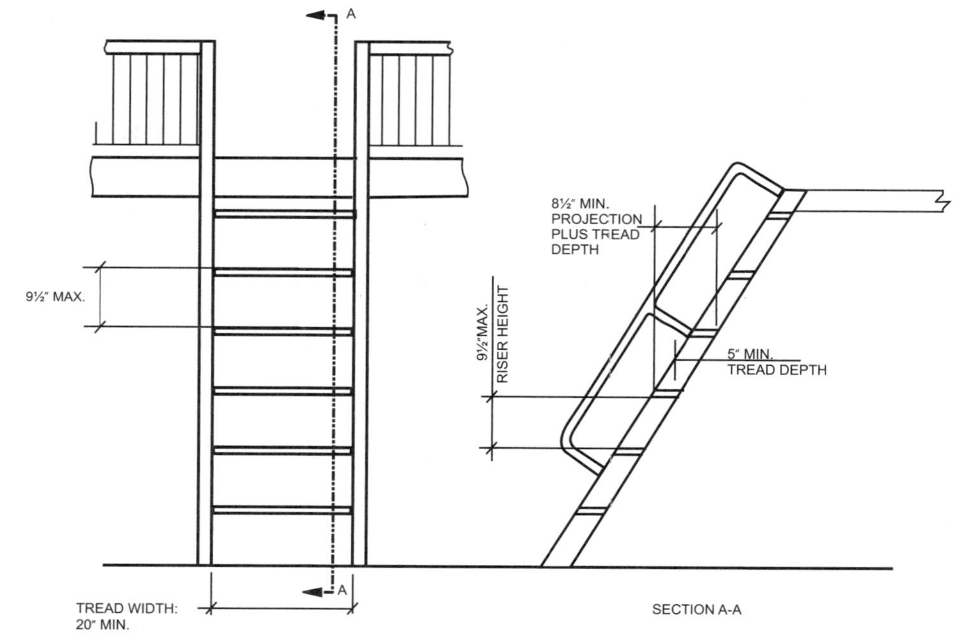

For SI: 1 inch = 25.4 mm.

Figure 1009.11
TYPICAL SHIP LADDER

1009.12 Handrails. *Stairways* shall have *handrails* on each side and shall comply with Section 1012. Where glass is used to provide the *handrail*, the *handrail* shall also comply with Section 2407.

Exceptions:

1. *Handrails* for *aisle stairs* are not required where permitted by Section 1028.13.

2. *Stairways* within dwelling units, *spiral stairways* and *aisle stairs* serving seating only on one side are permitted to have a *handrail* on one side only.

3. Decks, patios and walkways that have a single change in elevation where the landing depth on each side of the change of elevation is greater than what is required for a landing do not require *handrails*.

4. In Group R-3 occupancies, a change in elevation consisting of a single riser at an entrance or egress door does not require *handrails*.

5. Changes in room elevations of three or fewer risers within dwelling units and sleeping units in Group R-2 and R-3 do not require *handrails*.

❖ Handrails have four recognized functions in stairway use. First, they serve to guide persons in ascent and descent, especially important for those with low vision, and in cases of fire where vision might be obscured by smoke, handrails serve as guides directing the user along the path of egress travel. Second, they provide a tool for the user to exert stabilizing forces longitudinally (along the length of the rail), vertically and, most importantly, transversely (perpendicular) to the rail as the body transfers weight from side to side with each leg swing of the unique gait used on stairs. Third, they provide for pulling when arms are used to augment legs in ascent of steeper angles or when such climbing strategies result in more efficient use of the strengths of the user. Fourth, they are a tool that can be utilized to help in the arrest of a fall. In these capacities handrails serve to aid in the use of the stairway and are required on both sides of stairways in compliance with Section 1012 to allow passing users unencumbered access to a handrail. Finally, when glass is the material used to provide the handrail, it must comply with Section 2407.

Note that if the handrail extension is at a location that could be considered a protruding object, the handrail must return to the post at a height of less than 27 inches (686 mm) above the floor. Handrails along the stair runs are not considered protruding objects.

The exceptions state conditions where handrails are only required on one side or are not needed at all. Stepped aisles with assembly seating are addressed specifically in Section 1028 (see Exceptions 1 and 2). By nature or their construction, spiral stairways can only have a single handrail (see Exception 2). In accordance with Exceptions 2 and 5, within dwelling units, all stairways can have one handrail, and stairs

with three or fewer rises are not required to have any handrails. Since "Stair" is defined as one or more risers, Exceptions 3 and 4 are necessary. Exception 4 exempts the single step at the front or back door of a Group R-3 dwelling unit (i.e., townhouse). Decks, patios and walkways often move down with the grade. When there are single steps, either off the patio or deck to grade, or along the surface, a handrail is not required (Exception 3, see Figure 1009.12). Many of these exceptions dealing with residential units are consistent with the IRC.

Figure 1009.12
EXAMPLE OF SECTION 1009.12, EXCEPTION 3

1009.13 Stairway to roof. In buildings four or more stories above *grade plane*, one *stairway* shall extend to the roof surface, unless the roof has a slope steeper than four units vertical in 12 units horizontal (33-percent slope). In buildings without an occupied roof, access to the roof from the top story shall be permitted to be by an *alternating tread device*.

❖ Because of safety considerations, roofs used for habitable purposes such as roof gardens, observation decks, sporting facilities (including jogging or walking tracks and tennis courts) or other similar uses, must be provided with conventional stairways that will serve as required means of egress. Access by ladders or an alternating tread device for such uses is not permitted.

In buildings four or more stories high, roofs that are not used for habitable purposes must be provided with ready access by conventional stairways or by an alternating tread device (see Section 1009.10). If this stair is also to provide access to an elevator penthouse on the roof, see additional requirements in Section 1009.14. Two reasons for this are access for roof or rooftop equipment repair and fire department access during a fire event. Sloping roofs with a rise greater than 4 inches (102 mm) for every 12 inches (305 mm) in horizontal measurement (4:12) are exempt from the requirements of this section because of the steepness of the construction and the inherent dangers to life safety.

While it is not specifically required that roof access be through an exit stairway enclosure, since part of the intent is for fire department access to the roof, it is advisable. Section 1022.8 requires signage at the level of exit discharge indicating whether the stairway has roof access.

1009.13.1 Roof access. Where a *stairway* is provided to a roof, access to the roof shall be provided through a *penthouse* complying with Section 1509.2.

> **Exception:** In buildings without an occupied roof, access to the roof shall be permitted to be a roof hatch or trap door not less than 16 square feet (1.5 m²) in area and having a minimum dimension of 2 feet (610 mm).

❖ The purpose of the penthouse or stairway bulkhead requirement in this section is to protect the walking surface of the stairway to the roof. The exception provides for situations when roof access is only needed for service or maintenance purposes, and where the access may be permitted by alternatives such as alternating tread devices, ship ladders or ladders.

1009.13.2 Protection at roof hatch openings. Where the roof hatch opening providing the required access is located within 10 feet (3049 mm) of the roof edge, such roof access or roof edge shall be protected by *guards* installed in accordance with the provisions of Section 1013.

❖ While guards are required at the edge of a normally occupied roof by Section 1013, there is also a safety concern for roof areas that need to be accessed by service personnel, inspectors and emergency responders. This requirement for guards provides a minimum measure of safety when the roof access is close to the room edge. This is consistent with the requirements at mechanical equipment in Sections 1013.5 and 1013.6.

1009.14 Stairway to elevator equipment. Roofs and *penthouses* containing elevator equipment that must be accessed for maintenance are required to be accessed by a *stairway*.

❖ The requirement for a stair to the roof for maintaining elevator equipment correlates the code with ASME A17.1/CSA B44, *Safety Code for Elevators and Escalators*. This referenced standard (see Section 3001.2) has required stairs and a door to access elevator equipment since 1955. More specifically, Section 2.27.3.2.1 of A17.1 states the following: "a stairway with a swinging door and platform at the top level, conforming to 2.7.3.3 shall be provided from the top floor of the building to the roof level. Hatch covers as a means of access to the roofs shall not be permitted." Alternating tread devices or ladders are not permitted as an alternative to the stairway for access to the elevator penthouse. This provision is more specific; therefore, while not prohibiting using the same stairway for access to the roof and the elevator penthouse

(see Sections 1009.13 and 1009.13.1), access to that elevator penthouse must be via a stairway with door access, not an alternating tread device and hatch.

SECTION 1010
RAMPS

1010.1 Scope. The provisions of this section shall apply to *ramps* used as a component of a *means of egress*.

Exceptions:

1. Other than *ramps* that are part of the *accessible routes* providing access in accordance with Sections 1108.2 through 1108.2.4 and 1108.2.6, ramped *aisles* within assembly rooms or spaces shall conform with the provisions in Section 1028.11.

2. Curb *ramps* shall comply with ICC A117.1.

3. Vehicle ramps in parking garages for pedestrian *exit access* shall not be required to comply with Sections 1010.3 through 1010.9 when they are not an *accessible route* serving *accessible* parking spaces, other required accessible elements or part of an *accessible means of egress*.

❖ Ramps provide an alternative method of vertical means of access to or egress from a building. Ramps are required for access to building areas for mobility-impaired persons (see Chapter 11) and for small changes in floor elevations that are a safety hazard in themselves (see Section 1003.5). All ramps intended for pedestrian usage, whether required or otherwise provided, must comply with the requirements of this section. The code considers any walking surface that has a slope steeper than one unit vertical in 20 units horizontal (5-percent slope) to be a ramp (see the definition for "Ramp" in Section 1002.1).

Exception 1 clarifies that in assembly rooms with sloped floors, ramps other than the routes utilized to and from accessible wheelchair seating locations (see Section 1108.2) are permitted to be designed in accordance with the sloped aisle provisions for assembly with fixed seating in Section 1028.

Exception 2 references specific curb cut requirements found in Section 406 of ICC A117.1.

Exception 3 addresses parking garages. An accessible route is required to and from any accessible parking space, and all ramp provisions must be followed. However, ramps that provide access to and from nonaccessible spaces in the remainder of the parking garage need only comply with the provisions for slope and guard requirements. This permits nonaccessible portions of garages to be constructed as a continuous slope. Ramps that are strictly for vehicles, such as jump ramps, are not required to meet any of the ramp provisions.

1010.2 Slope. *Ramps* used as part of a *means of egress* shall have a running slope not steeper than one unit vertical in 12 units horizontal (8-percent slope). The slope of other pedestrian *ramps* shall not be steeper than one unit vertical in eight units horizontal (12.5-percent slope).

> **Exception:** *Aisle ramp* slope in occupancies of Group A or assembly occupancies accessory to Group E occupancies shall comply with Section 1028.11.

❖ Maximum slope is limited to facilitate the ease of ascent and to control the descent of persons with or without a mobility impairment. The maximum slope of a ramp in the direction of travel is limited to one unit vertical in 12 units horizontal (1:12) (see Figure 1010.2). Ramps in existing buildings may be permitted to have a steeper slope at small changes in elevation (see Section 3409.8.5 or IEBC Sections 308.8.5 and 605.1.4). An example of a ramp that is not part of a means of egress and, therefore, allowed to be a maximum slope of 1:8, is an industrial access ramp that provides access to a raised floor level around a piece of equipment.

The exception is a reference to the assembly seating requirements commonly found in sports facilities, theaters and lecture halls. However, the ramps that are part of an accessible route for ingress or egress are still required to meet the 1:12 maximum slope. Only the ramps used elsewhere in the assembly space may utilize the 1:8 maximum slope permitted in Section 1028.11.

1010.3 Cross slope. The slope measured perpendicular to the direction of travel of a *ramp* shall not be steeper than one unit vertical in 48 units horizontal (2-percent slope).

❖ The limitation of one unit vertical in 48 units horizontal on the slope across the direction of travel is to prevent a severe cross slope that would pitch a user to one side (see Figure 1010.2).

1010.4 Vertical rise. The rise for any *ramp* run shall be 30 inches (762 mm) maximum.

❖ Because pushing a wheelchair up a ramp requires a great deal of energy, landings must be situated so that a person can rest after each 30-inch (762 mm) elevation change (see Figure 1010.2).

1010.5 Minimum dimensions. The minimum dimensions of *means of egress ramps* shall comply with Sections 1010.5.1 through 1010.5.3.

❖ These minimum dimension requirements allow the ramp to function as a means of egress and an accessible route.

1010.5.1 Width. The minimum width of a *means of egress ramp* shall not be less than that required for *corridors* by Section 1018.2. The clear width of a *ramp* between *handrails*, if provided, or other permissible projections shall be 36 inches (914 mm) minimum.

❖ The requirements for the width of a means of egress ramp are 36 inches (914 mm) minimum, similar to that established by Section 1018.2 for corridors. Note that the clear width of 36 inches (914 mm) is required between the handrails and any other obstructions (i.e., handrail supports, curbs) for proper clearance for a person in a wheelchair. This is different from stairways where handrails are permitted to project into the required width. The 36-inch (914 mm) minimum clear width between handrails is consistent with ICC A117.1 and the new ADA/ABA Accessibility Guidelines.

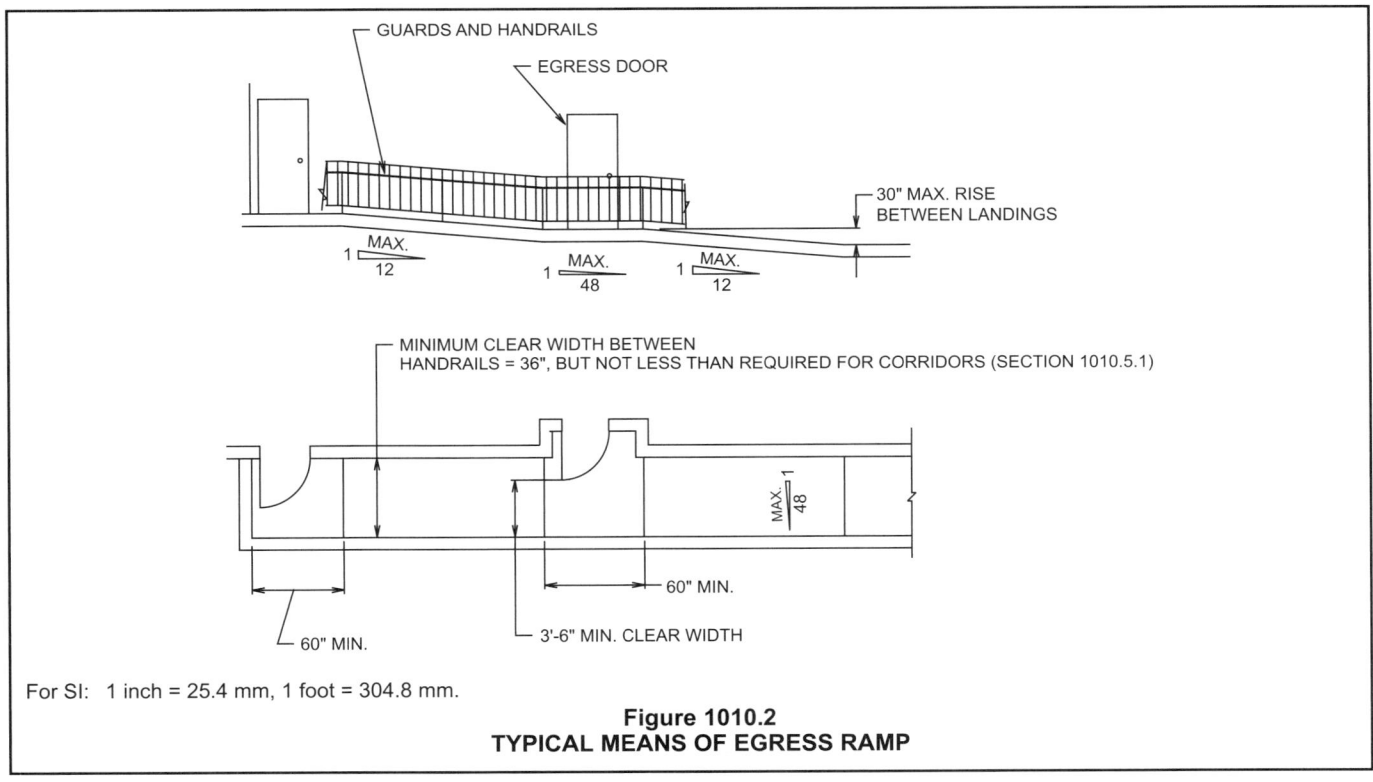

For SI: 1 inch = 25.4 mm, 1 foot = 304.8 mm.

Figure 1010.2
TYPICAL MEANS OF EGRESS RAMP

1010.5.2 Headroom. The minimum headroom in all parts of the *means of egress ramp* shall not be less than 80 inches (2032 mm).

❖ The requirement for headroom on any part of an egress ramp is identical to the requirement of a conventional (nonspiral) stairway (see Section 1009.2). General headroom heights along the means of egress are addressed in Section 1003.2.

1010.5.3 Restrictions. *Means of egress ramps* shall not reduce in width in the direction of egress travel. Projections into the required *ramp* and landing width are prohibited. Doors opening onto a landing shall not reduce the clear width to less than 42 inches (1067 mm).

❖ The purpose of not allowing ramps to reduce in width in the direction of egress travel is to prevent a restriction that would interfere with the flow of occupants out of a facility. This would include ramp landings in accordance with Section 1010.6.2. Handrails are the only exception in accordance with Sections 1010.5.1 and 1012.8.

Doors that open onto a ramp landing, including those at the top and bottom landings, must not reduce the clear width to less than 42 inches (1067 mm). This is a more restrictive provision than for corridors that would permit the reduction to one-half the required

width (see Section 1005.2). Since one of the purposes of a ramp is to accommodate persons with physical disabilities, it must provide the additional clear width for access by those confined to wheelchairs without the interference or potential blockage caused by the swing of a door (see Figures 1010.2 and 1010.5.3).

1010.6 Landings. *Ramps* shall have landings at the bottom and top of each *ramp*, points of turning, entrance, exits and at doors. Landings shall comply with Sections 1010.6.1 through 1010.6.5.

❖ Landings must be provided to allow users of a ramp to rest on a level floor surface and to adjust to the change in floor surface pitch.

Landings are required at the top and bottom of each ramp run (see Figure 1010.6). In addition, Section 1010.4 requires a landing every 30 inches (762 mm) of vertical rise of the ramp. The requirements for landings allow those occupants of the structure the ability to negotiate all changes in direction, and prepare themselves to either ascend or descend the ramp and to rest. If there is a door at the top or the bottom of the ramp, there are additional requirements in Section 1010.5.3 for door swing over the landing and Section 405 of ICC A117.1 for maneuvering space and turning space at the door.

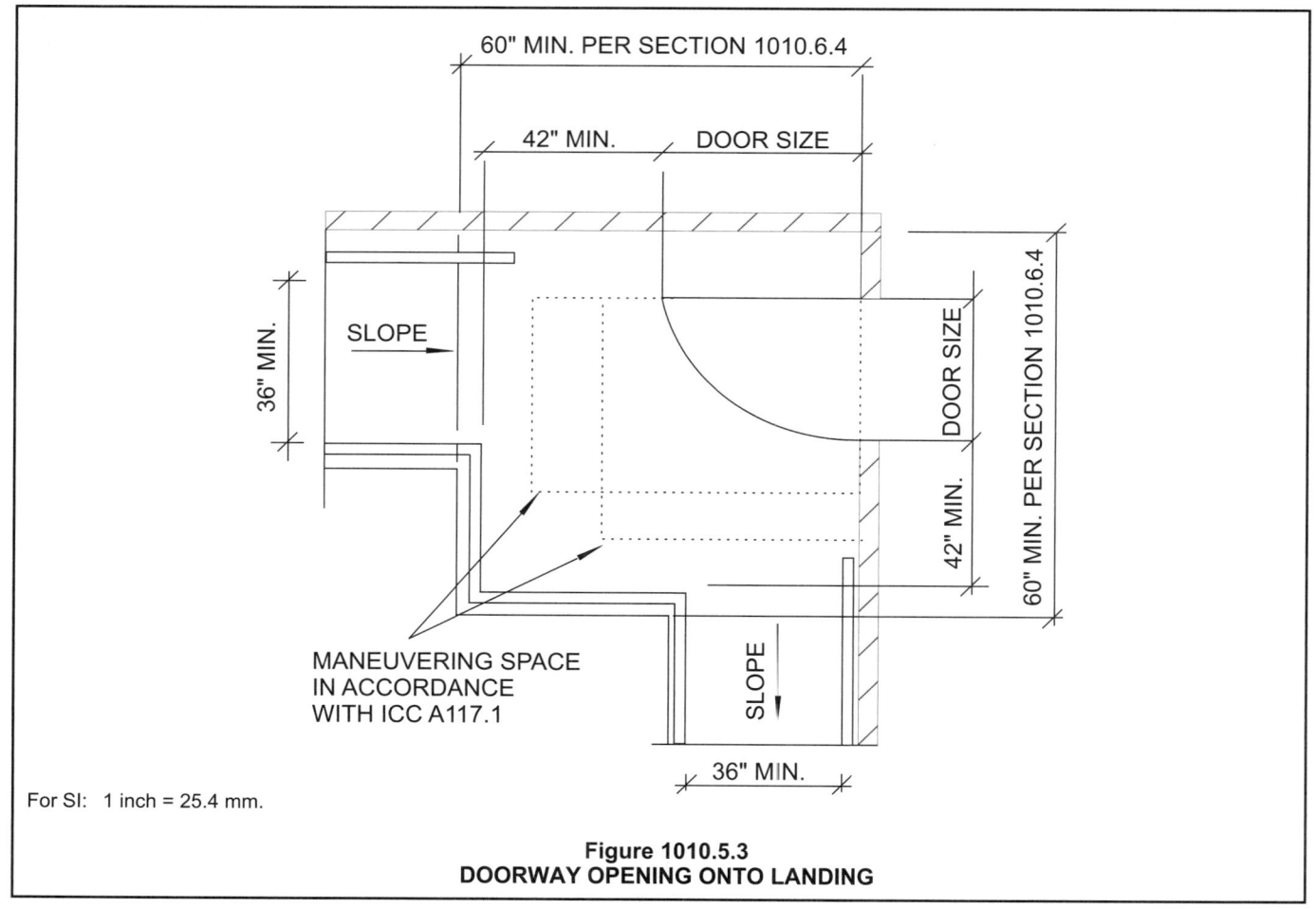

For SI: 1 inch = 25.4 mm.

**Figure 1010.5.3
DOORWAY OPENING ONTO LANDING**

1010.6.1 Slope. Landings shall have a slope not steeper than one unit vertical in 48 units horizontal (2-percent slope) in any direction. Changes in level are not permitted.

❖ Landings must be almost flat. This allows persons confined to a wheelchair to come to a complete stop without having to activate the brake or hold themselves stationary at the landing. The maximum slope or cross slope of the landing in any direction is 1:48 (see Figure 1010.2). This minimum slope is to allow for drainage to limit the accumulation of water on the landing surface.

1010.6.2 Width. The landing shall be at least as wide as the widest *ramp* run adjoining the landing.

❖ The width of all landings must be consistently as wide as the widths of the ramp runs leading to them. Means of egress ramps cannot be reduced in width in the direction of egress travel. This is also applicable to the landings connecting the ramp runs (see Figure 1010.6).

1010.6.3 Length. The landing length shall be 60 inches (1525 mm) minimum.

Exceptions:

1. In Group R-2 and R-3 individual dwelling and sleeping units that are not required to be *Accessible units*, *Type A units* or *Type B units* in accordance with Section 1107, landings are permitted to be 36 inches (914 mm) minimum.

2. Where the *ramp* is not a part of an *accessible route*, the length of the landing shall not be required to be more than 48 inches (1220 mm) in the direction of travel.

❖ The landings for ramps must be at least 60 inches (1524 mm) long (see Figure 1010.6). This allows persons confined to wheelchairs a sufficient distance to stop and rest along with any persons who may be as-sisting them. This requirement is directly applicable to straight-run ramps that may require an intermediate landing at every 30 inches (762 mm) of vertical rise (see Figure 1010.2). If the landing is also to be used to negotiate a change in the ramp's direction, Section 1010.6.4 is applicable. If a door overlaps the landing, Section 1010.5.3 is applicable.

The exceptions provide for smaller landings in dwelling and sleeping units and other locations where the ramp is not part of an accessible route. Exception 1 is consistent with the IRC. Exception 2 would be applicable in areas such as service ramps and ramps that served assembly seating areas that did not contain any wheelchair spaces.

1010.6.4 Change in direction. Where changes in direction of travel occur at landings provided between *ramp* runs, the landing shall be 60 inches by 60 inches (1524 mm by 1524 mm) minimum.

Exception: In Group R-2 and R-3 individual dwelling or sleeping units that are not required to be *Accessible units*, *Type A units* or *Type B units* in accordance with Section 1107, landings are permitted to be 36 inches by 36 inches (914 mm by 914 mm) minimum.

❖ When a change in direction is made in the ramp at a landing, the landing must be a square of at least 60 inches (1524 mm). This allows the person confined to a wheelchair enough room to negotiate the turn with minimal effort. The length of the landing may need to exceed 60 inches (1524 mm) to match the widths of the two ramp runs. In any case, the landing would still need to be 60 inches (1524 mm) wide (see Figures 1010.5.3 and 1010.6). If a door overlaps the landing, Section 1010.5.3 is applicable.

The exception provides for smaller landings in dwelling and sleeping units where the ramp is not part of an accessible route. This is consistent with requirements in the IRC.

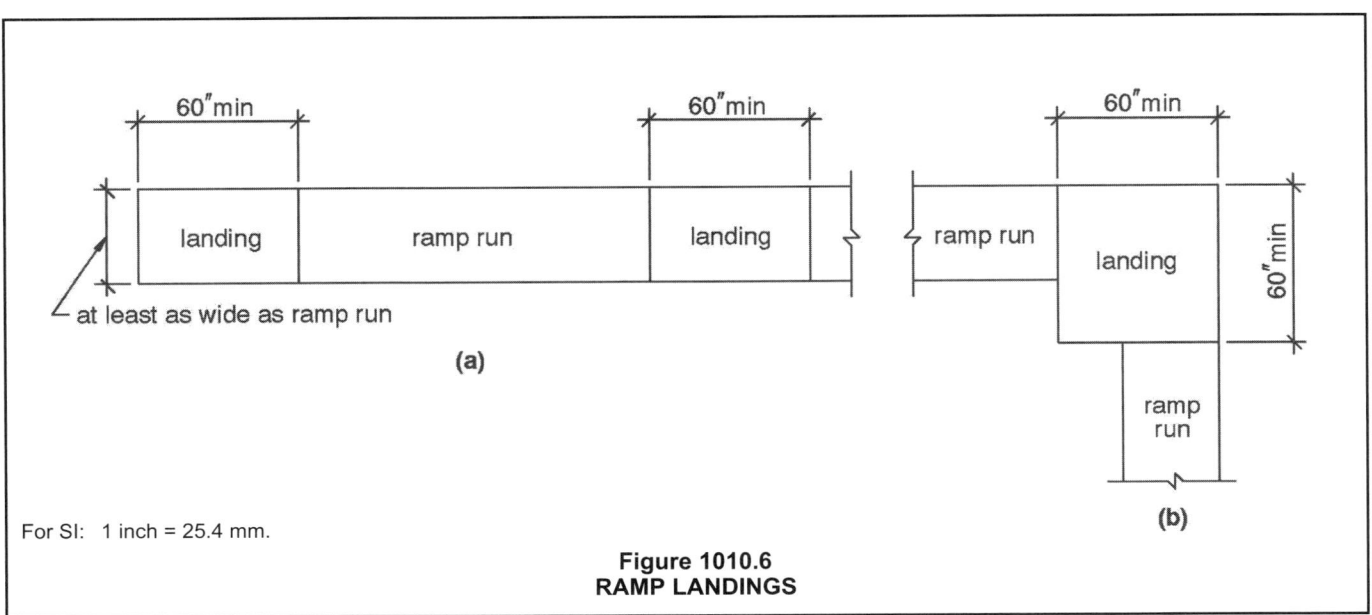

For SI: 1 inch = 25.4 mm.

**Figure 1010.6
RAMP LANDINGS**

1010.6.5 Doorways. Where doorways are located adjacent to a *ramp* landing, maneuvering clearances required by ICC A117.1 are permitted to overlap the required landing area.

❖ This section specifies that the area required for maneuvering to open the door and the area of the landing are allowed to overlap. It is not necessary to provide the sum of the two area requirements (see Figure 1010.5.3). Requirements for maneuvering space and turning space at the top and bottom of ramps are found in Section 405 of ICC A117.1.

1010.7 Ramp construction. All *ramps* shall be built of materials consistent with the types permitted for the type of construction of the building, except that wood *handrails* shall be permitted for all types of construction. *Ramps* used as an *exit* shall conform to the applicable requirements of Sections 1022.1 through 1022.6 for *exit enclosures*.

❖ Material requirements for the type of construction as required by Section 602 for floors are also the material requirements for ramp construction. The ramp, if used as an exit, must be enclosed and protected similar to an exit stairway within a vertical exit enclosure.

1010.7.1 Ramp surface. The surface of *ramps* shall be of slip-resistant materials that are securely attached.

❖ As the pace of exit travel becomes hurried during emergency situations, the probability of slipping on smooth or slick floor surfaces increases. To minimize the hazard, all floor surfaces in the means of egress are required to be slip resistant. The use of hard floor materials with highly polished, glazed, glossy or finely finished surfaces should be avoided. This is consistent with Section 1003.4.

Field testing and uniform enforcement of the concept of slip resistance is not practical. One method used to establish slip resistance is that the static coefficient of friction between leather [Type 1 (Vegetable Tanned) of Federal Specification KK-L-165C] and the floor surface is greater than 0.5. Laboratory test procedures such as ASTM D 2047 can determine the static coefficient of resistance. Bulletin No. 4 entitled "Sur-

faces" issued by the U.S. Architectural and Transportation Barriers Compliance Board (ATBCB) contains further information regarding slip resistance.

1010.7.2 Outdoor conditions. Outdoor *ramps* and outdoor approaches to *ramps* shall be designed so that water will not accumulate on walking surfaces.

❖ Outdoor ramps, landings and the approaches to the ramp must be sloped so that surfaces do not accumulate water so as to provide a safe path of egress travel at all times. While not specifically stated, any interior locations, such as near a pool, should also have the ramps designed to limit the accumulation of water in order to maintain slip resistance (see Sections 1003.4 and 1010.7.1).

Where exterior ramps are used in moderate or severe climates, there may also be a concern to protect the ramp from accumulations of snow and ice to provide a safe path of egress travel at all times, including winter. Maintenance of the means of egress in the IFC requires an unobstructed path to allow for full instant use in case of a fire or emergency (see IFC Section 1028.2). Typical methods for protecting these egress elements include roof overhangs or canopies, heated slab and, when approved by the building official, a reliable snow removal maintenance program.

1010.8 Handrails. *Ramps* with a rise greater than 6 inches (152 mm) shall have handrails on both sides. *Handrails* shall comply with Section 1012.

> **Exception:** *Handrails* for ramped *aisles* are not required where permitted by Section 1028.13.

❖ To aid in the use of a ramp, handrails are to be provided. Handrails are intended to provide the user with a graspable surface for guidance and support. All ramps with a vertical rise greater than 6 inches (152 mm) between landings are to be provided with handrails on both sides [see Figures 1010.8(1) and 1012.2]. General strength requirements for handrails are found in Section 1012 with a reference to Section 1607.7. Note that if the handrail extension is at a location that

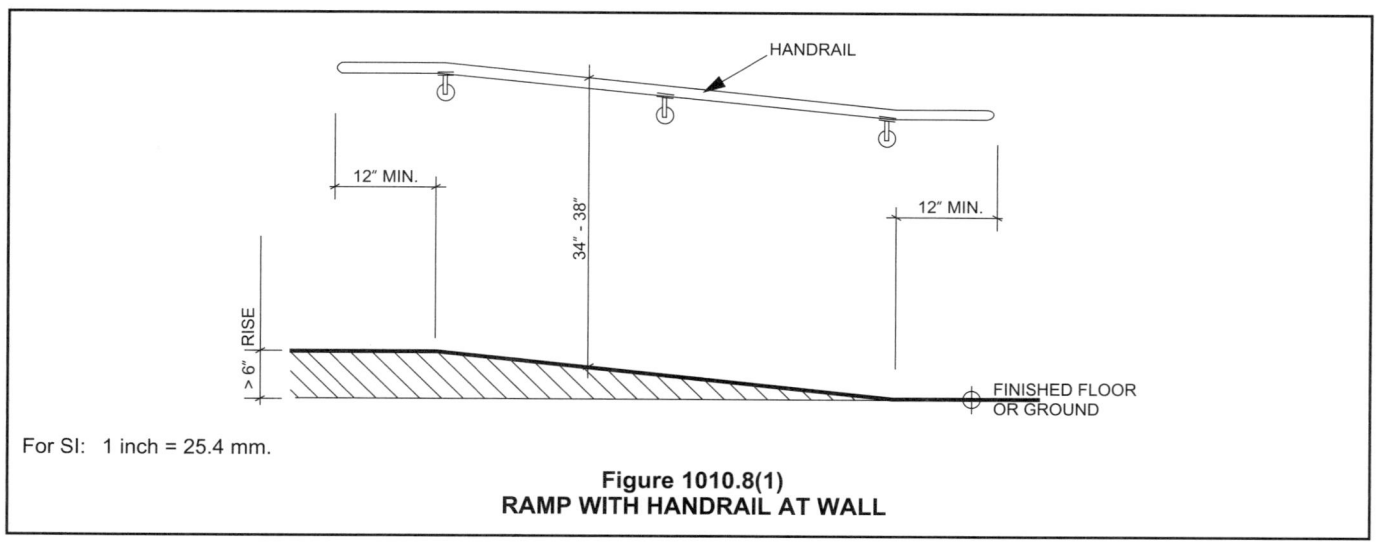

For SI: 1 inch = 25.4 mm.

Figure 1010.8(1)
RAMP WITH HANDRAIL AT WALL

could be considered a protruding object, the handrail must return to the post at a height of less than 27 inches (686 mm) above the floor. Handrails along the ramp runs are not considered protruding objects.

Depending on the configuration of the ramp and the adjacent walking surface, ramps may require a combination of handrails, edge protection and guards. See Figures 1010.8(1), 1010.8(2), 1010.8(3) and 1010.8(4) for an illustration of some of the alternatives.

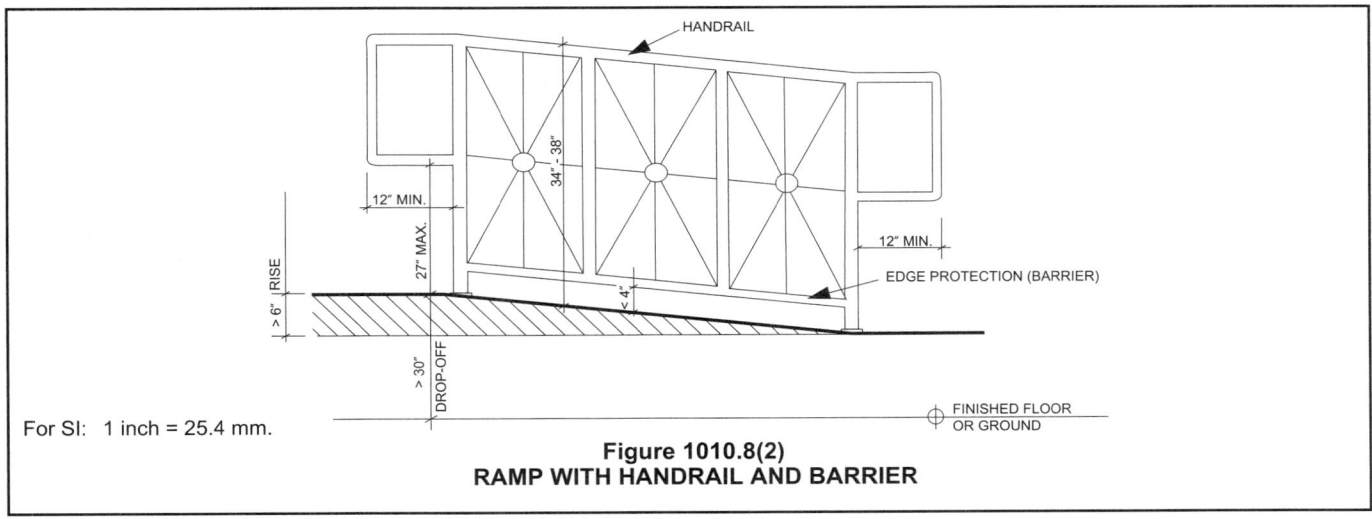

For SI: 1 inch = 25.4 mm.

Figure 1010.8(2)
RAMP WITH HANDRAIL AND BARRIER

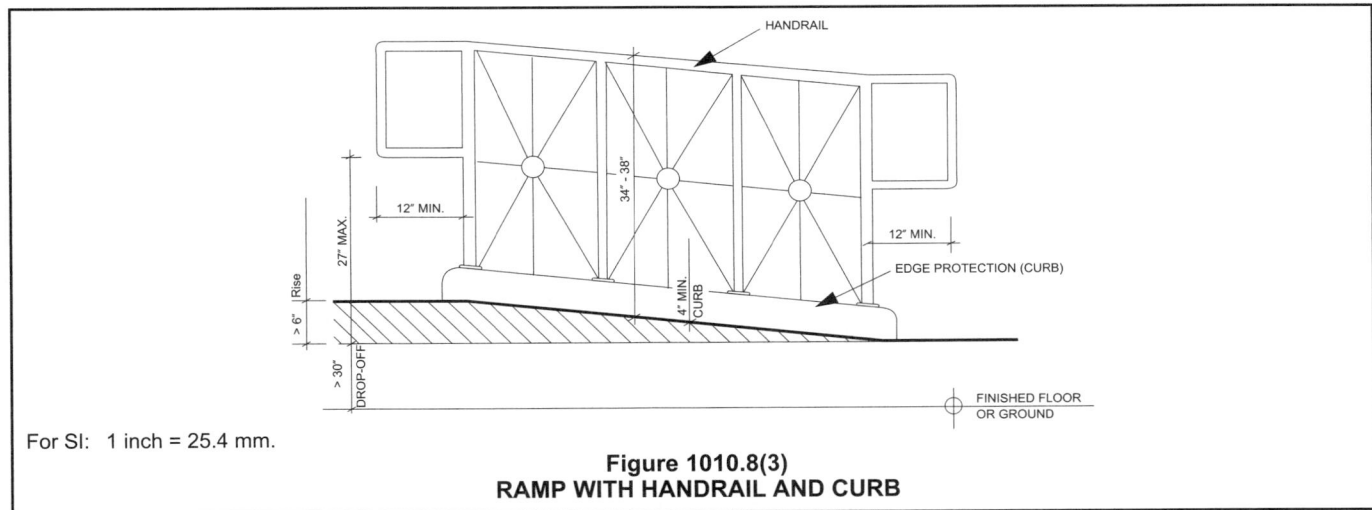

For SI: 1 inch = 25.4 mm.

Figure 1010.8(3)
RAMP WITH HANDRAIL AND CURB

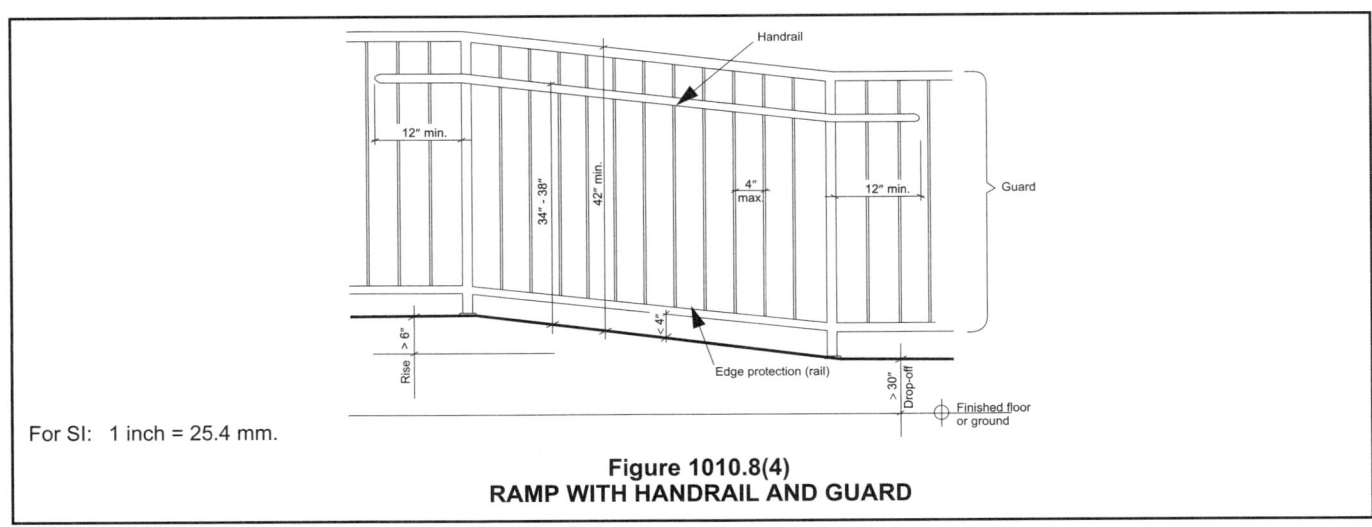

For SI: 1 inch = 25.4 mm.

Figure 1010.8(4)
RAMP WITH HANDRAIL AND GUARD

1010.9 Edge protection. Edge protection complying with Section 1010.9.1 or 1010.9.2 shall be provided on each side of *ramp* runs and at each side of *ramp* landings.

Exceptions:

1. Edge protection is not required on *ramps* that are not required to have *handrails*, provided they have flared sides that comply with the ICC A117.1 curb ramp provisions.

2. Edge protection is not required on the sides of ramp landings serving an adjoining *ramp* run or *stairway*.

3. Edge protection is not required on the sides of *ramp* landings having a vertical drop off of not more than $^1/_2$ inch (12.7 mm) within 10 inches (254 mm) horizontally of the required landing area.

4. In assembly spaces with fixed seating, edge protection is not required on the sides of *ramps* where the *ramps* provide access to the adjacent seating and *aisle accessways*.

❖ This section of the code now addresses the comprehensive requirements for edge protection for all ramps. It must be noted that edge protection is not the same as the requirements for guards. The presence of a guard does not necessarily provide adequate edge protection and the presence of adequate edge protection does not satisfy the requirements for a guard. Edge protection is necessary to prevent the wheels of a wheelchair from leaving the ramp surface or becoming lodged between the edge of the ramp and any adjacent construction. For example, a ramp may be located relatively adjacent to the exterior wall of a building. However, between the ramp edge and the exterior wall, there is a strip of earth for landscape purposes. Without adequate edge protection, persons confined to wheelchairs could possibly have their wheels run off the side of the ramp into the landscape, causing them to tip. These requirements are consistent with Section 405 of ICC A117.1 and those in the new ADA/ABA Accessibility Guidelines.

Exception 1 allows a ramp to have minimal edge protection as long as its vertical rise is 6 inches (152 mm) or less. The exception is predicated on the ramp not needing any handrails, which is established by the provisions of Section 1010.8. Such a ramp would only need flared sides or returned curbs. For specific details of these types of edge protection, the provisions of Section 406 of ICC A117.1 for curb ramps must be followed.

Exception 2 reiterates that edge protection is not literally required around each side of a ramp landing. Obviously, edge protection is not required along that portion of the landing that directly adjoins the next ramp run; it is only required along the unprotected sides of ramp landings.

Exception 3 states that edge protection is not required for those sides of a ramp landing directly adjacent to the ground surface that gently slopes away from the edge of the landing. If the grade adjacent to

the ramp landing slopes no more than $^1/_2$:10 (which equates to 1:20) away from the landing, additional edge protection is not required. Such a gradual slope would not be detrimental to persons confined to wheelchairs as they negotiate the ramp landing. Note that this exception is limited to landings, not the ramp surface itself. The ramp must meet the edge protection in Section 1010.9.1 or 1010.9.2.

Exception 4 allows for no edge protection along ramps that serve fixed seating on the side where the seats are located. If curbs were provided they would be a tripping hazard for people coming in and out of the aisle accessways between the seats.

Depending on the configuration of the ramp and the adjacent walking surface, ramps may require a combination of handrails, edge protection and guards. See Figures 1010.8(1), 1010.8(2), 1010.8(3) and 1010.8(4) for illustrations of some alternatives.

1010.9.1 Curb, rail, wall or barrier. A curb, rail, wall or barrier shall be provided to serve as edge protection. A curb must be a minimum of 4 inches (102 mm) in height. Barriers must be constructed so that the barrier prevents the passage of a 4-inch- diameter (102 mm) sphere, where any portion of the sphere is within 4 inches (102 mm) of the floor or ground surface.

❖ Edge protection for ramps and ramp landings may be achieved with a built-up curb or other barrier, such as a rail, wall or guard. The barrier must be located near the surface of the ramp and landing such that a 4-inch-diameter (102 mm) sphere cannot pass through any openings. An example of an effective barrier would be the bottom rail of a guard system. If the bottom rail is located less than 4 inches (102 mm) above the ramp and landing surface, edge protection has been provided. If a curb option is used, the curb must be a minimum of 4 inches (102 mm) high. The curb or barrier prevents the wheel of a wheelchair from running off the edge of the surface and provides people with visual disabilities a toe stop at the edge of the walking surface (see Figure 1010.9.1).

1010.9.2 Extended floor or ground surface. The floor or ground surface of the *ramp* run or landing shall extend 12 inches (305 mm) minimum beyond the inside face of a *handrail* complying with Section 1012.

❖ An alternative to providing some type of barrier at the edge of the ramp (see Section 1010.9.1) is to make the ramp surface wider than the handrails provided at either side. The combination of the wider surface and the handrail would assist in preventing a wheelchair or crutch tip from moving off the ramp during a temporary slip (see Figure 1010.9.1).

1010.10 Guards. *Guards* shall be provided where required by Section 1013 and shall be constructed in accordance with Section 1013.

❖ To protect the user from falls to surfaces below, guards are to be provided where the sides of a ramp or landing are more than 30 inches (762 mm) above the adjacent grade. Guards are to be constructed in accordance with Section 1013, including the minimum height of 42 inches (1067 mm) [see Figure 1010.8(4)].

Depending on the configuration of the ramp and the adjacent walking surface, ramps may require a combination of handrails, edge protection and guards. See Figures 1010.8(1), 1010.8(2), 1010.8(3) and 1010.8(4) for illustrations of some alternatives.

SECTION 1011
EXIT SIGNS

1011.1 Where required. *Exits* and *exit access* doors shall be marked by an *approved exit* sign readily visible from any direction of egress travel. The path of egress travel to *exits* and within *exits* shall be marked by readily visible *exit* signs to clearly indicate the direction of egress travel in cases where the *exit* or the path of egress travel is not immediately visible to the occupants. Intervening *means of egress* doors within *exits* shall be marked by *exit* signs. *Exit* sign placement shall be such that no point in an *exit access corridor* or *exit passageway* is more than 100 feet (30 480 mm) or the *listed* viewing distance for the sign, whichever is less, from the nearest visible *exit* sign.

Exceptions:

1. *Exit* signs are not required in rooms or areas that require only one *exit* or *exit access*.

2. Main exterior *exit* doors or gates that are obviously and clearly identifiable as *exits* need not have *exit* signs where *approved* by the *building official*.

3. *Exit* signs are not required in occupancies in Group U and individual sleeping units or dwelling units in Group R-1, R-2 or R-3.

4. *Exit* signs are not required in dayrooms, sleeping rooms or dormitories in occupancies in Group I-3.

5. In occupancies in Groups A-4 and A-5, *exit* signs are not required on the seating side of vomitories or openings into seating areas where *exit* signs are provided in the concourse that are readily apparent from the vomitories. Egress lighting is provided to identify each vomitory or opening within the seating area in an emergency.

❖ Where an occupancy has two or more required exits or exit accesses, the means of egress must be provided with illuminated signs that readily identify the location of, and indicate the path of travel to, the exits. The signs must be illuminated with letters reading "Exit." The illumination may be internal or external to the sign. The signs should be visible from all directions in the exit access route. In cases where the signs are not visible to the occupants because of turns in the corridor or for other reasons, additional illuminated signs must be provided indicating the direction of egress to an exit. "Exit" signs must be located so that, where required, the nearest one is within 100 feet (30 480 mm), of the sign's listed viewing distance. While not a referenced standard, UL 924 permits exit signs to be listed with a viewing distance of less than 100 feet (30 480 mm). When a sign is listed for a viewing distance of less than 100 feet (30 480 mm) the label on the sign will indicate the appropriate viewing distance. If such a sign is used, the spacing of the signs should be based on the listed viewing distance.

Typically, once an occupant enters an exit enclosure, exit signs are no longer needed; however, in buildings with more complicated egress layouts, it is possible that the direction for egress travel within the exit is not immediately apparent. For example, exit passageways can be part of the path of exit travel at the level of exit discharge or transfer floors. Evacuees may hesitate or be confused when the vertical travel becomes horizontal travel, which may result in a delay in evacuation. In these situations, exit signs may be needed within the exit enclosure (see Figure 1011.1).

The exceptions identify conditions where exit signs are not necessary since they would not increase the safety of the egress path.

For Exceptions 1 and 3, the assumption is that the occupants are familiar enough with the space to know the way out, which in most cases is also the way they come in.

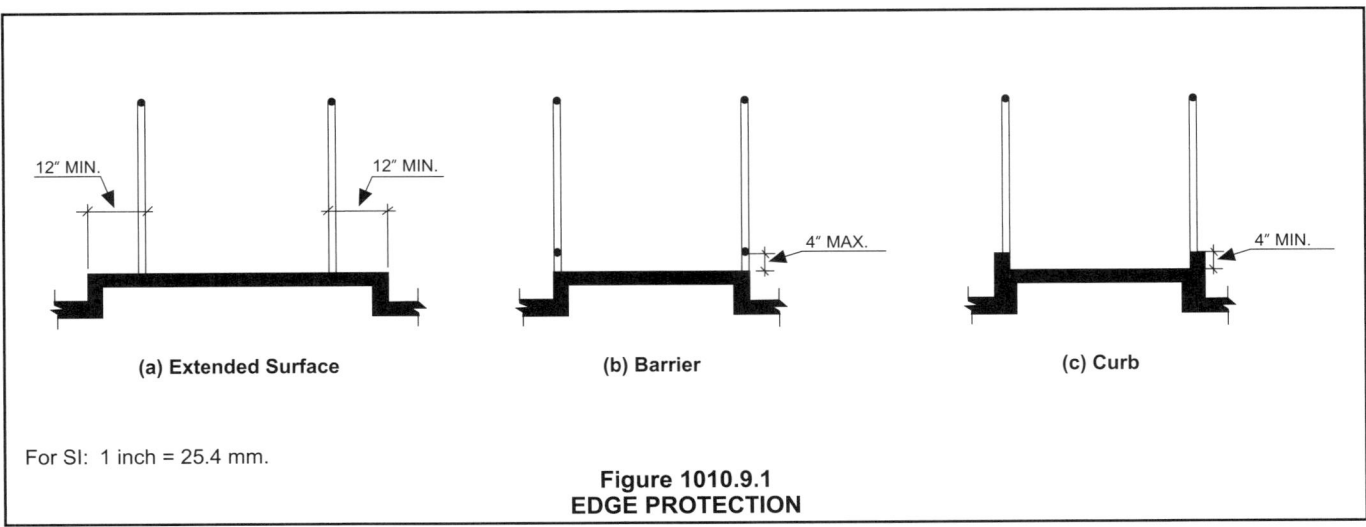

For SI: 1 inch = 25.4 mm.

(a) Extended Surface (b) Barrier (c) Curb

12" MIN. 12" MIN. 4" MAX. 4" MIN.

Figure 1010.9.1
EDGE PROTECTION

In accordance with Exception 2, when the exit is identifiable in itself and is the main exterior door through which the occupants would enter the building, "Exit" signs are not required. For example, a two-story Group B building has a main employee/customer entrance. The entrance consists of a storefront arrangement with glass doors and sidelights. The entrance is centrally located within the building. These main exterior exit doors can be quickly observed as being an exit and would not need to be marked with an "Exit" sign.

In accordance with Exception 4, "Exit" signs are not required in detainee living and sleeping room areas of Group I-3 buildings. In cases of emergency, occupants in Group I-3 are escorted by staff to the exits and to safety. The "Exit" signs also represent potential weapons when they are accessible to the residents.

In the Group A-4 and A-5 occupancies described in Exception 5, the egress path is obvious and thus exit signs are not needed. Additionally, because of the configuration of the vomitories, the "Exit" signs are not readily visible to the persons immediately adjacent to or above the vomitory.

1011.2 Illumination. *Exit* signs shall be internally or externally illuminated.

> **Exception:** Tactile signs required by Section 1011.3 need not be provided with illumination.

❖ This section simply provides the scope for illumination of regulated "Exit" signs. Exit signs must be illuminated so that they are readily apparent in situations where the lights may be off or the building has lost power.

Tactile (i.e., raised letters and Braille) exit signs are specifically addressed in Section 1011.3.

1011.3 Tactile exit signs. A tactile sign stating EXIT and complying with ICC A117.1 shall be provided adjacent to each door to an *area of refuge*, an exterior area for assisted rescue, an *exit stairway*, an *exit ramp*, an *exit passageway* and the *exit discharge*.

❖ This signage is needed to indicate which doors are serving as exits for those persons with visual impairments. Signs are needed on the required exits in the building, including at doors leading to exit stairway enclosures and exit passageways, within the exit enclosures leading to the outside and any exit doors that lead directly to the outside. While an area of refuge may be located within an exit stairway enclosure, Section 1007.3 also allows the area of refuge to be located immediately outside of the stairway enclosure. In this situation, tactile exit signage would be required both at the door leading into the area of refuge and, once in the enclosure, the door leading to the exit stairway. Exterior areas for assisted rescue are typically located immediately outside of an exit door (see Section 1007.7). This is not intended to preclude the tactile signage from including additional information as long as "Exit" is first. For example, labeling the door to the exit enclosure as "Exit stairway" would indicate to the visually impaired person that once they moved through the door they would be dealing with vertical travel. This could be considered an additional safety feature. This section is also referenced in Section 1110.3.

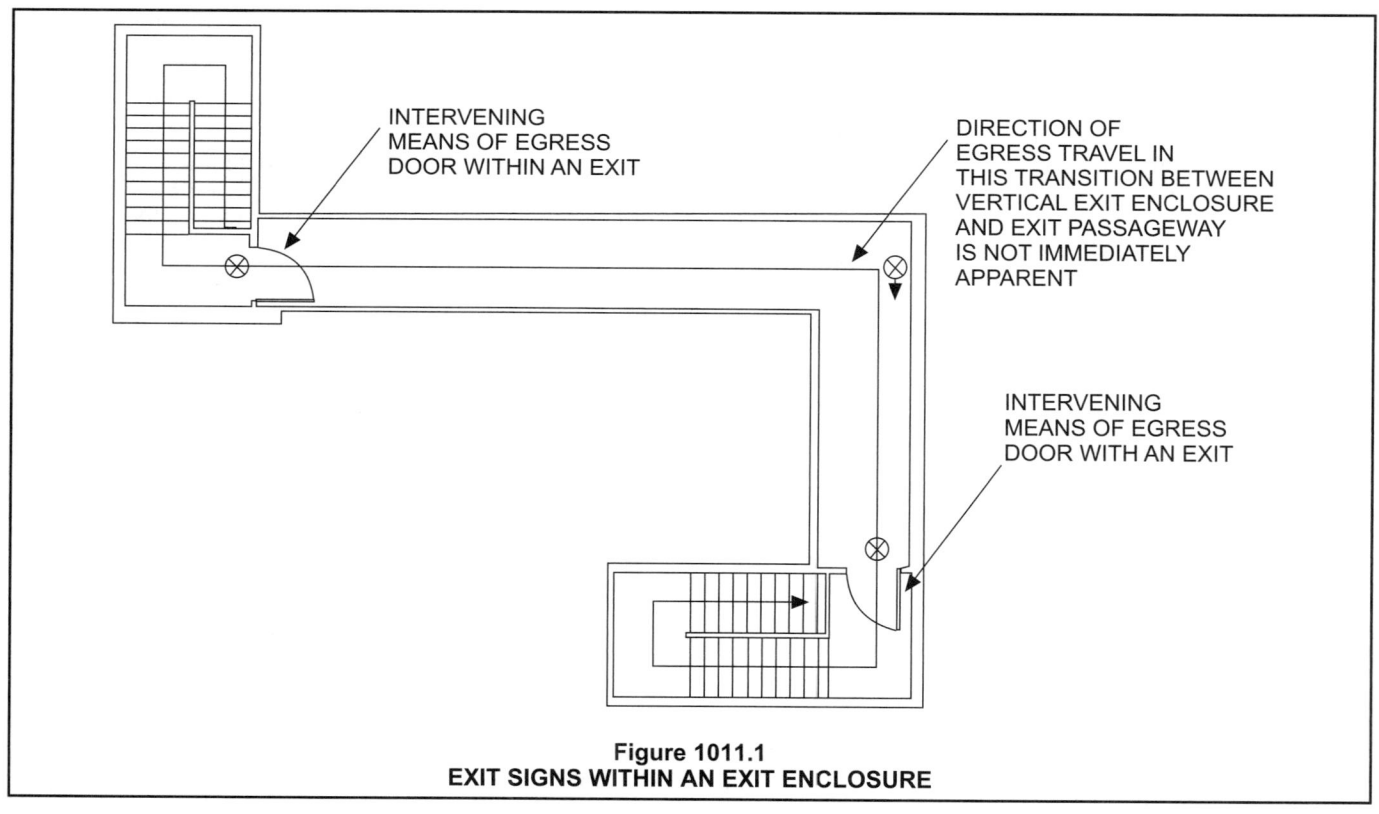

Figure 1011.1
EXIT SIGNS WITHIN AN EXIT ENCLOSURE

Tactile signage in accordance with Section 703 of ICC A117.1 includes both raised lettering and Braille. While not required to be illuminated, if the raised letters were also of high contrast, illumination of this sign would be advantageous for a person with partial sight.

1011.4 Internally illuminated exit signs. Electrically powered, *self-luminous* and *photoluminescent exit* signs shall be *listed* and labeled in accordance with UL 924 and shall be installed in accordance with the manufacturer's instructions and Chapter 27. *Exit* signs shall be illuminated at all times.

❖ All exit signage must be listed and labeled as indicated in UL 924, *Standard for Safety of Emergency Lighting and Power Equipment*. Listed "exit" signs are required by UL 924 to meet the same graphics, illumination and power sources defined in Sections 1011.5.1 through 1011.5.3 for externally illuminated signs. Internal illumination may be electrically powered or be of a self-luminous or photoluminescent product. Electrically powered would include LED, incandescent, fluorescent and electroluminescent types of signs. If a sign is photoluminescent, the "charging" source must be continually available (see the definitions in Section 1002.1 for "Photoluminescent" and "Self-luminous"). "Exit" signs must be illuminated at all times, including when the building may not be fully occupied. If a fire occurs late at night, there may be cleaning crews or persons working overtime in the building who will need to be able to find the exits. The reference to Chapter 27 is so the signs will be equipped with a connection to an emergency power supply.

1011.5 Externally illuminated exit signs. Externally illuminated *exit* signs shall comply with Sections 1011.5.1 through 1011.5.3.

❖ Externally illuminated exit signage must meet the graphic, illumination and emergency power requirements in the referenced sections. The requirements are the same as for internally illuminated signage.

1011.5.1 Graphics. Every *exit* sign and directional *exit* sign shall have plainly legible letters not less than 6 inches (152 mm) high with the principal strokes of the letters not less than $^3/_4$ inch (19.1 mm) wide. The word "EXIT" shall have letters having a width not less than 2 inches (51 mm) wide, except the letter "I," and the minimum spacing between letters shall not be less than $^3/_8$ inch (9.5 mm). Signs larger than the minimum established in this section shall have letter widths, strokes and spacing in proportion to their height.

The word "EXIT" shall be in high contrast with the background and shall be clearly discernible when the means of *exit* sign illumination is or is not energized. If a chevron directional indicator is provided as part of the *exit* sign, the construction shall be such that the direction of the chevron directional indicator cannot be readily changed.

❖ Every "Exit" sign and directional sign located in the exit access or exit route is required to have a color contrast vivid enough to make the signs readily visible, even when not illuminated. Letters must be at least 6 inches (152 mm) high and their stroke not less than $^3/_4$-inch

(19.1 mm) wide (see Figure 1011.5.1). The sizing of the letters is predicated on the readability of the wording from a distance of 100 feet (30 480 mm).

While red letters are common for exit signs, sometimes green on black is used in auditorium areas with low lighting levels, such as theaters, because that color combination tends not to distract the audience's attention. It is more important that the "Exit" sign be readily visible with respect to the background.

"Exit" signs may be larger than the minimum size specified; however, the standardized proportion of the letters must be maintained. Externally illuminated signage that is smaller could use the requirements in UL 924 for guidance; however, sign spacing would need to be adjusted, and alternative approval would be through the building official having jurisdiction.

A "chevron directional indicator" is the same as a directional arrow. The language is intended to be consistent with UL 924.

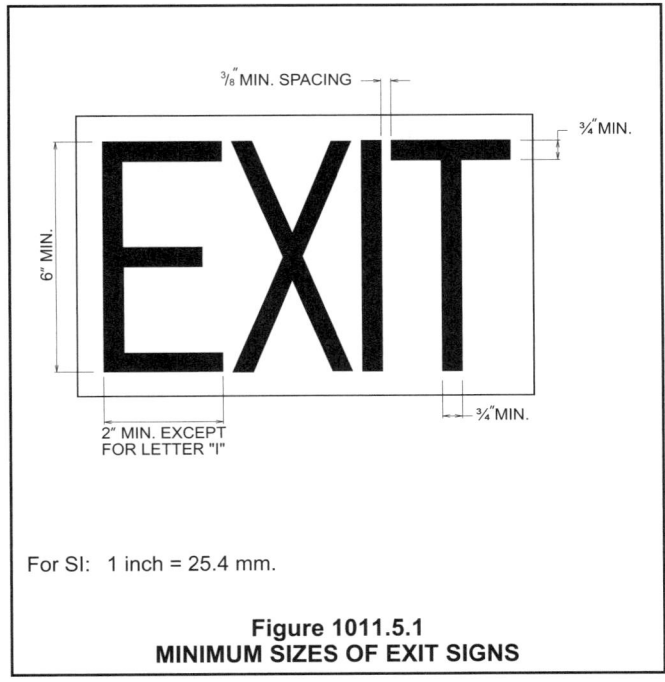

For SI: 1 inch = 25.4 mm.

Figure 1011.5.1
MINIMUM SIZES OF EXIT SIGNS

1011.5.2 Exit sign illumination. The face of an *exit* sign illuminated from an external source shall have an intensity of not less than 5 foot-candles (54 lux).

❖ Every "Exit" sign and directional sign must be continuously illuminated to provide a light intensity at the illuminated surface of at least 5 foot-candles (54 lux). It is not a requirement that the "Exit" signs be internally illuminated. An external illumination source with the power capabilities specified by Section 1011.5.3 is acceptable.

1011.5.3 Power source. *Exit* signs shall be illuminated at all times. To ensure continued illumination for a duration of not less than 90 minutes in case of primary power loss, the sign illumination means shall be connected to an emergency power system provided from storage batteries, unit equipment or an on-site generator. The installation of the emergency power system shall be in accordance with Chapter 27.

Exception: *Approved exit* sign illumination means that provide continuous illumination independent of external power sources for a duration of not less than 90 minutes, in case of primary power loss, are not required to be connected to an emergency electrical system.

❖ "Exit" signs must be illuminated on a continuous basis so that when a fire emergency occurs, occupants will be able to identify the locations of the exits. The reliability of the power sources supplying the electrical energy required for maintaining the illumination of exit signs is important. When power interruptions occur, "Exit" sign illumination must be obtained from an emergency power system. This does not imply that the sign must be internally illuminated. Whatever illumination system is used, whether internal or external, it must be connected to a system designed to pick up the power load required by the "Exit" signs after loss of the normal power supply.

Where self-luminous signs are used, connection to the emergency electrical supply system is not required.

The IFC requirements for emergency power for emergency egress lighting and exit signage in existing buildings only requires a 60-minute time duration. This is not a conflict, but rather recognition of the loss of battery storage capability over a length of time.

SECTION 1012
HANDRAILS

1012.1 Where required. *Handrails* for *stairways* and *ramps* shall be adequate in strength and attachment in accordance with Section 1607.7. *Handrails* required for *stairways* by Section 1009.12 shall comply with Sections 1012.2 through 1012.9. *Handrails* required for *ramps* by Section 1010.8 shall comply with Sections 1012.2 through 1012.8.

❖ Handrails are required at stairways and ramps. In all situations, they must be designed in accordance with the structural requirements in Section 1607.7.1. There are, however, distinct differences in how the handrail requirements are applied in stairways and ramps. Stairways and their handrails are not part of an accessible route and are not subject to ICC A117.1. The specific section references allow for this consideration. Handrails are also very distinct from guards, even though they are sometimes incorrectly called "guardrails." The "handrail" is the element that is used during vertical travel, for guidance, stabilization, pulling and as an aid in arresting a possible fall. Guards are located near the side of an elevated walking surface to minimize the possibility of a fall to a lower level and are discussed in Section 1013. In residential applications, however, the top rail of a guard when located at handrail height may also serve as a handrail.

1012.2 Height. *Handrail* height, measured above *stair* tread *nosings*, or finish surface of *ramp* slope, shall be uniform, not less than 34 inches (864 mm) and not more than 38 inches (965 mm). *Handrail* height of *alternating tread devices* and ship ladders, measured above tread *nosings*, shall be uniform, not less than 30 inches (762 mm) and not more than 34 inches (864 mm).

❖ It has been demonstrated that for safe use, the height of handrails must not be less than 34 inches (864 mm) nor more than 38 inches (965 mm) above the leading edge of stairway treads, landings or other walking surfaces (see Figure 1012.2). This requirement is applicable for all uses, including handrails within a dwelling unit.

An alternating tread device, defined as "a device that has a series of steps between 50 and 70 degrees from horizontal, usually attached to a center support rail in an alternating manner," is permitted for use in specific applications. Since an alternating tread device differs significantly from other types of stairways, several of the fundamental stairway requirements are not appropriate. The permitted range for handrail height for alternating tread devices allows for a lower height above the tread nosings. The minimum required height has been reduced from 34 inches (864 mm) to 30 inches (762 mm), with a new maximum permitted height of 34 inches (864 mm). The special features of an alternating tread device result in differences of handrail use, such as different arm posture, the hand gripping the handrail near a higher part of the body and the use of handrails under the arms for stabilization. Therefore, a lower handrail height is more appropriate.

1012.3 Handrail graspability. All required *handrails* shall comply with Section 1012.3.1 or shall provide equivalent graspability.

Exception: In Group R-3 occupancies; within dwelling units in Group R-2 occupancies; and in Group U occupancies that are accessory to a Group R-3 occupancy or accessory to individual dwelling units in Group R-2 occupancies; handrails shall be Type I in accordance with Section 1012.3.1, Type II in accordance with Section 1012.3.2 or shall provide equivalent graspability.

❖ The ability of grasping a handrail firmly and sliding the hand along the rail's gripping surface without meeting obstructions are important factors in the safe use of stairways and ramps. These properties are largely a function of the shape of the handrail. Handrails for stairways and ramps must meet the specifications of Section 1012.3.1 or be determined to have grasping properties and attributes equivalent to profiles allowed in Section 1012.3.1. Such determinations of equivalence are considered by local building officials as an option based on the profile presented and the building official's evaluation of its properties. A complete evaluation will consider the four basic functions of a handrail: guidance, stabilization, pulling and arresting a fall. The determination could be made by comparative use on stairs or ramps of properly mounted samples. Care should be given to understand the exception for the interference of handrail mounts on the gripping surfaces of smaller profiles in effectively evaluating their graspability. For a discussion of this, see Section 1012.4.

The exception allows for an alternative, Type II handrail, within residential units and their associated structures. A handrail on common stairways within an apartment building or on the steps to the front door of a townhouse could not use this exception. This is consistent with the IRC.

1012.3.1 Type I. *Handrails* with a circular cross section shall have an outside diameter of at least $1^{1}/_{4}$ inches (32 mm) and not greater than 2 inches (51 mm). If the *handrail* is not circular, it shall have a perimeter dimension of at least 4 inches (102 mm) and not greater than $6^{1}/_{4}$ inches (160 mm) with a maximum cross-section dimension of $2^{1}/_{4}$ inches (57 mm). Edges shall have a minimum radius of 0.01 inch (0.25 mm).

❖ Handrails have traditionally been regulated as either circular or noncircular rails. The noncircular rails have previously been limited to a maximum perimeter dimension of $6^{1}/_{4}$ inches (160 mm), with other limitations addressing minimum perimeter and maximum cross-sectional dimensions. These handrails shapes are now referred to as Type I handrails.

This class of handrails includes circular cross sections with an outside diameter of at least $1^{1}/_{4}$ inches

(32 mm) but not greater than 2 inches (51 mm). It limits the perimeter of the cross section such that the gripping surface incorporates the bottom of the rail. A handrail with either a very narrow or a large cross section is not graspable in a power grip by all able-bodied users and certainly not by those with hand-strength or flexibility deficiencies. Noncircular Type I cross sections can be approved by way of the alternative noncircular criteria in this section, but the bottom of the rail must be considered part of the suitable gripping surface. Edges must be slightly rounded and not sharp. An example is shown in Figure 1012.3.1.

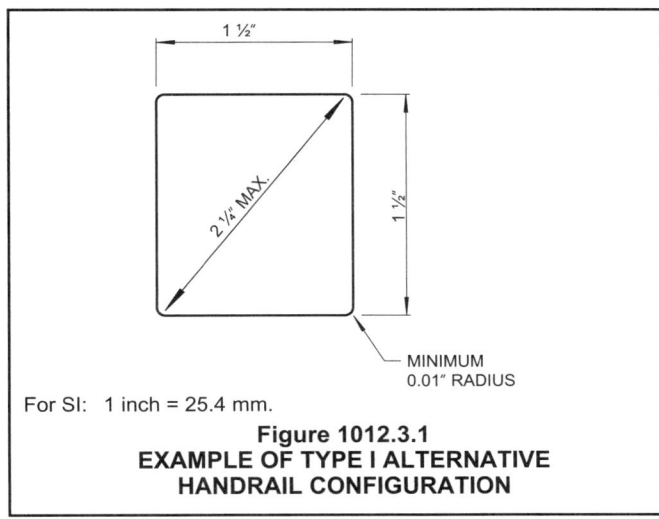

For SI: 1 inch = 25.4 mm.

Figure 1012.3.1
EXAMPLE OF TYPE I ALTERNATIVE
HANDRAIL CONFIGURATION

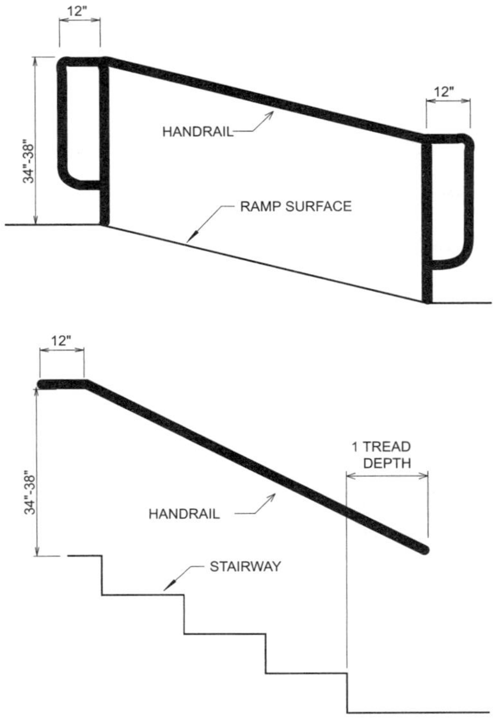

For SI: 1 inch = 25.4 mm.

Figure 1012.2
HANDRAIL REQUIREMENTS

1012.3.2 Type II. *Handrails* with a perimeter greater than $6^{1}/_{4}$ inches (160 mm) shall provide a graspable finger recess area on both sides of the profile. The finger recess shall begin within a distance of $^{3}/_{4}$ inch (19 mm) measured vertically from the tallest portion of the profile and achieve a depth of at least $^{5}/_{16}$ inch (8 mm) within $^{7}/_{8}$ inch (22 mm) below the widest portion of the profile. This required depth shall continue for at least $^{3}/_{8}$ inch (10 mm) to a level that is not less than $1^{3}/_{4}$ inches (45 mm) below the tallest portion of the profile. The minimum width of the *handrail* above the recess shall be $1^{1}/_{4}$ inches (32 mm) to a maximum of $2^{3}/_{4}$ inches (70 mm). Edges shall have a minimum radius of 0.01 inch (0.25 mm).

❖ Handrail profiles having a perimeter dimension greater than $6^{1}/_{4}$ inches (160 mm), identified as Type II handrails, are acceptable within dwelling units and their associated structures when complying with all of the specific dimensional requirements.

Research has shown that Type II handrails have graspability that is essentially equal to or greater than the graspability of handrails meeting the long-accepted and codified shape and size now defined as Type I.

The key features of the graspability of Type II handrails are graspable finger recesses on both sides of the handrail. These recesses allow users to firmly grip a properly proportioned grasping surface on the top of the handrail, ensuring that the user can tightly retain a grip on the handrail for all forces that are associated with attempts to arrest a fall.

This class of handrails incorporates a grip surface with controlled recesses for the purchase of the fingers and opposing thumb. These handrail shapes allow the use of a power-span grip that need not encompass the bottom surfaces of the rail allowing the design of taller cross sections that can eliminate the interference of mountings and provide a completely uninterrupted gripping surface for the user. The limits of the position and depth of the required recesses represent the minimum standard. Optimizing the design within the parameters with larger recesses and complete finger clearance from mountings will enhance the performance of the profile. Although this standard allows design flexibility, it is important to follow the specifications accurately. Each drawing in Figure 1012.3.2 illustrates the requirements of each sentence for clarity.

1012.4 Continuity. *Handrail* gripping surfaces shall be continuous, without interruption by newel posts or other obstructions.

Exceptions:

1. *Handrails* within dwelling units are permitted to be interrupted by a newel post at a turn or landing.

2. Within a dwelling unit, the use of a volute, turnout, starting easing or starting newel is allowed over the lowest tread.

3. *Handrail* brackets or balusters attached to the bottom surface of the *handrail* that do not project horizontally beyond the sides of the *handrail* within $1^{1}/_{2}$ inches (38 mm) of the bottom of the *handrail* shall not

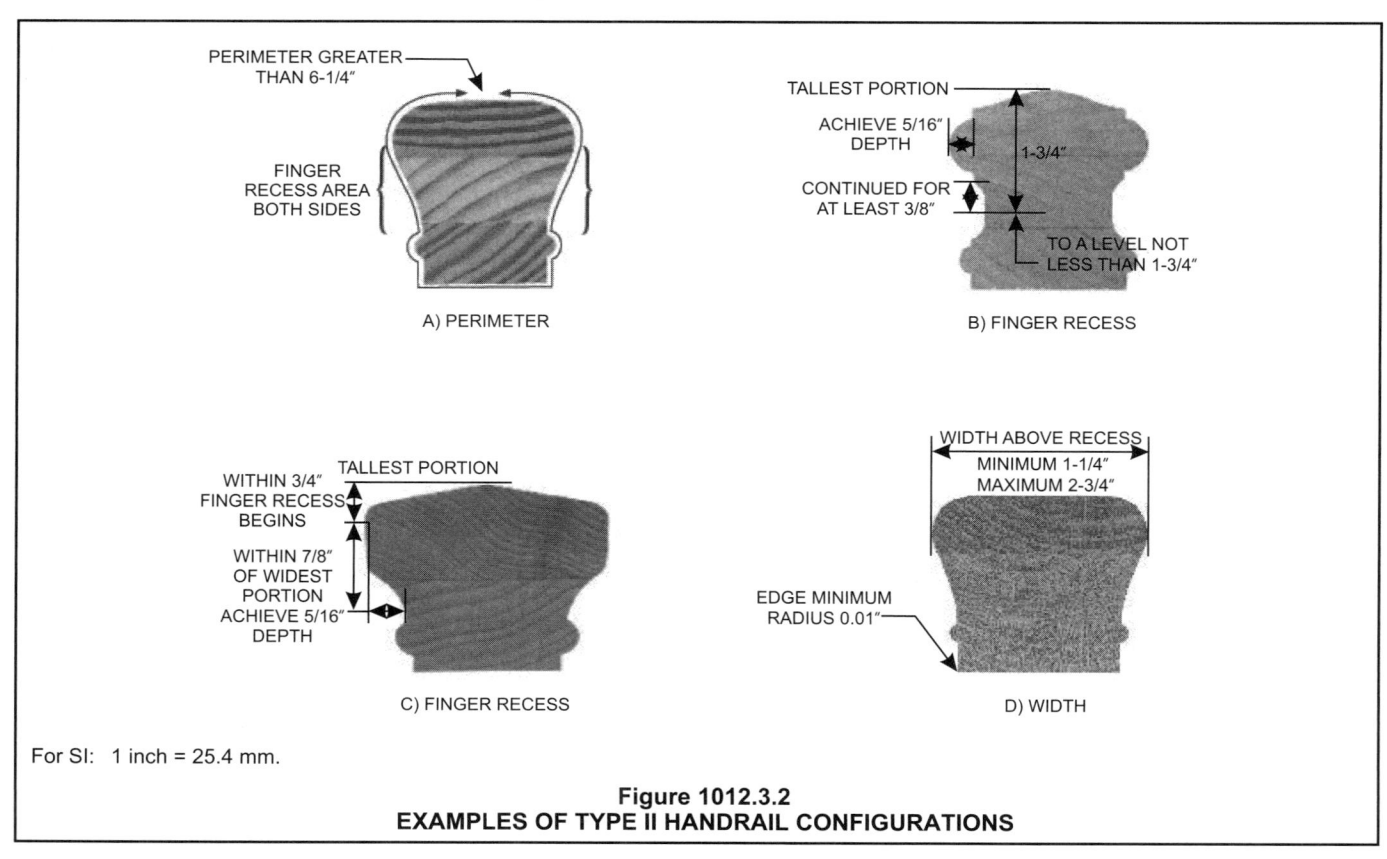

For SI: 1 inch = 25.4 mm.

Figure 1012.3.2
EXAMPLES OF TYPE II HANDRAIL CONFIGURATIONS

be considered obstructions. For each $^1/_2$ inch (12.7 mm) of additional *handrail* perimeter dimension above 4 inches (102 mm), the vertical clearance dimension of $1^1/_2$ inches (38 mm) shall be permitted to be reduced by $^1/_8$ inch (3 mm).

4. Where *handrails* are provided along walking surfaces with slopes not steeper than 1:20, the bottoms of the *handrail* gripping surfaces shall be permitted to be obstructed along their entire length where they are integral to crash rails or bumper guards.

❖ The degree of occupant safety as it relates to handrail use is a function of the features of handrail construction.

Handrails must be usable for their entire length without requiring the users to release their grasp. Typically, if using the handrail while traveling the means of egress, an individual's arm is extended to lead the body with the hand forming a loose grip on the top and sides of the rail. If handrails are to be of service to the occupants, they must be uninterrupted and continuous. Oversize newels, changes in the guard system, or excessive supports at the bottom of small perimeter handrails can cause interruption of the handrail, requiring the occupants to release their grip [see Figure 1012.4(1)]. Exception 1 allows the interruption of the handrail by a newel post at an the intersection of two handrail sections at a turn within a flight or at a landing; however, this exception is not applicable to curved or spiral stairs. Exception 2 provides for familiar and historical handrail details often combined with guards that have been used for years in dwelling units without substantiated incidence to the safety of the occupants. Exception 3 provides specifications for methods of handrail support to limit interruptions of the grip surfaces at

the bottom of smaller perimeter shapes that otherwise would deter or impede the user's ability to attain a stabilizing grip that is essential to safe stairway use. Larger handrail sizes permit shorter brackets since geometrically the finger clearance is still maintained [see Figure 1012.4(2)]. For example, a Type II handrail may elevate the fingertips completely above the area where supports are attached. Exception 4 allows for products that serve dual purposes, such as the bumper guard/handrail sometimes found in hospitals and nursing homes along corridors, to have a continuous bottom support. Since these are only permitted on slopes that are less than what is defined as a ramp, the handrails are more for assistance rather than to arrest a fall.

1012.5 Fittings. *Handrails* shall not rotate within their fittings.

❖ Fittings are those component pieces of a continuous handrail that are shaped or bent and attached to the longer sections of straight or curved handrail to provide for transition at changes in pitch, direction or to provide for termination of a continuous handrail. Fittings and handrails must be securely joined to ensure a stable handrail that does not allow any portion to rotate when grasped.

1012.6 Handrail extensions. *Handrails* shall return to a wall, *guard* or the walking surface or shall be continuous to the handrail of an adjacent *stair flight* or ramp run. Where *handrails* are not continuous between *flights*, the *handrails* shall extend horizontally at least 12 inches (305 mm) beyond the top riser and continue to slope for the depth of one tread beyond the bottom riser. At *ramps* where *handrails* are not continuous between runs, the *handrails* shall extend horizontally above the landing 12 inches (305 mm) minimum beyond the top and bottom of *ramp* runs. The extensions of *handrails* shall be in the same

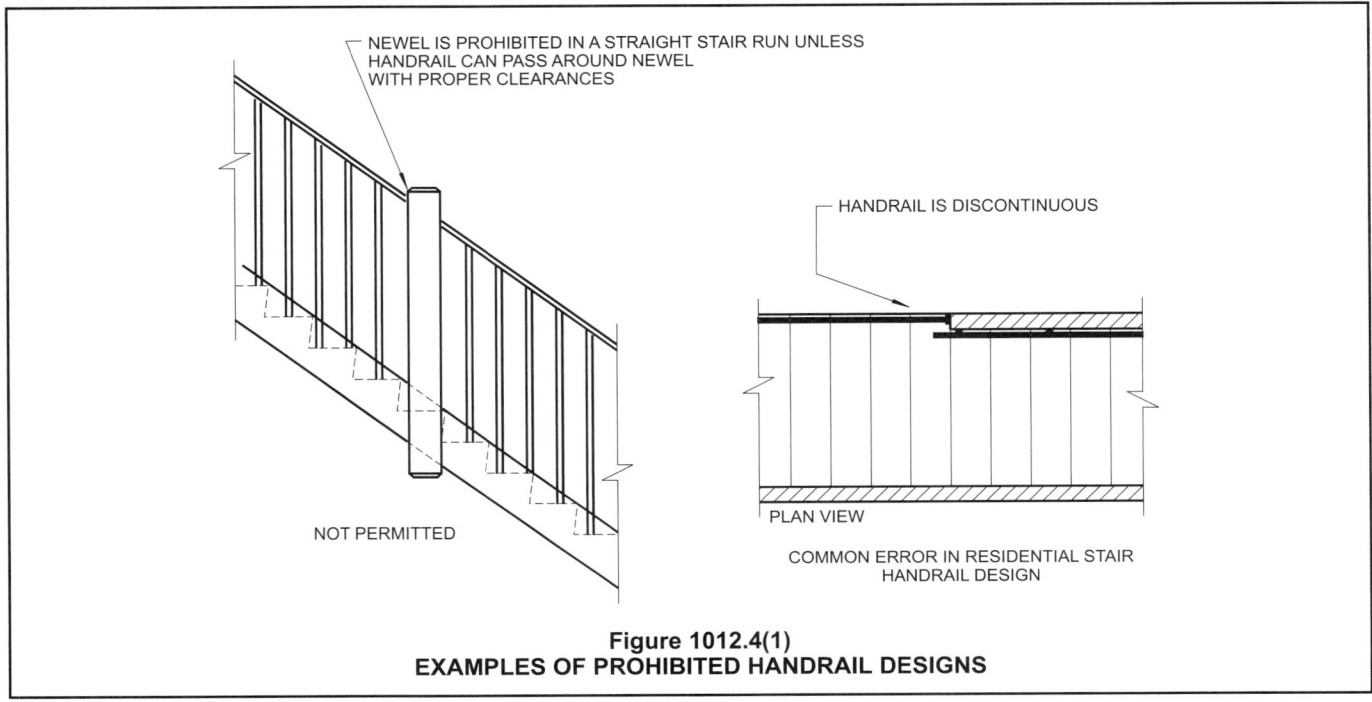

NEWEL IS PROHIBITED IN A STRAIGHT STAIR RUN UNLESS HANDRAIL CAN PASS AROUND NEWEL WITH PROPER CLEARANCES

NOT PERMITTED

HANDRAIL IS DISCONTINUOUS

PLAN VIEW

COMMON ERROR IN RESIDENTIAL STAIR HANDRAIL DESIGN

Figure 1012.4(1)
EXAMPLES OF PROHIBITED HANDRAIL DESIGNS

direction of the *stair flights* at *stairways* and the *ramp* runs at *ramps*.

Exceptions:

1. *Handrails* within a dwelling unit that is not required to be *accessible* need extend only from the top riser to the bottom riser.

2. *Aisle handrails* in Group A and E occupancies in accordance with Section 1028.13.

3. *Handrails* for *alternating tread devices* and ship ladders are permitted to terminate at a location vertically above the top and bottom risers. Handrails for *alternating tread devices* and ship ladders are not required to be continuous between *flights* or to extend beyond the top or bottom risers.

❖ The purpose of the handrail return requirements is to prevent a person from catching an article of loose clothing on it or being injured by falling onto the end of an extended handrail.

The length that a handrail extends beyond the top and bottom of a stairway, ramp or intermediate landing where handrails are not continuous to another stair flight or ramp run is an important factor for the safety of the users. An occupant must be able to grasp securely a handrail beyond the last riser of a stairway or the last sloped segment of a ramp. Handrails that bend around a corner do not provide this stability; therefore, the handrail must extend in the direction of the stair flight or ramp run. Where a user could keep his or her hand on the handrail, such as the continuous handrail at the landing of a switchback stairway or ramp, the handrail extensions are not required.

For stairways, handrails must be extended 12 inches (305 mm) horizontally beyond the top riser and sloped a distance of one tread depth beyond the bottom riser. For ramps, handrails must be extended 12 inches (305 mm) horizontally beyond the last sloped ramp segment at both the top and bottom locations. These handrail extensions are not only required at the top and bottom on both sides of stairways and ramps, but also at other places where handrails are not continuous, such as landings and platforms. These requirements are intended to reflect the current provisions of ICC A117.1 (see Figure 1012.2) and the new ADA/ABA Guidelines. Note that if the handrail extension is at a location that could be considered a protruding object, the handrail must return to the post at a height of less than 27 inches (686 mm) above the floor.

In accordance with Exception 1, handrail extensions are not required where a dwelling unit is not required to meet any level of accessibility (i.e. Accessible unit, Type A unit or Type B unit). Handrail extensions are permitted to end at a newel post or turnaround. Exception 2 provides for handrails along ramped or stepped aisles in assembly seating areas and educational facilities that provide seating similar to assembly seating configurations, such as in a lecture hall. It is necessary to have discontinuous handrails for assembly installations to provide for circulation of the occupants from the aisle to the seating areas; however, the handrail must still extend the full run of the aisle stairs or ramp. Exception 3 allows for the unique construction considerations for alternating tread devices and ship ladders. Again, usage of these devices is very limited. Since alternative tread devices and ship ladders are typically utilized as a safer alternative to a ladder, they are often located in tight spaces where traditional-type stairs cannot be used. With a much steeper angle than traditional stairs and differing usage, the practicality of the extension and continuity provisions was not deemed appropriate.

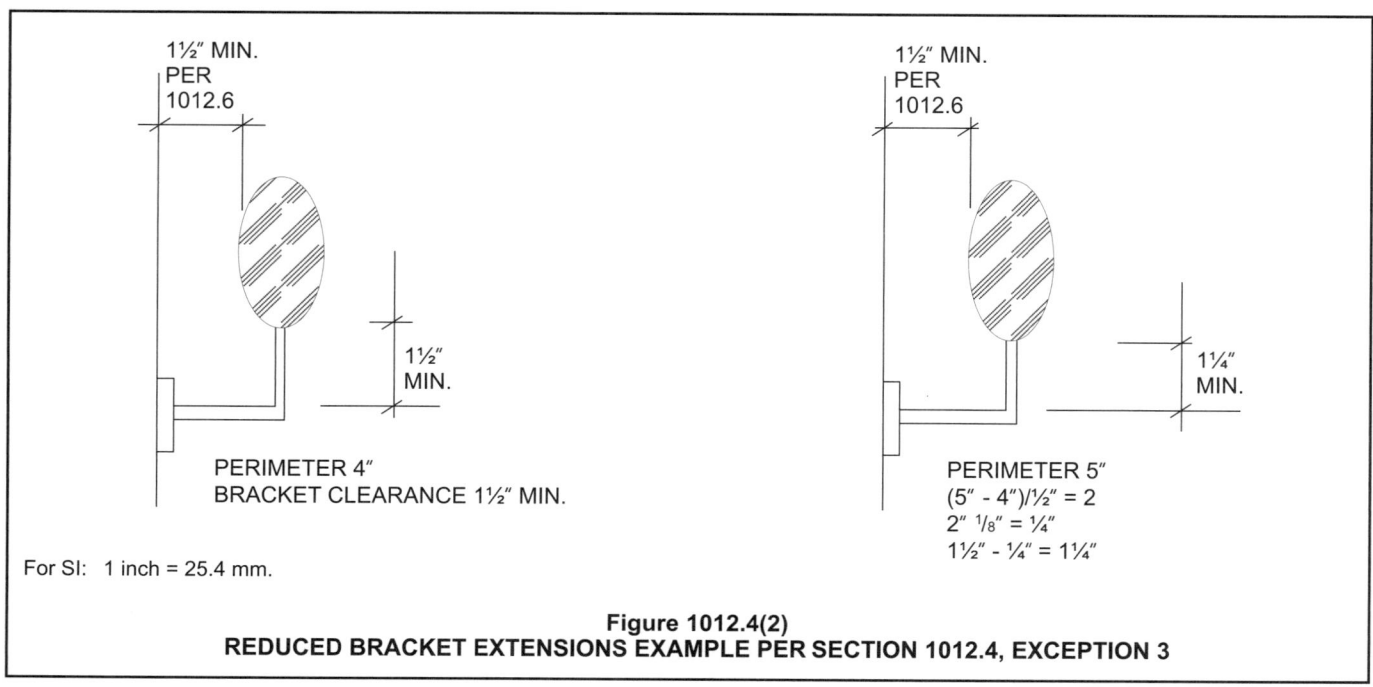

1½″ MIN.
PER 1012.6

1½″ MIN.

PERIMETER 4″
BRACKET CLEARANCE 1½″ MIN.

1½″ MIN.
PER 1012.6

1¼″ MIN.

PERIMETER 5″
(5″ - 4″)/½″ = 2
2″ ⅛″ = ¼″
1½″ - ¼″ = 1¼″

For SI: 1 inch = 25.4 mm.

Figure 1012.4(2)
REDUCED BRACKET EXTENSIONS EXAMPLE PER SECTION 1012.4, EXCEPTION 3

1012.7 Clearance. Clear space between a *handrail* and a wall or other surface shall be a minimum of 1¹/₂ inches (38 mm). A *handrail* and a wall or other surface adjacent to the *handrail* shall be free of any sharp or abrasive elements.

❖ A clear space is needed between a handrail and the wall or other surface to allow the user to slide his or her hand along the rail with fingers in the gripping position without contacting the wall surface, which could have an abrasive texture. In climates where persons may be expected to be wearing heavy gloves during the winter, an open design not adjacent to a wall or other surface where possible would be desirable at an exterior stairway, or a stairway directly inside the entrance to a building. [See Figures 1012.4(2) and 1012.8(2) for an illustration of handrail clearance.]

1012.8 Projections. On ramps, the clear width between *handrails* shall be 36 inches (914 mm) minimum. Projections into the required width of *stairways* and *ramps* at each *handrail* shall not exceed 4¹/₂ inches (114 mm) at or below the *handrail* height. Projections into the required width shall not be limited above the minimum headroom height required in Section 1009.2.

❖ Handrails may not project more than 4¹/₂ inches (114 mm) into the required width of a stairway, so that the clear width of the passage will not be seriously reduced [see Figure 1012.8(1)]. This is consistent with Section 1003.3.2. This projection may exist below the handrail height as well [see Figure 1012.8(2)].

1012.9 Intermediate handrails. *Stairways* shall have intermediate *handrails* located in such a manner that all portions of the *stairway* width required for egress capacity are within 30 inches (762 mm) of a *handrail*. On monumental *stairs*, *handrails* shall be located along the most direct path of egress travel.

❖ In order to always be available to the user of the stairway, the maximum distance to a handrail from within the required width must not exceed 30 inches (762 mm). People tend to walk adjacent to handrails, and if intermediate handrails are not provided for very wide stairways, the center portion of such stairways will normally receive limited use. More importantly, in emergencies, the center portions of wide stairways with handrails would be used more aptly to speed up egress travel rather than delay it by overcrowding at the sides with the handrails. This would especially be true under panic conditions. Without the requirement for intermediate handrails, the use of wide interior stairways could become particularly hazardous.

The distance to the handrail applies to the "required width" of the stairway. If a stairway is greater than 60 inches (1524 mm) in width, but only 60 inches (1524 mm) are required based on occupant load (see Section 1005.1), intermediate handrails are not required. Adequate safety is provided since every user is within 30 inches (762 mm) of a handrail.

The criteria for monumental stairways deal with the very wide stairway in relation to the required width. While handrails on both sides of the stairway may be

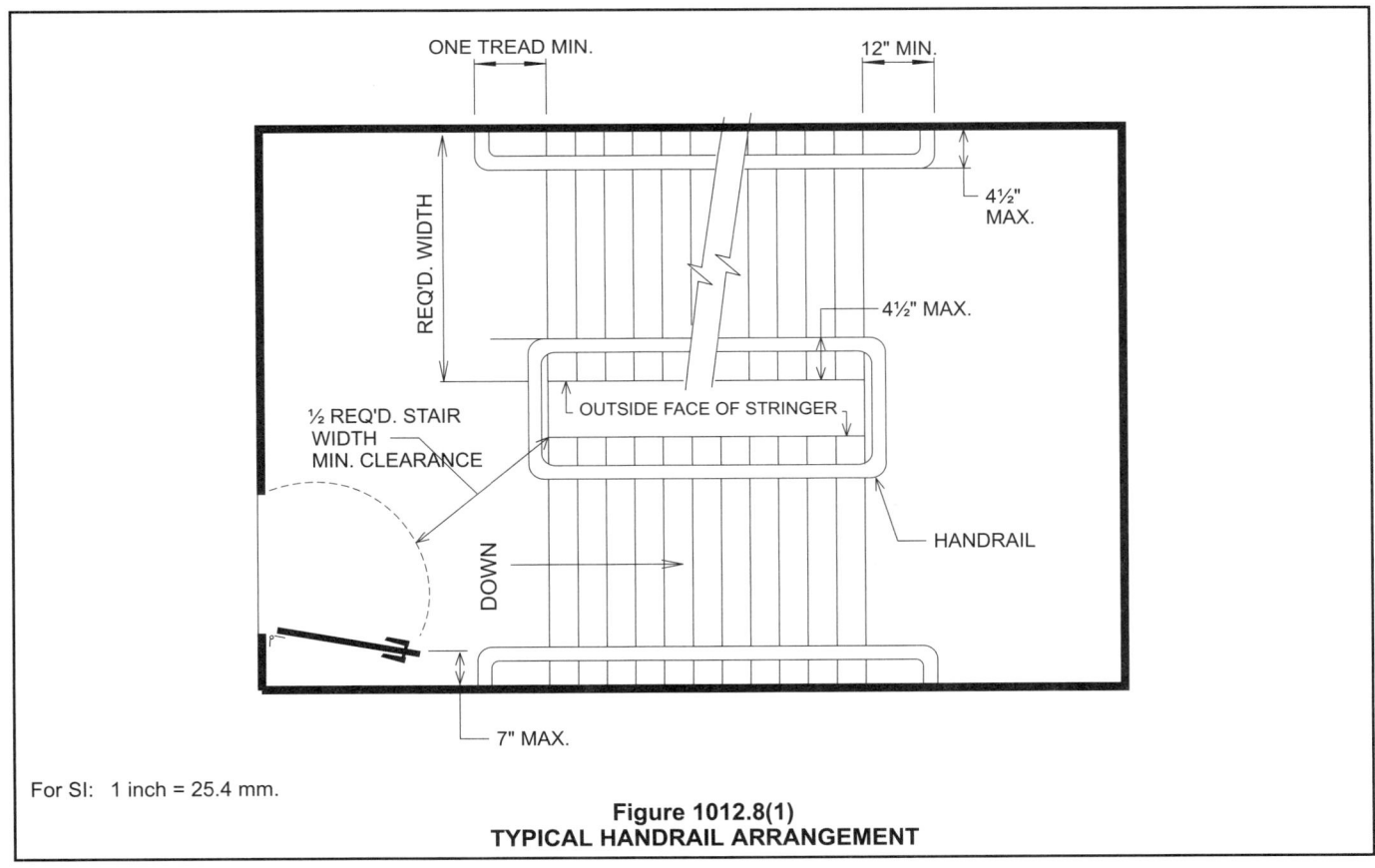

For SI: 1 inch = 25.4 mm.

Figure 1012.8(1)
TYPICAL HANDRAIL ARRANGEMENT

sufficient to accommodate the required width, the handrails may not be near the stream of traffic or even apparent to the user. In this case, the handrails are to be placed in a location more reflective of the egress path (see Figure 1012.9 for handrail locations for monumental stairs).

SECTION 1013
GUARDS

1013.1 Where required. *Guards* shall be located along open-sided walking surfaces, including *mezzanines*, *equip-* *ment platforms*, *stairs*, *ramps* and landings that are located more than 30 inches (762 mm) measured vertically to the floor or grade below at any point within 36 inches (914 mm) horizontally to the edge of the open side. *Guards* shall be adequate in strength and attachment in accordance with Section 1607.7.

Exception: *Guards* are not required for the following locations:

1. On the loading side of loading docks or piers.

2. On the audience side of stages and raised platforms, including steps leading up to the stage and raised platforms.

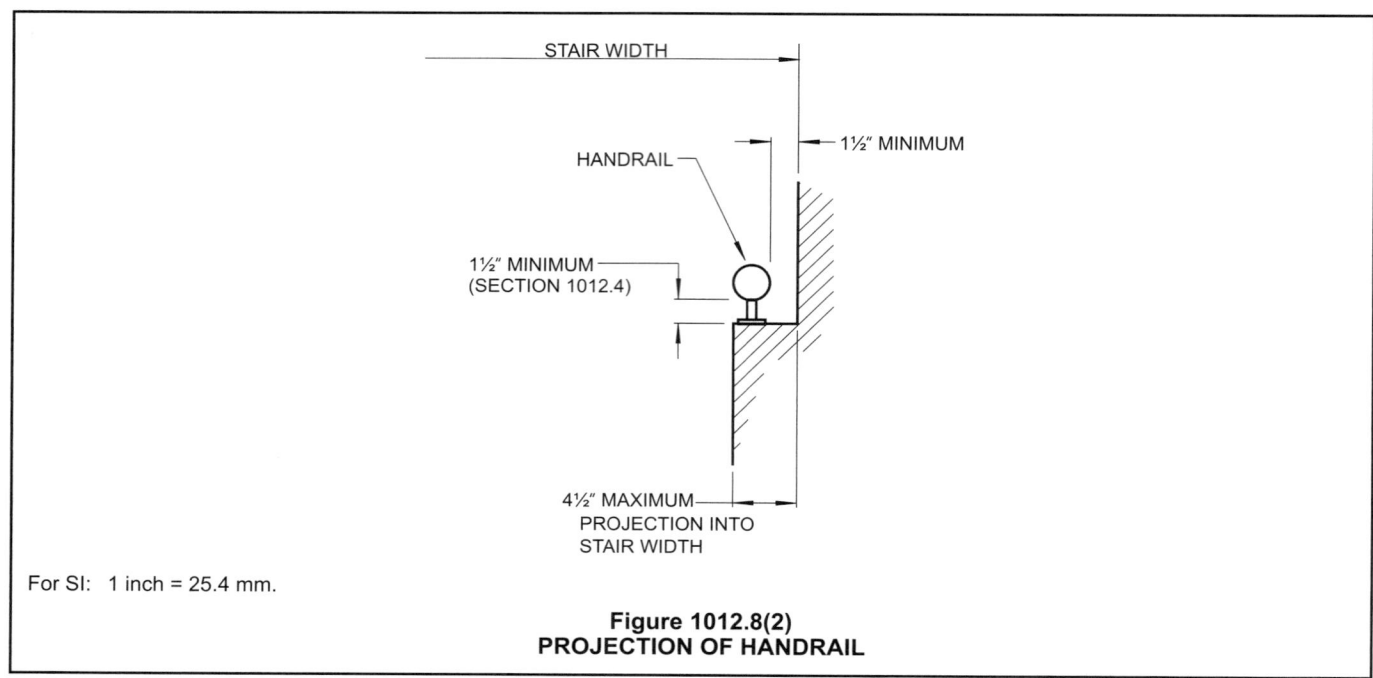

For SI: 1 inch = 25.4 mm.

Figure 1012.8(2)
PROJECTION OF HANDRAIL

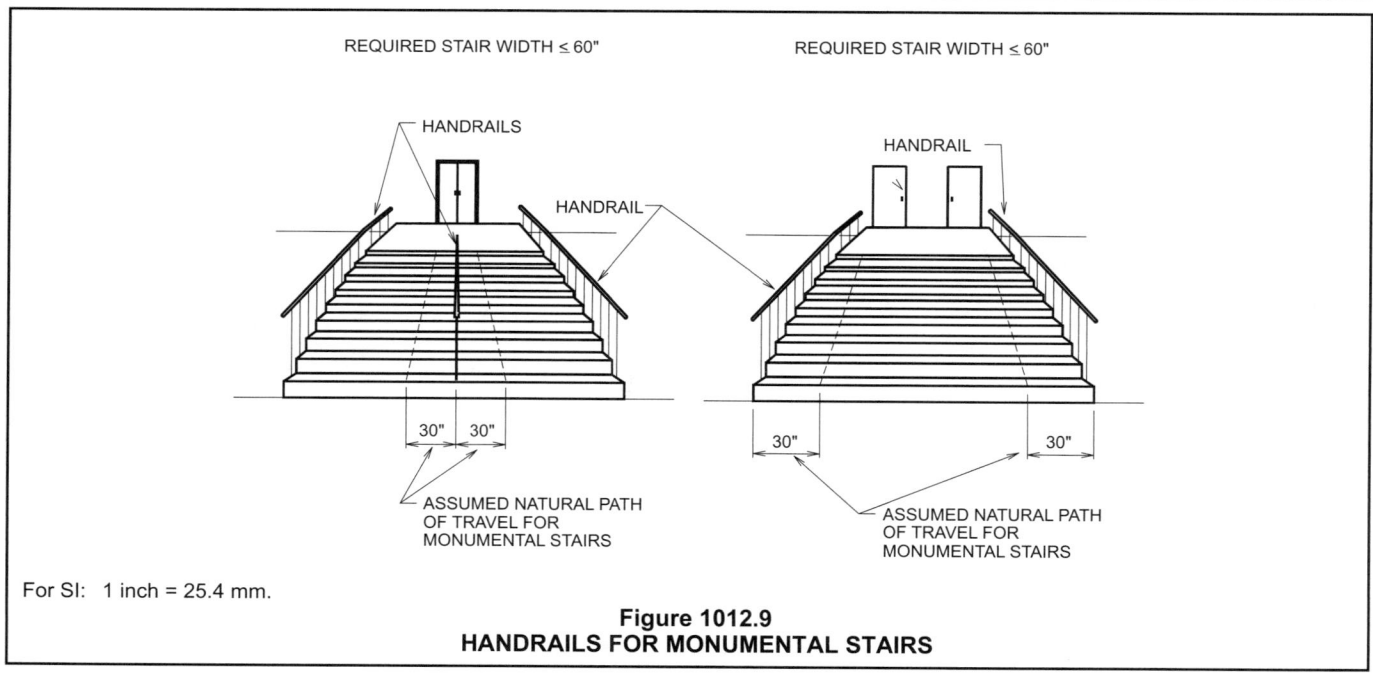

For SI: 1 inch = 25.4 mm.

Figure 1012.9
HANDRAILS FOR MONUMENTAL STAIRS

3. On raised stage and platform floor areas, such as runways, ramps and side stages used for entertainment or presentations.

4. At vertical openings in the performance area of stages and platforms.

5. At elevated walking surfaces appurtenant to stages and platforms for access to and utilization of special lighting or equipment.

6. Along vehicle service pits not accessible to the public.

7. In assembly seating where *guards* in accordance with Section 1028.14 are permitted and provided.

❖ Where one or more sides of a walking surface are open to the floor level or grade below, a guard system must be provided to minimize the possibility of occupants accidentally falling to the surface below [see Figure 1013.1(1)]. A guard is required only where the difference in elevation between the higher walking surface and the surface below is greater than 30 inches (762 mm). When the ground slopes away from the edge, the vertical distance from the walking surface to the grade or floor below must also be more than 30 inches (762 mm) on the lowest point within a 36-inch (914 mm) radius measured horizontally from the edge of the open-sided walking surface [see Figure 1013.1(2)].

The loads for guard design are addressed in Section 1607 and are typically 50 plf (222 N) along the top. If

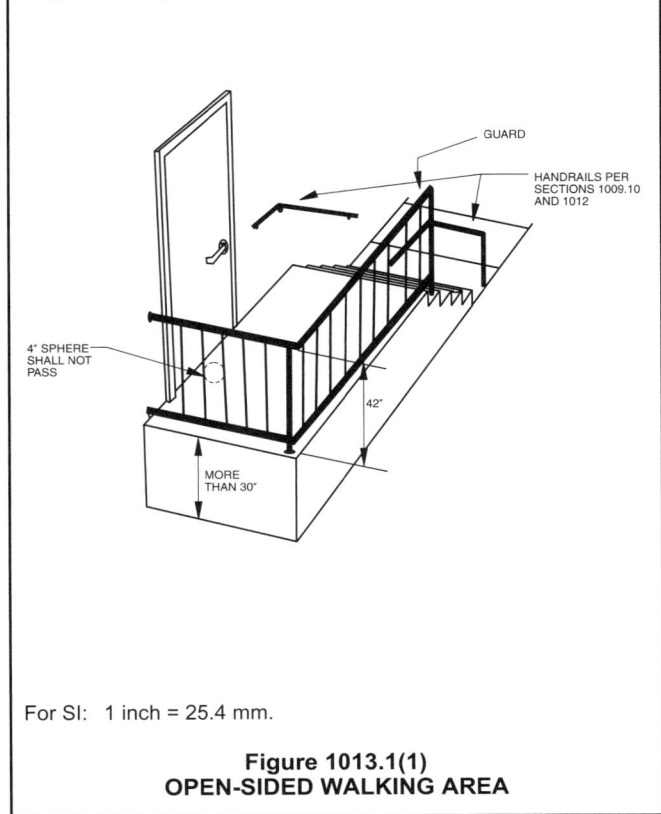

For SI: 1 inch = 25.4 mm.

Figure 1013.1(1)
OPEN-SIDED WALKING AREA

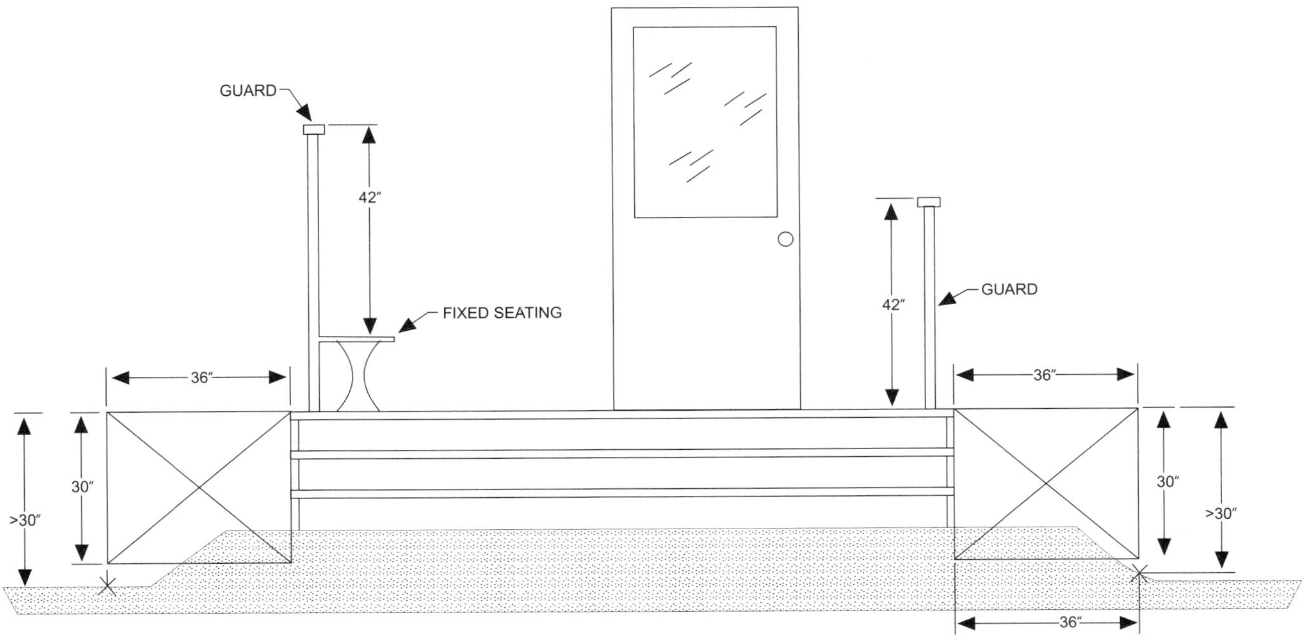

For SI: 1 inch = 25.4 mm.

Figure 1013.1(2)
DROP OFF AND GUARD HEIGHT MESAUREMENTS

glazing is used as part of a guard system, or windows are located adjacent to stairways or ramps, the guard must also comply with Section 1013.1.1 (see Sections 1013.1.1, 1607.7 and 2407).

Most of the exceptions identify situations when guards are not practical, such as along loading docks, stages and their approaches and vehicle service pits. Exception 7 references the lower guards permitted at locations where a line of sight for assembly spaces is part of the considerations.

1013.1.1 Glazing. Where glass is used to provide a *guard* or as a portion of the *guard* system, the *guard* shall also comply with Section 2407. Where the glazing provided does not meet the strength and attachment requirements of Section 1607.7, complying *guards* shall also be located along glazed sides of open-sided walking surfaces.

❖ The loads for guard design in Section 1013.1, which references Section 1607, are typically 50 plf (222 N) along the top. Two different situations are addressed with glazing: where glazing is installed in a guard on the side of a stairway, ramp or landing; or when a stairway, ramp or landing is adjacent to a window when the glazing has not been designed to resist the forces from a fall (see Sections 1607.7 and 2407 and Figure 1013.1.1).

Figure 1013.1.1
GUARD SYSTEM WITH GLAZING

1013.2 Height. Required *guards* shall be not less than 42 inches (1067 mm) high, measured vertically above the adjacent walking surfaces, adjacent fixed seating or the line connecting the leading edges of the treads.

Exceptions:

1. For occupancies in Group R-3, and within individual dwelling units in occupancies in Group R-2, *guards* on the open sides of *stairs* shall have a height not less than 34 inches (864 mm) measured vertically from a line connecting the leading edges of the treads.

2. For occupancies in Group R-3, and within individual dwelling units in occupancies in Group R-2, where the top of the *guard* also serves as a *handrail* on the open sides of *stairs*, the top of the *guard* shall not be less than 34 inches (864 mm) and not more than 38 inches (965 mm) measured vertically from a line connecting the leading edges of the treads.

3. The height in assembly seating areas shall be in accordance with Section 1028.14.

4. Along *alternating tread devices* and ship ladders, *guards* whose top rail also serves as a *handrail*, shall have height not less than 30 inches (762 mm) and not more than 34 inches (864 mm), measured vertically from the leading edge of the device tread *nosing*.

❖ Guards must not be less than 42 inches (1067 mm) in height as measured vertically from the top of the guard down to the leading edge of the tread along steps or to an adjacent walking surface for floors and ramps [see Figures 1013.1(1), 1013.2 and 1010.8(4)]. When there is a fixed seat next to a walking surface that requires a guard, the guard height is measured from the top of the seat surface [see Figure 1013.1(2)]. Experience has shown that 42 inches (1067 mm) or more provides adequate height for protection purposes. This puts the top of the guard above the center of gravity of the average adult. Note that with this height requirement, at locations where both a guard and handrail are required, the handrail cannot be at the top of the guard except as permitted in Exception 2.

The requirement for measuring the height of the guard from a fixed seat is due to concerns that people may stand on the seat, even if it is not a walking surface. Remember, the requirement for a guard is measured from the floor in accordance with Section 1013.1. This requirement is not intended to regulate such items as planters or loose furniture next the drop off.

Because of safety concerns, the designer may want to install a guard where there is a drop-off of less than 30 inches (762 mm). Decorative guards may be utilized to support handrails or serve as part of the edge protection along a ramp. When nonrequired guards/barriers are provided, the 42-inch (1067 mm) minimum height is not required.

Exceptions 1 and 2 are for nontransient residential occupancies and address guard heights only along the stairways. The handrail provisions allow some residential stairways to only have one handrail (see Section 1009.12). Exceptions 1 and 2 allow for a reduced guard height not only when the guard is also used as a handrail but also when it just serves the purpose of a guard along a stairway. The reduced allowable guard height along stairways is consistent with current construction practice.

Exception 3 references the lower guards permitted at locations where a line of sight for assembly spaces is part of the consideration.

Exception 4 permits a reduction in guard heights based on the limited used and unique design considerations for alternating tread devices and ship ladders (see Sections 1009.10 and 1009.11).

1013.3 Opening limitations. Required *guards* shall not have openings which allow passage of a sphere 4 inches (102 mm) in diameter from the walking surface to the required *guard* height.

Exceptions:

1. From a height of 36 inches (914 mm) to 42 inches (1067 mm), *guards* shall not have openings which allow passage of a sphere $4^3/_8$ inches (111 mm) in diameter.

2. The triangular openings at the open sides of a *stair*, formed by the riser, tread and bottom rail shall not allow passage of a sphere 6 inches (152 mm) in diameter.

3. At elevated walking surfaces for access to and use of electrical, mechanical or plumbing systems or equipment, *guards* shall not have openings which allow passage of a sphere 21 inches (533 mm) in diameter.

4. In areas that are not open to the public within occupancies in Group I-3, F, H or S, and for *alternating tread devices* and ship ladders, *guards* shall not have openings which allow passage of a sphere 21 inches (533 mm) in diameter.

5. In assembly seating areas, *guards* at the end of *aisles* where they terminate at a fascia of boxes, balconies and galleries shall not have openings which allow

passage of a sphere 4 inches in diameter (102 mm) up to a height of 26 inches (660 mm). From a height of 26 inches (660 mm) to 42 inches (1067 mm) above the adjacent walking surfaces, *guards* shall not have openings which allow passage of a sphere 8 inches (203 mm) in diameter.

6. Within individual dwelling units and sleeping units in Group R-2 and R-3 occupancies, *guards* on the open sides of *stairs* shall not have openings which allow passage of a sphere $4^3/_8$ (111 mm) inches in diameter.

❖ The basis for limiting openings in a guard to a 4-inch (102 mm) sphere is because of research that indicates that a 4-inch (102 mm) opening will prevent nearly all children 1 year in age or older from falling through the guard. The allowable opening increases to a $4^3/_8$-inch (203 mm) sphere at heights where falling through the guard is not an issue (Exception 1).

An exception to the 4-inch (102 mm) spacing requirement is that a 6-inch (152 mm) opening is allowed for openings formed by the riser, tread and bottom rail of guards at the open side of a stairway (Exception 2). This is because the geometry of the openings is such that the entire body cannot pass through the triangular opening. In the case of a standard stair, limiting such openings to a 4-inch (102 mm) sphere is impractical to achieve with a sloped bottom member in the guard (see Figure 1013.2).

Exceptions 3 and 4 address areas where the presence of small children is unlikely and often prohibited. Guards along walkways leading to electrical, mechanical and plumbing systems or equipment and in occupancies in Groups I-3, F, H and S may be constructed in such a way that a sphere 21 inches (533 mm) in diameter will not pass through any of the openings [see Figure 1013.3(1)]. This requirement allows the use of one horizontal intermediate member with the standard guard height of 42 inches (1067 mm).

Exception 5, for the guard infill near the top of the guard in assembly seating areas, is to reduce sightline

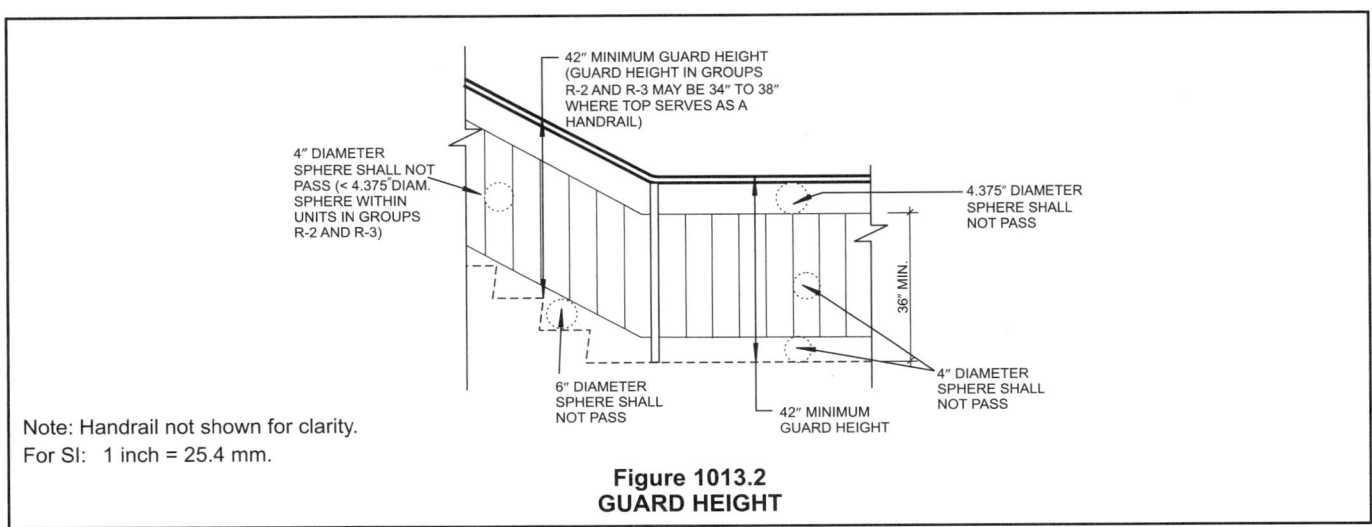

Note: Handrail not shown for clarity.
For SI: 1 inch = 25.4 mm.

Figure 1013.2
GUARD HEIGHT

problems.

Exception 6 is to allow a stairway within a residence that chooses to use the 7-inch rise/11-inch run (178 mm rise/279 mm run) stair configuration to have two spindles per stair tread instead of three spindles. Where the 7³/₄-inch rise/10-inch run (197 mm rise/254 mm run) configuration (see Section 1009.4.2, Exception 5) is utilized, the two spindles would meet the 4-inch (102 mm) maximum provision.

While many of the provisions in guards are to lessen the chances of falls, the opening limitations do not prohibit the use of horizontal guard members as long as they meet the maximum opening provisions. Good design can greatly reduce the opportunity for someone to "climb" the guard [see Figure 1013.3(2)].

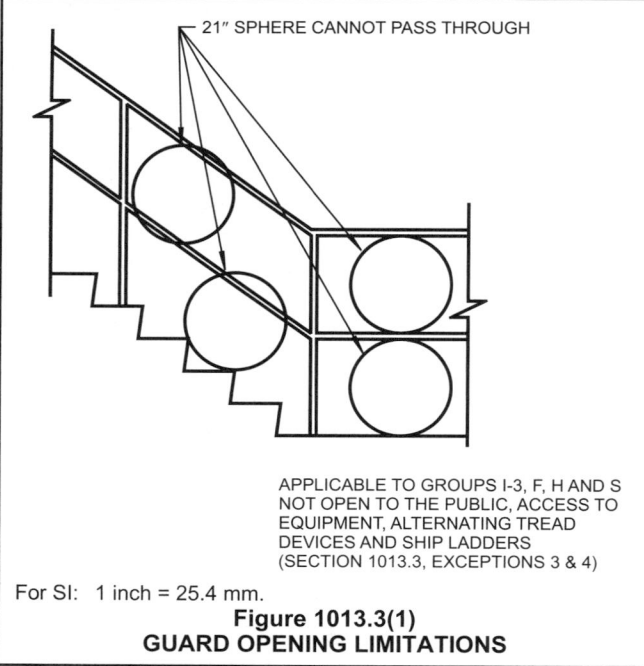

21″ SPHERE CANNOT PASS THROUGH

APPLICABLE TO GROUPS I-3, F, H AND S NOT OPEN TO THE PUBLIC, ACCESS TO EQUIPMENT, ALTERNATING TREAD DEVICES AND SHIP LADDERS (SECTION 1013.3, EXCEPTIONS 3 & 4)

For SI: 1 inch = 25.4 mm.

Figure 1013.3(1)
GUARD OPENING LIMITATIONS

For SI: 1 inch = 25.4 mm.

Figure 1013.3(2)
EXAMPLE OF GUARD WITH HORIZONTAL MEMBERS

1013.4 Screen porches. Porches and decks which are enclosed with insect screening shall be provided with *guards* where the walking surface is located more than 30 inches (762 mm) above the floor or grade below.

❖ Insect screening located on the open sides of porches and decks does not provide an adequate barrier to reasonably protect an occupant from falling to the surface below. Guards are required on the open sides of porches and decks where the floor is located more than 30 inches (762 mm) above the surface below. The guards must comply with the provisions of Section 1013. This provision does not reference the measurement method in Section 1013.1, but if a fixed seat is provided at the edge of the porch, the guard height would need to comply with Section 1013.2.

1013.5 Mechanical equipment. *Guards* shall be provided where appliances, equipment, fans, roof hatch openings or other components that require service are located within 10 feet (3048 mm) of a roof edge or open side of a walking surface and such edge or open side is located more than 30 inches (762 mm) above the floor, roof or grade below. The *guard* shall be constructed so as to prevent the passage of a sphere 21 inches (533 mm) in diameter. The *guard* shall extend not less than 30 inches (762 mm) beyond each end of such appliance, equipment, fan or component.

❖ The purpose of this requirement is to protect workers from falls off of roofs or from open-sided walking surfaces when doing maintenance work on equipment. The guard opening is allowed to be up to 21 inches (533 mm) since children are not likely to be in such areas. This requirement allows the use of horizontal intermediate members.

1013.6 Roof access. *Guards* shall be provided where the roof hatch opening is located within 10 feet (3048 mm) of a roof edge or open side of a walking surface and such edge or open side is located more than 30 inches (762 mm) above the floor, roof or grade below. The *guard* shall be constructed so as to prevent the passage of a sphere 21 inches (533 mm) in diameter.

❖ The code already requires guards around equipment on the roof; this section is intended to provide the same level of safety at the opening that the service personnel are using to access the roof (see Figure 1013.6).

SECTION 1014
EXIT ACCESS

1014.1 General. The *exit access* shall comply with the applicable provisions of Sections 1003 through 1013. *Exit access* arrangement shall comply with Sections 1014 through 1019.

❖ Sections 1014 through 1019 include the design requirements for exit access and exit access components. The general requirements that also apply to the exit access are in Sections 1003 through 1013.

1014.2 Egress through intervening spaces. Egress through intervening spaces shall comply with this section.

1. Egress from a room or space shall not pass through adjoining or intervening rooms or areas, except where such adjoining rooms or areas and the area served are accessory to one or the other, are not a Group H occupancy and provide a discernible path of egress travel to an *exit*.

    Exception: *Means of egress* are not prohibited through adjoining or intervening rooms or spaces in a Group H, S or F occupancy when the adjoining or intervening rooms or spaces are the same or a lesser hazard occupancy group.

2. An *exit access* shall not pass through a room that can be locked to prevent egress.

3. *Means of egress* from dwelling units or sleeping areas shall not lead through other sleeping areas, toilet rooms or bathrooms.

4. Egress shall not pass through kitchens, storage rooms, closets or spaces used for similar purposes.

    Exceptions:

    1. *Means of egress* are not prohibited through a kitchen area serving adjoining rooms constituting part of the same dwelling unit or sleeping unit.

    2. *Means of egress* are not prohibited through stockrooms in Group M occupancies when all of the following are met:

        2.1. The stock is of the same hazard classification as that found in the main retail area;

        2.2. Not more than 50 percent of the *exit access* is through the stockroom;

        2.3. The stockroom is not subject to locking from the egress side; and

        2.4. There is a demarcated, minimum 44-inch-wide (1118 mm) *aisle* defined by full- or partial-height fixed walls or similar construction that will maintain the required width and lead directly from the retail area to the *exit* without obstructions.

❖ This section allows adjoining spaces to be considered a part of the room or space from which egress originates, provided that there are reasonable assurances that the continuous egress path will always be available. The code does not limit the number of intervening or adjoining rooms through which egress can be made, provided that all other code requirements (i.e., travel distance, number of doorways, etc.) are met. An exit access route, for example, may be laid out such that an occupant leaves a room or space, passes through an adjoining space, enters an exit access corridor, passes through another room and, finally, into an exit [see Figure 1014.2(2)] as long as all other code requirements are satisfied.

The intent of Item 1 is not that the accessory space be limited to the 10-percent area in Section 508.2.1, but that the spaces be interrelated so that doors between the spaces will not risk being blocked or locked. For example, a conference room and managers' offices could exit through the secretary's office to reach the exit access corridor; or several office spaces could exit through a common reception/lobby area. Requir-

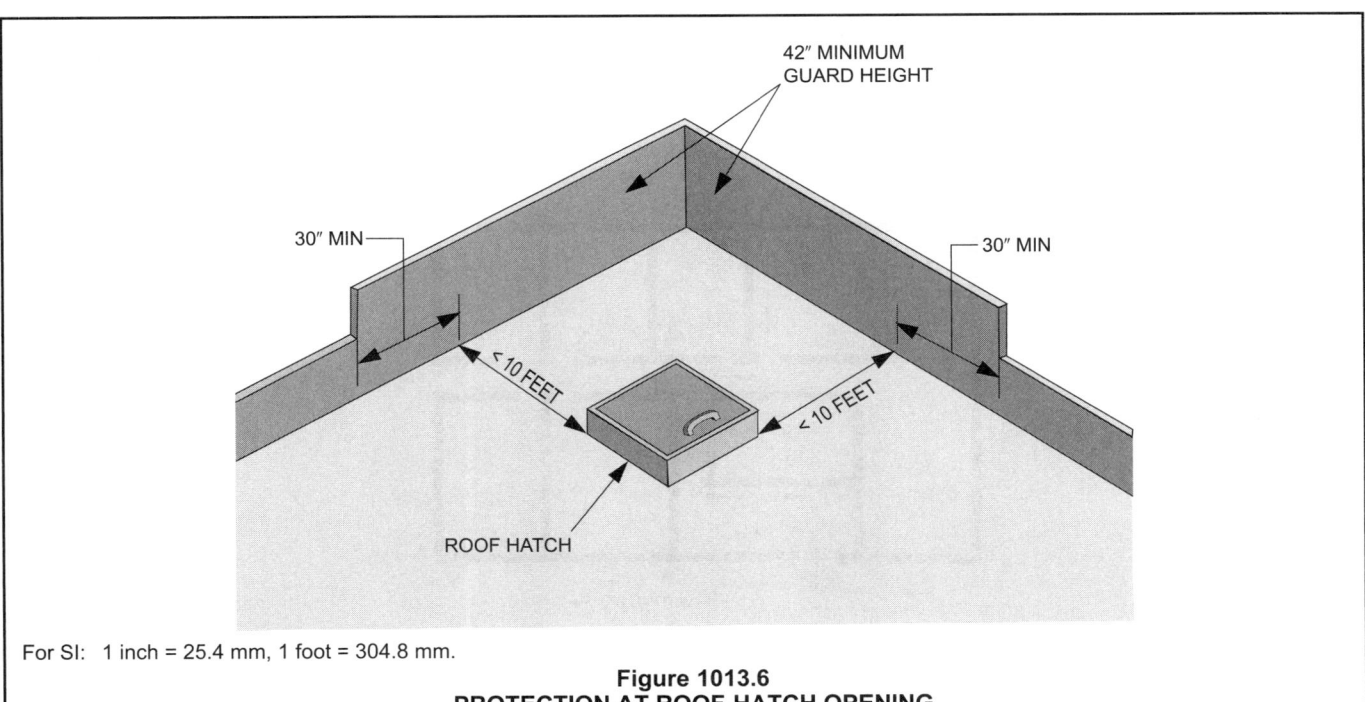

For SI: 1 inch = 25.4 mm, 1 foot = 304.8 mm.

Figure 1013.6
PROTECTION AT ROOF-HATCH OPENING

ing occupants to egress from an area and pass through an adjoining Group H that can be characterized by rapid fire buildup, or worse, places them in an unreasonable risk situation [see Figure 1014.2(1)]; therefore, this illustrated egress path would be prohibited. As an exception to Item 1, in facilities that may contain a Group H area, buildings of Group H, S or F can exit through adjoining rooms or spaces that have the same or less hazard. For example, a person exiting from a Group H storage room (see Section 415) could egress either through a similar Group H storage area or through the factory to get to an exit, but the person in the factory could not egress through the Group H storage rooms to get to the outside.

As expressed in Item 2, a common code enforcement problem is a locked door in the egress path. Twenty-five workers perished in September 1991 when they were trapped inside the Imperial Food Processing Plant in Hamlet, North Carolina, partially be-cause of locked exit doors. As long as the egress door is readily openable in the direction of egress travel without the use of keys, special knowledge or effort (see Section 1008.1.9.5), the occupants can move un-impeded away from a fire emergency. Relying on an egress path through an adjacent dwelling unit to be available at all times is not a reasonable expectation. Egress through an adjacent business tenant space can be unreasonable given the security and privacy measures the adjacent tenant may take to secure such a space. However, egress through a reception area that serves a suite of offices of the same tenant is clearly accessible and is permitted.

Item 3 is to address concerns for the path of egress travel within individual units. The concern once again is possible locking devices. Egress for one bedroom should not be through another bedroom or bathroom.

The concern in Item 4 is that kitchens, storage rooms and similar spaces may be subject to locking or

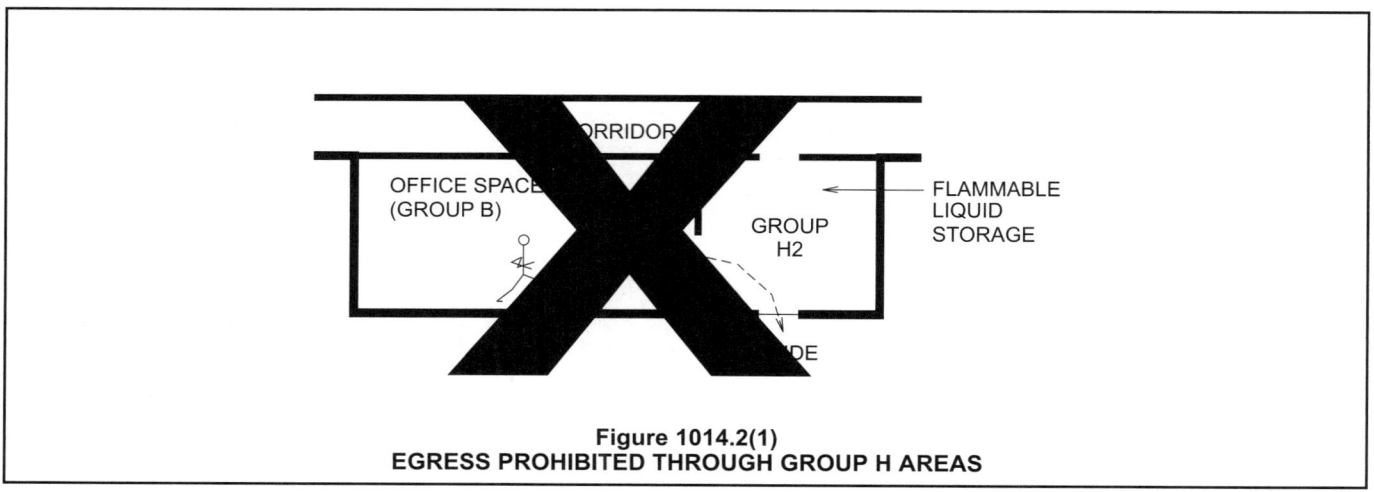

Figure 1014.2(1)
EGRESS PROHIBITED THROUGH GROUP H AREAS

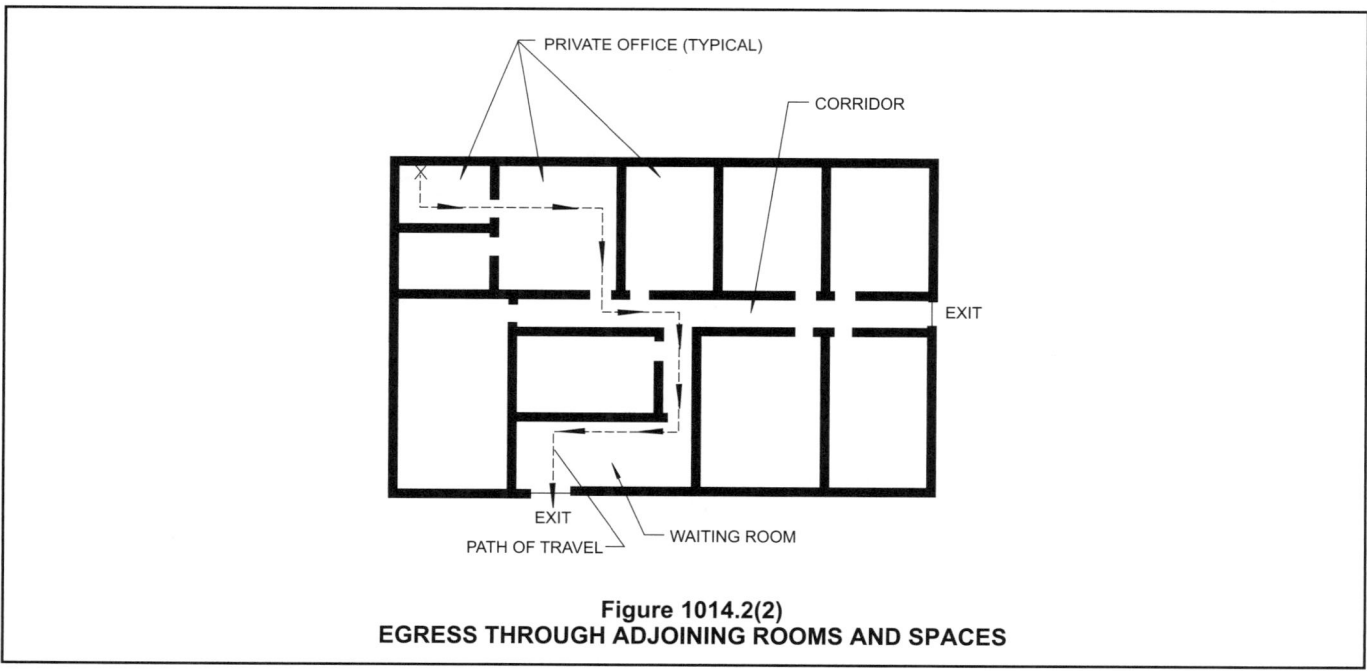

Figure 1014.2(2)
EGRESS THROUGH ADJOINING ROOMS AND SPACES

blockage of the exit access path. This is not a general provision for all Group S occupancies; therefore, it is not the intent of this provision to address the situation of egress for offices through an associated warehouse space. Item 4, Exception 1, does not apply this same prohibition to areas within dwelling or sleeping units. However, for other spaces, for example, a means of egress should not be through the working portions of a commercial kitchen behind a restaurant or the stock storage area of a storage room behind a mercantile occupancy. A definitive path must be available through the space. The four items listed in Item 4, Exception 2, are intended to provide measurable criteria to increase the likelihood that the exit access path of travel would always be available and identifiable. It is not acceptable to just mark the path on the floor. Whatever defines the route must permanently establish the egress path in a manner to maintain the minimum required unobstructed width.

1014.2.1 Multiple tenants. Where more than one tenant occupies any one floor of a building or structure, each tenant space, dwelling unit and sleeping unit shall be provided with access to the required *exits* without passing through adjacent tenant spaces, dwelling units and sleeping units.

> **Exception:** The *means of egress* from a smaller tenant space shall not be prohibited from passing through a larger adjoining tenant space where such rooms or spaces of the smaller tenant occupy less than 10 percent of the area of the larger tenant space through which they pass; are the same or similar occupancy group; a discernable path of egress travel to an *exit* is provided; and the *means of egress* into the adjoining space is not subject to locking from the egress side. A required *means of egress* serving the larger tenant space shall not pass through the smaller tenant space or spaces.

❖ Where a floor is occupied by multiple tenants, each tenant must be provided with full and direct access to the required exits serving that floor without passing through another tenant space. Tenants frequently lock the doors to their spaces for privacy and security. Should an egress door that is shared by both tenants be locked, occupants in one of the spaces could be trapped and unable to reach an exit. Therefore, an egress layout where occupants from one tenant space travel through another tenant space to gain access to one of the required exits from that floor is prohibited.

This limitation is so that occupants from all tenant spaces will have unrestricted access to the required egress elements while maintaining the security and privacy of the individual tenants. This limitation is based on one of the fundamental principles of egress: to provide a means of egress where all components are capable of being used by the occupants without keys, tools, special knowledge or effort (see Section 1008.1.9.5).

A common practice is to have a bank or small restaurant located within a large grocery store or department store. These can be separate tenants. In these situations, the small tenants are not open when the main store is closed. The intent of the exception is to allow those small tenants to egress through the large tenant. Since there may be times when the larger tenant is open and the smaller is closed (e.g., bank holidays), the larger tenant cannot exit through the smaller tenant.

1014.2.2 Group I-2. Habitable rooms or *suites* in Group I-2 occupancies shall have an *exit access* door leading directly to a *corridor*.

> **Exception:** Rooms with *exit* doors opening directly to the outside at ground level.

❖ The purpose of this section is to establish the means of egress requirements for rooms and suites that are unique to Group I-2 occupancies. Refer to Section 407 for general egress requirements unique to Group I-2 (e.g., corridors and smoke compartments). Group I-2 typically employs a "protect-in-place" strategy in which patients are relocated by staff to adjacent smoke compartments prior to evacuation. Direct access to the corridor system is a key component to staff access and patient movement. Patient treatment and sleeping rooms are also permitted to be arranged as suites (see Figure 1014.2.2 for an example plan of a patient suite). The figure illustrates a patient suite of more than 1,000 square feet (93 m²) where two remote egress doors are required. The criteria in this section recognizes the low patient-to-staff ratio of these facilities where the staff is directly responsible for the safety of the patients in the event of a fire.

"Habitable rooms" are not intended to include individual bathrooms, closets and similar spaces, as well as briefly occupied spaces, such as control rooms in radiology and small storage/supply rooms.

Health care suites allow for the elimination of the smoke partition between the patient rooms and the staff areas. Curtains or glass partitions are often desirable to allow for a more efficient operation and/or direct and immediate supervision. By considering the connected space an intervening room rather than a corridor, these provisions can be applied.

Traditionally, suites are used when patients require direct observation and immediate access as a component of their medical care. Examples of this include post anesthesia care units, critical care units, cardiac care areas, recovery rooms and surgical suites. While not prohibited in nursing homes, suite configurations are more commonly found in hospital situations. By the definition of the term "Suites" in Section 1002.1, these can be groups of sleeping rooms or treatment rooms and their associated support/staff spaces. Maximum area, number of exit access doors, travel distance within the suite and allowances for intervening rooms between the suite and the corridor are addressed in Section 1014.2.3 for patient sleeping rooms, and Section 1014.2.4 for patient treatment areas. The use of the term "Suite" is not intended to prohibit other groups of rooms that are not patient care areas, such as records, or the business offices found in hospitals.

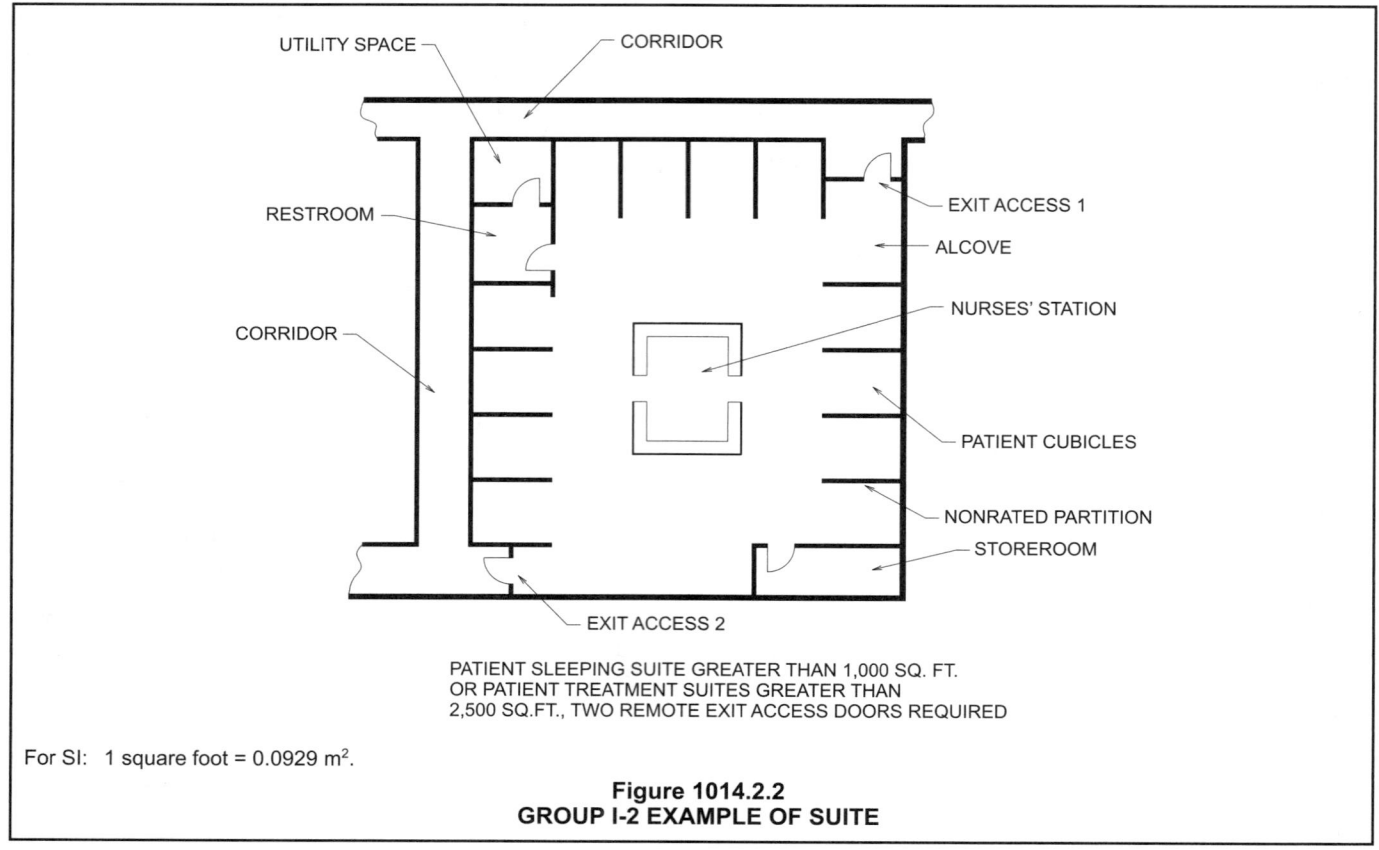

For SI: 1 square foot = 0.0929 m².

Figure 1014.2.2
GROUP I-2 EXAMPLE OF SUITE

1014.2.3 Suites in patient sleeping areas. Patient sleeping areas in Group I-2 occupancies shall be permitted to be divided into *suites* with one intervening room if one of the following conditions is met:

1. The intervening room within the *suite* is not used as an *exit access* for more than eight patient beds.

2. The arrangement of the *suite* allows for direct and constant visual supervision by nursing personnel.

❖ This group of sections deals with "suites" that include patient sleeping rooms. Staff awareness is key to recognizing a dangerous condition and beginning evacuation procedures. In accordance with Section 1014.2.2, "suites" must lead directly to a corridor in order to allow for quick evacuation. The term "Intervening rooms" means a room serving as part of the required means of egress from another room. Section 1014.2.3 allows for there to be one room between the patient sleeping rooms and the corridor in two different scenarios.

1. Where patient rooms within suites do not have direct and constant visual supervision by staff, the suite is limited to eight patient beds if they want to allow for one intervening room.

2. Where the suite can be arranged such that direct and constant supervision of patients can be provided, they can always have one intervening room.

To illustrate Option 1, consider a suite of rooms that serves for infant delivery. Birthing rooms are patient treatment rooms, but standard procedure is now for the patient to stay in that same room rather than being moved, so these rooms are also patient sleeping rooms. The rooms in the suite are typically divided by walls from a central nurse/doctor area so the staff has quick and direct access, but not direct and constant visual supervision; therefore, the suite is limited to eight patient beds. To illustrate Option 2, consider a critical care unit (CCU). Patient rooms are grouped around a central nursing station. These sleeping areas may be directly open to the nurse's station, or the intervening walls may have large windows or glass sliding doors that allow for direct and constant supervision. These patient sleeping rooms can also have one room between the sleeping room and each exit to the corridor, but the number of beds is not limited.

1014.2.3.1 Area. *Suites* of sleeping rooms shall not exceed 5,000 square feet (465 m²).

❖ The combined area of the patient sleeping rooms, the staff areas and any associated spaces (i.e., clean room, nutrition centers, storage room) and any intervening rooms (in accordance with Section 1014.2.3) are permitted to have a total area of 5,000 square feet (465 m²). This suite must be separated from other areas by smoke barriers in accordance with Section 1014.2.7, and meet the 100-foot (30 480 mm) travel distance for exit access in Section 1014.2.3.3.

1014.2.3.2 Exit access. Any patient sleeping room, or any *suite* that includes patient sleeping rooms, of more than 1,000 square feet (93 m²) shall have at least two *exit access* doors remotely located from each other.

❖ A suite that contains a patient sleeping room and an area of 1,000 square feet (93 m²) or less, can have only one exit access door to a corridor, provided that the 100-foot (30 480 mm) maximum travel distance (see Section 1014.2.3.3) is met. For suites that contain patient sleeping rooms and have an area of between 1,000 square feet and 5,000 square feet (93 m² and 465 m²), there must be at least two exit access doors to the corridor. These doors must meet the separation requirements in Section 1015.2.

1014.2.3.3 Travel distance. The travel distance between any point in a *suite* of sleeping rooms and an *exit access* door of that *suite* shall not exceed 100 feet (30 480 mm).

❖ The path of travel from the most remote corner of any patient sleeping room, through the suite, to the exit access door leading to the corridor, must not exceed 100 feet (30 480 mm). As a more specific provision, this would override the 75-foot (22 860 mm) common path of travel requirements in Section 1014.3 in a single-exit suite. In a two-exit suite, at least one of the exits must be within 100 feet (30 480 mm) of travel of every point in the suite. The path of travel measurement would include any travel through the intervening room permitted in Section 1014.2.3. Any travel distance within the suite is included in the overall exit access travel distance of 200 feet (60 960 mm) (see Table 1016.1). The 50-foot (15 240 mm) maximum travel distance out of each patient sleeping room (Section 1014.2.6) into the suite must also be met. To see how the travel distances work together, see Figure 1014.2.3.3.

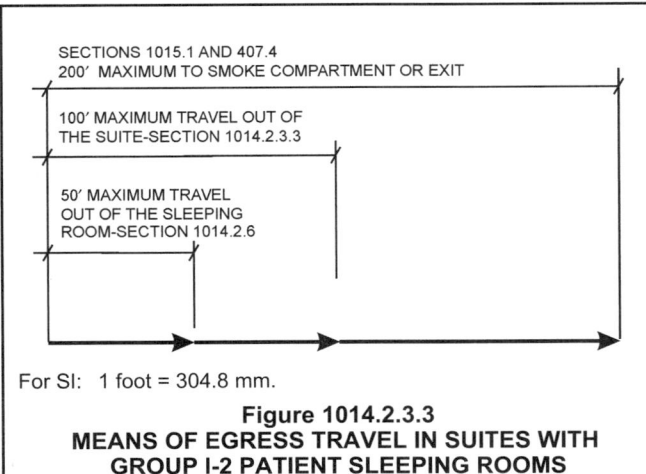

SECTIONS 1015.1 AND 407.4
200' MAXIMUM TO SMOKE COMPARTMENT OR EXIT

100' MAXIMUM TRAVEL OUT OF
THE SUITE-SECTION 1014.2.3.3

50' MAXIMUM TRAVEL
OUT OF THE SLEEPING
ROOM-SECTION 1014.2.6

For SI: 1 foot = 304.8 mm.

Figure 1014.2.3.3
MEANS OF EGRESS TRAVEL IN SUITES WITH
GROUP I-2 PATIENT SLEEPING ROOMS

1014.2.4 Suites in areas other than patient sleeping areas. Areas other than patient sleeping areas in Group I-2 occupancies shall be permitted to be divided into *suites*.

❖ According to the definition for "Suite" in Section 1002.1, suites covered by these provisions are limited to areas that include patient sleeping rooms or patient treatment rooms. This group of sections deals with suites for patient treatment rooms. Common examples include diagnostic or therapy areas where the function of treatment requires adjacency. This could also include in-patient or outpatient surgery suites, emergency treatment area suites or areas such as recovery rooms or dialysis treatment centers. Patients may be anesthetized, but staff/patient ratios are high and staff are trained in proper evacuation procedures.

1014.2.4.1 Area. *Suites* of rooms, other than patient sleeping rooms, shall not exceed 10,000 square feet (929 m²).

❖ The combined area of the patient treatment rooms, the staff areas and any associated spaces (i.e., clean room, nutrition centers and storage room) and any intervening rooms (in accordance with Sections 1014.2.4.3 and 1014.2.4.4) are permitted to have a total area of 10,000 square feet (929 m²). This suite must be separated from other areas by smoke barriers in accordance with Section 1014.2.7.

1014.2.4.2 Exit access. Any room or *suite* of rooms, other than patient sleeping rooms, of more than 2,500 square feet (232 m²) shall have at least two *exit access* doors remotely located from each other.

❖ A suite that contains a patient treatment room and an area of 2,500 square feet (232 m²) or less can have only one exit access door to a corridor, provided that the 100-foot (30 480 mm) maximum travel distance (see Section 1014.2.4.3) is met. As a more specific provision, this would override the 75-foot (22 860 mm) common path of travel requirements in Section 1014.3 in a single-exit suite. For suites that contain patient treatment rooms and have an area of between 2,500 square feet and 10,000 square feet (232 m² and 929 m²), there must be at least two exit access doors to the corridor. These doors must meet the separation requirements in Section 1015.2. In a two-exit suite, at least one of the exits must be within 100 feet (30 480 mm) of travel of every point in the suite (see Section 1014.2.4.3).

1014.2.4.3 One intervening room. For rooms other than patient sleeping rooms, *suites* of rooms are permitted to have one intervening room if the travel distance within the *suite* to the *exit access* door is not greater than 100 feet (30 480 mm).

❖ The term "intervening rooms" means a room serving as part of the required means of egress from another room. In a situation where the suite configuration would require exiting through a room adjacent to, but not part of the treatment room, such as the staff supervision area, patient preparation or waiting area, the travel distance is limited to 100 feet (30 480 mm) maximum. This would be measured from the most remote point in the patient treatment room, through the intervening room, to the exit access door to the corridor. Any travel distance within the suite is included in the overall exit access travel distance of 200 feet (60 960 mm) (see Table 1016.1). Many of the common suite layouts include a "long, skinny room" that resembles a corridor. If this room is not constructed to meet all of the corridor requirements, this is an intervening room. To see how the travel distances work together, see Figure 1014.2.4.3.

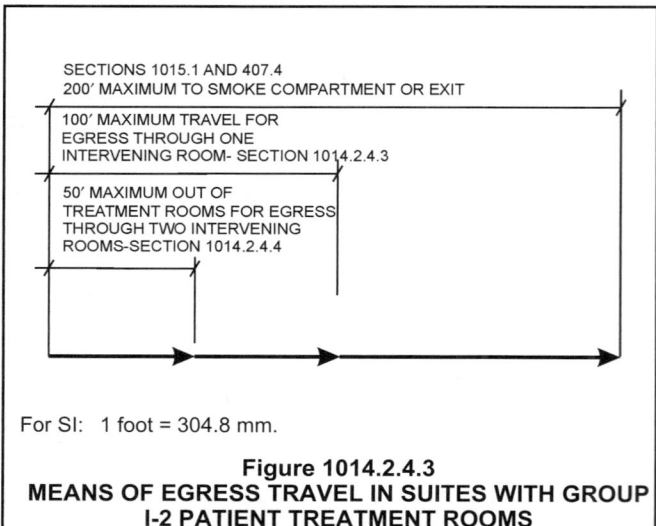

For SI: 1 foot = 304.8 mm.

Figure 1014.2.4.3
MEANS OF EGRESS TRAVEL IN SUITES WITH GROUP I-2 PATIENT TREATMENT ROOMS

1014.2.4.4 Two intervening rooms. For rooms other than patient sleeping rooms located within a *suite*, *exit access* travel from within the *suite* shall be permitted through two intervening rooms where the travel distance to the *exit access* door is not greater than 50 feet (15 240 mm).

❖ In a situation where the suite configuration would require exiting through two rooms adjacent to, but not part of the treatment room, the travel distance is limited to 50 feet (15 240 mm) maximum. This would be measured from the most remote point in the patient treatment room, through the intervening rooms, to the exit access door to the corridor. Any travel distance within the suite is included in the overall exit access travel distance of 200 feet (60 960 mm) (see Table 1016.1). An example of this might be an MRI suite where access to the treatment room was through both a waiting area and a patient prep/staff supervisory area.

1014.2.5 Exit access through suites. *Exit access* from all other portions of a building not classified as a *suite* in a Group I-2 occupancy shall not pass through a *suite*.

❖ Suites are intended to serve specific functions or patients. This section clarifies that the suite should not be used as part of the exit access of other areas within the building. It is intended to prevent corridors from discharging into a maze of intervening rooms prior to entering the exit.

1014.2.6 Travel distance. The travel distance between any point in a Group I-2 occupancy patient sleeping room and an *exit access* door in that room shall not exceed 50 feet (15 240 mm).

❖ This section limits the overall travel distance within an individual patient sleeping room to the exit access door in that room. This is a separate concept from the travel distance within a suite (see Section 104.2.3.3) and should be considered for each patient sleeping

room. To see how the travel distances work together, see Figure 1014.2.3.3.

1014.2.7 Separation. *Suites* in Group I-2 occupancies shall be separated from other portions of the building by a *smoke partition* complying with Section 711.

❖ This section clarifies the type of partition that should be used to separate the suite from the corridor, adjacent suite or any other portion of the building. This section requires separation by a smoke partition, similar to the construction of a corridor wall in Group I-2 (see Section 407.3).

1014.3 Common path of egress travel. In occupancies other than Groups H-1, H-2 and H-3, the *common path of egress travel* shall not exceed 75 feet (22 860 mm). In Group H-1, H-2 and H-3 occupancies, the *common path of egress travel* not exceed 25 feet (7620 mm). For *common path of egress travel* in Group A occupancies and assembly occupancies accessory to Group E occupancies having fixed seating, see Section 1028.8.

Exceptions:

1. The length of a *common path of egress travel* in Group B, F and S occupancies shall not be more than 100 feet (30 480 mm), provided that the building is equipped throughout with an *automatic sprinkler system* installed in accordance with Section 903.3.1.1.

2. Where a tenant space in Group B, S and U occupancies has an *occupant load* of not more than 30, the length of a *common path of egress travel* shall not be more than 100 feet (30 480 mm).

3. The length of a *common path of egress travel* in a Group I-3 occupancy shall not be more than 100 feet (30 480 mm).

4. The length of a common path of egress travel in a Group R-2 occupancy shall not be more than 125 feet (38 100 mm), provided that the building is protected throughout with an *approved automatic sprinkler system* in accordance with Section 903.3.1.1 or 903.3.1.2.

❖ Common path of egress travel does not apply to stories or spaces with one exit. See Section 1021.2 for travel limitations for one-, two- and three-story buildings and Section 1015.1 for spaces with one exit. The definition for "Common path" indicates the provisions are only applicable when access to two or more exits is required.

The common path of travel is the distance measured from the most remote point in a space to the point in the exit path where the occupant has access to two required exits in separate directions. The distance limitations are applicable to all paths of travel that lead out of a space or building where two exits are required. An illustration of this distance is found in Figure 1014.3. The illustration reflects two examples of a common path of travel where the occupants at points A and B are able to travel in only one direction before they reach a point at which they have a choice of two paths of travel to the required exits from the building. Note that from point A, the occupants have two available paths, but these merge to form a single path out of the space. This is also considered a common path of travel. The common path of travel is considered part of the overall travel distance limitations in Section 1016.1.

The reference to Section 1028.8 is for common path of travel requirements in assembly or education facilities where there is fixed seating, such as in a lecture room or sports facility.

The exceptions increase the allowable length of the common path of travel based on the installation of a sprinkler system, a low occupant load or for a Group I-3 occupancy.

SECTION 1015
EXIT AND EXIT ACCESS DOORWAYS

1015.1 Exits or exit access doorways from spaces. Two *exits* or *exit access doorways* from any space shall be provided where one of the following conditions exists:

Exception: Group I-2 occupancies shall comply with Section 1014.2.2 through 1014.2.7.

1. The *occupant load* of the space exceeds one of the values in Table 1015.1.

   Exception: In Group R-2 and R-3 occupancies, one *means of egress* is permitted within and from individual dwelling units with a maximum *occupant load* of 20 where the dwelling unit is equipped throughout with an *automatic sprinkler system* in accordance with Section 903.3.1.1 or 903.3.1.2.

2. The *common path of egress travel* exceeds one of the limitations of Section 1014.3.

3. Where required by Section 1015.3, 1015.4, 1015.5, 1015.6 or 1015.6.1.

Where a building contains mixed occupancies, each individual occupancy shall comply with the applicable requirements for that occupancy. Where applicable, cumulative *occupant loads* from adjacent occupancies shall be considered in accordance with the provisions of Section 1004.1.

❖ This section dictates the minimum number of paths of travel an occupant is to have available to avoid a fire incident in the occupied room or space. While providing multiple egress doorways from every room is unrealistic, a point does exist where alternative egress paths must be provided based on the number of occupants at risk, the distance any one occupant must travel to reach a doorway and the relative hazards associated with the occupancy of the space. Generally, the number of egress doorways required from any room or space coincides with the occupant load threshold criteria set forth for the minimum number of exits required in a building (see Section 1021.1).

The general exception allows for hospital and nursing home rooms to egress in accordance with the specific criteria in Sections 1014.2.2 through 1014.2.7. Group I-2 occupancies are not addressed in Table 1015.1.

In accordance with Item 1, the limiting criteria in Table 1015.1 for rooms or spaces permitted to have a single exit access doorway are based on an empirical judgment of the associated risks.

If the occupants of a room are required to egress through another room, as permitted in Section 1014.2, the rooms are to be combined to determine if multiple doorways are required from the combined rooms. For example, if a suite of offices shares a common reception area, the entire suite with the reception area must meet both the occupant load and the travel distance criteria.

It should be noted that where two doorways are required, the remoteness requirement of Section 1015.2 is applicable.

The exception for Item 1 allows for individual dwelling units to be considered a space with one means of

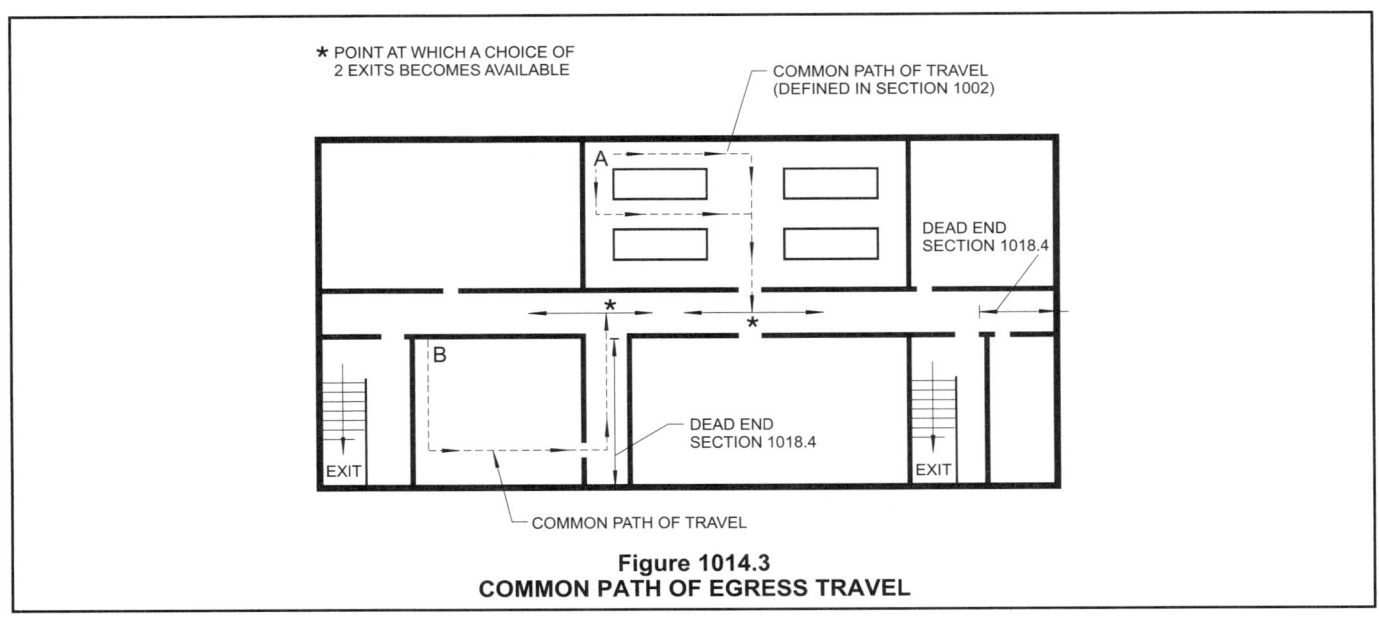

Figure 1014.3
COMMON PATH OF EGRESS TRAVEL

egress providing when the units combine they meet the provisions for buildings in Section 1021.2. If the building is sprinklered with an NFPA 13 or 13R system, the occupant load can be 20 (4,000 square feet apartment/200 square feet per occupant = 20 occupants), whereas if the unit complies with the table, the occupant load is limited to 10 people per unit (2,000 sq.ft. apartment/200 square feet per occupant = 10 occupants).

Item 2 sets the limits for a single means of egress based on travel distance. Where the common path of travel exceeds any one of the limits in Section 1014.3, two egress paths are required for safe egress from the space.

Item 3 addresses when two means of egress may be required in boiler, incinerator and furnace rooms; refrigerator machinery rooms or refrigerated rooms and spaces; and stage areas, including gridirons, catwalks and galleries.

TABLE 1015.1
SPACES WITH ONE EXIT OR EXIT ACCESS DOORWAY

OCCUPANCY	MAXIMUM OCCUPANT LOAD
A, B, Eª, F, M, U	49
H-1, H-2, H-3	3
H-4, H-5, I-1, I-3, I-4, R	10
S	29

a. Day care maximum occupant load is 10.

❖ The table represents an empirical judgment of the risks associated with a single means of egress from a room or space based on the occupant load in the room, the travel distance to the exit access door and the inherent risks associated with the occupancy (such as occupant mobility, occupant familiarity with the building, occupant response and the fire growth rate). The number 49 is for consistency with other thresholds, such as panic hardware (see Section 1008.1.10).

Group E has a reference to Note a. Day care occupancies are limited to a maximum of 10 occupants before two exits are required from a room. This limit is in consideration of needing a quick means of egress for toddlers and would be consistent with day care occupancies with children under 2¹/₂ years of age. This provision was the result of the passage of Code Change E10–04/05. The code change proponent's reason statement was mainly focused on infants and the exception in Section 308.5.2. That exception would allow day care facilities with up to 100 children 2¹/₂ years of age or less with rooms with direct access to the exterior to be classified as Group E. A concern is, with the facility now being classified as Group E, that each room, even though they have children 2¹/₂ years of age or less, could use the means of egress requirements for Group E for these rooms as opposed to Group I-4 for that space.

Since the occupants of Groups I and R may be sleeping and, therefore, not able to detect a fire in its early stages without staff supervision or room detectors, the number of occupants in a single egress room

or space is limited to 10. See the exception to Item 1 for Group R-2 and R-3 individual units.

Because of the potential for rapidly developing hazardous conditions, the single egress condition in Groups H-1, H-2 and H-3 is limited to a maximum of three persons. Because the materials contained in Groups H-4 and H-5 do not represent the same fire hazard potential as those found in Groups H-1, H-2 and H-3, the occupant load for spaces with one means of egress is increased.

Because of the reduced occupant density in Group S and the occupants' normal familiarity with the building, the single egress condition is permitted with an occupant load of 29.

1015.1.1 Three or more exits or exit access doorways. Three *exits* or *exit access doorways* shall be provided from any space with an *occupant load* of 501 to 1,000. Four *exits* or *exit access doorways* shall be provided from any space with an *occupant load* greater than 1,000.

❖ This section is correlated with Section 1021.1 for conditions where three or more exits are required from a story.

1015.2 Exit or exit access doorway arrangement. Required *exits* shall be located in a manner that makes their availability obvious. *Exits* shall be unobstructed at all times. *Exit* and *exit access doorways* shall be arranged in accordance with Sections 1015.2.1 and 1015.2.2.

❖ Exits need to be unobstructed and obvious at all times for the safety of occupants to evacuate the building in an emergency situation. This is consistent with the requirements in Section 1008.1 for exit or exit access doors to not be concealed by curtains, drapes, decorations or mirrors. Whether the doors from the space are exit access doors leading to a hallway or exit doors leading to an exit enclosure or directly to the outside, they must be located in accordance with the next two sections.

1015.2.1 Two exits or exit access doorways. Where two *exits* or *exit access doorways* are required from any portion of the *exit access*, the *exit* doors or *exit access doorways* shall be placed a distance apart equal to not less than one-half of the length of the maximum overall diagonal dimension of the building or area to be served measured in a straight line between *exit* doors or *exit access doorways*. Interlocking or *scissor stairs* shall be counted as one *exit stairway*.

Exceptions:

1. Where *exit enclosures* are provided as a portion of the required *exit* and are interconnected by a 1-hour fire-resistance-rated *corridor* conforming to the requirements of Section 1018, the required *exit* separation shall be measured along the shortest direct line of travel within the *corridor*.

2. Where a building is equipped throughout with an *automatic sprinkler system* in accordance with Section 903.3.1.1 or 903.3.1.2, the separation distance of the *exit* doors or *exit access doorways* shall not be less than one-third of the length of the

maximum overall diagonal dimension of the area served.

❖ This section provides a method to determine, quantitatively, remoteness between exits and exit access doors based on the dimensional characteristics of the space served. This measure has been common practice for some years with significant success. Very simply, the method involves determining the maximum dimension between any two points in a floor or a room (e.g., a diagonal between opposite corners in a rectangular room or building or the diameter in a circular room or building). If two doors or exits are required from the room or building (see Sections 1015.1 and 1021.1), the straight-line distance between the center of the thresholds of the doors must be at least one-half of the maximum dimension [see Figure 1015.2.1(1)].

While technical proof is not available to substantiate this method of determining remoteness, it has been found to be realistic and practical for building designs except for the common building with exits in a center core and office spaces around the perimeter.

If a scissor stairway is utilized, regardless of the sep-aration of the two entrances, the scissor stair may only be counted as one exit. Two stairways within the same enclosure could result in both stairways being unnavigable in an emergency if smoke penetrated the single enclosure (see the definition for "Scissor stairways" in Section 1002). Interlocking stairways that occur over the same building footprint but are within separate enclosures do not meet the definition for "Scissor stairways" and may count as two independent exits. Due to concern about smoke migration, careful review of the construction details and verification that they meet all the provisions for fire barriers and horizontal assemblies must be made. Of special concern would be the provisions for continuity, penetrations and joints.

Where one or more exit stairways are permitted to be unenclosed, the remoteness measurement shall begin at the center of the top riser of the unenclosed stairways. When enclosure is provided, distances are measured to the door of the enclosure.

In Exception 1, a method of permitting the distance between exits to be measured along a complying corri-

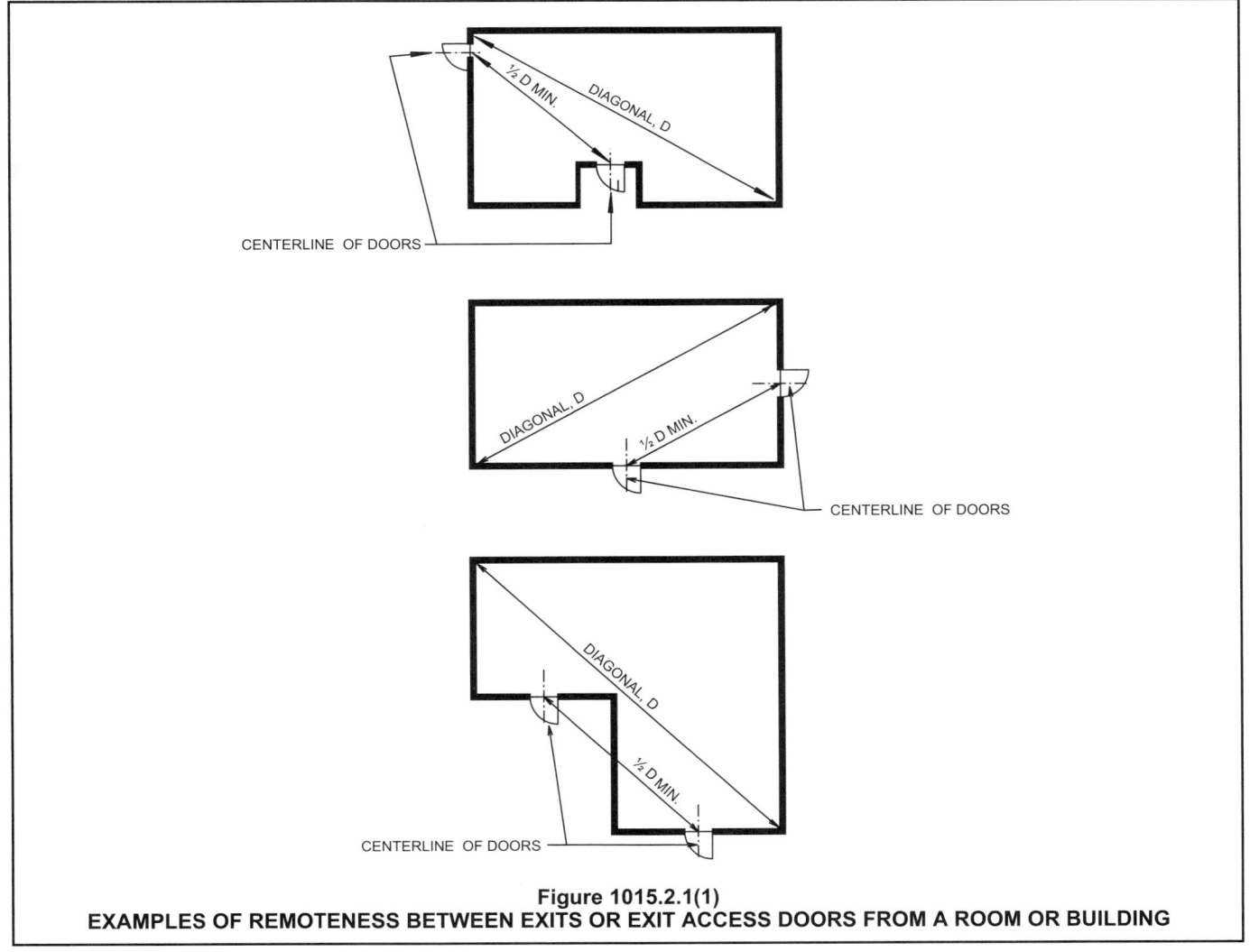

Figure 1015.2.1(1)
EXAMPLES OF REMOTENESS BETWEEN EXITS OR EXIT ACCESS DOORS FROM A ROOM OR BUILDING

dor connecting the exits has served to mitigate the disruption to this design concept [see Figure 1015.2.1(2)].

As reflected in Exception 2, the protection provided by an automatic sprinkler system can reduce the threat of fire buildup so that the reduction in remoteness to one-third of the diagonal dimension is not unreasonable, based on the presumption that it provides the occupants with an acceptable level of safety from fire [see Figure 1015.2.1(3)]. The automatic sprinkler system must be installed throughout the building in accordance with NFPA 13 or 13R. This reduced separation (one-third diagonal) may also be used when applying the requirements of Exception 1.

In applying the provisions of this section, it is important to recognize any convergence of egress paths that may exist. Figure 1015.2.1(4) illustrates that although the assembly room has remotely located exit access doors, the doors from the entire space do not meet the criteria of this section.

1015.2.2 Three or more exits or exit access doorways. Where access to three or more *exits* is required, at least two *exit* doors or *exit access doorways* shall be arranged in accordance with the provisions of Section 1015.2.1.

❖ When there are three or more required exits from a building or exit access doors from a room, they are to be analyzed identically to the method described in Section 1015.2.1. Two of the exits or exit access doors must meet the remoteness test and any additional exits or doors can be located anywhere within the floor plan that meets the code requirements, including independence, accessibility, capacity and continuity.

1015.3 Boiler, incinerator and furnace rooms. Two *exit access doorways* are required in boiler, incinerator and furnace rooms where the area is over 500 square feet (46 m²) and any fuel-fired equipment exceeds 400,000 British thermal units (Btu) (422 000 KJ) input capacity. Where two *exit access doorways* are required, one is permitted to be a fixed ladder or an *alternating tread device*. *Exit access doorways* shall be separated by a horizontal distance equal to one-half the length of the maximum overall diagonal dimension of the room.

❖ This section requires two exit access doorways for the specified mechanical equipment spaces because of the level of hazards in this type of occupancy. A fixed ladder or an alternating tread device is permitted for the occupants to egress where two doorways are required. The remoteness of the exit access doorways specified in this section is to give the occupants two paths of travel to exit the room so that if one doorway is not available, the alternate path can be used.

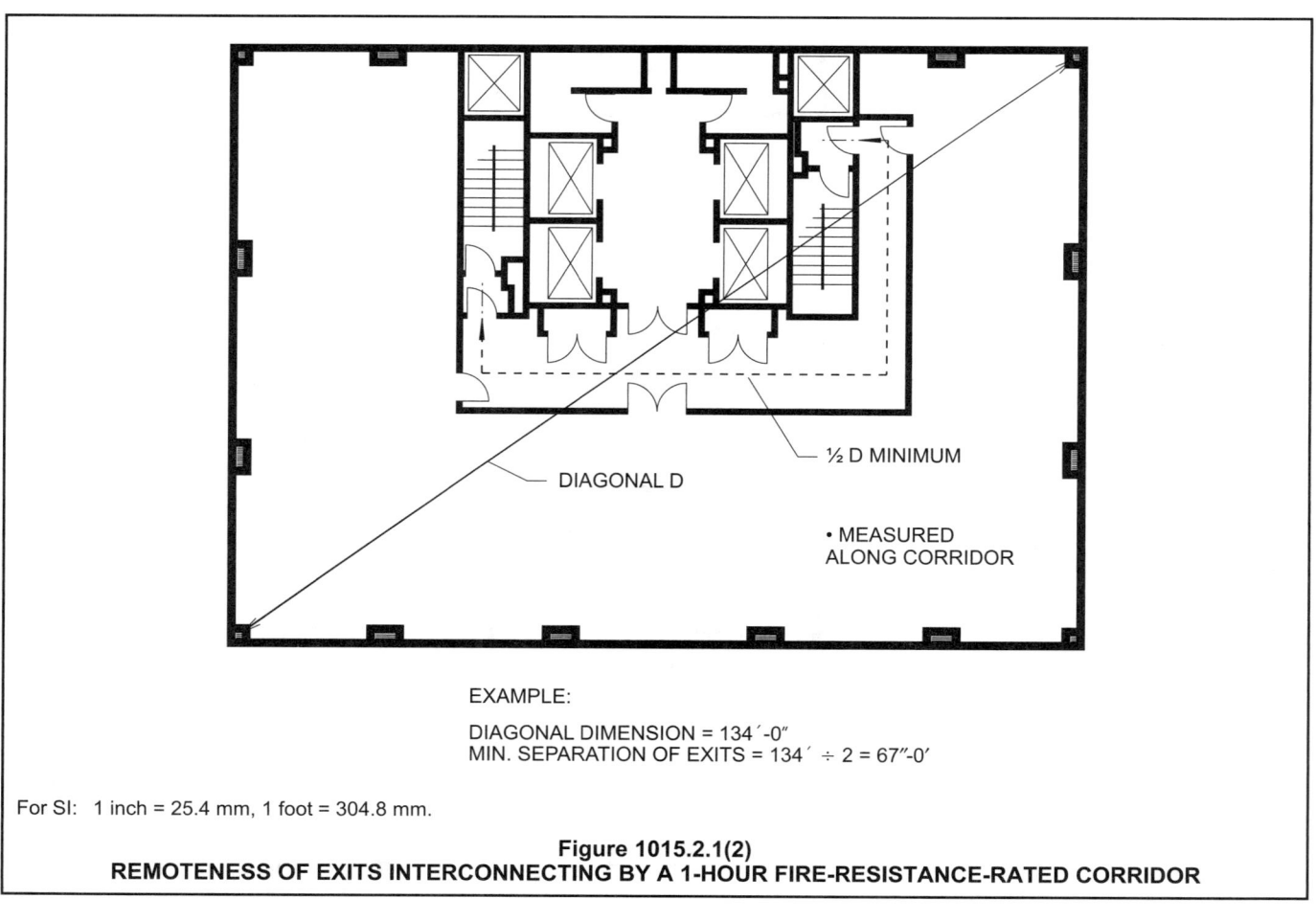

EXAMPLE:

DIAGONAL DIMENSION = 134´-0″
MIN. SEPARATION OF EXITS = 134´ ÷ 2 = 67″-0´

For SI: 1 inch = 25.4 mm, 1 foot = 304.8 mm.

Figure 1015.2.1(2)
REMOTENESS OF EXITS INTERCONNECTING BY A 1-HOUR FIRE-RESISTANCE-RATED CORRIDOR

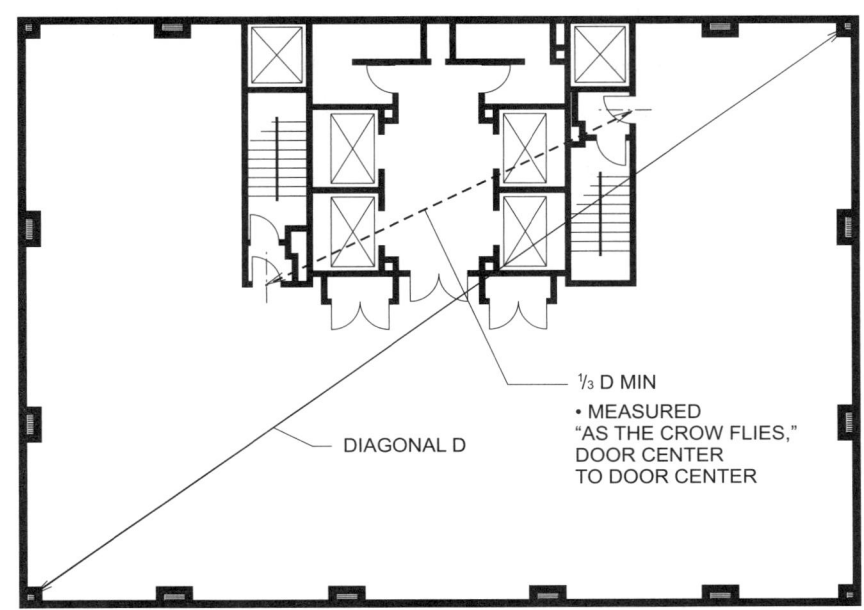

EXAMPLE:
DIAGONAL DIMENSION = 134'-0"
MIN. SEPARATION OF EXITS = 134 ÷ 3 = 44'-8"

For SI: 1 inch = 25.4 mm, 1 foot = 304.8 mm.

Figure 1015.2.1(3)
REMOTENESS OF EXITS IN A BUILDING WITH AN AUTOMATIC SPRINKLER SYSTEM

DOES NOT COMPLY!

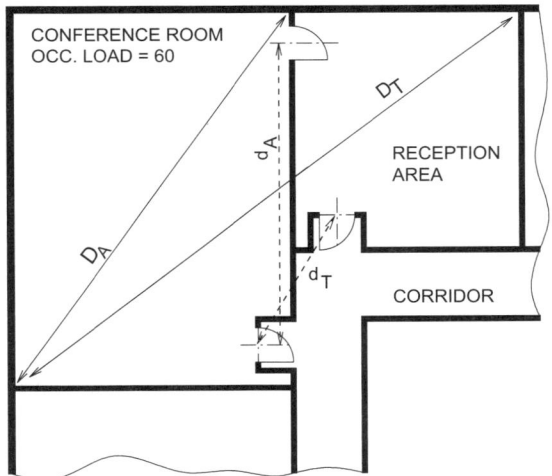

D_A = 42'-0"
d_A = 21'-0"
D_T = 58'-0"
d_T = 7'-0"

ALTHOUGH d_A EQUALS 1/2 D_A
THE OVERALL DESIGN FOR THE SPACE
IS NOT ACCEPTABLE BECAUSE d_T IS
LESS THAN 1/2 D_T.

For SI: 1 inch = 25.4 mm, 1 foot = 304.8 mm.

Figure 1015.2.1(4)
REMOTE LOCATION OF EXIT ACCESS DOORS
(ADJOINING ROOMS)

1015.4 Refrigeration machinery rooms. Machinery rooms larger than 1,000 square feet (93 m²) shall have not less than two *exits* or *exit access* doors. Where two *exit access doorways* are required, one such doorway is permitted to be served by a fixed ladder or an *alternating tread device*. *Exit access doorways* shall be separated by a horizontal distance equal to one-half the maximum horizontal dimension of room.

All portions of machinery rooms shall be within 150 feet (45 720 mm) of an *exit* or *exit access doorway*. An increase in travel distance is permitted in accordance with Section 1016.1.

Doors shall swing in the direction of egress travel, regardless of the *occupant load* served. Doors shall be tight fitting and self-closing.

❖ The reasons for these requirements are the same as for Section 1015.3. Travel distance is to be limited in accordance with Section 1016.1. The travel distance is to be increased in accordance with Section 1016.1. For example, the travel distance limit for a large refrigeration machinery room classified as Group F-1 that has a sprinkler system throughout the entire building in accordance with NFPA 13 would be 250 feet (76 200 mm) based on Table 1016.1. The 150-foot (45 720 mm) maximum distance to an exit or exit access doorway that is specified in this section would not apply in this example. The 150-foot (45 720 mm) travel distance is intended to be applied where a sprinkler system is not installed and to shorten the time that occupants would be exposed to the hazards within the machinery room.

1015.5 Refrigerated rooms or spaces. Rooms or spaces having a floor area larger than 1,000 square feet (93 m²), containing a refrigerant evaporator and maintained at a temperature below 68°F (20°C), shall have access to not less than two *exits* or *exit access* doors.

Travel distance shall be determined as specified in Section 1016.1, but all portions of a refrigerated room or space shall be within 150 feet (45 720 mm) of an *exit* or *exit access* door where such rooms are not protected by an *approved automatic sprinkler* system. Egress is allowed through adjoining refrigerated rooms or spaces.

Exception: Where using refrigerants in quantities limited to the amounts based on the volume set forth in the *International Mechanical Code*.

❖ The commentary for Sections 1015.1 to 1015.4 also applies to this section. The exception is intended to apply if Chapter 11 of the *International Mechanical Code®* (IMC®) does not require a refrigeration machinery room due to the small amount of refrigerant used (see the commentary for Section 1104 of the IMC for further explanation of the machinery room requirements).

1015.6 Stage means of egress. Where two *means of egress* are required, based on the stage size or *occupant load*, one *means of egress* shall be provided on each side of the stage.

❖ Two means of egress are required from stages in accordance with Section 410.5.3. The stage means of egress paths are to be separate so the two means of egress are independent.

1015.6.1 Gallery, gridiron and catwalk means of egress. The *means of egress* from lighting and access catwalks, galleries and *gridirons* shall meet the requirements for occupancies in Group F-2.

Exceptions:

1. A minimum width of 22 inches (559 mm) is permitted for lighting and access catwalks.

2. *Spiral stairs* are permitted in the *means of egress*.

3. *Stairways* required by this subsection need not be enclosed.

4. *Stairways* with a minimum width of 22 inches (559 mm), ladders or *spiral stairs* are permitted in the *means of egress*.

5. A second *means of egress* is not required from these areas where a means of escape to a floor or to a roof is provided. Ladders, *alternating tread devices* or *spiral stairs* are permitted in the means of escape.

6. Ladders are permitted in the *means of egress*.

❖ The purpose of this section is to specify the various options that are allowed for means of egress from theater lighting and access catwalks, galleries and gridirons. The requirements are consistent with the use being limited to service personnel (see also Section 410).

SECTION 1016
EXIT ACCESS TRAVEL DISTANCE

1016.1 Travel distance limitations. *Exits* shall be so located on each *story* such that the maximum length of *exit access* travel, measured from the most remote point within a *story* along the natural and unobstructed path of egress travel to an *exterior exit* door at the *level of exit discharge*, an entrance to a vertical *exit enclosure*, an *exit passageway*, a *horizontal exit*, an *exterior exit stairway* or an exterior *exit ramp*, shall not exceed the distances given in Table 1016.1.

Exceptions:

1. Travel distance in *open parking garages* is permitted to be measured to the closest riser of open *exit stairways*.

2. In outdoor facilities with open *exit access* components and open *exterior exit stairways* or *exit ramps*, travel distance is permitted to be measured to the closest riser of an *exit stairway* or the closest slope of the *exit ramp*.

3. In other than occupancy Groups H and I, the *exit access* travel distance to a maximum of 50 percent of the *exits* is permitted to be measured from the most remote point within a building to an *exit* using unenclosed *exit access stairways* or *ramps* when connecting a maximum of two stories. The two connected stories shall be provided with at least two *means of egress*. Such interconnected stories shall not be open to other stories.

4. In other than occupancy Groups H and I, *exit access* travel distance is permitted to be measured from the most remote point within a building to an *exit* using

unenclosed *exit access stairways* or *ramps* in the first and second stories above *grade plane* in buildings equipped throughout with an *automatic sprinkler system* in accordance with Section 903.3.1.1. The first and second stories above *grade plane* shall be provided with at least two *means of egress*. Such interconnected stories shall not be open to other stories.

Where applicable, travel distance on unenclosed *exit access stairways* or *ramps* and on connecting stories shall also be included in the travel distance measurement. The measurement along *stairways* shall be made on a plane parallel and tangent to the *stair* tread *nosings* in the center of the *stairway*.

❖ The length of travel, as measured from the most remote point within a structure to an exit, is limited to restrict the amount of time that the occupant is exposed to a potential fire condition [see Figure 1016.1(1)]. The route must be assumed to be the natural path of travel without obstruction. This commonly results in a rectilinear path similar to what can be experienced in most occupancies, such as a schoolroom or an office with rows of desks [see Figure 1016.1(2)]. The "arc" method, using an "as the crow flies" linear measurement, must be used with caution, as it seldom represents typical floor design and layout and, in most cases, would not be deemed to be the natural, unobstructed path.

The travel distance is measured from each and every occupiable point on a floor to the closest exit. While each occupant may be required to have access to a second or third exit, the travel distance limitation is only applicable to the distance to the nearest exit. In effect, this means that the distance an occupant must travel to the second or third exit is not regulated.

Travel distance is measured along the exit access path. Exit access travel distance may include travel on a stairway if it is not constructed to meet the definition of an exit (i.e., enclosure, discharge, etc.). An example of this would be an unenclosed exit access stairway from a mezzanine level or steps along the path of

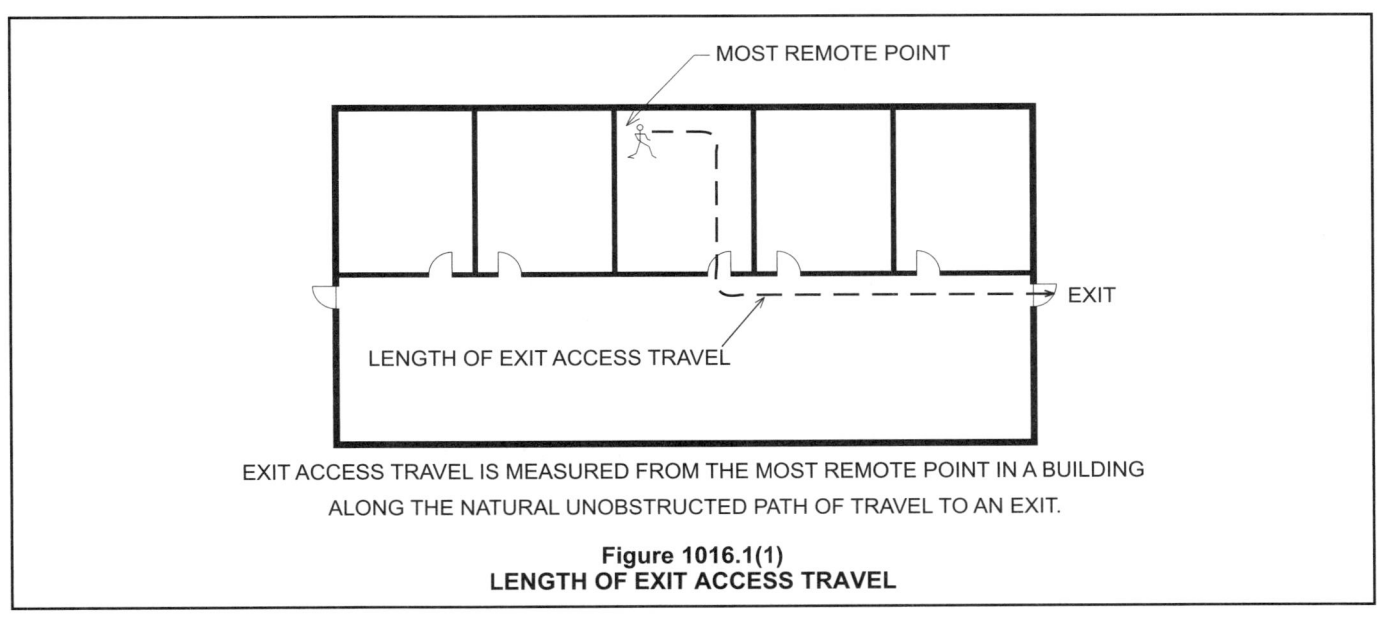

EXIT ACCESS TRAVEL IS MEASURED FROM THE MOST REMOTE POINT IN A BUILDING ALONG THE NATURAL UNOBSTRUCTED PATH OF TRAVEL TO AN EXIT.

Figure 1016.1(1)
LENGTH OF EXIT ACCESS TRAVEL

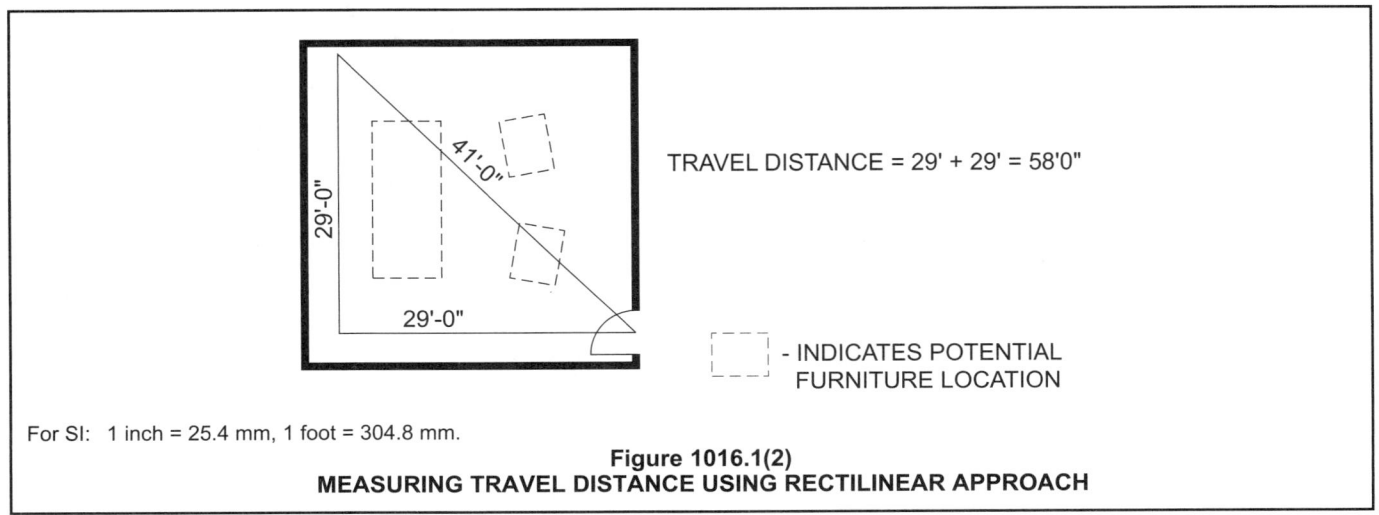

TRAVEL DISTANCE = 29' + 29' = 58'0"

☐ - INDICATES POTENTIAL FURNITURE LOCATION

For SI: 1 inch = 25.4 mm, 1 foot = 304.8 mm.

Figure 1016.1(2)
MEASURING TRAVEL DISTANCE USING RECTILINEAR APPROACH

travel in a split floor-level situation. When Section 1022.1 permits an exit stairway to be unenclosed, the travel distance would also include travel down the open stairway and to: a vertical exit enclosure; a horizontal exit; or an exit door to the outside. An example of this would be an open exit stairway within an individual dwelling unit (see Section 1022.1, Exception 3), or an open exit stairway from a small space not open to the public (see Section 1022.1, Exception 1).

Exceptions 1 and 2 provide for a travel distance terminating at the top of an open exit stair in an open parking structure, or an open exit stair or ramp in outdoor facilities (e.g., stadiums, exterior stairways from open balconies, observation decks and amusement structures) (see Section 1022.1, Exceptions 2 and 4). This is appropriate in view of the low hazard in these facilities.

Exceptions 3 and 4 address the special concerns for an open stairway between floors other than those covered by Exceptions 1 and 2 and the exterior stairway provisions in Section 1026. These stairways are considered exit access stairways, rather than exit stairways, so travel distance includes measuring down the stairs and to an exit door to the outside or the door to an enclosed exit stairway—similar to mezzanines. The measurement for the travel distance must be from the most remote point, down the exit access stairway and out of the building to the beginning of the exit discharge.

The distance of travel within an exit enclosure (e.g., vertical exit enclosure, exit passageway) and in the exit discharge portion of the means of egress is also not regulated. Section 1021.2 permits certain buildings to be provided with a single exit. In instances where there is a single exit, travel distances less than those permitted in Table 1016.1 apply (see Table 1021.2).

Exception 3 allows one story level of a required exit access stairway to be unenclosed. This exception may be used in either sprinklered or nonsprinklered buildings; however, it may not be used in Group H or I occupancies. Since this exception is limited to 50 percent of the egress stairways, it may not be used in a single-exit building. There are two common instances where this exception has been used. The first is for two-story buildings where one of the required stairways is required to be enclosed but the other is not. The second case is for a multistory building where one of the required means of egress stairways between levels may be open between two adjacent levels. In this situation, after moving down one level on an open exit access stairway, the required number of enclosed exit stairways must be available. Once the enclosed exit is entered, that level of protection must be maintained until the occupants reach the exit discharge. The most typical example of the second case is when a second means of egress is provided via exit access stairway for a tenant who does not have access to two enclosed exits on the upper floor because the second stair is only available through another tenant space.

Access to exits may not be through another tenant space in accordance with Sections 1014.2 and 1014.2.1. It should be noted that a stair shaft could not be discontinued at some point in the height of a multistory building and then continued one floor level lower, such that a person would have to come out of the stair enclosure and then back into it at the next level lower. This arrangement would not be allowed according to Sections 1020.1 and 1021.3, which require the level of protection of an exit to be continuous from the point of entry into the exit to the exit discharge. The open exit access stairway must meet the same separation requirements as enclosed exit stairways.

Exception 4 is limited to buildings sprinklered throughout by an NFPA 13 system and is only applicable to the stairway between the first and second floor. This exception is not available for Group H or I occupancies or upper floors in a multistory building. Literally, this exception would permit the second floor of any building to use two, three or four unenclosed stairways to serve as the required means of egress for that level. However, in most cases in a multistory building, the second level would utilize the enclosed exit stairways moving down from the upper floors. Sections 1019.1 and 1021.3 require the level of protection of an exit to be continuous from the point of entry, to the exit and to the exit discharge; therefore, the exit stairways coming down from the upper level could not be open between the second and first level. Therefore, this particular exception will most typically be utilized in two-story buildings. Open exit access stairways must meet the same remoteness requirements as enclosed exit stairways.

The last sentence in the section below the four exceptions emphasizes that exit access travel distance includes travel down the exit access stairways and to an exit—such as the outside, the door of an enclosed exit, through a horizontal exit, or the door leading to an exterior exit stairway.

TABLE 1016.1. See page 10-115.

❖ This table reflects the maximum distance a person is allowed to travel from any point in a building floor area to the nearest exit along a natural and unobstructed path. While quantitative determinations or formulas are not available to substantiate the tabular distances, empirical factors are utilized to make relative judgments as to reasonable limitations. Such considerations include the nature and fitness of the occupants; the typical configurations and physical conditions of each group; the level of fire hazard with respect to the specific uses of the facilities, including fire spread and the potential intensity of a fire. The inclusion of an automatic sprinkler system throughout the building can serve to control, confine or possibly eliminate the fire threat to the occupants so an increased travel distance is permitted. Increased travel distances are permitted when an automatic sprinkler system is installed in accordance with NFPA 13 or 13R.

When measuring travel distance, it is important to

consider the natural path of travel [see Figure 1016.1(1)]. In many cases, the actual layout of furnishings and equipment is not known or is not identified on the plans submitted with the permit application. In such instances, it may be necessary to measure travel distance using the legs of a triangle instead of the hypotenuse [see Figure 1016.1(2)]. Since most people tend to migrate to more open spaces while egressing, measurement of the natural path of travel typically excludes areas of the building within approximately 1 foot (305 mm) of walls, corners, columns and other permanent construction. Where the travel path includes passage through a doorway, the natural route is generally measured through the centerline of the door openings.

The "not permitted" in the table is an indication that this type of use must always be sprinklered. The common path of travel in Section 1014.3 is part of the overall exit access travel distance, with both starting at the same point. Common path of travel stops when the occupant has a choice of at least two exits, and overall travel distance stops when an occupant gets to the closest exit.

Note a is a reference to other travel distance limitations in the code. Notes b and c are simply a reference to the allowed type of sprinkler system, NFPA 13, 13R or 13D.

TABLE 1016.1
EXIT ACCESS TRAVEL DISTANCE[a]

OCCUPANCY	WITHOUT SPRINKLER SYSTEM (feet)	WITH SPRINKLER SYSTEM (feet)
A, E, F-1, M, R, S-1	200	250[b]
I-1	Not Permitted	250[c]
B	200	300[c]
F-2, S-2, U	300	400[c]
H-1	Not Permitted	75[c]
H-2	Not Permitted	100[c]
H-3	Not Permitted	150[c]
H-4	Not Permitted	175[c]
H-5	Not Permitted	200[c]
I-2, I-3, I-4	Not Permitted	200[c]

For SI: 1 foot = 304.8 mm.

a. See the following sections for modifications to exit access travel distance requirements:
   Section 402.4: For the distance limitation in malls.
   Section 404.9: For the distance limitation through an atrium space.
   Section 407.4: For the distance limitation in Group I-2.
   Sections 408.6.1 and 408.8.1: For the distance limitations in Group I-3.
   Section 411.4: For the distance limitation in special amusement buildings.
   Section 1014.2.2: For the distance limitation in Group I-2 hospital suites.
   Section 1015.4: For the distance limitation in refrigeration machinery rooms.
   Section 1015.5: For the distance limitation in refrigerated rooms and spaces.
   Section 1021.2: For buildings with one exit.
   Section 1028.7: For increased limitation in assembly seating.
   Section 1028.7: For increased limitation for assembly open-air seating.
   Section 3103.4: For temporary structures.
   Section 3104.9: For pedestrian walkways.

b. Buildings equipped throughout with an automatic sprinkler system in accordance with Section 903.3.1.1 or 903.3.1.2. See Section 903 for occupancies where automatic sprinkler systems are permitted in accordance with Section 903.3.1.2.

c. Buildings equipped throughout with an automatic sprinkler system in accordance with Section 903.3.1.1.

1016.2 Exterior egress balcony increase. Travel distances specified in Section 1016.1 shall be increased up to an additional 100 feet (30 480 mm) provided the last portion of the *exit access* leading to the *exit* occurs on an exterior egress balcony constructed in accordance with Section 1019. The length of such balcony shall not be less than the amount of the increase taken.

❖ This section allows an additional travel distance on exterior egress balconies since the accumulation of smoke is much less on a balcony. Note that the length of the increase is not to be more than the length of the exterior balcony. For example, if the length of the balcony is 75 feet (22 860 mm), the additional travel distance is limited to 75 feet (22 860). In order for the increase to apply, the exterior balcony must be located at the end of the path of egress travel and not in some other portion of the egress path.

SECTION 1017
AISLES

1017.1 General. *Aisles* serving as a portion of the *exit access* in the *means of egress* system shall comply with the requirements of this section. *Aisles* shall be provided from all occupied portions of the *exit access* which contain seats, tables, furnishings, displays and similar fixtures or equipment. *Aisles* serving assembly areas shall comply with Section 1028. *Aisles* serving reviewing stands, *grandstands* and *bleachers* shall also comply with Section 1028. The required width of *aisles* shall be unobstructed.

Exception: Doors complying with Section 1005.2.

❖ This section addresses aisles and aisle accessways for other than assembly areas that are covered in Section 1028. Aisle accessways for mercantile are addressed in Section 1017.3. Aisle accessways for tables and seating are covered by Section 1017.4.1.

"Aisle accessway" is defined in Section 1002 as "that portion of exit access that leads to an aisle." The term "Aisle" is defined in Section 1002 as "an exit access component that defines and provides a path of egress travel." Given the many possible configurations of fixtures and furniture, both permanent and moveable, the determination of where aisle accessways stop and aisles begin is often subject to interpretation.

Typically, the aisle accessways lead to the aisles, which in turn lead to the exits. Since the aisle serves as a path for means of egress similar to a corridor, the requirements for doors obstructing the aisle are the same (see Section 1005.2).

A cross reference back to Section 1005.2 from the exceptions for width in corridors (see Section 1018.3), aisles (see Section 1017.1), exit passageways (see Section 1023.2) and exit courts (see Section 1027.5.1) reinforces the fact that the protrusion limits provision is generally applicable for these types of confined routes.

1017.2 Aisles in Groups B and M. In Group B and M occupancies, the minimum clear *aisle* width shall be determined by Section 1005.1 for the *occupant load* served, but shall not be less than 36 inches (914 mm).

Exception: Nonpublic *aisles* serving less than 50 people and not required to be *accessible* by Chapter 11 need not exceed 28 inches (711 mm) in width.

❖ This requirement establishes aisle width criteria for Group B and M occupancies based on the occupant load served by the aisle. The reference to Section 1005.1 would trigger a requirement for aisles wider than 36 inches (914 mm) when the anticipated occupant load that the aisle served was larger than 180 (36 inches/0.2 = 180 occupants). The exception addresses aisles that may be found in an archival file room or stock storage racks.

If fixtures are permanent, such as in a typical grocery store or office cubicles in a business, the aisle provisions would be applicable throughout. In a situation where there were groups of displays separated by aisles, the area within the displays may be considered aisle accessways (see Section 1017.3).

1017.3 Aisle accessways in Group M. An *aisle accessway* shall be provided on at least one side of each element within the *merchandise pad*. The minimum clear width for an *aisle accessway* not required to be *accessible* shall be 30 inches (762 mm). The required clear width of the *aisle accessway* shall be measured perpendicular to the elements and merchandise within the *merchandise pad*. The 30-inch (762 mm) minimum clear width shall be maintained to provide a path to an adjacent *aisle* or *aisle accessway*. The common path of travel shall not exceed 30 feet (9144 mm) from any point in the *merchandise pad*.

Exception: For areas serving not more than 50 occupants, the common path of travel shall not exceed 75 feet (22 880 mm).

❖ The definition for "Merchandise pad" can be found in Section 1002. The idea is that the merchandise pad contains the movable displays and aisle accessways. Aisles and permanent walls or displays would define the extent of the merchandise pad. Large department stores will have multiple merchandise pads (see Figure 1017.3). In accordance with Section 105.2, Item 13, movable cases, counters and partitions not over 5 feet, 9 inches (1753 mm) in height do not require a building permit to move, add or alter. Every element within a merchandise pad must adjoin a minimum 30-inch-wide (762 mm) aisle accessway on at least one side. Travel within a merchandise pad is limited, with a maximum common path of travel of 30 feet (9144 mm). The common path of travel limitation is extended to 75 feet (22 m) in those areas serving a maximum occupant load of 50.

Figure 1017.3
AISLES AND AISLE ACCESSWAYS IN MERCANTILE

1017.4 Seating at tables. Where seating is located at a table or counter and is adjacent to an *aisle* or *aisle accessway*, the measurement of required clear width of the *aisle* or *aisle accessway* shall be made to a line 19 inches (483 mm) away from and parallel to the edge of the table or counter. The 19-inch (483 mm) distance shall be measured perpendicular to the side of the table or counter. In the case of other side boundaries for *aisle* or *aisle accessways*, the clear width shall be measured to walls, edges of seating and tread edges, except that *handrail* projections are permitted.

Exception: Where tables or counters are served by fixed seats, the width of the *aisle accessway* shall be measured from the back of the seat.

❖ Seating is often provided adjacent to aisles and aisle accessways. In measuring the width of an aisle or aisle accessway for movable seating, the measurement is taken at a distance of 19 inches (483 mm) perpendicular to the side of the table or counter. This 19-inch (483 mm) space from the edge of the table or counter to the line where the aisle or aisle accessway measurement begins is intended to represent the space occupied by a typical seated occupant. This dimension is also considered to be adequate to accommodate seats with armrests that are too high to fit under the table where fixed seats are used. The aisle width is permitted to be measured from the back of the seat based upon the exception. As indicated in Figure 1017.4, where seating abuts an aisle or aisle accessway, 19 inches (483 mm) must be added to the required aisle or aisle accessway width for seating on only one side and 38 inches (965 mm) for seating on both sides. When seating will not be adjacent to the aisles or aisle passageways, as is the case when tables are at an angle to the aisle or aisle accessway, the measurement may be taken to the edge of the seating, table, counter or tread.

Sections 1017.4.1 through 1017.4.3 address aisle accessways. For aisles, see Section 1028.

1017.4.1 Aisle accessway for tables and seating. *Aisle accessways* serving arrangements of seating at tables or counters shall have sufficient clear width to conform to the capacity requirements of Section 1005.1 but shall not have less than the appropriate minimum clear width specified in Section 1017.4.2.

❖ This section specifies two criteria for the determination of the required width of aisle accessways: the require-

AISLE ACCESSWAY
(12" MINIMUM)
AISLE WIDTH
(36" MINIMUM)
AISLE OR
AISLE ACCESSWAY
AISLE OR
AISLE ACCESSWAY
WIDTH
WIDTH + 19"
19"
19"
AISLE OR
AISLE ACCESSWAY
WIDTH
WIDTH
+ 38"
19"

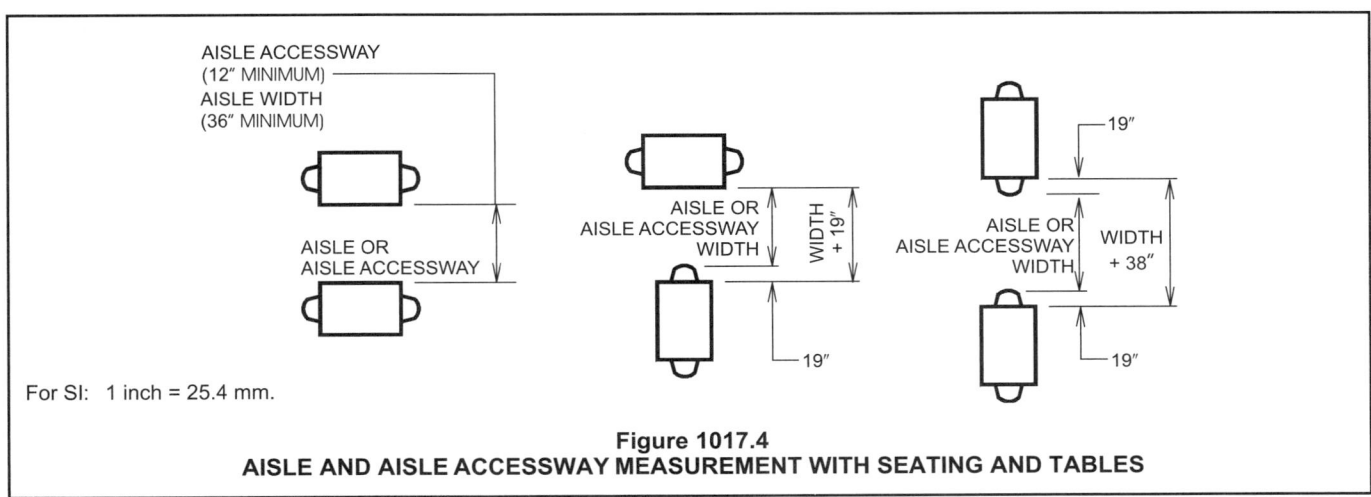

For SI: 1 inch = 25.4 mm.

Figure 1017.4
AISLE AND AISLE ACCESSWAY MEASUREMENT WITH SEATING AND TABLES

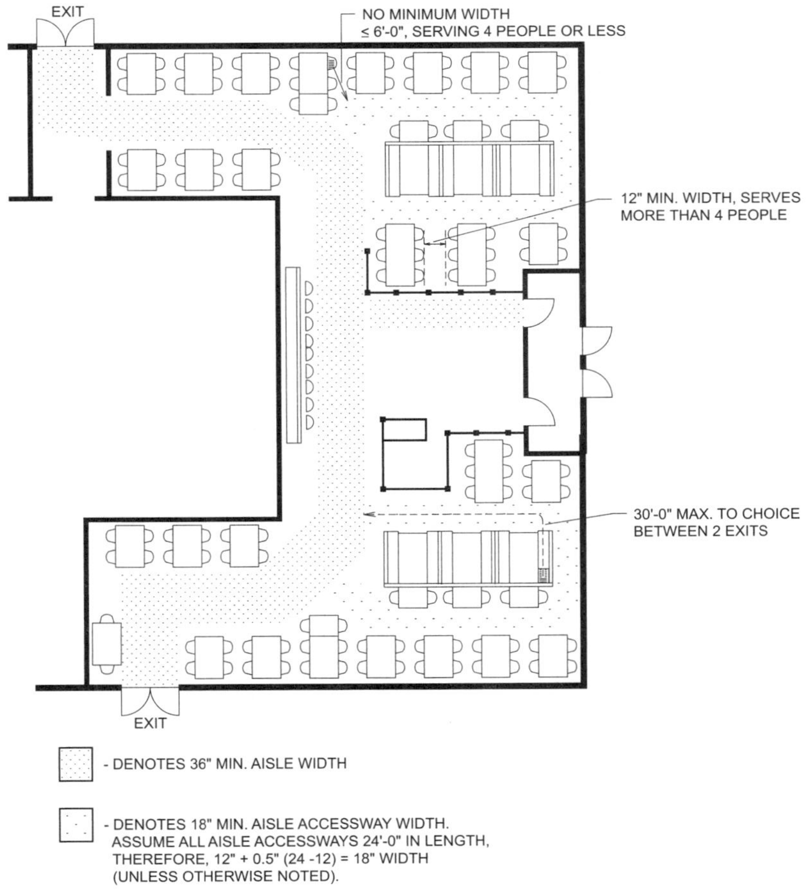

EXIT
NO MINIMUM WIDTH
≤ 6'-0", SERVING 4 PEOPLE OR LESS
12" MIN. WIDTH, SERVES MORE THAN 4 PEOPLE
30'-0" MAX. TO CHOICE BETWEEN 2 EXITS
EXIT
- DENOTES 36" MIN. AISLE WIDTH
- DENOTES 18" MIN. AISLE ACCESSWAY WIDTH. ASSUME ALL AISLE ACCESSWAYS 24'-0" IN LENGTH, THEREFORE, 12" + 0.5" (24 -12) = 18" WIDTH (UNLESS OTHERWISE NOTED).

For SI: 1 inch = 25.4 mm, 1 foot = 304.8 mm.

Figure 1017.4.1
ACCESS FOR TABLES AND SEATING

ments of Sections 1005.1 and 1017.4.2. The aisle accessway width is to be the wider of the two requirements.

The requirements of Sections 1017.4.1, 1017.4.2 and 1017.4.3 are illustrated in Figure 1017.4.1.

1017.4.2 Table and seating accessway width. *Aisle accessways* shall provide a minimum of 12 inches (305 mm) of width plus ¹/₂ inch (12.7 mm) of width for each additional 1 foot (305 mm), or fraction thereof, beyond 12 feet (3658 mm) of *aisle accessway* length measured from the center of the seat farthest from an *aisle*.

Exception: Portions of an *aisle accessway* having a length not exceeding 6 feet (1829 mm) and used by a total of not more than four persons.

❖ The relationship of tables and seating results in a situation in which it is difficult to determine which chairs are served by which aisle accessway; therefore, the width of the aisle accessway is a function of the distance from the aisle. The same minimum 12 inches (305 mm) is used and is increased ¹/₂ inch (12.7 mm) for each additional foot of travel beyond 12 feet (3658 mm).

Recognizing that the normal use of table and chair seating will require some clearance for access and service, the exception eliminates the minimum width criteria if the distance to the aisle [or an aisle accessway of at least 12 inches (305 mm)] is less than 6 feet (1829 mm) and the number of people served is not more than four. Therefore, the first 6 feet (1829 mm) are not required to meet any minimum width criteria. After the first 6 feet (1829 mm), the requirements for an aisle accessway will apply. The length of the aisle accessway is then restricted by Section 1017.4.3. When the maximum length of the aisle accessway is reached, either an aisle, corridor or exit access door must be provided (see Figure 1017.4.1).

1017.4.3 Table and seating aisle accessway length. The length of travel along the *aisle accessway* shall not exceed 30 feet (9144 mm) from any seat to the point where a person has a choice of two or more paths of egress travel to separate *exits*.

❖ At some point in the exit access travel, it is necessary to reach an aisle complying with the minimum widths of Section 1017.1. An aisle accessway travel distance is not to exceed 30 feet (9144 mm), which may represent a dead-end condition (see Figure 1017.4.1).

SECTION 1018
CORRIDORS

1018.1 Construction. *Corridors* shall be fire-resistance rated in accordance with Table 1018.1. The *corridor* walls required to be fire-resistance rated shall comply with Section 709 for *fire partitions*.

Exceptions:

1. A *fire-resistance rating* is not required for *corridors* in an occupancy in Group E where each room that is used for instruction has at least one door opening directly to the exterior and rooms for assembly purposes have at least one-half of the required *means of egress* doors opening directly to the exterior. Exterior doors specified in this exception are required to be at ground level.

2. A *fire-resistance rating* is not required for *corridors* contained within a dwelling or sleeping unit in an occupancy in Group R.

3. A *fire-resistance rating* is not required for *corridors* in *open parking garages*.

4. A *fire-resistance rating* is not required for *corridors* in an occupancy in Group B which is a space requiring only a single *means of egress* complying with Section 1015.1.

❖ The purpose of corridor enclosures is to provide fire protection to occupants as they travel the confined path, perhaps unaware of a fire buildup in an adjacent floor area. The base protection is a fire partition having a 1-hour fire-resistance rating (see Table 1018.1). The table allows a reduction or elimination of the fire-resistance rating depending on the occupant load and the presence of an NFPA 13 or 13R automatic sprinkler system throughout the building.

Section 709 addresses the continuity of fire partitions serving as corridor walls. In addition to allowing the fire partitions to terminate at the underside of a fire-resistance-rated floor/ceiling or roof/ceiling assembly, the supporting construction need not have the same fire-resistance rating in buildings of Type IIB, IIIB and VB construction as specified in Section 709. If such walls were required to be supported by fire-resistance-rated construction, the use of these construction types would be severely restricted when the corridors are required to have a fire-resistance rating. Section 407.3 requires that corridor walls in Group I-2 occupancies that are required to have a fire-resistance rating must be continuous to the underside of the floor or roof deck above or at a smoke-limiting ceiling membrane. Continuity is required because of the defend-in-place protection strategy utilized in such buildings. Requirements for corridor construction within Group I-3 occupancies are found in Section 408.8. Dwelling unit separation in Groups I-1, R-1, R-2 and R-3 is found in Section 420. Ambulatory health care provisions have special requirements in Section 422. For additional requirements for an elevator lobby that is adjacent to or part of a corridor, see the commentary to Section 708.14.

Exception 1 indicates a fire-resistance rating is not required for corridors in Group E when any room adjacent to the corridor that is used for instruction or assembly purposes has a door directly to the outside. Because these rooms are provided with an alternative egress path due to the requirement for exterior exits, the need for a fire-resistance-rated corridor is eliminated.

In accordance with Exception 2, a fire-resistance rating for a corridor contained within a single dwelling unit (e.g., apartment, townhouse) or sleeping unit (e.g., hotel guestroom, assistive living suite) is not required for practical reasons. It is unreasonable to expect fire doors and the associated hardware in homes and similar occupancies.

Given the relatively smoke-free environment of open parking structures, Exception 3 does not require rated corridors in these types of facilities.

If an office suite is small enough that only one means of egress is required from the suite, Exception 4 indicates that a rated corridor would not be required in that area. The main corridor that connected these suites to the exits would be rated in accordance with Table 1018.1.

TABLE 1018.1. See below.

❖ The required fire-resistance ratings of corridors serving adjacent spaces are provided in Table 1018.1. The fire-resistance rating is based on the group classification (considering characteristics such as occupant mobility, density and knowledge of the building as well as the fire hazard associated with the classification), the total occupant load served by the corridor and the presence of an automatic sprinkler system.

Where the corridor serves a limited number of people (see the second column in Table 1018.1), the fire-resistance rating is eliminated because of the limited size of the facility and the likelihood that the occupants would become aware of a fire buildup in sufficient time to exit the structure safely. The total occupant load that the corridor serves is used to determine the requirement for a rated corridor enclosure. Corridors serving a total occupant load equal to or less than that indicated in the second column of Table 1018.1 are not required to be enclosed with fire-resistance-rated construction. For example, a corridor serving an occupant load of 30 or less in an unsprinklered Group B occupancy is not required to be enclosed with fire-resistance-rated construction. This example is illustrated in Figure 1018.1.

The purpose of corridor enclosures is to provide fire protection to occupants as they travel the confined path, perhaps unaware of a fire buildup in an adjacent floor area. The base protection is a fire partition having a 1-hour fire-resistance rating. The table allows a re-

TABLE 1018.1
CORRIDOR FIRE-RESISTANCE RATING

OCCUPANCY	OCCUPANT LOAD SERVED BY CORRIDOR	REQUIRED FIRE-RESISTANCE RATING (hours)	
		Without sprinkler system	With sprinkler system[c]
H-1, H-2, H-3	All	Not Permitted	1
H-4, H-5	Greater than 30	Not Permitted	1
A, B, E, F, M, S, U	Greater than 30	1	0
R	Greater than 10	Not Permitted	0.5
I-2[a], I-4	All	Not Permitted	0
I-1, I-3	All	Not Permitted	1[b]

a. For requirements for occupancies in Group I-2, see Sections 407.2 and 407.3.
b. For a reduction in the fire-resistance rating for occupancies in Group I-3, see Section 408.8.
c. Buildings equipped throughout with an automatic sprinkler system in accordance with Section 903.3.1.1 or 903.3.1.2 where allowed.

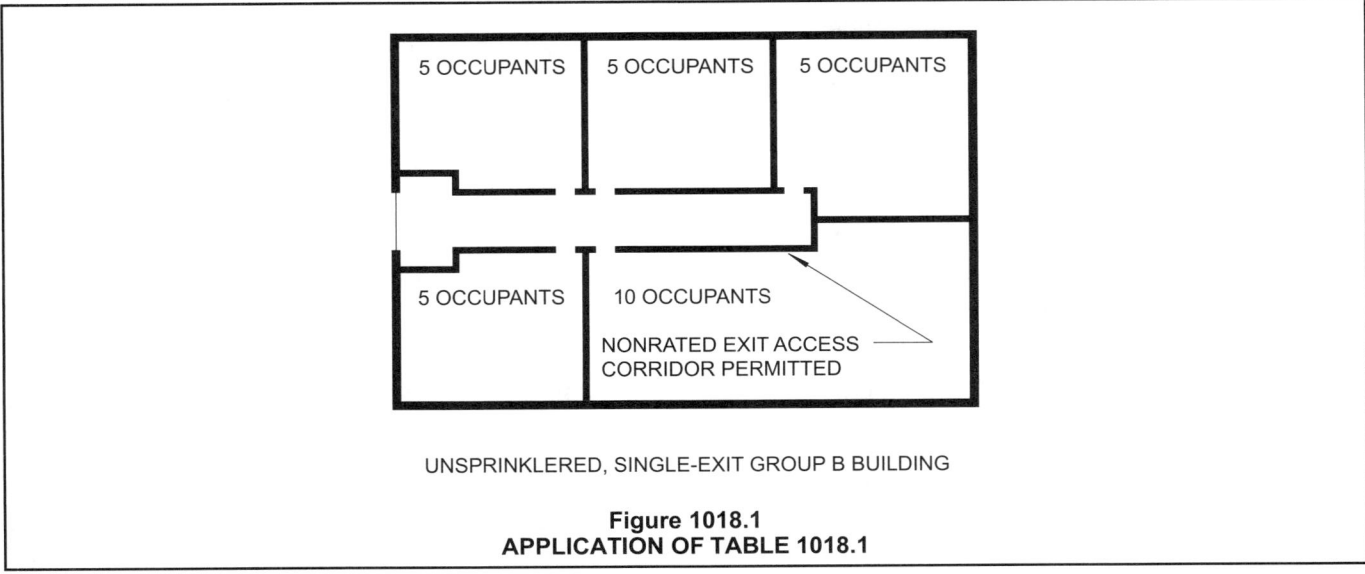

UNSPRINKLERED, SINGLE-EXIT GROUP B BUILDING

Figure 1018.1
APPLICATION OF TABLE 1018.1

duction or elimination of the fire-resistance rating depending on the occupant load and the presence of an NFPA 13 or 13R automatic sprinkler system throughout the building.

A common mistake is assuming a building is sprinklered throughout and utilizing the corridor rating reductions, when in fact certain requirements in NFPA 13 would not consider the building sprinklered throughout. For example, a health club installs a sprinkler system, but chooses to eliminate the sprinklers over the swimming pool in accordance with the exception in Section 507.3. Any corridors within the building that serve greater than 30 occupants must be rated because the building would not be considered sprinklered throughout in accordance with NFPA 13 requirements.

Note that because of the hazardous nature of occupancies in Groups H-1, H-2 and H-3, fire-resistance-rated corridors are required under all conditions. Regardless of the presence of a fire sprinkler system, a 1-hour-rated corridor enclosure is typically required in high-hazard occupancies. The only exception is for Group H-4 and H-5 occupancies. Occupancies that contain materials or operations constituting a health hazard do not pose the same relative fire or explosion hazard as Group H-1, H-2 or H-3 materials. As such, in Group H-4 or H-5, where the corridor serves a total occupant load of 30 or less, a fire-resistance-rated enclosure is not required. The "not permitted" in the third column is in coordination with Section 903.2.5, which requires all Group H buildings to be fully sprinklered.

The code acknowledges that an automatic sprinkler system can serve to control or eliminate fire development that could threaten the exit access corridor. Most occupancies where sleeping rooms are not present (Groups A, B, E, F, M, S and U) are permitted to have nonfire-resistance-rated corridors if a sprinkler system is installed throughout the building in accordance with NFPA 13.

In residential facilities, the response time to a fire may be delayed because the residents may be sleeping. With this additional safety concern, the requirements for corridors are more restrictive than nonresidential occupancies. If the corridor serves more than 10 occupants, it is required to be rated. If the building is sprinklered throughout with either an NFPA 13 or 13R system, then the rating of the corridor may be reduced to 1/2 hour. Note the exception for fire-resistance-rated corridors within an individual dwelling or sleeping unit in Section 1018.1. Also note that the reduction in the rating of the corridor walls is not permitted when an NFPA 13D sprinkler system is provided.

While all Group I facilities are supervised environments, the level of supervision in Group I-2 and I-4 occupancies would permit assisted evacuation by staff in an emergency; therefore, corridors are not required to be rated. Corridors in Groups I-2 are also regulated by Section 407.3. Due to the lower staff/resident ratio in Group I-1 and the limitation on free egress in Group

I-3, corridors must have a 1-hour fire-resistance rating (see Section 408.8 for a reduction in the corridors in Group I-3). The "not permitted" in the third column is in coordination with Section 903.2.5, which requires all Group I buildings to be fully sprinklered.

1018.2 Corridor width. The minimum *corridor* width shall be as determined in Section 1005.1, but not less than 44 inches (1118 mm).

Exceptions:

1. Twenty-four inches (610 mm)—For access to and utilization of electrical, mechanical or plumbing systems or equipment.

2. Thirty-six inches (914 mm)—With a required occupant capacity of less than 50.

3. Thirty-six inches (914 mm)—Within a dwelling unit.

4. Seventy-two inches (1829 mm)—In Group E with a *corridor* having a required capacity of 100 or more.

5. Seventy-two inches (1829 mm)—In *corridors* and areas serving gurney traffic in occupancies where patients receive outpatient medical care, which causes the patient to be not capable of self-preservation.

6. Ninety-six inches (2438 mm)—In Group I-2 in areas where required for bed movement.

❖ The corridor widths specified in Section 1018.2 and its exceptions are long established minimums originally derived from human dimensions, practical concerns, occupant load and psychological considerations. Additional corridor capacity, when necessary, is determined in accordance with Section 1005, Egress Width.

The number of occupants using a corridor for egress establishes the required capacity of a corridor, as well as for any specific portion of a corridor system. Portions of a corridor system may differ in width for a variety of reasons not related to code minimums. The designer and building official are expected to verify that corridor widths and corridor fire-resistance ratings are in accordance with Sections 1005 and 1018.

The required occupant capacity of a corridor is based on the total occupant load of the rooms and spaces served by the corridor as determined by Section 1004. Where a corridor is served by two exits in opposite directions, the corridor capacity is split to determine the minimum required width of those exits (exit door, exit stairway) at each end of the corridor. The total occupant load served by a corridor is not split to establish the corridor fire-resistance rating or capacity.

The width of passageways, aisles and corridors is a functional element of building construction that allows the occupants to circulate freely and comfortably throughout the floor area under nonemergency conditions. Under emergency situations, the egress passageways must provide the needed width to accommodate the number of occupants that must utilize the corridor for egress.

When the occupant load of the space exceeds 49, the minimum width of the passageway, aisle or corridor serving that space is required to be at least 44 inches (1118 mm) to permit two unimpeded parallel columns of users to travel in opposite directions. When the occupant load served is 49 or less, a minimum width of 36 inches (914 mm) is permitted and the users are expected to encounter some intermittent travel interference from fellow users, but the lower occupant load makes those occasions infrequent and tolerable. The 36-inch (914 mm) minimum width is also required within a dwelling unit.

Passageways that lead to building equipment and systems must be at least 24 inches (610 mm) in width to provide a means to access and service the equipment when needed. Due to the frequency of the servicing intervals and the limited number of occupants in these normally unoccupied areas, a reduced width is warranted. This minimum width criteria applies to many common situations, such as stage lighting and special-effects catwalks; catwalks leading to heating and cooling equipment as well as passageways providing access to boilers, furnaces, transformers, pumps, piping and other equipment.

Except for small buildings, Group E occupancies are required to have minimum 72-inch-wide (1829 mm) corridors where the corridors serve educational areas. This width is needed not only for proper functional use, but also because of the edge effect caused by student lockers and other boundary attractions and objects. Service and other corridors outside of educational areas, such as an administrative area, would be regulated consistent with their use. Note that Section 1018.3 would not allow the wall lockers to overlap the required corridor width.

In Group I-2 occupancies, where the corridor is utilized during a fire emergency for moving patients confined to beds, it is required to be at least 96 inches (2438 mm) in clear width. This width requirement is applicable to all areas where there are patient sleeping rooms, and may also be required in some of the treatment room areas where in-house patients will be brought in in beds or gurneys. This minimum width allows two rolling beds or gurneys to pass in a corridor and permits the movement of a bed/gurney into the corridor through a room door. In Group I-2 and ambulatory care center areas, where the movement of beds is not anticipated, such as administrative and some outpatient areas of a hospital or clinic, the corridor would not be required to be 96 inches (2438 mm) wide. The minimum width would be determined by one of the appropriate applicable criteria. For outpatient medical care, where the patient may not be capable of self-preservation, such as some outpatient surgery areas or dialysis treatment areas, the 72-inch-wide (1829 mm) corridor is required based on Exception 5. This would include Group I surgical areas, areas such as MRI suites or dialysis centers, emergency rooms or Group B ambulatory care centers.

1018.3 Corridor obstruction. The required width of *corridors* shall be unobstructed.

Exception: Doors complying with Section 1005.2.

❖ It is important to maintain required corridor width so that the path of travel to an exit is continually available and unobstructed. However, due to the fact that corridors tend to be lined with doors, there are allowances under Section 1005.2. In no case may the door block more than 50 percent of the corridor width. In addition, when fully open, the doors must not protrude more than 7 inches (178 mm). This is consistent with the provisions in aisles, corridors, stairways, ramps, exit passageways and exit discharge courts.

A cross reference back to Section 1005.2 from the exceptions for width in corridors (see Section 1018.3), aisles (see Section 1017.1), exit passageways (see Section 1023.2) and exit courts (see Section 1027.5.1) reinforces the fact that the protrusion limits provision is generally applicable for these types of confined routes.

1018.4 Dead ends. Where more than one *exit* or *exit access doorway* is required, the *exit access* shall be arranged such that there are no dead ends in *corridors* more than 20 feet (6096 mm) in length.

Exceptions:

1. In occupancies in Group I-3 of Occupancy Condition 2, 3 or 4 (see Section 308.4), the dead end in a *corridor* shall not exceed 50 feet (15 240 mm).

2. In occupancies in Groups B, E, F, I-1, M, R-1, R-2, R-4, S and U, where the building is equipped throughout with an *automatic sprinkler system* in accordance with Section 903.3.1.1, the length of the dead-end *corridors* shall not exceed 50 feet (15 240 mm).

3. A dead-end *corridor* shall not be limited in length where the length of the dead-end *corridor* is less than 2.5 times the least width of the dead-end *corridor*.

❖ The requirements of this section apply where a space is required to have more than one means of egress according to Section 1015.1.

Dead ends in corridors and passageways can seriously increase the time needed for an occupant to locate the exits. More importantly, dead ends will allow a single fire event to eliminate access to all of the exits by trapping the occupants in the dead-end area. A dead end exists if the occupant of the corridor or passageway has only one direction to travel to reach any of the building exits [see Figure 1018.4(1)]. While a preferred building layout would be one without dead ends, a maximum dead-end length of 20 feet (6096 mm) is permitted and is to be measured from the extreme point in the dead end to the point where the occupants have a choice of two directions to the exits. Having to go back only 20 feet (6096 mm) after coming to a dead end is not such a significant distance as to cause a serious delay in reaching an exit during an emergency situation.

A dead end results whether or not egress elements open into it. A dead end is a hazard for occupants who enter the area from adjacent spaces, travel past an exit into a dead end or enter a dead end with the mistaken assumption that an exit is directly accessible from the dead end.

Note that Section 402.4.5 deals with dead-end distances in a covered mall and assumes that, with a sufficiently wide mall in relation to its length, alternative paths of travel will be available in the mall itself to reach an exit (i.e., the mall is not to be confused as being a corridor).

Under special conditions, exceptions to the 20-foot (6096 mm) dead-end limitation apply.

Exception 1 is permitted based on the considerations of the functional needs of Group I-3 Occupancy

Conditions 2, 3 or 4, the requirements for smoke compartmentalization in Section 408.6 and the requirement for automatic sprinkler protection of the facility in Section 903.2.6.

Exception 2 recognizes the fire protection benefits and performance history of automatic fire sprinkler systems. While the degree of hazard in Group B, E, F, M, S and U occupancies does not initially require an automatic fire suppression system, the length of a dead-end corridor or passageway is permitted to be extended to 50 feet (15 240 mm) where an automatic fire sprinkler system in accordance with NFPA 13 is provided throughout the building. This exception is also permitted in Group I-1, R-1, R-2 and R-4 occupancies. In addition, these provisions are consistent with those in the IEBC and IFC in the regulation of dead-end corridors in existing buildings undergoing alterations.

Exception 3 addresses the condition presented by

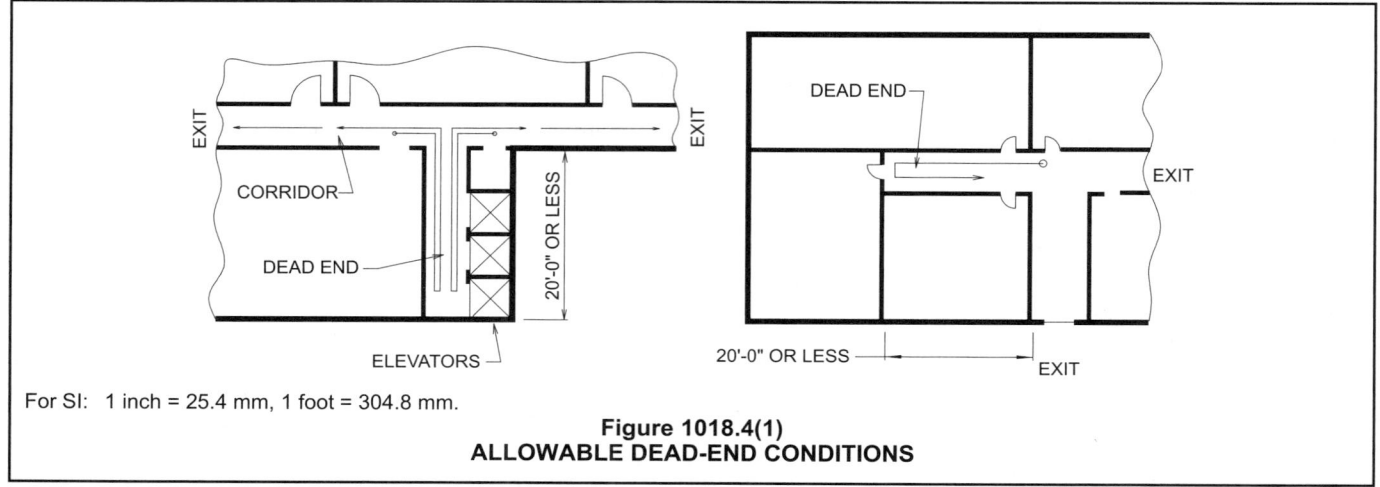

For SI: 1 inch = 25.4 mm, 1 foot = 304.8 mm.

Figure 1018.4(1)
ALLOWABLE DEAD-END CONDITIONS

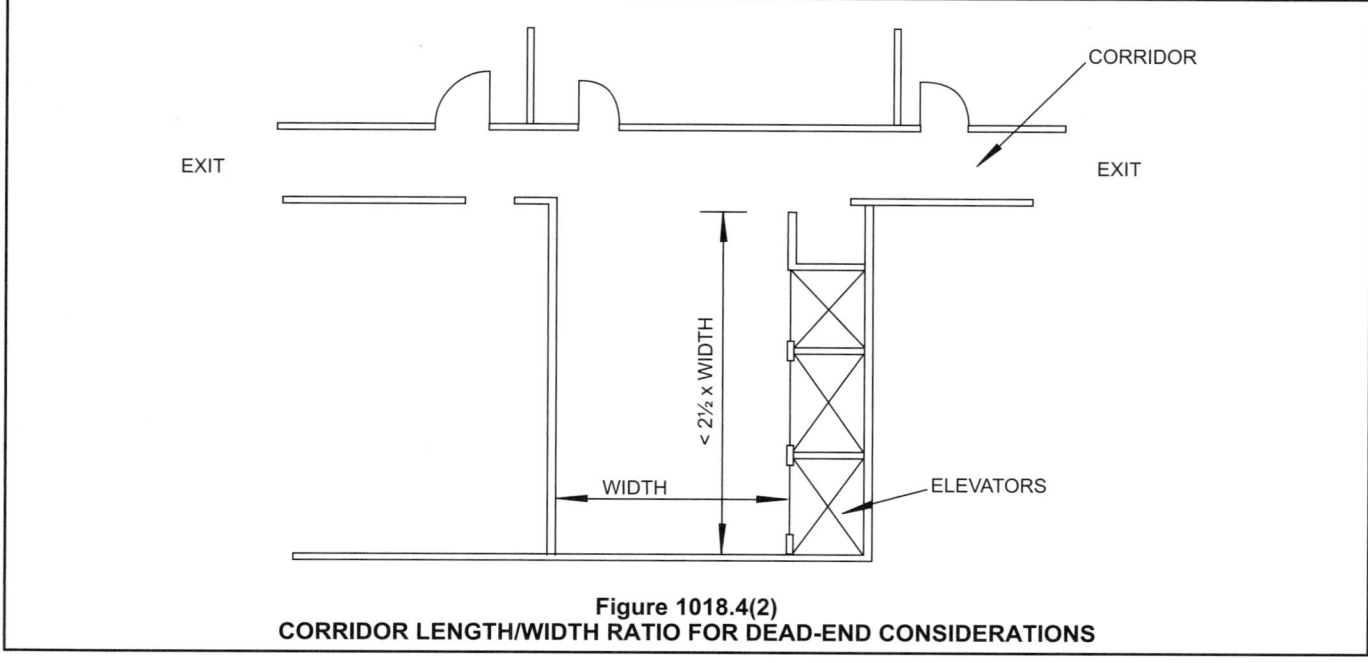

Figure 1018.4(2)
CORRIDOR LENGTH/WIDTH RATIO FOR DEAD-END CONSIDERATIONS

"cul-de-sac" elevator lobbies directly accessible from exit access corridors. In such an elevator lobby, lengths of 20 to 30 feet (6096 to 9144 mm) are common for three- or four-car elevator banks. Typically, the width of this elevator lobby is such that the possibility of confusion with a path of egress is minimized. Below the $2^1/_2$:1 ratio, the dead end becomes so wide that it is less likely to be perceived as a corridor leading to an exit. For example, based on the $2^1/_2$:1 ratio limitation, a 25-foot-long (7620 mm) dead end over 10 feet (3048 mm) in width would not be considered a dead-end corridor [see Figure 1018.4(2)]. For additional elevator lobby requirements, see the commentary to Section 708.14.1.

1018.5 Air movement in corridors. *Corridors* shall not serve as supply, return, exhaust, relief or ventilation air ducts.

Exceptions:

1. Use of a *corridor* as a source of makeup air for exhaust systems in rooms that open directly onto such *corridors*, including toilet rooms, bathrooms, dressing rooms, smoking lounges and janitor closets, shall be permitted, provided that each such *corridor* is directly supplied with outdoor air at a rate greater than the rate of makeup air taken from the *corridor*.

2. Where located within a dwelling unit, the use of *corridors* for conveying return air shall not be prohibited.

3. Where located within tenant spaces of 1,000 square feet (93 m²) or less in area, utilization of *corridors* for conveying return air is permitted.

4. Incidental air movement from pressurized rooms within health care facilities, provided that the *corridor* is not the primary source of supply or return to the room.

❖ Two of the most critical elements of the means of egress are the required exit stairways and corridors. Exit stairways serve as protected areas in the building that provide occupants with safe passage to the level of exit discharge. Corridors that provide access to the required exits frequently limit the direction of egress travel (e.g., travel forward or backward only). Since required exits and corridors are critical elements in the means of egress, the potential spread of smoke and fire into these spaces must be minimized. The scope of this section is corridors. For requirements for the exits, see Section 1022.

The use of these corridors as part of the air distribution system could render those egress elements unusable. The intent is to have positive pressure in the corridors. Therefore, any air movement condition that could introduce smoke into these vital egress elements is prohibited. It is not the intent of this section to prohibit the air movement necessary for ventilation and space conditioning of corridors, but rather to pre-

vent those spaces from serving as conduits for the distribution of air to, or the collection of air from, adjacent spaces. This restriction also extends to door transoms and door grilles that would allow the spread of smoke into a corridor. This limitation is not, however, intended to restrict slight pressure differences across corridor doors, such as a negative pressure differential maintained in kitchens to prevent odor migration into dining rooms. Note that air distribution via ducted systems located in or above corridors is acceptable since the corridor itself would not be functioning as a duct.

The four exceptions to this section identify conditions where a corridor can be utilized as part of the air distribution system. The exceptions apply only to exit access corridors, not to exit passageways.

Exception 1 addresses the common practice of using air from the corridor as makeup air for small exhaust fans in adjacent rooms. Where the corridor is supplied directly with outdoor air at a rate equal to or greater than the makeup air rate, negative pressure will not be created in the corridor with respect to the adjoining rooms and smoke would generally not be drawn into the corridor.

Regarding Exception 2, it is common practice to locate return air openings in the corridors of dwelling units and draw return air from adjoining spaces through the corridor. Such use of dwelling unit corridors for conveying return air is not considered to be a significant hazard and is permitted. Dwelling units are permitted to have unprotected openings between floors. Corridors in dwelling units that serve small occupant loads are short in length and are not required to be fire-resistance rated. For these reasons, the use of the corridor or the space above a corridor ceiling for conveying return air does not constitute an unacceptable hazard.

Exception 3 permits corridors located in small tenant spaces to be used for conveying return air based on the relatively low occupant load and the relatively short length of the corridor. These conditions do not pose a significant hazard. In the event of an emergency, the occupants of the space would tend to simply retrace their steps to the entrance.

Health care facilities require direct pressurization control of certain rooms to provide a clean and sterile environment for patients. For example, operating rooms and pharmacies are required to have positive air pressure in the room, resulting in a general air movement out of the room. This ensures that airborne contaminants do not infect a sterile procedure or supplies. Pressurization is achieved by supplying air at a greater or lesser rate than the return air. Exception 4 recognizes the need of infection control and clarifies that the corridor should not be the primary source of supply return. There should be supply and return air in the room.

1018.5.1 Corridor ceiling. Use of the space between the *corridor* ceiling and the floor or roof structure above as a return air plenum is permitted for one or more of the following conditions:

1. The *corridor* is not required to be of fire-resistance-rated construction;

2. The *corridor* is separated from the plenum by fire-resistance-rated construction;

3. The air-handling system serving the *corridor* is shut down upon activation of the air-handling unit *smoke detectors* required by the *International Mechanical Code*;

4. The air-handling system serving the *corridor* is shut down upon detection of sprinkler waterflow where the building is equipped throughout with an *automatic sprinkler system*; or

5. The space between the *corridor* ceiling and the floor or roof structure above the *corridor* is used as a component of an *approved* engineered smoke control system.

❖ This section identifies five different conditions where the space above the corridor ceiling is permitted to serve as a return air plenum only. Since a return air plenum operates at a negative pressure with respect to the corridor, any smoke and gases within the plenum should be contained within that space. Conversely, a supply plenum operates at a positive pressure with respect to the corridor, thus increasing the likelihood that smoke and gases will infiltrate the corridor enclosure. Where any one of the five conditions is present, the use of the corridor ceiling space as a return air plenum is permitted.

Where the corridor is permitted to be constructed without a fire-resistance rating (see Section 1018.1), Item 1 permits the space above the ceiling to be utilized as a return air plenum without requiring it to be separated from the corridor with fire-resistance-rated construction.

Item 2 is only applicable to corridors that are required to be enclosed with fire-resistance-rated construction. Compliance with this item requires the plenum to be separated from the corridor by fire-resistance-rated construction equivalent to the rating of the corridor enclosure itself. Therefore, the ceiling membrane itself must provide the fire-resistance rating required of the corridor enclosure. Section 709.4, Exception 3, is an example of this method of construction.

Items 3 and 4 recognize that the hazard associated with smoke spread through a plenum is minimized if the air movement is stopped.

It is not uncommon for an above-ceiling plenum to be utilized as part of the smoke removal system. This practice is permitted by Item 5. Due to the way these systems are designed, the higher equipment ratings and the power supply provisions, this is considered acceptable.

1018.6 Corridor continuity. Fire-resistance-rated corridors shall be continuous from the point of entry to an *exit*, and shall not be interrupted by intervening rooms.

Exception: Foyers, lobbies or reception rooms constructed as required for *corridors* shall not be construed as intervening rooms.

❖ This section requires the fire protection offered by a corridor to be continuous from the point of entry into the corridor and to an exit. This is to protect occupants from the accumulation of smoke or fire exposure and to allow for sufficient time to evacuate the building. Where a corridor is served by two or more exits, only one of the exits is required to be accessed directly from the corridor. Other exits may be accessed through intervening spaces in accordance with Section 1014.2, provided that there is an opening protective at the end of the corridor to separate the rated corridor from the intervening rooms. Thus, occupants will always have their protected path to an exit, and at the same time allows a reasonable degree of design freedom [see Figures 1018.6(1) and 1018.6(3)].

Note that when vertical exit enclosures are not required around an exit stairway or ramp (see Section 1022.1), such as within an open parking garage, the "point of entry to the exit" would be the top of the stairway or ramp. Since this connection would be an opening into the corridor, an opening protective would need to be provided at the transition between the protected corridor and the open exit stairway. Adequate landings meeting with the stairway and door provisions must be provided.

When a level is permitted to have open exit access stairways in accordance with Section 1016.1, Exceptions 3 and 4, the corridor protection would be required to continue down the exit access stairway and to an enclosed exit stairway or to an exit door leading to the outside. Since the stairway effectively becomes part of the corridor, doors would not be required at the top and bottom of the stairway as they are for vertical exit enclosures.

The exception allows a foyer, lobby or reception room to be located on the path of egress from a corridor or as part of the fire-resistance-rated corridor, provided the room has the same fire-resistance-rated walls and doors as required for the corridor. The use of the term should be viewed as limiting the types of uses that may occur within the protected corridor. Occupied spaces within the corridor should have very limited uses and hazards. Foyers and lobbies are included in this exception based on the low fire hazard of the contents in such rooms [see Figure 1018.6(2)].

A consideration must be the coordination of the requirements for fire-resistance-rated corridor continuity when connected to an elevator. When an elevator opens into a corridor that is required to be of fire-resistance-rated construction, the opening between the elevator shaft and the corridor must be protected to meet not only the shaft's fire protection rating but also the additional smoke and draft protection requirements necessary to limit the spread of smoke into the corridor. This additional smoke and draft control requirement is found in Section 715.4.3.1. Because elevator hoistway doors do not typically comply as smoke- and draft-control assemblies, they would not be able to open directly into a corridor that is required to have protected openings. The provisions in Section 708.14 waiving the requirements for an elevator lobby do not also waive the corridor opening protection requirements. Therefore, to maintain the integrity of the corridor, the elevator hoistway shaft doors opening into such rated corridors will need to be separated from the corridor by one of the following methods of protection:

1. A lobby needs to be provided with the appropriate doors (see Figure 1018.6.(4) and Section 708.14).

2. Additional doors must be provided at the hoistway (see Figure 1018.6(5) and Sections 715.4.3 and 3002.6).

3. An elevator shaft door meeting both the smoke and draft protection requirements for corridor

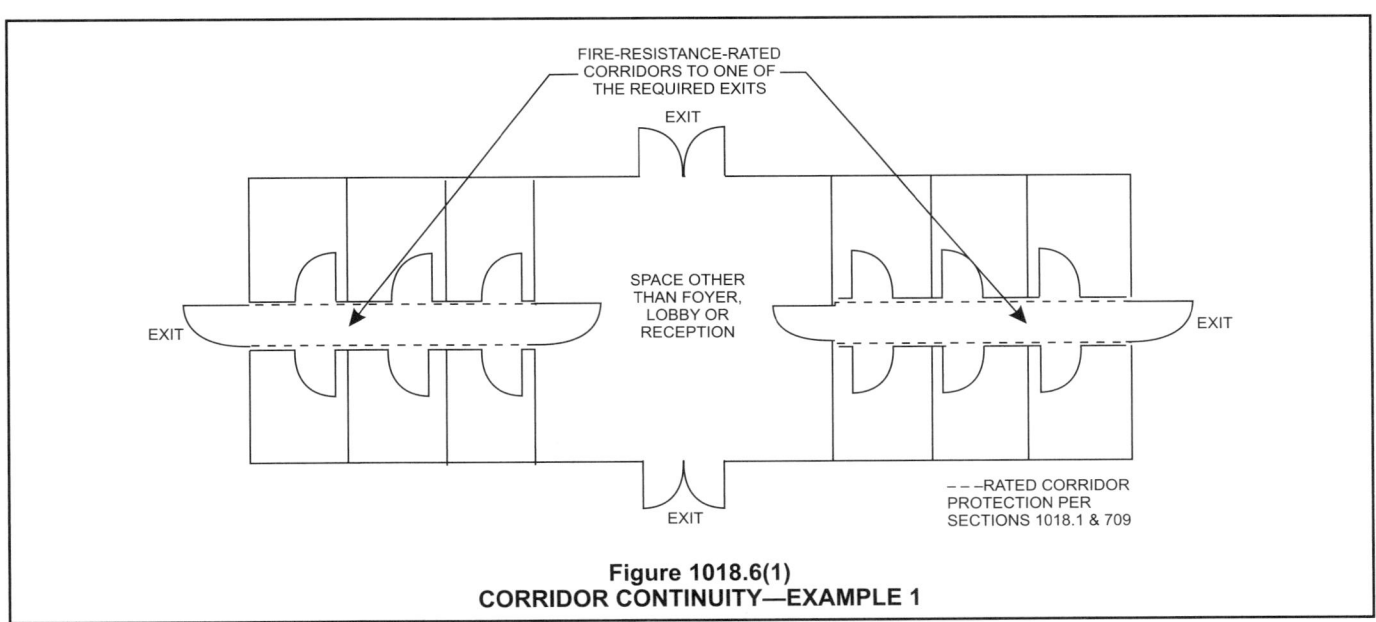

Figure 1018.6(1)
CORRIDOR CONTINUITY—EXAMPLE 1

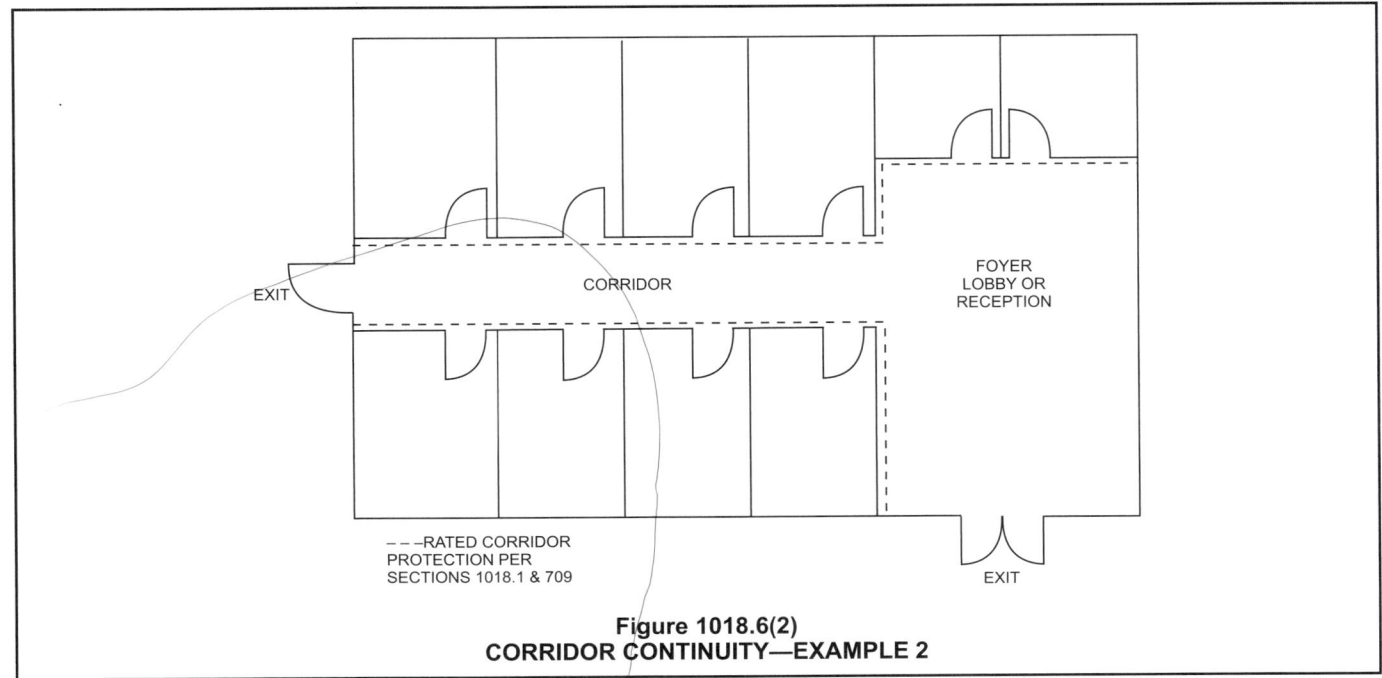

Figure 1018.6(2)
CORRIDOR CONTINUITY—EXAMPLE 2

doors in Section 715.4.3.1 as well as the appropriate fire protection rating of Table 715.4 for the shaft must be provided.

4. The corridor must be separated from the lobby [see Figure 1018.6(6)].

Option 4 is permitted when the other end of the fire-resistance-rated corridors lead directly to an exit. While many elevator hoistway shaft doors are tested and labeled for the 1-hour or 1¹/₂-hour fire-resistance rating (see Section 715.4), very few, if any of the doors

typically sold in the U.S. will also meet the smoke and draft requirements (see Section 715.4.3.1) that would allow them to open directly into a fire-resistance-rated corridor. Because of this, Items 1, 2 and 4 above will be the general methods for protecting such openings.

For requirements and additional explanation of the elevator lobby requirements for elevator lobbies that are adjacent to rated corridors, see Section 708.14. For requirements for exit enclosures, see Section 1022.

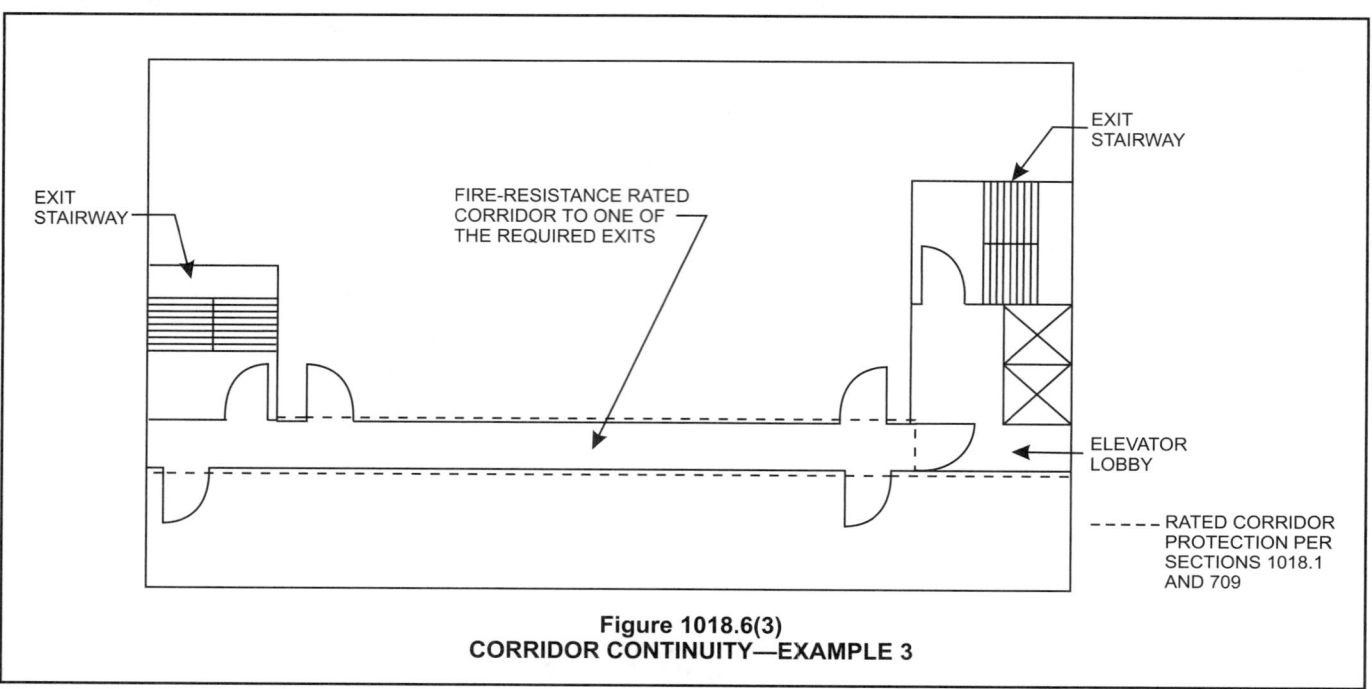

Figure 1018.6(3)
CORRIDOR CONTINUITY—EXAMPLE 3

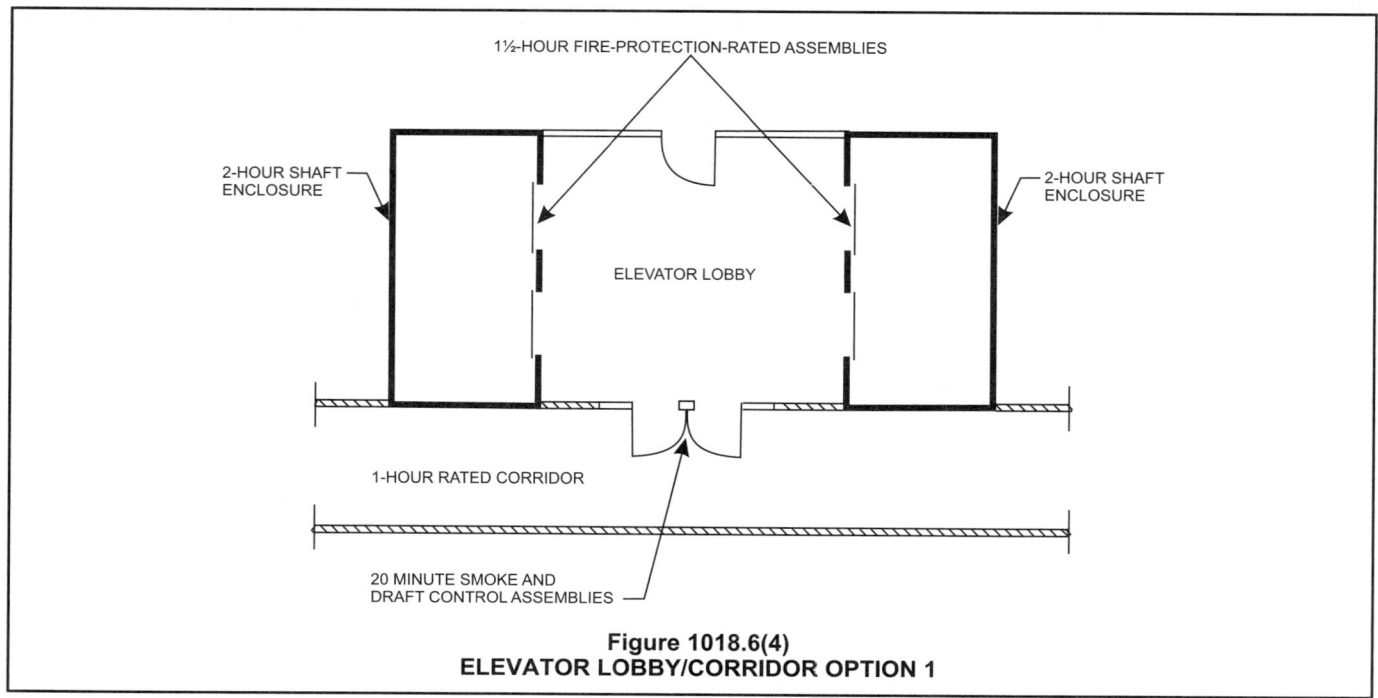

Figure 1018.6(4)
ELEVATOR LOBBY/CORRIDOR OPTION 1

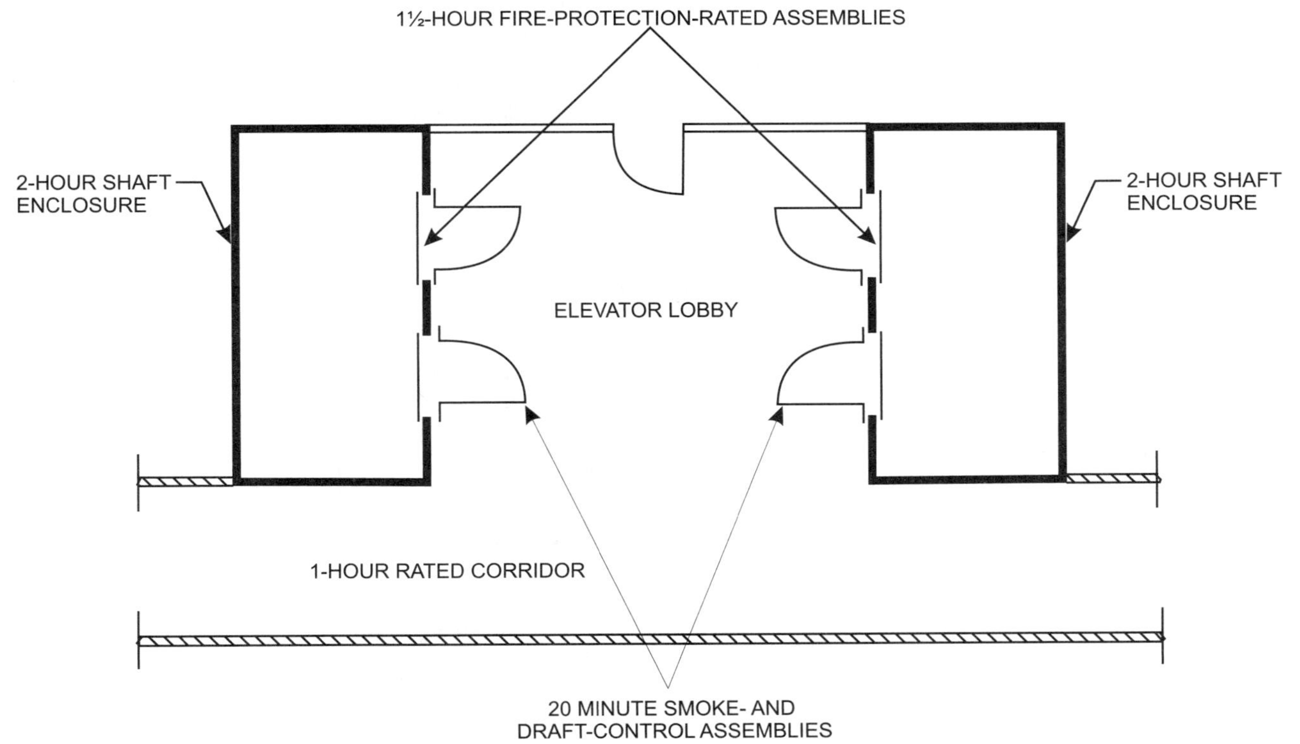

1½-HOUR FIRE-PROTECTION-RATED ASSEMBLIES

2-HOUR SHAFT
ENCLOSURE

2-HOUR SHAFT
ENCLOSURE

ELEVATOR LOBBY

1-HOUR RATED CORRIDOR

20 MINUTE SMOKE- AND
DRAFT-CONTROL ASSEMBLIES

Figure 1018.6(5)
ELEVATOR LOBBY/CORRIDOR OPTION 2

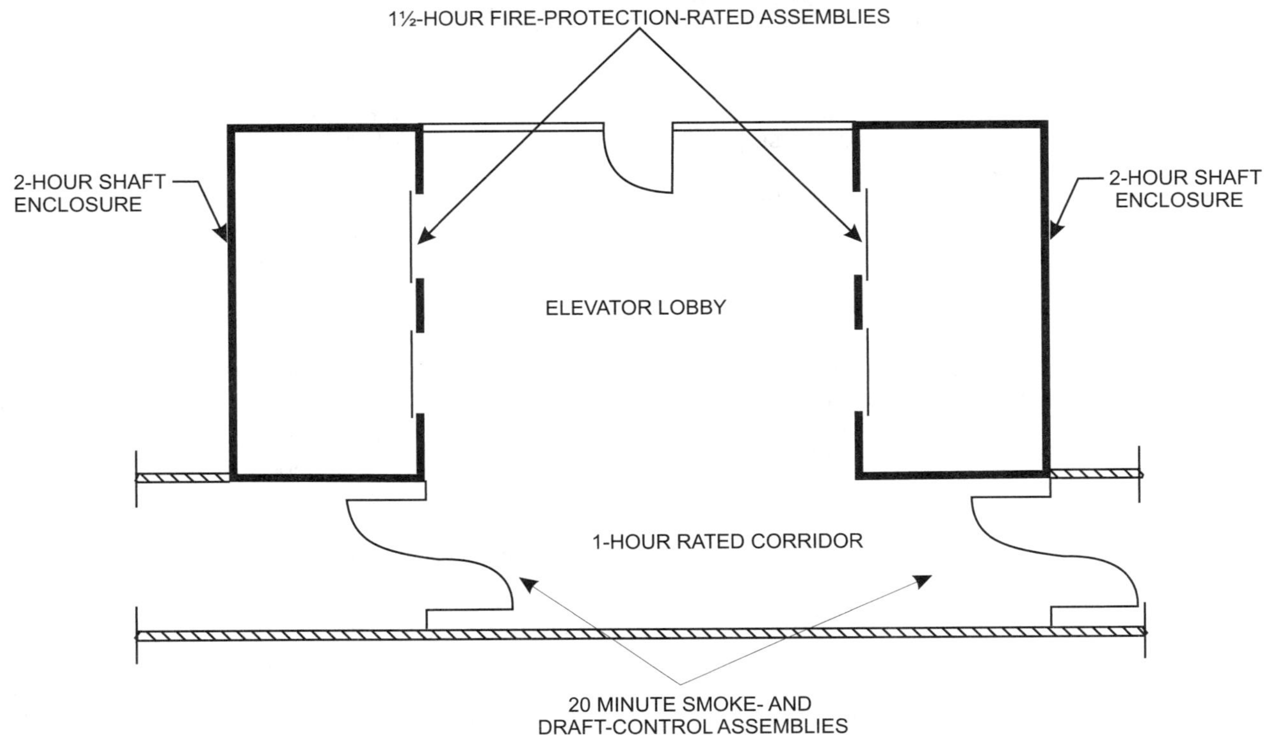

1½-HOUR FIRE-PROTECTION-RATED ASSEMBLIES

2-HOUR SHAFT
ENCLOSURE

2-HOUR SHAFT
ENCLOSURE

ELEVATOR LOBBY

1-HOUR RATED CORRIDOR

20 MINUTE SMOKE- AND
DRAFT-CONTROL ASSEMBLIES

Figure 1018.6(6)
ELEVATOR LOBBY/CORRIDOR OPTION 3

SECTION 1019
EGRESS BALCONIES

1019.1 General. Balconies used for egress purposes shall conform to the same requirements as *corridors* for width, headroom, dead ends and projections.

❖ This section regulates balconies that are used as an exit access element. Requirements are the same as exit access corridors, except for the enclosure.

Where exterior egress balconies are used in moderate or severe climates, there may also be a concern to protect the egress balcony from accumulations of snow and ice to provide a safe path of egress travel at all times, including winter. Maintenance of the means of egress in the IFC requires an unobstructed path to allow for full instant use in case of a fire or emergency. Typical methods for protecting these egress elements include roof overhangs or canopies, heated slab and, when approved by the building official, a reliable snow removal maintenance program.

1019.2 Wall separation. Exterior egress balconies shall be separated from the interior of the building by walls and opening protectives as required for *corridors*.

> **Exception:** Separation is not required where the exterior egress balcony is served by at least two *stairs* and a dead-end travel condition does not require travel past an unprotected opening to reach a *stair*.

❖ An exterior exit access balcony has a valuable attribute in that the products of combustion may be freely vented to the open air. In the event of a fire in an adjacent space, the products of combustion would not be expected to build up in the balcony area as would commonly occur in an interior corridor. However, there is still a concern for the egress of occupants who must use the balcony for exit access, and consequently, may have to pass the room or space where the fire is located. Therefore, an exterior exit access balcony is required to be separated from interior spaces by fire partitions, as is required for interior corridors. The other provisions of Section 1018 relative to dead ends and opening protectives also apply.

If there are no dead-end conditions that require travel past an unprotected opening and the balcony is provided with at least two stairways, then the wall separating the balcony from the interior spaces need not have a fire-resistance rating (see Figure 1019.2). Such an arrangement reduces the probability that occupants will need to pass the area with the fire to gain access to an exit.

1019.3 Openness. The long side of an egress balcony shall be at least 50 percent open, and the open area above the guards shall be so distributed as to minimize the accumulation of smoke or toxic gases.

❖ This section provides an opening requirement that is intended to preclude the rapid buildup of smoke and toxic gases. A minimum of one side of the exterior balcony is required to have a minimum open exterior area of 50 percent of the side area of the balcony. The side

openings are to be uniformly distributed along the length of the balcony.

Exterior egress balconies are required to be a minimum of 10 feet (3048 mm) away from the lot line or adjacent buildings (see Section 1027.3).

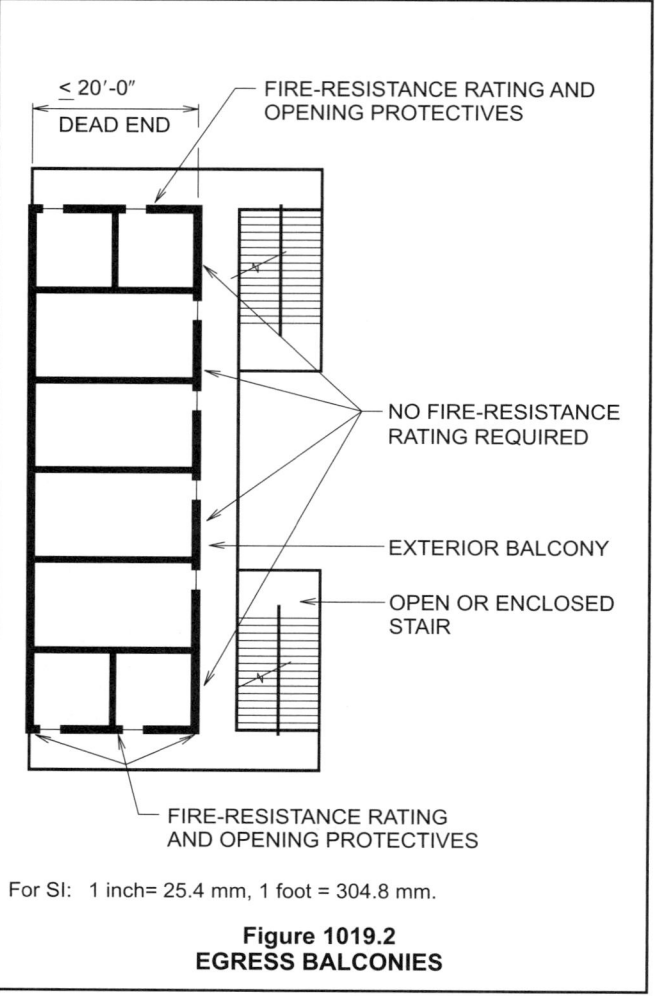

For SI: 1 inch= 25.4 mm, 1 foot = 304.8 mm.

Figure 1019.2
EGRESS BALCONIES

SECTION 1020
EXITS

1020.1 General. *Exits* shall comply with Sections 1020 through 1026 and the applicable requirements of Sections 1003 through 1013. An *exit* shall not be used for any purpose that interferes with its function as a *means of egress*. Once a given level of exit protection is achieved, such level of protection shall not be reduced until arrival at the *exit discharge*.

❖ The use of required exterior exit doors, exit stairways, exit passageways and horizontal exits for any purpose other than exiting is prohibited, because it might interfere with use as an exit. This is not intended to prohibit a door or stairway being used as part of normal circulation patterns, such as the exit doors also serving as entrances, or using the stairway to move between floors when there is not an emergency. However, these spaces must not include furniture, storage or work space. For example, the use of an exit stairway landing for storage, vending machines, copy machines, displays or any purpose other than for exiting is not permitted. Such a situation could not only lead to obstruction of the path of exit travel, thereby creating a hazard to life safety, but if the contents consist of combustible materials, then the use of the stairway as a means of egress could be jeopardized by a fire in the exit enclosure. The restriction does not apply solely to combustible contents, because of the potential for obstruction and difficulties in limiting the materials to noncombustible.

It is recognized that standpipe risers are provided within the stair enclosure and that vertical electrical conduit may be necessary for power or lighting. However, such risers must be located so as not to interfere with the required clear width of the exit. For example, a standpipe riser located in the corner of a stairway will not reduce the required clear width of the landing. This also holds true when the stairway landing is used as an area of refuge. The spaces for wheelchairs must not obstruct the general path of egress travel [see Figures 1007.3(1) and 1007.3(2)]. Electrical conduit and mechanical equipment are permitted when necessary to serve the exit enclosure.

Sections 1020 through 1026 apply to all exits but do not apply to elements of the means of egress that are not exits, such as exit access corridors and passageways or elements of the exit discharge. For exit discharge options for the enclosure, other than a door leading directly to the outside, see the requirements for exit passageways in Section 1023 or the options permitting usage of a lobby or vestibule in the exceptions to Section 1027.1.

1020.2 Exterior exit doors. Buildings or structures used for human occupancy shall have at least one exterior door that meets the requirements of Section 1008.1.1.

❖ The purpose of this section is to specify that at least one exterior exit door is required to meet the door size requirements in Section 1008.1.1. It is not the intent of this section to specify the number of exit doors required, which is addressed in Section 1021.1.

1020.2.1 Detailed requirements. Exterior *exit* doors shall comply with the applicable requirements of Section 1008.1.

❖ The purpose of this section is simply to provide a cross reference from the exit section to all of the detailed requirements for doors that are included in Section 1008.1 and all of its subsections. For example, the requirements for door operation on exterior exit doors are controlled by Section 1008.1.9.

1020.2.2 Arrangement. Exterior *exit* doors shall lead directly to the *exit discharge* or the *public way*.

❖ The exterior exit door is to be the entry point of the exit discharge or lead directly to the public way. When a person reaches the exterior exit door, he or she is directly outside, where smoke and toxic gases are not a health hazard. Additionally, this section will keep exterior doors at other locations, such as to an exit balcony, from being viewed as an exit.

SECTION 1021
NUMBER OF EXITS AND CONTINUITY

1021.1 Exits from stories. All spaces within each *story* shall have access to the minimum number of *approved* independent *exits* as specified in Table 1021.1 based on the *occupant load* of the *story*. For the purposes of this chapter, occupied roofs shall be provided with *exits* as required for stories.

Exceptions:

1. As modified by Section 403.5.2.

2. As modified by Section 1021.2.

3. *Exit access stairways* and *ramps* that comply with Exception 3 or 4 of Section 1016.1 shall be permitted to provide the minimum number of *approved* independent *exits* required by Table 1021.1 on each *story*.

4. In Group R-2 and R-3 occupancies, one *means of egress* is permitted within and from individual dwelling units with a maximum *occupant load* of 20 where the dwelling unit is equipped throughout with an *automatic sprinkler system* in accordance with Section 903.3.1.1 or 903.3.1.2.

5. Within a *story*, rooms and spaces complying with Section 1015.1 with *exits* that discharge directly to the exterior at the *level of exit discharge*, are permitted to have one *exit*.

❖ This section requires every floor of a building to be served by at least two exits (see Figure 1021.1). Similarly, every portion of a floor must also be provided with access to at least two exits. Where more than 500 occupants are located on a single floor, additional exits must be provided. For example, if a floor has an occupant load of 750, each occupant of that floor must have access to not less than three exits (see Table 1021.1).

This section also addresses the need for exits from roofs that are occupied by the public, such as rooftop decks or dining areas. It is important that exit enclosures serve the roof level in addition to the other floor levels.

The text references buildings and spaces with one means of egress (see Sections 1021.2 and 1015.1). Where a building requires more than one exit, that number, determined in accordance with Table 1021.1, is required from each floor level. This is true even if an individual floor level qualifies as a space with one means of egress in accordance with Table 1015.1. Table 1015.1 is intended to be applicable to rooms and spaces, but not to entire floor levels. One of the main concerns has been that vertical travel takes longer than horizontal travel in emergency exiting situations. However, if the single exit space can exit directly to the exterior rather than into an interior corridor, this provides a higher level of safety. This is the reasoning behind Exception 5. While the term "building" limits the area addressed to that bordered by exterior walls or fire walls, a common application of Exception 5 is on a tenant-by-tenant basis. For example, a strip mall may not meet the provisions for a building with one means of egress. However, assume a tenant meets the provisions for a space with one means of egress in accordance with Section 1015.1. This tenant could exist as either a stand-alone single-exit building or a single-exit tenant space that exits into an interior corridor. Is it not as safe to permit this tenant to exist as part of a larger building with the door exiting directly to the exterior?

While not specifically stated in this section, there is a situation where the single means of egress can be used from a multilevel space. Section 505 permits mezzanines to be considered part of the floor below for purposes of means of egress. When a mezzanine meets the occupant load in Table 1015.1 and the common path of travel distance (see Section 1014.3) measured from the most remote point to the bottom of the stairway, it can be considered a space with one means of egress.

Exception 4 is based on multiple years of practice within an individual dwelling or sleeping unit. In Group R-2 and R-3 buildings with multistory dwelling or sleeping units, the means of egress from a dwelling or sleeping unit is typically permitted to be from one level only. In a Group R-2 apartment- or townhouse-style building, if the unit has an occupant load of 20 or less, the building is sprinklered with an NFPA 13 or 13R system and the common path of travel from the most remote point on any level to the exit door from the unit itself is 125 feet (22 860 mm) maximum (see Section 1014.3, Exception 4), that unit may have only one means of egress. However, once the occupants exit the unit itself, they must be outside or the floor level must have access to two or more means of egress for all tenants, depending on the number required for the

building as a whole. While this exception also lists Group R-3, Section 1021.2 says all Group R-3 dwellings can have one exit regardless of occupant load or type of sprinkler system. Common path of travel is not applicable in single-exit buildings. The emergency escape and rescue opening addressed in Section 1029 does not count towards the required number of exits.

Exception 3 is to allow for open exit access stairways, as permitted in Section 1016.1, Exceptions 3 and 4, to count towards the required number of exits for that upper or lower level floor or space. For example, Section 1016.1, Exception 4 allows for a two-story office building to use two open exit access stairways provided that exit access travel distance includes travel down the stairway and to the exit door. This exception allows the two exit access stairways to meet the two-exit requirement in Table 1021.1 for that second level.

Exception 2 allows for the single-exit buildings in Section 1021.2.

The high-rise provisions in Section 403.5.2 have a requirement for an additional stairway in buildings 420 feet (128 m) or taller. Exception 1 is in recognition of that additional requirement.

TABLE 1021.1
MINIMUM NUMBER OF EXITS FOR OCCUPANT LOAD

OCCUPANT LOAD (persons per story)	MINIMUM NUMBER OF EXITS (per story)
1-500	2
501-1,000	3
More than 1,000	4

❖ Table 1021.1 specifies that the minimum number of exits available to each occupant of a floor is based on the total occupant load of that floor. This is so that at least one exit will be available in case of a fire emer-

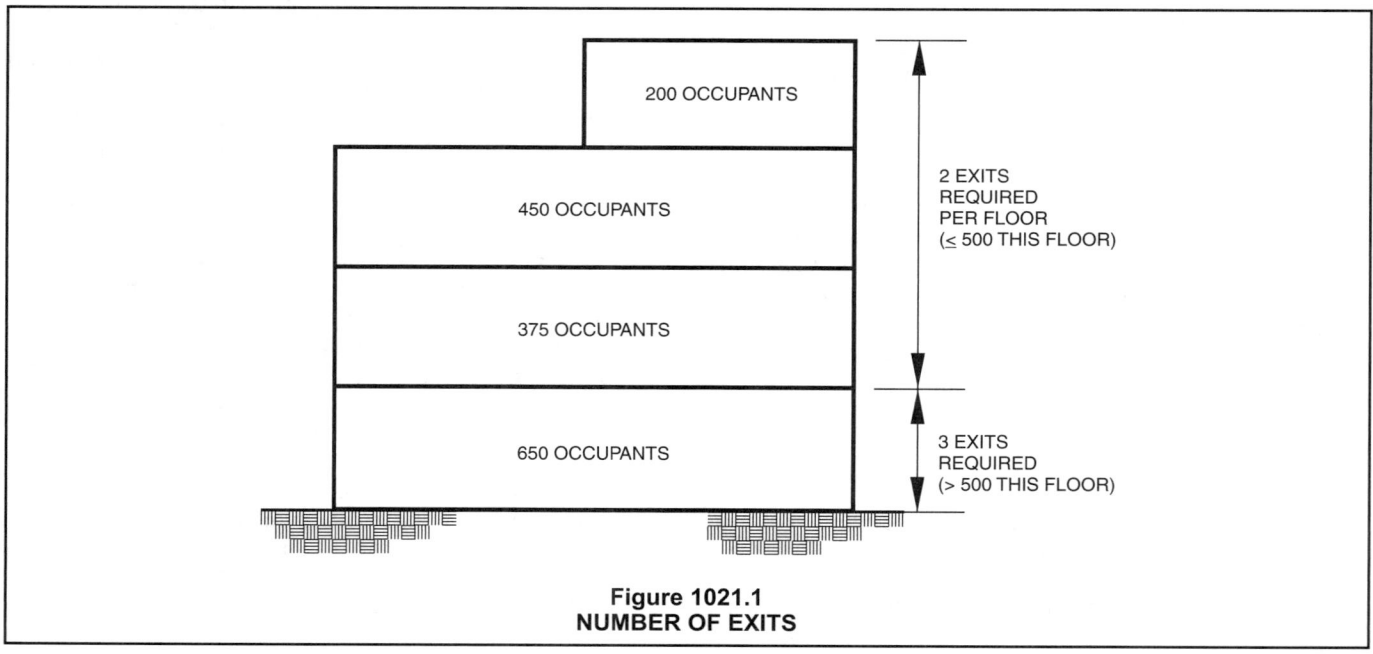

Figure 1021.1
NUMBER OF EXITS

gency and to provide increased assurance that a larger number of occupants can be accommodated by the remaining exits when one exit is not available. While an equal distribution of exit capacity among all of the exits is not required, a proper design would not only balance capacity with the occupant load distribution, but also consider a reasoned distribution of capacity to avoid a severe dependence on one exit.

1021.1.1 Exits maintained. The required number of *exits* from any *story* shall be maintained until arrival at grade or the *public way*.

❖ The need for exits to be independent of each other cannot be overstated. Each occupant of each floor must be provided with the required number of exits without having to pass through one exit to gain access to another. Each exit is required to be independent of other exits to prohibit such areas from merging downstream and becoming, in effect, one exit.

As discussed in conjunction with Section 1015.2, the location of two exits is to accomplish certain remoteness, and any additional exits are to be accessible from any point on the floor.

1021.1.2 Parking structures. Parking structures shall not have less than two *exits* from each parking tier, except that only one *exit* is required where vehicles are mechanically parked. Vehicle ramps shall not be considered as required *exits* unless pedestrian facilities are provided.

❖ At least two exits from each parking tier are required in parking structures, except where the resulting occupant load is minimal, such as where the vehicles are mechanically parked. A vehicle-only ramp may be considered as one of the required exits if pedestrian walkways are provided along the ramp. The ramps that are lined with parking spaces are not considered vehicle ramps. In open parking garages, according to Section 1022.1, Exception 4, the exit stairways are not required to be enclosed, since an open parking structure is designed to permit the ready ventilation of the products of combustion to the outside by exterior wall openings (see Section 406.3). Also, parking structures are characterized by open floor areas that allow the occupants to observe a fire condition and choose a travel path that would avoid the fire threat.

1021.1.3 Helistops. The *means of egress* from helistops shall comply with the provisions of this chapter, provided that landing areas located on buildings or structures shall have two or more *exits*. For landing platforms or roof areas less than 60 feet (18 288 mm) long, or less than 2,000 square feet (186 m²) in area, the second *means of egress* is permitted to be a fire escape, *alternating tread device* or ladder leading to the floor below.

❖ This section provides occupants of helistops with adequate exit facilities. The reduction in the exit requirements for small-area helistops is based on the low number of occupants that are associated with these facilities.

1021.2 Single exits. Only one *exit* shall be required from Group R-3 occupancy buildings or from stories of other build-ings as indicated in Table 1021.2. Occupancies shall be permitted to have a single *exit* in buildings otherwise required to have more than one *exit* if the areas served by the single *exit* do not exceed the limitations of Table 1021.2. Mixed occupancies shall be permitted to be served by single *exits* provided each individual occupancy complies with the applicable requirements of Table 1021.2 for that occupancy. Where applicable, cumulative *occupant loads* from adjacent occupancies shall be considered in accordance with the provisions of Section 1004.1. Basements with a single *exit* shall not be located more than one *story* below *grade plane*.

❖ Buildings with one exit are permitted where the configuration and occupancy meet certain characteristics so as not to present an unacceptable fire risk to the occupants. Buildings that are relatively small in size have a shorter travel distance and fewer occupants; thus, having access to a single exit does not significantly compromise the safety of the occupants since they will also be alerted to and get away from the fire more quickly. It is important to note that the provisions in Section 1021.2 apply to individual stories. Multiple single-exit spaces may exist in the same building, including those cases where differing occupancies exist. Therefore, Table 1021.2 can address mixed occupancy buildings.

Occupants of a story of limited size and configuration may have access to a single exit, provided that the building does not have more than one level below the first story above the grade plane. The limitation on the number of levels below the first story is intended to limit the vertical travel an occupant must accomplish to reach the exit discharge in a single-exit building. Taken to its extreme, the code would otherwise allow a single exit in a building with one or two stories above grade and an unlimited number of stories below grade. This would be clearly inadequate for those stories below grade.

Only one exit is required from a story where permitted by Table 1021.2 regardless of the number of exits required from other stories in the building. For example, a Group B occupancy on the second floor of a multistory building is only required to have one exit from the story provided its occupant load does not exceed 29 and the maximum travel distance to an exit does not exceed 75 feet (22 860 mm). The number of occupants and travel distances on the other stories do not affect the determination of the second story as a single-exit story. Other stories are also regulated independently as to number of exits.

Where multiple tenants or occupancies are located on a specific story, they are to be regulated independently for single-exit determination. The provisions can be applied to specific portions of the story, rather than the story as a whole. As an example, the second story of a building houses two office tenants, each with their own independent means of egress. Each tenant would be permitted a single, but separate, exit provided each had an occupant load of less than 30 and a travel distance not exceeding 75 feet (22 860 mm). This portion-by-portion philosophy also applies to a mixed occupancy condition provided each of the individual occupancies does not exceed the limitations of Table 1021.2 (see Figure 1021.2).

Group R-3 building occupancies are permitted to have a single exit since they do not have more than two dwelling units (see Section 310).

Table 1021.2. See below.

❖ Table 1021.2 lists the characteristics a building must have to be of single-exit construction, including occupancy, maximum height of building above grade plane, maximum occupants or dwelling units per floor and exit access travel distance per floor. The occupant load of each floor is determined in accordance with the provisions of Section 1004.1. The exit access travel distance is measured along the natural and unobstructed path to the exit, as described in Section 1016.1. If the occupant load is exceeded, as indicated

in Table 1021.2, two exits are required from each floor in the building. Likewise, if the travel distance or number of dwelling units is exceeded, as indicated in Table 1021.2, two exits are required from each floor of the building.

The exit enclosure required in a two-story, single-exit building is identical to any other complying exit (i.e., interior stairs, exterior stairs, etc.). Similarly, the fire-resistance rating required for opening protectives is identical to that required by Section 715.

Notes a and b are in reference to unique exit criteria for parking structures and air traffic control towers. Notes c and d provide for increased travel distance for Groups R-2, F, S and B with certain limitations. Note e specifies a maximum occupant load for day care facilities of Group E, but does not provide any additional criteria for travel distance; therefore, 75 feet (22 860 mm)

TABLE 1021.2
STORIES WITH ONE EXIT

STORY	OCCUPANCY	MAXIMUM OCCUPANTS (OR DWELLING UNITS) PER FLOOR AND TRAVEL DISTANCE
First story or basement	A, B[d], E[e], F[d], M, U, S[d]	49 occupants and 75 feet travel distance
	H-2, H-3	3 occupants and 25 feet travel distance
	H-4, H-5, I, R	10 occupants and 75 feet travel distance
	S[a]	29 occupants and 100 feet travel distance
Second story	B[b], F, M, S[a]	29 occupants and 75 feet travel distance
	R-2	4 dwelling units and 50 feet travel distance
Third story	R-2[c]	4 dwelling units and 50 feet travel distance

For SI: 1 foot = 304.8 mm.

a. For the required number of exits for parking structures, see Section 1021.1.2.

b. For the required number of exits for air traffic control towers, see Section 412.3.

c. Buildings classified as Group R-2 equipped throughout with an automatic sprinkler system in accordance with Section 903.3.1.1 or 903.3.1.2 and provided with emergency escape and rescue openings in accordance with Section 1029.

d. Group B, F and S occupancies in buildings equipped throughout with an automatic sprinkler system in accordance with Section 903.3.1.1 shall have a maximum travel distance of 100 feet.

e. Day care occupancies shall have a maximum occupant load of 10.

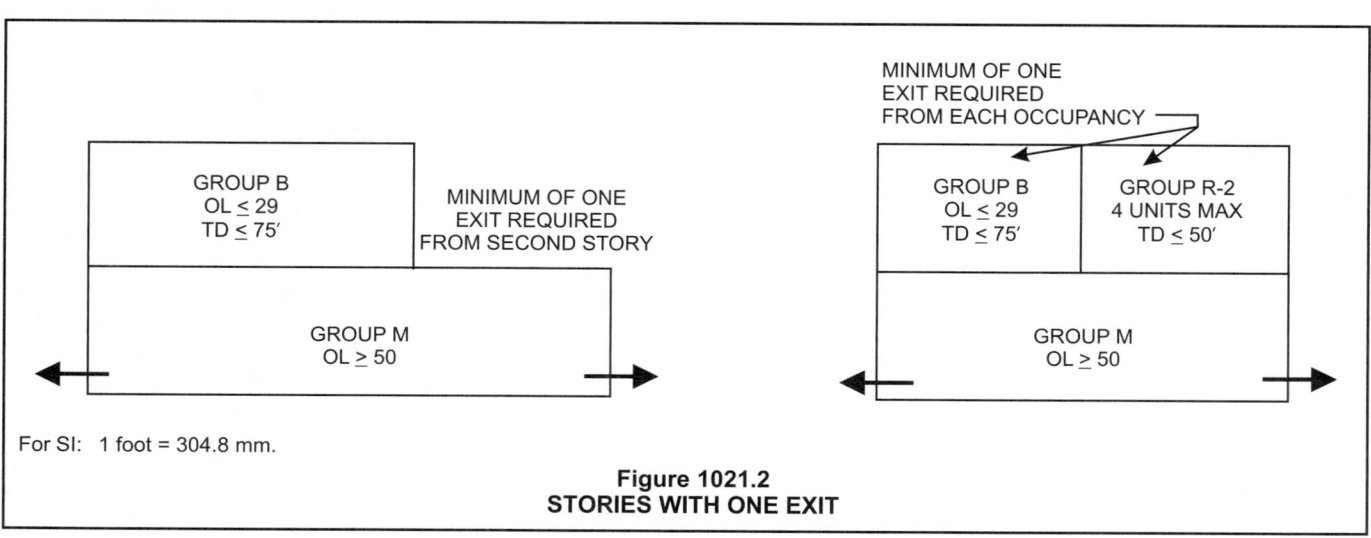

For SI: 1 foot = 304.8 mm.

Figure 1021.2
STORIES WITH ONE EXIT

maximum travel distance would still be applicable. The provision in Note e was the result of the passage of Code Change E10–04/05. The code change proponent's reason statement was mainly focused on infants and the exception in Section 308.5.2. That exception would allow day care facilities with up to 100 children $2^1/_2$ years in age or less with rooms with direct access to the exterior to be classified as Group E. A concern is, with the facility now being classified as Group E, that the building, even though children $2^1/_2$ years of age or less were present, could use the means of egress requirements for Group E for these buildings as opposed to Group I-4 for that space.

1021.3 Exit continuity. *Exits* shall be continuous from the point of entry into the *exit* to the *exit discharge*.

❖ This is consistent with the exit termination requirements in Section 1022.2 and exit discharge requirements in Section 1023.4. The intent of this section is to provide safety in all portions of the exit by requiring continuity of the fire protection characteristics of the exit enclosure. Exit passageways (see Section 1023) are a continuation of an exit enclosure. This would include, but not be limited to, the fire-resistance rating of the exit enclosure walls and the opening protection rating of the doors. Section 1027.1 would allow for an alternative for direct access to the outside via an intervening lobby or vestibule.

Horizontal exits (see Section 1025), while not providing direct access to the outside of the structure, do move occupants to another "building" by moving through a fire wall (see Section 1025 and the definition for "Area, building" in Section 502.1) into a refuge area protected by fire barriers and horizontal assemblies. Horizontal exits are commonly used in hospitals and jails for a defend-in-place type of protection.

1021.4 Exit door arrangement. *Exit* door arrangement shall meet the requirements of Sections 1015.2 through 1015.2.2.

❖ The intent of this section is to provide a cross reference to the exit door requirements for separation.

SECTION 1022
EXIT ENCLOSURES

1022.1 Enclosures required. *Interior exit stairways* and interior *exit ramps* shall be enclosed with *fire barriers* constructed in accordance with Section 707 or *horizontal assemblies* constructed in accordance with Section 712, or both. *Exit enclosures* shall have a *fire-resistance rating* of not less than 2 hours where connecting four stories or more and not less than 1 hour where connecting less than four stories. The number of stories connected by the *exit enclosure* shall include any basements but not any *mezzanines*. *Exit enclosures* shall have a *fire-resistance rating* not less than the floor assembly penetrated, but need not exceed 2 hours. *Exit enclosures* shall lead directly to the exterior of the building or shall be extended to the exterior of the building with an *exit passageway* conforming to the requirements of Section 1023, except as permitted in Section

1027.1. An *exit enclosure* shall not be used for any purpose other than *means of egress*.

Exceptions:

1. In all occupancies, other than Group H and I occupancies, a *stairway* is not required to be enclosed when the *stairway* serves an *occupant load* of less than 10 and the *stairway* complies with either Item 1.1 or 1.2. In all cases, the maximum number of connecting open stories shall not exceed two.

   1.1. The *stairway* is open to not more than one story above its *level of exit discharge*; or

   1.2. The *stairway* is open to not more than one story below its *level of exit discharge*.

2. *Exits* in buildings of Group A-5 where all portions of the *means of egress* are essentially open to the outside need not be enclosed.

3. *Stairways* serving and contained within a single residential dwelling unit or sleeping unit in Group R-1, R-2 or R-3 occupancies are not required to be enclosed.

4. *Stairways* in open parking structures that serve only the parking structure are not required to be enclosed.

5. *Stairways* in Group I-3 occupancies, as provided for in Section 408.3.8, are not required to be enclosed.

6. *Means of egress stairways* as required by Sections 410.5.3 and 1015.6.1 are not required to be enclosed.

7. *Means of egress stairways* from balconies, galleries or press boxes as provided for in Section 1028.5.1 are not required to be enclosed.

❖ This section requires that all interior exit stairways or ramps are to be enclosed with fire barriers and horizontal assemblies having a fire-resistance rating of at least 1 hour. The fire-resistance rating of the enclosure must be increased to 2 hours if the stairway or ramp connects four or more stories or if it penetrates a floor system with a fire-resistance rating of 2 hours or more (see Table 602 for Type I construction). Note that the criteria are based on the number of stories connected by the stairway or ramp and not the height of the building. Therefore, a building that has three stories located entirely above the grade plane and a basement would require an enclosure with a 2-hour fire-resistance rating if the stairway or ramp connects all four stories. The fire-resistant requirements for exit enclosures are consistent with those for shaft enclosures. Where the floor construction penetrated by the exit enclosure has a fire-resistance rating, the exit enclosure must have the same minimum rating. For example, an exit enclosure that penetrates a 2-hour floor assembly must have a minimum fire-resistance rating of 2 hours, regardless of the number of stories the enclosure connects. The fire-resistance rating of an exit enclosure need never exceed 2 hours. If the floor assembly penetrated requires a minimum 3-hour fire-resistance rating, the exit enclosure rating is only required to be 2-hour fire-resistance rated.

The enclosure is needed because an exit stairway or ramp penetrates the floor/ceiling assemblies between the levels, thus creating a vertical opening or shaft. In cases of fire, a vertical opening may act as a chimney, causing smoke, hot gases and light-burning products to flow upward (buoyant force). If an opening is unprotected, these products of combustion will be forced by positive pressure differentials to spread horizontally into the building spaces. There are exceptions for shaft protection around stairways and ramps that are not part of a means of egress in Section 708.2 or exit access stairways as permitted in Section 1016.1, Exceptions 3 and 4. Such stairways may include ornamental stairways within an atrium; or convenience stairways provided in excess of the minimum number required for egress and not designed as an exit; open exit access stairways between the first and second floor of an office building; or an open exit access stairway that serves as the second means of egress from a tenant space when one floor does not have access to the second exit due to tenant layouts.

The enclosure of interior stairways or ramps with construction having a fire-resistance rating is intended to prevent the spread of fire from floor to floor. Another important purpose is to provide a safe path of travel for the building occupants and to serve as a protected means of access to the fire floor by fire department personnel. For this reason, this section and Section 1022.3 limit the penetrations and openings permitted in a stairway enclosure. Exit passageways are considered an extension of the exit enclosure. While most exit enclosures have doors leading directly to the outside, there are limited allowances for the exit to discharge through a lobby or vestibule (see Section 1027).

It is important that an exit stairway or ramp not be used for any purpose other than as a means of egress. For example, there is a tendency to use stairway landings for storage purposes. Such a situation obstructs the path of exit travel and if the stored contents consist of combustible materials, the use of the stairway as a means of egress may be jeopardized, creating a hazard to life safety. However, the restriction on the use of an exit stairway or ramp is not limited to situations when the contents are combustible.

For travel distance measurements at the exit stairways in the exceptions, see the commentary to Section 1016.1. The exceptions in this section are only for the vertical exit enclosure. These stairways/ramps are still considered an exit for purposes of travel distance and number of exits. Any exit stairway or ramp must still comply with the exit discharge requirements of Section 1027 or go into an exit passageway in accordance with Section 1023.

This section allows exceptions to the enclosure requirements stated above. While not specifically mentioned as an exception, Section 1026 for exterior stairways is considered to provide an equivalent level of protection to enclosed interior stairways.

Exception 1 allows an open exit stairway from the first to the second floor or from the basement to the first floor where the stairway does not serve more than 10 occupants. An example of the application of this exception is a retail space that has a small office located on the second story, or a small mechanical or storage area in the basement. This exception allows the stair serving the small second floor or basement area to be unenclosed based on the small number of occupants served.

In Exception 2, stairways in occupancies in Group A-5 in which the means of egress is essentially open to the outside need not be enclosed because of the ability to vent the fire to the outside. The criteria specified in Section 1026 should be used to determine if the stairway is "essentially open."

In Exception 3, the exit stairways in the listed Group R occupancies are not required to have enclosures because of the small occupant load. This exception is limited to stairways located within the individual dwelling or sleeping units. Section 708.2, Exception 1, limits the open exit stairway within dwelling units to four stories or less.

In Exception 4, stairways located in open parking structures are exempt from the enclosure requirements because of the ease of accessibility by the fire services, the natural ventilation of such structures, the low level of fire hazard, the small number of people using the structure at any one time and the excellent fire record of such structures (see commentary, Section 1021.1.2).

In Exception 5, because of security needs in detention facilities, one of the exit stairs is permitted to be glazed in a manner similar to atrium enclosures. Specific limitations and requirements are discussed in Section 408.3.8.

Regarding Exception 6, due to the nature of stages and platforms, stage exit stairways from the side of a stage, spaces under the stage, fly galleries and gridirons are not required to be enclosed.

Exception 7 addresses the unique situation for large assembly seating areas. Since press boxes, galleries and balconies all have the same atmosphere and fire recognition as the rest of the seating, the stairways to these spaces are protected by the same system as the seating (i.e., smoke protected, open to the outside); therefore, it is logical to treat the access to these spaces the same as the seating bowl.

1022.2 Termination. *Exit enclosures* shall terminate at an *exit discharge* or a *public way.*

> **Exception:** An *exit enclosure* shall be permitted to terminate at an *exit passageway* complying with Section 1023, provided the *exit passageway* terminates at an *exit discharge* or a *public way.*

❖ This is consistent with the exit continuity requirements in Section 1021.3 and exit discharge requirements in Section 1023.4. The intent of this section is to provide safety in all portions of the exit by requiring continuity of the fire protection characteristics of the exit enclosure. Exit passageways (see Section 1023) are a continuation of an exit enclosure. This would include, but not be limited to, the fire-resistance rating of the exit enclosure walls and the opening protection rating of the doors. Section 1027.1 would allow for an alternative for direct access to the outside via an intervening lobby or vestibule.

Horizontal exits (see Section 1025), while not providing direct access to the outside of the structure, do move occupants to another "building" by moving through a fire wall (see Section 1025 and the definition for "Area, building" in Section 502.1) into a refuge area protected by fire barriers and horizontal assemblies. Horizontal exits are commonly used in hospitals and jails for a defend-in-place type of protection.

1022.2.1 Extension. Where an *exit enclosure* is extended to an *exit discharge* or a *public way* by an *exit passageway*, the *exit enclosure* shall be separated from the *exit passageway* by a *fire barrier* constructed in accordance with Section 707 or a *horizontal assembly* constructed in accordance with Section 712, or both. The *fire-resistance rating* shall be at least equal to that required for the *exit enclosure*. A *fire door assembly* complying with Section 715.4 shall be installed in the *fire barrier* to provide a *means of egress* from the *exit enclosure* to the *exit passageway*. Openings in the *fire barrier* other than the *fire door assembly* are prohibited. Penetrations of the *fire barrier* are prohibited.

> **Exception:** Penetrations of the *fire barrier* in accordance with Section 1022.4 shall be permitted.

❖ Once a person enters an exit enclosure, that same level of protection should be provided to them until they can leave the building. When a vertical exit enclosure connects to an exit passageway, either at the ground level or at an intermediate transition floor, the exit passageway must provide the same level of protection as the stairway, including fire-resistance of the walls, floor, ceiling and supporting construction and protection of any openings. At the junction between the exit enclosure and the exit passageway, there must be both a rated fire barrier and a fire door. This has the additional benefit of preventing any smoke that may migrate into the exit passageway from also moving up the exit stairway or ramp. Permitted penetrations are the same as those limitations set for the stairway enclosure. See Sections 1011.1 and 1024 for egress markings and signage within these types of spaces.

For the situation when an exit stairway is constructed as a smokeproof enclosure or a pressurized stairway, see Section 1022.9.1.

1022.3 Openings and penetrations. *Exit enclosure* opening protectives shall be in accordance with the requirements of Section 715.

Openings in *exit enclosures* other than unprotected exterior openings shall be limited to those necessary for *exit access* to the enclosure from normally occupied spaces and for egress from the enclosure.

Elevators shall not open into an *exit enclosure*.

❖ The only openings that are permitted in fire-resistance-rated exit stairways or ramp enclosures are doors that lead from normally occupied spaces into the enclosure and doors leading out of the enclosure to the outside. This restriction on openings essentially prohibits the use of windows in an exit enclosure except for those exterior windows that are not exposed to any hazards. This requirement is not intended to prohibit windows or other openings in the exterior walls of the exit enclosure. The verbiage "unprotected exterior openings" includes windows or doors not required to be protected by either Section 705.8 or 1022.6. The only exception would be window assemblies that have been tested as wall assemblies in accordance with ASTM E 119. The objective of this provision is to minimize the possibility of fire spreading into an exit enclosure and endangering the occupants or even preventing the use of the exit at a time when it is most needed. The limitation on openings applies regardless of the fire protection rating of the opening protective. The limitation on openings from normally occupied areas is intended to reduce the probability of a fire occurring in an unoccupied area, such as a storage closet, which has an opening into the stairway, thereby possibly resulting in fire spread into the stairway. Other spaces that are not normally occupied include, but are not limited to, toilet rooms, electrical/mechanical equipment rooms and janitorial closets. For connection between the vertical exit enclosure and an exit passageway, see Sections 1022.1 and 1022.9.1.

Elevators may not open into exit enclosures. The difficulty is to have elevator doors that can meet the opening protectives for a fire barrier, but still operate effectively as elevator doors. For additional information on elevator lobbies and doors, see the commentary for Sections 708.14 and 1018.6.

These opening limitations are very similar to those required for an exit passageway (see Section 1023.5).

1022.4 Penetrations. Penetrations into and openings through an *exit enclosure* are prohibited except for required *exit* doors, equipment and ductwork necessary for independent ventilation or pressurization, sprinkler piping, standpipes, electrical raceway for fire department communication systems and electrical raceway serving the *exit enclosure* and terminating at a steel box not exceeding 16 square inches (0.010 m²). Such penetrations shall be protected in accordance with Section 713. There shall be no penetrations or communication openings, whether protected or not, between adjacent *exit enclosures*.

❖ This section specifically lists the items that are allowed to penetrate a vertical exit enclosure. This is consistent for all types of exit enclosures, including stair or ramp vertical exit enclosures and exit passageways (see Section 1023.6). In general, only portions of the building service systems that serve the exit enclosure are allowed to penetrate the exit enclosure. As indicated in the commentary to Section 1020.1, standpipe systems are commonly located in the exit stair enclosures. If two exit enclosures are adjacent to one another, there must be no penetrations between them, thereby limiting the chances of smoke being in both stairwells. This requirement is not intended to prohibit windows in the exterior walls of stairways that are not required to be rated. This section and Section 1022.5 are meant to work together.

1022.5 Ventilation. Equipment and ductwork for *exit enclosure* ventilation as permitted by Section 1022.4 shall comply with one of the following items:

1. Such equipment and ductwork shall be located exterior to the building and shall be directly connected to the *exit enclosure* by ductwork enclosed in construction as required for shafts.

2. Where such equipment and ductwork is located within the *exit enclosure*, the intake air shall be taken directly from the outdoors and the exhaust air shall be discharged directly to the outdoors, or such air shall be conveyed through ducts enclosed in construction as required for shafts.

3. Where located within the building, such equipment and ductwork shall be separated from the remainder of the building, including other mechanical equipment, with construction as required for shafts.

In each case, openings into the fire-resistance-rated construction shall be limited to those needed for maintenance and operation and shall be protected by opening protectives in accordance with Section 715 for shaft enclosures.

Exit enclosure ventilation systems shall be independent of other building ventilation systems.

❖ The purpose of the requirements for the ventilation system equipment and ductwork is to maintain the fire resistance of the exit enclosure. The exit enclosure ventilation system is to be independent of other building systems to prevent smoke in the exit enclosure from traveling to other areas of the building. This section and Section 1022.4 are meant to work together.

1022.6 Exit enclosure exterior walls. *Exterior walls* of an *exit enclosure* shall comply with the requirements of Section 705 for *exterior walls*. Where nonrated walls or unprotected openings enclose the exterior of the *stairway* and the walls or openings are exposed by other parts of the building at an angle of less than 180 degrees (3.14 rad), the building *exterior walls* within 10 feet (3048 mm) horizontally of a nonrated wall or unprotected opening shall have a *fire-resistance rating* of not less than 1 hour. Openings within such *exterior walls* shall be protected by opening protectives having a *fire protection rating* of not less than $^3/_4$ hour. This construction shall extend vertically from the ground to a point 10 feet (3048 mm) above the topmost landing of the *stairway* or to the roof line, whichever is lower.

❖ This section does not require exterior walls of a stairway enclosure to have the same fire-resistance rating as interior walls; however, to minimize the potential fire spread into the stairway from the exterior, the issue of the exterior wall of the stairway and adjacent exterior walls of the building is addressed. Essentially, there are two alternatives where an exposure hazard exists: either (1) provide protection to the stairway by having a fire-resistance rating on its exterior wall or (2) provide

a fire-resistance rating to the walls adjacent to the stairway. The ratings apply for a distance of 10 feet (3048 mm) measured horizontally and vertically from the stairway enclosure where those walls are at an angle of less than 180 degrees (3.14 rad) from the enclosure's exterior wall [see Condition 1 in Figure 1022.6(1)]. When the adjacent exterior wall is protected in lieu of the stairway enclosure wall, the protection is to extend from the ground to a level of 10 feet (3048 mm) above the highest landing of the stairway. However, the protection is not required to extend beyond the normal roof line of the building.

The 180-degree (3.14 rad) angle criteria is based on the scenario where the exterior wall of the stair enclosure is in the same plane and flush with the exterior wall of the building [see Conditions 2 and 3 in Figure 1022.6(1)]. In this scenario, a fire would need to travel 180 degrees (3.14 rad) around in order to impinge on the stair. Based on studies of existing buildings, this 180-degree (3.14 rad) spread of fire does not appear to be a problem. This criteria is only applicable when the angle between the walls is 180 degrees (3.14 rad) or less.

As the fire exposure on the exterior is different than can be expected on the interior, the fire-resistance rating of the exterior wall is not required to exceed 1 hour, regardless of whether it is the stairway enclosure wall or the adjacent exterior wall, unless the exterior wall is required by other sections of the code to have a higher fire-resistance rating. The fire protection rating on any openings in the exterior wall of a stairway enclosure or adjacent exterior wall within 180 degrees (3.14 rad) is to be a minimum of $^3/_4$ hour [see Figure 1022.6(2)].

In a situation where the upper levels are smaller than lower levels, an interior stairway can end up having an exterior wall when it moves above the roof of the lower levels. In this situation, the question is the rating requirements for the exterior wall of the stairway over the roof. Therefore, the exterior wall of the stairway must meet the vertical opening provisions in Section 705.8.6.

1022.7 Discharge identification. A *stairway* in an *exit enclosure* shall not continue below its *level of exit discharge* unless an approved barrier is provided at the *level of exit discharge* to prevent persons from unintentionally continuing into levels below. Directional *exit* signs shall be provided as specified in Section 1011.

❖ So that building occupants using an exit stairway during an emergency situation will be prevented from going past the level of exit discharge, the run of the stairway is to be interrupted by partitions, doors, gates or other approved means. These devices help the users of the stairway to recognize when they have reached the point that is the level of exit discharge. Exit signs, including tactile, are to be provided for occupant guidance at the door leading to the way out (i.e., directly to the exterior, or via exit passageway, lobby or vestibule). Furthermore, signs are to be placed at each floor landing in all interior exit stairways connecting more than three floor levels that designate the level or story of the landings in accordance with Section 1022.8.

The code does not specify the type of material or construction of the barrier used to identify the level of exit discharge. The key issues to be considered in the selection and approval of the type of barrier to be used are: (1) will the barrier provide a visible and physical means of alerting occupants who are exiting under emergency conditions that they have reached the level of exit discharge and (2) is the barrier constructed of materials that are permitted by the construction type of the building? In an emergency situation, some occupants are likely to come in contact with the barrier during exiting before realizing that they are at the level of exit discharge. Therefore, the barrier should be constructed in a manner that is substantial enough to withstand the anticipated physical contact, such as pushing or shoving. It would be reasonable, as a minimum, to design the barrier to withstand the structural load requirements of Section 1607.5 for interior walls and partitions. The barrier could be opaque (such as gypsum wallboard and stud framing) or not (such as a wire grid-type material).

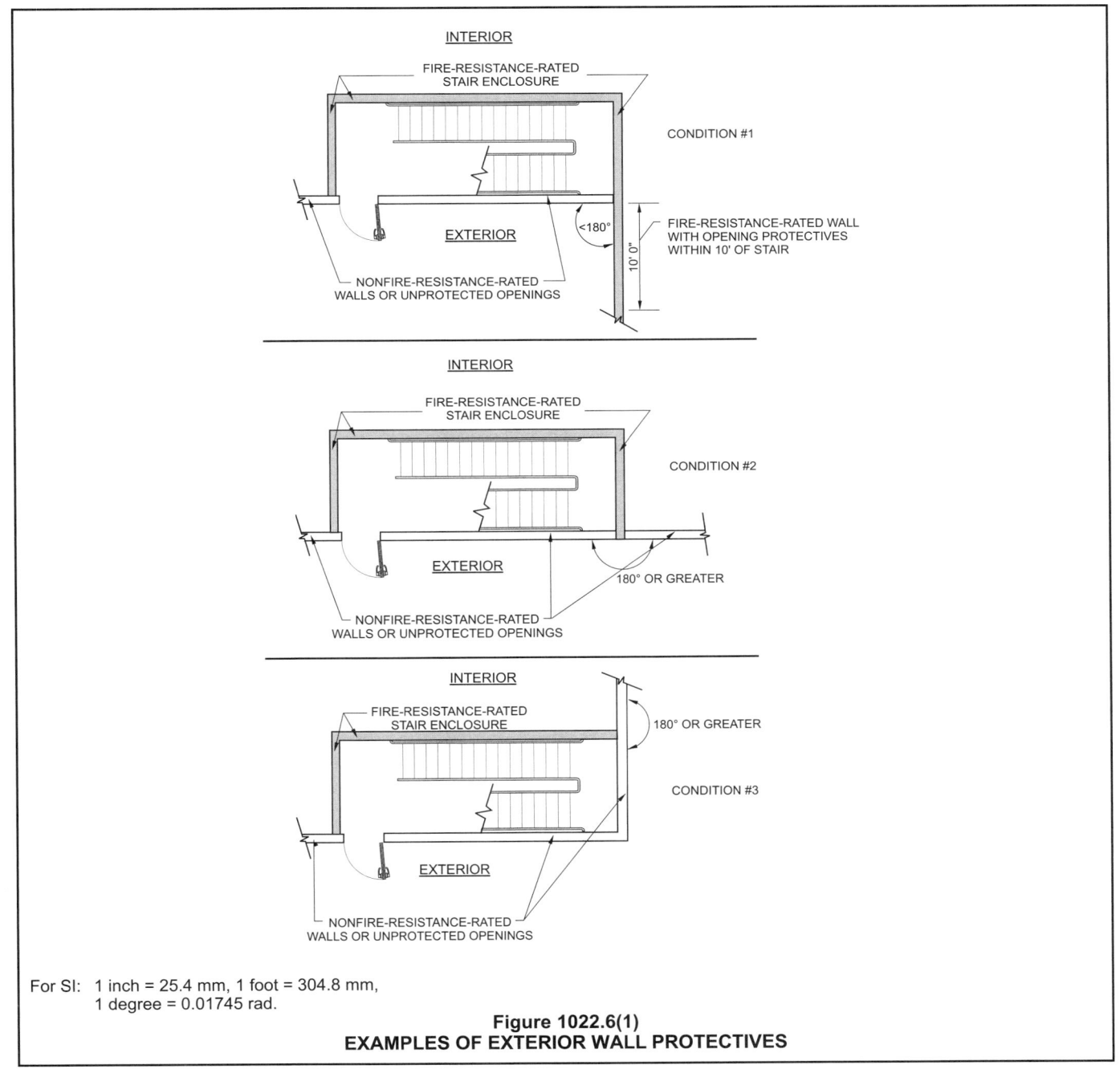

For SI: 1 inch = 25.4 mm, 1 foot = 304.8 mm,
1 degree = 0.01745 rad.

Figure 1022.6(1)
EXAMPLES OF EXTERIOR WALL PROTECTIVES

The use of signage only or relatively insubstantial barriers, such as ropes or chains strung across the opening, is typically not sufficient to prevent occupants from attempting to continue past the level of exit discharge during an emergency.

Figure 1022.7 is an example of one method of discharge identification.

1022.8 Floor identification signs. A sign shall be provided at each floor landing in *exit enclosures* connecting more than three stories designating the floor level, the terminus of the top and bottom of the *exit enclosure* and the identification of the *stair* or *ramp*. The signage shall also state the *story* of, and the

direction to, the *exit discharge* and the availability of roof access from the enclosure for the fire department. The sign shall be located 5 feet (1524 mm) above the floor landing in a position that is readily visible when the doors are in the open and closed positions. Floor level identification signs in tactile characters complying with ICC A117.1 shall be located at each floor level landing adjacent to the door leading from the enclosure into the corridor to identify the floor level.

❖ Signs are to be placed at each floor landing in all exit stairways connecting more than three stories. The signs are to designate the level or story of the landings above or below the level of exit discharge. The pur-

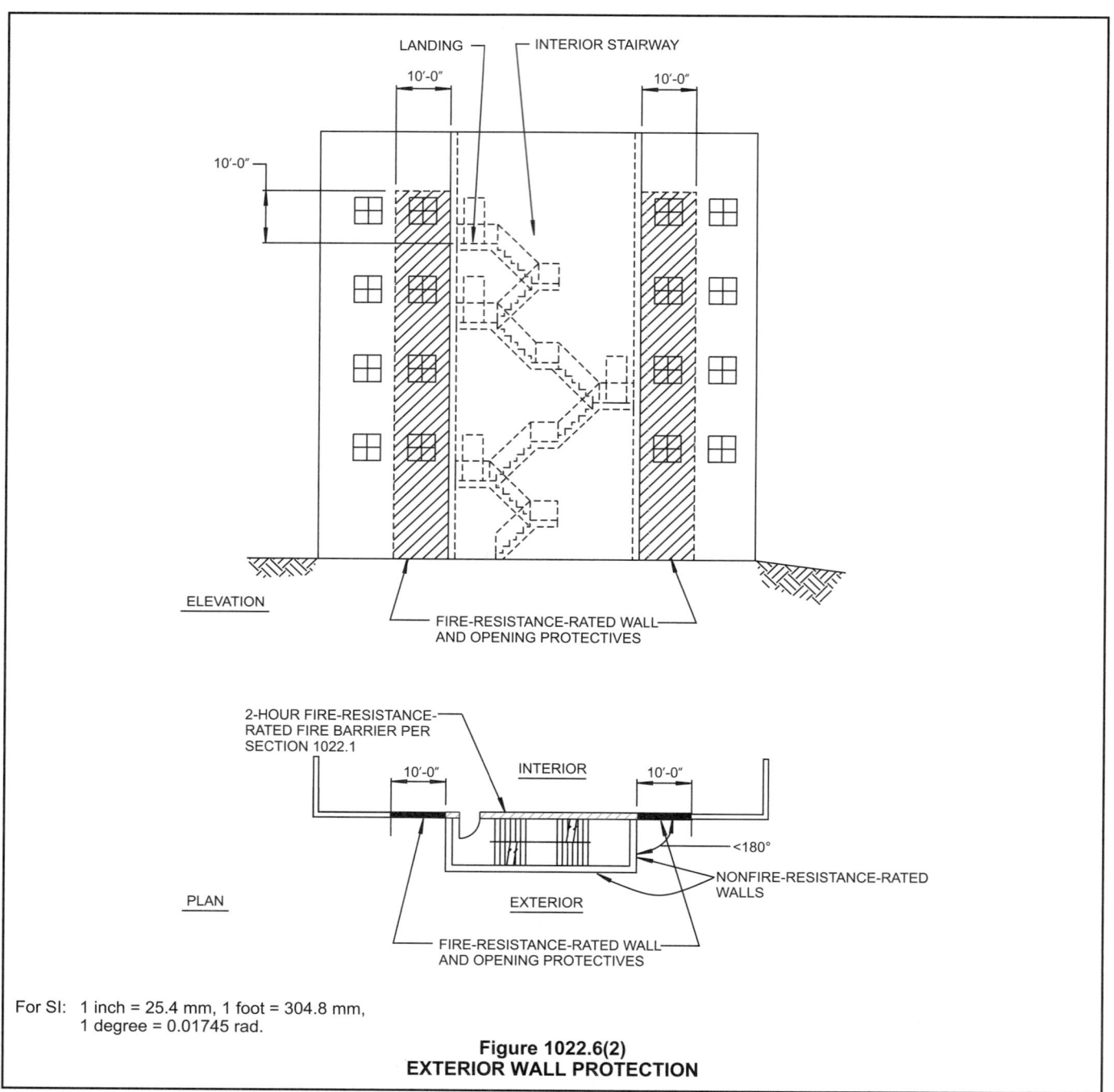

For SI: 1 inch = 25.4 mm, 1 foot = 304.8 mm,
1 degree = 0.01745 rad.

Figure 1022.6(2)
EXTERIOR WALL PROTECTION

pose is to inform the occupants of their location with respect to the level of exit discharge as they use the stairway to leave the building. More importantly, it allows the fire service to locate and gain quick access to the fire floor. At each level, the direction to the exit discharge is required to be indicated. The identification of the level that is the exit discharge also is to be indicated at each level. The identification of the roof access availability is for the fire department. Roof access is required by Section 1009.13. For visibility, the signs are required to be located approximately 5 feet (1524 mm) above the floor surface and to be visible when the stairway door is open. The need to designate levels remaining to reach the level of exit discharge may mean that the numbering is other than that designation used by building management. For example, a designation of P1, P2, P3, etc., would not be acceptable for stairways in the basement parking garage, since in themselves they do not designate the floor level below the level of exit discharge.

To aid people with vision impairments, the floor designation must also be available in both raised letters and Braille at each door. Tactile signage indicating the door leading to the exterior is covered in Section 1011.3.

1022.8.1 Signage requirements. *Stairway* identification signs shall comply with all of the following requirements:

1. The signs shall be a minimum size of 18 inches (457 mm) by 12 inches (305 mm).

2. The letters designating the identification of the stair enclosure shall be a minimum of $1^1/_2$ inches (38 mm) in height.

3. The number designating the floor level shall be a minimum of 5 inches (127 mm) in height and located in the center of the sign.

4. All other lettering and numbers shall be a minimum of 1 inch (25 mm) in height.

5. Characters and their background shall have a nonglare finish. Characters shall contrast with their background, with either light characters on a dark background or dark characters on a light background.

6. When signs required by Section 1022.8 are installed in interior *exit enclosures* of buildings subject to Section 1024, the signs shall be made of the same materials as required by Section 1024.4.

❖ The requirements for stairway identification signage will provide for a consistent approach. The intent is to make signs visible and immediately recognizable to occupants and emergency responders using the stairway. In addition, if the building is a high-rise and luminous egress path markings are required (see Section 1024.1), the stairway identification signage must also be self-luminous or photoluminescent. In order to also meet the contrast requirements in Item 5, typically the sign will be dark letters on a glow-in-the-dark background.

Figure 1022.7
EXAMPLE OF DISCHARGE IDENTIFICATION

1022.9 Smokeproof enclosures and pressurized stairways. In buildings required to comply with Section 403 or 405, each of the *exit enclosures* serving a *story* with a floor surface located more than 75 feet (22 860 mm) above the lowest level of fire department vehicle access or more than 30 feet (9144 mm) below the finished floor of a *level of exit discharge* serving such stories shall be a *smokeproof enclosure* or pressurized *stairway* in accordance with Section 909.20.

❖ While smokeproof enclosures and pressurized stairways for exit stairways can, at the designer's option, be used in buildings of any occupancy, height and area, this section specifically requires smokeproof enclosures or pressurized stairway to be provided when either of two conditions occur.

The first condition requires all exit stairways in buildings with floor levels higher than 75 feet (22 860 mm) above the level of exit discharge (i.e., high-rise) to be smokeproof enclosures or pressurized stairways. The reason for this provision is that in very tall buildings, often during fire emergencies, total and immediate evacuation of the occupants cannot be readily accomplished. In such situations, exit stairways become places of safety for the occupants and must be adequately protected with smokeproof enclosures or pressurization to provide a safe egress environment. In order to provide this safe environment, the enclosure must be constructed to resist the migration of smoke caused by the "stack effect." Stack effect occurs in tall enclosures such as chimneys, when a fluid such as smoke, which is less dense than the ambient air, is introduced into the enclosure. The smoke will rise due to the effect of buoyancy and will induce additional flow into the enclosure through openings at the lower levels.

The second condition applies when an occupiable floor level is located more than 30 feet (9144 mm) below the level of exit discharge (i.e., underground buildings) serving such floor levels. Stairways serving those levels are also required to be protected by smokeproof enclosures or pressurization because underground portions of a building present unique problems in providing not only for life safety but also access for fire-fighting purposes. The choice of a 30-foot (9144 mm) threshold for this requirement is intended to provide a reasonable limitation on vertical travel distance before the requirement applies.

Detailed system requirements for a smokeproof enclosure or pressurization are in Section 909.20. Note that Sections 403 and 405 each have exceptions for specific building types that are not to be regulated by those sections. Likewise, the requirements of this section do not apply to buildings identified in those exceptions.

1022.9.1 Termination and extension. A *smokeproof enclosure* or pressurized *stairway* shall terminate at an *exit discharge* or a *public way*. The *smokeproof enclosure* or pressurized *stairway* shall be permitted to be extended by an *exit passageway* in accordance with Section 1022.2. The *exit passageway* shall be without openings other than the *fire door assembly* required by

Section 1022.2 and those necessary for egress from the *exit passageway*. The *exit passageway* shall be separated from the remainder of the building by 2-hour *fire barriers* constructed in accordance with Section 707 or *horizontal assemblies* constructed in accordance with Section 712, or both.

Exceptions:

1. Openings in the *exit passageway* serving a *smokeproof enclosure* are permitted where the *exit passageway* is protected and pressurized in the same manner as the *smokeproof enclosure*, and openings are protected as required for access from other floors.

2. Openings in the *exit passageway* serving a pressurized *stairway* are permitted where the *exit passageway* is protected and pressurized in the same manner as the pressurized *stairway*.

3. The *fire barrier* separating the *smokeproof enclosure* or pressurized *stairway* from the *exit passageway* is not required, provided the *exit passageway* is protected and pressurized in the same manner as the *smokeproof enclosure* or pressurized *stairway*.

4. A *smokeproof enclosure* or pressurized *stairway* shall be permitted to egress through areas on the level of discharge or vestibules as permitted by Section 1027.

❖ The walls that comprise the smokeproof enclosure or pressurized stairway, which includes the stairway shaft and the vestibules, must be fire barriers having a fire-resistance rating of at least 2 hours. This level of fire endurance is specified because exit stairways in high-rise buildings serve as principal components of the egress system and as the source of fire service access to the fire floor. This supersedes any allowed reduction of enclosure rating, even if the stair from the level that is more than 30 feet (9144 mm) below exit discharge connects three stories or less.

The first two exceptions apply to openings in the exit passageway that are permitted, provided the exit passageway is protected and pressurized. If the exit stairway enclosure is connected to an exit passageway, Exception 3 allows there to not be a door between, since this would interfere with the pressurization of the stairway and passageway as a combined exit system. In accordance with Exception 4, 50 percent of the stairways in smokeproof enclosures or pressurized stairways can use the exit discharge exceptions in Section 1027 to egress through a lobby or vestibule.

1022.9.2 Enclosure access. Access to the *stairway* within a *smokeproof enclosure* shall be by way of a vestibule or an open exterior balcony.

Exception: Access is not required by way of a vestibule or exterior balcony for *stairways* using the pressurization alternative complying with Section 909.20.5.

❖ See Figures 1022.9.2(1) and 1022.9.2(2) for illustrations of access to the smokeproof stairway by way of a vestibule or an exterior balcony. The purpose of this requirement is to keep the enclosure clear of smoke. Where a pressurized stairway is used, these elements are not necessary.

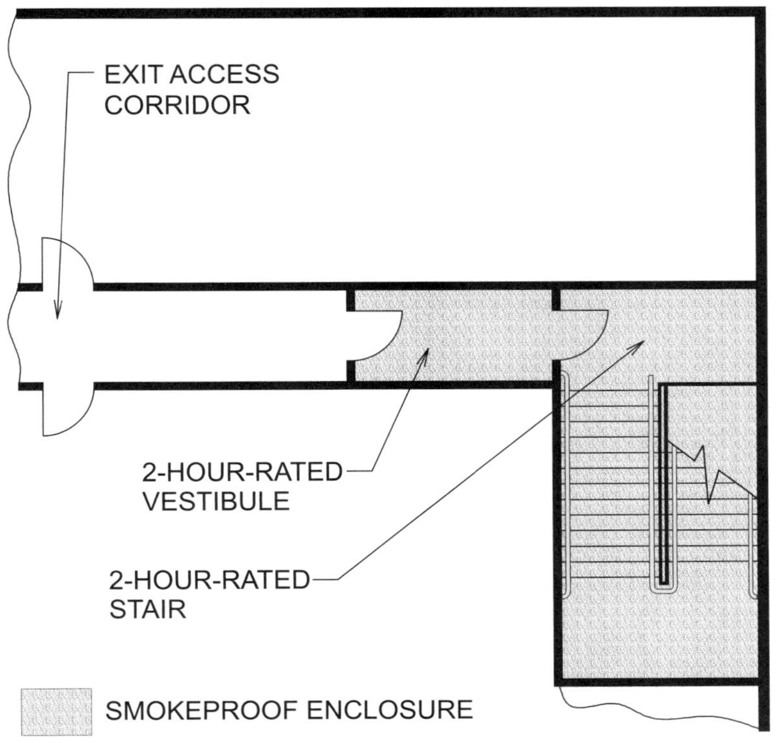

EXIT ACCESS
CORRIDOR

2-HOUR-RATED
VESTIBULE

2-HOUR-RATED
STAIR

SMOKEPROOF ENCLOSURE

Figure 1022.9.2(1)
TYPICAL SMOKEPROOF ENCLOSURE ENTRY

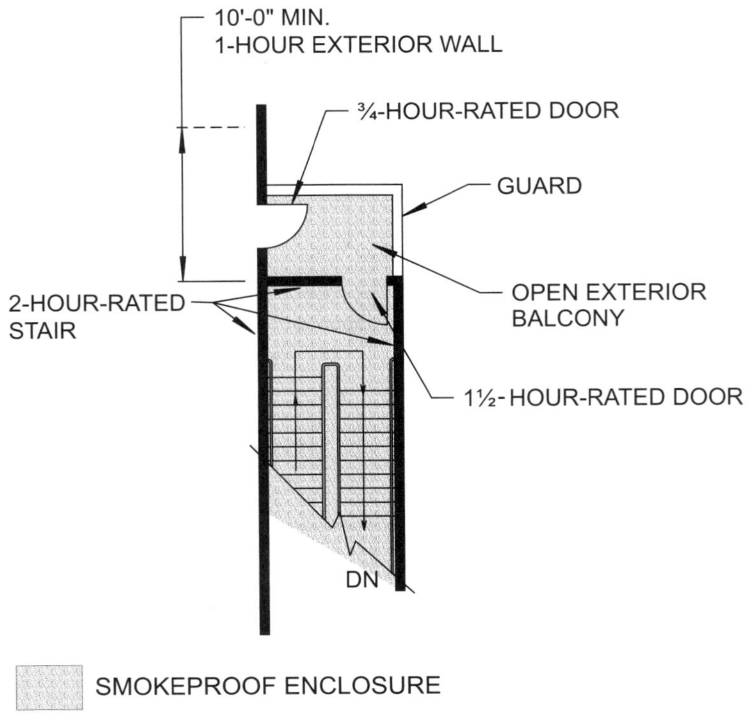

10'-0" MIN.
1-HOUR EXTERIOR WALL

¾-HOUR-RATED DOOR

GUARD

OPEN EXTERIOR
BALCONY

2-HOUR-RATED
STAIR

1½- HOUR-RATED DOOR

DN

SMOKEPROOF ENCLOSURE

For SI: 1 inch = 25.4 mm, 1 foot = 304.8 mm.

Figure 1022.9.2(2)
TYPICAL SMOKEPROOF ENCLOSURE WITH OPEN BALCONY ARRANGEMENT

SECTION 1023
EXIT PASSAGEWAYS

1023.1 Exit passageway. *Exit passageways* serving as an *exit* component in a *means of egress* system shall comply with the requirements of this section. An *exit passageway* shall not be used for any purpose other than as a *means of egress*.

❖ This section provides acceptable methods of continuing the protected path of travel for building occupants. The building designer or owner is given these different options for achieving this protected path of travel. See Figure 1023.1 for an illustration of an exit passageway arrangement. In the case of office buildings or similar types of structures, the exit stairways are often located at the central core or in line with the centrally located exit access corridors. Exit passageways may be used to connect the exit stair to the exterior exit door or to connect stairway enclosures when a stairway needs to shift over. Such an arrangement provides great flexibility in the design use of the building. Without the passageway at the grade floor or the level of exit discharge, the occupants of the upper floors or basement levels would have to leave the safety of the exit stairway to travel to the exterior doors. Such a reduction of protection is not acceptable (see Section 1027 for exit discharge alternatives).

The exit passageway is considered an exit enclosure, such as in a mall or unlimited area building. Therefore, exit passageways may also be used on their own in locations not connected with a stair enclosure. Sometimes on large floor plans, an exit passageway may be used to extend an exit into areas that would not otherwise be able to meet the travel distance requirements. Like vertical exit stairways, there is no travel distance limitation within exit passageways.

1023.2 Width. The width of *exit passageways* shall be determined as specified in Section 1005.1 but such width shall not be less than 44 inches (1118 mm), except that *exit passageways* serving an *occupant load* of less than 50 shall not be less than 36 inches (914 mm) in width. The required width of *exit passageways* shall be unobstructed.

Exception: Doors complying with Section 1005.2.

❖ The width of an exit passageway is to be determined in accordance with Section 1005.1, based on the number of occupants served in the same manner as for corridors. The greater of the minimum width or the width determined based on occupancy is to be used. In situations where the exit passageway also serves as an exit access corridor for the first floor, the corridor width must comply with the stricter requirement.

A cross reference back to Section 1005.2 from the exceptions for width in corridors (see Section 1018.3), aisles (see Section 1017.1), exit passageways (see Section 1023.2) and exit courts (see Section 1027.5.1) reinforces the fact that the protrusion limits provision is generally applicable for these types of confined routes.

1023.3 Construction. *Exit passageway* enclosures shall have walls, floors and ceilings of not less than 1-hour *fire-resistance rating*, and not less than that required for any connecting *exit enclosure*. *Exit passageways* shall be constructed as *fire barriers* in accordance with Section 707 or *horizontal assemblies* constructed in accordance with Section 712, or both.

❖ The entire exit passageway enclosure is to be fire-resistance rated as specified. The floors and ceilings are required to be rated in addition to the walls. When used separately, a minimum 1-hour fire-resistance rating is required. Where extending an exit enclosure, the rating must not be less than the exit enclosure so that the degree of protection is kept at the same level. Re-

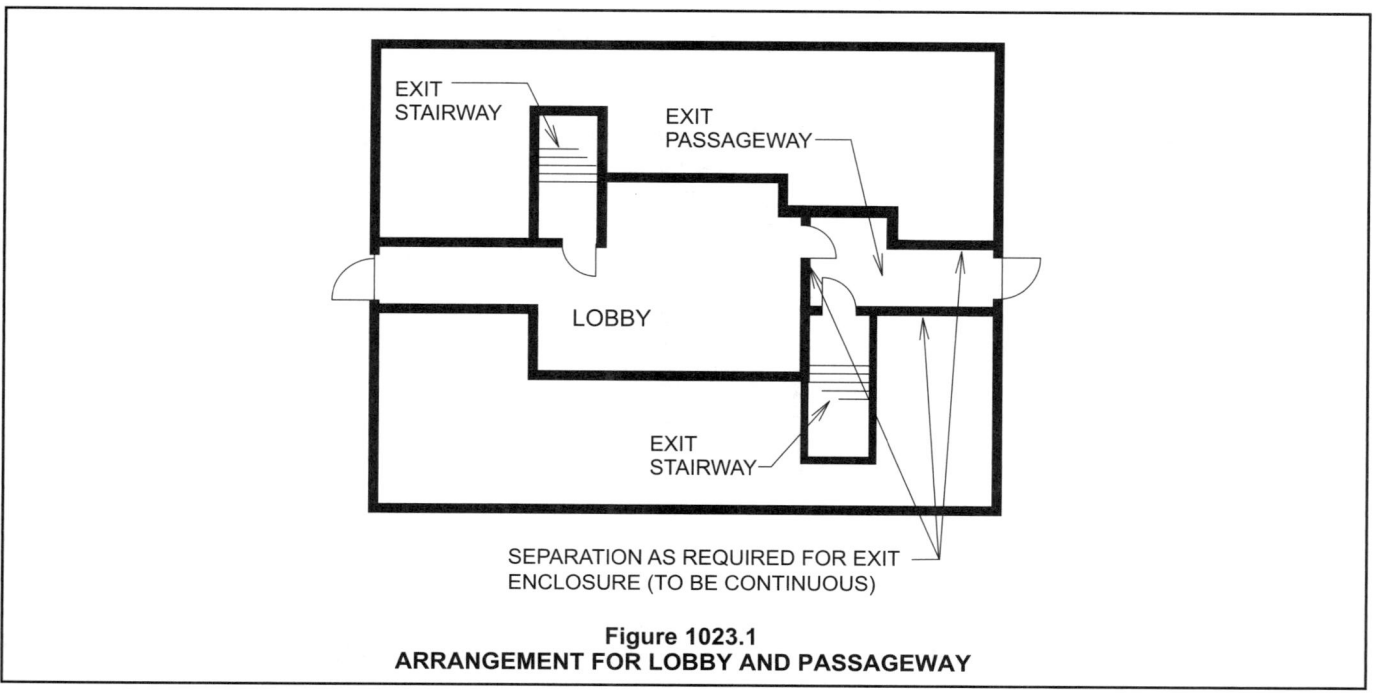

Figure 1023.1
ARRANGEMENT FOR LOBBY AND PASSAGEWAY

member that if the exit passageway extends over a lower level, all supporting construction is to have the same fire-resistance rating as the elements supported in accordance with the continuity requirements for fire barriers and horizontal assemblies (see Sections 707.5 and 712.4). The continuity requirements would also be a concern for the rated ceiling of the exit passageway. An alternative for the ceiling of the exit passageway could be a top shaft enclosure (see Section 708.12).

1023.4 Termination. *Exit passageways* shall terminate at an *exit discharge* or a *public way*.

❖ This is consistent with the exit continuity and enclosure requirements in Sections 1021.3 and 1022.2. The intent of this section is to provide safety in all portions of the exit by requiring continuity of the fire protection characteristics of the exit enclosure in combination with an exit passageway. This would include, but not be limited to, the fire-resistance rating of the exit enclosure walls and the opening protection rating of the doors. Section 1027.1 would allow for an alternative for direct access to the outside via an intervening lobby or vestibule.

Horizontal exits (see Section 1025), while not providing direct access to the outside of the structure, do move occupants to another "building" by moving through a fire wall (see Section 1025 and the definition for "Area, building" in Section 502.1) into a refuge area protected by fire barriers and horizontal assemblies. Horizontal exits are commonly used in hospitals and jails for a defend-in-place type of protection.

1023.5 Openings and penetrations. *Exit passageway* opening protectives shall be in accordance with the requirements of Section 715.

Except as permitted in Section 402.4.6, openings in *exit passageways* other than exterior openings shall be limited to those necessary for *exit access* to the *exit passageway* from normally occupied spaces and for egress from the *exit passageway* .

Where an *exit enclosure* is extended to an *exit discharge* or a *public way* by an *exit passageway*, the *exit passageway* shall also comply with Section 1022.2.1.

Elevators shall not open into an *exit passageway*.

❖ The requirements for exit passageways are very similar to those required for vertical exit enclosures (see Section 1022.3). The only openings that are permitted in fire-resistance-rated exit passageways are doors that lead either from normally occupied spaces or from the vertical exit enclosure. This restriction on openings essentially prohibits the use of windows in an exit passageway except for those exterior windows that are not exposed to any hazards by fire separation distance (see Section 705.8) or by adjacent exterior walls similar to what is specified for vertical exit enclosures (see Section 1022.6). The only exception would be window assemblies that have been tested as wall assemblies in accordance with ASTM E 119. The objective of this provision is to minimize the possibility of fire spreading

into an exit passageway and endangering the occupants or even preventing the use of the exit at a time when it is most needed. The limitation on openings applies regardless of the fire protection rating of the opening protective. The limitation on openings from normally occupied areas is intended to reduce the probability of a fire occurring in an unoccupied area, such as a storage closet, that has an opening into the exit passageway, thereby resulting in smoke spreading into the exit passageway. Other spaces that are not normally occupied include, but are not limited to, toilet rooms, electrical/mechanical equipment rooms and janitorial closets. Note that exit passageways prohibit elevators from opening directly into the passageway. There are some exceptions for these unoccupied spaces in exit passageways in covered malls (see Section 402.4.6).

The third paragraph addresses when the exit enclosure could transition from the vertical enclosure of a stairway or ramp to the exit passageway. While the exit passageway is an extension of the protection offered by the vertical exit, there must still be a door (i.e., opening protective) between the bottom of the stairway or ramp enclosure and the exit passageway. This is to prevent any smoke that may migrate into the exit passageway from also moving up the exit stairway or ramp.

Elevators may not open into exit passageways. The difficulty is to have elevator doors that can meet the opening protectives for a fire barrier, but still operate effectively. For additional information on elevator lobbies and doors, see the commentary for Sections 707.14 and 1018.6.

These opening limitations are very similar to those required for an exit enclosure (see Section 1022.3).

1023.6 Penetrations. Penetrations into and openings through an *exit passageway* are prohibited except for required *exit* doors, equipment and ductwork necessary for independent pressurization, sprinkler piping, standpipes, electrical raceway for fire department communication and electrical raceway serving the *exit passageway* and terminating at a steel box not exceeding 16 square inches (0.010 m²). Such penetrations shall be protected in accordance with Section 713. There shall be no penetrations or communicating openings, whether protected or not, between adjacent *exit passageways*.

❖ This section specifically lists the items that are allowed to penetrate an exit passageway. This is consistent for all types of exit enclosures, including stair or ramp vertical exit enclosures (see Section 1022.4) and exit passageways. In general, only portions of the building service systems that serve the exit passageways are allowed to penetrate the exit passageway. If two exit passageways are adjacent to one another, there must be no penetrations between them, thereby limiting the chances of smoke being in both passageways. This requirement is not intended to prohibit windows in the exterior walls of exit passageways that are not required to be rated.

SECTION 1024
LUMINOUS EGRESS PATH MARKINGS

1024.1 General. *Approved* luminous egress path markings delineating the exit path shall be provided in buildings of Groups A, B, E, I, M and R-1 having occupied floors located more than 75 feet (22 860 mm) above the lowest level of fire department vehicle access in accordance with Sections 1024.1 through 1024.5.

Exceptions:

1. Luminous egress path markings shall not be required on the *level of exit discharge* in lobbies that serve as part of the exit path in accordance with Section 1027.1, Exception 1.

2. Luminous egress path markings shall not be required in areas of *open parking garages* that serve as part of the exit path in accordance with Section 1027.1, Exception 3.

❖ Improved safety for individuals negotiating stairs during egress of a high-rise building is provided by improving the visibility of stair treads and handrails under emergency conditions. A second source of emergency power for exit illumination, exit signs and stair shaft pressurization systems in smokeproof enclosures is currently mandated for high-rise buildings. In the event of an emergency that disconnects utility power, the emergency power source should engage, causing the stair shaft to be illuminated and smoke free by the pressurization system. Unfortunately, such systems can fail under demand conditions. The provisions of Section 1024 add an additional level of safety to the egress path by requiring the installation of photoluminescent or self-illuminating marking systems which do not require electrical power and its associated wiring and circuits. An additional means for ensuring that occupants can safely egress a building via exit stairs is now available even if the emergency power supply and system fails to operate. The groups indicated have a high anticipated occupant load or occupants may not be as familiar with the space.

The exceptions indicate that if the exit stairway enclosure discharges through the lobby (see Section 1027.1, Exception 1), or the lowest level of an open parking garage (see Section 1027.1, Exception 3), the egress markings would not be required outside the stairway enclosure for the portion from the stairway enclosure to the door leading to the outside. If the exit stairway discharges through an exit passageway, or through a vestibule, the exit path markings must continue to the door leading to the outside.

1024.2 Markings within exit enclosures. Egress path markings shall be provided in *exit enclosures*, including vertical *exit enclosures* and *exit passageways*, in accordance with Sections 1024.2.1 through 1024.2.6.

❖ Egress path markings are required inside the exit stair/ramp enclosure for all floors. If the stairway connects to an exit passageway as part of the travel down to the level of exit discharge or on the level of exit dis-

charge, the path markings must also be continued in the exit passageway.

The subsections include marking the tread nosings, the surrounding edges of landings and any exit passageways, handrails and any protruding objects within the enclosure.

All exit path markings are required to be solid and continuous stripes. A key requirement for marking systems is that their design must be uniform. The placement and dimensions of markings must be consistent throughout the same exit enclosure. By specifying a standard marking dimension, the requirements will ensure that the marking is visible during dark conditions and provides consistent and standard application in the design and enforcement of exit path markings. Markings installed on stair steps, perimeter demarcation lines and handrails must have a minimum width of 1 inch (25 mm). For stair steps and perimeter demarcation lines, their maximum width cannot exceed 2 inches (51 mm). The provisions for stair steps, perimeter demarcation lines and handrails allow the width of the marking to be reduced to less than 1 inch (25 mm) when marking stripes are listed in accordance with UL 1994.

1024.2.1 Steps. A solid and continuous stripe shall be applied to the horizontal leading edge of each step and shall extend for the full length of the step. Outlining stripes shall have a minimum horizontal width of 1 inch (25 mm) and a maximum width of 2 inches (51 mm). The leading edge of the stripe shall be placed at a maximum of $^1/_2$ inch (13 mm) from the leading edge of the step and the stripe shall overlap the leading edge of the step by not more than $^1/_2$ inch (13 mm) down the vertical face of the step.

Exception: The minimum width of 1 inch (25 mm) shall not apply to outlining stripes *listed* in accordance with UL 1994.

❖ Stripes are required the full width of the stairway on all tread nosings and along the leading edge of stair landings. These demarcation lines serve to identify the transition from the stair steps to the landing, which is important to minimize the risk of a fall inside of a stairway enclosure that is not illuminated. In order to clearly identify the leading edge of the step, the front edge of the stripe must be within $^1/_2$ inch (13 mm) plus or minus of the leading edge or the tread (see Figure 1024.2.1).

The code does not specify any minimum slip-resistance requirements for luminous products installed on walking surfaces. However, Section 1003.4 does require all walking surfaces for means of egress to be slip resistant. Persons with vision impairments often rely on high contrast elements to delineate changes in elevation such as that required in Sections 1009.4.4 and 1028.11.2. In medium light conditions, the luminous materials may be hard to discern. Luminous materials installed adjacent to dark contrasting materials may help with both situations.

The provisions for stair steps, perimeter demarcation lines and handrails allow the width of the marking to be reduced to less than 1 inch (25 mm) when marking stripes are listed in accordance with UL 1994.

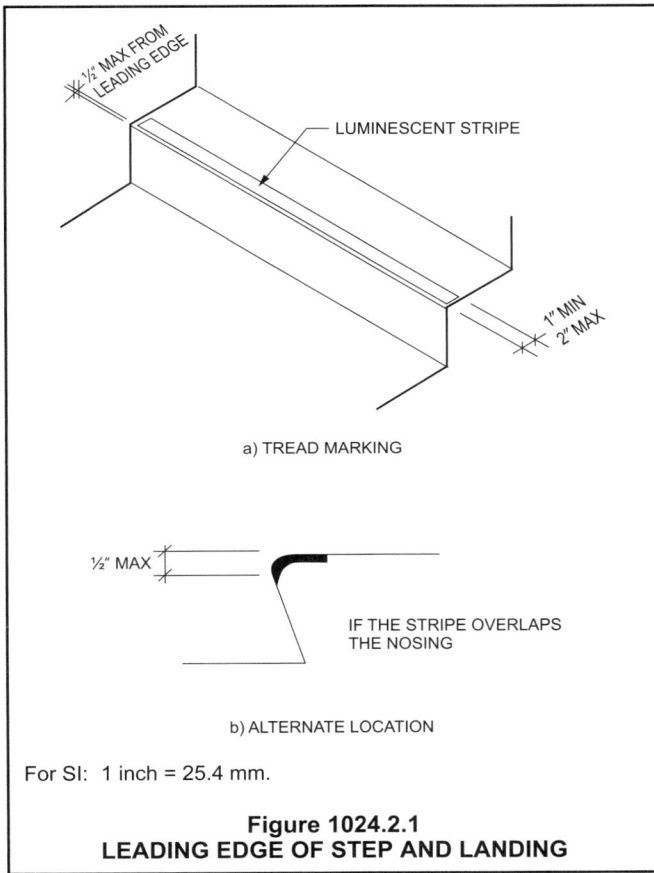

For SI: 1 inch = 25.4 mm.

Figure 1024.2.1
LEADING EDGE OF STEP AND LANDING

1024.2.2 Landings. The leading edge of landings shall be marked with a stripe consistent with the dimensional requirements for steps.

❖ The edge of the landing at the top of the steps must be marked in the same manner as the tread nosing so that a person can tell where the steps start. See the commentary to Section 1024.2.1 and Figure 1024.2.1.

1024.2.3 Handrails. All *handrails* and *handrail* extensions shall be marked with a solid and continuous stripe having a minimum width of 1 inch (25 mm). The stripe shall be placed on the top surface of the *handrail* for the entire length of the *handrail*, including extensions and newel post caps. Where *handrails* or *handrail* extensions bend or turn corners, the stripe shall not have a gap of more than 4 inches (102 mm).

Exception: The minimum width of 1 inch (25 mm) shall not apply to outlining stripes *listed* in accordance with UL 1994.

❖ The handrail must have a glow-in-the-dark stripe down the entire length. The 1-inch (25 mm) minimum width stripe needs to be on the top of the handrail both so it can charge and so that someone would be best able to see it in the dark. Bends or turns may result in a break in the marking which must not be more than 4 inches (102 mm) maximum (see Figure 1024.2.3).

The provisions for stair steps, perimeter demarcation lines and handrails allow the width of the marking to be reduced to less than 1 inch (25 mm) when marking stripes are listed in accordance with UL 1994.

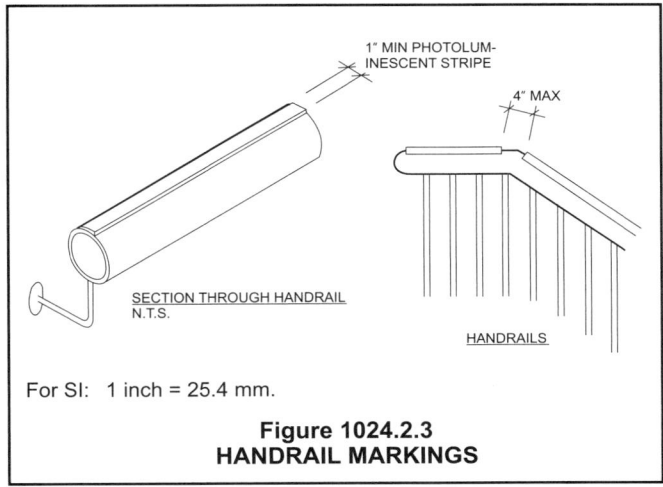

For SI: 1 inch = 25.4 mm.

Figure 1024.2.3
HANDRAIL MARKINGS

1024.2.4 Perimeter demarcation lines. *Stair* landings and other floor areas within *exit enclosures*, with the exception of the sides of steps, shall be provided with solid and continuous demarcation lines on the floor or on the walls or a combination of both. The stripes shall be 1 to 2 inches (25 mm to 51 mm) wide with interruptions not exceeding 4 inches (102 mm).

Exception: The minimum width of 1 inch (25 mm) shall not apply to outlining stripes *listed* in accordance with UL 1994.

❖ In addition to the leading edge of the landing, the landing must have a stripe all the way around the edge. If the enclosure includes any type of exit passageway, that corridor must also have a perimeter stripe. The stripe can be on the floor or on the wall at baseboard height (see Sections 1024.2.4.1 and 1024.2.4.2).

The provisions for stair steps, perimeter demarcation lines and handrails allow the width of the marking to be reduced to less than 1 inch (25 mm) when marking stripes are listed in accordance with UL 1994.

1024.2.4.1 Floor-mounted demarcation lines. Perimeter demarcation lines shall be placed within 4 inches (102 mm) of the wall and shall extend to within 2 inches (51 mm) of the markings on the leading edge of landings. The demarcation lines shall continue across the floor in front of all doors.

Exception: Demarcation lines shall not extend in front of *exit* doors that lead out of an *exit enclosure* and through which occupants must travel to complete the exit path.

❖ Stripes shall extend all the way around any stair landings. This section specifies the option of stripes on the floor (see Figure 1024.2.4.1). On a typical landing, the stripe will extend across the front of the door, indicating to someone moving down the stairway that they should continue (see Figure 1024.2.4.3). At the level of exit discharge the occupant has a different visual queue. The line should not extend in front of the door leading to the exterior (see Figure 1024.2.6). This should allow people to understand which door they need to move through to get to safety.

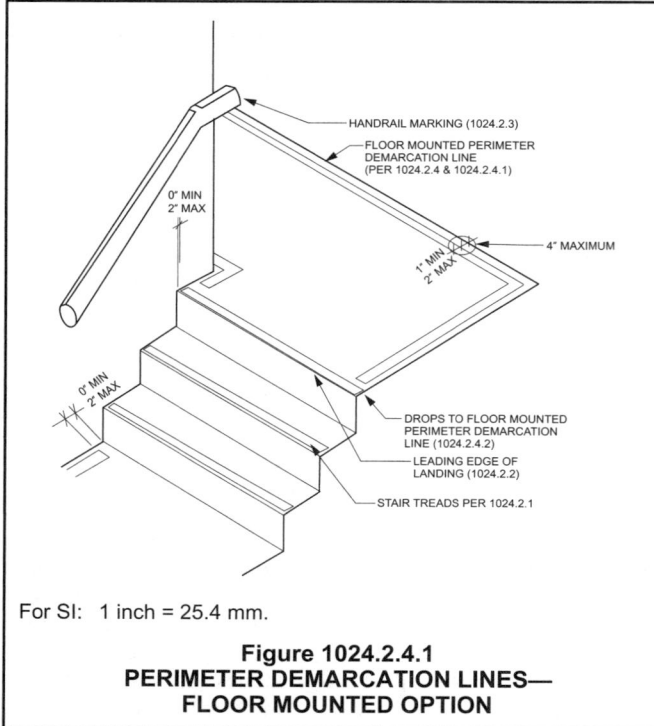

For SI: 1 inch = 25.4 mm.

Figure 1024.2.4.1
PERIMETER DEMARCATION LINES—
FLOOR MOUNTED OPTION

1024.2.4.2 Wall-mounted demarcation lines. Perimeter demarcation lines shall be placed on the wall with the bottom edge of the stripe no more than 4 inches (102 mm) above the finished floor. At the top or bottom of the *stairs*, demarcation lines shall drop vertically to the floor within 2 inches (51 mm) of the step or landing edge. Demarcation lines on walls shall transition vertically to the floor and then extend across the floor where a line on the floor is the only practical method of outlining the path. Where the wall line is broken by a door, demarcation lines on walls shall continue across the face of the door or transition to the floor and extend across the floor in front of such door.

> **Exception:** Demarcation lines shall not extend in front of *exit* doors that lead out of an *exit enclosure* and through which occupants must travel to complete the exit path.

❖ Stripes that surround the stairway landings can be on the floor or on the wall. This section specifies the option of stripes on the wall (see Figure 1024.2.4.2). See the commentary to Sections 1024.2.4 and 1024.2.4.1.

1024.2.4.3 Transition. Where a wall-mounted demarcation line transitions to a floor-mounted demarcation line, or vice versa, the wall-mounted demarcation line shall drop vertically to the floor to meet a complementary extension of the floor-mounted demarcation line, thus forming a continuous marking.

❖ While the perimeter stripes on landing can be on either the wall or the floor, when they transition from one to another, the lines should appear continuous. See Section 1024.2.4.2 for special requirements for when the

wall perimeter marking transitions to the floor stripe indicating leading edge of the landing.

When perimeter lines cross a door frame, the lines can transition or stay on the same plain (see Figure 1024.2.4.3).

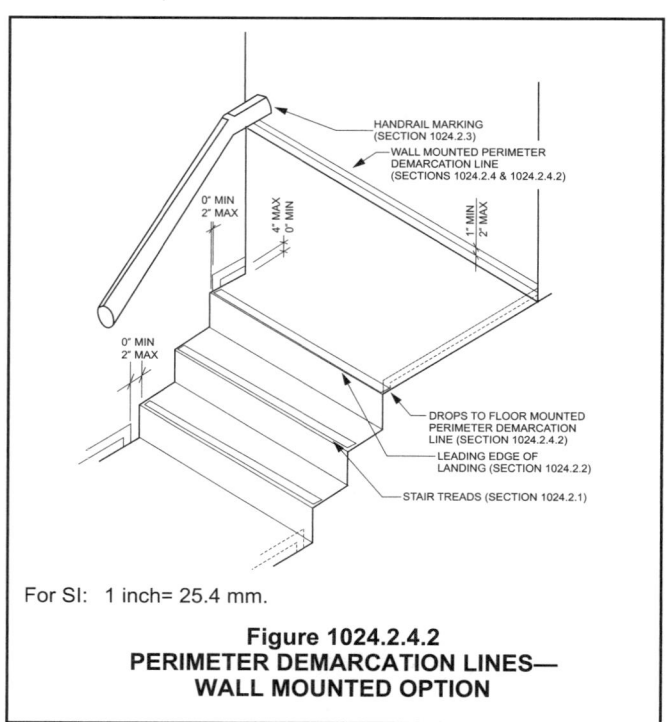

For SI: 1 inch= 25.4 mm.

Figure 1024.2.4.2
PERIMETER DEMARCATION LINES—
WALL MOUNTED OPTION

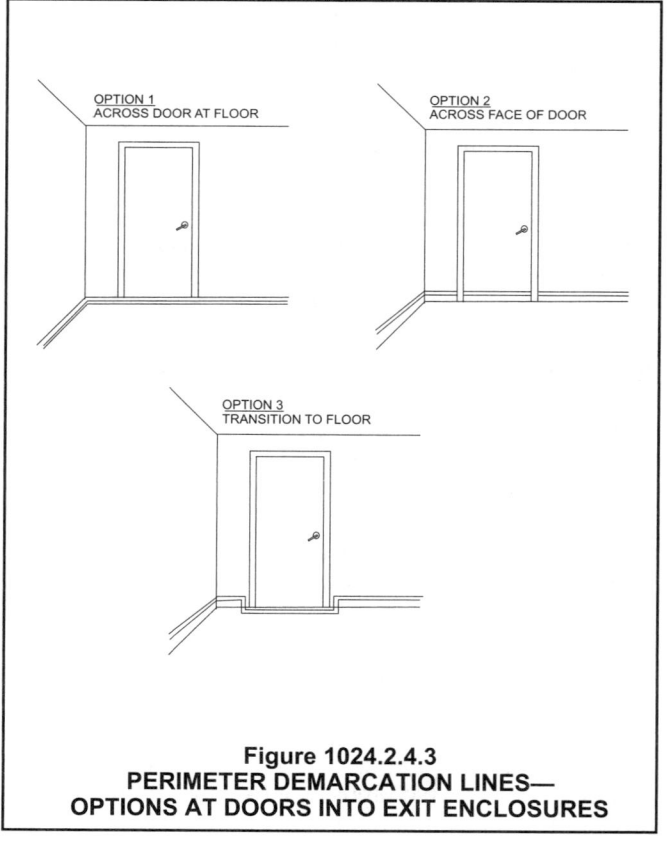

Figure 1024.2.4.3
PERIMETER DEMARCATION LINES—
OPTIONS AT DOORS INTO EXIT ENCLOSURES

1024.2.5 Obstacles. Obstacles at or below 6 feet 6 inches (1981 mm) in height and projecting more than 4 inches (102 mm) into the egress path shall be outlined with markings no less than 1 inch (25 mm) in width comprised of a pattern of alternating equal bands, of luminescent luminous material and black, with the alternating bands no more than 2 inches (51 mm) thick and angled at 45 degrees (0.79 rad). Obstacles shall include, but are not limited to, standpipes, hose cabinets, wall projections and restricted height areas. However, such markings shall not conceal any required information or indicators including, but not limited to, instructions to occupants for the use of standpipes.

❖ Any obstacles within the stairway must be marked with a dashed line of diagonal slashes (see Figure 1024.2.5). Which items must be marked as obstacles is consistent with the intent of protruding object provisions in Section 1003.3. However, there is a difference in the height of 6 feet, 6 inches (1981 mm) instead of 6 feet, 8 inches (2032 mm). Items permitted within the stairway enclosure are limited by Sections 1022.1 and 1022.4.

1024.2.6 Doors from exit enclosures. Doors through which occupants within an *exit enclosure* must pass in order to complete the exit path shall be provided with markings complying with Sections 1024.2.6.1 through 1024.2.6.3.

❖ Doors within an exit enclosure can lead directly to the exterior, to an exit passageway on the level of exit discharge, to a lobby or vestibule at the level of exit discharge (see Section 1027.1) or an exit passageway on a transition level. Doors at these locations should be marked as indicated in the following three subsections. Combined with the landing markings not extending across the bottom of this door, this will provide several visual cues to indicate that this is the door to continue through to get out of the building (see Figure 1024.2.6)

The NIST egress study of the World Trade Center indicated that transition floors can cause delays in egress because people hesitate when it is unclear about which way they should continue. Effectively marking these doors four ways will decrease that hazard. Which doors to mark is consistent with the exit sign requirements within exit enclosures in Section 1011.1.

1024.2.6.1 Emergency exit symbol. The doors shall be identified by a low-location luminous emergency exit symbol complying with NFPA 170. The exit symbol shall be a minimum of 4 inches (102 mm) in height and shall be mounted on the door, centered horizontally, with the top of the symbol no higher than 18 inches (457 mm) above the finished floor.

❖ The door shall include a low level exit symbol within 18 inches (457 mm) of the floor. For an example of the emergency exit symbol, see Figure 1024.2.6.1. The sign on the door can be just this symbol or it can also contain additional information such as a directional arrow or "EXIT."

For SI: 1 inch = 25.4 mm.

Figure 1024.2.6.1
EMERGENCY EXIT SYMBOL

1024.2.6.2 Door hardware markings. Door hardware shall be marked with no less than 16 square inches (406 mm²) of luminous material. This marking shall be located behind, immediately adjacent to or on the door handle and/or escutcheon. Where a panic bar is installed, such material shall be no less than 1 inch (25 mm) wide for the entire length of the actuating bar or touchpad.

❖ Door hardware locations must be clearly visible. If a panic bar is used, a stripe with a minimum width of 1 inch (25 mm) should be provided down the entire length of the activation bar/paddle [see Figure 1024.2.6(a)]. If lever hardware is used, a donut, square or rectangle with a minimum area of 16 square inches (10 322 mm²) should be provided behind the hardware [see Figure 1024.2.6(b)]. There is also the option of marking the door handle itself, but it would be difficult to get 16 square inches (10 322 mm²) of visible surface area. Plus, over time, a finish on the hardware has a greater chance of wearing off with normal use. The language does allow freedom for the designer to decide with wear and hardware options which configuration would give the best results.

1024.2.6.3 Door frame markings. The top and sides of the door frame shall be marked with a solid and continuous 1 inch to 2 inch (25 mm to 51 mm) wide stripe. Where the door molding does not provide sufficient flat surface on which to locate the stripe, the stripe shall be permitted to be located on the wall surrounding the frame.

❖ Doors must be marked along the sides and top with stripes similar to those provided on the stair nosing. Door frames come in a variety of shapes. If there is not space on the door for the marking stripe, the stripe can be around the perimeter of the door [see Figure 1024.2.6(a) and (b)].

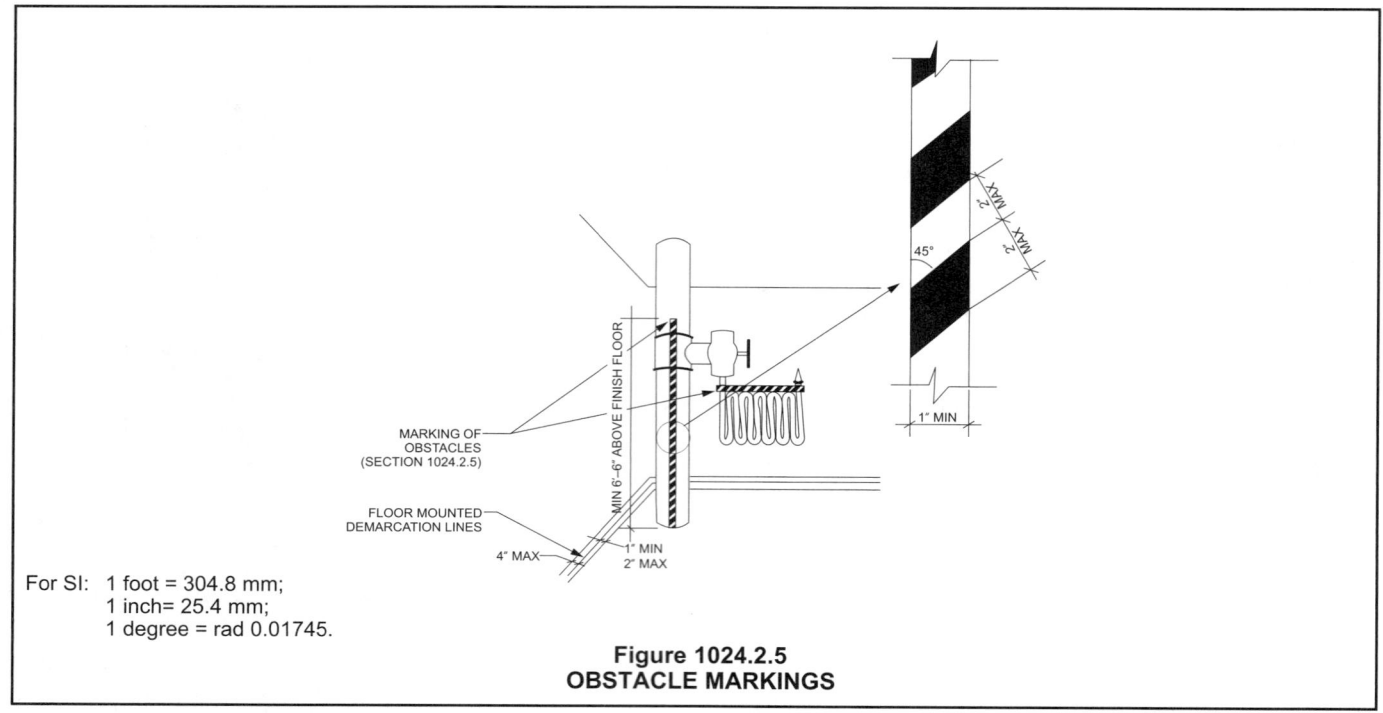

For SI: 1 foot = 304.8 mm;
          1 inch= 25.4 mm;
          1 degree = rad 0.01745.

Figure 1024.2.5
OBSTACLE MARKINGS

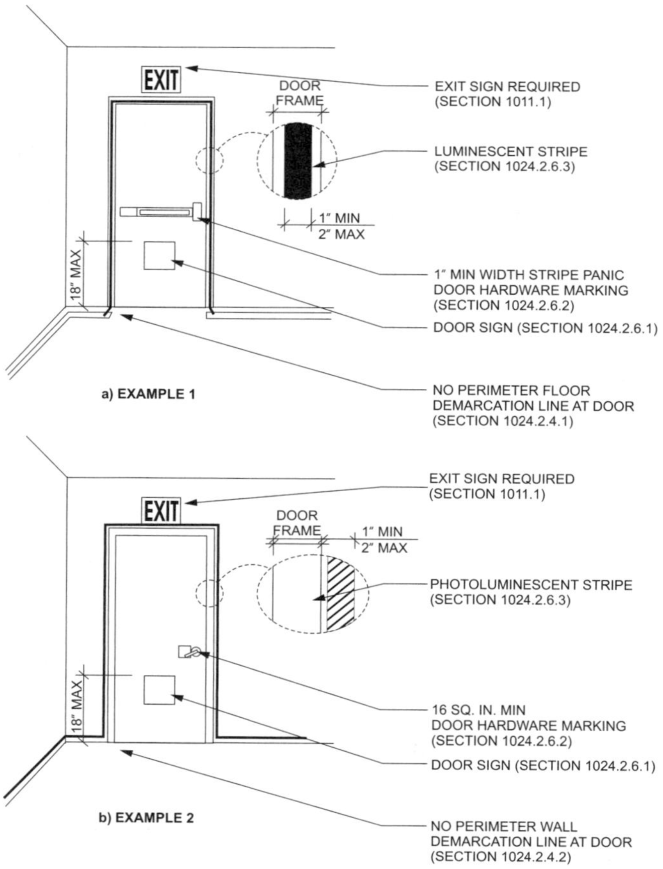

For SI: 1 inch = 25.4 mm;
          1 square inch = 645.16 mm.

Figure 1024.2.6
EXAMPLES OF EXIT DOOR TO EXTERIOR OPTIONS

1024.3 Uniformity. Placement and dimensions of markings shall be consistent and uniform throughout the same *exit enclosure*.

❖ All exit path markings are required to be solid and continuous stripes. A key requirement for marking systems is that their design must be uniform. The placement and dimensions of markings must be consistent throughout the same exit enclosure. By specifying a standard marking dimension, the requirements will ensure that the marking is visible during dark conditions and provides consistent and standard application in the design and enforcement of exit path markings. Markings installed on stair steps, perimeter demarcation lines and handrails must have a minimum width of 1 inch (25 mm). For stair steps and perimeter demarcation lines, their maximum width cannot exceed 2 inches (51 mm). The provisions for stair steps, perimeter demarcation lines and handrails allow the width of the marking to be reduced to less than 1 inch (25 mm) when marking stripes are listed in accordance with UL 1994.

1024.4 Self-luminous and photoluminescent. Luminous egress path markings shall be permitted to be made of any material, including paint, provided that an electrical charge is not required to maintain the required luminance. Such materials shall include, but are not limited to, *self-luminous* materials and *photoluminescent* materials. Materials shall comply with either:

1. UL 1994; or

2. ASTM E 2072, except that the charging source shall be 1 foot-candle (11 lux) of fluorescent illumination for 60 minutes, and the minimum luminance shall be 30 millicandelas per square meter at 10 minutes and 5 millicandelas per square meter after 90 minutes.

❖ Products utilized to meet the requirements for luminous egress path markings in high-rise buildings (see Sections 411.7 and 1024) or exit signs (see Section 1011.4) may be photoluminescent or self-luminous (see definitions in Section 1002.1). An example of photoluminescent material is paint or tape that is charged by exposure to light. When the lights are turned off, the product will "glow" in the dark. Self-luminous products do not need an outside light source to charge them like photoluminescent materials do.

A variety of materials can comply with the referenced standards for egress path markings—UL1994, and ASTM E 2072 and for signs—UL 924.

ASTM E 2072 allows the use of paints and coatings, which can be useful because it avoids a potential tripping hazard, especially in locations where the surface substrate may not be even. The luminescence of the selected marking system must provide an illumination of 1 foot-candle (11 lux) for 60 minutes, which is consistent with the requirement in Section 1006.2 for the illumination of walking surfaces. Section 1006.3 does require the emergency lighting system to have power for 90 minutes, however, due to normal battery consid-

erations, the IFC only requires a 60-minute duration in existing buildings.

1024.5 Illumination. *Exit enclosures* where photoluminescent exit path markings are installed shall be provided with the minimum *means of egress* illumination required by Section 1006 for at least 60 minutes prior to periods when the building is occupied.

❖ Analogous to rechargeable batteries, many photoluminescent egress path markings require exposure to light to perform properly. Thus, photoluminescent egress path markings must be exposed to a minimum 1 foot-candle (11 lux) of light energy at the walking surface for at least 60 minutes prior to the building being occupied. The charging rate for photoluminescent egress path markings is based on the wattage of lamps used to provide egress path illumination. Therefore it is important to verify that the specified lamps have sufficient wattage to meet the specified time period. This requirement may be a concern for buildings developed with the *International Energy Conservation Code®* (IECC®) or trying for LEED certification.

Note that this requirement does not apply to self-luminous materials since these materials operate independently of the external power source. See the definitions for "Photoluminescent" and "Self-luminous" in Section 1002.1.

SECTION 1025
HORIZONTAL EXITS

1025.1 Horizontal exits. *Horizontal exits* serving as an *exit* in a *means of egress* system shall comply with the requirements of this section. A *horizontal exit* shall not serve as the only *exit* from a portion of a building, and where two or more *exits* are required, not more than one-half of the total number of *exits* or total *exit* width shall be *horizontal exits*.

Exceptions:

1. *Horizontal exits* are permitted to comprise two-thirds of the required *exits* from any building or floor area for occupancies in Group I-2.

2. *Horizontal exits* are permitted to comprise 100 percent of the *exits* required for occupancies in Group I-3. At least 6 square feet (0.6 m²) of accessible space per occupant shall be provided on each side of the *horizontal exit* for the total number of people in adjoining compartments.

❖ Horizontal exits can provide up to 50 percent of the exits from an area of a building. A horizontal exit cannot serve as the only exit from a single exit space. The percentage is higher for Group I-2 and I-3 occupancies where the evacuation strategy is defend in place rather than direct egress (see Figure 1025.1 for a typical horizontal exit arrangement). Section 1025.4 allows for some areas to have all the exits from a space to be horizontal exits under specific conditions (see

Section 1025.4).

A horizontal exit may be an element of a means of egress when in compliance with the requirements of this section. The actual horizontal exit is the protected door opening in a wall, open-air balcony or bridge that separates two areas of a building. A horizontal exit is often used in hospitals and in prisons where it is not feasible or desirable that all occupants exit the facility (see Section 1002.1 for the definition of a "Horizontal exit").

Horizontal exits and their associated "refuge areas" are considered to provide the same or higher level of protection as an "area of refuge" for people who cannot use the egress system. Sections 1007.3 and 1007.4 allow for a horizontal exit or an area of refuge as alternatives. See these sections for exceptions for buildings with sprinkler systems and/or where the path of travel has protection from the accumulation of smoke (i.e., open parking garages, open air assembly seating, smoke-protected assembly seating).

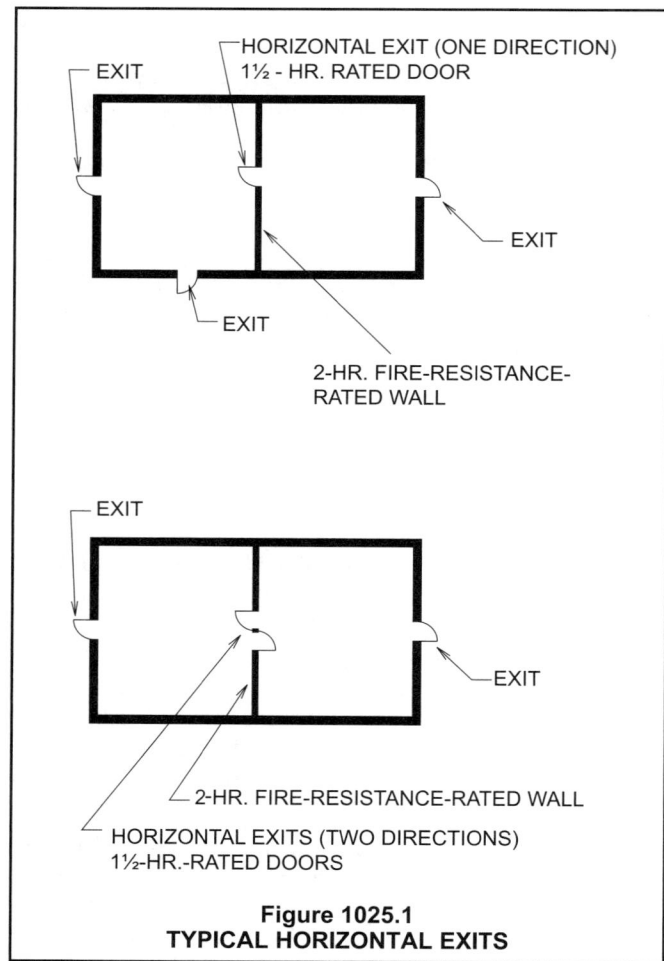

Figure 1025.1
TYPICAL HORIZONTAL EXITS

1025.2 Separation. The separation between buildings or refuge areas connected by a *horizontal exit* shall be provided by a *fire wall* complying with Section 706; or it shall be provided by a *fire barrier* complying with Section 707 or a *horizontal assembly* complying with Section 712, or both. The minimum

fire-resistance rating of the separation shall be 2 hours. Opening protectives in *horizontal exits* shall also comply with Section 715. Duct and air transfer openings in a *fire wall* or *fire barrier* that serves as a *horizontal exit* shall also comply with Section 716. The *horizontal exit* separation shall extend vertically through all levels of the building unless floor assemblies have a *fire-resistance rating* of not less than 2 hours with no unprotected openings.

> **Exception:** A *fire-resistance rating* is not required at *horizontal exits* between a building area and an above-grade *pedestrian walkway* constructed in accordance with Section 3104, provided that the distance between connected buildings is more than 20 feet (6096 mm).

Horizontal exits constructed as *fire barriers* shall be continuous from *exterior wall* to *exterior wall* so as to divide completely the floor served by the *horizontal exit*.

❖ The basic concept of a horizontal exit is that during a fire emergency, the occupants of a floor will transfer from one fire area to another. Separation between areas of a building can be accomplished by either a fire wall (see Section 706) a fire barrier (see Section 707) horizontal exits (see Section 712), or a combination thereof, with a fire-resistance rating not less than 2 hours. Any fire shutters or fire doors must have an opening protective of not less than $1^1/_2$ hours (see Table 715.4). Ducts and air transfer openings must comply with Section 716.

In buildings of Groups I-2 and I-3, it may also be desirable (while not mandatory) for the horizontal exit to serve as a smoke barrier. In such cases, the wall containing the horizontal exit must also comply with the requirements for a smoke barrier (see Section 710).

In order to decrease the amount of smoke able to migrate around the edges of a horizontal exit, the horizontal exit must extend from at least the floor to the deck above (i.e., fire barrier), as well as across the floor level from one side of the building to another. Moving up from floor to floor, there are two choices. One option is that the horizontal exit can extend vertically through all levels of the building (i.e., fire wall or fire barriers). The second option is to utilize fire barriers that are not aligned vertically (i.e., a combination of fire barriers and horizontal assemblies), but then the floor must have a 2-hour fire-resistance rating and no unprotected openings are permitted between any two refuge areas. The supporting construction would also have to be a minimum of 2 hours.

The exception is permitting a pedestrian walkway or sky bridge to act as a horizontal exit when buildings are at least 20 feet (6096 mm) apart.

1025.3 Opening protectives. *Fire doors* in *horizontal exits* shall be self-closing or automatic-closing when activated by a *smoke detector* in accordance with Section 715.4.8.3. Doors, where located in a cross-corridor condition, shall be automatic-closing by activation of a *smoke detector* installed in accordance with Section 715.4.8.3.

❖ For the safety of occupants using a horizontal exit, it is important that the doors be fire doors that are self-closing or automatic closing by activation of a smoke detector. Smoke detectors that initiate automatic closing should be located at both sides of the doors (see the commentary to Section 907.3 for an additional explanation of the installation requirements). Any openings in the fire barriers or fire walls used as horizontal exits must be protected in coordination with the rating of the wall. There is a reference to Section 715 for opening protectives.

1025.4 Capacity of refuge area. The refuge area of a *horizontal exit* shall be a space occupied by the same tenant or a public area and each such refuge area shall be adequate to accommodate the original *occupant load* of the refuge area plus the *occupant load* anticipated from the adjoining compartment. The anticipated *occupant load* from the adjoining compartment shall be based on the capacity of the *horizontal exit* doors entering the refuge area. The capacity of the refuge area shall be computed based on a net floor area allowance of 3 square feet (0.2787 m²) for each occupant to be accommodated therein.

> **Exception:** The net floor area allowable per occupant shall be as follows for the indicated occupancies:
>
> 1. Six square feet (0.6 m²) per occupant for occupancies in Group I-3.
>
> 2. Fifteen square feet (1.4 m²) per occupant for ambulatory occupancies in Group I-2.
>
> 3. Thirty square feet (2.8 m²) per occupant for nonambulatory occupancies in Group I-2.

The refuge area into which a *horizontal exit* leads shall be provided with *exits* adequate to meet the occupant requirements of this chapter, but not including the added *occupant load* imposed by persons entering it through *horizontal exits* from other areas. At least one refuge area *exit* shall lead directly to the exterior or to an *exit enclosure*.

> **Exception:** The adjoining compartment shall not be required to have a *stairway* or door leading directly outside, provided the refuge area into which a *horizontal exit* leads has stairways or doors leading directly outside and are so arranged that egress shall not require the occupants to return through the compartment from which egress originates.

❖ The building area on the discharge side of a horizontal exit must serve as a refuge area for the occupants of both sides of the floor areas connected by the horizontal exit. Therefore, adequate space must be available on each side of the wall to hold the full occupant load of that side, plus the number of occupants from the other side that may be required to use the horizontal exit. These refuge areas are meant to hold the occupants temporarily in a safe place until they can evacuate the premises in an orderly manner or, in the case of hospitals and like facilities, to hold bedridden patients and other nonambulatory occupants in a protected area until the fire emergency has ended. The size of the refuge area is based on the nature of the expected occupants. In the case of Group I-3, the area will be used to hold the occupants until deliberate egress can be accomplished with staff assistance or supervision. In other cases, it is assumed that the occupants simply wait in line to egress through the required exit facilities provided on the discharge side. Although similar language is used in describing the "area of refuge" for an accessible means of egress, Section 1007.6 specifies area requirements that are insufficient for use as a "refuge area" for a horizontal exit. Care must be taken when applying both principles to the same horizontal exit.

The 3-square-feet (0.28 m²) per occupant requirement is based on the maximum permitted occupant density at which orderly movement to the exits is reasonable. The 30-square-feet (2.8 m²) per hospital or nursing home patient requirement is based on the space necessary for a bed or litter. It should be noted that 30 square feet (2.8 m²) is not based on the total occupant load, as would be determined in accordance with Section 1004.1, but rather on the number of nonambulatory patients. The 15-square-feet (1.4 m²) requirement for occupancies in Group I-2 facilities is based on each ambulatory patient having a staff attendant.

In a single-tenant facility, any of the spaces that are constantly available (e.g., not lockable) can be used as places of refuge. However, in spaces housing more than one tenant, public refuge areas, such as corridors or passageways, must be provided and be accessible at all times. This requirement is necessary because if a horizontal exit connected two areas occupied by different tenants, the tenants could (for privacy and security purposes) render the necessary free access through the horizontal exit ineffective. When the horizontal exit discharges into a public or common space, such as a corridor leading to an exit, each tenant can obtain the desired security.

Note that the capacity of the exit (such as an exit stairway) from the refuge area to which the horizontal exit leads is required to be sufficient for the design occupant load in the area, and does not include those who come into the space from other areas via the horizontal exit. This is because the adjacent refuge area is of sufficient safety to house occupants during a fire or until the egress system is available.

The door through the horizontal exit and the second exit must meet the separation requirements in Section 1015.2.1. Measurement of the travel distance stops at the horizontal exit. There are not travel distance requirements for the distance from the horizontal exit to the vertical exit enclosure on the other side; however, the areas on each side need to be evaluated for all means of egress requirements individually.

When there is one horizontal exit and two fire compartments, at least one exit from each side of the horizontal exit must go directly to the outside of an exit stairway or ramp enclosure (see Figure 1025.1). The exception allows for a central building/fire area with access to two horizontal exits and no direct exterior exit door or exit stairway/ramp enclosure as long as the piece on each side has access to exterior exits or exit stairways/ramps (see Figure 1025.4).

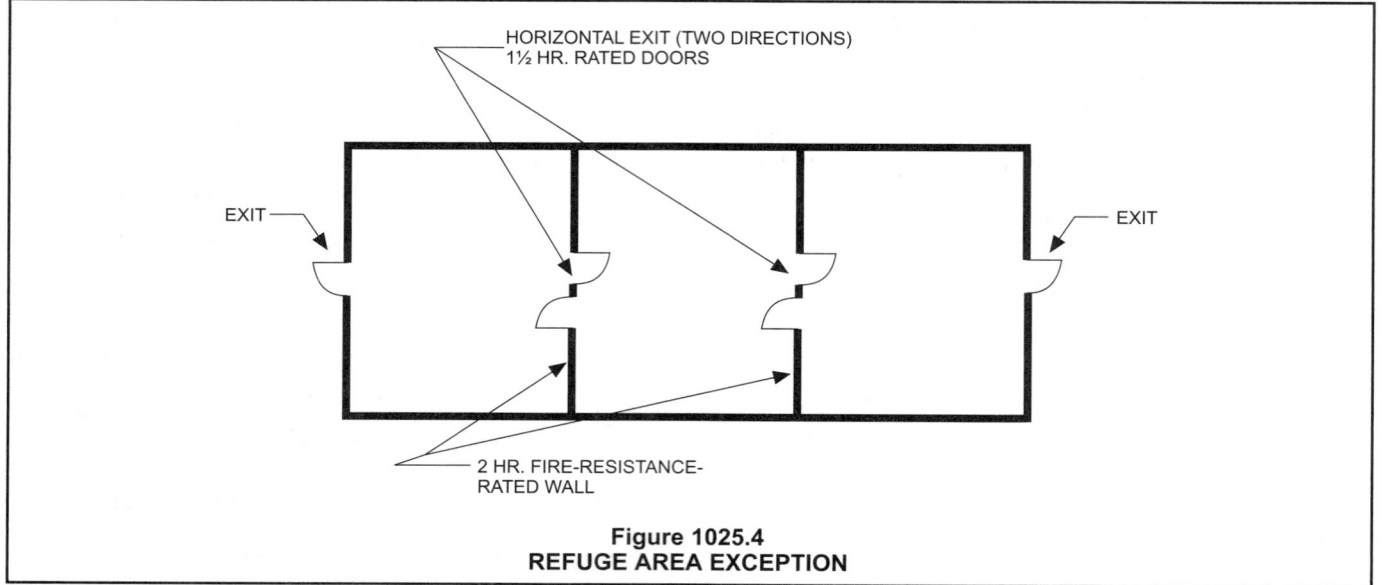

Figure 1025.4
REFUGE AREA EXCEPTION

SECTION 1026
EXTERIOR EXIT RAMPS AND STAIRWAYS

1026.1 Exterior exit ramps and stairways. *Exterior exit ramps* and *stairways* serving as an element of a required *means of egress* shall comply with this section.

> **Exception:** *Exterior exit ramps* and *stairways* for outdoor stadiums complying with Section 1022.1, Exception 2.

❖ Ramps and stairways can be exit access, exit or exit discharge elements. Exit access and exit discharge stairs and ramps typically involve a change of elevation of less than a story. Most exit ramps and exit stairways traverse a full story or more. This section addresses exterior ramps and exterior stairways that function as exit elements.

Exterior exit ramps and stairways are an important element of the means of egress system and must be designed and constructed so that they will serve as a safe path of travel. The general requirements in Section 1009 also apply to exterior stairways (for ramp provisions, see Section 1010).

The exception references Exception 2 in Section 1022.1, which allows exterior ramps and stairways that serve outdoor assembly facilities to be open. The openness criteria in Section 1026.2 would be an appropriate guideline for what is intended.

1026.2 Use in a means of egress. *Exterior exit stairways* shall not be used as an element of a required *means of egress* for Group I-2 occupancies. For occupancies in other than Group I-2, *exterior exit ramps* and *stairways* shall be permitted as an element of a required *means of egress* for buildings not exceeding six stories above *grade plane* or having occupied floors more than 75 feet (22 860 mm) above the lowest level of fire department vehicle access.

❖ This section specifies the conditions where an exterior ramp or stairway can be used as a required exit. Exterior exit stairways are not permitted for Group I-2 since

quick evacuation of nonambulatory patients from buildings is impractical. Some of the patients may not be capable of self-preservation and, therefore, may require assistance from the staff. The period of evacuation of nonambulatory patients could become lengthy.

With the exception of outdoor stadiums (see Section 1026.1), exterior ramps or stairways are not allowed to be a required exit in buildings that exceed six stories or 75 feet (22 860 mm) in height due to the hazard of using such a ramp or stairway in poor weather. Some persons may not be willing to use such a stair due to vertigo. When confronted with a view from a great height, vertigo sufferers can become confused, disoriented and dizzy. They could injure themselves, become disoriented or refuse to move (freeze). In a fire situation, they could become an obstruction in the path of travel, possibly causing panic and injuries to other users of the exit.

1026.3 Open side. *Exterior exit ramps* and *stairways* serving as an element of a required *means of egress* shall be open on at least one side. An open side shall have a minimum of 35 square feet (3.3 m²) of aggregate open area adjacent to each floor level and the level of each intermediate landing. The required open area shall be located not less than 42 inches (1067 mm) above the adjacent floor or landing level.

❖ An important factor in considering exterior exit ramps or stairways is that natural ventilation is assumed to occur. This is so that smoke will not be trapped above the ramp or stairway walking surfaces and obscure safe egress.

The exterior ramp or stairway must have at least one of its sides directly facing an outer court, yard or public way. This will allow the products of combustion escaping from the interior of the building to quickly vent to the outdoor atmosphere and let the building occupants egress down the exterior ramp or stairway. Since exterior ramps or stairways are occasionally partially enclosed within the building construction, minimum amounts of exterior openings are specified by the code.

The openings on each and every floor level and landing must total 35 square feet (3.3 m²) or greater. The opening is to occur higher than 42 inches (1067 mm) from the floor and intermediate landing levels. With a standard 8-foot (2438 mm) ceiling height minus the 3 foot, 6 inch (1067 mm) guard, and approximately an 8-foot-wide (2438 mm) opening, the result would be 4.5 feet x 8 feet = 36 square feet. The high openings dissipate the smoke buildup from the exterior ramp or stairway (see Figure 1026.3). The bottom edge of the opening is consistent with the height requirements for guards (see Section 1013.2).

1026.4 Side yards. The open areas adjoining *exterior exit ramps* or *stairways* shall be either *yards*, *courts* or *public ways*; the remaining sides are permitted to be enclosed by the *exterior walls* of the building.

❖ This section simply specifies the type of areas that the exterior opening of the exterior ramp or stair is to adjoin. These open spaces will enable the smoke to dissipate from the exterior ramp or stairway so it will be usable as a required exit. See Section 1026.3 for a discussion of the opening requirements. See Sections 1027.5 and 1206 for the minimum sizes of yards and courts.

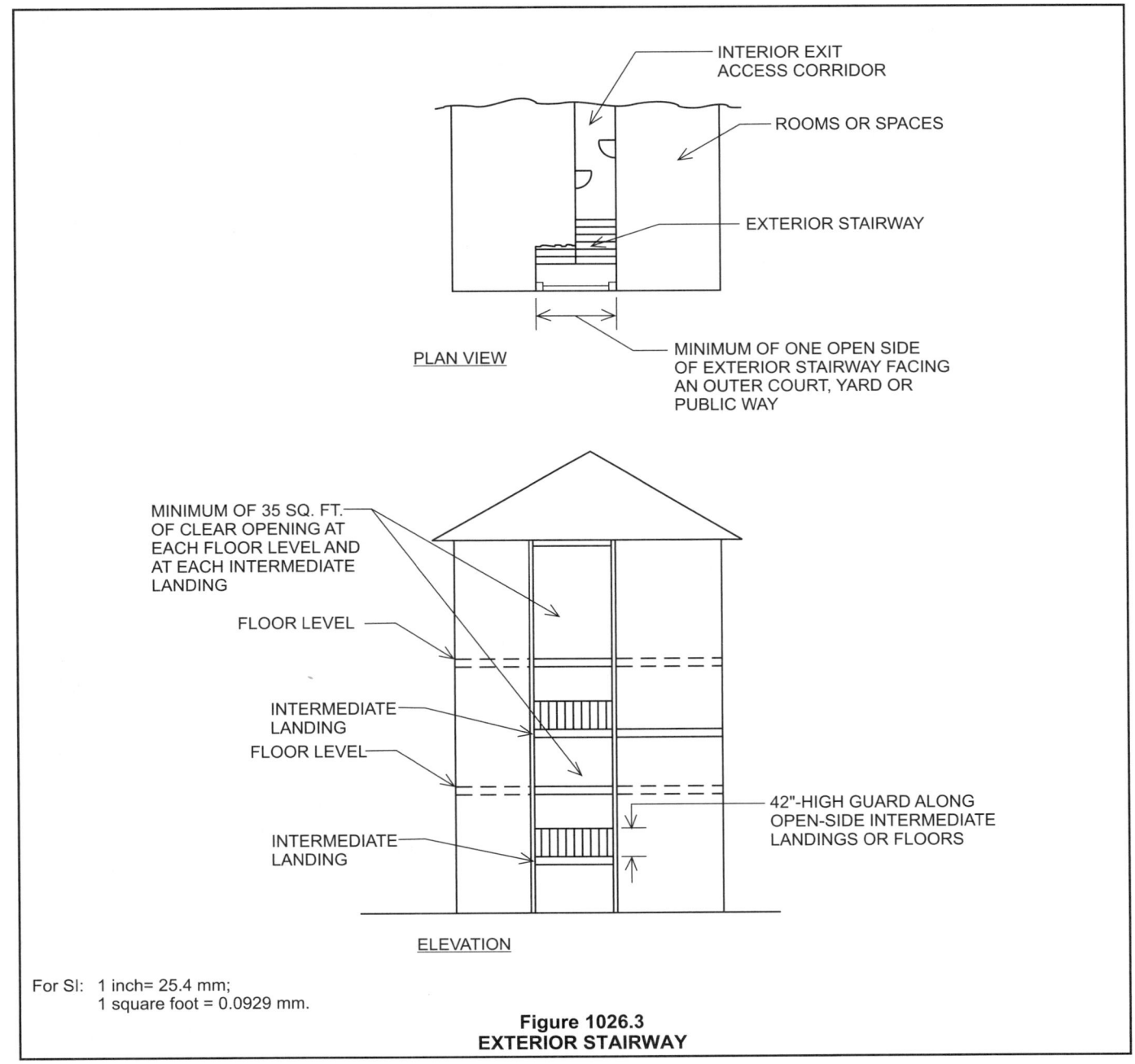

INTERIOR EXIT ACCESS CORRIDOR

ROOMS OR SPACES

EXTERIOR STAIRWAY

MINIMUM OF ONE OPEN SIDE OF EXTERIOR STAIRWAY FACING AN OUTER COURT, YARD OR PUBLIC WAY

PLAN VIEW

MINIMUM OF 35 SQ. FT. OF CLEAR OPENING AT EACH FLOOR LEVEL AND AT EACH INTERMEDIATE LANDING

FLOOR LEVEL

INTERMEDIATE LANDING

FLOOR LEVEL

INTERMEDIATE LANDING

42"-HIGH GUARD ALONG OPEN-SIDE INTERMEDIATE LANDINGS OR FLOORS

ELEVATION

For SI: 1 inch= 25.4 mm;
1 square foot = 0.0929 mm.

Figure 1026.3
EXTERIOR STAIRWAY

1026.5 Location. *Exterior exit ramps* and *stairways* shall be located in accordance with Section 1027.3.

❖ The location requirements of this section protect the users of the exterior exit ramp or stairway from the effects of a fire in another building on the same lot or an adjacent lot. The separation distance reduces the exposure to heat and smoke. If the stairway is closer than specified, then adjacent buildings' exterior walls and openings are to be protected in accordance with Section 705, so that the users of the exterior exit are protected. The reason for the required distance to a lot line is to provide for a future building that could be built on the adjacent lot. While buildings on the same lot can be considered one building for height and area limitations (see Section 503.1.2), they must be separated by a minimum of 10 feet (3048 mm) if there is a path for exit discharge between them.

1026.6 Exterior ramps and stairway protection. *Exterior exit ramps* and *stairways* shall be separated from the interior of the building as required in Section 1022.1. Openings shall be limited to those necessary for egress from normally occupied spaces.

Exceptions:

1. Separation from the interior of the building is not required for occupancies, other than those in Group R-1 or R-2, in buildings that are no more than two stories above *grade plane* where a *level of exit discharge* serving such occupancies is the first *story above grade plane*.

2. Separation from the interior of the building is not required where the *exterior ramp* or *stairway* is served by an exterior *ramp* or balcony that connects two remote *exterior stairways* or other *approved exits*, with a perimeter that is not less than 50 percent open. To be considered open, the opening shall be a minimum of 50 percent of the height of the enclosing wall, with the top of the openings no less than 7 feet (2134 mm) above the top of the balcony.

3. Separation from the interior of the building is not required for an *exterior ramp* or *stairway* located in a building or structure that is permitted to have unenclosed *interior stairways* in accordance with Section 1022.1.

4. Separation from the interior of the building is not required for *exterior ramps* or *stairways* connected to open-ended *corridors*, provided that Items 4.1 through 4.4 are met:

    4.1. The building, including *corridors*, *ramps* and *stairs*, shall be equipped throughout with an *automatic sprinkler system* in accordance with Section 903.3.1.1 or 903.3.1.2.

    4.2. The open-ended *corridors* comply with Section 1018.

    4.3. The open-ended *corridors* are connected on each end to an *exterior exit ramp* or *stairway* complying with Section 1026.

    4.4. At any location in an open-ended *corridor* where a change of direction exceeding 45 degrees (0.79 rad) occurs, a clear opening of not less than 35 square feet (3.3 m²) or an *exterior ramp* or *stairway* shall be provided. Where clear openings are provided, they shall be located so as to minimize the accumulation of smoke or toxic gases.

❖ Exterior exit ramps or stairways must be protected from interior fires that may project through windows or other openings adjacent to the ramp or stairway, possibly endangering the occupants using this means of egress to reach grade. The protection of an exterior ramp or stairway is to be obtained by separating the exterior exit from the interior of the building using walls having a fire-resistance rating of at least 1 hour with opening protectives. Consistent with the protection required in Sections 1022.1 and 1022.6 for interior exit stairways, the fire-resistance rating must be provided for a distance of 10 feet (3048 mm) horizontally and vertically from the ramp or stairway edges, and from the ground to a level of 10 feet (3048 mm) above the highest landing.

All window and door openings falling inside the 10-foot (3048 mm) horizontal separation distance as well as all window and door openings 10 feet (3048 mm) above the topmost landing and below the stairway must be protected with minimum ³/₄-hour fire protection-rated opening protectives [see Figure 1026.6(1)].

Openings within the width of the stairway must only be from normally occupied spaces. This is consistent with the requirements for vertical exit enclosures (see Sections 1022.3 and 1022.4).

Exception 1 indicates that opening protectives are not required for occupancies (other than Groups R-1 and R-2) that are two stories or less above grade when the level of exit discharge is at the lower story. The reason for this exception is that in cases of fire in low buildings, the occupants are usually able to evacuate the premises before the fire can emerge through exterior wall openings and endanger the exit ramp or stairways. In hotels and apartments, however, the occupants' response to a fire emergency could be significantly reduced because they may be either unfamiliar with the surroundings or sleeping.

Exception 2 allows the opening protectives to be omitted when an exterior exit access balcony is served by two exits and when the exits are remote from each other. Remoteness is regulated by Section 1015.2. This exception is applicable to all groups. In such instances, it is unlikely that the users of the exterior ramp or stairway will become trapped by fire, since they have the option of using the balcony to gain access to either of the two available exits, and the products of combustion will be vented directly to the outside (see Section 1019.1 regarding exterior balconies). At least one-half of the total perimeter of the exterior balcony must be permanently open to the outside. The requirement for at least one-half the height of that level to be open allows for columns, solid guards and decorative elements, such as arches. With the top of the opening at least 7 feet (2134 mm) above the walking surface, products of combustion can vent and allow passage below the smoke layer [see Figure 1026.6(2)].

Exception 3 exempts exterior ramps or stairways from protection when Section 1022.1 permits unenclosed interior exit stairways. This would be areas such as open parking garages, individual dwelling units in Groups R-2 and R-3, etc.

Exception 4 deletes the requirement for a separation between the interior of the building and the exterior wall area where an open-ended corridor (breezeway) interfaces with an exterior ramp or stairway. The separation is not needed as a result of the NFPA 13 or 13R sprinkler system in all areas of the building, including the open-ended corridor and the

exterior exit. The other characteristics of the open-ended corridor described in this exception are needed so that it is safe to be used in the event of a fire. The requirements for an exterior ramp or stairway at each end and the opening or an exterior ramp or stairway where the open-ended corridor has a change of direction of greater than 45 degrees (0.79 rad) are for adequate ventilation of the open-ended corridor. Exit access travel distance on an open-ended corridor is measured to the first riser of an exterior stair or the beginning slope of an exterior ramp.

Similar language is used in describing the exterior wall requirements for an exterior area for assisted rescue (see Section 1007.8). If this option is utilized at an exterior stairway or nonaccessible ramp, the exceptions in this section would not be permitted since occupants of the exterior area for assisted rescue could not immediately move away from the building.

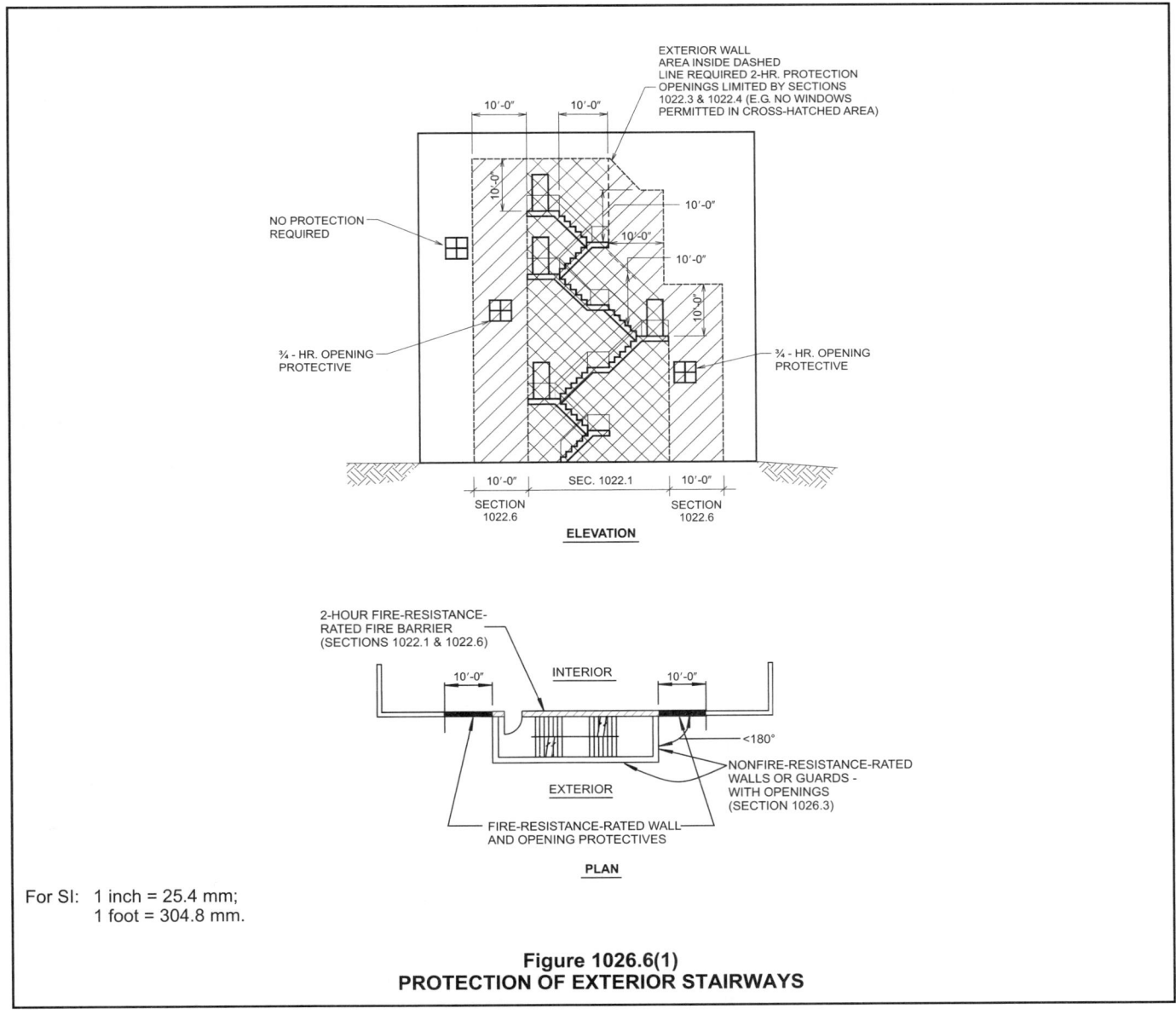

Figure 1026.6(1)
PROTECTION OF EXTERIOR STAIRWAYS

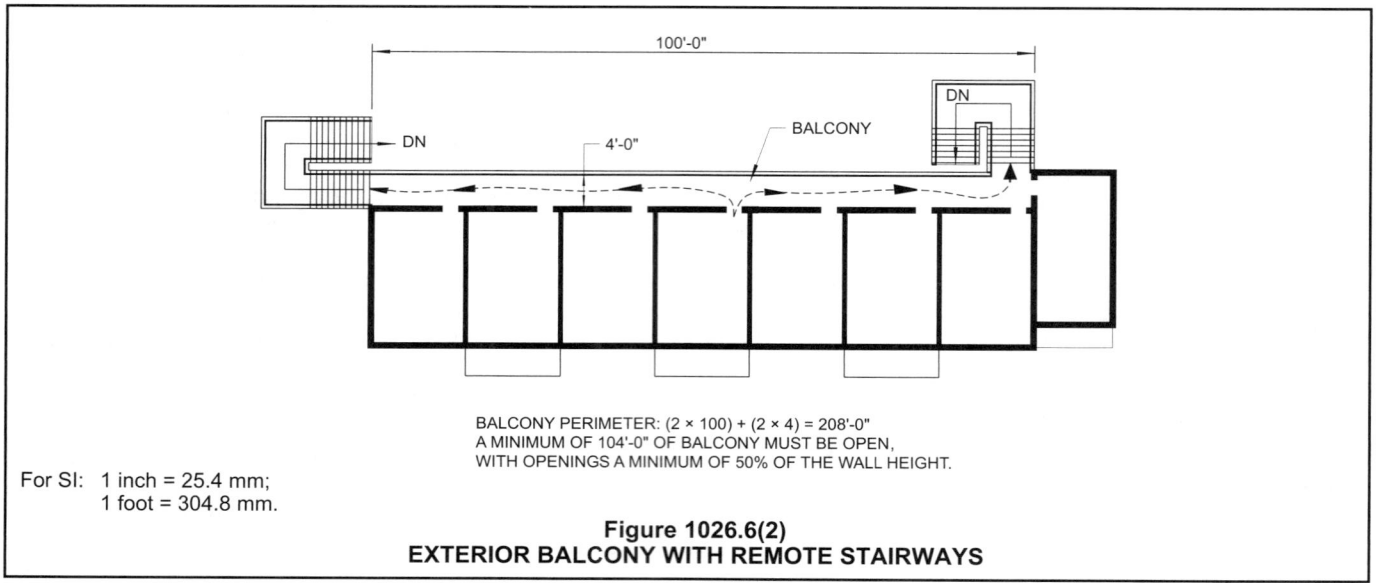

For SI: 1 inch = 25.4 mm;
1 foot = 304.8 mm.

BALCONY PERIMETER: (2 × 100) + (2 × 4) = 208'-0"
A MINIMUM OF 104'-0" OF BALCONY MUST BE OPEN,
WITH OPENINGS A MINIMUM OF 50% OF THE WALL HEIGHT.

Figure 1026.6(2)
EXTERIOR BALCONY WITH REMOTE STAIRWAYS

SECTION 1027
EXIT DISCHARGE

1027.1 General. *Exits* shall discharge directly to the exterior of the building. The *exit discharge* shall be at grade or shall provide direct access to grade. The *exit discharge* shall not reenter a building. The combined use of Exceptions 1 and 2 below shall not exceed 50 percent of the number and capacity of the required *exits*.

Exceptions:

1. A maximum of 50 percent of the number and capacity of the *exit enclosures* is permitted to egress through areas on the level of discharge provided all of the following are met:

    1.1. Such *exit enclosures* egress to a free and unobstructed path of travel to an exterior *exit* door and such *exit* is readily visible and identifiable from the point of termination of the *exit* enclosure.

    1.2. The entire area of the *level of exit discharge* is separated from areas below by construction conforming to the *fire-resistance rating* for the *exit enclosure*.

    1.3. The egress path from the *exit enclosure* on the *level of exit discharge* is protected throughout by an *approved automatic sprinkler system*. All portions of the *level of exit discharge* with access to the egress path shall either be protected throughout with an *automatic sprinkler system* installed in accordance with Section 903.3.1.1 or 903.3.1.2, or separated from the egress path in accordance with the requirements for the enclosure of *exits*.

2. A maximum of 50 percent of the number and capacity of the *exit enclosures* is permitted to egress through a vestibule provided all of the following are met:

    2.1. The entire area of the vestibule is separated from areas below by construction conforming to the *fire-resistance rating* for the *exit enclosure*.

    2.2. The depth from the exterior of the building is not greater than 10 feet (3048 mm) and the length is not greater than 30 feet (9144 mm).

    2.3. The area is separated from the remainder of the *level of exit discharge* by construction providing protection at least the equivalent of *approved* wired glass in steel frames.

    2.4. The area is used only for *means of egress* and *exits* directly to the outside.

3. *Stairways* in *open parking garages* complying with Section 1022.1, Exception 4, are permitted to egress through the *open parking garage* at their *levels of exit discharge*.

4. *Horizontal exits* complying with Section 1025 shall not be required to discharge directly to the exterior of the building.

❖ The exit discharge is the third piece of the means of egress system, which includes exit access, exit and exit discharge. The general provisions for means of egress in Sections 1003 through 1013 are applicable to the exit discharge. The basic provision is that exits must discharge directly to the outside of the building. The exit discharge is the path from the termination of the exit to the public way. When this is not practical, there are four alternatives: an exit passageway (see Section 1023), an exit discharge lobby (see Section 1027.1, Exception 1), an exit discharge passageway (see Section 1027.1, Exception 2) or a horizontal exit (see Sections 1025 and 1027.1, Exception 4). Open parking garages, since there is little to no accumulation of smoke in the building, are a special case, and are addressed in Exception 3.

While Exceptions 1 and 2 could be applicable for

exit passageways and exit ramps, most of the real-life application of the exceptions is for exit stairways. This commentary for Section 1027 will be limited to enclosed exit stairways. Up to 50 percent of the exit stairways in a building may use either Exception 1 or 2; therefore, neither exception is viable for a single-exit building. In a two- or three-exit building, either a lobby or a vestibule can be used for exit discharge for one of the exit stairways. In a four-exit building, two of the exit stairways can use either a lobby or a vestibule for exit discharge.

An interior exit discharge lobby is permitted to receive the discharge from an exit stairway in lieu of the stairway discharging directly to the exterior. A fire door must be provided at the point where the exit stairway discharges into the lobby. Without an opening protective between the stairway and a lobby, it would be possible for the stairway to be directly exposed to smoke movement from a fire in the lobby. The opening protective provides for full continuity of the vertical component of the exit arrangement. Additionally, in buildings where stair towers must be pressurized, pressurization would not be possible without a door at the lobby level.

An exit discharge lobby is the sole location recognized in the code where an exit element can be used for purposes other than pedestrian travel for means of egress. The lobby may contain furniture or decoration and nonoccupiable spaces may open directly into the lobby. The lobby, and all other areas on the same level that are not separated from the lobby by fire barriers consistent with the rating of the stair enclosure, must be sprinklered in accordance with an NFPA 13 or NFPA 13R system [see Figure 1027.1(1)]. If the entire level is sprinklered, no separation is required. In this case, the automatic sprinkler system is anticipated to control and (perhaps) eliminate the fire threat so as not to jeopardize the path of egress of the occupants. The lobby floor and any supporting construction must be rated the same as the stairway enclosure. If the lobby is slab on grade, this requirement is not applicable. This is consistent with the fundamental concept that an exit enclosure provides the necessary level of protection from adjacent areas. A path of travel through the lobby must be continually clear and available. The exit door leading out of the building must be visible and identifiable immediately when a person leaves the exit. This does not mean the exterior exit door must be directly in front of the door at the bottom of the stairway, but the intent is that it should be within the general range of vision. It should not be required that a person must turn completely around or go around a corner to be able to see the way out.

An exit is also allowed to discharge through a vestibule, provided it complies with the specified requirements of Exception 2. Note that a vestibule is not allowed to have any vending machines within the space or any storage rooms opening into the vestibule. The vestibule floor and any supporting construction must be rated the same as the stairway enclosure. If the vestibule is slab on grade, this requirement is not applicable. The walls of the vestibule itself must have something at least equivalent with wired glass in steel frames. This is typically an opening protective with a fire protection rating of at least 20 minutes. The size of the vestibule is limited so that it cannot be used for other activities, and the travel distance from the exit stairway to the exterior exit doorway is limited [see Figures 1027.1(2) and 1027.1(3)].

The reason that stairs in open parking garages are

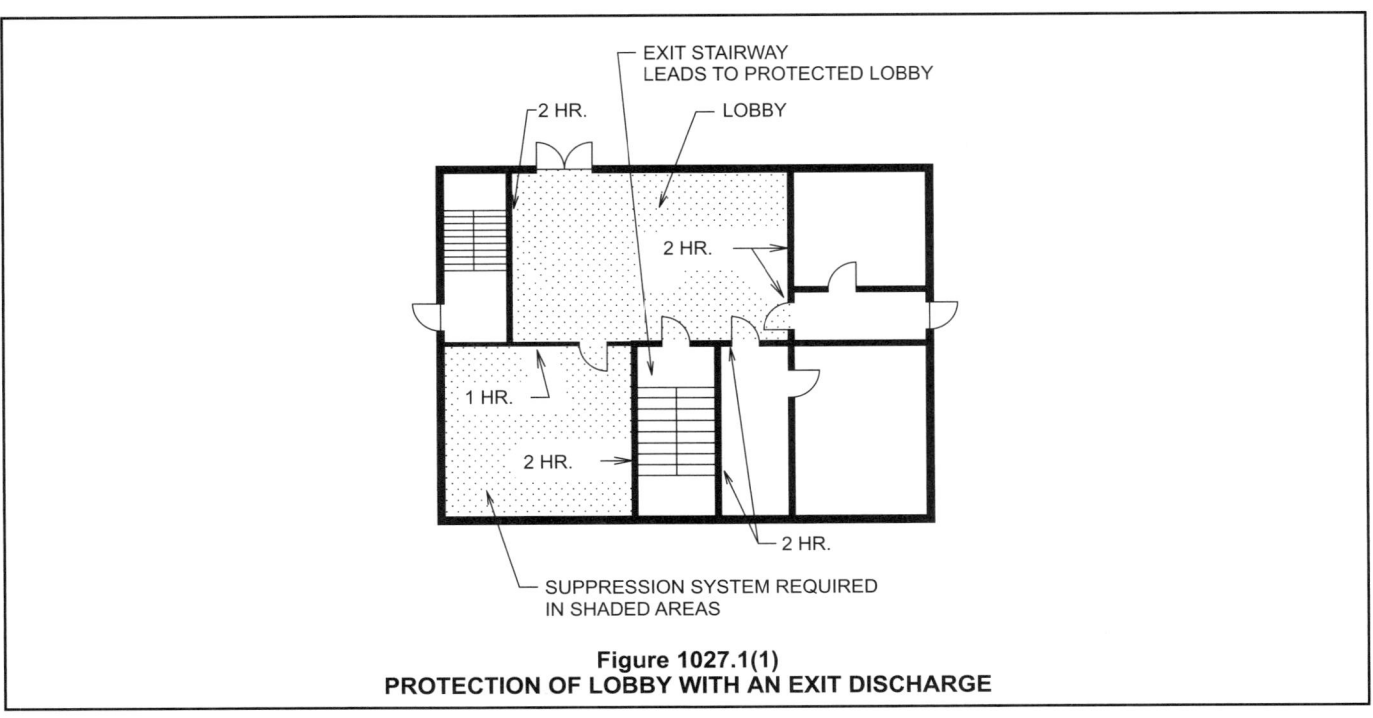

Figure 1027.1(1)
PROTECTION OF LOBBY WITH AN EXIT DISCHARGE

allowed to be unenclosed is because of the low hazard and free venting aspects. When an occupant leaves the stair in an open parking garage, typically he or she can move in several directions to reach the perimeter of the building. This, coupled with the open nature of the parking garage structure, allows occupants a safe path of travel from the bottom of the stairway and across the parking level to the exterior and public way without the additional protection of an exit passageway, exit lobby or exit vestibule. This path must meet the general means of egress requirements and may not be down a driveway unless a pedestrian path is also provided (see Section 1010.1).

A horizontal exit creates refuge areas, so this unique type of exit is not required to discharge directly to the outside. Many hospitals and correctional facilities use horizontal exits to "defend in place" rather than require an immediate building evacuation. There are exit stairways or exits available from the refuge areas, so occupants can move to the outside if they need to (see Section 1025 for additional information).

1027.2 Exit discharge capacity. The capacity of the *exit discharge* shall be not less than the required discharge capacity of the *exits* being served. "area of refuge"

❖ This section specifies the exit discharge capacity. The exit discharge is required to be designed for the required capacity of all the exits it serves. If the exit discharge serves two exits, it is to be designed for the sum of the occupants served by both exits. Note that the capacity of the exit discharge is not required to be the capacity of both exits, which is higher than the sum of the occupants served by both exits.

1027.3 Exit discharge location. Exterior balconies, *stairways* and *ramps* shall be located at least 10 feet (3048 mm) from adjacent *lot lines* and from other buildings on the same lot unless the adjacent building *exterior walls* and openings are protected in accordance with Section 705 based on *fire separation distance.*

❖ The location requirements of this section protect the users of the exterior exit ramp or stairway from the effects of a fire in another building on the same lot or an adjacent lot. The separation distance reduces the exposure to heat and smoke. If the stairway is closer than specified, then adjacent buildings' exterior walls and openings are to be protected in accordance with Section 705, so that the users of the exterior exit are protected. The reason for the required distance to a lot line is to provide for a future building that could be built on the adjacent lot. While buildings on the same lot can be considered one building for height and area limitations (see Section 503.1.2), they must be separated by a minimum of 10 feet (3048 mm) if there is a path for exit discharge between them.

While this section is referenced from exterior exit stairways and ramps (see Section 1026.5), it is not directly referenced from egress balconies (see Section 1019). For an illustration of how exterior egress balconies and exterior exit stairways work together, see Figure 1027.3.

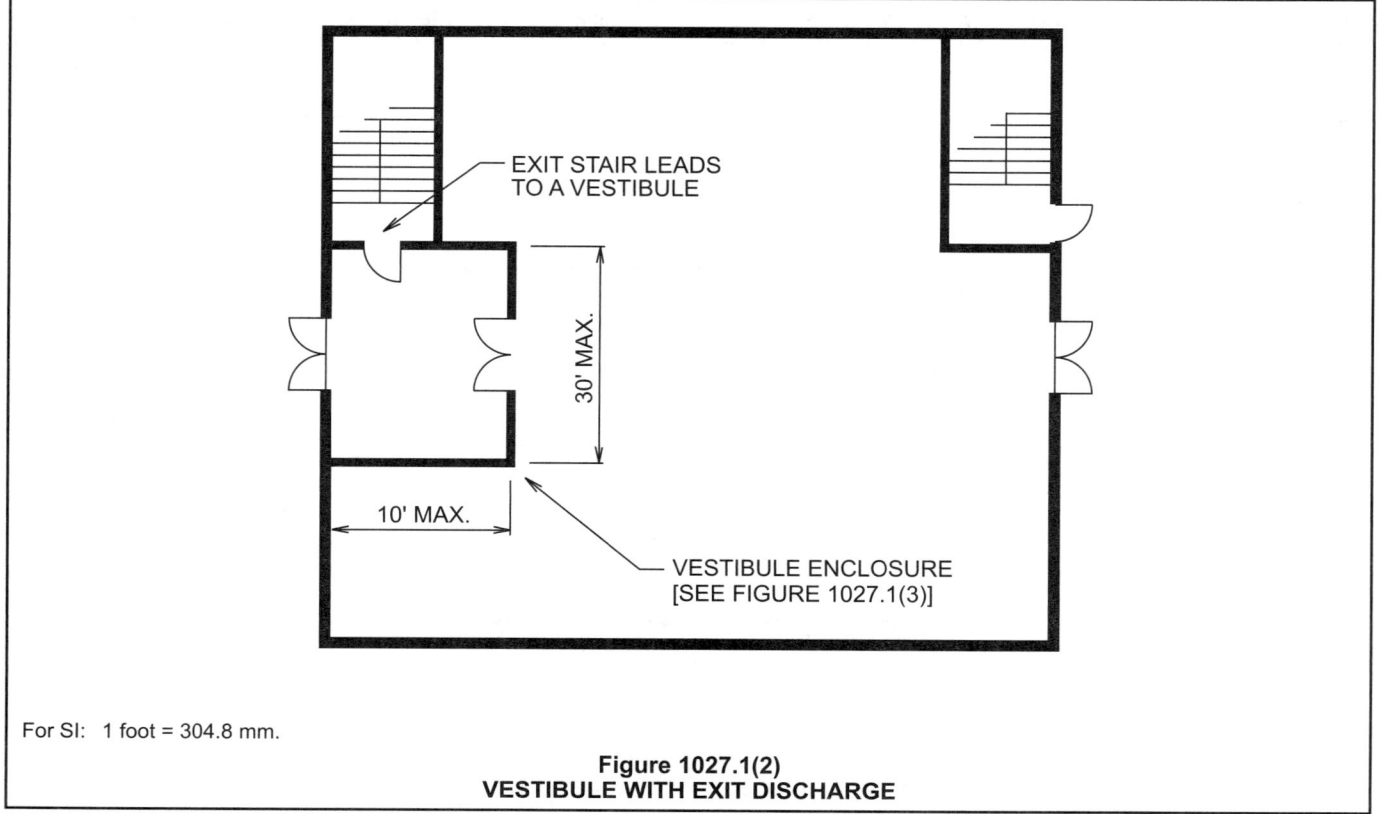

For SI: 1 foot = 304.8 mm.

Figure 1027.1(2)
VESTIBULE WITH EXIT DISCHARGE

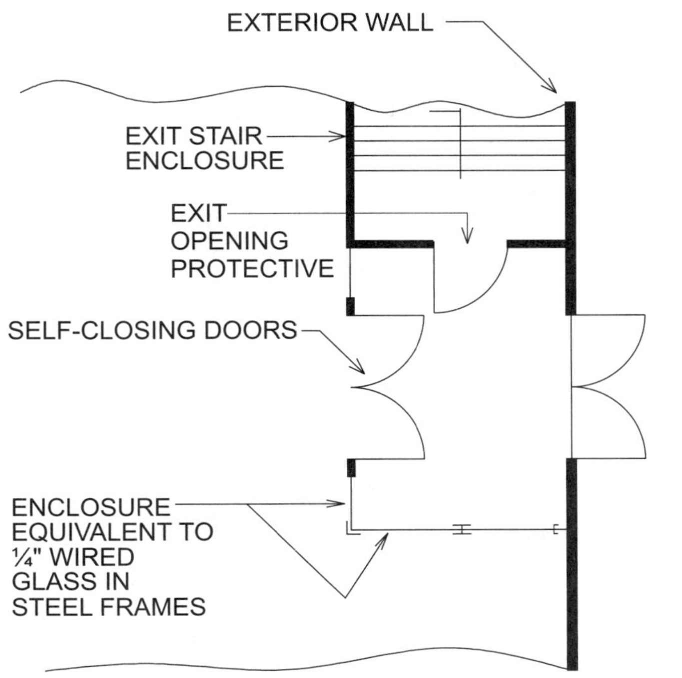

For SI: 1 inch = 25.4 mm.

Figure 1027.1(3)
SEPARATION DETAILS FOR VESTIBULE WITH EXIT DISCHARGE

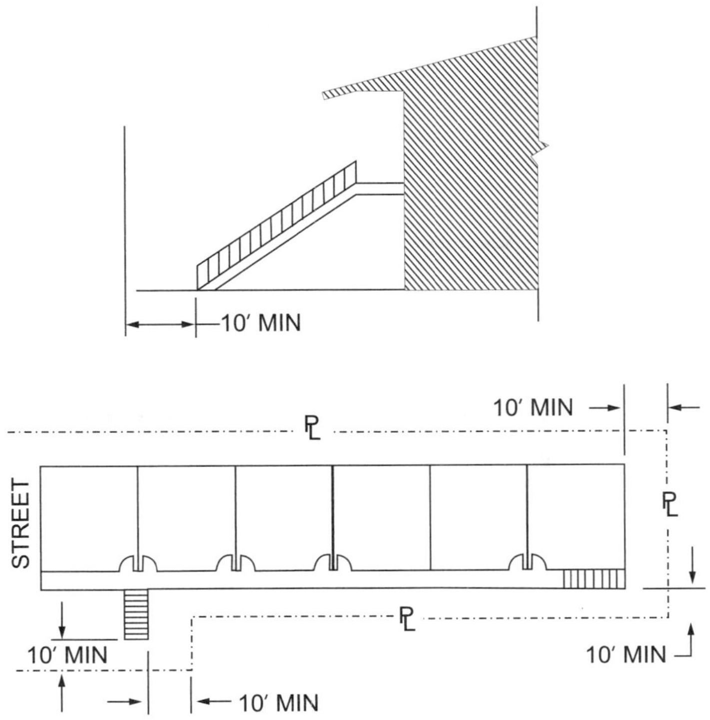

For SI: 1 foot = 304.8 mm.

Figure 1027.3
EXTERIOR BALCONY AND STAIRWAY ADJACENT TO LOT LINE

1027.4 Exit discharge components. *Exit discharge* components shall be sufficiently open to the exterior so as to minimize the accumulation of smoke and toxic gases.

❖ This section includes exit discharge components that currently only include egress courts. It should be noted that the general requirements in Sections 1003 through 1013 will still apply when those components are used in the exit discharge.

1027.5 Egress courts. *Egress courts* serving as a portion of the *exit discharge* in the *means of egress* system shall comply with the requirements of Section 1027.

❖ This section and the following subsections address the detailed requirements for egress courts. It is essential that exterior egress courts that serve occupants from an exit to a public way be sufficiently open to prevent the accumulation of smoke and toxic gases in the event of a fire.

See Figure 1027.5 for an illustration of an exit discharge, including an egress court. See Section 1206 for additional minimum width and openness requirements.

1027.5.1 Width. The width of *egress courts* shall be determined as specified in Section 1005.1, but such width shall not be less than 44 inches (1118 mm), except as specified herein. *Egress courts* serving Group R-3 and U occupancies shall not be less than 36 inches (914 mm) in width. The required width

of *egress courts* shall be unobstructed to a height of 7 feet (2134 mm).

Exception: Doors complying with Section 1005.2.

Where an *egress court* exceeds the minimum required width and the width of such *egress court* is then reduced along the path of *exit* travel, the reduction in width shall be gradual. The transition in width shall be affected by a guard not less than 36 inches (914 mm) in height and shall not create an angle of more than 30 degrees (0.52 rad) with respect to the axis of the *egress court* along the path of egress travel. In no case shall the width of the *egress court* be less than the required minimum.

❖ The width of an exterior court is to be determined in the same fashion as for an interior corridor. The width is not to be less than required to serve the number of occupants from the exit or exits and not less than the minimum specified in this section (see also Section 1206). A cross reference back to Section 1005.2 from the exceptions for width in corridors (see Section 1018.3), aisles (see Section 1017.1), exit passageways (see Section 1023.2) and exit courts (see Section 1027.5.1) reinforces the fact that the protrusion limits provision is generally applicable for these types of confined routes.

Many egress courts are significantly larger than required. Thus, the code allows such an egress court to decrease in width along the path of travel to the public way. The gradual transition requirement is so the flow

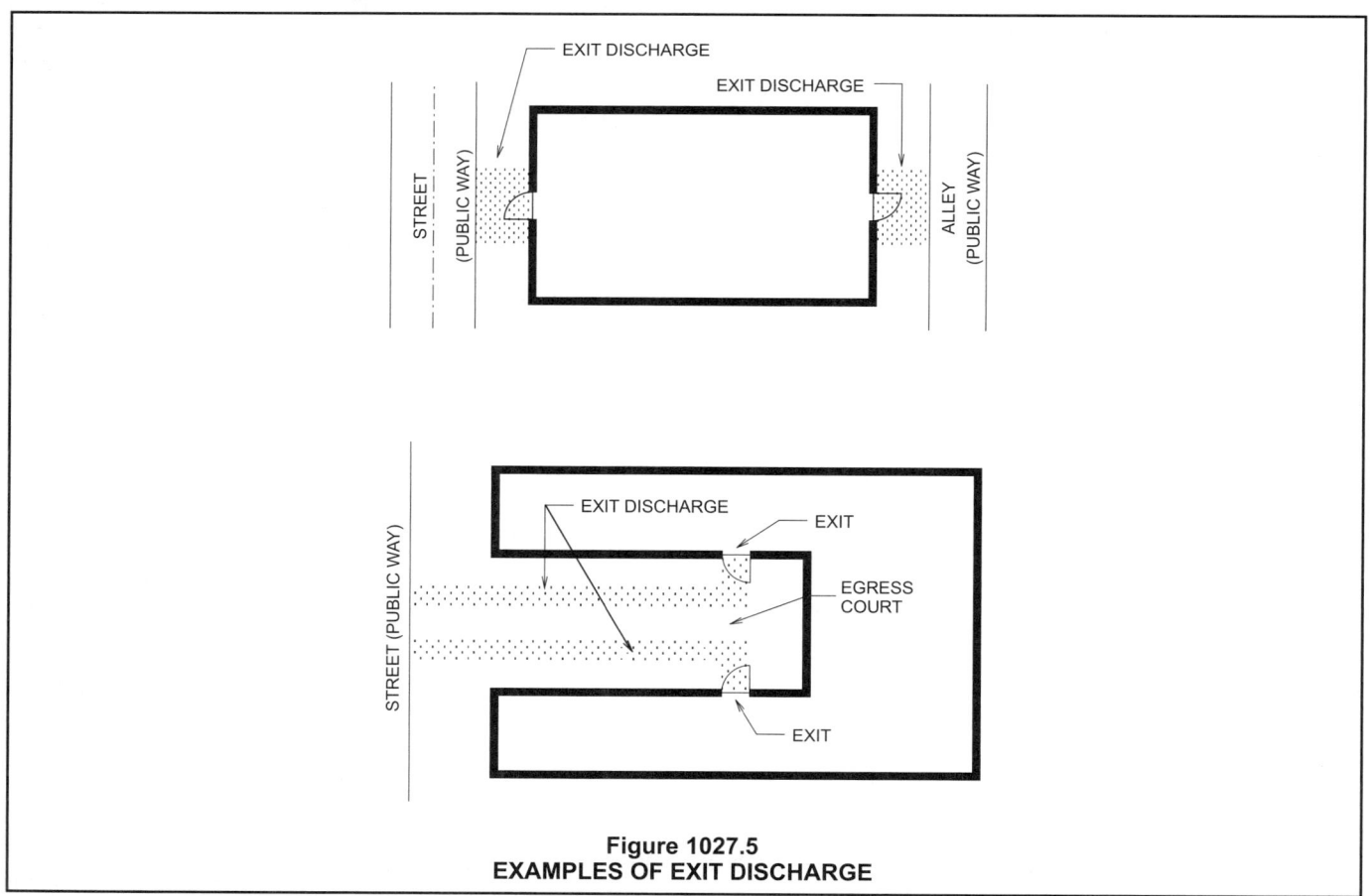

Figure 1027.5
EXAMPLES OF EXIT DISCHARGE

of the occupants will be uniform without pockets of congestion. The transition requirements should be applied to egress courts where a reduction results in a width that is near the minimum based on the number of occupants served. It is this condition where the uniform flow of occupants is essential.

1027.5.2 Construction and openings. Where an *egress court* serving a building or portion thereof is less than 10 feet (3048 mm) in width, the *egress court* walls shall have not less than 1-hour *fire-resistance-rated* construction for a distance of 10 feet (3048 mm) above the floor of the *court*. Openings within such walls shall be protected by opening protectives having a *fire protection rating* of not less than $^3/_4$ hour.

Exceptions:

1. *Egress courts* serving an *occupant load* of less than 10.

2. *Egress courts* serving Group R-3.

❖ The purpose of this section is to protect the occupants served by the egress court from the building that they are exiting from. If occupants must walk closely by the exterior walls of the exit court, the walls are required to have the specified fire-resistance rating and the openings are required to be protected as specified. This requirement is only for the first 10 feet (3048 mm) above the level of the egress court since the exposure hazard from walls and openings above this level is reduced.

The two exceptions provide for egress courts that serve a very low number of occupants and the specified residential occupancy where the protection requirement would be located.

1027.6 Access to a public way. The *exit discharge* shall provide a direct and unobstructed access to a *public way.*

Exception: Where access to a *public way* cannot be provided, a safe dispersal area shall be provided where all of the following are met:

1. The area shall be of a size to accommodate at least 5 square feet (0.46 m²) for each person.

2. The area shall be located on the same lot at least 50 feet (15 240 mm) away from the building requiring egress.

3. The area shall be permanently maintained and identified as a safe dispersal area.

4. The area shall be provided with a safe and unobstructed path of travel from the building.

❖ There are instances where the path of travel to the public way is not safe or not achievable due to site constraints or security concerns. The provisions in this section specify what would constitute a safe area to allow occupants of a building to assemble in an emergency. The 5 square feet (0.28 m²) would allow adequate space for standing persons as well as some space for persons in wheelchairs or on stretchers. Everyone who is expected to wait in this area for fire department assistance must be a minimum of 50 feet (15 240 mm) away from the building. This refuge must always remain open and not be used for parking, storage or temporary structures.

SECTION 1028
ASSEMBLY

1028.1 General. Occupancies in Group A and assembly occupancies accessory to Group E which contain seats, tables, displays, equipment or other material shall comply with this section.

❖ Although most of the provisions in Section 1028 focus on fixed seating auditoriums or theaters, this section also addresses loose seats, tables, displays, equipment, etc. For example, Section 1017.4 provides measurement criteria adjacent to seating at tables, but relies on Section 1028.9 to determine the required aisle width. Assembly spaces that contain elements that would affect the path of travel for the means of egress must comply with this section. Assembly spaces require special consideration due to the large occupant loads and possible low lighting (e.g., nightclubs, theaters). The group description of educational facilities allows for some accessory assembly spaces to be considered Group E for height and area limitation, type of construction, sprinkler requirements, etc. Not all assembly spaces associated with Group E are accessory. For additional explanations, see the commentary to Section 305.1. However, for evaluation of the occupant load and the means of egress in these spaces, they should be regulated based on their function, rather than their occupancy group.

1028.1.1 Bleachers. *Bleachers*, *grandstands* and *folding and telescopic seating*, that are not building elements, shall comply with ICC 300.

❖ On February 24, 1999, the Bleacher Safety Act of 1999 was introduced in the House of Representatives. The bill, which cites the ICC and the code, authorizes the Consumer Product Safety Commission (CPSC) to issue a standard for bleacher safety. This was in response to concerns relative to accidents on bleacher-type structures. As a result, the CPSC developed and revised the *Guidelines for Retrofitting Bleachers*. The ICC Board of Directors decided that a comprehensive standard dealing with all aspects of both new and existing bleachers was warranted and authorized the formation of the ICC Consensus Committee on Bleacher Safety. The committee is comprised of 12 members, including the requisite balance of general, user interest and producer interest.

ICC 300 was completed in December 2001, and submitted to ANSI on January 1, 2002. ICC 300 was reissued with some revisions in 2007. While the term "bleachers" is generic, the standard addresses all aspects of tiered seating associated with bleachers, grandstands and folding and telescopic seating. These types of seating are supported on dedicated structural systems, which in turn may sit on the ground or on a building floor system. Single seats or bench seats bolted down to a stepped floor are not considered a bleacher or grandstand and should comply with Section 1028. "Building element" is defined in Section 702.1, while "Bleachers," "Grandstands" and "Folding and telescopic seating" are defined in Section 1002.1. While ICC 300 is consistent and also relies on Chapter

10 of the code for some provisions, the standard addresses items specific to these types of seating arrangement. The bleacher standard references Chapter 11 of the code and ICC A117.1 for accessibility requirements.

1028.2 Assembly main exit. Group A occupancies and assembly occupancies accessory to Group E occupancies that have an *occupant load* of greater than 300 shall be provided with a main *exit*. The main *exit* shall be of sufficient width to accommodate not less than one-half of the *occupant load*, but such width shall not be less than the total required width of all *means of egress* leading to the *exit*. Where the building is classified as a Group A occupancy, the main *exit* shall front on at least one street or an unoccupied space of not less than 10 feet (3048 mm) in width that adjoins a street or *public way*.

> **Exception:** In assembly occupancies where there is no well-defined main *exit* or where multiple main *exits* are provided, *exits* shall be permitted to be distributed around the perimeter of the building provided that the total width of egress is not less than 100 percent of the required width.

❖ Assembly buildings as well as other buildings that include spaces that function as assembly spaces (i.e., the band classroom in a school, the training room in an office, the cafeteria in a large factory) present an unusual life safety problem that includes frequent high occupant densities and therefore large occupant loads and the opportunity for irrational mass response to a perceived emergency, i.e., panic. For this reason, the code requires a specific arrangement of the exits.

Studies have indicated that in any emergency, occupants will tend to egress via the same path of travel used to enter the room and building. Therefore, a main entrance to the building must also be designed as the main exit to accommodate this behavior, even if the required exit capacity might be more easily accommodated elsewhere. The main entrance (and exit) must be sized to accommodate at least 50 percent of the total occupant load of the structure and must front on a large, open space, such as a street, for rapid dispersal of the occupants outside the building. The remaining exits must also accommodate at least 50 percent of the total occupant load from each level (see Figure 1028.2). The total occupant load includes those within the theater seating area, the foyer and any other space (e.g., ticket booth, concession stand, offices, storage and the like). When the assembly space is within a mixed use building, the intent is that the main exits from the space comply with these provisions for one-half the capacity, but not necessarily that they lead directly to the outside. Egress requirements from the building would depend on how it was anticipated for the assembly space occupants to disperse. For example, an office building may have a large training/conference room where the path of exit access travel from the room goes out a main exit from the space and then disseminates into the general floor egress system. While the room exit access doors may need to meet the 50 percent criteria, the exception may be applicable once the occupants leave the room

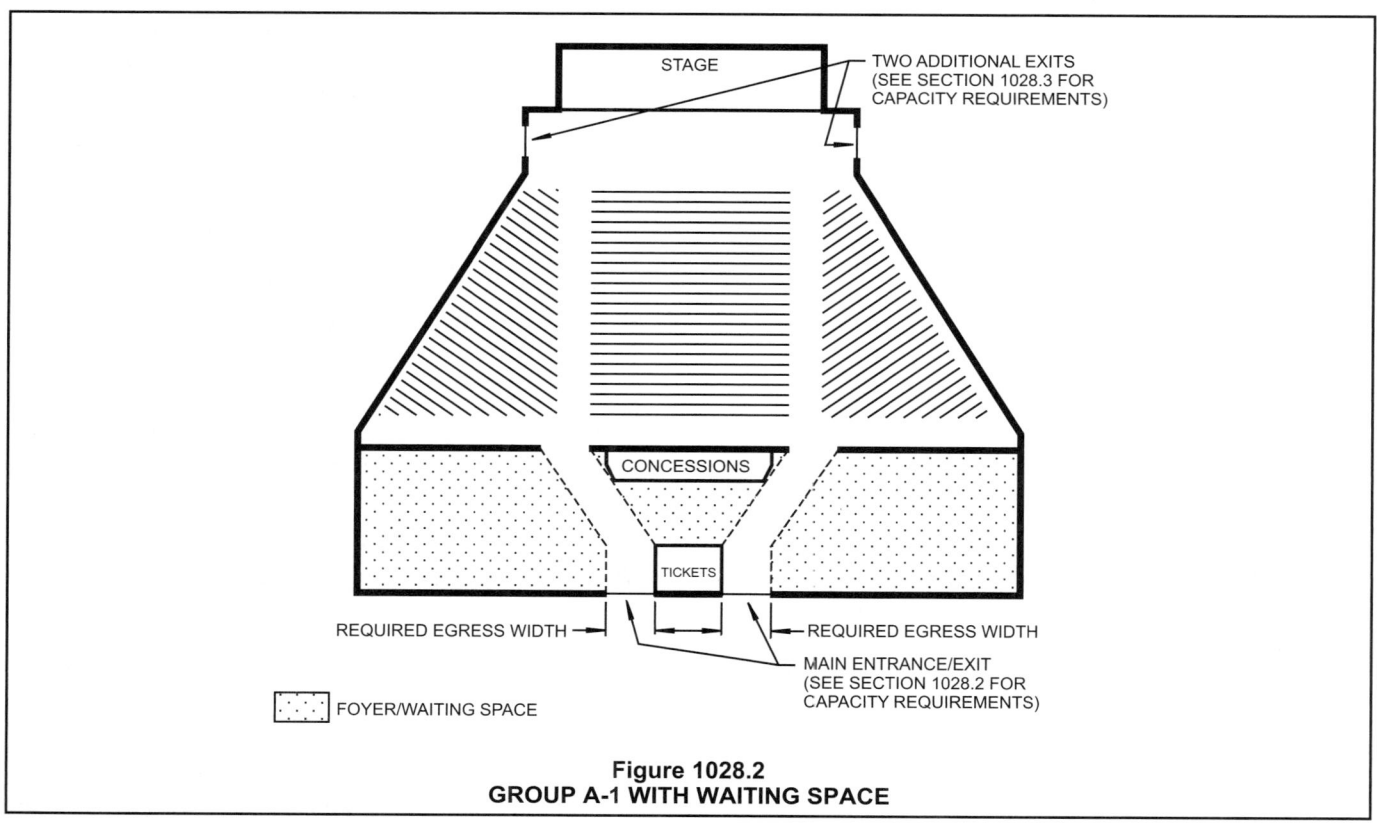

Figure 1028.2
GROUP A-1 WITH WAITING SPACE

and enter the general floor egress system.

The required width of the means of egress in places of assembly is more often determined by the occupant load than in most other occupancies. In other occupancies, the minimum required widths and the travel distances will often determine the required widths of the exits.

This section only requires the main exit to accommodate 50 percent of the occupant load when there is a single main entrance. Therefore, a large stadium or civic center, in which there are numerous entrances (and exits), need not comply with the main entrance criteria. This condition is addressed in the exception.

1028.3 Assembly other exits. In addition to having access to a main *exit*, each level in Group A occupancies or assembly occupancies accessory to Group E occupancies having an *occupant load* greater than 300, shall be provided with additional *means of egress* that shall provide an egress capacity for at least one-half of the total *occupant load* served by that level and comply with Section 1015.2.

> **Exception:** In assembly occupancies where there is no well-defined main *exit* or where multiple main *exits* are provided, *exits* shall be permitted to be distributed around the perimeter of the building, provided that the total width of egress is not less than 100 percent of the required width.

❖ This section provides for the egress of one-half of the total occupant load by way of exits other than the main exit that is described in Section 1028.2. The exception addresses assembly spaces, such as a school gymnasium or a large stadium or civic center in which there are numerous entrances (and exits), none of which are a main entrance or exit.

1028.4 Foyers and lobbies. In Group A-1 occupancies, where persons are admitted to the building at times when seats are not available, such persons shall be allowed to wait in a lobby or similar space, provided such lobby or similar space shall not encroach upon the required clear width of the *means of egress*. Such foyer, if not directly connected to a public street by all the main entrances or *exits*, shall have a straight and unobstructed *corridor* or path of travel to every such main entrance or *exit*.

❖ In theaters, people may be arriving for the next show while another group has yet to exit. This is extremely common in multiplex theater complexes. In every case, the main entrance (exit) and all other exits are to be constantly available for the entire building occupant load.

For example, because of the queuing of large crowds, particularly in theaters where a performance may be in progress and people must wait to attend the next one, standing space is often provided. For reasons of safety, such spaces cannot be located in or interfere with established paths of egress from the assembly areas. While a facility may choose to separate the route for means of egress using partitions or railings from the general lobby space to allow for easy traffic flow through the lobby to the street, it is not required to designate these areas (see Figure 1028.2).

1028.5 Interior balcony and gallery means of egress. For balconies, galleries or press boxes having a seating capacity of 50 or more located in Group A occupancies, at least two *means of egress* shall be provided, with one from each side of every balcony, gallery or press box and at least one leading directly to an *exit*.

❖ This section states the threshold where two means of egress are required based on the occupant load of the interior balcony, gallery or press box. Note that one of the means of egress must lead directly to an exit. However, Section 1028.5.1 does not require the stairways serving the balcony to be enclosed in an exit enclosure. These requirements will ensure that at least one path of travel is always available and occupants face a minimum number of hazards.

For balconies with 50 or fewer occupants, see Section 1028.8.

1028.5.1 Enclosure of openings. *Interior stairways* and other vertical openings shall be enclosed in an *exit enclosure* as provided in Section 1022.1, except that *stairways* are permitted to be open between the balcony, gallery or press box and the main assembly floor in occupancies such as theaters, places of *religious worship*, auditoriums and sports facilities. At least one *accessible means of egress* is required from a balcony, gallery or press box level containing accessible seating locations in accordance with Section 1007.3 or 1007.4.

❖ This section allows the stairways that lead from interior balconies, galleries and press boxes to be unenclosed from those spaces to the main floor where the interior balconies are within large assembly spaces, such as theaters, places of religious worship, auditoriums and sports arenas. Thus, vertical exit enclosures are not required for stairways or ramps leading to these facilities. When these areas contain accessible wheelchair spaces (see Sections 1104.3.2 and 1108.2), at least one means of egress must be accessible. While the section references only indicate exit stairways or elevators (see Sections 1007.3 and 1007.4), there are special allowances for the use of platform lifts (see Sections 1007.5 and 1109.7, Item 2) in assembly spaces to allow for dispersion of wheelchair spaces to a variety of locations. This is especially important in assembly spaces with sloped or tiered seating. Section 1007.5 states that if a platform lift is permitted as part of an accessible route, it should also be permitted as part of the accessible means of egress if it is provided with standby power.

1028.6 Width of means of egress for assembly. The clear width of *aisles* and other *means of egress* shall comply with Section 1028.6.1 where *smoke-protected seating* is not provided and with Section 1028.6.2 or 1028.6.3 where *smoke-protected seating* is provided. The clear width shall be measured to walls, edges of seating and tread edges except for permitted projections.

❖ The means of egress width for assembly occupancy is to be in accordance with this section and the referenced sections instead of the criteria specified in Section 1005.1. The width factors in Section 1028.6 and its subsections apply to those stairs, aisle steps, corridors, passageways and ramped surfaces that serve the assembly seating areas.

Different means of egress width criteria are also specified for assembly seating where smoke protection is provided versus areas it is not provided. The egress width for smoke-protected seating is allowed to be less than for areas where smoke protection is not provided, since the smoke level is required to be maintained at least 6 feet (1829 mm) above the floor of the means of egress, according to Section 1028.6.2.1.

1028.6.1 Without smoke protection. The clear width of the *means of egress* shall provide sufficient capacity in accordance with all of the following, as applicable:

1. At least 0.3 inch (7.6 mm) of width for each occupant served shall be provided on *stairs* having riser heights 7 inches (178 mm) or less and tread depths 11 inches (279 mm) or greater, measured horizontally between tread *nosings*.

2. At least 0.005 inch (0.127 mm) of additional *stair* width for each occupant shall be provided for each 0.10 inch (2.5 mm) of riser height above 7 inches (178 mm).

3. Where egress requires *stair* descent, at least 0.075 inch (1.9 mm) of additional width for each occupant shall be provided on those portions of *stair* width having no *handrail* within a horizontal distance of 30 inches (762 mm).

4. Ramped *means of egress*, where slopes are steeper than one unit vertical in 12 units horizontal (8-percent slope), shall have at least 0.22 inch (5.6 mm) of clear width for each occupant served. Level or ramped *means of egress*, where slopes are not steeper than one unit vertical in 12 units horizontal (8-percent slope), shall have at least 0.20 inch (5.1 mm) of clear width for each occupant served.

❖ This section prescribes the criteria needed to calculate the clear widths of aisles and aisle accessways in order to provide sufficient capacity to handle the occupant loads established by the "catchment areas" described in Section 1028.9.2. Clear width is to be measured to walls, edges of seating and tread edges.

The criteria for determining the required widths are based on analytical studies and field tests that used people to model egress situations [see Figures 1028.6.1(1) and 1028.6.1(2)].

Criterion 1 addresses the method for determining the required egress width for aisles and aisle accessways that are stepped. This method corresponds with the requirements of Table 1005.1 for egress width per occupant of stairways in an unsprinklered building.

Criterion 2 addresses the method for determining the additional stair width required for aisle and aisle accessway stairs with risers greater than 7 inches (178 mm).

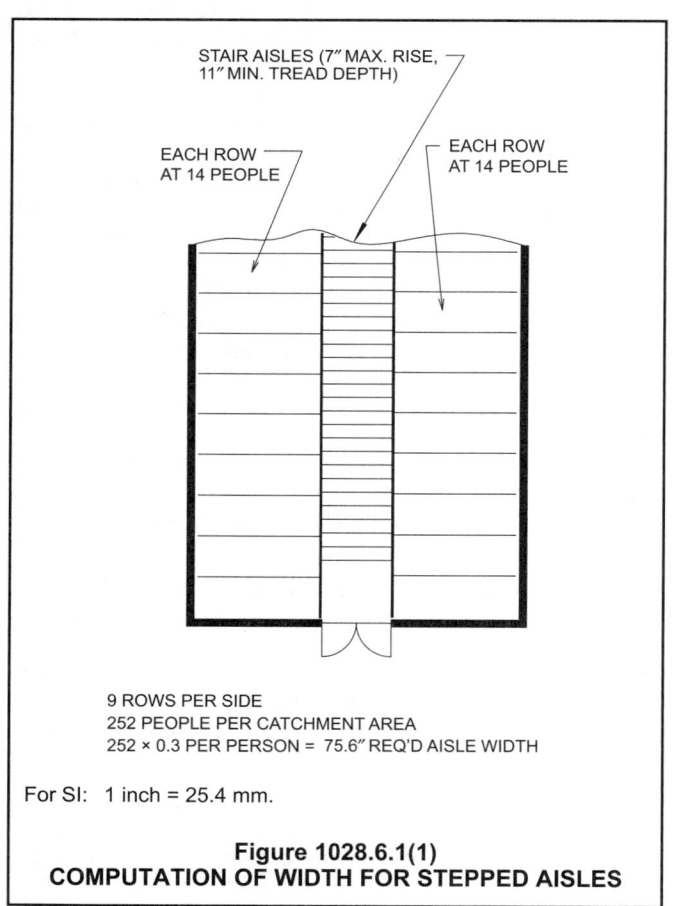

9 ROWS PER SIDE
252 PEOPLE PER CATCHMENT AREA
252 × 0.3 PER PERSON = 75.6″ REQ'D AISLE WIDTH

For SI: 1 inch = 25.4 mm.

Figure 1028.6.1(1)
COMPUTATION OF WIDTH FOR STEPPED AISLES

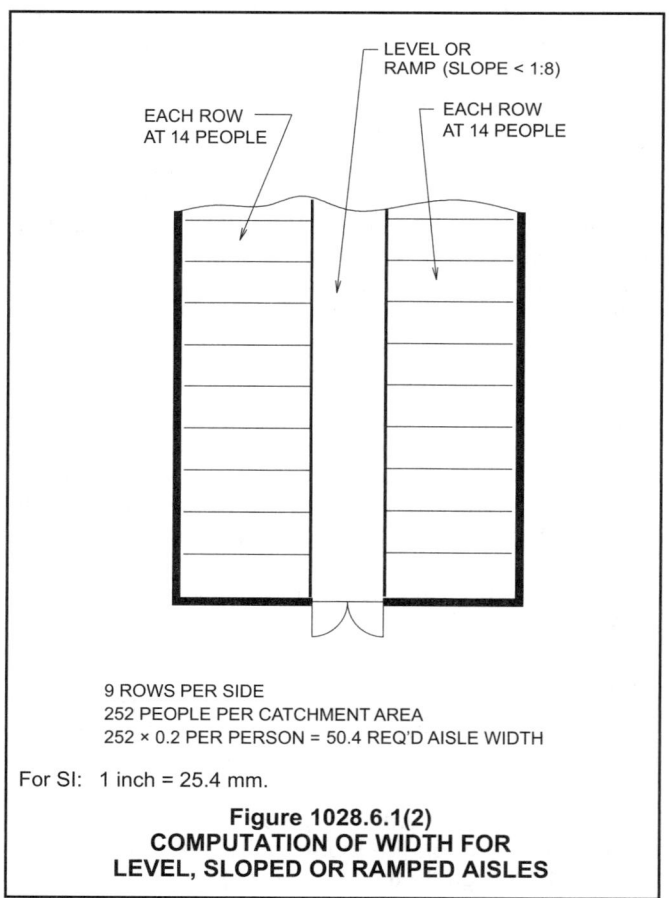

9 ROWS PER SIDE
252 PEOPLE PER CATCHMENT AREA
252 × 0.2 PER PERSON = 50.4 REQ'D AISLE WIDTH

For SI: 1 inch = 25.4 mm.

Figure 1028.6.1(2)
COMPUTATION OF WIDTH FOR
LEVEL, SLOPED OR RAMPED AISLES

Criterion 3 addresses the method for determining the additional stair width where a handrail is not located within 30 inches (762 mm).

Criterion 4 addresses the method for determining the required widths for level or ramped means of egress.

1028.6.2 Smoke-protected seating. The clear width of the *means of egress* for *smoke-protected assembly seating* shall not be less than the *occupant load* served by the egress element multiplied by the appropriate factor in Table 1028.6.2. The total number of seats specified shall be those within the space exposed to the same smoke-protected environment. Interpolation is permitted between the specific values shown. A life safety evaluation, complying with NFPA 101, shall be done for a facility utilizing the reduced width requirements of Table 1028.6.2 for *smoke-protected assembly seating*.

Exception: For an outdoor smoke-protected assembly with an *occupant load* not greater than 18,000, the clear width shall be determined using the factors in Section 1028.6.3.

❖ Special consideration is given to facilities with features that will prevent the means of egress from being blocked by smoke. Facilities to be considered smoke protected by Sections 1028.6.2.1 through 1028.6.2.3 are permitted increases in travel distance, egress capacity, longer dead-end aisles and increased row lengths. All of these result in an increase of allowable egress time. Typically, model codes based on research by Dr. John Fruin and others recognize the need for occupants exposed to the fire environment to evacuate to a safe area within 90 seconds of notification and to reach an area of refuge within 5 minutes. With the increases permitted for smoke-protected facilities, these times are effectively doubled since the time available for safe egress also increases.

The exception is a pointer to the specific criteria for outdoor seating areas. For outdoor stadiums with 18,000 seats or greater, use Table 1028.6.2.

TABLE 1028.6.2. See below.

❖ This section requires the egress component to be of adequate size to accommodate the occupant load.

The egress width per occupant for nonsmoke-protected seating is to be based on Section 1028.6.1 and is similar to the provisions in Table 1005.1. For smoke-protected seating, the egress width per occupant is based on Table 1028.6.2.

1028.6.2.1 Smoke control. *Means of egress* serving a *smoke-protected assembly seating* area shall be provided with a smoke control system complying with Section 909 or natural ventilation designed to maintain the smoke level at least 6 feet (1829 mm) above the floor of the *means of egress*.

❖ The means of egress and the assembly seating area are required to have some type of smoke control system that will prevent smoke buildup from encroaching on the egress path. This may be a mechanical smoke control system, designed in accordance with Section 909, or a natural ventilation system.

In either type of system, the major consideration is that a smoke-free environment be maintained at least 6 feet (1829 mm) above the floor of the means of egress for a period of at least 20 minutes.

1028.6.2.2 Roof height. A *smoke-protected assembly seating* area with a roof shall have the lowest portion of the roof deck not less than 15 feet (4572 mm) above the highest *aisle* or *aisle accessway*.

Exception: A roof canopy in an outdoor stadium shall be permitted to be less than 15 feet (4572 mm) above the highest *aisle* or *aisle accessway* provided that there are no objects less than 80 inches (2032 mm) above the highest *aisle* or *aisle accessway*.

❖ One element of a smoke-protected assembly seating facility is that the lowest portion of the roof is required to be at least 15 feet (4572 mm) above the highest aisle or aisle accessway. The objective of this provision is to have a minimum 6-foot (1829 mm) smoke-free height to accommodate safe egress through the area. The additional 9 feet (2743 mm) of height is to provide a volume of space that will act to dissipate smoke. The measurement of the height is shown in Figures 1028.6.2.2(1) and 1028.6.2.2(2).

TABLE 1028.6.2
WIDTH OF AISLES FOR SMOKE-PROTECTED ASSEMBLY

TOTAL NUMBER OF SEATS IN THE SMOKE-PROTECTED ASSEMBLY OCCUPANCY	INCHES OF CLEAR WIDTH PER SEAT SERVED			
	Stairs and aisle steps with handrails within 30 inches	Stairs and aisle steps without handrails within 30 inches	Passageways, doorways and ramps not steeper than 1 in 10 in slope	Ramps steeper than 1 in 10 in slope
Equal to or less than 5,000	0.200	0.250	0.150	0.165
10,000	0.130	0.163	0.100	0.110
15,000	0.096	0.120	0.070	0.077
20,000	0.076	0.095	0.056	0.062
Equal to or greater than 25,000	0.060	0.075	0.044	0.048

For SI: 1 inch = 25.4 mm.

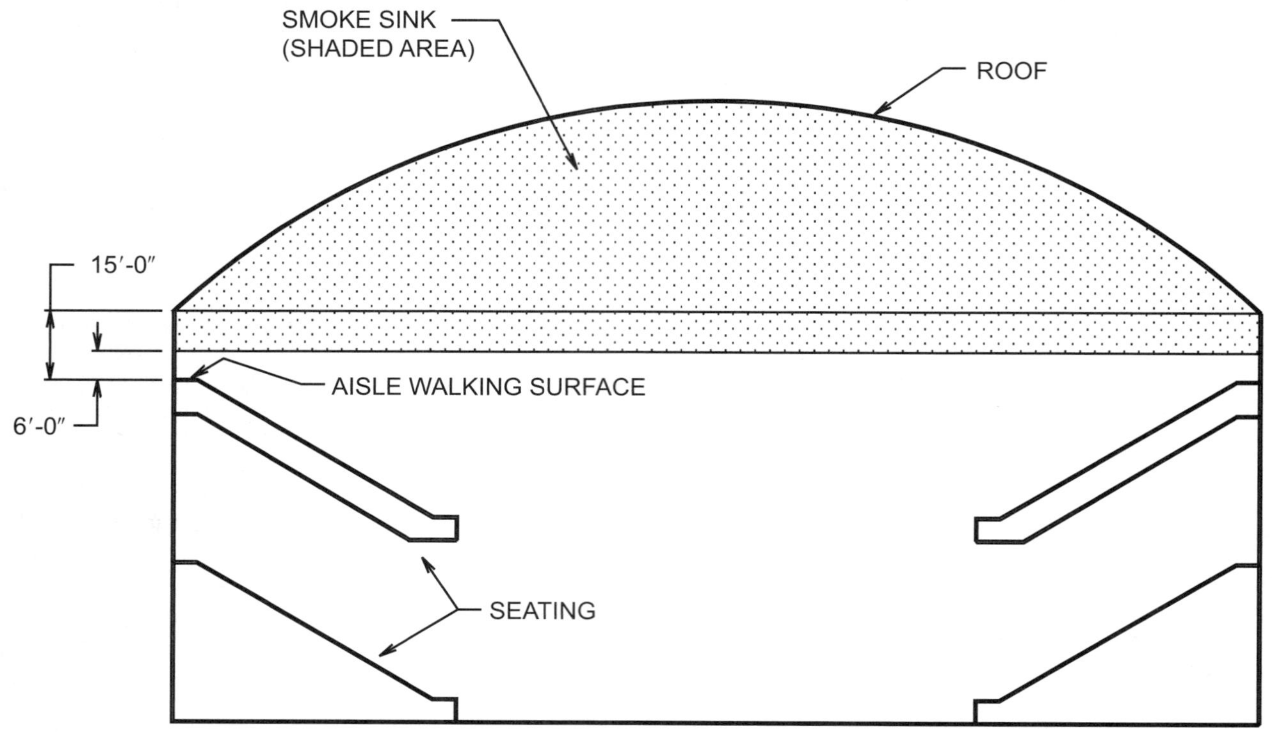

For SI: 1 inch = 25.4 mm, 1 foot = 304.8 mm.

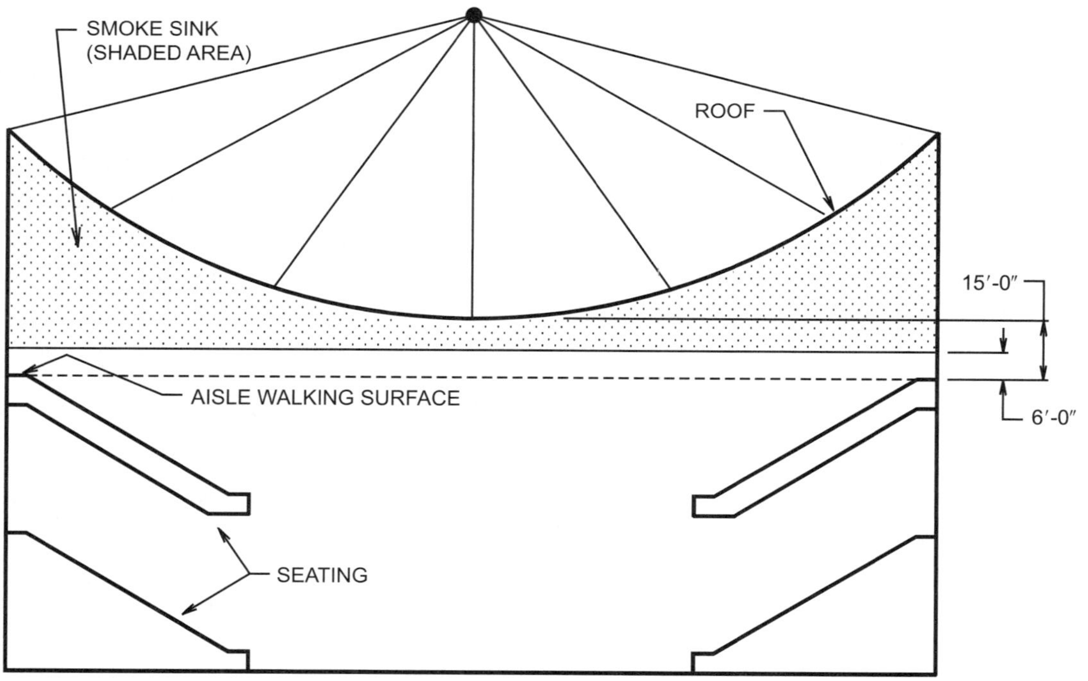

For SI: 1 inch = 25.4 mm, 1 foot = 304.8 mm.

Figure 1028.6.2.2(2)
ROOF HEIGHT (CONCAVE SUSPENDED ROOF)

　　　　　　　　　　　　　　　　　2009 INTERNATIONAL BUILDING CODE® COMMENTARY

1028.6.2.3 Automatic sprinklers. Enclosed areas with walls and ceilings in buildings or structures containing *smoke-protected assembly seating* shall be protected with an *approved automatic sprinkler system* in accordance with Section 903.3.1.1.

Exceptions:

1. The floor area used for contests, performances or entertainment provided the roof construction is more than 50 feet (15 240 mm) above the floor level and the use is restricted to low fire hazard uses.

2. Press boxes and storage facilities less than 1,000 square feet (93 m²) in area.

3. Outdoor seating facilities where seating and the *means of egress* in the seating area are essentially open to the outside.

❖ If there are areas in the smoke-protected assembly seating structure enclosed by walls and ceilings, the entire structure is to be provided with an automatic sprinkler designed to meet the requirements of NFPA 13. NFPA 13R systems are not acceptable for this use.

Exception 1 indicates that the area over the playing field or performance area is not required to be sprinklered if the use of the floor area is restricted. If the facility is used for conventions, trade shows, displays or similar purposes, sprinklers would be required throughout, since the occupancy would no longer be a low fire-hazard use. A characteristic of a low fire-haz-

ard occupancy is that the fuel load due to combustibles is approximately 2 pounds per square foot (9.8 kg/m²) or less.

In order for the contest, performance or entertainment area to be unsprinklered, the roof over that area must be at least 50 feet (15 240 mm) above the floor in addition to the floor area meeting the low fire-hazard criteria. The 50-foot (15 240 mm) criterion was selected because the response time for sprinklers at this height is extremely slow. It is estimated that the response time for standard sprinklers [50 feet (15 240 mm) above a floor with a fire having a heat release rate of 5 British thermal units (Btu) per square foot per second] exceeds 15 minutes. Therefore, it is not reasonable to install sprinklers at that height with little expectation of timely activation [see Figure 1028.6.2.3(1)]. Note that if this exception is utilized, the trade-offs for a fully sprinklered building, such as increased height and area limitations or decreased corridor ratings, are no longer permitted.

Exception 2 indicates that automatic sprinklers are not required in small spaces in buildings. Sprinklers are required in press box and storage areas of outdoor facilities when the aggregate area exceeds 1,000 square feet (93 m²). The primary reason for sprinklers in these areas is that both are anticipated to have a relatively large combustible load when compared to the main seating and participant areas. Additionally, in the case of storage areas, there is an increased potential

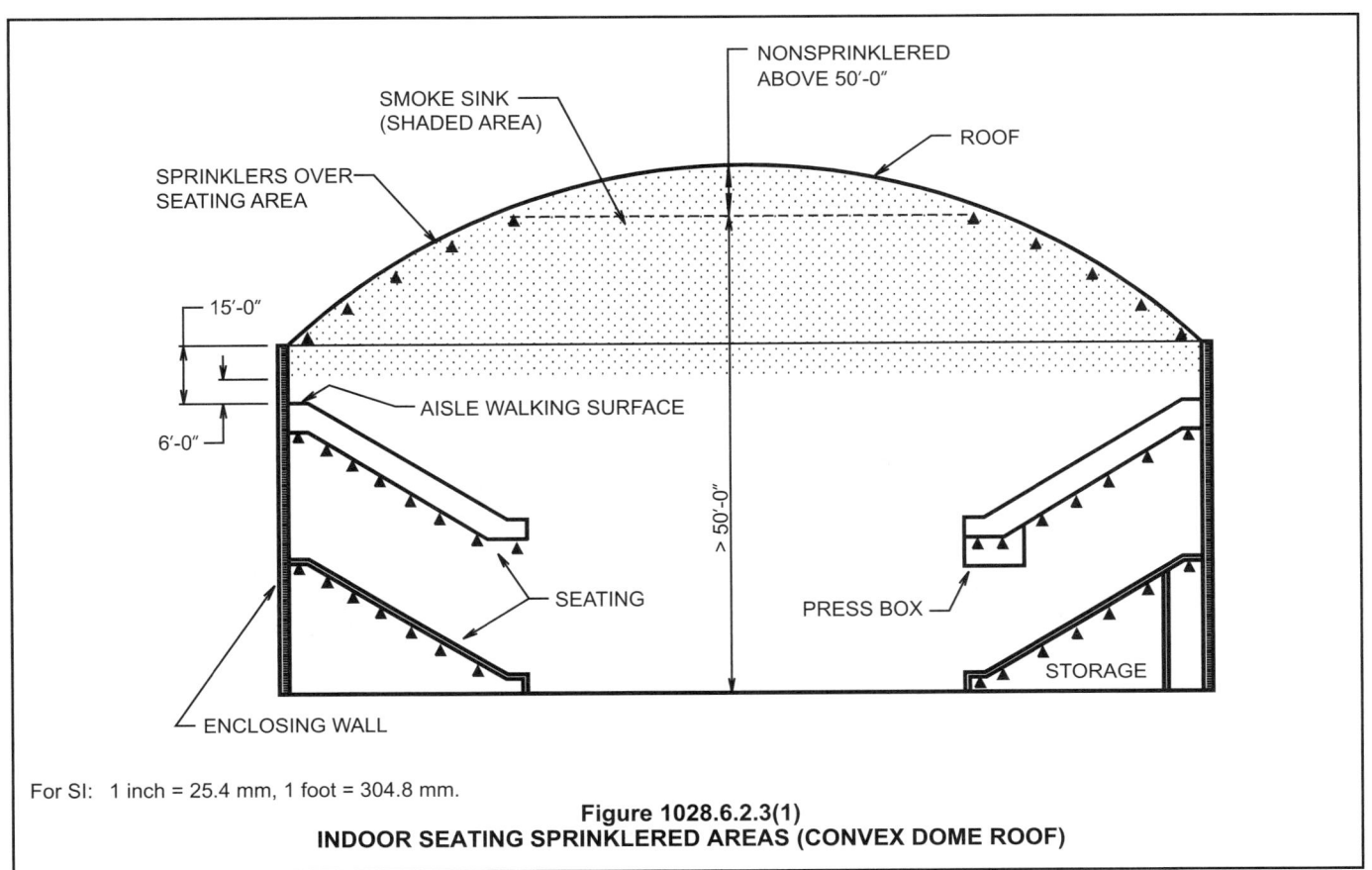

For SI: 1 inch = 25.4 mm, 1 foot = 304.8 mm.

Figure 1028.6.2.3(1)
INDOOR SEATING SPRINKLERED AREAS (CONVEX DOME ROOF)

for an undetected fire condition to occur [see Figure 1028.6.2.3(2)].

Exception 3 provides for outdoor seating facilities where smoke entrapment is not a safety concern.

1028.6.3 Width of means of egress for outdoor smoke-protected assembly. The clear width in inches (mm) of *aisles* and other *means of egress* shall be not less than the total *occupant load* served by the egress element multiplied by 0.08 (2.0 mm) where egress is by *aisles* and *stairs* and multiplied by 0.06 (1.52 mm) where egress is by *ramps*, *corridors*, tunnels or vomitories.

Exception: The clear width in inches (mm) of *aisles* and other *means of egress* shall be permitted to comply with Section 1028.6.2 for the number of seats in the outdoor smoke-protected assembly where Section 1028.6.2 permits less width.

❖ This section has the coefficients for the determination of the width of egress required for outdoor smoke-protected assembly areas. Note that the coefficients are significantly less when compared to the values in Section 1028.6.1 for assembly areas without smoke protection. The coefficients are also less than those for smoke-protected assembly seating in Table 1028.6.2 except for very large assembly areas. The exception in this section would apply where the coefficients in Table 1028.6.2 are less than those in this section.

Low coefficients are a result of the very low hazard of outdoor smoke-protected assembly areas.

Generally, an outdoor assembly area meets the smoke control requirements of Section 1028.6.1 by natural ventilation and does not require an automatic sprinkler system according to Section 1028.6.3, Exception 3.

1028.7 Travel distance. *Exits* and *aisles* shall be so located that the travel distance to an *exit* door shall be not greater than 200 feet (60 960 mm) measured along the line of travel in nonsprinklered buildings. Travel distance shall be not more than 250 feet (76 200 mm) in sprinklered buildings. Where *aisles* are provided for seating, the distance shall be measured along the *aisles* and *aisle accessway* without travel over or on the seats.

Exceptions:

1. *Smoke-protected assembly seating*: The travel distance from each seat to the nearest entrance to a vomitory or concourse shall not exceed 200 feet (60 960 mm). The travel distance from the entrance to the vomitory or concourse to a *stair*, *ramp* or walk on the exterior of the building shall not exceed 200 feet (60 960 mm).

2. Open-air seating: The travel distance from each seat to the building exterior shall not exceed 400 feet (122

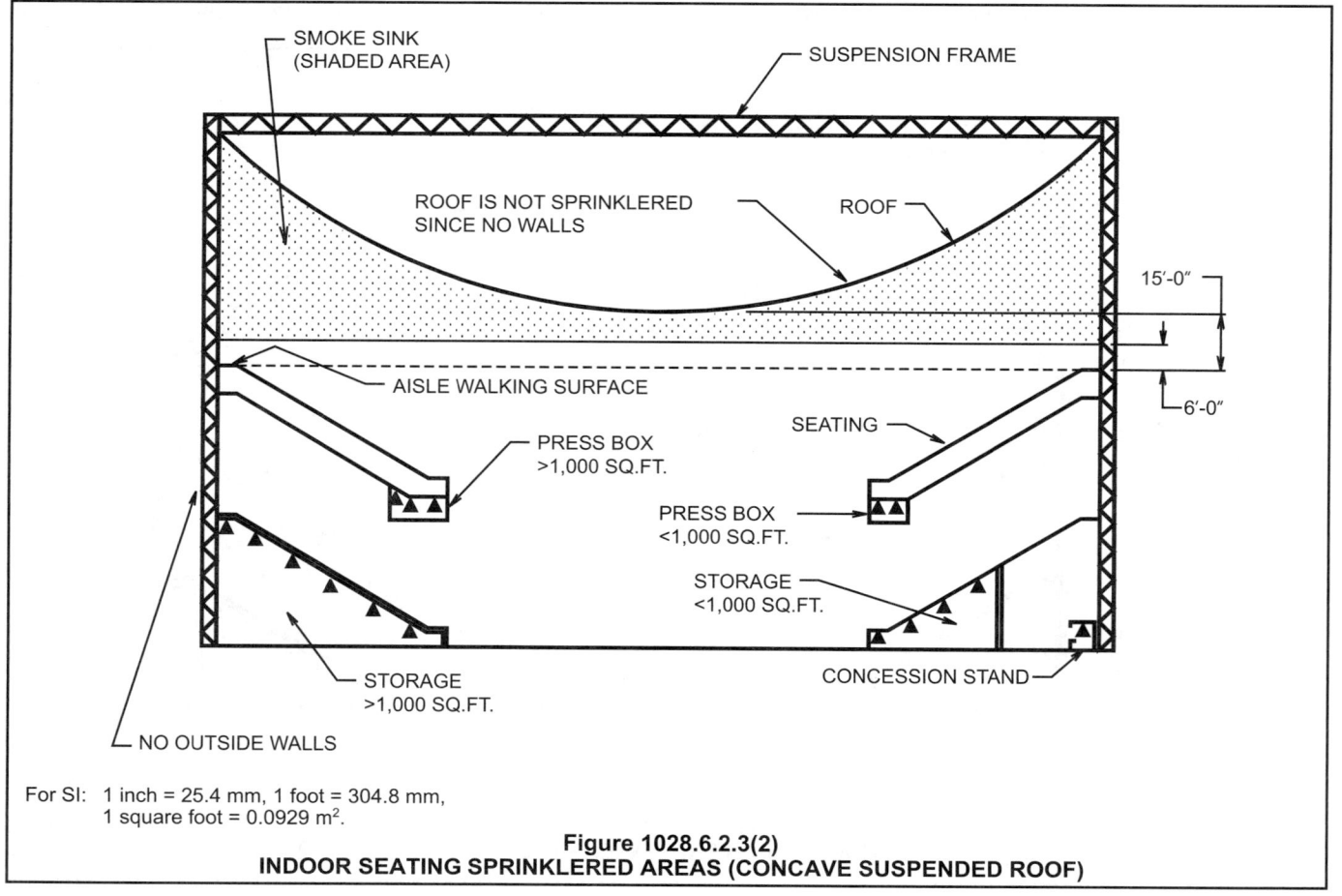

For SI: 1 inch = 25.4 mm, 1 foot = 304.8 mm, 1 square foot = 0.0929 m².

Figure 1028.6.2.3(2)
INDOOR SEATING SPRINKLERED AREAS (CONCAVE SUSPENDED ROOF)

m). The travel distance shall not be limited in facilities of Type I or II construction.

❖ This section includes the travel distance limits for an assembly occupancy, which are the same as those in Table 1016.1. The travel distance is to be measured in the same path as the occupants would normally take to exit the facility.

Exception 1 provides an extended travel distance for smoke-protected assembly seating that meets the requirements of Sections 1028.6.1 through 1028.6.3. Exception 2 applies to outdoor open-air seating areas where the smoke and fire hazard is very low. The Type I and II construction referred to in this exception is described in Section 602.

1028.8 Common path of egress travel. The *common path of egress travel* shall not exceed 30 feet (9144 mm) from any seat to a point where an occupant has a choice of two paths of egress travel to two *exits*.

Exceptions:

1. For areas serving less than 50 occupants, the *common path of egress travel* shall not exceed 75 feet (22 860 mm).

2. For *smoke-protected assembly seating*, the *common path of egress travel* shall not exceed 50 feet (15 240 mm).

❖ The maximum travel distance down a single access row of seating to a location where a patron would have two choices for a way out of the space is 30 feet (9144 mm). In smoke-protected seating, the common path of travel can be up to 50 feet (15 240 mm).

If the room or space (e.g., box, gallery or balcony) has 50 or fewer occupants, the travel distance can be increased to 75 feet (22 860 mm). For example, this allows for a path of travel from a box seat, out of the box and to a main aisle or even a corridor located outside the assembly room itself. When this section is referenced for accessible means of egress (see Section 1007.1, Exception 3), the utilization of Exception 1 would include the entire occupant load of the box, gallery or balcony, not just the number of wheelchair spaces and/or companion seats. Wheelchair spaces that are integrated into the general seating would have the same common path of travel distance of 30 feet (9144 mm) before the person needing the accessible route could choose two different paths for accessible means of egress. This provides the same level of protection for the persons in the accessible seating as provided for others within the space.

1028.8.1 Path through adjacent row. Where one of the two paths of travel is across the *aisle* through a row of seats to another *aisle*, there shall be not more than 24 seats between the two *aisles*, and the minimum clear width between rows for the row between the two *aisles* shall be 12 inches (305 mm) plus 0.6 inch (15.2 mm) for each additional seat above seven in the row between *aisles*.

Exception: For *smoke-protected assembly seating* there shall be not more than 40 seats between the two *aisles* and

the minimum clear width shall be 12 inches (305 mm) plus 0.3 inch (7.6 mm) for each additional seat.

❖ In establishing the point where the occupants of a row served by a single access aisle have two distinct paths of travel, the code allows one of those paths to be through the rows of an adjacent seating area or section. This requirement increases the row widths for the single-access seating section and the adjacent dual-access seating section. This allows the occupants to either travel down the single access aisle or readily traverse the oversized row widths to gain access to a second means of egress. This exception allows a greater number of seats spaced with a minimum clearance of 12 inches (305 mm) for smoke-protected assembly seating that complies with Sections 1028.6.2.1 through 1028.6.2.3 or Section 1028.6.3. For the base width requirements for single- and dual-access rows, see the commentary to Sections 1028.10 through 1028.10.2.

1028.9 Assembly aisles are required. Every occupied portion of any occupancy in Group A or assembly occupancies accessory to Group E that contains seats, tables, displays, similar fixtures or equipment shall be provided with *aisles* leading to *exits* or *exit access doorways* in accordance with this section. *Aisle accessways* for tables and seating shall comply with Section 1017.4.

❖ This section requires that each assembly area have designated aisles. For aisle accessway requirements, see Section 1028.10. Assembly area aisle accessways between tables and chairs are to comply with the width requirements in Section 1017.4.

1028.9.1 Minimum aisle width. The minimum clear width for *aisles* shall be as shown:

1. Forty-eight inches (1219 mm) for *aisle stairs* having seating on each side.

    Exception: Thirty-six inches (914 mm) where the *aisle* serves less than 50 seats.

2. Thirty-six inches (914 mm) for *aisle stairs* having seating on only one side.

3. Twenty-three inches (584 mm) between an *aisle stair handrail* or *guard* and seating where the *aisle* is subdivided by a *handrail*.

4. Forty-two inches (1067 mm) for level or ramped *aisles* having seating on both sides.

    Exceptions:

    1. Thirty-six inches (914 mm) where the *aisle* serves less that 50 seats.

    2. Thirty inches (762 mm) where the *aisle* does not serve more than 14 seats.

5. Thirty-six inches (914 mm) for level or ramped *aisles* having seating on only one side.

    Exceptions:

    1. Thirty inches (762 mm) where the aisle does not serve more than 14 seats.

2. Twenty-three inches (584 mm) between an *aisle stair* handrail and seating where an *aisle* does not serve more than five rows on one side.

❖ The clear widths of aisles and other means of egress established by the formulas given in Section 1028.6 must not be less than the minimum width requirements of this section. The development of minimum width requirements is based on the association of aisle capacity with the path of exit travel as influenced by the different features of aisle construction. The purpose is to create an aisle system that would provide an even flow of occupant egress. The minimum width of the aisles is also based on an anticipated movement of people in two directions. The exceptions are only intended to be applicable to the item directly above.

1028.9.2 Aisle width. The *aisle* width shall provide sufficient egress capacity for the number of persons accommodated by the catchment area served by the *aisle*. The catchment area served by an *aisle* is that portion of the total space that is served by that section of the *aisle*. In establishing catchment areas, the assumption shall be made that there is a balanced use of all *means of egress*, with the number of persons in proportion to egress capacity.

❖ The determination of required aisle and aisle accessway width is a function of the occupant load. In calculating the required widths, the assumption is that in a system or network of aisles and aisle accessways serving an occupied area, people will normally exit the area in a way that will distribute the occupant load throughout the system in proportion to the egress capacity of the aisles and aisle accessways. Each aisle and aisle accessway would take its tributary share (catchment area) of the total occupant load (see Figure 1028.9.2).

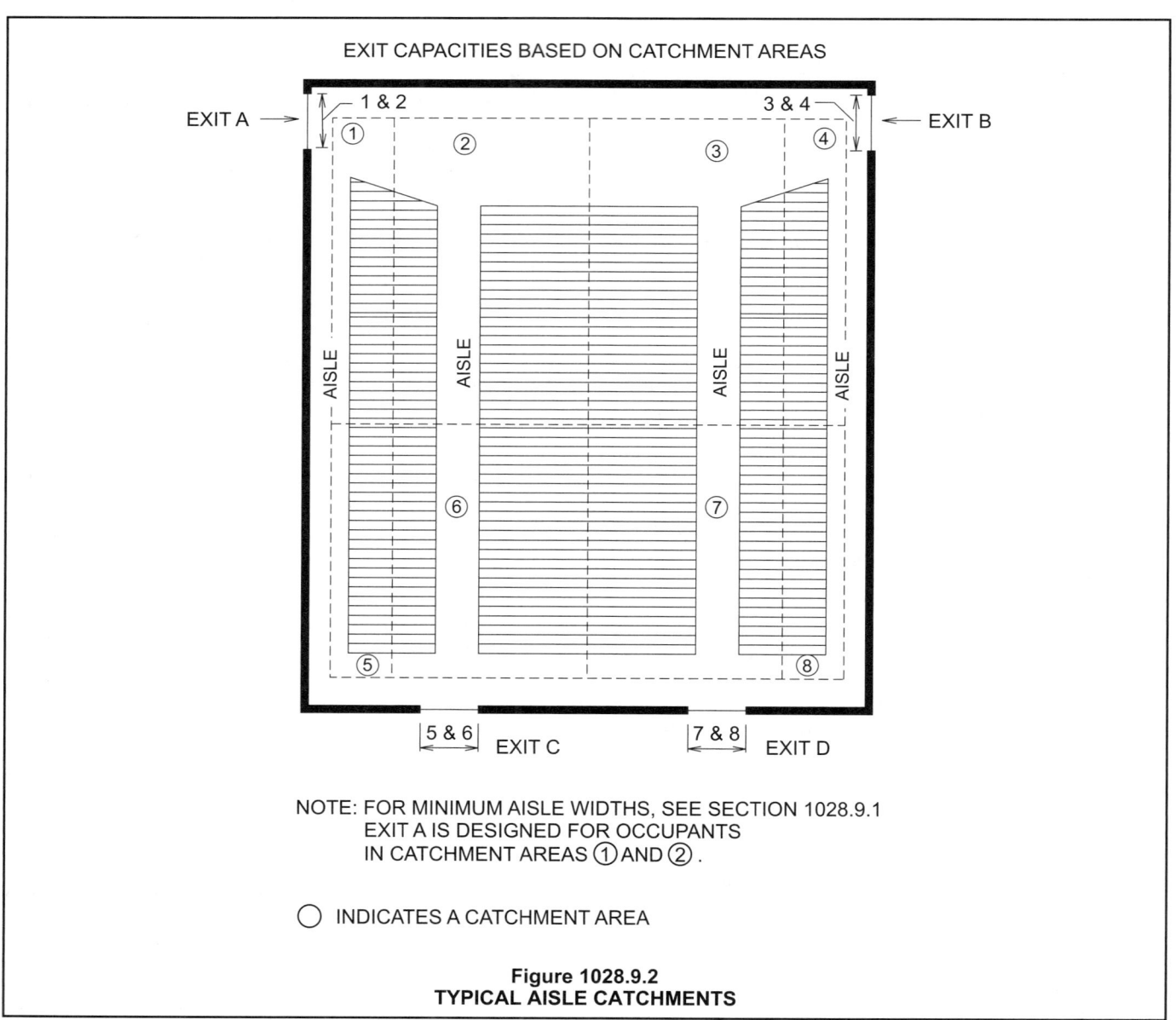

Figure 1028.9.2
TYPICAL AISLE CATCHMENTS

In addition to the provisions in this section, the requirement for the capacity of the main exit and other exits must also be considered (see Section 1028.2). While this section assumes an equal distribution, Section 1028.2 requires that where the facility has a main exit, the main exit and the access thereto must be capable of handling 50 percent of the occupant load.

1028.9.3 Converging aisles. Where *aisles* converge to form a single path of egress travel, the required egress capacity of that path shall not be less than the combined required capacity of the converging *aisles*.

❖ Where one or more aisles or aisle accessways meet to form a single path of egress travel, that path must be sized to handle the combined occupant capacity of the converging aisles and aisle accessways (see Figure 1028.9.3). The reason for this requirement is to maintain the natural pace of travel all the way through the aisle accessways or aisles to the exits and to minimize the queuing of occupants.

This section requires combining the required occupant capacity of converging aisles and aisle accessways, but not necessarily the required widths. For example, if two 48-inch (1219 mm) aisles converge, the result need not be a 96-inch (2438 mm) aisle unless the 48-inch (1219 mm) width of the aisles is required based on the requirements of Section 1028.6 for the actual occupant load served. However, if the 48-inch (1219 mm) width is not based on the occupant load but is required to comply with the minimum aisle width requirements of Section 1028.9.1, the resulting aisle width must be sized for the total occupant load served by the converging aisles, as determined by Section 1028.6, but not less than the minimum widths of Section 1028.9.1.

1028.9.4 Uniform width. Those portions of *aisles*, where egress is possible in either of two directions, shall be uniform in required width.

❖ Aisles that connect or lead to opposite exits must be of uniform width throughout their entire length to allow for exit travel in two directions without creating a traffic bottleneck (see Figure 1028.9.4).

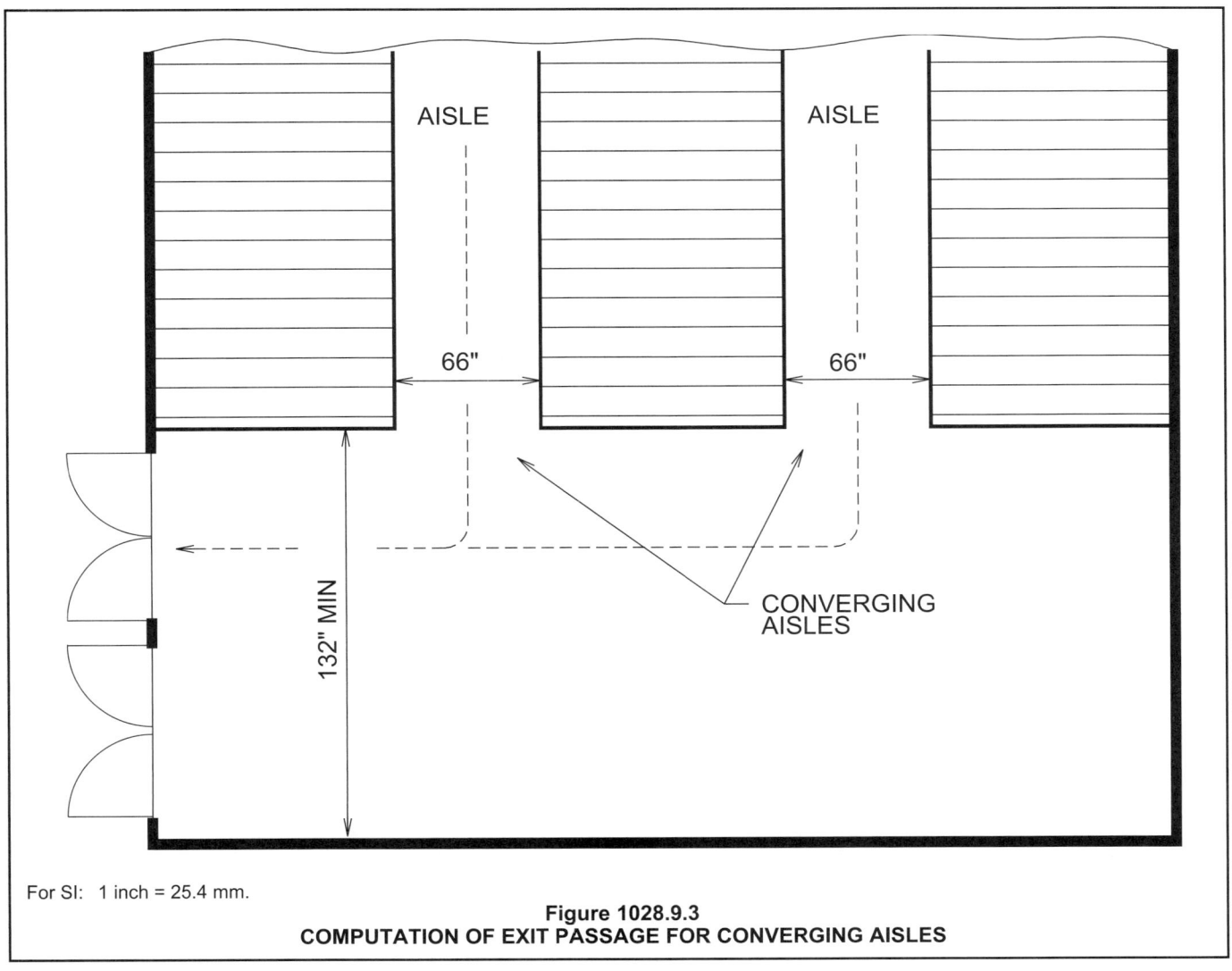

For SI: 1 inch = 25.4 mm.

Figure 1028.9.3
COMPUTATION OF EXIT PASSAGE FOR CONVERGING AISLES

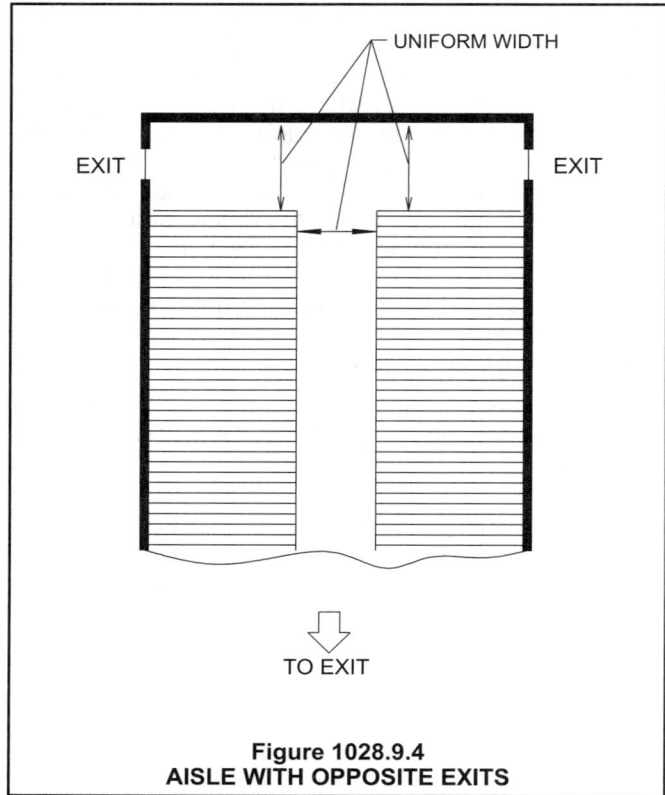

Figure 1028.9.4
AISLE WITH OPPOSITE EXITS

1028.9.5 Assembly aisle termination. Each end of an *aisle* shall terminate at cross *aisle*, foyer, doorway, vomitory or concourse having access to an *exit*.

Exceptions:

1. Dead-end *aisles* shall not be greater than 20 feet (6096 mm) in length.

2. Dead-end *aisles* longer than 20 feet (6096 mm) are permitted where seats beyond the 20-foot (6096 mm) dead-end *aisles* are no more than 24 seats from another *aisles*, measured along a row of seats having a minimum clear width of 12 inches (305 mm) plus 0.6 inch (15.2 mm) for each additional seat above seven in the row.

3. For *smoke-protected assembly seating*, the dead-end *aisle* length of vertical *aisles* shall not exceed a distance of 21 rows.

4. For *smoke-protected assembly seating*, a longer dead-end *aisle* is permitted where seats beyond the 21-row dead-end aisle are not more than 40 seats from another *aisle*, measured along a row of seats having an *aisle accessway* with a minimum clear width of 12 inches (305 mm) plus 0.3 inch (7.6 mm) for each additional seat above seven in the row.

❖ Both ends of a cross aisle must terminate at either an intersecting aisle, a foyer, a doorway or a vomitory (lane) that gives access to an exit(s). Each exception allows an aisle to have a dead-end of limited length. Exceptions 1 and 2 address dead end aisles in assembly

seating areas with or without smoke protection. Exceptions 3 and 4 address dead-end aisles only in smoke-protected assembly seating. In accordance with Exception 1, dead-end aisles (similar to corridors and passageways) that terminate at one end of a cross aisle or at a foyer, doorway or vomitory must not be more than 20 feet (6096 mm) in length. The intent of the row width requirements in the exceptions is to provide sufficient clear width between rows of seating to allow the occupants in times of emergency to pass quickly from a dead-end aisle to the aisle at the opposite end. In Exception 2, the 0.6-inch (15 mm) increase beyond seven seats is consistent with the minimum width determined in accordance with Section 1028.10.2 for single access rows. The code recognizes that one dead-end aisle may not be usable, thus creating a single access row condition. Exceptions 3 and 4 allow longer dead-end aisles for smoke-protected assembly seating that complies with Sections 1028.6.2.1 through 1028.6.2.3 or Section 1028.6.3 (see Figure 1025.9.5).

The overall purpose of this section is to provide aisle/seating arrangements that would allow the occupants to seek safe and rapid passage to exits in case of fire or other emergency.

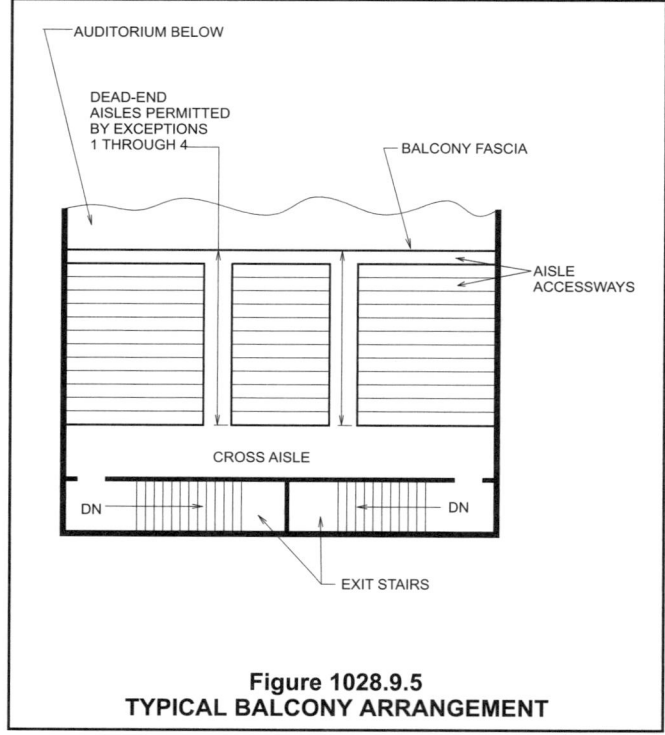

Figure 1028.9.5
TYPICAL BALCONY ARRANGEMENT

1028.9.6 Assembly aisle obstructions. There shall be no obstructions in the required width of *aisles* except for *handrails* as provided in Section 1028.13.

❖ Except for handrails, aisles are required to be clear of any obstructions so that the full width is available for egress purposes. Handrails are allowed to project into the required aisle width in the same manner as handrail projections in stairways.

1028.10 Clear width of aisle accessways serving seating. Where seating rows have 14 or fewer seats, the minimum clear *aisle accessway* width shall not be less than 12 inches (305 mm) measured as the clear horizontal distance from the back of the row ahead and the nearest projection of the row behind. Where chairs have automatic or self-rising seats, the measurement shall be made with seats in the raised position. Where any chair in the row does not have an automatic or self-rising seat, the measurements shall be made with the seat in the down position. For seats with folding tablet arms, row spacing shall be determined with the tablet arm in the used position.

> **Exception:** For seats with folding tablet arms, row spacing is permitted to be determined with the tablet arm in the stored position where the tablet arm when raised manually to vertical position in one motion automatically returns to the stored position by force of gravity.

❖ The requirements of this section are applicable to theater-type seating arrangements. This includes both "continental" and "traditional" seating arrangements. Theater-type seating is characterized by a number of seats arranged side by side and in rows. In this type of seating arrangement, the potential exists for a large number of occupants to be present in a confined environment where the ability of the occupants to move is limited. In order to egress, people are required to move within a row before reaching an aisle or aisle accessway, and the aisle or aisle accessway also limits its movement toward an exit. To provide adequate passage between rows of seats, this section requires that the clear width between the back of a row to the nearest projection of the seating immediately behind must be at least 12 inches (305 mm) (see Figure 1028.10). Where chairs are manufactured with automatic or self-lifting seats, the minimum width requirement may be measured with the seats in a raised position. These are commonly used in college lecture halls. When tablet arm chairs are used, the required width is to be determined with the tablet arm in its usable position. The exception allows for folding arms that fall back into the stored position when a person rises out of the seat.

Even if someone has raised the writing surface on both sides, when students egress down the row they would at most encounter one additional tablet to move out of the way. With these types of arms, the aisle accessway can be measured for the seat or arm as indicated in Figure 1028.10.

With respect to self-rising seats, ASTM F 851 provides one method of determining acceptability.

1028.10.1 Dual access. For rows of seating served by *aisles* or doorways at both ends, there shall not be more than 100 seats per row. The minimum clear width of 12 inches (305 mm) between rows shall be increased by 0.3 inch (7.6 mm) for every additional seat beyond 14 seats, but the minimum clear width is not required to exceed 22 inches (559 mm).

> **Exception:** For *smoke-protected assembly seating*, the row length limits for a 12-inch-wide (305 mm) *aisle accessway*, beyond which the *aisle accessway* minimum clear width shall be increased, are in Table 1028.10.1.

❖ Where rows of seating are served by aisles or doorways located at both ends of the path of row travel, the number of seats that may be used in a row may be up to, but not more than, 100 (continental seating) and the minimum required clear width aisle accessway of 12 inches (305 mm) between rows of seats must be increased by 0.3 inch (8 mm) for every additional seat beyond 14, but not more than a total of 22 inches (559 mm) (see Figure 1028.10.1). For example, in a row of 24 seats, the minimum clear width would compute to 15 inches (381 mm) [12 + (0.3 × 10)]. For a row of 34 seats, a clear width of 18 inches (457 mm) would be required. Increases in the clear width between rows of seats would occur up to a row of 46 seats. From 47 to 100 seats, a maximum clear width between rows of 22 inches (559 mm) would apply.

Since the row is to provide access to an aisle in both directions, the minimum width applies to the entire length of the row aisle accessway.

The exception allows more seats in a row with the minimum 12-inch (305 mm) seat spacing since safe egress time is extended for this condition.

For additional aisle accessway width requirements

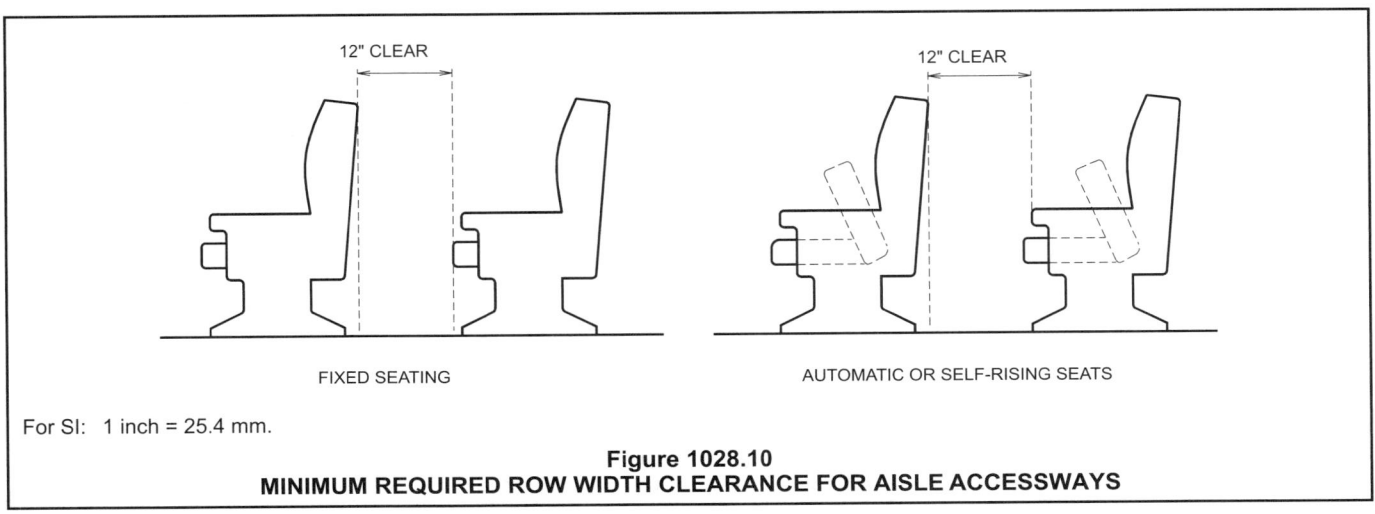

For SI: 1 inch = 25.4 mm.

Figure 1028.10
MINIMUM REQUIRED ROW WIDTH CLEARANCE FOR AISLE ACCESSWAYS

when one of the means of egress at the end of the single access row is through a dual access row, see Section 1028.8.1.

TABLE 1028.10.1
SMOKE-PROTECTED
ASSEMBLY AISLE ACCESSWAYS

TOTAL NUMBER OF SEATS IN THE SMOKE-PROTECTED ASSEMBLY OCCUPANCY	MAXIMUM NUMBER OF SEATS PER ROW PERMITTED TO HAVE A MINIMUM 12-INCH CLEAR WIDTH AISLE ACCESSWAY	
	Aisle or doorway at both ends of row	Aisle or doorway at one end of row only
Less than 4,000	14	7
4,000	15	7
7,000	16	8
10,000	17	8
13,000	18	9
16,000	19	9
19,000	20	10
22,000 and greater	21	11

For SI: 1 inch = 25.4 mm.

❖ Table 1028.10.1 recognizes the increased egress time available in smoke-protected assembly seating areas. Therefore, the table permits greater lengths of rows that have the minimum 12 inches (305 mm) of clear width. When a row exceeds the lengths identified in the table, the row width is to be increased in accordance with Section 1028.10.1 [0.3 inch (8 mm) per additional seat] for dual access rows and Section 1028.10.2 [0.6 inch (15 mm) per additional seat] for single access rows. The requirements of this table are based on the total number of seats contained within the assembly space.

1028.10.2 Single access. For rows of seating served by an *aisle* or doorway at only one end of the row, the minimum clear width of 12 inches (305 mm) between rows shall be increased by 0.6 inch (15.2 mm) for every additional seat beyond seven seats, but the minimum clear width is not required to exceed 22 inches (559 mm).

Exception: For *smoke-protected assembly seating,* the row length limits for a 12-inch-wide (305 mm) *aisle accessway,* beyond which the *aisle accessway* minimum clear width shall be increased, are in Table 1028.10.1.

❖ Where rows of seating are served by an aisle or doorway at only one end of a row, the minimum clear width of 12 inches (305 mm) between rows of seats must be increased by 0.6 inch (15 mm) for every additional seat beyond seven, but not more than a total of 22 inches (559 mm) (see Figure 1028.10.2). While this section does not specify the maximum number of seats permitted in a row, the 30-foot (9144 mm) common path of travel limitation (see Section 1028.8) essentially restricts the single access row to approximately 20 seats, based on an 18-inch (457 mm) width per seat. A row of 12 seats would compute to a required minimum width of 15 inches [12 + (0.5 × 5)]. Similarly, a row of 17 seats would require a clear width of 18 inches (457 mm) and so on. Since dual access is not provided, incremental increases would be permitted in the aisle accessway width as shown in Figure 1028.10.2. Incremental increases in the required width would occur up to the maximum number of seats, which is determined by the 30-foot (9144 mm) dead-end limitation.

The reason for increasing the row accessway widths incrementally with increases in the number of seats per row is to provide more efficient passage for the occupants who are using the aisle accessway. As a practical matter, where dual-access (see Section

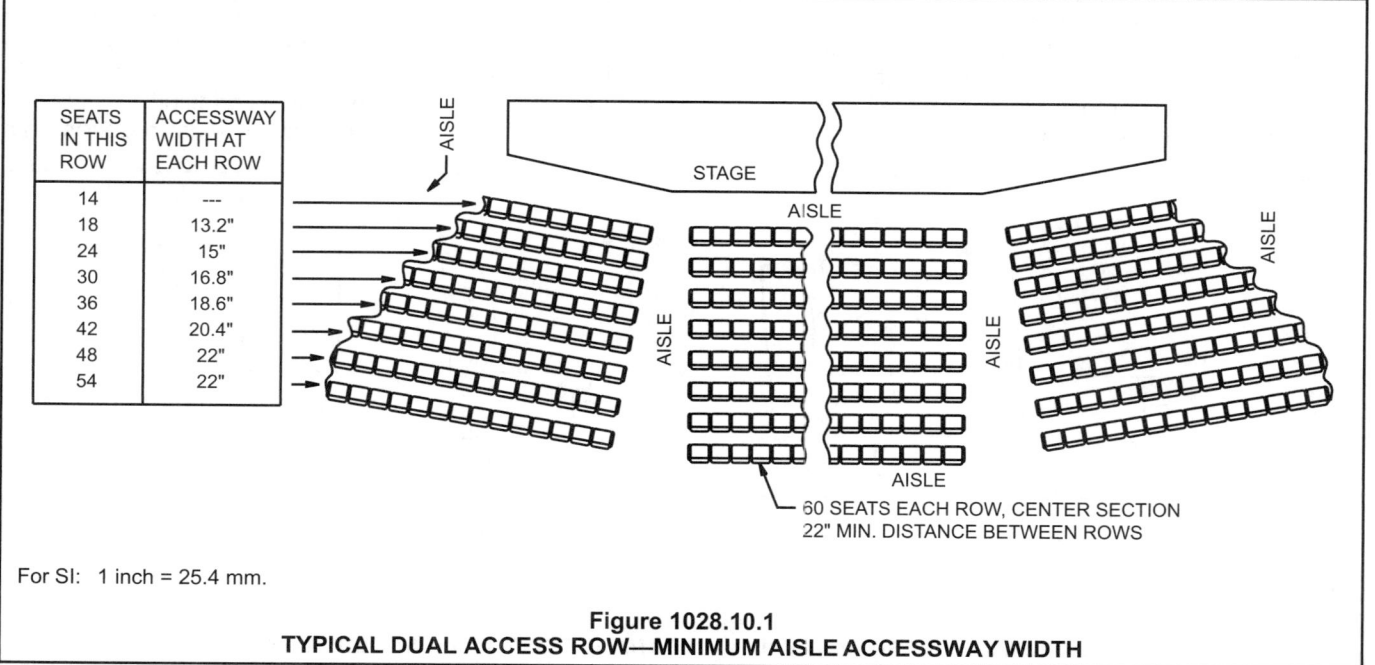

SEATS IN THIS ROW	ACCESSWAY WIDTH AT EACH ROW
14	---
18	13.2"
24	15"
30	16.8"
36	18.6"
42	20.4"
48	22"
54	22"

60 SEATS EACH ROW, CENTER SECTION
22" MIN. DISTANCE BETWEEN ROWS

For SI: 1 inch = 25.4 mm.

Figure 1028.10.1
TYPICAL DUAL ACCESS ROW—MINIMUM AISLE ACCESSWAY WIDTH

1028.10.1) and single-access seating arrangements are used together, the largest computed clear width dimension would normally be applied by the designer to both arrangements so that the rows of seats will be in alignment. For additional aisle accessway width requirements when one of the means of egress at the end of the single access row is through a dual access row, see Section 1028.10.1.

1028.11 Assembly aisle walking surfaces. *Aisles* with a slope not exceeding one unit vertical in eight units horizontal (12.5-percent slope) shall consist of a *ramp* having a slip-resistant walking surface. *Aisles* with a slope exceeding one unit vertical in eight units horizontal (12.5-percent slope) shall consist of a series of risers and treads that extends across the full width of *aisles* and complies with Sections 1028.11.1 through 1028.11.3.

❖ Assembly facilities such as theaters and auditoriums often require sloping or stepped floors to provide seated occupants with preferred sightlines for viewing presentations (for sightlines for wheelchair spaces, see Section 1108.2). Aisles must, therefore, be designed to accommodate the changing elevations of the floor in such a manner that the path of travel will allow occupants to leave the area at a rapid pace with minimal possibilities for stumbling or falling during times of emergency.

This section requires that aisles with a gradient of one unit vertical and eight units horizontal (12.5 percent slope) or less must consist of a ramp with a slip-resistant surface. Aisles with a gradient exceeding one unit vertical and eight units horizontal (12.5-percent slope) must consist of a series of treads and risers that comply with the requirements of Sections 1028.11.1 through 1028.11.3. Note that ramps that serve as part of an accessible route to and from accessible wheelchair spaces must comply with the more restrictive requirements for ramps in Section 1010 (see Section 1010.1, Exception 1).

While not specifically indicated for stepped aisles, such floor surfaces must also be slip resistant in accordance with Section 1003.4. Field testing and uniform enforcement of the concept of slip resistance is not practical. One method used to establish slip resistance is that the static coefficient of friction between leather [Type 1 (Vegetable Tanned) of Federal Specification KK-L-165C] and the floor surface is greater than 0.5. Laboratory test procedures can determine the static coefficient of resistance.

What must be recognized here is that stepped aisles are part of the floor construction and are intended to provide horizontal egress. Tread and riser construction for this purpose should not be compared to the requirements for treads and risers in conventional stairways that serve as means of vertical egress. Sometimes, because of design considerations, the gradient of an aisle is required to change from a level floor to a ramp and then to steps. In cases where there is no uniformity in the path of travel, occupants tend to be considerably more cautious, particularly in the use of stepped aisles, than they would normally be in the use of conventional stairways.

1028.11.1 Treads. Tread depths shall be a minimum of 11 inches (279 mm) and shall have dimensional uniformity.

Exception: The tolerance between adjacent treads shall not exceed $^3/_{16}$ inch (4.8 mm).

❖ Depths of treads are not to be less than 11 inches (279 mm) and uniform throughout each flight, except that a variance of not more than $^3/_{16}$ inch (4.8 mm) is permitted between adjacent treads to accommodate variations in construction. While this provision is the same as the limiting dimension for treads in interior stairways (see Section 1009.3), it rarely applies in the construction of stepped aisles. A more common form of stepped aisle construction is to provide a tread depth equal to the back-to-back distance between rows of seats. This way the treads can be extended across the full length of the row and serve as a supporting platform for the seats. Other arrangements might require two treads between rows of seats.

In theaters, for example, the back-to-back distance between rows of fixed seats usually ranges some-

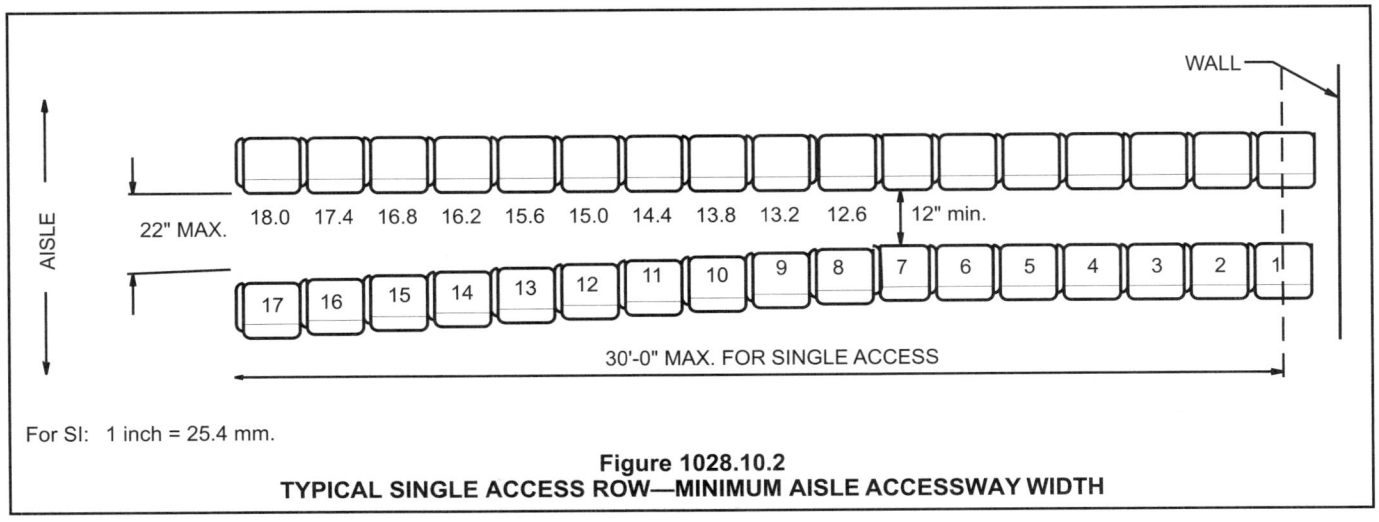

For SI: 1 inch = 25.4 mm.

**Figure 1028.10.2
TYPICAL SINGLE ACCESS ROW—MINIMUM AISLE ACCESSWAY WIDTH**

where between 3 and 4 feet (914 and 1219 mm), depending on seat style and seat dimensions as well as the ease of passage between the rows (see Figure 1028.11.1). The selection of single-tread or two-tread construction between rows of seats depends on the gradient and suitable riser height (see Section 1028.11.2), as needed for sightlines.

In comparing this section with Section 1028.11.2, it is significant to note the emphasis placed on the tread dimension. While not desirable, the code permits riser heights to deviate; however, tread dimensions must not vary beyond the 0.188-inch (4.8 mm) tolerance.

1028.11.2 Risers. Where the gradient of *aisle stairs* is to be the same as the gradient of adjoining seating areas, the riser height shall not be less than 4 inches (102 mm) nor more than 8 inches (203 mm) and shall be uniform within each *flight*.

Exceptions:

1. Riser height nonuniformity shall be limited to the extent necessitated by changes in the gradient of the adjoining seating area to maintain adequate sightlines. Where nonuniformities exceed 0.188 inch (4.8 mm) between adjacent risers, the exact location of such nonuniformities shall be indicated with a distinctive marking stripe on each tread at the *nosing* or leading edge adjacent to the nonuniform risers. Such stripe shall be a minimum of 1 inch (25 mm), and a maximum of 2 inches (51 mm), wide. The edge marking stripe shall be distinctively different from the contrasting marking stripe.

2. Riser heights not exceeding 9 inches (229 mm) shall be permitted where they are necessitated by the slope of the adjacent seating areas to maintain sightlines.

❖ In stepped aisles where the gradient of the aisle is the same as the gradient of the adjoining seating area, riser heights are not to be less than 4 inches (102 mm) nor more than 8 inches (203 mm) (see Figure 1028.11.2). For the safety of the occupants, risers should have uniform heights, where possible, throughout each flight of steps. However, nonuniformity of riser heights is permitted in cases where changes to the gradient in the adjoining seating area are required because of sightlines and other seating layout considerations.

Where variations in height exceed $^3/_{16}$ inch (4.8 mm) between adjacent risers, a distinctive marking stripe

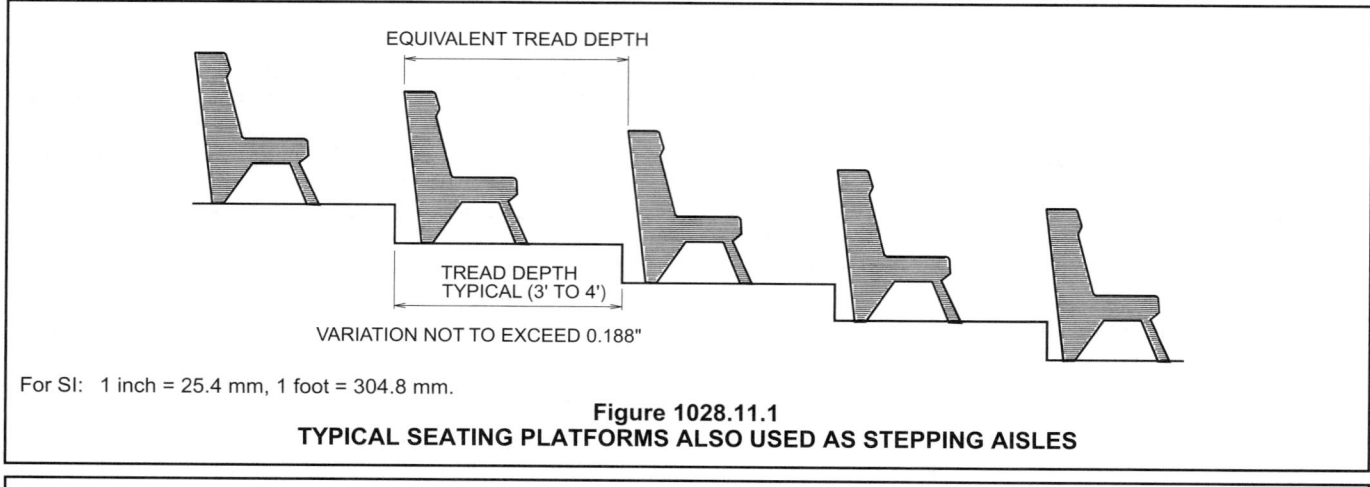

For SI: 1 inch = 25.4 mm, 1 foot = 304.8 mm.

Figure 1028.11.1
TYPICAL SEATING PLATFORMS ALSO USED AS STEPPING AISLES

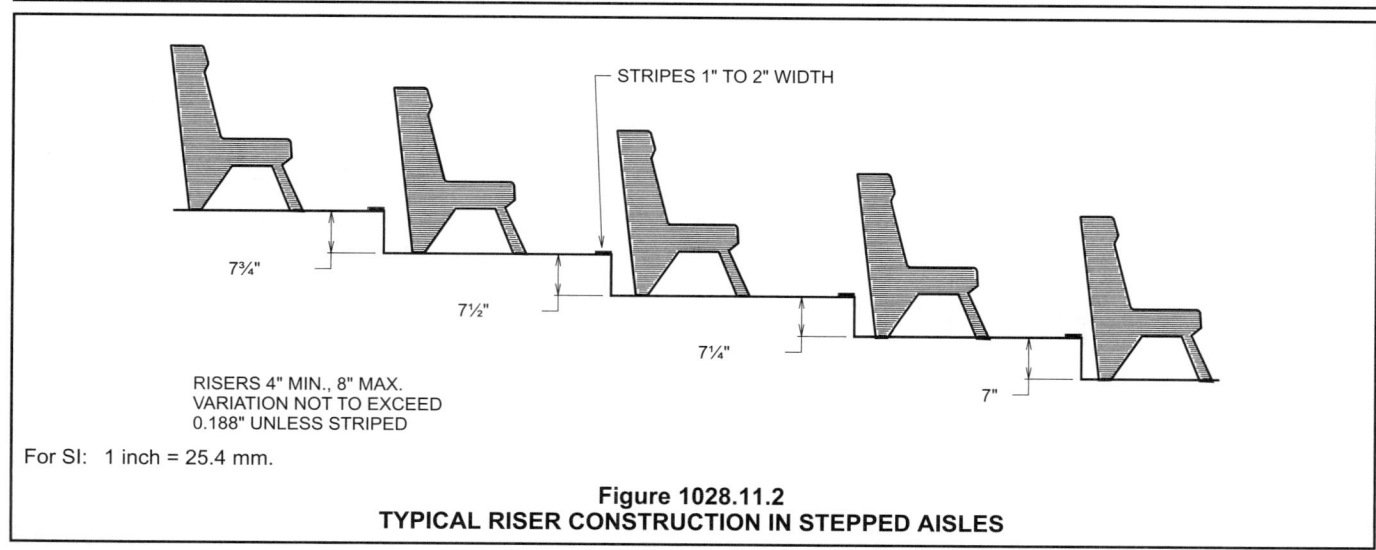

For SI: 1 inch = 25.4 mm.

Figure 1028.11.2
TYPICAL RISER CONSTRUCTION IN STEPPED AISLES

between 1 and 2 inches (25 and 51 mm) wide is to be located on the nosings of each tread where the variations occur as a visual warning to the occupants to be cautious. Frequently, this is done with "runway" lights. Note that this stripe must be different from the tread contrast marking stripes required in Section 1028.11.3. These stripes must be visible in lighted conditions; therefore, these stripes are not required to comply with the provisions for luminous tread markings in Section 1024.

In comparing this section with Section 1028.11.1, it is significant to note the emphasis placed on the tread dimension. While not desirable, the code permits riser heights to deviate; however, Section 1028.11.1 does not permit tread dimensions to vary beyond the $^3/_{16}$-inch (4.8 mm) tolerance.

1028.11.3 Tread contrasting marking stripe. A contrasting marking stripe shall be provided on each tread at the *nosing* or leading edge such that the location of each tread is readily apparent when viewed in descent. Such stripe shall be a minimum of 1 inch (25 mm), and a maximum of 2 inches (51 mm), wide.

Exception: The contrasting marking stripe is permitted to be omitted where tread surfaces are such that the location of each tread is readily apparent when viewed in descent.

❖ The exception provides for the omission of the contrasting marking stripe where the tread is readily apparent such as when aisle stair treads are provided with a roughened metal nosing strip or where lighted nosings occur. In this situation, the user is aware of the treads without the marking stripe. This stripe must be different from the marking stripe required for nonuniform risers in Section 1028.11.2, Exception 1.

These stripes must be visible in lighted conditions; therefore, these stripes are not required to comply with the provisions for luminous tread markings in Section 1024.

1028.12 Seat stability. In places of assembly, the seats shall be securely fastened to the floor.

Exceptions:

1. In places of assembly or portions thereof without ramped or tiered floors for seating and with 200 or fewer seats, the seats shall not be required to be fastened to the floor.

2. In places of assembly or portions thereof with seating at tables and without ramped or tiered floors for seating, the seats shall not be required to be fastened to the floor.

3. In places of assembly or portions thereof without ramped or tiered floors for seating and with greater than 200 seats, the seats shall be fastened together in groups of not less than three or the seats shall be securely fastened to the floor.

4. In places of assembly where flexibility of the seating arrangement is an integral part of the design and function of the space and seating is on tiered levels, a maximum of 200 seats shall not be required to be fastened

to the floor. Plans showing seating, tiers and *aisles* shall be submitted for approval.

5. Groups of seats within a place of assembly separated from other seating by railings, *guards*, partial height walls or similar barriers with level floors and having no more than 14 seats per group shall not be required to be fastened to the floor.

6. Seats intended for musicians or other performers and separated by railings, *guards*, partial height walls or similar barriers shall not be required to be fastened to the floor.

❖ The purpose of this section is to require that assembly seating be fastened to the floor where it would be a significant hazard if loose and subject to tipping over. The exceptions allow loose assembly seating for situations where the hazard is lower, such as floors where ramped or tiered seating is not used, where no more than 200 seats are used and for box seating arrangements where a limited number of seats are within railings, guards or partial height walls.

1028.13 Handrails. Ramped *aisles* having a slope exceeding one unit vertical in 15 units horizontal (6.7-percent slope) and *aisle stairs* shall be provided with *handrails* located either at the side or within the *aisle* width.

Exceptions:

1. *Handrails* are not required for ramped *aisles* having a gradient no greater than one unit vertical in eight units horizontal (12.5-percent slope) and seating on both sides.

2. *Handrails* are not required if, at the side of the *aisle*, there is a *guard* that complies with the graspability requirements of *handrails*.

3. *Handrail* extensions are not required at the top and bottom of *aisle stairs* and *aisle ramp* runs to *permit* crossovers within the *aisles*.

❖ For the safety of occupants, handrails must be provided in aisles where ramps exceed a gradient of one unit vertical in 15 units horizontal (6.67-percent slope) (see Figure 1028.13).

Exception 1 omits the handrail requirements where ramped aisles are not steep and seats are on both sides to reduce the fall hazard.

Exception 2 allows handrails to be omitted where there is a guard at the side of the aisle with a top rail that complies with the requirements for handrail graspability (see Section 1012.3). Note that the guard must meet the height and opening requirements specified in Section 1013 or 1028.14, as applicable.

While Section 1028.13.1 does allow for discontinuous handrails, and Exception 3 (as well as Section 1012.6, Exception 2) exempts handrail extensions, the handrail must extend the full run of the aisle stair. Stopping the handrail a couple of risers from the bottom of the stair flight would be considered a code violation.

1028.13.1 Discontinuous handrails. Where there is seating on both sides of the *aisle*, the *handrails* shall be discontinuous with gaps or breaks at intervals not exceeding five rows to facilitate access to seating and to permit crossing from one side of the *aisle* to the other. These gaps or breaks shall have a clear width of at least 22 inches (559 mm) and not greater than 36 inches (914 mm), measured horizontally, and the *handrail* shall have rounded terminations or bends.

❖ Where aisles have seating on both sides, handrails may be located at the sides of the aisles, but are typically located in the center of the aisle. The width of each section of the subdivided aisle between the handrail and the edge of seating is not to be less than 23 inches (584 mm) (see Section 1028.9.1, Item 3).

For reasons of life safety in fire situations and also as a practical matter in the efficient use of the facility, a handrail down the middle of an aisle should not be continuous along its entire length. Crossovers must be provided by means of gaps or breaks in the handrail installation. Such openings must not be less than 22 inches (559 mm) nor more than 36 inches (914 mm) wide, and must be provided at intervals not exceeding the distance of five rows of seats (see Figure 1028.13.1). All handrail terminations should be designed to have rounded ends or bends to avoid possible injury to the occupants (see Figure 1028.13).

1028.13.2 Intermediate handrails. Where *handrails* are provided in the middle of *aisle stairs*, there shall be an additional intermediate *handrail* located approximately 12 inches (305 mm) below the main *handrail*.

❖ Handrail installations down the middle of an aisle must be constructed with intermediate rails located 12 inches (305 mm) below and parallel to main handrails. This is to provide handholds for children, to prevent people from using the handrail like a gym apparatus and possibly injuring themselves and from ducking to get under the rail (see Figure 1028.13).

1028.14 Assembly guards. Assembly *guards* shall comply with Sections 1028.14.1 through 1028.14.3.

❖ This section establishes the scope of the assembly guard provisions.

1028.14.1 Cross aisles. Cross *aisles* located more than 30 inches (762 mm) above the floor or grade below shall have *guards* in accordance with Section 1013.

Where an elevation change of 30 inches (762 mm) or less occurs between a cross *aisle* and the adjacent floor or grade

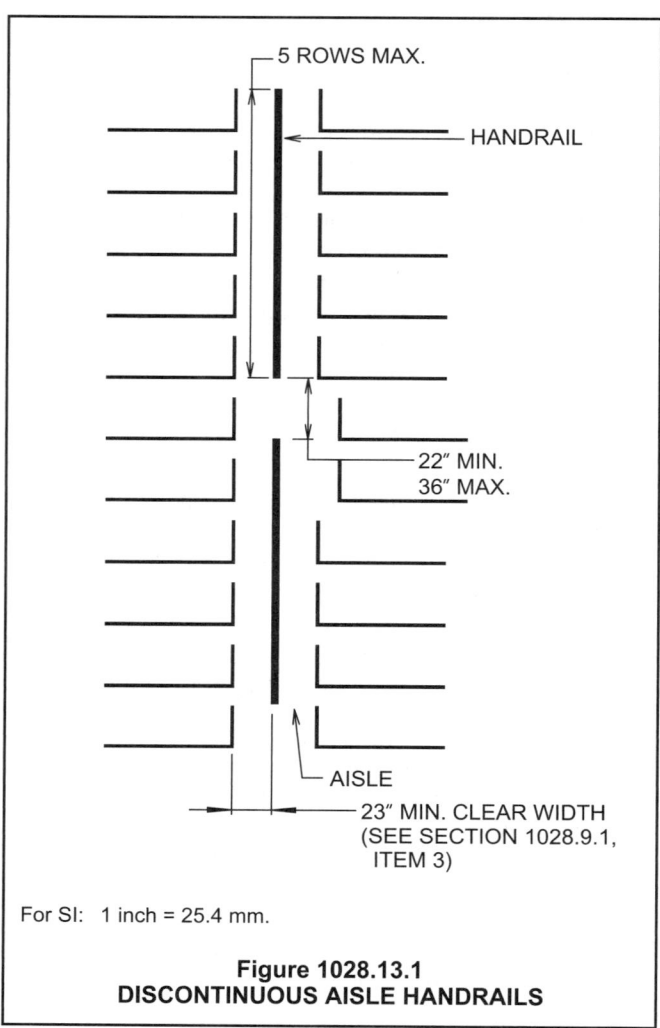

For SI: 1 inch = 25.4 mm.

Figure 1028.13.1
DISCONTINUOUS AISLE HANDRAILS

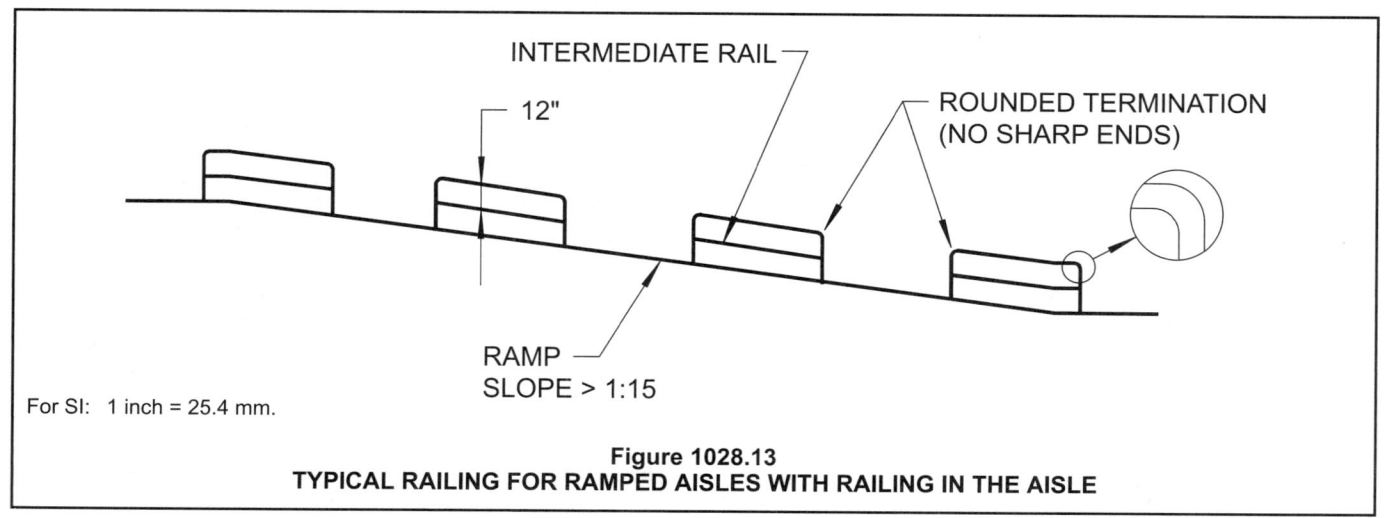

For SI: 1 inch = 25.4 mm.

Figure 1028.13
TYPICAL RAILING FOR RAMPED AISLES WITH RAILING IN THE AISLE

below, *guards* not less than 26 inches (660 mm) above the *aisle* floor shall be provided.

Exception: Where the backs of seats on the front of the cross *aisle* project 24 inches (610 mm) or more above the adjacent floor of the *aisle*, a *guard* need not be provided.

❖ The purpose of this section is to provide for occupant safety with guards along elevated cross aisles. The minimum height of the guard is a function of the cross-aisle elevation above the adjacent floor or grade below [i.e., 42 inches (1067 mm) high with more than a 30-inch (762 mm) drop-off and 26 inches (660 mm) high with a 30-inch (762 mm) or less drop-off]. When the backs of the seats adjacent to the cross aisle are a minimum of 24 inches (610 mm) above the floor level of the cross aisle, they will serve as the guard (see Figure 1028.14.2 for an illustration of the requirements in this section).

1028.14.2 Sightline-constrained guard heights. Unless subject to the requirements of Section 1028.14.3, a fascia or railing system in accordance with the *guard* requirements of Section 1013 and having a minimum height of 26 inches (660 mm) shall be provided where the floor or footboard elevation is more than 30 inches (762 mm) above the floor or grade below and the fascia or railing would otherwise interfere with the sightlines of immediately adjacent seating. At *bleachers*, a *guard* must be provided where required by ICC 300.

❖ This section specifies a height of 26 inches (660 mm) for guards within assembly seating areas other than at the end of aisles where a vertical 36-inch (914 mm) height is required. This is to provide a reasonable degree of safety while providing for sightlines within the viewing area. The guard opening configuration must comply with Section 1013.3 (see Figure 1028.14.2 for an illustration of the requirements in this section).

1028.14.3 Guards at the end of aisles. A fascia or railing system complying with the *guard* requirements of Section 1013 shall be provided for the full width of the *aisle* where the foot of the *aisle* is more than 30 inches (762 mm) above the floor or grade below. The fascia or railing shall be a minimum of 36 inches (914 mm) high and shall provide a minimum 42 inches (1067 mm) measured diagonally between the top of the rail and the *nosing* of the nearest tread.

❖ This section applies only at the end of aisles where the foot (the lower end) of the aisle is greater than 30 inches (762 mm) above the adjacent floor or grade below. The guard must satisfy both of the specified height requirements to provide safety for persons at the end of the aisle. The 36-inch (914 mm) minimum height is measured from the floor vertically to the top of the guard. The minimum 42-inch (1067 mm) diagonal dimension from the nosing of the nearest stair tread to the top of the fascia or guard is to provide sufficient height for a fall from the nearest stair tread (see Figure 1028.14.2 for an illustration of the requirements in this section).

1028.15 Bench seating. Where bench seating is used, the number of persons shall be based on one person for each 18 inches (457 mm) of length of the bench.

❖ The purpose of this section is to specify the length of bench for each occupant for bench and bleacher seating. This is commonly used to calculate the occupant load of bench or bleacher seating for egress purposes and is not intended to limit any individual to an 18-inch (457 mm) area. This is consistent with the fixed seating occupant loads indicated in Section 1004.7.

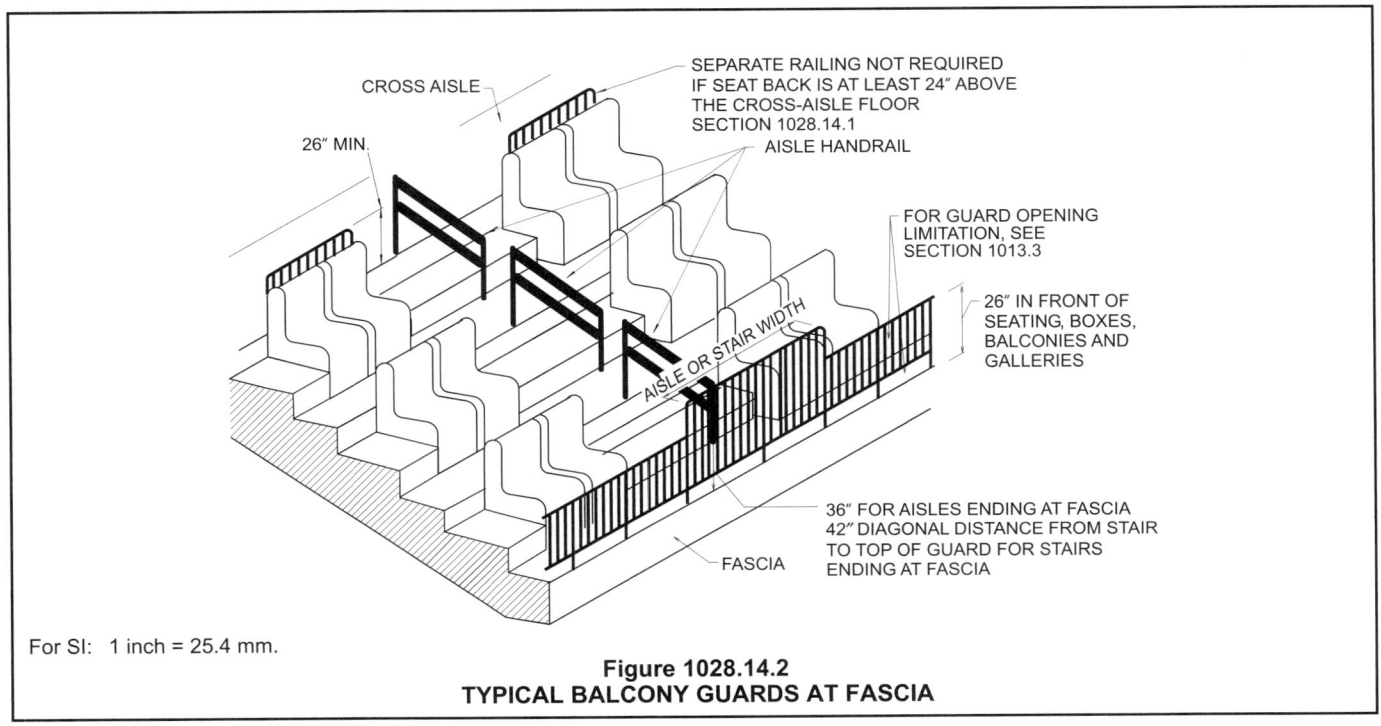

For SI: 1 inch = 25.4 mm.

Figure 1028.14.2
TYPICAL BALCONY GUARDS AT FASCIA

SECTION 1029
EMERGENCY ESCAPE AND RESCUE

1029.1 General. In addition to the *means of egress* required by this chapter, provisions shall be made for emergency escape and rescue in Group R and I-1 occupancies. Basements and sleeping rooms below the fourth *story above grade plane* shall have at least one exterior *emergency escape and rescue opening* in accordance with this section. Where basements contain one or more sleeping rooms, *emergency escape and rescue openings* shall be required in each sleeping room, but shall not be required in adjoining areas of the basement. Such openings shall open directly into a *public way* or to a *yard* or *court* that opens to a *public way*.

Exceptions:

1. In other than Group R-3 occupancies, buildings equipped throughout with an *approved automatic sprinkler system* in accordance with Section 903.3.1.1 or 903.3.1.2.

2. In other than Group R-3 occupancies, sleeping rooms provided with a door to a fire-resistance-rated *corridor* having access to two remote *exits* in opposite directions.

3. The *emergency escape and rescue opening* is permitted to open onto a balcony within an *atrium* in accordance with the requirements of Section 404, provided the balcony provides access to an *exit* and the dwelling unit or sleeping unit has a *means of egress* that is not open to the *atrium*.

4. Basements with a ceiling height of less than 80 inches (2032 mm) shall not be required to have emergency escape and rescue windows.

5. *High-rise buildings* in accordance with Section 403.

6. *Emergency escape and rescue openings* are not required from basements or sleeping rooms that have an *exit* door or *exit access* door that opens directly into a *public way* or to a *yard*, *court* or exterior *exit* balcony that opens to a *public way*.

7. Basements without *habitable spaces* and having no more than 200 square feet (18.6 m²) in floor area shall not be required to have emergency escape windows.

❖ This section requires emergency escape and rescue provisions in groups where occupants may be sleeping during a potential fire buildup, but are capable of self-preservation (Groups R and I-1). A basement and each sleeping room are to be provided with an exterior window or door that meets the minimum size requirements and is operable for emergency escape by methods that are obvious and clearly understood by all users. Sleeping rooms four stories or more above grade are not required to be so equipped, since fire service access at that height, as well as escape through such an opening, may not be practical or reliable. In accordance with Chapter 9, such buildings will also be equipped throughout with an automatic fire suppression system. The provision for basements is in recognition that such types of spaces typically only have a

single path of egress and often have no alternative routes available as other levels do.

It is important to note that this window is an element of escape and does not comprise any part of the means of egress unless it is a door with appropriate egress component characteristics.

Exception 1 assumes that the automatic sprinkler system can control fire buildup and reduce, if not eliminate, the need for an occupant to use an emergency escape window. The exception applies to buildings equipped throughout with an NFPA 13 or 13R sprinkler system.

Exception 2 allows another acceptable means of escape; that is, a door directly from the sleeping room to a corridor with exits in opposite directions, to substitute for the escape window.

Exception 3 provides for dwelling and sleeping units that have egress windows to a balcony that is within an atrium. The exception specifies that the dwelling or sleeping unit is to have another means of egress that does not pass through the atrium so that an independent route of egress is provided.

Exceptions 4 and 7 are intended to exempt basements that would not be likely to have sleeping rooms in them from the requirement to have emergency escape and rescue openings.

Exception 5 is in correlation with the exception for emergency escape windows in high-rise buildings addressed in Section 403.4.

The intent of Exception 6 is to permit sleeping rooms with a direct access to an exterior-type environment, such as a street or exit balcony, to not have an emergency escape window. The open atmosphere of the escape route would increase the likelihood that the means of egress be available even with the delayed response time for sleeping residents.

1029.2 Minimum size. *Emergency escape and rescue openings* shall have a minimum net clear opening of 5.7 square feet (0.53 m²).

Exception: The minimum net clear opening for *emergency escape and rescue* grade-floor *openings* shall be 5 square feet (0.46 m²).

❖ The dimensional criteria of the openings are intended to permit fire service personnel (in full protective clothing with a breathing apparatus) to enter from a ladder, as well as permit occupants to escape. The net clear opening area and minimum dimensions are intended to provide a clear opening through which an occupant can pass to escape the building or a fire fighter can pass to enter the building for rescue or fire suppression activities. Since the emergency escape windows must be usable to all occupants, including children and guests, the required opening dimensions must be achieved by the normal operation of the window from the inside (e.g., sliding, swinging or lifting the sash). It is impractical to assume that all occupants can operate a window that requires a special sequence of operations to achieve the required opening size. While most occupants are familiar with the normal operation by which to open the window, children and guests are frequently unfamiliar with special procedures necessary to remove the sashes. The time spent in comprehending the special operation unnecessarily delays egress from the bedroom and could lead to panic and further confusion. Thus, windows that achieve the required opening dimensions only through operations such as the removal of sashes or mullions are not permitted. It should be noted that the minimum area cannot be achieved by using both the minimum height and minimum width specified in Section 1029.2.1 (see Figure 1029.2).

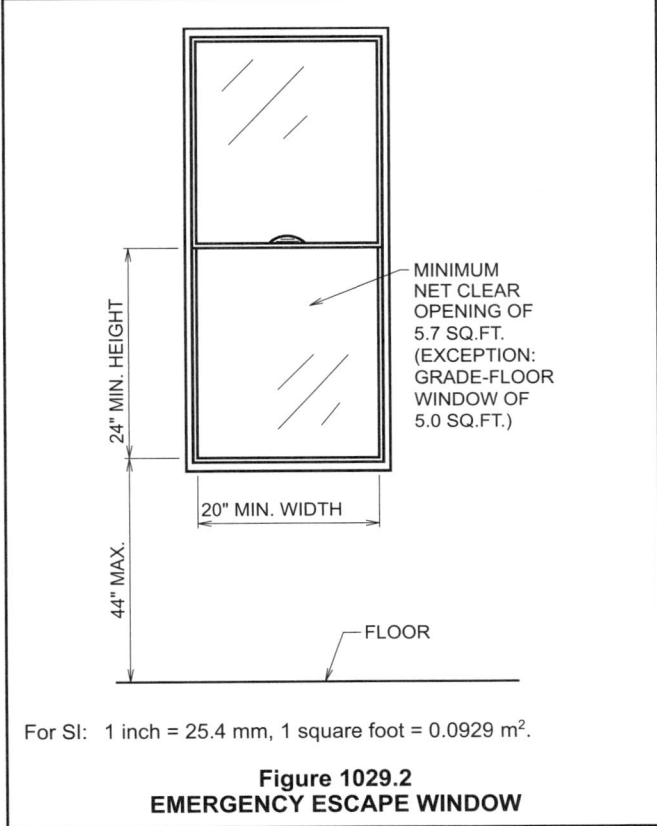

For SI: 1 inch = 25.4 mm, 1 square foot = 0.0929 m².

Figure 1029.2
EMERGENCY ESCAPE WINDOW

1029.2.1 Minimum dimensions. The minimum net clear opening height dimension shall be 24 inches (610 mm). The minimum net clear opening width dimension shall be 20 inches (508 mm). The net clear opening dimensions shall be the result of normal operation of the opening.

❖ Note that the minimum dimensions in this section and the minimum area requirements in Section 1029.2 both apply. Thus, a grade-floor window that is only 24 inches (610 mm) in height must be 30 inches (762 mm) wide to meet the 5-square-foot (0.46 m²) area requirement of Section 1029.2 for grade-floor window (see Figure 1029.2).

1029.3 Maximum height from floor. *Emergency escape and rescue openings* shall have the bottom of the clear opening not greater than 44 inches (1118 mm) measured from the floor.

❖ This section limits the height of the bottom of the clear opening to 44 inches (1118 mm) or less such that it

can be used effectively as an emergency escape (see Figure 1029.2). For a minimum sill height that may affect the emergency escape window, see Section 1405.13.2.

1029.4 Operational constraints. *Emergency escape and rescue openings* shall be operational from the inside of the room without the use of keys or tools. Bars, grilles, grates or similar devices are permitted to be placed over *emergency escape and rescue openings* provided the minimum net clear opening size complies with Section 1029.2 and such devices shall be releasable or removable from the inside without the use of a key, tool or force greater than that which is required for normal operation of the escape and rescue opening. Where such bars, grilles, grates or similar devices are installed in existing buildings, *smoke alarms* shall be installed in accordance with Section 907.2.11 regardless of the valuation of the *alteration*.

❖ If security grilles, decorations or similar devices are installed on escape windows, such items must be readily removable to permit occupant escape without the use of any tools, keys or a force greater than that required for the normal operation of the window.

Where bars, grilles or grates are placed over the emergency escape and rescue opening, it is important that they are easily removable. Thus, the requirements for ease of operation are the same as required for windows.

The smoke alarms that are required for existing buildings where such items are installed provides advance warning of a fire for safety purposes.

1029.5 Window wells. An *emergency escape and rescue opening* with a finished sill height below the adjacent ground level shall be provided with a window well in accordance with Sections 1029.5.1 and 1029.5.2.

❖ Emergency escape and rescue openings that are partially or completely below grade need to have window wells so that they can be used effectively (see Figure 1029.5).

1029.5.1 Minimum size. The minimum horizontal area of the window well shall be 9 square feet (0.84 m²), with a minimum dimension of 36 inches (914 mm). The area of the window well shall allow the *emergency escape and rescue opening* to be fully opened.

❖ This section specifies the size of the window well that is needed for a rescue person in full protective clothing and breathing apparatus to use the rescue opening. The required 9 square feet (0.84 m²) is the horizontal cross-sectional area of the window well. Thus, if the window well projects away from the plane of the window 3 feet (914 mm), the required dimension in the plane of the window along the wall is also 3 feet (914 mm) (see Figure 1029.5).

1029.5.2 Ladders or steps. Window wells with a vertical depth of more than 44 inches (1118 mm) shall be equipped with an *approved* permanently affixed ladder or steps. Ladders or rungs shall have an inside width of at least 12 inches (305 mm), shall project at least 3 inches (76 mm) from the wall and shall be spaced not more than 18 inches (457 mm) on center (o.c.) vertically for the full height of the window well. The ladder or steps shall not encroach into the required dimensions of the window well by more than 6 inches (152 mm). The ladder or steps shall not be obstructed by the *emergency escape and rescue opening*. Ladders or steps required by this section are exempt from the *stairway* requirements of Section 1009.

❖ This section specifies that a ladder or steps be provided for ease of getting into and out of window wells that are more than 44 inches (1118 mm) deep.

Usually ladder rungs are embedded in the wall of the window well. The 44-inch (1118 mm) dimension is the depth of the window well, not the distance from the bottom of the window well to grade. Thus, if the floor of a window well is 40 inches (1016 mm) below grade, but the wall of the window well projects above grade by 6 inches (152 mm), steps or a ladder are required since the vertical depth is 46 inches (1168 mm).

It is important that the ladder not obstruct the operation of the emergency escape window (see Figure 1029.5).

Bibliography

The following resource materials are referenced in this chapter or are relevant to the subject matter addressed in this chapter.

24 CFR, *Fair Housing Accessibility Guidelines* (FHAG). Washington, DC: Department of Housing and Urban Development, 1991.

36 CFR Parts 1190 and 1191 Final Rule, *The Americans with Disabilities Act (ADA) Accessibility Guidelines; Architectural Barriers Act (ABA) Accessibility Guidelines*. Washington, DC: Architectural and Transportation Barriers Compliance Board, July 23, 2004.

ANSI/NSPI-8, *1996 Model Barrier Code for Residential Swimming Pools, Spas, and Hot Tubs*. Alexandria, VA: Association of Spa and Pool Professionals, 1996.

Architectural and Transportation Barriers Compliance Board, 42 USC 3601-88, *Fair Housing Amendments Act* (FHAA). Washington, DC: United States Code, 1988.

ASME A17.1/CSA B44-2007, *Safety Code for Elevators and Escalators*. New York: American Society of Mechanical Engineers, 2007.

ASME A18.1-05, *Safety Standard for Platform Lifts and Stairway Chairlifts*. New York: American Society of Mechanical Engineers, 2005.

ASTM D 2047-99, *Test Method for Static Coefficient of Friction Polish-coated Floor Surfaces as Measured by the James Machine*. West Conshohocken, PA: ASTM International, 1999.

ASTM E 119-07, *Test Method for Fire Tests of Building Construction and Materials*. West Conshohocken, PA: ASTM International, 2007.

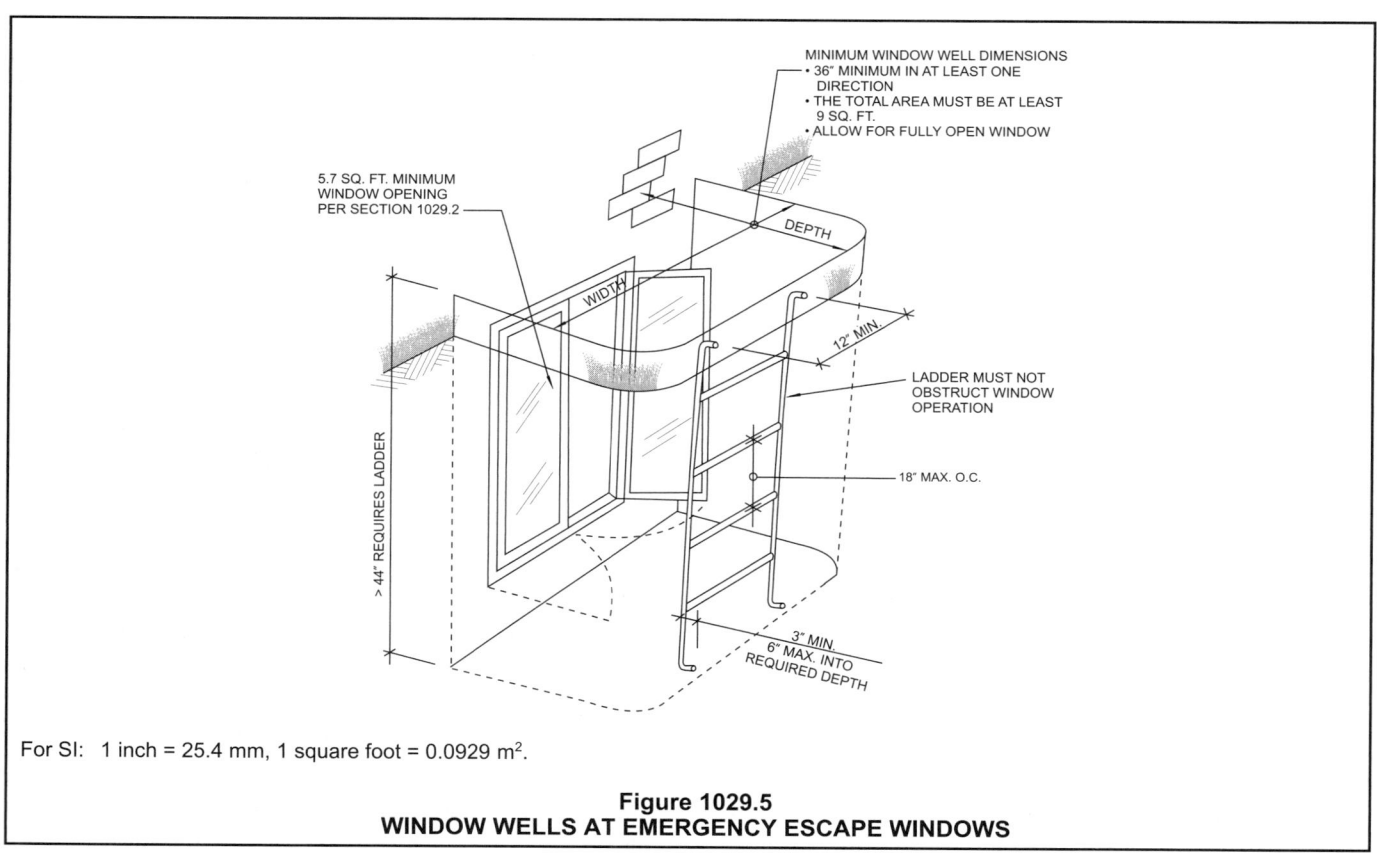

For SI: 1 inch = 25.4 mm, 1 square foot = 0.0929 m².

Figure 1029.5
WINDOW WELLS AT EMERGENCY ESCAPE WINDOWS

ASTM E 2072-04, *Standard Specification for Photoluminescent (Phosphorescent) Safety Markings*. West Conshohocken, PA: ASTM International, 2004.

ASTM F 851-87 (Reapproved 2000), *Test Method for Self-rising Seat Mechanisms*. West Conshohocken, PA: ASTM International, 2000.

DOJ 28 CFR, Part 36 (Appendix A)-91, *ADA Accessibility Guidelines for Buildings and Facilities*. Washington, DC: U.S. Department of Justice, 1991.

DOJ 28 CFR, Part 36-91, *Americans with Disabilities Act (ADA)*. Washington, DC: U.S. Department of Justice, 1991.

Final Report of the HUD Review of the Fair Housing Accessibility Requirements and the 2003 International Building Code (IBC). February 18, 2005 (Docket No FR-4943-N-02).

Final Report of the HUD Review of the Fair Housing Accessibility Requirements and the 2006 International Building Code (IBC). May 31, 2007 (Docket No FR-5136-N-01).

HUD 24 CFR, Part 100, *Federal Fair Housing Accessibility Guidelines*. Washington, DC: U.S. Department of Housing and Urban Development.

ICC 300-07, *ICC Standards on Bleachers, Folding and Telescopic Seating, and Grandstands*. Washington, DC: International Code Council, 2007.

ICC A117.1-03, *Accessible and Usable Buildings and Facilities*. Washington, DC: International Code Council, 2003.

IEBC-09, *International Existing Building Code*. Washington, DC: International Code Council, 2009.

IECC-09, *International Energy Conservation Code*. Washington, DC: International Code Council, 2009.

IFC-09, *International Fire Code*. Washington, DC: International Code Council, 2009.

IMC-09, *International Mechanical Code*. Washington, DC: International Code Council, 2009.

IRC-09, *International Residential Code*. Washington, DC: International Code Council, 2009.

NFPA 13-07, *Installation of Sprinkler Systems*. Quincy, MA: National Fire Protection Association, 2007.

NFPA 13D-07, *Installation of Sprinkler Systems in One- and Two-family Dwellings and Manufactured Homes*. Quincy, MA: National Fire Protection Association, 2007.

NFPA 13R-07, *Installation of Sprinkler Systems in Residential Occupancies Up to Four Stories in Height*. Quincy, MA: National Fire Protection Association, 2007.

NFPA 70-08, *National Electrical Code*. Quincy, MA: National Fire Protection Association, 2008.

NIST IR 4770-92, *Report on Staging Areas for Persons with Mobility Limitations*. Washington, DC: National Institute of Standards and Technology, 1992.

UL 10C-98, *Positive Pressure Fire Tests of Door Assemblies—with Revisions through November 2001*. Northbrook, IL: Underwriters Laboratories, Inc., 1998.

UL 305-07, *Panic Hardware*. Northbrook, IL: Underwriters Laboratories, Inc., 2007.

UL 924-06, *Standard for Safety Emergency Lighting and Power Equipment*. Northbrook, IL: Underwriters Laboratories Inc., 2006.

UL 1994-04, *Standard for Luminous Egress Path Marking Systems—with Revisions through February 2005*. Northbrook, IL: Underwriters Laboratories, Inc., 2005.

Chapter 11:
Accessibility

General Comments

Chapter 11 contains provisions that set forth requirements for accessibility of buildings and their associated sites and facilities for people with physical disabilities. Existing building criteria is addressed in Section 34011. Appendix E is included in the code to address accessibility for items in the new Americans with Disabilities Act Accessibility Guidelines (ADAAG) that were not typically enforceable through the standard traditional building code enforcement approach system (e.g., beds, room signage).

In July 2004, the United States Access Board published the new design guidelines under the name Americans with Disabilities Act and Architectural Barriers Act Accessibility Guidelines, otherwise known as the ADA/ABA Guidelines. For purposes of the following discussion, the 2004 document will be referred to as the "ADA/ABA Guidelines" while the original guidelines of 1991 are referred to as just "ADAAG."

The *International Residential Code*® (IRC®) references Chapter 11 for accessibility provisions; therefore, this chapter may be applicable to housing covered under the IRC (see commentary, Sections 1107.6 and 1107.6.3). Structures referenced to Chapter 11 from the IRC would be considered Group R-3.

Section 1101 contains the broad scope statement of the chapter and identifies the baseline criteria for accessibility as being in compliance with this chapter and ICC A117.1. ICC A117.1 *Accessible and Useable Buildings and Facilities* is the consensus national standard that sets forth the details, dimensions and construction specifications for accessibility.

Section 1102 contains definitions of terms that are associated with accessibility.

Section 1103 describes the applicability of the provisions of this chapter. Accessibility is broadly required in all buildings, structures, sites and facilities. Those specific circumstances in which accessibility is not required (or is limited) are set forth as exceptions.

Section 1104 contains the requirements for interior and exterior accessible routes. An accessible route is a key component of the built environment that provides a person with a disability access to spaces, elements, facilities and buildings. Note that ramps are addressed in Section 1010.

Section 1105 contains requirements for accessible entrances to buildings and structures. Note that requirements for accessible means of egress are addressed in Section 1007.

Section 1106 sets forth the requirements for accessible parking facilities and passenger loading zones.

Section 1107 contains various accessibility require-ments that are unique to occupancies that contain dwelling units and sleeping units and are applicable in addition to other general requirements of this chapter. Specific provisions unique to Group I and R occupancies are included. Requirements in this section are coordinated with the requirements found in the Fair Housing Accessibility Guidelines (FHAG).

Section 1108 contains various accessibility requirements that are unique to specific occupancies, other than Groups I and R, and are applicable in addition to other general requirements of this chapter. Specific provisions unique to assembly seating, performance areas, self-storage and judicial facilities are included.

Section 1109 contains various requirements that are applicable to features and facilities that are not occupancy related, including requirements for toilet and bathing facilities; sinks; kitchens; drinking fountains; elevators; lifts; storage facilities; detectable warnings; seating at tables, counters and work surfaces; service facilities; controls and operating mechanisms; fuel dispensing systems; and recreational facilities.

Section 1110 sets forth requirements for signage identifying certain required accessible elements.

History—Building Codes and Federal Laws

Access to buildings and structures for people with physical disabilities has been a subject that the building codes have regulated since the early 1970s. They have consistently relied on a consensus national standard, CABO/ANSI A117.1, as the technical basis for accessibility. The title of CABO/ANSI A117.1 is now ICC/ANSI A117.1 to reflect that the International Code Council® (ICC®) is the secretariat for this standard. Accessibility is not a new subject to the construction regulatory community. There has been a great deal of emphasis and awareness placed on the subject of accessibility through the passage of two federal laws. The Americans with Disabilities Act (ADA) and the Fair Housing Amendment Act (FHA) are federal regulations that affect building construction as it relates to accessibility.

ADA, signed into law in July 1990, is a very broad civil rights law designed to protect persons with disabilities. There are five sections of the ADA that address different aspects of civil rights for people with disabilities. Title I deals with employment and generally prohibits discrimination against disabled people in employment. Title II deals with public services, including access to all services and facilities receiving federal funding, and access to public transportation, including buses, rail lines and passenger transit facilities. Title III requires access to a large category of buildings and structures in a manner that is equal or comparable to that available to the general public. (Note that FHA is a companion document to

ADA. FHA deals with residential and institutional living arrangements that, for the most part, are not covered under ADA.) Title IV deals with accessible telecommunication facilities. Title V further defines miscellaneous issues dealing with ADA compliance.

Titles II and III of the ADA overlap requirements customarily regulated by building codes. Most Title II facilities may opt to follow either the FED-STD-795-88, Uniform Federal Accessibility Standards (UFAS) or ADAAG for compliance with accessibility requirements. This group includes anyone receiving federal funding, i.e., colleges, schools, park districts, local and state governments, hospitals, etc. Implementation of Title III of the ADA is based on the requirements of ADAAG. Both UFAS and ADAAG set forth scoping and technical requirements for building design and construction. The federal government is interested in bringing more uniformity to federal accessibility standards and will be looking at ways in which harmonization of all federal and private-sector standards can be accomplished. In many ways, UFAS and ADAAG are similar to a building code and cover many of the same matters that are dealt with in such codes. It is important to emphasize, however, that the ADA is written as civil rights legislation and not as a building code. The enforcement mechanism for civil rights laws is significantly different than that for traditional building regulation.

The ADA does not preempt the adoption or enforcement of accessibility-related codes by state and local governments. All buildings constructed within a jurisdiction must comply with locally adopted building codes and any applicable state codes, as well as ADAAG; however, as a federal law, state and local governments have neither the authority nor the responsibility to enforce ADA. Enforcement of ADA is the responsibility of the U.S. Department of Justice (DOJ). As previously described, ADA covers a large range of issues dealing with disabilities. Under Titles II and III, the scoping and technical requirements of UFAS or ADAAG deal directly with accessibility of buildings. Typically, enforcement of these regulations can only take place after the construction process. Any differences between the regulations must be ultimately reconciled by the building owner or designer. This creates a difficult situation for the building owners and designers who are faced with multiple sets of regulations. In an effort to alleviate this problem, some states have adopted ADAAG as the referenced standard for their accessibility laws. Currently, this creates difficulties in enforcement, since ADAAG is not written in language that is customarily used in building codes and relied on for enforcement by state and local enforcement officials. Fortunately, there are many similarities between the requirements of ADAAG and the code. In addition, compliance with the code typically takes place during the plan review and construction process. This is a less costly time to make required changes to a facility that are necessary for compliance with applicable requirements. Again, the building official is expected to enforce the code adopted by the state or local jurisdiction, and any applicable state laws. The building official

is neither required to nor responsible for interpretation or enforcement of ADA unless the state or local jurisdiction has referenced ADAAG in a state's accessibility laws. No attempt should be made to represent that ADAAG is being interpreted or enforced by the building official or the jurisdiction when ADAAG is not specifically adopted by the jurisdiction.

The concept of two completely separate and independent, broad-based sets of regulations that affect building design and construction may sound onerous, but the situation may not be quite as bad as it appears. The reason is that the code has a high level of consistency with ADAAG. Since the early 1990s, code change activity has incorporated the recommendations of the Council of American Building Officials' (CABO) Board for the Coordination of the Model Codes (BCMC) [succeeded in 1995 by the ICC Board for the Development of the Model Codes (BDMC)]. BDMC undertook the effort to review comprehensively all facets of the accessibility issue. This effort included a comparison of ADAAG with the provisions of the model codes in use at that time. BDMC's recommendations were ultimately reflected in the code. There are still differences between the code and ADAAG, but the differences have been minimized.

Substantial activity to revise the federal standards has taken place resulting in the ADA/ABA Guidelines (July 2004). Efforts for coordination with the federal accessibility requirements are ongoing. Representatives from interested accessibility groups, Department of Housing and Urban Development (HUD) and the United States Access Board have been attending and participating in the code change process for the code and ICC A117.1. Additionally, the ICC has participated in the public comment process on the development of federal regulations. The ICC has worked toward, and will continue to strive for, accessibility regulations that reflect the highest possible degree of consistency with federal regulations and, more importantly, reasonable and appropriate provisions to meet the needs of people with disabilities.

The FHA is federal legislation promulgated by HUD that extends fair housing protection against discrimination on the basis of family status or persons with disabilities. The standard for FHA is the FHAG, which sets forth accessibility scoping and technical requirements for a broad category of residential construction. The scope of ADAAG generally excludes occupancies that are covered by FHAG; however, FHA is being interpreted to overlap the scope of the coverage of ADA in some respects (e.g., dorms, nursing homes). Residential buildings that are covered by FHAG must also comply with the locally adopted building code. The same circumstances with respect to differences in the requirements and responsibility for enforcement described for ADA also exist with FHA.

Again, as a federal law, building officials have neither the responsibility nor the authority to enforce the requirements of FHAG. Inquiries regarding those laws should be referred to the DOJ or HUD. Until such time that their laws are coordinated with and consider traditional code enforcement mechanisms and methods, building officials are advised not to attempt to interpret or enforce these laws. Permit applicants should be advised that the work they propose has not been reviewed for compliance with FHAG.

Due to concerns about residential construction complying with all applicable codes and laws, the National Association Home Builders (NAHB), along with others, approached the model code groups about incorporating the FHAG requirements into the model code requirements. Since the FHA is a civil rights law rather than a building code, careful study was needed to interpret them into enforceable language that could be utilized in model codes. Through the efforts of the BDMC, recommendations to the model codes were proposed in 1994. The "adaptable" dwelling unit requirements were carried forward as a Type A unit, with the FHA requirements reflected in the Type B units.

In 2000, HUD reviewed the 2000 edition of the IBC and the referenced accessibility standard ICC/ANSI A117.1-1998 for compliance with FHAG. Based on HUD's report, a series of modifications were proposed as part of the 2000 code change cycle. The proposed modifications were accepted by the voting members and were incorporated into the 2001 Supplement to the IBC and eventually into the 2003, 2006 and 2009 editions of the code. As a result of the earlier code changes, HUD eventually issued press releases stating that the 2000 code with the 2001 Supplement and 2003 code (with the ICC/ANSI A117.1-1998 accessibility standard) and the 2006 IBC (with the ICC/ANSI A117.1-2003 accessibility standard) could be considered "safe harbor" for anyone wanting to comply with FHAG scoping and technical requirements. The original safe harbor status that was bestowed on the 2000 IBC and its accessory documents was identified in the preamble of HUD's Fair Housing Act Design Manual.

With the 2003 and the 2006 editions of the code being safe harbor documents, architects and developers could design and be reviewed for compliance with the code through the typical building code review process. At the same time, they would also be in compliance with the building requirements for the FHAG. At the time this commentary was written, HUD had not reviewed the 2009 edition of the code for safe-harbor status. Since the code is developed through a code change/public comment process, there have been changes to items in Chapter 11. This is typically a good way for new concerns, concepts or procedures to be incorporated into the requirements. It is the ICC's opinion that none of these changes has resulted in jeopardizing the safe-harbor status between FHA and code requirements for this edition.

Substantial activity to revise the federal standards has taken place resulting in the ADA/ABA Guidelines (July 2004). Efforts for coordination with the federal accessibility requirements are ongoing. Representatives from interested accessibility groups, HUD and the Access Board have been attending and participating in the code change process for the codes and ICC A117.1. Addition-ally, the ICC has participated in the public comment process on the development of federal regulations. The ICC has worked toward, and will continue to strive for, accessibility regulations that reflect the highest possible degree of consistency with federal regulations and, more importantly, reasonable and appropriate provisions to meet the needs of people with disabilities.

Philosophy

The fundamental philosophy of the code on the subject of accessibility is that everything is required to be accessible. This is reflected in the basic applicability requirement (see Section 1103.1). The code's scoping requirements then address the conditions under which accessibility is not required in terms of exceptions to this general mandate. In the early 1990s, building codes tended to describe where accessibility was required in each occupancy, and any circumstance not specifically identified was excluded. The more recent codes represent a fundamental change in approach. Now one must think of accessibility in terms of "if it is not specifically exempted, it must be accessible."

Another important concept is that of "mainstreaming." There are many accessibility issues that not only benefit people with disabilities, but also provide a tangible benefit to people without disabilities. This type of requirement can be set forth in the code as generally applicable without necessarily identifying it specifically as an accessibility-related issue. Such a requirement would then be considered as having been mainstreamed. For example, the limitation on objects protruding into corridors, aisles and passageways (see Section 1003.3) is intended to aid people with a vision impairment by reducing the potential for unintended contact that may cause injury. Clearly, this has an additional benefit to people without a vision impairment. A protruding object can be encountered by a sighted person who is simply not paying close attention. The concept of mainstreaming is responsive to the desire of people with disabilities to not be singled out and categorized separately from the remainder of society. It is therefore important to recognize that, while the provisions of Chapter 11 are specifically grouped and identified as accessibility related, they do not represent all of the issues that must be taken into consideration when evaluating accessibility in buildings.

There are many items that have some basis in accessibility requirements, but actually result in more "user-friendly" buildings for all of us. Physical disabilities can be permanent or temporary; can affect all age groups; can involve all levels of abilities and can range from persons with minor visual, hearing or mobility impairments to persons who are blind, deaf or confined to a wheelchair. Everyone will benefit from accessible features in buildings, either directly or indirectly through those they care for. With some foresight in designing our built environment, access into and throughout buildings can be better for everyone throughout their lifetimes.

Purpose

The purpose of this chapter is to set forth requirements for accessibility applicable to those elements of the built environment that are included within the scope of the code. There are two categories or types of requirements that must be addressed in order to accomplish accessibility: scoping and technical. Scoping requirements describe what and where accessibility is required, or how many accessible features or elements must be provided. For example, a requirement that at least one of the first 25 sleeping units in a hotel must be accessible is referred to as a scoping requirement because it describes what and how many accessible features are required. Another example of scoping is the exception for detached one- and two-family dwellings from accessibility requirements because it defines what is not required to be accessible. A technical requirement is intended to refer to a statement indicating how accessibility is to be accomplished. For example, a requirement indicating that in order for a parking space to be accessible, it must have a width of 96 inches (2438 mm) with an adjacent 60-inch (1524 mm) access aisle is considered a technical requirement. The provisions of Chapter 11 are scoping requirements in that they all address what and where accessibility is required or how many accessible elements are required. ICC A117.1 indicates how to make something accessible and is the consensus national standard that is referenced for establishing technical requirements. In addition to the provisions of Chapter 11, there are accessibility-related technical requirements elsewhere in the code. These are subject-matter-specific provisions that have been mainstreamed into the chapter or section of the code that deals with those subjects. As such, the code and its references regulate accessibility comprehensively by setting forth the scoping and technical requirements that establish the minimum level of accessibility required in the built environment.

SECTION 1101
GENERAL

1101.1 Scope. The provisions of this chapter shall control the design and construction of facilities for accessibility to physically disabled persons.

❖ This section establishes the scope of the chapter as providing for design and construction of facilities for accessibility to disabled persons. The scope is broadly inclusive of all aspects of construction that affect the ability of people with disablties to approach, enter and utilize a facility. The term "facility," as defined in Section 1102.1, includes not only buildings and structures, but also the site on which they are located. Features of a site, such as parking areas and paths of travel from a public way to a structure, affect accessibility and are, therefore, within the scope of Chapter 11. Chapter 11, in conjunction with mainstreamed provisions throughout the code, sets forth scoping requirements.

1101.2 Design. Buildings and facilities shall be designed and constructed to be *accessible* in accordance with this code and ICC A117.1.

❖ This section establishes the primary and fundamental relationship of ICC A117.1 to the code. The code text is intended to "scope" or provide thresholds for application of required accessibility features. The referenced standard contains technical provisions indicating how compliance with the code is achieved. In short, Chapter 11 specifies what, when and how many accessible features are required; the referenced standard indicates how to make that feature accessible. Compliance with both the code and the standard is required.

In accordance with Section 102.3, standards are utilized only to the extent that they are referenced. Note that ICC A117.1 includes technical criteria for several items that are not actually scoped in the code. Such items are not required to comply with ICC A117.1 criteria unless specifically scoped by the authority have jurisdiction. Some examples of items in ICC A117.1 that are not referenced by the IBC are:

Section 403.6 – Corridor handrails

Section 504 – Stairways

Section 704 – Telephones (scoped in IBC Appendix E)

Section 707 – Automatic Teller Machines (scoped in IBC Appendix E)

SECTION 1102
DEFINITIONS

1102.1 Definitions. The following words and terms shall, for the purposes of this chapter and as used elsewhere in the code, have the meanings shown herein:

❖ This section contains definitions of terms that are associated with the subject matter of this chapter. It is important to emphasize that these terms are not exclusively related to this chapter but are applicable everywhere the term is used in the code.

Definitions of terms can help in the understanding and application of the code requirements. The purpose for including these definitions within this chapter is to provide more convenient access to them without having to refer back to Chapter 2. For convenience, these terms are also listed in Chapter 2 with a cross reference to this section. Terms that are italicized provide a visual identification throughout the code that a definition exists for that term. The use and application of all defined terms, including those defined herein, are set forth in Section 201.

ACCESSIBLE. A *site*, building, *facility* or portion thereof that complies with this chapter.

❖ This definition identifies the fundamental concept of Chapter 11. Accessibility is deemed to be accomplished if a building, site or facility complies with the applicable provisions of the code and ICC A117.1. It is not the intent of the code to accommodate fully every type and range of disability; it would not be feasible to do so. The extent to which the code requires accessible features in the various occupancies covered by Chapter 11 (scoping) and the characteristics those features are required to meet through reference to ICC A117.1 (technical requirements) establish that which the code considers fully accessible.

ACCESSIBLE ROUTE. A continuous, unobstructed path that complies with this chapter.

❖ An accessible route enables a person with a disability to approach and utilize a facility's accessible fixtures and features. The design and construction of an accessible route is governed predominantly by provisions necessary for accessibility to a person using a wheelchair. There are typically more physical barriers in the built environment to people with a mobility impairment than in any other category of disability. An accessible route must also be safe and usable by people with other disabilities and, therefore, requirements are set forth in consideration of those needs. For example, there are restrictions on objects that protrude into an accessible route in consideration of a person with a visual impairment. Accessible routes are required for both ingress and egress (see Sections 1007 and 1104).

ACCESSIBLE UNIT. A *dwelling unit* or *sleeping unit* that complies with this code and the provisions for *Accessible units* in ICC A117.1.

❖ There are three levels of accessibility described in the IBC pertaining to dwelling units. Accessible units (always spelled with a capital "A"), Type A units and Type B units. Accessible units are required to be constructed as fully accessible, meaning all required features are present at first occupancy. Unlike Type A and Type B units, Accessible units have no features left as adaptable. Accessible units provide a "higher" level of accessibility than Type A and Type B units and are mandated in all Group I (as a percentage), in Group R-1 (per Table 1107.6.1.1), in most Group R-2 congregate living (as a percentage) and in Group R-4 (at least one unit). The technical criteria for Accessible dwelling units are identified in Section 1002 of the 2003 ICC A117.1, whereas in the 1998 ICC A117.1 they were spread through Chapters 1 through 9. Also see the

commentary for the definitions of "Dwelling unit" and "Sleeping unit" and to Section 1107.2.

CIRCULATION PATH. An exterior or interior way of passage from one place to another for pedestrians.

❖ Examples of circulation paths include sidewalks, walkways, corridors, aisles, courtyards, ramps, stairways and landings. While a stairway is never part of an accessible route, it can be part of a general circulation path for ambulatory persons (see Section 1009).

COMMON USE. Interior or exterior *circulation paths*, rooms, spaces or elements that are not for public use and are made available for the shared use of two or more people.

❖ Some buildings include areas that are restricted to employees only or where public access is limited. Common use spaces may be part of employee work areas but do not include public use spaces. Any space that is shared by two or more persons, such as copy areas, break rooms, toilet rooms or circulation paths, are common use areas. A grade school classroom would be another example of a common use space (see also the commentary for the definition of "Public-use areas" and "Employee work area").

DETECTABLE WARNING. A standardized surface feature built in or applied to walking surfaces or other elements to warn visually impaired persons of hazards on a *circulation path*.

❖ A detectable warning is a change in texture that is detectable by a person with a vision impairment. Specifications for a detectable warning surface are set forth in ICC A117.1. Detectable warnings are only required for transit platform edges so that users can be confident of the warning that is intended to be communicated and not be confused by multiple, different surfaces intended to convey the same warning (see commentary, Section 1109.9).

DWELLING UNIT OR SLEEPING UNIT, MULTISTORY. See definition for "*Multistory unit*."

❖ See the commentary for "Multistory unit."

DWELLING UNIT OR SLEEPING UNIT, TYPE A. See definition for "*Type A unit*."

❖ See the commentary for "Type A dwelling unit."

DWELLING UNIT OR SLEEPING UNIT, TYPE B. See definition for "*Type B unit*."

❖ See the commentary for "Type B dwelling unit."

EMPLOYEE WORK AREA. All or any portion of a space used only by employees and only for work. *Corridors*, toilet rooms, kitchenettes and break rooms are not *employee work areas*.

❖ An employee work area is different in an office versus on a factory line. An employee work area may expand past the station or desk where an employee performs his or her job. An employee work area could include common use spaces, but not public use spaces. Depending on the duties of the employee, it may also include copy areas, stock rooms, filing areas, an assembly line, etc. (see also the commentary for the definitions of "Common use" and "Public-use areas").

Note that not all employee only areas are considered part of employee work areas (i.e., bathrooms, corridors, breakrooms).

FACILITY. All or any portion of buildings, structures, *site* improvements, elements and pedestrian or vehicular routes located on a site.

❖ This term is intentionally broad and includes all portions within a site and all aspects of that site that contain features required to be accessible. This includes parking areas, exterior walkways leading to accessible features, recreational facilities, such as playgrounds and picnic areas, as well as any structures on the site (see also the commentary to the definition of "Site").

INTENDED TO BE OCCUPIED AS A RESIDENCE. This refers to a *dwelling unit* or *sleeping unit* that can or will be used all or part of the time as the occupant's place of abode.

❖ A unit that is a person's home, rather than a unit used for a more transient nature, is a place of abode. Fair housing regulations do not include a 30-day criteria for transient/nontransient, similar to what has been traditionally used by the building codes (see commentary, Section 1107); therefore, beach homes, timeshares, extended stay hotels, etc., may be included.

MULTILEVEL ASSEMBLY SEATING. Seating that is arranged in distinct levels where each level is comprised of either multiple rows, or a single row of box seats accessed from a separate level.

❖ Assembly rooms may include a sloped seating arrangement (i.e., either ramped or stepped) to improve the viewing of the event for the occupants. These spaces can be single- or multiple-level arrangements. For example, for an auditorium with a sloped floor, the entire main floor is a single level. A level can be a balcony or a separate section of seating in an arena or stadium, such as skyboxes. The upper-seating bowl in the coliseum is a separate level, as is the loge in the theater; however, it is not the intent of this provision that each row of seats be considered a separate level.

MULTISTORY UNIT. A *dwelling unit* or *sleeping unit* with *habitable space* located on more than one *story*.

❖ A multistory dwelling or sleeping unit has living, sleeping, eating, cooking or bathroom space on more than one floor level within the unit (see Section 1107.7.2). A residence with only a garage underneath or an unfinished basement would not be a multistory unit.

PUBLIC ENTRANCE. An entrance that is not a *service entrance* or a *restricted entrance*.

❖ A public entrance is one that provides access for the general public or employees, other than the service entrance or a restricted entrance (see the commentary for the definition of "Service entrance" and "Restricted entrance" and to Section 1105.1).

PUBLIC-USE AREAS. Interior or exterior rooms or spaces that are made available to the general public.

❖ This term is utilized to describe all interior and exterior spaces or rooms that may be occupied by the general public for any amount of time. Spaces that are utilized by the general public may be located in facilities that are publicly or privately owned. Examples include the lobby in an office building, a high-school gymnasium with assembly seating an open-air stadium, a multipurpose room, an exposition hall, a restaurant dining room, a health club, etc. (see also the commentary for the definitions of "Common use" and "Employee work area").

RESTRICTED ENTRANCE. An entrance that is made available for *common use* on a controlled basis, but not public use, and that is not a *service entrance*.

❖ The key to this provision is that the entrance has a controlled access or some type of limiting basis. This may be an entrance for jurors only at a courthouse, visitors only at a jail or employees only at a factory. A sports facility where there is control at the entrance for ticket holders only, or a building with locked entrances, is not typically considered a restricted entrance (see the commentary for the definitions of "Service entrance" and "Public entrance" and to Section 1105.1.3).

SELF-SERVICE STORAGE FACILITY. Real property designed and used for the purpose of renting or leasing individual storage spaces to customers for the purpose of storing and removing personal property on a self-service basis.

❖ A portion or space within these facilities can be rented by persons to store personal property. Movement of items into and out of the space is handled by the individual.

SERVICE ENTRANCE. An entrance intended primarily for delivery of goods or services.

❖ This entrance is utilized primarily for accepting or sending deliveries of goods and services. Often this entrance is directly associated with a loading dock, and is not considered a public or restricted entrance (see the commentary for the definitions of "Service entrance" and "Public entrance" and to Section 1105.1.5).

SITE. A parcel of land bounded by a *lot line* or a designated portion of a public right-of-way.

❖ A site, for purposes of accessibility requirements, is the same as that which is considered in the application of other code requirements. The property within the boundaries of the site is under the control of the owner. The owner can be held responsible for code compliance of the site and all facilities on it.

TYPE A UNIT. A *dwelling unit* or *sleeping unit* designed and constructed for accessibility in accordance with this code and the provisions for *Type A units* in ICC A117.1.

❖ A Type A unit has some elements that are constructed for accessibility [e.g., 32-inch (813 mm) clear width doors with maneuvering clearances] and some elements that are constructed as adaptable (e.g., blocking for future installation of grab bars). A Type A dwelling unit is designed and constructed to provide accessibility for wheelchair users throughout the unit, and as such, is considered more accessible than a Type B dwelling unit. The technical requirements for the interior of Type A units are in Section 1003 of ICC A117.1 (see commentary, Section 1107.2).

TYPE B UNIT. A *dwelling unit* or *sleeping unit* designed and constructed for accessibility in accordance with this code and the provisions for *Type B units* in ICC A117.1, consistent with the design and construction requirements of the federal Fair Housing Act.

❖ A Type B dwelling or sleeping unit is designed and constructed to provide a minimal level of accessibility, and as such, is considered less accessible than either an Accessible unit or a Type A unit. The requirements for Type B units are intended to be consistent with the Fair Housing Amendments Act (FHA). The technical requirements for the interior of Type B units are in Section 1004 of ICC A117.1 (see commentary, Section 1107.2).

WHEELCHAIR SPACE. A space for a single wheelchair and its occupant.

❖ A wheelchair space is a designated space for a person to be stationary in his or her wheelchair as part of a fixed assembly seating configuration. The wheelchair space must be sized in accordance with ICC A117.1 (see Section 1108.2.4 for wheelchair space dispersion requirements).

SECTION 1103
SCOPING REQUIREMENTS

1103.1 Where required. *Sites*, buildings, *structures*, *facilities*, elements and spaces, temporary or permanent, shall be *accessible* to persons with physical disabilities.

❖ This section establishes the broad principle that all buildings, structures and their associated sites and facilities are required to be accessible to persons with physical disabilities. This would include anyone who utilizes a space, including occupants, employees, students, spectators, participants and visitors. The approach taken by the code on the subject of accessibility is to require all construction to be accessible and then provide for the acceptable level of inaccessibility that is reasonable and logical. In codes created before the early 1990s, the approach was to list the conditions and occupancies to which the accessibility requirements applied; however, this is no longer practical, since the exceptions are far fewer than the circumstances to which accessibility applies. The 15 exceptions to this section, Sections 1103.2.1 to 1103.2.15, reflect the extent to which accessibility in construction is either exempt or reduced in scope.

1103.2 General exceptions. *Sites*, buildings, *structures*, *facilities*, elements and spaces shall be exempt from this chapter to the extent specified in this section.

❖ Accessibility is generally applicable to all building sites and all spaces and elements within the constructed facilities except as specifically exempted in the subsections to Section 1103.2.

1103.2.1 Specific requirements. *Accessibility* is not required in buildings and *facilities*, or portions thereof, to the extent permitted by Sections 1104 through 1110.

❖ This section provides a correlative reference to the various sections in Chapter 11 that identify when the intended number of accessible elements in various occupancies is less than 100 percent. The number of accessible fixtures and elements required by the referenced sections are deemed to provide adequate accessibility for those circumstances. For example, Section 1106 does not require all parking spaces in a parking facility to be accessible, Section 1107.6.1 does not require all sleeping units in Group R-1 to be fully accessible and Section 1109.8 does not require all storage to be accessible.

1103.2.2 Existing buildings. Existing buildings shall comply with Section 3411.

❖ The second section is a reference to the existing building criteria in Section 3411 for accessibility concerns in existing buildings. In accordance with Section 3401.4, the *International Existing Building Code®* (IEBC®) could also be used (see commentary, Section 3411).

1103.2.3 Employee work areas. Spaces and elements within *employee work areas* shall only be required to comply with Sections 907.5.2.3.2, 1007 and 1104.3.1 and shall be designed and constructed so that individuals with disabilities can approach, enter and *exit* the work area. Work areas, or portions of work areas, other than raised courtroom stations, that are less than 300 square feet (30 m²) in area and elevated 7 inches (178 mm) or more above the ground or finish floor where the elevation is essential to the function of the space shall be exempt from all requirements.

❖ This section states that elements within individual work stations are not required to be accessible, with the exception of visible alarms (see Section 907.5.2.3), accessible means of egress (see Section 1007) and circulation paths (see Section 1104.3.1). The assumption is that the employment nondiscrimination requirements of the Americans with Disabilities Act (ADA) will provide for "reasonable accommodations" to the disability of the employee at that station. In other words, employers will modify individual work stations for the specific requirements of the individual utilizing the space. An accessible route will be required to each work station. An example of this is an individual work station in a laboratory. Installing sinks and built-in counters at accessible levels (see commentary, Sections 1109.3 and 1109.10) could make the station impractical for use by a person without a disability. When a station is required to be adapted for an individual, it would be revised based on the individual's needs and abilities. An accessible route to each work station in the laboratory would be required so that access to and from that station would be available. Note that the 36-inch (914 mm) clear width for the accessible route is the same as the minimum required width of an exit access aisle.

There is an additional exception for work areas that need to be raised 7 inches (178 mm) or more above the floor and have an area of less than 300 square feet (14 m²). Examples would include a raised area around a metal stamping machine, a safety manager's observation station on a production line or the pulpit area in a church. Raised courtroom areas are specifically addressed in Section 1108.4.1.

1103.2.4 Detached dwellings. Detached one- and two-family *dwellings* and accessory structures, and their associated *sites* and facilities, are not required to be *accessible*.

❖ This section exempts detached one- and two-family dwellings from accessibility requirements. The key word here is "detached." For example, a structure containing four or more dwelling units, even if the dwelling units are separated by fire walls, would still be required to be accessible as indicated in Section 1107.

Although one- and two-family detached dwellings are typically regulated by the IRC, this exception in the code is still necessary. Single-family dwellings or duplexes that are four stories or higher would be designed and constructed under Group R-3 requirements of the code. The IRC references Chapter 11 of the code for accessibility requirements in Section R320. Those multiple-family structures with four or more dwelling units that qualify as townhouses or congregate residences with four or more sleeping units under the IRC are required to comply with the requirements for Group R-3 in Section 1107.6.3.

1103.2.5 Utility buildings. Occupancies in Group U are exempt from the requirements of this chapter other than the following:

1. In agricultural buildings, access is required to paved work areas and areas open to the general public.

2. Private garages or carports that contain required *accessible* parking.

❖ This section exempts Group U from accessibility requirements except as indicated, on the basis that such structures are a low priority when considering the need for accessibility. Areas in utility buildings that are required to be accessible would be paved work areas or areas open to the general public, such as if a farmer included a farmstand within his or her barn or buildings in which accessible parking spaces are located.

1103.2.6 Construction sites. Structures, *sites* and equipment directly associated with the actual processes of construction including, but not limited to, scaffolding, bridging, materials hoists, materials storage or construction trailers are not required to be *accessible*.

❖ This section exempts structures directly associated with the construction process because the need for accessibility on a continuous or regular basis in those circumstances is unlikely to arise. Note that all structures that may be involved during a construction project are not exempt—only those specifically involved in the ac-

tual process of construction. For example, if mobile units are brought into house classrooms during a school addition, these classrooms must be accessible.

1103.2.7 Raised areas. Raised areas used primarily for purposes of security, life safety or fire safety including, but not limited to, observation galleries, prison guard towers, fire towers or lifeguard stands, are not required to be *accessible* or to be served by an *accessible route*.

❖ If there is a reason to elevate an area for concerns about security or safety, these areas are not required to be accessible (see Figure 1103.2.7).

Figure 1103.2.7
EXAMPLE OF RAISED AREA EXCEPTION

1103.2.8 Limited access spaces. Nonoccupiable spaces accessed only by ladders, catwalks, crawl spaces, freight elevators or very narrow passageways are not required to be *accessible*.

❖ Nonoccupiable, limited access spaces are considered areas where work could not reasonably be performed by a person using a wheelchair. These areas are not required to be accessible.

1103.2.9 Equipment spaces. Spaces frequented only by personnel for maintenance, repair or monitoring of equipment are not required to be *accessible*. Such spaces include, but are not limited to, elevator pits, elevator *penthouses*, mechanical, electrical or communications equipment rooms, piping or equipment catwalks, water or sewage treatment pump rooms and stations, electric substations and transformer vaults, and highway and tunnel utility facilities.

❖ Spaces that only contain heating, ventilation and air conditioning (HVAC), electrical, elevator, communication and similar types of equipment are considered areas where work could not reasonably be performed by a person utilizing a wheelchair. These areas are not required to be accessible.

1103.2.10 Single-occupant structures. Single-occupant structures accessed only by passageways below grade or elevated above grade including, but not limited to, toll booths that are accessed only by underground tunnels, are not required to be *accessible*.

❖ This section addresses facilities such as toll booths, which are raised to facilitate access to higher vehicles as well as to provide protection from being struck by vehicles. Access to these spaces are often lanes by tunnel or bridge. Some of these types of facilities could also be covered by Section 1103.2.3.

1103.2.11 Residential Group R-1. Buildings of Group R-1 containing not more than five *sleeping unit*s for rent or hire that are also occupied as the residence of the proprietor are not required to be *accessible*.

❖ This section exempts small bed-and-breakfast or transient boarding house facilities that are also the home of the owner.

1103.2.12 Day care facilities. Where a day care facility (Groups A-3, E, I-4 and R-3) is part of a *dwelling unit*, only the portion of the structure utilized for the day care facility is required to be *accessible*.

❖ When an adult day care or child day care facility is part of a person's home, the accessibility requirements are only applicable to the portion of the home that constitutes the day care facility, not the home itself.

1103.2.13 Live/work units. In live/work units constructed in accordance with Section 419, the portion of the unit utilized for nonresidential use is required to be *accessible*. The residential portion of the live/work unit is required to be evaluated separately in accordance with Sections 1107.6.2 and 1107.7.

❖ Live/work units are dwelling units in which a significant portion of the space includes a nonresidential use operated by the tenant/owner. Although the entire unit is classified as a Group R-2 occupancy, for accessibility purposes it is viewed more as a mixed-use condition. The residential portion of the unit is regulated differently for accessibility purposes than the nonresidential portion.

The floor area of the dwelling unit that is intended for residential use is regulated under the provisions of Section 1107.6.2 for Group R-2 occupancies. The requirements for an Accessible unit, Type A unit or Type B unit would be applied based upon the specific residential use of the unit and the number of units in the structure. The exceptions for Type A and Type B units set forth in Section 1107.7 would also exempt such units where applicable. For example, a two story dwelling unit above the work unit in a non-elevator building would never be subject to Type B requirements due to the exemption in Section 1107.7.2. The code does not clarify if a two story live/work unit with the business on the entire first floor and the residence on the entire second floor would be considered a multi-story dwelling unit for purposes of the exception in Section 1107.7.2.

In the nonresidential portion of the unit, full accessibility would be required based upon the intended use. For example, if the nonresidential area of the unit is utilized for hair care services, all elements related to the

service activity must be accessible. This would include site parking where provided, site and building accessible routes, the public entrance, and applicable patron services. In essence, the work portion of the live/work unit would be regulated in the same manner as a stand-alone commercial occupancy.

1103.2.14 Detention and correctional facilities. In detention and correctional facilities, *common use* areas that are used only by inmates or detainees and security personnel, and that do not serve holding cells or housing cells required to be *accessible*, are not required to be *accessible* or to be served by an *accessible route*.

❖ Section 1107.5.5 addresses when sleeping units, special holding or housing cells and medical care units in detention and correctional facilities are required to be accessible. If the purpose of any common or shared space is to serve only the associated cells that are not required to be accessible (e.g., shared bathrooms or living space serving a specific group of cells), then those common or shared spaces are not required to be accessible.

1103.2.15 Walk-in coolers and freezers. Walk-in coolers and freezers intended for employee use only are not required to be *accessible*.

❖ Walk-in coolers and freezers usually have features that make accessibility difficult. For thermal efficiency they may have raised floors, special door seals, unconventional door-operating hardware, tight internal storage, wheeled racking systems, etc. For these and other reasons they are not required to have an accessible entry or to be accessible within.

SECTION 1104
ACCESSIBLE ROUTE

1104.1 Site arrival points. *Accessible routes* within the *site* shall be provided from public transportation stops; *accessible* parking; *accessible* passenger loading zones; and public streets or sidewalks to the *accessible* building entrance served.

Exception: Other than in buildings or facilities containing or serving *Type B units*, an *accessible route* shall not be required between *site* arrival points and the building or facility entrance if the only means of access between them is a vehicular way not providing for pedestrian access.

❖ The intent of this section is to require an accessible route from the point at which one enters the site to any buildings or facilities that are required to be accessible on that site. It is presumed that people with disabilities are capable of gaining access to the site from such locations as accessible parking, public transportation stops, loading zones, public streets or sidewalks.

The exception addresses vehicular routes that provide the only route between an arrival point and an accessible entrance. An accessible route for pedestrian access is not required except in buildings or structures having or serving Type B units. For example, if there is a bus stop at the front of an industrial complex, but the only route to the building entrance is via a long driveway, an accessible pedestrian route to that entrance from the bus stop is not required. For special considerations in residential developments, see the commentary to Sections 1107.4, 1107.7.4, 1107.7.5, 1109.14.1 and 1109.14.2.

1104.2 Within a site. At least one *accessible route* shall connect *accessible* buildings, *accessible* facilities, *accessible* elements and *accessible* spaces that are on the same *site*.

> **Exception:** An *accessible route* is not required between *accessible* buildings, *accessible* facilities, *accessible* elements and *accessible* spaces that have, as the only means of access between them, a vehicular way not providing for pedestrian access.

❖ Developments may include several buildings on the same site. The intent of this section is to require an accessible route to all facilities offered on a site. Often sites are designed such that the only way to reach a building or facility is by automobile. If there are multiple, separated parking areas serving one or more buildings on a site, an accessible route is required between all such parking facilities and the buildings they serve. If there is an exterior feature, such as a swimming pool, located on a site containing multiple buildings, an accessible route is required from each building to the swimming pool. The exception clarifies that an accessible route is not required where no pedestrian access is otherwise intended or provided to a particular building or feature on the site. For special considerations in residential developments with recreational facilities, see the commentary to Sections 1107.3 and 1109.14.

1104.3 Connected spaces. When a building or portion of a building is required to be *accessible*, an *accessible route* shall be provided to each portion of the building, to *accessible* building entrances connecting *accessible pedestrian walkways* and the public way.

Exceptions:

1. In assembly areas with fixed seating, an *accessible route* shall not be required to serve levels where *wheelchair spaces* are not provided.

2. In Group I-2 facilities, doors to *sleeping units* shall be exempted from the requirements for maneuvering clearance at the room side provided the door is a minimum of 44 inches (1118 mm) in width.

❖ This section requires that there be at least one route from an accessible entrance to all required accessible features within a building. If an area is addressed with a specific exception, then an accessible route is not required. For example, mechanical penthouses are exempt under Section 1103.2.9; therefore, an accessible route would not be required to this area.

Once someone gets into a space, they must also be able to evacuate in an emergency situation. This may or may not be via the same route. For accessible means of egress requirements, see Section 1007.

Exception 1 reemphasizes that levels in assembly seating that do not contain wheelchair seating loca-

tions are not required to be accessed by an accessible route; however, the types of services available in the facility must be considered when determining what services must be available to accessible seats. A route is required to services from accessible seats in accordance with Section 1108.2.1.

Exception 2 recognizes the practical implications of door maneuvering clearances on the room side at in-swinging doors in smaller sized patient sleeping rooms in hospitals and nursing homes. The maneuvering clearance is still required on the outside of the Accessible patient sleeping room door. Since the maneuvering clearance in ICC A117.1 is based on the size of the door, there is no credit provided for the wider 44-inch (1118 mm) door width (required in Section 1008.1.1 in Group I-2 for movement of bed) versus the standard 36-inch (914 mm) wide door. Without this exception a patient room would likely require extra width when a toilet room is located on the same wall as the door. Due to operational procedures and observation needs, the door position is normally open, therefore, the need for door maneuvering clearance on the interior side is reduced. At hospitals, nursing homes and rehabilitation centers, the patient room door is usually closed only when privacy is needed during procedures.

The percentage of Accessible units and Type B units in hospitals, nursing homes and rehabilitation centers is addressed in Sections 1107.5.2, 1107.5.3 and 1107.5.4. Maneuvering clearances are part of an accessible route, and accessible routes connect accessible elements; therefore, maneuvering clearances at patient sleeping rooms that are not Accessible or Type B units are not required since an accessible route is not required. For additional information on when Type B units are required in institutional facilities, see the commentary to Section 1107.

1104.3.1 Employee work areas. *Common use circulation paths* within *employee work areas* shall be *accessible routes*.

Exceptions:

1. *Common use circulation paths*, located within *employee work areas* that are less than 300 square feet (27.9 m²) in size and defined by permanently installed partitions, counters, casework or furnishings, shall not be required to be *accessible routes*.

2. *Common use circulation paths*, located within *employee work areas*, that are an integral component of equipment, shall not be required to be *accessible routes*.

3. *Common use circulation paths*, located within exterior *employee work areas* that are fully exposed to the weather, shall not be required to be *accessible routes*.

❖ This requirement for common use circulation paths within employee work areas is consistent with the exception in Section 1103.2.3. An accessible route is required to each employee work area. When employees share work areas, an accessible route must be available throughout that area. Note that the accessible route minimum width of 36 inches (914 mm) clear is consistent with the minimum means of egress pathways.

Exception 1 addresses the accessible route within small employee work areas. Shared work areas that are less than 300 square feet (28 m²) and confined by walls, partitions, permanently installed equipment, cabinets and counters are not required to have an accessible route through that particular area. The intent was to allow such areas as the last two workstations down an aisle, two or three workstations in a small office, the employee side of a beverage bar, portions of commercial kitchens, etc. An accessible route is required to these areas, but not necessarily throughout the area. Again, modifications would be performed at a later date based on employee needs.

Exception 2 permits nonaccessible areas around and through pieces of equipment. An example would be the shared work areas around a piece of assembly equipment in a factory where one or more persons are required to monitor and operate the machine and incoming or outgoing product.

Exception 3 is an exception for outdoor work areas. This would be applicable to landscapers, sewer workers, gravel pit crews, etc.

1104.3.2 Press boxes. Press boxes in assembly areas shall be on an *accessible route*.

Exceptions:

1. An *accessible route* shall not be required to press boxes in *bleachers* that have points of entry at only one level, provided that the aggregate area of all press boxes is 500 square feet (46 m²) maximum.

2. An *accessible route* shall not be required to free-standing press boxes that are elevated above grade 12 feet (3660 mm) minimum provided that the aggregate area of all press boxes is 500 square feet (46 m²) maximum.

❖ Press boxes are required to be served by an accessible route. If the occupant load is five or less, this could be provided by a platform lift (see Section 1109.7, Item 3).

Exception 1 is mainly applicable to press boxes located at the back of the bleacher seating in an outdoor sports facility. The "point of entry at only one level" refers to access to the press box directly from the bleacher seating (see Figure 1104.3.2).

Exception 2 is mainly applicable to the free-standing press box. If that press box is either less than 12 feet (3677 mm) above the ground or more than 500 square feet (46 m²) in area, it must be served by an accessible route.

For both exceptions, if one press box is provided, the total area of the press box must be less than 500 square feet (46 m²). If more than one press box overlooks the same playing field, the aggregate area of both press boxes must be less than 500 square feet

(46 m²). The aggregate area is not intended to be applicable to press boxes that happen to be on the same multiple-facility site. For example, if a high school has a football field with two press boxes, the area of those two press boxes must be added together; however, if the same high school also has a press box for the baseball field, the area of the press box in the baseball field would not be included with the area of the press boxes in the football field.

Figure 1104.3.2
EXAMPLE OF PRESS BOX ACCESSED FROM
BLEACHER SEATING

1104.4 Multilevel buildings and facilities. At least one *accessible route* shall connect each *accessible* level, including *mezzanines*, in multilevel buildings and facilities.

Exceptions:

1. An *accessible route* is not required to stories and *mezzanines* that have an aggregate area of not more than 3,000 square feet (278.7 m²) and are located above and below *accessible* levels. This exception shall not apply to:

    1.1. Multiple tenant facilities of Group M occupancies containing five or more tenant spaces;

    1.2. Levels containing offices of health care providers (Group B or I); or

    1.3. Passenger transportation facilities and airports (Group A-3 or B).

2. Levels that do not contain *accessible* elements or other spaces as determined by Section 1107 or 1108 are not required to be served by an *accessible route* from an *accessible* level.

3. In air traffic control towers, an *accessible route* is not required to serve the cab and the floor immediately below the cab.

4. Where a two-story building or facility has one *story* with an *occupant load* of five or fewer persons that does not contain *public use* space, that *story* shall not

be required to be connected by an *accessible route* to the *story* above or below.

5. Vertical access to elevated employee work stations within a courtroom is not required at the time of initial construction, provided a *ramp*, lift or elevator complying with ICC A117.1 can be installed without requiring reconfiguration or extension of the courtroom or extension of the electrical system.

❖ At least one accessible route is required between levels in a facility. This requirement does not mandate an elevator. The accessible route between levels can be via ramps, platform lifts (where permitted), limited use/limited access (LULA) elevators, passenger elevators, etc. The intent of the exceptions is to allow limited areas to be inaccessible without restricting access to services available to the general public.

Exception 1 addresses conditions under which it may not be practical or economical to provide an accessible route in multilevel buildings. The primary economic consideration is that, in the vast majority of circumstances, the means of providing an accessible route to floor levels above or below the entrance level of the building will be by an elevator. This exception applies to levels above and below the entry level that have an aggregate area of 3,000 square feet (279 m²) or less. For example, if a building had a floor area of 2,000 square feet (186 m²) above the entrance level and a floor area of 2,000 square feet (186 m²) below the entrance level, since the aggregate area of the basement and second floor is 4,000 square feet (372 m²), at least one of the two floor areas would be required to be connected to the entrance level by an accessible route. There are certain facilities that, despite being relatively small in size, are not exempt from the requirement for an accessible route. Due to the critical nature of the services provided, offices of health care providers, passenger transportation facilities and airports are not included in this exception. Multitenant mercantile occupancies (i.e., five or more tenants in the facility), which include facilities such as shopping malls, are not included in the exception because individual retail stores have limited types of commodities (i.e., shoes, eyeglasses, sporting goods, etc.), and the exception would mean that people with disabilities would not have access to the same range of goods that are available to the general public. In these cases, access must be provided such that people with disabilities may patronize all the establishments within the facility. The 3,000-square-foot (279 m²) exception can be applied to a mercantile facility with fewer than five tenants.

Exception 2 is a direct reference to the specific items addressed in Sections 1107 and 1108. It is not a general exception for Groups A, I, R and S. The purpose of Exception 2 is to address areas within specific occupancies where a single element would be repeated multiple times on a level. Where Sections 1107 and 1108 would allow construction of a floor level with no accessible features, this section is not intended to trigger an accessible route to that otherwise inaccessible level. For example, an accessible route would not be required to upper levels in a hotel with all public spaces and all required accessible sleeping units located on the ground floor. If an elevator is to be provided in a building that qualifies for this exception, and the elevator is not part of an accessible route, signage must be provided in accordance with Section 1110.2.

Exception 3 is specific to the unique situations found in air traffic control towers.

Exception 4 permits small nonpublic second floors to not require access. An example would be the second floor in a doctor's office that is used only for storage. The occupant load table in Section 1004.1.1 would limit this storage area to 1,500 square feet (139 m²) [i.e., 300 square feet (28 m²) per occupant × 5 occupants maximum = 1,500 square feet (28 m²)]. If the second floor also contained a mechanical room, that area would be exempt under Section 1103.2.9; however, if the doctor chose later to expand exam rooms into this second floor, he or she would have to provide an accessible route. Another example would be a second level in an airport that included only operational offices. In this case, the area would be limited by the occupant load table in Section 1004.1.1 to 500 square feet (46 m²) [i.e., 100 square feet (9 m²) per occupant × 5 occupants maximum = 500 square feet (46 m²)]. While the exception is specifically stated as applicable only to second-story buildings, it could be interpreted to apply to a one-story building with a basement level or a one-story building with a mezzanine.

Exception 5 is specific to raised platform areas used for employee work stations (judge, clerk, court reporter) in courtrooms. These areas are permitted to be constructed as adaptable for a future accessible route if and when a route is necessary for a mobility impaired employee. This exception does not apply to raised public areas in courtrooms such as witness stands or jury boxes. The assembly seating provisions in Section 1108.2 could be applied to raised, sloped or tiered visitor seating in the gallery areas of the courtrooms.

1104.5 Location. *Accessible routes* shall coincide with or be located in the same area as a general *circulation path*. Where the *circulation path* is interior, the *accessible route* shall also be interior. Where only one *accessible route* is provided, the *accessible route* shall not pass through kitchens, storage rooms, restrooms, closets or similar spaces.

Exceptions:

1. *Accessible routes* from parking garages contained within and serving *Type B units* are not required to be interior.

2. A single *accessible route* is permitted to pass through a kitchen or storage room in an *Accessible unit*, *Type A unit* or *Type B unit*.

❖ One of the objectives of accessibility requirements is to normalize, to the extent possible, the facilities provided for disabled persons and those comparable facilities that exist for people without disabilities. In addition, the intent of this section is to avoid the circumstance where an interior path between facilities is provided but the only accessible route between those same facilities is an exterior path.

When only one accessible route is provided, this section puts limits on the types of spaces through which that accessible route may pass so that it is readily available. Spaces, such as storage rooms, restrooms, closets and kitchens, can be subject to being locked or their availability may be restricted to authorized personnel; therefore, they would not serve as reliable accessible routes for all building occupants.

Exception 1 addresses individual dwelling units that include their own garages and residential facilities with shared garage levels. This exception is limited to accessible parking for Type B units. It is intended to allow a person to exit the garage and enter through the front door, instead of requiring access from the garage into the unit directly. Section 1106.2 requires that when parking is provided within a building, accessible parking must also be located within the building. If this exception is utilized in accordance with Sections 1104.1 and 1107.4, an accessible route must be provided from the accessible parking spaces to the front door.

Exception 2 allows a single accessible route to pass through a kitchen or storage room within a dwelling unit that is required to be an Accessible, Type A or Type B dwelling or sleeping unit. Given the space limitations and typical arrangements of dwelling units, it is not unreasonable to allow access, for example, through a kitchen, to reach an eating area or an exterior patio as the only means of access.

1104.6 Security barriers. Security barriers including, but not limited to, security bollards and security check points shall not obstruct a required *accessible route* or *accessible means of egress*.

> **Exception:** Where security barriers incorporate elements that cannot comply with these requirements, such as certain metal detectors, fluoroscopes or other similar devices, the *accessible route* shall be permitted to be provided adjacent to security screening devices. The *accessible route* shall permit persons with disabilities passing around security barriers to maintain visual contact with their personal items to the same extent provided others passing through the security barrier.

❖ This requirement provides guidance for when a security feature is required along an accessible route. An example would be the security checkpoints in airports. The intent is that the route for a person with mobility impairments should move through security checkpoints as close to the typical route as possible. It is recognized that some people using wheelchairs or with braces could not move through standard security barriers, such as metal detectors.

SECTION 1105
ACCESSIBLE ENTRANCES

1105.1 Public entrances. In addition to *accessible* entrances required by Sections 1105.1.1 through 1105.1.6, at least 60 percent of all *public entrances* shall be *accessible*.

> **Exceptions:**
>
> 1. An *accessible* entrance is not required to areas not required to be *accessible*.
>
> 2. Loading and *service entrances* that are not the only entrance to a tenant space.

❖ A facility is not accessible if the entrances into it are inaccessible. This section establishes a reasonable criteria for providing accessible entrances. A facility is not required to have all of its entrances accessible in order to provide reasonable accommodation to disabled persons. If a facility has multiple public entrances, as a minimum, it is not considered unreasonable to require at least 60 percent of the entrances to be accessible. In addition to the 60-percent accessible public entrances, entrances that have a specific function or provide access to only certain portions of the facility must be addressed (see commentary, Sections 1105.1.1 through 1105.1.6). Sections 1110.1 and 1110.2 require signage at all entrances if all entrances are not accessible.

Doors that are only for means of egress are not considered when determining the number of entrances required to be accessible. However, these doors may need to be on an accessible route for accessible means of egress requirements (see Section 1007). Depending on the arrangement, sometimes the entrance requirements are more restrictive, and sometimes the means of egress requirements are more restrictive. For example, take a small tenant space with two exterior doors—the front door is the public entrance and the back door is the service entrance. Section 1105.1 along with Exception 2 would require only the front door to be on an accessible route for ingress into the building. However, when a space has two means of egress required, Section 1007.1 would require both doors to be on an accessible route for means of egress purposes.

Exception 1 is self-evident in that if a facility or portion of a facility is not required to be accessible, then the entrances to such facilities or spaces are not required to be accessible. An example would be an exterior entrance to the sprinkler room (see Section 1103.2.9).

Exception 2 exempts loading and service entrances from the accessibility requirement on the basis that such entrances are unlikely to be used on a regular basis and may be raised to allow ruck unloading (see definition in Section 1102). Loading and service entrances are required to be accessible if they are the only means of access into a facility or tenant space. This is consistent with Section 1105.1.5.

The intent of the reference to Sections 1105.1.1 through 1105.1.6 is to provide reasonable and convenient availability of an accessible entrance from the accessible facilities provided on the site. All entrances that serve distinct arrival points on the site are required to be accessible. For example, a public entrance from a public parking structure must be accessible (see Section 1105.1.1), as well as the employees-only entrance adjacent to separate employee parking (see Section 1105.1.3). This should deter the formerly common practice of designating an entrance as "the handicapped entrance" (which is inappropriate) without regard to its location relative to such arrival points as accessible parking facilities, transportation facilities, passenger loading zones, taxi stands, public streets or sidewalks and tunnels or elevated walkways. This section also is intended to provide reasonable and convenient availability of an accessible entrance to the building's accessible vertical movement elements, such as the elevators. This would allow a person with disabilities to move readily throughout all levels of the building.

1105.1.1 Parking garage entrances. Where provided, direct access for pedestrians from parking structures to buildings or facility entrances shall be *accessible*.

❖ Occasionally, a parking garage is attached to an office building or mall. An accessible route must be provided from the accessible parking spaces in that parking garage to an accessible entrance into the office or mall portion of the building.

1105.1.2 Entrances from tunnels or elevated walkways. Where direct access is provided for pedestrians from a pedestrian tunnel or elevated walkway to a building or facility, at least one entrance to the building or facility from each tunnel or walkway shall be *accessible*.

❖ Elevated walkways are sometimes provided between adjacent buildings for easy access or protection from the weather for people who must commonly move between certain buildings. Tunnels may serve the same purpose. Washington, DC, has tunnels between many of its congressional buildings. Chicago has a "pedway" system connecting the commuter train stations with the basement level of many buildings in the downtown area. Minneapolis deals with the extreme winter weather conditions in that city by having a system of elevated walkways that connect many of the downtown buildings (see Figure 1105.1.2). When these types of walkways or tunnels are provided, at least one of the building entrances off of these walkways or tunnels must be an accessible entrance.

1105.1.3 Restricted entrances. Where *restricted entrances* are provided to a building or facility, at least one *restricted entrance* to the building or facility shall be *accessible*.

❖ Where access into a specific entrance is controlled (see definition, Section 1102), at least one of that type of entrance must be accessible. An entrance that is locked is not necessarily a restricted entrance. For example, an apartment building may have multiple entrances for tenants where they enter with their key; however, there may be only one entrance for visitors if they need to access a tenant intercom system to gain entrance. The one visitor entrance is required to be accessible.

Figure 1105.1.2
EXAMPLE OF ELEVATED WALKWAY ENTRANCE

1105.1.4 Entrances for inmates or detainees. Where entrances used only by inmates or detainees and security personnel are provided at judicial facilities, detention facilities or correctional facilities, at least one such entrance shall be *accessible*.

❖ Where access to a specific entrance is limited for security reasons to inmates or detainees and the staff supervising them, that entrance must be accessible. This type of situation could occur at courthouses, jails and police stations(see also Sections 1103.2.14, 1107.5.5 and 1108.4.2).

1105.1.5 Service entrances. If a *service entrance* is the only entrance to a building or a tenant space in a facility, that entrance shall be *accessible*.

❖ Loading and service entrances are required to be accessible if they are the only means of access into a facility or tenant space. Typically, loading and service entrances are exempted from the accessibility requirement on the basis that such entrances are unlikely to be used on a regular basis and may be raised to allow for truck unloading (see definition in Section 1102). This is consistent with Section 1105.1, Exception 2.

1105.1.6 Tenant spaces, dwelling units and sleeping units. At least one *accessible* entrance shall be provided to each tenant, *dwelling unit* and *sleeping unit* in a facility.

Exceptions:

1. An *accessible* entrance is not required to tenants that are not required to be *accessible*.

2. An *accessible* entrance is not required to *dwelling units* and *sleeping units* that are not required to be *Accessible units*, *Type A units* or *Type B units*.

❖ Each tenant space must have at least one accessible entrance. Dwelling and sleeping units that are Accessible, Type A or Type B units must have at least one accessible entrance. If a building is a single-tenant building, the 60-percent entrance requirement in Section 1105.1 is also applicable. If a space, whether a tenant, dwelling unit or sleeping unit, does not have accessibility requirements, then an accessible entrance is not required.

SECTION 1106
PARKING AND PASSENGER LOADING FACILITIES

1106.1 Required. Where parking is provided, *accessible* parking spaces shall be provided in compliance with Table 1106.1, except as required by Sections 1106.2 through 1106.4. Where more than one parking facility is provided on a *site*, the number of parking spaces required to be *accessible* shall be calculated separately for each parking facility.

> **Exception:** This section does not apply to parking spaces used exclusively for buses, trucks, other delivery vehicles, law enforcement vehicles or vehicular impound and motor pools where lots accessed by the public are provided with an *accessible* passenger loading zone.

❖ Accessible parking facilities are an important component of building and site accessibility. This section addresses the number of parking spaces required. The location of such parking spaces in relation to accessible building entrances is addressed in Section 1106.6. Parking facility requirements are primarily intended to facilitate accessibility for people with a mobility impairment.

This section is not intended to require that parking facilities be provided; rather, when parking is provided, the minimum number of accessible parking spaces specified herein must be provided. ICC A117.1 contains detailed requirements on the size and configuration of accessible parking spaces, including the required access aisle adjacent to the space, as well as the dimensional requirements for van-accessible parking spaces required by Section 1106.5.

Multitenant facilities, such as a shopping mall, or large facilities, such as hospitals or assembly occupancies, may have multiple parking lots or parking garages. If multiple parking lots, or a combination of parking garages and parking lots are provided for a facility, the number of spaces in each separate parking facility is utilized to determine the number of accessible spaces required. Accessible car and van parking would then be provided in each parking facility accordingly, except as otherwise permitted by Section 1106.6.

In certain types of lots it is not logical to require accessible parking spaces. The exception helps to clarify these situations. Parking lots that contain only parking for buses, trucks and delivery vehicles would not need accessible parking spaces. Lots where the parked cars are limited to vehicles used by law enforcement, impound lots or motor pools are also not required to provide accessible spaces. An accessible passenger drop-off is required to be provided on the site, so that when a person with a mobility impairment comes to that site, he or she has a place to exit the arrival vehicle safely and enter a vehicle coming from the lot.

TABLE 1106.1
ACCESSIBLE PARKING SPACES

TOTAL PARKING SPACES PROVIDED	REQUIRED MINIMUM NUMBER OF ACCESSIBLE SPACES
1 to 25	1
26 to 50	2
51 to 75	3
76 to 100	4
101 to 150	5
151 to 200	6
201 to 300	7
301 to 400	8
401 to 500	9
501 to 1,000	2% of total
1,001 and over	20, plus one for each 100, or fraction thereof, over 1,000

❖ The ratio of required number of accessible parking spaces is based on the accessible parking requirements of the Uniform Federal Accessibility Standards (UFAS). It does not reflect the demographic statistics on wheelchair usage that were used to scope other requirements in Chapter 11, because the majority of disabled parking permit and license plate holders in most states are ambulatory, mobility-impaired persons. The required ratios are intended to be responsive to the anticipated demand for all facilities, except as provided for in Sections 1106.2 through 1106.4, such that accessible parking spaces will be reasonably available on demand. Section 1110.1 states that signage is not required on the one required accessible parking space when the total number of parking spaces provided is four or less. This could be burdensome for the building tenant in that the accessible parking space, which is restricted for use only by authorized vehicles, could constitute anywhere from 25 to 100 percent of the available parking. This may unduly restrict the availability of parking for all other vehicles and patrons of the facility.

1106.2 Groups R-2 and R-3. At least 2 percent, but not less than one, of each type of parking space provided for occupancies in Groups R-2 and R-3, which are required to have *Accessible*, *Type A* or *Type B dwelling* or *sleeping units*, shall be *accessible*. Where parking is provided within or beneath a building, *accessible* parking spaces shall also be provided within or beneath the building.

❖ This section provides a separate criterion for the required number of accessible parking spaces for occupancies in Groups R-2 and R-3 that include Accessible, Type A or Type B units. The 2-percent requirement is based on HUD's Fair Housing Accessibility Guidelines (FHAG). Section 1107.7 identifies buildings where accessible units may not be required.

Where parking is provided within or beneath a building, accessible parking spaces also are to be provided within or beneath the building. If a combination of surface and covered parking is provided, accessible parking may be provided in both locations. This is intended to establish consistency in the type and location of parking spaces available to all people. If parking is provided in individual private parking garages, 2 percent of the parking garages would have to contain accessible parking spaces (see the exception to Section 1106.5). Note that parking provided for visitors should comply with the numbers indicated in Table 1106.1.

In a development, typically parking for dwelling units is considered on a site basis rather than a building by building basis. Accessible parking should be dispersed throughout the development so as to provide the best access possible. It is not the intent to require accessible parking spaces at the entrance to every building, or within every strip of parking garages. For example, it would not be logical to ask for a surface space and a garage space for each building in developments with multiple four-unit buildings.

1106.3 Hospital outpatient facilities. At least 10 percent, but not less than one, of patient and visitor parking spaces provided to serve hospital outpatient facilities shall be *accessible.*

❖ Certain facilities can be expected to have a higher demand for accessible parking spaces than that reflected in Table 1106.1. Medical outpatient facilities are one such facility. This section requires that at least 10 percent of the parking provided for patients and visitors of these types of facilities be accessible. Parking provided for the employees of these facilities should comply with Table 1106.1. If a hospital contains multiple clinics or types of facilities, the amount of parking dedicated to the outpatient facility is interpretive. Options for determining the number of accessible spaces could be based on the area percentage of the outpatient facilities to other facilities or the anticipated number of patients and visitors to these facilities in relation to other facilities.

1106.4 Rehabilitation facilities and outpatient physical therapy facilities. At least 20 percent, but not less than one, of the portion of patient and visitor parking spaces serving rehabilitation facilities specializing in treating conditions that affect mobility and outpatient physical therapy facilities shall be *accessible.*

❖ Medical facilities that specialize in treatment or other services for people with mobility impairments can be expected to have a higher demand for accessible parking spaces. In these cases, at least 20 percent of the parking spaces provided for visitors and patients are required to be accessible. Parking provided for the employees of these facilities should comply with Table 1106.1. If a hospital contains multiple types of facilities, the amount of parking dedicated to the outpatient facil-

ity is interpretive. Options for determining the number of accessible spaces could be based on the area percentage of the outpatient facilities to other facilities or the anticipated number of patients and visitors to these facilities in relation to other facilities.

1106.5 Van spaces. For every six or fraction of six *accessible* parking spaces, at least one shall be a van-accessible parking space.

Exception: In Group R-2 and R-3 occupancies, van-accessible spaces located within private garages shall be permitted to have vehicular routes, entrances, parking spaces and access aisles with a minimum vertical clearance of 7 feet (2134 mm).

❖ Vans that are specially equipped to accommodate a person using a wheelchair require larger parking spaces than the typical accessible parking space. This is because they are usually equipped with a mechanized loading/unloading platform that extends outward from the passenger side of the vehicle. ICC A117.1 contains the dimensional requirements for van-accessible spaces. In order to accommodate the growing usage of these specially equipped vans, this section requires that one in every six accessible parking spaces, or a fraction thereof, be of a size to accommodate these vans. The requirement for van-accessible spaces is not intended to require a greater number of accessible spaces than required by Table 1106.1. For example, if Table 1106.1 requires six accessible parking spaces, one of those six must be van accessible. If seven accessible spaces are required by Table 1106.1, two of those seven are required to be van accessible. On the other hand, if only one accessible parking space is required by Table 1106.1, that space would be a van-accessible space.

The exception modifies the height criteria for van-accessible spaces in ICC A117.1 for private garages serving dwelling units. This is a reduction from the 98-inch (2489 mm) height requirement in ICC A117.1. Since many vans are now converted by lowering the floor rather than raising the roof, the 82-inch (2083 mm) height should accommodate most private passenger vans with lifts.

1106.6 Location. *Accessible* parking spaces shall be located on the shortest *accessible route* of travel from adjacent parking to an *accessible* building entrance. In parking facilities that do not serve a particular building, *accessible* parking spaces shall be located on the shortest route to an *accessible* pedestrian entrance to the parking facility. Where buildings have multiple *accessible* entrances with adjacent parking, *accessible* parking spaces shall be dispersed and located near the *accessible* entrances.

Exceptions:

1. In multilevel parking structures, van-accessible parking spaces are permitted on one level.

2. *Accessible* parking spaces shall be permitted to be located in different parking facilities if substantially equivalent or greater accessibility is provided in

terms of distance from an *accessible* entrance or entrances, parking fee and user convenience.

❖ As previously stated, the majority of disabled parking permits and license plate holders in most states are ambulatory, mobility-impaired persons. Travel distance, as well as severe weather conditions encountered when traversing from the parking lot to the building entrance, are more difficult for persons with mobility impairments to deal with than the general population. The intent of this section is to locate the parking so that the people utilizing the accessible parking spaces have to travel a minimum distance to an accessible entrance (see Figure 1106.6). This requirement is stated in performance terms and requires a degree of subjective judgment on the part of both the designer and the building official in determining the appropriate location for accessible spaces that meets the intent of this section. If a facility has multiple accessible entrances, accessible parking spaces are to be dispersed consistent with the location of these entrances.

Exception 1 is intended to acknowledge a practical difficulty associated with multilevel parking structures. In many cases, a multilevel parking structure will serve accessible building entrances on more than one or all of its parking levels. Specially equipped vans for persons using wheelchairs may be modified by raising the roof of the vehicle in order to provide greater interior headroom. Consequently, such vans require greater vertical clearances. Typical parking structure design may not easily accommodate the necessary vertical clearance for accessible vans due to their low floor-to-ceiling heights. It would be impractical and economically unjustified to require parking structures to be designed solely for the purpose of enabling van-accessible spaces to be located on upper levels. Accordingly, the exception allows the required van-accessible parking spaces to be located on only one level of a multilevel parking structure. The route to and from the space, as well as at the van space and asso-

ciated access aisle, must meet the minimum clear height of 98 inches (2489 mm) as specified in Section 502.6 of ICC A117.1. This will usually be the entry level of the parking facility.

Exception 2 addresses sites where multiple parking facilities are provided to serve a single destination or facility. Since one or more of the parking lots or parking garages may be more attractive to users for various reasons, it is acceptable to locate the required accessible parking based on perceived user convenience and preferences, including distance, parking fees and amenities.

1106.7 Passenger loading zones. Passenger loading zones shall be designed and constructed in accordance with ICC A117.1.

❖ This section does not require passenger loading zones to be provided; however, where provided, they must be accessible in accordance with ICC A117.1. For locations where passenger loading zones are required, see Sections 1106.7.1 through 1106.7.4.

1106.7.1 Continuous loading zones. Where passenger loading zones are provided, one passenger loading zone in every continuous 100 linear feet (30.4 m) maximum of loading zone space shall be *accessible*.

❖ The intent of this section is to address the continuous loading zones found at most larger airports as well as some other types of transportation facilities, such as commuter train "kiss-n-ride" drop-off areas. At least one 20-foot (6096 mm) section for every 100 feet (30 480 mm) of passenger loading zone must meet the accessibility provisions in ICC A117.1.

1106.7.2 Medical facilities. A passenger loading zone shall be provided at an *accessible* entrance to licensed medical and long-term care facilities where people receive physical or medical treatment or care and where the period of stay exceeds 24 hours.

❖ The requirement for accessible passenger loading zones is applicable to both medical and long-term care facilities. The most common examples are hospitals and nursing homes; however, this could also be applicable to some rehabilitation and assisted living facilities (typically Group I-2). Medical and long-term care facilities typically have a higher percentage of people with mobility impairments living in the facility or visiting on a regular basis. Due to their mobility impairment, such people will be frequently driven by a friend or family member, and it is desirable and convenient to drop them off at the entrance rather than parking and requiring them to travel from the parking lot to the facility's entrance. As such, this section requires that a passenger loading zone be provided at no less than one accessible entrance to the facility.

1106.7.3 Valet parking. A passenger loading zone shall be provided at valet parking services.

❖ Valet parking is typically the public access point to a facility. When valet parking is provided, the passenger loading zone must be accessible so that it may be uti-

Figure 1106.6
EXAMPLE OF ACCESSIBLE PARKING SPACES

lized by persons requiring wheelchairs. This would include the access aisle and an accessible route from the access aisle to the front entrance. Note that providing valet parking services does not eliminate the need for accessible parking spaces. Some vehicles that are modified for persons with disabilities may not be drivable by a valet attendant. In addition, valet parking may only be provided during certain periods of time, such as during dinner hours at a restaurant.

1106.7.4 Mechanical access parking garages. Mechanical access parking garages shall provide at least one passenger loading zone at vehicle drop-off and vehicle pick-up areas.

❖ Mechanical parking garages allow a much higher density of vehicles in a given footprint due to the elimination of ramps and drive lanes. With respect to the customer interface, mechanical garages are similar to valet parking and, therefore, require similar scoping for accessibility.

SECTION 1107
DWELLING UNITS AND SLEEPING UNITS

❖ The development of this section was part of the coordination effort with the Fair Housing Amendments Act (FHA) and the *Fair Housing Accessibility Guidelines* (FHAG). For further explanation, see the general comments for this chapter.

One of the phrases developed as part of this coordination was "intended to be occupied as a residence." This defines a type of dwelling or sleeping unit that is a person's home or place of abode. The intent of this language is to clarify that, consistent with the FHA, all dwelling units or sleeping units that can or will be used as a place of abode—even for a short period of stay—must meet the requirements for Type B dwellings. There is a presumption in the IBC that if a building is nontransient (i.e., occupants stay more than 30 days), it is covered by the requirements in this section. In addition, in coordination with the FHA, occupancies that allow stays of fewer than 30 days may be required to meet Type B design and construction requirements.

Accordingly, this commentary will provide guidance for determining when short-term occupancies that might not typically be viewed as housing are covered by this section. Such short-term Group R occupancies may include residential hotels and motels; corporate housing; seasonal vacation units; timeshares; boarding houses; dormitories and migrant-farm worker housing. Also included are some occupancies in Group I, such as nursing homes, assisted living facilities, hospices and homeless shelters.

The key factor in determining whether any of these occupancies are subject to Type B requirements is whether the occupant will have the right to return to the property and whether he or she would have anywhere else in which to return. If it is intended that an occupant will have a right to return to the property and will not have anywhere else to return, the unit is "intended to be occupied as a residence" and must meet the

Type B requirements regardless of the length of stay. Thus, for example, homeless shelters where occupants have a right to return nightly must comply with Type B unit criteria even if the occupants stay only a few nights. Additionally, nursing homes in which a resident moves after vacating his or her primary residence are subject to Type B requirements.

If the occupants have a right to return to the property but also have another place in which to return to, the unit may still be subject to Type B criteria. Additional factors must be considered to determine whether the property is a short-term dwelling or sleeping unit that must meet Type B criteria, or a transient property that is not required to comply with Type B criteria. These factors must be considered by owners, builders, developers, architects and other designers to determine whether or not a building must be designed and constructed in accordance with Type B unit requirements.

For additional discussion on the elements related to hotels, motels, corporate housing and seasonal vacation units and timeshares, see the commentary to Section 1107.6.1 (Group R-1). With respect to most other occupancies, the following factors are also relevant to determine whether the units are subject to Type B requirements:

1. Whether the property is to be marketed as short-term housing;

2. Whether the terms and length of occupancy will be established through a written agreement;

3. How payment will be calculated (e.g., on a daily, weekly, monthly or yearly basis); and

4. What types of amenities and services are offered with the occupancy.

For example, an assisted living facility that provides sleeping units and medical services to its occupants, and bills them on a monthly basis is subject to Type B design and construction requirements. In addition, housing for migrant farm workers that is provided in conjunction with the worker's employment (whether or not rent is paid) and contains amenities for cooking and sleeping would be subject to Type B criteria.

1107.1 General. In addition to the other requirements of this chapter, occupancies having *dwelling units* or *sleeping units* shall be provided with *accessible* features in accordance with this section.

❖ There are two basic types of facilities that this section covers: dwelling units and sleeping units. A dwelling unit is defined in Section 202 as a single unit that contains permanent provisions for "living, sleeping, eating, cooking and sanitation." A sleeping unit is defined in Section 202 as a room in which people sleep, which can include some of the provisions found in a dwelling unit but not all. Occupancy of dwelling units or sleeping units can be transient or nontransient. Dwelling units are typically apartments, condominiums, detached homes or townhouses. Dwelling units can be located in hotels that offer cabins, suites or rooms with kitchen facilities. Bedrooms within dwelling units are not considered sleeping units.

A sleeping unit could be a typical hotel guestroom; a bedroom in a congregate residence, such as a dorm, sorority house, fraternity house, convent, monastery or boarding house; a nursing home room or a jail cell.

1107.2 Design. *Dwelling units* and *sleeping units* that are required to be *Accessible units, Type A units* and *Type B units* shall comply with the applicable portions of Chapter 10 of ICC A117.1. Units required to be *Type A units* are permitted to be designed and constructed as *Accessible units.* Units required to be *Type B units* are permitted to be designed and constructed as *Accessible units* or as *Type A units.*

❖ There are three levels of accessibility that can be required in a dwelling unit or sleeping unit: Accessible units, Type A units and Type B units.

An Accessible unit is constructed for full accessibility in accordance with Section 1002 of ICC A117.1. For example, grab bars are in place in the bathrooms, a clear floor space is provided for front approach at the kitchen sink and bathroom lavatories, 32-inch (813 mm) clear width doors with maneuvering clearances and lever hardware are provided, etc. None of the elements in the unit are constructed for adaptability. The requirements for an Accessible unit are more restrictive than either a Type A unit or a Type B unit.

A Type A unit has some elements that are constructed accessible [e.g., 32-inch (813 mm) clear width doors with maneuvering clearances and lever hardware] and some elements designed to be added or altered when needed (e.g., grab bars can be easily added in bathrooms since blocking in the walls is in place). Type A units follow the technical criteria in Section 1003 of ICC A117.1. This type of unit is less accessible than an Accessible unit and more accessible than a Type B unit.

The scoping or technical requirements for Type B units are consistent with the requirements for units required by the FHAG. A Type B unit is constructed to a lower level of accessibility than either an Accessible unit or a Type A unit. While a person who uses a wheelchair could maneuver in a Type B unit, the technical requirements are geared more towards persons with lesser mobility impairments. Type B units follow the technical requirements in Section 1004 of ICC A117.1. Areas of a Type B unit are allowed to be totally nonaccessible (e.g., sunken living room, extra bedrooms on a mezzanine level). Side approach is permitted to sinks in the kitchen and lavatories in the bathroom rather than planning for a front approach. Some elements are constructed with a minimal level of accessibility [e.g., doors within the unit have a 31³/₄-inch (806 mm) clear width but do not require maneuvering clearances], while some elements are designed to be altered when needed (e.g., blocking in the walls of the bathroom for future installation of grab bars).

This section also takes into consideration the fact that Accessible unit requirements are more stringent than Type A requirements, and Type A requirements

are more stringent than Type B requirements. Units are permitted to be constructed to a higher level of accessibility than required.

The technical criteria for each type of unit in ICC A117.1 is organized in the same order and section number for each element, so make comparisons between types easier. For example, looking at the door provisions in Sections 1002.5, 1003.5 and 1004.5 of ICC A117.1 would clarify that the maneuvering clearances are required at all doors that are part of the accessible route through an Accessible unit or a Type A unit, but for the Type B unit, the maneuvering clearance is only required on the outside of the front door to the unit, not within the unit. Another example would be the requirements for operable parts: Section 1002.9 and 1003.9 address requirements for plumbing fixture and appliance controls, while per Section 1004.9, Type B units do not require plumbing fixture and appliance controls to meet operable parts requirements.

1107.3 Accessible spaces. Rooms and spaces available to the general public or available for use by residents and serving *Accessible units, Type A units* or *Type B units* shall be *accessible. Accessible* spaces shall include toilet and bathing rooms, kitchen, living and dining areas and any exterior spaces, including patios, terraces and balconies.

Exceptions:

1. Recreational facilities in accordance with Section 1109.14.

2. In Group I-2 facilities, doors to *sleeping units* shall be exempted from the requirements for maneuvering clearance at the room side provided the door is a minimum of 44 inches (1118 mm) in width.

❖ Spaces available for use by residents or the general public that are associated with Accessible units, Type A units or Type B units are required to be accessible. This would include spaces within the unit and outside the unit in the same building, as well as facilities that may be outside the building somewhere else on the site. The intent is that a person with a disability could take full advantage of the amenities associated with his or her dwelling or sleeping unit. While it is fairly clear which elements within a unit must be accessible, the associated facilities outside the unit vary depending on the type of building and facilities provided. Public use spaces within a building would include lobby areas while spaces for residents could include shared storage areas or refuse rooms. Shared spaces in nursing homes or assisted living facilities might be dining rooms, therapy rooms and recreational rooms. In dormitories, residents may share study rooms, laundry rooms or mail rooms and facilities located remotely, such as cafeterias. Examples of shared spaces in congregate residences could include community kitchens, living rooms and shared bathrooms. Examples in apartment complexes could include grouped mailboxes, rental offices, exercise rooms and exterior spaces, such as community buildings and swimming pools.

Exception 1 allows for limited access to exterior recreational facilities when multiples of the same type of facility are provided. An example would be multiple tennis courts on the same site for use by residents in a complex (see Section 1109.14 for additional information). The net result is that a person with a disability, either a resident or guest, can access all Accessible, Type A and Type B units, as well as any associated facilities and public areas on the site.

Exception 2 recognizes the practical implications of door maneuvering clearances on the room side at in-swinging doors in smaller sized patient sleeping rooms in hospitals and nursing homes. The maneuvering clearance is still required on the outside of the Accessible patient sleeping room door. Since the maneuvering clearance in ICC A117.1 is based on the size of the door, there is no credit provided for the wider 44-inch (1118 mm) door width (required in Section 1008.1.1 in Group I-2 for movement of bed) versus the standard 36-inch (914 mm) door. Without this exception a patient room would likely require extra width when a toilet room is located on the same wall as the door. Due to operational procedures and observation needs, the door position is normally open, therefore, the need for door maneuvering clearance on the interior side is reduced. At hospitals, nursing homes and rehabilitation centers, the patient room door is usually closed only when visitors or caregivers are present.

The percentage of Accessible units and Type B units in hospitals, nursing homes and rehabilitation centers is addressed in Sections 1107.5.2, 1107.5.3 and 1107.5.4. Maneuvering clearances are part of an accessible route, and accessible routes connect accessible elements; therefore, maneuvering clearances at patient sleeping rooms that are not Accessible or Type B units are not required since an accessible route is not required. For additional information on when Type B units are required in institutional facilities, see the commentary to Section 1107. This is consistent with Section 1104.3, Exception 2.

1107.4 Accessible route. At least one *accessible route* shall connect *accessible* building or facility entrances with the primary entrance of each *Accessible unit, Type A unit* and *Type B unit* within the building or facility and with those exterior and interior spaces and facilities that serve the units.

Exceptions:

1. If due to circumstances outside the control of the owner, either the slope of the finished ground level between *accessible* facilities and buildings exceeds one unit vertical in 12 units horizontal (1:12), or where physical barriers or legal restrictions prevent the installation of an *accessible route*, a vehicular route with parking that complies with Section 1106 at each *public* or *common use* facility or building is permitted in place of the *accessible route*.

2. Exterior decks, patios or balconies that are part of *Type B units* and have impervious surfaces, and that

are not more than 4 inches (102 mm) below the finished floor level of the adjacent interior space of the unit.

❖ Section 1107.3 indicates which spaces associated with accessible dwelling and sleeping units (i.e. Accessible, Type A and Type B units) are required to be accessible. Section 1105 indicates which building entrances are required to be accessible. The intent of this section is twofold: that there will be at least one accessible route that connects required accessible building entrances with the entrance of all accessible dwelling and sleeping units, and that there also must be accessible routes connecting accessible unit entrances with all interior and exterior common and public use spaces and facilities that serve the accessible dwelling or sleeping unit. Some additional facilities that serve units that are not really considered spaces would be accessible parking (see Section 1106.1), passenger loading zones (see Section 1106.7) and public transportation stops located either on site or adjacent to the perimeter. In determining the site arrival points that serve an accessible unit, the designer should make a reasonable determination of how pedestrians and vehicles would arrive at the site.

The Department of Housing and Urban Development's (HUD) review of the 2003 and 2006 editions of the code found that it meets or exceeds the seven design and construction requirements of the Fair Housing Act (FHA) and are "safe harbor" documents for complying with the Fair Housing Accessibility Guidelines (FHAG). To read the final report of the HUD review, go to http://www.hud.gov/offices/fheo/disabilities/modelcodes/. HUD stated, "Having a more recent edition of the IBC recognized by HUD as a safe harbor will ensure that covered dwelling units will be built with the accessible features required by the Act." For additional information, see the commentary at the beginning of Chapter 11.

The following commentary concerning the accessible route serving dwelling units and sleeping units has been provided by HUD:

The five examples serve to illustrate a variety of conditions that may be encountered on nonlevel building sites where Exception 1 is applicable. These examples pertain to the accessible route between facilities on a site. In addition to the on-site accessible routes, the accessible route(s) between the site arrival point(s) and facilities on the site must be considered. Note that the exception in Section 1104.1 does not apply to facilities containing or serving Type B units.

The intent of this section is to ensure that there will be at least one accessible route that connects all accessible building and facility entrances with the entrance of all Accessible, Type A and Type B units. To qualify as an accessible route, that route must serve pedestrians (i.e., sidewalk or other walkway). People with disabilities who need the features of an Accessible, Type A or Type B dwelling or sleeping unit cannot use them if accessible routes are not provided from the entrances to buildings or facilities to the primary entrance of their dwelling or sleeping unit. There also must be accessible routes connecting accessible building or facility entrances with all interior and exterior spaces and facilities that serve such dwelling or sleeping units. For example, if a development has a recreational facility, such as a community center, persons with disabilities who need the features of an Accessible, Type A or Type B unit need an accessible route from their dwelling unit to that community center.

Exception 1 is intended to provide consistency with the federal Fair Housing Act (FHA), which recognizes that, in very rare circumstances, an accessible pedestrian route between an accessible entrance to an accessible unit or an accessible entrance to a building containing accessible units and an exterior public use or common use facility may be impractical because of factors outside the control of the owner. Section 1107.4 requires an accessible pedestrian route between covered dwelling units and public use or common use areas and facilities that are required to be accessible, except in rare circumstances that are outside the control of the owner where extreme terrain or impractical site characteristics result in a finished grade exceeding 8.33 percent or physical barriers or legal restrictions prevent the installation of an accessible pedestrian route. In these cases, Exception 1 allows access to be provided by means of a vehicular route leading from the accessible parking serving the accessible unit to the accessible parking serving the public use or common use facility. Accessible parking complying with Section 1106 must be provided in each parking area. If a building containing accessible units also contains accessible features that are required by other code provisions or federal, state or local laws, then Exception 1 may not apply at all.

It is important to understand that compliance with the accessible design and construction requirements of the FHA is a legal obligation applicable to all architects, engineers, builders, developers and others involved in the design and construction of housing. HUD's regulations implementing the FHA make it clear that the burden of showing the applicability of exceptions is the responsibility of those individuals and entities involved in the design and construction of such housing. In order to ensure compliance with FHA, architects, engineers, developers, builders and others who use the IBC must make accessibility a priority at the planning and design phase of Group I and R developments, including the siting of housing and public use or common use areas. To do this, at the initial stage of site planning and design for all sites, before considering whether Exception 1 applies, persons and entities involved in the design of covered residential occupancies must determine whether and how the exceptions of Sections 1107.7.4 and 1107.7.5 apply.

After careful site planning and design has been completed, the following factors may then be considered to determine whether it is outside the control of the owner to provide an accessible pedestrian route between a building/Type B dwelling unit entrance and a given public use or common use facility. Each such route must be analyzed individually. Exception 1 will only apply when at least one of the following factors is present:

1. Legal restrictions outside the control of the owner. These include setback requirements, tree-save ordinances, easements, environmental restrictions and other limitations that prevent installation of an accessible pedestrian route without violating the law.

2. Physical barriers outside the control of the owner. These include physical characteristics of the site, which are outside the control of the owner that prevent the installation of an accessible pedestrian route.

3. On sites that qualify for the exceptions of Sections 1107.7.4 and 1107.7.5, the presence of extreme terrain or other unusual site characteristics (e.g., flood plain, wetlands) outside the control of the owner that would require substantial additional grading to achieve a slope that will allow for an accessible pedestrian route.

In considering whether the additional grading is substantial enough to qualify for Exception 1, the extent to which the builder has elected to grade the site for other purposes unassociated with accessibility must be considered. If grading for those other purposes is extensive, then substantial additional grading would be necessary to provide the required accessible pedestrian route. If grading for other purposes is not extensive, and substantial additional grading is necessary to provide an accessible pedestrian route, then reliance on Exception 1 would be appropriate. Note that when determining whether the additional grading is substantial, one may not consider the grading that the builder must perform to provide accessible pedestrian routes from site arrival points to the accessible entrances of accessible dwelling or sleeping units. If none of the factors above are present, Exception 1 does not apply. If one or more of these factors is present, then the next step in determining whether Exception 1 applies (i.e., the vehicular route is the only feasible option) is to consider alternative locations and designs for buildings, facilities and accessible pedestrian routes connecting each accessible building and accessible dwelling unit entrance and each public use or common use area required to be accessible to ensure that there is no other way to provide the required accessible pedestrian routes. It is important to recognize that if a road sloping 8.33 percent or less can be provided, then an accessible pedestrian route would also be feasible and must be provided.

Following are some examples to illustrate the proper application of Exception 1:

Example 1: An undisturbed site has slopes of 8.33 percent or less between required accessible entrances to Type B dwelling units and public use or common use areas and there are no legal restrictions or other unique characteristics preventing the construction of accessible routes. For aesthetic reasons, the developer would like to create some hills or decorative berms on the site. Because there are no extreme site conditions (severe terrain or unusual site characteristics such as flood plains), and no legal barriers that prevent installation of an accessible pedestrian route between the buildings/Type B dwelling units and any planned public use or common use facilities, the developer will still be obligated to provide accessible pedestrian routes. Exception 1 to Section 1107.4 is inapplicable in this circumstance.

Example 2: A developer plans to construct several buildings with Type B units clustered in a level area of a site that has some slopes of 10 percent. A swimming pool and tennis court will be added on the two opposing sides of the site. The builder plans grading that will result in a finished grade exceeding a slope of 8.33 percent along the route between the Type B units and the swimming pool and tennis court. There are no physical barriers or legal restrictions outside the control of the owner or builder that prevent him or her from reducing the existing grade to provide an accessible pedestrian route between the Type B units and the pool and tennis court. Therefore, the builder's building plan would not be approved under the code because it is within the owner's control to have the final grading fall below 8.33 percent and meet the slope and other requirements for an accessible pedestrian route. Accessible pedestrian routes between the Type B units, pool and tennis court must be provided.

Example 3: A multiple-family housing complex is built on two sections of a large piece of property, which is divided by a wide stream running through protected wetlands. Both sections of the property are at the same relative elevation and have dwelling units with accessible routes from site arrival points; however, a combination clubhouse and swimming pool is located on one section of the property. Access to each section is provided by an existing public road outside the boundary of the site, which includes a bridge over the stream. Environmental restrictions prevent construction of any type of paved surface between the two sections within the boundary of the site. If environmental restrictions do not prevent the construction of an accessible pedestrian route, such as a boardwalk, through the wetlands connecting the two sections, then the accessible pedestrian route must be provided even if a road cannot be provided. If construction of any type of pedestrian route is prohibited, then

a vehicular route that utilizes the public road and bridge is permitted with parking complying with Section 1106 located at the clubhouse/swimming pool, even though the vehicular route relies on a public road instead of a road through the development.

Example 4: A narrow and deep site has a level section in the front taking up most of the site and another level section at the back that is located up a steep incline. The developer chooses to place all of the buildings/Type B dwelling units on the front section to provide for accessible routes from site arrival points to building entrances. After considering all options for siting buildings and facilities in different locations, including the priority of accessibility, the most feasible location for a planned swimming pool is at the top of the higher section to the rear of the property. Because of the narrowness of the site and the relative elevation of the upper level at the rear of the property, it is impracticable to construct an accessible pedestrian route to the pool; however, a road that slopes more than 8.33 percent can be provided. Under these circumstances, Exception 1 is applicable and access to the swimming pool on the upper level of the site may be provided by means of a vehicular route with parking complying with Section 1106 provided at the Type B dwelling units and the pool.

Example 5: A developer plans to build a multiple-family housing complex with nonelevator buildings on a site with hilly terrain. All of these buildings will have some Type B dwelling units. The developer plans to locate a tennis court on the site. There are gentle slopes exceeding 8.33 percent with existing trees between the entrances to the Type B units and the tennis court. There is also a tree-save ordinance in place. If the builder can grade the site to allow for an accessible pedestrian route to the tennis court without disturbing the trees in violation of the tree-save ordinance, then an accessible pedestrian route between the Type B units and the planned location of the tennis court must be provided. If, however, the grading necessary to reduce the slope of the site near the trees to provide an accessible route would cause tree loss or damage in violation of the ordinance, then the developer cannot grade without violating the tree-save ordinance. The developer must then consider whether the tennis court can be relocated so it is served by an accessible pedestrian route, and if so, the tennis court must be relocated. If the tennis court cannot be relocated so it can be served by an accessible pedestrian route, then the developer may provide a vehicular route from the Type B dwelling units to the tennis court with parking complying with Section 1106 at the Type B dwelling units and the tennis court. Note, however, that if the developer can provide an accessible pedestrian route from some of the buildings without violating the ordinance, the developer must do so, even if it is necessary to provide a vehicular route from other buildings. Additionally, if the grading and construction of the proposed vehicular route can be limited to 8.33

percent by design and would not violate the tree-save ordinance, it is likely that an additional accessible walkway adjacent to the vehicular route would also fall under the scope of work that would not violate the tree-save ordinance and, therefore, must be provided, eliminating the use of Exception 1.

This completes the guidance on this topic provided by HUD for inclusion in this commentary.

Exception 2 is intended for application to Type B units with either a private deck, patio or balcony within an individual unit or to a shared deck, patio or balcony in a "residents only" area of a congregate residence. Decks, patios or balconies located in areas such as the community building or swimming pool should be constructed accessible. When the outside walking surface is impervious to water, such as concrete, there may be a 4-inch (102 mm) step-down between the inside finished floor surface and the outside floor surface. The step-down is permitted to address the concern of water infiltration under the doors at such locations. With a maximum 4-inch (102 mm) difference in elevation, adaptation for accessibility is achievable with raised sleepers and decking. Normal door threshold criteria apply.

1107.5 Group I. *Accessible units* and *Type B units* shall be provided in Group I occupancies in accordance with Sections 1107.5.1 through 1107.5.5.

❖ This section introduces five subsections that set forth the threshold for accessibility in institutional occupancies.

Note that among the various categories of institutional facilities, only Accessible units and Type B units are scoped. There are no requirements for Type A units in institutional occupancies. The criteria for Accessible units and Type B units are specified in Chapter 10 of ICC A117.1.

1107.5.1 Group I-1. *Accessible units* and *Type B units* shall be provided in Group I-1 occupancies in accordance with Sections 1107.5.1.1 and 1107.5.1.2.

❖ Group I-1 is comparable in many respects to Group R-2 in that it is a residential setting. The difference is that Group I-1 is a supervised environment because of the reduced abilities of the occupants; physical, mental or both (see Section 308.2).

1107.5.1.1 Accessible units. At least 4 percent, but not less than one, of the *dwelling units* and *sleeping units* shall be *Accessible units.*

❖ The threshold for accessibility in apartments (Group R-2), in accordance with Section 1107.6.2.1.1, is that 2 percent of the units are required to be Type A units. It is anticipated that Group I-1 will experience a greater demand for accessible facilities because of the nature of the occupants and the frequency of occupant turnover. As such, this section requires that 4 percent of the sleeping units and their associated facilities (e.g.,

bathing and toilet facilities) must meet the criteria for Accessible units.

1107.5.1.2 Type B units. In structures with four or more *dwelling units* or *sleeping units intended to be occupied as a residence*, every *dwelling unit* and *sleeping unit intended to be occupied as a residence* shall be a *Type B unit.*

> **Exception:** The number of *Type B units* is permitted to be reduced in accordance with Section 1107.7.

❖ As discussed in the general comments to this section, most Group I-1 assisted living facilities are intended to serve as a person's place of residence. Since a Group I-1 starts at an occupant load of 16 residents, the Type B criteria for any units that are not Accessible units are almost always required. The exception is a general reference to Section 1107.7, which addresses situations where it is logical to back off on the requirements for Type B units within a structure.

1107.5.2 Group I-2 nursing homes. *Accessible units* and *Type B units* shall be provided in nursing homes of Group I-2 occupancies in accordance with Sections 1107.5.2.1 and 1107.5.2.2.

❖ The requirements for accessibility in Group I-2 are based on the anticipated frequency of usage by disabled persons. There are different thresholds established for different types of health care occupancies, based on the nature of their activity. A nursing home is anticipated to have a high percentage of residents who use wheelchairs or some type of mobility aid.

1107.5.2.1 Accessible units. At least 50 percent but not less than one of each type of the *dwelling units* and *sleeping units* shall be *Accessible units.*

❖ Nursing homes, which generally provide long-term care, are more likely to have patients with physical disabilities. As such, 50 percent of the patient sleeping rooms and their associated facilities (e.g., bathing and toilet facilities) must meet the criteria for Accessible units. Sections 1104.3.2, Exception 2, and 1107.3, Exception 2, exempt any 44-inch (1118 mm) doors from the maneuvering clearance on the inside of the patient sleeping room. Wider doors are required in Section 1008.1.1 for Group I-2 facilities where residents are moved while in their beds.

Housing the elderly should not automatically be classified as a "nursing home." Some residential care/assisted living facilities provide quarters for residents that more closely represent a residential setting (i.e., sleeping, eating, cooking and sanitation within the individual living accommodations) and the residents are more ambulatory than in a nursing home. Based on the anticipated frequency of residents needing wheelchairs, these facilities might be more appropriately classified within the 4- or 10-percent requirement for Accessible units (see Sections 1107.5.1.1 and 1107.5.3.1).

1107.5.2.2 Type B units. In structures with four or more *dwelling units* or *sleeping units intended to be occupied as a residence*, every *dwelling unit* and *sleeping unit intended to be occupied as a residence* shall be a *Type B unit*.

> **Exception:** The number of *Type B units* is permitted to be reduced in accordance with Section 1107.7.

❖ As discussed in the general comments to this section, most Group I-2 nursing homes are intended to serve as a person's place of residence. The Type B criteria for any units that are not Accessible units are almost always required. The exception is a general reference to Section 1107.7, which addresses situations where it is logical to back off on the requirements for Type B units within a structure.

1107.5.3 Group I-2 hospitals. *Accessible units* and *Type B units* shall be provided in general-purpose hospitals, psychiatric facilities, detoxification facilities and *residential care/assisted living facilities* of Group I-2 occupancies in accordance with Sections 1107.5.3.1 and 1107.5.3.2.

❖ The requirements for accessibility in Group I-2 are based on the anticipated disabled occupants. There are different thresholds established for different types of health care occupancies based on the nature of their services and level of care provided. The anticipated need in hospitals is partially based on patients needing varying levels of assistance due to temporary disabilities resulting from illness or operations.

1107.5.3.1 Accessible units. At least 10 percent, but not less than one, of the *dwelling units* and *sleeping units* shall be *Accessible units*.

❖ While the anticipated percentage of patients in general-purpose hospitals is typically higher than 10 percent, there is also an anticipation that many of the patients will be routinely assisted by staff. Section 1109.2, Exception 5, exempts the bathrooms associated with critical care or intensive care. Sections 1104.3.2, Exception 2 and 1107.3, Exception 2, exempt any 44-inch (1118 mm) doors from the maneuvering clearance on the inside of the Accessible patient sleeping room. Wider doors are required in Section 1008.1.1 for Group I-2 facilities where there is movement of patients in their beds. In the case of psychiatric facilities, detoxification facilities or residential care/assisted facilities, it is less likely that the typical patient will be physically disabled.

1107.5.3.2 Type B units. In structures with four or more *dwelling units* or *sleeping units intended to be occupied as a residence*, every *dwelling unit* and *sleeping unit intended to be occupied as a residence* shall be a *Type B unit*.

> **Exception:** The number of *Type B units* is permitted to be reduced in accordance with Section 1107.7.

❖ As discussed in the general comments to this section, most Group I-2 hospitals are not intended to serve as a person's place of residence, therefore, Type B units are not required in most hospitals. However, Group I-2 residential care/assisted living facilities are typically a person's place of residence, and the Type B criteria for

any units that are not Accessible units are required. Whether or not a Group I-2 psychiatric or detoxification facility is a person's place of abode is subject for interpretation. Please see the general discussion to this section for additional information. The exception is a general reference to Section 1107.7, which addresses situations where it is logical to back off on the requirements for Type B units within a structure.

1107.5.4 Group I-2 rehabilitation facilities. In hospitals and rehabilitation facilities of Group I-2 occupancies which specialize in treating conditions that affect mobility, or units within either which specialize in treating conditions that affect mobility, 100 percent of the *dwelling units* and *sleeping units* shall be *Accessible units*.

❖ The requirements for accessibility in Group I-2 are based on the anticipated frequency of usage by disabled persons. There are different thresholds established for different types of health care occupancies based on the nature of their activity. In rehabilitation facilities that specialize in treating conditions that affect mobility, the anticipated need is very high compared to other Group I-2 facilities. In such cases, it is realistic to presume that the majority, if not all, of the patients will have some degree of mobility impairment and will, thus, require accessible features. Accordingly, all patient rooms and their associated bathing and toilet facilities are required to be Accessible units. Sections 1104.3.2, Exception 2 and 1107.3, Exception 2, exempt any 44-inch (1118 mm) doors from the maneuvering clearance on the inside of the patient sleeping room. Wider doors are required in Section 1008.1.1 for Group I-2 facilities where there is movement of patients in their beds. While this facility may also be intended to be occupied as a residence, Accessible unit requirements are more restrictive than Type B requirements, so those needs are already addressed.

1107.5.5 Group I-3. *Accessible units* shall be provided in Group I-3 occupancies in accordance with Sections 1107.5.5.1 through 1107.5.5.3.

❖ The requirement for accessibility in Group I-3 facilities was taken from the ADA/ABA Guidelines. Section 1107.5.5.1 contains requirements for general housing cells. Sections 1107.5.5.2 and 1107.5.5.3 are additional requirements for special types of cells or holding areas. While jail cells may be considered a detainee's residence, since the FHA does not cover detention and correctional facilities, Type B units are not required in Group I-3 occupancies.

1107.5.5.1 Group I-3 sleeping units. In Group I-3 occupancies, at least 2 percent, but not less than one, of the *dwelling units* and *sleeping units* shall be *Accessible units*.

❖ The requirement for sleeping units or cells in Group I-3 is for 2 percent of the general housing cells and their associated facilities to be Accessible units. The current trend in jail design is to group housing cells into "pods." Since the administration of the correctional facility will determine where detainees will reside, there are no requirements for distribution of the Accessible units. A correctional facility may decide to locate all its Accessible cells in one pod due to suicide prevention and other safety concerns. Sections 1103.2.14 and 1109.10 have exceptions for spaces shared by detainees for areas within the jail where Accessible units are not provided.

1107.5.5.2 Special holding cells and special housing cells or rooms. In addition to the *Accessible units* required by Section 1107.5.5.1, where special holding cells or special housing cells or rooms are provided, at least one serving each purpose shall be an *Accessible unit.* Cells or rooms subject to this requirement include, but are not limited to, those used for purposes of orientation, protective custody, administrative or disciplinary detention or segregation, detoxification and medical isolation.

> **Exception:** Cells or rooms specially designed without protrusions and that are used solely for purposes of suicide prevention shall not be required to include grab bars.

❖ If special types of holding or housing cells are provided within a facility (e.g., orientation, protective custody, disciplinary detention, segregation, detoxification, medical isolation), at least one of each type must be an Accessible unit. A cell that is designed specifically for a suicide watch is not required to be an Accessible unit.

1107.5.5.3 Medical care facilities. Patient *sleeping units* or cells required to be *Accessible units* in medical care facilities shall be provided in addition to any medical isolation cells required to comply with Section 1107.5.5.2.

❖ When medical care facilities are provided within a detention or correctional facility, 10 percent of the units are required to be Accessible units. The intent is to be consistent with the hospital requirements found in Section 1107.5.3.1.

1107.6 Group R. *Accessible units, Type A units* and *Type B units* shall be provided in Group R occupancies in accordance with Sections 1107.6.1 through 1107.6.4.

❖ This section introduces four subsections that address accessibility requirements for occupancies in Group R. Criteria for Accessible units, Type A units and Type B units are specified in Chapter 10 of ICC A117.1.

The IRC references this chapter for accessibility provisions; therefore, this chapter may be applicable to housing covered under the IRC (see commentary, Section 1107.6.3). Structures referenced to Chapter 11 from the IRC would be considered Group R-3 for purposes of accessibility only.

1107.6.1 Group R-1. *Accessible units* and *Type B units* shall be provided in Group R-1 occupancies in accordance with Sections 1107.6.1.1 and 1107.6.1.2.

❖ The terms "dwelling unit" and "sleeping unit" are utilized instead of "guestrooms" so that the amenities in the units themselves are reflected. An exception applicable to small bed-and-breakfast style hotels is provided in Section 1103.2.11.

Hotels, motels, boarding houses and other short-term housing types are required to provide Accessible units. Where such facilities are also intended to be occupied as a residence, Type B unit criteria is

applicable. For additional discussion, please see the general comments at the beginning of this chapter.

The following factors should be considered where persons can stay more or less than 30 days, including hotels, motels, corporate housing and seasonal vacation units, in determining the applicability of Type B unit criteria:

1. Amenities included inside the units, including kitchen facilities;

2. Whether the property is to be marketed to the public as short-term housing;

3. Whether the terms and length of the occupancy will be established through a lease or other written agreement; and

4. How payment is calculated (e.g., on a daily, weekly, monthly or yearly basis).

If the amenities and operation of the units are closer to those of apartments than of hotels, they are subject to Type B requirements. For example, if a hotel is marketed as short-term housing, payment is made monthly, and if the units contain kitchens, the hotel would be subject to Type B unit criteria. For additional information see the commentary at the beginning of this chapter and Section 1107.

1107.6.1.1 Accessible units. In Group R-1 occupancies, *Accessible dwelling units* and *sleeping units* shall be provided in accordance with Table 1107.6.1.1. All R-1 units on a *site* shall be considered to determine the total number of *Accessible units. Accessible units* shall be dispersed among the various classes of units. Roll-in showers provided in *Accessible units* shall include a permanently mounted folding shower seat.

❖ The required number of Accessible dwelling or sleeping units in Group R-1 is indicated in Table 1107.6.1.1. The table requires Accessible units to include a variety of bathing options. Where roll-in-type showers are provided, this section requires that they must include folding shower seats mounted on the shower wall. These showers can then serve the dual purpose of roll-in and transfer showers. This would provide a higher level of accessibility than ICC A117.1, which allows roll-in showers to be constructed with or without a shower seat.

For sites where the hotel rooms are provided in multiple buildings, all the units on the site should be considered to determine the number of Accessible units. For example, if a hotel consists of several small buildings on a site, the same number of Accessible units would be required as if the hotel was constructed as a single structure.

Accessible units must be dispersed among the different types of units. This is not an automatic requirement to disperse the Accessible units to different floor levels. When different classes of dwelling or sleeping units are available, special amenities must also be made available in the Accessible units, including suites or larger rooms, kitchenettes, executive levels, etc. For example, if sleeping units with three options of single room, suit or kitchenettes are provided, and two Accessible units are required, the Accessible units must be provided in two of the three options. If no other types of amenities are provided on other levels, the hotel could provide the Accessible units on the main level. Some hotels chose to provide all Accessible units on the main level due to concerns for disabled guests' ease of escape from the building during an emergency, such as a fire. Keep in mind that services available to guests (i.e., pools, exercise rooms, laundry rooms) must also be on an accessible route for guests in the Accessible units.

TABLE 1107.6.1.1. See below.

❖ The number of Accessible dwelling and sleeping units required is coordinated with the ADA/ABA Guidelines requirements. The ADAAG Review Federal Advisory Committee reviewed all the information and data provided by the industry in support of a lower number of Accessible units being required. The committee determined that accessible features are useful to persons with disabilities beyond those who use wheelchairs, and that good design would lessen the "institutional" impression that may make Accessible units less attractive to nondisabled guests. To accommodate wheelchair users in relatively large facilities, this table requires a minimum number of units to be equipped with roll-in-type showers. In accordance with Section 1107.6.1.1, the roll-in showers must have seats, therefore, they can serve the dual purpose of transfer and roll-in shower. Currently ICC A117.1 permits seats in roll-in showers to be an option. The intent of the second column in this table is that different options (i.e., roll-in shower, transfer shower or bathtubs) are available in the Accessible units. It is not the intent of this column to prohibit roll-in showers with seats when there are only one or two Accessible units in the hotel.

1107.6.1.2 Type B units. In structures with four or more *dwelling units* or *sleeping units intended to be occupied as a residence*, every *dwelling unit* and *sleeping unit intended to be occupied as a residence* shall be a *Type B unit*.

Exception: The number of *Type B units* is permitted to be reduced in accordance with Section 1107.7.

❖ Most hotels and motels are not intended to be occupied as a residence; therefore, the criteria for Type B units would not be applicable. However, in certain situations, extended-stay hotels or similar corporate housing arrangements are required to meet Type B unit criteria. For additional information on how to determine if this section is applicable, see the comments at the beginning of this chapter and Sections 1107 and 1107.6.1. The Accessible sleeping units required by Section 1107.6.1.1 exceed Type B unit requirements. The exception is a general reference to Section 1107.7, which addresses situations where it is logical to back off on the requirements for Type B units within a structure.

1107.6.2 Group R-2. *Accessible units*, *Type A units* and *Type B units* shall be provided in Group R-2 occupancies in accordance with Sections 1107.6.2.1 and 1107.6.2.2.

❖ The terms "dwelling unit" and "sleeping unit" are utilized instead of "apartment," "dorm room," "sleeping accommodations," "suites," etc., so that the amenities in the units themselves are reflected and for consistent application within all Group R-2 facilities. Since these types of facilities are intended to be occupied as a residence, Type B unit criteria would be applicable. For a complete discussion, please see the comments to this chapter and this section.

Timeshare properties require a different analysis. Timeshare owners have an interest in the property, yet they typically stay less than 30 days. For a timeshare

TABLE 1107.6.1.1
ACCESSIBLE DWELLING UNITS AND SLEEPING UNITS

TOTAL NUMBER OF UNITS PROVIDED	MINIMUM REQUIRED NUMBER OF ACCESSIBLE UNITS WITHOUT ROLL-IN SHOWERS	MINIMUM REQUIRED NUMBER OF ACCESSIBLE UNITS WITH ROLL-IN SHOWERS	TOTAL NUMBER OF REQUIRED ACCESSIBLE UNITS
1 to 25	1	0	1
26 to 50	2	0	2
51 to 75	3	1	4
76 to 100	4	1	5
101 to 150	5	2	7
151 to 200	6	2	8
201 to 300	7	3	10
301 to 400	8	4	12
401 to 500	9	4	13
501 to 1,000	2% of total	1% of total	3% of total
Over 1,000	20, plus 1 for each 100, or fraction thereof, over 1,000	10 plus 1 for each 100, or fraction thereof, over 1,000	30 plus 2 for each 100, or fraction thereof, over 1,000

property to be subject to Type B requirements, the owners must have an ownership interest in the property itself, rather than in just coming to the area for a vacation without ties to a particular property. The following additional factors must be considered to determine whether timeshare units must meet Type B design and construction requirements:

1. Whether traditional rights of ownership are to be unrestricted (e.g., whether the timeshare owner has the right to occupy, alter or exercise control over a particular unit over a period of time);

2. The nature of the ownership interest conveyed (e.g., fee simple); and

3. The extent to which operations resemble those of a hotel, motel or inn (e.g., reservation, central registration, meals, laundry service).

If an owner's rights regarding the units are subject to few restrictions and the operation of the units is closer to that of condominiums/apartments than hotels, the units are subject to Type B requirements.

1107.6.2.1 Apartment houses, monasteries and convents. *Type A units* and *Type B units* shall be provided in apartment houses, monasteries and convents in accordance with Sections 1107.6.2.1.1 and 1107.6.2.1.2.

❖ Unless they receive some type of federal funding, apartments and condominiums are not covered by the current ADAAG or the ADA/ABA Guidelines. Convents and monasteries are associated with religious organizations, so they also are not covered by ADA requirements. Since all these facilities may serve as a person's residence, they are covered by FHAG. A convent or monastery cannot be sued under FHA for selectively limiting its residents; however, it must comply with the FHAG construction requirements for the residential units that it provides. The scope for Type B units is consistent with FHAG requirements.

1107.6.2.1.1 Type A units. In Group R-2 occupancies containing more than 20 *dwelling units* or *sleeping units*, at least 2 percent but not less than one of the units shall be a *Type A unit*. All R-2 units on a *site* shall be considered to determine the total number of units and the required number of *Type A units*. *Type A units* shall be dispersed among the various classes of units.

Exceptions:

1. The number of *Type A units* is permitted to be reduced in accordance with Section 1107.7.

2. *Existing structures* on a *site* shall not contribute to the total number of units on a *site*.

❖ To be able to better accommodate a person who uses a wheelchair, 2 percent of apartments and condominiums are required to provide Type A units within the development. Congregate residences, such as convents and monasteries, must also provide Type A units. (Other types of congregate living arrangements are addressed in Section 1107.6.2.2.) The housing indus-

try has been concerned that dwelling units equipped with the full range of features to accomplish accessibility will not be marketable to people without disabilities. The Type A requirements are intended to accomplish a middle ground that will satisfy both needs. Allowances made during the construction process for future conversion for accessibility, such as reinforcement in bathroom walls for future installation of grab bars, will allow a unit to be altered later at a considerably lesser cost.

Type A units are required when the site contains more than 20 dwelling units or sleeping units. So that there is a consistent number of Type A units in multibuilding sites based on the size of a development as a whole, all the buildings are added together to determine the number required. For example, a 300-unit building would require six Type A units, and a development with 75 buildings with four units per building would also require six Type A units. Exception 2 is in recognition that a development may be built in stages. Only the units that are being constructed as part of that development phase are considered in determining the number of Type A units.

The Type A units must be dispersed among the classes of units provided. For example, if one-, two- and three-bedroom units are available with the development and two Type A units are required, it is the designer's choice as to which two of the options to provide as Type A units. The designer, however, cannot choose to only provide the one-bedroom option with both Type A units. This is not intended to require Type A units to be provided in different buildings in a multibuilding site. Many times in multibuilding developments, there are shared facilities such as clubhouses or pools. The designer may choose to locate the Type A units in the building closest to those amenities for ease of access for the residents. This is acceptable as long as the dispersion by type requirement is met.

Exception 1 allows the number of Type A units to be reduced in accordance with Section 1107.7.5 when the building's first-floor elevation is required to be raised due to flood-plain regulations. This is the only exception that allows for a reduction in the number of Type A units required.

1107.6.2.1.2 Type B units. Where there are four or more *dwelling units* or *sleeping units intended to be occupied as a residence* in a single structure, every *dwelling unit* and *sleeping unit intended to be occupied as a residence* shall be a *Type B unit*.

Exception: The number of *Type B units* is permitted to be reduced in accordance with Section 1107.7.

❖ When four or more dwelling or sleeping units are provided in a single structure, those units must meet Type B criteria (note the use of the term "structure" instead of "building"). The criteria is applicable if four dwelling units are built together, regardless of fire walls [see Figures 1107.6.2.1.2(1) and 1107.6.2.1.2(2)]. The exception is a general reference to Section 1107.7, which addresses situations where it is logical to back off on the requirements for Type B units within a structure.

Type A units provided in accordance with Section 1107.6.2.1.1 exceed Type B unit requirements.

1107.6.2.2 Group R-2 other than apartment houses, monasteries and convents. In Group R-2 occupancies, other than apartment houses, monasteries and convents, *Accessible units* and *Type B units* shall be provided in accordance with Sections 1107.6.2.2.1 and 1107.6.2.2.2.

❖ This would include congregate housing types, such as dormitories, boarding houses, fraternities and sororities. Nontransient hotels and motels would typically be grouped with these requirements since they may also be operating as transient and nontransient. A building used for two purposes must be designed for the most restrictive provisions (see Section 302.1). Time shares should comply with Section 1107.6.2.1 or 1107.6.2.2, based on whichever it most closely represents (see also commentary, Section 1107.6.2).

Dormitories are typically located in universities. Since universities typically receive some type of federal funding, their dormitories are covered by ADA requirements. Boarding houses, fraternity houses and sorority houses have similar types of living arrangements; therefore, it seems logical that the same anticipated need should result in the same level of required Accessible units. Since all these facilities typically serve as a person's residence, they are covered by FHAG. The scope for Accessible units is consistent with ADA requirements for dormitories. The scope for Type B units is consistent with FHAG requirements.

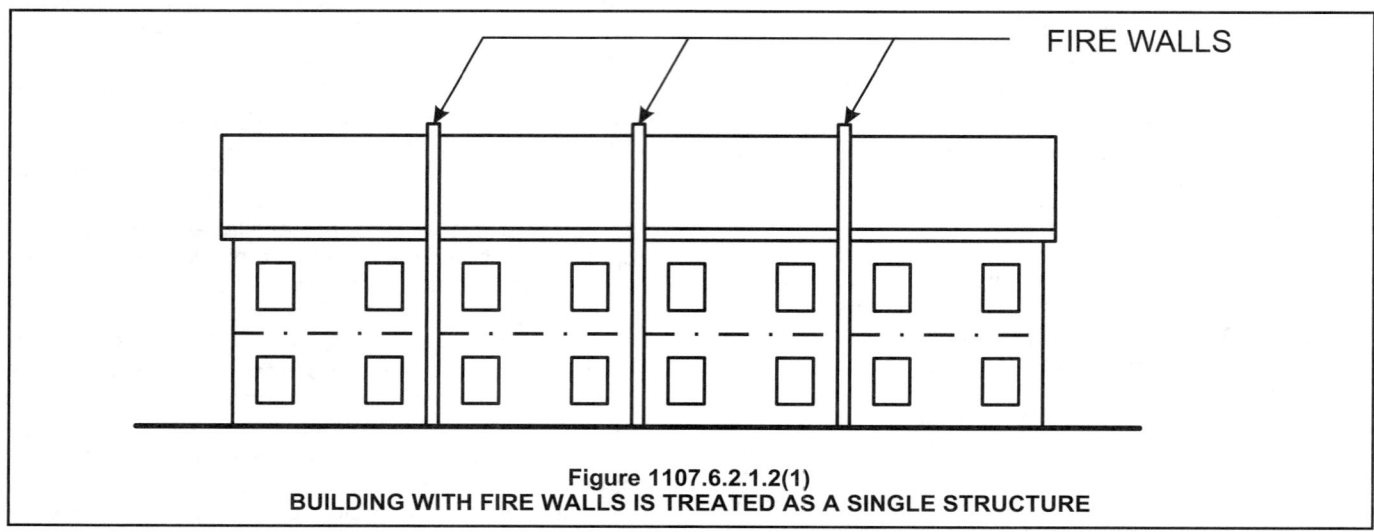

Figure 1107.6.2.1.2(1)
BUILDING WITH FIRE WALLS IS TREATED AS A SINGLE STRUCTURE

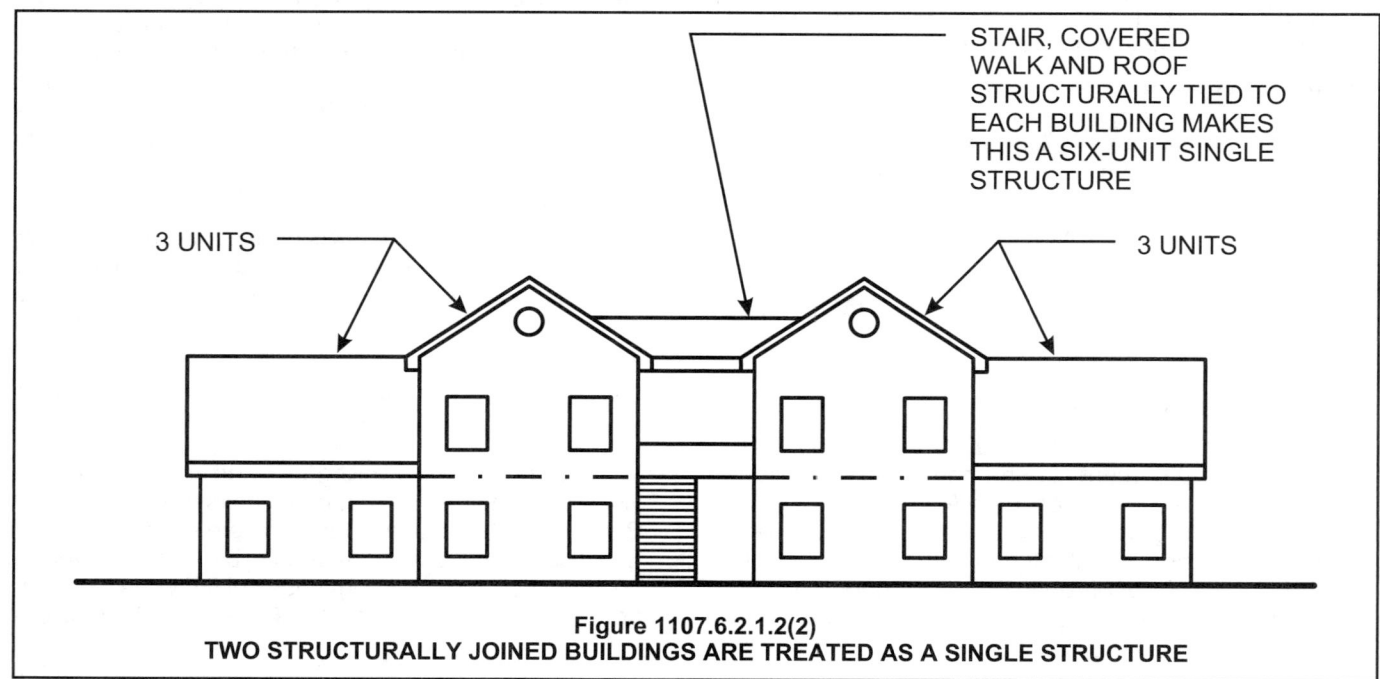

Figure 1107.6.2.1.2(2)
TWO STRUCTURALLY JOINED BUILDINGS ARE TREATED AS A SINGLE STRUCTURE

1107.6.2.2.1 Accessible units. *Accessible dwelling units* and *sleeping units* shall be provided in accordance with Table 1107.6.1.1.

❖ The number of Accessible units required in these Group R-2 facilities is the same as that required in hotels and motels. All associated and shared areas (e.g., bathrooms, kitchens, study rooms) must also be accessible in accordance with ICC A117.1. A certain number of the Accessible units must have associated bathrooms equipped with a roll-in shower.

1107.6.2.2.2 Type B units. Where there are four or more *dwelling units* or *sleeping units intended to be occupied as a residence* in a single structure, every *dwelling unit* and every *sleeping unit intended to be occupied as a residence* shall be a *Type B unit.*

> **Exception:** The number of *Type B units* is permitted to be reduced in accordance with Section 1107.7.

❖ When four or more sleeping units are provided in a single structure, those units must meet Type B criteria. The Accessible sleeping units required by Section 1107.6.2.2.1 exceed Type B unit requirements. The exception is a general reference to Section 1107.7, which addresses situations where it is logical to back off on the requirements for Type B units within a structure.

1107.6.3 Group R-3. In Group R-3 occupancies where there are four or more *dwelling units* or *sleeping units intended to be occupied as a residence* in a single structure, every *dwelling unit* and *sleeping unit intended to be occupied as a residence* shall be a *Type B unit.*

> **Exception:** The number of *Type B units* is permitted to be reduced in accordance with Section 1107.7.

❖ When four or more dwelling or sleeping units are provided in a single structure, those units must meet Type B criteria (note the use of the term "structure" instead of "building"). Since Group R-3 structures can be a series of one or two dwellings per building separated by fire walls (see Section 310.1), the provisions for Type B units would apply to groups of R-3 units. Transient congregate residences with 10 or fewer occupants and nontransient congregate residences with 20 or fewer occupants are permitted to comply with Group R-3 construction requirements (see Section 310.1). Small bed-and-breakfast-type hotels are exempted from accessibility under Section 1103.2.11. These congregate residences may be covered by ADAAG, so a designer/owner may want to consider providing Accessible units (see Sections 1107.6.1.1 and 1107.6.2.2.1) in facilities such as hotels, dormitories, sororities and fraternities. The criteria is applicable if four dwelling units are built together, regardless of fire walls. The exception is a general reference to Section 1107.7, which addresses situations where it is logical to back off on the requirements for Type B units within a structure.

Section 101.2, Exception 1 states that detached one- and two-family dwellings and townhouses that are both three stories or less and have an independent means of egress must comply with the IRC. Section 1103.2.4 exempts detached single-family homes and duplexes from accessibility requirements, but not townhouses. In the IRC, townhouses are further defined as extending from foundation to roof and open on at least two sides; therefore, the typical side-by-side townhouse is constructed using the IRC. Any configuration of townhouses that does not meet all four criteria has to be constructed under the IBC as Group R-2 or possibly Group R-3. If the structure is divided into one- or two-dwelling units per building with fire walls, the structure is a Group R-3 (see commentary, Section 310). Group I-1 and I-2 institutional-type facilities with five or fewer residents and Group R-4 facilities have the option of complying with the code or the IRC, since they often operate similar to a single-family home (see Sections 308.2, 308.3 and 310.1). Day care facilities within single-family homes are specifically addressed under Section 1103.2.12.

Section 320.1 of the IRC refers any structure with four or more sleeping units or dwelling units back to the code as Group R-3 buildings for accessibility requirements. Since these types of facilities typically serve as a person's permanent residence, the units must meet Type B criteria. The exception is a general reference to Section 1107.7, which addresses situations where it is logical to back off on the requirements for Type B units within a structure.

1107.6.4 Group R-4. *Accessible units* and *Type B units* shall be provided in Group R-4 occupancies in accordance with Sections 1107.6.4.1 and 1107.6.4.2.

❖ Group R-4 facilities are limited to between six and 16 residents. A Group R-4 is basically a small Group I-1 facility (see Sections 308.2 and 310.1). The Group R-4 occupancy was originally developed in response to some lawsuits filed under FHA concerning homes for mentally disabled adults. Under the past legacy codes, a structure where persons live in a supervised environment is a Group I occupancy. In neighborhoods zoned residential only, a variance is required if a group of mentally disabled adults wants to move into a single-family-style dwelling. Additionally, Group I facilities are typically required to be sprinklered throughout with an NFPA 13 system. This type of sprinkler system is not what is typically used in a single-family home. The 16-resident criteria is based on two things: 1) In the last census, 98 percent of the households in the United States that identified themselves as single family had 16 or fewer residents; and 2) In facilities where residents are capable of self-preservation, the number 16 also happens to be the limit of residents permitted in a building where an NFPA 13D sprinkler system can be utilized. Establishing these types of facilities as part of Group R eliminates potential conflict with the zoning issue. The limit on the number of residents and allowing the alternative sprinkler system addresses the discrimination in housing concerns and provides a reasonable level of sprinkler protection for the residents.

1107.6.4.1 Accessible units. At least one of the *dwelling* or *sleeping units* shall be an *Accessible unit.*

❖ The requirement for one sleeping unit and its associated facilities (e.g., bathing room) to meet Accessible unit criteria is consistent with Group I-1 requirements. All common rooms are required to be fully accessible. Section 1002 of ICC A117.1 provides the technical criteria for Accessible sleeping units.

1107.6.4.2 Type B units. In structures with four or more *dwelling units* or *sleeping units intended to be occupied as a residence*, every *dwelling unit* and *sleeping unit intended to be occupied as a residence* shall be a *Type B unit.*

> **Exception:** The number of *Type B units* is permitted to be reduced in accordance with Section 1107.7.

❖ Since these types of facilities typically serve as a person's permanent residence, the sleeping unit must meet Type B criteria. The Accessible sleeping unit required by Section 1107.6.4.1 exceeds Type B unit requirements. The exception is a general reference to Section 1107.7, which addresses situations where it is logical to back off on the requirements for Type B units within a structure. Since Group R-4 is a congregate residence, this section would be applicable when four or more sleeping units are provided.

1107.7 General exceptions. Where specifically permitted by Section 1107.5 or 1107.6, the required number of *Type A units* and *Type B units* is permitted to be reduced in accordance with Sections 1107.7.1 through 1107.7.5.

❖ Section 1107.5 or 1107.6 establish when Accessible, Type A or Type B units are expected to be provided. Section 1107.7 covers the general exceptions where it is reasonable and logical to not provide accessibility for some of the units. This would be consistent with the general provisions of Section 1103. Code users should start out with the assumption that everything, in this case Group I and R dwelling and sleeping units, is required to be accessible, and then back off from that level of accessibility when specific exceptions are indicated. Note that there are no exceptions for Accessible units. While Type A units are mentioned in Section 1107.7.1, the only exception for Type A units is found in Section 1107.7.5.

Sections 1107.7.1 through 1107.7.4 primarily deal with nonelevator buildings. If elevators are provided, except as addressed in Section 1107.7.3, all units in the building are required to meet Type B criteria or better. Providing access to the upper floors in these cases would require either an elevator or a ramp system (or multiple elevators or ramps), both of which would be unreasonable. These exceptions are intended to be consistent with the scope of FHAG. It should be noted that these are only exceptions to the requirements for Type B dwelling units.

1107.7.1 Structures without elevator service. Where no elevator service is provided in a structure, only the *dwelling units* and *sleeping units* that are located on stories indicated in Sections 1107.7.1.1 and 1107.7.1.2 are required to be *Type A units* and *Type B units*, respectively. The number of *Type A units* shall be determined in accordance with Section 1107.6.2.1.1.

❖ Only the units located on the floor levels defined in Sections 1107.7.1.1 and 1107.7.1.2 are required to contain Type A or Type B units when no elevator service is provided in the building. Floor levels that do not meet the criteria are not required to have an accessible route to that level, nor do the dwelling units or sleeping units on those levels have to meet Type A or Type B unit criteria. The building must still have the minimum number of Type A units required on the first floor.

1107.7.1.1 One story with Type B units required. At least one *story* containing *dwelling units* or *sleeping units intended to be occupied as a residence* shall be provided with an *accessible* entrance from the exterior of the structure and all units *intended to be occupied as a residence* on that *story* shall be *Type B units.*

❖ This section basically states that Type B units must be provided on at least one level of a building that is not equipped with elevator service. For example, on a flat site, a two-story structure with Type B dwelling units or sleeping units on the first floor would not require an accessible route to, or Type B units on, the second floor (see Figure 1107.7.1.1).

1107.7.1.2 Additional stories with Type B units. On all other stories that have a building entrance in proximity to arrival points intended to serve units on that *story*, as indicated in Items 1 and 2, all *dwelling units* and *sleeping units intended to be occupied as a residence* served by that entrance on that *story* shall be *Type B units.*

1. Where the slopes of the undisturbed *site* measured between the planned entrance and all vehicular or pedestrian arrival points within 50 feet (15 240 mm) of the planned entrance are 10 percent or less, and

2. Where the slopes of the planned finished grade measured between the entrance and all vehicular or pedestrian arrival points within 50 feet (15 240 mm) of the planned entrance are 10 percent or less.

Where no such arrival points are within 50 feet (15 240 mm) of the entrance, the closest arrival point shall be used unless that arrival point serves the *story* required by Section 1107.7.1.1.

❖ This section addresses the idea that a building could have two levels that would be provided with accessible routes. This could be a structure built into a hill or a building with the first level a few feet down and the second level a few feet up from grade level [see Figure 1107.7.1.2(1)].

The basic test for multiple accessible levels is to check for the slope between the building entrance and any arrival points within a 50-foot (15 240 mm) arc. If no arrival points (e.g., sidewalk, parking) are within that 50-foot (15 240 mm) arc, the closest arrival point should be used as a reference. If the slope between these two points is 10 percent or less both before and after grading of the site, than an accessible route is required to that level [see Figures 1107.7.1.2(2) and 1107.7.1.2(3)].

In the case of sidewalks, the closest point to the entrance will be where a public sidewalk entering the site intersects with the sidewalk to the entrance. In the case of resident parking areas, the closest point to the planned entrance will be measured from the entry point to the parking area that is located closest to the planned entrance. The measurement for elevation should be taken from the center of the entrance door to the top of the pavement at the arrival point.

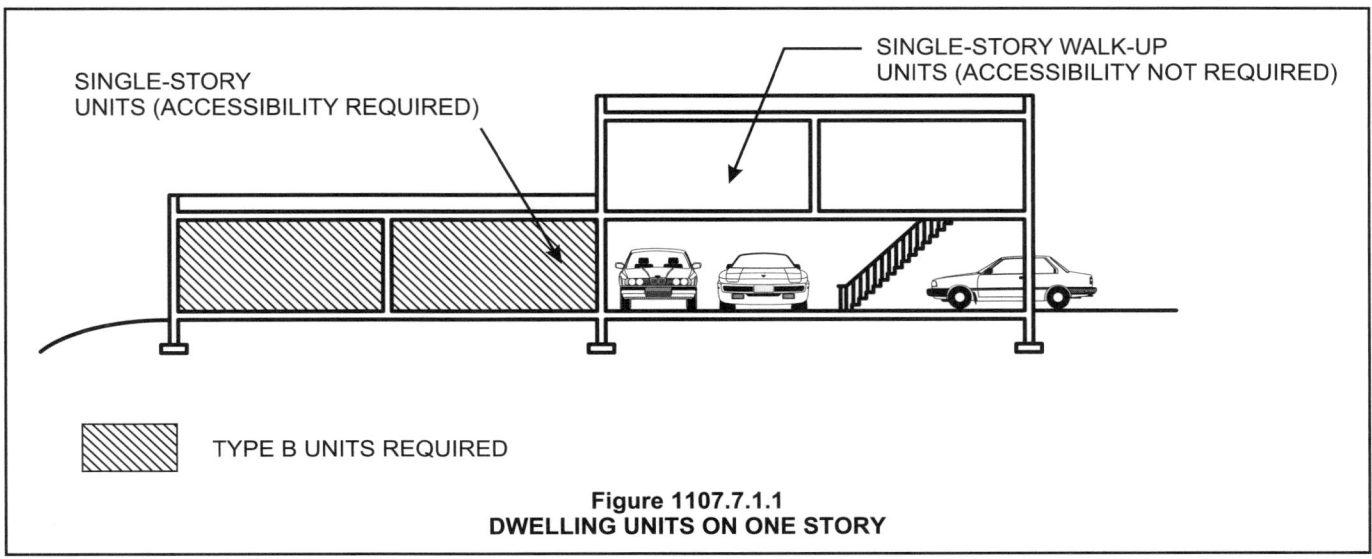

SINGLE-STORY
UNITS (ACCESSIBILITY REQUIRED)

SINGLE-STORY WALK-UP
UNITS (ACCESSIBILITY NOT REQUIRED)

TYPE B UNITS REQUIRED

Figure 1107.7.1.1
DWELLING UNITS ON ONE STORY

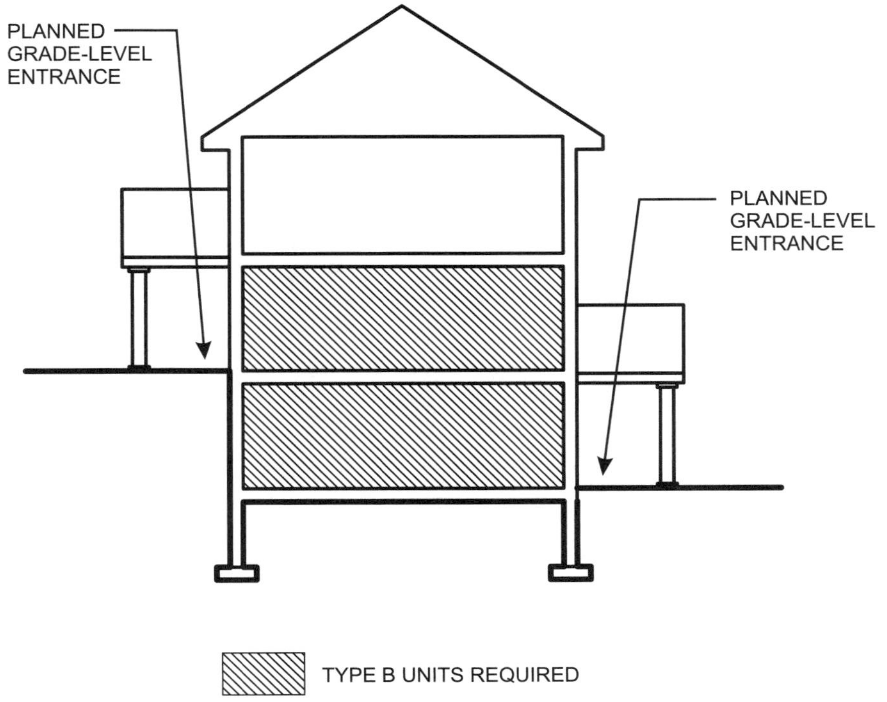

PLANNED
GRADE-LEVEL
ENTRANCE

PLANNED
GRADE-LEVEL
ENTRANCE

TYPE B UNITS REQUIRED

Figure 1107.7.1.2(1)
BUILDING HAS TWO ACCESSIBLE LEVELS, EACH WITH AN ACCESSIBLE ENTRANCE ON AN ACCESSIBLE ROUTE

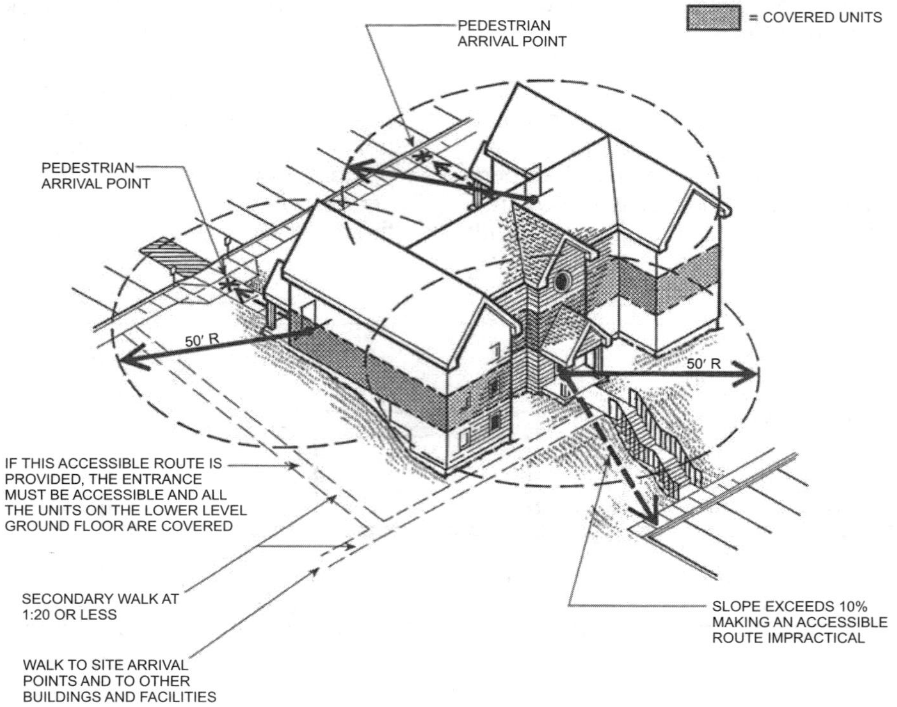

PEDESTRIAN
ARRIVAL POINT

= COVERED UNITS

PEDESTRIAN
ARRIVAL POINT

50' R

50' R

IF THIS ACCESSIBLE ROUTE IS
PROVIDED, THE ENTRANCE
MUST BE ACCESSIBLE AND ALL
THE UNITS ON THE LOWER LEVEL
GROUND FLOOR ARE COVERED

SECONDARY WALK AT
1:20 OR LESS

SLOPE EXCEEDS 10%
MAKING AN ACCESSIBLE
ROUTE IMPRACTICAL

WALK TO SITE ARRIVAL
POINTS AND TO OTHER
BUILDINGS AND FACILITIES

For SI: 1 foot = 304.8 mm.

Figure 1107.7.1.2(2)
SINGLE BUILDING WITH MULTIPLE COMMON ENTRANCES—LOWER GROUND FLOOR UNITS MAY BE COVERED

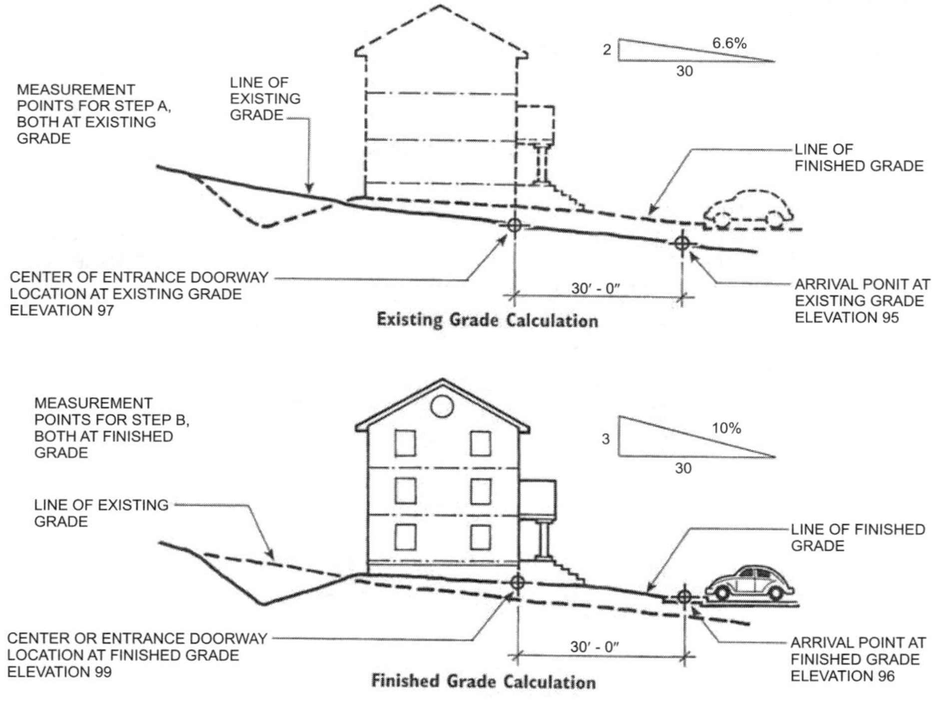

2 ◺ 6.6%
30

MEASUREMENT
POINTS FOR STEP A,
BOTH AT EXISTING
GRADE

LINE OF
EXISTING
GRADE

LINE OF
FINISHED GRADE

CENTER OF ENTRANCE DOORWAY
LOCATION AT EXISTING GRADE
ELEVATION 97

30' - 0"

ARRIVAL PONIT AT
EXISTING GRADE
ELEVATION 95

Existing Grade Calculation

3 ◺ 10%
30

MEASUREMENT
POINTS FOR STEP B,
BOTH AT FINISHED
GRADE

LINE OF EXISTING
GRADE

LINE OF FINISHED
GRADE

CENTER OR ENTRANCE DOORWAY
LOCATION AT FINISHED GRADE
ELEVATION 99

30' - 0"

ARRIVAL POINT AT
FINISHED GRADE
ELEVATION 96

Finished Grade Calculation

For SI: 1 foot = 304.8 mm.

Figure 1107.7.1.2(3)
EXAMPLE FOR GRADE CALCULATIONS

1107.7.2 Multistory units. A *multistory dwelling* or *sleeping unit* which is not provided with elevator service is not required to be a *Type B unit*. Where a *multistory unit* is provided with external elevator service to only one floor, the floor provided with elevator service shall be the primary entry to the unit, shall comply with the requirements for a *Type B unit* and a toilet facility shall be provided on that floor.

❖ Section 1107.7.2 addresses multistory dwelling units (see the definitions in Section 1102.1). For example, the typical townhouse scenario, where there is a series of two-story units adjacent to each other in a single structure, would not require Type B units. If multistory units are provided in a building containing elevators and access is provided at only one floor level, that level must be accessible and contain a toilet facility [see Figures 1107.7.2(1) and 1107.7.2(2)].

1107.7.3 Elevator service to the lowest story with units. Where elevator service in the building provides an *accessible route* only to the lowest *story* containing *dwelling* or *sleeping units intended to be occupied as a residence*, only the units on that *story* which are *intended to be occupied as a residence* are required to be *Type B units*.

❖ Section 1107.7.3 exempts dwelling units located on upper floors as long as the dwelling units on the first floor containing dwelling units are at least Type B units. For example, a three-story structure has a business on the first floor with apartments on the second and third levels. An accessible route is required to the second level, and all dwelling units on that level are required to be Type B units; however, an accessible route and Type B units are not required to the third level. (Note: If an elevator is utilized to provide access to only the lowest level containing dwelling or sleeping units, this building would not be considered an "elevator" building) [see Figures 1107.7.3(1) and 1107.7.3(2)].

1107.7.4 Site impracticality. On a *site* with multiple nonelevator buildings, the number of units required by Section 1107.7.1 to be *Type B units* is permitted to be reduced to a percentage which is equal to the percentage of the entire *site* having grades, prior to development, which are less than 10 percent, provided that all of the following conditions are met:

1. Not less than 20 percent of the units required by Section 1107.7.1 on the *site* are *Type B units*;

2. Units required by Section 1107.7.1, where the slope between the building entrance serving the units on that *story* and a pedestrian or vehicular arrival point is no greater than 8.33 percent, are *Type B units*;

3. Units required by Section 1107.7.1, where an elevated walkway is planned between a building entrance serving the units on that *story* and a pedestrian or vehicular arrival point and the slope between them is 10 percent or less are *Type B units*; and

4. Units served by an elevator in accordance with Section 1107.7.3 are *Type B units*.

❖ Section 1107.7.4 addresses multiple buildings on a sloping site. For example, if an apartment complex was built on a steep or hilly site, the number of Type B dwelling units required on the ground floor could be reduced due to the difficulty of providing accessible routes. A minimum of 20 percent of the total ground floor units on the site must be Type B units, regardless of site complications (see Figure 1107.7.4).

1107.7.5 Design flood elevation. The required number of *Type A units* and *Type B units* shall not apply to a *site* where the required elevation of the lowest floor or the lowest horizontal structural building members of nonelevator buildings are at or above the *design flood elevation* resulting in:

1. A difference in elevation between the minimum required floor elevation at the primary entrances and vehicular and pedestrian arrival points within 50 feet (15 240 mm) exceeding 30 inches (762 mm), and

2. A slope exceeding 10 percent between the minimum required floor elevation at the primary entrances and vehicular and pedestrian arrival points within 50 feet (15.24 m).

Where no such arrival points are within 50 feet (15.24 m) of the primary entrances, the closest arrival points shall be used.

❖ Residential structures in flood hazard areas must be elevated (see Section 1612). If, based on the required floor elevation, the criteria in either Item 1 or 2 are met, it is considered that an accessible route to the units is not feasible. This section applies to both Type A and Type B dwelling units [see Figures 1107.7.5(1) and 1107.7.5(2)].

The design flood elevation is the height at which the lowest floor level is required to be located by the local jurisdiction. Sometimes this is the same as the base flood elevation, but a community may choose to require additional "free board" to require a higher design flood elevation. This will reduce their potential flood losses and can increase the insurance rating for the area, thus possibly reducing the cost of both flood insurance and rebuilding costs in the event of a flood. Since this is an issue of public safety, there are reductions in the level of accessibility required in residential facilities without elevators that were required to be raised to meet flood provisions. Please note that the exception is based on the required design flood elevation. If someone would choose to construct the first-floor elevation higher than the design floor elevation, the measurement would still be from the design flood elevation rather than the actual floor elevation.

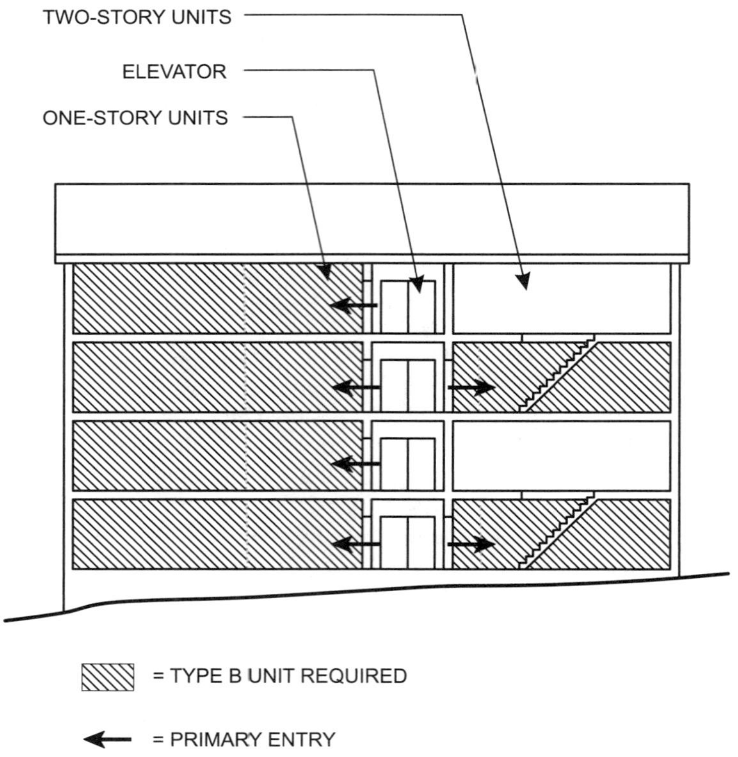

TWO-STORY UNITS

ELEVATOR

ONE-STORY UNITS

▨ = TYPE B UNIT REQUIRED

← = PRIMARY ENTRY

Figure 1107.7.2(1)
BUILDINGS WITH ELEVATOR(S): ALL SINGLE-STORY UNITS AND
THE PRIMARY ENTRY LEVEL OF MULTISTORY UNITS

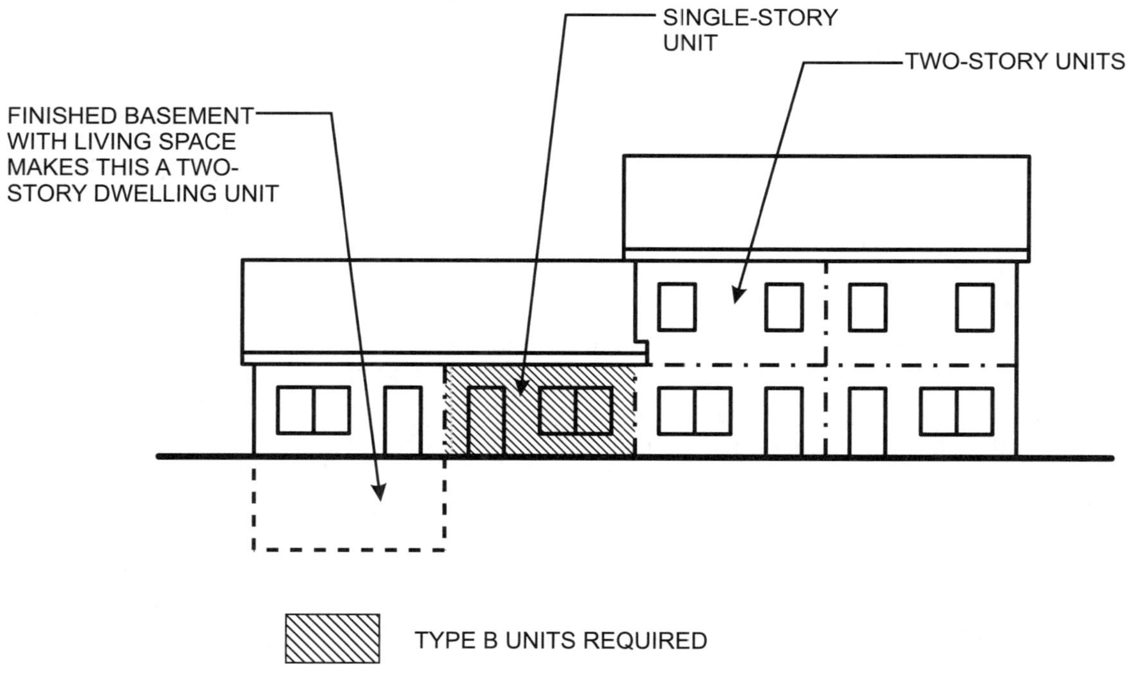

SINGLE-STORY
UNIT

TWO-STORY UNITS

FINISHED BASEMENT
WITH LIVING SPACE
MAKES THIS A TWO-
STORY DWELLING UNIT

▨ TYPE B UNITS REQUIRED

Figure 1107.7.2(2)
GROUND FLOOR UNITS IN BUILDINGS OF FOUR OR MORE UNITS

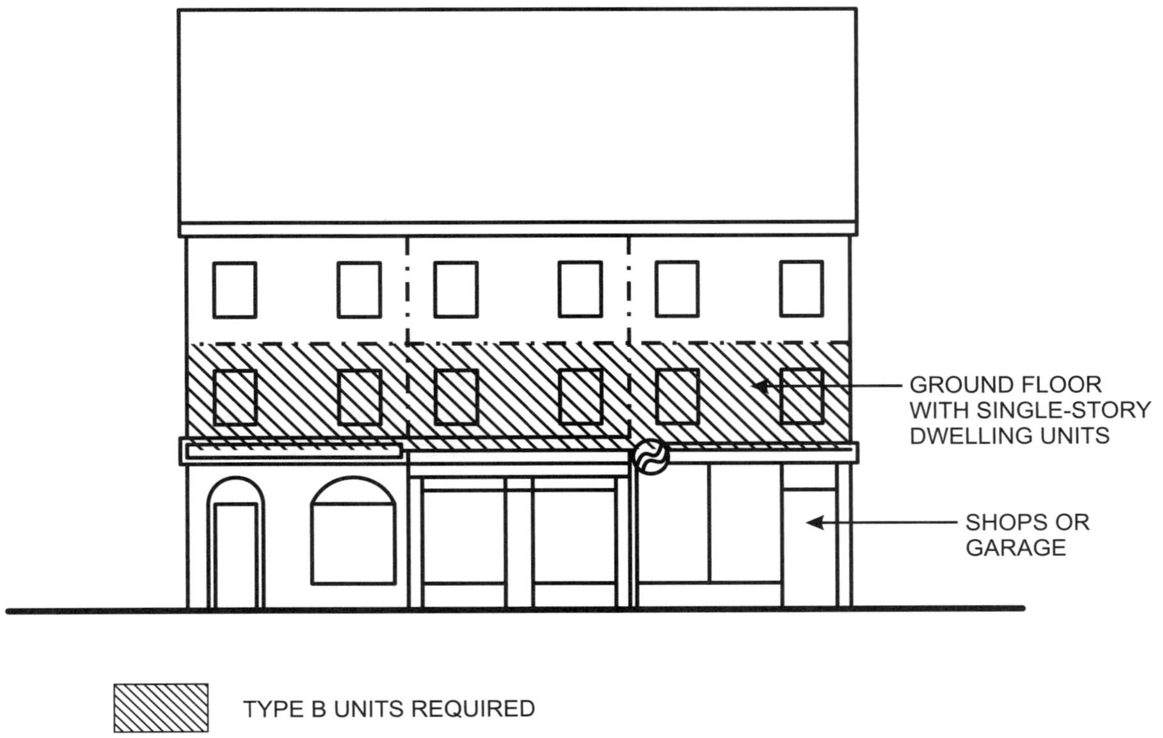

TYPE B UNITS REQUIRED

Figure 1107.7.3(1)
DWELLING UNITS OVER SHOPS AND GARAGES

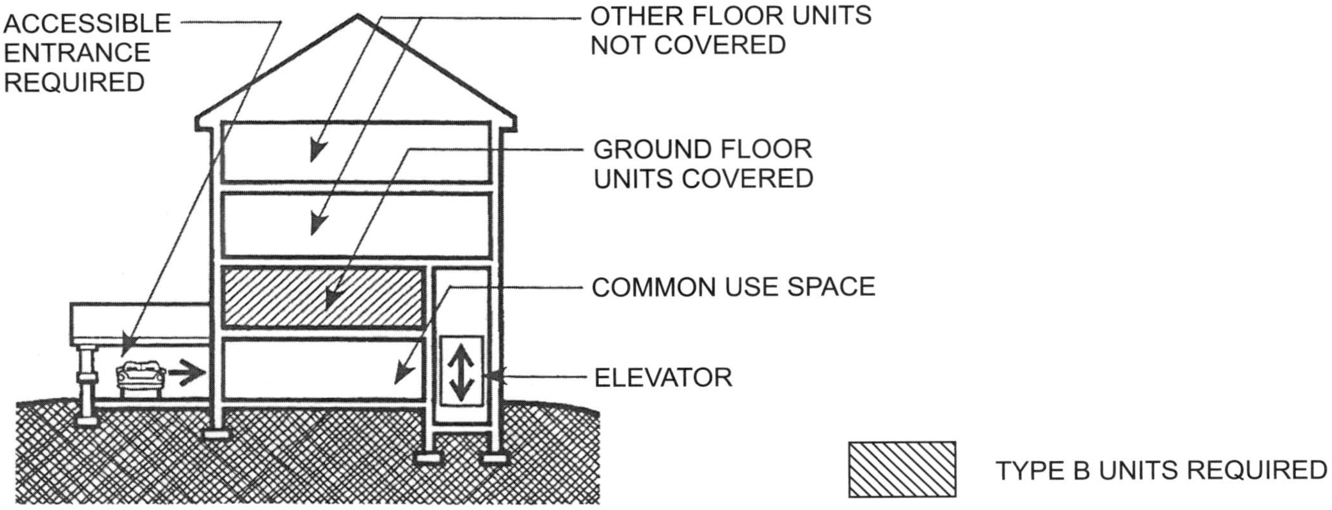

TYPE B UNITS REQUIRED

Figure 1107.7.3(2)
ELEVATOR TO FIRST FLOOR OF DWELLING UNITS ABOVE GRADE OR
ENTRANCE LEVEL DOES NOT MAKE AN ELEVATOR BUILDING

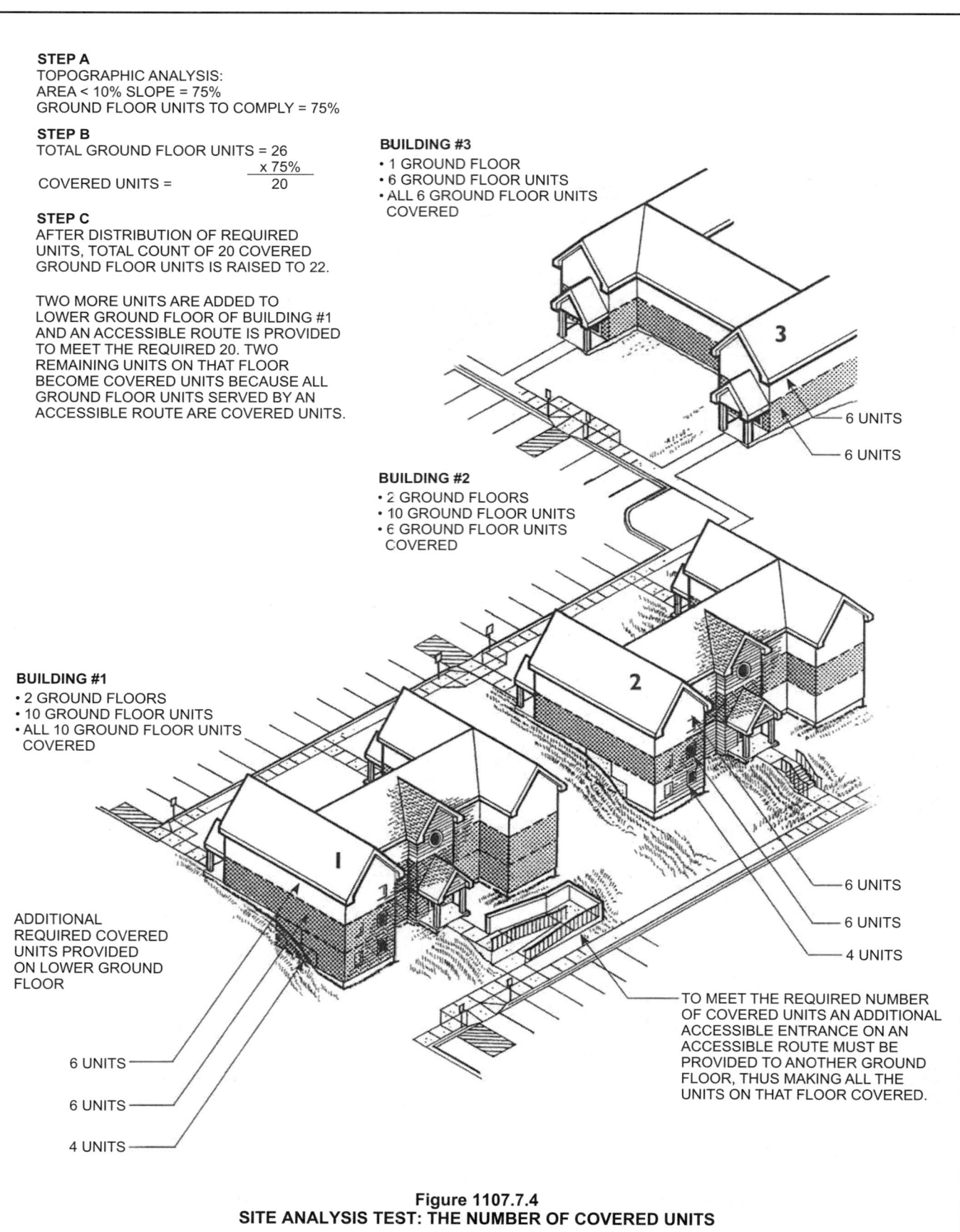

STEP A
TOPOGRAPHIC ANALYSIS:
AREA < 10% SLOPE = 75%
GROUND FLOOR UNITS TO COMPLY = 75%

STEP B
TOTAL GROUND FLOOR UNITS = 26
x 75%
COVERED UNITS =   20

STEP C
AFTER DISTRIBUTION OF REQUIRED
UNITS, TOTAL COUNT OF 20 COVERED
GROUND FLOOR UNITS IS RAISED TO 22.

TWO MORE UNITS ARE ADDED TO
LOWER GROUND FLOOR OF BUILDING #1
AND AN ACCESSIBLE ROUTE IS PROVIDED
TO MEET THE REQUIRED 20. TWO
REMAINING UNITS ON THAT FLOOR
BECOME COVERED UNITS BECAUSE ALL
GROUND FLOOR UNITS SERVED BY AN
ACCESSIBLE ROUTE ARE COVERED UNITS.

BUILDING #3
• 1 GROUND FLOOR
• 6 GROUND FLOOR UNITS
• ALL 6 GROUND FLOOR UNITS
  COVERED

3

6 UNITS
6 UNITS

BUILDING #2
• 2 GROUND FLOORS
• 10 GROUND FLOOR UNITS
• 6 GROUND FLOOR UNITS
  COVERED

BUILDING #1
• 2 GROUND FLOORS
• 10 GROUND FLOOR UNITS
• ALL 10 GROUND FLOOR UNITS
  COVERED

ADDITIONAL
REQUIRED COVERED
UNITS PROVIDED
ON LOWER GROUND
FLOOR

6 UNITS
6 UNITS
4 UNITS

6 UNITS
6 UNITS
4 UNITS

TO MEET THE REQUIRED NUMBER
OF COVERED UNITS AN ADDITIONAL
ACCESSIBLE ENTRANCE ON AN
ACCESSIBLE ROUTE MUST BE
PROVIDED TO ANOTHER GROUND
FLOOR, THUS MAKING ALL THE
UNITS ON THAT FLOOR COVERED.

Figure 1107.7.4
SITE ANALYSIS TEST: THE NUMBER OF COVERED UNITS

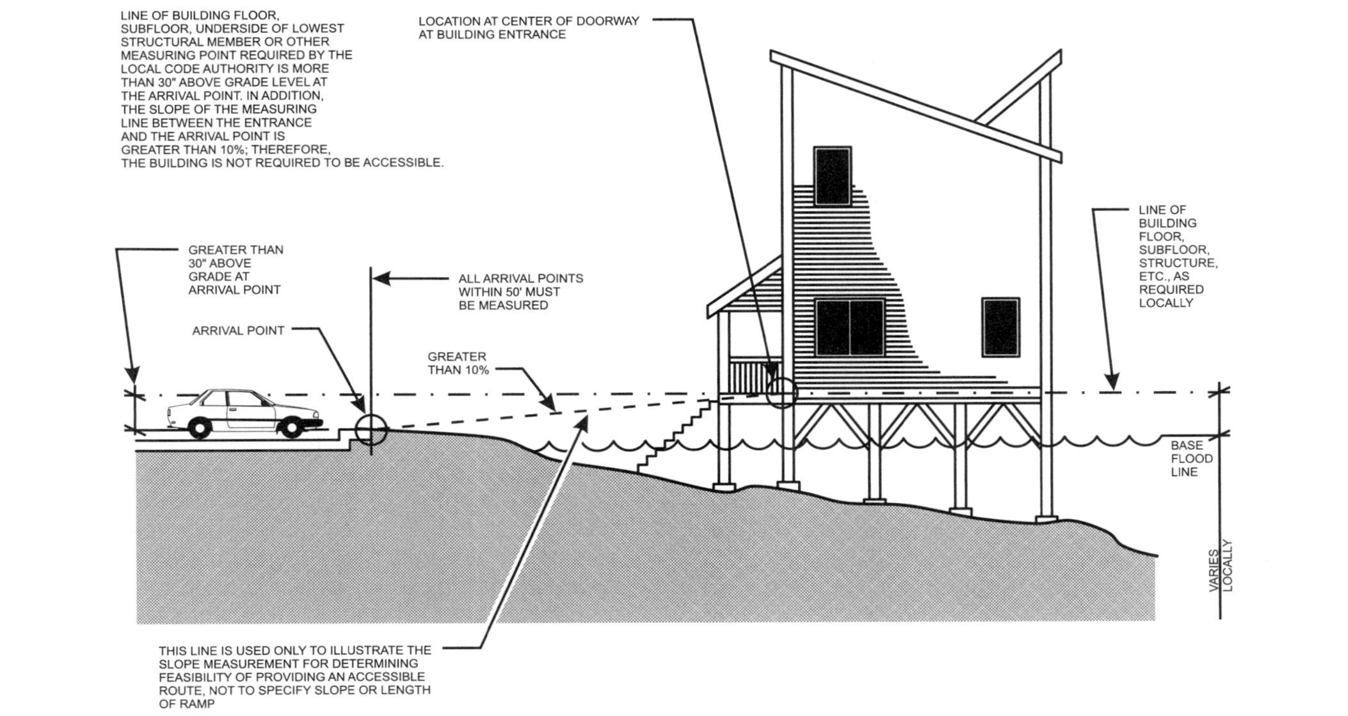

LINE OF BUILDING FLOOR, SUBFLOOR, UNDERSIDE OF LOWEST STRUCTURAL MEMBER OR OTHER MEASURING POINT REQUIRED BY THE LOCAL CODE AUTHORITY IS MORE THAN 30" ABOVE GRADE LEVEL AT THE ARRIVAL POINT. IN ADDITION, THE SLOPE OF THE MEASURING LINE BETWEEN THE ENTRANCE AND THE ARRIVAL POINT IS GREATER THAN 10%; THEREFORE, THE BUILDING IS NOT REQUIRED TO BE ACCESSIBLE.

LOCATION AT CENTER OF DOORWAY AT BUILDING ENTRANCE

LINE OF BUILDING FLOOR, SUBFLOOR, STRUCTURE, ETC., AS REQUIRED LOCALLY

GREATER THAN 30" ABOVE GRADE AT ARRIVAL POINT

ALL ARRIVAL POINTS WITHIN 50' MUST BE MEASURED

ARRIVAL POINT

GREATER THAN 10%

BASE FLOOD LINE

VARIES LOCALLY

THIS LINE IS USED ONLY TO ILLUSTRATE THE SLOPE MEASUREMENT FOR DETERMINING FEASIBILITY OF PROVIDING AN ACCESSIBLE ROUTE, NOT TO SPECIFY SLOPE OR LENGTH OF RAMP

For SI: 1 inch = 25.4 mm, 1 foot = 304.8 mm.

Figure 1107.7.5(1)
BUILDING MEETS BOTH ACCESSIBILITY CRITERIA FOR EXEMPTION

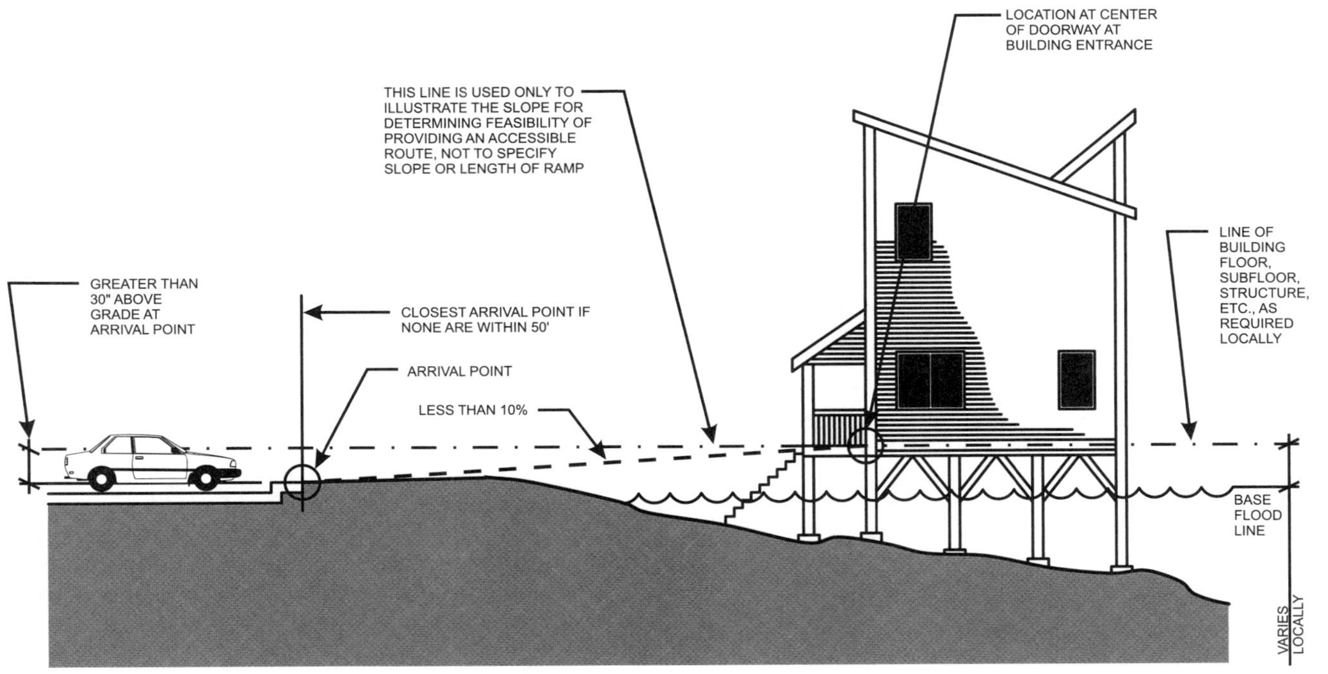

THIS LINE IS USED ONLY TO ILLUSTRATE THE SLOPE FOR DETERMINING FEASIBILITY OF PROVIDING AN ACCESSIBLE ROUTE, NOT TO SPECIFY SLOPE OR LENGTH OF RAMP

LOCATION AT CENTER OF DOORWAY AT BUILDING ENTRANCE

LINE OF BUILDING FLOOR, SUBFLOOR, STRUCTURE, ETC., AS REQUIRED LOCALLY

GREATER THAN 30" ABOVE GRADE AT ARRIVAL POINT

CLOSEST ARRIVAL POINT IF NONE ARE WITHIN 50'

ARRIVAL POINT

LESS THAN 10%

BASE FLOOD LINE

VARIES LOCALLY

For SI: 1 inch = 25.4 mm, 1 foot = 304.8 mm.

Figure 1107.7.5(2)
BUILDING MUST COMPLY WITH ACCESSIBILITY REQUIREMENTS

SECTION 1108
SPECIAL OCCUPANCIES

1108.1 General. In addition to the other requirements of this chapter, the requirements of Sections 1108.2 through 1108.4 shall apply to specific occupancies.

❖ The criteria provided herein are occupancy specific, and are intended to result in a reasonable level of accessibility in areas with assembly seating, self-service storage facilities and judicial facilities.

1108.2 Assembly area seating. Assembly areas with fixed seating shall comply with Sections 1108.2.1 through 1108.2.8. Dining areas shall comply with Section 1108.2.9. In addition, lawn seating shall comply with Section 1108.2.6.

❖ Sections 1108.2.1 through 1108.2.8 specifically address facilities with fixed seating utilized for purposes of viewing an event, typically facilities and spaces of occupancies of Groups A-1, A-3, A-4 and A-5. These criteria would also be applicable in assembly-type spaces with fixed seats that are located in buildings of other occupancies. The requirement for access to services; number and dispersion of wheelchair spaces and their associated companion seats; designated aisle seats; lawn seating; assistive listening systems and performance areas are addressed. The requirements for seating in areas for eating or drinking, typically Group A-2 spaces or facilities, and the dispersion of this seating are addressed in Section 1108.2.9.

1108.2.1 Services. If a service or facility is provided in an area that is not *accessible*, the same service or facility shall be provided on an *accessible* level and shall be *accessible*.

❖ This section establishes an important concept. The intent is that all types of services provided by a facility must be accessible. For example, a stadium may have a particular section of seating that is not accessible (assuming that the required number and dispersion of spaces is in compliance with the code). This section establishes that if a souvenir stand is located in or near that section, one of three circumstances would apply: an additional souvenir stand must be provided in the facility on an accessible level; the souvenir stand would have to be relocated to an accessible level or the section where the souvenir stand is located must be made accessible. This section also emphasizes that not only must that service or facility be located on an accessible level, but the service itself must also be accessible. Obviously, it would be inappropriate to locate a souvenir stand on an accessible level if the approach or entrance to that area is inaccessible.

1108.2.2 Wheelchair spaces. In theaters, *bleachers*, *grandstands*, stadiums, arenas and other fixed seating assembly areas, *accessible wheelchair spaces* complying with ICC A117.1 shall be provided in accordance with Sections 1108.2.2.1 through 1108.2.2.4.

❖ The intent of this section is to provide a reasonable number of spaces in an assembly occupancy with fixed seating to accommodate persons who use wheelchairs. Demographic statistics from the National Center for Health Statistics on the number of noninstitutionalized Americans who use wheelchairs indicate that these requirements are realistic. These required wheelchair spaces consist of an open, available floor space in which the wheelchair takes the place of the fixed seat that would otherwise occupy that space. Section 1108.2.3 is consistent with ICC A117.1 in requiring companion seating adjacent to each wheelchair space.

1108.2.2.1 General seating. *Wheelchair spaces* shall be provided in accordance with Table 1108.2.2.1.

❖ The number of required wheelchair spaces is indicated in Table 1108.2.2.1 and is based on the total number of fixed seats set up to view the same event, with the exception of box seats covered in Sections 1108.2.2.2 and 1108.2.2.3. For example, the fixed seating on all levels and all types provided in a sports stadium, excluding box seats, is used to calculate the number of required wheelchair spaces for the general seating. The requirements for box- or suite-type seating are calculated separately. Seating provided in different rooms, such as in a series of lecture halls in a university, must be calculated separately. Note that the percentage of wheelchair spaces is less for facilities with a capacity of more than 5,000. The industry has been able to provide statistics to show that the higher percentage of wheelchair spaces is not typically utilized in large facilities.

TABLE 1108.2.2.1
ACCESSIBLE WHEELCHAIR SPACES

CAPACITY OF SEATING IN ASSEMBLY AREAS	MINIMUM REQUIRED NUMBER OF WHEELCHAIR SPACES
4 to 25	1
26 to 50	2
51 to 100	4
101 to 300	5
301 to 500	6
501 to 5,000	6, plus 1 for each 150, or fraction thereof, between 501 through 5,000
5,001 and over	36 plus 1 for each 200, or fraction thereof, over 5,000

❖ This table sets forth the required number of wheelchair spaces based on the capacity of seating in the space containing the fixed seating. Any fixed seating with less than four seats is not required to provide wheelchair spaces. There is special criteria for wheelchair spaces in facilities with a seating capacity of over 500 and over 5,000. For example, six wheelchair spaces would be required for an assembly space with a seating capacity of 500. Seven wheelchair spaces would be required for a seating capacity of 501 through 650. Both of the last two rows use the term "or fraction thereof" to designate the next step up.

1108.2.2.2 Luxury boxes, club boxes and suites. In each luxury box, club box, and suite within arenas, stadiums and *grandstands*, *wheelchair spaces* shall be provided in accordance with Table 1108.2.2.1.

❖ When luxury boxes, club boxes or suites are provided, each luxury box, club box or suite must have an accessible route to that box and at least one wheelchair space and associated companion space. If the luxury box, club box or suite has more than 25 seats, the number of required wheelchair spaces is increased in accordance with Table 1108.2.2.1. For example, if a stadium has three luxury boxes with 51 seats in each, at least four wheelchair spaces and their associated companion seats must be provided in each box. While the total of all three luxury boxes (i.e., 153 occupants) would only require five total wheelchair spaces, these specific types of boxes must be calculated individually.

Luxury boxes are most commonly found in larger facilities, thus the reference to arenas, stadiums and grandstands. Boxes that are portions of balconies, such as in an opera house, or boxes separated by rails in general seating areas, such as behind home plate in baseball stadiums, are addressed in Section 1108.2.2.3.

1108.2.2.3 Other boxes. In boxes other than those required to comply with Section 1108.2.2.2, the total number of *wheelchair spaces* provided shall be determined in accordance with Table 1108.2.2.1. *Wheelchair spaces* shall be located in not less than 20 percent of all boxes provided.

❖ Examples of the boxes covered in this section are the seats separated by railings or low walls behind home plate in a baseball stadium [see Figure 1108.2.2.3(1)], or side balconies in opera houses [see Figure 1108.2.2.3(2)]. Luxury boxes are typically located in a separate level, and are often attached to some type of party room behind a group of seats.

When boxes other than luxury boxes, club boxes or

suites are provided, at least 20 percent of the boxes must have an accessible route to that box and at least one wheelchair space and associated companion space in that box. The total number of box seats is utilized to calculate the total number of required wheelchair spaces in accordance with Table 1108.2.2.1. For example, if a stadium has five boxes with 51 seats in each, the total of all five boxes (i.e., 255 occupants) would require five total wheelchair spaces. The wheelchair spaces could be located in one box (i.e., 20 percent minimum) or dispersed to all five boxes. Note that there is additional dispersion criteria for assembly seating based on lines of sight in ICC A117.1, Section 802.8, as well as the integration of wheelchair spaces indicated in Section 1108.2.3 and dispersion by level in Section 1108.2.4.

Figure 1108.2.2.3(2)
EXAMPLE OF BOX SEATING IN A THEATER

1108.2.2.4 Team or player seating. At least one *wheelchair space* shall be provided in team or player seating areas serving areas of sport activity.

> **Exception:** *Wheelchair spaces* shall not be required in team or player seating areas serving bowling lanes that are not required to be located on an *accessible route* in accordance with Section 1109.14.4.1.

❖ Accessible seating for sports teams and players is a logical code requirement that allows players to maintain team participation when injured. This requirement coordinates with the new ADA/ABA Guidelines.

The exception acknowledges that only 5 percent of lanes in bowling facilities are required to be provided with an accessible route. Since teams sit at their lanes while playing, this provides a balance between player seating and access to the playing field (see Section 1109.14.4.1).

Figure 1108.2.2.3(1)
EXAMPLE OF BOX SEATING IN A SPORTS VENUE

1108.2.3 Companion seats. At least one companion seat complying with ICC A117.1 shall be provided for each *wheelchair space* required by Sections 1108.2.2.1 through 1108.2.2.3.

❖ A companion seat allows a friend, relative, associate, etc., to accompany a person using a wheelchair at an event and to share in the experience with the same level of companionship as others attending the event. The ICC A117.1 standard permits two wheelchair spaces to be located next to each other; however, each wheelchair space must also be adjacent to at least one companion seat. Accessible seating areas are sometimes provided with nonfixed companion chairs allowing the users to adjust the adjacency of seats and wheelchair spaces to their liking. If fixed seats are provided, the companion seat must have shoulder alignment with the person using the wheelchair space.

1108.2.4 Dispersion of wheelchair spaces in multilevel assembly seating areas. In *multilevel assembly seating* areas, *wheelchair spaces* shall be provided on the main floor level and on one of each two additional floor or *mezzanine* levels. *Wheelchair spaces* shall be provided in each luxury box, club box and suite within assembly facilities.

Exceptions:

1. In multilevel assembly spaces utilized for worship services where the second floor or *mezzanine* level contains 25 percent or less of the total seating capacity, *wheelchair spaces* shall be permitted to all be located on the main level.

2. In *multilevel assembly seating* where the second floor or *mezzanine* level provides 25 percent or less of the total seating capacity and 300 or fewer seats, all *wheelchair spaces* shall be permitted to be located on the main level.

3. *Wheelchair spaces* in team or player seating serving areas of sport activity are not required to be dispersed.

❖ When facilities provide seating on multiple levels, wheelchair spaces must also be provided on multiple levels. For example, if a theater has a main level and a balcony level, wheelchair spaces must be provided on both. If a theater has a main level and two balcony levels, wheelchair spaces must be provided on the main level and at least one of the two balcony levels. While the term "level" is difficult to define in some assembly arrangements, the intent is to provide the wheelchair user with a choice of seating similar to what is available to all persons in the space. A "level" is not each time that there is a step-down to another row of seating. Subjective judgement on the part of both the designer and the building official is required for each individual facility.

There are three specific exceptions to this multilevel dispersion requirement.

Exception 1 is applicable to assembly spaces utilized for worship services. There is typically not a cost or line-of-sight issue in this type of facility, as there is in other assembly seating areas. The balcony is not re-

quired to contain accessible seating or have an accessible route, as long as accessible seating is provided on the main level and the balcony does not contain more than 25 percent of the seating.

Exception 2 is applicable to balconies in assembly facilities with a limited number of seats in the upper level. If accessible seating is provided on the main level and the balcony contains less than 25 percent of the seating (and less than 300 seats), then the balcony is not required to contain accessible seating or have an accessible route.

Exception 3 relieves the requirement for wheelchair space dispersion among the various team or player seating options that may be provided in a facility. However, there must be at least one wheelchair space for each team.

The dispersement requirements in the code must be combined with the additional wheelchair dispersement requirements in ICC A117.1, including side to side (horizontally) across a venue, front to back (variety of distances) within the seating, and by type (distinct service or amenities) of seating. The standard also looks at requirements for line of sight or seated or standing spectators. Wheelchair spaces and their associated companion seats may be grouped, but must, at a minimum, be distributed into the number of locations indicated in Table 802.10 of ICC A117.1.

For requirements for accessible means of egress from these wheelchair spaces, please see Sections 1007.1, Exception 3, 1028.5.1 and 1028.8.

1108.2.5 Designated aisle seats. At least 5 percent, but not less than one, of the total number of aisle seats provided shall be designated aisle seats and shall be the aisle seats located closest to *accessible routes*.

Exception: Designated aisle seats are not required in team or player seating serving areas of sport activity.

❖ Designated aisle seats allow some persons with mobility impairments to be able to easily access certain seats along aisles. Due to an impairment, it may be difficult or impossible for some to move further into the narrow space between rows. It is not the intent of the designated aisle seats to be transfer locations. The person using this space can be someone using a walker, crutches or cane or anyone who may have difficulty with steps; hence, the requirement that the designated aisle seats be close to accessible routes. When seats have armrests, ICC A117.1 requires each designated aisle seat to have a folding or retractable armrest allowing easier access or better fit.

1108.2.6 Lawn seating. Lawn seating areas and exterior overflow seating areas, where fixed seats are not provided, shall connect to an *accessible route*.

❖ Lawn seating on both flat and sloped sites is fairly common at less formal venues. It may be the only type of seating provided or one of several options. Sometimes lawns are overflow areas, with or without movable chairs, located beyond the fixed seating areas. In any case, these lawn areas must be located on an accessible route. Access onto the lawn seating depends on the ability of the person and/or assistance from companions (see Figure 1108.2.6).

Figure 1108.2.6
EXAMPLE OF LAWN SEATING

1108.2.7 Assistive listening systems. Each assembly area where audible communications are integral to the use of the space shall have an assistive listening system.

> **Exception:** Other than in courtrooms, an assistive listening system is not required where there is no audio amplification system.

❖ This section is intended to accommodate people with a hearing impairment. Assembly occupancies, such as stadiums, theaters, auditoriums, lecture halls and similar spaces, are required to make provisions for the use of assistive listening devices in accordance with Section 1108.2.7.1 and Table 1108.2.7.1. In these assembly areas, audible communication is often integral to the use and full enjoyment of the space. This requirement offers the possibility for individuals with hearing impairments to attend functions in these facilities without having to give advance notice and without disrupting the event in order to have a portable assistive listening system set up and made ready for use.

There are three primary types of listening systems available: induction loop, AM/FM and infrared. Each type of system has certain advantages and disadvantages that the designer should take into consideration when choosing the type of system that is most appropriate for the intended application. Signage notifying the general public of the availability of these systems must be provided in accordance with Section 1110.3.

If an audio amplification system is not provided within a facility, an assistive listening system is not re-

quired. Given the essential nature of the proceedings, this exception is not applicable to courtrooms. All courtrooms must have an assistive listening system.

1108.2.7.1 Receivers. Receivers shall be provided for assistive listening systems in accordance with Table 1108.2.7.1.

> **Exceptions:**
> 1. Where a building contains more than one assembly area, the total number of required receivers shall be permitted to be calculated according to the total number of seats in the assembly areas in the building, provided that all receivers are usable with all systems and if assembly areas required to provide assistive listening are under one management.
> 2. Where all seats in an assembly area are served by an induction loop assistive listening system, the minimum number of receivers required by Table 1108.2.7.1 to be hearing-aid compatible shall not be required.

❖ Table 1108.2.7.1 specifies the number of required receivers in an assembly occupancy. Exception 1 states that if a facility has more than one assembly space with an audio amplification system, such as a multiplex theater, then the total seating for all the spaces may be used to determine the number of receivers required.

Exception 2 states that since induction loop technology renders an entire space accessible to the assistive listening system, the hearing-aid compatible receivers are not necessary.

TABLE 1108.2.7.1. See page 11-42.

❖ Table 1108.2.7.1 specifies the number of required receivers in an assembly occupancy. At least 25 percent of the receivers provided (but not less than two receivers) must be hearing-aid compatible, as persons with hearing aids typically cannot use earpieces or head-phone-equipped receivers. When determining the general number of receivers required, the table is to be applied as though it read "2 plus 1 for each additional 25 over 50, or a fraction thereof." For example, where 51 to 75 total seats are provided, three assistive listening devices are required. Where 76 to 100 total seats are provided, four assistive listening devices are required.

The number of required receivers is based on the capacity of the assembly areas. Note that as the size of the assembly area increases, the number of required receivers also increases; however, the percentage of receivers to seats decreases. This is based on the actual usage in large assembly areas.

1108.2.7.2 Public address systems. Where stadiums, arenas and *grandstands* provide audible public announcements, they shall also provide equivalent text information regarding events and facilities in compliance with Sections 1108.2.7.2.1 and 1108.2.7.2.2.

❖ If stadiums, arenas or grandstands provide public announcements, the same information should be displayed on some type of electronic signage. Most stadiums, arenas and grandstands have electronic scoreboards that are capable of displaying text messages. If electronic signage is not provided, compliance with Sections 1108.2.7.2.1 and 1108.2.7.2.2 is not required.

1108.2.7.2.1 Prerecorded text messages. Where electronic signs are provided and have the capability to display prerecorded text messages containing information that is the same, or substantially equivalent to information that is provided audibly, signs shall display text that is equivalent to audible announcements.

Exception: Announcements that cannot be prerecorded in advance of the event shall not be required to be displayed.

❖ If prerecorded messages are part of the event, the same information should be displayed in a text format. Text display is not required for announcements that are not prerecorded. Note that text information is only required when electronic signage is provided.

1108.2.7.2.2 Real-time messages. Where electronic signs are provided and have the capability to display real-time messages containing information that is the same, or substantially equivalent, to information that is provided audibly, signs shall display text that is equivalent to audible announcements.

❖ If the electronic signage in the facility is capable of displaying real-time messages, then the same information being provided to the general audience through audible means should also be displayed in text. Note that text information is only required when electronic signage is provided.

1108.2.8 Performance areas. An *accessible route* shall directly connect the performance area to the assembly seating area where a *circulation path* directly connects a performance area to an assembly seating area. An *accessible route* shall be provided from performance areas to ancillary areas or facilities used by performers.

❖ Performance areas, such as stages, orchestra pits, band platforms, choir lofts and similar spaces, must be accessible. If there is a direct route from the seating to the performance area, there must also be an accessible route. For example, if steps are provided from the assembly seating area to the stage within the theater, then an accessible route (e.g., ramp or platform lift) to the stage must also be provided within the theater. An

accessible route must also be provided to any ancillary areas, such as greenrooms or practice/warm-up rooms. The intent is that a person with mobility impairments could participate in the event. This could include high school graduation with students coming from the audience up onto the stage to receive their diplomas; participating with the community band; playing in the orchestra for a performance; acting in a production or giving a speech.

1108.2.9 Dining areas. In dining areas, the total floor area allotted for seating and tables shall be *accessible*.

Exceptions:

1. In buildings or facilities not required to provide an *accessible route* between levels, an *accessible route* to a *mezzanine* seating area is not required, provided that the *mezzanine* contains less than 25 percent of the total area and the same services are provided in the *accessible* area.

2. In sports facilities, tiered dining areas providing seating required to be *accessible* shall be required to have *accessible routes* serving at least 25 percent of the dining area, provided that *accessible routes* serve *accessible* seating and where each tier is provided with the same services.

❖ Dining areas most frequently occur in Group A-2 (restaurants, cafeterias, portions of nightclubs, dinner theaters, etc.). The provisions of this section are intended to govern such areas, rather than the criteria specified in Sections 1108.2.2 through 1108.2.5. This section requires the total floor area allotted for seating and tables to be accessible, including dining areas that are raised or lowered by one or more risers. The two exceptions specifically address tiered dining areas in sports facilities and dining on mezzanine levels. Exception 1 is intended to acknowledge a practical and reasonable limitation in buildings with a mezzanine level that contain only seating and that seating provides less than 25 percent of the total seating for the dining area. In addition, all services provided in the mezzanine must also be available in an accessible area. Any mezzanine condition that does not meet the criteria of this exception does not qualify for the excep-

TABLE 1108.2.7.1
RECEIVERS FOR ASSISTIVE LISTENING SYSTEMS

CAPACITY OF SEATING IN ASSEMBLY AREAS	MINIMUM REQUIRED NUMBER OF RECEIVERS	MINIMUM NUMBER OF RECEIVERS TO BE HEARING-AID COMPATIBLE
50 or less	2	2
51 to 200	2, plus 1 per 25 seats over 50 seats*	2
201 to 500	2, plus 1 per 25 seats over 50 seats*	1 per 4 receivers*
501 to 1,000	20, plus 1 per 33 seats over 500 seats*	1 per 4 receivers*
1,001 to 2,000	35, plus 1 per 50 seats over 1,000 seats*	1 per 4 receivers*
Over 2,000	55, plus 1 per 100 seats over 2,000 seats*	1 per 4 receivers*

Note: * = or fraction thereof

tion and must be served by an accessible route. Nonqualifying examples are as follows: a dining mezzanine that contains 25 percent or more of the total dining area; a raised or depressed dining area that is not actually a mezzanine; and a mezzanine level that contains the only area where a specific service or amenity is provided, like a bar or private party room.

Exception 2 is for sports facilities. Some sports facilities also have accommodations for dining or picnicking while watching the event. For line-of-sight issues, the dining terraces are tiered. If the same services are available on the accessible level as any other level, only 25 percent of the dining area is required to be on an accessible route. Section 1109.7 allows platform lifts to provide access to these levels when the sporting event and dining facilities are outdoors, such as at a baseball park.

At this time the code does not contain specific information on tiered dining facilities that also have issue of line of sight for viewing an event, such as a dinner theater or dinner seating during a sport event such as at a racetrack. These types of spaces may want to look at the provisions for both dining (Section 1108.2.9) and seating for viewing an event (Section 1108.2.2) and develop a reasonable compromise.

1108.2.9.1 Dining surfaces. Where dining surfaces for the consumption of food or drink are provided, at least 5 percent, but not less than one, of the dining surfaces for the seating and standing spaces shall be *accessible* and be distributed throughout the facility and located on a level accessed by an *accessible route*.

❖ Section 1108.2.9.1 establishes the criteria for the percentage of spaces at tables, booths, bars and counters that will be used for eating or drinking that must be accessible. This criteria is consistent with Section 1109.10 for the required percentage of accessible built-in surfaces in all other occupancies. The accessible surfaces are also required to be distributed throughout the facility such that a comparable choice of locations and types (i.e., tables, booths, counters, etc.) is available (see Figure 1108.2.9.1). This requirement, in conjunction with Section 1108.2.9, provides a reasonable and appropriate degree of accessibility throughout dining areas. The result is that a person with a mobility impairment will be able to approach, enter and move about in virtually all portions of a dining area. The entire dining or drinking area must be accessible. In addition, 5 percent of the total spaces at dining surfaces provided must be accessible. Each seating location at a table or seat section of a bar is considered a dining surface. The issue of whether a portion of a bar or dining counter in a restaurant is required to be accessible is subjective. The assumption is that if other types of seating are provided adjacent to the counter, then services provided at the counter will also be available at the adjacent seating; therefore, if adequate accessible seating is available adjacent to the bar area, the bar is not required to be lowered. If the

Figure 1108.2.9.1
EXAMPLE OF DINING SEATING

bar is the only eating or dining surface in a restaurant or in a separate room in the restaurant, then a portion of the bar must be made accessible.

1108.3 Self-service storage facilities. *Self-service storage facilities* shall provide *accessible* individual self-storage spaces in accordance with Table 1108.3.

❖ This section addresses facilities that provide self-storage units or spaces. These types of facilities are often storage garages located in long rows. Some facilities also provide climate-controlled storage within large multistory warehouses. The key is that the storage is moved in and out and accessed by the renter of the space. The intent is to provide access for persons with disabilities to this service without requiring the entire facility to be accessible.

TABLE 1108.3
ACCESSIBLE SELF-SERVICE STORAGE FACILITIES

TOTAL SPACES IN FACILITY	MINIMUM NUMBER OF REQUIRED ACCESSIBLE SPACES
1 to 200	5%, but not less than 1
Over 200	10, plus 2% of total number of units over 200

❖ The minimum number of accessible spaces is based on the total number of self-storage spaces available in a facility.

1108.3.1 Dispersion. *Accessible* individual self-service storage spaces shall be dispersed throughout the various classes of spaces provided. Where more classes of spaces are provided than the number of required *accessible* spaces, the number of *accessible* spaces shall not be required to exceed that required by Table 1108.3. *Accessible* spaces are permitted to be dispersed in a single building of a multibuilding facility.

❖ Self-storage facilities may offer a variety of spaces, such as heated/nonheated, different sizes, etc. If the variety offered is greater than the number of accessible spaces required, the accessible spaces should be dispersed as much as possible. For example, if a facility offers the choice of a 10-foot by 10-foot (3048 mm by 3048 mm) unit and a 10-foot by 20-foot (3048 mm by 6096 mm) unit, but only one accessible space is required, then only one unit is required to be accessible. The choice of which unit is made accessible is up to the building owner or designer. When a facility has multiple buildings, accessible spaces are permitted to be located in a single building.

At this time the ICC A117.1 does not provide specific information on how to make a self-service storage unit accessible. Since overhead doors are not typically considered accessible, alternatives to be discussed between the owner/designer and code official are the possibility of man doors on the side of the unit or electric garage door openers.

1108.4 Judicial facilities. Judicial facilities shall comply with Sections 1108.4.1 through 1108.4.3.

❖ Accessibility in judicial facilities (e.g., courthouses) includes access to all public areas as well as special requirements for the courtrooms, holding cells and visitation areas. This section was added for coordination with new requirements for judicial facilities from the DOJ.

1108.4.1 Courtrooms. Each courtroom shall be *accessible* and comply with Sections 1108.4.1.1 through 1108.4.1.5.

❖ All courtrooms are required to be accessible for participants at every level. This would include participation as a employee of the court, counselor, litigant, juror, witness or observer. All courtrooms must be on an accessible route from both the public side and courtroom staff side. The waiting areas, lawyer meeting rooms and vestibule that are sometimes located immediately adjacent to the courtroom entrance must be accessible. Access must be readily available to the gallery as well as the witness stand, and jury box. Members of the jury must have an accessible route to the jury deliberation room. The witness box in a courtroom is often raised to have a line of sight between the judge and witness. It is not acceptable to ask someone to testify from an area outside the box; therefore, the witness box must always be accessible. Elements would include an accessible route to all portions of the courtroom unless limited by the following subsections which identify areas and features of courtrooms that have specific accessibility requirements.

An accessible route must also be available to the holding cells addressed in Section 1108.4.2.

The ICC was very proud to participate in a special Courthouse Access Advisory Committee that assisted the Access Board in developing recommendations for accessibility to courtrooms and courthouses. This report includes information on both requirements and recommended design practices. The committee was able to construct actual mock-ups to verify design configurations. The committee report can be located at http://www.access-board.gov/caac/.

1108.4.1.1 Jury box. A *wheelchair space* complying with ICC A117.1 shall be provided within the jury box.

Exception: Adjacent companion seating is not required.

❖ Jury boxes may be configured in numerous ways, including with multiple tiers. At least one wheelchair space within the "box" is required. A wheelchair space adjacent to (or outside of) the jury box could not be used to satisfy the requirement. Since only jury members are seated in the jury box, companion seating is not required.

1108.4.1.2 Gallery seating. *Wheelchair spaces* complying with ICC A117.1 shall be provided in accordance with Table 1108.2.2.1. Designated aisle seats shall be provided in accordance with Section 1108.2.5.

❖ Gallery seating in a courtroom is regulated similarly to other assembly seating areas with respect to wheelchair spaces, associated companion seating and designated aisle seats (see Figure 1108.4.1.2).

**Figure 1108.4.1.2
GALLERY SEATING**

1108.4.1.3 Assistive listening systems. An assistive listening system must be provided. Receivers shall be provided for the assistive listening system in accordance with Section 1108.2.7.1.

❖ Every courtroom must have an assistive listening system. Unlike other assembly occupancies, the requirement is not dependent upon an audio amplification system being provided. The assumption is that any person in a courtroom with a hearing impairment should be provided with an enhanced ability to hear the proceedings when desired. Issues of privacy may dictate the type of system provided within the courtrooms.

1108.4.1.4 Employee work stations. The judge's bench, clerk's station, bailiff's station, deputy clerk's station and court reporter's station shall be located on an *accessible route*. The vertical access to elevated employee work stations within a courtroom is not required at the time of initial construction, provided a *ramp*, lift or elevator complying with ICC A117.1

can be installed without requiring reconfiguration or extension of the courtroom or extension of the electrical system.

❖ Employee work stations in courtrooms are permitted to be constructed as adaptable for a future accessible route if and when a route is necessary for a mobility impaired employee. Depending on the type of courtroom (i.e., panel, trial, jury, traffic), employee work stations will vary. The judge is typically raised to establish a level of authority over the courtroom and to have his or her eyes approximately level with those of the standing lawyers. Associated clerk stations are raised to ease interaction with the judge. The vertical portion of the accessible route (ramp, platform lift, elevator) can be an adapting element and is expected to be predesigned and readily incorporated into the courtroom when needed. This option is consistent with the modification for employees based on the need of the employee expressed in Section 1103.2.3. This option does not apply to the raised public areas in courtrooms, namely, witness stands, jury boxes and galleries, which are required to be constructed accessible.

1108.4.1.5 Other work stations. The litigant's and counsel stations, including the lectern, shall be *accessible* in accordance with ICC A117.1.

❖ The well of the court typically contains the area for the litigant's and counselor's tables and a lectern for lawyers to address the judge and/or jury. Depending on the courtroom, this configuration may vary. The tables should be constructed appropriately for a front approach for work stations (see Figure 1108.4.1.5). The lectern can be adjustable so that it can work for lawyers seated in wheelchairs as well as those standing.

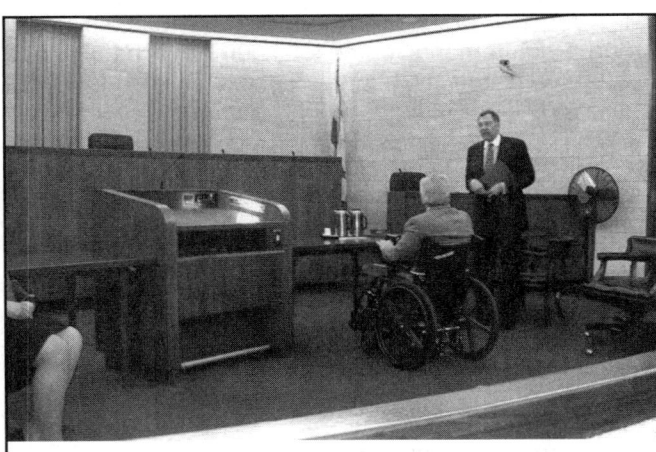

Figure 1108.4.1.5
COURTROOM WELL WORK STATIONS

1108.4.2 Holding cells. Central holding cells and court-floor holding cells shall comply with Sections 1108.4.2.1 and 1108.4.2.2.

❖ If holding cells are provided, Sections 1108.4.2.1 and 1108.4.2.2 indicate which cells shall be accessible.

1108.4.2.1 Central holding cells. Where separate central holding cells are provided for adult males, juvenile males, adult females or juvenile females, one of each type shall be *accessible*. Where central holding cells are provided and are not separated by age or sex, at least one *accessible* cell shall be provided.

❖ Some courthouse facilities have a central holding area for detainees that are in the courthouse for judicial proceedings. At least one central holding cell is required to be accessible. If separate holding cells are provided for males and females or to separate juvenile offenders from the adult offenders, at least one of each type is required to be accessible.

1108.4.2.2 Court-floor holding cells. Where separate court-floor holding cells are provided for adult males, juvenile males, adult females or juvenile females, each courtroom shall be served by one *accessible* cell of each type. Where court-floor holding cells are provided and are not separated by age or sex, courtrooms shall be served by at least one *accessible* cell. *Accessible* cells shall be permitted to serve more than one courtroom.

❖ Some courthouse facilities have a holding cell located in the area of the courtroom for detainees who are in the courthouse for judicial proceedings. For security reasons, these holding cells may be provided on each floor of the facility in order to have the detainees as close to the courtrooms they are to appear in as possible. At least one holding cell per floor is required to be accessible. One accessible holding cell could serve all the courtrooms on one floor. If separate holding cells are provided for males and females or to separate juvenile offenders from the adult offenders, at least one of each type is required to be accessible.

1108.4.3 Visiting areas. Visiting areas shall comply with Sections 1108.4.3.1 and 1108.4.3.2.

❖ Persons appearing in court proceedings may need to talk with their lawyer, social worker, parole officer, etc. Typically, these visitation areas include cubicles or counters in supervised areas.

1108.4.3.1 Cubicles and counters. At least 5 percent but no fewer than one of the cubicles shall be *accessible* on both the visitor and detainee sides. Where counters are provided, at least one shall be *accessible* on both the visitor and detainee sides.

Exception: This requirement shall not apply to the detainee side of cubicles or counters at noncontact visiting areas not serving *accessible* holding cells.

❖ If detainee/visitor counters or cubicles are provided, at least 5 percent must be accessible. Since it is not known if the person needing the accessible counter is on the visitor side or the detainee side, at least one counter location shall be accessible on both sides. In accordance with the exception, visiting areas that do not serve accessible holding cells would still require 5 percent of the counters or cubicles to be accessible on the visitor's side. This is consistent with the exception to certain areas of jails in accordance with Section 1103.2.14.

1108.4.3.2 Partitions. Where solid partitions or security glazing separate visitors from detainees, at least one of each type of cubicle or counter partition shall be *accessible*.

❖ Section 1108.4.3.1 requires 5 percent of cubicles or counters in courthouse visiting areas to be accessible in accordance with the built-in counter/work surface requirements. This section deals with the visiting situation where solid partitions of security glazing separate visitors from detainees. To facilitate communication, the system may include some type of handset, headphones or microphone. The system must consider in its design that either user may use a wheelchair or have difficulty bending or stooping. Any type of telephone or headphone system must have a volume control.

SECTION 1109
OTHER FEATURES AND FACILITIES

1109.1 General. *Accessible* building features and facilities shall be provided in accordance with Sections 1109.2 through 1109.14.

Exception: *Type A units* and *Type B units* shall comply with ICC A117.1.

❖ This section clarifies that the provisions of Section 1109 are applicable in addition to all other requirements of Chapter 11. For example, although Section 1109.2 sets forth accessibility requirements for toilet rooms and bathing facilities within a store, it is not intended to mean that these are the only accessibility requirements that are applicable to stores. The building features and facilities covered by this section are required to be accessible to the extent set forth herein. The exception is a reminder that Type A and Type B dwelling units should comply with ICC A117.1 within the individual dwelling units. Common and public areas associated with Group R-2 and R-3 occupancies should comply with Section 1109.

1109.2 Toilet and bathing facilities. Each toilet room and bathing room shall be *accessible*. Where a floor level is not required to be connected by an *accessible route*, the only toilet rooms or bathing rooms provided within the facility shall not be located on the inaccessible floor. At least one of each type of fixture, element, control or dispenser in each *accessible* toilet room and bathing room shall be *accessible*.

Exceptions:

1. In toilet rooms or bathing rooms accessed only through a private office, not for *common* or *public use* and intended for use by a single occupant, any of the following alternatives are allowed:

    1.1. Doors are permitted to swing into the clear floor space, provided the door swing can be reversed to meet the requirements in ICC A117.1;

    1.2. The height requirements for the water closet in ICC A117.1 are not applicable;

    1.3. Grab bars are not required to be installed in a toilet room, provided that reinforcement has been installed in the walls and located so as to permit the installation of such grab bars; and

    1.4. The requirement for height, knee and toe clearance shall not apply to a lavatory.

2. This section is not applicable to toilet and bathing rooms that serve *dwelling units* or *sleeping units* that are not required to be *accessible* by Section 1107.

3. Where multiple single-user toilet rooms or bathing rooms are clustered at a single location, at least 50 percent but not less than one room for each use at each cluster shall be *accessible*.

4. Where no more than one urinal is provided in a toilet room or bathing room, the urinal is not required to be *accessible*.

5. Toilet rooms that are part of critical care or intensive care patient sleeping rooms are not required to be *accessible*.

❖ This section generally requires toilet rooms and bathing rooms to be accessible. A person using a wheelchair must be able to approach and enter the room. Within the toilet room and/or bathing room, a minimum of one of each element or fixture provided is required to be accessible. Elements and fixtures include such things as water closets, lavatories, mirrors, towel dispensers, hand dryers and any other device that is installed and intended for use by the occupants of the room. Showers and bathtubs are both considered bathing fixtures, but not as different types of bathing fixtures. If a shower and a bathtub are both provided in the same bathing room, only one is required to be accessible. The designer or owner can choose which one. Requirements for the total number of bathing and toilet rooms are in Chapter 29 of the code [which is duplicated from Section 403 of the *International Plumbing Code*® (IPC®)]. Large assembly and mercantile occupancies must also include a family or assisted use toilet room in addition to their other toilet facilities (see commentary, Section 1109.2.1).

ICC A117.1 is the document referenced for all accessible toilet room requirements. The technical requirements in ICC A117.1 are based on allowing a person in a wheelchair to perform a side transfer [see Figure 1109.2(1)]. Maintaining a clear floor space at each fixture is important [see Figure 1109.2(2) for possible configurations].

Exception 1 addresses a condition in which a toilet room or bathing room is permitted to be adaptable rather than fully accessible. The intent is that if a toilet room is part of an individual office and serves only the occupant of that office, the adaptable toilet room can be readily modified to be fully accessible based on that individual's needs. Preplanning during construction and design, such as installing blocking for grab bars and arranging plumbing fixtures to have adequate clear floor space, will facilitate future alterations.

Exception 2 is intended to correlate with the provisions of Section 1107, which establish the minimum number of facilities required to be Accessible, Type A or Type B units. Without this exception, the code would literally be requiring accessible fixtures in inaccessible dwelling and sleeping units. It is important to note that the bathrooms associated with required Accessible, Type A and Type B dwelling and sleeping units must comply with the requirements in ICC A117.1, Chapter 10.

Exception 3 specifies that all toilet rooms that are clustered together need not be accessible. In such configurations, typically found in a doctor's office or drug test center, the requirement is reduced to a 50 percent minimum. If these toilet rooms are clustered in separate locations, such as in a multiclinic facility, the 50 percent minimum would be applied to each cluster. A single-occupant women's bathroom adjacent to a single-occupant men's bathroom is not considered a cluster since they each serve a different sex.

The IPC permits urinals to be substituted for water closets to a maximum of 67 percent in each toilet room (see IPC Section 419.2). Exception 4 states that if only one urinal is provided within a toilet room, that urinal is not required to be accessible. This situation would typically only occur in toilet rooms with one water closet compartment.

While Exception 2 would exempt all the nonaccessible patient rooms in a hospital from accessible bathroom requirements, the intent of Exception 5 is also to exempt the Accessible units that may be provided within the critical-care or intensive-care units from requiring accessible bathrooms. In critical-care or intensive-care units the patients are often too ill to use the bathroom without assistance; therefore, assistance is offered and expected in these areas to all patients. In addition, critical-care and intensive-care rooms often must be designed to maximize free space for equipment and personnel in case of emergency care situations.

There are allowances in the ICC A117.1 for a toilet room and water closet compartment specifically designed for children's use. This may be appropriate in areas such as day care facilities and certain areas of elementary schools.

1109.2.1 Family or assisted-use toilet and bathing rooms. In assembly and mercantile occupancies, an *accessible* family or assisted-use toilet room shall be provided where an aggregate of six or more male and female water closets is required. In buildings of mixed occupancy, only those water closets required for the assembly or mercantile occupancy shall be used to determine the family or assisted-use toilet room requirement. In recreational facilities where separate-sex bathing rooms are provided, an *accessible* family or assisted-use bathing room shall be provided. Fixtures located within family or assisted-use toilet and bathing rooms shall be included in determining the number of fixtures provided in an occupancy.

Exception: Where each separate-sex bathing room has only one shower or bathtub fixture, a family or assisted-use bathing room is not required.

❖ The requirements for family or assisted-use (referred to as "family" in the following commentary) toilet and bathing rooms were recommended for adoption into the codes by the Board for the Coordination of the Model Codes (BCMC), in conjunction with representatives of the United States Architectural and Transportation Barriers Compliance Board (Access Board). The primary issue relative to family toilet/bathing facilities is that some people with disabilities require assistance to utilize them. If that attendant is of the opposite sex, a toilet or bathing facility that can accommodate both persons is required. It is important to note that these provisions will typically not result in a substantial cost burden to the building owners, since the fixtures provided also count towards the minimum number required (see Section 2902.1.2 or IPC Section 403.1.2).

The family or assisted-use bathing room requirements, although beneficial to all occupancies and

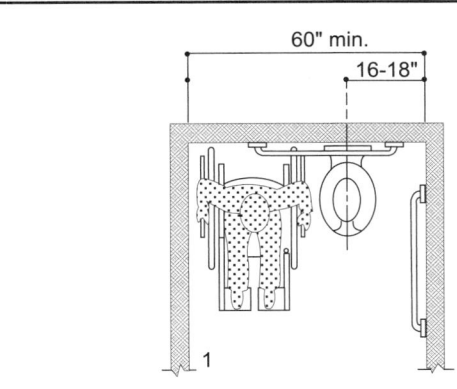

TAKES TRANSFER POSITION, REMOVES ARMREST, SETS BRAKE

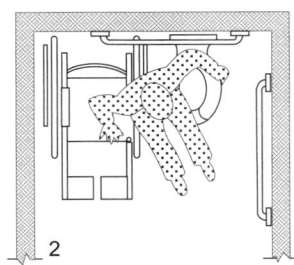

TRANSFERS

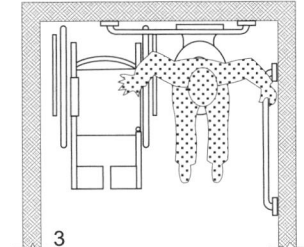

POSITIONS ON TOILET

For SI: 1 inch = 25.4 mm.

Figure 1109.2(1)
SIDE APPROACH WHEEL CHAIR TRANSFER

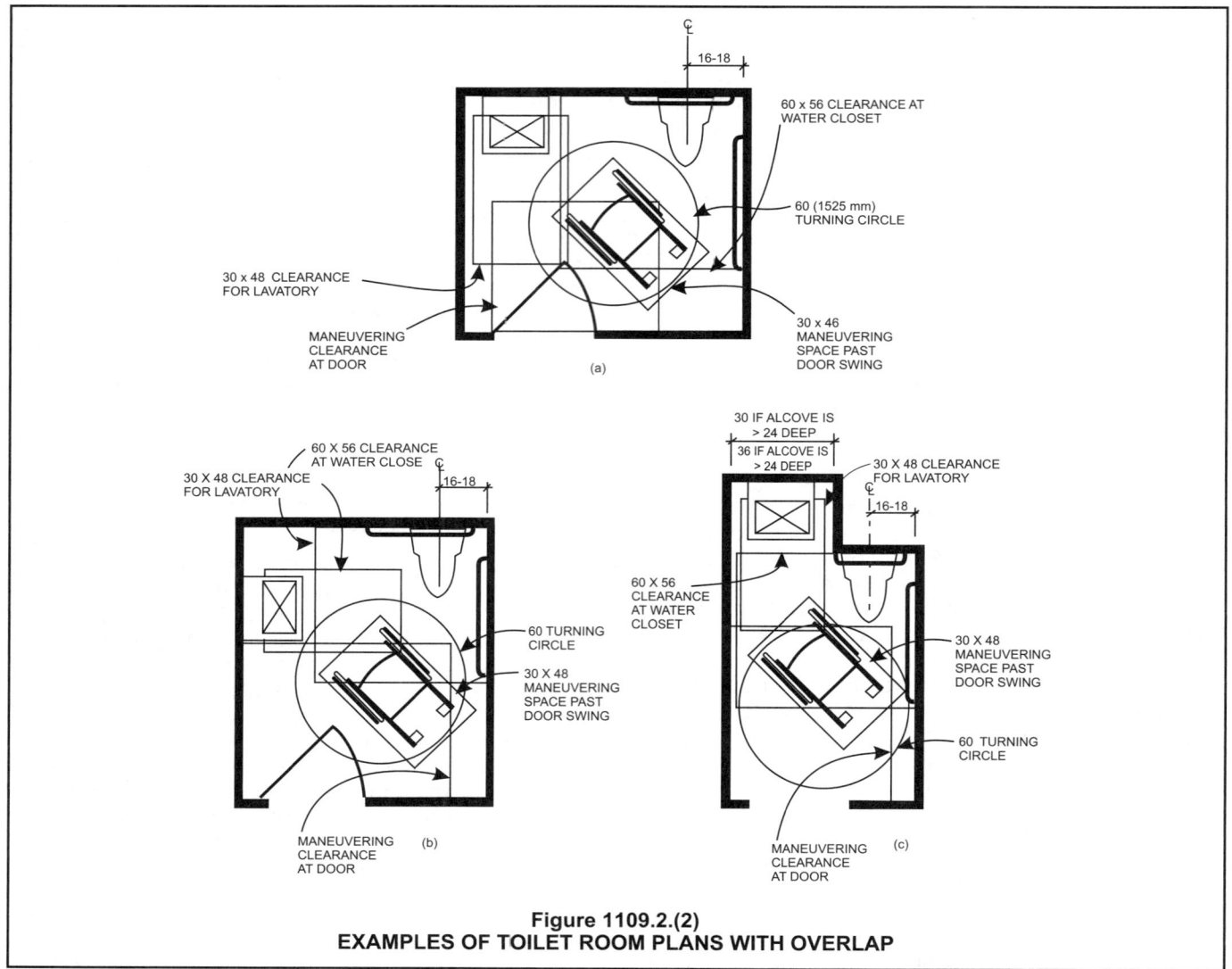

Figure 1109.2.(2)
EXAMPLES OF TOILET ROOM PLANS WITH OVERLAP

building sizes, have been limited to those structures that typically have large transient occupant loads, namely assembly and mercantile. The section also identifies how mixed-occupancy buildings are to be addressed relative to the calculation. The fixtures provided in family toilet and bathing rooms count towards the number of required fixtures for the occupancy. The number of fixtures to be located in such rooms is limited, based on the premise that these rooms are securable (see commentary, Section 1109.2.1.2). The provision for bathing facilities is primarily geared towards recreational facilities. It is only required where the designer has chosen to provide separate-sex facilities for bathing and has provided more than one tub or shower for each sex. The exception acknowledges that smaller bathing rooms such as those described can be utilized by a disabled person with assistance without much hardship to a nondisabled person.

Many facilities have been installing family bathrooms as part of their standard bathroom layouts prior to the code requirement, viewing them as "customer

friendly." For example, a person requiring assistance can also be a small child or an elderly adult. A parent shopping or attending an event with a child of the opposite sex can utilize this facility as well as an adult who may need assistance from a spouse.

Note that the requirement for family bathrooms does not exempt separate-sex bathrooms from providing accessible or ambulatory stalls. For a discussion of family bathrooms in existing buildings, see Section 3411.8.11.

1109.2.1.1 Standard. Family or assisted-use toilet and bathing rooms shall comply with Sections 1109.2.1.2 through 1109.2.1.7 and ICC A117.1.

❖ The facilities are required to comply with the provisions of ICC A117.1 as well as the additional criteria established in the following subsections.

1109.2.1.2 Family or assisted-use toilet rooms. Family or assisted-use toilet rooms shall include only one water closet and only one lavatory. A family or assisted-use bathing room in accordance with Section 1109.2.1.3 shall be considered a family or assisted-use toilet room.

> **Exception:** A urinal is permitted to be provided in addition to the water closet in a family or assisted-use toilet room.

❖ A family toilet room must include a lavatory and a water closet. The exception permits a urinal to also be installed in the family toilet room if desired. If a family bathing room is provided within a facility, it can serve as the required family toilet room.

1109.2.1.3 Family or assisted-use bathing rooms. Family or assisted-use bathing rooms shall include only one shower or bathtub fixture. Family or assisted-use bathing rooms shall also include one water closet and one lavatory. Where storage facilities are provided for separate-sex bathing rooms, *accessible* storage facilities shall be provided for family or assisted-use bathing rooms.

❖ A family bathing facility is required to have one shower or tub, one water closet and one lavatory. The shower can be a transfer type, roll-in type or a combination of the two. Accessible storage facilities, such as lockers, are also required if storage facilities are provided in the separate-sex bathing facilities. A family bathing room can also serve a dual purpose (bathing and toilet) as the required family toilet room (see commentary, Section 1109.2.1.2).

1109.2.1.4 Location. Family or assisted-use toilet and bathing rooms shall be located on an *accessible route*. Family or assisted-use toilet rooms shall be located not more than one *story* above or below separate-sex toilet rooms. The *accessible route* from any separate-sex toilet room to a family or assisted-use toilet room shall not exceed 500 feet (152 m).

❖ A one-story, 500-foot (1524 mm) limitation for access to customer toilet facilities is currently in Section 2902.3.2 (which is duplicated from Section 402.3.2 of the IPC). The distance to the customer toilet facilities is measured from the main entrance of a store or space to the toilet facility. The distance in this section is measured from the separate-sex facility to the family facility. The general travel distance requirement, in combination with this section, could result in a total travel distance for a family toilet room of two stories or 1,000 feet (3048 mm) maximum. An accessible route is required to provide access to the family facilities. Signage is required at both the separate-sex and family facilities in accordance with Sections 1110.1 and 1110.2.

1109.2.1.5 Prohibited location. In passenger transportation facilities and airports, the *accessible route* from separate-sex toilet rooms to a family or assisted-use toilet room shall not pass through security checkpoints.

❖ Security checkpoints in airports and similar facilities represent a potential delay, which may cause missed flights or connections. Due to security concerns, more and more facilities are adding security checkpoints. For example, many large sports facilities, courthouses and government buildings have checkpoints where the public move into certain portions of the facilities. While not required, due to the delay in moving through

a security system, a designer might want to follow this same guidance for location of the family bathrooms in other types of facilities.

1109.2.1.6 Clear floor space. Where doors swing into a family or assisted-use toilet or bathing room, a clear floor space not less than 30 inches by 48 inches (762 mm by 1219 mm) shall be provided, within the room, beyond the area of the door swing.

❖ The clear floor space provisions are intended to provide a room that is large enough to allow a person in a wheelchair to enter and close the door before utilizing the fixtures. This requirement is also in ICC A117.1 [see Figure 1109.2(2)].

1109.2.1.7 Privacy. Doors to family or assisted-use toilet and bathing rooms shall be securable from within the room.

❖ Since privacy while utilizing bathing and toilet facilities is an issue, the door to the facility must be securable from the inside. The securing mechanism must be within reach and not require any tight pinching, grasping or sharp turning of the wrist to operate.

  While it is not prohibited to lock bathrooms, this bathroom should be as easily accessed as the separate men's and women's bathroom. A person who needed this facility should not have to find someone with a key for the family bathroom if the men's and women's bathrooms are readily available. This bathroom does count towards the required fixture count (see Section 2902.1.2) and as such must be available.

1109.2.2 Water closet compartment. Where water closet compartments are provided in a toilet room or bathing room, at least one wheelchair-*accessible* compartment shall be provided. Where the combined total water closet compartments and urinals provided in a toilet room or bathing room is six or more, at least one ambulatory-*accessible* water closet compartment shall be provided in addition to the wheelchair-*accessible* compartment. Wheelchair-*accessible* and ambulatory-*accessible* compartments shall comply with ICC A117.1.

❖ There are different configurations of water closet compartments that facilitate different degrees of physical disability. The provisions of ICC A117.1 establish the configuration and dimensional requirements for various types of water closet compartments. A wheelchair-accessible compartment is one in which sufficient space is provided for the wheelchair to enter completely the water closet compartment [see Figures 1109.2.2(2) and 1109.2.2(3)]. The wheelchair user then transfers from the wheelchair to the water closet in order to utilize the fixture. It is important that the required clear floor space be maintained. The configuration did not intend for a lavatory to be provided within the minimum-sized compartment. If a lavatory is located within a compartment, it must meet the same provisions as a single-occupant toilet room.

  An ambulatory-accessible compartment is intended to facilitate use by a person with a mobility impairment that necessitates the use of a walking aid, such as a cane or walker [see Figure 1109.2.2(1)]. An ambulatory-accessible water closet compartment is not intended to be utilized by a person in a wheelchair. The

36-inch (914 mm) width is intended to allow standing persons to support themselves utilizing the grab bars on both sides. A wider compartment would not allow adequate bearing. Since these provisions are intended to address, within reason, the needs of both ranges of mobility impairment, this section requires a wheelchair-accessible compartment in all cases. In larger toilet rooms (i.e., those with a total of six or more water closet compartments and urinals), one ambulatory-accessible compartment is required in addition to the wheelchair-accessible compartment.

This section is not intended to increase the required number of fixtures beyond that required by the IPC. For example, if a toilet room contains 10 water closets, eight water closet compartments may be of conventional design, one water closet compartment must be wheelchair accessible and one must be ambulatory accessible.

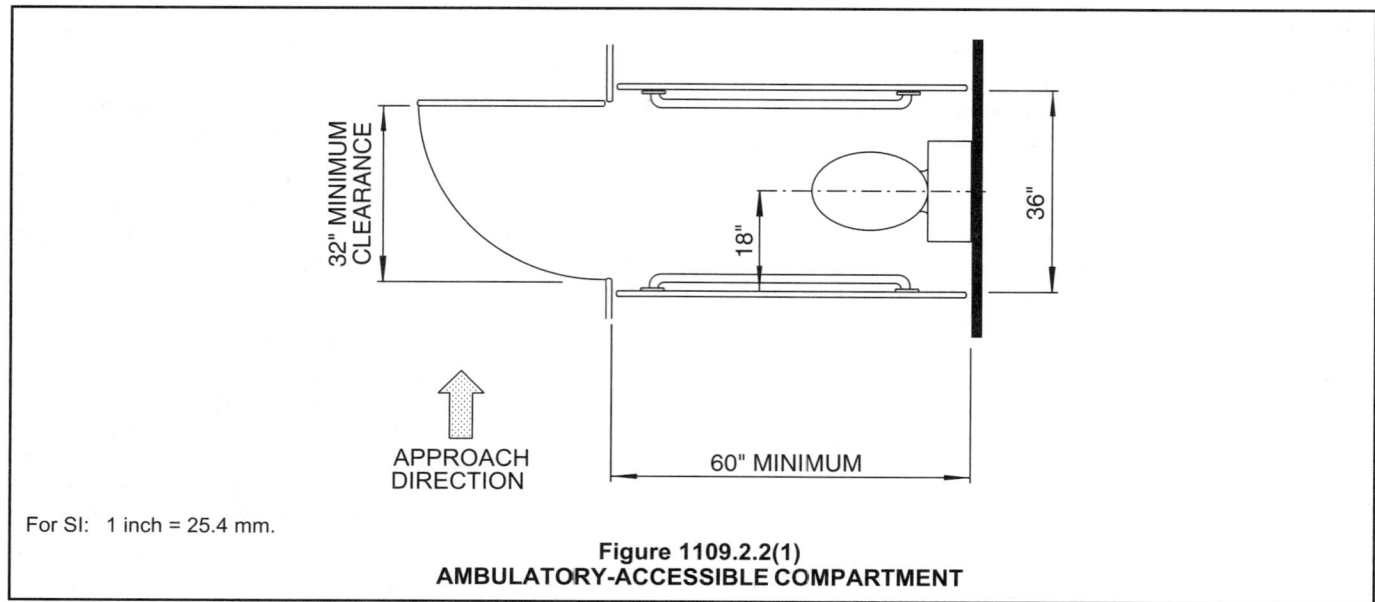

For SI: 1 inch = 25.4 mm.

Figure 1109.2.2(1)
AMBULATORY-ACCESSIBLE COMPARTMENT

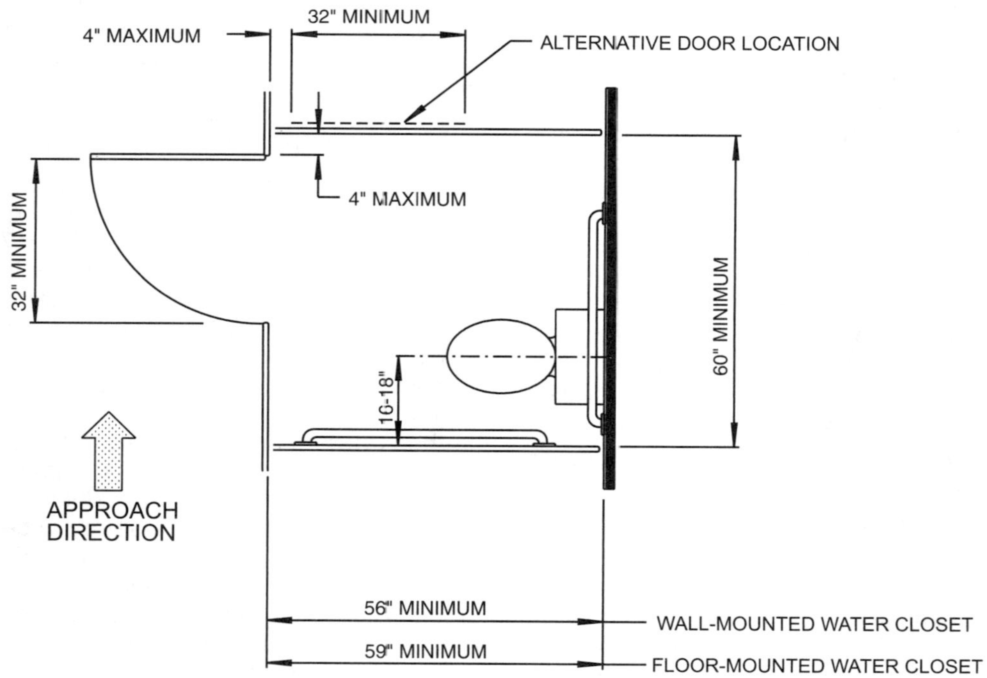

For SI: 1 inch = 25.4 mm.

Figure 1109.2.2(2)
WHEELCHAIR-ACCESSIBLE COMPARTMENT; OUT-SWINGING DOOR

1109.2.3 Lavatories. Where lavatories are provided, at least 5 percent, but not less than one, shall be *accessible*. Where the total lavatories provided in a toilet room or bathing facility is six or more, at least one lavatory with enhanced reach ranges in accordance with ICC A117.1, shall be provided.

❖ A lavatory is a type of sink for which the primary purpose is hand washing. This section provides scoping requirements for lavatories located in spaces not specifically addressed or exempted by Section 1109.2. If a lavatory is provided in toilet and bathing rooms specifically exempted by Section 1109.2, it is not required to be accessible. Lavatories are not as commonly part of an employee work station as sinks are, however, if they are they are exempted under Section 1103.2.3. An example of a lavatory that was part of an employee's work area would be the lavatories found in the exam rooms. The lavatories are there for the nurses and doctors to wash their hands before examining a patient.

When lavatories are provided in other areas, a minimum of 5 percent, but not less than one, is required to be accessible via a front approach with appropriate knee and toe clearances.

In large facilities, where the designer has located six or more lavatories within one toilet facility (similar to the ambulatory stall), one of the lavatories must also be provided with enhanced reach range for the faucet controls in accordance with Section 606.5 of ICC A117.1. This would require the faucets to be located on the side of the lavatory, or provide access to the side of one of the lavatories, or to have automatic controls. This is to address the needs of persons who may have a limited reach over the lavatory where they may have difficulty reaching the controls. The lavatory with enhanced reach range could be the same lavatory as the accessible lavatory, or it could be another lavatory.

There are allowances in ICC A117.1 for lavatories specifically designed for children's use. This may be appropriate in areas such as day care facilities and certain areas of elementary schools.

1109.3 Sinks. Where sinks are provided, at least 5 percent but not less than one provided in *accessible* spaces shall comply with ICC A117.1.

Exception: Mop or service sinks are not required to be *accessible*.

❖ A sink may have numerous functions, unlike a lavatory which is primarily provided for hand washing. Sinks in kitchens, kitchenettes and classrooms are examples that may have scoping and accessibility requirements different than those for lavatories. When sinks are provided, a minimum of 5 percent, but not less than one, is required to be accessible via a front approach with knee and toe clearances. There is a specific exception here for mop or service-type sinks. Service sinks include a wide variety of specialized sinks manufactured for multiple or specific functions. Service sinks are sized and positioned for their function which usually renders them incapable of being accessible. Also, when a sink is part of an individual work station, that sink is not required to be accessible in accordance with Section 1103.2.3.

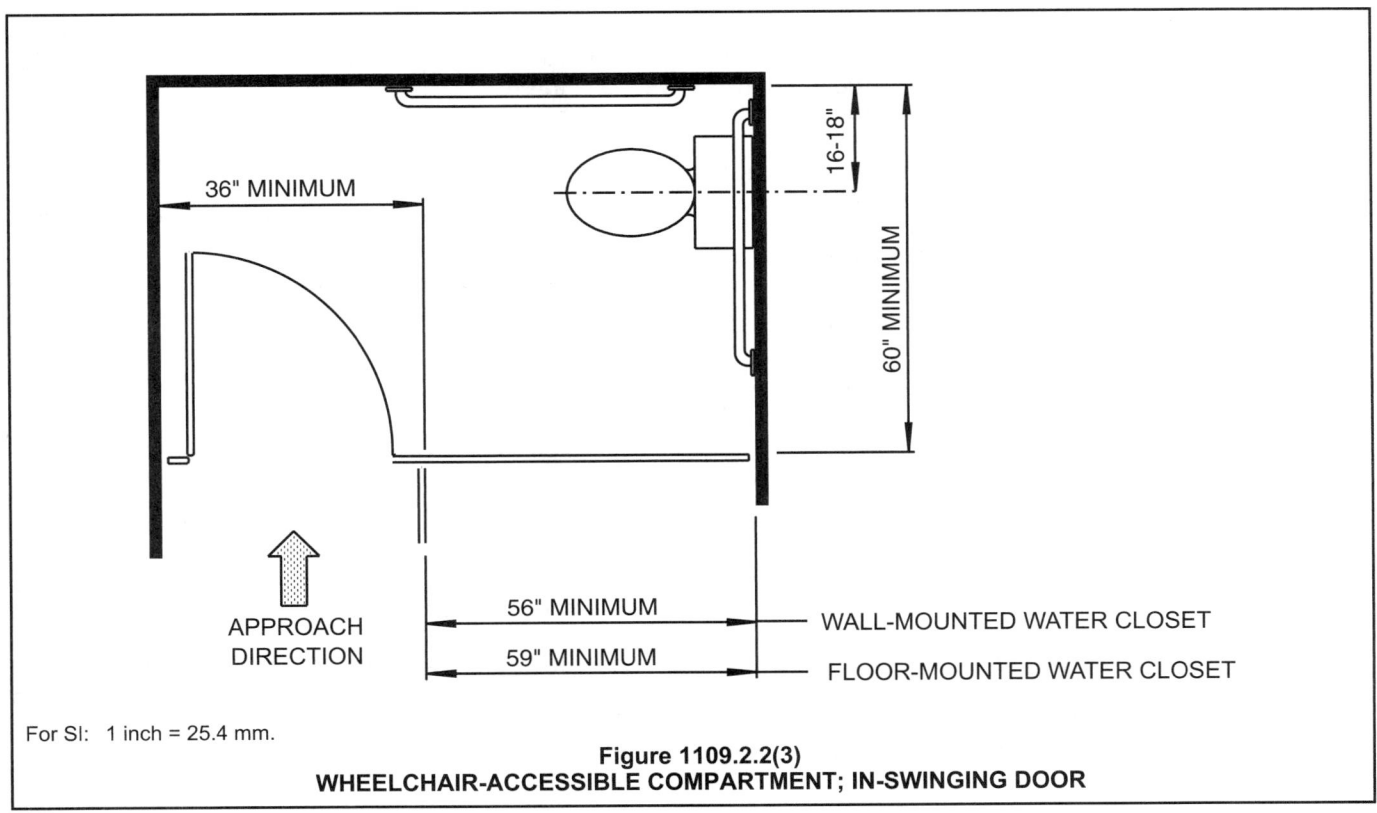

For SI: 1 inch = 25.4 mm.

Figure 1109.2.2(3)
WHEELCHAIR-ACCESSIBLE COMPARTMENT; IN-SWINGING DOOR

A forward approach to sinks, as required for a lavatory within a toilet room, is not always required (see commentary, Section 1109.4).

There are allowances in the ICC A117.1 for sinks specifically designed for children's use. This may be appropriate in areas such as day care facilities and certain areas of elementary schools.

1109.4 Kitchens and kitchenettes. Where kitchens and kitchenettes are provided in *accessible* spaces or rooms, they shall be *accessible* in accordance with ICC A117.1.

❖ Kitchens, kitchenettes and wet bars in accessible spaces, other than within Type A and Type B dwelling units, must be accessible. When a cooktop or conventional range is not provided, side approach to the sink is acceptable.

1109.5 Drinking fountains. Where drinking fountains are provided on an exterior site, on a floor or within a secured area, the drinking fountains shall be provided in accordance with Sections 1109.5.1 and 1109.5.2.

❖ This section establishes a reasonable threshold for the required number of accessible drinking fountains. It should be noted that this section does not require the installation of drinking fountains where none are required or provided. Chapter 29 (which is duplicated from Section 403 of the IPC) contains criteria indicating the number of drinking fountains that are required based on occupancy.

Current requirements for plumbing fixtures (see IBC Table 2902 and IPC Table 403) stipulate only one drinking fountain for many facilities or tenant spaces. The provisions in Section 1109.5.1 for a minimum of two drinking fountains would be more restrictive than the plumbing requirements for these facilities and spaces. Additionally, the plumbing requirements would allow for 50 percent of the drinking fountains to be substituted for bottled water coolers (see IPC 410.1). While the provisions in Sections 1109.5.1 and 1109.5.2 would not prohibit the substitution, the requirement for two drinking fountains would effectively negate this unless there were at least three drinking fountains required for a floor.

1109.5.1 Minimum number. No fewer than two drinking fountains shall be provided. One drinking fountain shall comply with the requirements for people who use a wheelchair and one drinking fountain shall comply with the requirements for standing persons.

Exception: A single drinking fountain that complies with the requirements for people who use a wheelchair and standing persons shall be permitted to be substituted for two separate drinking fountains.

❖ Where a single drinking fountain is provided or required by another code, this section mandates a minimum of two fixtures be provided: one for seated persons and one for standing persons. The seated and standing drinking fountains that serve a facility need not be provided at the same location in the facility. The

exception allows the use of a single fixture that accommodates both seated and standing persons. Technical criteria for both wheelchair accessible fountains and standing person fountains are located in Section 602 of ICC A117.1.

There are allowances in the ICC A117.1 for drinking fountains specifically designed for children's use. This may be appropriate for facilities such as day care facilities and certain areas of elementary schools.

1109.5.2 More than the minimum number. Where more than the minimum number of drinking fountains specified in Section 1109.5.1 are provided, 50 percent of the total number of drinking fountains provided shall comply with the requirements for persons who use a wheelchair and 50 percent of the total number of drinking fountains provided shall comply with the requirements for standing persons.

Exception: Where 50 percent of the drinking fountains yields a fraction, 50 percent shall be permitted to be rounded up or down, provided that the total number of drinking fountains complying with this section equals 100 percent of the drinking fountains.

❖ When an even number of drinking fountains is provided, half must accommodate seated persons and half must accommodate standing persons. The exception addresses when an odd number of drinking fountains is provided.

An example:
- Two drinking fountains are required by Section 1109.5.1.
- Seven drinking fountains are provided.
- Fifty percent of seven is three and one-half.
- Rounding up yields four; rounding down yields three.
- Therefore, there are two choices:
  - Provide four sitting and three standing fountains; or
  - Provide three sitting and four standing fountains.
- Both choices comply since the complying fixtures total 100 percent.

This logic applies whenever an odd number of three or more drinking fountains is provided, regardless of the total quantity.

1109.6 Elevators. Passenger elevators on an *accessible route* shall be *accessible* and comply with Section 3001.3.

❖ This section requires all passenger elevators on an accessible route to be accessible. The reference to Section 3001.3 is somewhat redundant, in that it requires all passenger elevators to conform to ICC A117.1. It is not the intent of this section to prohibit limited access/limited use elevators (LULA), private residence elevators or platform lifts (as permitted by Section 1109.7). LULAs and private residence elevators are recognized as a type of passenger lift by ASME A17.1.

Their limitations of use are controlled by that standard. For example, LULA maximum travel distance is 25 feet (7620 mm), and private residence elevators are limited to within or serving individual dwelling units.

It should also be noted that unlike plumbing fixtures, this section does not establish that a certain number of passenger elevators are required to be accessible while other passenger elevators may be of a design that is not fully accessible. All passenger elevators that are on an accessible route are required to be accessible. If a bank of passenger elevators is provided as part of an accessible route, all elevators in that bank must be accessible.

The reference to an accessible route is important because some areas in certain buildings may not be required to be accessible (see Sections 1103.2 and 1104.4). In addition, certain types of elevators (e.g., service elevator) may not be part of an accessible route.

Note that there are additional requirements for elevators in buildings four stories and higher in order to accommodate stretchers (see Section 3002.4) and when serving as part of an accessible means of egress in buildings five stories and taller (see Section 1007.2.1).

1109.7 Lifts. Platform (wheelchair) lifts are permitted to be a part of a required *accessible route* in new construction where indicated in Items 1 through 10. Platform (wheelchair) lifts shall be installed in accordance with ASME A18.1.

1. An *accessible route* to a performing area and speaker platforms in Group A occupancies.

2. An *accessible route* to *wheelchair spaces* required to comply with the *wheelchair space* dispersion requirements of Sections 1108.2.2 through 1108.2.6.

3. An *accessible route* to spaces that are not open to the general public with an *occupant load* of not more than five.

4. An *accessible route* within a *dwelling* or *sleeping unit*.

5. An *accessible route* to wheelchair seating spaces located in outdoor dining terraces in Group A-5 occupancies where the *means of egress* from the dining terraces to a *public way* are open to the outdoors.

6. An *accessible route* to jury boxes and witness stands; raised courtroom stations including judges' benches, clerks' stations, bailiffs' stations, deputy clerks' stations and court reporters' stations; and to depressed areas such as the well of the court.

7. An *accessible route* to load and unload areas serving amusement rides.

8. An *accessible route* to play components or soft contained play structures.

9. An *accessible route* to team or player seating areas serving areas of sport activity.

10. An *accessible route* where existing exterior *site* constraints make use of a ramp or elevator infeasible.

❖ This section indicates that platform (wheelchair) lifts are only permitted to be part of a required accessible route in new construction in limited situations. If a platform lift is permitted as part of the accessible route into a space, it can also serve as part of the accessible route used for means of egress out of the space (with the exception of Item 10) if it has standby power (see Section 1007.5). There are concerns about the potential for a large number of people needing to exit the building under the allowances for Item 10; therefore, an accessible means of egress must be via another route. Platform lifts may be used more extensively in existing construction (see Section 3411.8.3). This is in recognition that circumstances in existing buildings may make it impractical to accomplish accessibility by use of an elevator, in which case a platform lift is a reasonable alternative. Accessible means of egress is not required in existing buildings undergoing alterations (see Section 1007.1, Exception 1 or Section 3411.6, Exception 2); therefore, a platform lift as part of the accessible means of egress in these situations is not an issue.

A previous edition of the technical standard (i.e., ASME A17.1-1996) required key operation to platform lifts, which unnecessarily inhibited independent access by persons with physical disabilities. The requirements for platform lifts have been removed from the elevator standard and now have their own standard, ASME A18.1, *Safety Standard for Platform Lifts and Stairway Chairlifts*. This newer standard allows push-button operation of platform lifts, thus making independent access much easier.

The listed items indicate when a platform lift can be utilized as part of a required accessible route in new construction. Items 1 and 2 allow platform lifts to be utilized for access to performing (i.e., stages, orchestra pits) and viewing areas (i.e, wheelchair spaces in stepped seating areas) in assembly occupancies. Item 3 specifies that a platform lift may be used to provide access to a nonpublic area with five or less occupants, such as a projection booth. Item 4 allows for a platform lift within an individual dwelling or sleeping unit. Item 5 is intended to address an outdoor dining area in Group A-5, such as a picnic terrace in a baseball park.

Item 6 permits platform lifts to provide access to the raised and/or lowered areas typically found in courtroom settings. This is consistent with Section 1108.4.

Item 7 allows platform lifts for access to loading and unloading areas serving amusement rides. It does not attempt to address access to any other element of a ride.

Item 8 identifies children's play spaces, the entry to which may be raised or depressed, as an appropriate place for a platform lift. Such spaces are often associated with restaurants, shopping malls and preschools.

Item 9 acknowledges that team or player seating may be a small depressed area (ie., dugout) capable of being efficiently served by a wheelchair lift.

Item 10 recognizes that existing site constraints may make installation of a ramp or elevator infeasible. An example would be the situation of dealing with existing public sidewalks, easements and public ways in downtown urban areas. This situation would be most common in hilly areas where the street and sidewalk follows grade and the building's floor is level, resulting in steps up or down at entrances.

Note that a platform lift that was not part of the required accessible route could be used to facilitate access to any space. The governing factor in those situations would be the limitations of the product itself for capacity and travel distance. Item 4 permits platform lifts to be utilized within individual dwelling or sleeping units as part of a required accessible route. Single-family homes are not required to be accessible; however, persons with mobility impairments may choose to install stairway chairlifts in their private homes. Section 1009.1, Exception 4, specifies a minimum stairway width for platform lifts or chair lifts provided within dwelling units.

A platform lift is an electrically operated, mechanical device designed to transport a person who cannot use stairs over a short vertical distance. Platform lifts must be sized to accommodate a wheelchair user. Platform lifts can be used by wheelchair users and persons with limited mobility and are sometimes equipped with folding seats. A fold-down seat that moves up the stairway is not a platform lift, but rather a chair lift. A stairway

chair lift is not suitable for a person using a wheelchair and cannot serve as part of a required accessible route. A platform lift is most suitable for changes of elevation of one story or less where the installation of a ramp is not feasible [see Figures 1109.7(1) and 1109.7(2)].

While new technologies are expanding current options, there are two basic types of platform lifts: vertical lifts and inclined lifts. Vertical lifts are similar to elevators in that they travel only up and down in a fixed vertical space. Inclined platform lifts are usually installed in conjunction with a stairway and travel along the slope of the stairway. Inclined lifts are a design consideration for long flights of stairs where a vertical platform lift is not practical, where headroom is limited or where ceilings are low.

1109.8 Storage. Where fixed or built-in storage elements such as cabinets, shelves, medicine cabinets, closets and drawers are provided in required *accessible* spaces, at least one of each type shall contain storage space complying with ICC A117.1.

❖ Fixed or built-in storage elements are common in many occupancies, including both residential and many nonresidential occupancies, such as office buildings, recreational facilities, etc. Examples would include mailboxes, coat closets, storage closets, lockers, etc. The code does not require storage facilities, but when such facilities are provided, this section requires at least one of each type to contain storage space in accordance with ICC A117.1. Storage elements within Accessible, Type A and Type B units are specifically addressed under ICC A117.1 Chapter 10.

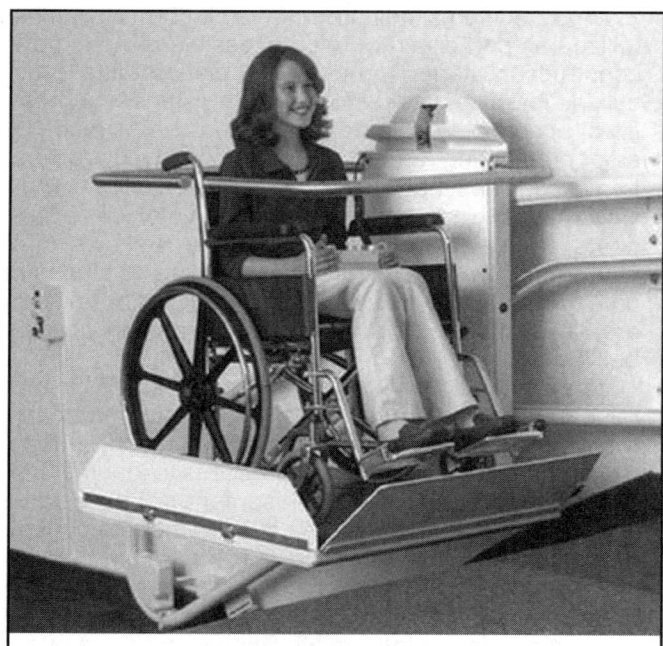

Figure 1109.7(1)
PLATFORM (WHEELCHAIR) LIFT
(Photo courtesy of Wheel-evator)

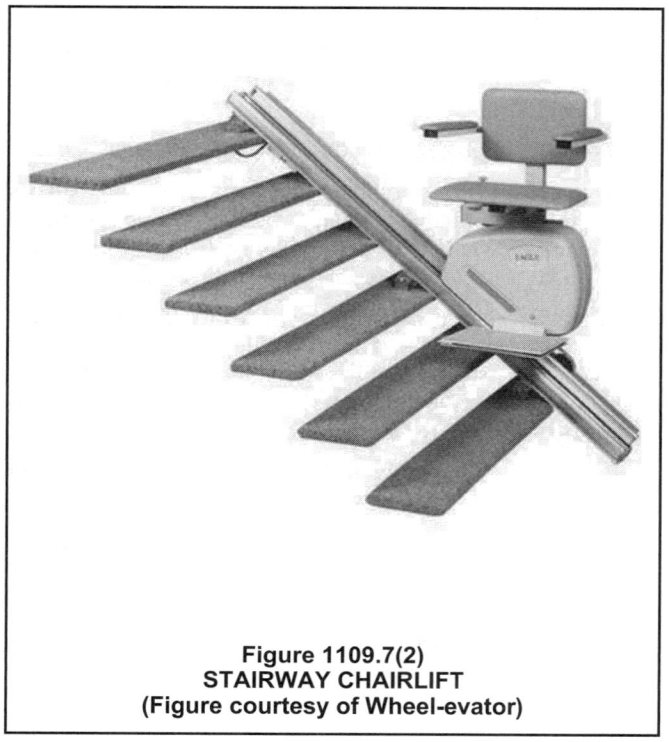

Figure 1109.7(2)
STAIRWAY CHAIRLIFT
(Figure courtesy of Wheel-evator)

1109.8.1 Lockers. Where lockers are provided in *accessible* spaces, at least five percent, but not less than one, of each type shall be *accessible*.

❖ Lockers are often provided in schools, health clubs, etc. At least 5 percent of the lockers in each location must have adequate clear floor space for wheelchair approach with exterior locks and latches and interior coat hooks and shelves within reach ranges. In accor-

Figure 1109.8.1(1)
EXAMPLE OF LOCKER AREA

Figure 1109.8.1(2)
EXAMPLE OF LOCKER ROOMS

dance with Section 803.4 of ICC A117.1, an accessible bench must be provided in all locker rooms so that people can transfer out of their chairs to change clothes. These benches should be located adjacent to the accessible lockers. In accordance with ICC A117.1 Section 903, acessible benches must be against a wall or have a back. If lockers are provided for just storage of items, such as in a school hallway, this is not considered a locker room like in a school gymnasium, so benches are not required.

1109.8.2 Shelving and display units. Self-service shelves and display units shall be located on an *accessible route*. Such shelving and display units shall not be required to comply with reach-range provisions.

❖ The intent of this section is that persons who utilize wheelchairs can see items on shelves in stores and libraries, but someone may be required to assist them in reaching the items.

1109.8.3 Coat hooks and shelves. Where coat hooks and shelves are provided in toilet rooms or toilet compartments or in dressing, fitting or locker rooms, at least one of each type shall be *accessible* and shall be provided in *accessible* toilet rooms without toilet compartments, *accessible* toilet compartments and *accessible* dressing, fitting and locker rooms.

❖ If amenities such as coat hooks or shelves are provided for use by the general public (in nonaccessible spaces), then those same amenities must be provided in an accessible space.

1109.9 Detectable warnings. Passenger transit platform edges bordering a drop-off and not protected by platform screens or *guards* shall have a *detectable warning*.

Exception: *Detectable warnings* are not required at bus stops.

❖ A detectable warning is a standardized feature built in or applied to walking surfaces to warn a visually impaired person of a hazard on or near his or her path of travel that may otherwise go unnoticed and could result in injury to that person. A typical example of the need for a detectable warning is at any walking surface where a siginifcant drop-off occurs that, if unnoticed, has the potential to cause injury should a person fall.

This section requires a detectable warning at the edges of passenger transit platforms where they border a drop-off. For example, loading platforms in both light- and heavy-rail transit stations where passengers await the arrival of trains are a serious potential hazard to a person with a vision impairment when there is no train at the station. There have been incidents where people with severe vision impairments have fallen off the transit platform and been killed or seriously injured. The presence of a detectable warning at the platform edge where it borders the drop-off would be encountered and recognized by a person with a vision impairment either through detection by a long cane or by foot contact with the detectable warning surface.

The exception for bus stops is established so as not to require a detectable warning at the curb between a sidewalk and the street. Detectable warnings at curbs are contradictory to the envisioned application of detectable warnings in the built environment. The curb itself is a recognizable and well-known cue to people with vision impairments to proceed with caution.

There is a great deal of controversy and general disagreement as to the benefit or advisability of detectable warnign surfaces throughout the built environment. The use of such warnings at transit starions is, to date, the only well documented use of detectable warnings, as discussed in studies such as "Tactile Warnings to Promote Safety in the Vicinity of Transit Platform Edges," conducted by the Urban Mass Transportation Administration, and "Pathfinder Tactile Tile Demonstration Test Project," conducted by the Metro-Dade Transit Agency. The Access Board will be looking into this issue as part of the development of Public Rights-of-Way Guidelines.

ICC A117.1 prescribes the type of surface that consitiutes a detectable warning in Section 705. One option for the detectable warning surface that is currently considered suitable consists of raised, truncated domes with a diameter between 0.9 inches (23 mm) and 1.4 inches (36 mm), a height of approximately 0.2 inch (5.1 mm) and center-to-center spacing of between 1.6 (41 mm) and 2.4 inches (61 mm). However, there is current and ongoing research into the use and application of truncated domes as a suitable detectable-wraning surface for various applications, including interior and exterior locations. ICC A117.1 requires that the type of detectable warning surface utilized must be standard throughout a building, facility, site or complex of buildings. If different types of detectable warning surfaces are utilized, their usefulness is diminished. The different messages they would convey to a person with a vision impairment would not be consistent and may be confusing and easily misinterpreted.

It is anticipated that future applications of detectable warnings will be considered when additional research and documentation of their usefulness and suitability is available. One issue, among many, is the durability of a dtecable warning surface in an exterior application and the potential difficulty that truncated domes may present to a person in a wheelchair, as well as to non-disabled persons who may have to negotiate the surface.

1109.10 Seating at tables, counters and work surfaces. Where seating or standing space at fixed or built-in tables, counters or work surfaces is provided in *accessible* spaces, at least 5 percent of the seating and standing spaces, but not less than one, shall be *accessible* . In Group I-3 occupancy visiting areas at least 5 percent, but not less than one, cubicle or counter shall be *accessible* on both the visitor and detainee sides.

Exceptions:

1. Check-writing surfaces at check-out aisles not required to comply with Section 1109.11.2 are not required to be *accessible*.

2. In Group I-3 occupancies, the counter or cubicle on the detainee side is not required to be *accessible* at noncontact visiting areas or in areas not serving *accessible* holding cells or *sleeping unit*s.

❖ Many occupancies, such as libraries, classrooms, restaurants and other public spaces, are equipped with fixed seating, tables or both (see Figure 1109.10). Correctional facilities often have cubicles or counters that are set up for visiting sessions between detainees and their lawyers or visitors. This section requires that at least 5 percent, but not less than one, be accessible. The threshold of 5 percent is consistent with the requirement for accessible seating in dining areas as specified in Section 1108.2.9.1. Fixed or built-in seating that is part of an individual work station is not required to be accessible (see Section 1103.2.3). These accessible table and work surfaces are expected to include adequate knee and toe clearances as specified in ICC A117.1. For requirements for points of sales or service that do not include a work station, see Section 1109.11.3. For example, the check-writing station in the center of the bank lobby is considered a work station; however, the teller window is typically considered a service counter. The accessible check-writing station would require knee and toe clearances, while the accessible teller window would not.

Note that the requirement addresses built-in counters, tables or work stations for standing or seating. Built-in surfaces where persons are typically standing are most often installed at a height of 36 to 48 inches (914 to 1219 mm). Depending on the overall configuration of the counters in the space, a portion of the counter for standing persons may need to be lowered to an accessible height of 28 to 34 inches (711 to 864 mm) and configured for forward approach.

The intent of Exception 1 is that if a check-writing surface is provided at a nonaccessible checkout aisle, this surface is not required to meet the counter and work surface requirements in ICC A117.1 (see Section 1109.11.2).

Exception 2 is consistent with the Group I-3 exception found in Section 1103.2.14. If Accessible units are not required by Section 1107.5.5, accessible elements are not required on the detainee side of the cubicles. Since the detainees may have a disabled person visiting them on the public side, at least 5 percent of the stations must be accessible.

There are allowances in the ICC A117.1 for work surfaces specifically designed for children's use. This may be appropriate in facilities such as day care facilities and certain areas of elementary schools, or areas frequented mostly by children, such as children's museums or the children's areas in libraries.

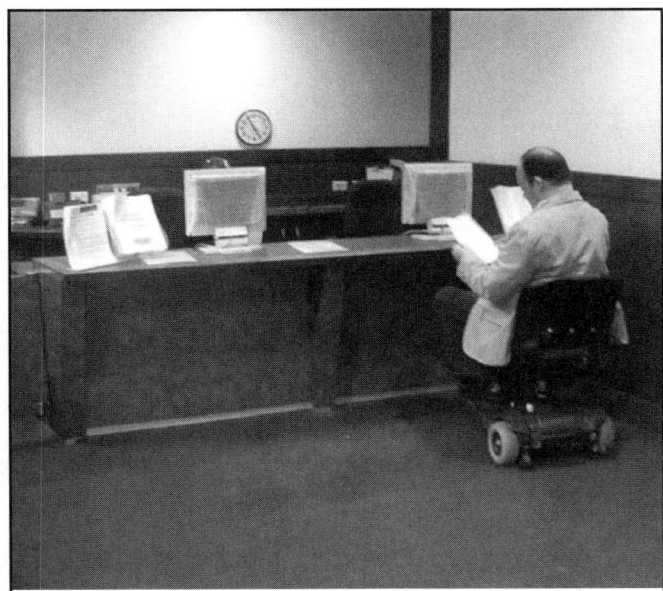

Figure 1109.10
EXAMPLE OF ACCESSIBLE WORK STATION

1109.10.1 Dispersion. *Accessible* fixed or built-in seating at tables, counters or work surfaces shall be distributed throughout the space or facility containing such elements and located on a level accessed by an *accessible route*.

❖ An appropriate distribution of seating and tables is required and is intended to be consistent with the distribution of all such seating and tables in the facility. For example, a library may contain several separate areas in which fixed seating or tables are installed. This provision would prohibit the location of all accessible seating and tables in a single specific area of the facility. However, this should not be construed to mean that every separate area in which fixed seating or tables are provided is required to contain an accessible seat or table. For example, if there are four areas in which fixed seating or tables are located, and two such seats or tables are required by this section to be accessible, the intent of this section is to require that one accessible seat or table be located in two of the four areas.

If the facility has floor levels exempted from an accessible route (see Section 1104.4), accessible work surfaces are not required to be located on the nonaccessible level.

1109.11 Service facilities. Service facilities shall provide for *accessible* features in accordance with Sections 1109.11.1 through 1109.11.5.

❖ This section introduces five subsections that address accessibility of dressing rooms; service counters and windows; check-out aisles; food service; queues and waiting lines, etc. These types of elements are broadly termed "service facilities" and are intended to permit usage by persons with physical disabilities.

1109.11.1 Dressing, fitting and locker rooms. Where dressing rooms, fitting rooms or locker rooms are provided, at least 5 percent, but not less than one, of each type of use in each cluster provided shall be *accessible*.

❖ This section establishes a reasonable minimum threshold for accessible dressing, fitting and locker rooms. This section does not require that dressing, fitting and locker rooms be provided, but when such facilities are provided, at least 5 percent of each group of rooms is required to be accessible. This section also requires an appropriate distribution of accessible facilities based on distinct and different functions of each group of facilities provided. For example, a department store may have one group of fitting rooms serving the menswear department and a separate group of fitting rooms serving the women's wear department. This section would not permit the location of all accessible fitting rooms in one department and no accessible fitting rooms in another department. This may result in a number of accessible fitting rooms greater than 5 percent of the total number of all fitting rooms in the facility. For example, if there are two dressing rooms in each of three different departments in a retail store, a total of three accessible fitting rooms would be required: one in the three areas. Where amenities such as coat hooks, lockers or shelves are provided in inaccessible dressing, fitting or locker rooms, they must also be provided in the accessible rooms (see Section 1109.8). In accordance with Section 803.4 of ICC A117.1, an accessible bench must be provided in all accessible dressing, fitting and locker rooms (see Figure 1109.11.1).

1109.11.2 Check-out aisles. Where check-out aisles are provided, *accessible* check-out aisles shall be provided in accordance with Table 1109.11.2. Where check-out aisles serve different functions, at least one *accessible* check-out aisle shall be provided for each function. Where check-out aisles serve different functions, *accessible* check-out aisles shall be provided in accordance with Table 1109.11.2 for each function. Where check-out aisles are dispersed throughout the building or facility, *accessible* check-out aisles shall also be dispersed. Traffic control devices, security devices and turnstiles located in *accessible* check-out aisles or lanes shall be *accessible*.

❖ This section applies to sales locations configured as lanes or aisles, often designed to accommodate shopping carts, such as are typically found in supermarkets, drugstores, discount retail stores, etc. If checkout aisles serve different functions, at least one check-out aisle that serves each function shall be accessible. If a facility offers checkouts at a variety of locations, the accessible locations must also be dispersed. Where security devices such as turnstiles and automatically activated gates are provided, they are required to be accessible. The required number of check-out aisles for each function is set forth in Table 1109.12.2. Signage is required in accordance with Section 1110.1.

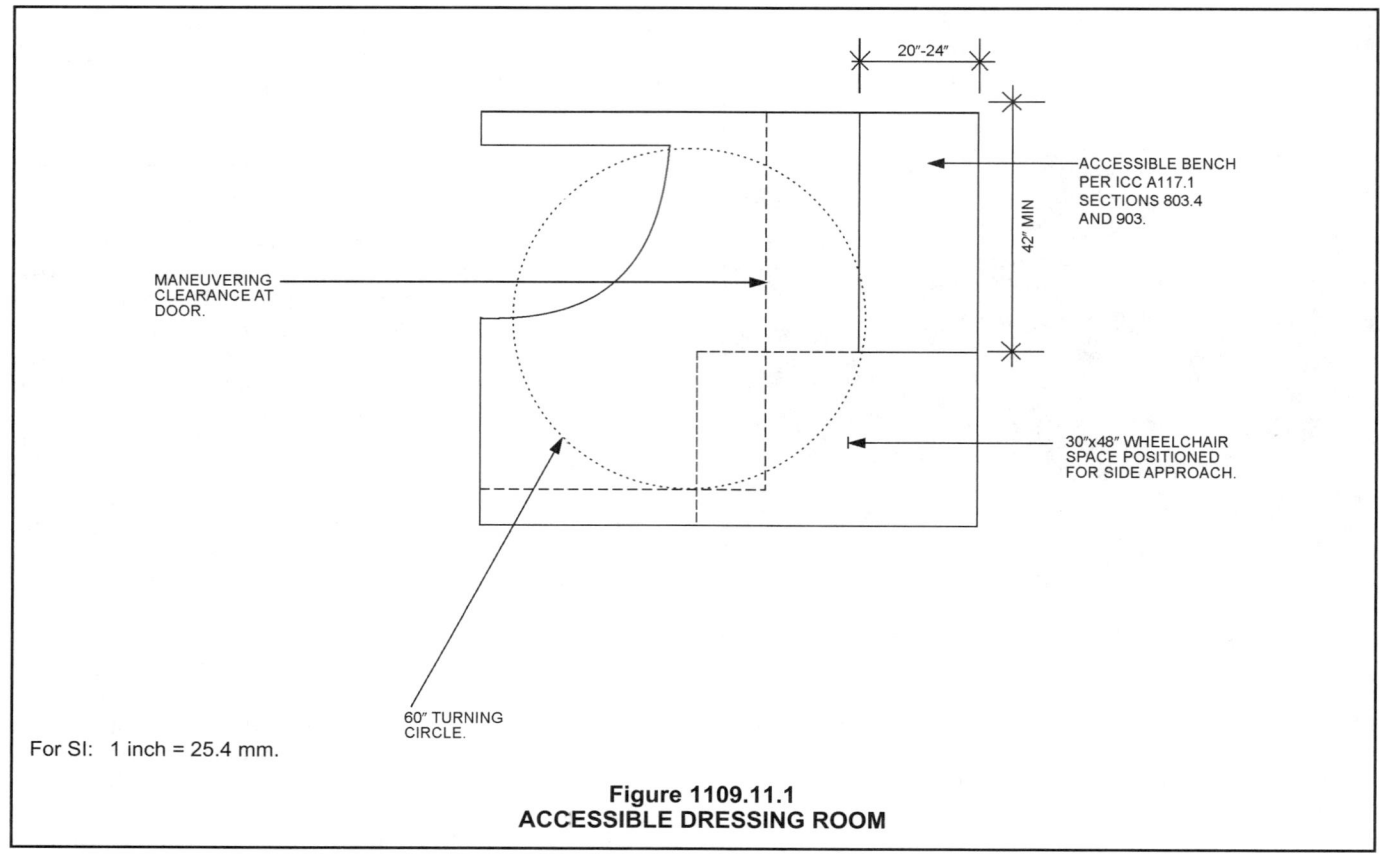

For SI: 1 inch = 25.4 mm.

Figure 1109.11.1
ACCESSIBLE DRESSING ROOM

TABLE 1109.11.2
ACCESSIBLE CHECK-OUT AISLES

TOTAL CHECK-OUT AISLES OF EACH FUNCTION	MINIMUM NUMBER OF ACCESSIBLE CHECK-OUT AISLES OF EACH FUNCTION
1 to 4	1
5 to 8	2
9 to 15	3
Over 15	3, plus 20% of additional aisles

1109.11.3 Point of sale and service counters. Where counters are provided for sales or distribution of goods or services, at least one of each type provided shall be *accessible*. Where such counters are dispersed throughout the building or facility, *accessible* counters shall also be dispersed.

❖ Wherever sales or service counters or windows are provided, at least one window or a portion of a counter is required to be accessible. For example, most hotel registration and check-out functions occur at a counter. The intent of this section is to require that one portion of the counter area be accessible. Accessibility is accomplished by providing a lower counter height to accommodate a person using a wheelchair and locating that counter on an accessible route. This is not necessarily intended to require that multiple lower counter heights be provided based on the different types of services that are offered. For example, this is not intended to require two separate, lower counter

heights: one for hotel registration and one for check-out. One lower counter height can be provided at which both functions are accomplished. In the above described situation, dispersion of the accessible counters would be required if the counters themselves were dispersed in separate locations.

The intent of the accessible service window is to permit the customer to interact with the service representative the same way they interact with the general public. An extra piece of counter stuck on the front of the high service counter [typically 42- to 48-inches (1067 to 1219 mm) high] may be appropriate as a temporary fix for barrier removal but is not acceptable in new construction. ICC A117.1 permits a maximum counter height of 36 inches (914 mm) at a service window (see commentary, Section 1109.10 and Figure 1109.11.3).

1109.11.4 Food service lines. Food service lines shall be *accessible*. Where self-service shelves are provided, at least 50 percent, but not less than one, of each type provided shall be *accessible*.

❖ All lines that provide access to food service must meet the accessible route provisions in ICC A117.1. Where self-service shelves are utilized, such as in a cafeteria, at least half of each type must be on an accessible route and within reach ranges.

**Figure 1109.11.3
SERVICE COUNTERS**

1109.11.5 Queue and waiting lines. Queue and waiting lines servicing *accessible* counters or check-out aisles shall be *accessible*.

❖ If patrons must form a line for service, then the waiting line must meet the accessible route provisions of ICC A117.1. When dealing with permanent crowd control barriers, special attention must be paid to turns around obstructions or dividers. Accessible waiting lines are required at locations where accessible counters, windows or check-out aisles are provided, such as banks, ticket counters, fast-food establishments and similar facilities. A separate waiting line for persons with disabilities is not a viable alternative.

1109.12 Controls, operating mechanisms and hardware. Controls, operating mechanisms and hardware intended for operation by the occupant, including switches that control lighting and ventilation and electrical convenience outlets, in *accessible* spaces, along *accessible routes* or as parts of *accessible* elements shall be *accessible*.

Exceptions:

1. Operable parts that are intended for use only by service or maintenance personnel shall not be required to be *accessible*.

2. Electrical or communication receptacles serving a dedicated use shall not be required to be *accessible*.

3. Where two or more outlets are provided in a kitchen above a length of counter top that is uninterrupted by a sink or appliance, one outlet shall not be required to be *accessible*.

4. Floor electrical receptacles shall not be required to be *accessible*.

5. HVAC diffusers shall not be required to be *accessible*.

6. Except for light switches, where redundant controls are provided for a single element, one control in each space shall not be required to be *accessible*.

7. Access doors or gates in barrier walls and fences protecting pools, spas and hot tubs shall be permitted to have operable parts of the release of latch on self-latching devices at 54 inches (1370 mm) maximum and 48 inches minimum above the finished floor or ground, provided the self-latching devices are not also self-locking devices, operated by means of a key, electronic opener, or integral combination lock.

❖ This section requires that any controls intended to be utilized by the occupants of the space be accessible when such controls are located in an accessible space, along accessible routes or as part of an accessible element. For example, the switch controlling the general lighting of a room that is required to be accessible also must be accessible. This is also true for most of the electrical outlets that are located in the room. Thermostats that are intended to be adjustable by the occupants must also be accessible. Another example would be the button or switch that activates an automatic hand dryer in a toilet room. If the fixture is mounted such that the outlet for the air is at an accessible height, the purpose is defeated if the switch to activate the device is inaccessible.

The exceptions listed are similar to the exceptions already located in ICC A117.1 for Accessible, Type A and Type B dwelling and sleeping units. Since the same problems exist in nonresidential facilities, the exceptions are appropriate.

Exception 1 deals with items that are not intended for use by the general occupants of the space. This would include controls such as thermostats that are intended to be adjustable only by authorized personnel. Another example would be the controls restricted to use by authorized personnel for a public address system in a meeting room.

Exception 2 deals with connections for a dedicated use. For example, the outlet behind the refrigerator, where it plugs in, is not required to be accessible.

Exception 3 is for kitchens and kitchenettes. If more than one electrical outlet is provided over a portion of the countertop, only one is required to be accessible.

Exception 4 permits floor receptacles not to be accessible, since by being in the floor they are out of the reach range.

Exception 5 is for HVAC diffusers. The typical locations for these elements are most often in the floor, high up on the walls or on the ceiling.

Exception 6 allows for redundant controls for everything but lighting. For example, a ceiling fan could be operated by a wall switch as well as by the chain on the fan itself.

Exception 7 addresses the operable parts on safety gates and doors serving pools, spas and hot tubs. The operable parts are required to be higher than normal to be above the reach of young children (see Section 1008.1.9.2).

1109.12.1 Operable window. Where operable windows are provided in rooms that are required to be *accessible* in accordance with Sections 1107.5.1.1, 1107.5.2.1, 1107.5.3.1,

1107.5.4, 1107.6.1.1, 1107.6.2.1.1, 1107.6.2.2.1 and 1107.6.4.1, at least one window in each room shall be *accessible* and each required operable window shall be *accessible*.

> **Exception:** *Accessible* windows are not required in bathrooms and kitchens.

❖ This section specifies when operable windows must be accessible, limiting the requirements to Accessible and Type A dwelling and sleeping units in Group I-1, I-2, R-1, R-2 and R-4 occupancies. There are no requirements for access to operable windows in Type B units. The concern is that such a requirement would force designers to opt for nonoperable windows when permitted by the code. If an operable window is required for natural ventilation or an emergency escape window within an Accessible or Type A unit, then that window must be accessible. Due to their typical locations within the space (e.g., over the sink in a kitchen, raised in a bathroom for privacy), operable windows in kitchens or bathrooms are exempted.

1109.13 Fuel-dispensing systems. Fuel-dispensing systems shall comply with ICC A117.1.

❖ With the exceptions specific to fuel-dispensing systems (i.e, gas pumps) in ICC A117.1, this section exempts gas and diesel fuel pumps at service stations from all accessibility provisions, except for the unobstructed reach range provisions in ICC A117.1. Basically, all the operable parts (e.g., pump controls, credit card readers, receipt dispenser, etc.) must be located between 15 inches and 48 inches (381 mm and 1219 mm) above the parking surface. Curbs must be considered when determining the height of controls. The intent is to address the needs of persons with limited reach range. Therefore, if the island the pump is on is wide enough to allow standing space in front of the pump, that is acceptable. A curb ramp is not required. Fuel-dispensing systems that are part of a work station are exempted by Section 1103.2.3.

1109.14 Recreational and sports facilities. Recreational and sports facilities shall be provided with *accessible* features in accordance with Sections 1109.14.1 through 1109.14.4.

❖ Recreational and sports facilities such as tennis courts, swimming pools, baseball fields, etc., may be part of development on a site such as a school or apartment complex. Group R-2 and R-3 facilities are addressed in Sections 1109.14.1 and 1109.14.2 so that equitable opportunity for use by all occupants of a multibuilding site would be evaluated. All other occupancies must comply with Section 1109.14.3.

1109.14.1 Facilities serving a single building. In Group R-2 and R-3 occupancies where recreational facilities are provided serving a single building containing *Type A units* or *Type B units*, 25 percent, but not less than one, of each type of recreational facility shall be *accessible*. Every recreational facility of each type on a site shall be considered to determine the total number of each type that is required to be *accessible*.

❖ Many multiple-family developments include common recreational facilities, such as a swimming pool, community meeting room, tennis court, playground, etc., which are available for use by residents of the complex. In some cases, there may be more than one type of recreational facility located at different points within the development. This section establishes the criterion that 25 percent, but not less than one, of each type of recreational facility be accessible. For example, if two separate tennis court areas are provided, and one contains two tennis courts and another contains seven tennis courts, then a total of three tennis courts must be accessible. In a single building site, the three accessible tennis courts could all be in one area or divided between the two areas. This satisfies the requirement that at least 25 percent of the total number of courts be made accessible.

1109.14.2 Facilities serving multiple buildings. In Group R-2 and R-3 occupancies on a single *site* where multiple buildings containing *Type A units* or *Type B units* are served by recreational facilities, 25 percent, but not less than one, of each type of recreational facility serving each building shall be *accessible*. The total number of each type of recreational facility that is required to be *accessible* shall be determined by considering every recreational facility of each type serving each building on the site.

❖ In a multibuilding site, the same criteria applies as for single building sites; however, in addition, if certain facilities only serve certain buildings, at least 25 percent of them in each area must be accessible. For example, if two separate tennis court areas are provided, each serving a specific building, and one contains two tennis courts and another contains seven tennis courts, then 25 percent at each location, for a total of three tennis courts, must be accessible. While the number is the same for the single building, the distribution requirement is different. In this case, one tennis court in the two-court area must be made accessible and two of the tennis courts in the seven-court area must be made accessible. This satisfies the requirement that at least 25 percent of the courts for each building are made accessible. If all tennis courts serve all the buildings, then the distribution could be the same as discussed in Section 1109.14.1.

1109.14.3 Other occupancies. All recreational and sports facilities not falling within the purview of Section 1109.14.1 or 1109.14.2 shall be *accessible*.

❖ When tennis courts, pools, baseball diamonds, playgrounds or other recreational facilities are provided in occupancies other than Group R-2 or R-3, all facilities must be accessible.

The Access Board has included requirements for many types of recreational facilities in the new ADA/ABA Guidelines. The 2009 edition of ICC A117.1, while not referenced by the 2009 edition of the code, does include a new chapter with technical information on accessibility for many types of recreational facilities. Either document would be a good resource when addressing accessibility issues for different types of recreational facilities.

1109.14.4 Recreational and sports facilities exceptions. Recreational and sports facilities required to be *accessible* shall be exempt from this chapter to the extent specified in this section.

❖ This section identifies requirements or exceptions for five specific types of recreational facilities and/or elements within those facilities. These items are intended to harmonize with the new ADA/ABA Guidelines.

1109.14.4.1 Bowling lanes. An *accessible route* shall be provided to at least 5 percent, but no less than one, of each type of bowling lane.

❖ Bowling lanes are typically depressed from the circulation and viewing areas in bowling facilities for sightline purposes. As with multiples in many nonsport facilities, only a percentage are required to be served by an accessible route.

1109.14.4.2 Court sports. In court sports, at least one *accessible route* shall directly connect both sides of the court.

❖ It is not permissible to have the only access to, for example, the far side of a court pass through the near side of the court. This requirement eliminates the possibility of having to travel through ongoing play on one side of a court to access the opposite side of the court.

1109.14.4.3 Raised boxing or wrestling rings. Raised boxing or wrestling rings are not required to be *accessible*.

❖ It is impractical and unnecessary for raised boxing and wrestling rings to be accessible. This exception recognizes a basic premise of accessibility in the built environment for reasonable and logical accommodation.

1109.14.4.4 Raised refereeing, judging and scoring areas. Raised structures used solely for refereeing, judging or scoring a sport are not required to be *accessible*.

❖ This exception is in step with several of the general exceptions in Section 1103.2 where small raised areas provided for specific purposes are exempted. As with any other exception, there is nothing that would prohibit the installation of specialized equipment to allow a disabled person to perform the particular function.

1109.14.4.5 Raised diving boards and diving platforms. Raised diving boards and diving platforms are not required to be *accessible*.

❖ As with boxing rings, it is impractical and unnecessary for diving boards and platforms to be accessible.

SECTION 1110
SIGNAGE

1110.1 Signs. Required *accessible* elements shall be identified by the International Symbol of Accessibility at the following locations:

1. *Accessible* parking spaces required by Section 1106.1 except where the total number of parking spaces provided is four or less.

2. *Accessible* passenger loading zones.

3. *Accessible* rooms where multiple single-user toilet or bathing rooms are clustered at a single location.

4. *Accessible* entrances where not all entrances are accessible.

5. *Accessible* check-out aisles where not all aisles are accessible. The sign, where provided, shall be above the check-out aisle in the same location as the check-out aisle number or type of check-out identification.

6. Unisex toilet and bathing rooms.

7. *Accessible* dressing, fitting and locker rooms where not all such rooms are *accessible*.

8. *Accessible areas of refuge* in accordance with Section 1007.9.

9. Exterior areas for assisted rescue in accordance with Section 1007.9.

❖ Identification of accessible elements can be accomplished by use of an International Symbol of Accessibility (see Figure 1110.1). These figures are international in that they are recognized throughout the world as that which identifies accessibility.

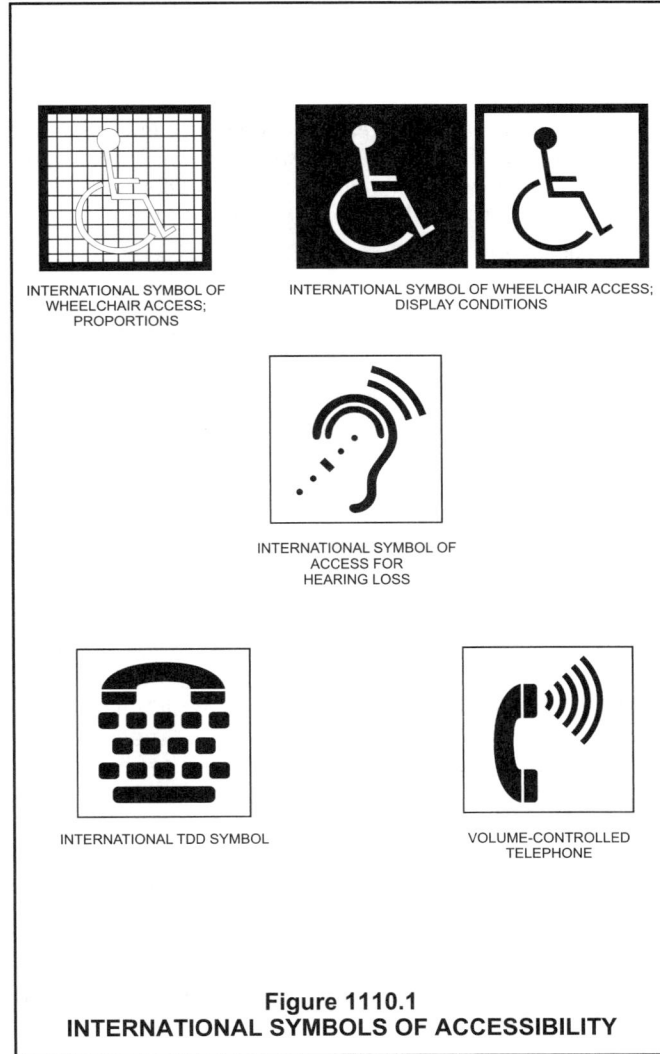

INTERNATIONAL SYMBOL OF WHEELCHAIR ACCESS; PROPORTIONS

INTERNATIONAL SYMBOL OF WHEELCHAIR ACCESS; DISPLAY CONDITIONS

INTERNATIONAL SYMBOL OF ACCESS FOR HEARING LOSS

INTERNATIONAL TDD SYMBOL

VOLUME-CONTROLLED TELEPHONE

Figure 1110.1
INTERNATIONAL SYMBOLS OF ACCESSIBILITY

There are nine specific circumstances in which required accessible elements are to be identified, as indicated in Items 1 through 9 of this section. Generally, these are locations in which not all of the facilities provided are accessible and, therefore, it is necessary to identify those that are accessible so that they can be readily recognized by the intended user. For example, Section 1106.1 specifies the required number of accessible parking spaces. If these are not identified by signage, it would be difficult and unnecessarily inconvenient for one to identify their location. (Note that the requirement for parking signage is limited to when more than four total parking spaces are provided. To require an accessible space to be reserved when it involves 25 to 100 percent of the parking provided is an undue hardship to the building tenant.) One of the concepts embodied in the code is to mainstream accessibility in recognition that many of the features that make facilities and elements accessible are also useful and of benefit to people without disabilities. Part of this principle includes the idea that if an element is universally usable by people both with and without disabilities, there is no need for signage specifically identifying the element as being accessible. Hence, the signage requirements generally address circumstances in which not all of the elements will be accessible.

1110.2 Directional signage. Directional signage indicating the route to the nearest like *accessible* element shall be provided at the following locations. These directional signs shall include the International Symbol of Accessibility:

1. Inaccessible building entrances.

2. Inaccessible public toilets and bathing facilities.

3. Elevators not serving an *accessible route*.

4. At each separate-sex toilet and bathing room indicating the location of the nearest family or assisted-use toilet or bathing room where provided in accordance with Section 1109.2.1.

5. At *exits* and *exit stairways* serving a required *accessible* space, but not providing an *approved accessible means of egress*, signage shall be provided in accordance with Section 1007.10.

❖ There are circumstances in which it is useful and necessary to locate directional signage at certain inaccessible elements to indicate the route to the nearest like accessible element. For example, not all building entrances are required to be accessible (see Section 1105). Should a person in a wheelchair happen to approach an inaccessible building entrance at an unfamiliar facility, it is appropriate to provide direction indicating where the nearest accessible entrance is located. The same circumstance presents itself at inaccessible public toilet and bathing facilities, at elevators that do not serve an accessible route and in assembly and mercantile occupancies where family-assisted use accessible toilet and bathing facilities are

required.

This requirement for directional signage works in conjunction with the fact that everything is initially assumed to be accessible; therefore, items are not required to be identified by the International Symbols of Accessibility in the interest of mainstreaming. See the general comments to this chapter and the commentary to Section 1110.1 for a further discussion of mainstreaming.

1110.3 Other signs. Signage indicating special accessibility provisions shall be provided as shown:

1. Each assembly area required to comply with Section 1108.2.7 shall provide a sign notifying patrons of the availability of assistive listening systems.

   Exception: Where ticket offices or windows are provided, signs are not required at each assembly area provided that signs are displayed at each ticket office or window informing patrons of the availability of assistive listening systems.

2. At each door to an *area of refuge*, an exterior area for assisted rescue, an egress *stairway*, *exit passageway* and *exit discharge*, signage shall be provided in accordance with Section 1011.3.

3. At *areas of refuge*, signage shall be provided in accordance with Section 1007.11.

4. At exterior areas for assisted rescue, signage shall be provided in accordance with Section 1007.11.

5. At two-way communication systems, signage shall be provided in accordance with Section 1007.8.2.

6. Within *exit enclosures*, signage shall be provided in accordance with Section 1022.8.

❖ It is considered desirable that all the requirements for accessible signage be in one location. Item 1 requires signage indicating that assistive listening systems are provided when they are required by Section 1108.2.7. Items 2 through 6 are references to signage criteria in specific sections that have been mainstreamed in Chapter 10.

Bibliography

The following resource materials are referenced in this chapter or are relevant to the subject matter addressed in this chapter.

24 CFR, *Fair Housing Accessibility Guidelines* (FHAG). Washington, DC: Department of Housing and Urban Development, 1991.

36 CFR Parts 1190 and 1191 Final Rule, *The Americans with Disabilities Act (ADA) Accessibility Guidelines; Architectural Barriers Act (ABA) Accessibility Guidelines*. Washington, DC: Architectural and Transportation Barriers Compliance Board, July 23, 2004.

42 USC 3601-88, *Fair Housing Amendments Act* (FHA). Washington, DC: United States Code, 1988.

"Accessibility and Egress for People with Physical Disabilities." CABO Board for the Coordination of the Model Codes Report, 1993.

ASME A17.1/CSA B44-07, *Safety Code for Elevators and Escalators*. New York, NY: American National Standards Institute, 2007.

ASME A18.1-05, *Safety Standard for Platform Lifts and Stairway Chairlifts*. New York, NY: American National Standards Institute, 2005.

DOJ 28 CFR, Part 36-91, *Americans with Disabilities Act* (ADA). Washington, DC: Department of Justice, 1991.

DOJ 28 CFR, Part 36-91 (Appendix A), *ADA Accessibility Guidelines for Building and Facilities* (ADAAG). Washington, DC: Department of Justice, 1991.

Fair Housing Act Design Manual. U.S. Department of Housing and Urban Development, Revised 1998.

FED-STD-795-88, *Uniform Federal Accessibility Standards*. Washington, DC: General Services Administration; Department of Defense; Department of Housing and Urban Development; U.S. Postal Service, 1988.

Final Report of the HUD Review of the Fair Housing Accessibility Requirements and the 2003 International Building Code (IBC). February 18, 2005 (Docket No FR-4943-N-02).

Final Report of the HUD Review of the Fair Housing Accessibility Requirements and the 2006 International Building Code (IBC). May 31, 2007 (Docket No FR-5136-N-01).

HUD 24 CFR, Part 100, *Fair Housing Accessibility Guidelines*. Washington, DC: U.S. Department of Housing and Urban Development.

ICC/A117.1-03, *Accessible and Usable Buildings and Facilities*. Washington, DC: International Code Council, 2003.

IEBC-09, *International Existing Building Code*. Washington, DC: International Code Council, 2009.

IPC-09, *International Plumbing Code*. Washington, DC: International Code Council, 2009.

IRC-09, *International Residential Code*. Washington, DC: International Code Council, 2009.

"Pathfinder Tactile Tile Demonstration Test Project." Metro-Dade Transit Agency, 1988.

"Tactile Warnings to Promote Safety in the Vicinity of Transit Platform Edges." Urban Mass Transportation Administration, 1987.

Vital Health Statistics Series 10, No. 135, Tables 1 & 2. 1977 National Health Interview Survey, National Center for Health Statistics.

Chapter 12:
Interior Environment

General Comments

Chapter 12 contains provisions governing the interior environment requirements of all buildings and structures intended for human occupancy.

Section 1201 identifies the scope of the chapter. Section 1202 provides definitions applicable to terms used in this chapter.

Section 1203 identifies special enclosed spaces where accumulated moisture must be removed by ventilation and establishes criteria (minimum openable area) for the only method of measuring compliance with the natural ventilation requirements for occupied spaces.

Section 1204 identifies the minimum space-heating requirement for interior spaces intended for human occupancy.

Section 1205 requires light for every room or space intended for human occupancy. The method of compliance is the choice of the designer, who may elect to provide artificial instead of natural light. Prescriptive requirements for stairway lighting in dwelling units are also included.

Section 1206 specifies the minimum requirements for courts and yards, including area width, accessibility for cleaning and the location of air intakes when natural light or natural ventilation is the chosen design option.

Section 1207 establishes the sound transmission control requirements for air-borne and structure-borne sound in residential buildings.

Section 1208 addresses the minimum ceiling height for all habitable and occupiable spaces along with other spaces as specified (i.e., toilet rooms and bathrooms). The minimum floor area for rooms in dwelling units is also specified.

Section 1209 provides minimal opening requirements for access to crawl spaces and attics.

Section 1210 contains requirements for toilet room surfaces and fixture surrounds.

The environmental and physiological justification for the code requirements relevant to light and ventilation of occupiable spaces is based on knowledge, technology and practices developed over centuries of building structures in which humans live and work. These design practices have been further validated by studies during the 19th and 20th centuries.

The greatest impact on these provisions, and those that have resulted in changes during the past 30 years, has been from the interest in energy conservation through construction practices, including the minimization of exterior wall openings.

The requirements related to interior sound transmission and control are considered an important aspect of human comfort in residential occupancies.

Purpose

The purpose of Chapter 12 is to establish minimum conditions for the interior environment of a building. The size of spaces, light, ventilation and noise intrusion are all addressed in order to define the acceptable conditions to which any occupant may be exposed. Design options of natural and mechanical systems are introduced and the criteria for performance are specified.

Even though it was not completely understood, the need for "fresh air" had been recognized for centuries. Designers of centuries-old adobe buildings in the southwestern United States, hide-covered Native American tepees of the plains and frame houses of early settlers in the eastern United States all relied upon the buoyancy of warm air, enabling it to rise and cooler air to flow in to replace it. Whether the design relied on solar energy, thermal mass or even wind velocity to cause the movement, it still reflected a natural movement and, as a result, has been termed "natural ventilation." Only recently have we begun to recognize the reasons for ventilation and the implications of failing to provide an adequate quantity and acceptable quality of air for all occupants. The expression "sick building syndrome" has crept into our vocabulary and reflects the increased understanding of the relationship between interior environment requirements and the physiological well-being of the occupants.

The other purpose of regulating the interior environment is psychological. Merely providing adequate conditions is not sufficient if the occupant does not perceive them as adequate. Minimum space requirements (floor area, yard dimensions or ceiling height) address the need to perceive adequate light, ventilation and space to promote psychological well-being. Regulation of sound transmission also bears directly on the psychological and long-term physical well-being of the occupant.

Finally, adequate lighting from natural sources also meets the physical and psychological needs of the occupants and contributes directly to their overall safety. Safe use of any building under ordinary and emergency conditions depends greatly on proper illumination of the space. This chapter references the *International Mechanical Code*® (IMC®) as the performance standard to which ventilation must be compared and the installation standard for mechanical systems used in buildings regulated by the code.

SECTION 1201
GENERAL

1201.1 Scope. The provisions of this chapter shall govern ventilation, temperature control, lighting, *yards* and *courts*, sound transmission, room dimensions, surrounding materials and rodent proofing associated with the interior spaces of buildings.

❖ This section identifies the scope of Chapter 12. The requirements of this chapter are intended to govern and regulate the need for light, ventilation, sound transmission control, interior space dimensions and materials surrounding plumbing fixtures in all buildings. It is the intent of the code that the user must comply with these regulations for all newly constructed buildings and structures and for all buildings and structures, or portions thereof, when there is to be a change of occupancy or within any additions to a building.

SECTION 1202
DEFINITIONS

1202.1 General. The following words and terms shall, for the purposes of this chapter and as used elsewhere in this code, have the meanings shown herein.

❖ Definitions of terms that are associated with the content of this section are contained herein. These definitions can help in the understanding and application of the code requirements. It is important to emphasize that these terms are not exclusively related to this section but are applicable everywhere the term is used in the code. The purpose for including these definitions within this section is to provide more convenient access to them without having to refer back to Chapter 2. For convenience, these terms are also listed in Chapter 2 with a cross reference to this section. The use and application of all defined terms, including those defined herein, are set forth in Section 201.

SUNROOM. A one-story structure attached to a building with a glazing area in excess of 40 percent of the gross area of the structure's *exterior walls* and roof.

❖ This terminology is provided in order to address separate requirements for sunrooms with regard to ventilation of adjoining spaces (see Section 1203.4.1.1). For the 2009 code the term "sunroom addition" was simplified to "sunroom." Provisions later in the chapter that still say sunroom addition. Those provisions are applicable regardless of whether the sunroom is an addition to an existing structure or is part of the original design and construction.

THERMAL ISOLATION. A separation of conditioned spaces, between a sunroom addition and a *dwelling unit*, consisting of existing or new wall(s), doors and/or windows.

❖ This terminology is required for the same reason provided in the definition of "Sunroom" (see above). Although the exact phrase "thermal isolation" is not used in the code, the phrase "thermally isolated" is found in a number of provisions. This definition provides guidance in applying this variation on the defined term.

SECTION 1203
VENTILATION

1203.1 General. Buildings shall be provided with natural ventilation in accordance with Section 1203.4, or mechanical ventilation in accordance with the *International Mechanical Code*.

❖ Every room or space must be provided with ventilation. The selection of natural versus mechanical ventilation on a room-by-room or space-by-space basis is the designer's prerogative. Certain conditions require mechanical exhaust even though natural ventilation systems have been selected by the designer. Section 1203.4.2 of the code, and Sections 401.6 and 403 of the IMC direct the treatment of those special situations. Existence of these conditions, however, does not require providing mechanical ventilation other than to address the specific condition. Other rooms or spaces not affected by those conditions may be served by natural systems.

1203.2 Attic spaces. Enclosed *attics* and enclosed rafter spaces formed where ceilings are applied directly to the underside of roof framing members shall have cross ventilation for each separate space by ventilating openings protected against the entrance of rain and snow. Blocking and bridging shall be arranged so as not to interfere with the movement of air. A minimum of 1 inch (25 mm) of airspace shall be provided between the insulation and the roof sheathing. The net free ventilating area shall not be less than $^1/_{300}$ of the area of the space ventilated, with 50 percent of the required ventilating area provided by ventilators located in the upper portion of the space to be ventilated at least 3 feet (914 mm) above eave or cornice vents with the balance of the required ventilation provided by eave or cornice vents.

❖ All attic spaces and each separate space formed between solid roof rafters are required to be cross ventilated where the ceiling is applied directly to the underside of the roof rafters. Care must be taken, however, to provide cross ventilation in a manner that does not introduce moisture to the attic area.

Snow infiltration can occur when the attic ventilation openings are not sufficiently protected against the entrance of snow or rain, or when more than 50 percent of the ventilation openings are located along the ridge or gable wall of the roof rather than at the eave. When the wind blows perpendicular to a roof ridge vent, a negative pressure builds up across the ridge that draws air out of the attic space through the attic vents. Cross flow of air through the attic can be achieved when outside air is drawn into the attic through the eave or cornice and exits through the ridge or gable vents. In order for this to occur, eave or cornice vents must be greater than or equal to the area of the ridge or gable vents.

Vents that permit snow or rain to infiltrate the attic are not permitted. While there is no specific test standard for this performance aspect of vents, snow infiltration through roof vents can be addressed by what is referred to as "balanced" venting. Balanced venting is providing at least 50 percent of the required ventilating area in the upper third of the space being ventilated (e.g., through ridge or gable vents). The balance of the required ventilation is provided by eave or cornice vents that are greater than or equal to the ventilation area provided by the ridge or gable vents. Most ridge vent manufacturers require slightly more ventilation area in the eaves than provided in the ridge vent to help prevent snow infiltration. If insufficient eave or cornice ventilation is provided, air will be drawn through the ridge or gable vents. Snow-laden air can enter the attic space through a ridge or gable vent; therefore, it is preferable to have both eave and ridge vents, with the eave vent area being greater than or equal to the ridge vent area (i.e., "balanced" venting).

If an adequate amount of ventilation area in the upper portion of the space is not provided and the ventilation area is provided mainly at the eave or cornice vents, then air will enter and leave the attic space at the eave, and very little cross flow of air will occur.

A test method that has been devised for ridge vent manufacturers involves the use of a 13-foot, 6-inch-diameter (4115 mm) propeller of a 2,650-horsepower aircraft engine wind generator. Snow is simulated with fine, soft wood sawdust, added to the airstream at about 5 pounds (2 kg) per minute. Using this method, the wind speed is varied, because it is not known at what speed the most snow infiltration will occur or the factors that will determine that each wind speed was sustained for a period of 5 minutes. The entire roof system, with all vents, must be installed in the test set-up in order to get a true measure of the potential (or lack thereof) for snow infiltration.

Attic ventilation openings cannot be placed in roof areas subject to snow drifts. These roof areas are subject to greater concentrations of snow, which could increase the chances of snow entering the attic through the ventilation openings. In addition, the ventilation openings are required to be covered with corrosion-resistant mesh or similar material in accordance with Section 1203.2.1. If roof spaces are not created (e.g., solid concrete roof sections), ventilation is not required, as there is no concealed space for condensation to accumulate.

The amount of area needed for ventilating a roof space is also established in this section.

The following example illustrates the calculation of required ventilation areas for an attic space [see Figure 1203.2(1)]:

Note: The area of the attic must include the area of the eave or soffit.

Attic ventilation area	=	$^1/_{300}$ of area
Required ventilating area	=	1,100/300
	=	3.67 sq. ft.
	=	528 sq. in.
Provide 50% by	=	528 × 0.5 ridge or roof vent
	=	264 sq. in.

Provide soffit ventilation	=	264 sq. in. total or 264/2(50)
	=	2.64 sq. in./ft.
Area of attic	=	1,100 sq. ft.

Common methods used to provide soffit ventilation include manufactured units, strips or soffit panels and holes or slots (with screening) that meet the criteria and are approved.

It is important to note that the distribution of ventilation openings should be uniform along the length of the soffits and ridge.

Common methods used to provide roof ridge ventilation include manufactured roof units, ridge vents and gable louvers.

Common methods used to provide soffit ventilation include manufactured units, strips or soffit panels and holes or slots (with screening) that meet the criteria and are approved.

Note that the net-free ventilation area of equipment used for soffit, roof and gable ventilation must be determined. One cannot simply calculate the ventilation area based on the opening created in the roof, wall or soffit. Products vary by manufacturer, and a review of the listing and specifications is necessary to verify actual or net-free ventilation areas.

Where an attic space is not created, but the ceiling membrane is applied directly to the bottom of the solid roof rafters, each rafter space is to be ventilated separately. In this type of installation, it is particularly important that cross ventilation is developed between each rafter space by providing vents at the ridge and eave [see Figure 1203.2(2)]. For small sections of roofs such as above dormers and for roofs that are flat or of

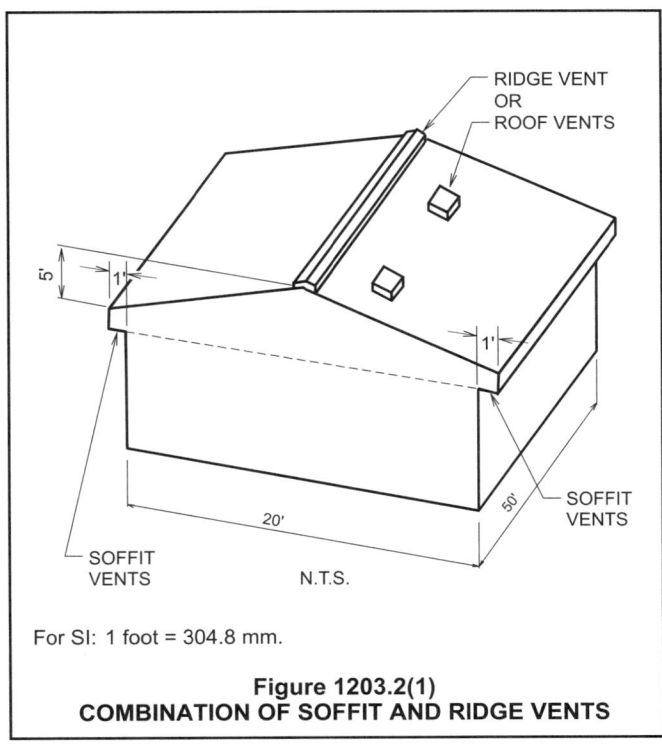

RIDGE VENT
OR
ROOF VENTS

SOFFIT VENTS

SOFFIT VENTS

5'

1'

1'

20'

50'

N.T.S.

For SI: 1 foot = 304.8 mm.

Figure 1203.2(1)
COMBINATION OF SOFFIT AND RIDGE VENTS

very low slope, the 30-inch (762 mm) height difference between the highest and lowest vent locations may not be readily achievable. The designer and building official need to ensure that the ventilation openings for such roofs will result in adequate ventilation.

1203.2.1 Openings into attic. Exterior openings into the *attic* space of any building intended for human occupancy shall be protected to prevent the entry of birds, squirrels, rodents, snakes and other similar creatures. Openings for ventilation having a least dimension of $^{1}/_{16}$ inch (1.6 mm) minimum and $^{1}/_{4}$ inch (6.4 mm) maximum shall be permitted. Openings for ventilation having a least dimension larger than $^{1}/_{4}$ inch (6.4 mm) shall be provided with corrosion-resistant wire cloth screening, hardware cloth, perforated vinyl or similar material with openings having a least dimension of $^{1}/_{16}$ inch (1.6 mm) minimum and $^{1}/_{4}$ inch (6.4 mm) maximum. Where combustion air is obtained from an *attic* area, it shall be in accordance with Chapter 7 of the *International Mechanical Code.*

❖ Ventilation openings that would permit the entrance of small animals into the structure must be protected in accordance with this section. Hardware cloth is a particular kind of metal wire cloth screening, and perforated vinyl is a plastic screening or grid with openings of similar dimensions. Metal or vinyl are specified because of their resistance to deterioration over time; therefore, whatever material is used, it must be nondeteriorating in addition to having the least dimension of the minimum and maximum openings of $^{1}/_{16}$ and $^{1}/_{4}$ inches (3.2 and 6.4 mm), respectively.

Combustion air is air supplied to the room where a fuel-burning appliance is located in order that combustion of the fuel can take place in a safe and complete manner. Chapter 7 of the IMC permits combustion air to be taken from an attic space that is ventilated by openings to the exterior, under certain conditions. Those conditions are based on the configuration of the attic space and the ventilation openings themselves (see Chapter 7 of the IMC and the IMC commentary for more information).

1203.3 Under-floor ventilation. The space between the bottom of the floor joists and the earth under any building except spaces occupied by basements or cellars shall be provided with ventilation openings through foundation walls or *exterior walls.* Such openings shall be placed so as to provide cross ventilation of the under-floor space.

❖ The intent of this section is to create an adequate flow of air through crawl spaces to achieve the ventilation goals of controlling temperature, humidity and accumulation of gases. The entire space must be properly ventilated by openings that are distributed to effect cross flow and include corner areas. Although the code does not specify the exact location of openings, an equal distribution of openings on at least three sides of a building, with at least one opening near each corner of the building, is typically sufficient.

Mechanical ventilating devices also can be installed to force air movement and ventilate the space, in which case the location and number of ventilation openings are less critical (see Section 1203.3.2, Exception 3). The amount of ventilation openings required can be drastically reduced if a vapor retarder is used on the ground surface in the crawl space, in accordance with Section 1203.3.2, Exception 2. Also, in accordance with Exception 3 of that section, when a vapor retarder is used, the installation of operable louvers (to close the openings in the coldest times of the year) is permitted.

1203.3.1 Openings for under-floor ventilation. The minimum net area of ventilation openings shall not be less than 1 square foot for each 150 square feet (0.67 m² for each 100 m²) of crawl-space area. Ventilation openings shall be covered for their height and width with any of the following materials, provided that the least dimension of the covering shall not exceed $^{1}/_{4}$ inch (6 mm):

1. Perforated sheet metal plates not less than 0.070 inch (1.8 mm) thick.

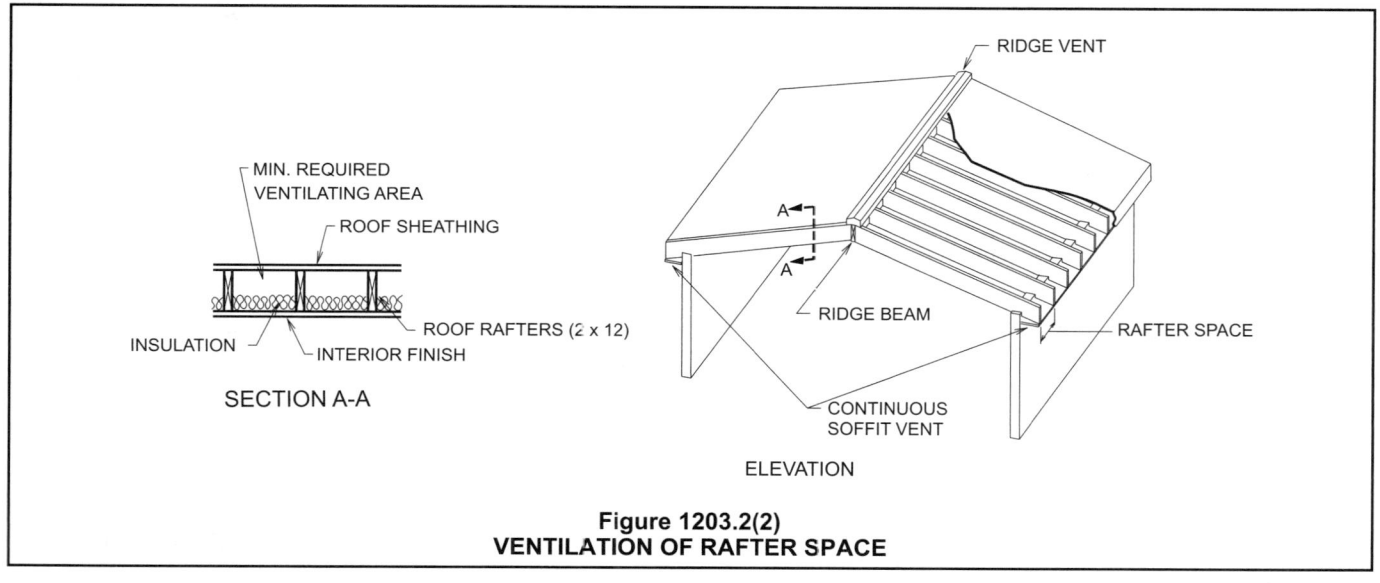

Figure 1203.2(2)
VENTILATION OF RAFTER SPACE

2. Expanded sheet metal plates not less than 0.047 inch (1.2 mm) thick.

3. Cast-iron grilles or gratings.

4. Extruded load-bearing vents.

5. Hardware cloth of 0.035 inch (0.89 mm) wire or heavier.

6. Corrosion-resistant wire mesh, with the least dimension not exceeding $^1/_8$ inch (3.2 mm).

❖ The following is an example of the area calculation: A rectangular building that is 60 feet (18 288 mm) long and 20 feet (6096 mm) wide has a plan area of 1,200 square feet (111.5 m²). The amount of ventilation opening required is 1,200/150 = 8 square feet (0.74 m²) by 144 square inches (92 903 mm²) per square foot = 1,152 square inches (0.74 m²). This is the total (aggregate) amount of ventilation opening that must be distributed among all the openings. This required amount of openings may be reduced by a factor of 10 if a vapor retarder is used on the ground surface in accordance with Section 1203.3.2, Exception 2.

The requirement for covering the openings with perforated plates, corrosion-resistant wire mesh or other covering is to keep small animals out. Six alternatives are given for this covering, and they all must have openings that have no dimension exceeding $^1/_4$ inch (6.4 mm).

1203.3.2 Exceptions. The following are exceptions to Sections 1203.3 and 1203.3.1:

1. Where warranted by climatic conditions, ventilation openings to the outdoors are not required if ventilation openings to the interior are provided.

2. The total area of ventilation openings is permitted to be reduced to $^1/_{1,500}$ of the under-floor area where the ground surface is covered with a Class I vapor retarder material and the required openings are placed so as to provide cross ventilation of the space. The installation of operable louvers shall not be prohibited.

3. Ventilation openings are not required where continuously operated mechanical ventilation is provided at a rate of 1.0 cubic foot per minute (cfm) for each 50 square feet (1.02 L/s for each 10 m²) of crawl space floor area and the ground surface is covered with a Class I vapor retarder.

4. Ventilation openings are not required when the ground surface is covered with a Class I vapor retarder, the perimeter walls are insulated and the space is conditioned in accordance with the *International Energy Conservation Code*.

5. For buildings in flood hazard areas as established in Section 1612.3, the openings for under-floor ventilation shall be deemed as meeting the flood opening requirements of ASCE 24 provided that the ventilation openings are designed and installed in accordance with ASCE 24.

❖ This section lists the locations and conditions where ventilation openings can be omitted entirely or the area of required openings can be reduced. Exception 1 could be used in extremely cold climates, where ventilation openings are a serious breach of the structure in terms of energy usage. It provides for ventilating the crawl space to the interior conditioned space of the building, which is heated and can accept moisture from the underground space without detrimental effects on the building structure.

The use of a Class I vapor retarder material on the ground surface inhibits the flow of moisture from the ground surface into the crawl space and thus reduces, if not virtually eliminates, the need for ventilation. Therefore, Exception 2 provides for a drastic reduction in the amount of ventilation openings required. While the vapor retarder may significantly reduce the moisture accumulation, ventilation openings are still required but may be equipped with manual dampers to permit them to be closed during the coldest weeks of the year in northern climates.

Exception 3 provides for the use of mechanical ventilation, such as an exhaust fan similar to a bathroom exhaust fan, to keep air moving through the crawl space. A Class I vapor retarder on the ground surface is also required when using Exception 3.

When a crawl space is provided with a Class I vapor retarder on the ground surface and is mechanically conditioned and insulated, it becomes like any other space in the conditioned structure and ventilation openings are not required in accordance with Exception 4. Requirements for insulating structures are found in the *International Energy Conservation Code®* (IECC®).

Section 1612 of the code requires buildings located in flood hazard areas to be constructed in accordance with ASCE 24. Exception 5 is necessary to coordinate the requirements for openings in ASCE 24 with this section. In most cases, the ventilation requirements of Section 1203.3 are not satisfied by installation of the flood openings because the requirements in ASCE 24 specify that flood openings are to be no more than 1 foot (305 mm) above the adjacent grade, while most airflow vents that meet the requirements of Section 1203.3 are installed immediately below the elevated floor. If there are a sufficient number of airflow vents located within 1 foot (305 mm) of the adjacent grade, they may satisfy both requirements. In either case, both airflow ventilation and flood opening requirements must be met for buildings and structures located in flood hazard areas. For further guidance, refer to FEMA publication FIA-TB-1, *Openings in Foundation Walls for Buildings Located in Special Flood Hazard Areas*.

1203.4 Natural ventilation. Natural ventilation of an occupied space shall be through windows, doors, louvers or other openings to the outdoors. The operating mechanism for such openings shall be provided with ready access so that the openings are readily controllable by the building occupants.

❖ This section provides the standard of natural ventilation for all occupied spaces. Openings to the outdoor air, such as doors, windows and louvers, provide natural ventilation. The section does not, however, state or intend that the doors, windows or openings actually be constantly open. The intent is that they be maintained in an operable condition so that they are available for use at the discretion of the occupant.

1203.4.1 Ventilation area required. The minimum openable area to the outdoors shall be 4 percent of the floor area being ventilated.

❖ This section specifies the ratio of openable doors, windows or openings to the floor space being ventilated but does not address the distribution around the space or location of these openings. It is the designer's prerogative to distribute openings in such a manner as to accomplish the natural ventilation of the space. When inadequate natural ventilation is provided, mechanical ventilation can supplement any inadequacy (see Chapter 4 of the IMC). The plan reviewer can determine compliance with this section. For example, in Figure 1203.4.1, the combined openable area (the net-free area of a door, window, louver, vent or skylight, etc., when fully open) of double-hung windows B and C is equal to 4 percent of the floor area [300 × 0.04

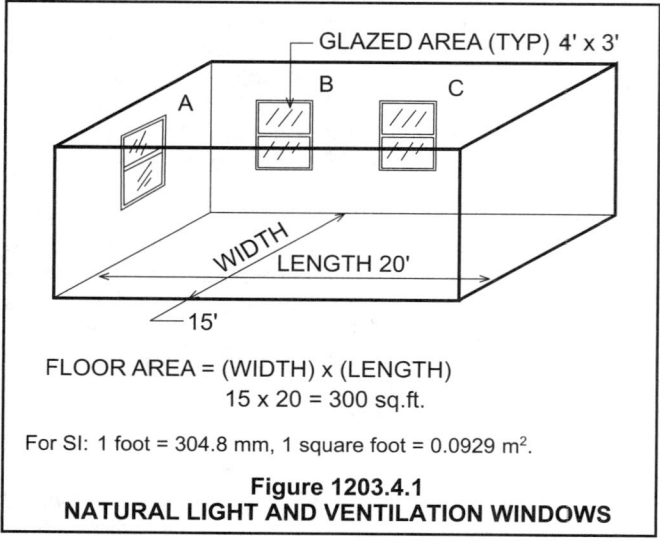

GLAZED AREA (TYP) 4' x 3'

A   B   C

WIDTH

LENGTH 20'

15'

FLOOR AREA = (WIDTH) x (LENGTH)
15 x 20 = 300 sq.ft.

For SI: 1 foot = 304.8 mm, 1 square foot = 0.0929 m².

Figure 1203.4.1
NATURAL LIGHT AND VENTILATION WINDOWS

= 12 square feet (1 m²)]. The openable area of window A is not required and need not open onto a court or yard complying with Section 1206.

1203.4.1.1 Adjoining spaces. Where rooms and spaces without openings to the outdoors are ventilated through an adjoining room, the opening to the adjoining room shall be unobstructed and shall have an area of not less than 8 percent of the floor area of the interior room or space, but not less than 25 square feet (2.3 m²). The minimum openable area to the outdoors shall be based on the total floor area being ventilated.

Exception: Exterior openings required for ventilation shall be permitted to open into a *thermally isolated* sunroom addition or patio cover provided that the openable area between the sunroom addition or patio cover and the interior room shall have an area of not less than 8 percent of the floor area of the interior room or space, but not less than 20 square feet (1.86 m²). The minimum openable area to the outdoors shall be based on the total floor area being ventilated.

❖ Adjacent spaces with large connecting openings may share sources of light and ventilation. This section deals with the natural ventilation of connecting interior spaces, and it is the designer's obligation to place openings between rooms with exterior openings and connecting spaces without exterior openings in such a manner as to accomplish natural ventilation of the connected space. For purposes of ventilation, this section establishes a minimum openness requirement for the common wall between a room with openings to the exterior and an interior room without openings to the exterior. The minimum amount of openness required in that common wall is 8 percent of the floor area of the interior room or 25 square feet (2.33 m²), whichever is greater. The openable area of the exterior openings in the "outer" room is required to be equal to or greater than 4 percent (in accordance with Section 1203.4.1) of the total combined floor areas served.

Figure 1203.4.1.1 shows a cut-away of an interior room (Room A) adjacent to a room with openings to the exterior (Space B). The openable area of exterior openings in Space B is required to be equal to or

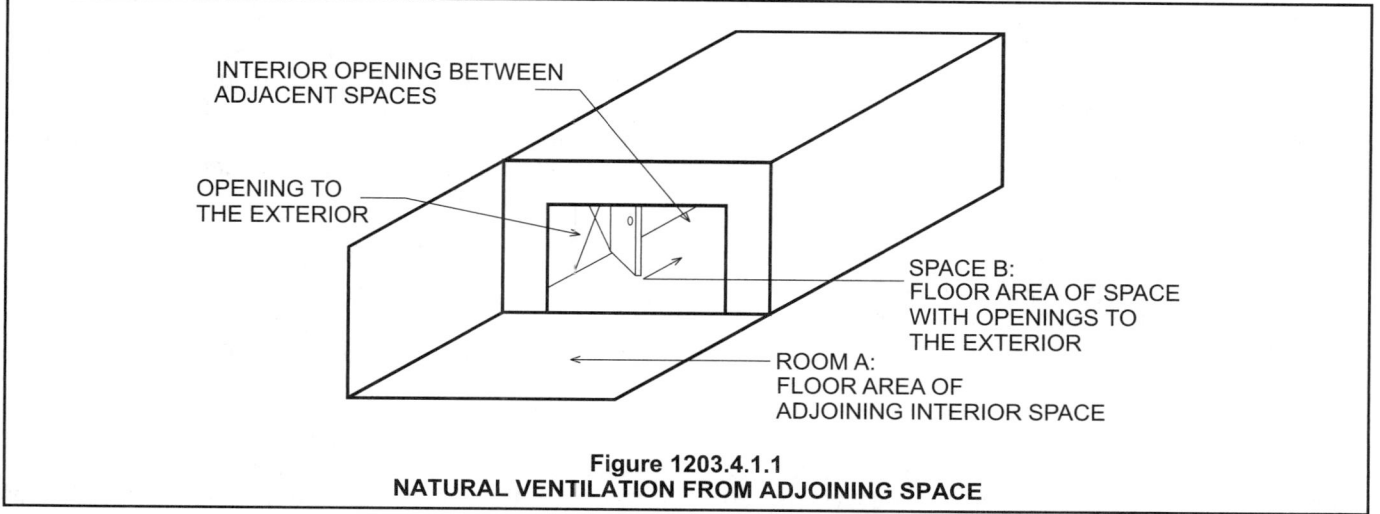

INTERIOR OPENING BETWEEN
ADJACENT SPACES

OPENING TO
THE EXTERIOR

SPACE B:
FLOOR AREA OF SPACE
WITH OPENINGS TO
THE EXTERIOR

ROOM A:
FLOOR AREA OF
ADJOINING INTERIOR SPACE

Figure 1203.4.1.1
NATURAL VENTILATION FROM ADJOINING SPACE

greater than 0.04 times the area of the entire space (floor area of Space A plus floor area of interior Space B). The opening in the wall between adjacent spaces must be a minimum of 25 square feet (2.33 m²), but not less than 0.08 times the floor area of interior Space A. Since the opening between the adjacent spaces must be unobstructed in accordance with this section, a door cannot be installed in the opening.

The exception deals with a very common circumstance, especially in residential construction. As long as the sunroom is large enough and is thermally isolated, the building owner need not move ventilation openings when installing an addition that falls within the definition of "Sunroom." Sunrooms can be part of initial construction.

1203.4.1.2 Openings below grade. Where openings below grade provide required natural ventilation, the outside horizontal clear space measured perpendicular to the opening shall be one and one-half times the depth of the opening. The depth of the opening shall be measured from the average adjoining ground level to the bottom of the opening.

❖ This section is applicable whenever occupied spaces below grade are dependent upon natural ventilation through structures like window wells. In order to provide adequate ventilation, this section sets the minimum horizontal clear space adjacent to the opening used for natural ventilation. Without this minimum horizontal area, there will be inadequate air movement through the opening.

As illustrated in Figure 1203.4.1.2, the opening area required for the story below grade intended for human occupancy is:

$$A = 0.04\ (L \times W)$$

The area of the window in the vertical plane (w × h) must equal or exceed the required opening area. Additionally, the horizontal dimension from the window to the well wall must equal one and one-half times the depth of the openable portion of the window at the lowest point. If the story below grade is not intended for human occupancy, ventilation is required to be provided in accordance with Section 1203.3 for underfloor spaces.

1203.4.2 Contaminants exhausted. Contaminant sources in naturally ventilated spaces shall be removed in accordance with the *International Mechanical Code* and the *International Fire Code*.

❖ Contaminants in the air are to be collected and exhausted by special means. Chapters 4 and 5 of the IMC specify areas or conditions that must be separately addressed. For example, there are many operations listed in Chapter 5 of the IMC that produce contaminants that cannot be properly or safely treated by natural means. Natural ventilation only anticipates normal occupancy by people and not the extra heat loads, dust, vapors and other contaminants generated by some activities.

1203.4.2.1 Bathrooms. Rooms containing bathtubs, showers, spas and similar bathing fixtures shall be mechanically ventilated in accordance with the *International Mechanical Code*.

❖ Chapter 4 of the IMC contains provisions for bathroom ventilation, requiring mechanical exhaust without recirculation of air at specific rates that depend on the occupancy group.

1203.4.3 Openings on yards or courts. Where natural ventilation is to be provided by openings onto *yards* or *courts*, such *yards* or *courts* shall comply with Section 1206.

❖ In order that adequate air movement will be provided through openings to naturally ventilated rooms, the openings must directly connect to yards or courts with the minimum dimensions specified in Section 1206.

1203.5 Other ventilation and exhaust systems. Ventilation and exhaust systems for occupancies and operations involving flammable or combustible hazards or other contaminant

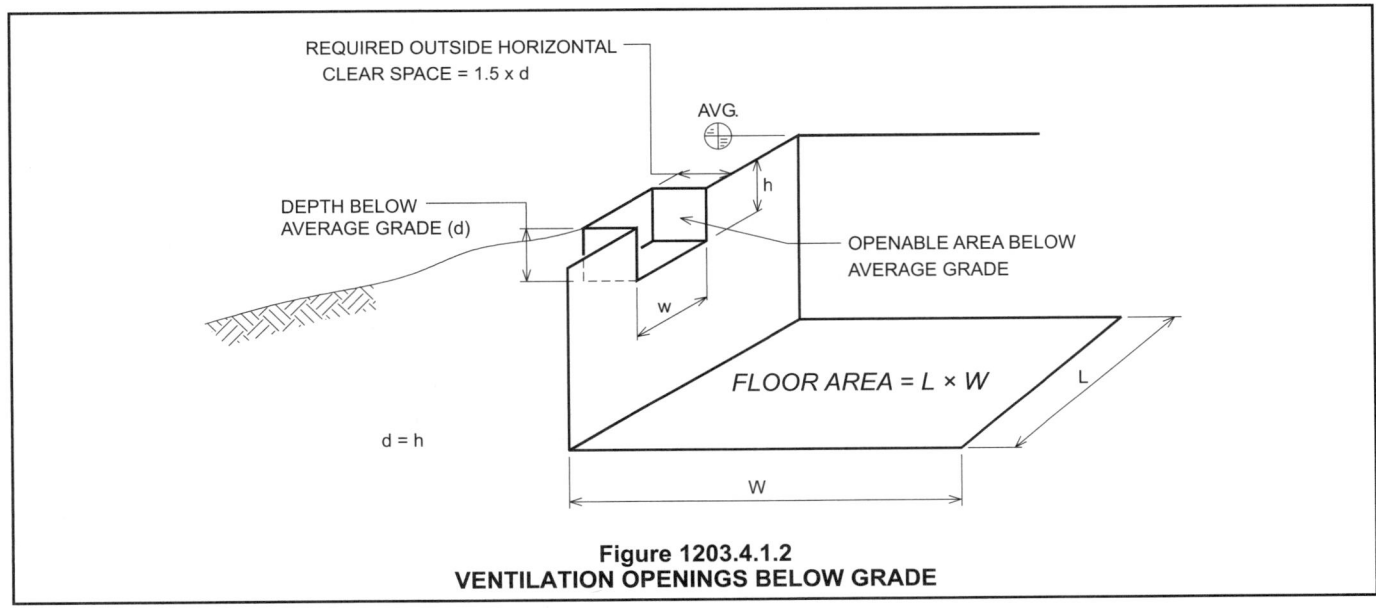

Figure 1203.4.1.2
VENTILATION OPENINGS BELOW GRADE

sources as covered in the *International Mechanical Code* or the *International Fire Code* shall be provided as required by both codes.

❖ Chapter 5 of the IMC contains specific provisions for hazardous exhaust systems in occupancies such as vehicle repair garages, aircraft fueling stations, dry cleaning plants, spray painting operations and hazardous production materials (HPM) facilities. Many of these provisions are duplicated from the *International Fire Code®* (IFC®), which also contains provisions for the handling and storage of hazardous materials.

SECTION 1204
TEMPERATURE CONTROL

1204.1 Equipment and systems. Interior spaces intended for human occupancy shall be provided with active or passive space-heating systems capable of maintaining a minimum indoor temperature of 68°F (20°C) at a point 3 feet (914 mm) above the floor on the design heating day.

Exception: Interior spaces where the primary purpose is not associated with human comfort.

❖ Heating facilities are required for comfort in all new construction. The systems may be either active (such as a forced-air furnace) or passive (such as solar systems), as long as the specified performance is achieved.

The outdoor design temperatures are taken from the ASHRAE *Handbook of Fundamentals* and are listed in Appendix D of the *International Plumbing Code®* (IPC®). Outdoor design temperatures provide a baseline from which heat load calculations are made. Heating system capacity is dependent upon the predicted outdoor temperatures during the heating season. As the outdoor temperature falls, the heat input to a building must increase to offset the increasing heat losses through the building envelope. Heating systems are designed to have the capacity to maintain the desired indoor temperature when the outdoor temperature is at or above the outdoor design temperature. When the outdoor temperatures are below the outdoor design temperature, the heating system will not be able to maintain a desired indoor temperature. It would be impractical, for example, to design a heating system based on the assumption that someday it might be -20°F (-29°C) outdoors if the outdoor temperature in that region rarely, if ever, dropped that low. In such a case, the heating system would be oversized and, thereby, less efficient and economical.

The winter outdoor design temperature is defined as follows: For 97.5 percent of the total hours in the northern hemisphere heating season, from December through February, the predicted outdoor temperatures will be at or above the values given in Appendix D of the IPC. It would be unreasonable to expect any heating system to maintain a desired indoor temperature when the outdoor temperature is below the design

temperature. When the 97.5-percent column in Appendix D of the IPC is used, it can be assumed that the actual outdoor temperature will be at or below the design temperature for roughly 54 hours of the total of 2,160 hours in the months of December through February (2.5 percent of 2,160 hours = 54).

The exception recognizes that not all interior spaces are associated with human comfort by the nature of their uses, such as a commercial cooler or freezer. These and similar spaces would not require heating systems.

SECTION 1205
LIGHTING

1205.1 General. Every space intended for human occupancy shall be provided with natural light by means of exterior glazed openings in accordance with Section 1205.2 or shall be provided with artificial light in accordance with Section 1205.3. Exterior glazed openings shall open directly onto a *public way* or onto a *yard* or *court* in accordance with Section 1206.

❖ This section establishes that an option can be exercised on a room-by-room or space-by-space basis. The option allows the designer to provide either natural light in accordance with this chapter or equivalent levels of artificial lighting.

1205.2 Natural light. The minimum net glazed area shall not be less than 8 percent of the floor area of the room served.

❖ This section establishes the minimum glazed area required based on the floor area served by the window. This is required only for spaces that are not provided with artificial light in accordance with Section 1205.3. It is the intent of the code to establish this ratio as the minimum glazed opening onto yards or courts, in accordance with Section 1205.1.

Early codes set this standard at 10 percent of the floor area served. This ratio was derived from certain architectural styles that yielded adequate light and ventilation; however, this is a more than adequate amount and has been reduced to the current levels because of energy conservation issues. Openings in excess of that minimum area are permitted to open onto areas other than a complying court or yard. In Figure 1203.4.1, the room dimensions are 15 feet by 20 feet (4572 mm by 6096 mm), or 300 square feet (27.9 m²) of area. If windows B and C are double hung, with a combined glazed area of 24 square feet (2.23 m²), they provide the minimum area required of 8 percent of the floor area (24/300 = 0.08). In this example, glazing unit A is not required for natural light; therefore, it need not face onto a required yard or court.

1205.2.1 Adjoining spaces. For the purpose of natural lighting, any room is permitted to be considered as a portion of an adjoining room where one-half of the area of the common wall is open and unobstructed and provides an opening of not less than one-tenth of the floor area of the interior room or 25 square feet (2.32 m²), whichever is greater.

Exception: Openings required for natural light shall be permitted to open into a *thermally isolated* sunroom addition or patio cover where the common wall provides a glazed area of not less than one-tenth of the floor area of the interior room or 20 square feet (1.86 m²), whichever is greater.

❖ In a case where a space (or room) has no glazed area open to the required courts or yards but is adjacent to one that does, it may "borrow" natural lighting from the adjacent space if: (1) the wall between the adjoining spaces is at least one-half open and unobstructed; (2) the opening equals at least 10 percent of the floor area of the interior space and (3) the opening is not less than 25 square feet (2.33 m²). The required glazed area facing the required court or yard must not be less than 8 percent of the total floor area of all rooms served. For example, in Figure 1205.2.1, the glazed area in Space B is required to be equal to or greater than 0.08 (floor area of Space A + floor area of Space B).

In the figure, the opening between the adjacent spaces must meet all three criteria: the wall must be at least half open and unobstructed, it must be a minimum of 25 square feet (2.33 m²) and it must be not less than one-tenth of the floor area of space A.

The exception deals with a very common circumstance, especially in residential construction. As long as the sunroom is large enough and is thermally isolated, the building owner need not move openings for lighting when installing an addition that falls within the definition of "Sunroom." Note that sunrooms can also be part of the initial construction of a building.

1205.2.2 Exterior openings. Exterior openings required by Section 1205.2 for natural light shall open directly onto a *public way*, *yard* or *court*, as set forth in Section 1206.

Exceptions:

1. Required exterior openings are permitted to open into a roofed porch where the porch:

    1.1. Abuts a *public way*, *yard* or *court*;

    1.2. Has a ceiling height of not less than 7 feet (2134 mm); and

    1.3. Has a longer side at least 65 percent open and unobstructed.

2. Skylights are not required to open directly onto a *public way*, *yard* or *court*.

❖ In order that enough light will be provided through openings to naturally lit rooms, the openings must open onto yards or courts with the minimum dimensions specified in Section 1206. Skylights admit light directly from above and, therefore, are not required to face a court or yard in accordance with Exception 2. Exception 1 gives the criteria by which a roofed porch may be located directly outside required openings without significantly obstructing the entrance of light to the space.

1205.3 Artificial light. Artificial light shall be provided that is adequate to provide an average illumination of 10 foot-candles (107 lux) over the area of the room at a height of 30 inches (762 mm) above the floor level.

❖ The section establishes the minimum required illumination for rooms without the minimum required natural light (see Figure 1205.3). Please note that Section 1006.2 requires 1 foot-candle (11 lux) of light at the walking surface of all means of egress.

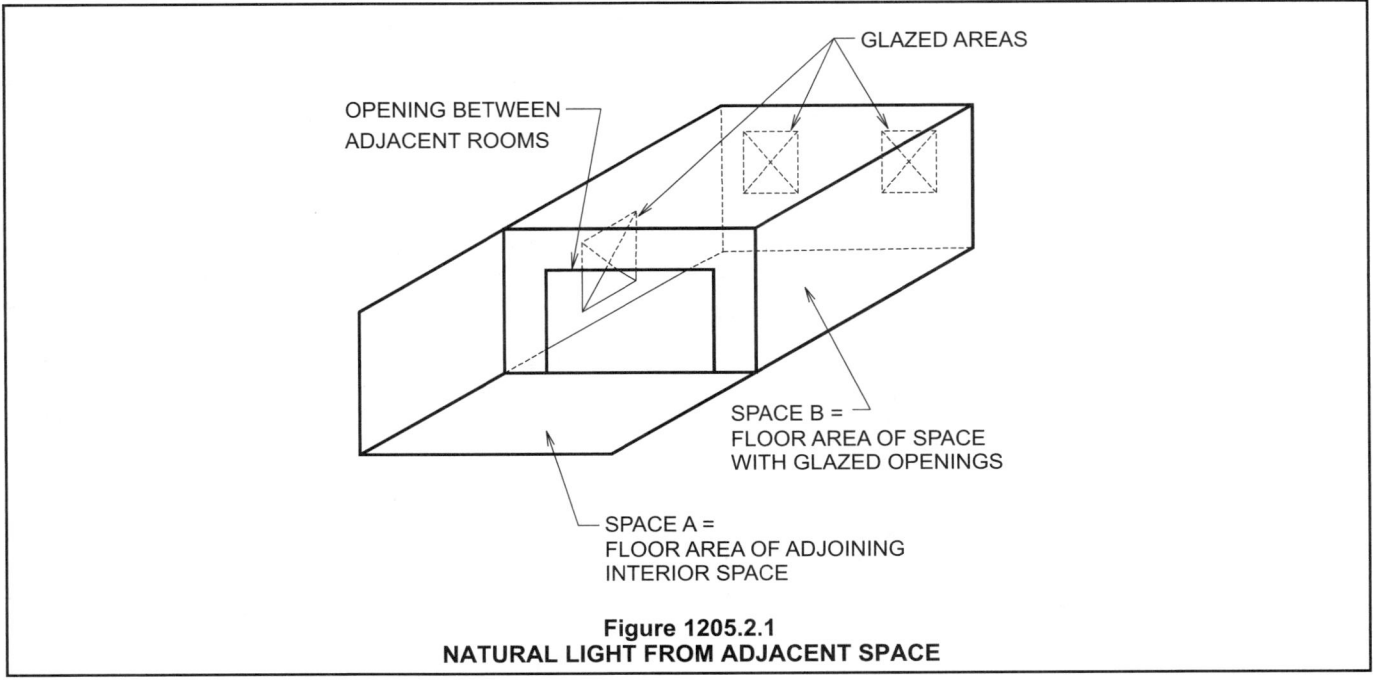

Figure 1205.2.1
NATURAL LIGHT FROM ADJACENT SPACE

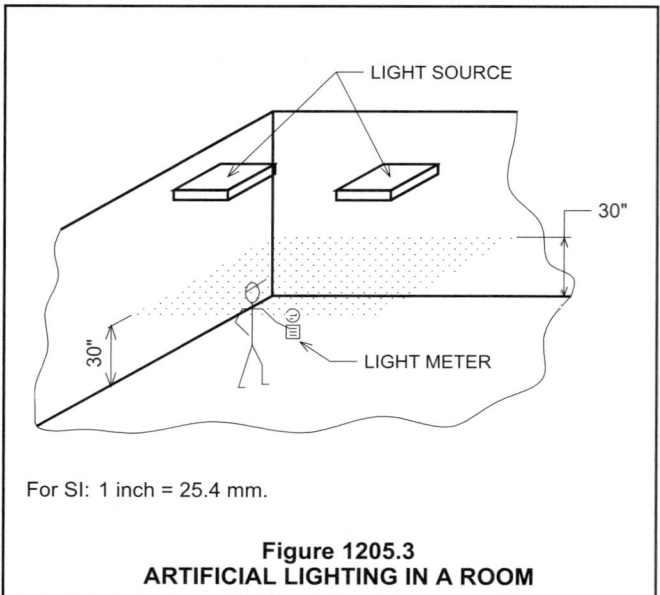

For SI: 1 inch = 25.4 mm.

Figure 1205.3
ARTIFICIAL LIGHTING IN A ROOM

1205.4 Stairway illumination. *Stairways* within *dwelling units* and *exterior stairways* serving a *dwelling unit* shall have an illumination level on tread runs of not less than 1 foot-candle (11 lux). *Stairs* in other occupancies shall be governed by Chapter 10.

❖ Illumination is essential for stairway safety during normal use, as well as during egress in an emergency. The lighting must be operable by switches in the vicinity of the stairway, located as required by the *National Electrical Code* (NEC). Emergency egress lighting, also referred to as "means of egress illumination," is required in occupancies other than dwelling units at a lower rate of illumination (see commentary, Sections 1006 and 1205.5).

1205.4.1 Controls. The control for activation of the required *stairway* lighting shall be in accordance with NFPA 70.

❖ The NEC provides for controls at the top and bottom of stairways within dwelling units, allowing an occupant to illuminate the stairways before traversing any stairways, regardless of the direction of travel. Illuminated switches, where required, allow an occupant to quickly find the switches when the stairways are dark.

Illumination controls for exterior stairways that are operable from the inside of a dwelling unit allow an occupant to safely egress by activating exterior stairway illumination prior to leaving the building. Exterior stairways must be provided with the minimum illumination level specified in Section 1205.4.

1205.5 Emergency egress lighting. The *means of egress* shall be illuminated in accordance with Section 1006.1.

❖ Means of egress illumination is required in all buildings to allow occupants enough light to negotiate the exit access (such as corridors) and exits (such as enclosed stairways) at all times the building is occupied (see commentary, Section 1006.1).

SECTION 1206
YARDS OR COURTS

1206.1 General. This section shall apply to *yards* and *courts* adjacent to exterior openings that provide natural light or ventilation. Such *yards* and *courts* shall be on the same property as the building.

❖ These provisions are intended to regulate those exterior areas of a building or structure that are supposed to supply required natural light or ventilation to interior spaces. These requirements are intended to increase the likelihood that the exterior walls are provided with enough adjacent open space to allow the required light and ventilating air to freely enter the exterior wall openings. These exterior areas are defined as "Courts" and "Yards." Courts and yards must be open, uncovered and on the same lot as the building. They may be either partly or wholly enclosed by the building. Requirements are provided in Section 1206, which regulates the minimum width, area, air intake and drainage of courts and yards. The requirements of Sections 1206.2 through 1206.3.3 do not apply if artificial ventilation and lighting is provided for the spaces opening onto the court or yard in accordance with Section 1203.1 or 1205.1. See Figure 1206.1 for examples of courts and yards.

1206.2 Yards. *Yards* shall not be less than 3 feet (914 mm) in width for buildings two *stories* or less above *grade plane*. For buildings more than two *stories above grade plane*, the minimum width of the *yard* shall be increased at the rate of 1 foot (305 mm) for each additional *story*. For buildings exceeding 14 *stories above grade plane*, the required width of the *yard* shall be computed on the basis of 14 *stories above grade plane*.

❖ A yard is distinguished from a court by the definitions in Chapter 2 (see commentary, Chapter 2). A court is bounded on at least three sides by exterior building walls or similar enclosing devices, whereas a yard is typically located between a building and a lot line and is open on at least two sides or ends (see Figure 1206.1).

The required width of a yard is measured perpendicular from the face of the wall to the opposing wall on the other side of the yard. A five-story building would be required to have a yard at least 6 feet (1829 mm) in width [3 feet (914 mm) plus 1 foot (305 mm) each for stories three through five]. A 20-story building is required to have a yard at least 15 feet (4572 mm) [3 feet (914 mm) plus 1 foot (305 mm) each for stories three through 14]. The last sentence of the section simply requires a minimum yard width of 15 feet (4572 mm) for all buildings over 14 stories in height. If the building is adjacent to a court rather than a yard, the requirements of Section 1206.3 apply. Neither Section 1206.2 nor 1206.3 is applicable if artificial lighting and ventilation is provided for spaces facing the yard or court in accordance with Sections 1203.1 and 1205.1. Note that the stories are measured above grade plane.

1206.3 Courts. *Courts* shall not be less than 3 feet (914 mm) in width. *Courts* having windows opening on opposite sides shall not be less than 6 feet (1829 mm) in width. *Courts* shall not be less than 10 feet (3048 mm) in length unless bounded on one end by a *public way* or *yard*. For buildings more than two *stories above grade plane*, the *court* shall be increased 1 foot (305 mm) in width and 2 feet (610 mm) in length for each additional *story*. For buildings exceeding 14 *stories above grade plane*, the required dimensions shall be computed on the basis of 14 *stories above grade plane*.

❖ "Courts" are defined in Chapter 2 as being bounded on no less than three sides by walls or other enclosing construction. A court adjacent to a five-story building would be required to have a width measured perpendicular from the wall facing the court of at least 6 feet (1829 mm) when required openings are on one wall and 9 feet (2743 mm) when required openings are on opposing walls [3 feet (914 mm) plus 1 foot (305 mm) each for stories three through five, or 6 feet (1829 mm) plus 1 foot (305 mm) each for stories three through five]. If the same court is bounded on all sides, the required minimum length would be 16 feet (4877 mm) [10 feet (3048 mm) plus 2 feet (610 mm) each for stories three through five]. The last sentence simply requires all buildings higher than 14 stories to have a minimum court width of 15 feet (4572 mm) without

opposing required openings, a minimum width of 18 feet (5486 mm) where required openings oppose each other and a minimum length of 34 feet (10 363 mm) if bounded on all sides [width equals 3 feet (914 mm) or 6 feet (1829 mm) plus 1 foot (305 mm) each for stories three through 14, and length equals 10 feet (3048 mm) plus 2 feet (610 mm) each for stories three through 14]. The requirements of this section are not applicable if artificial lighting and ventilation are provided for spaces facing the court in accordance with Sections 1203.1 and 1205.1. Note that the stories are measured above grade plane. See Figure 1206.3 for examples of court dimensions.

1206.3.1 Court access. Access shall be provided to the bottom of *courts* for cleaning purposes.

❖ Courts must be accessed for maintenance. Clearly, a court intended to be a source of ventilation air must be maintained in a manner conducive to its purpose.

1206.3.2 Air intake. *Courts* more than two *stories* in height shall be provided with a horizontal air intake at the bottom not less than 10 square feet (0.93 m²) in area and leading to the exterior of the building unless abutting a *yard* or *public way*.

❖ This section is applicable only to courts that are bounded on all four sides by walls or other construction. A fully bounded court takes on characteristics

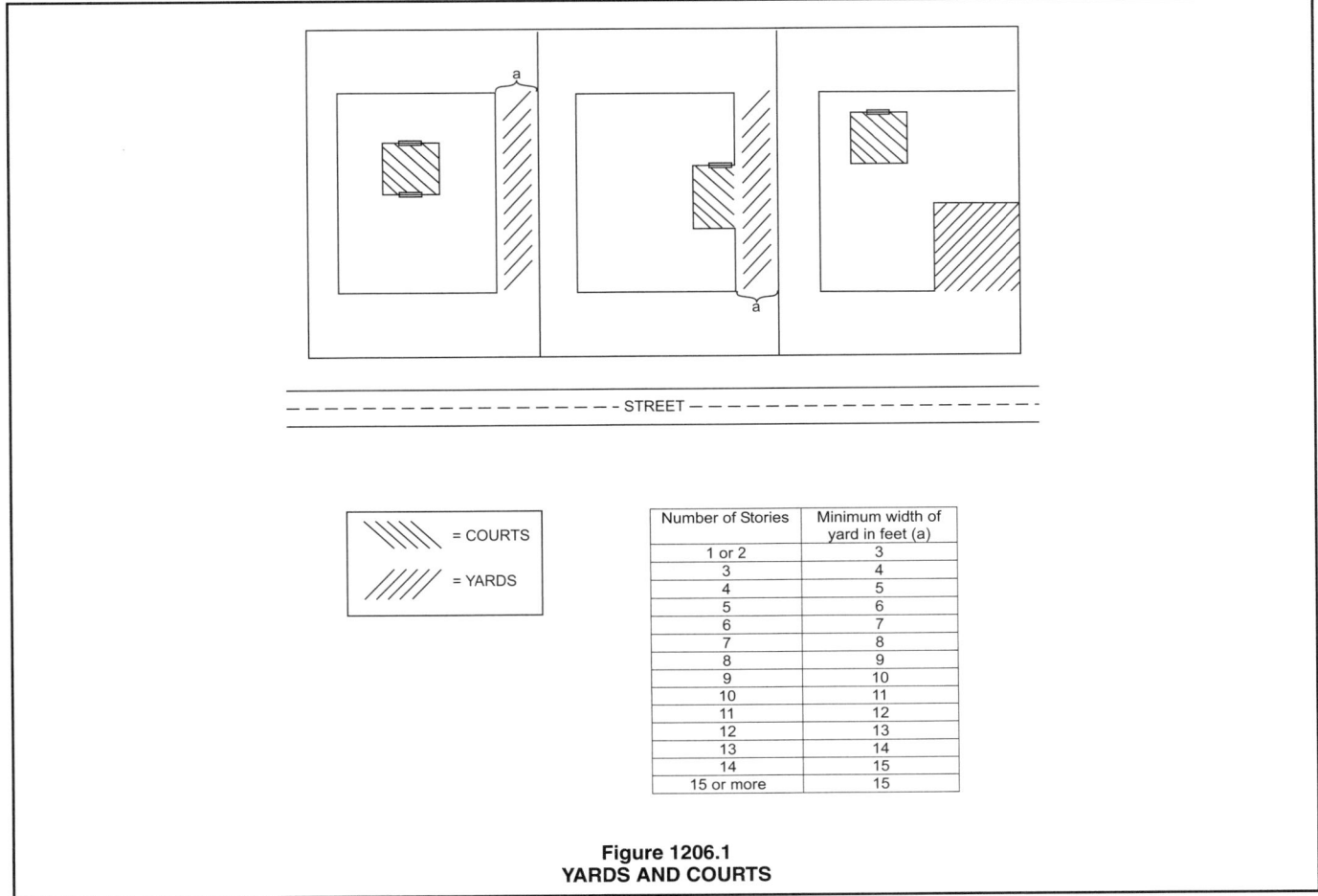

Number of Stories	Minimum width of yard in feet (a)
1 or 2	3
3	4
4	5
5	6
6	7
7	8
8	9
9	10
10	11
11	12
12	13
13	14
14	15
15 or more	15

Figure 1206.1
YARDS AND COURTS

similar to a chimney during summer weather when the building mass is heated from the daytime sun. In order for a fully bounded court to function as an efficient source of natural ventilation, the bottom of the court must have a source of fresh air (similar to a chimney). This source of fresh air is supplied through the required opening of 10 square feet (0.93 m²) connected directly to a street or yard. The requirements of this section are not applicable if artificial lighting and ventilation are provided for spaces facing the court in accordance with Sections 1203.1 and 1206.1.

1206.3.3 Court drainage. The bottom of every *court* shall be properly graded and drained to a public sewer or other approved disposal system complying with the *International Plumbing Code.*

❖ A court is an inherent water trap. A court that is not both graded and drained will accumulate water and remain in a saturated condition, which will promote an insanitary condition, including odors. Stagnant water is often a breeding ground for disease-carrying insects. Based on the design and nature of the soil after construction, paving the court may be the best solution to eliminate a problem.

SECTION 1207
SOUND TRANSMISSION

1207.1 Scope. This section shall apply to common interior walls, partitions and floor/ceiling assemblies between adjacent *dwelling units* or between *dwelling units* and adjacent public areas such as halls, *corridors*, *stairs* or service areas.

❖ Since noise transmission can be quantified and affects the quality of life, the code incorporates regulations that address noise transmission in multiple-family residential construction, wherein the occupants may have no control over noise. The regulated components of construction are those through which noise is primarily transmitted.

1207.2 Air-borne sound. Walls, partitions and floor/ceiling assemblies separating *dwelling units* from each other or from public or service areas shall have a sound transmission class (STC) of not less than 50 (45 if field tested) for air-borne noise when tested in accordance with ASTM E 90. Penetrations or openings in construction assemblies for piping; electrical devices; recessed cabinets; bathtubs; soffits; or heating, ventilating or exhaust ducts shall be sealed, lined, insulated or otherwise treated to maintain the required ratings. This requirement shall not apply to *dwelling unit*

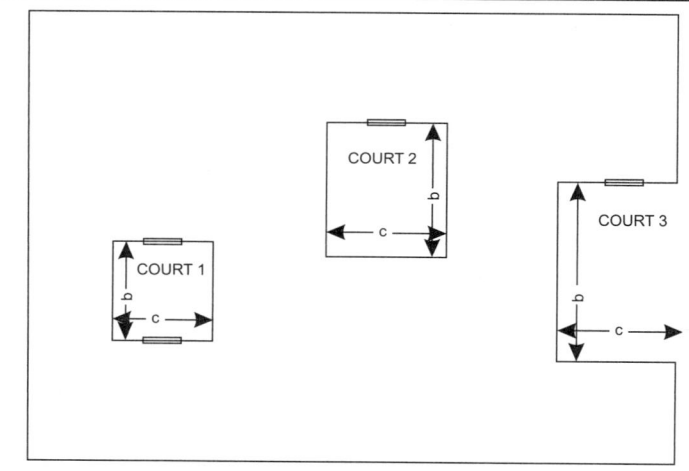

Number of Stories	Court 1		Court 2		Court 3	
	Width-b	Length-c	Width-b	Length-c	Width-b	Length-c
1 or 2	6	10	3	10	3	N/A
3	7	12	4	12	4	N/A
4	8	14	5	14	5	N/A
5	9	16	6	16	6	N/A
6	10	18	7	18	7	N/A
7	11	20	8	20	8	N/A
8	12	22	9	22	9	N/A
9	13	24	10	24	10	N/A
10	14	26	11	26	11	N/A
11	15	28	12	28	12	N/A
12	16	30	13	30	13	N/A
13	17	32	14	32	14	N/A
14	18	34	15	34	15	N/A
15 or more	18	34	15	34	15	N/A

N/A: Not regulated by Section 1206.3.

Figure 1206.3
DIMENSIONS OF COURTS

entrance doors; however, such doors shall be tight fitting to the frame and sill.

❖ The code requires common walls between dwelling units and between dwelling units and public areas to have a minimum sound transmission class (STC) of 50. The STC is a measure of an assembly's ability to resist sound transmission. The higher the number (rating), the higher the resistance (less sound transmission). Standard architectural wall construction assemblies have been tested for sound transmission ratings, and reference to the construction specifications will yield such information. Air-borne noise originates in the air, such as voice or music. For structure-borne sound, see the commentary to Section 1207.3.

As a rule, vertical assemblies meeting the requirements of this section consist of double walls or walls containing insulation similar to exterior walls.

1207.2.1 Masonry. The sound transmission class of concrete masonry and clay masonry assemblies shall be calculated in accordance with TMS 0302 or determined through testing in accordance with ASTM 90.

❖ The STC values found in the referenced standard TMS 0302 are derived from laboratory testing of masonry assemblies in accordance with ASTM E 90. This reference provides a quicker alternative for compliance with the STC ratings required by Section 1207.2 of the code for masonry and clay masonry assemblies.

1207.3 Structure-borne sound. Floor/ceiling assemblies between *dwelling units* or between a *dwelling unit* and a public or service area within the structure shall have an impact insulation class (IIC) rating of not less than 50 (45 if field tested) when tested in accordance with ASTM E 492.

❖ The impact insulation class (IIC) is a measure of an assembly's ability to resist sound transmission. The higher the number (rating), the higher the resistance (less sound transmission). Floors between dwelling units and those between dwelling units and public areas are required to have a minimum IIC rating of 50.

Usually, floor assemblies that are carpeted meet the minimum requirement for an IIC rating of 50. Other areas with hard-surfaced finishes may require additional treatment or insulation to comply with these requirements.

There are various resource documents containing STC ratings and IIC ratings, including the following: GA 600, *Fire Resistance Design Manual*, by the Gypsum Association; NCMA TEK 69A, *New Data on Sound Reduction with Concrete Masonry Walls*, by the National Concrete Masonry Association; and BIA TN 5A, *Sound Insulation—Clay Masonry Walls*, by the Brick Institute of America. These or any other similar

sources can be submitted as a basis for approval once it is demonstrated to the building official that the data are based on ASTM E 90 and E 492.

SECTION 1208
INTERIOR SPACE DIMENSIONS

1208.1 Minimum room widths. *Habitable spaces*, other than a kitchen, shall not be less than 7 feet (2134 mm) in any plan dimension. Kitchens shall have a clear passageway of not less than 3 feet (914 mm) between counter fronts and appliances or counter fronts and walls.

❖ This provision specifies the minimum horizontal dimensions required for all habitable rooms. Any room that functions as a living room, bedroom, dining room or any other similar habitable room must be sized such that a cylinder with a diameter of 7 feet (2134 mm) may be placed in it. Only kitchens are exempt from this requirement. This code provision allows the circulation of ventilation air through the space while maintaining reasonably sized living quarters for the occupants. "Habitable space" is defined in Chapter 2. Smaller spaces, such as bathrooms and hallways, do not need to meet the minimum room width requirements because they are not considered habitable rooms.

1208.2 Minimum ceiling heights. Occupiable spaces, *habitable spaces* and *corridors* shall have a ceiling height of not less than 7 feet 6 inches (2286 mm). Bathrooms, toilet rooms, kitchens, storage rooms and laundry rooms shall be permitted to have a ceiling height of not less than 7 feet (2134 mm).

Exceptions:

1. In one- and two-family *dwellings*, beams or girders spaced not less than 4 feet (1219 mm) on center and projecting not more than 6 inches (152 mm) below the required ceiling height.

2. If any room in a building has a sloped ceiling, the prescribed ceiling height for the room is required in one-half the area thereof. Any portion of the room measuring less than 5 feet (1524 mm) from the finished floor to the ceiling shall not be included in any computation of the minimum area thereof.

3. *Mezzanines* constructed in accordance with Section 505.1.

❖ Occupiable spaces or rooms (including habitable spaces) are required to have a specific minimum ceiling height. Bathrooms, toilet rooms, kitchens, storage rooms and laundry rooms are permitted to have a lower minimum ceiling height in accordance with this section. Ceiling height is one of the variables that affects the circulation of air in a space. Additionally, there is a psychological need for spaciousness in a living space or in one of the accessory spaces.

Figure 1208.2(1) illustrates the application of Exception 1 for beams and girders spaced no more than 4 feet (1219 mm) on center in one- and two-family dwellings.

Figure 1208.2(2) illustrates the application of Exception 2. Rooms with sloped ceilings are required to meet two distinct conditions. First, the area of the room having a floor-to-ceiling clearance of less than 5 feet (1524 mm) does not contribute to the minimum floor area required by Section 1208.3. Second, at least one-half of the actual total area of the room must meet the minimum ceiling height requirements of Section 1208.2 [see Figure 1208.2(2)].

Finally, Exception 3 is consistent with Section 505.1, which establishes the minimum ceiling height for mezzanines at 7 feet (2134 mm). In accordance with Section 505.2, mezzanines cannot exceed one-third of the area of the room in which they are located.

1208.2.1 Furred ceiling. Any room with a furred ceiling shall be required to have the minimum ceiling height in two-thirds of the area thereof, but in no case shall the height of the furred ceiling be less than 7 feet (2134 mm).

❖ This section only applies to rooms required to have a ceiling height of no less than 7 feet, 6 inches (2286 mm). In Figure 1208.2.1, floor area A_1 must be greater than or equal to two-thirds (length times width). Note that only those ceiling heights furred to a height of less than 7 feet, 6 inches (2286 mm) affect area A_1.

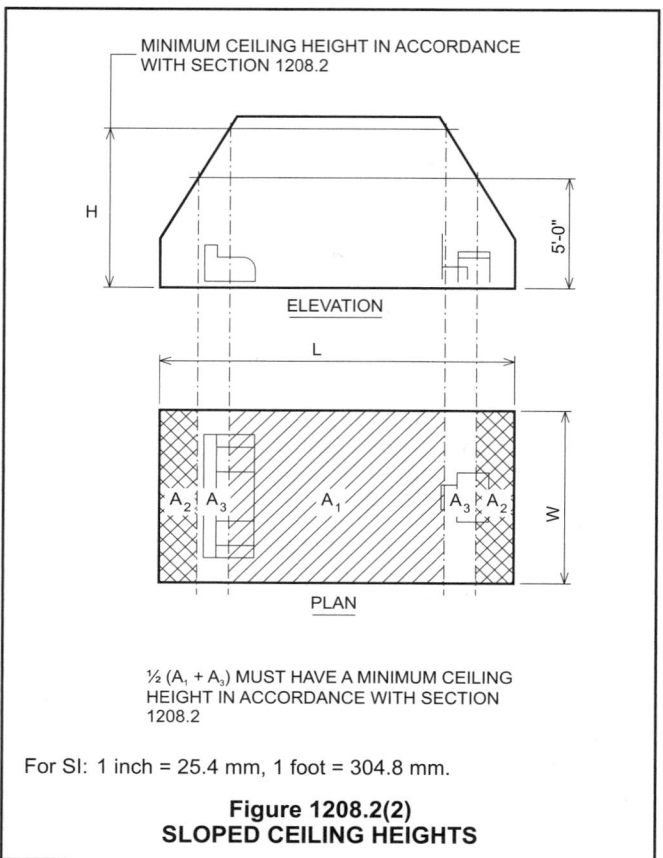

½ (A_1 + A_3) MUST HAVE A MINIMUM CEILING HEIGHT IN ACCORDANCE WITH SECTION 1208.2

For SI: 1 inch = 25.4 mm, 1 foot = 304.8 mm.

Figure 1208.2(2)
SLOPED CEILING HEIGHTS

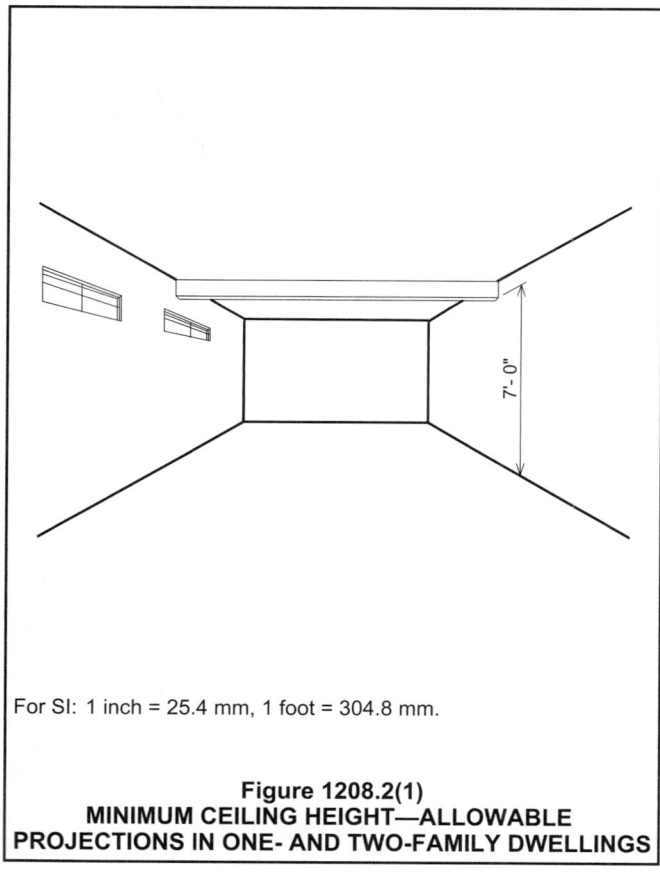

For SI: 1 inch = 25.4 mm, 1 foot = 304.8 mm.

Figure 1208.2(1)
MINIMUM CEILING HEIGHT—ALLOWABLE PROJECTIONS IN ONE- AND TWO-FAMILY DWELLINGS

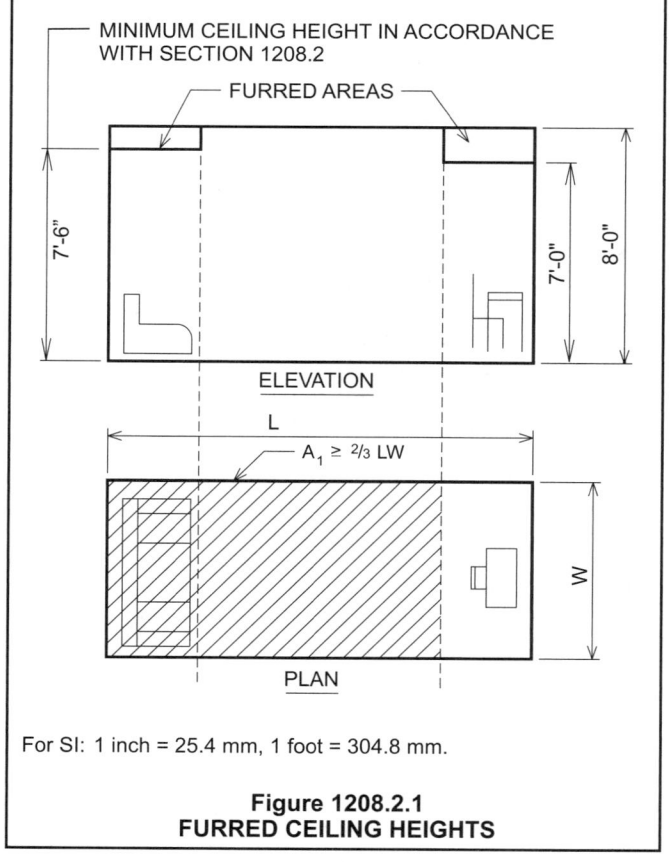

For SI: 1 inch = 25.4 mm, 1 foot = 304.8 mm.

Figure 1208.2.1
FURRED CEILING HEIGHTS

1208.3 Room area. Every *dwelling unit* shall have at least one room that shall have not less than 120 square feet (13.9 m²) of *net floor area*. Other habitable rooms shall have a *net floor area* of not less than 70 square feet (6.5 m²).

> **Exception:** Every kitchen in a one- and two-family *dwelling* shall have not less than 50 square feet (4.64 m²) of *gross floor area*.

❖ This section applies only to dwelling units. A minimum area of 120 square feet (13.9 m²) for at least one room of each dwelling unit and a minimum of 70 square feet (6.5 m²) for all other habitable rooms, except kitchens, are the minimum standards for habitable rooms. These minimums reflect the physiological requirements of light and ventilation and also preserve the individual's perception of space and the elements necessary for a psychological sense of well-being. The code does not regulate the minimum area of rooms and spaces in other than residential occupancies.

The exception permits smaller kitchens in one- and two-family dwelling units built under the code. The minimum required area of 50 square feet (4.65 m²) is still larger than the minimum required for dwellings built under the *International Residential Code®* (IRC®). Under the IRC, there is no minimum area or dimension required for kitchens.

1208.4 Efficiency dwelling units. An efficiency living unit shall conform to the requirements of the code except as modified herein:

1. The unit shall have a living room of not less than 220 square feet (20.4 m²) of floor area. An additional 100 square feet (9.3 m²) of floor area shall be provided for each occupant of such unit in excess of two.

2. The unit shall be provided with a separate closet.

3. The unit shall be provided with a kitchen sink, cooking appliance and refrigeration facilities, each having a clear working space of not less than 30 inches (762 mm) in front. Light and ventilation conforming to this code shall be provided.

4. The unit shall be provided with a separate bathroom containing a water closet, lavatory and bathtub or shower.

❖ Efficiency units are very small apartments consisting of one or two rooms and a bathroom. Efficiency units that comply with this section are not required to comply with the minimum area requirements in Section 1208.3; however, the total allowable number of occupants in the dwelling is limited, depending on the area of the unit in accordance with Item 1. The purpose of both efficiency units and this section is to provide for combined use of spaces in an economical or "efficient" manner without jeopardizing health or comfort. This is possible because of the occupant limitation.

In addition to the living room, the apartment must have a separate closet and bathroom. There are no minimum area requirements for these spaces; however, the fixture clearances in the bathroom must be as

required in the IPC. There is not a requirement for a separate kitchen, but if the required sink and appliances are located in the living room, they cannot encroach on the required minimum floor space.

SECTION 1209
ACCESS TO UNOCCUPIED SPACES

1209.1 Crawl spaces. Crawl spaces shall be provided with a minimum of one access opening not less than 18 inches by 24 inches (457 mm by 610 mm).

❖ The requirements of this section establish 18 inches by 24 inches (457 mm by 610 mm) as the minimum size opening for crawl spaces. If access is through a wall, the 18-inch (457 mm) minimum would be the height and the 24-inch (610 mm) minimum would be the width.

Items such as plumbing and wiring installations pass through crawl spaces at times. Required initial and periodic inspections, and repairs cannot be carried out without access to such crawl spaces.

1209.2 Attic spaces. An opening not less than 20 inches by 30 inches (559 mm by 762 mm) shall be provided to any *attic* area having a clear height of over 30 inches (762 mm). A 30-inch (762 mm) minimum clear headroom in the *attic* space shall be provided at or above the access opening.

❖ Access to the attic provides a convenient and nondestructive means for fire department personnel to visually check for an attic fire and, if need be, gain entry to the concealed spaces and suppress a fire. Access to attic spaces can be provided through the ceiling within each compartment that is created by draftstops or through openings within the draftstops themselves. Openings located within the draftstop are required to be self-closing and the opening protective must provide structural fire integrity (the ability to remain in place) similar to the draftstop. Access is required when the attic space has a clear height greater than 30 inches (762 mm) measured from the top of the ceiling joists (or top of the floor sheathing, if present) to the underside of the roof rafters.

1209.3 Mechanical appliances. Access to mechanical appliances installed in under-floor areas, in *attic* spaces and on roofs or elevated structures shall be in accordance with the *International Mechanical Code*.

❖ Access to mechanical appliances is needed to maintain and service the equipment. See Section 306 in the IMC for detailed requirements.

SECTION 1210
SURROUNDING MATERIALS

1210.1 Floors and wall base finish materials. In other than *dwelling units*, toilet, bathing and shower room floor finish materials shall have a smooth, hard, nonabsorbent surface. The intersections of such floors with walls shall have a smooth, hard, nonabsorbent vertical base that extends upward onto the walls at least 4 inches (102 mm).

❖ The purpose of this requirement is to provide non-absorbent surfaces that can be maintained in a sanitary condition. The 4-inch (152 mm) extension of the surface up the surrounding walls is so that the wall will not absorb moisture during cleaning and, thus, will be left in a clean condition. This provision does not require that the same material that is on the floor be extended up the wall. As long as the wall material is smooth, hard and nonabsorbent, and there is adequate seal between the materials to restrict moisture from getting behind the surface material and into the wall construction, the intent of the requirement should be met.

1210.2 Walls and partitions. Walls and partitions within 2 feet (610 mm) of urinals and water closets shall have a smooth, hard, nonabsorbent surface, to a height of 4 feet (1219 mm) above the floor, and except for structural elements, the materials used in such walls shall be of a type that is not adversely affected by moisture.

Exceptions:

1. *Dwelling units* and *sleeping units*.

2. Toilet rooms that are not accessible to the public and which have not more than one water closet.

Accessories such as grab bars, towel bars, paper dispensers and soap dishes, provided on or within walls, shall be installed and sealed to protect structural elements from moisture. For walls and partitions also see Section 2903.

❖ The walls and partitions near urinals and water closets need to have the surface specified in this section since they are subject to moisture. Exception 1 recognizes that water closet facilities in dwelling units and sleeping units are not exposed to as much use as those that serve the public and, thus, are easier to maintain. Exception 2 acknowledges that toilet fixtures that do not serve the public are also subject to less use. This section requires protection of the structural supports for accessories so that they will maintain their strength.

1210.3 Showers. Shower compartments and walls above bathtubs with installed shower heads shall be finished with a smooth, nonabsorbent surface to a height not less than 70 inches (1778 mm) above the drain inlet.

❖ The 70-inch (1778 mm) requirement in this section is based on the height of the shower compartment and walls that are exposed to significant moisture that would cause the surface to become insanitary over a long period of time.

1210.4 Waterproof joints. Built-in tubs with showers shall have waterproof joints between the tub and adjacent wall.

❖ The joint between the tub and wall must be sealed to prevent moisture from getting into the supporting floor and framing. Waterproof joints are also needed to keep the concealed area of the wall in a sanitary condition.

1210.5 Toilet rooms. Toilet rooms shall not open directly into a room used for the preparation of food for service to the public.

❖ The requirement that toilet rooms not open directly into rooms where food is prepared for the public is necessary to keep the food preparation area in a sanitary condition.

Bibliography

The following resource materials are referenced in this chapter or are relevant to the subject matter addressed in this chapter.

ASCE 24-05, *Flood Resistance Design and Construction Standard.* Reston, VA: American Society of Civil Engineers, 2005.

ASHRAE-2005, ASHRAE *Handbook of Fundamentals.* Atlanta: American Society of Heating, Refrigerating and Air-Conditioning Engineers, Inc., 2005.

ASTM E 90-04, *Test Method for Laboratory Measurement of Air-borne Sound Transmission Loss of Building Partitions and Elements.* West Conshohocken, PA: ASTM International, 2004.

ASTM E 492-04, *Test Method for Laboratory Measurement of Impact Sound Transmission Through Floor-ceiling Assemblies Using the Tapping Machine.* West Conshohocken, PA: ASTM International, 2004.

BIA TN 5A-83, *Sound Insulation—Clay Masonry Walls.* Reston, VA: Brick Institute of America, 1983.

FEMA FIA-TB 1, *Openings in Foundation Walls for Buildings Located in Special Flood Hazard Areas.* Washington, DC: Federal Emergency Management Agency, 2008.

GA 600-06, *Fire Resistance Design Manual.* Evanston, IL: Gypsum Association, 2006.

IECC-09, *International Energy Conservation Code.* Washington, DC: International Code Council, 2009.

IFC-09, *International Fire Code.* Washington, DC: International Code Council, 2009.

IMC-09, *International Mechanical Code.* Washington, DC: International Code Council, 2009.

IPC-09, *International Plumbing Code.* Washington, DC: International Code Council, 2009.

IRC-09, *International Residential Code.* Washington, DC: International Code Council, 2009.

NCMA TEK 69A-92, *New Data on Sound Reduction with Concrete Masonry Walls.* Herndon, VA: National Concrete Masonry Association, 1992.

NFPA 70-08, *National Electrical Code.* Quincy, MA: National Fire Protection Association, 2008.

TMS 0302-07, *Standard Method for Determining the Sound Transmission Class Rating for Masonry Walls.* Boulder, CO: The Masonry Society, 2007.

Chapter 13:
Energy Efficiency

General Comments

Chapter 13 provides for the design and construction of energy-efficient buildings and structures or portions thereof intended primarily for human occupancy by direct reference to the *International Energy Conservation Code®* (IECC®).

Purpose

The purpose of Chapter 13 is to provide minimum design requirements that will promote efficient utilization of energy in building construction. The requirements are directed toward the design of building envelopes with adequate thermal resistance and low air leakage, and toward the design and selection of mechanical, service water heating, electrical and illumination systems that will promote the effective use of depletable energy resources and encourage the use of nondepletable energy resources.

SECTION 1301
GENERAL

1301.1 Scope. This chapter governs the design and construction of buildings for energy efficiency.

❖ The scope of Chapter 13 is applicable to all buildings and structures, as well as their components and systems that are regulated by the IECC. The IECC thereby addresses the design of energy-efficient building envelopes, and the selection and installation of energy-efficient mechanical, service-water heating, electrical distribution and illumination systems and equipment for the effective use of energy in both residential and commercial buildings.

1301.1.1 Criteria. Buildings shall be designed and constructed in accordance with the *International Energy Conservation Code*.

❖ The energy conservation requirements of this chapter rely exclusively on the technical provisions of the IECC. Compliance with the IECC is the sole means of demonstrating compliance with the technical provisions of Chapter 13.

The climate basis for the 2009 IECC is derived from geographical zones that are based on multiple climate variables (so that both heating and cooling considerations are accommodated). Further, within the U.S., the zones are completely defined by political boundaries (county lines) so that code users will never have to choose from disparate climate data sources to determine local requirements. The climate zones were developed in an open process, in consultation with relevant standards committees of the American Society of Heating, Refrigerating and Air-conditioning Engineers (ASHRAE). The zones are designed to be an appropriate foundation for both residential and commercial codes, and may be useful in other contexts, as well.

The IECC is designed to increase consumer awareness of a home's energy features by making baseline requirements uniform within a jurisdiction and by requiring a disclosure of each house's *R*-values, *U*-factors, and heating, ventilating and air-conditioning (HVAC) efficiencies.

The IECC is designed, to the extent practicable, to incorporate aspects of the latest building science regarding energy efficiency and its effects on moisture control and durability. For example, the IECC contains provisions related to unvented crawl spaces, modifies vapor retarder requirements, requires sealing of air handlers in garages and limits worst-case glazing *U*-factors in locations where moisture condensation can be a serious problem.

Bibliography

The following resource materials are referenced in this chapter or are relevant to the subject matter addressed in this chapter.

IECC-09, *International Energy Conservation Code*. Washington, DC: International Code Council, 2009.

Chapter 14:
Exterior Walls

General Comments

Chapter 14 provides requirements for the materials and construction creating finished exterior surfaces of a building or structure. The chapter provides requirements resulting in minimum permitted weather resistance and fire performance.

In the past, there were relatively few materials available for application to exterior walls for protection against weather and exposure to other outside elements. With the development of new methods and materials over the years, architects and builders began to use a variety of materials for different appearances, im-proved insulating quality, sound transmission control and fire resistance. The code has developed prescriptive and performance regulations to control these aspects and the types and thickness of exterior wall coverings.

Purpose

The purpose of Chapter 14 is to provide the minimum requirements for exterior wall coverings. This chapter also includes the minimum regulations for materials, and the minimum thicknesses and installation requirements for exterior weather coverings and various wall veneers. Limitations on the use of combustible exterior wall finishes are also included.

SECTION 1401
GENERAL

1401.1 Scope. The provisions of this chapter shall establish the minimum requirements for exterior walls; *exterior wall* coverings; *exterior wall* openings; exterior windows and doors; architectural *trim*; balconies and similar projections; and bay and oriel windows.

❖ The requirements for exterior wall construction, including components such as openings, doors, windows, trim and balconies, are specified herein. Also specified are provisions intended to prevent damage to a building from the effects of weather and from moisture intrusion into or through the exterior walls.

SECTION 1402
DEFINITIONS

1402.1 General. The following words and terms shall, for the purposes of this chapter and as used elsewhere in this code, have the meanings shown herein.

❖ Definitions of terms can help in the understanding and application of the code requirements. The purpose for including these definitions within this chapter is to provide more convenient access to them without having to refer back to Chapter 2.

For convenience, these terms are also listed in Chapter 2 with a cross reference to this section. Terms that are italicized provide a visual identification throughout the code that a definition exists for that term.

The use and application of all defined terms, including those defined herein, are set forth in Section 201.

ADHERED MASONRY VENEER. Veneer secured and supported through the adhesion of an *approved* bonding material applied to an *approved* backing.

❖ This type of masonry veneer relies on the backing surface for both vertical and lateral load resistance. The components of adhered masonry veneer construction generally include the masonry veneer, the adhering material and the backing to which the veneer is attached. It should be noted that the term "approve" means components are subject to approval by the building official or the authority having jurisdiction, in accordance with Section 104.

ANCHORED MASONRY VENEER. Veneer secured with *approved* mechanical fasteners to an *approved* backing.

❖ This type of masonry veneer is generally supported from below and anchored to the sheathing, studs or other structural portion of the wall. Veneers provide little, if any, strength to the wall and are, therefore, considered to be nonstructural. Anchored masonry veneer is unique in that it is usually supported from below by a footing, lintels or shelf angles. Anchored masonry veneer must not, however, support loads other than its own dead loads, wind loads and seismic loads resulting from the dead load of the wall.

BACKING. The wall or surface to which the veneer is secured.

❖ The backing is the portion of a structure that provides support for the exterior veneer. The backing also typically resists lateral and transverse loads imposed by the veneer and may be load bearing or nonload bearing.

EXTERIOR INSULATION AND FINISH SYSTEMS (EIFS). EIFS are nonstructural, nonload-bearing, *exterior wall* cladding systems that consist of an insulation board attached either adhesively or mechanically, or both, to the substrate; an integrally reinforced base coat and a textured protective finish coat.

❖ EIFS is an exterior cladding that is specifically dealt with in Section 1408 of the code. A definition is necessary in order to ensure the proper application of these requirements.

EXTERIOR INSULATION AND FINISH SYSTEMS (EIFS) WITH DRAINAGE. An EIFS that incorporates a means of drainage applied over a *water-resistive barrier.*

❖ Although Exterior insulation and finish systems (EIFS) and Exterior insulation and finish systems (EIFS) with drainage are somewhat similar, the presence of the water-resistive barrier makes it necessary to treat this material differently in code regulations.

EXTERIOR WALL. A wall, bearing or nonbearing, that is used as an enclosing wall for a building, other than a *fire wall*, and that has a slope of 60 degrees (1.05 rad) or greater with the horizontal plane.

❖ A wall is defined as an exterior element that encloses a structure and that has a slope equal to or greater than 60 degrees (1.05 rad) from the horizontal plane. Exterior enclosing elements with slopes less than this are generally subjected to more severe weather exposure than vertical surfaces and thus may experience a greater amount of water intrusion. These sloped surfaces, which may include elements, such as inset windowsills, sloped parapets and other architectural elements, should be designed to resist water penetration in a manner similar to a roof.

EXTERIOR WALL COVERING. A material or assembly of materials applied on the exterior side of exterior walls for the purpose of providing a weather-resisting barrier, insulation or for aesthetics, including but not limited to, veneers, siding, exterior insulation and finish systems, architectural *trim* and embellishments such as cornices, soffits, facias, gutters and leaders.

❖ Materials such as wood, masonry, metal, concrete, structural glass and plastics are used for exterior wall coverings. It should be noted that exterior wall coverings that are combustible are to meet the combustible materials requirements of Section 1406.2.

EXTERIOR WALL ENVELOPE. A system or assembly of *exterior wall* components, including *exterior wall* finish materials, that provides protection of the building structural members, including framing and sheathing materials, and conditioned interior space, from the detrimental effects of the exterior environment.

❖ This definition is needed for the understanding and proper application of the weather protection requirements in Section 1403.2.

FIBER-CEMENT SIDING. A manufactured, fiber-reinforcing product made with an inorganic hydraulic or calcium silicate binder formed by chemical reaction and reinforced with discrete organic or inorganic nonasbestos fibers, or both. Additives that enhance manufacturing or product performance are permitted. Fiber-cement siding products have either smooth or textured faces and are intended for *exterior wall* and related applications.

❖ Fiber-cement siding is also a material used for weather-resistant siding. It is manufactured from fiber-reinforced cement, and the code permits its use as either panel siding or horizontal lap siding.

METAL COMPOSITE MATERIAL (MCM). A factory-manufactured panel consisting of metal skins bonded to both faces of a plastic core.

❖ These types of panels are sandwich construction composed of thin metal (usually aluminum or steel) sheets covering a solid plastic core. The panels typically have thicknesses that do not exceed 0.25 inch (6.4 mm). Metal composite material (MCM) panels are different from other types of sandwich panels in that they do not contain foam plastic cores and are not typically intended to provide thermal insulation.

METAL COMPOSITE MATERIAL (MCM) SYSTEM. An *exterior wall* covering fabricated using MCM in a specific assembly including joints, seams, attachments, substrate, framing and other details as appropriate to a particular design.

❖ This section provides a definition of the components that are typically part of an MCM installation. The components of the exterior wall covering system consist of framing members for the attachment and support of the MCM panels; the types of joints and seams used to maintain the weather resistance of the system and the means for attaching the entire system to the building substrate or structural frame. Figures 1402.1(1) through (3) illustrate the details of some types of MCM systems.

VENEER. A facing attached to a wall for the purpose of providing ornamentation, protection or insulation, but not counted as adding strength to the wall.

❖ Veneers are wall facings or claddings of various materials that are used for environmental protection or ornamentation on the exterior of walls. Veneers are nonstructural in that they do not carry any load other than their own weight.

VINYL SIDING. A shaped material, made principally from rigid polyvinyl chloride (PVC), that is used as an *exterior wall covering.*

❖ Vinyl siding is discussed in Sections 1404.9 and 1405.13. This material offers a low-maintenance option in a siding product that does not require regular painting. It is offered in a variety of colors and textures. Some vinyl siding is manufactured with some relief on the face of the product to emulate wood grain. It is important when applying the material to use an attachment method that allows the product to move as it expands and contracts. Static connections will cause the material to crack and warp.

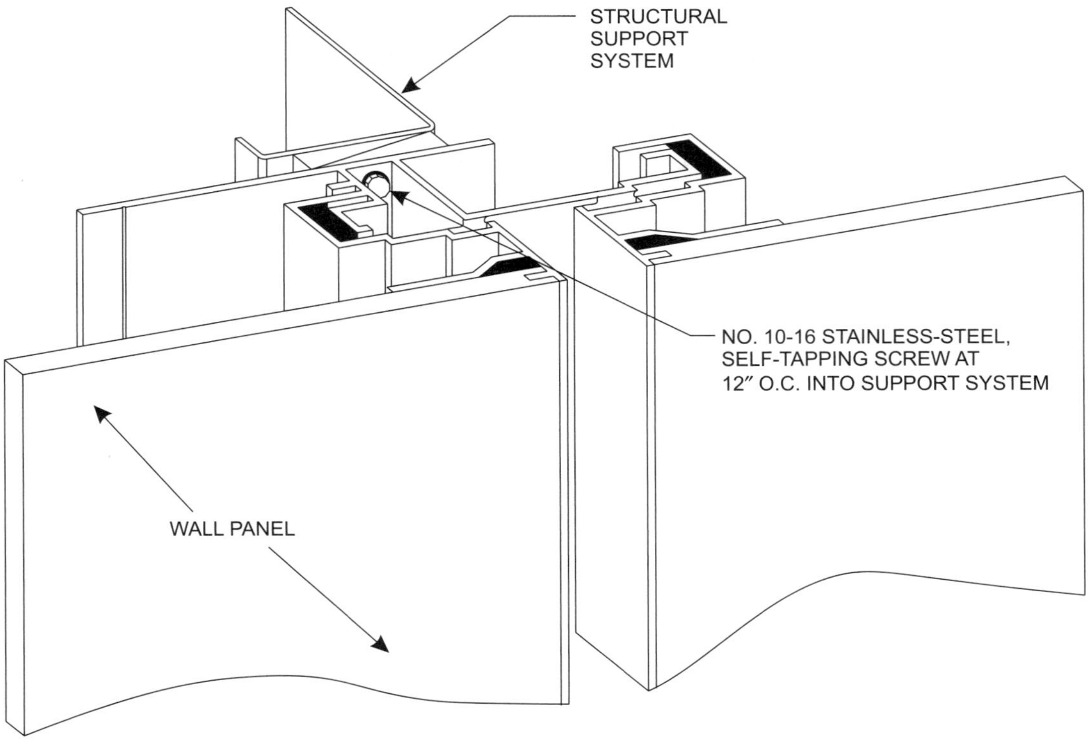

STRUCTURAL
SUPPORT
SYSTEM

NO. 10-16 STAINLESS-STEEL,
SELF-TAPPING SCREW AT
12″ O.C. INTO SUPPORT SYSTEM

WALL PANEL

Figure 1402.1(1)
CONTINUOUS EDGE GRIP SYSTEM

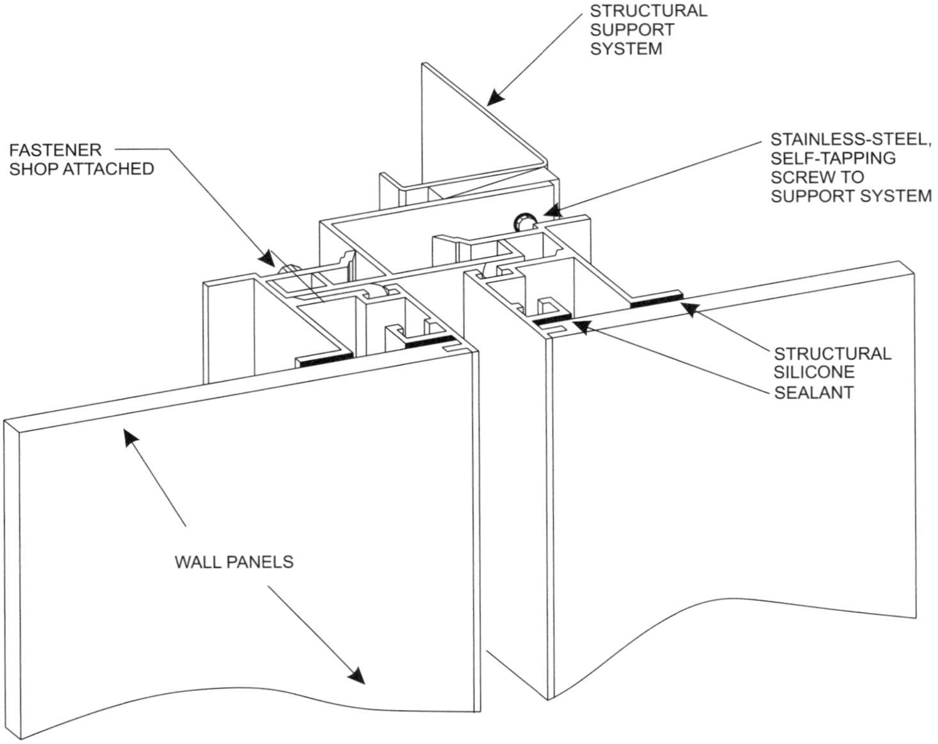

STRUCTURAL
SUPPORT
SYSTEM

FASTENER
SHOP ATTACHED

STAINLESS-STEEL,
SELF-TAPPING
SCREW TO
SUPPORT SYSTEM

STRUCTURAL
SILICONE
SEALANT

WALL PANELS

Figure 1402.1(2)
SECTION THROUGH VERTICAL JOINT (HORIZONTAL JOINT SIMILAR)

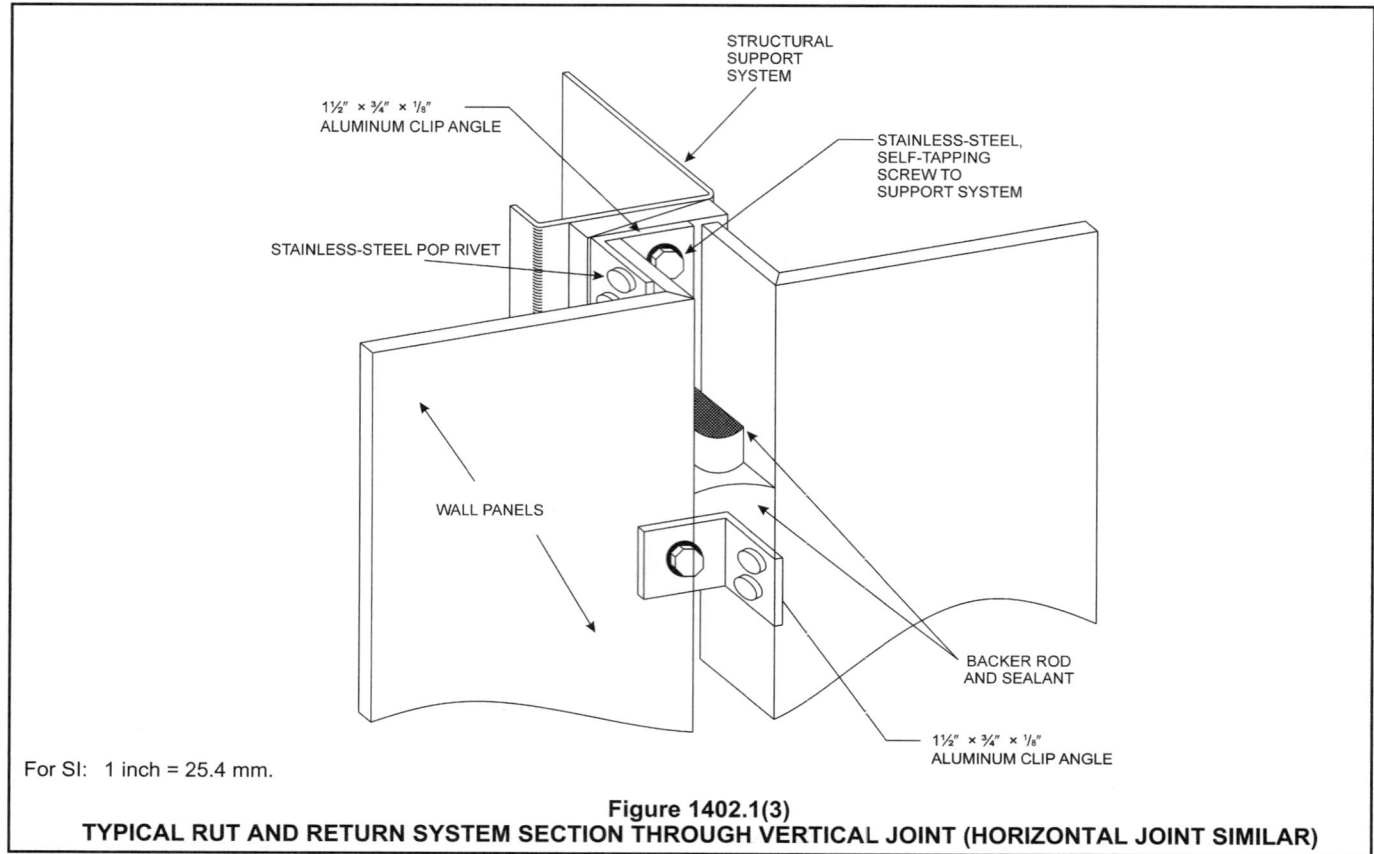

For SI: 1 inch = 25.4 mm.

Figure 1402.1(3)
TYPICAL RUT AND RETURN SYSTEM SECTION THROUGH VERTICAL JOINT (HORIZONTAL JOINT SIMILAR)

WATER-RESISTIVE BARRIER. A material behind an *exterior wall covering* that is intended to resist liquid water that has penetrated behind the exterior covering from further intruding into the *exterior wall* assembly.

❖ Protection of the building envelope from moisture intrusion is a primary concern. The ability of the water-resistive barrier to provide weather resistance and maintain the integrity of the building envelope is key to controlling water-based problems like mold, decay and deterioration of a structure. Water-resistive barriers are discussed in Section 1404.2.

SECTION 1403
PERFORMANCE REQUIREMENTS

1403.1 General. The provisions of this section shall apply to exterior walls, wall coverings and components thereof.

❖ The exterior walls of buildings provide support for the floor and roof systems; act as barriers against the outside environment; reduce heat loss from the interior and, in some instances, provide the sole resistance to such destructive forces as earthquakes and floods. This section provides references to other applicable sections of the code that are intended to detail specific requirements to accomplish these objectives. Additionally, this chapter contains detailed requirements related to the performance of exterior wall coverings and the exterior wall envelope.

1403.2 Weather protection. Exterior walls shall provide the building with a weather-resistant *exterior wall envelope*. The *exterior wall envelope* shall include flashing, as described in Section 1405.4. The *exterior wall envelope* shall be designed and constructed in such a manner as to prevent the accumulation of water within the wall assembly by providing a *water-resistive barrier* behind the exterior veneer, as described in Section 1404.2, and a means for draining water that enters the assembly to the exterior. Protection against condensation in the *exterior wall* assembly shall be provided in accordance with Section 1405.3.

Exceptions:

1. A weather-resistant *exterior wall envelope* shall not be required over concrete or masonry walls designed in accordance with Chapters 19 and 21, respectively.

2. Compliance with the requirements for a means of drainage, and the requirements of Sections 1404.2 and 1405.4, shall not be required for an *exterior wall envelope* that has been demonstrated through testing to resist wind-driven rain, including joints, penetrations and intersections with dissimilar materials, in accordance with ASTM E 331 under the following conditions:

    2.1. *Exterior wall envelope* test assemblies shall include at least one opening, one control joint, one wall/eave interface and one wall sill. All tested openings and penetrations shall be representative of the intended end-use configuration.

   2.2. *Exterior wall envelope* test assemblies shall be at least 4 feet by 8 feet (1219 mm by 2438 mm) in size.

   2.3. *Exterior wall envelope* assemblies shall be tested at a minimum differential pressure of 6.24 pounds per square foot (psf) (0.297 kN/m²).

   2.4. *Exterior wall envelope* assemblies shall be subjected to a minimum test exposure duration of 2 hours.

The *exterior wall envelope* design shall be considered to resist wind-driven rain where the results of testing indicate that water did not penetrate control joints in the *exterior wall* envelope, joints at the perimeter of openings or intersections of terminations with dissimilar materials.

  3. Exterior insulation and finish systems (EIFS) complying with Section 1408.4.1.

❖ All exterior walls of buildings must be protected against damage caused by precipitation, wind and other weather conditions.

The main text of this section prescribes three basic components of a weather-resistive exterior wall assembly: a water-resistive barrier installed over the building substrate; flashings at penetrations and terminations of the exterior wall finish and a means of draining moisture that may penetrate behind the finish back to the exterior.

Section 1404.2 is referenced for the requirements of the water-resistive barrier and Section 1405.3 is referenced for requirements for the flashings (see commentary, Sections 1404.2 and 1405.3). This section does not, however, contain a prescriptive requirement for the means of drainage to be provided. The method to provide the means of drainage is a performance criterion and must be evaluated based upon the ability to allow moisture that may penetrate behind the exterior wall covering to drain back to the exterior. This may be as complicated as a rain-screen, pressure-equalized type of exterior assembly or as simple as providing discontinuities or gaps between the surface of the substrate and the back side of the finish, such as through the use of noncorrodible furring.

For common types of construction, such as vinyl siding or brick veneer, the typical practice of installing building paper and weeps will comply with the intent of this section. Discontinuities between the exterior covering and substrate must be such that they encourage the flow of moisture via gravity or capillary action to a location where the water may exit, such as at flashings and weeps. The absence of a means of drainage may result in the accumulation of moisture that becomes trapped between the finish and the substrate. Over time, extended exposure to moisture may contribute to the degradation of the finish, building substrate or even the structural elements of the exterior wall.

Exception 1 states that where the exterior wall envelope is designed and constructed of concrete or masonry materials in accordance with the requirements of Chapters 19 and 21, respectively, the water-resistive barrier and means of drainage may be omitted. This is because the penetration of moisture behind the exterior wall finish is not detrimental to concrete and masonry substrates.

Exception 2 permits the use of exterior wall finishes that do not meet the prescriptive requirements of Section 1403.2, provided that the system, with penetration details, is tested for wind-driven rain resistance. The test specimen(s) must incorporate those penetration and termination details intended for use and the system will be limited to use with those details that successfully pass the test. The minimum panel size specified represents that which is commonly used in testing to ASTM E 331; however, this does not preclude the testing of larger panels if desired. The modifications to the test pressure differential and test duration are intended to represent more closely conditions that will be encountered in service. The pass/fail criteria is based upon the visual observation of moisture on the rear side of the wall assembly. In cavity-type assemblies, such as stud walls, this requires the observation of locations such as the rear face of the exterior wall sheathing and wall framing members for the presence of moisture. The test method is intended to assess the performance of the method(s) intended for use in sealing the interface between the termination of the exterior wall finish and the penetration items or abutting construction. The method is not necessarily intended to test the performance of the penetrating item.

Exception No. 3 relates to EIFS that is constructed in accordance with Section 1408.2. This material does not have the means for draining water that is called out in this section.

Walls designed and constructed in accordance with this chapter must also comply with the requirements of the *International Energy Conservation Code®* (IECC®).

1403.3 Structural. *Exterior walls*, and the associated openings, shall be designed and constructed to resist safely the superimposed loads required by Chapter 16.

❖ Exterior walls and their associated openings are required to resist all structural loads in accordance with the provisions of Chapter 16. This section is a correlative cross reference to emphasize the applicability of Chapter 16.

1403.4 Fire resistance. *Exterior walls* shall be fire-resistance rated as required by other sections of this code with opening protection as required by Chapter 7.

❖ The required fire-resistance rating of exterior walls is set forth in Tables 601 and 602. Table 602 is applicable to both load-bearing and nonload-bearing walls, since it addresses the prevention of fire spread from one building to an adjacent building. Load-bearing walls must comply with the greater of the requirements in Tables 601 and 602, based on the type of construction of the building. The size of openings in exterior walls is limited to prevent the spread of fire to other buildings. The commentary to Section 704 should also be reviewed when designing exterior walls. Trim on exterior walls is regulated by Section 1406.

The allowable size of openings in exterior walls is tabulated in Table 704.8. Section 714.3.7 specifies the required fire protection rating for opening protectives when required by Table 704.8. Where a fire-resistance rating is not required for the wall and unprotected openings are allowed, the glazing and the sash or frame may be of any material permitted by the code.

1403.5 Flood resistance. For buildings in flood hazard areas as established in Section 1612.3, *exterior walls* extending below the design flood elevation shall be resistant to water damage. Wood shall be pressure-preservative treated in accordance with AWPA U1 for the species, product and end use using a preservative *listed* in Section 4 of AWPA U1 or decay-resistant heartwood of redwood, black locust or cedar.

❖ Flood-resistant construction of exterior walls requires special consideration. Construction materials used in exterior walls of buildings that are located in flood hazard areas must be resistant to the effects of flooding. Some of the properties these materials must possess include resistance to prolonged contact with water, the ability to be cleaned and disinfected after the water recedes and negligible loss of physical properties after exposure to water. Additionally, systems must possess structural strength to resist the hydrodynamic and impact forces related to flooding. Wood construction materials in locations below the base flood elevation are required to be resistant to decay and exposure to water (see the commentary to Section 2303.1.8 for more discussion on preservative treatment of wood). Appendix G contains requirements specific to construction within areas that are prone to flooding.

1403.6 Flood resistance for high-velocity wave action areas. For buildings in flood hazard areas subject to high-velocity wave action as established in Section 1612.3, electrical, mechanical and plumbing system components shall not be mounted on or penetrate through exterior walls that are designed to break away under flood loads.

❖ ASCE 24 and Section 1612.3 provide for the option to design exterior walls to break away under flood loads. This section prohibits the installation of penetrations for electrical, plumbing or mechanical systems through such exterior walls. The breakaway provisions are a method of protecting the remaining building frame and building system. Installation of penetrations through these walls could serve in part to defeat that purpose or to reduce the ability of the breakaway system to perform as designed.

SECTION 1404
MATERIALS

1404.1 General. Materials used for the construction of exterior walls shall comply with the provisions of this section. Materials not prescribed herein shall be permitted, provided that any such alternative has been *approved*.

❖ Section 1404 contains performance requirements and detailed installation specifications for a number of common materials used in or on exterior walls of buildings. Reference is made to the applicable chapters that contain additional requirements. These chapters should be reviewed by the reader.

It is not the intent of the code to prevent the use of any material that is equivalent in performance to those specified in Section 1404.

1404.2 Water-resistive barrier. A minimum of one layer of No.15 asphalt felt, complying with ASTM D 226 for Type 1 felt or other *approved* materials, shall be attached to the studs or sheathing, with flashing as described in Section 1405.4, in such a manner as to provide a continuous *water-resistive barrier* behind the *exterior wall* veneer.

❖ Many exterior veneers provide weather resistance but may allow either penetration of water through joints or seams or the development of condensation to occur behind the veneer. To increase the weather resistance of the wall, a layer of asphalt felt or other approved material is required to be installed over the wall backing. The felt layer and the flashing provide a water-resistive barrier behind the exterior veneer. The water-resistive membrane must be attached to the studs or sheathing in a way that will not allow the penetration of water as a result of the attachment method.

1404.3 Wood. Exterior walls of wood construction shall be designed and constructed in accordance with Chapter 23.

❖ Chapter 23 contains general requirements regulating the use of wood in buildings; however, this section contains specific provisions for the minimum thicknesses of wood siding, wood backing surfaces and hardboard siding, as well as installation and nailing requirements for wood used as weatherboarding or a wall covering.

1404.3.1 Basic hardboard. Basic hardboard shall conform to the requirements of AHA A135.4.

❖ Basic hardboard is required to comply with AHA A135.4. This material is not intended for use as exposed siding or as a weather covering, but it is often a component of veneered wall construction.

1404.3.2 Hardboard siding. Hardboard siding shall conform to the requirements of AHA A135.6 and, where used structurally, shall be so identified by the *label* of an *approved* agency.

❖ Hardboard conforming to AHA A135.6 is manufactured in panels—usually 4 feet by 8 feet (1219 mm by 2438 mm) in size or in sizes simulating drop siding where the material is to be applied horizontally over structural sheathing. When used as a panel, hardboard siding is applied directly to the wood framing members as a backing material. Hardboard panels are not adequate to serve as a nailing base (spanning between studs) for coverings; however, they can be used to provide some lateral resistance to the wall [see Table 2308.9.3(6)]. Coverings are fastened directly at framing locations, through the panels.

1404.4 Masonry. Exterior walls of masonry construction shall be designed and constructed in accordance with this section and Chapter 21. Masonry units, mortar and metal accessories used in anchored and adhered veneer shall meet the physical requirements of Chapter 21. The backing of anchored and adhered veneer shall be of concrete, masonry, steel framing or wood framing.

❖ Chapter 21 contains the general requirements for the use of masonry in buildings. The material classification of masonry in this section includes materials such as concrete masonry; bricks of clay, shale or calcium silicate; clay-tile units; glazed masonry units; terra cotta; natural or cast stone and ceramic tile. Chapter 21 and this section contain provisions for the minimum thickness of masonry units, limitations on loading, heights of veneers, backing and attachment.

1404.5 Metal. Exterior walls of formed steel construction, structural steel or lightweight metal alloys shall be designed in accordance with Chapters 22 and 20, respectively.

❖ Chapters 20 and 22 contain the general requirements for use of metal in buildings. Chapters 20 and 22 as well as this section contain provisions for the minimum thickness of metal weather coverings, and requirements for protection from corrosion, installation and grounding.

1404.5.1 Aluminum siding. Aluminum siding shall conform to the requirements of AAMA 1402.

❖ Although aluminum siding is addressed in Chapter 20, it is more specifically addressed here as an exterior wall covering. Aluminum siding is required to conform to the referenced standard. The thickness, installation and protection provisions of this section are also applicable (see commentary, Sections 1405.11, 1405.11.1, 1405.11.2, 1405.11.3 and 1405.11.4).

1404.5.2 Cold-rolled copper. Copper shall conform to the requirements of ASTM B 370.

❖ This material is commonly used for roofing, flashing, gutters, downspouts and general sheet metal work. ASTM B 370 provides the specification for both copper sheet and copper strip materials.

1404.5.3 Lead-coated copper. Lead-coated copper shall conform to the requirements of ASTM B 101.

❖ This material is commonly used for roofing, flashing, gutters, downspouts and general sheet metal work. The lead coating is applied by a hot-dipped method. ASTM B 101 specifies that the lead must uniformly coat both surfaces, edges and ends of the copper.

1404.6 Concrete. Exterior walls of concrete construction shall be designed and constructed in accordance with Chapter 19.

❖ Chapter 19 contains the general requirements for use of concrete in buildings, including provisions for the minimum thickness of concrete units used as veneers and wall coverings and their installation and protection.

1404.7 Glass-unit masonry. Exterior walls of glass-unit masonry shall be designed and constructed in accordance with Chapter 21.

❖ Structural glass block is generally regulated by the provisions of Sections 2103 and 2104; however, structural glass veneers used as wall coverings must comply with the thickness requirement of Table 1405.2. Criteria for the support of glass unit masonry on wood frame construction are also contained in Exception 4 of Section 2304.12.

1404.8 Plastics. Plastic panel, apron or spandrel walls as defined in this code shall not be limited in thickness, provided that such plastics and their assemblies conform to the requirements of Chapter 26 and are constructed of *approved* weather-resistant materials of adequate strength to resist the wind loads for cladding specified in Chapter 16.

❖ This section provides for the use of plastic panels for exterior walls. The code uses the term "light-transmitting plastic" to apply to plastics used for exterior wall panels and roof panels, as well as for glazing and skylights. The panels must meet the requirements of Chapter 26, as well as certain requirements of Chapter 14 for weather resistance and Chapter 16 for structural capabilities.

1404.9 Vinyl siding. Vinyl siding shall be certified and labeled as conforming to the requirements of ASTM D 3679 by an *approved* quality control agency.

❖ Plastics are addressed in Chapter 26; however, polyvinyl chloride (PVC) siding is specifically addressed here as an exterior wall covering. PVC siding is required to conform to the provisions of ASTM D 3679. Installation of vinyl siding is prescribed by Section 1405.14 (see the commentary to Section 1405.14 for additional discussion on the installation of vinyl siding). Note that the product must be certified and labeled, which means that the manufacturer must have regular inspections by an approved quality control agency.

1404.10 Fiber-cement siding. Fiber-cement siding shall conform to the requirements of ASTM C 1186, Type A, and shall be so identified on labeling listing an *approved* quality control agency.

❖ Fiber-cement siding is also a material used for weather-resistant siding. It is manufactured from fiber-reinforced cement, and the code permits its use as either panel siding or horizontal lap siding. The material standard is ASTM C 1186. Note that the product is required to bear a label, which means that the manufacturer must have regular inspections by a third-party inspection quality control agency.

1404.11 Exterior insulation and finish systems. Exterior insulation and finish systems (EIFS) and exterior insulation and finish systems (EIFS) with drainage shall comply with Section 1408.

❖ Exterior insulation and finish systems (EIFS) and exterior insulation and finish systems (EIFS) with drainage are relatively complex exterior veneers that require great care and attention regarding the materials used and the installation details. This exterior veneer is now dealt with in Section 1408. All of the performance criteria, installation and design are covered in this section.

SECTION 1405
INSTALLATION OF WALL COVERINGS

1405.1 General. *Exterior wall coverings* shall be designed and constructed in accordance with the applicable provisions of this section.

❖ Exterior wall coverings used to provide weather protection to the structure are required to comply with all the applicable provisions contained in Section 1405.

1405.2 Weather protection. *Exterior walls* shall provide weather protection for the building. The materials of the minimum nominal thickness specified in Table 1405.2 shall be acceptable as *approved* weather coverings.

❖ The exterior walls of a structure must be designed and constructed to provide for the health and safety of the occupants and to protect the structure from the detrimental effects of weather exposure.

   This section introduces Table 1405.2, which is a tabulation of the minimum acceptable nominal thicknesses for a number of common weather coverings.

TABLE 1405.2. See next column.

❖ This table should be used in addition to all other applicable chapters and requirements of the code for the specific material listed. Testing and experience have determined that the minimum tabulated thicknesses will be durable and protect the building against the elements for relatively long periods of time when attached and maintained as required.

   Note a of Table 1405.2 allows material less than 0.5-inch-thick (12.7 mm) wood siding to be used, provided that it is installed over an approved sheathing conforming to the requirements of Section 2304.6.1 (see commentary, Section 2304.6.1).

   Note b of Table 1405.2 establishes that the minimum permitted nominal thickness is based on the narrowest solid thickness of material, exclusive of texture.

   Section 2304.6.1 allows fiberboard and particleboard for sheathing. Although permitted as sheathing, fiberboard and particleboard materials, and other similar materials, do not provide a nailing base for weatherboarding. When used over such materials, 0.5-inch (12.7 mm) wood siding is to be fastened to the wood-framing members or otherwise adequately secured. Wood siding having a thickness of less than 0.5 inch (12.7 mm) is not permitted to be fastened to foam plastic and other sheathing materials not having ade-

TABLE 1405.2
MINIMUM THICKNESS OF WEATHER COVERINGS

COVERING TYPE	MINIMUM THICKNESS (inches)
Adhered masonry veneer	0.25
Aluminum siding	0.019
Anchored masonry veneer	2.625
Asbestos-cement boards	0.125
Asbestos shingles	0.156
Cold-rolled copper[d]	0.0216 nominal
Copper shingles[d]	0.0162 nominal
Exterior plywood (with sheathing)	0.313
Exterior plywood (without sheathing)	See Section 2304.6
Fiber cement lap siding	0.25[c]
Fiber cement panel siding	0.25[c]
Fiberboard siding	0.5
Glass-fiber reinforced concrete panels	0.375
Hardboard siding[c]	0.25
High-yield copper[d]	0.0162 nominal
Lead-coated copper[d]	0.0216 nominal
Lead-coated high-yield copper	0.0162 nominal
Marble slabs	1
Particleboard (with sheathing)	See Section 2304.6
Particleboard (without sheathing)	See Section 2304.6
Precast stone facing	0.625
Steel (approved corrosion resistant)	0.0149
Stone (cast artificial)	1.5
Stone (natural)	2
Structural glass	0.344
Stucco or exterior cement plaster	
Three-coat work over:	
Metal plaster base	0.875[b]
Unit masonry	0.625[b]
Cast-in-place or precast concrete	0.625[b]
Two-coat work over:	
Unit masonry	0.5[b]
Cast-in-place or precast concrete	0.375[b]
Terra cotta (anchored)	1
Terra cotta (adhered)	0.25
Vinyl siding	0.035
Wood shingles	0.375
Wood siding (without sheathing)[a]	0.5

For SI: 1 inch = 25.4 mm.

a. Wood siding of thicknesses less than 0.5 inch shall be placed over sheathing that conforms to Section 2304.6.

b. Exclusive of texture.

c. As measured at the bottom of decorative grooves.

d. 16 ounces per square foot for cold-rolled copper and lead-coated copper, 12 ounces per square foot for copper shingles, high-yield copper and lead-coated high-yield copper.

quate strength as a nailing base spanning between studs or framing members.

1405.3 Vapor retarders. Class I or II vapor retarders shall be provided on the interior side of frame walls in Zones 5, 6, 7, 8 and Marine 4.

Exceptions:

1. Basement walls.

2. Below-grade portion of any wall.

3. Construction where moisture or its freezing will not damage the materials.

❖ New to the 2009 code are requirements for vapor retarders. The purpose of these requirements is to provide prescriptive methods for moisture control. This subject has been added to the code and removed from the IECC, because moisture control is not an energy conservation issue. The code now contains three different vapor retarder classes, based upon the vapor permeability of the material. These are defined in Chapter 2. The basic requirement is for Class I or II vapor retarders to be installed on the interior side of frame walls in Climate Zones 5, 6, 7, 8, and Marine 4. These climate zones are defined in the International Energy Conservation Code. Figure 301.1 and Table 301.1 of the IECC provide information regarding the climate zones for all geographic locations in the United States (see Commentary Figure 1405.3). These are reproduced here for convenience. Also, Tables 301.3(1) and 301.3(2) define international climate zones outside of the United States. These are reprinted in this commentary as well (see Commentary Table 1405.3). As can be seen by studying Figure 301.1, Climate Zones 5, 6, 7, 8 and Marine 4 are in the middle to northern portions of the continental United States, and Alaska. These are areas where colder temperatures can be expected in winter months, which cause moisture from the interior of the building to condensate in the exterior walls of the building. A Class I or II vapor retarder is therefore called for on the interior side of the exterior wall.

1405.3.1 Class III vapor retarders. Class III vapor retarders shall be permitted where any one of the conditions in Table 1405.3.1 is met.

❖ Wall assemblies can be designed and constructed to dry inwards, outwards and to both sides in all climate zones. This section allows more flexibility in the design and construction of moisture forgiving wall systems. These requirements recognize that many common materials function to various degrees to slow the passage of moisture. In many situations common materials such as the kraft facing on a fiberglass batt or latex paint may serve to retard moisture sufficiently. In par-

ticular, the "standard" sheet of polyethylene is usually not required as a vapor retarder in walls. This section therefore allows the use of Class III vapor retarders in lieu of the Class I or II otherwise called out in Section 1405.3. Classes of vapor retarders are defined in Chapter 2.

The Class III vapor retarders allow more moisture vapor to pass through them. These are allowed to be used with exterior wall assemblies described in Table 1405.3.1 that can be expected to dry. Section 1405.3.2 describes materials that are deemed to be Class III vapor retarders, and Section 1405.3.3 describes the attributes of cladding that make it "vented" as called out in Table 1405.3.1.

TABLE 1405.3.1
CLASS III VAPOR RETARDERS

ZONE	CLASS III VAPOR RETARDERS PERMITTED FOR:[a]
Marine 4	Vented cladding over OSB Vented cladding over plywood Vented cladding over fiberboard Vented cladding over gypsum Insulated sheathing with R-value ≥ R2.5 over 2×4 wall Insulated sheathing with R-value ≥ R3.75 over 2×6 wall
5	Vented cladding over OSB Vented cladding over plywood Vented cladding over fiberboard Vented cladding over gypsum Insulated sheathing with R-value ≥ R5 over 2×4 wall Insulated sheathing with R-value ≥ R7.5 over 2×6 wall
6	Vented cladding over fiberboard Vented cladding over gypsum Insulated sheathing with R-value ≥ R7.5 over 2×4 wall Insulated sheathing with R-value ≥ R11.25 over 2×6 wall
7 and 8	Insulated sheathing with R-value ≥ R10 over 2×4 wall Insulated sheathing with R-value ≥ R15 over 2×6 wall

For SI: 1 pound per cubic foot = 16 kg/m³.

a. Spray foam with a minimum density of 2 lbs/ft³ applied to the interior cavity side of OSB, plywood, fiberboard, insulating sheathing or gypsum is deemed to meet the insulating sheathing requirement where the spray foam R-value meets or exceeds the specified insulating sheathing R-value.

❖ This table describes the situations where it is permissible to use Class III vapor retarders instead of Class I or II. Class III vapor retarders allow more moisture vapor to pass through them. These are allowed to be used with exterior wall assemblies described in Table 1405.3.1 that can be expected to dry. The table provides types of wall assemblies that can be used with Class III vapor retarders in different climate zones. The climate zones are determined based upon Chapter 3 of the IECC.

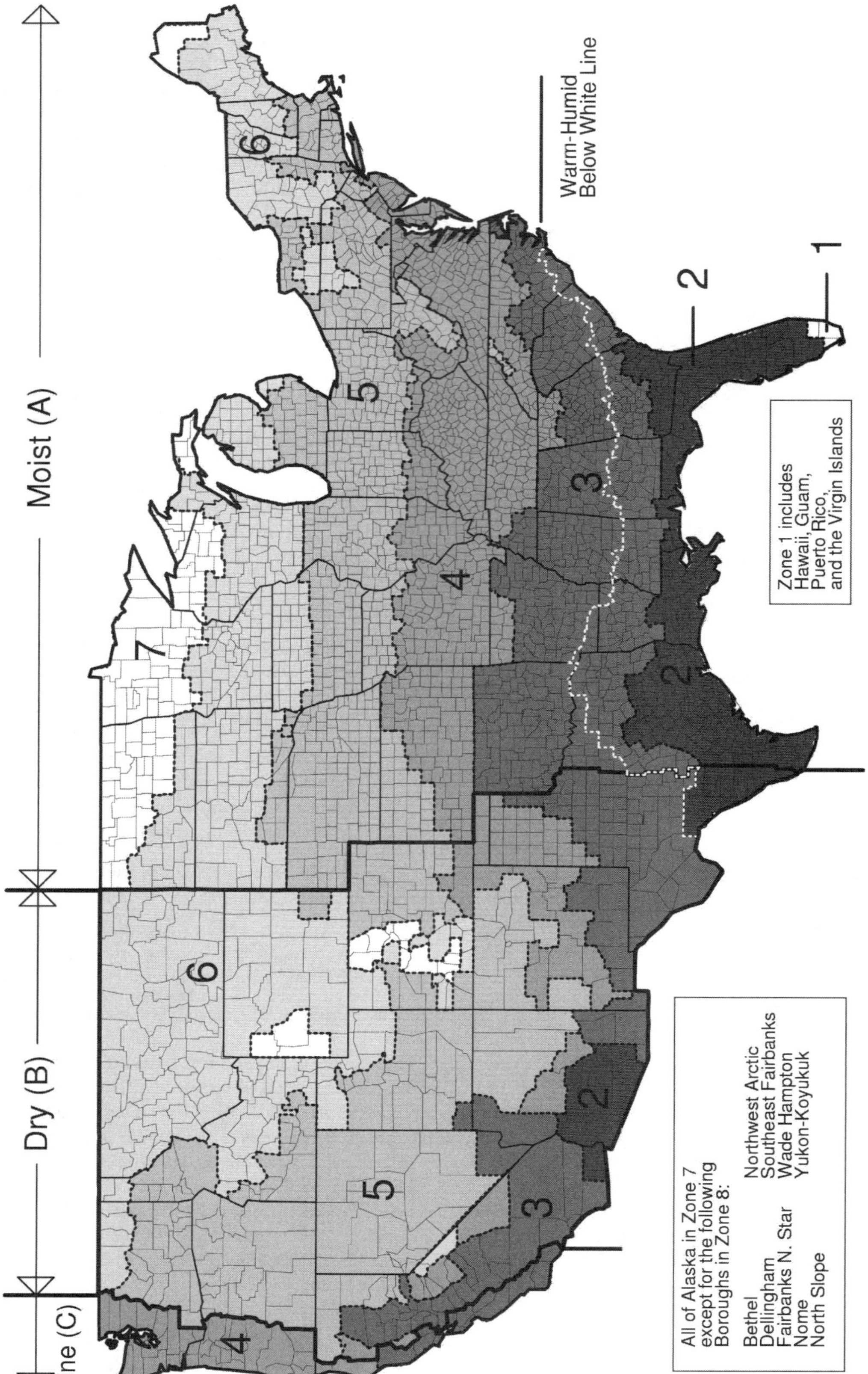
Marine (C)
Dry (B)
Moist (A)
Warm-Humid Below White Line
All of Alaska in Zone 7 except for the following Boroughs in Zone 8:
Bethel
Dellingham
Fairbanks N. Star
Nome
North Slope
Northwest Arctic
Southeast Fairbanks
Wade Hampton
Yukon-Koyukuk
Zone 1 includes Hawaii, Guam, Puerto Rico, and the Virgin Islands
1
2
3
4
5
6
7

FIGURE 1405.3
CLIMATE ZONES

Table 1450.3
CLIMATE ZONES, MOISTURE REGIMES, AND WARM-HUMID DESIGNATIONS BY STATE, COUNTY AND TERRITORY

Note: Table 301.1 in the 2006 edition has been replaced in its entirety. Margin lines are omitted for clarity.

Key: A – Moist, B – Dry, C – Marine. Absence of moisture designation indicates moisture regime is irrelevant.
Asterisk (*) indicates a warm-humid location.

US STATES

ALABAMA

3A Autauga*
2A Baldwin*
3A Barbour*
3A Bibb
3A Blount
3A Bullock*
3A Butler*
3A Calhoun
3A Chambers
3A Cherokee
3A Chilton
3A Choctaw*
3A Clarke*
3A Clay
3A Cleburne
3A Coffee*
3A Colbert
3A Conecuh*
3A Coosa
3A Covington*
3A Crenshaw*
3A Cullman
3A Dale*
3A Dallas*
3A DeKalb
3A Elmore*
3A Escambia*
3A Etowah
3A Fayette
3A Franklin
3A Geneva*
3A Greene
3A Hale
3A Henry*

3A Houston*
3A Jackson
3A Jefferson
3A Lamar
3A Lauderdale
3A Lawrence
3A Lee
3A Limestone
3A Lowndes*
3A Macon*
3A Madison
3A Marengo*
3A Marion
3A Marshall
2A Mobile*
3A Monroe*
3A Montgomery*
3A Morgan
3A Perry*
3A Pickens
3A Pike*
3A Randolph
3A Russell*
3A Shelby
3A St. Clair
3A Sumter
3A Talladega
3A Tallapoosa
3A Tuscaloosa
3A Walker
3A Washington*
3A Wilcox*
3A Winston

ALASKA

7 Aleutians East

7 Aleutians West
7 Anchorage
8 Bethel
7 Bristol Bay
7 Denali
8 Dillingham
8 Fairbanks North Star
7 Haines
7 Juneau
7 Kenai Peninsula
7 Ketchikan Gateway
7 Kodiak Island
7 Lake and Peninsula
7 Matanuska-Susitna
8 Nome
8 North Slope
8 Northwest Arctic
7 Prince of Wales-Outer Ketchikan
7 Sitka
7 Skagway-Hoonah-Angoon
8 Southeast Fairbanks
7 Valdez-Cordova
8 Wade Hampton
7 Wrangell-Petersburg
7 Yakutat
8 Yukon-Koyukuk

ARIZONA

5B Apache
3B Cochise
5B Coconino
4B Gila
3B Graham

3B Greenlee
2B La Paz
2B Maricopa
3B Mohave
5B Navajo
2B Pima
2B Pinal
3B Santa Cruz
4B Yavapai
2B Yuma

ARKANSAS

3A Arkansas
3A Ashley
4A Baxter
4A Benton
4A Boone
3A Bradley
3A Calhoun
4A Carroll
3A Chicot
3A Clark
3A Clay
3A Cleburne
3A Cleveland
3A Columbia*
3A Conway
3A Craighead
3A Crawford
3A Crittenden
3A Cross
3A Dallas
3A Desha
3A Drew
3A Faulkner
3A Franklin

4A Fulton
3A Garland
3A Grant
3A Greene
3A Hempstead*
3A Hot Spring
3A Howard
3A Independence
4A Izard
3A Jackson
3A Jefferson
3A Johnson
3A Lafayette*
3A Lawrence
3A Lee
3A Lincoln
3A Little River*
3A Logan
3A Lonoke
4A Madison
4A Marion
3A Miller*
3A Mississippi
3A Monroe
3A Montgomery
3A Nevada
4A Newton
3A Ouachita
3A Perry
3A Phillips
3A Pike
3A Poinsett
3A Polk
3A Pope
3A Prairie

(continued)

Table 1450.3—continued
CLIMATE ZONES, MOISTURE REGIMES, AND WARM-HUMID DESIGNATIONS BY STATE, COUNTY AND TERRITORY

3A Pulaski

3A Randolph

3A Saline

3A Scott

4A Searcy

3A Sebastian

3A Sevier*

3A Sharp

3A St. Francis

4A Stone

3A Union*

3A Van Buren

4A Washington

3A White

3A Woodruff

3A Yell

CALIFORNIA

3C Alameda

6B Alpine

4B Amador

3B Butte

4B Calaveras

3B Colusa

3B Contra Costa

4C Del Norte

4B El Dorado

3B Fresno

3B Glenn

4C Humboldt

2B Imperial

4B Inyo

3B Kern

3B Kings

4B Lake

5B Lassen

3B Los Angeles

3B Madera

3C Marin

4B Mariposa

3C Mendocino

3B Merced

5B Modoc

6B Mono

3C Monterey

3C Napa

5B Nevada

3B Orange

3B Placer

5B Plumas

3B Riverside

3B Sacramento

3C San Benito

3B San Bernardino

3B San Diego

3C San Francisco

3B San Joaquin

3C San Luis Obispo

3C San Mateo

3C Santa Barbara

3C Santa Clara

3C Santa Cruz

3B Shasta

5B Sierra

5B Siskiyou

3B Solano

3C Sonoma

3B Stanislaus

3B Sutter

3B Tehama

4B Trinity

3B Tulare

4B Tuolumne

3C Ventura

3B Yolo

3B Yuba

COLORADO

5B Adams

6B Alamosa

5B Arapahoe

6B Archuleta

4B Baca

5B Bent

5B Boulder

6B Chaffee

5B Cheyenne

7 Clear Creek

6B Conejos

6B Costilla

5B Crowley

6B Custer

5B Delta

5B Denver

6B Dolores

5B Douglas

6B Eagle

5B Elbert

5B El Paso

5B Fremont

5B Garfield

5B Gilpin

7 Grand

7 Gunnison

7 Hinsdale

5B Huerfano

7 Jackson

5B Jefferson

5B Kiowa

5B Kit Carson

7 Lake

5B La Plata

5B Larimer

4B Las Animas

5B Lincoln

5B Logan

5B Mesa

7 Mineral

6B Moffat

5B Montezuma

5B Montrose

5B Morgan

4B Otero

6B Ouray

7 Park

5B Phillips

7 Pitkin

5B Prowers

5B Pueblo

6B Rio Blanco

7 Rio Grande

7 Routt

6B Saguache

7 San Juan

6B San Miguel

5B Sedgwick

7 Summit

5B Teller

5B Washington

5B Weld

5B Yuma

CONNECTICUT

5A (all)

DELAWARE

4A (all)

DISTRICT OF COLUMBIA

4A (all)

FLORIDA

2A Alachua*

2A Baker*

2A Bay*

2A Bradford*

2A Brevard*

1A Broward*

2A Calhoun*

2A Charlotte*

2A Citrus*

2A Clay*

2A Collier*

2A Columbia*

2A DeSoto*

2A Dixie*

2A Duval*

2A Escambia*

2A Flagler*

2A Franklin*

2A Gadsden*

2A Gilchrist*

2A Glades*

2A Gulf*

2A Hamilton*

2A Hardee*

2A Hendry*

2A Hernando*

2A Highlands*

2A Hillsborough*

2A Holmes*

2A Indian River*

2A Jackson*

2A Jefferson*

2A Lafayette*

2A Lake*

2A Lee*

2A Leon*

2A Levy*

2A Liberty*

2A Madison*

2A Manatee*

2A Marion*

2A Martin*

1A Miami-Dade*

1A Monroe*

2A Nassau*

2A Okaloosa*

2A Okeechobee*

2A Orange*

2A Osceola*

2A Palm Beach*

2A Pasco*

(continued)

Table 1450.3—continued
CLIMATE ZONES, MOISTURE REGIMES, AND WARM-HUMID DESIGNATIONS BY STATE, COUNTY AND TERRITORY

2A Pinellas*	2A Charlton*	4A Gordon	3A Morgan	4A Union
2A Polk*	2A Chatham*	2A Grady*	4A Murray	3A Upson
2A Putnam*	3A Chattahoochee*	3A Greene	3A Muscogee	4A Walker
2A Santa Rosa*	4A Chattooga	3A Gwinnett	3A Newton	3A Walton
2A Sarasota*	3A Cherokee	4A Habersham	3A Oconee	2A Ware*
2A Seminole*	3A Clarke	4A Hall	3A Oglethorpe	3A Warren
2A St. Johns*	3A Clay*	3A Hancock	3A Paulding	3A Washington
2A St. Lucie*	3A Clayton	3A Haralson	3A Peach*	2A Wayne*
2A Sumter*	2A Clinch*	3A Harris	4A Pickens	3A Webster*
2A Suwannee*	3A Cobb	3A Hart	2A Pierce*	3A Wheeler*
2A Taylor*	3A Coffee*	3A Heard	3A Pike	4A White
2A Union*	2A Colquitt*	3A Henry	3A Polk	4A Whitfield
2A Volusia*	3A Columbia	3A Houston*	3A Pulaski*	3A Wilcox*
2A Wakulla*	2A Cook*	3A Irwin*	3A Putnam	3A Wilkes
2A Walton*	3A Coweta	3A Jackson	3A Quitman*	3A Wilkinson
2A Washington*	3A Crawford	3A Jasper	4A Rabun	3A Worth*
GEORGIA	3A Crisp*	2A Jeff Davis*	3A Randolph*	**HAWAII**
2A Appling*	4A Dade	3A Jefferson	3A Richmond	1A (all)*
2A Atkinson*	4A Dawson	3A Jenkins*	3A Rockdale	**IDAHO**
2A Bacon*	2A Decatur*	3A Johnson*	3A Schley*	5B Ada
2A Baker*	3A DeKalb	3A Jones	3A Screven*	6B Adams
3A Baldwin	3A Dodge*	3A Lamar	2A Seminole*	6B Bannock
4A Banks	3A Dooly*	2A Lanier*	3A Spalding	6B Bear Lake
3A Barrow	3A Dougherty*	3A Laurens*	4A Stephens	5B Benewah
3A Bartow	3A Douglas	3A Lee*	3A Stewart*	6B Bingham
3A Ben Hill*	3A Early*	2A Liberty*	3A Sumter*	6B Blaine
2A Berrien*	2A Echols*	3A Lincoln	3A Talbot	6B Boise
3A Bibb	2A Effingham*	2A Long*	3A Taliaferro	6B Bonner
3A Bleckley*	3A Elbert	2A Lowndes*	2A Tattnall*	6B Bonneville
2A Brantley*	3A Emanuel*	4A Lumpkin	3A Taylor*	6B Boundary
2A Brooks*	2A Evans*	3A Macon*	3A Telfair*	6B Butte
2A Bryan*	4A Fannin	3A Madison	3A Terrell*	6B Camas
3A Bulloch*	3A Fayette	3A Marion*	2A Thomas*	5B Canyon
3A Burke	4A Floyd	3A McDuffie	3A Tift*	6B Caribou
3A Butts	3A Forsyth	2A McIntosh*	2A Toombs*	5B Cassia
3A Calhoun*	4A Franklin	3A Meriwether	4A Towns	6B Clark
2A Camden*	3A Fulton	2A Miller*	3A Treutlen*	5B Clearwater
3A Candler*	4A Gilmer	2A Mitchell*	3A Troup	6B Custer
3A Carroll	3A Glascock	3A Monroe	3A Turner*	5B Elmore
4A Catoosa	2A Glynn*	3A Montgomery*	3A Twiggs*	6B Franklin

(continued)

Table 1450.3—continued
CLIMATE ZONES, MOISTURE REGIMES, AND WARM-HUMID DESIGNATIONS BY STATE, COUNTY AND TERRITORY

6B Fremont	4A Crawford	4A Madison	4A White	5A Howard
5B Gem	5A Cumberland	4A Marion	5A Whiteside	5A Huntington
5B Gooding	5A DeKalb	5A Marshall	5A Will	4A Jackson
5B Idaho	5A De Witt	5A Mason	4A Williamson	5A Jasper
6B Jefferson	5A Douglas	4A Massac	5A Winnebago	5A Jay
5B Jerome	5A DuPage	5A McDonough	5A Woodford	4A Jefferson
5B Kootenai	5A Edgar	5A McHenry		4A Jennings
5B Latah	4A Edwards	5A McLean	**INDIANA**	5A Johnson
6B Lemhi	4A Effingham	5A Menard	5A Adams	4A Knox
5B Lewis	4A Fayette	5A Mercer	5A Allen	5A Kosciusko
5B Lincoln	5A Ford	4A Monroe	5A Bartholomew	5A Lagrange
6B Madison	4A Franklin	4A Montgomery	5A Benton	5A Lake
5B Minidoka	5A Fulton	5A Morgan	5A Blackford	5A La Porte
5B Nez Perce	4A Gallatin	5A Moultrie	5A Boone	4A Lawrence
6B Oneida	5A Greene	5A Ogle	4A Brown	5A Madison
5B Owyhee	5A Grundy	5A Peoria	5A Carroll	5A Marion
5B Payette	4A Hamilton	4A Perry	5A Cass	5A Marshall
5B Power	5A Hancock	5A Piatt	4A Clark	4A Martin
5B Shoshone	4A Hardin	5A Pike	5A Clay	5A Miami
6B Teton	5A Henderson	4A Pope	5A Clinton	4A Monroe
5B Twin Falls	5A Henry	4A Pulaski	4A Crawford	5A Montgomery
6B Valley	5A Iroquois	5A Putnam	4A Daviess	5A Morgan
5B Washington	4A Jackson	4A Randolph	4A Dearborn	5A Newton
ILLINOIS	4A Jasper	4A Richland	5A Decatur	5A Noble
5A Adams	4A Jefferson	5A Rock Island	5A De Kalb	4A Ohio
4A Alexander	5A Jersey	4A Saline	5A Delaware	4A Orange
4A Bond	5A Jo Daviess	5A Sangamon	4A Dubois	5A Owen
5A Boone	4A Johnson	5A Schuyler	5A Elkhart	5A Parke
5A Brown	5A Kane	5A Scott	5A Fayette	4A Perry
5A Bureau	5A Kankakee	4A Shelby	4A Floyd	4A Pike
5A Calhoun	5A Kendall	5A Stark	5A Fountain	5A Porter
5A Carroll	5A Knox	4A St. Clair	5A Franklin	4A Posey
5A Cass	5A Lake	5A Stephenson	5A Fulton	5A Pulaski
5A Champaign	5A La Salle	5A Tazewell	4A Gibson	5A Putnam
4A Christian	4A Lawrence	4A Union	5A Grant	5A Randolph
5A Clark	5A Lee	5A Vermilion	4A Greene	4A Ripley
4A Clay	5A Livingston	4A Wabash	5A Hamilton	5A Rush
4A Clinton	5A Logan	5A Warren	5A Hancock	4A Scott
5A Coles	5A Macon	4A Washington	4A Harrison	5A Shelby
5A Cook	4A Macoupin	4A Wayne	5A Hendricks	4A Spencer
			5A Henry	

(continued)

Table 1450.3—continued
CLIMATE ZONES, MOISTURE REGIMES, AND WARM-HUMID DESIGNATIONS BY STATE, COUNTY AND TERRITORY

5A Starke	5A Clarke	6A Lyon	**KANSAS**	4A Harvey
5A Steuben	6A Clay	5A Madison	4A Allen	4A Haskell
5A St. Joseph	6A Clayton	5A Mahaska	4A Anderson	4A Hodgeman
4A Sullivan	5A Clinton	5A Marion	4A Atchison	4A Jackson
4A Switzerland	5A Crawford	5A Marshall	4A Barber	4A Jefferson
5A Tippecanoe	5A Dallas	5A Mills	4A Barton	5A Jewell
5A Tipton	5A Davis	6A Mitchell	4A Bourbon	4A Johnson
5A Union	5A Decatur	5A Monona	4A Brown	4A Kearny
4A Vanderburgh	6A Delaware	5A Monroe	4A Butler	4A Kingman
5A Vermillion	5A Des Moines	5A Montgomery	4A Chase	4A Kiowa
5A Vigo	6A Dickinson	5A Muscatine	4A Chautauqua	4A Labette
5A Wabash	5A Dubuque	6A O'Brien	4A Cherokee	5A Lane
5A Warren	6A Emmet	6A Osceola	5A Cheyenne	4A Leavenworth
4A Warrick	6A Fayette	5A Page	4A Clark	4A Lincoln
4A Washington	6A Floyd	6A Palo Alto	4A Clay	4A Linn
5A Wayne	6A Franklin	6A Plymouth	5A Cloud	5A Logan
5A Wells	5A Fremont	6A Pocahontas	4A Coffey	4A Lyon
5A White	5A Greene	5A Polk	4A Comanche	4A Marion
5A Whitley	6A Grundy	5A Pottawattamie	4A Cowley	4A Marshall
	5A Guthrie	5A Poweshiek	4A Crawford	4A McPherson
IOWA	6A Hamilton	5A Ringgold	5A Decatur	4A Meade
5A Adair	6A Hancock	6A Sac	4A Dickinson	4A Miami
5A Adams	6A Hardin	5A Scott	4A Doniphan	5A Mitchell
6A Allamakee	5A Harrison	5A Shelby	4A Douglas	4A Montgomery
5A Appanoose	5A Henry	6A Sioux	4A Edwards	4A Morris
5A Audubon	6A Howard	5A Story	4A Elk	4A Morton
5A Benton	6A Humboldt	5A Tama	5A Ellis	4A Nemaha
6A Black Hawk	6A Ida	5A Taylor	4A Ellsworth	4A Neosho
5A Boone	5A Iowa	5A Union	4A Finney	5A Ness
6A Bremer	5A Jackson	5A Van Buren	4A Ford	5A Norton
6A Buchanan	5A Jasper	5A Wapello	4A Franklin	4A Osage
6A Buena Vista	5A Jefferson	5A Warren	4A Geary	5A Osborne
6A Butler	5A Johnson	5A Washington	5A Gove	4A Ottawa
6A Calhoun	5A Jones	5A Wayne	5A Graham	4A Pawnee
5A Carroll	5A Keokuk	6A Webster	4A Grant	5A Phillips
5A Cass	6A Kossuth	6A Winnebago	4A Gray	4A Pottawatomie
5A Cedar	5A Lee	6A Winneshiek	5A Greeley	4A Pratt
6A Cerro Gordo	5A Linn	5A Woodbury	4A Greenwood	5A Rawlins
6A Cherokee	5A Louisa	6A Worth	5A Hamilton	4A Reno
6A Chickasaw	5A Lucas	6A Wright	4A Harper	5A Republic

(continued)

Table 1450.3—continued
CLIMATE ZONES, MOISTURE REGIMES, AND WARM-HUMID DESIGNATIONS BY STATE, COUNTY AND TERRITORY

4A Rice
4A Riley
5A Rooks
4A Rush
4A Russell
4A Saline
5A Scott
4A Sedgwick
4A Seward
4A Shawnee
5A Sheridan
5A Sherman
5A Smith
4A Stafford
4A Stanton
4A Stevens
4A Sumner
5A Thomas
5A Trego
4A Wabaunsee
5A Wallace
4A Washington
5A Wichita
4A Wilson
4A Woodson
4A Wyandotte

KENTUCKY

4A (all)

LOUISIANA

2A Acadia*
2A Allen*
2A Ascension*
2A Assumption*
2A Avoyelles*
2A Beauregard*
3A Bienville*
3A Bossier*
3A Caddo*
2A Calcasieu*
3A Caldwell*

2A Cameron*
3A Catahoula*
3A Claiborne*
3A Concordia*
3A De Soto*
2A East Baton Rouge*
3A East Carroll
2A East Feliciana*
2A Evangeline*
3A Franklin*
3A Grant*
2A Iberia*
2A Iberville*
3A Jackson*
2A Jefferson*
2A Jefferson Davis*
2A Lafayette*
2A Lafourche*
3A La Salle*
3A Lincoln*
2A Livingston*
3A Madison*
3A Morehouse
3A Natchitoches*
2A Orleans*
3A Ouachita*
2A Plaquemines*
2A Pointe Coupee*
2A Rapides*
3A Red River*
3A Richland*
3A Sabine*
2A St. Bernard*
2A St. Charles*
2A St. Helena*
2A St. James*
2A St. John the Baptist*
2A St. Landry*
2A St. Martin*

2A St. Mary*
2A St. Tammany*
2A Tangipahoa*
3A Tensas*
2A Terrebonne*
3A Union*
2A Vermilion*
3A Vernon*
2A Washington*
3A Webster*
2A West Baton Rouge*
3A West Carroll
2A West Feliciana*
3A Winn*

MAINE

6A Androscoggin
7 Aroostook
6A Cumberland
6A Franklin
6A Hancock
6A Kennebec
6A Knox
6A Lincoln
6A Oxford
6A Penobscot
6A Piscataquis
6A Sagadahoc
6A Somerset
6A Waldo
6A Washington
6A York

MARYLAND

4A Allegany
4A Anne Arundel
4A Baltimore
4A Baltimore (city)
4A Calvert
4A Caroline
4A Carroll

4A Cecil
4A Charles
4A Dorchester
4A Frederick
5A Garrett
4A Harford
4A Howard
4A Kent
4A Montgomery
4A Prince George's
4A Queen Anne's
4A Somerset
4A St. Mary's
4A Talbot
4A Washington
4A Wicomico
4A Worcester

MASSACHU-SETTS

5A (all)

MICHIGAN

6A Alcona
6A Alger
5A Allegan
6A Alpena
6A Antrim
6A Arenac
7 Baraga
5A Barry
5A Bay
6A Benzie
5A Berrien
5A Branch
5A Calhoun
5A Cass
6A Charlevoix
6A Cheboygan
7 Chippewa
6A Clare
5A Clinton

6A Crawford
6A Delta
6A Dickinson
5A Eaton
6A Emmet
5A Genesee
6A Gladwin
7 Gogebic
6A Grand Traverse
5A Gratiot
5A Hillsdale
7 Houghton
6A Huron
5A Ingham
5A Ionia
6A Iosco
7 Iron
6A Isabella
5A Jackson
5A Kalamazoo
6A Kalkaska
5A Kent
7 Keweenaw
6A Lake
5A Lapeer
6A Leelanau
5A Lenawee
5A Livingston
7 Luce
7 Mackinac
5A Macomb
6A Manistee
6A Marquette
6A Mason
6A Mecosta
6A Menominee
5A Midland
6A Missaukee
5A Monroe
5A Montcalm
6A Montmorency

(continued)

Table 1450.3—continued
CLIMATE ZONES, MOISTURE REGIMES, AND WARM-HUMID DESIGNATIONS BY STATE, COUNTY AND TERRITORY

5A Muskegon
6A Newaygo
5A Oakland
6A Oceana
6A Ogemaw
7 Ontonagon
6A Osceola
6A Oscoda
6A Otsego
5A Ottawa
6A Presque Isle
6A Roscommon
5A Saginaw
6A Sanilac
7 Schoolcraft
5A Shiawassee
5A St. Clair
5A St. Joseph
5A Tuscola
5A Van Buren
5A Washtenaw
5A Wayne
6A Wexford

MINNESOTA
7 Aitkin
6A Anoka
7 Becker
7 Beltrami
6A Benton
6A Big Stone
6A Blue Earth
6A Brown
7 Carlton
6A Carver
7 Cass
6A Chippewa
6A Chisago
7 Clay
7 Clearwater
7 Cook

6A Cottonwood
7 Crow Wing
6A Dakota
6A Dodge
6A Douglas
6A Faribault
6A Fillmore
6A Freeborn
6A Goodhue
7 Grant
6A Hennepin
6A Houston
7 Hubbard
6A Isanti
7 Itasca
6A Jackson
7 Kanabec
6A Kandiyohi
7 Kittson
7 Koochiching
6A Lac qui Parle
7 Lake
7 Lake of the Woods
6A Le Sueur
6A Lincoln
6A Lyon
7 Mahnomen
7 Marshall
6A Martin
6A McLeod
6A Meeker
7 Mille Lacs
6A Morrison
6A Mower
6A Murray
6A Nicollet
6A Nobles
7 Norman
6A Olmsted
7 Otter Tail

7 Pennington
7 Pine
6A Pipestone
7 Polk
6A Pope
6A Ramsey
7 Red Lake
6A Redwood
6A Renville
6A Rice
6A Rock
7 Roseau
6A Scott
6A Sherburne
6A Sibley
6A Stearns
6A Steele
6A Stevens
7 St. Louis
6A Swift
6A Todd
6A Traverse
6A Wabasha
7 Wadena
6A Waseca
6A Washington
6A Watonwan
7 Wilkin
6A Winona
6A Wright
6A Yellow
   Medicine

MISSISSIPPI
3A Adams*
3A Alcorn
3A Amite*
3A Attala
3A Benton
3A Bolivar
3A Calhoun

3A Carroll
3A Chickasaw
3A Choctaw
3A Claiborne*
3A Clarke
3A Clay
3A Coahoma
3A Copiah*
3A Covington*
3A DeSoto
3A Forrest*
3A Franklin*
3A George*
3A Greene*
3A Grenada
2A Hancock*
2A Harrison*
3A Hinds*
3A Holmes
3A Humphreys
3A Issaquena
3A Itawamba
2A Jackson*
3A Jasper
3A Jefferson*
3A Jefferson Davis*
3A Jones*
3A Kemper
3A Lafayette
3A Lamar*
3A Lauderdale
3A Lawrence*
3A Leake
3A Lee
3A Leflore
3A Lincoln*
3A Lowndes
3A Madison
3A Marion*
3A Marshall

3A Monroe
3A Montgomery
3A Neshoba
3A Newton
3A Noxubee
3A Oktibbeha
3A Panola
2A Pearl River*
3A Perry*
3A Pike*
3A Pontotoc
3A Prentiss
3A Quitman
3A Rankin*
3A Scott
3A Sharkey
3A Simpson*
3A Smith*
2A Stone*
3A Sunflower
3A Tallahatchie
3A Tate
3A Tippah
3A Tishomingo
3A Tunica
3A Union
3A Walthall*
3A Warren*
3A Washington
3A Wayne*
3A Webster
3A Wilkinson*
3A Winston
3A Yalobusha
3A Yazoo

MISSOURI
5A Adair
5A Andrew
5A Atchison
4A Audrain

(continued)

Table 1450.3—continued
CLIMATE ZONES, MOISTURE REGIMES, AND WARM-HUMID DESIGNATIONS BY STATE, COUNTY AND TERRITORY

4A Barry	4A Howard	4A Pulaski	3B Clark	4A Middlesex
4A Barton	4A Howell	5A Putnam	5B Douglas	4A Monmouth
4A Bates	4A Iron	5A Ralls	5B Elko	5A Morris
4A Benton	4A Jackson	4A Randolph	5B Esmeralda	4A Ocean
4A Bollinger	4A Jasper	4A Ray	5B Eureka	5A Passaic
4A Boone	4A Jefferson	4A Reynolds	5B Humboldt	4A Salem
5A Buchanan	4A Johnson	4A Ripley	5B Lander	5A Somerset
4A Butler	5A Knox	4A Saline	5B Lincoln	5A Sussex
5A Caldwell	4A Laclede	5A Schuyler	5B Lyon	4A Union
4A Callaway	4A Lafayette	5A Scotland	5B Mineral	5A Warren
4A Camden	4A Lawrence	4A Scott	5B Nye	
4A Cape Girardeau	5A Lewis	4A Shannon	5B Pershing	**NEW MEXICO**
4A Carroll	4A Lincoln	5A Shelby	5B Storey	4B Bernalillo
4A Carter	5A Linn	4A St. Charles	5B Washoe	5B Catron
4A Cass	5A Livingston	4A St. Clair	5B White Pine	3B Chaves
4A Cedar	5A Macon	4A Ste. Genevieve		4B Cibola
5A Chariton	4A Madison	4A St. Francois	**NEW HAMP-**	5B Colfax
4A Christian	4A Maries	4A St. Louis	**SHIRE**	4B Curry
5A Clark	5A Marion	4A St. Louis (city)	6A Belknap	4B DeBaca
4A Clay	4A McDonald	4A Stoddard	6A Carroll	3B Dona Ana
5A Clinton	5A Mercer	4A Stone	5A Cheshire	3B Eddy
4A Cole	4A Miller	5A Sullivan	6A Coos	4B Grant
4A Cooper	4A Mississippi	4A Taney	6A Grafton	4B Guadalupe
4A Crawford	4A Moniteau	4A Texas	5A Hillsborough	5B Harding
4A Dade	4A Monroe	4A Vernon	6A Merrimack	3B Hidalgo
4A Dallas	4A Montgomery	4A Warren	5A Rockingham	3B Lea
5A Daviess	4A Morgan	4A Washington	5A Strafford	4B Lincoln
5A DeKalb	4A New Madrid	4A Wayne	6A Sullivan	5B Los Alamos
4A Dent	4A Newton	4A Webster		3B Luna
4A Douglas	5A Nodaway	5A Worth	**NEW**	5B McKinley
4A Dunklin	4A Oregon	4A Wright	**JERSEY**	5B Mora
4A Franklin	4A Osage		4A Atlantic	3B Otero
4A Gasconade	4A Ozark	**MONTANA**	5A Bergen	4B Quay
5A Gentry	4A Pemiscot	6B (all)	4A Burlington	5B Rio Arriba
4A Greene	4A Perry		4A Camden	4B Roosevelt
5A Grundy	4A Pettis	**NEBRASKA**	4A Cape May	5B Sandoval
5A Harrison	4A Phelps	5A (all)	4A Cumberland	5B San Juan
4A Henry	5A Pike		4A Essex	5B San Miguel
4A Hickory	4A Platte	**NEVADA**	4A Gloucester	5B Santa Fe
5A Holt	4A Polk	5B Carson City (city)	4A Hudson	4B Sierra
		5B Churchill	5A Hunterdon	4B Socorro
			5A Mercer	

(continued)

Table 1450.3—continued
CLIMATE ZONES, MOISTURE REGIMES, AND WARM-HUMID DESIGNATIONS BY STATE, COUNTY AND TERRITORY

5B Taos	5A Orange	4A Burke	3A Jones	4A Wake
5B Torrance	5A Orleans	3A Cabarrus	4A Lee	4A Warren
4B Union	5A Oswego	4A Caldwell	3A Lenoir	3A Washington
4B Valencia	6A Otsego	3A Camden	4A Lincoln	5A Watauga
NEW YORK	5A Putnam	3A Carteret*	4A Macon	3A Wayne
5A Albany	4A Queens	4A Caswell	4A Madison	4A Wilkes
6A Allegany	5A Rensselaer	4A Catawba	3A Martin	3A Wilson
4A Bronx	4A Richmond	4A Chatham	4A McDowell	4A Yadkin
6A Broome	5A Rockland	4A Cherokee	3A Mecklenburg	5A Yancey
6A Cattaraugus	5A Saratoga	3A Chowan	5A Mitchell	
5A Cayuga	5A Schenectady	4A Clay	3A Montgomery	**NORTH DAKOTA**
5A Chautauqua	6A Schoharie	4A Cleveland	3A Moore	6A Adams
5A Chemung	6A Schuyler	3A Columbus*	4A Nash	7 Barnes
6A Chenango	5A Seneca	3A Craven	3A New Hanover*	7 Benson
6A Clinton	6A Steuben	3A Cumberland	4A Northampton	6A Billings
5A Columbia	6A St. Lawrence	3A Currituck	3A Onslow*	7 Bottineau
5A Cortland	4A Suffolk	3A Dare	4A Orange	6A Bowman
6A Delaware	6A Sullivan	3A Davidson	3A Pamlico	7 Burke
5A Dutchess	5A Tioga	4A Davie	3A Pasquotank	6A Burleigh
5A Erie	6A Tompkins	3A Duplin	3A Pender*	7 Cass
6A Essex	6A Ulster	4A Durham	3A Perquimans	7 Cavalier
6A Franklin	6A Warren	3A Edgecombe	4A Person	6A Dickey
6A Fulton	5A Washington	4A Forsyth	3A Pitt	7 Divide
5A Genesee	5A Wayne	4A Franklin	4A Polk	6A Dunn
5A Greene	4A Westchester	3A Gaston	3A Randolph	7 Eddy
6A Hamilton	6A Wyoming	4A Gates	3A Richmond	6A Emmons
6A Herkimer	5A Yates	4A Graham	3A Robeson	7 Foster
6A Jefferson	**NORTH CAROLINA**	4A Granville	4A Rockingham	6A Golden Valley
4A Kings	4A Alamance	3A Greene	3A Rowan	7 Grand Forks
6A Lewis	4A Alexander	4A Guilford	4A Rutherford	6A Grant
5A Livingston	5A Alleghany	4A Halifax	3A Sampson	7 Griggs
6A Madison	3A Anson	4A Harnett	3A Scotland	6A Hettinger
5A Monroe	5A Ashe	4A Haywood	3A Stanly	7 Kidder
6A Montgomery	5A Avery	4A Henderson	4A Stokes	6A LaMoure
4A Nassau	3A Beaufort	4A Hertford	4A Surry	6A Logan
4A New York	4A Bertie	3A Hoke	4A Swain	7 McHenry
5A Niagara	3A Bladen	3A Hyde	4A Transylvania	6A McIntosh
6A Oneida	3A Brunswick*	4A Iredell	3A Tyrrell	6A McKenzie
5A Onondaga	4A Buncombe	4A Jackson	3A Union	7 McLean
5A Ontario		3A Johnston	4A Vance	6A Mercer
				6A Morton

(continued)

**Table 1450.3—continued
CLIMATE ZONES, MOISTURE REGIMES, AND WARM-HUMID DESIGNATIONS BY STATE, COUNTY AND TERRITORY**

7 Mountrail	5A Crawford	5A Montgomery	3A Caddo	3A McCurtain
7 Nelson	5A Cuyahoga	5A Morgan	3A Canadian	3A McIntosh
6A Oliver	5A Darke	5A Morrow	3A Carter	3A Murray
7 Pembina	5A Defiance	5A Muskingum	3A Cherokee	3A Muskogee
7 Pierce	5A Delaware	5A Noble	3A Choctaw	3A Noble
7 Ramsey	5A Erie	5A Ottawa	4B Cimarron	3A Nowata
6A Ransom	5A Fairfield	5A Paulding	3A Cleveland	3A Okfuskee
7 Renville	5A Fayette	5A Perry	3A Coal	3A Oklahoma
6A Richland	5A Franklin	5A Pickaway	3A Comanche	3A Okmulgee
7 Rolette	5A Fulton	4A Pike	3A Cotton	3A Osage
6A Sargent	4A Gallia	5A Portage	3A Craig	3A Ottawa
7 Sheridan	5A Geauga	5A Preble	3A Creek	3A Pawnee
6A Sioux	5A Greene	5A Putnam	3A Custer	3A Payne
6A Slope	5A Guernsey	5A Richland	3A Delaware	3A Pittsburg
6A Stark	4A Hamilton	5A Ross	3A Dewey	3A Pontotoc
7 Steele	5A Hancock	5A Sandusky	3A Ellis	3A Pottawatomie
7 Stutsman	5A Hardin	4A Scioto	3A Garfield	3A Pushmataha
7 Towner	5A Harrison	5A Seneca	3A Garvin	3A Roger Mills
7 Traill	5A Henry	5A Shelby	3A Grady	3A Rogers
7 Walsh	5A Highland	5A Stark	3A Grant	3A Seminole
7 Ward	5A Hocking	5A Summit	3A Greer	3A Sequoyah
7 Wells	5A Holmes	5A Trumbull	3A Harmon	3A Stephens
7 Williams	5A Huron	5A Tuscarawas	3A Harper	4B Texas
OHIO	5A Jackson	5A Union	3A Haskell	3A Tillman
4A Adams	5A Jefferson	5A Van Wert	3A Hughes	3A Tulsa
5A Allen	5A Knox	5A Vinton	3A Jackson	3A Wagoner
5A Ashland	5A Lake	5A Warren	3A Jefferson	3A Washington
5A Ashtabula	4A Lawrence	4A Washington	3A Johnston	3A Washita
5A Athens	5A Licking	5A Wayne	3A Kay	3A Woods
5A Auglaize	5A Logan	5A Williams	3A Kingfisher	3A Woodward
5A Belmont	5A Lorain	5A Wood	3A Kiowa	**OREGON**
4A Brown	5A Lucas	5A Wyandot	3A Latimer	5B Baker
5A Butler	5A Madison	**OKLAHOMA**	3A Le Flore	4C Benton
5A Carroll	5A Mahoning	3A Adair	3A Lincoln	4C Clackamas
5A Champaign	5A Marion	3A Alfalfa	3A Logan	4C Clatsop
5A Clark	5A Medina	3A Atoka	3A Love	4C Columbia
4A Clermont	5A Meigs	4B Beaver	3A Major	4C Coos
5A Clinton	5A Mercer	3A Beckham	3A Marshall	5B Crook
5A Columbiana	5A Miami	3A Blaine	3A Mayes	4C Curry
5A Coshocton	5A Monroe	3A Bryan	3A McClain	5B Deschutes

(continued)

**Table 1450.3—continued
CLIMATE ZONES, MOISTURE REGIMES, AND WARM-HUMID DESIGNATIONS BY STATE, COUNTY AND TERRITORY**

4C Douglas
5B Gilliam
5B Grant
5B Harney
5B Hood River
4C Jackson
5B Jefferson
4C Josephine
5B Klamath
5B Lake
4C Lane
4C Lincoln
4C Linn
5B Malheur
4C Marion
5B Morrow
4C Multnomah
4C Polk
5B Sherman
4C Tillamook
5B Umatilla
5B Union
5B Wallowa
5B Wasco
4C Washington
5B Wheeler
4C Yamhill

PENNSYLVANIA
5A Adams
5A Allegheny
5A Armstrong
5A Beaver
5A Bedford
5A Berks
5A Blair
5A Bradford
4A Bucks
5A Butler
5A Cambria
6A Cameron

5A Carbon
5A Centre
4A Chester
5A Clarion
6A Clearfield
5A Clinton
5A Columbia
5A Crawford
5A Cumberland
5A Dauphin
4A Delaware
6A Elk
5A Erie
5A Fayette
5A Forest
5A Franklin
5A Fulton
5A Greene
5A Huntingdon
5A Indiana
5A Jefferson
5A Juniata
5A Lackawanna
5A Lancaster
5A Lawrence
5A Lebanon
5A Lehigh
5A Luzerne
5A Lycoming
6A McKean
5A Mercer
5A Mifflin
5A Monroe
4A Montgomery
5A Montour
5A Northampton
5A Northumberland
5A Perry
4A Philadelphia
5A Pike

6A Potter
5A Schuylkill
5A Snyder
5A Somerset
5A Sullivan
6A Susquehanna
6A Tioga
5A Union
5A Venango
5A Warren
5A Washington
6A Wayne
5A Westmoreland
5A Wyoming
4A York

RHODE ISLAND
5A (all)

SOUTH CAROLINA
3A Abbeville
3A Aiken
3A Allendale*
3A Anderson
3A Bamberg*
3A Barnwell*
3A Beaufort*
3A Berkeley*
3A Calhoun
3A Charleston*
3A Cherokee
3A Chester
3A Chesterfield
3A Clarendon
3A Colleton*
3A Darlington
3A Dillon
3A Dorchester*
3A Edgefield
3A Fairfield
3A Florence

3A Georgetown*
3A Greenville
3A Greenwood
3A Hampton*
3A Horry*
3A Jasper*
3A Kershaw
3A Lancaster
3A Laurens
3A Lee
3A Lexington
3A Marion
3A Marlboro
3A McCormick
3A Newberry
3A Oconee
3A Orangeburg
3A Pickens
3A Richland
3A Saluda
3A Spartanburg
3A Sumter
3A Union
3A Williamsburg
3A York

SOUTH DAKOTA
6A Aurora
6A Beadle
5A Bennett
5A Bon Homme
6A Brookings
6A Brown
6A Brule
6A Buffalo
6A Butte
6A Campbell
5A Charles Mix
6A Clark
5A Clay
6A Codington

6A Corson
6A Custer
6A Davison
6A Day
6A Deuel
6A Dewey
5A Douglas
6A Edmunds
6A Fall River
6A Faulk
6A Grant
5A Gregory
6A Haakon
6A Hamlin
6A Hand
6A Hanson
6A Harding
6A Hughes
5A Hutchinson
6A Hyde
5A Jackson
6A Jerauld
6A Jones
6A Kingsbury
6A Lake
6A Lawrence
6A Lincoln
6A Lyman
6A Marshall
6A McCook
6A McPherson
6A Meade
5A Mellette
6A Miner
6A Minnehaha
6A Moody
6A Pennington
6A Perkins
6A Potter
6A Roberts

(continued)

Table 1450.3—continued
CLIMATE ZONES, MOISTURE REGIMES, AND WARM-HUMID DESIGNATIONS BY STATE, COUNTY AND TERRITORY

6A Sanborn	4A Giles	4A Perry	3B Baylor	3B Crane
6A Shannon	4A Grainger	4A Pickett	2A Bee*	3B Crockett
6A Spink	4A Greene	4A Polk	2A Bell*	3B Crosby
6A Stanley	4A Grundy	4A Putnam	2A Bexar*	3B Culberson
6A Sully	4A Hamblen	4A Rhea	3A Blanco*	4B Dallam
5A Todd	4A Hamilton	4A Roane	3B Borden	3A Dallas*
5A Tripp	4A Hancock	4A Robertson	2A Bosque*	3B Dawson
6A Turner	3A Hardeman	4A Rutherford	3A Bowie*	4B Deaf Smith
5A Union	3A Hardin	4A Scott	2A Brazoria*	3A Delta
6A Walworth	4A Hawkins	4A Sequatchie	2A Brazos*	3A Denton*
5A Yankton	3A Haywood	4A Sevier	3B Brewster	2A DeWitt*
6A Ziebach	3A Henderson	3A Shelby	4B Briscoe	3B Dickens
	4A Henry	4A Smith	2A Brooks*	2B Dimmit*
TENNESSEE	4A Hickman	4A Stewart	3A Brown*	4B Donley
4A Anderson	4A Houston	4A Sullivan	2A Burleson*	2A Duval*
4A Bedford	4A Humphreys	4A Sumner	3A Burnet*	3A Eastland
4A Benton	4A Jackson	3A Tipton	2A Caldwell*	3B Ector
4A Bledsoe	4A Jefferson	4A Trousdale	2A Calhoun*	2B Edwards*
4A Blount	4A Johnson	4A Unicoi	3B Callahan	3A Ellis*
4A Bradley	4A Knox	4A Union	2A Cameron*	3B El Paso
4A Campbell	3A Lake	4A Van Buren	3A Camp*	3A Erath*
4A Cannon	3A Lauderdale	4A Warren	4B Carson	2A Falls*
4A Carroll	4A Lawrence	4A Washington	3A Cass*	3A Fannin
4A Carter	4A Lewis	4A Wayne	4B Castro	2A Fayette*
4A Cheatham	4A Lincoln	4A Weakley	2A Chambers*	3B Fisher
3A Chester	4A Loudon	4A White	2A Cherokee*	4B Floyd
4A Claiborne	4A Macon	4A Williamson	3B Childress	3B Foard
4A Clay	3A Madison	4A Wilson	3A Clay	2A Fort Bend*
4A Cocke	4A Marion		4B Cochran	3A Franklin*
4A Coffee	4A Marshall	**TEXAS**	3B Coke	2A Freestone*
3A Crockett	4A Maury	2A Anderson*	3B Coleman	2B Frio*
4A Cumberland	4A McMinn	3B Andrews	3A Collin*	3B Gaines
4A Davidson	3A McNairy	2A Angelina*	3B Collingsworth	2A Galveston*
4A Decatur	4A Meigs	2A Aransas*	2A Colorado*	3B Garza
4A DeKalb	4A Monroe	3A Archer	2A Comal*	3A Gillespie*
4A Dickson	4A Montgomery	4B Armstrong	3A Comanche*	3B Glasscock
3A Dyer	4A Moore	2A Atascosa*	3B Concho	2A Goliad*
3A Fayette	4A Morgan	2A Austin*	3A Cooke	2A Gonzales*
4A Fentress	4A Obion	4B Bailey	2A Coryell*	4B Gray
4A Franklin	4A Overton	2B Bandera*	3B Cottle	3A Grayson
4A Gibson		2A Bastrop*		

(continued)

**Table 1450.3—continued
CLIMATE ZONES, MOISTURE REGIMES, AND WARM-HUMID DESIGNATIONS BY STATE, COUNTY AND TERRITORY**

3A Gregg*	3B Kent	3A Morris*	3A Smith*	3A Young
2A Grimes*	3B Kerr	3B Motley	3A Somervell*	2B Zapata*
2A Guadalupe*	3B Kimble	3A Nacogdoches*	2A Starr*	2B Zavala*
4B Hale	3B King	3A Navarro*	3A Stephens	**UTAH**
3B Hall	2B Kinney*	2A Newton*	3B Sterling	5B Beaver
3A Hamilton*	2A Kleberg*	3B Nolan	3B Stonewall	6B Box Elder
4B Hansford	3B Knox	2A Nueces*	3B Sutton	6B Cache
3B Hardeman	3A Lamar*	4B Ochiltree	4B Swisher	6B Carbon
2A Hardin*	4B Lamb	4B Oldham	3A Tarrant*	6B Daggett
2A Harris*	3A Lampasas*	2A Orange*	3B Taylor	5B Davis
3A Harrison*	2B La Salle*	3A Palo Pinto*	3B Terrell	6B Duchesne
4B Hartley	2A Lavaca*	3A Panola*	3B Terry	5B Emery
3B Haskell	2A Lee*	3A Parker*	3B Throckmorton	5B Garfield
2A Hays*	2A Leon*	4B Parmer	3A Titus*	5B Grand
3B Hemphill	2A Liberty*	3B Pecos	3B Tom Green	5B Iron
3A Henderson*	2A Limestone*	2A Polk*	2A Travis*	5B Juab
2A Hidalgo*	4B Lipscomb	4B Potter	2A Trinity*	5B Kane
2A Hill*	2A Live Oak*	3B Presidio	2A Tyler*	5B Millard
4B Hockley	3A Llano*	3A Rains*	3A Upshur*	6B Morgan
3A Hood*	3B Loving	4B Randall	3B Upton	5B Piute
3A Hopkins*	3B Lubbock	3B Reagan	2B Uvalde*	6B Rich
2A Houston*	3B Lynn	2B Real*	2B Val Verde*	5B Salt Lake
3B Howard	2A Madison*	3A Red River*	3A Van Zandt*	5B San Juan
3B Hudspeth	3A Marion*	3B Reeves	2A Victoria*	5B Sanpete
3A Hunt*	3B Martin	2A Refugio*	2A Walker*	5B Sevier
4B Hutchinson	3B Mason	4B Roberts	2A Waller*	6B Summit
3B Irion	2A Matagorda*	2A Robertson*	3B Ward	5B Tooele
3A Jack	2B Maverick*	3A Rockwall*	2A Washington*	6B Uintah
2A Jackson*	3B McCulloch	3B Runnels	2B Webb*	5B Utah
2A Jasper*	2A McLennan*	3A Rusk*	2A Wharton*	6B Wasatch
3B Jeff Davis	2A McMullen*	3A Sabine*	3B Wheeler	3B Washington
2A Jefferson*	2B Medina*	3A San Augustine*	3A Wichita	5B Wayne
2A Jim Hogg*	3B Menard	2A San Jacinto*	3B Wilbarger	5B Weber
2A Jim Wells*	3B Midland	2A San Patricio*	2A Willacy*	**VERMONT**
3A Johnson*	2A Milam*	3A San Saba*	2A Williamson*	6A (all)
3B Jones	3A Mills*	3B Schleicher	2A Wilson*	**VIRGINIA**
2A Karnes*	3B Mitchell	3B Scurry	3B Winkler	4A (all)
3A Kaufman*	3A Montague	3B Shackelford	3A Wise	**WASHINGTON**
3A Kendall*	2A Montgomery*	3A Shelby*	3A Wood*	5B Adams
2A Kenedy*	4B Moore	4B Sherman	4B Yoakum	

(continued)

Table 1450.3—continued
CLIMATE ZONES, MOISTURE REGIMES, AND WARM-HUMID DESIGNATIONS BY STATE, COUNTY AND TERRITORY

5B Asotin	4A Boone	5A Summers	6A Kenosha	6A Wood
5B Benton	4A Braxton	5A Taylor	6A Kewaunee	
5B Chelan	5A Brooke	5A Tucker	6A La Crosse	**WYOMING**
4C Clallam	4A Cabell	4A Tyler	6A Lafayette	6B Albany
4C Clark	4A Calhoun	5A Upshur	7 Langlade	6B Big Horn
5B Columbia	4A Clay	4A Wayne	7 Lincoln	6B Campbell
4C Cowlitz	5A Doddridge	5A Webster	6A Manitowoc	6B Carbon
5B Douglas	5A Fayette	5A Wetzel	6A Marathon	6B Converse
6B Ferry	4A Gilmer	4A Wirt	6A Marinette	6B Crook
5B Franklin	5A Grant	4A Wood	6A Marquette	6B Fremont
5B Garfield	5A Greenbrier	4A Wyoming	6A Menominee	5B Goshen
5B Grant	5A Hampshire		6A Milwaukee	6B Hot Springs
4C Grays Harbor	5A Hancock	**WISCONSIN**	6A Monroe	6B Johnson
4C Island	5A Hardy	6A Adams	6A Oconto	6B Laramie
4C Jefferson	5A Harrison	7 Ashland	7 Oneida	7 Lincoln
4C King	4A Jackson	6A Barron	6A Outagamie	6B Natrona
4C Kitsap	4A Jefferson	7 Bayfield	6A Ozaukee	6B Niobrara
5B Kittitas	4A Kanawha	6A Brown	6A Pepin	6B Park
5B Klickitat	5A Lewis	6A Buffalo	6A Pierce	5B Platte
4C Lewis	4A Lincoln	7 Burnett	6A Polk	6B Sheridan
5B Lincoln	4A Logan	6A Calumet	6A Portage	7 Sublette
4C Mason	5A Marion	6A Chippewa	7 Price	6B Sweetwater
6B Okanogan	5A Marshall	6A Clark	6A Racine	7 Teton
4C Pacific	4A Mason	6A Columbia	6A Richland	6B Uinta
6B Pend Oreille	4A McDowell	6A Crawford	6A Rock	6B Washakie
4C Pierce	4A Mercer	6A Dane	6A Rusk	6B Weston
4C San Juan	5A Mineral	6A Dodge	6A Sauk	
4C Skagit	4A Mingo	6A Door	7 Sawyer	**US**
5B Skamania	5A Monongalia	7 Douglas	6A Shawano	**TERRITORIES**
4C Snohomish	4A Monroe	6A Dunn	6A Sheboygan	**AMERICAN**
5B Spokane	4A Morgan	6A Eau Claire	6A St. Croix	**SAMOA**
6B Stevens	5A Nicholas	7 Florence	7 Taylor	1A (all)*
4C Thurston	5A Ohio	6A Fond du Lac	6A Trempealeau	
4C Wahkiakum	5A Pendleton	7 Forest	6A Vernon	**GUAM**
5B Walla Walla	4A Pleasants	6A Grant	7 Vilas	1A (all)*
4C Whatcom	5A Pocahontas	6A Green	6A Walworth	
5B Whitman	5A Preston	6A Green Lake	7 Washburn	**NORTHERN**
5B Yakima	4A Putnam	6A Iowa	6A Washington	**MARIANA**
	5A Raleigh	7 Iron	6A Waukesha	**ISLANDS**
WEST VIRGINIA	5A Randolph	6A Jackson	6A Waupaca	1A (all)*
5A Barbour	4A Ritchie	6A Jefferson	6A Waushara	**PUERTO RICO**
4A Berkeley	4A Roane	6A Juneau	6A Winnebago	1A (all)*
				VIRGIN ISLANDS
				1A (all)*

1405.3.2 Material vapor retarder class. The *vapor retarder class* shall be based on the manufacturer's certified testing or a tested assembly.

The following shall be deemed to meet the class specified:

Class I: Sheet polyethylene, nonperforated aluminum foil

Class II: Kraft-faced fiberglass batts or paint with a perm rating greater than 0.1 and less than or equal to 1.0

Class III: Latex or enamel paint

❖ The vapor retarder class is defined in Section 202 of the code in terms of vapor permeability. The test method called out for determination of the perm value is ASTM E 96. However, this section provides a list of materials that can be used for each class of vapor retarder and deemed to comply with the test standard. No testing is required for these materials. All other materials are required to be tested.

1405.3.3 Minimum clear airspaces and vented openings for vented cladding. For the purposes of this section, vented cladding shall include the following minimum clear airspaces.

1. Vinyl lap or horizontal aluminum siding applied over a weather-resistive barrier as specified in this chapter.

2. Brick veneer with a clear airspace as specified in this code.

3. Other *approved* vented claddings.

❖ This section is intended to define what is intended by "vented cladding" called out in Table 1405.3.1. As can be seen, vented cladding is material commonly used in the code, including vinyl siding, aluminum siding, and brick with a clear airspace.

By "other approved vented claddings," the code intends materials that can be shown to allow drying of the components behind them in the same manner and at the same rate as the specific materials in Items 1 or 2.

1405.4 Flashing. Flashing shall be installed in such a manner so as to prevent moisture from entering the wall or to redirect it to the exterior. Flashing shall be installed at the perimeters of exterior door and window assemblies, penetrations and terminations of *exterior wall* assemblies, *exterior wall* intersections with roofs, chimneys, porches, decks, balconies and similar projections and at built-in gutters and similar locations where moisture could enter the wall. Flashing with projecting flanges shall be installed on both sides and the ends of copings, under sills and continuously above projecting *trim*.

❖ Water that enters the exterior wall can lead to the decay of wood and the degradation of other building materials that are sensitive to moisture. Some of the most common points of moisture penetration in an exterior wall are at intersections of windows, doors, chimneys and roof lines with other framing These potential points of moisture intrusion must be closed by corro-

sion-resistant flashings or by other acceptable practices [see Figures 1405.4(1)–1405.4(5)]. Materials such as asphalt-saturated felt or building paper are not acceptable for flashing at required locations, in part due to the fact that these materials have a limited ability to resist water penetration. The flashing installation locations stated in this section represent those that are most commonly encountered. As an alternative to the installation of flashings, testing of the specific penetrations, terminations and intersections of the exterior wall finish must be performed in accordance with Section 1403.2 (see commentary, Section 1403.2).

Improperly sloped flashings may enable the accumulation of moisture in the wall assembly, which may result in eventual degradation of the wall finish, building substrate or even the flashing material. Without the extension of the flashing beyond the face of the exterior wall finish, moisture may reenter the wall assembly directly below the flashing through capillary action as indicated in Figure 1405.4(5).

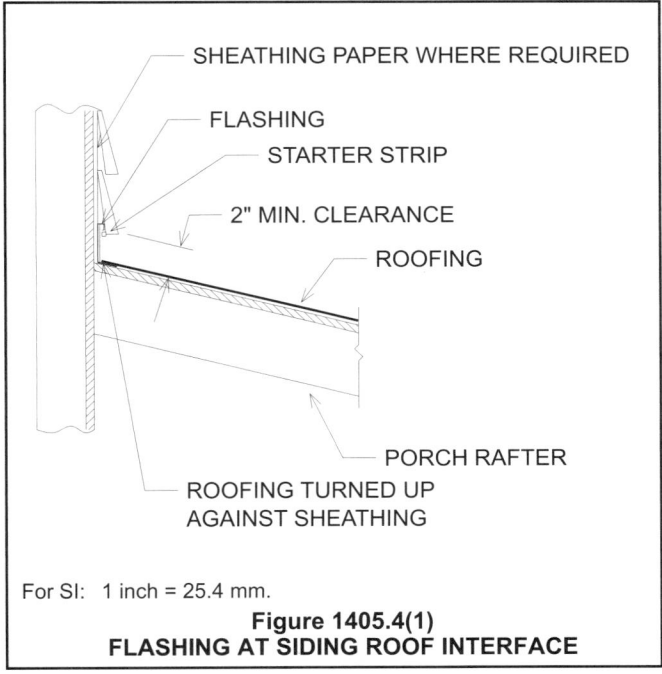

SHEATHING PAPER WHERE REQUIRED
FLASHING
STARTER STRIP
2" MIN. CLEARANCE
ROOFING
PORCH RAFTER
ROOFING TURNED UP AGAINST SHEATHING

For SI: 1 inch = 25.4 mm.

Figure 1405.4(1)
FLASHING AT SIDING ROOF INTERFACE

1405.4.1 Exterior wall pockets. In exterior walls of buildings or structures, wall pockets or crevices in which moisture can accumulate shall be avoided or protected with caps or drips, or other *approved* means shall be provided to prevent water damage.

❖ Changes in building lines and elevations, materials and tops of walls, etc., can produce wall pockets, crevices and other openings that allow the entry of rain, snow, ice and other sources of moisture. Where these openings are present, additional covers, caps, flashings, etc., must be provided to prevent the entry of moisture.

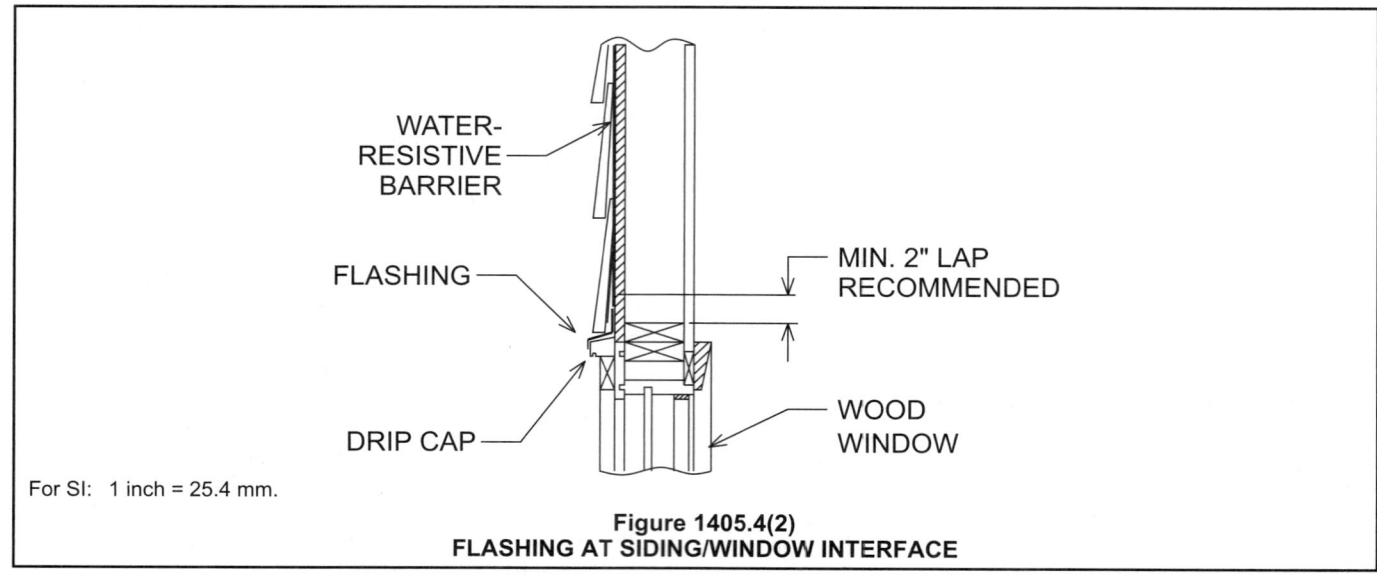

For SI: 1 inch = 25.4 mm.

Figure 1405.4(2)
FLASHING AT SIDING/WINDOW INTERFACE

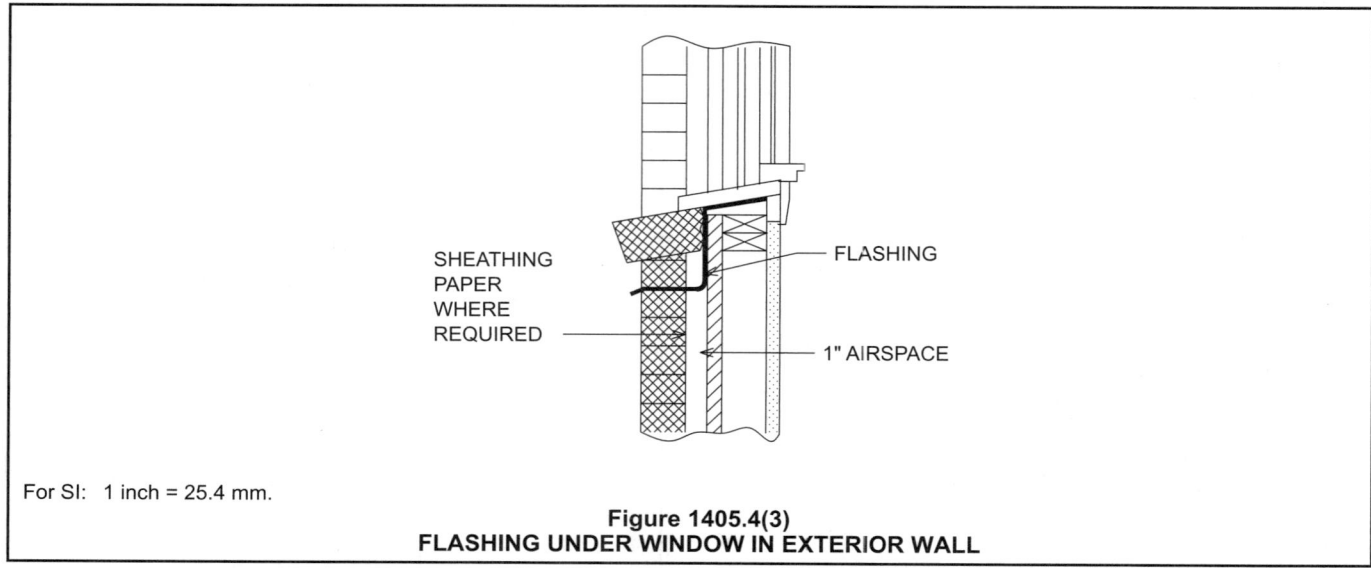

For SI: 1 inch = 25.4 mm.

Figure 1405.4(3)
FLASHING UNDER WINDOW IN EXTERIOR WALL

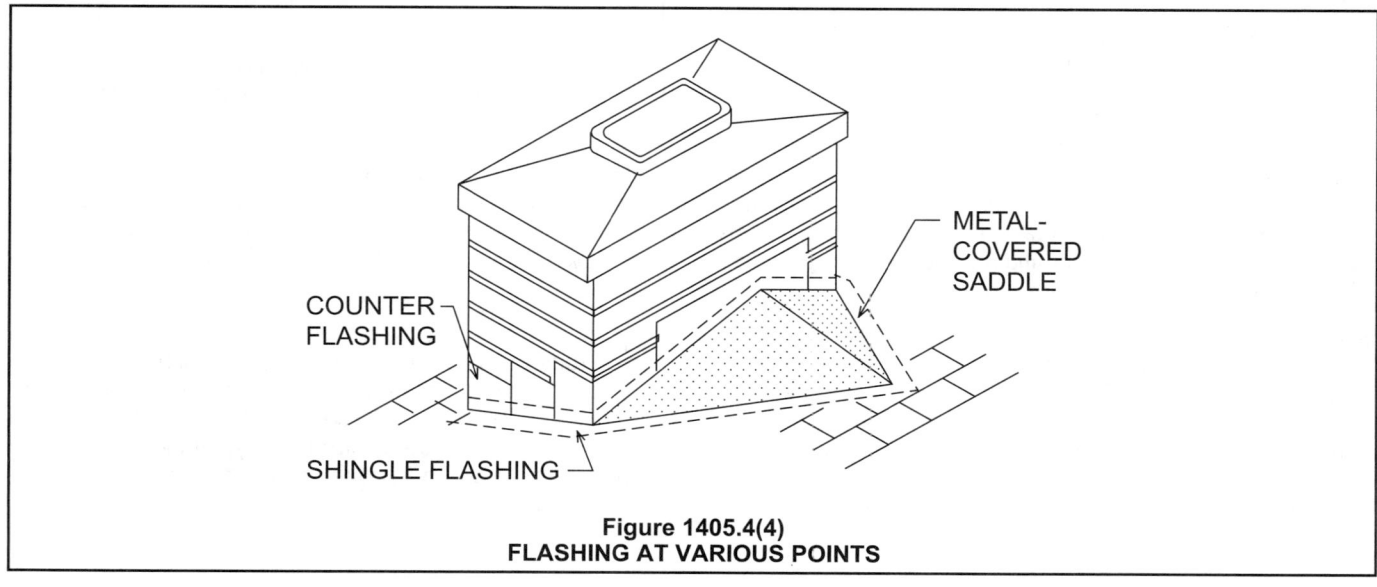

Figure 1405.4(4)
FLASHING AT VARIOUS POINTS

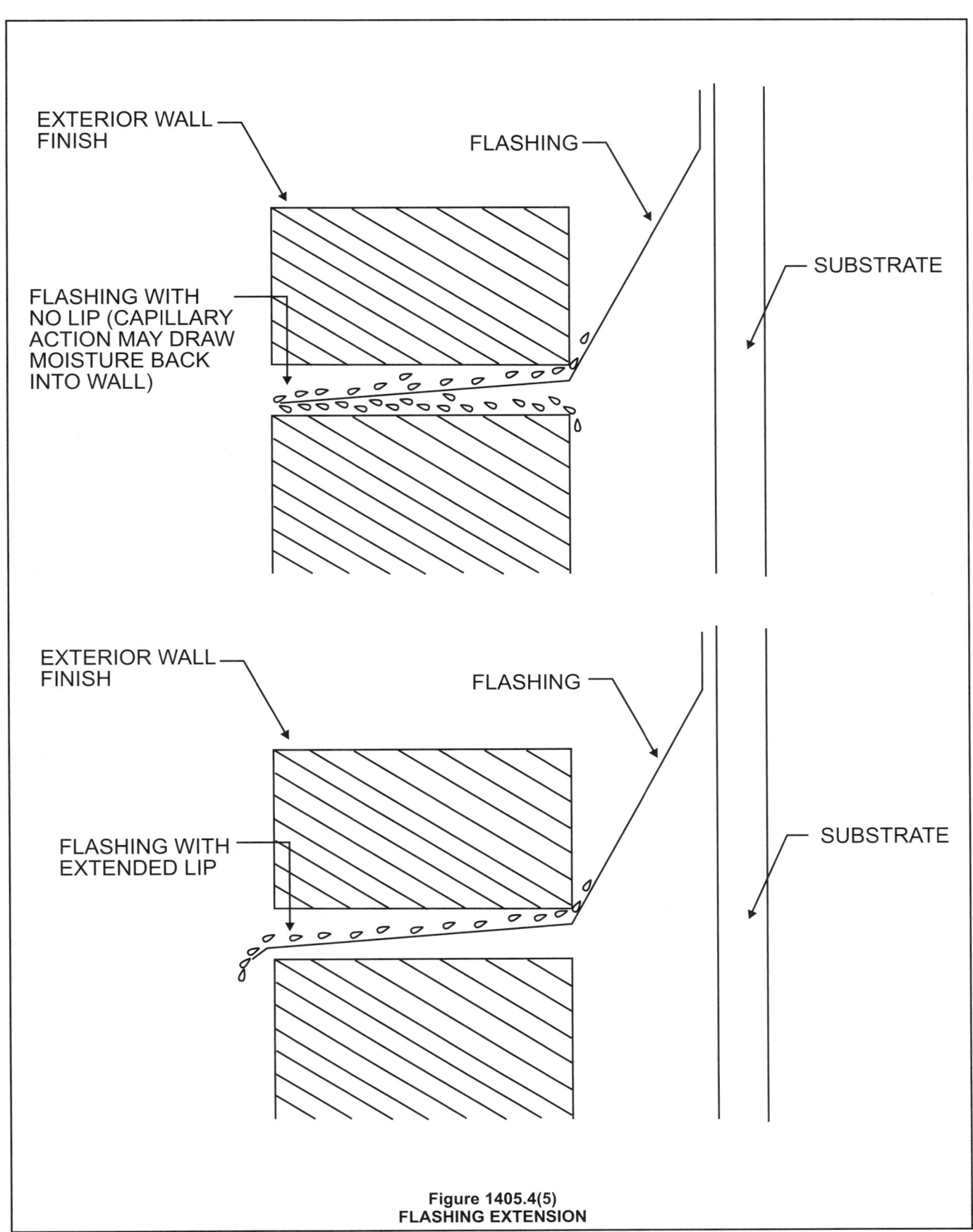

Figure 1405.4(5)
FLASHING EXTENSION

1405.4.2 Masonry. Flashing and weep holes in anchored veneer shall be located in the first course of masonry above finished ground level above the foundation wall or slab, and other points of support, including structural floors, shelf angles and lintels where anchored veneers are designed in accordance with Section 1405.6.

❖ Water can penetrate walls constructed of masonry materials from a number of sources. Some of the most common causes of water penetration are: the porous nature of many masonry materials; openings such as small cracks, that may occur between the masonry unit and mortar joint due to expansion and contraction of the wall over time; and the interface between the masonry and penetrations or dissimilar materials.

Walls constructed of anchored masonry veneer must be designed to accommodate water penetration and to provide a means to both stop the moisture penetration and for the moisture to leave the wall. This section stipulates the use of flashing to stop moisture penetration into the wall backing and small openings (weepholes) to allow the moisture to exit the wall. In order for these two components to function as intended, the weepholes must be installed immediately above the flashing so that moisture collected by the flashing can exit the wall, preventing the accumulation of water at the flashing location.

1405.5 Wood veneers. Wood veneers on exterior walls of buildings of Type I, II, III and IV construction shall be not less than 1 inch (25 mm) nominal thickness, 0.438-inch (11.1 mm) exterior hardboard siding or 0.375-inch (9.5 mm) exterior-type wood structural panels or particleboard and shall conform to the following:

1. The veneer shall not exceed 40 feet (1219 mm) in height above grade. Where fire-retardant-treated wood is used, the height shall not exceed 60 feet (1829 mm) in height above grade.

2. The veneer is attached to or furred from a noncombustible backing that is fire-resistance rated as required by other provisions of this code.

3. Where open or spaced wood veneers (without concealed spaces) are used, they shall not project more than 24 inches (610 mm) from the building wall.

❖ Wood veneers, which are combustible exterior materials, are permitted for use on buildings of other than combustible (Type V) construction where the minimum thickness requirements of this section and three construction criteria are met (see the commentary to Section 1406.2.1 for more discussion on wood veneers as a combustible material on exterior walls).

Item 1 limits the maximum height of the installation of the wood veneer and is coordinated with the provisions of Section 1406.2.2 (see the commentary to Section 1406.2.2 for more discussion on the limitations of use of combustible exterior materials). The increase of one additional story is permitted due to the limited combustible nature of fire-retardant-treated wood (FRTW) siding (see commentary, Section 2303.2). Item 2 requires the backing to be noncombustible, which is consistent with requirements of Chapter 6 for buildings of Type I, II, III and IV construction. The backing must also have the fire-resistance rating required by other provisions of the code, such as in Tables 601 and 602. Item 3 limits the maximum distance that wood veneers may project from the face of the building.

1405.6 Anchored masonry veneer. Anchored masonry veneer shall comply with the provisions of Sections 1405.6, 1405.7, 1405.8 and 1405.9 and Sections 6.1 and 6.2 of TMS 402/ACI 530/ASCE 5.

❖ Sections 6.1 and 6.2 of ACI 530/ASCE 5/TMS 402 are to be used for the design and detailing requirements for anchored masonry veneers. See the commentary for ACI 530/ASCE 5/TMS 402 for an explanation of the anchored masonry veneer requirements.

1405.6.1 Tolerances. Anchored masonry veneers in accordance with Chapter 14 are not required to meet the tolerances in Article 3.3 G1 of TMS 602/ACI 530.1/ASCE 6.

❖ The tolerances in the referenced standard apply to masonry units that are utilized in masonry wall construction and are not intended to be applied to masonry veneers.

1405.6.2 Seismic requirements. Anchored masonry veneer located in Seismic Design Category C, D, E or F shall conform to the requirements of Section 6.2.2.10 of TMS 402/ACI 530/ASCE 5. Anchored masonry veneer located in Seismic Design Category D shall also conform to the requirements of Section 6.2.2.10.3.3 of TMS 402/ACI 530/ASCE 5.

❖ This section requires that the seismic design and detailing of anchored masonry veneer be performed in accordance with referenced standard ACI 530 for veneers located in Seismic Design Categories C, D, E and F (see Chapter 16 for more information on seismic design categories). Section 6.2.2.10 of ACI 530 contains criteria for components such as masonry anchor spacing, joint reinforcement bedding and splicing. Masonry veneer that is located in Seismic Design Category D must conform to the additional requirements listed for Seismic Design Categories E and F because of the weight of the veneer materials. This is especially the case if the higher veneer is installed.

1405.7 Stone veneer. Stone veneer units not exceeding 10 inches (254 mm) in thickness shall be anchored directly to masonry, concrete or to stud construction by one of the following methods:

1. With concrete or masonry backing, anchor ties shall be not less than 0.1055-inch (2.68 mm) corrosion-resistant wire, or *approved* equal, formed beyond the base of the backing. The legs of the loops shall be not less than 6 inches (152 mm) in length bent at right angles and laid in the mortar joint, and spaced so that the eyes or loops are 12 inches (305 mm) maximum on center (o.c.) in both directions. There shall be provided not less than a 0.1055-inch (2.68 mm) corrosion-resistant wire tie, or *approved* equal, threaded through the exposed loops for every 2 square feet (0.2 m²) of stone veneer. This tie shall

be a loop having legs not less than 15 inches (381 mm) in length bent so that it will lie in the stone veneer mortar joint. The last 2 inches (51 mm) of each wire leg shall have a right-angle bend. One-inch (25 mm) minimum thickness of cement grout shall be placed between the backing and the stone veneer.

2. With stud backing, a 2-inch by 2-inch (51 by 51 mm) 0.0625-inch (1.59 mm) corrosion-resistant wire mesh with two layers of *water-resistive barrier* in accordance with Section 1404.2 shall be applied directly to wood studs spaced a maximum of 16 inches (406 mm) o.c. On studs, the mesh shall be attached with 2-inch-long (51 mm) corrosion-resistant steel wire furring nails at 4 inches (102 mm) o.c. providing a minimum 1.125-inch (29 mm) penetration into each stud and with 8d common nails at 8 inches (203 mm) o.c. into top and bottom plates or with equivalent wire ties. There shall be not less than a 0.1055-inch (2.68 mm) corrosion-resistant wire, or *approved* equal, looped through the mesh for every 2 square feet (0.2 m²) of stone veneer. This tie shall be a loop having legs not less than 15 inches (381 mm) in length, so bent that it will lie in the stone veneer mortar joint. The last 2 inches (51 mm) of each wire leg shall have a right-angle bend. One-inch (25 mm) minimum thickness of cement grout shall be placed between the backing and the stone veneer.

❖ This section permits direct anchorage of stone veneers to the wall backing by one of two specified methods. It should be noted that support from below is re-

quired to be provided in accordance with Section 1405.6. Since the framing is providing only lateral anchorage, thicknesses of up to 10 inches (254 mm) are permitted.

There are two methods of anchorage prescribed in this section, based upon the construction of the wall backing. The first method specifies the anchor type, size and spacing for concrete and masonry backing (see Figure 1405.7). The second method specifies the anchorage requirements for wood frame walls.

Both methods require that the anchorage ties be of corrosion-resistant material. Corrosion-resistant ties are necessary because of the potential for corrosion caused by both water penetration and an adverse reaction of the metal with the mortar or other masonry material.

According to the first method, the specified tie spacings, in conjunction with the cement grout, are intended to achieve lateral support points of the veneer at adequate spacings. Where lateral ties are not installed at the required locations, buckling of the masonry veneer under vertical dead loads and lateral loads is possible. It should be noted that the required 1 inch (25 mm) of mortar grout creates walls that are not considered cavity wall construction.

The second method requires that two layers of waterproof paper be installed over the stud construction to provide protection from moisture penetration.

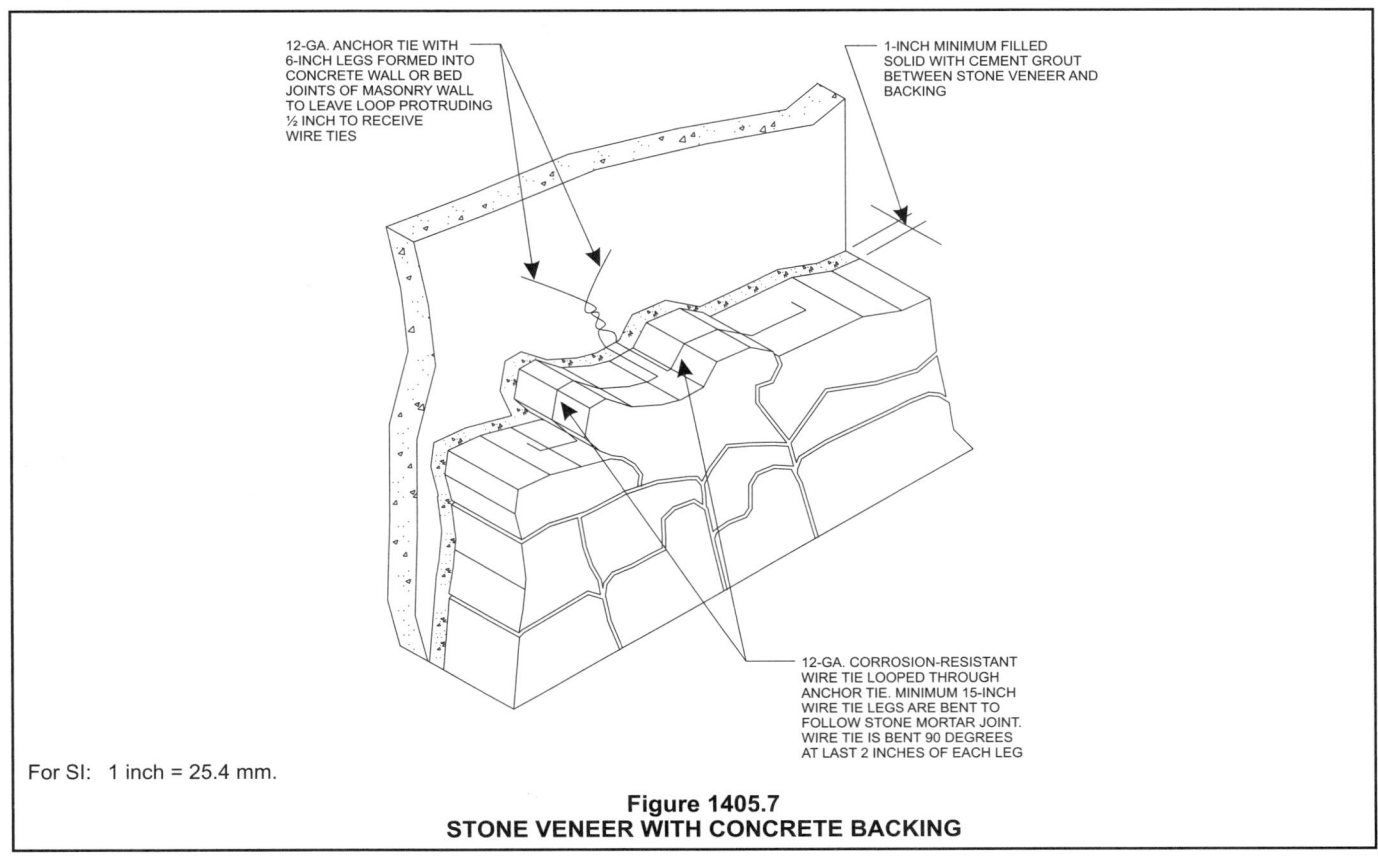

For SI: 1 inch = 25.4 mm.

Figure 1405.7
STONE VENEER WITH CONCRETE BACKING

1405.8 Slab-type veneer. Slab-type veneer units not exceeding 2 inches (51 mm) in thickness shall be anchored directly to masonry, concrete or stud construction. For veneer units of marble, travertine, granite or other stone units of slab form ties of corrosion-resistant dowels in drilled holes shall be located in the middle third of the edge of the units, spaced a maximum of 24 inches (610 mm) apart around the periphery of each unit with not less than four ties per veneer unit. Units shall not exceed 20 square feet (1.9 m²) in area. If the dowels are not tight fitting, the holes shall be drilled not more than 0.063 inch (1.6 mm) larger in diameter than the dowel, with the hole countersunk to a diameter and depth equal to twice the diameter of the dowel in order to provide a tight-fitting key of cement mortar at the dowel locations when the mortar in the joint has set. Veneer ties shall be corrosion-resistant metal capable of resisting, in tension or compression, a force equal to two times the weight of the attached veneer. If made of sheet metal, veneer ties shall be not smaller in area than 0.0336 by 1 inch (0.853 by 25 mm) or, if made of wire, not smaller in diameter than 0.1483-inch (3.76 mm) wire.

❖ Slab-type veneer refers to thin stone masonry veneers that are individually supported. These units do not bear on the successive units below, but are anchored to and supported vertically by the backup construction with corrosion-resistant dowels located at the edges of each unit. This type of veneer anchorage applies to the use of thin stone units that are typically not set in mortar, but are made water tight at the edges through the use of sealant materials. The strength of the veneer ties is required to be adequate to support twice the weight of the veneer in tension to provide for an adequate margin of safety, such that the veneer will remain in place in case it works loose of the grout and backing.

1405.9 Terra cotta. Anchored terra cotta or ceramic units not less than 1⅝ inches (41 mm) thick shall be anchored directly to masonry, concrete or stud construction. Tied terra cotta or ceramic veneer units shall be not less than 1⅝ inches (41 mm) thick with projecting dovetail webs on the back surface spaced approximately 8 inches (203 mm) o.c. The facing shall be tied to the backing wall with corrosion-resistant metal anchors of not less than No. 8 gage wire installed at the top of each piece in horizontal bed joints not less than 12 inches (305 mm) nor more than 18 inches (457 mm) o.c.; these anchors shall be secured to ¼-inch (6.4 mm) corrosion-resistant pencil rods that pass through the vertical aligned loop anchors in the backing wall. The veneer ties shall have sufficient strength to support the full weight of the veneer in tension. The facing shall be set with not less than a 2-inch (51 mm) space from the backing wall and the space shall be filled solidly with portland cement grout and pea gravel. Immediately prior to setting, the backing wall and the facing shall be drenched with clean water and shall be distinctly damp when the grout is poured.

❖ The minimum permitted 1⅝-inch (41 mm) thickness for terra cotta units is necessary in order for each unit to have an inherent level of structural integrity. In order to increase the mechanical anchorage of the backing, dovetail webs are required and are embedded in the

2-inch (51 mm) cement grout space. The strength of the veneer ties is required to be adequate to support the full weight of the veneer in tension in order to achieve an adequate margin of safety, such that the veneer will remain in place in case it works loose of the grout and backing (see Figure 1405.9).

1405.10 Adhered masonry veneer. Adhered masonry veneer shall comply with the applicable requirements of Section 1405.10.1 and Sections 6.1 and 6.3 of TMS 402/ACI 530/ASCE 5.

❖ Sections 6.1 and 6.3 of ACI 530/ASCE 5/TMS 402 are to be used for the design and detailing requirements for adhered masonry veneers. See the commentary for ACI 530/ASCE 5/TMS 402 for an explanation of the adhered masonry veneer requirements. Figures 1405.10(1) and 1405.10(2) illustrate two examples of an adhered masonry veneer.

1405.10.1 Interior adhered masonry veneers. Interior adhered masonry veneers shall have a maximum weight of 20 psf (0.958 kg/m²) and shall be installed in accordance with Section 1405.10. Where the interior adhered masonry veneer is supported by wood construction, the supporting members shall be designed to limit deflection to $\frac{1}{600}$ of the span of the supporting members.

❖ This section addresses masonry veneer that is supported on wood members. The provisions of Section 2104.1.6 are not intended to preclude masonry veneer on wood supports. In fact, Section 2304.12 contains provisions for the support of interior masonry veneer on wood-frame construction. The deflection limit is to minimize cracking of the veneer.

1405.11 Metal veneers. Veneers of metal shall be fabricated from *approved* corrosion-resistant materials or shall be protected front and back with porcelain enamel, or otherwise be treated to render the metal resistant to corrosion. Such veneers shall not be less than 0.0149-inch (0.378 mm) nominal thickness sheet steel mounted on wood or metal furring strips or approved sheathing on the wood construction.

❖ All metal veneers used as exterior weather coverings are required to be manufactured from materials that are inherently corrosion resistant or to be protected with other materials, such as coatings, to prevent corrosion.

For sheet steel, the minimum nominal thickness is 0.0149 inches (0.38 mm), not including the thickness of the corrosion-resistant coating. For aluminum siding, an 0.019-inch-thick (0.48 mm) value is the minimum. Aluminum siding must also conform to AAMA 1402. The minimum thicknesses for other weather coverings are obtained from Table 1405.2.

1405.11.1 Attachment. Exterior metal veneer shall be securely attached to the supporting masonry or framing members with corrosion-resistant fastenings, metal ties or by other *approved* devices or methods. The spacing of the fastenings or ties shall not exceed 24 inches (610 mm) either vertically or horizontally, but where units exceed 4 square feet (0.4 m²) in area there shall be not less than four attachments per unit. The metal attachments shall have a cross-sectional area not less than provided by W 1.7 wire. Such attachments and their supports shall be capable of resisting a horizontal force in accordance with the wind loads specified in Section 1609, but in no case less than 20 psf (0.958 kg/m²).

❖ This section prescribes minimum requirements for the attachment of metal veneers to the exterior wall con-struction. The attachment of the veneer may be ac-complished through the use of fasteners, metal ties, clips or other means. The means of attaching the pan-els and supporting members to which the metal ve-neer is fastened must be able to resist the design wind loads determined in accordance with Chapter 16, but in no case a horizontal load of less than 20 psf (0.958 kg/m²).

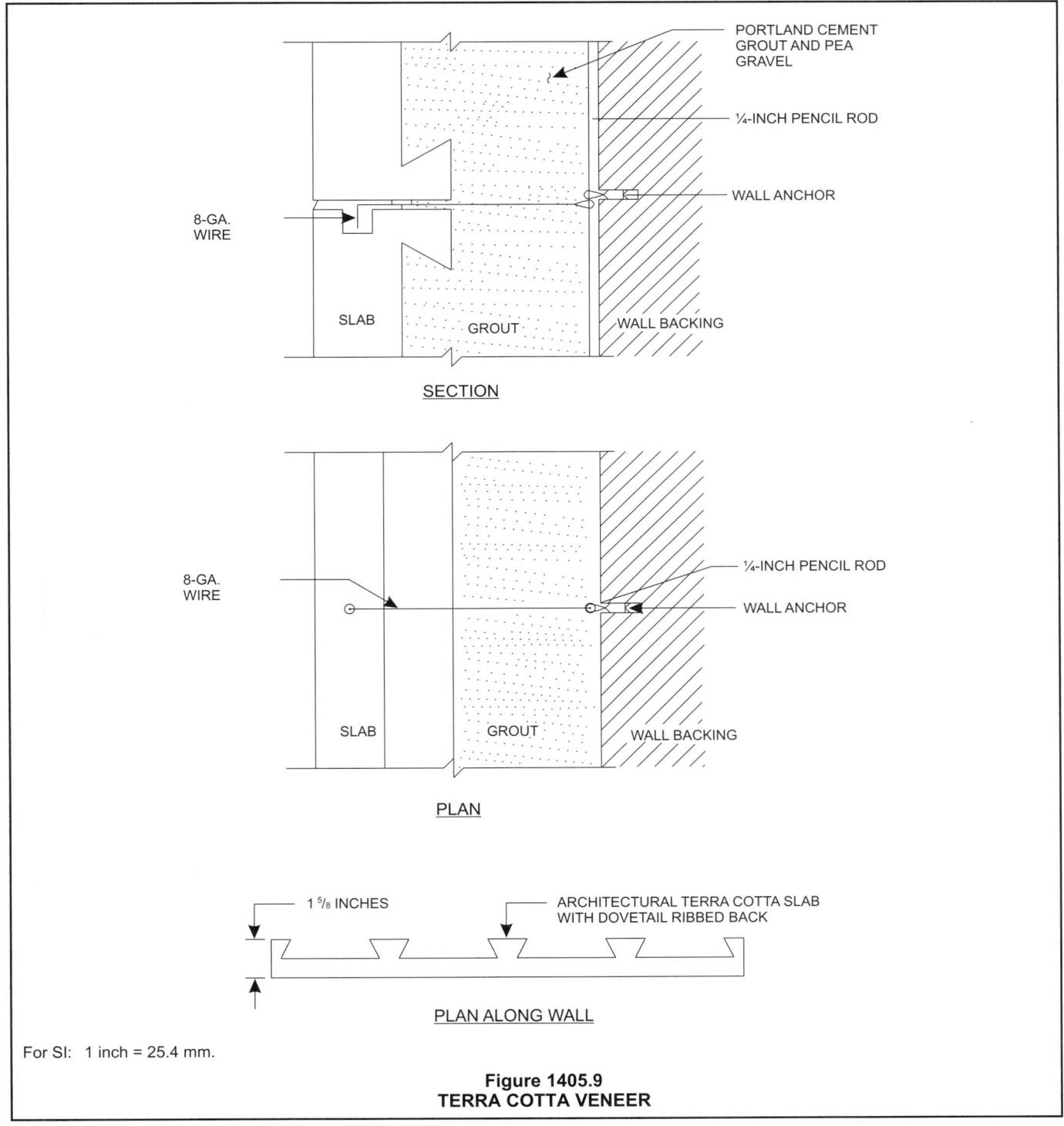

For SI: 1 inch = 25.4 mm.

**Figure 1405.9
TERRA COTTA VENEER**

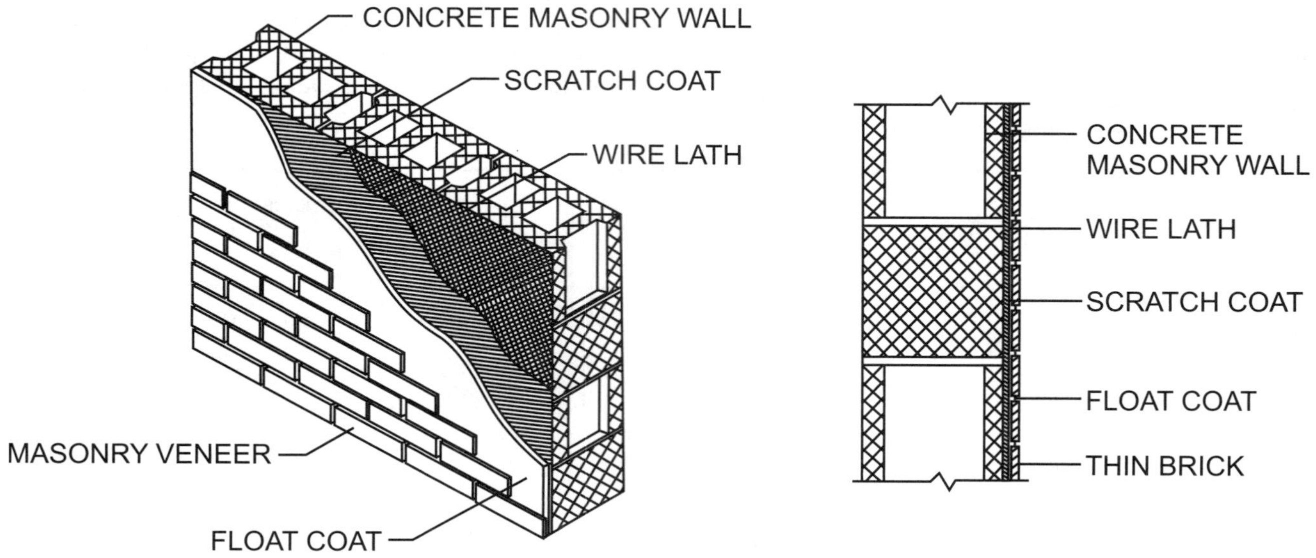

Figure 1405.10(1)
MASONRY VENEER ADHERED TO MASONRY BACKING
(Courtesy of Brick Industry)

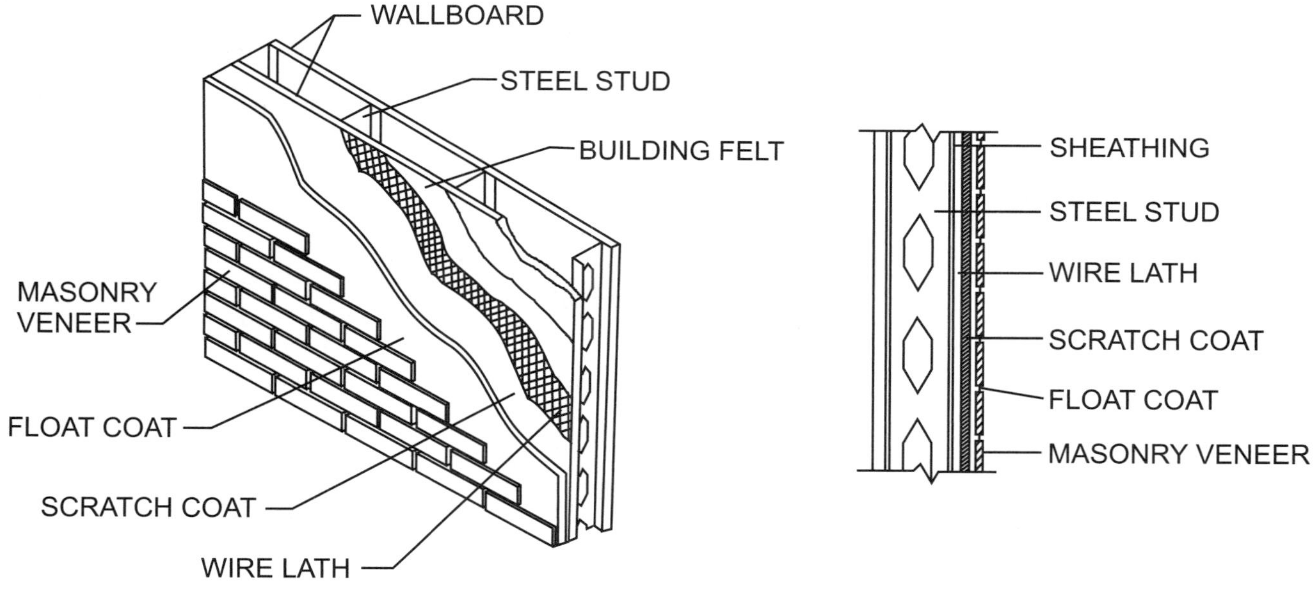

Figure 1405.10(2)
MASONRY VENEER ADHERED TO STEEL FRAME BACKING
(Courtesy of Brick Industry)

1405.11.2 Weather protection. Metal supports for exterior metal veneer shall be protected by painting, galvanizing or by other equivalent coating or treatment. Wood studs, furring strips or other wood supports for exterior metal veneer shall be *approved* pressure-treated wood or protected as required in Section 1403.2. Joints and edges exposed to the weather shall be caulked with *approved* durable waterproofing material or by other *approved* means to prevent penetration of moisture.

❖ Supports for metal veneers are required to be protected from corrosion induced by moisture. The corrosion resistance may be provided through a surface treatment applied to the metal or through the use of materials that are inherently corrosion resistant. Where wood supports are used, they must be protected through pressure treatment or covered with a minimum of one layer of water-resistive barrier as specified in Section 1404.2. Protection of joints and edges must be provided to prevent the entrance of moisture that may result in damage to the veneer, supports or building framing.

1405.11.3 Backup. Masonry backup shall not be required for metal veneer except as is necessary to meet the fire-resistance requirements of this code.

❖ The type of substrate that metal veneers may be installed over is not limited, except in cases where the wall is required to have a fire-resistance rating in accordance with other sections of the code.

1405.11.4 Grounding. Grounding of metal veneers on buildings shall comply with the requirements of Chapter 27 of this code.

❖ The need to provide electrical grounding for metal veneers has been widely discussed for more than 30 years. Examples of hazards have been demonstrated where electric services have been violated. Most metal-siding manufacturers have made provisions in their designs for grounding. This section references Chapter 27 without providing additional specifics regarding grounding of siding.

1405.12 Glass veneer. The area of a single section of thin exterior structural glass veneer shall not exceed 10 square feet (0.93 m²) where it is not more than 15 feet (4572 mm) above the level of the sidewalk or grade level directly below, and shall not exceed 6 square feet (0.56 m²) where it is more than 15 feet (4572 mm) above that level.

❖ Structural glass veneer is produced in the same manner as plate glass (see commentary, Chapter 24), with the addition of metallic oxides that are blended into the glass to provide opacity and color. To accommodate the forces due to expansion and contraction of differing materials in exterior installations, the maximum area of structural glass veneer is limited based upon the height above grade.

1405.12.1 Length and height. The length or height of any section of thin exterior structural glass veneer shall not exceed 48 inches (1219 mm).

❖ Where Section 1405.11 regulates the maximum area of any single section of structural glass veneer, this section limits the maximum dimension of any single section to 48 inches (1219 mm) when used in exterior application.

1405.12.2 Thickness. The thickness of thin exterior structural glass veneer shall be not less than 0.344 inch (8.7 mm).

❖ Structural glass veneers are produced in a range of thicknesses; however, the minimum thickness permitted for use as an exterior veneer is limited to $^{11}/_{32}$ inch (8.7 mm).

1405.12.3 Application. Thin exterior structural glass veneer shall be set only after backing is thoroughly dry and after application of an *approved* bond coat uniformly over the entire surface of the backing so as to effectively seal the surface. Glass shall be set in place with an *approved* mastic cement in sufficient quantity so that at least 50 percent of the area of each glass unit is directly bonded to the backing by mastic not less than $^1/_4$ inch (6.4 mm) thick and not more than $^5/_8$ inch (15.9 mm) thick. The bond coat and mastic shall be evaluated for compatibility and shall bond firmly together.

❖ Structural glass veneer is bonded to a solid backing, such as masonry, through the use of a mastic adhesive. The backing must be provided with a coating that will serve to seal it from moisture absorption, provide a uniform surface to which the mastic can adhere and provide a relatively level surface. The bond coat must be compatible for use with the backing material so that debonding will not occur over time. Some common types of bond coats are a float coat of cement mortar and portland cement stucco.

The glass veneer is attached to the bond coat by applying mastic adhesives. Mastics are characterized by their ability to remain pliable over time, rather than harden. The minimum and maximum thicknesses permitted for the mastic application are specified, since an application that is too thin would not provide an adequate bond and one that is too thick could sag over time. To minimize the possibility of debonding, it is necessary to determine that the materials intended for use as a mastic and as a bond coat are compatible with one another.

1405.12.4 Installation at sidewalk level. Where glass extends to a sidewalk surface, each section shall rest in an *approved* metal molding, and be set at least $^1/_4$ inch (6.4 mm) above the highest point of the sidewalk. The space between the molding and the sidewalk shall be thoroughly caulked and made water tight.

❖ To protect glass veneer from breakage when installed at sidewalk level, the veneer is required to be supported by a metal molding. The molding must be protected from corrosion and be constructed of a material that is compatible for contact with the glass. The molding is to be located a minimum of $1/4$ inch (6.4 mm) above the sidewalk to minimize the potential for cracking of the veneer due to expansion and contraction.

1405.12.4.1 Installation above sidewalk level. Where thin exterior structural glass veneer is installed above the level of the top of a bulkhead facing, or at a level more than 36 inches (914 mm) above the sidewalk level, the mastic cement binding shall be supplemented with *approved* nonferrous metal shelf angles located in the horizontal joints in every course. Such shelf angles shall be not less than 0.0478-inch (1.2 mm) thick and not less than 2 inches (51 mm) long and shall be spaced at *approved* intervals, with not less than two angles for each glass unit. Shelf angles shall be secured to the wall or backing with expansion bolts, toggle bolts or by other *approved* methods.

❖ Structural glass veneer installed more than 36 inches (914 mm) above sidewalk level is subject to impact from pedestrians and can pose a hazard if the veneer were to become dislodged from the backing. To minimize the potential for this occurring, it is necessary to provide additional supports that are attached to the backing. The supports are intended to supplement the mastic.

1405.12.5 Joints. Unless otherwise specifically *approved* by the *building official*, abutting edges of thin exterior structural glass veneer shall be ground square. Mitered joints shall not be used except where specifically *approved* for wide angles. Joints shall be uniformly buttered with an *approved* jointing compound and horizontal joints shall be held to not less than 0.063 inch (1.6 mm) by an *approved* nonrigid substance or device. Where thin exterior structural glass veneer abuts nonresilient material at sides or top, expansion joints not less than $1/4$ inch (6.4 mm) wide shall be provided.

❖ Glass has an expansion coefficient of roughly 4.5, which is less than most metals. To minimize the potential for cracking or chipping of the glass panels and the development of sharp edges due to differential movement, this section stipulates both the requirements for factory finishing of the glass veneer panel edges and the minimum requirements for joints between sections.

1405.12.6 Mechanical fastenings. Thin exterior structural glass veneer installed above the level of the heads of show windows and veneer installed more than 12 feet (3658 mm) above sidewalk level shall, in addition to the mastic cement and shelf angles, be held in place by the use of fastenings at each vertical or horizontal edge, or at the four corners of each glass unit. Fastenings shall be secured to the wall or backing with expansion bolts, toggle bolts or by other methods. Fastenings shall be so designed as to hold the glass veneer in a vertical plane independent of the mastic cement. Shelf angles providing both support and fastenings shall be permitted.

❖ While Section 1405.12.4.1 addresses the requirements for protection of pedestrians for installations of veneer that exceed 3 feet (914 mm) above sidewalk level, this section address installations that exceed 12 feet (3658 mm) above sidewalk level. Installations above 12 feet (3658 mm) are not subject to impact

from pedestrians but are subject to wind loads. Therefore, each veneer section must be secured in place by mechanical means, in addition to the mastic and shelf angles specified in Sections 1405.12.3 and 1405.12.6, respectively.

1405.12.7 Flashing. Exposed edges of thin exterior structural glass veneer shall be flashed with overlapping corrosion-resistant metal flashing and caulked with a waterproof compound in a manner to effectively prevent the entrance of moisture between the glass veneer and the backing.

❖ The presence of moisture behind the glass veneer can result in loss of strength of either the mastic, the bond coat or both. In addition, corrosion of the supports or fasteners can occur. This can increase the potential for failure of the structural glass veneer. This section addresses the requirements for flashing of structural glass veneers to prevent the penetration of moisture behind the veneers.

1405.13 Exterior windows and doors. Windows and doors installed in exterior walls shall conform to the testing and performance requirements of Section 1715.5.

❖ Windows and doors that are part of the exterior building envelope are to be tested for wind load resistance in accordance with the methods specified in Section 1714.5 (see commentary, Section 1714.5).

1405.13.1 Installation. Windows and doors shall be installed in accordance with *approved* manufacturer's instructions. Fastener size and spacing shall be provided in such instructions and shall be calculated based on maximum loads and spacing used in the tests.

❖ Windows and doors that are part of the exterior envelope are to be installed in accordance with the method in which they were tested (see Section 1405.13) and the window or door manufacturer's installation instructions.

1405.13.2 Window sills. In Occupancy Groups R-2 and R-3, one- and two-family and multiple-family dwellings, where the opening of the sill portion of an operable window is located more than 72 inches (1829 mm) above the finished grade or other surface below, the lowest part of the clear opening of the window shall be at a height not less than 24 inches (610 mm) above the finished floor surface of the room in which the window is located. Glazing between the floor and a height of 24 inches (610 mm) shall be fixed or have openings through which a 4-inch (102 mm) diameter sphere cannot pass.

> **Exception:** Openings that are provided with window guards that comply with ASTM F 2006 or F 2090.

❖ This requirement is intended to provide a level of protection to children to help keep them from falling from windows when the window sill is too close to the finished floor level. By raising the lowest operable portion of a window to 24 inches (610 mm) or more the sill height is above the center of gravity of smaller children. By restricting the sill position for the rough opening of the window any replacement windows that could be placed into the opening should also meet this code requirement.

1405.14 Vinyl siding. Vinyl siding conforming to the requirements of this section and complying with ASTM D 3679 shall be permitted on exterior walls of buildings located in areas where the basic wind speed specified in Chapter 16 does not exceed 100 miles per hour (45 m/s) and the *building height* is less than or equal to 40 feet (12 192 mm) in Exposure C. Where construction is located in areas where the basic wind speed exceeds 100 miles per hour (45 m/s), or building heights are in excess of 40 feet (12 192 mm), tests or calculations indicating compliance with Chapter 16 shall be submitted. Vinyl siding shall be secured to the building so as to provide weather protection for the exterior walls of the building.

❖ The installation of vinyl siding on Type V buildings is limited based upon the basic wind speed for a given location. For areas where the basic wind speed is greater than 100 miles per hour (44 m/s), wind load resistance testing and structural calculations supporting the ability of the siding to withstand the basic wind speed must be provided.

1405.14.1 Application. The siding shall be applied over sheathing or materials listed in Section 2304.6. Siding shall be applied to conform with the *water-resistive barrier* requirements in Section 1403. Siding and accessories shall be installed in accordance with *approved* manufacturer's instructions. Unless otherwise specified in the *approved* manufacturer's instructions, nails used to fasten the siding and accessories shall have a minimum 0.313-inch (7.9 mm) head diameter and $^1/_8$-inch (3.18 mm) shank diameter. The nails shall be corrosion resistant and shall be long enough to penetrate the studs or nailing strip at least $^3/_4$ inch (19 mm). Where the siding is installed horizontally, the fastener spacing shall not exceed 16 inches (406 mm) horizontally and 12 inches (305 mm) vertically. Where the siding is installed vertically, the fastener spacing shall not exceed 12 inches (305 mm) horizontally and 12 inches (305 mm) vertically.

❖ This section prescribes requirements for the installation of vinyl siding. The installation must also comply with ASTM D 4756, which is referenced in ASTM D 3679 (see commentary, Section 1404.9) and the manufacturer's instructions. These instructions are necessary for compliance with the performance requirements of this chapter. The prescriptive fastener requirements of this section correspond with the requirements contained in ASTM D 4756.

1405.15 Cement plaster. Cement plaster applied to exterior walls shall conform to the requirements specified in Chapter 25.

❖ The materials and installation of cement plaster (sometimes referred to as "stucco") are governed by the requirements of Chapter 25. It should be noted that cement plaster is the only plaster type that is permitted for exterior use. Gypsum plaster, which is also governed by Chapter 25, is generally subject to deterioration in the presence of moisture and is therefore not permitted for use in exterior applications.

1405.16 Fiber-cement siding. Fiber-cement siding complying with Section 1404.10 shall be permitted on exterior walls of Type I, II, III, IV and V construction for wind pressure resis-

tance or wind speed exposures as indicated by the manufacturer's listing and *label* and *approved* installation instructions. Where specified, the siding shall be installed over sheathing or materials *listed* in Section 2304.6 and shall be installed to conform to the *water-resistive barrier* requirements in Section 1403. Siding and accessories shall be installed in accordance with *approved* manufacturer's instructions. Unless otherwise specified in the *approved* manufacturer's instructions, nails used to fasten the siding to wood studs shall be corrosion-resistant round head smooth shank and shall be long enough to penetrate the studs at least 1 inch (25 mm). For metal framing, all-weather screws shall be used and shall penetrate the metal framing at least three full threads.

❖ Fiber cement siding is also a material used for weather-resistant siding. It is manufactured from fiber-reinforced cement, and the code permits its use as either panel siding or horizontal lap siding. Sections 1405.16.1 and 1405.16.2 states that the applicable material standard is ASTM C 1186. Note that the product is required to bear a label, which means that the manufacturer must have regular inspections by a third-party control agency.

1405.16.1 Panel siding. Fiber-cement panels shall comply with the requirements of ASTM C 1186, Type A, minimum Grade II. Panels shall be installed with the long dimension either parallel or perpendicular to framing. Vertical and horizontal joints shall occur over framing members and shall be sealed with caulking, covered with battens or shall be designed to comply with Section 1403.2. Panel siding shall be installed with fasteners in accordance with the *approved* manufacturer's instructions.

❖ This section specifies the material standard and type of material required for fiber cement siding used in a panel configuration. In addition specific details of installation necessary for fiber cement siding are called out.

1405.16.2 Lap siding. Fiber-cement lap siding having a maximum width of 12 inches (305 mm) shall comply with the requirements of ASTM C 1186, Type A, minimum Grade II. Lap siding shall be lapped a minimum of $1^1/_4$ inches (32 mm) and lap siding not having tongue-and-groove end joints shall have the ends sealed with caulking, covered with an H-section joint cover, located over a strip of flashing or shall be designed to comply with Section 1403.2. Lap siding courses shall be installed with the fastener heads exposed or concealed in accordance with the *approved* manufacturer's instructions.

❖ This section specifies the material standard and type of material required for fiber cement siding used in a lap configuration. In addition specific details of installation necessary for fiber cement siding are called out.

1405.17 Fastening. Weather boarding and wall coverings shall be securely fastened with aluminum, copper, zinc, zinc-coated or other *approved* corrosion-resistant fasteners in accordance with the nailing schedule in Table 2304.9.1 or the *approved* manufacturer's installation instructions. Shingles and other weather coverings shall be attached with appropriate standard-shingle nails to furring strips securely nailed to studs, or with *approved* mechanically bonding nails, except where sheathing is of wood not less than 1-inch (25 mm) nominal thickness or of wood structural panels as specified in Table 2308.9.3(3).

❖ The proper attachment of exterior weatherboard or cladding is critical in order to remain durable and stay in place under anticipated weather conditions. This section requires noncorrodible nails and other connectors. The requirements of Table 2304.9.1 address the attachment of wood structural panels and panels of particleboard and fiberboard. Lapped siding products of these materials are to be attached in accordance with the manufacturer's installation instructions.

The relatively narrow and variable widths of wood shingles and shakes make it necessary to provide a continuous nailing base. This is permitted to be wood sheathing, wood structural panels not less than $^5/_{16}$-inch (7.9 mm) thick or horizontal furring strips. Dimension lumber furring strips are required to be 1-inch (25 mm) minimum nominal thickness. Shingles are attached with special mechanically bonding nails.

Wood shingles and shakes are required to be fastened with at least No. 14 B&S-gage corrosion-resistant nails in accordance with Table 2304.9.1. The nail must be long enough to penetrate the framing, sheathing or furring strip at least $^3/_4$ inch (19.1 mm). In accordance with Sections 1508.8.5 (shingles) and 1508.9.6 (shakes), shingles and shakes must be fastened with not more than two fasteners per shingle or shake.

Wood siding must be attached to the sheathing or framing in accordance with Table 2304.9.1. The method of nailing for the various siding patterns is illustrated in Figure 1405.17. Wood siding, especially plain bevel siding, applied over foam plastic sheathing, re-

quires special attention to nailing. Cooperative research by the lumber and plastic industries has shown that poor performance may be expected if the nailing is inadequate. Since the foam plastic sheathing is not an adequate nail base, the siding is required to be nailed through the sheathing to the wood-framing members. Condensation can occur on the sheathing and the back of the siding, keeping the back wet. The weather face will remain dry. This differential moisture condition will cause the siding to "curl" or warp and pull improperly installed nails out of the framing members.

SECTION 1406
COMBUSTIBLE MATERIALS ON THE EXTERIOR SIDE OF EXTERIOR WALLS

1406.1 General. Section 1406 shall apply to *exterior wall coverings*; balconies and similar projections; and bay and oriel windows constructed of combustible materials.

❖ The requirements given in this section apply to exterior elements of combustible construction. The elements included are exterior wall veneers such as vinyl or wood siding, architectural trim, gutters, half-timbering, balconies, bay and oriel windows, plastic panels and foam plastics. The code requires walls to be noncombustible for buildings of construction Types I, II, III and IV, in order to limit the fuel load. Many common types of materials used for architectural enhancement of exterior walls are combustible. Therefore, the code provides limitations as well as requirements for flame spread and radiant heat exposure testing in order to mitigate the hazards associated with combusti-

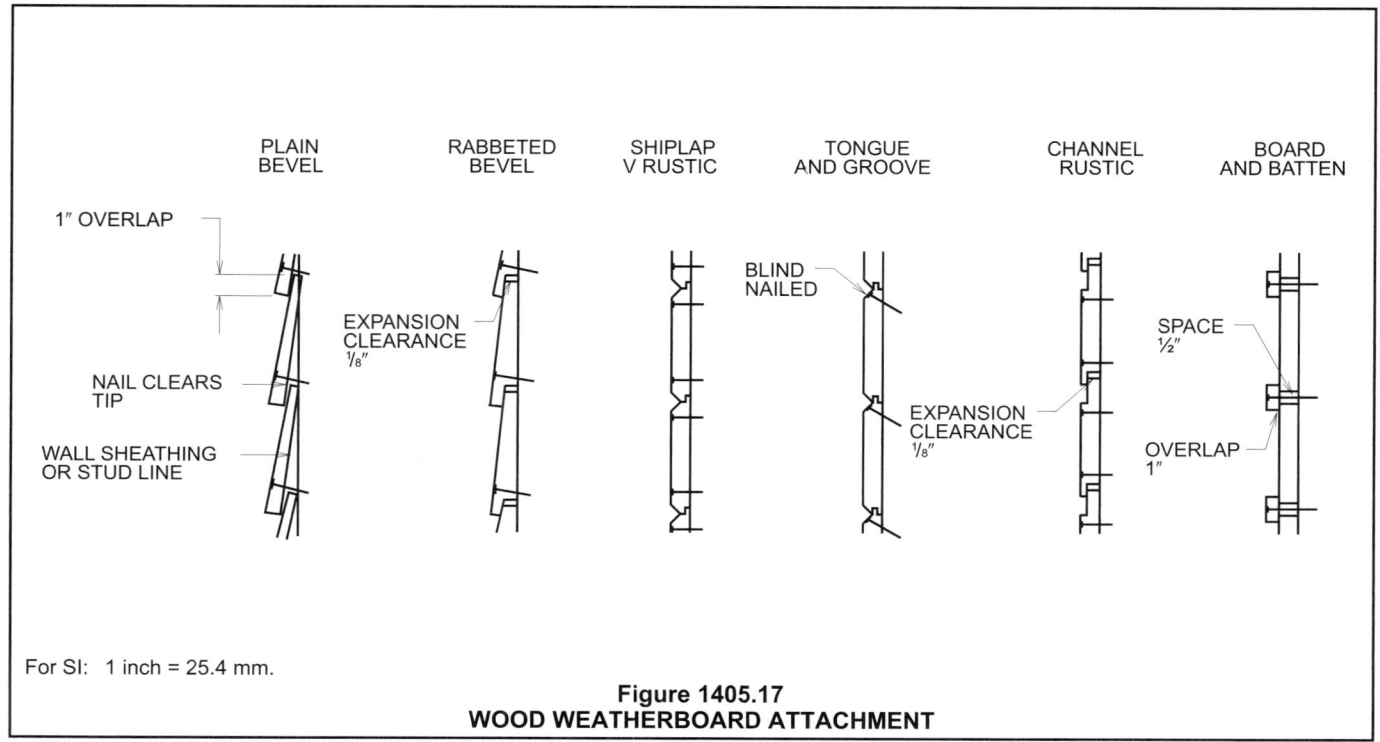

For SI: 1 inch = 25.4 mm.

Figure 1405.17
WOOD WEATHERBOARD ATTACHMENT

ble materials as permitted elements of noncombustible construction.

1406.2 Combustible exterior wall coverings. Combustible *exterior wall coverings* shall comply with this section.

Exception: Plastics complying with Chapter 26.

❖ The specific requirements relating to the combustible exterior wall finish include resistance to ignition through radiant heat exposure testing (see Section 1406.2.1) and limitations on the height and the amount of combustible materials.

The exception is provided for plastic materials that comply with the requirements of Chapter 26 for limitations on flame spread, ignition properties and amount of plastic.

1406.2.1 Ignition resistance. Combustible *exterior wall coverings* shall be tested in accordance with NFPA 268.

Exceptions:

1. Wood or wood-based products.

2. Other combustible materials covered with an exterior covering other than vinyl sidings listed in Table 1405.2.

3. Aluminum having a minimum thickness of 0.019 inch (0.48 mm).

4. *Exterior wall coverings* on *exterior walls* of Type V construction.

❖ In order to continue to provide the level of safety for adjoining property that is consistent with the fire safety provisions for opening limitations in exterior walls, radiant heat testing is required for combustible materials other than wood, using 1.25 W/cm² as a baseline exposure criterion.

This section requires that testing be performed in accordance with the NFPA 268 test method to measure the incident radiant heat that will cause a given exterior finish material to ignite.

There are four exceptions to the test requirements. Exception 1 is given because wood or wood-based products have been chosen as the baseline criterion for evaluation of other combustible materials. Testing in accordance with this section would be redundant since such materials are in compliance by virtue of being the baseline material.

Exception 2 relates to materials covered by one of the veneers listed in Table 1405.2. For example, the exception would apply to the building paper under vinyl siding, not the vinyl siding itself. Again, this is an exception to eliminate any confusion that might result from the presence of combustible components within a wall assembly that are not the exposed exterior finish.

Exception 3 is provided because, although most aluminum alloys are combustible, aluminum has a higher ignition temperature than wood. Although not tested, it is anticipated that the critical radiant heat flux would therefore also be higher.

Exception 4 is provided because its basic intent

relates to concerns over flame propagation on exteriors of buildings required to be of noncombustible construction.

1406.2.1.1 Fire separation 5 feet or less. Where installed on *exterior walls* having a *fire separation distance* of 5 feet (1524 mm) or less, combustible *exterior wall coverings* shall not exhibit sustained flaming as defined in NFPA 268.

❖ In order to qualify for use on walls with a fire separation distance as little as 5 feet (1524 mm), the material must not flame when exposed to a 12.5 kW/m² exposure.

1406.2.1.2 Fire separation greater than 5 feet. For fire separation distances greater than 5 feet (1524 mm), an assembly shall be permitted that has been exposed to a reduced level of incident radiant heat flux in accordance with the NFPA 268 test method without exhibiting sustained flaming. The minimum *fire separation distance* required for the assembly shall be determined from Table 1406.2.1.2 based on the maximum tolerable level of incident radiant heat flux that does not cause sustained flaming of the assembly.

❖ Materials with a lower tolerance to radiant heat (less than 12.5 kW/m²) can be used, but the trade-off is an increased fire separation distance to reduce the potential of ignition from radiant heat.

TABLE 1406.2.1.2
MINIMUM FIRE SEPARATION FOR COMBUSTIBLE VENEERS

FIRE SEPARATION DISTANCE (feet)	TOLERABLE LEVEL INCIDENT RADIANT HEAT ENERGY(kW/m²)	FIRE SEPARATION DISTANCE (feet)	TOLERABLE LEVEL INCIDENT RADIANT HEAT ENERGY(kW/m²)
5	12.5	16	5.9
6	11.8	17	5.5
7	11.0	18	5.2
8	10.3	19	4.9
9	9.6	20	4.6
10	8.9	21	4.4
11	8.3	22	4.1
12	7.7	23	3.9
13	7.2	24	3.7
14	6.7	25	3.5
15	6.3		

For SI: 1 foot = 304.8 mm, 1 Btu/H² × °F = 0.0057 kW/m² × K.

❖ The table provides an increased minimum fire separation distance when the resistance of a material to radiant heat is less than 12.5 kW/m² required for veneers on walls with a separation distance of 5 feet (1524 mm). The required resistance to radiant heat decreases with increasing separation distances. It is not anticipated that interpolation will be involved, since fire separation distances are not enforced in fractions of feet. For example, if testing indicated a tolerable level of radiant heat energy of 9.5 kW/m², the required minimum fire separation distance would be 10 feet (30 480 mm).

1406.2.2 Type I, II, III and IV construction. On buildings of Type I, II, III and IV construction, *exterior wall coverings* shall be permitted to be constructed of wood in accordance with Section 1405.5, or other equivalent combustible material, complying with the following limitations:

1. Combustible *exterior wall coverings* shall not exceed 10 percent of an *exterior wall* surface area where the *fire separation distance* is 5 feet (1524 mm) or less.

2. Combustible architectural *trim* shall be limited to 40 feet (12 192 mm) in height above grade.

3. Combustible *exterior wall coverings* constructed of *fire-retardant-treated wood* complying with Section 2303.2 for exterior installation shall not be limited in wall surface area where the *fire separation distance* is 5 feet (1524 mm) or less and shall be permitted up to 60 feet (18 288 mm) in height above grade regardless of the *fire separation distance*.

❖ All architectural trim and exterior veneers on buildings of Type I, II, III and IV construction that are more than 40 feet (12 192 mm) above grade are required to be noncombustible. This is primarily because of the difficulties associated with fire-fighter access to higher portions of a building and to minimize potential injury from flying brands at high elevation. Combustible architectural trim and exterior veneers are permitted below the 40-foot (12 192 mm) height even when the overall building height exceeds 40 feet (12 192 mm). Combustible trim and veneers are also permitted on buildings of Type V construction. It should be noted that although Table 503 limits buildings of Type V construction to 40 feet (12 192 mm) in height, such buildings may be increased in height when protected with an automatic sprinkler system in accordance with Section 504.2.

In the case of FRTW the material is unlimited in area regardless of the fire separation distance. The only qualification is that the material is limited to 60 feet (18 288 mm) in height above grade. FRTW is pressure treated with chemicals that make the wood more resistant to flame spread. Therefore, it can be expected to perform better than other combustible materials in the resistance to flame.

1406.2.3 Location. Where combustible *exterior wall covering* is located along the top of *exterior walls*, such *trim* shall be completely backed up by the *exterior wall* and shall not extend over or above the top of *exterior walls*.

❖ Combustible trim is not permitted to extend above the exterior wall to which it is attached. This is intended to limit the overall potential involvement of exterior combustible materials in a fire and to reduce the risk of fire spreading to or from such materials from the interior limits of the building. A typical example of architectural trim in this instance is a wood canopy attached to the exterior wall of a strip shopping center, with the canopy fully backed up by the wall.

1406.2.4 Fireblocking. Where the combustible *exterior wall covering* is furred from the wall and forms a solid surface, the distance between the back of the covering and the wall shall not

exceed $1^5/_8$ inches (41 mm). Where required by Section 717, the space thereby created shall be fireblocked.

❖ This section limits the area of concealed space behind combustible veneers installed on the exterior of buildings of Type I, II, III or IV construction. The intent of fireblocking is to limit the potential for fire to spread through the concealed spaces formed by combustible veneers or architectural trim. Fireblocking materials and methods are regulated by Section 717.

1406.3 Balconies and similar projections. Balconies and similar projections of combustible construction other than *fire-retardant-treated wood* shall be fire-resistance rated in accordance with Table 601 for floor construction or shall be of Type IV construction in accordance with Section 602.4. The aggregate length shall not exceed 50 percent of the buildings perimeter on each floor.

Exceptions:

1. On buildings of Type I and II construction, three stories or less above *grade plane*, *fire-retardant-treated wood* shall be permitted for balconies, porches, decks and exterior stairways not used as required exits.

2. Untreated wood is permitted for pickets and rails or similar guardrail devices that are limited to 42 inches (1067 mm) in height.

3. Balconies and similar projections on buildings of Type III, IV and V construction shall be permitted to be of Type V construction, and shall not be required to have a *fire-resistance rating* where sprinkler protection is extended to these areas.

4. Where sprinkler protection is extended to the balcony areas, the aggregate length of the balcony on each floor shall not be limited.

❖ Because these elements are, in a sense, an extension of floor construction, combustible appendages are required to afford the same required fire-resistance rating as required for floor construction in Table 602, unless the appendage is of FRTW or heavy timber construction (Type IV construction). As an additional safeguard against exterior fire spread, the aggregate length of combustible appendages must not exceed 50 percent of the building perimeter on each floor. Balconies, porches, decks, supplemental exterior stairs and similar appendages in buildings of Types I and II construction are required to be constructed of noncombustible materials in order to prevent fire involvement and fire spread up or along the exterior of a noncombustible building. In buildings of Types III, IV and V construction, the use of combustible materials for these elements is permitted.

Exception 1 permits balconies and similar appendages to be constructed of FRTW where buildings of Types I and II construction do not exceed three stories in height. This is due to limited combustibility of FRTW. The three-story limitation is similar to the provisions contained in Section 1406.2.2 for combustible trim.

Exception 2 permits the use of combustible guardrails and handrails for all types of construction in an attempt to alleviate the warping and maintenance problems associated with thin lumber members used for pickets and rails. These items need not be constructed of FRTW.

Exception 3 is applicable to buildings of Types III, IV and V construction. Balconies and similar projections need not be constructed of FRTW nor have a fire-resistance rating when the appendages are protected with an automatic sprinkler system. The presence of sprinkler protection, such as a dry pendent sprinkler, will also serve to limit fire spread from floor to floor.

Balconies, porches, decks and supplemental exterior stairways that are not attached to or supported by the building are separate structures and are to be built accordingly. Although the structural support may be independent, such structures could serve to assist vertical fire spread depending on the construction of the exterior wall and the presence or lack of opening protectives.

1406.4 Bay windows and oriel windows. Bay and oriel windows shall conform to the type of construction required for the building to which they are attached.

> **Exception:** *Fire-retardant-treated wood* shall be permitted on buildings three stories or less of Type I, II, III and IV construction.

❖ In all buildings of other than Type V construction, bay windows and similar appendages are required to be constructed of noncombustible materials (see Figure

1406.4). This is consistent with the requirements for exterior walls in Type I, II, III and IV construction in Sections 602.2 through 602.4. Based upon the limited combustibility of these elements of FRTW, the exception permits their use on buildings of Type I, II, III and IV construction when the elements are constructed of FRTW and the building does not exceed three stories.

SECTION 1407
METAL COMPOSITE MATERIALS (MCM)

1407.1 General. The provisions of this section shall govern the materials, construction and quality of metal composite materials (MCM) for use as *exterior wall coverings* in addition to other applicable requirements of Chapters 14 and 16.

❖ MCMs are panels with a solid plastic core that is encapsulated with an aluminum facer on both sides. MCMs are typically used in exterior applications as a weather covering; however, they do not provide insulation. Section 1407 regulates MCMs that are used as exterior wall coverings.

1407.1.1 Plastic core. The plastic core of the MCM shall not contain foam plastic insulation as defined in Section 2602.1.

❖ MCMs are intended to consist of metal skins covering a thin plastic core. The core is not intended to consist of foam plastic that is regulated under Section 2603. MCM panels are unique and perform significantly different from foam plastic insulation under fire conditions.

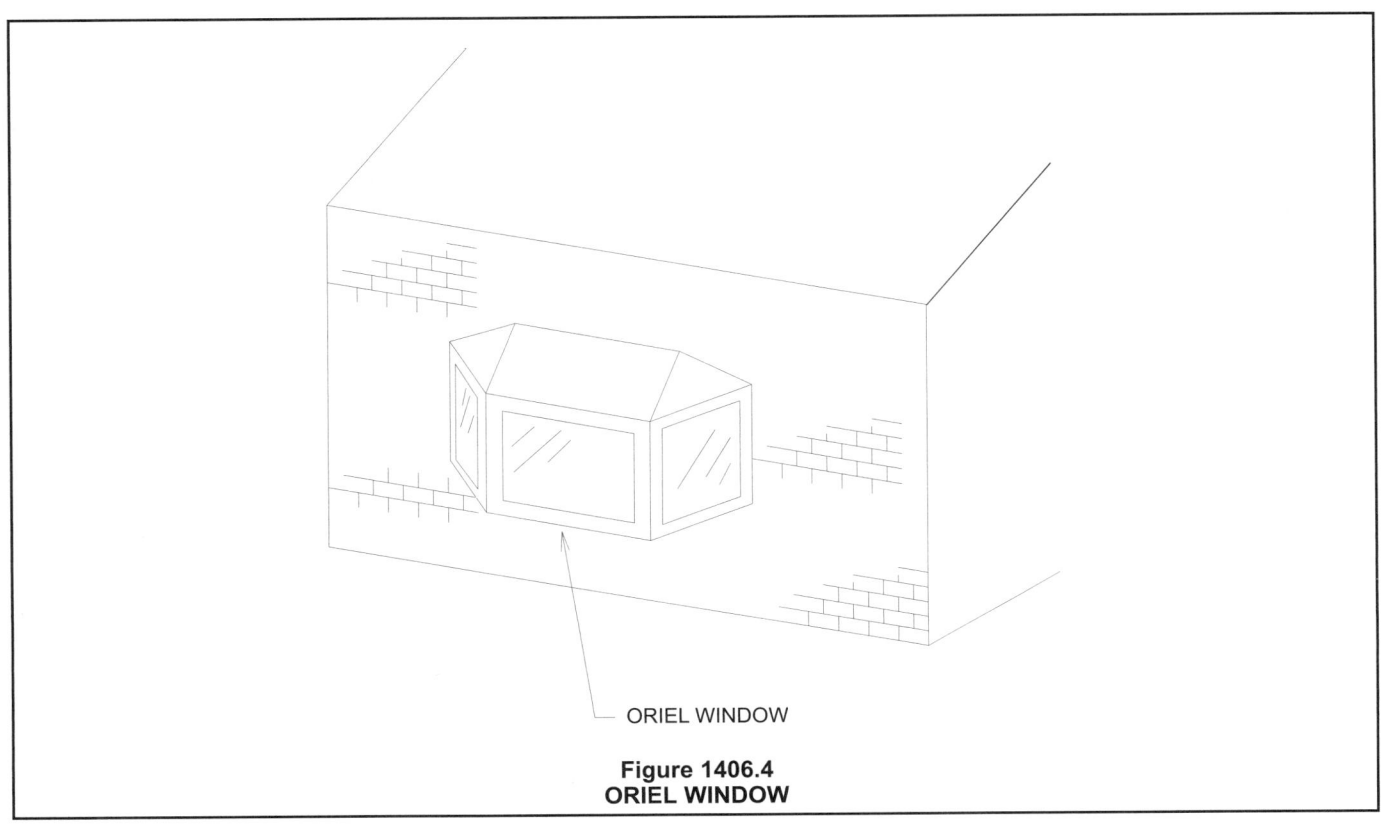

ORIEL WINDOW

Figure 1406.4
ORIEL WINDOW

1407.2 Exterior wall finish. MCM used as *exterior wall* finish or as elements of balconies and similar projections and bay and oriel windows to provide cladding or weather resistance shall comply with Sections 1407.4 through 1407.14.

❖ The plastic core of aluminum composite materials is a combustible material. This section provides a correlation with Sections 1406.2, 1406.3 and 1406.4 regulating the use of combustible materials on the exterior of buildings.

1407.3 Architectural trim and embellishments. MCM used as architectural *trim* or embellishments shall comply with Sections 1407.7 through 1407.14.

❖ In a manner similar to Section 1407.2, the provisions of this section provide a correlation with Section 1406.2.2 regulating the use of combustible trim on the exterior of buildings.

1407.4 Structural design. MCM systems shall be designed and constructed to resist wind loads as required by Chapter 16 for components and cladding.

❖ As previously stated in numerous locations throughout this chapter, exterior weather coverings must be designed and installed to resist the design wind loads required by Chapter 16. These provisions apply to the MCM panels, the supporting elements that are part of the MCM system and the attachment of the system to the building frame.

1407.5 Approval. Results of *approved* tests or an engineering analysis shall be submitted to the *building official* to verify compliance with the requirements of Chapter 16 for wind loads.

❖ The determination that an MCM system will be able to resist a specific wind load may be accomplished through wind load testing, structural engineering calculations or a combination of both. In all instances, documentation substantiating the system's ability to resist the design wind loads is required to be submitted to the building official.

1407.6 Weather resistance. MCM systems shall comply with Section 1403 and shall be designed and constructed to resist wind and rain in accordance with this section and the manufacturer's installation instructions.

❖ As with other exterior weather coverings, MCM systems must comply with the requirements of Section 1403 to prevent the entry of moisture into the building or structure. The manufacturer's installation instructions contain details that are specific to a weather-tight installation.

1407.7 Durability. MCM systems shall be constructed of *approved* materials that maintain the performance characteristics required in Section 1407 for the duration of use.

❖ All materials used as weather coverings are required to exhibit a certain level of durability to be effective in providing protection to the structure and its occupants. MCM panels are a composite construction and the bond between the aluminum facing and plastic core

must be maintained for the material to maintain the performance characteristics specified in Section 1407. For example, if the bond between the panel and core is not durable, debonding could occur. This could result in detrimental performance, such as possible water intrusion.

1407.8 Fire-resistance rating. Where MCM systems are used on exterior walls required to have a *fire-resistance rating* in accordance with Section 705, evidence shall be submitted to the *building official* that the required *fire-resistance rating* is maintained.

Exception: MCM systems not containing foam plastic insulation, which are installed on the outer surface of a fire-resistance-rated *exterior wall* in a manner such that the attachments do not penetrate through the entire *exterior wall* assembly, shall not be required to comply with this section.

❖ Evidence must be submitted to confirm that the installation of MCM systems will not reduce the fire-resistance rating of exterior walls that are required by Section 704.5 to have a rating. This data may consist of testing performed in accordance with ASTM E 119 of the MCM applied to the fire-resistance-rated wall assembly, as provided for in Section 703.2, or an alternative method, such as engineering analysis, as provided for in Section 703.3, respectively (see commentary, Sections 703.2 and 703.3).

1407.9 Surface-burning characteristics. Unless otherwise specified, MCM shall have a *flame spread index* of 75 or less and a smoke-developed index of 450 or less when tested in the maximum thickness intended for use in accordance with ASTM E 84 or UL 723.

❖ MCMs are required to have a basic flame spread rating of 75 and a smoke-developed rating of 450. This is consistent with the requirements for other types of plastics intended for use in building construction, including foam plastics. As indicated in Section 407.10.1, the flame spread rating required is more restrictive (less that 25) when MCMs are installed on buildings of Type I, II, III and IV construction.

1407.10 Type I, II, III and IV construction. Where installed on buildings of Type I, II, III and IV construction, MCM systems shall comply with Sections 1407.10.1 through 1407.10.4, or Section 1407.11.

❖ Due to the combustible core of MCM panels, the fire performance of these panels is regulated. Sections 1407.10.1 through 1407.10.4 stipulate the fire testing that is required to permit the use of these materials on buildings of other than Type V construction. It should be noted that the requirements of these sections are similar to the provisions contained in Chapter 26 for foam plastic materials. Where fire testing is specified, the panels are required to be tested in the maximum thickness intended for use. This is because the maximum thickness represents the greatest potential fuel load in a fire condition.

It should be noted that since MCM panels have an aluminum facing with a minimum thickness of 0.019 inch (0.48 mm), these materials are not required to undergo the ignition-resistance testing required by Section 1406.2.1.

1407.10.1 Surface-burning characteristics. MCM shall have a *flame spread index* of not more than 25 and a smoke-developed index of not more than 450 when tested as an assembly in the maximum thickness intended for use in accordance with ASTM E 84 or UL 723.

❖ The limitations on the maximum flame spread and smoke-developed indexes correspond with the requirements of Section 2603.5.4 for foam plastic materials used on the exterior of buildings of other than Type V construction (see commentary, Section 2603.5.4).

1407.10.2 Thermal barriers. MCM shall be separated from the interior of a building by an *approved* thermal barrier consisting of $^1/_2$-inch (12.7 mm) gypsum wallboard or equivalent thermal barrier material that will limit the average temperature rise of the unexposed surface to not more than 250°F (121°C) after 15 minutes of fire exposure in accordance with the standard time-temperature curve of ASTM E 119 or UL 263. The thermal barrier shall be installed in such a manner that it will remain in place for not less than 15 minutes based on a test conducted in accordance with UL 1715.

❖ The requirements for the installation of a thermal barrier to separate the MCM from the interior of the building are similar to the requirements contained in Section 2603.5.2. The thermal barrier requirements in this section are identical to the requirements of Section 2603.4 (see the commentary to Section 2603.4 for additional discussion on thermal barriers).

1407.10.3 Thermal barrier not required. The thermal barrier specified for MCM in Section 1407.10.2 is not required where:

1. The MCM system is specifically *approved* based on tests conducted in accordance with UL 1040 or UL 1715. Such testing shall be performed with the MCM in the maximum thickness intended for use. The MCM system shall include seams, joints and other typical details used in the installation and shall be tested in the manner intended for use.

2. The MCM is used as elements of balconies and similar projections, architectural *trim* or embellishments.

❖ The exemption of the installation of the thermal barrier required by Section 1407.10.2 may be accomplished in two ways. The first is through the use of large-scale fire testing of the MCM system. The system must be tested in a configuration that incorporates details that are representative of a typical installation. The second provision of this section exempts the installation of a thermal barrier where the MCM is limited to use as a component of balconies and similar projections or as architectural trim and embellishments.

1407.10.4 Full-scale tests. The MCM system shall be tested in accordance with, and comply with, the acceptance criteria of

NFPA 285. Such testing shall be performed on the MCM system with the MCM in the maximum thickness intended for use.

❖ The final data required to permit the use of MCM systems on buildings of Type I, II, III and IV construction are the result of full-scale testing of the system. This testing is intended to determine that the system will not have the tendency to propagate the spread of fire over the surface of the panels or through the panel core.

1407.11 Alternate conditions. MCM and MCM systems shall not be required to comply with Sections 1407.10.1 through 1407.10.4 provided such systems comply with Section 1407.11.1 or 1407.11.2.

❖ Section 1407.10 provides an alternative to compliance with the requirements of Section 1407.9 based upon a specific set of criteria. These criteria limit the amount of material and height above grade in a manner similar to the provisions of Section 1406 for combustible materials.

1407.11.1 Installations up to 40 feet in height. MCM shall not be installed more than 40 feet (12 190 mm) in height above grade where installed in accordance with Sections 1407.11.1.1 and 1407.11.1.2.

❖ The height limitation of 40 feet (12 192 mm) corresponds to the limitations of Section 1406.2.2 for architectural trim. In addition to the height limitation, the material must comply with the requirements of Sections 1407.10.1 and 1407.10.2.

1407.11.1.1 Fire separation distance of 5 feet or less. Where the *fire separation distance* is 5 feet (1524 mm) or less, the area of MCM shall not exceed 10 percent of the *exterior wall* surface.

❖ To limit the potential for the spread of fire between buildings that are in close proximity to one another, the maximum amount of combustible exterior veneer is limited in size. These limitations are identical to those of Section 1406.2.2 for combustible veneers.

1407.11.1.2 Fire separation distance greater than 5 feet. Where the *fire separation distance* is greater than 5 feet (1524 mm), there shall be no limit on the area of *exterior wall* surface coverage using MCM.

❖ Because of the height and area limitations of Section 1407.10, the MCM is permitted to have a higher flame spread than that required under Section 1407.9.1. The maximum flame spread of 75 is identical to the maximum flame spread permitted for foam plastics of Section 2603.3.

1407.11.2 Installations up to 50 feet in height. MCM shall not be installed more than 50 feet (15 240 mm) in height above grade where installed in accordance with Sections 1407.11.2.1 and 1407.11.2.2.

❖ The maximum installed height of MCM on buildings of Type I, II, III and IV construction may be increased up to 50 feet (15 240 mm) when the requirements of Sections 1407.10.2.1 and 1407.10.2.2 are complied with.

1407.11.2.1 Self-ignition temperature. MCM shall have a self-ignition temperature of 650°F (343°C) or greater when tested in accordance with ASTM D 1929.

❖ Keeping with the requirements in Chapter 26 for other plastic materials, the flame spread of the MCM is not permitted to exceed 75 and the smoke-developed index may not exceed 450. In addition to the limitations on flame spread and smoke development, the MCM must be tested for self-ignition in accordance with ASTM D 1929. These requirements are identical to those contained in Section 2606.4 for light-transmitting plastics (see commentary, Section 2606.4).

1407.11.2.2 Limitations. Sections of MCM shall not exceed 300 square feet (27.9 m²) in area and shall be separated by a minimum of 4 feet (1219 mm) vertically.

❖ The size limitations of this section are consistent with the provisions of Section 2605.2, Item 3 and Section 2606.7.3. The area limitation is intended to minimize the amount of concentrated combustible materials on the exterior of noncombustible buildings. The separation requirement is intended to minimize the potential for extensive flame propagation over the face of the wall.

1407.12 Type V construction. MCM shall be permitted to be installed on buildings of Type V construction.

❖ Because of the combustible nature of Type V construction, the fire testing requirements for MCM used as exterior veneers on these buildings is less severe than for other construction types. This is similar in concept to the provisions of Section 1406.

1407.13 Foam plastic insulation. MCM systems containing foam plastic insulation shall also comply with the requirements of Section 2603.

❖ The purpose of this section is to simply point the user of the code to Section 2603 for foam plastics when the MCM contains foam plastic components. The common use of foam plastic in MCM insulation would be the core of the material.

1407.14 Labeling. MCM shall be labeled in accordance with Section 1703.5.

❖ Because of the composite nature of MCMs and the proprietary nature of the panel cores, the fabrication of the panels must be subject to independent third-party verification. The verification of this inspection is evidenced by labeling of the product by the third-party agency. Refer to Section 1704.3 for more discussion on the labeling of products.

SECTION 1408
EXTERIOR INSULATION AND FINISH SYSTEMS (EIFS)

1408.1 General. The provisions of this section shall govern the materials, construction and quality of exterior insulation and finish systems (EIFS) for use as *exterior wall coverings* in addition to other applicable requirements of Chapters 7, 14, 16, 17 and 26.

❖ EIFS and EIFS with drainage are relatively complex exterior veneers that require great care and attention in the materials used and the installation details. This exterior veneer is now dealt with in this separate section. All of the performance criteria, installation, and design are covered in this section. EIFS is defined in Section 1402 of this code as wall cladding systems consisting of a foam plastic insulation material covered by a plastic reinforced portland cement base coat and a textured finish coat. The complexity of this material comes in the installation details. For the material to be able to perform as a watertight barrier, the various layers must be carefully applied to strict specifications.

1408.2 Performance characteristics. EIFS shall be constructed such that it meets the performance characteristics required in ASTM E 2568.

❖ ASTM E 2568 defines the necessary performance of EIFS systems. This standard contains basic test criteria for the materials and the entire system. These are generally tests that provide minimum performance levels for the system, including salt spray, accelerated weathering, ignition resistance, fire endurance and wind resistance.

1408.3 Structural design. The underlying structural framing and substrate shall be designed and constructed to resist loads as required by Chapter 16.

❖ This section addresses the structural design of the supporting members. The section underscores the fact that Section 1408 does not obviate the need to support the EIFS with a structural support system.

1408.4 Weather resistance. EIFS shall comply with Section 1403 and shall be designed and constructed to resist wind and rain in accordance with this section and the manufacturer's application instructions.

❖ The important aspect of this section is that, like all other exterior cladding materials, EIFS must be weather resistant as well. Section 1403 provides the performance criteria for exterior cladding.

1408.4.1 EIFS with drainage. EIFS with drainage shall have an average minimum drainage efficiency of 90 percent when tested in accordance the requirements of ASTM E 2273 and is required on framed walls of Type V construction and Group R1, R2, R3 and R4 occupancies.

❖ For wood-framed walls or any light-frame construction using wood sheathing or other water soluble type of material, it is essential to protect the supporting substrate from moisture. The EIFS system, like all other exterior veneers, must be able to allow water that gets behind the cladding to drain.

1408.4.1.1 Water-resistive barrier. For EIFS with drainage, the *water-resistive barrier* shall comply with Section 1404.2 or ASTM E 2570.

❖ The EIFS systems with drainage are required to have a barrier behind the cladding that serves as the "drainage plane" where all water penetration is stopped and directed downward and out of the wall. The section or referenced standard provide basic test criteria for the materials that would be used behind an EIFS system with drainage.

1408.5 Installation. Installation of the EIFS and EIFS with drainage shall be in accordance with the EIFS manufacturer's instructions.

❖ Manufacturers' instructions provide the proper installation parameters for these materials, including the proper mix proportions for the cement-based substrate, the correct sequence for installation and setting times, among others. The manufacturer's instructions form the job specification for this material.

1408.6 Special inspections. EIFS installations shall comply with the provisions of Sections 1704.1 and 1704.14.

❖ The installation of EIFS systems require attention to detail because the performance of the system can only be assured if the system is installed correctly. This sensitivity requires that special inspections be performed to ensure that ambient conditions were within acceptable range, installation sequences were followed, and all components meet the manufacturer's requirements.

Bibliography

The following resource materials are referenced in this chapter or are relevant to the subject matter addressed in this chapter.

AAMA 1402-86, *Standard Specification for Aluminum Siding, Soffit and Fascia*. Palatine, IL: American Architectural Manufacturers Association, 1986.

ACI 530/ASCE 5/TMS 402-99, *Building Code Requirements for Masonry Structures.* Farmington Hills, MI: American Concrete Institute, 1999.

ACI 530.1/ASCE 6/TMS 602-99, *Specifications for Masonry Structures.* Farmington Hills, MI: American Concrete Institute, 1999.

AHA A135.4-95, *Basic Hardboard*. Palatine, IL: American Hardboard Association, 1995.

AHA A135.6-98, *Hardboard Siding*. Palatine, IL: American Hardboard Association, 1998.

ASTM B 101-02, *Specification for Lead-coated Copper Sheet and Strip for Building Construction*. West Conshohocken, PA: ASTM International, 2002.

ASTM D 1929-96, *Test Method for Ignition Properties of Plastics*. West Conshohocken, PA: ASTM International, 1996.

ASTM D 3679-96a, *Specification for Rigid Poly (Vinyl Chloride)(PVC) Siding*. West Conshohocken, PA: ASTM International, 1996.

ASTM D 4756-96, *Standard Practice for Installation of Rigid Poly (Vinyl Chloride)*. West Conshohocken, PA: ASTM International, 1996.

ASTM E 84-98, *Test Methods for Surface-burning Characteristics of Building Materials*. West Conshohocken, PA: ASTM International, 1998.

ASTM E 96-95, *Standard Test Method for Water Vapor Transmission of Materials*. West Conshohocken, PA: ASTM International, 1995.

ASTM E 119-98, *Test Methods for Fire Tests of Building Construction Materials*. West Conshohocken, PA: ASTM International, 1998.

ASTM E 331-93, *Test Method for Water Penetration of Exterior Walls, Doors by Uniform Static Air Pressure Difference*. West Conshohocken, PA: ASTM International, 1993.

IECC-09, *International Energy Conservation Code*. Washington, DC: International Code Council, 2009.

NFPA 268-96, *Standard Test Method for Determining Ignitability of Exterior Wall Assemblies Using a Radiant Heat Energy Source*. Quincy, MA: National Fire Protection Association, 1996.

Chapter 15:
Roof Assemblies and Rooftop Structures

General Comments

Chapter 15 regulates the materials, design, construction and quality of roofs and roof structures for all buildings and structures. Requirements for weather protection, fire classification, flashings and insulation are provided to govern the roof slab or deck and its supporting members, in addition to the covering that is applied to the roof for weather resistance, fire resistance or appearance. Specific installation requirements for certain steep-slope and low-slope roof coverings are provided.

Section 1502 contains definitions of terms associated with the provisions in this chapter.

Section 1503 provides requirements for roof coverings to protect against the effects of weather.

Section 1504 provides performance requirements including referenced standards for evaluating the performance of new and innovative roof coverings.

Section 1505 provides a classification system to determine a roof covering's effectiveness against certain fire test exposures. Specific classifications of roof coverings are required based on the type of construction classification of the structure.

Section 1506 provides requirements for the application of roof covering materials.

Section 1507 provides requirements for the installation of roof coverings.

Section 1508 establishes requirements for the use of combustible roof insulation in all types of construction.

Section 1509 establishes minimum requirements for specific structures that can be constructed on and above the roof.

Section 1510 provides requirements for reroofing, including adequate structural supports, recovering of existing roof materials and replacement.

Purpose

The provisions in Chapter 15 for roof construction and covering are intended to provide a weather-protective barrier at the roof and, in most circumstances, a fire-retardant barrier to prevent flaming combustible materials, such as flying brands from nearby fires, from penetrating the roof construction. The chapter is essentially prescriptive in nature and is based on decades of experience with various traditional roofing materials. These prescriptive rules are very important for satisfactory performance of the roof covering even though the reason for a particular requirement may be lost. The provisions are based on an attempt to prevent past unsatisfactory performance of the various roofing materials and components.

SECTION 1501
GENERAL

1501.1 Scope. The provisions of this chapter shall govern the design, materials, construction and quality of roof assemblies, and rooftop structures.

❖ This section specifies the scope and applicability of the code for all roofs, roof assemblies and roof structures. Requirements are provided to regulate the materials, design and construction of roofs and rooftop structures such as penthouses, water tanks and other structures. See also the definition of "Roof assembly" in Section 1502. Other roof assemblies addressed elsewhere in the code and not addressed in this chapter include membrane structures (see Section 3102) and light-transmitting plastics (see Section 2606).

SECTION 1502
DEFINITIONS

1502.1 General. The following words and terms shall, for the purposes of this chapter and as used elsewhere in this code, have the meanings shown herein.

❖ Definitions of terms can help in the understanding and application of the code requirements. The purpose for including these definitions within this chapter is to provide more convenient access to them without having to refer back to Chapter 2.

For convenience, these terms are also listed in Chapter 2 with a cross reference to this section. Terms that are italicized provide a visual identification throughout the code that a definition exists for that term.

The use and application of all defined terms, including those defined herein, are set forth in Section 201.

AGGREGATE. In roofing, crushed stone, crushed slag or water-worn gravel used for surfacing for roof coverings.

❖ Aggregate is gravel, stone or slag used as a roof surfacing to provide a walking surface and protection to the roof covering. An aggregate typically is not used as a ballast material based on its small size.

BALLAST. In roofing, ballast comes in the form of large stones or paver systems or light-weight interlocking paver systems and is used to provide uplift resistance for roofing systems that are not adhered or mechanically attached to the roof deck.

❖ Ballast is a material, usually stone or concrete, that is used specifically to provide uplift resistance for roof coverings that are not otherwise attached to the structure.

BUILT-UP ROOF COVERING. Two or more layers of felt cemented together and surfaced with a cap sheet, mineral aggregate, smooth coating or similar surfacing material.

❖ Because of their low melting points and self-healing characteristics, built-up roofs are typically constructed of coal tar membranes that are commonly installed on lower slopes and dead-level roofs.

INTERLAYMENT. A layer of felt or nonbituminous saturated felt not less than 18 inches (457 mm) wide, shingled between each course of a wood-shake roof covering.

❖ According to Section 1507.9.5, interlayment is required to comply with ASTM D 226, Type I, which is commonly referred to as No. 15 asphalt felt.

MECHANICAL EQUIPMENT SCREEN. A partially enclosed *rooftop structure* used to aesthetically conceal heating, ventilating and air conditioning (HVAC) electrical or mechanical equipment from view.

❖ These are used mainly for appearance and to conceal HVAC equipment from view.

METAL ROOF PANEL. An interlocking metal sheet having a minimum installed weather exposure of 3 square feet (0.279 m²) per sheet.

❖ There are two general categories of metal roofing systems: architectural metal roofing and structural metal roofing. Architectural metal roofs are generally water-shedding roof systems and structural metal roofs have hydrostatic (water barrier) characteristics. The difference between a "metal roof panel" and a "metal roof shingle" is the weather exposure (i.e., that portion of the roofing exposed) (see Section 1507.4).

METAL ROOF SHINGLE. An interlocking metal sheet having an installed weather exposure less than 3 square feet (0.279 m²) per sheet.

❖ See the definition of "Metal roof panel."

MODIFIED BITUMEN ROOF COVERING. One or more layers of polymer-modified asphalt sheets. The sheet materials shall be fully adhered or mechanically attached to the substrate or held in place with an *approved* ballast layer.

❖ These are composite sheets consisting of copolymer-modified bitumen, often reinforced and sometimes surfaced with various types of films, foils and mats.

PENTHOUSE. An enclosed, unoccupied structure above the roof of a building, other than a tank, tower, spire, dome cupola or bulkhead.

❖ The definition of "Penthouse" is similar to the definition of "Mezzanine," in that specific area limitations are mandated. Any enclosed structure that is located above the surrounding roof surfaces can be considered a penthouse as long as it meets the criteria within Section 1509.2. By complying with these requirements, the penthouse is considered as part of the story below (see Section 1509.2) and, therefore, does not contribute to the height of the building either in number of stories above grade or in feet. If the proposed penthouse does not meet these requirements, it must be considered as an additional story of the building or structure.

POSITIVE ROOF DRAINAGE. The drainage condition in which consideration has been made for all loading deflections of the roof deck, and additional slope has been provided to ensure drainage of the roof within 48 hours of precipitation.

❖ The primary purpose of positive roof drainage is to allow for prompt roof drainage, thereby preventing the roof from being damaged or adversely impacting the structural support due to the additional load.

REROOFING. The process of recovering or replacing an existing roof covering. See "Roof recover" and "Roof replacement."

❖ This term refers to the process of covering an existing roof system with a new roofing system (see Section 1510).

ROOF ASSEMBLY. A system designed to provide weather protection and resistance to design loads. The system consists of a roof covering and roof deck or a single component serving as both the roof covering and the roof deck. A roof assembly includes the roof deck, *vapor retarder*, substrate or thermal barrier, insulation, *vapor retarder* and roof covering.

The definition of "Roof assembly" is limited in application to the provisions of Chapter 15.

❖ With respect to its application as it relates to Chapter 15, a roof assembly is an assembly of interacting roof components (including the roof deck) designed to weatherproof and, normally, to insulate a building's top surface.

ROOF COVERING. The covering applied to the roof deck for weather resistance, fire classification or appearance.

❖ This definition identifies the specific membrane of the entire roof system that provides weather protection and any required resistance to exterior fire exposure.

The code has specific performance and prescriptive requirements for this covering to provide a durable, weather-resistant surface for the entire structure. A roof covering is considered the membrane that provides the weather-resistance and fire-performance characteristics required by the code.

ROOF COVERING SYSTEM. See "Roof assembly."

❖ See the definition of "Roof assembly."

ROOF DECK. The flat or sloped surface not including its supporting members or vertical supports.

❖ A roof deck is the structural surface to which the roofing and waterproofing system (including insulation) is applied.

ROOF RECOVER. The process of installing an additional roof covering over a prepared existing roof covering without removing the existing roof covering.

❖ This term refers to the process of covering an existing roof system with a new roofing system.

ROOF REPAIR. Reconstruction or renewal of any part of an existing roof for the purposes of its maintenance.

❖ Roofs should be maintained for the purpose of protection. If a section is damaged, then it should be repaired immediately.

ROOF REPLACEMENT. The process of removing the existing roof covering, repairing any damaged substrate and installing a new roof covering.

❖ This definition refers to the process of covering an existing roof system with a new roofing system.

ROOF VENTILATION. The natural or mechanical process of supplying conditioned or unconditioned air to, or removing such air from, attics, cathedral ceilings or other enclosed spaces over which a roof assembly is installed.

❖ Ventilation of the attic prevents moisture condensation on cold surfaces, and, therefore, will prevent dry rot on the bottom surface of shingles or wood roof decks.

ROOFTOP STRUCTURE. An enclosed structure on or above the roof of any part of a building.

❖ This definition includes all appurtenances constructed and located above the surrounding roof surfaces. These items (water tanks or cooling towers) or architectural features (spires or cupolas) are regulated by specific code provisions.

SCUPPER. An opening in a wall or parapet that allows water to drain from a roof.

❖ These devices are for the purpose of allowing overflowing water to drain off the roof. These devices are commonly larger than the roof drain.

SINGLE-PLY MEMBRANE. A roofing membrane that is field applied using one layer of membrane material (either homogeneous or composite) rather than multiple layers.

❖ This is a flexible or semiflexible roof covering or waterproofing layer, whose primary function is the exclusion of water.

UNDERLAYMENT. One or more layers of felt, sheathing paper, nonbituminous saturated felt or other *approved* material over which a steep-slope roof covering is applied.

❖ According to Section 1507.2.3, underlayment is required to comply with ASTM D 226 or ASTM D 4869, Type I, which is commonly referred to as No. 15 asphalt felt.

SECTION 1503
WEATHER PROTECTION

1503.1 General. Roof decks shall be covered with *approved* roof coverings secured to the building or structure in accordance with the provisions of this chapter. Roof coverings shall be designed and installed in accordance with this code and the *approved* manufacturer's instructions such that the roof covering shall serve to protect the building or structure.

❖ The roof covering provides protection from water intrusion into a building. The roofing system has historically been one of the most problematic areas of a building. Without a properly constructed roof assembly, water may infiltrate the building causing damage to building materials and contents. A roof assembly, for the purposes of this chapter, is defined as including the roof deck and components above that make up the roof covering. For a roof covering to perform the way it must to protect the building, it must meet certain requirements. The code provides basic requirements for the construction of roof assemblies. The designer will typically include in his or her roof specification compatibility of materials, deck type, weather conditions, roof slope, structural loads, roof drainage, roof penetrations and future reroofing.

1503.2 Flashing. Flashing shall be installed in such a manner so as to prevent moisture entering the wall and roof through joints in copings, through moisture-permeable materials and at intersections with parapet walls and other penetrations through the roof plane.

❖ Flashing must be installed in a very specific manner so that moisture does not enter the building construction or get under the roof covering. As indicated, the flashing is installed anyplace the roof covering is interrupted or terminated.

1503.2.1 Locations. Flashing shall be installed at wall and roof intersections, at gutters, wherever there is a change in roof slope or direction and around roof openings. Where flashing is of metal, the metal shall be corrosion resistant with a thickness of not less than 0.019 inch (0.483 mm) (No. 26 galvanized sheet).

❖ Flashing consists of components that weatherproof or seal the roof system at discontinuities, such as perimeters, penetrations, walls, expansion joints, valleys, drains and other places, where the roof covering is interrupted or terminated. For example, membrane base flashing covers the edge of the field membrane, and cap flashings or counterflashings shield the upper edges of the base flashing.

Metal flashing must be no less than the specified minimum thickness and be corrosion resistant. "Corrosion resistance" is defined in Section 202 as "The ability of a material to withstand deterioration of its surface or its properties when exposed to its environment." This definition includes metallic and nonmetallic materials.

1503.3 Coping. Parapet walls shall be properly coped with noncombustible, weatherproof materials of a width no less than the thickness of the parapet wall.

❖ Coping is the covering piece on top of a wall which is exposed to the weather, usually made of metal, masonry or stone. It is typically sloped to shed water back onto the roof.

[P] 1503.4 Roof drainage. Design and installation of roof drainage systems shall comply with Section 1503 and the *International Plumbing Code.*

❖ Positive roof drainage is required for essentially two reasons: to prevent excessive ponding of water and the consequent rapid deterioration of the roof-covering material, including potential for early failure of the roof covering; and to prevent excessive ponding and the resulting failure of the roof structure or, alternatively, the need for increased strength of the roof system to support ponding. Section 1611.2 requires the structural design calculations to consider ponding and preclude a progressive deflection of the structural roof members.

1503.4.1 Secondary drainage required. Secondary (emergency) roof drains or scuppers shall be provided where the roof perimeter construction extends above the roof in such a manner that water will be entrapped if the primary drains allow buildup for any reason.

❖ This section requires all buildings to have some method for preventing the accumulation of unplanned excessive rainwater. A secondary drainage system is required where the building has parapet walls or other construction on the perimeter of the building that would cause ponding.

The intent here is to limit the amount of ponding water that will be placed on the roof by rainfall. If the building is designed so water cannot pond on the roof, such as roofs sloped toward the edge of the building, secondary drainage is not required. This simple concept should be carried to all portions of the roof, so that if any portion is designed so water can pond, secondary drains would be required.

Bear in mind that a roof could be intentionally designed to allow ponding to some design depth such as for a rainwater harvesting system.

1503.4.2 Scuppers. When scuppers are used for secondary (emergency overflow) roof drainage, the quantity, size, location and inlet elevation of the scuppers shall be sized to prevent the depth of ponding water from exceeding that for which the roof was designed as determined by Section 1503.4.1. Scuppers shall not have an opening dimension of less than 4 inches

(102 mm). The flow through the primary system shall not be considered when locating and sizing scuppers.

❖ This section provides performance requirements for the situation where wall scuppers are provided as a means of secondary roof drainage. This section also provides minimum prescriptive opening dimensions for scuppers regardless of the results of the performance design. The performance design typically takes into account the flow rate, the length of the scupper opening and the water pressure (head) on the scupper.

1503.4.3 Gutters. Gutters and leaders placed on the outside of buildings, other than Group R-3, private garages and buildings of Type V construction, shall be of noncombustible material or a minimum of Schedule 40 plastic pipe.

❖ All roofs are required to be designed and built with positive drainage (see Section 1507.10.1 for the slope requirements of built-up roofs). "Positive roof drainage" is defined in Section 1502. Ponding water has detrimental effects on a roof system, including roof surface and membrane deterioration due to debris accumulation; vegetation and fungal growth; deck deflections which can lead to structural damage; ice formation; tensile splitting; and the possibility of water entering the building.

1503.5 Roof ventilation. Intake and exhaust vents shall be provided in accordance with Section 1203.2 and the manufacturer's installation instructions.

❖ During cold weather, condensation is deposited on cold surfaces when, for example, warm, moist air rising from the interior of a building and through the attic comes in contact with the roof deck. Ventilation prevents moisture condensation on the cold surfaces and, therefore, will reduce or prevent dry rot on the bottom surface of shingles or wood roof decks.

1503.6 Crickets and saddles. A cricket or saddle shall be installed on the ridge side of any chimney or penetration greater than 30 inches (762 mm) wide as measured perpendicular to the slope. Cricket or saddle coverings shall be sheet metal or of the same material as the roof covering.

❖ A cricket or saddle is a raised roof substrate or structure constructed to channel or direct surface water around a chimney or other similar roof penetration.

SECTION 1504
PERFORMANCE REQUIREMENTS

1504.1 Wind resistance of roofs. Roof decks and roof coverings shall be designed for wind loads in accordance with Chapter 16 and Sections 1504.2, 1504.3 and 1504.4.

❖ It is important that roof coverings and roof decks remain intact and in place when subjected to high winds. Without an intact roof system, the building would be subjected either to water damage, which would reduce its structural stability, or to higher wind pressures for which the building is not designed. The following sections specify requirements for different roof coverings regarding wind loads. The roof decks are included in the requirements of Chapter 16.

1504.1.1 Wind resistance of asphalt shingles. Asphalt shingles shall comply with Section 1507.2.7.

❖ For the installation of asphalt shingles related to wind-resistance performance, this section refers to Section 1507.2.7 (see commentary, Section 1507.2.7).

1504.2 Wind resistance of clay and concrete tile. Wind loads on clay and concrete tile roof coverings shall be in accordance with Section 1609.5.

❖ To determine the minimum wind loads to use in the design and installation of clay and concrete roof tile, this section refers to Section 1609.5. When required by Table 1507.3.7, the tiles must be installed to resist the wind loads as calculated in accordance with Section 1609.5.

1504.3 Wind resistance of nonballasted roofs. Roof coverings installed on roofs in accordance with Section 1507 that are mechanically attached or adhered to the roof deck shall be designed to resist the design wind load pressures for components and cladding in accordance with Section 1609.

❖ This section addresses wind resistance of steep-slope systems that are mechanically attached by fasteners or adhered to the roof slab or deck by adhesives. Steep-slope roof coverings are those described in Section 1507 with a roof slope of 2:12 or greater. The basic wind speed as obtained from Chapter 16 for the building's location must be modified for the building's overall height above grade and exposure category, wind gust effect and wind importance factor. A structural design is required to demonstrate that the proposed fastening schedule will resist the wind uplift loads.

1504.3.1 Other roof systems. Roof systems with built-up, modified bitumen, fully adhered or mechanically attached single-ply through fastened metal panel roof systems, and other types of membrane roof coverings shall also be tested in accordance with FM 4474, UL 580 or UL 1897.

❖ Minimum requirements addressing the wind resistance of low-slope roof systems (typically less than 2:12) are established in this section. Besides compliance with the requirements of Section 1504.3, membrane-type roof coverings are required to be physically tested for compliance with one of three standards: FM 4474, UL 580 or UL 1897.

1504.3.2 Metal panel roof systems. Metal panel roof systems through fastened or standing seam shall be tested in accordance with UL 580 or ASTM E 1592.

Exception: Metal roofs constructed of cold-formed steel, where the roof deck acts as the roof covering and provides both weather protection and support for structural loads,

shall be permitted to be designed and tested in accordance with the applicable referenced structural design standard in Section 2209.1.

❖ This section specifies the test that metal panel roof systems are required to pass before installation. Standards and test methods have been established for the proper installation of metal panel roof systems whether fastened or standing seamed.

The exception recognizes metal decks that provide both weather protection and support for structural loads, including wind and live loads. Many of these metal decks are structural elements where stresses can be calculated just like the design approach used for the roof framing members that support the roof deck. The AISI specification referenced in Section 2209.1 recognizes a wide variety of roofing profiles where calculations are applicable. For roof profiles that cannot be directly calculated, the specification requires additional testing, including many of the tests currently referenced by the code for wind resistance. In addition, unlike many of the wind tests, the specification gives guidance on the selection of appropriate safety factors.

1504.4 Ballasted low-slope roof systems. Ballasted low-slope (roof slope < 2:12) single-ply roof system coverings installed in accordance with Sections 1507.12 and 1507.13 shall be designed in accordance with Section 1504.8 and ANSI/SPRI RP-4.

❖ Sections 1507.12 and 1507.13 prescribe the installation for ballasted low-slope, single-ply roof system coverings. These coverings are to be designed in accordance with Section 1504.8 and ANSI/SPRI RP-4, which is listed in Chapter 35.

1504.5 Edge securement for low-slope roofs. Low-slope membrane roof system metal edge securement, except gutters, shall be designed and installed for wind loads in accordance with Chapter 16 and tested for resistance in accordance with ANSI/SPRI ES-1, except the basic wind speed shall be determined from Figure 1609.

❖ This section references the appropriate standard for low-slope roofs, which is ANSI/SPRI ES 1, listed in Chapter 35. This standard is important in preventing failures for low-slope roof systems. Since the standard references ASCE 7-05, a reference to Figure 1609 is provided to clarify that the basic wind speeds must be determined using Chapter 16.

1504.6 Physical properties. Roof coverings installed on low-slope roofs (roof slope < 2:12) in accordance with Section 1507 shall demonstrate physical integrity over the working life of the roof based upon 2,000 hours of exposure to accelerated weathering tests conducted in accordance with ASTM G 152, ASTM G 155 or ASTM G 154. Those roof coverings that are subject to cyclical flexural response due to wind loads shall not demonstrate any significant loss of tensile strength for unreinforced membranes or breaking strength for reinforced membranes when tested as herein required.

❖ This section, which addresses the effects of weather on the performance of the roof covering, requires that weather testing be performed and includes performance criteria for those roof coverings that will have to perform while subjected to wind loads.

The ASTM test methods reference cover exposure to carbon-arc-type or xenon-arc-type light (with or without water), fluorescent ultraviolet-condensation-type rays and concentrated natural sunlight. In conjunction with these tests, roofing materials must not show any significant signs of failure caused by cyclical wind conditions. The tensile strength in unreinforced membranes and the breaking strength in reinforced membranes must endure such conditions without any significant strength reduction.

1504.7 Impact resistance. Roof coverings installed on low-slope roofs (roof slope < 2:12) in accordance with Section 1507 shall resist impact damage based on the results of tests conducted in accordance with ASTM D 3746, ASTM D 4272, CGSB 37-GP-52M or the "Resistance to Foot Traffic Test" in Section 5.5 of FM 4470.

❖ All roof-covering materials must withstand impact loads, such as hail and driving rainstorms. The referenced standards set criteria for the testing of certain roofing materials on low-slope roofs, such as plastic film and bituminous roofing systems. Each test uses similar devices to test the material. The tests consist of dropping a dart or missile, which weighs from $2^1/_2$ to 5 pounds (1.1 to 2.3 kg), approximately 4 to 5 feet (1219 to 1524 mm) and then evaluating the material for failure.

1504.8 Aggregate. Aggregate used as surfacing for roof coverings and aggregate, gravel or stone used as ballast shall not be used on the roof of a building located in a hurricane-prone region as defined in Section 1609.2, or on any other building with a mean roof height exceeding that permitted by Table 1504.8 based on the exposure category and basic wind speed at the site.

❖ This section recognizes that aggregate, gravel and stone used on roofs can be blown off of roofs that are subjected to hurricane-strength winds, or winds that are increased due to building height. Field assessments of damage to buildings caused by high-wind events have shown that gravel or stone blown from the roofs of buildings has exacerbated damage to other buildings due to breakage of glass. Once glass is broken, higher internal pressures are created within the building, which can result in structural damage to interior walls, and to the cladding of walls and roof surface subjected to negative external pressures. Even where the higher internal pressure is considered and the building designed accordingly, the breakage of windows will generally result in substantial wind and water damage to the building's interior and contents.

TABLE 1504.8
MAXIMUM ALLOWABLE MEAN ROOF HEIGHT PERMITTED FOR BUILDINGS WITH AGGREGATE ON THE ROOF IN AREAS OUTSIDE A HURRICANE-PRONE REGION

BASIC WIND SPEED FROM FIGURE 1609 (mph)[b]	MAXIMUM MEAN ROOF HEIGHT (ft)[a, c]		
	Exposure category		
	B	C	D
85	170	60	30
90	110	35	15
95	75	20	NP
100	55	15	NP
105	40	NP	NP
110	30	NP	NP
115	20	NP	NP
120	15	NP	NP
Greater than 120	NP	NP	NP

For SI: 1 foot = 304.8 mm; 1 mile per hour = 0.447 m/s.
a. Mean roof height as defined in ASCE 7.
b. For intermediate values of basic wind speed, the height associated with the next higher value of wind speed shall be used, or direct interpolation is permitted.
c. NP = gravel and stone not permitted for any roof height.

❖ This table contains criteria that defines locations where aggregate can be used on roofs of buildings outside a hurricane-prone region. The criteria include basic wind speed and exposure category.

SECTION 1505
FIRE CLASSIFICATION

1505.1 General. Roof assemblies shall be divided into the classes defined below. Class A, B and C roof assemblies and roof coverings required to be listed by this section shall be tested in accordance with ASTM E 108 or UL 790. In addition, *fire-retardant-treated wood* roof coverings shall be tested in accordance with ASTM D 2898. The minimum roof coverings installed on buildings shall comply with Table 1505.1 based on the type of construction of the building.

Exception: Skylights and sloped glazing that comply with Chapter 24 or Section 2610.

❖ The code designates the use of any particular classification of roof coverings based on the type of construction of the building. A minimum Class B roof covering is required for all roofs that have a minimum 1-hour fire-resistance rating in accordance with Table 602. Roofs without a required fire-resistance rating require a minimum Class C roof covering.

The exception clarifies that skylights and sloped glazing are not required to be tested for fire classification as roof coverings. Skylights and sloped glazing are considered to be individual assemblies that are required to meet specific requirements listed in other portions of the code. For example, the fire requirements for light-transmitting plastic skylights are covered in Section 2610, and the material and curbing requirements for glass skylights and sloped glazing is given in Chapter 24.

TABLE 1505.1[a, b]
MINIMUM ROOF COVERING CLASSIFICATION
FOR TYPES OF CONSTRUCTION

IA	IB	IIA	IIB	IIIA	IIIB	IV	VA	VB
B	B	B	C[c]	B	C[c]	B	B	C[c]

For SI: 1 foot = 304.8 mm, 1 square foot = 0.0929 m².

a. Unless otherwise required in accordance with the *International Wildland-Urban Interface Code* or due to the location of the building within a fire district in accordance with Appendix D.

b. Nonclassified roof coverings shall be permitted on buildings of Group R-3 and Group U occupancies, where there is a minimum fire-separation distance of 6 feet measured from the leading edge of the roof.

c. Buildings that are not more than two stories above grade plane and having not more than 6,000 square feet of projected roof area and where there is a minimum 10-foot fire-separation distance from the leading edge of the roof to a lot line on all sides of the building, except for street fronts or public ways, shall be permitted to have roofs of No. 1 cedar or redwood shakes and No. 1 shingles.

❖ See the commentary to Section 1505.1.

1505.2 Class A roof assemblies. Class A roof assemblies are those that are effective against severe fire test exposure. Class A roof assemblies and roof coverings shall be *listed* and identified as Class A by an *approved* testing agency. Class A roof assemblies shall be permitted for use in buildings or structures of all types of construction.

Exceptions:

1. Class A roof assemblies include those with coverings of brick, masonry or an exposed concrete roof deck.

2. Class A roof assemblies also include ferrous or copper shingles or sheets, metal sheets and shingles, clay or concrete roof tile or slate installed on noncombustible decks or ferrous, copper or metal sheets installed without a roof deck on noncombustible framing.

❖ Roof coverings that are effective against the severe fire test exposure of ASTM E 108 are classified as Class A. Exception 1 prescribes roof covering materials that are acceptable as Class A roof coverings without having to be tested in accordance with ASTM E 108, based on the past performance of these materials. Traditional Class A roof coverings include the following noncombustible coverings: brick, masonry or exposed concrete. Any assembly that is tested in accordance with ASTM E 108 requirements for severe fire exposure, and that is listed and labeled by an approved testing agency as a Class A roof covering, is allowed as a roof covering on any type of construction designated in the code. Similarly, Exception 2 prescribes roof covering materials that are acceptable as Class A roof coverings as long as they are installed over a noncombustible deck or, in some cases, without a deck but over noncombustible framing. Here again these specific installations would not require ASTM E 108 fire testing. Lastly, note that the materials and specific installations in both exceptions do not require listing or labeling by an approved agency.

1505.3 Class B roof assemblies. Class B roof assemblies are those that are effective against moderate fire-test exposure. Class B roof assemblies and roof coverings shall be *listed* and identified as Class B by an *approved* testing agency.

❖ Roof coverings that are effective against the moderate fire test exposure of ASTM E 108 are classified as Class B. Those roof coverings listed and identified as Class A by an approved testing agency are also allowed to be used where Class B is required. For buildings and structures of any type of construction designated in the code, Class A or B roof coverings are allowed.

1505.4 Class C roof assemblies. Class C roof assemblies are those that are effective against light fire-test exposure. Class C roof assemblies and roof coverings shall be *listed* and identified as Class C by an *approved* testing agency.

❖ Roof coverings that are effective against the light fire test exposure of ASTM E 108 are classified as Class C. There are no specific materials or products that automatically qualify as Class C roof coverings. However, note that those roof coverings listed and identified as Class A or B are allowed to be used where a Class C is required. Class C roof coverings that are listed and labeled by an approved testing agency may only be used on Types IIB, IIIB and VB construction.

1505.5 Nonclassified roofing. Nonclassified roofing is *approved* material that is not *listed* as a Class A, B or C roof covering.

❖ Nonclassified roofing is a roofing material that is approved by the building official that is not otherwise classified (see the definition of "Approved" in Section 202).

1505.6 Fire-retardant-treated wood shingles and shakes. *Fire-retardant-treated wood* shakes and shingles shall be treated by impregnation with chemicals by the full-cell vacuum-pressure process, in accordance with AWPA C1. Each bundle shall be marked to identify the manufactured unit and the manufacturer, and shall also be labeled to identify the classification of the material in accordance with the testing required in Section 1505.1, the treating company and the quality control agency.

❖ This section specifically identifies the particular pressure-impregnation method to be used to treat wood shakes and shingles. This method consists of placing bundles of the roofing materials inside a retort, which is a huge cylindrical pressure container. A vacuum is then created inside the retort, which in turn draws the air and moisture from the millions of cells per square inch in western red cedar. The fire-retardant chemical mixture is then injected at pressures reaching 150 psi (1034 kPa). The chemical is forced into every cell of the shakes and shingles—even the innermost layers. After the products are pressure treated, they are then placed in a special drying kiln where they undergo finishing with a thermal cure at temperatures of up to 2,000°F (1093°C). This section further identifies the test standard to be followed, the method of labeling and the information required per bundle of shingles in order for the building official to evaluate code compliance. This includes the testing requirements found in Section 1505.1.

1505.7 Special purpose roofs. Special purpose wood shingle or wood shake roofing shall conform with the grading and application requirements of Section 1507.8 or 1507.9. In addition, an underlayment of $5/_8$-inch (15.9 mm) Type X water-resistant gypsum backing board or gypsum sheathing shall be placed under minimum nominal $1/_2$-inch-thick (12.7 mm) wood structural panel solid sheathing or 1-inch (25 mm) nominal spaced sheathing.

❖ Informal tests of special purpose roofs have shown that they provide a significant increase in protection for combustible roof coverings, and again the protection offered is considered appropriate if applied in the manner stipulated.

SECTION 1506
MATERIALS

1506.1 Scope. The requirements set forth in this section shall apply to the application of roof-covering materials specified herein. Roof coverings shall be applied in accordance with this chapter and the manufacturer's installation instructions. Installation of roof coverings shall comply with the applicable provisions of Section 1507.

❖ This section establishes specifications and standards for materials and installation techniques in conjunction with recognized industry standards and acceptable roofing applications. By complying with the specified material requirements, slope limitations, underlayment requirements and fastening schedules, the proposed roof covering can be expected to perform as anticipated.

1506.2 Compatibility of materials. Roofs and roof coverings shall be of materials that are compatible with each other and with the building or structure to which the materials are applied.

❖ Material compatibility requires many considerations. For example, some roof coverings might not be compatible with certain roof deck material, certain sealants or solvents, and some are not compatible with oils. Additionally, it is necessary to have the correct combinations of materials to adequately respond to moisture, humidity and energy conservation for buildings of specific materials, in given climates.

1506.3 Material specifications and physical characteristics. Roof-covering materials shall conform to the applicable standards *listed* in this chapter. In the absence of applicable standards or where materials are of questionable suitability, testing by an *approved* agency shall be required by the *building code official* to determine the character, quality and limitations of application of the materials.

❖ This section acknowledges that there may be materials on the market for which the code does not currently

reference a test standard. In those cases, the building official will have to require testing by an approved lab to determine the character, quality and application limitations of the materials. The building official must also require testing when he or she questions the suitability of the materials.

1506.4 Product identification. Roof-covering materials shall be delivered in packages bearing the manufacturer's identifying marks and *approved* testing agency labels required in accordance with Section 1505. Bulk shipments of materials shall be accompanied with the same information issued in the form of a certificate or on a bill of lading by the manufacturer.

❖ Identification of the roofing materials is mandatory in order to verify that they comply with quality standards. In addition to bearing the manufacturer's label or identifying mark, prepared roofing and built-up roofing materials are required by the code to carry a label of an approved agency that inspects the material and finished products during manufacture.

SECTION 1507
REQUIREMENTS FOR ROOF COVERINGS

1507.1 Scope. Roof coverings shall be applied in accordance with the applicable provisions of this section and the manufacturer's installation instructions.

❖ This section requires that the installation be in accordance with the code and the manufacturer's installation instructions. Often other requirements must be met, such as UL or FM requirements. It is very possible that there will be times that these requirements conflict. Once adopted, the code is law. The other provisions are not law, and compliance with them does not necessarily mean that compliance with the code has been achieved. In cases where conflicts arise, the code requirements govern.

1507.2 Asphalt shingles. The installation of asphalt shingles shall comply with the provisions of this section.

❖ This section establishes the criteria for asphalt shingle roofs. Two general types of asphalt shingles specifically apply to this section: strip shingles (e.g., three-tab shingles) and individual interlocking shingles (e.g., T-lock shingles). Asphalt strip shingles, which are the most common, are produced in a variety of finished appearances, including three-tab, random or multiple-tab, no-cutout and laminated architectural.

Asphalt shingles are typically classified in two types: cellulose felt reinforced (i.e., organic shingles) and fiberglass mat reinforced (i.e., fiberglass shingles).

Although it is not specifically addressed in this chapter, the roofing industry generally recommends the attic space below asphalt shingle roofs be properly ventilated.

Additional information regarding asphalt shingle roofs is provided in the *NRCA Roofing Manual: Steep-slope Roof Systems* published by the National Roofing Contractors Association (NRCA).

1507.2.1 Deck requirements. Asphalt shingles shall be fastened to solidly sheathed decks.

❖ Solid sheathed roof decks typically include nominal 1-inch (25 mm) solid lumber or wood structural panels. It is recommended that solid sawn lumber, not more than 8 inches (203 mm) in width, be used. Wider pieces are more likely to warp and cup. The code is specific about the types of decking allowed depending on the material used. For example, wood roof decking is addressed in Chapter 23. Manufacturers may also require the application of their materials to specific roof deck materials for warranty of their products.

1507.2.2 Slope. Asphalt shingles shall only be used on roof slopes of two units vertical in 12 units horizontal (17-percent slope) or greater. For roof slopes from two units vertical in 12 units horizontal (17-percent slope) up to four units vertical in 12 units horizontal (33-percent slope), double underlayment application is required in accordance with Section 1507.2.8.

❖ Asphalt shingles are intended to be applied to a steep roof. A minimum slope is crucial in the performance of such shingles because it determines the surface drainage. Where the roof slope is less than 4:12, water drainage from the roof is slowed down and has a tendency to back up under the roofing and cause leaks. Also, the effect of ice dams at the eaves is more pronounced on low-slope roofs and, as a result, special precautions are necessary for satisfactory performance of the roofing materials.

1507.2.3 Underlayment. Unless otherwise noted, required underlayment shall conform to ASTM D 226, Type I, ASTM D 4869, Type I, or ASTM D 6757.

❖ Asphalt shingle roofs are required to be installed with a felt underlayment. Underlayment serves as secondary protection against wind-driven rain and other moisture penetrations, as well as offering some protection against moisture when individual shingles become damaged or dislodged in high winds. The code also specifies double coverage of the underlayment where the slope is between 2:12 and 4:12 (see Section 1507.2.2). The code requires that underlayment of one layer of No. 15 asphalt felt be provided under asphalt shingle roof coverings with a slope greater than 4:12.

Underlayment also serves as a separator sheet between the shingles and deck. As a separator, the underlayment provides some protection to the shingles from uneven roof deck edges and deck fasteners. Underlayment also helps eliminate a rectangular pattern of ridges in the membrane over insulation or deck joints.

Underlayment offers protection from resins and other chemicals contained in wood board sheathing.

Using the appropriate underlayments, decking materials and shingles will help in meeting the fire classification ratings.

Underlayment for asphalt shingle roofs is required to comply with ASTM D 226, Type I, or ASTM D 4869, Type I. Products complying with the "Type I" designa-

tion of these standards are commonly called No. 15 asphalt felt.

1507.2.4 Self-adhering polymer modified bitumen sheet. Self-adhering polymer modified bitumen sheet shall comply with ASTM D 1970.

❖ Self-adhering polymer modified bitumen sheets meeting ASTM D 1970 requirements are intended for use as underlayment for ice dam protection. These underlayments have an adhesive layer that is exposed by removal of a protective sheet. The top surface of the sheet is suitable to work on during the application of the exposed roofing.

1507.2.5 Asphalt shingles. Asphalt shingles shall comply with ASTM D 225 or ASTM D 3462.

❖ ASTM D 225 is a specification for asphalt roofing in shingle form, composed of single or multiple thicknesses of organic felt saturated and coated on both sides with asphalt and surfaced on the weather side with mineral granules. This standard is well established in the roofing industry. Generally, all organic shingles currently manufactured comply with this standard.

ASTM D 3462 is a specification for asphalt roofing in shingle form, composed of single or multiple thicknesses of glass felt impregnated and coated on both sides with asphalt, and surfaced on the weather side with mineral granules. The shingles may be either locked together during installation, or have a factory-applied self-sealing adhesive. Additionally, asphalt shingles meeting this standard must pass the Class A fire exposure test requirements of ASTM E 108 and the wind-resistance test requirements of ASTM D 7158 (see also Section 1507.2.7.1).

Many of the "20-year warranted" shingles and some "25-year warranted shingles" currently on the market do not comply with ASTM D 3462. Fiberglass shingles complying with ASTM D 3462 will typically be identified as such on the shingle bundle packaging or the manufacturer will supply a written certification of compliance with the standard.

1507.2.6 Fasteners. Fasteners for asphalt shingles shall be galvanized, stainless steel, aluminum or copper roofing nails, minimum 12 gage [0.105 inch (2.67 mm)] shank with a minimum $^3/_8$ inch-diameter (9.5 mm) head, of a length to penetrate through the roofing materials and a minimum of $^3/_4$ inch (19.1 mm) into the roof sheathing. Where the roof sheathing is less than $^3/_4$ inch (19.1 mm) thick, the nails shall penetrate through the sheathing. Fasteners shall comply with ASTM F 1667.

❖ The fasteners allowed are either of metal that is corrosion resistant or galvanized to be corrosion resistant. It is important to use fasteners that are long enough to penetrate the roof covering, flashing, underlayment and into the deck a minimum of $^3/_4$ inch (19.1 mm). When the deck is less than $^3/_4$ inch (19.1 mm) in thickness, the fastener must penetrate through the deck. It is recommended that the nail be at least $^1/_8$ inch (3.2 mm) through the deck.

1507.2.7 Attachment. Asphalt shingles shall have the minimum number of fasteners required by the manufacturer, but not less than four fasteners per strip shingle or two fasteners per individual shingle. Where the roof slope exceeds 21 units vertical in 12 units horizontal (21:12), shingles shall be installed as required by the manufacturer.

❖ This section clarifies that the manufacturers instructions must be consulted when installing asphalt shingles, unless the manufacturer's instructions contain fastening requirements less than those prescribed in this section. As a minimum, there needs to be four fasteners for each strip-type shingle and two fasteners for individual shingles.

For roof slopes greater than 21:12, the manufacturer should be consulted for installation requirements. Typically such an installation will include a 5-inch (127 mm) exposure, with each tab cemented in place using asphalt roofing cement compatible with the shingle.

1507.2.7.1 Wind resistance. Asphalt shingles shall be tested in accordance with ASTM D 7158. Asphalt shingles shall meet the classification requirements of Table 1507.2.7.1(1) for the appropriate maximum basic wind speed. Asphalt shingle packaging shall bear a label to indicate compliance with ASTM D 7158 and the required classification in Table 1507.2.7.1(1).

Exception: Asphalt shingles not included in the scope of ASTM D 7158 shall be tested and labeled to indicate compliance with ASTM D 3161 and the required classification in Table 1507.2.7.1(2).

❖ This section recognizes two test methods for the wind resistance of asphalt shingles: ASTM D 7158, which provides for a method of testing that is appropriate for sealed asphalt shingles, and ASTM D 3161, which provides for a method of testing that is appropriate for unsealed asphalt shingles. Each of these standards results in shingle "Classification" based on the ability of the asphalt shingle installation to resist wind uplift. Tables 1507.2.1(1) and 1507.2.1(2) contain minimum classification requirements based on the maximum basic wind speed as determined from Figure 1609.

TABLE 1507.2.7.1(1)
CLASSIFICATION OF ASPHALT ROOF
SHINGLES PER ASTM D 7158[a]

MAXIMUM BASIC WIND SPEED FROM FIGURE 1609	CLASSIFICATION REQUIREMENT
85	D, G or H
90	D, G or H
100	G or H
110	G or H
120	G or H
130	H
140	H
150	H

a. The standard calculations contained in ASTM D 7158 assume exposure category B or C and building height of 60 feet (18 288 mm) or less. Additional calculations are required for conditions outside of these assumptions.

TABLE 1507.2.7.1(2)
CLASSIFICATION OF ASPHALT SHINGLES PER ASTM D 3161

MAXIMUM BASIC WIND SPEED FROM FIGURE 1609	CLASSIFICATION REQUIREMENT
85	A, D or F
90	A, D or F
100	A, D or F
110	F
120	F
130	F
140	F
150	F

❖ Tables 1507.2.1(1) and 1507.2.1(2) contain minimum classification requirements based on the maximum basic wind speed as determined from Figure 1609 (see commentary, Section 1507.2.7.1).

1507.2.8 Underlayment application. For roof slopes from two units vertical in 12 units horizontal (17-percent slope) and up to four units vertical in 12 units horizontal (33-percent slope), underlayment shall be two layers applied in the following manner. Apply a minimum 19-inch-wide (483 mm) strip of underlayment felt parallel with and starting at the eaves, fastened sufficiently to hold in place. Starting at the eave, apply 36-inch-wide (914 mm) sheets of underlayment overlapping successive sheets 19 inches (483 mm), by fastened sufficiently to hold in place. Distortions in the underlayment shall not interfere with the ability of the shingles to seal. For roof slopes of four units vertical in 12 units horizontal (33-percent slope) or greater, underlayment shall be one layer applied in the following manner. Underlayment shall be applied shingle fashion, parallel to and starting from the eave and lapped 2 inches (51 mm), fastened sufficiently to hold in place. Distortions in the underlayment shall not interfere with the ability of the shingles to seal.

❖ Roofs of low slope, 2:12 to 4:12, shed water much more slowly than steeper roofs and therefore require greater protection from water backing up under the shingles. This is particularly important when considering wind-driven water and ice damming. For such slopes, two layers of underlayment are required. For greater slopes, only one layer of underlayment is necessary, except at eaves where ice dams are potentially a problem. Installation of underlayment material over a distorted surface can result in reduced wind resistance and weather (moisture) resistance, and poor aesthetics. Regardless of the slope, underlayment need only be fastened well enough to stay in place until the shingles are applied. The application of shingles will also serve to fasten the underlayment.

1507.2.8.1 High wind attachment. Underlayment applied in areas subject to high winds (greater than 110 mph in accordance with Figure 1609) shall be applied with corrosion-resistant fasteners in accordance with the manufacturer's instructions. Fasteners are to be applied along the overlap at a maximum spacing of 36 inches (914 mm) on center.

❖ The fasteners must be approved for asphalt shingle installation, and should be flat headed and corrosion resistant.

1507.2.8.2 Ice barrier. In areas where there has been a history of ice forming along the eaves causing a backup of water, an ice barrier that consists of at least two layers of underlayment cemented together or of a self-adhering polymer modified bitumen sheet shall be used in lieu of normal underlayment and extend from the lowest edges of all roof surfaces to a point at least 24 inches (610 mm) inside the *exterior wall* line of the building.

> **Exception:** Detached accessory structures that contain no conditioned floor area.

❖ Ice dams form when snow melts over the warmer parts of a roof and refreezes over the colder eaves. This ice formation acts like a dam and causes water to back up beneath the roof covering. The water will eventually leak causing damage to the structure, including the walls, ceilings and roof [see commentary, Figures 1507.8.4(1) and (2)]. In areas where this is prevalent, the installation of an ice barrier is required in accordance with this section.

There is an exception to this section that exempts accessory buildings from such restrictions as they are unheated structures where the need for protection against ice dams is unnecessary. The same exception is found in Sections 1507.5.4, 1507.6.4, 1507.7.4, 1507.8.4 and 1507.9.4.

1507.2.9 Flashings. Flashing for asphalt shingles shall comply with this section. Flashing shall be applied in accordance with this section and the asphalt shingle manufacturer's printed instructions.

❖ This section establishes the requirement for roof flashings for asphalt shingles to prevent leakage and further establishes performance criteria.

1507.2.9.1 Base and cap flashing. Base and cap flashing shall be installed in accordance with the manufacturer's instructions. Base flashing shall be of either corrosion-resistant metal of minimum nominal 0.019-inch (0.483 mm) thickness or mineral-surfaced roll roofing weighing a minimum of 77 pounds per 100 square feet (3.76 kg/m²). Cap flashing shall be corrosion-resistant metal of minimum nominal 0.019-inch (0.483 mm) thickness.

❖ Roof system edges must be weatherproofed or sealed where they intersect with other vertical components, such as walls or chimneys. Flashing is composed of two parts: the base and cap. For a general discussion of roof flashings, see the commentary to Section 1503.2.

1507.2.9.2 Valleys. Valley linings shall be installed in accordance with the manufacturer's instructions before applying shingles. Valley linings of the following types shall be permitted:

1. For open valleys (valley lining exposed) lined with metal, the valley lining shall be at least 24 inches (610 mm) wide and of any of the corrosion-resistant metals in Table 1507.2.9.2.

2. For open valleys, valley lining of two plies of mineral-surfaced roll roofing complying with ASTM D

3909 or ASTM D 6380 shall be permitted. The bottom layer shall be 18 inches (457 mm) and the top layer a minimum of 36 inches (914 mm) wide.

3. For closed valleys (valleys covered with shingles), valley lining of one ply of smooth roll roofing complying with ASTM D 6380, and at least 36 inches (914 mm) wide or types as described in Item 1 or 2 above shall be permitted. Self-adhering polymer modified bitumen underlayment complying with ASTM D 1970 shall be permitted in lieu of the lining material. Valleys are the internal angle formed by the intersection of two sloping roof planes. For asphalt shingle roofs, the valley protection is categorized as open or closed. An open valley is a method of construction in which the steep-slope roofing on both sides is trimmed along each side of the valley, exposing the valley flashing.

Closed valleys are those that are covered with shingles, and include methods known as closed-cut valleys and woven valleys.

Closed-cut valleys are a method of valley construction in which shingles from one side of the valley extend across the valley, while shingles from the other side are trimmed back approximately 2 inches (51 mm) from the valley centerline.

Woven valley is a method of valley construction in which shingles from both sides of the valley extend across the valley and are woven together by overlapping alternate courses as they are applied.

TABLE 1507.2.9.2. See page 15-12.

❖ These corrosion-resistant metals may be used in open-valley construction as valley lining where the lining is exposed.

1507.2.9.3 Drip edge. Provide drip edge at eaves and gables of shingle roofs. Overlap to be a minimum of 2 inches (51 mm). Eave drip edges shall extend 1/4 inch (6.4 mm) below sheathing and extend back on the roof a minimum of 2 inches (51 mm). Drip edge shall be mechanically fastened a maximum of 12 inches (305 mm) o.c.

❖ Drip edge is a metal flashing or other overhanging component applied at the roof edge intended to control the direction of dripping water and help protect the underlying building components. Drip edge has an outward projecting lower edge to direct the water away from the building. It is also used to break the continuity of contact between the roof perimeter and the wall components to help prevent capillary action.

Metal drip edge is a formed metal flashing that extends back from, and bends down over, the roof edge. Along roof gables, rakes and the eave, the edge drip is applied over the underlayment.

1507.3 Clay and concrete tile. The installation of clay and concrete tile shall comply with the provisions of this section.

❖ Additional information regarding clay and concrete tile roofs is provided in the *NRCA Roofing Manual: Steep-slope Roof Systems* published by the NRCA.

TABLE 1507.2.9.2
VALLEY LINING MATERIAL

MATERIAL	MINIMUM THICKNESS	GAGE	WEIGHT
Aluminum	0.024 in.	—	—
Cold-rolled copper	0.0216 in.	—	ASTM B 370, 16 oz. per square ft.
Copper	—	—	16 oz
Galvanized steel	0.0179 in.	26 (zinc-coated G90)	—
High-yield copper	0.0162 in.	—	ASTM B 370, 12 oz. per square ft.
Lead	—	—	2.5 pounds
Lead-coated copper	0.0216 in.	—	ASTM B 101, 16 oz. per square ft.
Lead-coated high-yield copper	0.0162 in.	—	ASTM B 101, 12 oz. per square ft.
Painted terne	—	—	20 pounds
Stainless steel	—	28	—
Zinc alloy	0.027 in.	—	—

For SI: 1 inch = 25.4 mm, 1 pound = 0.454 kg, 1 ounce = 28.35 g, 1 square foot = 0.093 m².

1507.3.1 Deck requirements. Concrete and clay tile shall be installed only over solid sheathing or spaced structural sheathing boards.

❖ The choice of deck material is critical to the life of the roof system. Decks must be capable of withstanding loads such as the weight of the tile and associated roof components and accessories. The weight of tile can increase significantly due to water absorption, depending on its porosity. It is important for designers to note that some types of roof tiles may develop increased porosity with age.

1507.3.2 Deck slope. Clay and concrete roof tile shall be installed on roof slopes of $2^{1}/_{2}$ units vertical in 12 units horizontal (21-percent slope) or greater. For roof slopes from $2^{1}/_{2}$ units vertical in 12 units horizontal (21-percent slope) to four units vertical in 12 units horizontal (33-percent slope), double underlayment application is required in accordance with Section 1507.3.3.

❖ Along with the manufacturer's installation specifications, this section gives specific limitations on the application of concrete and clay tile. Thus, tile is not allowed to be installed on slopes less than $2^{1}/_{2}$:12 regardless of the manufacturer's literature. Additionally, for low-sloped roofs, double underlayment is required.

1507.3.3 Underlayment. Unless otherwise noted, required underlayment shall conform to: ASTM D 226, Type II; ASTM D 2626 or ASTM D 6380, Class M mineral-surfaced roofing.

❖ Because of the long service life of many tile roofs, an underlayment should be chosen that will provide protection for a comparable period of time. ASTM D 226 includes the physical requirements for Type II asphalt-saturated felt, which is commonly called No. 30 asphalt felt.

ASTM D 2626 covers the felt base sheet with fine mineral surfacing on the top side, with or without perforations. It is intended that nonperforated felt be used in this application. The *NRCA Roofing Manual: Steep-*

slope Roof Systems has recommendations regarding the choice of underlayment for specific roof slopes and applications.

ASTM D 6380, Class M, designates mineral-surfaced roll-roofing products.

1507.3.3.1 Low-slope roofs. For roof slopes from $2^{1}/_{2}$ units vertical in 12 units horizontal (21-percent slope), up to four units vertical in 12 units horizontal (33-percent slope), underlayment shall be a minimum of two layers applied as follows:

1. Starting at the eave, a 19-inch (483 mm) strip of underlayment shall be applied parallel with the eave and fastened sufficiently in place.

2. Starting at the eave, 36-inch-wide (914 mm) strips of underlayment felt shall be applied overlapping successive sheets 19 inches (483 mm) and fastened sufficiently in place.

❖ The code requires that the underlayment be laid with two layers, which provides two thicknesses of the underlayment at any point.

1507.3.3.2 High-slope roofs. For roof slopes of four units vertical in 12 units horizontal (33-percent slope) or greater, underlayment shall be a minimum of one layer of underlayment felt applied shingle fashion, parallel to, and starting from the eaves and lapped 2 inches (51 mm), fastened only as necessary to hold in place.

❖ A single layer of underlayment is required on all roof slopes of 4:12 or greater. Reinforced underlayment is required on all roofs with spaced sheathing in order to avoid breakthrough of the tiles.

1507.3.4 Clay tile. Clay roof tile shall comply with ASTM C 1167.

❖ ASTM C 1167 covers clay tiles intended for use as a roof covering where durability and appearance are required to provide a weather-resistant surface of specified design. These tiles are made of clay, shale or other earthy substances and are heat treated at an elevated temperature. The process develops a fired bond between the particulate constituents to provide the strength and durability requirements of ASTM C 1167. The tiles are classified into three grades based on their resistance to weathering. Additionally, the standard covers performance characteristics, including durability, freezing and thawing, strength, efflorescence, permeability and reactive particulates.

1507.3.5 Concrete tile. Concrete roof tile shall comply with ASTM C 1492.

❖ ASTM C 1492 is the standard that covers the physical requirements for concrete roof tiles. The water absorption requirement in ASTM C 1492 for normal roof weight tiles is 12^1/$_2$ percent, which is lower than the 15 percent previously listed in the code. In addition, ASTM C 1492 adds clarity with regards to the water absorption requirements for medium- and lightweight roof tiles. ASTM C 140 is still referenced for the testing procedures in ASTM C 1492 even though it is no longer directly referenced in the code text of Section 1508.3.5.

1507.3.6 Fasteners. Tile fasteners shall be corrosion resistant and not less than 11 gage, ^{5}/$_{16}$-inch (8.0 mm) head, and of sufficient length to penetrate the deck a minimum of ^{3}/$_4$ inch (19.1 mm) or through the thickness of the deck, whichever is less. Attaching wire for clay or concrete tile shall not be smaller than 0.083 inch (2.1 mm). Perimeter fastening areas include three tile courses but not less than 36 inches (914 mm) from either side of hips or ridges and edges of eaves and gable rakes.

❖ Fasteners are to be corrosion resistant, such as hot-dipped galvanized steel, aluminum or stainless steel, needle or diamond pointed, with large flat heads.

1507.3.7 Attachment. Clay and concrete roof tiles shall be fastened in accordance with Table 1507.3.7.

❖ Refer to the commentary to Section 1504.2 for other information in the code related to the wind resistance of clay and concrete roof tile.

TABLE 1507.3.7. See page 15-14.

❖ The table covers wind speed, mean roof height and roof slope requirements for fastener installation. It also covers interlocking requirements for specific areas and criteria for roofs.

1507.3.8 Application. Tile shall be applied according to the manufacturer's installation instructions, based on the following:

1. Climatic conditions.
2. Roof slope.
3. Underlayment system.
4. Type of tile being installed.

❖ Clay and concrete tile come in two generic forms: roll tile and flat tile. Either one may be interlocking but must be applied according to the manufacturer's instructions.

1507.3.9 Flashing. At the juncture of the roof vertical surfaces, flashing and counterflashing shall be provided in accordance with the manufacturer's installation instructions, and where of metal, shall not be less than 0.019-inch (0.48 mm) (No. 26 galvanized sheet gage) corrosion-resistant metal. The valley flashing shall extend at least 11 inches (279 mm) from the centerline each way and have a splash diverter rib not less than 1 inch (25 mm) high at the flow line formed as part of the flashing. Sections of flashing shall have an end lap of not less than 4 inches (102 mm). For roof slopes of three units vertical in 12 units horizontal (25-percent slope) and over, the valley flashing shall have a 36-inch-wide (914 mm) underlayment of either one layer of Type I underlayment running the full length of the valley, or a self-adhering polymer-modified bitumen sheet complying with ASTM D 1970, in addition to other required underlayment. In areas where the average daily temperature in January is 25°F (-4°C) or less or where there is a possibility of ice forming along the eaves causing a backup of water, the metal valley flashing underlayment shall be solid cemented to the roofing underlayment for slopes under seven units vertical in 12 units horizontal (58-percent slope) or self-adhering polymer-modified bitumen sheet shall be installed.

❖ Flashings are required to maintain the integrity of weather-resistant roofs (see commentary, Section 1503.2).

1507.4 Metal roof panels. The installation of metal roof panels shall comply with the provisions of this section.

❖ There are two general categories of metal roofing systems: architectural metal roofing and structural metal roofing. Architectural metal roofs are generally water-shedding roof systems and structural metal roofs have hydrostatic (water barrier) characteristics.

Architectural metal roof systems are usually characterized by a flat pan with ^{3}/$_4$- to 1^1/$_2$-inch (19.1 to 38 mm) ribs on each side. The absence of intermediate ribs and massive side ribs gives a clean appearance, but does not provide the panels with the strength to be considered a structural panel. Architectural metal roofing systems are typically designed for steep slopes so that water will shed off the roof panels. This type of design does not require the seams to be water tight. Therefore, the slope for such nonwater-tight roof panels is required to be not less than 3:12. One exception is traditional flat seamed, soldered or welded metal roofing, which is acceptable on slopes less than 3:12. An example of roofing that has traditionally been used this way is copper. Architectural metal roofing systems require solid decking.

"Metal roof panels" are defined in Section 1502 as an interlocking metal sheet having a minimum installed weather exposure of 3 square feet (0.28 m²) per sheet. In contrast, metal roof shingles have less than 3 square feet (0.28 m²) of exposure.

TABLE 1507.3.7
CLAY AND CONCRETE TILE ATTACHMENT[a, b, c]

GENERAL — CLAY OR CONCRETE ROOF TILE			
Maximum basic wind speed (mph)	Mean roof height (feet)	Roof slope up to < 3:12	Roof slope 3:12 and over
85	0-60	One fastener per tile. Flat tile without vertical laps, two fasteners per tile.	Two fasteners per tile. Only one fastener on slopes of 7:12 and less for tiles with installed weight exceeding 7.5 lbs./sq. ft. having a width no greater than 16 inches.
100	0-40		
100	> 40-60	The head of all tiles shall be nailed. The nose of all eave tiles shall be fastened with approved clips. All rake tiles shall be nailed with two nails. The nose of all ridge, hip and rake tiles shall be set in a bead of roofer's mastic.	
110	0-60	The fastening system shall resist the wind forces in Section 1609.5.3.	
120	0-60	The fastening system shall resist the wind forces in Section 1609.5.3.	
130	0-60	The fastening system shall resist the wind forces in Section 1609.5.3.	
All	> 60	The fastening system shall resist the wind forces in Section 1609.5.3.	

INTERLOCKING CLAY OR CONCRETE ROOF TILE WITH PROJECTING ANCHOR LUGS[d, e] (Installations on spaced/solid sheathing with battens or spaced sheathing)				
Maximum basic wind speed (mph)	Mean roof height (feet)	Roof slope up to < 5:12	Roof slope 5:12 < 12:12	Roof slope 12:12 and over
85	0-60	Fasteners are not required. Tiles with installed weight less than 9 lbs./sq. ft. require a minimum of one fastener per tile.	One fastener per tile every other row. All perimeter tiles require one fastener. Tiles with installed weight less than 9 lbs./sq. ft. require a minimum of one fastener per tile.	One fastener required for every tile. Tiles with installed weight less than 9 lbs./sq. ft. require a minimum of one fastener per tile.
100	0-40			
100	> 40-60	The head of all tiles shall be nailed. The nose of all eave tiles shall be fastened with approved clips. All rake tiles shall be nailed with two nails The nose of all ridge, hip and rake tiles shall be set in a bead of roofers's mastic.		
110	0-60	The fastening system shall resist the wind forces in Section 1609.5.3.		
120	0-60	The fastening system shall resist the wind forces in Section 1609.5.3.		
130	0-60	The fastening system shall resist the wind forces in Section 1609.5.3.		
All	> 60	The fastening system shall resist the wind forces in Section 1609.5.3.		

INTERLOCKING CLAY OR CONCRETE ROOF TILE WITH PROJECTING ANCHOR LUGS (Installations on solid sheathing without battens)		
Maximum basic wind speed (mph)	Mean roof height (feet)	All roof slopes
85	0-60	One fastener per tile.
100	0-40	One fastener per tile.
100	> 40-60	The head of all tiles shall be nailed. The nose of all eave tiles shall be fastened with approved clips. All rake tiles shall be nailed with two nails The nose of all ridge, hip and rake tiles shall be set in a bead of roofer's mastic.
110	0-60	The fastening system shall resist the wind forces in Section 1609.5.3.
120	0-60	The fastening system shall resist the wind forces in Section 1609.5.3.
130	0-60	The fastening system shall resist the wind forces in Section 1609.5.3.
All	> 60	The fastening system shall resist the wind forces in Section 1609.5.3.

For SI: 1 inch = 25.4 mm, 1 foot = 304.8 mm, 1 mile per hour = 0.447 m/s, 1 pound per square foot = 4.882 kg/m².

a. Minimum fastener size. Corrosion-resistant nails not less than No. 11 gage with $^5/_{16}$-inch head. Fasteners shall be long enough to penetrate into the sheathing $^3/_4$ inch or through the thickness of the sheathing, whichever is less. Attaching wire for clay and concrete tile shall not be smaller than 0.083 inch.

b. Snow areas. A minimum of two fasteners per tile are required or battens and one fastener.

c. Roof slopes greater than 24:12. The nose of all tiles shall be securely fastened.

d. Horizontal battens. Battens shall be not less than 1 inch by 2 inch nominal. Provisions shall be made for drainage by a minimum of $^1/_8$-inch riser at each nail or by 4-foot-long battens with at least a $^1/_2$-inch separation between battens. Horizontal battens are required for slopes over 7:12.

e. Perimeter fastening areas include three tile courses but not less than 36 inches from either side of hips or ridges and edges of eaves and gable rakes.

1507.4.1 Deck requirements. Metal roof panel roof coverings shall be applied to a solid or closely fitted deck, except where the roof covering is specifically designed to be applied to spaced supports.

❖ The deck for metal roofing is required to be solid or closely fitted, except where the panels are specifically designed to be applied to spaced supports. Structural standing seam metal panel roof systems possess strength characteristics that allow them to span between structural supports, as is commonly used on preengineered metal buildings. The metal panel ribs are not seamed or interlocked like a true standing seam metal roof system.

1507.4.2 Deck slope. Minimum slopes for metal roof panels shall comply with the following:

1. The minimum slope for lapped, nonsoldered seam metal roofs without applied lap sealant shall be three units vertical in 12 units horizontal (25-percent slope).

2. The minimum slope for lapped, nonsoldered seam metal roofs with applied lap sealant shall be one-half unit vertical in 12 units horizontal (4-percent slope). Lap sealants shall be applied in accordance with the *approved* manufacturer's installation instructions.

3. The minimum slope for standing seam of roof systems shall be one-quarter unit vertical in 12 units horizontal (2-percent slope).

❖ Typically, structural metal roofing systems are designed to resist water at laps and seams with sealants applied in the seams. It is usually recommended that these types of roof systems have a slope of not less than $^1/_2$:12, although some manufacturers allow slopes as low as $^1/_4$:12. While the code requires a minimum slope of 3:12, a lower slope is permitted on standing-seam-type roof systems, which require only $^1/_4$:12. The term "standing seam" is used to refer to almost any roof panel with a raised vertical rib. Strictly speaking, it only indicates those metal panels that interlock or are seamed together vertically above the panel's pan.

1507.4.3 Material standards. Metal-sheet roof covering systems that incorporate supporting structural members shall be designed in accordance with Chapter 22. Metal-sheet roof coverings installed over structural decking shall comply with Table 1507.4.3(1). The materials used for metal-sheet roof coverings shall be naturally corrosion resistant or provided with corrosion resistance in accordance with the standards and minimum thicknesses shown in Table 1507.4.3(2).

❖ The requirements for metal roof coverings that incorporate supporting structural members into their design are contained in Chapter 22. Other metal-sheet roof coverings are required to comply with Table 1507.4.3(1).

Many of the materials listed in Table 1507.4.3(1) are inherently corrosion resistant, such as aluminum, copper and lead. For these corrosion-resistant materials a material standard is listed for the base material; however, for steel roofing products, coatings are added to the base material to provide the necessary corrosion resistance.

TABLE 1507.4.3(1). See page 15-16.

❖ This table provides specific guidance and appropriate metal roof coverings. The requirements vary based upon the type of metal used, such as aluminum, steel, copper, lead and zinc. In some cases, standards are referenced; in other cases, basic criteria is provided.

TABLE 1507.4.3(2). See page 15-16.

❖ Steel coating standards are located in Table 1507.4.3(2). Having this table separate from Table 1507.4.3(1) helps to clarify that these standards only address the process necessary to provide corrosion resistance for the base steel. The coating descriptions in Table 1507.4.3(2) are consistent with the applicable coating standards.

1507.4.4 Attachment. Metal roof panels shall be secured to the supports in accordance with the *approved* manufacturer's fasteners. In the absence of manufacturer recommendations, the following fasteners shall be used:

1. Galvanized fasteners shall be used for steel roofs.

2. Copper, brass, bronze, copper alloy or 300 series stainless-steel fasteners shall be used for copper roofs.

3. Stainless-steel fasteners are acceptable for all types of metal roofs.

❖ Fasteners used to attach metal roofing to their supporting construction are to have the manufacturer's approval for that purpose. If such approval is not provided, then galvanized fasteners are required for steel roofs; copper, brass, bronze, copper alloy or 300 series stainless steel fasteners are required for copper roofs and stainless-steel fasteners are permitted on any type of metal roof. The purpose of these provisions is to provide compatibility with roof materials and prevent corrosion.

1507.5 Metal roof shingles. The installation of metal roof shingles shall comply with the provisions of this section.

❖ See the commentary to Section 1507.4 for a general discussion about metal roof systems.

1507.5.1 Deck requirements. Metal roof shingles shall be applied to a solid or closely fitted deck, except where the roof covering is specifically designed to be applied to spaced sheathing.

❖ The manufacturer's instructions must be followed in addition to any restrictions required by the code.

1507.5.2 Deck slope. Metal roof shingles shall not be installed on roof slopes below three units vertical in 12 units horizontal (25-percent slope).

❖ Metal shingles cannot be installed on roof slopes less than 3:12.

1507.5.3 Underlayment. Underlayment shall comply with ASTM D 226, Type I or ASTM D 4869.

❖ A single layer of underlayment is required under all metal shingles other than flat metal shingles. The underlayment must meet the performance criteria of ASTM D 226, Type I, or ASTM D 4869.

1507.5.4 Ice barrier. In areas where there has been a history of ice forming along the eaves causing a backup of water, an ice barrier that consists of at least two layers of underlayment cemented together or of a self-adhering polymer-modified bitumen sheet shall be used in lieu of normal underlayment and extend from the lowest edges of all roof surfaces to a point at least 24 inches (610 mm) inside the exterior wall line of the building.

Exception: Detached accessory structures that contain no conditioned floor area.

❖ Ice dams form when snow melts over the warmer parts of a roof and refreezes over the colder eaves. This ice formation acts like a dam and causes water to back up beneath the roof covering. The water will eventually leak causing damage to the structure, including the walls, ceilings and roof [see commentary, Figures 1507.8.4(1) and (2)]. In areas where this is prevalent, the installation of an ice barrier is required in accordance with this section.

There is an exception to this section that exempts accessory buildings from such restrictions as they are unheated structures where the need for protection against ice dams is unnecessary.

1507.5.5 Material standards. Metal roof shingle roof coverings shall comply with Table 1507.4.3(1). The materials used for metal-roof shingle roof coverings shall be naturally corrosion resistant or provided with corrosion resistance in accordance with the standards and minimum thicknesses specified in the standards listed in Table 1507.4.3(2).

❖ ASTM A 653 regulates steel sheet for roofing and siding that is zinc coated on continuous lines and by the cut-length method. Material of this quality is furnished flat, in coils and cut lengths, and is formed in cut lengths. Roofing and siding includes corrugated, V-crimp, roll roofing and many special patterns. Corrugated roofing and siding sheet is produced in a number of corrugations, with variations of pitch and depth. ASTM A 755 covers steel sheet that is metallic coated by the hot-dipped process and prepainted by the coil-coating process with organic films for exterior-exposed building products of various qualities. Sheet material of this designation is furnished in coils, cut lengths and formed cut lengths. ASTM B 101 covers lead-coated copper sheets for architectural uses.

Many of the materials listed in Table 1507.4.3(1) are inherently corrosion resistant, such as aluminum, copper and lead. For these corrosion-resistant materials a material standard is listed for the base material; however, for steel roofing products, coatings are added to the base material to provide the necessary corrosion resistance.

Steel coating standards are located in Table 1507.4.3(2). Having this table separate from Table 1507.4.3(1) helps to clarify that these standards only address the process necessary to provide corrosion resistance for the base steel. The coating descriptions in Table 1507.4.3(2) are consistent with the applicable coating standards.

TABLE 1507.4.3(1)
METAL ROOF COVERINGS

ROOF COVERING TYPE	STANDARD APPLICATION RATE/THICKNESS
Aluminum	ASTM B 209, 0.024 inch minimum thickness for roll-formed panels and 0.019 inch minimum thickness for press-formed shingles.
Aluminum-zinc alloy coated steel	ASTM A 792 AZ 50
Cold-rolled copper	ASTM B 370 minimum 16 oz./sq. ft. and 12 oz./sq. ft. high yield copper for metal-sheet roof covering systems: 12 oz/sq. ft. for preformed metal shingle systems.
Copper	16 oz./sq. ft. for metal-sheet roof-covering systems; 12 oz./sq. ft. for preformed metal shingle systems.
Galvanized steel	ASTM A 653 G-90 zinc-coated[a].
Hard lead	2 lbs./sq. ft.
Lead-coated copper	ASTM B 101
Prepainted steel	ASTM A 755
Soft lead	3 lbs./sq. ft.
Stainless steel	ASTM A 240, 300 Series Alloys
Steel	ASTM A 924
Terne and terne-coated stainless	Terne coating of 40 lbs. per double base box, field painted where applicable in accordance with manufacturer's installation instructions.
Zinc	0.027 inch minimum thickness; 99.995% electrolytic high grade zinc with alloy additives of copper (0.08% - 0.20%), titanium (0.07% - 0.12%) and aluminum (0.015%).

For SI: 1 ounce per square foot = 0.0026 kg/m²,
1 pound per square foot = 4.882 kg/m²,
1 inch = 25.4 mm, 1 pound = 0.454 kg.

a. For Group U buildings, the minimum coating thickness for ASTM A 653 galvanized steel roofing shall be G-60.

TABLE 1507.4.3(2)
MINIMUM CORROSION RESISTANCE

55% Aluminum-zinc alloy coated steel	ASTM A 792 AZ 50
5% Aluminum alloy-coated steel	ASTM A 875 GF60
Aluminum-coated steel	ASTM A 463 T2 65
Galvanized steel	ASTM A 653 G-90
Prepainted steel	ASTM A 755[a]

a. Paint systems in accordance with ASTM A 755 shall be applied over steel products with corrosion resistant coatings complying with ASTM A 792, ASTM A 875, ASTM A 463 or ASTM A 653.

1507.5.6 Attachment. Metal roof shingles shall be secured to the roof in accordance with the *approved* manufacturer's installation instructions.

❖ Fasteners used to attach metal roofing to the roof are to have the manufacturer's approval for that purpose.

1507.5.7 Flashing. Roof valley flashing shall be of corrosion-resistant metal of the same material as the roof covering or shall comply with the standards in Table 1507.4.3(1). The valley flashing shall extend at least 8 inches (203 mm) from the centerline each way and shall have a splash diverter rib not less than $^3/_4$ inch (19.1 mm) high at the flow line formed as part of the flashing. Sections of flashing shall have an end lap of not less than 4 inches (102 mm). In areas where the average daily temperature in January is 25°F (-4°C) or less or where there is a possibility of ice forming along the eaves causing a backup of water, the metal valley flashing shall have a 36-inch-wide (914 mm) underlayment directly under it consisting of either one layer of underlayment running the full length of the valley or a self-adhering polymer-modified bitumen sheet complying with ASTM D 1970, in addition to underlayment required for metal roof shingles. The metal valley flashing underlayment shall be solidly cemented to the roofing underlayment for roof slopes under seven units vertical in 12 units horizontal (58-percent slope) or self-adhering polymer-modified bitumen sheet shall be installed.

❖ See the commentary to Section 1503.2.

1507.6 Mineral-surfaced roll roofing. The installation of mineral-surfaced roll roofing shall comply with this section.

❖ Mineral-surfaced roll roofing is an asphalt roll roofing material in some cases having a selvage edge, which is a specially defined edge (lined for demarcation) designed for some special purpose, such as overlapping or seaming. It is typically 36 inches (914 mm) in width and is surfaced with coarse mineral granules. In mineral-surfaced roll roofing, the selvage is not surfaced with coarse mineral granules to allow better adhesion of the overlapping sheet.

There are two general methods for applying roll roofing: the exposed nail method and the concealed nail method. Depending on the slope and nailing method used, roll roofing can be applied parallel (downslope roof edge) or perpendicular to the eave (parallel with the rake) (see Section 1507.6.1).

1507.6.1 Deck requirements. Mineral-surfaced roll roofing shall be fastened to solidly sheathed roofs.

❖ Prior to the application of the roll roofing on wood plank or plywood decks, the deck should be inspected for delamination of plywood, warped boards and proper nailing.

For roll roofing applied parallel to the eave, a minimum roof slope of 2:12 is recommended for application by the concealed-nail method, and 6:12 is recommended for application by the exposed-nail method.

For roll roofing applied parallel to the rake, a minimum roof slope of 3:12 is recommended for application by the concealed-nail method, and 4:12 is rec-

ommended for application by the exposed-nail method.

1507.6.2 Deck slope. Mineral-surfaced roll roofing shall not be applied on roof slopes below one unit vertical in 12 units horizontal (8-percent slope).

❖ It is not recommended that asphalt roll roofing be applied to deck slopes of less than 2:12, with one exception. Roll roofing with a 19-inch (483 mm) selvage edge (commonly referred to as double coverage or split-sheet) may be applied on a deck with a slope as low as 1:12; however, it should only be used on roofs that will drain by gravity. It is not recommended for decks that will allow puddling and require evaporation to dry the roof.

1507.6.3 Underlayment. Underlayment shall comply with ASTM D 226, Type I or ASTM D 4869.

❖ A single layer of underlayment is required under mineral-surfaced rolled roofing. The underlayment must meet the performance criteria of ASTM D 226, Type I, or ASTM D 4869.

1507.6.4 Ice barrier. In areas where there has been a history of ice forming along the eaves causing a backup of water, an ice barrier that consists of at least two layers of underlayment cemented together or of a self-adhering polymer-modified bitumen sheet shall be used in lieu of normal underlayment and extend from the lowest edges of all roof surfaces to a point at least 24 inches (610 mm) inside the exterior wall line of the building.

Exception: Detached accessory structures that contain no conditioned floor area.

❖ Ice dams form when snow melts over the warmer parts of a roof and refreezes over the colder eaves. This ice formation acts like a dam and causes water to back up beneath the roof covering. The water will eventually leak causing damage to the structure, including the walls, ceilings and roof [see commentary, Figures 1507.8.4(1) and (2)]. In areas where this is prevalent, the installation of an ice barrier is required in accordance with this section.

There is an exception to this section that exempts accessory buildings from such restrictions as they are unheated structures where the need for protection against ice dams is unnecessary.

1507.6.5 Material standards. Mineral-surfaced roll roofing shall conform to ASTM D 3909 or ASTM D 6380.

❖ ASTM D 6380 covers asphalt roll roofing (organic felt). ASTM D 3909 covers asphalt-impregnated and coated glass felt roll roofing surfaced on the weather side with mineral granules for use as a cap sheet in the construction of built-up roofs.

1507.7 Slate shingles. The installation of slate shingles shall comply with the provisions of this section.

❖ Slate is a dense, tough, durable natural rock or stone material that is practically nonabsorbent. This natural rock has cleavage planes that allow the slate to be easily split into relatively thin layers. Slate also possesses a natural grain that usually runs perpendicular to the cleavage. Slate is usually split so the length of the shingle runs in the direction of the grain.

Slate roofing has a long service life when consideration is given to all components of the roofing system. Some grades of slate combined with proper roof deck, underlayments, fastening and accessories have a service life in excess of 75 years.

1507.7.1 Deck requirements. Slate shingles shall be fastened to solidly sheathed roofs.

❖ Due to the long service life expected of a slate roof system, careful consideration should be given to the deck material. Deck material that can be expected to last as long as the service life of the roof should be chosen. Additionally, consideration should be given to the weight of the slate when choosing the deck. Slate is available in many different thicknesses and, therefore, the weight will vary.

1507.7.2 Deck slope. Slate shingles shall only be used on slopes of four units vertical in 12 units horizontal (4:12) or greater.

❖ This type of roof covering cannot be installed on roof slopes that are less than 4:12.

1507.7.3 Underlayment. Underlayment shall comply with ASTM D 226, Type I or ASTM D 4869.

❖ A single layer of underlayment is required under slate shingles. The underlayment must meet the performance criteria of ASTM D 226, Type I, or ASTM D 4869.

1507.7.4 Ice barrier. In areas where the average daily temperature in January is 25°F (-4°C) or less or where there is a possibility of ice forming along the eaves causing a backup of water, an ice barrier that consists of at least two layers of underlayment cemented together or of a self-adhering polymer-modified bitumen sheet shall extend from the lowest edges of all roof surfaces to a point at least 24 inches (610 mm) inside the exterior wall line of the building.

Exception: Detached accessory structures that contain no conditioned floor area.

❖ Ice dams form when snow melts over the warmer parts of a roof and refreezes over the colder eaves. This ice formation acts like a dam and causes water to back up beneath the roof covering. The water will eventually leak causing damage to the structure, including the walls, ceilings and roof [see commentary, Figures 1507.8.4(1) and (2)]. In areas where this is prevalent, the installation of an ice barrier is required in accordance with this section.

There is an exception to this section that exempts accessory buildings from such restrictions as they are unheated structures where the need for protection against ice dams is unnecessary.

1507.7.5 Material standards. Slate shingles shall comply with ASTM C 406.

❖ ASTM C 406 covers the material characteristics, physical requirements and sampling appropriate for the selection of slate as a roofing material.

1507.7.6 Application. Minimum headlap for slate shingles shall be in accordance with Table 1507.7.6. Slate shingles shall be secured to the roof with two fasteners per slate.

❖ Slate shingles are drilled or punched with holes for fasteners with consideration for proper headlap. Fastener material should be chosen based on the expected service life. Copper slating nails, stainless steel, aluminum-alloy, bronze or cut-brass roofing nails are often specified, depending on the particular project. Unprotected black-iron, electroplated and hot-dipped galvanized fasteners are usually not recommended.

TABLE 1507.7.6
SLATE SHINGLE HEADLAP

SLOPE	HEADLAP (inches)
4:12 < slope < 8:12	4
8:12 < slope < 20:12	3
slope ≥ 20:12	2

For SI: 1 inch = 25.4 mm.

❖ Headlap is the distance of overlap measured from the uppermost ply or course to the point that it overlaps the undermost ply or course.

1507.7.7 Flashing. Flashing and counterflashing shall be made with sheet metal. Valley flashing shall be a minimum of 15 inches (381 mm) wide. Valley and flashing metal shall be a minimum uncoated thickness of 0.0179-inch (0.455 mm) zinc-coated G90. Chimneys, stucco or brick walls shall have a minimum of two plies of felt for a cap flashing consisting of a 4-inch-wide (102 mm) strip of felt set in plastic cement and extending 1 inch (25 mm) above the first felt and a top coating of plastic cement. The felt shall extend over the base flashing 2 inches (51 mm).

❖ Flashing is required at all intersections between exterior walls and roofs. These are areas where rainwater can easily penetrate the building envelope (see Section 1503.2).

1507.8 Wood shingles. The installation of wood shingles shall comply with the provisions of this section and Table 1507.8.

❖ Wood shingles are defined by the roofing industry as sawed wood products featuring a uniform butt thickness per individual length.

TABLE 1507.8. See page 15-20.

❖ Wood shingles cannot be installed on roof slopes below 3:12. Single-layer underlayment is required at eaves, ridges, hips, valleys and all other changes of roof slope or direction to protect the roof from leakage caused by water backup. Each shingle must be securely fastened to the roof deck with a maximum of two fasteners. Care should be taken so that the fasteners do not cause splitting of the wood shingle. Fasteners are to be specified in the manufacturer's installation instructions.

1507.8.1 Deck requirements. Wood shingles shall be installed on solid or spaced sheathing. Where spaced sheathing is used, sheathing boards shall not be less than 1-inch by 4-inch (25 mm by 102 mm) nominal dimensions and shall be spaced on centers equal to the weather exposure to coincide with the placement of fasteners.

❖ Spaced sheathing is usually of 1-inch by 4-inch (25 mm by 102 mm) or 1-inch by 6-inch (25 mm by 152 mm) softwood boards. Note that solid sheathing is required for certain locations by Section 1507.8.1.1.

1507.8.1.1 Solid sheathing required. Solid sheathing is required in areas where the average daily temperature in January is 25°F (-4°C) or less or where there is a possibility of ice forming along the eaves causing a backup of water.

❖ Solid sheathing is usually softwood panels, which provide a smooth, even base for the roofing material and help stiffen the entire roof structure. Solid sheathing provides an extra degree of protection where ice damming is a possibility.

1507.8.2 Deck slope. Wood shingles shall be installed on slopes of three units vertical in 12 units horizontal (25-percent slope) or greater.

❖ Wood shingles cannot be installed on roof slopes below 3:12.

1507.8.3 Underlayment. Underlayment shall comply with ASTM D 226, Type I or ASTM D 4869.

❖ A single layer of underlayment is required under wood shingles. The underlayment must meet the performance criteria of ASTM D 226, Type I, or ASTM D 4869.

1507.8.4 Ice barrier. In areas where there has been a history of ice forming along the eaves causing a backup of water, an ice barrier that consists of at least two layers of underlayment cemented together or of a self-adhering polymer-modified bitumen sheet shall be used in lieu of normal underlayment and extend from the lowest edges of all roof surfaces to a point at least 24 inches (610 mm) inside the exterior wall line of the building.

Exception: Detached accessory structures that contain no conditioned floor area.

❖ Ice dams form when snow melts over the warmer parts of a roof and refreezes over the colder eaves. This ice formation acts like a dam and causes water to back up beneath the roof covering. The water will eventually leak causing damage to the structure, including the walls, ceilings and roof [see commentary, Figures 1507.8.4(1) and (2)]. In areas where this is prevalent, the installation of an ice barrier is required in accordance with this section.

There is an exception to this section that exempts accessory buildings from such restrictions they are unheated structures where the need for protection against ice dams is unnecessary.

1507.8.5 Material standards. Wood shingles shall be of naturally durable wood and comply with the requirements of Table 1507.8.5.

❖ Information about the installation of wood shingles is available in the *Design and Application Manual for New Roof Construction* published by the Cedar Shake and Shingle Bureau (CSSB). Additional information regarding wood shingle roofs is provided in the wood roofing section of the *NRCA Roofing Manual: Steepslope Roof Systems* published by the NRCA.

TABLE 1507.8.5
WOOD SHINGLE MATERIAL REQUIREMENTS

MATERIAL	APPLICABLE MINIMUM GRADES	GRADING RULES
Wood shingles of naturally durable wood	1, 2 or 3	CSSB

CSSB = Cedar Shake and Shingle Bureau

❖ The standards published by CSSB contain useful information on the grading rules for wood shingles. A third-party inspection agency's label is now required to document compliance of the shingle.

1507.8.6 Attachment. Fasteners for wood shingles shall be corrosion resistant with a minimum penetration of $^3/_4$ inch (19.1 mm) into the sheathing. For sheathing less than $^1/_2$ inch (12.7 mm) in thickness, the fasteners shall extend through the sheathing. Each shingle shall be attached with a minimum of two fasteners.

❖ Each shingle must be securely fastened to the roof deck with a minimum of two fasteners. Care should be taken so that the fasteners do not cause splitting of the wood shingle. Fasteners are to be as specified in the manufacturer's installation instructions.

1507.8.7 Application. Wood shingles shall be laid with a side lap not less than $1^1/_2$ inches (38 mm) between joints in adjacent courses, and not be in direct alignment in alternate courses. Spacing between shingles shall be $^1/_4$ to $^3/_8$ inches (6.4 to 9.5 mm). Weather exposure for wood shingles shall not exceed that set in Table 1507.8.7.

❖ Wood shingle exposure is specified in Table 1507.8.7. Depending on the grade of the material, total shingle length and slope of the roof deck, the table specifies the maximum length of exposure for weathering purposes.

TABLE 1507.8
WOOD SHINGLE AND SHAKE INSTALLATION

ROOF ITEM	WOOD SHINGLES	WOOD SHAKES
1. Roof slope	Wood shingles shall be installed on slopes of three units vertical in 12 units horizontal (3:12) or greater.	Wood shakes shall be installed on slopes of four units vertical in 12 units horizontal (4:12) or greater.
2. Deck requirement		
Temperate climate	Shingles shall be applied to roofs with solid or spaced sheathing. Where spaced sheathing is used, sheathing boards shall not be less than 1″ × 4″ nominal dimensions and shall be spaced on center equal to the weather exposure to coincide with the placement of fasteners.	Shakes shall be applied to roofs with solid or spaced sheathing. Where spaced sheathing is used, sheathing boards shall not be less than 1″ × 4″ nominal dimensions and shall be spaced on center equal to the weather exposure to coincide with the placement of fasteners. When 1″ × 4″ spaced sheathing is installed at 10 inches, boards must be installed between the sheathing boards.
In areas where the average daily temperature in January is 25°F or less or where there is a possibility of ice forming along the eaves causing a backup of water.	Solid sheathing required.	Solid sheathing is required.
3. Interlayment	No requirements.	Interlayment shall comply with ASTM D 226, Type 1.
4. Underlayment		
Temperate climate	Underlayment shall comply with ASTM D 226, Type 1.	Underlayment shall comply with ASTM D 226, Type 1.
In areas where there is a possibility of ice forming along the eaves causing a backup of water.	An ice barrier that consists of at least two layers of underlayment cemented together or of a self-adhering polymer-modified bitumen sheet shall extend from the eave's edge to a point at least 24 inches inside the exterior wall line of the building.	An ice barrier that consists of at least two layers of underlayment cemented together or of a self-adhering polymer-modified bitumen sheet shall extend from the lowest edges of all roof surfaces to a point at least 24 inches inside the exterior wall line of the building.
5. Application		
Attachment	Fasteners for wood shingles shall be hot-dipped galvanized or Type 304 (Type 316 for coastal areas) stainless steel with a minimum penetration of 0.75 inch into the sheathing. For sheathing less than 0.5 inch thick, the fasteners shall extend through the sheathing.	Fasteners for wood shakes shall be hot-dipped galvanized or Type 304 (Type 316 for coastal areas) with a minimum penetration of 0.75 inch into the sheathing. For sheathing less than 0.5 inch thick, the fasteners shall extend through the sheathing.
No. of fasteners	Two per shingle.	Two per shake.
Exposure	Weather exposures shall not exceed those set forth in Table 1507.8.7.	Weather exposures shall not exceed those set forth in Table 1507.9.8.
Method	Shingles shall be laid with a side lap of not less than 1.5 inches between joints in courses, and no two joints in any three adjacent courses shall be in direct alignment. Spacing between shingles shall be 0.25 to 0.375 inch.	Shakes shall be laid with a side lap of not less than 1.5 inches between joints in adjacent courses. Spacing between shakes shall not be less than 0.375 inch or more than 0.625 inch for shakes and taper sawn shakes of naturally durable wood and shall be 0.25 to 0.375 inch for preservative-treated taper sawn shakes.
Flashing	In accordance with Section 1507.8.8.	In accordance with Section 1507.9.9.

For SI: 1 inch = 25.4 mm, °C = [(°F) - 32]/1.8.

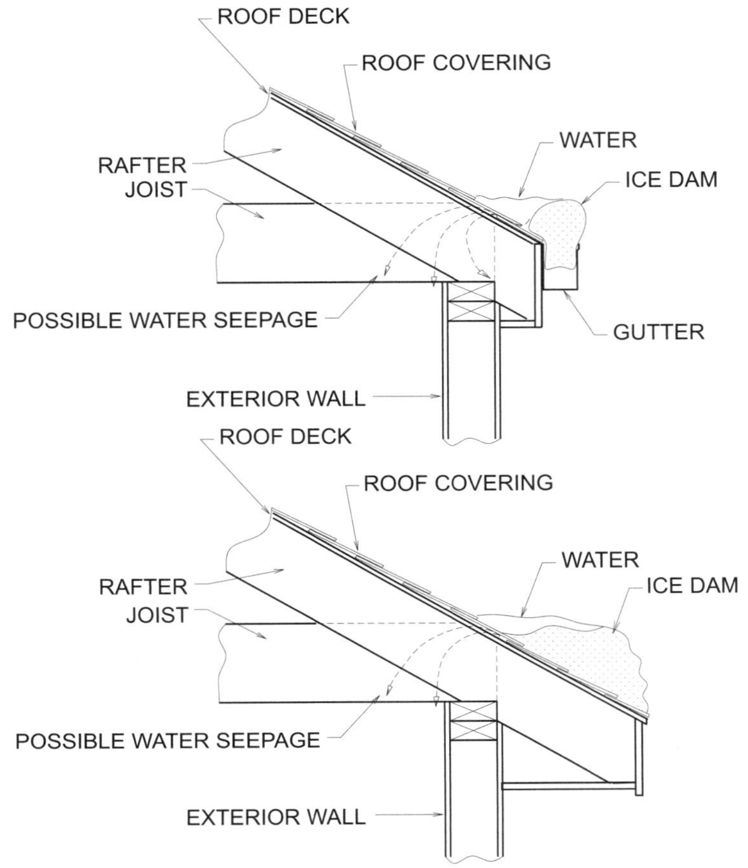

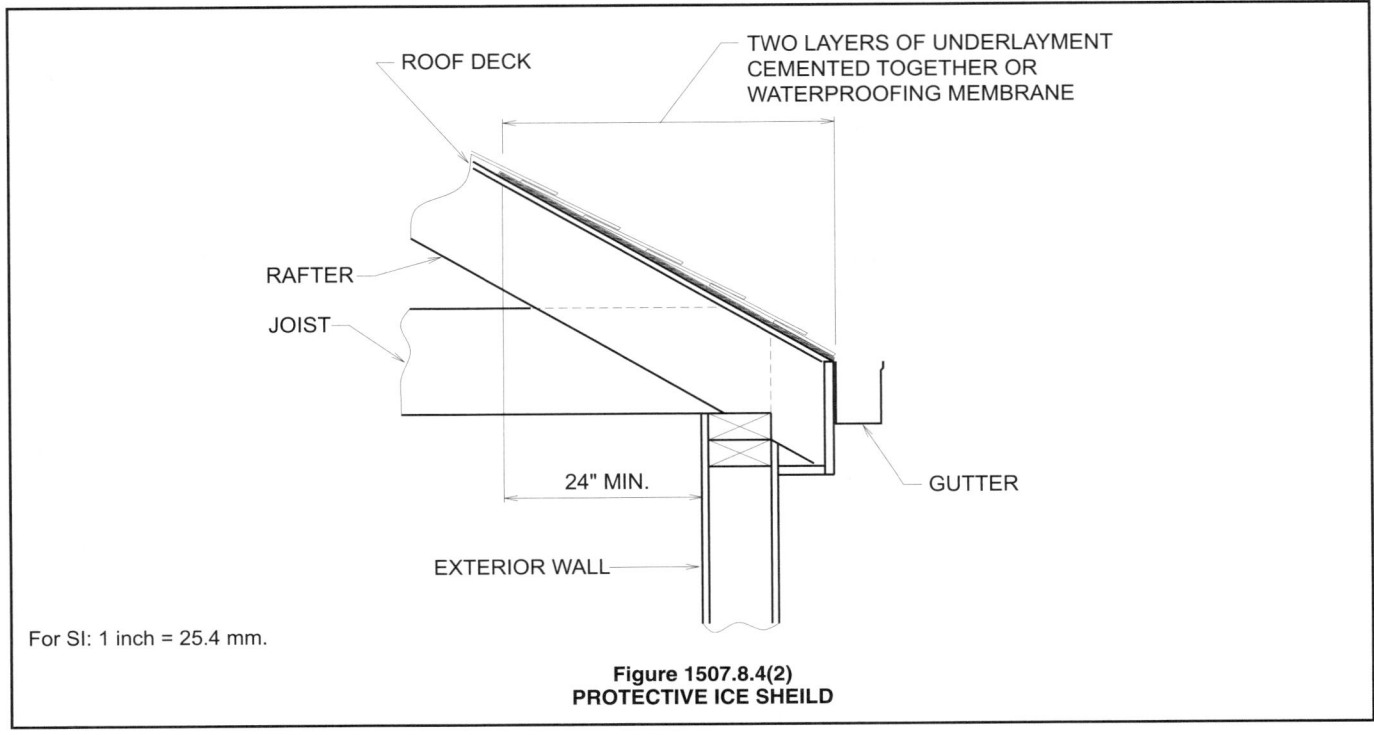

Figure 1507.8.4(1)
POSSIBLE WATER DAMAGE AT ICE DAMS

For SI: 1 inch = 25.4 mm.

Figure 1507.8.4(2)
PROTECTIVE ICE SHEILD

TABLE 1507.8.7
WOOD SHINGLE WEATHER EXPOSURE AND ROOF SLOPE

ROOFING MATERIAL	LENGTH (inches)	GRADE	EXPOSURE (inches) 3:12 pitch to < 4:12	EXPOSURE (inches) 4:12 pitch or steeper
Shingles of naturally durable wood	16	No. 1	3.75	5
	16	No. 2	3.5	4
	16	No. 3	3	3.5
	18	No. 1	4.25	5.5
	18	No. 2	4	4.5
	18	No. 3	3.5	4
	24	No. 1	5.75	7.5
	24	No. 2	5.5	6.5
	24	No. 3	5	5.5

For SI: 1 inch = 25.4 mm.

❖ The grade of the shingle will be listed on the label required for wood shingles. Each of the three different grades is available in three different lengths. For lower-sloped roofs (3:12 to 4:12), where wind uplift is more significant, the maximum weather exposure is always less than for steeper-sloped roofs (4:12 and greater).

1507.8.8 Flashing. At the juncture of the roof and vertical surfaces, flashing and counterflashing shall be provided in accordance with the manufacturer's installation instructions, and where of metal, shall not be less than 0.019-inch (0.48 mm) (No. 26 galvanized sheet gage) corrosion-resistant metal. The valley flashing shall extend at least 11 inches (279 mm) from the centerline each way and have a splash diverter rib not less than 1 inch (25 mm) high at the flow line formed as part of the flashing. Sections of flashing shall have an end lap of not less than 4 inches (102 mm). For roof slopes of three units vertical in 12 units horizontal (25-percent slope) and over, the valley flashing shall have a 36-inch-wide (914 mm) underlayment of either one layer of Type I underlayment running the full length of the valley or a self-adhering polymer-modified bitumen sheet complying with ASTM D 1970, in addition to other required underlayment. In areas where the average daily temperature in January is 25°F (-4°C) or less or where there is a possibility of ice forming along the eaves causing a backup of water, the metal valley flashing underlayment shall be solidly cemented to the roofing underlayment for slopes under seven units vertical in 12 units horizontal (58-percent slope) or self-adhering polymer-modified bitumen sheet shall be installed.

❖ Improper installation of flashing is the greatest cause of failures of roof covering systems. Whenever one plane of a roof intersects another plane, flashing is required where the planes intersect (see Section 1503.2).

1507.9 Wood shakes. The installation of wood shakes shall comply with the provisions of this section and Table 1507.8.

❖ This section establishes design and installation requirements for wood shakes, which are defined as roofing products split from logs and then shaped as required by the individual manufacturers. Wood

shakes must be labeled by an approved third-party inspection agency. A quality control program is required to contain a set of grading rules.

Information about the installation of wood shakes is available in the *Design and Application Manual for New Roof Construction* published by the CSSB. Additional information is provided in the wood roofing section of the *NRCA Roofing Manual: Steep-slope Roof Systems* published by the NRCA.

1507.9.1 Deck requirements. Wood shakes shall only be used on solid or spaced sheathing. Where spaced sheathing is used, sheathing boards shall not be less than 1-inch by 4-inch (25 mm by 102 mm) nominal dimensions and shall be spaced on centers equal to the weather exposure to coincide with the placement of fasteners. Where 1-inch by 4-inch (25 mm by 102 mm) spaced sheathing is installed at 10 inches (254 mm) o.c., additional 1-inch by 4-inch (25 mm by 102 mm) boards shall be installed between the sheathing boards.

❖ See the commentary to Section 1507.8.1.

1507.9.1.1 Solid sheathing required. Solid sheathing is required in areas where the average daily temperature in January is 25°F (-4°C) or less or where there is a possibility of ice forming along the eaves causing a backup of water.

❖ See the commentary to Section 1507.8.1.1.

1507.9.2 Deck slope. Wood shakes shall only be used on slopes of four units vertical in 12 units horizontal (33-percent slope) or greater.

❖ Wood shakes cannot be installed on roof slopes less than 4:12 to provide for adequate drainage of the roof.

1507.9.3 Underlayment. Underlayment shall comply with ASTM D 226, Type I or ASTM D 4869.

❖ A single layer of underlayment is required under wood shakes. The underlayment must meet the performance criteria of ASTM D 226, Type I, or ASTM D 4869.

1507.9.4 Ice barrier. In areas where there has been a history of ice forming along the eaves causing a backup of water, an ice barrier that consists of at least two layers of underlayment cemented together or of a self-adhering polymer-modified bitumen sheet shall be used in lieu of normal underlayment and extend from the lowest edges of all roof surfaces to a point at least 24 inches (610 mm) inside the exterior wall line of the building.

Exception: Detached accessory structures that contain no conditioned floor area.

❖ Ice dams form when snow melts over the warmer parts of a roof and refreezes over the colder eaves. This ice formation acts like a dam and causes water to back up beneath the roof covering. The water will eventually leak causing damage to the structure, including the walls, ceilings and roof [see commentary, Figures 1507.8.4(1) and (2)]. In areas where this is prevalent, the installation of an ice barrier is required in accordance with this section.

There is an exception to this section exempts accessory buildings from such restrictions as they are unheated structures where the need for protection against ice dams is unnecessary.

1507.9.5 Interlayment. Interlayment shall comply with ASTM D 226, Type I.

❖ A single layer of felt interlayment must be shingled between each course on all roof slopes.

1507.9.6 Material standards. Wood shakes shall comply with the requirements of Table 1507.9.6.

❖ Information about the installation of wood shingles is available in the *Design and Application Manual for New Roof Construction* published by CSSB. Additional information is provided in the wood roofing section of the *NRCA Roofing Manual: Steep-slope Roof Systems* published by the NRCA.

TABLE 1507.9.6
WOOD SHAKE MATERIAL REQUIREMENTS

MATERIAL	MINIMUM GRADES	APPLICABLE GRADING RULES
Wood shakes of naturally durable wood	1	CSSB
Taper sawn shakes of naturally durable wood	1 or 2	CSSB
Preservative-treated shakes and shingles of naturally durable wood	1	CSSB
Fire-retardant-treated shakes and shingles of naturally durable wood	1	CSSB
Preservative-treated taper sawn shakes of Southern pine treated in accordance with AWPA U1 (Commodity Specification A, Use Category 3B and Section 5.6)	1 or 2	TFS

CSSB = Cedar Shake and Shingle Bureau.
TFS = Forest Products Laboratory of the Texas Forest Services.

❖ Wood shakes must be labeled by an approved third-party inspection agency. A quality control program is required to contain a set of grading rules. Care must be taken to completely follow the manufacturer's installation instructions for each particular product. The building official must review and approve all installations with special note of any application limitations.

1507.9.7 Attachment. Fasteners for wood shakes shall be corrosion resistant with a minimum penetration of $^3/_4$ inch (19.1 mm) into the sheathing. For sheathing less than $^1/_2$ inch (12.7 mm) in thickness, the fasteners shall extend through the sheathing. Each shake shall be attached with a minimum of two fasteners.

❖ Each shake must be secured with a minimum of two fasteners. Fasteners are to be as specified in the manufacturer's installation instructions.

1507.9.8 Application. Wood shakes shall be laid with a side lap not less than $1^1/_2$ inches (38 mm) between joints in adjacent courses. Spacing between shakes in the same course shall be $^3/_8$ to $^5/_8$ inches (9.5 to 15.9 mm) for shakes and taper sawn shakes of naturally durable wood and shall be $^1/_4$ to $^3/_8$ inch (6.4 to 9.5

mm) for preservative taper sawn shakes. Weather exposure for wood shakes shall not exceed those set in Table 1507.9.8.

❖ Care must be taken to follow completely the manufacturer's installation instructions for each particular product. The building official must review and approve all installations with special note of any application limitations.

TABLE 1507.9.8
WOOD SHAKE WEATHER EXPOSURE AND ROOF SLOPE

ROOFING MATERIAL	LENGTH (inches)	GRADE	EXPOSURE (inches) 4:12 PITCH OR STEEPER
Shakes of naturally durable wood	18	No. 1	7.5
	24	No. 1	10[a]
Preservative-treated taper sawn shakes of Southern yellow pine	18	No. 1	7.5
	24	No. 1	10
	18	No. 2	5.5
	24	No. 2	7.5
Taper sawn shakes of naturally durable wood	18	No. 1	7.5
	24	No. 1	10
	18	No. 2	5.5
	24	No. 2	7.5

For SI: 1 inch = 25.4 mm.
a. For 24-inch by 0.375-inch handsplit shakes, the maximum exposure is 7.5 inches.

❖ The code provides specific weather exposure limitations for two different lengths of shakes. As with wood shingles, the longer the shake, the greater the exposure allowance, since wind uplift resistance is greater for the longer shakes. The exposure limitations are applicable regardless of the roof slope (4:12 or greater). The reduced weather exposure enhances the wind uplift resistance that these wood shakes will be able to provide.

1507.9.9 Flashing. At the juncture of the roof and vertical surfaces, flashing and counterflashing shall be provided in accordance with the manufacturer's installation instructions, and where of metal, shall not be less than 0.019-inch (0.48 mm) (No. 26 galvanized sheet gage) corrosion-resistant metal. The valley flashing shall extend at least 11 inches (279 mm) from the centerline each way and have a splash diverter rib not less than 1 inch (25 mm) high at the flow line formed as part of the flashing. Sections of flashing shall have an end lap of not less than 4 inches (102 mm). For roof slopes of three units vertical in 12 units horizontal (25-percent slope) and over, the valley flashing shall have a 36-inch-wide (914 mm) underlayment of either one layer of Type I underlayment running the full length of the valley or a self-adhering polymer-modified bitumen sheet complying with ASTM D 1970, in addition to other required underlayment. In areas where the average daily temperature in January is 25°F (-4°C) or less or where there is a possibility of ice forming along the eaves causing a backup of water, the metal valley flashing underlayment shall be solidly cemented to the roofing underlayment for slopes under seven units vertical in 12 units horizontal (58-percent slope) or self-adhering polymer-modified bitumen sheet shall be installed.

❖ Improper installation of flashing is the greatest cause of failure of roof covering systems. Whenever one plane of a roof intersects another plane, flashing is required where the planes intersect [see commentary, Section 1503.2 and Figures 1507.8.4(1) and (2)].

1507.10 Built-up roofs. The installation of built-up roofs shall comply with the provisions of this section.

❖ A built-up roof membrane is a continuous, semiflexible multiple-ply roof membrane, consisting of plies or layers of saturated felts, coated felts, fabrics or mats between which alternate layers of bitumen are applied. Generally, built-up roof membranes are surfaced with mineral aggregate and bitumen, a liquid-applied coating or granule-surfaced cap sheet.

1507.10.1 Slope. Built-up roofs shall have a design slope of a minimum of one-fourth unit vertical in 12 units horizontal (2-percent slope) for drainage, except for coal-tar built-up roofs that shall have a design slope of a minimum one-eighth unit vertical in 12 units horizontal (1-percent slope).

❖ Because of their low melting points and self-healing characteristics, coal tar membranes are commonly installed on lower slopes and dead-level roofs. One-quarter inch (6.4 mm) per foot (1:48) is the maximum recommended slope for a coal tar built-up roof.

1507.10.2 Material standards. Built-up roof covering materials shall comply with the standards in Table 1507.10.2.

❖ Asphalt and coal tar are the principal bitumens used for roofing purposes.

1507.11 Modified bitumen roofing. The installation of modified bitumen roofing shall comply with the provisions of this section.

❖ Modified bitumen roofing is a membrane-type roofing made up of composite sheets consisting of polymer-modified bitumen often reinforced and sometimes surfaced with various types of mats, films, foils and mineral granules. The bitumen is modified through inclusion of one or more polymers. There are two general types of polymer-modified bitumen membranes: those with the principal modifier being atactic polypropylene (APP), and those with bitumen modified with styrene butadiene styrene (SBS). They differ in characteristics, as well as application. For more information on the application of polymer-modified bitumen roofing membranes, refer to the *NRCA Roofing Manual: Membrane Roof Systems.*

1507.11.1 Slope. Modified bitumen membrane roofs shall have a design slope of a minimum of one-fourth unit vertical in 12 units horizontal (2-percent slope) for drainage.

❖ According to this section, all roofs must have a minimum slope of $^1/_4$:12, so that there is positive drainage of storm water to gutters, roof drains and other components of an approved storm sewer system in order to divert water away from the building or structure.

TABLE 1507.10.2
BUILT-UP ROOFING MATERIAL STANDARDS

MATERIAL STANDARD	STANDARD
Acrylic coatings used in roofing	ASTM D 6083
Aggregate surfacing	ASTM D 1863
Asphalt adhesive used in roofing	ASTM D 3747
Asphalt cements used in roofing	ASTM D 3019; D 2822; D 4586
Asphalt-coated glass fiber base sheet	ASTM D 4601
Asphalt coatings used in roofing	ASTM D1227; D 2823; D 4479
Asphalt glass felt	ASTM D 2178
Asphalt primer used in roofing	ASTM D 41
Asphalt-saturated and asphalt-coated organic felt base sheet	ASTM D 2626
Asphalt-saturated organic felt (perforated)	ASTM D 226
Asphalt used in roofing	ASTM D 312
Coal-tar cements used in roofing	ASTM D 4022; D 5643
Coal-tar saturated organic felt	ASTM D 227
Coal-tar pitch used in roofing	ASTM D 450; Type I or II
Coal-tar primer used in roofing, dampproofing and waterproofing	ASTM D 43
Glass mat, coal tar	ASTM D 4990
Glass mat, venting type	ASTM D 4897
Mineral-surfaced inorganic cap sheet	ASTM D 3909
Thermoplastic fabrics used in roofing	ASTM D 5665, D 5726

❖ The types of materials used in typical built-up roofing must comply with various standards as noted in the table. The following discusses some of those standards and what they address. Note that the titles and editions of all the standards are located in Chapter 35 of the code.

ASTM D 1863 covers the quality and grading of crushed stone, crushed slag and water-worn gravel suitable for use as aggregate surfacing on built-up roofs.

ASTM D 4601 covers asphalt-impregnated and coated glass fiber base sheet, with or without perforations, for use as the first ply of the built-up roofing. When not perforated, this sheet may be used as a vapor retarder under or between roof insulation with a solid top coating of asphaltic material.

ASTM D 2178 covers glass felt impregnated to varying degrees with asphalt, which may be used both with asphalts conforming to the requirements of ASTM D 312 in the construction of built-up roofs, and asphalts conforming to the requirements of ASTM D 449 in the membrane system of waterproofing.

ASTM D 2626 covers the base sheet with fine mineral surfacing on the top side, with or without perforations, for use as the first ply of a built-up roof. When not perforated, this sheet may be used as a vapor retarder under roof insulation.

ASTM D 226 covers asphalt-saturated organic felts, with perforations, that may be used with asphalts conforming to the requirements of ASTM D 312 in the construction of built-up roofs, and with asphalts conforming to the requirements of ASTM D 449 in the membrane system of waterproofing.

ASTM D 312 covers four types of asphalt intended for use in built-up roof construction. The specification is intended for general classification purposes only, and does not imply restrictions on the slope at which an asphalt must be used. There are four classification types. Type I includes asphalts that are relatively susceptible to flow at roof temperatures with good adhesive and self-sealing properties. Type II includes asphalts that are moderately susceptible to flow at roof temperatures. Type III includes asphalts that are relatively nonsusceptible to flow at roof temperatures for use in built-up roof construction on slope inclines from 8.3 to 25 percent. Type IV includes asphalts that are generally nonsusceptible to flow at roof temperatures for use in built-up roof construction on slope inclines from approximately 16.7 to 50 percent.

ASTM D 227 covers coal-tar saturated organic felt that may be used with coal-tar pitches conforming to the appropriate requirements of ASTM D 450 in the construction of built-up roofs and in the membrane system of waterproofing.

ASTM D 450 covers three types of coal-tar pitch suitable for use in the construction of built-up roofing, dampproofing and membrane waterproofing systems. Only Type I or III can be used in code-recognized roofing systems. Type I is suitable for use in built-up roofing systems with felts conforming to the requirements of ASTM D 227 or as specified by the manufacturer. Type III is suitable for use in built-up roofing systems, but has less volatile components than Type I.

ASTM D 3909 covers asphalt-impregnated and coated glass felt roll roofing surfaced on the weather side with mineral granules for use as a cap sheet in the construction of built-up roofs.

As noted, this list is not all-inclusive when compared to Table 1507.10.2, but includes materials commonly used as built-up roofing materials.

1507.11.2 Material standards. Modified bitumen roof coverings shall comply with CGSB 37-GP-56M, ASTM D 6162, ASTM D 6163, ASTM D 6164, ASTM D 6222, ASTM D 6223, ASTM D 6298 or ASTM D 6509.

❖ The materials and installation of modified bitumen roof coverings must be in accordance with the listed standards, which regulate modified bituminous roofing membranes, either prefabricated or reinforced. The standards are: CGSB 37-GP-56M, which regulates membrane, modified, bituminous, prefabricated and reinforced for roofing; ASTM D 6162, which regulates styrene butadiene styrene (SBS) modified bituminous sheet materials; ASTM D 6163, which regulates SBS using glass fiber reinforcements, ASTM D 6164, which regulates SBS using polyester reinforcements; ASTM D 6222, which regulates atactic polypropylene (APP)

modified bituminous sheet materials; ASTM D 6223, which regulates APP using polyester and glass fiber reinforcements; ASTM D 6298, which regulates SBS with a metal surface; and ASTM D 6509, which regulates APP using glass fiber reinforcements.

1507.12 Thermoset single-ply roofing. The installation of thermoset single-ply roofing shall comply with the provisions of this section.

❖ Thermoset is a material that solidifies or "sets" irreversibly when heated. Thinners evaporate during the curing process. Thermoset materials are those whose polymers are chemically cross-linked, commonly referred to as being "cured." Once they are fully cured, they can only be bonded to like material with an adhesive, as new molecular links may not be formed. There are four common subcategories of thermoset roof membranes: neoprene, chlorosulfonated polyethylene (CSPE), epichlorohydrin (DCH) and ethylene-propylene-diene monomer (or terpolymer) (EPDM).

1507.12.1 Slope. Thermoset single-ply membrane roofs shall have a design slope of a minimum of one-fourth unit vertical in 12 units horizontal (2-percent slope) for drainage.

❖ All roofs must have a minimum slope of $^1/_4$:12, so that there is positive drainage of storm water to gutters, roof drains and other components of an approved storm sewer system in order to divert water away from the building or structure.

1507.12.2 Material standards. Thermoset single-ply roof coverings shall comply with ASTM D 4637, ASTM D 5019 or CGSB 37-GP-52M.

❖ This section describes the standards to be used for materials and installation. The standards are: 1) ASTM D 4637, which regulates unreinforced and fabric-reinforced vulcanized rubber sheets made from EPDM or polychloroprene, and is intended for use in single-ply roof membranes exposed to weather; and 2) ASTM D 5019, which regulates reinforced nonvulcanized polymeric sheets made from chlorosulfonated polyethylene and polyisobutylene used in single-ply roof membranes; and 3) CGSB 37-GP-52M, which regulates sheet-applied elastomeric roofing membranes.

1507.12.3 Ballasted thermoset low-slope roofs. Ballasted thermoset low-slope roofs (roof slope < 2:12) shall be installed in accordance with this section and Section 1504.4. Stone used as ballast shall comply with ASTM D 448.

❖ The purpose of this section is to identify that this type of roof covering is allowed and to provide installation requirements. Further, this section specifies the minimum requirements for stone ballast to be used with this type of roofing system.

1507.13 Thermoplastic single-ply roofing. The installation of thermoplastic single-ply roofing shall comply with the provisions of this section.

❖ Thermoplastics are materials that soften when heated and harden when cooled. This process can be repeated, provided the material is not heated above the point at which decomposition occurs. Thermoplastic materials are distinguished from thermosets in that there is no chemical cross linking. Because of the materials' chemical nature, some thermoplastic membranes may be seamed by either heat (hot air) or solvent welding.

1507.13.1 Slope. Thermoplastic single-ply membrane roofs shall have a design slope of a minimum of one-fourth unit vertical in 12 units horizontal (2-percent slope).

❖ All roofs must have a minimum slope of $^1/_4$:12, so that there is positive drainage of storm water to gutters, roof drains and other components of an approved storm sewer system in order to divert water away from the building or structure.

1507.13.2 Material standards. Thermoplastic single-ply roof coverings shall comply with ASTM D 4434, ASTM D 6754, ASTM D 6878 or CGSB CAN/CGSB 37-54.

❖ This section establishes standards and performance criteria for the installation of thermoplastic single-ply roof coverings. Four standards are recognized: ASTM D 4434, ASTM D 6754, ASTM D 6878 and CGSB 37-GP-54M. These specifications cover flexible sheet made from poly (vinyl chloride) resin, ketone ethylene ester-based sheet roofing and thermoplastic polyolefin-based sheet roofing.

1507.13.3 Ballasted thermoplastic low-slope roofs. Ballasted thermoplastic low-slope roofs (roof slope < 2:12) shall be installed in accordance with this section and Section 1504.4. Stone used as ballast shall comply with ASTM D448.

❖ The purpose of this section is to identify that this type of roof covering is allowed and to provide installation requirements. Further, this section specifies the minimum requirements for stone ballast to be used with this type of roofing system.

1507.14 Sprayed polyurethane foam roofing. The installation of sprayed polyurethane foam roofing shall comply with the provisions of this section.

❖ Sprayed polyurethane foam is a foamed plastic material constructed by mixing a two-part liquid that is spray applied to form the base of an adhered roof system. The two-part component mixture reacts chemically and immediately expands when applied through special metering equipment to form a closed-cell foam. As the foam rises, it sets into a solid and a skin forms on the surface. The foam provides a thermal insulation and the skin provides a water-resistant surface.

1507.14.1 Slope. Sprayed polyurethane foam roofs shall have a design slope of a minimum of one-fourth unit vertical in 12 units horizontal (2-percent slope) for drainage.

❖ All roofs must have a minimum slope of $^1/_4$:12, so that there is positive drainage of storm water to gutters, roof drains and other components of an approved storm sewer system in order to divert water away from the building or structure.

1507.14.2 Material standards. Spray-applied polyurethane foam insulation shall comply with Type III or IV as defined in ASTM C 1029.

❖ ASTM C 1029 is the standard that regulates spray-applied rigid cellular polyurethane thermally insulated roof coverings. This specification covers the types and physical properties of spray-applied rigid cellular polyurethane intended for use as thermal insulation. Type III or IV foam insulation is required based on their higher compressive strengths. The operating temperatures of the surfaces to which the insulation is applied cannot be lower than -22°F (-30°C) or greater than 225°F (107°C).

1507.14.3 Application. Foamed-in-place roof insulation shall be installed in accordance with the manufacturer's instructions. A liquid-applied protective coating that complies with Section 1507.15 shall be applied no less than 2 hours nor more than 72 hours following the application of the foam.

❖ The roof system must be protected from ultraviolet light and degradation through the use of surfacing. Liquid-applied protected elastomeric coatings are the recommended surfacing for sprayed polyurethane foam roofing systems.

1507.14.4 Foam plastics. Foam plastic materials and installation shall comply with Chapter 26.

❖ Chapter 26 addresses foam plastic materials and installation.

1507.15 Liquid-applied coatings. The installation of liquid-applied coatings shall comply with the provisions of this section.

❖ Liquid-applied coatings are nonsynthetic roofing materials that are termed "cold applied" because hot bitumen is not employed in their application.

1507.15.1 Slope. Liquid-applied roofs shall have a design slope of a minimum of one-fourth unit vertical in 12 units horizontal (2-percent slope).

❖ All roofs must have a minimum slope of $^1/_4$:12, so that there is positive drainage of storm water to gutters, roof drains and other components of an approved storm sewer system in order to divert water away from the building or structure.

1507.15.2 Material standards. Liquid-applied roof coatings shall comply with ASTM C 836, ASTM C 957, ASTM D 1227 or ASTM D 3468, ASTM D 6083, ASTM D 6694 or ASTM D 6947.

❖ Liquid-applied roof coatings must be installed in accordance with the manufacturer's instructions. The coatings must comply with ASTM C 836, ASTM C 957, ASTM D 1227, D 3468 or ASTM D 6694. ASTM C 836 describes the required properties and test methods for a cold liquid-applied elastomeric membrane, one or two components, for waterproofing building decks subject to hydrostatic pressure in building areas to be occupied by personnel, vehicles or equipment. This specification only applies to a membrane system above which a separate wearing or traffic course will be applied. ASTM C 957 describes the required properties and test methods for a cold liquid-applied elastomeric membrane for waterproofing building decks not subject to hydrostatic pressure. The specification applies only to a membrane system that has an integral wearing surface. It does not include specific requirements for skid resistance or fire retardance, although both may be important in specific applications. ASTM D 1227 describes emulsified asphalt suitable for use as a protective coating for built-up roofs and other exposed surfaces with inclines of not less than 4 percent. ASTM D 3468 describes liquid-applied neoprene and chlorosulfonated polyethylene synthetic rubber solutions suitable for use in roofing and waterproofing. ASTM D 6694 describes liquid-applied silicone coatings used in spray on polyurethane foam roofing. ASTM D 6947 describes liquid- applied moisture cured polyurethane coatings used in spray polyurethane foam roofing systems.

1507.16 Roof gardens and landscaped roofs. Roof gardens and landscaped roofs shall comply with the requirements of this chapter and Sections 1607.11.2.2 and 1607.11.3.

❖ The purpose of this section is to require roof gardens and landscaped roofs to comply with the requirements for other roof systems contained in this chapter. Further, this section requires these types of roofs to meet the requirements for special purpose roofs as contained in Chapters 15 and 16.

SECTION 1508
ROOF INSULATION

1508.1 General. The use of above-deck thermal insulation shall be permitted provided such insulation is covered with an *approved* roof covering and passes the tests of FM 4450 or UL 1256 when tested as an assembly.

Exceptions:

1. Foam plastic roof insulation shall conform to the material and installation requirements of Chapter 26.

2. Where a concrete roof deck is used and the above-deck thermal insulation is covered with an *approved* roof covering.

❖ This section addresses two requirements for the use of above-deck thermal insulation. The first requirement is that an approved roof covering be used. The second requirement is that an under-deck flame spread test must be conducted and passed using either FM 4450 or the UL 1256 test methods. These tests were developed to evaluate the contribution of above-deck thermal insulations and their installation when exposed to a fire from below the deck. The primary construction assemblies that these tests have been developed for are roof assemblies that employ a steel roof deck with above-deck insulation materials.

Insulation may be used as building insulation, as well as a substrate to which the roof membrane is applied. In protected roof membrane systems, the insulation is applied over the membrane and is not expected to act as a substrate. The insulation should provide adequate support for the membrane and other associated rooftop materials, and permit limited rooftop traffic, such as for regular roof inspections and maintenance.

Rigid board roof insulations currently among the most commonly used for low-slope roofs are: cellular glass, glass fiber, mineral fiber, perlite, phenolic foam, polyisocyanurate foam, polystyrene foam, polyurethane foam, wood fiberboard and composite board.

Nonrigid roof insulations are generally used in steep-slope and some metal-roof assembly construction. Commonly made from cellulose, glass fiber and mineral fiber, these nonrigid insulations may be available in batts, blankets and loose-fiber forms.

Exception 2 recognizes that if a concrete deck is employed as part of the assembly it will provide protection to the above-deck insulation materials. This exception will apply to nonfoam plastic roof insulation materials, since foam plastic roof insulations are covered under Exception 1 and Chapter 26 of the code.

1508.1.1 Cellulosic fiberboard. Cellulosic fiberboard roof insulation shall conform to the material and installation requirements of Chapter 23.

❖ Chapter 23 addresses cellulosic fiberboard roof installation.

1508.2 Material standards. Above-deck thermal insulation board shall comply with the standards in Table 1508.2.

❖ The referenced material standards provide additional guidance on the physical properties of roof insulation when utilized as above-deck components of roof assemblies.

TABLE 1508.2
MATERIAL STANDARDS FOR ROOF INSULATION

Cellular glass board	ASTM C 552
Composite boards	ASTM C 1289, Type III, IV, V or VI
Expanded polystyrene	ASTM C 578
Extruded polystyrene board	ASTM C 578
Perlite board	ASTM C 728
Polyisocyanurate board	ASTM C 1289, Type I or Type II
Wood fiberboard	ASTM C 208

❖ This table incorporates industry-recognized material standards into the code for materials commonly used in above roof deck insulation practices.

SECTION 1509
ROOFTOP STRUCTURES

1509.1 General. The provisions of this section shall govern the construction of rooftop structures.

❖ This section identifies and establishes the criteria used in evaluating penthouse-type roof structures.

Other rooftop structures, such as water tanks, cooling towers, towers, spires, domes, cupolas, etc., are not to be considered as penthouses since the code has specific provisions for these structures (see Sections 1509.3, 1509.4 and 1509.5).

1509.2 Penthouses. A *penthouse* or *penthouses* in compliance with Sections 1509.2.1 through 1509.2.4 shall be considered as a portion of the *story* below.

❖ This section introduces the requirements necessary for a penthouse to be considered as a portion of the story directly below. Section 1509.2.1 contains height limitations, Section 1509.2.2 contains area limitations and Section 1509.2.3 contains use limitations.

1509.2.1 Height above roof. A *penthouse* or other projection above the roof in structures of other than Type I construction shall not exceed 28 feet (8534 mm) above the roof where used as an enclosure for tanks or for elevators that run to the roof and in all other cases shall not extend more than 18 feet (5486 mm) above the roof.

❖ The heights of roof structures are not limited provided the building is of Type I construction. For other types of construction, the roof structure is limited to 18 feet (3658 mm) in height unless it encloses tanks or elevators that extend to the roof, for which the maximum height is 28 feet (8534 mm). If the height exceeds that permitted, the roof structure should be counted as, and meet all the requirements of, a story.

1509.2.2 Area limitation. The aggregate area of penthouses and other rooftop structures shall not exceed one-third the area of the supporting roof. Such penthouses shall not contribute to either the *building area* or number of stories as regulated by Section 503.1. The area of the penthouse shall not be included in determining the *fire area* defined in Section 902.

❖ The area of the penthouse is also limited to one-third of the roof area. If the area of the penthouse exceeds one-third the roof area, regardless of height, then the penthouse must be considered a story.

1509.2.3 Use limitations. A *penthouse*, bulkhead or any other similar projection above the roof shall not be used for purposes other than shelter of mechanical equipment or shelter of vertical shaft openings in the roof. Provisions such as louvers, louver blades or flashing shall be made to protect the mechanical equipment and the building interior from the elements. Penthouses or bulkheads used for purposes other than permitted by this section shall conform to the requirements of this code for an additional *story*. The restrictions of this section shall not prohibit the placing of wood flagpoles or similar structures on the roof of any building.

❖ The use of the penthouse is also limited. Regardless of height or area, roof structures used for anything except sheltering mechanical equipment or vertical shaft openings must be considered as an additional story.

1509.2.4 Type of construction. Penthouses shall be constructed with walls, floors and roof as required for the building.

Exceptions:

1. On buildings of Type I construction, the exterior walls and roofs of penthouses with a *fire separation distance* of more than 5 feet (1524 mm) and less than 20 feet (6096 mm) shall be of at least 1-hour fire resistance-rated noncombustible construction. Walls and roofs with a *fire separation distance* of 20 feet (6096 mm) or greater shall be of noncombustible construction. Interior framing and walls shall be of noncombustible construction.

2. On buildings of Type I construction two stories above *grade plane* or less in height and Type II construction, the exterior walls and roofs of penthouses with a *fire separation distance* of more than 5 feet (1524 mm) and less than 20 feet (6096 mm) shall be of at least 1-hour fire-resistance-rated noncombustible or *fire-retardant-treated wood* construction. Walls and roofs with a *fire separation distance* of 20 feet (6096 mm) or greater shall be of noncombustible or *fire-retardant-treated wood* construction. Interior framing and walls shall be of noncombustible or fire retardant-treated wood construction.

3. On buildings of Type III, IV and V construction, the exterior walls of penthouses with a *fire separation distance* of more than 5 feet (1524 mm) and less than 20 feet (6096 mm) shall be at least 1-hour fire-resistance-rated construction. Walls with a *fire separation distance* of 20 feet (6096 mm) or greater from a common property line shall be of Type IV construction or noncombustible, or *fire-retardant-treated wood* construction. Roofs shall be constructed of materials and fire-resistance rated as required in Table 601 and Section 603, Item 25.3. Interior framing and walls shall be Type IV construction or noncombustible or *fire-retardant-treated wood* construction.

4. On buildings of Type I construction, unprotected noncombustible enclosures housing only mechanical equipment and located with a minimum *fire separation distance* of 20 feet (6096 mm) shall be permitted.

5. On buildings of Type I construction two stories or less above *grade plane* in height, or Type II, III, IV and V construction, unprotected noncombustible or *fire-retardant-treated wood* enclosures housing only mechanical equipment and located with a minimum *fire separation distance* of 20 feet (6096 mm) shall be permitted.

6. On one-story buildings, combustible unroofed mechanical equipment screens, fences or similar

enclosures are permitted where located with a *fire separation distance* of at least 20 feet (6096 mm) from adjacent property lines and where not exceeding 4 feet (1219 mm) in height above the roof surface.

7. Dormers shall be of the same type of construction as the roof on which they are placed, or of the exterior walls of the building.

❖ The general premise is that a penthouse be treated no differently than any other portion of the building as it is constructed of exterior walls, a floor and a roof. However, this section recognizes the reduced exposure of penthouses when the exterior wall is recessed from the exterior wall of the building.

Exceptions 1 through 6 allow for specific exceptions based on two thresholds of fire separation distance: greater than 5 feet (1524 mm) and up to 20 feet (6096 mm), and greater than 20 feet (6096 mm). See Section 702.1 for the definition of "Fire separation distance." The exterior walls of the penthouse, which are located within 5 feet (1524 mm) of the building's lot line, are subject to the same requirements as the exterior walls of the building. As such, the exterior walls are required to be of the same fire-resistance rating as the exterior wall of the story immediately below the penthouse.

If a penthouse is set back by a fire separation distance 5 feet (1524 mm) or more from the lot line, the risk of exposure to adjacent property from a fire within the penthouse, as well as fire exposure to the penthouse from adjacent property, is reduced. Accordingly, if a penthouse is set back 5 feet (1524 mm) or more from the building's lot line, the penthouse construction may be able to take advantage of one of the exceptions, typically based on the type of construction, resulting in a reduction in rating of the exterior wall of the penthouse. It should be noted that in many cases penthouses at least 20 feet (6096 mm) from the lot line are allowed further reductions.

Exception 7 recognizes typical dormer construction and requires dormers to be the same type of construction as the roof and/or exterior wall in which it is located.

1509.3 Tanks. Tanks having a capacity of more than 500 gallons (2 m³) placed in or on a building shall be supported on masonry, reinforced concrete, steel or Type IV construction provided that, where such supports are located in the building above the lowest *story*, the support shall be fire-resistance rated as required for Type IA construction.

❖ This section establishes the criteria used in evaluating rooftop tanks, such as water tanks.

1509.3.1 Valve. Such tanks shall have in the bottom or on the side near the bottom, a pipe or outlet, fitted with a suitable quick opening valve for discharging the contents in an emergency through an adequate drain.

❖ In the event of an emergency, there must be a means to drain the tank. This not only involves a quick-opening valve on the tank but also drains that are capable of handling the discharge. The drains must be sized and

installed in accordance with the *International Plumbing Code*® (IPC®).

1509.3.2 Location. Such tanks shall not be placed over or near a line of stairs or an elevator shaft, unless there is a solid roof or floor underneath the tank.

❖ Tanks are not permitted to be located above a stairway or elevator enclosure unless a solid roof or floor deck is provided underneath the tank. This is intended to provide a means of assurance that a shaft will not be affected by water from a leaking tank.

1509.3.3 Tank cover. Unenclosed roof tanks shall have covers sloping toward the outer edges.

❖ Roof tanks must be covered to prevent the accumulation of rain, ice and snow, which may cause an uncovered tank to overflow.

1509.4 Cooling towers. Cooling towers in excess of 250 square feet (23.2 m²) in base area or in excess of 15 feet (4572 mm) high where located on building roofs more than 50 feet (15 240 mm) high shall be of noncombustible construction. Cooling towers shall not exceed one-third of the supporting roof area.

> **Exception:** Drip boards and the enclosing construction of wood not less than 1 inch (25 mm) nominal thickness, provided the wood is covered on the exterior of the tower with noncombustible material.

❖ This section identifies and establishes criteria used in evaluating rooftop cooling towers.

1509.5 Towers, spires, domes and cupolas. Any tower, spire, dome or cupola shall be of a type of construction not less in *fire-resistance rating* than required for the building to which it is attached, except that any such tower, spire, dome or cupola that exceeds 85 feet (25 908 mm) in height above *grade plane*, exceeds 200 square feet (18.6 m²) in horizontal area or is used for any purpose other than a belfry or an architectural embellishment shall be constructed of and supported on Type I or II construction.

❖ This section identifies and establishes criteria used in evaluating rooftop structures other than penthouse-type structures, tanks and cooling towers. Examples of rooftop structures addressed in this section include towers, spires, domes and cupolas. Penthouse-type structures, tanks and cooling towers are addressed in Sections 1509.2, 1509.3 and 1509.4, respectively.

More specifically, when any of the structures addressed in this section exceed either 85 feet (25 908 mm) in height above grade plane or are greater than 200 square feet (18.6 m²) in area the construction type is limited to Type I or II. Section 1509.5.1 further restricts construction to noncombustible when the heights become more excessive (see commentary, Section 1509.5.1). These restrictions are related to the difficultly posed in fighting fires in such structures and also the danger to surrounding people, emergency responders and other buildings if such elements should fail during a fire.

1509.5.1 Noncombustible construction required. Any tower, spire, dome or cupola that exceeds 60 feet (18 288 mm) in height above the highest point at which it comes in contact with the roof, or that exceeds 200 square feet (18.6 m²) in area at any horizontal section, or which is intended to be used for any purpose other than a belfry or architectural embellishment, shall be entirely constructed of and supported by noncombustible materials. Such structures shall be separated from the building below by construction having a fire-resistance rating of not less than 1.5 hours with openings protected with a minimum 1.5-hour fire protection rating. Structures, except aerial supports 12 feet (3658 mm) high or less, flagpoles, water tanks and cooling towers, placed above the roof of any building more than 50 feet (15 240 mm) in building height, shall be of noncombustible material and shall be supported by construction of noncombustible material.

❖ These roof structures generally include communication towers, spires, cupolas and similar structures. Such structures must be constructed as required for the building type of construction classification. Because of the difficulty in fighting fires on tall buildings and structures, and generally the hazard associated with locating combustible materials on roofs of taller buildings, when the height of such structures exceeds 60 feet (18 288 mm) above the highest point at which it comes in contact with the roof, or is in excess of 200 square feet (19 m²) in area at any horizontal section, the structure and its supports must be of noncombustible construction. For similar reasons, most roof structures of any height or area placed on tall buildings [greater than 50 feet (15 240 mm) in height] are required to be constructed of noncombustible materials and be supported by noncombustible construction.

1509.5.2 Towers and spires. Towers and spires where enclosed shall have exterior walls as required for the building to which they are attached. The roof covering of spires shall be of a class of roof covering as required for the main roof of the rest of the structure.

❖ The towers addressed in this section are radio and television antenna towers, church spires and other towers and spires of similar nature. However, as with penthouses and roof structures, the code intends to obtain construction and fire resistance consistent with that of the building to which they are attached.

SECTION 1510
REROOFING

1510.1 General. Materials and methods of application used for recovering or replacing an existing roof covering shall comply with the requirements of Chapter 15.

Exception: Reroofing shall not be required to meet the minimum design slope requirement of one-quarter unit vertical in 12 units horizontal (2-percent slope) in Section 1507 for roofs that provide positive roof drainage.

❖ This section simply states that when a roof is replaced or recovered it must comply with Chapter 15 for the

materials and methods used with only one exception for low-sloped roofs. This section does not mandate that the entire roof be replaced but simply that the portion being replaced comply with Chapter 15.

For low-sloped roofs, the exception indicates that reroofing (i.e., recovering or replacement) is not required to meet the $^1/_4$:12 minimum slope requirement of Section 1507, provided that the roof has positive drainage. The term "positive drainage" is defined as the drainage condition in which consideration has been made for all loading deflections of the roof deck, and additional roof slope has been provided to ensure drainage of the roof area within 48 hours of rainfall.

1510.2 Structural and construction loads. Structural roof components shall be capable of supporting the roof-covering system and the material and equipment loads that will be encountered during installation of the system.

❖ The structural integrity of the roof must be maintained during reroofing operations, which can significantly contribute to the loading of the roof due to workers and material being present during this period of time. The roof support system must be able to structurally support all additional layers of roof covering material.

1510.3 Recovering versus replacement. New roof coverings shall not be installed without first removing all existing layers of roof coverings down to the roof deck where any of the following conditions occur:

1. Where the existing roof or roof covering is water soaked or has deteriorated to the point that the existing roof or roof covering is not adequate as a base for additional roofing.

2. Where the existing roof covering is wood shake, slate, clay, cement or asbestos-cement tile.

3. Where the existing roof has two or more applications of any type of roof covering.

Exceptions:

1. Complete and separate roofing systems, such as standing-seam metal roof systems, that are designed to transmit the roof loads directly to the building's structural system and that do not rely on existing roofs and roof coverings for support, shall not require the removal of existing roof coverings.

2. Metal panel, metal shingle and concrete and clay tile roof coverings shall be permitted to be installed over existing wood shake roofs when applied in accordance with Section 1510.4.

3. The application of a new protective coating over an existing spray polyurethane foam roofing system shall be permitted without tear-off of existing roof coverings.

❖ This section determines when all layers of previously installed roof covering systems must be removed prior to the installation of the new roof covering system. Note that roof coverings need to be removed down to the roof decking or sheathing.

When the existing roof or roof covering is water

soaked, it must be allowed to dry completely so as not to trap moisture beneath the new layer of covering. This could cause a rapid deterioration of the new covering material, as well as the existing sheathing. The existing covering is required to be removed if it cannot adequately dry out or if its physical properties have been permanently altered.

Wood shake, slate, clay, cement or asbestos-cement tile types of existing roof coverings historically do not make an adequate base for new roof coverings and could prevent the new covering from making a weather-tight seal. They could also allow penetration of water, snow, etc. These types of existing coverings must always be removed.

When the existing roof has two or more layers of any type of covering system, all layers need to be removed to enable the inspector and contractor to verify that the existing sheathing is not water damaged and still capable of providing an adequate nailing base.

Exception 1 states that new roofing systems that are designed to transmit all roof loads directly to the structural supports of the building do not necessitate that the existing roofing system be removed. Exception 2 allows certain roof covering, including metal, concrete panel and clay, to be placed over wood shingle and shake roofs only if any concealed combustible spaces are properly addressed in accordance with Section 1510.4 (see commentary, Section 1510.4). Exception 3 refers to the practice of "recoating" in which a new protective coating is placed over an existing spray polyurethane foam roofing system. Recoating can add many years to the effective life of a spray polyurethane foam roofing system without the associated downside of adding significant weight associated with other reroofing-type systems.

1510.4 Roof recovering. Where the application of a new roof covering over wood shingle or shake roofs creates a combustible concealed space, the entire existing surface shall be covered with gypsum board, mineral fiber, glass fiber or other *approved* materials securely fastened in place.

❖ "Roof recovering" is defined as the process of installing an additional roof covering over a prepared existing roof covering without removing the existing roof covering. Where recovering over wood shingles or shakes creates a combustible concealed space, the code requires that the entire surface of the wood shakes and shingles be covered with a material that will reduce the possibility of such materials adding fuel to a fire in such a space.

1510.5 Reinstallation of materials. Existing slate, clay or cement tile shall be permitted for reinstallation, except that damaged, cracked or broken slate or tile shall not be reinstalled. Existing vent flashing, metal edgings, drain outlets, collars and metal counterflashings shall not be reinstalled where rusted, damaged or deteriorated. Aggregate surfacing materials shall not be reinstalled.

❖ This section establishes requirements for materials to be reused for roof coverings. Historically, various types of materials have been removable without sub-

stantially damaging the material. Materials such as wood shingles and shakes, roll roofing and asphalt shingles are usually torn or cracked and cannot be reused. Fastener holes also violate the integrity of the material. Before reuse is allowed, materials such as slate, clay or cement tile should be examined thoroughly for cracks and deterioration.

1510.6 Flashings. Flashings shall be reconstructed in accordance with *approved* manufacturer's installation instructions. Metal flashing to which bituminous materials are to be adhered shall be primed prior to installation.

❖ Flashings to be reused or reconstructed must be in accordance with the manufacturer's installation instructions. Metal flashings that are to be reused for bituminous materials must be primed in accordance with the manufacturer's instructions.

Bibliography

The following resource material is referenced in this chapter or is relevant to the subject matter addressed in this chapter.

ASCE 7-05, *Minimum Design Loads for Buildings and Other Structures*. Reston, VA: American Society of Civil Engineers, 2005.

ASTM A 653/A 653M-07, *Specification for Steel Sheet, Zinc-coated or Zinc-coated (Galvanized or Zinc-iron Alloy-coated) by the Hot-dip Process*. West Conshohocken, PA: ASTM International, 2007.

ASTM A 755/A 755M-07, *Specification for Steel Sheet, Metallic-coated by the Hot-dip Process and Prepainted by the Coil-coating Process for Exterior Exposed Building Products*. West Conshohocken, PA: ASTM International, 2007.

ASTM B 101-02, *Specification for Lead-coated Copper Sheets and Strip for Building Construction*. West Conshohocken, PA: ASTM International, 2002.

ASTM C 140-07, *Test Method Sampling and Testing Concrete Masonry Units and Related Units*. West Conshohocken, PA: ASTM International, 2007.

ASTM C 406-06e01, *Specification for Roofing Slate*. West Conshohocken, PA: ASTM International, 2006.

ASTM C 836-06, *Specification for High-solids Content, Cold Liquid-applied Elastomeric Waterproofing Membrane for Use with Separate Wearing Course*. West Conshohocken, PA: ASTM International, 2006.

ASTM C 957-06, *Specification for High-solids Content, Cold Liquid-applied Elastomeric Waterproofing Membrane with Integral Wearing Surface*. West Conshohocken, PA: ASTM International, 2006.

ASTM C 1029-05a, *Specification for Spray-applied Rigid Cellular Polyurethane Thermal Insulation*. West Conshohocken, PA: ASTM International, 2005.

ASTM C 1167-03, *Specification for Clay Roof Tiles*. West Conshohocken, PA: ASTM International, 2003.

ASTM C 1492-03, *Standard Specification for Concrete Roof Tile.* West Conshohocken, PA: ASTM International, 2003.

ASTM D 225-04, *Specification for Asphalt Shingles (Organic Felt) Surfaced with Mineral Granules.* West Conshohocken, PA: ASTM International, 2004.

ASTM D 226-06, *Specification for Asphalt-saturated Organic Felt Used in Roofing and Waterproofing.* West Conshohocken, PA: ASTM International, 2006.

ASTM D 227-03, *Specification for Coal-tar-saturated Organic Felt Used in Roofing and Waterproofing.* West Conshohocken, PA: ASTM International, 2003.

ASTM D 312-00 (2006), *Specification for Asphalt Used in Roofing.* West Conshohocken, PA: ASTM International, 2006.

ASTM D 449-89 (1999)e1, *Specification for Asphalt Used in Dampproofing and Waterproofing.* West Conshohocken, PA: ASTM International, 1999.

ASTM D 450-07, *Specification for Coal-tar Pitch Used in Roofing, Dampproofing and Waterproofing.* West Conshohocken, PA: ASTM International, 2007.

ASTM D 1227-95 (2007), *Specification for Emulsified Asphalt Used as a Protective Coating for Roofing.* West Conshohocken, PA: ASTM International, 2007.

ASTM D 1863-05, *Specification for Mineral Aggregate Used on Built-up Roofs.* West Conshohocken, PA: ASTM International, 2005.

ASTM D 1970-01, *Specification for Self-adhering Polymer Modified Bituminous Sheet Materials Used as Steep Roof Underlayment for Ice Dam Protection.* West Conshohocken, PA: ASTM International, 2001.

ASTM D 2178-04, *Specification for Asphalt Glass Felt Used in Roofing and Waterproofing.* West Conshohocken, PA: ASTM International, 2004.

ASTM D 2626-04, *Specification for Asphalt-saturated and Coated Organic Felt Base Sheet Used in Roofing.* West Conshohocken, PA: ASTM International, 2004.

ASTM D 3161-06, *Standard Test Method for Wind-resistance of Asphalt Shingles (Fan-induced Method).* West Conshohocken, PA: ASTM International, 2006.

ASTM D 3462-07, *Specifications for Asphalt Shingles Made from Glass Felt and Surfaced with Mineral Granules.* West Conshohocken, PA: ASTM International, 2007.

ASTM D 3468-99 (2006)e01, *Specification for Liquid-applied Neoprene and Chlorosulfonated Polyethylene Used in Roofing and Waterproofing.* West Conshohocken, PA: ASTM International, 2006.

ASTM D 3909-97b (2004)e01, *Specification for Asphalt Roll Roofing (Glass Felt) Surfaced with Mineral Granules.* West Conshohocken, PA: ASTM International, 2004.

ASTM D 4434-06, *Specification for Poly (Vinyl Chloride) Sheet Roofing.* West Conshohocken, PA: ASTM International, 2006.

ASTM D 4601-04, *Specification for Asphalt-coated Glass Fiber Base Sheet Used in Roofing.* West Conshohocken, PA: ASTM International, 2004.

ASTM D 4637-04, *Specification for EPDM Sheet Used in Single-ply Roof Membrane.* West Conshohocken, PA: ASTM International, 2004.

ASTM D 4869-05e01, *Specification for Asphalt-saturated Organic Felt Shingle Underlayment Used in Roofing.* West Conshohocken, PA: ASTM International, 2005.

ASTM D 5019-07, *Specification for Reinforced Nonvulcanized Polymeric Sheet Used in Roofing Membrane.* West Conshohocken, PA: ASTM International, 2007.

ASTM D 6162-00A, *Specification for Styrene-butadiene-styrene (SBS) Modified Bituminous Sheet Materials Using a Combination of Polyester and Glass Fiber Reinforcements.* West Conshohocken, PA: ASTM International, 2000.

ASTM D 6163-00e01, *Specification for Styrene-butadiene-styrene (SBS) Modified Bituminous Sheet Materials Using Glass Fiber Reinforcements.* West Conshohocken, PA: ASTM International, 2000.

ASTM D 6164-05, *Specification for Styrene-butadiene-styrene (SBS) Modified Bituminous Sheet Metal Materials Using Polyester Reinforcements.* West Conshohocken, PA: ASTM International, 2005.

ASTM D 6222-02e01, *Specification for Atactic Polypropylene (APP) Modified Bituminous Sheet Materials Using Polyester Reinforcements.* West Conshohocken, PA: ASTM International, 2002.

ASTM D 6298-05, *Specification for Fiberglass Reinforced Styrene-butadiene-styrene (SBS) Modified Bituminous Sheets with a Factory-applied Metal Surface.* West Conshohocken, PA: ASTM International, 2005.

ASTM D 6380-03, *Standard Specification for Asphalt Roll Roofing (Organic) Felt.* West Conshohocken, PA: ASTM International, 2003.

ASTM D 6509-00, *Standard Specification for Atactic Polypropylene (APP) Modified Bituminous base Sheet Materials Using Glass Fiber Reinforcements.* West Conshohocken, PA: ASTM International, 2000.

ASTM D 6694-07, *Standard Specification for Liquid-applied Silicone Coating Used in Spray Polyurethane Foam Roofing.* West Conshohocken, PA: ASTM International, 2007.

ASTM D 6754-02, *Standard Specification for Ketone Ethylene Ester Based Sheet Roofing.* West Conshohocken, PA: ASTM International, 2002.

ASTM D 6878-06a, *Standard Specification for Thermoplastic Polyolefin Based Sheet Roofing.* West Conshohocken, PA: ASTM International, 2006.

ASTM D 6947-07, *Standard Specification for Liquid Applied Moisture Cured Polyurethane Coating Used in Spray Polyurethane Foam Roofing System.* West Conshohocken, PA: ASTM International, 2007.

ASTM D 7158-07, *Standard Test Method for Wind Resistance of Sealed Asphalt Shingles (Uplift Force/Uplift Resistance Method).* West Conshohocken, PA: ASTM International, 2007.

ASTM E 108-07a, *Standard Test Method for Fire Tests of Roof Coverings.* West Conshohocken, PA: ASTM International, 2007.

CGSB 37-GP-52M-(1984), *Roofing and Waterproofing Membrane, Sheet Applied, Elastomeric.* Ottawa, Ontario, Canada: Canadian General Standards Board, 1984.

CGSB 37-GP-54M-95, *Roofing and Waterproofing Membrane, Sheet Applied, Flexible, Polyvinyl Chloride.* Ottawa, Ontario, Canada: Canadian General Standards Board, 1995.

CGSB 37-GP-56M-80, *Membrane, Modified, Bituminous, Prefabricated and Reinforced for Roofing—with December 1985 Amendment.* Ottawa, Ontario, Canada: Canadian General Standards Board, 1980.

CSSB, *Design and Application Manual for New Roof Construction.* Sumas, WA: Cedar Shake and Shingle Bureau.

FM 4474-04, *Evaluating the Simulated Wind Uplift Resistance of Roof Assemblies using Static Positive and/or Negative Differential Pressures.* Johnson, RI: Factory Mutual Global Research Standards Laboratories Department, 2004.

IPC-09, *International Plumbing Code.* Washington, DC: International Code Council, 2009.

NRCA *Roofing Manual: Steep-slope Roof Systems—2009.* Rosemont, IL: National Roofing Contractors Association (NCRA).

NRCA *Roofing Manual: Membrane Roof Systems—2007.* Rosemont, IL: National Roofing Contractors Association (NCRA).

UL 580-06, *Test for Uplift Resistance of Roof Assemblies.* Northbrook, IL: Underwriters Laboratories, Inc.

UL 1256-02, *Fire Test of Roof Deck Construction—with Revisions through January 2007.* Northbrook, IL: Underwriters Laboratories, Inc., 2002.

UL 1897-04, *Uplift Tests for Roof Covering Systems.* Northbrook, IL: Underwriters Laboratories, Inc, 2004.

INDEX

Note: *This is taken from the index for the 2009 International Building Code. Volume I of the IBC Commentary only includes Chapters 1 through 15.*

A

H

Don't Miss Out On Valuable ICC Membership Benefits. Join ICC Today!

Join the largest and most respected building code and safety organization. As an official member of the International Code Council®, these great ICC® benefits are at your fingertips.

EXCLUSIVE MEMBER DISCOUNTS

ICC members enjoy exclusive discounts on codes, technical publications, seminars, plan reviews, educational materials, videos, and other products and services.

TECHNICAL SUPPORT

ICC members get expert code support services, opinions, and technical assistance from experienced engineers and architects, backed by the world's leading repository of code publications.

FREE CODE—LATEST EDITION

Most new individual members receive a free code from the latest edition of the International Codes®. New corporate and governmental members receive one set of major International Codes (Building, Residential, Fire, Fuel Gas, Mechanical, Plumbing, Private Sewage Disposal).

FREE CODE MONOGRAPHS

Code monographs and other materials on proposed International Code revisions are provided free to ICC members upon request.

PROFESSIONAL DEVELOPMENT

Receive Member Discounts for on-site training, institutes, symposiums, audio virtual seminars, and on-line training! ICC delivers educational programs that enable members to transition to the I-Codes®, interpret and enforce codes, perform plan reviews, design and build safe structures, and perform administrative functions more effectively and with greater efficiency. Members also enjoy special educational offerings that provide a forum to learn about and discuss current and emerging issues that affect the building industry.

ENHANCE YOUR CAREER

ICC keeps you current on the latest building codes, methods, and materials. Our conferences, job postings, and educational programs can also help you advance your career.

CODE NEWS

ICC members have the inside track for code news and industry updates via e-mails, newsletters, conferences, chapter meetings, networking, and the ICC website (www.iccsafe.org). Obtain code opinions, reports, adoption updates, and more. Without exception, ICC is your number one source for the very latest code and safety standards information.

MEMBER RECOGNITION

Improve your standing and prestige among your peers. ICC member cards, wall certificates, and logo decals identify your commitment to the community and to the safety of people worldwide.

ICC NETWORKING

Take advantage of exciting new opportunities to network with colleagues, future employers, potential business partners, industry experts, and more than 50,000 ICC members. ICC also has over 300 chapters across North America and around the globe to help you stay informed on local events, to consult with other professionals, and to enhance your reputation in the local community.

JOIN NOW! 1-888-422-7233, x33804 | www.iccsafe.org/membership

INTERNATIONAL CODE COUNCIL®

People Helping People Build a Safer World™

09-01547